The
NEW SOTHEBY'S WINE ENCYCLOPEDIA

The
NEW
SOTHEBY'S
WINE
ENCYCLOPEDIA

TOM STEVENSON &
ORSI SZENTKIRALYI, DJur

with a Foreword by
JAMIE RITCHIE

WASHINGTON, D.C.

Contents

Foreword

I BOUGHT AND READ MY FIRST COPY OF *THE SOTHEBY'S WINE ENCYCLOPEDIA* IN 1988 – when I really started to learn about wine – so I am delighted to introduce *The New Sotheby's Wine Encyclopedia* written by Tom Stevenson and published by National Geographic.

For me, the *Sotheby's Wine Encyclopedia* has always been the best overall book on wine, whether you use it as a reference or make a steady read-through, learning about the different grape varieties, countries, or regions. It provides the perfect amount of information in an easily readable and digestible format. The introductory pages that describe the world of wine, from growing and producing to buying and storing, provide the most accurate and concise descriptions that anyone needs. It is simply the best researched book about wine and fully lives up to the name "encyclopedia". When you use it to look up specific information, you invariably end up learning something beyond what you were searching for.

Interest in wine has been growing geographically, from Europe throughout the Americas and Asia, and demographically, as many younger people are becoming interested in the world's most complex beverage. Wine provides an opportunity of great discovery and joy every single day, whether we are tasting wines from a new producer, vintage, or region or enjoying the last bottle from a case of a mature vintage and reflecting on its evolution through the years. This book helps you explore new wines and return to old favorites with enthusiasm, optimism, and an eagerness for discovery. It gives us additional knowledge, so we can have a better ability to select the bottle of wine that will give us and our guests the most joy and pleasure.

There has never been a better time to buy wine. The advance in winemakers' technical knowledge of viticulture and vinification means that the quality and value we enjoy today is at least double what it was 10 or 20 years ago. In addition, there is a new generation of winemakers all around the world making exceptional wines, with a fresh approach to traditional grape varieties, sustainability, and the environment.

The market for wine is vibrant. Unlike Old Master paintings, we get a new supply of wine every year, and less of what has been consumed of the older vintages. With increasing geographic and demographic demand for fine wines that are limited in supply by the size of the vineyards and the appellations, such as the great wines of Burgundy or Bordeaux, it is logical that the price of these wines will continue to increase over time. This creates the opportunity for collectors to buy for both enjoyment and for investment.

One of the most interesting aspects of Sotheby's Wine business in the next 10 years will be the evolution of how wine is bought and sold, whether it be through the retail, auction, or exchange channels. The transition from brick and mortar to e-commerce is sure to make more wines available to more people, with more pricing transparency than ever before. It is our hope that the regulatory environment encourages a vibrant marketplace, without high tariffs and prohibitive regulation.

Wine is one of the great joys of the world. At Sotheby's Wine, we share this joy and are all lucky to work in an area which is also our passion. In ever more stressful times, everyone around the world is taking more time to enjoy and share experiences together – wine enhances all those experiences and *The New Sotheby's Wine Encyclopedia* is here to help in our pursuit of maximising that enjoyment.

Cheers!

Jamie Ritchie

Jamie Ritchie
Worldwide Head
SOTHEBY'S WINE

A New Edition for a New Decade

WELCOME TO *THE NEW SOTHEBY'S WINE ENCYCLOPEDIA*. Since the last edition was released almost 10 years ago, it has become ever more apparent that this encyclopedia needs a publisher whose name is every bit as iconic as Sotheby's itself. I found that publisher in the legendary National Geographic, and it was all rather serendipitous. Sean Moore, who was the Senior Editor on the second edition, moved to the United States, where he set up a book packaging company called Moseley Road in 2002. I happened to drop him an email one day, and he responded within less than two minutes saying that he just so happened to be thinking about the encyclopedia at the very moment my email popped up. Wasting no time, he asked me what I thought about National Geographic because he had an appointment with them the next day. I told him that I could not think of a more inspirational publisher for this encyclopedia, after which everything just tumbled into place. Furthermore, with Moseley Road playing an instrumental part in the encyclopedia's production and Sean at its helm, I had the support of one of only a handful of people in the entire world who understood first-hand the sheer volume of work involved.

When this encyclopedia was first published in 1988, it was just about possible for one person to visualize the global wine industry in the mind's eye, grasp all the threads, and draw them together to make one book. By 2011, however, when the last edition was published, the expansion of the world of wine had become so convoluted that it was no longer possible. I needed help. Thankfully I already had a first class network of contributors from the six annual editions of *Wine Report,* and I was happy to discover that some of them were available to work on the encyclopedia. I also needed someone special to provide day-to-day editorial control. When I say special, I mean somebody who has such a profound understanding of appellations that she would be able to maintain a disciplined overview of what is, after all, the raison d'être of this encyclopedia. While other books specialise in detailing the nooks and crannies of some vineyards, the ethos of Sotheby's has always been to map every appellation in the world, whether official or not, define them, and recommend their best producers. It is meant to be an encyclopedia to everything, providing information whittled down to a graspable level, not an in-depth, definitive study of a limited coverage.

I quickly realised that the ideal candidate for Managing Editor was already working with me at the Champagne & Sparkling Wine World Championships, where she has been the Tasting Quality Control Director since 2015. She was also one of less than half a dozen people in the world enrolled on both the Master Sommelier (MS) and Master of Wine (MW) programmes. I knew how bright she was and I knew she had great tenacity, but what made Orsi Szentkiralyi uniquely qualified for this encyclopedia was her Doctorate in Law specialising in appellations. I could not have attempted *The New Sotheby's Wine Encyclopedia* without such a brilliant team of contributors (credited in full, pp764–765), but even with their invaluable help, I would never have finished the project without Orsi. Even one of my grumpiest of contributors and oldest of friends told me this!

Tom Stevenson

WINE AS A CAREER CAME ABOUT LIKE A LIGHT-BULB MOMENT in my life: realising that it utilises a vast array of disciplines from art to science, I instantly knew I'd never be bored at work. For someone with a seemingly endless curiosity, who for a long time could not pick one from the many interesting subjects to focus on, it sounded like just the right thing. Little did I know that this choice meant a decade-long journey of intense studying, interspersed with gruelling exams. The further I went the harder it was to find the right resources to study from: most books were too superficial, factually wrong, outdated or just plain boring.

The copy of the *The Sotheby's Wine Encyclopedia* I bought in 2011 was a game changer to me. It was complete, in-depth, but also entertaining and easy to read; it could be used as a "one-stop shop" study material but also as a map source or a reference book if I just needed to check something.

It goes without saying how honoured I felt when Tom asked me to come on board for this new edition. It felt like a mammoth task, but also just the right challenge, and I cannot thank him enough for giving me this opportunity.

The decade passed between this edition and the last has brought immense changes to the industry: new regions and producers emerged; wine styles went in and out of fashion with the ever-evolving consumer preferences, shifting production volumes one way or another. Even the classic appellations are experiencing trying times today: climatic conditions are becoming more and more erratic and wines that are true to their own identity are harder and harder to make. To report on all of this would have been impossible without the contributing authors who are the very best of their own field, and their up-to-date knowledge and passion for the subject made my part of the job both educational and inspiring.

I was privileged to work alongside Sean Moore and Lisa Purcell at Moseley Road and to experience the professionalism at National Geographic while working on the brand-new maps and seeing the whole book come together. This book is great because of them.

I certainly hope that this new edition of the encyclopedia will bring readers from all over the world closer to the fascinatingly complex subject that is wine.

Orsi Szentkiralyi

Understanding the Fine Wine Market

The market for fine wine is vibrant and exciting and continues to to grow. In the following pages we reveal key aspects of this market: who is making fine wine and where; what wines are included in this market; who is buying fine wine; how fine wine changes hands; and how to collect it. We also illustrate why there has never been a better time to buy wine.

WHAT IS FINE WINE?

For our purposes, "fine wine" is defined as a wine that costs more than US$50 per bottle, is highly allocated, has ageing potential, and is in demand in the secondary market (see below).

THE FINE WINE MARKETPLACE

When talking about fine wine, there is a distinction between the primary and secondary markets. The primary market is the first sale of wine from the producer to the consumer handled through a distribution system that typically ends in a retail store or supermarket. The secondary market is the re-sale of wines by a consumer to other consumers through brokers, dealers, auction houses, and trade exchanges.

Fine wine is in global demand, and regulations govern individual markets on national and regional levels. Recent research estimates the total global wine market to have annual sales of US$200 billion, with $5 billion in fine wine.

CHANGING OF THE GUARD

Generational change is continually taking place in the wine industry, but the effect of this change is even greater today. As seasoned elders step aside to make way for the next crop of winemakers, fresh outlooks and new techniques are explored and put into practice. This creates or renews interest in different regions and producers, thereby increasing the range and styles of wines available in the market.

EXPANSION BEYOND TRADITIONAL BOUNDARIES

In this experimental spirit, and as a result of vastly increased land prices in their own vineyard areas, storied producers are exploring projects far beyond their traditional winemaking homes. Expansions into new geographical territories introduce less well-known regions to followers of well-known producers. Plus, expansion provides a hedge against climate change's impact in traditional regions and higher costs of acquiring vineyards, as land prices continue to rise.

SEARCHING FOR QUALITY AND VALUE

Increased prices for in-demand regions, particularly Burgundy and Champagne, have resulted in regional exploration at every tier of the distribution system, including consumers who are regularly discovering producers that are making the highest quality wines. On a global scale, winemaking knowledge and techniques have drastically improved. As a result, there are excellent, well-made wines from previously ignored areas providing delicious drinking with a dependable consistency, despite vintage variations.

There has never been a better time to buy wine than today: the quality that you get for US$20, or at any price point in any currency, is vastly better than it was 10 or 20 years ago.

DIVERSIFICATION IN THE SECONDARY MARKET

Bordeaux has been the backbone of the secondary market, by both volume and value, since Sotheby's began selling wine at auction in the United Kingdom in 1970. In recent years, however, it has ceded market share in terms of the value sold as demand for both Burgundy and spirits – specifically Scotch whisky and Japanese whisky – increase.

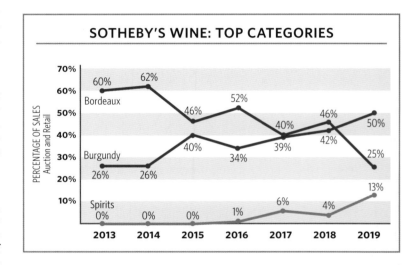

SOTHEBY'S WINE: TOP CATEGORIES

SOTHEBY'S WINE OWN LABEL COLLECTION
Sotheby's Wine partnered with well-known wineries to create a range of Sotheby's own label wines, which are available at their New York and Hong Kong retail locations.

SOTHEBY'S WINE: TOP PRODUCERS BY VALUE 2016–2019		
RANK	**PRODUCER**	**AGGREGATE SALES**
1	Domaine de la Romanée-Conti	$74 million
2	Pétrus	$21 million
3	Château Mouton Rothschild	$19 million
4	Château Lafite	$18 million
5	Domaine & Maison Leroy	$13 million
6	Domaine Coche-Dury	$10 million
7	Domaine Armand Rousseau	$10 million
8	Château Latour	$9 million
9	Château Haut Brion	$8 million
10	Château Margaux	$7 million

The "blue chip" wines considered suitable for investment still come from a relatively narrow band of regions, but the range of producers and vintages with potential for appreciation is widening. The appearance at auction of previously lesser-known producers from Bordeaux's right bank, Burgundy, the Northern Rhône, Piedmont, and Champagne is the result of a demand for options beyond the traditional supply.

In both the primary and secondary markets, the domination of the top producers has grown. In the wine world, where we once talked about growers, *domaines,* producers, and châteaux, now you frequently hear them referred to as "brands". Today producer Domaine de la Romanée-Conti (DRC) has the greatest share of the secondary fine wine market, representing 25 per cent of the value of all wine sold by Sotheby's Wine in 2019, as well as almost 45 per cent of the value of all Burgundy. The table above shows the shares represented by the top 10 wine producers sold by Sotheby's Wine from 2016 to 2019. Each year, the top producer lists are remarkably consistent.

Luxury groups continue to make acquisitions of the best wine producers (or vineyard sites) at a steady pace, and we foresee more and more investment groups acquiring family-run businesses over the next decades.

WHO IS BUYING?

Wealth creation is vital to the strength of the fine wine market, as with all luxury sectors. As wealth is being created by more people and, specifically, more young people around the world, the average age of wine buyers is decreasing, while the overall population of fine wine consumers increases. Technology and social media help to make wine education and trends instantly available everywhere, and as a result the demand for fine wine knows no geographical boundaries.

Asia

With the elimination of its wine tariffs in 2008, Hong Kong has emerged as the fine wine hub of Asia. Established US and UK merchants, brokers, and auction houses, as well as local Chinese businesses, sell wine from their bases in Hong Kong to consumers throughout the continent. Wines purchased in Hong Kong, regardless of the buyer's location, are typically also stored in Hong Kong. The steady growth in wealth creation throughout the region has resulted in Asia accounting for more than half of Sotheby's Wine's sales every year since 2013.

While the Asian market was once dominated by sales to residents of Hong Kong and mainland China, more recently consumers from countries such as Taiwan, Singapore, Thailand, Indonesia, Vietnam, and Japan have emerged. Now, Sotheby's Wine sells a higher percentage elsewhere in Asia than in Hong Kong.

Influencers

Tastemakers evolve, too. The dawn of social media has brought a changing of the influential guard. Where professional critics and publications were once the most trusted adjudicators, the opinions of sommeliers and crowd-sourced ratings, such as Vivino and Cellartracker, are now impacting the market.

WHERE TO BUY FINE WINE

The ways that we buy wine are also changing with the evolution of wines found in the marketplace and consumer behaviour. Both technology and regulations are driving factors in how people purchase wine.

Retail

Retailers are the most prominent source of current and recent releases of fine wines, as well as a good source of well-stored bottles with more age. Items are sold at a fixed price, and consumers can usually select the quantity they desire with immediate delivery. The newest vintage of Bordeaux – sold as En Primeur in Europe and as Futures in the United States – is traditionally sold through wine retailers. Even as specialty stores with highly curated selections become more common, resources such as the search engine Wine-Searcher offer buyers transparency about where to find wines at the most attractive prices in every location around the world.

The retail landscape today encompasses everything from grocery stores and national chains to more specialized fine wine merchants and brokers. Each one is an important part of the retail fabric: purchasing wine from producers, private collections, and wholesale channels. Some retailers operate without the traditional brick and mortar, listing wines online with a growing focus on e-commerce and selling through targeted email offers.

Beyond the fine and rare, there has been an interesting development of "private label" and "own label" wines that help consumers take the guesswork out of selecting their wines. Sotheby's Wine launched the Sotheby's Own Label Collection range of wines in 2019, which offers excellent value from classic regions at affordable prices, including Champagne, Burgundy, Bordeaux, Piedmont, and California (*see* photo on opposite page).

Auctions

For fine wine, auctions remain the most important and trusted source in the secondary market, providing the widest range of producers and vintages. Bidding methods have evolved, allowing buyers from all over the world to participate by watching live auctions and bidding over the phone, in addition to attending the sale in person or submitting absentee bids. At the same time, we are in a transition from live auctions to online-only and a combination of the two. The reputations and expertise of the very top auction houses provide assurance that the wines being sold are of highest quality, in superb condition, and of impeccable provenance. For sellers, auction houses offer the opportunity to maximize the value of their collections by selling it to the highest bidder – the truest method of setting a price.

Direct from producer and single-owner sales

More than ever, provenance is paramount when purchasing fine wine. The ultimate provenance, of course, is straight from the source. Auctions featuring wines directly from a producer's cellar consistently sell for a considerable premium, as collectors ascribe greater value to provenance and condition.

"Single Owner" sales offer wines entirely from one private collection. These are in very high demand. The individual collector has undertaken the onerous task of selecting, finding, purchasing, and storing wines over

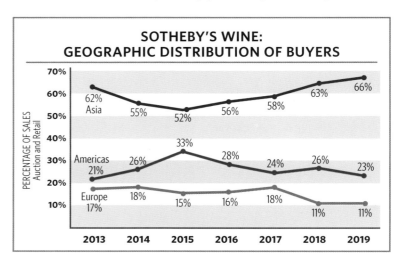

SOTHEBY'S WINE: MOST VALUABLE SINGLE-OWNER SALES

1. **The Classic Cellar of an American Collector**
2009–2013
$52 million

2. **Tran-scend-ent Wines**
2019
$30 million

3. **The Don Stott Cellar: 50 Years of Collecting Part I–IV**
2015–2017
$25 million

4. **The Cellar of William I Koch**
2016
$22 million

5. **The Philanthropist's Cellar**
2018
$16 million

6. **The Millennium Wine Cellar**
1999
$14 million

7. **The Ultimate Cellar**
2011
$12 million

8. **The Cellar from the Estate of Jerry Perenchio**
2018
$12 million

9. **The Ultimate Whisky Collection**
2019
$10 million

10. **The Bordeaux Collection from SK Networks**
2010
$10 million

Online auctions

Auctions listed solely online, rather than in printed catalogues or traditional salerooms, are without doubt the auction venue of the present and future. Online auctions allow for more variety in terms of the types, price ranges, and quantities of wines, and with transparent competitive bidding, prospective buyers know whether their bid is the highest throughout the timed sale.

Record sales and record prices

Prices for rare wines with unique verifiable provenance have been steadily increasing. Sotheby's Wine holds the auction records for highest price for a single bottle of wine, and highest price for a single bottle of spirits.

Sotheby's Wine also holds record for prices above $1 million with four record transactions listed in the chart below.

Direct to consumer

Increasingly, the direct-to-consumer (DTC) sales channel challenges the traditional systems of wine importation and distribution. Through email offers, winery subscriptions, or online sales platforms, selling wines directly to consumers eliminates whole tiers of the wine distribution system, delivering the highest margin to the winery. With this sales channel, producers have a direct relationship with the consumer, retain more control over the pricing, and can determine the allocations of their wines. This model is most widely used by the top wineries in California's Napa Valley and is now being adopted by wineries in Europe, particularly Burgundy.

Trading platforms and exchanges

Over the last 20 years, business-to-business (B2B) exchanges have been used by the wine industry to trade stock between themselves. The best

a long period of time. The wines in the collection usually reflect the personality of the collector and often have a particular focus on certain regions and producers. These single-owner auctions often become brands in their own right, garnering reputations among satisfied buyers. In 2019, at Sotheby's Wine, single-owner auctions accounted for more than 55 per cent of sales.

Lot 84 — Online

USD	450,000
EUR	388,594
GBP	340,574
CHF	446,119
JPY	50,618,673
HKD	3,532,737

SOTHEBY'S WINE: TOP WINE & SPIRITS WHICH SOLD FOR OVER US$1 MILLION

PRICE	DESCRIPTION	YEAR SOLD
$1.9 million	The Macallan Fine & Rare 60 Year Old, 1926 whisky *1 bottle sold at auction in London*	2019
$1.6 million	Romanée-Conti, Domaine de la Romanée-Conti Vertical 1992–2010 *14 bottles sold at auction in Hong Kong*	2014
$1.5 million	Domaine de la Romanée-Conti 2005 Assortment *7 Methuselahs (6 litres) sold at retail in Hong Kong*	2018
$1.1 million	A unique 50 case lot of Château Mouton Rothschild 1982 *50 cases (600 bottles) sold at auction in New York*	2006

JAMIE RITCHIE, WORLDWIDE HEAD OF SOTHEBY'S WINE, SELLING ROMANÉE-CONTI, DOMAINE DE LA ROMANÉE-CONTI 1945
Sotheby's top lot from the auction "Rare Domaine de la Romanée-Conti from the Personal Cellar of Robert Drouhin", sold for $558,000 – a world record for most-expensive single bottle of wine sold at auction. (*See also* photograph of bottle on Contents page.)

A SOTHEBY'S WINE ADVISOR DECANTS FOR AN EDUCATIONAL TASTING AT SOTHEBY'S WINE RETAIL STORE IN NEW YORK CITY
High-quality retailers can offer more than bottles of wine, with tastings and lectures that allow wine lovers to learn more about their favorite beverage.

known of these exchanges is Liv-ex. There have been a number of attempts to create peer-to-peer exchanges, but most have struggled to gain traction due to shipping and regulatory constraints. It is likely that consumer-to-business-to-consumer exchanges will gain in popularity, however, so long as they can aggregate enough property to constitute a market. Users of trading platforms can make offers on wines or receive bids for their own wines listed on the exchange. These exchanges also produce rich data on current market prices, bolster wine as an asset class for trading, and provide insights into the health of the wine market.

REGULATION

The sale and distribution of alcohol is heavily regulated throughout the world, with each country (and state) imposing restrictions, tariffs and taxes. These regulations fundamentally influence the global wine market, affecting the price of wine. The tightening and easing of regulations causes shifts in global demand.

With the 1920s Prohibition laws in the United States, the wine industry all but dried up, only to come back with force upon its repeal in 1933. More recently, the 1994 changes to regulations permitting wine to be sold at auction in New York altered the global secondary market, with North America overtaking Europe as the largest buyer of fine wine at auction. North Americans reigned as the least price-sensitive – and most eager – buyers until 2009, when Hong Kong's removal of all wine tariffs once again shifted the dynamic and tilted the wine auction market on its axis. Since 2009 Asian wine buyers have been the least price sensitive, and there has been a huge transfer of fine wine from both the United States and Europe to Asia. Trade wars, tariffs, and political decisions are a constant threat to the free movement of wine. We hope the future allows free access to wines for all who wish to buy it.

GUIDE TO BUYING AND COLLECTING FINE WINE
Retailers and brokers
Wine is a fragile, valuable, and increasingly in-demand beverage that is also a commodity. When buying fine and rare wine it is important to be selective in choosing sources. Only purchase wines from retail stores or brokers with stellar reputations; online forums and reviews can provide helpful feedback. Request provenance information and photographs for any mature, rare, and valuable bottles – keep in mind that if an offer seems too good to be true, it probably is. With resources such as Wine-Searcher, Cellar Tracker, and Vivino buyers have unprecedented access to information about fine wines. When possible, develop relationships with your wine retailer, broker, or specialist. In addition to providing expertise, establishing connections and letting them know your preferences can result in access to allocations and offers of hard-to-find wines.

Auction houses
Auction houses establish long-term relationships with both buyers and sellers and provide the service of efficiently transferring ownership from one to the other. Wine specialists at the top auction houses inspect every bottle offered for sale for condition and authenticity. Most houses provide background information about the collection's owner and provenance alongside the listing of wines, and they will provide photographs and further information upon request.

Provenance and authenticity
Be selective when purchasing fine and rare wines and only transact with well-known, reputable sources. Request documentation on provenance and obtain photographs clearly depicting the condition, the level in the bottle, label, and capsule and the color of the wine. If dealing in the ultra-rare and high value, employ an expert to authenticate wines before purchase. Counterfeit wines are an unfortunate reality, and prospective buyers must exercise vigilance.

Storage and insurance
For maximum enjoyment – and resale value – wine must be maintained in a temperature- and climate-controlled environment. If you cannot provide ideal home cellarage conditions, seek out professional storage for wines you plan to keep long-term. Bottles should lay on their sides to keep the cork in contact with the wine and avoid it drying out, and storage should be vibration free and away from direct sunlight.

Wine should be insured just like any other asset. Experts such as Sotheby's Wine can provide valuations to help you know what your wine collection is worth.

THE OUTLOOK
Wine and spirits, contemporary art, watches and jewelry – all are markets that share one important common denominator: they are produced every year, so there is new supply coming to the market regularly. With wine and spirits, the market receives a new vintage every year, so there is constant supply, but we must deduct the number of bottles of older vintages that are consumed. If demographic and geographic demand continues to increase, there is an economic imbalance and, over the long term, prices for fine wines will continue to increase.

At the same time, the quality of wine produced throughout the world will continue to improve, so even as fine and rare wines get more expensive, consumers at lower price points will still be able to drink better than ever before.

Commerce and discourse about wine will move increasingly to virtual realms, yet ultimately wine is a liquid asset meant to be shared and enjoyed. Armed with the information in this encyclopedia and the expert advisors at Sotheby's Wine, the journey will be rewarding – and fun.

TIPS FOR INVESTMENT

When buying wine as an investment, keep in mind the following tips:

- Vintages are important. Wines from the best years are more likely to appreciate over time. Although "sleepers" or off-vintages may surprise down the line and provide excellent value drinking, the safest are vintages that were highly regarded from the beginning.

- Try to buy in case quantities. If it is within your means, opt for full cases of wines – usually 6 or 12 standard bottles, or 3 or 6 magnums. Wines sold in case quantity often boast a per-bottle premium compared to single bottles or other quantities. The 12-bottle case used to be standard, but with the rising price per bottle, the 6-bottle quantity is equally considered a case.

- Variety is the spice of life. A thoughtfully chosen, well-rounded portfolio of wines of varying vintages, sizes, regions, and producers hedges against short-term trends in the market. Seek out the classics, but don't limit the potential for upside.

- Practice good record-keeping. When the time comes to sell your investment, potential auction houses or buyers like to see a paper trail. Save your invoices and shipping records.

PART ONE

TASTE *and* QUALITY

Factors Affecting Taste *and* Quality

THERE ARE SIX FUNDAMENTAL FACTORS THAT determine the quality and style of the wines produced in an area: location, climate, aspect, soil, viticulture and vinification, and grape varieties. In the introductory text for countries, regions, or districts, a "Factors Affecting Taste and Quality" panel provides a quick-reference guide to these six criteria. The first four listed (location, climate, aspect, and soil) are the constant factors that determine the ability to grow grapes anywhere; viticulture and vinification affect the potential quality of the specific grapes and the styles of wine they produce; whilst the choice of grape varieties decides the basic taste of the wine that can be made . . . which brings us to the winemaker, who is the idiosyncratic joker in the pack.

THE SINUOSITY OF TERRACED VINEYARDS IN THE DOURO VALLEY OF PORTUGAL IS A RESULT OF CAREFUL ATTENTION TO ASPECT AND CLIMATE
The hilly topography of this region means that growers must carefully plan the most advantageous aspects and elevations on which to plant their grapevines in order to get the best results. Terraced vineyards take advantage of the valley's continental climate that features hot and dry summers and cold winters.

Location

The location of a vineyard determines whether or not its climate is suitable for viticulture. The same grape grown in the same area can make two totally different wines, yet due to other factors affecting quality, different grapes grown continents apart may produce two wines that are very similar.

The vast majority of the world's wine-producing areas, including all of the classic and most successful wine regions, are located in the temperate zones, between 10°C (50°F) and 20°C (68°F), where *Vitis vinifera* is capable of producing premium-quality wine grapes. If you look at the Wine by Latitude map (*see* opposite page) you will notice that most of the vineyards in both hemispheres are found between 30° and 50° of latitude. These two latitudinal bands correspond approximately to the northern and southern temperate zones, but they can never represent precise parameters because altitude, prevailing winds, ocean currents, and the penetration of sea fog into coastal valleys influence the temperatures experienced beyond such simplicities as distance from the sun and the angle of its rays. Look at the isotherm map (*see* Wine by Temperate Zones on opposite page), and you can see the true extent of the northern and southern temperate zones.

According to wine-loving climatologist Dr Gregory Jones, whose studies have focused on a slightly different thermal zone (12 to 22°C, or 54 to 72°F), the line for the lower annual temperature has shifted 80 to 240 kilometres (50 to 150 miles) polewards in both hemispheres. This has been due to an increase of 1.7°C (3.1°F) in the average growing-season temperature in Europe and, over a similar time period (1948 to 2004), an increase of 1.5°C (2.7°F) in the western United States, driven mostly by increases in minimum temperatures. This might take the crisp, acidic edge off the coolest climate wines (in Germany's Mosel, for example), which, over the passage of time, will inevitably become fractionally softer than they used to be – but such generic changes to the potential taste profile of any given area will be partly hidden by the change in style of the wine producers themselves, most of whom are intentionally harvesting riper fruit than they used to. Sometimes this has been achieved by longer hang time (in other words, later picking), sometimes by the choice of an earlier ripening clone, or often both. Because many consumers have turned to – or have become conditioned by – softer, fatter, and sometimes sweeter wines, it is debatable if such differences are even noticed. The increase in minimum temperatures has also opened up greater quality potential for vineyards that were perhaps on the wrong side of the viticultural periphery (such as England) and will, if the current trend continues, offer serious opportunities to other areas that were once well outside the bounds of commercial viticulture (such as Scandinavia).

Varieties grown in vineyards that are located more polewards, or higher in latitude, exhibit different characteristics for the same variety in lower latitudes: sugar and thus potential alcohol decreases; the fruit style moves from full and lush through juicy to lean and crisp; the types of fruit flavour associated with the wine progresses from tropical through soft-orchard fruit to hard-orchard fruit in white wines; and from figs, prunes, and dried-fruits through black soft-fruits to red berries and then to cherries in red wines; while the body changes from full through medium to light. These are natural trends, and they can be modified, of course, albeit within certain limits, by the choice of site, clone, rootstock, training, date of picking, and vinification. All thing being equal, however, the difference in character will remain constant.

Most of the world's finest wines are produced in west coast locations, which tend to be cooler and less humid than east coast areas, due to ocean currents and their accompanying winds (*see* Ocean Currents on opposite page). Forests and mountain ranges protect the vines from wind and rain. A relatively close proximity to forests and large masses of water can influence the climate through transpiration and evaporation, providing welcome humidity in times of drought, although they can encourage rot. Thus, some factors can have both positive and negative effects.

KEY TO TASTE AND QUALITY SYMBOLS

How good a wine producer is and why a wine tastes the way it does are the twin pillars of this encyclopedia; thus, Factors Affecting Taste and Quality and the Star Rating System are regular features. The taste guides are found in Appellations, Producers, and

Wine Styles. Stars are given for a producer's general quality and do not necessarily apply to each wine that that a producer makes. When a producer listing has no stars, this indicates an acceptable but not outstanding standard.

TASTE GUIDE SYMBOLS

❦ Grape varieties used

♊ When to drink for optimum enjoyment; usually given as a range (eg 3–7 years) from the date of the vintage. "Upon purchase" is for wines that should be both bought and drunk while young. "Upon opening" is for fortified wines that can keep for many years but do not keep well once opened.

✓ Recommended producers, vineyards, and wines

♛ The 10 (or fewer) greatest wines for the type or style indicated

STAR RATING SYSTEM

☆ Open-stars indicate an intermediate rating equivalent to a half star.

★ Wines that excel within their class or style

★★ Very exceptional wines. The local equivalent of a Bordeaux super-second. *See* Micropedia.

★★★ The best wines that money can buy, without any allowances made for limitations of local style and quality

OTHER SYMBOLS

❿ A producer of organic wines

❷ A producer of biodynamic wines

❤ Wines of exceptional value for money, whether they are inexpensive or not

The following are seldom used, as there is a bias towards including producers than can be recommended.

❶ Wines that are inconsistent, too rustic, or judgement has been reserved for a stated reason

❌ Underperformers for their category or style

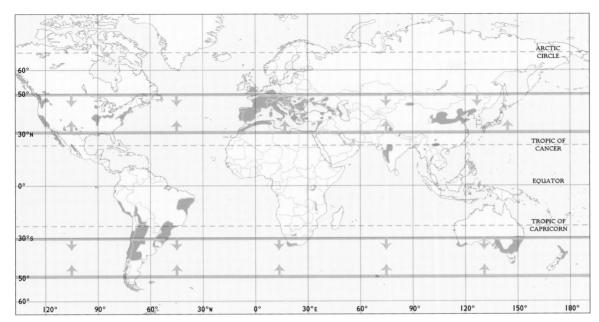

WINE BY LATITUDE
The most important areas of cultivation in both the Northern and Southern Hemispheres lie mainly between latitudes 30° and 50°.

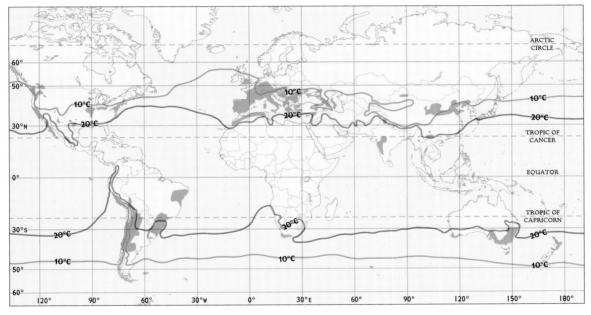

WINE BY TEMPERATURE ZONES
Vitis vinifera is best able to produce premium-quality wine grapes in areas where temperatures range between 10°C (50°F) and 20°C (68°F). In this map, with the boundaries adjusted for temperature rather than latitude, it is clear that these zones contain most of the world's vineyards. The factors that determine the widening and narrowing of this band include altitude and ocean currents (*see* below).

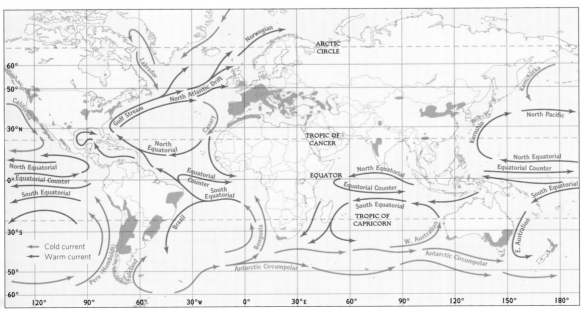

OCEAN CURRENTS
The ocean's currents link together to form the Global Ocean Conveyor Belt, which effectively controls the global climate and thereby ultimately dictates where grapevines can be cultivated. The currents work by a mechanism known as thermohaline circulation, which is driven by temperature (thermo) and salt (haline), although they are also tweaked by the pull of the moon (which should please the biodynamists). Compare all three maps on these pages, and it is easy to see how the Peru (or Humboldt) Current can push the 20°C (68°F) isotherm so far up the South American coast and how the Gulf Stream enables viticulture to survive in England.

Climate

Beyond the grape variety itself, climate or, more acutely, weather is the most important factor influencing the character and quality of wine. Climate determines if grapes should be grown, whereas weather dictates how they are grown. Climate is time-averaged weather, and weather is the fluctuation of climate at any given moment. Climate is thus predictable, whereas weather is effectively unpredictable beyond a week or two. In other words, climate is what the weather should be, while weather is what it actually is.

A grower must select a region with an amenable climate and hope that nature does not inflict too many anomalies. Although some vines survive under extreme conditions, most – and all classic – grapevines are confined to two relatively narrow climatic bands, as illustrated under Location (*see* pp18–19), and require a sympathetic combination of heat, sunshine, rain, and frost.

HEAT

Vines will not provide grapes suitable for winemaking if the annual mean temperature is less than 10°C (50°F). The ideal mean temperature is 14° to 15°C (57° to 59°F), with an average of no less than 19°C (66°F) in the summer and -1°C (30°F) in the winter. In order for the vines to produce a good crop of ripe grapes, the minimum heat-summation, measured in "degree-days" with an average of above 10°C (50°F) over the growing season, is 1,000° (using °C to calculate) or 1,800° (using °F to calculate). The chart at the top of the opposite column shows the degree-day totals over the growing season for a variety of viticultural regions from around the world.

WINTER SNOWS BLANKET THE PIEDMONT REGION IN NORTHERN ITALY
Ideal growing conditions for grapes require an annual mean temperature of at least 10°C (50°F). A winter average of no less -1°C (30°F) is also a critical factor.

AREA/REGION	DEGREE-DAYS, CELSIUS (FAHRENHEIT)
Trier, Mosel, Germany	945 (1,700)
Bordeaux, France	1,320 (2,375)
McLaren Vale, South Australia	1,350 (2,425)
Russian River, California, United States	2,000 (3,600)

SUNSHINE

Light is required for photosynthesis – the most important biological process of green plants –and there must be sufficient light for this, even in cloudy conditions. For vinegrowing, however, sunshine is needed more for its heat than its light. Approximately 1,300 hours is the minimum amount of sunshine required per growing season, but 1,500 hours is preferable.

AREA/REGION	HOURS OF SUNSHINE
Bordeaux, France	1,427 hours
Trier, Mosel, Germany	1,576 hours
McLaren Vale, South Australia	1,765 hours
Russian River, California, United States	2,100 hours

PUFFY CLOUDS FILTER THE SUN'S RAYS IN A VINEYARD IN SOUTHERN STEIERMARK (STYRIA), AUSTRIA
Grapevines need sufficient sunlight, but it is the heat of the sun more than its light that will make the difference in how well they grow.

CLIMATIC CONDITIONS

There are many factors that determine whether climatic conditions in a given area will be favourable for cultivating wine grapes.

FAVOURABLE
- A fine, long summer with warm, rather than hot, sunshine ensures that the grapes ripen slowly. Such weather leads to a good acid-sugar balance.
- A dry, sunny autumn is essential for ripening grapes and avoiding rot; but, again, it must not be too hot.
- The winter months from November to February (May to August in the Southern Hemisphere) are climatically flexible, with the vine able to withstand temperatures as low as -20°C (-4°F) and anything other than absolute flood or drought.
- Within the above parameters, the climate must suit the viticultural needs of specific grape varieties – for example, a cooler climate for Riesling, hotter for Syrah, and so on.

UNFAVOURABLE
- Major dangers are frost, hail, and strong winds, all of which can denude a vine and are particularly perilous when the vine is flowering or the grapes are ripening and at their most susceptible.

- Rain and/or cold temperatures during flowering may cause imperfect fertilization, which results in a physiological disorder called *millerandage*. The affected grapes contain no seeds and will be small and only partially developed when the rest of the cluster is fully matured.

- Persistent rain at, or immediately before, the harvest can lead to rot or dilute the wine, both of which can cause vinification problems.

- Sun is not often thought of as a climatic danger, but just as frost can be beneficial to the vine, so can sun be harmful. Too much encourages sap to go straight past the embryo grape clusters to the leaves and shoots. This causes a physiological disorder called *coulure*, which is often confused with *millerandage*. It is totally different, although both disorders can appear together due to the vagaries of climate. With *coulure*, either the flowers do not pollinate, so they do not become berries, or immature grapes or berries drop to the ground, and those that remain do not develop.

- Excessive heat during the harvest rapidly decreases the acid level of grape juice and makes the grapes too hot, creating problems during fermentation. It is especially difficult to harvest grapes at an acceptable temperature in very hot areas, such as South Africa. As a result, some wine estates harvest the grapes at night, when the grapes are at their coolest.

SHRIVELLED, BROWNING LEAVES SHOW EVIDENCE OF DAMAGE FROM AN EARLY-SPRING FROST
Frost damage primarily occurs in spring during a grapevine's early growth phase. During this phase, frost can kill the newly budding foliage or partially kill the vulnerable shoots and inflorescences, which can result in significant crop losses.

RAINFALL

A vine requires 68 centimetres (27 inches) of rain per year. Ideally, most of the rain should fall in the spring and the winter, but some is needed in the summer too. Vines can survive with less water if the temperature is higher, although rain in warm conditions is more harmful than rain in cool conditions. A little rain a few days before the harvest will wash the grapes of any sprays and is therefore ideal if followed by sun and a gentle, drying breeze. Torrential rain, however, can split berries and cause fungus. Below are annual rainfall figures for a variety of viticultural regions from around the world.

AREA/REGION	RAINFALL
McLaren Vale, South Australia	60 centimetres (24 inches)
Trier, Mosel, Germany	65 centimetres (26 inches)
Bordeaux, France	90 centimetres (36 inches)
Russian River, California, United States	135 centimetres (53 in ches)

FROST

Surprising as it may seem, some frost in a vineyard is desirable, providing it is in the winter, because it hardens the wood and kills spores and pests that the bark might be harbouring. Frost, however, can literally kill a vine, particularly at bud-break and flowering (*see* p 36).

VINTAGE

The anomalies of a vintage can bring disaster to reliable vineyards and produce miracles in unreliable ones. A vintage is made by weather, as opposed to climate. Although the climate may be generally good, uncommon weather conditions can sometimes occur. In addition to this, the vintage's annual climatic adjustment can be very selective; on the edge of a summer hailstorm, for example, some vineyards may be destroyed and produce no wine at all, while others are virtually unharmed and produce good wine. Vines situated between the two might be left with a partial crop of fruit that could result in wines of an exceptional quality if given a further two to three months of warm sunshine before the harvest, because reduced yields per vine produce grapes with a greater concentration of flavour.

Aspect

The aspect of a vineyard refers to its general topography – which direction the vines face, the angle and height of any slope, and so on – and how this interrelates with the climate.

There are few places in the world where winemaking grapes – as opposed to table grapes – are successfully grown under the full effect of a given region's prevailing climate. The basic climatic requirements of the vine are instead usually achieved by manipulating local conditions, keeping sunshine, sun strength, drainage, and temperature in mind.

SUNSHINE
In the Northern Hemisphere, south-facing slopes (and north-facing slopes in the Southern Hemisphere) attract more hours of sunshine and are therefore cultivated in cooler areas. In hotter regions, the slopes facing the opposite direction tend to be cultivated.

SUN STRENGTH AND DRAINAGE
Because of the angle, vines on a slope absorb the greater strength of the sun's rays. In temperate regions the sun is not directly overhead, even at noon, so its rays fall more or less perpendicular to a slope. Conversely, on flat ground the sun's rays are dissipated across a wider area, so their strength is diluted. (The plains are also susceptible to flooding and have soils that are usually too fertile, yielding larger crops of correspondingly inferior fruit.) Lake-valley and river-valley slopes are well suited for vines because rays are also reflected from the water.

A sloping vineyard also affords natural drainage. Hilltop vines are too exposed to wind and rain, however, and their presence, instead of a forest covering, deprives vines below of protection. Forested hilltops not only supply humidity in times of drought, but absorb the worst of any torrential rain that could wash away the topsoil below.

TEMPERATURE
Slopes are very desirable sites, but keep in mind that for every 100 metres (330 feet) above sea level, the temperature falls 1°C (1.8°F). This can result in an extra 10 to 15 days being needed for the grapes to ripen, and because of the extra time, the acidity will be relatively higher. A vineyard's altitude can thus be a very effective way of manipulating the quality and character of its crop. Riverside and lakeside slopes also have the advantages of reflected sunlight and the water acting as a heat reservoir, at night releasing heat that has been stored during the day. This not only reduces sudden drops in temperature that can be harmful, but also lessens the risk of frost. Depressions in slopes and the very bottoms of valleys collect cold air, are frost-prone, and slow growth, however.

VINES CLING TO THE STEEP RIVER BANKS OF THE REICHSBURG COCHEM ON THE MOSEL RIVER IN THE COCHEM-ZELL DISTRICT OF GERMANY
The hillsides of this serpentine river are blanketed with vineyards. With sharply rising banks, these vineyards are amongst the steepest in the world, with some planted at a precipitous 70-degree gradient. Inclines such as these call for nearly all of the grapes to be picked by hand, demonstrating how aspect can affect viticultural practises.

Soil

Topsoil is of primary importance to the vine because it supports most of its root system, including a majority of the feeding network. Subsoil always remains geologically true. Main roots penetrate several layers of subsoil, whose structure influences drainage, the root system's depth, and its ability to collect minerals.

The metabolism of the vine is well known, and the interaction between it and the soil is generally understood. The ideal medium in which to grow vines for wine production is one that has a relatively thin topsoil and an easily penetrable (and therefore well-drained) subsoil with good water-retaining characteristics. The vine does not like "wet feet", so drainage is vital, yet it needs access to moisture, so access to a soil with good water retention is also important. The temperature potential of a soil, its heat-retaining capacity, and its heat-reflective characteristics affect the ripening period of grapes: warm soils (gravel, sand, loam) advance ripening, while cold soils (clay) retard it. Chalk falls between these two extremes, and dark, dry soils are obviously warmer than light, wet soils. High-pH (alkaline) soils, such as chalk, encourage the vine's metabolism to produce sap and grape juice with a relatively high acid content. The continual use of fertilizers has lowered the pH level of some viticultural areas in France, and these are now producing wines of higher pH (less acidity).

THE MINERAL REQUIREMENTS OF THE VINE

Just as various garden flowers, shrubs, and vegetables perform better in one soil type as opposed to another, so too do different grape varieties. Certain minerals essential to plant growth are found in various soils. Apart from hydrogen and oxygen (which are supplied as water), the most important soil nutrients are nitrogen, which is used in the production of a plant's green matter; phosphate, which directly encourages root development and indirectly promotes an earlier ripening of the grapes (an excess inhibits the uptake of magnesium); potassium, which improves the vine's metabolism, enriches the sap, and is essential for the development of the following year's crop; iron, which is indispensable for photosynthesis (a lack of iron will cause chlorosis); magnesium, which is the only mineral constituent of the chlorophyll molecule (lack of magnesium also causes chlorosis); and calcium, which feeds the root system, neutralizes acidity, and helps create a friable soil structure (although an excess of calcium restricts the vine's ability to extract iron from the soil and therefore causes chlorosis).

GUIDE TO
VINEYARD SOILS

To the wine amateur, the details of geology are not always important; what matters is how soil affects the growth of vines. If one clay soil is heavier or more silty, sandy, or calcareous, that is relevant. But there is enough jargon used when discussing wine to think of mixing it with rock-speak.

Acid soil Any soil that has a pH of less than 7 (neutral). Typical acidic soils that are acidic due to their parent rock include brown or reddish-brown, sandy loams or sands, volcanic soils, and any igneous or silicate-rich soil. Neutral soils can become acidic from too much humus or acid rain. Acid soils are low in calcium and magnesium, with negligible amounts of soluble salts and reduced phosphorous availability.

Aeolian soil Sediments deposited by wind (eg loess).

Albariza White-surfaced soil formed by diatomaceous deposits, found in southern Spain.

Alberese A compact clay and limestone found in the Chianti region.

Albero Synonymous with *albariza*.

Albian A type of schist found in Maury, Roussillon.

Alkaline soil Any soil that has a pH of more than 7 (neutral). Typical alkaline soils include chalk and any calcareous soils.

Alluvial deposits (noun – alluvium) Material that has been transported by river and deposited. Most alluvial soils contain silt, sand, and gravel and are highly fertile.

Aqueous rocks One of the three basic rock forms (*see* Rock). Also called sedimentary or stratified.

Arenaceous rocks Formed by the deposits of coarse-grained particles, usually siliceous, and often decomposed from older rocks (eg, sandstone).

Arène A coarse, granitic sand ideally suited to the Gamay, *arène* is found in the Beaujolais region.

Argillaceous soils This term covers a group of sedimentary soils, commonly clays, shales, mudstones, siltstones, and marls.

Argovian marl A chalky, clay-like marl found in many parts of the Côte des Beaune.

Arkose A red, Triassic sandstone consisting of feldspar, quartz, and clay minerals, *arkose* is often found in the Côtes d'Auvergne and parts of Beaujolais (eg, St-Amour).

Aubuis Found in the Touraine district of the Loire and highly rated for Chenin Blanc in Vouvray and Montlouis, *aubuis* is a stony mix of permeable, fertile, calcareous clays that are said to be well suited to white grape varieties.

Barro A similar soil to *albariza* but brown in colour, sandier, and with less diatomaceous content. While Palomino grapes are grown on *albariza* soil, barro is reserved for Pedro Ximénez grapes.

BARRO
The rich brown soil of the Tierra de Barros wine region in Spain is conducive to the growth of Pedro Ximénez.

Basalt material This accounts for as much as 90 per cent of all lava-based volcanic rocks. It contains various minerals, is rich in lime and soda, but not quartz, the most abundant of all minerals, and it is poor in potash.

Bastard soil A *bordelais* name for medium-heavy, sandy-clay soil of variable fertility.

Bauxite As well as being a valuable ore mined for aluminium production, bauxite is found in limestone soils of Coteaux de Baux-de-Provence.

Block-like soil Referring to the soil structure, "block-like" indicates an angular or slanting arrangement of soil particles.

Boulbènes A *bordelais* name for a very fine siliceous soil that is easily compressed and hard to work. This "beaten" earth covers part of the Entre-Deux-Mers plateau in in Bordeaux.

Boulder *See* Particle size.

Calcareous clay An argillaceous soil with carbonate of lime content that neutralizes the clay's intrinsic acidity. Its low temperature also delays ripening, so wines produced on this type of soil tend to be more acidic.

Calcareous soil This label is for any soil, or mixture of soils, with an accumulation of calcium and magnesium carbonates. Essentially alkaline, it promotes the production of acidity in grapes, although the pH of each soil will vary according to its level of "active" lime. Calcareous soils are cool, with good water retention. With the exception of calcareous clays (*see* above), they allow the vine's root system to penetrate deeply and provide excellent drainage.

Carbonaceous soil Soil that is derived from rotting vegetation under anaerobic conditions. The most common carbonaceous soils are peat, lignite, coal, and anthracite.

CHALK
A French vineyard shows the pale colour of its chalky soil.

Chalk A type of limestone, chalk is a soft, cool, porous, brilliant-white, sedimentary, alkaline rock that encourages grapes with a relatively high acidity level. It also allows the vine's roots to penetrate and provides excellent drainage, while at the same time retaining sufficient moisture for nourishment. One of the few finer geological points that should be adhered to is that which distinguishes chalk from the numerous hard limestone rocks that do not possess the same physical properties.

Clay A fine-grained argillaceous compound with malleable, plastic characteristics and excellent water-retention properties. It is, however, cold, acid, offers poor drainage, and, because of its cohesive quality, is hard to work. An excess of clay can stifle the vine's root system, but a proportion of small clay particles mixed with other soils can be advantageous.

Clayey-loam A very fertile version of loam, but heavy to work under wet conditions, with a tendency to become waterlogged.

Coal Rarely seen as a vineyard soil, except for Chardonnay vines grown on one of the slagheaps at Haillicourt, near Béthune, in Pas-de-Calais.

Cobble See Particle size.

Colluvial deposits (noun – colluvium) Weathered material transported by gravity or hill-wash.

Crasse de fer Iron-rich hard-pan found in the Libournais area of France. Also called machefer.

Crystalline May be either igneous (eg granite) or metamorphic.

Dolomite A calcium-magnesium carbonate rock. Many limestones contain dolomite.

Entroques Type of hard limestone found in Burgundy (Montagny, for example).

Feldspar (or Felspar) One of the most common minerals, feldspar is a white- or rose-coloured silicate of either potassium-aluminium or sodium-calcium-aluminium and is present in a number of rocks, including granite and basalt.

Ferruginous clay Iron-rich clay.

Flint A siliceous stone that stores and reflects heat and is often associated with a certain "gun-flint" smell that sometimes occurs in wines, although this is not actually proven and may simply be the taster's auto-suggestion.

Gabbro A dark, coarse-grained igneous rock found in Muscadet.

Galestro Rocky, schistous clay soil commonly found in most of Tuscany's best vineyards.

Glacial moraine A gritty scree that has been deposited by glacial action.

Gore A pinkish, decomposed, granitic arenaceous soil found in Beaujoalais, St-Joseph, and Côtes Roannaise.

Gneiss A coarse-grained form of granite.

Granite A hard, mineral-rich rock that warms quickly and retains its heat. Granite contains 40 to 60 per cent quartz and 30 to 40 per cent potassium feldspar, plus mica or hornblende, and various other minerals. It has a high pH that reduces wine acidity. Thus, in Beaujolais, it is the best soil for the acidic Gamay grape. It is important to note that a soil formed from granite is a mixture of sand (partly derived from a disintegration of quartz and partly from the decomposition of feldspar with either mica or hornblende), clay, and various carbonates or silicates derived from the weathering of feldspar, mica, or hornblende.

Gravel A wide-ranging term that covers siliceous pebble of various sizes that are loose, granular, airy, and afford excellent drainage. Infertile, it encourages the vine to send its roots down deep in search of nutrients. Gravel beds above limestone subsoils produce wines with markedly more acidity than those above clay.

Greensand A dark greenish coloured, glauconite-rich sand of Cretaceous origin found in some vineyards in southeast England. Greensand is used as a water softener, which is ironic considering that it is found over chalk subsoil, known for its hard water.

Greywacke Argillaceous rocks that could have been formed as recently as a few thousand years ago by rivers depositing mudstone, quartz, and feldspar. Commonly found in Germany, South Africa, and New Zealand.

Gypsum Highly absorbent, hydrated calcium-sulphate that was formed during the evaporation of sea-water.

Gypsiferous marl A marly soil permeated with Keuper or Muschelkalk gypsum fragments, which improve the soil's heat-retention and water-circulation properties.

Hard-pan A dense layer of clay that forms if the subsoil is more clayey than the topsoil at certain depths. Hard-pans are impermeable to both water and roots, so they are not desirable too close to the surface but may provide an easily reachable water-table if located deep down. A sandy, iron-rich hard-pan known as iron-pan is commonly found in parts of Bordeaux.

Hornblende A silicate of iron, aluminium, calcium, and magnesium, it constitutes the main mineral found in basalt and is a major component of granite and gneiss.

Humus Organic material that contains bacteria and other micro-organisms that are capable of converting complex chemicals into simple plant foods. Humus makes soil fertile; without it, soil is nothing more than finely ground rock.

Igneous rock One of the three basic rock forms (see Rock), igneous rocks are formed from molten or partially molten material. Most igneous rocks are crystalline.

Iron-pan A sandy, iron-rich hard-pan.

Jory A volcanic soil, primarily basalt, which is in turn a hard and dense soil that often has a glassy appearance. One of the two primary soil types found in Oregon's Willamette Valley, particularly on the lower foothills, such as the Dundee Hills, where Pinot Noir excels.

GALESTRO
A vineyard worker in Pisa displays a handful of galestro soil. This rocky soil also appears in many Tuscan vineyards.

WENTWORTH-UDDEN SCALE OF PARTICLE SIZE

NAME		SIZE RANGE (millimetre/micrometre)	SIZE RANGE (approx inches)
Very coarse soil	Boulder	>256mm	>10.1in
	Cobble	64–256mm	2.5–10.1in
	Pebble	4–64mm	1.26–2.5in
Coarse soil	Gravel	2–4mm	0.079–0.157in
	Sand	62.5μm–2mm	0.0025–0.079in
Fine soil	Silt	3.9–62.5μm	0.00015–0.0025in
	Clay	0.98–3.9μm	3.8×10^{-5}–0.00015in

Keuper A term often used when discussing wines in Alsace, Keuper is a stratigraphic name for the Upper Triassic period and can mean marl (varicoloured, saliferous grey, or gypsiferous grey) or limestone (ammonoid).

Kimmeridgian soil A greyish-coloured limestone originally identified in, and so named after, the village of Kimmeridge in Dorset, England. A sticky, calcareous clay containing this limestone is often called Kimmeridgian clay.

Lacustrine limestone A freshwater limestone that forms at the bottom of lakes. Lacustrine-limestone soils have been found on Pelee Island and the Niagara district of Ontario, Yakima Valley in Washington, and Quincy in the Loire Valley.

Lignite The "brown coal" of Germany and the "black gold" of Champagne, this is a brown carbonaceous material intermediate between coal and peat. Warm and very fertile, it is mined and used as a natural fertilizer in Champagne.

Limestone Any sedimentary rock consisting essentially of carbonates. With the exception of chalk, few limestones are white; instead grey- and buff-coloured are probably the most common hues found limestone in wine areas. The hardness and water retention of this rock vary, but being alkaline limestone generally encourages the production of grapes with a relatively high acidity level. *See also* Lacustrine limestone.

Llicorella The Catalan name for a black slate and quartz soil found in Priorat, Spain.

Loam A warm, soft, crumbly soil with roughly equal proportions of clay, sand, and silt. It is perfect for large-cropping mediocre-quality wines but too fertile for fine wines.

Loess An accumulation of wind-borne, mainly silty material, that is sometimes calcareous but usually weathered and decalcified. Loess warms up relatively quickly and also has good water-retention properties.

Machefer *See Crasse de fer.*

Macigno Hard grey-blue sandstone found in the Chianti region.

Marl A cold, calcareous clay-like soil (usually 50 per cent clay content) that delays ripening and adds acidity to wine.

Marlstone Clayey limestone that has a similar effect to marl.

Metamorphic rock One of the three basic categories of rock (*see* Rock), this type is caused by great heat or pressure, often both.

Mica A generic name encompassing various silicate minerals, usually in a fine, decomposed-rock format.

Millstone Siliceous, iron-rich, sedimentary rock.

Moraine *See Glacial moraine.*

Mudstone A sedimentary soil similar to clay but without its plastic characteristics.

Muschelkalk Often used when discussing wines in Alsace, Muschelkalk is a stratigraphic name for the Middle Triassic period and can mean anything from sandstone (shelly, dolomitic, calcareous, clayey, pink, yellow, or millstone) to marl (varicoloured or fissile), dolomite, limestone (crinoidal or grey), and shingle.

Oolite A type of limestone.

Oolith A term used for small, round, calcareous pebbles that have grown through fusion of very tiny particles.

Palus A *bordelais* name for a very fertile soil of modern alluvial origin that produces medium-quality, well-coloured, robust wines.

Particle size The size of a rock determines its descriptive name. No handful of soil will contain particles of a uniform size, unless it has been commercially graded, of course, so all such descriptions can only be guesstimates, but it is worth noting what they should be, otherwise you will have nothing to base your guesstimates on. According to the Wentworth-Udden scale, they

are: boulder (greater than 256 millimetres), cobble (64 to 256 millimetres), pebble (4 to 64 millimetres), gravel (2 to 4 millimetres), sand (1/16 to 2 millimetres), silt (1/256 to 1/16 millimetre) and clay (smaller than 1/256 millimetre). Notice that even by this precise scale, Wentworth and Udden have allowed overlaps, thus a 1/16mm particle might either be sand or silt and, of course, sub-divisions are possible within each group, as there is such a thing as fine, medium, or coarse sand and even gritty silt (*see* chart at left).

Pebble *See* Particle size.

Pelite Fine-grained clayey-quartz sedimentary rock found in Banyuls in the Roussillon region of France.

Peperite Limestone or marly rock found on Madeira and along Idaho's Snake River Valley that has been ejected by volcanic activity and is literally "peppered" with tiny peppercorn-like grains of basalt.

Perlite A fine, powdery, light, and lustrous substance of volcanic origin with similar properties to diatomaceous earth.

Perruches Very stony, flinty clays combined with silica, perruches soils warm up quickly and are said to be why Sauvignon Blanc grapes grown on them have a flinty taste.

Phtanite Dark-coloured sedimentary rock bearing stratas of quartz crystals, found in Savennières and Coteaux du Layon.

Platy soil Referring to the soil structure, "platy" indicates a horizontal alignment of soil particles.

Porphyry A coloured igneous rock with high pH.

Precipitated salts A sedimentary deposit. Water charged with acid or alkaline material, under pressure of great depth, dissolves various mineral substances from rocks on the sea-bed, which are then held in solution. When the water flows to a place of no great depth or is drained away or evaporates, the pressure is reduced, the minerals are no longer held in solution and precipitate in deposits that may be just a few centimetres or several thousand metres deep. There are five groups: oxides, carbonates, sulphates, phosphates, and chlorides.

SILEX
This hard soil, a mix of flint, clay, and limestone, makes up the *terroir* of this Provençal vineyard. It also appears in the Loire Valley of France.

TERRA ROSSA
A vineyard in the Istria region of Croatia and Slovenia is planted in earthy red *terra rossa*. This type is also found in the wine regions of La Mancha, Spain, and Coonawarra, Australia.

Prism-like soil Referring to the soil structure, "prism-like" indicates a columnar or vertical arrangement of soil particles.

Pudding stones A term used for a large, heat-retaining conglomerate of pebbles.

Quartz The most common and abundant mineral, quartz is the crystalline form of silica. It is found in various sizes and in almost all soils, although sand and coarse silt contain the largest amount. Quartz has a high pH, which reduces wine acidity, but quartz that is pebble-sized or larger, stores and reflects heat, which increases alcohol potential.

Red earth *See Terra rossa.*

Rock A rock may be loosely described as a mass of mineral matter. There are three basic types of rock: igneous, metamorphic, and sedimentary (or aqueous or stratified).

Ruedas Red sandy-limestone soil found in the Montilla-Moriles region of Spain.

Ruffe A fine-grained, brilliant-red sandstone soil rich in iron-oxide, ruffe is found in parts of the Languedoc region of France, particulary the Vin de Pays des Coteaux de Salagou.

Safres A sandy-marl found in the southern Rhône Valley.

Saibro A decomposed red tufa soil that is highly regarded in Madeira.

Sand Tiny particles of weathered rocks and minerals that retain little water but constitute a warm, airy soil that drains well and is supposedly phylloxera-free.

Sandstone Sedimentary rock composed of sand-sized particles that have either been formed by pressure or bound by various iron minerals.

Sandy-loam Warm, well-drained, sand-dominated loam that is easy to work and suitable for early-cropping grape varieties.

Schist Heat-retaining, coarse-grain, laminated, crystalline rock that is rich in potassium and magnesium but poor in nitrogen and organic substances.

Scree Synonymous with colluvium deposits.

Sedimentary rock One of the three basic rock forms (see Rock), it includes arenaceous (eg, sandstone), argillaceous (eg, clay), calcareous (eg, limestone), carbonaceous (eg, peat, lignite, or coal), and siliceous (eg, quartz) and the five groups of precipitated salts, (oxides, carbonates, sulphates, phosphates, and chlorides). Sedimentary rocks are also called aqueous or stratified.

Shale Heat-retaining, fine-grain, laminated, moderately fertile sedimentary rock. Shale can turn into slate under pressure.

Shingle Pebble- or gravel-sized particle rounded by water-action.

Silex A hard, flint, clay and limestone rock famously promoted by Didier Dagueneau and others in Pouilly-Fumé in the Loire Valley.

Siliceous soil A generic term for acid rock of a crystalline nature. It may be organic (such as flint) or inorganic (quartz) and have good heat retention, but no water retention unless found in a finely ground form in silt, clay, and other sedimentary soils. Half of the Bordeaux region is covered with siliceous soils.

Silt A very fine deposit, with good water retention. Silt is more fertile than sand but is cold and offers poor drainage.

Slate Hard, often dark grey (but can be any colour between brown and bluish grey), fine-grain, plate-like rock formed under pressure from clay, siltstone, shale, and other sediments. It warms up quickly, retains its heat well, and is responsible for many fine wines, most notably from the Mosel.

Slaty-schist A sort of half-formed slate created under lower temperature and pressure than fully formed slate.

Spiroidal soil Referring to the soil structure, "spiroidal" indicates a granular or crumb-like composition of soil particles.

Steige A type of schist found on the north side of Andlau in Alsace, it has metamorphosed with the Andlau granite and is particularly hard and slaty. It has mixed with the granitic sand from the top of the Grand Cru Kastelberg and makes a dark, stony soil.

Stone This word should be used with rock types, such as limestone and sandstone, but is often used synonymously with pebble.

Stratified rock One of the three basic rock forms (see Rock); also called sedimentary or aqueous.

Terra rossa A red, clay-like, sometimes flinty sedimentary soil that is deposited after carbonate has been leached out of limestone. It is often known as "red earth".

Terres blanche Steep Kimmeridgian marls in Sancerre.

Tufa A limestone concretion that forms via water dripping through gaps in limestone, tufa is typical of the soil of Orvieto, Umbria, and is also found in Montalcino, Tuscany, as well as the Langhe region of Piedmont.

Tuff Rocks formed by fractured or water-bound material ejected by volcanic activity, tuff drains well and is found in Taburno, Campania, in Italy; Balatonfüred-Csopak, Balatonfelvidék and Balatonboglár around Lake Balaton in Hungary; and the Galilee region of Israel, particularly Upper Galilee and the Golan Heights.

Tuffeau A buff-coloured, sandstone-rich, otherwise chalky limestone as found in the Loire, particularly around Touraine, and used in the construction of many of its châteaux.

Volcanic soils Derived from two sources, volcanic soils are lava-based (the products of volcanic flow) and vent-based (material blown into the atmosphere). Some 90 per cent of lava-based rocks and soils are comprised of basalt, while others include andesite, pitchstone, rhyolite, and trachyte. Vent-based matter has either been ejected as molten globules, cooled in the air, and dropped to earth as solid particles (pumice), or as solid material and fractured through the explosive force with which it was flung (tuff).

Willakenzie A silty clay-loam colluvium, this is one of the two primary soil types found in Oregon's Willamette Valley.

Viticulture and Vinification

Winemakers get all the glory, but their success depends on the grapes they use. Vinification is an art and a science, but it is the art and science of damage limitation. Viticulture is all about preparation and providing the potential, which is why those who control the vineyards of successful wines are the true wine heroes.

It is always possible to raise the quality in the vineyard, often by a significant amount, but once the grapes are harvested, the potential of the wine is capped and cannot be raised. The winemaker's job is, therefore, one of containment and, when things do not go right, mounting a search-and-rescue mission. As the late Johnny Hugel once put it, "As soon as the grape is removed from the vine, it is exposed to the air, the clock starts ticking, and the winemaker's job becomes a race against time."

In the ideal *domaine,* where the grapes are carefully picked by hand, placed into small trays to prevent them from being crushed under their own weight, and the vines are close enough to the winery that all the trays are delivered to the press house within an hour (which is highly unlikely), the winemaker is already dealing with grapes that have effectively started to deteriorate. Even in a perfectly run ideal winery, the grapes probably represent no more than 95 to 97 per cent of their potential at the time of picking, and the most talented winemakers on earth would struggle to make a wine that goes into the bottle at anything close to 90 per cent, with most wines much closer to 80 per cent. By comparison, it is relatively easy to double the potential quality in the vineyard through the choice of grape variety and/or specific clones of that grape variety, the choice of rootstock, where the vines are planted, the planting density, how the vines are pruned and cared for, and when and how they are harvested. Instead of struggling to produce a 90 per cent potential wine, after improvements in the vineyard, a few years later the winemaker could put his or her feet up and churn out something closer to the equivalent of 180 per cent of the quality he or she used to deliver.

It is the variety of grape that determines the basic flavour of a wine, but it is the way the variety is grown that has the most profound effect on the quality of the wine.

PINOT NOIR GRAPES ON A CONVEYOR BELT BEGIN THEIR JOURNEY THROUGH THE WINEMAKING PROCESS
Once grapes are separated from their vines, it is a race against the clock to keep the effects of natural deterioration at bay. Viticultural and vinification decisions are based on minimizing any damage in order to produce the best wines possible from a given harvest.

Viticultural Management Systems

The management of vineyards is a science in itself, a branch of horticulture specifically devoted to grapevines. The duties of the viticulturalist include monitoring and controlling pests and diseases, fertilization, irrigation, canopy management, monitoring fruit development and characteristics, helping to decide the best time to harvest, and maintaining the vines during the fallow winter months. Viticulturists often work closely with winemakers because vineyard management is the foundation on which winemaking exists.

Beginning with the variety of grape and then looking at the site specifics (microclimate, aspect, slope, and desired style of wine), different trellising systems may be employed to gain maximum benefit from individual vines in specific circumstances. This is part of canopy management, giving the vine and – more important, the ripening grapes – access to maximum sun for photosynthesis and ripening.

The trellis supports the weight of the trunk and the fruit, and it can be as simple as a single wire in a high cordon or a number of wires, for example, a Lyre system. It is a major investment in time and materials, and mistakes at the outset will cost dearly further down the line.

CONVENTIONAL VITICULTURE

This involves all of the above, taking into account many centuries of grape-growing and winemaking experience to take the vine through its annual cycle of growth, reproduction, and harvest to make quality wines. A conventional vineyard has only one crop, the vine, and it is prized and protected above all else. Use of herbicides, insecticides, and pesticides is common to control the many pests and diseases that vines are susceptible to and to maximise yield and quality. There is no certification or labelling requirement for conventional wines (although many appellations have their own strict rules) and some in the "alternative" camp have suggested that "chemgro" would be appropriate.

Although there are only legislative limits in chemical usage in conventional farming, one should not forget that these sprays are also expensive. Any unnecessary excess cuts into the profit of the vineyard, so it would be foolish to imagine conventional viticulturalists as chemical-loving maniacs trying to poison all of us.

ORGANIC VITICULTURE

This looks at the vineyard holistically and rejects the use of certain synthetic chemicals and fertilisers, focussing on strengthening the vine naturally to withstand pests and disease and developing a healthy soil and balanced ecosystem in the vineyard.

Biodiversity is encouraged, and secondary crops, often planted between the vines, help to warn of disease and regulate water intake, offset weeds, and divert pests. Secondary crops also encourage a range of insects, birds, and small animals and make it harder for a single pest or disease to embed itself in the vineyard.

The organic movement is strictly regulated, and wines must be certified to claim organic status on the label. Many wineries are organic in practice but choose not to certify, however, because, apart from the cost, the stamp does not guarantee the wine's quality. It is confusing for consumers and divisive in itself.

THE LABEL PROCLAIMS THAT EMILIANA ADOBE RESERVA SAUVIGNON BLANC IS MADE FROM ORGANICALLY GROWN GRAPES
One of the decisions a viticulturist makes is just how much or how little chemicals he or she will use in the vineyard. Wine producers such as the Emiliana Vineyards in Chile opt for the latter, specialising in creating organically grown and eco-friendly wines. Depending on the country, there are various federal, state, and local organisations that certify the status of a given vineyard, whether it uses organic, biodynamic, or other environmentally sustainable practises.

Critics of the organic movement claim that just because something is naturally occurring, it does not make it better for the consumer's health (like copper sulphate in Bordeaux mixture), and the false idea that organic viticulture is spray-free and natural gives a distorted image to the consumer. Moreover, as chemicals permitted in organic vineyard management are often less potent than the conventional ones, they have to be used more frequently, exposing the workers and the environment to harmful effects more often.

A VINEYARD WORKER FILLS COW HORNS WITH MANURE TO MAKE BIODYNAMIC HORN MANURE
A soil spray made from cow manure is often used in biodynamic vineyards. Horn manure, commonly known as "500", is obtained by stuffing a cow's horn with high-quality manure and then burying it over the winter period, keeping it below the soil surface for six months. After it is dug up, it is mixed with water and then used as a growth-enhancing spray. Whether the practise works or not seems to be just as much a product of a winemaker's care as it is about a wine consumer's sensibilities.

SUSTAINABLE AGRICULTURE

Sustainability is not just a trendy buzz word – it goes to the very heart of the industry. But not everyone is convinced that the catch-all term has value or meaning when applied to viticulture. Sustainable agriculture values soil health and biodiversity as paramount, and in vineyard terms this can enhance the health of vineyards and the concept of *terroir*, as well as building consumer interest and evolving a better work environment.

In essence, this is organic viticulture without the certification, but with an added overlay of consideration for the needs of vineyard workers and minimising the input needed to maintain the vineyard management system.

BIODYNAMIC AGRICULTURE

Biodynamics is part of Rudolf Steiner's wider system of spiritual sciences, or anthroposophy. It is based on the belief that living systems have "formative forces" that start at the farthest reaches of the planetary system. In the vineyard this means planting by the phases of the moon and aligning with ancient Ley lines that connect paths of positive energy. A complex system of herbal sprays and composting techniques is applied in the vineyard.

There are strict standards laid down and certification and labelling is overseen by the Demeter Association. Estimates suggest around 5 per cent of the world's vineyards are certified organic or biodynamic, but it is possible to follow the principles without certifying.

A number of unconventional ceremonial practices, such as burying a cow horn filled with manure at the autumn equinox and digging it up in the spring, also make the biodynamic movement difficult to take seriously. It has been

called the "Hogwarts school of viticulture", and it is not always successful in damp, mildew-prone areas, which have seen significant crop failures.

There is very little scientific evidence to back up the system, but critics have largely been silenced by the quality of some wines produced, and these command a high price premium as a result – expensive to produce and expensive to drink. Some of the iconic (and most expensive) producers are farming their grapes biodynamically, whether they advertise it or not. We can safely say that the close-monitoring biodynamic methods require often result in earlier detection of pests and diseases, so gentler methods will suffice in defending the vines.

BEST PRACTICES PRODUCE BEST RESULTS

Conventional viticulture focuses on the plant as a priority, but organic, biodynamic, and sustainable systems prioritise the health of the soil and the vineyard as a whole and limiting harm to the environment and other living creatures.

Certification and labelling does help consumers to make informed choices, and the "alternative" movement has also captured the imagination of a younger generation of drinkers who are enjoying the artisanal nature of very individual wines rather than conventional, mass-made products.

Setting aside the airy-fairy stuff, it has been acknowledged that much of it is simply good farming practice, with an added feel-good factor. But in order to produce reliable, consistent quality levels and yield at an affordable price, the pragmatic approach – ever the remit of the farmer – may be the one that endures.

Vine Training

The grapevine is a naturally sprawling plant that, without training, would spend much of its energy crawling on the ground putting down roots, instead of bringing fruit.

The manner in which a grapevine is trained will guide the size, shape, and height of the plant towards reaping maximum benefits from the local conditions of aspect and climate. Vines can be trained high to avoid ground frost or low to hug any heat that may be reflected by stony soils at night. There may be a generous amount of space between rows to attract the sun and avoid humidity. On the other hand, vines may be intensively cultivated to form a canopy of foliage to avoid too much sun.

The fundamental reason for training and pruning a vine is to avoid phylloxera and to ensure that the purity of the fruiting stock is maintained. It is crucial to ensure that no cane ever touches the ground. Should one find its way to the soil, its natural inclination is to send out suckers that will put down roots. Within two or three years the majority of a grafted vine's above-ground network would be dependent not upon grafted roots, but upon the regenerated root system of the producing vine. Not only would this put the vine at the mercy of phylloxera, but that part of the vine still receiving its principal nourishment from the grafted hybrid rootstock would also send out its own shoots and, unchecked by any sort of pruning, these would produce hybrid fruit.

STYLES OF VINE TRAINING

Within the two basic systems of cane training and spur training, hundreds of different styles are employed, each developed for a reason, and all having their own advantages and disadvantages. In order to discern the style used in a particular vineyard, it is always best to look at the vines between late autumn and early spring when the branches are not camouflaged with leaves. In the illustrations below, which are not drawn to scale, vines are shown as they appear during their winter dormancy, with the following season's fruiting canes shown in green.

Bush vine – Spur-training system

The bush vine is an unsupported version of the Gobelet system (*see* box, below left). The term *bush vine* originated in Australia, where a few old vineyards – usually planted with Grenache – are still trained in this fashion. Bush vines are traditional in Beaujolais of France (where both supported and unsupported methods are referred to as Gobelet) and they are commonly found throughout the most arid areas of the Mediterranean. Because they are unsupported, the canes often flop downwards when laden with fruit, giving the vine a sprawling, straggly look. In the Beaujolais *crus,* the total number of canes on a vine is restricted to between three and five, but in other, less-controlled wine areas a bush vine may have as many as 10 canes. This method is only suitable for training low-vigour vines.

Bush vine

Chablis – Spur-training system

As the name implies, this style of vine training was originally developed in the Chablis district, although the method employed there now is, in fact, the Guyot Double. Champagne is the most important winemaking region to employ the Chablis system for training vines; there, it is used for more than 90 per cent of all the Chardonnay grown. Either three, four, or five

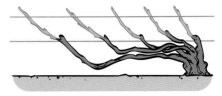
Chablis

permanent branches may be cultivated, each one being grown at yearly intervals. This results in a three-year-old vine (the minimum age for AOC Champagne) having three branches, a four-year-old having four branches, and so forth. The distance between each vine in the same row determines the eventual life of the oldest branch because, when it encroaches upon the next vine, it is removed and a new one cultivated from a bud on the main trunk. The Chablis spur-training system is, in effect, little more than a slanting bush vine unsupported by a central post.

BASIC SYSTEMS OF VINE TRAINING

There are two basic systems of vine training: cane training and spur training, of which there are many local variations. Cane-trained vines have no permanent branch because all but one of the strongest canes (which will be kept for next season's main branch) are pruned back each year to provide a vine consisting of almost entirely new growth. Apart from the trunk, the oldest wood on a cane-trained vine is the main branch and that is only ever one year old. This system gives a good spread of fruit over a large area, and allows easier regulation of annual production, because the number of fruiting buds can be increased or decreased. With spur training there is no annual replacement of the main branch, thus a solid framework is formed. It is easy, therefore, to know which basic training system has been applied to a vine simply by looking at the main branch. Even if you cannot recognize the specific style of training, if the main branch is thin and smooth, you will know that it has been cane trained, whereas if it is thick, dark, and gnarled, it has been spur trained.

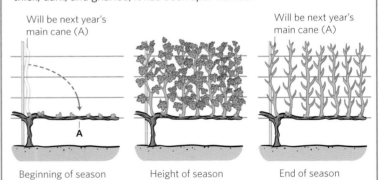

Will be next year's main cane (A)
Will be next year's main cane (A)

A

Beginning of season Height of season End of season

GUYOT – AN EXAMPLE OF CANE TRAINING
In the winter, the main horizontal cane is cut off and the spare cane (A) is bent horizontally to be tied to the bottom wire, where it will become next season's main cane. Another shoot close to the trunk will be allowed to grow as next season's spare cane.

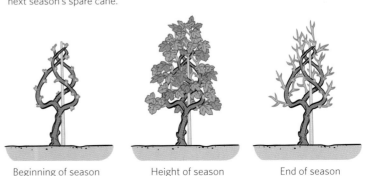

Beginning of season Height of season End of season

GOBELET – AN EXAMPLE OF SPUR TRAINING
The main canes on a spur-trained vine are all permanent, and will only be replaced if they are damaged. Only the year-old shoots are pruned back.

Cordon de Royat – Spur-training system

This is to Champagne's Pinot Noir what the Chablis system is to its Chardonnay and is nothing more complicated than a spur-trained version of Guyot Simple. There is even a double variant, although this is rarely cultivated purely for its own ends, its primary reason for existence being to replace a missing vine on its blind side. When, after their winter pruning, they are silhouetted against the sky, Cordon de Royat vines look very much like columns of gnarled old men in perfect formation; they all face forward, bent almost double, as if with one arm dug into the pit of the back and the other seeking the support of a stick.

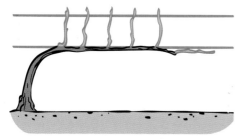

Cordon de Royat

Geneva Double Curtain – Spur-training system

This downward-growing, split-canopy system was developed by Professor Nelson Shaulis of the Geneva Experimental Station in New York state in the early 1960s to increase the volume and ripening of locally grown Concord grapes. Since then, GDC, as it is often referred to, has been adopted all over the world (particularly in Italy). However, unlike Concord varieties, classic *vinifera* vines have an upward-growing tendency, which makes the system more difficult to apply. Successful results are obtained by shoot positioning, which can either be accomplished via a movable wire (as in Scott Henry, *see opposite*), by hand, or even by machine. Yields from GDC are 50 per cent higher than those from the standard VSP trellis (*see opposite*), and the system offers increased protection from frosts (due to height above ground). It is ideal for full mechanization of medium- to high-vigour vineyards on deep fertile soils (low-vigour vines do not benefit).

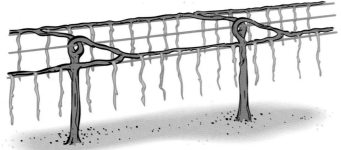

Geneva Double Curtain

Guyot – Cane-training system

Developed by Jules Guyot in 1860, both the Double and Simple forms shown here represent the most conservative style of cane training possible. It is the least complicated concept for growers to learn and, providing the number of fruiting canes and the number of buds on them are restricted, Guyot is the easiest means of restraining yields. Even when growers abuse the system, it is still the most difficult vine-training method with which to pump up production. This system is commonly used in Bordeaux, where the number of canes and buds are restricted by AOC rules (although, like most French bureaucratic systems, much depends on

Guyot Double

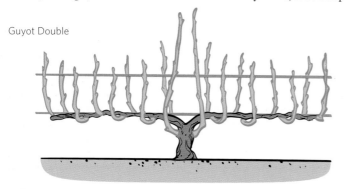

self-regulation – I have never seen an INAO inspector in the middle of a vineyard checking the variety of vines or counting them, let alone counting the number of canes or buds on canes!). Guyot is also used for some of the finest wines throughout the winemaking world, both Old and New.

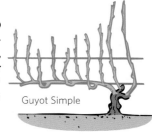

Guyot Simple

Lyre – Spur-training system

Also known as the "U" system, the canopy is divided, so as to allow a better penetration of light (thereby improving ripeness levels) and air (thereby reducing the incidence of cryptogamic disorders). Although it was developed in Bordeaux, the Lyre system is more common in the New World, where vine vigour is a problem, although not a major one. As with all split-canopy systems, the Lyre method is of no use whatsoever for low-vigour vineyards. Some growers have successfully adapted Lyre to cane training.

Lyre

Pendelbogen – Cane-training system

Also known as the European Loop, or Arc-Cane Training, this vine-training system, which is a variant of the Guyot Double (*see left*), is most popular in Switzerland and the flatter Rhine Valley areas of Germany and Alsace, although it can also be found in Mâcon, British Columbia, and Oregon. By bending the canes in an arch, Pendelbogen has more fruit-bearing shoots than the Guyot Double system, thereby providing higher yields. The arching does promote better sap distribution, which helps the produc-

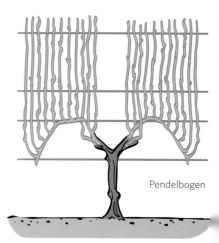

Pendelbogen

tion of more fruit, but it can also reduce ripeness levels, making the prime motive for adopting Pendelbogen one of economy, not of quality.

Scott Henry – Cane-training system

Developed by Scott Henry at the Scott Henry Vineyard in Oregon, this system effectively doubles the fruiting area provided by Guyot Double. It also offers a 60 per cent increase in fruiting area over the standard VSP Trellis system (*see right*). Scott Henry not only provides larger crops, but riper, better quality fruit – and because the canopy is split and, therefore, less dense, the wines are less herbaceous, with smoother tannins. Increased yields and increased quality may seem unlikely, but Kim Goldwater of Waiheke Island has records to prove it; and as Goldwater Estate is consistently one of New Zealand's best red wines, that is good enough for me. When I have asked growers who have tried the Scott Henry method why they have given it up, they invariably reply "Have you ever tried to grow vine shoots downwards? It doesn't work!". However, the downward-growing shoots actually grow upwards for most of the season. They are separated from the other upward-growing shoots by a moveable wire

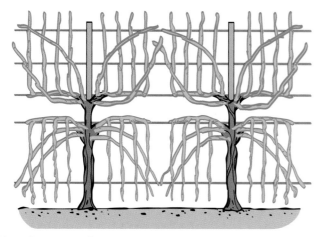

Scott Henry (cane-training)

that levers half the canopy into a downwards position. The secret of this system's success is to move the wire no earlier than two or three weeks before the harvest, which gives the canes no time to revert and, with the increasing weight of the fruit, no inclination. In areas where grazing animals coexist with vines (as sheep do in New Zealand, for instance), the fact that both halves of the canopy are a metre (two feet) or so above the ground most of the time allows under-vine weeds and water-shoots to be controlled without herbicides or manual labour, without fear of the crop being eaten in the process. Scott Henry is found mainly in the New World and is becoming increasingly popular.

Scott Henry – Spur-training system

When the Scott Henry cane-training system is adapted to spur training, each vine has two permanent spurs instead of four annual canes and produces either top canopies or bottom canopies, but not both. The vines are pruned at alternating heights to replicate the effect of cane training, and all canopies grow upwards until those on the lower vines are eventually levered downwards to the position shown here. The detailed structure of the Scott Henry system is almost impossible to identify when the vines are shrouded in foliage.

Scott Henry (spur-training)

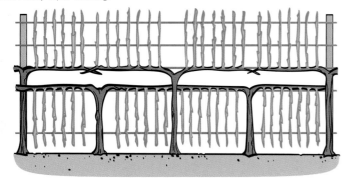

Minimal Pruning – Spur-training system

A wild, unruly mass that contains a central thicket of unpruned dead wood, which by definition has no disciplined form and is therefore impossible to illustrate. Some of the central thicket may be mechanically removed in the winter, but will still be a tangled mass. Initially, several canes are wrapped loosely around a wire, either side of the trunk, about 1.5 to 2 metres (5 to 6 feet) off the ground. The vine is then left to its own devices, although if necessary some of the summer shoots will be trimmed to keep the fruit off the ground. Some growers give up quite quickly because yields can initially be alarmingly high and as the volume increases, so the quality noticeably deteriorates. However, if they are patient, the vine eventually achieves a natural balance, reducing the length of its shoots and, consequently, the number of fruiting nodes on them. Although mature minimally pruned vines continue to give fairly high yields, the quality begins to improve after two or three years. By the sixth or seventh

year the quality is usually significantly superior to the quality achieved before minimal pruning was introduced – and the quantity is substantially greater. The ripening time needed by the grapes also increases, which can be an advantage in a hot climate, but disastrous in a cool one, particularly if it is also wet. Furthermore, after a number of years, the mass of old wood in the central thicket and the split-ends of machine-pruned cane ends surrounding it can make the vine vulnerable to various pests and diseases, especially rot and mildew, which is why minimally pruned vines in wet areas like New Zealand are more heavily pruned than they are in the hotter areas of Australia – such as Coonawarra and Padthaway – where minimal pruning first emerged and is still used to great effect.

Sylvos – Spur-training system

This is like Guyot Double, only the trunk is much longer – up to 2 metres (6 feet) – the main branches are permanent, and the fruiting canes are tied downwards, not upwards. Sylvos requires minimal pruning and is simple to maintain, lending itself to mechanization, but yields are low, unless pruned very lightly. *Vinifera* varieties do not like being forced

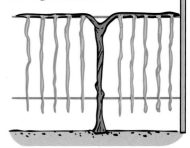

Sylvos

downwards, but shoot positioning has been introduced by growers in Australia (where it is called "Hanging Cane") and New Zealand. The system was originally conceived by Carlo Sylvos in Italy, where it is still very popular and is sometimes operated without a bottom wire, the canes falling downwards under their own weight. The main disadvantage is the dense canopy, which makes the vines prone to bunch-rot.

Sylvos (Hawke's Bay variant) – Spur-training system

This version of the Sylvos system was developed by Gary Wood of Montana Wines in the early 1980s on a Hawke's Bay vineyard belonging to Mark Read. The difference between this and the similar Scott Henry system is that it has two main spurs instead of four. With alternate fruiting canes on the same spur trained upwards then downwards, the canopy is more open, and grape clusters are farther apart. This reduces bunch-rot significantly and facilitates better spray penetration. Yields are increased by as much as 100 per cent. The only disadvantage is the longer ripening time needed, which is a risk in areas where late harvests have their dangers.

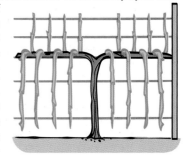

Sylvos (Hawke's Bay variant)

VSP Trellis – Cane-training system

The VSP (Vertical Shoot Positioned) Trellis is widely used, particularly in New Zealand, where it is commonly referred to as the "standard" trellising system. With the fruiting area contained within one compact zone on wires of a narrow span, it is ideally suited to mechanized forms of pruning, leaf removal, harvesting, and spraying, but it is prone to high vigour and shading. This is a very economic method and, when properly maintained, it is capable of producing good (but not top) quality wines. VSP is only suitable for low-vigour vines. A spur-trained version with just two main spurs is commonly encountered in France and Germany.

VSP Trellis

The Winery and Vineyard

Winemaking does not start in the vineyard; it begins long before the first vines are planted.

Millennia of knowledge and understanding of microclimatic conditions, not just of regions but also narrowed down to specific fields, with aspects and gradients observed, calculated, and recorded, combine to make the all-important decision of what and where to plant.

Here the viticulturist's skills come to the fore, and scientific advances are all harnessed to help add to the decision-making, often using high-tech computer systems and software.

GENTLE SLOPES

Slopes with a gradient of less than 15 per cent are thought to be ideal, specifically southwest facing in the Northern Hemisphere (or northeast below the Equator), to maximise ripening potential at the time-critical point of harvest. With a slope of greater than 15 per cent, vineyards can become hard to manage and soil erosion will become a major issue. Apart from important nutrients and topsoil getting washed down the slopes, the chemicals used in the vineyard may end up in rivers, further damaging the environment. With steep slopes, terracing can be used to minimise this impact, but they make working the vines much more labour-intensive, and time-saving equipment can rarely be used.

SOIL STRUCTURE

The structure of the soil is crucial to the growth of healthy vines. The topsoil is the primary support system for the roots, but roots can penetrate deep down into many layers of subsoil. The best soils have a thin topsoil and an accessible subsoil that has good water-retention abilities and drainage. Different types also have distinct heat-retention properties; for example, sand, gravel and loam soils retain heat better than clay, meaning that they can increase ripening speeds. The soil must have a good balance of nutrients; this can be encouraged through planting of cover crops, which can add nutrients and also combat erosion. Not surprisingly, certain grape varieties do better in certain soils depending on these characteristics, so what to plant is an important consideration at the outset when choosing a site.

WEATHER STATIONS

Weather stations, such as the one on the far right, are increasingly being used in vineyards to actively monitor weather conditions, as well as soil moisture and pests. The data from these stations is sent to a secure website in real time, allowing viticulturists to respond quickly and effectively to decisions on irrigation, vine growth, and any impending weather conditions that could impact the harvest. For example, when the grapes reach a certain level of ripeness this can start to attract pests like birds, and netting (as seen in the centre of the illustration) can be applied to keep them away.

PLANTING STRATEGIES

The right vineyard spacing and trellising very much depends on the anticipated vigour of the vines, which is determined by the grape variety and soil type. A viticulturist must leave enough space to ensure that at full growth a vine has minimal shading. Too much shade can lead to disease and uneven ripening. In general, more nutrient-rich soils translate into higher vigour vines that need wider spacing, whereas less nutrient-rich soils will have lower vigour and require less spacing. Trellising, such as vertical shoot positioning with wires, can be adapted for high- and low-vigour sites, but head-trained vines, with only a stake at the base and no wires, often require more spacing because their growth is less managed.

The decision of whether to do things manually or by machine applies to virtually everything in the vineyard. Machine planting is more expensive but allows for greater spacing precision and can reduce the time required

ANATOMY OF THE WINERY AND VINEYARD

1. Trucking in from nearby vineyards
2. River and contaminants
3. Vines on river slope
4. Trucking in from afar
5. Manual pruning
6. Machine pruning
7. Mechanical ploughing
8. Ploughing by horse
9. Manual and mechanical planting
10. Manual harvesting
11. Manual sorting
12. Protection from birds
13. Mechanical harvesting and sorting
14. Orders being shipped out
15. Loading orders
16. Warehouse for bottling, labelling, and packaging
17. Cellar door sales
18. *Barriques* for fermentation or ageing
19. Fermentation tanks
20. Sea pine rafters
21. Settling tanks
22. Black grape reception
23. White grape reception
24. Gravity-fed winery
25. Maceration
26. Terraced vineyards
27. Vertical-oriented vineyards
28. Checking ripeness
29. Weather station
30. Impact of soil structure

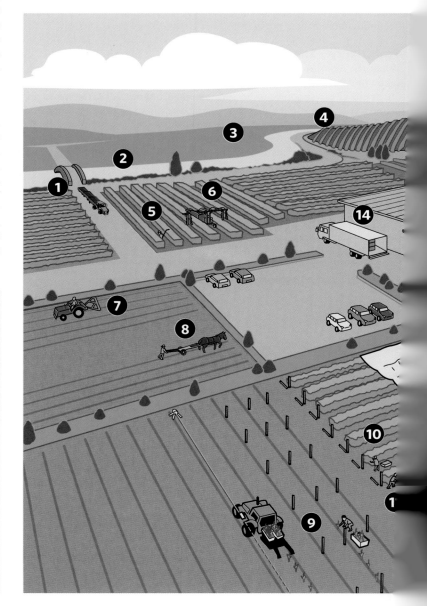

for planting from a few days to a few hours. To use a machine effectively, the land needs to be properly prepped, and it can only be used on bare soils that don't already have trellising or irrigation installed. Ploughing, which is typically done after the harvest or to prepare for the planting of new vines, was historically done with horses and is thought to reduce the need for herbicides. The advent of machinery to complete this task has allowed for more control and flexibility in timing; however, machines are heavier than horses and can compact the soil more. Few wineries still use horses for this.

PRUNING AND PICKING

Pruning is an essential viticultural activity. It determines yield, fruit positioning, and vine growth for the coming seasons. Good pruning practices ensure vine health and longevity, even affecting the ripening and quality of the fruit. Mechanical pruning exists, most but quality-oriented growers opt for manual pruning so that decisions can be made for each individual vine.

Although there is a romance to hand-picking, it is much more time-consuming and labour-intensive, whereas the use of machine pickers can speed up the process tremendously, operating round the clock if necessary. This is at the expense of having the human touch, however; for example, individual pickers can selectively choose ripe bunches and leave those that aren't ready, whereas a machine will simply pick everything: grapes, leaves, insects, and birds' nests. Hand-picking is also necessary in certain sites, such as terraced vineyards that machines can't easily access, and legislation may make hand-harvesting compulsory in certain areas.

WINEMAKING DECISIONS

Finally, once the grapes are harvested and brought to the winery there are many decisions a winemaker must make, including harvest date, length of maceration (for red wines), fermentation temperature, punchdowns versus pumpovers, use of oak versus steel tanks, wine-closure choice, and time of ageing in barrel (*barrique*), tank, or bottle before release. The hillside winery in this image has sea pine (*Pinus maritima*) rafters to guard against chlorine agents reacting with the wine and is gravity-fed, relying on the downhill flow to keep the process moving from start to finish. Not only is this a gentler process that is thought to preserve a wine's fruit flavours and an element of freshness, but it is also uses less electricity and is considered more sustainable.

Wineries can make wine using their own grapes, as well as those trucked in from neighbouring vineyards, giving greater scope for different wine styles. Additionally, having a welcoming cellar door that offers tasting and sales increases the options for selling wines directly to customers passing by. In many countries wine tourism is a lucrarive income generator for the wineries and surrounding areas.

After fermentation and ageing, the wines are bottled, labelled, and packaged in the warehouse for distribution worldwide.

We have come a long way from donkeys and wicker baskets, although some wineries still use traditional methods, and some winemakers have chosen to stay small and keep making quality artisanal products rather than scale up and lose the personal touch.

ANNUAL LIFE-CYCLE OF THE VINE

The calendar of events by which any well-established vine seeks to reproduce, through the production of grapes,
is outlined below, with a commentary on how the vine is cultivated to encourage the best grapes for winemaking.
The vine's year starts and finishes with the end and approach of winter.

| JAN | FEB | MAR | APR | MAY | JUN | JUL | AUG | SEP | OCT | NOV | DEC |

FEBRUARY Northern Hemisphere
AUGUST Southern Hemisphere

1. Weeping

Weeping is the first sign of the vine awakening after a winter of relative dormancy. When the soil at a depth of 25 centimetres (10 inches) reaches 10.2°C (50°F), the roots start collecting water and the sap in the vine rises, oozing out of the cane ends which were pruned in winter, in a manifestation called "weeping". This occurs suddenly, rapidly increases in intensity, and then decreases gradually. Each vine loses between half and five-and-a-half litres (10 pints) of sap. Weeping is the signal to prune for the spring growth. However, this poses a problem for the grower because the vine, once pruned, is at its most vulnerable to frost. But waiting for the danger of frost to pass wastes the vine's preciously finite energy and retards its growth, delaying the ripening of the fruit by as much as 10 days, and thus risking exposure of the fruit to autumn frosts later on.

WEEPING VINE
As the soil warms up the vine awakes, pushing sap out of its cane ends.

| JAN | FEB | MAR | APR | MAY | JUN | JUL | AUG | SEP | OCT | NOV | DEC |

MARCH to APRIL Northern Hemisphere
SEPTEMBER to OCTOBER Southern Hemisphere

2. Bud-break

In the spring, some 20 to 30 days after the vine starts to weep, the buds open. Different varieties bud-break at different times; there are early bud-breakers and late ones, and the same variety can bud-break at different times in different years due to climatic changes. Soil type can also affect the timing; clay, which is cold, will retard the process, while sand, which is warm, will promote it. Early bud-break varieties are susceptible to frost in northerly vineyards (southerly in the Southern Hemisphere), just as late-ripeners are vulnerable to autumn frosts. In the vineyard, pruning continues into March (September in the Southern Hemisphere). The vines are secured to their training frames, and the earth that was ploughed over the grafting wound to protect it in the winter is ploughed back, aerating the soil and levelling off the ground between the rows.

A BUD OPENS ON A VINE
Buds begin to open at a time determined by grape variety and climate.

| JAN | FEB | MAR | APR | MAY | JUN | JUL | AUG | SEP | OCT | NOV | DEC |

APRIL to MAY Northern Hemisphere
OCTOBER to NOVEMBER Southern Hemisphere

3. Emergence of shoots, foliage, and embryo bunches

Following bud-break, foliage develops and shoots are sent out. In mid-April (mid-October in the Southern Hemisphere), after the fourth or fifth leaf has emerged, tiny green clusters form. These are the flowers which, when they bloom, will develop into grapes. Commonly called embryo bunches, they are the first indication of the potential size of a crop. In the vineyard, spraying to ward off various vine pests, or cure diseases and other disorders, starts in May (November), and continues until the harvest. Many of these sprays are combined with systemic fertilizers to feed the vine directly through its foliage. These spraying operations are normally done by hand or tractor, but may sometimes be carried out by helicopter if the slopes are very steep or the vineyards too muddy to enter. At this time of year the vine can be affected by *coulure* or *millerandage* (*see* p22).

EMBRYO BUNCHES
Embryo bunches – the vine's flowers – form amidst the opening foliage.

| JAN | FEB | MAR | APR | MAY | JUN | JUL | AUG | SEP | OCT | NOV | DEC |

MAY to JUNE Northern Hemisphere
NOVEMBER to DECEMBER Southern Hemisphere

4. Flowering of the vine

The embryo bunches break into flower after the 15th or 16th leaf has emerged on the vine. This is normally about eight weeks after the bud-break and involves pollination and fertilization, and lasts for about 10 days. The weather must be dry and frost-free, but temperature is the most critical requirement. A daily average of at least 15°C (59°F) is needed to enable a vine to flower and between 20° and 25°C (68° and 77°F) is considered ideal. Heat summation, however, is more important than temperature levels, so the length of day has a great influence on the number of days the flowering will last. Soil temperature is more significant than air temperature, so the soil's heat-retaining capacity is a contributory factor. Frost is the greatest hazard, and many vineyards are protected by stoves or sprinkling systems.

A VINE FLOWER BEGINS TO BLOOM
The ten days of flowering forms a vulnerable period for a grapevine.

| JAN | FEB | MAR | APR | MAY | JUN | JUL | AUG | SEP | OCT | NOV | DEC |

JUNE to JULY Northern Hemisphere
DECEMBER to JANUARY Southern Hemisphere

5. Fruit set

After the flowering, the embryo bunches rapidly evolve into true clusters. Each fertilized berry expands into a grape, the first visible sign of the actual fruit that will produce the wine. This is called fruit set. The number of grapes per embryo bunch varies from variety to variety, as does the percentage that actually set into grapes. The panel below illustrates this. In the vineyard, spraying continues and summer pruning (cutting away some bunches) will concentrate the vine's energy on making fruit. In some vineyards this is the time for weeding, but in others the weeds are allowed to grow as high as 50 centimetres (20 inches) before they are mown and ploughed into the soil to break up the soil and provide the vines with excellent green manure.

CLUSTERS OF BERRIES TAKE SHAPE
Clearly recognizable grapes begin to form.

VARIETY	BERRIES PER EMBRYO BUNCH	GRAPES IN A RIPE CLUSTER	PERCENTAGE OF FRUIT SET
Chasselas	164	48	29%
Gewürztraminer	100	40	40%
Pinot Gris	149	41	28%
Riesling	189	61	32%
Sylvaner	95	50	53%

| JAN | FEB | MAR | APR | MAY | JUN | JUL | AUG | SEP | OCT | NOV | DEC |

AUGUST Northern Hemisphere
JANUARY Southern Hemisphere

6. Ripening of the grapes

As the grape develops its fleshy fruit, very little chemical change takes place inside the berry until its skin begins to turn a different colour – the process known as *véraison*. Throughout the grape's green stage, the sugar and acid content remains the same, but during August (January in the Southern Hemisphere), the ripening process begins in earnest – the skin changes colour, the sugar content dramatically increases, and the hard malic acid diminishes as the riper tartaric acid builds up. Although the tartaric acid content begins to decline after about two weeks, it always remains the primary acid.

It is at this stage that the grape's tannins are gradually hydrolysed. This is a crucial moment because only hydrolysed tannins are capable of softening as a wine matures. In the vineyard, spraying and weeding continue, and the vine's foliage is thinned to facilitate the circulation of air and thus reduce the risk of rot. Care must be taken not to remove too much foliage as it is the effect of sunlight upon the leaves, not the grapes, that causes the grapes to ripen.

GRAPES OF VARIEGATED COLOURS
In late summer the grapes begin to change colour – the sign of true ripening.

| JAN | FEB | MAR | APR | MAY | JUN | JUL | AUG | SEP | OCT | NOV | DEC |

AUGUST to OCTOBER Northern Hemisphere
FEBRUARY to MARCH Southern Hemisphere

7. Grape harvest

The harvest usually begins mid- to late September (mid- to late February in the Southern Hemisphere) and may last for a month or more, but, as is the case with all vineyard operations, the timing is earlier nearer to the equator and is dependent on the weather. Picking may, therefore, start as early as August (February) and finish as late as November (April). White grapes ripen before black grapes and must, in any case, be harvested a little bit earlier to achieve a higher acidity balance.

HAND-PICKING GRAPES
A vineyard's winemaker will determine the best time to pick the grapes.

| JAN | FEB | MAR | APR | MAY | JUN | JUL | AUG | SEP | OCT | NOV | DEC |

NOVEMBER to DECEMBER Northern Hemisphere
APRIL to MAY Southern Hemisphere

8. Grapes affected by *Botrytis cinerea*

In November the sap retreats to the protection of the vine's root system. As a result, the year-old canes begin to harden and any remaining grapes, cut off from the vine's metabolic system, start to dehydrate. The concentrated pulp that they become is subject to severe cold. This induces complex chemical changes in a process known as *passerillage*. In specialized sweet-wine areas the grapes are deliberately left on the vine to undergo this quality-enhancing experience and, in certain vineyards with suitable climatic conditions, growers pray for the appearance of *Botrytis cinerea,* or "noble rot".

"NOBLE ROT" COVERS SHRIVELLED GRAPES
These rotting grapes are soon to be harvested for sweet botrytised wine.

| JAN | FEB | MAR | APR | MAY | JUN | JUL | AUG | SEP | OCT | NOV | DEC |

DECEMBER to JANUARY Northern Hemisphere
MAY to JUNE Southern Hemisphere

9. *Eiswein* (Icewine)

In Germany, as well as Canada, it is possible to see grapes on the vine in December and even January. This is usually because the grower has hoped for *Botrytis cinerea,* or Edelfäule as it is called in Germany, but it has failed to occur on some grapes. Should frost or snow freeze the grapes, they can be harvested to produce *Eiswein* (called icewine in Canada), one of the world's most spectacular wines. As it is only the water that freezes, this can be skimmed off once the grapes are pressed in order to leave a super-concentrated unfrozen pulp that produces *Eiswein*.

GRAPES FROZEN ON THE VINE
Even snow-covered grapes that are still on the vine may yet become wine.

Grape Varieties

The grape variety used for a wine is the most influential factor in determining its taste. The factors that influence the inherent flavour of any grape variety are the same as those that determine the varietal taste of any fruit.

Of the 17 genera belonging to the Vitaceae (sometimes also called Vitidaceae and in older literatire, Ampelidacea), only the genus *Vitis* is important for winemaking. The list of winemaking species is just as limited: of the nearly 80 species in this genus, only *Vitis vinifera* (and its many cultivars) is generally grown for wine.

THE FAMILY VITACEAE

The Vitaceae, or the vine family, is a large and diverse group of plants ranging from the tiny pot-plant Kangaroo vine to the sprawling Virginia creeper. The Wine Vine Tree (*see* below) shows how *Vitis vinifera*, the classic winemaking species, relates to the rest of the vine family. Other species in this genus are used for rootstock.

Upon close examination, readers who possess earlier editions of this book might notice that I have slightly pruned the Wine Vine Tree. *Euvitis* (or *Euvites*) and *Muscadinia* used to be classified as sub-genera of *Vitis*, but *Muscadinia* is now considered to be its own genus because of the difference in the number of its chromosomes. Without this sub-genus, *Euvitis* becomes redundant and only *Vitis* remains. Furthermore, after several revisions, the number of genera in the botanical family Vitaceae has risen from 10 to 17, although Professor Markus Keller, author of *The Science of Grapevines* (Academic Press, 2010), which lists 17 genera, told me, "I may have to revise the number down to 15, 14, 12, even as low as 6, or perhaps up to 19 for the next edition, as the botanical classification in this area is in a constant state of flux!"

SIZE

The smaller the fruit, the more concentrated the flavour will be. Thus most classic grape varieties, such as Cabernet Sauvignon and Riesling, have small berries, although some varieties that rely more on elegance than power of concentration, such as the Pinot Noir, may yield large berries. Many varieties are known as *petit* or *gros* something, and it is usually the *petit* that is the better variety – Petit Vidure is Cabernet Sauvignon; Gros Vidure is Cabernet Franc.

SKIN STRUCTURE

The skin contains most of the aromatic characteristics with which we associate the varietal identity of any fruit. Its construction and thickness is, therefore, of paramount importance. For example, the thick-skinned Sauvignon Blanc produces an aromatic wine that, when ripe, varies in pungency from "peach" in a warm climate to "gooseberry" in a cool climate, and when underripe varies in herbaceousness, ranging from "grassy" to "elderflower" and even "cat's pee". Meanwhile the thin-skinned Sémillon produces a rather neutral wine, although its thin skin makes it susceptible to noble rot and is thus capable of producing one of the world's greatest botrytised sweet wines, with mind-blowing aromatics.

Skin Colour and Thickness

A dark-coloured, thick-skinned grape, such as Cabernet Sauvignon, produces very deep-coloured wines, while the lighter-coloured, thin-skinned grapes, such as Merlot, produce a less-intense colour.

ACID-SUGAR RATIO AND OTHER ELEMENTS

The grape's sugar content dictates the wine's alcohol level and the possibility of any natural sweetness; together with the acidity level, this determines the balance. The proportions of a grape's other constituents, or their products after fermentation, form the subtle nuances that differentiate the varietal characters. Although soil, rootstock, and climate have an effect on the ultimate flavour of the grape, the genetics of the vine dictate the end result.

ROOTSTOCK

Hundreds of rootstock varieties have been developed from various vine species, usually *Vitis berlandieri*, *Vitis riparia*, or *Vitis rupestris* because they are the most *phylloxera*-resistant. The precise choice of rootstock is dependent on its suitability to the vinestock on which it is to be grafted, as well as on its adaptability to the geographical location and soil type. The choice can increase or decrease a vine's productivity, and thus has a strong effect upon the quality of the wine produced from the grapes: generally, the lower the quantity, the higher the quality.

BOTANICAL FAMILY TREE FOR VITACEAE

GENUS *MUSCADINIA*

GENUS *VITIS* – 60 to 70 species, but only Vitis vinifera is important for winemaking varieties

VITACEAE FAMILY – Of the 17 genera belonging to the Vitaceae family (also called Ampelidaceae), only the genus *Vitis* is important for winemaking

Aligoté · Cabernet Franc · Cabernet Sauvignon · Chardonnay · Chenin Blanc · Gamay · Gewürztraminer · Grenache · Merlot · Muscat Blanc · Nebbiolo · Pinot Gris · Pinot Meunier · Pinot Noir · Riesling · Sangiovese · Sauvignon Blanc · Sémillon · Sylvaner · Syrah · Touriga Nacional · Viognier · Zinfandel · Others

Vitis rotundifolia · *Vitis munsoniana* · *Vitis popenoei* · *Vitis vinifera* · *Vitis riparia* · *Vitis labrusca* · *Vitis berlandieri* · *Vitis amurensis* · *Vitis rupestris* · *Vitis cariboea* · *Vitis argentifolia* · *Others*

THE WINE VINE TREE

The blue-green line running up from the ground traces the parentage of *Vitis vinifera*, the species from which all classic winemaking grapes come. Although *Vitis vinifera* is the most important species of the genus *Vitis*, wine is made from grape varieties from other species, notably *Vitis labrusca*, native to North America and, more recently, *Vitis amurensis*.

ABC OF GRAPE VARIETIES
INCLUDING SYNONYMS

It is estimated that there are more than 10,000 grape varieties in cultivation, but this could well include the same grapes under other names. With the relatively short and recent experience of DNA fingerprinting, it is thought that the real number might come down closer to 5,000.

Practicality dictates that the below list contains grape varieties that are mentioned in this book. For ease of use, the list of synonyms has been integrated with the grape variety entries, and all names have been listed alphabetically, regardless of colour, so you have only one section to search.

CROSSES AND HYBRIDS
A cross between grape varieties within one species is called a cross, and a cross between varieties from different species is a hybrid. Cross the same grape varieties more than once and the odds are that the new strains produced will not be the same. Thus *Sylvaner* x *Riesling* is the parentage not only of the Rieslaner but also of the Scheurebe, two totally different grapes. It is also possible to cross a variety with itself and produce a different grape. In the following glossary, the parentage of crosses and hybrids is always in *italics*.

Prior to our intervention in the process, most varieties originated by spontaneous crossing (most grapevines are hermaphrodite, but the flowers can be fertilized by the pollen from another variety to create a spontaneous cross) or as seedlings (when a grape drops to the ground and a seedling springs forth, it will never replicate the vine that produced it; thus, a new variety will always result).

CLONES AND CLONING
Within varietal limitations, intensive selection can produce a vine to suit specific conditions, such as to increase yield, resist certain diseases, or to thrive in a particular climate. Identical clones of this vine can then be replicated an infinite number of times by micro-biogenetic techniques. Clones are named by number and initial. For instance, "Riesling clone 88Gm" is the 88th clone of the Riesling variety produced at Geisenheim (Gm), the German viticultural research station. A "localized" clone is a vine that has evolved naturally under specific conditions within a particular environment. These may be named first by grape variety and then by locality and referred to as sub-varieties; however, the name of the original variety is often entirely forgotten with the passing of time, so that the variety acquires a new name altogether.

OVERVIEW OF GRAPE COLOURS

Ⓦ WHITE VARIETIES
Most white grapes actually range from a pale green (higher-acid skin sap) to an amber-yellow (lower acidity).

Ⓜ MID-COLOURED VARIETIES
Between black grapes and white grapes, there are numerous mid-coloured varieties. The French subdivide this category into *rouge, rosé,* and *gris,* but similar distinctions could be made within either black or white grape categories, which would confuse more than clarify; thus, all mid-coloured varieties here are classified as one.

Ⓑ BLACK VARIETIES
Black grapes vary from dark, ruddy red (higher-acid skin sap) to blue-black (lower acidity). While white wine can be made from most black grapes, because their juice is clear and uncoloured, red wine can be made only from black grapes, since it is the pigments in their skin, called anthocyanins, that give the wine its colour.

Ⓣ *TEINTURIER* VARIETIES
A *teinturier* variety literally means a "dyer" grape, so called because not only is its skin coloured, but so is its red-coloured juice. Without having to rely on highly phenolic anthocyanins, red wines produced from *teinturier* grapes can be deeply coloured with very little tannin, which can be an odd taste sensation unless crafted well (ie, blended with other grape varieties, aged in oak to extract wood tannins, and so on).

Note: The majority of varieties mentioned in the text below are cross-referenced, but readers will occasionally come across some references, particularly of varieties indicated as parents or distant relatives of various crosses and hybrids, that lead to dead ends. This is for purely practical purposes, as they would lead to ever more obscure entries, culminating in a far longer list of little more practical use.

Ⓢ GRAPE VARIETY SYNONYMS

Many varieties of grape are known by several different synonyms. The Malbec grape, for example, has at least 34 different names, including Pressac, Auxerrois, Balouzet, Cot, Estrangey, and Grifforin. The Chasselas officially has 213 synonyms. This would not be too confusing if the synonyms applied, uniquely, to the same grape variety, but unfortunately this is not the case. The Malbec is again a good example: a black grape, it is known as the Auxerrois in Cahors, but in Alsace and Chablis, the Auxerrois is a white grape, while in other parts of France, the Malbec is known as the Cahors.

Synonyms relating to localized clones or sub-varieties are often regarded as singularly separate varieties in their own right. The Italian Trebbiano, itself a synonym for the French Ugni Blanc, has many sub-varieties recognized by the Italian vine regulations. Also, many synonyms revolve around the name of another grape variety, although they are not necessarily related. Ampelographers distinguish between "erroneous" and "misleading" synonyms. The former refers to varieties that have mistakenly been given the name of another, totally different variety, whereas the latter refers to varieties whose names suggest, incorrectly, that they are related to another variety.

KEY

Ⓦ White grape variety

Ⓜ Mid-coloured grape variety

Ⓑ Black grape variety

Ⓣ *Teinturier* grape variety

Ⓢ Synonym

Ⓢ **Abbondosa** Synonym for Nuragus

Ⓑ **Abbuoto** Central Italian variety, incorrectly linked to the ancient Roman wine Caecuban (because this is a black grape, while Caecuban was a white renowned for turning "fire-coloured" as it aged), Abbuoto produces a deep-coloured, full-flavoured red with a tannic edge that repays ageing in oak and being blended with at least one

other variety to bring smoothness and intensity.

Ⓑ **Abouriou** This grape produces a Gamay-like red wine, as well as gallons of inexpensive rosé, in southwestern France.

Ⓢ **Acadie** Synonym for L'Acadie

Ⓑ **Ada Karasi** or **Adakarasi** A Turkish grape traditionally used for producing a soft, deep-coloured red.

Ⓑ **Agiorgitiko** An excellent indigenous Greek grape variety that is responsible for the rich and often oak-aged wines of Nemea, the Agiorgitiko (meaning "the grape of St George") is one of the most widely planted grape varieties in Greece. The best Agiorgitiko has a spicy richness to the fruit, and the greater the altitude at which it is grown, the more defined the spiciness becomes. This is a variety that can withstand the heat, but the cooler it is within the realms of a warm climate, the more it repays.

Ⓑ Aglianico This is the grape that made the famous Falernum of ancient Rome, the modern equivalent of which is typically dark and rustically rich. Many smoother examples are gradually being crafted by passionate winemakers, but there is never any getting away with the muscular structure of this grape. It is, after all, part of its attraction.

Ⓦ Airén This very ordinary-quality Spanish grape is the widest planted white wine grape in the world in terms of area. Grown for its good acidity in hot climates, it has a neutral character that makes it ideal for brandy and useful for tweaking white-wine blends in La Mancha.

Ⓦ Ak Dimrit or **Akdimrit** Technically a synonym for Dimrit Beyaz, but Ak Dimrit, literally meaning "white Dimrit", is the more common usage. Grown in the Central Anatolia and Mediterranean Coast regions. Also used for the production of *raki*.

Ⓦ Albana See box, below.

Ⓢ Albariño Synonym for Alvarinho

Ⓦ Albarola Native to northwestern Italy, this grape is mostly known as the principal variety for the Cinque Terre DOC, which is better known for its dramatic scenery than its slightly aromatic, light dry white wine.

Ⓢ Alcayata Synonym for Mourvèdre

Ⓑ Aleatico A Muscat-like variety grown principally in Southern Italy and, to a lesser extent, Tuscany, Aleatico usually produces vivid red sweet wine that may be fortified or not. This grape crops up far and wide, from Corsica to Elba, and California to Australia.

Ⓑ Alfrocheiro Preto This Portuguese grape is a blending component in the red wines of Alentejo, Dão, Estremadura and Terras do Sado.

Ⓣ Alicante Bouschet or **Alicante Henri Bouschet** The full name of this variety is Alicante Henri Bouschet, but it is invariably referred to simply

as Alicante Bouschet, even though other Alicante Bouschets exist, such as (Alicante Bouschet) Précoce, Tardif, and à Longues Grappes. Petit *Bouschet* x *Grenache* cross, this is a teinturier grape, with vivid, red juice. It was a favourite during Prohibition, when the rich colour of its juice enabled bootleggers to stretch wines with water and sugar. Alicante Bouschet is used primarily for blends and Port-style wines today and is seldom seen as a varietal table wine.

Ⓦ Aligoté A thin-skinned grape of unexceptional quality grown in Burgundy and Bulgaria. It makes tart wines of moderate alcoholic content, but in exceptionally hot years they can have good weight and richness. The variety's best wines come from certain Burgundian villages, especially Bouzeron, where the quality may be improved by the addition of a little Chardonnay.

Ⓦ Altesse The finest of Savoie's traditional varieties, the Altesse variety makes delightfully rich and fragrant wines.

Ⓑ Alvarelhão A minor Portuguese grape that is also one of Port's lesser varieties. A few parcels of this vine can be found in Galicia, Spain, and Victoria and New South Wales, Australia.

Ⓦ Alvarinho The classic Vinho Verde grape, en route to mainstream status due to its delicate, elegant, saline character. Rias Baixas produces some excellent examples under the Albariño synonym.

Ⓦ Amigne An old Swiss variety found in pockets of the Valais.

Ⓑ Ancellota An Italian variety used to deepen the colour of Lambrusca in Emilia-Romagna, Ancellota is also grown in Argentina, where Zucardi has successfully produced a pure varietal version for many years.

Ⓑ André This is a Blaufränkisch x St-Laurent cross that is primarily grown in Czechia's Moravia region. Also found in Germany's Saale-Unstrut region.

Ⓢ Ansolica Synonym for Inzolia

Ⓢ Ansonica Synonym for Inzolia

Ⓢ Aragnan Synonym for Picardan.

Ⓢ Aragones Synonym for Grenache and Tempranillo

Ⓢ Aragonêz Synonym for Tempranillo

Ⓑ Aramon This used to be the most widely planted vine in France. Although rapidly declining, there is still, even today, more than 3,000 hectares (7,415 acres) grown, mainly in the Hérault. Aramon produces an undistinguished wine, as does its rarely encountered siblings Aramon Blanc and Aramon Gris.

Ⓦ Arbane or **Arbanne** An ancient but minor Champagne grape variety that was supposed to be so noticeable in the tiniest proportion in Aube wines in the 19th century that when local winegrowers walked into a *cuverie* they would immediately detect it and exclaim, "Ah, Arbanne!" Currently undergoing a mini-revival in Champagne along with other permitted ancient varieties. Also found in the Vin de Pays des Coteaux de Coiffy between the Aube and Alsace. Arbanne Noir and Arbanne Rouge exist but are rarely encountered.

Ⓦ Arbois Authorized for various Loire wines, including Vouvray, where a handful of producers use its intrinsically high acidity to tweak the acidity of their sweeter wines in hotter years.

Ⓑ Arinarnoa This *Tannat* x *Cabernet Sauvignon* cross was created at Montpellier in 1956 but has only just started to be cultivated seriously beyond France and Lebanon since the late 1990s. It can now be found in Spain, Uruguay, Argentina, Brazil, and even China.

Ⓦ Arinto The Arinto is one of Portugal's potentially excellent white grapes. Its use in the small district of Bucelas is to make a crisp, lemony wine that ages well.

Ⓦ Arneis Literally meaning "little rascal", this Piedmontese grape was so named because of its erratic ripening. Once threatened by commercial extinction, this delicately aromatic variety is now fashionable, spreading to Victoria in Australia, Gisborne in New Zealand, and in California and Oregon.

Ⓦ Arrufiac Formerly a primary grape for Pacherenc du Vic-Bilh and Côtes de Saint-Mont, Arrufiac is high-yielding, rich in sugar, but low in character and finesse and has thus been reduced to secondary status.

Ⓢ Arvino Nero Synonym for Gaglioppo and Magliocco Dolce

Ⓑ Aspiran Bouschet or **Aspiran Noir** Once widely planted, the Aspiran is rarely found in France today, a little is used as a minor blending component in Minervois. However, Argentina has over 4,300 ha planted and the acreage is growing dynamically.

Ⓢ Asprinio Synonym for Greco di Tufo

Ⓢ Aspro Synonym for Xynisteri

Ⓦ Assyrtiko One of the better-quality indigenous varieties of Greece, the classic example of this variety is to be found on Santorini, although the super-*terroir* character of the island's volcanic-ash soil can be too pungent for unconditioned palates.

Ⓢ Aubaine Synonym for Chardonnay

Ⓑ Aubun This grape used to be a minor blending component in Cabardès and Côtes of Vivarais, but it was always more of a bland variety used to pad out a wine than provide any useful ability to tweak a blend. Consequently, it has been banned since 1995. Today it is permitted in the Languedoc but mainly grown in the Garde and the Aude.

Ⓦ Aurore The most useful character of this *Seibel 788* x *Seibel 29* hybrid is its very blandness, which has often been used to mute (as best as it can) the extrovert foxiness of Lambrusca wines.

Ⓢ Auvernat Synonym for Pinot Noir

Ⓢ Auvernat Gris Synonym for Pinot Gris or Pinot Meunier

Ⓦ Auxerrois At home in Alsace, Luxembourg, and England, Auxerrois makes a fatter wine than Pinot Blanc, so suits cooler situations. Its musky richness has immediate appeal, but it is inclined to low acidity and blowsiness in hotter climates.

Ⓢ Auxerrois Synonym for Malbec

Ⓢ Auxerrois Gris Synonym for Pinot Gris

Ⓢ Auxois Synonym for Chardonnay

Ⓦ Avesso A mildly aromatic, otherwise quite bland and lightweight Portuguese variety used for Vinho Verde.

Ⓢ Avillo Synonym for Picpoul Blanche

Ⓢ Axina de Margiai or **Axina 'e Pòberus** Synonym for Nuragus

Ⓑ Băbească or **Băbească Neagră** This variety is found mostly in the Nicoreşti area of Romania, where it is capable of producing a deep-coloured, spicy red wine.

Ⓦ Bacchus A (*Riesling* x *Sylvaner*) x *Müller-Thurgau* cross that is one of Germany's more superior crosses, Bacchus is refreshingly aromatic with zesty grapey fruit that can be a delight to drink when produced in cooler climes, such as Camel Valley in England.

Ⓑ Baco or **Baco Noir** See box, opposite.

Ⓢ Baco 1 Synonym for Baco Noir

Ⓑ Baga Principal grape variety for Bairrada, where it produces firm tannic reds, but the wines are developing more accessible fruit and softer tannins with new viticultural and winemaking practices.

Ⓢ Bangalore Blue Synonym for Isabella

Ⓦ ALBANA
This was the first Italian white grape to be accorded DOCG status back in 1987, yet it still struggles to deliver wines of an internationally acceptable standard. Producers such as Bissoni, Umberto Cesari, Leone Conti, Gallegati, and Fattoria Zerbina have all produced very good *passito* wines, but given the lengths that winemakers have to go to produce a true straw wine, they should be good. Albana is merely a high-yielding grape that makes a rather rustic, almost common dry wine.

Ⓜ **Barbarossa** or **Barbarossa Rosé** The name for a number of light-red grape varieties found in the Piedmont, Liguria, and Tuscany regions of Italy, as well as Corsica.

Ⓜ **Barbaroux** or **Barbaroux Rosé** This grape is found in Provence (AOCs of Cassis and Vin de Provence), where it either makes rosés or is used to lighten the reds that can get a bit heavy this far south in France.

Ⓑ **Barbera** *See box, right.*

Ⓦ **Baroque** or **Barroque** A herbaceous Colombard-like white wine grape from southwestern France, where it is primarily responsible for Tursan AOC.

Ⓑ **Bastardo** This is the classic Port grape and is identified as the Trousseau, an ancient variety once widely cultivated in the Jura, France.

Ⓢ **Beaunois** Synonym for Chardonnay

Ⓑ **Beichun, Beihong,** and **Beimei** Three of the more successful of the so-called "Bei" series of *Muscat Hamburg* x *Vitis amurensis* Ruprecht hybrids, which were bred in China to combat severe winter temperatures and yield extraordinarily high sugar content, these varieties now account for some 7,000 hectares (17,300 acres), mostly in the southern regions of the country.

Ⓑ **Béquignol** This old Bordeaux variety is rarely encountered in France today but is still widely grown in Argentina, where it produces soft, deeply coloured red wines.

Ⓢ **Bernarde** Synonym for Prié Blanc

Ⓦ **Bervedino** A minor grape used in the Colli Piacentini primarily as a major component in some of its *vin santo*, but also allowed as a secondary variety in Monterosso Val d'Arda, Bervedino makes a full yet soft wine with neutral fruit.

Ⓦ **Biancame** Central Italian grape responsible for Bianchello del Metauro DOC in the Marche and can be part of a blend or a standalone variety in Colli di Rimini DOC in Emilia-Romagna

Ⓢ **Bianchino** Synonym for Montuni

Ⓦ **Bianco d'Alessano** One of the latest-ripening varieties in CSIRO's (Australia's national science research agency) collection at Marbein in Victoria, demonstrating why this grape manages to hold on to decent acidity levels in even the hottest climes. In its native Apulia, Bianco d'Alessano is a secondary variety in the DOCs of Gravina, Lizzano, Locorotondo, and Martina Franca, but it enjoys success in the hot Riverina region of Australia.

Ⓦ **Biancolella** This Campanian variety is mostly grown on the island of Ischia but is also used as a component of DOCs Campi Flegrei and Penisola Sorrentina.

Ⓦ **Bical** A Central Portuguese grape, literally called "fly droppings" (Borrado des Moscas) in the Dão because of the brown spots the berries develop when ripe. It retains high natural acidity at moderate alcohol levels, making it well suited to white winemaking in hot areas.

Ⓢ **Bical Tinto** Synonym for Touriga Nacional

Ⓢ **Bidure** Synonym for Cabernet Franc and Cabernet Sauvignon

Ⓢ **Black Alicante** Synonym for Trincadeira

Ⓢ **Black Hamburg** Synonym for Schiava Grossa

Ⓢ **Black Malvoisie** Synonym for Cinsault

Ⓢ **Black Muscat** or **Black Muscat of Alexandria** Synonyms for Muscat of Hamburg

Ⓢ **Black Portugal** Synonym for Trincadeira

Ⓑ **Black Queen** Originally developed in Japan, this hybrid grape of complex parentage is popular in the small wine industries of Taiwan, Vietnam, Cambodia, and Thailand

Ⓢ **Black Spanish** Synonym for Jacquez

Ⓢ **Blanc Doux** Synonym for Sémillon

Ⓢ **Blanc Fumé** Synonym for Sauvignon Blanc

Ⓢ **Blanc de Morgex** Synonyms for the Prié Blanc

Ⓢ **Blanc Vert** Synonym for Sacy

Ⓢ **Blauburgunder** Synonym for Pinot Noir

Ⓢ **Blauer Portugieser** Synonym for Portugieser

Ⓑ **Blauer Wildbacher** Found in western Steiermark (Styria) in Austria and the Veneto region of Italy, the Blauer Wildbacher makes some nice fresh, crisp, dry rosé wines and even a few outlandish fizzy reds, but the light-bodied still reds are much less interesting.

Ⓑ **Blaufränkisch** A Central European variety that makes a relatively simple, light red wine. This grape is also grown in the Atlantic Northeast, where it is more commonly referred to as Lemberger (a name that is also used elsewhere, erroneously, as a synonym for Gamay), and goes by many synonyms in Europe.

Ⓑ **Bobal** The dominant variety in Utiel-Requena in Spain, where it makes fruity, medium-bodied, easy-drinking reds.

Ⓢ **Boal** or **Boal Branco** Synonym for Malvasia Fina

Ⓦ **Bombino** or **Bombino Bianco** Widely planted throughout central and southern Italy, where it has often been mistaken for Trebbiano. (The Trebbiano referred to as Abruzzese, Bianco di Chieti, Campolese, d'Abruzzo, di Avessano [sic], di Macerata, di Teramo, d'Ora, d'Oro, and Dorato di Teramo are all Bombino.) This is a tough-skinned grape that is highly resistant to parasites, diseases, and adverse climatic conditions. The Bombino is most at home in Emilia-Romagna, where it is known as the Pagadebit. There it produces a firmer-structured, more alcoholic wine than elsewhere and is sold in dry, semi-sweet, still and *frizzante* styles under the Pagadebit di Romagna DOC. This grape also plays a useful role as a blending component in the DOCs of Biferno and Pentro di Iserinia in Moloise, and Gravina, Leverano, Locorotondo, and San Severo in Puglia.

Ⓑ **Bombino Nero** Less widely planted and less well regarded than its white sibling, this grape is a secondary variety for the DOC of Lizzano (Apulia).

Ⓢ **Bonarda** Synonym for Bonarda variants, a confusing mess of incorrect synonyms and other varieties, such as Douce Noire, Corbeau Noir, Croatina, and Uva Rara.

Ⓦ **Bonarda Bianca** The white Bonarda is seldom seen outside of Argentina, where it is merely used as a bulk blending variety.

Ⓢ **Bonarda di Borgomasino** Synonym for Bonardina

Ⓢ **Bonarda di Gattinara** Synonym for Croatina and Uva Rara

Ⓢ **Bonarda a Grandes Grappes, Bonarda dell'Astigiano, Bonarda del Monferrato, Bonarda di Asti, Bonarda di Chieri, Bonarda di Gattinara, Bonarda di Piemonte, Bonarda du Piémont, Bonarda Nera,** or **Bonarda Nero** Synonyms for Bonarda Piemontese

Ⓢ **Bonarda Macun** or B**onarda Rotunda** Synonyms for Durasa

Ⓢ **Bonarda Novarese** Synonym for Uva Rara

Ⓦ **Bonarda di Piava** A minor blending component occasionally encountered in the Veneto region of Italy.

Ⓑ **Bonarda Piemontese** Traditionally used to support Nebbiolo in such DOC wines as Bramaterra, Gattinara, and Ghemme, it also makes pure varietal wine under the Piemonte DOC. But this native Italian grape is cultivated on a far greater scale in Argentina, where it is most successful in Syrah-Bonarda blends.

Ⓢ **Bonarda di Rovescala, Bonarda Grossa,** or **Bonarda Pignola** Synonyms for Croatina

Ⓑ **BARBERA**
A prolific Italian variety of black grape grown in Piedmont, the Barbera makes light, fresh, fruity wines that are sometimes very good. Because of its refreshing acidity, this grape can perform exceptionally well in warmer, New World climes.

Ⓑ **BACO OR BACO NOIR**

One of the more successful red hybrids, particularly when grown in Ontario or the Atlantic Northeast, where this grape can be relied upon to produce a decent medium-bodied red and can also churn out the odd nugget of gold that shows surprising smoky-spicy finesse. There are legions of Baco varieties, all produced in the 19th century by François Baco, but the very first Baco (*Folle Blanche* x *Riparia Grand Glabre*), aka Baco Noir or Baco 1, is the Baco that most hybrid winemakers and wine writers refer to. Most of the other Baco varieties are referred to simply by numbers. A few have their own primary names, such as Petit Boue, Baco, Bakouri, Caperan, Cazalet, Celine, Douriou, Estellat, Olivar, Rescape, and Totmur, but they were all originally registered as numbered Baco hybrids.

ⓑ CABERNET FRANC
Grown all around the world, but especially in Bordeaux, Cabernet Franc fares best as Bouchet in St-Emilion and at Pomerol, across the Dordogne River, where Cabernet Sauvignon is less well represented. It is grown under neutral conditions, and while it might not be easy to distinguish any significant varietal differences between the two Cabernets – suited, as they are, to different situations – the Cabernet Franc tends to produce a slightly earthy style of wine that is very aromatic but has less fine characteristics on the palate when compared to the Cabernet Sauvignon.

ⓑ Bondola An indigenous Swiss variety that can be used for rather rustic reds on its own, the Bondola is at its best when blended with Freisa and other local varieties in wines.

ⓢ Bordelais Synonym for Baroque

ⓢ Borrado des Moscas Synonym for **Bical**

ⓑ Bouchalès More commonly known as Prolongeau in parts of Bordeaux, Bouchalès is the primary name for this grape, which used to be a permitted variety in Blaye and basic Bordeaux appellations. It is now almost extinct, but Cuvée A & A from Château de la Vieille Chapelle is a stunning revival made from recently identified 100-year-old vines. This wine is like a dense *garagiste* Merlot, and for legal reasons it claims to be Merlot, but it's pure Bouchalès.

ⓢ Bouchet Synonym for Cabernet Franc and Cabernet Sauvignon

ⓦ Bourboulenc Found throughout southern France but particularly in the Rhône, where it is an authorized variety for several AOCs, most famously Châteauneuf-du-Pape, Bourboulenc is a late-ripening variety. This gives it the ability to retain good acid levels, which is the reason why it is so useful for tweaking blends. This is also one of the varieties James Busby took to Australia in 1832 after touring French and Spanish vineyards.

ⓦ Bouvier A modest-quality variety in Central Europe and one which, as Ranina, produces the "Tiger's Milk" wine of Slovenia.

ⓢ Bouviertraube Synonym for Bouvier

ⓢ Bovale Synonym for Carignan

ⓑ Brachetto This Muscat-like variety produces the highly fragrant, sweet, and sumptuous Brachetto d'Acqui DOC, the red wine equivalent of a top Asti.

ⓑ Braquet or **Braquet Noir** This grape is grown in a small area of Provence, where it is one of the primary varieties authorized for the production of red and rosé wines

under the Bellet AOC. A Braquet Blanc and Braquet Gris exist, but Braquet Blanc is also a synonym for Jurançon Blanc.

ⓢ Braquet des Jardins Synonym for Brachetto

ⓢ Brown Muscat Synonym for Muscat Blanc à Petits Grains and its red mutation in Australia

ⓑ Brugnola A close relative of Nebbiolo, this is a secondary variety for Valtellina Rosso DOC in Lombardy and for Bosco Eliceo DOC in Emilia-Romagna, Brugnola produces a firm but not full-bodied red wine with high acid levels.

ⓑ Brun Argenté This grape is allowed for the AOCs of Côtes du Rhône, Côtes du Rhône Villages, Gigondas, Rasteau, and Vacqueyras. The decrees for these appellations use the synonyms of both Camarèse and Vaccarèse in all cases. As these are uniquely synonyms for Brun Argenté, not any other variety, this grape is in fact authorized twice.

ⓢ Brunello Synonym for Sangiovese

ⓑ Brun Fourca or **Brun Fourcat** One of the secondary varieties used for red and rosé wines under the Palette AOC of Provence.

ⓢ Bual Synonym for Malvasia Fina

ⓢ Burgundi Kék and **Burgundi Nagyszemu** Synonyms for Gamay

ⓢ Burgundské Modré Synonym for Pinot Noir

ⓢ Burgundske Sede Synonym for Pinot Gris

ⓢ Buzetto Synonym for Ugni Blanc

ⓢ Cabernelle Synonym for Carménère

ⓑ Cabernet Franc See box, above.

ⓑ Cabernet Gernischt Synonym for Carménère

ⓢ Cabernet Gris Synonym for Cabernet Franc

ⓑ Cabernet Pfeffer Minor California variety. There are various opinions as to the grape's origins in the United States, the most plausible being a late-19th-century cross made by William Pfeffer.Others claim it is identical to the Bordeaux variety Gros Verdot.

ⓑ Cabernet Sauvignon The noblest variety of Bordeaux, the Cabernet Sauvignon, rich in colour, aroma, and depth, is vitally important to the classic Médoc wines. Many of its classical traits have been transplanted as far afield as California, Chile, and Australia. The complexities that this grape can achieve transcend simplistic comparisons to cedar, blackcurrants, or violets. The Cabernet Sauvignon's parents are now known to be Cabernet Franc and Sauvignon Blanc.

ⓢ Calabrese Synonym for Nero d'Avola

ⓑ Calitor or **Calitor Noir** One of the minor Provençal grapes that is steadily being grubbed up, due to its high yields, light colour, and neutral character. Calitor is also one of the varieties James Busby took to Australia in 1832, although only the odd vine survives there today. Calitor Blanc and Calitor Gris exist. Calitor is also a synonym for Brachetto.

ⓦ Camaralet or **Camaralet de Lasseube** A secondary variety for the AOCs of Béarn and Jurançon in southwest France. It produces only female flowers, therefore it needs another variety co-planted for pollination.

ⓢ Camaraou Synonym for Camaralet

ⓑ Camarate, Câmarate, or **Camarate Tinto** It is planted throughout Portugal, particularly in the Bairrada, where it is the second-most-planted red variety, but also in the Lisboa, Tejo and Dão regions.

ⓢ Camarèse Synonym for Brun Argenté

ⓑ Campbell Early Sounding more like a potato than a grape, this *Moore Early* x *(Belvidere x Muscat Hamburg)* hybrid is grown in very marginal wine-grape regions, such as South Korea, where it represents 75 per cent of the total vineyard area, and Japan's Miyazaki Prefecture, where the conditions are not unlike those of Florida. Under its synonym Island Belle, this variety is also grown on Stretch Island, Washington State.

ⓑ Canaiolo or **Canaiolo Nero** This native Tuscan grape is a secondary variety for wines such as Carmignano, Chianti, and Vino Nobile di Montepulciano but is capable of producing a soft, fruity red wine in a pure varietal format. Also found in Umbria, where it is traditionally blended with Sangiovese.

ⓢ Cannamelu Synonym for Guarnaccia Bianca or Tintora (red)

ⓢ Cannonadu, Cannonao, Cannonatu, Cannonau, and **Canonau** Synonyms for Grenache

ⓢ Cape Riesling Although not commercialized as such, the Crouchen is still commonly referred to in South Africa as Cape Riesling.

ⓜ Cardinal This American-bred cross is still grown in France, Spain, Morocco, Australia, and Argentina, although it has a neutral flavour and low acids and is mainly used as a table grape.

ⓑ Carignan or **Carignan Noir** A Spanish grape grown extensively in southern France and California. One of its synonyms – Mataro – is a common name for the Mourvèdre in Australia. A Carignan Blanc and Gris also exist.

ⓢ Carignane, Carignano, and **Cariñena** Synonyms for Carignan

ⓑ Carménère An old *bordelais* variety that was almost extinct, until it was discovered in Chile posing as Merlot, and then in Franciacorta, where it was pretending to be Cabernet Franc (interestingly, one of Carménère's synonyms is Grosse Vidure, shared with Cabernet).

ⓑ Carnelian A *Grenache* x *(Carignan* x *Cabernet Sauvignon)* cross developed in the 1930s by Professor Olmo for use in the hot continental climate of California's Central Valley. This grape produces rather rustic reds that do not compare to those of Ruby Cabernet or Rubired, two other Olmo crosses. The potential quality shows promise in Western Australia where it was planted mistaken for Sangiovese.

ⓦ Carricante The major grape in Etna DOC, Carricante usually produces a neutral-flavoured white wine unless planted at higher altitudes, where in the hands of a gifted winemaker such as Planeta, Carricante can have excellent acidity, with stunning minerality and finesse. Although various synonyms include the name Catarratto, Carricante is not related to the true Catarratto, which is in fact a secondary variety in the Etna DOC.

ⓑ Casavecchia Minor grape of uncertain parentage, only found in Campania.

ⓑ Cascade This hybrid is a prolific producer of light coloured, undistinguished red wine. It was used to breed L'Acadie, another hybrid grown successfully in Canada. Cascade's extreme sensitivity to virus diseases outweighs its early-ripening qualities and now the acreage is diminishing.

ⓢ Casculho or **Cascudo** Minor grape for Douro table wines. It can be also a synonym for Camarate, which is a different variety.

ⓢ Castelão de Cova da Beira Synonym for Trincadeira Preta

ⓢ Castelão Nacional and **Castelão Real** Synonym for Camarate

ⓑ Castets Originally from southwest France, having possibly travelled over the Pyrenees from Spain, this variety is almost extinct and rarely encountered beyond a few parcels in Palette AOC of Provence.

Ⓦ **CHARDONNAY**
The greatest non-aromatic dry white wine grape in the world, despite the proliferation of cheap, identikit Chardonnay wines that are churned out globally, hence its "ABC" reputation. This classic variety is responsible for producing the greatest white Burgundies and is one of the three major grape types used in the production of Champagne.

Ⓢ **Catalan** Synonym for Carignan or Mourvèdre

Ⓦ **Catarratto** This the secondary variety in the wines of Etna and is far more widely planted across Sicily than Carricante, the primary grape in that DOC. It is also used in Marsala production but the variety is capable of giving good quality varietal wine too.

Ⓜ **Catawba** A *labrusca* x *vinifera* cross that was once so famous that Henry Longfellow felt compelled to write an ode to it (*see* Ohio, p 628), this grape has the *labrusca*'s characteristically foxy aroma. It is still grown in New York State, Michigan, and Ontario, Canada, but plantings are on a decline due to its late ripening and unsuitability to make high-quality dry wines.

Ⓦ **Cayetana Blanca** Mediocre Iberian variety grown mainly in Spain and Portugal, used to make brandy and neutral dry whites.

Ⓢ **Cercial** Portuguese variety of *Sercial* x *Malvasia Fina* parentage, not to be confused the Madeira variety Sercial.

Ⓜ **Cereza** High-yielding Argentinian varietal that can be quite aromatic but lacks fruit and body, thus is mostly used as a blending component.

Ⓑ **Cesanese** This variety is found in Lazio, where it makes numerous medium-bodied red wines of no special quality in all styles, from dry to sweet, and still to sparkling.

Ⓑ **César** A minor grape variety of moderate quality that is still used in some areas of Burgundy, most notably for Bourgogne Irancy.

Ⓑ **Chambourcin** The rot-resistance of this grape is formidable, enabling it to yield half-decent Beaujolais-like wines even when contending with bracing sea breezes and lashing rain. This might not be my favourite grape, but I have tasted drinkable Chambourcin from as far afield as Illinois, Kansas, Kentucky, Michigan, Missouri, Nebraska, Pennsylvania, and Virginia in the United States; in Canada; in the Granite Belt and Hastings Valley in Australia; and even Madagascar in the Indian Ocean.

Ⓦ **Chardonel** Chardonel is a Seyval x Chardonnay cross made by Cornell University in 1953, but not released until 1990. So far the wines produced have been more bland than even the most anonymous Chardonnay.

Ⓦ **Chardonnay** *See* box, above.

Ⓦ **Chasan** Created by Paul Truel at INRA's Domaine de Vassal in 1958, this *Palomino* x *Chardonnay* cross is slightly aromatic with low acidity. Chasan is best bottled and consumed early, when a bare prickle of carbonic gas will enhance its freshness and crispness.

Ⓦ **Chasselas** or **Chasselas Blanc** Responsible for the best-forgotten Pouilly-sur-Loire wines (not to be confused with Pouilly-Blanc Fumé), this variety is at its modest best in Alsace and Switzerland's Valais (where it is known as the Fendant). Primarily a good eating grape. There are a dozen or more very minor true sub-varieties of Chasselas, but Chasselas Doré and Chasselas Vert are just synonyms, even though they are officially listed as separate varities in several appellations. Chasselas has no fewer than 318 synonyms.

Ⓢ **Chasselas Doré, Chasselas Roux,** or **Chasselas Vert** Synonyms for Chasselas Blanc

Ⓦ **Chenel** A South African cross of Chenin Blanc and Ugni Blanc

Ⓦ **Chenin Blanc** *See* box, below right.

Ⓢ **Chiavennasca** Synonym for Nebbiolo

Ⓑ **Ciliegiolo** A secondary variety for various wines in Tuscany, Umbria, and the Marches, the Ciliegiolo is also known as the Sangiovese Polveroso. DNA research suggests it is a parent to Sangiovese.

Ⓑ **Cinsault** A prolific grape variety found mainly in Southern Rhône and Languedoc-Roussillon vineyards, where it makes robust, well-coloured wines. It is best blended, as happens at Châteauneuf-du-Pape, for example.

Ⓢ **Cinsaut** Synonym for Cinsault

Ⓦ **Clairette** or **Clairette Blanche** A sugar-rich, intrinsically flabby grape best known for its many wines of southern France. It is the Muscat though, not the Clairette, which is chiefly responsible for the "Clairette de Die" in the Rhône. A pink-skinned variant also exists.

Ⓢ **Clairette à Grains Ronds** Synonym for Ugni Blanc or Bourboulenc

Ⓦ **Clairette à Gros Grains** A variant of Clairette Blanche bred by Antoine Besson and grown in parts of Provence.

Ⓢ **Clairette de Trans** Synonym for Clairette Blanche

Ⓢ **Clare Riesling** Synonym for Crouchen

Ⓢ **Clevner** Synonym for Pinot Blanc

Ⓦ **Cococciola** Found mainly in Abruzzo and Puglia, this grape yields straightforward, fruity, and herbal wines.

Ⓦ **Coda di Volpe** A minor Campanian grape variety whose main claim to fame is as the primary variety for Vesuvio Lacryma Christi.

Ⓦ **Códega** This Portuguese variety is grown in Transmontano, where it is usually blended into the wines of Chaves and Valpaços. Any black grape claiming to be Códega will in fact be Malvasia Grossa.

Ⓢ **Colombar** Synonym for Colombard

Ⓦ **Colombard** This produces thin, acidic wine ideal for the distillation of Armagnac and Cognac. It has also adapted well to the hotter winelands of California and South Africa, where its high acidity is a positive attribute. It can produce fresh, lively, everyday-drinking white wines with zesty fruit in Aquitaine, France.

Ⓢ **Colombier** Synonym for Colombard or Sémillon

Ⓑ **Colorino** Widely grown in Tuscany, particularly in Valdarno, Val d'Elsa, and the Val di Pesa, where this variety provides very loose bunches of small grapes that are easily aerated and thus, with their thick skins, are resistant to rot. This enables the production of deep-coloured wines in poor years when Sangiovese and other varieties with tightly packed bunches struggle to bring much colour to a wine, hence the Colorino name.

Ⓦ **Completer** A rare Swiss variety of ancient origin, this grape should be called the Malanstraube because it originates from the Malans area of Graubünden, where it makes distinct, complex wines.

Ⓑ **Complexa** A red Portuguese variety used to make Madeira. It was created as a crossing of Castelao, Muscat Hamburg, and Tintinha (Petit Bouschet)

Ⓑ **Concord** The widest-cultivated variety in North America outside of California, this *Vitis labrusca* variety has an extremely pronounced foxy flavour.

Ⓑ **Cornalin** or **Cornalin d'Aoste** This old Italian variety is virtually extinct in its native Valle d'Aosta and rarely encountered elsewhere, except for a few patches in the Valais, where it produces a rich and concentrated red wine. Although also known as the Humagne Rouge, this variety is not related to the Humagne Blanc.

Ⓑ **Cornifesto** A Portuguese variety grown in the Douro region for port and table wines.

Ⓦ **Cortese** A widely planted Piemontese variety that makes soft-textured, dry white wines that may or may not be slightly *frizzantino* when young. The best known example is perhaps Gavi DOCG.

Ⓑ **Corvina** This variety is widely planted in the Veneto region, where it is the mainstay of Valpolicella and is blended into many other wines. Too easily capable of light-coloured, lightweight reds with simple tart cherry and almond fruit with a touch of bitterness, Corvina needs to be restricted in the vineyard and nurtured in the winery to produce something special.

Ⓑ **Corvinone** This is not identical to Corvina, but instead a separate variety used in Valpolicella DOC and the Amarone and Recioto DOCGs, as a substitute of Corvina.

Ⓢ **Côt** Synonym for Malbec

Ⓦ **CHENIN BLANC**
A variety that acquired its name from Mont-Chenin in the Touraine district in about the 15th century but can be traced back to Anjou, around AD 845. The grape has a good acidity level, thin skin, and a high natural sugar content, making it very suitable for either sparkling or sweet wines, although some dry wines, notably Savennières, are made from it.

⒝ Counoise The colour and spice of Counoise can often play an instrumental, if minority, role in some wines of Châteauneuf-du-Pape, where it is one of the 13 permitted varieties. Also widely grown throughout the southern Rhône, neighbouring Languedoc, and Provence. Beware that Counoise is also a synonym for Aubun, a completely different variety.

⒮ Courbu Blanc and **Courbu Noir** Secondary varieties in many wines of southwestern France, the black-skinned variety is rarely encountered.

⒝ Criolla There is no such thing as true Criolla. This is not so much a variety as a loose family of grapes that has evolved throughout South America since the Spanish missionaries arrived in the wake of the conquistadors. The first grape variety the missionaries brought with them was, unsurprisingly, the so-called Mission Grape (aka Criolla Chica, Listán Prieto or, in Chile, the Pais). This was crossed either spontaneously or, in some cases, possibly by design, with other varieties brought from Spain (primarily Muscat d'Alexandrie and Tempranillo) to form many localized Criolla, such as Criolla Grande. Much research into South America's Criolla varieties has been conducted, but not enough to give a complete picture. They all have a propensity to produce a fairly similar style of rustic red wines that may or may not have a slight aromatic character.

⒮ Criolla Chica Synonym for Listán Prieto or Mission.

Ⓜ Criolla Grande An Argentine descendant of Mission Grape, this mildly aromatic *Mission Grape x Muscat d'Alexandrie* cross is the second-most widely cultivated mid-coloured grape in the world (after Pinot Gris).

⒮ Crna Moravka Synonym for Blaufränkisch

⒝ Croatina A northern Italian variety that has found itself at home in Lombardy and Piedmont, the Croatina is capable of producing well-coloured, soft, and fruity red wines. It has been confused with the Bonarda and even Nebbiolo in the past. The Oltrepò Pavese Bonarda DOC is 100 per cent Croatina

Ⓦ Crouchen Widely cultivated, particularly in Australia and South Africa, this variety used to be widely but incorrectly sold as Clare or Cape Riesling. Recent efforts to standardize varietal labelling to fall in line with EU regulations have reduced this practice, but it is still referred to by those traditional but false Riesling names.

⒮ Cruina Synonym for Corvina

Ⓦ Csabagyöngye This *Madeleine Angevine x Muscat Fleur d'Orange* cross was created in Hungary in 1904 and is grown in small pockets in Hungary, France, and Argentina.

⒮ Cynthiana Synonym for Norton

Ⓦ Damaschino This Sicilian grape has a neutral character and is allowed as a secondary variety in Alcamo DOC but usually ends up in Marsala blends. A little Damaschino is grown in the Barossa Valley of Australia, where it is known as Farana and was previously mistaken for Ugni Blanc, and in Provence, where it is called Mayorquin.

Ⓣ Deckrot This *Pinot Gris x Teinturier Färbertraube* cross is a *teinturier* grape mainly grown in Germany, where it is used to add colour to local red wines.

Ⓦ Delaware This American hybrid of uncertain parentage was developed in Frenchtown, New Jersey, and propagated in Delaware, Ohio, in the mid-19th century. Although grown in New York State and Brazil, it is far more popular in Japan.

Ⓦ Devin A *Gewürztraminer x Veltliner Rotweiss* cross grown in Slovakia for sweet and dry wines.

Ⓦ Dimyat An underrated Bulgarian variety grown in the Danube Plain and the Thracian Lowlands regions, where the Dimyat can produce fresh, tangy white wines. Dimyat Tcherven is a Russian mid-coloured variety.

⒝ Dimrit Turkish variety producing unexciting wines. Dimrit is a common synonym of many, mainly unrelated black and white grapes too.

⒝ Dolcetto Underrated variety found in northwestern Italy, where it is often over-cropped, but can be delightful when yields are restricted.

⒮ Dona Branca Synonym for Roupeiro

Ⓦ Donzelinho Branco The true Donzelinho Branco is a white Port variety, but it is also a synonym for Rabigato.

⒮ Donzelinho de Castelo, Donzelinho de Portugal, Donzelinho do Castello, or **Donzelinho Macho Synonyms** for Donzelinho Tinto

⒮ Donzelino de Castille Synonym for Trousseau

Ⓜ Donzelinho Roxo or **Donzelinho Rosa** A minor Port variety.

⒝ Donzelinho Tinto A Port variety and minor grape for Douro DOC wines.

⒝ Dornfelder *See box, below left.*

Ⓦ Drupeggio Central Italian grape, also known as Canaiolo Bianco. A minor component of the red and rosé Barco Reale de Carmignano DOC.

Ⓣ Dunkelfelder The parentage of this *teinturier* grape is uncertain. The dark wines produced from this variety are typically all colour and no flavour, without the tannin structure expected of any red wine, unless tannin-adjusted.

⒝ Duras A Gaillaçoise grape grown throughout southwest France as a secondary yet useful blending component, but at home in the Gaillac AOC, where it is the primary variety, albeit still blended. The Duras produces medium-sized clusters of small grapes with high sugar levels, but its rustic peppery fruit needs support.

Ⓦ Durella This Italian variety has good but not exceptional acidity and is used mostly to produce Monti Lessini Durello DOC, a sometimes good but never exceptional sparkling wine in the Veneto.

⒝ Dureza Apart from a few vines in the experimental vineyards of Domaine de Vassal in Montpellier, this minor Ardèche variety is for all practical purposes extinct but deserves to be preserved for no other reason than it was one of the parents that gave birth to the Syrah.

⒮ Durif Technically the primary name, this variety is, however, better known as Petite Sirah.

⒮ Early Burgundy Synonym for Abouriou

⒮ Edeltraube Synonym for Gewürztraminer and Savagnin Blanc

Ⓦ Ehrenfelser A *Riesling* x unknown *vinifera* cross, the Ehrenfelser variety has turned out to be a cul-de-sac in British Columbia, Canada, where it remains the preserve of a few die-hards winemakers.

Ⓦ Elbling This variety was once held in high esteem in Germany and France. The major Mosel grape in the 19th century, it is now mostly confined to the Ober-Mosel where its very acid, neutral flavour makes it useful for German *Sekt*.

Ⓦ Emerald Riesling A *Muscadelle du Bordelais* x *Garnacha* cross, this grape

was developed for cultivation in California by Professor Olmo of UC-Davis fame as the sister to his Ruby Cabernet cross.

Ⓦ Emir This variety comes from the Central Anatolian region of Turkey, where it is often blended with the Narince grape. It produces light, fresh, delicate wines.

Ⓦ Encruzado An important Dão variety, this is an under-rated grape with fine fruit and excellent acidity.

Ⓦ Erbaluce This Italian variety is at its best around Caluso in Piedmont, where it truly favours the production of *passito* wines.

⒮ Ermitage Synonym for Marsanne

Ⓦ Esgana Cão Better known as Madeira's classic Sercial grape, Esgana Cão (meaning "dog strangler"!) is the official primary name for this variety, which was once erroneously reputed to be a distant relative of the Riesling. This grape produces a lean, dry white wine with relatively high acidity and emphatic minerality, which translates into the driest and most assertive of Madeira's fortified wine styles.

⒮ Etraire Synonym for Persan-

⒝ Etraire de la Duï This Savoie grape originated in the Isère *département* and is cultivated throughout the region, where it likes clayey-limestone soils and has the potential to produce well-coloured, full-bodied red wines.

Ⓦ Ezerjó A distinctive Hungarian variety that is also grown in northern Serbia, Austria, Bulgaria, China, Czechia, Romania, Slovakia, and Ukraine, but is at its best in the Mór district west of Budapest, where Ezerjó (which means "a thousand good things") is crisp and racy with excellent minerality. Also used for cheaper sweet wines.

⒮ Faber Synonym for Faberrebe

Ⓦ Faberrebe A Weißburgunder x Müller-Thurgau cross grown in Germany, where it produces a fruity wine with a distinctive light Muscat aroma.

Ⓦ Falanghina The short name of two varieties, both widely planted throughout Campania: Falanghina Flegrera and Falanghina Beneventana. Falanghina produces round and fruity wines in still and fizzy formats. Current regulations do not make a distinction between the two varieties.

Ⓦ Favorita In Portugal this is the offspring of *Tamares* x *Muscat d'Alexandrie,* but in Italy, where the Favorita name is far more famous, particularly in Piedmont, it is a synonym for Vermentino.

⒮ Fehérburgundi Synonym for Pinot Blanc

⒮ Feinburgunder Synonym for Chardonnay

⒝ DORNFELDER
A Helfensteiner x Heroldrebe cross that produces light-to-medium-bodied reds with a malleable fruit character, Dornfelder is grown mostly in the Pfalz, Rheinheßen, and Württemberg regions of Germany, but can also be found in England and other marginal grape-growing areas of the world.

Ⓢ Fendant Synonym for Chasselas

Ⓑ Fer This vine has very hard wood that is notoriously difficult to prune, hence its name (meaning "iron"). The Fer is grown throughout southwest France, where it is often blended with Tannat, Malbec, or Cabernet grapes.

Ⓦ Fernão Pires Widely grown throughout Portugal, this grape variety produces fresh, floral-grapey style wines with citrus-Muscat type fruit and good acidity. A mid-coloured sibling called Fernão Pires Rosado exists.

Ⓦ Fetească Alba A Moldovan variety that is so widely cultivated in Romania that many Romanian winemakers consider it to be native to that country, the Fetească Alba is an under-rated grape that has the potential to produce distinctive dry white wines.

Ⓦ Fetească Regală The most widely planted variety of the Fetească family. It produces gently aromatic, fresh wines, some of which are suited to oak ageing.

Ⓑ Fetească Negră Less prolific than Fetească Alba, this variety is even more underrated.

Ⓦ Fiano This Campanian variety has been known since ancient Roman times, when it was called *Vitis apiana*. Fiano wines have a distinctive floral aroma and honeyed flavour, best illustrated by the top wines of Fiano di Avellino DOCG. There is no black variant; Fiano Rosso is a synonym for Aglianico.

Ⓜ Flora This California *Sémillon* x *Gewürztraminer* cross was bred by the famed Professor Olmo and has a tendency to produce overly exotic, often extremely blowsy wines, but it works well in a sparkling wine like Schramsberg's Crémant Demi-Sec, or in a sweet wine like Brown Brothers' Orange Muscat & Flora. If a red-skinned Flora is encountered, it will not be any relation to Flora but will in fact be a *V vinifera* x *labrusca* hybrid that was bred in 1850 by AM Sprangler of Philadelphia.

Ⓦ Folle Blanche Traditionally used for the distillation of Armagnac and Cognac, the Folle Blanche grape also produces the Gros Plant wine of the Loire Valley. It can produce fresh, lively, everyday-drinking white wines with zesty fruit in the Aquitaine, France.

Ⓑ Fortana Originating from Emilia-Romagna, this high-yielding grape can be found around Ravenna and Ferrara. Bosco Eliceo DOC permits varietal bottlings but a small percentage can be added the Lambrusco DOC too.

Ⓦ Francavidda A southern Italian variety capable of producing a delicate, dry white made in the Ostuni DOC, Apulia.

Ⓢ Francavilla Synonym for Francavidda

Ⓢ Franken or **Frankenriesling** Synonyms for Sylvaner

Ⓢ Frankenthaler Synonym for Schiava Grossa

Ⓢ Frankinja Crna or **Frankinja Modra** Synonyms for Blaufränkisch

Ⓢ Frankisch Synonym for Savagnin Blanc

Ⓢ Frankovka, Frankovka Modrá, Frankova, or **Frankovka Modrá** Synonyms for Blaufränkisch

Ⓑ Frappato Sicilian grape, yielding light-bodied, fruity, likeable wines as a varietal or playing a balancing role in blends.

Ⓑ Freisa The Freisa d'Asti has been known for centuries and was a favourite of King Victor Emmanuel, who drank these fruity reds as dry, still wines, although much of the production today is sweet and fizzy. Freisa is produced in Chieri and other areas of Piedmont, where it typically has an aroma of raspberry and rose petals. This variety, which is descended from Nebbiolo, is also grown in Veneto and Switzerland.

Ⓢ French Colombard Synonym for Colombard

Ⓦ Friulano This is better known as the Sauvignonasse, which literally means "Sauvignon-like" or "Sauvignon-ish", but it's not very Sauvignon-ish and is not, in fact, any relation to Sauvignon Blanc, even though another one of its synonyms is Sauvignon Vert. The high-yielding Sauvignonasse is perhaps most infamously grown in Chile, where it used to masquerade as Sauvignon Blanc until the mid-1990s, even though the wines were terrible. It was also once quite widely grown in California, as was the Sauvignon Vert, but as a synonym for Muscadelle, not Sauvignonasse. Although the Sauvignonasse and the Friulano are one and the same variety, when grown in Friuli-Venezia Giulia the Friulano (previously more commonly known as the Tocai Friulano) produces an altogether more enjoyable wine in both pure and blended formats. It is now turning up all over the place, and winemakers are making a real effort to produce wines of genuine quality and interest, such as the Friulano made by Channing Daughters in New York State. One day someone is going to make the following advice redundant by producing a truly excellent Sauvignonasse, but for now it is possible to say that wherever you come across Sauvignonasse it should be avoided, but wherever this grape is known as Friulano, you stand a good chance of finding something decent to drink.

Ⓢ Friularo Synonym for Raboso

Ⓢ Fromenteau Blanc Synonym for Roussanne and Savagnin Blanc

Ⓢ Fromenteau Gris Synonym for Pinot Gris

Ⓑ GAMAY OR GAMAY NOIR
The mass-produced wine of this famous grape from Beaujolais has a tell-tale "peardrop" aroma, indicative of its carbonic-maceration style of vinification. These wines should be drunk very young and fresh, although traditionally vinified wines from Beaujolais' 10 classic *crus* can be aged like other red wines and, after 10 or 15 years, develop Pinot Noir traits. This may be because the grape is an ancient, natural clone of Pinot Noir. In France, Gamay Beaujolais is the synonym for true Gamay.

Ⓢ Fromenteau Rouge Synonym for Gewürztraminer

Ⓢ Fromentot Synonym for Pinot Gris

Ⓢ Frontignac Synonym for Muscat Blanc à Petits Grains

Ⓑ Frühburgunder A lighter-bodied, delicately aromatic mutation of Pinot Noir that evolved in and around the village of Bürgstadt in the Franken, the Frühburgunder nearly became extinct in the 1970s because of its variable yields, but dedicated growers have deliberately restricted yields to coax some exquisitely elegant wines from this grape, which is at its best in southern Baden, the Pfalz, and the Ahr.

Ⓜ Frühroter Veltliner Not black as such, this grape has a darker skin than the more widely encountered Roter Veltliner, but like that grape, the Frühroter is not a true Veltliner and can in fact trace its parentage back to Silvaner and Roter Veltliner itself. This variety has in excess of 130 synonyms, many of which are incorrectly attributed to Malvasia.

Ⓢ Fruttana Synonym for Fortana

Ⓑ Fuella A secondary variety in the blends of Lavilledieu in southwest France and Bellet in Provence.

Ⓢ Fumé Blanc Synonym for Sauvignon Blanc

Ⓑ Fumin A minor variety that produces a robust red on its own but can be a useful blending component, adding structure to wines such as Torrette in the Valle d'Aosta DOC.

Ⓦ Furmint This strong, distinctively flavoured grape is the most important variety used to make Tokaji in Hungary. The Furmint's potential, however, is not only for botrytised sweet wine. If its best clones are sought out and their wines crafted by talented winemakers, Furmint variety could do for Hungarian wines what Grüner Veltliner has done for Austrian dry white wines. Although it has a very different varietal character, this grape has a similar aptitude to Grüner Veltliner in its ability to produce both profoundly complex, oak-fermented wines and a leaner, more mineral style.

Ⓑ Gaglioppo An ancient variety, the Gaglioppo is the most widely planted black grape in Calabria, where it is often blended with the Greco Noir to produce full, fruity, *chiaretto*-style wines. Its *Sangiovese* x *Mantonico di Bianco* parentage was only discovered by DNA fingerprinting, yet one of its synonyms is Mantonico Nero, indicating that local growers had an inkling of its origins long before the term DNA had been invented.

Ⓢ Gaioppo Synonym for Gaglioppo

Ⓦ Galbena This Moldovan variety is often found in Romania. It's full name is Galbena de Odobesti, and it is known by 33 different synonyms. But not all Galbena is Galbena: the name is also an incorrect synonym for Furmint and Fetească Regală.

Ⓢ Galbena Ourata or **Galbenă Uriasa** Synonyms for Galbena

Ⓢ Gallioppa or **Galloppo** Synonyms for Lacrima

Ⓢ Gamaret This Swiss *Gamay Noir* x *Reichensteiner* cross was first released in 1970, but serious commercial cultivation did not take place until the 1990s. As a pure varietal, Gamaret is typically light, yet reasonably well structured, with spicy red fruits. At its best, Gamaret has more fruit and body.

Ⓑ Gamay or **Gamay Noir** See box, above.

Ⓢ Gamay Beaujolais Synonym for Abouriou and the true Gamay

Ⓢ Gamay Blanc or **Gamay Blanc à Feuille Rond** Synonyms primarily for Melon de Bourgogne, Chenin Blanc, and Chardonnay

Ⓣ Gamay de Bouze or **Gamay Teinturier de Bouze** The widest-planted Gamay Teinturier, this variety is believed to have originated in Bouze-lès-Beaune in Burgundy, although most of its 310 hectares (765 acres) of plantings are now located in the Loire. Both Gamay de Bouze and Gamay de Chaudenay were permitted for most of the Loire VDQS appellations, and they are still allowed in the production of various *vins de pays* including Vin de Pays d'Urfé and Vin de Pays du Val de Loire (formerly Jardin de la France). Wines from *teinturier* grapes are supposed to have less finesse than red wines that derive their colour from the skins, yet Domaine de la Charmoise Les Cépages Oubliés from Henri Marionnet is a pure varietal Gamay de Bouze that can be very good indeed.

Ⓣ Gamay de Chaudenay or **Gamay Teinturier de Chaudenay** Almost as widely planted as Gamay de Bouze, approximately 250 hectares (615 acres) of this *teinturier* grape are currently grown in France. For a pure Gamay de Chaudenay, try Clos de la Bruyère Elément Terre from Julien Courtois, a near neighbour of Henri Marionnet. *See* Gamay de Bouze.

Ⓢ Gamay Noir or **Gamay Noir à Just Blanc** Synonym for Gamay

Ⓣ Gamay Teinturier Synonym for any coloured-juice Gamay, especially Gamay de Bouze and Gamay de Chaudenay, but also including Gamay Castille, Gamay Teinturier Fréaux, and Gamay Teinturier Mouro.

Ⓣ Gamay Teinturier Fréaux This is not a distinct variety but a *teinturier* mutation of the white-fleshed Gamay.

Ⓢ Gamza Synonym for Kadarka

Ⓦ Garganega The principal grape used in the production of Soave. At its best, this grape is capable of producing both dry and sweet wines that are surprisingly rich for their delicate balance.

Ⓢ Garnacha or **Garnacha Noir** Although this is the primary name for this variety, Garnacha is better known as Grenache beyond the boundaries of Spain.

The Grenache variety is grown in southern France, where it is partly responsible for the famous wines of Châteauneuf-du-Pape, Tavel, and many others. It is the mainstay of Rioja in Spain, makes Port-style and light rosé wines in California, and is also grown in South Africa. Grenache wines are rich, warm, and alcoholic – sometimes too much so – and require blending with other varieties. The true Grenache has nothing to do with the Grenache de Logroño of Spain, which is, in fact, the Tempranillo or Tinto de Rioja. Some sources say the Alicante (a synonym of the Grenache) is the Alicante Bouschet (or plain Bouschet in California), but this too is misleading.

Ⓢ Garnacha Blanca Synonym for Grenache Blanc

Ⓦ Gewürztraminer See box, below.

Ⓑ Girò A minor Italian grape mostly confined to Sardinia, where it is also known as the Girò Rosso di Spagna, reflecting the belief of some that it might have Spanish origins.

Ⓦ Glera This is the grape we have all come to know as Prosecco, but the name was changed in 2010 to Glera, when the sparkling wine of the same name and grape variety was elevated from DOC to DOCG status. Prosecco is now officially accepted exclusively as a wine appellation, and some ampelographers record Glera as this variety's primary name. This vine is thought to have originated in the village of Prosecco (now a coastal suburb of Trieste), where it was known as Glera, after which it was transplanted to the Colli Euganei in Veneto, where it was called Serprina. Only when it reached the Conegliano-Valdobbiadene region, north of Venice and south of the Dolomites, did this variety assume its Prosecco identity, having been named as such by Francesco Malvolti in 1772. Malvolti alleged that the Prosecco had been growing in the Conegliano-Valdobbiadene area since AD 700, however, and further claimed that it was the grape used for the ancient Roman wine known as *Pucinum,* although there is no documentation explaining how he arrived at either of these wild conclusions. The historical waters surrounding the origin of this vine are also muddied by the fact that a grape known as the Prosecco Nostrale is none other than the Malvasia Bianca Lunga (the Malvasia of Chianti), so Prosecco has been consistently mistaken for Malvasia over the ages, and vice versa, meaning claims of origin could easily be misleading. The Glera grape makes a very uninspiring still wine, except as a *vin de paille,* and in terms of sparkling wine gains nothing from time on yeast or, indeed, post-disgorgement ageing. It works best when tank-fermented and sold as young and as fresh as possible, as DOCG Prosecco is.

Ⓦ Godello Galician grape capable of making excellent, age-worthy wines. Its cultivation has spread to other regions in Spain (Bierzo DO, Castilla y León) and to Portugal, where it's known under many synonyms, such as Gouveio, Agodenho or Verdelho do Dão.

Ⓦ Goldburger Developed in 1922 by Professor Fritz Zweigelt, this *Welschriesling* x *Orangetraube* is primarily used in Austrian white wine blends.

Ⓢ Gordo Blanco Synonym for Muscat d'Alexandrie

Ⓦ Gouais or **Gouais Blanc** This mundane vine produces sour grapes that do not make a very pleasant wine. In the Middle Ages, Gouais vines were planted around a vineyard so that its sour grapes might deter animals and grape thieves. With Pinot Fin Teinturier, another unremarkable grape, the Gouais has, however, given birth to numerous classic varieties, including one of the world's greatest grapes – Chardonnay.

Ⓢ Gouveio or **Gouveio Real** Synonyms for Godello

Ⓑ Graciano An important variety used in the production of Rioja, where a small amount lends richness and fruit to a blend, Graciano is still a seriously underrated grape.

Ⓢ Grande Vidure Synonym for Carménère

Ⓣ Grand Noir de la Calmette This *Aramon x Petit Bouschet teinturier* grape was developed in 1855 at Domaine de la Calmette by Henri Bouschet of Alicante Bouschet fame.

Ⓦ Grasă or **Grasă de Cotnari** A very old Romanian variety often used for sweet wines due to its susceptibility to noble rot, this grape is often confused with the Hungarian Furmint.

Ⓢ Graševina Synonym for Welschriesling

Ⓢ Grauburgunder Synonym for Pinot Gris

Ⓢ Grauklevner Synonym for Pinot Gris

Ⓢ Gray Riesling Synonym for Trousseau Gris

Ⓢ Grecanico Synonym for Garganega

Ⓢ Grechetto and **Greco** These are often used synonymously for each other, which is incorrect as they are totally different varieties. There is a raft of synonyms with a place name attached that can either be true Grechetto or Greco but may also be synonyms for Malvasia, Trebbiano (Ugni Blanc) or other grapes.

Ⓑ Grechetto or **Grechetto Blanco** The true white Grechetto is grown further north than the Greco, mainly in Umbria, where it forms part of the blend for Orvieto and numerous other wines

Ⓢ Greco Nero The shared name of a number of different grapes in Calabria: Castiglione, Nerello, Magliocco Dolce, Greco Nero di Verbicaro, and Greco Nero di Sibari.

Ⓦ Greco or **Greco di Tufo** The true white Greco is more famous than the black "Grecos", even though it often plays no more than a supporting role in many Italian wines. The two most

At its most clear-cut and varietally distinctive in Alsace (where it is never spelled with an umlaut), this variety produces very aromatic wines that are naturally low in acidity. The Gewürztraminer's famous spiciness is derived from terpenes, which are found in the grape's skins. The grapes have to be ripe, otherwise the wine will just have a soft rose-petal character, and the wine must not be acid-adjusted, otherwise the spice will lack breadth and potential spicy complexity. All classic Gewürztraminer wines need bottle age to reveal their true spice-laden aromas.

famous exceptions are Greco di Tufo DOCG, which can be a delightfully delicate dry white wine when made by the likes of Feudi di San Gregorio, and Greco di Bianco DOC, which is a *passito* wine of mostly uninspiring quality, except in the hands of the gifted Umberto Ceratti, when it can be absolutely vivacious.

Ⓦ Green Hungarian The primary name for this Hungarian variety is Putzscheere, but it is better known as Green Hungarian. Small pockets of this vine are to be found in Hungary, Romania, and California

Ⓑ Grenache or **Grenache Noir** See box, opposite page, above right.

Ⓦ Grenache Blanc This is the white Grenache variant that is widely planted in France and Spain. It is an ancient Spanish variety with the potential to produce a good-quality, full-bodied wine.

Ⓜ Grenache Gris Of all the Spanish Garnacha variants, this is grown more in France than Spain, or anywhere else for that matter. Grenache Gris can be highly aromatic, with a touch of spice.

Ⓑ Grignolino This Italian variety originates from the Asti region and typically produces lightly tannic red wines with a slightly bitter aftertaste under the Grignolino d'Asti DOC.

Ⓦ Grillo The original Marsala variety, Grillo is still considered the best for that fortified wine, but it is also found throughout the rest of Sicily, where it is either blended or, mostly, makes an undistinguished soft, dry white wine, although the best can have a gentle, floral aroma, with tropical fruits on the palate and a touch of spice on the finish.

Ⓦ Gringet A minor white variety found in the Savoie, where it contributes to the blend of still and, mostly, sparkling wines.

Ⓢ Gris de Salces Synonym for Ondenc

Ⓑ Grolleau or **Grolleau Noir** A prolific grape with a high natural sugar content, it is important for the bulk production of Anjou rosé, but rarely interesting in terms of quality. Both Grolleau and Groslot names are used throughout the Loire – for instance, it's Grolleau in Anjou, but Groslot in the Fiefs Vendée, but both are pronounced the same, and both are the same variety. Internationally, Grolleau is the accepted standard, with Groslot its most important synonym. Grolleau Blanc and Grolleau Gris exists but is rarely encountered and they are just a colour mutation of the black variant.

Ⓑ Groppello At home in Breganze on the opposite bank of Lake Garda to Valpolicella, Groppello is definitely better suited to the pure varietal format. When blended with Sangiovese, Marzemino, or Barbera, as often happens, the results seldom

equal, let alone exceed, the sum of their parts. The original localized clone, the Groppello Gentile, is considered superior to the Groppello di Revòor Groppello Mocasina, both of which are also grown in the Lake Garda area.

Ⓢ Groslot Synonym for Grolleau

Ⓢ Gros Manseng Grown in southwestern France, where with the Petit Manseng it contributes to many wines, dry and sweet, but is most famous for producing the legendary and succulently sweet Jurançon Moelleux. Curiously, the Gros Manseng grapes are as small as, if not smaller than, those of the Petit Manseng, and in terms of wine character they are very similar, producing wines with a certain nervosity of fruit. There is far more Gros Manseng cultivated in France than Petit Manseng, and late-harvest Gros Manseng tends to be more passerillé, whereas the smaller percentage of Petit Manseng will attract more noble rot.

Ⓢ Gros Plant Synonym for Folle Blanche

Ⓢ Grosse Clairette Synonym for Bourboulenc, Vermentino, or Picardan

Ⓢ Gross Vernatsch (GroßVernatsch) or **Grossvernatsch (Großvernatsch)** Synonyms for Schiava Grossa

Ⓢ Grünedel, Grüner Silvaner, Grünfrankisch, and **Grünling** Synonyms for Sylvaner

Ⓦ Grüner Veltliner See box, below right.

Ⓢ Guarnaccia If the grapes are black, this is a synonym for Perricone, Tintora, and Vernaccia Nera. But if the grapes are white, it is either a separate variety (see below) or a synonym for Vernaccia Bianca.

Ⓦ Guarnaccia or **Guarnaccia Bianca** The true Guaranaccia is a white variety originating from Southern Italy where it is also called Coda di Volpe Bianca ("white foxtail") and thought to be the grape of the famous Roman Falerno wine.

Ⓢ Gutedel Synonym for Chasselas

Ⓦ Gutenborner This *Müller-Thurgau* x *Chasselas* cross is grown in Germany and England. It produces grapes with intrinsically high sugar levels but makes rather neutral wines.

Ⓢ Gwäss Synonym for Gouais

Ⓢ Hanepoot Synonym for Muscat d'Alexandrie

Ⓦ Hárslevelű This Hungarian grape is the second most important Tokaji variety. It produces full, rich, and powerfully perfumed wines.

Ⓢ Heunisch Weiss Synonym for Gouais

Ⓦ Hondarribi Zuri The main white grape in the Basque DOs; however, quite confusingly it refers to three distinct varieties: Courbu Blanc, Crouchen and Noah (a hybrid).

Ⓦ Humagne Blanc or **Humagne Blanche** This is no relation to the Humagne Rouge, which is also grown in the Valais of Switzerland, where this variety originated, the Humagne Blanc has lost its *Capsicum* aroma now that it is being grown at lower yields. Never alcoholic, the wines from this grape can be deceptively full and rich, with hints of exotic fruit on the palate and sometimes a touch of spice on the finish. A mid-coloured Humagne Gris exists.

Ⓢ Humagne Rouge Synonym for Cornalin

Ⓢ Hunter Riesling Although not commercialized as such, Sémillon is still commonly referred to in Australia as Hunter Riesling.

Ⓦ Huxelrebe A *Chasselas* x *Muscat Courtillier* cross that is grown in Germany and England, and is capable of producing good-quality wine. At its best, the Huxelrebe has a herbaceous-grapefruit bite with a hint of elderflower that can go very "cat's pee" in cold or wet years.

Ⓦ Impigno This Italian variety makes light, delicate dry white in Ostuni DOC, where it is blended with Francavidda.

Ⓦ Incrocio Bruni 54 A *Verdicchio* x *Sauvignon Blanc* cross, grown on a couple of hectares in the Marche.

Ⓢ Incrocio Manzoni 2.15 A *Prosecco* x *Cabernet Sauvignon* cross, grown only in Treviso, Italy.

Ⓢ Incrocio Terzi 1 A *Barbera* x *Cabernet Franc* cross, cultivated in Lombardia.

Ⓦ Inzolia This Marsala grape is also capable of producing surprisingly light and delicate dry wines in Sicily. Also a blending component for Elba DOC in Tuscany, where this variety is known as Ansonica.

Ⓦ Irsai Olivér This Hungarian **Pozsonyi** x *Perl Von Csaba* cross was created in 1930 and produces a very fresh, aromatic wine with grapey fruit, but it is intrinsically too soft without a bit of judicious blending or being bottled with a bit of residual gas to provide a crispy-crunchy impression that lifts the finish.

Ⓢ Iskendiriye Misketi Synonym for Muscat d'Alexandrie

Ⓑ Isabella One of the oldest American hybrids, probably a natural cross of a *V labrusca* and a *vinifera* species. Today it is widely grown in Brazil, India (known as Bangalore Blue), Moldova, and Ukraine.

Ⓢ Italianski Rizling or It**aliansky Rizling** Synonyms for Welschriesling

Ⓦ Jacquère or **Jacquère Blanche** The work-horse grape of the Savoie, the Jacquère is subject to rot, has a neutral flavour and high acidity.

Ⓣ Jacquez This grape is having something of a minor revival in Texas, where it is known and sold as Black Spanish (or Lenoir in other parts of the United States). It was widely grown in the 1830s to make a Port-like and communion wine. It is believed to be a *Vitis bourquiniana (aestivalis* x *vinifera)* hybrid. The Black Spanish in Texas can be quite dense but tastes a bit diffused and sometimes earthy.

Ⓢ Johannisberg Riesling Synonym for Riesling

Ⓑ Joubertin A minor Savoie grape, of which there is very little left in France but still a few hectares in Argentina.

Ⓦ Juhfark Hungarian variety grown almost exclusively in the Somló region, where it gives age-worthy, firm, mineral wines.

Ⓦ Jurançon Blanc This minor variety is found in southwest France, where it is authorized for Pineau des Charentes AOC, Vin de Pays du Comté Tolosan, and Armagnac.

Ⓑ Jurançon Noir A natural **Folle Blanche** x *Côt* cross, giving light reds and rosés, mostly in The Tarn and the Gers.

Ⓑ Kadarka One of Hungary's major grape varieties, also grown throughout the Balkans. It makes pleasant, light, and fruity wine.

Ⓢ Kalabaki or **Kalambaki** Synonyms for Limnio

Ⓑ Kalecik Karasi Central Turkish variety producifresh, fruity, light reds.

Ⓦ Kanzler A *Müller-Thurgau* x *Sylvaner* cross, producing a good Sylvaner substitute in the Rheinheßen.

Ⓦ GRÜNER VELTLINER
This is the most important wine grape in Austria, where it commonly produces fresh, well-balanced wines, with a light, fruity, sometimes slightly spicy, flavour. Top-quality Grüner Veltliner from the Wachau can have a penetrating ground white pepperiness. Some of the very top-quality wines have become too fat and heavy, whereas others are the equal of great Chardonnays. A number of unoaked Grüner Veltliners from the finest sites can have a beautiful minerality that sometimes verges on Riesling in character.

Ⓜ KOSHU

Although commonly viewed as a Japanese variety, the Koshu originated in Asia Minor, moving to China via the Silk Route, before Buddhists took it to Japan, where it has been cultivated in the Yamanashi district since at least the eighth century. Until the turn of the millennium, this grape was used only to make sweet wine or disappeared into modern Japanese blends, but producers have started to hone a pure Koshu in its own inimitable dry style, which would be an acquired taste for most Western palates but is revered and very much in vogue on its home market.

Ⓑ Kara Dimrit Technically synonyms for Dimrit Beyaz, but Kara Dimrit (or Karadimrit), literally meaning "red Dimrit", is the more common usage. Grown in the Central Anatolia and Mediterranean Coast regions. Also used for the production of *raki*.

Ⓢ Kékfrankos Synonym for Blaufränkisch

Ⓦ Kéknyelű Rare Hungarian variety that produces only female flowers, relying on another variety for pollination. It is grown mainly in the Badacsony area, giving elegant, complex, age-worthy wines.

Ⓦ Kerner A *Schiava Grossa* x *Riesling* cross that produces wines with a high natural sugar content and good acidity, but a very light aroma. Like Müller-Thurgau, Kerner has been planted in too many former Riesling vineyards to have much sympathy for it.

Ⓢ Klevener de Heiligensteiner Synonym for Savagnin Rosé

Ⓢ Klevner Synonym for Pinot Blanc

Ⓦ Knipperlé Alsatian variety, now almost exclusively grown on Alois Raubal's estate near Vienna.

Ⓣ Kolor This variety og *teinturier* grape is a *Pinot Noir* x *Teinturier Färbertraube* cross that was developed at the Freiburg Research Institute in Germany.

Ⓜ Koshu See box, above.

Ⓦ L'Acadie Specifically created at Vineland in Ontario for the harsh winters and short, hot summers of Nova Scotia, this hybrid is named after 17th-century Acadian French who settled in the Annapolis Valley of that province. L'Acadie has proved some degree of competence as a very minor blending component in some sparkling-wine *cuvées* produced by Benjamin Bridge in Nova Scotia's Gaspareau Valley.

Ⓑ Lacrima or **Lacrima di Morro d'Alba** Primarily known for producing a soft, medium-bodied red wine in the Marche province of Ancona, under the DOC Lacrima di Morro d'Alba. Lacrima (or Lacryma) Christi

del Vesuvio is not made from the Lacrima grape. Various localized clones, both black and white exist, but beware: Lacrima is also a synonym for Greco di Tufo, Gaglioppo, and Sangiovese.

Ⓢ Lafnetscha Old Swiss variety grown on a few hectares in the Haut-Valais today.

Ⓑ Lagrein An ancient Alto Adige variety, Lagrein was well known to Pliny, who called it Lageos, and is now known to be a grape of complex pedigree. This underrated grape has distinctively rich and chunky fruit with relatively high acid levels when young, but it is capable of developing a silky-smooth finesse after a few years in bottle. Also good for rosé and used as a blending component throughout the region.

Ⓑ Lambrusco See box, opposite page, above left.

Ⓢ Languedocien Synonym for Piquepoul

Ⓦ Lauzet A minor white variety that is used as a blending component in southwest France.

Ⓦ Leányka Hungarian variety producing mostly bland, soft, unexciting wines in the northeast of the country.

Ⓢ Lemberger Synonym for Blaufränkisch

Ⓦ Len de l'El A flavoursome, naturally sugar-rich grape that is used in Gaillac.

Ⓑ Léon Millot One of the better hybrid grapes, this *Millardet et Grasset* x *Goldriesling* cross (exactly the same parentage as the Maréchal Foch) has produced some honourable reds, most notably in Ontario and the US Eastern Seaboard and Great Lakes states.

Ⓑ Liatiko An important Greek grape, Liatiko is the primary variety for the wines of Dafnés and Sitia on Crete (Kríti).

Ⓢ Limberger Synonym for Blaufränkisch

Ⓑ Limnio An ancient variety known to Hesiodos and Polydeuctes as

Limnia, this grape originated on the island of Limnos, where its importance has diminished in recent times. Limnio is today found in northern Greece, in Rapsani, Halkidiki, and Mount Athose, where it is more of a blending component than a stand-alone variety. A late-ripening grape, Limnio has a tendency to show herbaceous pyrazines when not fully ripe but has the potential to provide good colour and body.

Ⓑ Limniona A near-extinct variety from Thessalia, now slowly regaining popularity due to its age-worthy, spicy, deep-coloured wines.

Ⓢ Lipovina Synonym for Hárslevelű

Ⓢ Listán Synonym for Palomino Fino

Ⓑ Listán Negro This grape, which is found mostly in the Canary Islands, has no links to Palomino Negro, but is a distinct variety of uncertain parentage. Whatever its upbringing, Listán Negro has so far made only modest, everyday red wines at best.

Ⓑ Listán Prieto Also known as the Mission Grape, this is the first variety the Spanish missionaries planted in the wake of the conquistadors, and it is still widely cultivated throughout South America and Mexico, where it is mostly used for rustic reds consumed by locals or lost in cheap mega-blends.

Ⓦ Liza A Serbian *Kunleany* x *Pinot Gris* hybrid developed in the 1970s to provide extreme winter hardiness, thanks to the Kunleany, which is itself part Amurensis.

Ⓢ Lladoner or **Lledoner Pelut** Synonyms for Grenache

Ⓦ Loureiro Considered one of the best two Vinho Verde grapes with Alvarinho, Loureiro is much higher cropping, but it was compared to Muscat in the 19th century and, at lower yields, has the potential to provide fine floral aromatics.

Ⓢ Macabeu or **Maccabeu** Synonym for Macabéo

Ⓦ Macabéo This is a Spanish variety used to "lift" a sparkling Cava blend and give it freshness. Bearing the name of Viura, it is also responsible for some of the best fresh, unoaked white Rioja.

Ⓦ Maceratino An old Marches variety, Maceratino produces at best a light, delicate dry white wine, but more often something far more neutral. Due to its variability, this grape has historically been mistaken for Greco and Verdicchio.

Ⓢ Mâconnais Synonym for Altesse and Chardonnay

Ⓦ Madeleine Angevine This is a *Précoce de Malingre* x *Madeleine Royale* cross that is grown quite successfully in England, where it produces a characteristically light-bodied, aromatic wine in some of the country's most northerly vineyards.

Ⓢ Madiran Synonym for Tannat

Ⓑ Magliocco Dolce Calabrian variety, along with the rare Magliocco Canino. In addition, Magliocco is also a synonym of Gaglioppo.

Ⓑ Malbec This grape is traditionally used in Bordeaux blends in order to provide colour and tannin. It is also grown in the Loire, Cahors, and Mediterranean regions, among many others, and was the grape responsible for the "black wine of Cahors" – a legendary name, if not wine, in the 19th century. Cahors wine, however, is now made from a blend of grapes and is an infinitely superior wine to its predecessor.

Ⓢ Malmsey or **Malvagia** Synonyms for Malvasia Bianca and Pinot Gris

Ⓢ Malvasia A large and confusing family of true Malvasias and wannabees, further complicated by the fact that Malvasia is a common synonym for Pinot Gris and an occasional synonym for Branco sem Nome, Códega, Marufo (black variety), Roupeiro, Tamares, Torrontés, Trincadeira Preta (black variety), and Teneron.

Ⓢ Malvasia Babosa (Portuguese); **Malvasia Bianca di Basilicata** (Italian); **Malvasia Bianca di Piemonte** (Italian, aka Malvasia Greca, Moscato Greco Nell'Astiniano); **Malvasia Bianca Lunga** (Italian, aka Malvasia Bianca di Bari, Malvasia Bianca di Toscana, Malvasia Cannilunga, Malvasia Cannilunga di Novoli, Malvasia de Chianti, Malvasia del Chianti, Malvasia di Arezzo, Malvasia di Brolio, Malvasia di San Nicandro, Malvasia di Trieste, Malvasia Lunga, Malvasia Piccola Lunga, Malvasia Pugliese Bianca, Malvasia Toscana, Malvasia Trevigna Verace); **Malvasia Branco de São Jorge, Malvasia Candida** (Italian via Crete, aka Malvasia Bianca di Candia, Malvasia de Madere, Malvasia Fina de Madere, Malvazija Kanida); **Malvasia de Lanzarote** (Spanish); **Malvasia de Oerias** (Portuguese); **Malvasia de Porto** (Portuguese); **Malvasia de Setúbal** (Portuguese); Malvasia del Lazio (Italian, aka Malvasia Col Puntino, Malvasia Gentile, Malvasia Nostrale, Malvasia Puntinata); **Malvasia di Candia Aromatica** (aka Malvasia Bianca Aromatica, Malvasia di Alessandria, Malvasia di Candia, Malvasia di Candia a Sapore Moscato); **Malvasia di Candia a Sapore Semplice, Malvasia di Napoli** (Italian); **Malvasia di Sardegna** (Italian, aka Malvasia de Sitges, Malvasia delle Lipari, Malvasia di Bosa, Malvasia di Lipari. Malvasia di Ragusa, Malvasia di Sardegna, Malvasia di Sitjes [sic], Malvasija de Doubrovnik Blanche, Malvoisie de Liparie, Malvoisie de Sitges, Malvoisie Doubrovatchka); **Malvasia Istriana** (Italian, aka

ⓑ LAMBRUSCO
This covers a multitude of sub-varieties, all of which essentially produce simplistic wine, primarily in Emilia-Romagna, where it is fizzed up, cherry-coloured, and cherry-flavoured. Lambrusco Salamino is by far the most widely planted, followed by Marani, Gasparossa, and then, at a slightly lower level, Foglia Frastagliata and Sorbara. Beyond this, there are small pockets of numerous other sub-varieties, mostly black but occasionally mid-coloured, such as Pedúncolo Rossa and Sorbara a Foglia Rossa. To add to the confusion, each sub-variety also has several synonyms.

Malvasia d'Istria, Malvasia del Carso, Malvasia del Lazio, Malvasia di Ronchi, Malvasia Friulana, Malvasia Istriana, Malvasia Nostrale, Malvasia Puntinata, Malvasika Istarska Bijela, Malvazija Istarska, Malvazija Istarska Bijela, Malvoisie de l'Istrie); **Malvasia Moscatel Fonte Grande** (Portuguese, aka Fonte Grande); **Malvasia Nostrana** (Italian); **Malvasia Parda** (Italian). These are not synonyms but are all recognized as singular localized clones of Malvasia Bianca (with their own synonyms in parentheses).

ⓦ Malvasia Bianca The true Malvasia Bianca is an Italian variety of possibly Greek origin. This grape is widely grown in Italy, France, and Spain. There are also fairly sizeable plantations in Brazil and the United States, with pockets in most other wine countries. The wines produced are typified by their full body and distinctive, almost musky aroma, but the most famous Malvasia is, of course, Malmsey Madeira.

ⓢ Malvasia a Bonifacio Synonym for Vermentino

ⓢ Malvasia Branca In addition to being a synonym for Malvasia Bianca, this is also a synonym for Códega.

ⓢ Malvasia di Casorzo (Italian, aka Malvasia Casorzo, Malvasia Nera di Casale, Malvasia Nera di Casorzo, Malvasia Nera di Piemonte, Moscatellina); **Malvasia di Castelnouvo del Bosco, Malvasia di Schierano** (Italian, aka Malvasia a Grappolo Corto, Malvasia di Casorzo, Malvasia di Castelnouvo); **Malvasia Nera di Basilicata** (Italian); **Malvasia Nera di Basilicata** (Italian); **Nera di Bolzano** (Italian); **Malvasia Nera di Brindisi** (Italian, aka Malvasia di Bitonto, Malvasia di Trani, Malvasia Nera di Bari, Malvasia Nera di Candia, Malvasia Nera di Lecce); **Malvasia Nera Lunga** (Italian, aka Moscatella

Nell'Alessandrina); and **Malvasia Preta** (Portuguese, aka Moireto do Dão). These are not synonyms but are all recognized as singular localized clones of Malvasia Nera.

ⓢ Malvasia Comun, Malvasia de la Rioja, Malvasia de Rioja, Malvasia de Roja, Malvasia Fina, Malvasia Grossa, Malvasia Riojana, Malvazija Bela, and **Malvaziya Fina** Synonyms for Malvasia Bianca

ⓢ Malvasia Corada or **Malvasia Fina de Douro** Synonyms for Vital

ⓢ Malvasia Fina The richest and fattest of Madeira's four classic grape varieties under the synonym Boal, it is also grown for wines produced on the Portuguese mainland in the DO appellations of Carcavelos and Lagos, and the IPRs of Chaves and Valpaços in the Trás-os-Montes region. In addition to being a synonym for Malvasia Bianca, this is also a synonym for Arinto, Douradinha, and Trebbiano Toscano.

ⓜ Malvasia Fina Roxa (Portuguese, aka Assario Roxo, Malvasia Roxa). This is a pink-skinned mutation of Malvasia Fina

ⓢ Malvasia Grossa Synonym for Malvasia di Sardegna and Códega and Vermentino

ⓢ Malvasia Grosso Synonym for Códega

ⓢ Malvasia de Manresa Synonym for Garganega

ⓑ Malvasia Nera Although not as widely cultivated as the white-skinned cultivars, this Italian variety is the most aromatic of all Malvasia variants.

ⓢ Malvasia Nera Agglomerata, Malvasia Nera di Bari, Malvasia Nera di Candia, Malvasia Odorossima, or **Malvoisie Noire Musquée** Synonyms for Malvasia Nera

ⓢ Malvasia Rei Synonym for Palomino

ⓜ Malvasia Rosa A mutation of Malvasia di Candia Aromatica, Malvasia Rosa was found by Mario Fregoni in 1967.

ⓢ Malvazia Krasnaja Synonym for Frühroter Veltliner

ⓢ Malvoisie Synonym for Malvasia and Pinot Gris

ⓢ Malvoisie Rosé Synonym for Frühroter Veltliner

ⓢ Mammola or **Mammolo** Synonyms for Sciaccarello

ⓑ Mandilaria A tannic Greek variety that is grown on Crete, Paros, and Rhodes, where it is either softened by other varieties or used to provide the backbone of a blend.

ⓑ Manseng Noir Little-known black-skinned variety unrelated to the Gros and Petit Manseng, Manseng Noir is also grown in southwest France, where it contributes to the fresh and fruity red-wine blends of Béarn.

ⓢ Mantonico Nero Synonym for Gaglioppo

ⓑ Maratheftiko An indigenous Cypriot grape, this underrated variety has the capability to produce wines of good colour with a rich, spicy black-fruit flavour.

ⓑ Maréchal Foch One of the better hybrid grapes, this *Millardet et Grasset* x *Goldriesling* (exactly the same parentage as the Léon Millot) cross has produced some honourable reds, most notably in Ontario and the the Eastern Seaboard and Great Lakes state of the United States.

ⓑ Mare's Nipple This Turkish variety was presented to Emperor Tai-Tsung of China in 674, when it was recorded that the purple grapes hung in bunches up to 60 centimetres (24 inches) long and made a "fiery" wine.

ⓢ Maria Gomes Synonym for Fernão Pires

ⓦ Mariensteiner A *Sylvaner* x *Rieslaner* cross that is grown in Germany and is generally considered superior to the Sylvaner.

ⓑ Marquette A relatively recent (released in 2006), cold-tolerant hybrid capale of giving decent wines, showing success in North America.

ⓦ Marsanne This grape makes fat, rich, full wines and is one of the two

major varieties used to produce the rare white wines of Hermitage and Châteauneuf-du-Pape.

ⓑ Marselan See box, below.

ⓑ Marsigliana A minor blending grape in Calabria. A Marsigliana Bianco exists, but this is also a synonym for a Lebanese grape also known as Afus Ali.

ⓢ Maru Synonym for Mare's Nipple

ⓑ Marufo Undistinguished Iberian blending grape giving light, thin wines.

ⓑ Marzemino This grape variety was probably born in northern Italy. By the 18th century it had achieved sufficient fame for Mozart to use it in his opera Don Giovanni as a preliminary to the seduction of Zerlina. A Marzemino cluster typically provides large, loosely bunched grapes that make aromatic, early-drinking red wines, either as a pure varietal in Trentino or blended in several other DOCs.

ⓢ Mataro Synonym for Mourvèdre and Carignan

ⓦ Mauzac A late-ripener with good natural acidity, grown in southwest France, Mauzac is flexible in the wines it produces but is used for sparkling wine and is particularly suitable for sweet sparklers in Limoux.

ⓜ Mauzac Rosé The mid-coloured Mauzac Blanc mutation is mostly restricted to Gaillac sparkling wines and vins de pays in the Comté Tolosan.

ⓑ Mavro The most planted grape on Cyprus, making low-acid, simple red wines.

ⓑ Mavrodaphne The most famous example of this grape is Mavrodaphne of Patras, a rich, sweet, red liqueur wine with a velvety smooth, sweet-oak finish. There is an affinity between Mavrodaphne and oak that reveals itself even when this variety is playing a supporting role, as it does in Gentilini's Syrah.

ⓑ Mavrotragno Greek variety from Santorini, used to make sweet wines and fresh, structured, spicy dry reds.

ⓑ Mavroud or **Mavrud** The Mavroud variety probably at its best in Assenovgrad, Bulgaria, where it produces dark, dry, plummy-spicy red wine that can age well.

ⓑ MARSELAN
One of the better new varieties, this *Cabernet Sauvignon* x *Grenache Noir* cross was created in 1961 and has become relatively widely planted in France since the new millennium, despite being restricted to *Vins de Table* (now *Vins de France*) and *Vins de Pays*. It has raspberry-blackcurrant fruit with a touch of smokiness and makes a straightforward yet substantial wine.

Ⓑ MERLOT
Merlot produces nicely coloured wines, soft in fruit but capable of great richness. It is invaluable in Bordeaux, bringing fruity lusciousness and a velvet quality to wines that might otherwise be rather hard and austere. It is the chief grape in Château Pétrus, which is the top name in Pomerol and one of the most well-regarded wines in the world.

Ⓜ Mayolet A minor blending component in the Valle d'Aosta.

Ⓢ Mayorquin Synonym for Damaschino.

Ⓢ Mazuelo Synonym for Carignan.

Ⓑ Melnik Technically Shiroka Melnishkais, the primary name for this Bulgarian variety, but most wine professionals, including those in the Bulgarian wine industry, refer to this grape simply as Melnik. It makes a well-coloured, smooth and richly flavoured red wine that may be soft or tannic, depending on how it was made.

Ⓦ Melon or **Melon de Bourgogne** This variety was transplanted from Burgundy to Nantais where it replaced less hardy vines after the terrible winter of 1709. Most famous for its production of Muscadet. When fully ripe, it makes very flabby wines, lacking in acidity, although curiously it can be successful in California's warmer climes.

Ⓑ Mencia Iberian variety, mainly grown in Castilla y León in Spain and in the Dão in Portugal, where it is called Jaen.

Ⓢ Menu Pineau or **Menu Pineau de Vouvray** Synonyms for Arbois

Ⓑ Mérille A minor blending component in Bergerac and the Frontonnais, Mérille is also a synonym for Bouchalès.

Ⓑ Merlot See box, above.

Ⓦ Merlot Blanc An undistinguished variety, this grape was banned in the wines of Blaye and Bourg in 1997 but is still allowed for the generic white-wine appellations of Bordeaux and Bordeaux Supérieur.

Ⓢ Meslier Synonym for Luglienga Bianca, Meslier Saint Francois (spontaneous *Gouais* x *Chenin Blanc* cross), Roublot, and Welschriesling. *See* Petit Meslier.

Ⓢ Meunier Synonym for Pinot Meunier

Ⓦ Mila A Serbian *Kunleany* x *Muscat Ottonel* hybrid developed in 1977 to provide extreme winter hardiness, thanks to the Kunleany, which is itself part Amurensis.

Ⓑ Milgranet A secondary variety authorized for the red wine blends of Lavilledieu in southwest France but almost extinct. An even more rarely encountered white Milgranet exists.

Ⓣ Millefleurien A localized *teinturier* clone of Gamay de Fréaux, this grape is restricted to a few vineyards around the village of Millefleur in Auvergne.

Ⓦ Misket This can either be a singular variety of Ukrainian Muscat or a generic synonym for Muscat, mostly Bulgarian.

Ⓜ Misket Cherven The Misket Cherven, or Red Miske,t is the most widely planted Misket variety in Bulgaria, where it produces fresh, light, aromatic white wines.

Ⓑ Misket Dunavski (*Chaouch Blanc* x *Muscat Hamburg*); **Misket Ran** (*MAI 3* x *Cardinal*); **Misket Rusenski** (*Muscat Hamburg* x *Cardinal*), **Ran Hamburgski Misket** (*Yulski Biser* x *Muscat Hamburg*). Black-skinned Misket crosses developed and grown in Bulgaria.

Ⓦ Misket Kailachki (*Muscat Hamburg* x *Villard Blanc hybrid*); **Misket Markovski** (*[Terra Promesa* x *Muscat Ottonel]* x *Muscat Ottonel*), **Misket Plovdivski** (*[Caush* x *Marsilsko Ranno]* x **Italia**); **Misket Sandanski** (*Siroka Melniska* x *[Tamyanka* x *Cabernet Sauvignon]*); **Misket Trakijski** (*Dimyat* x *Perl Von Csaba*); **Misket Varnenski** (*Dimyat* x *Riesling*); **Misket Vratchanskii** (*Corna Alba* x *Muscat Blanc à Petits Grains*). White-skinned Misket crosses developed and grown in Bulgaria.

Ⓢ Mission Grape Synonym for Listán Prieto

Ⓢ Modrý Portugal Synonym for Portugieser or Blauer Portugieser

Ⓦ Molette The true Molette is a minor white variety rarely encountered outside of the Savoie region of eastern France. If the grapes are black, then Molette will be a synonym for Mondeuse.

Ⓑ Molinara A minor blending component in northeast Italy.

Ⓢ Monastrell Synonym for Mourvèdre

Ⓑ Mondeuse This variety may have originally hailed from Friuli in northeastern Italy, where it is known as the Refosco. It is now planted as far afield as the Savoie in France, parts of the United States, including California, and in Switzerland, Italy, Argentina, and Australia, where it is often an important constituent of the fortified Port-type wines.

Ⓦ Mondeuse Blanche A minor white-wine variety rarely encountered outside the Savoie region of eastern France, the greatest claim to fame this grape has is as one of the two parents of the classic Syrah (the other parent being the equally nondescript Dureza!).

Ⓑ Monica This Sardinian variety produces a fragrant red wine that may be dry or sweet, fortified or not.

Ⓑ Montepulciano A late-ripening variety that performs best in the Abruzzi region of Italy, where its wines are very deep in colour, and can either be full of soft, fat, luscious fruit, or made in a much firmer, more tannic, style.

Ⓢ Montonico Nero Synonym for Gaglioppo

Ⓢ Montù Synonym for Montuni

Ⓦ Montuni This variety is found in Emilia-Romagna, where it makes light, dry, or sweet still or sparkling white wine with a slightly bitter finish.

Ⓢ Morellino Synonym for Sangiovese

Ⓑ Moreto or **Moreto doAlentejo** A minor but widely planted Portuguese variety, Moreto is only an average Port grape but is more useful when blended into numerous unfortified dry red wines.

Ⓢ Moreto do Dão Synonym for Camarate and Malvasia Preta

Ⓢ Moreto Mortagua Synonym for Trincadeira Preta

Ⓢ Morillon Synonym for Pinot Noir

Ⓢ Morillon Taconé Synonym for Pinot Meunier

Ⓦ Morio-Muskat This *Sylvaner* x *Muscat Blanc à Petit Grains* cross is widely grown in the Rheinpfalz and Rheinheßen of Germany, producing perfumed, grapey wines.

Ⓢ Morrastel Synonym for Graciano

Ⓢ Moscadello Synonym for Muscat Blanc à Petits Grains and Muscat Selvatico

Ⓢ Moscatel A generic Spanish synonym for almost any Muscat variety

Ⓦ Moscatel Galego A good quality white Port grape, Moscatel Galego is also known as Moscatel do Douro. A black variant also exists. Also a synonym for Muscat Blanc à Petits Grains and Muscat Noir à Petits Grains.

Ⓢ Moscatel Gordo or **Moscatel Gordo Blanco** Synonyms for Muscat d'Alexandrie

Ⓢ Moscatel Grano Menudo or **Moscatel Menudo Bianco** Synonyms for Muscat Blanc à Petits Grains

Ⓢ Moscatel de Málaga or **Moscatel Romano** Synonyms for Muscat d'Alexandrie

Ⓢ Moscatel Rosé Synonym for Muscat Rosé à Petits Grains

Ⓢ Moscatel Samsó or **Moscatel de Setúbal** Synonyms for Muscat d'Alexandrie

Ⓢ Moscato A generic Italian synonym for almost any Muscat variety

Ⓢ Moscato d'Asti or **Moscato di Canelli** Synonyms for Muscat Blanc à Petits Grains

Ⓢ Mourisco A confusing mishmash of varieties and synonyms clarified below. In Spain, rather than Portugal, when the Mourisco name is unqualified by anything else and the grapes are white, it will be a synonym for Palomino.

Ⓢ Mourisco Arsello, Mourisco Branco, and **Mourisco Portalegre** Synonyms for Cayetana Blanca

Ⓢ Mourisco de Braga Synonym for Mourisco de Semente

Ⓢ Mourisco du Douro, Mourisco Tinto, and **Mourisco Vero** Synonyms for Marufo

Ⓑ Mourisco de Semente A minor Port variety that is not as highly regarded as Mourisco de Trevões,

Ⓦ MUSCAT À PETITS GRAINS
There are two versions of this variety – the Muscat Blanc à Petits Grains and the Muscat Rosé à Petits Grains – and some vines that seem to produce a motley crop somewhere between the two, with berries of both colours on every branch. Where one is cultivated, the other is often growing close by but is seldom intermingled because the Blanc ripens one week later than the Rosé. The two greatest products of the Petits Grains are the dry wines of Alsace and the sweet, lightly fortified Muscat de Beaumes de Venise, although in the production of the former the variety is giving way to the Muscat Ottonel.

these grapes are fat, low in colour and acid, but rich in sugar, limiting their use to tawny style.

Ⓑ Mourisco de Trevões A minor Port variety but with grapes that are slightly smaller and more highly regarded than those of the Mourisco de Semente.

Ⓑ Mourvèdre Although Monastrell is its primary name, this variety is better known outside Spain as Mourvèdre – an excellent-quality grape that has been used more than other lesser varieties in Châteauneuf-du-Pape. The Mourvèdre is grown extensively throughout southern France and, under the name of Mataro, has become one of Australia's most widely cultivated black grapes, although it is a declining force in southern California.

Ⓦ Mtsvane Kakhuri Ancient Georgian grape capable of producing well-structured, citrusy wines.

Ⓑ Mouyssaguès A rarely encountered blending variety from southwest France.

Ⓢ Müller Rebe, Müller Schwarzriesling, or **Müllerrebe** Synonyms for Pinot Meunier

Ⓦ Müller-Thurgau A *Riesling* x *Madeleine Royale* cross, bred at Geisenheim in 1882 by Professor Hermann Müller. It is more prolific than the Riesling, has a typically flowery bouquet, and good fruit, but lacks the Riesling's characteristic sharpness of definition. Apart from once being the mot planted grape in Germany (due to the large quantities used in Liebfraumilch), it is widely cultivated in English and New Zealand vineyards, although in both cases the acreage devoted to it has shrunk markedly since the 1990s.

Ⓢ Munica Synonym for Monica

Ⓢ Muristrellu Synonym for Monastrell

Ⓦ Muscadelle This is a singular variety that has nothing to do with the Muscat family, although it does have a musky aroma and there is, confusingly, a South African synonym for the Muscat – the Muskadel. In Bordeaux, small quantities of this grape add a certain lingering "after-smell" to some of the sweet wines, but the Muscadelle is at its sublime best in Australia, where it is called the Tokay and produces a rich and sweet "liqueur wine".

Ⓢ Muscadet Synonym for Melon de Bourgogne

Ⓢ Muscadine A generic term for any variety belonging to *Vitis rotundifolia* (eg, Scuppernong).

Ⓑ Muscardin One of Châteauneuf-du-Pape's 13 permitted varieties, Muscardin produces a light-coloured red wine with good aromatic qualities. It is an authorized grape for many other wines throughout the Rhône but is not widely planted.

Ⓢ Muscat Synonym for every Muscat variety and sub-variety

Ⓦ Muscat d'Alexandrie An extremely important grape in South Africa, where it makes mostly sweet, but some dry, wines. In France, it is responsible for the fortified wine Muscat de Rivesaltes (a very tiny production of unfortified dry Muscat is also made in Rivesaltes), and the grape is also used for both wine and raisin production in California.

Ⓑ Muscat Bailey or **Muscat Bailey-A** A complex *Bailey* x *Muscat Hamburg* Japanese hybrid, where Bailey is the *American Big Extra* x *Triumph* hybrid, Big Extra is *Big Berry* x *Triumph*, Big Berry is from *Vitis labrusca* (one of the 17 genera belonging to the genus *Vitis*). and Triumph is *Concord* x *Chasselas Musque*. This convoluted cross was developed to withstand Japan's wet and windy climate, which it achieves. In so doing, though, it has resulted in wine that is more Muscat than red, which would be fine in itself, but to deliver a Japanese red wine that can be compared to the tannin structure of Western models, it is often blended with Merlot or Cabernet, and the effect is mesmerizingly weird, with two taste profiles that mix no better than oil and water!

Ⓦ Muscat Fleur d'Oranger A spontaneous cross of *Muscat Blanc à Petits Grains* x *Chasselas Blanc* occasionally encountered in California and Australia, where it is known as Orange Muscat and has produced some high-quality, early-drinking wines.

Ⓢ Muscat Gordo Blanco Synonym for Muscat d'Alexandrie

Ⓢ Muscat de Hambourg Synonym for Muscat Hamburg

Ⓑ Muscat Hamburg This *Schiava Grossa* x *Muscat d'Alexandrie* cross is a better table grape than wine grape but is widely planted and used for dessert wines.

Ⓦ Muscat Ottonel A *Chasselas* x *Muscat d'Eisenstadt* cross, originally from France. One of the mysteries of wine-grape history, this was first thought to be a cross of East European origin because of its proliferation in Romania and Hungary, and the fact that one of its two parents, Muscat d'Eisenstadt, obviously had Austrian connotations (Eisenstadt being the state capital of Burgenland). The name, though, turned out to be misleading, because Muscat d'Eisenstadt, a black table-grape variety, was developed by Jean-Pierre Vibert of Angers in France. It has a very clean, fresh, grapey-Muscatty aroma, and because of its relative hardiness, it continues to replace the more difficult-to-grow but potentially more intense, complex, and age-worthy Muscat à Petits Grains in Alsace.

Ⓑ NEBBIOLO
Famous for its production of Barolo, this grape is also responsible, totally or in part, for the other splendid wines of Piedmont in Italy, such as Gattinara, Barbaresco, Carema, and Donnaz. It often needs to be softened by the addition of Bonarda grapes, which have the same role as Merlot grapes in Bordeaux. In fact, Merlot is used in Lombardy to soften Nebbiolo for the production of Valtellina and Valtellina Superiore wines, of which Sassella can be the most velvety of all.

Ⓦ Muscat à Petits Grains See box, opposite page, bottom right.

Ⓢ Muscat Rosé à Petits Grains See Muscat à Petits Grains.

Ⓦ Muskat Moravské A *Muscat Ottonel* x *Prachttraube 23/33* cross developed and grown in Czechia.

Ⓢ Muskateller or **Muskotály** Synonyms for Muscat Blanc à Petits Grains

Ⓢ Nagy-Burgundi or **Nagyburgundi** Synonym for Pinot Noir

Ⓢ Napa Gamay Synonym for Valdiguié

Ⓦ Narince Used for rustic local white wines, often fizzy, sometimes aged in oak, in the Central Anatolian region of Turkey, where it is often blended with the Emir grape.

Ⓦ Nasco This Sardinian variety makes finely scented, delicate, dry and sweet white wines that can also be *liquoroso*.

Ⓢ Nascu Synonym for Nasco

Ⓢ Naturé Synonym for Savagnin

Ⓑ Nebbiolo See box, above.

Ⓢ Nebbiolo di Gattinara Synonym for Croatina

Ⓑ Négrette Known for its deep, dark colour, Négrette is grown in the Pays Nantais, Fiefs Vendée, Frontonnais, and Haute Garonne regions. There is also a little bit planted in the San Benito, California, where Wild Horse Winery has produced a pure varietal wine for many years.

Ⓑ Negroamaro This Apulian grape makes dark, easy-drinking red wines that used to be sold in bulk for blending but are starting to acquire their own reputation in the wines of Alezio, Brindisi, Leverano, Lizzano, Matino, Nardo, and Salice Salentino, not to mention contributing to many more.

Ⓑ Nerello Mascalese Interesting black grape widely grown on Sicily, where it is one of the varieties used for Marsala but has more individual potential than that and produces light-coloured, elegant reds and fruity rosés

Ⓢ Nerino Synonym for Sangiovese

Ⓑ Nero d'Avola Widely grown in Italy, especially on Sicily, where it is probably a native variety. Used

primarily as a blending variety in the Sicilian DOCs of Erice, Riesi and Cerasuolo DOCG. It can be made into a varietal wine in Menfi and Vittoria DOCs.

Ⓑ Nero di Troia This Apulian grape variety has no connection with the town of Troia in Apulia's northern province of Foggia, but refers to ancient Troy, whence it originated. It was brought to the region by the first Greeks to settle in the Taranto area and is best exemplified today by the medium-bodied, ruby-coloured, everyday-quality red wines produced under the Rosso Bartletta DOC, although it plays a good supporting role throughout Apulia.

Ⓦ Neuburger This Austrian *Roter Veltliner* x *Sylvaner* cross excels on chalky soil, where it can make full-bodied wines with a typically nutty flavour in all categories of sweetness. Also grown in Czechia.

Ⓑ Neyret This variety is used as a blending component in the northwest of Italy.

Ⓦ Niagara This *Concord* x *Cassady* is a native American *V labrusca* x *labrusca* cross and can be ruinously foxy. It is planted in the United States, Canada, Brazil, and Japan.

Ⓢ Niedda Synonym for Monica

Ⓢ Nielluccio Synonym for Sangiovese.

Ⓦ Noblessa This is a low-yielding *Madeleine Angevine* x *Sylvaner* cross grown in Germany, which produces grapes with a high sugar level.

Ⓦ Nobling A *Sylvaner* x *Chasselas* cross grown in Baden, Germany, its grapes have high sugar and acidity levels.

Ⓑ Nocera A minor Sicilian grape that contributes to the wines of Faro DOC.

Ⓑ Norton At one time, Norton and Cynthiana were thought to be different varieties, although they have since been identified as one and the same. Norton yields very small grapes with a high pip ratio, necessitating *déstelage* to avoid excessively harsh tannins in the wine. Acidity is high, with an unusually high proportion of malic acidity, and the wine has the unnerving ability seemingly to "stain"

Ⓜ PINOT GRIS
The most widely cultivated and fastest-expanding mid-coloured grape in the world, this variety is undoubtedly at its best in Alsace, where it can produce succulent, rich, and complex wines of great quality and a spiciness seldom found elsewhere. When coupled with an acidity that is lower than that of Pinot Blanc, an absence of spicy varietal definition will invariably be due to high yields. This is why Pinot Grigio in northeast Italy generally lacks the intensity of Alsace Pinot Gris, although both use the same clones.

glass. That said, the best examples are good, and it is the only wild grape capable of producing fine wine.

Ⓦ Nosiola At one time the least known of Trentino's traditional grapes because it is mostly used for *vin santo*, Nosiola has made some interesting dry white varietal wines in recent years.

Ⓑ Notar Domenico or **Notardomenico** Rarely encountered beyond its native land in Upper Salento, this loose-bunched variety is used as a blending component only.

Ⓦ Nuragus A Sardinian grape that produces dry, semi-sweet, and *frizzante* white wines under the Cagliari DOC.

Ⓢ Nzolia Synonym for Inzolia

Ⓑ Oeillade A minor grape variety in the southern Rhône and Languedoc regions, where it is now mostly used for blending into *vins de pays*.

Ⓢ Oesterreicher Synonym for Sylvaner

Ⓢ Olaszrizling Synonym for Welschriesling

Ⓢ Olivese Synonym for Greco di Tufo

Ⓦ Ondenc A grape once widely planted in southwest France and popular in Bergerac, Ondenc is now grown more in Australia than in France. Its acidity makes it useful for sparkling wines.

Ⓢ Oporto Synonym for Portugieser

Ⓦ Optima Developed in 1970, this (*Riesling* x *Sylvaner*) x *Müller-Thurgau* cross is already widely grown in Germany because it ripens even earlier than the early-ripening Müller-Thurgau.

Ⓢ Orange Muscat Synonym for Muscat Fleur d'Oranger

Ⓦ Orion This *Optima* x *Villard Blanc* cross is perhaps the least inspiring of four recent hybrids that have been allowed by the EU for the wine production wine because the percentage of non-*vinifera* is too tiny to be of any significance.

Ⓢ Oriou Synonym for Petit Rouge

Ⓢ Ormeasco Synonym for Dolcetto

Ⓦ Ortega A *Müller-Thurgau* x *Siegerrebe* cross grown in both Germany and England, its aromatic grapes have naturally high sugar and make a pleasantly fragrant and spicy wine.

Ⓦ Ortrugo A minor variety used in Emilia-Romagna for still and sparkling wine under the Colli Piacentini DOC.

Ⓢ Ottavianello Synonym for Cinsault

Ⓢ Pagadebit, Pagadebiti, or **Pagedebito** Synonym for Bombino Bianco and (if a black variety) Plavac Mali

Ⓢ Paien Synonym for Gewürztraminer

Ⓢ Pais Synonym for the Mission Grape

Ⓦ Palomino or **Palomino Fino** The classic Sherry grape variety, Palomino occupies more than 90 per cent of all the vineyards in Jerez, where its combination of low alcohol and acidity is perfect for that particular fortified-wine style. Known elsewhere in Spain as Listan, Palomino is used as a minor component in cheap Australian and South African blends (both unfortified and fortified) and has even produced Ecuador's first drinkable dry white wine.

Ⓜ Pamid This is the most widely cultivated of Bulgaria's indigenous mid-coloured grape varieties. It makes light and fruity quaffing wine.

Ⓢ Pampanino Synonym for Pampanuto

Ⓦ Pampanuto An undistinguished variety grown in Apulia.

Ⓢ Pansá Bianca Synonym for Xarel-lo

Ⓦ Parellada The major white grape variety of Catalonia, used for still wines and sparkling Cava, in which it imparts a distinctive aroma and is used to soften the firm Xarel-lo grape.

Ⓦ Pascal Blanc A minor blending variety in the Rhône, Pascal Blanc is being phased out of the Côtes du Ventoux and Côtes du Vivarais by 2014 but is still allowed in Rasteau and Cassis.

Ⓢ Passale Synonym for Monica

Ⓦ Passerina High-yielding variety grown in central Italy.

Ⓢ Pâté Noir Synonym for Salvador

Ⓦ Pecorino An old Italian variety, it used to be blending component for undistinguished white wines in Italy's Marches, but today it also produces firm, savoury, mineral varietal wines in the Offida DOC.

Ⓢ Pedepalumb Synonym for Piedirosso

Ⓑ Pedro Ximénez or **PX** Some unfortified wines are produced from this classic Sherry grape, but they lack acidity and character to stand on their own and are therefore swallowed up by such mega-sized blends that they are lost. Although its primary use in Sherry is as a sweetening agent (typically with 360 grams of residual sugar per litre), there has been a certain vogue for limited releases of very old PX sweetening wines, sold as the ultimate dessert wine. Huge, dark, deep, and powerfully rich, these special PX wines are piled high with complex, raisiny Muscovado flavours that can be compared in quality, weight, and intensity with only some of the oldest and rarest Australian liqueur Muscats.

Ⓦ Perdernã Average-quality blending component for white Port.

Ⓢ Per'e Palumme Synonym for Piedirosso

Ⓑ Periquita Widely grown throughout Portugal, from Lisbon to the Douro, this grape can be far too astringent if not harvested on tannin ripeness but has the potential to make great red wines, as well as being a good-quality Port variety.

Ⓦ Perle A *Savagnin Rosé* x *Müller-Thurgau* cross, this grape can survive winter temperatures as low as -30°C (-22°F), and produces a light, fragrant, and fruity wine, but in low yields.

Ⓢ Perle Von Csaba This *Madeleine Angevine* x *Muscat Fleur d'Orange* cross was created in Hungary in 1904 and is grown in small pockets in Hungary, France, and Argentina.

Ⓑ Perricone This Sicilian variety has started to shine since the introduction of the Sachia Perricone IGT, when growers started cropping at lower

levels and winemakers took more care in their vinification, including using oak barrels. With a little extra ripeness, the red fruit flavours of cherry, redcurrant, and raspberry typically found in Perricone can gravitate to denser, more black-fruit notes without being too alcoholic.

Ⓦ Perrum Synonym for Pedro Ximénez

Ⓑ Persan Minor red variety grown in the Savoie region.

Ⓣ Petit Bouschet This *Aramon Noir* x *Teinturier du Cher* hybrid is another *teinturier* cross bred by Louis Bouschet at Domaine de la Calmette by Henri Bouschet of Alicante Bouschet fame. Petit Bouschet is also a synonym for Cabernet Sauvignon and Alicante Bouchet.

Ⓑ Petit-Brun A minor blending component for the red and rosé wines of the Palette AOC in Provence.

Ⓢ Petit Cabernet Synonym for Cabernet Sauvignon

Ⓦ Petite Arvine This old Valais variety makes a rich, dry white, with a distinctive grapefruit character and good acidity in Switzerland, where it also adapts well to sweeter, late-harvest styles. In the Valle d'Aosta, Italy, the Petite Arvine produces a lighter, more fragrant dry white.

Ⓑ Petite Sirah An ancient cross between Peloursin and Syrah, the Petite Sirah seems more at home in California and Australia than its native France, where it has often been mistaken for Syrah. A synonym for Durif.

Ⓢ Petite-Vidure Synonym for Cabernet Sauvignon

Ⓦ Petit Manseng Grown in southwestern France, where with the Gros Manseng it is known primarily for producing the legendary Jurançon Moelleux. Less widely grown than the Gros Manseng, the Petit

Ⓑ PINOT MEUNIER
An important variety in Champagne, where vinified white gives more upfront appeal of fruit than the Pinot Noir when young and is therefore essential for young Champagnes. Its characteristics are more immediate in appeal but less fine than those of the Pinot Noir. The Pinot Meunier is extensively cultivated in the Marne Valley area of Champagne, where its resistance to frost makes it the most suitable vine. Its synonym Plant de Brie illustrates just how far west of Champagne (as we know it today) the old vineyards used to extend. Research carried out in 2008 clearly indicates that not all Pinot Meunier kept in nursery collections is even from the same varietal family, as some bore close resemblance to Pinot Noir, while others did not and, to quote, "several Meunier samples from different collections showed different genotypes" (in other words, they were different varieties). It is now found all over the place and can make very Pinot Noir-like red wine.

Manseng curiously does not have smaller grapes, although the berry size is still very small. Another curiosity is that Petit Manseng is extremely resistant to grey rot yet susceptible to noble rot, whereas late-harvested Gros Manseng grapes tend to be *passerillé*.

Ⓦ **Petit Meslier** An ancient Champagne variety and spontaneous cross of *Gouais* x *Savagnin Blanc,* the Petit Meslier is one of the permitted varieties for Vin de Pays des Coteaux de Coiffy and is making a small revival in Champagne itself.

Ⓢ **Petit Pineau** Synonym for Arbois

Ⓑ **Petit Rouge** A low-yielding variety in the Valle d'Aosta, where it is capable of producing deep, dark, highly perfumed red wines. There have been moves to change the name of this grape to Oriou (one of its many synonyms) because of the misconception that the "Petit" refers to its quality, even though it refers to the size of berry, and most classic varieties have smaller, rather than larger, grapes.

Ⓑ **Petit Verdot** A grape that has been used to good effect in Bordeaux because it is a late ripener, bringing acidity to the overall balance of a wine. Today it is successfully grown in Australia, Argentina, South Africa and the United States.

Ⓦ **Petra** A Serbian *Kunbarat* x *Pinot Noir* hybrid developed in the 1970s to provide extreme winter hardiness, thanks to the Kunbarat, which is itself part Amurensis.

Ⓦ **Phoenix** One of four recent hybrids that have been allowed by the EU for the production of wine because the percentage of non-vinifera is too tiny to be of any significance. A *Bacchus* x *Villard Blanc,* Phoenix has a slightly muted Bacchus aroma with a touch of herbaceous character in the fruit.

Ⓦ **Picardan** or **Picardan Blanc** Although the primary name according to some ampelographers is Araignan, this variety is best known as the Picardan, one of Châteauneuf-du-Pape's 13 permitted varieties, even if it plays only an occasional minor role in some Châteauneuf-du-Pape blanc. Picardan is also one of the Bourboulenc's synonyms, and Grosse Clairette is a synonym for both Bourboulenc and Picardan, illustrating why these two singular varieties have often been confused for one another. Although a Picardan Noir exists, it is not related, but Picardan Noir is also a synonym for both Bouchalès and Cinsault.

Ⓦ **Picolit** Typically produces grossly over-rated and horrendously over-priced sweet wine in Italy.

Ⓦ **Picpoul, Picpoul Blanche,** or **Picpoul de Pinet** Widely planted in the southern Rhône and Languedoc,

Picpoul is one of the 13 varieties authorized for Châteauneuf-du-Pape and produces a delicate dry white under the Coteaux du Languedoc – Picpoul de Pinet AOC. It is also a synonym for Folle Blanche.

Ⓑ **Picpoul Noir** High sugar levels make this a useful blending component for the wines of Corbières, Coteaux du Tricastin, Côtes-du-Rhône, Côtes-du-Ventoux, Languedoc, Minervois, and Vacqueyras. Picpoul Noir is also a synonym for Calitor.

Ⓑ **Piedirosso** Old Campanian variety, grown around Napoli producing soft but fresh Gamay-like varietal wines and used in belnds with Aglianico.

Ⓢ **Pigato** Synonym for Vermentino

Ⓢ **Pignatello** Synonym for Perricone

Ⓑ **Pignerol** Minor Provençal grape used as a blending component in the red and rosé wines of Bellet AOC.

Ⓑ **Pignola Valtellina** or **Pignola Valtellinese** Blending component in the Valtellina Superiore DOCG.

Ⓦ **Pignoletto** This is ancient variety from Emilia-Romagna, which makes fresh, crisp, and light still and sparkling varietal wines.

Ⓑ **Pineau d'Aunis** This grape is best known for its supporting role in the production of Rosé d'Anjou. A *teinturier* variant also exists.

Ⓢ **Pineau Blanche de Loire** or **Pineau de la Loire** Synonyms for Chenin Blanc

Ⓢ **Pineau de Saumur** Synonym for Grolleau

Ⓦ **Pinella** Although only a minor Veneto variety, Pinella has its own varietal appellation under the Colli Euganei DOC.

Ⓑ **Pinotage** A *Pinot Noir* x *Cinsault* cross developed in 1925, it occupies an important position in South African viticulture, where its rustic

and high-toned wine is greatly appreciated. If this variety has the potential that some South Africans believe, then they should actively encourage its spread throughout as many wine areas across the world, and benefit from as wide experimentation, as possible.

Ⓢ **Pinot Aigret** Synonym for Rufete

Ⓢ **Pinot d'Anjou** or **Pinot d'Aunis** Synonyms for Pineau d'Aunis

Ⓢ **Pinot Beurot** or **Pinot Burot** Synonyms for Pinot Gris

Ⓢ **Pinot Bianco** Synonym for Pinot Blanc

Ⓦ **Pinot Blanc** A variety that is perhaps at its best in Alsace, where it is most successful, producing fruity, well-balanced wines with good grip and alcohol content. It also has excellent potential in northern Italy as Pinot Bianco. Plantings are gradually diminishing worldwide.

Ⓢ **Pinot Blanc Cramant** or **Pinot Blanc Chardonnay** Synonyms for Chardonnay

Ⓢ **Pinot Blanco** Synonym for Chenin Blanc

Ⓢ **Pinot Blanc Vrai** Synonym for Pinot Blanc

Ⓢ **Pinot Branco** Synonym for Pinot Blanc

Ⓢ **Pinot Chardonnay** Synonym for Chardonnay

Ⓢ **Pinot Droit** or **Pinot Fin** Synonyms for Pinot Noir

Ⓢ **Pinot d'Evora** Synonym for Carignan

Ⓣ **Pinot Fin Teinturier** or **Pinot Teinturier** This Pinot Noir with its coloured juice has so far produced undistinguished wine, but with Gouais Blanc, another unremarkable variety, it has fathered one of the world's greatest grapes, the Chardonnay.

Ⓢ **Pinot Fleri** Synonym for Mourvèdre

Ⓢ **Pinot Giallo** Synonym for Chardonnay

Ⓢ **Pinot Grigio** Synonym for Pinot Gris

Ⓜ **Pinot Gris** *See* box, opposite page, top left.

Ⓢ **Pinot Liébault** This is a clone selection of Pinot Noir, believed to be more productive than the original.

Ⓢ **Pinot de la Loire** Synonym for Chenin Blanc

Ⓑ **Pinot Meunier** *See* box, opposite page, bottom right.

Ⓢ **Pinot Nero** Synonym for Pinot Noir

Ⓑ **Pinot Noir** *See* box, above.

Ⓢ **Pinot Noir Précoce** Synonym for Frühburgunder

Ⓢ **Pinot St George** Synonym for Négrette

Ⓢ **Pinot Vache** Synonym for Mondeuse

Ⓢ **Pinot Verdet** Synonym for Arbois

Ⓢ **Plant d'Arbois** Synonym for Poulsard Blanc

Ⓢ **Plant d'Arles** Synonym for Cinsault

Ⓢ **Plant d'Aunis** Synonym for Pineau d'Aunis

Ⓢ **Plant de Brie** Synonym for Pinot Meunier

Ⓢ **Plant Dore, Plant Doré**, or **Plant Fin** Synonyms for Pinot Noir

Ⓢ **Plant Noir** Synonym for Mondeuse

Ⓢ **Ploussard** Synonym for Poulsard Blanc and Poulsard Noir

Ⓢ **Portugal** Synonym for Tinto Amarella

Ⓑ **Portugieser** One of the most widely planted black varieties in Germany, the Portugieser is thought to have originated in the Danube district of Austria, not in Portugal as its name suggests. It makes a very ordinary, extremely light red wine, so

Ⓑ **PINOT NOIR**

This is one of the classic varieties of Champagne, although its greatest fame lies in Burgundy. The Pinot Noir has 258 official synonyms, and there are innumerable Pinots that are not related. In a study by Regner et al in 2000, it was concluded that "results suggest that Pinot was derived from a cross of *Schwarzriesling* x *Traminer*", which would make Meunier an even earlier heterogenously distinct variety, rather than (as previously thought) a variant of Pinot Noir, which itself is widely regarded as one of the very oldest identifiable grapevines. Regner's data is not inconsistent with the proposed parentage, but neither is it compelling, and to make matters worse, a subsequent scientific paper (Salmaso et al) claimed that Regner et al "proposed Riesling Renano and Traminer as the progenitors of Pinot Nero", when clearly they had not. Don't take any notice of claims that the Riesling has anything to do with the Pinot Noir's parentage. In the right place, under ideal climatic conditions, the Pinot Noir can produce the richest, most velvet-smooth wines in the world. Depending on climate and ripeness, its flavour can range from cherries to strawberries. Great Pinot Noir is also made in California (notably Russian River and Santa Barbara), Oregon, and Central Otago, New Zealand. While individual producers in the New World are making progress in trying to replicate the extraordinary balance of weight, finesse, and complexity that this grape can achieve in the truly great *grands crus* of Burgundy, there is a feeling that most of even the best Pinot Noir wines produced outside Burgundy are all barely more than "varietal wines" and that, when this is taken to its extreme, they become "fruit-bombs". Having stated that, it should be remembered that the worst *grand cru* Burgundies do not even qualify as "fruit-bombs"; thus, even in the greatest *terroirs,* the Pinot Noir is extremely unforgiving.

it is often vinified white in bad years to blend with white wines that are too acidic. In addition to Germany and Austria, the Portugieser is also found in Italy, France, Czechia, Hungary, Romania, and Croatia.

Ⓑ Poulsard or **Poulsard Noir** Grown in the Jura, where it contributes to various appellations, Poulsard Noir is the most commonly encountered Poulsard and has the capability of producing complex Burgundian-style wines, but it seldom does and probably needs slightly warmer climes to excel. A more aromatically spicy sub-variety exists, the Poulsard Noir Musquet, which was originally and erroneously thought to be a Muscat variety, when it was known as the Muscat Pelosard Noir. A little Poulsard Noir Musquet is grown in Hungary.

Ⓦ Poulsard Blanc This is a rarely encountered, undistinguished white version of the Jura's more famous Poulsard Noir.

Ⓜ Poulsard Rosé or **Poulsard Rouge** Rarely encountered, undistinguished mid-coloured version of the Jura's more famous Poulsard Noir.

Ⓦ Praça This is an average-quality white Port variety.

Ⓑ Premetta Minor Italian variety found in the Valle d'Aosta where it produces slightly tannic, bright cherry-red wines.

Ⓦ Prié Blanc This Italian variety produces light, delicate, sometimes sparkling white wine and is almost exclusively found in the Valle d'Aosta, although it is also grown in the Swiss Valais.

Ⓢ Primitivo Synonym for Zinfandel

Ⓢ Procanico Synonym for Ugni Blanc

Ⓢ Prolongeau Synonym for Bouchalès

Ⓢ Prosecco Synonym for Glera

Ⓢ Prugnolo Synonym for Sangiovese

Ⓢ Pugnet Synonym for Nebbiolo

Ⓢ Pulciano Synonym for Grechetto

Ⓢ PX Synonym for Pedro Ximénez

Ⓦ Rabaner This *Riesling (clone 88Gm)* x *Riesling (clone 64Gm)* cross has the dubious honour of being the variety that most resembles the Müller-Thurgau.

Ⓦ Rabigato This is the main white Port grape variety. It is also known as Rabo da Ovelha, or "ewe's tail", and is grown in the Douro Valley for unfortified wines under the name Donzelinho Branco.

Ⓦ Rabo de Ovelha Portuguese variety, producing light, gently aromatic, undistinguished wines.

Ⓢ Rabo de Ovelha Tinto Synonym for Negramoll

Ⓑ Raboso This underrated variety is indigenous to the Veneto region, where it can make excellent-value red wines that are full of sunny fruit. There are two distinct, localized clones, Raboso del Piave and Raboso Veronese. Raboso del Piave is the most widely planted and is most often used as a varietal wine. Raboso Veronese is the more productive; and while it does not have the same richness or positive character, it is a very effective blending component. Raboso Veronese is also found in Argentina. This clone should not be confused with Raboso Piava, which is a synonym for Durella, or Raboso Piave, the synonym for Béquignol Noir.

Ⓦ Raffiat This variety is a minor, undistinguished blending component in southwest France.

Ⓢ Ragusan Synonym for Greco di Tufo

Ⓑ Ramisco Famously grown ungrafted in trenches dug out of the sandy dunes of Sintra, Portugal, where it produces Colares, which is as tough as old boots.

Ⓢ Räuschling Synonym for Elbling

Ⓦ Ratinho A minor Port variety, capable of adding some fragrance and structure in tiny quantities.

Ⓢ Red Muscadel Synonym for Muscat Rosé à Petits Grains

Ⓑ Refosco This ancient Italian variety is a native of the Friuli-Venezia Giulia region of northern Italy, where it makes dark, spicy red wines. There are six distinctly individual varieties called Refosco, of which Refosco dal Peduncolo Rosso is the oldest, most highly regarded, and most widely cultivated, followed by Refosco Nostrano, aka Refosco di Faedis. The other cultivars are, Refosco Botton, Refosco d'Istria, Refosco Guarnieri, and Refosco Rauscedo (which is almost extinct). With just three exceptions, any other Refosco will probably be a synonym of one of these clones and most likely will refer to Refosco dal Peduncolo Rosso, which has the most synonyms. The three exceptions are Refosco as a synonym for Mondeuse and Terrano, and any white Refosco, which would merely be an incorrect synonym for the Ezerjó of Hungary.

Ⓑ Regent One of four recent hybrids that have been allowed by the EU for the production of wine because the percentage of non-*vinifera* is too tiny to be of any significance. A *Diana* x *Chambourcin* cross, Regent is supposed to provide a wine with good body and a touch of spice, although even the best of the samples I have tasted have been uninspiring. It is surprisingly widely planted in Germany and is also finding favour in Switzerland and the UK.

Ⓦ Regner The parents of this *Luglienca Bianca* x *Gamay* cross are a curious combination. Why anyone would consider crossing a table grape with the red wine grape of Beaujolais to create a German white wine variety is a mystery. Predictably it produces sugar-rich grapes and mild, Müller-Thurgau-like wines.

Ⓦ Reichensteiner Affectionately known as the "Rick Steiner" at Camel Valley, a Cornish vineyard that has long supplied the restaurants of local celebrity chef Rick Stein, this *Müller-Thurgau* x *Madeleine Angevine* x *Calabreser-Fröhlich* cross is grown in Germany and England. Its sugar-rich grapes produce a mild, delicate, somewhat neutral, Sylvaner-like wine.

Ⓢ Rhine Riesling Synonym for Riesling

Ⓦ Ribolla or **Ribolla Gialla** An old variety grown in the Friuli-Venezia Giulia region of northern Italy and in Slovenia around Gorizia, where it is the most important grape. Ribolla Nera exists, but this name is also a synonym for the Schioppettino.

Ⓦ Rieslaner A *Riesling* x *Sylvaner* cross mainly grown in Franken, Germany, where it produces sugar-rich grapes, and it is best suited to make sweet wines.

Ⓦ Riesling See box, below left.

Ⓢ Riesling Italico Synonym for Welschriesling

Ⓢ Riesling Renano Synonym for Riesling

Ⓢ Riesling x Sylvaner or **Rivaner** Synonyms for Müller-Thurgau

Ⓦ Rkatsiteli An old Georgian variety, this grape is also grown in Moldova, Bulgaria, and in very small parcels in Australia, Russia, Italy, China, and elsewhere.

Ⓦ Robola Confined to the Ionian islands, this is a good-quality Greek grape that is at its best when its minerality is encouraged.

Ⓦ Roditis This grape is used as a supporting variety in the making of Retsina, a use only eclipsed by its suitability for the distilling pot.

Ⓢ Rolle Synonym for Vermentino

Ⓦ Romorantin An obscure variety that is confined to the Loire Valley, Romorantin can trace its ancestry back to a natural *Gouais Blanc* x *Pinot Fin Teinturier* cross and is thus a sibling of Auxerrois, Chardonnay, Gamay, Melon de Bourgogne, Sacy, et al. This grape is capable of producing a delicate, attractive, and flowery wine if it is not overcropped.

Ⓑ Rondo One of four recent hybrids that have been allowed by the EU for the production of wine because the percentage of non-*vinifera* is too tiny to be of any significance. A *Zarya Severa* x *St-Laurent* cross with both Saperavi and Amurensis in the parentage of Zarya Severa, this grape is more useful (disease resistance, winter hardiness, and good colour) than outstanding.

Ⓢ Rossanella Synonym for Molinara and Aleatico

Ⓢ Rossara Synonym for Molinara and Schiava Lombarda

Ⓦ Rossese A group of different grape varieties, some black, some white. There is one recognized local clone of the dark-skinned variant, Rossese di Campochiesa in Liguria, but there is another dark-skinned variety called Rossese di Dolceacqua close to the French border, which is identical to Tibouren. Rossese Bianco (if the grape is white) is the name for four distinct varieties (Rossese Bianco from Liguria and another one from Piemonte, Rossese Bianco di San Biagio, Rossese Bianco di Monforte), but is also the synonym for Grillo.

Ⓑ Rossignola A minor blending component for Valpolicella.

Ⓑ Rossola Nera High-acid blending variety from Valtellina.

Ⓑ Rotberger A *Schiava Grossa* x *Riesling* cross. The parents seem an

Ⓦ RIESLING
The classic German grape produces a zesty, citrous, intensely flavoured wine of great mineral complexity, length, and longevity. When grown on certain soils, the terpenes in Riesling benefit from bottle-age and can, after several years, develop a so-called petrolly bouquet. Alsace and Austria make most (not all) of the best dry Riesling. In Australia, the wine from this grape has, by and large, a simplistic lime-fruit character that is prone to going petrolly in a relatively short while but lacks the finesse and complexity of truly classic petrolly aromas. Great Australian Rieslings do exist, however, made in traditional areas such as Grosset's Polish Hill in the Clare Valley, Holmoak in Tasmania, and Frogmore Creek FGR (for a delicate, Mosel-like balance of sweetness), also in Tasmania.

B SAGRANTINO
An Umbrian grape that was almost extinct until its revival in the 1980s, which directly led to the 1992 upgrading of Montefalco Sagrantino to DOCG. Traditionally this variety makes both dry and sweet passito reds. The name is supposed to have derived from sagra or "festival", suggesting that the wines it yields were originally reserved for feast days.

odd couple, but the offspring is surprisingly successful, producing some excellent rosé wines.

Ⓜ Roter Veltliner Not related to the Grüner Veltliner, although both are Austrian varieties. No coloured Veltliner variant has yet been identified. Roter veltliner produces full-bodied, spicy whites that age well for a few years.

Ⓦ Rotgipfler This Austrian variety makes a robust, full-bodied, spicy white wine of not dissimilar character to the Zierfandler, which is often made into a dry style, although the semi-sweet Rotgipfler of Gumpoldskirchen is probably the most famous rendition of this grape.

Ⓦ Roupeiro Technically, its primary name is Síria, but this grape is better known as Roupeiro. It is grown all over Portugal but is considered most noble in Alentejo, where it produces soft, full, well-rounded white wines.

Ⓦ Roussanne One of two major varieties used to produce the rare white wines of Hermitage and Châteauneuf-du-Pape in France's Rhône Valley, this grape makes the finer, more delicate wines, while those made from the Marsanne tend to be fatter and richer.

Ⓦ Roussette Commonly associated with Savoie, the true Roussette is, however, the Roussette of the Rhône, which is a good-quality grape that produces a rich, full-bodied wine. A Savoie wine that states Roussette on the label (eg, Roussette de Savoie) will definitely have Altesse inside the bottle, another good-quality grape, but lighter in body with higher acidity and more minerality.

Ⓦ Roussette d'Ayze The only Roussette in Savoie, the Roussette d'Ayze, is an undistinguished grape used as a blending component in very ordinary sparkling wine.

Ⓣ Royalty This Alicante Ganzin x Trousseau Noir hybrid is a teinturier grape that was developed in California, where it is grown to no great success in the hot Central Valley.

Ⓑ Rubin This Bulgarian Nebbiolo x Syrah cross was developed in 1944 and produces interesting, spicy-cherry fruit that even has a touch of

cracked black pepper but is perhaps a tad too alcoholic to rival either of its very classy parents.

Ⓣ Rubired Possibly the most successful teinturier grape bred so far, Rubired was once restricted to Port-style fortified wines (useful for its high ratio of acid to colour), but this Alicante Ganzin x Tinta Cão hybrid is being increasingly used in unfortified red wines in California and Australia.

Ⓑ Ruby Cabernet This Cabernet Sauvignon x Carignan cross was produced by Professor Olmo of UC Davis. Its origination date is often misquoted, so, to clarify, this grape was bred in 1936, first fruited in 1940, and made its first wine in 1946. Although one of the more successful Cabernet Sauvignon x Carignan crosses, this variety nevertheless inherited far more of the Carignan's genes than the Cabernet's, like all the rest. The Ruby Cabernet is highly resistant to heat and drought, but its most useful attribute is probably the ability to survive high winds, when the metabolic system of most other vines shuts down. Primarily grown in California, but with a fairly significant hectarage in South Africa and Australia and just a smattering in Spain and Argentina, the Ruby Cabernet typically provides a rather one-dimensional red wine with characteristic cherry fruit

Ⓑ Ruchè Best known for producing Ruchè di Castagnole Monferrato DOC, the Ruché is a rather mysterious grape of unknown but presumed Piedmontese origin that generally makes a light ruby-coloured wine with an aromatic twist, although through reduced yields, much darker, richer wines not dissimilar to Nebbiolo, only more aromatic, can be achieved.

Ⓑ Rufete A high-yielding, low-quality Portuguese grape for red wine, due to its low acidity. It is, however, a good blending component in the production of Port.

Ⓢ Ruländer Synonym for Pinot Gris

Ⓢ Rulandské Bílé Synonym for Pinot Blanc

Ⓢ Rulandské Modrý Synonym for Pinot Noir

Ⓢ Rulandské Šedé Synonym for Pinot Gris

Ⓢ Ryzlink Rýnský Synonym for Riesling

Ⓢ Ryzlink Vlašský Synonym for Welschriesling

Ⓦ Sacy A minor grape variety that produces bland "stretching" wine and is grown in small quantities in the Chablis district. Its high acidity could make it very useful in the production of sparkling wines.

Ⓑ Sagrantino See box, left.

Ⓣ Salvador This French teinturier grape is a Munson Noir (aka Jaeger 70) x vinifera hybrid that is mostly planted in California's Central Valley, with small amounts found in Australia, Bulgaria, Brazil, India, Mexico, and Romania. The juice of the Salvador is coloured and pulpy, but not gelatinous as some sources suggest.

Ⓦ Samarrinho A good-quality Port variety.

Ⓑ Sangiovese See box, above left.

Ⓢ Sangiovese Polveroso Bonechi Synonym for Ciliegiolo

Ⓢ Sangioveto Synonym for Sangiovese

Ⓢ Sankt Laurent Synonym for St-Laurent

Ⓣ Saperavi The origins of this Georgian teinturier grape are unclear, but its once fearsome wine is gradually becoming clearer. Symphony Wines in Australia's King Valley also produces a Saperavi, and with a nice touch of irony, David Nelson, an Australian winemaker, and Lado Uzunashvili, a Georgian-Australian winemaker, have done much to improve the Saperavi in Georgia since 2001. Recognized localized clones are Saperavi Atenis,

Saperavi Bezhashvilis, Saperavi Budeshuriseburi, Saperavi Grdzelmtevana, Saperavi Guriis, Saperavi Kartlis, Saperavi Pachkha, and Saperaviseburi.

Ⓢ Saperavi Ochanuri Not a Saperavi; synonym for Otskhanuri Sapere, a minor Georgian grape

Ⓢ Sauvignon Synonym for Sauvignon Blanc

Ⓦ Sauvignon Blanc See box, below.

Ⓜ Sauvignon Gris An extremely muted sibling with none of Sauvignon Blanc's gooseberry character, a little Sauvignon Gris is still found in the Saint-Bris AOC of Burgundy, but there is not much left in its native Bordeaux these daysIn the 19th century, it was transplanted from Bordeaux to Chile, where it has miraculously survived. Miraculously because, although no vinifera variety likes "wet feet", the sensitivity of Sauvignon Gris to wet soils is almost at a phobic level and, until relatively recently, all Chilean vineyards existed only because they were flood-irrigated. Most Sauvignon Gris these days seems to come from Chile, where it grows particularly well in the Leyda Valley. It is being replanted in France and has even started to appear in Marlborough, the New World heartland of Sauvignon Blanc.

Ⓢ Sauvignon Rosé Synonym for Sauvignon Gris

Ⓢ Sauvignon Vert Synonym for Muscadelle and Friulano

Ⓢ Sauvignonasse Synonym for Friulano

Ⓦ Savagnin Blanc This grape is a non-aromatic Traminer Weiss (or Weiß) and subjected to ageing under flor, the Savagnin Blanc is responsible for the Sherry-like vin jaune of the Jura, of which the best known is Château Chalon.

Ⓦ SAUVIGNON BLANC
Sauvignon Blanc is at its best defined in New Zealand, particularly Marlborough, and is still a long way from achieving a similar level of quality and consistency in its home location of the Loire Valley. It is improving in Bordeaux, where it is also used in Sauternes and Barsac blends. Some exciting wines have begun to emerge from the Cape winelands, South Africa (ignoring those that are artificially flavoured), but California remains hugely disappointing, even though its vines are the same Sauvignon Blanc clone as New Zealand's.

W SÉMILLON

In Sauternes and Barsac, this is the grape susceptible to "noble rot". Some say its aroma is reminiscent of lanolin, but as pure lanolin is virtually odourless, the comparison hardly conveys the Sémillon's distinctive bouquet. For dry wine, this grape is at its best in Australia, particularly the Hunter Valley, where its lime fruit takes to oak like a duck to water, whereas bottle-aged Sémillon can be sublime even after several decades.

M Savagnin Rosé This is the non-aromatic Traminer Rosé (from which the highly aromatic Gewürztraminer evolved). It grows primarily in the village of Heiligenstein, Alsace, where it is known as the Klevener de Heiligenstein. It was almost extinct in the 1950s, with just 0.15 hectares, or barely more than one-third of an acre) surviving, but with renewed interest this has grown to more than 40 hectares (100 acres).

W Savatiano Although used primarily for retsina, the highly drought-resistant Savatiano is grown all over Greece as white wine insurance against a lack of rainfall. Pure non-resinated Savatiano is not unknown, but due to its low acidity it usually forms part of a blend.

W Scheurebe A cross of Riesling and an unknown variety, this is one of the best of Germany's new varieties. When ripe, it makes very good aromatic wines, but, if it is harvested too early, very herbaceous cat's pee.

B Schiava The so-called Schiava family is only just being classified. DNA research has determined that 22 Schiava cultivars can in fact be grouped into four genetically unrelated groups according to their geographical origin: Valtellina, Bergamo, Brescia, and Alto Adige in Italy. It was concluded that the term *Schiava* (or "slave") referred to a similar cultivation practice in contiguous regions rather than any common genetic background.

S Schiava di Bergamo Synonym for Schiava Lombarda

W Schiava Bianca Least encountered of the Schiava family (and not DNA tested), this minor variety is an occasional blending component in Lombardy and the Alto Adige.

S Schiava di Como or **Schiava di Varese** Synonyms for Schiava Lombarda

B Schiava Gentile Supposedly the best Schiava, this is the most widely cultivated variety in the Alto Adige, where it produces mostly light-coloured, light-bodied red of undistinguished quality, sometimes with an almond character to the fruit and a light bitterness on the finish. The best Schiava Gentile wines are Alto Adige DOC, especially Colli di Bolzano and Santa Maddalena, yet they are merely soft, fruity, and early-drinking. Usually much better if bolstered by other varieties such as Lagrein or Pinot Nero.

M Schiava Grigia Seldom encountered mid-coloured variant of the Schiava family Most Schiava Grigia that does exist is to be found in the Alto Adige.

B Schiava Grossa The least appreciated of the Schiava clan, most Schiava Grossa is cultivated in the Alto Adige.

B Schiava Lombarda One of the lesser Schiava varieties, Schiava Lombarda is most often used as a blending component in the wines of Bergamasca and Valcalepio Rosso.

B Schiava Meranese Synonym for Schiava Lombarda

S Schilcher Synonym for Blauer Wildbacher

B Schioppettino An ancient Friulian variety, Schioppettino was nearly extinct until it suddenly became fashionable in the 1980s. When not harvested at full ripeness, Schioppetino can be just another medium-bodied wine hinting unimpressively at cherries. But when it is harvested ripe, this grape produces much rounder wines, with blacker fruit and a fine, cracked-pepper aroma.

M Schönburger A *Spätburgunder* x (*Chasselas Rosé* x *Muscat Hamburg*) cross, this grape is grown in Germany and England. It produces sugar-rich grapes that make wine with good aromatic qualities but low acidity.

S Schwartzriesling Synonym for Pinot Meunier

B Sciaccarello A Corsican variety of typically Italian origin, the Sciaccarello grape has aromatics and acidity that are comparable to those of the Tibouren in Provence. It is, in fact, the same grape as the Mammolo (Tuscany), which is traditionally blended with Sangiovese. In Corsica, the Sciaccarello is blended into the AOCs of Ajaccio, Patrimonio, and Vin de Corse Porto Vecchio, adding a distinctive floral aroma to the wines. This variety is seldom encountered outside Corsica, Tuscany, and a few other isolated parts of Italy, but as the Mammolo, the odd few vines can be found in Mudgee, Australia.

W Scuppernong This is not a *Vitis vinifera*, but a native American *Vitis rotundifolia*, thus a so-called Muscadine. Scuppernong's large and cherry-like grapes produce an extraordinary, disturbingly exotic, and most unusual wine. Yet it was from these grapes that "Captain" Paul Garrett first made his notorious, best-selling Virginia Dare wine from a string of East Coast wineries in the 1900s. Unless conditioned from birth to Scuppernong grapes, the extremely foxy wine they make will at best be off-putting to most wine consumers today. A black-skinned Scuppernong exists, but it is less widely cultivated.

W Sémillon See box, above.

S Sercial Synonym for Esgana Cão

S Serène Synonym for Syrah

S Serprina or **Serprino** Synonym for Glera

W Seyval Blanc This *Seibel 5656* x *Rayon d'Or* hybrid is the most successful of the many Seyve-Villard crosses. It is grown primarily in New York State and England, where it produces the better wine Although seldom accomplished, this grape can produce high-quality sparkling wine.

S Shiraz Synonym for Syrah

S Shiroka Melnishka Ioza Synonym for Melnik

M Siegerrebe This is a *Madeleine Angevine* x *Savagnin Rosé* cross, a grape that was once widely grown in Germany. Also found in England and elsewhere, Siegerrebe is primarily grown in cool, potentially wet climates for its early ripening capability and high sugar yield. It can have very good aromatic character but unfortunately has very low acidity.

S Silvaner or **Silvain Vert** Synonyms for Sylvaner

S Síria Synonym for Roupeiro

S Sousão or **Souzão** Synonym for Vinhão and Donzelinho Tinto

S Spätburgunder Synonym for Pinot Noir

S Spergola Synonym for Sauvignon Blanc

S St Emilion Synonym for Ugni Blanc

B St-Laurent This is a central European variety widely planted in Austria, Hungary, and Czechia, where it is appreciated for its soft tannins and Pinot-like character.

W St Pepin A relatively recent *Elmer Swenson 114* x *Seyval Blanc* hybrid, the St Pepin is primarily cultivated in the Atlantic Northeast and other marginal viticultural areas of the United States, where this grape benefits from early ripening and is capable of surviving winter temperatures as low as -32°C (-25°F). St Pepin typically produces fruity, but unexciting semi-sweet white wines but can be stretched to some intriguing limits.

S Steen Synonym for Chenin Blanc

S Sultana or **Sultaniye** Synonyms for Thompson Seedless

B Susumaniello This historic grape is currently underrated and plays only a minor blending role for the wines in Apulia. It could have much greater potential.

W Sylvaner or **Sylvaner Blanc** See box, below.

M Sylvaner Rouge A few hectares of this rare red mutation exist in Germany and other Central European countries under a number of synonyms: Zierfandler, Cirfandli, Österreicher Rot, among others.

W Symphony Plantations of this *Muscat d'Alexandrie* x *Grenache Gris* cross rapidly increased in California

W SYLVANER OR SYLVANER BLANC

Originally from Austria, this variety is widely planted throughout Central Europe. It is prolific, early maturing, and yields the dry wines of Franken and Alsace. It is also widely believed to be the Zierfandler of Austria. Sylvaner has a tart, earthy, yet neutral flavour, which takes on a tomato-like richness in the bottle. This grape is now known to be the offspring of the Österreichisch Weiss and Traminer. Angelo Puglisi of Ballandean Estate in the Granite Belt of Queensland has consistently produced the world's greatest Sylvaner with cane cutting and late harvests.

Ⓑ SYRAH

Because the name of this variety derived from Shiraz (which Australia has adopted), the capital of Fars, the idea arose that it must have originated in Persia. We now know it is pure French in its breeding, being the progeny of two minor varieties, the Mondeuse Blanche and Dreza. The Syrah makes fashionably dark-coloured red wines in many countries. The wines from this variety are generally perceived to fall into one of two basic styles: the classic French (sometimes referred to as the north European) style, with its black, cracked-peppercorn fruit and a certain astringency in its formative years; or the bigger, brasher, distinctively oaky, New World classic, as epitomized by Barossa Shiraz.

in the mid-1990s, althouth it was developed by Professor Olmo as long ago as 1940. The wine it makes is usually off-dry with a distinctly flowery-grapey Muscat aroma.

Ⓑ Syrah *See box, above.*

Ⓢ Szürkebarát Synonym for Pinot Gris

Ⓢ Tămaioasă or **Tamianka** Synonyms for Muscat Blanc

Ⓑ Tannat This grape originated from the Basque region and has the potential to produce deeply coloured, tannic wines of great longevity (although there are certain modern methods of vinification that often change the traditional character of Tannat wines). The variety's best-known wines are the attractive red Madiran and Irouléguy. A little Tannat wine is used for blending purposes in and around the town of Cahors.

Ⓑ Tarrango This *Touriga Nacional x Thompson Seedless* cross is a high-yielding Australian variety that is capable of producing fresh, fruity, Beaujolais-style wines.

Ⓦ Tazzelenghe This grape variety takes its name from the Friulian dialect word for *tazzalingua*, Italian for "tongue cutter," and refers not to the tingle of any residual carbonic gas, but to the wine's acidity.

Ⓑ Tempranillo *See box, right.*

Ⓑ Téoulier A minor blending component in the reds and rosés of Provence.

Ⓑ Teroldego This grape is thought to have originated in the Rotaliano area of the Alto Adige, where today it produces a surprisingly dark, full, and distinctive wine for a naturally high-yielding vine. In youth, the wines can be chewy, with a tannic raspberry richness, but this softens with age, eventually attaining a silky-violety finesse.

Ⓢ Terrano Synonym for Refosco d'Istria

Ⓦ Terrantez or **Terrantez da Madeira** This is a white grape variety, although you might be forgiven for thinking otherwise if you have a beautiful brick red 20-year-old Terrantez Madeira in your hand. Although virtually extinct on the island of Madeira today, it is possible to find numerous old releases, and the occasional pure Terrantez is still produced from time to time. When produced as Madeira, this grape is highly perfumed, providing a rich, powerfully flavoured, tangy-sweet wine with a typically drier finish.

Ⓦ Terret Blanc This is used as a minor blending component in Languedoc, even for rosés and reds, for which this grape has traditionally been used as an ameliorator.

Ⓢ Terret-Bourret Synonym for Terret Gris

Ⓜ Terret Gris An excellent and interesting blending component in the Languedoc, where old vines of this variety are capable of producing full-bodied wines with slightly spiced aromatic notes.

Ⓑ Terret Noir The least encountered of the Terret varieties, Terret Noir is one of the 13 permitted varieties for Châteauneuf-du-Pape, despite there being less than half a hectare (1.25 acres) planted in the entire region. It has a spicy-floral aroma, relatively high acidity, and has not yet been fully exploited.

Ⓦ Thompson Seedless I suspect that this variety crops up in many more wines than we might be led to expect, but perhaps its greatest claim to fame was the first *cuvées* of Omar Khayyam, the Indian sparkling wine. Those earliest wines were produced in the 1980s by fizzmaking genius Raphael Brisbois, who crafted such a good sparkler in its day that its marketeers (not Brisbois himself) had little trouble convincing wine journalists it was made from

Chardonnay. Not bad for a grape better known as sultana, not merely because it is something we imagine more suited to Granny's fruitcake than any wine product, but also because its firm flesh and lack of seeds make the sultana a difficult grape to press.

Ⓑ Tibouren This is an aromatic black-skinned grape that produces lightly coloured but relatively acidic red wine, better suited to the production of rosé.

Ⓢ Tignolo Synonym for Sangiovese

Ⓦ Timorasso Minor, old Piemontese variety, highly regarded for its gently aromatic, textural wines.

Ⓢ Tinta Amarela or **Tinta Amarella** Synonyms for Trincadeira Preta

Ⓢ Tinta Bairrada Synonym for Baga

Ⓢ Tinta Barca Synonym for Tinta da Barca and Touriga Franca

Ⓑ Tinta da Barca A seriously underrated old Port variety that produces dark, rich, concentrated wine. This grape is also known as Tinta Barca.

Ⓑ Tinta Barroca The wine produced by this top-quality Port grape is quite precocious and therefore useful for younger-drinking Ports or to dilute Port blends that are too tannic and distinctive.

Ⓢ Tinta Cão Synonym for Tinto Cão

Ⓑ Tinta Carvalha A minor blending component for both fortified and unfortified wines in the Douro, Tinta Carvalha is considered to be too light in body and colour to be much use in Port blends, but it is favoured by some Portuguese-owned producers, who traditionally make a much lighter style of Port. When they get the blend right, there is nothing lacking or insubstantial in those wines – just pure finesse.

Ⓢ Tinta Fina Synonym for Alicante Bouschet, Baga, and Grand Noir

Ⓑ Tinta Francisca Two different grape varieties are known by this name. First, there is the seldom-encountered minor Port grape variety of undistinguished quality, which gets its name from historic stories of its French origin. Second is the French *teinturier* grape with many synonyms, two of which are, intriguingly, Portugal and Oporto, but it is not the same as the true Oporto variety called Tinta Francisca, the grapes of which have a clear juice.

Ⓑ Tinta Grossa A popular variety grown in the Vidigueira region, whereas elsewhere in the Alentejo it is merely used as blending fodder. Tinta Grossa is also a synonym for Tinta Barroca and Marufo.

Ⓢ Tinta Martins Synonym for Tinto Martins

Ⓑ TEMPRANILLO

This is the most important variety in Rioja, where it is traditional to blend the grapes, although many pure Tempranillo wines of excellent quality are made. It is capable of producing long-lived wines of some finesse and complexity. It is also an important variety on the other side of the Iberian peninsula, where it is known as Tinta Roriz in the north and Aragonêz in the south. Tempranillo could do well in parts of Portugal, but its role is usually confined to that of a blending component, and we have to look to the New World before we find any non-Spanish wine areas taking this grape seriously. It is found in California, and a small amount grows in Oregon and even New Mexico, but it is in Australia where it produces by far the greatest number of pure varietal wines. It is also grown in Egypt, without much luck, and Thailand, where it is more successful.

Ⓑ TROUSSEAU OR TROUSSEAU NOIR
Under the Portuguese name of Bastardo and various other synonyms, this variety is 10 times more commonly encountered on the Iberian Peninsula than it is in its native France. It is high yielding, with very good sugar and acid levels, making it an extremely useful blending component, as which (as Bastardo) it is still most famously used in fortified Port wines. Bastardo is also a Madeira grape, and although rarely found today, it was possible to find pure Bastardo varietal Madeiras at one time. As for still wines, this remains a seriously under-rated variety.

Ⓢ **Tinta do Minho** Synonym for Donzelinho Tinto

Ⓑ **Tinta Negra** (formely Tinta Negra Mole) This Madeira grape was planted as an improvement over the illegal hybrids it replaced. An improvement it was, but it is nonetheless a low-quality variety that is several classes below the truly classic grapes of Madeira. Cheap Madeira made from this grape is sold in bulk to France, Germany, and Belgium.

Ⓢ **Tinta Roriz** Synonym for Tempranillo

Ⓢ **Tintinha** Synonym for Petit Bouschet

Ⓑ **Tinto Cão** This is a top-quality Port variety, a small proportion of which can add finesse and complexity to a blend, but this is not easy to pull off in the Douro Valley. Tinto Cão enjoys a cooler environment than most other authorized Port grapes, it requires training on wires to produce a decent crop, and it is a late ripener in one of the world's hottest wine valleys. This variety is also excellent for unfortified Douro wines, either as part of a blend or as a pure varietal.

Ⓑ **Tinto Martins** A seldom encountered minor Port grape variety of undistinguished quality.

Ⓢ **Tocai Friulano** Synonym for Friulano

Ⓢ **Tocai Rosso** Synonym for Grenache

Ⓢ **Tokay** Synonym for Muscadelle and Pinot Gris

Ⓦ **Torrontés** Various different naturalized clones of this lightly aromatic grape are found throughout South America, particularly in Argentina, but the most predominant are firstly the Torrontés Riojano, then Torrontés Sanjuanino. DNA fingerprinting has revealed the parentage of both to be the Mission Grape and Muscat d'Alexandrie.

Ⓢ **Touriga** Synonym for Touriga Nacional

Ⓑ **Touriga Brasileira** A minor Port variety of undistinguished quality

Ⓑ **Touriga Franca** or **Touriga Francesa** This classic Port grape is no relation of the Touriga Nacional. Its wine is less concentrated but of fine quality.

Ⓑ **Touriga Nacional** The finest Port grape in the entire Douro, it produces fantastically rich and tannic wine, with masses of fruit, and is capable of great longevity and complexity.

Ⓢ **Tramin** or **Traminer** Synonyms for Savagnin Blanc

Ⓢ **Traminer Aromatico, Traminer Aromatique, Traminer Musqué, Traminer Parfumé, Traminer Rosé, Traminer Rosso, Traminer Roz,** or **Traminer Rozovy** Synonyms for Gewürztraminer

Ⓦ **Traminette** A *Joannes Seyve 23–416* x *Gewürztraminer* hybrid that was developed by Cornell University and released in 1996. Since then, it has been planted here and there throughout the Eastern Seaboard and Great Lakes states. Generally perceived as a cool-climate grape, its high acidity does, however, raise the question of whether it might not fare much better in the warmer climes of California.

Ⓢ **Tramini** or **Tramini Piros** Synonyms for Gewürztraminer

Ⓦ **Trebbiano** or **Trebbiano Toscano** Although the primary name, this variety is better known outside Italy as Ugni Blanc. There are a large number of Trebbiano Somethings, which may be synonyms of other cultivars or the names of genetically distinct varieties.

Ⓢ **Trebbiano Bianco di Chieti, Trebbiano CampoleseTrebbiano di Avezzano, Trebbiano di Macerata, Trebbiano d'Ora,** and **Trebbiano Dorato di Teramo** Not true Trebbiano, these are all synonyms for Bombino Bianco.

Ⓢ **Trebbiano Casale, Trebbiano di Parenzo, Trebbiano di Piemonte,**

Trebbiano di Spagna, Trebbiano Dorato, Trebbiano Dorato Citta' Sant Angelo, Trebbiano Emiliano, Trebbiano Giallo, Trebbiano Modenese, Trebbiano Romagnolo, and **Trebbiano Spoletino** All singular varieties of the Trebbiano family. *See* Ugni Blanc.

Ⓢ **Trebbiano di Lugana** Synonym for Verdicchio

Ⓢ **Tresallier** Synonym for Sacy

Ⓑ **Tressot** An ancient spontaneous *Duras* x *Petit Verdot* cross that is almost extinct, except for a few vines growing in the Yonne district of France, where its wines are typically thin, weak, and without any merit.

Ⓑ **Tribidrag** Synonym for Zinfandel and Primitivo

Ⓦ **Trincadeira das Pratas** The white Trincadeira is less widely planted than the black Trincadeira and not so highly regarded.

Ⓑ **Trincadeira** or **Trincadeira Preta** One of the best Port grapes, this variety was imported by James Busby into Australia, where it is better known as Tinta Amarella and used to soften wine blends. In the Douro itself, this variety is also known as Tinta Amarela (with just one "l"). When yields are restricted, Trincadeira has the potential to produce wines of great finesse and fragrance.

Ⓑ **Triomphe** or **Triomphe d'Alsace** An unexceptional hybrid that was bred but outlawed in Alsace, this *Millardet et Grasset* x *Knipperlé* cross is often purchased by non-French growers who think they are growing a true Alsace variety.

Ⓢ **Trollinger** Synonym for Schiava Grossa

Ⓑ **Trousseau** or **Trousseau Noir** *See* box, above.

Ⓢ **Trousseau Blanc** Synonym for Ondenc

Ⓜ **Trousseau Gris** This grape is now more widely grown in Portugal, Argentina, California, and New Zealand than in its traditional home of the northern Jura, France. It is yet another grape that has been erroneously tagged with the Grey Riesling name in the New World but does not resemble classic Riesling in the slightest.

Ⓦ **Ugni Blanc** A variety that usually makes light, even thin, wines that have to be distilled, the Ugni Blanc is ideal for making Armagnac and Cognac. There are a few exceptions, however, but most wines are light, fresh, and quaffing at their very best. Known as the Trebbiano in Italy, the name Ugni Blanc is also incorrectly used as a synonym for the Spanish Viura (Macabéo) and Portuguese Douradinha.

Ⓜ **Ugni Rosé** or **Ugni Rouge** Although more often encountered in Italy as the Trebbiano Rosa or

Trebbiano Rosso, a little Ugni Rosé is, however, found in Provence, primarily as a minor blending component in the Palette AOC.

Ⓢ **Ull de Llebre** Synonym for Tempranillo

Ⓢ **Uva d'Aceto** Synonym for Fortana

Ⓢ **Uva Asprina** Synonym for Greco di Tufo

Ⓢ **Uva Canina** Synonym for Sangiovese

Ⓢ **Uva d'Oro** Synonym for Fortana

Ⓑ **Uva Rara** Widely cultivated in the Novarese province northwest of Milan, where it is a major variety in the red wines of Colline Novarese DOC. Uva Rara's medium-bodied, soft, fruity wines benefit from the structure and added complexity of the Nebbiolo grape.

Ⓢ **Uva di Spagna** Synonym for Carignan

Ⓢ **Uva di Troia** Synonym for Nero di Troia

Ⓢ **Vaccarèse** Synonym for Brun Argenté

Ⓑ **Valdiguié** This is now an undistinguished variety in France, where it was once planted widely. In warmer climes such as California, this grape comes into its own under the synonym Napa Gamay, producing a nice quaffing red that is far fruitier than most of the true Gamay produced in Beaujolais.

Ⓢ **Varnenski Misket** Synonym for Misket Varnenski

Ⓢ **Veltlinske Zelene** Synonym for Grüner Veltliner

Ⓦ **Verdeca** At its most expressive in the Locorotondo DOC, where, with Bianco d'Alessano, it produces light, fresh, fruity, dry white wines of good everyday quality, the Verdeca grape has often been mistaken for Verdicchio in the past.

Ⓦ **Verdejo** Northern Spanish variety producing excellent, gently aromatic, fresh varietal wines in Rueda DO.

Ⓦ **Verdelho Branco** A successful grape variety for making white wines in Australia in recent years and possibly connected to the Verdello of Italy, Verdelho is best known, however, as a classic grape variety grown on the Portuguese island of Madeira for the island's fortified wine. (A Verdelho *tinto* also exists on the island but is almost extinct.)

Ⓦ **Verdesse** Grown in the Savoie region of eastern France, this variety has good acidity and a crisp, light, fresh, and floral aroma.

Ⓢ **Verdet** Synonym for Arbois

Ⓦ **Verdicchio** or **Verdicchio Bianco** As well as being used to make Verdicchio wine, this grape is also employed for blending.

Ⓢ **Verdicchio Femmina, Verdicchio Peloso, Verdicchio Tirolese,** or **Verdicchio Verde** Not Verdicchio; these are all synonyms for Verdeca.

Ⓦ **VIOGNIER**
This individual, shy-bearing variety used to be confined to a tiny part of the Rhône Valley, where it produced the famous wines of Condrieu and Château Grillet. Although this variety crept out to Australia in the 1970s, the real explosion did not occur until the 1990s in Languedoc-Roussillon and California. At its best, Viognier has a lush, aromatic quality, with a distinctive peachy character, but it is all too often over-oaked. Without the searing effect of the Gewürztraminer's spicy backbone, this variety's intrinsically low acidity produces a wine that can be too soft and exotic for those with palates that prefer higher-acid, more mineral styles.

Ⓢ **Verdicchio Marina, Verdicchio Marino, Verdicchio Sirolese,** or **Verdicchio Tirolese** Not Verdicchio; these are all synonyms for Maceratino.

Ⓢ **Verdicchio Nera** Synonym for Greco Nero

Ⓢ **Verdurino** Synonym for Bianco d'Alessano

Ⓦ **Verduzzo** The Verduzzo is widely grown in the Friuli-Venezia Giulia region of Italy, where it plays an important blending role in many dry white wines. This grape variety responds particularly well to dessert-styles wines.

Ⓦ **Vermentino** The Vermentino variety is cultivated throughout the Rhône, Provence, and Corsica, where it makes unpretentious, highly drinkable, refreshing dry white wines.

Ⓢ **Vernaccia di Austera, Vernaccia di San Vero Milis,** or **Vernaccia Solarussa** Synonyms for Vernaccia di Oristano

Ⓑ **Vernaccia Nera** One of the primary varieties for Ischia DOC, where this grape is known as Guarnaccia, which produces a robust, fruity red wine.

Ⓦ **Vernaccia di Oristano** No relation to Vernaccia di San Gimignano, this is the most widely cultivated of all so-called Vernaccia grapes, with most of it growing between Cabras and Baratili on the island of Sardinia, where it produces various types of wine, the most famous of which is unfortified, Sherry-like wine aged under *flor* in chestnut barrels.

Ⓦ **Vernaccia di San Gimignano** This variety is no relation to Vernaccia di Oristano. It is a Tuscan grape that produces deliciously crisp and vibrantly fruity DOCG wines.

Ⓢ **Vernaccia di Toscana** Synonym for Vernaccia di San Gimignano

Ⓢ **Vernaccia Trentina** Synonym for Bianchetta Trevigiana

Ⓢ **Vernatsch** Synonym for Schiava Grossa

Ⓦ **Vespaiolo** Pure varietal wines are produced from this grape in the Breganze DOC.

Ⓑ **Vespolina** This Piedmontese variety is a natural offspring of the Nebbiolo grape. In addition to pure varietal wines made under the Colline Novaresi DOC, it may be blended with Nebbiolo in wines such as Bramaterra and Ghemme.

Ⓦ **Vidal Blanc** This *Ugni Blanc* x *Rayon d'Or* cross produces a better wine in Ontario and New York State than it does in England (whereas Seyval Blanc grown in England is superior to that grown in either New York State or Ontario). In recent times, Vidal Blanc has become famed for its prolific production of Ice Wine in Ontario, although most have noticeably high volatile acidity.

Ⓢ **Vidure** Synonym for Cabernet Sauvignon

Ⓑ **Vien de Nus** An old Valle d'Aosta variety, Vien de Nus is a blending component in the wines of Arnad-Montjovat, Enfer d'Arvier, Torrette, and, primarily, Nus itself.

Ⓦ **Villard Blanc** This *Seibel 6468* x *Subéreux* hybrid is in decline, but stil one of the widest cultivated of the Seyve-Villard crosses in France. Its slightly bitter, iron-rich wine cannot be compared with the attractive wine of the Seyve-Villard 5276, commonly known as Seyval Blanc.

Ⓣ **Vinhão** In the Minho, this Portuguese teinturier variety is rated as the best grape for red Vinho Verde, but it is probably better known in the Douro, where, as Sousão, it donates colour and acidity to a Port blend.

Ⓦ **Viognier** *See* box, left.

Ⓦ **Viosinho** Considered one of the best white Port varieties for body and flavour, Viosinho is also used as a blending component in the various still wines of Douro, Planalti Mirandês, and Trás-os-Montes, but it can be heavy in an unfortified format, unless grown at high altitudes.

Ⓦ **Vital** A not-so-vital, characterless grape, despite being an authorized variety for many Portuguese wines.

Ⓢ **Viura** Synonym for Macabéo

Ⓢ **Weissburgunder** or **Weißburgunder** Synonym for Pinot Blanc

Ⓦ **Welschriesling** No relation whatsoever to the true Riesling, this variety is still grown in Austria, Hungary, Italy, and Brazil, producing ordinary medium-dry to medium-sweet white wines.

Ⓢ **Wildbacher** Synonym for Blauer Wildbacher

Ⓢ **Wrotham Pinot** Synonym for Pinot Meunier

Ⓦ **Würzer** This German *Savagnin Rosé* x *Müller-Thurgau* cross was one of the last developed by George Scheu, the breeder of Scheurebe. Early ripening with a strong spicy aroma but highly susceptible to rot and on the decline.

Ⓦ **Xarel-lo** A Spanish grape variety vital to the sparkling Cava industry, Xarel-lo makes firm, alcoholic wines.

Ⓑ **Xinomavro** or **Xynomavro** Most commonly called Xinomavro. *Xyno* means "acid and *mavr* "black", an indication of how dark and long-lived the wines of this excellent Greek variety can be. It is a late ripener, which helps it retain excellent acidity levels at harvest. Naoussa is the heartland for this variety, but it makes beautiful wine in other parts of Greece and could potentially excel throughout the New World, should any Greek ex-pats want to give it a try.

Ⓦ **Xynisteri** This Cypriot grape is the most widely planted on the island, and its wines are best drunk young if at all. Also spelled Xinisteri, and also known as Aspro.

Ⓢ **Zagarese** Synonym for Zinfandel

Ⓢ **Zierfandler** Synonym for Sylvaner

Ⓜ **Zierfandler Rot** This is a minor Austrian variety capable of producing a full-bodied, flavourful dry white wine. Also a synonym for Sylvaner Rouge.

Ⓦ **Zierfandler Weiss** A minor Hungarian variety, listed by some ampelographers as a separate singular variety that is not related to the Austrian Zierfandler. However, with no DNA evidence one way or the other, and with much Austrian and Hungarian vinestock sharing a very close history, the jury remains out on this one.

Ⓑ **Zinfandel** *See* box, below.

Ⓢ **Zingarello** Synonym for Zinfandel

Ⓦ **Zlata** A Serbian *Irsai Oliver* x *Kunleany* hybrid developed in the 1970s to provide extreme winter hardiness, thanks to the Kunleany, which is itself part Amurensis.

Ⓢ **Zöld Szilváni** Synonym for Sylvaner

Ⓑ **Zweigelt** This *St-Laurent* x *Blaufränkisch* cross is grown primarily in its native Austria but is also widely grown in Hungary and Czechia, with pockets also found in Germany, Canada, and England. The norm for this high-yielding variety is rather light and lacklustre, but when yields are restricted, the wines can be big and soft, with peppery fruit.

Ⓑ **ZINFANDEL**
Once thought to be the only indigenous American *Vitis vinifera* grape, Zinfandel has now been positively identified by Isozyme "fingerprinting" as the Primitivo grape of southern Italy. The origins of this grape, however, are Croatian, where it is known as the Crljenak Kastelanski. Depending on the vinification method used, Zinfandel can produce many different styles of wine – from rich and dark, to light and fruity or nouveau style. They can be dry or sweet; white, rosé, or red; dessert or sparkling.

Harvesting and Pressing

Deciding when to pick is one of the most crucial choices a grower has to make each year. It varies according to the grape variety, the location of the vineyard, and the style of wine that is to be made. One of the impacts of climate change is that ripening curves started to move away from the usual, adding even more uncertainty to this risky process.

White wine generally benefits from the extra acidity of earlier harvested grapes, but they also need the varietal aroma and richness that can only be found in ripe grapes. It is essential to get the balance right. Red wine requires relatively less acidity, although it is still important, and profits from the increased colour, sugar, and tannin content of later harvested grapes.

Growers must also take the vagaries of the weather into account. Those who have the nerve to wait for perfectly ripe grapes every year can produce exceptional wines in even the poorest vintages, but they also run the risk of frost, rot, rain, or hail damage, which can destroy an entire year's income. Those who never take a chance may harvest healthy grapes, but in poor years they risk an unripe crop and mediocre wine.

MECHANICAL HARVESTING

The subject of mechanical harvesting is a contentious one. The advantages of mechanization are dramatically reduced labour costs and a quick harvest of the entire crop at optimum ripeness, but the vineyard has to be adapted to the machine chosen, and the reception and fermentation facilities must be enlarged to cope with the greater amounts and quicker throughput, which is costly. Disadvantages relate to compacting the soil, the efficiency of the machinery, and the quality of the wine.

As the machines beat the vine trunks with rubber sticks, the grapes drop onto a conveyor along with leaves and other matter, most of which is ejected as the fruit is sorted on its way to the hold. Apart from the waste that inevitably remains with the harvested grapes – which is becoming less as machines become more sophisticated – the main disadvantage of mechanical harvesting is the inability of a machine to distinguish between the ripe and the unripe or to sort the healthy from the diseased or plain rotten (the first thing to drop from any plant when shaken). This can be achieved manually, though, and the latest machines are capable of sorting the fruit. From a practical point of view, it would seem that this method of harvesting is better for red wine than it is for white wine, particularly sparkling, because it splits the grapes. This encourages oxidation and a loss of aromatics and, when harvesting black grapes for white sparkling wine, results in an undesirable coloration of the juice.

It is well known that machine harvesting is widely employed in New World countries like Australia, but few realize just how prevalent it is in France, where more than 60 per cent of the wines are machine harvested.

PRESSING

In the race against time to get the grapes from the vineyard to the winery, everything possible should be done to cosset the fruit. Ideally, the winery and the vineyards are close, and the grapes are transported in stackable plastic crates that are small enough to prevent damage from their own weight.

When grapes arrive at the winery, white varieties are destemmed and immediately pressed, whereas black grapes are crushed and kept in contact with their skins while they ferment (to extract colour and tannins). Some aromatic white varieties might also be crushed and kept in contact with their skins, but they will be pressed immediately prior to fermentation. Two famous exceptions are for Champagne (where the grapes, both white and black, are not destemmed before pressing) and Beaujolais

A WINERY WORKER CHECKS THE STATUS OF THE GRAPES IN A PNEUMATIC PRESS
Pressing takes place after crushing, which bursts the grapeskins so that the inner solids mingle with the skins and stems, absorbing the flavours, colours, and tannins that give a wine its distinct characteristics. The free-run juice released turns into a liquid called must. Pressing then takes place, separating the must from the berry's fibres and other solids.

A MECHANICAL HARVESTER OFFLOADS A BATCH OF GRAPES IN A TUSCAN VINEYARD
Mechanical harvesters allow for the most efficient and economical removal of grapes from the vines. As more and more wineries around the world turn to machine harvesting, grape-growers and viticultural researchers continue to focus on ways to refine machinery and adapt vineyards to be more compatible with mechanization.

Nouveau (for which whole clusters are neither destemmed nor crushed but kept under carbon dioxide pressure in a vat, where they undergo an aerobic intracellular fermentation). Some red wines are made with a percentage of whole-cluster fermentation, while others are made without the free-run juice (juice that runs freely from the grapes under their own weight as they are piled into a press), which gives more colour and structure to a wine because of the higher ratio of skins to remaining juice.

There are various types of press. First, they break down into continuous and batch presses. A continuous press usually operates on the principle of an Archimedes screw, which applies increasingly higher pressure as the grapes move through the press to exit in a never-ending bone-dry fibrous sausage of skins, seeds, and stalks. This is the harshest of the pressing processes, and as such it is primarily used for high-volume, low-cost wines. Batch presses are the most common form of press found in the majority of wineries and are essentially divided into two types: hydraulic plate presses (horizontal and vertical) and pneumatic (horizontal). The former category is most effectively illustrated by Vaslin, a firm whose name is synonymous with hydraulic presses. Hydraulic presses have their uses, particularly the vertical bunch presses, and the technology is much better than it was in the 1970s, when Vaslin horizontal hydraulic presses were everywhere. A horizontal press consists of a perforated cylinder containing plates either end. Grapes are loaded through a closeable hatch, and the press rotates to remove any free-run juice, after which the two metal plates move towards each other, squeezing the grapes between them, with the juice exiting through the holes in the perforated cylinder. The vertical hydraulic press, or basket press, is the oldest design of press known but is particularly useful in Champagne and for limited-production wines (because this type of press can be produced in the smallest sizes).

Pneumatic presses are the softest pressing and consist of a central or side rubber bladder that is inflated by compressed air, squashing the grapes against the perforated sides of the cylindrical press, where the juice escapes through holes. Top-quality pneumatic presses include those by Bücher, Willmes, and Magnum. The latest development in pneumatic presses involves flooding it with an inert gas such as nitrogen, carbon dioxide, or argon.

However softly grapes are pressed, there is no getting away from the fact that it is a destructive process traditionally carried out in an oxygen-rich environment; as such, it represents the first crisis point for any winemaker anxious to avoid the onset of oxidation. Pressing grapes in an inert-gas environment seeks to eliminate much of this risk. On the other hand, there are times when winemakers deliberately oxidize grape juice prior to fermentation (for either stylistic reasons or technical reasons).

Vinification

Methods of wine production can vary greatly not just from country to country, but from region to region and, quite commonly, even from grower to grower within the same village. Although the internationalization of winemaking techniques is a topic that is always much discussed, these techniques must be adapted to the individual means, needs, and goals of the winemaker.

In winemaking, much depends upon whether or not traditional values are upheld or innovations are sought and, for the latter, whether or not the technology is available. Whatever the winemaker decides, certain principles will, essentially, remain the same. These are described below, followed by sections on styles of wine and the processes common or unique to each one.

THE DIMINISHING QUALITY FACTOR

The quality of the grapes when they are harvested represents the maximum potential of any wine that can be made from them. A winemaker will never be able to transfer 100 per cent of this inherent quality to the wine, however, because deterioration sets in from the moment a grape is disconnected from the vine's metabolism. Furthermore, the very process of turning grapes into wine is necessarily a destructive one, so every action taken by the winemaker, however quality-conscious he or she may be, will inevitably erode some of the wine's potential. Winemakers, therefore, can only attempt to minimize the loss of potential quality.

It is relatively easy to retain approximately 80 per cent of the potential quality of a wine, but very difficult to preserve every percentile point after that. It is also relatively easy to double or even triple the basic grape quality

by better selection of vineyard sites, improved training methods, the use of superior clones, correct rootstock, and a reduction in yields. As a result, research has long since swung from the winery back to the vineyard.

That said, oenological practices are still important and how they are employed will have a profound effect not only on quality, but also on the style of the wine produced.

TONY RYNDERS, FORMER WINEMAKER OF DOMAINE SERENE, SAMPLES PINOT NOIR GRAPES DURING A WILLAMETTE VALLEY HARVEST
The decisions a winemaker makes both in the vineyard and in the winery determine the quality and character of the wines they produce. Winemaking is an idiosyncratic undertaking, and two winemakers working with the same raw materials can come up with very different end results.

HUMAN IMPACT

The OIV's (International Office of Vine and Wine) *terroir* definition places an emphasis on the grape-grower's and winemaker's practices, making the human element central to the typicity of the resulting wine. The decisions they make while following experience, instinct, and their appellations' rules deeply rooted in tradition plays a major part in the wine's identity.

Time and again I have seen that neighbouring winemakers can make wines of widely varying quality using virtually the same raw product and technology. Chemical analyses of the wines in question may be virtually indistinguishable, yet one wine will have all the definition, vitality, and expression of character that the other lacks. Why does this occur? Because it is always the winemakers with passion who are able to produce the finer, more characterful wines. Many inferior wines are made by the misuse of up-to-date technology, and I have seen dedicated winemakers produce absolutely spellbinding wines using totally inadequate or inferior equipment. If the test is between passionless, high-tech wine and wine made by an ill-equipped genius, the genius always wins, but employing a few caring oenologists (not all are) in the largest modern winery can make all the difference. Some winemakers have even been known to sleep by their vats during a particularly

difficult fermentation, so that they are on hand to make any adjustments immediately if anything goes wrong. From the grower who never hesitates to prune the vine for low yields yet always agonizes over the optimum time to harvest, to the winemaker who literally nurses the wines through each and every stage of fermentation and maturation and bottles at precisely the right time and at exactly the correct temperature, the human element is most seriously the joker in the pack when it comes to factors affecting the taste and quality of the wines produced.

THE JOKER AT WORK

There are many examples of the human joker changing the taste and quality of wine for either good or bad, but few practices have had such a widespread negative impact on modern winemaking as picking grapes by so-called physiological ripeness. This pseudo-scientific term (*see* Alsace, p270) has assumed an almost religious reverence among winemakers, turning former dry white-wine areas into preserves of increasingly sweet concoctions, and where there were once red wines brimming with vibrant fruit and finesse, we now have dark, dense wines full of nothing but dead fruit, the size and power of which we are expected to find awesome.

Other notable bad jokes include putting the best Chardonnay in new oak throughout the New World, which only hides the grape's best attributes, and using leftover wines for so-called unoaked Chardonnay, when logic tells us it should be the opposite way around: put the lesser wines in the oak to hide their imperfections, and leave the best Chardonnays unoaked to reveal every facet of their beauty. Whether oak or other winemaking techniques are used – lees stirring, acidifying, malolactic, carbonic maceration, or whatever – their tell-tale aromas should not be noticeable. At worst, they should play no more than a minor supporting role in a seamless presentation of the wine. It should be the wine, not the technique used to make the wine, that leaps out of the glass as soon as you pick it up. Yet some of these jokers deliberately show off their techniques, creating not complexity or finesse but the vinous equivalent of a painting by numbers; they make it so obvious how their wines are made. The good jokers are those winemakers who are seeking elegance and finesse over weight and alcohol – those who aspire to make wines with extraordinary great length, not due to their sheer size but because they know that the real skill is in achieving a balance required to provide length, no matter how light in weight the wine might be.

PRINCIPLES OF VINIFICATION

With modern technology, good everyday-drinking wines can be made any-where that grapes are grown. When such wines are not made, the reason is invariably a lack of equipment and expertise or an absence of self-respect. Finer-quality wines require vineyards that have a certain potential and win-emakers with a particular talent. When not even good everyday-drinking wines are made from fine wine vineyards, it is usually due to a combination of excessive yields and poor winemaking, and there is no excuse for either.

FERMENTATION

The biochemical process that transforms fresh grape juice into wine is called fermentation. Yeast cells excrete enzymes that convert natural fruit sugars into almost equal quantities of alcohol and carbonic gas. This process ceases when the supply of sugar is exhausted or when the alcoholic level reaches a point that is toxic for the yeast enzymes (usually 15 to 16 per cent, although certain strains can survive at 20 to 22 per cent). Traditionally, winemakers racked their wine from cask to cask (*see* box, opposite) until they were sure that fermentation had stopped, but there are now many other methods that halt fermentation artificially. These can involve the use of heat, sulphur diox-ide, centrifugal filtration, alcohol, pressure, or carbonic gas.

Heat

There are various forms of pasteurization (for table wines) and flash-pasteurization (for finer wines) and chilling operations that are used to stabilize wine. These operate on the basis that yeast cells are incapacitated at temperatures above 36°C (97°F) or below -3°C (26°F) and that yeast enzymes are destroyed above 65°C (149°F). Flash-pasteurization subjects wines to a temperature of about 80°C (176°F) for between 30 seconds and 1 minute, whereas fully fledged pasteurization involves lower tempera-tures of 50 to 60°C (122 to 140°F) for a longer period.

YEAST: THE FERMENTER

The yeasts used for fermentation may be divided into two categories: cultured yeasts and natural yeasts.

Cultured yeasts are nothing more than thoroughbred strains of natural wine yeasts that have been raised in a laboratory. They may be used because the juice has been cleansed of all organisms, including its yeasts, prior to fermentation or because the winemaker prefers their reliability or for a specific purpose, such as withstanding higher alcohol levels or the increased osmotic pressure that affects bottle-fermented sparkling wines.

Natural yeasts are to be found adhering to the pruina, a waxy substance that covers the skin of ripe grapes and other fruits. By the time a grape has fully ripened, the coating of yeasts and other micro-organisms, commonly referred to as the "bloom", contains an average of 10 million yeast cells, although only 1 per cent – or just 100,000 cells – are so-called wine-yeasts. A yeast cell is only microscopic, yet under favourable conditions it has the ability to split 10,000 sugar molecules every second during fermentation.

Addition of sulphur dioxide or sorbic acid

Dosing with one or more aseptic substances will kill off the yeasts.

Centrifugal filtration or filtration

Modern winery equipment is now capable of physically removing all the yeasts from a wine, either by filtration (simply pouring the wine through a medium that prevents certain substances passing through) or by

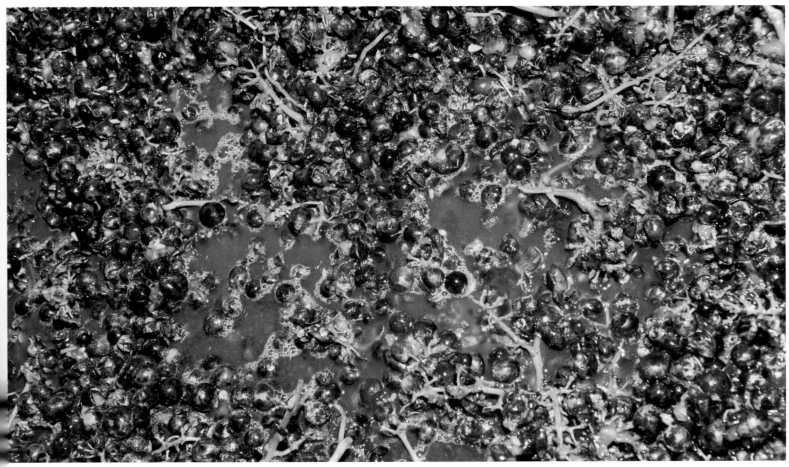

THE FRESHLY CRUSHED GRAPES FORM MUST, WHICH CONTAINS THE SKINS, SEEDS, AND STEMS
A winemaker must decide when to drain off the juice from the solids (called pomace) because this will determine the final characteristics of a wine. Once drained, yeast will be added to the juice to begin the fermentation process. Pomace is then often used in the vineyard as fertilizer for the next year's crop or used to make brandies, such as grappa.

centrifugal filtration (a process that separates unwanted matter from wine – or grape juice, if used at an earlier stage – by centrifugal force).

Addition of alcohol

Fortification raises the alcohol content to a level toxic to yeast.

Pressure

Yeast cells are destroyed by pressure in excess of eight atmospheres (the pressure inside a Champagne bottle is around six atmospheres).

Addition of carbonic gas

Carbonic gas is also a yeast destroyer. Yeast cells are destroyed in the presence of 15 grams per litre or more of carbonic gas (CO_2). (There are just over 10 grams per litre in a fully sparkling wine.)

THE USE OF SULPHUR

Sulphur is used in winemaking from the time the grapes arrive at the winery until just before the wine is bottled. It has several properties, including antioxidant and aseptic qualities, that make it essential for commercial winemaking. To some extent, all wines are oxidized from the moment the grapes are pressed and the juice is exposed to the air, but the rate of oxidation must be controlled. This is where sulphur is useful, because it has a chemical attraction for the tiny amounts of oxygen that are present in wine. One molecule of sulphur will combine with two molecules of oxygen to form sulphur dioxide (CO_2), or fixed sulphur. Once it is combined with the sulphur, the oxygen is neutralized and can no longer oxidize the wine. More oxygen will be absorbed by wine during the vinification process, of course, and there will also be a small head of air between the wine and the cork after bottling. It is for this reason that wines are bottled with a set amount of free sulphur (the amount of the total sulphur content that is not fixed). Occasionally a winemaker claims that sulphur is completely super-fluous to the winemaking process, but whereas low-sulphur regimes are actually to be encouraged, wines produced without it are usually dire or have a very short shelf-life.

One famous wine that claimed not to use any sulphur was so long-lived that I had a bottle independently analysed, only to find that it did contain sulphur. The quantity was small, but far too significant to have been created during fermentation (which is possible in tiny amounts). It was, therefore, an example of how effective a low-sulphur regime can be. Methods of reducing the level of SO_2 are well known, the most important being a very judicious initial dosage because a resistance to sulphur gradually builds up and, as a result, later doses always have to be increased.

Some wines can be over-sulphured and, although they are less common than they used to be, they are by no means rare. Over-sulphured wines are easily recognizable by their smell, which ranges from the slight whiff of a recently ignited match (which is the clean smell of free sulphur) to the stench of bad eggs (which is H_2S, where the sulphur has combined with hydrogen – literally the stuff of stink-bombs). When H_2S reacts with ethyl alcohol or one of the higher alcohols, foul-smelling compounds called mercaptans are formed. They can smell of garlic, onion, burnt rubber, or stale cabbage, depending on the exact nature of the compound. Mercaptans are extremely difficult for the winemaker to remove and can ruin a wine, which illustrates just how important it is to maintain a low-sulphur regime.

A WINERY WORKER AT CHÂTEAU PURCARI IN MOLDOVA POURS RED WINE INTO STAINLESS STEEL VATS
Stainless steel is the cornerstone of modern winemaking technology. Winemakers often prefer this type of container because it is easier to maintain than oak barrels and can be used for both fermentation and ageing and allows a grape's varietal characteristics to remain pure.

MALOLACTIC FERMENTATION

Malolactic fermentation is sometimes known as the secondary fermentation, but this is an inappropriate description. The malolactic, or "malo" (as it is sometimes called), is a biochemical process that converts the "hard" malic acid of unripe grapes into two parts "soft" lactic – or "milk" – acid (so-called because it is the acid that makes milk sour) and one part carbonic gas. Malic acid is a very sharp-tasting acid, which reduces during the fruit's ripening process. A significant quantity persists in ripe grapes, however, and, although reduced by fermentation, also in wine.

The quantity of malic acid present in a wine may be considered too much and the smoothing effect of replacing it with just two-thirds the quantity of the much weaker lactic acid is often desirable. This smoothing effect is considered vital for red wine, beneficial for fuller, fatter, more complex whites, and optional for lighter, crisper whites and certain styles of sparkling wine.

To ensure that the malo can take place, it is essential that specific bacteria are present. These are found naturally on grape skins among the yeasts and other micro-organisms, but commercially prepared bacteria may also be used. To undertake their task, they require a certain warmth, a low level of sulphur, a pH of between 3 and 4, and a supply of various nutrients found naturally in grapes.

STAINLESS STEEL OR OAK?

The use of stainles steel and oak containers for fermentation and maturation is not simply dependent on the cost (*see* left). The two materials produce opposing effects upon wine, so the choice is heavily dependent upon whether the winemaker wants to add character to a wine or keep its purity.

A stainless steel vat is a long-lasting, easy-to-clean vessel made from an impervious and inert material that is ideally suited to all forms of temperature control. It has the capacity to produce the freshest wines with the purest varietal character. An oak cask has a comparatively limited life, is not easy to clean (it can never be sterilized), makes temperature control very difficult, and is neither impervious nor inert. It allows access to the air, which encourages a faster rate of oxidation but also causes evaporation, which concentrates the flavour. Vanillin, the essential aromatic constituent of vanilla pods, is extracted from the oak by

THE COST OF NEW OAK

Two hundred 225-litre (49-gallon) oak casks with a total capacity of 450 hectolitres (9,900 gallons) cost between 4 and 10 times the cost of a single 450-hectolitre (9,900-gallon) stainless steel vat. After two years of much higher labour costs to operate and maintain the large number of small units, the volume of wine produced in the oak casks is 10 per cent less because of evaporation, and the winemaker faces the prospect of purchasing another 200 casks.

FROM GRAPE TO GLASS

Virtually every ingredient of a fresh grape can be found in the wine it makes, although additional compounds are produced when wine is made and any sedimented matter is disposed of before it is bottled. The most significant difference in the two lists below is the disappearance of fermentable sugar and the appearance of alcohol, although the constituents will vary according to the variety and ripeness of the grape and the style of wine produced.

The individual flavouring elements in any wine represent barely 2 per cent of its content. We can determine with great accuracy the amount and identity of 99 per cent of these constituents, but the mystery is that if we assembled them and added the requisite volume of water and alcohol, the result would taste nothing like wine, let alone like the specific wine we would be trying to imitate.

THE "INGREDIENTS" OF FRESH GRAPE JUICE

Percentage by volume			
73.5	Water		
25	Carbohydrates, of which:		
		20%	Sugar (plus pentoses, pectin, inositol)
		5%	Cellulose
0.8	Organic acids, of which:		
		0.54%	Tartaric acid
		0.25%	Malic acid
		0.01%	Citric acid (plus possible traces of succinic acid and lactic acid)
0.5	Minerals, of which:		
		0.25%	Potassium
		0.05%	Phosphate
		0.035%	Sulphate
		0.025%	Calcium
		0.025%	Magnesium
		0.01%	Chloride
		0.005%	Silicic acid
		0.1%	Others (aluminium, boron, copper, iron, molybdenum, rubidium, sodium, zinc)
0.13	Tannin and colour pigments		
0.07	Nitrogenous matter, of which:		
		0.05%	Amino acids (arginine, glutamic acid, proline, serine, threonine, and others)
		0.005%	Protein
		0.015%	Other nitrogenous matter (humin, amide, ammonia, and others)
Traces	Vitamins (thiamine, riboflavin, pyridoxine, pantothenic acid, nicotinic acid, and ascorbic acid)		

THE "CONTENTS" OF WINE

Percentage by volume			
86	Water		
12	Alcohol (ethyl alcohol)		
0.4	Glycerol		
	Organic acids, of which:		
		0.20%	Tartaric acid
		0.15%	Lactic acid
		0.05%	Succinic acid (plus traces of malic acid citric acid)
0.2	Carbohydrates (unfermentable sugar)		
0.2	Minerals, of which:		
		0.075%	Potassium
		0.05%	Phosphate
		0.02%	Calcium
		0.02%	Magnesium
		0.02%	Sulphate
		0.01%	Chloride
		0.005%	Silicic acid
		Traces	Aluminium, boron, copper, iron, molybdenum, rubidium, sodium, zinc
0.1	Tannin and colour pigments		
0.045	Volatile acids (mostly acetic acid)		
0.025	Nitrogenous matter, of which:		
		0.01%	Amino acids (arginine, glutamic acid, proline, serine, threonine, and others)
		0.015%	Protein and other nitrogenous matter (humin, amide, ammonia, and others)
0.025	Esters (mostly ethyl acetate, but traces of numerous others)		
0.004	Aldehydes (mostly acetaldehyde, some vanillin, and traces of others)		
0.001	Higher alcohols (minute quantities of amyl plus traces of isoamyl, butyl, isobutyl, hexyl, propyl, and methyl may be present)		
Traces	Vitamins (thiamine, riboflavin, pyridoxine, pantothenic acid, nicotinic acid, and ascorbic acid)		

oxidation and, with various wood lactones and unfermentable sugars, imparts a distinctive sweet and creamy vanilla nuance to wine. This oaky character takes on a smoky complexity if the wine is allowed to go through its malolactic fermentation in contact with the wood and becomes even more complex, and certainly better integrated, if the wine has undergone all or most of its alcoholic fermentation in cask. Oak also imparts wood tannins to low-tannin wine, absorbs tannins from tannic wine, and can exchange tannins with some wines. Oak tannins also act as catalysts, provoking desirable changes in grape tannins through a complex interplay of oxidations.

There has also been a move over the last decade or so for winemakers to choose concrete tanks for fermentation and ageing, which purportedly offer the best qualities of both steel and oak.

POST-FERMENTATION PROCEDURES
Numerous procedures can take place in the winery after fermentation and, where applicable, malolactic fermentation, have ceased. The five most basic procedures are racking, fining, cold stabilization, filtration, and bottling.

Racking
Draining the clear wine off its lees, or sediment, into another vat or cask is known as "racking" because of the different levels, or racks, on which the wine is run from one container into another. In modern vinification, this operation is usually conducted several times during vat or cask maturation. The wine gradually throws off less and less of a deposit. Some wines, such as Muscadet *sur lie,* are never racked.

Fining

After fermentation, wine may look hazy to the eye. Even if it does not, it may still contain suspended matter that threatens cloudiness in the bottle. Fining usually assists the clarification of wine at this stage. In addition, special fining agents may be employed to remove unwanted characteristics. When a fining agent is added to wine, it adheres to cloudy matter by physical or electrolytic attraction, creating tiny clusters (known as colloidal groups), which drop to the bottom of the vat as sediment. The most commonly encountered fining agents are egg white, tannin, gelatine, bentonite, isinglass, and casein. Winemakers have their preferences and individual fining agents also have their specific uses: positively charged egg white fines out negatively charged matter, such as unwanted tannins or anthocyanins, while negatively charged bentonite fines out positively charged matter, such as protein haze and other organic matter. Making wines suited to certain dietary requirements, such as veganism can also determine the type of fining agent used.

Cold stabilization

When wines are subjected to low temperatures, a crystalline precipitate of tartrates can form a deposit in the bottle. Should the wine be dropped to a very low temperature for a few days before bottling, this process can be accelerated, rendering the wine safe from the threat of a tartrate deposit in the bottle. For the past 25 years, cold stabilization has been almost obligatory for cheap commercial wines, and it is now increasingly used for those of better quality as well. This trend is a pity because the crystals are, in fact, entirely harmless and their presence is a completely welcome indication of a considerably more natural, rather than heavily processed, wine.

Filtration

Various methods of filtration exist, and they all entail running wine through a medium that prevents particles of a certain size from passing through. Filtration has become a controversial subject in recent times, with some critics claiming that anything that removes something from wine must be bad. Depending on who you listen to, this "something" is responsible for a wine's complexity, body, or flavour. Yet, although it is undeniable that filtration strips something from a wine, if it is unfiltered, it will throw a much heavier deposit and do so relatively quickly. The mysterious "something" is, therefore, purged from all wines at some time or other, whether they are filtered or not – and whether the critics like it or not. Filtration, like so many things, is perfectly acceptable if it is applied in moderation. The fact that many of the world's greatest wines are filtered is a testament to this.

I prefer less or no filtration, as do most quality-conscious winemakers. This is not because of any romantic, unquantifiable ideal; it is simply because I prefer wine to be as unprocessed and as natural as possible. This is a state that can only be achieved through as much of a hands-off approach as the wine will allow. Generally, the finer the wine, the less filtration required, as consumers of expensive wines expect sediment and are prepared to decant. Delicate reds, such as Pinot Noir, should be the least filtered of all, as I swear they lose fruit just by looking at them, and they certainly lose colour – that, at least, is quantifiable. No wine with extended barrel-ageing should ever require filtration – just a light, natural fining.

Each filtration is expensive and time-consuming, thus even producers of the most commercial, everyday wines (which even filtration critics accept must be filtered) should keep these operations to a minimum. The principle

FILLED BOTTLES TRAVEL ALONG A CONVEYOR BELT, READY FOR SEALING AND LABELLING
Bottling, the last phase of the winemaking process, is straightforward. First, a bottle is washed and sterilised and then dried out with an air shot. During the vacuum and sparge stage, nitrogen is added to protect the wine's quality and purity. The bottle is then filled with the wine, and another small quantity of nitrogen is added in the small empty space at the top of the bottle to help prevent any oxidation. The bottle is then corked or capped and sealed with a protective sleeve called a capsule. The last step will be labelling.

means of achieving this is by ensuring the best possible clarification by set-tling and racking. Fining (*see* opposite page) should always take precedence, as it is both kinder on the wine and much cheaper than filtering. There are four basic types of filtration: earth, pad, membrane, and crossflow.

Earth filtration
This system is primarily used after racking for filtering the wine-rich lees that collect at the bottom of the fermentation tank. A medium, usu-ally kieselguhr (a form of diatomaceous earth), is continuously fed into the wine and used with either a plate and frame filter or a rotary drum vacuum filter. Both types of filter are precoated with the medium, but in a plate and frame filter, the wine and medium mix is forced, under pres-sure, through plates, or screens, of the medium, in a manner similar to that of any other filter. For the rotary drum vacuum filter, however, the precoat adheres to the outside of a large perforated drum by virtue of the vacuum that is maintained inside. The drum revolves through a shallow bath into which the wine is pumped and literally sucks the wine through the precoat into the centre, where it is piped away. The advantage of this system is that on one side of the shallow bath there is a scraper that con-stantly shaves the coating on the drum to the desired thickness. The medium thus falls into the wine, with which it mixes, and is then sucked back on to and through the drum. It is a continuous process and a very economical one, as the amount of medium used is limited. In a plate and frame filter, the medium is enclosed and eventually clogs up, requiring the operation to be stopped, and the equipment to be dismantled and cleaned before the process can resume.

Pad filtration
Also called sheet filtration, this requires the use of a plate and frame filter with a variable number of frames into which filter pads or sheets can slide. Before it was outlawed in the 1970s, these used to be made of asbestos. They now contain numerous filtration mediums, ranging from diatoma-ceous earth to regular cellulose pads, the latter of which are the most com-monly used medium. Special filter formats include active carbon (to remove unwanted colour, which is frowned upon for all but the most com-mercial, high-volume wines) and electrostatically charged pads. These are designed to attract any matter that is suspended in the wine, most of which will possess a negative or positive charge. It is claimed that these electro-statically charged pads are more effective in filtering out matter than the same pads without a charge.

Membrane filtration
This is also called millipore filtration because the membranes contain microscopic holes capable of removing yeasts and other micro-organisms. These holes account for 80 per cent of the sheet's surface, so the through-put of wine can be extremely fast, provided it has undergone a light pre-fil-tration. Both filtration and pre-filtration can now, however, be done at the same time with new millipore cartridge filters. These contain two or more membranes of varying porosity, thus the coarser ones act as a screen for those with the most minuscule holes. Many producers will not put a full-bodied wine through a membrane filter because it filters down to such a microscopic level that it is thought to remove too much of the body.

Crossflow filtration
Originally designed to purify water, a crossflow filter varies from the oth-ers (with the exception of the rotary drum vacuum filter) because it is self-cleaning and never clogs up. The wine flows across the membrane, not into it, so only some of it penetrates. Most of it returns to the chamber from which it came and, because it flows very fast, takes with it any matter filtered out by the membrane.

BOTTLING
When visiting larger producers, automated bottling lines are the wine journalist's bête noire. They get faster and more complex each year and, having invested vast sums in the very latest bottling line, it is understand-able that proprietors are eager to show off their new high-tech toy. Yet, as John Arlott once told me, as he resolutely refused to set foot in the bottling

FORTIFIED AND AROMATIZED WINES

FORTIFIED WINES
Any wine, dry or sweet, red or white, to which alcohol has been added is classified as a fortified wine, whatever the inherent differences of vinification may be. Still wines usually have a strength of 8.5 to 15 per cent alcohol and fortified wines a strength of 17 to 24 per cent. The spirit added is usually, but not always, brandy made from local wines. It is totally neutral, with no hint of a brandy flavour. The amount of alcohol added, and exactly when and how it is added, is as critical to the particular character of a fortified wine as is its grape variety or area of production. *Mutage,* early fortification, and late fortification are all methods that may be used to fortify wines.

Mutage
This is the addition of alcohol to fresh grape juice, which prevents fermentation and produces fortified wines, known as *vins de liqueurs* in France, such as Pineau des Charentes in the Cognac region, Floc de Gascogne in Armagnac, Macvin in the Jura, and Ratafia in Champagne.

Early Fortification
This is the addition of alcohol after fermentation has begun, and it is often done in several small, carefully measured, timed doses spread over several hours or even days. The style of fortified wine being made will dictate exactly when the alcohol is added, and the style itself will be affected by the variable strength of the grapes from year to year. On average, however, alcohol is added to Port after the alcohol level has reached 6 to 8 per cent and added to the *vins doux naturels* of France, such as Muscat de

Beaumes de Venise, at any stage between 5 and 10 per cent.

Late Fortification
This is the addition of alcohol after fermentation has ceased. The classic drink produced by this method is Sherry, which is always vinified dry, with any sweetness added afterwards.

AROMATIZED WINES
With the exception of Retsina, the resinated Greek wine, aromatized wines are all fortified. They also all have aromatic ingredients added to them. The most important aromatized wine is vermouth, which is made from neutral white wines of two to three years of age, blended with an extract of wormwood (vermouth is a corruption of the German *Wermut,* meaning "wormwood"), vanilla, and various other herbs and spices. Although the earliest example of vermouth dates back to 16th-century Germany and the first commercial vermouth (Punt e Mes, created by Antonio Carpano of Turin) to 1786, the notion of adding herbs and spices to wine was prevalent in Roman times and goes back at least as far as ancient Egypt. Italian vermouths are produced in Apulia and Sicily and French vermouths in Languedoc and Roussillon. Chambéry is a delicate generic vermouth from the Savoie and Chambéryzette is a red-pink version flavoured with alpine strawberries, but such precise geographical aromatized wines are rare. Most, in fact, are made and sold under internationally recognized brands such as Cinzano and Martini. Other well-known aromatized wines include Amer Picon, Byrrh, Dubonnet (both red and white), Punt e Mes, St-Raphael, and Suze.

hall of Piper-Heidsieck, "I have not written a single word about bottling lines in my life and I'm not going to start now". They make a very dull experience and inevitably break down just as one's host boasts that it is the fastest bottling line in the world, but their smooth operation most days of the week is essential if the wine is to remain as fresh as possible in the bottle. All that readers need to know is that the bottles should be sterile, that fully automated lines cork, capsule, label, and box the wines, and that there is a device to detect any impurities before the bottles are boxed. All European systems either print on the label or laser-print directly on to the bottle, a lot number that identifies the date each batch was bottled. This means that in an emergency, a specific batch can be recalled, rather than the entire production having to be cleared from every wholesaler and retailer stocking a particular line.

Red Winemaking

The art of making red wine has been around for thousands of years. Through the centuries, the art has been refined into a science, and although winemaking is a natural process, winemakers guide it with different techniques to bring out their preferred characteristics.

On arrival at the winery, the grapes are usually crushed and destemmed, although it was once accepted practice to leave the stems on for a more tannic wine. Stem tannins are too harsh, however, and fail to soften as the wine matures. The modern winemaker can include a small quantity of stems if the grape variety requires extra structure or if the vintage needs firming up.

FERMENTATION

After the grapes are destemmed and lightly crushed, they are pumped into a vat, where fermentation may begin as early as 12 hours or as late as several days later. Even wines that will be cask-fermented must start off in vats, whether they are old-fashioned oak *foudres* or modern stainless steel tanks. This is because they must be fermented along with a *manta*, or cap, of grapeskins. To encourage fermentation, the juice may be heated and selected yeast cultures or partially fermented wine from another vat added. During fermentation, the juice is often pumped from the bottom of the vat to the top and sprayed over the *manta* to keep the juice in contact with the grapeskins. This ensures that the maximum colour is extracted. Other methods involve the *manta* being pushed under the fermenting juice with poles. Some vats are equipped with crude but effective grids that prevent the *manta* from rising; others rely on the carbonic gas that is given off during fermentation to build up pressure, releasing periodically and pushing the *manta* under the surface. Another system keeps the *manta* submerged in a "vinimatic", a sealed, rotating stainless-steel tank, based on the cement-mixer principle.

The higher the temperature during fermentation, the more colour and tannin will be extracted; the lower the temperature, the better the bouquet, freshness, and fruit will be. The optimum temperature for the fermentation of red wine is 29.4°C (85°F). If it is too hot, the yeasts produce certain substances (decanoic acid, octanoic acids, and corresponding esters) that inhibit their own ability to feed on nutrients and cause the yeasts to die. It is, however, far better to ferment hot fresh juice than to wait two weeks (which is normal in many cases) to ferment cooler but stale juice. The fuller, darker, more tannic, and potentially longer-lived wines remain in contact with the skins for anything between 10 and 30 days. Lighter wines, on the other hand, are separated from the skins after only a few days.

VIN DE GOUTTE AND VIN DE PRESSE

The moment the skins are separated from the juice, every wine is divided into two parts: free-run wine, or *vin de goutte,* and press wine, or *vin de presse.* The free-run juice runs out of the vat when the tap is opened. The remains – the *manta* of grapeskins, pips, and other solids – are put into a press to extract the very dark, extremely tannic "press wine". The free-run wine and the press wine are then pumped into separate vats or casks, depending on the style of the wine being made. These wines then undergo their malolactic conversion separately and are racked several times, fined, racked again, blended, then fined and racked once more before bottling.

CARBONIC MACERATION

There are several variations of carbonic maceration (known in French as *macération carbonique),* a technique used almost exclusively for making red wine, involving an initial fermentation under pressure of carbonic gas. The traditional method, dating back at least 200 years, was to put the uncrushed grapes in a closed container where, after a while, a natural fermentation would take place inside the grapes. When the grapes eventually exploded, filling the container with carbonic gas, a normal fermentation continued, and the grapes macerated in their own skins. Today the grapes are often placed into vats filled with carbonic gas from a bottle. Carbonic maceration produces light wines with good colour, soft fruit, and a "peardrop" aroma.

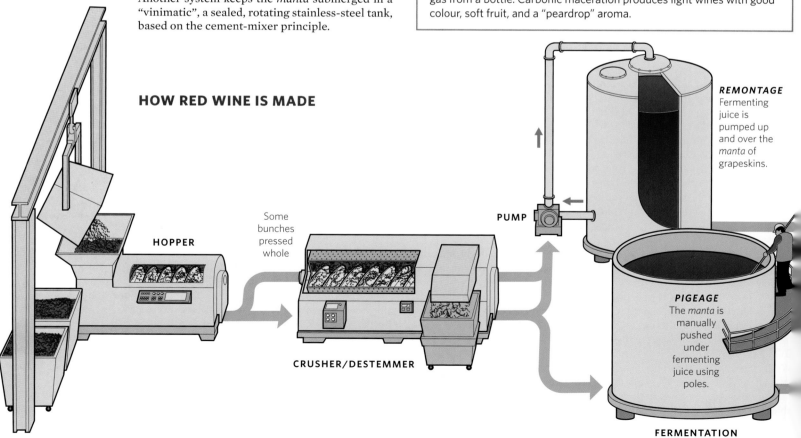

HOW RED WINE IS MADE

HOPPER

Some bunches pressed whole

CRUSHER/DESTEMMER

PUMP

REMONTAGE
Fermenting juice is pumped up and over the *manta* of grapeskins.

PIGEAGE
The *manta* is manually pushed under fermenting juice using poles.

FERMENTATION

THE WINEMAKING FACILITIES OF RIOJA PRODUCER MARQUÉS DE RISCAL INCLUDE BOTH STEEL AND OAK FOR FERMENTATION AND AGEING
The fermentation tanks are all small enough to ferment wines on a block-by-block basis. Note how they stand well clear of the ground to facilitate drainage by gravity.

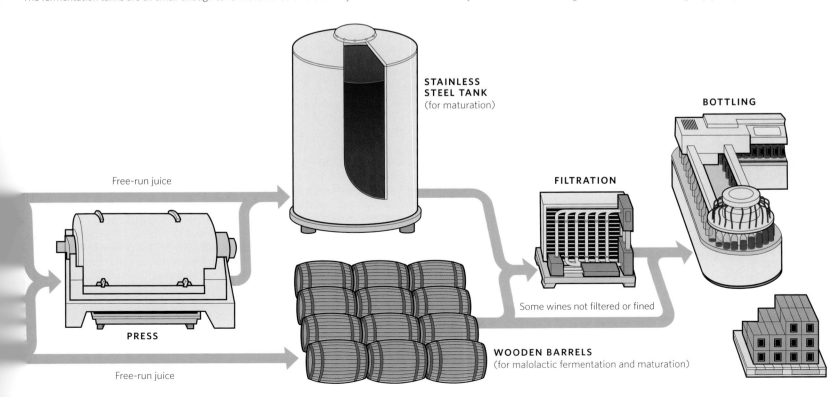

Free-run juice

STAINLESS STEEL TANK
(for maturation)

BOTTLING

FILTRATION

PRESS

Some wines not filtered or fined

WOODEN BARRELS
(for malolactic fermentation and maturation)

Free-run juice

ORSI VIGNETO SAN VITO SUI LIEVITI ORGANIC ORANGE WINE SITS ON ICE FOR A WINE-TASTING EVENT
The Georgian wine *qvevri* inspired many winemakers to try their hand at an orange wine. These days, most orange winemaking takes place in the Friuli-Venezia Giulia region of northeastern Italy, along the border of Slovenia. Italian winemaker Josko Gravner, who first attempted an orange wine in the late 1990s, launched the orange wine trend.

ORANGE WINE

These days, the hippest colour on the wine shelf – orange – is fast becoming a style category of its own, even though it has been around for thousands of years. With more obvious structure, tannins, and complexity than many whites, orange wine has been described as a "white wine with red wine characteristics". This is because red winemaking techniques are employed to make orange wine, extracting colour and flavour during pressing and fermentation.

Orange is also the emblematic wine of a modern generation of winemakers who take a low-intervention approach, allowing nature to set the pace in both the vineyard and the winery, in the main avoiding commercial yeasts, temperature control, filtering, and preservatives along the way. They reached into the past for ideas and found inspiration in *qvevri*-fermented Georgian amber wines, which have been made since ancient times.

No oranges are harmed in the making of these often (but not always) natural wines. The term refers only to the colour of the liquid, which can vary in intensity from dark pink through gold and pale amber to tawny. The hue is the result of extended maceration – steeping together juice, stems, pips, and skins of white grapes for anything from a week to a year – so that compounds such as carotenoids and natural phenols are released during fermentation to give an orange glow to the wine as well adding structure and complexity and health-giving antioxidants.

In the winery, grape selection is very important to avoid mould taint and green tannins, but once pressed they are often left to ferment naturally. Sulphur is usually added only at bottling (if at all), and many orange wines are unfiltered, giving them a recognisable cloudiness in the glass.

Low-intervention orange wines are fermented dry and often have an earthy, savoury character, while on the nose they can have aromas of dried fruit, exotic tea, or cider. Do not look for varietal typicity, because as it usually falls victim to the long maceration, but some aromatic varieties can give some attractive character.

Fermentation can be in steel, oak, concrete, or clay, and some oxygen is preferred – there is an argument that exposure to oxygen during winemaking prevents "bottle shock" (a temporary condition in which a wine's flavours are muted or disjointed). Fermentation of modern orange wines is often not temperature controlled and should not be confused with other styles of macerated white wines in which the skin contact takes place before controlled fermentation.

Orange wines are made all over the world, so there is no formal designation or appellation for them, and all are bottled as white wines, although some countries have lately put rules in place. In South Africa, where winemakers have enthusiastically embraced the orange movement (in particular in the Swartland region with Chenin Blanc), orange wines must have a minimum of 96 hours on skins and may only have a maximum sulphites content of 40 milligrammes per litre total.

Drinkers who are new to orange wines are encouraged to approach them with both caution and an open mind. Turn off expectations of mainstream tasting notes for white wine, and be prepared to be pleasantly surprised by an earthy new flavour experience. Orange wines have a layered complexity when made well, suited perfectly for matching with food, especially ingredients with umami influences such as mushrooms, asparagus, and washed-rind cheeses.

Rosé Winemaking

With the exception of pink Champagne, most of which is made by blending white wine with red, all quality rosés are produced by limited maceration or bleeding.

The pink-tinted – and too often watery wine – made from black grapes now enjoys year-round popularity with a large and loyal following that appreciates its mass appeal as an easy-going drink that one can consume without thinking too much about it. Sadly, the great rosés of the world are usually overlooked in the current trend of pale plonk.

All black grape varieties can be used to make rosé, on their own or in a blend. The acid and tannin structure and colour intensity of each variety will largely influence the resulting wine.

There are several ways to make rosé wines, all resulting in slightly different styles. The common idea behind them is to limit the colour intensity to a lighter or deeper shade, depending on the intention of the winemaker and the style of the region of origin. There are three main methods of producing a rosé: limited maceration, *saignée* (bleeding), and blending.

LIMITED MACERATION

Limited maceration is the most common method used in the production of rosé. Black-skinned grapes are gently crushed and the skins left in contact with the juice for a short time (generally from 2 to 24 hours) to allow colour to infuse before the juice is drained for fermentation. This is the most common way of rosé winemaking today, and it allows greater control over the resulting colour. It also gives the wine more structure and, in some cases, even ageability.

SAIGNÉE (BLEEDING)

This is an often misunderstood concept that means quite the opposite of what it sounds like, depending on where you are. A true-bled, or *saignée*, rosé is made from the juice that issues from black grapes pressed under their own weight. In most areas the slightly tinted juice of the black grapes is bled off to the desired colour level to concentrate the remaining must and make denser, more intense red wines. For making a rosé Champagne, however, saignée is understood as free-run juice being bled off *after* white must had been drawn from the black grapes, making this rosé a by-product of white winemaking (from black grapes of course).

ROEDERER'S "SECRET RECIPE" FOR ROSÉ CHAMPAGNE

Louis Roederer's method of creating its pink champagne is a unique combination of maceration, *saignée*, and co-fermentation that suits the house's style perfectly.

Ripe Pinot Noir grapes are cold-soaked in Chardonnay juice for a week to extract the colour without any crushing of the berries, which would add undersireable tannins. The coloured juice is drawn off, the remaining Pinot Noir berries are gently pressed, and the two batches combined. Acidic Chardonnay juice is added (between 20 and 50 per cent) to fix the colour by adjusting pH and adjusting the acidity of the Pinot Noir that was picked for ripeness.

BLENDING

The use of this technique is legally limited to Champagne in the European Union and to some New World countries, where regulations are less strict. The blending method is just what it sounds like – it involves simply mixing fermented white and red wine to create the pinkish hue.

To make rosé or pink Champagne, a small per cent of red wine is added to the white base wine (which can be made from either white Chardonnay, or black Pinot Noir and Pinot Meunier grapes) before the second fermentation to achieve the desired shade.

LIGHTSTRIKE

A word of caution is necessary when buying rosé: winemakers love to show off the delicate hues of their pink wines by bottling them in clear glass bottles (an obvious marketing strategy), but light, and in particular blue and ultraviolet light, can damage the amino acids in the wine. This can cause the wines to lose their delicate aromas and, in worse cases, to smell like cooked cabbage or a dirty dishcloth.

Lightstrike in wines is well known – cellars are kept dark and red wines are bottled in green glass to protect them – but we now know that light can damage a wine in as little as one hour, and clear glass has the least protection against destructive wavelengths. This makes wine displayed on supermarket and corner shop shelves particularly vulnerable and also puts bottles on brightly lit bar and restaurant shelves at risk. Yet, clear glass (and thinner, lighter glass to take account of environmental concerns) is still too often the winemaker's package of choice, even for premium white and sparkling wines, making the issue of lightstrike one for the consumer.

A DISPLAY SHOWS THE VARIATION OF ROSÉ COLOURS
For a majority of rosé, the colour produced is rarely a true pink, but instead will have hints of orange, blue, or purple.

White Winemaking

Until fairly recently, it could be said that two initial operations distinguished the white winemaking process from the red: first, an immediate pressing to extract the juice and separate the skins, and second, the purging, or cleansing, of this juice.

For white wines of expressive varietal character, the grapes are now often crushed and then macerated (sometimes in a vinimatic) for 12 to 48 hours to extract the aromatics that are stored in the skins. The juice that is run off and the juice that is pressed out of the macerated pulp then undergoes cleansing and fermentation like any other white wine.

With the exception of wines macerated in a vinimatic, the grapes are either pressed immediately on arrival at the winery or lightly crushed and then pressed. The juice from the last pressing is murky, bitter, and low in acidity and sugar, so only the first pressing, which is roughly equivalent to the free-run juice in red wine, together with the richest elements of the second pressing, should be used for white-wine production. Once pressed, the juice is pumped into a vat, where it is purged, or cleansed, which in its simplest form means simply leaving the juice to settle so that particles of grape-skin and any other impurities fall to the bottom. This purging may be helped by chilling, adding sulphur dioxide, and, possibly, a fining agent. Light filtration and centrifugation may also be applied during this process.

After cleansing, the juice is pumped into the fermenting vat, or directly into barrels if the wine is to be cask-fermented. The addition of selected yeast cultures occurs more often in the production of white wine because of its limited contact with the yeast-bearing skins and the additional cleansing that reduces the potential amount of wine yeasts available. Selected cultured yeast strains may also provide some desirable aromatic profiles.

CHARDONNAY GRAPES ARE LOADED INTO A CRUSHER AND DESTEMMER
White wine grapes can be crushed and destemmed first or be pressed whole.

The optimum temperature for fermenting white wine is 18°C (64°F), although many winemakers opt for between 10°C and 17°C (50°F and 63°F), and it is actually possible to ferment wine at temperatures as low as 4°C (39°F). At the lower temperatures, more esters and other aromatics are created, less volatile acidity is produced, and a lower dose of sulphur dioxide is required; on the other hand, the resulting wines are lighter in body and contain less glycerol.

With acidity an essential factor in the balance of fruit and, where appropriate, sweetness in white wines, many products are not permitted to undergo malolactic conversion and are not bottled until some 12 months after the harvest. Oak-matured wines – which, incidentally, always undergo malolactic conversion – may be bottled between 9 and 18 months, but wines that are made especially for early drinking are nearly always racked, fined, filtered, and bottled as quickly as the process will allow in order to retain as much freshness and fruitiness as possible.

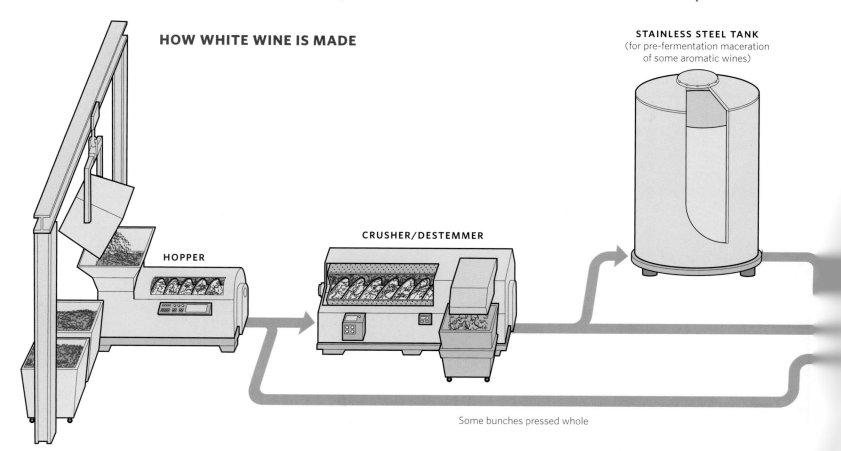

HOW WHITE WINE IS MADE

STAINLESS STEEL TANK
(for pre-fermentation maceration of some aromatic wines)

HOPPER

CRUSHER/DESTEMMER

Some bunches pressed whole

WHITE WINEMAKING TECHNIQUES

Some of the following winemaking techniques can be used for red wine as well as white, but without the benefit of a red wine's lengthy maceration on skins, a white wine inevitably needs more techniques for those who want to tweak the outcome. The higher the quality and the more specific the *terroir*, however, the less hands-on a winemaker needs to be to express the grapes grown and a sense of place.

Acidification Unless you do this all the time, acidification is not easy to get right, but when performed proficiently, it can be the saving of what would otherwise be a flabby wine. It also increases perceived freshness and reduces SO_2 requirements. Even the greatest wines have benefited from acidification.

Back-blending A judicious addition of fresh, sterile-filtered juice after fermentation can increase freshness and fruitiness with minimal increase in residual sweetness.

Blanc de noirs A red wine is only red because it has been macerated on its own skins. With very rare exceptions (*teinturier* grapes, for example), black varieties have clear juice, so making white wine from them is no big deal. It is not new either – the *champenois* have been doing it for centuries, and *blanc de noirs* Champagnes are consistently the best rendition of this style. Other *blanc de noirs* wines can and do work, but usually only when produced by someone who has been inspired to make such a wine. Far too many *blanc de noirs* are produced from grapes that, for one reason or another, were harvested before the skins had developed sufficient pigments to make a half-decent red wine, and the white wine alternative is often disappointing.

Cool fermentation The lower the temperature, the more freshness, less volatile acidity (VA), and less need for SO_2. But for every degree of temperature below 18°C (64°F), there is a trade-off between increasing these advantages and reducing fruit and body. At the lowest temperatures, the amount of amyl acetate produced becomes noticeable, though, as the aromas go from banana through peardrop to nail varnish.

Enzymes The use of enzymatic preparations to affect the biochemical pathways in white winemaking has been on the increase since the mid-1990s. Enzymes are used for improving the pressing or maceration (by reducing the risk of earthy or grassy notes) and increasing the filterability of musts and wines (by increasing

clarification), but most attention has been directed towards the flavour-altering capabilities of these preparations, particularly
for aromatic grape varieties such as Riesling, Gewürztraminer, and Muscat. Lafazym Arom, for example, is supposed to increase various terpenoid compounds (*see* Micropedia), but whether it and other similar enzymatic preparations have a significant effect on the right terpenes is open to question. It is difficult to judge the effect of these enzymes on either Gewürztraminer or Muscat, but there are reports that they give an inappropriate Turkish-delight floral aroma to Riesling.

Lees stirring What the French call *bâtonnage*, lees stirring has been taken over by certain New World winemakers who hamfistedly stir the lees in their barrel-fermented Chardonnays too frequently and for too long, until the wine becomes dominated by overtly rich, dairy-like aromas. *Bâtonnage* was developed to prevent the formation of hydrogen sulphide (*see* Egg entry in the Taste and Aroma Chart, as well as the Micropedia). When practised with a light hand, it should merely endow a wine with a barely perceptible umami-like (*see* p93) amplitude that is more textural than taste, with no vulgar tell-tale aromas.

Malolactic fermentation Thanks to the diacetyl, this can produce anything from big and buttery to "baby sick", but malolactic should never be noticeable on the nose – just a creamy texture at the back of the palate. *See also* Non-malolactic.

Micro-oxygenation The oxidation of wine is not only inevitable but desirable, to smooth out the tannins and open up the fruit during winemaking, or simply as part of the maturation process in barrel or bottle. Only the rate of oxidation determines a fault. The role of artificial micro-oxygenation (the introduction of a constant stream of minuscule oxygen bubbles into a wine) has replaced the traditional methods of exposing wine to air (such as that deliberately practised when racking a wine) and has become ubiquitous in the technological environment of premium winemaking.

Non-malolactic For crisp grapes (such as Sauvignon Blanc, Riesling) and aromatic varieties (such as Riesling and Gewürztraminer), malolactic is the last

thing needed. By avoiding it, the natural acidity is enhanced, and any minerality of fruit is retained.

Oak ageing It is always preferable to barrel-ferment, but ageing part or all of a wine in oak barrels following fermentation in stainless steel tanks is cheaper. If this can be achieved with a low-key malolactic in the barrel, then the oak-aged wine will be better integrated than a wine that has gone through malolactic in tank.

VA lift Tweaking the volatile acidity in a wine to lift the fruity notes of its aroma and intensify fruitiness on the palate must be undertaken with great care, as a touch too much can reveal acrid notes of acetic acid (vinegar).

Wild-yeast ferment It's debatable how many claimed wild-yeast ferments have been fermented by wild yeasts to any significant degree, unless the fermentations have been carried out in a closed and controlled environment, as research has shown that the yeast that gets into every nook and cranny of the cleanest press-house and winery (the true terror yeast) will always take over the fermentation after the first two or three degrees of alcohol. True wild-yeast ferments supposedly run the risk of excessive volatile acidity and are famed for their "funky" flavours.

Winter bottling When wines are bottled in the winter, they retain more of their residual carbonic gas, which can heighten the sense of acidity without being *pétillant* as such. This practice best suits crisper, non-malolactic, unoaked, and more mineral styles of white wine. For cheaper wines, a similar effect can be created all year around by sparging the wines with CO_2 as they are bottled.

Yeast manipulation Although everything ultimately depends on the initial ingredients in the grape juice being fermented, many of those ingredients are neutral precursors to very specific aromas that can only be activated by certain biochemical processes that yeast cells are capable of initiating. Bearing that in mind – and the fact that yeasts are responsible for more than 40 per cent of all the aromatic compounds found in wine – through those biochemical processes, it becomes clear that the choice of whether to use a proprietary can have a profound effect on the type of wine produced. For example, Zymaflore VL3 from Laffort promotes the production of 4MMP (the Sauvignon Blanc varietal aroma), 3MH (citrus), and 3MHA (passion fruit).

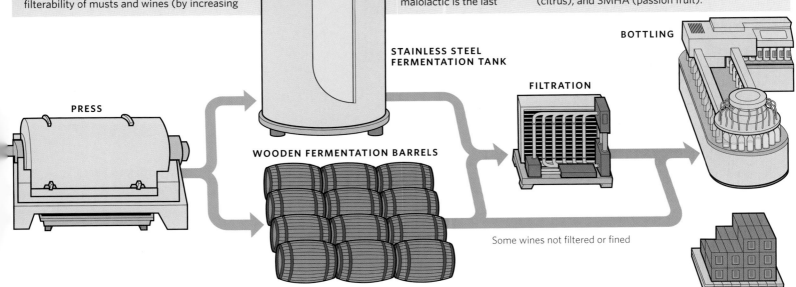

PRESS

STAINLESS STEEL
FERMENTATION TANK

WOODEN FERMENTATION BARRELS

FILTRATION

BOTTLING

Some wines not filtered or fined

Sparkling Winemaking

Sparkling wine is a drink of celebration, brought out to mark some of life's most joyous events. Yet, it is not a simple thing to get the right amount of fizz and the perfect bubble size that make this wine so special.

When grape juice is fermented, sugar is converted into alcohol and carbonic gas. For still wines, the gas is allowed to escape, but if it is prevented from doing so – by putting a lid on a vat or a cork in a bottle – it will remain dissolved in the wine until the lid or cork is removed, when the gas will be released to rush out of the wine in the form of bubbles. The production of all natural sparkling wines is based on this essential principle, using one of four different methods: *méthode champenoise,* bottle-fermented, *méthode rurale,* and *cuve close.* They can also be made by injecting the bubbles.

MÉTHODE CHAMPENOISE

Also referred to as *méthode traditionnelle* or *méthode classique* (France), *metodo classico* (Italy), and *Cap Classique* (South Africa), this term indicates a sparkling wine that has undergone a second fermentation in the bottle in which it is sold. A label may refer to a wine being "Individually fermented in this bottle", which is the beautifully simple American equivalent of *méthode champenoise.*

After the first fermentation, which might or might not include malolactic or partial malolactic, the *assemblage* takes place. This usually occurs in the first few months of the year following the harvest. At its most basic, and without taking into consideration any particular house style, it is a matter of balancing the characteristics of different wines produced in different places (far and wide or even merely different parts of a single vineyard) and, possibly, different grape varieties. For non-vintage Champagne and increasingly for other serious sparkling wines, the *assemblage* also involves the use of reserve wines. These are, quite literally, wines that are kept in reserve from previous years. The idea that reserve wines are primarily used to make *cuvées* conform to a particular house style no matter what the quality or character of the base-wine year is a misconception. The job of reserve wine is, if anything, closer to that of the *dosage* in that both make a sparkling wine easier to drink at a younger age. Reserve wines also provide a certain richness, fullness, and mellowed complexity, which is why it is generally considered that the more reserve wines are added, the better. Unless kept on their yeast, however, reserve wines can dilute the autolysis process (*see* below); thus, there may come a point when potential finesse is traded for instant complexity.

THE SECOND FERMENTATION AND *REMUAGE*

The second fermentation is the essence of the *méthode champenoise* and is the only way to produce a fully sparkling wine. After the blended wine has undergone its final racking, the *liqueur de tirage,* or bottling liquor, is added. This is a mixture of still Champagne, sugar, selected yeasts, yeast nutrients, and a clarifying agent. The amount of sugar added depends on the degree of effervescence required and the amount of natural sugar in the wine. The wines are bottled (usually in May or thereabouts) and capped with a temporary closure. The second fermentation can take between 10 days and three months, after which the bottles can be transferred to *pupitres* to undergo *remuage.* A *pupitre* consists of a pair of heavy, hinged, rectangular boards, each containing 60 holes, which have been cut at an angle to enable the bottle to be held by the neck in any position between horizontal and vertical (neck pointing downwards). *Remuage* is a method of riddling the bottles to loosen sediment, encouraging it to move to the neck of the bottle, where it collects. By hand this takes about eight weeks, although many companies now have computerized 504-bottle pallets that perform the task in just eight days. A technique involving the use of porous yeast capsules, which trap the sediment inside them, reduces *remuage* to a mere eight seconds, but this is seldom used. After *remuage,* the wine undergoes a period of ageing *sur point* (in a fully inverted position) before the sediment is removed.

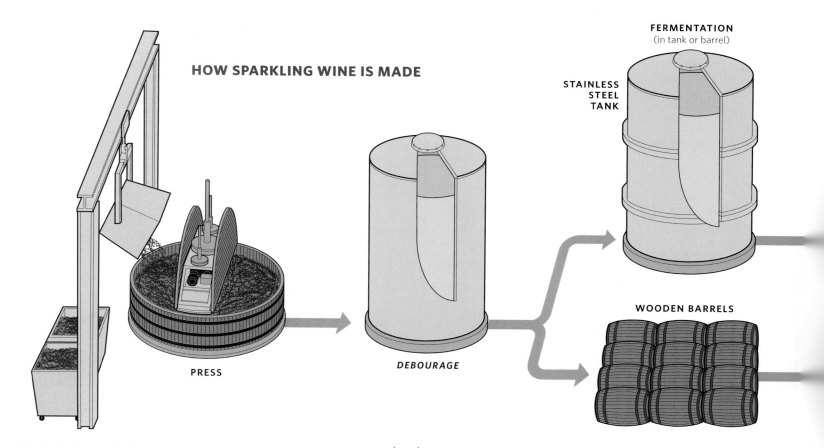

HOW SPARKLING WINE IS MADE

PRESS

DEBOURAGE

FERMENTATION
(in tank or barrel)

STAINLESS STEEL TANK

WOODEN BARRELS

OTHER SPARKLING WINEMAKING METHODS

Bottle-fermented This refers to a wine produced through a second fermentation in a bottle but (and this is the catch) not necessarily in the bottle in which it is sold. It may have been fermented in one bottle, transferred to a vat, and, under pressure at -3°C (26°F), filtered into another bottle. This is also known as the "transfer method".

Carbonation This is the cheapest method of putting bubbles into wine and simply involves injecting it with carbon dioxide. Because this is the method used to make fizzy soft drinks, it is incorrectly assumed that the bubbles achieved through carbonation are large and short-lived. They can be, and fully sparkling wines made by this method will indeed be cheapskates, but modern carbonation plants have the ability to induce the tiniest of bubbles, even to the point of imitating the "prickle" of wine bottled *sur lie*.

Cuve close, Charmat method, or **tank method** This is used for the bulk production of inexpensive sparkling wines that have undergone a second fermentation in large tanks before being filtered and bottled under pressure at -3°C (26°F). Contrary to popular belief, there is no evidence to suggest this is an intrinsically inferior way of making sparkling wine. It is only because it is a bulk-production method that it tends to attract mediocre base wines and encourage a quick throughput. I genuinely suspect that a *cuve close* produced from the finest base wines of Champagne and given the autolytic benefit of at least three years on its lees before bottling might well be indistinguishable from the "real thing".

Méthode rurale This refers to the precursor of *méthode champenoise*, which is still used today, albeit only for a few obscure wines. It involves no second fermentation, with the wine being bottled before the first alcoholic fermentation is finished.

BOTTLES OF CHAMPAGNE REST BOTTOMS-UP IN *PUPITRES* IN THE CAVES OF HOUSE TAITTINGER IN REIMS, FRANCE
For the riddling, or *remuage*, process, the bottles are placed at a 45-degree angle with the cork pointing down. A winery worker called a riddler will then twist the bottle a few degrees every day so that any sediment from the bottom or sides collects at the neck.

AGEING ON LEES AND AUTOLYSIS

Ageing a sparkling wine on its lees improves its quality because the sediment contains dead yeast cells that break down by autolysis, which contributes to the wine's inimitable "Champagny" character. The optimum duration of yeast ageing depends on the quality of the wine and how it has been processed. Although most Champagnes require at least three years and few sparkling wines produced outside of Champagne benefit from as much as three years, there are many exceptions. Some Champagnes are at their best between 18 months and two years on yeast, while a number of serious sparkling wines can improve for up to a decade. After a sparkling wine's specific optimum time on lees, the complexity stops building, and the wine simply remains fresher than earlier-disgorged wine, because its toasty and biscuity aromas, which require oxygen molecules to evolve, are suppressed in the reductive environment on yeast.

DISGORGEMENT AND DOSAGE

Disgorgement is the removal of the sediment from the wine, which is normally collected in a plastic pot. The normal modern method is known as *dégorgement à la glace,* which involves the immersion of the bottle neck in a shallow bath of freezing brine. This causes the sediment to adhere to the base of the plastic pot, enabling the bottle to be turned upright without disturbing the sediment. When the crown-cap is removed, the semi-frozen sediment is ejected by the internal pressure of the bottle. Only a little wine is lost, as the wine's pressure is reduced by its lowered temperature.

Before corking, bottles are topped up to their previous level, and *liqueur d'expédition* is added. In all cases except for *nature* and some *extra brut,* this *liqueur* includes a small amount of sugar: the younger the wine, the greater the *dosage* of sugar required to balance its acidity. A *brut* champagne may have between 0 and 12 grams of residual sugar per litre (the upper limit was 15 grams per litre until 2009), but it should have at least 6 grams if it is likely to be cellared, unless the producer intends for it to develop oxidatively.

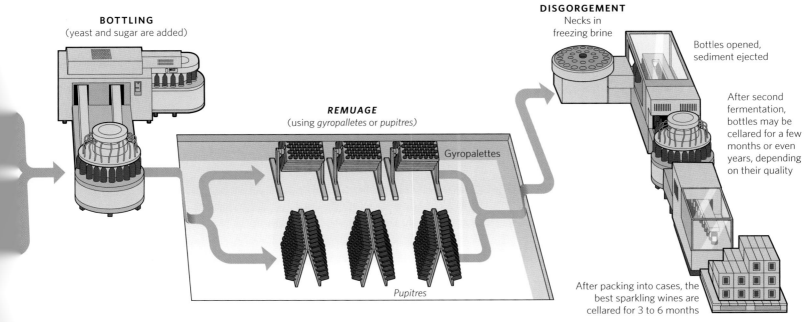

BOTTLING
(yeast and sugar are added)

REMUAGE
(using *gyropalletes* or *pupitres*)

Gyropalettes

Pupitres

DISGORGEMENT
Necks in freezing brine

Bottles opened, sediment ejected

After second fermentation, bottles may be cellared for a few months or even years, depending on their quality

After packing into cases, the best sparkling wines are cellared for 3 to 6 months

The Choice of Oak

"Why oak?" I have seen the question only once in print, yet it is the most fundamental question that could possibly be asked: why, out of all the woods around the world, is oak and, to any significant degree, only oak, used for barrelmaking?

The answer is that other woods are either too porous or contain overpowering aromatic substances that unpleasantly taint the wine. It is not entirely true to say that only oak is used in winemaking; chestnut, for example, is occasionally found in the Rhône and elsewhere, but it is so porous and so tannic that it is usually lined with a neutral substance, rendering the wood no different from any other lined construction material (such as concrete). A beech variety called *rauli* used to be popular in Chile until its winemakers, suddenly exposed to international markets, soon discovered they had become so used to the wood that they had not realized that it gave their wines a musty joss-stick character. Large, redwood tanks are still used in California and Oregon, but they are not greatly appreciated and, as the wood cannot be bent very easily, it is not practical for small barrels. Pine has a strong resinous character that the Greeks seem to enjoy, although they have had 3,000 years to acquire the taste. Most tourists try Retsina, but very few continue to drink it by choice when they return home. Moreover, it is made by adding resin, with no direct contact with the wood and, apart from an oddity called "Tea Wine", produced on La Palma in the Canary Islands, no wine to my knowledge is produced in barrels made of pine. Eucalyptus also has a resinous affect, acacia turns wine yellow, and hardwoods other than oak are often impossible to bend and contain aromatic oils that are undesirable.

White oak, on the other hand, is easily bent, has a low porosity, acceptable tannin content, and mild, creamy aromatic substances that either have an intrinsic harmony with wine or, like the Greeks, we have grown accustomed to the effect.

LARGE OR SMALL?

The size of the cask is critical to its influence because the smaller it is, the larger the oak-to-wine ratio, and the greater the oaky flavour it imparts. A 200-litre *barrique* has, for example, one-and-a-half times the internal surface area of oak for every litre of wine as a 500-litre cask. Traditional sizes for *barriques* range from 205 litres in Champagne, to 225 litres in Bordeaux and Spain, 228 litres in Burgundy, and 300 to 315 litres in Australia and New Zealand.

AN OAK BARREL

The staves of a barrel are held together by metal hoops, which are sometimes positioned at slightly different distances, depending on the traditions of the cooper, or *tonnelier,* in question. There may be a red colour between the two innermost hoops, but this is merely where some winemakers wish to conceal their own dribble marks around the bung by

dyeing the entire middle area with wine, which can look very impressive. When fermenting white wines, the bung is always uppermost and, even with a good ullage, or space, the hole may be left open during the most tumultuous period, but will be sealed with an air-lock valve when the fermentation process settles down and will remain closed during the malolactic fermentation. After racking, when all wines undergo several months of maturation, the barrels are filled to the very top, and positioned so that the tightly sealed bung is to one side, visually reminding cellar-workers that the casks are full.

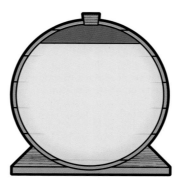

Fermenting position   Maturation position

Square barrels were even developed to increase the ratio of oak-to-wine and, although treated as a novelty, were actually more practical and economical than normal casks. More practical because they make more efficient use of storage space, and more economical because their straight sides could be reversed to create a new oak barrel from an old one. A couple of firms even built square stainless-steel tanks with oak panels made from oak staves that could be replaced, reversed, and adjusted in size to give different oak-to-wine ratios.

CHIPS OFF THE OLD BLOCK

Using old oak barrels for the finest *barrique*-fermented or *barrique*-matured wines is a question of style rather than a consideration of cost. But for less-expensive wines, it is almost entirely a question of economics, as barrels can double, for example, the cost of a *vin de pays*.

The use of new oak *barriques* for just a small percentage of a wine blend can add a certain subliminal complexity to it, although not the overt oakiness that so many people find attractive, yet so few are willing to pay very much for. Oak chips or shavings are the answer. Although generally believed to be a fairly recent phenomenon, the use of oak chips was sufficiently widespread by 1961 to warrant statutory controls in the United States. In fact, as a by-product of barrelmaking, today's ubiquitous chip probably has an equally long, if somewhat more covert, history. Oak chips have been one of the most potent weapons in the New World's armoury, producing relatively inexpensive, but distinctly premium-quality, wines to conquer international markets. This has been particularly evident in Australia, where flying winemakers have not only perfected oak chip wines, but by the early 1990s had exported the techniques to virtually every winemaking country in the world. Some experiments have demonstrated that wine matured with oak chips used in old barrels is "virtually indistinguishable" from the same wine stored in new oak. The range of oak

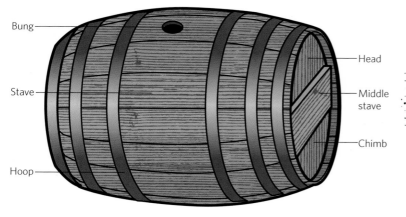

Bung

Stave

Hoop

Head

Middle stave

Chimb

AMERICAN OAK DUST
Light toast

FRENCH OAK CHIPS
Medium toast

AMERICAN OAK CHIPS
High toast

chip products is now very comprehensive, covering the entire range of oak varieties and different toast levels. Some are even impregnated with malolactic bacteria. If a wine label mentions oak, but not *barriques*, barrels, or casks, it is probably a clue that oak chips have been employed in the winemaking and, if the wine is cheap, you can bet on it.

ANYONE FOR TOAST?

Toasting is one operation in the barrelmaking process that has a very direct effect on the taste of the wine. In order to bend the staves, heat is applied in three stages: warming-up (pre-*chauffrage*), shaping (*cintrage*), and toasting (*bousinage*), each of which browns, or chars, the internal surface of the barrel. However, it is only the last stage – *bousinage* – that determines the degree of toasting. During toasting, furanic aldehydes (responsible for "roasted" aromas) reach their maximum concentration, the vanilla aroma of vanillin is heightened, and various phenols, such as eugenol (the chief aromatic constituent of oil of cloves), add a smoky, spicy touch to the complexity of oak aromas in wine.

There are three degrees of toasting: light, medium, and heavy. A light toasting is used by winemakers who seek the most natural oak character (although it is not as neutral as using staves that have been bent with steam); medium varies between a true medium, which suits most red wine demands, and the so-called medium-plus, which is the favourite for fermenting white wines; the third, a heavy toast, dramatically reduces the coconutty-lactones and leaves a distinctly charred-smoke character that

A BARRELMAKER CONTROLS THE FLAMES DURING *CINTRAGE*
A barrelmaker, or *tonnelier*, fires the inner sides of the staves in order to shape them. The final firing operation puts a light, medium, or heavy toast on their inner surface.

WINE NOTES

• Contrary to popular belief, the finer the oak's grain, the greater its porosity. This might sound counter-intuitive, but the warmer the weather, the quicker the growth – and the quicker the growth, the wider the grain. Although summer growth is quicker, it is also harder and more dense than spring growth, due to its hotter, drier growing environment. In cooler climates, where there is not as much summer growth, the incidence of less dense, tighter-grained, but more porous spring growth is significantly higher. The tighter the grain, the greater the micro-oxidation and, therefore, the more softening the effect on the tannins in a wine; thus, the tighter the grain, the greater its winemaking reputation. Most European oak is much tighter-grained than American oak, and Tronçais – which has the narrowest trunks and greatest proportion of spring wood to summer wood of all French forests – has the tightest grain, hence the greatest reputation.

can be overpowering unless used only as a small component in a blend. Furthermore, with time, the high carbon content of heavily toasted barrels can leach the colour out of some wines, so they tend to be used for white wines (often big, brash Chardonnays), although heavy toast is best suited to maturing Bourbon whiskey.

THE DIFFERENT OAKS

Both American and European oaks are used for winemaking. The aromatics of fast-growing, wide-grained American white oak, *Quercus alba,* are more pungent, while there are more tannins and finer texture found in slow-growing, tight-grained European brown oaks, *Quercus robur* (syn pedunculate oak) and *Quercus sessilis* (syn *Quercus petraea* and *Quercus rouvre*). Much of the appealing, if obvious, coconut character in American oak is also due to the very different barrelmaking techniques used in the United States.

Unlike European oak, American is sawn, not split. This ruptures the wood cells, releasing aromatic substances, especially vanillin and up to seven different lactones, which together explain the coconut aroma. American oak is also kiln-dried, which concentrates the lactones, while European oak is seasoned outside for several years, a process that leaches out some of the most aromatic substances, and reduces the more aggressive tannins. The whole process tends to accentuate the character of American oak, while subduing that of European oak.

Many French winemakers consider American oak vulgar. Even so, a little-known fact is that in ultra-conservative Bordeaux, quite a number of *châteaux* are experimenting with it. Coconut-flavoured claret is by no means a foregone conclusion, but oak is expensive, a little sawn *Quercus alba* goes a long way, and a small percentage of American oak, either as barrels or mixed staves, could, in fact, significantly reduce the percentage of new oak needed each year.

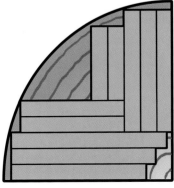

EUROPEAN SPLIT VERSUS AMERICAN SAWN
After the bark has been stripped from a log destined for barrelmaking, it will be split (as is done in Europe) or sawn into quarters (as it is done in the United States). The examples above clearly show that it is more economical to saw staves from a quarter, as opposed to splitting them. This is the major reason why American oak barrels cost half the price of European ones.

FRENCH OAK "APPELLATIONS"

Some winemakers swear by a particular variety of French oak, but most are suspicious of barrels that claim to be from one specific forest. Tronçais, for example, is one of the most famous oak forests, but the Tronçais forest supplies just a tiny percentage of all French, oak and if every barrel claiming to be from there actually were, the forest would have disappeared long ago.

Perhaps the Bordelais, who are the most experienced in buying new French oak, realized this centuries ago, as they have traditionally purchased barrels made from "mixed staves" (in other words, various forests).

Most winemakers buy from a particular cooper because they like their barrels. This usually has less to do with their construction than with the seasoning of the wood and the toasting of the barrels. Most barrels are not only made on a mix-and-match basis, but the mix of staves is likely to vary from year to year. This is because oak is sold by auction, and coopers buy on quality, not name. It is more important who owns the forest than where it is located, because known sources, such as the *Office National des Forêts*, which is the only purveyor of *Haute Futaie* oak and which only sells *Haute Futaie* oak, can guarantee a consistency that a private proprietor, even with good wood, cannot. Therefore, some companies, such as Seguin Moreau, follow a policy of buying only *Haute Futaie* oak.

REGIONAL OAK VARIETIES

For winemaking, the tighter the size of the oak grain, the better. The slower the tree grows, the tighter the grain – thus, cooler-climate European oak is older and tighter-grained than warmer-climate American oak. Of the European oaks, forest oak (Quercus sessilis) is preferred to solitary oak trees (Quercus robur) because its branches start higher and the trunk is longer and straighter. Solitary oaks grow faster and have a larger grain because they tend to grow in fertile soil where there is more water. Quercus sessilis is also preferred to Quercus robur because it is four times richer in aromatic components.

AMERICAN SOURCES

AMERICAN
Quercus alba

This oak covers most of the eastern United States. Some winemakers think that trees from Minnesota and Wisconsin are the best, while others find them too tannic and consider Appalachian oak, particularly from Pennsylvania, to be superior. Other popular oaks are found in Ohio, Kentucky, Mississippi, and Missouri. All are white oaks, fast-growing, wide-grained, with lower tannin (except for Oregon) than any European brown oaks, but with higher, sweeter, more coconutty aromatics. *Quercus alba* is favoured for traditional Rioja, Australian Shiraz, and California Zinfandel.

OREGON
Quercus gariana

Although a white oak, *Quercus gariana* has a significantly tighter grain than *Quercus alba*. Relatively few barrels were traditionally made from Oregon oak, but its usage is increasing as research claims that it has characteristics similar to European oak. As such, Oregon oak is now hand split rather than sawn, and being treated more similarly to European oak.

AMERICAN OAK (LEFT) AND FRENCH OAK (RIGHT)

FRENCH SOURCES

ALLIER
Quercus sessilis

Tight-grained with well-balanced, medium tannin and aromatics, Allier is highly regarded.

ARGONNE
Quercus sessilis

Tight-grained with low aromatics and tannin, this oak was used for Champagne before the advent of stainless-steel. Now seldom used.

BOURGOGNE
Quercus sessilis

Tight-grained with high tannin and low aromatics, most of this oak goes to Burgundian cellars.

LIMOUSIN
Quercus robur

Wider-grained with high tannin and low aromatics, this oak used to be favoured for Chardonnay, but is most widely used for brandy.

NEVERS
Quercus sessilis

Tight-grained with well-balanced, medium tannin and aromatics, Nevers is highly regarded.

TRONÇAIS
Quercus sessilis

The tightest-grained and, with Vosges, the highest tannin content, this oak grows in the Forest of Tronçais in the Allier *départment*. It is highly suitable for long-term maturation, owing to its understated aromatics, and it has been long a sought-after wood.

VOSGES
Quercus sessilis

The tight grain, very high tannin content, and understated but slightly spicy aromatics make Vosges an especially well-balanced oak for winemaking. It is underrated, particularly in its home region of Alsace, where even though few winemakers use *barriques*, those who do ironically seem to experiment with virtually every French forest except the one that is actually on their own doorstep. Vosges is especially popular in California and New Zealand, where some winemakers think it is similar to Allier and Nevers. Vosges deserves to receive greater recognition.

OTHER EUROPEAN SOURCES

BALKAN
Quercus robur

Often called Slavonian or Yugoslav oak, the grain is tight, with medium tannin and low aromatics. Balkan oak is popular for large oval casks, particularly in Italy.

HUNGARIAN
Quercus sessile

Hungarian oak is tight-grained with high aromatics and lower tannins. Hand-split Hungarian staves and barrels are becoming increasingly popular in the New World and Europe alike, due to their pricing and suitability for larger oak regimes.

PORTUGUESE
Quercus gariana

Cooperage oak (*Quercus gariana*) is far less of a commercial concern in Portugal than stunted cork-industry oak (*Quercus suber*), but the former's medium-grain wood has good aromatic properties, making it preferable in Portugal itself, where it is much cheaper than French oak.

RUSSIAN
Quercus sessilis

Tight-grained with low aromatics and easy to confuse with French oak under blind conditions, this was the major oak in Bordeaux during the 19th century and up until the 1930s. Thus all the great old vintages owe something to it. Thanks to investment from Seguin Moreau, which set up a cooperage in the Adygey region near the Black Sea in the late 1980s, French producers have begun to use Russian oak again.

Barrel Makers

There are millions of barrels produced every year around the world, with approximately half constructed from French oak. Some French-coopered barrels are made from American and Eastern European oak, just as some American coopers build French oak barrels in the Unites States. There has also been an increase in so-called hybrid barrels, produced from both French and American oak.

Since the turn of the 21st century, many French *tonnelleries* have stopped selling their French oak barrels based on the name of individual forests, opting instead to offer oak selected on the basis of grain tightness. Although they swear that grain is the ultimate arbiter of quality for oak and claim that selling by grain provides better consistency of product and greater ease of verification, this practice is more likely to be the result of a crackdown by the French authorities investigating oak-origin fraud within individual *tonnelleries*. (It is true that a simple inspection of the croze can show the tightness of grain, whereas who can tell what forest a particular barrel comes from without expensive analysis?)

GRAIN OF TRUTH

Selling solely by grain is wrong on many levels. Discarding such famous names as Allier, Limousin, Nevers, Tronçais, and Vosges – which have, over centuries, created the enviable reputation of French oak around the world – in favour of something that is as generic as "grain" is madness. It is like abandoning famous wine names, such as Bordeaux and Burgundy, in favour of a generic descriptor like "body". Despite the French wine industry's many faults, that just wouldn't happen, and a small number of *tonneliers,* such as Vincent Damy (now at Billon) and Jérôme Damy, refuse to abandon these world-famous forest names. Those barrelmakers who are adamant about the primacy of grain point out that the grain of oak varies from tree to tree within the same forest. They are probably right, but so what? What is to stop them from specifying the forest and grain? Or just specifying the forest and a variation of grain? Or specifying the grain first and naming the mix of forests used? Maybe they are missing a marketing opportunity; maybe there is a niche in the market for an Allier-Nevers blend or a Limousin-Tronçais-Vosges blend, just as there is for Cabernet-Shiraz or a Grenache-Syrah-Mourvèdre? This would at least keep the famous names of French forests alive. What it all boils down to is transparency and the oak equivalent of what Australian winemakers call "truth in labelling". Any *tonnellerie* found guilty of oak-origin fraud ought to be prosecuted – it's as simple as that. Yet some of the most vociferous proponents of not using forest names claim full traceability of forest origin. If that is so, why not proudly stamp it on every barrel?

A SELECTION OF THE BEST AND MOST WIDELY ENCOUNTERED

BARRELMAKERS

ADOUR

Chemin de la Riberotte, Cognac, France

Established in 1928, Adour produces barrels from both French and American oak, including Gascogne oak from the local region, which is especially suited to brandies. Adour makes its barrels through either steam or fire bending, with ageing from 24 to 36 months in their *parc à bois*. They also have presence in the United States as Adour USA Tonnellerie.

AQUITAINE

Cazouls-Les-Beziers, Languedoc, France

Established at Bellebat in the Entre-Deux-Mers, Tonnellerie d'Aquitaine has been supplying First Growth Bordeaux *châteaux* since 1860. Traditional Bordeaux and Burgundian barrels and puncheons are produced in French oak only, all air-seasoned for two to three years. Aquitaine recommends Allier and Tronçais for ageing Pinot Noir, Chardonnay, or Syrah 15 to 24 months; Nevers and Bertranges for ageing Pinot Noir, Merlot, Cabernet Sauvignon, or Cabernet Franc 12 to 18 months; Vosges for ageing Syrah or Grenache 9 to 16 months; and Belleme, Jupilles, and Berce for ageing white wines (Sauvignon Blanc, Chardonnay, or Viognier) and lighter reds (Sangiovese, Syrah, or Merlot) 8 to 12 months in barrel.

BERNARD

Lignières Sonneville, Cognac, France

This third-generation *cognaçais tonnellerie* was established in 1936, and its modern, computer-controlled facility still uses conventional brazier firing to toast the barrels. Hogsheads and puncheons are produced in addition to the classic Bordeaux and Burgundian barrels that are traditional in these parts. Winemakers who use these barrels talk about their exceptional consistency of performance.

BERTHOMIEU

La Charité-sur-Loire, Loire, France

All French oak barrels are constructed from a blend of Allier, Centre-France, Nevers, and Vosges oaks. (Berthomieu also offers American oak and hybrid barrels.) This *tonnellerie* has

developed a unique toasting system that combines steaming with firing to provide an exceptionally deep heat penetration by opening the pores of the staves. Winemakers like the elegant results using Berthomieu barrels, appreciating the sweetness and mouthfeel when vinifying whites, and roasted coffee bean and floral components when ageing reds.

BILLON

Beaune, Burgundy, France

Founded in 1947 and run today by Vincent Damy, a third-generation *tonnelier*, Billon is a sister company to Damy under Les *Tonnelleries* de Bourgogne, which is owned by Vincent's father, Jacques Damy. Billon specializes in oak from specific and hard-to-source forests such as Bertranges, Cher, Citeaux, Jupilles, and Jura. Its barrels are often used for Pinot Noir in New Zealand, where winemakers have begun using Billon for Chardonnay and Syrah. Billon seasons its oak in the open for two to three years, dependent on climatic conditions. Winemakers like the sweet spiciness that Billon barrels bring to their wines. Billon also produces barrels in acacia and chestnut.

TW BOSWELL

See World Cooperage

BOUTES

Beychac et Caillau, Bordeaux, France

Since 1880, this family-owned *tonnellerie* in Entre-Deux-Mers has specialized in Allier oak, air-seasoned in Allier itself and toasted longer and at lower temperatures than most other barrels. They offer two ranges, the Classique and Evolution. The Classique range includes the conventional "Tradition" (18 to 24 months seasoning and manual toasting) and "Selection" (minimum of three years seasoning, with computer-controlled toasting). The Evolution range includes the somewhat unconventional "Grande Réserve" barrels (which appeal to winemakers who want new barrels but only for micro-oxygenation, not the aromas of new oak), the "Coeur" (which selects oak that has had limited exposure to sun or rain) and "Soleil" (which selects oak that has had the most exposure to rain, sun, and wind).

CANTON

Lebanon, Kentucky, United States

Originally established in 1933 in Ohio, Canton moved to its current cooperage in Kentucky in 1983 and was subsequently acquired by Chêne & Cie, thus becoming a sister cooperage to France's Taransaud. Their Grand Cru barrel, introduced in 2002, was the first barrel to be made from American white oak and air seasoned for three years (this was then followed by the Grand Cru Limited Edition, seasoned for four years, and most recently Canton Five, seasoned for five years). They work specifically with American oak and, since 2015, have partnered with American Forests to plant trees for reforestation.

CADUS

Ladoix-Serrigny, Burgundy, France

Founded in 1996, Cadus was conceived by Vincent Bouchard of Bouchard Cooperage, an international barrel broker, with Maison Jadot and the Vicard *tonnellerie* as its major shareholders. Since 2013 one of the main independent French stave mills became a major shareholder in the business, ensuring a secure supply of quality French oak. They offer three main ranges of oak: the made-to-order "Origine" range, the "Sensoriel" range crafted for product performance, and the premium "C by Cadus" for the most exceptional wines. All oak is air-seasoned for a minimum of 30 months, with toasting strictly controlled by time and temperature. Winemakers find that Cadus barrels shine the spotlight on the fruit in their wines, ramping up the richness on the palate.

DAMY

Meursault, Burgundy, France

A father-and-son handcrafted operation, this *tonnellerie* was founded in 1946 and is now managed by third-generation Jérôme Damy, who has moved the business into a new, ultra-modern facility. Damy offers a choice of oak from Burgundy, Allier, Vosges, Cher, Centre de France, Limousin, Nevers, and Tronçais, as well as American and Hungarian oak. Its barrels are highly regarded in France for Chardonnay, which is why they are to be found in more than 80 per cent of the cellars in the Côte de Beaune, especially Puligny-Montrachet and Meursault. In New Zealand, Damy is widely used for Pinot Noir. This *tonnellerie* developed the "Long Toast" for preserving grape aromas in Chardonnay and delicate red wines. Winemakers enjoy the soft, floral effect of Damy oak, which always supports and never dominates.

DARGAUD & JAEGLE
Romanèch-Thorins, Burgundy, France

Founded in 1921, this family-owned *tonnellerie* in modern premises does not buy trees or split wood, believing that quality control can be executed only at stave stage. All Dargaud & Jaegle's barrels are therefore made from bought-in staves. The spiciness, purity of fruit, and roundness that these barrels bring to the palate makes Dargaud & Jaegle a favourite of many white-wine specialists.

DEMPTOS
Saint-Caprais-de-Bordeaux, France

Founded in 1825, Demptos remained a family-owned business for more than 160 years, opening up a subsidiary cooperage in the Napa Valley in 1982, just seven years prior to a takeover by François Frères. In 1999 French stave mill Merranderie Sogibois joined Demptos, followed by Tonnellerie Maury in 2014. Demptos's top-of-the-range "Réserve" oak barrel is produced from the company's finest tight-grained oak; its "Essencia" barrel is comprised exclusively of selected high norisoprenoidal (most aromatic) oak; and each barrel in the Demptos D'Collection is supposed to enhance a specific grape variety.

ERABLE
See Saury

EUROPEAN COOPERS
Bátaapáti, Szekszárd, Hungary

Established in 1997 with Antinori the major shareholder, European Coopers uses only Hungarian oak sourced from the Zemplen Hills in the Tokaj region and the Mecsek Hills near Pécs. The oak is aged for three years and toasted by hand using traditional fire pots. These barrels are sold under the Kádárok brand and are said to have less oak impact than French barrels, tending to highlight natural fruit aromas and soften the harsher phenolics.

FRANÇOIS FRÈRES
Saint Romain, Burgundy, France

Established in 1910 by Joseph François, this family-run Burgundian *tonnellerie* is now under the control of fourth-generation Jérôme François, whose empire includes either exclusive or partial ownership of eight cooperages and three stave mills. All toasting is strictly controlled by time and temperature. These barrels are definitely not shy in showing their presence in a wine, but many winemakers adore the integration of rich, smoky-creamy aromatics that François Frères barrels can add to their wines. Categorized around the length of time the barrels are seasoned, "Classique" is aged 18 to 24 months, "Exclusif" for 30 to 36 months, "Rare" for 40 to 48 months, and "Privilege" for three years. The "Horizon" is made from Hungarian oak. François Frères also owns Demptos.

GILLET
Saint Romain, Burgundy, France

This artisanal *tonnellerie* was established in 1966 by Claude Gillet, who did not even have his own workshop until 1978. Gillet sources French oak from Allier, Bertranges, Centre de France, Châtillonnais, Limousin, Nevers, and Vosges but also uses American oak from Missouri and Eastern European oak from Romania. Most barrels are multi-sourced and sold by grain, traditional size, and shape, but some, such as the Bordeaux Château Ferrés, are made exclusively from Allier. Understandably for a Burgundian *tonnellerie*, Gillet's barrels were once used only for Chardonnay and Pinot Noir, but New World producers are now having success with Sauvignon Blanc, Sémillon, Shiraz, and Viognier. Gillet barrels are generally thought to be very elegant, with low-key extraction of flavours, yet capable of providing a sweet, floral kiss to a wine.

KÁDÁROK
See European Coopers

LAGLASSE
Varize, Lorraine, France

Established in 1986 by Alain Laglasse, this small, family-owned business specializes in neutral barrels and alternative oak usage. It moved to new premises near Metz in 1998 and supplies Billon and Damy with some of their staves. Laglasse uses only Vosges oak seasoned for one year.

LEROI
See Saury

MERCIER
Barbezieux, Charente-Maritime, France

Strategically located between Bordeaux and Cognac, Mercier was established in 1960. Traditional Bordeaux and Burgundy barrels are made from oak that has been air-seasoned for a minimum of two years. Some winemakers find these barrels too dominant, while those who have worked the longest with Mercier believe that the oak is heavy-handed only in the first few months, after which the wines settle down beautifully, particularly the more powerful reds. Mercier offers an acacia barrel, as well as a number of alternative oak products, including chips, staves, and powders.

MERCUREY
Mercurey, Burgundy, France

A Burgundian *tonnellerie* with a *champenois* heart, this business was established in 1992 by Nicolas Tarteret, the owner of a sawmill at Aix-en-Othe, west of Troyes, the ancient capital of Champagne. This is not far from Estissac, where Nicolas's father Philippe Tarteret set up the first sawmill in Champagne. Just two years after the business was founded on two barrels a day, Bruno Lorenzon of Domaine Lorenzon was so impressed by the quality that he personally developed export markets for Mercurey barrels in Australia, Canada, New Zealand, and South Africa. All oak is sourced within 100 kilometres (60 miles) of the company's sawmill in Aix-en-Othe and air-seasoned for at least two years. Mercurey offers a number of special toasts, including the "Chalk Selection", which involves a light toasting of specially selected oak grown on chalky soils.

MILLET
Galgon, Bordeaux, France

This small, family-owned *tonnellerie* was founded in 1952 in Puisseguin by Guy Millet, whose son Dominique took over in 1995 and moved the business to new premises five years later. As might be expected, these barrels are made very much in the Bordeaux tradition and are sought out by winemakers for Merlot and Cabernet in various countries.

MISTRAL BARRELS
See TN Coopers

MORLIER
Saint-Estèphe, Bordeaux, France

A true artisanal *tonnelier*, Morlier provides barrels to some of the most famous Bordeaux *châteaux*, including at least one First Growth (Latour).

NADALIÉ
Ludon-Médoc, Bordeaux, France

Family-owned for five generations, Nadalié has supplied Bordeaux *châteaux* since 1910 and now produces the full range of traditional Bordeaux and Burgundian barrels. Nadalié uses French, Hungarian, and American oak (from its own Megnin Mills in Pennsylvania) and has operated its own cooperage in Calistoga, California, since 1980, the oldest French-style manufacturing cooperage in the Americas. The company also owns Tonnellerie Marsannay, which offers barrels from specific forests (Allier, Bertranges, Jupilles, Nevers, Tronçais, and Vosges), and has operations in Chile and Australia.

ODYSÉ
See TN Coopers

OREGON BARREL WORKS
McMinnville, Oregon, United States

Established in 1996 by Rick DeFerrari – the first (and still the only) person to hand-split Oregon oak for making barrels – Oregon Barrel Works makes hand-split barrels, seasoned for three years in the open and coopered in traditional Bordeaux or Burgundian style. To further refine the Oregon oak taste profile, Oregon Barrel Works initiated a water-immersion programme in 2006, whereby the wood is rinsed three times prior to coopering, and has combined this with a new slow-toasting process. In 2007 Oregon Barrel Works introduced a range of French oak barrels from specific vineyards (Bertranges, Centre de France, Chatillon, Fontainebleau, Nevers, and Vosges), all seasoned three years in the open in France but coopered in McMinnville, where they are marketed under the Tonnellerie DeFerrari trade name.

QUINTESSENCE
Beychac et Caillau, Bordeaux, France

In addition to traditional Bordeaux and Burgundian barrels made from French, Slovakian, and American (Missouri Ozark) oak, this *tonnellerie* also produces a "Fleur de Quintessence" constructed from extra-fine-grain Tronçais and Jupilles oak. Their "Hydro-Collection" is a series of water-immersed barrels that rely on water to accelerate tannin extraction.

RADOUX
Jonzac, Charente-Maritime, France

Robert Radoux established himself as an artisanal *tonnelier* at Jonzac, north of Bordeaux, in 1947, with a business that was developed into Tonnellerie Radoux in 1982 by his son Christian. The latter opened his own stave mill in 1987, also setting up cooperages in Stellenbosch, South Africa, that same year and in Rioja, Spain, and Santa Rosa, California, in 1994. Radoux joined the Francois Frères family in 2012. Radoux introduced crosscut staves for more extraction and developed a device called Oakscan, which measures tannin levels in staves and is used on every stave produced by the company to sort them by tannin. Radoux's standard French barrels are air-seasoned for one to two years and sold by grain, but bespoke barrels can be seasoned longer and produced from a specific forest.

ROUSSEAU
Couchey, Burgundy, France

Rousseau's barrels can be seen when visiting many Burgundy cellars, particularly in the Côte de Nuits. They offer three ranges: "Expert", "Traditional", and the "Confidential Series". Each range has barrels recommended for either whites or reds (or both) based on the desired characteristics in the final wine. Single-forest barrels are made from Bertranges, Châtillon, and Tronçais. Winemakers like the rich, creamy sweetness that Rousseau barrels bring to white wines (some believe this enhances tropical-fruit aromas) and the rich dark-chocolate aromas in red wines.

SAINT MARTIN
Pecarrere, Buzet, France

This business was established in 1945 as an artisanal *tonnelier* for Armagnac producers but segued into the Bordeaux market by the 1950s and started exporting in 1997. All oak is from Allier, Vosges, and other selected French forests and is air-seasoned for a minimum of two years (three for limited production of pure Tronçais).

SAURY
Brive, Corrèze, France

Established in 1873 and still family-owned, this firm purchased two other *tonnelleries* in 2006, both in Cognac: Leroi (dating back to 1773 and formerly belonging to Martell, to which Leroi still sells 80 per cent of its 13,000-barrel production) and Erable. These additions made Saury the fifth-largest cooperage in the world. All oak is air-seasoned for two to three years. Saury also produces water-immersion barrels, which are rinsed prior to cooperage and aptly branded "Immersion". Some winemakers of well-structured reds, such as Syrah and Zinfandel, like the way that Saury barrels have minimal impact upon varietal structure.

SEGUIN MOREAU

Chagny, Burgundy, France

Tonnellerie Moreau was established in 1838; Tonnellerie Seguin was established in 1870. After Rémy Martin became the sole owner of Seguin in 1972, Moreau was purchased, and the two companies were merged. Today, Seguin Moreau has two stave mills, one in France and one in the United States, plus three cooperages: one in Charente, one in Burgundy, and one in Napa. Now the largest individual barrelmaker in the world, Seguin Moreau makes some beautiful barrels, which can be seen when visiting the cellars of some of the world's greatest wines (such as Châteaux Cheval Blanc, Haut Brion, Latour, Margaux, Mouton Rothschild, and Yquem, Domaine Ramonet in Burgundy, and Krug in Champagne).

As might be expected, Seguin Moreau uses oak from a large range of sources – first and foremost France (from where they assemble a classic blend of *Haute Futaie* forests, including Allier, Tronçais, Nevers, and Centre de France), then Eastern Europe (Hungary, Slovakia, and Romania), Russia (Republic of Adygea), and the United States (the Appalachians and the Midwest forests of Minnesota, Iowa, and Missouri). The company makes all the classic barrel shapes, sizes, and types, all toasted for various durations at different temperatures. On these barrels you might see one of the following: "SVT" (Sélection Vendanges Tardives), made from ultra-tight-grain oak that has been seasoned in the open for 18 to 36 months and designed with sweet wines in mind; "SC" (Sélection Cabernet), very tight grain, for ageing all red Bordeaux varietals over extended periods; "SR" (Sélection Rouge), described as a workhorse for all red wines requiring 15 to 18 months in barrel; "SB" (Sélection Blanc), restricted to a medium-long toasting for white-wine ageing requiring oak character; "ST" (Sélection Terroir), designed to have minimum oak impact, allowing an expression of fruit and *terroir* over 8 to 12 months ageing in barrel; "BGC" (Bourgogne Grand Cru), made from very tight-grain oak and designed with ultra-premium white and red wines in mind, with up to 18 months in barrel; "CP" (Chagny Pinot), toasted in Burgundian style to enhance sweet oak tannins in red wine varietals not requiring long ageing; "CW" (Chagny White), the same Burgundian-style toasting but on oak with a medium- to semi-tight-grain oak, for faster expression of oak on white wines, particularly Chardonnay, where less barrel ageing is required; and "Grand Domaine", which is the red-wine equivalent of "CW". Seguin Moreau is famous – some might say infamous – for its U-Stave system, whereby the inside of the barrel consists of a continuous series of grooves, 10 millimetres wide and 5 millimetres deep, that increases the ratio of oak surface to wine by 75 per cent. No winemaker ever uses U-Stave barrels for 100 per cent of any wine, but some find the odd barrel can create an extremely useful blending component.

SIRUGUE

Nuits-Saint-Georges, Burgundy, France

Sirugue was established at Morey-Saint-Denis in 1903 by Félix Sirugue, whose father Victor Sirugue was the *tonnelier* for Maison Thomas-Massot at Gevrey Chambertin, later moving to Nuits-Saint-Georges. In the 1950s, when the introduction of cement tanks dramatically reduced the demand, Félix's son Emile looked abroad for sales; thanks to James Zellerbach, the American industrialist who made the first barrel-fermented California Chardonnay (Hanzell 1957 Chardonnay), this *tonnellerie* opened up sales of French oak to the United States. Still family-owned and run by Félix's grandchildren, Sirugue very traditionally produces just four types of barrel, although kosher versions are available. Winemakers who use Sirugue find that these barrels have typically subtle smoky vanillin aromas, amplifying the fruit in white wines and adding spice to red wines.

SUD-OUEST

Brens, Gaillac, France

As part of the Sylvaboise group since 1988, Sud-Ouest has exclusive access to three sawmills and two stave mills. In addition to the traditional Bordeaux and Burgundian barrels of various sizes produced in either French or American oak, Sud-Ouest also offers the "Alliance F" (which is composed of French staves with American heads) and "Alliance A" (American staves with French heads) and is one of the few *tonnelleries* to produce acacia barrels.

SYLVAIN

Saint Denis de Pile, Bordeaux, France

This business was established in 1957 by Gérard Sylvain, who opened a small barrel-repair workshop in Libourne. Under Gérard's son, Jean-Luc, the company moved to a full-production facility in Saint Denis de Pile, just north of the St-Émilion district. Sylvain uses exclusively French oak from 150 to 200-year-old trees in several forests, including Tronçais, Bercé, Fontainebleau, and Haguenau. It produces Bordeaux barrels (including "Château Tradition" and "Château Ferré") in six types: "Grand Réserve" (for very fine wines); "Réserve" (for long, traditional ageing); "Sélection" (fine grain, recommended for powerful, tannic wines, might include some *Quercus robur)*; "Signature" (extra-fine grain for increased aromatics); "Biodynamie" (made according to biodynamic principles); and "Blanc" (a fine-grain barrel developed for the balance and elegance of white wine).

TARANSAUD

Cognac, France

Established at Juillac-le-Coq in 1932 by Roger Taransaud, whose family have been *tonneliers* since 1672, the business moved to Cognac in 1937. Taransaud was taken over in 1972 by Cognac producer Hennessy, which sold the firm on to its current owners, the Pracomtal family, who built a new cooperage. The Pracomtal family owns the Chêne group, which also includes Jacques Garnier (a *tonnellerie* in Charente-Maritime), Canton (a Kentucky-based cooperage combining American oak and "French know-how"), and Thalés (reconditioned barrels), and co-owns the Kádár Hungary cooperage. All oak is air-seasoned for between two and three years, depending on the thickness of the staves (12 months for every 10 mm). This is very much a red-wine tonnelerie. Winemakers who use Taransaud barrels often say they like the smoky, roasted-coffee bean aromas they impart. Taransaud has been doing extensive research into the interaction between the wine and oak, offering "CHENOX", a device that allows winemakers to measure the level of dissolved oxygen in real-time in the barrel without having to open the bung in order to inform winemaking choices.

TERES TONNELLERIE

Skutare, Southern Region, Bulgaria

Bulgaria represents the fourth-largest reserves of oak forests in Europe, and Teres Tonnellerie, established as Bulgarian Oak in 2005, is the largest family-run *tonnellerie* in Bulgaria. All Bulgarian oak is air-seasoned for between two and three years and slow-toasted over open fire. The main distinction amongst their three lines of barrels (Select, Choice, and Premium) is the length of air-drying (a minimum of 24, 30, and 36 months respectively).

TN COOPERS

Santiago, Chile

Previously known as Tonelería Nacional (and Mistral Barrels in the United States), Chilean-owned TN Coopers works with French, American, and European oak, with headquarters in Santiago, Chile and outposts in Argentina, the United States, South Africa, and Italy. They use a patented software to maintain a consistency of flavors and aromas in their barrels. In addition, they offer three types of toasting: their traditional "Mistral" toasting through direct-flame contact, "Odysé" through convection toasting, and "Ambrosia" through slow-convection toasting. The convection toasting adds an element of sweetness to the wine. Each of these toasting styles is available across their lines of barrels and alternative oak products, including powders, chips, viniblocks, and zigzags. In 2001 Mistral reinvented the zigzag barrel renovation system, whereby 24 to 28 narrow new oak staves can be inserted through (and removed from) the bung-hole. Other zigzag systems require the removal of the head for insertion and removal, which is hardly winemaker-friendly, so this was a particularly welcome reinvention.

VICARD

Cognac, France

This sixth-generation *tonnellerie* was founded in Cognac by Paul Vicard, whose son Jean was named Best Craftsman in France in 1965. Vicard is the largest single-site cooperage in France, spread over 12 hectares (30 acres), including a 6-hectare (15-acre) wood yard that is due to be expanded to 8 hectares (20 acres). The company uses French, Eastern European, and American oak, with the American oak sourced exclusively from Missouri but aged in Cognac, where the climate is better suited to seasoning in the open. Toasting is carried out by a fully automated, patented state-of-the-art process to guarantee a specific temperature, duration, and percentage of humidity during firing. Vicard owns three stave mills (Merrains de France, Merrains de Cognac, and Merrains du Périgord), and its barrels are best suited to red wines, particularly Merlot, Cabernet Sauvignon, and Cabernet Franc. Vicard conducts extensive research and innovation projects, including the launch of "molecular toasting", perfected in 2011, which uses radiant heat for a uniform, reproducible, and precise toast.

WORLD COOPERAGE

Napa, California, United States

Established in 1912 by Thomas Walton Boswell, World Cooperage (also known as Cooperages 1912 and Tonnellerie du Monde) has since 2017 been under the guidance of fourth-generation Brad Boswell. World Cooperage owns the Quintessence *tonnellerie* in Bordeaux, which is supplied by its own sawmill at Monthureux-sur-Saône the heart of the Vosges region. World Cooperage is still owned by the Boswell family, who make and sell barrels under two brands: World Cooperage and the upmarket TW Boswell range. The company's traditional barrels are made from extra-fine-grain French or fine-grain or extra-fine-grain American white oak (Missouri Ozarks), plus hybrids of American oak staves and French heads. All their oak is air-seasoned a minimum of two years, with a limited availability of 36 months. The TW Boswell range consists of traditional barrels, made from French, Slovakian, and American (Missouri Ozark) oak. All TW Boswell oak is air-seasoned for 30 months, with a limited availability of 36 months. All oak can be ordered as fine or "Special Reserve", which is made from extra-fine grain.

SYLVAIN BARRELS FILL THE CELLAR IN THE ZEVENWACHT WINE ESTATE LOCATED IN CAPE TOWN, SOUTH AFRICA
This Bordeaux-based barrelmaker sources its oak exclusively from French forests, including Tronçais. Its completed barrels are sold to wineries across the globe.

Corks and Closures

"Put a cork in it" is urban slang, a rude way to tell someone to seals their lips and stop talking. And that is the purpose of the universal cork – to provide a seal. The need for a wine cork goes back to the days when amphorae being shipped to ancient Rome needed something that would keep the draft out and the wine in, but the "corks" were little more than reed bungs and rags.

Winemakers require a reliable seal for their precious bottles before sending them out into the world. Wine drinkers, however, have a different set of requirements, including how easy the bottle is to open, how easy it is to get the cork back into the bottle afterwards, and whether or not the contents are drinkable. A range of modern bottle closures from natural, technical, glass, and plastic corks to screw caps and crown caps allow the winemaker the final say in how the wine will reach the consumer. They must consider oxygen transfer rates (OTR), permeability, reactivity, taint, and contamination, as well as economy of production and environmental considerations. There has been much effort going into alternative closures but, as can be seen below, no Holy Grail has been found yet.

NATURAL CORK

Natural cork has been used to seal glass wine bottles since at least the 17th century, and the practise began in Cataluña in Spain, where the cork oak (*Quercus suber*) grows. This evergreen oak is now cultivated in Spain, Portugal, Algeria, Morocco, France, Italy, and Tunisia. The bark of cork oaks can be stripped without damaging the tree: a tree remains standing while large sections of its outer bark are cut and peeled from it. Once harvested, the cork oak will regenerate its outer bark. Once a tree reaches 25 years of age, its bark can be stripped once every 7 to 12 years, and with a lifespan of up to 200 years, a single cork oak can be harvested more than 16 times, making it an environmentally friendly, renewable source.

When the Spanish Civil War interrupted Catalan supplies, Algeria – which was then still a French wine-producing colony – took up production, but when its own economic stability became a concern in the 1960s Portugal stepped up production and with the help of EU grants now produces around 34 per cent of global production.

A WORKER UNLOADS A MULE'S CORK HARVEST IN LAS BATUECAS – SIERRA DE FRANCIA NATURE RESERVE IN SPAIN
The extraction of cork from the trees in the region takes place every seven to eight years. The source of cork, the *Quercus suber* tree, grows in Mediterranean woodlands and forests.

A CORK FROM A BOTTLE OF DOM PÉRIGNON CHAMPAGNE ASSUMES THE CHARACTERISTIC MUSHROOM SHAPE OF A YOUTHFUL VINTAGE
A Champagne cork before insertion has the cylindrical shape typical of all wine corks. The mushroom head is achieved when the cork is rammed into the bottle. The cork of a more mature Champagne will have shrunk and straightened through age.

TECHNICAL CORKS

The industry's love affair with cork has been a turbulent one, because cork has both strengths (recognition, tradition, compressibility, elasticity, and relative impermeability) and weaknesses (the greatest being its susceptibility to 2-, 4-, 6-trichloroanisole – TCA – cork taint and the unpredictable nature of how each bottle evolves under cork closure), which has driven the design and engineering of cork replacements in recent decades.

Technical cork makers take natural cork, break it down into tiny particles, and "wash" it with proprietary cleaning agents to eliminate TCA. It is then reconstructed using adhesives and/or compression. Although popular in Europe and North America, technical corks were initially slow to catch on because of their visual associations with cheaper, poor-quality agglomerate corks. Both main manufacturers, Diam/Diamant and Amorim (NDTech), promise taint-free closures with different OTR options for both still and sparkling wines.

The Diamant closures proved particularly reliable in preventing sparkling wines from random oxidation and spoilage, hence their popularity in high-quality areas like the United Kingdom and the Trento DOC in Italy.

OTHER OPTIONS

Invented in Australia, Zork offers a "peel and reseal" option for sparkling (SPK) and still (STL) wines using low-density polyethylene with a tear-off plastic seal. Acquired by US packaging giant Scholle in 2011, Zork is a popular option at the lower end of the sparkling wine market.

Screw caps have also evolved in recent years, and although they are mainly preferred where the wines are not considered suitable for long-term bottle ageing, some top-end wines are now under screw cap. For this technology, different bottles are required, ones with a thread on the outside. The seal is provided by the liner inside the top, and several options with different rates of permeability and OTR are available. As an industrial, rather than natural product, screw caps offer consistency and reliability as a closure, but critics have identified reductive issues in screw-capped wines. as well as oxidation where the seals have been damaged during bottling.

Plastic corks, either moulded, extruded, or co-extruded, are a cheaper option, and most producers now offer a range of OTR options for in-bottle ageing, although this is not considered a feasible option for long-term ageing. Visually they can be indistinguishable from natural cork and have the benefit of being recyclable, although critics flag up their inflexibility and problems with extraction and reinsertion.

Vino-Lok is a Czech-made glass stopper that has become increasingly popular in the last few years. It consists of a glass cork and plastic ring that creates the seal between the bottle neck and the stopper. The company claims that the technical performance of the Vino-Lok is at least comparable to other closure systems, but the main attraction for the consumer is definitely the sleek, clean design and the resealability. The fashion-conscious French are increasingly making Vino-Lok their stopper of choice, particularly for Provence rosés, where the clean, transparent design conveys purity and class, while the cork does not need to keep the wine fresh for decades.

Crown Caps are similar to bottle tops found on beer and soft-drink bottles. They sit on the outside of the bottle, rather than inside the neck, and like screw caps, the lining inside the top provides the seal. There is evidence to suggest they are at least as effective as screw caps and are inexpensive, easy to remove, and recyclable. With a decorative foil capsule, it is almost impossible to distinguish a crown cap from a cork closure on the shelf.

No single solution will please everyone. Each closure provides something different, and technology is moving fast to provide the solution to most problems and to keep natural cork at the top of the traditional closure wish list for the time being.

ALUMINIUM CAPS SCREW-OFF WINE BOTTLES
Often associated with cheap wines, screw caps have a few advantages, such as ease of opening, but their production and use can have negative environmental impacts.

Storing and Serving Wine

Many wine purchasers worry about how to properly store and serve wine, but unless you are buying rare and expensive vintages, both storing and serving are far easier tasks than many a wine snob would have you think.

More than 95 per cent of wine is ready to drink straight from the shelf, and a majority of it is actually consumed within 24 hours of purchase, so for most people correct wine storage is not an issue. How concerned should you be about how to serve a particular wine? Not very conerned – not all (but certainly most) of how and why we drink wines the way we do can be put down to conditioning, and much of that is over-fussy.

STORAGE

If you do not have a cellar – and most people do not – then there really is no need to store wine. Yet, if you want to keep a few bottles for convenience, common sense will tell you to place it somewhere relatively cool and dark. There is also a wide range of wine coolers available with the capacity between a dozen and a few hundred bottles that maintain perfect storage conditions for the wine lover who does not want to or cannot invest in an entire wine room or underground cellar.

On the other hand, if you are determined to build up a proper cellar of wine (and for the enthusiast, there is nothing more enjoyable), then there are some very important factors to consider, principally temperature and light.

TEMPERATURE

The perfect storage temperature for wine is purportedly 11°C (52°F), but anything between 5°C and 18°C (40°F and 65°F) will in fact suffice for most styles of wines, providing there is no great temperature variation over a relatively short period of time. Higher temperatures increase the rate of oxidation in a wine, therefore a bottle of wine stored at 18°C (65°F) will gradually get "older" than the same wine stored at 11°C (52°F); however, a constant 15°C (59°F) is far kinder to a wine than erratic temperatures between 5°C and 18°C (40°F and 65°F). Changes in temperature cause the cork to shrink and expand, which can loosen the closure's grip on the inner surface of the bottle's neck, rendering the wine liable to oxidation.

LIGHT

All wines are affected negatively by the ultraviolet end of the light spectrum, but delicate sparkling and rosé wines are particularly sensitive to short wavelength light. Brown or dead-leaf coloured wine bottles offer more natural protection from ultraviolet light than those that are made of traditional green glass – but dark green is better than light green, whereas blue and clear are the most vulnerable (which is why Louis Roederer Roederer Cristal is wrapped in protective yellow-orange cellophane).

OTHER FACTORS

A certain humidity (between 60 and 70 per cent) is essential to keep the cork moist and flexible. This is one reason why long-term storage in a domestic refrigerator should be avoided – the refrigeration process dehumidifies. Several days in a refrigerator is okay, but much longer than this, and the cork will start to dry out.

The position in which a wine bottle is stored is also extremely important: most wines should be stacked on their sides to keep their corks moist, and therefore fully swollen and airtight. Exceptions to this rule are sparkling wines and any wine that has been sealed with a screw cap. Champagne and any other sparkling wine may be safely stored in an upright position because the carbonic gas (CO_2) trapped in the space between the top of the wine and the base of the cork provides more than enough. Screw caps require no moistening, of course.

SERVING WINE

Whites have traditionally been served chilled and red at room temperature, or *chambré*. At higher temperatures, the odorous compounds found in all wines are more volatile, so the practice of serving full-bodied red wines *chambré* has the effect of releasing more aromatics into the bouquet. The days, however, the widespread use of refrigerators and central heating means that white wines are frequently served too cold and red wines too warm. *Chambré* is not the cozy 19 to 21°C of today's homes but the chilly 15 to 18°C of medieval French rooms. Once poured, the wine warms up in the glass quickly, so generally it's always better to serve a wine too cold than too warm.

One major effect of chilling wine is that more carbonic gas is retained at lower temperatures. This enhances the crispness and freshness and tends to liven the impression of fruit on the palate. It is thus beneficial to chill a youthful white wine but absolutely vital to serve a sparkling wine well chilled, because this keeps it bubbling longer.

Over-chilling wine kills its flavour and aroma, and it also makes the cork difficult to remove, as the wax on a cork adheres to the bottle. Over-warm wine, on the other hand, is bland to taste. The rough guide below is more than you need to know. I prefer not to complicate life with specific temperatures and simply think in terms of "putting a chill on" white or rosé wines and "taking the chill off" red wines if necessary.

WINE TYPE	SERVING TEMPERATURE
Sparkling (red, white, and rosé)	4.5 to 7°C (40 to 45°F)
White	7 to 10°C (45 to 50°F)
Rosé and light-bodied red	10 to 12.5°C (50 to 55°F)
Medium-bodied red	12.5 to 15.5°C (55 to 60°F)
Full-bodied red	15.5 to 18°C (60 to 65°F)

RAPID CHILLING AND INSTANT *CHAMBRÉ*

It is fine to chill wine in a refrigerator for a few days but not long term, because the cork may stick. Ten or 15 minutes in the deep-freeze has never done a wine any harm. The belief that this practice "burns" a wine is unfounded; the cold creeps evenly into the bottle by exactly the same principle as with rapid-chill sheaths.

Unlike cooling, warming a wine by direct heat is not an even process; whether standing a bottle by a fire or putting it under a hot tap, some of the wine gets too hot, leaving the rest too cold. Either just wait for the wine to come to the correct temperature in the room or pour it into a decanter that has been warmed up under hot tap water (making sure the water stream stays on the outside of the decanter.

DECANTING

With increasing age, many wines – especially reds – throw off a natural sediment of tannins and colouring pigments that collect in the base or along the sides of the bottle. Most sediment is loose and fine, but you may encounter a thin film of dark-coloured sediment adhering to the inside of the bottle. Often this is just a patch, with the shoulder being a favoured location for some reason, but occasionally it can be found to cover almost the entire bottle. The degree of adhesion varies but can at times be very stubborn, as anyone who has tried to rinse out such bottles will testify. The technical description of this bloom is an insoluble complex polymer of pigmented tannins and protein. It is not known why it affects some wines and not others, but it does appear to be more commonly found in high-pH red wines from exceptional vintages or hotter climes generally.

OPENING AND SERVING A BOTTLE OF CHAMPAGNE

- Remove the foil from the bulbous top end of the neck. Quite often there is a little foil tail sticking out, which you merely pull. Failing this, you may have to look for the circular imprint of the end of the wire cage, which will have been twisted, folded upwards, and pressed into the neck. When you find this, simply pull it outwards – this will rip the foil, enabling you to remove a section from just below the level of the wire cage.
- Holding the bottle upright at an angle, keep one hand firmly on the cork to make sure it will not surprise you by shooting out, untwist the wire with the other hand (six turns anti-clockwise), and loosen the bottom of the wire cage so that there is a good space all round. A good tip is

not to remove the wire cage, not only because that is when most bottles fire their corks unexpectedly, but also because it acts as a good grip, which you need when a Champagne cork is stuck tight.

- Transfer your grip on the cork to the other hand, which should completely enclose cork and cage, and, holding the base of the bottle with your other hand, twist both ends in opposite directions. As soon as you feel pressure forcing the cork out, try to hold it in, but continue the twisting operation until, almost reluctantly, you release the cork from the bottle. The mark of a professional is that the cork comes out with a sigh, not a bang.

- With just a little bit of practice, it will soon become clear that pouring from a Champagne bottle, particularly a magnum, is easier to achieve with just one hand, your thumb in the punt and your fingers spread beneath the bottle for support. The correct way to pour any sparkling wine is directly into a standing glass, not to tilt the glass as if you are pouring a lager. Recent research and common sense tell us that the "lager method" will preserve the carbonic gas content of a Champagne, but the mousse of a youthful non-vintage needs a little taming, and the softest, silkiest mousse will always enhance the more complex character of a mature vintage Champagne.

OPENING

HOLDING

POURING

Both red and white wines, particularly white, can also shed a crystalline deposit due to a precipitation of tartrates. The precipitations can appear on the end of the cork that is in contact with the wine, too. Although all these deposits are harmless, their appearance is distracting, and decanting will be necessary to remove them.

Preparing the bottle and pouring the wine
Several hours prior to decanting, move the bottle into an upright position. Doing so allows the sediment lying along the side of the bottle to fall to the bottom. Cut away the foil under the second lip of the bottle's neck, about half a centimetre under the top. This could well reveal a penicillin growth or, if the wine is an old vintage, a fine black deposit, neither of which will have had contact with the wine, but to avoid any unintentional contamination when removing the cork it is wise to wipe the lip of the bottle neck and the top of the cork with a clean, damp cloth. Insert a corkscrew, and gently withdraw the cork. Place a clean finger covered in tissue inside the top of the bottle and carefully remove any pieces of sediment, cork, or any tartrate crystals adhering to the inside of the neck, and then wipe the lip of the bottle neck with a clean, dry cloth.

Lift the bottle slowly in one hand and the decanter in the other, and bring them together over a light source, such as a candle or torch, to reveal any sediment as the wine is poured. Aim to pour the wine in a slow, steady flow so that the bottle does not jerk and wine does not "gulp for air". Such mishaps will disturb the sediment, spreading it through a greater volume of liquid. Wine that contains a lot of sediment naturally cannot be fully emptied into the decanter, pouring should stop when the deposit becomes visible in the neck through the light.

Filtering dregs
Personally, I flout tradition by pouring cloudy dregs through a fine-grade coffee filter paper. I always attempt to decant the maximum volume, thereby filtering the minimum, and no one has ever been able to tell the

difference under blind conditions. Once the wine is decanted, the next question is whether to serve it in the decanter or to rinse out the bottle, allow it to stand upside down to drip dry, and then refill the original bottle. If the label is in good condition, I often serve the wine in its own bottle. I think the original bottle is the best possible vessel from which to pour a wine, and it allows people to see what they are drinking. Older, delicate red wines, however, might not benefit from too much sloshing around, so they should stay in the decanter.

ALLOWING WINE TO BREATHE
Wine "feeds" on the small amount of air trapped inside the bottle between the wine and the cork and on the oxygen naturally absorbed by the wine itself. It is during this slow oxidation that various elements and compounds are formed or changed in a complex chemical process known as maturation. Allowing a wine to breathe is, in effect, creating a rapid, but less sophisticated, maturation. Breathing may be beneficial to certain still wines for several reasons, only some of which are known. The only generalization that tends to hold true is that breathing is likely to improve young, full-bodied, tannic red wines.

The fact that restaurants invariably decant older vintages has given rise to a belief that older vintages must be allowed to breathe. But this is misleading – and such actions can be detrimental to the wine. Good sommeliers will always decant any wine that has a sediment (and should, if they knows their wines, suggest decanting younger wines that are known to benefit from breathing), but any breathing of older vintages just because they have a sediment will not be intentional. Indeed, the older the wine, the more delicate it is, and the more likely it is that it will be on its last legs and should be consumed as soon after opening as possible. While it is true that some extremely venerable wines can build in the glass over many hours, you will always have that option however soon after opening you taste it, whereas if you leave it to breathe regardless, such a narrow window of opportunity will be lost forever.

Wine Accessories

Along with drinking, collecting, savouring, and storing wines come many other purchase decisions. The range of accessories associated with wine is designed to target the enthusiast, the gift giver, the lover, the friend, and everyone in between.

So assuming you have the most basic of twist-pull corkscrews, a small knife to cut through the foil or capsule, and a glass to pour it into, what else does the wine lover need? To be honest, the answer to that is nothing, *nada, rien*. Yet the wine world "support industry" is awash with options . . . for opening the bottle and for decorating it with fancy stoppers and oenology-themed gadgets before the wine has even been poured.

From KitchenCraft to Wusthof, Carrol Boyes to Le Creuset and of course, Coravin systems that cost thousands, bear in mind that a conventional waiters' friend corkscrew will do the same job for less than £5. The best type will have a two-step lever, resulting in a gentler pull action. It is virtually indestructible and once you get a little practice, it is one of the easiest to use.

The Ah-So, or cork-fork, is also known as the butler's friend. It is particularly useful when opening aged wines with older corks because it greatly reduces the risk of the cork breaking or crumbling. The two long prongs are inserted vertically into the cork and then levered into place before twisting the handle to extract the cork. When demonstrated, this is known as the "Ah-So" moment, explaining how the opener got its name.

And while we're on the subject of openers, let's not forget the sabres beloved of the *sabrageurs* – there's no greater wow factor than whipping out your sword to ceremonially lop the top off a bottle of champagne. But a good sabre will set you back around £1,000, not to mention the tuition to handle it relatively safely and effectively (never mind the occasional "bad results" on expensive bottles).

A SERVER OPENS A BOTTLE OF WINE WITH A WAITER'S FRIEND CORKSCREW
This type of opener is surely one of the most efficient and affordable tools to have in your wine accessories kit.

DECANTER OPTIONS

Carafes and decanters range from the sedate and traditional to elegant long-necked trumpets. Their functional purpose, however, is one or more of three things: to separate the sediment from the wine, to aerate the wine by helping it "open up", and to bring wine in bottles that are too cool to the correct temperature. Everything else is only aesthetics.

Decanters come with a plethora of optional add-ons, such as built-in marble aerators and cleaning kits, from brushes and metal beads to drying racks and microfibre polishing cloths.

The easiest way to keep them clean and shiny, however, is to fill them with filtered (soft) water, and drop in an effervescent unflavoured denture-cleaning tablet, leaving it to soak overnight. The wine stains will be dissolved while the glass remains unscratched.

While simply popping a bottle of white or sparkling wine in the fridge a couple of hours before opening (or 30 minutes in the case of a red) will chill a bottle perfectly adequately, enthusiasts have been persuaded to invest in cooling devices that keep the bottles at a constant temperature. Although they surely perform the job impeccably, the fun and discovery of how the wine's aromas change at different temperatures over the consumption of the bottle would also be lost. A simple ice bucket or a chilled cooling sleeve will do the job just fine.

OTHER GADGETS

Electric aerators that churn air through the liquid as it is poured (ostensibly saving time on decanting and airing), will set you back anything from £200 to £600, even more if they are also works of art like those from Carrol Boyes, which could hold pride of place on a bar or mantlepiece.

Having got the bottle open, chilled or not, red, white, or rosé, the device called Clef du Vin claims to tell you whether or not it's ageable. Designed and patented by sommelier Franck Thomas, the key uses a metal alloy to augment and speed up the process of oxidation that a wine may go through over time in the cellar. As the instrument is made mostly out of copper, however, you may save some money by using a (well-cleaned) copper penny for this experiment.

Wine fripperies rival the golf world for adding clutter, but not necessarily form or function, to your leisure pursuits. None of them will make you a more-informed wine drinker, any more than putter socks will get you that hole-in-one.

THE *PORRÓN*, PART DECANTER, PART DRINKING VESSEL, IS A TRADITIONA WINE PITCHER, ORIGINATING IN CATALUNYA, SPAIN
The *porrón* is shaped so that the wine stored inside has minimal contact with the air. It is meant to allow wine drinkers to share a communal quaff, while not letting the vessel touch their lips. It does, however, require practice to ensure that the wine pours into the mouth rather than spilling down one's clothing.

WINE PRESERVATION SYSTEMS

There are those who boast that there is no such thing as leftover wine and those who only sip a single glass for whom the idea of polishing off a bottle in a single sitting is anathema. Increasingly we are being guided to limit our alcohol intake and practice moderation for both physical and mental well-being, and this means wine drinkers will often want to extend the life of a wine past a single drinking occasion.

Once opened, a bottle of wine typically starts to lose its freshness and vibrancy within one to two days if red and perhaps a little longer, three to five days, if white (and stored in the refrigerator). Yet, if you want to enjoy that bottle of wine a few days later (or simply want to guarantee the same drinking experience of that first day it was opened), there are many home preservation systems on the market these days that claim to do just that.

There are two primary ways that these systems work: pump out the air in the bottle or cover the wine with a protective layer of inert gasses. In both methods, the objective is the same: to reduce the wine's contact with oxygen. Oxygen is the culprit for tiring out a wine or making it fall flat, dull, and, ultimately, unpleasurable to drink.

The Vacu Vin

Such home preservation systems can vary in price from a couple of pounds to a couple of hundred, with varying success rates. One of the first products on the market – and perhaps one of the most affordable – is the Vacu Vin. With this system, rather than the original cork, you insert a rubber stopper into an open bottle, connect the associated vacuum device, and then pump up and down until it clicks. The *click* indicates that the oxygen has been removed. The wine can then be stored upright and, when you'd like to drink that next glass, you simply remove the rubber stopper and pour yourself a glass (and can then can re-attach and pump again if you still have more wine to save).

Vacu Vin claims to preserve a wine's freshness for three to five days, and many home wine drinkers swear by it, although the data is mixed as to whether or not this system really is any better than simply sticking the cork back in the bottle and plopping the wine in the fridge (putting an opened bottle – red or white – in the fridge will slow down the oxidation process). A study conducted by *The Wall Street Journal* and Portland State University found that the vacuum at most removes 70 to 75 percent of the oxygen, but that 25 percent of that vacuum is lost within two hours (and virtually all is lost overnight). Still, at its minimal cost, using such a vacuum product can't hurt, but only possibly help.

Other preservation options

Another affordable alternative is the wine spray, which is a mixture of carbon dioxide, nitrogen, and argon gas that is sprayed into the open bottle, forming a protective layer between the wine and the oxygen. Using one of these, re-corking the bottle, and then putting it in the refrigerator can, in theory, preserve a wine for up to a week. Unsurprisingly, however, users have mixed results with these devices. Although some insist that this does extract a bit more life out of a wine, others say this can actually have a negative effect on the aroma and taste. Needless to say, this system is only effective if the bottle stands undisturbed, because the protective layer of gas must remain in full contact with the surface of the wine.

Perhaps the most expensive option, prized by wine lovers and sommeliers alike, is the Coravin, which will set you back a few hundred pounds, but allows you to drink a glass and preserve the remaining wine without ever actually opening the bottle. The system features a device with a pin-sized needle coated with Teflon that is inserted into the cork, extracting a single glass and inserting an inert argon gas into the bottle, then allowing the cork to miraculously seal itself back up, as if the bottle had never been opened. (In recent years a special cap for screw-cap bottles was developed, with a silicone membrane on top, which allows the needle to access the wine, although the bottle has to be opened to change the cap.)

Argon gas, which is also used in the wine sprays discussed above, is heavier than oxygen and thus has long been considered one of the best ways to prevent oxidation in wine. It is both an inert and non-toxic gas and has been shown to serve as a protective layer for wine from oxygen without having any reaction or effect on the wine itself.

By this logic, it's no wonder the Coravin has presented itself as the Holy Grail of wine preservation systems; not only does it make use of the reliable argon gas properties, but the bottle never even actually has to be opened as well. Surely this should mean that wines can then be preserved for years even once a glass has been extracted?

Alas, there's no perfect solution. and even the Coravin has its criticisms. For one thing, in 2014 the company faced a serious product recall when a number of wine bottles using the Coravin system exploded, sometimes injuring their owners. In response to this, the company sent repair kits to all device owners, and it has since then included neoprene sleeves with the product to contain the bottle and any possible breakage (though this is still a rare occurrence).

What's more, although the Coravin indeed works well for the first glass or two, some argue it's foolish to continue using for longer than this because it uses exponentially larger amounts of argon the more headspace there is in the bottle. Inevitably, despite argon's non-interaction with the wine, this will have an impact on the taste. Still, many home wine drinkers continue to swear by it, as do many sommeliers who argue that it has revolutionised the by-the-glass programme in their bars and restaurants (it's not unusual to see a list of "Coravin selections" nowadays on a wine list, offering expensive by-the-glass pours of high-priced wines it would have been unheard of to pour by the glass in the past).

This list barely even skims the surface of the never-ending list of devices out there claiming to prolong the life of your unfinished wine. Most rely on some version of vacuum-pumping or inserting an inert gas, though the craziest we've heard of is putting marbles in your semi-filled bottle to fill the space oxygen would otherwise take up.

When in doubt, if simply finishing the bottle isn't an option, and none of these options seem worth the expense, there's still nothing wrong with the old-fashioned trick of sticking the cork back in and popping the bottle into the fridge. After all, if you cannot finish a bottle of wine over three or four days, it may not be worth preserving.

AN ELECTRIC CORKSCREW, AERATOR, AND VACUUM STOPPER
Think before you buy. A whole industry exists to market wine tools. Despite the plethora of options hawked to wine lovers, to enjoy a good bottle, you need little more than a simple corkscrew, knife, and glass.

Wine Glasses

Over the past 30 years, wine glasses have moved from regional tradition to technical shapes, with today's emphasis clearly placed on physical and aesthetic qualities that have been engineered to suit specific styles of wine.

In the 19th and early 20th centuries, the so-called traditional styles of wine glass evolved on a regional basis, but these shapes were primarily for recognition purposes, in much the same way that the different regional shapes of wine bottle were developed. Their suitability to the wines in question was not considered to be as relevant as the ability to discern what glass a particular wine should be poured into at the table; consequently, most traditional wine glasses turned out to be the wrong shape and size, with rims that were much too thick. The first manufacturer to design stemware (as wine glasses have been collectively known since the 1920s) to suit the technical demands and aromatic virtues of specific styles of wine was Riedel. Although established in 1756, this family-owned company only set the wine world alight in 1973, with its Sommeliers collection of 10 specially designated wine glasses. Riedel had created some media attention with its Burgundy Grand Cru glass in 1958 and Bordeaux Grand Cru in 1961, capable of holding well over a bottle of wine each, and although they formed part of the Sommeliers collection, it was the size, precision, and design of the other eight glasses that grabbed the most attention 12 to 15 years later.

By the mid-1990s, much of the wine trade had come to accept the importance of technical wine glasses to showcase their best wines, even if it took much longer to filter through to consumers. Much but not all of the wine trade – there were still some very important annual trade tastings and wine competitions where tasters were expected to assess wines in small *copita* glasses, a tradition that lingered on far too long. The big sea-change came in 2004, when *Decanter* magazine launched its Decanter World Wine Awards (DWWA), the world's largest wine competition. *Decanter* was determined to do everything absolutely right from the very start, and this included the use of Riedel glasses, a massive undertaking considering the unprecedented scale of the DWWA. And it got the message across. Globally. Just two years later, I judged at The National Wine Show of Australia and was delighted to see the *copitas* had been replaced with decent-sized stemware. When I launched the Champagne & Sparkling Wine World Championships (CSWWC), I followed *Decanter*'s lead and used exactly the same Riedel glass, the Vinum Chianti (6416/15), but whereas that makes

RIEDEL SOMMELIERS HERITAGE
The hand-blown Sommelier range of stemware is one of the best, with just the right combination of overall size, shape, and rim thicknes.

an excellent universal glass, the CSWWC is a niche competition, and I soon realised that we needed something more specialised for tasting exclusively sparkling wine. After a lot of experimentation, I finally settled for Riedel's Veritas Champagne glass.

Aesthetics are always a serious consideration for any hedonistic pursuit, particularly one as subjective as wine and should not be discounted, but the design of technical tasting glasses goes beyond beauty and personal taste. The proportions of a glass are essential. The bowl needs to be of a minimum generous size to project the aromas. Do not accept smaller glasses for Port, Sherry, or dessert wine. That is a hangover from the 19th century that curiously persists even in top-quality stemware ranges. If anything, such wines require even larger bowls to cope with the increased volume and complexity of aromas, but whatever size a glass is, never fill it more than an inch or two to three centimetres. The fineness of the glass at the rim and the thinness of the stem, with no obvious joints (thus either hand-blown or latest generation machine-drawn), play an important role in the appreciation of wine, as if the elegance of design is directly transferred from glass to wine.

If you need convincing about the importance of stemware in the appreciation of wine, pour a *cru classé* Bordeaux into any precision-made wine glass, whether manufactured by Riedel, Schott Zwiesel, or whomever, and pour the same wine into a small, squat Paris goblet with a bulbous, rolled rim. Swirl, sniff, sip, and judge for yourself.

GABRIEL-GLAS
Swiss wine critic René Gabriel founded the firm after realising how many great wines are served in substandard glassware. Moulded in one piece to increase stability and designed in a slightly outward-curving side, these affordable and versatile glasses offer a surprisingly good tasting experience.
Particularly recommended: *Gold Edition*

LEHMANN
Founded in 1970 in Reims, France, Lehmann combines simplicity and functionality with aesthetics to create glasses that don't just look good but work well too. Their collaboration with wine professionals (notably Gerard Basset and Philippe Jamesse) resulted in some impressive designs. The Jamesse Prestige line of champagne glasses offers generous breathing space for the wine, helping it open up and release subtle aromas.
Particularly recommended: *Jamesse Prestige Grand Champagne 45cl*

RIEDEL
Established in Bohemia in 1756, Riedel became the first manufacturer of technical wine glasses in 1973, with designs that have been both brilliant and innovative, but for all the wrong reasons. They were intended to deliver wine to a certain part of the tongue in the belief that, in hitting or missing the dedicated taste receptors for sweetness, sourness, acidity, and bitterness, Riedel glasses would manipulate how we perceive a wine in a very fundamental way. We now know, of course, that separate taste zones do not exist, but the so-called tongue taste map (*see* box, opposite page) was the accepted truth in 1973, when these glasses were launched. Nonetheless, in pursuing non-existent taste zones, Riedel somehow managed to design the most brilliant stemware on the planet. The company continues to offer a ridiculous number of ranges, but over the last 10 years I have whittled down my favourites to just three each from two ranges: Superleggero, an expensive, but not the most expensive, hand-blown range for special occasions and entertaining; and Veritas for everyday use, which is expensive for machine-blown, but latest-generation technology with machine-pulled stems delivers stemware that is both exquisite and robust. The three styles I use are: Champagne Wine Glass (not Flute) for not just sparkling wine, but also overlaps into dry, lean, mineral white wines (machine-pulled); Oaked Chardonnay for fuller whites, fuller reds, and fortified and dessert wines; and Old World Syrah (Veritas) or Hermitage/Syrah (Superleggero) for more precision reds, rosé, and a crossover between mineral and fuller whites.
Particularly recommended: *Veritas (Champagne Wine Glass, Oaked Chardonnay, Old World Syrah), Superleggero (Champagne Wine Glass, Oaked Chardonnay, Hermitage/Syrah)*

ROSENTHAL

Established at Selb, Germany, in 1879, Rosenthal leans towards dinner-ware and has produced several inappropriate wine-glass designs along-side some truly serious stemware, particularly in its Studio Line. The Fuga part of this line includes the top-of-the-range Bordeaux Grand Cru and Burgundy Grand Cru, which are too large and vulgar, but the Fuga Glossy Red & White Wine Young make fine workhorse all-rounders.
Particularly recommended: *Fuga Glossy (Red Wine Young, White Wine Young)*

SCHOTT ZWEISEL

Established in 1872, my initial attraction to Schott Zweisel was its The First range, which had been designed in collaboration with the 2004 world champion sommelier Enrico Bernardo, but over the years my preference has changed to hand-blown, lead-free Zweisel 1872 Titaniun Crystal range. Particularly recommended: Titanium Crystal Sauvignon Blanc, Titanium Crystal Beaujolais, Titanium Crystal Fino Chianti
Particularly recommended: *Titanium Crystal Sauvignon Blanc, Titanium Crystal Beaujolais, Titanium Crystal Fino Chianti*

SPIEGELAU

This firm's history goes way back to at least 1521, when the German glassmaking factory of Spiegelau was mentioned in a will. The company was taken over by Nachtmann in 1990 and has been part of the Riedel group since 2004, when Nachtmann was purchased by Riedel. These glasses are often viewed as Riedel's cheaper secondary brand, but Riedel produces its own cheaper ranges and purchased Spiegelau for its quality, not its price. It also took it over because Nachtmann was Riedel's biggest competitor, not to erase it, but to increase its market share. But perhaps the biggest advantage for Riedel was Nachtmann developed the world's first machine-pulled stem, which provides the feel of hand-blown stem-ware at machine-blown prices.
Particularly recommended: *Highline Red Wine, Grand Palais Bordeaux Pokal*

STÖLZLE

Established in 1433, German glassmaker Stölzle produces a number of different ranges, but the only one that rocks my boat these days is the hand-blown QI range.
Particularly recommended: *Q1 Bordeaux, QI Burgundy*

ZALTO

The Denk'Art range of Austrian glassmaker Zalto stole the show at *Stern* magazine's 2009 invitation taste-off between 10 stemware manufactur-ers. Forget the firm's claim that the success of these glasses is all down to the slope of their bowls, with angles of 24, 48, or 72 degrees that replicate the triumvirate of angles upon which the Earth tilts on its axis, thus giving them almost magical qualities. These glasses are just exceptionally well designed, with stems so thin that swirling a glass of wine can feel as if the stem is bending, although obviously it's not. Some might find the straight sides of the angular design more proficient than elegant, but there is no denying that technical proficiency.
Particularly recommended: *Denk'Art Bordeaux*

ZALTO DENK'ART BURGUNDY, BORDEAUX, AND WHITE WINE
These mouth-blown glasses from Austrian maker Zalto have won design awards. A well-proportioned swell of the bowl allows a red wine glass to be filled halfway and still have plenty of room for heftier red wine aromatics to accumulate.

THE DEBUNKED TONGUE TASTE MAP

The concept whereby the original four basic tastes were detected by receptors found only in specific parts of the tongue, the so-called tongue taste map (at right), has now been thoroughly debunked.

The origins of the tongue taste map can be traced back to a PhD thesis written by D P Hanig and published in *Philosophische Studien* in 1901, but he did not conceive it. The graphs in Hanig's raw data reveal variations in sensitivity to the four basic tastes around the tongue, but these variations were so tiny as to be insignificant for any practical purposes.

Furthermore, Hanig's data did not reveal distinct areas, as they overlapped each other. The subtlety of this variation and the overlapping of zones was lost in 1942, when Hanig's paper was translated at Harvard by Edwin Garrigues Boring, upon whose work the tongue taste map was established. It was not until 1974 that Boring's error was uncovered by Virginia Collings, who re-examined Hanig's original data. The fact that this was not mentioned by Amerine and Roessler in their otherwise seminal work *Wines: Their Sensory Evaluation* some two years later, illustrates how long it took the wine trade to abandon the tongue taste map, despite the work of other scientists specializing in basic tastes, especially Linda Bartoshuk since 1993.

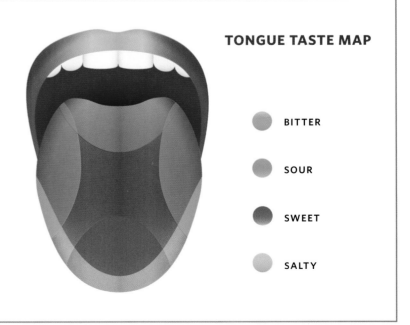

TONGUE TASTE MAP

● BITTER

● SOUR

● SWEET

● SALTY

The Taste of Wine

The difference between tasting and drinking is similar to test-driving a car you might buy and the relish of driving it afterwards. One is a matter of concentration, as you seek out merits and faults, while the other is a more relaxed and enjoyable experience. Almost anyone can learn to taste wine.

When tasting a wine, it is important to eliminate all distractions, especially comments made by others; it is all too easy to be swayed. The wine should be tasted, and an opinion registered before any ensuing discussions. Even at professionally led tastings, the expert's job is not to dictate but to educate, putting into perspective other people's natural responses to smells or tastes through clear and concise explanation. The three "basics" of wine-tasting are sight, smell, and taste, known as "eye", "nose", and "palate".

THE SIGHT, OR "EYE", OF A WINE

The first step is to assess the wine's limpidity, which should be perfectly clear. Many wines throw a deposit, but this is harmless if it settles to yield a bright and clear wine. If it is cloudy or hazy, the wine should be discarded. Tiny bubbles that appear on the bowl or cling persistently to the edge of the glass are perfectly acceptable in a few wines, such as Muscadet *sur lie* and Vinho Verde, but probably indicate a flaw in most other still wines, particularly if red and from classic Old World regions.

HOW WE TASTE SMELLS

All tastes other than sweetness, sourness, bitterness, saltiness, and umami are detected as aromas by receptors on the olfactory bulb, which transmits the data directly to the brain, where we are fooled into believing that the aromas are in fact complex tastes, or flavours. That explains why we lose so much of our ability to taste when our nose is blocked by a cold. The perception of all aroma-based flavours is influenced by one or more of the five basic tastes, as well as temperature and touch. The warmer the wine, the more aromas are given off, and the more flavour we perceive (thus melting ice cream has more flavour than frozen ice cream). The sense of touch should not be underestimated, as texture can be very influential. The smoother something is, the creamier it tastes. Imagine a smooth, creamy toffee; now imagine a toffee made with exactly the same ingredients but grainy in texture – not so creamy, is it? A watery wine gives us an entirely different taste sensation compared to a more viscous wine with theoretically the same aromas and basic tastes. Add some fizziness, and the taste parameters change even further, as will the strength of *mousse* and the size of bubble.

WHAT IS A "SUPERTASTER"?

The term *supertaster* was coined by Linda Bartoshuk at Yale Medical School in 1991 to refer to those who are supersensitive to various tastes, particularly bitterness. They are supersensitive because they have more taste buds than the rest of us, but it is more of a super-burden than a super-power. Around 50 per cent of the population are normal tasters, 25 per cent are so-called non-tasters, and 25 per cent are supertasters. Women are more likely to be supertasters than men, so although 25 per cent of the population are supertasters, by gender this breaks down as 35 per cent of women and 15 per cent of men. Also, Asians are more likely to be supertasters than Caucasians. Everything tastes bland to non-tasters, so they smother their food in sauces and spices. Supertasters are particularly sensitive to bitterness and dislike strong, bitter foods like raw broccoli, grapefruit juice, coffee, and dark chocolate. So, is a supertaster a super wine taster? No – the normal tannin content of all the world's greatest red wines would be far too bitter for a supertaster to bear, let alone enjoy.

The next step is to swirl the wine gently around the glass. So-called "legs" or "tears", thin sinewy threads of wine that run down the side of the glass, may appear. Contrary to popular belief, they are not indicative of high glycerol content, but are simply the effect of alcohol on wine's viscosity, or the way the wine flows. The greater the alcohol content, the less free-flowing, or more viscous, the wine actually becomes.

The colour of wine

Natural light is best for observing a wine's colour, the first clue to its identity once its condition has been assessed. Look at the wine against a white background, holding the glass at the bottom of the stem and tilting it away from you slightly. Red wines vary in colour from *clairet,* which is almost rosé, to tones so dark and opaque that they seem black. White wines range from a colourless water-white to deep gold, although the majority are a light straw yellow colour. For some reason, there are very few rosé wines that are truly pink in colour, the tonal range extending from blue-pink through purple-pink to orange-pink. Disregard any impression about a wine's colour under artificial lighting because it will never be true – fluorescent light, for example, makes a red wine appear brown.

Factors affecting colour

The colour and tonal variation of any wine, whether red, white, or rosé, is mainly determined by the grape variety. It is also influenced by the ripeness of the actual grapes, the area of production, the method of vinification, and the age of the wine. Dry, light-bodied wines from cooler climates are the lightest in colour, while fuller-bodied or sweeter-styled wines from hotter regions are the deepest. Youthful red wines usually have a purple tone, whereas young white wines may hint of green, particularly if they are from a cooler climate. The ageing process involves a slow oxidation that has a browning effect similar to the discoloration of a peeled apple that has been exposed to the air.

THE SMELL, OR "NOSE", OF A WINE

Whenever an experienced taster claims to be able to recognize in excess of a thousand different smells, many wine lovers give up all hope of acquiring even the most basic tasting skills. Yet they should not be discouraged. Almost everybody can detect and distinguish more than a thousand different smells, the majority of which are ordinary, everyday odours. Ask anyone to write down all the smells they can recognize, and most will be able to list several hundred without really trying. Yet a far greater number of smells are locked away in our brains waiting to be triggered.

The wine-smelling procedure is quite simple: give the glass a good swirl, put your nose into the glass, and take a deep sniff. While it is essential to take a substantial sniff, it is not practicable to sniff the same wine over and over again. This is because each wine activates a unique pattern of nerve ends in the olfactory bulb; these nerve ends are like small candles that are snuffed out when activated and take a little time to reactivate. As a result, subsequent sniffs of the same smell can reveal less and less, yet it is perfectly feasible to smell different smells, therefore different wines, one after the other.

THE TASTE, OR "PALATE", OF A WINE

As soon as one sniffs a wine the natural reaction is to taste it, but do this only after all questions concerning the nose have been addressed. The procedure is simple, although it may look and sound rather strange to the uninitiated. Take a good mouthful and draw air into the mouth through the wine; this makes a gurgling sound, but it is essential to do it in order to magnify the wine's volatile characteristics in the back of the throat.

The tongue itself reveals very little – just five basic tastes: sweetness, sourness (or acidity), bitterness, saltiness, and umami. The concept whereby these basic tastes are perceived by receptors in specific parts of the tongue (such as sweetness on the tip, acidity on the sides, etc) has now been thoroughly debunked. Apart from these five basic taste perceptions, we smell tastes rather than taste them. Any food or drink emits odorous vapours in the mouth that are automatically conveyed to the roof of the nasal passages.

Here the olfactory bulb examines, discerns, and catalogues them – as they originate from the palate, the natural inclination is to perceive them as tastes. For many of us, it is difficult to believe that we taste with an organ located behind the eyes at the top of the nose, but when we eat ice cream too quickly, we painfully experience precisely where the olfactory bulb is, as the chilly ice cream aromas literally freeze this acutely delicate sensory organ (*see* How We Taste Smells, opposite page). The texture of a wine also influences its taste; the prickly tactile sensation of carbon dioxide, for example, heightens our perception of acidity, while increased viscosity softens it.

THE COLOUR OF WINE IS ONE OF THE FACTORS DETERMINING ITS TASTE
Each wine style, whether red, white, or rose, has a range of colours. Red wine can range from pale, almost pink, to deep, nearly black. The depth of colour can tell you where the grapes may have been grown and hence, what to expect from its flavour. Be sure to evaluate its colour under natural lighting.

QUALITY AND TASTE: WHY OPINIONS DIFFER

Whether you are a novice or a Master of Wine, it is always personal preference that is the final arbiter when you are judging wine. The most experienced tasters can often argue endlessly over the relative merits and demerits of certain wines.

We all know that quality exists and more often than not agree which wines have it, and yet we are not able to define it. Lacking a solid definition, most experienced tasters would happily accept that a fine wine must have natural balance and finesse and show a definite, distinctive, individual character within its own type or style.

If we occasionally differ on the question of the quality of wine, should we disagree on what it tastes like? We may love or hate a wine, but surely the taste we perceive is the same? Surprisingly, the answer is no, it isn't.

Conveying specific taste characteristics from the mind of one person to that of another is difficult enough, whether one is writing a book or simply discussing a wine at a tasting. Much of this difficulty lies in the words we choose, but the problem is not confined to semantics. Even in a world of perfect communication, conveying impressions of taste would still be an inexact art because of the different threshold levels at which we pick up fundamental tastes and smells and because of the various tolerance levels at which we enjoy them. Because individuals require different quantities of acidity, tannin, alcohol, sugar, esters, and aldehydes in a wine before actually detecting them, the same wine has, literally, a different taste for each of us. In the unlikely event of people having the same threshold for every constituent and combination of constituents, disagreement would probably ensue because we also have different tolerance levels; therefore, some of us would enjoy what others dislike because we actually like the tastes and smells they dislike. Thresholds and tolerance levels vary enormously; the threshold for detecting sweetness, for example, varies by a factor of five, which explains the "sweet tooth" phenomenon, and there are an infinite number of tolerance levels. Apply this to every basic aroma and flavour, and it is surprising that we agree on the description of any wine.

WINE NOTES

- An average human being has about 10,000 taste buds, and within each of those there are around 100 taste receptors, making a total of around one million taste receptors. Most of these are on the tongue, but some are also present on the soft or upper palate, the insides of the cheeks and lips, and around the back of the throat and upper throat.

VISUAL EXAMINATION
Unless it has an extreme colour or hue, the appearance of a wine is the least interesting aspect of a tasting note for most readers, which is why most authors use colour descriptions sparingly. The eye, however, is one of the most important sensory organs for professional tasters, because even the most subtle shade or nuance can provide numerous clues to the wine's identity.

NOSING A WINE
As we smell most flavours, rather than taste them, a good sniff tells us a lot about a wine. But refrain from continuously sniffing, because this will dull your sense of smell. Take one good sniff, then pause for thought. Do not rush on to tasting the wine. Remember that no smells are specific to wine – they can all be compared to something familiar, but you need time to work out what they are.

WHAT IS UMAMI?

The term *umami* (meaning "deliciousness") was coined in 1908, when Professor Kikunae Ikeda identified it as the fifth basic taste, isolating monosodium glutamate (MSG) as the active chemical compound responsible. Ikeda immediately patented the manufacturing process for extracting MSG from wheat flour, and the first commercially available MSG became available under the trade name Ajinomoto (meaning "at the origin of flavour") in 1909. It was later discovered that MSG was not exclusively responsible for the umami taste but worked in a synergistic fashion with other substances, such as disodium inosinate and disodium guanylate, which occur naturally in meat, fish, vegetables, and dairy products.

Although MSG is tasteless, it increases the flavour of these other substances by six- to eightfold, resulting in the so-called umami taste.

The concept of umami has been accepted throughout much of Asia since the early 20th century, but it did not achieve scientific credence until 2002, when Charles Zuker and Nick Ryber identified taste receptors on the tongue that were capable of detecting amino acids. One amino acid we detect in this way is glutamic acid, and MSG is a salt of glutamic acid. It is found in protein-rich foods such as meat (particularly bacon), cheese (particularly Parmesan), soy sauce, mushrooms, broths, and stocks. In the 1950s, when scientists began to analyse the Eastern concept of umami, various

English-language descriptors were applied, such as "amplitude", "mouth-fullness", and "bloom". Perhaps the most appropriate one-word translation is "moreish", but a more complete description would be "a comforting and fulfilling brothy savouriness". The difficulty in assigning a simple yet precise, distinctive, stand-alone taste descriptor (such as sweetness, sourness, acidity, or bitterness) is indicative of how umami is not an obvious basic taste. From a totally unscientific perspective, the richness and complexity of umami makes it an ideal candidate for the sort of taste we smell. Even from a scientific standpoint, umami hardly fits in with the singularity of hypothesis that scientists normally apply to basic tastes.

MAGNIFYING THE TASTE OF A WINE
The tongue discerns only sweetness, sourness, bitterness, saltiness, and umami (*see* box above and the Micropedia). Every other "taste" we smell. By drawing air through a mouthful of wine, the volatilized aromas are taken into the back of the throat, where they are picked up by the olfactory bulb, which automatically analyses them and transmits the information to the brain as various so-called flavours.

SPITTING OUT
When tasting a large number of wines, each mouthful should be ejected after assessment to prevent alcohol affecting the ability to taste. Yet some wine will remain, even after spitting out, coating the inner surface of the mouth, where it is absorbed directly into the bloodstream. Contrary to popular belief, the more you taste, the better you taste; but it is a race between the wine sharpening the palate and the alcohol dulling the brain.

How to Assess Wine

As the ability to make wines of almost every type, style, and depth of colour has spread across the planet since the mid-1980s, it is becoming increasingly difficult to pinpoint a certain wine's origin in blind-tasting conditions.

Signature wine styles of famous regions are routinely copied worldwide today, and international trends bring once-unique flavour profiles ever closer together. Yet copycats are not the only threat to successful blind tasting: changing climatic conditions make it difficult for classic wine regions to stay true to their own identity, too. These challenges make arriving at a correct answer far more difficult, and it requires much wider tasting experience than ever before. Nonetheless, without understanding the underlying principles expressed below, it would be impossible to determine a wine's origins. Even when authenticity is losing ground to international trends, the logical process of questions and answers required to identify a wine under blind-tasting conditions remains very much the same.

SIGHT

Look at the colour: is it deep or pale? Is the colour vivid and youthful, or is there browning that might suggest its age? What does the rim, or meniscus, indicate? Does it retain the intensity of colour to the rim of the glass, which suggests a quality product or warmer clime, or does it fade to an unimpressive, watery finish? Some grape varieties tend to be lighter, others are usually inky dark, but much depends on the winemaking, so don't jump to a conclusion too early.

The colour alone is not a sure way of identifying the wine, but it certainly helps to confirm it once the other elements are assessed.

A SAMPLE TASTING

This guide provides a few simulated examples from a range of options in the complex business of tasting. It demonstrates that it is possible to approach the task systematically and rationally. When tasting, it is important to keep your options open until you have assessed the sight, smell, and taste. At each stage you should be seeking to confirm at least one of the possibilities that has arisen during the previous stage. Be confident, and do not be afraid to back your own judgement – it is the only way to learn.

SIGHT

The pale ruby is very light, so it is probably made from a thin-skinned grape such as Nebbiolo, Pinot Noir, or Grenache. There is a very fine sediment in the glass suggesting some age and potentially a traditional region of origin.

SMELL

The high-toned Pinot aroma dismisses the other options. But what is it exactly? The subtle baking spice character on the nose indicates some oak usage, but it does not pinpoint the origin: we must taste the wine to find out.

TASTE

Definitely Burgundian with a very distinctive, piquant Pinot Noir taste. The subtle integration of oak and the bright acidity, along the fine-grained tannin structure confirms the region. The perfumed, elegant character suggests a good quality village wine from Chambolle Musigny.

CONCLUSION

- **Grape variety** Pinot Noir
- **Region** Chambolle Musigny, Burgundy
- **Age** 4 to 5 years old
- **Comment** Good quality from a respected producer

SIGHT

Water-white with the palest straw tinge – there's little to say about this wine's appearance. There is, however, a certain viscosity detected on the glass wall, and a few stray bubbles suggest youth.

SMELL

The aromas are jumping out of the glass! Fruit salad with green apples, pineapple, and ripe citrus, along with a certain grassy herbaceousness, tinned peas, and a touch of sweatiness point us straight to Sauvignon Blanc.

TASTE

Bright, fresh, but ripe and almost tropical, this is a classic example of Marlborough Sauvignon.

CONCLUSION

- **Grape variety** Sauvignon Blanc
- **Region** Marlborough, New Zealand
- **Age** 1 year
- **Comment** A well-made, high-volume wine

SIGHT

Medium garnet colour with a ruby rim. The colour is neither too deep nor too light, but there is an attractive brightness to it. Bordeaux? Rioja? Maybe Chianti? It doesn't give much away.

SMELL

An immediately oaky nose, with lots of vanilla and coconut, bringing American oak to mind. But the plum fruit and the tarry, gently floral aromas do not suggest a New World wine. It must be Rioja.

TASTE

Balanced and complex, with silky tannins and fresh cherry and red plum notes. The oak spice is present, but well integrated into the palette of violets, tar, and old-leather character. This wine is ready to drink but has a long life ahead of it.

CONCLUSION

- **Grape variety** Tempranillo
- **Region** Rioja
- **Age** About 5 years old
- **Comment** Reserva quality, traditional producer

SIGHT

Golden straw colour, with lazy droplets on the glass indicating viscosity. It could be a youthful sweet wine or something ripe and luscious.

SMELL

The ripe apple and Meyer lemon notes are jumping out of the glass, layered with sweet vanilla and coconut from the new oak. This is definitely from the New World, and given the style, it could be Chardonnay from California or Chenin Blanc from South Africa.

TASTE

This is unapologetically full-bodied, creamy and ripe: a Chardonnay from a great growing region. The popcorn and salted caramel notes, along with the ripe apple flavours, are typical in Napa Valley, where new oak treatment is standard for expensive Chardonnays.

CONCLUSION

- **Grape variety** Chardonnay
- **Region** Napa Valley
- **Age** 2 to 3 years old
- **Comment** Top grower, great year

SMELL

If the first impression is very heady, is the wine fortified? (Classic fortified wines, such as Port, Sherry, and Madeira, do have easily recognizable characteristics, but it can still be difficult to distinguish between a robust wine with a naturally high alcohol level produced in a hot country and a fortified wine.) Does the wine have any intense distinctive aromas, or are they obscure or bland or simply reticent? Does the wine smell as youthful or as mature as it appears to the eye? Is it smooth and harmonious, suggesting the wine is ready to drink? If so, should it be drunk? If it is not ready, can you estimate when it will be? Is there a recognizable grape variety aroma? Are there any creamy coffee, coconut, or vanilla hints to suggest that it has been fermented or aged in new oak? If so, which region ages such wine in oak? Is it a simple wine or is there a degree of complexity? Are there any hints as to the area of production? Is its quality obvious or do you need confirmation on the palate?

TASTE

This should reflect the wine's smell and confirm any judgements you have already made. Should. But human organs are fallible, not least so the brain, so keep an open mind. Be prepared to accept contradiction as well as confirmation. Ask yourself all the questions you asked on the nose, but before you do, ask what your palate tells you about the acidity, sweetness, and alcoholic strength.

If you are tasting a red wine, its tannin content can be revealing. Tannin is derived from the grape's skin, and the darker and thicker it is and the longer the juice macerates with the skins, the more tannin there will be in the wine. A great red wine will contain so much tannin that it will literally pucker the mouth, while early-drinking wines will contain little.

If you are tasting a sparkling wine, on the other hand, its *mousse*, or effervescence, will give extra clues. The strength of the *mousse* will determine the style – whether it is fully sparkling, semi-sparkling, or merely *pétillant* – and the size of the bubbles will indicate the quality: the smaller they are, the better.

CONCLUSION

As you think about the visual, aromatic, and taste profile, try to draw conclusions regarding the climatic conditions the wine may have come from, what grape(s) may have been used, and whether it was made in a bold and expressive New World style or a more reserved Old World fashion. Is it fresh and youthful or mature and mellow? Does it offer complex aromas and a long finish, or is it straightforward and simple? Wise tasters do not risk their credibility by having a stab at anything more specific, such as the producer or vineyard unless he or she is 100 per cent sure. In the Master of Wine examination, marks are given for correct rationale, even if the conclusion that is drawn is wrong. Wine tasting is not a matter of guessing, it is about deduction, and getting it wrong should be encouraged because that is the only way to learn.

SIGHT
Stunning colour, a distinctive old gold hue immediately suggests a full, rich, and probably sweet wine. The viscosity is high, lazy tears run down the wall of the glass.

SMELL
This has the amazingly full, rich, and opulent nose of a botrytised wine. Anyone who dislikes sweet wine should smell a wine like this before giving up on it altogether.

TASTE
Everything is here from peaches, pineapple, and cream to the honeyed aromatics of a fairly mature wine. Only a classic Sauternes can have such intense flavours, yet possess great finesse.

CONCLUSION
- **Grape variety** Mostly Sémillon
- **Region** Sauternes
- **Age** About 15 years old
- **Comment** *Premier cru*, great vintage

SIGHT
The brick-red colour and watery, browning meniscus immediately suggest a thin-skinned grape and an aging wine. It can be a Pinot, a Nebbiolo, or even something obscure like a Nerello Mascalese.

SMELL
A lovely, complex aroma of violets, forest floor after rain, red morello cherries, and a lifted tone of balsamic – a typical Italian nose. The leather, cedar wood, and truffle character just adds to the complexity. It is not as aged as the colour would suggest, so the grape is probably Nebbiolo that tends to turn brown early.

TASTE
The palate perfectly reflects the nose: complex and delightful with firm tannins and vibrant acidity. This is definitely Piemonte and based on the power, balance, and overall quality of the wine, a single-vineyard wine from Barolo.

CONCLUSION
- **Grape variety** Nebbiolo
- **Region** Barolo, Piemonte
- **Age** 5 to 7 years old
- **Comment** Single vineyard from a good vintage

SIGHT
The distinctive yellow-gold colour retains its intensity to the rim. Various possibilities: a sweet wine, a full-bodied dry wine, a mature wine, or something obscure like Retsina. If none of those, it could be a Gewürztraminer.

SMELL
Gewürztraminer! Full, rich, and spicy, the aroma hits you between the eyes, and the first instinct is to think of Alsace. Usually you will be right, but bear in mind the possibility of a top grower in Germany or Austria. If the aroma were muted it might be Italian, and if exotic, Californian or New Zealand. This however, seems to be a classic example of a ripe Alsace vintage of perhaps four years of age.

TASTE
A rich-flavoured wine; full, fat, and fruity with well-developed spice and a soft, succulent finish. Evidently made from very ripe grapes.

CONCLUSION
- **Grape variety** Gewürztraminer
- **Region** Alsace
- **Age** About 4 to 5 years old
- **Comment** Very good quality

SIGHT
Intense, almost black colour that is virtually opaque, with a purple rim variation. Obviously from a thick-skinned grape variety, such as the Syrah, which has ripened under a very hot sun. Australia's Barossa or France's Rhône Valley? California?

SMELL
As intense on the nose as on the eye. Definitely Syrah, and judging by its spicy aroma with hints of dark chocolate and cassis jam with a eucalyptus-minty finish, we are in Australia. The Rhône and California can now be ruled out. More massive than complex, it must be from a larger area with a hot climate.

TASTE
Powerful and warming, the spicy-fruit flavour is rich with blackberries, blackcurrants, plums, and cinnamon and layered with vanilla oak. Drinking well, with rounded tannins and a velvety texture. This is a high-quality Australian Shiraz from a notable region of origin, such as the Barossa Valley.

CONCLUSION
- **Grape variety** Shiraz
- **Region** Barossa Valley
- **Age** About 3 years old
- **Comment** Good quality

Tastes and Aromas

A taste chart is a useful mind-jogging aid for identifying elusive aromas and flavours that you may have encountered in a wine but cannot quite put a name to.

Even seasoned wine tasters experience this lapse. Logically, though, anything the brain recognizes must be well known to it, and the odds are it is an everyday aroma or flavour, rather than something obscure. Therefore, it is not the aroma itself that is elusive, merely its name. This is not surprising, since we all have the sensory profile of more than a thousand everyday aromas locked away in our brains; the difficulty lies in accessing the information. I realized this long ago, which is why my personal tasting books always have a list of mind-jogging aromas and flavours. When I am on my travels and find that I cannot immediately identify a flower, fruit, or spice, I run my finger down the list until my brain connects with the aroma I am trying to discern.

HOW TO USE THE CHART

If you know the category (flower, fruit, spice etc) of the aroma you are looking for, use the Mind-Jogging Chart below to pinpoint the specific aroma you are trying to identify. Take the glass in one hand, swirl the wine, and take a sniff. If it is a flavour you are seeking, take a sip, and then explore the Guide to Tastes and Aroma that starts on the next page.

MIND-JOGGING CHART

Fruity	Citrus	Orange
		Lime
		Lemon
		Grapefruit
		Tangerine
	Berry	Blackberry
		Raspberry
		Strawberry
		Redcurrant
		Blackcurrant/cassis
		Blueberry
	Orchard/stone fruit (pitted fruit)	Cherry
		Apricot
		Peach
		Apple
		Pear
		Quince
		Plum
	Tropical fruit	Banana
		Melon
		Pineapple
		Passionfruit
		Mango
		Papaya
	Dried fruit	Fig
		Prune
		Raisin
	Other	Foxy/jelly
		Amylic/peardrop
Vegetative	Fresh	Cut green grass
		Bell pepper
		Mint
		Nettle
	Canned/cooked	Green beans
		Asparagus
		Artichoke
	Dried	Hay/Straw
		Tea
		Tobacco
Chemical	Petroleum	Tar
		Plastic
		Kerosene/petrol
		Diesel
	Sulphur	Rubbery
		Bad eggs
		Natural gas
		Truffle
		Garlic
		Skunk
		Cabbage
		Burnt match
		Sulphur dioxide
		Wet dog
	Pungent	Nail varnish
		Vinegar
		Garlic

Woody	Burnt	Coffee
		Smoky
		Burnt toast
	Resinous	Vanilla
		Eucalyptus
		Cedar
		Oak
	Phenolic	Bacon
		Phenolic
		Medicinal
Microbiological	Yeasty	Christmas cake
		Acacia
	Lactic	Creamy
		Buttery
		Yoghurt
		Sauerkraut
		Sweaty
	Other	Horsey
		Mousey
Caramel	Caramel	Honey
		Butterscotch
		Butter
		Soy sauce
		Chocolate
		Molasses
Earthy	Undergrowth	Dusty
		Truffle
		Mushroom
		Forest floor
	Mouldy	Mouldy
		Musty/Mildew
Floral	Floral	Elderflower
		Orange blossom
		Rose
		Violet
		Geranium
Spicy	Spicy	Cloves
		Black pepper
		Ginger
		Aniseed
		Liquorice
	Nutty	Almond
		Walnut
		Hazelnut
Oxidative	Oxidative	Dried fruits
		Biscuity (cookie-like)
	Oxidizing	Sherry
	Oxidized	Vinegar

THE ORIGIN OF EVERYDAY AROMAS IN WINE

Although no wines actually contain fruits (other than grapes, of course), flowers, vegetables, herbs, spices et al, it is perfectly reasonable to use their aromas and flavours when describing wines. To the uninitiated, it might sound rather fanciful to say that a wine is buttery, but diacetyl, which is used as an artificial flavouring to make margarine smell and taste buttery, is created naturally in wine as a by-product of the malolactic process. Wines, in fact, contain varying amounts of many chemical compounds that can be linked directly to a vast number of characteristic aromas or flavours.

Some of the compounds involved can evoke different aromas depending on the levels found and the presence of other compounds that can also exert an influence; and various unrelated compounds can induce a very similar aroma. The amount involved can be minuscule; a strong presence of the aromatic compounds responsible for peas and bell peppers or capsicums of the green variety can be detected, for example, at levels of one part in 100 billion!

Do not get carried away in the search for these aromas and flavours. It is far more important to concentrate on just one or two descriptors than to record a fruit cocktail or potpourri of aromas and flavours. When you read elaborate descriptions (not too many in this book, I hope), just ask yourself what such concoctions would actually smell like and whether it would be possible to discern any of their component parts.

GUIDE TO TASTES AND AROMAS

Whether we perceive any of a wine's characteristics as aromas or flavours, technically they are all aromas (see p90). Textural and tactile impressions made in the mouth and true tastes sensed by the tongue (sweetness, acidity, or sourness, bitterness, saltiness, and umami), also influence our perception, however. Where specific chemical compounds are known to be responsible for an aroma or flavour, they are mentioned, and are italicized so that those who are interested can identify the possible cause, while those who are not can skim across without interruption to the text.

Acacia This is the flowery autolytic aroma on a recently disgorged sparkling wine. It can be found in other white wines (*paratolylmethyl ketone*).

Almond Considered part of the varietal character of Gamay, fresh and toasted almond is also commonly found in all types of wine (*acetoin, acetophenone*), especially when aged in oak (*benzaldehyde*), whereas a bitter almond aroma (also *benzaldehyde*) is often detected in sparkling wines and red carbonic-maceration wines.

Aniseed This is characteristic of a Bas Rhin Riesling but can be found in almost any wine red or white (*anethole*).

Apple Apple is a white-wine aroma that ranges from green apple (*malic acid*) in under-ripe wines to soft, red-apple flavours in riper wines, where 50-odd known compounds might or might not be responsible.

Apple, bruised See Oxidative.

Apple blossom This aroma is typical of youthful Riesling (*p-ansic acid, amyl acetate*).

Apple peel This is a pithy apple character (*ethyl hexanoate, n-hexyl, n-butanoate, hexyl hexanoate*).

Apricot A pithy apricot character is less ripe and more bottle-aged than peachiness, which is a finer, juicier, more succulent fruitiness. Apricot is often found in Loire or German whites (*4-decanolide, amyl propoanate*).

Asparagus Asparagus is common in Sauvignon Blanc made from over-ripe grapes or kept too long in bottle. Some people adore this style, but most do not. It can develop into canned peas aroma (*isobutyl or segbutyl*). In a wine without any pyrazine character, it will be *dimethyl disulphide*, otherwise definitely *2-Methoxy-3-isobutylpyrazine*.

Baby sick This classic aroma informs of excessive malolactic (*diacetyl or lactic acid*).

Banana This flavour is found in cool-fermented whites and reds made by carbonic maceration (*amyl acetate or isoamyl acetate*, also known as "banana oil" and "pear oil", which, in excess, can lead to a nail-varnish aroma). A more profound banana character found on the aftertaste of Alsace Pinot Gris or Gewürztraminer is the precursor to bottle-aged spiciness.

Band-Aid This sticking-plaster aroma is one of the four olfactory defects known collectively as phenol off-flavour, or "poff" (*vinyl-4-phenol*).

Barnyard See Horsey.

Beetroot This fruity-vegetal earthiness may be found in some red wines, mostly Pinot Noir grown in unsuitable areas, and aged too long in bottle, or Cabernet Franc (*geosmin*).

Bell Pepper or **Capsicum** This can be found in a slightly grassy-herbaceous Sauvignon, a Loire Cabernet Franc, or a Cabernet Sauvignon from high-vigour vines. It used to be a big problem in New Zealand (*2-Methoxy-3-isobutylpyrazine*).

Bilberry See Blueberry.

Biscuity (cookie-like) Found in fine-quality, well-matured Champagnes, biscuitiness is the post-disgorgement bottle-aroma that typifies Pinot Noir, although many pure Chardonnay Champagnes develop a creamy-biscuitiness (*acetal, acetoin, diacetyl, benzoic aldehyde*, and *undecalactone*).

Blackberry This is detectable in ripe Pinot Noir in the black-fruit stage beyond strawberry (*ethyl caprylate, ethyl hexanoate, ethyl butyrate*, and *amyl propionate*).

Blackcurrant Characteristic of classic Cabernet, blackcurrant is also found in grapes such as Syrah and Carmenère, particularly when bottle-aged (*ethyl acetate, ethyl formate, various acids*, and *esters*).

Blueberry Also known as bilberry or whortleberry, this fruit gives an aroma that is much softer, more perfumed, and less intensely flavoured than blackcurrant (possibly *hexanoate*).

Bready The second stage of autolysis, as the flowery acacia-like aromas take on more substance and a certain creaminess (*diacetyl, undecalactone*, or *paratolylmethyl ketone*).

Broad bean (fava bean) A typical Sauvignon Blanc aroma between bell pepper and fresh green pea. It is also found in Cabernet Sauvignon (*2-Methoxy-3-isobutylpyrazine*).

Bubblegum Found in cool-fermented whites and reds made by carbonic maceration (*amyl acetate* or *isoamyl acetate*, also known as "banana oil" or "pear oil", which, in excess, can lead to a nail-varnish aroma).

Burnt Match A burnt match aroma is the clean, if somewhat choking, whiff produced by free sulphur. This is not a fault as such in a young or recently bottled wine, and it can be dispersed by swirling the wine around in the glass (*sulphur dioxide*).

Burnt Rubber This is a serious wine fault created by the reaction between ethyl alcohol and hydrogen sulphide, another wine fault, which produces a foul-smelling compound called *ethylmercaptan*.

Butter This characteristic is usually found in Chardonnay, and is caused by diacetyl, an artificial flavouring that is used by the food industry, but it is also produced naturally during the malolactic process (also *undecalactone*). It is inappropriate for classic sparkling wine, so the *champenois* utilize special low-diacetyl-forming bacteria.

Butterscotch Butterscotch is produced when very ripe, exotically fruity white wines are aged in well-toasted new oak *barriques*, and is most commonly found in New World wines (*cyclotene, diacetyl, maltol*, or *undecalactone*).

Cabbage or **Cauliflower** The presence of cabbage or cauliflower usually denotes a Chardonnay wine or a wine from the Pinot family. Some people think mature unfiltered Burgundy should have this aroma, or even one that is farmyardy or evocative of manure (*methylmercaptan*).

Candle wax Candle wax is a more accurate descriptor for Sémillon wines than the more commonly employed lanolin, since lanolin possesses no smell, even though it has a connotation of one (*aprylate, caproate,* or *ethyl capryate*).

Caramel This may be either a mid-palate flavour in young wines aged in new *barriques* or, as in tawny port, an aftertaste achieved through considerable ageing in used barrels (*cyclotene* or *maltol*).

Cardboard This characteristic can literally be produced by storing glasses in a cardboard box. It may also be caused by heavy-handed filtration or by leaving a wine to mature for too long in old wood. The glue-ridden smell of wet cardboard can be either *TCA* or *mercaptan.*

Carnation This is one of the four olfactory defects known collectively as phenol off-flavour, or "poff". Although perceived as a defect in wines from most grape varieties, it is said to contribute in a positive sense the varietal aroma of Gewürztraminer (*vinyl-4-guaiacol*).

Cat's pee The is the elderflower aroma taken to the extreme in Sauvignon Blanc, probably due to unripe grapes (*4-Mercapto-4-methylpentan-2-one*).

Cedary oak This is not as silly as it sounds. Various oak lactones have a cedary aroma, particularly in American oak, but this is also a common descriptor for Bordeaux, so it is assumed it can come from French oak, too.

Cheese Occasionally, a wine can have a clean cheese aroma (Emmental or blue-veined being most common), but a strong cheesy smell will be the result of a bacterial fault (*ethyl butryrate* or *S-ethythioacetate*).

Cherry Tart, red cherries are classic in cool-climate Pinot Noir, while black cherries can be part of the complexity of a great Cabernet or Syrah (*cyanhydrin benzaldehyde*).

Chocolate or **chocolate-box** This is the aroma or flavour typical of youngish Cabernet Sauvignon or Pinot Noir wines, when they are rich and soft with a high alcohol content and low acidity level. It may also be detected as part of the complexity of a mature wine or be due to various oak lactones.

Cinnamon Part of the aged complexity of many fine red wines, especially Rhône, cinnamon is also found in oak-aged whites, particularly Gewürztraminer or those made from botrytised grapes (*cinnamic aldehyde*).

Citrus fruit This is a very common taste descriptor in France, where agrume is attributed to the fresh complexity of many young white wines. It is often more complex than a specific citrus fruit aroma (one or more of *limonene, citonellol, linalool*).

Clementine *See* Mandarin.

Clove Clove is found in wines that have been matured or aged in new oak *barriques,* which gain this aroma during the process of being toasted (*see* pp 76–77). In addition, clove is found in Gewürztraminer from certain *terroirs,* such as Soultzmatt and Bergbieten in Alsace (eugenol or eugenic acid from oak if not an aromatic variety).

Coconut This characteristic is another found in great old Champagne. Pungent coconutty aromas are also produced by various wood lactones that are most commonly found in American oak, notably *Cis G-Octa Lactone* (could also be *capric acid*).

Coffee A sign of a great old Champagne, maybe 20 or 50 years old or more, coffee is now increasingly found on the finish of inexpensive red wines made with medium- or high-toast American oak chips.

Currant leaf Although associated with Sauvignon Blanc, this herbaceous character can be found in any wine made from under-ripe grapes or grapes from high-vigour vines. Can be green and mean on the finish.

Dried fruit The aroma of sultanas or currants is most commonly found in Italian Recioto or Amarone wines, and sometimes a yeast-complexed aroma in a Pinot Noir-dominant Champagne, whereas the aroma of raisins is characteristic of fortified Muscat.

Earthy Wines can have an earthiness on the palate that some people incorrectly attribute to the *terroir,* but this undesirable taste is unclean and not expressive of origin (*geosmin*).

Egg, hard-boiled or **rotten** Sulphur is added to wine in order to prevent oxidation, which it does by fixing itself to any oxygen that is present in the wine but, if it fixes with hydrogen, it creates hydrogen sulphide, which smells of hard-boiled or rotten eggs.

Elderflower Found in wines made from aromatic grape varieties, elderflower is good only when the aroma is clean and fresh, and the fruit ripe, but can verge on cat's pee when the grapes are unacceptably under-ripe (*pyrazines*).

Eucalyptus This aroma is noticeable in many Australian Cabernet Sauvignon and Shiraz wines, and could originate from leaves falling off eucalyptus trees into grape-pickers' baskets (*eucalyptol*).

Fennel A softer, more subtle, slightly herbal rendition of aniseed (*anethole*).

Fig A fig-like aroma is sometimes a characteristic of potential complexity in a youthful Chardonnay and may be found in combination with nuances of apple or melon (*ethyl propionate, isobutyl acetate*).

Flinty This is a subjective connotation for the finest Sauvignon Blanc.

Floral This generic flowery aroma is usually light and fresh (*linalool, 2-phenylethanol, methyl 2-methylpropanoate,* various *aldehydes*).

Flowery-fruity In terms of development, this could be viewed as a few months' more bottle-age than "floral" (*ß-damascenone, ethyl hexanoate*).

Foxy Foxy is the term used to describe the very distinctive, cloyingly sweet, and perfumed character of certain indigenous American grape varieties (*methyl anthranilate* or *ethyl anthranilate*).

Garlic When garlic is present as taste as well as or rather than aroma, the fault is definitely mercaptan (*4-methyl-thiobutan-1-ol*).

Gasoline *See* Petrol.

Geranium Commonly a sweet wine fault (*2-ethoxyhexa-3, 5-diene*), but also the sign of an Asti that is too old (*geraniol* degradation), it is always distinctive (also *glycyrrhizin* or *hexanedienol*).

Gingerbread Found in mature Gewürztraminer of the highest quality, when the true spiciness of this variety is mellowed by bottle-age (*citronellal, citronellol, eugenol, geranial, geraniol, linalool, nerol, myrcene*).

Gooseberry The classic aroma of a truly ripe, yet exceedingly fresh, crisp, and vibrant Sauvignon Blanc, gooseberry is most widely found in white wines from New Zealand, particularly Marlborough (*4-MMP, aka 4-mercapto-4-methylpentan-2-one*).

Grape Few wines are actually grapey, but grapiness is found in cheap German wines, young Gewürztraminer, and Muscat or Muscat-like wines (*ethyl caprylate, ethyl heptanoate,* and *ethyl perargonate*).

Grapefruit Grapefruit is found in the Jurançon Sec and Alsace Gewürztraminer, German or English Scheurebe and Huxelrebe, and Swiss Arvine wines (*3-mercaptohexanol* or a combination of *terpenes,* such as *linalool* and *citronellal*).

Grass, freshly cut Can be aggressive, but if fresh, light, and pleasant, it is a positive attribute of deliberately early-picked Sauvignon Blanc or Sémillon grapes (*methoxy-pyrazine hexenal* or *hexanedienol*).

Hay Like dull, flat, or oxidized grassiness, the hay characteristic can be found in sparkling wines that have undergone a slight oxidation prior to their second fermentation (*linalool oxides*).

Hazelnut Discovered as recently as 2000, roasted hazelnut from time in oak is 2-acetylthiazole. This characteristic is part of the complexity of mature white Burgundy and Champagne (*undecalactone, 4-methylthiazole, trimethylpyrazine,* or *diacetyl*).

Herbaceous Overt herbaceousness is a sign of under-ripeness or an over-vigorous canopy, although an understated herbaceousness can add an attractive dimension to some early-drinking wines (various *pyrazines*).

Honey Almost every fine white wine becomes honeyed with age, but particularly great Burgundy, classic German Riesling, and botrytised wines (*phenylethylic acid*).

Horsey Once thought to be part of a wine's complexity, then believed to be a mercaptan fault, the horsey (or barnyard or sweaty-saddle) aroma is now known to be a *Brettanomyces* fault and one of the four olfactory defects known collectively as phenol off-flavour, or "poff" (*ethyl-4-phenol*).

Jam Any red wine can be jammy, but Grenache has a particular tendency towards raspberry jam, while

Pinot Noir has a distinct tendency to evoke strawberry jam. A jammy flavour is not typical of a really fine wine, but can be characteristic of a wine that is upfront and lip-smacking.

Kerosene See Petrol.

Lavender Lavender is often found with lime on Australian wines, particularly Riesling, Muscat, or sparkling wines, and occasionally in German Riesling and even Vinho Verde.

Leather A dry, almost tactile impression of leather can be a complex element of many high-quality wines and should not be confused with the more pungent sweaty-saddle aroma (see Horsey). Leather can be a pure alcohol aroma "peeping" through the fruit.

Lemon Not as distinctive in wine as freshly cut lemon would suggest, many young white wines have simple, ordinary, almost mild lemony fruit or acidity (limonene or citronellal).

Lilac The lilac characteristic is found in some herbaceous reds and peppery Rhônes. It can be overpowering in Muscadine (α-terpineol).

Lily of the Valley This floral characteristic is found in New World Gewürztraminer (linalool).

Lime A truly distinctive aroma and flavour found in good-quality Australian Sémillon and Riesling; in the latter, it often turns to lavender in bottle (limonene, citronellal, or linalool).

Lime tree (Linden) Youthful Riesling typically has this characteristic (hotrienol).

Linden See Lime tree.

Liquorice This can be part of the complexity of red, white, and fortified wines of great concentration, particularly those that are made from late-harvested or sun-dried grapes (geraniol or glycyrrhizin).

Lychee Fresh lychee is depicted as the classic varietal character of Gewürztraminer, but is not as widely encountered as commonly imagined, whereas tinned lychee is commonly found in precocious white wines from off-vintages. Fresh lychee aroma is the product of at least 12 different compounds, the most dominant of which is cis-rose oxide.

Macaroons The almondy-coconutty taste of macaroons is a typical characteristic of a great old Champagne, being similar to a coconutty taste, but sweeter and more complex (undecalactone or capric acid).

Malt More at home as a beer aroma, malt is unwelcome rather than unpleasant in wine. The most common occurrence of malt is in a sparkling wine that has either had too long on its lees or has undergone a less-than- ideal autolysis (3-methylbutanal, 3-methylbutanol involving lactobacillus, 3-hydroxy-2- methyl-4-pyrone, aka maltol, gives the most distinctive maltiness and can develop into a blue- cheese aroma).

Magnolia Although characteristic of a Muscadine grape variety actually called Magnolia, this distinctly floral aroma can also be found in wines made from other grapes (geranyl acetone).

Mandarin or **clementine** A softer yet zestier and more defined aroma than simple orange, mandarin or clementine can be found in Muscat or Riesling of exceptional complexity and finesse (ethyl octanoate, ethyl decanoate, isoamyl alcohol, ethyl hexanoate, and isoamyl acetate).

Manure This aroma is a very extreme form of the "farmyardy" aroma, which some people (not this author) believe to be characteristic of great Pinot Noir. Certain New World winemakers try to emulate this aroma in their wines, but it is probably a fixed-sulphur fault, and quite possibly a mercaptan.

Melon A characteristic of young, cool-fermented, New World Chardonnay, melon may be found in combination with nuances of apple or fig (limonene, citronellal, or linalool).

Mint Although it is occasionally found in Bordeaux, mint is actually far more redolent in full-bodied New World reds, particularly Californian Cabernet (especially Napa) and Coonawarra Shiraz (l-carvone, menthol, menthone, menthyl acetate).

Mouse The mousey aroma was once thought to be a Brettanomyces fault but is now known to be a lactobacillus fault that can only occur when lysine is present in the wine (acetyl-tetrahydrpyridines).

Muscat The fresh, grapey aroma of the Muscat grape is an aromatic characteristic that can occasionally be found in wines made from other varieties, but it usually defines a Muscat wine and is the only grapey-grape aroma in the winemaking world (essentially a combination of geraniol, linalool, and nerol).

Mushroom A beautifully clean mushroom aroma is an indication of Pinot Meunier in a fine old Champagne, but if the aroma is musty, it will be a contamination fault such as infected staves or a corked wine.

Musty This has a drier perception than "mouldy" – the difference as in a dry, musty church and a damp, mouldy cellar. Although both could be due to infected staves, "mouldy" can also come from grapes affected by botrytis, although it is more likely to be grey rot than brown (2,4,6-trichloroanisole, or TCA).

Nail varnish or **polish** Nail varnish is the pungent, peardrop aroma produced by intensive carbonic maceration. It is found in the worst Beaujolais Nouveau (amyl acetate or isoamyl acetate, otherwise known as "banana oil" or "pear oil").

Nuts This ranges from the generic nuttiness of mature Burgundy, and the walnuts or hazelnuts in Champagne blanc de blancs, to the almondy fruit of young Italian red (acetoin, diacetyl, or undecalactone).

Nuttiness See Oxidative.

Oak Used on its own, the term oak is so generic as to be almost meaningless and should be qualified to indicate the type of oakiness, such as coffee-oak, creamy oak, lemony oak, spicy oak, sweet oak and vanilla oak (various oak lactones). See also Cedary oak.

Onion This aroma indicates a very unpleasant fault and can range from rubbery onion (ethanethiol) to pungent raw onion, garlic, or burnt rubber onion (diethyl disulphide).

Orange A good blind-tasting tip is that orange can be found in Muscat, but never Gewürztraminer. It is also found in some fortified wines and Ruby Cabernet (limonene, citronellal, or linalool).

Orange blossom This is typical of youthful dry Muscat (anisic acid, limonene, citronellol).

Oxidative This characteristic can range from bruised apple to Sherry (acetaldehyde) in a still wine and nutty in a sparkling wine. "Oxidative" can be positive for those who like the style, forming part of the complexity. But when the word "oxidized" is used in a description of a wine, it is only in a negative sense.

Paraffin See Petrol.

Passion fruit An intense New Zealand Sauvignon Blanc aroma, a touch of passion fruit adds to the complexity of a wine, but too much can push the varietal aroma into sweaty-armpit mode (mercaptohexanol).

Peach Found in ripe Riesling and Muscat, very ripe Sauvignon Blanc, true Viognier, Sézannais Champagne, New World Chardonnay, and botrytised wines (piperonal or undecalactone).

Peanut At low levels and with a pleasant popcorn hint, this characteristic is found in some Banyuls, Port, and Bordeaux, but a more pungent peanut aroma comes from ladybird or Asian ladybug taint, usually machine-harvested, where just one ladybird in a vat can be responsible for the peanut off-odour (adaline, adalinine, and 2-isopropyl-3-methoxypyrazine).

Pear Pear is found in cool-fermented whites and reds made by carbonic maceration (amyl acetate or isoamyl acetate, also known as "banana oil" or "pear oil", which, in excess, can lead to a nail-varnish aroma).

Peas, canned Common in Sauvignon Blanc made from over-ripe grapes or kept too long in bottle. Some people adore this style, but most do not. Can develop from an asparagus aroma (isobutyl or segbutyl). See Asparagus.

Peas, fresh The fresh green-pea aroma found in Sauvignon Blanc is closer to broad bean than it is to tinned peas or asparagus (2-methoxy-3-isobutylpyrazine).

Pebble, wet Not an aroma or taste as such, this sensation is reminiscent of the salivating effect produced when sucking a wet pebble and is indicative of Sauvignon Blanc of exceptional finesse.

Peppercorn Many young reds have a basic peppery character, but Syrah evokes the distinctive fragrance of crushed black peppercorns (rotundone), while for top-quality Grüner Veltliner, it is ground white pepper.

Petrol The so-called petrol (kerosene, gasoline, or paraffin) aroma is a well-known varietal characteristic of a classic, racy Riesling wine of some maturity. It does not literally smell or taste of petrol, but it is an instantly recognizable aroma descriptor that experienced tasters use without pretension. The active chemical compound responsible for the petrol aroma has been identified as trimethyldihydronaphthalene, or TDN for short. TDN develops during the bottle-ageing process through the degradation of beta-carotene, an antioxidant that is itself derived from lutein, another antioxidant. The ratio of beta-carotene to lutein is higher in Riesling than in any other white grape variety. Studies show that the lower the pH of a wine, the higher its potential for developing TDN, thus its propensity to develop in warmer climes. The longer it takes for the petrol aromas to emerge, the more finesse they have. Interestingly, cork absorbs 40 per cent of TDN; ergo, screw caps preserve petrol aromas.

Pineapple Pineapple is found in very ripe Chardonnay, Chenin Blanc, and Sémillon, especially in the New World, and almost any botrytized wine. It implies good acidity for the ripeness (*ethyl caprylate* or *butyl butyrate*).

Plastic or **polythene** This not uncommon off-odour is possibly caused by *benzohiazole* or unsaturated *olefins*.

Popcorn This can be a malolactic aroma, but *see also* Peanut.

Potato peelings More earthy and less fruity than beetroot, potato peelings is found in a wide range of red wines, and could be an indication of infected staves or corkiness (*geosmin*).

Quince or **quince jelly** This is a classic reductive aroma (*dimethyl sulphide, dimethyl disulphide*).

Raisin The raisin aroma is commonly found in fortified wines, particularly Muscat (*ß-damascenone*).

Raspberry This is sometimes found in Grenache, Loire Cabernet, Pinot Noir, and Syrah (evolving into blackcurrant in bottle) (*ethyl acetate, ethyl formate, various acids and esters*).

Redcurrant Youthful Pinot Noir, probably cool-climate grown, commonly has redcurrant aromas, but it is also sometimes found in young New World Cabernet Sauvignon and even Merlot.

Rhubarb Some yeasts (eg, Lalvin 71B-1122) tend to produce a rhubarb-like fruitiness, particularly in wines that have undergone carbonic maceration (*3-mercaptohexanol*).

Rose Rose petals can be found in many wines, particularly delicate Muscats and understated Gewürztraminers (*damascanone, diacetyl, geraniol, irone, nerol*, or *phenylethylic acid/rose oxide*).

Rubber This is a fixed-sulphur fault (carbon disulphide) or a mercaptan (*ethanethiol*), whereas burnt rubber can be either a sulphur fault (*diethyl sulphide*) or a mercaptan (*thiophen-2-thiol;* although at low concentrations, this can give a wonderful roast-coffee aroma).

Sauerkraut The lactic smell of a wine that has undergone excessive malolactic, sauerkraut is actually even less acceptable in wine than the sour milk or sour cream aroma (*diacetyl* or *lactic acid*).

Sherry This is the tell-tale sign of excessive acetaldehyde, which could turn a wine into vinegar unless it is Sherry or another type of fortified wine, which will be protected by its high alcohol content.

Skunk Once smelled, never forgotten, the pungent, highly resinous skunk aroma is a methyl mercaptan fault (*dimethyl-ethanethiol*).

Smoke Smoke is a complexity that might be varietal, as in the case of Syrah, but can be induced by stirring less during barrel fermentation, suggesting that the wine has not been racked, fined, or filtered (*guaiacol* or *4-ethyl-guaiacol*).

Smoky-spicy Although this can be part of a wine's complexity, it may also be one of the four olfactory defects known collectively as phenol off-flavour, or 'poff' (*ethyl-4-guaiacol*).

Soapy A youthful soapiness in white wines, particularly Riesling, can indicate potential complexity, but too distinctive a soapiness is a fault (*caprylate, caproate, ethyl caprate*).

Sour milk or **sour cream** The lactic smell of a wine that has undergone excessive malolactic, the sour-milk character may develop into a more pronounced sauerkraut aroma (*diacetyl* or *lactic acid*); *see also* Sauerkraut.

Spicy Many wines have a hint of spiciness, which is more exotic than peppery, but, after a few years' bottle-age, the spiciness of Gewürztraminer should almost burn the palate.

Stable *See* Horsey.

Stagnant water This is a mercaptan fault (*methanethiol*).

Strawberry Succulent, ripe strawberry fruit is found in classic Pinot Noir from a warm climate or top vintage. It is also found in Loire Cabernet (*ethyl acetate, ethyl formate, various acids, and esters*).

Summer fruits This aroma can be either a more complex or less distinct medley of raspberry, strawberry, blackberry, and blackcurrant (*ethyl acetate, ethyl butyrate*).

Sweaty An unattractive sweaty-armpit aroma can be the result of many compounds found in wine, but the best documented example is when passionfruit in Sauvignon Blanc goes sweaty (*mercaptohexyl acetate*).

Sweaty saddle *See* Horsey.

Tar Like liquorice with a touch of smoke, a tarry aroma in some full-bodied reds, typically Barolo and northern Rhône, could indicate a wine that has not been racked, fined, or filtered.

Toast Toastiness is commonly associated with Chardonnay and mature Champagne, particularly *blanc de blancs,* but it can be found in many wines. Toastiness can either be a slow-developing bottle-aroma or an

instant gift of new oak (*furanic aldehydes*). Current theory among research chemists is that the toastiness in Chardonnay wines is technically a fixed-sulphur fault, although it is a fault that many wine lovers have come to enjoy.

Tobacco The tobacco aroma is often found in mature reds, particularly Bordeaux, or big New World reds (*3-oxy-a-ionol, ß-damascenone, hydroxy-ß-damascenone*).

Toffee This is less creamy and more oxidative than caramel (various oak *lactones*).

Tomato We tend to think of tomato as a vegetable, but it is really a fruit and, although not a common feature in wines, it is found in bottle-aged Sylvaner and, with blood-orange, in Ruby Cabernet. Also found in Merlot, Pinot Noir, and various Italian wines (*trans-2-pentanal*).

Tomato leaf This characteristic is the distinctive aroma of deliberately oxidized Sauvignon Blanc juice prior to fermentation (*2-isobutylthiazole*).

Tropical fruit Usually found in New World whites, particularly Chardonnay, tropical-fruit aromas can also be found in Viognier, exotic Riesling, or Old World wines, such as Champagne from Sézanne or mature bottles of exceptionally hot vintages (*ß-damascenone*).

Truffle While truffle is often cited as part of the profound complexity of a great wine, if it is noticeable, it is a fixed-sulphur fault (*dimethyl sulphide*).

Vanilla Probably vanillin from new oak, although vanillin is also found in cork. Various vanillin-based compounds also have vanilla aromas, and a hint of vanilla can be due to unrelated compounds present in wines that have seen no oak or cork. DDMP has a vanilla-sugar aroma and is a product of Maillard reactions during the toasting of oak barrels (*vanillin, vanillyl acetate, acetovanillone, ethyl vanillate, methyl vanillate, guaiacol, furylacetone, 4-ethylguiacol, 2,3-dihydro-2,5-dihydroxy-6-methyl-4-H-pyran-4-one, aka DDMP*).

Vegetal Although unattractive sounding, vegetal can be either negative or positive (*trans-2-hexenol*).

Vinegar This is a classic volatile-acidity fault (*acetic acid*).

Violet Violets can often be found as part of the finesse on the finish of Cabernet-based red wines, notably Bordeaux, especially from Graves. It is possibly more tactile-based than a volatile aroma (*ionones*).

Wet dog or **wet wool** These aromas are heat-generated volatile sulphur faults involving the retro-Michael reaction of *methional*, which is thermally unstable and evolves rapidly into *acrolein* and *methanethiol*, which are responsible for the so-called wet-dog and wet-wool aromas and a stronger cooked-cauliflower smell.

Whortleberry *See* Blueberry.

Yeast-complexed Yeasty aromas are not welcome and are rarely encountered, but yeast-complexed aromas are, especially in Champagne, where they give a chewy creaminess to the fruit and encourage many different aromatic characters, such as dried fruit, developing into Christmas cake for yeast-complexed Pinot Noir.

FAULT FINDING

SYMPTOM	CAUSE	REMEDY
Bits of floating cork	You are the cause, and this is not a corked wine! Tiny bits of cork have become dislodged when opening the bottle.	This is not a fault, so drink the wine or give it away. Pour the glass through a strainer to remove the cork bits.
Sediment	All wines shed a deposit in time; most are drunk before they do.	Decant the bottle.
Coating on the inside of the bottle	Full-bodied reds from hot countries or exceptionally hot vintages in cool-climate countries are prone to shedding a deposit that adheres to the inside of the bottle.	You can check to see whether the wine pours clear. However, there is bound to be some loose sediment and, even when there is not, it is always safest to decant.
Cloudy haze	If it is not sediment, it will not drop out and is either a metal or protein haze.	Seek a refund. Home winemakers can try bentonite but, although this removes a protein haze, it could make a metal haze worse!
A film or slick on the surface	This is an oil slick caused by glasses or a decanter. Either they have not been properly rinsed or a minuscule amount of grease has come from the glass-cloth used to polish them.	Use detergent to clean glasses and rinse them thoroughly in hot water. Never use a glass-cloth for anything other than polishing glasses and never dry them with general-purpose tea towels.
Still wine with tiny bubbles clinging to the glass	An unwanted second fermentation or malolactic can make some still wines as fizzy as Champagne. White wines meant to be consumed very young (Vinho Verde or some Sauvignon Blancs) are sometimes bottled with a squirt of CO_2 that creates small bubbles.	If the wine is really fizzy, then take it back, but if the fault is just a spritz or prickle, swirl the glass or use a Vacu-vin to suck the gas out.
Asparagus or canned-peas aromas	Sauvignon Blanc from over-ripe grapes or kept too long in bottle.	No technical fault – buy a more recent vintage next time.
Cabbage, cauliflower, farmyardy, or manure aromas	Technically a fault (methylmercaptan), but half the wine trade would argue it is part of the complexity of some wines, particularly Burgundy.	Some retailers will refund you, but those who have personally selected this "traditional" style may not.
Currant leaf aromas or flavours	Caused by under-ripe grapes or high-vigour vines, this may be deliberate if wine is from a hot country and green on the finish.	Not a fault as such, although it is not exactly good winemaking so, if you cannot force yourself to drink the wine, throw it away.
Bubblegum, peardrops, or nail-varnish aromas	Produced by cool-fermentation in white wines and carbonic maceration in reds, and found in the worst Beaujolais Nouveau.	This is not a fault, so drink the wine or give it away.
Burnt match aromas (or a tickle in the nose or throat)	The clean whiff of free sulphur, which protects the wine, as opposed to fixed-sulphur faults, which have a pungent stench.	Swirl the glass or pour the wine vigorously into a jug and back into the bottle to disperse the aroma through aeration.
Burnt rubber or skunk	A serious wine fault created when ethyl alcohol reacts with hydrogen sulphide, a fixed-sulphur fault, to form a foul-smelling compound called ethylmercaptan.	Take the wine back for a full refund.
Cardboard aromas	This can be due to storing glasses in a cardboard box, heavy-handed filtration, or leaving a wine too long in old wood.	If the wine is still cardboardy in a clean, untainted glass, you could seek a refund, but may have to put it down to experience.
Cheese aromas	Occasionally a wine has a clean cheese aroma (Emmental or blue-veined being most common), but a real cheesy smell will be a bacterial fault (ethyl butryrate or S-ethythioacetate).	Take the wine back for a full refund.
Earthy aromas	Unclean, but not exactly a known fault.	As the wine was probably purchased by someone who thought it had a goût de terroir, you are unlikely to get a refund.
Hard-boiled or rotten-egg aromas	Sulphur is added to wine to prevent oxidation by fixing itself to any oxygen, but if it fixes with hydrogen it creates hydrogen sulphide, which is the stuff of stink bombs.	Theoretically, if you put a brass or copper object into the wine, the smell should drop out as a very fine brown sediment, but frankly it is quicker and easier to ask for a refund.
Geranium aromas	In a sweet wine, this is a sorbic acid and bacterial infection fault. In Asti or any other Muscat wine, the geraniol that gives the wine its classic flowery-peach character has degraded with age.	Take the wine back for a full refund.
Maderized (in any wine other than Madeira)	Maderization is undesirable in any ordinary, light table wine. Such a wine will have been affected by light or heat, or both.	Take the wine back for a full refund, unless you have kept it under bad conditions yourself.
Mushroom aromas	A clean mushroom aroma indicates Pinot Meunier in a fine old Champagne but, if musty, it will be a contamination fault.	If a contamination fault, take the wine back for a full refund.
Musty aromas, as in an old church	The wine is corked (or at least it is suffering from a corky taint).	Smell the wine an hour later: a corked wine will get worse and you should seek a refund. Harmless bottle mustiness disappears.
Mousy aromas	Caused by Brettanomyces yeast and malolactic bacteria, this is feared in the New World, but some Old World wineries consider this as an addition to a wine's complexity.	Give it to someone you do not like.
Onion or garlic aromas	A serious wine fault created when ethyl alcohol reacts with hydrogen sulphide, a fixed-sulphur fault, to form a foul-smelling compound called ethylmercaptan.	Take the wine back for a full refund.
Sauerkraut aromas	The smell of a wine that has undergone excessive malolactic, this is more unacceptable than sour milk or sour cream aromas.	Take the wine back for a full refund.
Sherry aromas (in any wine other than Sherry)	Excessive acetaldehyde: the wine is oxidized. An ordinary wine with excessive acetaldehyde turns into vinegar, but Sherry and other fortified wines are protected by a high alcohol content.	Take the wine back for a full refund.
Vinegar aromas	The distinctive aroma of acetic acid: the wine has oxidized.	Use it for salad dressing.

Wine with Food

There is only one golden rule when you are selecting a wine to accompany food: the more delicately flavoured the dish, the more delicate the wine should be, whereas richer-flavoured foods can take richer-flavoured wines.

The old wine trade maxim "buy on apple, sell on cheese" vividly illustrates that wine and food combinations range can have positive or negative effects. A raw apple highlights the smallest defect in a wine, whereas cheese is wine-friendly.

Wine should not overwhelm the food, and food should not overwhelm the wine: the most successful combinations enhance each other. To find the best combinations for you, you should begin with classic food-and-wine pairings and experiment from there. It took hundreds of years to whittle down the best food and wine pairings, so it makes sense to take advantage of that groundwork – not necessarily to accept every combination as if it should be carved in stone, but as a starting point from which to ask yourself whether you might prefer something lighter, fuller, drier, sweeter, more mineral, less buttery, oaky, or whatever

A WHITE WINE ACCOMPANIES MUSHROOM RISOTTO
For rice or pasta dishes, key the wine to the other ingredients. A classic *risotto ai funghi*, a typical dish of northern Italy, pairs well with a white from the Piedmont area.

SOME CLASSIC
FOOD AND WINE PAIRINGS

APERITIFS

SHERRY

Sherry is a traditional aperitif, but its use has been abused. If the first course is sufficiently well-flavoured or includes Sherry as an ingredient, then it can be an admirable choice. Mostly, however, even the lightest *fino* will have too much alcohol and flavour.

WINES

Perfect all-purpose aperitifs are light-bodied, dry, still, or sparkling white wines, although a rosé may be suitable, especially if the first wine served with the meal is also a rosé or a light red. Excellent choices for a white wine aperitif include a well-selected Mâcon, a light but serious Muscadet, a muted Loire Sauvignon or Chenin, and a Pinot Blanc or Sylvaner from Alsace (although bone-dry Muscat is a classic aperitif in the region itself). Almost any dry white from northeastern Italy, aromatic or relatively neutral, would do. Crisp dry whites of distinctive minerality, such as Albariño or Assyrtiko, are perfect. A fresh, young Australian Sémillon would be an ideal lead-in to a finer example with some decent bottle-age for the first course. A fresh, young, genuinely dry Mosel or Rhine would fulfill the same function as a Riesling of some maturity. Any light-bodied Chardonnay, Chenin, or Colombard from anywhere would also work well. The list is endless. If the choice is a rosé, choose one from the same area and preferably grape as the first wine of the meal, being careful to avoid any in clear glass bottles, unless you have personally transported them from the darkness of the producer's cellar to the darkness of your own. The aperitif par excellence in every conceivable situation is, of course, Champagne or other fine-quality sparkling wines from around the world.

♛ **Classic** *Champagne*
✔ **Budget choice:** *Cava Brut*

STARTERS

ARTICHOKE

If served with just butter or a Hollandaise sauce, a sparkling *blanc de blancs*, a light but slightly assertive Po:illy-Fumé, or an Assyrtiko from northern Greece would be ideal.

♛ **Classic:** *Champagne Blanc de Blancs*
✔ **Budget choice:** *Côtes de Gascogne*

ASPARAGUS

There are two perfect accompaniments for this vegetable: a young Muscat d'Alsace, which is a classic combination in Alsace, and a Sauvignon Blanc, the varietal aroma of which shares several pyrazines with asparagus. Marlborough Sauvignon is ace with green asparagus cooked *al dente*, while a more delicate Pouilly-Fumé *blanc de blancs* medium-weight white Burgundy and Californian or Pacific Northwest Chardonnay also work well.

♛ **Classic:** *Sauvignon Blanc*
✔ **Budget choice:** *Côtes de Gascogne*

AVOCADO

Defying all logic, the perfect accompaniment to avocado with vinaigrette is a (rare) dry Gewürztraminer d'Alsace, which should not work with such a naturally low acidity, but the opposite end of the scale, such as a light, mineral, acidic, dry white, has a workmanlike affinity.

♛ **Classic:** *Gewürztraminer d'Alsace*
✔ **Budget choice:** *Muscadet sur lie*

EGG, RICE, AND PASTA DISHES

A sparkling wine is the perfect foil to omelettes, quiches, soufflés, eggs cooked *en cocotte*, coddled, scrambled, or poached. The texture of these dishes is cut by the wine's acidity and effervescence, although a crisp, dry white of expressive minerality also works extremely well. Where the dish involves a runny egg yolk, all glasses on the table will have to be replaced, as the egginess does not go away and will spoil whatever is consumed from those vessels afterwards. Even (or, especially) the water tumblers must be removed. Savoury mousses or mousselines, rice or pasta dishes, whether hot or cold, vegetable or meat, fish or fowl, should be partnered according to the main ingredient or flavour, and the ideal choice could cover the entire spectrum of wine colours and styles.

♛ **Classic:** *Champagne*
✔ **Budget choice:** *Saumur Brut or Saumur Rouge*

FOIE GRAS

This is classic with Sauternes or Tokaji, but today such wines are considered too sweet and overwhelming to begin a meal. A Pinot Gris Vendange Tardive is the wine of choice in Alsace, and lighter equivalents like Jurançon can also be fabulous, but for something dry, go for a fine, mature vintage Champagne.

♛ **Classic:** *Sauternes*
✔ **Budget choice:** *Jurançon*

SALADS

Plain green salads with the barest touch of dressing need little more than a light, dry, white, such as a Muscadet, unless there is a predominance of bitter leaves, in which case something more assertive, but just as light, such as an Assyrtiko from northern Greece. A youthful Champagne is the best accompaniment to salads that include warm ingredients, especially if they include *lardons* and a soft, poached egg. Salad Niçoise and rosé is a safe bet.

♛ **Classic:** *Champagne*
✔ **Budget choice:** *Cava Brut*

SOUPS

Champagne is ideal with most purée, *velouté*, and cream soups, especially the more delicately flavoured recipes. It is virtually essential with a chilled soup, such as Vichyssoise, but Gazpacho requires a Sherry (anything from a *fino* to an authentically dry *oloroso* depending on the recipe). Most fine-quality sparkling wines (Trentodoc, Franciacorta, English sparkling wine, and the like) can match a shellfish bisque, but a rosé style makes a particularly picturesque partner. Rich-flavoured soups can take fuller wines, often red: a good game soup, for example, can respond well to the heftier reds of the Rhône, Bordeaux, Burgundy, and Rioja or their New World equivalents. A Lambrusco cuts through the texture of a genuine minestrone, and its cherry flavour matches the soup's rich, tomato tang.

♛ **Classic:** *Champagne*
✔ **Budget choice:** *Prosecco*

TERRINES AND PÂTÉS

Whether meat or vegetable, fish or fowl, a pâté should be partnered according to its main ingredient or flavour. Look in the appropriate entries and choose a similar wine. The following pairings are for a typical coarse pâté, which requires more acidity than most reds can provide, or some effervescence to cut across the fatty texture.

♛ **Classic:** Cru *Beaujolais*
✔ **Budget choice:** *Dry Lambrusco*

SEAFOOD

CAVIAR

The saltiness of caviar makes it wine-friendly, but even the cheapest caviar is expensive, so it makes sense to insist on the perfect accompaniment and only the effervescence of a sparkling wine can flow around and keep separate each individual egg. Champagne is the classic partner.

♛ **Classic:** *Prestige or vintage Champagne*
✔ **Budget choice:** *Fine but less expensive sparkling wine or mineral water*

FISH, GRILLED

Freshly caught and plain grilled is the most delicate and unadulterated way of serving fish. It preserves all the freshest and most delicate flavours of the fish and, as such, deserves an equally delicate, dry, crisp white wine, whether still or sparkling, to ensure that none of the natural flavours of the fish are masked.

♛ **Classic:** *Chablis*
✔ **Budget choice:** *Crémant de Luxembourg*

FISH, PAN-FRIED

A pan-fried fish dish requires a slightly fuller Chardonnay than that required for a plain-grilled fish dish or a more distinctive sparkling wine.

♛ **Classic:** *Puligny-Montrachet*
✔ **Budget choice:** *Unoaked New World Chardonnay*

FISH, SERVED WITH A SAUCE

A fish dish served with a cream sauce or *beurre noisette* requires a wine with more acidity or effervescence than normal. If the sauce is very rich, consider a wine with a greater intensity of flavour. Unless you have a palate that can withstand or even prefer red wine with fish, a fish dish that has been poached in red wine or served with a red wine sauce will be one of the very few situations where a red wine will not just be an acceptable accompaniment, but will in fact be the most preferable.

♛ **Classic:** *Champagne*
✔ **Budget choice:** *Lambrusco*

FISH STEWS

As indicated above, although red wine and fish react violently in the mouth for most people, fish dishes cooked in red wine present no such problems. The same goes for big, rich Mediterranean fish stews. Not just those that have red wine as an ingredient, but strangely all recipes that are dominated by tomato, which normally repel reds.

♛ **Classic:** *Provence*
✔ **Budget choice:** *Lambrusco*

MACKEREL

An assertive Loire Sauvignon is needed for mackerel, although a richer, oak-fermented Sauvignon from the New World would be preferable with smoked mackerel.

♛ **Classic:** *Sancerre*
✔ **Budget choice:** *Sauvignon de Touraine*

RIVER FISH

Generally, most river fish dishes go well with a fairly assertive rosé wine – the fish and the wine having a complementary earthiness – but a modest French-style Sauvignon can be just as effective. Graves wines and Champagne are especially successful with pike. Champagne and Montrachet are classic with salmon or salmon-trout, but any fine sparkling wine will also be excellent. Riesling is almost obligatory with trout, particularly when cooked *au bleu* (rapidly, in stock with plenty of vinegar).

♛ **Classic:** *Montrachet*
✔ **Budget choice:** *Haut Poitou Sauvignon*

SARDINES

Vinho Verde is the ideal wine with sardines, especially in Portugal, if freshly caught and cooked on the beach.

♛ **Classic:** *Albariño*
✔ **Budget choice:** *Vinho Verde*

SHELLFISH, CRUSTACEAN

Choose a top estate Muscadet, a Loire Sauvignon or Chenin, or a Mosel Riesling with simple prawn and shrimp dishes; a fine Sancerre or Pouilly Fumé with crayfish (or Saumur Blanc as an alternative to Sauvignon); and a Grand Cru Chablis or fine Champagne with crab or lobster. A fuller Chardonnay if potted.

♛ **Classic:** *Grand Cru Chablis*
✔ **Budget choice:** *Muscadet*

SHELLFISH, MOLLUSC

Chablis and Champagne are classic choices to accompany oysters and scallops. Choose a wine with more acidity or effervescence if the dish includes a cream sauce.

♛ **Classic:** *Champagne*
✔ **Budget choice:** *Guinness*

SMOKED FISH

Drink an oak-fermented or cheaper, oak-matured version of the wine you would drink with an unsmoked dish of the same fish.

♛ **Classic:** *Krug*
✔ **Budget choice:** *Oaked New World Chardonnay*

TUNA

If tuna is served rare like a steak, it is almost meaty in texture as well as flavour, and it demands an assertive Loire red or rosé, but it can be magnificent with Champagne. If the tuna comes in a tin, it will probably end up in an overpriced salad or sandwich, so why bother?

♛ **Classic:** *Champagne Rosé*
✔ **Budget choice:** *Cava Rosé*

MEAT DISHES

BEEF

Claret is the classic accompaniment to roast beef, but almost any red wine should work. Choose a younger, perhaps lighter and livelier style if the meat is served cold. A good New World Cabernet Sauvignon would do just as well – and might even be preferable – for steaks that are charred on the outside and pink in the middle. For pure beef burgers, an unpretentious, youthful red of almost any origin is a suitable choice for all beef dishes to some degree or another. Contrary to popular belief, many white wines, particularly fine, sparkling wines, have the flexibility to go with most beef dishes.

♛ **Classic:** *Cru Classé Bordeaux*
✔ **Budget choice:** *Inexpensive Cabernet Sauvignon*

DUCK

Roast duck is very versatile, but the best accompaniments include certain *cru* Beaujolais, such as Morgan or Moulin-à-Vent, fine red and white Burgundy, especially from the Côte de Nuits, and mature Médoc. Barbera is brilliant, too. With cold duck, a lighter *cru* Beaujolais (Fleurie) should be considered. The classic Franglais dish of the 1960s, Duck à l'Orange, repels most red wines and many whites, leaving Champagne as the perfect partner. In fact, a fine quality sparkling wine from any origin makes the ideal accompaniment to any duck dish, from a simple rare breast to any variant, with or without a sauce, because it has the effervescence and acidity to cut across the crispy fat and the

hidden sweetness of the *brut*-style to balance any sweetness that may be present.

♛ **Classic:** *Champagne*
✔ **Budget choice:** *Prosecco*

GAME

For lightly hung winged game, the wines that are suitable are the same as for poultry. If it is mid-hung, try a fullish *cru* Beaujolais like a Brouilly or even a Morgon, whereas well-hung game birds require a full-bodied red Bordeaux or Burgundy. Pomerol is the classic choice with well-hung pheasant. Treat lightly hung ground game in the same way as lamb, mid-hung meat in the same way as beef. Well-hung ground game can take big reds, such as Hermitage, Côte Rôtie, or an Australian Shiraz.

♛ **Classic:** *Pomerol*
✔ **Budget choice:** *Australian Shiraz*

GOOSE

Champagne is the classic choice, but other options are numerous, and it is a hard choice between a Loire red such as Chinon, Bourgueil, or Anjou Rouge and going white with a Vouvray or Riesling. The most important characteristic is plenty of acidity for this fatty bird. If the goose is served in a fruity sauce, stick to the white wines, where a little sweetness in the wine will do no harm.

♛ **Classic:** *Champagne*
✔ **Budget choice:** *Beaujolais Villages or South African Chenin Blanc*

HAM, BACON

Ham can react in the mouth with some red wines in a similar way to white fish and red wine, particularly if it is unsmoked, but young Beaujolais and light reds from almost anywhere are usually safe bets. Rosés and crisp, dry mineral whites are good, too. The more smoked, cured or crispy, the more ideal sparkling wine will be.

♛ **Classic:** *Champagne*
✔ **Budget choice:** *Albariño*

CIOPPINO PAIRS BEAUTIFULLY WITH RED WINE
Fish stews, such as the Italian-American Cioppino made with a variety of shellfish in a tomato-based wine sauce, call for reds such as Sangiovese or Pinot Noir. The similar French dish, Bouillabaisse, is based more on vegetable stock and is traditionally served with a dry rosé.

THINGS TO THINK ABOUT

Keep in mind a few useful tips when choosing the wine to go with a dish.

- **Two ways of choosing** In a restaurant, you are more likely to choose a wine to match the food, whereas when arranging a meal for others at home, there will be times when you know the wines you want to serve and will have choose the food to match the wine.

- **White wine flexibility** In general terms, white wine is more flexible than red. Any dish that ideally goes with a specific red wine can be acceptably accompanied by a white wine, especially if chosen for its structure and style. The reverse cannot be said for red.

- **Free pass** Don't know what to serve with a particular dish? When in doubt, serve sparkling wine, the most versatile of all food wines.

- **Matching the flavour** Do not try to match a wine with the main ingredient (meat, fish, or whatever); instead match to the strongest flavour, whether that is a sauce, garnish, spice, or what have you.

- **Saltiness in food** Salt reduces the perception of bitterness and astringency in wine, while it can increase sweetness or fruitiness. Saltiness is wine-friendly.

- **Sweetness in food** This reduces the perception of sweetness in wine, making it less fruity, while it increases sourness, bitterness, and astringency.

- **Acidity or tartness in food** Both reduce the perception of acidity in wine, while they increase bitterness and sweetness.

- **Umami in food** This increases the perception of bitterness in wine (which is why matching wine to Chinese takeaways loaded with MSG is so difficult).

- **Spiciness in food** Spiciness exaggerates tannins and bitterness and lessens fruitiness in wine (a squeeze of something citrus on a dish just before it is served can neutralise the negative effects of spiciness).

- **Rare or well-done** The rarer a steak is, the more tannin it can take, but that is not an excuse for harsh tannins.

- **Oaked with smoked** Whether fish, meat, cheese, or whatever, if you enjoy a particular wine with a certain food, you will find that an oak-fermented or oak-aged variant of that wine will best accompany a smoked version of the food in question.

LAMB

Although claret is as classic with lamb as it is with beef, Burgundy works at least as well, particularly when the meat is a little pink and the pinker it is, the more tannins in the wine it can take. It is well known in the wine trade that the flavour of lamb brings out every nuance of flavour in the finest of wines, which is why it is served more often than any other meat when a merchant is hosting a meal. As with beef, it is possible to drink almost any red wine, although this should, if anything, be a little lighter.

♔ **Classic:** *Cru Classé Bordeaux*
✓ **Budget choice:** *Inexpensive Pinot Noir*

OFFAL

Kidneys go with full, well-flavoured, but round wines, such as a mature Châteauneuf-du-Pape or Rioja. A Graves with some bottle age would be a good choice for white wine drinkers. Lamb and calf livers go well with a Syrah that is very good, but not too heavy, while chicken livers require something richer, such as Gigondas or Zinfandel. Pig and ox livers are much coarser in texture and flavour, requiring a full, robust, but not too fussy red. Either red or dry-to-medium-dry white wine may be served with sweetbreads, which are rich yet delicate. Lamb and calf sweetbreads take well to fine Saint-Émilion or Saint-Julien, if in a sauce, or a good white Burgundy if pan-fried.

♔ **Classic:** *Châteauneuf-du-Pape*
✓ **Budget choice:** *Gigondas*

PORK, POULTRY, AND VEAL

These meats are capable of accompanying a diverse range of wines. They combine beautifully with wines of all colours and many styles. Dry white wines, whether light or full, rich and oak-fermented, or crisp and mineral go equally well, as do reds, from light and assertive through rich to soft. Burgundy is perfect, from Mâcon to Côte de Beaune or Côte Chalonnaise to Côtes de Nuits. Montrachet, the king of white wines, is the perfect indulgence with belly-pork, the cheapest of cuts, when it is slow-cooked and served with crackling. Any decent Pinot Noir or Chardonnay will do, Old World or New. Champagne is brilliant, yet so is a good Lambrusco. Riesling is another perfect match. For chops, cutlets, or escalopes, grilled, pan-fried, or in a cream sauce, it is advisable to choose something with a higher acidity balance or some effervescence. Beaujolais is perhaps the best choice for roast pork, when served cold.

♔ **Classic:** *Montrachet*
✓ **Budget choice:** *dry of off-dry Lambrusco*

REGIONAL CUISINES

CHINESE DISHES

Much depends upon whether the dishes are from one of the authentic Chinese cuisines (such as Anhui, Fujian, Guangdong, Hunan, Jiangsu, Shandong, Sichuan, and Zhejiang) or American-Chinese, French-Chinese, Vietnam-Chinese, or whatever, but a fine sparkling wine is the safest bet with most dishes. Ginger is a common ingredient in many dishes, and that always repays a fine Alsace Gewürztraminer, while a good-quality German Riesling Kabinett is also generally very useful, particularly when partnering dishes in Black Bean or Oyster sauces. Spare ribs require a *Spätlese*, but anything from Westernised Chinese cuisine, especially sweet and sour, might be loaded with monosodium glutamate (MSG) and will be better with water. Where chillies and other hotter spices dominate, an Alsace Gewürztraminer with some natural sweetness will work. For duck or goose dishes, try a good-quality Vouvray *demi-sec*, still or sparkling. If delicate or relatively bland, nutty ingredients dominate, such as water chestnuts, bamboo shoots, or cashew nuts, a soft white such as a light Australian Chardonnay might suffice. The best match is probably spring rolls with champagne.

♔ **Classic:** *Champagne*
✓ **Budget choice:** *Lager or iced water*

INDIAN DISHES

Indian cuisines are also multifarious, but the dishes share a softness in texture when compared to other Asian cuisines, which are often crispy, yet Champagne or any other fine-quality sparkling wine is also the best all-rounder here. A light and slightly tannic red wine can go well with a number of Indian dishes, such as chicken tikka, korma, pasanda, tandoori, and even rogan josh, but you need something fresh and white, such as a Côtes de Gascogne or one of the lighter New Zealand Sauvignon Blancs, to accompany a vegetable tikka or Madras curry and something fruitier, like German Riesling Kabinett for a vindaloo.

♔ **Classic:** *Champagne*
✓ **Budget choice:** *Lager or Iced water*

JAPANESE DISHES

Sake is, of course, the classic accompaniment to almost all Japanese food, and because it is so undervalued, it is also the budget choice. Champagne or any fine sparkling wine will be perfect with sushi, nigiri, maki, and chirashi, as will be a crisp, dry white of notable minerality. The same goes for dashi, miso, butajiru, kenchin-jiru, noppei-jiru, and other soups and broths. Sauvignon Blanc is a better choice for the seaweed wrapped sashimi, but Riesling works too, particularly a Mosel Kabinett. Kasujiru demands sake, of course.

♔ **Classic:** *Sake*
✓ **Budget choice:** *Sake*

MEXICAN DISHES

True Mexican cuisine evolved from Spanish cuisine, but was restricted to local products like corn, beans (*frijoles de olla*), and spices. It is very innovative and continues to change. When in doubt, try Champagne or any other fine sparkling wines. Try Lambrusco, a fruity rosé, or a *crémant* with various tortillas, tacos, and burritos; Carmenère with barbecued, spiced-meat dishes and chorizo; fullish rosé with masa dough dishes; Grüner Veltliner with empanadas and pico de gallo salads; Sauvignon Blanc with chiles rellenos; and Cava with various Mexican rice dishes. Where sauces dominate, Sauvignon Blanc with green chilli and tomatillo salsa; Lambrusco with red chile and ranchero; Albariño with guacamole; and a Palo Cortado with molé.

♔ **Classic:** *Champagne*
✓ **Budget choice:** *Lambrusco*

THAI DISHES

Much spicier than Chinese, Thai's more intricate mix of flavours is often highlighted by lemongrass, basil, and fresh chili. Whitbier is a clever choice, but for the very hottest chilli-charged dishes, forget beer or wine, and stick to iced water. With low-key chilli, you can get away with a Marlborough Sauvignon Blanc or an English sparkling wine. Choose something fresh, but softer, such as Colombard, Franciacorta, or Cava for dishes that include coconut milk. When lemon grass, lime, and other zesty ingredients dominate, an un-oaked Australian Sémillon will be a wise choice.

♔ **Classic:** *Whitbier*
✓ **Budget choice:** *Iced water*

DESSERTS

CAKES AND GATEAUX

Many cakes, sponges, and gâteaux do not require wine, but Verdelho with Madeira cake is a Victorian classic. Banyuls enhances coffee-flavoured cakes, while Samos and various other fortified Muscats are very good when almonds or walnuts are present. A sweet sparkling Vouvray will go well with fruit-filled, fresh-cream gâteaux and fruit-flavoured cheese-cakes. Iberian Moscatels and Australian fortified wines are superb with Christmas or plum pudding, while Asti is ideal with mince pies and generally ends a formal afternoon tea better than Champagne. Pairing wine with chocolate is an age-old problem, and the darker and richer the chocolate is, the more difficult the quest becomes. Some people believe Banyuls and chocolate to be the perfect combination, and others consider that Maury, Port, Recioto della Valpolicella, and even Brachetto d'Acqui can work too, but the truth is that although the most successful accompaniment does not clash, it does not work any wonders either. For chocolate, stick to iced water.

♔ **Classic:** *Verdelho*
✓ **Budget choice:** *Iced water*

CRÈME BRULÉE, CRÈME CARAMEL

Something rich, sweet, and luxurious is required to accompany crème brulée or crème caramel. A botrytised Riesling would be too tangy, but a top Tokaji, Sauternes, or Samos Nectar would be excellent, as would a Pinot Gris SGN or a great Malmsey.

♔ **Classic:** *Cru Classé Sauternes*
✓ **Budget choice:** *Australian Liqueur Muscat*

FRESH FRUIT

A fine *demi-sec* sparkling wine is a good all-round choice, as indeed is water, but a late-harvest Riesling is the ideal partner to a fresh, plump, and juicy peach. Asti, Californian Muscat Canelli, and a well-selected Clairette de Die may also partner peaches, especially if served with strawberries and raspberries. A rich, fresh, and sweet Australian sparkling Shiraz or Quady Elysium Black Muscat works for red fruits alone and a fine, sweet Lambrusco for cherries.

♛ **Classic:** *Rheingau Auslese*
✓ **Budget choice:** *Iced water*

FRUIT TARTS AND PASTRIES

Coteaux du Layon with an apple tart or pie is as good as it gets. It also makes a good choice for peach, pear, plum, or banana pies, tarts, and other pastries. A botrytised Riesling is the preferred choice, however, for peach tarts or pastries, and a fresh Muscat de Beaumes de Venise has a lighter touch with more delicately flavoured pear-based pastries, while an Alsace Pinot Gris VT is the preference for plum, and a Barsac for banana. If any of the dishes include a caramelised ingredient, opt for a noticeably cask-fermented version of the appropriate wine.

♛ **Classic:** *Coteaux du Layon*
✓ **Budget choice:** *Samos*

ICE CREAM

The absolute classic accompaniment to ice cream is, of course, PX (Pedro Ximénez Sherry), but whether you should drink it or pour it over the desert like a sauce is a matter of personal preference. When ice cream is just part of a dessert, it is the other ingredients that should be the focus of what the wine choice should be. Ice cream on its own rarely calls for any accompaniment, and even iced water rarely works because it can seem warm by comparison. Muscat de Beaumes de Venise with Victorian Brown Bread Ice Cream is a partnership made in heaven.

♛ **Classic:** *PX*
✓ **Budget choice:** *Moscatel de Valencia*

MERINGUE-BASED DISHES

For meringue served as part of a fruit Vacherin or Pavlova, try a still or sparkling Moscato, Californian Muscat Canelli, a botrytised Riesling, and, particularly, a top Mosel *Beerenauslese* (because the fruit needs the acidity and the meringue needs the sweetness). The fruitiness of Vouvray *moelleux* would work better than a good Sauternes. For meringue desserts with nutty, coconutty, or biscuity ingredients, a Pinot Gris d'Alsace SGN, Torcolato from Veneto, Malmsey Madeira, or Samos Grand Cru would be equally successful, but with caramel, salted or otherwise, it would need Samos Nectar or a mature Malmsey. A Champagne *demi-sec* is classic with lightly poached meringue served as floating islands or snow eggs, but a *demi-sec* Crémant de Luxembourg works at least as well, if not better. An upper-level Mosel *Auslese* is the perfect partner to lemon meringue pie – a dessert that demands luxury, acidity, and a vibrant sweetness, but a lighter Coteaux du Layon would be preferable for a key lime pie.

♛ **Classic:** *Mosel* Auslese
✓ **Budget choice:** *Moscatel de Valencia*

MOUSSES, POSSETS, AND OTHER CREAMS

With vanilla, coffee, or nut-flavoured creams, a Muscat de Beaumes de Venise or Bual will work well. Most fruit-based creams need Riesling, although not too botrytised, but an authentic lemon posset (a cold-set dessert made from thickened cream and flavored with lemon) is simply too acid, even for Riesling, so don't try, because the posset itself will be the palate cleanser.

♛ **Classic:** *Bual*
✓ **Budget choice:** *Iced water*

CHEESES AND CHEESE DISHES

BLUE-VEINED CHEESES

A good blue cheese is best partnered by a sweet wine, which will impart a piquant sensation not dissimilar to that found in sweet and sour dishes. Many dessert wines will suffice. Hard blues, such as Stilton and Blue Cheshire, go best with Port.

♛ **Classic:** *Port*
✓ **Budget choice:** *Moscatel de Valencia*

SOFT AND SEMI-SOFT MILD CHEESES

Fine, *brut*-style sparkling wines and dry white wines, mineral or fruity, have a much greater affinity for soft cheeses, such as an Italian Bel Paese, French Port Salut, or Dutch Gouda, than do reds.

♛ **Classic:** *Champagne*
✓ **Budget choice:** *Crémant du Luxembourg*

SOFT AND SEMI-SOFT STRONG CHEESES (COW)

A semi-soft Munster demands a strong Gewürztraminer d'Alsace and the most decadent way to wash down a perfectly ripe soft cheese like Brie de Meaux or Brie de Melun is with a 20-year-old vintage Champagne. Washed-skin cheeses (that have been bathed in water, brine, or alcohol while ripening) are best with a rich white Burgundy but can also handle a Gewürztraminer d'Alsace.

♛ **Classic:** *Gewürztraminer d'Alsace*
✓ **Budget choice:** *Inexpensive sparkling wine*

HARD CHEESES (COW)

Hard cheeses can take a red wine, certainly more easily than either soft or semi-soft cheeses can, but they make an infinitely better match with dry white wine, either still or sparkling, from myriad grape varieties. Dry white wines are ideal with Caerphilly, for example, while off-dry styles are perfect with Wensleydale served with a slice of homemade apple pie. Mature Cheddar and other well-flavoured, hard cheeses, such as Parmesan or Manchego, will well repay everything from crisp, dry mineral whites and Gewürztraminer d'Alsace to Champagne.

♛ **Classic:** *Champagne*
✓ **Budget choice:** *Mâcon Villages*

GOAT AND SHEEP CHEESES

These little cheeses require an assertive, dry white wine of notable minerality, such as Sauvignon Blanc or Assyrtiko. Most so-called hard goat cheese is springy, not truly hard, but rarely encountered, authentically hard goat cheese, such as Penard Ridge, can be accompanied by almost any dry white wine, still or sparkling.

♛ **Classic:** *Champagne*
✓ **Budget choice:** *Albariño*

CHEESE FONDUE

It is possible to drink a wide range of dry white and sparkling wines with fondue, but it often pays to serve a wine from the same area of origin as the dish, thus Fendant from the Valais with a Swiss fondue, or Apremont or Crépy with fondue Savoyarde.

♛ **Classic:** *Fendant*
✓ **Budget choice:** *Côte de Gascogne*

CHEESE SOUFFLÉ

A cheese soufflé requires a good sparkling wine, preferably Champagne. If it is a very rich soufflé, such as soufflé Roquefort, then the wine must have the power to match it, such as a *blanc de noirs*.

♛ **Classic:** *Champagne*
✓ **Budget choice:** *Inexpensive sparkling wine*

MADEIRA WINE COMPLEMENTS A PORTUGUESE *BOLO DEL MEL* CAKE
When travelling, it never hurts to try out regional specialties. Here in the town of Funchal, diners are served the local wine with *bolo del mel*, a honey and nut cake from the Madeira Islands.

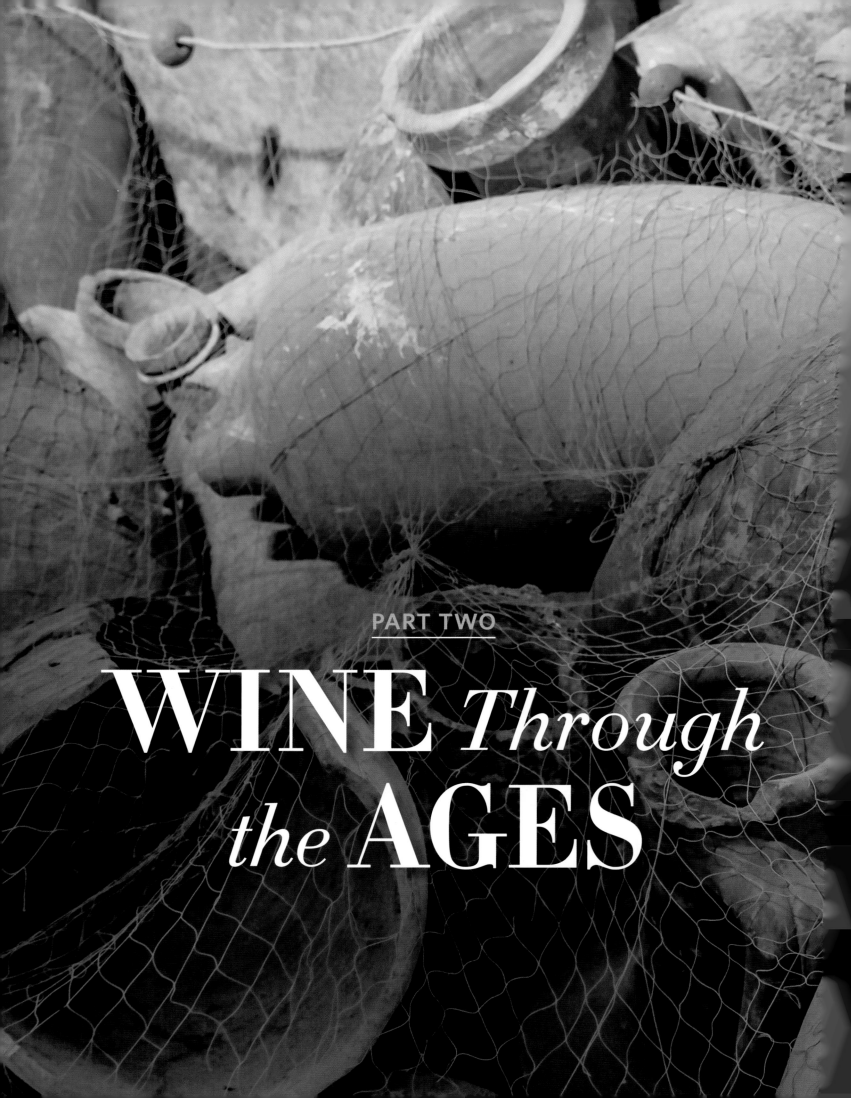

PART TWO

WINE *Through* the AGES

A Chronology *of* Wine

In a work that casts its net as far and wide
as this book, it would be impossible to provide
anything approaching a comprehensive historical
coverage without doubling its size and price.
Where space is limited, the usual solution is to
opt for a very selective history, choosing just two
or three pivotal topics to summarize. But when an
entire book devoted to the history of any subject
ends up being a compromise, such an approach
becomes so eclectic as to be meaningless.
The chronological format is also a compromise –
indeed, it is by definition a compromise – but
that is its advantage. It was conceived to be a
compromise. Instead of choosing two or three
pivotal topics to summarize, why not try to cover
as many topics as possible in as few words as
possible? A Chronology of Wine is the sort of
history that anyone would want to dip in and out
of, and by interspersing important historical
events with the odd bizarre wine-related fact,
I hope I have made it a fun read, too. Just as
the Micropedia has taken more than two decades
to evolve from its more modest origin as a
glossary of tasting and technical
terms, so it is expected that this chronology
will grow over forthcoming revisions.

**EPIPHANY OF DIONYSUS MOSAIC DEPICTS THE GREEK GOD
OF THE GRAPE HARVEST, WINEMAKING, AND WINE**
Uncovered in the the Villa of Dionysus (2nd century AD) in Dion, Greece,
the mosaic is one of many tributes to Dionysus (or Dionysos) produced in
the ancient world. Ancient Greeks saw wine (and the vines and grapes from
which it came) as a gift from the gods. The cult of Dionysus, rather than
portraying him as a god of drunkenness, saw him as centred on correct
consumption, which could bring joy and ease suffering.

Notes:
1. Where dates are approximate periods or carbon dated, the oldest possible date is given.
2. Geographical locations preceding events are the modern-day equivalents.

500 million BC	A distant relative of *Vitis vinifera* is speculated to be *Ampelopsis,* a climbing vine that existed when Earth had only one land mass, the supercontinent known as Pangaea.
200 million BC	Postulated appearance of *Vitis vinifera sylvestris,* the Eurasian wild grape, bearing fruit that would be black or dark red. All wine grapes we know today are descended from this single wild-grape variety. White grapes are highly unlikely at this juncture.
50 million BC	The oldest fossilized remains of *Vitis vinifera sylvestris* date from this period.
2.5 million BC	Several pockets of *Vitis vinifera sylvestris* survive during the Quaternary Ice Age, including, it is believed, in an area somewhere between the Black Sea and southern Caspian Sea, from where modern viticulture eventually spreads.
200,000 BC	Postulated appearance of the first white mutations of *Vitis vinifera sylvestris.*
21,000 BC	**Israel:** Remains of *Vitis vinifera sylvestris* carbon dated to this period are found at the Ohalo II site on the southwest shore of the Sea of Galilee, in Israel's Rift Valley.
8500 BC	**Armenia, Azerbaijan, Georgia, Turkey:** The first *vinifera* (*Vitis vinifera vinifera*) is domesticated by the earliest agricultural societies of Neolithic humans for fresh and dried grapes. • As the vine spreads through the Fertile Crescent (Mesopotamia and Levant) and beyond, it adapts to local conditions, from which distinctly separate grape varieties will emerge.
7000 BC	**China:** Residue of the oldest-known alcoholic product in carbon-dated storage jar found at Jiahu, a Neolithic village in the Yellow River Valley, fermented from rice, honey, and either grapes or hawthorn, possibly both.
6700 BC	**Middle East:** Desertification commences due to changes in the Earth's orbit and a tilt of its axis. Prior to this, the lands into which *vinifera* spread were green and lush.
6000 BC	**Georgia:** Wine residue in a carbon-dated storage jar found at Shulveri was the oldest-known wine in 2003.
5400 BC	**Iran:** Wine residue in a carbon-dated storage jar found at Hajji Firuz was the oldest-known wine in 1995.
5000 BC	**Egypt:** Evidence for the domestication of grapevine from tablets and papyri and Egyptian tombs
4000 BC	**Armenia:** In January 2011 the world's oldest "winery" was discovered in the cave complex at the Areni-1 archaeological site. A shallow trough about 1 metre (3 feet) across had been positioned to drain into a deep vat. This is thought to be a *lagar* or treading grapes. *Vinifera* grape seeds, remains of pressed *vinifera* grapes, fermentation jars, a cup, and a drinking bowl were also discovered inside the cave. **Lebanon:** Estimated arrival of viticulture in the northern part of Canaan that is now Lebanon. Wine plays an important role in the culture of both the Canaanites and the Phoenicians, their successors.
3150 BC	**Egypt:** More than 700 wine jars buried in the U-j tomb of Scorpion I, the earliest-known king of Egypt's Protodynastic Period, at Abydos. The wine is laced with pine resin, figs, and various herbs, including balm, coriander, mint, and sage.
3100 BC	**Egypt:** The word for wine exists in ancient Egyptian as *irp* (written using the eye symbol "ir" followed by the mat symbol "p"). • The world's earliest "wine labels" are stamped into clay on sealed wine jars later found in the tomb of King Den at Abydos, certifying that the wine they contained was produced at a vineyard dedicated to Horus. • By this time, wine jars are standardized in shape and volume (10, 20, and 30 litres), sealed with a wet-clay stopper that is spread down over the shoulder of the vessel to create a tight seal. • Tomb paintings show vines grown in raised troughs to avoid wasting precious irrigation resources and trained into arbours for easy harvesting.
3000 BC	**Iran:** Wine residue in a carbon-dated storage jar found at Godin Tepi was the oldest-known wine in 1991.
2550 BC	**Egypt:** A wine known as Chassut Red produced by vintner Sekem-Ka is reputed not to be ready to drink until it has aged 100 years.
2500 BC	**Egypt:** Illustration in a tomb at Saqqara depicts an early Egyptian bag press being squeezed by twisting poles. • Estimated earliest use of cork stoppers.

2400 BC	**Egypt:** Various mosaics from the Fourth Dynasty show grapes and wine production.
2137 BC	**China:** The earliest-recorded eclipse occurs, according to *The Shu King.* This volume also documents the execution of two royal astronomers, Hsi and Ho, for failing to predict the eclipse because they had drunk too much rice wine.
2000 BC	**Greece:** Viticulture is introduced by Phoenician and Egyptian traders to the Cycladic and Minoan (Cretan) civilizations of ancient Greece.
1600 BC	**Iran:** Glassmaking emerges in Mesopotamia.
1520 BC	**Egypt:** Painted scenes in the tomb of Amenhotep I depict the entire winemaking process, from harvesting and treading the grapes, through fermentation and storage in sealed jars with a hole drilled in the top (obviously to facilitate the escape of carbon dioxide, since these holes are later stopped with clay), which are then carried individually in net slings and transported by boat.
1450 BC	**Egypt:** Tomb illustration of wine being siphoned off its sediment and blended in preparation for, it is supposed, a banquet, though it could equally be for rebottling into fresh wine jars.
1345 BC	**Egypt:** Of the 26 wines later buried with Tutankhamun, five are from this vintage, including a "sweet wine" from the House of Aton on the Western River, produced by chief vintner Aperershop.
1344 BC	**Egypt:** The first white wine is made near Alexandria, its residue later found in one of the amphorae in Tutankhamun's tomb.
1330 BC	**Egypt:** Paintings in the tomb of Parennefer, adviser to Akhenaten, illustrate a primitive but effective form of air conditioning, because wine jars are stored in wet sand and fanned to keep cool. Although the reason for this practice is unknown, it would have slowed the fermentation, resulting in fruitier wines.
1200 BC	**Greece:** Greek viticulture is at its peak. Vines are trained in parallel rows, just as they are today, with care taken to ensure proper spacing between each vine. At least six different methods of pruning and training are employed, the choice depending on the variety of grape, type of soil, and wind strength. The classic wines of ancient Greece were great indeed, relative to their era, and worthy of note in the writings of Hippocrates, Homer, Plato, Pliny, Virgil, and many others.
1153 BC	**Egypt:** Rameses III is buried with more than 20,000 jars of wine. He had previously listed 513 vineyards as belonging to the temple of Amun-Ra.
1000 BC	**Italy:** Viticulture is introduced to southern Italy by the Greeks.
800 BC	**France, Portugal, Spain:** Viticulture is introduced by Phoenician traders.
750 BC	**Tunisia:** Viticulture and winemaking are believed to have commenced here when Phoenician traders founded the city of Carthage, which Greco-Roman legend puts at 814 BC; however, no Phoenician artefact older than 750 BC has been found west of Phoenicia's Levant origins.
700 BC	**Spain:** Archeological evidence found of Phoenician winemaking in Valdepeñas
650 BC	**Greece:** It is customary to drink wine mixed with water. Under the first Greek laws to be codified, Zaleucus decrees that anyone caught drinking undiluted wine shall be put to death.
620 BC	**Greece:** Under Draco's "Draconian law", anyone caught stealing grapes will face the death penalty.
600 BC	**Bulgaria, Moldova, Romania, Ukraine:** Vineyards are established by the Greeks. **France:** The Phocaeans (Greeks from the Ionian city of Phocaea in Anatolia, now Turkey) establish the port of Marseilles and introduce viticulture to the Gauls (ancient French).
500 BC	**Algeria, Morocco:** Viticulture probably established by the Carthaginians (Phoenician descendants). **France, Portugal, Spain:** Viticulture has been introduced by the Romans by this date. Although some vines were planted much earlier by Phoenicians, it was the Romans who established vineyards farther inland and started flourishing wine industries in most of Europe's classic wine regions. Initially, this was not for commercial purposes but merely to sustain the Roman army.
450 BC	**Greece:** The world's first wine laws regulating quality are introduced.
425 BC	**Greece:** The world's first wine law setting the earliest harvesting date is introduced.
424 BC	**Greece:** Herodotus, the world's first true historian, writes about casks of "palm wood filled with wine" being moved by boat to Babylon, invoking the possibility that the art of cooperage might have been invented by the ancient Mesopotamians, not the Celts as is generally accepted.

400 BC	**Spain:** Viticulture established on Ibiza by the Carthaginians (Phoenician descendants).
340 BC	**Greece:** Aristotle describes the "Black Wine" of Lesbos as tasting of oregano and thyme.
300 BC	**Portugal, Spain:** Viticulture is introduced here at Tarragona, Rioja, Duero, and Douro.
250 BC	**Iran, Syria:** Glassblowing is invented.
128 BC	**China:** *Vinifera* vines are introduced by General Chang, who plants seeds at the Imperial Palace in Chang An (now called Xian).
121 BC	**Italy:** This legendary Roman vintage is named the Opimian vintage, after Lucius Opimius, Rome's consul at the time, and is noted for a large harvest of exceptionally high quality, now believed to be due to the grapes shrivelling on the vine. Falernian is the greatest wine of this vintage, and although there are several documented examples of Falernian Opimian being drinkable at the age of 100 years or more, this would be contested by Cicero, who in 46 BC declared that, at 75 years old, "the Opimian is already too old to drink".
70 BC	**Italy:** An ancient Roman wine called Caecuban and produced on the Lazio coast was considered to be a First Growth, thus on a par with Falernum and Setinum.
50 BC	**France:** With all of Gaul under Roman occupation, the Romans discover the barrel and the art of cooperage, a Celtic invention (but see 424 BC).
19 BC	**Israel:** A large Roman two-handled wine amphora is sent from Italy to King Herod. It is found in an ancient garbage dump near the synagogue of Masada in 1996.
AD 43	**England:** The Romans brought the vine to England, and by this date every important villa had its own vineyard.
AD 65	**France, Italy, Spain:** The 12-book set *De Re Rustica* is completed by Lucius Junius Columella; it is the most comprehensive documented record of Roman viticulture that has survived to this day. Columella tells us that "many vineyards" yielded as much as 15 *cullei* per *jugera* (300 hl/ha), although some modern references give the impression that only 3 *cullei* per *jugera* (60 hl/ha) were the norm, but that was the yield at which Columella considered a vineyard to be unprofitable and recommended uprooting. Columella indicates that there were already many distinctly different varieties of grape and demonstrates that there was a keen and widespread sense of matching the most appropriate vine to a given *terroir* among farmers, who "assign to a fat and fertile land a vine that is slender and not too productive by nature; to lean land, a prolific vine; to heavy soil, a vigorous vine that puts forth much wood and foliage; to loose and rich soil, one that has few canes". Farmers were also aware of how to avoid rot: "He will know that it is not proper to commit to a damp place a vine with thin-skinned fruit and exceptionally large fruit, but one with fruit that is tough-skinned, small, and full of seeds."

PAGE FROM A 1491 GERMAN EDITION OF *DE RE RUSTICA*
Columella's work, translated as *On Rural Affairs* or *On Farming*, covers early viticulture.

AD 79	**Italy:** Pompeii is laid to waste by the eruption of Vesuvius. The town is preserved forever, which is how we know it boasted more than 160 bars selling wine and food. With a mere 10,000 inhabitants and bars less than 10 metres (30 feet) apart in places, Pompeii demonstrates that dining out is not merely a modern trend.
100	**Italy:** The city of Rome swells to 1 million inhabitants and has to import wine from Gaul and Iberia, despite the expansion of Italian vineyards.
250	**France, Italy:** As the flow of wine from France to Rome increases, the Romans realize the advantages of wooden casks over amphorae for transporting wine. From this juncture, casks become the vessel of choice for wine, and cooperage spreads throughout the Roman Empire.
651	**France:** The first conclusive mention of vines growing in Champagne in a letter from Grimoald the Elder to the Benedictine Bishop Remaclus, revealing that King

651 (cont'd)	Sigebert III of Austrasia had given Remaclus a vineyard at Terune (now known as Terron-sur-Aisne).
674	**China:** Chinese wine is first documented (but *see* 128 BC).
760	**France:** Saint Fulrade is reputed to have brought the Pinot Noir grape to the village of St-Hippolyte from Italy (but *see* 1894).
802	**Western and Central Europe:** Charlemagne bans "leaded wine". The addition of lead was to make a sour wine taste sweet, a hangover from Roman times, when wines were sweetened with concentrate, and lead vessels were preferred for boiling down the concentrate because copper vessels left a bitter taste.
862	**Germany:** Staffelter Hof, originally an abbey and the oldest-known winery in the world, is established in Kröv.
1001	**United States:** The first native American vines are identified by a German called Tyrker, who accompanied Viking Leif Ericson on a voyage of discovery to Newfoundland, Canada, and Northeast America almost 500 years before Christopher Columbus. According to the *Grœnlendinga Saga*, a tale first documented in the *Flateyjarbók* manuscript between 1387 and 1394, Tyrker found what he described as *vinber*, or wine berries, having told Ericson, "I was born where there is no lack of either grapes or vines." Ericson gave Newfoundland the name "Vinland." This was written nearly 400 years after the alleged event, but it was more than 100 years before Columbus "discovered" America, and in 1075 – more than 400 years before Columbus – Adam of Bremen wrote in *Descriptio Insularum Aquilonis* that the name comes from the huge amount of grapes growing there. Furthermore, a Viking settlement has now been unearthed in Newfoundland - not conclusive but convincing.
1086	**England:** The *Domesday Book* records some 40 significantly sized vineyards in England.
1125	**France:** Count of Champagne Thibaut II or Thibaut the Great, the second-most powerful man in France, establishes the great Champagne Fairs, which spread the fame of Champagne's wine throughout the growing markets of Europe.
1141	**Italy:** The Ricasoli family produced the first known Chianti at Castello di Brolio.
1152	**England, France:** The marriage between king-to-be Henry II and Eleanor of Aquitaine starts a special relationship between Bordeaux and England that continues to this day.
1224	**Cyprus:** The poem "Battle of Wines", written by Henry d'Andeli, tells of the first wine competition, organised by King Philip Augustus of France (1165-1223), who appointed an English priest to taste more 70 wines from France, Spain, Germany, and Cyprus. The priest classifies 49 wines that pleased him as "Celebrated", 15 that displeased him as "Excommunicated", and the overall winner the supreme title of "Apostle" (the sweet Commandria from Cyprus). *See also* 800 BC.
1232	**France:** This is such a hot year in Alsace that it is possible to fry eggs on paving stones.
1255	**France:** The crop in Alsace is so huge that, after filling all available casks, excess wine had to be used for mixing mortar.
1268	**Italy:** The *"nibiol"* vineyards of the Rivoli estate in Turin are believed to be the first mention of Nebbiolo. *See also* 1292.
1292	**Italy:** First recorded spelling of Nebbiolo is documented in Alba.
1304	**Italy:** First mention of the Muscat grape is in agricultural treatise from Bologna under the Latin name *Muscatellus*.
1307	**England:** There are claims that cork was shipped to England as early as this, possibly directly to Edward II from his cousin, King Diniz of Portugal.
1315	**France:** It rains continuously between May and November in Alsace.
1330	**France:** The first documented mention of Chardonnay is by the Cistercian monks, who planted it (and no other variety) at Clos de Vougeot in Burgundy.
1335	**Italy:** A document written by Francesco Scacchi dating from this year is later claimed, in the 20th century, as proof that the world's first sparkling wine was made in Italy. The document turns out to be no more than a warning of the ill effects of a spoiled wine that has either not stopped fermenting or has accidentally started to re-ferment.
1375	**France:** The first documented mention of Pinot Noir is in Burgundy.
1381	**England:** Richard II places a toll of "two roundlets of wyne" on any ship passing the Tower of London. In the late-14th century, a *roundlet* or *rundlet* was 15 imperial gallons.

1385	**Italy:** Giovanni di Pietro Antinori of Florence branches out from his family's lucrative silk and wool business to become a winemaker.
1418	**Portugal:** The island of Madeira is discovered by João Gonçalves Zarco.
1435	**Germany:** The first Riesling vines to be documented are purchased by Klaus Kleinfisch, the administrator of Rüsselsheim Castle, which is on the opposite bank of the River Main to the villages of Hochheim and Florsheim in the Rheingau but, ironically, no longer makes any wine. Widespread claims that the "storage inventory of the counts of Katzenelnbogen lists the purchase of six barrels of Riesling from a Rüsselsheim vintner" are false.
1455	**Portugal:** Alves da Mosto writes about the vines on Madeira producing "more grapes than leaves" with "clusters of extraordinary size".
1458	**France:** Grapes in Alsace ripen by 21 May.
1470	**Germany:** French monks take the Pinot Noir to the Rheingau.
1477	**France:** The first varietal wine, Riesling (then spelled Rissling), appears in Alsace.
1478	**England:** According to Shakespeare, the Duke of Clarence was drowned in a barrel of Malmsey in the Tower of London. There is no evidence of Madeira being shipped to England as early as this (see 1537), although it was obviously well known by the Bard's time.
1484	**France:** The Alsace crop is so large that prices plummet, and 50 litres of wine cost as little as one egg.
1487	**Germany:** A Prussian royal decree requires wines to be preserved with sulphur burned on wood chips.
1490	**Spain:** Wine areas are extensively planted during the supposedly abstemious rule of the Moors.
1500	**France:** Muscat and Traminer are mentioned in Alsace.
1505	**France:** The greatest vintage of the 16th century.
1515	**Portugal, France:** The first Madeira – dry and unfortified – is shipped to Francis I in France.
1521	**Mexico:** The first vines on the American continent are planted by the Spanish.
1524	**Mexico:** Hernando Cortés, the governor of New Spain, orders every Spaniard holding a grant of land to plant annually 1,000 vines per each native forced labourer for a period of five years.
1531	**France:** Later claims that a document of this date detailing that Benedictine monks of St-Hilaire in Limoux produced intentionally sparkling wine are untrue (see 1544).
1535	**Canada:** French explorer Jacques Cartier discovers the first native Canadian vines when he finds a large island in the St Lawrence River that is totally overrun by wild vines (Île d'Orléans).
1537	**Portugal, England:** The first documented shipment of Madeira to England mentions a dry unfortified wine called "malvoisie of the Isle of Madeer".
1540	**Germany:** The "Great Sun Year" withered vines along the Rhine under relentless heat and drought.
1544	**France:** Upon later inspection, the document that many sources refer to as detailing that Benedictine monks of St-Hilaire in Limoux produced sparkling wine dates from this year, not 1531, and contains no mention of any effervescence, only that the bottles used were a different shape and the price the wine achieved was higher. Irrationally (but not unprofitably), this is what local Limoux producers have convinced themselves is proof of a sparkling wine.
1548	**Chile:** The first vines are planted by Spanish conquistador Francisco Pizarro.
1554	**Argentina:** The first vines are planted at Santiago del Estero by Father Juan Cidrón.
1560	**Spain:** The first Spanish wine laws are introduced (for Rioja).
1564	**Canada:** It is theorized that Jesuit missionaries would have planted vines upon arrival in Canada. **United States:** French Huguenots produce the first American wine from Scuppernong grapes at Fort Caroline, 21 kilometres (13 miles) northeast of what will become Jacksonville, Florida. Florida is, ironically, destined to become the least viticultural state in the country.
1585	**United States:** On his voyage to Roanoke, Virginia, Sir Walter Raleigh observes "Here were great store of great red grapis veri pleasant."
1587	**Spain, England:** Sherry shoots to fame in England following the daring raid on Cádiz by Sir Francis Drake, who sets alight the Spanish fleet and makes off with 3,000 casks of it.
1590	**Hungary:** A Tokaji from this vintage is possibly the earliest documented noble-rot wine, according to a report dated 1856 (but see 1606). **Italy:** The Sangiovese is first documented as "Sangiogheto" in Tuscany.
1593	**Mexico:** The first of two claims of being the oldest winery in the Americas. Marquis of Aguayo claims that Captain Francisco de Urdiñola founded the Urdiñola winery in 1593, four years before Casa Madero (see 1597). Apparently this winery in Parras, Coahuila, later became known as Del Rosario before assuming the name of Marquis of Aguayo. It is still in the same location. No verification, but its wines have 1593 prominently displayed on the label.
1595	**Mexico:** Philip II of Spain prohibits any further planting of vines in New Spain (the colonies) to protect exports of Spanish wines.
1597	**Mexico:** Casa Madero also claims to be the coldest winery in the Americas, having been established by a royal decree of Philip II dated 19 August 1597 to Don Lorenzo Garcia of the San Lorenzo Hacienda (I can verify the existence of this document). Casa Madero is still in existence today. See also 1593.
1599	**England:** Cork stoppers are used in wine bottles in England by this date at the latest, according to As You Like It, which Shakespeare is supposed to have written in 1599 or 1600, although it was not published until 1623 (see 1685).
1606	**Hungary:** The oldest-known existence of a noble-rot wine is a Tokaji from this vintage. A Warsaw wine merchant known as Fukier (called Fugger in the 17th century and now a famous restaurant) still possessed 328 bottles when the city was invaded in World War II. But when Warsaw was "liberated" by the Russians, they found only 300 bottles of 1668 and a thousand or so from the vintages of 1682, 1737, 1783, and 1811 (but see 1590).
1615	**England:** A ban on charcoal burning is introduced to preserve forests for the English fleet. Coal-fired glass is thus found to be significantly stronger than wood-fired glass, as favoured by the French.
1622	**Italy:** Earliest documented mention of intentional sparkling wine production in De Salubri Potu Dissertatio by Francesco Scacchi.
1623	**United States:** The first European vines in New Hampshire are planted at the mouth of the Piscataqua River by Ambrose Gibbons but did not survive, and it would be almost another 250 years before any are successfully grown in this state.
1626	**France:** The Trimbach family travel from Switzerland to Alsace, where they establish their world-famous winery.
1630	**Hungary:** Szepsi Laczkó Máté describes the method for producing Tokaji wine from botrytised grapes.
1631	**France:** The greatest vintage of the 17th century.
1638	**Portugal:** The first Port house is established: C N Kopke & Co.
1642	**United States:** The first vines in what will become New York State are planted by the Dutch.
1644	**France:** First mention of Pinot Gris (as Grauklevner) in Alsace.
circa 1650	**France:** Cabernet France and Sauvignon Blanc naturally cross in Bordeaux to create the Cabernet Sauvignon variety.
1655	**South Africa:** The first vines are planted by Van Riebeek.
1657	**United States:** The first attempt at Prohibition dates back this far, when the General Court of Massachusetts makes it illegal to sell strong liquor "whether known by the name of rumme, strong water, wine, brandy, etc".
1658	**United States:** In the last year of his life, Oliver Cromwell offers by an Act of Assembly 4,535 kilogrammes (10,000 pounds) of tobacco to whoever "shall first make two tunne of wine raised out of a vineyard made in this colony". This prize was never claimed, and it quietly dropped in 1685, when it was eventually realized that winegrowing would never succeed.
1659	**South Africa:** First wine made by Van Riebeek.
1662	**England:** Earliest description of what would eventually become known as the Méthode Champenoise, the first recorded use of a source of yeast added to induce a second fermentation, in Some Observations concerning the Ordering of Wines by Dr Christopher Merret, six years before Dom Pérignon takes up his post as winemaker at the abbey of Hautvillers in Champagne. **United States:** Charles Calvert, governor of Maryland, plants 100 hectares (240 acres) of native vines; the wine was reportedly "as good as the best Burgundy", encouraging him to plant a further 40 hectares (100 acres) three years later.
1663	**England, France:** Samuel Pepys drinks "Ho Bryan" (Château Haut-Brion), the first branded or single-estate wine documented since Roman times.

1668   **France:** Dom Pérignon takes up his post as cellarmaster at the abbey of Hautvillers.

1673   **England:** The world's first wine auction is held.

1676   **England:** Sparkling Champagne is so popular in England that it is mentioned in English literature (*The Man of Mode* by Sir George Etheridge) 42 years before it was documented in France and 53 years before the first Champagne house is established.
France: A Fiscal Order issued on 15 February makes it illegal to transport Champagne (which was not yet sparkling) in bottles.

1679   **England:** On 14 March Samuel Pepys takes a ride in his coach to Hyde Park for "the first time this year, taking two bottles of champagne in my way".
South Africa: Simon van der Stel complains about the "revolting sourness" of South African wines.

1680   **Netherlands:** Anton Leeuwenhoek is the first person to observe yeast under a microscope (*see* 1662 and 1857).
South Africa: Simon van der Stel plants 100,000 vines in Constantia.

1685   **France:** Dom Pérignon is believed to have brought cork from Spain.

1688   **South Africa:** French Huguenots emigrate to the Cape and establish vineyards in the Franschhoek Valley.

1699   **Americas:** Carlos II, King of Spain, bans the production of wine throughout his colonies, except when making wine for the Church.

1703   **England, Portugal:** The Methuen Treaty, a military and commercial treaty between the two countries, ensures the supply of Port and corks to England.

1709   **France:** Louis XIV orders the frozen vineyards of the Loire-Atlantique to be replanted with Melon Blanc (Muscadet).

1716   **Italy:** The country's first wine laws are introduced (Chianti).

1718   **France:** The first documented mention of sparkling Champagne in France refers to its initial appearance some 20 years earlier (*Manière de cultiver la Vigne et de faire le Vin de en Champagne*, written by an anonymous author believed to be Jean Godinot).

1724   **France:** The first claim is made that it was "Dom Pérignon who found the secret of making white sparkling wine".

1728   **France:** A decree of the Council of State dated 25 May revokes the Fiscal Order of 1676 banning the transport of Champagne in bottle, allowing the shipping of Champagne in basketed quantities of 50 or 100 bottles.

1729   **France:** Ruinart, the first Champagne house, is founded.

1750   **Spain:** The world's first cork-stopper factory opens in Anguine.

1761   **South Africa:** Constantia wines are exported to Europe.

1766   **England:** Wine from this date is included in the very first Christie's auction (*see* 1965 and 1966).

1768   **England:** The Great Vine (Black Hambourg, aka Schiava Grossa) is planted by Capability Brown at Hampton Court. By 2010 it fills an entire greenhouse from a single trunk and requires a large field next to the greenhouse to remain fallow to sustain its roots.

1769   **United States:** The first vines planted in California by Franciscan monks in San Diego are the so-called Mission grapes (Criolla), brought from Mexico, but there is no documented evidence of wine produced until 1782.

1775   **Germany:** The practice of late harvesting is born at Schloss Johannisberg following the delay of a messenger from the prince-bishop of Fulda. Since 1718 it had been customary not to start the harvest at Schloss Johannisberg until it was announced in writing by the prince-bishop, but the messenger sent out in 1775 was delayed for 14 days. When the messenger finally turned up, many of the grapes were rotten, and it was feared the wine would be ruined. Instead, it was so special that Schloss Johannisberg started to produce late-harvest (*Spätlese*) wines on a regular basis. Just three years later, Thomas Jefferson writes about how Schloss Johannisberg stands out: "It has none of the acid of Hochheim and other Rheniss grapes . . . 1775 is the best!"

1782   **United States:** California wine is first documented. It was produced at San Juan Capistrano by Fathers Pablo de Mugártegui and Gregorio Amurrió, from vines transported by Don José Camacho, commander of the supply ship *San Antonio*.

1783   **Germany:** The country's first sparkling wines are made in the Rhine (*see* 1791).

1786   **United States:** The first commercial vineyard and winery in America is established near Philadelphia by Frenchman Peter Legaux of the Pennsylvania Vine Company.

1788   **Australia:** The first vineyard is planted at Farm Cove in New South Wales by the first governor, Captain Arthur Phillip, with vines brought from Rio de Janeiro and the Cape of Good Hope. • Nine vines are planted by William Bligh of HMS *Bounty* at the eastern end of Adventure Bay on Bruny Island. This is much farther south than Tasmanian vines grow today; consequently, it is no surprise that they had not survived when Bligh returned in 1791 (*see* 1823).
South Africa: The dessert wines of Constantia achieve legendary status throughout Europe.

1789   **England:** Just 21 years after Capability Brown planted the Great Vine at Hampton Court, William Speechly in *Treatise on the Culture of the Vine* describes a vine "believed to have been planted the Carmelite friars prior to the dissolution, which would have made it 250 years old" at Northallerton in Yorkshire, significantly farther north than any commercial vineyard in existence today. He noted that the "circumference of the trunk or stem a little above the surface of the ground is three feet eleven inches". The last known photograph of this vine was taken in 1956, although it is thought to have lingered on until the 1970s. In 1956 it would have been well over 400 years old, making it at least as old as the current oldest vine in the world (in Maribor, Slovenia).

1790s   **United States:** President George Washington serves Champagne at dinners held in Philadelphia (then the nation's capital) before the Champagne houses start shipping to America.

1791   **Germany:** The first mention of a sparkling German wine is recorded in *Intelligenziablatt* by Johann Funcke, who states that sparkling wine has been made in the Rhine since 1783.

1793   **France:** The temperature recorded in Alsace on 21 June is -20°C (-4°F)

1794   **Germany:** The first recorded *Eiswein* is produced in Franconia, although both Pliny and Martial document wines produced from frozen grapes in Roman times.

1799   **Ukraine:** The country's first sparkling wine is supposedly produced by Peter Simon Pallas, a German professor of natural history, in the Crimean town of Sudak.

1801   **France:** Thanks to Jean-Antoine Chaptal, the amount of sugar added to wine to increase its potential alcoholic strength (chaptalization) or sparkling (as in Champagne) can now be accurately quantified.

1803   **Germany:** Napoleon conquers the Rhine region, taking its vineyards from the Church and dividing them between several families, who are told to produce the lightest, most flavourful white wines in the world.

1804   **United States:** First vines planted in Indiana, 12 years before Indiana becomes a State.

1806   **France:** One year after the death of her husband, the widow Clicquot invents the precursor to the *pupitre*, a kitchen table with holes cut in its top, and begins *remuage* (riddling) experiments.

1808   **United States:** The country's first temperance organization is formed. Although the Temperance Society of Moreau and Northumberland sought to eliminate the sale and drinking of spirits (except for medical purposes), in its earliest days it only wanted to temper the consumption of other less alcoholic beverages, not ban them – so they said.

1810   **France:** Veuve Clicquot starts recording in detail the blend of every Champagne produced.

1811   **Canada:** Johann Schiller establishes the first Canadian winery in the area of Mississauga now known as Cooksville, Ontario.

1813   **United States:** The Massachusetts Society for the Suppression of Intemperance seeks to discourage, not prohibit, "hard liquor" (spirits).

1816   **United States:** The first "dry legislation" is enacted, prohibiting Sunday sales of any form of alcohol in Indiana.

1817   **United States:** Governor Sola reports 53,687 grape-bearing vines growing in Los Angeles.

1818   **France:** Antoine Müller, the *chef de caves* at Champagne Veuve Clicquot, perfects *remuage* by taking the Grande Dame's kitchen table one huge step closer to being a fully fledged *pupitre* by cutting the holes at an angle of 45 degrees.

1820   **France:** The first sparkling Burgundy is produced by J Laussere of Nuits-St-George.

1822   **Australia:** The first Australian wine documented is made in New South Wales by Gregory Blaxland from vines planted at Ermington, a few kilometres down river from Captain Phillip's second vineyard (*see* 1788).
France: Louis Pasteur is born in the small wine village of Dole in the Jura region (*see* 1857 and 1864).

1822
(cont'd)

South Africa: Of just over 22 million vines growing in the Cape vineyards, 21 million are the "green grape", aka Sémillon.

United States: Postulated earliest date by which George Gibbs could have imported Zinfandel vines from the Schönbrunn collection in Vienna, Austria, which logically should have included Crljenak Kastelanski (Zinfandel) because Schönbrunn was supposed to have housed every variety grown in the Austro-Hungarian Empire. Many of the vines received by Gibbs were not named, though, and Crljenak Kastelanski was not amongst those that were.

1823

Australia: The first vineyard in Tasmania is planted by Bartholomew Broughton at Prospect Farm (but *see* 1788 and 1956).

1824

Australia: In this year, some believe, James Busby plants the first Hunter Valley vineyard at Kirkton, run by his brother-in-law William Kelman. Other sources suggest the vineyard is established in 1825 or 1830 (the Kelman Winery celebrated its 100th anniversary in 1930). Certainly, Busby arrived with cuttings in 1824, and there is no record of their being lost or destroyed.

France: The first precise method of measurement of alcohol is developed by Joseph-Louis Gay-Lussac.

1825

Slovakia: The country's first sparkling wine is produced by Messrs Hubert and Habermann near Pozsony (today Bratislava), then part of Hungary.

United States: Traders working for the Hudson's Bay Company plant the first vines in what will become Washington State at Fort Vancouver on the Columbia River.

1826

Germany: George Kessler at Esslingen is generally accepted as the first German sparkling-wine producer (but *see* 1783 and 1791).

1827

France: In *Oenologie Française*, Jean-Alexandre Cavoleau claims that Hermitage is purchased by the Bordeaux wine trade to blend with its own wines. • Champagne corking machine invented.

1828

England: *The Operative Chemist* by Samuel Frederick Gray contains the earliest mention of oak chips: "One of the most common ingredients used for flavouring wines is oak chips; and from this the wretched Lisbon wines acquire the little taste they have."

1829

Australia: The first vineyard in Western Australia is planted by Thomas Waters, a veteran winemaker from South Africa, at Olive Farm in the Swan Valley.

Switzerland: The country's first sparkling wine is produced by Louis-Edouard Mauler at the Abbey of St-Pierre in Môtiers, Neufchâtel.

1830

Australia: The first Cabernet Sauvignon, Riesling, and Verdelho in Australia are planted by William Macarthur at Camden Park near Penrith in New South Wales.

United States: The "Black Zinfardel" catalogued by nurseryman William Prince, owner of the Linnaean Botanic Gardens on Long Island, is presumably Zinfandel, although the etymological origin of Zinfandel is unknown. (It could have been confused in cataloguing with Zierfandler, a white Austrian variety.) The "Black Zinfardel" is later sold as Zinfindal and Zinfendel (*see* 1846).

1833

United States: The first commercial winery in California is established by *bordelais* Jean-Louis Vignes.

1834

Australia: The first vineyard in Victoria is planted by William Ryrie in the Yarra Valley.

France: Phylloxera is found on European oak trees by French entomologist Boyer de Fonscolombe, who classifies and names the genus, but the deadly native American *Phylloxera vastatrix* (hereafter referred to in this chronology simply as phylloxera) is not yet identified.

1836

Australia: The first vineyard in South Australia is planted by John Barton Hack at Chichester Gardens, North Adelaide.

France: Earliest use of the most basic *liqueur de tirage* to induce an intentional second fermentation, 174 years after Dr Christopher Merret recorded its use in London (*Traité sur le Travail des Vins Blancs Mousseux* by Jean-Baptiste François)

United States: The first vines in Napa are planted by George Yount (who gave his name to Yountville) in a vineyard he named Napanook, now part of Dominus Estate.

1838

United States: Tennessee becomes the first state to prohibit alcohol.

1841

Australia: The first vines in the Barossa are planted by Joseph Gilbert at Pewsey Vale.

France: Adolphe Jacquesson invents wire cages to replace string *ficelage* for securing Champagne corks. This invention was patented on 5 July 1844 and later adapted by René Lebegue, whose idea it was to fit the wire cage with a ring that untwists for uncorking.

1842

Austria: The first Austrian sparkling wine is produced by Robert Schlumberger in Vienna.

England: The original screw-type Kilner jar, the earliest precursor to the screwcap, is invented by John Kilner of Yorkshire.

United States: The first American sparkling wine is produced by Nicholas Longworth in Cincinnati.

1843

Australia: The country's first sparkling wine is produced by James King Irrawing in the Hunter Valley (*see also* 1883 and 1920).

United States: Portland, Maine, is the first American city to go dry.

1846

United States: The first documented use of the Zinfandel spelling is by viticulturist J F Allen.

1851

Spain: The first Spanish sparkling wine is produced by Antoni Gali Comas.

United States: Maine is the first American state to go dry. The Maine Law, as it becomes known, is the model for other states to outlaw the sale of alcohol.

1852

United States: This is when, according to Arpad Haraszthy, Zinfandel was introduced to California by his father, the legendary Ágoston Haraszthy, who first planted it at Las Flores, south of San Francisco, and took it with him wherever he went. Arpad wrote this account 30 years after the event, however, while just six years after, in 1858, Ágoston failed to mention Zinfandel in his famous *Report on Grapes and Wine of California*. Nor did he mention the variety later that year, when he wrote a long letter to the State Agricultural Society describing his activities at Buena Vista Farm (*see* 1866). • Those who debunk Haraszthy consider that either Captain Frederick W Macondray or William Robert Prince was the first to bring Zinfandel to California. Certainly Macondray was one of two exhibitors of Zinfandel at the Mechanics' Institute Fair in San Francisco in 1857, and the vines had to have been established at least two years prior to that. Interestingly, Macondray and Haraszthy were neighbours.

1853

France: English Baron Nathaniel de Rothschild purchases Château Mouton in Bordeaux for 1.125 million gold francs.

1854

United States: The first vines are planted in Oregon in the Rogue River Valley.

1855

France: Bordeaux's famed classification is first published, developed specifically for the very first Universal Exposition. It was based on price, not on reputation or any other preconceived quality.

1856

United States: The boom in the economy and population created by the Gold Rush of the previous 10 years fuels demand for wine and provides both the capital and the entrepreneurs to kick-start the California wine industry, increasing the number of vines grown from 1.5 million in 1856 to 8 million in 1862. • San Francisco vintners start exporting California wines to England, Germany, Russia, and China.

1857

France: Louis Pasteur discovers the role of yeast in the fermentation process (but *see* 1680 Netherlands).

United States: Ágoston Haraszthy establishes Buena Vista Winery in Sonoma, California, planting 165 different varieties of vine imported from Europe.

1858

United States: It is speculated that phylloxera reached California by this date (but *see* 1873). • John L Mason patents what could be described as an American version of the Kilner jar (*see* 1842).

1860

France: Before anyone has ever heard about phylloxera, Léo Laliman, a viticulturist in Bordeaux, recommends grafting classic *vinifera* vines on to American rootstock as a protection against oidium (powdery mildew).

ÁGOSTON HARASZTHY
The "Father of California Viticulture"

1861

England: Britain introduces the Single Bottle Act allowing small grocer shops to sell wine by the bottle.

United States: Mary Todd Lincoln, the wife of Abraham Lincoln, starts the practice of serving American wines at the White House. • Samuel A. Whitney patented a bottle with an internal neck thread and a threaded glass-stopper.

1862

France: Unknown to anyone at the time, phylloxera-infested vines are planted in a *clos* in Roquemaure (now part of AOC Lirac) in the Rhône (*see* 1874 Veuve Borty's confession).

1863

England: Phylloxera is identified at Kew Gardens on vines imported from America.

1864

France: Louis Pasteur discovers microbial growths in wine and invents pasteurization as a means of combating spoiled wine.

1866 **France:** Louis Pasteur is the first to recognise the importance of tannins in the maturation of red wine.
Germany: Phylloxera discovered in the Botanische Gärten, Bonn (*see also* 1881).
United States: The first documented evidence of Zinfandel as a wine comes from Thomas Hart Hyatt, the editor of *California Rural Home Journal,* who describes a Buena Vista wine made "mostly from Mission, Zinfindal [sic], and Black St Peter" (*see* 1852).

1868 **Argentina:** Professor Pouet introduces Cabernet Sauvignon, Malbec, and Merlot from Bordeaux.
France: Professor Jules-Émile Planchon of the Montpellier School of Agriculture (hereafter referred to simply as Montpellier) identifies phylloxera as the cause of vineyards dying in the Rhône and deduces that infection spreads via cuttings imported from America by *vignerons* in search of hardy vines to experiment with following the havoc wreaked by oidium (powdery mildew) in the 1840s and 1850s (*see* 1860).

1869 **France:** Baron Paul Thénard tries to kill phylloxera with carbon bisulphide-filled trenches. He not only succeeds in killing the bugs but also kills the vines! • Louis Faucon, the owner of a small phylloxera-infested vineyard on a tributary of the Rhône, suffers one calamity after another when his vineyard is severely flooded but is delighted to discover, once the flood has receded, that all the phylloxera has drowned. His vines immediately pick up, and later that year, Faucon enjoys a good harvest (*see* 1874 Faucon).

1870 **France:** The French government offers a prize of 20,000 francs to the inventor of a cure for phylloxera. • The first varietal Sylvaner wine is produced in Alsace (the year that Germany assumed sovereignty).
United States: Thomas Munson, whose destiny it is to save the vineyards of France, becomes only the second person to graduate from a new college that would eventually become the University of Kentucky. • California overtakes Missouri and Ohio to become the largest winegrowing state in the United States.

1871 **France:** Phylloxera spreads throughout the entire Rhône Valley and reaches Bordeaux, Languedoc, and Provence.
Portugal: Phylloxera reaches Portuguese vineyards.
Turkey: Phylloxera reaches Turkish vineyards. Although mostly used for table grapes, the first vineyard infected had been planted with Bordeaux vines (imported by Kösé Riza Efendi).

1872 **Austria:** Phylloxera reaches Austrian vineyards (Klosterneuberg).
France: The first experiments to graft classic *vinifera* varieties onto American rootstock are conducted at Montpellier.
Ukraine: Phylloxera reaches Ukraine vineyards.

1873 **France:** Planchon travels to America to research phylloxera.
Switzerland: Phylloxera reaches Swiss vineyards.
United States: The first phylloxera identified in California is found in a Sonoma vineyard (but *see* 1858).

1874 **France:** Veuve Borty, the widow of a wine merchant in Roquemaure (*see* 1862), confesses to Professor Planchon that in 1862 her late husband had been given a gift of some "exotic" vines by a friend who had brought them with him from New York. Her husband had planted them in a walled garden, but the following year he noticed that Grenache and Alicante vines in the same garden had started to wither; then, in 1864 and 1865, they heard about vines in surrounding vineyards mysteriously dying. • Louis Faucon presents his paper *Sur la Maladie de la Vigne et sur Son Traitement par le Procédé de la Submersion* to the Académie des Sciences in Paris. He recommends flooding a vineyard for at least 40 days, which is discovered to be effective, but only temporarily, since it does not prevent reinfection. Flooding is also very expensive, requiring an enormous infrastructure of dikes, gates, pipes, steam-powered pumps, and of course an abundant supply of water. Even then, not all vineyards can be flooded. Most of the greatest sites are all on slopes to some extent. Furthermore, growers are acutely aware that *vinifera* vines do not like "wet feet"; consequently, it is debatable how many times a vineyard might withstand flooding, if indeed it was capable of flooding and its owners could afford multiple remedies. • The French government increases its prize for a cure for phylloxera from 20,000 francs (*see* 1870) to 300,000 francs and commissions Montpellier to evaluate all submissions.
Switzerland: Phylloxera reaches Switzerland (Baron Rothschild's estate at Pregny-Chambésy in the canton of Geneva).

1875 **France:** Phylloxera reaches Corsica, western Loire, and southern Burgundy. • Professor Millardet of Bordeaux starts work on hybrid rootstocks (*see* 1887).
Serbia: Phylloxera is identified at Pancsova (now Pancevo), just east of Belgrade.
Spain: Phylloxera reaches Spanish vineyards.

1875 **United States:** The first screw-sealed bottle is invented by William T Fry of
(cont'd) Brooklyn, New York, who develops a male-threaded *collerette* that encircles the neck of the bottle, on to which a female-threaded top is screwed.

1877 **Australia:** Phylloxera reaches Australia.
France: Of the 696 remedies for phylloxera submitted in response to the French government's prize of 300,000 francs, Montpellier considers 317 worthy of evaluation and puts them to the test in a nearby infested vineyard (Las Sorres). Only two remedies have any effect on phylloxera – potassium sulphide dissolved in human urine, and the direct application of sulphide – but neither method is fully effective, and what control they do achieve is only temporary, since neither solution prevents reinfestation. • The first reference in Champagne to the addition of a liqueur to promote a second fermentation is made by Professor Robinet in Epernay, 215 years after Merret's paper to the Royal Society (*see* 1662). • A sparkling Sauternes from Messrs Normandin Sparkling Sauternes Manufactory, in the tiny Charentes village of Châteauneuf, receives a gold medal at the Concours Régional d'Angoulême.
United States: French scientist Pierre Viala is tasked with saving Cognac from phylloxera and visits Thomas Munson in Denison, Texas, where the two agree on the most appropriate American rootstock to graft with the region's vines. Munson starts shipping cuttings to France and gears up production for future shipments.

1878 **France:** Phylloxera has invaded 39 *départements*, killing off 370,000 hectares (914,300 acres) of vineyard and infecting another 250,000 hectares (617,800 acres). • The first-ever instances of downy mildew in Europe appear in France, having crossed the Atlantic on American vines imported to defeat phylloxera. • Lermat-Robert et Cie of Bordeaux enters a sparkling Barsac at the Paris Exhibition. • Henry Vizetelly tastes "sparkling Chambertin, Romanée, and Vougeot of the highest order" at the Paris Exhibition.
United States: The first true screwcap (a moulded metal cap fitting over a screw moulded into the glass neck of the bottle) is patented by August Voege of Brooklyn, New York, in June 1878, and a variant is patented by John K Chase of New York in September 1878. Both are described as an "improvement in screwcaps for bottles", indicating that cruder screwcaps were in prior use.

1879 **France:** State-licensed nurseries are established to mass produce rootstocks.
Italy: Phylloxera reaches Italian vineyards.

1880 **France:** All vineyard regions with the exception of Champagne and Alsace (then part of Germany) are infested by phylloxera. • The addition of a *liqueur de tirage* starts to become widespread in Champagne. Prior to this, most Champagne was rendered sparkling by bottling before the first fermentation had ceased, a haphazard judgment that often resulted in a massive loss of production from exploding bottles. • The first sparkling wine in Alsace (then part of Germany) is produced by a small grower in Guebwiller called Dirler.
Italy: Phylloxera reaches Sicily.
Slovenia: Phylloxera reaches Pisece and Bizeljsko in the Bizeljsko-Sremic area.
United States, France: Swiss-born Hermann Jaeger, the father of direct-producing hybrid vines and owner of a nursery vineyard at New Switzerland, in the southwest Ozark region of Missouri, has shipped many of the "millions upon millions of American cuttings" that had been imported by France to replant phylloxera-infested vineyards. • US state legislation requires the University of California to commence research and instruction in viticulture and winemaking. • The first mention of the use of agglomerated cork is in a US patent for protective headgear issued to Abraham Moses.

1881 **Australia:** Australia's first "sparkling burgundy" is rosé coloured and produced from Pinot Noir grapes by Auguste d'Argent at the Victorian Champagne Company in Melbourne (*see* 1920).
Croatia: Phylloxera reaches Croatia.
France: At the International Phylloxera Congress held in Bordeaux, the consensus is that the only feasible long-term solution for vineyards continuing to grow classic *vinifera* varieties is to graft them on to American rootstock. Direct-producing hybrid vines also present a feasible long-term solution, but whatever anyone thinks about the wines they produce, everybody agrees that they cannot replicate the classic quality, character, and style of famous French wines.
Germany: Phylloxera reaches German vineyards.

1882 **Germany:** Carl Wienke invents the Waiter's Friend, which becomes the workhorse corkscrew for sommeliers all over the world and remains so for more than 100 years.
Serbia: Phylloxera reaches Serbia (Pančevo, Bannet).

1883 **Algeria:** Aware of the devastation caused by phylloxera in France and the plans to build the vineyards of Algeria into a second viticultural France, the colonial government imposes a strict law requiring growers to report any potential

1883 (cont'd)	symptoms. • The French are the first to experiment with cool fermentation, in a bid to improve the odds for winemaking in such a hot country (but *see* 1330 BC). **Australia:** Roseworthy College is established for agricultural studies but does not yet include winemaking (*see also* 1843 and 1920). **Bulgaria:** Phylloxera reaches Bulgaria (Vidin). **United States:** The first commercial winery in Texas, the Val Verde Winery, is established at Del Rio by Italian immigrant Frank Qualia.
1884	**Algeria:** In a further bid to protect Algeria from phylloxera, all imports of vines, fruits, and fresh vegetables are banned. **France:** The refrigerated bath required for *disgorgement à la glace* is invented by a Belgian, Armand Walfart; the system is first used in the same year by Champagne Henri Abelé.
1885	**Algeria:** Phylloxera is discovered at Sidi-Bel-Abbès and Tlemcen. **Israel:** The country's first commercial winery, Carmel Winery, is established by Baron Edmond James de Rothschild at Zikhron Ya'akov. **New Zealand:** Phylloxera reaches New Zealand.
1886	**France:** Jules Émile Planchon, Professor of Pharmacy at Montpellier University is the first person to identify that it was an insect ravaging French vines at this time. He is also an entomologist and thought it looked very similar to *Phylloxera quercus*, which lives on the underside of oak leaves, so he named it *Phylloxera vastatrix* (meaning Phylloxera "the devastator"). **South Africa:** Phylloxera reaches South Africa.
1887	**France:** The first completely successful rootstock, 41B, a Chasselas x Berlandieri cross, is launched at the Mâcon Wine Congress by its breeder, Professor Millardet.
1888	**Peru:** Phylloxera reaches Peruvian vineyards. **United States:** Thomas Munson is made a Chevalier of the Légion d'Honneur by the French government for his part in saving the French wine industry by supplying millions of American vines for grafting purposes. The French send a delegation to Denison, Texas, to present Munson with this award (*see also* 1870 United States and 1877 United States).
1889	**Austria:** Phylloxera has destroyed more than 42,000 hectares (104,000 acres) of vines. **England:** Most references credit Dan Rylands of Hope Glass Works in Barnsley as the inventor of the screwcap at this date, but the Americans got there more than a decade earlier (*see* 1878). **Greece:** Phylloxera reaches Greece.
1890	**Australia:** The first vines are planted in Coonawarra by John Riddoch at Yallum. • The first vines are planted in the Margaret River Valley by the Cullen family at Bunbury. **Czechia:** Phylloxera reaches Šatov, in the Znojmo district of Moravia. **France:** Bordeaux and Burgundy commence a complete replanting of their vineyards with grafted vines. **India:** This is the earliest date by which phylloxera is speculated to have reached India. This country's emerging British-built wine industry was killed off by phylloxera in the 1990s.
1891	**United States:** The crown cap (aka crown cork, and originally called the Crown Cork Tin Bottle Cap) is invented by William Painter of Maryland and sold by his employee, King Camp Gillette, who would go on to invent the disposable safety razor.

AN 1890 *PUNCH* CARTOON DEPICTS A PHYLLOXERA LOUSE GUZZLING FINE WINE
The caption reads "The phylloxera, a true gourmet, finds out the best vineyards and attaches itself to the best wines". Late-19th-century vineyards experienced the devastating phylloxera epidemic, which destroyed most of the vine for wine grapes in Europe, most notably in France.

1892	**China:** First *vinifera* vines are taken to China by Chinese businessman Zhang Bishi. **Israel:** Phylloxera destroys the vines at Carmel, the country's only vineyard (*see* 1885).
1893	**France:** Wine production exceeds pre-phylloxera level for the first time.
1894	**France:** Phylloxera finally reaches Champagne. **Italy:** First mention of Pinot Nero (aka Pinot Noir) in Italy being grown in Trentino and the Alto Adige.
1895	**Algeria:** Only the Alger *département* is free of phylloxera. **Portugal:** All Portuguese vineyard areas are infested with phylloxera. The northern areas are worst affected, destroying 32,000 hectares (79,000 acres) in the Douro region alone. **United States:** Mass production of wine bottles becomes possible following the invention of an automatic glass-blowing machine by Michael Owens of the Libby Glass Company in Toledo, Ohio (*see* 1903).
1898	**Australia:** Experimental cool fermentation is undertaken. **Greece:** Phylloxera reaches Thessaloniki.
1901	**Australia:** The first documented mention of Malbec is by Isaax Himmelhoch, who has planted it at his Grodno vineyard at Liverpool, near Sydney.
1903	**Algeria:** Despite phylloxera penetrating almost all of Algeria, the French manage to increase Algerian vineyards from just 15,000 hectares (37,000 acres) in 1878 to 167,000 hectares (412,600 acres) by 1903. **United States:** The first fully automatic bottle-making machine, "The Owens Machine", goes operational (*see* 1895).
1904	**France:** Phylloxera reaches Alsace (then part of Germany).
1907	**France:** The Charmat Method is invented by Eugène Charmat.
1908	**United States:** The viticultural department of the University of California moves to Davis.
1909	**Croatia:** Phylloxera is discovered in Dalmatia. **United States:** Charles McManus is said to have invented agglomerate cork closures.
1910	**China:** A French priest converts a Beijing church graveyard into a vineyard and small winery called Shangyi (*see* 1949 and 1987).
1911	**France:** Louis Étienne Ravaz suggests that the ratio of fruit weight harvested to wood cane weight pruned to be the key to consistent productivity and fruit quality. The Ravaz Index is still used today.
1912	**Macedonia:** Phylloxera is found in Macedonian vineyards. **South Korea:** Phylloxera reaches Korea (table grapes).
1914	**France:** Phylloxera has destroyed 1 million hectares (2.47 million acres) of French vineyards since 1875.
1920	**Australia:** It is unknown when Australia's "sparkling burgundy" evolved from a rosé-coloured Pinot Noir *cuvée* with a touch of sweetness to a seriously sweet, deep-coloured, pure Shiraz fizz, but it is believed that Great Western (which had been producing the lighter style since 1890) was the instigator and that the change in grape variety and style began in or around 1920 (*see also* 1843 and 1883). **United States:** More than 30 states are already totally dry by the time the Eighteenth Amendment to the Constitution is officially ratified, enforcing Prohibition throughout the entire United States. Despite this, 100 wineries continue production under strict federal control for medicinal purposes, for sacramental wines, and to make "salted wine" for cooking.
1921	**Luxembourg:** The country's first sparkling wine is produced by Bernard-Massard.
1925	**South Africa:** The Pinotage grape is developed by Professor Perold by crossing Pinot Noir and Cinsault.
1930s	**Australia:** Sydney Hamilton of Hamilton Ewells is probably the first to experiment with temperature control in winemaking. Prior to the stainless steel era, he immersed refrigerated copper coils in wooden fermentation tanks. In Australia, temperature-controlled white winemaking started in the late 1940s and temperature-controlled red winemaking in the early 1950s. By contrast, the earliest cooling in France was at Bordeaux properties like Château Dauzac, which relied on blocks of ice, and temperature-controlled winemaking did not become mainstream until the 1970s, well after the introduction of stainless-steel fermentation tanks.
1932	**Australia:** Cold stabilisation first used to prevent build-up of tartrate crystals in bulk wine exports.

1933 **United States:** National Prohibition ends. • E & J Gallo is established; it will become the world's largest wine producer from the 1960s until Constellation Brand is formed (*see* 2003).

1934 **United States:** The Wine Institute is incorporated, and California introduces state wine standards. • Frank Schoonmaker and Tom Marvel publish *The Complete Wine Book*, in which they urge American winemakers to plant better vine varieties, adopt better winemaking standards, and sell honest, informatively labelled wines, rather than relying on borrowed European names.

1935 **France:** Introduction of the AOC system.
Morocco: Phylloxera reaches Morocco.

1936 **Australia:** Roseworthy College launches the nation's first winemaking course.
Tunisia: Phylloxera reaches Tunisia and destroys all vines there.

1939 **United States:** Frank Schoonmaker starts sourcing and selling varietal wines from California, New York, and Ohio. Growers in all these states, but particularly California, start to replant their vineyards with classic *vinifera* varieties as soon as it is evident that Schoonmaker's wines are attracting a significant premium. The New World varietal wine trend begins.

1944 **France:** In his capacity as the Vichy mayor of Paris, Pierre Taittinger, founder of Champagne Taittinger and the far-right Jeunesses Patriotes, persuades General von Choltitz, the German governor of Paris, to defy Hitler's order to destroy the city.
United States: The University of California at Davis publishes a map of California with five climate zones based on heat summation, identifying which grape varieties are most suitable for each zone.

1945 **South Africa:** Nederburg Première Cuvée is the country's first *cuve close* sparkling wine (*see* 1971).

1947 **France:** Abandoned vineyards in the Médoc are quarried for gravel, destroying any possible use as vineyards in the future.

1948 **Italy:** The first vintage of Sassicaia, the first of the so-called super-Tuscans, is harvested (*see also* 1971).

1949 **China:** The Shangyi in Beijing is taken over by the Chinese government (*see* 1910 and 1987).

1950s **United States, Australia:** Small oak *barriques* are increasingly and eventually become the hallmark of New World wines.

1951 **Australia:** The first (experimental) vintage of the legendary Penfold's Grange is harvested.
France: The *champenois* are granted a reduction in AOC Champagne from 46,000 to 34,000 hectares to cope with a drop in sales and build-up of stocks.

1953 **France:** Baron Philippe de Rothschild begins a 20-year battle to upgrade his Château Mouton from its 1855 classification of Second Growth to First Growth (*see* 1973).

1954 **United States:** The first mechanical harvester, the so-called "Cutterbar Harvester", which cut and collected whole clusters of grapes, is developed by Lloyd Lamouria at UC Davis Agricultural Engineering Department.

1955 **United States:** The bag-in-box system concept was invented for the safe transportation of battery acid in 1947 by William R. Scholle, who patents the design in 1955. *See also* 1965 Australia and 1967 Australia.

1956 **Australia:** The first vines in Tasmania's modern viticultural era are planted by Jean Miguet at La Provence (now called, for legal reasons, Providence).

1957 **United States:** When James Zellerbach - industrialist, financier, head of the Marshall Plan mission in Italy, and US ambassador to Italy - makes the first Chardonnay at his Hanzell vineyard in Sonoma, he leaves nothing to chance, building a small but high-tech, gravity-fed winery full of gleaming, temperature-controlled, stainless-steel vats equipped with inert-gas blanketing that is way ahead of its time. He is also the first to insist on French oak barrels. When the Hanzell 1957 Chardonnay is released and found by critics to have a "complex French flavour", other wineries start ordering French oak. While some eventually go back to American oak for most or all of their wines, the French oak standard has been set for premium Chardonnay in California. • Ukrainian viticulturist Konstantin Frank does something that Americans thought impossible: he successfully grows *vinifera* grapes in New York State.

1958 **United States:** The University of California issues a report on suitable rootstock, recommending the high-yielding AxR#1 as "the nearest approach to an all-purpose stock", despite admitting that it had "only moderate phylloxera resistance". The clock for the second coming starts ticking.
Russia: A large part of the Fukier Tokaji collection (*see* 1606), which had been

1958 **(cont'd)** missing since the beginning of World War II, turns up at an extraordinary tasting laid on by Marshall Zhukov for his personal friend General Alan Shapley, deputy commander of the US Marine Corps. The vintages include 1811, 1783, 1737, 1682, and 1668 but unfortunately no 1606.

1959 **France:** Le Bouchage Mechanique develops a screwcap bottle and closure called Stelcap with a cork and paper liner specifically for wine (First Generation screwcaps).

1960 **Argentina:** When Moët & Chandon establishes Bodegas Chandon, which will make and sell Champaña, it is the beginning of the end of Champagne's integrity in its defence against others abusing its appellation (*see* 1993 and 2007).

1961 **France:** This is the greatest postwar Bordeaux vintage. • By the 1980s, the market performance of this vintage of Château Palmer will have led to the creation of the "super-second" concept, whereby some high-performing châteaux that are not First Growths start achieving prices close to First Growth prices and, as Palmer demonstrates, super-seconds do not necessarily have to be Second Growths.

1963 **Australia:** The first Cabernet Sauvignon is planted in the Hunter Valley.
Portugal: This is the greatest postwar Port vintage.

1964 **France, Australia:** Peter Wall of Yalumba winery in Australia commissions (*see* 1959) Le Bouchage Mechanique to adapt its Stelcap to modern Australian wine requirement (*see* 1965).

1965 **Australia:** Tom Angove patents first bag-in-box for wine. *See also* 1955 United States and 1967 Australia.
France, Australia: Le Bouchage Mechanique patents new screwcap called Stelvin, developed at the behest of Yalumba (see 1964), which have a synthetic liner and a long tail that remains on the bottle (Second Generation screwcaps).
England: Christie's makes plans to reopen a wine department.
United States: The first commercial use of mechanical harvesting takes place in New York State, but experimental machines have been tested in California and elsewhere since the early 1950s.

1966 **England:** Michael Broadbent MW is hired to head Christie's new wine department, exactly 200 years after its first wine auction (*see* 1766).
United States: Robert Mondavi opens his Napa Valley winery.

1967 **Australia:** The country's first vines of the modern era are planted at Vasse Felix by Dr Tom Cullity. • Inventor Charles Malpas and Penfolds Wines patent a plastic, air-tight tap welded to an aluminium film bladder, making bag-in-box wine more convenient for consumers.
United States, Italy: Dr Austin Goheen of UC Davis visits Bari in Puglia and is so reminded of Zinfandel when tasting the local Primitivo that he takes back cuttings to grow side by side with Zinfandel. They appear to be identical.

1968 **France:** Vins de Pays regulations are established.

1969 **Armenia:** Phylloxera reaches Armenian vineyards.
United States: Schramsberg's first rosé sparkling wine is produced from 100% Gamay (a grape that was permitted in the Aube region of Champagne until 1949).

1970 **United States:** First European (*Vitis vinifera*) grapes are planted in Michigan at Tabor Hill.

1971 **Australia:** The first Chardonnay is planted in the Hunter Valley. • The first vines in the Canberra District are planted by two doctors, Dr Edgar Riek (at Cullarin, on the northwest shore of Lake George) and Dr John Kirk (at Clonakilla Vineyard, in Murrumbateman).
England: Hugh Johnson's *The World Atlas of Wine* is published, establishing for the very first time a sense of place for everyone who drinks wine. People like Frank Schoonmaker have been trying to get across such a notion since the 1930s, but until Johnson the most this amounted to was a line drawing of minimal detail. Now wine lovers can sit back in an armchair, sip a delicious wine, and find out precisely where it came from on a beautiful, full-colour, highly detailed map.
Italy: The first vintage of Tignanello, the second so-called super-Tuscan (*see* 1948).
South Africa: Kaapse Vonkel is widely cited in South Africa itself as the country's first sparkling wine produced by the *méthode champenoise*, yet one of the most reliable of those sources, *The Complete Book of South African Wine*, also states in the very next paragraph, "In earlier days in the Cape, all sparkling wines were made this way, but with the perfection of tank fermentation, bottle fermentation all but disappeared." The first South African sparkling wine was probably produced in the 19th century, but exactly when and by whom is not yet known. *See also* 1945.

1973 | **France:** The first Vins de Pays are available to purchase.
United States: Domaine Chandon establishes the first French sparkling-wine venture in California, at Yountville in Napa County.

1974 | **Crete:** Phylloxera eventually reaches Crete.

1975 | **Australia:** Roseworthy Agricultural College lecturer Dr Richard Smart introduces the concept of vineyard region homoclime matching, arguing that if climate and, especially, temperature are so important in affecting the composition of grapes and, thus, the style and quality of wine, then why not, for example, look in Australia for similar climates (homoclimes) to those of famous French regions?
England: Colin Parnell and Tony Lord launch *Decanter* magazine.
Italy: Talento Metodo Classico sparkling wines are established.
United States: Zinfandel and Primitivo are declared to be one and the same following isozyme "fingerprinting" by Wade Wolfe, a PhD student at UC Davis, where both varieties have been growing side by side since 1967.

1976 | **France, United States:** Steven Spurrier holds the famous Judgment of Paris wine tasting that pits the best California wines against the best French wines.
England: The first English sparkling wine is made by two different producers in the same year: Felsted and Pilton Manor.
United States: The concept of AVAs (American Viticultural Areas) is considered by the ATF (Bureau of Alcohol, Tobacco and Firearms).

1977 | **Australia:** Andrew Pire introduces a humidity index to a homoclime-like analysis of Australian wine regions for his PhD thesis at the University of Sydney.
Germany: Atypical Ageing (ATA) first noticed.

1978 | **United States:** Robert Parker launches *The Wine Advocate*, introducing his 100-point rating system, and the wine world will never be the same again. • The regulations for AVAs are established and differ from the French AOC concept in that, although there must be some topographical and climatical homogeneity of growing zone, there is no attempt to control what grape varieties are grown, how vines should be cultivated, or what style of wine must be produced.

1979 | **United States:** In *The Myth of the Universal Rootstock,* Lucie T Morton wonders why the AxR#1 rootstock became the rootstock of choice for the California wine industry, when the French, who developed it, had long since concluded that AxR#1 and all other *vinifera rupestris* rootstocks are dangerously low in resistance. • Marvin Shanklin buys *Wine Spectator* and turns it into the most successful wine magazine in the world. *See also* 1976.

1980 | **Australia:** Dr Richard Smart and Dr Peter Dry of Roseworthy Agricultural College publish their climate classification for Australian viticultural regions, establishing the concept of homoclimes in the public domain. This also includes the first strictly defined meaning of water deficit (calculated over the growing season as the difference between rainfall and estimated vineyard water use).
China: The Tianjin Winery and Rémy Martin launch the Dynasty brand in the first joint Franco-Chinese wine venture.
United States: The first AVA is established in Augusta, Missouri.

1981 | **United States:** The Napa Valley is the second AVA to be established.

1982 | **India:** Raphael Brisbois of Piper Heidsieck Technology advises Sham Chougule to establish his Champagne Indage (later changed to Chateau Indage) sparkling-wine vineyard and winery of the modern Indian viticultural era at Narayangaon, in the Sahyadri Mountains north of Puna.

1983 | **Germany:** The Charta organization promoting dry Riesling is founded in the Rheingau.
United States: A new biotype of phylloxera hits Napa and Sonoma vineyards grafted on to AxR#1 rootstock.

1984 | **Cyprus:** Phylloxera reaches Cypriot vineyards.
England: Robert Joseph and Charles Metcalfe establish the International Wine Challenge (IWC), the first truly international wine competition. Despite being set in a country with no wine-producing industry of its own to speak of, the IWC attracts thousand of wines from countries whose producers had been reluctant to let judges in other wine-producing countries make qualitative assessments about their competitors' wines.

1985 | **Australia:** Domaine Chandon establishes its sparkling-wine venture in Victoria's Yarra Valley (*see* 1989).
Austria: The "anti-freeze scandal" is misreported by the media as a health scare. It was a scandal, and its greedy perpetrators did untold damage to the reputation of the Austrian wine industry for the next 10 years or so, but it was not a health scare. Ethylene glycol is anti-freeze and highly lethal, but the illegal additive used was diethylene glycol, which is less toxic than alcohol. The more diethylene glycol added, the more alcohol is displaced, and the less toxic the wine becomes.

1985
(cont'd) | **England:** A bottle of "Lafitte 1787" engraved with "Th. J." and believed to belong to Thomas Jefferson, is sold at auction for a world-record price of £105,000. *See also* 2008.
United States: The Bureau of Alcohol, Tobacco, Firearms, and Explosives rules that Zinfandel may not be used as a synonym for Primitivo, declaring there is not enough evidence to prove Zinfandel and Primitivo are the same. • The first Rhône varieties are planted in California (by Phelps).

1986 | **France, England:** In *Champagne* (Sotheby's Publications), Tom Stevenson reveals the so-called *"sur-lattes* scandal": how the use of *boues de villes* (refuse!) is turning the Côte de Blancs into the Blue Hills of Champagne, and the first discussion about a future expansion of the vineyards.
Italy: At least 22 Italians die and 90 others are hospitalized after drinking wine contaminated with methanol. Six suspects are named, and millions of gallons of wine are disposed of.
United States: Dutch-born yeast authority Herman Jan Phaff (1913-2001) postulates that the natural habitat of ancestral strains of *Saccharomyces cerevisiae* (wine yeasts) hundreds of millions of years ago appears to be the bark and sap exudate of oak trees. In which case, the habit of grapevines climbing trees may have encouraged the introduction of yeasts to the bloom of grapes, providing the first link in what has become the natural affinity between oak and wine (corks and barrels coming much later, of course).

1987 | **China:** The Shangyi winery in Beijing is renamed the Beijing Friendship Winery and, in partnership with Pernod-Ricard, launches the Dragon Seal brand.

1988 | **England:** Stuart and Sandy Moss, two Americans from Chicago, purchase Nyetimber, ignore all advice, and kick-start a golden age for English sparkling wine.

1989 | **Australia:** By Domaine Chandon's third vintage, Tony Jordan has already surpassed the quality established by Domaine Chandon's California operation, even though the latter had a 12-year head-start (*see* 1985).

1990 | **France/Australia:** Péchiney (which bought out Le Bouchage Mechanique) introduces screwcaps with a Saranex liner (Third Generation screwcaps).

1991 | **Australia:** Dr Richard Smart and colleagues offer a worldwide homoclime-matching service, using computers and international climate databases.
England, Chile: In the British publication *Wines & Spirits,* Tom Stevenson exposes that not all Chilean Sauvignon Blanc is Sauvignon Blanc, and that Chile sells eight times more Sauvignon Blanc wine than the highest possible yield of its claimed plantation of Sauvignon Blanc could possibly produce. French ampelographers later confirm that most Chilean Sauvignon Blanc is not Sauvignon Blanc.
United States: The French Paradox gets its first mention on the CBS show *60 Minutes,* hosted by Morley Shafer. Within weeks, US sales of red wine have increased 44 per cent, and Gallo's Hearty Burgundy is put on allocation.

1992 | **Germany:** Hans-Peter Frericks sues wine collector Meinhard Görke (aka Hardy Rodenstock) over alleged fake wine. The Munich state court finds that Görke has "adulterated the wine or knowingly offered adulterated wine". Görke appeals, and the pair file criminal complaints against each other for defamation. The action for fraud and the two defamation cases are eventually settled out of court.

1993 | **France, South America:** Moët & Chandon admits under pressure that there is no law that requires their subsidiaries to use "Champagne" or "Champaña" on its sparkling wines produced in Argentina or Brazil, despite claiming this in numerous court battles around the world (*see* 1960 and 2007). • Château Couhins-Lurton becomes the first Bordeaux *cru classé* to be sealed with a screwcap.

1994 | **Chile:** One French ampelographer invited to Chile, Jean-Michel Boursiquot, discovers that Chilean Merlot is in fact Carménère (which has since become the country's flagship wine).
France: This is the first vintage when some *bordelais* harvest by tannin ripeness.
United States: Using DNA typing, Professor Carole Meredith, a plant geneticist at UC Davis, establishes conclusively that the grape varieties Zinfandel and Primitivo are genetically the same.

1995 | **France:** Michel Bettane, one of the country's top wine writers, declares, "Today, *appellation contrôlée* guarantees neither quality nor authenticity."

1997 | **Azerbaijan:** Phylloxera reaches Nagorno-Karabakh.
China: The country's first sparkling wines are produced by Frenchman Denis Degache using 1993 reserve wines.
England: Nyetimber 1992 is served at Queen Elizabeth II's Golden Anniversary Lunch, becoming the catalyst for a new world-class sparkling wine industry.

**1997
(cont'd)**

Finland: Some 2,000 bottles of Heidsieck & Co Monopole 1907 Goût Américain is recovered from the wreck of the Jönköping, which was sunk by charges placed by the crew of a German U-boat. After 80 years in a totally reductive environment, the quality of all the bottles is stunning. *See also* 2010.

United States: An estimated 6,700 hectares (16,500 acres) of infected vineyards in Napa and 4,000 hectares (10,000 acres) in Sonoma have been uprooted and replanted since the 1983 phylloxera attack. • Professor Carole Meredith uses DNA "fingerprinting" to determine that only four of the seven Petite Sirah vines in the UC Davis collection are in fact Petite Sirah. Two are Durif and one is pure Syrah.

HEIDSIECK CHAMPAGNE CORK
Recovered from the wreck of the Swedish *Jönköping*, which had been sunk off the Finnish coast, the bottles of Heidsieck Champagne were surprisingly well-preserved.

1998

France, England: Tom Stevenson publishes a photo-facsimile of Christopher Merret's paper (*see* 1662) in *Christie's World Encyclopedia of Champagne & Sparkling Wine* (Absolute Press 1998), which becomes the only wine book to warrant a leader in any national newspaper (*The Guardian*).

1999

France: The so-called "Franco-Iranian Wine Incident". Iranian President Mohammad Khatami is due to make a state visit to France but insists on no wine being served at the proposed state dinner, not only to Iranian dignitaries, but to their French hosts as well. Protocol demands that a state visit requires a state banquet, but President Jacques Chirac declares that a French state dinner without French wine is impossible, so he cancels dinner and downgrades the Khatami's presence to an "official visit".

Israel: A team led by Robert Ballard (who located the *Titanic*) discover two Phoenician ships from Tyre that had sunk in 750 BC at a depth of 460 metres (1,500 feet) 50 kilometres (30 miles) off the coast of Israel and were carrying amphorae filled with wine.

2000

United States: The 1997 Plumpjack Reserve from Napa Valley is sold in two formats, one sealed with a cork, as per normal, and the other sealed with a screwcap. Both are exactly the same wine, but the screwcap is $10 per bottle more expensive. A point is being made.

2001

Australia, New Zealand: Dr Richard Smart identifies homoclimes of Martinborough and Marlborough in New Zealand and uses the same principle to search for vineyard locations at Tamar Ridge Estate in Tasmania. Typically, homoclime searches are initially conducted using only monthly maximum and minimum temperatures over the growing season, leaving other data layers – such as rainfall, humidity, sunshine, and calculated moisture stress – to be factored in later for the searches that are most likely to bear fruit. Other data, such as soils, may also be added.

2002

France: François Mauss, head of the Grand Jury of European Tasters, describes much of Beaujolais as "not proper wine" and its producers of "consciously commercializing a *vin de merde*".

United States: Two American scientists, Charles Zuker and Charles Ryber, confirm the Japanese concept of umami, a satisfying savouriness that is the fifth of the five basic tastes recognized by the palate, alongside sweetness, sourness, saltiness, and bitterness. • The first world class Gewürztraminer produced outside of Alsace is made in Michigan of all places, by Bryan Ulbrich under the Peninsula Cellars Manigold label.

2003

United States, Australia: E & J Gallo is overtaken as the world's largest wine producer when Constellation Brands is formed by merging Canandaigua, the second-largest wine producer in the United States, and BRL Hardy, the second-largest wine producer in Australia.

2004

England: Researchers demonstrate that drinkers of alcohol are more likely to get gout, and that beer drinkers are more prone than wine drinkers.

United States: Following the release of the American comedy film *Sideways*, US sales of Pinot Noir increase, while sales of Merlot drop. • Michigan fizz fanatic Larry Mawby launches a sparkling wine called Sex, after submitting as a joke an application for the name and being totally shocked when the federal authorities officially gave him permission (but *see* 2005).

2005

China: Phylloxera is suddenly rampant in Jiading and Shanghai.

EU: The use of oak chips is legal for the first time in Europe. *See also* 1828.

United States: Larry Mawby's Sex may have got past the normally puritanical US federal authorities (*see* 2004), but the sparkling wine is refused entry into the UK by the Portman Group, the wine trade's self-regulatory body, which deems it "sexually suggestive", even though there is no illustration or text to suggest that it is meant to be anything other than sex as in "gender." The Portman Group is perfectly happy, however, to allow many emphatically suggestive labels such as Old Tart, which is illustrated and contains descriptive text full of *double entendres*.

2006

Australia: Cullen, in the Margaret River, becomes the country's first carbon-neutral winery.

United States: Billionaire Bill Koch sues millionaire Meinhard Görke (aka Hardy Rodenstock) over a bottle of "Lafitte 1787" engraved with "Th. J." (supposedly standing for Thomas Jefferson), claiming that he has scientific proof that the wine is a fake and that the initials have been engraved by an electronic tool. A default judgment was entered against Görke in May 2010 because, as a German citizen, he refused to participate (*see also* 1992), but in March 2011 the US District Court of New York threw out Koch's case – not because the bottle was not a fake, but because Koch knew it was a fake when he bought it.

2007

France, South America: More than a decade after Moët & Chandon admitted there is no legal requirement for its subsidiaries to prostitute its own Champagne on bottles of South American sparkling wine, this unprincipled practice has stopped (*see* 1960 and 2007).

2008

England: Prince Charles converts his 38-year-old Aston Martin DB6 (a 21st-birthday present from the Queen) to run on bioethanol fuel distilled from surplus English wine; it now gets 1.6 kilometres (1 mile) for 4.5 bottles of wine.

United States: The so-called "Thomas Jefferson Lafitte 1787" (*see* 1985) is dated to be circa 1962 according to one tritium test and two carbon-14 tests, taken after investigations by David Molyneaux-Berry MW, the FBI, a former Inspector at Scotland Yard, and a former MI5 agent in Germany! • The film *Bottle Shock* is released. This comedy drama is loosely based on the Judgment of Paris tasting (*see* 1976).

2010

Finland: A cache of 168 bottles of Champagne from the 1830s is recovered from an unidentified shipwreck in Finland's Åland archipelago (identified as Champagne Juglar, a brand belonging to Champagne Jacquesson since 1829, and Champagne Veuve Clicquot), they are amongst the oldest Champagnes ever tasted. Described by Richard Juhlin in the auction catalogue as "floral and citrus", Tom Stevenson is quoted by *Decanter* magazine, "I cannot tell a lie: even the palate was laced with the stench of horse manure!". Juhlin later confesses that even the best bottles had the smell of "Brie de Meaux", while the worst had a "stench of manure, swamp, and rotten eggs". Why such a difference compared to the Heidsieck Monopole 1907 recovered from the Jönköping (*see* 1997)? Three factors determine the evolution of shipwrecked Champagne: heat, light, and time. The Heidsieck was found at a depth of 64 metres (210 feet) and a temperature of 2 to 4°C (35 to 39°F), whereas the Åland wreck was located at a shallower depth of 48 metres (160 feet), where the temperature was slightly higher, 4 to 6°C (39 to 43°F), where fractionally more light penetrated, and which had had twice the time (160 years) for the slow processes to do their worst.

France: An extortion attempt that threatened to kill-off the vineyards of Domaine de la Romanee-Conti is later revealed in *Shadows in the Vineyard* (Maximillian Potter, 2014).

2011

Austria: In February of this year, in the village of Sankt Georgen am Leithagebirge, the only known example of a vine of a variety once misleadingly called Grün Muskateller (but now known simply as St Georgen) is mutilated by an unknown vandal. The vine survives and because of its importance as one of the parents of the Grüner Veltliner, and its age, which is thought to be over 500 years (in which case it is considerably older than the "official" oldest vine in Maribor, Slovenia, *see* 1789), the Austrian government designates this vine as a protected natural monument.

2012

United States: Robert Parker sells controlling interest in *The Wine Advocate* to a group led by Soo Hoo Khoon Peng for a reported US$15 million.

2019

United States: Robert Parker, arguably the most powerful and influential wine critic of all-time, formally retires.

2020

France: Georges Duboeuf, the "King of Beaujolais", dies.

2020

Australia: Horrific wildfires, which started in 2019, devastate the country. With thousands of hectares of vines lost and extensive property damage, it will be years before the Australian wine industry can recover.

PART THREE

A WORLD
of WINE

The Wines of the World

CONSUMERS AROUND THE WORLD HAVE CONTINUED their gradual move to upmarket, quality wines, but with today's technology, even the cheapest wines should be drinkable. A decade ago consumers were switching from white wine to red, but they have since begun to swing back to white. Furthermore, rosé is no longer the forgotten wine style of sporadic and limited success, but a permanent force to be reckoned with. Sparkling wine is another style that has bucked the trend of declining sales and has been accepted for what it is by more-recent generations of wine consumers, who see it not merely as a celebratory drink, but as a fine wine in its own right and, most important, deserving its place at the table with other fine wines. Bottles are getting lighter to reduce the carbon footprint, but clear-glass bottles are so prolific that their intrinsic capacity to induce foul light-strike aromas now represents the single greatest danger to the quality of wine. Screwcaps are slowly gaining ground everywhere except New Zealand and Australia, where they have long since taken over, having eliminated cork taint in the process.

A SELECTION OF BOTTLE CAPS SHOW THE DIVERSE DESIGNS OF INTERNATIONAL WINEMAKERS
Screwcap closures are often replacing traditional corks in the wine industry across all styles and price points.

Global Production and Consumption

At a recent average of just over 270 million hectolitres, world wine production has dropped by 4 per cent in the past 10 years, but 2018 saw a one-off record of 293 million hectolitres and, with global wine consumption averaging 246 million hectolitres, every year sees an enormous surplus production.

Almost every country produces more wine than it can consume and export, but Europe accounts for 54 per cent of the world's wine production, including more than 80 per cent of its surplus. This problem was created by the Common Agricultural Policy (CAP), which encouraged deliberate overproduction throughout the second half of the 20th century. The wine regime under CAP was altered in 1999, 2008, and 2013 to combat this practice and during the first 15 years of the 21st century, the three countries with the largest surplus production significantly reduced the area under vine (Spain by 17 per cent, France by 13 per cent, and Italy by 17 per cent), but this has had very little impact on either surplus production or subsidies paid out. Nearly 20 years after the change in CAP, wine subsidies between 2014 and 2018 amounted to more than 6 billion euros, according to a recent European Commission report.

TOP 20 WINEGROWING COUNTRIES					
	COUNTRY	HECTARES		COUNTRY	HECTARES
1	Spain	969,000	11	Iran*	153,000
2	China	875,000	12	India*	151,000
3	France	793,000	13	Moldova	147,000
4	Italy	705,000	14	Australia	146,000
5	Turkey*	448,000	15	South Africa	126,000
6	United States	439,000	16	Uzbekistan	111,000
7	Argentina	218,000	17	Greece	106,000
8	Chile	212,000	18	Germany	103,000
9	Portugal	192,000	19	Brazil	85,000
10	Romania	191,000	20	Hungary	68,000

*Mostly table grapes

A SHOP AT THE JEAN LESAGE INTERNATIONAL AIRPORT IN QUEBEC CITY DISPLAYS CANADIAN DOMAINE DE LAVOIE SPARKLING WINES
Statistics such as per capita wine consumption can be complicated by the fact that there is no way to distinguish between sales to visitors to a country and wine bought and consumed by the local population.

BEWARE OF ALL-CONSUMING FACTS

Statistics for per capita wine consumption are notoriously unreliable. Even the supposedly same statistic (ie, per capita consumption for the same country in the same year) can vary in different documents and tables from the same organization. There are many anomalies at work. The Luxembourgois, to quote one major reference, "leave everyone else standing" when it comes to wine consumption, but the Belgians and Germans use Luxembourg as a giant superstore, buying up wine, cigarettes, and petrol in bulk because of the significantly lower taxes. The trade is so vast and Luxembourg's population so small that this cross-border trading distorts the imbibing prowess of the Luxembourgois, so precisely where they fit in the per capita scheme of things is debatable. By contrast, the population of France is so large that the trade makes no difference, and the Luxembourg trade would, in any case, be offset by cross-Channel transactions. Holiday resort islands, principalities, and tax havens with small populations, such as the Turks and Caicos Islands, Andorra, and the Cayman Islands, are prone to similar distortions, with much of the wine enjoyed by tourists rather than locals. Such tables therefore generally provide a good overview, rather than a detailed picture.

INTERNATIONALISATION?

Some critics are obsessed by what they see as a growing internationalisation of wine. This is often ascribed to the spread of such ubiquitous grapes as Chardonnay and Cabernet Sauvignon, especially into new and emerging regions and often at the expense of indigenous varieties. It is also attributed to a perceived homogenization in style due to technology, flying winemakers, and schools of oenology, where winemakers are all taught to sing from the same sheet. Anyone who has to taste thousands of wines for a living gets "Chardonnayed out" far too frequently these days, but I believe there are good reasons not to be so pessimistic.

Those ubiquitous grapes

They were once a necessary evil. It was not that long ago when it was a relief to taste a fresh young Chardonnay in parts of Languedoc-Roussillon, let alone Spain, southern Italy, Sardinia, Greece, and much more obscure wine-producing regions of the world. As recently as the early 1980s, it was almost impossible to discern whether many such areas had any potential whatsoever. Was the dire quality in these places caused by what were (then) unknown grape varieties, by the way that the vines were grown and harvested, a lack of quality winemaking experience, or the crude conditions under which the wines were produced? For anyone setting out to improve quality in these viticultural backwaters, the obvious place to start was in the vineyard, and, equally obviously, Chardonnay was the best choice.

Chardonnay might get a hard time in the press these days, but this was not always the case. If readers consult books and articles written in the 1980s, they will discover Chardonnay to be universally revered as one of the greatest white wine grapes in the world. Furthermore, its ability to transplant varietal characteristics to myriad soils and climates throughout the world was considered uncanny and, if anything, served to confirm its nobility. Jancis Robinson waxed lyrical about Chardonnay in her seminal work *Vines, Grapes and Wines* (1986), and even today only the most prejudiced wine drinker would deny that the best white Burgundies rank among the very greatest of dry white wines. If a winemaker in a lowly area of southern France, an Aegean island, or parts of China could not produce a half-decent wine out of the world's greatest, best-travelled grape variety, the vineyards in question would have been written off long ago.

PER CAPITA CONSUMPTION IN LITRES PER PERSON PER YEAR

	COUNTRY	2006	2007	2008	2009	2010	2011	2012	2013	2014	2015	2016	10-YEAR AVERAGE
1	Luxembourg (1)	67.6	61.8	66.5	62.9	64.5	59.8	60.5	58.2	58.9	56.6	54.2	1
2	Portugal (4)	53.5	50.3	50.6	49.9	52.2	53.1	55.2	46.2	47.8	53.8	52.5	3
3	France (2)	65.7	63.6	60.6	59.1	56.9	54.8	54	53.2	52.5	51.9	51.2	2
4	Slovenia (17)	31.1	39.5	45.9	42.9	43.3	43.2	40.9	35.8	36.8	42.5	45.8	6
5	Italy (3)	53.9	52.4	51.2	44.9	45.8	42.8	42	40.5	38	41.7	43.6	4
6	Seychelles (40)	5.7	5.7	5.7	7.1	12.9	12.9	14.3	14.3	27.1	32.9	41.4	34
7	Switzerland (5)	44.7	45.8	45.1	44	44.1	42.6	41.3	41.5	40.3	39.6	37.7	5
8	Croatia (9)	36.3	38.5	39.2	39.4	41.8	42.1	39.8	37.6	32.8	31.3	33.6	7
9	Austria (11)	35.1	34.9	34	33.7	33.5	35.2	36.4	38.1	40.5	31.9	32.4	9
10	New Caledonia (7)	37.7	35.6	37.2	34.7	34.2	35.8	34	34	35.5	37.1	32.4	8
11	Belgium (15)	33	32.7	33.4	31.6	30.4	31.9	31.2	31.2	28.6	31.7	31.9	12
12	Sao Tome And Príncipe (8)	36.7	35.6	30	25	31	43	43	37.3	37.3	32.7	31.8	10
13	Namibia (37)	9.7	9.5	4.3	5.3	5.2	4.5	9.9	21.4	31.1	35.6	30.3	36
14	Sweden (28)	19.4	23.3	26.2	27.7	28.4	29.1	29.1	29.7	29.2	28.9	29.2	18
15	Argentina (6)	38.3	38	35.9	34.3	31.9	31.7	32.1	32.7	31	31.6	28.7	11
16	Germany (18)	28.9	29.7	29.7	28.9	28.9	28.1	28.9	28.9	28.7	28.9	28.4	14
17	Hungary (13)	33.9	33.2	35.7	30.8	21.4	25.1	23.7	24.4	27.5	26.4	27.7	16
18	Australia (19)	28.8	28.8	29.3	29.1	30.1	29.7	29.3	28.6	28.4	28.3	27.5	15
19	Antigua And Barbuda (30)	18.3	12.9	11.4	10	8.6	8.6	15.7	24.3	27.1	20	26.3	32
20	New Zealand (20)	26.1	27.5	25.9	27	26.5	26.8	25.8	25.7	24.8	26	25	21
21	Spain (10)	35.3	33.7	30.9	28.4	27.3	24.9	24.7	24.7	24.8	24.9	24.9	17
22	Ireland (26)	20.6	20.9	21.1	18.7	19.7	21	22.6	23.8	23.2	23	24.1	25
23	Netherlands (21)	26.1	26.4	26.2	26.1	25.4	23.8	22.7	23.8	24.4	24.5	24.1	22
24	United Kingdom (22)	25.4	27.1	26.4	24.5	24.7	24.5	24.2	23.9	23.4	23.6	23.8	23
25	Greece (14)	33.2	34.1	33	31.1	33.4	29.4	31.7	31.1	27.4	25.1	23.7	13
26	Romania (25)	22.1	31.2	31.1	34.9	29.4	24.2	25.3	26.9	27.7	23.7	22.8	19
27	Uruguay (16)	32.1	31.3	30.6	28.7	27.2	26.2	24.2	25.7	24.1	23.7	22.5	20
28	Czech Republic (35)	14.8	20	22.1	22.3	22.1	22.2	19.9	17.3	17.9	20.9	21.3	29
29	Montenegro (24)	23.9	16.2	24.4	26.8	26.8	20.2	20	24.1	17.5	16.5	20.4	26
30	Malta (12)	34.7	23.2	18.9	20.9	18.6	16.7	18.1	17.2	16.7	18.1	19.5	28
31	Estonia (39)	9.2	10.2	12.8	12.5	12.2	14	15.5	17	17.8	18	18.2	38
32	Norway (32)	16.8	17.3	17.5	17.9	18.4	18.4	18.3	18.3	18.4	18.3	17.7	30
33	Chile (29)	19.2	23.7	18.3	24	23.5	22.5	23.3	21.5	21.2	18.3	17.1	27
34	Cape Verde (31)	17.7	19	15.6	13.4	15.2	13.2	12.3	10	11.9	14.3	17	37
35	Bulgaria (36)	11.4	12.3	12.4	13.6	9.3	7.9	16.1	12.7	15	16.3	16.8	39
36	Canada (34)	14.9	14.8	15.3	15.7	16	16.3	16.6	16.5	15.5	16	16.4	35
37	Cyprus (27)	19.5	18.6	16.7	17.7	17.3	15.1	15.5	17.5	17.2	17.2	16.3	31
38	Iceland (33)	15.7	16.3	16.3	22.4	18.8	14.6	15	15.4	15.4	16.2	16.3	33
39	Georgia (23)	24.1	31.3	33.7	27.8	26.8	27.1	23.4	19.3	19.5	17.4	15.9	24
40	Barbados (38)	9.6	9.1	11.4	9.6	10	9.1	9.1	10	10	12.6	14.8	40

Sources: Australian Wine Bureau; Wine Institute (USA); CIES; OIV; IVV; TDA; Statec.
* See Beware of All-Consuming Facts, facing page

WORLD WINE PRODUCTION IN THOUSAND HECTOLITRES

	COUNTRY	2006	2007	2008	2009	2010	2011	2012	2013	2014	2015	2016	10-YEAR AVERAGE
1	Italy (2)	52,036	45,981	46,970	47,314	48,525	42,772	45,616	54,029	44,229	49,996	50,920	1
2	France (1)	52,127	45,672	42,654	46,269	44,381	50,757	41,548	42,134	46,534	46,977	45,367	2
3	Spain (3)	38,273	36,408	35,913	36,093	35,353	33,397	31,123	45,308	39,494	37,703	39,670	3
4	United States (4)	19,440	19,870	19,340	21,965	20,887	19,140	21,650	24,366	23,098	21,731	23,715	4
5	China (7)	11,900	12,500	12,600	12,800	13,000	13,426	16,065	13,693	13,496	13,345	13,217	6
6	Australia (6)	14,263	9,620	12,448	11,784	11,420	11,180	12,259	12,310	11,863	11,912	13,100	7
7	South Africa (8)	9,398	9,783	10,165	9,986	9,327	9,725	10,569	10,982	11,460	11,231	10,531	8
8	Chile (10)	8,448	8,227	8,683	10,093	8,844	10,464	12,554	12,821	9,896	12,866	10,143	9
9	Argentina (5)	15,396	15,046	14,676	12,135	16,250	15,473	11,778	14,984	15,197	13,362	9,447	5
10	Germany (9)	8,916	10,261	9,991	9,228	6,906	9,132	9,012	8,409	9,202	8,819	9,013	10
11	Russian Federation (12)	6,280	7,280	7,110	3,911	7,605	6,960	6,237	5,064	5,112	5,594	6,646	12
12	Portugal (11)	7,542	6,073	5,689	5,894	7,148	5,622	6,327	6,231	6,206	7,048	6,010	11
13	Romania (13)	5,014	5,289	5,159	6,703	3,287	4,058	3,311	5,113	3,750	3,628	3,267	13
14	New Zealand (20)	1,332	1,476	2,052	2,050	1,900	2,350	1,940	2,484	3,204	2,347	3,139	19
15	Hungary (15)	3,100	3,222	3,460	3,065	1,646	2,508	1,765	2,644	2,427	2,572	2,545	16
16	Greece (14)	3,938	3,511	3,869	3,366	2,950	2,750	3,115	3,343	2,800	2,501	2,490	14
17	Austria (17)	2,256	2,628	2,993	2,352	1,737	2,814	2,125	2,392	1,999	2,268	1,953	18
18	Brazil (16)	2,372	3,502	3,683	2,720	2,459	3,460	2,967	2,710	2,607	2,699	1,257	15
19	Bulgaria (19)	1,757	1,796	1,617	1,246	1,030	1,098	1,337	1,755	833	1,367	1,206	21
20	Ukraine (18)	2,160	2,516	2,651	3,181	3,002	3,170	2,400	2,827	1,507	1,117	1,123	17
21	Switzerland (24)	1,011	1,039	1,072	1,112	1,030	1,119	1,004	839	934	850	1,077	24
22	Georgia (25)	950	1,475	1,105	1,043	1,034	1,108	830	997	920	1,196	884	23
23	Japan (27)	900	630	660	730	750	790	800	800	850	741	789	27
24	Republic of North Macedonia (28)	615	1,075	924	955	772	665	781	1,024	510	797	785	25
25	Croatia (22)	1,237	1,365	1,276	1,424	1,433	1,409	1,293	1,248	842	992	760	22
26	Uruguay (26)	939	884	852	724	769	901	962	667	670	658	755	26
27	Peru (29)	600	610	620	515	520	630	650	700	730	747	750	28
28	Serbia (21)	1,292	1,670	1,929	2,392	1,200	1,949	1,581	1,200	734	758	648	20
29	Czech Republic (34)	432	820	840	570	564	650	470	501	536	819	631	29
30	Algeria (23)	1,050	520	500	588	475	480	492	498	520	515	574	30
31	Canada (31)	527	502	543	454	507	614	620	697	530	560	548	32
32	Slovenia (32)	527	664	555	593	570	638	485	526	494	617	497	31
33	Turkey (38)	253	244	393	476	601	571	546	604	604	617	496	33
34	Turkmenistan (33)	438	444	390	402	390	390	390	390	442	452	455	35
35	Mexico (30)	568	630	557	353	346	385	394	418	397	425	404	34
	Other countries	3,520	3,260	3,465	3,215	3,125	3,360	3,484	3,647	3,494	3,737	3,706	
	World total	**280,807**	**266,493**	**267,404**	**267,701**	**261,743**	**265,915**	**258,480**	**288,355**	**268,121**	**273,564**	**268,518**	

Source: COPS; CIES; Wine Institute (USA); TDA; and individual national statistics
Note: The ranking for the previous period appears in parantheses after country name

Decline of indigenous varieties

Far from being under the threat of decline, these varieties owe their survival to the likes of Chardonnay and Cabernet Sauvignon muscling in on their areas of production. Had the so-called international varieties not achieved a certain success, these areas would have been grubbed up long ago (between 1980 and 1994, the area under vine throughout the world plummeted from over 10 million hectares to less than 8 million, this loss almost entirely stemming from the Old World). Having demonstrated a suitability for viticulture via yardsticks like Chardonnay, local winemakers started to apply the standards learned to the indigenous varieties that had once been the underperformers. In the New World, of course, there are virtually no indigenous grapes suited to winemaking, but again the experience with benchmark international varieties has encouraged growers in new and emerging wine areas to experiment with other European varieties. In both Old and New Worlds, there is too much ordinary Chardonnay being made, but it is only too much for critics and wine enthusiasts, because if it were too much for most consumers, its producers would go out of business and the wines would not exist.

Wine lovers of a certain age like to think that 1990 was not that long ago and that the wine world was already dominated by classic grape varieties. Yet, as the table below illustrates, many of the most widely planted varieties back then were from a bygone era, and although there has not been much movement in the Top 10 over the last decade, Airén is still a major force.

10 MOST WIDELY CULTIVATED GRAPES			
	1990	2010	2020	2020 HECTARES (Acres)
1	Airén	Cabernet Sauvignon	Cabernet Sauvignon	340,000 (840,000)
2	Garnacha Tinta	Merlot	Merlot	266,000 (657,300)
3	Rkatsiteli	Airén	Tempranillo	231,000 (570,800)
4	Sultana	Tempranillo	Airén	218,000 (538,700)
5	Ugni Blanc/ Trebbiano	Chardonnay	Chardonnay	211,000 (518,900)
6	Mazuelo	Syrah	Syrah	190,000 (470 000)
7	Merlot	Grenache Noir	Grenache Noir	163,000 (402,780)
8	Cabernet Sauvignon	Sauvignon Blanc	Sauvignon Blanc	121,000 (299 000)
9	Monastrell	Ugni Blanc/ Trebbiano	Pinot Noir	115,000 (285,000)
10	Bobal	Pinot Noir	Ugni Blanc/ Trebbiano	111,000 (274,300)

Homogenization of style

Technology is what you make of it. It can enable vast volumes of technically correct wines to be produced with more fruit and fresher flavours, thus enticing greater numbers of consumers to migrate from popular to finer wines, and that's not a bad thing. When the best technology is combined with the best traditions, it can also improve the quality and expressiveness of highly individual wines, and that's not bad either.

What about so-called flying winemakers? Are they responsible for homogenization in style? I cannot imagine how they can be, as they make less than 1 per cent of the world's wine. What flying winemakers have done is to leave a certain level of cleanliness and competence in their wake, and to show winemakers in underperforming wineries what can be achieved.

WORLD'S LARGEST WINE GROUPS (BY PRODUCTION)	
WINE GROUP	CASES
E & J Gallo	104 million*
Castel Group	53 million
The Wine Group	53 million
Treasury Wine Estates	35 million**
Viña Concha y Toro	35 million
Accolade Wines	34 million
Constellation Brands	31 million*
FeCoVitA	30 million
Grupo Peñaflor	22 million

* See also p 571 for US figures
** See also p 573 for US figures

WHITE AIRÉN GRAPES RIPEN IN A SPANISH VINEYARD
This variety of *Vitis vinifera* is native to Spain. Although it has dropped from its former first-place position, it is still one of the most widely cultivated grapes in the world.

They have also created a reverse flow in terms of where winemakers go to seek practical experience. Over the past 30 years, New World winemakers had gone to Europe to pick up tips on fruit handling and winemaking techniques; now the most inquisitive Europeans go to Australia, New Zealand, or California to learn.

As for schools of oenology, it used to be de rigueur to have the likes of Bordeaux, Dijon, or Montpellier on a winemaker's CV, but now the university faculties at Roseworthy (Australia) and Davis (USA) are the fashion, as much in Europe as the New World itself. There might be a danger that these schools are producing technically correct oenologists trained not to take risks, rather than winemakers whose passion is intimately entwined with the sort of risk-taking that can result in something truly special. Ultimately, however, it is better for a winemaker to understand what he or she is doing, though even that statement has its exceptions, with many a self-taught winemaker who has never opened a book on oenology producing some of the world's most stunning wines.

The Wines of France

FRENCH WINES ARE REGARDED AS THE BEST in the world. Well, according to the French they are, but a thread of this belief is even shared by France's fiercest New World competitors. Although the winemakers of Australia and California no longer try to copy famous French wine styles, they still consider them benchmarks. The great French wine regions are a fortunate accident of geography, climate, and *terroir*. Unlike Italy, the most famous wine regions of France are limited in number and simple to locate in the mind's eye, each with its own distinct identity. No other winemaking country in the world has such a wide range of cool climates; this factor has enabled France to produce the every imaginable type of classic wine style. Over many centuries of trial and error, the French have discovered that specific grapes are suited to certain soils, and through this, distinctive regional wine styles have evolved, so that every wine drinker knows what to expect from each bottle. Enshrining this in *Appellation d'Origine Contrôlée* laws has been the key to success for French wines, however bureaucratically frustrating and controlling that system might be.

THE SPECTACULAR GLASS TOWER OF LA CITÉ DU VIN EVOKES THE MOVEMENT OF WINE SWIRLING IN A GLASS
The Cité du Vin in the Bassins à Flot district of Bordeaux opened its doors in 2016. It is the world's largest museum dedicated to wine and celebrates its production and history from ancient times to the present day.

France

Although the success of French wine has been built on deservedly famous regions that have been enshrined by Appellation d'Origine Contrôlée (AOC) laws, the unwillingness and, at times, practical impossibility to police this system in any meaningful way gradually debased historic reputations at the precise point in history when New World producers had been eager to establish their own wines. This led to competition on export markets in the first decade of the new millennium, when the French simply could not compete on price, particularly against Australian wines.

France has a total of 793,000 hectares (1,959,582 acres) under vine, including those for Cognac, Armagnac, other brandies, vermouth, vinegar, and grape juice, producing an average of just over 49 million hectolitres (544 million cases) of wine across this spectrum every year. Since 2000 French wine production has dropped by 25 per cent in an attempt to reduce surplus production and encourage consumers to drink less wine, but of better quality.

ORIGIN OF APPELLATIONS

The fame of individual French wines goes back centuries, of course, but the first attempt to establish a French appellation by law (albeit rather rudimentary and in retrospect conspicuously incomplete) was for Champagne, and that was as recent as 1905. It was, however, the more adequate Law of 6 May 1919 that today is considered to be the foundation of the *Appellations d'Origine* system, although the creation of appellations did not commence until 1935 when the *Comité National des Appellations d'Origine* (CNAO) was founded. Although the 1919 law is general in nature, covering all agricultural products, there is a special mention for wines and spirits and, because of earlier, unique laws protecting Champagne (not just in 1905, but also 1908 and three separate laws in 1911), Articles 16 to 24 of

the 1919 law refer to Champagne alone. This is not the earliest law concerning protection of provenance, but it is probably the most important, and as such it forms Title II of Book VII of the *Code de Propriété Intellectuelle,* where it is recognised as a special form of trademark law.

In 1935, when CNAO (it was not known as the *Institut National des Appellations d'Origine,* or INAO, until 1947) was established, France became the first nation to set up a countrywide system for controlling the origin and authenticity of its wines, although individual winegrowing areas outside France had formalised their own quality controls much earlier (Chianti in 1716 and Rioja in 1560, for example). The task of CNAO (or INAO as it later became) was to devise and enforce regulations governing the production of each appellation. It did this, essentially, by taking stock of how each wine was made (grape varieties, viticultural and winemaking habits, and other factors) and authorised almost every practise according to "honest and traditional local usage", one of the basic principles underlying the ethos of the French AOC system. The first *Vins de Pays* were authorised five years earlier, in 1930, and in 1949 *Vins Délimité de Qualité Supérieur* (VDQS) were established as a halfway house for *Vins de Pays* aspiring to become AOCs. Below *Vins de Pays,* wine was sold as *vins de table* and could not indicate grape varieties, vintage, or provenance on its label.

THE LANDSCAPE OF THE MONBAZILLAC WINE REGION IS PAINTED LUSH GREEN AT THE HEIGHT OF SUMMER
After its formation in 1935, the CNAO had by the next year received and approved six applications for protected designations of origin. On 15 May 1936, then President Albert Lebrun signed the decree granting AOC status for Arbois, Tavel, Cognac, Cassis, Monbazillac, and Châteauneuf-du-Pape. Located in the wider region of Bergerac in southwestern France, Monbazillas AOC is for sweet botrytised wines that are neither as famous nor as pricey as those made in neighbouring Sauternes.

WINE NOTES

- There are more than 87,400 vineyard owners in France, of which 68,500 make and sell wine, but only 46,600 are producers of AOC/AOP wines.
- 74 per cent of all French vineyards are machine harvested.
- 9 per cent of all French vineyards are farmed organically or biodynamically.
- 51 per cent of all French wine is processed by 774 *cave coopératives*.
- Merlot is the most planted grape variety of any colour in France.
- Ugni Blanc is the widest planted white grape variety in France.

WINE REGIONS OF FRANCE
The coloured areas on this map identify the 10 main wine-producing regions of France, where the areas of *Appellation d'Origine Contrôlée,* which cover 474,000 hectares (1.2 million acres), are concentrated. The country has more than 800,000 hectares of vineyards in total, however, and many good, everyday-drinking wines are made in other parts of the country. *See also Vin de Pays* maps A (p331) and B, (p332).

Wine regions map of France with labelled regions: *Alsace and Lorraine* p268, *Champagne* p249, *Jura, Savoie, and Bugey* p309, *Burgundy* p215, *Loire Valley* p284, *Rhône Valley* p299, *Bordeaux,* p140, *Southwest France* p311, *Languedoc–Roussillon* p318, *Provence and Corsica* p326.

FRENCH WINE PRODUCTION BY TYPE

MILLIONS OF HECTOLITRES

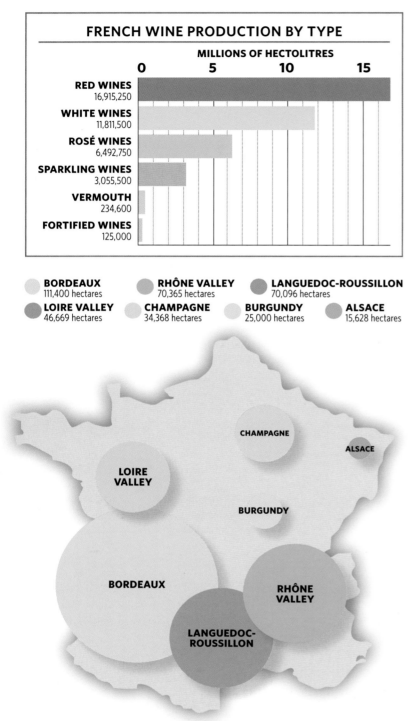

RED WINES 16,915,250	
WHITE WINES 11,811,500	
ROSÉ WINES 6,492,750	
SPARKLING WINES 3,055,500	
VERMOUTH 234,600	
FORTIFIED WINES 125,000	

- **BORDEAUX** 111,400 hectares
- **RHÔNE VALLEY** 70,365 hectares
- **LANGUEDOC-ROUSSILLON** 70,096 hectares
- **LOIRE VALLEY** 46,669 hectares
- **CHAMPAGNE** 34,368 hectares
- **BURGUNDY** 25,000 hectares
- **ALSACE** 15,628 hectares

MAJOR AOC/AOP WINE-PRODUCING REGIONS
From this map, it is easy to discern the relative size of production of the seven most important wine regions of France, which in total represent 373,526 hectares, or 84.4 per cent of the country's AOC/AOP vineyards.

BREAKDOWN OF FRENCH WINE PRODUCTION MEASURED IN HECTOLITRES

MILLIONS OF HECTOLITRES

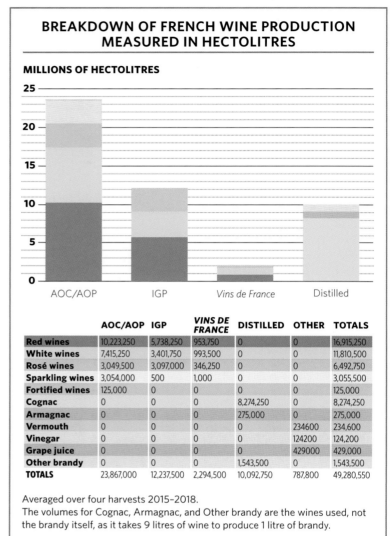

	AOC/AOP	IGP	VINS DE FRANCE	DISTILLED	OTHER	TOTALS
Red wines	10,223,250	5,738,250	953,750	0	0	16,915,250
White wines	7,415,250	3,401,750	993,500	0	0	11,810,500
Rosé wines	3,049,500	3,097,000	346,250	0	0	6,492,750
Sparkling wines	3,054,000	500	1,000	0	0	3,055,500
Fortified wines	125,000	0	0	0	0	125,000
Cognac	0	0	0	8,274,250	0	8,274,250
Armagnac	0	0	0	275,000	0	275,000
Vermouth	0	0	0	0	234600	234,600
Vinegar	0	0	0	0	124200	124,200
Grape juice	0	0	0	0	429000	429,000
Other brandy	0	0	0	1,543,500	0	1,543,500
TOTALS	23,867,000	12,237,500	2,294,500	10,092,750	787,800	49,280,550

Averaged over four harvests 2015–2018.
The volumes for Cognac, Armagnac, and Other brandy are the wines used, not the brandy itself, as it takes 9 litres of wine to produce 1 litre of brandy.

There are now in excess of 490 AOCs (some authorities state there are only 360 AOCs, but there are almost 500 if you count the individual designations within some multi-AOCs, such as the Mâcon with village AOC – which is a completely separate to Mâcon-Villages AOC – and encompasses 41 different geographically based subappellations. If wine sold as Mâcon Azé cannot legally be sold Mâcon Burgy or Mâcon Fuissé or Mâcon Uchizy, they must all be separate appellations.). The 490 AOCs cover 442,562 hectares (1,093,614 acres) and produce just under 24 million hectolitres (267 million cases) of wine. The AOC laws regulate the grape varieties used, viticultural practices, harvest and yield restrictions, minimum alcohol content, and winemaking techniques for each area. These rules even restrict the number of fruiting buds on a branch, but you will never see INAO inspectors checking what varieties are growing in the middle of a vineyard, let alone counting all of its fruiting buds. The system is obviously deficient, as evidenced by the glut of very poor-quality AOCs that are available at French supermarkets. The best AOC wines have always been superb, but without some sort of overhaul of the system, the onus remains very much on the knowledge of consumers to discern the reputation of individual producers. One fact of life every French wine lover should understand about the French appellation system in general and AOCs in particular is that it was not set up to guarantee quality, and it most certainly does not protect the consumer. Its historic purpose has always been to protect producers from fraudulent competition. Not a bad thing, for sure, but as the late René Renou (director of INAO, 2000–2006) told *The Wine Spectator* in 2004 "You'd be surprised how many AOCs have rules that are empty of meaning, consumers can't trust our AOCs".

RESETTING THE SYSTEM

Renou's comment was merely an echo of Alain Berger (director of INAO, 1990–1996), who had declared a decade earlier that "One can find on the market scandalously poor products with the *appellation contrôlée* halo". Renou was not content merely to confirm Berger's opinion, however; he was determined to do something about it. When he proposed a super-appellation for wines of "demonstrably superior quality", he shook the tradition and thus the very foundation of French culture. This is a country that produces both the best and the worst wines in the world because of its fixation with tradition. The French are always willing to preserve the best traditions, but unfortunately they are also loath to banish the worst.

Some ultra-traditionalists believed that any fundamental shake-up to a system that gave birth to wine regimes throughout Europe would threaten the moral authority of French wine, thus Renou's proposals were met with hostility in most quarters, while a few realists believed it to be an absolute necessity.

Those who supported Renou were, essentially, a quality-conscious minority, while those who outrightly opposed him included the worst of the worst and knew they could rely on the vast majority, who simply felt so safe with their traditions that they did not wish to rock the boat.

Initially, there was good news and bad news. The good news was that the Burgundians were for it. The bad was that the *bordelais* were not. Even the good news turned out to be bad, however, when the supposedly pro-Renou Jean-François Delorme, President of the Bureau Interprofessionnel des Vins de Bourgogne (BIVB), commented: "Of course, all of Burgundy's AOCs would be super-AOCs".

If every single bottle of Burgundy was the quality of the greatest wines of, say, Leroy, Dujac, or Georges Roumier, then no one would argue, but much too much of it is no better than *vin ordinaire,* while – as with every wine region – some wines can be almost undrinkable. So no, a pan-Burgundian application for super-AOCs would not be very super at all. Being no fool,

Renou allowed his historic proposal to "slip out" in London, where he could rely on its radical aims receiving widespread enthusiastic approval from blatantly objective third-party sources.

After presenting the plan in detail to the INAO's 80-member National Committee, he fleshed out more details about his super-appellation, which he called *Appellation d'Origine Contrôlée d'Excellence* (AOCE), and threw another spanner in the traditionalist works by introducing the novel concept of *Site et Terroir d'Excellence* (STE) for stand-out individual vineyards in relatively lesser appellations.

On past experience, the odds always were that a compromise would be sought. That the aims would become so fudged and the regulations so watered-down that they would not only be simply ineffective but would also probably be detrimental to the future of French wine. And that is precisely what has happened – but these new appellations and the reasons why they were rejected need exploring in more detail so that readers can at least understand Renou's good intentions and appreciate how catastrophically the French missed an opportunity to regain both moral and commercial dominance over the rest of the wine world. And why the growers and government minister involved deserve to reap whatever ill fortune they have sown.

PATRONS SIP WINE AND OTHER DRINKS ON THE TERRACE OF A PARISIAN CAFE
Red leads the other wine styles in total production figures, with about 70 per cent of it being consumed domestically – by natives and tourists alike. Although less white wine is produced than red in France, a greater proportion of it goes to export markets, with more than half of it consumed outside the country.

(Protected Geographical Indication) or IGP (*Indication Géographique Protegée*) in French to encompass *Vins de Pays*. AOP and IGP came into force on 1 August 2009. Reports that *Vin de Table* would be scrapped were misleading. It is more accurate to state that the name changed to *Vin de France* and the wines under this designation became able to mention the vintage and grape variety.

With the rejection of Renou's revolutionary proposals, Gaymard watered-down compromise of "tighter controls", the EU's new labelling regulations, and stiff competition, not so much at home, but on export markets, where wines from the New World, particularly Australia, were literally forcing French wines off the shelf, France desperately needed a flexibility in wine production and labelling that would enable it to fight back not only on quality, but on price too. The INAO therefore called for new *cahier des charges*, the basic rules upon which appellation are regulated, as a chance to start over again. INAO took the opportunity of the EU's new wine labelling regulations to shake-up the system and hopefully make it easier to compete with the New World. All appellations were thus

10 MOST WIDELY PLANTED GRAPE VARIETIES IN FRANCE BY COLOUR

For the second decade of the 21st century, there was no change whatsoever for black grape varieties in terms of positioning and very little as far as surface area is concerned, but the big differences for white grape varieties is the increase in Colombard and Vermentino, and the drop in Melon de Bourgogne of Muscadet fame. Previous period ranking appears in parentheses after the varietal name.

	2019	10 MOST WIDELY CULTIVATED GRAPES	
	2019	**BLACK GRAPES**	**2010**
1	115,723	Merlot (1)	114,675
2	85,555	Grenache (2)	90,991
3	67,152	Syrah (3)	67,382
4	47,718	Cabernet Sauvignon (4)	54,434
5	32,683	Cabernet Franc (6)	36,302
6	32,633	Carignan (5)	47,721
7	32,269	Pinot Noir (7)	30,086
8	24,192	Gamay (8)	29,698
9	19,705	Cinsault (9)	19,505
10	10,690	Meunier (10)	11,087
	513,632	**Total France**	**540,200**
	2019	**WHITE GRAPES**	**2010**
1	89,864	Ugni Blanc (1)	83,445
2	43,673	Chardonnay (2)	45,243
3	31,558	Sauvignon Blanc (3)	27,931
4	11,316	Colombard (7)	8,173
5	10,305	Sémillon (5)	11,566
6	10,280	Chenin Blanc (6)	9,825
7	8,713	Melon de Bourgogne (4)	12,305
8	7,547	Muscat Blanc à PG (8)	7,671
9	6,853	Viognier (9)	4,823
10	6,704	Vermentino (15)	3,089
	273,201	**Total France**	**201,300**

given the chance to turn the clock back. They were created merely as an official endorsement of local tradition and some of those of local traditions were less principled than others. Over the years these imperfect rules have been added to and, in some, case, bent in order to placate local politics. This was why each appellation was given the opportunity to go back to the basic EU wine regime requirements and to build in any new rules as they see fit. Some appellations were carrying so much extra baggage that they appeared to glad to simplify the day-to-day lives of everyone involved, but in the end, relatively few appellations made full use of this opportunity.

One small but positive outcome was the phasing out of VDQS, however. The final vintage under which a wine may proclaim VDQS status on label was 2010, and the category was officially abolished in 2011, when most had already been elevated to AOC status. Few AOCs took advantage of new *cahier des charges* to give their wines any significant advantage when fighting New World wines on export markets. About the only noticeable difference was that AOP has replaced AOC on many, but certainly not all, labels. AOP is not a direct replacement for AOC (or any other national designation), but it can be used on the label, just as VQPRD could have been, and very occasionally was. VQPRD was perhaps avoided because it could have been mistaken for VDQS, but the similarity of AOP to AOC has meant that it has been far more widely adopted. The rules have always enabled *Vin de Pays* to fight New World brands on a more level playing field, and the much shorter IGP classification has generally been regarded as fresher and easier to sell, making it one of the two categories to benefit most from the new regulations. The other is, of course, *Vin de France*. There was never any chance of *vin de table* making much of an impact on global markets. The name had such a low-quality connotation that its producers might as well have labelled it *vin de merde*. Without any indication of grape variety or vintage, there was little inclination to produce anything special, but *Vin de France* has become somewhat cool, and its intrinsic ability to blend from all over the country makes it the most flexible wine in the French armoury to do battle with the cheapest New World brands.

IGP

There are currently 177 appellations, covering 193,358 hectares (477,807 acres), and producing an annual average of just over 12 million hectolitres (133 million cases). Although legally established in 1930, *vins de pays* as we know them date from 1968 (and IGP as we know them now) did not become a marketable reality until 1973, when the rules for production were fixed. They were created, quite simply, by authorizing some *vins de table* to indicate their specific geographic origin and, if desired, the grape varieties used, but the effect was more profound than that. For a long while, it was even obligatory for these wines to carry the term *"Vin de Table"* in addition to their *vin de pays* appellation, but this regulation was discontinued in 1989 and *Vin de Table* itself was replaced by *Vin de France* in 2009. Allowing producers of potentially superior table wine areas to declare, for the first time, the origin and content of their wines gave back a long-lost pride to the best growers, who started to care more about how they tended their vines and made their wines and began to uproot low-quality varieties, replant with better ones, and restrict yields.

There are three categories of IGP, each with its own controls, but all regulated by the *Office National Interprofessionnel des Vins* (better known as ONIVINS), which merged with ONIFLHOR to become VINIFLHOR and then AgriFranceMer before it was eventually subsumed by INAO in 2016. Although every IGP must have a specified origin on the label, a wide range of grape varieties can be used and high yields are allowed. A certain flexibility on choice of grape varieties allows innovation, a character-building trait that has been denied winemakers under the AOC regime, but the high yields permitted have also caused many an unprincipled producer to jump on the relatively lucrative IGP bandwagon, thus the quality can vary greatly. This is, nevertheless, a most interesting category of wines, often infiltrated by foreign winemakers producing some of the most exciting, inexpensive, and upfront-fruity wines.

Many IGP are, of course, bland and will never improve, but some will make it up the hierarchical ladder to full AOC status. Certain individual producers, such as the Guiberts of Mas de Daumas Gassac and the Dürrbachs of Domaine de Trévallon, could not care less about AOC status,

FRANCE "SHOT IN FOOT"

A Bordeaux campaign to "drink less, drink better" was banned in 2004, after the National Association for the Prevention of Alcoholism (NAPA) took the Conseil Interprofessionnel des Vins de Bordeaux (CIVB) to court. Any rational person would have to question the motives of an organization that is so opposed to people drinking less, but better wine that it would drag them into a court of law. If NAPA seeks prohibition, it should do so openly. Furthermore, any rational person would also have to question the motives behind a law under which a court is obliged to declare that "drink less, drink better" is illegal because it "incites people to purchase wine". Why should inciting people to purchase less, better-quality wine be against the law? One of the Bordeaux posters was also declared illegal because it depicted someone drinking wine. According to this warped law, it is legal to display an alcoholic product on a poster, but not legal to display that product being consumed. What are people supposed to think they should do with such a product after purchasing it? The law in question is the Loi Evin, enacted in January 1991, and named after Claude Evin, a minister in the Rocard government. And so, finally, we must question this man's motives. Why, when Evin presented his proposal to the French Parliament, did he do so under the banner of "In the case of alcohol, only abuse is dangerous"? Certainly not to mislead Parliament. The National Assembly was informed precisely how far beyond the curbing of abuse this law extended, and was thus as guilty as the infamous Evin. So, if the minister was not hiding the draconian powers of his legislation behind such a misleading slogan from Parliament, who was he hiding them from? The French people?

Other French governments had inherited the Loi Evin, but had done nothing to curb its excesses until July 2004, when to get around its advertising restrictions, Hervé Gaymard, the French agricultural minister, announced plans to reclassify wine as a "food". This was quickly followed by a report requested by Prime Minister Raffarin on how to boost the "struggling wine industry" through advertising "without undermining the principles of consumer protection" in the Loi Evin. In fact, the Loi Evin has no principles of consumer protection – it has no principles, period.

A softening of the legislation has been widely discussed in France ever since it was enacted, but nothing ever happens. For example, in 2019, a group of 105 French MPs from the ruling party wanted to relax the law "in a supervised manner", but it was rejected by a health minister, who wanted to extend its restrictive powers.

---

as they happen to be making wines under the less draconian IGP system. They already have reputations as good as those of the finest French appellations and demand even higher prices than some expensive AOC wines. As a general rule, it is much wiser to pay a premium for an IGP wine than buy any "bargain" priced AOC/AOP. For a description of the various IGP categories, a map showing the areas producing IGP, and profiles of every IGP along with their finest producers, *see* pp331-341.

VIN DE FRANCE

Formerly *Vin de Table* (technically *Vin de Table Français*), this category encompasses 31,900 hectares (just over 76,000 acres) of unclassified vineyards in France, producing an annual average of just under 2.3 million hectolitres (more than 25 million cases) of *Vin de France,* (although this also includes a variable volume of grape juice, grape concentrate, and wine vinegar). This compares with a production of 34 million hectolitres (510 million cases) in the mid-1990s and is evidence of the ruthless culling of low-quality vineyards to reduce surplus production. Part of the reason for the change of name, it is claimed, was to avoid continuing the confusion in the United States, where a table wine is synonymous with a non-fortified wine. Historically, this came about because non-fortified wines were served at the table and thus applied to all such wines, whether lowly or a great growth from Burgundy or Bordeaux, as can be gleaned by reading almost any wine book from the 19th and early-20th centuries. It was never a derogatory term until the French made it so, and it is perhaps more derogatory in the United Kingdom than the United States.

Vin de France, like *Vin de Table*, might be at bottom of the quality ladder, but its wines do not have to be, and since the change from *Vin de Table* to *Vin de France*, the quality of many of these wines has soared. Just as the less strict regulations give IGP the flexibility to produce some outstanding wines that would not normally be permitted in their area of origin, so the even less strict *Vin de France* regulations give *Vin de France* even more flexibility to pull something really special out of the bag. We saw the possibility of such wines even under the old *Vin de Table* regulations, when Zind (a mouthwatering blend of Auxerrois, Pinot Blanc, and Chardonnay from three exceptional *lieux-dits* produced by Zind Humbrecht in Alsace) and historical wines such as Libre Expression (effectively a late-harvest version of Rivesaltes from Domaine Cazes that is dry and not fortified) and Muscat de Petit Grains Passerillé Vendange d'Octobre (a top-class oak-aged dessert wine produced from shrivelled grapes grown on old vines) commanded higher prices and exuded more quality and individual style than most AOC wines produced in the same area.

MERLOT GRAPES RIPEN IN A SAINT-ÉMILION VINEYARD
Merlot, a dark blue wine grape variety has long been associated with Bordeaux, where its was first mentioned in the late 18th century. The name "Merlot" derives from the French dialectical *merle* or *merlau,* the local name of blackbirds in the region. The Merlot is now the most commonly grown grape variety in France.

FRENCH LABEL LANGUAGE

Année Literally "year; this will refer to the vintage.

AOC or AOP *Appellation d'Origine Contrôlée* or *Appellation d'Origine Protegée,* the top official classification

Barriques Small oak barrels of approximately 225 litres capacity

Blanc White

Blanc de blancs Literally "white of whites"; a white wine made from white grapes, a term that is often, but not exclusively, used for sparkling wines

Blanc de noirs Literally "white of blacks"; a white wine made from black grapes, a term that is often, but not exclusively, used for sparkling wines. In the New World, the wines usually have a tinge of pink, often no different from a fully-fledged rosé, but a classic *blanc de noirs* should be as white as possible without artificial means.

Brut Normally reserved for sparkling wines, *brut* literally means "raw" or "bone dry", but in practice there is always some sweetness and so can at the most be termed only "dry".

Cave coopérative Cooperative cellars where members take their grapes at harvest time, and the wines produced are marketed under one label. Basic products will often be blended, but top-of-the-range wines can be from

very specific locations. A lot of *cooperative* wines can be ordinary, but the best *cooperatives* usually offer exceptional value.

Cépage This means "grape variety" and is sometimes used on the label immediately prior to the variety, thus *"Cépage Mauzac"* means that the wine is made from the Mauzac grape variety.

Châtaignier Chestnut (occasionally used to make barrels)

Château Literally "castle" or "mansion", but whereas many wines do actually come from magnificent edifices that could truly be described as *châteaux,* many that claim this description are merely modest one-storey villas and some are no more than purpose-built *cuveries,* while a few are actually tin sheds. The legal connotation of a *château*-bottled wine is the same as for any domaine-bottled wine.

Chêne Oak (commonly used to make barrels)

Clairet A wine that falls somewhere between a dark rosé and a light red wine

Climat A single plot of land with its own name, located within a specific vineyard

Clos Synonymous with *climat,* except this plot of land is either enclosed by walls, or was once

Côte or **côtes** Literally "slope(s)" or "hillside(s)" of one contiguous slope or hill, but in practical terms often no different from *coteau* and *coteaux*

Coteau or **coteaux** Literally "slope(s)" or "hillside(s)" in a hilly area that is not contiguous, but in practical terms often no different from *côte* and *côtes*

Crémant Although traditionally ascribed to a Champagne with a low pressure and a soft, creamy *mousse,* the term has now been phased out in Champagne as part of the bargain struck with other French sparkling wines that have agreed to drop the term *méthode champenoise.* In return they have been exclusively permitted to use this old Champagne term to create their own appellations, such as Crémant de Bourgogne or Crémant d'Alsace.

Cru or **crû** Literally means a "growth" or "plot of land"; this term usually implies that it has been officially designated in some way, such as a *cru bourgeois, cru classé, premier cru,* or *grand cru.*

Cru bourgeois An officially designated growth below *cru classé* in the Médoc.

Cru classé An officially classified vineyard in the Médoc and Provence

Cuve A vat. *Cuve* should not be confused with *cuvée.*

Cuve close A sparkling wine that has undergone second fermentation in a vat. *Cuve close* is synonymous with Charmat or Tank Method.

Cuvée Originally the wine of one *cuve,* or vat, but now refers to a specific blend or product that, in current commercial terms, will be from several vats, this term is used most commonly, but not exclusively, in Champagne.

Dégorgé or **dégorgement** Found on some sparkling wines, this term will be followed by the date when the wine was disgorged of the yeast sediment that has built up during its second fermentation in bottle.

Demi-muid A large oval barrel, usually made of oak, with a capacity of 300 litres (600 litres in Champagne)

Demi-sec Literally "semi-dry", but actually tastes quite sweet

Domaine Closer to "estate" than "vineyard", as a domaine can consist

of vineyards scattered over several appellations in one region. The wine in the bottle must come from the domaine indicated on the label. Some former *"châteaux"* have changed to "domaine" in recent years.

Doux A sweet wine

Encépagement The proportion of grape varieties in a blend or *cuvée*

Elevage par X or **eleveur X** Refers to the traditional function of a *négociant* (X), who has not made the wine initially, but purchased it, after which the *négociant* will age, possibly blend, bottle, and sell the wine

Elevée en … Aged in ….

Elevée et mise en bouteille par X The wine was aged and bottled by X, but none or only part of the wine was produced by X.

Foudre A large wooden cask or vat

Fûts Barrels, but not necessarily small ones

Grand cru Literally "great growth"; this term indicates the wine has come from a truly great vineyard (which, to put it in context, a lousy producer can make a hash of) in regions such as Burgundy and Alsace, where its use is strictly controlled. In Champagne, *grand cru* is much less meaningful, as it applies to entire villages, which might contain most of the best vineyards in the region, but where lesser vineyards also exist.

Grand vin This term has no legal connotation whatsoever, but when used properly in Bordeaux, *grand vin* applies to the main wine sold under the *château*'s famous name and will have been produced from only the finest barrels. Wines excluded during this process go into second, third, and sometimes fourth wines that are sold under different labels.

IGP (Indication Géographique Protegée) This is the French equivalent of the EU category Protected Geographical Indication. This has replaced *Vins de Pays.*

Lie French for "lee": *sur lie* refers to a wine kept in contact with its lees.

Lieu-dit A named site; this term is commonly used for a wine from a specific, named site.

Médaille Medal: *médaille d'or* is gold medal, *médaille d'argent* is silver, and *médaille de bronze* is bronze.

AN ARCH LEADS INTO THE *GRAND CRU* VINEYARD OF CLOS DE LA PUCELLE
Producers with vineyards that are enclosed by fences or walls will often have "clos" as part of their names, even if the plots are no longer enclosed.

Méthode ancestrale A sweet sparkling wine from Limoux that is produced by a variant of *méthode rurale*, a process that involves no secondary fermentation, the wine being bottled before the first alcoholic fermentation has finished

Méthode champenoise Contrary to popular belief, this term is not completely banned in the EU: it is expressly reserved for Champagne, the producers of which seldom bother to use it (cannot say "never" because I saw it once). *Méthode champenoise* is the process whereby an effervescence is produced through a secondary fermentation in the same bottle in which it is eventually sold.

Méthode classique One of the legal terms for a sparkling wine made by the *méthode champenoise*

Méthode deuxième fermentation, traditionnelle, méthode traditionnelle classique These are legal terms for a sparkling wine made by the *méthode champenoise* (*méthode deuxième fermentation* is used only in Gaillac).

Millésime or **millésimé** Refers to the vintage year

Mis en bouteille au domaine or **à la propriété** or **le château** This means "bottled at the domaine/property/*château*" indicated on the label (the wine must have been made exclusively from grapes grown there).

Mis en bouteille par ... Bottled by ... (not necessarily made by)

Mistelle Fresh grape juice that has been muted with alcohol before any fermentation can take place

Moelleux Literally "soft" or "smooth"; this term usually implies a rich, medium-sweet style in most French areas, except the Loire, where it is used to indicate a truly rich, sweet botrytis wine, thereby distinguishing it from *demi-sec*. This term was due to be obligatory in Alsace for wines that have a minimum residual sugar of 12 grams per litre (9 g/l for Riesling), but it was dropped after being vetoed by growers.

Monopole With one exception, this means that the wine comes from a single vineyard (usually a *grand cru*) that is under the sole ownership of the wine producer in question. The exception is Champagne Heidsieck & Co Monopole, which is just a trading name.

Mousse The effervescence of a sparkling wine

Mousseux Sparkling, found only on cheap fizz

Négociant Literally a "trader" or "merchant"; this name is used by wine producers whose business depends on buying in raw materials (the name is derived from the traditional practice of negotiating with growers to buy grapes and/or wine, and with wholesalers and retailers to sell wine).

Négociant-éleveur Wine firm that buys in ready-made wines for *élevage*

Négociant-propriétaire A *négociant* who owns vineyards and probably offers a range of its own produced wines

Non filtré This is a wine that has not been filtered, thus it is liable to drop a sediment much earlier than most other wines, and should always be decanted.

Oeil de perdrix Literally "partridge eye"; refers to a rosé-coloured wine.

Perlant Very slightly sparkling, even less so than *pétillant*

Pétillance, pétillant A wine with enough carbonic gas to create a light sparkle

Premier cru Literally a "first growth"; the status of a *premier cru* is beneath that of *grand cru* but only of relevance in areas where the term is controlled, such as Burgundy and Alsace. In Champagne, *premier cru* is much less meaningful, as it applies to entire villages, of which there are far too many and in which much lesser vineyards also exist.

Produit Product, as in Product of ...

Propriétaire-récoltant Grower-producer whose wine will be made exclusively from his own property

Récolte Refers to the vintage year

Réserve Ubiquitous and meaningless

Rosé Pink

Rouge Red

Rubis Sometimes seen on rosé wines, particularly, but not exclusively, on darker styles

Saignée A rosé wine produced by "bleeding" off surplus liquid from either the press or the fermenting vat

Sec Dry, but in a sparkling wine this will have more sweetness than a *brut*.

Sélection de grains nobles A rare, intensely sweet wine in Alsace made from selected botrytised grapes

Sélection par X Selected, but not produced or bottled by X

Supérieur As part of an appellation (for example, Mâcon Supérieur as opposed to Mâcon plain and simple); this refers not so much to a superior quality than – possibly – a superior alcoholic strength (the minimum criteria being 0.5 per cent ABV higher than the minimum for the same appellation without the *Supérieur* appendage).

Sur lie Refers to wines, usually Muscadets, that have been kept on their lees and have not been racked or filtered prior to bottling. This enhances the fruit of the normally bland Melon de Bourgogne grape used in Muscadet, often imparting a certain liveliness.

Tête de cuvée No legal obligation, but this term strongly suggests that a wine has been produced exclusively from the first flow of juice during the pressing, which is the highest in quality with the most acids, sugars, and minerals.

Tonneau A large barrel, traditionally four times the size of a *barrique*

Trie Usually used to describe the harvesting of selected overripe or botrytised grapes by numerous sweeps (or *tries*) through the vineyard

Union coopérative A regional collective of local cooperatives

Vendange tardive Late harvest, implying a sweet wine

Vieilles vignes Supposedly from old vines, implying lower yields and higher concentration. No legal parameters apply, but most reputable producers would not dream of using the term unless the vines are at least 35 to 50 years old.

Vieillissement Ageing

Vigne Vine

Vigneron Literally a vineyard worker, but often used to denote the owner-winemaker

Vignoble Vineyard

Vin Wine

Vin de l'année Synonymous with *vin primeur*

Vin doux naturel A fortified wine, such as Muscat de Beaumes de Venise, that has been muted during fermentation, after it has achieved between 5 and 8 per cent alcohol

Vin de France This has replaced the old *vin de table* terminology and the wines are a lot better now that they can mention grape varieties and vintage, which has given producers something worthwhile to work on and market. *See* p137.

Vin de glace French equivalent of *Eiswein* or icewine

Vin gris A delicate, pale version of rosé

Vin jaune The famous "yellow wine" of the Jura derives its name from the honey-gold colour that results from its deliberate oxidation beneath a Sherry-like *flor*. The result is quite similar to an aged *fino* Sherry, although *vin jaune* is not fortified.

Vin de liqueur A fortified wine that is muted with alcohol before fermentation can begin.

Vin mousseux Literally means "sparkling wine" without any connotation of quality one way or the other, but because all fine sparkling wines in France utilize other terms, it should be taken to mean a cheap, low-quality fizz.

Vin nouveau Synonymous with *vin primeur* – as in Beaujolais Nouveau

Vin de paille Literally "straw wine"; this is a complex sweet wine produced by leaving late-picked grapes to dry and shrivel while hanging from rafters over straw mats.

Vin primeur A young wine sold within weeks of the harvest and made to be drunk within the year. Beaujolais Primeur is the most commonly known example, but numerous other AOCs are allowed to be sold in this fashion, as is every IGT.

Vin de table Means "table wine". This term has been replaced by *Vin de France* (see pp136–137).

Vin d'une nuit A rosé or very pale red wine that is allowed contact with its grapeskins for one night only

Bordeaux

Thanks to its château-based classification system, established more than 150 years ago, Bordeaux has gained an iconic status that can be intimidating to wine lovers. But it would be a mistake to define the region by this one date in history, or by the perceived high prices of the classified growths; it is so much more complex and interesting than that.

In the southwest of France the Bordeaux wine region straddles the 45th parallel, about 50 kilometres (31 miles) inland from the Atlantic coast in the Département de la Gironde. With the city at their heart, the vineyards reach about 80 kilometres (50 miles) up the Médoc peninsula to the north, 80 kilometres (50 miles) to the southeast, 100 kilometres (62 miles) to the east towards the Dordogne, and just 20 kilometres (12 miles) to the west around the airport, where a few token vines are planted to welcome visitors.

This situation and the proximity to the ocean, where the warm Gulf Stream runs along the coast, give a temperate oceanic climate. This also brings rain: between 70 to 80 centimetres (28 to 32 inches) per year. It is an unpredictable climate; no two years seem to have the exact same weather pattern, no two vintages are ever quite the same, hence the importance of the vintage effect.

Bordeaux has been through a series of booms and busts over the centuries. In 56 BC, Burdigala, as Bordeaux was then known, became part of the Roman Empire, the Romans bringing their knowledge of viticulture. The first Bordeaux boom started in the Middle Ages when Eleanor, the Duchess of Aquitaine, married Henry Plantagenet, bringing the region under English rule. Booms have usually come from a foreign impetus: Dutch and German markets grew in the 17th century and Asian in the 21st century. Bordeaux's prosperity and its openness to foreign influence are due in no small part to both the importance of its port and international trade.

Bordeaux was France's largest port in the 17th century; it was the gateway to the trade route between Northern Europe and the South of France via the Canal du Midi and later the Canal de la Gironde linking the Atlantic to the Mediterranean. This key position favoured an international trade for Bordeaux wines that continues to this day: Bordeaux now exports about 42 per cent of its wine production for a value of almost 2 billion euros a year.

The size of Bordeaux means it produces a diversity of wines from everyday drinking to top-end classified growths of world renown. The vast majority of Bordeaux wine produced is everyday drinking wine, accessible by both price and style. The expensive classified, or equivalent, wines account for less than five per cent.

THE MODERN BORDEAUX-STYLE BOTTLE SHAPE
A modern Bordeaux bottle has straight sides and high, distinct shoulders. It is usually used to bottle heavier red wines. Any wine produced in the Bordeaux region is a Bordeaux, but over 90 per cent are reds made with Cabernet Sauvignon and Merlot.

PRESENT-DAY BORDEAUX

Recent years have seen a marked reduction in the size of the Bordeaux vineyard area thanks to a programme of removing land not deemed to be of sufficient potential. In 2006 a Vin de Pays de l'Atlantique was created, encompassing the Gironde, where Bordeaux is made, along with neighbouring departments. This was the first *vins de pays* in the region. A *vins de pays* allows more flexibility in the choice of grape varieties, yields, and labelling. Together with a campaign for improving quality, this reduced the surface area under vine in Bordeaux from about 123,000 hectares (303,940 acres) in the 1990s to 111,400 hectares (275,275 acres) in 2017.

Bordeaux wine remains a big global business; it generates around four billion euros turnover per year and employs 55,000 people in the region, directly or indirectly, which accounts for 20 per cent of total jobs in French viticulture. Over the past 20 years, the number of growers has almost halved, numbering 6,568 in 2016. Between them, they produce about 750 million bottles per year representing 15 per cent of total French wine production. The average property size is 17.5 hectares (about 40 acres). Although this has increased dramatically in the last 20 years from 7 hectares (17 acres), it is still far from the image of large properties that many people hold of Bordeaux. Although there are fewer than 7,000 vineyard owners, there are more than 12,000 château names appearing on Bordeaux labels. Some producers own a number of different châteaux, but only one château name per property may be used; a second name is allowed only if it was in use prior to 1983.

LA PLACE DE BORDEAUX: CHÂTEAUX, BROKERS, AND MERCHANTS

Since the Middle Ages, when Bordeaux was under the English crown, Bordeaux wines have been destined for export rather than French consumption. Wines were sold after the harvest in barrels to *négociants* (merchants) who undertook ageing and blending in their cellars on the waterfront port of Bordeaux. Most wines were shipped in barrels from the château to merchants; these merchants were responsible for the wines' *élevage*, known as *négociants-éleveurs,* and they did the bottling in Bordeaux. Brokers acted as intermediaries between the vineyards (château) and the *négociants,* recording wines according to their *cru,* or growth (their geographical origin), and the prices they fetched; thus the reputations of individual properties became established. The 19th century saw the rise of the *négociant* in Bordeaux. Many *négociants* were of Northern European origin: English, Scottish, Irish, Dutch, or German trading partners established in the port.

THE BORDEAUX WINE TRADE AND THE RISE OF CHÂTEAU BOTTLING

Although the majority (60 per cent) of bottling now takes place at the château (or vineyard), the *négociant* system, known as La Place de Bordeaux, continues to account for 70 per cent of the turnover of the Bordeaux wine business. Three hundred Bordeaux *négociants* buy and sell wines in both bottle and in bulk to 170 different countries worldwide, and 82 brokers are involved in over 80 per cent of the transactions between owner-seller and *négociant*-buyer.

The first property to bottle its wines at the château was Château Mouton Rothschild as early as the 1920s. Most of the other properties did not follow this habit until the 1970s. Château bottling was an expensive step. It meant a change in a cash flow model. Shipping wines in barrels immediately following the harvest meant that payment would be received in time to finance the next harvest. Given the average barrel ageing for most wines is between 12 to 18 months, château bottling therefore requires a cellar large enough to store two harvests' worth of production and enough finance to wait until 18 months after the harvest for payment.

The development of the *en primeur* system came from merchants paying for the wines in the spring following the harvest but not taking delivery until a year to 18 months later. Only a small proportion of Bordeaux wines are sold in this way. Depending upon the perceived quality of the vintage, up to 20 per cent of the value of the production of the crop will be purchased *en primeur.* The decision whether to pay in advance to secure supply will be made following the tastings of the new vintage held in the first week of April, followed by the châteaux announcements of their release price. Every year the international trade forecasts the demise of this system, but it seems to continue, if only for about 5 per cent of the volume of trade in an average year.

In today's market the role of the *négoce* has changed, its *élevage* role is primarily restricted to branded wines. The distinction between a château and *négociant* is less clear, as châteaux can and do open *négociant* houses, and *négociant* houses purchase vineyards.

There are currently 300 *négociants* in Bordeaux, with the top nine companies representing 50 per cent of the turnover, each of them with a turnover of more than 75 million euros.

COOPERATIVES

Many smaller properties do not have a winemaking facility at the vineyard but make their wines at a local cooperative. There are 33 wine cooperatives representing 23 per cent of Bordeaux's production and 40 per cent of the producers.

VOLUME VERSUS REPUTATION

As the table below illustrates, the famous appellations represent a relatively small amount of the *Bordelais vignoble,* and the classified growths only about 5 per cent, although they represent about 20 per cent of the value. Yet it could be argued that the reputation of Bordeaux has been built upon these *crus classés.*

TYPE	2017 PERCENTAGES FOR BORDEAUX PRODUCTION
Bordeaux red	85%
Dry white, including Crémant	10%
Rosé and clairet, including Crémant Rosé	4%
Sweet white	1%

DISTRICTS	SURFACE AREA Expressed as a percentage of Bordeaux vineyards
Bordeaux (including Bordeaux Supérieur red, rosé, and clairet)	47%
Médoc, Graves, and Pessac-Léognan (all ten appellations)	18%
Saint-Émilion, Pomerol, and Fronsac (known as Libournais, or "right bank")	11%
Côtes	12%
Dry white	9%
Sweet white	3%

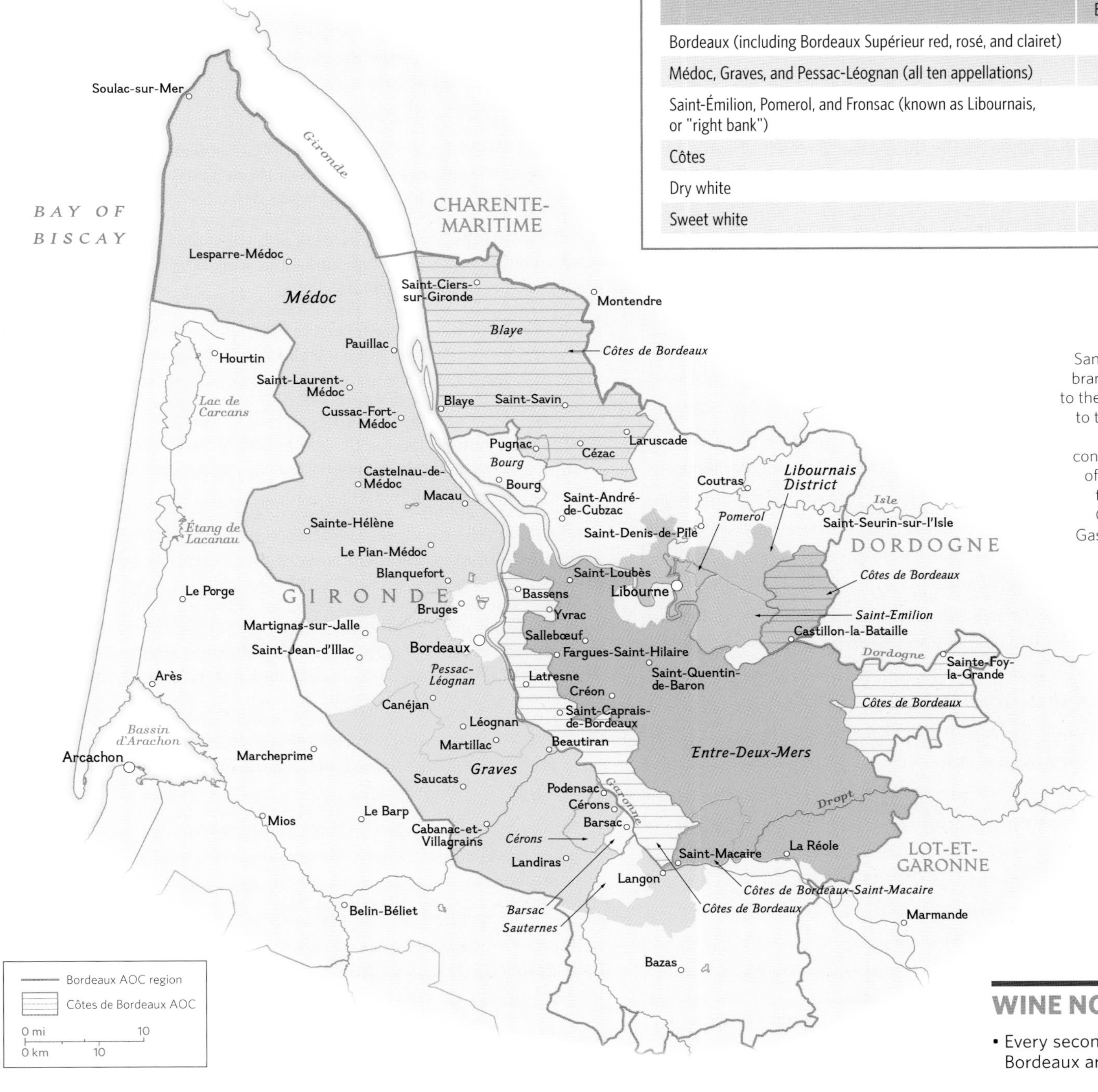

WINE REGIONS OF BORDEAUX
(see also p131)
Sandwiched between the brandy regions of Cognac to the north and Armagnac to the south, the Gironde *département*, which constitutes the AOC area of Bordeaux, consists of the former province of Guyenne and a part of Gascony called Bazedais.

Bordeaux AOC region

Côtes de Bordeaux AOC

0 mi   10
0 km   10

WINE NOTES

• Every second, 12 bottles of Bordeaux are sold worldwide.

THE CLASSIFICATION OF BORDEAUX WINES

Of all the Bordeaux classifications that exist, it is the 1855 Classification that is meant whenever anyone refers to "The Classification". It was created to present the most famous Bordeaux wines at the Paris Universal Exhibition. The Chamber of Commerce gave the responsibility of drawing it up to the Bordeaux trade brokers, or *courtiers*. These brokers kept records of the sales prices of wines from the top Bordeaux properties, so they effectively transcribed the price hierarchy of châteaux that had been established over more than 200 years of trading and awarded a classification based on these sales prices. It was submitted as a "complete list of classified red Bordeaux wines, as well as our great white wines". The original classification (*see* below and opposite page) give the 19th-century names in the original form as listed by the brokers on 18 April 1855 (Château Cantemerle was added in September of the same year).

The 1855 Classification officially includes wines from the Graves, Médoc, and Sauternes (Barsac being a village of the Sauternes appellation). There are now 61 red wines (60 from the Médoc, one from Graves) with five levels of classification. There are 27 sweet white wines with two levels of classification plus *première cru supérieur* awarded uniquely to Château d'Yquem, the highest-ranking property in the 1855 Classification. The Graves and the St-Émilion classifications were introduced a century later. Graves was first classified in 1953 and revised in 1959 and St-Émilion was first classified in 1955 and has been revised several times since (*see* p190).

THE CLASSIFICATION OF THE RED WINES OF THE MÉDOC

They say *terroir* never changes, but that is to ignore the human impact – so much has changed since 1855. Nearly all châteaux have increased their vineyards, either by expansion or by purchasing a neighbouring property – quite often both – meaning some châteaux have had their vineyards split

BARRELS IN STORAGE IN CHÂTEAU MOUTON ROTHSCHILD CELLAR
Château Mouton Rothschild was in the vanguard of château bottling with its abilty to store two harvests' worth of wine.

or totally absorbed. Furthermore, the varieties grown have changed, shrinking from maybe as many as a dozen varieties to just Cabernet Sauvignon, Merlot, Cabernet Franc, Petit Verdot, and, very occasionally, Malbec and Carmenère. Go back to the Médoc in 1855, and there are many places you would not recognise. The 1855 Classification was based on sales price, so it is interesting to see where these châteaux would figure in a price-based classification today. Every year the wine-trading website Liv-ex produces an up-to-date price listing of the wines of Bordeaux based on their sales price. This includes not only the red wines of the Médoc and Graves but also the right bank and some second wines of the properties that now have a strong following.

The Graves has its own classification that was introduced almost 100 years later in 1954 and reconfirmed in 1959. This includes just 16 châteaux: 13 red and 9 white wines. Some are classified for their white and not their red or vice versa and a few for both; it is the wine and not the property that is classified. Unlike the five levels in the 1855 Classification, there is only one level in the Graves; you either have it or you don't. These classified wines are all found to the north in what is now the Pessac-Léognan appellation. The Pessac-Léognan appellation was created in 1987 so the classification predates the appellation.

These wines are often overshadowed by the reputation of their Médoc neighbours to the north, possibly due to the 100-year head start, and consequently tend to be excellent value wines, accessible both in style and price (*see* "From Grape to Glass" for the list of Graves *cru classé*, p178).

At the same time as the Graves decided to classify their wines, the right bank also woke up to the advantage of having a classification and St-Émilion was classified in 1954. The St-Émilion classification can be considered more democratic. Whereas the 1855 Classification has never really changed – the exceptions being Château Cantemerle, which was added in at the last minute, and Château Mouton Rothschild, which was promoted from a second growth to a first in 1973 – St-Émilion re-classifies its châteaux almost every 10 years.

To qualify to be considered for classification, the wine must be an Appellation St-Émilion *Grand Cru*. There are three categories of classification in St-Émilion: *Grand Cru classé*, *Première Grand Cru classé B*, and *Première Grand Cru classé* A. The latest classification dates from 2012 and includes 64 classified growths (*Grand Cru classés*) and 18 first growths (*Première Grand Cru classés*) of which four are As: Château Angelus, Château Ausone, Château Cheval Blanc, and Château Pavie. (*See* The Classification of St-Émilion, p190)

CRU BOURGEOIS: A CLASS STRUGGLE

It's all change as far as the *bourgeois* growths are concerned. The original *Cru Bourgeois* Classification of 1932 had been carried out by Bordeaux wine brokers, as had the famous 1855 Classification. It recognised 444 Médoc properties divided into three hierarchal levels: *crus bourgeois*

THE 1855 CLASSIFICATION OF THE WHITE WINES OF THE GIRONDE

PREMIER CRU SUPÉRIEUR (Superior First Growth)
Yquem, Sauternes

PREMIERS CRUS (First Growths)
Latour Blanche, Bommes
(now Château La Tour Blanche)
Peyraguey, Bommes (now two properties: Château Lafaurie-Peyraguey and Château Clos Haut-Peyraguey)
Vigneau, Bommes (now Château Rayne-Vigneau)
Suduiraut, Preignac
Coutet, Barsac
Climens, Barsac
Bayle, Sauternes (now Château Guiraud)
Rieusec, Sauternes (now Château Rieussec, Fargues)
Rabeaud, Bomme (now two properties: Château Rabaud-Promis and Château Sigalas-Rabaud)

DEUXIÈMES CRUS (Second Growths)
Mìrat, Barsac (now Château Myrat)
Doisy, Barsac (now three properties: Château Doisy-Daëne, Château Doisy-Dubroca, and Château Doisy-Védrines)
Pexoto, Bommes (now part of Château Rabaud-Promis)
D'arche, Sauternes (now Château d'Arche)
Filhot, Sauternes
Broustet Nérac, Barsac (now two properties: Château Broustet and Château Nairac)
Caillou, Barsac
Suau, Barsac
Malle, Preignac (now Château de Malle)
Romer, Preignac (now two properties: Château Romer and Château Romer-du-Hayot, Fargues)
Lamothe, Sauternes (now two properties: Château Lamothe and Château Lamothe-Guignard)

THE 1855 CLASSIFICATION OF THE RED WINES OF THE GIRONDE

The 1855 Classification is replicated below, and the number before each producer indicates how they would be classed today according to the data provided by Benjamin Lewin MW in his book *What Price Bordeaux?* (Vendange Press, 2009). Zero implies a classification below Fifth Growth. In the second part of the box are the 14 unclassified wines, including second wines, that would qualify for *cru classé* status.

PREMIERS CRUS (First Growths)
- (1) Château Lafite, Pauillac (now Château Lafite-Rothschild)
- (1) Château Margaux, Margaux
- (1) Château Latour, Pauillac
- (1) Haut-Brion, Pessac (Graves)

SECONDS CRUS (Second Growths)
- (1) Mouton, Pauillac (now Château Mouton Rothschild and a first growth since 1973)
- (3) Rauzan-Ségla, Margaux
- (5) Rauzan-Gassies, Margaux
- (2) Léoville, St-Julien (now three properties: châteaux (2) Léoville-Las-Cases, (3) Léoville Poyferré, and (3) Léoville-Barton)
- (5) Vivens Durfort, Margaux (now Château Durfort-Vivens)
- (3) Gruau-Laroze, St-Julien (now Château Gruaud-Larose)
- (3) Lascombe, Margaux (now Château Lascombes)
- (3) Brane, Cantenac (now Château Brane-Cantenac)
- (2) Pichon Longueville, Pauillac (now two properties: Château Pichon-Longueville-Baron and Château Pichon-Longueville-Comtesse-de-Lalande)
- (2) Ducru Beau Caillou, St-Julien (now Château Ducru-Beaucaillou)
- (2) Cos Destournel, St-Estèphe (now Château Cos d'Estournel)
- (2) Montrose, St-Estèphe

TROISIÈMES CRUS (Third Growths)
- (4) Kirwan, Cantenac
- (4) Château d'Issan, Cantenac
- (4) Lagrange, St-Julien
- (3) Langoa, St-Julien (now Château Langoa-Barton)
- (4) Giscours, Labarde
- (4) St-Exupéry, Margaux (now Château Malescot-St-Exupéry)
- (0) Boyd, Cantenac (now two properties: (0) Château Boyd-Cantenac and (5) Château Cantenac Brown)
- (2) Palmer, Cantenac
- (4) Lalagune, Ludon (now Château La Lagune)
- (0) Desmirail, Margaux
- (4) Dubignon, Margaux (no longer in existence, but some of these original vineyards now belong to (4) Château Malescot-St-Exupéry, (2) Château Palmer, and (1) Château Margaux)
- (3) Calon, St-Estèphe (now Château Calon-Ségur)
- (5) Ferrière, Margaux
- (0) Becker, Margaux (now Château Marquis d'Alesme-Becker)

QUATRIÈMES CRUS (Fourth Growths)
- (4) St-Pierre, St-Julien (now Château St-Pierre-Sevaistre)
- (3) Talbot, St-Julien
- (4) Du-Luc, St-Julien (now Château Branaire-Ducru)
- (4) Duhart, Pauillac (at one time Château Duhart-Milon Rothschild, but now Château Duhart-Milon, although still Rothschild-owned)
- (0) Pouget-Lassale, Cantenac (now Château Pouget)
- (0) Pouget, Cantenac (now Château Pouget)
- (0) Carnet, St-Laurent (now Château La Tour-Carnet)
- (5) Rochet, St-Estèphe (now Château Lafon-Rochet)
- (4) Château de Beychevele, St-Julien (now Château Beychevelle)
- (4) Le Prieuré, Cantenac (now Château Prieuré-Lichine)
- (5) Marquis de Thermes, Margaux (now Château Marquis-de-Terme)

CINQUIÈMES CRUS (Fifth Growths)
- (3) Canet, Pauillac (now Château Pontet-Canet)
- (0) Batailley, Pauillac (now two properties: (0) Château Batailley and (5) Château Haut-Batailley)
- (3) Grand Puy, Pauillac (now Château Grand-Puy-Lacoste)
- (0) Artigues Arnaud, Pauillac (now Château Grand-Puy-Ducasse)
- (2) Lynch, Pauillac (now Château Lynch-Bages)
- (0) Lynch Moussas, Pauillac
- (5) Dauzac, Labarde
- (5) Darmailhac, Pauillac (now Château d'Armailhac)
- (5) Le Tertre, Arsac (now Château du Tertre)
- (5) Haut Bages, Pauillac (now Château Haut-Bages-Libéral)
- (0) Pédésclaux, Pauillac (now Château Pédésclaux)
- (0) Coutenceau, St-Laurent (now Château Belgrave)
- (0) Camensac, St-Laurent
- (5) Cos Labory, St-Estèphe
- (4) Clerc Milon, Pauillac
- (0) Croizet-Bages, Pauillac
- (0) Cantemerle, Macau

The following 14 wines would appear on today's classification but did not exist in 1855.
- (3) Forts de Latour
- (3) Pavillon Rouge de Margaux
- (3) Clos du Marquis
- (3) Carraudes de Lafite
- (4) Haut Marbuzet
- (4) Sociando Mallet
- (5) Tourelle Longueville
- (5) Alto Ego de Palmer
- (5) Phélan Ségur
- (5) Gloria
- (5) Pibran
- (5) Siran
- (5) Labégorce
- (5) Ormes de Pez

exceptionnel (six properties), *crus bourgeois supérieur* (99 properties), and *crus bourgeois* (339 properties).

Although this worked well enough for three decades, there were only 94 *cru bourgeois* properties remaining by the 1960s; the others having been grubbed up, abandoned, or consolidated. Furthermore, although authorised by the Bordeaux Chamber of Commerce, this classification had never been submitted for ministerial ratification, thus the term *cru bourgeois* was not recognised by the French government and therefore had little legal meaning. In 1962 the surviving properties established the *Syndicat des Crus Bourgeois* and set about establishing a new classification, which made an initial assessment of 101 *crus bourgeois* in 1966, and this was expanded to 124 in 1978.

Although the 94 surviving properties had widened their scope to include other châteaux, the classification did exclude a number of highly regarded châteaux. Additionally, those selected as the very best (*cru bourgeois exceptionnel*) and generally superior (*cru bourgeois supérieur*) were no longer allowed to use those terms, as the Common Market regulations of 1979 permitted only the words *cru bourgeois*. Being French, they used the terms anyway. They were not the only ones flouting the law, however, because many of those demoted in 1978 from one of the higher ranks they had previously enjoyed continued to use their 1932 status, while the term *cru bourgeois* was often used for second wines, unclassified properties, and even châteaux outside of the Médoc in Blaye, Bourg, and Sauternes.

It was obvious that the legal situation had to be rectified, but it was not until November 2000 that a new *cru bourgeois* law could be agreed and ratified by ministerial decree. This legislation required the classification to be reassessed every 12 years, and the first such reassessment took place in June 2003, when just 247 of the 490 châteaux that applied were successful. Such a high failure rate indicated a level of seriousness. No classification will ever be perfect, as there will always be deserving producers who are excluded – not simply due to a flaw in the system or errors of judgment, but because they might simply have abstained from the process. Overall, however, the judging panel on this occasion got it right. Inevitably, some château owners objected to the outcome.

The objections achieved critical mass in 2007, when, after an acrimonious series of legal challenges and counter-challenges, a total of 77 aggrieved châteaux appeared before Jean-Pierre Valeins, the magistrate of Bordeaux's administrative court of appeal. Although Valeins admitted that it was not a case of who was right and who was wrong, he had no hesitation in annulling the classification – a decision that was subsequently endorsed by the French government. The magistrate revealed that 4 of the 18 members on the classification panel had direct ties to a number of châteaux included in the 2003 classification. Yet his decision essentially came down to the fact that, in a classification, all the wines are judged in relation to one another.

In November 2009 the French government ratified a decree for the latest incarnation of *cru bourgeois*. This annual scheme was called the Reconnaissance system, whereby the term *cru bourgeois* – and *cru bourgeois* only – will be awarded to châteaux on an exclusively vintage-by-vintage basis. In the first of a two-step process open to all properties in the Médoc, the *cuverie* of all applicants is subjected to a quality-control check (a process that will be repeated every five years), and any successful château may submit its wine for annual blind assessment by an independently audited panel of impartial tasters. The wines are tasted two years after they were made, at the time of bottling. Compared to the 247 successful applicants out of the 490 châteaux that applied for *cru bourgeois* status in 2003, the number qualifying for *cru bourgeois* in 2008, the first Reconnaissance vintage, was 243 out of 290 submitted, which might make the cynic think that the French were back to their old tricks.

You can see, however, that this is not so when you look a bit deeper. The first thing that should give readers a clue is that the 243 successful applicants in 2008 were almost the same as the 247 in 2003. Most of the other 240-odd châteaux either accepted that it would not be worth the effort and cost of applying, or their *cuveries* failed the quality-control check. There are also some highly regarded properties that previously held the status of *cru bourgeois exceptionnel* or *cru bourgeois supérieur* and did not see the value in applying for what was effectively a lower-grade title for the 2008 vintage. The very best of those are probably right and can no doubt make a better go of it by building their own brand. Those that cannot might return to the fold.

The Cru Bourgeois Reconnaissance in September 2017 for the 2015 vintage (a particularly good vintage) included 271 properties from across seven of the eight Médoc appellations. (There is currently no St-Julien member). There is no denying that in a group of almost 300 properties, there will be a variation in quality. Some vineyards have elected not to be part of the classification due to this variation, asking how they can differentiate themselves in such a large group.

To answer this criticism the French public authorities have approved the process that will allow a return to the historical three-tier hierarchy. A supervised independent jury will assess a hierarchy of the quality, consistency, and ageing capacity of the wines, judged by a blind tasting of several vintages. Criteria will also include traceability with the authentication of each bottle and respect for the environment by the vineyards, with

RECENT BORDEAUX VINTAGES

2018 A difficult start, saved by summer sun. Very wet conditions in the first months of the year, with disease pressure from mildew and rot, followed by hailstorms in May and July. Flowering, however, largely escaped the adverse weather, and the sun came out in June to stay right until October, providing near-perfect ripening conditions with just enough water to avoid stress on the vines. Those that weren't hit by disease and hail had very good yields and fruit quality, but balance will be important, as alcohol levels are typically higher in reds and the acidity lower in whites. Botrytis onset in Sauternes and Barsac was delayed due to the sunny and dry weather, and hail damage was considerable to some estates.

2017 A particularly painful year for many producers. A historically damaging frost in April and the hail in August meant several vineyards lost most or all of their production. The effect was uneven: vineyards closest to the water on the left bank and higher ground escaped lightly and others lost 100 per cent. Total Bordeaux production was down by almost 40 per cent.

An early season meant early bud break, especially early for Merlot, making it all the more vulnerable when the frost fell. The sunny August was good for growers with grapes left and gave an early and quick harvest, producing some good wines. The reds are fresh, with lovely fruit and supple, elegant tannins. Sémillon was particularly successful for sweet whites with great concentration and an elegant balance between sugar and acidity.

2016 Quality and quantity. Winter rains boosted water reserves after a dry 2015, and despite a damp spring flowering went well before a warm summer. Yield was the highest for a decade, despite green harvesting that helped quality. The rain in September saved the grapes from excess water stress and the cooler evenings helped the elegance, giving great colour with lovely fruit ripe tannins.

The whites planted on cooler soils have a lovely fresh expression, and the summer weather allowed grapes to ripen perfectly ready for botrytis triggered by the mid-September rains.

2015 An outstanding vintage in terms of quality and quantity for all varieties and colours. Perhaps favouring right bank Merlots, but excellent across the region with perfect growing conditions and no rot to affect good yields.

A fine April and the warmest June for over a century, followed by a beautiful summer. Water stress slowed maturity, but August rains saved the vintage. A cool and sunny September and October meant pickers could wait for perfect ripeness but, with cooler nights, not at the expense of freshness.

Early noble rot gave a large and early crop of rich and elegant sweet white wines; 80 per cent of the crop was harvested by the end of September. The dry white wines are also very good. The red wines have the elegance of a great Bordeaux with good ageing potential, but they should drink earlier than the 2010s thanks to more rounded tannins.

2014 A relief after the challenging 2013 vintage. Things were not looking good after a dreary summer, but all was saved by an outstanding September and October, especially favourable to Cabernet.

After a cool start, things started to warm up in late July, but this caused storms and some hail on the right bank and in the Entre-Deux-Mers before becoming cooler again. The damp conditions increased the risk of disease, necessitating expensive treatments. By the end of August it was looking as bad, if not worse, than 2013. The weather picked up in September, and the good weather continued through October, ensuring ripening in extremis. A relatively late harvest in fine weather meant picking took place in excellent weather conditions for all varieties, with both good ripeness and acidity.

The cool summer was great for dry whites, giving ripe grapes such high acidity that some wine makers used malolactic fermentation (very rare in Bordeaux). The warm, dry days of September delayed noble rot in Sauternes and Barsac until the mid-October showers, but then more hot weather with misty mornings gave ideal conditions for berries to "roast" and for selective picking, although yields were low.

The 2014 red wines are better than 2013, deep in colour, fruity, relatively full bodied, with the Cabernets, both Franc and Sauvignon, really profiting from the Indian summer. These fresh, fruit-driven reds are perfect for medium-term drinking.

2013 An unloved vintage by growers. A difficult growing season and a small harvest. Low yields reinforced by strict selection meant there wasn't much wine but, given the low demand, that's perhaps not a bad thing.

Wet winter weather continued through spring causing late flowering and shot berries. April was warmer and drier for bud burst, but frost hit Sauternes, Blaye, the northern Médoc, around Libourne, the Entre-Deux-Mers, and Graves. June was the coldest and wettest since the dreadful 1992 vintage.

Much hoped-for hot and dry summer weather arrived in July. Wine growers started to be cautiously optimistic, but to no avail. Storms arrived on the back of the hot weather bringing heavy rain, strong winds, and some hail. On 2 August hail wiped out about 80 per cent of over 10,000 hectares (24,700 acres) of Entre-Deux-Mers.

inspections carried out at the properties throughout the classification period. The new classification system will come into effect from 2020.

The 2003 classification was the last to include a hierarchy. Only nine *crus bourgeois exceptionnel* were designated, none of which have participated in the Cru Bourgeois Reconnaissance. Perhaps they will consider rejoining the party now the hierarchy is to be re-instigated?

CRUS ARTISANS

The other historic classification in the Médoc that is on the move is *Les Crus Artisans,* a less well-known classification. This group includes fewer properties, and they tend to be smaller (between 1 and 5 hectares, or 2.5 to 12.5 acres), so they are not always easy to find on export markets. Despite this being a historical term used as early as 1868 in the Cocks and Féret *Bordeaux and Its Wines,* the first official *cru artisan* classification dates from 2006 when 44 properties were classified, with a planned reclassification every 10 years. The latest classification was announced in May 2018, not exactly on schedule. This classification will hold for five years, from the 2017 until the 2021 vintage. Seven out of eight of the Médoc appellations are included (Pauillac is not represented) for a total of 36 properties, although the surface area remains the same as in 2006 due to continued consolidation of smaller vineyards.

An artisan winemaker is defined as a producer responsible for the entire production process: vineyard work, vinification, aging of the wine, bottling, packaging, and sales. The disappearance of many of the *cru artisans* since

WHAT MAKES A GREAT VINTAGE?

When assessing Bordeaux vintages, it is worth looking at the conditions that can lead to a great vintage.
- Warm and dry spring weather to ensure quick flowering and fruit-set leading to good pollination and even ripening
- A warm, dry July giving a gradual onset of water stress to slow down and then finally stop vine growth by the start of *véraison* (colour change)
- A warm, dry (but not excessively so) August and September to allow full ripening of the various grape varieties
- Dry and medium-warm weather at harvest to pick the grapes at optimum ripeness without the risk of dilution from rain or rot

If only it was like this every year . . .

2006 – due to some owners retiring and properties being bought up by larger neighbours – underlines the problems that these small, family-run properties are facing, even in some of Bordeaux's more prestigious appellations.

These classifications can have a role to play in helping to keep these small producers in business, raising awareness of their very existence to the trade and consumers alike.

A wet, mild September and October caused grey rot to spread dramatically both before and during the harvest, especially on thinner-skinned Merlot. Late-ripening plots and Cabernet Sauvignon vines were least affected. A lot of green selection was needed on already low-yield vines and picking dates were based more on rot risk than ripeness.

It was possible to make some deeply coloured, fruity wines on the right *terroir,* with meticulous green pruning, low yields, quick picking, precise sorting, and precision winemaking. Quite a challenge!

The cool and wet weather was better for dry white wines and the wet, mild autumn was favourable to early widespread development of *Botrytis cinerea,* so there were some lovely sweet whites, aromatic but less sweet than 2009 and 2011. For the reds 2013 was unquestionably the most difficult vintage in the past thirty years.

2012 A good, if not great vintage that was written off by many as, at best, a classic vintage. Yet upon release it turned out to be a wine that is already drinking very nicely indeed due to the soft tannins making for early drinking.

A cool wet June meant a slow start to the season. The sun shone from mid July through August, almost too dry in some places, but the tricky start meant that ripening was uneven. The late harvest favoured earlier-ripening Merlot, as cool wet weather in mid-October made it difficult for many Cabernets to reach full maturity, added to which there was a lot of mildew, especially tricky to manage for organic vineyards.

The best wines come from properties that select grapes through severe green harvest and picking, concentrating on the ripe fruit. Yields were low but there are some lovely wines.

The dry whites were picked before the change of weather, but the sweet wines had a more difficult vintage with noble rot arriving late September, hampered by rain in October and finished off by rain in November.

2011 With early drought, storms, slow ripening, and a very heterogeneous harvest, 2011 was a vintage where skills in the vineyard were as important as in the cellars. Given these difficulties, the 2011s showed better than expected when tasted *en primeur.* A good but not great vintage, back to a classic Bordeaux style characterised by freshness and tannins, rather than the ripe fruit we had become used to in the hot vintages of 2009 and 2010.

A warm and dry start with great fruit set but the hot June – over 40°C (104°F) – damaged vines on drier soils, through drought and sun burn. Difficult and uneven ripening meant lots of harlequin bunches (bunches with different-coloured grapes) that needed eliminating at green harvest.

Results were uneven across the region. Limestone subsoils, such as St-Émilion, had sufficient water reserves to resist drought so tannins could ripen. Pessac-Léognan, Graves, and Sauternes enjoyed better rainfall than the Médoc and St-Émilion. Early August rains didn't help much, but an early harvest in a hot dry September reduced mildew risk. It was a long harvest waiting for full ripeness but keeping an eye out for rot.

St-Estèphe was badly hit by hail in September, meaning an early harvest, especially of remaining fruit to avoid any rot developing on damaged fruit.

Selection was the key to good wine in 2011. Châteaux that had invested some of their profits from the successful 2009s and 2010s in selection machines and optical sorters put them to good use.

There are some excellent wines, if not across the board. Merlots on cooler limestone and clay are coloured and fresh and the Cabernet Francs give a complimentary structure. Left bank Cabernets have good structure, but volumes are small.

A very good vintage for dry whites, and an excellent one for Sauternes, despite some hail damage in April. Early September rains and morning mists gave good noble rot, followed by hot weather giving great concentration and a quick harvest.

2010 This year has had the bad luck to be a very good – possibly very great – vintage that follows on the heels of the extraordinary 2009. The potential greatness of 2010 started early, however, when it suffered from *coulure* at fruit set (as did 1961), and the style of the wine will forever be marked by its very dry summer, with hardly any rainfall in August. This resulted in a small crop, 20 to 30 per cent down across the region, of small berries with very thick skins and (because of beneficial rain at the start of September) good phenolic ripeness. The 2010s are bigger and heftier than the 2009s, but at the time of publication they do not appear to be as fine or as consistent across the appellations. Thus, I would suggest that, although probably a great vintage, 2010 is not as great as 2009.

2009 One of the greatest Bordeaux vintages on record. The 2009 reds on both banks (although a tad stronger on the left bank) are at least as good as the 2005s but bigger in every sense, with more fruit, more tannin, and more acidity. The greatest 2009s are totally *terroir*-driven, while the greatest 2005s are more man-made. This is the 1961 of modern times. With a vintage this great, you do not need many words, as the wines will do all the talking. Sauternes and Barsac have been surprisingly underrated by some, but have no fear: this is one of the all-time great botrytis vintages.

ARE CLASSIFICATIONS STILL RELEVANT?

Historically, with the exception of Pomerol, regions that have never had a classification remain the less well-known wines of Bordeaux. It could be said that Graves and St-Émilion are still playing catch-up in the notability stakes with the Médoc, which has had a 100-year head start. This is one argument that underlies the importance the classifications have had in building the reputation of the Bordeaux vineyards.

To judge their importance, however, look at the weight they carry in the market. All the classifications together represent fewer than 500 out of about 7,500 wineries. Only 5 per cent of the volume of production is classified (or equivalent), but it does represent 20 per cent of the value. *Cru bourgeois* and St-Émilion continue to move with the times, or at least reclassify regularly. Médoc, Graves, and Sauternes don't – or do they?

The way these top wines of Bordeaux are sold is a unique system of future sales, or *"en primeur"*. Wine made in the autumn will be presented to the trade for tasting the following spring, despite the fact that these wines will remain in the tender loving care of the winemakers until they are bottled 18 months later.

Following these spring tastings, once the trade and the critics have had their say, prices for this new vintage are fixed. Interested parties must pay in full to reserve their allocation. This is in the hope that when the wine becomes physical (bottled and delivered), they will not only be assured of the supply of the desired wines but they will also have purchased them at a preferential price.

It could be argued that the real classification is the one that appears every year – the price that the market will pay for these wines. That was, after all, the basis of the original 1855 Classification.

THE GRAPE VARIETIES OF BORDEAUX

Bordeaux makes blends. There are single-varietal wines in Bordeaux, but they are rare. The red and white grape varieties permitted to comply with Bordeaux appellation regulation are limited.

RIPE CABERNET SAUVIGNON GRAPES
These world-renowned grapes are thought to be the classic Bordeaux variety, even though Merlots are more widely planted. Yet it is these complex black grapes that often give Bordeaux wine blends their structure and backbone.

Red varieties

The majority – 88 per cent – of Bordeaux production is red (of which about 6 per cent is rosé). Contrary to what one might expect, it is Merlot, not Cabernet Sauvignon, that is the most widely planted grape variety in Bordeaux. Cabernet Sauvignon represents less than 24 per cent of black grapes cultivated in Bordeaux, whereas Merlot accounts for more than 63 per cent. It is nearer to the truth, therefore, to say that Cabernet Sauvignon gives backbone to Merlot, rather than to suggest that Merlot softens Cabernet Sauvignon (which is the old adage).

Although the Médoc region is defined in the minds of many by Cabernet Sauvignon, even here Merlot makes up almost 50 per cent of blends, although this varies from appellation to appellation as we will see.

Pétrus (*see* pp 198, 201), one of the most expensive wines in the world, is now 100 per cent Merlot. Cabernet Sauvignon is a classic grape, quite possibly the greatest red wine grape in the world, but its importance for Bordeaux is often overstated.

White varieties

Sémillon is the most widely planted white grape variety in Bordeaux (almost 50 per cent). It is significant both in terms of its extent of cultivation and quality. This grape variety is susceptible to botrytis, the "noble rot" that results in classic Sauternes and Barsac. It is therefore considered to be the world's greatest sweet wine grape. Sémillon also accounts for a large part of the blends of most of the fine dry white wines of Bordeaux. Sauvignon Blanc plays the supporting role in the production of sweet wines and is also used for dry wines.

Varietal contributions to the blend – red

Cabernet Sauvignon is the most complex and distinctive of all black Bordeaux grapes. It has a firm tannic structure, yet with time reveals a powerful, rich, and long-lasting flavour. Wines from this grape can have great finesse; their bouquets often possess a blackcurrant or violet character when young, evolving to complex aromas of tobacco and cedarwood with age.

Cabernet Franc has similar characteristics to Cabernet Sauvignon, but may also have a leafy, sappy, or earthy taste, depending on where it is cultivated. It shines when grown on the limestone and clay soils of St-Émilion.

Merlot is soft, silky, and often opulent. It is a grape that charms and can make wines with lots of juicy-rich and spicy fruit and surprising power when grown on the clay soils of Pomerol. Merlot-dominated blends are often more accessible young than the Cabernet-dominated blends.

There are small amounts of Malbec, whose thick skin is also rich in colour, and some Carmenère. Petit Verdot is a late-ripener with a naturally high acidity and beautiful dark colour. With improved agricultural techniques and better understanding of *terroir*, Petit Verdot is becoming more popular in the Médoc, where its fresh acidity and dark colour are prized in blends.

Varietal contributions to the blend – white

Sauvignon Blanc in Bordeaux is delicate and easy to drink. It does not have the same bite as it does in the Loire vineyards of Sancerre or Pouilly-Fumé, but the varietal character is pronounced, with a crisper, fresher style characterised by citrus and floral aromas.

Sauvignon Gris is a greyish pink-skinned mutation of Sauvignon Blanc, yielding a slightly softer, but deeper-flavoured wine. Although a minor blending grape in Bordeaux (only 2 per cent of total plantings), it recently gained popularity as a varietal wine in the New World. (The Bordeaux appellation legislation only mentions Sauvignon as a primary grape and does not distinguish between the two variants.)

The white Sémillon grape provides a lovely round mouthfeel; it makes succulent sweet white wines that are capable of great longevity, as well as dry whites. Its lower acidity is perfectly complimented by the aromatic and more acidic Sauvignon Blanc, producing dry white wines of character. It lends itself well to the complexity brought by barrel fermentation and ageing.

Muscadelle is a richly scented grape that adds aromatic lift to the dry blends, where it plays a minor role.

Early harvesting, pre-fermentation maceration on the grape skins to draw out the aromatics, and longer, cooler fermentation in stainless steel have all combined to produce a range of dry white wines, with excellent examples in the Graves, but also the Entre-Deux-Mers and some notable dry whites even in the classic red region of the Médoc. A lot of this improvement is due to the work of the late Denis Dubourdieu, known in Bordeaux as the Pope of White Whites. With his team at the University of Bordeaux he was responsible for identifying the molecules responsible for the aromatic characteristics of Sauvignon Blanc and conducting research into what soils and agricultural and winemaking techniques allow the best expression and the best preservation of these aromatics in the finished wine.

WHY BORDEAUX BLENDS?

Although grape varieties allowed in Bordeaux are controlled by the appellation rules, it is the soils in the Bordeaux region that explain the importance of having a range of grape varieties. In this temperate and humid climate, soil plays an important role in the choice of grape variety; for example Cabernet Sauvignon will struggle to ripen unless planted on the warmer, better drained gravel soils found mainly on the left bank (Médoc and Graves). The blend is not a fixed parameter. A cooler year might damage the future yield of the early-flowering Merlot. From one year to the next, the raw material that a winemaker has to work with will vary. A vineyard is a continually changing puzzle. A vine will normally be cultivated for about 70 years before being uprooted. The land will then lie fallow for a couple of years before being replanted with young vines. Any grapes produced on these vines cannot be used for appellation wine for the first three years. After this, for the next few years they will tend to bear a lot of fruit but rarely of a very great quality.

Vineyard renewal is managed on a rotation basis to ensure that there is a balance between the ages of the vines. This is often dictated by cash flow as well as estate management, because uprooting and replanting is an expensive operation both in investment and in lost production. Consequently, with each vintage the vineyard will have a slightly different age profile and a different raw material for the winemaker to assess. Every year he will have to rethink the blend.

A blend implies a choice, a selection, and it is one of the keys in defining the style of a wine. The blend is not made from picking all the grapes available in the vineyard and putting them in a vat to ferment together to make the wine. Blending gives the winemaker the possibility of offering several different wines to the consumer. Many properties blend a second wine. A second wine is not a dustbin blend for everything that doesn't make it into the first wine; this wine also carries the name of the château and represents the brand. A second wine is often blended as a more accessible wine, in style as well as in price point, often from younger vines, producing less concentrated juice or perhaps with a higher percentage of Merlot. Wine not deemed suitable for either blend will be sold in bulk not carrying the château name.

TERROIR AND PRECISION VITICULTURE

Bordeaux has always talked up *"terroir"*, but in recent years a deeper understanding of what this really means has seen an emphasis away from what is happening in the cellar to focus more on what is happening in the vineyard. Châteaux now proudly show their vineyard maps detailing soil types and the grape varieties planted on them. Each plot within the vineyard is treated as a separate entity with its different soils, different slopes and exposure, different varietals, different age of vines, and the like. This all contributes to making each plot of the vineyard quite different from its neighbour. They may, for example, require diverse agricultural operations throughout the year, including planting density, canopy management, crop thinning, and different picking dates to ensure the best maturity.

A better understanding and measurement of ripeness has seen a significant advance in Bordeaux over the past 20 years. Instead of uniquely measuring physiological ripeness (the balance of sugar and acidity) the ripeness of tannins is now accurately assessed. This is especially important in Bordeaux where the red grapes have a high tannin concentration, where tannin ripeness often comes later than physiological ripeness. A better understanding and treatment of diseases such as mildew and oidium means vines are healthier, creating more photosynthesis from a greater healthy leaf surface and better maturity. Awareness of the precise phenolic content also allows the winemaker to fine-tune maceration and fermentation for each batch of grapes.

PRECISION WINEMAKING

Identifying and separating these plots in the vineyards results in as many wines as there are plots of land. As the knowledge of *terroir* becomes more and more precise, the plots become smaller and more numerous and so do the vats. In the not-so-distant past there would have been a few enormous vats in the wine cellars, now there are as many vats as there are plots of land, allowing winemaker to vinify grapes from each plot separately should they so desire. The winemaker can then apply the latest winemaking knowledge available in a much more precise manner. Because grapes from each plot are dissimilar, they will also need a different approach in vinification, including such parameters as how long to macerate skins, pips, and juice together; pump over frequency; vinification temperature; and other factors. All of this gives a wider palette to work from when blending – there is more to the Bordeaux blend that just the grape varieties.

Traditionally winemaking in Bordeaux took place in wooden or concrete vats, but these started to be replaced with stainless steel in the 1960s. The first property to do so was Château Haut-Brion. Today in the wine cellars of Bordeaux you will see a mix of oak, cement, and stainless steel vats of many different shapes and sizes, with sometimes a mix of all three, allocating different grape varieties to different materials. There is much experimentation now with amphora, cement and wooden eggs, and other containers, encouraged by the plot-by-plot approach.

The 1990s saw a big technical revolution in winemaking in Bordeaux, in part thanks to the "garage wine" revolution in the right bank – St-Émilion and Pomerol. Here smaller properties, and a high percentage of Merlot, lead to more experimentation: lower yields, later picking dates, and precise selection – all practises that are now common across Bordeaux.

THE ENVIRONMENTAL CHALLENGE

Bordeaux vineyards, like other agricultural sectors in France, have recently come under harsh criticism for their pesticide and herbicide use.

Consumers are concerned about residues in the final product and the negative effect on the environment, but winemakers, vineyard workers, and the populations surrounding the vineyards are also concerned about the more immediate effects of the treatments themselves.

Bordeaux suffers from a particularly damp climate thanks to its proximity to the Atlantic Ocean. This means that diseases such as mildew and oidium are rife. Organic treatment of these diseases is particularly difficult because organic sprays are washed away by rain and must be reapplied – a task made even more difficult on heavy clay soils when they are wet.

Only 8 per cent of Bordeaux vineyards currently adhere to an organic certification, although many more use organic methods, eschewing certification to enable them to treat with non-organic methods if weather conditions call for it. Others feel that relying on copper sulphate, a treatment allowed in organic production, goes against their philosophy.

Le Conseil Interprofessionnel du Vin de Bordeaux (CIVB) invests about 1.2 million euros annually into research on reducing chemical use, including researching disease-resistant strains of grape varieties, treatments that stimulate the natural vine defences, and obtaining a more intimate understanding of vine disease to avoid blanket treatments. It also encourages producers to transition to environmentally friendly farming.

As a result, between 2014 and 2016 sales of carcinogenic, mutagenic, reprotoxic (CMR) pesticides in the Gironde were down 50 per cent, and herbicides sales fell by 35 percent. At the same time organic products represented 35 per cent of vineyard supplies sales in volume.

Wine appellation body *Les Organismes de Défense et de Gestion* (ODG), representing over 80 per cent of the Bordeaux vineyards, voted to change their appellations specifications to include agri-environmental measures. This is a big deal, as non-adherence by vineyards will result in the loss of appellation status and wine being sold as Wine Without Geographical Indication (VSIG). Change is happening, however slowly, in the right direction, despite the climatic challenges the growers continue to face.

PRIMEURS, OR BORDEAUX FUTURES

The classified growths (and equivalent) represent approximately 5 per cent of volume of Bordeaux production but 20 per cent of its value – figures that vary slightly from year to year.

The trade tastes the wines in the spring following the harvest, at the *Union des Grands Crus* (UGC) trade tastings while the wines are still in barrel. They will not be bottled for at least another year to 18 months. At these tastings, the local and international trade and press assess the individual château's wines and the vintage as a whole. Following the feedback, the château – in conjunction with the *négociants* and brokers – fix a price for the wines that will then be offered via the brokers to the *négociants*. The *négociants* will then add their margin and offer the wines to their clients as futures. The wines will only become "physical" 18 months down the road.

Traditionally, payment for these wines is split: one-third at reservation, one-third after about six months, and the final third before delivery when the wines are bottled and delivered, usually in the late summer or autumn two years after harvest.

The system of *en primeur*, or futures purchase, does not work for all châteaux, nor in all vintages, and the percentage of wines offered *en primeur* varies from year to year. It is deemed to be of greatest interest for the top properties for which there is a high demand and also in top vintages for the same reason. Merchants, both in Bordeaux and overseas, will only put money up front if they feel that the product will be in short supply or if they feel there is a possibility of price appreciation.

For the châteaux the advantage of the *en primeur* system is, of course, cash flow. Being able to sell your crop in June before physical delivery in over a year's time is more advantageous than waiting until the following summer, even if the price may be a little lower.

This system has historical origins. Up until the 1970s it was common practice for all châteaux to deliver their wines in barrels to the *négociants* in the early spring following the harvest and to receive payment then in time to finance the next harvest. The *négociant* would then undertake the ageing, blending, and bottling. This system changed relatively recently. The first producer to systematically undertake château-bottling was Philippe de Rothschild at Château Mouton Rothschild, starting with the 1924 vintage, introducing at the same time an artist's interpretation for the label.

The wine scandal of the 1970s (when a few *négociants* were found to have adulterated some wines) precipitated the adoption of château-bottling by most of the top properties soon afterwards, and now 60 per cent of Bordeaux wine is currently château-bottled. Aging the wines on site meant, however, that the châteaux needed more labour and storage facilities (the Great Barrel Hall at Mouton was built in 1926 to house the wines kept back until bottling), but also had an 18-month cash-flow delay. This was compensated for by the introduction of the *en primeur* campaign.

In more difficult vintages some top growths systematically sell their production *en primeur* very quickly with all the allocations being take up, whereas others may struggle.

This may be a reflection of the quality – or perceived quality – and value for money of the product, but it is also a reflection of the brand image of the château. The *négociants* will be more interested in taking up their allocations if the brand image is strong, and they can therefore rely on the pull of the marketplace. Château Latour stepped out of the *en primeur* campaign with the 2012 vintage, which raised a few eyebrows and questions about the permanence of the system. To be clear, Latour has announced that they will be selling their wines once they become physical and when they consider them ready for drinking. They wil, however, continue to sell through the La Place de Bordeaux. We said the major advantage of *en primeur* sales for a château is cash flow at the expense of margin. For Latour, controlling margins might be a higher priority than cash flow. Other properties may be on standby to see how this experiment goes. Probably only five to six other top châteaux may be in a position to follow this lead, so overall the knock-on effect is likely to be very limited; some *négociants* declined the latest physical offer of Latour – we shall see.

BARREL AGEING

After fermentation and prior to bottling, the majority of red Bordeaux is matured in 225-litre (59 gallons) Bordeaux oak casks called *barriques*. Some wines may undergo their second fermentation (also known as malolactic fermentation) in vats and others in these oak barrels. The barrels are then used for ageing – the idea being that the oak is more quickly absorbed, giving a more flattering aspect when it comes to tasting *en primeur*. It would seem, however, that in the long term this makes little difference. The duration of barrel ageing and the percentage of new oak casks used will depend on the quality and structure of the wine, and this will vary according to the vintage. The bigger the wine, the more new oak it can take and the longer maturation it will need. Many of the top-growths wines spend 18 to 24 months in a high percentage of new oak to reach maturity. Other fine-quality Bordeaux wines do not need so much time, perhaps only 12 to 18 months, and do not benefit from 100 per cent new oak; between 30 and 50 per cent may be enough.

If you get the chance, put your nose into a new oak cask before it has been used. The wonderful creamy-smoky, vanilla and charcoal aroma is the very essence of what should come through when a fine wine has been properly matured in oak.

THE APPELLATIONS OF GENERIC
BORDEAUX

BORDEAUX AOC

This appellation could be viewed as the first step on the Bordeaux ladder of quality. Wines carrying the generic appellation may come from any AOC vineyard in the entire Gironde. Some of the most interesting wines hail from classic areas, where the more specific appellation is confined to a precise style: such as a red Bordeaux produced by a château in Sauternes. If the wine is a brand, it should be ready to drink.

RED Most are dry luncheon claret styles, made for early drinking and usually softened by a high Merlot content.

🍷 *Cabernet Sauvignon, Cabernet Franc, Carmenère, Merlot, Malbec, Petit Verdot*

🍷 *1–5 years*

WHITE This is by far the most variable appellation category, with many dull wines. They are normally dry, but the appellation allows residual sweetness. They may be sold from 1 December without any mention of primeur or nouveau.

🍷 *Sémillon, Sauvignon, Muscadelle, plus up to 30% in total of Merlot Blanc, Colombard, Mauzac, Ondenc, Ugni Blanc*

🍷 *1–2 years*

ROSÉ Many Bordeaux and Bordeaux Supérieur properties produce rosé and clairet, alongside their red wines. There is a move away from the darker to the lighter style of rosé. Rosé and clairet production is often higher in lighter vintages where the juice is run off to concentrate the must of red wines. Now with its increased popularity, plots of land are specifically designated for rosé production, picked earlier at higher acidities and made from a direct press rather than a saignée method. Clairet remains popular in the region although not always easy to find in export markets.

🍷 *Cabernet Sauvignon, Cabernet Franc, Carmenère, Merlot, Malbec, Petit Verdot*

🍷 *Immediately*

BORDEAUX SUPÉRIEUR AOC

Technically superior to Bordeaux AOC by only half a degree of alcohol for rouge and two degrees for blanc, yet most of these wines do seem to have a greater consistency of quality, and therefore value. All generics are variable, but this one is less so than most. A Bordeaux Supérieur can feature the words "Grand Vin de Bordeaux" on the label, a Bordeaux AOC may not.

RED These dry wines vary a lot in texture and weight but are generally fuller and richer than most red wines using the basic Bordeaux appellation.

🍷 *Cabernet Sauvignon, Cabernet Franc, Carmenère, Merlot, Malbec, Petit Verdot*

🍷 *2–6 years*

WHITE Dry or sometimes off-dry, light- to medium-bodied white wines, they are little seen.

🍷 *Sémillon, Sauvignon, Muscadelle, plus Merlot Blanc up to 30% in total of Colombard, Mauzac, Ondenc, Ugni Blanc; the proportion of Merlot Blanc must not exceed 15%*

🍷 *1–2 years*

CÔTES DE BORDEAUX AOC

This relatively new combined appellation (March 2009) can be traced back as far as 1975, when growers felt they stood a better chance of survival together than apart, following the economic depression of the early 1970s. Up until recently many of the small, family-owned and -run properties were not big enough to market their wines under their own brand. They relied on cooperatives or the merchants. Now a younger generation of *vignerons* have

BORDEAUX'S GENERIC BRANDS

Think of Bordeaux, and you probably have an image of a "château". You'd be right. More than 60 per cent of Bordeaux wine is bottled at the property where the grapes are grown and the wine is made, easily identified with "Mise en Bouteille au Château" on the label. Most of the remaining 40 per cent is bottled by Bordeaux merchants (négociants) under their own labels, or those of their clients, and sold as branded wines. These wines have not always enjoyed a great brand image, often considered to be blended from leftover wines that châteaux choose to not bottle under their own name, but this sector of Bordeaux has recently undergone a renaissance.

A branded wine offers large volumes of a consistent style, year in year out, something that wine produced at a single property cannot. Some clients are looking for just this: reassurance that every time they reach for a bottle it will taste exactly the same. Not every wine enthusiast's dream perhaps, but it appeals to an important part of the market.

Brands can do this as the wines are sourced from across the Bordeaux vineyard. Bordeaux is big enough to offer négociants plenty of choice of raw material from different soils and grape varieties, giving them the flexibility to create a consistent signature style.

Up until the 1970s it was rare for châteaux to bottle their own wine; négociants took delivery of the wine in the spring following the harvest ageing, and blending and bottling all took place in their cellars, along the Quai des Chartrons in Bordeaux. Philippe de Rothschild pioneered the change to château bottling in the 1920s, determined to control quality from start to finish, so it is ironic that Bordeaux's best-known and largest brand, Mouton Cadet, was created by him in 1930. Mouton Cadet produces over 12 million bottles a year (total Bordeaux production is about 800 million bottles a year).

Branded wines are generally blended from wines purchased in bulk from different sources; these may be individual producers or cooperatives. Less than 1 per cent of Bordeaux harvest is bought and sold as grapes; it is the finished wine that is traded, a very different model to many other regions.

Négociants are building closer relationships with producers rather than relying on brokers to source on the open Bordeaux market. This includes working directly with the in-house winemakers who act as consultants to suppliers from field to cellar, ensuring the supply, the quality, and consistency they are looking for with each vintage in exchange for contracts to guarantee wine purchasing.

You could consider these wines as the training wheels of Bordeaux; allowing people to discover Bordeaux with no fear of mispronunciation, challenging appellations, or classifications in markets where the complexity of Bordeaux can be a barrier for consumers.

MOUTON CADET
Created in 1930 Mouton Cadet was originally the cadet (little brother) of the Grand Vin: Château Mouton Rothschild. The brand, as we know it today, was born after the Second World War, when the blend grew to include wines from other Bordeaux appellations. Now all the wines blended for Mouton Cadet are sourced through exclusive contracts with almost 500 producer partners (about half of which are in the Côtes de Bordeaux appellations). Annual production is over 12 million bottles.

The Merlot-dominated Bordeaux Rouge is the famous flagship, but the complete range includes whites and rosés, including a pure Bordeaux Sauvignon and an Ice Rosé with a screw-cap closure. The more concentrated Réserve range is destined for wine shops and restaurants; sold in heavier bottles, it features Médoc, St-Émilion, and Graves white and red, as well as Bordeaux white and red appellations.

DOURTHE NO.1
Probably the best, largest-selling branded Bordeaux available, Dourthe No.1 celebrated its 30th birthday in 2018. It started as a dry white in the mid 1980s, a red was introduced in 1993, and a rosé in 2005. Current sales reach nearly 2 million bottles of which 50 per cent are the Bordeaux Rouge, 40 per cent Bordeaux Sauvignon Blanc, and 10 per cent Bordeaux Rosé. A more exclusive Dourthe La Grande Cuvée range also offers a Médoc and a St-Émilion. It is worth looking out for Essence de Dourthe, available since 2005 and only in good vintages. It is a blend of wines uniquely sourced from plots in their own vineyards across Bordeaux.

MICHEL LYNCH
Produced by Jean-Michel Cazes of Château Lynch Bages in an unashamedly up-front, fruity style. Red, white, and rosé are all thoroughly recommended.

PREMIUS
Produced by Yvon Mau, the step up between the premium Premius and Premius Exigence is about as big as it can get for two wines sharing the same brand. The stricter selection of grapes and no-expense-spared handling shows through in the richer, oak-layered fruit. The line now includes red, white, rosé, and Sparkling Crémant de Bordeaux white and rosé (700,000 bottles worldwide).

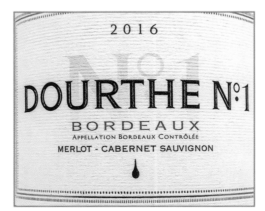

DOURTHE NO.1
This company celebrated 30 years in 2018.

SIRIUS
From the respected Sichel family, who also own Château d'Angludet in Margaux and are a major shareholder in classified-growth Château Palmer. Sirius Bordeaux is available in white and red.

DOMAINES BARON DE ROTHSCHILD
Mouton Cadet is not the only Rothschild brand. Domaine Baron de Rothschild (Lafite) market three brands aimed at different price points and markets to differentiate themselves from their English neighbours: Saga, Legende, and Réserve; they include a St-Émilion, a Médoc, and a Pauillac, as well as Bordeaux red and white.

CORDIER
The presentation of their Bordeaux Blanc and Bordeaux Rouge has been updated with a resolutely modern bottle and label that reflect the modern style of the wines. The red is a Merlot-Cabernet Sauvignon blend, with no oak, all up-front, fresh fruit, and the white is 100 per cent Sauvignon Blanc. Both are easy, early-drinking wines.

LA COUR PAVILLON
One of the oldest Bordeaux brands previously made at Château Loudenne is now sourced from other producers across Bordeaux by La Maison Montagnac. Not so easy to find as it once was but still worth looking out for.

THOMAS BARTON RÉSERVE
from Barton & Guestier
As well as a range under the Barton & Guestier name, this merchant house produces this Thomas Barton Médoc. Packaged in a traditional heavy bottle, it is an excellent example of a Bordeaux brand.

CLARENCE DILLON WINES
Clarence Dillon Wines is part of Domaine Clarence Dillon, owners of Château Haut-Brion, La Mission Haut-Brion, and Quintus in St-Émilion. They launched into the négociant business in 2005 with their Clarendelle range. Unashamedly trading on the name of the Haut-Brion brand, with the strap line "Inspired by Haut-Brion", the range includes a red, white, and rosé Bordeaux and an "amber wine" that is in fact a Monbazillac, not a Bordeaux wine. Good value basic Bordeaux.

GINESTET
Maison Ginestet owns several successful brands, the most well-known and long surviving being Ginestet Rouge, easily identified by the eagle crest embossed on the bottle shoulder. In addition to this reliable and easy-drinking Merlot-led blend, they also produce a Bordeaux Blanc (a Sauvignon-Sémillon blend) and a rosé.

BERRY BROTHERS AND RUDD
Many merchants outside of Bordeaux also produce their own Bordeaux blends. The most famous in the UK is Good Ordinary Claret from Berry Brothers and Rudd. It is an excellent value as is the rest of the range of Bordeaux appellations made for them exclusively from some really good properties across Bordeaux.

stepped in, taking more control of their destiny. What difference has this made? Having a common identity means a larger production to promote under a common logo and a bigger budget to do it with. The initial four appellations (Côtes de Bordeaux Saint-Macaire is part of Côtes de Bordeaux Cadillac) were joined by Côtes de Bordeaux Sainte-Foy in 2016.

Although the Côtes de Bordeau–named districts are diverse in the type and style of wine they produce, the combined Côtes de Bordeaux appellation plain and simple may be used only for red wines. For Blaye – Côtes de Bordeaux, *see* p205; Côtes de Bordeaux Cadillac, p209; Côtes de Bordeaux Castillon, p186, 188; Côtes de Bordeaux Francs; p186, 188; Côtes de Bordeaux Sainte-Foy p209; and Côtes de Bordeaux Saint-Macaire p209.

RED In theory, these could be a blend of any two or more of four appellations, so are impossible to describe in any generic terms at this juncture.

 Cabernet Sauvignon, Cabernet Franc, and Merlot, plus secondary varieties Malbec, Carmenère, and Petit Verdot

CRÉMANT DE BORDEAUX AOC

This was introduced in 1990 to replace the old Bordeaux Mousseux AOC (which was phased out in 1995). Changing the appellation has done nothing to change the product because, like its predecessor, Crémant de Bordeaux is merely a modest and inoffensive fizz. It lacks the spirit and expressiveness to stand out from the sea of far cheaper, but equally boring, sparkling wines that exist almost everywhere. I have tasted much better examples from areas far less suited to sparkling wine than Bordeaux.

CHÂTEAU LAMOTHE
The château was renovated in the 19th century, but this property has produced estate-bottled wine since the 16th.

SPARKLING WHITE These wines varys from dry to sweet and light- to medium-bodied, butthey are almost always bland.

 Sémillon, Sauvignon, Muscadelle, Ugni Blanc, Colombard, Cabernet Sauvignon, Cabernet Franc, Carmenère, Merlot, Malbec, Petit Verdot

1–2 years

SPARKLING ROSÉ The authorities should have taken advantage of the introduction of a new appellation to allow the inclusion of white grapes for this style, as this would potentially have improved the quality.

 Cabernet Sauvignon, Cabernet Franc, Carmenère, Merlot, Malbec, Petit Verdot

2–3 years

THE WINE PRODUCERS OF
BORDEAUX, BORDEAUX SUPÉRIEUR, AND CÔTES DE BORDEAUX

CHÂTEAU DES ARRAS
St-André-de-Cubzac
Bordeaux Supérieur AOC
☆ V

Deep-coloured wines with good structure and lots of chunky fruit.

CHÂTEAU BAUDUC
Creon
Bordeaux Supérieur AOC

Englishman Gavin Quinney produces a range of modern Bordeaux wines here that are fruit forward, excellent value, and sold directly from his own website.

CHÂTEAU BEAULIEU
Le Gay
Bordeaux Supérieur AOC
★

Brilliant value until 2005, then a step up in quality. This is one to watch – definitely a future star. But make sure you get in before the price shoots up!

CHÂTEAU CARSIN
Rions
Bordeaux Blanc AOC
and Bordeaux Rosé AOC

Finnish-owned property, influenced by the New World. The reds are under the Côtes de Bordeaux Cadillac appellation.

CHÂTEAU LA CLYDE
Tabanac
Côtes de Bordeaux AOC
★ V

These aromatic, deep-coloured, ruby-red wines show good spice and fruit. The white has finesse and balance.

CHÂTEAU LA COMBE-CADIOT
Blanquefort
Bordeaux Supérieur AOC
☆ V

Well-coloured wines with a big bouquet; oaky with delicious fruit.

CHÂTEAU COURTEY
St-Macaire
Bordeaux AOC
☆ V

Intensely flavoured old-style wines, with quite a remarkable bouquet.

CHÂTEAU DINTRANS
Ste-Eulalie
Bordeaux AOC

This château produces attractive, nicely coloured, fruity red wines.

CHÂTEAU GOUMIN
Dardenac
Bordeaux AOC

Goumin is another successful André Lurton château. It produces up to 10,000 cases of pleasant, soft, fruity red wine and 5,000 cases of white wine that is slightly fuller than other similar Lurton products.

CHÂTEAU GRAND VILLAGE
Mouillac
Bordeaux Supérieur AOC

Rich and easy Merlot-dominated, oak-aged red, and a good second wine under the "Beau Village" label.

CHÂTEAU GREE LAROQUE
St-Ciers-d'Abzac
Bordeaux Supérieur AOC
★

Elegantly rich wines, with succulent, beautifully focused fruit supported by fine-grained tannins.

CHÂTEAU GROSSOMBRE
Branne
Bordeaux Supérieur AOC
☆ V

From the André Lurton stable, a beautifully concentrated red, one of Bordeaux's great bargains. No longer produces white.

CHÂTEAU DE HAUX
Haux
Bordeaux Supérieur AOC
☆ V

These red and white wines are gorgeously ripe and ready to drink, absolutely fresh, and very elegant. Probably the top-performing château in the area for both red and white.

CHÂTEAU JEAN FAUX
Ste-Radegonde
Bordeaux Supérieur AOC
★

An ancient wine property that once belonged to the Knights Templar, Château Jean Faux has been brought back to life by wine-loving tonnelier Pascal Collotte, with the advice of top consultant Stéphane Derenoncourt.

CHÂTEAU LAFITTE
Camblanes-et-Meynac
Côtes de Bordeaux AOC

Nothing like the real thing (the famous *premier cru*), of course, but the wine is decent, well structured, and capable of improving with age – a cheap way to get a Château Lafitte on the table, even if it is not the Château Lafite. Michel Rolland is the consulting oenologist.

CHÂTEAU LAMOTHE

Haux

Côtes de Bordeaux AOC
and Bordeaux Blanc AOC

Some exceptionally good wines have been produced in recent years at this château. Its name from "La Motte", a rocky spur that protects the vineyard.

CHÂTEAU LAROCHE BEL AIR

Baurech

Côtes de Bordeaux AOC

Absolutely delicious-drinking reds under the basic Château Laroche label and an even better selection of oak-aged reds under Laroche Bel Air. Belongs to the Jean Merlaut collection.

CHÂTEAU LATOUR LAGUENS

St-Martin-du-Puy

Bordeaux Supérieur AOC

From this ancient château, parts of which date back to the 14th century, 10,000 cases of attractive, well-balanced, smooth red Bordeaux Supérieur are produced every year. This château's technically sound wines often win prizes and enable its devotees to claim that they can afford to drink Château Latour every day, at minimum expense.

CHÂTEAU LAVILLE

St-Loubès

Bordeaux Supérieur AOC

These are rich, tannic, and powerfully structured wines with spicy fruit.

CHÂTEAU LESTRILLE

Saint Germain du Puch

Bordeaux Supérieur,
Blanc, Rosé, and Clairet AOC

Excellent example of the new Bordeaux produced by dynamic winemakers Estelle and Jean-Louis Roumage, who head a long-held family property that offers a range of affordable fruit-driven everyday Bordeaux wines using sustainable agriculture.

CHÂTEAU DE LUGAGNAC

Pellegrue

Bordeaux Supérieur AOC

This romantic 15th-century château makes attractive, firm, and fleshy reds that can frequently show some finesse.

CHÂTEAU MACHORRE

St-Martin-de-Sescas

Bordeaux Supérieur AOC

☆

This very respectable red is sold under the Bordeaux appellation.

Other wines: *Cuvée La Villa, Gravière de la Madeleine*

CHÂTEAU MARJOSSE

Branne Tizac-de-Curton

Bordeaux AOC and Bordeaux Blanc AOC

☆

This is Pierre Lurton's home, where he makes lush and up-front reds with creamy-silky fruit and a beautiful dry white.

CHÂTEAU MÉAUME

Coutras

Bordeaux Supérieur AOC

Alan Johnson-Hill used to be a UK wine merchant before settling north of Pomerol. Since then he has gained a reputation for cleverly tailoring this red Bordeaux to young British palates.

CHÂTEAU MORILLON

Monségur

Bordeaux Supérieur AOC

☆

These are rich, fat, and juicy wines.

CHÂTEAU PÉNIN

Génissac

Bordeaux Supérieur AOC

★☆

With their dark fruits and dark colour, the quality wines from this château have been nothing short of *cru classé* standard since the turn of the century.

CHÂTEAU PEY LA TOUR

Bordeaux AOC, Bordeaux Supérieur AOC,
and Bordeaux Blanc AOC

Salleboeuf

★

This large property at the heart of the Entre-Deux-Mers region is owned by the wine merchant Dourthe, who bring their investment and *cru classé* winemaking skills to this property, producing consistently excellent wines for their price point. The Réserve is always a deeply impressive, almost pure varietal Merlot wine.

CHÂTEAU LE PIN BEAUSOLEIL

St-Ciers-d'Abzac

★

Originally named Le Pin, this ancient property was purchased in 1997 by Arnaud Pauchet, who renovated and renamed it. He soon started making a name for his wine, with the help of consultant Stéphane Derenoncourt. He sold it in 2002 to Dr Michael Hallek, who retained Derenoncourt as a consultant.

STÉPHANE DERENONCOURT
Acclaimed French *vigneron* Stéphane Derenoncourt works as a consultant for numerous estates in Bordeaux.

CHÂTEAU PLAISANCE

Capian

Bordeaux Blanc AOC

This château makes excellent, slightly rich Bordeaux Blanc Sec from 50-year-old vines and ferments it in oak.

CHÂTEAU DE PLASSAN

Tabanac

Bordeaux Blanc AOC

☆

The basic Bordeaux Blanc Sec has fresh, swingeing Sauvignon fruit; the more expensive Bordeaux Blanc Sec, fermented and aged in oak, has lovely creamy fruit and a fine, lemony- vanilla finish.

CAVE DE QUINSAC

La Tresne Cooperative

Bordeaux Supérieur,
Clairet, Rosé, Blanc AOC

Look for delicately coloured, light-bodied, rosé-style wines sold as clairet.

CHÂTEAU DE REIGNAC

Saint Loubès

Bordeaux Supérieur AOC
and Bordeaux Supérieur Blanc AOC

This Entre-Deux-Mers property promotes itself as a Bordeaux *non cru classé*, laying claim to its high quality. Extensive investment by the Vatelot family resulted in glowing reviews and the wine shines in its category. The top *cuvée*, Balthus, is possibly the most expensive Bordeaux Supérieur on the market. They also produce a dry Bordeaux Blanc.

CHÂTEAU REYNIER

Grezillac

Bordeaux Supérieur, Blanc, and Rosé AOC

Marc Lurton (nephew to André Lurton) and his wife, Agnes, produce excellent value Bordeaux and Bordeaux Supérieur reds and a delightful white and a rosé. They also sell under the Château de Bouchet label and make an organic Red Bordeaux Supérieur near St-Émilion called Château Tour de Boisset.

CHÂTEAU REYNON

Bordeaux Blanc AOC

☆

This is a star-performing château that produces *cru classé*-quality dry white wine. It is now run by the family of the late Denis Dubourdieu, pioneer of quality dry white Bordeaux. All wines here can be relied upon, right down to the second wine Le Clos de Reynon. It also produces excellent red Côtes de Bordeaux Cadillac AOC.

CHÂTEAU ROC-DE-CAYLA

Targon

Bordeaux AOC

Easy-drinking, well-balanced reds with good fruit and some finesse.

LE ROSÉ DE CLARKE

Castelnau-de-Médoc

Bordeaux Rosé AOC

This wine has all the fragrance expected from a classic dry rosé.

ROSÉ DE LASCOMBES

Margaux

Bordeaux Rosé AOC

Refreshing, fruity rosé of excellent character, quality, and finesse.

CHÂTEAU SAINTE-COLOMBE

Bordeaux AOC

☆

This château was purchased in 1999 by Gérard Perse of Pavie; since 2008 the grapes have been blended with grapes from the second wine of Pavie to make l'Esprit de Pavie, a Bordeaux appellation.

CHÂTEAU DE SEGUIN

La Tresne

Bordeaux Supérieur AOC

Look out for the Cuvée Prestige, which is rich and smoother than the basic Château de Seguin.

CHÂTEAU THIEULEY

Créon

Bordeaux Blanc AOC
and Bordeaux Rosé AOC

☆

These are medium-bodied, elegant reds that possess more than just a hint of cask ageing. Also fine, fresh, floral, and fruity white wines and a fresh rosé and a clairet. They also produce Clos Sainte-Anne in Côtes de Bordeaux red and Côtes de Bordeaux Liquoreux.

CHÂTEAU TIMBERLAY

St-André-de-Cubzac

Deep-coloured, full of flavour, but not without a certain elegance.

CHÂTEAU TOUR-DE-L'ESPÉRANCE

Galgon

Bordeaux Supérieur AOC

☆

This is soft and smooth wine, full of fat, ripe, and juicy fruit, yet not without finesse.

CHÂTEAU DE TOUTIGEAC

Targon

Bordeaux AOC and Bordeaux Blanc AOC

☆

Château de Toutigeac is a well-known property that produces full, rich red wine. It is made for early drinking and is the best the château produces.

CHÂTEAU DE LA VIEILLE CHAPELLE

Lugnon-et-l'Île-du-Carney

Bordeaux Supérieur, Blanc, and Rosé AOC

☆

Try La Cuvée A&A, which is the world's very last pure Bouchalès (Prolongeau).

CHÂTEAU DE LA VIEILLE TOUR

St-Michel-Lapujade

Bordeaux Supérieur AOC

☆

Consistently rich and smooth, even in notoriously harsh vintages.

The Médoc

In the Medoc, the style of wine alters more radically over short distances than in any other French wine district.

The Médoc takes its name from the Latin phrase *medio aquae* – "between the waters" – referring to the Gironde estuary and the Atlantic Ocean. It is a long, thin strip of prized vines, extending northwest from the city limits of Bordeaux to the Pointe de Grave at the top of the Peninsula. It is here where the vast majority of the most famous châteaux of Bordeaux are found, and yet this was the last major district of Bordeaux to be cultivated. While winemaking in the Libournais district of St-Émilion began as early as the Roman occupation, it was another thousand years before scattered plots of vines spread along the Médoc. At that time the marshland of the Médoc was difficult to cross and impossible to cultivate. Dutch engineers drained the marshland in the 17th century to clear it for agricultural use.

Once drained, the soils were seen to be a majority of gravel, similar to the Graves to the South, where the most expensive wines of the day were being produced. This was a time of great prosperity, thanks to the success of the port of Bordeaux. Merchants and other notables rushed north to buy up land and plant large estates. Under their commercial management they quickly became successful, and the wines were shipped by their owners to the thirsty Northern European markets. This success was crowned by the 1855 Classification, and they have remained a benchmark style for Bordeaux wines to this day.

The term *Médoc* can be confusing – it is used generically to describe the whole of the Médoc peninsula, but it in fact includes eight different appellations, including the Médoc AOC. This appellation is the largest, covering more than 30 per cent of the surface area under vines. Legally all eight appellations of the Médoc may apply for the Médoc AOC, but to the north of St-Estèphe there is a specific area that produces exclusively Médoc AOC. There has been a recent increase in vine planting alongside older vineyard plots and a technical revolution in winemaking. Historically the presence of many small estates created an important cooperative movement here that allowed vineyards to survive the crisis of the 1930s. Of the 440 vine growers, 214 belong to the cooperative cellars, making up approximately a quarter of the production.

There are no classified growths within the appellation, but there are over 100 *cru bourgeois* and 9 *crus artisans*. The average percentage of Merlot in the blends is around 50 per cent due to the cooler sandy and clay soils in amongst the gravel. The best vineyards are closest to the estuary, where this gravel is concentrated. Some leading lights include Château Loudenne, La Tour de By, Castera, Greysac, Les Ormes Sorbet, and Patache d'Aux.

THE MÉDOC STYLE: VARIATIONS ON A THEME

There are eight appellations within the Médoc region. The two regional appellations are Médoc to the north and Haut-Médoc to the south. The Haut-Médoc appellation is farther upstream of the Gironde. This long stretch of vineyards (60 kilometres, or 37 miles) reaches from the north of Bordeaux up to just beyond St-Estèphe and is interspersed with the "village appellations" – from north to south they are: St-Estèphe, Pauillac, St-Julien, Moulis-en-Médoc, Listrac-Médoc, and Margaux. The quality of wine from this appellation has been noted since the 1800s. Although the classified growths of Bordeaux only represent about 5 per cent of the production of Bordeaux, it is this classification that has put these properties on the map.

A variety of different *terroirs* and microclimates over this large area explain the diversity of Haut-Médoc wines, but they share the characteristic of being well balanced and full-bodied without being too powerful. Traditional wisdom has it that vineyards with a view of the water make the best wines. The most prestigious properties lie closest to the Garonne and the Gironde. These soils have the deepest and whitest gravel, giving the best conditions for Cabernet Sauvignon to ripen, producing wines with

FACTORS AFFECTING TASTE AND QUALITY

LOCATION
The Médoc lies on the left bank of the Gironde estuary, stretching northwest from Bordeaux in the south to Soulac in the north.

CLIMATE
Two large masses of water – the Atlantic and the Gironde – act as a heat regulator and help provide a microclimate ideal for viticulture. The Gulf Stream generally gives the Médoc mild winters, warm summers, and long, sunny autumns. The district is protected from westerly and northwesterly winds by the continuous coastal strip of pine forest that runs roughly parallel to the Médoc.

ASPECT
Undulating hillsides with knolls and gentle slopes characteristise the Médoc. The best vineyards can "see the river", and all areas of the Haut-Médoc gradually slope from the watershed to the Gironde. Marshy areas, where vines cannot be grown, punctuate most communes.

SOIL
Similar topsoils lie over different subsoils. Topsoils are typically outcrops of gravel, consisting of sand mixed with siliceous gravel of varying particle size. Subsoils may contain gravel and reach a depth of several metres, or may consist of sand, often rich in humus, and some limestone and clay.

VITICULTURE & VINIFICATION
Only red wines can use the Médoc appellation, although a small amount of prestigious whites are produced here under the Bordeaux appellation. You will find stainless steel, oak, and cement vats, and sometimes all three in the same cellar. Vats are becoming smaller and smaller as producers concentrated on precision viticulture and plot-by-plot fermentation.

GRAPE VARIETIES
Primary varieties: Cabernet Sauvignon, Merlot, Cabernet Franc
Secondary varieties: Petit Verdot, Carmenère, Malbec

CHÂTEAU KIRWAN
A bottle of Bordeaux decorates the gardens of the Château Kirwan in the Médoc.

power, elegance and ageing potential. The best are well suited to ageing. Of the 301 wine growers 5 are *crus classés*, 88 are *crus bourgeois*, 17 are *crus artisans*, and 63 belong to one of the four cooperatives.

The wines of Margaux are soft and velvety and full of charm, although they are very much vins de garde and will improve well with age. The wines of St-Julien are elegant with a very pure flavour. They have the delicate touch of Margaux, yet lean closer to Pauillac in body. The wines of Pauillac are powerful, the archetypical expression of Bordeaux left bank dominated by Cabernet Sauvignon.

St-Estèphe has traditionally been known for its rustic charm with a few classic wines, but investment and technology is changing the robustness of these spicy wines to richness. Beyond St-Estèphe lies the commune of St-Seurin-de-Cadourne, whose wines are entitled to use the Haut-Médoc appellation, after which the appellation becomes Médoc AOC.

THE FIGHT FOR GRAVEL

The best soils for vine-growing also happen to be the most suitable for gravel quarrying. After the war, in the absence of any legislation, gravel quarrying started in abandoned vineyards. Once the gravel was gone, the opportunity to reclaim the area as a vineyard was lost. There is plenty of gravel in the Gironde estuary itself, but it is more profitable to take it from an open pit. Quarrying companies will continue to plunder the Médoc's finite resources until the government agrees to protect them, but past administrations have shown little interest.

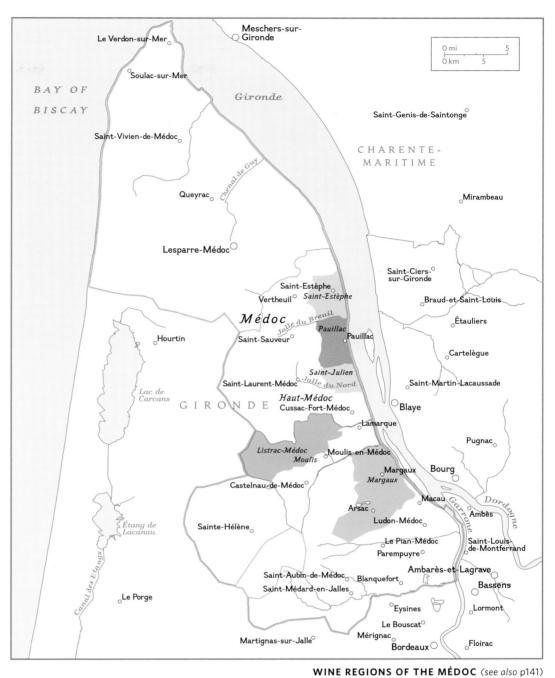

WINE REGIONS OF THE MÉDOC (see also p141)
The Médoc, a narrow strip of land between the Gironde estuary and the Atlantic Ocean, stretches northwards from the city of Bordeaux to the Pointe de Grave. The climate is Bordeaux's mildest, moderated by both the estuary and the ocean.

PROPORTION OF AOC AREA UNDER VINE REPRESENTED BY *CRUS CLASSÉS*

APPELLATION	UNDER VINE		CRUS CLASSÉS		REPRESENTS*
	HA	(ACRES)	HA	(ACRES)	
Médoc	5,560	(13,739)	-	-	No crus classés
Haut-Médoc	4,316	(10,665)	505	(1,248)	12% of AOC, 13.5% of crus classés
Listrac	787	(1,945)	-	-	No crus classés
Moulis	610	(1,507)	-		No crus classés
St-Estèphe	1,229	(3,037)	318	(786)	26% of AOC, 8.5% of crus classés
Pauillac	1,212	(2,995)	1,126	(2,782)	93% of AOC, 30% of crus classés
St-Julien	910	(2,249)	789	(1,950)	87% of AOC, 21% of crus classés
Margaux	1,512	(3,736)	994	(2,456)	66% of AOC, 26% of crus classés
TOTAL	1,6136	(39,873)	3,732	(9,222)	23% of all Médoc AOCs

(*Not 100% because Château Haut-Brion was included in the 1855 Classification)

DISTRIBUTION OF MÉDOC *CRUS CLASSÉS* THROUGHOUT THE APPELLATIONS

APPELLATION	GROWTHS					
	1ST	2ND	3RD	4TH	5TH	TOTAL
Haut-Médoc	0	0	1	1	3	5
St-Estèphe	0	2	1	1	1	5
Pauillac	3	2	0	1	12	18
St-Julien	0	5	2	4	0	11
Margaux	1	5	10	3	2	21
TOTAL	4	14	14	10	18	60

THE APPELLATIONS OF
THE MÉDOC

DECANTING MÉDOC WINES
Médoc wines are justifiably lauded, such as a Lynch-Bages from Pauillac and a Léoville-Las Cases from St-Julien.

HAUT-MÉDOC AOC

This AOC encompasses the Médoc's four finest communes – Margaux, St-Julien, Pauillac, and St-Estèphe – as well as the less well-known Listrac and Moulis communes. Outside these appellations lie a variety of different *terroirs* and microclimates over a large area that explain the diversity of Haut-Médoc wines, but the wines share the characteristic of being well balanced; full-bodied without being too powerful. The best are well suited to ageing. Of the 301 wine growers 5 are *crus classées*, 88 are *crus bourgeois*, 17 are *crus artisans*, and 63 belong to one of the 4 cooperatives.

RED These dry wines have a generosity of fruit tempered by a firm structure, and are medium- to full-bodied.

🍷 *Cabernet Sauvignon, Cabernet Franc, Merlot, Malbec, Petit Verdot, Carménère*

🥂 *6–15 years (crus classés); 5–8 years (others)*

LISTRAC-MÉDOC AOC

Listrac has had a chequered past. Already famous in the 18th century it was one of the largest vine-growing villages in the Médoc peninsula in 1913, but the crisis of the 1930s changed this. In 1935 a handful of winegrowers created a cooperative, Le Cave Grand-Listrac, to face the economic difficulties of the time. Today there are almost 40 members and they export all around the world. ften overlooked, mainly due to a lack of Classified growths and a reputation for somewhat rustic wines, recent investment has brought many of these wines rightly back into the spotlight, although they still remain excellent value for money.

RED These dry, medium- to full-bodied wines have the fruit and finesse of St-Julien combined with the firmness of St-Estèphe. The most successful wines tend to have a large proportion of Merlot, which enjoys the Haut-Médoc's clay soil.

🍷 *Cabernet Sauvignon, Cabernet Franc, Carménère, Merlot, Malbec, Petit Verdot*

🥂 *5–10 years*

MARGAUX AOC

The best Margaux are potentially the greatest wines in the whole of Bordeaux, but this is an appellation that covers five communes encompassing a great diversity of soil, and some of its wines not unnaturally have a tendency to disappoint. Margaux benefits enormously from having a namesake château, which is unique in Bordeaux, and the fact that this property sets the most extraordinarily high standards has done no harm to the reputation and price of these wines generally. The phenomenal success of Château Margaux has, however, unfairly raised expectations of many lesser-quality châteaux in the area, but those critics who widely accuse proprietors of sacrificing quality for quantity could not be further from the truth. There are individual châteaux that overproduce and therefore fail to achieve their full potential, but excessive volume is not typically the problem with this appellation, since it has the lowest yield per hectare of the four famous Médoc AOCs.

RED Exquisite, dry, medium-bodied, and sometimes full-bodied wines that can be deep-coloured and fabulously rich, yet they have great finesse and a silky finish.

🍷 *Cabernet Sauvignon, Cabernet Franc, Carménère, Merlot, Malbec, Petit Verdot*

🥂 *5–20 years (crus classés); 5–10 years (others)*

MÉDOC AOC

The appellation is the largest of all eight of the Médoc appellations, covering over 30% of the surface area under vines. Legally all the eight appellations of the Médoc Peninsula from the north of the city of Bordeaux almost to the Pointe de Grave, may apply for the (AOC) Médoc, but to the north of Saint-Estèphe there is a specific area, which produces exclusively Médoc AOC. It is here where the majority of Médoc wines come from. Its vineyards have undergone a rapid and extensive expansion since the mid-1970s. Of the 440 vine growers 214 belong to the cooperative cellars making up a quarter of the production. There are no classified growths within the appellation but there are over 100 *cru bourgeois* and 9 *crus artisans*.

RED The best of these dry, medium-bodied wines are similar to good Haut-Médocs, but the style is less sophisticated.

🍷 *Cabernet Sauvignon, Cabernet Franc, Carménère, Merlot, Malbec, Petit Verdot*

🥂 *4–8 years*

MOULIS AOC or
MOULIS-EN-MÉDOC AOC

Moulis is said to take its name from the windmills built on its higher land (the word *moulin* in French for "mill"). The Moulis appellation is a narrow band four and a half miles long running perpendicular to the Gironde Estuary with a diversity of *terroirs*: pure Garonne and Pyrenean gravel, clay, and limestone. Of the 46 winegrowers in the appellation, 39 remain independent, with important investment from more prestigious neighbours. The wines tend to be more approachable and rounder than their Listrac neighbours. Like Listrac, Moulis has no *cru classé* châteaux, despite adjoining Margaux, the appellation that has the highest number of such properties in the Médoc.

RED These are dry, medium-bodied, sometimes full-bodied, wines with more power than those of Margaux, but far less finesse.

🍷 *Cabernet Sauvignon, Cabernet Franc, Carménère, Merlot, Malbec, Petit Verdot*

🥂 *12 years*

PAUILLAC AOC

The commune of Pauillac vies with Margaux as the most famous appellation, but it is without doubt the most rock solid and consistent of the Bordeaux AOCs, while its *premiers crus* of Latour, Lafite, and Mouton make it the most important.

RED Dark and virtually opaque, great Pauillac is a dry, powerfully constructed wine, typically redolent of blackcurrants and new oak. It might be unapproachable when young, but it is always rich with fruit when mature. Although it does not have the grace of great Margaux, Pauillac brings power and style together to produce wines of incomparable finesse for their size.

🍷 *Cabernet Sauvignon, Cabernet Franc, Carménère, Merlot, Malbec, Petit Verdot*

🥂 *9–25 years (crus classés); 5–12 years (others)*

ST-ESTÈPHE AOC

The potential of the St-Estèphe appellation is exemplified by Cos d'Estournel, which is one of the best *deuxièmes crus* in the Médoc, but its strength lies in its range of *crus bourgeois*. The area under vine is slightly less than that of Margaux, which has the largest area, but St-Estèphe has far more unclassified châteaux, and even the best wines are wonderfully cheap.

RED If Pauillac is the stallion of the four famous appellations, St-Estèphe must be the shire horse. These dry, full-bodied wines are big and strong, yet not without dignity. St-Estèphe demands affection and, with the rich fruit of a sunny year, deserves it. These most enjoyable sweet-spice and cedary wines can have lots of honest, chunky fruit. Cos d'Estournel is the thoroughbred of the commune.

🍷 *Cabernet Sauvignon, Cabernet Franc, Carménère, Merlot, Malbec, Petit Verdot*

🥂 *8–25 years (crus classés); 5–12 years (others)*

ST-JULIEN AOC

St-Julien is the smallest of the four famous appellations and the most intensively cultivated, with almost 50 per cent of the commune under vine. There are no first growths, but there are as many as five seconds, and the standard and consistency of style of St-Julien wines are very high. This AOC overlaps part of the commune of Pauillac, and, historically, Châteaux Latour and Pichon-Longueville-Comtesse-de-Lalande could as easily have become St-Julien AOC as they did Pauillac AOC.

RED These are dry, medium-bodied, sometimes full-bodied, wines that have purity of style, varietal flavour, and can be long-lived. Well balanced and elegant, these wines fall somewhere between the lushness that is typical of Margaux and the firmer structure of Pauillac.

🍷 *Cabernet Sauvignon, Cabernet Franc, Carménère, Merlot, Malbec, Petit Verdot*

🥂 *6–20 years (crus classés); 5–12 years (others)*

THE WINE PRODUCERS OF
THE MÉDOC

CHÂTEAU D'AGASSAC
Haut-Médoc AOC
Cru Bourgeois
★☆

This estate's first wine is made from their oldest vines grown in deep gravel *terroir*. The wine is matured for 12 to 15 months, with 60 per cent in new oak.

RED Dark-coloured, plummy wine, with a lot of soft, ripe fruit.

🍷 *Cabernet Sauvignon 50%, Merlot 47%, Cabernet Franc 3%*

🍷 *4–10 years*

Second wine: *Château Pomiès-Agassac*

Other wines: *Château Pomiès-Agassac Tête de Cuvée, L'Agassant d'Agassac, Précision d'Agassac*

CHÂTEAU D'AURILHAC
Haut-Médoc AOC
Cru Bourgeois
★

A relative newcomer that has quickly developed a cult following. The grapes are machine harvested, and the wine is matured in wood for 12 months, with 35 per cent new oak.

RED A flashy, huge, dark, and dense wine with masses of fruit to balance the ripe tannins and extrovert oak.

🍷 *Cabernet Sauvignon 50%, Merlot 44%, Cabernet Franc 3%, Petit Verdot 3%*

🍷 *5–15 years*

Second wine: *Château La Fagotte*

CHÂTEAU BEAUMONT
Haut-Médoc AOC
Cru Bourgeois
☆

A large property that consistently produces wines of good quality. This wine is matured in wood for 12 months, with 35 per cent new oak.

RED These are aromatically attractive wines with elegant fruit and supple tannin.

🍷 *Cabernet Sauvignon 50%, Merlot 47%, Petit Verdot 3%*

🍷 *4–8 years*

Second wine: *Château d'Arvigny*

Other wines: *Tours de Beaumont*

CHÂTEAU BEL-AIR LAGRAVE
Moulis AOC
☆ V

This growth was classified *cru bourgeois* in 1932, but not included in the Syndicat's 1978 list. The wine is matured in wood for 18 to 20 months, with 70 per cent new oak.

RED These vividly coloured wines have a fine bouquet and a firm tannic structure.

🍷 *Cabernet Sauvignon 50%, Merlot 47%, Petit Verdot 3%*

🍷 *8–20 years*

Second wine: *Château Peyvigneau*

Other wines: *Château Haut Franquet, Châteaux Maleterre*

CHÂTEAU BELGRAVE
Haut-Médoc AOC
5ème Cru Classé
★☆

Situated on a good gravel bank behind Château Lagrange. The wine is matured in wood for 12 to 14 months, with up to 35 per cent new oak.

RED A good balance of blackcurrant fruit and ripe acidity, with much more supple tannin structure than used to be the case and vanilla overtones of new oak.

🍷 *Merlot 50%, Cabernet Sauvignon 46%, Petit Verdot 4%*

🍷 *8–16 years*

Second wine: *Diane de Belgrave*

CHÂTEAU BEL-ORME-TRONQUOY-DE-LALANDE
Haut-Médoc AOC
★

This property has a confusingly similar name to Château Tronquoy-Lalande, St-Estèphe. The chateau is owned and the wines made by the Quié family of Château Rauzan Gassies, classified growth of Margaux. This wine is matured in wood for 12 months.

RED These are round, charming wines.

🍷 *Merlot 65%, Cabernet Sauvignon 35%*

🍷 *7–15 years*

CHÂTEAU BERNADOTTE
Haut-Médoc AOC
Cru Bourgeois
★☆

Consistently performing above its class, this château is situated on fine, gravelly ground that once had the right to the Pauillac appellation and formed part of a *cru classé*. The quality started to improve from 1996 when the redoubtable Madame May-Éliane de Lencquesaing, who also owned Pichon-Longueville-Comtesse-de-Lalande, purchased it. In 2007 Champagne Louis Roderer become the owners when they purchased Pichon Longueville Comtesse de Lalande. Hong Kong businessman Antares Cheng bought the property in 2012 and continues to produce wines with the same quality with Hubert de Bouard as a consultant. This wine is now matured in wood for 14 to 16 months, with 30 per cent new oak.

RED These wines are very stylish, with lush Cabernet fruit backed up by the creamy richness of new oak.

🍷 *Cabernet Sauvignon 58%, Merlot 41%, Petit Verdot 1%*

🍷 *6–12 years*

Second wine: *Château Le Fournas Bernadotte*

Other wines: *Désirée, Rosé de Bernadotte.*

CHÂTEAU BISTON-BRILLETTE
Moulis AOC
★☆ V

This top-quality Moulis property ages its wines in wood for 12 to 15 months, with up to 35 per cent new oak.

RED Wines that are very rich in colour and fruit with a full, spicy-cassis character and a supple tannin structure.

🍷 *Cabernet Sauvignon 55%, Merlot 45%*

🍷 *5–15 years*

Second wine: *Château Biston*

CHÂTEAU BLAIGNAN
Médoc AOC
Cru Bourgeois
★

The area around Château Blaignan has had vineyards since the 14th century. Its wine achieved *cru bourgeois* status in 1932, under the name "Château Taffard de Blaignan". Today the 97-hectare (240-acre) property is owned by CA Grands Crus. The wine matures for 18 months in vats and oak barrels.

RED Full bodied with good fruit, this is an approachable wine in its youth.

🍷 *Cabernet Sauvignon 50%, Merlot 41%, Cabernet Franc 9%*

🍷 *3–12 years*

Other wine: *Château Prieuré de Blaignan*

CHÂTEAU BOUQUEYRAN
Moulis AOC
☆ V

A big improvement in quality and value since this 13-hectare (32-acre) property was leased by Philippe Porcheron of nearby Château Rose Saint-Croix and Château Marajollia in Margaux. Wines are matured in wood for 16 months with up to 50 per cent new oak. La Fleur de Bouqueyran is a superior *cuvée*.

RED Lovely deep-coloured, deep-flavoured wines of not inconsiderable style and finesse.

🍷 *Merlot 57%, Cabernet Sauvignon 43%*

🍷 *5–10 years*

Second wine: *Fleur de Bouqueyran, Château Rose Cantegrit*

CHÂTEAU LE BOURDIEU VERTHEUIL
Haut-Médoc AOC

Sitting between Vertheuil and St-Estèphe, this château was classified *cru bourgeois* in 1932, but not included in the Syndicat's 1978 list. This wine is matured in wood for 12 months, with 30 per cent new oak.

RED Well-coloured, full-bodied with robust character not lacking in charm.

🍷 *Merlot 55%, Cabernet Sauvignon 40%, Petit Verdot 5%*

🍷 *7–15 years*

Second wine: *Château Haut-Brignays*

Other wines: *Château Victoria, Chateau Picourneau*

CHÂTEAU BRANAS GRAND POUJEAUX
Moulis AOC
★☆ V

Bought by Justin Onclin in 2002, *negociant* and owner of Château Villemaurine in St-Émilion. These excellent and rapidly improving wines are aged for wood for 15 to 18 months, with 30 per cent new oak.

RED An increase in Merlot gives this wine plenty of accessible fruit, charming aromatic properties, and increasing finesse.

🍷 *Merlot 50%, Cabernet Sauvignon 45%, Petit Verdot 5%*

🍷 *5–12 years*

Second wine: *Les Eclats de Branas Grand Poujeaux*

CHÂTEAU BRILLETTE
Moulis AOC
Cru Bourgeois
★☆ V

This château's name reputedly derives from its glinting, pebbly soil. The wine is matured in wood for 12 months, with 40 per cent new oak.

RED These are attractively coloured wines of full but supple body, with delightful summer-fruit and vanilla aromas. Easily equivalent to *cru classé* quality.

🍷 *Merlot 54%, Cabernet Sauvignon 35%, Cabernet Franc 7%, Petit Verdot 4%*

🍷 *5–12 years*

Second wine: *Haut Brillette*

CHÂTEAU D'AGASSAC
This Haut-Médoc wine estate has a history dating back to the 13th century.

CHÂTEAU CAMBON-LA-PELOUSE
Haut-Médoc AOC
Cru Bourgeois
☆

This wine is matured in wood for 12 months, with 45 per cent new oak.

RED Soft, medium- to full-bodied wines with fresh and juicy flavours.

 Merlot 52%, Cabernet Sauvignon 44%, Petit Verdot 4%

🍷 *3–8 years*

Other wines: *Château Trois Moulins, L'Aura*

CHÂTEAU CAMENSAC
Haut-Médoc AOC
5ème Cru Classé
★ **V**

Situated behind Château Belgrave, this property was purchased and renovated in the mid-1960s by the Forner brothers, who are of Spanish origin, and later established Marquès de Cáceres in Rioja. Camensac began making wine equivalent to its classification in the late 1970s and since 1995 has been performing beyond its class. It is matured in wood for 18 months, with 40 per cent new oak.

RED Well-structured wine, with a medium weight of fruit and a certain finesse.

🍷 *Cabernet Sauvignon 60%, Merlot 40%*

🍷 *8–20 years*

Second wine: *Le Second de Camensac*
Other wines: *La Closerie de Camensac*

CHÂTEAU CANTEMERLE
Haut-Médoc AOC
5ème Cru Classé
★ ☆ **V**

Managed for 10 years by Cordier, who oversaw the replanting of vines and cellar improvements. This 90-hectare (222-acre) property is now independently managed and sold through the Bordeaux marketplace. The wine is normally matured in wood for 16 months (12 in barrel), with one-third new oak. It is currently performing above its classification.

RED Deliciously rich wines of fine colour, creamy-oaky fruit, beautiful balance, and increasing finesse.

🍷 *Cabernet Sauvignon 64%, Merlot 27%, Cabernet Franc 5%, Petit Verdot 4%*

🍷 *8–20 years*

Second wine: *Villeneuve Les Allées de Cantemerle*

CHÂTEAU CAP-LÉON-VEYRIN
Listrac AOC
☆

Originally just called Château Cap-Léon. The vines of this property are planted in two plots of clay-gravel soil over marl.

RED Deep-coloured, full-bodied, richly flavoured wines with high extract levels and a good balance of tannin.

🍷 *Merlot 60%, Cabernet Sauvignon 35%, Petit Verdot 5%*

🍷 *5–12 years*

CHÂTEAU COUFRAN
This Bordeaux goes well with classic meat dishes and also hearty pasta and cheeses.

CHÂTEAU LA CARDONNE
Médoc AOC
Cru Bourgeois
★ **V**

Formerly owned by the Rothschilds of Lafite, who expanded and renovated the property. This wine is matured in wood for 12 months, with 50 per cent new oak.

RED Attractive, medium-bodied wines with a good, grapey perfume and a silky texture, made in an elegant style.

 Merlot 50%, Cabernet Sauvignon 45%, Cabernet Franc 5%

🍷 *6–10 years*

CHÂTEAU CARONNE-STE-GEMME
Haut-Médoc AOC
Cru Bourgeois
★ ☆ **V**

This property is situated south of Château Lagrange – a superb island of vines on a gravel plateau. Matured in wood for 12 months, with 25 per cent new oak.

RED Full-bodied wines rich in flavour with undertones of creamy oak and a supple tannin structure.

🍷 *Cabernet Sauvignon 60%, Merlot 37%, Petit Verdot 3%*

🍷 *8–20 years*

Second wine: *Château Labat*

CHÂTEAU CASTÉRA
Médoc AOC
Cru Bourgeois
V

The original château was reduced to ruins by the Black Prince in the 14th century. This wine is matured in wood for 12 months, with one-third new oak.

RED Soft-textured, medium-bodied wines best drunk relatively young.

 Merlot 65%, Cabernet Sauvignon 25%, Cabernet Franc 5%, Petit Verdot 5%

🍷 *4–8 years*

Second wine: *Château Bourbon La Chapelle*
Other wines: *Marquis de Castera, Apostrophe, Alexandrin, Perle Rosé (Bordeaux Rosé)*

CHÂTEAU CHANTELYS
Médoc AOC
Cru Bourgeois
☆ **V**

Owner Christine Courrian Braquissac brings a gentle touch to the naturally firm wines of this district.

RED Well-coloured, medium-bodied, gently rich-flavoured wines of some elegance.

🍷 *Cabernet Sauvignon 60%, Merlot 30%, Petit Verdot 10%*

🍷 *3–8 years*

Second wine: *Les Iris de Chantelys*

CHÂTEAU CHARMAIL
Haut-Médoc AOC
Cru Bourgeois
★ **V**

The wines from this château have improved dramatically since its excellent 1996 vintage and continue to perform well in blind tastings. This wine is matured in wood for 12 months, with 30 per cent new oak.

RED Rich, spicy, and long, with well-rounded, ripe tannins.

🍷 *Merlot 48%, Cabernet Sauvignon 30%, Cabernet Franc 20%, Petit Verdot 2%*

🍷 *3–7 years*

Second wine: *Les Tours de Charmail*

CHÂTEAU CHASSE-SPLEEN
Moulis AOC
★★ **V**

The property belongs to the Merlaut Villars family, who also own the *cru classé* Château Haut-Bages-Libéral and the excellent unclassified growth of Château la Gurgue in Margaux. It has been under the management of Celine Villars since 2000 (who also manages Château Camensac). The wine is matured in wood for 18 months with 40 per cent new oak and is usually of *cru classé* quality. Certainly, it well deserves being recently classified as one of only nine *crus bourgeois exceptionnels*.

RED Full-bodied wines of great finesse, vivid colour, with a luxuriant, creamy-rich flavour of cassis and chocolate with warm, spicy-vanilla undertones. Easily equivalent to *cru classé* quality.

🍷 *Cabernet Sauvignon 52%, Merlot 9.5%, Petit Verdot 4.5%, Cabernet Franc 4%*

🍷 *8–20 years*

Second wine: *L'Oratoire de Chasse-Spleen, Heritage de Chasse-Spleen, Blanc de Chasse Spleen (b-Bordeaux Blanc) Gressier Grand Poujeaux (Moulis)*

CHÂTEAU CISSAC
Haut-Médoc AOC
Cru Bourgeois
V

Château Cissac is always good value, especially in hot years. It is fermented in wood and matured in cask with no *vin de presse*. This wine is matured in wood for 16 months, with 30 per cent new oak.

RED These are deep-coloured, well-flavoured, full-bodied wines made in a *vin de garde* style.

🍷 *Cabernet Sauvignon 70%, Merlot 22%, Petit Verdot 8%*

🍷 *8–20 years*

Second wine: *Les Reflets du Cissac*
Other wines: *Château de Breuil*

CHÂTEAU CITRAN
Haut-Médoc AOC

This substantial-sized property was once run by Château Coufran, then passed into Japanese ownership under the Fujimoto company, which invested in improvements in the vineyard and winery. In 1997 it was taken back into French ownership under the auspices of Groupe Taillan and is run personally by Céline Villars. This wine is matured in wood for 15 months, with 30 per cent new oak.

RED A once solid, if plodding, Médoc of robust character, the style since the new millennium has shown a true plumpness of fruit, with not inconsiderable finesse.

🍷 *Merlot 58%, Cabernet Sauvignon 47%, Cabernet Franc 5%*

🍷 *5–15 years*

Second wine: *Moulins de Citran*

CHÂTEAU LA CLARE
Médoc AOC
Cru Bourgeois
☆ **V**

Owned by Jean Guyon of Château Rollan de By fame, this well-established property is receiving renewed attention of late. Approximately 30 per cent of the wine is matured in wood for 12 months, with 60 per cent new oak.

RED A rich, nicely coloured, medium-bodied wine with some spicy finesse.

🍷 *Cabernet Sauvignon 45%, Merlot 40%, Cabernet Franc 5%*

🍷 *4–8 years*

CHÂTEAU CLARKE
Listrac AOC
★ **V**

This estate's vines were dug up and its château pulled down in 1950. All was abandoned until 1973, when it was purchased by Baron Edmond de Rothschild.

He completely restored the vineyard and installed an ultra-modern winery. Since the 1981 vintage, it has become one of the Médoc's fastest-rising stars. The wine is fermented in stainless steel and matured in wood for 12 to 16 months, with up to 70 per cent new oak.

RED Well-coloured wines have a good measure of creamy-smoky oak, soft fruit, and increasing finesse.

🍷 *Merlot 70%, Cabernet Sauvignon 30%*

🍷 *7–25 years*

Second wine: *Château Granges des Domaines Edmond de Rothschild, Le Merle Blanc de Chateau Clarke (Bordeaux Blanc)*

CHÂTEAU COUFRAN
Haut-Médoc AOC
☆

These wines are matured in wood for 12 months, with 25 per cent new oak.

RED Frank and fruity, this medium- to full-bodied wine has a chunky, chocolaty flavour, which is dominated by Merlot.

🍷 *Merlot 85%, Cabernet Sauvignon 15%*

🍷 *4–12 years*

Second wine: *Numéro 2 de Couffran, Château la Rose-Maréchale*

CHÂTEAU DUTRUCH GRAND POUJEAUX
Moulis AOC
★

Dutruch is one of the best Grand Poujeaux satellite properties. It also makes two other wines from the specific-named plots "La Bernède" and "La Gravière". Matured in wood for 12 months, with 30 per cent new oak.

RED These are fine, full-bodied wines of excellent colour, fruit, and finesse.

🍷 *Merlot 50%, Cabernet Sauvignon 45%, Petit Verdot 5%*

🍷 *7–15 years*

Other wines: *La Bernède-Grand-Poujeaux, La Gravière-Grand-Poujeaux*

CHÂTEAU FONRÉAUD
Listrac AOC
Cru Bourgeois

This splendid château has south-facing vineyards situated on and around a knoll, This wine is matured in wood for 12 to 18 months, with one-third new oak.

RED Attractive medium-to-full-bodied wines of good fruit and some style.

🍷 *Cabernet Sauvignon 52%, Merlot 44%, Petit Verdot 4%*

🍷 *6–12 years*

Second wine: *Legende de Fonréaud, Clos les Demoiselles, Le Cygne de Fonreaud Bordeaux Blanc)*

CHÂTEAU FOURCAS DUPRÉ
Listrac AOC
★

A charming house with vineyards situated on gravel over iron-pan soil, which can excel in hot years. This wine is matured in wood for 12 months, with one-third new oak.

RED The good colour, bouquet, and tannic structure of these wines is rewarded with rich fruit in good years.

🍷 *Cabernet Sauvignon 44%, Merlot 44%, Cabernet Franc 10%, Petit Verdot 2%*

🍷 *6–12 years*

Second wine: *Hautes Terres de Fourcas Dupré*

CHÂTEAU FOURCAS HOSTEN
Listrac AOC
★

Under the ownership of the Momméja brothers since 2010, the property has undergone a complete renovation, with replanted vines and updated cellars. The attention to detail in both agriculture and winemaking clearly shows in the quality of these wines, which are matured in wood for 12 months, with one-third new oak.

RED Deeply coloured and full-bodied wines, rich in fruit and supported by a firm tannic structure, although the style is becoming more supple and can even be quite fat.

🍷 *Cabernet Sauvignon 44%, Merlot 55%, Cabernet Franc 1%*

🍷 *8–20 years*

Second wine: *Les Cèdres d'Hosten, Le Blanc de Fourcas Hostens*

CHÂTEAU LES GRANDS CHÊNES
Médoc AOC
☆

The quality here has been excellent since the mid-1990s, but took another step up in 1999, when Bernard Magrez purchased the château. The wine is matured in wood for 16 months, with one-third new oak.

RED Lush and charming, with opulent fruit and fine acidity balance.

🍷 *Cabernet Sauvignon 70%, Merlot 29%, Cabernet Franc 1%*

🍷 *6–15 years*

CHÂTEAU GRESSIER GRAND POUJEAUX
(See also Château Chasse-Spleen)
Moulis AOC
Cru Bourgeois
★

This château was classified *cru bourgeois* in 1932, but not in 1978. Owned and made by the Villars team (who also own Château Chasse-Spleen), it has in recent years produced successful wines that compare well with good *crus classés*. This wine is matured in wood for 24 months, with one-third new oak.

RED Full-bodied wines with plenty of fruit and flavour. Well worth laying down.

🍷 *Cabernet Sauvignon 50%, Merlot 50%*

🍷 *6–12 years*

CHÂTEAU GREYSAC
Médoc AOC
Cru Bourgeois
☆

Another excellent Médoc from the Jean Guyon, Château Rollan de By stable. This

wine is matured in wood for 12 months, with 50 per cent new oak.

RED Stylish, medium-bodied wines with silky-textured, ripe-fruit flavours.

🍷 *Merlot 65%, Cabernet Sauvignon 29%, Cabernet Franc 3%, Petit Verdot 3%*

🍷 *6–10 years*

CHÂTEAU HANTEILLAN
Haut-Médoc AOC
Cru Bourgeois
★

Under the keen direction of Catherine Blasco, this large property in the Haut-Médoc produces a consistently fine wine. Blasco's achievements have also been consolidated by the new winemaking facilities. This wine is matured in wood for 14 to 16 months, with 40 per cent new oak.

RED The wine has a fine colour, spicy bouquet with underlying vanilla-oak tones, ripe fruit, and supple tannins.

🍷 *Cabernet Sauvignon 50%, Merlot 5%, Petit Verdot 5%*

🍷 *6–12 years*

Second wine: *Château Larrivaux Hanteillan*

CHÂTEAU LACOMBE-NOILLAC
Haut-Médoc AOC
Cru Bourgeois
☆

This property is the home of the most northerly vines in the Médoc. The wine is matured in wood for six months, with 15 per cent new oak.

RED Elegant, medium-bodied of surprising style and finesse for the location.

🍷 *Cabernet Sauvignon 58%, Merlot 32%, Cabernet Franc 6%, Petit Verdot 4%*

🍷 *4–10 years*

Second wine: *Rives de Gravelongue*
Other wines: *Château Les Traverses La Franque*

CHÂTEAU LA LAGUNE
Haut-Médoc AOC
3ème Cru Classé
★☆

This property was purchased by the Frey family in 2000 and is managed by Caroline Frey, who also runs the Rhône house Paul Jaboulet Aîné Hermitage. The immaculate vineyard of this fine château is the first *cru classé* encountered after leaving Bordeaux and is situated on sand and gravel soil. The château and the cellars were completely renovated in 2003. This wine is matured in wood for 16 to 18 months, with 50 per cent new oak. Certified organic since 2016.

RED These wines are deep-coloured with complex cassis and stone-fruit flavours intermingled with rich, creamy-vanilla oak nuances. They are full-bodied but supple.

🍷 *Cabernet Sauvignon 60%, Merlot 30%, Petit Verdot 10%*

🍷 *5–30 years*

Second wine: *Moulin de Lagune*
Other wines: *Mademoiselle L*

CHÂTEAU DE LAMARQUE
Haut-Médoc AOC
★

This large and impressive vineyard continues to improve under the ownership of Pierre-Gilles and Marie-Hélène Gromand-Brunet d'Évry. The wine is matured in wood for 16 to 18 months, with one-third new oak.

RED The supple style of a modern Médoc, with plenty of real fruit flavour, and an enticingly perfumed bouquet.

🍷 *Cabernet Sauvignon 46%, Merlot 25%, Cabernet Franc 24%, Petit Verdot 5%*

🍷 *5–12 years*

Second wine: *D de Lamarque*
Other wines: *Château Cap de Haut*

CHÂTEAU LAMOTHE-CISSAC
Haut-Médoc AOC
Cru Bourgeois
★

An up-and-coming wine from one of Bordeaux's oldest properties, Lamothe-Cissac is matured in wood for 12 months, with 20 per cent new oak. It has recently started to outperform Cissac.

RED Classically proportioned, Cabernet-dominated wines of excellent potential longevity.

🍷 *Cabernet Sauvignon 58%, Merlo 35%, Petit Verdot 5%, Cabernet Franc 2%*

🍷 *4–16 years*

CHÂTEAU LANESSAN
Haut-Médoc AOC
★☆

When the owner forgot to submit samples – or could not be bothered – this château lost the chance to be included in the 1855 Classification. It was later classified *cru bourgeois* in 1932, but not included in the Syndicat's 1978 list. Since 2009 Paz Espajo has managed the property for the owners, the Bouteiller family. She has overseen investments in replanting the vines and refurbishing the cellars, bringing this property to the level of quality its location merits. The vines are planted on beautiful gravel outcrops in the north of the Haut-Médoc on the boundary of St-Julien. The grapes are machine harvested, and the wines are matured in wood for 15 months, with no claim of any new oak.

RED Big, intensely flavoured wines of deep, often opaque, colour and a quality that closely approaches that of a *cru classé*, requiring a similar minimum ageing before its fruit shows through.

🍷 *Cabernet Sauvignon 60%, Merlot 35%, Petit Verdot 4%, Cabernet Franc 1%*

🍷 *7–20 years*

Second wine: *Les Calèches de Lanessan*
Other wines: *Château de Ste-Gemme, Voyage de Lanessan*

CHÂTEAU LAROSE-TRINTAUDON
Haut-Médoc AOC
Cru Bourgeois
★☆

This is the largest estate in the Médoc. It 1986 the insurance company Allianz bought the domain. Their continuing investment ensures the standard of these wines, which are matured in wood for 12 months with 25 per cent new oak, and it is as high as it has ever been.

RED Medium-bodied, and sometimes full-bodied, wines with an elegantly rich flavour of juicy summer fruits, vanilla, and truffles, backed up by supple tannins.

 Cabernet Sauvignon 60%, Merlot 40%

6–15 years

Second wine: *Les Hauts de Trintaudon*
Other wines: ✔ *Château Larose-Perganson (cru bourgeois), Château Arnaud (cru bourgeois)*

CHÂTEAU LESTAGE-DARQUIER
Moulis AOC
★ ❷

This is the least encountered of the many Poujeaux châteaux (formerly sold as Château Lestage-Darquier-Grand-Poujeaux), but well worth digging out. This wine is matured in wood for 9 to 12 months, with 100 per cent year-old oak.

RED Densely coloured wines, rich in bouquet and fruit, with a powerful structure.

 Cabernet Sauvignon 50%, Merlot 46%, Cabernet Franc 2%, Petit Verdot 2%

8–20 years

CHÂTEAU LIVERSAN
Haut-Médoc AOC
Cru Bourgeois
★ ❷

Prince Guy de Polignac purchased the estate of Château Liversan in 1984, when it was inexorably linked with Champagne Pommery, but it was first leased and then sold to the owners of Patache d'Aux. Both properties are now under the ownership of the Antoine Moueix group. The vineyard is on fine, sandy gravel over a limestone subsoil, just 3 kilometres (almost 2 miles) from Lafite and Mouton Rothschild. The wine is fermented in stainless steel and matured in wood for 12 months, with 25 per cent new oak.

RED Rich and flavourful wines, of full body and some style. They are gaining in class with each vintage.

 Merlot 50%, Cabernet Sauvignon 44%, Cabernet Franc 4%, Petit Verdot 2%

7–20 years

Second wine: *Les Charmes des Liversan*

CHÂTEAU LOUDENNE
Médoc AOC
Cru Bourgeois
☆ ❷

This pink-washed, Chartreuse-style château, with its lawns running down to the Gironde, once belonged to W and A Gilbey, who ran it in a style that harked back to the last days of the British Empire. Since 2013 it belongs to the Chinese group Moutai, joined by Camus Cognac as minority shareholders in 2016. The property is now run by Camus, who are doing everything it takes to bring it back to its former glory, replanting with some Petit Verdot and some Sauvignon Gris for their white. Loudenne was the very first Médoc property to produce a

white wine in 1880 (a quarter Sémillon and three-quarters Sauvignon), aged for 8 months in oak. Still being produced, it is reminiscent of a white Graves. The red wine is matured in wood for 12 to 15 months, with 25 per cent new oak.

RED Full-bodied wines with a spicy-black-currant bouquet, sometimes silky and hinting of violets, with underlying vanilla oak, a big mouthful of rich and ripe fruit, excellent extract, and great length.

 Cabernet Sauvignon 50%, Merlot 50%, Cabernet Franc 7%, Malbec 2%, Petit Verdot 1%

5–15 years

Second wine: *Loudenne des Rives*
Other wines: *Loudenne les Roses, Blanc de Loudenne*

CHÂTEAU MALESCASSE
Haut-Médoc AOC
Cru Bourgeois
☆ ❷

Philippe Austruy (owner of La Commanderie de Peyrassol in Provence and La Quinta da Côrte in the Douro) bought the property in 2012. He completely renovated the cellars and the chateau and opened the doors of the château to the public. The wine is matured in wood for 16 months, with 30 per cent new oak.

RED Fruit-forward, elegant, good-value claret, getting better with each vintage.

 Cabernet Sauvignon 39%, Merlot 53%, Petit Verdot 9%

4–8 years

Second wine: *Le Moulin Rose de Château Malescasse*

CHÂTEAU DE MALLERET
Haut-Médoc AOC
Cru Bourgeois
☆ ❷

Château de Malleret is a vast estate, which incorporates a stud farm with two training race tracks and stables for both hunting and racing. The vineyard boasts 60 hectares (148 acres). This wine is matured in wood for 12 months, with up to 30 per cent new oak. Stephen

Derenoncourt has been consulting since 2014 in the brand-new cellar furnished with cement and oak vats.

RED Delightful wines of good bouquet, medium body, and juicy-rich fruit. Improving. Vinified in concrete and oak

 Cabernet Sauvignon 60%, Merlot 36%, Petit Verdot 4%

5–12 years

Second wine: *Le Baron*
Other wines: *Château Barthez, Le Margaux*

CHÂTEAU MAUCAILLOU
Moulis AOC
★ ☆ ❷

Consistently one of the best-value wines produced in Bordeaux. This wine is matured in wood for 16 months, with up to 60 per cent new oak.

RED Deep-coloured, full-bodied wine with masses of velvety-textured fruit, beautiful cassis and vanilla flavours, and supple tannins.

 Cabernet Sauvignon 55%, Merlot 36%, Petit Verdot 7%, Cabernet Franc 2%

6–15 years

Second wine: *No 2 de Maucaillou*
Other wines: *Le Haut-Médoc de Maucaillou (Haut-Médoc AOC)*

CHÂTEAU MAUCAMPS
Haut-Médoc AOC
Cru Bourgeois
★ ❷

Situated between Macau itself and the *cru classé* Cantemerle to the south, this château makes superb use of its 15 hectares (37 acres) of fine, gravelly vineyards. This wine is matured in wood for 16 months, with up to 40 per cent new oak.

RED Always a deep-coloured, full-bodied wine with plenty of fruit flavour supported by supple tannins.

Cabernet Sauvignon 60%, Merlot 35%, Petit Verdot 5%

5–12 years

Second wine: *Clos de May*
Other wines: *Château Priban, Château Dasvin Bel-Air (Cru Bourgeois)*

CHÂTEAU MAUVESIN
Moulis AOC

Lilian and Michel Sartorius Barton of Leoville and Langoa Barton purchased this property in 2011. The wines are made by the eighth-generation Melanie Sartorius-Barton and are fermented in stainless steel in a brand-new fermentation cellar and aged for 12 months in one-third new oak in the fully renovated barrel cellar.

 Merlot 46%, Cabernet Sauvignon 36%, Cabernet Franc 6%, Petit Verdot 2%

Second wine: *L'Impression de Mauvesin Barton*
Other wines: *Le Haut-Médoc de Mauvesin Barton*

CHÂTEAU LE MEYNIEU
Haut-Médoc AOC
★ ❷

This property is under the same ownership as Château Lavillotte and Domaine de La Ronceray in St-Estèphe. The deep-coloured wine is not filtered before it is bottled. Aged for 18 months in 30 per cent new oak.

RED This is a deep, dark, brooding wine of dense bouquet and solid fruit which promises much for the future.

Cabernet Sauvignon 60%, Merlot 35%, Cabernet Franc 5%

7–15 years

CHÂTEAU MOULIN-À-VENT
Moulis AOC
Cru Bourgeois
❍

One-third of the property of Château Moulin-à-Vent overlaps the commune of Listrac, but its appellation is still Moulis. This wine is matured in wood for 20 months, with 25 per cent new oak. Organic agriculture.

RED Medium-bodied wines with an elegant bouquet and a full flavour.

 Merlot 60%, Cabernet Sauvignon 15%, Petit Verdot 15%, Cabernet Franc 10%

7–15 years

CHÂTEAU NOAILLAC
Médoc AOC
Cru Bourgeois
☆ ❷

The wine is matured in wood for up to 12 months, with 30 per cent new oak.

RED Deliciously fruity style, underpinned by a discreet use of oak.

Cabernet Sauvignon 45%, Merlot 55%, Petit Verdot 5%

3–8 years

Second wine: *Château Moulin-de Noaillac*
Other wines: *Noaillac Prestige, La Rosé Noaillac*

CHÂTEAU LOUDENNE
Located in the Haut-Médoc, the elegant 17th-century château, with its distinctive pink-washed walls is the centerpiece of this vineyard's magnificent grounds.

A CHÂTEAU POTENSAC WINE CRATE IS CREATIVELY USED AS A BIKE BASKET
Part of Domaines Delon, Château Potensac's wines received *cru bourgeois exceptionnel* status in 2003.

CHÂTEAU LES ORMES SORBET

Médoc AOC
Cru Bourgeois
★

Owned by the Boivert family since 1764. The wine today is matured in wood for 16 to 18 months, with one-third new oak.

RED Once reputed for its characterful wines of substantial body, dense fruit, and positive flavour, this château has made an increasingly opulent style, with fine aromatics, since the early 1990s, fully justifying its *cru bourgeois supérieur* in the 2003 classification.

🍷 Cabernet Sauvignon 60%, Merlot 30%, Petit Verdot 5%

🍷 4–15 years

Second wine: *Château de Conques*

CHÂTEAU PALOUMEY

Haut-Médoc AOC
Cru Bourgeois
☆

An old property that enjoyed fame in the 19th century but was grubbed up and forgotten until Martine Cazeneuve replanted it in 1990. The property is now run by her son, Pierre. The wine is matured in wood for 12 to 15 months, with one-third new oak.

RED Deliciously rich and ripe, fruit-driven style that is underpinned by a nicely restrained use of oak.

🍷 Cabernet Sauvignon 55%, Merlot 40%, Cabernet Franc 5%

🍷 4–8 years

Second wine: *Les Ailes de Paloumey, Plume de Paloumey, Château Haut-Carmaillet*

Other wines: *Château la Bessan (Margaux), Château la Carricq (Moulis)*

CHÂTEAU PATACHE D'AUX

Médoc AOC
Cru Bourgeois
☆

This old property once belonged to the Aux family, descendants of the counts of Armagnac, but was purchased by a syndicate headed by Claude Lapalu in 1964. In 2016, Antoine Moueix properties purchased the estate. Although Patache d'Aux is always reliable, it has performed particularly well since its stunning 1990 vintage. The wine is matured in wood for 12 months, with 25 to 30 per cent new oak.

RED Stylish, highly perfumed, medium-bodied wine with very accessible fruit.

🍷 Cabernet Sauvignon 60%, Merlot 30%, Cabernet Franc 7%, Petit Verdot 3%

🍷 4–8 years

Second wine: *Le Relais de Patache d'Aux*

Other wines: *Les Chevaux de Patache d'Aux, La Patache*

CHÂTEAU PEYRABON

Haut-Médoc AOC
Cru Bourgeois
☆

Virtually unknown until 1998, when it was purchased by Millesima (formerly Les Vins des Grands Vignobles). The wine is matured in wood for 14 months, with 25 per cent new oak.

RED Sturdy style, but not lacking in fruit, and usually very good value.

🍷 Cabernet Sauvignon 50%, Merlot 33%, Cabernet Franc 15%, Petit Verdot 2%

🍷 5–10 years

Second wine: *Château Pierbone*

Other wines: *Château Le Fleur Peyrabon (Pauillac)*

CHÂTEAU PLAGNAC

Médoc AOC
Cru Bourgeois

Owned by Domaines Cordier since 1972, this property has produced consistently good-value red Bordeaux since the end of that decade. The wine is matured in wood for 12 months with 25 per cent new oak.

RED Full-bodied and full-flavoured, with some breed; lots of up-front Merlot fruit and a smooth finish.

🍷 Cabernet Sauvignon 65%, Merlot 35%

🍷 4–10 years

Second wine: *Château Haut de Plagnac*

CHÂTEAU PONTEY

Médoc AOC
Cru Bourgeois
★☆

Owned by the Bordeaux *négociant* Quancard family, this château occupies an excellent location on a gravel plateau. The wines are matured in wood for 12 months, with one-third new oak.

RED These wines have always been cleverly constructed and brimming with lush, oaky fruit, but they are even more lush than ever.

🍷 Merlot 55%, Cabernet Sauvignon 45%

🍷 3–12 years

Second wine: *Château Vieux Prezat*

Other wines: *Château Pontey Caussan*

CHÂTEAU POTENSAC

Médoc AOC
★☆

This property is under the same ownership as Château Léoville-Las-Cases in St-Julien, and its wines often aspire to *cru classé* quality, fully justifying the *cru bourgeois exceptionnel* classification in 2003. The wine is fermented in stainless steel, then matured in wood for 18 months, with 20 per cent new oak.

RED Classy, full-bodied wines of a lovely brick-red colour, with lots of fruit and underlying chocolate and spice flavours.

🍷 Merlot 49.6%, Cabernet Sauvignon 33.7%, Cabernet Franc 15.7%, Petit Verdot 1%

🍷 6–15 years

Second wine: *Chapelle de Potensac*

CHÂTEAU POUJEAUX

Moulis AOC
★★

After Chasse-Spleen, this château produces the best wine in Moulis and is easily the equivalent of a good *cru classé*. Hence it was no surprise that it was classified as one of only nine *crus bourgeois exceptionnels* in 2003. Aged for 12 months in 30 per cent new oak.

RED Full-bodied and deep-coloured wine with a very expansive bouquet and creamy-rich, spicy fruit.

🍷 Cabernet Sauvignon 50%, Merlot 40%, Cabernet Franc 5%, Petit Verdot 5%

🍷 10–25 years

Second wine: *La Salle-de-Château-Poujeaux*

CHÂTEAU PREUILLAC

Médoc AOC
Cru Bourgeois
★

Purchased in 1998 by Yvon Mau, who has sold his eponymously named *négociant* business to Freixenet in a bid to build up a portfolio of estate properties. A lot of money and attention was devoted to renovating the property, reducing yields, and updating the vinification facilities, which is why director Jean-Christophe Mau was upset not to have been included in the 2003 *cru bourgeois* classification. The grapes are hand-picked then sorted on tables, and the wines are fermented in oak, then matured in wood for 12 to18 months, with one third new oak.

RED The creamy-cassis and chocolaty fruit in these wines is a testament to Mau's efforts. Preuillac used to be rather rustic, but now shows more finesse with each and every vintage.

🍷 Merlot 45%, Cabernet Sauvignon 40%, Cabernet Franc 5%, Petit Verdot 1%

🍷 4–12 years

Second wine: *Esprit de Preuillac*

Other wines: *Émotion Preuillac*

CHÂTEAU RAMAGE-LA-BATISSE

Haut-Médoc AOC
Cru Bourgeois
★

This property has surpassed itself since the late 1980s, making wines of a remarkable quality-price ratio that are matured in wood for 18 months, with 50 per cent new oak.

RED Rich, well-flavoured, oaky wines that are immediately attractive in light years and ridiculously inexpensive *vins de garde* in the best vintages. Not unlike a poor man's Lynch-Bages.

🍷 *Cabernet Sauvignon 51%, Merlot 41%, Cabernet Franc 6%, Petit Verdot 2%*

🍷 *7–15 years*

Second wine: *Clos de Ramage*
Other wines: *Château du Terrey, La Bastide de Ramage, Chateau Tourteran (Cru Bourgeois)*

CHÂTEAU ROLLAN DE BY
Médoc AOC
★☆

Château Rollan de By was the first vineyard bought by Jean Guynon in 1989; it is now the flagship for his collection of properties, which covers 185 hectares (457 acres) in the Northern Médoc under "Domaines Rollan de By". The wine is matured in wood for 12 months, with 60 per cent new oak. It has had a string of blind-tasting victories. The Château Haut-Condissas is a super-selection, with 100 per cent new oak.

RED Lots of up-front fruit, but long and classy, with plenty of finesse. Not big and, surprisingly, not over-oaked.

🍷 *Merlot 70%, Cabernet Sauvignon 10%, Cabernet Franc 10%, Petit Verdot 10%*

🍷 *4–12 years*

Other wines: *Château Haut-Condissas, Chateau Tour Seran (Cru Bourgeois), Château la Clare, Château la Rose de By, Château Greysac, Château de By, Château Monthil, Le Blanc de Château Rolland de By (Bordeaux Blanc), Château Rolland de By Rosé.*

CHÂTEAU ST-PAUL
Haut-Médoc AOC
Cru Bourgeois
☆ Ⓥ

This vineyard is pieced together from parcels previously owned by two St-Estèphe châteaux. This wine is matured in wood for 12 months, with 25 per cent new oak.

RED Plump, fruit-driven reds of good class and finesse, with a long finish.

🍷 *Merlot 50%, Cabernet Sauvignon 40%, Cabernet Franc 8%, Petit Verdot 2%*

🍷 *4–12 years*

CHÂTEAU SÉNÉJAC
Médoc AOC
☆ Ⓥ

The wine is matured in wood for 12 to 15, with 30 per cent new oak.

RED A firm, full-flavoured wine of excellent longevity, Sénéjac is not, however, for the faint-hearted, especially in its rather stern youthful years.

🍷 *Cabernet Sauvignon 51%, Merlot 37%, Cabernet Franc 8%, Petit Verdot 4%*

🍷 *5–15 years*

Second wine: *Compte de Senejac*

CHÂTEAU SIGOGNAC
Médoc AOC
Ⓥ

This property is run by Louis and Marion Allard. The wine is matured in wood for 12 months, with 20 per cent new oak.

RED Consistently good value, a lunchtime claret of some elegance.

🍷 *Merlot 35%, Cabernet Sauvignon 30%, Cabernet Franc 30%, Petit Verdot 5%*

🍷 *3–8 years*

Second wine: *Benjamin de Sigognac*
Other wines: *Château la Croix de Chevalier, l'Enclos de Sigognac (100% Merlot), La Jonquille (100% Semillon), Blanc de l'Estuaire, Rosé de l'Estuaire*

CHÂTEAU SOCIANDO-MALLET
Haut-Médoc AOC
★★ Ⓥ

This property has been making a name for itself since 1970, when Jean Gautreau raised standards to near *cru classé* quality. The quality of Château Sociando-Mallet continued to increase throughout the 1990s, when between 80 and 100 per cent new oak became the norm. Its owners did not bother to submit wines for the 2003 reclassification, because it already achieves a higher price than a number of *cru classé* wines, but it is undoubtedly at least *cru bourgeois exceptionnel* in quality, if not name.

RED These are powerfully built wines that are rich in colour and extract. Often totally dominated by vanilla oak in their youth, they are backed up with plenty of concentrated cassis fruit.

🍷 *Merlot 54%, Cabernet Sauvignon 42%, Cabernet Franc 4%*

🍷 *10–25 years*

Second wine: *La Demoiselle de Sociando-Mallet*
Other wines: *Jean Gautreau (100% Cabernet Sauvignon)*

CHÂTEAU SOUDARS
Médoc AOC
☆ Ⓥ

Château Soudars was formed by the combination of several parcels of rock-strewn land that had become overgrown with brambles until Eric Miailhe took over the property in 1973 and spent several years clearing it. The wine is fermented in stainless steel and matured in wood for 12 months, with up to 40 per cent new oak.

RED Excellent, well-coloured wines of good structure and accessible fruit.

🍷 *Merlot 50%, Cabernet Sauvignon 49%, Cabernet Franc 1%*

🍷 *5–10 years*

CHÂTEAU LE TEMPLE
Médoc AOC
Cru Bourgeois
☆ Ⓥ

Château le Temple is starting to live up to its reputation from the early 1900s, when it was known as the "Lafite of the Bas-Médoc". This wine is matured in wood for 12 months, with 22 per cent new oak.

RED Increasingly lush and fruit-dominant, without losing its classic, tannin structure.

🍷 *Cabernet Sauvignon 60%, Merlot 35%, Petit Verdot 5%*

🍷 *7–15 years*

Second wine: *Château Balirac*

CHÂTEAU LA TOUR DE BY
Médoc AOC
★ Ⓥ

The tower of Tour de By was once a lighthouse. The wine is of very good quality; it is matured in wood for 12 months, with up to 30 per cent new oak.

RED These deeply coloured, full-bodied, richly flavoured wines have good spicy fruit, backed up by a firm tannic structure.

🍷 *Cabernet Sauvignon 60%, Merlot 35%, Petit Verdot 5%*

🍷 *6–12 years*

Second wine: *Château La Roque-de-By*
Other wines: *Hértiage Marc Pagès, Les Tourterelles, Le Rosé de la Tour de By.*

CHÂTEAU LA TOUR-CARNET
Haut-Médoc AOC
4ème Cru Classé
★

This charming, 13th-century miniature moated castle has a well-kept vineyard. Its wines used to be lacklustre, but have been transformed by owner Bernard Magrez of Pape-Clément, who purchased the property in 1999. The wine is matured in wood for 12 to 18 months, with 30 per cent new oak.

RED Much riper, more opulent fruit, with some lush new oak in a supporting role, this wine has the richness of flavour it used to lack, with improvements noticeable from one year to the next.

🍷 *Merlot 56%, Cabernet Sauvignon 40%, Petit Verdot 3%, Cabernet Franc 1%*

🍷 *5–15 years*

Second wine: *Les Pensées de la Tour Carnet*
Other wines: *Le Médoc de La Tour Carnet, Le Blanc de la tour Carnet*

CHÂTEAU TOUR HAUT-CAUSSAN
Médoc AOC
Cru Bourgeois
★ ☆

Well known for its landmark windmill, this property has been in the hands of the Courrian family, since 1877. It is now run by Philippe Courrian, who also makes wine in Corbières. The grapes are all hand-harvested, and the wine is matured in wood for 12 months, with 25 per cent new oak.

RED Rich, well-coloured wines with a great concentration of fruit and nicely integrated, creamy oak.

🍷 *Cabernet Sauvignon 50%, Merlot 50%*

🍷 *4–10 years*

Other wines: *Château Landotte*

CHÂTEAU TOUR SAINT-BONNET
Médoc AOC
Cru Bourgeois
★ Ⓥ

Situated on fine, gravelly ridges, this property was known as Château la Tour Saint-Bonnet-Cazenave in the 19th century. This wine is matured in wood for 18 months, with 50 per cent new oak.

RED Firm, full-flavoured, well-coloured wines of consistent quality.

🍷 *Cabernet Sauvignon 45%, Merlot 45%, Malbec 5%, Petit Verdot 5%*

🍷 *7–15 years*

Second wine: *Château La Fuie Saint-Bonnet*

CHÂTEAU VERDIGNAN
Haut-Médoc AOC
★ Ⓥ

Since 1972 this has been the property of the Miailhe family, who continue to improve the quality. The wine is fermented in stainless steel and matured in wood for 12 to 14 months, with 25 per cent new oak.

RED Medium-bodied, fruity wines, made in a soft and silky style.

🍷 *Cabernet Sauvignon 72%, Merlot 25%, Cabernet Franc 3%*

🍷 *5–10 years*

Second wine: *No 2 de Verdignan*
Other wines: *Château Plantey-de-la-Croix*

CHÂTEAU VILLEGEORGE
Haut-Médoc AOC
★ Ⓥ

This château was classified *cru bourgeois* in 1932. Lucien Lurton purchased it in 1973, but he then resigned from the Syndicat, so the château was not included in its 1978 list – though it was superior to a few that were. The present incumbent, Marie-Louise (Lucien's daughter) has continued improving the quality, achieving *supérieur* status in the 2003 reclassification. The wine is matured in wood for 10 to 16 months, with up to 30 per cent new oak.

RED Full-bodied wines, with a lovely colour, increasingly lush and opulent fruit, understated, creamy oak, and an increasing spicy finesse.

🍷 *Cabernet Sauvignon 63%, Merlot 37%*

🍷 *6–12 years*

Second wine: *L'Etoile de Villegeorge*
Other wines: *Château la Tour de Bessan (Margaux)*

CHÂTEAU LA TOUR-HAUT-CAUSSAN
This Médoc château's iconic windmill graces its wine labels.

The Médoc:
Saint-Estèphe

At the heart of the peninsula, close to the Gironde, St-Estèphe is an historic terroir with proof of vine production dating back to the Romans. Some of the great wine estates were created as early as the 14th century.

As with the Haut-Médoc, St-Estèphe châteaux cover the full range of producers: 74 properties include five *crus classés*, 19 *crus bourgeois*, 1 *cru artisan*, 32 other independents, and 17 vineyards that work with the cooperative.

The wines of St-Estèphe have always been well structured, with natural longevity, but they now have more lushness of fruit, which allows the wines to be accessible when relatively young. It was once essential to buy only the greatest vintages and wait 20 years or more before drinking them. But the increasing use of the Merlot grape, as well as Cabernet Sauvignon and Cabernet Franc, and a tendency to favour vinification techniques that extract colour and fruit in preference to the harsher tannins provide richer, fruitier, and eminently drinkable wines in most vintages.

With only five *crus classés* covering a mere 8.5 per cent of the commune, St-Estèphe is a rich source of undervalued clarets, where the prices paid by wine drinkers are unlikely to be sent soaring by wine investors. Enthusiasts rather than speculators will benefit from the fact that no fewer than four of the Médoc's nine *crus bourgeois exceptionnels* from the 2003 classification are to be found in this commune.

CHÂTEAU COS D'ESTOURNEL

St-Estèphe might lack *crus classés,* but it is not lacking in class. If it had only one *cru classé* – the stunning, stylish Château Cos d'Estournel – St-Estèphe would still be famous. Since 2000 it has been owned by Michel Reybier, but its reputation soared when Bruno Prats took control in 1971. This success can be put down to maximizing the true potential of its exceptional *terroir,* a superb, south-facing ridge of gravel with perfect drainage. Those vineyards on heavier soil with less gravel and more clay tend to produce more rustic wines.

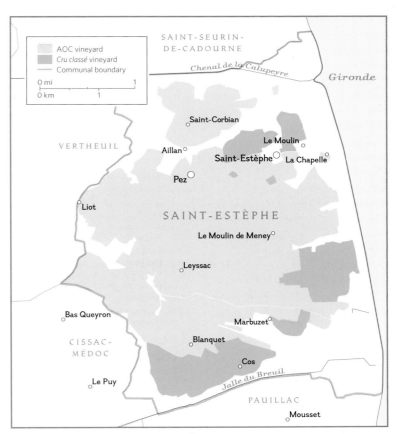

VINE-GROWING AREAS OF ST-ESTÈPHE (see also p141)
Of the Haut-Médoc's four best-known communes, St-Estèphe is the most northerly, although the actual AOC area covers only part of the commune.

MODERN ST-ESTÈPHE

Most wines from St-Estèphe have always been well structured, with natural longevity, but they now have more lushness of fruit, which allows the wines to be accessible when relatively young. It was once essential to buy only the greatest vintages and wait 20 years or more before drinking them. The increasing use of the Merlot grape, as well as Cabernet Sauvignon and Cabernet Franc, and a tendency to favour vinification techniques that extract colour and fruit in preference to the harsher tannins, provide richer, fruitier, and eminently drinkable wines in most vintages.

CHÂTEAU COS D'ESTOURNEL
The wine produced at this château was classified as one of 15 *Deuxièmes Crus* (Second Growths) in the original Bordeaux Wine Official Classification of 1855.

FACTORS AFFECTING TASTE AND QUALITY

LOCATION
St-Estèphe is the most northerly of the four classic communes of the Médoc. It is situated 18 kilometres (11 miles) south of Lesparre, bordering the Gironde.

CLIMATE
As for the Médoc (see p152).

ASPECT
In St-Estèphe, the undulating relief rolling down to the estuary has excellent drainage. A topsoil of quartz and large pebbles acts as a temperature and humidity regulator.

SOIL
The topsoil of the St-Estèphe wine region is gravelly and more fertile than in communes farther south, with the subsoil exposed in parts, consisting of clay beds, stony-clay, and limestone over iron-pan.

VITICULTURE & VINIFICATION
Only the red wines have the right to the appellation in this commune. With increasing emphasis placed on the Merlot grape, which can now account for up to 50 per cent of the vines cultivated in some châteaux, reduced use of *vin de presse,* and improved vinification techniques, these wines are becoming far more accessible in less sunny years. During the vinification, all grapes must be destalked, and duration of skin contact averages three weeks. Maturation in cask currently varies between 15 and 24 months.

GRAPE VARIETIES
Primary varieties: Cabernet Franc, Cabernet Sauvignon, Merlot
Secondary varieties: Carmenère, Malbec, Petit Verdot

THE WINE PRODUCERS OF
SAINT-ESTÈPHE

CHÂTEAU ANDRON-BLANQUET
★ V

Under the same ownership as Château Cos Labory, the vineyards of this property are situated above the gravel crest of *cru classé* châteaux that overlook Château Lafite-Rothschild in Pauillac. This wine is matured in wood for 12 months, with 30 per cent new oak.

RED An exceptionally well-made wine that consistently rises above its *petit château* status. Fermented and matured in cask, it has good fruit and a distinctive style.

🍇 *Cabernet Sauvignon 60%, Merlot 25%, Cabernet Franc 15%*

🍷 *4–10 years*

Second wine: *Château St-Roch*

CHÂTEAU BEAU-SITE
★ V

This property, owned by Bordeaux wine Merchant Borie Manoux, should not be confused with Château Beau-Site Haut-Vignoble. The wine is matured in wood for 16 to 18 months, with 50 per cent new oak.

RED A stylish, medium-bodied, sometimes full-bodied, wine that often has an elegant finish reminiscent of violets.

🍇 *Cabernet Sauvignon 70%, Merlot 30%*

🍷 *3–10 years*

CHÂTEAU LE BOSCQ
★ ☆ V

This property has always produced good wine, but quality increased dramatically in the 1980s. It was taken over by Dourthe-Kressman in 1995.

RED Superbly aromatic, almost exotic, full-bodied wine that is elegant and rich with the flavour of summer fruits and is nicely backed up with new oak.

🍇 *Cabernet Sauvignon 35%, Merlot 6%, Petit Verdot 7%, Cabernet Franc 2%*

🍷 *5–12 years*

Second wine: *Héritage de Le Boscq*

CHÂTEAU CALON-SÉGUR
3ème Cru Classé
★ ☆ V

From the Gallo-Roman origins of this château grew the community of St-Estèphe. The first wine estate in the commune, it used to boast "Premier Cru de St-Estèphe" on its label until other producers objected. The heart dominates the label since the 18th century when the then owner Marquis de Segur declared that he made wine at Lafite and at Latour but his heart remained at Calon. Since 2012 this property is owned by French insurance company Suravenir, Jean-Pierre Moueix, which has a minority sharing-holding consult. The wine is matured in wood for 18 to 20 months, with 100 per cent new oak.

RED Full, fruity, well-structured wine that has a creamy, rich flavour. It is of consistently good quality and improves well in bottle.

🍇 *Cabernet Sauvignon 56%, Merlot 35%, Cabernet Franc 7%, Petit Verdot 2%*

🍷 *3–20 years*

Second wine: *Marquis de Calon Ségur*
Other wines: *Le Saint Estèphe de Calon Segur*

CHÂTEAU CAPBERN (GASQUETON)
★ V

The property is under the same ownership as Château Calon-Ségur. Since the 2013 vintage the chateau has dropped Gasqueton from its name. The vineyards are found north and south of the village of St-Estèphe. This wine is matured in wood for 18 to 20 months, with 60 per cent new oak.

RED Medium-weight, ripe, and fruity wine of consistent quality.

🍇 *Cabernet Sauvignon 52 %, Merlot 46%, Cabernet Franc 2%*

🍷 *4–12 years*

CHÂTEAU CHAMBERT-MARBUZET
★ ☆ V

Technically faultless, hand-harvested wine produced in limited quantities from the sister château of Haut-Marbuzet. Many would rate it easily equivalent to a *cru classé*. This wine is matured in wood for 18 months, with 50 per cent new oak.

RED Aromatically attractive, medium-bodied, sometimes full-bodied, wine. It is rich, ripe, and fruity, with plenty of caramel-oak and sufficient tannin to age well.

🍇 *Cabernet Sauvignon 70%, Merlot 30%*

🍷 *3–10 years*

CHÂTEAU COS D'ESTOURNEL
2ème Cru Classé
★★★ V

This was one of the very first super-seconds to emerge, and this was the achievement of one man, Bruno Prats, although he would claim it to be teamwork. In 1998 Prats was forced by French tax laws to sell out to Groupe Taillan, who in 2001 sold it on to Michel Reybier, a Geneva-based food manufacturer and owner of the luxury hotel group La Reserve. Cos d'Estournel has no château as such, merely a bizarre façade to the winery with huge, elaborately carved oak doors that once adorned the palace of the Sultan of Zanzibar. The wine is fermented in stainless steel in the ultra-modern, gravity-fed cellar renovated in 2000, but all of it is matured in cask for 18 to 24 months, with 100 per cent new oak for big years and up to 70 per cent for lighter vintages.

RED A rich, flavoursome, and attractive wine of full body, great class, and distinction; without doubt the finest wine in St-Estèphe. It is uniquely generous for the appellation and capable of amazing longevity, even in the poorest years. This is a complex wine with silky fruit and great finesse.

🍇 *Cabernet Sauvignon 74%, Merlot 23%, Cabernet Franc 2%, Petit Verdot 1%*

🍷 *8–20 years*

Second wine: *Les Pagodes de Cos*
Other wines: *Cos d'Estournel Blanc, (Appellation Bordeaux Blanc) Goulée by Cos d'Estournel (Appellation Médoc)*

CHÂTEAU COS LABORY
5ème Cru Classé
★

Until the late 19th century, this property formed part of Château Cos d'Estournel. During the 1920s, it was purchased by distant cousins of Madame Audoy, the current owner. The wine is matured in wood for 15 to 18 months, with 50 per cent new oak.

RED These wines used to be merely light and elegant with a certain degree of finesse, even when at their best. However, recent vintages have displayed a very welcome change to a distinctly fuller, fruitier, and fatter style.

🍇 *Cabernet Sauvignon 60%, Merlot 35%, Cabernet Franc 5%*

🍷 *5–15 years*

Second wine: *Charme de Cos Labory*

CHÂTEAU LE CROCK
★ ☆ V

Under the same ownership as Château Léoville Poyferré of St-Julien, this property was promoted to *supérieur* status in the 2003 reclassification. This hand-harvested wine is matured in wood for 16 to18 months, with 25 per cent new oak.

RED These dark-coloured, substantial wines have surged in quality since 1995 under the guidance of Michel Rolland.

🍇 *Cabernet Sauvignon 53%, Merlot 33%, Cabernet Franc 9%, Petit Verdot 5%*

🍷 *6–15 years*

Second wine: *Château La Croix de Saint-Estèphe*

CHÂTEAU DOMEYNE
☆ V

This property was not classified *cru bourgeois* in 1932, nor was it listed by the Syndicat in 1978, but it certainly should have been. This wine is matured in wood for 12 months, with 25 per cent new oak.

RED These are typically deep-coloured, rich-flavoured wines that have an excellent marriage of fruit and oak. They are smooth and well-rounded wines that can be drunk while fairly young.

🍇 *Cabernet Sauvignon 60%, Merlot 40%*

🍷 *3–8 years*

CHÂTEAU HAUT-MARBUZET
★★ V

One of several properties belonging to Henri Duboscq. These wines receive 15 to 18 months in 100 per cent new oak – extremely rare even for *cru classé* châteaux.

RED These full-bodied, deep-coloured wines are packed with juicy fruit, backed up by supple tannin. They are marked by a generous buttered-toast and creamy-vanilla character.

🍇 *Merlot 50%, Cabernet Sauvignon 40%, Cabernet Franc 5%, Petit Verdot 5%*

🍷 *4–12 years*

Second wine: *Chateau MacCarthy*
Other wines: *Château Chambert Marbuzet, Château Tour de Marbuzet*

CHÂTEAU LA HAYE
Cru Bourgeois
★ V

New equipment thanks to the new ownership since 2012, 12 to 18 months in 46 per cent new oak casks every year, and a fair proportion of old vines combine to produce some exciting vintages at this property.

ST-ESTÈPHE *CRU CLASSÉ* STATISTICS

CRUS CLASSÉS IN AOC ST-ESTÈPHE
Five châteaux (by number: 8% of *crus classés* in the Médoc) with 309ha (763ac) of vineyards (by area: 8.5% of *crus classés* in the Médoc and 26% of this AOC)

1ERS CRUS CLASSÉS
None

2ÈMES CRUS CLASSÉS
Two châteaux (by number: 14% of *2ème crus classés* in the Médoc) with 195ha (482ac) of vineyards (by area: 19% of *2ème crus classés* in the Médoc)

3ÈMES CRUS CLASSÉS
One château (by number: 7% of *3ème crus classés* in the Médoc) with 55ha (136ac) of vineyards (by area: 9% of *3ème crus classés* in the Médoc)

4ÈMES CRUS CLASSÉS
One château (by number: 10% of *4ème crus classés* in the Médoc) with 41ha (101ac) of vineyards (by area: 7% of *4ème crus classés* in the Médoc)

5ÈMES CRUS CLASSÉS
One château (by number: 5.5% of *5ème crus classés* in the Médoc) with 18ha (44ac) of vineyards (by area: 2% of *5ème crus classés* in the Médoc)

RED Always limpid, this medium-bodied, sometimes full-bodied, wine is rich in colour and flavour, well balanced, and lengthy, with vanilla-oak evident on the finish.

🍷 *Cabernet Sauvignon 52%, Merlot 40%, Petit Verdot 8%*

🍷 *5–8 years*

Second wine: *Le Cedre de Château La Haye*
Other wines: *Château Bel Air, Majesty de Château la Haye*

CHÂTEAU LAFON-ROCHET
4ème Cru Classé
★

When Guy Tesseron purchased this vineyard, which is situated on the borders of Pauillac, in 1959, he embarked on a project to increase the proportion of Cabernet Sauvignon grapes used in the wine. This proved to be a mistake for Lafon-Rochet's *terroir,* however, and has been rectified in recent years. Now managed by his grandson Basile Tesseron, the wine produced here is matured in wood for 16 to 18 months, with up to 50 per cent new oak.

RED More fruit and finesse from the mid-1990s onwards.

🍷 *Cabernet Sauvignon 57%, Merlot 37%, Petit Verdot 4%, Cabernet Franc 2%*

🍷 *5–12 years*

Second wine: *Pellerins de Lafon Rochet*

CHÂTEAU LAVILLOTTE
★☆ 🅥

This star-performing *petit château* gives good value. This wine is matured in wood for 16 months, with up to 40 per cent new oak.

RED These are dark-coloured wines with a deep and distinctive bouquet. Smoky, full-bodied, intense, and complex.

🍷 *Cabernet Sauvignon 72%, Merlot 25%, Petit Verdot 3%*

🍷 *5–12 years*

Other wines: *Domaine de Ronceray, Château Meynieu (Haut-Médoc AOC)*

CHÂTEAU LILIAN LADOUYS
Cru Bourgeois
★

Just a few hundred metres from Lafite, this property has its first mention in the 16th century. Purchased by the Lorenzetti family in 2008 and extended to 80.5 hectares (200 acres) by the acquisition of Châteaux Clauzet and Tour de Pez, the wines now have a larger proportion of Cabernet in the blend. Matured in 40 per cent new oak for 14 months.

RED Modern style wine with some elegance from Cabernet Sauvignon

🍷 *Cabernet Sauvignon 59%, Merlot 37%, Petit Verdot 4%*

🍷 *5–12 years*

Second wine: *La Devise de Lilian Ladouys*

CHÂTEAU DE MARBUZET
Cru Bourgeois
☆

Under the same ownership of the Prats family, formerly as of Cos d'Estournel, Marbuzet used to include the wines rejected from the grand vin of that "super-second", serving as a second wine. Since 1994, when Pagodes de Cos was created, all the wine from this château has been produced exclusively from its own vineyard. The wine is matured in wood for 12 to 14 months in used second-fill Cos d'Estournel barrels.

RED Elegant, medium-bodied, and sometimes full-bodied, wines, well balanced with good fruit and a supple finish.

🍷 *Cabernet Sauvignon 46%, Merlot 42%, Petit Verdot 12%*

🍷 *4–10 years*

LE MARQUIS DE SAINT-ESTÈPHE
☆ 🅥

This wine is produced by the conscientious Cave Coopérative Marquis de Saint-Estèphe, who mature it in wood for 12 months, with 30 per cent new oak.

RED A consistently well-made, good-value, usually medium-bodied (although sometimes full-bodied) wine, with increasingly more noticeable oak.

🍷 *Cabernet Sauvignon 50%, Merlot 50%*

🍷 *3–6 years*

Other wines: *Leo de Prades*

CHÂTEAU MEYNEY
Cru Bourgeois
★ 🅥

One of the oldest château in St-Estèphe, situated next to Château Montrose overlooking the Gironde Estuary. It has been owned by CA Grands Crus since 2004. Consistent in managing to produce fine wines in virtually every vintage.

RED These wines used to be big, beefy, chunky, and chewy, and required at least 10 years in bottle. They have changed, and for the better, acquiring a silky-textured finesse and ageing gracefully without so many years in bottle.

🍷 *Cabernet Sauvignon 60%, Merlot 30%, Petit Verdot 10%*

🍷 *5–25 years*

Second wine: *Prieuré de Meyney*

CHÂTEAU MONTROSE
2ème Cru Classé
★★ 🅥

This "youngest" of the *cru classé* vineyards, this property grew out of an inconsequential plot of vines retained by M Dumoulin, the former owner of Calon-Ségur, when he sold that château in 1824. By 1855, Montrose had grown to 96 hectares (237 acres), as Dumoulin bought and exchanged parcels of land from and with his neighbours. Despite the new-found importance of Montrose, Calon-Ségur was still considered by locals to be its superior, thus there was much surprise when Montrose was classified as a *deuxième cru classé,* above *troisième cru* Calon-Ségur. Martin and Olivier Bouygues purchased the property in 2006 and have completely renovated the vineyard and the cellars. The monumental cellars completed in 2013, alongside their objective of being carbon neutral, they have obtained the Haute Valeur Environemental certification. The wines are matured in wood for 18 months, with 60 per cent new oak.

RED The inhibiting factor used to be its "stemmy" tannins. A vintage of exceptional richness and fatness was required to overcome the aggressive character produced by these tannins. The excellent 1994 gave hope that this château had started to harvest the grapes when they are tannin-ripe and was applying more specific maceration techniques. Performance since the mid-1990s has clearly demonstrated that this wine is a true *deuxième cru classé.* The property is now producing wines of exceptional elegance, whilst losing none of the power of the appellation.

🍷 *Cabernet Sauvignon 60%, Merlot 32%, Cabernet Franc 6%, Petit Verdot 2%.*

🍷 *8–25 years*

Second wine: ✓ *La Dame de Montrose*
Other wines: *Tertio de Montrose (before 2017 called Le Saint-Estèphe de Montrose)*

CHÂTEAU LES ORMES DE PEZ
★★ 🅥

Owner Jean-Michel Cazes of Pauillac's Château Lynch-Bages installed new stainless steel vats in 1981 and raised the quality of these wines from good to sensational. Matured in wood for 12 to 15 months, with 50 per cent new oak, this affordable wine is equivalent to a good *cru classé.*

RED Dark and fruity, yet capable of ageing with a herbal complexity.

🍷 *Cabernet Sauvignon 54%, Merlot 37%, Cabernet Franc 7%, Petit Verdot 2%*

🍷 *3–15 years*

CHÂTEAU DE PEZ
★★ 🅥

This property was purchased by Louis Roederer Champagne in 1995, also owners of Château Pichon Comtesse in Pauillac. The wines are fermented in wooden vats, then matured in small casks for 12 months, with 30 per cent new oak. These wines are easily the equivalent of *cru classé* quality.

RED Consistently one of the best *bourgeois* growths in the entire Médoc. A medium-bodied wine, it has a rich fruit flavour and good tannic structure and can mature into a sublime, cedary wine.

🍷 *Cabernet Sauvignon 52%, Merlot 43%, Petit Verdot 3%, Cabernet Franc 2%*

🍷 *6–20 years*

CHÂTEAU PHÉLAN-SÉGUR
★☆ 🅥

Purchased in 1984 by the Gardinier family. After 30 years of continual investment and improvement, they sold the property to Philippe Van de Vyvere in 2018. He has continued with the same technical team but is investing in a new cellar, increasing the number of vats to allow for an even more precise vinification. Michel Rolland consults at the property. Now one of the best-value wines in the Médoc, this wine is matured in wood for about 12 months, with up to 50 per cent new oak.

RED Increasingly stylish wines of good colour and a certain plumpness of fruit, without loss of true St-Estèphe structure. Unlike most other success stories in the commune, this has not been achieved by increasing Merlot; the reverse, in fact.

🍷 *Cabernet Sauvignon 58%, Merlot 39%, Cabernet Franc 1.5%, Petit Verdot 1.5%*

🍷 *5–10 years*

Second wine: *Franck Phélan*
Other wines: *La Croix Bonis*

CHÂTEAU POMYS
☆ 🅥

Classified *cru bourgeois* in 1932, but was not included in the 1978 list, Château Cos d'Estournel bought the château in 2017 – it was once the home of Louis Gaspard Estournel. It has re-opened as a hotel called La Maison d'Estournel in 2019.

CHÂTEAU TOUR DE PEZ
★☆ 🅥

Huge investments made since this château changed hands in 1989 paid off in 2003 when it was promoted to *crus bourgeois supérieur.* This wine is matured in wood for 12 to 14 months, with 30 per cent new oak.

RED Consistently elegant, medium-bodied wine with good, plump, fleshy fruit.

🍷 *Merlot 60%, Cabernet Sauvignon 35%, Petit Verdot 5%*

🍷 *3–7 years*

Second wine: *Les Hauts de Pez*

CHÂTEAU TRONQUOY-LALANDE
☆ 🅥

Bought by the Bouygues brothers, owner of neighbouring Chateau Montrose in 2006, who have invested in new a winemaking facility and aging cellars. The wines are aged in 40 per cent oak for a year.

RED This wine can be dark and tannic, but as from the 1996 vintage it has displayed more fruit and finesse thanks to a dominance of Merlot in the blend.

🍷 *Merlot 49%, Cabernet Sauvignon 45%, Petit Verdot 6%*

🍷 *3–7 years*

Second wine: *Tronquoy de Sainte-Anne*
Other wines: *Chateau Tronquoy Lalande Bordeaux Blanc*

The Médoc:
Pauillac

Pauillac is the archetypal Bordeaux wine, what many people think of when they think of Bordeaux, or at least the Médoc.

It is most famous for the three *premiers crus* of Latour, Lafite, and Mouton. If the wine is allowed to evolve slowly, it achieves an astonishing degree of finesse for its weight.

Pauillac is, however, an appellation of quite surprising contrasts. There are 54 wine growers, including 18 *crus classés,* three-quarters of the Médoc's *premiers crus,* and two-thirds of the region's *cinquièmes crus.* There are only three *cru bourgeois.* There are 21 growers that belong to the one cooperative and another 22 independent winemakers that are worth searching out.

Thanks to a series of undulating gravel outcrops with excellent natural drainage, Cabernet Sauvignon is at its most majestic in Pauillac, and though the much-vaunted blackcurrant character of this grape is much in evidence in Pauillac, it is always beautifully balanced by a tannic structure. Powerful when young, the aromas of berry fruits (blackcurrant, raspberry) or flowers (violets, roses, irises) infuse the wines Of Pauillac. They improve with age to become rich and complex, with characteristic aromas of cedar and often graphite freshness.

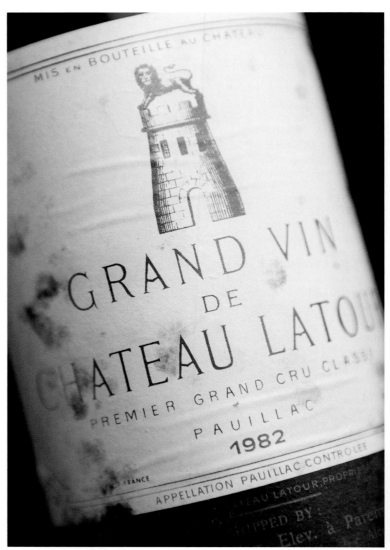

CHÂTEAU LATOUR, VINTAGE 1982
Due to radical changes in vinification techniques introduced in the 1960s, Château Latour became the most consistent of Bordeaux's great *premiers crus classés*. It produces the archetypal Pauillac wine, which balances weight with finesse.

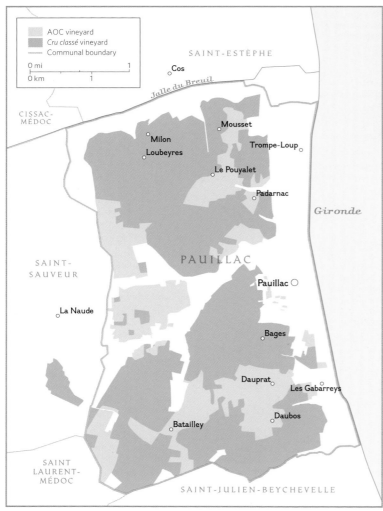

VINE-GROWING AREAS OF PAUILLAC (see also p141)
Blessed with three *premiers crus,* Lafite-Rothschild and Mouton Rothschild in the north, and Latour to the south, Pauillac is sandwiched between St-Estèphe and St-Julien.

PAUILLAC *CRU CLASSÉ* STATISTICS

CRUS CLASSÉS IN AOC PAUILLAC
18 châteaux (by number: 30% of *crus classés* in the Médoc) with 1,126ha (2,782ac) of vineyards (by area: 30% of *crus classés* in the Médoc and 93% of this AOC)

1ERS CRUS CLASSÉS
Three châteaux (by number: 75% of *1ers crus classés* in the Médoc) with 286ha (707ac) of vineyards (by area: 78% of *1ers crus classés* in the Médoc)

2ÈMES CRUS CLASSÉS
Two châteaux (by number: 14% of *2èmes crus classés* in the Médoc) with 109ha (269ac) of vineyards (by area: 11% of *2èmes crus classés* in the Médoc)

3ÈMES CRUS CLASSÉS
None

4ÈMES CRUS CLASSÉS
One château (by number: 10% of *4èmes crus classés* in the Médoc) with 108ha (267ac) of vineyards (by area: 18% of *4èmes crus classés* in the Médoc)

5ÈMES CRUS CLASSÉS
12 châteaux (by number: 67% of *5èmes crus classés* in the Médoc) with 634ha (1,567ac) of vineyards (by area: 66% of *5èmes crus classés* in the Médoc)

Note: Only Margaux has more *cru classé* châteaux than Pauillac, and no communal AOC has a greater concentration of *cru classé* vines.

FACTORS AFFECTING TASTE AND QUALITY

LOCATION
Pauillac is sandwiched between St-Estèphe to the north and St-Julien to the south.

CLIMATE
As for the Médoc (see p152).

ASPECT
Pauillac consists of two large, low-lying plateaux, one to the northwest of the town of Pauillac, the other to the southwest. Exposure is excellent, and both drain down gentle slopes, eastwards to the Gironde, westwards to the forest, or north and south to canals and streams.

SOIL
Pauillac's two plateaux are massive gravel beds, reaching a greater depth than any found elsewhere in the Médoc. The water drains away before the iron-pan subsoil is reached. St-Sauveur consists of shallow sand over a stony subsoil to the west, and gravel over iron-pan (or more gravel) in the centre and south.

VITICULTURE & VINIFICATION
Only red wines have the right to the Pauillac appellation. Some *vin de presse* is traditionally used by most châteaux. Skin contact duration averages between three and four weeks, and maturation in cask currently varies between 18 and 24 months.

GRAPE VARIETIES
Primary varieties: Cabernet Franc, Cabernet Sauvignon, Merlot
Secondary varieties: Carmenère, Malbec, Petit Verdot

***ROUTE DES CHÂTEAUX* IN PAUILLAC**
The *Routes des Châteaux*, or "castle road", is a wine route that winds its way through the Médoc region, including through the vineyards of Pauillac.

THE WINE PRODUCERS OF
PAUILLAC

CHÂTEAU D'ARMAILHAC
5ème Cru Classé
★☆

Baron Philippe de Rothschild purchased Château Mouton d'Armailhac in 1933. In 1956 he renamed it Château Mouton-Baron Philippe,. In 1975 he changed it again, calling it Mouton-Baronne-Philippe in honour of his late wife. In 1991 it reverted to d'Armailhac, but without the Mouton tag; the baron believed the wine to be in danger of assuming second wine status due to the overwhelming prestige of Mouton Rothschild. This property borders that of Mouton, and one of Baron Philippe's reasons for acquiring it was to provide an easier, more impressive access to the famous *premier cru*. The wines, which are matured in wood for 16 months, with one third new oak, are produced with the same care and consideration. Despite this attention, I have criticized the "austere, light and attenuated style" of Château d'Armailhac in the past, which "proves that even money cannot buy *terroir*", but within limits improvements can always be made, and by tweaking harvesting and vinification techniques, owners Baron Philippe de Rothschild SA produced the greatest wine this property has ever known in 2000. That was an extraordinary vintage, but the style shift and giant stride in quality continue.

RED Since 2000 much riper tannins and a more supple structure has given this wine more velvety fruit than previous vintages had established. That was the start of a new era.

 Cabernet Sauvignon 52%, Merlot 36%, Cabernet Franc 10%, Petit Verdot 2%

🍷 *6–15 years*

CHÂTEAU BATAILLEY
5ème Cru Classé
★☆ V

This château responds well to sunny years and produces underrated and undervalued wine. The 1985 and 1986 were possibly the best bargains in Bordeaux. The wine is matured in wood for 16 to 18 months, with 55 per cent new oak.

RED This wine has sometimes been rustic and too assertive in the past, but now shows its class with fine, succulent fruit supported by a ripe tannic structure and a complex creamy-oak aftertaste.

 Cabernet Sauvignon 70%, Merlot 25%, Cabernet Franc 3%, Petit Verdot 2%

🍷 *10–25 years*

Second wine: *Lions de Batailley*

CHÂTEAU CLERC MILON
5ème Cru Classé
★

This property was purchased by Baron Philippe de Rothschild in 1970. After more than a decade of investment and quite a few disappointing vintages along the way, it came good in 1981, achieved sensational quality in 1982, and now consistently performs well above its classification. This wine, which is matured in wood for 18 months with one-third new oak, is one worth watching.

RED A deep-coloured, medium-bodied, sometimes full-bodied, wine with cassis-cum-spicy-oak aromas and rich berry flavours well balanced by ripe acidity.

🍷 *Cabernet Sauvignon 50%, Merlot 37%, Cabernet Franc 10%, Petit Verdot 2%, Carménère 1%*

🍷 *10–20 years*

Second wine: *Pastourelle de Clerc Milon*

CHÂTEAU COLOMBIER-MONPELOU
☆ V

In the 1874 edition of *Bordeaux and Its Wines*, this property, also belonging to Baron Philippe de Rothschild, was described as a *"quatrième cru."* Of course, it was never officially classified as such, but its recent *cru bourgeois supérieur* achievement is about right. This wine is matured in wood for 16 months, with 40 per cent new oak.

RED Rich, spicy fruit with fine Cabernet character, backed up by good, ripe tannic structure and vanilla-oaky undertones.

🍷 *Cabernet Sauvignon 55%, Merlot 35%, Cabernet Franc 5%, Petit Verdot 5%*

🍷 *5–12 years*

Other wines: *Château Grand Canyon, Château Pey la Rose, Château Coubersant, Château de Puy la Rose*

CHÂTEAU CORDEILLAN-BAGES
★

Jean-Michel Cazes of Château Lynch-Bages was the driving force behind the group of growers from the Médoc who renovated Château de Cordeillan, turning it into an extensive complex comprising a hotel, restaurant, and wine school. With only a very small production from just 2 hectares (5 acres) of vineyards, this hand-harvested wine is produced by the Lynch-Bages winemaking team and is now uniquely available for sale at the hotel and restaurant.

RED As dark and as dense as might be expected from a wine produced with the Lynch-Bages influence. Smoky oak, tobacco plant, and violet aromas weave their way through the chocolaty Cabernet fruit.

🍷 *Cabernet Sauvignon 80%, Merlot 20%*

🍷 *3–8 years*

CHÂTEAU CROIZET-BAGES
5ème Cru Classé
★ V

Under the same ownership as Château Rauzan-Gassies of Margaux and situated on the Bages plateau, Croizet-Bages is a classic example of a "château with no château". Its wine is matured in wood for 12 to 14 months in 55 per cent new oak. Much improved as of the mid-1990s, then stepped up a gear at the turn of the millennium under the management of Jean-Philippe Quié.

RED Not one of the most deeply coloured Pauillacs, this medium-bodied wine has always been easy-drinking but has been more stylish since 1999, gaining both in gravitas and finesse.

🍷 *Cabernet Sauvignon 58%, Merlot 39%, Petit Verdot 3%*

🍷 *6–12 years*

Second wine: *Alias Croizet Bages*

CHÂTEAU DUHART-MILON-ROTHSCHILD
4ème Cru Classé
★☆

Another "château with no château", Duhart-Milon was purchased by the Lafite branch of the Rothschild family in 1962. Its wines prior to this date were almost entirely Petit Verdot, and so only in abnormally hot years did it excel with this late-ripening grape, which is traditionally cultivated for its acidity. Interestingly, in the near-tropical heat of 1947, Duhart-Milon managed to produce a wine that many considered to be the best of the vintage. The Rothschilds expanded these vineyards bordering Lafite and replanted them with the correct combination of varieties to suit the *terroir*. In 1994 Charles Chevalier arrived, and since 1996 he has moved these wines up a gear. Since his retirement in 2018 Eric Kohler has stepped into his shoes, managing all the Bordeaux properties of the group. The wine is matured from 10 to 18 months in wood, with 50 per cent new oak.

RED These wines are elegantly perfumed, deliciously rich in creamy-oaky fruit, and have exceptional balance and finesse.

🍷 *Cabernet Sauvignon 67%, Merlot 33%*

🍷 *8–16 years*

Second wine: *Moulin de Duhart*

CHÂTEAU FONBADET
Cru Bourgeois
★

Surrounded by classified growths, this growth was classified *cru bourgeois* in 1932, but was not included in the Syndicat's 1978 list, although it is superior to a few that were, and achieved due recognition in the 2003 reclassification. Many of the vines are in excess of 80 years old, with an average age of 50 years for the entire 20-hectare (50-acre) vineyard. This hand-harvested wine is matured in wood for 16 months, with 50 per cent new oak. Great-value Pauillac.

RED This typical Pauillac red wine has a deep, almost opaque colour, an intense cassis, cigar-box, and cedarwood bouquet, a concentrated spicy, fruit flavour with creamy-oak undertones, and a long finish.

🍷 *Cabernet Sauvignon 60%, Merlot 20%, Cabernet Franc 15%, Malbec and Petit Verdot 5%*

🍷 *6–15 years*

Other wines: *Château Haut-Pauillac, Château Padarnac, Château Montgrand-Milon, Château Tour du Roc-Milon, Château Pauillac*

CHÂTEAU GRAND-PUY-DUCASSE
5ème Cru Classé
★☆ V

Under the ownership of CA Grands Crus, the wine arm of Credit Agricole, the same owners as La Tour de Mons, Château Meyney, and Château Blaignan, this property produces an undervalued wine that comes from various plots scattered across half the commune. One of

the best-value, improving *cru classé* wines available, this wine is matured in wood for 16 to 18 months, with up to 30 to 40 per cent new oak.

RED Well-balanced, relatively early-drinking, medium-bodied, sometimes full-bodied, wine of classic Pauillac cassis character and more suppleness than is usual for this commune.

🍷 *Cabernet Sauvignon 60%, Merlot 40%*

🍷 *5–10 years*

Second wine: *Prélude à Grand-Puy Ducasse*

CHÂTEAU GRAND-PUY-LACOSTE
5ème Cru Classé
★★ V

Under the skilful guidance of François-Xavier Borie, now seconded by his daughter, Emmeline. The wine is matured in wood for 16 to 18 months, with 75 per cent new oak.

RED Deep-coloured with complex cassis, cigar-box spice, and vanilla bouquet, with lots of fruit, length, and finesse.

🍷 *Cabernet Sauvignon 75%, Merlot 20%, Cabernet Franc 5%*

🍷 *10–20 years*

Second wine: *Lacoste-Borie*

CHÂTEAU HAUT-BAGES-LIBÉRAL
5ème Cru Classé
★

Under the same ownership as Château Chasse-Spleen in Moulis and the excellent unclassified Château la Gurgue in Margaux, this dynamic property is currently producing sensational wines. They are matured for 16 months in wood, with up to 40 per cent new oak.

RED Dark, full-bodied wines with masses of concentrated spicy-cassis fruit, great tannic structure, and ripe vanilla oak. In a word – complete.

🍷 *Cabernet Sauvignon 80%, Merlot 17%, Petit Verdot 3%*

🍷 *8–20 years*

Second wine: *La Chapelle de Bages, La Fleur de Haut-Bages Libéral, Le Pauillac de Haut-Bages Libéral*

Other wines: *Le Haut-Médoc de Haut-Bages Libéral, Le Rosé de Vertheuil*

CHÂTEAU HAUT-BATAILLEY
5ème Cru Classé
★ V

Under joint ownership of the Borie and Des Brest families since the 1930s, this producer saw a marked improvement in quality under the management of François-Xavier Borie of Grand-Puy-Lacoste. The vineyard was purchased by the Cazes family in 2017 and now comes under the same management as neighbouring Château Lynch-Bages. The wine is matured in wood for 12 months, with 60 per cent new oak.

RED Haut-Batailley is well-coloured and medium-bodied and shows more elegance and finesse than Batailley, although it can lack the latter's fullness of fruit.

🍷 *Cabernet Sauvignon 70%, Merlot 25%, Cabernet Franc 5%*

🍷 *7–15 years*

Second wine: *Verso (since 2018 vintage)*

CHÂTEAU LAFITE-ROTHSCHILD
1er Cru Classé
★★★

From 1994 this famous château, the vineyard of which includes a small plot in St-Estèphe, had been run along traditional lines and with fastidious care by Charles Chevalier for Baron Éric de Rothschild of the French branch of the Rothschilds. Since their retirement in 2018, Eric Kohler has stepped into the shoes of Chevalier and Saskia de Rothschild into those of her father, Eric. The St-Estèphe portion of the Lafite vineyard is allowed to bear the Pauillac appellation, having been part of the Lafite-Rothschild estate for several hundred years. A change of style occurred in the mid-1970s, when the decision was taken to give the wines less time in cask, Under Chevalier they went into hyper-drive and are often the very best of the *premiers crus*. Lafite is arguably the best performing of all *premiers crus*. This hand-harvested wine is matured in wood for 18 to 20 months, with 100 per cent new oak.

RED Not the biggest of the *premiers crus*, but Lafite is nevertheless textbook stuff: a rich delicacy of spicy fruit flavours, continuously unfolding, supported by an array of creamy-oak and ripe tannins; a wine with incomparable finesse.

🍷 *Cabernet Sauvignon 70%, Merlot 25%, Cabernet Franc 3%, Petit Verdot 2%*

🍷 *25–50 years*

Second wine: ✔ *Carruades de Lafite*

CHÂTEAU LATOUR
1er Cru Classé
★★★

When temperature-controlled, stainless-steel vats were installed in 1964, then owner the British Pearson Group was accused by the French of turning a Bordeaux *premier cru* into a dairy. These French critics conveniently ignored the fact that Château Haut-Brion had done the same three years earlier. Allied Lyons (then owner of Harveys Bristol Cream) bought the property in 1989, but in 1993 they needed to liquidate various shareholdings to finance takeovers and sold Latour to François Pinault, the French industrialist. In 1998 Pinault appointed Frédéric Engerer, a graduate of one of France's top business schools, as the new young president of Latour. He has revolutionized working practices in both the vineyard and winery. Hélène Genin has been the technical director of the property since 2014. This hand-harvested wine is matured in wood for 18 months, with 100 per cent new oak.

RED Despite its close proximity to the neighbouring commune of St-Julien, Latour is the archetypal Pauillac. Its ink-black colour accurately reflects the immense structure and hugely concentrated flavour of this wine. If Lafite is the ultimate example of finesse, then Latour is the

ideal illustration of how massive a wine can be while still retaining great elegance.

🍷 *Cabernet Sauvignon 76 %, Merlot 22%, Cabernet Franc and Petit Verdot 2%*

🍷 *15–60 years*

Second wine: ✔ *Les Forts de Latour*
Other wines: *Pauillac de Latour*

CHÂTEAU LYNCH-BAGES
5ème Cru Classé
★★ V

This château is sited on the edge of the Bages plateau, a little way out of Pauillac. It is on the southern fringe of the small town of Bages. Jean-Michel Cazes produces wines that some people describe as "poor man's Latour (or Mouton)". Well, that cannot be such a bad thing, but if I were rich, I would drink as many *cinquièmes crus* from this château as *premiers crus* from elsewhere. Since 1980 the successes in off-vintages have been extraordinary, making it more consistent than some *deuxièmes crus*. The 2019 vintage will be the first realised in a brand-new and sorely need new cellar. This hand-harvested wine is matured in wood for 18 months, with 75 per cent new oak.

RED An intensely deep-purple-coloured wine of seductive character that is packed with fruit and has obvious class. It has a degree of complexity on the nose, a rich, plummy flavour, supple tannin structure, and a spicy blackcurrant and vanilla aftertaste.

🍷 *Cabernet Sauvignon 70%, Merlot 25%, Cabernet Franc 4%, Petit Verdot 2%*

🍷 *8–30 years*

Second wine: *Echo de Lynch-Bages*
Other wines: *Blanc de Lynch-Bages (rare "Médoc" white wine)*

CHÂTEAU LYNCH-MOUSSAS
5ème Cru Classé
☆

Owned by the third generation, Chantal and Philippe Castéja, of Borie-Manoux, this property has been renovated and the wines could well improve. The wine is matured in wood for 14 to 24, with 55 to 60 per cent new oak.

RED After a false start in the mid-1980s, when an exceptionally stylish 1985 turned out to be exactly that (an exception), this château continued to produce light, rather insubstantial wines of no specific character or quality. Yet with better selection, riper fruit, less time in cask, and more than twice the amount of new oak Lynch-Moussas used to receive, it has made significantly better wines since the mid-1990s.

🍷 *Cabernet Sauvignon 75%, Merlot 25%*

🍷 *4–8 years*

Second wine: *Les Hauts de Lynch-Moussas*

CHÂTEAU MOUTON ROTHSCHILD
1er Cru Classé
★★★

The famous case of the only wine ever to be officially reclassified since 1855, Baron Philippe de Rothschild's plight ended with Mouton's status being justly raised to *premier cru* in 1973. Through promotion of Mouton's unique character, he was probably responsible for elevating the Cabernet Sauvignon grape to its present high profile. Part of his campaign to keep this château in the headlines was the introduction of a specially commissioned painting for the label of each new vintage. The hand-harvested wine is matured in wood for up to 22 months, with 100 per cent new oak.

RED It is difficult to describe this wine without using the same descriptive terms as those used for Latour, but perhaps the colour of Mouton reminds one more of damsons and the underlying character is more herbal, sometimes even minty. And although it ages just as well as Latour, it becomes accessible slightly earlier.

🍷 *Cabernet Sauvignon 81%, Merlot 15%, Cabernet Franc 3%, Petit Verdot 1%*

🍷 *12–60 years*

Second wine: *Le Petit Mouton de Mouton Rothschild*

Other wines: *Aile d'Argent (another rare "Médoc" white)*

CHÂTEAU PEDESCLAUX
5ème Cru Classé

Little-seen *cru classé* produced from two very well-situated plots of vines, one bordering Lynch-Bages, the other between Mouton Rothschild and Pontet-Canet. In 2013 Jacky Lorenzetti (owner of *cru bourgeois* Château Lilian Ladouys in St-Estèphe) purchased this property, along with a share of Château d'Issan, *cru classé* of Margaux. Emmanuel Cruse manages both *crus classés*. Under the new ownership investment has been made in a spectacular new gravity-fed cellar, changing the profile of the wine to a more powerful expression of Pauillac. The wine is matured in wood for 18 months, with 70 per cent new oak.

RED Full, firm, traditional style of Pauillac that is slow-maturing and long-lasting.

🍷 *Cabernet Sauvignon 65%, Merlot 25%, Cabernet Franc 5%, Petit Verdot 5%*

🍷 *15–40 years*

Second wine: *Fleur de Pedesclaux*

CHÂTEAU PIBRAN
★

This growth was classified *cru bourgeois* in 1932, but was not in the Syndicat's 1978 list, although it is superior to a few that were. The vineyard is owned by AXA Millésimes, owners of Pichon-Baron, and the wines are made by the same team.

RED This wine has bags of blackcurrant fruit yet retains the characteristic Pauillac structure and firm finish.

🍷 *Cabernet Sauvignon 54%, Merlot 45%, Cabernet Franc 1%*

🍷 *6–16 years*

Second wine: *Château Tour Pibran*

CHÂTEAU PICHON LONGUEVILLE BARON or CHÂTEAU LONGUEVILLE AU BARON DE PICHON-LONGUEVILLE
2ème Cru Classé
★★☆

The smaller of the two Pichon vineyards and in the past the less inspiring, although many experts reckoned the *terroir* of Pichon-Baron to be intrinsically superior to that of its neighbour, the star-performing Pichon-Comtesse. Indeed, I once even suggested that Madame de Lencquesaing, then owner of Pichon-Comtesse, should buy this château and cast her seemingly irresistible spell on what would inevitably be the still greater *terroir* of the two properties combined. Whether she had such an ambition, or even the cash to consider it, AXA Millésimes got there first and has been remarkably successful at conjuring the most incredible quality from Pichon-Baron. Essentially by reducing Cabernet Sauvignon in favour of Merlot and almost tripling the amount of new oak used, AXA has ensured that Pichon-Baron now lives up to its potential. The wine is matured in wood for 15 to 18 months, with 80 per cent new oak.

RED Intensely coloured, full-bodied wine with concentrated spicy-cassis fruit backed up by supple tannins, which are rich and heady with smoky-creamy oak and complexity. True super-second quality.

🍷 *Cabernet Sauvignon 65%, Merlot 30%, Cabernet Franc 3%, Petit Verdot 2%*

🍷 *8–25 years*

Second wine: *Les Tourelles de Longueville, Les Griffons de Pichon-Baron*

Other wines: *Château Pibran*

CHÂTEAU PICHON LONGUEVILLE COMTESSE-DE-LALANDE
2ème Cru Classé
★★☆

There is a limit to the quality of any wine and this is determined by the potential quality of its grapes. But at Pichon-Comtesse (as it is known), the formidable French winemaker Madame May-Éliane de Lencquesaing demanded the maximum from her *terroir* – and consistently got it. In 2007 she sold Pichon-Comtesse to the Rouzaud family, owners of Louis Roederer Champagne. Since 2012 Nicolas Glumineau, previously from Château Montrose, has managed the property. Substantial investments have been made in the new cellar in 2013. The wine is matured in wood for 18 months, with 60 per cent new oak.

RED This temptress is the Château Margaux of Pauillac. It is silky textured, beautifully balanced, and seductive. A wine of great finesse, even in humble vintages.

🍷 *Cabernet Sauvignon 65%, Merlot 25%, Cabernet Franc 7%, Petit Verdot 3%*

🍷 *10–30 years*

Second wine: *Réserve de la Comtesse*

Other wines: *Les Gartieux*

CHÂTEAU PONTET-CANET
5ème Cru Classé
★☆ Ⓑ

The reputation of this château suffered for decades in the last century, but many wine enthusiasts thought the situation would be reversed when Guy Tesseron purchased the property in 1975. The 1985 vintage gave a glimmer of hope, but it was not until 1995 that the breakthrough occurred, and the 1998 vintage was nothing less than outstanding. Guy's son Alfred Tesseron now owns and manages the estate with the assistance of technical director Jean-Michel Comme. Under Alfred, the property has shifted to biodynamic farming, producing the first *grand cru classé* to be 100 per cent biodynamic. Certified biodynamic since 2010, Château Pontet-Canet has dramatically changed the yields and the quality of the wines. The biodynamic philosophy spills over into the cellars, reducing the percentage of new oak used for aging to 50 percent, with 35 percent of the ageing taking place in amphora-inspired concrete dolia tanks, partly made from the soils of the property.

RED These wines are fruity and graceful, with a rich, smooth, oaky touch.

🍷 *Cabernet Sauvignon 62%, Merlot 32%, Cabernet Franc 4%, Petit Verdot 2%*

🍷 *6–12 years*

Other wines: *Les Hauts de Pontet-Canet*

CHÂTEAU LAFITE-ROTHSCHILD
An illustration of Château Lafite-Rothschild published in *L'Illustration, Journal Universel* in 1868. The property was one of only four wine-producing châteaux of Bordeaux originally awarded First Growth status in the 1855 Classification. It still holds that classification, and its red wines are consistently ranked amongst the finest – and most expensive – in the world.

The Médoc:
Saint-Julien

The commune's fame is disproportionate to its size – it is smaller than any of the other three classic Médoc appellations: St-Estèphe, Pauillac, and Margaux.

St-Julien has no *premiers crus* – nor even *cinquièmes crus* – although some châteaux sometimes produce wines that are undeniably of *premier cru*. The concentration of its 11 *crus classés* in the middle of the classification is St-Julien's real strength, enabling this commune justly to claim that it is the most consistent of the Médoc appellations: these quintessential clarets have a vivid colour, elegant fruit, superb balance, and great finesse.

It is perhaps surprising that wines from 16 hectares (40 acres) inside St-Julien's borders may be classified as AOC Pauillac, particularly in view of the perceived difference in style between these appellations. This illustrates the "grey area" that exists when communal boundaries overlap the historical borders of great wine estates, and highlights the existence and importance of blending, even in a region reputed for its single-vineyard wines. We should not be too pedantic about communal differences of style: if these borders of the Médoc followed local history, Château Latour, the most famous château of Pauillac, would today be in St-Julien – and that makes me wonder how the wines of this commune might be described under such circumstances.

CHÂTEAU BEYCHEVELLE
An arrow points the way to an autumn wine tasting at this estate.

ST-JULIEN PROFILE

APPELLATION AREA
Covers part of the commune of St-Julien only

SIZE OF COMMUNE
1,554ha (3,840ac)

AOC AREA UNDER VINE
910ha (2,249ac), 59% of commune

SURFACE AREA OF *CRUS CLASSÉS*
789ha (1,950ac), 51% of commune, 87% of AOC

SPECIAL COMMENTS
Some 16ha (40ac) of St-Julien are classified as AOC Pauillac

▢	AOC vineyard
▣	*Cru classé* vineyard
—	Communal boundary

0 mi   ½
0 km   ½

VINE-GROWING AREAS OF ST-JULIEN
(*see also* p141)
St-Julien lies south of Pauillac in the centre of the Médoc. Many distinguished châteaux are sited here.

FACTORS AFFECTING TASTE AND QUALITY

LOCATION
St-Julien lies in the centre of the Haut-Médoc, 4 kilometres (2.5 miles) south of Pauillac.

CLIMATE
As for the Médoc (see p152).

ASPECT
The gravel crest of St-Julien slopes almost imperceptibly eastwards towards the village and drains into the Gironde.

SOIL
Fine, gravel topsoil of good-sized pebbles in vineyards within sight of the Gironde. Farther inland, the particle size decreases and the soil begins to mix with sandy loess. The subsoil consists of iron-pan, marl, and gravel.

VITICULTURE & VINIFICATION
Only red wines have the right to the appellation. All grapes must be destalked. Some *vin de presse* may be used according to the needs of the vintage. Skin contact duration averages two to three weeks and most châteaux allow 18 to 22 months' maturation in cask.

GRAPE VARIETIES
Primary varieties: Cabernet Franc, Cabernet Sauvignon, Merlot
Secondary varieties: Carmenère, Malbec, Petit Verdot

CHÂTEAU GRUAUD-LAROSE
The stately château of this estate sits in the middle of a large, manicured park.

THE WINE PRODUCERS OF
SAINT-JULIEN

CHÂTEAU BEYCHEVELLE
4ème Cru Classé
★★

The immaculate and colourful gardens of this château never fail to take away the breath of passers-by. Beychevelle also boasts one of the most famous legends of Bordeaux: its name is said to be a corruption of *basse-voile*, the command to "lower sail". This arose because the Duc d'Épernon, a former owner who was also an admiral of France, apparently required the ships that passed through the Gironde to lower their sails in respect. His wife would then wave her kerchief in reply. This story, however, is not true. I prefer the story of the sailors who lowered their trousers and revealed their sterns, which shocked the duchess but made her children laugh. The wines are matured in wood for 18 months with 50 per cent new oak.

RED Medium- to full-bodied wines of good colour, ripe fruit, and an elegant oak and tannin structure. They can be quite fat in big years.

🍷 *Cabernet Sauvignon 52%, Merlot 40%, Cabernet Franc 5%, Petit Verdot 3%*

🍷 *12–20 years*

Second wine: ✔ *Amiral de Beychevelle*
Other wines: *Les Brulières de Beychevelle*

CHÂTEAU BRANAIRE-DUCRU
4ème Cru Classé
★

The vineyards of this château are situated farther inland than those of Beychevelle and Ducru-Beaucaillou, and its soil contains more clay and iron-pan than theirs, so the wine here is fuller and can be assertive, although never austere. It is matured in wood for 18 months, with up to 50 per cent new oak, and it is remarkably consistent.

RED This is a quite full-bodied wine, which is richly flavoured and can show a certain chocolate character in big years. It has a distinctive bouquet that sets it apart from other St-Juliens.

🍷 *Cabernet Sauvignon 75%, Merlot 20%, Cabernet Franc 5%*

🍷 *2–25 years*

Second wine: *Duluc*

CHÂTEAU LA BRIDANE

The owners have maintained a vineyard here since the 14th century.

RED Attractive, fruity, medium-bodied wine that is easy to drink.

🍷 *Cabernet Sauvignon 55%, Merlot 45%*

🍷 *3–6 years*

CHÂTEAU DUCRU-BEAUCAILLOU
2ème Cru Classé
★★☆

One of the super-seconds, the quality of this classic St-Julien château, the flagship property of the Borie empire, is legendary. In both good years and bad it remains remarkably consistent, and, although relatively expensive, the price fetched falls short of those demanded by *premiers crus*, making it a relative bargain. The wine is matured in wood for 18 months, with 100 per cent new oak.

RED This wine has a fine, deep colour that can belie its deft elegance of style. There is richness of fruit, complex spiciness of oak, great finesse, and exquisite balance.

🍷 *Cabernet Sauvignon 65%, Merlot 25%, Cabernet Franc 5%, Petit Verdot 5%*

🍷 *15–30 years*

Second wine: *La Croix Ducru Beaucaillou*
Other wines: *Le Petite Caillou, Château Lalande-Borie*

CHÂTEAU DU GLANA
☆

This property is under the same ownership as Château Plantey in Pauillac and Château La Commanderie in St-Estèphe.

RED Du Glana is normally an unpretentious and medium-weight wine, but it excels in really hot years, when it can be deliciously ripe and juicy. The wine is aged for 12 months in 40 per cent new oak.

🍷 *Cabernet Sauvignon 65%, Merlot 35%*

🍷 *3–6 years*

CHÂTEAU GLORIA
★

Gloria excites opposite passions in wine drinkers: some consider it the equal of several *cru classé* wines – even superior in some cases – while others believe it earns an exaggerated price based on the reputation of merely a handful of vintages. Certainly the owners saw nothing to be gained by entering this wine for consideration in the 2003 reclassification, although it should have easily made *cru bourgeois supérieur* status had they done so. Henri Martin created the 50-hectare (124-acre) property by buying up plots from grands *crus classsés* over a number of years. In 2016 Château Gloria completes an all-encompassing renovation of its entire winemaking facilities, vat rooms, cellars, and other areas. The wine is matured in wood for 14 months, with 40 per cent new oak.

RED A deep plum-coloured, full-bodied wine with masses of fruit and a rich, exuberant character.

🍷 *Cabernet Sauvignon 65%, Merlot 25%, Cabernet Franc 5%, Petit Verdot 5%*

🍷 *12–30 years*

Second wine: *Chateau Peymartin*
Other wines: *Chateau Haut-Beychevelle-Gloria*

CHÂTEAU GRUAUD-LAROSE
2ème Cru Classé
★★

This large property produces consistently great wines of a far more solid structure than most other St-Julien wines. Anyone who has tasted the supposedly mediocre 1980 Sarget de Gruaud-Larose (made from the wines rejected from the grand vin) will realize the true potential of Château Gruaud-Larose in any year. If anything, the quality of Gruaud-Larose is still improving. The wines are aged between 18 to 24 months in 80 per cent new oak.

RED Full-bodied, rich, and plummy wine with masses of fruit. Its spicy blackcurrant flavour is supported by a structure of ripe tannins.

🍷 *Cabernet Sauvignon 61%, Merlot 29%, Cabernet Franc 7%, Petit Verdot 3%*

🍷 *10–40 years*

Second wine: ✔ *Sarget de Gruaud-Larose*

CHÂTEAU LAGRANGE
3ème Cru Classé
★★☆

When the Ban de Vendanges was held at this Japanese-owned château in 1986, everyone realized that the company Suntory was not simply content to apply state-of-the-art technology; it seriously intended to make Lagrange the best-quality wine in St-Julien. It might well have succeeded in this ambition. The formidable Bordeaux oenologist Professor Peynaud dubbed Lagrange a "dream estate" and describes its vinification centre as "unlike any other in the whole of Bordeaux". Each vat is, according to Peynaud, a "wine-making laboratory". The wine spends 21 months in wood, with up to 60 per cent new oak.

ST-JULIEN *CRU CLASSÉ* STATISTICS

CRUS CLASSÉS IN AOC PAUILLAC
Eleven châteaux (by number: 18% of *crus classés* in the Médoc) with 789ha (1,950ac) of vineyards (by area: 22% of *crus classés* in the Médoc and 87% of this AOC)

1ERS CRUS CLASSÉS
None

2ÈMES CRUS CLASSÉS
Five châteaux (by number: 36% of *2èmes crus classés* in the Médoc) with 399ha (986ac) of vineyards (by area: 39% of *2èmes crus classés* in the Médoc)

3ÈMES CRUS CLASSÉS
Two châteaux (by number: 14% of *3èmes crus classés* in the Médoc) with 138ha (341ac) of vineyards (by area: 20% of *3èmes crus classés* in the Médoc)

4ÈMES CRUS CLASSÉS
Four châteaux (by number: 40% of *4èmes crus classés* in the Médoc) with 249ha (615ac) of vineyards (by area: 42% of *4èmes crus classés* in the Médoc)

5ÈMES CRUS CLASSÉS
None

RED A deeply coloured wine with intense spicy-fruit aromas. It is full-bodied, silky-textured, and rich, with an exquisite balance and finish.

 Cabernet Sauvignon 67%, Merlot 28%, Cabernet Franc and Petit Verdot 5%

🍷 8–25 years
Second wine: ✓ *Les Fiefs de Lagrange*
Other wine: *Les Arums de Lagrange, Bordeaux Blanc*

CHÂTEAU LALANDE-BORIE
★ Ⓥ

Under the same ownership as the illustrious Château Ducru-Beaucaillou, Lalande-Borie is an inexpensive introduction to the wines of St-Julien.

RED These are well-coloured wines, dominated by rich, blackcurranty Cabernet Sauvignon flavours. Some vintages are fat and juicy, while others are more ethereal and tannic.

🍷 Cabernet Sauvignon 65%, Merlot 25%, Cabernet Franc 10%

🍷 5–10 years

CHÂTEAU LANGOA-BARTON
3ème Cru Classé
★ Ⓥ

This beautiful château was known as Pontet-Langlois until 1821, when it was purchased by Hugh Barton, grandson of "French Tom" Barton, the founder of Bordeaux *négociant* Barton & Guestier, and is now run by Anthony Barton and his daughter Lilian. Both Langoa-Barton and Léoville-Barton are made here using very traditional techniques. The wine is matured in wood for 18 to 24 months, with 60 per cent new oak.

RED Attractive, easy-drinking wine with good fruit and acidity. It is lighter than the Léoville and can sometimes taste a little rustic in comparison, but has gained a degree of extra elegance and finesse in recent years.

🍷 Cabernet Sauvignon 57%, Merlot 34%, Cabernet Franc 9%

🍷 10–25 years
Second wine: ✓ *Lady Langoa*

CHÂTEAU LÉOVILLE BARTON
2ème Cru Classé
★★☆ Ⓥ

A quarter of the original Léoville estate was sold to Hugh Barton in 1826, but the château remained in the hands of the Léoville estate and is now called Château Léoville-Las Cases (*see below*). This wine is made by Anthony Barton at Langoa-Barton (*see left*). It is matured in wood for 24 months, with a minimum of one-third new oak. Although it is the better of the two Barton estates, it has been considered significantly beneath the standard set by Léoville-Las Cases – since the late 1980s, however, it has performed equally as well. A great château in ascendancy.

RED Excellent wines of great finesse and breeding; they are darker, deeper, and richer than the Langoa-Barton, which is itself of very good quality. With maturity, a certain cedarwood complexity develops in this wine and gradually overwhelms its youthful cassis and vanilla character.

🍷 Cabernet Sauvignon 74%, Merlot 23%, Cabernet Franc 3%

🍷 15–30 years
Second wine: *La Réserve de Léoville Barton*

CHÂTEAU LÉOVILLE-LAS CASES
2ème Cru Classé
★★☆

The label reads "Grand Vin de Léoville du Marquis de Las Cases", although this wine is commonly referred to as "Château Léoville-Las Cases". This estate represents the largest portion of the original Léoville estate. This is a great wine, and it certainly qualifies as one of the super-seconds, while Clos du Marquis is one of the finest second wines available and probably the best value St-Julien. The grand vin spends 18 months in wood, with 80 per cent new oak.

RED Dark, damson-coloured, full-bodied, and intensely flavoured wine, complex, classy, and aromatically stunning. A skilful amalgam of power and finesse.

 Cabernet Sauvignon 65 61%, Merlot 19 21%, Cabernet Franc 13 16%, Petit Verdot 3%

🍷 15–35 years
Second wine: *Le Petit Lion du Marquis de Las Cases*
Other wines: ✓ *Clos du Marquis, Potensac (Médoc AOC)*

CHÂTEAU LÉOVILLE POYFERRÉ
2ème Cru Classé
★☆

This property once formed a quarter of the original Léoville estate and probably suffers from being compared to the other two châteaux, whose properties were also part of Léoville – Léoville-Barton and Léoville-Las Cases. Yet in the context of St-Julien as a whole, it fares very well, and since 1982 it has had some extraordinary successes. Didier Cuvelier has presided over this property since 1979, and since the involvement of Michel Rolland from the mid-1990s, quality has gone up another gear. Wine is matured in wood for 18 to 20 months, with 80 per cent new oak.

RED This wine has always been tannic, but it is now much fuller in fruit, richer in flavour, and darker in colour, with oaky nuances.

🍷 Cabernet Sauvignon 61%, Merlot 27%, Petit Verdot 8%, Cabernet Franc 4%

🍷 12–25 years
Second wine: *Pavillon de Léoville Poyferré*
Other wines: *Château Moulin-Riche*

CHÂTEAU MOULIN-DE-LA-ROSE
★ Ⓥ

This vineyard is well situated, being surrounded by *crus classés* on virtually all sides. Its wine is fermented in stainless steel and aged in cask for 20 months, with one-third new oak.

RED This attractively aromatic wine is unusually concentrated and firm for a minor St-Julien, but rounds out well after a few years in bottle.

🍷 Cabernet Sauvignon 65%, Merlot 35%

🍷 6–12 years

CHÂTEAU ST-PIERRE
4ème Cru Classé
★☆ Ⓥ

This property was bought in 1982 by Henri Martin, who owns the *bourgeois* growth Château Gloria. The wine is matured in wood for between 14 to 16 months, with 50 per cent new oak.

RED Once an astringent, coarse wine; now ripe, fat, and full of cedarwood, spice, and fruit.

 Cabernet Sauvignon 75%, Merlot 15%, Cabernet Franc 10%

🍷 8–25 years
Second wine: *Esprit de Saint-Pierre*

CHÂTEAU TALBOT
4ème Cru Classé
★☆ Ⓥ

Named after the English commander who fell at the Battle of Castillon in 1453, this property remains under Cordier family ownership, while its sister château Gruaud-Larose now belongs to Groupe Taillan. To contrast the style of these two St-Juliens is justifiable, but to compare their quality is not: Château Talbot is a great wine and closer to the style of a classic St-Julien, but intrinsically it does not have the quality nor the consistency of Château Gruaud-Larose. Talbot is matured in wood for between 15 and 18 months, with one-third new oak.

RED A graceful wine, medium-bodied, with elegant fruit, gently structured by ripe oak tannins and capable of considerable finesse.

🍷 Cabernet Sauvignon 66%, Merlot 30%, Petit Verdot 4%

🍷 8–30 years
Second wine: ✓ *Connétable Talbot*
Other wines: *Caillou Blanc*

CHÂTEAU TEYNAC
☆ Ⓥ

This fine gravel vineyard once formed part of *cru classé* Château St-Pierre.

RED This is a well-balanced, medium- to full-bodied wine with good spice and a firm tannin structure.

🍷 Cabernet Sauvignon 70%, Merlot 28%, Petit Verdot 2%

🍷 6–10 years Ⓥ

ENTRANCE GATE TO CHÂTEAU LÉOVILLE-LAS CASES
The lion sitting atop the archway inspired the name of this estate's second wine.

The Médoc:
Margaux

The Margaux appellation, stretching over five communes, is the largest and most southerly village appellation in the Médoc. It is the most diverse with the full range of wines from the premier grand cru classé *to the* cinquième, *as well as the famous* crus bourgeois *and* crus artisans.

The other three great Médoc AOCs – Saint-Estèphe, Pauillac, and Saint-Julien – are connected in one unbroken chain of vineyards, but Margaux stands alone to the south, with its vines spread across five communes – Labarde, Arsac, and Cantenac to the south; Margaux in the centre; and Soussans to the north. Margaux and Cantenac are the most important communes and, of course, Margaux contains the *premier cru* of Château Margaux itself. Cantenac has a slightly larger area under vine and no fewer than eight classified growths, including the star performer Château Palmer. It also has more *cru classé* châteaux than any other Médoc appellation, including an impressive total of 10 *troisièmes crus*. Margaux wines are often described as feminine, supple, and elegant but with great aging potential, thanks to their tannic structure.

Dominated by Garonne gravel on the central plateau about 6.5 kilometres (4 miles) long and 2.25 kilometres (1.25 miles) wide. To the southeast, this plateau overlooks the low-lying land by the estuary. Its east side is marked by gentle, dry valleys carving a succession of outcrops.

CHÂTEAU MARGAUX
A woodcut from an extract from *Richesse Gastronomique de France* (published in *L'Illustration: journal universel,* Paris, 1868) shows how this illustrious château appeared in the 19th century. Chateau Margaux wines are lauded worldwide.

The river deposited the gravel of Margaux in the early Quaternary Period. It is on this ancient layer of gravel over a terrace of limestone and clay marl subsoil that the best Margaux *crus* are to be found. It is also here that the two rivers, the Dordogne and the Garonne join, mixing dark, sandy soil from the Massif Central to the gravel at the heart of the appellation.

AN OUTSTANDING WINE
If the massive Pauillac wines of Château Latour and Château Mouton are an object lesson in how it is possible to bombard the senses with power and flavour, and yet retain quite remarkable finesse, then the exquisite wines of Margaux at their very best are perfect proof that complexity does not necessarily issue from an intense concentration of flavour.

This is not to suggest that Margaux wines do not actually possess some concentration; indeed, Château Margaux has a remarkable concentration of flavour, and it remains the quintessential wine of this appellation.

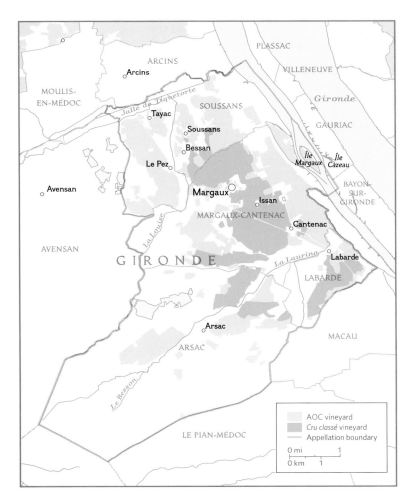

FACTORS AFFECTING TASTE AND QUALITY

LOCATION
Some 28 kilometres (17 miles) northwest of Bordeaux, in the centre of the Haut-Médoc, encompassing the communes of Cantenac, Soussans, Arsac, and Labarde, in addition to Margaux itself.

CLIMATE
As for the Médoc (*see* p152).

ASPECT
One large, low-lying plateau centring on Margaux, plus several modest outcrops that slope west towards the forest.

SOIL
Shallow, pebbly, siliceous gravel over a gravel subsoil interbedded with limestone.

VITICULTURE & VINIFICATION
Only red wines have the right to the Margaux appellation. All grapes must be destalked. On average, between 5 and 10 per cent *vin de presse* may be used in the wine, according to the needs of a particular vintage. Skin contact duration averages 15 to 25 days, with the period of maturation in cask currently varying between 12 and 24 months.

GRAPE VARIETIES
Primary varieties: Cabernet Franc, Cabernet Sauvignon, Merlot
Secondary varieties: Carmenère, Malbec, Petit Verdot

VINE-GROWING AREAS OF MARGAUX (*see also* p141)
Of the classic Médoc appellations, Margaux – the most famous – stands alone to the south and can boast more *cru classé* châteaux than any of the others.

MARGAUX

CHÂTEAU D'ANGLUDET
★☆

This château is owned by the Sichel family, who are also part-owners of the star-performing Château Palmer. Since the late 1980s this château has established itself as *cru classé* quality. The wine is matured in wood for 12 months, with up to one-third new oak.

RED Vividly coloured, medium- to full-bodied wines with excellent fruit, finesse, and finish – classic Margaux.

🍷 *Cabernet Sauvignon 46%, Merlot 41%, Petit Verdot 10 13%*

🍷 *10–20 years*

Second wine: *Réserve d'Angludet*

CHÂTEAU D'ARSAC
Cru Bourgeois

Until recently, this was the only property in Arsac not to benefit from the Margaux appellation. Purchased by wine merchant Philippe Raoux in 1986, this estate has expanded its vineyards from just over 3 hectares (7.5 acres) to 160 (395 acres), of which 54 (133 acres) are now classified as Margaux. The wines are matured in wood for 12 to 18 months, with 20 per cent new oak.

RED These are deep-coloured, full-bodied wines.

🍷 *Cabernet Sauvignon 60%, Merlot 40%*

🍷 *7–15 years*

Second wine: *Château Le Monteil-d'Arsac (Haut-Médoc)*

Other wines: *Château d'Arsac Blanc, The Winemaker's Collection*

CHÂTEAU BOYD-CANTENAC
3ème Cru Classé

Producing traditional-style wines from old vines, Château Boyd-Cantenac is a property at the heart of the Margaux appellation owned by the Guillemet family since the 1930s. The wine is matured in wood for 16 months, with 85 per cent new oak.

RED Full-bodied, firm wine of good colour that needs a long time in bottle to soften. The mediocre 1980 was particularly successful.

🍷 *Cabernet Sauvignon 60%, Merlot 25%, Petit Verdot 9%, Cabernet Franc 6%*

🍷 *12–20 years*

Second wine: *Jacques Boyd, Josephine Boyd, La Croix de Boyd-Cantenac*

Other wines: *Clos Maucaillou (Bordeaux Supérieur)*

CHÂTEAU BRANE-CANTENAC
2ème Cru Classé
★☆ⓥ

This property is a superb plateau of immaculately kept vines situated on gravel over limestone and is owned and run by Henri Lurton. The wine is matured in wood for 18 months, with up to 70 per cent new oak.

RED These stylish wines have a smoky-cream and new-oak bouquet, deliciously rich fruit, and finesse on the palate. They are top-quality wines, velvety and beautifully balanced.

🍷 *Cabernet Sauvignon 55%, Merlot 40%, Cabernet Franc 4.5%, Carmenère 0.5%*

🍷 *8–25 years*

Second wine: *Le Baron de Brane*

Other wines: *Margaux de Brane*

CHÂTEAU CANTENAC BROWN
3ème Cru Classé
☆

Ever since I drank a 50-year-old half-bottle of 1926 Cantenac Brown in splendid condition, I have had a soft spot for this château, which has, frankly, been disproportionate to the quality of its wines. Despite heavy investment after being purchased by AXA in 1989, these wines have not noticeably improved. That, however, did not stop Syrian-born, British-based property billionaire Simon Halabi from snapping up this château for a rumoured £50 million ($86 million) in 2006. At the time, he insisted that the aim was to invest further in Cantenac Brown, to make it one of the best wines in Margaux. The wine is aged in 50 to 70 per cent new oak for 12 to 15 months.

RED A similar weight to Brane-Cantenac, but with a less velvety and generally more rustic style.

🍷 *Cabernet Sauvignon 65%, Merlot 30%, Cabernet Franc 5%*

🍷 *10–25 years*

Second wine: *Brio de Cantenac Brown*

Other wines: *Château Brown Lamartine, Alto de Cantenac Brown*

CHÂTEAU CHARMANT
☆

This property was not classified as a *cru bourgeois* in 1932, nor listed by the Syndicat in 1978, but it certainly deserves recognition today. The wine is aged in 25 per cent new oak for 12 months.

RED An elegant wine with plenty of fruit and a soft finish. It makes delightful drinking when young.

🍷 *Merlot 50%, Cabernet Sauvignon 30%, Cabernet Franc 15%, Petit Verdot 5%*

🍷 *3–8 years*

Other wines: *Château Galiane*

CHÂTEAU DAUZAC
5ème Cru Classé
★ⓑ

Now owned by MAIF, the quality of the wines from this château has steadily increased since the mid-1990s. Since the management of Laurent Fortin, the property has invested heavily in sustainable and biodynamic practises. The wine, which is matured in wood for 12 to 18 months with 60 per cent new oak, is steadily improving.

RED Ruby-coloured, medium-bodied, round, and attractively fruity wines that are easy to drink.

🍷 *Cabernet Sauvignon 69%, Merlot 29%, Petit Verdot 2%*

🍷 *6–12 years*

Second wine: *Labastide Dauzac,*

Other wines: *Aurore de Dauzac, Le Haut-Médoc de Dauzac, D de Dauzac (Bordeaux)*

CHÂTEAU DESMIRAIL
3ème Cru Classé
★

A "château with no château" (because the building that was its château was purchased by Château Marquis d'Alesme-Becker), Desmirail has been on the ascent since its purchase by the Lurton family, but it still has a way to go before it becomes a true *troisième cru*. It is now owned and run by Denis Lurton. The wine is matured in wood for 12 months, with 25 to 50 per cent new oak.

RED A medium-bodied wine that is nicely balanced, with gentle fruit flavours and supple tannins. It is well made and gradually gaining in finesse.

🍷 *Cabernet Sauvignon 70%, Merlot 29%, Petit Verdot 1%*

🍷 *7–15 years*

Second wine: *Initial de Desmirail*

Other wines: *Domaine de Fontarney, Origine Desmirail, Le Haut-Médoc de Desmirail, Le Rosé de Desmirail*

CHÂTEAU DEYREM-VALENTIN
☆

This château was classified *cru bourgeois* in 1932, but was not included in the Syndicat's 1978 list, although it is supe-

rior to a few that were. Its vineyards adjoin those of Château Lascombes. Christelle Sorge is the fifth generation of the family to manage the property and with the help of Hubert de Bouard as a consultant, is producing an elegant expression of Margaux. The wine spends 12 to 16 months in oak, of which one-third is new.

RED Honest, medium-bodied, fruity wine of some elegance.

🍷 *Cabernet Sauvignon 50%, Merlot 49%, Petit Verdot 5 1%*

🍷 *4–10 years*

Other wines: *Château Soussans, Château Valentin (Haut-Médoc)*

CHÂTEAU DURFORT-VIVENS
2ème Cru Classé
★☆ⓥⓑ

This property is owned by Gonzague Lurton, who is married to Claire Villars, the administrator of Haut-Bages Libéral and Ferrière, as well as Acaibo in Sonoma, California. Château Durfort-Vivens has become one of the best-value *cru classé* wines of Margaux since the mid-1990s. Fermented in oak and cement tanks, the wine is matured in wood for 18 to 20 months, with up to 50 per cent new oak. The property has been certified biodynamic since 2016.

RED Supple and polished at its best, but it can be a bit chewy and firm when young. The 1985 was particularly rich and impressive.

🍷 *Cabernet Sauvignon 80%, Merlot 18%, Cabernet Franc 2%*

🍷 *10–25 years*

Second wine: *Vivens, Le Relais de Durfort-Vivens*

Other wines: *Jardins de Durfort (exclusive for the Chinese market)*

MARGAUX *CRU CLASSÉ* STATISTICS

CRUS CLASSÉS IN AOC MARGAUX
Twenty-one châteaux (by number: 35% of *crus classés* in the Médoc) with 994ha (2,456ac) of vineyards (by area: 35% of *crus classés* in the Médoc and 66% of this AOC)

1ERS CRUS CLASSÉS
One château (by number: 25% of *1ers crus classés* in the Médoc) with 79ha (195ac) of vineyards (by area: 24% of *1ers crus classés* in the Médoc)

2ÈMES CRUS CLASSÉS
Five châteaux (by number: 36% of *2èmes crus classés* in the Médoc) with 310ha (766ac) of vineyards (by area: 31% of *2èmes crus classés* in the Médoc)

3ÈMES CRUS CLASSÉS
Ten châteaux (by number: 71% of *3èmes crus classés* in the Médoc) with 408ha (1,008ac) of vineyards (by area: 60% of *3èmes crus classés* in the Médoc)

4ÈMES CRUS CLASSÉS
Three châteaux (by number: 30% of *4èmes crus classés* in the Médoc) with 110ha (272ac) of vineyards (by area: 18.5% of *4èmes crus classés* in the Médoc)

5ÈMES CRUS CLASSÉS
Two châteaux (by number: 11% of *5èmes crus classés* in the Médoc) with 86ha (212ac) of vineyards (by area: 9% of *5èmes crus classés* in the Médoc)

CHÂTEAU FERRIÈRE
3ème Cru Classé
★ 🅥 🅑

When managed by Château Lascombes this was little more than second-label status, but it has gained in both exposure and quality since this property was purchased by the Villars family. It is now owned and managed by Claire Villars. Certified organic since 2013 and biodynamic since 2017.

RED Quick-maturing wine of medium weight and accessible fruit.

🍇 *Cabernet Sauvignon 64%, Merlot 30%, Petit Verdot 4%, Cabernet Franc 2%*

🍷 *4–8 years*

Second wine: *Les Remparts de Ferrière, La Dame de Ferrière*

CHÂTEAU GISCOURS
3ème Cru Classé
★☆

This property is situated in the commune of Labarde. It was purchased in 1952 by the Tari family, who restored the château, the vineyard, and the quality of its wine to their former glory. In 1995 the vines were sold to Eric Albada Jelgerms, a Dutch businessman who continued to improve the wines. Since his death in 2018 the property continues to be managed by Alexander Van Beck who has been with the property since 1997. The wine is matured in wood for 15 to 18 months, with 50 per cent new oak.

RED Vividly coloured wine, rich in fruit and finesse. Its vibrant style keeps it remarkably fresh for years.

🍇 *Cabernet Sauvignon 60%, Merlot 32%, Petit Verdot 5%, Cabernet Franc 3%*

🍷 *8–30 years*

Second wine: *La Sirène de Giscours*
Other wines: *Château Duthil (Haut Médoc), Le Haut-Médoc de Giscours, Le Rosé de Giscours*

CHÂTEAU LA GURGUE

This property was classified *cru bourgeois* in 1932, but not included in the 1978 list, although it is superior to a few that were. Owned and managed by Claire Villars Lurton, alongside *crus classés* Haut-Bages-Libéral, Château Ferrière, and Château Chasse-Spleen. The wine is aged in 50 per cent new oak for 12 months.

RED Soft, elegant, medium-bodied wine of attractive flavour and some finesse, improving since 2000.

🍇 *Cabernet Sauvignon 50%, Merlot 45%, Petit Verdot 5%*

🍷 *4–12 years*

CHÂTEAU D'ISSAN
3ème Cru Classé
★★

This beautiful 17th-century château is frequently cited as the most impressive in the entire Médoc, and its remarkable wines, matured in wood for 18 months, with up to 50 per cent new oak, are consistently just as spectacular.

RED This wine really is glorious! Its luxuriant bouquet is immediately seductive; its fruit is unbelievably rich and sumptuous. A great wine of great finesse.

🍇 *Cabernet Sauvignon 60%, Merlot 40%*

🍷 *10–40 years*

Second wine: *Blason d'Issan*
Other wines: *Haut-Médoc d'Issan, Moulin d'Issan (Bordeaux Supérieur)*

CHÂTEAU KIRWAN
3ème Cru Classé
★☆ 🅥

Château Kirwan is a well-run and improving property owned by the Schyler family, also owners of Bordeaux *négociant* Schröder & Schÿler with Eric Boissenot consulting. Continued investment, including the new concrete cellars built in 2017 have seen the wines improve. The wine is matured in wood for 18 to 24 months, with up to 50 per cent new oak.

RED Deep-coloured, full-bodied, rich and concentrated wines that are well made and gaining in generosity, riper tannins, and new oak influence with each passing vintage.

🍇 *Cabernet Sauvignon 50%, Merlot 35%, Cabernet Franc 10%, Petit Verdot 5%*

🍷 *10–35 years*

Second wine: *Les Charmes de Kirwan*

CHÂTEAU LABÉGORCE
★ 🅥

This château was classified *cru bourgeois* in 1932, but not included in the 1978 list. Since Hubert Perrodo, a wine-loving oil tycoon, purchased Labégorce in 1989, the quality and price of its wines have increased steadily. After his death in 2006 the property has been managed by his daughter Nathalie, who continues to invest and improve the quality. Matured in wood for 15 months, with up to 40 per cent new oak. Labégorce has purchased neighbouring Château Labégorce-Zédé, which ceased production after the 2008 vintage, when its 22 hectares (54 acres) were absorbed by this château.

RED Well-coloured wine with good balance of concentration and finesse.

🍇 *Cabernet Sauvignon 50%, Merlot 45%, Cabernet Franc 3%, Petit Verdot 2%*

🍷 *5–15 years*

Second wine: *Zédé de Labégorce*

CHÂTEAU LASCOMBES
2ème Cru Classé
★★

The wines of this large, 120-hectare (297-acre) property, have always been good, yet they improved dramatically under René Vanatelle. It was Vanatelle who introduced a true second wine, Chevalier de Lascombes. Dominique Befve, formerly of Château Lafite, has managed the property since 2001 when the Colony Capital group purchased it with René Vatelot. They invested heavily in replanting and a new cellar. He remains in charge under the new ownership of the MACSF insurance group since 2011. The wine is matured in wood for 18 to 20 months, with one-third new oak.

RED Full-bodied, rich, and concentrated wine with ripe fruit, a lovely cedarwood complexity, and supple tannin.

🍇 *Cabernet Sauvignon 55%, Merlot 40%, Petit Verdot 5%*

🍷 *8–30 years*

Second wine: *Chevalier Lascombes*
Other wines: *Haut-Médoc de Lascombes*

CHÂTEAU MALESCOT ST-EXUPÉRY
3ème Cru Classé
★

English-owned until 1955, when it was purchased by Roger Zuger, whose brother owned Château Marquis d'Alesme-Becker. The wine is matured in wood for 13 months, with 80 per cent new oak.

RED Richer, more complex wines have been produced since 1996.

🍇 *Cabernet Sauvignon 50%, Merlot 35%, Cabernet Franc 10%, Petit Verdot 5%*

🍷 *8–25 years*

Second wine: *La Dame de Malescot*

CHÂTEAU MARGAUX
1er Cru Classé
★★★ 🅥

This is the most famous wine in the world and, since its glorious rebirth in 1978, the greatest. Its quality may occasionally be matched, but it is never surpassed. Purchased in 1977 for 72 million francs by the late André Mentzelopoulos, who spent an equal sum renovating it, this fabulous jewel in the crown of the Médoc is now run by his daughter, Corinne Mentzelopoulos. Both Château Margaux and its second wine, "Pavillon Rouge", are vinified in oak vats and matured for 18 to 24 months in 100 per cent new oak.

RED If finesse can be picked up on the nose, then the stunning and complex bouquet of Château Margaux is the yardstick. The softness, finesse, and velvety texture of this wine belies its depth. Amazingly rich and concentrated, with an elegant, long, and complex finish supported by ripe tannins and wonderful smoky-creamy oak aromas. This is as near perfection as we will ever get.

🍇 *Cabernet Sauvignon 75%, Merlot 20%, Cabernet Franc 4.5%, Petit Verdot 0.5%*

🍷 *15–50 years*

Second wine: *Pavillon Rouge du Château Margaux*
Other wines: *Pavillon Blanc du Château Margaux, Margaux du Château Margaux*

CHÂTEAU MARQUIS D'ALESME
3ème Cru Classé

Like Château Malescot St-Exupéry, this was English-owned until purchased by Jean-Claude Zuger, who also purchased the maison of neighbouring Desmirail to act as its château. Purchased by Hubert Perrodo of Château Labégorce, just before his untimely death in a skiing accident in 2006, it is now managed by his daughter Nathalie with Marjolaine de Cornack as technical director in the new gravity-fed cellars. The wine is aged for 18 months in 65 per cent new oak.

MARGAUX PROFILE

APPELLATION AREA
Covers parts of the communes of Arsac, Cantenac, Labarde, Margaux, and Soussans.

TOTAL SIZE OF ALL FIVE COMMUNES
7,512ha (18,562ac)

AOC AREA UNDER VINE
1,512ha (3,736 ac (20% of communes)

TOTAL SURFACE AREA OF *CRUS CLASSÉS*
994ha (2,456ac) (13% of communes, 56% of AOC)

RED The wines have improved immensely under the new team, presenting a classic and generous Margaux elegance.

🍇 *Cabernet Sauvignon 63%, Merlot 30%, Petit Verdot 5%, Cabernet Franc 2%*

🍷 *8–20 years*

CHÂTEAU MARQUIS D'ALESME-BECKER

Same wine as Château Marquis d'Alesme, but the new owners have removed the Becker name from recent vintages. *See* Château Marquis d'Alesme.

CHÂTEAU MARQUIS-DE-TERME
4ème Cru Classé

Situated next to Château Margaux, this once majestic estate developed the reputation for producing tight, tannic, one-dimensional wines, but its quality has picked up since the late 1970s and has been performing extremely well since 1983. The wine is matured in wood for 24 months, with one-third new oak.

RED Appears to be developing a style that is ripe and rich, with definite and delightful signs of new oak. The 1984 was quite a revelation.

🍇 *Cabernet Sauvignon 60%, Merlot 30%, Petit Verdot 7%, Cabernet Franc 3%*

🍷 *10–25 years*

Second wine: *La Couronne de Marquis de Termes*
Other wines: *Fleur de Marquis de Terme, M de Marquis de Terme, Le 9 de Marquis de Terme*

CHÂTEAU MARSAC SÉGUINEAU

This property in the heart of the Marsac plateau was classified *cru bourgeois* in 1932, but it was not included in the 1978 list, although it is superior to a few that were. The vineyards of this château include some plots that originally belonged to a *cru classé*. Along with its neighbour, La Tour de Mons, it belongs to CA Grands Crus, the winemaking group of the Credit Agricole Bank. The wine is aged for 12 months in 25 per cent new oak.

RED Medium- to full-bodied wines of good bouquet and a soft style.

🍷 Merlot 65%, Cabernet Sauvignon 23%, Cabernet Franc 12%

🍷 5–12 years

Second wine: *Château Gravières-de-Marsac*

CHÂTEAU MONBRISON
★ **V**

Purchased by American Robert Davis in the 1930s on his marriage to Kathleen Johnston, the vineyard was uprooted in the 1930s and replanted by their daughter in the 1960s. It thrived in the 1970s and 1980s thanks to the talent of Jean-Luc Vonderheyden, who pioneered green harvesting in the Médoc. Upon his untimely death in 1992, his brother Laurent took over and the property, which continues to go from strength to strength. The wines now vie with those of true *cru classé* standard.

RED This château's second label offers a brilliant selection of beautifully deep-coloured wines with spicy-oak, super-rich juicy fruit, and a fine structure of supple tannin.

🍷 Cabernet Sauvignon 60%, Merlot 28%, Cabernet Franc 7%, Petit Verdot 5%

🍷 8–15 years

Second wine: *Bouquet de Monbrison*
Other wines: *Haut-Médoc de Monbrison*

CHÂTEAU PALMER
3ème Cru Classé
★★ ☆ **B**

Only Château Margaux outshines this property, jointly owned by Dutch (Malher-Besse) and British (the Sichel family) interests. Château Palmer 1961 and 1966 regularly fetch prices at auction that equal those fetched by the *premiers crus*. A true super-second since the introduction of Alter Ego, a sort of super-premium second wines, in 1998. The property is certified biodynamic since 2017. The wine is matured in wood for 20 to 22 months, with up to 50 per cent new oak.

RED Almost opaque-coloured, with masses of cassis fruit and an exceedingly rich, intense, and complex construction of creamy, spicy, cedarwood, and vanilla flavours.

🍷 Cabernet Sauvignon 47%, Merlot 47%, Petit Verdot 6%

🍷 12–35 years

Second wine: *Alter Ego*

CHÂTEAU PONTAC-LYNCH
Cru Bourgeois
★ **V**

This property was classified *cru bourgeois* in 1932, but not included in the Syndicat's 1978 list, although it is superior to a few that were. It has been in the official *cru bourgeois* selection from 2008 to 2015. The vineyards are well situated and surrounded by *crus classés*. In conversion to organic agriculture since 2017.

RED Richly perfumed, deeply coloured, full-bodied wines of good structure.

🍷 Merlot 45%, Cabernet Sauvignon 40%, Cabernet Franc 10%, Petit Verdot 5%

🍷 6–15 years

Second wine: *Pontac-Phénix (Haut-Médoc)*
Other wines: *La Quintessance de Pontac-Lynch*

CHÂTEAU POUGET
4ème Cru Classé

Under the same ownership as Boyd-Cantenac. The wine is matured in wood for 18 months, with 70 per cent new oak.

RED Well-coloured, full-bodied wine with good depth of flavour. Good, but not great, and could be more consistent.

🍷 Cabernet Sauvignon 60%, Merlot 30%, Petit Verdot 10%

🍷 10–25 years

Second wine: *Antoine Pouget*

CHÂTEAU PRIEURÉ-LICHINE
4ème Cru Classé
★

The late Alexis Lichine purchased Château Prieuré in 1951 and added his name to it. To develop the small rundown vineyard, he bought various prized plots of vines from Palmer, Kirwan, Giscours, Boyd-Cantenac, Brane-Cantenac, and Durfort-Vivens – some 77.5 hectares (190 acres). The composite classification must be higher than its official status – the wines certainly are. Lichine's son Sacha ran the property until 1999, when he sold it. The current owners, the Ballande family, have steadily improved the quality under top consultant Stéphane Derenoncourt. The

SIGNBOARDS IN MARGAUX
In this fertile wine country, vineyards can cluster near one another. On this route, Château Marquis-De-Terme, Château Rauzan-Gassies, and Château Rauzan-Ségla share a neighbourhood.

wines are matured in wood for 16 months, with 50 per cent new oak.

RED Well-coloured, full-bodied wines, plummy and rich, with good blackcurrant fruit supported by supple tannins and a touch of vanilla-oak.

🍷 Cabernet Sauvignon 50%, Merlot 45%, Petit Verdot 5%

🍷 7–20 years

Second wine: *Château de Clairefont, Confidences de Prieuré Lichine*
Other wines: *Le Blanc de Prieuré Lichine (Bordeaux Blanc), Le Clocher de Prieuré Lichine (Haut Médoc)*

CHÂTEAU RAUZAN-GASSIES
2ème Cru Classé
☆

Until the French Revolution of 1789, this property and Château Rauzan-Ségla were one large estate. The globe-trotting Professor Peynaud was brought in to steer this wine back on course in the early 1980s, followed by Jean-Louis Camp, formerly of Loudenne. The property is now run by the sister and brother team of Anne-Francoise and Jean-Philippe Quié whose family have owned the property since 1946. The wine is matured in wood for 12 to 14 months, with 55 per cent new oak, although there is little evidence of it on the palate.

RED The 1996 and 1998 suggested an upturn in quality, and this has been partly realized in vintages.

🍷 Cabernet Sauvignon 58%, Merlot 40%, Petit Verdot 2%

🍷 7–15 years

Second wine: *Gassies*

CHÂTEAU RAUZAN-SÉGLA
2ème Cru Classé
★★

The quality of this once-disappointing château began to lift in the 1980s due to significant investment in the property from its owner, the Bordeaux *négociant* house of Eschenauer, which also instigated a far stricter selection of the *grand vin*. In 1994 Rauzan-Ségla was sold to Chanel, the under-bidder for Latour (sold by Allied-Lyons to the French industrialist François Pinault). Since then key personnel from the *premier cru* were brought in to keep the improvements in full swing. The wine is matured in wood for 18 months, with 65 per cent new oak and is currently one of Bordeaux's top-performing *deuxièmes crus*.

RED In classic years, this wine is deep and dark, with a powerful tannic construction, and more than enough intensely flavoured fruit to match. Lesser vintages are dark for the year, but much more lush, with softer tannins.

🍷 Cabernet Sauvignon 62%, Merlot 36%, Cabernet Franc 1%, Petit Verdot 1%

🍷 15–30 years

Second wine: *Ségla*

CHÂTEAU SIRAN
★ ☆ **V**

The vineyard is well situated, with immaculately manicured vines that bor-

der those of châteaux Giscours and Dauzac. The wine is matured in wood for 12 to 18 months, with one-third new oak, in air-conditioned cellars. Owned by the Miailhe de Burgh family, who are direct descendants of Sir Patrick Sarsfield, the daring Irish Jacobite who gave William of Orange such a bloody nose.

RED Stylish, aromatic wines of good body, creamy-spicy fruit, length, and obvious class. Easily equivalent to *cru classé* quality.

🍷 Merlot 46%, Cabernet Sauvignon 40%, Petit Verdot 13%, Cabernet Franc 1%

🍷 8–20 years

Second wine: *S de Siran*
Other wines: *St-Jacques de Siran (Bordeaux Supérieur), Bel Air de Siran (Haut Médoc)*

CHÂTEAU TAYAC

As Bernard Ginestet, whose family owned Château Margaux for 40 years, once wrote, "this is one of the largest of the smaller properties, and one of the smallest of the larger".

RED Firm, medium- to full-bodied wines of good character, although somewhat rustic; they tend to be coarse in lesser years.

🍷 Cabernet Sauvignon 60%, Merlot 35%, Petit Verdot 5%

🍷 6–12 years

Second wine: *Château Grand Soussans*
Other wines: *Château Tayac Cuvée Larauza, Château Tayac Cuvée Labory de Tayac, Château Tayac, Cuvée les Gravières*

CHÂTEAU DU TERTRE
5ème Cru Classé
★ ☆

An underrated *cru classé*, this château has well-situated vineyards. It is under the same ownership as Giscours since 1998. The wine is matured in wood for 15 to 17 months, with 45 per cent new oak.

RED Although the scent of violets is supposed to be common to Margaux wines, this is one of the few in which I pick it up. The wine is medium- to full-bodied, rich in fragrant fruit, and has excellent balance, with obvious class.

🍷 Cabernet Sauvignon 43%, Merlot 33%, Cabernet Franc 19%, Petit Verdot 5%

🍷 8–25 years

Second wine: *Les Hauts du Tertre*

CHÂTEAU LA TOUR DE MONS

These wines are aged in wood for 12 months, with 20 per cent new oak and have improved enormously since the late 1980s. Easily equivalent to *cru classé* quality.

RED As richly flavoured as ever, but without the tannins or acidity that used to be this wine's pitfall.

🍷 Merlot 56%, Cabernet Sauvignon 38%, Petit Verdot 6%

🍷 10–30 years

Second wine: *Marquis de Mons*

Graves, Cérons, Sauternes, and Barsac

The finest red Graves wines are produced in Pessac-Léognan, good red and improving dry white wines in the centre of Graves and the great sweet wines of Sauternes and Barsac in the south. The emphasis in production is on classic red wines.

The silky-smooth red wines of the Graves district have been famous since the Middle Ages, when they were protected by local laws that punished those who dared to blend them with other Bordeaux wines. Château Haut-Brion was the only red wine outside the Médoc to be classified in 1855, and such was its reputation that it was placed alongside the *premiers crus* of Latour, Lafite, and Margaux. Beneath Haut-Brion, there are a few great wines equivalent in quality to *deuxième* or *troisième cru*, but only a few.

The relative lack of superstars in Graves is offset by a higher base quality of wine and greater consistency of performance in the red wines at least. There are 43 communes in this appellation. Much the best are Léognan, Talence, and Pessac, after which Martillac and Portets are the most outstanding, followed by Illats and Podensac. All the greatest wines are therefore in the north of the Graves district, amid the urban sprawl of Bordeaux,

and this presents something of a problem. The once-peaceful left bank of the Garonne is slowly and inexorably disappearing. As the city bursts outwards, more rural vineyards are encircled by the concrete jungle, and many quite simply vanish. How many Bordeaux aficionados who fly directly to the airport in Mérignac stop to consider the cost of such progress? In 1908 there were 30 winemaking properties in the commune of Mérignac; today there is just one – Château Picque-Caillou. The conurbated communes of Cadaujac, Gradignan, Léognan, Martillac, Mérignac, Pessac, Talence, and Villenave d'Ornon have lost 214 wine châteaux over the same period.

THE PROBLEM OF WHITE GRAVES SOLVED

The quality and reputation of the red wines have always been well established, but white Graves had a serious identity problem that came to a crisis point in the mid-1980s. Although fine white Graves were being produced, most of it was in the northern communes, but they were tarred with the same brush as the worst white wines from farther south. It was not simply a north-south divide; there was also an identity problem – should they be making rich, oak-aged blends or light and fluffy Sauvignon Blanc? Paradoxically, the worst wines came from some of the best properties in the north, produced by winemakers who either did not know how to, or did not care to, clean up their act, as they continued to sell tired, over-sulphured, oxidized, and flabby wines on the back of their decaying reputations.

An official north-south divide, however, proved to be the solution for, since 1987, when the Pessac-Léognan AOC was introduced, things have never looked better for Graves. The Pessac-Léognan appellation is a single appellation for both red and white wines from the communes Cadaujac, Canéjan, Gradignan, Léognan, Martillac, Mérignac, Pessac, St-Médard-d'Eyrans, Talence, and Villenave d'Ornon. This has had the effect of giving the northern châteaux the official quality recognition they both wanted

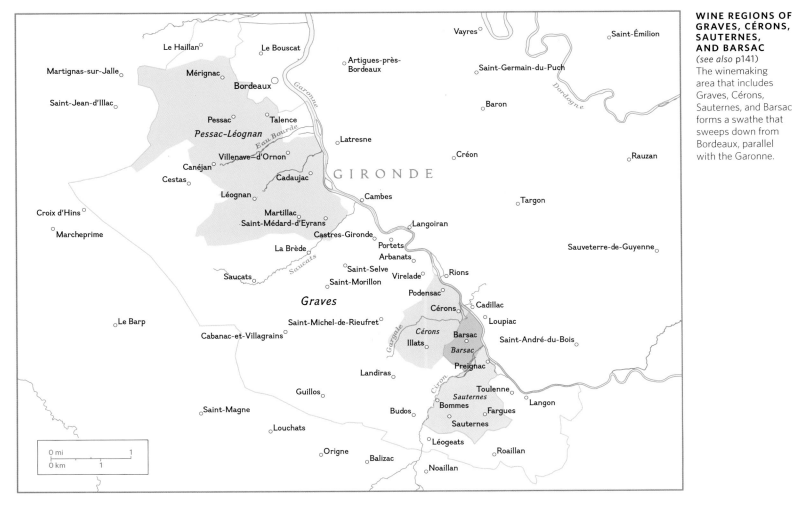

WINE REGIONS OF GRAVES, CÉRONS, SAUTERNES, AND BARSAC
(see also p141*)*
The winemaking area that includes Graves, Cérons, Sauternes, and Barsac forms a swathe that sweeps down from Bordeaux, parallel with the Garonne.

RIPE GRAPES HARVESTED IN TRACTOR, BARSAC
Machine harvesting causes the berries to fall off the stem, leaving the stem attached to the vine. Once in the bins, high-speed fans remove leaves and other debris.

and deserved. It was a bit slow to start off – after all, Pessac-Léognan hardly trips off the tongue, and there were worries about its marketability. There is still a tendency to put Graves on labels and use Pessac-Léognan to qualify the wine as if it were a higher classification of Graves, which for all practical purposes it is.

Once the châteaux realized that foreign markets were picking up on the superior connotation of Pessac-Léognan, use of the appellation soon became widespread. Whether by their own volition or due to peer pressure, many of the underperformers have become the most quality-conscious châteaux in the appellation, and this has spurred producers in the south to improve their wines. They do not like being considered inferior, and as they intend to prove they are not, the consumer can only gain.

CÉRONS

This is an area situated within the boundaries of Graves. It is the stepping stone between dry white Graves and sweet white Sauternes and Barsac. The châteaux of Cérons have been given the official right to make both red and white Graves, Graves Supérieur (which may be dry but is usually sweet), and, of course, the sweet wine of Cérons – a wine that has enjoyed a modest reputation for nearly 200 years. In fact, only 20 per cent of the production in this area is sold as Cérons, because the appellation covers three communes: those of Illats, Podensac, and Cérons itself. Many of the vineyards comprise scattered plots, some of which are partially planted with acacias.

SAUTERNES AND BARSAC

The gap between ordinary sweet white wines and the great wines of Sauternes and Barsac is as wide as that between sweet and dry wines. What creates this gap is something called "complexity" – to find out what that is, sample the aroma of a glass of mature Sauternes. The wines produced in Sauternes are not only the world's most luscious, but also the most complex wines. I have seen hardened men who resolutely refuse to drink anything sweeter than lemon juice go weak at the knees after one sniff of Château Suduiraut, and I defy the most stubborn and bigoted anti-sweet-wine drinker not to drool over a glass of Château

---

FACTORS AFFECTING TASTE AND QUALITY

LOCATION
The left bank of the Garonne river, stretching southeast from just north of Bordeaux to 10 kilometres (6 miles) east of Langon. Cérons, Sauternes, and Barsac are tucked into the southern section of the Graves district.

CLIMATE
Very similar to the Médoc, but fractionally hotter and with slightly more rainfall. In Sauternes and Barsac it is mild and humid, with an all-important autumnal alternation of misty mornings and later sunshine, the ideal conditions for noble rot.

ASPECT
The suburbs of Bordeaux sprawl across the northern section of this district, becoming more rural beyond Cadaujac. Graves has a much hillier terrain than the Médoc, with little valleys cut out by myriad streams that drain into the Garonne. Some of the vineyards here are quite steep. The communes of Sauternes, Bommes, and Fargues are hilly, but Preignac and Barsac on either side of the Ciron – a small tributary of the Garonne – have gentler slopes.

SOIL
Travelling south through the district, the gravelly topsoil of Graves gradually becomes mixed with sand, then with weathered limestone, and eventually with clay. The subsoil also varies, but basically it is iron-pan, limestone, and clay, either pure or mixed. Cérons has a stony soil, mostly flint and gravel, over marl; there is reddish clay-gravel over clay, or gravelly iron-pan in Sauternes, and clay-limestone over clay-gravel in Fargues. The gravel slopes of Bommes are sometimes mixed with heavy clay soils, while the plain is sandy clay with a reddish clay or limestone subsoil. Preignac is sand, gravel, and clay over clay-gravel in the south, becoming more alluvial over sand, clay, and limestone closer to Barsac. Where the classified growths of Barsac are situated, the soil is clay-limestone over limestone, elsewhere the topsoil mingles with sandy gravel.

VITICULTURE & VINIFICATION
Some châteaux add a certain amount of *vin de presse* to the red wine. The *cuvaison* varies between 8 and 15 days, although some Graves châteaux permit 15 to 25 days. Maturation in cask is generally between 15 and 18 months. The sweet white wines of Sauternes and Barsac are made from several *tries* of late-harvested, overripe grapes which, ideally, have noble rot. Destalking is usually unnecessary. The fermentation of grape juice so high in sugar content is difficult to start and awkward to control, but it is usually over within two to eight weeks. The exact period of fermentation depends upon the style desired. Many of the best wines are matured in cask for one and a half to three and a half years.

GRAPE VARIETIES
Primary varieties: Cabernet Franc, Cabernet Sauvignon, Merlot, Sauvignon Blanc, Sémillon
Secondary varieties: Malbec, Muscadelle, Petit Verdot

---

d'Yquem 1967. Astonishingly, there are dissenters, but for me Yquem is by far the best wine of these two appellations, Sauternes and Barsac. The battle for second place is always between the soft, luscious style of Suduiraut, and the rich, powerful character of Rieussec, with Climens, Nairac, and the non-classified growths of Gilette and de Fargues in close pursuit. Guiraud has the potential to go right to the top, and with so many châteaux seriously improving, they could all end up chasing each other for the number two spot.

The "noble rot"

Yquem might be the ultimate sweet white wine, but many other great wines are made in these two small areas tucked away in the Bordeaux backwaters. What gives all these wines their hallmark of complexity is, literally, a lot of rot – namely "noble rot", or the fungal growth *Botrytis cinerea*. The low-lying hills of Sauternes and, to a lesser extent, of Barsac, together with a naturally warm but humid climate, provide a natural breeding ground for botrytis, the spores of which are indigenous to the area. They remain dormant in the vineyard soil and on vine bark until they are activated by suitable conditions – alternating moisture and heat (the early-morning mist being followed by hot, mid-morning sunshine). The spores latch on to the skin of each grape, replacing its structure with a fungal growth and feeding on moisture from within the grape. They also devour five-sixths of the grape's acidity and one-third of its sugar, but as the amount of water consumed is between one-half and two-thirds, the effect is to concentrate the juice into a sticky, sugar-rich pulp. A healthy, ripe grape with a potential of 13 per cent alcohol is thus converted into a mangy-looking mess with a potential of between 17.5 per cent and 26 per cent. The spread of botrytis through a vineyard is neither orderly nor regular, and the harvest may take as long as 10 weeks to complete, with the pickers making various sorties, or *tries,* through the vineyard. On each *trie,* only the affected grapes should be picked, but care must be taken to leave some rot on each bunch to facilitate its spread. The longer the growers await the miraculous noble rot, the more the vines are prone to the ravages of frost, snow, hail, and rain, any of which could destroy an entire crop.

The viticulture of Sauternes and Barsac is the most labour-intensive of any region. The yield is very low, officially a maximum of 25 hectolitres per hectare (112 cases per acre), about half that in the Médoc, and the levels achieved in the best châteaux are much lower, around 15 to 20 hectolitres per hectare (67 to 90 cases per acre). At Yquem it is even less, the equivalent of one glass per vine. On top of all this, the vinification is, at the very least, difficult to handle, and maturation of a fine sweet wine demands a good proportion of very expensive new oak.

Variations in character

Not all the sugar is used up during fermentation, even when a wine of perhaps 14 to 15 per cent alcohol is made. The remaining unfermented sugar, often between 50 and 120 grams per litre, gives the wine its natural sweetness. Unlike Sauternes' German counterparts, however, its alcohol level is crucial to its character. Its strength, in harmony with the wine's sweetness, acidity, and fruit, give it a lusciousness of concentration that simply cannot be matched. Yet its complexity is not the effect of concentration, although an increased mineral level is no doubt an influence. Complexity is created by certain new elements that are introduced into the grape's juice during the metabolic activities of its botrytis – glycerol, gluconic acid, saccharic acid, dextrin, various oxidizing enzymes, and an elusive antibiotic substance called "botrycine".

It is easy to explain how these components of a botrytised wine that form its inimitably complex character can vary. When tasting wine from different *tries* at the same château, the intensity of botrytised character varies according to the "age" of the fungus when the grapes are harvested. Wines made from the same percentage of botrytised grapes collected at the beginning and end of the harvest are noticeably mute compared to those in the middle when the rot is at its most rampant. If it is not surprising that youthful *Botrytis cinerea* has an undeveloped character, the same cannot be said of late-harvested. Many people believe that the longer botrytis establishes itself, the more potent its effect, but this is not true.

THE VENERABLE CHÂTEAU D'YQUEM IN THE SAUTERNES REGION
This château in Aquitaine has a long history, passing from the English crown to the French in the 15th century. Today this wine estate is renowned for the painstaking care it takes in the selection and quality of its harvest. It is also known for its often-delectable sweet Sauternes.

FROM GRAPE TO GLASS

The only Graves property to be classified in 1855 was Château Haut-Brion. The Syndicat, for the defence of the Graves appellation, wanted to create its own classification but was prevented from doing so until the 1921 law was changed in 1949. The first classification was not made until 1953, and this itself was later modified in 1959. Distinction is made between red wines and white wines, but no attempt at ranking between the various growths is made – they all have the right to use the term *cru classé*. It can be seen from the 535 hectares (1,320 acres) of classified properties listed below that this represents not much more than 13 per cent of the total 4,350 hectares (10,745 acres) of delimited vineyards planted in the Graves and Pessac-Léognan appellation.

RED WINES	COMMUNE	AREA CURRENTLY UNDER VINE	
		HECTARES	(ACRES)
Château Bouscaut	Cadaujac	34ha	(84ac)
Château Carbonnieux	Léognan	50ha	(124ac)
Domaine de Chevalier	Léognan	50ha	(124ac)
Château de Fieuzal	Léognan	58ha	(143ac)
Château Haut-Bailly	Léognan	30 ha	(74ac)
Château Haut-Brion	Pessac	50ha	(124ac)
Château La Mission-Haut-Brion	Pessac	25ha	(62ac)
Château Latour-Haut-Brion	Talence	La Tour Brion is now part of Haut-Brion (see above)	
Château La Tour-Martillac	Martillac	40ha	(99ac)
Château Malartic-Lagravière	Léognan	53ha	(131ac)
Château Olivier	Léognan	45ha	(111ac)
Château Pape-Clément[1]	Pessac	53ha	(131ac)
Château Smith-Haut-Lafitte[1]	Martillac	67ha	(166ac)
TOTAL AREA UNDER VINE:		**555 HECTARES**	**(1,373 ACRES[2])**

NOTES

1 These châteaux also produce a dry white wine, but only the red is classified as *cru classé*.

2 These figures are not precise conversions of the hectare totals, but are column totals and differ because of rounding up.

WHITE WINES	COMMUNE	AREA CURRENTLY UNDER VINE	
		HECTARES	(ACRES)
Château Bouscaut	Cadaujac	10ha	(25ac)
Château Carbonnieux	Léognan	42ha	(104ac)
Domaine de Chevalier	Léognan	5ha	(12ac)
Château de Fieuzal	Léognan	10ha	(25ac)
Château Couhins-Lurton[3]	Villenave	13ha	(32ac)
Château Haut-Brion	Pessac	3ha	(7ac)
Château La Tour-Martillac	Martillac	10ha	(25ac)
Château La Mission Haut-Brion	Talence	4ha	(10ac)
Château Laville-Haut-Brion	Talence	now La Mission Blanc (see above)	
Château Malartic-Lagravière	Léognan	7ha	(17ac)
Château Olivier	Léognan	10ha	(25ac)
TOTAL AREA UNDER VINE:		**114 HECTARES**	**(282 ACRES[2])**

3 This château also produces a red wine, but only the white is classified as *cru classé*.

The rewards, the reality, and the future

A good Sauternes is the most arduous, expensive, and frustrating wine in the world to produce – and what is the winemaker's reward? Very little, I'm afraid. Apart from Château d'Yquem – not only the greatest Sauternes but some would argue the greatest wine per se – the wines of this region fail to realize their true worth. This is predictable in a world where the trend is towards lighter and drier styles of wine and may have a positive short-term effect for Sauternes aficionados, for it means a cheaper supply of their favourite wine. In the long term, however, this is not a positive way to operate, and some proprietors simply cannot afford to go on. The Comte de Pontac uprooted all the vines at his *deuxième cru* Château de Myrat in Barsac, and even the ever-optimistic Tom Heeter, former owner of Château Nairac, once said, "You have to be at least half-crazy to make a living out of these wines." We certainly do not deserve the luscious wines of Sauternes and Barsac if we continue to ignore them, but if the authorities had more sense and the owners more business acumen, these wines could literally become "liquid gold".

The only way ahead

The vineyards of Sauternes and Barsac should also be allowed to sell red and dry white wines under the Graves appellation. If this is a right accorded to modest Cérons, why not to its illustrious neighbours? Many châteaux already make red and dry white wines, but they are sold under the cheaper "Bordeaux" appellation. Tom Heeter was right, the proprietors must be half-crazy, because their motivation for producing these alternative products is to subsidize the cost of making their botrytised wine, when they should be trying to supplement their income. Given the incentive of a superior appellation, the châteaux should concentrate on making the finest red and dry white wines every year. Only when conditions appear favourable should some of the white grape crop be left on the vine, with fingers crossed for an abundance of *Botrytis cinerea*. Instead of these châteaux investing in new oak for modest vintages, they should utilize the casks for the red and the dry white. The result would be a tiny amount of the world's most luscious wine, maybe 3 or 4 years in 10. It would no longer be necessary to attempt the impossible task of selling an old-fashioned image to young wine drinkers; the limited supply would outstrip the current demand. After 30 years of watching this area's vain attempts to win over popular support for its wines, I have come to accept the view of Comte Alexandre de Lur-Saluces, proprietor of Château d'Yquem. When asked to justify the price of Yquem, he simply said his wines are not made for everyone; they are made for those who can afford them.

THE APPELLATIONS OF
GRAVES, CÉRONS, SAUTERNES, AND BARSAC

BARSAC AOC

The commune of Barsac is one of five that have the right to the Sauternes appellation. (The others are Preignac, Fargues, Bommes, and Sauternes itself.) Some generic wines sold in bulk may take advantage of this, but all individual properties are sold as Barsac. The wine must include overripe botrytised grapes harvested in *tries*.

WHITE Luscious, intensely sweet wines similar in style to Sauternes, but perhaps lighter in weight, slightly drier, and less rich. As in Sauternes, 1983 is one of the best vintages of the 20th century.

🍷 Sémillon, Sauvignon Blanc, Muscadelle

🍷 6–25 years for most wines; between 15–60 years for the greatest

CÉRONS AOC

These inexpensive wines from an area adjacent to Barsac are the best value-for-money sweet wines in Bordeaux. They must include overripe botrytised grapes harvested in *tries*.

WHITE Lighter than Barsac, but often just as luscious, the best of these wines can show true botrytis complexity.

🍷 Sémillon, Sauvignon Blanc, Muscadelle

🍷 6–15 years for most wines

GRAVES AOC

This appellation begins at the Jalle de Blanquefort, where the Médoc finishes and runs for 60 kilometres (37 miles) along the left bank of the Garonne. Almost two-thirds of the wine is red and is consistently high in quality and value.

RED I was brought up on the notion that with full maturity a Graves reveals itself through a certain earthiness of character. Experience has taught me the opposite. The biggest Graves from hot years can have a denseness that may combine with the smoky character of new oak to give the wine a roasted or tobacco-like complexity, but Graves is intrinsically clean. Its hallmark is its vivid fruit, clarity of style, silky texture, and hints of violets.

🍷 Cabernet Sauvignon, Cabernet Franc, Merlot Secondary grape varieties: Malbec, Petit Verdot

🍷 6–15 years

WHITE This is the disappointing half of the appellation: light- to full-bodied, from pure Sauvignon to pure Sémillon (with all proportions of blends in between, flabby to zingy, and unoaked to heavily oaked). Pay strict attention to the château profiles on the following pages. These wines may be sold from 1 December following the harvest without any mention of *primeur* or *nouveau*.

🍷 Sémillon, Sauvignon Blanc, Muscadelle

🍷 1–2 years for modest wines; 8–20 years for the best

GRAVES SUPÉRIEUR AOC

Some surprisingly good would-be Barsacs lurk under this appellation that is rarely seen yet accounts for more than a fifth of all white Graves produced.

WHITE This wine can be dry, but most is a sweet style, similar to Barsac.

🍷 Sémillon, Sauvignon Blanc, Muscadelle

🍷 6–15 years

PESSAC-LÉOGNAN AOC

Introduced in September 1987, this appellation covers the 10 best communes that have the right to the Graves AOC, and it is not by chance that it also encompasses 55 of the best estates, including all the *crus classés*. The technical requirements are similar to Graves except that the Carmenère may be used for red wines; white wines must contain at least 25 per cent Sauvignon Blanc and a slightly stricter yield. If you are not sure which château to buy in the Graves, it is worth remembering this appellation and paying a premium for it.

RED Soft, silky reds of great violety elegance, and not lacking either concentration or length. Most have been aged in a percentage of new oak, which adds a smoky or tobacco-like complexity.

🍷 Cabernet Sauvignon, Cabernet Franc, Merlot, Malbec, Petit Verdot, Carmenère

🍷 6–20 years

WHITE The serious styles are invariably oaked these days, with oodles of flavour, often tropical and fruity, with a firm acid structure. These wines may be sold from 1 December following the harvest without any mention of primeur or nouveau.

🍷 A minimum of 25% Sauvignon, plus Sémillon, Muscadelle

🍷 Usually 3–8 years, but up to 20 years for the best

SAUTERNES AOC

The much hillier communes of Bommes, Fargues, and Sauternes produce the richest of all Bordeaux's dessert wines, while the châteaux in the lower-lying, flatter Preignac make wines very close in style to Barsac. The wine must include overripe botrytised grapes harvested in *tries*.

WHITE Golden, intense, powerful, and complex wines that defy the senses and boggle the mind. They are rich in texture, with masses of rich, ripe, and fat fruit. Pineapple, peach, apricot, and strawberry are some of the lush flavours that can be found, and the creamy-vanilla character of fruit and new oak matures into a splendid honeyed sumptuousness that is spicy and complex. Above all, these wines are marked by the distinctive botrytis character.

🍷 Sémillon, Sauvignon Blanc, Muscadelle

🍷 10–30 years for most wines; between 20 and 70 years for the greatest

THE 17TH-CENTURY CHÂTEAU DE CÉRONS IN THE GRAVES REGION
Set in the heart of the village of Cérons, an imposing manor house situated on a terrace overlooking the Garonne River is home to this vineyard.

GRAVES AND CÉRONS

CHÂTEAU D'ARCHAMBEAU
Illats
★☆

Sited in Podensac, one of the communes of Cérons, this fine property is owned by Dr Jean Dubourdieu, nephew of Pierre Dubourdieu of Doisy-Daëne, a *deuxième cru* in Barsac. He produces a fine-quality, fragrant, and attractively aromatic red wine, which has the typical silky Graves texture. The deliciously fresh, crisp, and fruity dry white Graves is better than some efforts by certain *cru classé* châteaux. His soft, fruity Cérons is *moelleux* with the emphasis more on perfume than richness.

Other wines: *Château Mourlet, Château La Citadelle, Château Moulin de Ségurat*

CHÂTEAU LA BLANCHERIE
La Brède
Ⓞ★

This fresh and lively dry white Graves is cool fermented and has plenty of juicy fruit flavour balanced with ripe acidity. The red wine of La Blancherie is a medium- to full-bodied wine that is matured in casks and has an engaging, spicy bouquet and a rich, fruity flavour.

CLOS BOURGELAT
Cérons
AOC Cérons
★☆ Ⓥ

These botrytised wines have great aroma, finesse, and complexity. (Also produce dry white and red Graves)

CHÂTEAU BOUSCAUT
Cadaujac
Cru Classé (red and white)
★☆

Belongs to Laurent and Sophie Cogombles (née Lurton). The red wine is matured in wood for 18 months, with 25 per cent new oak. The white wine is fermented and matured for up to six months in 100 per cent new oak.

RED Until the 1980s this wine was big, tough, and tannic with little charm. Recent vintages have shown increasing suppleness, but the wine still struggles to find form. The second wine, Château Valoux, is a really excellent wine for its class.

🍷 *Merlot 55%, Cabernet Sauvignon 35%, Cabernet Franc 5%, Malbec 5%*

🍷 *8–20 years*

Second wine: *Château Valoux*

WHITE This dry, medium-bodied white wine has exotic fruit flavours supported by gentle oak.

🍷 *Sémillon 70%, Sauvignon 30%*

🍷 *5–10 years*

CHÂTEAU CARBONNIEUX
Léognan
Cru Classé (red and white)
★☆

This is the largest wine estate in Graves. The white wine, the better known of the two styles, is fermented and aged in oak barrels and vats with lees stirring.

🍷 *Sauvignon 60%, Sémillon 40%*

RED I frankly did not care for this wine until the splendid 1985 vintage, which seduced me with its creamy-oak nose, silky-textured fruit, and supple tannin.

🍷 *Cabernet Sauvignon 55%, Merlot 35%, Cabernet Franc Malbec 7%, and Petit Verdot 3%*

🍷 *6–18 years*

WHITE Once solid and uninspiring, this wine has really come into its own since the early 1990s. From this time Château Carbonnieux has been lush and creamy with well-integrated new oak and not a little finesse.

🍷 *Sauvignon 60%, Sémillon 40%*

🍷 *2–5 years*

Second wine: *Château La Tour Léognan, La Croix Carbonnieux*

CHÂTEAU LES CARMES-HAUT-BRION
Pessac
★

From 1584 until the French Revolution in 1789, this property belonged to the Carmelite White Friars, or Carmes, hence the name. This soft, Merlot-dominated wine has always been in the shadow of its more famous neighbour, Haut-Brion, and always will, but there has been a noticeable shift upwards in quality, especially since the purchase by Patrice Pichet in 2010 and his huge investment in vines and the cellar.

CHÂTEAU DE CÉRONS
Cérons
AOC Cérons
★ Ⓥ

Recently taken over by Caroline and Xavier Perromat, this beautiful historic property has undergone an amazing renewal. It is now producing elegant white and red Graves, as well as the consistently superb, white botrytised wines.

Other wines: *De Calvimont (dry white Graves, red Graves, and Cérons), Château du Mayne (Graves), La Quille, (Graves red and white and Bordeaux Rosé AOC)*

CHÂTEAU DE CHANTEGRIVE
Podensac
☆ Ⓥ

This château produces a substantial quantity of an excellent, soft, and fruity red Graves (50-50 Cabernet Sauvignon and Merlot) that is matured in casks for a further 12 months with 50 per cent new oak. It also produces an elegant, aromatic, cool-fermented dry white Graves entirely from the first pressing (Sémillon 50 per cent and Sauvignon 50 per cent). The proprietor also owns Château d'Anice. Smaller production of Cuvée Henri Leveque in red and the oak-fermented and aged Cuvée Caroline in white.

Other wines: *La Grive Rosé, La Rose Nouet*

DOMAINE DE CHEVALIER
Léognan
Cru Classé (red and white)
★★☆ Ⓥ

One of the top three Graves after Haut-Brion, this extraordinary property gives me more pleasure than any other in this AOC. It utilizes the most traditional methods to produce outstanding red and dry white wine. Fermenting red wine at a temperature as high as 32°C (89°F) might encourage some problems elsewhere, but under the meticulous care of those at the Domaine de Chevalier, this practice, designed to extract the maximum tannins and colouring material, is a positive advantage. The red wine is matured in wood for up to 24 months, with 40 to 60 per cent new oak. The white wine is fermented and matured in wood for 18 months, with up to 30 per cent new oak. *Vigneron* Stéphane Derenoncourt has been consulting on the red wines produced here since 2003.

RED Deep-coloured, medium-to-full or full-bodied wines, stunningly rich in fruit and oak, with intense cedarwood and tobacco overtones, yet subtle, seductive, and full of finesse. These are wines of great quality, longevity, and complexity.

🍷 *Cabernet Sauvignon 63%, Merlot 30%, Cabernet Franc 5%, Petit Verdot 2%*

🍷 *15–40 years*

WHITE Even better than the red, but produced in frustratingly small quantities, this star-bright, intensely flavoured dry wine is almost fat with exotic fruit and epitomizes finesse.

🍷 *Sauvignon 70%, Sémillon 30%*

🍷 *8–20 years*

Second wine: *L'Esprit de Chevalier*

CLOS FLORIDÈNE
Pujols-sur-Ciron
★★ Ⓥ

Owned by the family of the late Denis Dubourdieu, Bordeaux's white wine revolutionary, the property produces a sensational dry white Graves (Sémillon 70 per cent, Sauvignon 30 per cent) from this small estate. The red Clos Floridène (Cabernet Sauvignon 56 per cent, Merlot 43 per cent, Muscadelle 1 per cent) possesses an extraordinary combination of rich fruit and elegant new oak and is the equivalent of a top *cru classé*.

Second wine: *Drapeaux de Floridène*

CHÂTEAU COUHINS
Villenave-d'Ornon
Cru Classé (white only)

The Institut National de La Récherche Agronomique (INRA) and Lucien Lurton share this estate. INRA produces a separate wine, which is cool fermented with no maturation in wood. Château Couhins also produces a red Pessac-Léognan), but it is not a *cru classé*.

WHITE Clean, crisp, and fruity dry white wines that are well made.

🍷 *Sauvignon 90%, Sauvignon Gris 10%*

🍷 *2–4 years*

CHÂTEAU COUHINS-LURTON
Villenave-d'Ornon
Cru Classé (white only)
★☆ Ⓥ

The highest-performing half of the Couhins estate owned by André Lurton. The wine is fermented and matured in 100 per cent new oak.

WHITE Delicious dry wines that have all the advantages of freshness and fruitiness, plus the complexity of oak. Surprisingly fat for pure Sauvignon. Château also produces a red Pessac-Léognan; Cabernet Sauvignon25%, 75% Merlot, but it is not a *cru classé*.

🍷 *Sauvignon 100%*

🍷 *3–8 years*

CHÂTEAU DE CRUZEAU
St-Médard-d'Eyrans
☆ Ⓥ

Situated on a high, south-facing crest of deep, gravel soil, this property belongs to André Lurton, owner of Château Couhins-Lurton, the high-performance white Graves *cru classé*. De Cruzeau makes 18,000 cases of full-bodied red Graves (Cabernet Sauvignon 55 per cent, Merlot 45 per cent) that is ripe and velvety with a spicy-cedarwood complexity. This château also produces around 5,000 cases of a fine-quality white Pessac-Léognan (Sauvignon 100 per cent) that after some five years of maturation develops an intense citrus bouquet and flavour.

CHÂTEAU FERRANDE
Castres

A large property that, like so many in Graves, makes better red wine than white. The red wine (Cabernet Sauvignon 35 per cent, Merlot 35 per cent, Cabernet Franc 30 per cent) is a consistently good-quality, chocolaty Graves that is matured in wood for 15 to 18 months, with 10 to 15 per cent new oak. The dry white Graves (Sémillon 60 per cent, Sauvignon 35 per cent, Muscadelle 5 per cent) is somewhat less inspiring.

CHÂTEAU DE FIEUZAL
Léognan
Cru Classé (red only)
★★

This property occupies the highest and best exposed gravel crest in the commune. The vineyard and the château are immaculate, which is reflected in the style of its wines.

RED A deeply coloured, full-bodied, rich, stylish wine with typical Graves silky texture and ample finesse. De Fieuzal also produces a rich, exotic, and oaky dry white wine that is not *cru classé*, yet is one of the finest white Graves (Pessac-Léognan) produced.

Cabernet Sauvignon 42%, Merlot 50%, Petit Verdot 5 %, Cabernet Franc 3%

12–30 years

Second wine: *L'Abeille de Fieuzal*

GRAND ENCLOS DU CHÂTEAU DE CÉRONS

Cérons

★☆ V

Although historically part of Château de Cérons, this is not under the same ownership. The original estate belonged to the Marquis de Calvimont, but was split in two by the route from Bordeaux to Spain, which was constructed in 1875. The marquis then sold the property in three separate lots, one of which was called Grand Enclos. Today the property belongs to Giorgio Cavana. The white wines of Grand Enclos are equally as rich and potentially as complex as those of Château de Cérons itself (which is so-called because it retains the marquis' château). The proprietor also makes a dry white wine and a red Graves at nearby Château Lamouroux.

CHÂTEAU HAURA

Illats

Another property from the Denis Dubourdieu stable. Produces an excellent red Graves and a tiny amount of AOC Cerons.

CHÂTEAU HAUT-BAILLY

Léognan

Cru Classé (red only)

★★☆ V

This château's well-kept vineyard is located in an excellent gravel crest bordering the eastern suburbs of Léognan. The late US banker Bob Wilmers purchased it in 1998, and he kept the same team on, notably Veronique Saunders, the granddaughter of the previous owners. Consistent investment in vines and cellars brought this wine up to a level its *terroir* deserves. Unusually for Pessac-Léognan, the property only produces red wine. This red Graves is matured in wood for up to 18 months, with 50 per cent new oak.

RED The class of fruit and quality of new oak is immediately noticeable on the creamy-ripe nose of this wine. Always an elegant and stylish expression of Pessac-Léognan.

Cabernet Sauvignon 60%, Merlot 34%, Cabernet Franc 3%, Petit Verdot 3%

12–25 years

Second wine: *Le Parde Haut-Bailly* (since 2018 this wine has been rechristened Haut-Bailly II and the third wine, previously labelled Pessac-Léognan de Haut-Bailly as HB de Haut-Bailly)

Other wines: *Château Le Pape, Rosé de Haut-Bailly (certain vintages)*

CHÂTEAU HAUT-BRION

Pessac

Cru Classé (red and white)

★★★

In 1663 this famous château was mentioned in Pepys's diary as "Ho Bryan". It

has been under American ownership since 1935, when it was purchased by banker Clarence Dillon. The parent company is called Domaine Clarence Dillon, and Dillon's great-grandson Prince Robert of Luxembourg is the president. Jean-Philippe Delmas is the technical director. The red wine is fermented in stainless steel and matured in wood for 24 to 27 months, with 100 per cent new oak. The white wine is fermented and matured in 100 per cent new oak. The second wine used to be Bahans Haut-Brion but is now sold as Le Clarence Haut-Brion.

RED This supple, stylish, medium- to full-bodied wine has a surprisingly dense flavour for the weight and a chocolaty-violet character. The ideal commercial product, it develops quickly and ages gracefully.

Cabernet Sauvignon 44%, Merlot 45%, Cabernet Franc 9%

10–40 years

Second wine: *Le Clarence Haut-Brion*

WHITE This is not one of the biggest white Graves, but it is built to last. It is sumptuous, oaky, and teeming with citrus and exotic fruit flavours.

Sauvignon 48%, Sémillon 52%

5–20 years

Second wine: *Le Clarté de Haut-Brion*

CHÂTEAU LANDIRAS

Landiras

Under the ownership of Michel Pélissié since 2007. The property produces a red and a white Graves under the label Château Landiras and a *cuvée* Jeanne de Lestonnac.

Other wines: *Château La Ouarde, Château Peyron Bouché*

CHÂTEAU LARRIVET-HAUT-BRION

Léognan

★★

Originally called Château Canolle, the name was at one point changed to Château Haut-Brion-Larrivet. Larrivet is a small stream that flows through the property, and *Haut-Brion* means "high gravel", referring to the gravel plateau west of Léognan on which the vineyard is situated. Château Haut-Brion took legal action over the re-naming, and since 1941 the property and its wines have been known as Château Larrivet-Haut-Brion. The red wine (Cabernet Sauvignon 50 per cent, Merlot 45 per cent, Cabernet Franc 5 per cent), which is matured in wood for 18 months with 25 per cent new oak, is certainly *cru classé* standard, being a well-coloured and full-bodied Graves with good flavour, spicy-cedarwood undertones, and a firm tannic structure. The white wine (Sauvignon Blanc 80 per cent, Sémillon 20 per cent) has leapt in quality since 1996.

CHÂTEAU LAVILLE-HAUT-BRION

Talence

Cru Classé (white only)

★★

Since 1983, this small vineyard has been owned by Clarence Dillon, American proprietor of Château Haut-Brion. This

"château with no château" is thought of as the white wine of La Mission. The wine is fermented and matured in cask. Since the 2009 vintage this label has become extinct, and the wine is sold as Château La Mission-Haut-Brion Blanc, but the profile remains for readers who encounter vintages in this name.

WHITE Until 1982 the style was full, rich, oaky, and exuberant, tending to be more honeyed and spicy with a floral finesse since 1983. Both styles are stunning and complex.

Sémillon 85%, Sauvignon 14%, Muscadelle 1%

6–20 years

Other wines: *It shares a second wine with Château Haut-Brionn: la Clarté de Haut-Brionn*

CHÂTEAU LA LOUVIÈRE

Léognan

★★ V

Part of André Lurton's Graves empire, this château has made a smart about-turn since 1985 as far as the quality of its red wine goes. A string of dull, lifeless vintages has come to an end with the beautiful, deep, and vividly coloured wines of the years 1985 and 1986. There was another step up in quality in the mid-1990s, since when this has been a truly splendid, full-bodied red Graves that is rich in spicy-blackcurrant fruit and new oak (Cabernet Sauvignon 60 per cent, Merlot 40 per cent). The 100 per cent Sauvignon Blanc white wines of Château La Louvière have always been excellent, but even here there has been a gigantic leap in quality. These are exciting and complex wines that deserve to be among the very best *crus classés*.

Second wine: *"L" de Louvière (white and red and in certain vintages : 1989, 1996, 2003, and 2010), Les Lions de Château La Louvière (a semi-sweet white Graves Supérieur)*

CHÂTEAU MAGENCE

St-Pierre-de-Mons

☆ V

A good property making 5,000 cases of a supple, well-perfumed, red wine (Cabernet Sauvignon 50 per cent, Cabernet Franc 5 per cent, Merlot 45 per cent) and 10,000 cases of attractive, aromatic, cool-fermented dry white Graves (Sauvignon 80 per cent, Sémillon 20 per cent).

CHÂTEAU MALARTIC-LAGRAVIÈRE

Léognan

Cru Classé (red and white)

★★

This 50-hectare (124-acre) vineyard forms a single block around the château. Purchased by the Bonnie family in 1996 it is enjoying a well-deserved resurgence consistently producing much higher-quality wines since the 1980s. The red wine is fermented in a mixture of stainless steel and oak vats at a low temperature (16°C/61°F), and matured in wood for 20 to 22 months, with 50 per cent new oak. The white wine is matured in 50 per cent new oak for 7 to 8 months.

RED Rich, garnet-coloured with an opu-

lent sweet-oak nose, penetrating flavour, and supple tannin structure.

Cabernet Sauvignon 45%, Merlot 45%, Cabernet Franc 8%, Petit Verdot 2%

7–25 years

WHITE Barrel fermented and aged on their lees, the wines have a balance of fresh minerality and citrus notes.

Sauvignon 80%, Sémillon 20%

5–12 years

Second wine: *La Réserve de Malartic (red and white)*

CHÂTEAU LA MISSION-HAUT-BRION

Pessac

Cru Classé (red only)

★★☆

Under the ownership of Henri Woltner, this was the pretender to the throne of Graves. Little wonder, then, that Clarence Dillon of Haut-Brion snapped it up when the opportunity arose in 1983. The red wine is matured in wood for 20 to 24 months, with 100 per cent new oak.

RED Despite different winemaking techniques, Dillon's La Mission is no less stunning than Woltner's. Both styles are deeper, darker, and denser than any other wine Graves can manage. They are essentially powerful yet elegant wines that merit some bottle-age.

Cabernet Sauvignon 45%, Merlot 45%, Cabernet Franc 10%

15–45 years

WHITE This was sold as Château Laville-Haut-Brion until 2009, when it metamorphosed into this château's white sibling – and what a vintage to choose for the change, effectively doubling the price of this wonderful but already extremely expensive wine. Clear-cut floral minerality at the front of the palate; deep, rich, and honeyed on the finish.

Second wine: *La Chapelle de la Mission-Haut-Brion (red), La Clarté (white), which is a joint second white wine with Château Haut-Brion White*

CHÂTEAU OLIVIER

Léognan

Cru Classé (red and white)

☆

In the hands of the de Bethmann family since the 1800s, this beautiful moated medieval château sits on some of the best *terroir* of Pessac-Léognan. Under the new management of Laurent Lebrun, new plots have been planted and a new cellar created and the improvement in the wine is palpable. The red wine is matured in wood for 12 months in one-third new oak; the white wine up to 10 months, with 35 per cent new oak.

RED The fruit is now easier-drinking and the oak, which used to be aggressive, more supple and creamy.

Cabernet Sauvignon 60%, Merlot 40%

WHITE This wine began to sparkle as early as 1985, with some quite outstanding vintages in the 1990s, since when has been an added freshness, real fruit flavour, and some positive character developing.

HAND-PICKING BOTRYTISED GRAPES IN THE SAUTERNES REGION
This warm and wet region is famous for its susceptibility to "noble rot", essential for the production of its world-famous sweet wines.

🍷 *Sémillon 75%, Sauvignon 23%, Muscadelle 2%*

🍷 *3-7 years*

Second wine: *Le Dauphin d'Olivier*

CHÂTEAU PAPE-CLÉMENT
Pessac
Cru Classé (red only)
★★

After a disastrous period in the 1970s and early 1980s, Pape-Clément began to improve in 1985 and 1986, due to stricter selection of the *grand vin* and the introduction of a second wine. Some critics rate these two vintages highly, and they were very good wines, but when examined in the context of the enormous potential of this vineyard, my brain tells me they were not at all special, even if my heart wants them to be. The trio of 1988, 1989, and 1990 wines turned out to be the best this château has produced since 1953, although they are nowhere near as great and still not special at the very highest level of Graves wine. However,

even in the string of lesser vintages Bordeaux experienced in the early 1990s, Pape-Clément managed to produce good wines, and with 1995, 1996, 1998, and 1999, it has truly regained the reputation of its former glory years. The red wine from this château is matured in wood for 18 months, with a minimum of 70 per cent new oak.

RED Medium-bodied wines of excellent deep colour, a distinctive style, and capable of much finesse.

🍷 *Cabernet Sauvignon 55%, Merlot 40%, Cabernet Franc 5%*

Second wine: *Le Clémentin (white and red)*

Note *This château also produces a little non-cru classé white Graves, made from Sémillon 40%, Sauvignon 50%, and Muscadelle 10%. Vinified and aged for 16 months in oak barrels on the lees except for 15% aged in concrete eggs.*

CHÂTEAU RAHOUL
Portets

The wine produced by Château Rahoul is not quite as exciting as it was in the 1980s,

when the property was home to – although never owned by – the maestro Peter Vinding-Diers (*see* Château Landiras). Now under the ownership and management of Bordeaux merchant Dourthe, both red and white wines are still reliable sources of very good value oak-aged Graves.

Second wine: *L'Orangerie de Rahoul*
Other wines: *Château la Garance*

CHÂTEAU RESPIDE-MÉDEVILLE
Toulenne
★ ⓥ

Following in the footsteps of her father, Christian, Julie Médeville and her husband, Xavier, continue to produce fine examples of red and the white Graves using the best of modern vinification combined with new oak. The red is a well-coloured wine with rich, ripe fruit, some spice, and a creamy, new-oak aftertaste, good for early drinking. The white is a rich, creamy-vanilla concoction with soft, succulent fruit and a fat finish.

CHÂTEAU DU ROCHEMORIN
Martillac
☆ ⓥ

Originally called La Roche Morine ("the Moorish rock"), this estate has a history that extends at least as far back as the eighth century when Bordeaux was defended by the Moors from attacking Saracens. Another château belonging to André Lurton, Rochemorin produces a fine, elegant, fruity red Graves that is well balanced and has a good spicy finish (Cabernet Sauvignon 65 per cent, Merlot 35 per cent). Rochemorin also produces a very clean dry white 100 per cent Sauvignon Blanc Pessac-Léognan

Other wines: *Château Coucheroy (white and red)*

CHÂTEAU DE ROQUETAILLADE LA GRANGE
Mazères
★☆ ⓥ

This is a very old property, that produces some 12,000 cases of an attractive, well-coloured red Graves that has an aromatic bouquet and a delicious spicy-cassis flavour. This wine is made from Merlot 40 per cent, Cabernet Sauvignon 25 per cent, Cabernet Franc 25 per cent, Malbec 5 per cent, and Petit Verdot 5 per cent. Its firm, tannic structure means it matures gracefully over 15 or more years. The white wine, which is made from Sémillon 80 per cent, Sauvignon 20 per cent, is less successful.

Second wine: *Château de Carolle*
Other wines: *Château de Roquetaillade-le-Bernet*

Note *The Guignard family now own eight vineyards across the Graves, including these three.*

CHÂTEAU DU SEUIL
Cérons
ⓞ

Up-and-coming Graves property producing fine, elegant reds and fruity, oak-aged whites, both proving to be of

increasingly excellent value. In certified organic production since 2012.

Other wines: *Heritage du Seuil Graves (red and white), Château du Seuil Cerons, Domaine du Seuil (in Cotes de Bordeaux and in Bordeaux Blanc)*

CHÂTEAU SMITH-HAUT-LAFITTE
Martillac
Cru Classé (red only)
★☆ ⓥ

The reputation of these wines began to soar under consultant Michel Rolland, but the quality has stepped up another gear since 2001, when Stéphane Derenoncourt took over. The red wine matures in wood for 18 months, with 50 per cent new oak.

RED These wines are now in a richer style with creamy-oak undertones and up-front fruit.

🍷 *Cabernet Sauvignon 65%, Merlot 30%, Cabernet Franc 4%, Petit Verdot 1%*

🍷 *8-20 years*

Second wine: *Les Hauts-de-Smith-Haut-Lafitte*
Other wines: *Le Petit Haut Lafitte*

Note *A white Pessac-Léognan is also made. Sauvignon Blanc 90%, Sauvignon Gris 5%, Sémillon 5% barrel-fermented and aged on the lees in 50% new oak for 12 months. It is not a cru classé, yet ironically now considered one of the finest white wines in Pessac-Léognan.*

CHÂTEAU LA TOUR-HAUT-BRION
Talence
Cru Classé (red only)

The final vintage under this label was in 2005 and the wine is now included in the production of Château la Mission Haut-Brionn and its second wine La Chapelle de La Mission Haut-Brionn.

CHÂTEAU LA TOUR-MARTILLAC
Martillac
Cru Classé (red and white)
ⓞ★☆

The estate takes its name from the 12th-century tower that stands in the main courtyard of the château. Its red wine is matured in wood for 16 months with 40% new oak. The white is fermented in oak, and matured for 15 months on the lees, 11 of which in oak.

RED Not big or bold wines with immediate appeal; the reds are elegant with some finesse. The fruit in recent vintages has tended to be a bit plumper, but in bottle these wines develop creamy-oak flavour.

🍷 *Cabernet Sauvignon 55%, Merlot 40%, Petit Verdot 5%*

🍷 *8-20 years*

Second wine: *La Grave-Martillac*
Other wines: *Lacrox-Martillac, Château Langlet*

WHITE The stunning 1986 vintage heralded a new era of exciting dry whites. This is very fresh, elegant wine, the fruit gently balanced by complex nuances of oak.

🍷 *Sémillon 40 %, Sauvignon 60%*

🍷 *4-8 years*

THE WINE PRODUCERS OF
SAUTERNES AND BARSAC

CHÂTEAU D'ARCHE

Sauternes
2ème Cru Classé
★

This property dates from 1530. It was known as Cru de Bran-Eyre until it was bought by the Comte d'Arche in the 18th century. It has been inconsistent. The wine sees up to 50 per cent new oak.

WHITE The successful Château d'Arche is an elegantly balanced wine that is more in the style of Barsac than Sauternes. It is sweet, rich, and has complex botrytis flavours, which often puts it on par with a *premier cru,* although it is less plump than most Sauternes. Easily equivalent to a classed growth in quality, and the Crème de Tête is even better.

🍷 *Sémillon 80%, Sauvignon 15%, Muscadelle 5%*

🍶 *8–25 years*

Second wine: *Prieuré d'Arche*

Other wines: *d'Arche-Lafaurie (dry white)*

CHÂTEAU BASTOR-LAMONTAGNE

Preignac
☆ V

A large property that deserves *deuxième cru* status. The wine is matured in wood for up to 16 months, with 30 per cent new oak. Lighter years such as 1980, 1982, and 1985 lack botrytis but are successful in an attractive mellow, citrus style. Big years such as 1983 lack nothing: the wines are full, rich, and stylish with concentrated botrytis flavour and ample class. The years 1989, 1990, 1996, 1997, 1998, 1999, and 2001 all very successful.

Second wine: *Les Remparts du Bastor-Lamontagne*

CHÂTEAU BROUSTET

Barsac
2ème Cru Classé
★ V

The wine produced at Château Broustet is matured in wood for 22 months in 40 per cent new oak.

WHITE Château Broustet can be a delightful wine, with a fruit-salad-and-cream taste, a very elegant balance, and some spicy-botrytis complexity.

🍷 *Sémillon 80%, Sauvignon 15%, Muscadelle 5%*

🍶 *8–25 years*

Second wine: *Château de Ségur*

CHÂTEAU CAILLOU

Barsac
2ème Cru Classé
★☆ V

This château gets its name from the *cailloux,* the stones that are brought to the surface during ploughing. These stones have been used to enclose the entire 15-hectare (37-acre) vineyard. This is not one of the better-known *deuxièmes crus,* but it consistently produces wines of a very high standard, and so deserves to be.

WHITE The white wine is a rich, ripe, and spicy-sweet Barsac with concentrated botrytis flavours underscored by refined oak. It is not the fattest of Barsacs, but made in the richer rather than lighter style.

🍷 *Sémillon 90%, Sauvignon 10%*

🍶 *8–30 years*

Second wine: *Les Tourelles, Les Erables*

Other wines: *Le Vin Sec du Caillou Sec (dry white)*

CHÂTEAU DE LA CHARTREUSE

Preignac
★ V

This is the same stunning wine as Château Saint-Amande, but under a different exclusive label. *See also* Château Saint-Amande.

CHÂTEAU CLIMENS

Barsac
1er Cru Classé
★★☆

Under the ownership of Bérénice Lurton, this property has long been considered one of the top wines of both appellations. The wine is matured in wood for 20 to 24 months with up to 35 to 45 per cent new oak. In 2010 the whole property was converted to biodynamic viticulture.

WHITE The fattest of Barsacs, yet its superb acidity and characteristic citrus style give it an amazingly fresh and zippy balance. This wine has masses of creamy-ripe botrytised fruit supported by good cinnamon and vanilla-oak flavours.

🍷 *Sémillon 100%*

🍶 *10–40 years*

Second wine: *Les Cyprès de Climens*

CHÂTEAU CLOS HAUT-PEYRAGUEY

Bommes
1er Cru Classé
★★

Bernard Margez added this property, originally part of Château Lafaurie-Peyraguey, to his portfolio in 2012. The wine is fermented and matured in wood for 18 months, with up to 25 per cent new oak.

WHITE This wine now flaunts a positively eloquent bouquet and has a rich flavour with complex botrytis creamy-oak nuances – very stylish.

🍷 *Sémillon 95%, Sauvignon 5%*

🍶 *8–25 years*

Second wine: *La Symphonie de Haut Peyraguey, Château Haut-Bommes*

CHÂTEAU COUTET

Barsac
1er Cru Classé
★★

This château is usually rated a close second to Climens, but is capable of matching it in some vintages and its occasional production of tiny quantities of *tête de cuvée* called "Cuvée Madame" often surpasses it. It is fermented and matured for 18 months in cask with 100 per cent new oak.

WHITE This wine has a creamy vanilla-and-spice bouquet, an initially delicate richness that builds on the palate, good botrytis character, and oaky fruit.

🍷 *Sémillon 75%, Sauvignon 23%, Muscadelle 2%*

🍶 *8–25 years (15–40 years for Cuvée Madame)*

Second wine: *La Chartreuse de Coutet*

Other wines: *Cuvée Madame de Château Coutet (the Reserve wine), Opalie de Château Coutet (dry white)*

CHÂTEAU DOISY-DAËNE

Barsac
2ème Cru Classé
★★ V

Managed by Fabrice and Jean-Jacques, sons of the late Denis Dubourdieu. They ferment this wine in oak until the desired balance of alcohol and sweetness is achieved, and then mature it in 25 per cent new oak for 10 months before finishing in stainless steel for 9 months. The wine also undergoes various low-sulphur techniques. The result is a wine equal to a Barsac *premier cru.*

WHITE This is a wine of great floral freshness and elegance, with a delightful honeyed fragrance of deliciously sweet fruit, delicate botrytis character, hints of creamy oak, and perfect balance.

🍷 *Sémillon 86%, Sauvignon 14%*

🍶 *8–20 years*

Other wines: *L'Extravagant (certain vintages), Le Grand Vin Sec de Doisy-Daëne*

CHÂTEAU DOISY-VÉDRINES

Barsac
2ème Cru Classé
★☆ V

This is the original and largest of the three Doisy châteaux. It is owned by the Castéja family, who also own Bordeaux *négociant* house Joanne. The wine is matured in wood for 18 months with one-third new oak.

WHITE This wine was somewhat lacklustre until 1983, since when it has explod-

AN AERIAL VIEW OF CHÂTEAU DE MALLE GIVES A GLIMPSE OF IT EXTENSIVE HOLDINGS
With vineyards spanning both the Sauternes and the Graves AOCs, this estate produce Sauternes whites and Graves whites and reds.

ed with character. Rich, ripe, and oaky, with a concentrated botrytis complexity.

 Sémillon 80%, Sauvignon 15%, Muscadelle 5%

🍷 8–25 years

Second wine: *Château Petit Védrines*

CHÂTEAU DE FARGUES
Fargues
★★

The eerie ruin of the 14th-century Château de Fargues is the ancestral home of the Lur-Saluces family and is now under renovation. The small production of ultra-high-quality wine is made by essentially the same fastidious methods as Yquem, including fermentation and maturation in 50 per cent new oak (depending on the vintage). It is powerful and viscous, very rich, succulent, and complex, with a fat, toasty character (Sémillon 80 per cent, Sauvignon 20 per cent). Easily equivalent to a classed growth.

CHÂTEAU FILHOT
Sauternes
2ème Cru Classé
☆

The beautiful Château Filhot was built between 1780 and 1850. This splendid château has a potentially great vineyard, which has not lived up to its potential. In 1995 the cellar was modernised introducing temperature control and 20 to 24 months of mainly oak ageing with 30 per cent new oak bringing more complexity and freshness to the wines.

WHITE At best these are well-made, simply fruity, and sweet.

🍷 Sémillon 60%, Sauvignon 36%, Muscadelle 4%

Other wines: *Gold Reserve*

CHÂTEAU GILETTE
Preignac
★

Julie Gonet Medeville and her husband continue the "Antiquaire de Sauternes" tradition of her father, Christian Médeville, ageing the wines of the property (made from Sémillon 90 per cent, Sauvignon 8 per cent, Muscadelle 2 per cent) in cement vats under anaerobic conditions for an amazing 20 years before bottling and selling it. The Crème de Tête is *premier cru* quality with a powerful bouquet and intense flavour of liquorice and peaches and cream, followed by a long barley-sugar aftertaste. The Crème de Tête deserves ★★, but I am less impressed with Château Gilette's regular bottlings (if, indeed, any bottling at this property can be so described!).

CHÂTEAU GUIRAUD
Sauternes
1er Cru Classé
★★ⓓ

This property has been on the up since 1981, when the Narby family of Canada purchased it, replanting, re-equipping the winery, and renovating the château. Only Yquem is on as high ground as Guiraud, and as drainage is a key factor affecting the quality of the greatest Sauternes, where heavy clay soils domi-

nate, the potential for this wine is very exciting. Under the majority ownership of Robert Peugeot since 2006 with co-owners Olivier Bernard of Domaine de Chevalier and Stephan von Neipperg of Château Canon-la-Gaffelière, and estate manager Xavier Planty, the property has been under organic agriculture since 2011. The wine is matured in wood for 18 to 24 months with around 90 per cent new oak.

WHITE After two dismal decades, great Sauternes arrived at this château with the classic 1983 vintage, the first true botrytis wine under Narby's ownership. Guiraud is now plump with Sémillon fruit and fat with botrytis character. It also has the highest percentages of Sauvugnon amongst the *crus classés* of Sauternes and Barsac. A deliciously sweet wine with luxuriant new oak, complexity, and considerable finesse.

🍷 Sémillon 65%, Sauvignon 35% (one of the highest percentages of Sauvignon amongst the Crus Classés of Sauternes and Barsac)

🍷 12–35 years

Second wine: *Le Petit Guiraud (since 2016)*
Other wines: *Vin Blanc Sec "G" (dry white)*

CHÂTEAU HAUT-BOMMES
Bommes

Part of the Clos Haut-Peyraguey estate, owned by Bernard Magrez since 2012. Just five hectares (12 acres) of 95 per cent Sémillon and 5 per cent Sauvignon barrel fermented and aged in 50 per cent new oak.

CHÂTEAU LES JUSTICES
Preignac
★ ⓥ

Under the same ownership as the star-performing Château Gilette, a tiny 8.5-hectare (21-acre) property, near the mist-giving Ciron river. Here the wines enjoy only two years ageing in vats. Les Justices is a consistent wine of excellent quality that is riper and fruitier than Gilette and the equivalent of a *deuxième cru*.

CHÂTEAU LAFAURIE-PEYRAGUEY
Bommes
1er Cru Classé
★★☆ ⓓ

In 2014 Silvio Denz, the Swiss businessman who also owns Château Faugères and Château Péby Faugères in Saint-Émilion, purchased this property. He has invested in the vineyard and cellar to bring this property back to its first growth status. He has also converted the château into a five-star Lalique hotel, a boost for the Sauternes region. Its wine is fermented and matured in wood from 18 to 20 months in between 40 to 70 per cent new oak, depending upon the vintage.

WHITE The combination of botrytis and oak is like pineapples and peaches and cream in this elegant wine that keeps fresh and retains an incredibly light colour in old age.

🍷 Sémillon 93%, Sauvignon 6%, Muscadelle 1%

🍷 8–30 years

CHÂTEAU LAMOTHE-DESPUJOLS
Sauternes
2ème Cru Classé

In 1961 the Lamothe vineyard was split in two. The section belonging to Jean Despujols has been the most disappointing half up until the 1985 vintage, but it has really come into its own since 1990.

WHITE Fuller, richer, and sweeter than previously expected, and in an oily, fuller-bodied style, with overtly attractive tropical fruit character.

🍷 Sémillon 85%, Sauvignon 10%, Muscadelle 5%

CHÂTEAU LAMOTHE-GUIGNARD
Sauternes
2ème Cru Classé
☆ ⓥ

The Guignards are really trying to achieve something with their section of the Lamothe vineyard. The wine is matured in wood for 24 months with 20 per cent new oak.

WHITE Rich, spicy, and concentrated wines of full body and good botrytis character.

🍷 Sémillon 90%, Muscadelle and Sauvignon 10%

🍷 7–20 years

CHÂTEAU LIOT
Barsac
☆ ⓥ

This wine is elegant, with light but fine botrytis character and the creamy vanilla of new oak – probably the equivalent of a *deuxième cru* in quality and is excellent value. Owner Jean-Nicol David also produces at Château Saint-Jean des Graves a dry white Graves, a Graves Supérieures, and a fruity red Graves.

Second wine: *Château du Levant*

CHÂTEAU DE MALLE
Preignac
2ème Cru Classé

The vineyard is spread across on the Sauternes and Graves appellations and produces white and red Graves, as well as Sauternes. This vineyard does not shine every year, but when it does, it can be superb value.

WHITE These are firm, well-concentrated wines often influenced more by *passerillage* than botrytis. Delicious, rich, and luscious.

🍷 Sémillon 75%, Sauvignon 22%, Muscadelle 3%

🍷 7–20 years

Second wine: *Les Fleurs du Malle*
Other wines: *M de Malle (dry white Graves), Château Pessan (Graves)*

CHÂTEAU MÉNOTA
Barsac
ⓥ

This quaint old property – with its historic towers and ramparts – has exported its wines to England since the 16th century. Château Ménota produces very good

Barsac, despite the unusually high proportion of Sauvignon Blanc (60 per cent).

CHÂTEAU NAIRAC
Barsac
2ème Cru Classé
★☆

Tom Heeter established the practice of fermenting and maturing his wine in up to 100 per cent new oak – Nevers for vanilla and Limousin for backbone – and his perfectionist ex-wife, Nicole Tari, has continued this format with great success. The property is now run by their son Nicolas, who continues to improve the quality.

WHITE These are rich and oaky wines that require ample ageing to show true finesse. With enough bottle maturity, the tannin and vanilla harmonize with the fruit, and the rich botrytis complexity emerges.

🍷 Sémillon 90%, Sauvignon 6%, Muscadelle 4%

🍷 8–25 years

CHÂTEAU RABAUD-PROMIS
Bommes
1er Cru Classé
★☆

The wines of this once-grand property used to be awful. It was sad to see the vineyard, château, and wine so neglected. What a joy to witness such a dramatic change. It began with the 1983; and the vintages are now something special.

WHITE A lovely gold-coloured wine with full, fat, and ripe botrytis character on the bouquet and palate.

🍷 Sémillon 80%, Sauvignon 18%, Muscadelle 2%

🍷 8–25 years

CHÂTEAU RAYMOND-LAFON
Sauternes

Owner Pierre Meslier was *régisseur* at Yquem. In the mid-1980s the wine was overrated and overpriced, and although things improved since the 1989 and 1990 vintages, it still performed at only *deuxième cru* level and was still overrated and overpriced. Now that the style has been cleaned up and plumped out, Raymond-Lafon (Sémillon 80 per cent, Sauvignon 20 per cent) is a nice Sauternes, but not superior to Rieussec or Climens.

CHÂTEAU RAYNE-VIGNEAU
Bommes
1er Cru Classé
★

The quality of Rayne-Vigneau had plummeted to dismal depths until 1985. It has a higher Sémillon content than the statistics would suggest, due to the 5,000 cases of dry Sauvignon Blanc that are sold as "Rayne Sec". This property is considered by many to be on some on the finest *terroir* of the appellation. After years of underperformance, the property was purchased in by the Credit Agricole, who invested heavily. Now owned by Tresor du Patrimoine Group, the investment and resulting improvement in quality continue. The wine is now matured in wood for 24 months with 50 per cent new oak.

WHITE Château Rayne-Vigneau is now a very high-quality wine that has an elegant peachy ripeness to its botrytis character.

🍷 Sémillon 74%, Sauvignon 24%, Muscadelle 2%

🍷 8–25 years

Second wine: Madame de Rayne

Other wines: Gold de Rayne Vigneau (100% Sémillon, in certain vintages), Le Sec de Rayne Vigneau (100% dry Sauvignon, in certain vintages)

CHÂTEAU RIEUSSEC
Fargues
1er Cru Classé
★★☆

This fine property has increased in quality since its acquisition by Domaines Rothschild in 1984. They have reduced yields and increased barrel ageing. None of the Sauvignon produced here is used for Château Rieussec (it goes in the "R"), effectively making the wine 96 per cent Sémillon. It is barrel-fermented and cask-matured for 18 to 26 months with 50 per cent new oak.

WHITE One of the richest and most opulent of Sauternes, with intense pineapple fruit and a heavily botrytised character.

🍷 Sémillon 90%, Sauvignon 7%, Muscadelle 3%

🍷 12–35 years

Second wine: Carmes de Rieussec

Other wines: "R" de Château Rieussec (dry white)

CHÂTEAU DE ROLLAND
Barsac
☆

The 15th-century chateau of this vineyard is one of the oldest in Barsac.

WHITE Fresh and elegant, with an emphasis on fruit. The Guignard family, who also own the excellent Château de Roquetaillade-La-Grange at Mazères in Graves, produces an improving Barsac.

🍷 Sémillon 80%, Sauvignon 15%, Muscadelle 5%

🍷 5–8 years

Second wine: Comte de Rolland

Other wines: Château de Rolland (Graves dry white, Bordeaux dry white, and red Graves), Château de Rolland Les Allouettes (Graves dry white), Château de Rolland Les Astérides (red Bordeaux), Rubis d'Automne (100% Merlot Rosé)

CHÂTEAU ROMER
Fargues
2ème Cru Classé

The original Romer estate was divided in 1881, and at just 5 hectares (13 acres) this is the smallest part. I have never come across the wine. Owned and managed by the Francois Janoueix group.

🍷 Sémillon 100%

CHÂTEAU ROMER-DU-HAYOT
Fargues
2ème Cru Classé

Monsieur André du Hayot owns these 10 hectares (25 acres) of vines on a fine clayey-gravel crest that was once part of the original Romer estate. The wines are little seen, but represent very good value.

WHITE Fresh, not oversweet, fruit-salad and cream style, with light botrytis character and an elegant balance.

🍷 Sémillon 70%, Sauvignon 25%, Muscadelle 5%

🍷 5–12 years

CHÂTEAU ROUMIEU
Barsac
★

This property, which borders the classified growths of Climens and Doisy-Védrines, has produced luscious sweet wines of a richer than normal style in some vintages

🍷 Sémillon 89%, Sauvignon 10%, Muscadelle 1%

CHÂTEAU ROUMIEU-LACOSTE
Barsac
★

Hervé Dubourdieu makes a consistently fine Barsac here (Sémillon 80 per cent, Sauvignon 20 per cent) with good botrytis concentration.

CHÂTEAU SAINT-AMANDE
Preignac
★

An elegant and stylish wine (Sémillon 67 per cent, Sauvignon 33 per cent) that is very attractive when young, yet some vintages have potentially excellent longevity and are often equivalent to a classed growth in quality. Part of the production of this property is sold under the Château de la Chartreuse label.

Second wine: Château de la Chartreuse

CHÂTEAU SIGALAS-RABAUD
Bommes
1er Cru Classé
★☆

This 14-hectare (35-acre) property is the jewel of the original estate. Owned by the de Lambert family since the 19th century and now co-owned and managed by Laure de Lambert Compeyrot, the sixth generation. Under her guidance, the quality of the wine continues to improve, spending 16 to 24 months ageing in new oak. She has introduced two dry white wines and a non-sulphured sweet wine.

WHITE A stylish wine with an elegant botrytis bouquet and deliciously fresh fruit on the palate.

🍷 Sémillon 86%, Sauvignon 14%

🍷 6–15 years

Second wine: Le Lieutenant de Sigalas

Other wines: Le Demoiselle de Sigalas (dry white), La Semillante de Sigalas (dry white), Le 5 (non-sulphured sweet white)

CHÂTEAU SIMON
Barsac
2ème Cru Classé

A combination of modern and traditional methods produces a mildly sweet wine from Sémillon 90 per cent, Sauvignon 8 per cent, and Muscadelle 2%. Most Sauternes and Barsacs are aged in Nevers or Limousin oak. Sometimes Allier is used, but at Simon they mature the wine in Merrain oak for two years.

CHÂTEAU SUAU
Barsac
2ème Cru Classé

The vineyard belongs to Roger Biarnès, who makes the wine at his Château Navarro in Cérons because the original château is under different ownership. Also, this is not the same property as Château Suau at Capian, on the opposite bank of the Garonne, almost directly north.

WHITE This attractive, fresh, and fragrantly fruity wine has a gentle citrus-and-spice botrytis complexity. Improvement noted in the 2005 vintage.

🍷 Sémillon 80%, Sauvignon 10%, Muscadelle 10%

🍷 6–12 years

CHÂTEAU SUDUIRAUT
Preignac
1er Cru Classé
★★☆

This splendid 17th-century château, with its picturesque parkland, effectively evokes the graceful beauty found in its luscious wines. Suduiraut's superb 100-hectare (245-acre) vineyard enjoys a good susceptibility to noble rot, and adjoins that of Yquem. The wines went through an inconsistent patch in the 1980s, but have improved dramatically under the watchful eye of the AXA insurance group. The wines are fermented and matured in cask for 18 months, with 50 per cent new oak.

WHITE Soft, succulent, and sublime, this is an intensely sweet wine of classic stature. It is rich, ripe, and viscous, with great botrytis complexity that benefits from good bottle-age.

🍷 Sémillon 90%, Sauvignon 10%

🍷 8–35 years

Second wine: Castelnau de Suduiraut

Other wines: Lions de Suduiraut (Sauternes), S de Suduiraut, Le Blanc Sec de Suduiraut (dry whites)

CHÂTEAU LA TOUR BLANCHE
Sauternes
1er Cru Classé
★☆

This property was placed at the head of the premiers crus in 1855, when only Yquem was deemed superior, but until relatively recently it failed to live up to this reputation. Even at its lowest ebb, few critics would have denied that these vineyards possessed great potential, but the wines were ordinary. This was made all the more embarrassing by the fact that the state-owned La Tour Blanche was a school of agriculture and oenology that was supposed to teach others how to make Sauternes. This depressing situation began to change in the mid-1980s when the château started increasing the proportion of Sémillon, picking much riper grapes and implementing stricter selection in both vineyard and chais. Fermentation is in wood (with up to 90 per cent new oak, but averaging 25 per cent in most years). The results were particularly exciting in the 1990s, such as the excellent 1990 (its greatest ever)

WHITE These are now so rich they are almost fat and bursting with plump, ripe, juicy fruit and oodles of complex botrytis character.

🍷 Sémillon 83%, Sauvignon 12%, Muscadelle 5%

🍷 8–20 years

Second wine: Les Charmilles de la Tour Blanche

Other wines: Les Brumes de la Tour Blanche Sauternes, Duo de la Tour Blanche (dry white), Les Jardins de La Tour Blanche Bordeaud (red, white, and rosé)

CHÂTEAU D'YQUEM
Sauternes
1er Cru Supérieur
★★★

This most famous of all châteaux, classified at the highest level of 1er Cru Supérieur in 1855, belonged to the English crown from 1152 to 1453. It then passed into the hands of Charles VII of France. In 1593 Jacques de Sauvage acquired tenant's rights to the royal property, and in 1711 his descendants purchased the fiefdom of Yquem. In 1785 it passed into the hands of the Lur-Saluces family. The tries tradition was kept alive at Yquem when it was long forgotten by other noble châteaux. Like Pétrus, one of Yquem's "secrets" is its pickers. They are all skilled; they know what to pick and, just as important, what to leave. The gap between tries can vary from three days to several weeks. Housing and feeding 120 pickers for several weeks of inactivity is not cheap.

The property has been run with passionate care by succeeding generations, although LVMH (which owns Moët & Chandon) purchased a majority share in 1999 (after three years of acrimony between members of the Lur-Saluces family).

In 1972 the harvest consisted of 11 tries spread over 71 days. In that year no wine was sold as Château d'Yquem, as it was the case with the 2012 vintage. This is not to say that Yquem's fastidious attention to selection and quality does not pay off in some poor vintages. But in good years, because of the strict selection in the vineyard, the amount of wine that is finally used is as high as 80 to 90 per cent. The wines are fermented and matured for up to 20 to 22 months with 100 per cent new oak. Other terroirs in Sauternes and Barsac are potentially comparable, but, no matter how conscientious their owners, none makes the same sacrifices as Yquem. This is reinforced by its unique situation on the high outcrop dominating the appellation, at the heart of all the other first growths of Sauternes.

WHITE This wine represents the ultimate in richness, complexity, and class. No other botrytis wine of equal body and concentration has a comparable finesse, breeding, and balance. Some of the characteristic aromas and flavours include peach, pineapple, coconut, nutmeg, and cinnamon, with toasty-creamy vanilla and caramel flavours of new oak.

🍷 Sémillon 80%, Sauvignon 20%

🍷 20–60 years

Other wines: "Y" de Château d'Yquem (dry white)

The Libournais and Fronsadais

The ancient port town of Libourne, on the banks of the Dordogne River, is at the centre of the 10 appellations that make up this Libournais region of Bordeaux. Dominated by the Merlot grape, the vineyards here produce deep-coloured, silky- or velvety-rich wines of classic quality in the St-Émilion and Pomerol regions, in addition to wines of modest quality, but excellent value and character, in the "satellite" appellations that surround them.

St-Émilion is the largest appellation of the Libournais, accounting for almost 45 per cent of the total area. The size of St-Émilion – and its long history – go a long way to explain its dominance.

The map of the region shows the Barbanne River running to the north of St-Émilion and Pomerol, separating them from the satellite appellations of St-Émilion and Lalande-de-Pomerol. Fronsac and Canon Fronsac lay to the northwest, on the opposite bank of the Isle, a tributary of the Dordogne. Carry on across the Isle and you are back in limestone country. Fronsac and Canon-Fronsac enjoy spectacular views across the Dordogne valley from the top of the *tertre*, or hill, of Fronsac. The vines enjoy the southwesterly-facing slopes and excellent drainage. It is a mystery why these wines are not better known in export markets. Their style resembles St-Émilion, and the soils at the heart of the appellation are very similar to the classified growth of St-Émilion. It wasn't always the case. In the 17th century Cardinal Richelieu owned vines here, and his wine was served at court. So popular were these wines that rumour has it Château Saint-Martin may then have changed its name to Château Canon (now a St-Émilion first growth) to profit from the popularity associated with the name.

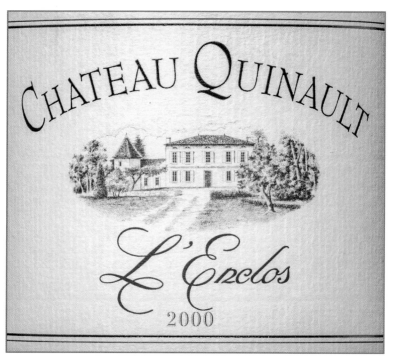

CHÂTEAU QUINAULT L'ENCLOS RED BORDEAUX
This château followed in the footsteps of Le Pin, producing a *"vins de garage"* from grapes grown on a small plot within their appellation.

SERENDIPITY OUT OF MISFORTUNE

In the mid-1950s, many Libournais wines were harsh, and even the best AOCs did not enjoy the reputation they do today. Most growers believed that they were cultivating too much Cabernet Sauvignon and Malbec for their particular *terroir* and decided that they should plant more vines of Cabernet Franc. A few growers argued for the introduction of Merlot, which was allowed by the regulations, because it would give their wines the suppleness they desired. Even if they could have agreed on united action, changing the encépagement of an entire district would have been a very long-term project, as well as being extremely expensive. In 1956, however, frost devastated the vineyards, forcing the Libournais growers into action. With poor, short crops inevitable for some years to come, prices soared, enabling them to carry out the massive replanting that, ironically, they could not have afforded prior to the crisis. This devastation led to the wholesale cultivation of Merlot and Cabernet Franc, which established a completely different style of wines, providing the catalyst for the spectacular post-war success of St-Émilion and Pomerol.

CÔTES DE BORDEAUX CASTILLON AND CÔTES DE BORDEAUX FRANCS

Technically part of the Côtes de Bordeaux family, Castillon and Francs are the immediate easterly neighbours of St-Émilion and enjoy the same clay and limestone *terroir* that is at the heart of St-Émilion. Dominated by Merlot, the wines are similar in style – a similarity that is reinforced by the fact that many winemakers from St-Émilion have invested in the region because the land is considerably more affordable that St-Émilion. As well as their investment they have bought their know-how. Côtes de Bordeaux Castillon is one of those overlooked appellations of Bordeaux that represent such great value for money, sharing the same excellent clay and limestone *terroir* and often the same excellence in winemaking.

THE *GARAGISTE* EFFECT

In 1979 Jacques Thienpont, whose family owns neighbouring Vieux Château Certan in Pomerol, purchased a small 1.6-hectare (4-acre) vineyard called Le Pin, named after a lone pine tree, for one million francs. The grapes had previously been used in other blends and may well have been destined for Vieux Château Certan, but Jacques decided to make his own wine. He made the tiny production in what was effectively a garage, under the dilapidated house. After Robert Parker praised the 1982 vintage, bringing it to the attention of wine lovers, the quality of the wine combined with the tiny production made this *"vin de garage"* so popular it became one of the most expensive wines of Bordeaux, often out-pricing neighbouring Pétrus, as well as first growths.

By the late 1980s Le Pin's extraordinary financial success had become so obvious to a number of neighbouring *vignerons* that they attempted to replicate it with small plots of land in their own appellation: Jean-Luc Thunevin with Château de Valandraud (first vintage 1991), Gérard and Dominique Bécot with Château La Gomerie (first vintage 1995), von Neipperg with Château La Mondotte (first vintage 1996), Francis Gaboriaud with Château L'Hermitage (first vintage 1997), Alain and Françoise Raynaud with Château Quinault L'Enclos (first vintage 1997), and Jonathon Maltus with Le Dôme (first vintage 1998).

It was not just about size. There was a dynamism in St-Émilion and Pomerol that led winemakers to experiment with new techniques such as lower yields, severe selection, cold-soak pre-fermentation, malolactic fermentation in oak, and a high percentage of new oak. All these techniques may seem mainstream now but were very experimental at the time.

Why on the right bank? Apart from the success of Le Pin, properties are smaller and often family owned, allowing a greater freedom and independence that also comes from a more recent commercial history. There were also some very dynamic and innovative winemakers and oenologists here, including leading lights Michel Roland, Jean-Claude Berrouet, and Jean-Luc Thunevin. The higher percentages of Merlot in the blends compared to the left bank wines also lent themselves to more experimentation perhaps?

The term *vin de garage* was first coined by French wine writer Michel Bettane based on Le Pin's garage-cellar. It created a certain resistance amongst the (mainly left bank) old guard as being all about "cooking" rather than *terroir*. All these techniques are now quite commonplace, however; although an excess of concentration, new oak (sometimes 200 per cent), and extraction have now been dialled back in favour of a more traditional elegance.

Due to their continued quality and very limited supply, some of these wines still command good prices (especially Le Pin). Le Pin has had a brand-new cellar since 2012.

THE SATELLITE APPELLATIONS OF ST-ÉMILION AND POMEROL
There are four other appellations that carry the name St-Émilion: Montagne-St-Émilion, Lussac-St-Émilion, Puisseguin-St-Émilion, and St-Georges-St-Émilion. Collectively these are known as the satellites, as they all circle off to the north of St-Émilion, each carrying the name of the village that is at their centre.

These appellations all have a similar blend: Merlot dominated with some complimentary Cabernet Franc, occasionally some Cabernet Sauvignon, and now and again a little Malbec. This is due to a similarity in the *terroir* of clay and limestone and their hilly topography. The mosaic of *terroirs* is reinforced by a much greater variation in elevation compared to the left bank: from almost sea level up to 100 metres (328 feet) compared to the high point of Listrac of 47 metres (155 feet). It really is very beautiful, well worth going off the St-Émilion-Pomerol beaten track to discover

Montagne-St-Émilion is the largest of these appellations at just over 1,500 hectares, or 3,707 acres, (the two appellations of St-Émilion and St-Émilion Grand Cru total over 5,000 hectares (12,355 acres). Located to the north of St-Émilion, it is surrounded by two small tributaries of the Isle River: the Barbanne and the Lavié, bringing more sandy-clay soils to the west.

Only slightly smaller, Lussac-St-Émilion consists of terraces of plateaux and valleys that form a natural southern-facing amphitheatre. The south of the appellation has clay and limestone slopes with more silt deposits in the north left by the receding Isle River.

The Puisseguin-St-Émilion appellation is smaller, just over 700 hectares (1,730 acres) to the east of Lussac and Montagne. It has a more homogeneous *terroir* of some of the hardest limestone in the region, known as *astride* due to the fossilised starfish often found in the rock. This limestone retains moisture resulting in a cool subsoil that lends elegance and freshness to the wines.

St-Georges-St-Émilion is the smallest appellation, just under 200 hectares (495 acres). It is tucked in between St-Émilion and Montagne-St-Émilion on an almost entirely clay-limestone soil promontory.

Given the underlying similarity between these appellations, the role of the individual winemakers really makes a difference.

There is a feeling here of working together, rather than seeing their neighbours as competition. Montagne and St-Georges have their own *syndicat*, or wine bureau, but since 2007 St-Émilion and St-Émilion Grand Cru, Lussac-St-Émilion, and Puisseguin-St-Émilion have joined forces to create "Le Conseil des Vins de St-Émilion".

To the north of Pomerol lies the lesser-known satellite commune of Lalande-de-Pomerol. It is overshadowed by its illustrious neighbour; despite having a good complexity of soils with river born gravels, it is considered a lighter and perhaps more rustic expression of the area. It is now coming into its own, thanks to influential and talented winemakers from St-Émilion and Pomerol producing excellent (and still affordable wines) from this appellation.

CHÂTEAU DE LUSSAC, ST-ÉMILION AOC
A display of this St-Émilion satellite's estate wines, including its eponymous wine and its second wine, Le Libertin de Lussac.

WINE REGIONS OF THE LIBOURNAIS DISTRICT (*see also p141*)
This great red wine area includes St-Émilion, Pomerol, and their satellites. In this district, the Merlot grape reigns supreme, with its succulent fruit essential to the local style.

THE GENERIC APPELLATIONS OF THE
LIBOURNAIS AND FRONSADAIS

CANON FRONSAC AOC

Fronsac AOC and Canon Fronsac AOC will no doubt be the next wines to be "discovered" by budget-minded Bordeaux drinkers. The best of these wines are Canon Fronsac AOC. With lower yields and stricter selection, these wines could equal all but the best of St-Émilion and Pomerol.

RED Full-bodied, deep-coloured, rich, and vigorous wines with dense fruit, fine spicy character, plenty of finesse, and good length.

🍷 *Cabernet Franc, Cabernet Sauvignon, Malbec, Merlot, Petit Verdot, Carménère*

🍷 *7–20 years*

CÔTES DE BORDEAUX CASTILLON AOC

This is an attractive hilly area squeezed between St-Émilion, the Dordogne River, and the Dordogne département. Its wine has long been appreciated for quality, consistency, and value. These wines used to be sold as Bordeaux and Bordeaux Supérieur wine until the 1989 vintage, when Côtes-de-Castillon received its own AOC status.

RED Firm, full-bodied, fine-coloured wines with dense fruit and finesse.

🍷 *Cabernet Franc, Cabernet Sauvignon, Carménère, Malbec, Merlot, Petit Verdot*

🍷 *5–15 years*

CÔTES DE BORDEAUX FRANCS AOC

This forgotten area's vineyards are contiguous with those of Puisseguin-St-Émilion and Lussac-St-Émilion and have a very similar clay-limestone over limestone and iron-pan soil. The Bordeaux Supérieur version of these wines differs only in its higher alcohol level.

RED These are essentially robust, rustic, full-bodied wines that are softened by their high Merlot content.

🍷 *Cabernet Franc, Cabernet Sauvignon, Malbec, Merlot*

🍷 *5–10 years*

WHITE Little-seen dry wines of clean, fruity character.

🍷 *Sauvignon Blanc, Sémillon, Muscadelle*

🍷 *5–10 years*

CÔTES DE BORDEAUX FRANCS LIQUOREUX AOC

This style of Bordeaux-Côtes-de-Francs wine must by law be naturally sweet and made from overripe grapes that possess at least 238 grams of sugar per litre. The wines must have a minimum level of 14.5 per cent alcohol and 51 grams of residual sugar per litre.

WHITE Rich, genuinely *liquoreux* wines; only tiny amounts are made.

🍷 *Sauvignon Blanc, Sémillon, Muscadelle*

🍷 *5–15 years*

FRONSAC AOC

This generic appellation covers the communes of La Rivière, St-Germain-la-Rivière, St-Aignan, Saillans, St-Michel-de-Fronsac, Galgon, and Fronsac.

RED These full-bodied, well-coloured wines have good chunky fruit and a fulsome, chocolaty character. Not quite the spice or finesse of Canon Fronsac, but splendid value.

🍷 *Cabernet Franc, Cabernet Sauvignon, Malbec, Merlot, Petit Verdot, Carménère*

🍷 *6–15 years*

LALANDE-DE-POMEROL AOC

This good-value appellation covers the communes of Lalande-de-Pomerol and Néac. No matter how good they are, even the best are but pale reflections of classic Pomerol.

RED Firm, meaty Merlots with lots of character but without Pomerol's texture and richness.

🍷 Cabernet Franc, Cabernet Sauvignon, Malbec, Merlot, Petit Verdot

🍷 *7–20 years*

LUSSAC-ST-ÉMILION AOC

A single-commune appellation 9 kilometres (5.5 miles) northeast of St-Émilion.

RED The wines produced on the small gravelly plateau to the west of this commune are the lightest but have the most finesse. Those produced on the cold, clayey lands to the north are robust and earthy, while those from the

clay-limestone in the southeast have the best balance of colour, richness, and finesse.

🍷 *Cabernet Franc, Cabernet Sauvignon, Carménère, Malbec, Merlot, Petit Verdot*

🍷 *5–12 years*

MONTAGNE-ST-ÉMILION AOC

This appellation includes St-Georges-St-Émilion, a former commune that is today part of Montagne-St-Émilion. St-Georges-St-Émilion AOC, and Montagne-St-Émilion AOC are the best of all the appellations that append "St-Émilion" to their names.

RED Full, rich, and intensely flavoured wines that mature well.

🍷 *Cabernet Franc, Cabernet Sauvignon, Malbec, Merlot, Petit Verdot*

🍷 *5–15 years*

POMEROL AOC

The basic wines of Pomerol fetch higher prices than those of any other Bordeaux appellation. The average Merlot content of a typical Pomerol is around 80 per cent.

RED It is often said that these are the most velvety-rich of the world's classic wines, but they also have the firm tannin structure that is necessary for successful long-term maturation and development. The finest also have surprisingly deep colour, masses of spicy-oak complexity, and great finesse.

🍷 *Cabernet Franc, Cabernet Sauvignon, Malbec, Merlot, Petit Verdot*

🍷 *5–10 years (modest growths), 10–30 years (great growths)*

PUISSEGUIN-ST-ÉMILION AOC

This commune has a clay-limestone topsoil over a stony subsoil, and the wines it produces tend to be more rustic than those of the Montagne-St-Émilion AOC.

RED These are rich and robust wines with a deep flavour and lots of fruit and colour, but they are usually lacking in finesse.

🍷 *Cabernet Franc, Cabernet Sauvignon, Carménère, Malbec, Merlot, Petit Verdot*

🍷 *5–10 years*

ST-ÉMILION AOC

These wines must have a minimum of 11 per cent alcohol, but in years when chaptalisation (the addition of sugar to grape juice to increase alcohol content) is allowed, there is also a maximum level of 13.5 per cent.

RED Even in the most basic St-Émilions the ripe, spicy-juiciness of the Merlot grape should be supported by the firmness and finesse of the Cabernet Franc. The great châteaux achieve this superbly: they are full, rich, and concentrated, chocolaty, and fruit-cakey.

🍷 *Cabernet Franc, Cabernet Sauvignon, Carménère, Malbec, Merlot, Petit Verdot*

🍷 *6–12 years (modest growths), 12–35 years (great growths)*

ST-GEORGES-ST-ÉMILION AOC

Along with the Montagne region of Montagne-St-Émilion, this is the best parish of the outer areas.

RED These are deep-coloured, plummy wines with juicy, spicy fruit and good supporting tannic structure.

🍷 *Cabernet Franc, Cabernet Sauvignon, Malbec, Merlot, Petit Verdot*

🍷 *5–15 years*

VINEYARDS OF FRONSAC AOC LINE THE DORDOGNE
The Dordogne snakes through this eastern region of Bordeaux. Merlot wines dominate this appellation.

The Libournais and Fronsadais:
Saint-Émilion

The Romans were the first to cultivate the vine in St-Émilion, a small area that has exported its wines to various parts of the world for over 800 years. In the first half of the 20th century it lapsed into obscurity, but in the last 50 years it has risen like a phoenix.

St-Émilion as we know it is a phenomenon of the post-war era, but there are many reminders of this wine's ancient past – from the famous Château Ausone, named after the Roman poet Ausonius, to the walled hilltop village of St-Émilion itself, which has survived almost unchanged since the Middle Ages. In contrast, the Union des Producteurs, the largest single-appellation coopérative in France, is a graphic illustration of the best in modern, technologically sophisticated wine production. Today, there are almost a thousand *crus* within 10 kilometres (6 miles) of the village of St-Émilion that may use this appellation.

THE APPEAL OF ST-ÉMILION WINES

For those who find red wines too harsh or too bitter, St-Émilion, with its elegance and finesse, is one of the easiest with which to make the transition from white to red. The difference between the wines of St-Émilion and those of its satellites is comparable to the difference between silk and satin, whereas the difference between St-Émilion and Pomerol is like the difference between silk and velvet: the quality is similar, but the texture is not – although, of course, we must be humble about categorizing such complex entities as wine areas. It could justifiably be argued that the *graves* (gravelly terrain) that produces two of the very best St-Émilions – Château Cheval-Blanc and Château Figeac – has more in common with Pomerol than with the rest of the appellation.

FACTORS AFFECTING TASTE AND QUALITY

LOCATION
St-Émilion is on the right bank of the Dordogne, 50 kilometres (80 miles) east of Bordeaux.

CLIMATE
The climate is less maritime and more continental than that of the Médoc, with a greater variation in daily temperatures – there is also slightly more rain during spring and substantially less during summer and winter.

ASPECT
The village of St-Émilion sits on a plateau where vines grow at an altitude of 25 to 100 metres (80 to 330 feet). These vineyards are quite steep, particularly south of the village where two slopes face each other. The plateau heads eastwards as hilly knolls. North and west of the village, the vineyards are on flatter ground.

SOIL
St-Émilion's soil is extremely complex (see The Question of Quality, below) and is part of the area known as "Pomerol-Figeac graves" that encompasses the châteaux Cheval Blanc and Figeac.

VITICULTURE & VINIFICATION
Some of the *vin de presse,* usually the first pressing only, is considered necessary by many châteaux. Skin-contact usually lasts for 15 to 21 days but may last up to four weeks. Some wines spend as little as 12 months in cask, but the average is nearer to 15 to 18 months.

GRAPE VARIETIES
Primary varieties: Cabernet Franc, Cabernet Sauvignon, Merlot
Secondary varieties: Carmenère, Malbec

THE QUESTION OF QUALITY

The diverse nature of St-Émilion's soil has led to many generalizations that attempt to relate the quantity and character of the wines produced to the soils from which they come. Initially the wines were lumped into two crude categories, *côtes* ("hillside" or "slope") and *graves* ("gravelly terrain"). The term *côtes* was supposed to describe fairly full-bodied wines that develop quickly; the term *graves,* fuller, firmer, and richer wines that take longer to mature.

The simplicity was appealing, but it ignored the many wines produced on the stretch of deep sand between St-Émilion and Pomerol and those of the plateau, which has a heavier topsoil than the *côtes.* It also failed to distinguish between the eroded *côtes* and the deep-soiled bottom slopes. Most important, it ignored the fact that many châteaux are spread across more than one soil type (*see* the list of classified growths, next page) and that they have various other factors of *terroir,* such as aspect and drainage, which affect the character and quality of a wine (*see* Soil Types of St-Émilion, p191).

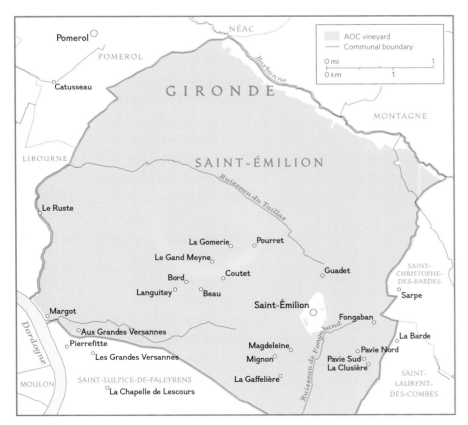

A WINE SHOP IN ST-ÉMILION
Oenophiliac tourists flock to this medieval town that gives its name to the well-respected red wine region.

VINE-GROWING AREAS OF SAINT-ÉMILION (*see also* p141)
Most of the highest-classified châteaux are located on the slopes around Saint-Émilion itself or to the northwest, where they neighbour the vineyards of Pomerol.

THE CLASSIFICATION OF ST-ÉMILION

St-Émilion could be considered as the most democratic of the Bordeaux classifications, because, unlike its 1855 counterpart in the Medoc, Graves, and Sauternes, since its creation in 1955 it is reassessed every 10 years (or thereabouts). The classification was revised in 1969, 1985 (some six years late), and again in 1996, 2006, and most recently 2012.

Inclusion in the running for this classification is not automatic, properties must apply, and about 100 participated with the objective of winning either the "Grand Cru Classé" status or the coveted "Premier Grand Cru Classé'.

The 2006 classification kicked out 11 properties and brought in another four, but 4 of the 11 demoted properties took exception and made a legal fight. (Despite having signed the application saying they would abide by the classifications findings.) They took the appellation to court and successfully had the classification results annulled.

A sense of compromise was reached in 2008, allowing those properties that were justly recompensed for their efforts to keep their upgrade and those who were pushed down a level allowed to stay at their status as per the previous 1996 classification, up to and including the 2011 vintage.

The 2006 classification therefore counted 72 properties and covered 16 per cent of the surface area of the St-Émilion vineyard.

The first qualifier for candidates to this classification is that they must produce wines in the St-Émilion Grand Cru appellation rather than Saint-Émilion AOC, with the quality restrictions (lower yield, longer ageing, and winemaking and bottling with the St-Émilion appellation region) that accompany this higher-quality appellation. Out of the 5,500 hectares (13,590 acres) and 770 properties concerned over two-thirds is currently under this more prestigious Grand Cru appellation.

To avoid another legal challenge, the 2012 classification was managed from Paris by the government agency L'Insitut National des Appellations d'Origines (INAO) and was piloted by wine professionals, none of whom were from the Bordeaux region. Assessment was controlled by an independent quality-control organisation.

They blind-tasted 10 vintages (1999 to 2008, for 2012 classification) or 15 vintages if a promotion from grand cru classé to premier grand cru classé was envisaged. The tasting results contributed 50 per cent of the final decision for grand cru classé and 30 per cent for premier grand cru classé, with the balance accounted for by criteria such as quality of technical equipment, hygiene of the winery, and historic reputation of the château and its wine.

The 2012 classification includes 82 properties, 64 Grands Crus Classés, and 18 Premiers Grands Crus Classés.

Until 2012 there were only ever two Premier Grand Cru Classé (A)s: Château Ausone and Château Cheval Blanc. There are now four, as Château Angelus and Château Pavie joined the elite club.

2012 LIST OF THE CRUS CLASSES OF THE APPELLATION D'ORIGINE CONTRÔLEE SAINT-ÉMILION GRAND CRU

PREMIERS GRANDS CRUS CLASSÉS:
1. Château Angélus (A)
2. Château Ausone (A)
3. Château Cheval Blanc (A)
4. Château Pavie (A)
5. Château Beauséjour (héritiers Duf-fau-Lagarrosse)
6. Château Beau-Séjour-Bécot
7. Château Bélair-Monange (merged with other Premier Grand Cru Classé Château Magdeleine in 2012)
8. Château Canon
9. Château Canon-la-Gaffelière
10. Château Figeac
11. Clos Fourtet
12. Château la Gaffelière
13. Château Larcis Ducasse
14. La Mondotte
15. Château Pavie Macquin
16. Château Troplong Mondot
17. Château Trottevieille
18. Château Valandraud

GRANDS CRUS CLASSÉS:
(In alphabetical order)
19. Château l'Arrosée
20. Château Balestard la Tonnelle
21. Château Barde-Haut
22. Château Bellefont-Belcier
23. Château Bellevue
24. Château Berliquet
25. Château Cadet-Bon
26. Château Cap de Mourlin
27. Château le Chatelet
28. Château Chauvin
29. Château Clos de Sarpe
30. Château la Clotte
31. Château la Commanderie
32. Château Corbin
33. Château Côte de Baleau
34. Château la Couspaude
35. Château Dassault
36. Château Destieux
37. Château la Dominique
38. Château Faugères
39. Château Faurie de Souchard
40. Château de Ferrand
41. Château Fleur Cardinale
42. Château La Fleur Morange
43. Château Fombrauge
44. Château Fonplégade
45. Château Fonroque
46. Château Franc Mayne
47. Château Grand Corbin
48. Château Grand Corbin-Despagne
49. Château les Grandes Murailles
50. Château Grand Mayne
51. Château Grand-Pontet
52. Château Guadet
53. Château Haut-Sarpe
54. Clos des Jacobins
55. Couvent des Jacobins
56. Château Jean Faure
57. Château Laniote
58. Château Larmande
59. Château Laroque
60. Château Laroze
61. Clos la Madeleine
62. Château la Marzelle
63. Château Monbousquet
64. Château Moulin du Cadet
65. Clos de l'Oratoire
66. Château Pavie Decesse
67. Château Peby Faugères
68. Château Petit Faurie de Soutard
69. Château de Pressac
70. Château le Prieuré
71. Château Quinault l'Enclos
72. Château Ripeau
73. Château Rochebelle
74. Château Saint-Georges-Cote-Pavie
75. Clos Saint-Martin
76. Château Sansonnet
77. Château la Serre
78. Château Soutard
79. Château Tertre Daugay
80. Château la Tour Figeac
81. Château Villemaurine
82. Château Yon-Fig

SOIL TYPES OF ST-ÉMILION

In the numerous *premier grand cru classé* and *grand cru classé* vineyards spread over St-Émilion a number of different soil types occur. With soil and *terroir* changing dramatically from vineyard to vineyard, the wines produced within possess characteristics unique to each.

ANCIENT SAND

This is a thick blanket of large-grain sand over a subsoil of *molasse*. The bulk of this sand extends northeast from the village of St-Émilion towards Pomerol (*see* map on p189). Although this area seems to have a gentle slope all round it, and the sand is very permeable, the *molasse* below is flat and impermeable. The water collects, saturating root systems and increasing soil acidity. Some châteaux benefit greatly from underground drainage pipes.

BOTTOM SLOPES

The gentler bottom slopes of the côtes have a deep, reddish brown, sandy-loam topsoil over yellow sand subsoil.

CÔTES

The lower-middle to upper slopes of the côtes have a shallow, calcareous, clay-silty-loam topsoil with a high active lime content. It is quite sandy on the middle slopes, and the topsoil thins out higher up. The subsoil is mostly molasse, not the impermeable type found under the ancient sand and graves, but a weathered, absorbent molasse of limestone or sandstone.

GRAVES

In the Graves, deep gravel topsoil with a subsoil of large-grain sand lies over a very deep, hard, and impermeable sedimentary rock called *molasse*. The gravel here is similar to that found in the Médoc.

SANDY-GRAVEL

This is sandy and sandy-gravel topsoil over sandy-gravel, ferruginous gravel, and iron-pan.

ST-CHRISTOPHE PLATEAU

This area is clay-limestone and clay-sand topsoil over limestone and *terra rossa* subsoil (a red, clay-like, limestone soil).

ST-ÉMILION PLATEAU

This area is covered in shallow clay-limestone and clay sand, shell debris, and silt topsoil over eroded limestone subsoil.

AN AERIAL VIEW OF THE ST-ÉMILION LANDSCAPE SHOWS THE NUMEROUS PLOTS UNDER VINE IN THIS REGION
Vineyards chequer the area around the town of St-Émilion. The variation of soil types in this region of Bordeaux results in the diverse wine styles produced here.

SAINT-ÉMILION

CHÂTEAU ANGÉLUS
Premier Grand Cru Classé (A)
★★

This is a large property with a single plot of vines on the south-facing *côtes*. At one time the château produced wines in the old "farmyard" style, That ended with the 1980 vintage, because, under the guidance of Hubert de Boüard de Laforest, quality constantly improved, and the wine was promoted to Premier Grand Cru Classé (A) in 2012. The wine is aged in new oak, barrels, and *foudres* for 18 to 22 months. The property is under conversion to organic agriculture.

RED This is a soft, silky, and seductive wine. The luxury of new oak is having a positive effect on the quality, character, and ageing potential of this wine.

🍷 Merlot 53%, Cabernet Franc 46%, Petit Verdot 1%

🍷 7–20 years

Second wine: *Carillon de l'Angélus*
Other wines: *No 3 d'Angélus*

CHÂTEAU AUSONE
Premier Grand Cru Classé (A)
★★★

After gifted winemaker Pascal Delbeck took control of Château Ausone in 1975, this prestigious property produced wines of stunning quality, and it now deserves its superstar status. In 1997 Alain Vauthier took control and along with his daughter Pauline now makes the wine. He has increased panting density and introduced a second wine. The vineyard of Château Ausone has a privileged southeast exposure, and its vines are fairly established at an average of 50 years of age. They are capable of yielding very concentrated wines, which are then matured in wood for up to 24 months, with 100 per cent new oak.

RED These rich, well-coloured wines have opulent aromas and scintillating flavours. They are full in body, compact in structure, and refined in character, with masses of spicy-cassis fruit and creamy-oak undertones. These wines are the quintessence of class, complexity, and finesse.

🍷 Cabernet Franc 65%, Merlot 35%

🍷 15–45 years

Second wine: *Chapelle d'Ausone*
Other wines: *Moulin Saint Georges*

CHÂTEAU BALESTARD LA TONNELLE
Grand Cru Classé
★

The label of this wine bears a 15th-century poem by François Villon that cites the name of the château. The Capdemourlin family of neighbouring Château Cap De Mourlin own and run this property. One-third of the wine is matured in 50 per cent new oak for up to 15 to 18 months, and the remainder in one-year-old oak.

RED The gentle, ripe aromas of this wine belie its staunchly traditional style.

It is a full-bodied wine of great extract, tannin, and acidity that requires time to soften, but it has masses of fruit and so matures gracefully.

🍷 Merlot 70%, Cabernet Franc 25%, Cabernet Sauvignon 5%

🍷 10–30 years

Second wine: *Chanoine de Balestard*

CHÂTEAU BEAUSÉJOUR
Premier Grand Cru Classé (B)
★★

These little-seen wines consistently underwhelmed critics until the 1980s, since when Château Beauséjour began to produce darker, fuller wines with more class, but I was not that impressed until the stellar class 2000 vintage. The property is by Duffau–Lagarosse, and since 2009 Nicolas Thienpont has signed the wines.

RED Lovely rich wine, with concentrated fruit, and great potential longevity from 2000 onwards.

🍷 Merlot 80 %, Cabernet Franc 20 %

🍷 7–15 years

Second wine: *Croix de Beauséjour héritiers Duffau-Lagarrosse*

CHÂTEAU BEAU-SÉJOUR BÉCOT
Premier Grand Cru Classé (B)
★★

Since 1979 this property has almost doubled in size by merging with two *grands crus classés*, Château la Carte and Château Trois Moulins. In 1985 Beau-Séjour Bécot was the only *premier grand cru classé* to be demoted in the revision of the St-Émilion classification. The demotion was not due to its quality or performance, which were consistently excellent, but because of its expansion. In 1996 this château was promoted back to a Premier Grand Cru Classé (B). The wine is fermented in stainless steel and matured in wood for 18 months, with 70 per cent new oak. The third generation of the family are now in charge with Juliette Bécot and her husband, Julien, at the helm.

RED Once lightweight and high-tone, this wine is now full, rich, and truly characterful. The silky Merlot fruit develops quickly but is backed up with creamy new oak.

🍷 Merlot 80%, Cabernet Franc 15%, Cabernet Sauvignon 5%

🍷 7–25 years

Other wines: *La Gomerie (100% Merlot Saint-Émilion Grand Cru), vinified separately from neighbouring vines since 2016. Beau-Séjour-Bécot*

CHÂTEAU BÉLAIR-MONANGE
Premier Grand Cru Classé (B)
★★

Le Château Bélair was purchased by Jean-Pierre Moueix in 2008 and, the name changed to Château Bélair-Monange with the 2008 vintage. In 2012

Bélair-Monange and neighbouring Château Magdelaine were confirmed as *premier grands crus classés* and were amalgamated into Château Bélair-Monange. The wine is matured in wood for 16 to 18 months. Up to half is aged in new oak. This is one of the very best *premier grands crus*.

RED This is a deep-coloured, full-bodied wine with a rich flavour of plums, chocolate, black cherries, and cassis. It has great finesse.

🍷 Merlot 90%, Cabernet Franc 10%

🍷 10–35 years

Second wine: *Annonce de Bélair-Monange*
Other wines: *Haut Roc Blanquant*

CHÂTEAU BELLEVUE
★☆

This small property was originally called Fief-de-Bellevue and belonged to the Lacaze family from 1642 to 1938. Stripped of its *grand cru classé* status in 2006, a 50 per cent share of the property was purchased by the neighbours at Château Angélus in 2007. Now made by the Angélus team of Hubert de Boüard de Laforest and Emmanuelle d'Aligny-Fulchi, it was reinstated as a *grand cru classé* in 2012.

RED Beautifully ripe fruit, sleek, with great finesse, and remarkably evocative of its *terroir*.

🍷 Merlot 100%

🍷 5–10 years

CHÂTEAU BERGAT
★

Emile Castéja (of wine merchant Borie-Manoux) and the Preben Hansen family jointly own this small vineyard. It was included in Château Trotte Vieille, a *premier grand cru classé* and another Casteja family St-Émilion estate, after the 2012 Saint-Émilion Classification. Its final vintage was 2011.

CHÂTEAU BERLIQUET
Grand Cru Classé
★☆

The only property upgraded to *grand cru classé* in 1985. In 2017 the Chanel Group, owners of the neighbouring *premier grand cru classé* Château Canon, bought it. The vineyard and cellars are undergoing a complete renovation. The wine is in wood for 15 to 18 months, with 50 per cent new oak.

RED These are deep, dark, and dense wines with spicy-cassis fruit and good vanilla oak.

🍷 Merlot 70%, Cabernet Franc 25%, Cabernet Sauvignon 5%

🍷 10–30 years

CHÂTEAU CADETBON
Grand Cru Classé

This property was demoted from *grand cru classé* in 1985, reinstated in 1996, then demoted again in 2006! It was one of the

properties that challenged the classification, resulting in the re-do of 2012. Since 2003 under the wine wizardry of oenology consultants Stéphane Derenoncourt and Frederic Massie, the wine began stepping up another gear, and in 2012 it was reinstated as a *grand cru classé*. In 2017, the property embarked on a program of conversion to biodynamic agriculture.

RED Until 2003 this wine was not the richest on the block, but it had an ampleness that made for easy, early drinking. Since 2004 there has been increasing richness, with a very fine, elegant tannin structure.

🍷 Merlot 80%, Cabernet Franc 20%

🍷 7–20 years

CHÂTEAU CADET-PIOLA
★

This small property was purchased by La Mondiale insurance company in 2009. Following the 2012 classification the vines were included within the vineyard of neighbouring Château Soutard, also owned by La Mondiale. Its final vintage was 2011.

CHÂTEAU CANON
Premier Grand Cru Classé (B)
★★

Eric Fornier sold Château Canon to the Wertheimer brothers, owners of Chanel, in 1996. Under the management of John Kolasa, who also managed their Margaux property, Château Rauzan-Ségla, they invested heavily in the vineyard, which was completely replanted, and in the cellars. In 2000 the Wertheimers bought the nearby property Curé Bon with INAO agreeing to incorporate 3.5 hectares (8.6 acres) into Canon from the 2000 vintage. In 2011 they purchased neighbouring Château Matras. Most of these vines now go into the second wine (previously Clos Canon), which is vinified in the renovated cellars of Matras and now labelled as Croix Canon. Nicolas Audebert took over in 2014. The grand vin, fermented in stainless steel vats and matured in wood for 18 months with 50 per cent new oak, is one of the best of St-Émilion's *premier grands crus classés*.

RED These wines have a deep purple colour, an opulent cassis bouquet and are very rich and voluptuous on the palate with masses of juicy Merlot fruit and spicy-complexity.

🍷 Merlot 70%, Cabernet Franc 30%

🍷 8–30 years

Second wine: *Croix Canon*

CHÂTEAU CANON-LA-GAFFELIÈRE
Premier Grand Cru Classé
★☆

Owned since 1985 by Comte von Neipperg, Château Canon-la-Gaffeliére is one of the oldest properties in St-Émilion. Its wines were fermented in stainless steel vats until wooden vats were installed in 1997, and the quality soared the very next year. Malolactic fermentation is carried out in barrel, and the wines are matured for 15 to 18 months, with up to 80 per cent new oak. They now practice organic agriculture.

RED This wine has been on form, really plump and displaying vivid, concentrated, spicy fruit and a creamy-oak finish.

🍷 *Merlot 50%, Cabernet Franc 40%, Cabernet Sauvignon 10%*

🕒 *8–20 years*

Other properties: *La Mondotte (Premier Grand Cru Classé), Clos de l'Oratoire (St-Émilion Grand Cru Classé), Château Peyreau (St-Émilion Grand Cru), Château d'Aguilhe (Côtes de Bordeaux Castillon AOC), Clos Marsalette (Pessac-Léognan)*

CHÂTEAU CAP DE MOURLIN
Grand Cru Classé
★

In 1983 there were two versions of this wine, one bearing the name of Jacques Capdemourlin and another of Jean Capdemourlin. Jacques amalgamated the two properties in 1983, building new winemaking facilities and a barrel-ageing cellar. The wine is matured in new, French oak barrels for up to 15 to 18 months, with 50 per cent new oak.

RED An attractive, well-made, medium-bodied wine, with exquisitely fresh dark fruit, and a smooth finish.

🍷 *Merlot 65%, Cabernet Franc 25%, Cabernet Sauvignon 10%*

🕒 *6–15 years*

Second wine: *Capitan de Mourlin*

CHÂTEAU LA CARTE
Grand Cru Classé

Since 1980, the vineyards of Château La Carte have been merged with those of *premier grand cru classé* Château Beau-Séjour-Bécot.

CHÂTEAU CHAPELLE-MADELEINE

Classified *grand cru classé* until 1996. Since 1971 these vineyards have been merged with those of *premier grand cru classé* Château Ausone.

CHÂTEAU CHAUVIN
Grand Cru Classé
★

Purchased by Sylvie Cazes in 2014 and reconfirmed as a *grand cru classé* in the 2012 classification. This property's wine is matured in wood for 15 to 18 months, with 60 per cent new oak. It is difficult to find, but its quality makes it deserving of better distribution.

RED When on form, Château Chauvin can have excellent colour, an aromatic bouquet, full body, and plummy fruit. More hits than misses these days, as the wines are made with an increasing degree of finesse.

🍷 *Merlot 75%, Cabernet Franc 20%, Cabernet Sauvignon 5%*

🕒 *4–10 years*

Second wine: *Folie de Chauvin*

CHÂTEAU CHEVAL BLANC
Premier Grand Cru Classé (A)
★★★

The unusual aspect of this great wine is its high proportion of Cabernet Franc, which harks back to the pre-1956 era. Switching to a majority of Merlot vines was advantageous for most Libournais properties, but keeping a majority of Cabernet Franc was even better for Château Cheval Blanc. Since 1998 the property has been in the hands of Baron Albert Frère and Bernard Arnault. The wine is matured in wood for 16 to 18 months with 100 per cent new oak.

RED These wines have all the sweet, spicy richness one expects from a classic St-Émilion property situated on graves.

🍷 *Cabernet Franc 52%, Merlot 43%, Cabernet Sauvignon 5%*

🕒 *12–40 years*

Second wine: *Le Petit Cheval*
Other wines: *Le Petite Cheval Blanc (Bordeaux Blanc)*

CLOS FOURTET
Premier Grand Cru Classé (B)
★☆ 🆅

Previously called Château Clos Fourtet, this property had an inconsistent record but improved in the 1990s. Philippe Cuvelier, also owner of Château Poujeaux, bought the estate in 2001, with Stéphane Derenoncourt as consultant. The wines are matured in wood for 14 to 18 months, using 60 to 80 per cent new oak in the underground limestone caves on the edge of the town of St-Émilion.

RED Opulent and medium-bodied with silky Merlot fruit, gaining in complexity and finesse.

🍷 *Merlot 83%, Cabernet Franc 9%, Cabernet Sauvignon 8%*

🕒 *6–12 years*

Second wine: *La Closerie de Fourtet*
Other wines: *Château Poujeaux (since 2008), Château Les Grandes Murailles (since 2013), Château Côte de Baleau (since 2013)*

CHÂTEAU CLOS DES JACOBINS
Grand Cru Classé
★

In 2004 Magali and Thibault Decoster purchased Clos des Jacobins, along with Château de La Commanderie followed by Château de Candale and Château Roc de Candale in 2017. The wines are aged for 18 months in 75 per cent new oak.

RED These are rich, fat wines, bursting with chocolate and black-cherry flavours.

🍷 *Merlot 80%, Cabernet Franc 18%, Cabernet Sauvignon 2%*

🕒 *8–25 years*

CLOS DE L'ORATOIRE
Grand Cru Classé
★★

Clos de l'Oratoire belongs to Stephan von Neipperg, who has developed a cult following for the property's wine. The wine is matured in wood for 18 months, with up to 80 per cent new oak.

BELL SCULPTURE IN THE GARDEN OF CHÂTEAU ANGÉLUS
This estate's wine label has always featured a bell on a light background, and over the years the bell has become the emblem of the property.

RED These fine, full-flavoured wines tend to have great concentration and style.

🍷 *Merlot 80%, Cabernet Franc 20%*

🕒 *7–15 years*

CLOS ST-MARTIN
Grand Cru Classé
★

The smallest classified growth in St-Émilion, Clos St-Martin is owned and managed by Sophie Fourcade, with Michel Rolland as consultant winemaker. The wine is aged for 20 months in new oak barrels.

RED Vivid colour with ripe Merlot fruit, silky texture, and elegant style.

🍷 *Merlot 75%, Cabernet Franc 15%, Cabernet Sauvignon 10%*

🕒 *6–15 years*

CHÂTEAU LA CLOTTE
Grand Cru Classé
❶

This property is one of the oldest Bordeaux vineyards on the right bank. Since 2014 Château La Clotte has been owned by the Vautier family of Château Ausone and managed by Pauline Vauthier. The wine is aged in 50 per cent new oak barrels for up to 15 months.

RED This estate can make attractive and elegant wines with lots of soft, silky fruit that are a match for its peers.

🍷 *Merlot 80%, Cabernet Franc 10%, Cabernet Sauvignon 10%*

🕒 *5–20 years*

CHÂTEAU LA CLUSIÈRE

A small enclave within the property of Château Pavie and under the same ownership as Pavie and Château Pavie Decesse. After 2001 the vineyards were officially merged into Château Pavie.

CHÂTEAU CORBIN
Grand Cru Classé
☆

The Corbin estate is divided into five separate properties bordering the Pomerol district, with soils more reminiscent of Pomerol than the classic clay and limestone traditionally associated with St-Émilion: the soil is a mixture of sand, clay, and the famous iron-oxide that gives those truffle aromas to older Pomerols. The estates that now carry the family name are Château Corbin, Château Grand Corbin, Château Grand Corbin Despagne, and Château Corbin Michotte, all *grands crus classés* in the 2012 classification, and *grand cru* Grand Corbin Manuel. Château Corbin has been in the Cruse family for four generations and is currently owned and made by Anabelle Cruse. The wine is fermented in cement tanks and one-third is matured in 50 per cent new oak.

RED Deep-coloured, full-bodied, and deliciously rich, but rather rustic for a classified growth.

🍷 *Merlot 80%, Cabernet Franc 20%*

🕒 *6–12 years*

Second wine: *Divin de Corbin, XX de Corbin*

CHÂTEAU CORBIN MICHOTTE

Château Corbin Michotte is the smallest of the Corbin estates at only 7 hectares (17 acres). This wine is fermented in stainless steel and 60 per cent is matured in new oak, the remainder aged in tanks.

RED A dark, deeply flavoured, full-bodied wine that has rich, juicy Merlot fruit and some finesse.

🍷 Merlot 65%, Cabernet Franc 30%, Cabernet Sauvignon 5%

🍷 6–15 years

Second wine: *Les Abeilles*

CHÂTEAU CÔTE DE BALEAU
Grand Cru Classé
★

This property was unjustly demoted from its *grand cru classé* status in the 1985 revision and reinstated in 2012. This wine estate is wned by the Cuvelier family of Clos Fourtet, who also own Château Grandes Murailles. This wine is aged in wood and 50 per cent of the barrels are renewed every year.

RED Full, rich, and well-balanced wines that have good fruit, some fat, and an attractive underlying vanilla character.

🍷 Merlot 80%, Cabernet Franc 15%, Cabernet Sauvignon 5%

🍷 4–12 years

CHÂTEAU LA COUSPAUDE
Grand Cru Classé
★

This property was demoted from its *grand cru classé* status in 1985 but, following a string of good vintages, was promoted back to its original classification in 1996. Since then the quality at Château la Couspaude has continued to improve. The wine is matured in wood, with 100 per cent new oak.

RED This wine has lots of upfront, juicy-Merlot fruit with an increasing amount of finesse.

🍷 Merlot 75%, Cabernet Franc 20%, Cabernet Sauvignon 5%

🍷 3–7 years

Second wine: *Château Saint Hubert*

CHÂTEAU COUTET

This property was demoted from its *grand cru classé* status in 1985, and the owners have decided to no longer participate in the classification. It has a record of producing finer wines than La Couspaude (*see* previous entry), but unfortunately has the same lack of consistency. The vineyard is certified organic since 2012 and is aged in 20 per cent new oak.

RED Most vintages have a light but elegant style, with a firm tannin structure.

🍷 Cabernet Franc 30%, Merlot 60%, Cabernet Sauvignon 3%, Malbec 7%

🍷 4–8 years

Second wine: *Château Belles Cîmes*

COUVENT DES JACOBINS
Grand Cru Classé
★

The wine from the young vines of this property is not included in its *grand vin* but is used to make a second wine called "Château Beau Mayne". One-third of the production is matured in wood, with 50 to 55 per cent new oak for 12 to 15 months in limestone cellars in the centre of the village of St-Émilion. The property is under conversion to organic agriculture.

RED The delicious, silky-seductive fruit in this consistently well-made wine is very stylish and harmonious.

🍷 Merlot 80%, Cabernet Franc 12% Petit Verdot 8%

Other wines: *Caliceum (made from old vines)*

CHÂTEAU CROQUE MICHOTTE

Grand cru classé until 1996, this property certainly deserves its former status. The wine is fermented in stainless steel and matured in 60 per cent new oak for between 18 and 24 months, with up to one-third new oak.

RED A delightful and elegant style of wine, brimming with juicy, soft, and silky Merlot fruit.

🍷 Merlot 74%, Cabernet Franc 25%, Cabernet Sauvignon 1%

🍷 5–12 years

CHÂTEAU CURÉ BON LA MADELEINE
Grand Cru Classé

Surrounded by *premier grands crus classés* such as Ausone, Belair, and Canon, this property has had an excellent record, but was absorbed by Château Canon in the summer of 2000.

CHÂTEAU DASSAULT
Grand Cru Classé
★☆

This property was promoted to *grand cru classé* in the 1969 revision of the 1954 Saint-Émilion classification. It has an excellent record and more than deserves its classification. The wine is fermented in a mixture of barrels, cement, and stainless steel and matured in 65 per cent new oak for 12 months. With its beautifully understated Lafite-like label, Dassault's presentation is perfect.

RED Supremely elegant wines that always display a delicate marriage of fruit and oak in perfect balance, with fine acidity and supple tannin.

🍷 Merlot 70%, Cabernet Franc 23%, Cabernet Sauvignon 7%

🍷 8–25 years

Second wine: *Le D de Dassault*
Other wines: *Château la Fleur, Château Faurie de Souchard*

CHÂTEAU LA DOMINIQUE
Grand Cru Classé
★★

This property, one of the best of the *grands crus classés,* is situated close to

Château Cheval Blanc on the graves in the extreme west of St-Émilion and enjoys a mix of the clay and gravel soil of Pomerol and the clay and limestone of St-Émilion. In 2012 a brand-new cellar with stainless-steel vats was opened with a fashionable restaurant on the rooftop. It is matured in wood for 16 months, with 60 per cent new oak.

RED Very open and expressive wines that are plump and attractive, full of ripe, creamy fruit with elegant underlying oak.

🍷 Merlot 81%, Cabernet Franc 16%, Cabernet Sauvignon 3%

🍷 8–25 years

Second wine: *Relais de la Dominique*

CHÂTEAU FAURIE DE SOUCHARD

Demoted from *grand cru classé* in 2006 and reinstated in 2012, the property was purchased by Dassault in 2013.

RED Recent vintages have increased in concentration and colour, and the wine is now aged in 75 per cent new oak

🍷 Merlot 70%, Cabernet Franc 25%, Cabernet Sauvignon 5%

🍷 4–7 years

Second wine: *Pavillon de Faurie de Souchard*

CHÂTEAU DE FERRAND
Grand Cru Classé

Owned by the family of Marcel Bich, this property situated on the limestone plateau above the Dordogne valley has been beautifully restored and are making wines of a quality meriting its *cru classé* status. The wine is aged for 14 to 16 months in 50 per cent new French oak barrels.

🍷 Merlot 75%, Cabernet Franc 15%, Cabernet Sauvignon 10%.

🍷 12–30 years

Second wine: *Le Différent de Château de Ferrand*

CHÂTEAU-FIGEAC
Premier Grand Cru Classé (B)
★★★

Some critics suggest that the unusually high proportion of Cabernet Sauvignon in the *encépagement* (varietal blend) of this great château is wrong, but owner Thierry de Manoncourt refuted this. After his death in 2010, his son-in-law Eric d'Aramon managed the property. In 2013 Jean-Valmy Nicolas took over as director, working closely with the family and overseeing considerable investment, including a new cellar that will be ready for the 2020 harvest. This château belongs with the elite of Ausone and its *graves* neighbor Cheval Blanc. The wine is matured in wood for 18 to 20 months, with 100 per cent new oak.

RED Impressively ripe, rich, and concentrated wines with fine colour, a beautiful bouquet, stunning creamy-ripe fruit, great finesse, and a wonderful spicy complexity.

🍷 Merlot 30%, Cabernet Franc 35%, Cabernet Sauvignon 35%

🍷 12–30 years

Second wine: *Petit-Figeac, La Grange Neuve de Figeac*

Other wines: *Château La Fleur Pourret, Château de Millery*

CHÂTEAU FONPLÉGADE
Grand Cru Classé

Since 2004 Château Fonpleagde has been owned by Denise and Stephen Adams, Americans with a wine interest in Napa. They have invested heavily in the property, both the vines and the cellars. The wine is matured in wood for 12 to 18 months, in 60 per cent new oak. Certified biodynamic since 2013.

RED A spectacular improvement in both quality and consistency since 2004. These are elegant wines with length and polished tannins.

🍷 Merlot 95%, Cabernet Franc 5%

🍷 5–12 years

Second wine: *Fleur de Fonplégade*
Other wines: *Château de l'Enclos, Pomerol*

CHÂTEAU FONROQUE
Grand Cru Classé

Located just northwest of St-Émilion itself, this secluded property was certified organic in 2006 and biodynamic 2008. In 2001 Alain Moueix inherited it and though under the ownership of the Guillard family since 2017, Moueix continues to manage it. The wine is matured in 30 per cent new oak.

RED This is a deep-coloured, well-made wine with a fine plummy character that shows better on the bouquet and the initial and middle palate than on the finish.

🍷 Merlot 80%, Cabernet Franc 20%

🍷 6–15 years

Second wine: *Château Cartier*

CHÂTEAU LA GAFFELIÈRE
Premier Grand Cru Classé (B)
★☆

This property has belonged to the Malet-Roquefort family for three centuries. The cellar was completely re-structured in 2013 to concentrate on plot-by-plot vinification. After a string of aggressive, ungenerous vintages, Château la Gaffelière has produced increasingly excellent wines since the mid-1980s. The wine is matured in wood for 14 to 16 months, with 50 per cent new oak.

RED These wines are concentrated and tannic, but they now have much more finesse, fat, and mouth-tingling richness than previously.

🍷 Merlot 75%, Cabernet Franc 25%

🍷 12–35 years

Second wine: *Clos la Gaffelière*
Other wines: *Château Armens, Château Chapelle d'Alienor, Château La Connivence (Pomerol)*

CHÂTEAU LA GOMERIE
Grand Cru
★★

Owned by Gérard and Dominique Bécot of Château Beau-Séjour-Bécot fame, I rather get the impression that the tiny production of this 100 per cent Merlot vinified in 100 per cent new oak is the vinous equivalent of sticking one finger up to the authorities. The unfair demotion of Beau-Séjour-Bécot in 1985 (rectified in 1996) was a cruel and unjustified blow to the Bécots, who had improved the quality of their wine. By producing an unclassified super-premium St-Émilion that demands and receives a higher price than its own *premier grand cru classé,* the Bécots have demonstrated that the classification is meaningless. La Gomerie was included in the blend of Château Beau-Séjour-Bécot from 2012 until 2015, but since 2016 it is produced once again under a separate label.

RED Masses of rich, ripe Merlot fruit dominate this wine, despite its 100 per cent new oak. A stunning wine that deserves its cult following.

🍷 *Merlot 100%*

🍷 *4–18 years*

CHÂTEAU GRAND CORBIN
Grand Cru Classé
★

Since the 2012 classification, Château Haut Corbin and Château Grand Corbin, which both belong to the Mutuelles d'Assurance du Bâtiment et des Travaux Publics have merged into a single property. This wine is fermented in a mix of wood and concrete tanks and aged in 50 per cent new oak.

RED A dark, deeply flavoured, full-bodied wine that has rich, juicy Merlot fruit and some finesse.

🍷 *Merlot 70%, Cabernet Franc 25%, Cabernet Sauvignon 5%*

🍷 *6–15 years*

Second wine: *Les Charmes de Grand Corbin*

ENTRANCE TO CHÂTEAU LA GAFFELIÈRE
The estate's signboard proudly proclaims the chateau's *premier grand cru classé* status.

CHÂTEAU GRAND-CORBIN-DESPAGNE
Grand Cru Classé
★

This part of the Corbin estate was bought by the Despagne family – hence the name. It was demoted in the 1996 St-Émilion Classification but reinstated in 2006 under the management of the genial François Despagne, seventh generation of the family to run the vineyard. The vineyard has been farmed sustainably under the Terravitis certification since 2005 and certified organic since 2010 – no pesticides have ever been used. The wine is fermented in a mix of stainless steel and traditional cement vats and matured in wood for up to 18 months, with some new oak.

RED A well-coloured wine of full and rich body with plenty of creamy fruit and oak, supported by supple tannin.

🍷 *Merlot 65%, Cabernet Franc 24%, Cabernet Sauvignon 1%*

🍷 *7–25 years*

Second wine: *Petit Corbin Despagne*

Other wines : *Château Le Chemin à Pomerol, Château Ampélia en Castillon – Côtes de Bordeaux, Château Reine Blanche Saint-Émilion Grand Cru*

CHÂTEAU GRAND MAYNE
Grand Cru Classé
★☆

This château ferments its wine in a mixture of oak and stainless steel vats and ages it in wood with 70 per cent new oak.

RED This is a firm, fresh, and fruity style of wine that had a rather inconsistent reputation until the 1990s when the wines have had much more richness than in previous years.

🍷 *Merlot 80%, Cabernet Franc 15%, Cabernet Sauvignon 5%*

🍷 *4–10 years*

Second wine: *Filia de Grand Mayne*

CHÂTEAU GRANDES MURAILLES
Grand Cru Classé
★

This property was demoted from its *grand cru classé* status in 1985, unjustly I think. The wines produced here were better and more consistent than those of many châteaux that were not demoted. Château Grandes Murailles was, however, promoted back to its previous status in the 1996 classification. Since 2013 the Cuvelier family of Château Clos Fourtet have owned the property. The wine is fermented in stainless steel vats and matured in 100 per cent new oak.

RED These elegant, harmonious wines have good extract and a supple tannin structure that quickly softens. They are a delight to drink when relatively young, although they also age gracefully.

🍷 *Merlot 100%*

🍷 *5–20 years*

CHÂTEAU GRAND-PONTET
Grand Cru Classé
★☆

The wine is matured in wood for 12 to 18 months with 70 per cent new oak.

RED After a string of very dull vintages, this property is now producing full-bodied wines of fine quality and character. They are fat and ripe, rich in fruit and tannin, with delightful underlying creamy-oak flavours.

🍷 *Merlot 70%, Cabernet Franc 15%, Cabernet Sauvignon 15%*

🍷 *6–15 years*

CHÂTEAU GUADET
Grand Cru Classé
☆

This property deserves its status. The wines are matured in wood for 18 to 20 months, using up to 40 per cent new oak.

RED These are wines that show the silky charms of Merlot very early, and then tighten up for a few years before blossoming into finer and fuller wines.

🍷 *Merlot 75%, Cabernet Franc 25%*

🍷 *7–20 years*

CHÂTEAU HAUT-CORBIN

Merged into Château Grand Corbin after the 2012 classification. The 2011 was the last vintage.

CHÂTEAU HAUT-SARPE
Grand Cru Classé
★

Although not one of the top performers, this château certainly deserves its status. The property belongs to the other right bank *negociant* Janoueix. The wine is matured in wood for 18 months, using 60 per cent new oak.

RED Elegant, silky, and stylish medium-bodied wines that are best appreciated when young.

🍷 *Merlot 70%, Cabernet Franc 30%*

🍷 *4–8 years*

CHÂTEAU JEAN FAURE
Grand Cru Classé

This property was demoted from its *grand cru classé* status in 1985 but included in the 2012 classification. Situated near Pomeol, it is marked by its high percentage of Cabernet Franc. The wine is matured in wood for 18 months, with 50 per cent new oak. Certified organic since 2017.

RED These wines have good colour and easy, attractive, supple fruit.

🍷 *Cabernet Franc 50%, Merlot 45%, Malbec 5%*

🍷 *3–8 years*

CHÂTEAU LAMARZELLE
Grand Cru Classé
☆

This property preserved its classification in 1996, while Château Grand Barrail Lamarzelle-Figeac was demoted – only to follow suit in 2006 and then be reinstated in 2012.

RED Forward, fruity wines

🍷 *Merlot 70%, Cabernet Franc 20 %, Cabernet Franc 10%*

🍷 *3–7 years*

CHÂTEAU LANIOTE
Grand Cru Classé
☆

An old property that incorporates the "Holy Grotto" where St-Émilion lived in the eighth century. The wine is fermented in concrete and aged in 40 to 50 per cent new oak for 12 to 16 months.

RED Stylish medium-bodied wines, with plenty of fresh, elegant fruit.

🍷 *Merlot 80%, Cabernet Franc 15%, Cabernet Sauvignon 5%*

🍷 *6–12 years*

CHÂTEAU LARCIS DUCASSE
Grand Cru Classé
★☆

This property, whose vineyard is situated on the Côte de Pavie, matures its wine in 75 per cent new oak for 24 months.

RED Full, rich wines, particularly in the best years.

🍷 *Merlot 83%, Cabernet Franc 17%*

🍷 *4–8 years*

CHÂTEAU LARMANDE
Grand Cru Classé
★☆

Owned by Château Soutard, consistently one of the best *grands crus classés* in St-Émilion. The wine is fermented in stainless stee and matured in wood for 12 to 18 months, with 60 per cent new oak.

RED Superb wines typified by their great concentration of colour and fruit. They are rich and ripe with an abundancy of creamy-cassis and vanilla flavours that develop into a cedarwood complexity.

🍷 *Merlot 65%, Cabernet Franc 30%, Cabernet Sauvignon 5%*

🍷 *8–25 years*

CHÂTEAU LAROQUE
Grand Cru Classé
★

Made a *grand cru classé* in 1996, and re-confirmed in the recent 2012 classification.

RED As smooth and fruity as might be expected, with good tannic edge and increasing oak influence.

🍇 *Merlot 87%, Cabernet Franc 11%, Cabernet Sauvignon 2%*

🍷 *4–16 years*

Second wine: *Les Tours de Laroque*

CHÂTEAU LAROZE
Grand Cru Classé
★ Ⓥ

This 19th-century château was named Laroze after a "characteristic scent of roses" was said to be found in its wines. The wine is matured in 65 per cent new oak for 16 months.

RED The wine does have a soft and seductive bouquet, although I have yet to find "roses" in it. It is an immediately appealing wine of some finesse that is always a delight to drink early.

🍇 *Merlot 68%, Cabernet Franc 26%, Cabernet Sauvignon 6%*

🍷 *4–10 years*

Second wine: *La Fleur Laroze, Lady Laroze*

CHÂTEAU MAGDELAINE

Merged with Château Bélair-Monange. The 2012 classification confirmed both as *premier grands crus classés*. The 2011 was the last vintage under the name of Château Magdelaine.

CHÂTEAU MATRAS

Purchased by Château Canon in 2011, the vineyard is now part of Château Canon, with most of the vines going into the second wine of the property; Croix Canon. The cellars and an old chapel have been completely renovated, and it is the religious connotation that inspired the name.

CHÂTEAU MONBOUSQUET
Grand Cru Classé
★☆

This was hypermarket owner Gérard Perse's first venture into wine, and he openly admits that he bought the property for its beauty, rather than out of any detailed analysis of its viticultural potential. That said, with the help of consultant Michel Rolland, he took this château to unbelievable heights and, having done so, set about analysing what could and should be purchased in St-Émilion from a purely viticultural perspective. He set his sights on Pavie and Pavie Decesse, which he purchased in 1998 and 1997. Perse sold a portion of Château Monbousquet to an anonymous French pension fund in 2013.

RED Voluptuous, velvety, and hedonistic, these wines lack neither complexity nor finesse, but they are so delicious to drink that their more profound qualities easily slip by. Or should that be slip down?

🍇 *Cabernet Sauvignon 10%, Cabernet Franc 30%, Merlot 60%*

🍷 *4–15 years*

Second wine: *Angelique de Monbousquet*
Other wines: *Monbousquet Blanc (Bordeaux AOC)*

CHÂTEAU MONDOTTE
Premier Grand Cru Classé
★★ Ⓞ

Since the 1996 vintage Stephan von Neipperg's *vin de garage* has been classified as a *premier grand cru classé* and by 2012 had surpassed the quality and price of his excellent *grand cru classé* Château Canon-la-Gaffelière. Low yield, 100 per cent oak for an average of 18 months, and ludicrous prices. Certified organic in 2014.

RED Extraordinary colour, density, and complexity for a wine that is not in the slightest bit heavy and makes such charming and easy drinking.

🍇 *Cabernet Franc 25%, Merlot 75%*

🍷 *5–20 years*

CHÂTEAU MOULIN DU CADET
Grand Cru Classé
★ Ⓞ Ⓥ

The property belongs to the Lefévère family, also owners of Château Sansonnet. The wine is aged in 60 per cent new oak for 12 to 18 months.

RED These wines have good colour, a fine bouquet, delightfully perfumed Merlot fruit, excellent finesse, and some complexity. They are not full or powerful, but what they lack in size, they more than make up for in style.

🍇 *Merlot 100%*

🍷 *6–15 years*

CHÂTEAU PAVIE
Premier Grand Cru Classé (A)
★★ Ⓥ

Gérard Perse bought his top-performing château 1998, along with Pavie Decesse and La Clusière. Perse used to own a group of hypermarkets, but his love of fine wine began to take over his working life when in 1993 he purchased Monbousquet. The link between Pavie under its previous owners, the Valette family, and Perse is Michel Rolland, who remained consultant. Pavie has produced some of the greatest wines of St-Émilion in recent decades, resulting in the promotion to Premier Grand Cru Classé (A) in 2012. The wine is matured in wood for 18 to 32 months, with 70 to 100 per cent new oak.

RED Great, stylish wines packed with creamy fruit and lifted by exquisite new oak. Fabulous concentration since 1998, without losing any finesse, although I will keep an open mind about the 2003 for a decade or so!

🍇 *Merlot 65%, Cabernet Franc 25%, Cabernet Sauvignon 10%*

🍷 *8–30 years*

Second wine: *Aromes de Pavie*
Other wines: *Château Pavie Decesse, Château Bellevue Mondotte, Clos*

Lunelles (Côtes de Bordeaux Castillon), Esprit de Pavie (Bordeaux AOC)

CHÂTEAU PAVIE DECESSE
Grand Cru Classé
★☆

This property was under the same ownership as Château Pavie when the Valette family were the owners and still is under Gérard Perse. Although it is not one of the top *grands crus classés*, it is consistent and certainly worthy of its status. The wine is aged for 18 to 24 months in new oak.

RED There was a huge sea change in colour, quality, and concentration as of the 1998 and 1999 vintages.

🍇 *Merlot 90%, Cabernet Franc 10%*

🍷 *6–12 years*

CHÂTEAU PAVIE MACQUIN
Premier Grand Cru Classé (B)
★☆

This property was named after Albert Macquin, a local grower who pioneered work to graft European vines on to American rootstock. These wines noticeably improved in the 1990s when Nicolas Thienpont of Vieux Château Certan began overseeing the production. The wine is aged for 16 to 20 months in 60 per cent new oak.

RED Much richer, with more fruit and new oak in recent years.

🍇 *Merlot 84%, Cabernet Franc 14%, and Cabernet Sauvignon 2%*

🍷 *4–8 years*

Second wine: *Les Chênes de Maquin*

CHÂTEAU PETIT FAURIE DE SOUTARD
Grand Cru Classé
★☆

Some poor wines produced in the mid-1990s and at the turn of the millennium resulted in demotion from *grand cru classé* in 2006, but great improvements from 2003 resulted in reinstatement in 2012. Half of its production is matured in new oak for up to a year.

RED This wine has soft, creamy aromas on the bouquet, some concentration of smooth Merlot fruit on the palate, a silky texture, and a dry, tannic finish. It is absolutely delicious when young but gains a lot from a little bottle-age.

🍇 *Merlot 65%, Cabernet Franc 30%, Cabernet Sauvignon 5%*

🍷 *3–8 years*

CHÂTEAU PIPEAU
Grand Cru
★ Ⓥ

The wine of Château Pipeau is matured in wood for 18 months, with 100 per cent new oak.

RED Rich, stylish and quite striking wines that are full of fruit, underpinned by creamy-smoky oak, and easier to drink younger than most St-Émilions.

🍇 *Merlot 90%, Cabernet Franc 5%, Cabernet Sauvignon 5%*

🍷 *4–10 years*

CHÂTEAU LE PRIEURÉ
Grand Cru Classé
★☆

Dating back to 1696, since 2018 this property is under the same ownership as Château Siaurac in Lalande-de-Pomerol, the Artemis group. The wine produced here is matured in wood for 18 to 24 months, with 40 per cent new oak.

RED Light but lengthy wines of some elegance that are best enjoyed when young and fresh.

🍇 *Merlot 80%, Cabernet Franc 20%*

🍷 *4–8 years*

Second wine: *Le Délice du Prieuré*

CHÂTEAU QUINAULT L'ENCLOS
Grand Cru Classé
★ Ⓥ

Château Quinault l'Enclos is one of the wines that surfed on the "garage wine" phenomena following its purchase by Dr Alain Raynaud in 1997, who then sold it to Château Cheval Blanc in 2008. The property is within the neighbouring town of Libourne and has a unique warmer microclimate with a lot of gravel in the soil. A new cellar was built in 2017. This wine is matured in wood for 18 months, with 50 per cent new oak.

RED Exceptionally concentrated and complex.

🍇 *Merlot 75 %, Cabernet Franc 15 %, Cabernet Sauvignon 15%*

🍷 *6–20 years*

CHÂTEAU QUINTUS

This wine estate was created in 2011 when the Dillon family (of Château Haut-Brion) bought Château Tertre Daugay and renamed it Château Quintus. In 2016 the Dillon family bought neighbouring *grand cru classé* Château l'Arrosée and integrated into the vineyard, which now dominates the limestone outcrop to the southwest of the St-Émilion plateau. The wine produced here aged in 40 per cent new oak for 12 months.

🍇 *Merlot 75%, Cabernet Franc 25%, Cabernet Sauvignon 3%*

Second wine: *Le Dragon de Quintus*
Other wines: *Le Saint Émilion de Quintus*

CHÂTEAU ST-GEORGES (CÔTE PAVIE)
Grand Cru Classé
★

Owned by Jacques Masson, this small property's vineyard is well situated on the Côte Pavie, lying close to those of Château Pavie and Château la Gaffelière. Stéphane Derenoncourt has consulted for the property since 2016. The wine is fermented in stainless steel and matured in one-third new oak for 14 months.

RED This is a delicious medium-bodied wine with plump, spicy-juicy Merlot fruit, made in an attractive early-drinking style that does not lack finesse.

🍇 *Merlot 80%, Cabernet Franc 20%*

🍷 *4–8 years*

CHÂTEAU SANSONNET
Grand Cru Classé

The property has been owned by the Lefévère family since 2009. Even before their ownership much improvement had been made, resulting in its inclusion in the 2012 classification.

RED Supple and attractive.

🍷 *Merlot 85%, Cabernet Franc 15%*

🍷 *3–7 years*

Other wines: *Château Moulin du Cadet, Château Soutard-Cadet, Château Harmonie*

CHÂTEAU LA SERRE
Grand Cru Classé

This is another property that has seen tremendous improvement in quality. It occupies two terraces on St-Émilion's limestone plateau, one in front of the château and one behind. The wine is fermented in lined concrete tanks and matured in wood for 12 to 18 months with 50 per cent new oak.

RED This wine initially charms, then goes through a tight and sullen period, making it reminiscent of Château Guadet St-Julien. Their styles, however, are very different. When young, this is quite a ripe and plump wine, totally dominated by new oak. In time, the fruit emerges to form a luscious, stylish wine of some finesse and complexity.

🍷 *Merlot 80%, Cabernet Franc 20%*

🍷 *8–25 years*

Second wine: *Menuts de la Serre*

CHÂTEAU SOUTARD
Grand Cru Classé
★ ☆ V

The large and very fine château on this estate was built in 1740 for the use of the Soutard family in summer. Vines have grown here since Roman times. In 2006 French insurance company La Mondial bought the property from the de Lineris family. La Mondial already owned neighbouring Château Larmande and Château Grand Faurie la Rose. They have since invested heavily in spectacular gravity-fed cellars, in replanting, and in renovating the Château. The wine is matured in wood for 18 months, with up to one-third new oak casks.

RED This dark, muscular, and full-bodied wine is made in true *vin de garde* style, which means it improves greatly while ageing. It has great concentrations of colour, fruit, tannin, and extract. With time it can also achieve great finesse and complexity.

🍷 *Merlot 63%, Cabernet Franc 28%, Cabernet Sauvignon 7%*

🍷 *12–35 years*

Second wine: *Jardins de Soutard*

THE BEST OF THE REST

With more than a thousand châteaux in this one district, it not practical to feature every recommendable St-Émilion wine. This is a list of the best of the rest: châteaux that consistently make wine that stands out for either quality or value, sometimes both. Those marked with a star (*) sometimes produce wines that are better than many *grands crus classés.*

*Château Bellefont-Belcier †
*Château La Bienfaisance
*Château Cantenac (since 2002)
*Château Carteau Côtes Daugay
*Château Le Castelot (organic)
*Château Le Châtelet †
Château Cheval Noir
Clos Cantenac
*Château la Commanderie †
(see Clos des Jacobins)
*Château Destieux†
*Château Faugères
*Château Ferrand-Lartique
(since 2003)
*Château la Fleur (see Château Dassault)
*Château Fleur Cardinale †
*Château Fleur-Cravignac
*Château la Fleur Morange †
Château la Fleur Pourret
*Château Fombrauge †
Château Grand Barrai Lamarzelle-Figeac
*Château la Grave Figeac
*Château Haut Brisson
*Château Haut-Pontet

Château Lapelletrie
*Château Laroque †
*Lucia
*Château Magnan la Gaffelière
*Château Moulin-St-Georges (see Ausone)
Château Patris
*Château Petit-Figeac (see Château Figeac)
Château Petit-Gravet
*Château Petit Val
Château Peyreau (see Château Canon La Gaffeliere)
*Château Pindefleurs
*Château Plaisance
Château de Pressac †
Château Puy Razac
Château Roc Blanquant – 3rd wine of Belair Monagne (see above)
*Château Rol Valentin (since 1999)
*Château Teyssier
Château Tour St-Christophe

† *Grand cru classé in 2012*

CHÂTEAU TERTRE DAUGAY
★ ☆

The Dillon family, owners of Château Haut-Brion and La Mission Haut-Brion, bought this property from Comte Léo de Malet-Roquefort in 2011 and renamed it Quintus. The final vintage was 2010.

CHÂTEAU TERTRE-RÔTEBOEUF
★★

François Mitjavile's cult wine is yet more proof that the only important classification is made by the consumer. They choose not to be part of the Saint Émilion Classification. The wine is aged in 100 per cent new oak.

RED Huge, oaky, complex, and cultish: the Leonetti of St-Émilion!

🍷 *Cabernet Franc 20%, Merlot 80%*

🍷 *5–20 years*

CHÂTEAU LA TOUR FIGEAC
Grand Cru Classé
★ ☆

This property was attached to Château Figeac in 1879, and today it is one of the best of the *grands crus classés.* Since 1994 it has been managed by Otto Rettenmaier, who was one of the early converts to organic and biodynamic agriculture. The wine is matured in wood for 13 to 15 months, with 50 per cent new oak.

RED These are fat and supple wines with a very alluring bouquet and masses of rich, ripe cassis fruit gently supported by smoky-creamy oak.

🍷 *Merlot 65%, Cabernet Franc 35%*

🍷 *4–8 years*

Second wine: *L'Esquisse de Château la Tour Figeac*

CHÂTEAU LA TOUR DU PIN FIGEAC
★ ☆

This property was sold by Antoine Moueix to neighbouring Cheval Blanc in 2006. Part of the vineyard has now been included in the Cheval Blanc vineyard, and part is used to make the new Petit Cheval Blanc white wine from the property. The label no longer exists.

CHÂTEAU TRIMOULET
Grand Cru Classé
★

This is an old property overlooking St-Georges–St-Émilion. The wine is matured in wood for 12 months, with 100 per cent new oak.

RED This well-coloured wine has an overtly ripe and fruity aroma, lots of creamy-oaky character, a fruit flavour, and supple tannin.

🍷 *Merlot 60%, Cabernet Franc 20%, Cabernet Sauvignon 20%*

🍷 *7–20 years*

CHÂTEAU TROPLONG MONDOT
Premier Grand Cru Classé (B)
★★

This property was promoted to Grand Cru Classé (B) in 2006 and reconfirmed in 2012 thanks to the hard work of Christine Valette. Sadly Christine died in 2014, and French reinsurance company SCOR purchased the property in 2017. The wine is matured in wood for 18 months with 65 per cent new oak.

RED A powerful, big-boned expression of St-Émilion with lots of ripe black fruit and structured tannins; in the best years it is destined to age well.

🍷 *Merlot 90%, Cabernet Sauvignon 8%, Cabernet Franc 2%*

🍷 *8–15 years*

Second wine: *Mondot*

CHÂTEAU TROTTEVIEILLE
Premier Grand Cru Classé (B)
★ ☆

Owned by Bordeaux *négociant* Borie-Manoux, this property has the reputation of producing a star wine every five years or so interspersed by very mediocre wines indeed. It has, however, made very consistent since 1985 and now makes true *premier grand cru classé* quality wine every year. The wine is matured in wood for 12 to 18 months, with up to 100 per cent new oak.

RED The wine has fabulous Merlot-fruit richness with new oak and the power of a true *premier grand cru classé.*

🍷 *Merlot 55%, Cabernet Franc 40%, Cabernet Sauvignon 5%*

🍷 *8–25 years (successful years only)*

CHÂTEAU VALANDRAUD
Premier Grand Cru Classé
★★

Owned by Jean-Luc Thunevin, Château Valandraud is one of St-Émilion's best-known vins de garage, although it has grown from its original small plot of 0.6 hectares (1.5 acres) to 8.8 hectares (22 acres), making it more of a *parc de stationnement.* The wines are aged for 18 to 30 months in 100 per cent new oak.

RED Wines that even Gary Figgins of Leonetti in Washington state might complain were too oaky!

🍷 *Merlot 65%, Cabernet Franc 25%, Cabernet Sauvignon 5%, Malbec 4%, Carmenère 1%*

🍷 *6–25 years*

CHÂTEAU VILLEMAURINE
Grand Cru Classé

Since 2005 Château Villemaurine has belonged to the Onclin family, Bordeaux wine merchants and owners of Château Branas Grand Poujeaux. The wine is matured in 70 to 90 per cent new oak for 16 to 18months. Demoted from *grand cru classé* in 2006, it was reinstated in 2012.

RED These are full-bodied wines of excellent, spicy Merlot fruit, good underlying oak, and firm structure.

🍷 *Merlot 80%, Cabernet Sauvignon 20%*

🍷 *8–25 years*

Second wine: *Les Angelots de Villamaurine*

Other wines: *Clos Larcis*

The Libournais and Fronsadais:
Pomerol

The most velvety and sensuous clarets are produced in Pomerol, yet the traveller passing through this small and rural area, with its dilapidated farmhouses at every turn, few true châteaux, and no really splendid ones, must wonder how this uninspiring area can produce such magnificently expensive wines.

The prosperity of recent years has enabled Pomerol's properties to indulge in more than just an extra lick of paint, but renovation can only restore, not create, and Pomerol essentially remains an area with an air of obscurity. Even Château Pétrus, which is the greatest growth of Pomerol and produces what for the last 20 years has been consistently the world's most expensive wine, is nothing more than a simple farmhouse.

There has been no attempt to publish an official classification of Pomerol wines, but Vieux Château Certan was considered to be its best wine in the 19th century. Nowadays, however, Pétrus is universally accepted as the leading growth. It commands prices that dwarf those of wines such as Mouton and Margaux, so could not be denied a status equivalent to that of a *premier cru*. Indeed, Le Pin has become considerably more expensive than Pétrus itself. If, like the 1885 classification, we classify Pomerol according to price, next would come Lafleur, followed by a group including L'Évangile, La Fleur-Pétrus, La Conseillante, Trotanoy, and a few others, all of which can cost as much as a Médoc First Growth. It is difficult to imagine, but Pomerol was ranked as an inferior sub-appellation of St-Émilion until it obtained its independent status in 1900. Even then it was a long hard struggle, since even Pétrus did not become sought after until the mid-1960s.

FACTORS AFFECTING TASTE AND QUALITY

LOCATION
Pomerol is a small rural area on the western extremity of the St-Émilion district, just northeast of Libourne.

CLIMATE
The same as for St-Émilion (*see* p 189).

ASPECT
This modest mound, with Château Pétrus and Château Vieux Certan situated at its centre, is the eastern extension of the Pomerol-Figeac *graves* (gravelly terrain). The vines grow on slightly undulating slopes that, over a distance of 2 kilometres (1.2 miles), descend from between 35 and 40 metres above sea level (115–130 feet) to 10 metres (33 feet).

SOIL
Pomerol's soil is sandy to the west of the national highway and to the east, where the best properties are situated on the sandy-gravel soil of the Pomerol-Figeac graves. The subsoil consists of an iron-pan known as *crasse de fer* or *machefer*, with gravel in the east and clay in the north and centre. The château of Pétrus lies in the very centre of the Pomerol-Figeac *graves*, on a unique geological formation of sandy-clay over *molasse* (sandstone).

VITICULTURE & VINIFICATION
Some of Pomerol's châteaux use a proportion of *vin de presse* according to the requirement of the vintage. At Pétrus, the *vin de presse* is added earlier than is normal practice, in order to allow it to mature with the rest of the wine – this is believed to reduce harshness. The duration of skin-contact is usually between 15 and 21 days, but is sometimes as brief as 10 days or as long as four weeks. The wines stay in cask for between 18 and 20 months.

GRAPE VARIETIES
Primary varieties: Cabernet Franc, Cabernet Sauvignon, Merlot
Secondary varieties: Malbec

VINE-GROWING AREAS OF POMEROL (*see also* p141)
The sleepy area of Pomerol and Lalande-de-Pomerol fans out above the riverside town of Libourne. None of the so-called châteaux is particularly imposing: among the most attractive are Château Nénin and Vieux Château Certan.

THE CHÂTEAU CLINET PROPERTY SIGN GREETS VISITORS TO THE WINE ESTATE
One of the oldest vineyards in Pomerol, this estate has been planting vines since the late 18th century.

LALANDE-DE-POMEROL

AOC vineyard
Communal boundary

0 mi 1/2
0 km 1/2

Barbanne

Chevrol

NÉAC

POMEROL

Néac

Pomerol

LIBOURNE

Ruisseau du Mauvais Temps

Dordogne

Catusseau

La Bordette

Libourne

SAINT-ÉMILION

L'Épinette

Ruisseau du Taillas

THE WINE PRODUCERS OF
POMEROL

CHÂTEAU BEAUREGARD
★☆

An American architect who visited Pomerol after World War I built a replica of Beauregard called Mille Fleurs on Long Island, New York. Quality changed dramatically at Beauregard in 1985, two years before the arrival of Michel Rolland, who followed 1985's superb wine with others and was generally responsible for turning this château around. The property was purchased by the Cathiard (Smith Haut Lafitte) and Moulin (Galleries Lafayette) families in 2014. They have overseen a dramatic renovation of the cellars with new concrete vats. The property was certified organic in 2014. The wine is matured in wood for 15 to 20 months with 60 per cent new oak.

RED Firm, elegant, and lightly rich wine with floral-cedarwood fruit.

🍷 *Merlot 70%, Cabernet Franc 25 %, Cabernet Sauvignon 5%*

🍷 *5–10 years*

Second wine: *Le Benjamin de Beauregard*
Other wines: *Chateau Pavillon Beauregard, Lalande de Pomerol*

CHÂTEAU BONALGUE

This small property lies on gravel and sand northwest of Libourne and belongs to the Libournais *négociant* Audy. The wine is matured in 50 per cent new oak for 18 months.

RED This medium- to full-bodied wine has always been of respectable quality with a frank attack of refreshing fruit flavours, a supple tannin structure, and a crisp finish.

🍷 *Merlot 90%, Cabernet Franc 10%*

🍷 *5–10 years*

CHÂTEAU LE BON PASTEUR
★

This is the personal property of wine consultant Michel Rolland, where he puts into action his signature winemaking techniques of low yields, ripe fruit, and 100 per cent new oak ageing. The wine is matured in 100 per cent new oak wood for 15 to 18 months.

RED These intensely coloured, full-bodied, complex wines are packed with cassis, plum, and black-cherry flavours.

🍷 *Merlot 80%, Cabernet Franc 20%*

🍷 *8–25 years*

CHÂTEAU BOURGNEUF
☆ **V**

This property is situated close to Château Trotanoy. It has an honourable, if not exciting, record and a 10-hectare (25-acre) vineyard.

RED Made in a quick-maturing style with soft fruit and a light herbal finish.

🍷 *Merlot 90%, Cabernet Franc 10%*

🍷 *4–8 years*

CHÂTEAU LA CABANNE
★

This is a fine estate producing increasingly better wine. The wine is matured in wood for 18 months, with one-third new oak.

RED These medium-bodied, and sometimes full-bodied, wines have fine, rich, chocolaty fruit.

🍷 *Merlot 92%, Cabernet Franc 8%*

🍷 *7–20 years*

Second wine: *Domaine de Compostelle*

CHÂTEAU CERTAN DE MAY DE CERTAN
★★

This can be a confusing wine to identify because the "De May de Certan" part of its name is in very small type on the label and it is usually referred to as "Château Certan de May". It is matured in wood for 16 to 20 months, with 80 per cent new oak.

RED This is a firm and tannic wine that has a powerful bouquet bursting with fruit, spice, and vanilla.

🍷 *Merlot 70%, Cabernet Franc 25%, Cabernet Sauvignon 5%*

🍷 *15–35 years*

CHÂTEAU CLINET
★☆

This wine, which is matured in wood with 60 to 75 per cent new oak, has undergone a revolution in recent years. It used to disappoint those looking for the typically fat, gushy-juicy style of Pomerol, and critics often blamed this on the wine's high proportion of Cabernet Sauvignon. The 1985 vintage was more promising (a touch plumper than previous vintages, with more juicy character), so Clinet's previous lack of typical Pomerol character was evidently not entirely due to the blend of grape varieties, although the vineyard has since undergone a radical change in varietal proportions. The excellent 2009 vintage really saw this wine take off, and the 2010, 2015, and 2016 are not too far behind.. Clinet has been under the ownership of the Laborde family since 1999 and the management of Ronan Laborde since 2003.

RED Château Clinet is now producing exceedingly fine, rich, ripe wine with ample, yet supple tannin structure mixed with oaky tannins to produce a creamy-herbal-menthol complexity.

🍷 *Merlot 88%, Cabernet Sauvignon 12%*

🍷 *7–20 years*

CLOS DU CLOCHER
★☆ **V**

This belongs to the Libournais *négociant* Audy. The wine is aged in oak barrels: two-thirds new oak and one-third one-year-old casks.

RED These are deliciously deep-coloured, attractive, medium-bodied, sometimes full-bodied, wines that have plenty of plump, ripe fruit, a supple structure, intriguing vanilla undertones, and plenty of finesse.

🍷 *Merlot 70%, Cabernet 30%*

🍷 *8–20 years*

Second wine: *Esprit de Clocher*
Other wines: *Château Monregard-Lacroix*

CLOS L'ÉGLISE
★☆

There are several "Église" properties in Pomerol. The wine from this one is matured in wood for 24 months, with some new oak.

RED A consistently attractive wine with elegant, spicy Merlot fruit and firm structure; it is eventually dominated by violet Cabernet perfumes.

🍷 *Merlot 80%, Cabernet Franc 20%*

🍷 *6–15 years*

Second wine: *Esprit de l'Église*

CLOS RENÉ
★ **V**

This property is situated just south of l'Enclos on the western side of the N89. The wine is matured in wood for 24 months with up to 15 per cent new oak. An underrated wine, it represents good value.

RED A splendid spicy-blackcurrant bouquet, plenty of fine plummy fruit on the palate, and a great deal of finesse. These wines are sometimes complex in structure, and are always of excellent quality.

🍷 *Merlot 70%, Cabernet Franc 20%, Malbec 10%*

🍷 *6–12 years*

Other wines: *Moulinet-Lasserre*

CHÂTEAU LA CONSEILLANTE
★★☆

If Pétrus is rated a "megastar", this property must be rated at least a "superstar" in the interests of fairness. The wine is matured in wood for 18 months, with 50 to 80 per cent new oak.

RED This wine has all the power and concentration of the greatest wines of Pomerol, but its priorities are its mind-blowing finesse and complexity.

🍷 *Cabernet Franc 20%, Merlot 80%*

🍷 *10–30 years*

Second wine: *Duo de Conseillante*

VINTAGE SIGNS OF BORDEAUX REGION WINE BRANDS AT A PARIS FLEA MARKET
Its neighbour Margaux produces one of the world's most famous wine, but Pomerol has a rich history of fine wine, especially clarets.

CHÂTEAU LA CROIX
★

Under the ownership of Jean-Philippe Janoueix, the same ownership as Château La Croix-St-Georges.

RED These attractive wines are quite full-bodied, yet elegant and quick-maturing, with fine, spicy Merlot fruit.

🍇 *Merlot 90%, Cabernet Franc 10%*

🍷 *5–10 years*

Second wine: *Le Gabachot*

CHÂTEAU LA CROIX DE GAY
★

This property is situated in the north of Pomerol on sandy-gravel soil, and the wine is matured in wood for 18 months, with up to 50 per cent new oak.

RED Used to be somewhat lightweight, but attractive, with easy-drinking qualities, the fruit in Croix de Gay has plumped-up in recent vintages.

🍇 *Merlot 95%, Cabernet Franc 5%*

🍷 *4–8 years*

Other wines: *Château le Commandeur, Vieux-Château-Groupey*

CHÂTEAU LA CROIX-ST-GEORGES
★

Under the ownership of Domaines Jean-Philippe Janoueix, this wine is matured in wood for 18 months, with 50 per cent new oak.

RED Rich, soft and seductive.

🍇 *Merlot 91%, Cabernet Franc 9%*

🍷 *4–10 years*

CHÂTEAU DU DOMAINE DE L'ÉGLISE

This is the oldest estate in Pomerol and now owned by the Casteja family of *négociant* house Borie Manou. The wine is matured in wood for 18 to 24 months, with 60 per cent new oak.

RED This is another attractive, essentially elegant wine that is light in weight and fruit.

🍇 *Merlot 95%, Cabernet Franc 5%*

🍷 *4–8 years*

CHÂTEAU L'ÉGLISE-CLINET
★★

The wine produced here is matured in wood for 15 to 17 months with 50 to 80 per cent new oak. Under the winemaking skill of Denis Durantou since 1983, they have become a Pomerol reference. Quality in overdrive since the 1990s.

RED Deep-coloured wines with a rich, seductive bouquet and a big, fat flavour bursting with spicy blackcurrant fruit and filled with creamy-vanilla oak complexity.

🍇 *Merlot 85%, Cabernet Franc 14%, Malbec 1%*

🍷 *8–30 years*

Second wine: *La Petite l'Église*

CHÂTEAU L'ENCLOS
★ Ⓥ

The vineyard is situated on an extension of the sandy-gravel soil from the better side of the N89. The wine is matured in wood for 12 to 18 months, with around one-third new oak.

RED These are deliciously soft, rich, and voluptuous wines, full of plump, juicy Merlot fruit and spice.

🍇 *Merlot 80%, Cabernet Franc 19%, Malbec 1%*

🍷 *7–15 years*

Second wine: *Le Petit Enclos*

CHÂTEAU L'ÉVANGILE
★★☆

Situated close to two superstars of Pomerol, Vieux Château Certan and Château la Conseillante, this château, owned by the Rothschilds of Lafite Rothschild, produces stunning wines that are matured in wood for 18 months with 70 per cent new oak.

RED Dark but not brooding, these fruity wines are rich and packed with summer fruits and cedarwood. More Merlot and new oak since 1998 has given the wine more generosity.

🍇 *Merlot 80%, Cabernet Franc 20%*

🍷 *8–20 years*

Second wine: *Blason de l'Evangile*

CHÂTEAU FEYTIT-CLINET

From 2000 this property has belonged exclusively to the Chasseuil family and is made by Jérémy Chasseuil. It is seeing huge improvement in recent vintages under his management. Some vines are over 70 years old. The wine is matured in 70 per cent new oak for 18 to 22 months.

RED Consistently well-coloured and stylish wines that are full of juicy plum and black cherry flavours.

🍇 *Merlot 90%, Cabernet Franc 10%*

🍷 *7–15 years*

Second wine: *Les Colombiers de Feytit-Client*

CHÂTEAU LA FLEUR-DE-GAY
★☆

Owned by the Raynaud family, owners of la Croix de Gay, with the ubiquitous Michel Rolland consulting, this wine is matured in wood for 18 months, with 100 per cent new oak.

RED Big, concentrated fruit underpinned by firm tannic structure.

🍇 *Merlot 85%, Cabernet Franc 15%*

🍷 *8–20 years*

CHÂTEAU LA FLEUR-PÉTRUS
★★

Owned by Libournais *négociant* J-P Moueix, Château La Fleur-Pétrus, producer of one of the best Pomerols, is situated close to Château Pétrus, but on soil that is more gravelly. Four hectares (10 acres) of Château le Gay were purchased in 1994 and incorporated in this property, fattening out the style. The wine is matured in 50 per cent new oak for 16 to 18 months.

RED Although recent vintages are relatively big and fat, these essentially elegant wines rely more on exquisiteness than richness. They are silky, soft, and supple.

🍇 *Merlot 91%, Cabernet Franc 6%, Petit Verdot 3%*

🍷 *6–20 years*

CHÂTEAU LE GAY
★ Ⓥ

The wine is matured in 100 per cent new oak for 18 months.

RED Firm and ripe, this big wine is packed with dense fruit and coffee-toffee oak.

🍇 *Merlot 90%, Cabernet Franc 10%*

🍷 *10–25 years*

Second wine: *Manoir de Gay*

CHÂTEAU GAZIN
★☆ Ⓥ

This château's record was disappointing until the stunning 1985 vintage, and it has been on a roll ever since, having abandoned harvesting by machine, introduced new, thermostatically controlled vats, and employed various quality-enhancing practices, not the least being a second wine, which enables stricter selection of grapes. The wine is matured in wood for 18 months with up to 50 per cent new oak.

RED Marvellously ripe and rich wine with plump fruit. It should have a great future.

🍇 *Merlot 90%, Cabernet Franc 3%, Cabernet Sauvignon 7%*

🍷 *8–20 years*

Second wine: *Hospitalet de Gazin*

CHÂTEAU LA GRAVE

The gravelly vineyard of this property has an excellent location. The Trignant de Boisset element of this château's name has been dropped. It is owned by Christian Moueix and farmed by J-P Moueix. The wine is matured in wood for 16 to 18 months with 40 per cent new oak.

RED Supple, rich and fruity, medium-bodied wines of increasing finesse.

🍇 *Merlot 85%, Cabernet Franc 15%*

🍷 *7–15 years*

CHÂTEAU HOSANNA
★☆ Ⓥ

This property was called Château Certan-Marzelle until 1956 and was purchased in 1999 by Jean-Paul Moueix, with Christian Moueix in control. Almost immediately 4 hectares (10 acres) of less well-positioned vineyards were sold to Nénin and strict selection imposed since the 1999 vintage. The wine is matured in wood for 16 to 18 months, with 50 per cent new oak.

RED These ripe, voluptuous wines become darker and denser since 1999.

🍇 *Merlot 70%, Cabernet Franc 30%*

🍷 *8–20 years*

CHÂTEAU LAFLEUR
★★☆

This property has a potential for quality second only to Château Pétrus itself, but it has a very inconsistent record. The quality and concentration of the wines have soared since 1985.

RED This is a well-coloured wine with a rich, plummy-porty bouquet, cassis fruit, a toasty-coffee oak complexity, and great finesse.

🍇 *Cabernet Franc 50%, Merlot 50%*

🍷 *10–25 years*

Second wine: *Les Pensées de Lafleur*

CHÂTEAU LAFLEUR-GAZIN
★

This property has been run by the J-P Moueix team on behalf of its owners since 1976. It produces a wine that is matured in 30 to 40 per cent new oak for 16 to 18 months.

RED Well-made wines of good colour and bouquet, supple structure, and some richness and concentration.

🍇 *Merlot 85%, Cabernet Franc 15%*

🍷 *6–15 years*

CHÂTEAU LAGRANGE
★ Ⓥ

Not to be confused with its namesake in St-Julien, this property belongs to the firm J-P Moueix. The wine is aged in wood for 16 to 18 months, in 40 per cent new oak.

RED The recent vintages of this full-bodied wine have been very impressive, with an attractive and accessible style.

🍇 *Merlot 95%, Cabernet Franc 5%*

🍷 *8–20 years*

CHÂTEAU LATOUR À POMEROL
★★ Ⓥ

Château Latour à Pomerol now belongs to the last surviving sister of Madame Loubat, Madame Lily Lacoste (owner of one-third of Château Pétrus) and has been farmed by J-P Moueix since 1962. The wine is matured in wood for 16 to 18 months with 40 per cent new oak.

RED These deep, dark wines are luscious, voluptuous, and velvety. They have a great concentration of fruit and a sensational complexity of flavours.

🍇 *Merlot 90%, Cabernet Franc 10%*

🍷 *12–35 years*

CHÂTEAU MAZEYRES
★

This 25.5-hectare (63-acre) property is managed by Alain Moueix and is certified biodynamic since 2018. 30 per cent of the wine is aged in new oak barrels, 50 per cent in barrels of one year, and 20 per cent in small concrete eggs

RED These elegant wines are rich, ripe, and juicy, and have silky Merlot fruit and some oaky finesse.

🍇 *Merlot 71%, Cabernet Franc 24.3% Petite Verdot 2.6%*

🍷 *5–12 years*

Second wine: *Le Seuil de Mazeyres*

CHÂTEAU MOULINET
★ Ⓥ

This large estate belongs to Armand Moueix. The wine is matured in wood for 18 months, with 40 per cent new oak.

RED Attractively supple with a light, creamy-ripe fruit and oak flavour.

🍷 *Merlot 5%, Cabernet Franc 15%*

🍾 *5–10 years*

Second wine: *Clos Sainte Anne*

CHÂTEAU NÉNIN
★

This large and well-known property between Catussau and the outskirts of Libourne. The wine has had a disappointing record, but there has been a noticeable improvement since this property was purchased by Jean-Hubert Delon in 1997, who added 4 hectares (10 acres) from Certan-Giraud, replanted, and built a new vat room. The wines are aged in 30 per cent new oak barrels for 18 months.

RED No similarity to Nénin of the past, vintages from 1998 onwards have been increasingly full, deep, and concentrated, with a lush, opulent fruit, and a silky finish.

🍷 *Merlot 76%, Cabernet Franc 23%, Cabernet Sauvignon 1%*

🍾 *5–18 years*

Second wine: ✔ *Fugue de Nenin*

CHÂTEAU PETIT-VILLAGE
★★

This property borders Vieux Château Certan and Château La Conseillante, and it therefore has the advantage of a superb *terroir* and a meticulous owner in Axa Millésimes. The result is a wine of superstar quality, even in poor years. Petit-Village is matured in wood for about 18 months with at least 50 per cent of the casks made from new oak.

RED These wines seem to have everything. Full and rich with lots of colour and unctuous fruit, they have a firm structure of ripe and supple tannins and a luscious, velvety texture. Classic, complex, and complete.

PÉTRUS, VINTAGE 2009
Known as the finest wine of the Pomerol appellation, Pétrus consistently ranks among the world's most expensive wines.

🍷 *Merlot 75%, Cabernet Franc 18%, Cabernet Sauvignon 7%*

🍾 *8–30 years*

Second wine: *Le Jardin de Petit Village*

CHÂTEAU PÉTRUS
★★★

The Libournais *négociant* Jean-Pierre Moueix was in technical control of this estate from 1947 until his death at the age of 90 in March 2003. Before the previous owner, Madame Loubat, died in 1961, she gave one share of Pétrus to Monsieur Moueix. She had no children, just a niece and a nephew who were not on the best of terms, so Madame Loubat wisely gave Moueix the means of ensuring that family disagreements would not be able to harm the day-to-day running of Pétrus. In 1964 the nephew sold his shares to Jean-Pierre Moueix. In 2001 his eldest son, Jean-François, became the sole owner.

RED The blue smectite clay on which the vines are planted allows the Merlot grapes to develop very high tannins and keep a good level of acidity. Both assure the impressive longevity of the wine, yet the very gentle extraction of only the ripest tannins allows Pétrus to be very pleasant to taste even very young.

🍷 *Merlot 100%*

🍾 *20–50 years*

CHÂTEAU LE PIN
★★

In 1979 by Jacques Thienpont purchased this tiny property neighbouring the family vineyard of Vieux Château Certan. The wine rose to fame with the 1982 vintage and has since sold at prices over and above the first growths. A small modern winery was added in 2012. The yield is very low, and the wine is fermented in stainless steel and matured in wood for 18 months with 100 per cent new oak.

RED These oaky wines are very full-bodied and powerfully aromatic with a sensational spicy-cassis flavour dominated by decadently rich, creamy-toffee, toasty-coffee oak.

🍷 *Merlot 100%*

🍾 *10–40 years*

Other wines: There is no real second wine, but a Trilogie – a blend of three different vintages – is made from declassified lots.

CHÂTEAU PLINCE
★ Ⓥ

This property is owned by the Moreau family. The wine is matured in vats for 6 months and in wood for 12 to 18 months, with one-third new oak.

RED These wines are fat, ripe, and simply ooze with juicy Merlot flavour. Although they could not be described as aristocratic, they are simply delicious.

🍷 *Merlot 72%, Cabernet Franc 23%, Cabernet Sauvignon 5%*

🍾 *4–8 years*

CHÂTEAU LA POINTE
Ⓥ

In 2008 Eric Monneret took over the management of this important Pomerol property, changing the planting and agricultural practises and renovating the vat room and cellar. The wines age for 12 months in 50 per cent new oak.

RED The wines now show lovely fruit, with plummy-chocolaty overtones.

🍷 *Merlot 90%, Cabernet Franc 10%, Malbec 5%*

🍾 *5–12 years*

Second wine: *Ballade de La Pointe*

CHÂTEAU ROUGET
☆ Ⓥ

One of the oldest properties in Pomerol. The wine is matured in wood for 18 months in one-third new oak

RED Château Rouget produces excellent red wines with a fine bouquet and elegant flavour. Fat and rich, with good structure and lots of ripe fruit, they are at their most impressive when mature.

🍷 *Merlot 85%, Cabernet Franc 15%*

🍾 *10–25 years*

Second wine: *Le Carillon de Rouget*

CHÂTEAU DE SALES
☆ Ⓥ

With 47.6 hectares (117 acres) under vine out of 90 hectares (222 acres), this is the largest property in the Pomerol appellation situated in the very northwest of the district. Despite an uneven record, it has demonstrated its potential and inherent qualities on many occasions. The wine is matured in wood for 12 months with 15 to 20 per cent new oak. Since 2017 it has been under the management of Vincent Montigaud, and changes in the cellar and the vines have seen their wines go from strength to strength.

RED When successful, these wines have a penetrating bouquet and a palate jam-packed with deliciously juicy flavours of succulent stone-fruits such as plums, black cherries, and apricots.

🍷 *Merlot 73%, Cabernet Franc 12%, Cabernet Sauvignon 15%*

🍾 *7–20 years*

Second wine: *Château Chantalouette*

CHÂTEAU DU TAILHAS

The wines of this château are matured in wood for 18 months with 50 per cent new oak.

RED Consistently attractive, with silky Merlot fruit and creamy oak.

🍷 *Merlot 80%, Cabernet Franc 10%, Cabernet Sauvignon 10%*

🍾 *5–12 years*

CHÂTEAU TAILLEFER

This potentially excellent property belongs to Claire and Antoine, children of Bernard Moueix. The wines are matured in wood for between 18 and 22 months, with the addition of one-third new oak.

RED These wines are attractively light and fruity, revealing the potential of the iron pan soil.

🍷 *Merlot 75%, Cabernet Franc 25%*

🍾 *4–8 years*

Second wine: *Chateau Fontmarty (not every vintage)*

CHÂTEAU TROTANOY
★★☆

Some consider this property, purchased by J-P Moueix in 1953, to be second only to Château Pétrus, although in terms of price L'Évangile and Lafleur have overtaken it. The wine is matured in wood for up to 16 to 18 months, with 50 per cent new oak.

RED This inky-black, brooding wine has a powerful bouquet and a rich flavour, which is supported by a firm tannin structure and a complex, creamy-toffee, spicy-coffee oak character.

🍷 *Merlot 90%, Cabernet Franc 10%*

🍾 *15–35 years*

VIEUX CHÂTEAU CERTAN
★★☆

This was once regarded as the finest-quality growth in Pomerol, and Vieux Château Certan remains one of Bordeaux's great wines. This wine is matured in wood for 18 to 24 months with up to 60 75 to 100 per cent new oak. The property is owned by the Thienpont family and managed by Alexandre Thienpont.

RED This is an attractive, garnet-coloured, full-bodied wine that has a smouldering, smooth, and mellow flavour. It displays great finesse and complexity of structure.

🍷 *Merlot 70 %, Cabernet Franc 25%, Cabernet Sauvignon 5%*

🍾 *12–35 years*

Second wine: *La Gravette de Certan*

CHÂTEAU LA VIOLETTE
☆

In the same way as Château Laroze in St-Émilion is said to be named for aroma of roses, so this château is named after its aroma of violets – or so the story goes. It is located in Catussau, and its vineyards are scattered about the commune. The wine, matured in 100 per cent new oak for 12 to 18 to months, can be inconsistent, but I think it has great potential.

RED I have not tasted this wine as frequently as I would like, but I can enthusiastically recommend the best vintages, when they have a rich and jubilant flavour of Merlot fruit, which is ripe and fat.

🍷 *Merlot 100%*

🍾 *5–15 years*

CHÂTEAU VRAY CROIX DE GAY

A small property on good gravelly soil next to Château le Gay, it is co-owned by Artemis Domanies (owners of Château Latour in Pauillac) along with Château le Prieuré in Saint-Émilion and Château Siaurac in Lalande de Pomerol. The wine is matured in wood in 18 months.

RED The wine can be full, rich, chocolate and black cherry flavoured, with the best vintages showing more fat and oak.

🍷 *Merlot 84%, Cabernet Franc 16%*

🍾 *5–10 years*

SAINT-ÉMILION AND POMEROL SATELLITES

DOMAINE DE L'A
Côtes de Bordeaux Castillon AOC
★☆

The home of ace consultant Stéphane Derenoncourt, and unlike plumbers and decorators, he does not neglect his own property. Sleek and rich, early drinking, yet benefits from several years' additional ageing, and it's biodynamic. A future star!

CHÂTEAU D'AIGUILHE
Côtes de Bordeaux Castillon AOC
☆

Purchased in 1998 by Stephan von Neipperg, who has restored the property. Super since 1999. The wine is aged in 50 per cent new oak for 15 to 18 months.

Merlot 80%, Cabernet Franc 20%,
Second wine: *Seigneurs d'Aiguilhe*
Other wines: *Le Blanc d'Aiguilhe*

CHÂTEAU DES ANNEREAUX
Lalande-de-Pomerol AOC

Attractive, fruity, medium-bodied wines of some elegance. Organic agriculture.

CHÂTEAU BARRABAQUE
Canon-Fronsac AOC
☆

Situated on the mid-*côte,* this 80 per cent Merlot has really shone since the late 1990s. Excellent Cuvée Prestige is the *crème de la crème* here.

CHÂTEAU BEL-AIR
Puisseguin-St-Émilion AOC

This property makes generous, fruity, early-drinking wines.

CHÂTEAU DE BEL-AIR
Lalande-de-Pomerol AOC
★

One of the best of the appellation, this property has fine, sandy gravel. Aged in 40 per cent new oak with François Despagne of Château Corbin Despagne as consultant oenologist.

Merlot 73%, Cabernet Franc 19%, Cabernet Sauvignon 8%

CHÂTEAU BELAIR-MONTAIGUILLON
St-Georges-St-Émilion AOC
★

Consistently rich, deliciously fruity. The best wine selected from old vines and matured in cask, including some new oak, is sold as Château Belair-St-Georges.

CHÂTEAU DE BELCIER
Côtes de Bordeaux Castillon AOC

This property produces fruity red wines.

Merlot 57%, Cabernet Franc 28%, Cabernet Sauvignon 9%, Malbec 6%

CHÂTEAU CALON
Montagne-St-Émilion AOC and St-George-St-Émilion AOC
☆

This château is under the same ownership as the *grand cru classé* Château Corbin-Michotte. The wine is good quality, with a juicy style and very Merlot in character. Part of this vineyard falls within the St-George-St-Émilion area, and this produces the best wine.

CHÂTEAU CANON
Canon-Fronsac AOC
★☆

This is one of several Fronsadais properties formerly owned, and still managed by, J-P Moueix. It was sold to Jean Galand in September 2000. It produces one of the best wines in this appellation from 100 per cent Merlot (40 years old).

CHÂTEAU CAP DE MERLE
Lussac-St-Émilion AOC
☆

Château Cap de Merle is wine guru Robert Parker's best Lussac performer for the 1981, 1982, and 1983 vintages. These wines today remain consistently good value.

CHÂTEAU CARLES
Fronsac AOC
★

This producer's primary wine is attractive and juicy, but it is the 1.5 blockbusting 95 per cent Merlot selection, sold as Château Haut Carles, that really stands out here.
Other wines: *Château Haut Carles*

CHÂTEAU CASSAGNE-HAUT-CANON
Canon-Fronsac AOC
★

Château Cassagne-Haut-Canon produces a selection of rich, full, fat, fruitcake-flavoured wines that are especially attractive when they are still young.

CHÂTEAU LES CHARMES-GODARD
Côtes de Bordeaux Francs AOC

White Côtes de Bordeaux Francs produced by Nicolas Thienpont on the same site of his Côtes de Bordeaux Francs red Château Peygueraud and Chateau la Prade and Côtes de Bordeaux Castillon Château Alcée. Thienpont also owns Pavie Macquin, Larciss Ducasse, Berliquet, and Beauséjour Duffau Lagarrosse in neighbouring St-Émilion.

CHÂTEAU DE CLOTTE
Côtes de Bordeaux Castillon AOC
★

This property has the right to both the Côtes de Bordeaux Castillon and Côtes de Bordeaux Francs appellations, but uses only the former.

CHÂTEAU DU COURLAT
Lussac-St-Émilion AOC

These are spicy-tannic wines with good fruit flavours.

CHÂTEAU COUSTOLLE VINCENT
Côtes Canon-Fronsac AOC

Château Coustolle Vincent's wines are well-flavoured, and matured in up to 20 per cent new oak.

CHÂTEAU DALEM
Fronsac AOC
★☆

These soft and velvety wines develop quickly but have finesse and are among the very best of their appellation.

CHÂTEAU DE LA DAUPHINE
Fronsac AOC
☆

Owned by Moueix until 2000 when Jean Halley, the CEO of and major shareholder in Carrefour, bought the property and invested heavily in renovating the cellars and vineyard. In the hands of the Labrune family since 2015, La Dauphine was certified organic in 2015. One of the most dynamic properties of the Fronsac appellation.
Second wine: *La Delphis de la Dauphine*

CHÂTEAU DURAND LAPLAIGNE
Puisseguin-St-Émilion AOC
★

The excellent-quality wine produced by Château Durand Laplaigne is grown using clay-and-limestone soil, with a strict selection of grapes, and modern vinification techniques.

CHÂTEAU LA FLEUR DE BOÜARD
Lalande-de-Pomerol AOC
★ (V)

Since 1998, this château has been owned by Hubert de Boüard de Laforest, co-owner of Château Angelus. It uses all the latest winemaking technology and could be considered a testing site before they use the same techniques at Angelus. It puts out powerful, fruit-driven wines, some of the best in the appellation.
Second wine: *Le Lion de la Fleur Bouard*
Other wines: *Le Plus de la Fleur Bouard*

CHÂTEAU FONTENIL
FRONSAC AOC
★☆

This rich, velvety 90 per cent Merlot high-flyer has lashings of new oak and is from Michel Rolland's own property.

CHÂTEAU GABY
Canon Fronsac AOC

An excellent example of Canon Fronsac wine produced on beautiful slopes above the Dordogne river.
Second wine: *Princesse Gaby*
Other wines: *Gaby Cuvée*

CHÂTEAU GRAND-BARIL
Montagne-St-Émilion AOC

Attractive, fruity wine made by the agricultural school in Libourne.

CHÂTEAU GRAND ORMEAU
Lalande-de-Pomerol AOC
★

A rich and lusciously fruity wine that is matured in 50 per cent new oak, Grand Ormeau is very classy for its appellation.
Second wine: *Chevalier d'Haurange*

CHÂTEAU GUIBEAU-LA FOURVIEILLE
Puisseguin-St-Émilion AOC
★☆

Much investment has gone into this property, the wines of which are now considered to be the best in Puisseguin. Under organic production.
Second wine: *Château Guibeau*
Other wines: *Château Guibeau Grand Reserve*

CHÂTEAU HAUT-CHAIGNEAU
Lalande-de-Pomerol AOC
★ (V)

Look out for Château Le Sergue, which is selected from this property's best wines and matured in 80 per cent new oak. A poor man's Mondotte?
Other wines: *Château Le Sergue*

CHÂTEAU HAUT-CHATAIN
Lalande-de-Pomerol AOC
★☆ (V)

Fat, rich, and juicy wines with definite hints of new-oak vanilla are made by Château Haut-Chatain.

CHÂTEAU LES HAUTS-CONSEILLANTS
Lalande-de-Pomerol AOC

Owned by *négociant* company J-B Audy, Château les Hauts-Conseillants is another fine Néac property.

CHÂTEAU HAUT-TUQUET
Côtes de Bordeaux Castillon AOC

This wine is consistently good.

CHÂTEAU JEANDEMAN
Fronsac AOC

This château produces fresh, fruity wine with good aroma.

CHÂTEAU JUNAYME
Canon-Fronsac AOC
★ (V)

Well-known wines of finesse.

CHÂTEAU DES LAURETS
Puisseguin-St-Émilion AOC

The appellation's largest property, owned by Edmond de Rothschild.
Other wines: *Château des Laurets Baron*

CHÂTEAU DE LUSSAC
Lussac-St-Émilion AOC

Château de Lussac produces well-balanced, early-drinking wine.

CHÂTEAU DU LYONNAT
Lussac-St-Émilion AOC

The appellation's largest property.
Other wines: *La Rose Peruchon, Chateau Lyonnat Emotion*

CHÂTEAU MAISON BLANCHE
Montagne-St-Émilion AOC

Château Maison Blanche produces attractive wine that is easy to drink. Under Biodynamic production.
Second wine: *Les Piliers de Maison Blanche*

CHÂTEAU MAQUIN-ST-GEORGES
St-Georges-St-Émilion AOC

This wine is 80 per cent Merlot.

CHÂTEAU MAUSSE
Canon-Fronsac AOC

Wines with good aroma and flavour produced on one of the Guy Janoueix collection of vineyards.

CHÂTEAU MAYNE-VIEIL
Fronsac AOC

Easy-drinking wines with good Merlot spice and fruit.

CHÂTEAU MAZERIS
Canon-Fronsac AOC

This is a 350-year-old estate producing right bank–style Merlot.

CHÂTEAU MONCETS
Lalande-de-Pomerol AOC

This Néac property makes a fine, rich, and elegant Pomerol look-alike.
Other wines: *Château de Chambrun, Château de Chambrun "Le Bourg", Château La Bastidette (Montagne-St-Émilion)*

CHÂTEAU MOULIN HAUT-LAROQUE
Fronsac AOC

Well-perfumed, quite fat wines with lots of fruit and good tannin.
Second wine: *Chateau Hervé Laroque*

CHÂTEAU LA PAPETERIE
Montagne-St-Émilion AOC

They produce wines with a rich nose and a big fruit-filled palate.

CHÂTEAU DU PONT DE GUESTRES
Lalande-de-Pomerol AOC

This château produces full, ripe, fat wines of good quality.

CHÂTEAU LA PRADE
Côtes de Bordeaux Francs AOC

☆

From the Nicolas Thienpont stable (*see* Chateau les Charmes).

CHÂTEAU LE PUY
Côtes de Bordeaux Francs AOC and Bordeaux Supérieur AOC

A large range of small *cuvées* produced under biodynamic agriculture.

CLOS PUY ARNAUD
Côtes de Bordeaux Castillon AOC

★

Owner-winemaker Thierry Valette has produced increasingly concentrated, yet extremely approachable wines since 2000.

CHÂTEAU PUYCARPIN
Côtes de Bordeaux Castillon AOC

This property produces an excellent range of red wines from biodynamic viticulture.

CHÂTEAU PUYGUERAUD
Côtes de Bordeaux Francs AOC and Bordeaux Supérieur AOC

★

Aromatically attractive wines with good colour and supple fruit. One of the leading wines of the Côtes de Bordeaux Francs appellation. Produced by Nicolas Thienpont. (*See* Château La Prade and Château Les Charmes).
Other wines: *Georges* (80% Malbec), *Château Lauriol, Château Puygueraud Blanc* (100% Sauvignon)

CHÂTEAU RICHELIEU
Fronsac AOC

Excellent since 2003. The 2005 was stunning vintage, especially the wonderfully lush and satisfying prestige *cuvée* "La Favorite de Richelieu."
Second wine: *Trois Musketeers*

CHÂTEAU DE LA RIVIÈRE
Fronsac AOC

Magnificent wines that are built to last: they are rich, tannic, and fruity and matured in up to 40 per cent new oak.
Second wine: *Les Sources de Chateau la Rivière*
Other wines: *Aria, la Blanc de Château la Rivière, Le Rosé de Château la Rivière*

CHÂTEAU ROCHER-BELLEVUE
Côtes de Bordeaux Castillon AOC

Organic and biodynamic wine from the Vignobles Gabriel & Co family. A good St-Émilion look-alike that regularly wins medals.
Other wines: *La Palène, Coutet-St-Magne*

CHÂTEAU ROUDIER
Montagne-St-Émilion AOC

Quality wines that are well-coloured, richly flavoured, finely balanced, and long and supple.

CHÂTEAU LA ROUSSELLE
Fronsac AOC

This château has produced lovely rich, ripe, elegant wines since 2003.

CHÂTEAU SIAURAC
Lalande-de-Pomerol AOC

Fine, firm, and fruity wines. The property is co-owned by Artémis Domaines, owner of Château Latour (*see also* Château Vray Croix de Gay in Pomerol).
Second wine: *Plaisir de Siaurac*

CHÂTEAU STE-COLOMBE
Bordeaux AOC

This château was purchased in 1999 by Gérard Perse of Pavie. Since 2008 the grapes have been blended with grapes from the second wine of Pavie to make l'Esprit de Pavie, a Bordeaux appellation.

CHÂTEAU ST-GEORGES
St-Georges-St-Émilion AOC

An impressive property that dominates this smallest of the Saint-Émilion Satellite appellations. Super quality wine of great finesse.

CHÂTEAU DE LA RIVIÈRE LABEL
One of the older estates in the Fronsac, they produces both red and white wine.

CHÂTEAU TOUMALIN
Canon-Fronsac AOC

Nathalie and Xavier Miravete bought Château Toumalin in 2008. They produce fresh, fruity wine.

CHÂTEAU TOUR-DU-PAS-ST-GEORGES
St-Georges-St-Émilion AOC

An excellent and inexpensive entrée into the world of *premier cru* claret.

CHÂTEAU DES TOURELLES
Lalande-de-Pomerol AOC

Fine wines with vanilla undertones from the Bernard Janoueix portfolio of right bank vineyards.

CHÂTEAU TOURNEFEUILLE
Lalande-de-Pomerol AOC

This rich, long-lived wine is the best of the appellation. The owners also make wine at Château Lécuyer in Pomerol and Château La Révérence in Saint-Émilion.

CHÂTEAU DES TOURS
Montagne-St-Émilion AOC

☆ V

The largest property in the appellation. The wine is big, full, and fleshy, yet soft and easy to drink.

CHÂTEAU LA VALADE
Fronsac AOC

Elegant, aromatic, and silky-textured wines, which are made exclusively from Merlot grapes.

CHÂTEAU LA VIEILLE CURE
Fronsac AOC

★☆

Very fresh and velvety with delightfully floral, summer fruit aromas, these wines are at the very top of their appellation. The 1998 is stunning.

VIEUX-CHÂTEAU-ST-ANDRÉ
Montagne-St-Émilion AOC

★☆ V

A soft, exciting wine, full of cherry, vanilla, and spice flavours.

CHÂTEAU LA VILLARS
Fronsac AOC

Soft, fat, and juicy wines of excellent quality, one-third of which are matured in new oak.

CHÂTEAU VRAY-CANON-BOYER
Canon-Fronsac AOC

This château produces a fruity, medium-bodied wine that is attractive for early drinking from 90 per cent Merlot grapes.

Bourg and Blaye

Ninety-five per cent of the wine produced in Bourg and Blaye is good-value red. Tiny Bourg makes more wine than its five-times-larger neighbour, Blaye, and most of the vines grown in Blaye come from a cluster of châteaux close to the borders of Bourg.

As one would expect of an area that has supported a settlement for 400,000 years, Bourg has a close-knit community. Comparatively recently, the Romans used neighbouring Blaye as a *castrum,* a fortified area in the defence system that shielded Bordeaux. According to some sources, the vine was cultivated in Bourg and Blaye as soon as the Romans arrived. Vineyards were certainly flourishing here long before those of the Médoc, just the other side of the Gironde.

Bourg is a compact, heavily cultivated area with pretty hillside vineyards at every turn. The vine is less important in Blaye, which has other interests, including a caviar industry based at its ancient fishing port where sturgeon is still a major catch. The south-facing vineyards of Blaye are mostly clustered in the countryside immediately bordering Bourg and, despite the similarity of the countryside, traditionally produce slightly inferior wines to those of Bourg. The D18 motorway appears to be a barrier beyond which the less intensely cultivated hinterland takes on a totally different topography, where the more expansive scenery is dotted with isolated forests.

THE POTENTIAL OF BOURG AND BLAYE

To the Romans, these south-facing vineyards overlooking the Gironde seemed the ideal place to plant vines. Indeed, the quality achieved today in these vineyards would have surpassed the most optimistic hopes of those past masters of the vine. By today's expectations, however, Bourg and Blaye have been relegated to a viticultural backwater. This is a pity, as there are some exciting quality wines being made here. Perhaps when the world has woken up to the gems in Canon-Fronsac, more curious consumers might turn their attention to the better producers in this district. As soon as wine lovers are prepared to pay higher prices for these wines, so its proprietors will be able to restrict yields, improve vinification techniques, and indulge in a percentage of new oak.

FACTORS AFFECTING TASTE AND QUALITY

LOCATION
The vineyards fan out behind the town of Bourg, which is situated on the right bank of the confluence of the Dordogne and the Garonne, some 20 kilometres (12.5 miles) north of Bordeaux. Blaye is a larger district that unfolds beyond Bourg.

CLIMATE
These two areas are less protected than the Médoc from westerly and northwesterly winds and have a higher rainfall.

ASPECT
Bourg is very hilly with vines cultivated on steep limestone hills and knolls up to a height of 80 metres (260 feet). In the southern section of Blaye, the country is rich and hilly, with steep slopes overlooking the Gironde that are really just a continuation of those in Bourg. The northern areas are gentle and the hills lower, with marshes bordering the viticultural areas.

SOIL
In Bourg the topsoil is clay-limestone or clay-gravel over a hard limestone subsoil, although in the east the subsoil sometimes gives way to gravel and clay. The soil in Blaye is clay or clay-limestone over hard limestone on the hills overlooking the Gironde, getting progressively sandier going east.

VITICULTURE & VINIFICATION
There are many grape varieties here, some of which are far too inferior or unreliable to contribute to the quality of these wines, particularly the whites. Bourg produces the best reds and Blaye the best whites, but there is relatively little white wine made in either appellation – even Blaye is 90 per cent red and Bourg is in excess of 99 per cent red. Very few *petits châteaux* in both areas can afford the use of casks, let alone new ones, and much of the wine in Bourg is made by one of its five *coopératives.*

GRAPE VARIETIES
Primary varieties: Cabernet Franc, Cabernet Sauvignon, Merlot, Sauvignon Blanc, Sémillon
Secondary varieties: Béguignol, Chenin Blanc, Colombard, Folle Blanche, Malbec, Merlot Blanc, Muscadelle, Petit Verdot, Prolongeau (Bouchalès), Ugni Blanc

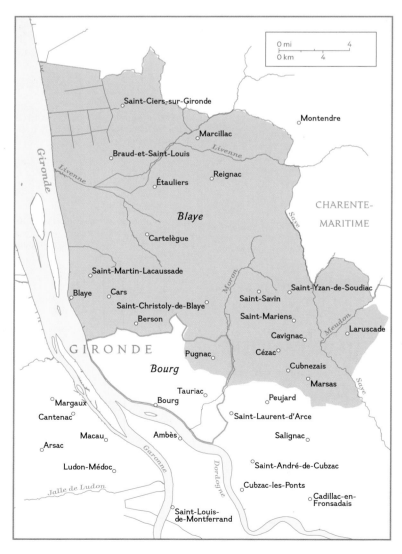

HAND-HARVESTED GRAPES
Picking grapes by hand is always the preferred method, especially for small smaller vineyards that cannot risk losing any fruit.

VINE-GROWING AREAS OF BOURG AND BLAYE
(see also p141)
Most of the best growths of Bourg and Blaye are clustered behind the respective ports that give this wine-producing area its name. Bourg, the smaller area, has a higher concentration of vineyards and generally produces the better wines.

AN AERIAL VIEW OF THE TOWN OF BLAYE SHOWS VINEYARDS NESTLING IN THE RAMPARTS OF THE RUINS OF THE CITADEL OF BLAYE
The 17th-century Citadel of Blaye once guarded the ancient fishing port of Blaye against the approach of marauders from the Atlantic.

THE APPELLATIONS OF
BOURG AND BLAYE

The appellations in this area are very confusing for the consumer, with a needless array of designations. There is really no reason why just two AOCs – Côtes de Blaye and Côtes de Bourg – could not be used for all the wines produced here.

BLAYE AOC

Blaye, or Blayais, is a large and diverse appellation of variable quality.

RED A few properties cultivate obscure varieties, such as Prolongeau and Béguignol. Many properties utilize the more prestigious sounding Premières Côtes de Blaye AOC, hence hardly anyone bothers to sell their wine under this plain appellation.

🍷 *Cabernet Sauvignon, Cabernet Franc, Merlot, Malbec, Prolongeau, Béguignol, Petit Verdot*

🍷 *3–7 years*

WHITE Since 1997 the Ugni Blanc variety has dominated the appellation, with the Merlot Blanc and Folle Blanche varieties banned. The allowable ripeness level has been lowered from 170 to 153 grams of sugar, with no more than 4 grams residual allowed in the finished wine, and a maximum of 12.5 per cent alcohol imposed, thus ensuring fresher, crisper wines. These wines may be sold from 1 December following the harvest without any mention of *primeur* or *nouveau*.

🍷 *Ugni Blanc, plus up to 10% in total of Folle Blanche, Colombard, Chenin Blanc, Sémillon, Sauvignon Blanc, Muscadelle*

🍷 *1–2 years*

BLAYE – CÔTES DE BORDEAUX AOC

The boundary and technical requirements for this appellation are precisely the same as for the Premières Côtes de Blaye. The wines may be red or dry white.

BOURG AOC

Also called Bourgeais, this appellation, which covers both red and white wines, has fallen into disuse because the growers prefer to use the Côtes de Bourg AOC, which is easier to market but conforms to the same regulations. *See also* Côtes de Bourg AOC.

BOURGEAIS AOC

See Bourg AOC

CÔTES DE BLAYE AOC

Unlike the Bourg and Côtes de Bourg appellations, which cover red and white wines, Côtes de Blaye is white only. Blaye, however, may be red or white.

WHITE As much white Côtes de Blaye is produced as basic Blaye. The wines are similar in style and quality.

🍷 *Merlot Blanc, Folle Blanche, Colombard, Chenin Blanc, Sémillon, Sauvignon Blanc, Muscadelle*

🍷 *1–2 years*

CÔTES DE BOURG AOC

Bourg is one-fifth the size of Blaye, yet it traditionally produces a greater quantity and, more important, a much finer quality of wine than that produced at Blaye.

RED Excellent-value wines of good colour, full of solid, fruity flavour. Many are very stylish indeed.

🍷 *Cabernet Sauvignon, Cabernet Franc, Merlot, Malbec*

🍷 *3–10 years*

WHITE A very small quantity of this light, dry wine is produced and sold each year. It may be sold from 1 December following the harvest without any mention of *primeur* or *nouveau*.

🍷 *Sémillon, Sauvignon Blanc, Muscadelle, Merlot Blanc, Colombard, plus up to 10% Chenin Blanc*

🍷 *1–2 years*

PREMIÈRES CÔTES DE BLAYE AOC

This appellation covers the same area as Blaye and Côtes de Blaye, but only classic grapes are used here, and the minimum alcoholic strength is higher. The Premières Côtes De Blaye area has very good potential for producing quality wines.

RED There are one or two excellent properties that use a little new oak.

🍷 *Cabernet Sauvignon, Cabernet Franc, Merlot, Malbec*

🍷 *4–10 years*

WHITE Dry, light-bodied wines that may have a fresh, lively, grapey flavour.

🍷 *Sémillon, Sauvignon Blanc, Muscadelle*

🍷 *1–2 years*

THE WINE PRODUCERS OF

BOURG

CHÂTEAU DE BARBE
Villeneuve

Château de Barbe is a property that makes substantial quantities of light-styled, Merlot-dominated, gently fruity red wines, which are easy to drink.

CHÂTEAU BÉGOT
Lansac

This property produces some 5,000 cases of agreeably fruity red wine, which is best drunk young.

CHÂTEAU DU BOUSQUET
Bourg-sur-Gironde
☆

This large, well-known château produces some 40,000 cases of red wine that offers excellent value for the money, The wine is fermented in stainless steel and aged in oak; it has a big bouquet, and a smooth feel.

CHÂTEAU BRULESCAILLE
Tauriac

Château Brulescaille's vineyards are very well-sited and produce agreeable wines for early drinking.

CHÂTEAU CONILH-LIBARDE
Bourg-sur-Gironde

This small vineyard overlooking Bourg-sur-Gironde and the river makes soft, fruity red wines.

CHÂTEAU CROUTE-COURPON
Bourg-sur-Gironde

A small but recently enlarged estate, it produces honest, fruity red wines.

CHÂTEAU EYQUEM
Bayon-sur-Gironde

Owned by the serious winemaking Bayle-Carreau family, which also owns several other properties, this wine is not normally, however, purchased for its quality. It is enjoyable as a light luncheon claret, but the real joy is in the spoof of serving a red "Yquem".

CHÂTEAU FALFAS
Bourg-sur-Gironde
 ☆ V

Owned by John and Vonique Cochran, whose biodynamic wines are consistently rich, mid-term drinkers of no little finesse.

CHÂTEAU FOUCHÉ
Bourg-sur-Gironde

This wine is firm, yet has fat, juicy fruit and a smooth finish.

CHÂTEAU GÉNIBON-BLANCHEREAU
Bourg-sur-Gironde

This small vineyard produces attractive wines that have all the enjoyment upfront and are easy to drink.

CHÂTEAU GRAND-LAUNAY
Teuillac
☆ V

This property has been developed from the vineyards of three estates: Domaine Haut-Launay, Château Launay, and Domaine les Hermats. Mainly red wine is produced, although a very tiny amount of 100 per cent Sauvignon Gris white is also made. The top *cuvée* is sold as Château de Launay Reserve. All wines are organic.

CHÂTEAU DE LA GRAVE
Bourg-sur-Gironde
☆ V

An important property situated on one of the highest points of Bourg-sur-Gironde, it produces a large quantity of light, fruity red wine and a very tiny amount of white with a high percentage of Colombard.

CHÂTEAU GUERRY
Tauriac
☆ V

Some 10,000 cases of really fine wood-aged red wines are produced at this château. The wines have good structure, bags of fruit, and a smooth, elegant flavour.

CHÂTEAU GUIONNE
Lansac

These are easy-drinking wines, full of attractive Merlot fruit, good juicy flavour, and some finesse. A little white wine of some interest and depth is also made.

CHÂTEAU HAUT-MACÔ
Tauriac

This rustic red wine is full of rich, fruity flavours, and good acidity. The proprietors also own a property called Domaine de Lilotte in Bourg-sur-Gironde producing attractive, early-drinking red wines under the Bordeaux Supérieur appellation.
Other wines: *Les Bascauds*

CHÂTEAU HAUT-ROUSSET AND CHÂTEAU ROUSSET
Bourg-Sur-Gironde

This fairly large property produces some 12,000 cases of decent, everyday-drinking red wine and 1,000 cases of white. The red wines from a small vineyard close by are sold under the Château la Renardière label.

CHÂTEAU DE LIDONNE
Bourg-sur-Gironde
 ★☆

This very old property produces an excellent-quality red wine, powerfully aromatic and full of Cabernet character. Its name comes from the 15th-century monks who looked after the estate and offered lodgings to passing pilgrims: *Lit-Donne,* or "Give Bed".

CHÂTEAU PEYCHAUD
Teuillac
☆ V

These fruity red wines are easy to drink when young. A little white is also made. It is under the same ownership as Château Peyredoulle and Château le Peuy-Saincrit.

CHÂTEAU ROC DE CAMBES
Bourg-sur-Gironde
 ★☆

You can expect to pay five times the price of any other Côtes de Bourg for this blockbuster, which is made by François Mitjavile, the owner of Château Tertre-Rôteboeuf in the commune St Laurent-des-Combes, a few kilometers southeast of the picturesque town of St-Émilion. Some vintages are worth the extra money, however.

CHÂTEAU ROUSSET
Samonac
(*See* Château Haut-Rousset and Château Rousset above.)

CHÂTEAU SAUMAN
Villeneuve

These immaculate vineyards produce a good-quality red wine for medium-term maturity. The château has been in the ownership of the Sinan-Braud family for over five generations. The proprietor also owns the red-wine producing Domaine du Moulin de Mendoce in the same commune.

CHÂTEAU TOUR-DE-TOURTEAU
Samonac
 ★ V

This property was once part of Château Rousset (*see* above); however, the wines are definitely bigger and richer than those of Rousset.

VINEYARDS SPILL OVER THE GENTLE SLOPES ALONG THE BLAYE AND BOURG WINE TRAIL
With increasing interest in this region's wines, the Blaye and Bourg Wine Trail is seeing more tourist traffic. Along this route are found many family-owned properties, and wine tastings and tours are often run by the winemakers themselves.

THE WINE PRODUCERS OF
BLAYE

CHÂTEAU BARBÉ
Cars
☆

This château produces well-made and overtly fruity red and white wines. Château Barbé is one of several properties owned by the Bayle-Carreau family (see also La Carelle and Eyquem in Bourg).

CHÂTEAU BOURDIEU
Berson

This old and well-known property produces Cabernet-dominated red wines of a very firm structure that receive time in oak. Seven hundred years ago this estate was accorded the privilege of selling "clairet", a tradition it maintains today by ageing the blended production of various vineyards in oak.

CHÂTEAU LA CARELLE
St-Paul

More than 11,000 cases of agreeable red wines are made at this property. The owner also runs Châteaux Barbé.

CHÂTEAU CHARRON
St-Martin-Lacaussade
☆

These very attractive, well-made, juicy-rich, Merlot-dominated red wines are matured in oak, some of which is new. A small amount of white wine is also made.

CHÂTEAU CRUSQUET-DE-LAGARCIE
Cars
★☆

A tremendously exciting, richly styled red wine: deep coloured, bright, big, full of fruit, vanilla, and spice. A small amount of dry white wine is sold as "Clos-des-Rudel" and an even smaller quantity of sweet white wine as "Clos-Blanc de Lagarcie".

CHÂTEAU L'ESCADRE
Cars
☆

These elegant red wines are well-coloured, full, and fruity. They can be enjoyed young, but also improve with age. A small amount of fruity white wine is produced. The Carreau family owns eight properties of over 82 hectares (203 acres) of the Blaye-Côtes de Bordeaux AOC.

Second wine: *Château la Croix*

CHÂTEAU GIGAULT CUVÉE VIVA
Mazion
★☆

This producer offers amazing richness and opulence of fruit for the price.

CHÂTEAU SEGONZAC CELLAR
The wine of this château is matured in small oak barrels for 12 months.

DOMAINE DU GRAND BARRAIL
Plassac
☆

This château makes a fine-quality red wine that attracts by its purity of fruit. A little white wine is also produced. The proprietor also owns Château Gardut-Haut-Cluzeau and Domaine du Cavalier in Cars.

CHÂTEAU LES GRANDS MARÉCHAUX
St-Girons d'Aiguevives
★

Bags of black fruit for your bucks.

CHÂTEAU DU GRAND PIERRE
Berson
★

This property can produce tremendous value medium- to full-bodied red wine with sweet, ripe fruit. Fresh, zesty, dry white wine of agreeable quality is also made here.

CHÂTEAU HAUT BERTINERIE AND BERTINERIE
Cubnezais
★☆

The consistent class of this wine, with its silky fruit and beautifully integrated oak, makes it stand out above the rest as the best-value dry white currently made in Bordeaux.

CHÂTEAU DE HAUT-SOCIONDO
Cars

Agreeably light and fruity red and white wines are made here.

CHÂTEAU LES JONQUEYRES
St-Paul-de-Blaye
★☆

This château produces lush Merlot-dominated reds with lots of well-integrated creamy oak. The wines are impeccably produced.

CHÂTEAU LACAUSSADE SAINT MARTIN "TROIS MOULINS"
St-Martin Lacaussade
★

Owned by Jacques Chardat, but managed and marketed by Vignobles Germain, it produces both sumptuous red and classy dry white that benefit from 100 per cent new oak.

CHÂTEAU MARINIER
Cézac
☆

Twice as much red wine as white is produced at this property. The red is agreeably fruity, but the white is much the better wine: smooth, well balanced, lightly rich, and elegant. Red and rosé wines are also produced under the Bordeaux appellation.

CHÂTEAU MENAUDAT
St-Androny
☆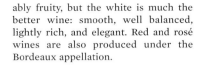

These are extremely attractive, full, and fruity red wines.

CHÂTEAU LES MOINES
Blaye

A red-only château, producing a light- to medium-bodied, fresh and fruity wine for easy drinking.

CHÂTEAU MONCONSEIL-GAZIN
Plassac

This property produces Blaye – Côtes de Bordeaux red and white, as well as a small production of the more concentrated, new oak-aged Blaye and a rosé in "Vin de France".

CHÂTEAU LES PETITS ARNAUDS
Cars
☆

This property, run for seven generations by the Carreau family, produces attractively aromatic red wines, which are pleasingly round and fruity. Dry white Blaye and a *moelleux* white owned and made by the Carreau family.

CHÂTEAU PEYREDOULLE
Berson
☆

This 16th-century property, situated on a clay and limestone *terroir*, gets its name from its vineyard soil: *Peyredoulle* means "hard rock" in the local dialect. Château Peyredoulle produces mainly good-quality red wine, although it does make some white wine. The proprietors also own Château Peychaud in the Bourgeais commune of Teuillac and the Bordeaux AOC of Château le Peuy-Saincrit.

CHÂTEAU PEYREYRE
St-Martin-Lacaussade
★

This château produces rich-flavoured reds of some finesse. Bordeaux white and rosé is also made.

CHÂTEAU SEGONZAC
St-Genès-de-Blaye
★

Château Segonzac's easy-drinking red wines are light, well made, fresh, firm, and agreeably fruity.

Entre-Deux-Mers

Entre-Deux-Mers, which literally means "between two seas" is situated between the Dordogne and Garonne rivers. It is Bordeaux's largest district, and produces inexpensive dry white wines and an increasing volume of excellent value-for-money red wines that are entitled to the Bordeaux, Bordeaux Supérieur, and Premières Côtes de Bordeaux appellations.

Technological progress in winemaking occurred earlier and more quickly in Entre-Deux-Mers than in any other district of Bordeaux. As early as the 1950s and 1960s, there was a grass-roots viticultural movement to drop the traditional low vine-training systems and adopt the revolutionary "high-culture" system that contrasted with the methods common throughout Bordeaux. These new vineyard techniques were followed in the 1970s by a widespread adoption of cool-fermentation techniques. With fresh, light, and attractively dry white wines being made at many châteaux, the major export markets suddenly realized it would be easier to sell the name Entre-Deux-Mers rather than continue to sell what had become boring Bordeaux Blanc. This was even better if the wine could boast some sort of individual *petit château* personality.

THE "HIGH-CULTURE" SYSTEM OF VINE TRAINING

Entre-Deux-Mers in the late 1940s and early 1950s was a sorry place. The wines were sold in bulk, ending up as anonymous Bordeaux Blanc, and much of the decline in the Bordeaux region was centred on this district. But the new post-war generation of winegrowers was not content with this state of affairs. Although times were difficult and the economy was deteriorating, the young, technically minded *vignerons* realized that the district's compressed *boulbènes* soil, which was choking the vines, could not be worked by their ancestors' outdated methods, and they therefore took a considerable financial risk to rectify the situation. They grubbed up every other row of vines, thus increasing the spacing between the rows, and trained the plants on a "high-culture" system similar to that practised farther south in Madiran and Jurançon (also in Austria where it was originally conceived and was called the Lenz Moser system). This system allowed machinery to work the land and break up the soil. It also increased the canopy of foliage, which intensified chlorophyll assimilation and improved ripening.

COOL FERMENTATION

In the 1970s university-trained personnel at the well-funded Entre-Deux-Mers *coopératives* invested in temperature-controlled stainless steel vats and led the way in cool-fermentation techniques. Prior to this, fermentation temperatures were often in excess of 28°C (83°F), but it was soon discovered that the lower the temperature, the more aromatic compounds were released. They determined that fermentation could take place at temperatures as low as 4°C (39°F), although the risk of stuck-fermentation (when the

WINE REGIONS OF ENTRE-DEUX-MERS
(see also p141)
The varied countryside of this district spreads out between the rivers Dordogne and Garonne as their paths diverge. The Premières Côtes form a narrow strip along the south side.

fermentation process stops) was greater at such low temperatures. It soon became clear that the ideal fermentation temperature was somewhere between 10°C (50°F) and 18°C (64°F). This increased the yield of alcohol and important aromatic and flavour compounds. It also reduced both the loss of carbonic gas and the presence of volatile acidity and required less sulphur dioxide. In the mid-1980s it was confirmed that 18°C (64°F) is the optimum temperature for fermentation. Lower temperatures also produce amylic aromas, which in small quantities are fine, but the lower the temperature, the more the wine is dominated by these nail-varnish aromas.

BORDEAUX HAUT-BENAUGE AOC

Situated above the Premières Côtes, opposite Cérons, this area corresponds to the ancient and tiny county of Benauge. To claim this appellation, as opposed to Entre-Deux-Mers-Haut-Benauge, the grapes are restricted to the three classic varieties and must be riper, with a minimum sugar level of 195 grams per litre instead of 170 grams per litre. The yield is 10 per cent lower, and the minimum alcoholic level is 1.5 per cent higher.

WHITE Dry, medium-sweet, and sweet versions of this light-bodied, fruity wine may be made.

🍷 *Sémillon, Sauvignon Blanc, Muscadelle*

🍷 *1–3 years for dry and medium-sweet wines; 3–6 years for sweet*

CADILLAC AOC

Of the trio of sweet-wine areas on the right bank of the Garonne, Cadillac is the least known. It encompasses 21 communes, 16 of which form the canton of Cadillac, yet very little wine is produced under this appellation – just one-fifth of that made in Loupiac, or one-tenth of that made in Ste-Croix-du-Mont. The regulations state that the wines must be made from botrytised grapes harvested in successive *tries*, but there is little evidence of this in the wines, which at best have the character of *passerillage*. The *terroir* could produce wines of a much superior quality, but it would be costly to do so, and sadly, this appellation does not fetch a high enough price to justify the substantial investment needed.

WHITE Attractive honey-gold wines with fresh, floral aromas and a semi-sweet, or sweet, fruity flavour.

🍷 *Sémillon, Sauvignon Blanc, Muscadelle*

🍷 *3–8 years*

CÔTES DE BORDEAUX CADILLAC AOC

This is a bit of a misnomer. The boundaries for the other three Côtes de Bordeaux appellations are exactly the same as the AOC with which they are prefixed (Côtes de Blaye for Blaye, Côtes de Castillon for Castillon, and Côtes de Francs for Francs), but Cadillac Côtes de Bordeaux encompasses not only Cadillac but also Premières Côtes de-Bordeaux, Côtes de Bordeaux St-Macaire, Loupiac, and Ste-Croix-du-Mont. Despite the proliferation of wine styles within these AOCs, the wines for Cadillac Côtes de Bordeaux must be red only.

CÔTES DE BORDEAUX STE-FOY AOC

Until relatively recently, Ste-Foy-Bordeaux was known primarily for its white wines, but it now produces as much red as white. There is a high proportion of "organic" winemakers in this area.

RED Ruby-coloured, medium-bodied wines made in a soft, easy-drinking style.

🍷 *Cabernet Sauvignon, Cabernet Franc, Merlot, Malbec, Petit Verdot*

🍷 *3–7 years*

WHITE Mellow, semi-sweet wines of uninspiring quality, and fresh, crisp dry white wines that have good aroma and make attractive early drinking. These wines may be sold from 1 December following the harvest without any mention of *primeur* or *nouveau*.

🍷 *Sémillon, Sauvignon Blanc, and Muscadelle, and up 10 per cent in total of Merlot Blanc, Colombard, Mauzac, and Ugni Blanc*

🍷 *1–3 years*

CÔTES DE BORDEAUX ST-MACAIRE AOC

These little-seen wines come from an area at the eastern extremity of the Premières Côtes de Bordeaux. Of the 2,300 hectares (6,000 acres) of vineyards that may use this appellation, barely 30 hectares (75 acres) bother to do so.

ROAD THROUGH LOUPIAC VINEYARDS
This appellation is known for its excellent sweet wines.

WHITE Medium-bodied, medium-sweet, or sweet wines that are attractive in a fruity way, but unpretentious.

🍷 *Sémillon, Sauvignon Blanc, Muscadelle*

🍷 *1–3 years*

ENTRE-DEUX-MERS AOC

This is the largest district in the region, and after the generic Bordeaux Blanc, it is its greatest-volume white-wine appellation. Entre-Deux-Mers has a growing reputation for exceptional-value wines of a high technical standard.

WHITE Crisp, dry, light-bodied wines that are fragrant, aromatic, and usually predominantly Sauvignon Blanc. These are clean, cool-fermented wines. They may be sold from 1 December following the harvest without any mention of *primeur* or *nouveau*.

🍷 *At least 70 per cent Sémillon, Sauvignon Blanc, and Muscadelle, plus a maximum of 30 per cent Merlot Blanc and up to 10 per cent in total of Colombard, Mauzac, and Ugni Blanc*

🍷 *1–2 years*

ENTRE-DEUX-MERS-HAUT-BENAUGE AOC

These wines are drier than those of the Bordeaux-Haut-Benauge appellation, and their blends may include a greater number of grape varieties, although the same nine communes comprise both appellations. The wines comply with the less rigorous regulations of Entre-Deux-Mers and, consequently, this AOC produces four times the volume of wine than does Bordeaux-Haut-Benauge. Entre-Deux-Mers-Haut-Benauge has so far produced only dry wines, with the exception of 1983, when a luscious vintage arrived that was easy to make into sweet wines.

WHITE These dry wines are very similar to those of Entre-Deux-Mers.

🍷 *At least 70 per cent Sémillon, Sauvignon Blanc, and Muscadelle, plus a maximum of 30 per cent Merlot Blanc and up to 10 per cent in total of Colombard, Mauzac, and Ugni Blanc*

🍷 *1–3 years*

GRAVES DE VAYRES AOC

An enclave of gravelly soil on the left bank of the Dordogne, this appellation produces a substantial quantity of excellent-value red and white wines.

RED These are well-coloured, aromatic, medium-bodied wines with fragrant, juicy-spicy, predominantly Merlot fruit. They are richer than those found elsewhere in Entre-Deux-Mers.

🍷 *Cabernet Sauvignon, Cabernet Franc, Carmenère, Merlot, Malbec, Petit Verdot*

🍷 *4–10 years*

WHITE Mostly dry and off-dry styles of fresh, fragrant, and fruity wines made for early drinking. Occasionally, sweeter styles are made. These wines may be sold from 1 December following the harvest without any mention of *primeur* or *nouveau*.

🍷 *Sémillon, Sauvignon Blanc, and Muscadelle plus a maximum of 30 per cent Merlot Blanc*

🍷 *1–3 years*

LOUPIAC AOC

This appellation is located on the right bank of the Garonne, opposite Barsac. It is by far the best sweet wine appellation in Entre-Deux-Mers and its wines are always excellent value. According to the regulations, Loupiac must be made with the "assistance" of overripe botrytised grapes and, unlike Cadillac, these wines often have the honeyed complexity of "noble rot". The best Loupiac wines come from vineyards with clay-and-limestone soil.

WHITE Luscious medium- to full-bodied wines that are sweet or intensely sweet, honey-rich, and full of flavour. They can be quite complex and in suitable years have evident botrytis character.

🍷 *Sémillon, Sauvignon Blanc, Muscadelle*

🍷 *5–15 years (25 in exceptional cases)*

PREMIÈRES CÔTES DE BORDEAUX AOC

This 60-kilometre (37-mile) strip of southwest-facing slopes covers 170 hectares (420 acres) of vines scattered through 37 communes, each of which has the right to add its name to this appellation. They are: Bassens, Baurech, Béguey, Bouliac, Cadillac, Cambes, Camblanes, Capian, Carbon Blanc, Cardan, Carignan, Cenac, Cenon, Donzac, Floirac, Gabarnac, Haux, Langoiran, Laroque, Lestiac, Lormont, Monprimblanc, Omet, Paillet, Quinsac, Rions, Sémens, St-Caprais-de-Bordeaux, St-Germain-de-Graves, St-Maixant, Ste-Eulalie, Tabanac, Le Tourne, La Tresne, Verdelais, Ville-nave de Rions, and Yvrac.

RED The best red wines come from the northern communes. These well-coloured, soft, and fruity wines are a cut above basic Bordeaux AOC.

🍷 *Cabernet Sauvignon, Cabernet Franc, Carmenère, Merlot, Malbec, Petit Verdot*

🍷 *4–8 years*

WHITE Since the 1981 harvest, no dry wines have been allowed under this generally unexciting appellation. They must have at least some sweetness, and most are in fact semi-sweet. Simple, fruity wines, well made for the most part, but lacking character.

🍷 *Sémillon, Sauvignon Blanc, Muscadelle*

🍷 *3–7 years*

STE-CROIX-DU-MONT AOC

This is the second-best sweet-white appellation on the right bank of the Garonne, and it regularly produces more wine than Barsac. Like Loupiac wines, these wines must be made with the "assistance" of overripe botrytised grapes. They have less honeyed complexity of "noble rot" than Loupiac wines, but often have more finesse.

WHITE Fine, viscous, honey-sweet wines that are lighter in body and colour than Loupiac wines. Excellent value when they have rich botrytis character.

🍷 *Sémillon, Sauvignon Blanc, Muscadelle*

🍷 *5–15 years (25 in exceptional cases)*

THE WINE PRODUCERS OF

THE WINE PRODUCERS OF
ENTRE-DEUX-MERS

CHÂTEAU ARNAUD-JOUAN

Cadillac

Côtes de Bordeaux Cadillac AOC
and Cadillac AOC

☆

This large, well-situated vineyard makes interesting, attractive wines.

DOMAINE DU BARRAIL

Monprimblanc

Côtes de Bordeaux Cadillac AOC
and Cadillac AOC

☆ V

Both the red Côtes de Bordeaux and sweet white Cadillac produced at this property are worth watching.

CHÂTEAU DE BEAUREGARD

Entre-Deux-Mers AOC and Bordeaux-Haut-Benauge AOC

The red and white wines are well-made. The red has good structure, but is softened by the spice of the Merlot.

CHÂTEAU BEL-AIR

Vayres

Graves de Vayres AOC

Most of the wines made here are red and well-coloured. They are aromatic wines of a Cabernet character.

CHÂTEAU BIAC

Langoiran

Côtes de Bordeaux Cadillac AOC
and Cadillac AOC

An historic property with spectacular views across the Garonne River, owned by the Asseily family since 2006 when the property underwent a transformation in both planting and wine making. The property produces three red wines: Château Biac and B de Biac, both in the Côtes de Bordeaux Cadillac AOC; a tiny production of Felix de Biac, a Vin de France; and an excellent sweet Cadillac, Secret de Biac.

CHÂTEAU BIROT

Béguey

Côtes de Bordeaux Cadillac AOC
and Cadillac AOC

☆ V

Popular for its easy-drinking whites, this property also produces well-balanced red wines of some finesse.

CHÂTEAU LA BLANQUERIE

Mérignas

Entre-Deux-Mers AOC

Dry white wine with a Sauvignon character and a fine finish.

BORDEAUX VINEYARDS LINE THE RIVER GARONNE
This river is one of the two bodies of water that give this region its name and forms its southern boundry. The other, the Dordogne, form the area's northern boundary.

CHÂTEAU BONNET

Grézillac

Entre-Deux-Mers AOC

★☆

This large top-performing Entre-Deux-Mers château is owned by André Lurton. It produces crisp, fresh, and characterful white wines, elegant rosé, and soft, fruity, extremely successful (Bordeaux Supérieur) red.

Other wines: *Tour-de-Bonnet, Château Bonnet Reserve*

CHÂTEAU BRÉTHOUS

Camblanes-et-Meynac

Premières Côtes de Bordeaux AOC, Côtes de Bordeaux Cadillac AOC, and Cadillac AOC

★ V O

The red wines are forward and attractive, yet well-structured, while the whites are succulent and sweet. In organic production.

CHÂTEAU CANET

Guillac

Entre-Deux-Mers AOC

★ V

These excellent white wines from this property are clean and crisp, with good fruit and an elegant balance.

CHÂTEAU CARBONNEAU

Pessac-sur-Dordogne

Côtes de Bordeaux Ste-Foy AOC

The chateau produces a full range of wines, including their classic Bordeaux blend, La Verrière, which is a small production of oak-aged wine worth seeking out, as is the Cabernet Franc–dominated Sequoia.

CHÂTEAU CAYLA

Rions

Côtes de Bordeaux AOC

★ V

The reds are elegant and accessible with just a touch of new oak.

DOMAINE DE CHASTELET

Quinsac

Premières Côtes de Bordeaux AOC
and Cadillac AOC

★ V

Domaine de Chastelet produces red wine that is delicious, yet firm and complex, with very well-balanced blackcurrant fruit flavours and a hint of vanilla oak.

CLOS JEAN

Loupiac

Loupiac AOC

The wines produced by Clos Jean are of similar quality to those of Château du Cros, but more refined and ethereal in character.

CHÂTEAU DU CROS

Loupiac

Loupiac AOC

★ V

The fine, fat, succulent sweet wines of Château du Cros are among the best of the appellation.

CHÂTEAU FAUGAS

Côtes de Bordeaux Cadillac AOC

Well-balanced reds with attractive berry-fruit flavours.

CHÂTEAU FAYAU

Cadillac

Côtes de Bordeaux Cadillac AOC and Cadillac AOC

Part of the Medeville group of chateaux in the southern Graves and Entre deux Mers, this château produces succulent sweet wines in addition to red, clairet, and dry white wines.

Other wines: *Clos des Capucins*

CHÂTEAU FONGRAVE

Gornac

Entre-Deux-Mers AOC

These dry white wines have a fresh and tangy taste. Also produces a red Bordeaux.

Other wines: *Silence by Fongrave Cotes de Bordeaux Saint Macaire*

CHÂTEAU GOUDICHAUD

Vayres

Graves de Vayres AOC

This property also extends into St-Germain-du-Puch in Entre-Deux-Mers, where it produces some very respectable wines.

CHÂTEAU DU GRAND MOUËYS

Capian

Côtes de Bordeaux Cadillac AOC

Côtes de Bordeaux Blanc
and Bordeaux Rosé

★ V

Excellent-value reds for medium-term ageing are currently made at this château.

Second wine: *Les Templiers*

CHÂTEAU HOSTENS PICANT

Grangeneuve Nord

Côtes de Bordeaux Ste-Foy AOC

★ V

Since the current owners Nadine and Yves Picant purchased this property in 1986, broke off relations with the local cooperative, and built their own winery, this château has not looked back. The top wine, LVCVLLVS Cuvée d'Exception, is ★☆ quality, but the "basic" *grand vin* is excellent, and the refreshing dry white and rosé (sold as the seldom-seen Bordeaux Clairet AOC) offer deliciously good value.

SÉMILLON GRAPES
Sémillon is a common white variety in the blends produced in the Entre-Deux-Mers.

CHÂTEAU DU JUGE

Cadillac

Côtes de Bordeaux Cadillac AOC
and Cadillac AOC

Part of the Jean Medeville stable of chateaux. Respectable red and dry white wines are produced at Château du Juge, and in some years a little sweet white wine of high quality is also made. Both red and white wines are extraordinarily good value.

CHÂTEAU LAPÉYÈRE

Côtes de Bordeaux Cadillac AOC

Well-structured red wine, dominated by Cabernet Sauvignon.

CHÂTEAU LAMOTHE

Haux

Côtes de Bordeaux AOC
and Bordeaux Blanc AOC

Some exceptionally good wines have been produced in recent years at this château, which derives its name from "La Motte", a rocky spur that protects the vineyard.

CHÂTEAU LAROCHE BEL AIR

Baurech

Côtes de Bordeaux AOC

Absolutely delicious-drinking reds under the basic Château Laroche label, and an even better selection of oak-aged reds under Laroche Bel Air. Belongs to the Jean Merlaut collection.

CHÂTEAU LAURETTE

Cadillac

Ste-Croix-du-Mont AOC

☆

This property is under the same ownership as Château Lafue, and runs along similar lines.

CHÂTEAU LOUBENS

Cadillac

Ste-Croix-du-Mont AOC

★

This château produces rich, liquorous, superbly balanced sweet white wines. Dry white wines are sold as "Fleur Blanc", and a little red wine is also made.

Other wines: *Fleur Blanc de Château Loubens*

CHÂTEAU LOUPIAC-GAUDIET

Loupiac

Loupiac AOC

Fine, honey-rich sweet wines hinting of crystallized fruit are produced here.

CHÂTEAU LOUSTEAU-VIEIL

Cadillac

Ste-Croix-du-Mont AOC

★

This property produces richly flavoured, high-quality sweet wines.

CHÂTEAU DES MAILLES

Cadillac

Ste-Croix-du-Mont AOC

Some outstanding sweet wines are produced at Château des Mailles, alongside Bordeaux red, white, and rosé. Since 2015 Laurence, the fifth generation of the Larrieu family, has taken over the property and the winemaking, improving the quality across the board.

CHÂTEAU LA MAUBASTIT

Côtes de Bordeaux Ste-Foy AOC

Some 5,000 cases of white and 2,000 of red, both organic wines, are sold under the Bordeaux appellation.

CHÂTEAU MORLAN-TUILIÈRE

St-Pierre-de-Bat

Entre-Deux-Mers-Haut-Benauge AOC
and Bordeaux Haut-Benauge AOC

One of the best properties of the area, producing a vibrant, crystal-clear Entre-Deux-Mers-Haut-Benauge, a Bordeaux Supérieur in the *moelleux* style, a fairly full-bodied red Bordeaux AOC, a clairet, and a Cremant de Bordeaux. Organic production.

CHÂTEAU MOULIN DE LAUNAY

Soussac

Entre-Deux-Mers AOC

Despite the vast quantity produced, the dry white wine is crisp and fruity and of a very fine standard. A little red is also produced.

Other wines: *Plessis, Château Tertre-de-Launay, Château de Tuilerie, Château la Vigerie*

CHÂTEAU MOULIN DE ROMAGE

Les Lèves et Thoumeyragues

Côtes de Bordeaux Ste-Foy AOC

This château produces equal quantities of organic red and white.

CHÂTEAU PETIT-PEY

St-André-du-Bois

Côtes de Bordeaux St-Macaire AOC

Good, sweet white St-Macaire and agreeably soft red Bordeaux AOC are made at this property.

CHÂTEAU PEYREBON

Grézillac

Entre-Deux-Mers AOC

This *cru bourgeois* property, owned by Patrick Bernard of the wine merchant Millésima, produces red and white wine in almost equal quantities. The dry white is fine and flavoursome.

CHÂTEAU PEYRINES

Mourens

Entre-Deux-Mers-Haut-Benauge AOC
and Bordeaux Haut-Benauge AOC

★☆

The vineyard of this château has an excellent southern exposure and produces fruity red and white wines.

CHÂTEAU DE PIC

Le Tourne

Côtes de Bordeaux Cadillac AOC

The basic red of the property is a lovely, creamy-sweet, easy-drinking, fruity wine. A superb oak-aged red under the Cuvée Tradition label is also made.

CHÂTEAU PICHON-BELLEVUE

Vayres

Graves de Vayres AOC

The red wines from this estate are variable, but the dry white wine are delicate and refined.

CHÂTEAU PLAISANCE

Capian

Côtes de Bordeaux Cadillac AOC

The Cuvée tradition, in which the wine is aged in oak and is unfiltered, gives rich, ripe fruit with supple tannin structure and smoky oak. (*See also* Bordeaux and Bordeaux Supérieur, p148.)

Other wines: *De l'Esplanade, Château Florestin*

CHÂTEAU DE PLASSAN

Tabanac

Côtes de Bordeaux Cadillac AOC

The basic red has a lot of character, with cherry-minty undertones in riper years. The fuller, more complex *cuvée spéciale* is worth paying for, however, particularly if you want a wine to accompany food. (*See also* Bordeaux and Bordeaux Supérieur, p148.)

CHÂTEAU PONTETTE-BELLEGRAVE

Vayres

Graves de Vayres AOC

This property has a reputation for subtly flavoured, dry white wines and also produces a red Graves de Vayres and a Bordeaux Rosé.

CHÂTEAU PUY BARDENS

Cambes

Premières Côtes de Bordeaux AOC

This top-performing château produces reds with sweet, ripe, fat fruit and a soft, velvety finish.

CHÂTEAU LA RAME

Cadillac

Ste-Croix-du-Mont AOC

One of the top wines of the appellation, La Rame can have fruit, with cream and honey flavours.

CHÂTEAU REYNON

Béguey

Côtes de Bordeaux Cadillac AOC,
Cadillac AOC, and Bordeaux Blanc AOC

This property produces a superb, oak-aged Côtes de Bordeaux Cadillac red wine, a dry white, and a Cadillac. (*See also* Bordeaux and Bordeaux Supérieur, p148.)

CHÂTEAU RICAUD

Loupiac

Loupiac AOC

Château Ricaud wines are once again the best in the Loupiac appellation. They suffered a significant decline under the previous proprietor, but now under the management of Dourthe display great class. As well as the Loupiac, they produce a red Côtes de Bordeaux Cadillac, a Bordeaux Supérieur, and a Bordeaux Blanc.

CHÂTEAU DE LA SABLIÈRE-FONGRAVE

Gornac

Entre-Deux-Mers-Haut-Benauge AOC
and Bordeaux Haut-Benauge AOC

Sold as a Bordeaux Supérieur, the red wine of this château is fairly robust and requires time in bottle to soften. A much better-quality dry white is produced and sold under the Entre-Deux-Mers appellation.

CHÂTEAU TANESSE

Langoiran

Côtes de Bordeaux Cadillac AOC,
Cadillac AOC, and Bordeaux Blanc AOC

Previously a Cordier property, Château Tanesse now belongs to the Gonfrier family, who own and manage several properties across Bordeaux. Farmed under the "Haut Valeur Environemental" certification, they produce a red Côtes de Bordeaux Cadillac, a fine-quality Sauvignon-style dry white, and a sweet Cadillac.

CHÂTEAU THIEULEY

La Sauve

Entre-Deux-Mers AOC

☆

Château Thieuley is a family property run by the Courselle sisters. They produce two reds, two whites, a delightful light rosé and a clairet under the Château Thieuley label, a red Côtes de Bordeaux and a Bordeaux liquoreux under the Clos Sainte Anne label, a red Bordeaux under the Saint Genès label, a Crémant de Bordeaux, and two "Vins de France" from Syrah and Chardonnay.

CHÂTEAU TOUR PETIT PUCH

St-Germain-du-Puch

Graves de Vayres AOC

☆

Attractively coloured, well-made wines, with a touch of spice.

Burgundy

Burgundy has always been famous for its vineyards, from Chablis down to Beaujolais. The quality of the wines has improved in every corner of the region, but nature has reduced the yield during the past 10 to 15 years. The number of great bottles produced has thus become limited and therefore even more sought after. The monks who were working on the terroir concept for centuries would hardly believe what heights the fame of Burgundy has reached today.

Before anyone gets frightened of the high prices and decides not to fall in love with Burgundy, let me quote one of the favourite maxims of the local *vignerons:* "We have wine for everyone". Certainly fine Burgundy is not cheap, and the quality does not always match the price tag. Cheap-and-good Burgundy exists, however, but one has to explore for it, has to find it, he or she has to be adventurous, taking risks and tasting wines from all over the *vignoble* . . . and someday, without any disappointment I hope, the consumer can decide if the *Bourguignons* really have wine for everyone. I trust they can be justified.

Say "Burgundy" and most people think of the famous wines of the Côte de Nuits and Côte de Beaune: Chambertin, Musigny, Montrachet. Yet Burgundy stretches from Chablis, which is close to the Aube vineyards of Champagne, down to Beaujolais, which is in the Rhône *département*. In fact, the Côte de Nuits and Côte de Beaune account for less than 10 per cent of Burgundy, while even a much-reduced Beaujolais represents a third of the region's entire production. Problems with the quality of so much Beaujolais led to the *vin de merde* story by *Lyon Mag* in 2001, but few people outside of Beaujolais's own backyard would have heard of this if 56 cooperatives had not decided to sue the local rag. Within a short time, wine lovers all over the world discovered that Lyon Mag had described 1.1 million cases of Beaujolais as *vin de merde* because no one wanted it; consequently, it had to be distilled (*see* p246). Since the case came to light,

BURGUNDY AT A GLANCE

DISTRICT	PERCENTAGE OF ALL BURGUNDY PRODUCTION	HECTOLITRES (CASES)	PERCENTAGE BY CATEGORY		
			RED/ ROSÉ	WHITE	(GRANDS CRUS)
Generic AOCs	18	369,633 (4,004,358)	56	44	–
Crémant AOCs	6	153,.950 (1,667,791)	9	91	–
Chablis	12	238,334 (2,581,952)	–	100	2,00%
Côte de Nuits	4	71,679 (776.523)	96	4	11,00%
Hautes-Côtes de Nuits	1	28,000 (303,333)	97	3	–
Côtes de Beaune	8	144,506 (1,565,482)	63	37	4,00%
Hautes-Côtes de Beaune	1	39,000 (422,500)	98	2	–
Côte Chalonnaise	3	79,899 (865,783)	52	48	–
Mâconnais	15	343,086 (3,716,765)	15	85	–
Beaujolais	33	662,457 (7,176,618)	98	2	–
TOTAL	100	2,130,544[1] (23,080.893[1])	58.5[2]	41.5[2]	–

[1] Totals might not tally with each other due to rounding of numbers above.

[2] Average percentage for region

A SOMMELIER PRESENTS A BOTTLE OF DOMAINE COFFINET-DUVERNAY CHASSAGNE-MONTRACHET *1ER CRU* LA MALTROIE
The Montrachet *grand cru* vineyard in the Côte de Beaune subregion of Burgundy is widely considered the world's best for dry white Chardonnays. Chardonnays are cheese-friendly wines and pair well with creamy or buttery cheeses like brie.

the figures speak for themselves, as production has plummeted by more than 50 per cent at the bottom of the Beaujolais market. The best Beaujolais today is better than it has ever been, but because it became impossible to get rid of the cheapest and nastiest Beaujolais, producers began to sell it as Bourgogne AOC. That was perfectly legal, but it began damaging the generic reputation of Burgundy itself, prompting calls to remove Beaujolais from the Burgundy region altogether. Even though it is made from an entirely different grape variety grown way down in the Rhône, that was never going to happen, but the pressure it brought to bear on the situation did force through changes in the AOCs. Starting with the 2011 vintage, wines from the Beaujolais district can no longer be sold under the Bourgogne AOC and must be sold as Coteaux Bourgignons AOC (formerly called Bourgogne Ordinaire or Bourgogne Grand Ordinaire). This appellation is open to wines from all over the Burgundy region but, in practice, will primarily be produced from Gamay grown in Beaujolais. Additionally, a new generic appellation, Bourgogne Côte d'Or AOC, was created in 2017 for wines exclusively from the Côte d'Or.

Burgundy still produces the world's greatest Chardonnay and Pinot Noir wines and the only Gamay wines ever to achieve classic status, but it is increasingly debased by a growing number of lacklustre, sometimes quite disgusting, mass-market wines that rely solely upon the reputation of the famous Burgundian names, which their producers then abuse to sell low-quality wines at high prices.

Burgundy, or *Bourgogne* as it is known in French, is an area rich in history, gastronomy, and wine, but unlike the great estates of Bordeaux, the finest Burgundian vineyards are owned by a proliferation of smallholders. Prior to 1789, the church owned most of the vineyards in Burgundy, but these were seized and broken up as a direct result of the Revolution, which was as much anti-church as anti-aristocracy. In Bordeaux, although some of the large wine estates were owned by the aristocracy, many were owned by the *bourgeoisie*, who, because of their long association with the English, were anti-papist and so escaped the full wrath of the Revolution. In Burgundy the great vineyards were further fragmented by inheritance laws, which divided the plots into smaller and smaller parcels. Consequently, many *crus*, or growths, are now owned by dozens of individual growers. The initial effect of this proprietorial carve-up was to encourage the supremacy of *le négoce*. Few commercial houses had been established prior to the mid-18th century because of the difficulty of exporting from a landlocked area, but with better transport and no opposition from land-owning aristocracy, merchant power grew rapidly. A network of brokers evolved in which dealers became experts on very small, localized areas.

As ownership diversified even further, it became a very specialized, and therefore rewarding, job to keep an up-to-date and comprehensive knowledge of a complex situation. The brokers were vital to the success of a *négociant*, and the *négoce* himself was essential to the success of international trade and therefore responsible for establishing the reputation of Burgundy.

THE ROLE OF THE *NÉGOCIANT*

Until the early 1980s, virtually all Burgundy would be sold through *négociants* and, although many of them had their own vineyards, these wines were seldom domaine-bottled. Faced with the rise of domaine-bottled Burgundies from small growers, most of the old-fashioned merchants were devoured by Boisset (*see* box below), which has become the largest merchant of its kind. Boisset has shown a determination to join the ranks of the few traditional *négociants* who can be said to be top performing. These currently include Bouchard Père, Drouhin, Louis Jadot, and Leroy, who have not only enlarged their own domaines but have also taken an increasingly proactive role in the vineyards of their suppliers, tending now to buy in grapes or must rather than wines. Until this new strategy evolved, the purchase of wine rather than grapes was the major difference between the old-style *négociant* and the new-wave growers-cum-merchants. The best of the latter include the likes of Jean-Marc

Boillot, Michel Colin, Bernard Morey, Sauzet, Anne Gros, Méo-Camuzet, and many more who tend to buy grapes to expand their range of single-vineyard wines, in contrast to the old-style merchant, who still churns out examples from nearly every village under the Burgundian sun.

The decline of the old-style merchant is clearly highlighted by domaine-bottling statistics. Today, as much as 90 per cent of all *grand cru* wines (and 50 per cent of all *premiers crus*) are domaine bottled, although – tellingly – only 24 per cent of the entire production of Burgundy is domaine bottled.

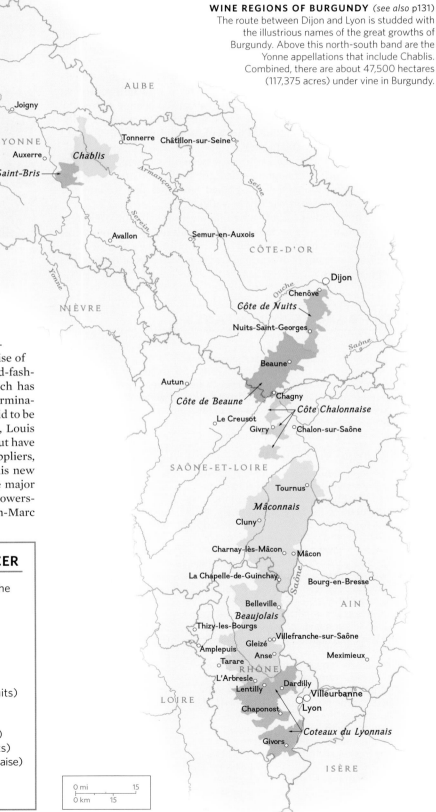

WINE REGIONS OF BURGUNDY (*see also* p131)
The route between Dijon and Lyon is studded with the illustrious names of the great growths of Burgundy. Above this north-south band are the Yonne appellations that include Chablis. Combined, there are about 47,500 hectares (117,375 acres) under vine in Burgundy.

BOISSET: BURGUNDY'S LARGEST WINE PRODUCER

The super-*négociant* Boisset has holdings throughout Burgundy. The following list includes its latest Burgundian acquisitions.

- Jean-Claude Boisset (Côte de Nuits)
- Bouchard Aîné & Fils (Côte de Beaune)
- F Chauvenet (Côte de Nuits)
- Château du Grand Talencé (Beaujolais)
- Jaffelin (Côte de Beaune)
- Benoît Lafont (Beaujolais)
- Mommessin (Beaujolais)
- J Moreau & Fils (Chablis)
- Morin Père & Fils (Côte de Nuits)
- Joseph Pellerin (Beaujolais)
- Pierre Ponnelle (Côte de Beaune)

- Ropiteau Frères (Côte de Beaune)
- Thomas-Bassot (Côte de Nuits)
- Thorin (Beaujolais)
- Charles Vienot (Côte de Nuits)
- Domaine de la Vougeraie (Côte de Nuits)
- Labouré Roi (Côte de Nuits)
- Louis Bouillot (Côte de Nuits)
- Antonin Rodet (Côte Chalonnaise)
- Chateau de Pierreux (Beaujolais)

BURGUNDY'S RICH DIVERSITY

The wines of Burgunday are as diverse as the region itself. Chablis in the north of Burgundy produces the crispest white Chardonnay wines in the world. It is also geographically closer to Champagne, of which it was once a part, than to the rest of Burgundy. After travelling more than 100 kilometres (60 miles) southwest from Champagne, we reach the great Burgundy districts of the Côte d'Or: the Côte de Nuits is encountered first, followed by the Côte de Beaune.

If you associate Côte de Nuits with "night" or "darkness" and Côte de Beaune with "bone-white", then you will easily remember which area is most famous for which wine: for although both *côtes* each make excellent red and white wines, most of the greatest red Burgundies come from the Côte de Nuits, whereas most of the greatest white Burgundies come from the Côte de Beaune.

The Côte Chalonnaise region – probably the least-known, but certainly the best-value wine district of Burgundy – produces similar, if somewhat less classic, red and white styles to those in the Côte de Beaune. Softer still are the primarily white wines of the Mâconnais district, where Pouilly-Fuissé AOC rules supreme. (Note: This wine should never be confused with Pouilly Fumé AOC in the Loire.)

Still in Burgundy, but farther south, is the Beaujolais region, which is in the Rhône *département,* although its soft, light, fluffy, fruity red wines are far removed from the archetypal full-bodied red wines that we immediately think of as coming from the Rhône Valley.

Beaujolais can be delicious, but unfortunately most of it is not. Whereas Beaujolais Nouveau was never meant to be considered as a serious wine, it is now considered a joke, and far too many of even the best *cru* Beaujolais wines are simply overpriced. Yet that is all part of Burgundy's rich tapestry of wines.

A TINY CHURCH SITS AMIDST VINEYARDS OF THE CORE D'OR
The Côte D'Or, or "Golden Coast", encompasses both the Côte de Nuits and the Cote De Beaune, two of Burgundy's finest wine-producing regions

RECENT BURGUNDY VINTAGES

2018 The year ended in a beautiful harvest. Fine grape maturity with nice acidity that made a good balance in the wine. As of publication, no 2018 bottles have been released, but the growers have every reason to smile. Quality comes down to the yields in Beaujolais. Those that chose not to take advantage of the potential generosity of the vintage made wines of the highest quality. But of course not everybody did that.

2017 After long years of short crops, 2017 was able to bring higher volumes that could please the demand for Burgundy. The whole of the region was hit by spring frost that did not impact the quantity, except in Chablis, where the crop was down by about 20 to 25 per cent. In every area the wines show fine balance, full of floral and fruity aromas. It is rather a red wine vintage, but the whites show comparable quality to 2014 on the Côte de Beaune. Some great, great wines were produced from lower yields in Beaujolais, and outside of some important places that were badly hailed (Fleurie and Moulin-à-Vent – but not only) there is a high-achieving consistency.

2016 Reds are extremely fruity; the tannins are silky and the structure precise. Most have huge potential to brilliantly age. The whites are fruity, rather crisp, having nice balance and length. The Côtes did an excellent job. Great quality in the Maconnais, too, with precision and elegance. Sadly, the quantities are far behind the average. The fruit is fresher and nicer in Beaujolais than in 2015, but it is a much less consistent vintage due to a poor start to the season and many areas suffering from frost; it was also higher yielding.

2015 Harvest this year started at the very end of August. The red wines are really fruity, deep in colour, and surprisingly accessible from the very beginning. A great red wine vintage. The whites of the whole of Burgundy have nice complexity and ripeness but lack the freshness I do love. In some of the southern areas the high alcohol has a bit of unpleasant impact on the balance of the wines. Beaujolais produced a once-in-a-generation vintage, by all accounts – with depth, consistency, and concentration – wines to age as long as you like.

2014 The white Burgundies are really well balanced and harmonious in 2014, no matter from which area they come from. I sampled brilliant wines in Chablis and in Macon, too. On the Côte de Beaune they are fresh, vivid, yet complex, and they show great ageing potential. Everything is here to offer one of the greatest white vintages of Burgundy. Most of the reds have nice balance, and there is much elegance here, too. They will age gracefully. Generally a lighter – more airy – but very delicious vintage in Beaujolais.

2013 The wines from Chablis show rich minerality and ageing potential. In the Maconnais and the Côte Chalonnaise, the specialty of the year showed a fine minerality with lots of citrus-driven aromas and freshness. The Côte de Beaune produced plenty of unforgettable bottles. The year is especially rich in white fruit and freshness; nevertheless, they are perfectly balanced. Many reds of the Côte de Nuits suffer from under-ripeness. In

the other red wine areas, such as the Côte de Beaune and the Côte Chalonnaise, the results are a lot more promising. This is a white wine vintage, however. A difficult vintage in Beaujolais, with more herbs in the aromas and an acidity that was not always balanced.

2012 Chablis and Auxerrois produced complex and dense wines with hints of citrus fruit. The Côte de Beaune suffered from hail, but the outcome was excellent: wines with real depth and concentration as well as long aftertaste. Out of the two southern regions the Côte Chalonnaise enjoyed better growing conditions. The wines there are complex and rich in 2012. On the Côte de Nuits they are rather on par with those of 2010, but as a famous grower said "2012 has more depth". I would definitely add that 2012 is a bit less precise. Some outstanding wines in the Beaujolais. No mistake, this is a vintage that can offer both concentration and clarity – but it is also a vintage of inconsistencies and multiple challenges in the growing season. The best are first class, but many are not.

2011 The September rain slowed down the picking. As a result the hands around the sorting table had to be very active to discard the unripe berries. I tasted plenty of bottles of 2011 – in both colours – showing vegetal, green tobacco leaf-like aromas. If I had to drop a vintage in the decade, I would definitely choose 2011. In Beaujolais this was a vintage the growers thought they would only witness once in a generation: a benchmark year.

THE GENERIC APPELLATIONS OF

BURGUNDY

BOURGOGNE AOC

Many writers consider Bourgogne AOC to be too basic and boring to warrant serious attention, but this should be the most instructive of all Burgundy's appellations. If a producer cares about the quality of the *Bourgogne*, how much more effort does that producer put into making higher-quality wines? Finding a delicious, easy-to-drink *Bourgogne* is delightful. Light-red/dark-rosé wines may be sold as Bourgogne Clairet AOC, but the style is outmoded and the appellation rarely seen.

RED The only *Bourgogne* worth seeking out is that with the flavour and aroma of pure Pinot Noir.

🍷 *Pinot Noir, Pinot Gris, plus Pinot Blanc and Chardonnay, in the Yonne district César, and Gamay (if from one of the original nine crus Beaujolais – but not Régnié)*

🍷 *2–5 years*

WHITE There are a lot of boring whites made under this appellation. Unless you have access to something more interesting like the wine from J-F Coche-Dury or another top grower, it is safer to buy an inexpensive Mâcon AOC. These wines may be sold as *primeur* or *nouveau* as from the third Thursday of November following the harvest.

🍷 *Chardonnay, Pinot Blanc, Pinot Gris*

🍷 *1–4 years*

ROSÉ The wines produced under this appellation are acceptable, but they are never special – it is the least exciting category of Bourgogne.

🍷 *Pinot Noir, Pinot Gris, Pinot Blanc, Chardonnay, plus in the Yonne district César*

🍷 *1–4 years*

✓ *Bourgogne Rouge Robert Arnoux • Ghislaine Barthod • Bertagna • Jean-Marc Boillot • Pascal Bouley • Carré-Courbin • Sylvain Cathiard • Château de Chamilly • Philippe Charlopin • Christian Clerget • J-F Coche-Dury • De la Combe • Joseph Drouhin • Alex Gambal • Anne Gros • Henri Jayer • Michel Juillot • Pierre Labet • Labouré-Roi • Michel Lafarge • Marie-Hélène Laugrotte • Dominique Laurent • Olivier Leflaive • Lucien Lemoine • Leroy ⓑ • Hubert Lignier • Catherine et Claude Maréchal (Cathérine, Gravel) • Jean-Philippe Marchand • Mauperthuis (Grande Réserve) • Olivier Merlin (Les Cras) ⓞ • Denis Mortet • Lucien Mouzard • De Perdrix • Des Pitoux • De la Pousse d'Or • Daniel Rion & Fils • Michel & Patrice Rion • Nicolas Rossignol • Emmanuel Rouget • Tollot-Beaut • Vallet Frères • A & P de Villaine ⓞ • De la Vougeraie ⓑ (Terres de Familles)*

Bourgogne Blanc: *Jean-Baptiste Béjot • Simon Bize & Fils • Jean-Marc Boillot • Jean-Marc Brocard (Cuvée Jurassique) • J-F Coche-Dury • Coste-Caumartin • Alex Gambal • Patrick Javillier • François Jobard • Labouré-Roi • Michel Lafarge • Hubert Lamy • Leflaive ⓑ • Olivier Leflaive • Lucien Lemoine • Lorenzon • Catherine et Claude Maréchal • Bátrice & Gilles Mathias • Olivier Merlin ⓞ • Pierre Morey ⓑ • Antonin Rodet • Georges Roumier • Michel Rouyer (Domaine du Petit-Béru) • Tollot-Beaut • Henry de Vézelay • A & P de Villaine ⓞ*

Bourgogne Rosé: *Abbaye du Petit Quincy • Philippe Defrance • Elise Villiers*

Note: For other local generic wines *see:*

Chablis: *Bourgogne Chitry AOC, Bourgogne Coulanges-la-Vineuse AOC, Bourgogne Côtes d'Auxerre AOC, Bourgogne Côte Saint-Jacques AOC, Bourgogne Épineuil AOC, Bourgogne Vézelay AOC*

Côte de Nuits and Hautes-Côtes de Nuits: *Bourgogne Hautes-Côtes de Nuits AOC, Bourgogne Le Chapitre AOC, Bourgogne Montrecul AOC*

Côte de Beaune and Hautes-Côtes de Beaune: *Bourgogne La Chapelle Notre-Dame AOC, Bourgogne Hautes-Côtes de Beaune AOC*

Côte Chalonnaise: *Bourgogne Côte Chalonnaise AOC, Bourgogne Côtes du Couchois AOC*

BOURGOGNE ALIGOTÉ AOC

The finest Bourgogne-Aligoté wines come from the village of Bouzeron in the Mercurey region, which has its own appellation (*see also* Bouzeron AOC, p237). With the exception of the wines below, the remaining Aligoté can be improved by adding *crème de cassis*, a local blackcurrant liqueur, to create an aperitif known as a "Kir".

WHITE These are dry wines that are usually thin, acid, and not very pleasant: bad examples are even worse and are becoming widespread. When good, however, Aligoté can make a refreshing change from Burgundy's ubiquitous Chardonnay wines. Even among the top producers, however, this grape seems to inflict its own inconsistency. Any one of the following recommended producers could make the greatest Aligoté of the vintage one year, then turn out something very ordinary the next year. Bourgogne Aligoté may be sold as *primeur* or *nouveau* from the third Thursday of November following the harvest.

🍷 *Aligoté, and a maximum of 15% Chardonnay*

🍷 *1–4 years*

✓ *D'Auvenay ⓑ • Bersan • Marc Brocard • Arnaud Ente • Naudin-Ferrand • De la Folie • Alex Gambal • Ghislaine et Jean-Hugues Goisot • François Jobard • Daniel Largeot • Catherine et Claude Maréchal • Edmond Monot • Alice et Olivier Moor • Jacky Renard • De la Sarazinière (Clos des Bruyères) • Thévenot-le-Brun & Fils • Vignerons des Terres Secrètes (Château des Moines) • A & P de Villaine ⓞ*

BOURGOGNE CÔTE D'OR AOC

The producers of the Regional Appellation "Bourgogne", located in Côte de Beaune and Côte de Nuits, have been able to use the additional mention "Bourgogne Côte d'Or", which thus becomes a Bourgogne with additional geographical denomination since 2017. This name is reserved for red and white still wines produced within the 40 villages in Côte de Beaune and Côte de Nuits.

BOURGOGNE GAMAY AOC

Created in 2011, this new appellation preserves Gamay under the Bourgogne AOC label but will be restricted to vineyards within the 9 of the 10 *crus* Beaujolais.

BOURGOGNE GRAND-ORDINAIRE AOC

See Coteaux Bourguignons AOC

BOURGOGNE MOUSSEUX AOC

Since December 1985 this appellation has been limited to, and remains the only outlet for, sparkling red Burgundy.

SPARKLING RED Once a favourite fizzy tipple in the pubs of pre-war Britain. This wine's sweet flavour is very much out of step with today's sophisticated consumers.

🍷 *Pinot Noir, plus up to 5% Pinot Gris, Pinot Blanc, and Chardonnay*

🍷 *Upon purchase*

GAMAY GRAPES
A fairly new appellation, Bourgogne Gamay incorporates 25 villages of the Mâconnais and Beaujolais. A majority of the *crus* Beaujolais retain the right to use the label AOC Bourgogne, with the restriction that if the wine contains more than 30 per cent Gamay, it must be labeled AOC Bourgogne Gamay.

CHÂTEAU DE MONTIGNY-SUR-AUBE
Estate owner Marie-France Ménage-Small produces Crémants de Bourgogne at this impressive château, which dates back to the 12th century.

BOURGOGNE ORDINAIRE AOC

See Coteaux Bourguignons AOC

BOURGOGNE PASSE-TOUT-GRAINS AOC

Made from a mélange of Pinot Noir and Gamay grapes, *passe-tout-grains* is the descendant of an authentic peasant wine. A grower would fill his vat with anything growing in his vineyard and ferment it all together. Thus *passe-tout-grains* once contained numerous grape varieties. The Pinot Noir and Gamay varieties were, however, the most widely planted, and the wine naturally evolved as a two-grape product. Up until 1943, a minimum of one-fifth Pinot Noir was enforced by law; now the minimum is one-third.

RED Many *passe-tout-grains* used to be drunk too early, as the better-quality examples require a few years of bottle-ageing to show the aristocratic influence of their Pinot Noir content. With an increase in Pinot Noir production and more modern vinification techniques, more producers have begun making softer, less rustic *passe-tout-grains*, which are easier to drink when young. They remain relatively modest wines.

🍷 *A minimum of 30% Pinot Noir, plus a minimum of 15% Gamay and a combined maximum of 15% Chardonnay, Pinot Blanc, and Pinot Gris*

🍷 *2–6 years*

ROSÉ This dry, pink version is worth trying.

🍷 *A minimum of 30% Pinot Noir, plus a minimum of 15% Gamay and a combined maximum of 15% Chardonnay, Pinot Blanc, and Pinot Gris*

🍷 *1–3 years*

✔ *Robert Chevillon • Edmond Cornu & Fils • Michel Lafarge • Laurent Ponsot (former Volpato old vines) • Lejeune • Daniel Rion & Fils*

COTEAUX BOURGUIGNONS AOC

The name of this appellation has altered since the 2011 vintage, when no wines made in the Beaujolais district could be declassified into Bourgogne AOC but had to be sold under the lower-quality Bourgogne Ordinaire or Bourgogne Grand-Ordinaire. This was as a result of complaints from Côte d'Or producers that inferior Gamay sold under Burgundy's name was damaging their reputation. Forced to accept this demotion, but not wanting to be lumbered with the "ordinary" or "very ordinary" tag, Beaujolais producers settled for Coteaux Bourguignons as a mutually acceptable compromise. Although this appellation is open to wines from all over the Burgundy region, in all likelihood the reds will be primarily produced from Gamay grown in Beaujolais.

RED Rather fruity wines that have some ability to age.

🍷 *A minimum of 85% Gamay, Pinot Noir, and (in the Yonne département only) César, plus Chardonnay, Pinot Blanc, and Pinot Gris, with a combined maximum of 10% Gamay de Bouze and Gamay de Chaudenay*

WHITE A lighter and fruity style of Bourgogne Blanc that should be drunk young.

🍷 *Aligoté, Chardonnay, Melon de Bourgogne, Pinot Blanc, Pinot Gris*

ROSÉ Interesting rosés of Coteaux Bourguignons can be tasted from the different departments. Worth trying.

🍷 *A minimum of 85% Gamay, Pinot Noir, and (in the Yonne département only) César, plus Chardonnay and Pinot Blanc, but only if part of a field blend with the black varieties*

CRÉMANT DE BOURGOGNE AOC

The Crémant de Bourgogne appellation was created in 1975 to supersede the Bourgogne Mousseux AOC, which failed to inspire a quality image because the term *mousseux* also applied to cheap sparkling wines. The Bourgogne Mousseux appellation is now reserved for red wines only. The major production centres for Crémant de Bourgogne are the Yonne, Chatillonais, Region de Mercurey, and the Mâconnais. There are already many exciting examples of these wines, and the quality is certain to improve as more producers specialize in cultivating grapes specifically for sparkling wines, rather than relying on excess or inferior grapes, as was the traditional practice in Burgundy.

SPARKLING WHITE These sparkling wines are dry but round, and the styles range from fresh and light to rich and toasty.

🍷 *A minimum of 30% of Pinot Noir, Pinot Gris, Pinot Blanc, Chardonnay, plus Sacy, Aligoté, Melon de Bourgogne, and a maximum of 20% Gamay*

🍷 *3–7 years*

SPARKLING ROSÉ Until now the best pink Crémant produced outside of Champagne has come from Alsace. Good examples are made in Burgundy, but have not yet realized their potential.

🍷 *A minimum of 30% of Pinot Noir, Pinot Gris, Pinot Blanc, Chardonnay, plus Sacy, Aligoté, Melon de Bourgogne, and a maximum of 20% Gamay*

🍷 *2–5 years*

✔ *Caves de Bailly • André Bonhomme • Paul Chollet • André Delorme • Cave de Lugny • Roux Père • Cave de Viré • Veuve Ambal*

The Chablis District

No doubt Chablis is one of the best-known white wine areas in the world. Its soil and climatic conditions are identical with those of the southern end of Champagne. Although the wines made here undergo different vinification methods and have varying styles, they attract a wide range of aficionados.

Like Champagne, Chablis owes much of its success to a cool and uncertain northern climate that puts viticulture on a knife edge. This is a source of constant worry – and not a little diabolical wine – but when everything comes together just right, Chablis can produce the most electrifying Chardonnay in the world.

Known as the "Golden Gate", this area has the advantage of being the inevitable first stop for anyone visiting the Burgundy region by car, whether directly from Paris or via Champagne. Situated in the Yonne *département,* much of which once formed part of the ancient province of Champagne, Chablis gives the distinct impression of an area cut off not simply from the rest of Burgundy but also from the rest of France. Indeed, the great *négociants* of the Côte d'Or rarely visit Chablis and have never made any significant penetration into what appears to be a closed-shop trade.

THE VARYING STYLES OF CHABLIS

The traditional description of Chablis is of a wine of clear, pale colour with a green hue around the rim. It is very straightforward, with a pronounced steely character, very direct attack, and a high level of acidity that needs a few years to round out. This description, however, rarely applies today, as much has changed in the way these wines are made at both ends of the quality spectrum.

Thirty-five years ago, most Chablis did not undergo malolactic fermentation. The wines that resulted had a naturally high acidity, and were hard, green, and ungenerous in their youth, although they often matured into wines of incomparable finesse. Most Chablis wines now undergo malolactic fermentation and cold stabilization, which is used to precipitate tartrates (although some wines fermented or matured in small oak casks do not), making the wine fuller, softer, and rounder.

At the top end of the market, there are two distinctly different schools. Some wines are fermented in stainless steel and bottled early to produce the most direct and upfront style, while others are fermented in wood and matured in casks with an increasing amount of new oak. Writers often describe the unoaked, stainless-steel-fermented Chablis as "traditional", but these vats were introduced in the 1960s, so it cannot be a well-established tradition. The oak barrel is much older, of course, and thus far more traditional, but what the critics really mean is that new oak has never been a feature of Chablis winemaking. The crisp, clean style of Chablis fermented in stainless steel is therefore closer to the original style: traditional by default. Obviously, the most authentic style of Chablis is the wine that is made in old or, more accurately, well-used casks. The traditional Chablisienne cask, known as a *feuillette,* is only half the size of a normal Burgundian barrel, and thus has twice the effect, but not being new oak, this would be an oxidative effect, not creamy-vanilla or other aromatics. Yet the more rapid oxidative effect of the *feuillette* does explain why the wines were traditionally bottled early, retaining the minerality of the fruit and invariably imparting a slight spritz, further separating the style of these wines from Chardonnay produced in the Côte d'Or.

FACTORS AFFECTING TASTE AND QUALITY

LOCATION
Chablis is isolated halfway between Beaune and Paris, 30 kilometres (19 miles) from the southernmost vineyards of Champagne, but 100 kilometres (60 miles) from the rest of Burgundy.

CLIMATE
This Chablis district has a semi-continental climate with minimal Atlantic influence, which results in a long, cold winter, a humid spring, and a fairly hot, very sunny summer. Hail storms and spring frosts are the greatest hazards.

ASPECT
All the *grands crus* are located on one stretch of southwest-facing slopes just north of the commube of Chablis itself, where the vineyards are at a height of between 150 and 200 metres (490 to 660 feet). Apart from the southwest-facing slopes of Fourchaume and Montée de Tonnerre, the *premier cru* slopes face southeast.

SOIL
This area is predominantly covered with calcareous clay, and the traditional view is that of the two major types, Kimmeridgian and Portlandian, only the former is suitable for classic Chablis, but this is neither proven nor likely. Geologically they have the same Upper Jurassic origin. Any intrinsic geographical differences should be put down to aspect, microclimate, and the varied nature of the sedimentary beds that underlie and interbed with the Kimmeridgian and Portlandian soils.

VITICULTURE & VINIFICATION
The vineyards in Chablis have undergone rapid expansion, most particularly in the generic appellation and in the *premiers crus,* both of which have doubled in size since the early 1970s. Mechanical harvesting has now found its way to the *grands crus* slopes of Chablis, but smaller producers still pick by hand. Most Chablis is fermented in stainless steel, but oak barrels are making a comeback, although too much new oak fights against the lean, austere intensity of the Chardonnay vines grown in this district.

GRAPE VARIETIES
Primary varieties: Chardonnay
Secondary varieties: Aligoté, César, Gamay, Melon de Bourgogne, Pinot Beurot (Pinot Gris), Pinot Blanc, Sauvignon Blanc, Sacy, Tressot

RAOUL GAUTHERIN & FILS CHABLIS GRAND CRU VAUDÉSIRS
The best glass of Chablis should display the classic dry, clean, and green flavour.

What makes the divide between oaked and unoaked Chablis even wider is the fact that the leaner, more mineral style of wine produced in this district can fight against the effects of new oak, whereas the fatter, softer, more seductive wines of the Côte d'Or embrace it with open arms. Recognizing that some people enjoy new oak characteristics, the recommendations in this book include producers of the best oaky Chablis. The trend for new oak peaked sometime in the late 1990s, however. Nowadays, even producers known for the oakiest Chablis have been holding back, to promote the minerality meant to be expressive of the *terroir*.

There has always been a certain inconsistency about the Chablis district – which is only to be expected given its uncertain climate – and this has never deterred the devotees of its wines. Things have gone from bad to worse over the last 20 to 25 years, however, and it is not the area's weather that has always been to blame – it is the increasing yields harvested by greedy producers and sloppy winemaking. There are still great joys to be had with the best and most passionately produced of Chablis, from the lowliest appellation to the greatest *grands crus*, but wine buyers must be increasingly vigilant.

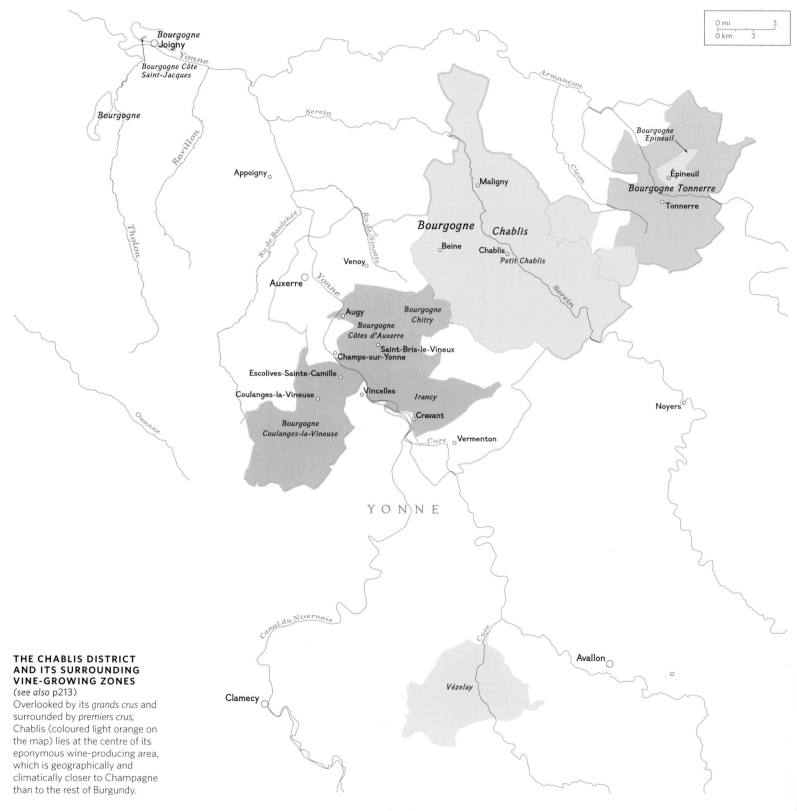

THE CHABLIS DISTRICT AND ITS SURROUNDING VINE-GROWING ZONES
(see also p213)
Overlooked by its *grands crus* and surrounded by *premiers crus*, Chablis (coloured light orange on the map) lies at the centre of its eponymous wine-producing area, which is geographically and climatically closer to Champagne than to the rest of Burgundy.

OTHER WINES OF YONNE

Other than Chablis, the two best-known wines of the Yonne are the reds of Bourgogne Irancy AOC and the white Saint-Bris AOC, made from the Sauvignon Blanc grape – a trespasser from the Loire. Other grapes peculiar to the Yonne are the black Césars and Tressots and the white Sacy. None are permitted in any Burgundian appellation other than AOC Bourgogne from Yonne, and they are not even widely cultivated here. César is the most interesting of these varieties, albeit rather rustic. It is a low-yielding vine that produces a thick, dark, tannic wine, though Simonnet-Febvre makes one of the better examples. César grows best at Irancy and can make a positive contribution when carefully blended with Pinot Noir. Just 5 or 10 per cent of César is required, but a few growers use as much as 20 per cent, and this tends to knock all the elegance out of the light-bodied local Pinot Noir. Tressot is thin, weak, and without any merit, but this blank canvas, together with the advantage of its typically high yield, made it the obvious partner to the César in bygone times. Occasionally encountered, it usually tastes like a thin, coarse Beaujolais. This district's other viticultural oddity is the steadily declining Sacy, a high-yielding white grape that produces acidic, neutrally flavoured wines best used in Crémant de Bourgogne, although most of it has traditionally been sold to Germany for making *Sekt*.

THE APPELLATIONS OF
THE CHABLIS DISTRICT

BOURGOGNE CHITRY AOC

A true single-village appellation with regional classification defined in 1993 in that only vines in Chitry qualify. This village neighbours St-Bris-le-Vineux to the northeast, and its vineyards comprise 27 hectares (67 acres) of black grapes and 40 hectares (100 acres) of white grapes. Generally, the reds are the least interesting, whereas the whites can be modestly successful, with their attractive, fresh-lemony fruit and an occasional aromatic hint on the nose and finish. Chalmeau is the best performer here.

🍷 *Pinot Noir, Pinot Blanc, Pinot Gris, César, Tressot, Chardonnay*

🍾 *Upon purchase*

✓ *Patrick, Elodie & Christine Chalmeau (blanc only)*
 • *Griffe (blanc only)* • *Simonnet-Febvre* • *Marcel Giraudon (rouge only)*

BOURGOGNE CÔTE SAINT-JACQUES AOC

One of four local Bourgogne appellations created in the 1990s (*see also* Bourgogne La Chapelle Notre Dame, p230; Bourgogne Le Chapitre, p223, and Bourgogne Montrecul, p223). Côte Saint-Jacques overlooks Joigny, which has the most northerly vineyards in Burgundy. Vines were first permitted to grow here for basic Bourgogne AOC in 1975, but just a few hectares were planted. Currently the largest vineyard owner is Alain Vignot, whose father pioneered winemaking in the area. At the present time, the area barely manages to produce a light-bodied white wine, although Vignot makes a *vin gris* from Pinot Noir and Pinot Gris, which Joigny was famous for in pre-phylloxera times. What they need here is the true Auxerrois of Alsace (which has no connection with nearby Auxerre). The reasoning would be exactly the same as for Alsace, where more Auxerrois is used to make so-called Pinot Blanc wine the farther north the vineyards are situated. In the Côte de Nuits it would be too fat and spicy, but in Joigny it would simply bring some generosity to the wines.

RED Michel Lorain's Clos des Capucins outclasses the rest of this appellation with its good colour and truly expressive fruit.

🍷 *Pinot Noir, Chardonnay, Pinot Blanc, Pinot Gris, César, Tressot*

🍾 *1–3 years*

WHITE Not worth buying.

🍷 *Chardonnay, Pinot Blanc*

🍾 *Upon purchase*

ROSÉ There are a handful of pure Pinot Gris wines made elsewhere in Burgundy, but they are usually white wines, whereas the tradition here is to make a rosé or *vin gris* from this variety, either pure (Alain Vignot) or blended.

🍷 *Pinot Noir, Chardonnay, Pinot t Blanc, Pinot Gris, César, Tressot*

🍾 *Upon purchase*

✓ *Calmus Pére & Fils* • *Christophe Lepage* • *Alain Vignot (rouge and rosé only)*

BOURGOGNE CÔTES D'AUXERRE AOC

Overlapping the Saint-Bris appellation, Côte d'Auxerre covers various parcels of vines scattered throughout the hillsides overlooking Augy, Auxerre-Vaux, Quenne, Champs-sur-Yonne, St-Bris-le-Vineux, and in part of Vincelottes that does not qualify for Irancy.

RED With the exception of Bersan or Guilhem & Jean-Hugues Goisot in a good vintage, these wines are invariably disappointingly light and dilute.

🍷 *Pinot Noir, Chardonnay, Pinot Blanc, Pinot Gris, César, Tressot*

🍾 *Upon purchase*

WHITE Guilhem & Jean-Hugues Goisot make a range of different Côtes d'Auxerre, and their Corps de Garde and Gondonne *cuvées* usually outclass the rest of the competition, although Bailly-Lapierre gives Goisot a good run for its money in some years.

🍷 *Chardonnay, Pinot Blanc*

🍾 *Upon purchase*

ROSÉ Light, fresh, easy-drinking rosé that is better than a lot of supposedly finer AOCs.

🍷 *Pinot Noir, Pinot Blanc, Chardonnay, Pinot Gris, César, Tressot*

🍾 *Upon purchase*

✓ *Bailly-Lapierre* • *Bersan* • *Guilhem & Jean-Hugues Goisot* • *Simonnet-Febvre* • *Tabit & Fils*

BOURGOGNE COULANGES-LA-VINEUSE AOC

The far-flung borders of Coulanges-la-Vineuse encompass no fewer than six communes in addition to its own: Charentenay, Escolives-Sainte-Camille, Migé, Mouffy, Jussy, and Val-de-Mercy. Vines cover almost 120 hectares (300 acres), of which 20 hectares (50 acres) are planted with white varieties.

RED Most of these wines are primarily Pinot Noir with a dash of César. Most are simply frank and fruity, though the best can be quite rich, with truly expressive fruit.

🍷 *Pinot Noir, Chardonnay, Pinot Blanc, Pinot Gris, César, Tressot*

🍾 *1–3 years*

WHITE Not worth buying.

🍷 *Chardonnay, Pinot Blanc*

🍾 *Upon purchase*

ROSÉ Not tasted.

🍷 *Pinot Noir, Pinot Blanc, Chardonnay, Pinot Gris, César, Tressot*

🍾 *Upon purchase*

✓ *Du Clos du Roi (rouge only)* • *Jean-Luc Houblin (rouge only)* • *Jean-Pierre Maltoff (rouge only)* • *Alain Rigoutat (rouge only)*

BOURGOGNE ÉPINEUIL AOC

A true single-village appellation, Épineuil consists of 85 hectares (210 acres) of hillside vineyards on the banks of the Armançon River, overlooking Tonnerre, northeast of Chablis. Planted mostly with Pinot Noir, there are just 6 hectares (15 acres) of white grapes.

RED Most are light and undistinguished, but Domaine Dominique Gruhier/Domaine de l'Abbaye du Petit Quincy makes a much darker version, especially its Côte de Grisey and Côte de Grisey Cuvée Juliette, which many consider to be the finest wines of Épineuil. Neither light nor dark, the ruby-coloured wine produced by the local *coopérative*, La Chablisienne, is probably the most consistent wine in the appellation.

🍷 *Pinot Noir, Chardonnay, Pinot Blanc, Pinot Gris, César, Tressot*

🍾 *1–5 years*

WHITE Few wines deliver more than a straightforward fresh, crisp, dry white of fairly neutral character.

🍷 *Chardonnay, Pinot Blanc*

🍾 *Upon purchase*

ROSÉ At its best, this can be a deliciously fresh, easy-drinking rosé.

🍷 *Pinot Noir, Pinot Blanc, Chardonnay, Pinot Gris, César, Tressot*

🍾 *Upon purchase*

✓ *Dominique Gruhier/Abbaye du Petit Quincy* • *Savary* • *Dampt Fréres (white only)* • *Dominique Gruhier* • *Alain Mathias*

BOURGOGNE TONNERRE AOC

This appellation was created in 2006 for pure Chardonnay wine from the Armançon Valley, immediately east of Chablis. Vines have grown here since the 9th century and had, by the start of the 19th century, extended to more than 2,000 hectares (5,000 acres). These vineyards were destroyed by phylloxera in the 1890s and not replanted until 1987, when local grower Bernard Dampt planted the first Tonnerre vines of modern times.

WHITE Lighter and more ethereal than Chablis, with flashes of brilliance when the vintage is right.

🍷 *Chardonnay*

🍾 *1–2 years*

✓ *Dampt Fréres* • *Alain Mathias*

VÉZELAY AOC

Some 70 hectares (172 acres) of vines of the Bourgogne Vézelay AOC within the area of of Vézelay, Asquins, Saint-Père, and Tharoiseau were promoted to Village AOC status by the INAO in 2017.

WHITE The top-performing white wines from these appellations are superior to the lower end of Chablis, which is relatively much more expensive. They are invariably pure Chardonnay, yet rarely taste like Chablis, being softer and smoother.

🍷 *Chardonnay*

🍾 *1–3 years*

✓ *La Croix Montjoie* • *Elise Villiers*

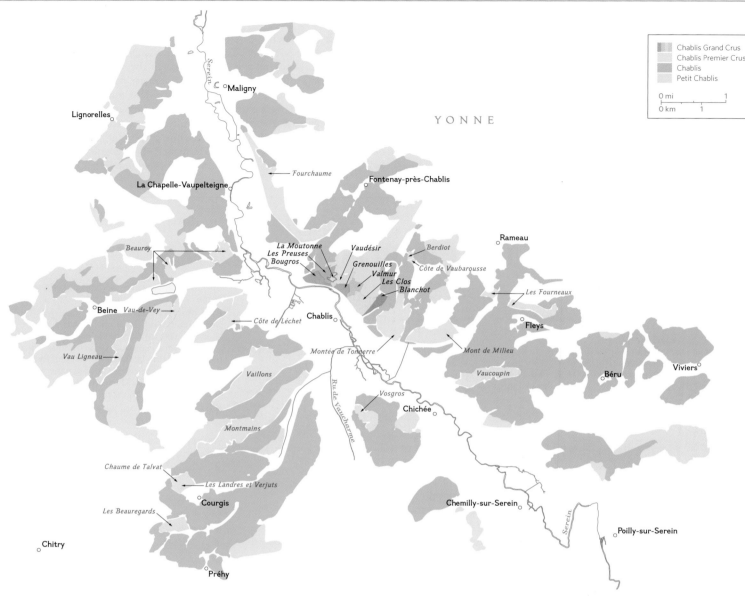

YONNE

Maligny

Lignorelles

Fourchaume

La Chapelle-Vaupelteigne

Fontenay-près-Chablis

Beauroy

La Moutonne
Les Preuses
Bougros

Vaudésir

Berdiot

Rameau

Grenouilles
Valmur
Les Clos
Blanchot

Côte de Vaubarousse

Les Fourneaux

Beine *Vau-de-Vey*

Côte de Léchet

Chablis

Fleys

Montée de Tonnerre

Mont de Milieu

Vau Ligneau

Vaillons

Vaucoupin

Béru

Viviers

Vosgros

Chichée

Montmains

Chaume de Talvat

Les Landres et Verjuts

Chemilly-sur-Serein

Les Beauregards

Courgis

Poilly-sur-Serein

Chitry

Préhy

Legend:
- Chablis Grand Crus
- Chablis Premier Crus
- Chablis
- Petit Chablis

0 mi — 1
0 km — 1

THE *GRANDS* AND *PREMIERS CRUS* OF THE CHABLIS DISTRICT

This map clearly illustrates that all 7 *grands crus* are huddled together on one contiguous southwest-facing slope of exceptional situation and aspect, whereas the 17 *premiers crus* are scattered across the surrounding area on isolated slopes of excellent, but secondary, exposure.

CHABLIS AOC

With careful selection, basic Chablis can be a source of tremendous-value, classic 100 per cent Chardonnay wine, particularly in the best vintages. The appellation covers a relatively large area, however, with many vineyards that do not perform well, and there are far too many mediocre winemakers. Basic Chablis needs to come from the most favourable locations, where the grower restricts the yield and the winemaker selects only the best wines; it is not an appellation in which shortcuts can be taken. Cheap Chablis can be dire, even in superior appellations; better therefore to pay for a top *cuvée* of basic Chablis than to be seduced by a cut-price *premier cru*. (*See also* Chablis Premier Cru and Petit Chablis AOC.)

WHITE When successful, these wines have the quintessential character of true Chablis – dry, clean, green, and expressive, with just enough fruit to balance the "steel".

🍷 *Chardonnay*

🍷 2-6 years

✔ Barat • Baillard • Billaud-Simon • Jean-Marc Brocard • *La Chablisienne* • Chateau de Béru • Jean Collet & Fils • *Jean Defaix* • René & Vincent Dauvissat • Jean-Paul & Benoit Droin (especially *Vieilles Vignes*) • Gérard Duplessis

• William Fèvre de la Maladière • Jean-Pierre & Corinne Grossot • *Laroche* • Olivier Leflaive • Christian Moreau Pére • *Christian Moreau Pére & Fils* • Louis Moreau (especially Domaines de Biéville and Cèdre Doré) • *Sylvain Mosnier* (*Vieilles Vignes*) • Gilbert Picq (*Vieilles Vignes*) • *François Raveneau*

CHABLIS GRAND CRU AOC

The seven *grands crus* of Chablis are all located on one hill that overlooks the town of Chablis itself. They are Blanchot, Bougros, Les Clos, Grenouilles, Les Preuses, Valmur, and Vaudésir. One vineyard called La Moutonne is not classified in the appellation as a *grand cru,* but the authorities permit the use of the coveted status on the label because it is physically part of other *grands crus*. In the 18th century La Moutonne was in fact a 1-hectare (2.5-acre) *climat* of Vaudésir, but under the ownership of Louis Long-Depaquit, its wines were blended with those of three other *grands crus* (namely, Les Preuses, Les Clos, and Valmur). This practice came to a halt in 1950 when, in a bid to get La Moutonne classified as a separate *grand cru*, Long-Depaquit agreed to limit its production to its current location, which cuts across parts of Vaudésir and Les Preuses. Its classification never actually took place, but the two *grands crus* that it overlaps are probably the finest of all.

WHITE Always totally dry, the *grands crus* are the biggest, richest, most complex of all Chablis and should always boast great minerality. Many are overwhelmed by new oak, however. Their individual styles depend very much on how the winemaker vinifies and matures the wine, but when well made they are essentially as follows: Blanchot has a floral aroma and is the most delicate of the *grands crus* (Michel Laroche's Réserve de l'Obédiance is the greatest Chablis *grand cru* I have tasted, taking richness to the very limit, while not losing sight of true Chablis structure, crispness, minerality, and finesse); Bougros has the least frills of all the *grands crus*, but is vibrant with a penetrating flavour; Les Clos is rich, luscious, and complex with great mineral finesse and beautiful balance; Grenouilles should be long and satisfying, yet elegant, racy, and aromatic; Les Preuses gets the most sun and is vivid, sometimes exotic, quite fat for Chablis, yet still expressive and definitely complex, with great finesse; Valmur has a fine bouquet, rich flavour, and smooth texture; Vaudésir has complex, intense flavours that display great finesse and spicy complexity; and La Moutonne is fine, long-flavoured, and wonderfully expressive.

🍷 *Chardonnay*

🍷 6–20 years

✓ Blanchot Billaud-Simon (Vieilles Vignes) • Laroche • Raveneau • Vocoret & Fils; Bougros William Fèvre • Joseph Drouhin • De la Maladière • Michel Laroche; Les Clos Billaud-Simon • La Chablisienne • René & Vincent Dauvissat • Jean-Paul & Benoit Droin • Joseph Drouhin • Caves Duplessis • William Fèvre • Michel Laroche • De la Maladière • Domaines des Malandes • Louis Michel & Fils • Pinson • Raveneau • Servin; Grenouilles La Chablisienne (Château Grenouilles) • Louis Michel; La Moutonne Long-Depaquit; Les Preuses Billaud-Simon • Joseph Drouhin • René & Vincent Dauvissat • William Fèvre; Valmur Jean Collet • Jean-Paul & Benoit Droin • William Fèvre • Olivier Leflaive • Raveneau; Vaudésir Billaud-Simon • La Chablisienne • Jean-Paul Droin • Joseph Drouhin • William Fèvre • Michel Laroche • Des Malandes • Louis Michel Fils

CHABLIS PREMIER CRU

This appellation includes the following *premiers crus*: Les Beauregards, Beauroy, Berdiot, Chaume de Talvat, Côte de Jouan, Côte de Léchet, Côte de Vaubarousse, Fourchaume, Les Fourneaux, Montée de Tonnerre, Montmains, Mont de Milieu, Vaillons, Vaucoupin, Vau-de-Vey (or Vaudevey), Vau Ligneau, and Vosgros. Unlike the *grands crus*, the 17 *premiers crus* of Chablis are scattered among the vineyards of 15 surrounding communes, and the quality and style is patchy. Montée de Tonnerre is the best *premier cru* throughout the producers and across the many vintages. One of its *lieux-dits*, Chapelot, is considered by many to be the equivalent of a *grand cru*. After Montée de Tonnerre, Côte de Léchet, Les Forêts (which is a *climat* within Montmains), Fourchaume, Mont de Milieu, and Vaillons vie for second place.

WHITE Dry wines that can vary from light-bodied to fairly full-bodied, but should always be finer and longer-lasting than wines of the basic Chablis appellation, although without the concentration of flavour expected from a *grand cru*.

🍷 Chardonnay

🍾 4–15 years

✓ Les Beauregards Jean-Marc Brocard; Beauroy Sylvain Mosnier; Berdiot None; Chaume de Talvat None; Côte de Jouan Michel Colbois; Côte de Léchet Jean-Paul Droin • Jean Defaix • Sylvain Mosnier; Côte de Vaubarousse None; Fourchaume Billaud-Simon • La Chablisienne • De Chantemerle • Jean-Paul Droin • Gérard Duplessis • Jean

Durup • William Fèvre (Côte de Vaulorent) • Lamblin & Fils • Michel Laroche (Vieilles Vignes) • Des Malandes • Louis Michel • Savary • Verget; Les Fourneaux Jean-Pierre & Corinne Grossot • Louis Moreau; Montée de Tonnerre Billaud-Simon • Jean-Paul Droin (Vieilles Vignes) • Caves Duplessis • William Fèvre • Louis Michel • Raveneau (including Chapelot and Pied d'Aloue lieux-dits) • Guy Robin & Fils; Montmains La Chablisienne • René & Vincent Dauvissat (La Forest [sic] lieu-dit) • Jean-Paul & Benoit Droin • Caves Duplessis • Des Malandes • Des Marronniers • Louis Michel • Pinson • Raveneau (Butteaux lieu-dit) • Guy Robin & Fils (Butteaux lieu-dit) • Robert Vocoret & Fils (La Fôret lieu-dit); Mont de Milieu Barat • Billaud-Simon (especially Vieilles Vignes) • La Chablisienne • Jean Collet • Jean-Pierre & Corinne Grossot • De Meulière • Pinson; Vaillons Barat • Billaud-Simon • René & Vincent Dauvissat (Séchet lieu-dit) • Jean Defaix • Jean-Paul & Benoit Droin • Gérard Duplessis • Michel Laroche (Vieilles Vignes) • François Raveneau • Verget; Vaucoupin Jean-Pierre & Corinne Grossot; Vau-de-Vey Jean Durup • Michel Laroche; Vau Ligneau Thierry Hamelin • Louis Moreau; Vosgros Jean-Paul & Benoit Droin • Gilbert Picq & ses Fils

IRANCY AOC

Irancy was promoted from Bourgogne Irancy to full village AOC status in its own right in 1999. This red-wine-only appellation encompasses 160 hectares (395 acres) in Irancy and the neighbouring villages of Cravant and Vincelottes.

RED Irancy is supposed to be the "famous" red wine of Chablis, but it is not really that well known and is not even the best local red wine. Dominique Gruhier's Domaine de l'Abbaye du Petit Quincy / Côte de Grisey Cuvée Juliette is consistently superior to the best Irancy, though it possesses a certain richness of fruit that most examples of this modest appellation lack. Original Irancy was a pure César wine: Anita and Jean-Pierre Colinot make the most interesting example (Les Mazelots César).

🍷 Pinot Noir, Pinot Gris, César

🍾 1–3 years

✓ Benoît Cantin (Cuvée Emiline) • Jean-Marc Brocard • Anita & Jean-Pierre Colinot (Palotte, Les Mazelots, Les Mazelots César, Les Bessys, Côte-de-Moutier) • Vincent Dauvissta • Christophe Ferrari Domaine Saint-Germain (rouge only) • Félix

PETIT CHABLIS AOC

A depreciatory appellation that covers inferior soils and expositions within the same area as generic Chablis, with the exception of Ligny-le-Châtel, Viviers, and Collan. Although I have found four more reliable producers than last time, this appellation should be downgraded to VDQS or uprooted. The rumour is that it will be phased out, but do not misunderstand: this does not mean that they will uproot the vines, simply that this inferior land will instead produce Chablis, albeit of a *petit vin* quality.

WHITE Apart from the occasional pleasant surprise, most if the whites are mean and meagre dry wines of light to medium body. The producers below provide most of the pleasant surprises.

🍷 Chardonnay

🍾 2–3 years

✓ Jean-Marc Brocard • Jean Durup Chateau de Maligny • Duplessis • Thierry Hamelin • Savary

SAINT-BRIS AOC

This wine is as good as most Sauvignon Blanc AOCs, and considerably better than many other white AOCs made from lesser grape varieties. Twenty-five years ago I did not believe it possible that Sauvignon de St-Bris would ever overcome Burgundy's Chardonnay chauvinism to rise to the ranks of Appellation Contrôlée, but in 2003, Sauvignon de St-Bris VDQS was promoted to Saint-Bris AOC (retrospectively applied to the 2001 vintage).

WHITE Fine wet-grass or herbaceous aromas, full smoky-Sauvignon flavours, and a correct, crisp, dry finish. Made from the rare Sauvignon Gris, Jean-Hugues & Guilhem Goisot's Cuvée du Corps de Garde Gourmand Fié Gris is the best, most consistent, and most unusual wine in this appellation, while the same producer's "straight" Sauvignon is easily the top thoroughbred Sauvignon Blanc.

🍷 Sauvignon Blanc, Sauvignon Gris

🍾 2–5 years

✓ Jean-Marc Brocard • Christophe Ferrari Domaine Saint Germain • Félix • Jean-Hugues & Guilhem Goisot • La Chablisienne • Simonnet Febvre

GRAND CRU VINEYARDS LINE THE HILLSIDES OF THE CHABLIS WINE DISTRICT
A stretch of land north of Chablis is home to the district's *grand cru* vineyards, including Bougros, Les Preuses, Vaudésir, Grenouilles, Valmur, Les Clos, and Blanchots.

The Côte de Nuits and Hautes Côtes de Nuits

The Côte de Nuits is considered to be the capital of Pinot Noir, the land of the finest expression of the queen of black grapes. This is definitely a red-wine area (with one or two extraordinary whites), and it is home to 22 of Burgundy's 23 red grands crus.

The Côte D'Or, or "golden slope", is the departmental name for both the Côte de Nuits and the Côte de Beaune. Firmness, weight, and finesse are the key words to describe the wines produced here, and these characteristics intensify as the vineyards progress north. A string of villages with some of the richest names in Burgundy – Gevrey-Chambertin, Morey-St-Denis, Chambolle-Musigny, Vosne-Romanée, and Nuits-Saint-Georges – these slopes ring up dollar signs in the minds of merchants throughout the world. Ironically, the most famous appellations can also produce some of Burgundy's worst wines if they are not in the right hands.

CHÂTEAU DE GEVREY-CHAMBERTIN
Located in one of the finest wine regions of Burgundy, the château dates to earlier than the 13th century. It once served as a priory belonging to the Abbey of Cluny and has been the property of the Masson family since 1858. With 2 hectares (5 acres) of associated vineyard, it is open to tourists for wine tastings and visits.

FACTORS AFFECTING TASTE AND QUALITY

LOCATION
The Côte de Nuits is a narrow, continuous strip of vines stretching from Dijon to just north of Beaune, with the Hautes Côtes de Nuits in the southwestern hinterland.

CLIMATE
Semi-continental climate with minimal Atlantic influence, which results in a long, cold winter, a humid spring, and a fairly hot, very sunny summer. Hail is its greatest natural hazard, and heavy rain is often responsible for diluting the wines and causing rampant rot.

ASPECT
A series of east-facing slopes which curve in and out to give some vineyards northeastern, some southeastern aspects. The vines grow at an altitude of between 225 and 350 metres (740 to 1,150 feet) and, apart from Gevrey-Chambertin and Prémeaux-Prissey, those vineyards that have the right to the village and higher appellations rarely extend eastwards beyond the RN 74 road.

SOIL
Vineyards here sit in a subsoil of sandy-limestone, which is exposed in places, but usually covered by a chalky scree mixed with marl and clay particles on higher slopes and richer alluvial deposits on lower slopes. Higher slopes sometimes have red clay.

VITICULTURE & VINIFICATION
The vines of the Côte de Nuits are trained low to benefit from heat reflected from the soil at night. For red wines, the grapes are almost always de-stemmed and the juice is kept in contact with the skins for between eight and ten days. Less than three per cent of the wine produced is white, but this is mostly high quality and traditionally cask fermented. The best wines are matured in oak.

GRAPE VARIETIES
Primary varieties: Chardonnay, Pinot Noir
Secondary varieties: Aligoté, Gamay, Melon de Bourgogne, Pinot Beurot (Pinot Gris), Pinot Blanc

VINE-GROWING ZONES OF THE CÔTE DE NUITS AND HAUTES-CÔTES DE NUITS (*see also* p213)
The best vineyards of the Côte de Nuits form a tighter, more compact strip than those of the Côte de Beaune (*see also* p228), and the wines produced are tighter, with more compact fruit. The Hautes Côtes de Nuits is in the southwest.

THE CÔTE DE NUITS AND HAUTES-CÔTES DE NUITS

Note: Each *grand cru* of Côte de Nuits has its own appellation and is listed individually below. The *premiers crus* do not, however, and are therefore listed under the appellation of the village in which the vineyards are situated. *Premiers crus* that are virtually contiguous with *grand cru* vineyards are in italics; they are possibly superior to those that do not share a boundary with one or more *grand crus*, and generally superior to those that share boundaries with village AOC vineyards.

BONNES MARES AOC
Grand Cru

Bonnes Mares is the largest of the two *grands crus* of Chambolle-Musigny. It covers 13.5 hectares (34 acres) in the north of the village, on the opposite side to Musigny, the village's other *grand cru,* and extends a further 1.5 hectares (3.5 acres) into Morey-St-Denis.

RED A fabulous masculine style with sheer depth of flavour give something rich and luscious, yet complex and complete.

🍷 Pinot Noir, plus up to 15% combined total of Pinot Gris, Pinot Blanc, Chardonnay

🍷 12–25 years

✓ D'Auvenay ❸ • Bruno Clair • Dujac • R Groffier • Dominique Laurent • J F Mugnier • Georges Roumier • Nicolas Potel • Comte Georges de Vogüé • de La Vougeraie

BOURGOGNE LE CHAPITRE AOC

One of the Côte de Nuits' two Bourgogne *lieux-dits* created in 1993, La Chapitre is located at Chenove, between Marsannay and Dijon. According to Anthony Hanson (in his book Burgundy), the wines of Chenove once fetched higher prices than those of Gevrey (Chenove's most famous vineyard is Clos du Roi, of which Labouré-Roi took management in 1994), but the only example I have tasted (Domaine Bouvier) was not special.

BOURGOGNE HAUTES-CÔTES DE NUITS AOC

A source of good-value wines, these vineyards have expanded since the 1970s, and the quality is improving noticeably. Half-red/half-rosé wines may be sold as Bourgogne Clairet Hautes-Côtes de Nuits AOC, but the style is outmoded and the appellation rarely encountered.

RED Medium-bodied and medium- to full-bodied wines with good fruit and some true Côte de Nuits character. The wines from some growers have fine oak nuances.

🍷 Pinot Noir, Pinot Blanc, Pinot Gris, Chardonnay

🍷 4–10 years

WHITE Just about 20 per cent of the production is dry white. Most have a good weight of fruit, but little finesse.

🍷 Chardonnay, Pinot Blanc, Pinot Gris

🍷 1–4 years

ROSÉ Little-seen, but those that have cropped up have been dry, fruity, and delicious wines of some richness.

🍷 Pinot Noir, Pinot Blanc, Pinot Gris

🍷 1–3 years

✓ Bertagna • J-C Boisset • Bizot • Dufouleur • Giboulot • Anne Gros • Michel Gros • Robert Jayer-Gilles • Thibault Liger-Belair • Naudin-Ferrand • Thévenot-le-Brun & Fils

BOURGOGNE MONTRECUL or MONTRE-CUL or EN MONTRE-CUL AOC

The other Bourgogne *lieu-dit* in this district is actually located on the outskirts of Dijon. Some might think this is a wine to moon over, but the only wine I have tasted (again Domaine Bouvier) was, again, not special. If no one is going to produce something exciting under these Bourgogne *lieux-dits*, why bother with them in the first place?

CHAMBERTIN AOC
Grand Cru

This is one of the nine *grands crus* of Gevrey-Chambertin. All of them (quite legally) add the name Chambertin to their own and one, Clos de Bèze, actually has the right to sell its wines as Chambertin.

RED Always full in body and rich in extract, Chambertin is not, however, powerful like Corton, but graceful and feminine with a vivid colour, stunning flavour, impeccable balance, and lush, velvety texture.

🍷 Pinot Noir, Pinot Gris, Pinot Blanc, Chardonnay

🍷 12–30 years

✓ Bouchard Père & Fils • Dujac • Leroy ❸ • Denis Mortet • Jacques Prieur • Henri Rebourseau • Armand Rousseau • Jean Trapet Père & Fils

CHAMBERTIN-CLOS DE BÈZE AOC
Grand Cru

Another Gevrey-Chambertin *grand cru.* The wine may be sold simply as Chambertin, the name of a neighbouring *grand cru,* but Chambertin may not call itself Clos de Bèze.

RED This wine is reputed to have a greater finesse than Chambertin but slightly less body, tending to have aromas of oriental spices. It is just as sublime.

🍷 Pinot Noir, Pinot Gris, Pinot Blanc, Chardonnay

🍷 12–30 years

✓ Bouchard Père & Fils • Burguet • Bruno Clair • Faiveley • R Groffier • Louis Jadot • Dominique Laurent • Raphet • Armand Rousseau

CHAMBOLLE-MUSIGNY AOC

This village is very favourably positioned, with a solid block of vines nestled in the shelter of a geological fold.

RED Many of these medium-bodied to fairly full-bodied wines have surprising finesse and fragrance for mere village wines.

🍷 Pinot Noir, Pinot Gris, Pinot Blanc, Chardonnay

🍷 8–15 years

✓ Ghislaine Barthod • Sylvain Cathiard (Les Clos de L'Orme) • Dujac Père & Fils • J F Mugnier • Géantet Pansiot • Perrot-Minot • Daniel Rion & Fils • Georges Roumier • Comte Georges de Vogüé

CHAMBOLLE-MUSIGNY PREMIER CRU AOC

This appellation includes the following *premiers crus:* Les Amoureuses, Les Baudes, Aux Beaux Bruns, Les Borniques, Les Carrières, Les Chabiots, Les Charmes, Les Châtelots, La Combe d'Orveau, Aux Combottes, Les Combottes, Les Cras, Derrière la Grange, Aux Echanges, Les Feusselottes, Les Fuées, Les Grands Murs, Les Groseilles, Les Gruenchers, Les Hauts Doix, Les Lavrottes, Les Noirots, Les Plantes, and Les Sentiers. The outstanding *premier cru* is Les Amoureuses, with Les Charmes and Les Cras as very respectable podium finishers.

RED The best have a seductive bouquet and deliciously fragrant flavour.

🍷 Pinot Noir, Pinot Gris, Pinot Blanc, Chardonnay

🍷 10–20 years

✓ Amiot-Servelle (Amoureuses) • Ghislaine Barthod • Jean-Claude Boisset • Jean-Jacques Confuron • R Groffier • Dominique Laurent • Leroy ❸ • Denis Mortet • J F Mugnier • Mugneret-Gibourg • Perrot-Minot • Laurent Ponsot • Michelle & Patrice Rion • Georges Roumier • Comte Georges de Vogüé

CHAPELLE-CHAMBERTIN AOC
Grand Cru

This is one of the nine *grands crus* of Gevrey-Chambertin, and it is comprises two *climats* called En la Chapelle and Les Gémeaux.

CONFRÉRIE DES CHEVALIERS DU TASTEVIN

After the three terrible vintages of 1930, 1931, and 1932, and four years of worldwide slump following the Wall Street Crash of 1929, Camille Rodier, and Georges Faiveley formed the Confrérie des Chevaliers du Tastevin to revive Burgundy's fortunes. They named the brotherhood after the traditional Burgundian *tastevin,* a shallow, dimpled, silver tasting-cup with a fluted edge.

The first investitures took place on 16 November 1934, in a cellar in Nuits-St-Georges; the Confrérie now boasts thousands of members in numerous foreign chapters and averages 20 banquets a year at Château du Clos de Vougeot. Until the late 1980s, the distinctive Tastevinage label was a useful means of quickly identifying wines within basic appellations that had risen above their modest status. The selection gradually became less rigorous, though, with some decidedly underperforming wines slipping through. In recent years, however, there appears to be more consistency, once again making the Tastevinage label worth watching out for.

SINGERS ENTERTAIN AT A CHEVALIERS DU TASTEVIN DINNER
This is one of the three Grand Meals, or *Les Trois Glorieuses,* held in the Chateau du Clos de Vougeot, around the time of the Hospices de Beaune wine auction (*see* p233).

RED The lightest of all the *grands crus*, with a delightful bouquet and flavour.

🍷 *Pinot Noir, Pinot Gris, Pinot Blanc, Chardonnay*

🍾 *8–20 years*

✓ *Louis Jadot • Jean Trapet Père & Fils • Cécile Tremblay*

CHARMES-CHAMBERTIN AOC
Grand Cru

This is the largest Gevrey-Chambertin *grand cru*, and it is part of the vineyard is known as Mazoyères, from which Mazoyères-Chambertin has evolved. Mazoyéres can be bottled as Charmes.

RED Soft, sumptuous wines with ripe-fruit flavours and pure Pinot character, although some slightly lack finesse.

🍷 *Pinot Noir, Pinot Gris, Pinot Blanc, Chardonnay*

🍾 *10–20 years*

✓ *Denis Bachelet • Camille Giroud • Confuron-Cotétidot • Claude Dugat • Bernard Dugat-Py • Dujac • Frédéric Magnien • Géantet-Pansiot • Vincent Girardin • Perrot-Minot • Sérafin Père & Fils • Joseph & Philippe Roty • Christophe Roumier • Armand Rousseau • Taupenot-Merme • De la Vougeraie Z*

CLOS DE BÈZE AOC

This is an alternative appellation for Chambertin-Clos de Bèze. *See* Chambertin-Clos de Bèze AOC.

CLOS DES LAMBRAYS AOC
Grand Cru

This vineyard was classified as one of the four *grands crus* of Morey-St-Denis only as recently as 1981, although the owner used to put *"grand cru classé"* (illegally) on the label.

RED The vineyard was replanted later on and now produces fine, elegant wines with silky fruit of a good, easily recommendable quality.

🍷 *Pinot Noir, Pinot Gris, Pinot Blanc, Chardonnay*

🍾 *10–20 years*

✓ *Des Lambrays*

CLOS DE LA ROCHE AOC
Grand Cru

Covering an area of almost 17 hectares (42 acres), Clos de La Roche is twice the size of the other *grands crus* of Morey-St-Denis.

RED A deep-coloured, rich, and powerfully flavoured *vin de garde* with a silky texture. Many consider it the greatest *grand cru* of Morey-St-Denis.

🍷 *Pinot Noir, Pinot Gris, Pinot Blanc, Chardonnay*

🍾 *10–20 years*

✓ *Jean-Claude Boisset • Dujac • Leroy Ⓑ • Hubert Lignier • Michel Magnien & Fils • Ponsot • Armand Rousseau*

CLOS ST-DENIS AOC
Grand Cru

This is the *grand cru* that the village of Morey attached to its name when it was the best growth in the village, a position now contested by Clos de la Roche and Clos de Tart.

RED Strong, fine, and firm wines with rich liquorice and berry flavours that need time to come together.

🍷 *Pinot Noir, Pinot Gris, Pinot Blanc, Chardonnay*

🍾 *10–25 years*

✓ *Bertagna • Philippe Charlopin-Parizot • Dujac • Louis Jadot • Michel Magnien & Fils • Ponsot • Nicolas Potel*

CLOS DE TART AOC
Grand Cru

This is one of the four *grands crus* of Morey-St-Denis. It is entirely owned by Francois Pinault. In addition to Clos de Tart itself, a tiny part of the Bonnes Mares *grand cru* also has the right to this appellation.

RED This *monopole* yields wines with a penetrating Pinot flavour, to which Mommessin adds such a spicy-vanilla character from 100 per cent new oak that great bottle maturity is required for a completely harmonious flavour.

🍷 *Pinot Noir, Pinot Gris, Pinot Blanc, Chardonnay*

🍾 *15–30 years*

✓ *Clos de Tart*

A STONE GATE MARKS THE CLOS DE LA ROCHE *GRAND CRU* VINEYARD NEAR MOREY-ST-DENIS
Located in the northern part of the commune of Morey-St-Denis in the Côte-d'Or *département*, Clos de la Roche is a red wine appellation, with Pinot Noir its main grape variety. The largest landholder here is Domaine Ponsot, which owns three of the original four hectares of the vineyard, which is now more than four times that size.

CLOS DE VOUGEOT AOC
Grand Cru

The only *grand cru* of Vougeot, it is a massive 50-hectare (123-acre) block of vines with no fewer than 85 registered owners. It has been described as "an impressive sight, but a not very impressive site". This mass ownership situation has often been used to illustrate the classic difference between Burgundy and Bordeaux, where an entire vineyard belongs to one château, and so the wine can be blended to a standard quality and style every year.

RED With individual plots ranging in quality from truly great to very ordinary, operated by growers of varying skills, it is virtually impossible to unravel the intrinsic characteristics of this *cru*. Its best wines, however, have lots of silky Pinot fruit, an elegant balance, and a tendency towards finesse rather than fullness.

🍷 *Pinot Noir, Pinot Gris, Pinot Blanc, Chardonnay*

🍾 *10–25 years*

✓ *Arnoux-Lachaux • Joseph Drouhin • Faiveley • Jean Grivot • Anne Gros • Michel Gros • Alain Hudelot-Noëllat • Louis Jadot • Dominique Laurent • Leroy* Ⓑ *• Méo-Camuzet • Mugneret-Gibourg • Nicolas Potel • Château la Tour • De la Vougeraie* Ⓑ

CLOS VOUGEOT AOC

See Clos de Vougeot AOC

CÔTE DE NUITS-VILLAGES AOC

This AOC covers the wines produced in one or more of five communes: Fixin and Brochon, situated in the north of the district, and Comblanchien, Corgoloin, and Prissy in the south.

RED Firm, fruity, and distinctive wines made in true, well-structured Côte de Nuits style.

🍷 *Pinot Noir, Pinot Gris, Pinot Blanc, Chardonnay*

🍾 *6–10 years*

WHITE Very little is made, and I have never encountered it.

🍷 *Chardonnay, Pinot Blanc, Pinot Gris*

✓ *De l'Arlot • Sylvie Esmonin • Gachot-Monot • Naudin-Ferrand • Daniel Rion & Fils (Le Vaucrain)*

ECHÉZEAUX AOC
Grand Cru

This 30-hectare (74-acre) vineyard is the larger of the two *grands crus* of Flagey-Echézeaux and is composed of 11 *climats* owned by no fewer than 84 smallholders.

RED The best have a fine and fragrant flavour that relies more on delicacy than power, but too many deserve no more than a village appellation.

🍷 *Pinot Noir, Pinot Gris, Pinot Blanc, Chardonnay*

🍾 *10–20 years*

✓ *Arnoux-Lachaux • Albert Bichot (Clos Frantin) • Confuron-Cotétidot • Dujac • Faiveley • Forey Père & Fils • Vincent Girardin • Jean Grivot • A-F Gros • Louis Jadot • Robert Jayer-Gilles • Anne Gros • Mongeard-Mugneret • Mugneret-Gibourg • De la Romanée-Conti • Emmanuel Rouget • Cécile Tremblay*

FIXIN AOC

Fixin is one of the oldest villages along the Côte d'Or, with evidence of inhabitation tracing back to AD 830.

RED Well-coloured wines that can be firm, tannic *vins de garde* of excellent quality and even better value.

🍷 *Pinot Noir, Pinot Gris, Pinot Blanc, Chardonnay*

🍾 *6–12 years*

WHITE Rich, dry, and concentrated wines that are rare, but exciting and well worth seeking out. Bruno Clair shows what Pinot Blanc can produce when it is not over-cropped.

🍷 *Chardonnay, Pinot Blanc, Pinot Gris*

🍾 *3–8 years*

✓ *Bart • Berthaut • Bruno Clair • Pierre Gelin • Joliet Père & Fils • Mongeard-Mugneret • Philippe Naddef*

FIXIN PREMIER CRU AOC

This appellation includes the following *premiers crus*: Les Arvelets, Clos du Chapitre, Aux Cheusots, Les Hervelets, Le Meix Bas, La Perrière, Queue de Hareng (located in neighbouring Brochon), En Suchot, and Le Village. The best premier crus are La Perrière and Clos du Chapitre. Clos de la Perrière is a monopoly owned by Philippe Joliet that encompasses En Souchot and Queue de Hareng as well as La Perrière itself.

RED Splendidly deep in colour and full in body with masses of blackcurrant and redcurrant fruit, supported by a good tannic structure.

🍷 *Pinot Noir, Pinot Gris, Pinot Blanc, Chardonnay*

🍾 *10–20 years*

WHITE I have not encountered any, but it would not be unreasonable to assume that it might be at least as good as a basic Fixin *blanc*.

✓ *Berthaut • Pierre Gelin • Joliet Pére & Fils • Mongeard-Mugneret*

GEVREY-CHAMBERTIN AOC

Famous for its *grand cru* of Chambertin, the best growers also produce superb wines under this appellation. Some vineyards overlap the village of Brochon.

RED These well-coloured wines are full, rich, and elegant, with a silky texture and a perfumed aftertaste reminiscent of the pure fruit of Pinot Noir.

🍷 *Pinot Noir, Pinot Gris, Pinot Blanc, Chardonnay*

🍾 *7–15 years*

✓ *Denis Bachelet • Burguet • Charlopin-Parizot • Confuron-Cotétidot • Drouhin-Laroze • Bernard Dugat-Py • Bruno Clair • Dujac Père & Fils • Labouré-Roi • Marc Roy (Vieilles Vignes, Clos Priéur) • Denis Mortet • Géantet Pansiot • Armand Rousseau*

GEVREY-CHAMBERTIN PREMIER CRU AOC

This appellation includes the following *premiers crus*: Bel Air, La Bossière, Les Cazetiers, Champeaux, Champitennois, Champonnet, Clos du Chapitre, Cherbandes, Au Closeau, Combe au Moine, Aux Combottes, Les Corbeaux, Craipillot, En Ergot, Etournelles (or Estournelles), Fonteny, Les Goulots, Lavaut (or Lavout St-Jacques), La Perrière, Petite Chapelle, Petits Cazetiers, Plantigone (or Issarts), Poissenot, Clos Prieur-Haut (or Clos Prieure), La Romanée, Le Clos St-Jacques, and Les Varoilles.

RED These wines generally have more colour, concentration, and finesse than the village wines, but, with the possible exception of Clos St-Jacques, do not quite match the *grands crus*.

🍷 *Pinot Noir, Pinot Gris, Pinot Blanc, Chardonnay*

🍾 *10–20 years*

✓ *Burguet • Bruno Clair • Drouhin-Laroze • Claude Dugat • Bernard Dugat-Py • Dujac Pére & Fils • Frédéric Esmonin • Sylvie Esmonin • Faiveley • Jean-Claude Fourrier • Heresztyn-Mazzini • Louis Jadot • Denis Mortet • Philippe Naddef • Géantet Pansiot • Michelle & Patrice Rion • Joseph & Philippe Roty • Armand Rousseau • Sérafin Pére & Fils • Jean Trapet Père & Fils • de La Vougeraie*

GRANDS ECHÉZEAUX AOC
Grand Cru

The smaller and superior of the two *grands crus* of Flagey-Echézeaux, this area is separated from the upper slopes of the Clos de Vougeot by a village boundary.

RED Fine and complex wines that should have a silky bouquet, often reminiscent of violets. The flavour can be very round and rich but is balanced by a certain delicacy of fruit.

🍷 *Pinot Noir, Pinot Gris, Pinot Blanc, Chardonnay*

🍾 *10–20 years*

✓ *Drouhin • Gros Frère & Sœur • François Lamarche • Mongeard-Mugneret • De la Romanée-Conti*

LA GRANDE RUE AOC
Grand Cru

The newest *grand cru* of Vosne-Romanée, La Grande Rue was generally considered to have the best potential of all the *premiers crus* in this village. Officially upgraded in 1992, although the quality of wine produced by its sole owner, François Lamarche, was erratic to say the least. The *terroir* is, however, undeniably superior for a *premier cru* and as the domaine has been on form since its promotion. Perhaps the cart should come before the horse sometimes.

RED When they get this right, the wine is well coloured with deep, spicy-floral, black cherry fruit.

🍷 *Pinot Noir, Pinot Gris, Pinot Blanc, Chardonnay*

🍾 *7–15 years*

✓ *Lamarche*

GRIOTTES-CHAMBERTIN AOC
Grand Cru

The smallest of the nine *grands crus* of Gevrey-Chambertin.

RED The best growers produce deep-coloured, delicious wines with masses of soft-fruit flavours and all the velvety texture that could be expected of Chambertin itself.

🍷 *Pinot Noir, Pinot Gris, Pinot Blanc, Chardonnay*

🍾 *10–20 years*

✓ *Joseph Drouhin • Claude Dugat • Duroche • Jean-Claude Fourrier • Louis Jadot • Ponsot • Joseph & Philippe Roty*

LATRICIÈRES-CHAMBERTIN AOC
Grand Cru

One of the nine *grands crus* of Gevrey-Chambertin, situated above the Mazoyères climat of Charmes-Chambertin. A tiny part of the adjoining *premier cru*, Aux Combottes, also has the right to use this AOC.

RED Solid structure and a certain austerity connect the two different styles of this wine (early drinking and long maturing). They sometimes lack fruit and generosity, but wines from the top growers recommended below are always the finest to be found.

🍷 *Pinot Noir, Pinot Gris, Pinot Blanc, Chardonnay*

🍾 *10–20 years*

✓ *Drouhin-Laroze • Faiveley • Leroy* Ⓑ *• Ponsot • Jean Trapet Père & Fils*

MARSANNAY AOC

This village, situated in the very north of the Côte de Nuits, has long been famous for its rosé and has recently developed a reputation for its red wines. In May 1987 it was upgraded to full village AOC status from its previous appellation of Bourgogne Marsannay. There is talk of a classification of Marsannay that would result in a new Marsannay *premier cru* AOC, which should go ahead in 2020. This would be at least as justifiable as Maranges *premier cru* in the Côte de Beaune, not so much in the sense that either village has vineyards of truly *premier cru* quality, but it is at least useful in highlighting the better vineyards.

RED Firm and fruity wines with juicy redcurrant flavours, hints of liquorice, cinnamon and, if new oak has been used, vanilla.

🍷 *Pinot Noir, Pinot Gris, Pinot Blanc, Chardonnay*

🍾 *4-8 years*

WHITE Mostly light and tangy, early-drinking wines, although the top wines made in a suitable vintage, by the likes of Charles Audoin and Louis Jadot, are a lot richer, creamier and offer serious ageing potential.

🍷 *Chardonnay, Pinot Blanc, Pinot Gris*

🍾 *2-5 years*

ROSÉ These dry wines are rich, rather than light and fragrant. Packed with ripe fruit flavours that can include blackberry, blackcurrant, raspberry, and cherry, they are best consumed young, although people do enjoy them when the wine has an orange tinge and the fruit is over-mature.

🍇 Pinot Noir, Pinot Gris, Pinot Blanc, Chardonnay

🍷 1–3 years

✓ Charles Audoin • René Bouvier • Coillot Pére & Fils (Les Boivins) • Louis Jadot • Denis Mortet (Les Longeroies) • Géantet Pansiot (Champ Perdrix)

MAZIS-CHAMBERTIN AOC
Grand Cru

Sometimes known as Mazy-Chambertin, this is one of the nine *grands crus* of Gevrey-Chambertin. Some say this is the vineyard that can produce wines close to Chambertin in terms of quality.

RED These complex wines have a stature second only to Chambertin and Clos de Bèze. They have a fine, bright colour, super-silky finesse, and a delicate flavour that lasts.

🍇 Pinot Noir, Pinot Gris, Pinot Blanc, Chardonnay

🍷 10–20 years

✓ D'Auvenay 🅱 • Bernard Dugat-Py • Frédéric Esmonin • Faiveley • Guillon & Fils • Harmand-Geoffroy • Leroy 🅱 • Bernard Maume • Philippe Naddef • Joseph & Philippe Roty • Armand Rousseau

MAZOYÈRES-CHAMBERTIN AOC

This is an alternative appellation for Charmes-Chambertin. *See* Charmes-Chambertin AOC.

MOREY-ST-DENIS AOC

This excellent little wine village tends to be overlooked. The fact that it is situated between two world-famous places, Gevrey-Chambertin and Chambolle-Musigny, coupled with the fact that Clos St-Denis is no longer considered to be Gevrey-Chambertin's top *grand cru*, does little to promote the name of this village.

RED The best of these village wines have a vivid colour, a very expressive bouquet, and a smooth flavour with lots of finesse. A Morey-St-Denis from a domaine such as Dujac can have the quality of a top *premier cru*.

🍇 Pinot Noir, Pinot Gris, Pinot Blanc, Chardonnay

🍷 8–15 years

WHITE Dujac produces an excellent Morey-St-Denis blanc, but more interesting, although less consistent, is Ponsot's Monts Luisants blanc. Although Monts Luisants is a *premier cru*, the upper section from which this wine comes is part of the village appellation (the southeastern corner is classified *grand cru* and sold as Clos de la Roche). When Ponsot gets the Monts Luisants blanc right, it can be a superbly fresh, dry, buttery-rich wine, and some writers have compared it to a Meursault.

🍇 Chardonnay, Pinot Blanc, Pinot Gris

🍷 3–8 years

✓ Pierre Amiot • Philippe Charlopin-Parizot • Dujac • Bruno Clair • Hubert Lignier • Michel Magnien • Perrot-Minot • Ponsot • Forey Pére & Fils

MOREY-ST-DENIS PREMIER CRU AOC

This appellation includes the following *premiers crus*: Clos Baulet, Les Blanchards, La Bussière, Les Chaffots, Aux Charmes, Les Charrières, Les Chénevery, Aux Cheseaux, Les Faconnières, Les Genevrières, Les Gruenchers, Les Millandes, Monts Luisants, Clos des Ormes, Clos Sorbè, Les Sorbès, Côte Rôtie, La Riotte, Les Ruchots, and Le Village.

RED These wines should have all the colour, bouquet, flavour, and finesse of the excellent village wines plus an added expression of *terroir*. The best *premiers crus* are: Clos des Ormes, Clos Sorbè, and Les Sorbès.

🍇 Pinot Noir, Pinot Gris, Pinot Liébault

🍷 10–20 years

WHITE The only white Morey-St-Denis I know is from the upper section of Monts Luisants belonging to Ponsot (*see* Morey-St-Denis AOC). To my knowledge, no other white Morey-St-Denis *premier cru* is made.

🍇 Chardonnay

✓ Pierre Amiot • Dujac • Heresztyn-Mazzini • Hubert Lignier • Frédéric Magnien (Les Ruchots) • Perrot-Minot • Georges Roumier

MUSIGNY AOC
Grand Cru

Musigny is the smaller of Chambolle-Musigny's two *grands crus*. It covers some 10 hectares (25 acres) on the opposite side of the village to Bonnes Mares.

RED These most stylish of wines have a fabulous colour and a smooth, seductive, and spicy bouquet. The velvet-rich fruit flavour constantly unfolds to reveal a succession of taste experiences.

🍇 Pinot Noir, Pinot Gris, Pinot Blanc, Chardonnay

🍷 10–30 years

WHITE Musigny blanc is a rare and expensive dry wine produced solely at Comte Georges de Vogüé. It combines the steel of a Chablis with the richness of a Montrachet, although it never quite achieves the quality of either.

🍇 Chardonnay

🍷 8–20 years

✓ Joseph Drouhin • Louis Jadot • Leroy 🅱 • J-F Mugnier (Château de Chambolle-Musigny) • Georges Roumier • Comte Georges de Vogüé

NUITS AOC

See Nuits-St-Georges AOC

NUITS PREMIER CRU AOC

See Nuits-St-Georges Premier Cru AOC

NUITS-ST-GEORGES AOC

More than any other, the name of this town graphically projects the image of full flavour and sturdy structure for which the wines of the Côte de Nuits are justly famous.

RED These are deep-coloured, full, and firm wines, but they can sometimes lack the style and character of wines from Gevrey, Chambolle, and Morey.

🍇 Pinot Noir, Pinot Gris, Pinot Blanc, Chardonnay

🍷 7–15 years

WHITE I have not tasted this wine. (*See* white Nuits-St-Georges Premier Cru.)

✓ De l'Arlot • R Chevillon • J Chauvenet • Jean-Jacques Confuron • Confuron-Cotétidot • Thibault Liger-Belair • Odoul-Coquard • A Michelot • Des Perdrix • Daniel Rion & Fils

NUITS-ST-GEORGES PREMIER CRU AOC

This appellation includes the following *premiers crus*: Les Argillats, Aux Boudots, Aux Bousselots, Les Cailles, Les Chaboeufs, Aux Chaignots, Chaine-Carteau (or Chaines-Carteaux), Aux Champs Perdrix, Aux Cras, Les Crots, Les Damodes, Les Didiers, Château Gris, Les Hauts Pruliers, Aux Murgers, En la Perrière Noblet (or En la Perrière Noblot), Les Perrières, Les Porets, Les Poulettes, Les Procès, Les Pruliers, La Richemone, La Roncière, Rue de Chaux, Les St-Georges, Clos St-Marc (or Aux Corvées), Les Terres Blanches, Aux Thorey, Les Vallerots, Les Vaucrains, and Aux Vignerondes. The *premiers crus* in the village of Prémeaux-Prissey are Les Argillières, Clos Arlot, Clos des Corvées, Clos des Corvées Pagets, Les Forêts (or Clos des Forêts St-Georges), Les Grandes Vignes, Clos de la Maréchale, and Aux Perdrix.

RED These wines have a splendid colour, a spicy-rich bouquet, and a vibrant fruit flavour which can be nicely underpinned with vanilla.

🍇 Pinot Noir, Pinot Gris, Pinot Blanc, Chardonnay

🍷 10–20 years

WHITE Henri Gouges's La Perrière is dry, powerful, almost fat, with a spicy-rich aftertaste. The vines used for this wine were propagated from a mutant Pinot Noir that produced bunches of both black and white grapes.

Gouges cut a shoot from a white-grape-producing branch of the mutant vine in the mid-1930s, and there is now just under half a hectare (just over one acre) of the vines planted by Gouges, none of which have ever reverted to producing black grapes.

🍇 Chardonnay, Pinot Blanc, Pinot Gris

🍷 5–10 years

✓ Ambroise • De l'Arlot • Arnoux-Lachaux • Bouchard Père & Fils • Jean Chauvenet (Les Perrières) • Robert Chevillon • Jean-Jacques Confuron • Confuron-Cotétidot • Dubois & Fils • Faiveley • Forey Père & Fils • Henri Gouges • Jean Grivot • Anne Gros • Alain Hudelot-Noëllat • Robert Jayer-Gilles • Dominique Laurent • Philippe & Vincent Léchenaut • Leroy 🅱 • Thibault Liger-Belair • Alain Michelot • Mugneret-Gibourg • Gérard Mugneret • Gilles Remoriquet • Daniel Rion & Fils • Michelle & Patrice Rion • Patrice Rion • Antonin Rodet • Jean Tardy • de la Vougeraie 🅱

RICHEBOURG AOC
Grand Cru

One of the six *grands crus* at the heart of Vosne-Romanée's vineyards. It is for red wine only.

RED This is a gloriously rich wine that has a heavenly bouquet and is full of velvety and voluptuous fruit flavours.

🍇 Pinot Noir, Pinot Gris, Pinot Blanc, Chardonnay

🍷 12–30 years

✓ Jean Grivot • Anne Gros • Alain Hudelot-Noëllat • Leroy 🅱 • Méo-Camuzet • De la Romanée-Conti

LA ROMANÉE AOC
Grand Cru

This vineyard is owned by the Liger-Belair family, but the wine was matured, bottled, and sold by Bouchard Père & Fils until 2002, when Louis-Michel Liger-Belair took over. Less than 1 hectare (2.5 acres), it is the smallest *grand cru* of Vosne-Romanée.

RED This is a full, fine, and complex wine that might not have the voluptuous appeal of a Richebourg, but recent vintages are very promising for a real *grand vin*.

🍇 Pinot Noir, Pinot Gris, Pinot Blanc, Chardonnay

🍷 12–30 years

✓ Comte Liger-Belair

ROMANÉE-CONTI AOC
Grand Cru

This Vosne-Romanée grand cru is under 2 hectares (5 acres) in size and belongs solely to the famous Domaine de la Romanée-Conti.

RED As the most expensive Burgundy in the world, this wine must always be judged by higher standards than all the rest. Yet I must admit that I never fail to be amazed by the stunning array of flavours that continuously unfold in this fabulously concentrated and utterly complex wine.

🍇 Pinot Noir, Pinot Gris, Pinot Blanc, Chardonnay

🍷 15–35 years

✓ De la Romanée-Conti

ROMANÉE-ST-VIVANT AOC
Grand Cru

The largest of the six *grands crus* on the lowest slopes and closest to the village.

RED This is the lightest of the fabulous *grands crus* of Vosne-Romanée, but what it lacks in power and weight it makes up for in finesse.

🍇 Pinot Noir, Pinot Gris, Pinot Blanc, Chardonnay

🍷 10–25 years

✓ Arnoux-Lachaux • Sylvain Cathiard • Jena-Jacques Confuron • Joseph Drouhin • Follin-Arbelet • Alain Hudelot-Noëllat • Dujac Pére & Fils• Leroy 🅱 • De la Romanée-Conti

VINEYARDS SURROUND THE COMMUNE OF VOSNE-ROMANÉE
The *grand cru* vineyards of the Vosne-Romanée AOC supply grapes for some of Burgundy's most sought-after wines. The appellation encompasses both the commune of Vosne-Romanée and its neighbour Flagey-Échezeaux. This is a red wine appellation, and Pinot Noir must be the main grape.

RUCHOTTES-CHAMBERTIN AOC

Grand Cru

This is the second-smallest of the nine *grands crus* of Gevrey-Chambertin. It is situated above Mazis-Chambertin and is the last grand cru before the slope turns to face north.

RED Normally one of the lighter Chambertin look-alikes, but top growers like Mugneret-Gibourg, Roumier, and Rousseau seem to achieve a bagful of added ingredients in their splendidly rich wines.

🍷 *Pinot Noir, Pinot Gris, Pinot Blanc, Chardonnay*

🍷 *8–20 years*

✓ *Frédéric Esmonin • Mugneret-Gibourg • Christophe Roumier • Armand Rousseau*

LA TÂCHE AOC

Grand Cru

One of Vosne-Romanée's six *grands crus,* this fabulous vineyard belongs to the world-famous DRC (Domaine de la Romanée-Conti), which also owns the Romanée-Conti *grand cru.*

RED This wine is indeed extremely rich and very complex, but it does not in comparative terms quite have the richness of Richebourg nor the complexity of Romanée-Conti. It does, however, have all the silky texture anyone could expect from the finest of Burgundies, and no other wine surpasses La Tâche for finesse.

🍷 *Pinot Noir, Pinot Gris, Pinot Blanc, Chardonnay*

🍷 *12–30 years*

✓ *De la Romanée-Conti*

VOSNE-ROMANÉE AOC

This is the most southerly of the great villages of the Côte de Nuits. Some Vosne-Romanée vineyards are in neighbouring Flagey-Echézeaux. This village includes the *grand cru* Romanée-Conti vineyard.

RED Sleek, stylish, medium-bodied wines of the purest Pinot Noir character and with the silky texture so typical of the wines of this village.

🍷 *Pinot Noir, Pinot Gris, Pinot Blanc, Chardonnay*

🍷 *10–15 years*

✓ *Arnoux-Lachaux • Sylvain Cathiard • Bruno Clavelier • Confuron-Cotétidot • Bernard Dugat-Py (Vieilles Vignes) • Forey Père & Fils • Jean Grivot • Anne Gros • Michel Gros • Méo-Camuzet • Mugneret-Gibourg • Perrot-Minot • Emmanuel Rouget • Fabrice Vigot*

VOSNE-ROMANÉE PREMIER CRU AOC

This appellation includes the following *premiers crus:* Les Beaux Monts, Les Brûlées, Les Chaumes, La Combe Brûlées, La Croix Rameau, Cros-Parantoux, Les Gaudichots, Les Hauts Beaux Monts, Aux Malconsorts, Les Petits Monts, Clos des Réas, Aux Reignots, and Les Suchots. The *premier cru* vineyards Les Beaux Monts Bas, Les Beaux Monts Hauts, En Orveaux, and Les Rouges du Dessus are actually in Flagey-Echézeaux.

RED These are well-coloured wines with fine aromatic qualities that are often reminiscent of violets and blackberries. They have a silky texture and a stylish flavour that is pure Pinot Noir. The best *premiers crus* to look out for are: Les Brûlées, Cros-Parantoux, Les Petits-Monts, Aux Reignots, Les Suchots, and Les Beaumonts (an umbrella name for the various Beaux Monts, high and low).

🍷 *Pinot Noir, Pinot Gris, Pinot Blanc, Chardonnay*

🍷 *10–20 years*

✓ *De l'Arlot • Arnoux-Lachaux • Albert Bichot (Clos Frantin) • Sylvain Cathiard • Confuron-Cotétidot • Jean Grivot • Alain Hudelot-Noëllat • S Javouhey • Lamarche • Leroy ᗷ • Comte Liger-Belair • Méo-Camuzet • Gérard Mugneret • Perrot-Minot • Nicolas Potel • Daniel Rion & Fils • Michelle & Patrice Rion • Emmanuel Rouget*

VOUGEOT AOC

The modest village wines of Vougeot are a relative rarity. This appellation, covering less than 5 hectares (12 acres), is less than one-tenth of the area encompassed by the Clos de Vougeot *grand cru* itself.

RED These are seldom-seen, fine-flavoured, and well-balanced wines that are overpriced due to their scarcity. It is better to buy a *premier cru*.

🍷 *Pinot Noir, Pinot Gris, Pinot Blanc, Chardonnay*

🍷 *8–20 years*

WHITE An even scarcer rarity thant the red.

✓ *Bertagna • Mongeard-Mugneret*

VOUGEOT PREMIER CRU AOC

This appellation includes the following *premiers crus:* Les Crâs, Clos de la Perrière, Les Petits Vougeots, La Vigne Blanche. The *premiers crus* are located between Clos de Vougeot and Musigny and, from a *terroir* point of view, should produce much better wines. Because Clos de Vougeot is a red-only AOC, white wines that originate from within the walls of that enclosed vineyard can be sold as white Vougeot Premier Cru.

RED This wine can be nicely coloured, medium bodied, and have an attractive flavour with a good balance and a certain finesse.

🍷 *Pinot Noir, Pinot Gris, Pinot Blanc, Chardonnay*

🍷 *10–20 years*

WHITE A clean, rich, and crisp wine of variable quality, called Clos Blanc de Vougeot, is produced by L'Héritier Guyot from the *premier cru* of La Vigne Blanche.

🍷 *Chardonnay, Pinot Blanc, Pinot Gris*

🍷 *4–10 years*

✓ *Bertagna (Clos de la Perrière monopole) • Chauvenet-Chopin • Lamarche • De la Vougeraie ᗷ (Clos du Prieuré monopole)*

The Côte de Beaune and Hautes Côtes de Beaune

If the Côte de Nuits is the land of Pinot Noir, then the Côte de Beaune, with seven of Burgundy's eight white grands crus, is really Chardonnay country. The wines produced here are the richest, longest-lived, most complex, elegant, and stylish white wines in the world. The reds are also fine – Pinot Noir in the Côte de Beaune is renowned for its softness and finesse, characteristics that become more evident as one progresses south through the region.

DOMAINE BONNEAU DU MARTRAY IN PERNAND-VERGELESSES
These *grand cru* vineyards fall in the white wine Corton-Charlemagne AOC. The appellation is named after the Holy Roman Emperor Charlemagne, who once owned the hill of Corton on which the vineyards now rest.

Entering the Côte de Beaune from the Nuits-St-Georges end, the most immediately visible viticultural differences are its more expansive look and the much greater contrast between the deep, dark, and so obviously rich soil found on the inferior eastern side of the RN 74 road and the scanty patches of pebble-strewn thick drift that cover the classic slopes west of the road.

It is often said that the slopes of the Côte de Beaune are gentler than those of the Côte de Nuits; however, there are many parts that are just as sheer, although the best vineyards of the Côte de Beaune are located on the middle slopes, which have a gentler incline. The steeper upper slopes produce good but generally lesser wines in all cases, except the vineyards of Aloxe-Corton – which are, anyway, more logically part of the Côte de Nuits than the Côte de Beaune.

Côte de Beaune-Villages AOC

0 mi — 3
0 km — 3

VINE-GROWING ZONES AND VILLAGE APPELLATIONS OF THE CÔTE DE BEAUNE AND HAUTES CÔTES DE BEAUNE
(see also p213)
Most of the village and Hautes-Côtes appellations are clustered between Nuits-Saint-Georges in the north and Chagny in the south. The Côte d'Or is a hilly ridge that follows the trajectory of the Autoroute du Soleil.

THE APPELLATIONS OF
THE CÔTE DE BEAUNE & HAUTES CÔTES DE BEAUNE

Notes: 1. All the *grands crus* of Côte de Beaune have their own separate appellations and are therefore listed individually below. The *premiers crus* do not, however, and are therefore listed under the appellation of the village in which the vineyards are situated. *Premiers crus* that are virtually contiguous with *grand cru* vineyards are italicized; they are possibly superior to those that do not share a boundary with one or more *grands crus*, and generally superior to those that share boundaries with village AOC vineyards.

2. Where white grapes are permitted in red and rosé wines, they must come from a co-planted vineyard in small proportions (5 to 15 per cent of the vines).

3. Leflaive plain and simple refers to Domaine Leflaive, while Olivier Leflaive is the *négociant*.

ALOXE-CORTON AOC

This village is more Côte de Nuits than Côte de Beaune, as its 99 per cent red-wine production suggests.

RED These deeply coloured, firm-structured wines with compact fruit are reminiscent of reds from northern Côte de Nuits. They are excellent value for money.

🍷 *Pinot Noir, Pinot Gris, Pinot Blanc, Chardonnay*

🍾 *10–20 years*

WHITE Very little Aloxe-Corton *blanc* is made, but Comte Senard makes a lovely buttery-rich, concentrated pure Pinot Gris wine (which, although it is definitely a white wine, makes it red according to the regulations).

🍷 *Chardonnay, Pinot Blanc*

🍾 *4–8 years*

✓ *Bruno Clair • Cornu & Fils • Follin-Arbelet • Naudin-Ferrand • Comte Senard • Tollot-Beaut*

ALOXE-CORTON PREMIER CRU AOC

This appellation includes the following *premiers crus*: Les Chaillots, Les Fournières, Les Guérets, *Clos des Maréchaudes, Les Maréchaudes,* Les Meix (or Clos du Chapitre), *Les Paulands, Les Valozières, and* Les Vercots. La Coutière, *La Maréchaude, Les Moutottes, Les Petites Lolières,* and *La Toppe au Vert* are *premier cru* vineyards of Aloxe-Corton in Ladoix-Serrigny.

RED Can have an intense bouquet and a firm, spicy-fruit flavour. The best *premiers crus* are Les Fournières, Les Valozières, Les Paulands, Les Maréchaudes, and Les Vercots.

🍷 *Pinot Noir, Pinot Gris, Pinot Blanc, Chardonnay*

🍾 *10–20 years*

WHITE I have never encountered any.

✓ *Cachat-Ocquidant (Les Marechaudes) • Capitain-Gagnerot • Antonin Guyon • Thibault Liger-Belair • Tollot-Beaut & Fils*

AUXEY-DURESSES AOC

Auxey-Duresses is a beautiful village, set in an idyllic valley behind Monthélie and Meursault.

RED The redshere are attractive wines that are not very deep in colour, but they possess a softness of fruit and a little finesse.

🍷 *Pinot Noir, Pinot Gris, Pinot Blanc, Chardonnay*

🍾 *6–12 years*

WHITE Medium-bodied wines with a full, spicy-nutty flavour, like that of a modest Meursault.

🍷 *Chardonnay, Pinot Blanc*

🍾 *3–7 years*

✓ *Robert Ampeau & Fils • d'Auvenay ❸ • Jean-Pierre Diconne • Alain Gras • Louis Jadot • Henri Latour & Fils • Olivier Leflaive • Leroy ❸ • Prunier (all and sundry) • Roy • Taupenot-Merme*

AUXEY-DURESSES PREMIER CRU AOC

This appellation includes the following *premiers crus:* Bas des Duresses, Les Breterins, La Chapelle, Climat du Val, Clos du Val, Les Duresses, Les Écussaux, Les Grands-Champs, and Reugne.

FACTORS AFFECTING TASTE AND QUALITY

LOCATION
The Côte de Beaune abuts the Côte de Nuits on its southern tip and stretches almost 30 kilometres (18 miles) past the town of Beaune to Cheilly-lès-Maranges. Its vines are contiguous, although those of the Hautes-Côtes de Beaune in the western hinterland are divided into two by the Côte de Beaune vineyards of St-Romain.

CLIMATE
This area has a slightly wetter, more temperate climate than the Côte de Nuits, and the grapes tend to ripen a little earlier. Hail is still a potential hazard and quite often early spring frosts bring bitter days to the area.

ASPECT
Composed of a series of east-facing slopes, up to 2 kilometres (just over 1 mile) wide, which curve in and out to give some vineyards northeastern aspects, and others southeastern aspects. Here the vines grow at an altitude of between 225 and 380 metres (740 to 1,250 feet) on slightly less steep slopes than those of the Côte de Nuits. South of Beaune, no vines with the right to the village (and higher) appellations extend past the RN 74 road on to the flat and fertile ground beyond.

SOIL
The vineyards are composed of a limestone subsoil with sporadic beds of oolitic ironstones with flinty clay and calcareous topsoils. Light-coloured marl topsoil is found in the vineyards of Chassagne and Puligny.

VITICULTURE & VINIFICATION
The vines are trained low to benefit from heat reflected from the soil at night. In the south of the district, the system employed is similar to that used in parts of Champagne and slightly different from elsewhere on the Côte de Beaune. For red wines, the grapes are almost always de-stemmed and the juice kept in contact with the skins for between 8 and 10 days. Classic white wines are cask-fermented and the best wines, both red and white, matured in oak. The flavour of the Pinot Noir grape can easily be overwhelmed by oak, so it always receives less new-oak maturation than Chardonnay does.

GRAPE VARIETIES
Primary varieties: Chardonnay, Pinot Noir
Secondary varieties: Aligoté, Gamay, Melon de Bourgogne, Pinot Beurot (Pinot Gris), Pinot Blanc, Tressot

A WINE MERCHANT TEMPTS PASSERS-BY WITH A DISPLAY OF ITS WARES NEAR THE MARCHÉ AUX VINS IN BEAUNE
The walled town of Beaune is surrounded by the vineyards of the AOC of the same name, as well as the Beaune Premiers Crus AOC. Both appellations focus on Pinot Noir, Pinot Gris, Pinot Blanc, and Chardonnay varieties.

RED The *premiers crus* provide nicely coloured soft wines with good finesse. The best have fine redcurrant Pinot character with the creamy-oak of cask maturity.

🍷 *Pinot Noir, Pinot Gris, Pinot Liébault*

🍾 *7–15 years*

WHITE Excellent value, smooth, and stylish wines in the Meursault mould.

🍷 *Chardonnay, Pinot Blanc*

🍾 *4–10 years*

✓ *Comte Armand • Jean-Pierre Diconne • Michel Prunier • Vincent Prunier*

AUXEY-DURESSES-CÔTES DE BEAUNE AOC

This is an alternative appellation for red wines only. *See* Auxey-Duresses AOC.

BÂTARD-MONTRACHET AOC
Grand Cru

This *grand cru* is situated on the slope beneath Le Montrachet and overlaps both Chassagne-Montrachet and Puligny-Montrachet.

WHITE Full-bodied, intensely rich wine with masses of nutty, honey-and-toast flavours. It is one of the best dry white wines in the world.

🍷 *Chardonnay*

🍾 *8–20 years*

✓ *Henri Boillot • Bouchard Père & Fils • Jean-Noël Gagnard • Vincent Girardin • Louis Latour • Leflaive ⓑ • Marc Morey • Pierre-Yves Colin-Morey ⓑ • Michel Niellon • Paul Pernot • Ramonet • Antonin Rodet • Étienne Sauzet*

BEAUNE AOC

Beaune gives its name to village wines and *premiers crus*, but not to any *grands crus*.

RED These soft-scented, gently fruity wines are consistent and good value.

🍷 *Pinot Noir, Pinot Gris, Pinot Blanc, Chardonnay*

🍾 *6–14 years*

WHITE An uncomplicated dry Chardonnay wine with a characteristic soft finish.

🍷 *Chardonnay, Pinot Blanc*

🍾 *3–7 years*

✓ *Bouchard Père & Fils • Darviot-Perrin • Dominique Laurent (Vieilles Vignes) • Albert Morot • Tollot-Beaut • De la Vougeraie Z*

BEAUNE PREMIER CRU AOC

This appellation includes the following *premiers crus*: Les Aigrots, Aux Coucherias, Clos de la Féguine, Aux Cras, Clos des Avaux, Les Avaux, Le Bas des Teurons, Les Beaux Fougets, Belissand, Les Blanches Fleurs, Les Boucherottes, Les Bressandes, Les Cents Vignes, Champs Pimont, Les Chouacheux, L'Écu (or Clos de l'Écu), Les Epenottes (or Les Epenotes), Les Fèves, En Genêt, Les Grèves, Clos Landry (Clos Ste-Landry), Les Longes, Le Clos des Mouches, Le Clos de la Mousse, Les Marconnets, La Mignotte, Montée Rouge, Les Montrevenots, En l'Orme, Les Perrières, Pertuisots, Les Reversées, Clos du Roi, Les Seurey, Les Sizies, Clos Ste-Anne (or Sur les Grèves), Les Teurons, Les Toussaints, Les Tuvilains, La Vigne de l'Enfant Jésus, and Les Vignes Franches (or Clos des Ursules)

RED The best *crus* of this appellation are medium-bodied with a delightfully soft rendition of Pinot fruit and lots of finesse. Look out for Faiveley's Clos de L'Écu, a *monopole climat*. Faiveley should make some sort of quality statement here.

🍷 *Pinot Noir, Pinot Gris, Pinot Blanc, Chardonnay*

🍾 *10–20 years*

WHITE These wines have lovely finesse, and they can display a toasty flavour more common to richer

growths. Domaine Jacques Prieur's Champs Pimont usually stands out.

🍷 *Chardonnay, Pinot Blanc*

🍾 *5–12 years*

✓ *Arnoux Père & Fils • Jean-Claude Boisset • Bouchard Père & Fils • Chanson Père & Fils (Clos de Fèves, Clos des Mouches) • Joseph Drouhin • Camille Giroud • Louis Jadot • Jaffelin • Michel Lafarge • Louis Latour • Rapet Père & Fils • Albert Morot • Jacques Prieur • Tollot-Beaut*

BIENVENUES-BÂTARD-MONTRACHET AOC
Grand Cru

This is one of Puligny-Montrachet's four *grands crus*.

WHITE Not the fattest dry wines, but they have great finesse, immaculate balance, and some of the nuttiness and honey-and-toast flavours expected in all Montrachets.

🍷 *Chardonnay*

🍾 *8–20 years*

✓ *Jacques Carillon • Vincent Girardin • Louis Latour • Leflaive ⓑ • Paul Pernot • Bachelet-Ramonet • Étienne Sauzet*

BLAGNY AOC

A red-only appellation from Blagny, a tiny hamlet shared by the communes of Meursault and Puligny-Montrachet.

RED These rich, full-flavoured, Meursault-like red wines are underrated.

🍷 *Pinot Noir, Pinot Gris, Pinot Blanc, Chardonnay*

🍾 *8–15 years*

✓ *Robert Ampeau & Fils • Lamy-Pillot*

BLAGNY PREMIER CRU AOC

This appellation includes the following *premiers crus*: La Garenne (or Sur la Garenne), Hameau de Blagny (in Puligny-Montrachet), La Jeunelotte, La Pièce sous le Bois, Sous Blagny (in Meursault), Sous le Dos d'Ane, and Sous le Puits.

RED These rich wines have even more grip and attack than basic Blagny.

🍷 *Pinot Noir, Pinot Gris, Pinot Blanc, Chardonnay*

🍾 *10–20 years*

✓ *Robert Ampeau • François Jobard • Larue (Sous le Puits) • Leflaive ⓑ • Matrot*

BLAGNY-CÔTE DE BEAUNE AOC

An alternative appellation for Blagny. *See* Blagny AOC.

BOURGOGNE LA CHAPELLE NOTRE-DAME AOC

One of three Bourgogne *lieux-dits* created in 1993. La Chapelle Notre Dame is located in Serrigny, which is just east of Ladoix. Red, white, and rosé may be produced according to the same grapes and rules as for Bourgogne AOC (see p215). P Dubreuil-Fontaine has produced a wine from this *lieu-dit* for many years. It has always been decent, not special, but capable of improving in bottle. Nudant also makes a light, fresh red from this appellation.

BOURGOGNE HAUTES-CÔTES DE BEAUNE AOC

This appellation is larger and more varied than Hautes-Côtes de Nuits. Half-red / half-rosé wines may be sold as Bourgogne Clairet Hautes-Côtes de Beaune AOC, but the style is outmoded and the appellation rarely encountered.

RED Ruby-coloured, medium-bodied wines with a Pinot perfume and a creamy-fruit finish.

🍷 *Pinot Noir, Pinot Gris, Pinot Blanc, Chardonnay*

🍾 *4–10 years*

WHITE Not very frequently encountered.

🍷 *Chardonnay, Pinot Blanc*

🍾 *1–4 years*

ROSÉ Pleasantly dry and fruity wines with some richness and a soft finish.

🍷 *Pinot Noir, Pinot Gris, Pinot Blanc, Chardonnay*

🍾 *1–3 years*

✓ *Contat-Grange • Lucien Jacob • Didier Montchovet • Claude Nouveau • Naudin-Ferrand*

CHARLEMAGNE AOC
Grand Cru

This white-only *grand cru* of Aloxe-Corton overlaps Pernand-Vergelesses and is almost, but not quite, identical to the *grand cru* of Corton-Charlemagne.

CHASSAGNE-MONTRACHET AOC

These village wines have a lesser reputation than those of Puligny-Montrachet.

RED Firm, dry wines with more colour and less softness than most Côte de Beaune reds.

🍷 *Pinot Noir, Pinot Gris, Pinot Blanc, Chardonnay*

🍾 *10–20 years*

WHITE These whites are an affordable introduction to the great wines of Montrachet.

🍷 *Chardonnay, Pinot Blanc*

🍾 *5–10 years*

✓ *Marc Colin & Fils • Fontaine-Gagnard • Jean-Noël Gagnard • Marquis de Laguiche • Château de la Maltroye • Bernard Morey • Jean-Marc Morey*

CHASSAGNE-MONTRACHET PREMIER CRU AOC

This appellation includes the following *premiers crus*: Abbaye de Morgeot, Les Baudines, *Blanchot Dessus*, Les Boirettes, Bois de Chassagne, Les Bondues, La Boudriotte, Les Brussonnes, En Cailleret, La Cardeuse, Champ Jendreau, Les Champs Gain, La Chapelle, Clos Chareau, Chassagne, Chassagne du Clos St-Jean, Les Chaumées, Les Chaumes, Les Chenevottes, Les Combards, Les Commes, Ez Crets, Ez Crottes, *Dent de Chien*, Les Embrazées, Les Fairendes, Francemont, La Grande Borne, La Grande Montagne, Les Grandes Ruchottes, Les Grands Clos, Guerchère, Les Macherelles, *La Maltroie*, Morgeot, Les Murées, Les Pasquelles, Petingeret, Les Petites Fairendes, Les Petits Clos, Clos Pitois, Les Places, Les Rebichets, En Remilly, La Romanée, La Roquemaure, Clos St-Jean, Tête du Clos, Tonton Marcel, Les Vergers, *Vide Bourse*, Vigne Blanche, Vigne Derrière, and En Virondot.

RED These wines have the weight of a Côte de Nuits and the softness of a Côte de Beaune.

🍷 *Pinot Noir, Pinot Gris, Pinot Blanc, Chardonnay*

🍾 *10–25 years*

WHITE These are flavoursome dry wines, but they are lacking the finesse of those in the neighbouring appellation of Puligny.

🍷 *Chardonnay, Pinot Blanc*

🍾 *6–15 years*

✓ *Guy Amiot & Fils • Roger Belland (Morgeot Clos Pitois monopole) • Blain-Gagnard • Marc Colin & Fils • Pierre-Yves Colin-Morey • Vincent Dancer • Colin-Deléger & Fils • Fontaine-Gagnard • Jean-Noël Gagnard • Gagnard-Delagrange • Vincent Girardin • Gabriel et Paul Jouard • Vincent et François Jouard • Louis Latour • Olivier Leflaive • Lamy-Pillot • Château de la Maltroye • Bernard Moreau & Fils • Marc Morey • Michel Niellon • Roux*

CHASSAGNE-MONTRACHET-CÔTE DE BEAUNE AOC

This is an alternative appellation for red wines only. *See* Chassagne Montrachet AOC.

CHEVALIER-MONTRACHET AOC
Grand Cru

This is one of Puligny-Montrachet's four *grands crus*.

WHITE Fatter and richer than Bienvenues-Bâtard-Montrachet, this wine has more explosive flavour and precision than Bâtard-Montrachet.

🍷 *Chardonnay*

🥂 *10–20 years*

✓ *D'Auvenay* ⓑ *• Bouchard Père & Fils • Vincent Dancer • Georges Déléger • Vincent Girardin • Louis Jadot • Leflaive* ⓑ *• Marc Morey • Michel Niellon • Pierre-Yves Colin-Morey*

CHOREY-LÈS-BEAUNE AOC

This satellite appellation of Beaune produces exciting, underrated wines.

RED Although next to Aloxe-Corton, Chorey has all the soft and sensuous charms that are quintessentially Beaune.

🍷 *Pinot Noir, Pinot Gris, Pinot Blanc, Chardonnay*

🥂 *7–15 years*

WHITE Less than 1 per cent of the wines produced in this village are white.

✓ *Arnoux Père & Fils • Jaffelin • Maillard • Tollot-Beaut*

CHOREY-LÈS-BEAUNE CÔTE DE BEAUNE AOC

This is an alternative appellation for red wines only. *See* Chorey-lès-Beaune AOC.

CORTON AOC

Grand Cru

This is one of the *grands crus* of Aloxe-Corton (it extends into Ladoix-Serrigny and Pernand-Vergelesses). Corton is the only *grand cru* in the Côte de Beaune that includes red and white wines and thus parallels the Côte de Nuits *grand cru* of Musigny. The following 25 *climats* may hyphenate their names (with or without the prefix) to the Corton appellation: Basses Mourottes, En Charlemagne, Hautes Mourottes, La Vigne-au-Sai, Le Charlemagne, Le Clos du Roi, Le Corton, Le Meix Lallemand, Le Rognet et Corton, Les Bressandes, Les Chaumes, Les Chaumes et la Voierosse, Les Combes, Les Fiètres, Les Grandes Lolières, Les Grèves, Les Languettes, Les Maréchaudes, Les Meix, Les Moutottes, Les Paulands, Les Perrières, Les Pougets, Les Renardes, Les Vergennes.

So it is possible to get a red Corton-Charlemagne. Beware: Corton is one of the least dependable *grands crus* for quality. When it is great, it is wicked, but production is vast (3,500 to 3,800 hectolitres per year) – over twice the volume of the huge, over-producing Clos de Vougeot, and more than one-quarter of all the red *grand cru* wine produced in both the Côte de Nuits and Côte de Beaune.

RED These wines may sometimes appear intense and broody in their youth, but, when fully mature, a great Corton has such finesse and complexity that it can stun the senses.

🍷 *Pinot Noir, Pinot Gris, Pinot Blanc, Chardonnay*

🥂 *12–30 years*

WHITE This is a medium-bodied to full-bodied wine with a fine, rich flavour.

🍷 *Chardonnay*

🥂 *10–25 years*

✓ *Arnoux Père & Fils • Bouchard Père & Fils • Chevalier Père & Fils (Rognet) • Vincent Girardin • Antonin Guyon • Louis Jadot • Leroy* ⓑ *• Chandon de Briailles • Faiveley • Michel Juillot • De La Romanée Conti • Jacques Prieur • Rapet Père & Fils • Comte Senard • Tollot-Beaut*

CORTON-CHARLEMAGNE AOC

Grand Cru

This famous *grand cru* of Aloxe-Corton extends into Ladoix-Serrigny and Pernand-Vergelesses. Although the best Corton-Charlemagne is incomparable, there is so much of it produced that consumers can pay a lot of money for relatively disappointing wines. At an average of 2,280 hectolitres, the production of Corton-Charlemagne represent more than two out of every three bottles of all the

grand cru white wines produced in both the Côte de Nuits and Côte de Beaune.

WHITE At its natural best, this is the most sumptuous of all white Burgundies. It has a fabulous concentration of rich, buttery fruit flavours, a dazzling balance of acidity, and delicious overtones of vanilla, honey, and cinnamon.

🍷 *Chardonnay, Pinot Blanc*

🥂 *5–25 years*

✓ *Bertrand Ambroise • Bonneau du Martray • Bouchard Père & Fils • J-F Coche-Dury • Joseph Drouhin • Genot-Boulanger • Vincent Girardin • Pierre-Yves Colin-Morey • Louis Jadot • Patrick Javillier • Michel Juillot • Louis Latour • Olivier Leflaive • Pavelot • Jacques Prieur • Rapet Père & Fils • Remoissenet Père & Fils • Christophe Roumier*

CÔTE DE BEAUNE AOC

Wines that are entitled to the actual Côte de Beaune appellation are restricted to a few plots on the Montagne de Beaune above Beaune itself.

RED These are fine, stylish wines that reveal the purest of Pinot Noir fruit. They are produced in the soft Beaune style.

🍷 *Pinot Noir, Pinot Gris, Pinot Blanc, Chardonnay*

🥂 *10–20 years*

WHITE Little-seen, dry basic Beaune.

🍷 *Chardonnay, Pinot Blanc*

🥂 *3–8 years*

✓ *Emmanuel Giboulot • Chantal Lescure • Lycée Viticole de Beaune*

CÔTE DE BEAUNE-VILLAGES AOC

Although AOC Côte de Nuits-Villages covers red and white wines in a predominantly red-wine district, AOC Côte de Beaune-Villages applies only to red wines in a district that produces the greatest white Burgundies.

RED Excellent-value fruity wines, made in true soft Beaune style.

🍷 *Pinot Noir, Pinot Gris, Pinot Blanc, Chardonnay*

🥂 *7–15 years*

✓ *Michel Delorme • Andre Morey • Bachey-Legros*

CRIOTS-BÂTARD-MONTRACHET AOC

Grand Cru

The smallest of Chassagne-Montrachet's three *grands crus* with its 1.57 hectare (3.87 acres) surface.

WHITE This wine has some of the weight of its great neighbours and a lovely hint of honey-and-toast richness, but it is essentially the palest and most fragrant of all the Montrachets.

🍷 *Chardonnay*

🥂 *8–20 years*

✓ *D'Auvenay* ⓑ *• Roger Belland • Blain-Gagnard • Vincent Girardin • Hubert Lamy*

LADOIX AOC

Parts of Ladoix-Serrigny have the right to use the Aloxe-Corton Premier Cru appellation or the *grands crus* of Corton and Corton-Charlemagne. Ladoix AOC covers the rest of the wine produced in the area.

RED Many wines are rustic versions of Aloxe-Corton, but there are some fine grower wines that combine the compact fruit and structure of a Nuits with the softness of a Beaune.

🍷 *Pinot Noir, Pinot Gris, Pinot Blanc, Chardonnay*

🥂 *7–20 years*

WHITE Just about 25 per cent of the production is white, and it is not very well distributed.

🍷 *Chardonnay, Pinot Blanc*

🥂 *4–8 years*

✓ *Capitain-Gagnerot • Edmond Cornu & Fils • François Gay & Fils • Chevalier Pére & Fils • Henri Naudin-Ferrand*

LADOIX PREMIER CRU AOC

This appellation includes the following *premiers crus*: Basses Mourottes, Bois Roussot, Les Buis, Le Clou d'Orge, La Corvée, Les Gréchons et Foutriéres, *Hautes Mourottes*, Les Joyeuses, La Micaude, En Naget, and *Le Rognet et Corton*. These *premier cru* vineyards were expanded from 14 to 24 hectares (35 to 59 acres) in 2000.

RED These wines are decidedly finer in quality and deeper in colour than those with the basic village appellation.

🍷 *Pinot Noir, Pinot Gris, Pinot Blanc, Chardonnay*

🥂 *7–20 years*

WHITE Rarely seen.

✓ *Ambroise • Capitain-Gagnerot • Edmond Cornu & Fils • Faiveley • Henri Naudin-Ferrand • Nudant • G & P Ravaut*

LADOIX-CÔTE DE BEAUNE AOC

This is an alternative appellation for red wines only. *See* Ladoix AOC.

MARANGES AOC

In 1989 Maranges AOC replaced three separate appellations: Cheilly-lès-Maranges AOC, Dézize-lès-Maranges AOC, and Sampigny-lès-Maranges AOC. This trio of villages had previously shared the once moderately famous *cru* of Marange, which is on a well-exposed hillside immediately southwest of Santenay. The red wines may also be sold as Côte de Beaune AOC or Côte de Beaune-Villages AOC. Production used to be erratic, with most of the wine wine (including much of that qualifying for *premier cru*) sold to *négociants* for blending into Côte de Beaune-Villages, but some dedicated growers managed to get the name around.

RED Wines with a very pure Pinot perfume, which are developing good colour and body.

🍷 *Pinot Noir, Pinot Gris, Pinot Liébault*

🥂 *2–7 years*

WHITE Rarely produced in any of the three villages and never in Dézize-lès-Maranges, as far as I am aware, although it is allowed.

✓ *Chevrot & Fils • Roger Belland • Jaffelin • Vincent Girardin • Bachelet-Monnot • Claude Nouveau*

MARANGES CÔTE DE BEAUNE AOC

This is an alternative appellation for red wines only. *See* Maranges AOC.

MARANGES PREMIER CRU AOC

This appellation includes the following *premiers crus:* Clos de la Boutière, La Croix Moines, La Fussière, Clos de la Fussiére, Le Clos des Loyères, Le Clos des Rois, Les Clos Roussots. Some of these *climats* officially designated as *premiers crus* are a bit of a puzzle. Clos de la Boutière, for example, was originally just plain old La Boutière, and Les Clos Roussots once adjoined a *premier cru* called Les Plantes de Marange, but was never classified as one itself. According to maps prior to the merging of the three appellations, Les Plantes de Marange, Maranges, and En Maranges were the authentic names of the most important vineyards in the area classified as *premier cru*, but they have since adopted less repetitive *lieux-dits*, and such revisionism is probably justified from a marketing aspect.

RED The best examples of these wines are well coloured, with a good balance of fruit, often red fruits, and a richer, longer finish than those wines bearing the basic Maranges appellation.

🍷 *Pinot Noir, Pinot Gris, Pinot Liébault*

WHITE Rarely encountered.

🍷 *Chardonnay, Pinot Blanc*

✓ *Vincent Bachelet • Chevrot & Fils • Roger Belland • Bachelet-Monnot • Claude Nouveau*

MEURSAULT AOC

The greatest white Côte de Beaune is either Montrachet or Corton-Charlemagne, yet Meursault is probably better known and is certainly more popular.

RED This is often treated as a novelty, but it is a fine wine in its own right, with a firm edge.

🍷 *Pinot Noir, Pinot Gris, Pinot Blanc, Chardonnay*

🥂 *8–20 years*

WHITE Even the most basic Meursault should be deliciously dry with a nutty-buttery-spice quality added to its typically rich flavour.

🍷 *Chardonnay, Pinot Blanc*

🥂 *5–12 years*

✔ *Robert Ampeau & Fils • d'Auvenay ❽ • Ballot-Millot • Bouchard Père & Fils • Boyer-Martenot • Alain Coche-Bizouard • J-F Coche-Dury • Vincent Dancer • Jean-Philippe Fichet • Henri Germain (Limozin) • Albert Grivault • Patrick Javillier • Antoine Jobard • Rémy Jobard • Comte Lafon ❽ • Michelot • François Mikulski • Pierre Morey ❽ • Guy Roulot*

MEURSAULT PREMIER CRU AOC

This appellation includes the following *premiers crus*: Aux Perrières, Les Bouchères, Les Caillerets, Les Charmes-Dessous (Charmes), Les Charmes-Dessus (Charmes), Les Chaumes de Narvaux (Genevrières), Les Chaumes des Perrières (Genevrières), Les Cras, Les Genevrières-Dessous (Genevrières), Les Genevrières-Dessus (Genevrières), Les Gouttes d'Or, La Jeunelotte, Clos des Perrières, Les Perrières-Dessous (Perrières), Les Perrières-Dessus (Perrières), La Pièce sous le Bois, Les Plures (Santenots), Le Porusot, Les Porusot-Dessous (Porusot), Le Porusot-Dessus (Porusot), Clos des Richemont (Les Cras), Les Santenots Blancs (Santenots), Les Santenots du Milieu (Santenots), Sous Blagny, and Sous le Dos d'Âne.

RED Finer and firmer than the basic village wines, these reds need plenty of time to soften.

🍷 *Pinot Noir, Pinot Gris, Pinot Blanc, Chardonnay*

🥂 *10–20 years*

WHITE Great Meursault should always be rich. Their various permutations of nutty, buttery, and spicy Chardonnay flavours may often be submerged by the honey, cinnamon, and vanilla of new oak until considerably mature.

🍷 *Chardonnay, Pinot Blanc*

🥂 *6–15 years*

✔ *D'Auvenay ❽ • Ballot-Millot • Yves Boyer-Martenot • Michel Bouzereau & Fils • J-F Coche-Dury • Vincent Dancer • Henri Germain & Fils • Vincent Girardin • Albert Grivault (Clos des Perrières) • Patrick Javillier • Antoine Jobard • Comte Lafon ❽ • Vincent Latour • Jean Latour Labille • Leroy ❽ • Martelet de Cherisey (Meursault-Blagny) • Matrot • Mazilly Père & Fils (Les Meurgers) • Michelot • Pierre Morey ❽ • Alain Patriarche (Les Grands Charrons) • Jacques Prieur • François Mikulski • Remoissenet Père & Fils • Guy Roulot*

MEURSAULT-BLAGNY PREMIER CRU AOC

This appellation includes the following *premiers crus*: La Jeunelotte, La Pièce sous le Bois, Sous Blagny, Sous le Dos d'Âne. An alternative appellation for Meursault wines from vineyards in the neighbouring village of Blagny. The wines must be white, otherwise they claim the Blagny Premier Cru appellation. *See* Meursault Premier Cru AOC.

MEURSAULT-CÔTE DE BEAUNE AOC

Alternative red-wine-only appellation for Meursault. *See* Meursault AOC.

MEURSAULT-SANTENOTS AOC

This is an alternative appellation for Meursault Premier Cru that comes from a part of the Volnay-Santenots appellation. *See* Volnay-Santenots AOC.

MONTHÉLIE AOC

Monthélie's wines, especially the *premiers crus*, are probably the most underrated in Burgundy.

RED These excellent wines have a vivid colour, expressive fruit, a firm structure, and a lingering, silky finish.

🍷 *Pinot Noir, Pinot Gris, Pinot Blanc, Chardonnay*

🥂 *7–15 years*

WHITE Relatively little white wine is produced.

🍷 *Chardonnay, Pinot Blanc*

🥂 *3–7 years*

✔ *Eric Boigelot • Denis Boussey • Eric Boussey • J-F Coche-Dury • Paul Garaudet • Comte Lafon ❽ • Monthélie-Douhairet Porcheret • Remi Jobard*

MONTHÉLIE PREMIER CRU AOC

This appellation includes the following *premiers crus*: Le Cas Rougeot, Les Champs Fulliots, Les Duresses, Le Château Gaillard, Clos des Toisiéres, Le Clos Gauthey, Le Clou des Chênes, Les Clous, Le Meix Bataille, Les Riottes, Sur la Velle, La Taupine, Les Vignes Rondes, Le Village, and Les Barbières.

RED Monthélie's *premiers crus* are hard to find, but worth the effort.

🍷 *Pinot Noir, Pinot Gris, Pinot Blanc, Chardonnay*

🥂 *8–20 years*

WHITE Paul Garaudet's delicately perfumed Champs-Fulliot blanc is one of the few white wines that come from this appellation.

✔ *Eric Boigelot • Denis Boussey (Les Champs Fulliots) • Changarnier • Gérard Doreau • Paul Garaudet • Château de Monthélie ❽*

MONTHELIE-CÔTE DE BEAUNE AOC

This is an alternative appellation for red wines only. *See* Monthélie AOC.

MONTRACHET AOC

Grand Cru

Many consider Montrachet to be the greatest dry white wine in the world. Growers are convinced the quality of Montrachet is not only exceptional, but the *terroir* is also capable of producing the same high-class grape even in lesser vintages.

WHITE When it is fully mature, Montrachet has the most glorious and expressive character of all dry white wines. Its honeyed, toasty, floral, nutty, creamy, and spicy aromas are simply stunning.

🍷 *Chardonnay*

🥂 *10–30 years*

✔ *Guy Amiot & Fils • Amiot-Bonfils • Bouchard Père & Fils • Pierre-Yves Colin-Morey • Marc Colin • Louis Jadot • Comte Lafon ❽ • Leflaive ❽ • Marquis de Laguiche (made and sold by Joseph Drouhin) • Jacques Prieur • De la Romanée-Conti*

PERNAND-VERGELESSES AOC

This village, near Aloxe-Corton, is the most northerly appellation of the Côte de Beaune.

RED With the exception of the silky wines recommended below, too many of these are rustic and overrated and would be better off in a *négociant* Côte de Beaune-Villages blend.

🍷 *Pinot Noir, Pinot Gris, Pinot Blanc, Chardonnay*

🥂 *7–15 years*

WHITE Although this village is famous for its Aligoté, growers can also produce smooth and well-balanced wines of Chardonnay.

🍷 *Chardonnay, Pinot Blanc*

🥂 *4–8 years*

✔ *Denis Père & Fils • P Dubreuil-Fontaine • Denis Père & Fils • Olivier Leflaive*

THE HOSPICES DE BEAUNE, ALSO KNOWN AS HOTEL-DIEU DE BEAUNE
The wines produced from the vineyards of this former charitable almshouse are sold at an annual charity wine auction held every November.

THE HOSPICES DE BEAUNE LABEL

This distinctive label indicates that the wine comes from vineyards belonging to the Hospices de Beaune, a charitable institution that has cared for the sick and poor of Beaune since 1443. Half a millennium of gifts and legacies has seen the accumulation of vineyards that now total some 62 hectares (153 acres) of *premiers crus* and *grands crus*.

After some criticism of the unreliability of certain wines, a new *cuverie* was built at the rear of the famous Hôtel-Dieu, known for its magnificent Flemish-crafted roof. All the wines are now matured in new oak casks. There remain, however, the variations that result from the different *élevages* of the various casks of the same *cuvée*. At the most innocent level,

these may result from one *négociant* giving his wine more or less cask-maturation than another, and temperature and humidity levels can also radically change the wine's alcohol and extract content. The cellar management of less scrupulous firms might also be a consideration.

Since 1859 these wines have been sold by auction and, because of the publicity gained by this annual event, the prices fetched are now generally much higher than the going rate. We consumers must be prepared to pay a relatively higher price for these wines and help the cause.

After the wines have been auctioned, they become the full responsibility of the purchaser.

THE HOSPICES DE BEAUNE *CUVÉES*

RED WINES

- **Cuvée Clos des Avaux**
  AOC Beaune *premier cru*
  Unblended Les Avaux

- **Cuvée Baronne du Bay**
  AOC Corton *grand cru*
  Unblended Clos du Roi

- **Cuvée Billardet**
  AOC Pommard
  A blend of Les Noizons,
  Les Arvelets, Les Cras

- **Cuvée Jean-Luc Bissey**
  AOC Echezeaux *grand cru*
  Unblended Les Echezeaux
  Dessus

- **Cuvée Blondeau**
  AOC Volnay *premier cru*
  A blend of Les Champans,
  En Taille Pieds, Les Ronceret,
  Les Mitans

- **Cuvée Boillot**
  AOC Auxey-Duresses *premier cru*
  Unblended Les Duresses

- **Cuvée Brunet**
  AOC Beaune *premier cru*
  A blend of Les Bas des Teurons,
  Les Bressandes, Les Cents
  Vignes

- **Cuvée Suzanne Chaudron**
  AOC Pommard
  A blend of Pommard

- **Cuvée Madeleine Collignon**
  AOC Mazis-Chambertin *grand cru*
  Unblended Mazis-Chambertin

- **Cuvée Raymond Cyrot**
  AOC Pommard
  A blend of Pommard and
  Pommard *premier cru*

- **Cuvée Cyrot-Chaudron**
  AOC Beaune *premier cru*
  Unblended Les Montrevenots

- **Cuvée Cyrot-Chaudron**
  AOC Clos de La Roche *grand cru*
  Unblended Les Froichots

- **Cuvée Pierre Floquet**
  AOC Beaune *premier cru*
  Unblended Les Gréves

- **Cuvée Georges Kritter**
  AOC Clos de La Roche *grand cru*
  Unblended Clos de la Roche

- **Cuvée Dames de la Charité**
  AOC Pommard *premier cru*
  A blend of Les Petits Epenots, Les
  Rugiens, La Refène, Les Combes
  Dessus

- **Cuvée Dames Hospitalières**
  AOC Beaune *premier cru*
  A blend of Les Bressandes, La
  Mignotte, Les Teurons

- **Cuvée Maurice Drouhin**
  AOC Beaune *premier cru*
  A blend of Les Avaux, Les
  Grèves, Les Boucherottes,
  Champs Pimont

- **Cuvée Charlotte Dumay**
  AOC Corton *grand cru*
  A blend of Renardes, Les
  Bressandes

- **Cuvée Forneret**
  AOC Savigny-lès-Beaune
  premier cru
  A blend of Les Vergelesses, Bas
  Vergelesses

- **Cuvée Fouquerand**
  AOC Savigny-lès-Beaune
  A blend of Les Talmettes, Aux
  Gravains, Les Serpentières

- **Cuvée Christine Friedberg**
  AOC Santenay
  Blended Santenay

- **Cuvée Gauvain**
  AOC Volnay *premier cru*
  Unblended Les Santenots
  du Millieu

- **Cuvée Arthur Girard**
  AOC Savigny-lès-Beaune
  premier cru
  A blend of Les Peuillets, Les Bas
  Marconnets

- **Cuvée Dom Goblet**
  AOC Pommard *premier cru*
  Unblended Les Petits-Epenots

- **Cuvée Guigone De Salins**
  AOC Beaune *premier cru*
  A blend of Les Bressandes,
  En Seurey, Champs Pimont

- **Cuvée Hugues Et Louis Bétault**
  AOC Beaune *premier cru*
  A blend of Les Grèves,
  La Mignotte, Les Aigrots

- **Cuvée Lebelin**
  AOC Monthélie
  Unblended Les Duresses

- **Cuvée Jehan De Massol**
  AOC Volnay-Santenots
  premier cru
  Unblended Les Santenots

- **Cuvée Général Muteau**
  AOC Volnay *premier cru*
  A blend of Volnay-le-Village,
  Carelle sous la Chapelle,
  En Cailleret Dessus, Les Frémiets,
  Taille Pieds

- **Cuvée Docteur Peste**
  AOC Corton *grand cru*
  A blend of Les Bressandes,
  Les Chaumes et la Voierosse,
  Les Grèves

- **Cuvée Rameau-Lamarosse**
  AOC Pernand-Vergelesses
  premier cru
  Unblended Les Basses
  Vergelesses

- **Cuvée Nicolas Rolin**
  AOC Beaune *premier cru*
  A blend of Les Cents Vignes,
  Les Grèves, En Genêt, Les
  Teurons, Les Bressandes

- **Cuvée Rousseau-Deslandes**
  AOC Beaune *premier cru*
  A blend of Les Cent Vignes,
  Les Montrevenots, La Mignotte

WHITE WINES

- **Cuvée De Bahèzre De Lanlay**
  AOC Meursault-Charmes
  premier cru
  A blend of Les Charmes Dessus,
  Les Charmes Dessous

- **Cuvée Baudot**
  AOC Meursault-Genevrières
  premier cru
  A blend of Genevrières Dessus,
  Les Genevrières Dessous

- **Cuvée Philippe Le Bon**
  AOC Meursault-Genevrières
  premier cru
  A blend of Genevrières Dessus,
  Les Genevrières Dessous

- **Cuvée Jean-Brocard**
  AOC Chablis *premier cru*
  Unblended Cote de Lechet

- **Cuvée Paul Chanson**
  AOC Corton-Vergennes *grand cru*
  Unblended Corton-Vergennes

- **Cuvée Bernard Clerc**
  AOC Puligny-Montrachet
  Unblended Les Reuchaux

- **Cuvée Dames de Flandres**
  AOC Bâtard-Montrachet
  grand cru
  Unblended Bâtard-Montrachet

- **Cuvée Goureau**
  AOC Meursault
  A blend of Les Peutes Vignes,
  Les Grands Charrons, Les Cras

- **Cuvée Albert Grivault**
  AOC Meursault-Charmes
  premier cru
  Unblended Les Charmes Dessus

- **Cuvée Jehan Humblot**
  AOC Meursault *premier cru*
  Unblended Les Porusots

- **Cuvée Loppin**
  AOC Meursault
  A blend of Les Criots, Les Cras

- **Cuvée Joseph Menault**
  AOC Saint-Romain
  A blend of Le Village Haut,
  Sous la Velle

- **Cuvée Docteur Peste**
  AOC Corton *grand cru*
  A blend of Chaumes, Voierosses,
  Les Grèves, Les Bressandes

- **Cuvée Françoise Poisard**
  AOC Pouilly-Fuissé
  A blend of Les Plessys,
  Les Robées, Les Chevrières

- **Cuvée du Roi Soleil**
  AOC Corton-Charlemagne
  grand cru
  Unblended Les Renardes

- **Cuvée Françoise-De-Salins**
  AOC Corton-Charlemagne
  grand cru
  Unblended Le Charlemagne

- **Cuvée Suzanne Et Raymond**
  AOC Beaune *premier cru*
  Unblended Les Montrevenots

PERNAND-VERGELESSES-CÔTE DE BEAUNE AOC

This is an alternative appellation for red wines only. *See* Pernand-Vergelesses AOC.

PERNAND-VERGELESSES PREMIER CRU AOC

This appellation includes the following *premiers crus:* En Caradeux, Creux de la Net, Les Fichots, Vergelesses, Ile des Vergelesses, Clos Berthet, Sous Frétille, and Village de Pernand.

RED These wines repay their cost, keeping until the fruit develops a silkiness that hangs gracefully on the wine's structure and gives Pernand's *premiers crus* the class its village wines lack.

🍷 *Pinot Noir, Pinot Gris, Pinot Blanc, Chardonnay*

🍷 *10–20 years*

WHITE The Beaune firm of Chanson Père & Fils produces a consistent wine of medium body, which is dry but mellow. Pavelot, however, is the best white *premier cru* from this village that I have tasted.

🍷 *Chardonnay, Pinot Blanc*

🍷 *4–8 years*

✓ *Delarche • P Dubreuil-Fontaine • Roger Jaffelin & Fils (Creux de la Net) • Pavelot • Rapet Père & Fils • Rollin Père & Fils*

POMMARD AOC

A very famous village with a "reborn" image built up by a group of dedicated and skilful winemakers.

RED The "famous" dark, alcoholic, and soupy wines of Pommard are now mostly a thing of the past, having been replaced by exciting fine wines.

🍷 *Pinot Noir, Pinot Gris, Pinot Blanc, Chardonnay*

🍷 *8–16 years*

✓ *Robert Ampeau & Fils • Comte Armand • Billard-Gonnet • Jean-Marc Boillot • Bernard & Louis Glantenay • Leroy* **B** *• Louis Boillot • Hubert de Montille • Parent • Château de Pommard • Vaudoisey-Creusefond (Croix Blanche)*

POMMARD PREMIER CRU AOC

This appellation includes the following *premiers crus:* Les Arvelets, Les Bertins, Clos Blanc, Les Boucherottes, La Chanière, Les Chanlins-Bas, Les Chaponnières, Les Charmots, Les Combes-Dessus, Clos de la Commaraine, Les Croix Noires, Derrière St-Jean, Clos des Epeneaux, Les Fremiers, Les Grands Epenots, Les Jarolières, En Largillière (or Les Argillières), La Clos Micot, Les Petits Epenots, Les Pézerolles, La Platière, Les Poutures, La Refène, Les Rugiens-Bas, Les Rugiens-Hauts, Les Saussilles, Clos de Verger, and Le Village.

RED The best *crus* are the various climats of Les Rugiens (deep and voluptuous) and Les Epenots (soft, fragrant, and rich).

🍷 *Pinot Noir, Pinot Gris, Pinot Blanc, Chardonnay*

🍷 *10–20 years*

✓ *Comte Armand (Les Petits Epenots Z) • Denis Carré • Billard-Gonnet • Jean-Marc Boillot • De Courcel (Les Vaumuriens) • Vincent Dancer • Lejeune • Leroy* **B** *• Olivier Leflaive • Francois Mikulski • Moissenet Bonnard (Les Epenots) • Hubert de Montille • Pierre Morey* **B** *• Parent • Nicolas Potel • De la Pousse d'Or • De la Vougeraie* **B**

PULIGNY-MONTRACHET AOC

One of two Montrachet villages producing some of the greatest dry whites in the world.

RED Although some fine wines are made, Puligny-Montrachet rouge demands a premium for its scarcity.

🍷 *Pinot Noir, Pinot Gris, Pinot Blanc, Chardonnay*

🍷 *10–20 years*

WHITE Basic Puligny-Montrachet from a top grower is a very high-quality wine: it wil be full bodied, fine, and steely, requiring a few years to develop a nutty honey-and-toast flavour.

🍷 *Chardonnay, Pinot Blanc*

🍷 *5–12 years*

✓ *Robert Ampeau • Jean-Marc Boillot • Jacques Carillon • Francois Carillon • Des Lambrays (Clos du Cailleret) • Leflaive* **B** *• Olivier Leflaive • Étienne Sauzet*

PULIGNY-MONTRACHET PREMIER CRU AOC

This appellation includes the following *premiers crus:* Le Cailleret (or Demoiselles), Les Chalumeaux, Champ Canet, Champ Gain, Au Chaniot (Les Folatiéres), Clavaillon, Les Combettes, Ez Folatières (Les Folatiéres), La Garenne (or Sur la Garenne), Clos de la Garenne, Hameau de Blagny, La Jaquelotte (or Champ Canet), Clos des Meix, Les Perriéres (Clos de la Mouchère), Peux Bois (Les Folatiéres), Les Pucelles, Les Referts, En la Richarde (Les Folatiéres), Sous le Courthil (Les Chalumaux), Sous le Puits, and La Truffière.

RED A lot less than one per cent of the production is red wine. It is difficult to encounter any.

🍷 *Pinot Noir, Pinot Gris, Pinot Blanc, Chardonnay*

WHITE A *premier cru* Puligny by a top grower such as Étienne Sauzet is one of the most flavour-packed taste experiences imaginable.

🍷 *Chardonnay, Pinot Blanc*

🍷 *7–15 years*

✓ *Guy Amiot & Fils • Robert Ampeau • D'Auvenay* **B** *• Jean-Marc Boillot • Jean-Claude Boisset • Michel Bouzereau & Fils (Les Champs Gains) • Jacques Carillon • Joseph Drouhin • Louis Jadot (La Garenne) • Leflaive* **B** *• Jean Chartron • Olivier Merlin* **O** *• Martelet de Cherisey • Hubert de Montille • Marc Morey • Jacques Prieur • Étienne Sauzet*

PULIGNY-MONTRACHET-CÔTE DE BEAUNE AOC

This is an alternative appellation for red wines only. *See* Puligny-Montrachet AOC.

ST-AUBIN AOC

This underrated village has many talented winemakers and is an excellent source for good-value wines.

RED Delicious, ripe but light, fragrant, and fruity red wines that quickly develop a taste of wild strawberries.

🍷 *Pinot Noir, Pinot Gris, Pinot Blanc, Chardonnay*

🍷 *4–8 years*

THE COURTYARD OF THE CHÂTEAU DE POMMARD LEADS TO ITS ENCLOSED VINEYARD
Established in 1726, the property features two châteaux, gardens, and a walled vineyard, Clos Marey-Monge, which encompasses approximately 20 hectares (50 acres).

WHITE Super-value white wines – a sort of "Hautes-Côtes Montrachet".

🍷 *Chardonnay, Pinot Blanc*

🥂 *3-8 years*

✓ *Jean-Claude Bachelet • Françoise & Denis Clair • De Villaine • Marc Colin & Fils • Hubert Lamy*

ST-AUBIN-CÔTE DE BEAUNE AOC

This is an alternative appellation for red wines only. *See* St-Aubin AOC.

ST-AUBIN PREMIER CRU AOC

This appellation includes the following *premiers crus:* Le Bas de Gamay à l'Est, Bas de Vermarain à l'Est, Les Castets, Les Champlots, Es Champs, Le Charmois, La Chatenière, Les Combes, Les Combes au Sud, Les Cortons, En Créot, Derrière chez Edouard, Derrière la Tour, Echaille, Les Frionnes, Sur Gamay, Marinot, En Montceau, Les Murgers des Dents de Chien, Les Perrières, Pitangeret, Le Puits, En la Ranché, En Remilly, Sous Roche Dumay, Sur le Sentier du Clou, Les Travers de Marinot, Vignes Moingeon, Village, and En Vollon à l'Est. The best of these *premiers crus* are Les Frionnes and Les Murgers des Dents de Chien, followed by La Chatenière, Les Castets, En Remilly, and Le Charmois.

RED Very appealing strawberry and oaky-vanilla wines that are delicious young, yet improve further with age.

🍷 *Pinot Noir, Pinot Gris, Pinot Blanc, Chardonnay*

🥂 *5-15 years*

WHITE These dry wines are often superior to the village wines of Puligny-Montrachet, and they will always be much cheaper.

🍷 *Chardonnay, Pinot Blanc*

🥂 *4-10 years*

✓ *Guy Amiot & Fils • Jean-Claude Bachelet • Marc Colin & Fils • Vincent Girardin • Hubert Lamy • Lamy-Pillot • Patrick Miolane • Bernard Moreau • Marc Morey & Fils • Henri Prudhon & Fils • Gérard Thomas & Filles*

ST-ROMAIN AOC

A little village amid picturesque surroundings in the hills above Auxey-Duresses.

RED Good-value, medium-bodied, rustic reds that have a good, characterful flavour.

🍷 *Pinot Noir, Pinot Gris, Pinot Blanc, Chardonnay*

🥂 *4-8 years*

WHITE Fresh and lively, light- to medium-bodied dry white wines of an honest Chardonnay style.

🍷 *Chardonnay, Pinot Blanc*

🥂 *3-7 years*

✓ *Ambroise Bertrand • De Chassorney • Joseph Drouhin • Alain Gras • Rapet Francois & Fils*

ST-ROMAIN-CÔTE DE BEAUNE AOC

This is an alternative appellation for red wines only. *See* St-Romain AOC.

SANTENAY AOC

This most southerly village appellation of the Côte d'Or (but not of the Côte de Beaune) is a reliable source of good-value Burgundy.

RED These wines are fresh and frank, with a clean rendition of Pinot Noir fruit supported by a firm structure.

🍷 *Pinot Noir, Pinot Gris, Pinot Blanc, Chardonnay*

🥂 *7-15 years*

WHITE Only two per cent of Santenay is white, but some good buys can be found among the top growers.

🍷 *Chardonnay, Pinot Blanc*

🥂 *4-8 years*

✓ *Guy Amiot & Fils • Roger Belland • Marc Colin & Fils • Vincent Girardin • Alain Gras • Prieur-Brunet*

SANTENAY PREMIER CRU AOC

This appellation includes the following *premiers crus:* Beauregard, Beaurepaire, La Comme, La Comme Dessus (Beauregard), Clos Faubard, Les Fourneaux, Grand Clos Rousseau, Les Gravières, La Maladière, Clos des Mouches, Passetemps, Clos Rousseau, and Les-Gravières-Clos de Tavannes. The best are Clos de Tavannes, Les Gravières, La Maladière, and La Comme Dessus.

RED In the pure and frank mould of Pinot Noir wines, but with an added expression of *terroir*.

🍷 *Pinot Noir, Pinot Gris, Pinot Blanc, Chardonnay*

🥂 *6-15 years*

WHITE Rarely encountered.

🍷 *Chardonnay, Pinot Blanc*

🥂 *5-10 years*

✓ *Roger Belland • Françoise & Denis Clair • Vincent Girardin • Alain Gras (also sold as René Gras-Boisson) • Lecquin-Colin (Vieilles Vignes) • Bernard Moreau • Lucien Muzard & Fils • De la Pousse d'Or • Jean-Marc Vincent (Beaurepaire, Passetemps)*

SANTENAY-CÔTE DE BEAUNE AOC

This is an alternative appellation for red wines only. *See* Santenay AOC.

SAVIGNY AOC

See Savigny-lès-Beaune AOC

SAVIGNY PREMIER CRU AOC

See Savigny-lès-Beaune Premier Cru AOC

SAVIGNY-CÔTE DE BEAUNE AOC

This is an alternative appellation for red wine only. *See* Savigny-lès-Beaune AOC.

SAVIGNY-LÈS-BEAUNE AOC

This village has gifted winemakers producing very underrated and undervalued wines. Their popularity seems to be growing these days, however.

RED Delicious, easy-to-drink, medium-bodied wines that are very soft and Beaune-like in style.

🍷 *Pinot Noir, Pinot Gris, Pinot Blanc, Chardonnay*

🥂 *7-15 years*

WHITE Some excellent dry wines with good concentration of flavour, a smooth texture, and some finesse, but they are difficult to find.

🍷 *Chardonnay, Pinot Blanc*

🥂 *4-10 years*

✓ *Robert Ampeau • Simon Bize & Fils • Bouchard Père & Fils • Chandon de Briailles • Philippe Delagrange (blanc) • Pierre Guillemot • Lucien Jacob • Catherine & Claude Maréchal • Parent • Jean-Marc Pavelot • Jean Marc Naudin • Rollin Père & Fils • Tollot-Beaut*

SAVIGNY-LÈS-BEAUNE PREMIER CRU AOC

This appellation includes the following *premiers crus:* Aux Clous, Aux Fournaux, Aux Gravains, Aux Guettes, Aux Serpentières, Les Hauts Marconnets (or Les Marconnets), Basses Vergelesses, Bataillère (or Aux Vergelesses), Champ Chevrey (or Aux Fournaux), Les Charnières, Hauts Jarrons (or Les Hauts Jarrons) Bas Marconnets (or Les Marconnets), Les Jarrons (La Dominode), Les Lavières, Les Narbantons, Petits Godeaux, Les Peuillets, Redrescut (or Redrescul), Les Rouvrettes, and Les Talmettes.

RED These wines have a very elegant, soft, and stylish Pinot flavour that hints of strawberries, cherries, and violets. The best are: Les Lavières, La Dominode, Aux Vergelesses, Les Marconnets, and Aux Guettes.

🍷 *Pinot Noir, Pinot Gris, Pinot Blanc, Chardonnay*

🥂 *7-20 years*

WHITE For whites, Domaine des Terregelesses produces a splendidly rich, dry Les Vergelesses.

🍷 *Chardonnay, Pinot Blanc*

🥂 *5-15 years*

✓ *Simon Bize & Fils • Bouchard Père & Fils • Chandon de Briailles • Chanson Père & Fils • Bruno Clair • Maurice Ecard • François Gay & Fils • A-F Gros (Clos des Guettes) • Louis Jadot • Leroy 🅑 • Albert Morot (La Bataillère aux Vergelesses) • Olivier (Les Peuillets) • Pavelot • Des Terregelesses (Les Vergelesses) • Tollot-Beaut*

SAVIGNY-LÈS-BEAUNE-CÔTE DE BEAUNE AOC

This is an alternative appellation for red wines only. *See* Savigny-lès-Beaune AOC.

VOLNAY AOC

Volnay ranks in performance with such great *crus* as Gevrey-Chambertin and Chambolle-Musigny. It is the most southerly red-wine-only appellation in the Côte d'Or, and the only great wine village located above its vineyards.

RED These wines are not cheap, but they are firm and well coloured with more silky finesse than should be expected from a village appellation.

🍷 *Pinot Noir, Pinot Gris, Pinot Blanc, Chardonnay*

🥂 *6-15 years*

✓ *Marquis d'Angerville • Michel Lafarge • Régis Rossignol • Bitouzet Prieur*

VOLNAY PREMIER CRU AOC

This appellation includes the following *premiers crus:* Les Angles, Les Aussy, La Barre, Bousse d'Or (or Clos de la Bousse d'Or), Les Brouillards, En Cailleret, Les Caillerets, Cailleret Dessus (part of which may be called Clos des 60 Ouvrées), Carelles Dessous, Carelle sous la Chapelle, Clos de la Caves de Ducs, En Champans (or Champans), Chanlin, En Chevret, Clos de la Chapelle, Clos des Chênes, Clos de Ducs, Clos du Château des Ducs, Frémiets (or Clos de la Rougeotte), La Gigotte, Les Grands Champs, Lassolle, Les Lurets, Les Mitans, En l'Ormeau, Pitures Dessus, Pointes d'Angles, Robardelle, Le Ronceret, Taille Pieds, En Verseuil (or Clos du Verseuil), and Le Village.

RED No *grands crus*, but its silky-smooth and fragrant *premiers crus* are great wines, showing tremendous finesse. The best are: Clos des Chêne, Taille Pieds, Bousse d'Or, Clos de Ducs, the various climats of Cailleret, Clos des 60 Ouvrées, and En Champans.

🍷 *Pinot Noir, Pinot Gris, Pinot Blanc, Chardonnay*

🥂 *8-20 years*

✓ *Marquis d'Angerville • Lucien Boillot • Jean-Marc Bouley (Les Carelles) • Vincent girardin • Louis Jadot • Michel Lafarge • Comte Lafon 🅑 • Olivier Leflaive • Leroy 🅑 • Hubert de Montille • Nicolas Potel • De la Pousse d'Or • Rossignol-Changarnier*

VOLNAY-SANTENOTS PREMIER CRU AOC

This confusing appellation is in Meursault, not Volnay, although it does run up to the boundary of that village. It dates back to the 19th century, when the Meursault lieu-dit of Les Santenots du Milieu became famous for its red wines. White wines cannot be called Volnay-Santenots and must be sold as Meursault or, if produced from the two-thirds of this vineyard furthest from the Volnay border, they may be sold as Meursault Premier Cru AOC or Meursault-Santenots. The right to the Volnay-Santenots appellation was accorded by Tribunal at Beaune in 1924.

RED Not often found, these are similar to Volnay with good colour and weight, but can lack its silky elegance.

🍷 *Pinot Noir, Pinot Gris, Pinot Blanc, Chardonnay*

🥂 *8-20 years*

✓ *Lucien Boillot • Comte Lafon 🅑 • Leroy 🅑 • Matrot • François Mikulski • Jacques Prieur • Ropiteau Fréres*

The Côte Chalonnaise

The Côte Chalonnaise might be a simple district in wine terms. It has just five appellations, two of which are exclusively white and three that are red or white, but the quality of wines is very good. The time when more public attention is paid to the region seems to be coming. A nice bottle of Côte Chalonnaise is outstanding value for money.

Once the forgotten area of Burgundy, the Côte Chalonnaise, or Région de Mercurey, as it is sometimes called, was long perceived as too serious for its own good. Because its flavour-some reds and buttery whites have more in common with the wines of the Côte de Beaune than elsewhere, merchants categorized them as "inferior" or "pretentious".

Perhaps the Côte Chalonnaise would not have been forgotten had merchants thought of it more as a superior Mâconnais than as an inferior Côte de Beaune. Over the past 20 years, however, as merchants across the world have become more willing to seek out lesser-known wines, the area has built up a reputation as one of Burgundy's best sources of quality wines.

EXCELLENT *NÉGOCIANTS*

The Burgundy drinker is blessed with a fine choice of *négociants* in the Côte Chalonnaise. The list of excellent *négociants* includes Chandesais, Delorme, and Faiveley. There is also a good *coopéra-tive* at Buxy and an increasing number of talented growers. This area produces fine Crémant de Bourgogne and, in Bouzeron, has the only single-village appellation of Aligoté wine.

VINE-GROWING ZONES OF THE CÔTE CHALONNAISE
(*see also* p213)
The vine-growing zones form three separate "islands" west of Chalon between the Côte de Beaune to the north and the Mâconnais to the south.

[Map of the Côte Chalonnaise showing places: CÔTE-D'OR, Santenay, Chagny, Bouzeron, Rully, Chassey-le-Camp, Rully, Saint-Maurice-lès-Couches, Nyon, Couches, Bourgogne Côtes du Couchois, Le Breuil, Mercurey, Saint-Martin-sous-Montaigu, Bourgogne Côte Chalonnaise, Saint-Denis-de-Vaux, Dracy-le-Fort, Orbise, Chalon-sur-Saône, Givry, Jambles, Givry, Sainte-Hélène, Saint-Désert, Moroges, SAÔNE-ET-LOIRE, Bissey-sous-Cruchaud, Buxy, Montagny-lès-Buxy, Saint-Germain-lès-Buxy, Montagny, Saint-Micaud, Jully-lès-Buxy, Saint-Vallerin, Grosne]

0 mi — 3
0 km — 3

FACTORS AFFECTING TASTE AND QUALITY

LOCATION
These three islands of vines that make up the Côte Chalonnaise, are situated to the west of Châlon-sur-Saône, 350 kilometres (217 miles) south east of Paris, between the Côte de Beaune in the north and the Mâconnais in the south.

CLIMATE
The climate of the Côte Chalonnaise is lightly drier than that of the Côte d'Or, with many of the best slopes protected from the worst ravages of frost and hail.

ASPECT
This is a disjointed district in which the great plateau of the Côte d'Or peters out into a complex chain of small hills with vines clinging to the most favourable slopes, at an altitude of between 230 and 320 metres

(750 to 1,050 feet), in a far more sporadic fashion than those in the Côte d'Or.

SOIL
Limestone subsoil are covered with clay-sand topsoils that are sometimes enriched with iron deposits. At Mercurey there are limestone ooliths mixed with iron-enriched marl.

VITICULTURE & VINIFICATION
The wines are produced in an identical way to those of the Côte de Beaune, with no exceptional viticultural or vinification techniques involved (*see* p229).

GRAPE VARIETIES
Primary varieties: Chardonnay, Pinot Noir
Secondary varieties: Aligoté, Pinot Blanc, Pinot Gris

VINEYARDS CHEQUER THE HILLSIDES OF MONTAGNY-LÈS-BUXY
This commune is part of the Montagny AOC, a white wine appellation with 49 *premier cru* vineyards. Chardonnay is the appellation's main grape variety.

THE APPELLATIONS OF
THE CÔTE CHALONNAISE

BOURGOGNE CÔTE CHALONNAISE AOC

Beginning with the 1990 vintage, basic Bourgogne made exclusively from grapes harvested in the region may bear this specific appellation.

RED A & P de Villaine's Digoine has outstanding fruit and finesse for such a modest appellation. Only Michel Goubard's Mont-Avril comes close in richness and can be truly excellent, but is not in the same class. Both are terrific value.

🍷 *Pinot Noir, Pinot Gris, Chardonnay, Pinot Blanc, César*

🍷 *1–3 years*

WHITE Normally softer and fuller than a Mâcon, but not particularly rich in fruit, except for A & P de Villaine's Clous, which is the best white wine in the appellation, and Venot's La Corvée, the number two.

🍷 *Chardonnay, Pinot Blanc, Pinot Gris*

🍷 *Upon purchase*

ROSÉ Seldom encountered, and not special, although no reason why it should not be successful.

🍷 *Pinot Noir, Pinot Gris, Chardonnay, Pinot Blanc*

🍷 *Upon purchase*

✓ *René Bourgeon • Caves de Buxy • André Delorme • Michel Derain • Michel Goubard (Mont Avril) • Guy Narjoux • Venot (La Corvée) • A & P de Villaine* 🅞

BOURGOGNE CÔTES DU COUCHOIS, OR BOURGOGNE COUCHOIS, AOC

Although the Côtes du Couchois extends across five communes (Dracy-Lès-Couches, Saint-Jean-de-Trézy, Saint-Maurice-Lès-Couches, Saint-Pierre-de-Varennes, and Saint-Cernin-du-Plain) in addition to Couches itself, at just 20 hectares (50 acres) this appellation is nowhere near as large as the Côte Chalonnaise. The wines I have tasted (Domaine des Trois Monts, Bichot's Château de Dracy, and Serge Prost), although restricted in number and style (red only, although white and rosé are also permitted), have not been special.

BOUZERON AOC

In 1979 Bouzeron became the only *cru* to have its own appellation specifically for the Aligoté grape, as Bourgogne Aligoté Bouzeron AOC, and in 1998 it achieved full village status.

WHITE This excellent and interesting dry wine is the fullest version of Aligoté available. In weight, fruit, and spice, its style is nearer to Pinot Gris than to Chardonnay. The A & P de Villaine Bouzeron s a class apart.

🍷 *Aligoté*

🍷 *1–5 years*

✓ *Bougeot • Ancien Carnot (sold by Bouchard Père & Fils) • Chanzy Frères • André Delorme • A & P de Villaine Õ*

GIVRY AOC

Underrated wines from just south of Mercurey.

RED Light- to medium-bodied, soft, and fruity wine with delightful nuances of cherry and redcurrant.

🍷 *Pinot Noir, Pinot Gris, Chardonnay*

🍷 *5–12 years*

WHITE Just 10 per cent of Givry is white – a deliciously clean, dry Chardonnay that can have an attractive spicy-buttery hint on the aftertaste.

🍷 *Chardonnay, Pinot Blanc, Pinot Gris*

🍷 *3–8 years*

✓ *René Bourgeon • Chofflet-Vaudenaire • Michel Derain • Didier Erker (En Chenèvre) • Mme du Jardin • Joblot • Louis Latour • François Lumpp • Parize Père & Fils (Champ Nalot) • Jean-Paul Ragot • R Remoissenet & Fils • Baron Thénard*

GIVRY PREMIER CRU AOC

This appellation includes the following *premiers crus:* Clos de la Barraude, Les Berges, Bois Chevaux, Bois Gauthier, Clos de Cellier aux Moines, Clos Charlé, Clos du Cras Long, Les Grandes Vignes, Grand Prétants, Clos Jus, Clos Marceaux, Marole, Petit Marole, Petit Prétants, Clos St-Paul, Clos St-Pierre, Clos Salomon, Clos de la Servoisine, Vaux, Clos du Vernoy, En Vignes Rouge, and Le Vigron.

RED Best examples are medium bodied, soft, rich, and fruity with delightful nuances of cherry and redcurrant.

🍷 *Pinot Noir, Pinot Gris, Pinot Blanc*

🍷 *5–12 years*

WHITE Deliciously clean, dry Chardonnay that can have an attractive spicy-buttery hint on the aftertaste.

🍷 *Chardonnay, Pinot Blanc, Pinot Gris*

🍷 *3–8 years*

✓ *René Bourgeon • Chofflet-Vaudenaire • Michel Derain • Mme du Jardin • Joblot • Louis Latour • François Lumpp • Parize & Fils • Jean-Paul Ragot • R Remoissenet & Fils • Baron Thénard*

MERCUREY AOC

These wines, including the *premiers crus*, account for two-thirds of the production of the entire Côte Chalonnaise.

RED Medium-bodied with excellent colour and fine varietal character; exceptional quality-for-price ratio.

🍷 *Pinot Noir, Pinot Gris, Chardonnay*

🍷 *5–12 years*

WHITE Dry wines that combine the lightness and freshness of the Mâconnais with some of the fatness and butteriness of the Côte de Beaune.

🍷 *Chardonnay, Pinot Gris*

🍷 *3–8 years*

✓ *Brintet (white Vieilles Vignes) • Château de Chamilly • Louis Desfontaine • Lorenzon • Antonin Rodet*

MERCUREY PREMIER CRU AOC

This appellation includes the following *premiers crus:* La Bondue, Les Byots, La Cailloute, Champs Martins, La Chassière, Le Clos, Clos des Barraults, Clos Château de Montaigu, Clos l'Evêque, Clos des Myglands, Clos du Roi, Clos Tonnerre, Clos Voyens (or Les Voyens), Les Combins, Les Crêts, Les Croichots, Les Fourneaux (or Clos des Fourneaux), Grand Clos Fortoul, Les Grands Voyens, Griffères, Le Levrière, Le Marcilly (or Clos Marcilly), La Mission, Les Montaigus (or Clos des Montaigus), Les Naugues, Les Petits Voyens, Les Ruelles, Sazenay, Les Vasées, and Les Velley.

RED These *premiers crus* have increased in number from 5 to 27 and in area from 15 to over 100 hectares (37 to 250 acres).

🍷 *Pinot Noir, Pinot Gris, Chardonnay*

🍷 *5–15 years*

WHITE I have encountered white Mercurey only at the basic village level.

✓ *Brintet • Faiveley • Jeannin-Nastet • Émile Juillot • Michel Juillot • Lorenzon • Guy Narjoux • François Raquillet • Antonin Rodet (Château de Chamirey)*

MONTAGNY AOC

As all the vineyards in this AOC are *premiers crus*. The only wines that appear under the basic village appellation are those that fail to meet the technical requirement of 11.5 per cent alcohol before chaptalization.

WHITE These dry white wines are good-value, fuller versions of the white Mâcon type.

🍷 *Chardonnay*

🍷 *3–10 years*

✓ *Maurice Bertrand & François Juillot • Caves de Buxy • Faiveley • Louis Latour • Antonin Rodet*

MONTAGNY PREMIER CRU AOC

This appellation includes the following *premiers crus:* Les Bassets, Les Beaux Champs, Les Bonnevaux, Les Bordes, Les Bouchots, Le Breuil, Les Burnins, Les Carlins, Les Champs-Toiseau, Les Charmelottes, Les Chandits, Les Chazelles, Clos Chaudron, Le Choux, Les Clouzeaux, Les Coères, Les Combes, La Condemine, Cornevent, La Corvée, Les Coudrettes, Les Craboulettes, Les Crets, Creux des Beaux Champs, L'Epaule, Les Garchères, Les Gouresses, La Grand Pièce, Les Jardins, Les Las, Les Males, Les Marais, Les Marocs, Les Monts Cuchots, Le Mont Laurent, La Mouillère, Moulin l'Echenaud, Les Pandars, Les Pasquiers, Les Pidans, Les Platières, Les Resses, Les St-Mortille, Les St-Ytages, Sous les Roches, Les Thilles, La Tillonne, Les Treufferes, Les Varignys, Le Vieux Château, Vignes Blanches, Vignes sur le Clou, Les Vignes Couland, Les Vignes Derrière, Les Vignes Dessous, La Vigne Devant, Vignes Longues, Vignes du Puits, Les Vignes St-Pierre, and Les Vignes du Soleil. With every one of its 60 vineyards classified as *premier cru*, Montagny is unique among the villages of Burgundy.

WHITE These delicious dry wines have a Chardonnay flavour that is more akin to that of the Côte de Beaune than it is to Mâconnais.

🍷 *Chardonnay*

🍷 *4–12 years*

✓ *Stéphane Aladame • Arnoux Père & Fils • Caves de Buxy • Château de Davenay (blanc) • Maurice Bertrand & François Juillot • Château de la Saule • Jean Vachet*

RULLY AOC

The Côte Chalonnaise's northernmost appellation produces wines that are closest in character to those from the southern Côte de Beaune.

RED These delightfully fresh and fruity wines of light to medium body and some finesse are uncomplicated when young but develop well.

🍷 *Pinot Noir, Pinot Gris, Chardonnay*

🍷 *5–12 years*

WHITE Serious dry wines that tend to have a crisper balance than wines made farther south in Montagny, although a few can be quite fat.

🍷 *Chardonnay, Pinot Gris*

🍷 *3–8 years*

✓ *Chanzy Frères • André Delorme • Joseph Drouhin • Raymond Dureuil-Janthial • de Folie • des Fromanges • H et P Jacqueson • de la Renard • Antonin Rodet (Château de Rully)*

RULLY PREMIER CRU AOC

This appellation includes the following *premiers crus:* Agneux, Bas de Vauvry, La Bressaude, Champ-Clou, Chapitre, Clos du Chaigne, Clos St-Jacques, Les Cloux (or Cloux), Ecloseaux, La Fosse, Grésigny, Margotey (or Margoté), Marissou, Meix-Caillet, Mont-Palais, Moulesne (or Molesme), Phillot, Les Pieres, Pillot, Préau, La Pucelle, Raboursay (or Rabourcé), Raclot, La Renarde, and Vauvry.

RED Fine-quality, medium-bodied wines with a silky texture added to the summer-fruit flavour.

🍷 *Pinot Noir, Pinot Gris, Chardonnay*

🍷 *5–15 years*

WHITE Generally finer, fuller, and richer dry wines, many with excellent finesse.

🍷 *Chardonnay, Pinot Gris*

🍷 *4–12 years*

✓ *Belleville • Jean-Claude Brelière • Michel Briday • André Delorme • de Folie • Vincent Girardin • H et P Jacqueson • Laborde-Juillot • Albert Sounit*

The Mâconnais

The Mâconnais produces "the world's greatest value pure Chardonnay wines". Although they never quite match the quality produced in the Cote d'Or, their bottles are brilliant value for money. The region vinifies three times more Chardonnay than the rest of Burgundy put together. A new, truly deserved reclassification in Pouilly-Fuissé – upgrading some 22 climats to premier cru rank – is in progress, and should be settled before the 2020 vintage.

The Mâconnais is an ancient viticultural area that was renowned as long as 1,600 years ago when Ausonius, the Roman poet of St-Émilion, mentioned its wines. Today, it makes sense to couple it with the Beaujolais because, although Chardonnay is the dominant white grape in both districts (as it is in the rest of Burgundy), the Gamay is the dominant black grape, which it is not elsewhere; this forms a link between the two. The Mâconnais can be seen as essentially a white-wine producing area, while the Beaujolais is almost entirely red.

MÂCON *ROUGE* – A RELIC OF THE PAST?

Although the Mâconnais is a white-wine district in essence, some 25 per cent of the vines planted here are in fact the purple-skinned Gamay and a further 7.5 per cent are Pinot Noir. The Gamay does not, however, perform very well on the limestone soils of the Mâconnais, and despite the smoothing effect of modern vinification techniques, these wines will always be of rustic quality with a characteristic hard edge.

Theoretically, basic Mâcon *rouge* can be made from either Gamay or Pinot Noir or the blend of the two varieties. In practice, however, growers use Gamay. Yet a red wine from Mâcon followed by the name of a village has to be 100 per cent Gamay. There are a few areas of granite in the very south of the district where good Gamay can be produced with some consistency, and a few villages are even beginning to make something drinkable on limestone – albeit usually only where there is clay or other soils to affect the pH. Gone, too, is the Mâcon Supérieur AOC, which had dropped into almost total disuse.

FACTORS AFFECTING TASTE AND QUALITY

LOCATION
Situated halfway between Lyon and Beaune, the vineyards of the Mâconnais adjoin those of the Côte Chalonnaise to the north and overlap with those of Beaujolais to the south.

CLIMATE
The climate is similar to that of the Côte Chalonnaise, but with a Mediterranean influence gradually creeping in towards the south, so occasional storms are more likely.

ASPECT
The soft, rolling hills in the north of the Mâconnais, which are a continuation of those in the Côte Chalonnaise, give way to a more closely knit topography, with steeper slopes and sharper contours becoming increasingly prominent as one travels farther south into the area that overlaps the Beaujolais.

SOIL
The topsoil of the Mâconnais consists of scree and alluvium or clay and clay-sand and covers a limestone subsoil.

VITICULTURE & VINIFICATION
Some exceptional wines (the Vieilles Vignes Château de Fuissé made in Pouilly-Fuissé, for example) can stand a very heavy oak influence, but most of the whites are fermented in stainless steel and bottled very early to retain as much freshness as possible. The reds are vinified by carbonic maceration, either fully or in part.

GRAPE VARIETIES
Primary varieties: Chardonnay, Gamay
Secondary varieties: Aligoté, Melon de Bourgogne, Pinot Beurot (Pinot Gris), Pinot Blanc, Pinot Noir (restricted to generic Bourgogne AOC)

THE VINE-GROWING AREA OF THE MÂCONNAIS *(see also p213)*
Concentrated to the west of the river Saône, the famous appellations of the Mâconnais spread out over the area to the northwest and interlace with those of Beaujolais to the south.

DOMAINE ROGER LUQUET SAINT-VÉRAN VERS LES MONTS, VINTAGE 2011
The Saint-Véran AOC is a subregion of the Mâconnais , but this white wine appellation straddles both the Mâconnais and the Beaujolais appellations and is split into two "islands" by the Pouilly-Fuissé appellation.

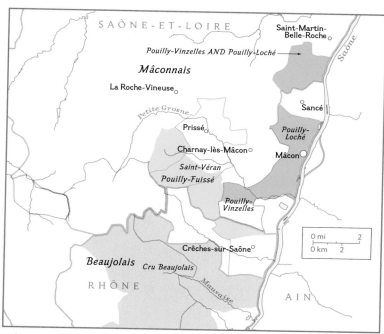

MÂCONNAIS OR BEAUJOLAIS?
The Mâcon and Beaujolais appellations overlap, with parts of Mâcon, Saint-Véran, and Pouilly-Fuissé creeping into red-wine country. Almost half of Saint-Véran is in Beaujolais, going as far as the famous *cru* Beaujolais of Saint-Amour, where the lacework of vineyards dictate whether a grower can produce one or more of Beaujolais, Beaujolais Villages, Saint-Amour, Mâcon, and Saint-Véran appellations.

THE APPELLATIONS OF

THE MÂCONNAIS

MÂCON AOC

Most wines from this district-wide appellation are produced in the area north of the Mâcon-Villages area.

RED Better vinification techniques have improved these essentially Gamay wines, but Gamay is still a grape that does not like limestone.

🍷 *Gamay, Pinot Noir*
🍶 *2–6 years*

WHITE These are basic-quality Chardonnay whites wines that are also fresh, frank, tasty, and dry; they are easy to quaff and superb value. They may be sold as *primeur* or *nouveau* from the third Thursday of November following the harvest.

🍷 *Chardonnay*
🍶 *1–4 years*

ROSÉ These lightweight wines have an attractive, pale raspberry colour and light fruit flavour, and are more successful than their counterparts. They may be sold as *primeur* or *nouveau* from the third Thursday of November after the harvest.

🍷 *Gamay, Pinot Noir*
🍶 *1–3 years*

✓ *Patrick Bénas • Stephane Brocard (rouge) • Cave de Lugny • Jean-Francois Gonon (rouge) • Guffens-Heynen • Héritiers du Comte Lafon* ⑧ *• Jean Thévenet (Bongran* ⑩*) • Louis Latour • Saumaize-Michelin (Les Bruyères rouge) • Valette*

MÂCON-VILLAGES AOC

Of 96 villages within the 90 communes in the basic Mâcon appellation, this superior white-only appellation encompasses no fewer than 84 villages, of which 11 overlap the Beaujolais appellation (Ancelle, Chaintré, Chânes, Chasselas, Crêches-sur-Saône, La Chapelle-de-Guinchay, Leynes, Pruzilly, Romanèche-Thorins, St-Amour-Bellevue, and St-

Vérand). *See also* the following appellation for individual Mâcon village denominations.

WHITE These are some of the world's most delicious, thirst-quenching, easy-drinking, dry Chardonnay wines. They also represent tremendous value. These wines may be sold as *primeur* or *nouveau* commencing from the third Thursday of November following the harvest.

🍷 *Chardonnay*
🍶 *1–4 years*

✓ *Mâcon-Villages (without a named village) Cave des Vignerons de Buxy (Clos de Mont-Rachet) • Jean-Francois Gonon • Héritiers du Comte Lafon* ⑧ *• Jean Thévenet • Vignerons des Terres Secrètes*

MÂCON WITH VILLAGE DENOMINATIONS

These AOCs have been consolidated from 41 white-only denominations to just 27, and most of these appellations now allow red and rosé wines, as well as white. The white wines can now be made only from Chardonnay (with Pinot Blanc no longer allowed), while the red and rosé wines must be pure Gamay (no Pinot Noir or Pinot Gris). All recommendations are for white wines only, unless specifically stated otherwise.

MÂCON AZÉ AOC

Red, white, and rosé wines exclusively from the village of Azé, which has a good *coopérative*. Azé has a reputation for fatness and consistency.

✓ *Cave Coopérative d'Azé (Cuvée Jules Richard) • De la Bruyère*

MÂCON BRAY AOC

Red, white, and rosé wines grown in the villages of Blanot, Bray, Chissey-lès-Mâcon, and Cortambert. Reds have occasionally done well here.

✓ *De Thalie*

MÂCON BURGY AOC

Red, white, and rosé wines exclusively from the village of Burgy. The whites from Domaine de Chervin can show stunning elegance.

✓ *De Chervin*

MÂCON BUSSIÈRES AOC

Red, white, and rosé wines exclusively from the village of Bussières, where Domaine de la Sarazinière manages to craft reds of extraordinary elegance for Gamay (especially Les Devants) but truly excels with brilliantly pure whites (Le Pavillon and Cuvée Claude Seigneuret).

✓ *Héritiers du Comte Lafon* ⑧ *• De la Sarazinière*

MÂCON CHAINTRÉ AOC

Red, white, and rosé wines grown in the villages of Chaintré, Chânes, and Crêches-sur-Saône, which all overlap the Beaujolais district. Indeed, Chaintré is home to Georges Duboeuf's older brother Roger, but he produces strictly white wines. This village is also one of four that form the appellation of Pouilly-Fuissé.

✓ *CCV de Chaintré • Roger Duboeuf • Valette*

MÂCON CHARDONNAY AOC

Red, white, and rosé wines grown in the villages of Chardonnay, Ozenay, Plottes, and in part of Tournus. These wines have a certain following due to some extent, no doubt, to the novelty of their name. Nevertheless, the *coopérative* does produce fine wines.

✓ *Cave de Lugny (L'Originel)*

MÂCON CHARNAY-LÈS-MÂCON AOC

Excellent red, white, and rosé wines exclusively from the village of Charnay-lès-Mâcon, east of Pouilly-Fuissé, where even Gamay can show well in some hands (Jean-Paul Brun).

BERZÉ-LE-CHÂTEL IN SAÔNE-ET-LOIRE
Built on a rocky spur overlooking the Mâconnais vineyards, the massive fortifications of Berzé-le-Châtel, the seat of the first feudal barony of Mâcon, boast 13 medieval towers.

✓ *Jean-Paul Brun (Terres Dorées) • Jean Manciat • Manciat-Poncet • Didier & Catherine Tripoz (Clos des Tournons)*

MÂCON CRUZILLE AOC

This appellation is for red, white, and rosé wines grown in the villages of Grevilly, Martailly-lès-Brancion, and in part of Cruzille. This village denomination is in the extreme north of the appellation area.

✓ *Bret Brothers* **B** *• Guillot-Broux*

MÂCON DAVAYÉ AOC

Red, white, and rosé wines exclusively from the village of Davayé. Some excellent whites are made in this village, which is also part of Saint-Véran AOC.

✓ *Des Deux Roches • Lycée Viticole de Davayé*

MÂCON FUISSÉ AOC

These top-quality white wines are grown exclusively in the village of Fuissé, which unsurprisingly is also one of four communes that form the Pouilly-Fuissé AOC.

✓ *Bret Brothers • Christophe Cordier • De Fussiacus • Auvigue • Robert-Denogent*

MÂCON IGÈ AOC

Red, white, and rosé wines exclusively from the village of Igè, a village denomination that is seldom seen, but has a good reputation.

✓ *Fichet • Les Vignerons d'Igé*

MÂCON LOCHÉ AOC

This is an AOC for white wines only, grown exclusively in the village of Loché, an associated commune of Mâcon itself, which also has the right to the Pouilly-Loché and Pouilly-Vinzelles AOCs.

✓ *Caves des Grands Crus Blancs • Marcel Couturier*

MÂCON LUGNY AOC

Red, white, and rosé wines grown in the villages of Bissy-la-Mâconnaise, Lugny, Saint-Gengoux-de-Scissé, and in part of Cruzille. The white wines of Mâcon Lugny are probably the most ubiquitous of all Mâcon village wines that have a right to their own appellation.

✓ *Louis Latour • Cave de Lugny • Jean Rijckaert*

MÂCON MANCEY AOC

Red, white, and rosé wines grown in the villages of Boyer, La Chapelle-sous-Brancion, Etrigny, Jugy, Laives, Mancey, Montceaux-Ragny, Nanton, Royer, Sennecey-le-Grand, and Vers et Tournus. Mancey itself has something of a reputation for red wine, and the local cooperative's Les Essentielles Vieilles Vignes is probably the most consistent.

✓ *Cave des Vignerons de Mancey (Les Essentielles)*

MÂCON MILLY-LAMARTINE AOC

Red, white, and rosé wines grown in the villages of Berzé-la-Ville, Berzé-le-Châtel, Milly-Lamartine, and Sologny. The white wines from this top Mâcon appellation enjoy the best reputation, particularly those by Cordier and Lafon from the Clos du Four. The red wines from the local *coopérative*, Vignerons des Terres Secrètes, can be good despite its limestone soil, but their white Mâcon Milly-Lamartine is far better.

✓ *Christophe Cordier (Clos du Four) • Héritiers du Comte Lafon (Clos du Four)* **B** *• Vignerons des Terres Secrètes*

MÂCON MONTBELLET AOC

These white wines of this appellation are grown exclusively in the village of Montbellet and develop well after a couple of years in bottle.

✓ *Jean Rijckaert (En Pottes Vieilles Vignes)*

MÂCON PÉRONNE AOC

Red, white, and rosé wines grown in the villages of Péronne, Saint-Maurice-de-Satonnay, and in part of Mâcon Clessé. These wines tend to be fuller on the nose and stronger on the palate than those of most villages, but with less initial charm.

✓ *Cave de Lugny*

MÂCON PIERRECLOS AOC

Red, white, and rosé wines exclusively from the village of Pierreclos. Although there are some good reds produced on granitic soils in this village, the white wines truly excel, and Guffens-Heynen is by far the best producer.

✓ *Des Deux Roches (rouge) • Guffens-Heynen • Henri de Villamont • Vignerons des Terres Secrètes (rouge) • Jean Thévenet*

MÂCON PRISSÉ AOC

This appellation is for red, white, and rosé wines exclusively from the village of Prissé, which is also part of the Saint-Véran appellation. Look for the excellent minerality in the whites.

✓ *Vignerons des Terres Secrètes • Thibert Père & Fils*

MÂCON LA ROCHE-VINEUSE AOC

Red, white, and rosé wines grown in the villages of Chevagny-lès-Chevrières, Hurigny, and La Roche-Vineuse. These under-rated wines are produced on west- and south-facing slopes north of Pouilly-Fuissé, the potential of which is pushed to the very limits by Olivier Merlin, one of Mâcon's greatest winemakers.

✓ *Olivier Merlin ❶ • Vignerons des Terres Secrètes*

MÂCON SOLUTRÉ-POUILLY AOC

White wines only, grown exclusively in the village of Solutré-Pouilly, which is one of the four communes of Pouilly-Fuissé and also forms part of the Saint-Véran area. Its wines, therefore, have a choice of three appellations. Robert-Denogent is outstanding.

✓ *Auvigue • Drouhin • Robert-Denogent • des Gerbeaux*

MÂCON SAINT-GENGOUX AOC

Far from the madding crowd in the northwestern corner of the Mâconnais, the red, white, and rosé wines from this very large village appellation are made from vines grown in the villages of Ameugny, Bissy-sous-Uxelles, Bonnay, Bresse-sur-Grosne, Burnand, Champagny-sous-Uxelles, Chapaize, Cortevaix, Curtil-sous-Burnand, Lournand, Malay, Massy, Saint-Gengoux-le-National, Saint-Ythaire, Salornay-sur-Guye, Savigny-sur-Grosne, Sigy-le-Châtel, and La Vineuse. Rarely encountered up to now, but the excellent *coopérative* at Buxy produces a good white from Saint-Gengoux-le-National itself, where their members have vines on Mont Goubot, which is one of the highest points in the village.

✓ *Cave des Vignerons de Buxy (Buxynoise)*

MÂCON SERRIÈRES AOC

Only red and rosé wines, grown exclusively in the village of Serrières. Only red wines encountered so far, and nothing has stood out.

MÂCON UCHIZY AOC

Red, white, and rosé wines exclusively from the village of Uchizy, which is adjacent to Chardonnay and usually produces good-quality, thirst-quenching white wines.

✓ *Bret Brothers ❸ • Héritiers du Comte Lafon ❸ • Raphaël Sallet (Clos des Ravières) • Mallory & Benjamin Talmard • Vignerons des Terres Secrètes*

MÂCON VERGISSON AOC

White wines only, grown exclusively in the village of Vergisson, which is one of the four villages with the right to the Pouilly-Fuissé AOC. These wines are richer than most and can be aged a few years in bottle without losing their freshness and appeal.

✓ *Daniel Barraut • Michel Forest*

MÂCON VERZÉ AOC

Red, white, and rosé wines exclusively from the village of Verzé, where it is the white wines that truly excel, exhibiting some of the purest minerality in the Mâconnais.

✓ *Des Gandines • Leflaive • Nicolas Maillet • Vignerons des Terres Secrètes*

MÂCON VINZELLES AOC

White wines only, grown exclusively in the village of Vinzelles, which also has the right to the Pouilly-Vinzelles AOC. These are excellent, fresh, vibrant wines.

✓ *Bret Brothers (Domaine Soufrandière) Z*

POUILLY-FUISSÉ AOC

This pure Chardonnay wine should not be confused with Pouilly-Fumé, the Sauvignon Blanc wine from the Loire. This appellation covers a wide area of prime vineyards spread over four villages (Chaintré, Fuissé, Solutré-Pouilly, and Vergisson), but there is considerable variation. A reclassification of the AOC Pouilly-Fouissé has been going on for more than 10 years now. As a result there should be 22 *climats* bearing the *premier cru* labels hopefully very soon. The following *climats (lieux-dits)* may be attached to the Pouilly-Fuissé AOC: Les Chevrières, Les Vignes Blanches, Aux Chailloux, and Les Crays.

WHITE These dry wines range from typical Mâcon blanc style, through slightly firmer versions, to the power-packed, rich oaky flavours of Michel Forest and Vincent's Château Fuissé Vieilles Vignes. Although these last two are widely regarded as the finest in Pouilly-Fuissé, they are by no means typical of Mâcon, leaning more towards the Côte Chalonnaise, sometimes even the Côte de Beaune, and ardent admirers of the more traditional, light, and fluffy Mâcon will not even touch them. On the other hand, wines from producers such as Guffens-Heynen can be just as rich and intense as either Forest or Château Fuissé, but without any oak whatsoever.

🍷 *Chardonnay*

🍷 *3–8 years*

✓ *Auvigue • Daniel et Martine Barraud • Chateau de Beauregard • Christophe Cordier • Nadine Ferrand (Prestige) • Des Gerbeaux (Cuvée Jacques Charvet) • Guffens-Heynen • Bruno Jeandeau (Terre Jeanduc) • Dominique Lafon • Roger Lassarat • Olivier Merlin ❶ • Catherine & Pascal Rollet (Clos de la Chapelle) • Jean Rijckaert • Thibert Père & Fils (Vignes de la Côte) • Pierre Vessigaud (Vieilles Vignes) • Robert Denogent • Valette • Vincent & Fils (Château de Fuissé Vieilles Vignes)*

POUILLY-LOCHÉ AOC

One of Pouilly-Fuissé's two satellite appellations. Good value. Two *climats (lieux-dits)* may be attached to the Pouilly-Loché AOC: Les Mûres and Aux Barres.

WHITE This village may produce Mâcon-Loché AOC, Pouilly-Loché AOC, or Pouilly-Vinzelles AOC. The dry wines of this village are more of the Mâcon style, whatever the AOC.

🍷 *Chardonnay*

🍷 *1–4 years*

✓ *Bret Brothers • Christophe Cordier • Louis Jadot (Château de Loché) • Tripoz (Clos des Rocs)*

POUILLY-VINZELLES AOC

One of Pouilly-Fuissé's two satellite appellations. Some of these wines have the potential to challenge Pouilly-Fuissé. Two *climats (lieux-dits)* may be attached the Pouilly-Vinzelles AOC: Les Quarts and Les Longeays.

WHITE More the Mâcon-type of Pouilly-Fuissé, for similar reasons to those of Pouilly-Loché.

🍷 *Chardonnay*

🍷 *1–4 years*

✓ *Bret Brothers • Thibert Père & Fils (Thibert-Parisse) • Valette*

SAINT-VÉRAN AOC

This appellation overlaps the Mâconnais and Beaujolais districts, and is itself bisected by the Pouilly-Fuissé AOC, with two villages to the north (Davayé and Prissé) and five to the south (Chânes, Chasselas, Leynes, Saint-Amour, and Saint-Vérand). Saint-Véran was named after Saint-Vérand, but the "d" was dropped in deference to the growers in certain other villages, who it was feared would not support the new appellation if they felt their wines were being sold under the name of another village. This appellation was introduced in 1971 to provide a more suitable outlet for white wines produced in Beaujolais than the Beaujolais Blanc appellation.

WHITE Excellent value, fresh, dry, and fruity Chardonnay wines that are very much in the Mâcon-

Villages style. Vincent, the proprietor of Château Fuissé, produces an amazingly rich wine that is far closer to Pouilly-Fuissé than to Macon-Villages, with hints of oak and honey.

🍷 *Chardonnay*

🍷 *1–4 years*

✓ *Daniel et Martine Barraud • Christophe Cordier • Pierre Janny • Roger Lassarat • Olivier Merlin ❶ • Paquet • Jean Rijckaert • Vignerons des Terres Secrètes*

VIRÉ-CLESSÉ AOC

This appellation was created in February 1999, by combining the Mâcon-Viré and Mâcon-Clessé sub-appellations and recognizing the wines from Viré-Clessé in their own right, above and beyond those of Mâcon-Villages. The appellation was retrospectively applied to the 1998 vintage, but producers could continue using either of the original two AOCs up to and including the 2002 vintage. Viré was always the most ubiquitous of the Mâcon-Villages wines, but it was also the most consistent and one of the best, while Clessé showed the most finesse. The regulations are slightly stricter in terms of yield, with an increased minimum natural sugar content for wines bearing one of the following *climats (lieux-dits)*: La Montagne, La Bussière, En Collonge, and Quintaine.

🍷 *Chardonnay*

🍷 *1–4 years*

✓ *All those recommended under Mâcon-Viré and Mâcon-Clessé are potential recommendations here, but those that have actually excelled are: André Bonhomme • Bret Brothers (Sous Les Plantes) • Héritiers du Comte Lafon ❸ • Jean Rijckaert (L'Epinet) • Jean Thévenet (Emilian Gillet) • Cave de Viré (Cuvée Spéciale)*

LA ROCHE DE SOLUTRÉ
The dramatic shape of the La Roche de Solutré (Rock of Solutré) towers over the village's vineyard. Solutré is one of the 42 villages of the Mâcon-Villages appellation and is a commune of Pouilly-Fuissé.

The Beaujolais

Beaujolais is currently the subject of many positive reviews and column inches, whilst at the same time it is also seeing a groundswell of interest – youthful interest – in its wines. Can there be a better position for a region to find itself in? Well, there remains one fly in this ointment, relatively speaking: the market for these wines remains in the doldrums – not in terms of volume, but in terms of price.

Despite the last 10 years having delivered a string of impressive vintages – with impressive wines to match – the market pricing lags significantly behind sentiment, and that pricing was created by an older sentiment, one that was created by producers riding, indeed relying on, a tsunami of poor wine sold under the name of Beaujolais Nouveau. Older people remember great Gamay. Younger people love the dynamism and diversity of styles coming from the region today, but all that the people in between can recollect is the awful Beaujolais Nouveau of the 1990s.

There is no "one truth" as to what Beaujolais is today, but all the categories are interesting for inquisitive consumers. Beaujolais Nouveau is resurgent – for quality if not yet market sentiment. Beaujolais Blanc is growing very quickly and likewise is the market for Beaujolais of all labels made with low or "zero-added" sulfur.

THE VINE-GROWING AREA OF THE BEAUJOLAIS *(see also p213)*
Forming the southernmost part of the Burgundy region, the Beaujolais area is planted almost entirely with Gamay vines, the best vineyards being on granite soil.

NEW BEAUJOLAIS NOUVEAU

Beaujolais Nouveau became official in 1951, but the template for this wine had its genesis in the wine bars of Paris in the years before. In essence, this was a juicy, young, fruity wine – with fermentations overwhelmingly of carbonic maceration emphasising these elements. After a slow start, the sales of this category grew in popularity at an astounding rate. Beaujolais Nouveau became a cash cow, and as the volumes skyrocketed, attention to quality apparently faded – unsurprisingly in concert with fading market sentiment for the wine. It seems incontrovertible that the combination of market dominance of this category of Beaujolais over the more classic *crus,* coupled with years of poor quality Nouveau led to the significant decline in market sentiment for all the wines from the entire region. For most consumers, Nouveau *was* Beaujolais. Nothing exemplifies this more than a quick comparison of the bulk price of Saint-Amour (one of the *cru* Beaujolais) in the 1970s when its price was roughly equivalent to that of Chambolle-Musigny (from the Côte d'Or). Today's bulk price for Saint-Amour is hardly two euros per litre. In a nutshell, this is the moribund commercial state of Beaujolais today.

Yet something had to change, and change it has. The general quality of Beaujolais Nouveau is now unrecognisable from what was produced in the 1990s, and there are even three new categories: Beaujolais-Villages Nouveau *rouge,* Beaujolais-Villages Nouveau rosé, and Beaujolais Nouveau rosé. The rosés are generally excellent, whereas the Beaujolais-Villages Nouveau, empirically, is a strange beast – a

GEORGES DUBŒUF BEAUJOLAIS NOUVEAU ROSÉ, VINTAGE 2019
The region has become inextricably linked with "nouveau" wines, whose reputation suffered after the uninspiring 1990s vintages, though recent examples are far superior.

FACTORS AFFECTING TASTE AND QUALITY

LOCATION
Beaujolais, the most southerly of Burgundy's districts, is located in the Rhône *département*, which is 400 kilometres (250 miles) southeast of Paris.

CLIMATE
Beaujolais has an essentially sunny climate tempered by the Atlantic and the Mediterranean, as well as by continental influences. Although the annual rainfall and temperature averages are ideal for winegrowing, they are subject to sudden stormy changes due to the influence of the Mediterranean – most of the recent vintages have seen localised summer storms of hail.

ASPECT
A series of east-facing slopes, up to 2 kilometres (just over 1 mile) wide, which curve in and out to give some vineyards northeastern aspects, and others southeastern aspects. Here the vines grow at an altitude of between 225 and 380 metres (740 to 1,250 feet) on slightly less steep slopes than those of the Côte de Nuits. South of Beaune, no vines with the right to the village (and higher) appellations extend past the RN 74 road on to the flat and fertile ground beyond.

SOIL
The northern Beaujolais, which encompasses the famous *crus* and those communes entitled to the Beaujolais-Villages AOC, is an area renowned for its granite-based soil, the only type on which the Gamay has so far excelled. Topsoils are often schistous or made up of decomposed granite mixed with sand and clay. The south is essentially limestone based, and this is a problem for Gamay, which accordingly produces much lighter wines that lack the class of those in the north. There has been significant growth in the planting of Chardonnay here and, most recently, plantations of Pinot Noir.

VITICULTURE & VINIFICATION
In recent years there has been a groundswell of young *vignerons* looking to work organically in the vines, but the steepness of many slopes coupled with the sandiness of the soil has limited their aspirations, so the application of herbicide remains an important soil-management tool in Beaujolais. There is much market interest in low-sulfur/no-added sulfur/natural wines. The latter "banner" garnering mixed sentiment and having no clear definitions, so I will restrict commentary to the more consistent quality delivered by the first two. The low-sulfur wines are typically produced with no added sulfur during both vinification and *elevage,* but these wines will usually receive a small dose of sulfur at bottling time, whereas the no-added sulfur wines also eschew sulfur at bottling time. Aromatically these low-sulfur wines have similar traits: roundness and less direct focus – though producers such as Yvon Metras and Camille & Mathieu Lapierre regularly seem to deliver more clarity than most. The Lapierre domaine is also one of the few to openly acknowledge the extra sensitivity of low-sulfur wines to storage temperature, so produce both low-sulfur and no-added sulfur variants of all their *cuvées* – the back label of the "no-added" wines recommending storage below 14°C (57.2°F). The aim of the Lapierres is to match the *cuvées* to the individual logistics of their buyers.

GRAPE VARIETIES
Primary varieties: Gamay
Secondary varieties: Aligoté, Chardonnay, Melon de Bourgogne, Pinot Beurot (Pinot Gris), Pinot Blanc, Pinot Noir (restricted to Côteaux Bourguignons AOC)

Nouveau wine that needs cellaring. In the ripe vintages of 2017 to 2019 this Beaujolais-Villages Nouveau was more concentrated, structured, and intense than its cousin with a Beaujolais Nouveau label and was rarely ready to drink on the third Thursday of November.

There has been a groundswell of new producers and delicious wines in the category of Nouveau in the last 10 years, but look at supermarket shelves outside of France for Beaujolais Nouveau and you will struggle to find the wines. Too little, too late?

WHITE BEAUJOLAIS
From virtually nothing 15 years ago, the region's current production of whites from Chardonnay has much more visibility. Whilst amounting to only 2.5 per cent of the region's total output, this was still more than 19,000 hectolitres in 2018. Of those 2018 whites 72 per cent were produced under the AOC Beaujolais category, with the remaining volume attributed to AOC Beaujolais-Villages. Production of whites is largely in the south of Beaujolais, where you will find more "argilo-calcaire" soils – particularly in the Pierres Dorées area. Although much of this white wine is destined for the production of Crémant, many domaines today feel the need to have a white wine to "complete their range", even if their holdings are on granitic soils. Many also feel the need to label this wine (legally) as Bourgogne Chardonnay and for two reasons: first, they maintain that the Bourgogne label is easier to sell (is this because the customer might assume that it comes from the Côte d'Or?), and second, because producers worry that if they stop using the Bourgogne label, then their "right" to use that label will be taken away. As a still wine, there is usually much richness and a little mineral rigour – you would rarely mistake it for a Mâcon and certainly not a white of the Côte d'Or or Chablis. Good whites exist from the likes of Domaine Robert Perroud, Domaine Longère. and Château Thivin, but you must be prepared to search hard to find them.

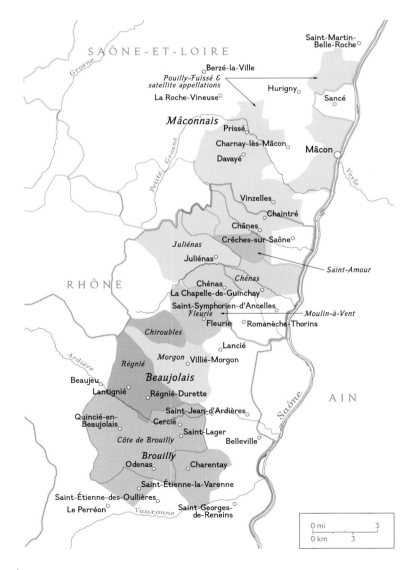

CRUS BEAUJOLAIS VINEYARDS
All the best growths of Beaujolais are found in the northeast corner of the region, where they overlap with some of the finest wines of Mâcon.

THE *VIN DE MERDE* CASE

After 100,000 hectolitres of Beaujolais had to be distilled in 2001, a French wine critic by the name of François Mauss claimed in *Lyon Mag,* a small French magazine, that poor-quality Beaujolais of this ilk was "not proper wine, but rather a sort of lightly fermented and alcoholic fruit juice". No wonder they could not sell it. Mauss blamed the need for distillation on the craze for Beaujolais Nouveau, which is rushed to market barely two months after the harvest. Beaujolais producers, he claimed, had ignored all warning signs that consumers were no longer willing to buy such wine, which he branded as *vin de merde,* or "shit wine", and the proverbial *merde* hit the fan.

Even though this description was not the assessment of *Lyon Mag* per se, but rather a quote from someone being interviewed, and despite the fact that the publication balanced these comments with those of a Beaujolais representative defending the wine, an association of 56 Beaujolais cooperative producers decided to sue the magazine, They sued, not for libel, but instead under a rarely used French law that protects products from being denigrated. A cynic might ask why these producers did not feel able to sue for libel, but even more disturbing is the fact that a modern-day court in a supposedly civilized society could actually find what this magazine published to be illegal in any way. Yet it did, and more disturbingly, in 2003 the court (in Villefranche-sur-Saône, the heart of Beaujolais country) considered this quote to be so serious a wrongdoing that it ordered *Lyon Mag* to pay 350,000 euros, which for such a small, employee-owned publication would have put it out of business. On appeal, this was reduced to 113,000 euros and, after a backlash of bad publicity around the world, the Beaujolais cooperatives decided not to pursue any damages, just their costs. In 2005, the highest court of appeal reversed the decision against the magazine, finding no cause of action against the publication, and the Beaujolais cooperatives were ordered to pay 2,000 euros in court costs to *Lyon Mag.*

If the association of cooperative producers started this case to protect the reputation of Beaujolais Nouveau, it clearly achieved the opposite result: its reputation had been dealt a more severe blow by the worldwide coverage of "The Shit Wine Case". If the cooperatives don't like wine critics describing Beaujolais Nouveau as crap, they should make it delicious and fruity to drink. Is that too much to ask?

A WINDMILL STANDS IN THE VINEYARDS OF MOULIN-À-VENT
With a well-deserved reputation as one of the most noteworthy of the Beaujolias appellations, Moulin-à-Vent is known as the "King of Beaujolais". These darker, heavier wines set themselves apart from the lighter Nouveaus.

THE APPELLATIONS OF
THE BEAUJOLAIS

BEAUJOLAIS AOC

This generic Beaujolais appellation accounts for half the wine produced in the district and more than half of this is sold as Beaujolais "Primeur". The basic quality of these wines means that they cannot be bought from the great *négociants* of the Côte d'Or, although *cru* Beaujolais can be.

RED Due to their method of vinification, most of these wines have a pear drop or bubble-gum character to their fruitiness. The best of them also have a delightful freshness and frankness that beg for the wine to be consumed in large draughts.

🍷 *Gamay, Gamay de Bouze, Gamay de Chaudenay, Melon, Aligoté, Chardonnay, Pinot Noir, Pinot Gris*

🍸 *1–3 years*

WHITE The production of white wine, made from Chardonnay, has jumped significantly in the past 10 years. in 2018, 5.5 per cent of AOC Beaujolais was declared as white – more than 14,000 hectolitres. The largest outlet is for making Crémant, but bottlings under the Beaujolais AOC or with a Bourgogne Chardonnay label are everywhere today.

🍷 *Chardonnay, Aligoté*

🍸 *1–3 years*

ROSÉ Fresh, "pretty", and fruity.

🍷 *Gamay, Gamay de Bouze, Gamay de Chaudenay, Melon, Aligoté, Chardonnay, Pinot Noir, Pinot Gris*

🍸 *1–3 years*

✔ *Jean-Paul Brun (Terres Dorées) • Domaine Monternot – Les Jumeaux • Domaine Girin • Claire et Fabien Chasselay*

BEAUJOLAIS NOUVEAU AOC

See Beaujolais Primeur AOC.

BEAUJOLAIS PRIMEUR AOC

At its height, more than half of all the Beaujolais produced was sold as *vin de primeur,* but production has dropped to one-third following the *vin de merde* saga (*see* p246). Swinging from the vine one moment, this wine is subjected to intensive carbonic maceration in order to hit the shelf just a few weeks later (usually the third Thursday of November). Sold mostly as Beaujolais Primeur in France but better known on export markets as Beaujolais Nouveau, this is supposed to be a fun wine, but by the 1990s this merrymaking had worn a little thin.

BEAUJOLAIS SUPÉRIEUR AOC

Only 1 per cent of all Beaujolais wines carry this appellation. It is exceedingly rare to find a Beaujolais Supérieur that is superior in anything – other than strength – to basic Beaujolais AOC, because this appellation merely indicates that the wines contain an extra 1 per cent alcohol. Red and rosé Beaujolais Supérieur may be sold as *primeur* or *nouveau* from the third Thursday of November following the harvest.

RED These are by no means superior to Beaujolais AOC – buy basic Beaujolais for fun or *cru* Beaujolais for more serious drinking.

🍷 *Gamay, Gamay de Bouze, Gamay de Chaudenay, Melon, Aligoté, Chardonnay, Pinot Noir, Pinot Gris*

🍸 *3–8 years*

WHITE Barely 5 per cent of vineyards in this tiny appellation produces white wine. Fine as it may be, it has no intrinsic superiority over the quaffing quality of basic Beaujolais Blanc.

🍷 *Chardonnay, Aligoté*

🍸 *1–3 years*

ROSÉ I have not encountered any pink versions of this appellation.

✔ *Cave Beaujolais du Bois-d'Oingt • Cave Saint-Vérand*

BEAUJOLAIS-VILLAGES AOC

The 30 villages that are allowed to add their names to the Beaujolais AOC (*see* Beaujolais [village name] AOC), plus the *crus* Beaujolais, also have the right to this appellation, and they must use it if the wine is a blend of wines from two or more villages.

RED Good examples of these wines are well coloured with a rich Gamay flavour and should exhibit more concentration and tension when compared the more basic Beaujolais AOC.

🍷 *Gamay, Gamay de Bouze, Gamay de Chaudenay, Melon, Aligoté, Chardonnay, Pinot Noir, Pinot Gris*

🍷 *3–8 years*

WHITE Of the white Beaujolais produced in 2018, 28 per cent was declared as Beaujolais-Villages at harvest time – 5,400 hectolitres. As with Beaujolais Blanc, it is difficult to estimate how much is actually sold under that declared label, as it may have been used either to make Crémant or take a Bourgogne Chardonnay label. A wine of richness, but do not look for tension.

🍷 *Chardonnay, Aligoté*

🍷 *1–3 years*

ROSÉ This is an attractive wine, and accounts for a little over 2 per cent of Beaujolais' production in 2018. These wines may also be sold as *primeur* or *nouveau* from the third Thursday of November following the harvest.

🍷 *Gamay, Gamay de Bouze, Gamay de Chaudenay, Melon, Aligoté, Chardonnay, Pinot Noir, Pinot Gris*

🍷 *1–3 years*

BEAUJOLAIS [VILLAGE NAME] AOC

Of the 30 villages that may add their names to this appellation, very few do. One reason is that all or part of 7 of these villages (asterisked – * – below) qualify for one of the superior *cru* Beaujolais appellations, and it makes no sense to use a less famous name to market the wines. Another is that 8 of the villages are entitled to the Mâcon-Villages AOC (marked "M"), and four of these are also within the St-Véran AOC (*see also* p241), marked "S-V", which overlaps with the Mâconnais and Beaujolais; some of these villages, of course, produce more white wine than red. These village names are also under-exploited because they are either unknown or more suggestive of Mâcon than Beaujolais. It is thus easier to sell them as nothing more specific than Beaujolais-Villages. The wines may be sold as *primeur* or *nouveau* from the third Thursday of November following the harvest.

The following is a complete list of villages that may use the appellation: Arbuisonnas; Les Ardillats; Beaujeu; Blacé; Cercié*; Chânes (M, S-V); La Chapelle-de-Guinchay* (M); Charentay; Denicé; Durette; Emeringes*; Jullié*; Lancié; Lantignié; Leynes (M, S-V); Marchampt; Montmelas; Odenas; Le Perréon; Pruzilly* (M); Quincié*; Rivolet; (M); St-Étienne-des-Ouillères; St-Étienne-la-Varenne*; St-Julien; St-Lager; St-Symphorien-d'Ancelles (M); Salles; Vaux; and Vauxrenard.

RED Good examples should be richly flavoured Gamay wines of similar quality to non-specific Beaujolais-Villages but with more personality.

🍷 *Gamay, Pinot Noir, Pinot Gris, Gamay de Bouze, Gamay de Chaudenay, Aligoté, Chardonnay, Melon*

🍷 *3–8 years*

WHITE Chardonnay is the principle white grape, but Aligoté is allowed as an "accessory" grape – ie with a maximum percentage. But no 100 per cent Aligoté wine is allowed today with either a Beaujolais or Beaujolais Villages label. That said, older planted parcels – those pre-dating 2004 – could theoretically produce a 100 per cent Aligoté wine – but with a Bourgogne Aligoté label, rather a Beaujolais label.

🍷 *Chardonnay, Aligoté*

🍷 *1–3 years*

ROSÉ Rare but growing, they can often surprise in quality, particularly when offered as *primeur* or *nouveau*.

🍷 *Gamay, Pinot Noir, Pinot Gris, Gamay de Bouze, Gamay de Chaudenay, Aligoté, Chardonnay, Melon*

🍷 *1–3 years*

BROUILLY AOC
Cru Beaujolais

The largest and most southerly of the 10 *cru* villages, this is the only one, with Côte de Brouilly, to permit grapes other than Gamay.

RED Most Brouilly are serious wines. They are not quite as intense or detailed as Côte de Brouilly wines, but they are full, rich, and fruity. They can be quite tannic.

🍷 *Gamay, Chardonnay, Aligoté, Melon de Bourgogne*

🍷 *2–7 years (4–12 years for vin de garde styles produced in the very best vintages)*

✓ *Georges Duboeuf (Château de la Pérrière, Domaine de Lafayette, Pisse Vieille) • Collin Bourisset • Pierre Marie Chermette • Château de la Térrière • Robert Perroud*

CHÉNAS AOC
Cru Beaujolais

The smallest of the *cru* Beaujolais, situated on the slopes above Moulin-à-Vent – herein lies a clue – most of the best land of Chénas carries the label Moulin à Vent. These slopes used to be occupied by oak trees, and so its name derives from *chêne*, the French for "oak" – unfortunately some producers take that link for granted and produce vanilla-infused wines.

RED Although most Chénas cannot match the power of the wines from neighbouring Moulin-à-Vent, they are nevertheless in the full and generous mould.

🍷 *Gamay, Aligoté, Chardonnay, Melon de Bourgogne*

🍷 *3–8 years (5–15 years for vin de garde styles produced in the very best vintages)*

✓ *Paul-Henri & Charles Thillardon • Château Bonnet (Pierre-Yves Perrachon) • Céline et Nicolas Hirsch • Cave de le Château de Chénas (Coeur de Granit)*

CHIROUBLES AOC
Cru Beaujolais

Situated higher in the hills, similar to the Haute Combe of Juliénas or Fleurie's Madone. Chiroubles can produce beautiful fragrance, though the steep hills here are hard to work and don't suit organic viticulture – at least not at the price the wines can achieve in the market.

THE LEGEND OF PISSE VIEILLE

The vineyard of Pisse Vieille in Brouilly amuses English-speaking consumers, who are dismayed by those writers who dare only to print the vineyard's story in French. It goes like this:

One day, an old woman (*vielle* in French) called Mariette went to confession. The priest was new to the village and unaware of its dialect. He also did not know that Mariette was hard of hearing. When he heard her confession, he merely said *"Allez! Et ne péchez plus!"* (*"Go! And do not sin again!"*). Mariette misheard this as *"Allez! Et ne piché plus!"*, which in the dialect meant *"Go! And do not piss again!"*, *piché* being the local form of *pisser*. Being a devout Catholic, Mariette did exactly as she was told. When her husband asked what terrible sin she had committed she refused to tell and, after several days, he went to ask the new priest. When he found out the truth he hurried home, and as soon as he was within shouting distance, began yelling *"Pisse, vieille!"* (*"Piss, old woman!"*).

RED These light to full-bodied wines have a perfumed bouquet and a deliciously delicate, crushed-grape flavour. They are charming to drink when young, but exceptional examples can improve with age.

🍷 *Gamay, Aligoté, Chardonnay, Melon de Bourgogne*

🍷 *1–8 years (5–15 years for vin de garde styles produced in the very best vintages)*

✓ *Fabien Collonge • Gilles Paris • Vignerons de Bel Air • Patrick Bouland • Louis Tête*

VINEYARDS STRETCH OUT BENEATH MONT BROUILLY
The Brouilly AOC, in central-north Beaujolais, contain significant plantings of the Gamay grape variety.

CÔTE DE BROUILLY AOC
Cru Beaujolais

A glance at the map of the Côte de Brouilly appellation would imply that this is an island in the middle of the Brouilly appellation – but it's more delineated than that; the Côte de Brouilly is actually a large round hill, of all orientations to the sun, that looks down upon the appellation of Brouilly. The Côte de Brouilly also has its own distinct form of blue-green granite that sets it apart from the rest of Brouilly, with finer, less ripe, better delineated flavours.

RED A fine Côte de Brouilly is full, rich, and flavoursome. Its fruit should be vivid and intense, with none of the earthiness that may be found in a Brouilly.

🍷 *Gamay, Pinot Noir, Pinot Gris, Aligoté, Chardonnay, Melon de Bourgogne*

🍷 *3–8 years (5–15 years for vin de garde styles produced in the very best vintages)*

✓ *Jean-Paul Brun (Terres Dorées) • Robert Perroud • Château des Ravatys • Laurent Martray • Château Thivin (all cuvées) • Georges Duboeuf (Domaine du Riaz)*

COTEAUX DU LYONNAIS AOC

This is not part of the true Beaujolais district, but it falls within its sphere of influence and certainly utilizes classic Beaujolais grapes. In May 1984, this wine was upgraded from VDQS to full AOC status.

RED Light-bodied wines with fresh Gamay fruit and a soft balance. May be sold as *primeur* or *nouveau* from the third Thursday of November following the harvest.

🍷 *Gamay, Gamay de Bouze, Gamay de Chaudenay*

🍷 *2–5 years*

WHITE These are fresh and dry Chardonnay wines that are softer than a Mâconnais and lack the definition of a Beaujolais Blanc. These wines may be sold as *primeur* or *nouveau* from the third Thursday of November following the harvest.

🍷 *Chardonnay, Aligoté, Pinot Blanc*

🍷 *1–3 years*

ROSÉ I have not encountered these wines.

✓ *Ris Descotes (blanc) • de Prapin (blanc) • Clos du Saint-Marc*

FLEURIE AOC
Cru Beaujolais

The evocatively-named Fleurie (or "flowery") is one of the more expensive of the *crus* and its finest wines are the quintessence of classic Beaujolais.

RED The wines of Fleurie quickly develop a fresh, fragrant, and fittingly floral style. Not as light and delicate as some writers suggest, their initial charm belies a positive structure and a depth of fruit that can sustain the wines for many years.

🍷 *Gamay, Aligoté, Chardonnay, Melon de Bourgogne*

🍷 *2–8 years (4–16 years for vin de garde styles produced in the very best vintages)*

✓ *Château des Bachelards • Yvon Metras • Château de Fleurie Reserve (Jean Loron) • Gilles Paris • Jules Desjourneys • Clos de la Roilette (Alain et Christie Coudert)*

JULIÉNAS AOC
Cru Beaujolais

Situated in the hills above St-Amour, Juliénas is named after Julius Caesar and, according to local legend, was the first Beaujolais village to be planted. It is probably the most underrated of the 10 *crus* Beaujolais.

RED The spicy-rich, chunky-textured fruit of a youthful, lower slope, Juliénas or the airy, faintly spiced, higher slope Juliénas – both will develop a classy, satin-smooth appeal if given time to develop in bottle.

🍷 *Gamay, Aligoté, Chardonnay, Melon de Bourgogne*

🍷 *3–8 years (5–15 years for vin de garde styles produced in the very best vintages)*

✓ *Vincent Audras • Jaques Charlet, Julénas Clos des Poulettes (Jean Loron) • Château des Capitain (Georges Duboeuf) • Domaine du Bois du Chat (Jérémy Bally) • Michel & Sylvain Tête*

MORGON AOC
Cru Beaujolais

Just as the Côte de Brouilly is a finer form of Brouilly, so the wines of Mont du Py in the centre of Morgon are far more powerful than those of the surrounding vineyards in this commune.

RED Although these wines are variable in character and quality, the best of them rank with those of Moulin-à-Vent as the most sturdy of all Beaujolais. They have a singularly penetrating bouquet and very compact fruit – they resemble the Cortons of the Côte de Beaune.

🍷 *Gamay, Aligoté, Chardonnay, Melon de Bourgogne*

🍷 *4–9 years (6–20 years for vin de garde styles produced in the very best vintages)*

✓ *Château Bellevue • Daniel Bouland • Jean-Marc Burgaud • Louis-Claude Desvignes • Mee Godard • Gilles Paris • Jules Desjourneys • Camille et Mathieu Lapierre •*

MOULIN-À-VENT AOC
Cru Beaujolais

Because of its sheer size, power, and reputation for longevity, Moulin-à-Vent is known as the "King of Beaujolais". The powerful character of Moulin-à-Vent has been attributed to the high manganese content of its soil. The availability of manganese to the vine's metabolic system depends on the pH of the soil, and in the acid, granite soil of Beaujolais, manganese is all too readily available. For a healthy metabolism, however, the vine requires only the tiniest trace of manganese, so its abundance at Moulin-à-Vent could be toxic (to the vine that is, not the consumer!), may well cause chlorosis, and would certainly affect the vine's metabolism. This naturally restricts yields and could alter the composition of the grapes produced.

RED These well-coloured wines have intense fruit, excellent tannic structure, and, in many cases, a spicy-rich oak flavour.

🍷 *Gamay, Aligoté, Chardonnay, Melon de Bourgogne*

🍷 *4–9 years (6–20 years for vin de garde styles produced in the very best vintages)*

✓ *Domaine Rochgrès (Maison Albert Bichot) • Château des Jacques • Paul Janin • Château du Moulin-à-Vent • Le Nid*

RÉGNIÉ AOC
Cru Beaujolais

The growers claim that this village was the first to be planted with vines in Beaujolais, but so do the growers of Juliénas. Régnié was upgraded to full *cru* Beaujolais status in December 1988. Too many half-hearted efforts nearly sank the ship while it was being launched, but the best growers are steadfastly carving a reputation for this still-fledgling *cru* Beaujolais. Given the relative youth of this label in the ranks of the *crus*, it is no surprise that this remains the most variable of all *crus* – producing great to modest wines.

RED There are two distinct styles of red wine here: one is light and fragrant, the other much fuller and more meaty, but all the best examples are fruity and supple, showing a fresh, invigorating aroma.

🍷 *Gamay, Aligoté, Chardonnay, Melon de Bourgogne*

🍷 *2–7 years (4–12 years for vin de garde styles produced in the very best vintages)*

✓ *Gilles Copéret • Raphaël Chopin • Château de la Terrière • Julien Sunier*

ST-AMOUR AOC
Cru Beaujolais

This is the most northerly of the 10 *crus*, and it is more famous for its Mâconnais wines than for its *cru* Beaujolais. Despite its modern-day connotations, St-Amour has nothing to do with love – its name derives from St-Amateur, a Roman soldier who was converted to Christianity andis believed to have founded a monastery in the locality.

RED Charming wines of fine colour, seductive bouquet, and soft, fragrant, fruity flavour. Whilst the best will repay ageing, too many are bottled early to cash in on St.Valentine's day drinking.

🍷 *Gamay, Aligoté, Chardonnay, Melon de Bourgogne*

🍷 *2–8 years (4–12 years for vin de garde styles produced in the very best vintages)*

✓ *Château des Bachelards (Fleurie) • Domaine de la Pirolette • Domaine du Paradis • Château de Saint*

WORKERS HARVEST WINE GRAPES IN THE VINEYARDS OF FLEURIE
As this appellation's name suggests, the wines of the Fleurie AOC are fragrantly floral-tinged Beaujolais reds.

Champagne

In a world where technology is capable of producing high-quality, classic brut *sparkling wine almost anywhere, it is a testament to its* terroir *that Champagne is still the clear leader, but the gap has closed significantly and the* champenois *cannot afford to be complacent.*

It is Champagne's *terroir* that sets it apart from the rest of the sparkling wine world. Nowhere else can Chardonnay, Pinot Noir, or Meunier grapes ripen at just 9.7 per cent ABV with as much as 53 per cent tartaric acid. It is the lean structure of such low-alcohol, high-acid wines that makes Champagne ideal for bottle fermentation, and it is the fresh, crisp, undeveloped flavours of its relatively simple base wines that enables Champagne to seamlessly soak up the slowly evolving yeast-complexed aromas of autolysis. Of all the facors contributing to Champagne's *terroir*, location and climate are by far the most important. The height and slope of Champagne's vineyards, its soil, subsoil, and everything else might be important, but by comparison to location and climate, they are merely the icing on the cake.

A SPECIFIC WINE, NOT A STYLE

Champagne is not a generic term that can be used for any sparkling wine: it is the protected name of one specific sparkling wine produced exclusively from grapes grown within the legally defined area of northern France historically known as Champagne. In Europe and various countries throughout the world, strict laws ensure the authenticity of the name "Champagne", but this principle is not respected everywhere. The most blatant abuse in the developed world is in the Unites States. Most producers of world-class American sparkling wines want to be recognised for what they are and where they come from, thus steer clear of muddying the waters, but the ethical (not legal) misuse of "Champagne" is rife amongst the cheapest US fizz. However, even they are not entirely to blame because for many years some of the most powerful Champagne houses hypocritically sold their own-produced South American sparkling wines as "Champaña" or "Champahna". If it's good enough for the *champenois* . . .

CLASSIFICATION OF CHAMPAGNE VINEYARDS

In 1911 Champagne's vineyards were quality-rated on a village-by-village basis using a percentile system known as the *Échelle des Crus,* which determined the difference in price for grapes grown in each village. With just 11 villages rated 100 per cent, the lowest-rated at 22.55 per cent, and the margin between some of the rankings as negligible as 0.02 per cent, this classification was truly percentile in its span and demonstrably fastidious about the smallest differential in quality between one growth and another, thus initially proved to be a fair and robust pricing system.

Long before its eventual demise in 2003, however, the *Échelle des Crus* had gone through so many reclassifications that the lowest rated had soared from 22.5 to 80 per cent, the smallest differential in quality had grown 50-fold, and the system was regarded as little more than a politically biased shopping-list.

The *Échelle de 1911* effectively classified 11 *grands crus* (100 per cent) and, surprisingly, just 7 *premiers crus* (90 to 99 per cent), although neither of those terms were indicated at that juncture because the right to use the terms *grand cru* and *premier cru* in Champagne was not be granted until 1952. By 1985 there were 17 *grands crus,* and by 2003 there were 42 *premiers crus.*

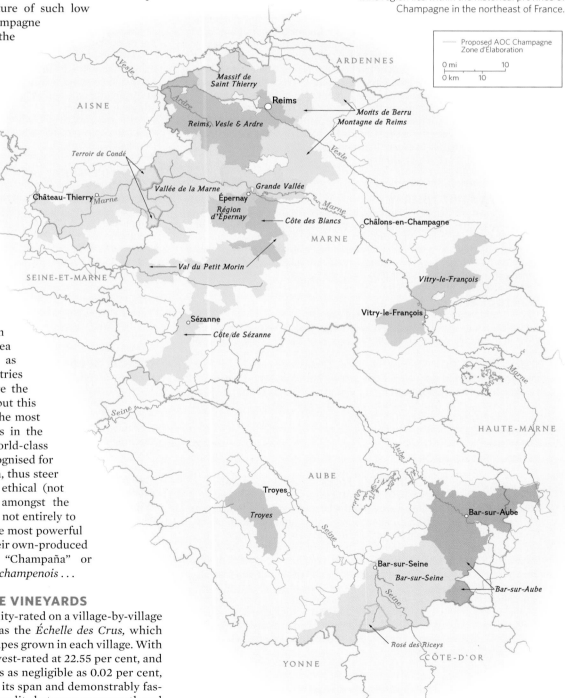

CHAMPAGNE AREAS UNDER VINE
(*see also* p131)
Located in the northeast of France, this world-renown wine region lies within the historical province of Champagne in the northeast of France.

Proposed AOC Champagne Zone d'Élaboration

0 mi   10
0 km   10

DISTRICTS OF CHAMPAGNE

MONTAGNE DE REIMS
• *40% Pinot Noir, 34% Meunier, 26% Chardonnay*
Widely taken to be synonymous with Grande Montagne: a hilly, vine-clad outcrop with a densely forested top that rises to some 270 metres (885 feet) above the surrounding plains between Reims (to the north) and Épernay (to the south), the Montagne de Reims is by its broadest definition, a much larger, S-shaped district, curling around the north of Reims itself, from Monts de Berru to the Massif de Saint Thierry, through the Vesle and Ardre valleys (also known as the Petite Montagne) and the Grande Montagne itself and ending with the vineyards of Bouzy, to the northeast of Aÿ.

SUBDISTRICTS
GRANDE MONTAGNE
• *56% Pinot Noir, 30% Chardonnay, 14% Meunier*
This subdistrict is Champagne's place par excellence for the Pinot Noir, with no fewer than 10 of the region's 17 *grands crus* and 12 of its *premiers crus*. The Grande Montagne can be divided yet again into the northern, eastern, and southern montagnes. Vines on the northern montagne face north and would not ripen grapes but for the convection of warm air from a thermal zone that builds up above the montagne during the day. They yield some of the darkest-coloured Pinot Noir grapes, producing bigger-bodied wines than those from the southern montagne, although southern montagne wines often have a deeper flavour, finer aroma, greater finesse, and, because of the higher chalk content, greater minerality. The eastern montagne is mostly planted with Chardonnay, the use of which is the secret to success for some of the greatest and most age-worthy *blanc de blancs* in Champagne.
Primary style: *Rich and defiantly Pinot*
Best villages: *Ambonnay, Bouzy, Trépail (91% Chardonnay), Verzenay, Verzy, Villers-Marmery (98% Chardonnay)*

MASSIF DE SAINT-THIERRY
• *54% Meunier, 29% Pinot Noir, 17% Chardonnay*
A minor area that some consider to be part of the more informal Petite Montagne district, but it is geologically separate. This is limestone country, not chalk, and the most renown it has ever achieved dates back to the early-19th century for *vin rouge,* not Champagne.
Primary style *Light, non-descript blending fodder*
Best villages *Prouilly, Saint-Thierry, Trigny*

MONTS DE BERRU
• *92% Chardonnay, 7% Meunier, 1% Pinot Noir*
An isolated outcrop of vines a few kilometres east of Reims, the vines grow in three villages perched on the corners of a pure chalk, triangular elevation. It is quite extraordinary to stand in the northern *montagne* and scan the horizon to realise that there are not only vineyards much farther north, but that they are Chardonnay, not Pinot Noir. An even smaller and more isolated outcrop of vines to the east of Mont de Berru is an area known as Moronvilliers, which have been added to the Mont Berru subdistrict for bureaucratic purposes.
Primary style *Light, fresh, and uplifting*
Best villages *Nogent-l'Abesse*

REIMS, VESLE & ARDRE
• *62% Meunier, 25% Pinot Noir, 13% Chardonnay*
The Vesle and the Ardre valleys should to be regarded as separate subdistricts. The Ardre valley is the heart of the so-called Petite Montagne, where 13 *premiers crus* and the best vineyards in this district are found. They provide an excellent source of Meunier, although Pinot Noir and Chardonnay plantations have increased of late, often at the expense of the best Meunier *terroirs.* The city of Reims itself should be separated from the suburbs of Reims, which might consist of villages on the periphery of Reims, but their vineyards are farther out and part of the Vesle valley, Massif de Saint-Thierry, and the lower slopes of the Montagne de Reims. Within the city limits there are no fewer than 54 hectares (133 acres), just over 22 hectares (54 acres) of which belong to Pommery's famous Clos Pompadour (produced exclusively in magnums). Just 1.5 kilometres (1 mile) west of Clos Pompadour is Clos Lanson (first vintage was 2006). A much larger expanse of 28 hectares (69 acres) can be found located either side of the aptly named Allée du Vignoble in Murigny, owned by Taittinger.
Primary style *Fresh and fruity, with some individual wines standing out*
Best villages *Ecueil, Jouy-les-Reims, Sacy, Villedommange*

VALLÉE DE LA MARNE
• *60% Meunier, 23% Pinot Noir, 17% Chardonnay*
The river Marne enters AOC Champagne at Vitry-le-François in the very east of the region, flowing northwards to Châlons-en-Champagne, then west through the very heart of the Champagne region before exiting at Saâcy-sur-Marne, just 36 kilometres (22 miles) from the outskirts of Paris. It is generally considered to be the home of Meunier because of that grape's hardier attributes and greater propensity for heavier soils than Pinot Noir. The vineyards directly facing the Marne have a southern aspect on the right bank and northern on the left.

SUBDISTRICTS
GRANDE VALLÉE
• *65% Pinot Noir, 20% Chardonnay, 15% Meunier*
The Grande Vallée supposedly starts at Tours-sur-Marne (even though its vines are contiguous with those of Bouzy and are thus unarguably part of the Montagne de Reims) and stretches as far west as Cumières. These right bank–only vineyards might contain only two *grands crus,* but one (Aÿ-Champagne) is historically the most famous of all, and this tiny stretch of the river boasts eight *premiers crus,* all of which well and truly deserve the distinction, including one (Mareuil-sur-Aÿ) that many believe to be *grand cru* quality. Although this is Pinot country (especially in Aÿ-Champagne and Mareuil-sur-Aÿ), it can also provide some stunning Meunier.
Primary style *Flashes of intensity and immediacy in general, while Aÿ-Champagne probably produces the most profound Pinot in the region*
Best villages *Aÿ-Champagne, Cumières, Dizy, Hautvillers, Mareuil-sur-Aÿ*

RÉGION D'ÉPERNAY (AKA COTEAUX SUD D'ÉPERNAY)
• *46% Meunier, 42% Chardonnay, 12% Pinot Noir*
The most famous property in this area south of Épernay is Taittinger's Château la Marquetterie at Pierry (Folies de La Marquetterie was launched in 2006). La Marquetterie gained its name because of its reputation for cultivating alternating plots of black and white grapes due, it was supposed, to the differences in soil composition. This has led to the notion that the entire Région d'Épernay possesses an equally complex geology, but it is in fact similar to that of the Côte des Blancs, only with a more sandy clay mix. Unlike Reims, there are no vineyards within Épernay's city limits.
Primary style *Flashes of intensity and immediacy in general, while Aÿ-Champagne probably produces the most profound Pinot in the region*
Best villages *Aÿ-Champagne, Cumières, Dizy, Hautvillers, Mareuil-sur-Aÿ*

TERROIR DE CONDÉ
• *65% Pinot Noir, 20% Chardonnay, 15% Meunier*
The most confusing of all subdistricts because even local growers do not understand why some villages are included while others are not. The Terroir de Condé consists of Barzy-sur-Marne, Passy-sur-Marne, Trélou-sur-Marne, and Baulne-en-Brie, yet only the last of these is legitimately part of the canton of Condé-en-Brie (the name under which this subdistrict used to be referred to), while the first three villages are not only miles apart, but they are located on the right bank of the Marne, whereas Baulne-en-Brie is on the left bank.
Primary style *Capable but disparate* terroirs *that lack a coherent identity*
Best villages *Trélou-sur-Marne*

RIVE DROITE
• *70% Meunier, 18% Pinot Noir, 12% Chardonnay*
Not shown on the map because of it artificially segregates the so-called Vallée de la Marne Ouest, simply because that area crosses the Marne border into the Aisne and Seine-et-Marne, which are an anathema to certain old-school idealists. That said, there are villages within this subdistrict that can rival the Grand Vallée.
Primary style *Floral aromas and precocious fruit, but with more body and potential longevity than normally found outside the Grande Vallée itself*
Best villages *Damery, Ste-Gemme, Venteuil*

RIVE GAUCHE
• *76% Meunier, 12% Chardonnay, 12% Pinot Noir*
Also not shown on the map. It makes no geographical, geological, or topological sense for the left bank of the Marne to stretch so far south through the hinterland that encompasses the upper reaches of the

Surmelin, yet officially it does. Although these north-facing vineyards have twice the chalk content of those on the sunnier, south-facing right bank, they are much less contiguous and are not as well regarded.
Primary style Capable but disparate terroirs *that lack a coherent identity.*
Best villages Leuvrigny

VALLÉE DE LA MARNE OUEST
• 69% Meunier, 16% Chardonnay, 15% Pinot Noir

Also not shown on the map, the Vallée de la Marne Ouest comprises all the vines of the Aisne (aka Champagne Axonais) and, the most westerly of all Champagne vineyards, the Seine-et-Marne.
Primary style Fruity, early-drinking, becoming rustic as it ages
Best villages Charly-sur-Marne, Chézy-sur-Marne, Esômes-sur-Marne, Mont-Saint-Pierre, Sauluchery, Villers-Saint-Denis

CÔTE DES BLANCS
• 85% Chardonnay, 8% Meunier, 7% Pinot Noir

Illogically, this district is composed of five different and widespread subdistricts, including the world-famous Côte des Blancs. If the authorities wish to bring together all the primary but diverse Chardonnay areas under one umbrella name, it would make more sense to call it Les Côtes des Blancs plural, but Côte des Blancs singular it is.

SUBDISTRICTS
CÔTE DES BLANCS
• 97% Chardonnay, 2% Pinot Noir, 1% Meunier

The name of this district is derived from its almost exclusive cultivation of white Chardonnay grapes. More than in any other area of Champagne it is noticeable that the tops of hills and spurs are cultivated with vines, whereas elsewhere they are usually left au naturel with forest or scrub to prevent erosion. However, it is the vines on the mid-slopes that always produce the finest wines (followed by top and lower slopes, then brow, with the plains below well and truly last in line, particularly those vineyards that have crept onto the "wrong side" of the D9). The Côte des Blancs contains six of Champagne's 17 *grands crus* and four of its 42 *premiers crus*.
Primary style Cramant and Avize are aromatically the most floral of all Côte des Blancs with intrinsically citrus fruit, whereas Oger and, particularly, Le Mesnil-sur-Oger, have the greatest minerality and potential longevity.
Best villages Cramant, Le Mesnil-sur-Oger

TROYES
• 91% Chardonnay, 9% Pinot Noir, 0% Meunier

A stone's throw west of Troyes, the ancient capital of Champagne, is Montgueux, where a substantial island of vines exists on pure Turonian chalk. This will be substantially expanded under the new boundary proposals, when the right to grow vines will be extended to 11 neighbouring villages.
Primary style Muscular minerality.
Best villages Montgueux

CÔTE DE SÉZANNE (AKA SÉZANNAIS)
• 75% Chardonnay, 19% Pinot Noir, 6% Meunier

All the villages of the Sézannais are located within the Marne *département* with the exception of Villenauxe-la-Grande, which is in the Aube. Unlike the Côte des Blancs, small amounts of Meunier are planted everywhere, while Pinot Noir represents a significant minority in all but one village (Chantemerle). For many years, this subdistrict has inexplicably remained an unknown corner of Champagne, even though *blanc de blancs* was on the rise and its more opulent style so well suited to the new generation of wine drinkers.
Primary style Extremely aromatic with lush tropical fruit that can be almost musky
Best villages Barbonne-Fayel (including Pinot Noir), Bethon, Villenauxe-la-Grande (including Pinot Noir)

VAL DU PETIT MORIN
• 52% Chardonnay, 38% Meunier, 10% Pinot Noir

Not a truly specialist Chardonnay subdistrict. Indeed, many more of these vineyards were planted with Meunier as recently as the 1990s. Formerly known as the Region de Congy or Region de Congy-Villevenard, the Val du Petit Morin comprises 20 villages between the Côte des Blancs and the Côte de Sézanne. Bizarrely it includes two *premier cru* growths east of Vertus (Voipreux and Villeneuve-Renneville), the vines of which are contiguous with the Côte des Blancs. Coligny (Val-des-Marais) and Etrechy are also classified as *premiers crus*. Apart from Voipreux and Villeneuve-Renneville, clays and argillaceous alluvium dominate the Val du Petit Morin, and outcrops of chalk are relatively few.
Primary style Mixed
Best villages Coligny (Val-des-Marais), Etrechy, Villeneuve-Renneville, Voipreux

VITRY LE FRANÇOIS (AKA VITYRAT)
• 98% Chardonnay, 1% Pinot Noir, 1% Meunier

This is where the Canal de la Marne au Rhin joins the Marne river, thus technically part of the Vallée de la Marne, although its remoteness on the eastern periphery of the Champagne region, its pure chalk soil, and the production of almost exclusively Chardonnay grapes are all good reasons not to confuse the issue by classifying it as part of that Meunier-dominated, clay-soil district. Vitry le François is a rapidly expanding subdistrict, having started with just 3 hectares (7.5 acres) in 1970, the vineyards had grown to 170 hectares (420 acres) by the 1990s and are now in excess 457 hectares (1,130 acres), with five additional villages to receive planation rights under the so-called expansion. The wines produced here are increasingly highly regarded by the region's biggest blenders, although Vitry le François has yet to produce singular Champagnes of truly exceptional quality.
Primary style Fine minerality
Best villages Bassu, Bassuet, Vavray-le-Grand, Vitry-en-Perthois

AUBE
The Aube also includes the most southerly village of the Côte de Sézanne and the entire Troyes subdistrict (*see above*).

CÔTE DES BAR
• 85% Pinot Noir, 11% Chardonnay, 3% Meunier, 1% Other

Generally warmer and sunnier than the rest of Champagne, giving the wines a certain plumpness of fruit, which can be a positive boon in cooler vintages, but can be a disadvantage in years like 2003. This is, however, relative, as the entire district is more northerly than Chablis, which itself is more northerly than Sancerre, thus it is definitively a northern zone of France where crisp whites are the norm and reds are difficult but not impossible to produce. Although planted primarily with Pinot Noir, this district should be Chardonnay country. It used to be extensively planted with Gamay until a law was passed requiring that variety to be pulled up. Local growers opted for Pinot Noir as the obvious black grape upgrade. Nobody questioned why they were growing black grapes or, indeed, why Gamay had been planted in the first place. It was not as if the geology of the Côte des Bar was Gamay-loving granite. No, it is Kimmeridgian, just like nearby Chablis. Pinot Noir might have been a logical black-grape upgrade, but Chardonnay ought to have been the obvious choice and, if the decision had been made any time from the 1960s onwards, it would have been.

SUBDISTRICTS
BAR-SUR-AUBOIS (AKA BARSURAUBOIS)
• 80% Pinot Noir, 13% Chardonnay, 7% Meunier

The smaller of the Aube's two subdistricts, the vineyards here are more scattered than they are in Bar-sur-Seine. The northeastern corner of the Barsuraubois subdistrict located in the Haut-Marne *département* and the number of villages here will increase under the so-called expansion.
Primary style Mellow-classic
Best villages Baroville, Bligny, Colombé-le-Sec, Urville

BAR-SUR-SEINE (AKA BARSÉQUANAIS)
• 88% Pinot Noir, 11% Chardonnay, 1% Meunier

The larger of the Côte des Bar's two subdistricts, the Barséquanais has 50 per cent more vineyards than the Barsuraubois. It also has a greater concentration of more contiguous vineyards and houses UCAVIC, the massive but dynamic super-cooperative that is better known by its primary brand Devaux. The vines are generally more protected than they are in the Barsuraubois, with slightly cooler summers and not quite as cold or as exposed in the winter. The subdistrict also encompasses a third and completely separate AOC called Rosé des Riceys, a dark-coloured still rosé made exclusively from Pinot Noir.
Primary style Fruity-classic.
Best villages Avirey-Lingey, Les Riceys

CHAMPAGNE'S SO-CALLED EXPANSION

The decision by Champagne to increase its area under vine has been rightly condemned. Not because of any expansion per se, but because of its timing. The future expansion of AOC Champagne has been the subject of heated internal debate since at least the 1970s, but its support would wax and wane as sales and stocks went up or down. The worst possible moment to announce an expansion would be to wait until almost every square inch of AOC land had already been cultivated, when sales were at an all-time high (despite efforts to curb demand by increasing prices) and when, as a result, stocks were draining fast. Because it would just look so greedy. And yet that is exactly what the *champenois* did when they took the decision in 2003 and announced it in 2007. The Champagne industry compounded the issue by failing to point out that it was not an expansion outwards, but instead a consolidation inwards. Nor was it quick to point out that in 1951, when sales were down and stocks piling up, they requested a reduction in the size of AOC Champagne. The expansion involves all geographical AOC zones of Champagne.

FUTURE ZONES

The geographical framework by which the Champagne (and indeed every French wine) appellation is constructed consists of three zones: the Zone de l'Élaboration, the Zone de Production, and the Zone Parcellaire de Production de Raisins. The Zone de l'Élaboration is the outer limits of the region, where grapes, bulk wines, and bottled wines may be freely transported. It is also that part of the region where – and only where – it is legal to vinify the wines of that specific appellation. The Zone de Production is located within the Zone de l'Élaboration and consists of the communes where vines may be cultivated, but only within a lacework of delimited areas representing a fraction of the surface area. The Zone Parcellaire de Production de Raisins is that lacework of delimited areas. To provide real-life examples, the appellations mapped in this book are primarily the Zones de l'Élaboration, since they usually represent just one contiguous block or perhaps a few blocks at most and thus are visually easy to grasp.

Under the proposed expansion, the Zone de l'Élaboration is set to increase from 637 communes to 675; not through simple growth, but by removing 117 villages and adding 157. This has nothing to do with increasing vineyards and everything to do with the logistics of moving grapes to *pressoirs* and wines to *cuveries*. The Zone de Production currently consists

WINE NOTES

- There are almost 12,000 different brands of Champagne, including over 3,000 "buyer's own-brands".

- Of the 36,604 owners of AOC Champagne vineyards, most are absentee landlords, with only 15,800 who work the land and who are collectively known as "growers".

- Just 3,200 growers actually make and sell their own Champagne (and many of these also sell a proportion of their harvest to houses).

- There are at least 1,500 growers who appear to sell their own Champagne, but it is in fact identical to the Champagnes sold by fellow members of the same cooperatives (see *Récoltants-Coopérateurs* under Types of Producer, p258).

- Champagne's 15,800 growers own 90 per cent of the vineyards, but sell only 18 per cent of all the Champagne produced (the cooperatives selling another 9 per cent).

- Champagne's 264 houses own just 12 per cent of the vineyards, but account for 73 per cent of Champagne sales.

- One-third of all Champagnes consumed in France are cooperative Champagnes.

- Champagne is unique among AOC wines because it does not have to indicate *"Appellation Contrôlée"* on the label.

- There are 250 million bubbles in a bottle of Champagne, according to photo-analysis, not 49 million as hypothesized by scientist Bill Lembeck (because he did not take into account that the bubbles decrease in size as pressure decreases).

- The pressure inside a Champagne bottle is equal to that inside a double-decker bus tyre.

- As the wine ages, minuscule amounts of oxygen enter a Champagne bottle through the cork, against the internal pressure, due to a principle known as exchange of gases.

- Champagne's pure white chalk subsoil is part of a seabed that dried up 65 million years ago and each cubic inch was formed by the slow accumulation of one trillion dead coccoliths, calcareous sea organisms.

THE *GRAND CRU* AND *PREMIER CRU* VILLAGES OF CHAMPAGNE
Within the three important districts surrounding Épernay, 17 villages have *grand cru* status and 44 the status of *premier cru*.

Vesle

Bezannes
Les Mesneux
Reims
Coulommes-la-Montagne
Vrigny
Cormontreuil
Trois-Puits
Montbré
Pargny-Grigny
Puisieulx
Jouy-lès-Reims
Sacy
Verzenay
Villedommange
Taissy
Beaumont-sur-Vesle
Villers-aux-Noeuds
Sillery
Écueil
Montagne de Reims
Villers-Allerand
Sermiers
Ludes
Verzy
Chamery
Villers-Marmery
Rilly-la-Montagne
Mailly-Champagne
Billy-le-Grand
Chigny-les-Roses
Louvois
Trépail
Champillon
Avenay-Val-d'Or
Tauxières-Mutry
Vaudemange
Mutigny
Bouzy
Ambonnay
Cumières
Hautvillers
Marne
Dizy
Tours-sur-Marne
Vallée de la Marne
Aÿ-Champagne
Mareuil-sur-Aÿ
Épernay
Bisseuil
Pierry
Chouilly
Oiry
Cuis
Côte des Blancs
Cramant
Avize
Grauves
Oger
Somme-Soude
Le Mesnil-sur-Oger
Villeneuve-Renneville
Vertus
Étréchy
Voipreux
Bergères-lès-Vertus
Coligny
(Val-des-Marais)

Grand Cru commune
Premier Cru commune
Commune boundary
0 mi   4
0 km   4

of 319 villages, but is set to rise to 362, with the expansion initially planning to drop 2 and add 40, after which a further 5 villages were added in 2007.

The Zone Parcellaire de Production de Raisins currently extends to 35,208 hectares (87,000 acres) – of which 34,368 (84,925 acres) were planted in 2018 – and the latest semi-official guestimate is that this will increase by some 5,000 hectares (12,355 acres) to around 40,490 hectares (100,050 acres), although in the real world of a parcel-by-parcel revision, the actual figure will depend on the experts' review, the consultations that follow, and the legal challenges that will inevitably erupt in its wake!

CURRENT TRENDS

The major changes in recent years have been:

- **Rise of grower Champagnes.** Talk often lags behind the reality. From the amount of attention that grower Champagnes attract, readers might be forgiven for thinking that their sales are soaring, but they have actually plummeted by more than 26 per cent over the last 10 years and today represent less than 5 per cent of all Champagnes exported. Growers, however, often show more respect for Champagne as a wine, whereas the houses as an industry (not individually) have a tendency to treat it more as luxury goods than fine wine.

PRESSING GRAPES IN A *PRESSOIR COQUARD*
Nearly a third of all Champagne grapes are pressed using this traditional method.

FACTORS AFFECTING TASTE AND QUALITY

LOCATION
This most northerly of the AOC wine regions of France lies some 145 kilometres (90 miles) northeast of Paris and is separated from Belgium by the forested hills of the Ardennes. Four-fifths of the region is in the Marne, and the balance is spread over the Aube, Aisne, Seine-et-Marne, and the Haute-Marne.

CLIMATE
This cold and wet northern climate is greatly influenced by the Atlantic, which has a cooling effect on its summer and makes the seasons more variable but buffered by the Oceanic-Continental corridor. Its position at the northern maritime edge of Europe's winemaking belt stretches the duration of the vine's growth cycle to the limit, making frost a major problem during spring and autumn, with rain threatening rot at harvest-time.

ASPECT
Gently rolling east- and southeast-facing slopes of the Côte des Blancs at altitudes of 120 to 200 metres (380 to 640 feet). On the slopes of the Montagne de Reims (a plateau), the vines grow at altitudes similar to those on the Côte. The best valley vineyards lie in sheltered situations on the right bank of the Marne.

SOIL
The vineyards of the Côte des Blancs, Montagne de Reims, Marne Valley, Vityrat, and Côte de Sézanne are all situated on a porous chalk subsoil up to 300 metres (960 feet) thick, which is covered by a thin layer of drift derived in various proportions from sand, lignite, marl, loam, clay, and chalk rubble, with the topsoil growing deeper with more marl, clay, and sand in the Vallée de la Marne. Champagne's pure-white, chalk subsoil is highly porous, draining almost instantly, which avoids flooding, and yet possesses amazing water-retention at depth, where every cubic metre stores 300 to 400 litres (317–422 gallons) of water, making it the world's largest reservoir, which can be drawn from as deep as 500 metres (1,640 feet) by joint capillary action of both chalk and vines in times of drought. The chalk's high active-lime content encourages the vines to produce grapes that have relatively high acid when they become ripe.

VITICULTURE & VINIFICATION
Vine training systems are restricted to: Cordon de Royat (favoured for Pinot Noir, but may be used for any variety, anywhere), Chablis (favoured for Chardonnay, but may be used for any variety, anywhere), Guyot Single or Double (may be used for any variety, but not allowed in *premieres* or *grands crus)*, and Vallée de la Marne (only for Meunier, not allowed in *premieres* or *grands crus)*. Grapes must be picked in whole bunches, thus until machinery is developed to achieve this, it effectively disallows mechanical harvesting; hence the entire Champagne crop is gathered by hand. Traditionally picking starts mid-September, but since 2003 August harvests every few years have become a fact of Champagne life. Contrary to popular belief, the grapes are not early-picked, but are on average picked two weeks later than Bordeaux. The current 30-year average is for grapes to ripen in Champagne with 9.7% ABV and 53% tartaric acid. Rain invariably interrupts the flowering, resulting in at least two crops, although the second, known in *champenois* as the *bouvreu*, rarely ripens and, as the name indicates, is usually left for the birds. The once-traditional practice of sorting grapes in the vineyards has made a comeback in recent years. Almost one-third of Champagne grapes are still pressed using the traditional, vertical Champagne press or *pressoir Coquard,* although pneumatic presses and Coquard's new inclined stainless-steel press (PAI) have gained ground. Whole-bunch pressing is required by law. Increasing use is being made of mostly stainless-steel vats with temperature-controlled conditions for the first fermentation, but a few houses and many growers still ferment part or all of their wines in cask, with the younger generation employing a percentage of new oak. Bottling for the second fermentation always used to be in the spring following the harvest, but some producers have started to delay this, with varying degrees of success, until summer, autumn, and even the following spring.

The second fermentation gives the wine its sparkle, and this almost always takes place in the bottle in which it is sold (exceptions being for quarter bottles and, for some producers, bottles larger than Jeroboams). Add the 1 to 1.5% of alcohol of chaptalisation needed to bring these wines up to bottling strength and a further 1.5% of alcohol from the second fermentation, and we see the reason for Champagne's "classic lean structure" – because it has the fruit and body of 9.7% ABV grapes but the alcoholic structure of a 12 or 12.5% wine. A non-vintage Champagne must not be sold until at least 15 months after 1 January following the harvest (of the youngest wines in the blend), although most non-vintage Champagnes now spend 24 months on yeast, with a number much higher than that. Vintaged Champagne must have at least 36 months on yeast, and the actual vintage sold will give a good indication of time on yeast. Yeast-complexed fruit aromas derived from autolysis further modify Champagne's classic lean structure. The introduction of Jetting and DIAM MytiK Diamant® corks heralds a new era of consistent TCA-free post-disgorgement development.

GRAPE VARIETIES
Primary varieties: Chardonnay, Meunier, Pinot Noir
Secondary varieties: Arbanne, Petit Meslier, Pinot Blanc, Pinot Gris

- **August harvests.** These are a new and worrying feature. There was not one August harvest throughout the entire 20th century, yet 2003 was the first of six August harvests within the last 15 years. Champagne's famous "lean structure" is best achieved in a sunny September or October, when the vine's metabolism is temporarily halted each night by the sudden drop in temperature, preserving acidity, but on clammy nights in August no such diurnal difference exists, and acids plummet.

- **The lowering of *dosage*.** This has been the knee-jerk reaction to a sudden appearance of August harvests and to riper grapes from traditional September harvests. Some producers have been more creative, combining smaller drops in *dosage* with the use of some non-malolactic wines, a change of priorities for reserve wines (balancing older reserves with younger, fresher, crisper reserves) and other potential remedies, while others chose to include low- or no-sugar *dosage* with a low- or no-SO2 approach that has been trending amongst younger producers.

- **The cult for "oxidative" Champagnes.** This coincided with the trend for low- or no-*dosage* and low- or no-SO2 and makes as much sense as "reductive" sherries. The magnums from some self-proclaimed producers of oxidative-style Champagne are not oxidative, therefore the wine itself was clearly not made in that style, despite claims to the contrary, and oxidation process in the bottles that are oxidative evidently occurs after disgorgement. What would cause that? Low or no SO2 at the point of disgorgement? Oxidation that sets in quickly enough after disgorgement to be able to claim the *faux* style of "oxidative" can only deteriorate further.

- **Champagne rosé.** This is now a major style, whereas throughout all but the last three years of the 20th century, it was of distinctly ephemeral popularity, averaging less than 3 per cent of production, whereas today it represents almost 10 per cent. It was never taken seriously by the *champenois* because it was seen as a fad, the popularity of which would barely last two or three years before consumers moved on, leaving those who had geared up production with surplus stocks that would turn orange before they could be sold. In the run-up to millennium celebrations, however, Champagne rosé took off, eventually becoming the locomotive for increasing the sales of rosé for almost all appellations and origins, both still and sparkling. Instead of wondering whether sales would last two or three years, the *champenois* are now into their third decade and consequently have been able to plan ahead, ring-fencing the right Pinot Noir clones in the most suitable locations to provide the ideal red wine or grapes best-suited to maceration. Quality is zooming!

- **Proliferation of clear-glass bottles.** Any wine in a clear-glass bottle exposed to artificial light for 60 minutes starts building stinky light-strike aromas. This 60 minutes can be accumulative, and in sunlight it is much quicker (*see* Micropedia). While many of Champagne's competitors have quickly learned that clear-glass bottle is a fault waiting to happen, the greatest sparkling wine appellation in the world is willing to sacrifice its hard-earned reputation because the marketing people want these bottles for *blanc de blancs* and rosé *cuvées*. Any *chef de caves* or CEO who uses marketing as an excuse to over-rule quality should be sacked.

- **Jetting and MytiK Diamant® corks.** These are the most important quality-driven inventions in the history of sparkling wine – most others have been about speed, volume, or efficiency. Jetting can either expel all oxygen before the final cork is inserted or guarantee exactly the same, minuscule amount of oxygen in every bottle, thereby eliminating or controlling oxidation, ensuring consistency of post-disgorgement development, if the closure also provides consistency. MytiK Diamant® corks are the only closures at the moment that provide precisely the same oxygen ingress (as opposed to possible variations of up to a thousandfold for agglomerated and natural corks) and a guarantee of no TCA taint. Jetting and MytiK Diamant® corks working together will transform the consistency of all Champagnes, particularly the greatest and oldest. No longer will we have to make the excuse that there is no such thing as a great wine, only great bottles!

- **Increased sales of magnums.** Sales of magnums have increased by more than 50 per cent since the beginning of the new millennium. This underlines the fact that more Champagne is being cellared – although there is nothing to stop readers from opening a magnum rather than a bottle on a regular basis and making it last twice as long. If the wine is even better on the second day, it indicates that it will improve if cellared even longer.

IS HOMEGROWN BEST?

Is it better for a Champagne producer to grow its own grapes? Does this guarantee a better quality and greater consistency of style? As the table below demonstrates, the answer is "maybe" to the first question and "no" to the second. Own grapes do not guarantee a better quality or a greater consistency of style. Own grapes make it cheaper and easier to control the raw material and achieve these ends, but there are more than enough cooperatives and growers making poor Champagne exclusively from their own vineyards to realize that they do not guarantee quality. Gosset, which does not own a single vine, and Alfred Gratien, which owns just one hectare, demonstrate that quality and consistency can be achieved exclusively from purchased grapes. The vineyards belonging to Charles & Piper Heidsieck represent a minuscule 5 per cent, yet the quality and consistency is stunning.

VINEYARDS BELONGING TO THE HOUSES

PRODUCER	VINEYARDS OWNED (hectares)	PROPORTION OF PRODUCTION	CHAMPAGNES MADE FROM OWN GRAPES
Billecart-Salmon	80	40%	Clos Saint-Hilaire (100%)
Bollinger	178	60%	Vieilles Vignes Françaises (100%), Grande Année (80-90%)
Cattier	34	20%	Clos du Moulin (100%)
Deutz	42	18%	Entire range (18%)
Drappier	57	30%	Grande Sendrée (100%)
J Dumangin	5.5	39%	Trio des Ancêtres (100%)
Henri Giraud	12	34%	Entire range (34%)
Gosset	0	0%	None
Alfred Gratien	1	Effectively 0%	Effectively none
Charles & Piper Heidsieck	65	5%	Entire range (5%)
Henriot	30	26%	Entire range (26%)
Jacquesson	33	80%	Toulette, Tue Boeuf, Chemin des Conges (100%)
Krug	21	35%	Clos du Mesnil, Clos d'Ambonnay (100%)
Lanson	57	12%	Clos du Lanson (100%)
AR Lenoble	18	44%	Les Aventures (100%)
Lombard et Medot	5.5	3%	Les Correttes, Les Marquises, Chemin de Flavigny (100%)
Nicolas Maillart	8	57%	Les Chaillots Gillis, Les Francs du Pied (100%)
Marguet	8	91%	Le Parc, La Grande Ruelle, Les Bermonts and Les Crayères (100%)
GH Martel	200	17%	Clos de Château de Bligny (100%)
Moët & Chandon	1,200	20%	Dom Pérignon (100%), Vintage *cuvées* (50-75%)
Mumm	218	25%	Entire range (25%)
Joseph Perrier	21	23%	Entire range (23%)
Perrier-Jouët	65	20%	Entire range (20%)
Bruno Paillard	34	60%	Entire range (60%)
Philipponnat	20	30%	Clos des Goisses (100%)
Pol Roger	92	45%	Entire range (45%)
Pommery	2,800	55%	Les Clos Pompadour (100%)
Roederer	237	50%	Vintage *cuvées* (100%)
Taittinger	288	45%	Entire range (45%)
Veuve Clicquot	390	20%	Grande Dame (up to 100%)

LARGEST PRODUCERS AND THE BRANDS THEY OWN

GROUP	TOTAL PRODUCTION (BOTTLES)	BRANDS
LVMH	65 million	Krug, Mercier, Moët & Chandon, Ruinart, Veuve Clicquot
Lanson-BCC	24 million	Besserat de Bellefon, Boizel (also Camuset, Kremer, Montoy, Veuve Borodin, Veuve Delaroy), Alexandre Bonnet (Ferdinand Bonnet, Petrot Bonnet), Maison Burtin (owns 200 low-profile brands), Chanoine, Lanson, Philipponnat (Abel Lepitre, Henri Peyraud, de Saint Marceaux, Saint Rémi), De Venoge
Vranken Monopole	15 million	Barancourt, Charbaut, Collin, Demoiselle, Diamant, Germain, Heidsieck & Co Monopole, Charles Lafitte, Pommery, Vranken
Pernod Ricard	14.5 million	Mumm, Perrier-Jouët
Laurent-Perrier	14 million	Beaumet, De Castellane (also Chaurey, Freminet, Ettore Bugatti, Jacques Cattier, A Mérand), Delamotte, Jeanmaire, Laurent-Perrier (also Lemoine), Oudinot, Salon
CV de Chouilly	11 million	Mostly Nicolas Feuillatte, plus St Nicholas, St Maurice, Desroches, Camille d'Haubaine, Henri Macquard, Philippe de Nantheuil
Martel-Rapeneau	10 million	Château de Bligny, De Cazenove (also Baudry, J Lanvin, Magenta, Marguerite Christel), Mansard-Baillet (Mansard), G H Martel (also Balahu, Comte de Lamotte, P L Martin, Mortas, De Noiron, Charles Orban, Marcel Pierre, Rapeneau, Charles du Roy), Vieille Français
EPI	9 million	Charles Heidsieck, Piper-Heidsieck
Alliance-Champagne	7 million	Jacquart, Montaudon
Duval-Leroy	6 million	Duval-Leroy (also Baron de Beaupré, E Michel, Henri de Varlane, Paul Vertay)
Thienot	6 million	Canard-Duchêne, Joseph-Perrier, Malard (also Georges Goulet, Gobillard), Marie Stuart, Alain Thienot (also Billiard, Castille, Petitjean)
Taittinger	6 million	Taittinger, Irroy, Saint-Eyremond
Roederer	4 million	Deutz, Louis Roederer (also Théophile)

MOULIN DE VERZENAY (VERZENAY WINDMILL), GRANDE MONTAGNE
Located in the commune of Verzenay, the Moulin de Verzenay looks out over the city of Reims, where GH Mumm & Cie, which owns the vineyards is based, One of the largest producers in Champagne, Mumm owns around 220 hectares (540 acres) of vines, mostly in top *grand cru* vineyards. These holdings account for a quarter of Mumm's production.

SIGN INDICATING GROWER CHAMPAGNE AT CHARTOGNE-TAILLET IN MERFY
Chartogne-Taillet displays a Les Champagnes de Vignerons ("Vinegrower Champagnes") sign, which many *récoltant-manipulant* producers of Grower Champagne exhibit at their facilities.

TYPES OF PRODUCERS

The label on a bottle of Champagne will include some initials, usually found at the bottom and preceding a small code or number. These initials are the key to the type of producer who has made the Champagne.

NM = *Négociant-Manipulant*
Traditionally referred to as a "house", a *négociant-manipulant* is a producer that is allowed to purchase grapes and *vins clairs* in large volumes from growers, cooperatives, and other houses. A few houses, such as Bollinger and Louis Roederer, own sufficient vineyards to supply as much as two-thirds of their own needs, whereas some own none or hardly any at all (Alfred Gratien).

RM = *Récoltant-Manipulant*
Otherwise known as a grower-producer, or grower for short. In principle, a grower is not allowed to buy grapes or *vins clairs,* as the Champagne sold under its label is supposed to be 100 per cent from its own vineyards. However, growers are legally permitted to purchase grapes or wines from each other, up to a maximum of 5 per cent of their total production. The reasoning behind this is to enable those with exclusively Chardonnay vines to buy Pinot Noir for the production of rosé.

CM = *Coopérative-Manipulant*
A cooperative of growers that makes and sells Champagne under one or more brands that it owns.

RC = *Récoltant-Coopérateur*
A grower who delivers grapes to a cooperative and, in part or full payment, receives back ready-made Champagne, which they sell under their own name. Although more than 4,000 RC brands are registered, they are seldom encountered even in France, and many of these cooperative clones are still sold with RM numbers. Moves to rectify this are ongoing.

SR = *Société de Récoltants*
A publicly registered firm set up by two or more growers – often related – who share premises to make and market their Champagne under more than one brand.

ND = *Négociant-Distributeur*
A company that buys and sells Champagne but does not make it.

MA = *Marque d'Acheteur*
A brand name owned by the purchaser, such as a restaurant, supermarket, or wine merchant.

RECENT CHAMPAGNE VINTAGES

2018 An August harvest. Although there will be some truly great Champagnes released by a few producers, this is a vintage that has been saved, not a vintage that is intrinsically great. All the hype suggesting otherwise emanates from a sense of relief, following a runaway ripening process that has caused widespread anxiety over acid levels. However, after harvesting a huge crop of healthy grapes that must be sold, the *champenois* suffered short-term memory loss and nare a word was heard about low acidity again. In fact, it was perfect! Yet the total acidity for the region as a whole was the lowest since the pan-European drought year of 2003 (and that was the lowest ever), while malic acidity in 2018 was the lowest on record. While everyone should make a small amount of pure vintage for future reference in an era of increasingly warmer, earlier-picked years, most of these wines should be used for non-vintage and to flush out 2017s and 2016s from the recent reserves.

2017 An August harvest. Meteorologically the opposite of 2016, yet it resulted in almost equally difficult conditions in the vineyard and disappointing quality. The only advantage this poor vintage has over 2016 is the Chardonnay, which fared better

2016 Famously described by Régis Camus of Piper-Heidsieck as a "nightmare vintage", with the gods throwing everything at the vineyards in Biblical proportions. When the white-skinned Chardonnay turns out to be the hardest hit of all varieties, the year has to be bad in a strange way.

2015 An August harvest. Low acidity, but exceptionally good pH for such an early crop. These fine, nicely structured wines could provide Champagne with the first August harvest of true vintage quality.

2014 A large crop of variable quality (poor to average) that could have been a lot worse, had it not been saved by a dry and sunny September.

2013 Picking of this large crop continued into October, producing an excellent quality of classic style, long and linear fruit, promising to be a lovely slow-developing vintage, reminiscent of 1979s.

2012 A small crop of truly great quality that could challenge 2008 for the title of the greatest vintage of the first two decades of the new millennium. Both 2008 and 2012 have a very fresh, direct, and invigorating style that is exquisitely poised with the crispest acidity, great linearity and boundless energy. Both vintages have a remarkable purity of fruit borne of crops that have been rescued from the jaws of disaster by a sunny September. Only time will tell which is best.

2011 An August harvest. Indeed, this was one of the earliest harvests in history, and it followed an early spring and a hot summer. Exceptions will always exist, but in general terms, 2011 is bit like an undercharged 2016, and that was bad enough.

COOPERATIVE BRANDS AT A GLANCE

COOPERATIVE	PRIMARY LOCATION	MEMBER COOPERATIVES	VINEYARDS OWNED (hectares)	BOTTLES SOLD UNDER OWN BRAND	BRANDS
CVRB (Baroville)	Côte des Bar	1	112	0.18 million	Barfontarc
CCM (Mardeuil)	Vallée de la Marne	1	89	0.55 million	Beaumont des Crayères
CRVC (Reims)	Region-wide	1	900	0.5 million	Castelnau (also De Montpervier)
CVGJR (Janvry)	Reims, Vesle & Ardre	1	115	0.4 million	Ch. De l'Auche (also Charles de Courcelles, Hubert de Lossey, Philippe de Morney, Prestige des Sacres)
URC (Colombé-le-Sec)	Côte des Bar	1	120	0.04 million	Charles Clément
CVF (Fontette)	Côte des Bar	1	330	0.32 million	Charles Collin
Charles Heston (Villers-Franqueux)	Massif de Saint-Thierry	1	115	0.07 million	Charles Heston
CVCA (Ville-sur-Arce)	Côte des Bar	1	310	1.5 million	Chassenay d'Arce (also Montaubret, Alice Bardot, d'Armanville, Lelieur, Baron de Villeboerg and member of Union Auboise)
CVNB (Neuville-sur-Seine)	Côte des Bar	1	170	0.5 million	Clérambault
COGEVI (Aÿ-Champagne)	Vallée de la Marne	1	650	0.5 million	Collet
Coopérative Vini-cole de Cuis	Côte des Blancs	1	40	0.05 million	De Blémond
CUV (Vandières)	Vallée de la Marne	1	154	0.1 million	De l'Argentaine
ACVPC (Aÿ-Champagne)	Vallée de la Marne	1	150	0.6 million	De la Brèche
Union Champagne (Avize)	Côte des Blancs	15	2,300	0.8 million	De Saint Gall (also René Florancy, Lechere, Chevalier de Melline, Pierre Vaudon, Veuve de Medts, Christian Martial)
Union Auboise (Bar-sur-Seine)	Côte des Bar	12	800	0.75 million	Devaux (also Léonce d'Albe, Nicole d'Aurigny)
Dom Caudron (Passy-Grigny)	Vallée de la Marne	1	130	0.7 million	Domaine Caudron
CVM (Mancy)	Region d'Épernay	1	113	1.2 million	Esterlin (also d'Alencourt, Victor Lejeune)
UVCB (Bethon)	Côte de Sézanne	1	120	0.15 million	G. Gruet et Fils
Les Coteaux du Landion (Meurville)	Côte des Bar	1	180	0.03 million	Gaston Cheq
GVV (Vincelles)	Vallée de la Marne	1	123	0.5 million	H. Blin
Alliance-Champagne	Region-wide	4	2,600	7 million	Jacquart (also Montaudon)
CCBN (Bethon)	Côte de Sézanne	1	145	0.35 million	Le Brun de Neuville
UPRM (Le Mesnil-sur-Oger)	Côte des Blancs	1	310	0.1 million	Le Mesnil
CVRC (Grauves)	Côte des Blancs	1	60	0.1 million	Le Royal Coteaux
Mailly Grand Cru	Montagne de Reims	1	70	0.5 million	Mailly Grand Cru
CCR (Les Riceys)	Côte des Bar	1	95	0.06 million	Marquis de Pomereuil
CVMM (Monthe-lon)	Region d'Épernay	1	73	0.02 million	Mont Hau-ban
CVC or CVCNF (Chouilly)	Côte des Blancs	82	2,100	11 million	Nicolas Feuillatte (also St Nicholas, St Maurice, Desroches, Camille d'Haubaine, Henri Macquard, Philippe de Nantheuil)
Grands Terroirs de Champagne (Reims)	Montagne de Reims	1	390	0.5 million	Palmer
COVAMA (Châ-teau-Thierry)	Vallée de la Marne	1	700	0.7 million	Pannier (De Brienne)
La Goutte d'Or (Oger)	Côte des Blancs	1	120	0.5 million	Paul Goerg (also Napoleon, Ch. & A. Prieur)
CVA (Ambonnay)	Montagne de Reims	1	147	0.08 million	Saint-Réol
CVAV (Avize)	Côte des Blancs	1	10	0.07 million	Sanger
Coopérative de Pierry	Region d'Épernay	1	85	0.05 million	Vincent d'Astrée (also member of Alliance-Champagne)

THE WINE STYLES OF
CHAMPAGNE

Note: Although all Champagnes are infinitely superior in magnum format, if a magnum is recommended, the recommendation is exclusively for magnums.

NON-VINTAGE BRUT

No Champagne may be bottled before 1 January following harvest and cannot be sold until 15 months after the date of bottling, with at least 12 months on yeast. Non-vintage *brut* accounts for 80 per cent of all Champagne sold. The base wine for a non-vintage blend, to which reserve wines may be added, will always be from the last harvest. Most producers use between 10 and 15 per cent reserves from the previous two or three years, although some utilize as much as 40 per cent, while a few will add much less reserve wine in volume, but from a greater number of much older vintages. In recent years, when trying to combat the fatter base wines resulting from the increasing frequency of warmer years, there has been a demand at the *assemblage* for a radically different type of reserve wine: one that adds freshness and crispness rather than an instant mellowness and complexity. The introduction of strategic reserves (the so-called *réserve personnelle* or *réserve qualitative individuelle*, as opposed to reserve wines per se) involves holding them *en blocage*, but with the legal right to refresh on a rolling basis, thus provide the perfect tool with which to manipulate their true reserve wine stock. In the not too distant future, it is likely that each producer will achieve its own balance between relatively small volumes of mature reserves and considerably larger volumes of, quite literally, immature reserves, and that the base wine itself will gravitate from the most recent year to a blend of the three most recent years. Some are already doing this. The top wines here include some special and premium *cuvées,* but for real, deep-pocket, truly deluxe *cuvées,* see the best non-vintage prestige or deluxe *cuvées. See also* Multivintage Brut.

♛ *Deutz • Devaux (Cuvee D, magnum) • Charles Heidsieck • Laurent-Perrier (La Cuvée) • Mumm • Piper-Heidsieck • Pommery (magnum) • Louis Roederer • Ruinart ("R") • Taittinger*

VINTAGE BRUT

No vintage Champagne may be sold until at least 36 months after the date of bottling. There is a law that restricts each producer from making or selling more than 80 per cent of any year's harvest as vintage Champagne. This is to ensure that at least 20 per cent of the highest-quality crops are conserved for the future blending of non-vintage *cuvées,* the foundation of the entire wine industry. The sad truth is there is no danger of any producer selling even a tiny fraction of 80 per cent as vintage Champagne, because this category represents less than 2 per cent of Champagne sales, year on year. Whether to declare any year a vintage is up to individual producers. Before the recent warming of Champagne's climate, there would be three to four true vintage-quality years every 10 years, yet most producers averaged seven vintages every 10 years. Vintage Champagne is the product of stricter selection during the *assemblage,* even in the greatest vintage years. By having a large enough palette of raw material to choose from and applying ever more strict selection, it is possible to produce "vintage quality" Champagne in even the poorest years. The three primary differences of vintage Champagne are a different structure, more autolytic finesse, and a slower development in bottle. Because of the stricter selection employed at *assemblage,* most of the base wines chosen have far less chaptalisation, and many have none at all. Since no reserve wines are allowed, vintage-base wines possess more protein (which drops out as reserve wines age) and this helps the autolytic process, endowing the wines with an acacia-like floweriness and other precursors to yeast-complexed aromas.

Finally, not only is it aged longer than a non-vintage, but without a myriad of reserve wine components ageing at different rates, it evolves at a significantly slower rate.

♛ *Barons de Rothschild • Cédric Bouchard • Charles Heidsieck • Lanson (magnum) • Mumm • Piper-Heidsieck • Louis Roederer • Taittinger • Thiénot (magnum) • Veuve Clicquot (magnum)*

BLANC DE BLANCS

Literally meaning "white of whites", this wine is produced entirely from white Chardonnay grapes and possesses the greatest ageing potential of all Champagnes. A *blanc de blancs* may be made in any district of Champagne, but the most famous come from the Côte des Blancs, and the best of those are made between Cramant and Le Mesnil-sur-Oger. It is the most linear of Champagnes, with the longest, most focused profile, relying on intensity without weight.

♛ **Non-vintage:** *Barons de Rothschild • Pierre Gimonnet (Cuis) • Guiborat Fils (Prisme) • Charles Heidsieck • Henriot • Lanson (Extra Age) • Palmer • Perrier-Jouët • Piper-Heidsieck (Essential) • Ruinart*

Vintage: *Ayala • Billecart-Salmon (Cuvée Louis Salmon) • Gaston Chiquet (Réserve d'Aÿ, magnums) • Deutz • Pierre Gimonnet • Le Mesnil (Prestige) • Mumm (RSRV) • Pol Roger (magnum) • Louis Roederer • Vilmart & Cie*

Prestige or Deluxe: *Boizel (Joyeau Chardonnay) • Guy Charlemagne (Le Mesnillésime) • Deutz (Amour de Deutz) • Charles Heidsieck (Cuvée des Millénaires) • Henriot (Cuve 38) • Perrier-Jouët (Belle Epoque) • Ruinart (Dom Ruinart) • De Saint Gall (Orpale) • Salon ("S") • Taittinger (Comtes de Champagne)*

Single-vineyard: *Agrapart (Venus) • Cédric Bouchard (La Haute-Lemblé) • Guy Charlemagne (Coulmets) • Ulysse*

DOM PÉRIGNON CHAMPAGNE *BRUT*, VAINTAGE 2005
"Brut" on the label indicates that this is a dry to very dry Champagne.

LEVELS OF CHAMPAGNE SWEETNESS

The sweetness of a Champagne can be accurately indicated by its residual sugar level, measured in grams per litre. The percentage *dosage*, which some books refer to, is not an accurate indicator because the liqueur added in the *dosage* itself can and does vary in sweetness. The legal levels changed in 2009 and are as follows.

BRUT NATURE: 0 to 3 grams per litre
(Absolutely bone dry!)

EXTRA BRUT: 0 to 6 grams per litre
(Bone dry)

BRUT: 0 to 12 grams per litre*
(Dry to very dry, but should never be austere)

EXTRA SEC OR EXTRA DRY: 12 to 17 grams per litre
(A misnomer – dry to medium-dry)

SEC OR DRY: 17 to 32 grams per litre
(A bigger misnomer – medium to medium-sweet)

DEMI SEC: 33 to 50 grams per litre
(Definitely sweet, but not true dessert sweetness)

DOUX: 50+ grams per litre
(Very sweet; this style was favoured by the tsars but is no longer commercially produced)

* Although it appears as if the top end of the *brut dosage* has decreased from 15 to 12 grams, the regulations permit a 3-gram variation. Officially, this allows for the odd bottle that might still have a little residual sugar after the second fermentation, but in practice, nothing has changed: the top end of *brut* is still 15g.

Collin (Les Enfers, Les Pierrières, Les Roises) • Guiborat Fils (Le Mont Aigu) • Krug (Clos du Mesnil) • Larmandier-Bernier (Vieille Vigne du Levant) • Jacques Lassaigne (Le Cotet, Clos Sainte-Sophie) • Egly-Ouriet (Les Crayères) • Pierre Peters (Les Chétillons)

BLANC DE NOIRS

Unlike New World *blanc de noirs* (which literally translate as "white of blacks"), these Champagnes must be made entirely from black grapes. A classic *blanc de noirs* is 100 per cent Pinot Noir, but Pinot Meunier may also be used, either in its pure form or blended with Pinot Noir. Another departure from New World *blanc de noirs* is that in Champagne it is meant to be white – and the whiter it is, the better it is deemed to be – whereas New Word *blanc de noirs* is not only expected to show some colour, it is often impossible to see any difference between it and the myriad shades of a rosé. The first *blanc de noirs* to gain any fame in Champagne or elsewhere was Bollinger's Vieilles Vignes Françaises, a unique, pure Pinot Noir Champagne traditionally made from overripe grapes grown on two tiny plots of ungrafted vines in Aÿ. It remains the most famous *blanc de noirs* on the market, with a hefty price tag to match. Few other *blanc de noirs* can be compared to the build and weight of the Beast of Bollinger, although the nearest is, unsurprisingly, Krug's even more expensive Clos d'Ambonnay. Relatively few houses make a *blanc de noirs*, but single-vineyard renditions have become increasingly fashionable amongst *terroir*-minded growers.

Non-vintage: *Paul Bara (Comtesse Marie de France)* • *Alexandre Bonnet (Cuvée Noir)* • *Gremillet* • *Palmer* • *Jacques Rousseaux (Grande Réserve)* • *François Secondé (La Loge)* • *De Venoge (Princes)*

Vintage: *Gremillet* • *Lombardi* • *Serge Mathieu* • *Thiénot (Cuvée Garance)*

Prestige or Deluxe: *Bollinger (Vieilles Vignes Françaises)* • *Canard-Duchêne (Charles VII)*

Single-vineyard: *Cédric Bouchard (Bechalin, Presle, Les Ursules, Val Vilaine)* • *Deutz (Côte Glacière, Meurtet)* • *J Dumangin (Hippolyte Pinot Noir, Achille Meunier)* • *Serge & Olivier Horiot (Contrées Val Bazot)* • *Jacquesson (Vauzelle Terme)* • *Krug (Clos d'Ambonnay)* • *Leclerc-Briant (Les Basses Prières)* • *Marguet (La Grande Ruelle)* • *Pierre Paillard (Les Maillerettes)* • *Philipponnat (Les Cintres)*

Pure Meunier: *Xavier Leconte (Coeur d'Histoire)* • *Moussé Fils (Les Vignes de Mon Village)* • *Eric Taillet (Bansionensi)*

EXTRA SEC

Rarely encountered except on supermarket-quality Champagne, fooling all those who like to "talk dry but drink sweet" with the "extra dry" misnomer on a Champagne with 12 to 17 grams of residual sugar per litre (up to 20 grams, if the margin of error is taken into account).

EXTRA BRUT

The best Champagnes in this subsection of the *brut* category can be wonderfully bracing, but with just 0 to 6 grams of residual sugar, cellaring is not advised unless in magnum. The top end of the scale (4 to 6 grams) represents a good *dosage* for late-released mature vintages.

SEC

With a residual sugar greater than 17 grams per litre but less than 32 grams, the lower end of this style (say, 17 to 24 grams) can make a fabulous gastronomic experience when accompanying dishes with a distinct hint of sweetness, such as *foie gras* or a game dish with a fruit component.

RICH or RICHE

Under the old Champagne laws, these terms were strictly used as an official alternative designation for *demi-sec*, but they are currently used for anything between *sec* and *doux*.

DEMI SEC

Although this style of Champagne may contain between 33 and 50 grams of residual sugar per litre, most are closer to the minimum these days and, as such, fall between two stools: too sweet for conventional use at 33 grams, while even at 50 grams, a *demi-sec* Champagne would struggle with many desserts. Much *demi-sec*, therefore, is used to dispose of inferior-quality Champagne, hidden beneath a veneer of sugar and sold to consumers who do not know better and could not care less as it is quaffed at Christmas or the New Year. If you are a consumer who does know better, however, and has found a good use for *demi-sec* Champagne, the following are seriously produced using high-quality base wines.

Non-vintage: *Billecart-Salmon* • *Henriot* • *Mumm* • *Pol Roger (Rich)* • *Piper-Heidsieck (Sublime)* • *Taittinger (Nocturne)*

Vintage: *Deutz*

DOUX

According to André Simon, *doux* represented as much as 60 per cent of all Champagne shipments until 1960, and half of that had 100 grams of residual sugar, but then *brut* caught on, and within 20 years sweet Champagnes had disappeared. Only in recent years have they started to make something of a comeback, but the *champenois* have forgotten the second-most essential *dosage* ingredient for this style (*esprit de cognac*), and the results have been predictably disappointing. About the best so far has been Veuve Clicquot's Rich and Rich Rosé, both of which contain 60 grams of residual sugar. They work as *doux*, but they are in fact designed for mixology – to add ice, orange peel, or other garnishes or extras.

ROSÉ

The first conventional Champagne rosé was an *Oeil de Perdrix Mousseux* that was made by Ruinart and sent on 14 March 1764 to Baron von Welzel, the "chief cupbearer" to the Grand Duke of Mecklenburg-Strelitz, whereas the first blended Champagne rosé was produced by Clicquot in 1777. Conventional rosé exclusively involves the maceration of skins and juice to extract pigments, but because this region has a long history of blending white wine with a little red to achieve a pink colour, and most Champagne rosé has always been produced by this method, it was allowed by AOC regulations, which merely codify traditional practices within a legal framework. Thus for many years, Champagne rosé was the only rosé in Europe allowed to blend white wines with red. Elsewhere it was illegal. With increased EU integration, however, the right to use this anomalous method inevitably had to be granted to the producers of all other sparkling rosé wines in Europe. Is it an inferior method? No. Nearly all of the greatest and most expensive Champagne rosé *cuvées* are made by blending, but the red wine used has to be special.

Until the late 1990s, Champagne rosé was highly cyclical, enjoying short bursts of popularity between decades of minimal interest. Under such circumstances, there was no motivation to fine tune the production of red wines best suited to colour a rosé, and most Champagne rosés relied on Coteaux Champenois purchased on the internal market through a *courtier*. Demand for the rosé style has been non-stop since 1997, however, and it is now a permanent feature in the range of every producer. This has enabled the *champenois* to ring-fence the most appropriate plots and perfect techniques to craft deeply coloured red wines that are extremely soft, opulent, and aromatically expressive, yet with very little depth of red wine flavour and absolutely no firmness. If sold as a red wine, they would fall over pretty quickly, but as a colouring agent, such wines are perfect.

It is possible to make equally great Champagne rosé through maceration, but because of the region's location and climate, and the fact that its grapes are picked at sparkling wine ripeness, not red wine ripeness, the tannins are harsh and not ripe, easily rendering the wines far too rustic to achieve any of the elegance or finesse expected from this style of Champagne and, in the worst-case scenario, leaving them prone to oxidative aromas. Some producers, such as Louis Roederer, have demonstrated that it can be done on a consistent basis, but it's not as simple as just macerating the juice. Roederer, for example, cold-soaks Pinot Noir grapes in Chardonnay juice for one week, never crushing the black berries and with absolute fermentation, which would only extract tannins. The coloured Chardonnay juice is then drained off, after which the Pinot Noir grapes are allowed to be pressed, and the two are combined with 20 to 50 per cent acidic Chardonnay juice. This is because the Pinot Noir has been picked for ripeness, and needs additional, natural acidity for balance. The acidity also helps stabilize the colour. The temperature is then gently increased to encourage the first fermentation, with approximately 20 per cent fermented in oak in large, used French oak *foudres*. So no shortcuts there. Beware of claims of *saignée*. True *saignée*, as the name (literally "to bleed") suggests, is the coloured, free-run juice that naturally bleeds from black grapes before they are pressed – no maceration is involved, other than that passively caused by bunches weighing down on themselves in the press. Yet, if you read the back label of many Champagnes claiming to use the *saignée* method, you will see that they also claim to be produced by maceration. *Saignée* has become little more than an inaccurate marketing term.

Non-vintage: *Barnaut (Authentique)* • *Deutz* • *Devaux (Cuvée D, magnum)* • *Charles Heidsieck* • *Perrier-Jouët (Blason)* • *Ruinart (magnum)* • *Thiénot* • *Veuve Clicquot (magnum)*

Vintage: *Deutz* • *Charles Heidsieck* • *Pol Roger* • *Louis Roederer*

Prestige or Deluxe: *Billecart-Salmon (Cuvée Elisabeth)* • *Deutz (Amour de Deutz)* • *Dom Pérignon* • *Perrier-Jouët (Belle Epoque)* • *Louis Roederer (Cristal)* • *Rare* • *Ruinart (Dom Ruinart)* • *Veuve Clicquot (Grande Dame)*

Single-vineyard: *Veuve Fourny (Les Rougesmonts)* • *Philipponnat (Clos des Goisses Juste)*

BRUT NATURE

The first non-*dosage* Champagne to be sold was Laurent-Perrier's Grand Vin Sans Sucre in 1889. In the modern era, these wines became fashionable in the early 1980s, when consumers began seeking lighter, drier wines. It was a trend that was driven by critics with too little Champagne experience, who had enjoyed the privilege of tasting wonderful old vintages straight off their lees, leading them to believe that the *dosage* was "make-up" and a Champagne without any *dosage* was somehow intrinsically superior. A *brut nature* can be the most wonderfully fresh and invigorating style of Champagne, but as a rule, it is best consumed when fresh and full of vitality. I recommend cellaring *brut nature* only in magnum format or larger, only if sealed with DIAM MytiK Diamant® cork, and only if it has received at least 20g/L SO2 at disgorgement. Even then, what the ageing process brings is more about feeding curiosity than providing an improvement. Few *bruts nature* have sufficient consistency to be recommended. Most of the best are to be encountered *en passant* and should be grabbed while they can. Don't make the mistake of leaving it until another day because, for that Champagne, another day might never come.

Non-vintage: *Guy Charlemagne*

Vintage: *Louis Roederer (Philippe Starck)*

Single-vineyard: *Agrapart (Venus)*

MULTIVINTAGE BRUT

This term is not legally defined, enabling it to become the most abused description in the Champagne industry, thanks to those producers who have a problem with the term *non-vintage*. They seem to think it is a negative term, when in fact it simply means what it says: "a wine without a vintage". In French, this is even more explicit, as the equivalent term *sans-année* literally means "without vintage". Instead of being ashamed of creating something that is greater than the sum of its parts by way of the art of *assemblage*, they should be proud of *sans-année*. Even more important, they should not mislead consumers by using the term *multivintage* for a non-vintage that is not a blend of exclusively vintage years. By this definition Krug Grande Cuvée is not a multivintage, although it is a great prestige *cuvée*. If a Champagne that claims to be a multivintage contains any wine from years that have not sold or set aside to be sold as a vintage by the producer in question, it could lead to a prosecution under various sales description laws

PRESTIGE *CUVÉE* CRISTAL CHAMPAGNE, VINTAGE 2007
Although Louis Roederer's Cristal was first produced in 1876, it was initially made exclusively for Tsar Alexander II and was not sold to the public until 1945, making Dom Pérignon, launched in 1936, the very first commercial available prestige *cuvée*.

worldwide. To prevent this, the term *multivintage* should be enshrined in law, making it obligatory to indicate the vintage years on the back label.

♛ *Cattier (Clos du Moulin) • Laurent-Perrier (Grand Siècle, magnum) • Moët & Chandon (MCIII)*

PRESTIGE or DELUXE CUVÉE

A prestige *cuvée* is typically made from a producer's own vineyards, with a blend that is often restricted to *grand cru* grapes, although the very greatest examples (Cristal and Dom Pérignon) also happen to contain a small amount of *premier cru* grapes. If vintage Champagne is the product of stricter selection during the *assemblage*, then the selection for prestige *cuvées* is several orders higher. They are invariably produced by the most traditional methods (fermented in wood, sealed with a cork and agrafe, rather than a crown-cap, then hand-disgorged), aged for longer than the regular vintage *cuvée*, and sold in specially shaped bottles at very high prices. Some are clearly over-priced, while others are over-refined, having so much mellowness that all the excitement has been lost, but a good number are truly exceptional Champagnes and worth every penny.

Cristal and Dom Pérignon still dominate the prestige *cuvée* market today, followed by Perrier-Jouët Belle Epoque, with all the others way behind. As the Champagne world is flooded with premium and other even more special *cuvées*, I have kept my top selections to absolute, slam-dunk prestige *cuvées*.

♛ **Non-vintage:** *Krug (Grande Cuvée) • Laurent-Perrier (Grand Siècle) • Moët & Chandon (MCIII)*
Vintage: *Deutz (Amour de Deutz) • Dom Pérignon • Krug (Vintage) • Perrier-Jouët (Belle Epoque) • Rare • Louis Roederer (Cristal) • Veuve Clicquot (Grande Dame)*

SINGLE-VINEYARD

The first single-vineyard to be commercialised was Clos de la Chapitre in the 1860s, when Amédée Tarin sold it as Clos du Mesnil and produced it in two styles: a *blanc de blancs* and a

rosé. It was not widely known until well after Krug purchased it in 1972 and immediately replanted the *clos*. When Krug launched its first vintage (1979) of Clos du Mesnil in 1986, the market slowly started to understand what a single-vineyard Champagne was. It took the fame of a great producer like Krug to promote the concept because Philipponnat had been producing its now-famous Clos des Goisses since as early as 1935, and it took a while before the market realised that Drappier's Grande Sendrée was a single-vineyard Champagne, even though 1975 was its first vintage.

Leclerc-Briant announced its trio of Les Authentiques in 1994, and the fanfare of Moët's ill-fated Trilogie des Grands Crus in 2001 certainly put single-vineyards on the map. In 2003 Billecart-Salmon launched the 1995 vintage of Clos Saint-Hilaire, and Taittinger quietly released its Les Folies de la Marquetterie in 2004. The houses had woken up to the concept of single-vineyard *cuvées* by the turn of the millennium, yet the growers, who were ideally placed to exploit Champagne's *terroir*, were still fast asleep. Having seen, however, how these wines have gained traction, demand very high prices, and generate greater profit margins, the current generation of growers has not been as slow as its predecessors, crafting a raft of their own single-vineyard Champagnes. There are now over 250, and that number is still climbing.

♛ **Non-vintage:** *Cédric Bouchard (Bechalin, Presle, Les Ursules, Val Vilaine) • Cattier (Clos du Moulin) • Ulysse Collin (Les Enfers, Les Pierrières, Les Roises) • Egly-Ouriet (Les Crayères) • Jacques Lassaigne (Le Cotet) • Taittinger (Les Folies de la Marquetterie)*
Vintage: *Krug (Clos d'Ambonnay, Clos du Mesnil) • Philipponnat (Les Cintres, Clos des Goisses, Clos des Goisses Juste Rosé) • Pommery (Clos Pompadour) • Cédric Bouchard (La Haute-Lemblé) • Guy Charlemagne (Coulmets) • Deutz (Côte Glacière, Meurtet) • Larmandier-Bernier (Vieille Vigne du Levant) • Jacques Lassaigne (Clos Sainte-Sophie) • Pierre Peters (Les Chétillons)*

COTEAUX CHAMPENOIS AOC

This appellation covers still wines (red, white, and rosé)

produced in Champagne. They used to be low in quality, but high in price. Now they are even more expensive, yet they can be positively exciting in quality. Although still variable due to Champagne's location, Coteaux Champenois can now soar to greater heights, thanks to climate change and the current generation of Champagne growers who want to explore the full and varied potential of their *terroir*.

♛ *Bérêche (Blanc Les Monts Fournois, Les Montées) • Bollinger (La Côte aux Enfants) • Dehu (La Rue des Noyers) • Charles Heidsieck • Jacques Lassaigne (Haut Revers du Chutat)*

RATAFIA AOC

This *vin de liqueur* is made in the same fashion as Pineau des Charentes AOC (*see* p284), by adding neutral spirits to unfermented grape juice. The only ratafia I have found outstanding in 40 years of tasting remains Janisson Baradon 2005 Single-Cask, which, although brown in colour and not particularly limpid, possessed an incredible taste of pure chocolate and was a delight to drink.

ROSÉ DES RICEYS AOC

This is not part of the Coteaux Champenois AOC but is a totally separate appellation. This pure Pinot Noir, still, pink wine is made in the commune of Les Riceys in the Aube *département* and is something of a legend locally. Its fame dates back to the 17th century and Louis XIV, who is said to have served it as often as he could. Traditionally, Rosé des Riceys is rather deep coloured and must be made by partial *macération carbonique*. It has been such a long time since I tasted the greatest Rosé des Riceys of my life, a 15-year-old vintage 1971 from Horiot Père & Fils, that I have begun to wonder just how great it really was. I am in print as having waxed lyrical since day one, however, so I suppose I will have to trust the palate of my former self. Since there is a resurgence of interest in Rosé des Riceys by the current generation of growers, I live in reasonable hope of encountering further examples of such a high order.

THE WINE PRODUCERS OF
CHAMPAGNE

HENRI ABELÉ
Avize
★

Owned by Freixenet, which was overtaken by Henkell in 2018, this is the second foothold the German giant now has in Champagne (Alfred Gratien being the other, of course).

✔ *Entire range (preferably in magnum)*

ACE OF SPADES
See Armand de Brignac

AGRAPART
Avize
★☆

Malolactic can show through at times, but there is no doubting the creamy-walnut complexity of Agrapart's single-vineyard Vénus, although its Mineral and L'Avizoise *cuvées* consistently outperform it.

✔ *Entire range (preferably in magnum)*

APOLLONIS
Le Mesnil-sur-Oger
☆

Capable of fresh and grippy Meunier, with nice energy and life.

✔ *Cuvee Authentic Meunier*

ARMAND DE BRIGNAC
Chigny-les-Roses
★

Also known as Ace of Spades, this Champagne was launched by Cattier in 2006, but will be forever associated with Shawn Carter, aka Jay-Z, who has owned its distribution rights in the US since 2014, having promoted this Champagne as early as 2006 in his video of "Show Me What You Got", a single from his 2006 "comeback" album *Kingdom Come*. Always young, fresh, soft, and fruity, but each *cuvée* is a serious step up in quality in magnum (equivalent two stars). Best consumed on purchase or within three years, although the *blanc de blancs* possesses a longer potential longevity.

✔ *Entire range (magnum)*

AYALA & CO
Aÿ-Champagne
★☆

Owned by Bollinger, Ayala aims lower than its famous parent brand, but hits its more modest target more accurately and more consistently. Specialises the low-/no-*dosage* style, although *brut* still represents the majority of sales.

✔ *Entire range*

PAUL BARA
Bouzy
★☆

Grower Champagnes that are always made with the pure expression of the richness of Bouzy. Comtesse Marie de France is the ultimate standout here.

✔ *Millésimé • Spécial Club • Comtesse Marie de France*

BARNAUT
Bouzy
★☆

Philippe Secondé is on form and improving. In 2016, his Authentique Rosé won the first (and, at the time of writing, only) Grower Champagne trophy awarded by Champagne & Sparkling Wine World Championships.

✔ *Entire range*

BARONS DE ROTHSCHILD
Reims
★★

These well-made Champagnes are sold with the heavy premium of the Rothschild imprimatur, but currently outperform well-established Champagne brands, warranting a significant hike in its star rating. All three wine branches of the Rothschild dynasty put their historical differences to one side to form this joint venture in 2005 in collaboration with the Goutte d'Or cooperative in Vertus, where the wines are made. Vintaged *cuvées* are particularly impressive.

✔ *Entire range*

BEAUMONT DES CRAYÈRES
Mardeuil
☆

The vineyards of this cooperative are mostly of modest origin, but they are tended like gardens, without resort to machines or chemical treatments to produce richly flavoured Champagne with a light *dosage*.

✔ *Fleur de Rosé • Fleur de Prestige*

BÉRÊCHE
Ludes
★

Some *cuvées* do not cellar well, but the oxidative character that can blight these Champagnes does appear to be diminishing, and the Coteaux Champenois can be outstanding.

✔ *Brut Réserve, Les Monts Fournois, Les Montées*

BESSERAT DE BELLEFON
Épernay
★

The quality has steadily risen under Lanson-BCC, particularly in magnum. On their own, Brut Bleu and Grand Cru Blanc de Blancs in magnum are two-star quality. The primary criticism here is too much clear glass.

✔ *Brut Bleu (magnum) • Grand Cru Blanc de Blancs (magnum)*

BILLECART-SALMON
Mareuil-sur-Aÿ
★★

With the exception of the Cuvée Sous Bois and Clos Saint-Hilaire, which can be oaky and oxidative, the style of this beautiful boutique Champagne house remains true to its original form of freshness, purity, and bright fruit, with 100

years of fame for its pale and delicate rosé. The non-vintage *cuvées* are tremendous value, while the vintaged *cuvées* can be absolutely majestic.

✔ *Entire range*

CHÂTEAU BLIGNY
Bligny
★

Part of the family-owned *négociant* group, GH Martel, who run this property as a *récoltant-manipulant*. Improving quality.

✔ *Blanc de Blancs • Clos de Château Bligny*

BOIZEL
Épernay
★★★

Part of Lanson-BCC and still run by its former sole owners, the warm and welcoming, customer-friendly Roques-Boizel family, who remain major shareholders in Lanson-BCC itself. Freshness and fruit are the hallmarks here, with vintaged *cuvées* possessing exceptional length and intensity.

✔ *Blanc de Blancs • Rosé • Grand Vintage • Joyau de France (brut, Chardonnay, rosé)*

BOLLINGER
Aÿ-Champagne
★★

Despite the high rating (given for intrinsic, not lasting, quality), the problem here is the same one Bollinger has always had: no SO_2 at time of disgorgement, which has been the case since the days of Madame Lily Bollinger. When such wines are first released, they can shine brightly and beautifully for a limited period of time, during which they will be fabulous. They soon deteriorate if cellared, however, unless a collector likes Bollinger's downward slope of bruised apple fruit with oxidative and aldehydic aromas. Although intentional, the previous *chef de caves*, Mathieu Kauffmann, attempted to combat this vulnerability through the introduction of a new bottle shape based on an original 1829 design. This has a narrower neck that is supposed to reduce oxygen ingress by 10 per cent, but the first release was more oxidative than the last release in the old bottle. Kauffmann also had plans to install jetting in 2012, but they were shelved, and in 2013 he became first *chef de caves* in Bollinger's history to resign. He did so "on friendly terms", but with immediate effect and has had impeccably sealed lips ever since. His successor, Gilles Descôtes, who had been with Bollinger 10 years when he took over and was thus well-versed in his employer's idiosyncratic sulphur regime, immediately hired Denis Bunner, an oenologist with the CIVC, to oversee the introduction of jetting. This was not simply to reduce the amount of oxygen entering the bottle during disgorgement, but also for consistency: to guarantee that each bottle would have exactly the same oxygen ingress at this vital juncture. Some of Bollinger's *cuvées* might shine brightly for longer than others, depending on

how much free SO_2 there is left over prior to disgorgement, but they would have to be seriously over-sulphured in the first place to compensate for the oxidative shock of disgorgement and provide protection going forward. Common sense dictates drink on purchase.

✔ *Entire range (magnums only)*

ALEXANDER BONNET
Les Riceys
★☆

Over the last 20 years or so, since purchased by what is now Lanson-BCC, Alexander Bonnet has developed from being a useful source of Aube vineyards (47 hectares, or 116 acres) and BOBs, including some very fine *blanc de noirs*, to a very well-respected producer in its own right.

✔ *Cuvée Noir • Cuvée Noir Rosé • Blanc de Noirs • Perle Rosée*

CÉDRIC BOUCHARD
Celles-sur-Ource
★★

This is one artisanal producer who manages to get away with no *dosage* more times than not. With very few vines – and those pruned to reduce yields – the production is tiny and prices are high. All the wines are from single-vineyards, with a style that is seductively soft and vibrantly fruity, but that description does no justice to the nuanced complexity and finesse of each *cuvée*. These wines benefit from decanting and not being served too chilled. All these *cuvées* are beautifully presented under the Jeanne de Roses label.

✔ *Entire range*

RAYMOND BOULARD
Cauroy-les-Hermonville
★

Run by Francis Boulard, whose style tends to be lifted fruit with oxidative complexity. Avoid the *brut nature*, which usually needs the *dosage* it lacks.

✔ *Cuvée Rosé Brut*

CHÂTEAU DE BOURSAULT
Boursault
★

Madame Clicquot's historic château is owned by the Fringhian, whose Champagnes now outperform the vineyard's former 84 per cent *échelle*, probably due to the fact they are a genuine *clos* enclosed within the château's walls.

✔ *Rosé*

EMMANUEL BROCHET
Villers-aux-Noeuds
★☆

With just 2.5 hectares (6 acres) and the same *cuvées* not necessarily produced every year, the tiny production from this highly regarded grower is much sought after. Very pure and precise, with great minerality of fruit.

✔ *Entire range*

CANARD-DUCHÊNE
Ludes
★

Veuve Clicquot owned Canard-Duchêne from 1978 until 2003, when it sold it to the Alain Thiénot group. The problem here is inconsistency, with far too many occasions when even the best *cuvées* are oxidative or have VA issues. When these Champagnes are on form, however, they are very fine indeed and represent stunning value for money.

✓ *Charles VII Blanc de Blancs • Charles VII Blanc de Noirs • Léonie • Vintage*

CASTELNAU
Reims

Produced by CRVC (*Coopérative Régionale des Vins de Champagne*), which was the original creative force behind the Jacquart brand prior to the formation of Alliance Champagne. CRVC severed all ties with Jacquart in 2006, however, and since then De Castelnau (simply Castelnau since 2017) became its primary focus. Magnums are outstanding.

✓ *Blanc de Blancs • Vintage • Hors Catégorie*

CATTIER
Chigny-les-Roses

The star here is Clos du Moulin, Cattier's great single-vineyard Champagne, which would be a two-star Champagne in its own right. Clos du Moulin is a blend of three vintage years (indicated on the back label). Too many clear glass bottles. *See also* Armand de Brignac.

✓ *Blanc de Noirs • Clos du Moulin*

CLAUDE CAZALS
Reims
☆

It was the late Claude Cazals, who, with Jacques Ducoin, co-invented the first true precursor to the *gyropalette*, filing for a patent in 1968. It was also at Champagne Claude Cazals where Laurent Fresnet, the recently appointed *chef de caves* at GH Mumm, cut his teeth in the 1990s. I would like to see more freshness and elegance in these wines.

CHARLES DE CAZANOVE
Reims
★

Established by Charles de Bigault de Cazanove in Avize in 1843 (on the site of a former Champagne producer founded in 1811), Champagne de Cazanove enjoyed considerable fame in Paris at the end of the 19th century. In 1958 the De Cazanove family sold the business to the Martini group, which sold it to Moët in 1979. Retaining the vineyards, Moët disposed of the brand, cellars, and stocks to the Lombard family in 1985, who sold it to GH Martel in 2004. On the whole, Lombard improved the quality, but was not consistent and sometimes prone to heavy-handed malolactic. Under GH Martel, consistency has improved, and the production is well made and good value, especially the Stradivarius range. Stradivarius was originally a Chardonnay-dominated prestige *cuvée* launched by Thierry Lombard, but the current owners have fleshed out what was an exceptional Champagne de Cazanove *cuvée* into a full range.

✓ *Entire Stradivarius range*

CHANOINE
Reims
★

This second-oldest Champagne house Chanoine was established in 1730, just one year after Ruinart, but had disappeared until it was relaunched by Philippe Baijot and Bruno Paillard in 1991. Now part of Lanson-BCC, these Champagnes can be excellent quality and value, if they make it out of the winery alive. Unfortunately, many of the best Tsarine and Tsarina *cuvées* are produced in clear glass bottles, and if they are not compromised by light-struck aromas before shipment, they all too often are when put on display in retail outlets. *Caveat emptor.*

CHAPUY
Oger
★

Apart from the new Livrée Noire label, the presentation of these Champagnes still harks back to the 1950s, but the Champagnes inside are fresher and more up-to-date, particularly the Blanc de Blancs Grand Cru Millésimé.

✓ *Brut Réserve Grand Cru • Brut Blanc de Blancs Millésimé • Livrée Noire Cuvée Prestige*

GUY CHARLEMAGNE
Le Mesnil-sur-Oger
★★

Rich and exuberant in youth, these great Champagnes tighten up in bottle before emerging with even greater intensity after several years of cellaring under ideal conditions. Coulmets and Le Mesnillésime are the stars here.

✓ *Entire range*

CHARTOGNE-TAILLET
Merfy
★☆

Elisabeth and Philippe Chartogne work hard to squeeze every morsel of potential from their vines, and their current focus on single-vineyard *cuvées* is part of that effort, but they need to dial down the oak. Notwithstanding the issue of oak being too dominant, there are some fine Champagnes produced here.

✓ *Entire range*

CHAUVET
Tours-sur-Marne
★☆

A small, quality-conscious house situated opposite Laurent Perrier, Chauvet is owned by the Chauvet-Paillard family and run by Jean François Paillard, who is a cousin of Pierre Paillard of Bouzy and Bruno Paillard of Reims and a nephew of Antoine Gosset, the former owner of Champagne Gosset in Aÿ. Clean, precise, and well-made wines created by possibly the most unknown or under-rated producer in Champagne.

✓ *Entire range*

GASTON CHIQUET
Dizy
★☆

Owners Antoine and Nicolas Chiquet are cousins of brothers Jean-Hervé and Laurent Chiquet of Champagne Jacquesson. Apart from the occasional, malolactic-dominant aroma, this environmentally friendly, family-owned *récoltant-manipulant* produces lush, lovely, creamy, fruit-driven Champagne of exceptional value. Even when they make something tongue-in-cheek, like the sub-entry-level Insolent *cuvée*, it turns out well. Why Insolent? Because it is made from 100 per cent *vins de tailles*, something no other *champenois* would dare to admit, and to achieve a pale, bright colour, it has to be completely carbon-filtered: also something no other *champenois* would dare to admit. If that is not insolent enough, its very fruity primary aromas are the nearest you will find to Prosecco in Champagne. In terms of pure quality, Réserve d'Aÿ stands head and shoulders above a super range of boutique beauties.

✓ *Entire range*

ULYSSE COLLIN
Congy
★★

The star player in Champagne's little-known district of Val du Petit Morin, Olivier Collin (son of Ulysse) has always produced big, Burgundian-styled Chardonnay and bold Pinot Noir Champagnes. He can overdo the oak, and there has been a tendency towards oxidation, but the oak is always the right type, and he fills his Champagnes to the brim with majestic fruit. As Collin's single-vineyard Champagnes have gained precision over the years, so the oxidative character has started to fall away. Completely in many cases. And we see he is producing a quality that is well beyond the norm for the Val du Petit Morin.

✓ *Entire range*

HAUTVILLERS CAVE WITH CHAMPAGNE BOTTLES IN *GYROPALETTES*
The *gyropalette* was invented by Claude Cazals and Jacques Ducion. It is used to make sparkling wine by the traditional method, in which the second fermentation takes place in the bottle. Rather than cellar workers performing the tasks manually, the *gyropallete* twists the bottles, shakes them slightly, and moves them progressively to a vertical position with the neck pointing down. This process, called *remuage,* removes the yeast sediment left in the bottle after the fermentation.

DEHU PÈRE ET FILS
Fossoy
★☆

The top wine amongst these Meunier-dominated Champagnes has to be the 100 per cent Meunier single-vineyard La Rue des Noyers.

✓ *La Rue des Noyers*

DELAMOTTE PÈRE & FILS
Le Mesnil-sur-Oger
☆

Part of the Laurent-Perrier group and located next door to Champagne Salon, which also belongs to Laurent-Perrier, Delamotte is a small *négociant* house that has been privately owned by the Nonancourt family (the owners of Laurent-Perrier) since the end of World War I, when it was purchased by Marie-Louise de Nonancourt, the sister of Victor and Henri Lanson. Delamotte is, technically, the origin of Champagne Lanson.

✓ *Blanc de Blancs (NV & Vintage)*

A&J DEMIÈRE
Fleury-la-Rivière

They produce a wonderfully rich and toasty 2009 Cuvée Lysandre Blanc de Meuniers, with fine, high-acid yet mellow fruit, and it is really quite long.

PAUL DÉTHUNE
Ambonnay
★☆

As one of the more consistent growers, Paul Déthune always makes good vintage and rosé, but is best known for his prestige *cuvée,* the luxuriously rich, big, deliciously creamy Princesse des Thunes, which is made from an *assemblage* of mature vintages.

✓ *Entire range*

DEUTZ
Aÿ-Champagne
★★☆

Smart and sleek in a style totally different from Louis Roederer, its parent company, Deutz will never be the cheapest Champagne, but it is the highest quality and will always be great value. The entry-level *brut* and *brut rosé* are amongst the best and most consistent in the region, with Amour de Deutz, Côte Glacière, and Meurtet, the true standouts.

✓ *Entire range*

DEVAUX
Bar-sur-Seine
★☆ ⓥ

This is super-value Champagne, especially the utterly delicious *cuvées* of rosé in magnum, which are brimming with bold, ripe, juicy fruit.

✓ *Entire range*

DOM PÉRIGNON
Épernay
★★★

The first prestige *cuvée* to be commercialised is part of Moët & Chandon, even though they attempt to market it as a distinctly separate brand. This is impossible while it mentions Moët & Chandon on the label, has the same matriculation number (NM-549-002), and does not have its own winery. Dom Pérignon as a Champagne would never have existed but for Champagne Mercier (which did not belong to Moët at the time) and an English journalist by the name of Laurence Venn. The brand itself came into Moët's possession as part of Francine Durand-Mercier's dowry when she married Comte Paul Chandon-Moët in 1927, while the concept of a prestige *cuvée* was the brainchild of Venn, who was the UK marketing consultant to the Syndicat de Grandes Marques de Champagne. During the Great Depression, some Champagne houses went broke, while other sold off their vineyards to survive. The common strategy was to cut costs and slash prices, but when they consulted Venn at a meeting of the *syndicat* in 1932, he proposed the opposite. He recommended the production of a luxury *cuvée* of exceptional quality that should be sold in a replica of an original 18th-century Champagne bottle at more than twice the price of the most expensive vintage Champagne ever sold. They thought he was mad and asked how they could be expected to sell a luxury *cuvée* at an unprecedented high price when they evidently could not even sell cut-price Champagne? Venn pointed out that this could be achieved by targeting the British aristocracy, one of the few sectors of the market capable of such extravagances during hard times. This advice was soundly rejected by everyone in the *syndicat* with the exception of Robert-Jean de Vogüé of Moët & Chandon. De Vogüé took Venn to dinner that evening and picked his brains, and so Cuvée Dom Pérignon was born in 1936, 15 years after its first vintage. The current *chef de caves,* Vincent Chaperon, has big shoes to fill, following in the footsteps of his mentor, Richard Geoffroy, who did more than any of his predecessors to "push the envelope" of the Dom Pérignon phenomenon by creating the mystique of *plenitude* and releasing longer-aged P2 and P3 *cuvées.*

✓ *Entire range*

DOSNON
Avirey-Lingey
★

Although some *cuvées* can be under-dosaged, all the Champagnes from this small *négociant* are impeccably produced. The Récolte Noire, a pure Pinot Noir *blanc de noirs,* is the standout.

✓ *Récolte Noire* • *Alliae*

DOYARD
Vertus
★

With a choice of 11 hectares (27 acres) of prime sites, mostly *grand cru* villages, the style is rich, with some wines partially fermented in well-integrated oak.

✓ *Entire range*

DRAPPIER
Urville
★☆ ⓥ

The low-sulphur regime here was once Drappier's greatest asset, and it consistently delivered brilliant, ultra-fruity Champagnes that quickly (but not too rapidly) acquired a mellow biscuity complexity. Now, however, it has been taken its low-sulphur strategy to the extreme, so unless you catch a *cuvée* very recently released and still shining brightly, you need to buy exclusively in larger format bottles, which, thankfully, Drappier specialises in.

✓ *Entire range (magnum and larger)*

J DUMANGIN
Chigny-les-Roses
★ ⓥ

Run by Gilles Dumangin, a lovely guy who should target TOWIE-lovers because he learned English in Essex and speaks their language.

✓ *Hippolyte Pinot Noir* • *Achille Meunier*

DUVAL-LEROY
Vertus
★☆ ⓥ

Duval-Leroy has achieved a great consistency of quality for its fresh, light, and elegant house style. With single-vineyard *cuvées,* dabbling with oak, and a seriously high-class prestige *cuvée,* there have been interesting choices from this forward-thinking producer, including collaboration with organic and biodynamic growers. Femme is always stunningly complex, but amazingly fresh and long-lived in magnum, yet invariably oxidative in 75cl bottles.

✓ *Entire range*

EGLY-OURIET
Ambonnay
★☆

A grower with a cult following. Some *cuvées* can still look quite dark, and when they do, the oak has a touch of volatility to it; overall these Champagnes are a lot fresher than they used to be, allowing the complexity to shine through rather than become confused by oxidative notes, which are far fewer than they used to be.

✓ *Les Crayères* • *Grand Cru Blanc de Noirs Vieilles Vignes*

NICOLAS FEUILLATTE
Chouilly
★ ⓥ

The quality of this giant cooperative has not increased dramatically but has spread horizontally across the range, which can now be considered satisfactory, but there is still inconsistency, with the odd *cuvée* occasionally either disappointing or surprisingly good.

✓ *Cuvée Spéciale* • *Palmes d'Or* • *Palmes d'Or Rosé*

FLUTEAU
Gyé-sur-Seine
★ ⓥ

Thierry Fluteau and his American wife, Jennifer, tend to release their Champagnes one year ahead of most others, when they are very fruity and easy to drink. Vintaged *cuvées* are worth keeping and should be purchased in magnums, which are at their beautiful best in their seventh or eighth year.

✓ *Entire range*

GARDET & CIE
Chigny-les-Roses
★ ⓥ

Now part of Groupe Prieux, which also owns Ployez-Jacquemart. Gardet has always prided itself on the longevity of its wines, but I prefer the vintage almost before it is even released.

✓ *Brut Tradition* • *Prestige Charles Gardet Millésime*

GATINOIS
Aÿ-Champagne
★

Pinot-dominated, pure Aÿ grower Champagnes of class and intensity.

✓ *Entire range*

GEOFFROY
Aÿ-Champagne
★ ⓥ

Non-malolactic Champagnes of fine precision and capable of great complexity, with useful information on the back label, including the *assemblage* years, disgorgement date, and *dosage.*

✓ *Entire range*

PIERRE GIMONNET
Cuis
★★ ⓥ

Connected through marriage to the Larmandier family, Didier and Olivier Gimonnet own an impressive 28 hectares (69 acres) of vineyards and produce Champagne Larmandier Père et Fils, as well as Champagne Pierre Gimonnet. The quality here is exceptionally high, and prices are very reasonable. Uniquely, some reserve wines are kept in bottles, which is similar to Bollinger's famous habit, but not identical, because Bollinger will keep some of their reserve wines in magnums.

✓ *Entire range*

HENRI GIRAUD
Aÿ-Champagne

Having been one of the first to acknowledge the extraordinary quality of Henri Giraud's 1993 Grand Cru Fût de Chêne Brut, I am disappointed that he has not been able to replicate that quality. He promotes the production of oak from local Argonne forests, often far too enthusiastically for all but those who enjoy chewing through a plank. Whether it is egg-shaped, concrete tanks or *terroir*-specific use of terracotta eggs, there is no shortage of marketing coming from this producer.

PAUL GOERG
Vertus
★ ⓥ

This small but superior cooperative produces very fine and elegant, yet rich, concentrated Champagnes.

✓ *Entire range*

GOSSET
Aÿ-Champagne
★☆ ⓥ

Established in 1584, Gosset is the oldest house in the region, but for still wines, not sparkling. This non-malolactic range used to be one of my favourites, but there have been too many instances where the *dosage* is too low for balance of these intense Champagnes, and oxidative aromas are becoming increasingly commonplace, particularly in the top *cuvées,* which seem to try too hard. I much prefer the mid-range non-vintage and the straight vintage.

✓ *Grande Réserve Brut • Grand Rosé • Grand Millésime*

HENRI GOUTORBE
Aÿ-Champagne
★

With 15 hectares (37 acres) of vineyards and half a million bottles in stock, Henri Goutorbe is one of the larger wine growers. It is also one of the largest *pépiniéristes* (vine nurseries) in Champagne and owns a boutique hotel in Aÿ. At their best, these are rich, classic, and well-structured Champagnes that become very satisfying with age.

✓ *Cuvée Millésime Grand Cru Aÿ • Spécial Club*

ALFRED GRATIEN
Épernay
★★

This house is part of Gratien, Meyer, Seydoux & Cie, who also sell a range of Loire traditional-method wines under the Gratien & Meyer brand. Alfred Gratien is one of the most traditionally produced Champagnes available. The old vintages never fail to amaze; they are brilliant in quality and retain a remarkable freshness for decades. The non-vintage wine is beautifully fresh, with mature reserves coming through on the palate, making it a wine to drink now, yet one that will improve with age.

✓ *Brut Classique • Blanc de Blancs • Brut Millésime*

GREMILLET
Balnot sur Laignes
★

Small-but-growing house producing fresh and easy-drinking, fruity-style Champagnes.

✓ *Entire range*

GUIBORAT FILS
Cramant
★★

Richard Fouquet is the fifth generation at this estate of 8 hectares (19 acres) in Chouilly, Cramant, Mardeuil, and Oiry. He is a talented grower whose Champagnes are beautifully fresh and brimming with elegance, energy, and minerality.

✓ *Entire range*

HEIDSIECK & CO MONOPOLE
Épernay
☆

Part of the Vranken empire, this was to be his jewel in the crown until the Belgian financial wizard got his hands on Pommery, so now it plays a definite second fiddle, but that can an advantage in terms of value.

✓ *Rosé Top • Gold Top*

CHARLES HEIDSIECK
Reims
★★★

The superb quality of Charles Heidsieck in the modern era was masterminded by the late Daniel Thibault, one of Champagne's truly greatest winemakers, an asset that was totally wasted by inept marketing under Rémy-Cointreau ownership. That quality legacy was ably maintained by his successor, Régis Camus, and, for a tragically short while, by Thierry Roset, whose unexpected passing away led to the arrival of the talented and charming Cyril Brun. For the last 10 years or so, Charles Heidsieck (and Piper) has been under family ownership of the EPI group, which brought in a young and dynamic CEO by the name of Stephen Leroux, who has repositioned this brand where it belongs, in the top tier of Champagne producers.

✓ *Entire range*

HENRIOT
Reims
★★

Henriot has been under family ownership since 1994, when Joseph Henriot left Clicquot and relaunched this house, where Laurent Fresnet was *chef de caves* until the end of 2019 (when he took over at GH Mumm). The only weak link, the old and often oxidised Enchanteleurs, has been transmuted into the much fresher and finer Hemera (Greek goddess of daylight), which together with Cuvée 38, has given this house a new lease on life across its entire range.

✓ *Entire range*

PIERRE ET FRANCOIS HURÉ
Ludes
★☆

If the Champagne has to be an oak style, then the Huré brothers' precision use of this material is for me. I just love the 4 Elements Pinot Meunier.

✓ *Entire 4 Elements range*

OLIVIER HORIOT
Les Riceys Bas
★☆

The increasingly more refined Rosé des Riceys, made exclusively from and by this small estate, represent the bulk of production here, but the Champagne should not be overlooked, particularly the single-vineyard *blanc de noirs* Contrées Val Bazot, which clearly stands out.

✓ *Entire range*

JACQUART
Reims

The Jacquart brand is fully owned by Alliance Champagne, a commercial grouping of three cooperatives: COGEVI (aka Collet), COVAMA (aka Pannier), and Union Auboise (aka Vve Devaux). The quality of Jacquart used to be consistent, fresh and under-rated, but lost its way in recent years, becoming rather oxidative. Following the resignation of Floriane Eznack, we will have to wait and see what happens under a new *chef de caves.*

JACQUESSON & FILS
Dizy
★★

This small, family-owned house is run by the Chiquet brothers, two of the most charming people you are likely to meet. Jacquesson has pioneered various quality-enhancing practices in the vineyards that have been adopted elsewhere, while in the winery it has turned Champagne's traditional non-vintage concept on its head by not attempting any consistency in style. Rather than smooth out the differences presented by each harvest, the new concept is to produce the best possible non-vintage Champagne each year and to number the *cuvées* (starting with No.733 for the 2005-based *cuvée*), so they can be cellared, followed, and compared. Increasingly oxidative aromas are a concern.

✓ *Entire range (magnums)*

JACQUINOT
Épernay

Jean-Manuel Jacquinot's biggest claim to fame is his association with Nyetimber in England, but he has not produced a memorable Champagne of his own since the 1990 Symphony Brut Grande Réserve, and that was just good, not exceptional.

JANISSON
Verzenay
★

Owner Manuel Janisson usually manages to produce an excellent Pinot-dominated style, although oxidative notes have been noted at times. He also works with Claude Thibault in Virginia, USA, to produce Thibault-Janisson sparkling wines.

✓ *Grand Cru Brut • Grand Cru Blanc de Blancs • Grand Cru Blanc de Noirs*

JANISSON-BARADON
Épernay
★

From egg-shaped concrete tanks to 7 Cépages, made from all seven permitted varieties, and the only ratafia I have found outstanding in 40 years of tasting (2005 Single-Cask), Janisson Baradon has proved to be an interesting and rewarding producer.

✓ *7 Cépages • Toulette • Tue Boeuf • Chemin des Conges • Grand Cru Vintage*

KRUG & CO
Reims
★★★

Krug puts quality first and makes Champagne in its own individual style, regardless of popular taste or production costs. This sort of quality is not equalled by any other Champagne house, although it is not to everyone's taste. Clos d'Ambonnay is a great Champagne, but it is not as great as some Krug-worshipping critics made out when they saw the eye-watering $3,000 price tag. Although it is a great Champagne, I personally do not find it as great as Clos du Mesnil (it depends on the year), and Clos du Mesnil is not as great as Krug Vintage, yet for one bottle of Clos d'Ambonnay, you can

buy a case of Krug Vintage in magnum. The price asked is the price of rarity, not intrinsic quality, but for those who are Krugists and have deep enough pockets, it will not matter, even if, like me, they prefer the Krug Vintage.

✓ *Entire range*

BENOÎT LAHAYE
Bouzy
★

There are some strangely exotic grape varieties growing here, but legal ones too! I have been impressed by the *brut nature* in magnum.

✓ *Entire range*

LAHERTE
Chavot
☆

Less oxidative aromas these day, but magnums are the best bet.

✓ *Entire range*

LANSON-BCC

Formerly known as BCC, the Lanson-BCC group is one of the industry's youngest and most dynamic companies, comprising Lanson, Boizel, Chanoine (including Tsarine), Philipponnat, De Venoge, Besserat de Bellefon, Alexander Bonnet (including Ferdinand Bonnet), and Maison Burtin (including hundreds of *sous marques*). Bruno Paillard is the chairman of BCC, and the Paillard family are, directly and indirectly, the largest shareholders in this publicly quoted company, although Champagne Bruno Paillard is not part of the Lanson-BCC group itself. Philippe Baijot (former chairman of Lanson) and the Roques-Boizel family (*see* Boizel) are the two other major shareholders, each of them owning more than the total combined public shareholding.

LANSON PÈRE & FILS
Reims
★★

Purchased in 1991 by Marne et Champagne and now owned by Lanson-BCC, this house famously eschewed malolactic, excelling at classic, slow-maturing, biscuity vintage Champagnes. Indeed, Lanson was the only large-volume *grande marque* to remain true to Champagne's historic non-malolactic style, although this strict precept and, consequently, the style itself has softened since the arrival of Hervé Dantan, the new *chef de caves* in 2015. With the exception of the Gold Label Vintage, partial malolactic is now used judiciously whenever Hervé believes there is a need. The Gold Label, however, remains true to the origins of this house, as it is the only way to accurately depict the differences between vintages (malolactic is the great leveller). The real stars here are the Vintage Collection, available only in magnum.

✓ *Entire range*

LARMANDIER-BERNIER
Vertus
★★

Run by Pierre Larmandier, a cousin of Larmandier Père & Fils (whose wines are made by Didier Gimonnet), Pierre owns 15 hectares (37 acres) in Avize, Chouilly, Cramant, Oger, and Vertus, from which he produces Champagnes of exquisite minerality. Vieille Vigne du Levant (previously sold as Vieille Vigne de Cramant) is the standout here.

✓ *Entire range*

J LASSALLE
Chigny-les-Roses

With a substantial 20 hectares (49 acres), this serious-sized *récoltant-manipulant* makes delicious Champagnes that seem to get fresher, finer, and more elegant with each vintage.

✓ *Entire range*

JACQUES LASSAIGNE
Montgueux
★★

Unique, deeply resonating Chardonnay of Montgueux, these stylishly presented Champagnes are highly enjoyable on purchase, when they shine at their brightest.

✓ *Le Cotet • Haut Revers du Chutat • Clos Sainte-Sophie*

LAURENT-PERRIER
Tours-sur-Marne
★★ⓥ

This house also owns Château Malakoff (producer of Beaumet, Jeanmaire, and Oudinot brands), De Castellane, Delamotte, and Salon. *Chef de caves* Michel Fauconnet retired at the end of 2019, and Dominique Demarville, formerly of Veuve Clicquot, took over at the beginning of 2020. The greatest wine here is Cuvée Grand Siècle in magnum (which really is a totally different animal to the 75cl), but the cash cow has always been Cuvée Rosé Brut, with its famous *Financial Times* pink label. The introduction of numbered iterations of Grand Siècle now allows collectors to cellar and compare each and every blend of this great Champagne. With the focus of attention on Grand Siècle and Cuvée Rosé Brut, however, the vintage is overlooked and under-rated. Furthermore, the remodelled entry-level non-vintage, La Cuvée Brut, has consistently impressed.

✓ *Entire range (except Ultra Brut)*

XAVIER LECONTE
Troissy
★☆

Coeur d'Histoire might not be Xavier Leconte's best Champagne, but it is extremely good and at the time of writing the best example of a pure Meunier *cuvée* I have ever tasted with its floral aromas edging onto red-berry fruit.

✓ *Entire range*

LECLERC BRIANT
Épernay
★☆Ⓑ

The spirit of Pascal Leclerc-Briant still lingers here, despite the fact that he sadly passed away as long ago as 2010. This is thanks to the biodynamic legacy left by such a larger-than-life character and Hervé Jestin, Champagne's most famous biodynamic consultant, who has taken over as *chef de caves*. Leclerc-Briant is now owned by Mark Nunnelly and his wife, Denise Dupré, who seem to understand intimately that Champagne is a long-term, slow-burn investment. This is rare for new investors to this industry, but perhaps less surprising considering the investment expertise of Mark Nunnelly, who was formerly the managing director of Bain Capital in Boston. Nunnelly has also purchased and totally renovated the Royal Champagne hotel and (now) Michelin-starred restaurant, which sits high above the town of Épernay. Not a bad purchase for someone more used to putting together multibillion-dollar deals to buy companies such as Dunkin' Donuts and Domino's Pizza.

✓ *Entire range*

AR LENOBLE
Damery
★☆

With 18 hectares (44 acres) of vineyards in Damery, Chouilly, and Bisseuil, brother and sister Antoine and Anne Malassagne create what they openly claim to be an oxidative style, yet the 2008 Grand Cru Blanc de Blancs Chouilly was so fresh, clean, and precise, without a hint of oxidative aroma, that it not only won a gold medal at the Champagne & Sparkling Wine World Championships in 2017, but was also awarded the Chairman's Trophy. This was to illustrate that the so-called "oxidative style" is mostly a myth. The wine in magnums is exactly the same as the wine in 75cl bottles, so it is evidently not oxidative in and of itself. Early-onset oxidation can only be due to what happens next, namely disgorgement and specifically low or no SO2. With half the ratio of oxygen to wine, magnums are far more robust to resist, thus we can see the wine in its true, unadulterated form for much longer.

✓ *Magnums*

LAURENT LEQUART
Passy-Grigny
★

Although not always consistent, Lequart's 10 hectares (24 acres) in Passy-Grigny can produce some of the best Meunier in Champagne, particularly his Reserve Extra Brut.

✓ *Entire range*

LILBERT-FILS
Cramant
★

This small grower consistently produces firm *blancs de blancs* that show finesse and age gracefully.

✓ *Entire range*

LOMBARDI
Balnot-sur-Laignes
★☆

Superb vintaged *blanc de noirs*.

✓ *Entire range*

LVMH

Short for Louis Vuitton-Moët-Hennessy, a multinational luxury goods conglomerate. Christian Dior, which was a subsidiary of Moët & Chandon in the 1970s, now owns 42 per cent of LVMH and controls 59 per cent of its voting rights. LVMH is run by Bernard Arnault, who is the richest man in Europe. Arnault owns 100 per cent of Christian Dior. LVMH owns the largest group in the Champagne industry and was formed as recently as 1987, when Louis Vuiton, which owned Veuve Clicquot at the time, merged with Moët-Hennessy, which owned Moët & Chandon, Mercier, and Ruinart. In 1999, these houses were joined by Krug and, more recently, efforts have been made to spin-off Dom Pérignon as a brand in its own right. The guiding principle for the growth of LVMH's Champagne group has always been to maximise brand value, and to achieve this it has bought up distressed houses whenever they have come on the market, retained the vineyards and sold on the brands. This has seen the group grow from 600 hectares (1,482 acres) with sales of 30 million bottles to the current 1,700 hectares (4,200 acres) and sales of 65 million bottles, discarding along the way brands such as Canard-Duchêne (which was effectively the second label of Veuve Clicquot), Lanson, Pommery, De Cazanove, and Montaudon. In addition to Champagne, sparkling wines have been produced under the Chandon brand in Argentina (1959), Australia

HAND-HARVESTING IN CRAMANT, CÔTE DES BLANCS
Cramant is one of the five *grand cru* villages of the Côte des Blancs. Here the subsoil is composed entirely of belemnite chalk, which many believe produces the best expression of Champagne. The chalk bed that surfaces in Côte des Blancs is part of the Paris Basin, which is a contiguous seam of chalk that extends westward, beneath Paris, to emerge as the White Cliffs of Dover in the UK.

(1986), Brazil (1973), California (1973), China (2013), India (2014), and for a short while Spain (1987–2003), while other wineries belonging to LVMH include Ao Yun (China), Bodega Numanthia (Spain), Cape Mentelle (Australia), Château Cheval Blanc (Bordeaux), Château d'Yquem (Bordeaux), Cheval des Andes (Argentina), Clos des Lambrays (Burgundy), Cloudy Bay (New Zealand), Newton Vineyard (California), and Terrazas de los Andes (Argentina).

NICOLAS MAILLART
Mailly-Champagne

Recommended for lovers of oaked Champagne, but it is less obvious in the non-vintage Platine Brut and, my favourite, the vintaged *brut* in magnum.

✓ *Entire range*

MAILLY GRAND CRU
Mailly-Champagne

A large range of richly flavoured mono-*cru* Champagnes.

✓ *Entire range*

HENRI MANDOIS
Pierry

The house style of Henri Mandois is for elegant Champagnes that have a very satisfying length of attractive, creamy fruit and a fine balance. Jimmy Boyle's favourite Champagne.

✓ *Entire range*

VINEYARDS OF MOËT & CHANDON IN AŸ-CHAMPAGNE
Cultivating Pinot Noir, Pinot Meunier, and Chardonnay varieties, the extensive vineyards of this house spread over all five of the main areas of Champagne – Montagne de Reims, Côte des Blancs, Vallée de la Marne, Sézanne, and Aube.

MARGAINE
Villers-Marmery

From 6.5 hectares (16 acres) of vineyards in Villers-Marmery, one of the very best villages for Chardonnay on the Montagne de Reims, Arnaud Margaine produces top-quality, rich yet crisp *blanc de blancs* of great longevity.

✓ *Entire range*

MARGUET
Ambonnay

Benoît Marguet worked with Hervé Jestin to produce Sapience, a prestige *cuvée* produced from vines located just above Krug's Clos d'Ambonnay, but my favourite here so far is a single-vineyard *blanc de noirs* from La Grande Ruelle, also from Ambonnay.

GH MARTEL
Épernay

Primary brand of the Martel Group. The wines are clean, fresh, and good value.

✓ *Entire range*

MARTEL GROUP
Épernay

After GH Martel was acquired by the Rapeneau family in 1979, they effectively marginalised their own brand, Champagne Rapeneau & Cie, which had been established by Ernest Rapeneau in 1901, and concentrated their efforts on Martel, gradually building up a group that now consists of Château de Bligny, Charles de Cazanove, Mansard, P Louis Martin, and Charles Orban. Other labels include Balahu, Comte de Lamotte, Maxim's, Mortas, Comte de Noiron, Marcel Pierre, Charles du Roy, and, of course, Ernest Rapeneau. All the Champagne are at least clean, fresh, and good value. They are all worthy of one star and can occasionally produce a stunner. In total the Martel Group owns 200 hectares (494 acres) of vineyards, producing eight million bottles a year and making this one of the few cash-rich businesses in Champagne.

SERGE MATHIEU
Avirey-Lingey

A small grower in the Aube, it consistently produces excellent Champagnes that are beautifully focussed, have much finesse, and a real richness of fruit for such light and elegantly balanced wines. Isabelle Mathieu and her husband, Michel Jacob, should receive an award for the cartoon on their website, www.champagne-serge-mathieu.fr.

✓ *Entire range*

MERCIER
Épernay

Mercier, established by Eugène Mercier in 1858, was the original owner of the Dom Pérignon brand, although this house never utilized it and gave it to Moët as part of Francine Durand-Mercier's dowry when she married Comte Paul Chandon-Moët in 1927. Moët returned the favour by purchasing Champagne Mercier some 40 years later. Exporting of Mercier is tolerated, rather than pursued, which is why it is seldom seen outside of France, where it remains the best-selling brand of Champagne, despite Moët being number one globally. Mercier is generally fuller and fatter than Moët, with unashamed Aube fruit ripeness. Generally it is less elegant too, although some of the vintages can excel. The range has much diminished in recent years.

✓ *Entire range*

LE MESNIL
Le Mesnil-sur-Oger
★☆ Ⓥ

If only this cooperative's basic non-vintage *blanc de blancs,* by far the most commonly encountered, were as stunning as the three strict selections recommended below.

✓ *Cuvée Prestige • Sublime • Cuvée Heritage*

MOËT & CHANDON
Épernay
★★

The largest Champagne house by a mile, Moët is also the leading Champagne company in the LVMH group. Never has the quality been higher or more consistent than under the watchful eye of *chef de caves* Benoît Gouez, who took over in 2005. Part of the reason for the exceptional consistency has been jetting (Moët were first) and, since 2006, the use of DIAM Mytik Diamant® for all *cuvées*, which has to be the most strategically acute judgement taken by any *chef de caves* in the first 12 months of tenure. Furthermore, these technologies can only preserve what already exists. They cannot improve mediocre wines, so it also highlights the quality that Benoît has delivered from the get-go. Moët's space-age facility at Mont Aigu on the outskirts of Oiry is the most impressive winery I have seen. Not only has it been sustainably built, but it is also beautifully designed, with 95 per cent of the building underground, yet so light and airy that when Moët's workers do a stint there, they don't want to leave. There are no pipes on the floor or, indeed, anywhere. The wines are moved automatically and monitored by oenologists, who seem to float above the tanks, plugging their laptops into terminals, where they can determine everything they want to know, right down to the O_2 content of juice on arrival.

✓ *Entire Range*

PIERRE MONCUIT
Le Mesnil-sur-Oger

Sometimes oxidative, while other times a bit heavy-handed with the malolactic, but when a wine shines it is fresh and crisp, with lively-rich fruit that is satisfyingly tasty, and usually matures into a creamy-biscuit style.

✓ *Grand Cru Millésime • Cuvée Nicole Moncuit Vieille Vigne*

MONTAUDON
Reims

Made by Jacquart, but not oxidative and thus much preferable.

✓ *Entire range*

MOUSSÉ FILS
Cuisles

My absolute favourite is a 100 per cent Meunier *blanc de noirs* called Les Vignes de Mon Village, which has a vanilla aroma, a lovely, voluminous *mousse* supporting fresh, fleshy, nicely mellowed fruit, with some fresh, youthful elements, fine acidity, and length.

✓ *Entire range*

MOUTARD
Buxeuil

This very inconsistent but pioneering producer almost single-handedly preserved the tradition of making Champagne with ancient varieties such as Arbanne (called Arbane here). But except for the occasional *cuvée*, they have been unattractive, which is such a pity from a producer I feel grateful for. Worse still, François Moutard is starting to dabble in Champagnes without sulphur. Oh, dear! Maybe one of his sons and nephews he has brought into the business has a flair for winemaking, and he can keep the ideas coming, but allow someone else full control of the production? If not, then Moutard is of a size that should be able to afford an outside professional as *chef de caves*.

MOUTARDIER
Le Breuil

Englishman Jonathan Saxby has long gone. He had married into this rural Champagne-producing family and seemed to be gradually increasing quality, but nothing has grabbed me about these wines in a while.

GH MUMM & CIE
Reims
★★

With Perrier-Jouët, this house is part of Pernod-Ricard's Champagne group, which masterfully hired a young Dominique Demarville to clean up the quality of Mumm Champagne in the 1990s. Demarville was so successful that he was headhunted by Veuve Clicquot, and his assistant, Didier Mariotti, took over Mumm. At the end of 2019, Demarville moved from there to Laurent-Perrier, and Mariotti took his place at Veuve Clicquot, but in the meantime Mariotti continued increasing quality here, so that Mumm's Cordon Rouge was not only clean, but became one of the best non-vintage *cuvées,* regularly winning medals at the Champagne & Sparkling Wine World Championships. The only problem here, strangely, is Cuvée R Lalou, which started with such promise as a true gastronomic *cuvée* but has shown signs of oxidation.Let's wait and see what happens when the new *chef de caves,* Laurent Fresnet (ex-Henriot) releases the next vintage. Although there is no complaint about the quality of the *cuvées* in Mumm's RSRV range, in their consolidated, dumpy bottles, there has been confusion over what they represent: RSRV Blanc de Blancs (a repackaging of the famous Mumm de Cramant, which has always been from a single year and now indicates that vintage); RSRV Blanc de Noirs (repackaging of Mumm de Verzenay); and RSRV 4.5 (repackaging of Brut Sélection, previously marketed as Grand Cru). To add to the confusion there are now two new Champagnes Cuvée 4 and Cuvée 6, which have no connection to RSRV 4.5 or, indeed, the RSRV range at all; they are in the traditional Champagne bottle shape and simply refer to the number of years on yeast. Super Champagnes, lousy marketing.
✓ *Entire range*

FRANCIS ORBAN
Leuvrigny
★☆

A promising grower with 7.5 hectares (18.5 acres) in the heartland of Meunier. Mostly fine and fresh, the pure Meunier non-vintage rosé has been outstanding, with its lovely peach colour, vanilla-dusted aromas, and clean, crisp, yeast-complexed fruit.
✓ *Entire range*

BRUNO PAILLARD
Reims
★☆

Bruno Paillard opts for elegance rather than body or character, and his Première Cuvée is not just elegant, but one of the most consistent non-vintage Champagnes

L'AVENUE DE CHAMPAGNE IN ÉPERNAY
This famous street in Épernay, the "Capital of Champagne", is home to many leading champagne houses, such as Perrier-Jouët, Mercier, Moët & Chandon, Boizel, de Venoge, Vranken, Pol Roger, and GH Martel. It is now a UNESCO World Heritage site.

on the market. The *blanc de blancs* and vintage are the standouts here.
✓ *Entire range*

PIERRE PAILLARD
Bouzy
★☆

Expect classy Champagnes of richness and elegance from a grower who practices biodynamics but has not so far been interested in becoming certified.
✓ *Entire range*

PALMER & CO
Reims
★★☆

This producer is a great source of vintage Champagnes, particularly vintage *blanc de blancs* in magnum, which is often pure Montagne de Reims and will age gracefully for 30 years or more. Although *blanc de blancs* is the signature style here, Palmer also excels at the opposite end of the spectrum, with an amazing *blanc de noirs*. The only disappointment here is the oxidative prestige *cuvée,* Amazone, an anomaly in Palmer's world of pristine Champagne. This cooperative is one of my top half-dozen Champagne producers.
✓ *Millésime • Blanc de Blancs (NV & Vintaged Magnum) • Blanc de Noirs*

PANNIER
Château-Thierry
★

The past 10 years have seen a slow but steady increase in quality here, yet this phenomenon has come with a decrease in consistency. Some *cuvées* and vintages can certainly slug it out with the best, but most cannot, and I like that sense of the unknown, where there is a chance of occasionally encountering a superstar *cuvée*.

✓ *Brut Tradition • Brut Vintage • Blanc de Noirs • Egérie de Pannier*

JOSEPH PERRIER
Châlons-sur-Marne
★

Not as consistent as it used to be, with oxidative notes creeping into some its *cuvées*.

PERRIER-JOUËT
Épernay
★★

Belle Époque might be the headline grabber here, but the entire range is a joy, from the entry-level non-vintage (particularly in magnum) up. It's a pity that Perrier-Jouët has stopped producing the vintaged Grand Brut (1998 was the last), but it became necessary as soon as sales of Belle Époque took off in the mid-1990s. Belle Époque in magnum is, of course, superior, but Belle Époque in jeroboam is in a totally different league altogether, thanks to being vinified to a lower pressure. The non-vintage *grand brut* is extraordinary value. Hervé Deschamps, who has been *chef de caves* since 1993, retires in 2020, when he will be replaced by Séverine Frerson, formerly of Charles Heidsieck.
✓ *Entire range*

PHILIPPONNAT
Mareuil-sur-Aÿ
★★

Under the day-to-day control of Charles Philipponnat, a direct descendant of the founder, this house is a key player in the Lanson-BCC group. During Charles's tenure, the house style here has moved quite dramatically from a classic Champagne to a more powerfully structured, Pinot Noir-influenced food wine. Clos des Goisses (individually rated three stars) remains

Philipponnat's jewel in the crown, and although its individual plot spin-offs are fascinating, it is interesting to note that they have not (yet?) equalled Clos des Goisses itself, even if the plots concerned are supposedly superior. Clos des Goisses is a very special vineyard and one of the few that could – and should – be produced every year.
✓ *Entire range*

PIPER-HEIDSIECK
Reims
★★☆

Under the same family ownership (the EPI group) as Charles Heidsieck, Piper was always viewed as good value, but lesser quality, volume-selling *grande marque* compared to its sibling. When the legendary *chef de caves* Daniel Thibault passed away, however, he was replaced by Régis Camus, who left the day-to-day winemaking of Charles Heidsieck in the safe hands of Thierry Roset and performed a miracle by lifting Piper to the quality of Charles Heidsieck. Indeed, Piper would deserve a three-star rating, had its range still included Champagne Rare, but that has been spun-off as its own brand. Régis retired to become face of Champagne Rare and handed the reins of Piper over to Séverine Frerson, who he had mentored since 2002. Yet, she was head-hunted by Perrier-Jouët after fewer than four months as *chef de caves,* and the position has since been filled by Emilien Boutillat, ex-Cattier. He not only has to prove himself, but he also has to do that without Champagne Rare, creating a new prestige *cuvée* to fill the gap and digging deep into his Armand de Brignac experience to market it.
✓ *Entire range*

LOUIS ROEDERER AND TAITTINGER CHAMPAGNES
Both of these Reims Champagne houses are lauded for their ability to consistently produce the finest quality of Champagne, worthy of their high star rankings.

PLOYEZ-JACQUEMART

Ludes

★

This producer is owned by Groupe Prieux since 2004, but still run by the founder's granddaughter, Laurence Ployez-Krommydas. The potential longevity of these Champagnes has reduced as the *dosage* has dropped, but they are still elegantly rich and well made.

✔ *Entire range*

POL ROGER & CO

Épernay

The new *chef de caves* here since 2017 is Damien Cambres, who was in charge of La Goutte d'Or cooperative in Vertus, where he produced Champagne Paul Goerg. We must wait to see his first *cuvées*, which I hope will put to bed the premox rather than oxidative style that has been worry for quite a while now. Either Dominique Petit, the previous *chef de caves,* changed his winemaking habits of a lifetime and lowered the sul-

phur, which seems an unlikely and a risky step to take towards the end of his tenure or, as I hope, the enormous heat generated and stored while renovating the *cuverie* and cellars came back to bite these wines. The premox problem seems to have abated in recent years, but it has not disappeared, and some wines seem to be recovering more quickly than others, which leads me to suspect residual heat as the cause. This problem is not something that anyone at Pol Roger wants to admit even exists. My personal attachment to this house has been clear from the dedication in *Champagne* (Sotheby's Publications) in 1986, but I have always been totally honest and cannot turn a blind eye. One thing is sure and that is no renovations that require the excavation of the scale I saw should ever be undertaken with winemaking and stock still in place. The end result is excellent, but Pol Roger should have moved their entire production and storage to temporary premises before the first jackhammer let rip.

POMMERY

Reims

★★

I get a lot of stick for recommending Pommery NV. Let's face it – I get a lot of stick for recommending Pommery full stop. Come to that, I get a lot of stick for recommending any large house, and that stick comes from people who believe that large means bland, and only small is beautiful. Yet, all sizes of Champagne producer can produce a good, bad, or indifferent quality, and readers who understand the technical processes involved will know that larger producers have an advantage when producing quality sparkling wine. That does not mean that small producers cannot be beautiful. They can, but face challenges that larger producers do not, so it is only the really talented and dedicated small producers who can be consistently beautiful. Any Pommery doubters should first open and taste together both a 75cl bottle and a magnum of NV (Brut Royal, Apanage, whatever), then taste the difference,

after which they should serve the magnum blind to their other "small is beautiful" friends. If they cannot taste the difference between bottle and magnum, and if they do not appreciate the magnum, then fine, carry on the damnation, but take the plunge first. Pommery is one of the few *grande marque* houses that claim 100 per cent *grand cru* for all of its standard vintage Champagne.

✔ *Brut Royal (magnum)* • *Brut Apanage (magnum)* • *Brut Apanage Blanc de Blancs* • *Grand Cru Millésimé* • *Les Clos Pompadour*

JÉRÔME PRÉVOST

Gueux

Like the Grand Old Duke of York, when Jérôme Prévost is up, he is up, and should be drunk down, not kept. Jérôme Prévost can produce some great Meunier Champagnes, but enjoying them depends not on having a cellar, but buying the moment you taste a *cuvée* that is singing and then drinking on purchase!

RARE
Reims
★★★

Piper-Heidsieck has spun-off Champagne Rare à la Dom Pérignon as its own brand and faces not only the same difficulties as Dom Pérignon in that the Rare bottle bears the name Piper-Heidsieck in small print and its matriculation number (NM-211-001), but it also has to overcome the practicalities of establishing Rare as a brand in its own right. Dom Pérignon works, despite the difficulties of carrying Moët's name and matriculation number, because it sounds like the name of a producer and even teetotallers have heard of Dom Pérignon, but Rare? It might be fine as the name of a *cuvée*, but what does it mean as a brand? The quality, class, and longevity are, however, nothing short of superb.

✓ *Entire range*

LOUIS ROEDERER
Reims
★★★

This family-owned *grande marque* has always been one of the most consistent, greatest-quality Champagne houses, but under Jean-Baptiste Lécaillon, the *chef de caves* and vice-president, Louis Roederer has won an astonishing 50 gold medals at the Champagne & Sparkling Wine World Championships, and no less than 25 Best in Class awards and 22 trophies, including the Supreme World Champion trophy on an unprecedented three occasions.

✓ *Entire range*

JACQUES ROUSSEAUX
Verzenay
☆

Brother and sister Eric and Céline Rousseaux produce a voluptuous and beautifully textured Grande Réserve Blanc de Noirs, although it is distinctly oaky.

✓ *Grande Réserve*

RUINART
Reims
★★☆

Under Frédéric Panaiotis, the *chef de cave* since 2007, this entire range has freshened up, and the non-vintage *blanc de blancs* in magnum has become one of the greatest *blanc de blancs* in the world, with or without a vintage. The huge improvement in both the quality (particularly the intensity) of the vintaged blend and the non-vintage R de Ruinart cannot be overstated. Fred's main task now is to bring intensity where there is weight to Dom Ruinart Blanc de Blancs and thus increase the elegance and finesse without dismantling the DNA of this classic icon. A big ask, but his first vintage, the 2007, gives a small hint of what is in store.

✓ *Entire range*

DE SAINT GALL
Avize
★☆

Champagne de St-Gall is the Union Champagne Cooperative's primary brand, with Cuvée Orpale its prestige *cuvée*. The whole range is fresh, crisply fruity, and offers excellent value, while Cuvée Orpale is a *blanc de blancs* of excellent longevity that is occasionally quite superb.

✓ *Entire range*

SALON
Le Mesnil-sur-Oger
★★

Not the special wine it used to be. Very few great vintages since the 1982 and no megastars since 1971, 1947, and 1928.

✓ *"S"*

FRANÇOIS SECONDÉ
Sillery
★☆

The only producer of pure Sillery Champagne, and La Loge Blanc de Noirs Sillery is a truly exceptional pure Pinot *cuvée*.

✓ *Entire range*

JACQUES SELOSSE
Avize

I preferred the Champagnes of Anselme's father. Yes, I have tasted some incredible Champagnes from Anselme Selosse, although the last of those was before he opted to leave a headspace in his barrels to allow a yeast-flor to develop. No, none of Anselme's Champagnes have improved with age, in my opinion. I would have been happy – deliriously happy – to have drunk the entire contents of any of those fabulous Champagnes at the time of tasting, however, and that is my advice to readers: see it, taste it, buy it, drink it! Now is the time for Anselme's son, Guillaume . . . let's give him some time.

PIOT SEVILLANO
Vincelles
★

Capable of fine, rich Meunier with smoky-floral complexity and lovely sapid fruit.

✓ *Provocante*

DE SOUSA
Avize

Erick de Sousa has always been a very dedicated grower, but he shines more brightly in the vineyard than the winery, where the results can often be too oxidative or estery.

ERIC TAILLET
Baslieux-sous-Châtillon

Producer of one of the best pure Meunier *cuvées* I have tasted recently (Bansionensi Extra Brut), and one of the worst (Le Bois de Binson).

TAITTINGER
Reims
★★☆

In December 2019 Pierre-Emmanuel Taittinger stepped down from the role of president, the position of which went to his daughter, Vitalie, who will be assisted by her brother, Clovis. The year before saw another change, when Loïc Dupont, Taittinger's long-serving *chef de caves* retired, and Alexandre Ponnavoy, who had been his deputy since 2015, was appointed as his replacement. Ponnavoy worked at Louis Roederer, including some time at Roederer Estate in California, before joining the Station Oenotechnique de Champagne as a consultant in 2007. Taittinger is currently at the very top of its game in almost all of its *cuvées*, but whereas the iconic Comtes de Champagne Blanc de Blancs is so close to perfection (and definitely a three-star Champagne on its own), the red wine used for Comtes de Champagne Rosé still needs some work, so all eyes will be on what Ponnavoy does there, while maintaining the quality and consistency everywhere else.

✓ *Entire range*

TARLANT
Oeuilly
★☆

This grower has lots of ambition and a heart in the right place, so it is good to see his wines improve year on year. The future is bright, providing he does not get too carried away with Georgian clay amphorae. Dabble by all means, but do not lose the plot.

✓ *Entire range*

THIÉNOT
Reims
★

A very savvy financier who has quietly built up a significant group (including Joseph Perrier, Canard-Duchêne, George Goulet, and Marie Stuart), Alain Thiénot also produces good-value, fruit-driven Champagnes under his own label, ranging from a successful pouring quality non-vintage, through good-quality vintage *brut* and often surprisingly excellent rosé, to the trio of top *cuvées* named after family members: Alain, Garance, and Stanislas, all of which are occasionally outstanding, but can sometimes disappoint.

✓ *Entire range*

DE VENOGE
Épernay
★

Part of the Lanson-BCC group, this remains a good-value, clean, creamy-fruity, and well-made brand. De Venoge's former prestige Cuvée des Princes, a vintaged Champagne that had a brilliant track record despite its tacky, tear-shaped bottle, has been spun off into a non-vintage premium range (all recommended despite the ridiculous bottle), while a new duo of *brut* and rosé prestige Champagne have been introduced as Louis XV in a clear version of the tear-shaped bottle (which not only looks even more kitsch, but renders the Champagne vulnerable to light-strike).

✓ *Entire range*

JL VERGNON
Le Mesnil-sur-Oger
★☆

A *récoltant-manipulant* beginning in 1985, JL Vergnon became a *négociant-manipulant* in 2012. This allows the purchase of grapes to grow the brand, although volumes have remained relatively small, even by grower standards.

✓ *Entire range*

VEUVE CLICQUOT–PONSARDIN
Reims
★★☆

Chef de caves Dominique Demarville has now left, but before his departure he reduced the number of vintages released to just 3 in 10, ostensibly to make vintage Champagne special again, but in truth to give Yellow Label non-vintage blends access to some of the very best wines, which would otherwise have been reserved for the vintage. Since 2008 he fermented a small but growing proportion of both vintage and non-vintage wines in oak for the first time since 1961. The volume is so small that the oak cannot possibly be noticed, but the micro-oxygenation involved does produces a textural enhancement that is not dissimilar to the creaminess created by an aromatically low-key malolactic, a process that will have to be reduced to combat increased ripeness from climate change. Another change inspired by Demarville was to push the proportion of Pinot Noir in the prestige *cuvée* Grande Dame up to 92 per cent in 2008 (a Champagne to die for in magnum!). Demarville was the first Veuve Clicquot *chef de caves* to leave by his own volition, rather than through retirement, thus his tenure of less than 14 years was short for this house, but he made a significant difference, thus will not go unnoticed. Didier Mariotti, who worked with Demarville at Mumm, has replaced him at Veuve Clicquot.

✓ *Entire range*

VEUVE FOURNY
Vertus
★★

A small *négociant* house capable of producing great quality Champagne.

✓ *Blanc de Blancs Vertus Brut Nature • Blanc de Blancs Vertus Brut*

VILMART
Rilly-la-Montagne
★★

The "poor man's Krug", as somebody once wrote.

✓ *Entire range*

VRANKEN-POMMERY
Épernay

Owned by Belgian Paul Vranken, who undoubtedly has one of the cleverest commercial brains in the region, the Vranken-Pommery group includes lacklustre brands such as Barancourt, Bissinger & Co, Charbaut, Germain, and Vranken (which includes Diamant and Demoiselle), plus, of course, Heidsieck & Co Monopole and Pommery. Vranken-Pommery also owns subsidiaries in France (Provence: Château La Gordonne; Sables de Camargue: Domaine Royal de Jarras); Portugal (Port: Rozès and São Pédro; Douro: Terras do Grifo); UK (Louis Pommery England); and USA (Louis Pommery California).

Alsace

After more than a quarter of a century of increasing sweetness, in which wines with no indication of sweetness on the label can turn out to be anything from medium dry to distinctly sweet, the reputation for crisp, dry white food wines that made Alsace so desirable has flown out of the window. Vendanges Tardives (VT) and Sélection de Grains Nobles (SGN) should of course have some sweetness, and the label reveals this, of course, but all other Alsace wine should be dry to one degree or another.

Unfortunately, apart from classic dry wines still produced by a few traditionalists, the white wines of Alsace are no longer dry. Fortunately, however, some of the few are globally famous, including but not restricted to Hugel, JosMeyer, Kientzler, Muré, Trimbach, and Domaine Weinbach. There will be some residual sugar in the odd wine from the most traditional of producers, particularly Gewürztraminer and Pinot Gris, but by and large consumers can count on finding mostly crisp, dry whites.

Sweetness is an immediate problem that this region has to face, and this is covered later in greater detail (*see* Disappearing Dry Wines, p270), but there is another, more long-term issue that has to be addressed at some time or another, and that is the multiplicity of wines. All 9-plus varietal styles are not a hindrance in themselves; indeed, they make Alsace such a fascinating region. Yet, at a producer level, such multiplicity makes it impossible for individual wineries to express themselves, unless each one wants to be a Jack of All Trades and Master of None. There needs to be more focus. Not everywhere in Alsace is suitable for all 9 grape varieties (or 13 in the case of Marcel Deiss). In a region that prides itself on its complex mosaic of *terroirs*, producers should be carving out their own niche as Riesling specialists, Pinot Gris specialists, Muscat specialists etc, with maybe two or three other varieties that are also expressive of their particular *terroir*. Not growing all 9 varieties and producing several different *cuvées* of each, let alone VT, SGN, and Crémant.

MIXED BLESSINGS

A fascinating mixture of French and German characteristics pervades this northeastern fragment of France, cut off from the rest of the country by the barrier of the Vosges mountains and separated from neighbouring Germany by the mighty Rhine. Most of the lower foothills of the Vosges had been cleared and cultivated with vines by the second century AD, and from the seventh century, when the Frisians navigated the Rhine in Alsace for the first time, the fame of these wines began to spread throughout Central Europe. In the ninth century, Ermoldus Nigellus, a Roman poet from Aquitaine, compared the wines of Sigolsheim to those of Falernum, the most famous of antiquity.

VINEYARDS LINE THE LOWER SLOPES OF THE VOSGES MOUNTAIN RANGE IN KAYSERSBERG
The mountains serve to protect the vineyards of Alsace from the elements. Here, scattered plots fill the hillsides from the ruins of the Château de Kaysersberg keep to the town below. The Vallée de Kaysersberg is filled with vineyards and is home to many important wineries.

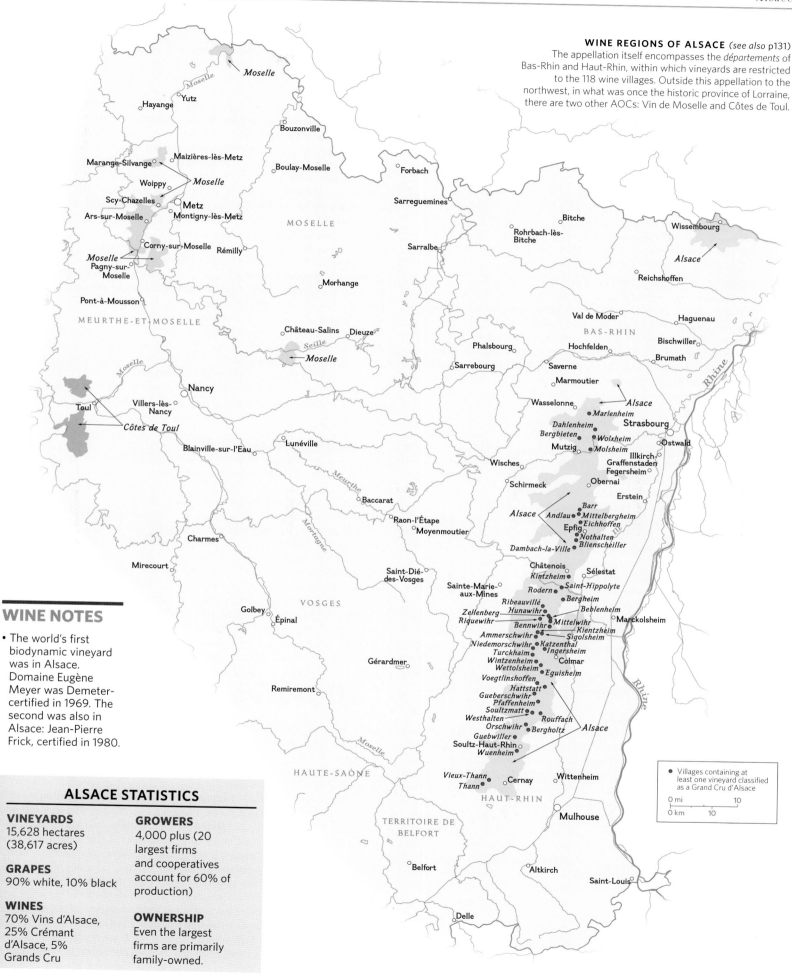

The appellation itself encompasses the *départements* of Bas-Rhin and Haut-Rhin, within which vineyards are restricted to the 118 wine villages. Outside this appellation to the northwest, in what was once the historic province of Lorraine, there are two other AOCs: Vin de Moselle and Côtes de Toul.

Moselle
Hayange Yutz
Bouzonville
Maizières-lès-Metz
Marange-Silvange
Boulay-Moselle
Forbach
Moselle
Woippy
Sarreguemines
Bitche
Rohrbach-lès-Bitche
Wissembourg
Scy-Chazelles Metz
Ars-sur-Moselle Montigny-lès-Metz
Alsace
Corny-sur-Moselle Rémilly
Sarralbe
Reichshoffen
Moselle
Pagny-sur-Moselle
Pont-à-Mousson
Morhange
Val de Moder
Haguenau
MEURTHE-ET-MOSELLE
MOSELLE
BAS-RHIN
Bischwiller
Château-Salins Dieuze
Phalsbourg
Hochfelden
Brumath
Toul
Seille
Moselle
Sarrebourg
Saverne
Alsace
Nancy
Marmoutier
Villers-lès-Nancy
Wasselonne
Marlenheim
Strasbourg
Côtes de Toul
Dahlenheim
Bergbieten *Wolxheim*
Blainville-sur-l'Eau
Lunéville
Mutzig *Molsheim* Illkirch-
Wisches
Graffenstaden
Fegersheim
Schirmeck
Obernai
Baccarat
Erstein
Raon-l'Étape
Barr
Alsace *Andlau* *Mittelbergheim*
Moyenmoutier
Epfig *Eichhoffen*
Charmes
Nothalten
Blienscheiller
Dambach-la-Ville
Mirecourt
Saint-Dié-des-Vosges
Châtenois Sélestat
Kintzheim
Sainte-Marie-aux-Mines *Rodern* *Saint-Hippolyte*
Bergheim
Golbey
VOSGES
Ribeauvillé *Beblenheim*
Zellenberg *Hunawihr*
Épinal
Riquewihr *Mittelwihr* Marckolsheim
Bennwihr
Kientzheim
Ammerschwihr *Sigolsheim*
Niedemorschwihr *Katzenthal*
Turckheim *Ingersheim*
Wintzenheim Colmar
Wettolsheim
Voegtlinshoffen *Eguisheim*
Gérardmer
Hattstatt
Gueberschwihr
Pfaffenheim
Remiremont
Soultzmatt
Westhalten *Rouffach*
Orschwihr *Bergholtz*
Guebwiller *Alsace*
Soultz-Haut-Rhin
Wuenheim
Moselle
Vieux-Thann Wittenheim
Thann Cernay
HAUTE-SAÔNE
HAUT-RHIN
Mulhouse
TERRITOIRE DE BELFORT
Belfort
Altkirch
Saint-Louis
Delle
Rhine

• Villages containing at least one vineyard classified as a Grand Cru d'Alsace

0 mi ————— 10
0 km ————— 10

WINE NOTES

• The world's first biodynamic vineyard was in Alsace. Domaine Eugène Meyer was Demeter-certified in 1969. The second was also in Alsace: Jean-Pierre Frick, certified in 1980.

ALSACE STATISTICS

VINEYARDS
15,628 hectares (38,617 acres)

GRAPES
90% white, 10% black

WINES
70% Vins d'Alsace, 25% Crémant d'Alsace, 5% Grands Cru

GROWERS
4,000 plus (20 largest firms and cooperatives account for 60% of production)

OWNERSHIP
Even the largest firms are primarily family-owned.

GRANDS CRUS OF ALSACE

There are 51 *grands crus,* and they range in size from 3.2 hectares, or 7.9 acres, (Kanzlerberg) to 80.3 hectares, or 198 acres, (Schlossberg). My personal view is that there were probably no more than 20 true *grands crus* and that many of those that were eventually classified are bloated versions of their original selves (Brand is the most obvious and easily provable example, but it is by no means the only one). It would have been more logical to start with a much larger number of *lieux-dits,* promote the best to *premier cru* status 10 to 15 years later, then promote the best of those to *grand cru* after another similar period. That ship has sailed, however, and there are 51 Alsace *grands crus,* and they are what they are, but in true "Benjamin Button" going-backwards tradition, Alsace is now considering *premiers crus.* The Alsace Vintners Association asked their membership which of 400 as-yet-unclassified *lieux-dits* deserved *premier cru* status, and, not unsurprisingly, received 150 applications. The process is ongoing. It will happen.

By 1400 there were as many as 430 wine-growing villages (only 118 today), and Alsace reached its viticultural pinnacle in the 15th and 16th centuries, when its wines rivalled those of Bordeaux as the most famous in Europe. After the Treaty of Westphalia put an end to the Thirty Years' War in 1648 and gave the French sovereignty over Alsace, royal edicts were issued in 1662, 1682, and 1687, proffering free land to anyone willing to restore it to full productivity. Swiss, Germans, Tyroleans, and Lorrainers poured into Alsace, infusing the region with a unique combination of cultures that define Alsace to this day. In 1871, at the end of the Franco-Prussian War, the region came under German control again and remained so until the end of World War I, when the French regained sovereignty. Alsace then began to reorganize the administration of its vineyards in line with the new French AOC system, but in 1940 Germany reclaimed the province from France before the process could be completed. Only after World War II, when Alsace reverted to France once more, was the quest for AOC status resumed, finally being realized in 1962.

The vineyards of Alsace are dotted with medieval towns of cobbled streets and timbered buildings, reflecting – as do the wines – the region's myriad Gallic and Prussian influences. The grapes are a mixture of German, French, and the exotic, with the German Riesling and Gewürztraminer (written without an umlaut in Alsace), the French Pinot Gris, and the decidedly exotic Muscat comprising the four principal varieties. Sylvaner, another German grape, also features to some extent, while other French varieties include Pinot Noir, Pinot Blanc, Auxerrois, and Chasselas. Although Gewürztraminer is definitely German (and fine examples are still to be found in the Pfalz, its area of origin), only in Alsace is it quite so spicy. And only in Alsace do you find spicy Pinot Gris, a grape that is neutral elsewhere. Even the Pinot Blanc may produce spicy wines in Alsace, although this is normally due to the inclusion of the fatter, slightly spicy Auxerrois grape.

DISAPPEARING DRY WINES

There has been an explosion in the numbers of producers making and selling VT and SGN wines (from just two or three producers prior to the decree to hundreds now) but the volume of these late-harvest wines remains minute at just 2 per cent of Alsace wines as a whole. The vast bulk of Alsace wines have traditionally been dry, but they have become increasingly sweet over the last 25 years or so, and some are very sweet indeed. So much so that it has become almost impossible for the wine-drinking public to tell whether the wine they are buying is dry or sweet. This is not because Alsace producers deliberately chose to create sweeter wines; it is merely the regrettable by-product of good intentions and worshipping the false concept of "physiological ripeness".

Grammatically, "physiological ripeness" is a tautological nonsense: the ripening process is intrinsically a physiological one. It is thus as pointless as calling time "chronological time". The use of this term infers that other, less apt definitions of ripeness exist, which is why it was latched onto by an increasing number of growers with an almost religious fervour. They believe that "physiological ripeness" involves a certain skin colour, stem maturity, tannin ripeness, aroma development, berry shrivel, pulp texture, and seed ripeness. The truth is, however, that all these factors, together with other, more conventional parameters, such as sugar, acidity, and pH, are merely indicators of the *progress* of grape ripeness, not some magical, singular definition of ripeness itself. As the late Dr Tony Jordan was fond of saying "the only ripeness that matters is flavour ripeness", and that begins when tartaric acid starts to dominate. The decision when to pick after this point is a subjective judgment that is based purely on the type and style of wine to be produced, providing that inclement weather does not force the winemaker's hand.

There are some very famous winemakers in Alsace who maintain that Riesling grapes grown on limestone or calcareous-clay do not attain "physiological ripeness" until they are much richer in sugar than the same variety grown on other soils, making it impossible to produce a balanced dry wine on limestone or calcareous-clay soils. Well, Trimbach's Clos Ste-Hune and Cuvée Frédéric-Émile are both grown on calcareous-clay, and both are beautifully balanced and totally dry. If believers in "physiological ripeness" consider Clos Ste-Hune grapes to be "physiologically unripe", then I would submit that as good evidence that all dry white wines should be made from unripe grapes.

No one is saying no to sweet wine in Alsace. No one should dictate the style of wine any producer wants to make. If producers want to make an entire range of wine with high alcohol and residual sugar, they are entitled to do so, but they should not pretend they are forced to do this because of so-called physiological ripeness. In recent years, some of these producers have moved their argument from physiological ripeness to climate change. Apparently, Alsace has become too hot to produce dry white wines, which might resonate with some, but it is – again – easily disproved by those who manage to make classic, crisp, dry white wines. It might be warmer than it used to be, and it is certainly drier, if reduced botrytis levels are any indication, but Alsace is still white wine country, and, if the eventual product is made with skill, it remains one of the world's greatest dry white wine regions.

UNINTENTIONAL BEGINNINGS

The loss of dry white wine in Alsace was unplanned but, with hindsight, it was inevitable and unfortunately was the result of very good intentions. In 1982 Alsace received a rebuke from Brussels for gross over-production, when it averaged 127 hectolitres per hectare (hl/ha). One year later, the first 25 *grand cru* sites were announced, and, understandably, a lot of producers saw this as an opportunity to reduce yields and raise profiles. The maximum *grand cru* yield had been set at 70 hl/ha in 1975 with the trial run for Grand Cru Schlossberg, but the hugely respected Léonard Humbrecht called for this to be halved. In the early 1970s Humbrecht had famously replanted the abandoned steep slopes of the Rangen in Thann, and his pursuit of low yields, not just in the *grands crus,* but across the entire Domaine Zind Humbrecht, was admired by all. As I wrote in *The Wines of Alsace* (Faber & Faber, 1993) "Few people command the respect that Léonard Humbrecht does. He is a legend throughout Alsace, not just among other growers, but merchants too. There is not one *négociant* too large or too lofty to acknowledge his tireless enthusiasm and almost evangelical crusade for lower yields". As other growers started to follow his lead of higher-density planting and, in particular, limiting yields through severely strict pruning, the official maximum reduced to 65 hl/ha, and then 55 hl/ha (they average 45 hl/ha today, with many famous names significantly lower than that).

At very much the same time as all this was brewing, the late Johnny Hugel (who incidentally chaired the first committee to consider Alsace Grand Cru and from that experience became vociferously opposed to the *grand cru* system) was working hard to introduce the concept of two speciality wines: VT and SGN. He had a devil of a job persuading growers that they could achieve the potential level of ripeness he wanted to set, and at the very top end (SGN) he had to settle for 220 grams per litre (Riesling and Muscat) and 243 grams per litre (Pinot Gris and Gewürztraminer). Later, as growers who were already pruning hard on the *grand cru* sites, thanks to Léonard Humbrecht, began to focus on VT and SGN styles, the game was afoot to produce higher and higher residual sugar. When they could not prune any harder, they began thinning the bunches, with yields plummeting lower and lower. It was not long before the amount of residual sugar *after* fermentation of some Alsace SGNs began to exceed the amount of sugar required for the Sauternes appellation *before* fermentation, and all those growers who opposed Johnny now wished they had allowed the levels to be set much higher. To achieve this, much of the crop will be removed to make standard

VINES LINE THE STEEP SLOPES OF THE RANGEN IN THAN
The vineyards of this area gained prominence in the 1970s when Léonard Humbrecht planted vines on the abandoned slopes with a goal to improve quality through lower yields.

FACTORS AFFECTING TASTE AND QUALITY

LOCATION
The northeast corner of France, flanked by the Vosges mountains and bordered by the Rhine and Germany's Black Forest. Six rivers rise in the fir-capped Vosges, flowing through the 97-kilometre (60-mile) strip of vineyards to feed the River Ill.

CLIMATE
Protected from the full effect of Atlantic influences by the Vosges Mountains, these vineyards are endowed with an exceptional amount of warm sunshine (800 hours per year) and a very low rainfall (55 to 60 centimetres, or 23 to 24 inches per year), as the rain clouds tend to shed their load on the western side of the Vosges, when they climb over the mountain range, creating warm and dry *foehn* winds that sweep along the plains.

ASPECT
The vineyards nestle on the lower east-facing slopes of the Vosges at a relatively high altitude of about 180 to 360 metres (600 to 1,200 feet), and at an angle ranging between 25 degrees on the lower slopes and 65 degrees on the higher ones. The best vineyards are generally found on mid-slopes with a south or southeast aspect,

but many good growths are also located on north- and northeast-facing slopes. In many cases, the vines are cultivated on the top as well as the sides of a spur, but the best sites are always protected by forested tops, with the vines kept well away from the trees. Some vineyards on the plains, however, can yield very good-quality wines where the soils are favourable.

SOIL
Alsace has the most complex geological and solumological situation of all the great wine areas of France. The three basic morphological and structural areas are: the siliceous edge of the Vosges; limestone hills; and the hydrous alluvial plain, which in total consist of 13 geological formations and in excess of 20 soil combinations. The siliceous soils include: colluvium and fertile sand over granite, stony-clay soil over schist, gneiss, various fertile soils over volcanic sedimentary rock, and poor, light, sandy soil over sandstone. The limestone soils include: dry, stony, and brown alkaline soil over limestone, brown sandy calcareous soils over sandstone and limestone, heavy fertile soils over clay-and-limestone, and brown alkaline soil over chalky marl. The hydrous

alluvial soils include sandy-clay and gravel over alluvium, brown decalcified loess, and dark calcareous soils over loess.

VITICULTURE & VINIFICATION
The vines are trained high on wires to avoid spring ground frost. There are a high number of organic and biodynamic producers. Classic wines are generally fermented as dry as possible, although this has been ignored by an increasing number of producers who harvest at such ripe levels that the wines regularly contain residual sugar and are sometimes sweeter than *vendanges tardives*. Although very few have a strict no-malolactic policy (Trimbach being one of the exceptions), the general trend is to avoid it if at all possible. In some years, however, it happens spontaneously during the fermentation process and is widespread, although if handled correctly, the results are more creamy-tactile than buttery aromas.

GRAPE VARIETIES
Primary varieties: Gewürztraminer, Muscat (Blanc/Rosé à Petits Grains, and Ottonel), Pinot Gris, Riesling
Secondary varieties: Auxerrois, Chardonnay, Chasselas, Klevener de Heiligensteiner (Savagnin Rosé), Pinot Blanc, Pinot Noir, Sylvaner

wines, while the remaining bunches continue to shrivel to VT and SGN levels, but making "dry" wine from the first harvest of vines trained and pruned for sweet wine hardly ever results in the lighter, finer structure required.

Whatever excuse any individual producer might make, the increase in alcohol and residual sugar levels in non-VT and non-SGN wines is the inevitable consequence of lower yields, particularly in the *grand cru* vineyard. This in itself would not be a problem if consumers could determine whether a wine was dry and that was set to happen starting from the 2016 vintage, when new labelling regulations were due to come into force. But INAO was already considering new regulations for VT and SGN and claimed they were unable to work on two things at once, and here, four years later, consumers are still waiting for INAO's bureaucrats to pull their collective finger out. It is not as if Alsace is requesting anything controversial. Its association of growers were consulted and gave almost unanimous support (just two objections) and the regulation INAO has been asked to authorise is, in fact, word-for-word in line with EU regulations: to impose an obligation on the labelling of Alsace to include the descriptor "Dry" or "*Sec*" on any Alsace white wine with less than 4 (g/L) of sugar, or 7g/L of sugar, if the total acidity is 9 grams or more (expressed as tartaric). Watch this space . . .

SEEING RED

Although Alsace is definitely still white wine country (90 per cent of production is white if Crémant d'Alsace is included), and crisp dry whites are easily possible for those with the skill and desire to make the classic style, whether entry-level or the greatest wine this region produces, it is indeed warmer than it used to be. Even if it is insufficiently warmer to prevent the production of crisp dry white wine, it should make it easier to ripen Pinot Noir and make increasingly finer reds. Forty years ago most Pinot Noir *rouge* was barely darker than a Pinot Noir rosé. In fact many were simply labelled Pinot Noir, leaving consumers to guess the style for themselves. Some still are. Thirty years ago the colour started to deepen as the demand from local *winstubs* ("wine rooms") for a regional red to sell by the *pichet* ("pitcher") increased, yet many of these were red in colour only. They seldom went through malolactic fermentation because they were made by producers with a white wine mentality and a fear that MLF bacteria would take up lodgings in their cellars and affect their white wines. Twenty years ago there was a move to making serious red wine from Pinot Noir, but far too many were over-extracted and over-oaked. Ten years ago, a few producers were starting to get things right (Blanck, Deiss, Hugel, Muré, and Weinbach), but in the scheme of things they were failures at this style compared to their German cousins across the Rhine. Now any producer with half a reputation has to be serious about Pinot Noir, but the fact of the matter is that not everywhere is suitable for this grape variety, and until it is recognised as a *grand cru* variety, we will never see the true potential of red wine in Alsace from the Pinot Noir, except from those producers who continue to make those wines in all but name, using the initial of their *grand cru* as a clue for consumers.

RECENT ALSATIAN VINTAGES

2018 Following a year's worth of rain in the first seven months, the weather turned dry and hot, continuing into and through the harvest, which enjoyed an Indian summer of sunny days and chilly nights. Early flowering and an early harvest produced intensely coloured Pinot Noir and the highest volumes of VT and SGN for 30 years. Fresh and fruity varietal wines, but for ready drinking, not keeping, due to low acidity.

2017 Together with 2011, 2009, 2007, and 2003, this was one of the five earliest harvest in the last 40 years, illustrating the warmer climate since the new millennium. Deeply coloured Pinot Noir with silky tannins. Large volume of VT. A difficult year for Gewürztraminer.

2016 A difficult year due to persistent rain and below-average temperatures. Grapes harvested were clean, but variable in ripeness. Pinot Gris is the stand-out variety.

2015 A hot June and July, with temperatures exceeding 40ºC (104ºF), drought, storms, and hail. Rain in mid-August was welcome and the vines responded amazingly. The third small harvest in a row, but with very clean grapes, high-quality grapes. All varieties performed exceptionally well, with good volumes of VT and SGN.

2014 Problematic weather (hot drought conditions at the beginning, fairly cool summer) with Suzukii fruit-flies adding to the difficulties, required swift and strict sorting. Those who succeeded were repaid with less richness, but high freshness and acidity. The second small harvest running, it was the Pinot varieties (Blanc, Gris, and Noir) that fared best, and Gewürztraminer was surprisingly good, with particularly spicy aromas.

2013 One of those years when growers have to decide whether they should wait for additional ripeness or pick at each and every opportunity the weather allowed. In most cases it was those who picked early who were the most successful and, in such circumstances, it was the intrinsically early-picked Crémant d'Alsace that fared best.

2012 A less-than-perfect year climatically produced beautifully balanced wines of the highest quality from Crémant d'Alsace to all varietals, with Riesling standing out due to its longer than usual ripening period.

2011 From -20ºC (-4ºF) the preceding winter to high spring temperatures and drought, followed by heavy rainfall and cool temperatures in the summer, this was by no means an easy year. Winds eliminated the possibility of rot, however, and had the secondary effect of concentrating the berries, like a wind-induced *passerillage*. After an Indian summer, sorting of the grapes and chilling the must and controlling the fermentation, the quality was surprisingly good for all varieties.

THE *GRANDS CRUS* OF

ALSACE

The original *grand cru* legislation was introduced in 1975 and used Schlossberg as its first example. Apart from Schlossberg, the first *grands crus* did not contain the name of any growth and some producers blended from far and wide. It was not until 1983 that the first list of 25 *grand cru* sites appeared and, three years later, a further 23 were added. By 2001 there were 50, and in 2006 Kaefferkopf became the 51st and final *grand cru*. In 2011 the term *grand cru* was upgraded from a designation of status within the AOC of Alsace or Vin d'Alsace to an appellation in and of itself, through the publication of 51 individual AOCs.

The exclusion of Kaefferkopf until 2006 was indicative of what was wrong with the declared strategy of this *grand cru* system in the first place: to limit its status to pure varietal wines and to restrict those varietal wines to just four "noble" varieties (Muscat, Riesling, Pinot Gris, and Gewürztraminer). Not only did this rob us of the chance to

enjoy the finest-quality Pinot Noir, Pinot Blanc, Sylvaner, and Chasselas, but to deny Pinot Noir, one of the greatest wine grapes in the world, as a "noble" variety can only be described as perverse. Furthermore – and this is where Kaefferkopf comes in – restricting *grand cru* status to pure varietal wines excluded Kaefferkopf, which was not only a truly great *cru* of Alsace, but also one that was historically famous for its production of classic blends. In 2005, however, after Marcel Deiss had forced through the acceptance of his *complantation* wines (field blends) for Altenberg de Bergheim Grand Cru, *grand cru* status for Kaefferkopf was inevitable. In 2006, Sylvaner was allowed for Zotzenberg.

In 2004, the growers of Zotzenberg successfully won the right for its Sylvaner to be granted *grand cru* status, which set a precedent that could and should be followed by other *grands crus* for Pinot Noir, but so far this status has not been conferred on the noblest of grapes. In the past

INAO has made the excuse that there are insufficient plantings of Pinot Noir for it to be a *grand cru* in its own right and for other lowly varieties, INAO has imposed a planting deadline (Altenberg de Bergheim, for example), but when it comes to the iconic wines of Alsace in a warming world, INAO should look forwards, not backwards, and encourage future plantings. With the input of Burgundy's top consultants and with the right clones growing on the warmest sites, the high-quality red wines produced should receive the premium expected for a *grand cru* wine, and that should fund future growth and improvement.

ALTENBERG DE BERGBIETEN
Bergbieten (29.07 hectares)

This is an exceptional growth, but not a truly great one. Its gypsum-permeated, clayey-marl soil is best for Gewürztraminer, which has a very floral character with

A SAMPLING OF WHITE WINES SET OUT FOR A TASTING IN RIBEAUVILLÉ
Alsace is best known for it wonderful whites. Here the selection includes the most famous. From left: Pinot Blanc, Riesling, Pinot Gris, Gewürztraminer, and Muscat

immediate appeal, yet can improve for several years in bottle. The wines are allowed to be blended from different varieties. The local *syndicat* has renounced the PLC, limiting the yield to a maximum of 55 hectolitres per hectare. It has also outlawed all chaptalisation and has increased the minimum ripeness of Riesling and Muscat from 11 to 12 per cent and that of Pinot Gris and Gewürztraminer from 12.5 to 14 per cent.

♛ *Loew* ⓑ • *Frédéric Mochel* • *Roland Schmitt*

ALTENBERG DE BERGHEIM
Bergheim (35.06 hectares)

This Altenberg has been a true *grand cru* since the 12th century. Its calcareous clay soil is best suited to Gewürztraminer, which is tight and austere in youth, but gains in depth and bouquet with ageing. This *cru* now permits wines made from a blend of grapes that extends beyond the four original varieties (50 to 70 per cent Riesling, 10 to 25 per cent Pinot Gris, 10 to 25 per cent Gewürztraminer, and, if planted before 2005, up to 10 per cent Pinot Blanc, Pinot Noir, Muscat, and Chasselas).

♛ *Marcel Deiss* ⓑ • *Gustave Lorentz* • *Sylvie Spielmann* ⓞ

ALTENBERG DE WOLXHEIM
Wolxheim (31.2 hectares)

Although appreciated by Napoleon, this calcareous clay *cru* cannot honestly be described as one of the greatest growths of Alsace, but it does have a certain reputation for its Riesling.

♛ *Dagobert* • *Dischler* • *Lissner* • *Muhlberger* • *Zoeller*

BRAND
Turckheim (57.95 hectares)

This *cru* might legitimately be called Brand New, the original Brand being a tiny *cru* of little more than three hectares. In 1924 it was expanded to include surrounding sites: Steinglitz, Kirchthal, Schneckenberg, Weingarten, and Jebsal, each with its own fine reputation. By 1980 it had grown to 30 hectares (74 acres) and it is now almost double that. This confederation of *lieux-dits* is one of the most magnificent sites in the entire region, and the

quality of the wines consistently excites me – great Riesling, Pinot Gris, and Gewürztraminer. The local *syndicat* has reduced chaptalisation of Riesling and Muscat from the allowable 1.5 to 1 per cent and that of Pinot Gris and Gewürztraminer from 1.5 to 0.5 per cent. It has also increased the minimum ripeness of Riesling and Muscat from 11 to 11.5 per cent and that of Pinot Gris and Gewürztraminer from 12.5 to 13.5 per cent.

♛ *Albert Boxler* • *Dopff Au Moulin* • *JosMeyer* ⓑ • *Armand Hurst* • *Zind Humbrecht* ⓑ

BRUDERTHAL
Molsheim (18.4 hectares)

Riesling and Gewürztraminer – reputedly the best varieties – occupy most of this calcareous clay *cru*. The best Riesling are fruity with an elegance when genuinely dry. Pinot Gris can also be impressive. The local *syndicat* has renounced all chaptalisation. It has also increased the minimum ripeness of Riesling and Muscat from 11 to 12 per cent and that of Pinot Gris and Gewürztraminer from 12.5 to 13 per cent.

♛ *Alain Klingenfus* • *Robert Klingenfus* • *Gérard Neumeyer*

EICHBERG
Eguisheim (57.62 hectares)

This calcareous clay *cru* has the lowest rainfall in Colmar and produces very aromatic wines of exceptional delicacy, yet great longevity. Famous for Gewürztraminer, which is potentially the finest in Alsace, Eichberg is also capable of making superb, long-lived Riesling and Pinot Gris. Léon Beyer does not sell *grand cru* as such, but its Cuvée des Comtes d'Eguisheim is 100 per cent pure Eichberg. The local *syndicat* has reduced chaptalisation of Riesling from 1.5 to 1 per cent and outlawed it completely for Pinot Gris and Gewürztraminer. It has also increased the minimum ripeness of Riesling from 11 to 11.5 per cent and Pinot Gris and Gewürztraminer from 12.5 to 13 per cent. The chaptalisation and minimum ripeness of Muscat remain the same.

♛ *Charles Baur* ⓞ • *Emile Beyer* • *Paul Ginglinger* ⓞ • *Gruss* • *Albert Hertz* ⓑ

ENGELBERG
Dahlenheim and Scharrachbergheim (14.8 hectares)

One of the least-encountered *grands crus* of Alsace, this vineyard gets long hours of sunshine and is supposed to favour Gewürztraminer and Riesling.

♛ *Bechtold* ⓑ • *Loew* ⓑ • *Pfister* • *Roland Schmitt*

FLORIMONT
Ingersheim and Katzenthal (21 hectares)

Mediterranean flora abounds on the sun-blessed, calcareous clay slopes of this *cru* – hence its name, meaning "hill of flowers". The excellent microclimate here produces some stunning Riesling and Gewürztraminer.

♛ *François Bohn* • *Justin Boxler* • *Jean Geiler* • *René Meyer* • *Bruno Sorg*

FRANKSTEIN
Dambach-la-Ville (56.2 hectares)

Frankstein is not so much one vineyard as four separate spurs. The warm, well-drained, granite soil of this *cru* is best suited to the production of delicate, racy Riesling and elegant Gewürztraminer.

♛ *Charles Frey* ⓑ • *Kamm* • *Schaeffer-Woerly*

FROEHN
Zellenberg (14.6 hectares)

This *cru* sweeps up and around the southern half of the hill upon which the small wine-making town Zellenberg is situated. The marly-clay soil suits Muscat, Gewürztraminer, and Pinot Gris varieties – in that order – and the wines are typically rich and long lived. The local *syndicat* has outlawed all chaptalisation and increased the minimum ripeness of Riesling and Muscat from 11 per cent to 12 per cent and that of Pinot Gris and Gewürztraminer from 12.5 to 13.5 per cent.

♛ *Becker* ⓞ

FURSTENTUM

Kientzheim and Sigolsheim (30.5 hectares)

This estate is best for Riesling, although the vines have to be well established to take full advantage of the calcareous soil. Gewürztraminer can also be fabulous – in an elegant, more floral, less spicy style – and Pinot Gris excels even when the vines are very young. The local *syndicat* has reduced all chaptalisation from the allowable 1.5 to 0.5 per cent. It has also increased the minimum ripeness of Riesling and Muscat from 11 to 11.5 per cent and Pinot Gris and Gewürztraminer from 12.5 to 13 per cent.

♔ *CV Bestheim* • *Paul Blanck* ❶ • *Bott-Geyl* ❸ • *Albert Mann* ❸ • *Weinbach* ❸

GEISBERG

Ribeauvillé (8.53 hectares)

Geisberg has been well documented since as long ago as 1308 as Riesling country par excellence. The calcareous, stony-and-clayey sandstone soil produces fragrant, powerful, and long-lived wines of great finesse. The local *syndicat* has reduced the chaptalisation of Riesling, Pinot Gris, and Gewürztraminer from 1.5 to 0.5 per cent. It has also increased the minimum ripeness of Riesling from 11 to 11.5 per cent and that of Pinot Gris and Gewürztraminer from 12.5 to 13 per cent. The chaptalisation and minimum ripeness of Muscat remain the same.

♔ *Kientzler* • *F.E. Trimbach*

GLOECKELBERG

Rodern and St-Hippolyte (23.4 hectares)

This clay-granite *cru* is known for its light, elegant, and yet persistent style of wine, with Gewürztraminer and Pinot Gris the most successful varieties. The local *syndicat* has halved the PLC, thereby reducing the absolute maximum yield allowed to 60.5hl/ha.

♔ *Bott-Frères* • *Fernand Engel* ❶ • *Koberlé-Kreyer*

GOLDERT

Gueberschwihr (45.35 hectares)

Dating back to the year 750, and recognized on export markets as long ago as 1728, Goldert derives its name from the colour of its wines, the most famous of which is the golden Gewürztraminer. The Muscat grape variety also excels on the calcareous clay soil, and whatever the varietal, the style of wine is always rich and spicy, with a luscious creaminess.

♔ *Ernest Burn* • *Bernard Humbrecht* • *Zind Humbrecht* ❸

HATSCHBOURG

Hattstatt and Voegtlinshoffen (47.36 hectares)

Gewürztraminer from the south-facing slope, with its calcareous marl soil, excels, but Pinot Gris and Riesling are also excellent.

♔ *André Hartmann* • *Lucien Meyer*

HENGST

Wintzenheim (75.78 hectares)

Hengst Gewürztraminer is a very special, complex wine, seeming to combine the classic qualities of this variety with the orange zest and rose-petal aromas more characteristic of the Muscat grape variety. But Hengst is a very flexible *cru*. Besides Gewürztraminer, its calcareous marl soil also produces top-quality Muscat, Riesling, and Pinot Gris. No one does it better than JosMeyer. The local *syndicat* has renounced all chaptalisation. It has also increased the minimum ripeness of Riesling and Muscat from 11 to 12 per cent and that of Pinot Gris and Gewürztraminer from 12.5 to 13 per cent.

♔ *Paul Buecher* • *Barmès-Buecher* ❸ • *JosMeyer* ❸ • *Hubert Krick* • *Albert Mann* ❸ • *Zind Humbrecht* ❸ • *CV Turckheim*

KAEFFERKOPF

Ammerschwihr (71.65 hectares)

The saga of Kaefferkopf's classification finally ended in 2006, when it became the 51st *grand cru*. The first list of potential *grands crus* was published in 1975, and although Kaefferkopf had been the very first named site in Alsace to have its boundaries delimited (in 1932 by a Colmar tribunal) and was also the only named site to be recognized in the original AOC Alsace of 1962, it was not included in the 50 *grands crus* delimited between 1983 and 1992. Kaefferkopf's traditional practice of blending two or more grape varieties was in direct conflict with the pure varietal connotation of AOC Alsace Grand Cru. Furthermore, it was claimed that the original 1932 delimitation covered an area of geologically diverse soils. The concept of *grand cru* wine consisting of more than one variety was accepted in principle by 2005, however, following the changes to Altenberg de Bergheim, and the question of geological uniformity, so often spouted by pedants of the *grand cru* system, can only be true to the subsoil, not the topsoil, and the subsoil across the entire area originally classified by the tribunal is pure granite. All in all, some 70 hectares have been delimited, compared to the 67.81 hectares delimited in 1932. This not only includes vineyards that were not previously classified as Kaefferkopf; it also excludes some vineyards that were. (Their owners do have the right to use the Kaefferkopf name for the next 25 years, however!) Authorized grape varieties are Gewürztraminer, Riesling, and Pinot Gris (each of which may be either a single variety or part of a blend) plus Muscat (only as part of a blend). Blends must be 60 to 80 per cent Gewürztraminer, 10 to 40 per cent Riesling, with optional up to 30 per cent Pinot Gris and up to 10 per cent Muscat.

♔ *Jean-Baptiste Adam* ❸ • *Pierre Adam* • *Marcel Freyburger* • *Horcher* • *Keuhn* • *Alfred Meyer* • *Maurice Schoech* • *Etienne Simonis* ❸

YIELD AND RIPENESS CRITERIA

	AOC ALSACE		CRÉMANT		COMMUNALE		LIEU-DIT		GRAND CRU		VT		SGN	
	Maximum Yield	Minimum Ripeness	Maximum Yield	Minimum Ripeness	Maximum Yield	Minimum Ripeness	Maximum Yield	Minimum Ripeness	Maximum Yield	Minimum Ripeness	Maximum Yield	Minimum Ripeness	Maximum Yield	Minimum Ripeness
Auxerrois	100hl/ha	10.0%	80hl/ha	9.0%	72hl/ha	10.5%	68hl/ha	10.5%	-	-	-	-	-	-
Chardonnay	-	-	80hl/ha	9.0%	-	-	-	-	-	-	-	-	-	-
Chasselas	100hl/ha	9.5%	-	-	72hl/ha	9.5%	68hl/ha	10.5%	-	-	-	-	-	-
Gewürztraminer	80hl/ha	11.5%	-	-	72hl/ha	12.0%	68hl/ha	12.0%	55hl/ha	12.5%	N/A	16.0%	N/A	19.0%
Klevener de Heiligenstein	-	-	-	-	72hl/ha	11.0%	-	-	-	-	-	-	-	-
Muscat	90hl/ha	9.5%	-	-	72hl/ha	10.5%	68hl/ha	10.5%	55hl/ha	11.0%	N/A	14.5%	N/A	17.0%
Pinot Blanc	100hl/ha	10.0%	80hl/ha	9.0%	72hl/ha	10.5%	68hl/ha	10.5%	-	-	-	-	-	-
Pinot Gris	80hl/ha	11.5%	80hl/ha	9.0%	72hl/ha	12.0%	68hl/ha	12.0%	55hl/ha	12.5%	N/A	16.0%	N/A	19.0%
Pinot Noir	-	-	80hl/ha	9.0%	-	-	-	-	-	-	-	-	-	-
Pinot Noir (rosé)	75hl/ha	10.0%	80hl/ha	9.0%	72hl/ha	10.0%	68hl/ha	10.0%	-	-	-	-	-	-
Pinot Noir (red)	60hl/ha	11.0%	-	-	60hl/ha	11.5%	60hl/ha	11.5%	-	-	-	-	-	-
Riesling	90hl/ha	10.0%	80hl/ha	9.0%	72hl/ha	10.5%	68hl/ha	10.5%	55hl/ha	11.0%	N/A	14.5%	N/A	17.0%
Sylvaner	100hl/ha	9.5%	-	-	72hl/ha	10.5%	68hl/ha	10.5%	-	-	-	-	-	-

Only four varieties qualify for Alsace Grand Cru, unless other varieties are expressly allowed for an individual *grand cru,* in which case they will be indicated under that specific *grand cru* profile.

No maximum yields for either VT or SGN because its academic if the grapes manage to achieve such high ripeness levels.

KANZLERBERG
Bergheim (3.23 hectares)

Although this tiny *cru* adjoins the western edge of Altenberg de Bergheim, the wines of its gypsum-permeated, clayey-marl *terroir* are so different from those of the Altenberg growth that the vinification of the two sites has always been kept separate. The wines have the same potential longevity, but the Kanzlerbergs are fuller and fatter. Kanzlerberg has a reputation for both Riesling and Gewürztraminer, but their ample weight can be at odds with their varietal aromas when young, and both wines require plenty of bottle age to achieve their true finesse. The local *syndicat* has renounced the PLC, limiting the yield to an absolute maximum of 55hl/ha. It has also outlawed all chaptalisation and increased the minimum ripeness of Riesling and Muscat from 11 to 12 per cent and that of Pinot Gris and Gewürztraminer from 12.5 to 14 per cent.

♛ *Gustave Lorentz* • *Sylvie Spielmann* ⓪

KASTELBERG
Andlau (5.82 hectares)

One of the oldest vineyards in Alsace, Kastelberg has been planted with vines since the Roman occupation. Situated on a small hill next to Wiebelsberg, Andlau's other *grand cru*, the very steep, schistous *terroir* has long proved to be an excellent site for racy and delicate Riesling, although the wines can be very closed when young and require a few years to develop their lovely bottle aromas. Kastelberg wines remain youthful for 20 years or more, and even show true *grand cru* quality in so-called "off" years.

♛ *Moritz* • *Guy Wach*

KESSLER
Guebwiller (28.53 hectares)

Though a *cru* more *premier* than grand (the truly famous sites of Guebwiller being Kitterlé and Wanne, the latter not classified), the central part of Kessler is certainly deserving of *grand cru* status. Here, the vines grow in a well-protected, valley-like depression, one side of which has a very steep, south-southeast facing slope. Kessler is renowned for its full, spicy, and mellow Gewürztraminer, but Riesling can be much the greater wine. The local *syndicat* has renounced the PLC, limiting the yield to an absolute maximum of 55hl/ha. It has also outlawed all chaptalisation and increased the minimum ripeness of Riesling and Muscat from 11 to 11.5 per cent and that of Pinot Gris and Gewürztraminer from 12.5 to 13.5 per cent.

♛ *Dirler-Cadé* Ⓑ • *Schlumberger*

KIRCHBERG DE BARR
Barr (40.63 hectares)

The true *grands crus* of Barr are Gaensbroennel and Zisser, but these have been incorporated into the calcareous marl *terroir* of Kirchberg, which is known for its full-bodied yet delicate wines that exhibit exotic spicy fruit, a characteristic that applies not only to Pinot Gris and Gewürztraminer but also to Riesling.

♛ *Jean & Hubert Heywang* • *Klipfel* • *André Lorentz*

KIRCHBERG DE RIBEAUVILLÉ
Ribeauvillé (11.4 hectares)

One of the few *lieux-dits* that has regularly been used to market Alsace wine over the centuries. It is famous for Riesling, which typically is firm, totally dry, and long lived, developing intense petrolly characteristics with age. Kirchberg de Ribeauvillé also produces great Muscat with a discreet, yet very specific, orange-and-musk aroma, excellent acidity, and lots of finesse. The local *syndicat* has reduced the chaptalisation of Riesling, Pinot Gris, and Gewürztraminer from 1.5 to 0.5 per cent. It has also increased the minimum ripeness of Riesling from 11 to 11.5, and that of Pinot Gris and Gewürztraminer from 12.5 to 13 per cent. The chaptalisation and minimum ripeness of Muscat remain the same.

♛ *Kientzler* • *Jean Sipp* • *Louis Sipp* ⓪

KITTERLÉ
Guebwiller (25.79 hectares)

Of all the grape varieties that are grown on this volcanic sandstone *terroir*, it is the crisp, petrolly Riesling that shows greatest finesse. Gewürztraminer and Pinot Gris are also very good in a gently rich, supple, and smoky-mellow style. The local *syndicat* has renounced the PLC, limiting the yield to an absolute maximum of 55hl/ha. It has also outlawed all chaptalisation and has increased the minimum ripeness of Riesling and Muscat from 11 to 11.5 per cent and that of Pinot Gris and Gewürztraminer from 12.5 to 13.5 per cent.

♛ *Schlumberger*

MAMBOURG
Sigolsheim (61.85 hectares)

The reputation of this *cru* has been documented since 783, when it was known as the "Sigolttesberg". A limestone coteau, with calcareous clay topsoil, Mambourg stretches for well over 1 kilometre (3/4 mile), penetrating further into the plain than any other spur of the Vosges foothills. Its vineyards, supposed to be the warmest in Alsace, produce wines that tend to be rich and warm, mellow and liquorous. Both the Gewürztraminer and the Pinot Gris have plenty of smoky-rich spice in them. The local *syndicat* has reduced chaptalisation of Riesling, Pinot Gris and Gewürztraminer from 1.5 to 0.5 per cent. It has also increased the minimum ripeness of Riesling from 11 to 12 per cent and that of Pinot Gris and Gewürztraminer from 12.5 to 13 per cent.

♛ *CV Bestheim* • *Jean Biecher* • *François Bléger* • *J-M & F Bernhard* • *Marcel Deiss* Ⓑ • *Charles Sparr* • *André Thomas*

MANDELBERG
Mittelwihr and Beblenheim (22 hectares)

Mandelberg – "almond tree hill" – has been planted with vines since Gallo-Roman times and used as an appellation since 1925. Its reputation has been built on Riesling, although today more Gewürztraminer is planted; high-quality Pinot Gris and Muscat are also produced. The local *syndicat* has renounced the PLC, limiting the yield to an absolute maximum of 55hl/ha. It has also outlawed chaptalisation for Pinot Gris and Gewürztraminer, increasing the minimum ripeness for these two grapes from 12.5 to 13 per cent. The minimum ripeness for Riesling and Muscat remains at 11 per cent and the possibility of chaptalisation is retained exclusively for these two varieties.

♛ *Becker* ⓪ • *Bott-Geyl* Ⓑ • *Philippe Gocker* • *André Stentz* ⓪

MARCKRAIN
Bennwihr and Sigolsheim (53.35 hectares)

Markrain is the east-facing slope of the Mambourg, which overlooks Sigolsheim. The soil is a limestone-marl with oolite pebbles interlayering the marl beds. It is mostly planted with Gewürztraminer, with some Pinot Gris and a few Muscat vines. The Bennewihr *coopérative* usually makes a decent Pinot Gris under the Bestheim label. The local *syndicat* has reduced the chaptalisation of Pinot Gris and Gewürztraminer from an allowable 1.5 to 1 per cent and increased the minimum ripeness of these grapes from 12.5 to 13 per cent. The chaptalisation and minimum ripeness of Muscat and Riesling remain the same.

♛ *Bouxhof* • *Michel Fonné* • *Etienne Simonis* Ⓑ

MOENCHBERG
Andlau and Eichhoffen (11.83 hectares)

Moenchberg – "monk's hill"– was owned by a Benedictine order until 1097, when it was taken over by the inhabitants of Eichhoffen. With its clayey-marl soil, excellent exposure to the sun, and very hot, dry microclimate, this *cru* has built up a reputation for firm, intensely fruity, and very racy Riesling. Pinot Gris can be excellent. Not to be confused with the equally excellent *grand cru* Muenchberg of Nothalten.

♛ *Rémy Gresser* • *Maurer*

A RETRO-STYLE BICYCLE DECORATES A POST ALONGSIDE A VINEYARD NEAR MITTELWIHR
Located in the Mandelberg wine region, Mittelwihr was nearly destroyed during World War II. Since then, vineyards have been revitalized and are thriving.

MUENCHBERG
Nothalten (17.7 hectares)

This sunny vineyard belonging to the abbey of Baumgarten, whose monks tended vines in the 12th century, nestles under the protection of the Undersberg, a 900-metre (2,950-feet) peak in the Vosges mountains. The striking style of Muenchberg's wines is due in part to the special microclimate it enjoys, and also to the ancient, unique, and pebbly volcanic sandstone soil. The wines are now allowed to be blended from different varieties. The local *syndicat* has renounced the PLC, limiting the yield to an absolute maximum of 55hl/ha. It has also outlawed all chaptalisation and has increased the minimum ripeness of Riesling and Muscat from 11 to 12 per cent and that of Pinot Gris and Gewürztraminer from 12.5 to 14 per cent.

♛ *Pierre Koch* • *Armand Landmann* • *André Ostertag* Ⓑ • *Philippe Sohler*

OLLWILLER
Wuenheim (35.86 hectares)

Ollwiller's annual rainfall is one of the lowest in France, with Riesling and Gewürztraminer faring best on its clayey-sand soil – although it is not one of the greatest *grands crus,* and the recommended wines are rated in that context.

♛ *CV Vieil Armand*

OSTERBERG
Ribeauvillé (24.6 hectares)

This stony-clay growth abuts Geisberg, another Ribeauvillé *grand cru,* and makes equally superb Riesling country. The wines age very well, developing the petrolly nose of a fine Riesling. Trimbach owns vines here, and together with a contiguous plot with vines in Geisberg (both of which overlook the winery) produce its superb Riesling Cuvée Frédéric Émile. Much of Trimbach's Pinot Gris Réserve Personnelle also comes from this *grand cru*.

♛ *Joggerest* • *Kientzler* • *CV de Ribeauvillé* • *Louis Sipp* ⓪

PFERSIGBERG
Eguisheim and Wettolsheim (74.55 hectares)

Also spelled Pfersichberg and Pfirsigberg. A calcareous sandstone soil well known for its full, aromatic, and long-lived Gewürztraminer – although Pinot Gris, Riesling, and Muscat also fare well here. The wines all share a common succulence of fruit acidity and possess exceptional aromas. Léon Beyer's Cuvée Particulière is not sold as a *grand cru,* but 100 per cent Pfersigberg. The local *syndicat* has reduced chaptalisation of Riesling from 1.5 to 1 per cent and outlawed it completely for Pinot Gris and Gewürztraminer. It has also increased the minimum ripeness of Riesling from 11 to 11.5 per cent and that of Pinot Gris and Gewürztraminer from 12.5 to 13 per cent. The chaptalisation and minimum ripeness of Muscat remain the same.

♛ *Charles Baur* ❶ • *Emile Beyer* • *Paul Ginglinger* ❶ • *Jean-Luc Meyer* • *Bruno Sorg*

PFINGSTBERG
Orschwihr (28.15 hectares)

All four *grand cru* varieties grow well on the calcareous-marl and clayey-sandstone of this *cru*, producing wines of typically floral aroma, combined with rich, honeyed fruit. The local *syndicat* has renounced all chaptalisation. It has also increased the minimum ripeness of Riesling and Muscat from 11 to 12 per cent and that of Pinot Gris and Gewürztraminer from 12.5 to 13.5 per cent.

♛ *François Braun* • *Bernard Haegelin* ❸ • *François Schmitt* • *Albert Ziegler*

PRAELATENBERG
Kintzheim and Orschwiller (18.7 hectares)

Although Praelatenberg dominates the north side of the village of Orschwiller, virtually all the *cru* actually falls within the boundary of Kintzheim, 1.5 kilometres (1 mile) away. The locals say that all four varieties grow to perfection here, but Pinot Gris used to be the best, but has been toppled by Riesling in recent years. Muscat can also be exceptionally fine. The local *syndicat* has reduced all chaptalisation from an allowable 1.5 to 1 per cent. It has also increased the minimum ripeness of Riesling and Muscat from 11 to 11.5 per cent and that of Pinot Gris and Gewürztraminer from 12.5 to 13 per cent.

♛ *Allimant-Laugner* • *Engel Frères* ❶ • *Les Faîtières*

RANGEN
Thann and Vieux-Thann (22.13 hectares)

The 15th-century satirist Sebastian Brant, writing about the little-known travels of Hercules through Alsace, reveals that the mythical strongman once drank so much Rangen that he fell asleep. So ashamed was he on waking that he ran away, leaving behind his bludgeon – the club which today appears on Colmar's coat of arms.

So steep that it can be cultivated only when terraced, Rangen's volcanic soil is very poor organically, but extremely fertile minerally. It also drains very quickly, and its dark colour makes it almost too efficient in retaining the immense heat that pours into this sweltering suntrap. This fierce heat and rapid drainage, however, are essentially responsible for the regular stressing of the vine, which is what gives the wines their famed power and pungency. Rangen produces great wines even in the poorest years, making it a true *grand cru* in every sense.

♛ *Schoffit* • *Wolfberger* • *Zind Humbrecht* ❸

ROSACKER
Hunawihr (26.18 hectares)

First mentioned in the 15th century, this *cru* has built up a fine reputation for Riesling, but one wine – Trimbach's Clos Ste-Hune – is, almost every year, far and away the finest Riesling in Alsace. Occasionally other producers make an exceptional vintage that may challenge it, but

THE CENTRAL STREET OF MEDIEVAL RIQUEWIHR VILLAGE LEADS TO THE DOLDER WATCHTOWER
One of "Les Plus Beaux Villages de France", Riquewihr is typical of many of the wine towns along the Alsace Wine Route. Oenophiles can sample the wines of the region as the stop in colourful towns such as this, as well as Eguisheim, Colmar, Kaysersber, Rosheim, Rosheim, Ribeauvillé, Obernai, Dambach-la-Ville, Kintzheim, and Bergheim.

none has consistently matched Clos Ste-Hune's excellence. Trimbach makes no mention of Rosacker on its label because the family believes, as do a small number of internationally known producers, who avoid using the term, that much of Rosacker should not be classified as *grand cru* (although Trimbach used to sell Clos Ste-Hune *grand cru* in the 1940s). Rosacker's calcareous and marly-clay soil is rich in magnesium and makes fine Gewürztraminer as well as top Riesling. The local *syndicat* has reduced chaptalisation of Riesling, Pinot Gris, and Gewürztraminer from 1.5 to 1 per cent. It has also increased the minimum ripeness of Riesling from 11 to 11.5 per cent and that of Pinot Gris and Gewürztraminer from 12.5 to 13 per cent. The chaptalisation and minimum ripeness of Muscat remain the same.

♛ *L'Agapé • Eblin-Fuchs* Ⓑ *• Mader • CV de Hunawihr • F.E. Trimbach (Clos Ste-Hune)*

SAERING
Guebwiller (26.75 hectares)

This vineyard, first documented in 1250 and marketed since 1830, is situated below Kessler and Kitterlé, Guebwiller's other two *grands crus*. Like Kessler, this *cru* is more *premier* than *grand* (yet still better than many of the *grands crus*). The floral, fruity, and elegant Riesling is best, especially in hot years, when it becomes exotically peachy, but Muscat and Gewürztraminer can also be fine.

♛ *Dirler-Cadé* Ⓑ *• Rominger* Ⓑ *• Schlumberger*

SCHLOSSBERG
Kientzheim and Kaysersberg (80.28 hectares)

The production of Schlossberg was controlled by charter in 1928, and in 1975 it became the first Alsace *grand cru*. Although its granite *terroir* looks to be shared equally by the two sites, less than half a hectare belongs to Kaysersberg. Schlossberg is best for Riesling, but Gewürztraminer can be successful in so-called "off" vintages. The wine is full of elegance and finesse, whether produced in a classic, restrained style, as Blanck often is, or with the more exuberant fruit that typifies Weinbach. The local *syndicat* has reduced the chaptalisation of Pinot Gris and Gewürztraminer from an allowable 1.5 to 0.5 per cent and increased the minimum ripeness of these grapes from 12.5 to 13 per cent. The chaptalisation and minimum ripeness of Riesling and Muscat remain the same.

♛ *Paul Blanck* Ⓞ *• Joseph Fritsch • Albert Mann* Ⓑ *• Weinbach* Ⓑ *• F.E. Trimbach*

SCHOENENBOURG
Riquewihr (53.4 hectares)

This vineyard has always had a reputation for producing great Riesling and Muscat, although modern wines show Riesling to be supreme, with Pinot Gris vying with Muscat for the number two spot. Schoenenbourg's gypsum-permeated, marly-and-sandy soil produces very rich, aromatic wines in a *terroir* that has potential for VT and SGN.

♛ *L'Agapé • Dopff Au Moulin • Dopff & Irion • Armand Shreyer*

SOMMERBERG
Niedermorschwihr and Katzenthal (28.36 hectares)

Known since 1214, the fame of this *cru* was such that a strict delimitation was in force by the 17th century. Situated in the foothills leading up to Trois-Épis, its granite soil is supposed to be equally excellent for all four classic varieties, although Riesling stands out in my experience. Sommerberg wines are typically aromatic, with an elegant succulence of fruit.

♛ *Albert Boxler*

SONNENGLANZ
Beblenheim (32.8 hectares)

In 1935, two years after Kaefferkopf was defined by tribunal at Colmar, Sonnenglanz received a similar certification, but unlike at Kaefferkopf, its producers failed to exploit the appellation until 1952, when the local cooperative was formed. Once renowned for its Sylvaner, the calcareous clay soil of Sonnenglanz is best suited to Gewürztraminer and Pinot Gris, which can be very ripe and golden in colour.

♛ *Bott-Geyl* Ⓑ

SPIEGEL
Bergholtz and Guebwiller (18.26 hectares)

Known for only 50 years or so, this is not one of the great *grands crus* of Alsace. Its sandstone and marl *terroir* can produce fine, racy Riesling with a delicate bouquet, however, and good, though not great, Gewürztraminer and Muscat. The local *syndicat* has renounced all chaptalisation of Pinot Gris and Gewürztraminer, but permissible chaptalisation of Riesling and Muscat remains at 1.5 per cent. It has also increased the minimum ripeness of Pinot Gris and Gewürztraminer from 12.5 to 13.5 per cent, but that of Riesling and Muscat remains 11 per cent.

♛ *Dirler-Cadé* Ⓑ *• Eugène Meyer* Ⓑ

SPOREN
Riquewihr (23.7 hectares)

Sporen is one of the truly great *grands crus*, its stony, clayey-marl soil producing wines of remarkable finesse. Historically, this *terroir* is famous for Gewürztraminer and Pinot Gris, which occupy virtually all of its vineyard today, but it was also traditional to grow a mix of varieties and vinify them together to produce a classic non-varietal wine, such as Hugel's Sporen Gentil, which was capable of ageing 30 years or more (and in a totally different class from that firm's own Gentil, which does not come from Sporen and is a blend of separately vinified wines).

♛ *Bott Geyl • Dopff Au Moulin • Meyer-Fonné • Vieille Forge*

STEINERT
Pfaffenheim and Westhalten (38.9 hectares)

Pinot Gris is the king of this stony, calcareous *cru*, although historically Schneckenberg (now part of Steinert) was always renowned for producing a Pinot Blanc that tasted more like Pinot Gris. The Pfaffenheim cooperative still produces a Pinot Blanc Schneckenberg (although not a *grand cru*, of course), and its stunning Pinot Gris steals the show. Steinert's Pinot Blanc-cum-Gris reputation illustrates the exceptional concentration of these wines. Gewürztraminer fares best on the lower slopes, Riesling on the higher, more sandy slopes. The local *syndicat* has reduced chaptalisation of Riesling and Muscat from the allowable 1.5 to 1 per cent and that of Pinot Gris and Gewürztraminer from 1.5 to 0.5 per cent. It has also increased the minimum ripeness of Riesling and Muscat from 11 to 12 per cent and that of Pinot Gris and Gewürztraminer from 12.5 to 13.5 per cent.

♛ *Pierre Frick* Ⓑ *• Moltès • Rieflé • Pierre Paul Zink*

STEINGRUBLER
Wettolsheim (22.95 hectares)

Although this calcareous-marl and sandstone *cru* is not one of the great names of Alsace, I have enjoyed some excellent Steingrubler wines, particularly Pinot Gris. This can be very rich, yet show great finesse. Certainly Steingrubler is one of the better lesser-known *grands crus*, and it could well be a great growth of the future. The local *syndicat* has reduced all chaptalisation from an allowable 1.5 to 1 per cent. It has also increased the minimum ripeness of Riesling and Muscat from 11 to 11.5 per cent and that of Pinot Gris and Gewürztraminer from 12.5 to 13 per cent.

♛ *Barmès-Buecher* Ⓑ *• Robert Dietrich • Albert Mann* Ⓑ *• Jean-Paul Schaffhauser • Wunsch & Mann*

STEINKLOTZ
Marlenheim (40.6 hectares)

Steinklotz ("block of stone") was part of the estate of the Merovingian king, Childebert II, from which Marlenheim derives its reputation for Pinot Noir, but since it has flown the *grand cru* flag, Steinklotz is supposed to be good for Pinot Gris, Riesling, and Gewürztraminer. Local growers are determined to have their Pinot Noir recognized, however, and will be applying for *grand cru* status for these wines.

♛ *Helfrich • Xavier Muller*

VORBOURG
Rouffach and Westhalten (73.61 hectares)

All four varieties excel in this calcareous sand-stone *terroir*, whose wines are said to develop a bouquet of peaches, apricots, mint, and hazelnut – but Riesling and Pinot Gris fare best. Muscat favours warmer vintages, when its wines positively explode with flavour, and Gewürztraminer excels in some years but not in others. Vorbourg catches the full glare of the sun from dawn to dusk and so is also well suited to Pinot Noir, which is consequently heavy with pigment. The local *syndicat* has renounced all chaptalisation. It has also increased the minimum ripeness of Riesling and Muscat from 11 to 12 per cent and that of Pinot Gris and Gewürztraminer from 12.5 to 13.5 per cent.

♛ *De l'École • Muré* Ⓑ

WIEBELSBERG
Andlau (12.52 hectares)

This vineyard has very good sun exposure, and its siliceous soil retains heat and drains well. Riesling does well, producing wines that can be very fine and floral, slowly developing a delicate, ripe-peachy fruit on the palate.

♛ *Rémy Gresser • Guy Wach*

WINECK-SCHLOSSBERG
Katzenthal and Ammerschwihr (27.4 hectares)

Granite vineyards enjoy a sheltered microclimate that primarily favours Riesling, followed by Gewürztraminer. The wines are light and delicate, with a fragrant aroma.

♛ *Jean-Baptiste Adam* Ⓑ *• Jean-Marc Bernhard • Paul Blanck* Ⓞ *• Jean-Paul Ecklé • Vignoble Klur* Ⓑ *• Jean Geiler • Meyer-Fonné • Vincent Spannagel*

WINZENBERG
Blienschwiller (19.2 hectares)

Locals claim that this *cru* is cited in "old documents" and that Riesling and Gewürztraminer fare best in its granite vineyards. The Riesling I have encountered has been light and charming, but not special. Gewürztraminer has definitely been much the superior wine, showing fine, fresh aromas with a refined spiciness of some complexity.

♛ *Hubert Metz • Hubert Meyer • Straub*

ZINNKOEPFLÉ
Soultzmatt and Westhalten (71.03 hectares)

The hot, dry microclimate of Zinnkoepflé gives rise to a rare concentration of Mediterranean and Caspian fauna and flora near its exposed summit. The heat and the arid, calcareous sandstone soil are what gives it its reputation for strong, spicy, and fiery styles of Pinot Gris and Gewürztraminer. The Riesling is a delicate and most discreet wine, but this is deceptive, and it can, given decent bottle-age, be just as powerful. Mention should be made of Seppi Landmann, who has produced some stunning Zinnkoepflé, but also far too many ordinary wines, and a few absolute bummers, thus does not get a wholehearted recommendation below.

♛ *Léon Boesch* Ⓑ *• Agathe Bursin • René Fleck • Jean-Marie Haag • Paul Kubler • Moltès • Gérard Nicollet • Schlegel-Boeglin*

ZOTZENBERG
Mittelbergheim (36.45 hectares)

First mentioned in 1364, when it was known as Zoczenberg, the wines of this calcareous clay *terroir* have been sold under its own *lieu-dit* since the beginning of the 20th century. It is historically the finest site in Alsace for Sylvaner, and in 2004 became the first *grand cru* to be officially recognized for this supposedly lowly grape variety. Gewürztraminer and Riesling show a creamy richness of fruit.

♛ *Boeckel • Armand Gilg • Alfred Wantz • Wittmann*

THE APPELLATIONS OF

ALSACE AND LORRAINE

ALSACE AOC

This appellation covers all the wines of Alsace (with the exception of Alsace Grand Cru and Crémant d'Alsace), but 95 per cent of the wines are often sold according to grape variety. These are: Pinot (which may also be labelled Pinot Blanc, Clevner, or Klevner), Pinot Gris, Pinot Noir, Riesling, Gewürztraminer, Muscat, Sylvaner, Chasselas (which may also be labelled Gutedel), and Auxerrois. This practice effectively creates nine "varietal" AOCs under the one umbrella appellation, and these are listed separately.

ALSACE COMMUNALE AOC

Introduced in 2011, this extended the 12 geographical designations already allowed to each and every one of the 118 wine villages in Alsace. The existing geographical designations remain in force and are restricted as follows.

DESIGNATION	RESTRICTIONS
Bergheim	White wines only – from Bergheim and Ribeauvillé
Blienschwiller	White wines only – from Blienschwiller
Coteaux du Haut Koenigsbourg	White wines only – from Chatenois, Kintzheim, Orschwiller, and Saint Hippolyte
Côtes de Barr	White wines only – from Barr
Côte de Rouffach	Red and white wines – from Pfaffenheim, Rouffach, and Westhalten
Klevener de Heiligenstein	White wines only – from Bourgheim, Gertwiller, Goxwiller, Heiligenstein, and Obernai
Ottrott	Red wines only – from Ottrott and Obernai
Rodern	Red wines only – from Rodern et Saint Hippolyte
Saint Hippolyte	Red wines only – from Saint Hippolyte and Orschwiller
Scherwiller	White wines only – from Scherwiller
Val Saint Grégoire	White wines only – from Turckheim, Zimmerbach, Walbach, and Wihr-au-Val (Rhin)
Vallée Noble	White wines only – from Westhalten and Soultzmatt

ALSACE LIEU-DIT AOC

Introduced in 2011, this allows the use of any authentic *lieu-dit* (named site), of which there are in excess of 400, although not all are yet in commercial use.

ALSACE GRAND CRU AOC

The current production of *grand cru* wine is approximately 5 per cent of the total volume of AOC Alsace. Because every *cru*, or growth, makes a wine of a specific character, it is impossible to give a generalized description. Where other grapes and blends apply, these are detailed under the *grand cru* in question.

WHITE See The *Grands Crus* of Alsace, p270.

🍷 *Muscat, Riesling, Gewürztraminer, Pinot Gris*

ALSACE PREMIER CRU AOC

This has been in INAO's In-Tray for several years, but it won't be attended to until after the Dry/*Sec* designation has been processed, which should have been completed in time for the 2016 harvest, but they do not want to begin that process until after the latest new rules for VT and SGN have been agreed.

ALSACE SÉLECTION DE GRAINS NOBLES AOC

These rare and sought-after wines are made from botrytis-affected grapes. Unlike Sauternes, however, Alsace is no haven for "noble rot", which occurs haphazardly and in much reduced concentrations. The wines are, therefore, produced in tiny quantities and sold at very high prices. The *sauternais* are often amazed not only by the high sugar levels (while chaptalisation is not permitted in Alsace, it has become almost mandatory in Sauternes), but also by how little sulphur is used, highlighting why SGN has become one of the world's greatest dessert wines.

WHITE Now made with less alcohol and higher sugar than when first introduced, these wines possess even more finesse than before. While Gewürztraminer is almost too easy to make, Pinot Gris offers the ideal balance between quality and price; just a couple of Muscat have been produced, and Riesling SGN is in a class of its own. Check the appropriate varietal entry for the best producers.

🍷 *Gewürztraminer, Pinot Gris, Riesling, Muscat*

🍷 *5–30 years*

ALSACE VENDANGE TARDIVE AOC

This is a subordinate designation that may be appended either to the basic appellation or to Alsace Grand Cru AOC. Its production is controlled, and the regulations are far stricter than for any AOC in France. VT is far less consistent in quality and character than SGN. This is because some producers make these wines from grapes that have the correct minimum sugar content, but that were picked with the rest of the crop, not late-harvested as such. Many VT therefore lack the true character of a late-harvested wine, which is only brought about by the complex changes that occur inside a grape that has remained on the vine until November or December. As the leaves begin to fall and the sap retreats to the protection of the root system, the grapes, cut off from the vine's metabolic system, start dehydrating. The compounds that this process (known as *passerillage*) produces are in turn affected by the prevailing climatic conditions. *Passerillé* grapes that have endured progressively colder temperatures (the norm) and those that have enjoyed a late Indian summer (not uncommon) will produce entirely different wines.

WHITE Whether dry, medium-sweet, or sweet, this relatively full-bodied wine should always have the true character of *passerillage* – although sometimes this will be overwhelmed by botrytis. Gewürztraminer is the most commonly encountered variety, but only the best have the right balance; Pinot Gris and Riesling both offer an ideal balance between quality, availability, and price. Muscat is almost as rare for VT as it is for SGN, as it tends to go very flabby when overripe.

🍷 *Gewürztraminer, Pinot Gris, Riesling, Muscat*

🍷 *5–20 years*

AUXERROIS AOC

Theoretically this designation does not exist, but Auxerrois is one of the varieties permitted for the production of Pinot wine and makes such a distinctly different product that it has often been labelled separately. This practice, currently on the increase, is "officially tolerated".

WHITE Fatter than Pinot Blanc, with a more buttery, honeyed, and spicy character to the fruit, the greatest asset of Auxerrois is its natural richness and immediate appeal. Inclined to low acidity, it can easily become flabby and so musky it tastes almost foxy, but the best Auxerrois can give Pinot Gris a run for its money.

🍷 *Auxerrois*

🍷 *Up to 5 years*

✓ *Paul Blanck ❶ • JosMeyer ❸ • Rieffel ❶ (Klevner Vieilles Vignes is pure Auxerrois) • Jean Rapp • Armand Hurst • Antoine Stoffel*

CHASSELAS AOC

Rarely seen, but enjoying something of a revival among a few specialist growers.

WHITE The best Chasselas wines are not actually bottled but sit in vats waiting to be blended into anonymous *cuvées* of Edelzwicker. They are neither profound nor complex, but teem with fresh, delicate fruit and are an absolute joy to drink. Even the best taste better before they are bottled than after, however. The fruit is so fragile that it needs a lift to survive the shock of being bottled, and it would probably benefit from being left on lees and bottled very cold to retain a bit of tongue-tingling carbonic gas.

🍷 *Chasselas*

🍷 *Upon purchase*

✓ *Albert Boxler • Paul Blanck ❶ • Pierre Frick ❸ • Kientzler • Schoffit*

CLASSIC ALSACE BLENDS

Despite the varietal wine hype, Alsace is more than capable of producing the finest-quality classic blends, but their number is small and dwindles every year, because so few consumers realize their true quality. It is difficult for producers to make potential customers appreciate why their blends are more expensive than ordinary Edelzwicker, but classic Alsace blends should no more be categorized with Edelzwicker than *crus classés* compared with generic Bordeaux. This category focuses attention on the region's top-performing blends which, whether or not they fetch them, deserve *grand cru* prices. Unlike the blending of Edelzwicker, for which various wines are mixed together – and where there is always the temptation to get rid of unwanted wines – the different varieties in most classic Alsace blends always come from the same vineyard and are traditionally harvested and vinified together, known as *complantation*, or a field-blend.

WHITE Most of these wines improve with age but go through phases when one or other grape variety dominates, which is interesting to observe and should help you to understand why you prefer to drink a particular blend. Depending on the amounts involved, Gewürztraminer typically dominates in the young wine, followed by Pinot Gris then, many years later, Riesling, but other varieties may also be involved when overripe.

🍷 *Chasselas, Sylvaner, Pinot Blanc, Pinot Gris, Pinot Noir, Auxerrois, Gewürztraminer, Muscat Blanc à Petits Grains, Muscat Rosé à Petits Grains, Muscat Ottonel, Riesling*

✓ *Marcel Deiss ❸ (Alsace Blanc) • Muller-Koeberlé (Langenberg Clos des Aubépines) • CV Ribeauvillé (Clos du Zahnacker) • Rominger ❸ (Ozmose)*

CLEVNER AOC

Commonly used synonym under which Pinot is marketed, sometimes spelled Klevner, but not to be confused with Klevener. *See* Pinot AOC.

CRÉMANT D'ALSACE AOC

Although small growers like Dirler had made Vin Mousseux d'Alsace as early as 1880, it was not until 1900 that Dopff Au Moulin created a sparkling wine industry on a commercial scale, and 1976 before an AOC was established. Except for small growers, the grapes for most crémant is grown on the plains. The harvest usually begins some 10 days before Alsace varietal wines and will typically be in August or September compared to September or October. The quality is good, but there is a lot of room for improvement.

GEWÜRZTRAMINER GRAPES RIPEN ON THE VINE
This aromatic wine grape has a pink to red skin colour. It is considered one of the four "noble" varieties, along with Muscat, Riesling, and Pinot Gris. For many wine drinkers, Gewürztraminer (usually spelled without the umlaut over the "u" in France) is the one most associated with the white wines of Alsace.

SPARKLING WHITE The Pinot Blanc has perfect acidity for this sort of wine; however, it can lack sufficient richness, and after intensive tastings I have come to the conclusion that the Pinot Gris has the just right acidity and richness.

🍇 *Pinot Blanc, Pinot Gris, Pinot Noir, Auxerrois, Chardonnay, Riesling*

🍷 *5–8 years*

SPARKING ROSÉ These delightful wines can have a finer purity of perfume and flavour than many pink Champagnes.

🍇 *Pinot Noir*

🍷 *3–5 years*

✓ *Bernard Becht • Dopff Aux Moulin • Joseph Gruss (Rosé, Prestige Magnum) • Haailler (Rosé) • Klein-Brand • Meyer-Fonné (Brut Extra Magnum) • CV Pfaffenheim • Rieflé • Wolfberger (Chardonnay, Pinot Gris)*

EDELZWICKER AOC

This appellation (*Edel* = noble + *Zwicker* = blend) is reserved for wines blended from two or more of the authorized grape varieties, and it was indeed once noble. However, since the banning of AOC Zwicker, which was never meant to be noble, and due to the fact that there has never been a legal definition of which varieties make a blend noble, producers simply renamed their Zwicker blends Edelzwicker. Consequently, this appellation has become so tarnished that many producers prefer to sell their cheaper AOC Alsace wines under brand names or simply AOC Alsace, rather than put the debased Edelzwicker name on the label. *See also* Classic Alsace Blends.

GEWÜRZTRAMINER AOC

No other wine region in the world has managed to produce Gewürztraminer with any real spice, which is probably why this is usually the first Alsace wines people taste. Its voluptuous, up-front style is always immediately appealing.

WHITE The fattest and most full-bodied of Alsace wines. Classic renditions of this grape typically have the aroma of banana when young and take three to four years in bottle to build up a pungent spiciness of terpene-laden aromas, often achieving a rich gingerbread character when mature.

🍇 *Gewürztraminer*

🍷 *3–10 years (20–30 years for great examples)*

✓ *Yves Amberg* **B** *• Arbogast • Barmès-Buecher* **B** *• Charles Baur* **O** *• Léon Beyer • Paul Blanck* **O** *• Bott-Geyl* **B** *• Albert Boxler • Burghart-Spettal • Clos des Terres Brunes • Fernand Engel* **O** *• Jean Geiler • Hartweg • Hugel & Fils • Jean Geiler • Paul Ginglinger* **O** *• Pierre-Henri Ginglinger* **O** *• Hertzog • JosMeyer* **B** *• Kientzler • Georges Klein • Meyer-Fonné • Muré* **B** *• Ostertag* **B** *• Saint Rémy* **B** *• Schlumberger • François Schmitt • Schoffit*

• Aline & Rémy Simon • Vignobles Reinhart • Jean Sipp • Bruno Sorg • Sylvie Spielmann **O** *• Antoine Stoffel • F.E. Trimbach • Weinbach* **B** *• Zind Humbrecht* **B** *• Zink*

VT: *Jean-Baptiste Adam • Bader • Léon Baur • Burghart-Spettal • Ernest Burn • Hugel & Fils • Kientzler • Baron Kirmann • Ostertag* **B** *• Seilly • Weinbach* **B** *• Zind Humbrecht*

SGN: *Hubert Beck • Léon Beyer • Paul Blanck* **O** *• Albert Boxler • Ernest Burn • Dirler-Cadé* **B** *• Hugel & Fils • JosMeyer* **B** *• Albert Mann* **B** *• Muré* **B** *• Jean-Paul Schauffhauser • Schlumberger • Aimé Stentz* **O** *• F.E. Trimbach • Weinbach* **B** *• Welty • Zind Humbrecht* **B**

GUTEDEL AOC

Synonym under which Chasselas may be marketed. *See* Chasselas AOC.

KLEVENER DE HEILIGENSTEIN AOC

Nothing to do with Klevner (a synonym for Pinot Blanc), Klevener de Heiligenstein is an oddity in Alsace because Klevener is in fact Savagnin Rosé, a grape variety that is native to the Jura farther south and not found anywhere else in Alsace. Klevener is also the only grape confined by law to a fixed area within Alsace (the village of Heiligenstein).

WHITE Dry, light-bodied wines of a subdued, spicy aroma, and delicate, fruity flavour. Sadly, no wines worth recommending.

🌿 *Savagnin Rosé*

🍷 *2-4 years*

✓ *Charles Boch • Paul Dock • Jean & Hubert Heywang • Daniel Ruff • Wittmann • Zeyssolf*

KLEVNER AOC
See Pinot Blanc AOC

MUSCAT AOC

The best Muscat wine, some growers believe, is made from the Muscat d'Alsace, a synonym for both the white and pink strains of the rich, full Muscat à Petits Grains. Others growers are convinced that the lighter, more floral Muscat Ottonel is best. A blend of the two is probably preferable. These wines are better in average years, or at least in fine years that have good acidity, rather than in truly great vintages. Served in Alsace as an aperitif and as a classic accompaniment to asparagus, especially the white asparagus for which Alsace is known.

WHITE These are dry, aromatic wines with fine floral characteristics that often smell of orange-flower water and taste of peaches. A top-quality Muscat that is expressive of its *terroir* is a great wine.

🌿 *Muscat Blanc à Petits Grains, Muscat Rosé à Petits Grains, Muscat Ottonel*

🍷 *Upon purchase*

✓ *Becker* 🅞 *• Kress-Bléger • Bott Frères • Ernest Burn • Dirler-Cadé* 🅑 *• Fleck • Michel Fonné • Pierre Henri Ginglinger* 🅞 *• Maurice Griss • Bernard Humbrecht • JosMeyer* 🅑 *• René Kientz • Kientzler • André Lorentz • Frédéric Mochel • Muré* 🅑 *• Schoffit • Bruno Sorg • Weinbach* 🅑 *• Zind Humbrecht* 🅑

VT: *Barmès-Buecher* 🅑 *• Muré* 🅑

SGN: *Jean-Baptiste Adam • Claude Bleger • Fernand Engel* 🅞 *• Albert Mann* 🅑

PINOT AOC

Not necessarily pure Pinot Blanc, this white wine may be made from any of the Pinot grape varieties, including Auxerrois (often confused with Pinot Blanc, but it is in fact a completely separate variety). Most Pinot wines are a blend of Pinot Blanc and Auxerrois; the further north the vines are cultivated, the more Auxerrois is used to plump out the Pinot Blanc.

WHITE Some Pinot wines are occasionally spineless, but lacklustre examples are not as common as they used to be, as it is the plump and juicy *cuvées* that have made this the fastest-growing category of Alsace wine.

🌿 *Pinot Blanc, Auxerrois, Pinot Noir (vinified white), Pinot Gris*

🍷 *2-4 years*

✓ *Jean-Baptiste Adam • André Blanck • Agathe Bursin • CV Cléebourg • Pierre Frick* 🅑 *• JosMeyer* 🅑 *• Koeberlé-Kreyer • Marcel Litchlé • Albert Mann* 🅑 *• CV Pfaffenheim • Edmond Schueller • Sylvie Spielmann* 🅞 *• Antoine Stoffel • Weinbach* 🅑 *• Zind Humbrecht* 🅑

PINOT BLANC AOC

This designation should only be used if the wine is made from 100 per cent Pinot Blanc. See also Pinot AOC.

PINOT GRIS AOC

This designation is now the most common way of marketing wine from the rich Pinot Gris grape. For hundreds of years, this variety has been known locally as Tokay d'Alsace. Legend had it that the Pinot Gris was brought back to Alsace from Hungary by Baron Lazare de Schwendi in the 1560s, and because Tokay or Tokaji was the only famous Hungarian wine, it was assumed that this was the grape Tokay was made from, and the name stuck. But they got it all wrong: this grape came to Alsace from Burgundy; Schwendi had nothing to do with it, and Pinot Gris has nothing to do with Tokay/Tokaji, of course. In the 1980s, the Hungarians wanted their name back, and initially the French agreed, but after counter arguments from Alsace

producers, alleging more than 400 years use of the name, the French reneged, and agreed instead to change the name from Tokay d'Alsace to Tokay-Pinot Gris, with vague promises that they might drop the Tokay bit at some time in the future. That did not happen, but the fall of communism did, and with Hungary's entry to the European Union, the French were forced to honour their partner's historic appellation, so all bottles had to be labelled Pinot Gris, with no reference to Tokay, starting with 2006 vintages.

WHITE This full-bodied, off-dry wine is decadently rich, but has excellent acidity, and its fullness of flavour never tires the palate. A young Pinot Gris can taste or smell of banana, sometimes be smoky, with little or no spice, but as it matures it increasingly develops a smoky-spice, toasty-creamy richness, finally achieving a big, honeyed walnut-brazil complexity with good bottle age. Top Alsace Riesling can be much finer, but the variety is so sensitive to soil conditions and handling that the quality is nowhere near as consistent as Pinot Gris across the board.

🌿 *Pinot Gris*

🍷 *5-10 years*

✓ *Jean-Baptiste Adam • Pierre Adam • Laurent Bannwarth* 🅑 *• Becker* 🅞 *• Paul Blanck* 🅞 *• François Bohn • Albert Boxler • Henri Ehrhart • Bernard Haegelin* 🅑 *• Hartweg • Victor Hertz • CV de Hunawihr • JosMeyer* 🅑 *• Kientzler • Henri Klée • Koehly • Albert Mann* 🅑 *• Jean-Luis & Fabien Mann* 🅑 *• Marzolf • Muller-Koeberle • André Ostertag* 🅑 *• Clos Sainte-Apolline • Schaller • Spitz & Fils • Shoepfer • Antoine Stoffel • Guy Wack • Wassler • Bernard Weber • Weinbach* 🅑 *• Xavier Wymann • Zind Humbrecht* 🅑 *• Zink*

VT: *François Bléger • Hugel & Fils • Vignobles Reinhart • Zind Humbrecht* 🅑

SGN: *Jean-Baptiste Adam • Léon Beyer • Albert Boxler • Fernand Engel* 🅞 *• Hugel & Fils • Albert Mann* 🅑 *• Scheidecker • F.E. Trimbach • Weinbach* 🅑 *• Zind Humbrecht* 🅑 *• Pierre Paul Zink*

PINOT NOIR AOC

Not so long ago, Pinot Noir d'Alsace was synonymous in style with rosé, but the trend has swung hard over towards a true red wine. After a steep learning curve, during which many wines were overextracted, lacked elegance, were prone to rapid oxidation, and bore the most ungainly caramelized characteristics, Alsace winemakers have now managed to master the handling of oak and red wine techniques. Alsace producers are now allowed to use a fatter bottle of almost Burgundian proportions for Pinot Noir instead of the Flûte d'Alsace, the use of which has been enshrined in law since 1959.

RED Most are unsatisfactory (still over-extracted, too tannic, or too oaky – sometimes all three!), but Marcel Deiss's Burlenberg is in a different class and is comparable to a good Burgundy. Why is it that although both Alsace and Germany started at the same time (mid-1980s) to develop Pinot Noir as a serious red wine style (as opposed to rosé), only Germany has succeeded? Over the last five years or so there have been some definite improvements, resulting in many more Pinot Noir that live up to their name, although how much is down to the efforts of producers or the warmer climate is difficult to discern. When Pinot Noir is recognised as a *grand cru* grape, it will be a shot in the arm for Alsace red wine. Until then, even the top producers below will not always succeed with every *cuvée* in every vintage: caveat emptor.

🌿 *Pinot Noir*

🍷 *2-6 years (12 years for exceptional* cuvées*)*

ROSÉ At its best, this dry, light-bodied wine has a deliciously fragrant aroma and flavour, which is reminiscent of strawberries, raspberries, or cherries. Seldom seen on export markets, but still commonly encountered by the *pichet* in local *winstubs*.

🌿 *Pinot Noir*

🍷 *1-2 years*

✓ *L'Agapé • Francis Beck • Becker* 🅞 *• Paul Blanck* 🅞 *• Bohn • Camille Braun* 🅞 *• Brobecker • Ernest Burn • Marcel Deiss* 🅑 *• Charles Fahrer • Fritz-Schmitt • Gsell • Huber & Bléger • Hugel & Fils • Bruno Hunold • Albert Klée • René & Michel Koch • Kress-Bléger • Pierre-Yves Meyer • Moltés • Muré • De La Vieille Forge • Vignobles Reinhart • Aimé Stentz* 🅑 *• Vincent Stoeffler • François Schwach • Wassler • Weinbach* 🅑 *• Welty*

RIESLING AOC

Of all Alsace grape varieties, Riesling is the most susceptible to differences in soil: clay soils give fatness and richness; granite Riesling is also rich but with more finesse; limestone adds obvious finesse but less richness; and volcanic soil gives a well-flavoured, spicy style.

WHITE In youth, fine Rieslings can show hints of apple, fennel, citrus, and peach, but they can be so firm and austere that they give no hint of the beautiful wines into which they can evolve.

🌿 *Riesling*

🍷 *4-20 years*

✓ *L'Agapé • Ansen • Anstotz • Arbogast • Bader • Baumann • A.L. Baur • Francis Beck • Becker • Emile Beyer • Léon Beyer • Paul Blanck • Léon Boesch* 🅑 *• Bohn • Dischler • Eblin-Fuchs • Fernand Engel* 🅞 *• Fahrer-Ackermann • Les Faîtières • René Fleck • Jean Geiler • Paul Ginglinger* 🅞 *• Willy Gisselbrecht • Rémy Gresser • Jean-Marie Haag • Hartweg • Horcher • CV de Hunawihr • Bernard Humbrecht • JosMeyer • Georges Klein • Albert Klur • Albert Maurer • Alfred Meyer • Jérôme Meyer • Kamm • Karcher • Mader • Muré • Joseph Moellinger • Château d'Orschwihr • André Ostertag* 🅑 *• Rentz • CV de Ribeauvillé • Christophe Rieflé • Roi Dagobert • Rolly Gassmann • Eric Rominger • Robert Roth • Jean-Paul Schaffhauser • Joseph Scharsch • Scheidecker • Scherb • Schlegel-Boeglin • Schlumberger • François Schmitt • Roland Schmitt • Schoffit • Bernard Schwach • J-L Schwarz • Aline & Rémy Simon • Etienne Simonis* 🅑 *• Jean Sipp • Louis Sipp* 🅞 *• Philippe Sohler • Bruno Sorg • Vincent Spannagel • F.E Trimbach • CV Turckheim • Ville de Colmar • Vorburger* 🅞 *• Guy Wach • Weinbach* 🅑 *• Wunsch & Mann • Fernand Ziegler • Zind Humbrecht* 🅑 *• Zoeller*

VT: *Becker* 🅞 *• Dopff Au Moulin • Dopff & Irion • Hugel & Fils • JosMeyer* 🅑 *• Kientzler • Lichtlé • Vignobles Reinhart • Schlumberger • Weinbach* 🅑 *• Zind Humbrecht* 🅑

SGN: *Hugel & Fils • Fernand Engel* 🅞 *• Frey-Sohler • Kientzler • Rolly Gassmann • Joseph Rudlof • F.E. Trimbach • Weinbach* 🅑 *• Zind Humbrecht* 🅑

SYLVANER AOC

Hugh Johnson once described the Sylvaner as "local tapwine", and this is how it should be served – direct from the stainless steel vat, with all the zip and zing of natural carbonic gas (normally filtered out during the bottling process). Like Chasselas, the fruit rarely survives the shock of bottling and like Muscat, Sylvaner does not suit heat, which is why it is best to buy in cooler years.

WHITE Sylvaner is an unpretentious, dry, light- to medium-bodied wine, with fragrance rather than fruitiness. It is generally best drunk young, but, like the Muscat, exceptionally long-living examples can always be found.

🌿 *Sylvaner*

🍷 *Upon purchase*

✓ *Charles Baur* 🅞 *• Boeckel • Agathe Bursin • Paul Ginglinger* 🅞 *• JosMeyer* 🅑 *• Loew* 🅑 *• Roland Schmitt • Wantz • Weinbach* 🅑 *• Zind Humbrecht* 🅑 *• Valentin Zusslin* 🅑

VIN D'ALSACE AOC

Alternative designation for Alsace AOC, which see.

THE APPELLATIONS OF
LORRAINE

CÔTES DE TOUL AOC

Part of the once-flourishing vineyards of Lorraine, these *côtes* are located in eight communes west of Toul (Blénod-lès-Toul, Bruley, Bulligny, Charmes-la-Côte, Domgermain, Lucey, Mont-le-Vignoble, and Pagney-derrière-Barine), in the *département* of Meurthe-et-Moselle, and were elevated to AOC status in 2003. Although the best are merely ready-drinking country wines, Laroppe easily outclasses the competition.

RED The Pinot Noir is the most successful, and the wine is usually sold as a pure varietal. It can have surprisingly good colour for wine from such a northerly region, and good cherry-Pinot character.

🍇 *Pinot Meunier, Pinot Noir*

🍷 *1–4 years*

WHITE These wines represent less than 2 per cent of the appellation, producing just 76 hectolitres (844 cases). Nevertheless, the Auxerrois is the best grape variety, with its fatness making it ideal for such a northerly area with a calcareous soil.

🍇 *Aligoté, Aubin, Auxerrois*

🍷 *1–3 years*

ROSÉ Most Côtes de Toul is made and sold as *vin gris*. This pale rosé is delicious when it is still youthful.

🍇 *Gamay, Pinot Meunier, Pinot Noir, plus a maximum of 15 per cent Aligoté, Aubin, and Auxerrois*

🍷 *Upon purchase*

✓ *De L'Ambrosie • Demange • Vincent Laroppe • Lelièvre • de la Linotte • Migot • Régina • CV du Toulois*

MOSELLE AOC

Although many restaurants list German Mosel wines as "Moselle", the river and the wine it produces are called Mosel in Germany, only becoming Moselle when crossing the border into France (and Luxembourg). There were 323 hectares of vineyards here in 1898, but the catastrophic infestation of phylloxera caused almost total devastation, and by 1985 there were just 3 hectares of vineyard left in Moselle. There has been some revitalisation, and there are now 65 hectares planted, 50 of which are classified AOC. Gamay is limited by law to a maximum of 30 per cent of the surface area of this appellation. Gamay, Auxerrois, and various Pinot varieties are mostly grown in the south, whereas Müller-Thurgau dominates the farther north you go. A maximum of 12.5 per cent alcohol is applied to prevent producers from over-chaptalising.

RED Château de Vaux produces a surprisingly good Pinot Noir.

🍇 *Gamay, Pinot Meunier, Pinot Noir*

🍷 *Upon purchase*

WHITE There has been an improvement in these wines, particularly the Pinot Gris.

🍇 *Auxerrois, Müller-Thurgau, Pinot Blanc, Pinot Gris, Riesling, Gewürztraminer*

🍷 *Upon purchase*

✓ *Les Béliers • Buzéa • Legrandjacques • Oury-Schreiber • Sontag • Du Stromberg • Château de Vaux*

THE *ROUTE DES VINS ALSACE* WINDS THROUGH THE COMMUNE OF HUSSEREN-LES-CHÂTEAUX
Starting near Strasbourg in the north and ending just south of Colmar, the 170-kilometre (105-mile) Alsace Wine Route leads travellers through viticultural villages, wine-producing towns, and rows and rows of vineyards. Its rich blend of German and French influences shows in its medieval half-timbered houses, ancient castles, and stately churches.

The Loire Valley

In winemaking terms, the Loire Valley is best imagined as a long ribbon with crisp white wines at either end and fuller wines of all types in the middle. It is the home of Sauvignon Blanc, the only wine area in the world that specializes in Cabernet Franc, and, in truly great vintages, makes some of the most sublime and sumptuous botrytised Chenin Blanc wines.

The Loire is the longest river in France. From its source in the Cévennes Mountains, it flows 1,000 kilometres (620 miles) through 12 *départements*. The variations in soil, climate, and grape varieties found along its banks and those of its tributaries are reflected in the wide range of wines grown in the four major wine-producing districts. Red, white, and rosé and still, *pétillant*, and fully sparkling wines are produced in 87 different appellations, ranging in style from bone dry to intensely sweet.

THE LOIRE'S MOST IMPORTANT GRAPE

The Chenin Blanc grape produces four distinctly different styles of wine – dry, semi-sweet, sweet, and sparkling. This is due to traditional practices that have been forced on growers by the vagaries of climate. This grape has abundant natural acidity and, if it receives enough sun, a high sugar content. But the Loire is considered a northern area in viticultural terms, and growers must contend with late frosts, cold winds, and variable summers. Given a sunny year, the grower's natural inclination is to make the richest wine possible with this sweet and tangy grape, but in difficult vintages, only a medium or a dry style can be achieved. Advances in winery technology and vinification means that today top growers are producing dry Chenins that exhibit excellent purity of fruit and concentration.

Chenin Blanc has been given pride of place in Anjou, Saumur, and Vouvray. It is a grape that should be harshly pruned, and yields must be carefully managed to prevent over-cropping. Fortunately, there are now many quality-conscious Loire growers, who restrict their yields and allow the grapes to ripen to full phenolic maturity. It is from such top-performing growers as Château Pierre-Bise in the Coteaux du Layon and Savennières, François Chidaine and Huet in Vouvray, and Jacky Blot from Domaine de la Taille aux Loups in Montlouis that we get Chenin Blanc wines that dazzle our palates.

FROM SOURCE TO SEA

For some inexplicable reason, most references restrict the Loire to its four major districts of Pays Nantais, Anjou-Saumur, Touraine, and Central Vineyards. Although some books incorporate the outlying areas of Fiefs Vendéens and Haut-Poitou, few mention the appellations of the Upper Loire, yet the wines of Côtes du Forez, Côte Roannaise, Côtes d'Auvergne, St-Pourçain, and Châteaumeillant are equally legitimate members of the Loire's family of wines. Nowadays the most famous appellations and some of the lesser-known ones could and should rank among the most exciting wines of the world.

RECENT LOIRE VINTAGES

2018 As with the rest of France, the June heat wave pushed vine growth on at a tremendous rate, but unlike most other regions, the drought had minimal effect. The harvest in September was carried out under blue skies and high temperatures, and the result was fantastic for every appellation, particularly for those that decreased yields from the outset by rubbing off buds and those that grass between the vines. Crispness, freshness, and good fruit will be the hallmark of most 2018s.

2017 April frosts caused major crop loss, particularly in Anjou (Savennières up to 95% lost), Touraine, and Vouvray. Near perfect vintage conditions during summer meant that by harvest time full maturity had been achieved in all appellations with classic Loire freshness and great concentration in the finished wines.

2016 Late April frosts caused significant damage across the region, especially in Muscadet (up to 60%) and Pouilly Fumé, where some sectors reported losses of 80%. A cool August and perfect September led to a late harvest that produced some crisp, fresh fruity wines at best. The exceptional Cabernet Francs are delicious, ripe, and fruity.

2015 As with many other regions across France and Europe, the Loire enjoyed a warm, dry summer. Muscadet suffered a little rain at harvest time, causing a small amount of rot. Sauvignon yields were generally down on average, but overall quality of reds, whites, and rosés is excellent.

2014 This year was a stand-out vintage for Cabernet Franc, but Melon B, Sauvignon Blanc, and Chenin Blanc (from bone dry to sweet) also produced top-quality wines. The only negative aspect was – once again – reduced yields in most areas.

2013 A June hailstorm virtually wiped out the entire crop in Vouvray and Montlouis. Overall quality for whites is decent with good acidity although cooler conditions during summer and up until harvest meant that the reds struggled to reach full maturity, leading to leafy green characters in some of the red appellations.

2012 Sancerre and Pouilly Fumé performed very well with good ripeness levels and normal yields thanks to great weather conditions. Cabernet Francs suffered from dilution due to the rains at harvest time, however. Muscadet, despite a testing growing season, produced superb quality wines, although yields were severely restricted. For Chenin, 2012 was a vintage which favoured dry wines over sweet.

2011 A really tough vintage with grey rot affecting many varietals across the region. Chenin Blanc seemed to fare better even though not exceptional.

WINE REGIONS OF THE LOIRE VALLEY (see also p131)

The longest river in France claims a larger number of appellations than any other classic wine region, commencing with Muscadet on the Atlantic side, and travelling upriver to Anjou–Saumur, Touraine, the Central Vineyards, and the little-known Upper Loire appellations.

WINE NOTES

- The Loire is the last wild river in Europe and was designated a World Heritage Site by the United Nations Educational, Scientific, and Cultural Organization (UNESCO).
- As a wild river, the water level can vary by several metres within a few days, and its islands slowly move from year to year.
- The Loire Valley is the second-largest sparkling wine region in France (although Crémant d'Alsace is the second-largest individual appellation).

EURE-ET-LOIR
Châteaudun
SARTHE
Loir
Touraine
La Flèche
Loir
Fleury-les-Aubrais
Saint-Jean-de-la-Ruelle
Saint-Jean-de-Braye
Orléans
Olivet
LOIRET
Vendôme
Beaugency
Central Vineyards
LOIR-ET-CHER
Gien
Blois
Cosson
Briare
Beuvron
Aubigny-sur-Nère
Anjou–Saumur
Baugé-en-Anjou
Salbris
Cosne-Cours-sur-Loire
MAINE-ET-LOIRE
Saint-Cyr-sur-Loire
Saint-Pierre-des-Corps
Tours
Saint-Avertin
Romorantin-Lanthenay
Loire
Saumur
Anjou–Saumur
Doué-la-Fontaine
Chinon
INDRE-ET-LOIRE
Vienne
Indrois
Touraine
Vierzon
Central Vineyards
NIÈVRE
Bourges
CHER
Nevers
Indre
Issoudun
Loire
Touraine
INDRE
Châtellerault
Claise
Châteauroux
Haut-Poitou
DEUX-SÈVRES
Clain
Le Poinçonnet
Saint-Amand-Montrond
Parthenay
Neuville-de-Poitou
Jaunay-Clan
Cher
Thouet
Poitiers
VIENNE
Argenton-sur-Creuse
Allier
Yzeure
Moulins
Saint-Pourçain
ALLIER
Montluçon
Bouble
Guéret
CREUSE
Cusset
Vichy
Côte Roannaise
Riorges
Roanne
Riom
Gerzat
Thiers
Clermont-Ferrand
Chamalières
Pont-du-Château
Beaumont
Cournon-d'Auvergne
LOIRE
Lignon
PUY-DE-DÔME
Montbrison
Côtes d'Auvergne
Issoire
Allier
Côtes du Forez
Loire

SANCERRE VINEYARD IN EARLY SPRING

Known as the "Garden of France", the Loire Valley produces a rich diversity of wine styles, including red, white, and rosé and sparkling wines of varying levels of fizz. Sancerre itself is famous for its whites but does make wines of all three colours.

THE LOIRE VALLEY

CÔTES D'AUVERGNE AOC

Upgraded from VDQS to AOC in 2010, Côtes d'Auvergne is located south of Saint-Pourçain and west of Côtes du Forez on the edge of the Massif Central. This is the most remote of all the outer areas that fall officially within the Loire region. The red and rosé wines from certain villages are superior and have therefore been given the right to use the following communal appellations: The red-only Côtes d'Auvergne-Boudes; Côtes d'Auvergne-Chanturgues; Côtes d'Auvergne-Châteaugay; and Côtes d'Auvergne-Madargues; and the rosé-only Côtes d'Auvergne-Corent.

RED The best of these dry, light-bodied, and fruity wines carry the Chanturgues appellation. Most are made from Gamay, a grape that has traditionally been grown in the area, and are very much in the style of Beaujolais.

🍷 *Gamay, Pinot Noir*

🍾 *1–2 years*

WHITE These dry, light-bodied wines made from Chardonnay have been overlooked but can be more accessible than the same variety grown in more classic areas of the Loire, thus they are surely a marketable commodity in view of Chardonnay's upmarket status.

🍷 *Chardonnay*

🍾 *1–2 years*

ROSÉ These are dry, light-bodied wines with an attractive cherry flavour. The best of them are from the Corent appellation.

🍷 *Gamay, Pinot Noir*

🍾 *Within 1 year*

✓ *Vignoble de l'Arbe Blanc ⓝ • Pradier • Gilles Persilier*

CÔTES DU FOREZ AOC

Similar to Beaujolais, this wine is produced in the Loire *département* adjacent to Lyon, and was upgraded from VDQS to AOC in 2000. The wines are improving through the efforts of the cooperative and a few quality-conscious growers, but they could be better.

RED These are dry, light-bodied wines with some fruit. They are best drunk young and at a cool temperature.

🍷 *Gamay*

🍾 *Upon purchase*

ROSÉ These simple, light-bodied, dry rosés make attractive, unpretentious picnic wines.

🍷 *Gamay*

🍾 *Upon purchase*

✓ *Gilles Bonnefoy (La Madone) ⓝ • Les Vignerons Foreziens • Odile Verdier et Jacky Logel ⓝ*

CÔTE ROANNAISE AOC

Red and rosé wines are made from a localized Gamay clone called Gamay Saint-Romain. These grapes are grown on south- and southwesterly-facing slopes of volcanic soil on the left bank of the Loire some 40 kilometres (25 miles) west of the Mâconnais district of Burgundy. This appellation was promoted from VDQS to full AOC status in 1994.

RED Some of these dry, medium- to full-bodied wines are produced using a form of carbonic maceration, and a few are given a little maturation in cask. The result can vary between well-coloured wines that are firm and distinctive and oaky and fruity Beaujolais-type versions for quaffing when they are young.

🍷 *Gamay*

🍾 *1–5 years*

ROSÉ These crisp, fruity, well-made wines are dry and medium bodied.

🍷 *Gamay*

🍾 *2–3 years*

✓ *Alain Baillon (Cuvée Monplaisir) • Paul et Jean-Pierre Benetère (Vieilles Vignes) • Des Pothiers ⓝ • De la Rochette (Vieilles Vignes) • Robert Serol ⓝ*

CRÉMANT DE LOIRE AOC

Of all the Loire's sparkling wines, this is probably the most underrated, yet it has the greatest potential because it can be blended from the wines of Anjou-Saumur and Touraine and from the widest range of grape varieties.

SPARKLING WHITE The better-balanced *cuvées* of these dry to semi-sweet, light- to medium-bodied wines are normally a blend of Chenin Blanc in the main, with a good dash of Cabernet Franc and Chardonnay. The best Chardonnay clones are yet to be established in the Loire, but at least Chardonnay is widely utilized for sparkling wines, whereas Pinot Noir is not – this is a mystery, as it is a proven variety in the region.

🍷 *Chenin Blanc, Cabernet Franc, Cabernet Sauvignon, Pineau d'Aunis, Pinot Noir, Chardonnay, Arbois, Grolleau Noir, and Grolleau Gris*

🍾 *1–3 years*

SPARKLING ROSÉ The best of these light- to medium-bodied wines are *brut*, and usually contain a high proportion of Cabernet Franc and Grolleau Noir grapes. Cabernet Franc makes the most distinctive wine. A pure Pinot Noir crémant rosé would be interesting to taste.

🍷 *Chenin Blanc, Cabernet Franc, Cabernet Sauvignon, Pineau d'Aunis, Pinot Noir, Chardonnay, Arbois, Grolleau Noir and Gris*

🍾 *Most are best consumed upon purchase, although some benefit if kept 1–2 years.*

✓ *Baumard • Yves Lambert • Château de l'Aulée • Château de Montgueret • De Nerleux • Château de Putille • CV de Saumur*

HAUT-POITOU AOC

Haut-Poitou became an AOC in 2010, after the VDQS designation was dropped. Situated 80 kilometres (50 miles) southwest of Tours, the Poitiers district produces wines that have achieved a remarkable reputation despite its hot, dry climate and flat land that is more suited to arable farming than to viticulture.

RED This is a wine to watch: although the really successful reds have until now been in short supply and confined mainly to Cabernet, there is a feeling that a general breakthrough is imminent. The quality of the entire appellation is likely to rise significantly, and there is promise of some exciting reds.

🍷 *A minimum of 60% Cabernet Franc, with Pinot Noir, Gamay, Gamay de Bouze, Gamay de Chaudenay, and Merlot.*

🍾 *Within 3 years*

WHITE Dry, light- to medium-bodied wines. Those made from pure Sauvignon are softer and more floral than most of their northern counterparts, yet they retain the freshness and vitality that is so important to this grape variety.

🍷 *Sauvignon Blanc, with Sauvignon Gris up to 40%*

🍾 *Within 1 year*

ROSÉ These dry, light- to medium-bodied wines are fresh and fruity. They must be a blend of all three permitted grape varieties.

🍷 *A minimum of 40% Cabernet Franc with Pinot Noir, and Gamay, at least 20% each*

🍾 *Within 3 years*

✓ *Ampelidae ⓝ • La Tour Beaumont • CV Haut-Poitou*

PINEAU DES CHARENTES AOC

This *vin de liqueur* is produced in the Cognac region, which is between the Loire and Bordeaux (thus not really the Loire Valley, but it has to go somewhere!).

WHITE Few manage to rise above the typically dull, oxidative *vin de liqueur* style that goes rancid with age.

🍷 *Cabernet Franc, Cabernet Sauvignon, Colombard, Folle Blanche, Jurançon Blanc, Meslier Saint-François, Merlot, Merlot Blanc, Montils, Sauvignon Blanc, Sémillon, Ugni Blanc*

🍾 *Upon opening*

ROSÉ Like the whites, only some of these wines manage to rise above the unexciting, oxidative *vin de liqueur* style.

🍷 *Cabernet Franc, Cabernet Sauvignon, Malbec, Merlot*

🍾 *Upon opening*

✓ *Barbeau & Fils (Tres Vieux Rosé Grand Réserve) • Chais du Rouissoir (Rosé) • Vinet (Félix-Marie de la Villière Blanc)*

ROSÉ DE LOIRE AOC

This dry rosé wine was introduced in 1974 to exploit the international marketing success of Rosé d'Anjou and to take advantage of the trend for drier styles. The result has been very disappointing on the whole, although the few producers (recommended below) who have really tried to demonstrate that very good quality dry rosé can be made throughout the Loire.

ROSÉ Dry, light- to medium-bodied rosé from the Loire that could (and should) be the most attractive wine of its type – a few growers try, but most do not. These wines may be sold from 1 December following the harvest without any mention of *primeur* or *nouveau*.

🍷 *Pineau d'Aunis, Pinot Noir, Gamay, Grolleau, and Cabernet Franc and Cabernet Sauvignon at least 30%*

🍾 *Upon purchase*

✓ *De Chanteloup • Des Rochelles • Cady • Château de Passavant*

ST-POURÇAIN AOC

The wines of Saint-Pourçain have a long and impressive history for what was an obscure VDQS until as recently as 2009, with vineyards first planted not by the Romans, as elsewhere in the Loire Valley, but by the Phoenicians. The area covers 19 communes southeast of the Bourges appellations of the Central Vineyards in the Allier *département*. The growers are quite ambitious, and many people think that these wines have a particularly promising future. There are 500 hectares (1,235 acres) of vineyards.

RED Dry, light- to medium-bodied wines, but they can vary from very light Beaujolais look-alikes to imitations of Bourgogne Passetoutgrains, depending on the grape varieties in the blend.

🍷 *Gamay, Pinot Noir*

🍾 *1–2 years*

WHITE These are dry, light- to medium-bodied wines. The Tresallier grape (which is known as the Sacy in Chablis), when blended with Chardonnay and Sauvignon, produces a crisp, toasty, full-flavoured wine that does have some merit. Very little is known about the Saint-Pierre-Doré, except that locals say it makes a "filthy wine", according to Master of Wine Rosemary George in French Country Wines.

🍷 *Chardonnay 50–58%, plus Tresallier 20–40% and Sauvignon Blanc up to 10%*

🍾 *1–2 years*

ROSÉ Crisp, dry, light- to medium-bodied wines that have a fragrance that is reminiscent of soft summer fruits. The rosés are generally more successful than the red wines of the area, but both styles are particularly refreshing and thirst quenching.

🍷 *Gamay*

🍾 *1–2 years*

✓ *De Bellevue • Jallet (Les Ceps Centenaires) • CV de Saint-Pourçain (Cuvée Réserve) • Cave Touzain*

Pays Nantais

Nantais is Muscadet country. The Côtes de Grandlieu, granted its own appellation in 1994, produces the richest Muscadet, while the Sèvre et Maine district and the Coteaux de la Loire to the north produce wines with extra acidity.

Southeast of Nantes lie the vineyards of Muscadet. The best are those of the Sèvre et Maine district, named after two rivers, which is much hillier than the surrounding countryside and protected from northwesterly winds by Nantes itself. Sèvre et Maine accounts for one-quarter of the general appellation area, yet 85 per cent of all Muscadet produced. Only in unusually hot or dry years, when they contain extra natural acidity, can Muscadet grapes grown farther north in the Coteaux de la Loire sometimes surpass those of Sèvre et Maine.

THE MUSCADET GRAPE AND ITS WINES

It is uncertain when the Muscadet grape, also known as the Melon de Bourgogne (now legally known as Melon Blanc) and the Gamay Blanc, was first planted in the area. There is a plaque at Château de la Cassemichère that claims that the first Muscadet vine was transplanted there from Burgundy in 1740. But Pierre Galet, the famous ampelographer (vine botanist), tells us that "following the terrible winter of 1709, Louis XIV ordered that the replanting of the frozen vineyards of Loire-Atlantique be with Muscadet *blanc*".

The wine produced from the Muscadet grape is neutral in flavour and bears no hint of the muskiness that some believe its name implies. Traditionally, it is harvested early to preserve acidity, and yet, in doing so, the grower risks making a wine that lacks fruit. But if the wine is left in contact with its sediment and bottled *sur lie*, this enhances the fruit, adds a yeasty roundness, and, by retaining more of the carbonic gas created during fermentation, imparts a certain liveliness and freshness.

DES CLAVIÈRES MUSCADET SÈVRE ET MAINE
Pays Nantais is known for its wines made from the Muscadet grape, and the appellation of Muscadet Sèvre et Maine grows a vast majority of the variety.

VINE-GROWING AREAS OF THE PAY NANTAIS
(see also p283)
The finest wines in the Pays Nantais are produced to the east of Nantes in Sèvre et Maine and Coteaux de la Loire.

FACTORS AFFECTING TASTE AND QUALITY

LOCATION
The Pays Nantais lies in the coastal area and the westernmost district of the Loire Valley, with vineyards occupying parts of the Loire-Atlantique and the Maine-et-Loire *départements*.

CLIMATE
Mild and damp, but winters can be harsh and spring frosts troublesome. Summers are generally warm and sunny, although they are sometimes rainy.

ASPECT
Some of the vineyards are found on the flat land around the mouth of the Loire southwest of Nantes. There are rolling hills in the Sèvre et Maine and Coteaux de la Loire, with the best vineyards on gentle riverside slopes. A number of the smaller valleys are actually too steep for viticulture, and the vines in these areas occupy the hilltops.

SOIL
The best vineyards of the Sèvre et Maine are light and stony, with varying proportions of sand, clay, and gravel above a granitic, schistous, and volcanic subsoil that is rich in potassium and magnesium. The Coteaux de la Loire is more schistous, while the Côtes de Grandlieu is schistous and granitic, and vineyards in the generic Muscadet appellation have sand with silt. These soils provide good drainage, essential here.

VITICULTURE & VINIFICATION
The Muscadet is a relatively frost-resistant, early-ripening grape that adapts well to the damp conditions of the Pays Nantais. It is harvested early (early to mid-September) to preserve its acidity, although a number of growers have started to reduce yields and pick riper. The best Muscadet is left in vat or barrel on its sediment – *sur lie* – until it is bottled. This imparts a greater depth and fruitiness, as well as a faint prickle of natural carbonic gas.

GRAPE VARIETIES
Primary varieties: Gros Plant (Folle Blanche), Muscadet/Melon B
Secondary varieties: Cabernet Franc, Cabernet Sauvignon, Chardonnay, Chenin Blanc, Gamay, Gamay de Bouze, Gamay de Chaudenay, Groslot Gris (Grolleau Gris), Négrette, Pinot Noir

COTEAUX D'ANCENIS AOC

These varietal wines, which come from the same area as Muscadet Coteaux de la Loire, were promoted to full AOC status in 2009.

RED Bone-dry to dry, and light- to medium-bodied wines that include Cabernets, made from both Cabernet Franc and Cabernet Sauvignon grapes. Surprisingly, they are not as successful as the juicy Gamay wines that represent no less than 80 per cent of the total production of this appellation.

🍷 *Cabernet Sauvignon, Cabernet Franc, Gamay, and Gamay de Chaudenay and Gamay de Bouze up to a combined total of 5%*

🍷 *2 years*

WHITE Dry to medium-dry, light-bodied wines. The Pinot Gris, also sold as "Malvoisie", is not as alcoholic as its Alsatian cousin, yet can possess a light richness that will linger in the mouth. The Chenin Blanc, known locally as "Pineau de la Loire", is rarely very special.

🍷 *Chenin Blanc, Pinot Gris*

🍷 *12–18 months*

ROSÉ Bone-dry to dry, light- to medium-bodied wines, some of which are fresh, firm, and lively. Gamay is the most popular grape variety.

🍷 *Cabernet Sauvignon, Cabernet Franc, Gamay, and Gamay de Chaudenay and Gamay de Bouze up to a combined total of 5%*

🍷 *2 years*

✓ *CV des Terroirs de la Noëlle • Guindon • Du Haut Fresne*

FIEFS VENDÉENS AOC

Fiefs Vendéens was categorised as *vin de pays* until 1984, and VDQS until 2009, when it was upgraded to full AOC status. The regulations controlling the grape varieties permitted for this appellation are unique. They determine the proportion of each variety that must be cultivated in the vineyard, yet they do not limit the percentages of grapes contained in the final blend; thus blends and pure varietals are allowed. Five communes may use their name on the label: Brem, Chantonnay, Mareuil, Pissotte and Vix.

RED The communes of Vix and Mareuil-sur-Lay-Disais produce the best wines. They are dry, medium-bodied, and firm, but not long-lived. They can have a grassy character, derived from the Cabernet Franc, which is the predominant grape grown in both these villages.

🍷 *A minimum of 50% Gamay and Pinot Noir, plus Cabernet Franc, Cabernet Sauvignon, Négrette, and Gamay de Chaudenay up to a maximum of 15%*

🍷 *Within 18 months*

WHITE Bone-dry to dry, light-bodied wines that, apart from those of Vix and Pissotte, are of limited quality.

🍷 *A minimum of 50% Chenin Blanc, plus Sauvignon Blanc and Chardonnay. A maximum of 20% Melon de Bourgogne in the communes of Vix and Pissotte and a maximum of 30% Groslot Gris in the coastal vineyards around Les Sables d'Olonne.*

🍷 *Upon purchase*

ROSÉ Dry, light- to medium-bodied wines. The best wines of Vix and Mareuil-sur-Lay-Disais are soft, delicate, and underrated.

🍷 *A minimum of 50% Gamay and Pinot Noir, plus Cabernet Franc, Cabernet Sauvignon, Négrette, and Gamay de Chaudenay to a maximum of 15%. A maximum of 30% Groslot Gris in the coastal vineyards around Les Sables d'Olonne.*

🍷 *Within 18 months*

✓ *Coirer • Vignobles Mourat • Saint Nicolas* ⓑ

GROS PLANT DU PAYS NANTAIS AOC

Gros Plant is the local synonym for the Folle Blanche – one of the grapes used to make Cognac. At one time, it was thought that Gros Plant du Pays Nantais would be the only VDQS to opt for IGT (*vin de pays*) rather than AOC, but a change of heart in 2010 led to its application for, and receipt of, full AOC status.

WHITE Gros Plant is normally so dry, tart, and devoid of fruit and body that it seems tough and sinewy to taste. I would rather drink lemon juice than 99 per cent of the Gros Plant that I have had, but if yields are limited and the wine bottled *sur lie*, it can have sufficient depth to match its inherent bite.

🍷 *A maximum of 20% Gros Plant, Montils, and Colombard*

🍷 *Usually upon purchase*

✓ *De la Boitaudière (Sur Lie) • Guy Bossard* ⓑ *(De l'Ecu) • La Haut-Vrignais (Sur Lie)*

MUSCADET AOC

This basic appellation covers the whole Muscadet area, yet the wines produced under it account for only 10 per cent of the total production.

WHITE Bone-dry, light-bodied wines that – with very few exceptions – are ordinary at best, and they often lack balance. These wines may be sold as *primeur* or *nouveau* as from the third Thursday of November following the harvest.

🍷 *Muscadet/Melon B*

🍷 *Upon purchase*

✓ *De la Chauvinière*

MUSCADET DES COTEAUX DE LA LOIRE AOC

The Coteaux de la Loire is the most northerly wine area on the French coast, above which it is almost impossible to grow grapes of sufficient ripeness for winemaking.

WHITE Bone-dry, light-bodied wines of variable quality, usually lacking in fruit, but can be the best balanced of all Muscadets in very hot years.

🍷 *Muscadet/Melon B.*

🍷 *Upon purchase*

✓ *Guindon • Du Moulin Giron • De la Varenne*

MUSCADET CÔTES DE GRANDLIEU AOC

Delimited in 1994, this area west of Sèvre et Maine once represented 73 per cent of the basic Muscadet appellation. The wines now fetch a nice premium above that received when they were merely perceived as generic Muscadets and, though some deserve elevated price and status, many plainly do not.

WHITE Bone-dry, light-bodied wines that initially displayed considerable variation in quality, but the best are now showing a fine, floral-minerality of fruit.

🍷 *Muscadet/Melon B.*

🍷 *Upon purchase*

✓ *Du Fief Guerin • Les Hautes Noëlles* ⓞ *• Château des Herbauges*

MUSCADET DE SÈVRE ET MAINE AOC

Classic Muscadet from a small area containing most of the best wines. Some 45 per cent of this appellation is bottled and sold as sur lie, having remained in contact with its sediment for at least one winter before bottling. Legally, Muscadet sur Lie cannot be bottled until the 1st of March following the harvests.

WHITE Bone-dry to dry, light-bodied wines. The best should have fruit, acidity, and elegance, but although they can be reminiscent of a modest white Burgundy and exceptional wines can survive considerable ageing.

🍷 *Muscadet/Melon B.*

🍷 *2 years, although some may last 3–4 years*

✓ *Guy Bossard* ⓑ *(De l'Ecu) • Michel Bregeon • De la Chauvinière • Coing de Saint-Fiacre • Famille Lieubeau • Les Frères Couillaud • Gadais Père & Fils • Grand Fief de la Cormeraie • La Haute Févrie • De la Louvetrie • Des Petites Cossardières • De la Quilla • Château des Roi • Abbaye de Ste Radegonde • Sauvion* ⓞ *• Château la Tarcière*

MUSCADET SUR LIE AOC

Since 1994, the term *Muscadet Sur Lie* may only be applied to one of the three sub-appellations (Coteaux du Loire, Côtes de Grandlieu, and Sèvre et Maine) and may not be used on any wines bearing the generic Muscadet AOC. Quite what the logic is to this is uncertain, because Gros Plant VDQS is permitted to use *sur lie*, and it is even more inferior than the generic Muscadet appellation. The lesser the wine, the more need for a *sur lie* boost, thus rather than limiting its use, more emphasis should be placed on stricter controls.

At the moment Muscadet *sur lie* must remain in contact with its sediment for one winter and may not be bottled before the 1st of March following the harvest. The wine must also be bottled directly off its lees, and must not be racked or filtered. There is, although, still no regulation on the size and type of vessel in which the wine should be kept *sur lie*. Some growers would like the term applied only to wines kept in wooden barrels, arguing that the effect of keeping a wine in contact with its lees in huge vats is negligible, but, at the very least, vats over a certain size should be equipped with paddles to circulate the lees.

MUSCADET AOC CRU COMMUNAL

Since 2007 a new category of top-quality Muscadet has been created under the *cru communal* designation. These wines require a minimum of 24 months aging *sur lie* (18 months *sur lie* for Le Pallet) before being bottled. This aging process enhances the complexity and mineral concentration of the wines offering a style not dissimilar to Chablis.

CRU COMMUNAL MUSCADET

There have always been rare examples of exceptionally ripe Muscadet, which tend to age like a Burgundy after a few years in bottle. More recently, however, a new high-quality style of wine picked at optimum maturity has been created under the Muscadet *cru communal* designation. Three communes – Le Pallet, Clisson, and Gorges – were ratified as *cru communal* zones in 2011, and they were joined by Goulaine, Château-Thébaud, Monnières-Saint-Fiacre, Mouzillon-Tillières, La-Haye-Fouassière, Vallet, and Champtoceaux in 2019. These *crus communaux* are predicated upon the subsoil of their *terroirs* and the bedrock on which they lie; for example, Clisson is on a granite bedrock covered in coarse soils of sand and pebbles, and Gorges is on clay and quartz, with a mainly gabbro subsoil.

Anjou-Saumur

Anjou-Saumur is a microcosm of the entire Loire Valley, with almost every grape available in the Loire producing virtually every style of wine imaginable – from dry to sweet, red through rosé to white and still wines to sparkling varieties.

Saumur is the Loire Valley's sparkling-wine centre; it is where tourists flock in the summer, visiting the numerous cellars hewn out of the solid tufa subsoil. The magnificent white tufa-stone castle that overlooks the town was built in the 14th century. It is regarded as one of the finest of all the Loire châteaux, and it is used by the Confrérie des Chevaliers du Sacavins (one of Anjou's several wine fraternities) for various inaugural ceremonies and celebrations.

THE WINES OF ANJOU

Rosé represents about 35 per cent of Anjou's total wine output, even though it is on the decline: the figure for rosé was 55 per cent in the late 1980s. Although rosé remains this district's most popular wine, it has a down-market image and is essentially a blend of minor grapes; thus, its commercial success has not propelled a specific variety to fame. Anjou's most celebrated grape is the Chenin Blanc used to make white

wines. This vine has been cultivated in the area for well over a thousand years. It has many synonyms, from "Pineau de la Loire" to "Franc-blanc", but its principal name, Chenin Blanc, stems from Mont-Chenin in 15th-century Touraine. Under other names it can be traced as far back as the year 845, to the abbey of Glanfeuil (south of the river in the Anjou district). The distinctive tang of the Chenin Blanc grape comes from its inherently high tartaric acid content and this, combined with a naturally high extract, makes for citrusy and pear fruit–style whites; it also helps that the grapes are grown on the four sun-blessed, southeasterly-facing slopes of Savennières.

Anjou growers tended to pick as late as possible. This invites the risk of rain, but by going over the vines several times in the time-honoured tradition of *tries*, picking only the ripest and healthiest grapes on each and every sweep of the vineyard, a miraculous wine may be made. Although this is a time-consuming, labour-intensive operation, the unique quality of overripe grapes can result in the most succulent and immaculately balanced of sweet wines; these treasures are vinous investments that are capable of great maturity and can achieve wonderfully complex honeyed characteristics. More recently, some growers have started to pick selected vineyards earlier, before the onset of botrytis, in order to craft elegant dry Chenins packed with minerality.

VINE-GROWING AREAS OF ANJOU-SAUMUR
(see also p283)
Boasting sparkling wine and more from Saumur and a range of wines from Angers's environs, Anjou-Saumur produces most types of wine found in the Loire as a whole.

VINES OF CHENIN BLANC FORM NEAT ROWS IN A COTEAUX DU LAYON VINEYARD
This area has produced sweet wines since the 14th century. The wines of this appellation are all made from Chenin Blanc grapes.

THE SPARKLING SAUMUR INDUSTRY

With the rapid growth of the Champagne market in the 19th century, producers in the Loire began to copy effervescent winemaking practices. Saumur eventually turned into the largest French sparkling-wine industry outside Champagne. In many parts of the Loire the Chenin Blanc grape has the perfect acidity required for a quality sparkling wine, although devotees of the true yeasty character of Champagne can find its bouquet sweet and aromatic, maintaining that its flavour is too assertive to be properly transmuted by the traditional method. The wines are hugely popular, however, and the mixture of Chardonnay and other neutral varieties can greatly improve the overall blend. The most ardent admirer of Champagne has been known to fall prey to the charms of a superior pure Chenin Blanc bubbly from this region, and even some critics have been known to become besotted by Bouvet-Ladubay's luxuriously ripe, oak-fermented Trésor.

THE REGION'S RED WINES

It is in Anjou, especially south of Saumur, that the Cabernet Franc emerges as the Loire's best red wine grape. Beyond neighbouring Touraine, however, its cultivation rapidly diminishes. The Loire is the largest wine region in France, yet surprisingly it boasts just three classic red wines – Saumur-Champigny, Bourgueil, and Chinon – to which we might now add Anjou-Villages. It is no coincidence that most of the vineyards producing these wines are clustered together in a compact area around the confluence of the Vienne and the Loire – two rivers that long ago established the gravel terraces so prized for growing Cabernet Franc today.

CHÂTEAU DE SAUMUR
Vineyards surround this 14th-century castle, one of the finest in the Loire. The castle towers above the bustling town of Saumur, where the buildings are distinguished by the brilliant white tufa-stone typical of the area.

FACTORS AFFECTING TASTE AND QUALITY

LOCATION
West-central district with mostly left-bank vineyards situated between Angers and Saumur.

CLIMATE
A gentle Atlantic-influenced climate with light rainfall, warm summers, and mild autumns, but frost is a problem in Savennières.

ASPECT
Soft, rolling hills that hold back the westerly winds. The best sites are the south-facing rocky hillsides of Savennières and the steep-sided valley of the River Layon.

SOIL
In the west and around Layon, the soil is schist with a dark, shallow topsoil that stores heat well and helps ripen the grapes, but some colder clay-soil areas produce heavier wines. The chalk-tufa soil in the east of the district around Saumur produces lighter wines, while the shale and gravel in Saumur-Champigny favours Cabernet Franc.

VITICULTURE & VINIFICATION
The Chenin Blanc is a particularly slow-ripening grape that is often left on the vine until November, especially in the Coteaux du Layon. The effect of the autumn sun on the dew-drenched, overripe grapes can encourage "noble rot", particularly in Bonnezeaux and Quarts-de-Chaume. In good years, pickers go through the vineyards several times, selecting only the ripest or most rotten grapes – a tradition known as *tries*. Most wines are bottled in the spring following the vintage, but wines produced from such richly sweet grapes take at least three months to ferment and, for this reason, might not be bottled until the following autumn.

GRAPE VARIETIES
Primary varieties: Cabernet Franc, Chenin Blanc, Gamay, Grolleau (Grolleau Noir)
Secondary varieties: Cabernet Sauvignon, Chardonnay, Côt (Malbec), Pineau d'Aunis, Sauvignon Blanc

THE GENERIC AND OUTLYING APPELLATIONS OF
ANJOU-SAUMUR

ANJOU AOC

The Anjou district encompasses the vineyards of Saumur; thus Saumur may be sold as Anjou, but not vice versa. The red wines are by far the best, the whites the worst, and the rosé wines, although waning in popularity, remain the most famous. Because *mousseux* has "cheap fizz" connotations, the wines officially designated as Anjou Mousseux appellation are often marketed simply as "Anjou".

RED The district's dry, medium- to full-bodied wines are made mostly from pure Cabernet Franc or with a touch of Cabernet Sauvignon. These delightful wines are best drunk young, although the odd oak-aged wine of surprising complexity can be found.

🍷 *Cabernet Franc, Cabernet Sauvignon, Pineau d'Aunis*

🍾 *1–3 years*

✓ *Des Baumard • Cady Y• Château de Fesles • Richard Leroy* Ⓞ *• Le Logis du Prieuré • Des Forges • Château Pierre-Bise • Saint Arnoul • De la Sansonnière* Ⓑ *• Château Soucherie (Champ aux Loups) • Des Rochelles* Ⓞ

WHITE Although these wines vary from dry to sweet and from light- to full-bodied types, there are too many aggressively acid-dry or simply mediocre medium-sweet Chenin Blanc wines in the appellation. Some improvement has been made by growers maximizing the 20 per cent Chardonnay and Sauvignon allowance, while Jacques Beaujeu and a few others use oak to smooth out Chenin's jagged edges, and Ogereau even ferments en barrique. These wines may be sold from 1 December following the harvest without any mention of *primeur* or *nouveau*.

🍷 *A minimum of 80% Chenin Blanc, and a maximum of 20% Chardonnay and Sauvignon Blanc*

🍾 *Upon purchase*

✓ *Philippe Delesvaux* Ⓞ *• Les Grandes Vignes (Varenne de Combre)* Ⓞ *• Ogereau • Château de Passavant* Ⓞ *• Château Pierre-Bise • Des Sablonnettes* Ⓞ *• Pithon-Paillé* Ⓞ *• Du Regain • Des Forges • Richou • Cady* Ⓞ

ROSÉ There is nothing intrinsically wrong with a wine that happens to be pink with some sweetness, although you would be forgiven for thinking this is exactly why some critics turn up their noses at these wines. I prefer dry rosés, but I enjoy medium-sweet rosés when they are very fresh and fruity. The trouble with so many of these medium-sweet, light- to medium-bodied, coral-pink wines is that, though they can be delicious in the early spring following the vintage, an alarming number quickly tire in the bottle. The moral is, therefore, that even when you have found an Anjou rosé you like, never buy it by the case. These wines may be sold as *primeur* or *nouveau* from the third Thursday of November following the harvest or from 1 December without any mention of *primeur* or *nouveau*.

🍷 *Predominantly Grolleau, with varying proportions of Cabernet Franc, Cabernet Sauvignon, Pineau d'Aunis, Gamay, Malbec*

🥂 *Upon purchase*

✓ *CV de la Loire (Elysis) • Château La Varière*

ANJOU COTEAUX DE LA LOIRE AOC

This rare, white-only appellation is situated southwest of Angers. Production is small and will dwindle even further as vineyards are replanted with Cabernet for the increasingly popular Anjou Rouge appellation. The minimum natural richness of the grapes at harvest is 221 grams per litre (the same as Sauternes), with a minimum residual sugar in the finished wine of 34 grams per litre.

WHITE Originally legally defined in 1946 as a traditionally sweet wine.

🍷 *Chenin Blanc*

🥂 *Within 1 year*

✓ *Musset Roullier*

ANJOU GAMAY AOC

Gamay is only allowed in Anjou AOC wines if the name of the grape is added to the appellation on the label.

RED These consist of dry to medium-dry, light-bodied wines that are rarely of great interest. These wines may be sold as *primeur* or *nouveau* from the third Thursday of November following the harvest.

🍷 *Gamay*

🥂 *Upon purchase*

✓ *De la Charmoise • Roy René (La Creusette)*

ANJOU MOUSSEUX AOC

This traditional-method wine is softer, but less popular than its Saumur equivalent, although it may come from the communes within Saumur itself.

SPARKLING WHITE These dry to sweet, light- to medium-bodied wines desperately need a change of regulation to allow a little Chardonnay in the blend. Chardonnay's fatter, more neutral character would enable producers to make a more classic, less frivolous style of sparkling wine.

🍷 *A minimum of 60% Chenin Blanc plus Cabernet Sauvignon, Cabernet Franc, Malbec, Gamay, Grolleau, Pineau d'Aunis*

🥂 *1–2 years*

SPARKLING ROSÉ If you want to know what Anjou Rosé tastes like with bubbles, try this light- to medium-bodied wine, which is mostly sold as *demi-sec*.

🍷 *Cabernet Sauvignon, Cabernet Franc, Malbec, Gamay, Grolleau, Pineau d'Aunis*

🥂 *Upon purchase*

ANJOU ROSÉ AOC

See Anjou AOC

ANJOU-VILLAGES AOC

This superior, red-wine-only appellation was first delimited in 1986, but it did not come into effect until 1991. If you buy from the best growers, you will get some of the finest red wines that the Loire has to offer.

RED The very best wines can be deeply coloured with a creamy raspberry aroma and flavour.

🍷 *Cabernet Franc, Cabernet Sauvignon*

🥂 *2–6 years*

✓ *Patrick Baudouin ❶ • De la Bergerie ❶ • Château Pierre-Bise • Des Griottes ❶ • L & F Martin • De la Motte •*

Ogereau • De la Poterie • De Putille • Des Quattre Quarres • Michel Robineau • Des Saulaies • Château La Varière

ANJOU BRISSAC AOC

This village was singled out under the Anjou-Villages appellation in 1998 and backdated for wines from 1996 onwards. It covers the area of Brissac-Quincé and nine surrounding communes.

✓ *Daviau (Château de Brissac) • De Haute-Perche • Du Prieuré • Richou • Des Rochelles (La Croix de Mission) • Château La Varière*

BONNEZEAUX AOC

With grapes grown on three south-facing river slopes of the commune of Thouarcé in the Coteaux du Layon, this is one of the undisputed great sweet wines of France. The grapes must be harvested in tries with the pickers collecting only the ripest, often botrytis-affected fruit, which can take up to two weeks. The minimum natural richness of the grapes at harvest is 238 grams per litre (compared to 221 grams per litre for Sauternes), and the minimum residual sugar is 51 grams.

WHITE Intensely sweet, richer, and more full-bodied than Quarts-de-Chaume, the other great growth of the Layon Valley, this wine can have pineapple and liquorice fruit when young, often achieving a beautiful honeyed-vanilla complexity with age.

🍷 *Chenin Blanc*

🥂 *Up to 20 years or more*

✓ *De la Petite-Croix • Philippe Delesvaux ❶ • Château de Fesles • Godineau/Des Petit Quarts • Claude Robin (Floriane) • De la Sansonnière ❸ • Des Rochelles • Château La Varière*

CABERNET D'ANJOU AOC

The appellation of Cabernet D'Anjou includes Saumur, and it was a *saumurois* named Taveau who, in 1905, was the first person to make an Anjou Rosé from Cabernet grapes. Despite its classic Cabernet content and an extra degree of natural alcohol, this is not as superior to Anjou Rosé as it should be because bulk sales at cheap prices have devalued its reputation.

ROSÉ Good examples of these medium to medium-sweet, medium-bodied wines produced by the best domaines have a clean and fruity character with aromas of raspberries. These wines may be sold as *primeur* or *nouveau* from the third Thursday of November following the harvest or from 1 December without any mention of *primeur* or *nouveau*.

🍷 *Cabernet Franc, Cabernet Sauvignon*

🥂 *Upon purchase*

✓ *La Croix des Loges • CV de la Loire (Elysis) • Des Petite Grouas • De Preville*

COTEAUX DE L'AUBANCE AOC

These wines are made from old vines grown on the schistous banks of the River Aubance. To be entitled to this appellation growers must use grapes that are well ripened and harvested by *tries*, a labour-intensive system that is not cost-effective, thus until recently most growers produced Cabernet d'Anjou. In the early 2000s a new generation of growers transformed this appellation, such that in 2003 the minimum natural sugar of the grapes at harvest was increased from 204 to 230 grams per litre (compared with 221 grams per litre for Sauternes) or 294 grams per litre for Coteaux l'Aubance "Sélection de Grains Nobles" as indicated on the label. The minimum residual sugar was also increased from 17 to 34 grams per litre.

WHITE A few growers still make this rich and semi-sweet, medium- to full-bodied wine of excellent longevity and exceptional quality.

🍷 *Chenin Blanc*

🥂 *5–10 years*

✓ *De Bablut ❶ • Daviau • De Montgilet (Clos Prieur) • Richou*

COTEAUX DU LAYON AOC

This appellation, which overlaps Anjou Coteaux de la Loire in the northwest and Saumur in the southeast, has been famous for its sweet white wines since the 1300s. In favourable sites the vines are sometimes attacked by "noble rot", but in all cases the grapes must be extremely ripe and harvested by tries to a minimum of 12 per cent alcohol from a maximum 30 hectolitres per hectare. Due to the relatively low price this appellation commands, harvesting by tries is feasible only for the top domaines. The minimum natural richness of the grapes at harvest is 221 grams per litre (the same as Sauternes) or 323 grams per litre for Coteaux du Layon "Sélection de Grains Nobles". The minimum residual sugar is 34 grams per litre.

WHITE This appellation produces green-gold- to yellow-gold-coloured, soft-textured, sweet, medium- to full-bodied wines, rich in fruit and potentially long lived.

🍷 *Chenin Blanc*

🥂 *5–15 years*

✓ *Patrick Baudouin ❶ • Baumard (le Paon) • Château Pierre-Bise • Des Forges • Philippe Delevaux ❶ • Dhomme • Château de Fesles (Château de la Roulerie) • De la Bergerie • Ogereau (Prestige) • Du Petit Val (Simon) • Du Portaille (Planche Mallet) • Du Regain (Le Paradis) • Des Sablonnettes (La Bohème) ❶*

COTEAUX DU LAYON PREMIER CRU CHAUME AOC

After much legal wrangling, there are now only two Chaume appellations: this one and Quarts-de-Chaume. Coteaux du Layon-Chaume and Chaume have both been rescinded, replaced in 2011 (a 2009 legislation for using simply the name "Charme" in the AOC was overturned). The geographical difference between Coteaux du Layon Premier Cru Chaume and Quarts-de-Chaume is that the latter is restricted to just three *lieux-dits:* Les Quarts, Les Roueres, and Le Veau in the commune of Rochefort-sur-Loire, whereas Chaume may come from selected parcels all over the commune of Rochefort-sur-Loire. The minimum natural sugar of grapes destined for the appellation is 272 grams per litre, rather than 238 grams for the other communal appellations, while the minimum residual sugar in the finished wine is 80 grams per litre, compared to just 34 grams for Coteaux du Layon (and indeed for Quarts-de-Chaume).

WHITE These are big, rich, lusciously sweet wines that will make their mark in years to come.

🍷 *Chenin Blanc*

🥂 *5–15 years*

✓ *Château Pierre-Bise • Michel Blouin • Cady • Château de Fesles (Château de la Roulerie Aunis) • Des Forges (Les Onnis) • Château de la Guimonière • L & F Martin • Du Petit Metris • Du Rocher • De la Soucherie*

COTEAUX DU LAYON VILLAGES AOC

Historically, these six villages have consistently produced the cream of all the wines in the Coteaux du Layon and thus have the right to add their names to the basic appellation. In minimum natural sugar of the grapes at harvest is 238 grams per litre (compared to 221 grams per litre for Sauternes) or 323 grams per litre for Coteaux du Layon Villages "Sélection de Grains Nobles". The minimum residual sugar is 34 grams per litre.

WHITE These sweet wines range from medium to full bodied. According to the Club des Layon Villages, Beaulieu has a soft, light aroma; Faye has a scent reminiscent of brushwood; Rablay is big, bold, and round; Rochefort is full-bodied, tannic, and matures well; St-Aubin has a delicate aroma that develops; and St-Lambert is robust yet round.

🍷 *Chenin Blanc*

🥂 *5–15 years*

✓ *Beaulieu (or Beaulieu-sur-Layon) • Château de Breuil • Château Pierre-Bise • Faye (or Faye-d'Anjou) • Château de Fresne (Clos de Cocus) • Richard Leroy • Rablay (or Rablay-sur-Layon) Des Sablonnettes ❶ • Rochefort (or Rochefort-*

POURING A GLASS OF ROSÉ D'ANJOU
Once a marketing miracle, these medium-sweet pink wines (also called Anjou Rosé or Rosé Anjou) sell less well to today's more sophisticated consumers.

sur-Loire)• De la Motte • St-Aubin (or Saint-Aubin-de-Luigné) Des Barres, Cady 🅞, Jo Pithon, De la Roche Moreau • St-Lambert (or Saint-Lambert-du-Lattay) Ogereau, Pithon-Paillé Y, Roy René (Les Cartelles), Michel Robineau

COTEAUX DE SAUMUR AOC

After the ban in 1985 on the use of the term *méthode champenoise* for wines produced or sold in the European Union (*see* p139), there were moves to develop this little-used appellation as the principal still wine of the Saumur district in order to promote Saumur AOC as an exclusively sparkling wine. That idea has, it seems, ground to a halt.

WHITE These relatively rare, semi-sweet, medium- to full-bodied white wines are richly flavoured and worth seeking out.

🍇 Chenin Blanc

🍷 5–10 years

✓ Champs Fleuris (Sarah) • De Nerleux • De Saint Just (La Valboisière)

QUARTS-DE-CHAUME AOC

These wines are made from grapes grown on the plateau behind the village of Chaume in the Coteaux-du-Layon commune of Rochefort-sur-Loire. The vineyards of Quarts-de-Chaume were once run by the abbey of Ronceray, whose landlord drew a quarter of the vintage as rent. The minimum natural sugar of the grapes at harvest is 298 grams per litre, and the minimum residual sugar is 34 grams per litre. See also Chaume AOC.

WHITE These are semi-sweet to sweet, medium- to full-bodied wines. Although harvested by tries and produced in the same manner as Bonnezeaux, Quarts-de-Chaume comes from a more northerly area and as a result is slightly lighter in body. It also tends to have a touch less sweetness.

🍇 Chenin Blanc

🍷 Up to 15 years or more

✓ Des Baumard • Château de Bellerive • De Laffourcade • Château de la Roche Moreau • Château Pierre-Bise • Château La Varière

ROSÉ D'ANJOU AOC

See Anjou AOC

SAUMUR AOC

Saumur, situated within the borders of the Anjou appellation, is regarded as the pearl of Anjou. Its wine may be sold as Anjou, but Anjou does not automatically qualify as Saumur. Like Anjou, its white wines are variable, yet its red wines are excellent. The Cabernet de Saumur AOC was changed to Saumur Rose in 2016

RED These fine, bone-dry to dry, medium- to full-bodied wines are often similar to the red wines of Anjou, although they can vary from light and fruity to deep-coloured and tannic.

🍇 A minimum of 70% Cabernet Franc, and up to 30% Cabernet Sauvignon and Pineau d'Aunis

🍷 1–10 years according to style

✓ Du Bois Mignon (La Belle Cave) • Du Collier • De Fiervaux • De Château-Gaillard 🅑 • Filliatreau 🅞 • Guiberteau 🅞 • Des Hautes Vignes • Langlois-Château (Vieilles Vignes) • De Nerleux • De la Paleine • Château de Passavant 🅞 • Des Raynières • CV de Saumur (Reserve) • Château Yvonne 🅞

WHITE Varying from bone-dry to sweet and from light- to full-bodied, these wines have a style more akin to Vouvray than Anjou, due to the limestone and the tufa soil. In poor-to-average years, however, a Saumur is easily distinguished by its lighter body, leaner fruit, and a tartness of flavour that can sometimes have a metallic edge on the aftertaste. These wines may be sold as from 1 December following the harvest without any mention of *primeur* or *nouveau*.

🍇 Chenin Blanc

🍷 Up to 3 years

✓ Du Collier • Guiberteau (Le Clos) 🅞 • Château de Hureau • Langlois-Château (Vieilles Vignes) • De la Paleine • Des Roches Neuves • Clos Rougeard 🅞 • CV de Saumur • Château de Villeneuve

ROSÉ This is a delicate, medium-sweet, light- to medium-bodied wine with a hint of straw to its pink colour and a distinctive raspberry aroma. These wines may be sold as *primeur* or *nouveau* from the third Thursday of November following the harvest or from 1 December without any mention of *primeur* or *nouveau*.

🍇 Cabernet Franc, Cabernet Sauvignon

🍷 Upon purchase

SPARKLING WHITE Although the production per hectare of this regions' sparkling wine is one-third more than for its Anjou equivalent, Saumur is – or at least should be – better in quality and style due to its Chardonnay content and the tufa-limestone soil. Most wines are made in a true, bone-dry, *brut* style, although the full gamut is allowed and wines up to *demi-sec* sweetness are relatively common. The vast majority of these wines have a tart greengage character, lack finesse, and do not pick up bottle aromas. The wines indicated below have an elegance sadly lacking in most Saumur and possess gentler, more neutral fruit, which will benefit from a little extra time in bottle, although after a while all these wines tend to age rather than mature gracefully. The creamy-rich *barrique*-fermented Bouvet Trésor is the one significant exception; not only is it just the best sparkling wine in the Loire, it can be compared to very good-quality Champagne, although it is in a very different style.

🍇 Chenin Blanc, plus Chardonnay and Sauvignon Blanc to a maximum of 20%, and Cabernet Sauvignon, Cabernet Franc, Malbec, Gamay, Grolleau, Pineau d'Aunis, and Pinot Noir up to 60%

🍷 3–5 years

✓ De Beauregard • De la Bessière • Bouvet-Ladubay (Mlle Ladubay, Trésor) • De Brizé • Gratien & Meyer (Flamme) • CV de Saumur (Spéciale)

SPARKLING ROSÉ An increasing number of pink Saumurs are pure Cabernet Franc, and an increasing number are showing very well. The aggressive potential of this grape, however, can quickly turn a thrilling raspberry-flavoured fizz into something hideous. Pure Cabernet Sauvignon rosés can be much smoother, less overt, and not as intrinsically Saumur as a Cabernet Franc *cuvée*.

🍇 Cabernet Sauvignon, Cabernet Franc, Malbec, Gamay, Grolleau, Pineau d'Aunis, Pinot Noir

🍷 Upon purchase

✓ Bouvet-Ladubay (Trésor) • Gratien & Meyer (Flamme)

SAUMUR PUY-NOTRE-DAME AOC

A superior red-only appellation, producing age-worthy wines from a minimum 85 per cent Cabernet Franc, with only Cabernet Sauvignon allowed in the blend. Yields are limited to 50 hectolitres per hectare, and the wine can only be sold from 1 June following the harvest.

🍇 Cabernet Franc, Cabernet Sauvignon

🍷 5–10 years

SAUMUR-CHAMPIGNY AOC

Many people believe that the vineyards southeast of Saumur, entitled to add the village name of Champigny to their appellation, produce the best red wines in the Loire.

RED The vineyards of this appellation produce bone-dry to dry, full-bodied wines with a distinctive deep colour and full and fragrant raspberry aromas, often tannic and long-lived.

🍇 Cabernet Franc, Cabernet Sauvignon, Pineau d'Aunis

🍷 5–10 years

✓ Du Bois Moze Pasquier • La Bonnelière 🅞 • Des Champs Fleuris • Filliatreau 🅞 • Château du Hureau • René-Noël Legrand • Petit Saint Vincent • De Rocfontaine (Vieilles Vignes) • Des Roches Neuves 🅑 • Clos Rougeard 🅞 • De Targé • Château de Villeneuve • Château Yvonne 🅞

SAUMUR MOUSSEUX AOC

This is the technically correct appellation for all fully sparkling white and rosé Saumur wines made by the traditional method, but producers have shied away from the down-market term *mousseux*, selling the wines simply as Appellation Saumur Contrôlée. There is no allowance for this in the regulations, but it is a widespread practice. A significant amount of red traditional method wine is also produced, but this cannot claim AOC status. See also Saumur AOC.

SAVENNIÈRES AOC

When this small portion of Anjou Coteaux de la Loire produced only sweet wines, the AOC regulations set a correspondingly low maximum yield. This concentrates the wines on four southeast-facing slopes of volcanic debris that produce the world's greatest dry Chenin Blanc.

WHITE Bone-dry to dry wines of great mineral intensity, Savennières can be some of the longest-lived dry white wines in the world. Most critics believe that the single greatest Savennières is Nicolas Joly's Clos de la Coulée de Serrant and, although many wine lovers agree that it is indeed one of the greatest wines of the Loire Valley, some believe that Baumard's Clos du Papillon (not to be confused with Clos du Papillon from other growers) consistently displays greater elegance and finesse. Over the past 15 years or so, a few producers have resurrected the semi-sweet style that used to be more popular in Savennières in the first half of the 20th century.

🍇 Chenin Blanc

🍷 5–8 years (10–15 years for Clos de la Coulée de Serrant)

✓ Des Baumard • Du Closel (Clos du Papillon) • Château de Fesles (Château de Varennes) • De Forges • Nicolas Joly (Clos de la Coulée de Serrant) 🅑 • De la Monnaie • Château Pierre-Bise • Pierre Soulez

SAVENNIÈRES COULÉE-DE-SERRANT AOC

One of just two single-vineyard designations authorized for Savennières, Coulée-de-Serrant is just 7 hectares (17 acres) and a mono-*cru*. It is solely owned by Nicolas Joly of Château de la Roche-aux-Moines. Many consider this to be the single-greatest Loire dry white wine. See also Savennières AOC.

SAVENNIÈRES ROCHE-AUX-MOINES AOC

The second and largest of the two single-vineyard designations authorized for Savennières, Roche-aux-Moines is 17 hectares (42 acres) and owned by three producers: Nicolas Joly of Château de la Roche-aux-Moines, Pierre Soulez of Château de Chamboureau, and Madame Laroche of aux Moines. See also Savennières AOC.

Touraine

Wines bearing Touraine appellations are prolific, and the better wines are capable of offering great value. Most consumers are familiar with the workhorse Sauvignon de Touraine, but there is much interest to be found in the best wines from Vouvray and the less well-known but equally elite wines of Montlouis, Bourgueil, and Chinon.

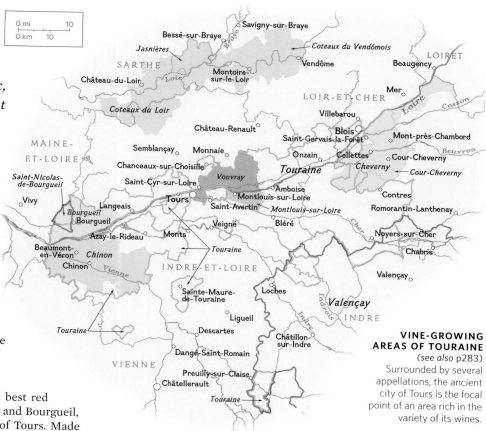

VINE-GROWING AREAS OF TOURAINE
(*see also* p283)
Surrounded by several appellations, the ancient city of Tours is the focal point of an area rich in the variety of its wines.

The wine-growing district around Tours dates back to Roman times, as does the town itself. The Cabernet Franc, known locally as Breton, was flourishing in the vineyards of the abbey of Bourgueil 1,000 years ago and, as recently as 500 years ago, the Chenin Blanc – Touraine's second-most widely planted white grape – acquired its name from Mont Chenin in the south of the district.

TOURAINE'S WINE REGIONS

With the possible exception of Saumur-Champigny, the best red wines in the Loire come from the appellations of Chinon and Bourgueil, which face each other across the Loire River, just west of Tours. Made predominantly from Cabernet Franc, good vintages aged in oak may be complex and comparable to claret, while the more everyday wines have the aromas of fresh-picked raspberries and can be drunk young and cool. To the east, Vouvray and Montlouis produce complex and alluring dry whites in most vintages and rich, sweet, long-lived *moelleux* wines from overripe Chenin Blanc grapes in sunny years. North of Tours, Jasnières produces wines from the same grape, but the dry style is distinctly different. Jasnières is a singular white sub-appellation within a wider red, white, and rosé AOC called the Coteaux du Loir. "Loir" is not a typographical error for "Loire", but a tributary of the great river. Using grapes grown on the banks of the Loir, Coteaux du Vendômois produces the full spectrum of still-wine styles, as does Cheverny to the east, including a distinctive dry white wine from the obscure Romorantin grape. Touraine Sauvignon Blanc makes an attractively priced, unassuming alternative to Sancerre, and the fruity Gamay makes easy-drinking reds and rosés. Here Chenin Blanc, also widely planted as in Anjou-Saumur, is more usually used to produce drier styles of Chenin with between 8 and 10 grams per litre of residual sugar, whereas the tradition had been to produce naturally sweet wines in great years when these grapes are full of sugar. The surplus of less-than-overripe grapes, like that in Anjou-Saumur, is traditionally utilized for sparkling wines.

FACTORS AFFECTING TASTE AND QUALITY

LOCATION
East-central district with most of its vineyards in the *département* of Indre-et-Loire, but they extend into Loir-et-Cher, Indre, and Sarthe.

CLIMATE
Touraine falls under some Atlantic influence, but the climate is less maritime than in the Nantes district and Anjou-Saumur. Protected from northerly winds by the Coteaux du Loir. Warm summer, low October rainfall.

ASPECT
Attractively rolling land, flatter around Tours itself, hillier in the hinterland. Vines are planted on gently undulating slopes, which are often south facing, at between 40 and 100 metres (130 to 330 feet) above sea level.

SOIL
Clay and limestone over tufa subsoil east of Tours around Vouvray and Montlouis. Tufa is chalk boiled by volcanic action. It is full of minerals, retains water, and can be tunnelled out to make large, cool cellars for storing wine. Sandy-gravel soils in low-lying Bourgueil and Chinon vineyards produce fruity, supple wines; the slopes, or *coteaux*, of sandy-clay produce firmer wines.

VITICULTURE & VINIFICATION
White-wine fermentation takes place at low temperatures and lasts for several weeks for dry wines, several months for sweet wines. The reds undergo malolactic fermentation. Some Bourgueil and Chinon is aged for up to 18 months in oak casks before bottling.

GRAPE VARIETIES
Primary varieties: Cabernet Franc, Chenin Blanc, Grolleau (Grolleau Noir), Sauvignon Blanc
Secondary varieties: Arbois, Cabernet Sauvignon, Chardonnay, Gamay, Gamay Teinturier, Côt (Malbec), Meslier, Pineau d'Aunis, Pinot Noir, Romorantin

FRESHLY HARVESTED CABERNET FRANC GRAPES
The finest red wine in this region is made from this black-skinned French variety.

THE APPELLATIONS OF
TOURAINE

BOURGUEIL AOC

Most of the vines of this appellation are grown on a sand-and-gravel plateau, or *terrasse*, by the river. The wines have a pronounced fruity character, and they are delicious to drink when they are less than six months old. Those grown on the south-facing clay and tufa slopes, or *coteaux*, ripen up to 10 days earlier and produce more full-bodied, longer-lived wines.

RED Bone-dry to dry, medium-bodied, lively wines, full of soft-fruit flavours, which are often aged in cask. They are very easy to quaff when less than six months old; many close up when in bottle and need time to soften. Wines from the *terrasse* vineyards are best drunk young, while those from the coteaux repay keeping.

🍷 *Cabernet Franc, with Cabernet Sauvignon up to 10%*

🍷 *Within 6 months or after 6 years*

ROSÉ Bone-dry to dry, light- to medium-bodied wines that are very fruity with aromas of raspberries and blackberries and good depth of flavour. They deserve to be better known.

🍷 *Cabernet Franc, with Cabernet Sauvignon up to 10%*

🍷 *2–3 years*

✓ Yannick Amirault (Les Quartier Vieilles Vignes) • De la Butte • Breton ❶ • Des Chesnaies (Prestige) ❶ • De la Chevalerie (Busardières) ❶ • De la Cotelleraie • Vignoble de la Grioche (Santenay) • Frédéric Mabileau ❶ • Des Ouches • Du Petit Bondiau (Le Petit Mont)

CHEVERNY AOC

Upgraded from VDQS to full AOC status in 1993. These good-value, crisp, and fruity wines deserved their promotion and should now be better known. Cheverny is usually made and marketed as a pure varietal wine, the most interesting of which, Romorantin, has been awarded its own AOC for the best vineyards around Cour-Cheverny itself. See also Cour-Cheverny AOC.

RED Dry, light- to medium-bodied wines. Gamay Teinturier de Chaudenay, which has coloured juice, is no longer permitted.

🍷 *Pinot Noir 60–84%, Gamay 16–40%, plus Côt (Malbec) up to 5%*

🍷 *1–2 years*

WHITE Dry, light-bodied, modest wines with a fine, flowery nose, delicate flavour, and crisp balance. Now primarily a Sauvignon Blanc wine.

🍷 *Sauvignon Blanc/Gris 60–84%, plus of Chardonnay, Chenin Blanc, and Orbois 16–40% each*

🍷 *1–2 years*

ROSÉ Only small quantities are produced, but the wines are agreeably dry and light-bodied and very consistent in quality. Since acquiring full AOC status, Cabernet Franc, Cabernet Sauvignon, and Malbec have been allowed for the production of these wines, which has boosted the quantity even further.

🍷 *Pinot Noir 60–84%, plus Gamay 16–40% and Côt (Malbec) up to 5%*

🍷 *1–2 years*

✓ De la Desoucherie • De la Gaudronnière (Tradition) • Du Moulin ❶ • Du Salvard • Sauger & Fils

CHINON AOC

The appellations of Chinon and Bourgueil produce the best red wine in Touraine using the Cabernet Franc grape, known locally as Breton. Chinon wines are generally more structured and *terroir*-driven than those of Bourgueil, but those from the tufa hill slopes have greater depth of fruit and flavour and age well.

RED Bone-dry to dry, medium-bodied wines that are lively and structured with good fruit concentration.

Most growers use small oak casks for ageing and produce wines of very good quality.

🍷 *Cabernet Franc, with Cabernet Sauvignon up to 10%*

🍷 *2–3 years*

WHITE A tiny production of clean, dry, light- to medium-bodied wines that are strangely aromatic for Chenin Blanc with an intriguing perfumed aftertaste.

🍷 *Chenin Blanc 100%*

🍷 *1–2 years*

ROSÉ These are dry, fairly light-bodied, smooth, and fruity wines which, like Bourgueil rosés, are very easy to drink and deserve to be better known.

🍷 *Cabernet Franc, with Cabernet Sauvignon up to 10%*

🍷 *2–3 years*

✓ Philippe Alliet • Château de l'Aulée (Artissimo) • Bernard Baudry • Breton ❶ • Château de Coulaine ❶ • Couly-Dutheil (Clos de l'Echo) • Charles Joguet ❶ • De la Noblaie ❶ • Philippe Pichard • Jean-Marie Raffault (Le Puy) • De Saint-Louand (Réserve de Trompegueux) • Serge et Bruno Sourdais

COTEAUX DU LOIR AOC

This is an area that had extensive vineyards in the 19th century. Production has since declined, and these generally unexciting wines come from the Loir, which is a tributary of, and not to be confused with, the Loire.

RED These are dry, medium-bodied wines that can have a lively character and good extract in sunny years.

🍷 *A minimum of 65% Pineau d'Aunis, with up to 30% each of Gamay, Cabernet Franc, and Côt (Malbec)*

🍷 *1–2 years*

WHITE Bone-dry to dry, light-bodied wines that are high in acidity and can be mean and astringent.

🍷 *Chenin Blanc 100%*

🍷 *As early as possible*

ROSÉ Dry, fairly light-bodied wines, a few of which are fruity and well balanced.

🍷 *A minimum of 65% Pineau d'Aunis, plus Gamay, Côt (Malbec), and Grolleau.*

🍷 *Within 1 year*

✓ De Bellivière ❶ • Jean-François Maillet (Réserve d'Automne)

COTEAUX DU VENDÔMOIS AOC

Situated on both banks of the Loir, upstream from Jasnières, this steadily improving wine was upgraded to full AOC status in 2001, when the *encépagement* changed.

RED Dry, fairly light-bodied wines that are full of soft-fruit flavours and very easy to drink.

🍷 *A minimum of 50% Pineau d'Auni, plus Pinot Noir and Cabernet Franc 10–40% , and a maximum of 20% Gamay*

🍷 *1–2 years*

WHITE Dry, fairly light-bodied wines which, when made from pure Chenin Blanc, have a tendency to be very astringent. Growers who blend Chardonnay with Chenin Blanc produce better-balanced wines.

🍷 *Chenin Blanc, with Chardonnay up to 20%*

🍷 *Within 1 year*

GRIS (ROSÉ) Fresh and fragrant, Coteaux du Vendômois is one of the most appealing, yet little seen, of the Loire's dry rosés.

🍷 *Pineau d'Aunis 100% (Gamay is no longer permitted)*

🍷 *1–2 years*

✓ Patrice Colin • Du Four à Chaux • CV de Villiers-sur-Loir

COUR-CHEVERNY AOC

Cheverny gained this special single-village appellation exclusively for Romorantin when it was upgraded in 1993 from a VDQS to an AOC, but it lost its right to produce sparkling wine. Romorantin is grown in the best sites around Cour-Cheverny. See also Cheverny AOC.

WHITE These wines can range from dry to sweet, but all are light bodied and modest, with a fine, flowery nose, delicate flavour, and crisp balance.

🍷 *Romorantin*

🍷 *1–2 years*

✓ De la Desoucherie • De la Gaudronnière (Le Mur de Gaudronnière) • Du Moulin ❶

JASNIÈRES AOC

This is the best area of the Coteaux du Loir – the wines produced in the Jasnières appellation can, in hot years, achieve a richness that compares well with those of Savennières in Anjou (*see p290*).

WHITE Medium-bodied wines that can be dry or sweet. They are elegant and age well in good years, but they can be unripe in poor years.

🍷 *Chenin Blanc 100%*

🍷 *2–4 years*

✓ De Bellivière ❶ • J Martellière (Poète)

MONTLOUIS-SUR-LOIRE AOC

All Montlouis appellations were renamed Montlou-is-sur-Loire AOC in November 2003. Like Vouvray, Montlouis produces wines that can be dry, medium-dry, or sweet depending on the vintage and, like those of its more famous neighbour, some of the greatest wines of Montlouis are the sweetest, most botrytis-rich wines, often sold as *moelleux*. The quality of dry whites has improved immeasurably in recent years, however, so that some Loire enthusiasts believe the very best can now rival the top whites of Burgundy. The wines are very similar in style to those of Vouvray, but while Montlouis is frequently underrated, Vouvray is often overrated. Certain critics rank Moyer's 1959 and the 1947 Montlouis alongside even the greatest vintages of Château d'Yquem Sauternes.

WHITE Light- to medium-bodied wines that can be dry or sweet. They are softer and more forward than the wines of Vouvray but can have the same honeyed flavour in fine years. Sweet Montlouis is aged in cask, but the best medium-dry styles are clean-fermented in stainless steel. More recently, some of the top producers of the appellation are fermenting bone-dry Chenins in barrel at very low temperatures with no malolactic fermentation. The results are truly amazing!

🍷 *Chenin Blanc 100%*

🍷 *1–3 years for medium-dry, up to 10 years for sweeter wines*

✓ Laurent Chatenay ❶ • François Chidaine ❸ • Cossais (Le Volagre) • De la Croix Melier • Alex Mathur (Dionys) • Dominique Moyer • De la Taille au Loups ❶

MONTLOUIS-SUR-LOIRE MOUSSEUX AOC

In poor vintages the grapes are used to make traditional method sparkling versions of Montlouis. The medium-dry (*demi-sec*) styles of the AOC are very popular in France.

SPARKLING WHITE These light- to medium-bodied wines can be *brut, sec, demi-sec,* or *moelleux*. The last two of these styles are made only in years that are particularly sunny.

🍷 *Chenin Blanc 100%*

🍷 *Upon purchase*

✓ Alain Joulin • Thierry et Daniel Mosny • De la Taille au Loups ❶

A CLUSTER OF CHÂTEUX PERCH ON THE HILLSIDE NEAR CHINON ON THE BANKS OF THE VIENNE
Along the *Val de Loire* wine route travellers will come across a 19th-century manor house set in a working organic vineyard in the medieval town of Chinon.

MONTLOUIS-SUR-LOIRE PÉTILLANT AOC

Gently effervescent, Montlouis Pétillant is one of the most successful, yet least encountered and under-rated, slightly sparkling French white wines.

SEMI-SPARKLING WHITE Light- to medium-bodied wines that can be dry or sweet. Very consistent in quality, with a rich, fruity flavour balanced by a delicate mousse of fine bubbles.

🍷 *Chenin Blanc 100%*

🍾 *Upon purchase*

✓ *Levasseur • Dominique Moyer*

ST-NICOLAS-DE-BOURGUEIL AOC

This is a commune with its own appellation in the northwest corner of Bourgueil. The soil is sandier than that of surrounding Bourgueil, and the wines are lighter but certainly equal in terms of quality. These are some of the finest red wines in the Loire.

RED Bone-dry to dry, medium-bodied wines that age well and have greater finesse than the wines of Bourgueil.

🍷 *Cabernet Franc, with Cabernet Sauvignon up to 10%*

🍾 *After 5–6 years*

ROSÉ A small amount of dry, medium-bodied rosé with firm, fruity flavour is produced.

🍷 *Cabernet Franc, with Cabernet Sauvignon up to 10%*

🍾 *Upon purchase*

✓ *Yannick Amirault • Max Cognard (Estelle, Malagnes) • De la Cotelleraie • Sebastian David • Frédéric Malibeau (Coutures, Eclipse) • Jacques et Vincent Malibeau (La Gardière Vieilles Vignes) • Clos des Quarterons (Vieilles Vignes) • Du Rochouard (Pierre du Lane)*

TOURAINE AOC

A prolific appellation with sparkling wines in dry and medium-dry styles, plus sweet white, red, and rosé still wines from all over Touraine. Most are pure varietal wines and the label should indicate which grape they have been made from. See also Touraine Mousseux AOC.

RED Dry, light- to medium-bodied. Ususally Côt and Cabernet Franc–based blends. If "Gamay" is indicated on the label, it must be at least 85 per cent Gamay. Those sold as *primeur* or *nouveau* from the third Thursday of November following the harvest must be made from 100 per cent Gamay.

🍷 *Côt (Malbec), Cabernet Franc, Gamay, Cabernet Sauvignon, Pinot Noir*

🍾 *Within 3 years*

WHITE These are bone-dry to dry, medium-bodied wines; they are fresh, aromatic, and fruity. Good Touraine Sauvignon is better than average Sancerre. These wines may be sold from 1 December following the harvest without any mention of *primeur* or *nouveau*.

🍷 *Sauvignon Blanc, with Sauvignon Gris up to 20%*

✓ *1–2 years*

ROSÉ Dry, light- to medium-bodied wines that must be blends in which none of the components exceed 70 per cent. Those based on Pineau d'Aunis are drier and subtler than Anjou rosé. They may be sold as *primeur* or *nouveau* from the third Thursday of November following the harvest.

🍷 *Cabernet Franc, Cabernet Sauvignon, Côt (Malbec), Gamay, Grolleau, Grolleau Gris, Pinot Noir, Pinot Meunier, Pinot Gris, Pineau d'Aunis*

🍾 *1–2 years*

✓ *De la Charmoise (also those simply labelled Henry Marionnet) • Des Corbillères • Grange Tiphaine 🅞 • De la Garrelière 🅑 • Jacky Marteau • De Bellevue • De Beauséjour • Jean-François Mérieau • Dominique Percereau • De la Puannerie 🅞 • Du Pre Baron • Clos Roche Blanche 🅞 • Les Vaucorneilles (Gamay)*

TOURAINE AMBOISE AOC

Modest white wines and light reds and rosés are produced by a cluster of eight villages surrounding and including Amboise. The vines are grown on both sides of the Loire adjacent to the Vouvray and Montlouis areas.

RED These are dry, light-bodied wines that are mostly blended. Those containing a high proportion of Malbec are the best.

🍷 *Gamay, Cabernet Franc, Cabernet Sauvignon*

🍾 *2–3 years*

WHITE These bone-dry to dry, light-bodied Chenin Blancs are usually uninspiring; the rosés are superior.

🍷 *Chenin Blanc 100%*

🍾 *Upon purchase*

ROSÉ These dry, light-bodied, well-made wines are mouth-watering.

🍷 *Cabernet Franc, Cabernet Sauvignon, Côt (Malbec), Gamay*

🍾 *Within 1 year*

✓ *Dutertre • Guy Saget • Lionel Truet (Grande Foucaudière)*

TOURAINE AZAY-LE-RIDEAU AOC

Good-quality wines from eight villages on either side of the Indre River, a tributary of the Loire.

WHITE Delicate, light-bodied wines that are usually dry but may be *demi-sec*.

🍷 *Chenin Blanc 100%*

🍾 *1–2 years*

ROSÉ Attractive, refreshing, truly dry (maximum of 3 grams per litre residual sugar) wines that are coral pink and often have a strawberry aroma.

🍷 *Grolleau 50–65%, plus Côt (Malbec), Gamay, Cabernet Franc, and Cabernet Sauvignon (the last two must not total more than 10% of the entire blend)*

🍾 *1–2 years*

✓ *Château de l'Aulée • Nicolas Paget • Pibaleau Père & Fils 🅞*

TOURAINE CHENONCEAUX AOC

Created in 2011 as a "super-Touraine". The best vineyards are situated overlooking the river Cher.

WHITE Fresh, crisp wines with good ripe stone fruit character, best enjoyed when young.

🍷 *Sauvignon Blanc 100%*

🍾 *Upon purchase*

RED Dry, medium-bodied with supple texture.

🍷 Côt (Malbec), with Cabernet Franc 35–50%

🍷 3–5 years

TOURAINE MESLAND AOC

Wines from the vineyards of Mesland and the five surrounding villages on the right bank of the Loire. The reds and rosés of this appellation are definitely well worth looking out for.

RED These dry, medium- to full-bodied wines are the best of the AOC and can be as good as those of Chinon or Bourgueil.

🍷 A minimum of 60% Gamay, plus Cabernet Franc and Côt (Malbec) 10–30%

🍷 1–3 years

WHITE Dry, light-bodied wines with a high acidity that is only tamed in the best and sunniest years.

🍷 A minimum of 60% Chenin Blanc, plus a maximum of 30% Sauvignon Blanc and a maximum of 15% Chardonnay

🍷 1–2 years

ROSÉ These dry, medium-bodied wines have more depth and character than those of Touraine Amboise.

🍷 A minimum of 80% Gamay, plus Cabernet Franc

🍷 1–3 years

✔ De la Briderie **B** • Château Gaillard **B**

TOURAINE MOUSSEUX AOC

Very good-value traditional method white and rosé wines.

SPARKLING WHITE These light- to medium-bodied wines are made in dry and sweet styles, and the quality is consistent due to the large production area, which allows for complex blending.

🍷 Chenin Blanc, but may also include Orbois up to 20% and Chardonnay and a combined maximum of 30% Cabernet, Pinot Noir, Pinot Gris, Pineau d'Aunis, Côt (Malbec), and Grolleau

🍷 Upon purchase

SPARKLING ROSÉ Light- to medium-bodied wines that are attractive when brut, though a bit cloying if sweeter.

🍷 Cabernet Franc, Cabernet Sauvignon, Côt (Malbec), Gamay, Grolleau, Grolleau Gris, Meunier, Pineau d'Aunis, Pinot Gris, Pinot Noir

🍷 1–2 years

✔ Blanc Foussy (Robert de Schlumberger, Veuve Oudinot) • Jean-Pierre Laissement (Rosé) • Monmousseau (JM)

TOURAINE OISLY AOC

Oisly is a white-only appellation, thought to be the birthplace of the Sauvignon grape.

WHITE Sauvignon Blanc 100%

🍷 Upon purchase

VALENÇAY AOC

Situated in the southeast of Touraine around the river Cher, these vineyards produce well-made, attractive wines. Promoted from VDQS to AOC in 2003, when the *encépagement* changed.

RED Dry, light-bodied, fragrant wines that can be very smooth and full of character. The Malbec, known locally as Côt, is particularly successful.

🍷 Gamay 30–60%, with Pinot Noir up to 20% and Côt (Malbec), plus Cabernet Franc up to 20%

🍷 1–2 years

WHITE Dry, light-bodied wines that are improved by the addition of Chardonnay.

🍷 A minimum of 70% Sauvignon Blanc, plus Orbois, Sauvignon Gris, and Chardonnay

🍷 1–2 years

ROSÉ Dry to medium-dry, light-bodied wines that can be full of ripe soft-fruit flavours.

🍷 Gamay 30–60%, Pinot Noir and Côt (Malbec) up to 20% in total, plus an optional maximum of 30% Pinot d'Aunis, with or without Cabernet Franc and Cabernet Sauvignon, which together must not exceed 20% of the entire blend

🍷 Upon purchase

✔ Chantal et Patrick Gibault • Jean-François Roy **O** • Nathalie Lafond • Bertrand Minchin

VOUVRAY AOC

These white wines may be dry, medium-dry, or sweet depending on the vintage. In sunny years, the classic Vouvray that is made from overripe grapes affected by the "noble rot" is still produced by some growers. In cooler years, the wines are correspondingly drier and more acidic, and greater quantities of sparkling wine are produced.

WHITE At its best, sweet Vouvray can be the richest of all the Loire sweet wines. In good years, the wines are very full bodied, rich in texture, and have the honeyed taste of ripe Chenin Blanc grapes.

🍷 Chenin Blanc, but may also contain Orbois

🍷 Usually 2–3 years; the sweeter wines can last up to 50 years

✔ Des Aubuisières • Champalou (Le Clos du Portail Sec) • Philippe Foreau (Clos Naudin) • De la Haute Borne (Tendre) • Huet **B** • Clos de Nouys • Pichot • De la Taille au Loups • Vigneau Chevreau **O** • Vignobles Robert (La Sablonnière)

VOUVRAY MOUSSEUX AOC

These sparkling wines are made from overripe grapes. In years when the grapes do not ripen properly they are converted into sparkling wines using the traditional method and blended with reserve wines from better years.

SPARKLING WHITE Medium- to full-bodied wines made in both dry and sweet styles. They are richer and softer than sparkling Saumur but have more edge than sparkling Montlouis.

🍷 Chenin Blanc, Arbois

🍷 Non-vintage 2–3 years, vintage brut and sec 3–5 years, vintage demi-sec 5–7 years

✔ Marc Brédif • Champalou • Champion • Clos de l'Epinay (Tête de Cuvée) • Philippe Foreau • Domaine de la Galinière (Cuvée Clément) • Sylvain Gaudron • Huet **B** • Laurent et Fabrice Maillet • Château de Moncontour (Cuvée Prédilection)

VOUVRAY PÉTILLANT AOC

These are stylish and consistent semi-sparkling versions of Vouvray, but very little is produced.

SEMI-SPARKLING WHITE Medium- to full-bodied wines made in dry and sweet styles. They should be drunk young.

🍷 Chenin Blanc, Orbois

🍷 Upon purchase

✔ Gilles Champion • Jean-Charles Cathelineau • Huet **B**

A WINE PRESS DECORATES THE ROADSIDE IN THE COMMUNE OF VOUVRAY
An old wine press overflowing with colourful flowers cheers passers-by in this area of Touraine. The best Vouvray wines are sweet whites.

Central Vineyards

In this district of scattered vineyards, all the classic wines are dry variations of the Sauvignon Blanc, but there are some discernible differences between the best of them – the concentration of Sancerre, the elegance of Pouilly Fumé, the fresh-floral character of Menetou-Salon, the lightness of Reuilly, and the purity of Quincy.

CAVES LA PERRIÈRE, SANCERRE
At Domaine de la Perrière the cellars are located in a vast natural cave dating back to the Tertiary Era, which was extended by monks in the Middle Ages. Visitors to the estate can tour the caves, which feature a winemaking museum and wine-tasting area.

The Central Vineyards are so called because they lie in the centre of France, not the centre of the Loire Valley. This is a graphic indication of how far the Loire Valley extends and, though it might not be a surprise to discover the vineyards of Sancerre are quite close to Chablis, it does take a leap of the imagination to accept that they are nearer to the Champagne region than to Tours. And who could discern by taste alone that Sancerre is equidistant between the production areas of such diverse wines as Hermitage and Muscadet?

Best known of all the towns in this district is Orléans, famous for its liberation by Joan of Arc from the English in 1429. The other important town is Bourges, which is situated in the south between the wine villages of Reuilly, Quincy, and Menetou-Salon, and was once the capital of the Duchy of Berry.

THE REGION'S SAUVIGNON BLANC WINES

The Sauvignon Blanc is to Central Vineyards what Muscadet is to the Pays Nantais. It produces the classic wine of the district, which, like Muscadet, also happens to be both white and dry. But two dry white wines could not be more different in style and taste. In the best Muscadet *sur lie* there should be a yeasty fullness, which can sometimes be misread as the Chardonnay character of a modest Mâcon. In Central Vineyard Sauvignons, however, whether they come from Sancerre or Pouilly – or even from one of the lesser-known, but certainly not lesser-quality, villages around Bourges – the aroma is so striking it sometimes startles.

The rasping dryness of the wine's flavour catches the breath and can come from only one grape variety.

A BURGUNDIAN INFLUENCE

Historically, this district was part of the Duchy of Burgundy, which explains the presence of Pinot Noir vines. After the scourge of phylloxera (a tiny, sap-sucking insect that feeds on vines), the area under vine shrank. Those areas brought back into production were mostly replanted with Sauvignon Blanc, which began to dominate the vineyards, although isolated spots of Pinot Noir were maintained. In the hands of a talented *vignerons*, some of the best of these wines today can be as good as Burgundy, and although perhaps less structured and more delicate they possess a purity of fruit that seduces.

FACTORS AFFECTING TASTE AND QUALITY

LOCATION
The most easterly vineyards of the Loire are situated in the centre of France, chiefly in the *départements* of Cher, Nièvre, and Indre.

CLIMATE
More continental than areas closer to the sea. The summers are shorter and hotter, and the winters are longer and colder. Spring frosts and hail are particular hazards in Pouilly. Harvests are irregular.

ASPECT
This region features chalk hills in a quiet, green landscape. Vines occupy the best sites on hills and plateaux. At Sancerre they are planted on steep, sunny, sheltered slopes at an altitude of 200 metres (660 feet).

SOIL
The soils of Central Vineyards are dominated by limestone or clay topped with gravel and flinty pebbles. When mixed with chalk-tufa, gravelly soils produce lighter, finer styles of Sauvignon wines; when combined with Kimmeridgian clay, the result is firmer and more strongly flavoured.

VITICULTURE & VINIFICATION
Some of the vineyard slopes in Sancerre are very steep, so cultivation and picking are done by hand. Most Central Vineyard properties are small and use the traditional wooden vats for fermentation, but some growers have stainless steel tanks.

GRAPE VARIETIES
Primary varieties: Pinot Noir, Sauvignon Blanc
Secondary varieties: Cabernet Franc, Chasselas, Gamay, Pinot Blanc, Pinot Gris

VINE-GROWING AREAS OF THE CENTRAL VINEYARDS (see also p283)
The most easterly of the Loire's vineyards, and the most central in France, Central Vineyards are famous for wines made from the Sauvignon Blanc grape.

THE APPELLATIONS OF
CENTRAL VINEYARDS

CHÂTEAUMEILLANT AOC

This appellation, which borders the Cher and Indre, around Bourges, was upgraded from the defunct VDQS to AOC in 2010.

RED The Gamay grape usually dominates the red wines, which should be drunk as young as possible.

🍷 *A minimum of 60% Gamay, plus Pinot Noir*

🍸 *6–12 months*

ROSÉ The best Châteaumeillant wines. They are fresh, grapey, and delicately balanced.

🍷 *A minimum of 60% Gamay, plus Pinot Noir, and Pinot Gris up to 15%*

🍸 *6–12 months*

✓ *Du Chaillot • CV Châteaumeillant • Valérie et Frédéric Dallot • De Pavillon*

COTEAUX DU GIENNOIS AOC

This appellation could boast nearly a thousand growers at the turn of the 20th century when it was 40 times its current size. Promoted from VDQS to AOC in 1998.

RED Dry, light-bodied red wines.

🍷 *Gamay, Pinot Noir – since 1992, neither variety may exceed 80% of the total blend.*

🍸 *1–2 years*

WHITE These are also dry, medium-bodied, aromatic white wines that bear many of the characteristics of their more famous neighbours in Sancerre and Pouilly-Fumé, but at a fraction of the price. A great value alternative for excellent quality Central Loire Sauvignon.

🍷 *Sauvignon Blanc*

🍸 *1–2 years*

ROSÉ These are pale salmon-coloured, light-bodied wines that can have a fragrant citrus character.

🍷 *Gamay, Pinot Noir – since 1992, neither variety may exceed 80% of the total blend.*

🍸 *1 year*

✓ *De Villargeau • Vignobles Bethier • Quintin Frères*

MENETOU-SALON AOC

This underrated appellation covers Menetou-Salon and the nine surrounding villages.

RED These are dry, light-bodied, crisp, and fruity wines with fine varietal aroma. They are best drunk young, although some oak-matured examples can age well.

🍷 *Pinot Noir*

🍸 *2–5 years*

WHITE Bone-dry to dry wines. They are definitely Sauvignon in character, but the flavour can have an unexpected fragrance.

🍷 *Sauvignon Blanc*

🍸 *1–2 years*

ROSÉ These are extremely good-quality, dry, light-bodied aromatic wines, full of straightforward fruit.

🍷 *Pinot Noir*

🍸 *Within 1 year*

✓ *Georges Chavet & Fils • Bertrand Minchin • Henry Pellé*

POUILLY BLANC FUMÉ AOC

See Pouilly Fumé AOC

POUILLY FUMÉ AOC

This used to be the world's most elegant Sauvignon Blanc wine, but too many wines of ordinary quality have debased this once-great appellation. In Pouilly-sur-Loire and its six surrounding communes, only pure Sauvignon wines have the right to use "Fumé" in the appellation name, a term that evokes the wine's gunsmoke character.

WHITE Great Pouilly Fumé is rare but when found, its crisp, gooseberry flavour will retain its finesse and delicacy in even the hottest years.

🍷 *Sauvignon Blanc*

🍸 *2–5 years*

✓ *Caillbourdin • Jean-Claude Châtelain • Didier Dagueneau • André & Edmond Figeat • Michel Redde • Château de Tracy • Bel Air • Caves des Vins de Pouilly-Fumé • Tinel-Blondelet • Des Rabichattes • Marchand*

POUILLY-SUR-LOIRE AOC

This wine comes from the same area as Pouilly Fumé, but it is made from the Chasselas grape. Chasselas is a good dessert grape, but it makes very ordinary wine.

WHITE Dry, light-bodied wines. Most are neutral, tired, or downright poor.

🍷 *Chasselas*

🍸 *Upon purchase*

✓ *Guy Saget • Bel Air*

QUINCY AOC

These vineyards, on the left bank of the Cher, are situated on a gravelly plateau. Although located between two areas producing red, white, and rosé wines, Quincy produces only white wine, from Sauvignon Blanc.

WHITE Bone-dry to dry, quite full-bodied wines in which the varietal character of the Sauvignon Blanc is evident. There is a purity that rounds out the flavour and seems to remove the rasping finish usually expected in this type of wine.

🍷 *Sauvignon Blanc*

🍸 *1–2 years*

✓ *Bailly • De La Commanderie • Mardon • De Puy-Ferrand • Philippe Portier • Silice de Quincy*

REUILLY AOC

Due to the high lime content in the soil, Reuilly produces wines of higher acidity than those of neighbouring Quincy.

RED These are dry, medium-bodied wines. Some are surprisingly good, although often tasting more of strawberries or raspberries than the more characteristic redcurrant flavour associated with Pinot Noir.

🍷 *Pinot Noir*

🍸 *2–5 years*

✓ *De Seresnes • Jean-Michel Sorbe*

WHITE These are bone-dry to dry, medium-bodied wines of good quality with more of a grassy than a gooseberry flavour, yet possessing a typically austere dry finish.

🍷 *Sauvignon Blanc*

🍸 *1–2 years*

ROSÉ This bone-dry to dry, light-bodied wine is a pure Pinot Gris wine, although it is simply labelled Pinot.

🍷 *Pinot Noir, Pinot Gris*

🍸 *2–5 years*

✓ *Aujard • Pascal Desroches (Clos des Varennes) • Nathalie Lafond • Matthieu Mabillot • Valéry Renaudat • De Seresnes • Jean-Michel Sorbe*

SANCERRE AOC

This appellation is famous for its white wines, but originally its reds were better known. Recently the reds and rosés have developed greater style.

RED These wines have been more variable in quality than the whites, but the consistency is improving rapidly. They are dry, light- to medium-bodied wines, with a pretty floral aroma and a delicate flavour.

🍷 *Pinot Noir*

🍸 *2–3 years*

✓ *Lucien Crochet • André Dezat*

WHITE Classic Sancerre should be bone dry, highly aromatic, and have an intense flavour – sometimes tasting of gooseberries or even peaches in a great year. Yet too many growers overproduce, never get the correct ripeness, and make the most miserable wines.

🍷 *Sauvignon Blanc*

🍸 *1–3 years*

ROSÉ Attractive, dry, light-bodied rosés with strawberry and raspberry flavours.

🍷 *Pinot Noir*

🍸 *Within 18 months*

✓ *Henri Bourgeois • François Cotat • Lucien Crochet • Matthieu Delaporte • André Dezat • Fouassier (Clos Paradis, Les Grands Groux) ❶ • Jean-Max Roger • Vignobles Berthier • Alphonse Mellot ❸ • Vacheron • Vincent Pinard • Claude Riffault • Merlin-Cherrier • Jean-Philippe Agisson • Reverdy-Ducroux*

ORLÉANS AOC

Centred around the city of Orléans, these wines have held AOC status since September 2006, with 13 communes entitled to the appellation. Although the region has made wine for centuries, during the 20th century winemaking declined here, and today only one-third of the appellation is now worked.

RED Dry, medium-bodied, fresh, and fruity wines that are given a short maceration, producing a surprisingly soft texture.

🍷 *Pinot Meunier 70–90% , plus Pinot Noir and Cabernet Franc*

🍸 *1–2 years*

WHITE Very small quantities of interesting wines are made from Chardonnay, known locally as Auvernat Blanc. Dry, medium-bodied, and surprisingly smooth and fruity.

🍷 *A minimum of 60% Chardonnay (Auvernat Blanc), plus Pinot Gris (Auvernat Gris)*

🍸 *1–2 years*

ROSÉ The local speciality is a dry, light- to medium-bodied rosé known as Meunier Gris – an aromatic *vin gris* with a crisp, dry finish.

🍷 *A minimum of 60% Pinot Meunier, plus Pinot Noir and Pinot Gris*

🍸 *Within 1 year*

✓ *Clos St-Fiacre • Vignerons de la Grand'Maison*

ORLÉANS-CLÉRY AOC

This relatively recent red-only appellation encompasses five communes and has about 28 hectares (69 acres) under vine. It champions the Cabernet Franc grape, with up to 25 per cent Cabernet Sauvignon allowed in the blend until 2020, from vineyards already established by 2006.

RED These are simple, fruity wines with some typical blackcurrant leaf character from Cabernet Franc. They are meant to be enjoyed young.

🍷 *Cabernet Franc (up to 25% Cabernet Sauvignon allowed until 2020)*

🍸 *2–5 years*

✓ *Jacky Legroux • CV de la Grand Maison*

The Rhône Valley

Famous for its full, fiery, and spicy-rich red wines, the Rhône Valley also produces lesser quantities of rosé and white and even some sparkling and fortified wines. Essentially red-wine country, and great red-wine country at that, the Rhône has also experienced a kind of revolution in white wine production. There has been a growing number of exotic, world-class whites in various appellations since the late 1980s, when just a few began to emerge in Châteauneuf-du-Pape.

WINE REGÔNS OF THE RHÔNE VALLEY
(*see also* p131)
Viticulturally, the Rhône Valley covers a large area of southern-central France – from Chasse-sur-Rhône and Vienne in the north to the heart of Provence in the south. There are about 70,365 hectares (173,875 acres) under vine.

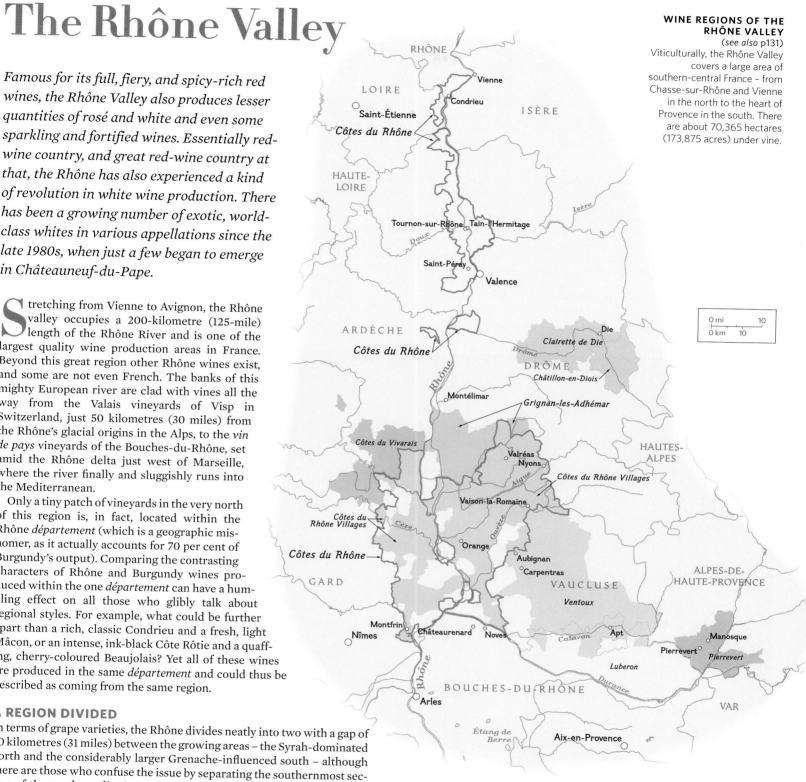

Stretching from Vienne to Avignon, the Rhône valley occupies a 200-kilometre (125-mile) length of the Rhône River and is one of the largest quality wine production areas in France. Beyond this great region other Rhône wines exist, and some are not even French. The banks of this mighty European river are clad with vines all the way from the Valais vineyards of Visp in Switzerland, just 50 kilometres (30 miles) from the Rhône's glacial origins in the Alps, to the *vin de pays* vineyards of the Bouches-du-Rhône, set amid the Rhône delta just west of Marseille, where the river finally and sluggishly runs into the Mediterranean.

Only a tiny patch of vineyards in the very north of this region is, in fact, located within the Rhône *département* (which is a geographic misnomer, as it actually accounts for 70 per cent of Burgundy's output). Comparing the contrasting characters of Rhône and Burgundy wines produced within the one *département* can have a humbling effect on all those who glibly talk about regional styles. For example, what could be further apart than a rich, classic Condrieu and a fresh, light Mâcon, or an intense, ink-black Côte Rôtie and a quaffing, cherry-coloured Beaujolais? Yet all of these wines are produced in the same *département* and could thus be described as coming from the same region.

A REGION DIVIDED

In terms of grape varieties, the Rhône divides neatly into two with a gap of 50 kilometres (31 miles) between the growing areas – the Syrah-dominated north and the considerably larger Grenache-influenced south – although there are those who confuse the issue by separating the southernmost section of the northern district and calling it the Middle Rhône. The north and south differ not only in terrain and climate, but also contrast socially, culturally, and gastronomically. Climate change is beginning to show stresses in the south, with rising levels of alcohol, though three of the last four vintages are very highly rated.

WINE FRAUD

In June 2017 the directors of Rhône merchant Raphaël Michel were arrested on criminal charges for fraudulent labelling of Côtes-du-Rhône wines. More than 66 million fake bottles (15 per cent of production) were produced between October 2013 and June 2016. There are some cases working through the French courts, but the matter seems to have blown over.

RHÔNE AT ITS BEST

The best Rhône vintages in recent times have broadly followed Bordeaux, being 2009, 2010, 2015, 2016, and 2018. There are variances between the Northern and Southern Rhône, with some divergent climatic factors at play, and the red versus white wines perform differently overall within each of the two main regions. Along with the above-mentioned vintages, the best of the past includes 2005, 2001, 1999, 1998, 1995, 1990, 1989, 1978, 1961, and 1959.

RECENT RHÔNE VINTAGES

2018 After a mild and wet spring bud-break came with warmth but some storms shortly after flowering causing some crop loss particularly in the south. Summer was sunny and hot, bringing a welcome shower in August. Harvest began in late August to early September in favourable conditions producing a good quantity and quality for most growers. Both red and white wines look set to age particularly well.

2017 Spring frost and rain caused trouble with mildew particularly in the south and for Condrieu in the north. Flowering was erratic and followed by an early dry season resulting in smaller berry clusters with reduced yields and a harvest of concentrated and intense wines. The best of the reds should age well, and producers were generally happy with quality if not quantity.

2016 A good start to the year with a warm early spring followed by cool then dry conditions from July through to a warm September and fine October for harvest. Well-balanced, fresh, and pristine intense reds and elegant white wines.

2015 A wet winter and spring that cleared by June offering no further rain until needed in mid-August. A splendid summer bringing cooler nights followed by a fine September producing a ripe and healthy vintage claimed by some to be as good as 1961. The reds are outstanding in all appellations, though high in alcohol in the south. The whites are very good, except those from the warmest sites in Condrieu.

2014 A cool and wet summer made it difficult to ripen the reds, and September hail in the north prompted some growers to pick prematurely. The reds lack phenolic ripeness and alcohol, making them light and grassy. The whites being naturally picked earlier before rain and hail were ripe with retained acidity and have been particularly acclaimed from Condrieu and St Péray.

2013 Spring was gloomy and then followed by record May rainfall that hit flowering especially in the south. After a cool June and wet July some beautiful weather in August and September salvaged ripening of the reduced crop. Reds are lighter but refreshing, and the whites fared extremely well to produce some concentrated, fresh, and aromatic wines among which Condrieu and Hermitage stand out as excellent.

2012 Some intermittent rain in spring compromised flowering, so there was a reduced crop. Warm and sunny summer with good harvest conditions produced a very good vintage of matured ripe fruit and soft tannins retaining sufficient acidity.

THE CHAPOUTIER CROZES-HERMITAGE VINEYARDS GROW ON THE SOUTHWEST SIDE OF THE STEEP GRANITE HILLS IN TAIN L'HERMITAGE
Winery and *négociant* Chapoutier, or Maison M. Chapoutier, is known for its acclaimed Hermitage wines and for including Braille on all its wine labels since 1996.

The Northern Rhône

The Northern Rhône is dominated by the ink-black wines of Syrah, the Rhône's only truly classic black grape. A small amount of white wine is also produced and, in the south of the district, at Saint-Péray, Châtillon-en-Diois, and Die, sparkling wines are made.

I t just might be the gateway to the south, but the Northern Rhône region has more in common with its neighbour to the north – Burgundy – than it does with the rest of the Rhône, even though its wines cannot be compared with those from any other area. Indeed, it would be perfectly valid to isolate the north as a totally separate region called the Rhône, which would, therefore, allow the Southern Rhône to be more accurately defined as a high-quality extension of the Midi.

THE QUALITY OF THE NORTHERN RHÔNE

The classic wines of Hermitage and Côte Rôtie stand shoulder to shoulder with the *crus classés* of Bordeaux in terms of pure quality, and the greatest Hermitage and Côte Rôtie deserve the respect given to *premiers crus* such as Latour, Mouton, or Lafite. Cornas can be even bigger and darker than Hermitage or Côte Rôtie, and the top growers can rival the best of wines from its better-known neighbours. The fine, dry white wines of Condrieu and Château Grillet are unique in character, yet the presence of such a style in this part of France is not as surprising as that of the sparkling white wines of St-Péray, Châtillon-en-Diois, and Die, particularly the latter, which Francophiles would describe as a superior sort of Asti.

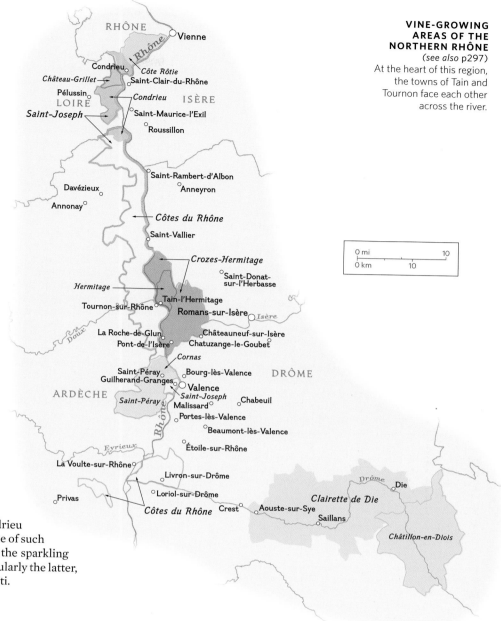

VINE-GROWING AREAS OF THE NORTHERN RHÔNE
(*see also* p297)
At the heart of this region, the towns of Tain and Tournon face each other across the river.

THE APPELLATIONS OF
THE NORTHERN RHÔNE

ARDÈCHE IGP
See 334.

BRÉZÈME CÔTES DU RHÔNE AOC

A regulatory curiosity, Brézème is not one of the 20 Côtes du Rhône Villages that may attach their name only to the end of that appellation but is an anomalous village that through a peculiarity in the AOC laws is allowed to put its name before the basic appellation (although some prefer to place Brézème after, in much larger letters). The story dates back to the mid-1800s, when the wines from Brézème almost rivalled those of Hermitage and sold for nearly as much. In 1943, when Brézème was first allowed to add its own name to the Côtes du Rhône AOC, the vineyards amounted to a mere 10 hectares (25 acres). Shortly afterwards, most of the vineyards were abandoned and the village was all but forgotten. By 1961 there was barely 1 hectare (2.47 acres) under cultivation, yet remarkably a few producers continued to make Brézème. According to Eric Texier, one of the most respected growers was a Monsieur Pouchoulin, who was known as the "Grandfather of Brézème". Texier, a former

nuclear engineer, took over Pouchoulin's vines, which are all Syrah and range from 60 to 100 years old. Thanks to Pouchoulin and others like him, the Brézème AOC was extended to include 84 hectares (208 acres) and, with renewed interest in the wine, plantings have increased by a further 22 hectares (55 acres), with more vines going in each year.

RED This pure Syrah wine has been likened to a Crozes-Hermitage, but some *cuvées* (Lombard's Grand Chêne and Eugène, for example) are aged in a percentage of new oak and are not at all like a Northern Rhône wine.

🍷 *The full spectrum of Côtes du Rhône grapes is allowed, but Syrah is the only red wine grape grown.*

🍷 *3–10 years*

WHITE Texier produces a pure Viognier and a pure Roussanne.

🍷 *The full spectrum of Côtes du Rhône grapes is allowed, but only Marsanne, Roussanne, and Viognier are grown.*

🍷 *3–10 years*

✓ Jean-Marie Lombard • Eric Texier (Syrah, Viognier)

CHÂTEAU GRILLET AOC

Château Grillet was under ownership of the Neyret-Gachet from 1830 until 2011, when it joined Château Latour in the Domaine Artémis portfolio of luxury brands collected by François Pinault, the third-richest man in France. Château Grillet is one of a very select club of single-estate appellations in France, the others being Coulée de Serrant in Savennières, Clos de Tart (also Pinault's) in Burgundy's Côte de Nuit, and its most famous *grand cru*, Romanée-Conti. Visited by Thomas Jefferson in 1787, it historically produced one of the world's great white wines, but suffering from under investment in the late 20th century, Château Grillet did not achieve its potential in the last few decades. Its reputation is being restored by team Artémis, with vineyard and winery improvements clearly evident in the 2015 vintage. Prices have risen commensurately, however, trebling the cost per bottle on release.

WHITE This is regaining its reputation as a wine of great finesse and complex character. Vines average 45 years old and undergo 18 months ageing in oak barrels. The best wines from different parcels are then blended

FACTORS AFFECTING TASTE AND QUALITY

LOCATION
The narrow strip of vineyards that belong to the Northern Rhône commences at Vienne, just south of Lyon, and extends southwards to Valence. Southeast from Valance up the Drôme tributary in the foothills of the Alps are France's highest-altitude vineyards, mostly producing sparkling wine.

CLIMATE
The general effect of the Mediterranean is certainly felt in the Northern Rhône, but its climate has a distinctly continental influence. This results in the pattern of warmer summers and colder winters, which is closer to the climate of southern Burgundy to the north than to that of the Southern Rhône. The climatic factor that the area does have in common with the southern half of the Rhône is the mistral, a bitterly cold wind that can reach up to 145 kilometres (90 miles) per hour and is capable of denuding a vine of its leaves, shoots, and fruit. As a result, many mistral-prone vineyards are protected by poplar and cypress trees. The wind can, however, have a welcome drying effect in humid harvest conditions. The average summertime temperature is just one degree Celsius lower than in the south, which means that, for now at least, viticulture in the north has not suffered to the extent as has the south from rising temperatures associated with climate change (see Climate Change under Southern Rhône, p302).

ASPECT
The countryside is generally less harsh than that of the southern Rhône, with cherry, peach, chestnut, and other deciduous trees in evidence. The valley vineyards are cut far more steeply into the hillsides than they are in areas farther south.

SOIL
The Northern Rhône's soil is generally light and dry, granitic and schistous. More specifically, it is made up of granitic-sandy soil on the Côte-Rôtie (calcareous-sandy on the Côte Blonde and ruddy, iron-rich sand on the Côte Brune); granitic-sandy soil with a fine overlay of decomposed flint, chalk, and mica, known locally as arzelle at Hermitage and Condrieu; heavier soil with patches of clay in Crozes-Hermitage; granitic sand with some clay between St-Joseph and St-Péray, getting stonier towards the southern end of the region, with occasional outcrops of limestone; and limestone and clay over a solid rock base in the area that surrounds Die.

VITICULTURE & VINIFICATION
Unlike the Southern Rhône, most Northern Rhône red wines are produced entirely or predominantly from a single grape variety, the Syrah, despite the long list of grapes that are occasionally used (see Secondary varieties, below). Viticultural operations are labour intensive in the northern stretch of the district and, owing to the cost, the vineyards were once under the threat of total abandonment. Since that threat, we pay much more for Côte Rôtie, but at least it has remained available. Vinification techniques are very traditional and, when wines are aged in wood, there is less emphasis on new oak than is given in Bordeaux or Burgundy, although chestnut casks are sometimes used.

GRAPE VARIETIES
Primary varieties: Syrah, Viognier, Marsanne, Roussanne
Secondary varieties: Aligoté, Bourboulenc, Calitor, Camarèse (Brun Argenté), Carignan, Chardonnay, Cinsault, Clairette, Counoise, Gamay, Grenache, Mauzac, Mourvèdre, Muscardin, Muscat Blanc à Petits Grains, Pascal Blanc, Picardan, Picpoul, Pinot Blanc, Pinot Noir, Terret Noir, Ugni Blanc, Vaccarèse (Brun Argenté)

---

for the main wine. Since 2011 a second wine, Pontcin, with the Côte du Rhone appellation has been produced, which for the 2013 and 2014 vintages represent better value than Château Grillet label.

🍷 *Viognier*

🥂 *4–15 years*

CHÂTILLON-EN-DIOIS AOC
This wine was raised to full AOC status in March 1975, yet in five decades this appellation has failed to make a mark and remains obscure. Grapes must be hand-picked.

RED Light in colour and body, thin in fruit, with little discernible character.

🍷 *Gamay, plus up to 25% Syrah and Pinot Noir*

WHITE Sold as pure varietal wines, the light and fresh, gently aromatic Aligoté is as good as the richer, fuller, and rather angular Chardonnay.

🍷 *Aligoté, Chardonnay*

ROSÉ I have not encountered any.

🍷 *Gamay, plus up to 25% Syrah and Pinot Noir*
✓ *Didier Cornillon (Clos de Beylière)*

CLAIRETTE DE DIE AOC
Despite the appellation name, the Clairette component of this wine may not exceed 25 per cent, because Muscat á Petits Grains is the primary grape. It is produced in the ancestral method with one fermentation only, as opposed to the two required for Crémant de Die and any other *méthode traditional* sparkling wine. Similar to the fashionable *pétillant naturel*, for the *méthode dioise ancestrale* the wine first undergoes a long, cold part-fermentation in bulk and when bottled must contain a minimum of 55 grams of residual sugar per litre and no *liqueur de tirage*. Fermentation continues inside the bottle, but when disgorged the wine must still retain at least 35 grams of residual sugar per litre. The wine undergoes *transvasage* (meaning it is filtered into a fresh bottle) and is corked without any addition of a *liqueur d'expédition* (see also Crémant de Die AOC). Production of both Die sparkling wine AOCs are dominated by Cave de Die Jalliance cooperative, representing around 75 per cent of growers with annual output of eight million bottles.

SPARKLING WHITE A very fresh, deliciously fruity, gently sparkling wine of at least *demi-sec* sweetness, with a ripe, peachy flavour.

🍷 *At least 75% Muscat à Petits Grains, plus Clairette*

🥂 *Upon purchase*
✓ *Buffardel Frères • CV du Diois (Clairdie, Clairdissime)*

COLLINES RHODANIENNES IGP
See p335.

COMTÉS RHODANIENS IGP
See p335.

CONDRIEU AOC
This is the greatest white-wine appellation in the entire Rhône Valley, with more up-and-coming young talent among its producers than anywhere else in the world. Initially the trend was to produce sweet and medium-sweet wines, but Condrieu has tended to be a distinctly dry wine for a couple of decades now. When the vintage permits, late-harvested sweeter wines are produced, and these styles are very much part of Condrieu's recent revival as one of the undisputed great white wines of the world.

WHITE These pale-gold-coloured wines were once essentially dry but had such an exotic perfume that you thought they were sweet when you first breathed in their heavenly aroma, but now so many of them have a real touch of sweetness. A great Condrieu has a beguiling balance of fatness, freshness, and finesse that produces an elegant, peachy-apricoty coolness of fruit on the finish. Do not be fooled into cellaring these wines – the dry ones in any case – as their greatest asset is the freshness and purity of fruit, which can only be lost over the years. Most of the top Condrieu are spoilt by too much new oak. Even those recommended below are too oaky from a classic point of view.

🍷 *Viognier*

🥂 *4–8 years*

✓ *Cuilleron (Les Ayguets, Les Chaillets, Vertige) • Delas (Clos Boucher) •Yves Gangloff • Guigal (La Dorianne) • Stéphane Montez (Les Grandes Chaillées, Domaine du Monteillet) • Remi Neiro • Alain Paret (Lys de Volan) • Georges Vernay • François Villard (Le Grand Vallon)*

CORNAS AOC
The sun-trap vineyards of Cornas produce the best value of all the Rhône's quality red wine. Yet you have to buy it as soon as it is released, because it is always sold far too young, so there is none left when it starts to show its true potential.

RED Ink-black, full-bodied, strong-flavoured, pure Syrah wines with lots of blackcurrant and blackberry fruit, lacking only a little finesse if compared to a great Hermitage or Côte Rôtie.

🍷 *Syrah*

🥂 *7–20 years*
✓ *Thierry Allemand • Auguste Clape • Jean-Luc Colombo • Du Coulet ❽ • Courbis (Les Eygats) • Yves Cuilleron • Eric & Joël Durand • Guillaume Gilles • Paul Jaboulet Aîné • Patrick Lesec • Johann Michel • Vincent Paris • Tardieu-Laurent (Vieilles Vignes) • Du Tunnel*

CÔTE RÔTIE AOC
The terraces and low walls of the "burnt" or "roasted" southeast-facing slopes of Côte Rôtie must be tended by hand, but the reward is a wine of great class that vies with Hermitage as the world's finest example of Syrah.

RED Garnet-coloured with full body, fire, and power, made fragrant by adding of Viogniers. Long-living and complex with nuances of violets and spices and great finesse.

🍷 *Syrah, plus up to 20% Viognier*

🥂 *10–25 years*
✓ *Patrick & Christophe Bonnefond • Clusel-Roch ❾ • Yves Cuilleron • Delas (Seigneur de Maugiron) • Duclaux • Des Entrefaux • Jean-Michel Gerin (Champin le Seigneur) • E Guigal • Paul Jaboulet Aîné • Joseph Jamet • Patrick Jasmin • Stéphane Montez • Michel Ogier • Stéphane Pichat • René Rostaing • Jean-Michel Stéphan • Vidal-Fleury (La Chatilonne) • François Villard (La Brocarde)*

COTEAUX DE DIE AOC

This appellation was created in 1993 to soak up wines from excess Clairette grape production that were formerly sold under previous sparkling-wine appellations.

WHITE Perhaps it is unfair to prejudge a wine that has had so little time to establish itself, but it is hard to imagine a quality wine made from this grape. In effect, Coteaux de Die is the still-wine version of Crémant de Die. It is easy to see why such a bland dry wine needs bubbles to lift it, yet so few make the transition successfully.

🍇 *Clairette*

🍷 *Upon purchase*

✓ *Jean-Claude Raspail*

CÔTES DU RHÔNE AOC

Although generic to the entire region, relatively little Côtes du Rhône (CDR) is made in this district, although 47 communes north of Montélimar are be entitled to do so. The only difference in technical criteria from those in the south is that Syrah may replace Grenache as the principal red wine variety.

CRÉMANT DE DIE AOC

This dry sparkling wine was introduced in 1993 to replace Clairette de Die Mousseux, which was phased out by January 1999. Like the old Clairette de Die Mousseux appellation, Crémant de Die AOC must be made by the traditional method; however, whereas the former (like Clairette de Die) could include up to 25 per cent Muscat à Petit Grains, the Crémant de Die must be made entirely from the Clairette variety. A maximum of 15 grams per litre of residual sugar is permitted for Crémant, whereas Clairette has at least 35 grams per litre.

SPARKLING WHITE Despite the reduction to only two sparkling wine AOCs distinguished by method, *cépage*, and levels of residual sugar, both Crémant de Die and Clairette de Die are regarded as neutral and lacklustre. Mostly sold within France as inexpensive sparkling wine, they have yet to more than dent export markets as was hoped for by simplifying the AOCs.

🍇 *Clairette*

🍷 *1–3 years*

✓ *Carod*

CROZES-HERMITAGE AOC

Crozes-Hermitage (also known as Crozes-Ermitage) is made from a relatively large area surrounding Tain, with the cooperative Cave de Tain responsible for 40 per cent, so the quality is very variable. Yet good Crozes-Hermitage will always be a great bargain. The area has provided affordable land for independent producers, as well as for those established Northern Rhône names wanting to expand production.

RED These well-coloured, full-bodied wines are similar to Hermitage, but they are generally less intense and have a certain smoky-rustic-raspberry flavour that only deepens into blackcurrant in the hottest years. The finest wines do, however, have surprising finesse, and make fabulous bargains.

🍇 *Syrah, plus up to 15% Roussanne and Marsanne*

🍷 *6–12 years (8–20 years for top wines and great years)*

WHITE These dry white wines are improving and gradually acquiring more freshness, fruit, and acidity.

🍇 *Roussanne, Marsanne*

🍷 *1–3 years*

✓ *Aléofane • Belle • Les Bruyères• Chapoutier ❸ (Les Varonnières) • Yann Chave (Tête de Cuvée) ⓞ • Les Chenets • CV des Clairmonts (Pionniers) • du Colombier (Gaby) • Combier ⓞ (Clos des Grives) • Emmanuel Darnaud • des Entrefaux • Fayolle • Alain Graillot • Hauts-Chassis • Paul Jaboulet Aîné (Thalabert) • les Lises • des Remizières (Emilie) • David Reynaud ⓞ • Gilles Robin • Marc Sorrel*

DRÔME IGP

See p337.

HERMITAGE AOC

Hermitage (also known as Ermitage or L'Ermitage) is one of the great classic French red wines, produced entirely from Syrah grapes in virtually all cases, although a small amount of Marsanne and Roussanne may be added. The vines are grown on a magnificent south-facing slope overlooking Tain. Traditional producers, like Jean-Louis Chave and Jaboulet, blend from different *climats* on Hermitage Hill, whereas some source enough fruit from each plot to vinify separately, which from west to east are: la Varogne, les Bessards, l'Hermite, le Meal, les Gréffieux, Beuame, Maison Blanche, Rocoule, Péléat, les Diogniéres, les Murets, lhomme, la Criox, and Torres et les Garennes.

RED These wines have a deep and sustained colour, a very full body, and lovely, plummy, lip-smacking, spicy, silky, violety, blackcurrant fruit. A truly great Hermitage has boundless finesse, despite the weighty flavour.

🍇 *Syrah, plus up to 15% Roussanne and Marsanne*

🍷 *12–30 years*

WHITE Big, rich, dry white wines with a full, round, hazelnut and dried-apricot flavour. They have continued to improve since the millennium but remain a niche curiosity, representing less than a quarter of the production from a location famed since Roman times. M Chapoutier is particularly recommended for white, as well as red.

🍇 *Roussanne, Marsanne*

🍷 *6–12 years*

✓ *M Chapoutier ❸ (Ermite, Méal, L'Orée, Pavillon) • Jean-Louis Chave1 • Yann Chave ⓞ • Emmanuel Darnaud • Delas (Les Bassards) • Alain Graillot • E Guigal (Ex Voto) • Paul Jaboulet Aîné (La Chapelle) • des Remizières1 (Emilie) • Marc Sorrel (Gréal) • Tardieu-Laurent (Vieilles Vignes)*

HERMITAGE VIN DE PAILLE AOC

In 1974 Gérard Chave made a *vin de paille* for "amusement". Chapoutier has since made several vintages of Hermitage *vin de paille* on a commercial basis – or as near commercial as you can get with this style of wine – and other producers include Michel Ferraton, Jean-Louis Grippat, and Guigal. Even the local *coopérative* has churned some out. At one time all Hermitage *blanc* was, in effect, *vin de paille*, a style that dates back in this locality to at least 1760. (Marc Chapoutier apparently drank the last wine of that vintage in 1964 when it was 204 years old.) The traditional *vin de paille* method in Hermitage is not as intensive as the one revived relatively recently in Alsace. There the grapes are dried out over straw beds until more than 90 per cent of the original juice has evaporated before they are pressed – even Chave's legendary 1974 was made from grapes bearing as much as a third of their original juice, but some of the results have been equally stunning.

WHITE Some *vins de paille* are rich and raisiny, while others (the best) have a crisp, vivid freshness about them, with intense floral-citrus aromas and a huge, long, honeyed aftertaste. After his famed 1974 vintage, Chave made a 1986 and at least a couple more vintages since. His *vin de paille* is reputed to be the greatest but is practically unobtainable and very expensive, even if made in the most commercial quantity. Jean-Louis Grippat made a *vin de paille* in 1985 and 1986, neither of which I have had the privilege to taste, but John Livingstone-Learmonth describes them as "not quite like late-harvest Gewurztraminer" in *The Wines of the Rhône*.

🍇 *Roussanne, Marsanne*

🍷 *Up to 30 years*

✓ *Chapoutier ❸*

SAINT-JOSEPH AOC

At last, some truly exciting wines are being produced under this appellation. Saint-Joseph has now replaced Cornas as the bargain seeker's treasure trove. Jean-Louis Chave produces a Saint-Joseph Domaine wine from land which, like in Hermitage, has been in his family since the 15th century, as well as the Offerus *cuvée* which includes some grapes purchased from his neighbours in Mauves and elsewhere, who are more focussed on farming apricots.

RED The best wines are dark, medium- to full-bodied, with intense blackberry and blackcurrant fruit aromas and plenty of soft fruit, whereas mediocre Saint-Joseph remains much lighter-bodied with a pepperiness more reminiscent of the Southern Rhône.

🍇 *Syrah, plus up to 10% Marsanne and Roussanne*

🍷 *3–8 years*

WHITE At their best, clean, rich, and citrus-resinous dry wines. Courbis is particularly recommended for white, as well as red.

🍇 *Marsanne, Roussanne*

🍷 *1–3 years*

✓ *Aléofane • De Champal • Chapoutier ❸ (Les Granits) • Jean-Louis Chave • Courbis (Les Roys) • Pierre Finon • Pierre Coursodon (Le Paradis Saint-Pierre) • Yves Cuilleron • Pierre Gaillard • Gonan1 (Les Oliviers) • E Guigal (Vignes de la Hospices) • Stéphane Montez • Vincent Paris • Tardieu-Laurent (Vieilles Vignes) • Du Tunnel • François Villard (Reflet)*

SAINT-PÉRAY AOC

This white-wine-only village appellation is situated immediately south of Cornas. Even smaller than Hermitage but producing similar volumes. The area is gradually increasing its vineyard areas and reputation though little more than 10 per cent is exported.

WHITE These wines are influenced by winds coming off the Massif Central to the north. There is a good topography with complex soils of sedimentary limestone, as well as granite, sand, and alluvial clay deposited by alpine glaciers. Marsanne usually dominates the still wines, which are pale in colour, with subtle fresh aromas and floral, mineral notes when young, tending to honeyed quince, baked apple, acacia, spice, nuts, and dried apricot with age. The emphasis is on purity and elegance, though Alain Voge, Bernard Gripa, and Du Tunnel produce some wines with more power and opulence from older Roussanne vines.

🍇 *Marsanne, Roussanne*

🍷 *1–3 years*

✓ *J-F Chaboud • Chapoutier • Bernard Gripa • Jean Lionnet • Du Tunnel • Alain Voge*

SPARKLING WHITE Traditional-method sparklers have been made here since the 1820s and were popular in the royal courts of Europe throughout the 19th century up until the 1960s. Richard Wagner consumed 100 bottles of sparkling St-Péray from Chapoutier while composing Parsifal, which he (allegedly) failed to pay them for.

🍇 *Marsanne, Roussanne*

✓ *Biguet • Chapoutier (La Muse de RW) • Du Tunnel • Alain Voges (Les Bulles D'Alain)*

THE RUINS OF CHÂTEAU DE CRUSSOL
The château overlooks the Rhône Valley and Saint-Péray, a white and sparkling wine appellation.

The Southern Rhône

The mellow warmth of the Grenache is found in most Southern Rhône wines, yet this region is, in fact, a blender's paradise, with a choice of up to 23 different grape varieties.

The Southern Rhône is a district dominated by herbal scrubland, across which blows a sweet, spice-laden breeze. This is a far larger district than the slender northern *côtes,* and its production is, not unnaturally, much higher. Allowing the north a generous 10 per cent of the generic Côtes du Rhône appellation, the Southern Rhône still accounts for a staggering 95 per cent of all the wines produced in the region.

WINES OF THE MIDI OR PROVENCE?

At least half of the Southern Rhône is in what was once called the Midi, an area generally conceded to cover the *départements* of the Aude, Hérault, and Gard, or more broadly the Languedoc. This is never mentioned by those intent on marketing the Rhône's image of quality, because the Midi was infamous for its huge production of *vin ordinaire.* The Rhône River marks the eastern border of the Midi, and its most famous appellations are geographically part of Provence. Yet, viticulturally, these areas do not possess the quasi-Italian varieties that dominate the vineyards of Provence and might, therefore, be more rationally defined as a high-quality extension of the Midi. It is telling that Costières de Nîmes defected from Eastern Languedoc in 2004 forming a southwestern extension of the Southern Rhône, claiming stronger synergies in terms of climate, soil, topography, and tradition with the latter rather than the former region.

CLIMATE CHANGE

Rising temperatures to a point have helped increase quality, but the tendency for higher alcohol wines, particularly made from Grenache, could become problematic. Alcohol levels above 15 per cent are against consumer trends, and appellation rules in the south require Grenache as the major component. Earlier picking and techniques to reduce alcohol can adversely impact flavour, and if irrigation becomes necessary, this will add to the difficulties. Time will tell how winemaking responds to these challenges.

FACTORS AFFECTING TASTE AND QUALITY

LOCATION
The Southern Rhône starts at Viviers, south of Valence, and runs south to Avignon.

CLIMATE
The Southern Rhône's climate is unmistakably Mediterranean, and its vineyards are far more susceptible to sudden change and abrupt, violent storms and are getting warmer than those of the Northern Rhône.

ASPECT
The terrain in the south is noticeably Mediterranean, with olive groves, lavender fields, and herbal scrub amid rocky outcrops.

SOIL
The limestone outcrops that begin to appear in the south of the Northern Rhône become more prolific and are often peppered with clay deposits, while the topsoil is noticeably stonier. Châteauneuf-du-Pape is famous for its creamy-coloured drift boulders, which vary according to location. Stone-marl soils persist at Gigondas, and there is weathered-grey sand in Lirac, Tavel, and Chusclan. The soils also incorporate limestone rubble, clay-sand, stone-clay, calcareous clay, and pebbles.

VITICULTURE & VINIFICATION
The vines are traditionally planted at an angle leaning into the wind so the mistral can blow them upright when they mature. The south is a district where blends reign supreme, but pure varietals are gaining ground. Traditional vinification methods are used on some estates, but modern techniques are common.

GRAPE VARIETIES
Primary varieties: Carignan, Cinsault, Grenache, Mourvèdre, Muscat Blanc à Petits Grains, Muscat Rosé à Petits Grains
Secondary varieties: Aubun, Bourboulenc, Calitor, Camarèse (Brun Argenté), Clairette, Clairette Rosé, Counoise, Gamay, Grenache Blanc, Grenache Gris, Macabéo, Marsanne, Mauzac, Muscardin, Oeillade, Picpoul Blanche, Picpoul Noir, Pinot Blanc, Pinot Noir, Rolle (Vermentino), Roussanne, Syrah, Terret Noir, Ugni Blanc, Vaccarèse (Brun Argenté), Viognier

VINE-GROWING AREAS OF THE SOUTHERN RHÔNE
(see also p297)
The wider southern part of the Rhône Valley area stretches its fingers down towards Provence and eastwards to the Alps.

THE APPELLATIONS OF
THE SOUTHERN RHÔNE

ARDÈCHE IGP
See p334.

BEAUMES-DE-VENISE AOC
Although primarily famous for its delectable, sweet Muscat wine, Beaumes-de-Venise also produces pleasant red wines, which were sold as Beaumes-de-Venise Côtes du Rhône until achieving single *cru* status in 2005. Dry white and rosé wines are produced under the generic AOC.

RED Good peppery-raspberry fruit flavour; the best have a certain plumpness and a touch of oak.

🍷 *At least 50% Grenache, plus 25–50% Syrah, with an optional choice of any other CDR grape varieties (a maximum of 10% white grapes)*

🍷 *2–7 years*

✓ *Cassan • CV de Beaumes-de-Venise (Chapelle Notre Dame d'Aubune, Les Garrigues d'Eric Beaumard) • Durban*

CAIRANNE AOC
When Vacqueyras was promoted to single *cru* status in 1990, Cairanne was clearly the top-performing village remaining in the Côtes du Rhône Villages appellation, and after launching an application in 2008 had to wait until 2016 for INAO's approval and see another four villages receive individual recognition before finally being awarded *cru* status in its own right in 2018.

RED An excellent source of rich, warm, and spicy red wines that seem to integrate better with oak than other Côtes du Rhône wines.

🍷 *At least 50% Grenache, plus 25–50% Syrah, with an optional choice of any other CDR grape varieties (a maximum of 10% white grapes)*

🍷 *2–7 years.*

✓ *d'Aeria • De l'Ameillaud • Daniel et Denis Alary • Brusset • Des Escaravailles • Les Grand Bois (Mireille) • Les Hautes Cances • De l'Oratoire St-Martin • Marcel Richaud (L'Ebrescade)*

WHITE Approximately 5% of Carianne production is white with citrus, white-fruit, and floral notes.

🍷 *At least 80% of a white blend has to be made up from Grenache Blanc, Clairette Blanche, Roussanne, Marsanne, Bourboulenc Blanc, and Viognier.*

🍷 *1–4 years*

CHÂTEAUNEUF-DU-PAPE AOC
The name Châteauneuf-du-Pape dates from the time of the Avignon Papacy in the 14th century. The appellation is well known for its amazingly stony soil, which at night reflects the heat stored during the day, but the size, type, depth, and distribution of the stones vary enormously, as does the aspect of the vineyards. These variations, plus the innumerable permutations of the 13 grape varieties that may be used, account for the diversity of its styles. In the early 1980s some growers began to question the hitherto accepted concepts of *encépagement* and vinification; winemaking in Châteauneuf-du-Pape is still in an evolutionary state. The steady decline of the traditionally dominant Grenache has speeded up as more growers are convinced of the worth of the Syrah and Mourvèdre. The Cinsault and Terret Noir are still well appreciated, and the Counoise is beginning to be appreciated for its useful combination of fruit and firmness. The use of new oak is under experimentation, and it already seems clear that it is better suited to white wine than red.

The regulations for this appellation have a unique safeguard designed within them to ensure that only fully ripe grapes in the healthiest condition are utilized in the production of its wines: between 5 and 20 per cent of the grapes harvested within the maximum yield for this AOC are rejected and may be used to make only *vin de table*. This process of exclusion is known as *le rapé*.

RED Due to the variations of *terroir* and almost limitless permutations of *encépagement*, it is impossible to describe a typical Châteauneuf-du-Pape. There are nonetheless two categories: the traditional, full, dark, spicy, long-lived style and the modern, easy-drinking Châteauneuf-du-Pape, the best of which are unashamedly upfront and brimming with lip-smacking, juicy fruit. Both are warmer and spicier than the greatest wines of Hermitage and Côte Rôtie.

🍷 *Grenache, Syrah, Mourvèdre, Picpoul, Terret Noir, Counoise, Muscardin, Vaccarèse, Picardan, Cinsault, Clairette, Roussanne, Bourboulenc*

🍷 *6–25 years*

WHITE Early harvesting has reduced sugar levels and increased acidity, and modern vinification techniques have encouraged a drop in fermentation temperatures, so these wines are not as full as those previously produced. They can still be very rich, albeit in a more opulent, exotic-fruit style, with a much fresher, crisper finish. The very best white Châteauneuf-du-Pape is generally agreed to be Château de Beaucastel Vieilles Vignes. Benedetti is particularly recommended for white, as well as red.

🍷 *Grenache, Syrah, Mourvèdre, Picpoul, Terret Noir, Counoise, Muscardin, Vaccarèse, Picardan, Cinsault, Clairette, Roussanne, Bourboulenc*

🍷 *1–3 years (4–6 years in exceptional cases)*

✓ *Paul Autard (Côte Ronde) • Barville • Château de Beaucastel ⓞ • De Beaurenard (Boisrenard) • Benedetti • Bois de Boursan (Felix) • Bousquet des Papes • La Boutinière • Clos du Caillou • Les Cailloux (Centenaire) • Réserve des Célestins • Chapoutier ⓑ (Barbe Rac, Croix de Bois) • De la Charbonnière • Gérard Charvin • De Cristia • Font de Michelle • Château de la Gardine • Grand Veneur • De la Janasse • Mark Kreydenweiss ⓑ • Lafond Roc-Epine • Patrick Lesec • De Marcoux ⓑ • Font de Michelle (Etienne Gonet) • Monpertuis • Château Mont-Redon • De la Mordorée (Reine des Bois) • De Nalys • Château la Nerthe ⓞ • De Panisse • Clos des Papes • Du Pegaü • Château Rayas • Roger Sabon • Clos St-Jean • Clos St Michel (Grand Clos) • De la Solitude • Tardieu-Laurent (Vieilles Vignes) • Pierre Usseglio • De la Vieille Julienne • Du Vieux Télégraphe • De Villeneuve ⓑ*

CLAIRETTE DE BELLEGARDE AOC
On the edge of the Carmargue, this small appelation of 40 hectares (99 acres) southeast of Nîmes has warm stoney soils suited to Clairette Blanche. AOC status since 1949, there are 15 growers.

WHITE Dry whites are produced here.

🍷 *100% Clairette Blanche*

🍷 *1 – 3 years*

COSTIÈRES DE NÎMES AOC
The most souterhly AOC in Rhône was previously considered part of Langeudoc and is also known as the Rhône Delta. Mediterranean climate.

RED and ROSÉ The wines of this AOC are mainly red (60 per cent) and rosé (35 per cent).

🍷 *A minimum of 60% from Grenache, Mourvèdre, or Syrah; Carignan and Cinsault also allowed*

WHITE These contain a minimum of 11.5 per cent alcohol and have citrus, white-fruit, and floral notes.

🍷 *Grenache Blanc, Roussanne, Marsanne, Bourboulenc, Clairette Blanche, Vermentino and Viognier*

🍷 *1 – 3 years*

COTEAUX DE PIERREVERT AOC
This appellation consists of some 400 hectares (990 acres) and was upgraded to AOC status in 1998.

RED Uninspiring wines with little original character to commend them.

🍷 *Carignan, Cinsault, Grenache, Mourvèdre, Oeillade, Syrah, Terret Noir*

🍷 *2–5 years*

WHITE Unspectacular, light, dry white wines with more body than fruit.

🍷 *Clairette, Marsanne, Picpoul, Roussanne, Ugni Blanc*

🍷 *1–3 years*

ROSÉ Well-made wines with a blue-pink colour and a crisp, light, fine flavour.

🍷 *Carignan, Cinsault, Grenache, Mourvèdre, Oeillade, Syrah, Terret Noir*

🍷 *1–3 years*

✓ *La Blaque (Réserve) • Vignobles de Régusse (Bastide des Oliviers)*

CÔTES DU RHÔNE AOC
A generic appellation that covers the entire Rhône region, although the vast majority of wines are actually produced in the Southern Rhône. There are some superb Côtes du Rhônes, but there are also some disgusting wines. The quality and character varies to such an extent that it would be unrealistic to attempt any generalized description. The red and rosé may be sold as *primeur* or *nouveau* from the third Thursday of November following the harvest, although the reds can be sold only unbottled as *vin de café*. Red, white, and rosé wines can go on sale from 1 December without mention of *primeur* or *nouveau*.

RED The red wines are the most successful of this basic appellation and often represent the best choice when dining in a modest French restaurant with a restricted wine list. Top *négociants* like Guigal consistently produce great-value Côtes du Rhône, but those below are all capable of very special quality.

🍷 *A minimum of 40% Grenache (except in the Northern Rhône, where Syrah may be considered the principal variety), plus Syrah and/or Mourvèdre, with up to 30% in total of any of the following: Camarèse, Carignan, Cinsault, Clairette Rosé, Counoise, Grenache Gris, Muscardin, Vaccarèse, Picpoul Noir, Terret Noir, and a maximum of 5% in total of any of the following white varieties: Clairette, Grenache Blanc, Marsanne, Picpoul, Roussanne, Bourboulenc, Ugni Blanc, Viognier*

🍷 *3–10 years*

WHITE These wines have improved tremendously in the past 5 to 10 years and continue to do so in encouraging fashion.

🍷 *Bourboulenc, Clairette, Grenache Blanc, Marsanne, Roussanne, and Viognier, and up to 20% Ugni Blanc and/or Picpoul*

🍷 *1–3 years*

ROSÉ Many of the best rosés are superior in quality to those from more expensive Rhône appellations.

🍷 *A minimum of 40% Grenache (except in the Northern Rhône, where Syrah may be considered the principal variety), plus Syrah and/or Mourvèdre, with up to 30% in total of any of the following: Camarèse, Carignan, Cinsault, Clairette Rosé, Counoise, Grenache Gris, Muscardin, Vaccarèse, Picpoul Noir, Terret Noir, and a maximum of 5% in total of any of the following white varieties: Clairette, Grenache Blanc, Marsanne, Picpoul, Roussanne, Bourboulenc, Ugni Blanc, Viognier*

🍷 *1–3 years*

✓ *Max Aubert (de la Présidente) • Castan • Coudoulet de Beaucastel • Cuilleron-Gaillard-Villard • Fond Croze • Château de Fontsegune • Grand Veneur • Clos de l'Hermitage • Paul Jaboulet Aîné • Clos des Magnaneraie • Clos Martin • Château de Montfaucon • Château Mont-Redon • Rigot • Des Roches Fortes (Prestige) • Xavier Vignon*

CÔTES DU RHÔNE VILLAGES AOC

Compared with the generic Côtes du Rhône, the Villages wines generally have greater depth, character, and quality. The area covered by the appellation is entirely within the Southern Rhône. If the wine comes from one commune only, then it has the right to append that name to the Côtes du Rhône appellation. Gigondas, Cairanne, Chusclan, and Laudun were the original Côtes du Rhône Villages, out of which some have attained their own AOCs (Gigondas in 1971, Vacqueyras in 1990, Beaumes-de-Venise and Vinsobres in 2005, Rasteau in 2010, and Cairanne in 2018). Other villages have been added, and the latest are Gadagne in 2012; Suze-la-Rousse, Sainte Cecile, and Vaison La Romaine with effect for the 2016 harvest onwards; and Saint-Andéol in 2018. The wines recommended below are for the Côtes du Rhône Villages appellation without any specified village. Following in this list you will find separate entries for each of the 21 villages that are currently allowed to add their name to this AOC.

RED These wines are mostly excellent.

🍷 *A minimum of 50% Grenache, plus a minimum of 20% Syrah and/or Mourvèdre, and a maximum of 20% in total of Camarèse, Carignan, Cinsault, Clairette Rosé, Counoise, Grenache Gris, Muscardin, Vaccarèse, Picpoul Noir, Terret Noir*

🍷 *3–10 years*

WHITE These wines are improving – Vieux Manoir du Frigoulas is the best.

🍷 *Bourboulenc, Clairette, Grenache Blanc, Marsanne, Picpoul, Roussanne, Viognier*

🍷 *1–3 years*

ROSÉ These wines can be very good.

🍷 *A minimum of 50% Grenache, plus a minimum of 20% Syrah and/or Mourvèdre, and up to 20% in total of Camarèse, Carignan, Cinsault, Clairette Rosé, Counoise, Grenache Gris, Muscardin, Vaccarèse, Picpoul Noir, Terret Noir, and a maximum of 20% in total of any of the following white varieties: Bourboulenc, Clairette, Grenache Blanc, Marsanne, Picpoul, Roussanne, Viognier; and a maximum of 20% Ugni Blanc and/or Picpoul*

🍷 *1–3 years*

✔ *Les Aphillanthes ❶ • Bouche • Le Clos du Caillou • Château la Gardine • Château d'Hugues (L'Orée des Collines) • De la Janasse (Terre Argile) • Mas de Libian ❶ • du Petit-Barbras (Le Chemin de Barbras Sélection) • La Réméjeanne (Eglantiers) • Dominique Rocher (Monsieur Paul) • St-Anne • Saint Estève d'Uchaux (Vieilles Vignes) • CV de St-Hilaire d'Ozilhan (Saveur du Temps) • Viret (Emergence)*

CHUSCLAN
CÔTES DU RHÔNE VILLAGES AOC
Red and rosé only

Chusclan may also be made from vines growing in Bagnols-sur-Cèze, Cadolet, Orsan, and St-Etienne-des-Sorts. These villages are situated just north of Lirac and Tavel, two famous rosé appellations, and make an excellent rosé. Most of the wines are red, however, and are produced in a good, quaffing style.

✔ *CV de Chusclan (Esprit de Terroir) • Château Signac*

GADAGNE
CÔTES DU RHÔNE VILLAGES AOC

To give this village its full name of Châteauneuf-de-Gadagne begins to explain how it has long been overshadowed by the more famous Châteauneuf on the other side of Avignon. It has been a village label since 1997, with the right to use its own name since 2012. Red wine only from Grenache, with Syrah and Mouvèdre.

✔ *Domaine de Bois de Saint-Jean • Chateau de Fontsegugne • Clos des Saumanes*

LAUDUN CÔTES DU RHÔNE VILLAGES AOC
Red, white, and rosé

Laudun can be made from vines grown outside the village in St-Victor-Lacoste and Tresques. Excelling in fine, fresh, and spicy red wines, Laudun also makes the best white wines in the CDRV appellation and a small amount of quite delightful rosé wines.

✔ *Château St-Maurice*

MASSIF D'UCHAUX
CÔTES DU RHÔNE VILLAGES AOC
Red only

A CDRV-named village since 2005. Its vines grow on sandstone hillsides at 100 to 280 metres (330 to 920 feet) at Massif d'Uchaux and five surrounding communes, including Mondragon, the first wines of which were recorded in 1290.

✔ *Château de Fonsalette • Château du Grand Moulas*

PLAN DE DIEU
CÔTES DU RHÔNE VILLAGES AOC
Red only

A CDRV-named village since 2005, Plan de Dieu extends over four communes, where grapes grow on gravelly terraces under the shadow of Mont Ventoux, an area that was originally planted with vines by the Knights Templar.

✔ *Bernard Latour*

PUYMERAS
CÔTES DU RHÔNE VILLAGES AOC
Red only

A CDRV-named village since 2005. The wines are made from grapes grown in Puymeras and four surrounding communes, including two in the neighbouring *département* of Drôme.

ROAIX CÔTES DU RHÔNE VILLAGES AOC
Red, white, and rosé

Neighbouring the vineyards of Séguret, Roaix produces a similar dark-coloured red wine that requires two or three years in bottle to mellow. A little rosé is also produced.

✔ *CV de Roaix-Séguret*

ROCHEGUDE
CÔTES DU RHÔNE VILLAGES AOC
Red, white, and rosé

Only the red wine of Rochegude may claim the CDRV appellation. The local *coopérative* wine is good quality, well coloured, soft, and plummy. The white and rosé are sold as generic Côtes du Rhône.

✔ *CV de Rochegude*

ROUSSET-LES-VIGNES
CÔTES DU RHÔNE VILLAGES AOC
Red, white, and rosé

The neighbouring villages of Rousset-les-Vignes and St-Pantaléon-les-Vignes possess the most northerly vineyards of this appellation, close to the verdant Alpine foothills. With the coolest climate of all the Côtes du Rhône villages, the wines are light, but soft and quaffable.

✔ *La Bouvade*

SABLET CÔTES DU RHÔNE VILLAGES AOC
Red, white, and rosé

This village's soft, fruity, and quick-maturing red and rosé wines are consistent in quality and are always good value.

✔ *De Piaugier (Montmartel)*

SAINT-ANDÉOL
CÔTES DU RHÔNE VILLAGES AOC
Red only

The village most recently (April 2018) granted the right to append its own name to Côte de Rhône Village for wines grown and produced in the communes of Bourg-Saint-Andéol, Saint-Just-d'Ardèche, Saint-Marcel-d'Ardèche, and Saint-Martin-d'Ardèche.

✔ *Domaine Sainte Andéol*

SAINTE-CÉCILE
CÔTES DU RHÔNE VILLAGES AOC
Red only

Sainte-Cécile-le-Vignes was raised to named village status with effect from the 2016 harvest omward, and the AOC comprises five communes in the Vaucluse *département*, along with the commune of Sainte-Cécile-le-Vignes itself. The others are Suze-la-Rousse, Tulette, Sérignan-du-Comtat, and Travaillan.

✔ *Domaine de la Berthete*

SAINT-GERVAIS
CÔTES DU RHÔNE VILLAGES AOC
Red, white, and rosé

The valley vineyards of St-Gervais are not those of the great river itself, but belong to the Céze, one of its many tributaries. The red wines are deliciously deep and fruity, and the whites are fresh and aromatic with an excellent, crisp balance for wines from such a southerly location.

✔ *Clavel (L'Étoile du Berger) • CV de St-Gervais (Prestige) • Sainte-Anne*

SAINT-MAURICE
CÔTES DU RHÔNE VILLAGES AOC
or ST-MAURICE-SUR-EYGUES
CÔTES DU RHÔNE VILLAGES AOC
Red, white, and rosé

Light, easy-drinking red and rosé wines. Production is dominated by the local *coopérative*.

✔ *CV des Coteaux St-Maurice*

SAINT-PANTALÉON-LES-VIGNES
CÔTES DU RHÔNE VILLAGES AOC
Red, white, and rosé

The neighbouring villages of St-Pantaléon-les-Vignes and Rousset-les-Vignes possess the most northerly vineyards of this appellation. With the coolest climate of all the CDRV, the wines are light, but soft and quaffable. Production is monopolized by the local *coopérative*, which has yet to make anything special.

SÉGURET
CÔTES DU RHÔNE VILLAGES AOC
Red, white, and rosé

Séguret produces red wine that is firm and fruity with a good, bright, deep colour. A little quantity of white and rosé is also made, but they seldom excel.

✔ *De l'Amauve • De Mourchon (Grande Réserve) • Eric Texier*

SINARGUES
CÔTES DU RHÔNE VILLAGES AOC
Red only

A CDRV-named village since 2005, for the most southerly-named village in this appellation, encompassing vines grown in Domazan, Estézargues, Rochefort du Gard, and Saze to the east of Avignon

SUZE-LA-ROUSSE
CÔTES DU RHÔNE VILLAGES AOC
Red only

Suze-la-Rousse covers 2,600 hectares (6,425 acres) in four communes and earned named village status in 2017. The town is overlooked by a medieval fortress housing the University of Wine, which was established in 1978. Varied calcareous soils with stony marl, clay, and sandy loam produce powerful, warm red wines.

VAISON LA ROMAINE
CÔTES DU RHÔNE VILLAGES AOC
Red only

In northern Vaucluse, the vineyards of Vaison la Romaine are composed of 70 per cent Grenache and 30 per cent Syrah. The wines have intense fruity aromas with spicy notes balancing freshness and generosity.

VALREAS CÔTES DU RHÔNE VILLAGES AOC

Red, white, and rosé

These are fine red wines with plenty of fruit flavour. A little rosé is also made.

✓ Clos Petite Bellane • Comte Louis de Clermont-Tonnerre • Des Grands Devers • Le Val des Rois

VISAN CÔTES DU RHÔNE VILLAGES AOC

Red, white, and rosé

These red wines have good colour and true *vin de garde* character. Fresh, quaffing white wines are also made.

✓ De la Coste Chaude • Olivier Cuilleras

CÔTES DU VIVARAIS AOC

The *côtes* of Vivarais look across the Rhône river to the *coteaux* of Tricastin. Its best *cru* villages (Orgnac, St-Montant, and St-Remèze) may add their names to the Côtes du Vivarais.

RED These light, quaffing reds are by far the best wines in the district.

🍷 A minimum of 90% in total of Syrah (minimum 40%) and Grenache (minimum 30%), with Cinsault and Carignan optional

🍾 1–3 years

WHITE These wines were always rather dull and disappointing but have improved in recent years.

🍷 At least two varieties from Clairette, Grenache Blanc, and Marsanne, with no variety accounting for more than 75%

🍾 1–3 years

ROSÉ Pretty pink, dry wines that can have a ripe, fruity flavour.

🍷 At least two varieties from Syrah, Grenache, and Cinsault, with no variety accounting for more than 80%

🍾 1–3 years

✓ Les Chais du Vivarais

DRÔME IGP

See p337.

DUCHÉ D'UZÈS AOC

This appellation, situated at the extreme south of the Rhône Valley at the crossroads of the Cévennes, the Languedoc, and Provence, has a Mediterranean climate. It was granted AOC status in 2013.

RED Fruity, approachable wines.

🍷 A minimum of 40% Syrah, and 20% Grenache, plus Carignan, Cinsaut, and Mourvèdre

🍾 1–3 years

WHITE The best whites have soft white peach and gentle floral character balanced by refreshing acidity.

🍷 A minimum of 40% Viognier and a minimum 30% Grenache Blanc, plus a total minimum of 20% Roussanne, Marsanne, and Vermentino. Clairette and Ugni Blanc are also permitted.

🍾 Upon purchase

ROSÉ Pretty pink, dry wines that can have a ripe, fruity flavour.

🍷 A minimum of 50% Grenache and a minimum of 20% Syrah, plus Carignan, Cinsaut, and Mourvèdre

🍾 Upon purchase

GIGONDAS AOC

Gigondas produces some of the most underrated red wines in the Rhône Valley.

RED The best have an intense black-red colour with a full, plummy flavour.

🍷 A maximum of 80% Grenache, plus at least 15% Syrah and Mourvèdre, and a maximum of 10% (in total) of Clairette, Picpoul, Terret Noir, Picardan, Cinsault, Roussanne, Marsanne, Bourboulenc, Viognier, Counoise, Muscardin, Vaccarèse, Pinot Blanc, Mauzac, Pascal Blanc, Ugni Blanc, Calitor, Gamay, and Camarèse

🍾 7–20 years

ROSÉ Good-quality, dry rosé wines.

🍷 A maximum of 80% Grenache and a maximum of 25% in total of Clairette, Picpoul, Terret Noir, Picardan, Cinsault, Roussanne, Marsanne, Bourboulenc, Viognier, Counoise, Muscardin, Vaccarèse, Pinot Blanc, Mauzac, Pascal Blanc, Ugni Blanc, Calitor, Gamay, and Camarèse

🍾 2–5 years

✓ Des Bosquets • La Bouissière • Brusset (Le Grand Montmirail, Les Hauts des Montmirail) • du Cayron • de la Gardette • Les Goubert (Florence) • Grapillon d'Or (Elevé en Vieux Foudres) • De Longue-Toque • Gabriel Meffre (Château Raspail) • De Montvac • Moulin de la Gardette • L'Oustau Fauquet (Secret de la Barrique) • Les Paillières • La Roubine • Saint Cosme • Saint-Damien (Les Souteyrades) • De St-Gayan (Fontmaria) • Sainte Anne • Santa Duc • La Tourade

GRIGNAN-LES-ADHÉMAR AOC

This was Coteaux du Tricastin until 2010, when producers changed the name for fear of bad publicity following a uranium leak from the Tricastin nuclear facility. The Comte de Grignan, from a local land-owning family with ancestry dating from the first crusade, was a 17th-century military leader under Louis XIV.

RED Very good wines, especially the deeply coloured, peppery Syrahs that are a delight after a few years in bottle. These wines may be sold as *primeur* or *nouveau* from the third Thursday of November following the harvest.

🍷 A maximum combined or individually of 80% Grenache and Syrah, plus Carignan, Cinsault, Marsanne, Marselan, Mourvèdre, Roussanne, or Viognier

🍾 2–9 years

WHITE A wider variety of grapes has enabled some producers to make richer, crisper, and more interesting wines.

🍷 Grenache Blanc, Clairette, Bourboulenc, Marsanne, Roussanne, Viognier

ROSÉ A small production of fresh and fruity dry rosé that occasionally yields an outstandingly good wine. Made using *saignée* method of bleeding juice from a vat in which red wine is being made, these wines may be sold as *primeur* or *nouveau* as from the third Thursday of November following the harvest.

🍷 A maximum combined or individually of 80% Grenache and Syrah, plus Carignan, Cinsault, Marsanne, Marselan, Mourvèdre, Roussanne, or Viognier

🍾 Upon purchase

✓ Bour • Des Estubiers • De Grangeneuve • De Montine • Saint-Luc

LIRAC AOC

This appellation was once the preserve of rosé, but the production of red is now on the increase.

RED In really great years, the Syrah and Mourvèdre can dominate despite the quite small quantities used, which produces a more plummy wine with silky-spicy finesse.

🍷 A minimum of 40% Grenache and 25% in total of Syrah and Mourvèdre, plus up to 10% Carignan and no limit on the amount of Cinsault (although this used to be restricted to 20%, and I suspect this relaxation is due to a clerical error when the regulations were rewritten in 1992)

🍾 4–10 years

WHITE A fragrant, dry white wine, the best of which has improved since 1992, when Marsanne, Roussanne, and Viognier were permitted for use in this appellation. The cheapest wines have declined in quality because the amount of Clairette allowed has been doubled. Macabéo and Calitor are no longer permitted.

🍷 A maximum of 60% each of Bourboulenc, Clairette, or Grenache Blanc, plus a maximum of 25% each of Ugni Blanc, Picpoul, Marsanne, Roussanne, and Viognier (but these secondary varieties must represent no more than 30% of the blend in total)

🍾 1–3 years

ROSÉ Production is declining in favour of red wine, but these dry rosés can have a delightful summer-fruit flavour that is fresher than either Tavel or Provence.

🍷 A minimum of 40% Grenache and 25% in total of Syrah and Mourvèdre, plus a maximum of 10% Carignan, no limit on the amount of Cinsault (but see red wine grapes, above), and up to 20% in total of Bourboulenc, Clairette, Grenache Blanc, Ugni Blanc, Picpoul, Marsanne, Roussanne, and Viognier

🍾 1–3 years

✓ Château d'Aquéria • Lafond Roc-Epine • Maby (La Fermade) • De la Mordorée

LUBERON AOC

When the wine of Luberon was promoted to full AOC status in 1988, much of the credit had to go to Jean-Louis Chancel and his vineyards at Château Val-Joanis. With its 16th-century Provençal house and gardens, this property has been described as the jewel in Luberon's crown.

RED These are bright, well-coloured wines, with plenty of fruit and character, improving with every vintage. The Val-Joanis Syrah Reserve Les Griottes has the quality

DOMAIN TOURBILLON, LUBERON
Wine-grower Benjamin Tourbillon's contemporary structure pays homage to the region's well-known and well-loved wines, including Chateauneuf-du-Pape, Gigondas, and Côtes-du-Rhône.

and density of a fine Hermitage. I'm not suggesting that it can be compared to the very best Hermitage, but it can stand shoulder to shoulder with many good examples in elegant rather than weightier styles.

🍷 *A blend of two or more varieties to include a minimum of 60% Grenache and Syrah (in total), of which Syrah must represent at least 10%; a maximum of 40% Mourvèdre; 20% each of Cinsault and Carignan; and 10% (either singly or in total) of Counoise, Pinot Noir, Gamay, and Picpoul*

🍷 *3–7 years*

WHITE White Luberon has no established style or reputation, so it seems bureaucratic nonsense to be so precise about the percentage of each grape variety that may or may not be included in its production. It showed promise in the late 1980s when Luberon was merely a VDQS, but even then the regulations were unnecessarily complicated. Since its upgrade to full AOC status, the regulations have become even more confusing, and I doubt if it is a coincidence that Luberon *blanc* has completely lost its way over the same period.

🍷 *A blend of two or more varieties, which can include Grenache Blanc, Clairette, Bourboulenc, Vermentino, and a maximum of 50% Ugni Blanc, and up to 20% (of the blend) of Roussanne and/or Marsanne*

🍷 *1–3 years*

ROSÉ These attractively coloured, fresh, fruity wines are much better quality than most Provence rosé.

🍷 *A blend of two or more varieties to include a minimum of 60% Grenache and Syrah (in total), of which Syrah must represent at least 10%; a maximum of 40% Mourvèdre; 20% each of Cinsault and Carignan; 20% in total of Grenache Blanc, Clairette, Bourboulenc, Vermentino, Ugni Blanc, Roussanne, and Marsanne; plus 10% (either singly or in total) of Counoise, Pinot Noir, Gamay, and Picpoul*

🍷 *Upon purchase*

✔ *De la Bastide de Rhodarès • Château la Canorgue • De la Citadelle • Château Constantin-Chevalier (Fondateurs) • De Fontenille (Prestige) • Cellier de Marrenon (Grand Luberon) • De Mayol • De la Royere • Château Thouramme* ⓞ *• Château Val Joanis*

MUSCAT DE BEAUMES-DE-VENISE AOC

These wines are the most elegant of the world's sweet fortified Muscat wines. Very little sweet Muscat was made before World War II. When the AOC was granted in 1945, the wine was classified as a *vin doux naturel*. The process by which this is made entails the addition of pure grape spirit in an operation called *mutage*, after the must has achieved 5 per cent alcohol by natural fermentation. The final wine must contain at least 15 per cent alcohol, plus a minimum of 110 grams per litre of residual sugar. The Coopérative des Vins and Muscats, which was established in 1956, accounts for a formidable 90 per cent of the wine produced. It is often said that this wine is always non-vintage, but in fact 10 per cent is sold with a vintage on the label.

WHITE/ROSÉ The colour varies between the rare pale gold and the common light apricot-gold, with an aromatic bouquet more akin to the perfume of dried flowers than fruit. You should expect Muscat de Beaumes-de-Venise to have hardly any acidity, which surprisingly does not make it cloying, despite its intense sweetness, but does enable it to be one of the very few wines to partner ice cream successfully.

🍷 *Muscat Blanc à Petits Grains, Muscat Rosé à Petits Grains*

🍷 *1–2 years*

✔ *De Coyeux • De Durban • Paul Jaboulet Aîné • Gabriel Meffre (Laurus) • Pigeade • De St-Saveur • Vidal-Fleury*

RASTEAU AOC

This village is best known for its *vin doux natural*, particularly its *rancio* style (*see* below), yet it produces nearly four times as much natural, unfortified dry red (in particular), white, and rosé as it does Rasteau of CDRV-level quality. The primarily Grenache-based *vin doux natural* appellation was created in the early 1930s and was promoted to

full AOC status in 1944; as such, it was the first of the Rhône's two *vin doux naturel* appellations. Earning AOC status for its still red wines in 2010, wines classified according to the original fortified appellation are distinguished by the prefix VDN.

RED (NOT FORTIFIED) Deep-coloured, full, and rich, with a spicy warmth.

🍷 *A minimum of 50% Grenache, plus a minimum of 20% Mourvèdre and Syrah, with an optional choice of any other CDR grape varieties (a maximum of 5% white grapes)*

🍷 *2–10 years*

✔ *De Beaurenard (Argiles Bleus) • Bressy-Masson • Des Coteaux des Travers (Cuvée Prestige) • La Courançonne (Magnificat) • Des Escaravailles (La Ponce) • La Soumade (Confiance) • Tardieu-Laurent*

RED (FORTIFIED) Most are rich, sweet, coarse, grapey-flavoured concoctions with plenty of grip, a rather awkward spiry aroma, and a pithy, apricot-skin aftertaste. Some are not even that sweet. Domaine de Beaurenard is in a different league, being smooth, rich, and truly sweet, but beautifully balanced and not in the slightest cloying, making it more like a vintage Port. The term *grenat* might be seen on some Rasteau Vin Doux Naturel, indicating a more reductive-style red that must not be bottled any later than 1 March of the second year, whereas *tuilé* (literally "tile-coloured") is an oxidative style that is orange-red in colour.

🍷 *A minimum of 90% Grenache, plus up to 10% in total of any CDR-permitted grape varieties*

🍷 *1–5 years*

✔ *De Beaurenard*

WHITE/TAWNY/ROSÉ (FORTIFIED) This wine can be white, tawny, or rosé, depending on the technique used and on the degree of ageing. It does not have the grip of the red, but has a mellower sweetness. Wines labelled *ambré* will be oxidative in style and tawny coloured.

🍷 *A minimum of 90% Grenache Gris or Blanc, plus up to 10% in total of any CDR-permitted grape varieties*

🍷 *1–5 years*

RANCIO (FORTIFIED) These fortified wines are similar to those above, except they must be stored in oak casks "according to local custom", which effectively means exposing the barrels to sunlight for a minimum of two years. This allows the wines to develop their distinctive *rancio* character.

🍷 *A minimum of 90% Grenache (Noir, Gris, or Blanc), plus up to 10% in total of any CDR-permitted grape varieties*

TAVEL AOC

Tavel is the most respected of all French dry rosé wines, but only the very best domaines live up to its reputation. In order to retain its characteristic freshness, Tavel's alcoholic level is restricted to a maximum of 13 per cent.

ROSÉ Some properties still cling to the old-style vinification methods, and, frankly, this means the wines are too old before they are sold. The top domaines in the appellation make clean-cut wines with freshly scented aromas and fine fruit flavours, which are invariably good food wines.

🍷 *Grenache, Cinsault, Clairette, Clairette Rosé, Picpoul, Calitor, Bourboulenc, Mourvèdre, and Syrah (none of which may account for more than 60% of the blend), plus a maximum of 10% Carignan*

🍷 *1–3 years*

✔ *Château d'Aquéria • Mireille Petit-Roudil • De la Mireille Petit-Roudil (Dame Rousse)*

VACQUEYRAS AOC

Formerly one of the single-village wines under the AOC Côtes du Rhône-Villages, Vacqueyras was elevated to full AOC status in 1990, without any mention of Côtes du Rhône or Côtes du Rhône-Villages, and is now theoretically on a par with Gigondas. The vines are spread over two communes, Vacqueyras and Sarrians, growing in the foothills of the Dentelles de Montmirail.

RED The best are dark, rich, and robust, with a warm, black-pepper spiciness.

🍷 *A minimum of 50% Grenache, plus a maximum of 20% in total of Syrah and/or Mourvèdre, plus a maximum of 10% in total of any of the following: Camarèse, Carignan, Cinsault, Clairette Rosé, Counoise, Grenache Gris, Muscardin, Vaccarèse, Picpoul Noir, Terret Noir*

🍷 *4–12 years*

WHITE Most tend to be either flabby or so dominated by modern, cool vinification methods that they are simply fresh and could come from anywhere.

🍷 *Grenache Blanc, Clairette, and Bourboulenc, plus a maximum of 50% in total of Marsanne, Roussanne, and Viognier*

🍷 *2–3 years*

ROSÉ Can be lovely, fresh, and fruity.

🍷 *A minimum of 60% Grenache, plus a minimum of 15% Mourvèdre and/or Cinsault, with an optional maximum of 10% of any of the following: Camarèse, Carignan, Cinsault, Clairette Rosé, Counoise, Grenache Gris, Muscardin, Vaccarèse, Picpoul Noir, Terret Noir*

🍷 *2–3 years*

✔ *Des Amouriers • De la Brunely (Elevé en Fût de Chêne) • De la Charbonnière • La Fourmone • Les Grands Cypres • De la Monardière (Vieilles Vignes) • De Montvac • Le Sang des Cailloux (Lopy) • Tardieu-Laurent (Vieilles Vignes) • Xavier Vignon (Povidis)*

VAUCLUSE IGP

See p341.

VINSOBRES AOC

Formerly a Côtes du Rhône Villages, Vinsobres has been an appellation in its own right since 2006 and now has full *cru* status. The vineyards are primarily devoted to Grenache (72 per cent) and Syrah (18 per cent), although there are small amounts of Carignan, Cinsault, and Mourvèdre.

RED Deep ruby in colour, with full, rich, spicy fruit. The quality has improved over the past few years.

🍷 *A minimum of 50% Grenache, plus a minimum of 20% Mourvèdre and Syrah, with an optional choice of any other CDR grape varieties (a maximum of 5% white grapes)*

🍷 *3–10 years*

✔ *L'Ancienne École • Des Ausellons • du Corinçon • du Moulin • Perrin • Domaine Jaume*

VENTOUX AOC

The limestone subsoil of this AOC produces a relatively lighter wine.

RED These fresh and fruity, easy-to-drink reds are the best wines in this AOC. The wines may be sold as *primeur* or *nouveau* from the third Thursday of November following the harvest.

🍷 *Grenache, Syrah, Cinsault, and Mourvèdre, plus a maximum of 30% Carignan and a maximum of 20% in total of Picpoul Noir, Counoise, Clairette, Bourboulenc, Grenache Blanc, Roussanne, and – until the 2014 vintage – Ugni Blanc, Picpoul Blanc, and Pascal Blanc*

🍷 *2–5 years*

WHITE The little white produced is seldom of interest. These wines may be sold as *primeur* or *nouveau* from the third Thursday of November following the harvest.

🍷 *Clairette, Bourboulenc, plus a maximum of 30% in total of Grenache Blanc, Roussanne, and – until the 2014 vintage – Ugni Blanc, Picpoul Blanc, and Pascal Blanc*

ROSÉ Fresh character and deliciously delicate fruit. They may be sold as *primeur* or *nouveau* from the third Thursday of November following the harvest.

🍷 *Same as red-wine grape varieties*

🍷 *Upon purchase*

✔ *La Ferme Saint Pierre (Roi Fainéant) • Le Murmurium • Martinelle • Château Talaud • Les Terrasses d'Eole (Lou Mistrau) • Château Valcombe • Vindémio • De la Verrière • Xavier Vignon (Xavier)*

The Jura, Savoie, and Bugey

The Jura wine region breaks all the winemaking rules with its idiosyncratic range of wines, yet these have caught the attention of wine enthusiasts worldwide; Savoie and Bugey are not far behind with wines from a myriad of unusual grapes.

Authentic, characterful, and downright obscure are some of the labels given to wines from the tiny vineyard areas on the foothills of the Jura and Alps mountains of eastern France. Yet in a short decade, from around 2009, Jura wines especially, with selected *cuvées* from Savoie and Bugey too, have sneaked onto the must-have wine lists of the hipster crowd, somms, and even established wine merchants in Tokyo, Stockholm, New York, and beyond.

In Jura, Savagnin – now known genetically to be the ancestor of many better-known varieties – was originally identified as the grape responsible for the rare and distinctive *vin jaune* (yellow wine), released only after six years barrel-ageing under a veil of yeast. It is not only these intense nutty, sherry-like, and extraordinarily long-lived wines that are made from this grape, but now lemony-fresh examples made more conventionally also prove its distinction; there are also Jura Chardonnays, which increasingly give top white Burgundies a run for their money, adding a distinctive *terroir* flavour of their own. Today, quality examples of all manner of white wine styles from the sparkling Crémant du Jura to the sweet *vin de paille*, with oxidative or classy barrel-fermented examples in between, demonstrate that Jura, with fewer than 2,000 hectares (4,942 acres) planted, is a world-class wine region. About one-third of the vineyards is planted with red grapes, the local Poulsard and Trousseau both considered equally seriously as Pinot Noir. These reds may look lightweight, but they have hidden depths, and the world can't find enough of them right now.

Bugey vineyards and the western vineyards of Savoie lie on the southern Jura foothills, while the main Savoie vineyard areas hug the lower slopes of the Prealps, with views of the towering, perennially snowy Alpine peaks. Savoie wines were all lapped up by skiers until recently, but a new generation of hard-working growers has emerged, many of them cultivating organically and earning a better return by selling their fresh and intense

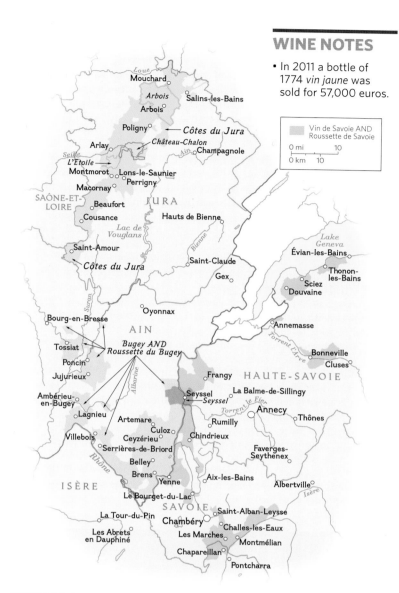

WINE REGIONS OF THE JURA, SAVOIE, AND BUGEY (*see also* p131)
Jura is a small wine region in eastern France sandwiched between Burgundy in the west and Switzerland in the east. Savoie is a mountainous region on the western edge of the Alps. Bugey lies between the Jura and the Savoie at the southern end of the Jura mountain chain; the Rhône river edges the region's southern border and loops to form the eastern border with Savoie.

RECENT JURA, SAVOIE, AND BUGEY VINTAGES

2018 A spectacular vintage for all three regions in terms of quality across all wine styles, with quantity generally good. Vignerons wreathed with smiles.

2017 A difficult vintage in the Jura, with half the normal volume due to frost damage, and variable quality. Better in Savoie and Bugey, but still challenging.

2016 An up-and-down vintage ended up with decent wines, if not spectacular. Late season varieties did best: Savagnin and Trousseau in the Jura; Mondeuse and Jacquère in Savoie.

2015 After a very warm summer, white wines lack a little acidity in all regions, but the reds are superb, notably Trousseau in the Jura and Mondeuse in Savoie and Bugey.

Earlier excellent vintages for Jura *vin jaune*: 2014 (not yet released), 2011, 2010, 2009, 2007, 2005, 2002, 1999

TALISSIEU VINEYARDS CASCADE DOWN THE JURA FOOTHILLS
The commune of Talissieu is located in the Bugey wine region, which is the smallest and possibly the least-known of France's many vine-growing areas.

wines beyond the Alpine boundaries. Consider the Jacquère, Altesse, Mondeuse, and Persan varieties, grown almost nowhere else in the world; these four indigenous varieties among dozens of others provide extra interest. Southern Bugey shares some of these varieties and is also attracting a younger generation of quality-focused growers, although Chardonnay dominates for now, while in northern Bugey the main focus is on making the frothy and light semi-sweet ancestral-method Cerdon, another wine that hipsters have taken to heart; it's the original pink Pet-Nat.

THE IMPACT OF CLIMATE CHANGE AND ORGANICS

In decades gone by there were years when ripening was a challenge in these areas, with their mountain-influenced weather, but climate change has made a real impact here with significant temperature increases, particularly in the period from March to October. Together with better vineyard practices and lower yields, ripening is no longer a problem in any of these regions. The unpredictability of the weather and extreme events that climate change has brought with it, however, have led to more violent rain or hailstorms and more deadly spring frosts, the latter often after the vine has had an early burst of growth.

Jura now has 20 per cent of its vineyards certified organic, and the small Bugey area with just 450 hectares (1,111 acres) is not far behind proportionately. Savoie, the largest of these three regions, with about 2,200 hectares (5,436 acres) under vine, only has 10 per cent organic at the time of writing, but conversions are increasing apace. With organics comes, typically, more dedication to making fine-quality wines, which often need patience to be at their best. But in too many vintages the vagaries of the weather have reduced yields so much that the most sought-after organic wineries simply cannot keep up with worldwide demand. The big worry is that interest in these fascinating but tiny wine regions (between them accounting for only half of one per cent of French wine volumes) could wither on the vine, so to speak, if enough good producers cannot supply their wines to customers on a consistent basis. It would be sad if the more adventurous wine drinkers gave up on Jura and Savoie wines so soon due to the supply constraints.

FACTORS AFFECTING TASTE AND QUALITY

LOCATION
Jura: East of Burgundy and west of the Jura Mountains and the Swiss border.
Savoie/Bugey: The scattered vineyards of Savoie stretch from south of Lac Léman (Lake Geneva) almost to Grenoble in the Alpine foothills; Bugey lies to the south of Jura, east of Beaujolais.

CLIMATE
Jura: The climate is northern continental, with hot summers and cold winters; much higher levels of rainfall than in Burgundy, and there is greater risk of hailstorms and frost.
Savoie/Bugey: Also continental, but hotter summers than in Jura. Some areas influenced by proximity of Lac Léman and Lac du Bourget, as well as the Rhône River. Also prone to high summer rainfall and increasingly violent hailstorms. Frost risk on lower slopes.

ASPECT
Jura: The vineyards are situated on the lower, southwesterly slopes of the first plateau that leads up to the Jura Mountains. The vines grow at an altitude of 250 to 450 metres (820 to 1,476 feet).
Savoie/Bugey: Vineyards in Bugey and western Savoie lie on the southern extremities of the Jura Mountains; most Savoie vineyards lie on the foothills of the Prealps, with the best ones facing south or southeast. Altitudes range from 250 to 500 metres (820 to 1,640 feet).

SOIL
Jura: Predominantly clay-limestone, known as marl, which appears in various colours; clay dominates over limestone in most areas. Topsoils vary greatly, with some warmer sandstone and gravels in Arbois.
Savoie/Bugey: Limestone dominates, with, in particular, limestone scree on the steepest slopes; glacial movements have created highly varied topsoils, with alluvial cones, sandstone, and clay deposits. Apremont and Abymes lie on the rocky, limestone deposits from the 1248 landslide of Mont Granier.

VITICULTURE & VINIFICATION
Jura: High rainfall and clay-dominant soils make vineyards hard to manage, both for weed control and disease, but nevertheless in 2019 about 20 per cent are under organic viticulture. The technique of making *vin jaune* is specific to Jura: late-picked Savagnin is fermented normally and the wine is transferred to old Burgundy barrels, incompletely filled. These are stored in ventilated cellars with temperature changes promoting growth of a flor-type yeast, which forms on the wine's surface. The barrels are not topped up, and to achieve the appellation *vin jaune,* the wine must not be bottled until six years and three months after harvest. Much is withdrawn early to be sold simply as Savagnin or as a blend with Chardonnay. Many Chardonnays and some Savagnins are made in a non-oxidative manner (tanks/barrels topped up – called ouillé in Jura) as in the rest of the world, some unoaked, some fermented and aged in oak, Burgundy-style, but rarely in new oak. Poulsard reds are often made with no or low SO2 additions, in a semi-carbonic maceration method; otherwise all reds are made traditionally, some with older oak ageing.
Savoie/Bugey: Working often on steep slopes means a slower uptake of organic methods in Savoie, with about 10 per cent in 2019; levels are higher in Bugey. Winemaking is traditional but with relatively little oak ageing, almost no new oak barrels. In Bugey, much sparkling wine is made including the ancestral-method Cerdon.

GRAPE VARIETIES
Jura: Chardonnay (Melon), Pinot Noir, Poulsard (Ploussard), Savagnin (Naturé), Trousseau
Savoie/Bugey: Altesse, Chasselas, Gamay, Gringet, Jacquère, Velteliner Rouge Précoce (Malvoisie), Chardonnay, Molette, Mondeuse (Noire), Mondeuse Blanche, Persan, Pinot Noir, Poulsard, Roussanne, Verdesse.

CHATEAU D'ARLAY *VIN JAUNE,* **VINTAGE 2001**
From the Côtes du Jura AOC, Château D'Arlay makes a fine *vin jaune;* nonetheless, Château-Chalon AOC produces the most famous of all the *vins jaunes* of Jura. Best drunk very old, this wine is made from 100 per cent Savagnin grapes and takes its name from its deep honey-gold colour.

THE JURA, SAVOIE, AND BUGEY

ARBOIS AOC

This is the largest appellation of the Jura, with vineyards in and around the town of Arbois, including in the village of Montigny-les-Arsures. It is used for all Jura wine styles as shown below. Pupillin is a single-commune appellation made to the same specification as Arbois AOC.

RED Poulsard is light coloured, but packs a punch; deeper Trousseau wines are more earthy and fuller, and Pinot offers rounder, fuller wines.

🍷 *Trousseau, Poulsard (Ploussard), Pinot Noir*

🍷 *1–8 years*

WHITE Varies hugely according to vinification and maturation methods. Chardonnays express the *terroir* well; topped-up (*ouillé*) Savagnin is lemony and fresh; oxidative wines may be blends or only Savagnin; the latter is effectively declassified *vin jaune*.

🍷 *Savagnin, Chardonnay*

🍷 *1–10 years*

ROSÉ Not to be confused with pale-coloured Poulsard reds, sometimes erroneously classified as rosé; these were traditionally firm and dry, but a paler more modern rosé is increasingly made.

🍷 *Poulsard, Trousseau, Pinot Noir*

🍷 *1–2 years*

✓ Jérôme Arnoux • Caveau de Bacchus • De la Borde ⓓ • Philippe Bornand ⓓ • Dugois • Michel Gahier • Hughes-Béguet ⓓ • Frédéric Lornet • Maison Overnoy ⓓ • Du Pélican ⓓ • De la Pinte ⓓ • Ratte ⓓ • De la Renardière ⓓ • Maison Rijckaert • De Saint-Pierre ⓓ • André & Mireille (Stéphane) Tissot ⓓ • De la Touraize ⓓ • De la Tournelle ⓓ • Gérard Villet ⓓ

VIN JAUNE This white wine, a Jura specialty, is aged in old Burgundy barrels under a layer of yeast, similar to Sherry's *flor*. The barrels are not topped up, and the wine may not be released until the beginning of its seventh year, leading to an increase in alcohol and oxidative character. The best are bone dry with complex flavours ranging from walnuts to curry spices and have incredible longevity. Arbois *vin jaune* tends to be the strongest and nuttiest in flavour due to the most extreme temperature variations of the ageing cellars.

🍷 *Savagnin*

🍷 *8–50+ years*

✓ Dugois • Fruitière Vinicole d'Arbois • Frédéric Lornet • De la Pinte ⓓ • Jacques Puffeney • Rolet • André & Mireille (Stéphane) Tissot ⓓ • De la Tournelle

VIN DE PAILLE Arbois *vin de paille* is made from grapes that are dried in warm attics to concentrate the juice, followed by a long fermentation and minimum three years ageing (18 months in oak). With a minimum 14 per cent alcohol, these vary from medium to very sweet, with an old-gold colour, the best being complex, showing honey and spiced dried fruits; some producers allow slight oxidation, adding a nutty character.

🍷 *Poulsard, Trousseau, Savagnin, Chardonnay*

🍷 *4–50+ years*

✓ Désiré Petit & Fils • Frédéric Lornet • André & Mireille (Stéphane) Tissot ⓓ

BUGEY AOC

The appellation covers wine made in the historic Bugey area of the Ain department and is divided into a northern part between Bourg-en-Bresse and Nantua and a southern part in the loop of the Rhône River with the area capital of Belley. There are three *crus*: Cerdon (semi-sweet ancestral-method sparkling rosé only), Manicle (red Pinot Noir and white Chardonnay only), and Montagnieu (red Mondeuse and sparkling white wines from predominantly Chardonnay, Altesse, and Mondeuse).

RED Light, potentially juicy wines that must be single varietal wines. They range from the fruity Gamay to the rich Mondeuse. Manicle must be made exclusively from Pinot Noir, while Montagnieu must be Mondeuse.

🍷 *Gamay, Mondeuse, Pinot Noir*

🍷 *1–8 years*

WHITE Dry to off-dry, fresh, light, and delicately fruity. Manicle must be made exclusively from Chardonnay.

🍷 *Mainly Chardonnay, plus Aligoté, Altesse, Jacquère, Mondeuse Blanche, and Pinot Gris*

🍷 *1–3 years*

ROSÉ Light and refreshing dry wines.

🍷 *Mainly Gamay and Pinot Noir, plus Mondeuse, Pinot Gris, and Poulsard*

🍷 *1–3 years*

SPARKLING WHITE These wines may be blends, especially the very fine Montagnieu, but are often pure Chardonnay – they are good value dry traditional-method wines.

🍷 *Mostly Chardonnay, Jacquère, Molette, with some Aligoté, Altesse, Gamay, Mondeuse (Noire and Blanche), Pinot Gris, and Poulsard*

🍷 *1–2 years*

✓ Maison Angelot • Maison Bonnard ⓓ • Caveau Bugiste (Manicle red and white) • Maison Yves Duport ⓓ • Franck Peillot • Thierry Tissot ⓓ

SPARKLING ROSÉ There is some traditional-method sparkling rosé, but much outside of Cerdon is made as a non-AOC ancestral-method wine. Cerdon is made predominantly from Gamay, sometimes blended with Poulsard by the *méthode ancestrale*, with a minimum of 22 grams of residual sugar per litre (usually closer to 45 grams). It is a fresh, light, grapey-aromatic sparkling wine with soft mousse.

🍷 *Mainly Gamay and Pinot Noir, plus Mondeuse, Pinot Gris, and Poulsard*

🍷 *1–2 years*

✓ (All Cerdon) Philippe Balivet ⓓ • Raphaël-Bartucci • Patrick Bottex • La Dentelle Lingot • Martin • Renardat-Fâche ⓓ • Rondeau

CHÂTEAU-CHALON AOC

The AOC Château-Chalon is exclusively for Jura's *vin jaune* wines made from Savagnin grown in a 50-hectare (123-acre) area around the village of the same name, and it is arguably the most legendary exponent. Château-Chalon is made in the same way as described above for Arbois *vin jaune* and subject to the same rules. Other wines made from grapes grown here must be labelled as Côtes du Jura AOC.

VIN JAUNE Compared to Arbois *vins jaunes*, Château-Chalon wines are often more elegant and less nutty, with complex flavours including dried fruit, peat, and curry spices.

🍷 *Savagnin*

🍷 *10–100 years*

✓ Baud Génération 9 • Berthet-Bondet ⓓ • Philippe Butin • Chevassu-Fassenet • Macle ⓓ • Jean-Luc Mouillard • André et Mireille (Stéphane) Tissot ⓓ

CÔTES DU JURA AOC

The Côtes du Jura is a generic appellation used for all Jura wine styles as shown below and covers the whole Jura appellation, including the AOC Arbois, L'Etoile, and Château-Chalon. The quantity sold under this appellation, however, is less than AOC Arbois and most comes from the area to the south.

RED These wine are usually lighter in colour and body than those from AOC Arbois.

🍷 *Poulsard, Trousseau, Pinot Noir*

🍷 *1–8 years*

WHITE See AOC Arbois

🍷 *Savagnin, Chardonnay*

🍷 *1–10 years*

ROSÉ See AOC Arbois.

🍷 *Poulsard, Trousseau, Pinot Noir*

🍷 *1–3 years*

✓ Badoz • Baud Génération 9 • Buronfosse ⓓ • Champ Divin ⓓ • Ganevat ⓓ • Grand • Labet ⓓ • Macle ⓓ • Marnes Blanches ⓓ • Pignier ⓓ • Maison Rijckaert • André & Mireille (Stéphane) Tissot ⓓ

VIN JAUNE See Arbois *vin jaune* above. Like Arbois *vins jaunes*, styles vary, with those from Arlay in particular more reminiscent of Château-Chalon AOC.

🍷 *10–50+ years*

✓ Château d'Arlay • Badoz • Caves Jean Bourdy ⓓ • Philippe Butin • Ganevat • Labet ⓓ • Marnes Blanches ⓓ • Pêcheur • Pignier ⓓ

VIN DE PAILLE See Arbois *vin de paille*.

🍷 *Poulsard, Trousseau, Savagnin, Chardonnay*

🍷 *4–50+ years*

✓ Château d'Arlay • Berthet-Bondet ⓓ • Caves Jean Bourdy ⓓ • Labet ⓓ • Pignier ⓓ

CRÉMANT DE SAVOIE AOC

Technically part of the main Savoie AOC, eventually meant to become a separate AOC, Crémant de Savoie was introduced from the 2014 vintage. It follows the rules for all Crémants in France (and Luxembourg) including that grapes must be hand harvested and the wines are made by the traditional method, with minimum nine months in bottle on the second fermentation lees and are sold before 12 months after harvest. Local grapes, Jacquère and Altesse, must account for a minimum 60 per cent with Jacquère itself accounting for 40 per cent minimum of the total blend; red grapes may not exceed 20 per cent. Due to the range of grapes used, the styles vary greatly from light, crisp 100 per cent Jacquère examples to more complex ones when blended with Altesse, Chardonnay, and/or Pinot Noir. Rosé is not yet allowed.

🍷 *Jacquère and Altesse with Chardonnay, Aligoté, Velteliner Rouge Précoce, Chasselas (Haute-Savoie only), Mondeuse Blanche, Gamay, Mondeuse Noire,*

🍷 *2–5 years*

✓ Blard • Giachino ⓓ • De l'Idylle • André et Michel Quenard • Jean-François Quénard • Saint-Germain ⓓ • Jean Vullien

CRÉMANT DU JURA AOC

This appellation was introduced in 1995 and like Crémant de Savoie above follows the rules for all Crémants. White and rosé are permitted: the white must be minimum 70 per cent Chardonnay, Pinot Noir and/or Trousseau; the rosés must have a minimum of 50 per cent red grapes. The whites can be *brut* or *demi-sec*, though most are *brut*, ranging from dry to off-dry, crisp with appley and stony flavours and soft *mousse*.

🍷 *Chardonnay, Pinot Noir, Poulsard, Trousseau, Savagnin*

🍷 *2–6 years*

✓ Fruitière Vinicole d'Arbois • Champ Divin ⓓ • Grand • De Montbourgeau • Désiré Petit & Fils • Rolet (Coeur de Chardonnay) • Pignier ⓓ • André & Mireille (Stéphane) Tissot ⓓ

L'ÉTOILE AOC

These wines are possibly named after the star-shaped fossils found in the local marl or from the five hilltops that form a star shape around the village of the same name. Only white wines may be made.

WHITE Most, including Chardonnay, are made in the oxidative style, but the best show just a delicate touch.

🍷 *Chardonnay, Savagnin*

🍾 *1–8 years*

✓ *Baud Génération 9 • Joly • De Montbourgeau • Philippe Vandelle*

VIN JAUNE See Arbois *vin jaune* above. *L'Etoile vin jaune* can be more delicate and stony due to higher levels of limestone in the area.

🍷 *Savagnin*

🍾 *10–100 years*

✓ *Château de l'Étoile • Joly • De Montbourgeau*

VIN DE PAILLE See Arbois *vin de paille.*

🍷 *Chardonnay, Poulsard, Savagnin*

🍾 *10–50+ years*

✓ *Château de l'Étoile • De Montbourgeau*

MACVIN AOC or MACVIN DU JURA AOC

Macvin, a *vin de liqueur*, is made by adding *marc* to grape juice, preventing fermentation – the *marc* is brandy distilled from the residue of skins and pips from the same producer and itself must already have been aged minimum 14 months in barrel. The blend is then aged in barrel for at least 10 months. The result, which can be white, rosé, or red, is unsurprisingly grapey and spirity and is served locally as an aperitif or over ice cream.

🍷 *Chardonnay, Savagnin, Poulsard, Trousseau, Pinot Noir*

🍾 *Non-vintage, but ages surprisingly well.*

✓ *Marnes Blanches ⓪ • De Montbourgeau • De la Renardière ⓪ • Jacques Tissot*

ROUSSETTE DU BUGEY AOC

A specific designation for whites from 100 per cent Altesse covering the whole AOC Bugey area with two geographic designations *(crus):* the tiny Virieu-le-Grand and much better known Montagnieu. Although not much Altesse is grown, it is increasing.

WHITE From dry to off-dry wines, high quality, herbal and steely wines, especially good from Montagnieu.

🍷 *Altesse*

🍾 *1–10 years*

✓ *Patrick Charlin (Montangieu) • Maison Yves Duport (Montangieu) ⓪ • Franck Peillot (Montangieu) • Des Plantaz*

ROUSSETTE DE SAVOIE AOC

A specific designation for whites from 100 per cent Altesse covering the whole AOC Savoie area. The following four specific geographic denominations *(crus)* may add their name to this appellation: Frangy, Marestel, Monterminod, and Monthoux. Plantings of the grape have been increasing due to the high potential quality.

WHITE A huge range of styles from dry to off-dry (sometimes with a little botrytis, or "noble rot"), these wines balance crisp acidity with fine herbal and stony flavours with tangy fruit. Potentially very long lived.

🍷 *Altesse*

🍾 *1–15 years*

✓ *Blard & Fils • Eugène Carrel et Fils (Marestel) • De Chevillard ⓪ • Dupasquier (Marestel) • Maison Philippe Grisard • Edmond Jacquin & Fils (Marestel) • Château de Lucey • Bruno Lupin (Frangy) • Jean Perrier et Fils (Monterminod) • Orchis ⓪ • Saint-Germain ⓪*

SEYSSEL AOC

This was the first AOC in Savoie and made famous by the Royal Seyssel brand (once owned by Varichon et Clerc) for sparkling wines. The appellation struggles to survive, and still white wines dominate.

WHITE Usually from 100 per cent Altesse but no longer allowed to be labelled Roussette de Seyssel; too often marred by residual sugar, whereas old vines in the area give the potential for classy, long-lived wines. There is a separate Seyssel AOC for Molette, always stated on the label, but they are too often thin and acidic.

🍷 *Altesse, Molette*

🍾 *1–3 years*

✓ *Clos de l'Arvière • De la Brune (Molette Vieilles Vignes) • Gérard Lambert • Maison Mollex*

SPARKLING WHITE After Crémant de Savoie AOC was created, the minimum level of Molette in Seyssel was to be increased to 70 per cent, but in practice most producers do not have enough Molette, and this is likely to reduce. Aged for much longer on second fermentation lees, the full, yeasty nose, fine *mousse*, and elegant flavour, makes the Royal Seyssel a yardstick for other producers.

🍷 *Altesse, Molette*

🍾 *1–3 years*

✓ *Clos de l'Arvière • Gérard Lambert (Royal Seyssel) • Maison Mollex*

SAVOIE/VIN DE SAVOIE AOC

This generic appellation covers all the wine styles in the several "islands" of Savoie vineyards, the only exceptions being Roussette de Savoie and Crémant de Savoie, see above. The following geographic designations *(crus)* have the right to add their name to the appellation providing the specified grapes are used: Abymes (white only, mainly Jacquère), Apremont (white only, mainly Jacquère), Arbin (red from Mondeuse), Ayze or Ayse (white or sparkling white from Gringet), Chautagne (white, mainly Jacquère and red from Gamay, Pinot, or Mondeuse), Chignin (white, mainly Jacquère, and red from Gamay, Pinot, or Mondeuse), Chignin-Bergeron (white from Roussanne), Crépy (white from Chasselas), Cruet (white only, mainly Jacquère), Jongieux (white, mainly Jacquère, and red from Gamay, Pinot, or Mondeuse), Marignan (white from Chasselas), Marin (white from Chasselas), Montmélian (white, mainly Jacquère), Ripaille (white from Chasselas), Saint-Jean-de-la-Porte (red Mondeuse), and Saint-Jeoire-Prieuré (white, mainly Jacquère).

RED Climate change and better viticulture and winemaking practices mean Mondeuse is a real star here: relatively light alcohol, berry fruit, and peppery spice.

The increasing range of Persan is very good too. Gamay and Pinot Noir are less interesting. Blends are few.

🍷 *Gamay, Mondeuse, Pinot Noir, plus Cabernet Franc, and Cabernet Sauvignon, Persan, and (in the Isère département) Etraire de la Dui, Joubertin, and Servanin*

🍾 *1–8 years*

WHITE The finest and longest-lived are the dry, but rich, apricotty Chignin-Bergeron from Roussanne and the Altesse wines (bottled usually as Roussette de Savoie AOC, see above). However, Jacquère produces increasingly good and more delicate steely Alpine dry whites. Ayze wines are much sought after, and whites from rare grapes are increasing.

🍷 *Aligoté, Altesse, Chardonnay, Jacquère, Mondeuse Blanche, Veltliner Rouge Précoce, plus (in the Haute-Savoie département) of Chasselas, Gringet, Roussette d'Ayze, and (in the Isère département) Marsanne and Verdesse*

🍾 *1–3 years*

ROSÉ Attractive, light, and fruity, dry rosés made for early drinking.

🍷 *See reds above*

🍾 *1–3 years*

✓ *De la Baraterie ⓪ • Curtet ⓪ • Dupasquier • Maison Philippe Grisard • De L'Idylle • Saint-Germain*

Best single-*cru* wines: Abymes: *Giachino ⓪;* **Apremont:** *Blard & Fils, Dupraz ⓪, Giachino ⓪, Jean Masson & Fils, Cellier du Palais;* **Arbin:** *Genoux, Louis Magnin, Fabien Trosset;* **Ayze:** *Belluard;* **Chignin:** *Berthollier ⓪, André & Michel Quenard, Jean-François Quénard;* **Chignin-Bergeron:** *Adrien Berlioz ⓪, Berthollier Y, Louis Magnin Y, Partagé-Gilles Berlioz ⓪ André & Michel Quenard, Jean-François Quénard , Jean Vullien;* **Jongieux:** *Eugène Carrel & Fils, Edmond Jacquin & Fils;* **Marignan:** *La Tour de Marignan ⓪;* **Marin:** *Delalex;* **St-Jean-de-la-Porte:** *De Chevillard, Des Côtes Rousses*

SPARKLING WHITE Since creating the new Crémant de Savoie AOC, the only Savoie AOC sparkling whites made are under the Ayze (or Ayse) geographic denomination *(cru)*, which produces some fine, dry examples.

🍷 *Gringet, plus optional Altesse and Roussette d'Ayze.*

🍾 *2–8 years*

✓ *Belluard ⓪ • Famille Montessuit*

SPARKLING ROSÉ Attractive, early-drinking wines, with a soft mousse and delicate fruit.

🍷 *Aligoté, Altesse, Chardonnay, Gamay, Jacquère, Mondeuse (Blanche and Noire), Pinot Noir, and Veltliner Rouge Précoce, plus (in the Haute-Savoie département) Chasselas*

🍾 *1–3 years*

CHÂTEAU DE RIPAILLE, THÔNON LES BAINS
Located on the edge of Lac Léman, this chateau was built in the 15th century by Amédée VIII. As well as serving as a residence of the dukes of Savoy, it is a former Carthusian monastery. Today, it is not technically a monopole but produces for the vast majority of the small but highly acclaimed *cru* of the Vin de Savoie Ripaille AOC.

Southwest France

This region encompasses numerous small, scattered areas that combine to produce an impressively wide range of excellent-value wines with diverse, but quite discernible, stylistic influences from Bordeaux, Spain, Languedoc-Roussillon, and the Rhône.

At the heart of the region lies Gascony, the great brandy district of Armagnac. It was from here that d'Artagnan set out in around 1630 to seek fame and fortune in the King's Musketeers. The narrow tracks upon which his eventful journey began still wind their lonely way around wooded hills and across bubbling brooks. Apart from brightly hued fields of cultivated sunflowers, surprisingly little has changed since Alexandre Dumas painted such a vivid and colourful picture of these parts, for they remain sparsely populated to this day. Time passes slowly even in the towns, where the main square is usually deserted all day long, except during the five o'clock rush hour, which lasts for all of 10 minutes.

THE DIVERSITY OF THE APPELLATIONS

The Southwest does not yet have a single wine of truly classic status, but it arguably offers more value for money and is a greater source of hidden bargains than any other French region. From the succulent, sweet Jurançon *moelleux* and Monbazillac to the fine wines of Bergerac, Buzet, and Marmandais, the revitalized "black wines" of Cahors, the up-and-coming Fronton, the tannic Madiran, and the highly individual Irouléguy of the Basque Country, this part of France represents tremendous potential for knowing wine drinkers.

A VINEYARD WORKER IN THE JURANÇON WINE REGION CAREFULLY SNIPS A BUNCH OF MANSENG GRAPES
Located in the foothills of the Pyrenees, the Jurançon AOC is known for its sweet white wines made from both Gros and Petit Manseng, as well as the Courbu variety.

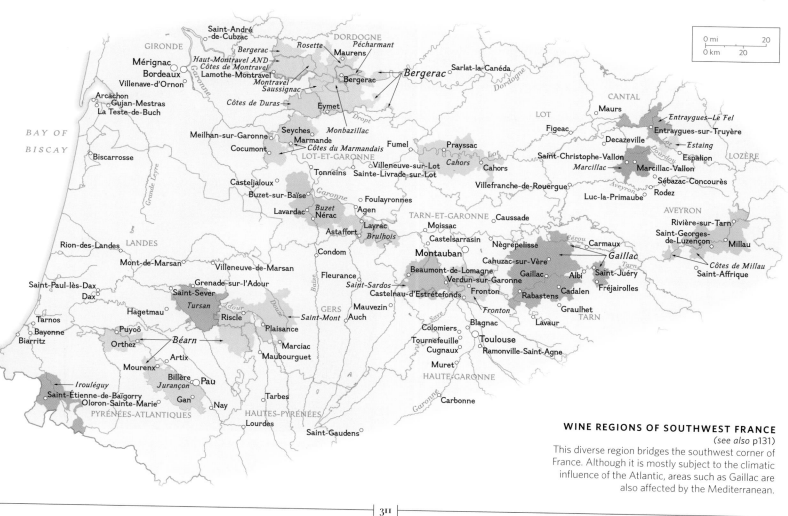

WINE REGIONS OF SOUTHWEST FRANCE
(see also p131)
This diverse region bridges the southwest corner of France. Although it is mostly subject to the climatic influence of the Atlantic, areas such as Gaillac are also affected by the Mediterranean.

RENAISSANCE OF CAHORS

As mentioned above as the "black wine" of Cahors, the primary grape and driving force of these wines is Côt (perhaps better known as Malbec). Either as a pure varietal wine or as a blending component (with Tannat and Merlot in Cahors) Côt gives an inky intense colour and characteristic plum flavours. In the mid-19th century Côt was taken from France and, most likely via Chile, arrived in Mendoza. Free from the disease pressure that could affect its French counterpart, with an enviously consistent climate and irrigation control, Côt flourished. Argentina certainly made Côt their own, with their Malbec wines characterized by deep colour, intense fruity flavors, velvety texture, and approachable nature. With the renaissance and varietal name recognition of "Malbec" (indeed many Cahors producers are now labelling their wine as such) engendered by the Argentines, Cahors winemakers have seen the opportunity to make Côt with a more approachable tannin structure and earlier drinking style, and rather than needing a knife and fork to get through a tannic wine when opening a recent vintage, there are now Cahors wines that can be enjoyed younger. Some winemakers still cleave to the old Cahors style, but now there is a choice for consumers, and with large tracts of excellent wine land yet to be cultivated around Cahors there is also room for expansion.

RECENT SOUTHWEST FRANCE VINTAGES

2018 A wet start gave way to a long, warm summer, producing concentrated reds and rich and powerful sweet whites.

2017 A decent vintage for both reds and whites.

2016 An exceptional year for the reds, particularly Cahors; hot but not excessively hot days and cooler nights giving concentrated, complex, fruitful wines with crisp acidity.

2015 June and July were warm and dry months with daily maximum temperatures ranging between 32°C to 40°C (86°F to 104°F) providing an exceptional vintage for both reds and whites.

2014 A decent vintage with some better conditions for botrytis late in the season.

FACTORS AFFECTING TASTE AND QUALITY

LOCATION
This is the southwest corner of France, bordered by Bordeaux, the Atlantic Ocean, the Pyrenees, and the vineyards of Languedoc-Roussillon.

CLIMATE
The climate of southwestern France is Atlantic-influenced, with wet winters and springs, warm summers, and long, sunny autumns. The vineyards of Cahors, Fronton, and Gaillac are subject to the greater heat but more changeable characteristics of the Mediterranean.

ASPECT
Mostly east- and east-through-to-south-facing slopes, affording protection from the Atlantic, in a varied countryside that can range from rolling and gently undulating to steep and heavily terraced hills.

SOIL
This collection of diverse areas has, not unexpectedly, a number of different soils: sandy-and-calcareous clay over gravel in the best vineyards of Bergerac; sandy soils on the Côte de Duras; calcareous and alluvial soils in the *côtes* of Buzet and Marmandais; gravel-clay and gravel crests over marly bedrock in the hilly hinterland of Cahors, and alluvial soils peppered with pebbly quartz, limestone, and gravel over a calcareous bedrock in the Lot Valley; limestone, clay-and-limestone, and gravel at Gaillac; sandy soils in Madiran, Tursan, and Irouléguy; and stony and sandy soils in Jurançon..

VITICULTURE & VINIFICATION
The viticultural traditions and vinification techniques of Bergerac, Buzet, Marmandais, and, to some extent, Cahors, are similar to those of Bordeaux. Other districts of this composite region have very much their own distinctive, individual practices: Béarn, Gaillac, and Jurançon produce almost every style of wine imaginable by many vinification techniques, among them the *méthode rurale* (known locally as the *méthode gaillaçoise*). Although the winemaking technique in these areas is generally very modern, the Basque district of Irouléguy remains stoutly traditional, allowing only the introduction of Cabernet Sauvignon and Cabernet Franc to intrude upon its set ways.

GRAPE VARIETIES
Abouriou, Arrufiac, Auxerrois (Malbec), Baroque, Cabernet Franc, Cabernet Sauvignon, Camaralet, Castets, Chardonnay, Chenin Blanc, Cinsault, Clairette, Colombard, Courbu Blanc, Courbu Noir, Duras, Fer, Folle Blanche, Fuella, Gamay, Gros Manseng, Jurançon Noir, Lauzet, Len de l'El, Manseng Noir, Mauzac, Mauzac Rosé, Mérille, Merlot, Milgranet, Mouyssaguès, Muscadelle, Négrette, Ondenc, Petit Manseng, Picpoul, Pinot Noir, Raffiat, Rousselou, Sauvignon Blanc, Sémillon, Syrah, Tannat, Ugni Blanc, Valdiguié

CAHORS BLACK WINES FROM CHÂTEAU DU CÈDRE AND CLOS TRIGUEDINA
The inky depth of colour gives these wines their nickname. Cahors black wines are made with Malbec (known locally as Côt) as the dominant grape variety. Either Malbec or Cot will appear on the label.

THE APPELLATIONS OF
SOUTHWEST FRANCE

BÉARN AOC

This modest AOC shines in the local Basque area with wines made in Bellocq, Lahontan, Orthez, and Saliès.

RED Fresh, light, and fruity wines with a good balance, but lacking depth.

🍷 *A minimum of 50% Tannat, plus Cabernet Franc, Cabernet Sauvignon, Fer, Manseng Noir, and Courbu Noir*

🍷 *1–4 years*

WHITE Light, dry, and aromatic wines.

🍷 *Petit Manseng, Gros Manseng, Courbu Blanc, Lauzet, Camaralet de Lasseube, Petit Courbu, Raffiat de Moncade (a minimum of 50% from the 2019 harvest onwards), and Sauvignon Blanc*

🍷 *1–2 years*

ROSÉ Simple, fruity, dry rosés with a fresh floral aroma.

🍷 *Tannat, Cabernet Franc, Cabernet Sauvignon, Fer, Manseng Noir, and Courbu Noir*

🍷 *1–2 years*

✓ *Bouscassé ⓝ (Rosé) • CV de Jurançon (Larribère)*

BERGERAC AOC

Adjoining Bordeaux, Bergerac produces wines that are sometimes mistaken for the modest appellations of its more famous neighbour. Its wines were shipped to London as early as 1250. Supplies dried up after the Hundred Years' War, which ended with the Battle of Castillon, and Castillon-la-Bataille marks the ancient English-French boundary between Bergerac and Bordeaux.

RED The best reds have a good garnet or ruby colour, fine fruit, and an elegant balance.

🍷 *Cabernet Sauvignon, Cabernet Franc, Merlot, Malbec, Fer, and Mérille*

🍷 *2–8 years*

WHITE Dry Bordeaux-style wines.

🍷 *Sémillon, Sauvignon Blanc, Sauvignon Gris, Muscadelle, Ondenc, Chenin Blanc, and Ugni Blanc (with the proviso that the quantity of Sauvignon Blanc and Sauvignon Gris used is at least equal)*

🍷 *1–3 years*

ROSÉ These are light, easy, and attractive dry wines.

🍷 *Cabernet Sauvignon, Cabernet Franc, Merlot, Malbec, Fer, and Mérille*

🍷 *1–3 years*

✓ *de l'Ancienne Cure • Julienne (Casnova des Conti) • Château Les Justices • Château Laulerie • Château Miaudoux (Inspiration) • Château Le Payral (Héritage) • Château Pion • Julien de Savignac (Rosé) • Château Thénac • Château Tour des Gendres • Les Verdots (Les Verdots selon David Fourtout)*

BRULHOIS AOC

Previously known as Côtes du Brulhois and elevated to AOC in 2011, this appellation encompasses vineyards along the Garonne immediately west of Buzet.

RED Decent, if unexciting, Bordeaux-like wine, though more rustic in style.

🍷 *Cabernet Franc, Cabernet Sauvignon, Fer, Merlot, Malbec, Tannat, and Abouriou*

🍷 *2–4 years*

ROSÉ A fresh, easy-to-drink dry wine.

🍷 *Cabernet Franc, Cabernet Sauvignon, Fer, Merlot, Malbec, Tannat, and Abouriou*

🍷 *1–3 years*

✓ *Château Grand Chêne (Elevé en Fûts de Chêne)*

BUZET AOC

Formerly known as Côtes de Buzet, this super-value Bordeaux satellite is located on the northern edge of the Armagnac region.

RED The best are always very good, with considerable finesse and charm.

🍷 *Merlot, Cabernet Sauvignon, Cabernet Franc, and Malbec, with a maximum of 10% from Abouriou and Petit Verdot*

🍷 *3–10 years (15 in exceptional cases)*

WHITE These dry whites are the least-interesting wines in this appellation. They may be sold from 1 December following the harvest without any mention of *primeur* or *nouveau*.

🍷 *Sémillon, Sauvignon Blanc, Sauvignon Gris, and Muscadelle, with a maximum of 10% from Colombard, Gros Manseng, and Petit Manseng*

🍷 *1–3 years*

ROSÉ Ripe, fruity, dry rosés. They may be sold from 1 December following the harvest without any mention of *primeur* or *nouveau*.

🍷 *Merlot, Cabernet Sauvignon, Cabernet Franc, Malbec, Abouriou, and Petit Verdot*

🍷 *1–4 years*

✓ *Baron d'Albret • CV de Buzet • Château Sauvagnères*

CAHORS AOC

The once-famous "black wine" of Cahors got its name from the Malbec grape, which, prior to phylloxera, produced dark, inky-coloured wines. But the vines did not graft well to the earliest American rootstocks and Cahors fell into decline. Compatible rootstocks were developed, and with the introduction of Merlot and Tannat, Cahors has started to improve. The sheer number of good producers now makes Cahors one of the most reliable red-wine appellations in France. The style used to require some ageing before being drinkable but a more modern approach taken by some producers now allow this to be consumed earlier. Vieux Cahors must be aged in oak for at least three years.

RED Most Cahors wines have a deep colour with a blackcurrant tinge. They are full of fruit and have a good, plummy, Bordeaux-like taste, with a silky texture and a distinctive violet-perfumed aftertaste.

🍷 *A minimum of 70% Malbec, plus a maximum of 30% (in total) of Merlot and Tannat*

🍷 *3–12 years (20 in exceptional cases)*

✓ *Château de Clasou • Château la Caminade (Commandery, Esprit) • Château du Cèdre • Château de Chambert • Cosse-Maisonneuve ⓝ • de la Coustarelle • Croix du Mayne (Elevé en Fûts de Chêne) • CV Côtes d'Olt • Clos de Gamot • Château Gautoul • des Grauzils • Château Haut Monplaisir (Pur Plaisir) • Clos d'Un Jour • Château Lacapell Cabanac ⓝ (Prestige Elevé en Fûts de Chêne) • Château Lagrezette • Château Lamartine (Expression, Particulière) • Primo Palatum • du Prince • Château La Reyne (Vente d'Ange) • des Saverines • Clos Triguedina*

CORRÈZE AOC

Includes a sub-zone "Coteaux de la Vézère" for wines from three municipalities of the département: Allassac, Donzenac, and Voulezac, and also a straw wine *vin de paille*. Achieved AOC status in 2017.

RED Expressive red fruit aromas with spicy notes.

🍷 *Corrèze AOC, vin de paille: Cabernet Franc, with optional Merlot and Cabernet Sauvignon; Coteaux de la Vézère: 100% Cabernet Franc*

🍷 *Dry 1–3 years, vin de paille 2–6 years*

WHITE The Chenins are aromatic and sometimes offer honey notes; the *vin de paille* is overripe with dried fruit.

🍷 *As part of Coteaux de la Vézère: Chenin Blanc; Vin de paille: Chardonnay, Sauvignon Blanc*

🍷 *Dry 1–3 years, vin de paille 2–6 years*

COTEAUX DU QUERCY AOC

Located just south of the Cahors area, and a *vin de pays* since 1976. In 2011 Coteaux du Quercy received full AOC status.

RED Richly coloured, full-bodied wines, the best of which have plenty of Morello cherry and black-berry fruit.

🍷 *Between 40% and 60% Cabernet Franc, plus optional Malbec, Merlot, Tannat, and Gamay.*

🍷 *1–4 years*

ROSÉ Soft, easy-drinking, fresh, and fruity rosé, with a maximum of 3 grams per litre residual sugar, thus truly dry.

🍷 *Between 40% and 60% Cabernet Franc, plus optional Malbec, Merlot, Tannat, and Gamay.*

🍷 *Upon purchase*

✓ *de Merchien*

CÔTES DE BERGERAC AOC

Geographically, there are no *côtes*; this appellation carries an extra half degree of alcohol minimum over Bergerac AOC, and the *cépage* for reds is more restrictive.

RED Should be richer than Bergerac AOC.

🍷 *Cabernet Sauvignon, Cabernet Franc, Merlot, and Malbec*

🍷 *3–10 years*

WHITE Dry Bordeaux-style wines.

🍷 *Sémillon, Sauvignon Blanc, Sauvignon Gris, Muscadelle, Ondenc, Chenin Blanc, and Ugni Blanc (with the proviso that the quantity cannot exceed the total of Sauvignon Blanc and Sauvignon Gris used)*

🍷 *1–3 years*

✓ *Château La Bard Les Tendoux (Elevé en Fûts de Chêne) • Château Les Mailleries (Dany Moelleux) • Château Masburel • Marlene & Alain Mayet (Révelation du Bois de Pourquie) • Château Tour des Verdots*

CÔTES DE DURAS AOC

An appellation of increasing interest.

RED Light Bordeaux-style wines.

🍷 *Cabernet Sauvignon as well as Franc, Merlot, amd Malbec*

🍷 *2–3 years*

WHITE Clean, crisp, and dry wines, except for those designated *moelleux*, which must have a minimum of 12 grams per litre of residual sugar.

🍷 *Sauvignon Blanc and Gris, Sémillon, Muscadelle, Mauzac, Chenin Blanc, Ondenc, and up to 25% Ugni Blanc/ Colombard (provided that the quantity of Sauvignon Blanc or Gris used is at least equal)*

🍷 *1–3 years*

ROSÉ These attractively coloured, dry, crisp, fruity rosés are firm and fresh.

🍷 *Cabernet Sauvignon, Cabernet Franc, Merlot, and Malbec*

🍷 *1–3 years*

✓ *des Allergrets (Elevé en Fûts de Chêne) • Duc de Berticot (Elevé en Fûts de Chêne) • Château la Grave Bechade (Alexandre Elevé en Fûts de Chêne) • Château Laplace • Château Moulière • Château La Petite Bertrande (Elevé en Fûts de Chêne) • du Petit Malromé (Sarah Elevé en Fûts de Chêne)*

CÔTES DU MARMANDAIS AOC

This successful Bordeaux imitation is situated on the border of Bordeaux itself; its vines grow on either side of the Garonne, on the left bank, Côtes de Cocumont, and the right bank, Côtes de Beaupuy. Few French people outside the region know anything about these wines, but the English have imported them since the 14th century.

RED Fresh, clean, and impeccably made wines.

🍷 *A maximum of 85% (in total) of Cabernet Franc, Cabernet Sauvignon, and Merlot, plus up to 50% (in total) of Abouriou, Malbec, Fer, Gamay, and Syrah*

🍷 *2–5 years*

WHITE These dry white wines are soft and delicious.

🍷 *Sauvignon Blanc and Gris, plus a maximum of 30% combined of Ugni Blanc and Sémillon*

🍷 *1–2 years*

ROSÉ Ripe and dry wines.

🍷 *A maximum of 85% (in total) of Cabernet Franc, Cabernet Sauvignon, and Merlot, plus up to 50% (in total) of Abouriou, Malbec, Fer, Gamay, and Syrah*

🍷 *1–2 years*

✓ *CV de Cocumont (Beroy) • Château La Gravette (Elevé en Fûts de Chêne) • Elian da Ros*

CÔTES DE MILLAU AOC

This appellation gained full AOC status in 2010.

RED The best *cuvées* usually contain a significant proportion of Syrah. *Vin primeur* is a speciality.

🍷 *A minimum of 30% Gamay and Syrah, between 10 to 30% Cabernet Sauvignon plus Fer and Duras*

WHITE Cannot be compared to the acidity found in the best Loire Chenin Blanc but are fresh and good for quaffing.

🍷 *A minimum of 50% Chenin Blanc, plus a minimum of 10% Mauzac*

ROSÉ Fresh and delicately fruity.

🍷 *A minimum of 50% Gamay, plus optional Syrah, Cabernet Sauvignon, Fer, and Duras*

✓ *CV des Gorges du Tarn (Maitre des Sampettes) • du Vieux Noyer*

CÔTES DE MONTRAVEL AOC

This wine must have a minimum of 25 grams and a maximum of 54 grams per litre of residual sugar in order to meet the requirements of the appellation. Any red wines produced here are sold as Bergerac AOC.

WHITE These fat, fruity wines are usually produced in a *moelleux* style.

🍷 *A minimum of 30% Sémillon, Sauvignon Blanc, and Gris, Muscadelle and a maximum of 10% Ondenc*

🍷 *3–8 years*

✓ *Château du Bloy • Château Lespinassat (Vieilles Vignes) • Château Masburel*

ENTRAYGUES–LE FEL AOC

This area overlaps the Aveyron and Cantal *départements* in the northeast of the region, but only a tiny amount is made. The wines are rarely encountered and never exported. Promoted to AOC in 2010.

RED Light, rustic wines that are best consumed locally because they need all the local ambience they can get.

🍷 *A minimum of 50% Fer, plus a minimum of 10% Cabernet Franc and Sauvignon and a maximum of 15% Mouyssaguès and Négret de Banhars*

🍷 *1–2 years*

WHITE Light, dry, crisp, and unpretentious wines.

🍷 *A minimum of 90% Chenin Blanc, plus optional Mauzac and Saint-Côme*

🍷 *Upon purchase*

ROSÉ Light, fresh, and very fruity, with an off-dry finish.

🍷 *A minimum of 50% Fer, plus a minimum of 10% Cabernet Franc and Sauvignon and a maximum of 15% Mouyssaguès and Négret de Banhars*

🍷 *1–2 years*

✓ *Jean-Marc Viguer*

ESTAING AOC

This area is contiguous with the southern tip of Entraygues–Le Fel AOC, and production is even more minuscule.

RED These wines are light bodied, attractive, and fruity.

🍷 *A minimum of 50% Fer and Gamay, plus a minimum of 10% Cabernet Franc or Sauvignon and a maximum of 30% Abouriou, Castet, Duras, Merlot, Mouyssaguès, Négret de Banhars, and Pinot Noir*

🍷 *1–3 years*

WHITE These are unpretentious, dry white wines with a crisp flavour and a rustic, tangy style.

🍷 *A minimum of 50% Chenin Blanc, a minimum of 10% Mauzac, and a maximum of 30% Saint-Côme*

🍷 *1–2 years*

ROSÉ These pleasant dry wines are probably the most interesting in the appellation.

🍷 *A minimum of 50% Fer and Gamay, plus a minimum of 10% Cabernet Franc or Sauvignon and a maximum of 30% Abouriou, Castet, Duras, Merlot, Mouyssaguès, Négret de Banhars, and Pinot Noir*

🍷 *Upon purchase*

✓ *CV d'Olt (Prestige)*

FLOC DE GASCOGNE AOC

This *vin de liqueur* is produced in the Armagnac region.

WHITE Few manage to rise above the typically dull, oxidative *vin de liqueur* style that goes *rancio* with age.

🍷 *A minimum of 70% Colombard, Gros Manseng, and Ugni Blanc, plus Baroque, Folle Blanche, Petit Manseng, Mauzac, Sauvignon Blanc and Gris, and Sémillon; no single varietal greater than 50%, however*

🍷 *Upon opening*

ROSÉ Few manage to rise above the typically dull, oxidative *vin de liqueur* style that goes *rancio* with age.

🍷 *A maximum of 50% Tannat, plus Cabernet Franc, Cabernet Sauvignon, Fer, Malbec, and Merlot*

🍷 *Upon opening*

✓ *CV du Muscat de Lunel (Prestige) • Saint Pierre de Paradis (Vendange d'Automne)*

FRONTON AOC

Previously Côtes du Frontonnais and renamed Fronton in 2008, this AOC is situated just west of Gaillac.

RED These medium- to full-bodied wines have excellent colour and violet-perfumed fruit.

🍷 *A minimum of 50% Négrette, a maximum of 40% Syrah, minimum of 25% Malbec and 25% Fer, a maximum of 25% Cabernet Franc and Cabernet Sauvignon combined, a maximum of 15% Gamay, plus a maximum of 5% combined Cinsault and Mérille*

🍷 *2–8 years*

ROSÉ Overtly fruity wines.

🍷 *A minimum of 50% Négrette, a maximum of 40% Syrah, a minimum of 25% Malbec and 25% Fer, a maximum of 25% Cabernet Franc and Cabernet Sauvignon combined, a maximum of 15% Gamay, plus a maximum of 5% combined Cinsault and Mérille*

🍷 *1–3 years*

✓ *Château Baudare • Château Bellevue la Forêt • Château Bouissel • Château Devès (Allegro) • Château Laurou (Elevé en Fûts de Chêne) • Château Plaisance ❶ • Château le Roc*

GAILLAC AOC

These vineyards, among the oldest in France, have only recently begun to make their mark. In order to emphasize local styles, there has been a concerted move away from classic grapes towards different native varieties. Sweet wines are available under the Gaillac Doux and Gaillac Vendanges Tardive appellations. The sparkling wines of Gaillac are now under the Gaillac Mousseux

BLACK DURAS GRAPES AWAIT HARVESTING
The Duras variety is found in just a few places, including the Côtes de Millau, Gaillac, and Vins d'Estaing AOCs. Duras is usually blended with other traditional varieties, such as Fer and Négrette, to make robust reds with peppery notes.

AOC and Gaillac Méthode Ancestrale (can also include *doux*) AOC appellations.

RED Wines made mostly in the fresh, soft-but-light carbonic-maceration style.

🍷 *A minimum of 60% Duras, Fer, Gamay, and Syrah, of which Duras and Fer must represent at least 40% of the entire blend; furthermore, there must be a minimum of 10% each of Duras and Fer, plus optional Cabernet Franc, Cabernet Sauvignon, Merlot, Gamay, and (not greater than 10%) Prunelard. Those red wines sold as primeur must be 100% Gamay.*

🍷 *1–3 years*

WHITE Dry and fresh.

🍷 *A minimum of 50% Len de l'El, Mauzac, Mauzac Rosé, and Muscadelle, plus Sauvignon Blanc and Ondenc*

🍷 *Upon purchase*

ROSÉ Easy to drink, light, fresh, and dry rosés.

🍷 *A minimum of 60% Duras, Fer, Gamay, and Syrah, of which Duras and Fer must represent a minimum of 40% of the entire blend; furthermore, there must be a minimum of 10% each of Duras and Fer, plus optional Cabernet Franc, Cabernet Sauvignon, Merlot, Gamay, and (a maximum of 10%) Prunelard*

🍷 *1–2 years*

✓ *Causse Marines (Délires d'Automne) • Manoir de l'Émeille (Tradition) • d'Escausses (La Vigne l'Oubli) • de Gineste (Aurore, Grand Cuvée) • de Larroque (Privilège d'Autan) • Château de Lastours • Château Lecusse (Spéciale) • Château Montels • Paysels (Tradition) • Robert Plageoles • René Rieux (Concerto) • Rotier • des Terrisses (Saint-Laurent) • Château La Tour Plantade*

GAILLAC DOUX AOC

These are naturally sweet wines that must contain a minimum of 45 grams per litre of residual sugar.

WHITE Sweet to very-sweet wines of ripe-peach, or richer, character.

🍷 *A minimum of 50% of Len de l'El, Mauzac, Mauzac Rosé, and Muscadelle, plus Sauvignon Blanc and Ondenc*

🍷 *5–15 years*

✓ *Barreau (Caprice d'Automne) • Château Palvie (Les Secrets) • Peyres-Combe (Flaveurs d'Automne) • Robert Plageoles • Rotier (Renaissance) • Sanbatan (Muscadelle) • de Vayssette (Maxime) • Les Vergnades*

GAILLAC MOUSSEUX AOC

This is a sparkling wine made by the traditional method.

SPARKLING WHITE These white wines are fresh and fragrant with a fine sparkle.

🍷 *A minimum of 50% of Len de l'El, Mauzac, Mauzac Rosé, and Muscadelle, plus Sauvignon Blanc and Ondenc*

🍷 *1–3 years*

SPARKLING ROSÉ Attractive, fresh, fruity wines.

🍷 *A minimum of 60% Duras, Fer, Gamay, and Syrah, of which Duras and Fer must represent a minimum of 40% of the entire blend; furthermore, there must be a minimum of 10% each of Duras and Fer, plus optional Cabernet Franc, Cabernet Sauvignon, Merlot, Gamay, and (not greater than 10%) Prunelard*

🍷 *1–2 years*

✓ *Manoir de l'Emeille • René Rieux*

GAILLAC MÉTHODE ANCESTRALE AOC

Sparkling wines made by the *méthode ancestrale*, involving just one fermentation, with no addition of a *liqueur de tirage*. The wine is bottled before the fermentation stops and no *liqueur d'expédition* is added prior to distribution, thus any residual sweetness is entirely from the original grape sugars. Styles include *brut* and *demi-sec*. A *doux* is also available but is governed by stricter rules and is given its own appellation (*see* next column).

HARVESTERS PICK FROM GOBELET-TRAINED VINES IN THE DOMAINE PLAGEOLES VINEYARD
This Gaillac producer farms Mauzac Vert, Mauzac Noir, Ondenc, Duras, Musscadelle, and Prunelart grapes, with a focus on cultivating varieties indigenous to the region.

SPARKLING WHITE These very fresh, fragrant, and grapey wines have a fine natural sparkle.

🍷 *Mauzac, Mauzac Rosé*

🍷 *1–3 years*

SPARKLING ROSÉ These wines are attractive, fresh, and deliciously fruity.

🍷 *A minimum of 60% Duras, Fer, Gamay, and Syrah, of which Duras and Fer must represent a minimum of 40% of the entire blend; furthermore, there must be a minimum of 10% each of Duras and Fer, plus optional Cabernet Franc, Cabernet Sauvignon, Merlot, Gamay, and (a maximum of 10%) Prunelard*

🍷 *1–2 years*

GAILLAC MÉTHODE ANCESTRALE DOUX AOC

This is a sparkling wine made by the *méthode ancestrale* (a minimum of 7 per cent alcohol) with at least 50 grams of residual natural sugar per litre.

SPARKLING WHITE Not as exotic as, say, Clairette de Die Méthode Dioise Ancestrale, but delicious, grapey, and fragrant all the same.

🍷 *Mauzac, Mauzac Rosé*

🍷 *1–3 years*

SPARKLING ROSÉ I have never come across a Gaillac Doux rosé, but if the white is anything to go by, it would be an interesting wine.

🍷 *A minimum of 60% Duras, Fer, Gamay, and Syrah, of which Duras and Fer must represent a minimum of 40% of the entire blend; furthermore, there must be a minimum of 10% each of Duras and Fer, plus optional Cabernet Franc, Cabernet Sauvignon, Merlot, Gamay, and (a maximum of 10%) Prunelard*

🍷 *1–2 years*

GAILLAC PREMIÈRES CÔTES AOC

These are dry white wines that come from 11 communes. The grapes must be riper than for ordinary Gaillac AOC, with a minimum alcohol of 11 per cent not surpassing 4 grams per litre of residual sugar.

WHITE A minimum of 50% Len de l'El, Mauzac, Mauzac Rosé, and Muscadelle, plus Sauvignon Blanc and Ondenc

✓ *Robert Plageoles • Château de Salettes*

GAILLAC VENDAGE TARDIVE AOC

These are late-harvest sweet wines that must contain a minimum of 100 grams per litre of residual sugar.

WHITE Sweet to very-sweet wines of ripe-peach, or richer, character.

🍷 *A minimum of 50% of Len de l'El and Ondenc, plus Mauzac, Mauzac Rosé, and Muscadelle*

🍷 *5–15 years*

HAUT-MONTRAVEL AOC

This wine must have a minimum of residual sugar of 85 grams per litre to meet the requirements of the appellation. Any red wines produced here are sold as Bergerac AOC.

WHITE Fat, fruity, and *moelleux* wines.

🍷 *A minimum of 50% Sémillon, with Sauvignon Blanc and Gris and Muscadelle plus a maximum of 10% Ondenc*

🍷 *3–8 years*

✓ *Château Puy-Servain*

IROULÉGUY AOC

The local *coopérative* dominates this Basque appellation, which makes some of the most distinctive red wines in Southwest France. Surprisingly, however, the production of rosé outweighs that of red.

RED These deep, dark-coloured, tannic wines have a rich and mellow flavour, with a distinctive earthy-and-spicy aftertaste.

🍷 *A minimum of 50% Tannat and Cabernet Franc (neither of these two exceeding 90%), with Cabernet Sauvignon*

🍸 *4–10 years*

WHITE These modest dry whites are the least interesting of this appellation.

🍷 *Courbu, Petit Courbu, Gros Manseng, and Petit Manseng*

ROSÉ This salmon-coloured, very fruity dry rosé is best drunk very young and fresh.

🍷 *A minimum of 90% Tannat, Cabernet Sauvignon, or Cabernet Franc, with Courbu, Petit Courbu, Gros Manseng, and Petit Manseng*

🍷 *Upon purchase*

✓ *Arretxa* ❶ • *Etxegaraya* • *Henri Mina*

JURANÇON AOC

The sweet version of this wine from the Pyrénées-Atlantiques was used at Henri de Navarre's christening in 1553 and is often sold today as *vendanges tardives*. Most production is, however, tart, dry, and nervy, and sold as Jurançon Sec (*see* next column).

WHITE The best wines have a fine, spicy, and tangy bouquet and flavour, and can hint of pineapples and peaches, candied peel, and cinnamon.

🍷 *A minimum of 50% Petit Manseng and Gros Manseng, with Courbu, Petit Courbu, Camaralet, and Lauzet. Those wines labelled* vendanges tardives *must be made exclusively from Petit Manseng and Gros Manseng.*

🍸 *5–20 years*

✓ *Bru-Baché* • *Cauhapé* • *Clos Lapeyre* ❶ • *Primo Palatum* • *Clos Thou* ❶ • *Clos Uroulat*

JURANÇON SEC AOC

This wine has to meet the same requirements as Jurançon AOC, but less residual sugar is allowed (a maximum of 4 grams per litre), and the grapes may be less ripe. It is best drunk young.

WHITE If any wine in the world habitually has a certain nervousness, it has to be Jurançon Sec. Most lack any individual character that would make them stand out from other dry whites, and none has the complexity and richness of Jurançon's late-harvest wines, but the best can have an intensity that hints of grapefruit.

🍷 *A minimum of 50% Petit Manseng and Gros Manseng, with Courbu, Petit Courbu, Camaralet, and Lauzet*

🍷 *2–5 years*

✓ *Capdevieille (Brise Océane)* • *Cauhape (Sève d'Automne)* • *du Cinquau* • *CV de Jurançon (Grain Sauvage)* • *Lapeyre* ❶ • *Larredya* • *Nigri* • *Primo Palatum*

MADIRAN AOC

This may be one of the most individually expressive appellations in Southwest France, but there are as many disappointments as successes. Many *domaines* are trying new oak, and there is a trend towards more Cabernet Franc.

RED You literally have to chew your way through the tannin in these dark, rich, and meaty wines when young.

🍷 *A range of 60 to 80% Tannat, plus Cabernet Franc, Cabernet Sauvignon, and Fer*

🍷 *5–15 years*

✓ *Berthoumieu* • *Château Bouscassé* • *Château Labranche-Laffont* • *Château Laffitte-Teston* • *Laffont* • *Château Montus* • *Plaimont (Arte Benedicte)* • *Primo Palatum* • *Château de Viela*

MARCILLAC AOC

This rarely encountered wine from the northeastern borderlands was upgraded to Marcillac AOC in 1990, when greater focus was placed on the local Fer and Gamay, while Jurançon Noir, Mouyssaguès, and Valdiguié were disallowed.

RED Rough and rustic when young, these wines soften with age.

🍷 *A minimum of 90% Fer, plus Cabernet Sauvignon, Merlot, and Prunelard*

🍸 *3–6 years*

ROSÉ Full, ripe, and attractive dry rosés that have expressed more individual style since the minimum Fer content has tripled.

🍷 *A minimum of 90% Fer, plus Cabernet Sauvignon, Merlot, and Prunelard*

🍸 *1–3 years*

✓ *CV de Ladrecht*

MONBAZILLAC AOC

An excellent value Sauternes-style appellation at the heart of Bergerac, these wines date back to 1080, when vines were planted by the Abbey of St-Martin on a hill called Mont Bazailhac.

WHITE These intensely sweet, rich wines are of a very high quality.

🍷 *Sémillon, Sauvignon Blanc and Gris, Muscadelle with a maximum 10% of Chenin Blanc, Ondenc and Ugni Blanc*

🍸 *7–20 years*

✓ *de l'Ancienne Cure (Abbaye, L'Extase)* • *Château Caillavel* • *Château Fonmorgues* • *Grande Maison* ❶ *(Cuvée du Château)* • *Château Tirecul La Gravière*

CHATEAU DE MONBAZILLAC, IN THE DORDOGNE REGION, IS HOME TO THE APPELLATION COOPERATIVE WINERY
The best-known property in the appellation, this château dates to about 1550, although Monbazillac's vineyards can be traced back to 11th-century Benedictine monasteries. Monzabazillas is famous for sweet wines produced from botrytised Sémillon, Sauvignon Blanc, and Muscadelle grapes.

MONTRAVEL AOC

The largest of the three Montravel appellations and the only one where the white wine can (and must) be dry.

RED Elegant, ruby-coloured reds that are very much in the Bergerac style.

🍷 *A minimum of 50% Merlot, plus Cabernet Franc, Cabernet Sauvignon, and Malbec*

🍾 *2–8 years*

WHITE These are dry, crisp, and aromatic Sauvignon-dominated wines.

🍷 *A minimum of 25% Sémillon and 25% Sauvignon Blanc or Gris, plus Muscadelle; optional Ondenc may not exceed 10%*

🍾 *1–2 years*

✓ *Château du Bloy (Le Bloy) • Château Laulerie • Moulin Caresse (Cent pour 100) • Château Pagnon • Château Pique-Sègue • Portes du Bondieu • Château Puy-Servain (Marjolaine)*

PACHERENC DU VIC-BILH AOC

New oak is being increasingly used in this white-only appellation, which covers the same area as Madiran.

WHITE Exotic floral aromas, a fruit salad of flavours, Pacherenc du Vic-Bilh is made dry (sold as *sec* with a maximum 4 grams per litre), medium-sweet (*moelleux*), and sweet (*doux*) styles, with both sweeter styles being upwards of 45 grams per litre.

🍷 *A minimum of 60% from Courbu, Petit Courbu, and Gros Manseng and Petit Manseng (with no single variety being more than 80%), optionally Arrufiac and (a maximum of 10%) Sauvignon Blanc*

🍾 *3–7 years*

✓ *Château Bouscassé • Berthoumieu • Château Labranche-Laffont • Château de Viela*

PÉCHARMANT AOC

These are the finest red wines of Bergerac; all but one (Saint-Saveur) of the communes in this appellation are also within the Rosette AOC.

RED All the characteristics of the Bergerac, but with a greater concentration of colour, flavour, and tannin.

🍷 *Cabernet Franc, Cabernet Sauvignon, Merlot, Malbec with no single variety exceeding 65% of the blend.*

🍾 *4–12 years*

✓ *Château Beauportail • Château de Biran (Prestige de Bacchus) • Château Champarel • Château Hugon (Elevé en Fûts de Chêne)*

ROSETTE AOC

A white-wine-only appellation. The wines must contain 25 to 51 grams per litre of residual sugar. Any red wines are sold as Bergerac AOC.

WHITE These wines are soft and delicately fruity with a sweet, waxy flavour.

🍷 *Between 15% and 70% Sauvignon Blanc and Gris, with Sémillon and Muscadelle*

🍾 *4–8 years*

✓ *Château de Contancie*

SAINT-MONT AOC

Situated within the Armagnac region, the vineyards of this appellation, which achieved AOC status in 2011, extend northwards from Madiran. Wine production is dominated by the local *coopérative*.

RED Well-coloured wines of good flavour and medium body. They are not dissimilar to a lightweight Madiran, although they could become deeper, darker, and more expressive if the *coopérative* were prepared to reduce yields and produce better-quality wine.

🍷 *A minimum of 60% Tannat, plus a minimum of 20% from Cabernet Sauvignon and Fer, with optional Cabernet Franc and Merlot*

🍾 *2–5 years*

WHITE Fruity, dry wines with a tangy finish.

🍷 *A minimum of 50% Gros Manseng, with 20% from Arrufiac and Petit Courbu; Courbu and Petit Manseng are optional*

🍾 *1–2 years*

ROSÉ Dry wines with a clean, fruity flavour.

🍷 *A minimum of 60% Tannat, plus a minimum of 20% from Cabernet Sauvignon and Fer, with optional Cabernet Franc and Merlot, and a maximum of 10% white varieties*

🍾 *1–3 years*

✓ *Producteurs Plaimont • Château Saint-Go (Elevé en Fûts de Chêne)*

SAINT-SARDOS AOC

Situated within the Lot-et-Garonne *département* along the Garonne southwest of Montauban, Saint-Sardos achieved AOC status in 2011 and is an appellation covering red and rose wines.

RED Spicy, hints of floral, fruit and liquorice.

🍷 *A minimum of 40% Syrah, a minimum of 20% Tannat, and a maximum of 15% Merlot. Cabernet Franc may also be used.*

🍾 *2–5 years*

ROSÉ Rounded, soft and vibrant, with red fruits and appealing floral flavours.

🍷 *A minimum of 40% Syrah, a minimum of 20% Tannat, and a maximum of 15% Merlot. Cabernet Franc may also be used.*

🍾 *1–3 years*

SAUSSIGNAC AOC

The wines of this sweet-wine appellation must have a minimum of 68 grams of residual sugar per litre. Any red wines are sold as Bergerac AOC.

WHITE The best can be very rich, fat, and full.

🍷 *Sémillon, Sauvignon Blanc and Gris, Muscadelle with a maximum of 10% Chenin Blanc, Ondenc, and Ugni Blanc*

🍾 *5–15 years*

✓ *Château Le Chabrier ⊙ • Château Eyssards (Flavie) • Château Grinou ⊙ (Elevé en Fûts de Chêne) • Château La Maurigne (La Maurigne) • Château Miaudoux*

TURSAN AOC

The reds rely on the Tannat, the same primary grape used in Madiran; the whites are essentially Baroque, a variety more at home in Tursan.

RED Rich and chewy, or finer-flavoured and aromatic, depending on the dominant grape.

🍷 *A minimum of 70% Cabernet Franc and Tannat, plus optional Cabernet Sauvignon, Merlot, and Fer; with each participating variety present at a minimum of 20%*

WHITE The *coopérative* traditionally makes a full-bodied white with a solid, somewhat rustic, rich flavour, but it is gradually being influenced by the growing reputation of the wine made under the relatively new Château de Bachen label, Baron de Bachen, which is far more aromatic and elegant.

🍷 *A minimum of 70% Baroque and Gros Manseng, plus optional Chenin Blanc, Petit Manseng, Raffiat de Moncade Sauvignon Blanc, and Sauvignon Gris. Each participating variety must be present as a minimum of 20%.*

🍾 *2–5 years*

ROSÉ These are unpretentious dry wines with good, juicy, fruit flavour.

🍷 *A minimum of 70% Cabernet Franc and Tannat, plus optional Cabernet Sauvignon, Merlot, and Fer with each participating variety present at a minimum of 20%.*

🍾 *1–3 years*

✓ *Michel Guérard (Baron de Bachen) • CV de Landais (Haute Carte)*

AN ADVERTISING BOTTLE OF TURSAN WINE IS MOUNTED ON A WALL IN SAINT-SEVER
The vineyards covered by the lesser-known Tursan AOC lie on the western edge of a cluster of better-known appellations, such as Armagnac and Madiran.

MORE GREEN WINES

In addition to the producers recommended in this directory that are either biodynamic or organic, there are also the following. No negative inference of quality should be taken from the fact that they are not featured among the other recommended producers. There are a number that have been recommended in other editions, and still make some fine wines, but have been culled to make room for others.

BIODYNAMIC
de Beaulande (Nastringues), de Lafage (Montpezat de Quercy), Chateau Laroque (St Antoine de Breuilh), de Souch (Jurançon), Château Vent d'Autun (Saint Matre)

ORGANIC
Chateau Haut Pontet (Vélines), Chateau Larchère (Pomport), de Matens (Gaillac), du Petit Malromé (St Jean de Duras), Château Richard (Monestier), Clos Julien (St Antoine de Brieuil), Château du Seigneur (Bergerac), Tinou (Lasseube), de La Tronque (Castelnau de Montmirail), Domaine des Vignals (Cestayrols), Château Le Clou (Pomport), de Durand (St Jean de Duras), de L'Antenet (Puy L'Évêque), Château le Barradis (Monbazillac), Baulandes (Vélines), Bois Moisset (Montans), Cailloux (Pomport), Pajot (Eauze), Theulet Marsalet (Monbazillac)

Languedoc-Roussillon

Dollar for dollar, the swelling numbers of top-quality red wines from Languedoc-Roussillon provide more value and excitement than any of the famous wine regions of France.

As well as being the largest wine region of France, Languedoc-Roussillon is also without doubt the most exciting, with a dynamism that other regions, which perhaps have more established reputations, often lack. It has attracted newcomers from outside the wine industry, for whom everything is possible. Once the main source of *le gros rouge*, anonymous red wine without any distinguishing characteristics – as well as making a sizeable contribution to the European wine lake – Languedoc-Roussillon now produces a host of individual and characterful wines that are often unbounded by appellation requirements. Traditionally the Languedoc has always been grouped with Roussillon, but the two regions are quite different. Roussillon covers the department of the Pyrenées-Orientales and was once part of Spain. They speak Catalan there, whereas the language of the Languedoc is Occitan. Roussillon has a strong tradition for *vins doux naturels,* fortified wines based on Grenache Noir, Blanc, and Gris. In the Languedoc the *vins doux naturels* are much less important, with just four small appellations based on Muscat, but there is an enormous variety of different table wines. These are mainly red, but white wines are also growing in quality and importance. And, surprisingly, given the popularity of Côtes de Provence rosé, the Languedoc makes more rosé than Provence.

LAND OF OPPORTUNITY

The region stretches from the fortified city of Carcassonne northeastward past Narbonne and Montpellier to Nîmes and to the estuary of the Rhône Valley and southeastward down to the border with Spain. Malepère, Cabardès, and Limoux are where Mediterranean influences meet Atlantic ones, producing *bordelais* as well as Mediterranean varieties. The vineyards continue east, with Corbières and Minervois on either side of the Aude river, and form a large amphitheatre in the hills, with St Chinian, Faugères, the new appellations of Terrasses du Larzac, and Pic St Loup. Nearer the coast are La Clape, with wonderful saline flavours in its white wines, and Picpoul de Pinet, with its fresh acidity making it the perfect accompaniment to oysters. The boundary between the Languedoc and the Rhône Valley is blurred with an overlap of appellations and IGPs (*Indication Géographique Protégée*) outside Nîmes.

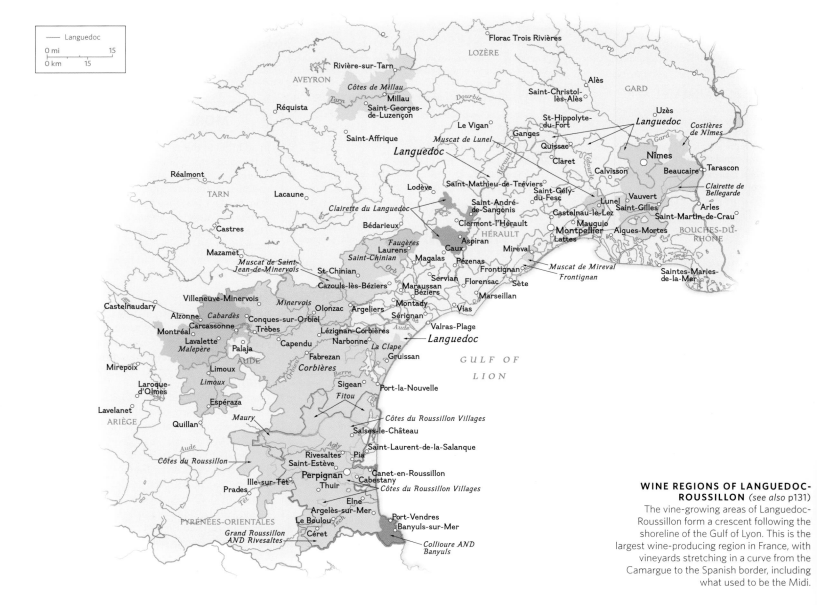

WINE REGIONS OF LANGUEDOC-ROUSSILLON (*see also p131*) The vine-growing areas of Languedoc-Roussillon form a crescent following the shoreline of the Gulf of Lyon. This is the largest wine-producing region in France, with vineyards stretching in a curve from the Camargue to the Spanish border, including what used to be the Midi.

RECENT LANGUEDOC-ROUSSILLON VINTAGES

2018 A difficult year. The unusually wet spring resulted in widespread mildew; the hot summer rectified some of the problems, so quality turned out well, but yields were low.

2017 Spring frosts, rarely seen in the Languedoc, curtailed yields in some areas. A summer drought also limited the crop. Quality was good but quantity low.

2016 A very dry summer made for a small crop of beautifully balanced wines. Hail in the Pic St-Loup decimated some vineyards.

2015 A very balanced year with rain at the right time. Summer temperatures were not too high, and the quality is excellent.

2014 – A challenging year with a warm spring making for an early bud break. Hail in La Clape and the Minervois. The first two weeks of September were warm and sunny, but then came the rain, which caused problems for those favouring a later harvest. The best wines are balanced with freshness and fruit.

Outsiders have had a huge impact on the industry in this region. In the last century, Australians played their part as flying winemakers and even buying vineyards. Today the outsiders come not just from elsewhere in France but from all over, with investment by Russians, the occasional Chinese entrepreneur, and most European countries. The attraction is the quality of the wine and the sense that nothing is impossible; land values are not yet extortionate, and the quality of the lifestyle along with the mild Mediterranean climate all add to the appeal.

The appellations are in a state of flux, however, with Languedoc at the bottom of the pyramid, then broader appellations such as St Chinian, Minervois, and Corbières, with various *crus*, such as Boutenac, La Livinière, and Roquebrun to complete the pyramid. Currently they are shifting sands. All the appellations are blends of two or more varieties, whereas the pure varietal wines are generally reserved for IGPs or indeed *vin de France*.

Two examples to illustrate the complexity: the recently recognised appellation of the Terrasses du Larzac has welcomed 25 new wine producers since 2011, mainly outsiders who have developed vineyards and created new estates. And the diversity of grape varieties in the Languedoc is such that the IGP Pays d'Oc, which covers the entire region, allows for 58 different grapes, with Albariño being the latest addition.

VINEYARDS GROW OUTSIDE THE WALLED CITY OF CARCASSONNE
In the western part of the Languedoc-Roussillon wine region, this medieval city forms the basis of the Cite de Carcassonne IGP. From the crenellated ramparts of the fortification, visitors can take in sweeping views of the vineyards surrounding the city.

FACTORS AFFECTING TASTE AND QUALITY

LOCATION
This region is a crescent of vineyards situated in southern France between the Rhône Valley to the east and the Pyrenees to the southwest, forming an amphitheatre around the Mediterranean coastline.

CLIMATE
The Mediterranean-influenced climate is generally well suited to the cultivation of the vine, although it is subject to occasional stormy weather. Two winds dominate: the cold and parching *mistral*, which blows down from the heights of the Alpine glaciers and the Massif Central, and the wet and warm *marin*, which comes in from the sea and can cause rot at harvest time. There are many microclimates in this collection of isolated vine-growing areas.

ASPECT
Famous for its unending tracts of flat, *vin ordinaire* vineyards that stretch across the vast plains. The best vineyard sites of Languedoc-Roussillon, however, mostly occupy south-, southeast-, and east-facing *garrigues* (scrublands) and hillsides or nestle beneath protective overhanging cliffs; there is a trend towards planting at higher altitudes.

SOIL
In general terms, the plains and valleys have rich alluvial soils, while the hillsides are schist or clay and limestone and the *garrigues*, or scrublands, are comprised of stony, carbonaceous soils over fissured limestone. Specific situations vary enormously, however.

VITICULTURE & VINIFICATION
This was once the great *vin ordinaire* region of France, where everything was mechanised and the vines of the plain were farmed like wheat or corn. Now there is a growing trend towards developing single-domaine vineyards that have potentially expressive *terroirs*, where classic varieties are grown and various traditional methods are combined with modern techniques. Limoux still practises the ancient *méthode rurale* for making sparkling wine – in Blanquette Méthode Ancestrale AOC (*see* opposite) – although the majority of the wines are *méthode champenoise* and sold either as Crémant or Blanquette de Limoux, depending on the grape variety blend,

GRAPE VARIETIES
Aspiran Gris, Aspiran Noir, Aubun, Bourboulenc (known erroneously as Malvoisie in Clape), Cabernet Sauvignon, Cabernet Franc, Carignan Noir, Carignan Blanc, Cinsaut, Clairette Blanche, Counoise, Fer, Grenache Noir, Grenache Blanc, Grenache Gris, Lledoner Pelut, Macabeo, Marsanne, Merlot, Mourvèdre, Muscat d'Alexandrie, Muscat Blanc à Petits Grains (*see* individual appellations for local synonyms), Oeillade, Piquepoul Blanc, Piquepoul Noir, Rolle (Vermentino), Roussanne, Syrah. Terret Blanc, Terret Noir, Tourbat (or Malvoisie du Roussillon)

THE APPELLATIONS OF
LANGUEDOC-ROUSSILLON

Note: In the following entries, a wine described as a *vin doux naturel* (or VDN) is made from very ripe grapes and fortified with pure grape spirit after its fermentation has reached 5 or 6 per cent. It has the natural sweetness of the grape. To be labelled "Rancio", a VDN must be stored in oak casks "according to local custom", and possibly also glass jars or demijohns, or *bonbonnes*, which often means exposing the barrels to direct sunlight, for a minimum of two years. This imparts the distinctive *rancio* flavour that is so prized in Roussillon. Depending on the colour of the original wine, the wine technique used, and how long it has been aged, the style of the wine produced varies; it can be red, white, rosé, or tawny.

BANYULS AOC

The most southerly appellation in France, with vineyards in four villages, notably Banyuls and Collioure, as well as Port-Vendres and Cerbère. The vines are grown on precipitous slopes of schist where man and mule have great difficulty maintaining a foothold; mechanisation is out of the question, yields are extremely low, and ripeness very high. You will often see *"rimage"* on labels; this word is derived from the Catalan *raïm*, or "grape", and describes a vintage wine, as opposed to a *rancio*.

RED TRADITIONNEL or **RIMAGE** This is the deepest and darkest of all VDNs. A rich, sweet, red Banyuls (without too much barrel-age) has a chocolaty, bottled-fruitiness, which is the nearest France gets to the great wines of Portugal's Douro region. It lacks the fire of a great port but has its own immense charm. After 15 to 20 years in bottle, a great Banyuls develops a curious but wonderful complexity that falls somewhere between the porty-plummy, dried-fruit spice of a mature vintage port and the coffee-caramel, nutty-raisiny smoothness of a fine old tawny.

🍷 *Grenache Noir, (plus Grenache Gris for Traditionnel) Carignan, Cinsaut, Cunois, Mourvedre, and Syrah*

🥂 *10–40 years*

WHITE, ROSÉ, AMBRÉ, TUILÉ Like all VDNs that may be made in red, white, and rosé style, they can all turn tawny with time, particularly *rancio* wines.

🍷 *All wines: Grenache Blanc, Grenache Gris, Macabeo, Tourbat, and a maximum of 20% of Muscat Blanc à Petits Grains and Muscat d'Alexandrie, and for white/ambré, a maximum 10% (in total) of Marsanne, Roussanne, Vermentino, and Carignan Blanc. For Tuile wines, the grapes of Banyuls Traditionnel also apply.*

🥂 *10–20 years*

✔ *CV l'Étoile (Extra Vieux) • Du Mas Blanc • De la Rectorie • De la Tour Vieille • Du Traginer ⊙ (Rimage Mis Tardive)*

BANYULS GRAND CRU AOC

The requirements for the Banyuls Grand Cru AOC are the same as for Banyuls AOC, but a minimum of 75 per cent Grenache Noir is required. The wine is matured in oak for at least 30 months. The wines are similar in character to those of the basic, although not at all ordinary, Banyuls appellation, but in terms of classic port styles, they veer more towards the tawny than vintage.

✔ *L'Abbé Rous (Christian Reynal) • L'Étoile (Réserve) • Cellier des Templiers*

BANYULS GRAND CRU "RANCIO" AOC
See Banyuls Grand Cru AOC

BANYULS "RANCIO" AOC
See Banyuls AOC

BLANQUETTE DE LIMOUX AOC

The individuality of Blanquette de Limoux as opposed to Crémant de Limoux is due to a high proportion of the local grape variety Mauzac, which can have an appealing rusticity and charm, that distinguishes it from more conventional sparkling wines.

SPARKLING WHITE These wines may have the distinctive aroma of fresh-cut grass, with fresh acidity.

🍷 *90% Mauzac, plus Chardonnay and/or Chenin Blanc*

🥂 *1–3 years (up to 12 for vintages)*

✔ *Antech (Tradition, Grande Réserve, Elégance) • Robert (Dame Robert, Maistre Blanquetïers) • Sieur d'Arques (Aimery Princesse) • Château Rives-Blanques. Delmas*

BLANQUETTE MÉTHODE ANCESTRALE AOC

Formerly called Vin de Blanquette, but still produced by the old *méthode rurale*.

SPARKLING WHITE These honeyed sweet sparkling wines are a hedonist's dream and should be far more commercially available. Some producers are attempting to streamline and modernise the ancient *méthode rurale*, thereby retaining the historical originality of Limoux.

🍷 *Mauzac*

✔ *Domaine Taudou, J. Laurens • Sieur d'Arques*

CABARDÈS AOC

An obscure appellation north of Carcassonne, these wines were once sold as Côtes du Cabardès et de l'Orbiel VDQS, but this hardly tripped off the tongue, thus the shortened form was in common usage long before the appellation was promoted to full AOC status in 1999. This is where the Atlantic meets the Mediterranean, combining the grape varieties of both regions.

RED The best wines have elegant fruit and a leaner, more Bordeaux-like balance than most of the warmer, spicy-ripe southern French reds.

🍷 *A minimum of 40% Grenache and Syrah, plus a minimum of 40% Cabernet Franc, Cabernet Sauvignon, and Merlot. The rest can be Cinsaut, Côt, and Fer.*

🥂 *3–8 years*

ROSÉ A rich, fruity, well-coloured rosé.

🍷 *A minimum of 40% Grenache and Syrah, plus a minimum of 40% Cabernet Franc, Cabernet Sauvignon, and Merlot*

🥂 *Between 2–3 years*

✔ *Domaine Cabrol (Vent d'Est) • Château de Pennautier (L'Esprit) • Château Ventenac. Domaine de Cazaban • Domaine Guilhem Barré*

CABRIÈRES AOC

This is a single commune of steep schist slopes close to the appellation of Clairette du Languedoc, making red, white, and rosé. The initial historic reputation of Cabrières was based on rosé.

✔ *Caves de l'Estabel, Mas Coris, Gérard Bertrand*

CLAIRETTE DU LANGUEDOC AOC

The oldest white appellation of the Languedoc may be dry, semi-sweet or *moelleux*, with 40 g/l residual sugar, as well as *rancio*, with three years ageing in barrel, and occasionally you may find it fortified as a *vin de liqueur*.

WHITE The dry white wine is fuller and balanced, and it has surprising ageability. *Rancio* is ripe and sweet, with refreshing acidity, and the *vin de liqueur is* rich and intense.

🍷 *Clairette Blanche*

🥂 *1–3 years or longer for naturally fermented wines, 8–20 years for fortified wines and rancio wines*

✔ *Domaine la Croix-Chaptal. Caves de l'Estabel, Domaines Paul Mas*

CLAIRETTE DU LANGUEDOC "RANCIO" AOC
See Clairette du Languedoc AOC

LA CLAPE AOC

An appellation since 2015 for red and white wines (any rosé is AOC Languedoc) coming from vineyards on a limestone outcrop, the Massif de la Clape, which was once an island before the river Aude changed its course. Bourboulenc is the defining grape variety, accounting for at least 40 per cent of the blend, or Bourboulenc with Grenache Blanc must make up 60 per cent, with Marsanne, Roussanne, Vermentino, Piquepoul, and Viognier, which is limited to 10 per cent. One of the most original white wines of the Languedoc, white La Clape has a wonderful salinity as so many of the vineyards are close to the sea, and it can age remarkably well. Red la Clape must be a blend of at least two varieties, namely Syrah, Mourvèdre, Grenache Noir, accounting for a minimum of 70 per cent of the blend, with Carignan and Cinsaut allowed to make up the rest.

✔ *Château d'Anglès • Château de Camplazens • Château l'Hospitalet • Château Négly (La Brise Marine, La Falaise) • Château Pech Redon • Château Rouquette-sur-Mer (Henry Lapierre) Domaine Sarrat de Goundy, Château Mire l'Etang*

COLLIOURE AOC

The red table wine twin of Banyuls produced in the same vineyard area.

RED These deep, dark, and powerful wines have a full and concentrated fruit flavour, with a soft, spicy aftertaste.

🍷 *A minimum of 60% (in total) of Grenache Noir, Mourvèdre, and Syrah (no variety may exceed 90%), plus Cinsaut, and a maximum of 30% Carignan.*

🥂 *2–10 years*

WHITE White Collioure was recognised in 2002.

🍷 *A minimum 70% in total of Grenache Blanc and/or Grenache Gris, plus a maximum of 30% in total (and individually a maximum of 15%) of Tourbat (Malvoisie de Roussillon), Macabeo, Marsanne, Roussanne, and Vermentino.*

🥂 *1–3 years*

ROSÉ From the same grape varieties as red Collioure, but rarely produced. Grenache Gris allowed up to 30 per cent.

🍷 *Same as for red*

🥂 *Upon purchase*

✔ *Du Mas Blanc • Celliers des Templiers (Abbaye de Valbonne) • Madeloc (Magenca) • De la Rectorie • De la Tour Vieille (Puig Oriol) • Du Traginer (Octobre) • Vial-Magnères • Coume del Mas • Les Clos de Paulilles*

CORBIÈRES AOC

When this appellation was elevated to full AOC status in December 1985, its area of production was practically halved, and has since decreased still further to a current 10,600 hectares (57,000 acres). The top estates may use carbonic maceration, followed by 12 months or so in oak and the results can be stunning. There is such a great diversity of *terroirs* in the appellation, that at one time there were moves to recognise as many as 11 different subzones. However, only one has been officially recognised, Boutenac, with its own AOC status (see Corbières-Boutenac). Currently the others in the pipeline are Alaric, (northerly zone around the dramatic Montagne d'Alaric, just west of Lézignan, with vines growing on well-drained slopes of gravel over limestone); Durban wedged between the two hilly halves of Fitou, this zone is cut off from any Mediterranean influence; Lagrasse protected limestone valley vineyards immediately west of Boutenac; and Corbières-Maritimes, which covers the coastal area around Sigean and the lagoons.

RED These wines have an excellent colour, a full, spicy-fruity nose, and a creamy-clean, soft palate that often hints of cherries, raspberries, and vanilla. Carignan, Syrah, Grenache Noir, Mourvèdre, with a minimum of two varieties, of which the principal grape is between 40 to 80 per cent of the blend. Carignan is the key variety. Cinsaut, Piquepoul Noir, and Terret Noir are also permitted, and Marselan is under consideration.

🍾 *A minimum of 50% Grenache, Lledoner Pelut, Mourvèdre, and Syrah, plus Carignan, Picpoul Noir, Terret, and a maximum of 20% Cinsaut (Macabeo and Bourboulenc are no longer allowed).*

🍷 *2–5 years (3–8 years in exceptional cases)*

WHITE Soft, dry wines that have acquired a more aromatic character, as white wines have improved all over the Languedoc There have been some successful experiments with oak fermentation. Two varieties minimum. Bourboulenc, Macabeo, Marsanne, Roussanne, Vermentino, Grenache Blanc.

🍷 *1–3 years*

ROSÉ The best of these dry wines have an attractive colour and are quite full-bodied and vinous, with fresh fruit. Two varieties minimum. Grenache Noir, Grenache Gris, Syrah, Cinsaut, Carignan.

🍷 *1–3 years*

✓ *Clos de L'Anhel • Clos Canos (Les Cocobirous) • CV du Mont Ténarel d'Octaviana (Sextant) • Les Clos Perdus • CV Castelmaure et d'Embrès • Domaine Ste Croix • Domaine la Cendrillon • Château Trillol . Domaine d'Aussières • Château la Baronne • Château Vieux Moulin*

CORBIÈRES-BOUTENAC AOC

This appellation in the central-northern district of Corbières is the first subzone to achieve its own AOC status, in 2005, and it encompasses nine villages in addition to Boutenac itself (Fabrezan, Ferrals-les-Corbières, Lézignan-Corbières, Luc-sur-Orbieu, Montséret, Ornaisons, Saint-André-de-Roquelongue, Saint-Laurent-de-la-Cabrerisse, and Thézan-des-Corbières). All the grapes must be manually harvested.

RED More black fruit than most Corbières.

🍾 *A minimum of 30% Carignan, with Grenache Noir, Mourvèdre, and Syrah*

🍷 *3–10 years*

✓ *Château la Voulte-Gasparets (Romain Pauc) • Château les Aiguilloux • Château de Luc • Château Ollieux-Romanis • Château Ste Estève*

CÔTES DU ROUSSILLON AOC

Covering a substantial part, 102 villages, of the department of the Pyrénées-Orientales, in the foothills of the Pyrenees. Recognised as an appellation in 1977.

RED The best of these wines has a good colour and a generosity of southern fruit, with some vanilla and spice.

🍾 *A blend of at least two of the following varieties, none more than 70%: Carignan (50% maximum), Cinsaut, Grenache Noir, Lledoner Pelut; Mourvèdre, and Syrah (25% maximum)*

🍷 *3–8 years*

WHITE As elsewhere in the south, the white wines have improved enormously, benefiting from improved methods in the cellar, but they still remain a very small part of the appellation.

🍾 *A blend of two varieties, and with a maximum 80% for the most important. A minimum of 50% of Grenache Blanc and Macabeo and Tourbat (or Malvoisie du Roussillon); plus Grenache Gris, Marsanne, Roussanne, and Vermentino. A maximum of 10% Viognier and Carignan Blanc can be added.*

🍷 *1–2 years*

ROSÉ Fresh and attractive dry wines.

🍾 *As for red*

🍷 *1–2 years*

✓ *Mas Amiel • Mas des Baux (Soleil) • De Casenove • CV Catalans • Mas Crémat (Dédicace, La Llose) • Cazes Frères •*

Clos des Fées • Jaubert-Noury (Château Planères) • Lafage (Le Vignon) • Du Mas Rous (Elevé en Fûts de Chêne) • Piquemal • Rey (Les Galets Roulés) • Sarda-Malet • Château de Corneilla • Mas Bécha • Domaine Modat • Domaine de la Perdrix • Domaine de la Préceptorie • Domaine la Soulane • Domaine Vaquer

CÔTES DU ROUSSILLON VILLAGES AOC

This appellation encompasses exclusively red wines from 51 villages, limited by the Aude department in the north and the Têt valley in the south, in the best area of the Côtes du Roussillon. Some producers make both Côtes du Roussillon and Côtes du Roussillon Villages.

RED Just as good value as basic Côtes du Roussillon, but the best can have even more character and finesse.

🍾 *A blend of at least two of the following varieties: Carignan, Grenache Noir, Lledoner Pelut, Syrah, and Mourvèdre. None should exceed 70%; Syrah and Mourvèdre are limited to a maximum of 30% and Carignan is limited to a maximum of 60%. Not to be sold until the middle of February following the harvest.*

🍷 *3–10 years*

✓ *Clot de L'Oum (Saint Bart Vieilles Vignes) • Clos des Fées • Força Real (Hauts de Força) • Mas de Lavail (Tradition) • Château Planèzes (Elevé en Fûts de Chêne) • Des S chistes (Tradition) Domaine la Toupie • Domaine de Rancy • Domaine Fontanel • Domaine de l'Edre • Domaine des Chênes*

CÔTES DU ROUSSILLON VILLAGES CARAMANY AOC

This good-value red wine conforms to the requirements of Côtes du Roussillon Villages, except for a minimum Syrah content and the stipulation that the Carignan must be vinified by carbonic maceration. This seemingly strange criterion stems from the fact that, in 1964, the local *coopérative* claimed to be the first in France to' use this technique. From the villages of Cassagnes and Bélesta as well as Caramany.

RED These are ripe spicy wines, characteristic of the appellation and often, but not always, are vinified by carbonic maceration,

🍾 *As for Côtes du Roussillon Villages, a minimum of two varieties, with the most important not exceeding 70%. Syrah must not exceed 40%. Carignan, a maximum of 60%, with obligatory carbonic maceration.*

🍷 *3–15 years*

✓ *Vignerons Catalans • CV Carmany*

CÔTES DU ROUSSILLON VILLAGES LATOUR-DE-FRANCE AOC

This is a fine-value red wine that conforms to the requirements of Côtes du Roussillon Villages. Virtually the entire production of Latour-de-France, produced by the village cooperative, used to be sold to, and through, the national French wine-shop group Nicolas, which may explain why this village was known better than any other. The neigh-

WORKERS LINE UP TO LOAD MERLOT GRAPES INTO A TRAILER IN AN AUDE VINEYARD
The gatherers strap large containers onto their backs and tip the hand-picked harvest into the trailer's shute attachment. The Aude department in southwest France is host to several AOC wines: La Clape, Quatourze, Corbières, Fitou, Minervois, Limoux, Côtes de Malepère, and Cabardès.

DOMAINE DE LA PLAINE AGES ITS SWEET MUSCAT WINE IN ANTIQUE DEMIJOHN BOTTLES
This maker of Muscat dessert wines is located within the jurisdiction of the Muscat de Frontignan AOC. All of Domaine de la Plaine's wines are made with Muscat Blanc à Petits Grains.

bouring villages of Cassagnes, Montner, Estagel, and Planèzes are also included in Latour-de-France. A minimum of two varietals, with no more that 60 per cent Carignan allowed in the vineyard.

RED Full in colour and body, these fine-value wines have a fruity flavour.

🍷 3–15 years

COTES DU ROUSSILLON VILLAGES LES ASPRES AOC

This is a red-wine-only designation for Côtes du Roussillon Villages, from 19 villages recognised in 2017. Permitted grape varieties are Carignan, Grenache, Syrah, and Mourvèdre, with the blend having to be of a minimum of three varieties, with the most important two not exceeding 90 per cent. Syrah, plus Mourvèdre (a minimum of 25 per cent, together or separately). Individually Syrah, Mourvèdre, and Grenache Noir must not each exceed 50 per cent. Carignan Noir, a maximum of 25 per cent.

CÔTES DU ROUSSILLON VILLAGES LESQUERDE AOC

Restricted to vines growing in the villages of Lansac, Lesquerde, and part of Rasiguères. A blend of at least two varietals, with the most important two totalling at least 70 per cent. Syrah up to 30 per cent, Carignan with obligatory car-

bonic maceration (*see* Côtes du Roussillon Caramany) must represent no more than 60 per cent of the vineyard. Not to be sold until the middle of February following the harvest.

RED Rich, well-coloured wines that deserve their own village designation.

🍷 As for Côtes du Roussillon Villages except that there must be a minimum of 30% Syrah and Mourvèdre

🍷 3–15 years

CÔTES DU ROUSSILLON VILLAGES TAUTAVEL AOC

Restricted to vines growing in the villages of Tautavel and Vingrau. It conforms to the requirements of Côtes du Roussillon Villages. A blend of at least two varietals, with the most important not exceeding 70 per cent. Grenache Noir and Lledoner Pelut at a minimum of 20 per cent, separately or together; Syrah plus Mourvèdre at a minimum of 30 per cent; Carignan Noir at a maximum of 60 per cent. Not for sale until the October of the year following the harvest.

RED Rich, well-coloured wines that deserve their own village designation.

🍷 3–15 years

✓ Des Chênes (La Carissa) • Fontanel (Prieuré Elevé en Fûts de Chêne)

CRÉMANT DE LIMOUX AOC

This sparkling wine was created in 1989 in recognition of the successful introduction of Chardonnay and Chenin Blanc into the vineyards of Limoux. Whereas Blanquette de Limoux is predominantly Mauzac, Crémant de Limoux depends on Chardonnay and Chenin Blanc, and also Pinot Noir. The choice was not merely about a name, but what direction and style the wine should follow. Most producers make both, and some also make still wine.

SPARKLING WHITE Chardonnay tends to be the main base, with just enough Mauzac retained to assure a certain style, and Chenin Blanc is used as a natural form of acid adjustment. The wines are generally more refined than Blanquette de Limoux, and the best have a finesse that the more traditional wines cannot quite match.

🍷 A maximum of 90% in total of Chardonnay and Chenin Blanc, including at least 20% of the latter, plus a maximum of 20% in total of Mauzac and Pinot Noir, with the latter restricted to no more than 10%

✓ CV Sieur d'Arques • Antech • J Laurens (Graimenous) Domaine Mouscaillo • Domaine Monsieur S • Domaine les Hautes Terres • Jean-Louis Denois .

FAUGÈRES AOC

A compact appellation covering seven villages and a couple of hamlets, recognised in 1982 for red and rosé, with white wine following in 2004. Schist is the determining feature of Faugères, but the name stems not only from the eponymous village but also from the previous success of Fine de Faugères. Neighbouring St Chinian also has some schist, as does Cabrières, but the wines are quite different, with an elegant freshness and spice.

RED Wines with good colour, and rich spicy flavours of the south. Sunshine in a glass, with Syrah, Mourvèdre, Grenache Noir, Cinsaut, and Carignan. Mourvèdre is obligatory in the vineyard, with a minimum of two varieties in the wine.

🍷 3–10 years

WHITE Fresh, exotic fruits with a touch of zest.

🍷 A minimum of two grape varieties, including at least 30% Roussanne, plus Grenache Blanc, Marsanne, Vermentino, and Clairette.

🍷 1–3 years

ROSÉ Small production of attractively coloured, ripe, and fruity dry rosés.

🍷 Cinsaut tends to dominate any blend, with Syrah, Grenache Noir, Mourvèdre, and Carignan the other options.

🍷 1–2 years

✓ Mas d'Alezon, Domaine Jean-Michel Alquier, Domaine des Trinités, Domaine Léon Barral, Domaine de Cébène, Domaine Ollier-Taillefer • Domaine de l'Ancienne Mercerie • Domaine de Sarabande . Domaine St Antonin • Château la Liquière • Domaine du Météore

FITOU AOC

The earliest red and red-only appellation of the Languedoc lies in the shadow of its larger neighbour Corbières. The appellation divides into two, côté montagne, with vineyards in the hills, and côté mer, with vineyards by the sea, covering nine villages in total. The two areas are quite different in character. Wine growers in the hilly vineyards make Corbières as well, and Fitou could almost be seen as a cru of Corbières. There is also an overlap with Rivesaltes.

RED These wines have a fine colour and a spicy warmth of Grenache Noir that curbs and softens the concentrated fruit and tannins of low-yielding Carignan.

🍷 A minimum of two varieties, with a minimum of 30% Carignan, plus Mourvèdre and Syrah accounting for a minimum of 10%, and at least 30% Grenache Noir and Lledoner Pelut, and no Cinsaut.

🍷 3–6 years (4–10 years in exceptional cases)

✓ Domaine Bertrand Bergé • Château de Nouvelles • De la Rochelière (Noblesse du Temps) • De Roland • Domaine Lerys • Mas des Capices • Domaine Jones • Château Champs des Soeurs

FRONTIGNAN AOC
See Muscat de Frontignan AOC

GRÈS DE MONTPELLIER AOC

New red-wine-only appellation as from the 2002 vintage, Grès de Montpellier effectively encompasses a wide area around the city of Montpellier, including the village designations of La Méjanelle, St-Christol, St-Drézéry, and St-Georges-d'Orques. Vérargues is no longer recognised as a separate area.

✓ *Château St Martin de la Garrigue • Château du Grès-St-Paul*

LANGUEDOC AOC

This appellation has replaced Coteaux du Languedoc as the bottom of the pyramid of Languedoc appellations, and it covers vineyards in four departments, the Aude, Hérault, and parts of the Gard. Illogically, Languedoc AOC wine can also be produced in Roussillon, where there is no other appellation. There are wide variations in style and flavour here, but they do have a consistent level of quality. Better wines will have a more precise appellation and definition; however, the wines of the village of Langlade, with a certain reputation, do not fit into any other appellation.

RED Full and honest red wines that make excellent everyday drinking.

🍷 *These wines are always a blend, allowing for Grenache Noir, Lledoner Pelut, Mourvèdre, Syrah Carignan Cinsaut, with other local varieties, such as Counoise, Terret Noir, Piquepoul Noir.*

🍷 *1–4 years*

WHITE Getting better by the day, some wonderfully fresh, aromatic, dry white wines are being made by the appellation's younger vignerons, sometimes with a little oak and not infrequently from very old vines.

🍷 *Bourboulenc, Clairette, Grenache Blanc, Grenache Gris, Piquepoul, Marsanne, Roussanne, Vermentino, Macabeo, Terret Blanc, and Carignan Blanc.*

ROSÉ These dry rosé wines have good fruit, and the best of them deserve as good a reputation as the rosés of Provence. And far more rosé is produced in the Languedoc than in Provence.

🍷 *As for red*

🍷 *1–2 years*

✓ *Terre3 des Dames • Roc d'Anglade*

LANGUEDOC (VILLAGE NAME) AOC

Except where stated, the wines bearing the names of the following areas within the appellation conform to the requirements of Languedoc AOC.

LIMOUX AOC

In 1992, Limoux became the first French AOC to insist upon barrel fermentation for its white wines and consequently established a reputation for the Chardonnay-dominated wines of this appellation. Red Limoux followed in 2004. Limoux Blanc from just Mauzac grapes has existed since 1959.

RED Some elegant wines are produced here.

🍷 *45 to 70% Merlot, plus Cabernet Franc, Cabernet Sauvignon – 35% maximum with Grenache Noir, Malbec, and Syrah making a combined minimum of 20%. There must be at least three varieties, no two of which may exceed 90% of the blend. Any Pinot Noir, which performs exceptionally well in Limoux, is an IGP Haute Vallée de l'Aude.*

🍷 *3–6 years*

WHITE Usually a single variety, of either Chardonnay, Chenin Blanc, or Mauzac. Chardonnay inevitably dominates, but Chenin Blanc and Mauzac can both offer unexpected pleasure, and a blend of the three is also possible, as in Château Rives-Blanques Trilogie. This producer is also one of the very few to make a pure Mauzac Occitanie and its Chenin Blanc Dédicace is one of my favourite whites of Limoux.

🍷 *Mauzac, Chardonnay, and Chenin Blanc*

🍷 *1–2 years*

✓ *D'Antugnac (Gravas) • Rives-Blanques (Dédicace) • Sieur d'Arques (Toques et Clochers) Domaine Mouscaillo, Domaine les Hautes Terres*

MALEPÈRE AOC

This is where the Atlantic meets the Mediterranean, on the slopes of the Massif de la Malepère, south of Carcassonne, and is the one appellation of the Languedoc that does not require the inclusion of any southern grape varieties. Good red and rosé wines, no white.

RED Well-coloured wines of medium to full body, with a certain bordelais structure.

🍷 *A minimum of 50% Merlot, a minimum of 20% Malbec, as well as optional Cabernet Franc, Cabernet Sauvignon, Cinsaut, Grenache Noir, and Lledoner Pelut.*

🍷 *3–7 years*

ROSÉ Attractive dry wines.

🍷 *A minimum of 50% Cabernet Franc, with Cinsaut and Grenache Noir, plus optional Cabernet Sauvignon, Malbec, and Merlot.*

🍷 *1–3 years*

✓ *Château Guilhem (Prestige) • Château Guiraud • Domaine Rose et Paul*

MAURY AOC

From four communes, centred on the small village of Maury, as well as Tautavel, St Paul de Fenouillèdes, and Rasiguères, where schist is the dominant soil. Like all fortified wines from Roussillon, Maury comes mainly from Grenache Noir, Gris, and Blanc, with Carignan Noir, Syrah, and Macabeo, and for white wines Grenache Blanc and Gris, Macabeo, Tourbat, Muscat d'Alexandrie, and Muscat a Petits Grains.

RED (GRENAT), WHITE, ROSÉ (TUILÉ), TAWNY (AMBRÉ), HORS D'AGE, RANCIO Characterful wines, with wonderful tangy, toasty, flavours and nutty-raisiny richness, especially the *rancio* styles that are aged in barrel for several years.

🍷 *Grenat and Tuilé – Grenache Noir, with Carignan, Syrah, and Macabeo also allowed. White and ambré also allow for Grenache Gris and Blanc, plus Macabeo, Muscat à Petits Grains, and Muscat d'Alexandrie. Thirty months ageing for Tuilé; Hors d'Age 5 years ageing.*

🍷 *10–30 years*

✓ *Mas Amiel • Jean-Louis Lafage (Prestige Vieilli en Fûts de Chêne) • CV de Maury (Solera 1928 Cask No.886) • Pouderoux • La Préceptorie de Centernach • La Coume du Roy • Domaine Fontanel*

MAURY SEC AOC

An appellation created in 2011 for red table wine from the same area as the VDN. The grape varieties permitted here, with a minimum of two in the blend, are Grenache Noir, Carignan, Mourvèdre, Syrah, and Lledoner Pelut, grown on schist. Grenache Noir accounts for a minimum of 60 per cent of the blend and 80 per cent of the vineyard. At least six months ageing. Most producers make both dry and sweet wines.

LA MÉJANELLE AOC

This appellation covers an area very close to the city of Montpellier. The one surviving producer is fighting a rearguard action against property developers. His *terroir* includes galets roulées similar to those found in Châteauneuf-du-Pape.

✓ *Château de Flaugergues*

MINERVOIS AOC

North of Corbières, this rocky area has the typically hot and arid air of southern France. It was recognised as an appellation in 1985, and the subzone of La Liviniere became a cru in 1999. Cazelles and Laure are two other *crus* in the pipeline.

RED Characterful reds with the spicy flavours of the *garrigues* of the south.

🍷 *Grenache Noir, Lledoner Pelut, Mourvèdre (minimum 20%), Syrah Carignan, Cinsaut, Piquepoul Noir, Terret Noir, and Aspiran.*

🍷 *1–5 years*

WHITE A mere 3 per cent of the appellation is white. A dry, fruity wine that has benefited from improvements in the cellar, so that it is generally fresh and lightly aromatic.

🍷 *Bourboulenc, Grenache Blanc, Macabeo, Marsanne, Roussanne, Vermentino, Clairette, Muscat à Petits Grains, Piquepoul Blanc, and Terret Blanc*

🍷 *Within 1 year*

ROSÉ Just 13 per cent of the appellation. Most of these are good-value wines with a pretty pink colour and a dry, fruity flavour.

🍷 *Grenache Noir, Lledoner Pelut, Mourvèdre, and Syrah, plus Carignan, Cinsaut, Picpoul Noir, and Terret Noir.*

🍷 *Within 1 year*

✓ *Des Aires Hautes • Borie de Maurel (Sylla) • Hegarty Chamans • Château Maris • Château d'Oupia (Oppius) • Château Saint-Jacques d'Albas • La Tour Boisée • Clos du Gravillas • Domaine Anne Gros • Château la Grave • Domaine Pierre Cros.*

MINERVOIS LA LIVINIÈRE AOC

Located in the central-northern zone of Minervois, La Livinière was recognised as a cru in 1999. The vines grow on the best areas of the Petit Causse, the sheltered limestone cliffs that skirt the foot of the Montagne Noir at a height of 140 metres (460 feet) and centred on the village of La Livinière. The cru has aspirations to be an independent appellation.

RED There must be 15 months ageing before sale, and where there is sufficient finesse to match the structure of these wines, they represent some of Minervois's best vins de garde.

🍷 *A minimum of 60% Syrah, Grenache Noir and Mourvèdre, or a minimum of 40% Syrah and Mourvèdre, with the rest being Carignan, Cinsaut, Aspiran, and Piquepoul.*

🍷 *2–10 years*

✓ *Clos de L'Escandil) • Château Faiteau • Clos Centeilles • Domaine Borie de Maurel • Domaine Ste Eulalie • Château Maris .*

MONTPEYROUX AOC

This village is located on clay and limestone hills next to the village of St-Saturnin. This is a village that has attracted more than its fair share of good wine growers, producing spicy warm red wines. Any white or rosé is AOC Languedoc or an IGP.

✓ *Domaine d'Aupilhac • Domaine Alain Chabanon • Villa Dondona • Domaine du Joncas • Mas d'Amélie*

MUSCAT DE FRONTIGNAN AOC

It is claimed that the Marquis de Lur-Saluces visited Frontignan in 1700 and that inspired him to make sweet wines at Château d'Yquem in Sauternes. Muscat de Frontignan is no longer botrytised – it is a fortified wine, a *vin doux naturel* (VDN), and you can also find late harvest wines from dried *passerillé* grapes, in an attempt to vary the range, given the fall in popularity of these luscious dessert wines.

WHITE The VDNs are golden in colour and rich, grapey and sweet, and the best are utterly delicious. They have a honeyed aftertaste with a slightly fatter style than those of Beaumes de Venise

🍷 *Muscat à Petits Grains*

🍷 *1–3 years*

✓ *Château de la Peyrade • Mas de Madame • Château Stony • CV de Frontignan .*

MUSCAT DE LUNEL AOC

Situated on limestone terraces northeast of Montpellier, this undervalued VDN approaches Frontignan for pure quality.

WHITE Lighter than Frontignan, these wines nevertheless have fine, fragrant Muscat aromas of great delicacy and length.

✓ *Muscat Blanc à Petits Grains*

🍷 *1–3 years*

✓ *Domaine le Clos de Bellevue • Château Grès St Paul • CV du Muscat de Lunel*

MUSCAT DE MIREVAL AOC

This is a little-seen VDN appellation.

WHITE Light and sweet wines that may have a better balance and (relatively) more acidity than those from neighbouring Frontignan.

🍷 *Muscat Blanc à Petits Grains*

🍷 *1–3 years*

✓ *Domaine de la Rencontre • Domaine du Mas Neuf*

MUSCAT DE RIVESALTES AOC

The town of Rivesaltes gives its name to two VDN (*vin doux naturel*) wines, namely Muscat de Rivesaltes and Rivesaltes, which cover 90 villages.

WHITE Rich, ripe, grapey-raisiny wines that are very consistent in quality.

🍷 *Muscat Blanc à Petits Grains, Muscat d'Alexandrie. Minimum 100gm/l sugar.*

🍷 *1–3 years*

✓ *Cazes Frères* ❸ *• Des Chênes • Lafage • Laporte • Château de Nouvelles • Sarda-Malet • Des Schistes • Arnaud de Villeneuve*

MUSCAT DE ST-JEAN-DE-MINERVOIS AOC

This tiny appellation, sandwiched between St Chinian and Minervois produces a delicious *vin doux naturel* that is sadly often overlooked. It deserves a wider reputation.

WHITE These golden wines have a balanced sweetness and an apricot flavour.

🍷 *Muscat Blanc à Petits Grains*

🍷 *1–3 years*

✓ *De Barroubio • Clos du Gravillas*

PÉZENAS AOC

Recognised as a cru of the Languedoc in 2007 and covering villages around the historic town of Pézenas. Prieuré de Saint-Jean-de-Bébian was a pioneering estate in the 1980s, but others have followed its example.

✓ *Prieuré de Saint-Jean-de-Bébian • Mas Gabriel • Domaine les Aurelles • Domaine le Conte de Floris • Domaine Ste Cécile du Parc • Domaine Monplézy*

PICPOUL-DE-PINET AOC

An appellation in its own right since 2013 and covering six communes, this white-wine-only appellation must be made, as the name suggests, from 100 per cent Piquepoul Blanc, from villages around the little town of Pinet. The wine is fresh and salty, with supple acidity and is perfect with oysters.

✓ *Cave Cooperative les Costières (Beauvignac) • Cave de l'Ormarine (Carte Noire, Duc de Morny) • Château St Martin de la Garrigue • Domaine la Croix Gratiot • Domaine Félines-Jourdan*

PIC-ST-LOUP AOC

An appellation since 2017 for red and rosé wines from 15 communes in Hérault and two in Gard, all in the vicinity of the Pic St-Loup, including some high-altitude locations that must rank among the coolest vineyards in southern France. Syrah must account for at least 50 per cent of the blend, with a minimum of two varieties. Grenache Noir, Mourvèdre, Carignan, and Cinsaut are the other options, as well as Counoise and Morastel. The wine must not be sold until 1 July following the vintage.

✓ *Domaine de l'Hortus • Mas Bruguière • Clos Marie • Château Lascaux • Château Cazeneuve • Bergerie du Capucin • Château de Lancyre • Domaine Clavel*

QUATOURZE AOC

A tiny area outside the town of Narbonne where one producer is stalwartly defending the reputation of Quatourze. The soil is galets roulées and the conditions quite different from neighbouring Corbières and la Clape.

✓ *Château Notre-Dame de Quatourze*

RIVESALTES AOC

This appellation covers 86 communes, in the Pyrenées-Orientales and nine in the Aude, so that it accounts for half of the *vin doux naturel* produced in France. The descriptions in the appellation include ambré for an oxidative style of tawny coloured Rivesaltes; tuilé for an oxidative red; and grenat for a reductive-style red, as well as rosé. Hors d'age requires a minimum of five years' ageing, for an Ambré or a Tuilé wine. It remains one of the great fortified wines of France, with the best examples rivalling old tawny port. Sadly these wines are often misunderstood and overlooked.

GRENAT From 100 per cent Grenache Noir. The warm, brick-red glow of these wines belies their astringent-sweet, chocolate, and cherry-liqueur flavour and drying, tannic finish. Bottled within two years of the harvest.

🍷 *3–40 years*

TUILÉ Up to 50 per cent Grenache Noir plus Grenache Blanc, Grenache Gris, Macabeo, Tourbat (also called Malvoisie du Roussillon), Muscat à Petits Grains, and Muscat d'Alexandrie. Aged for at least 30 months before sale.

AMBRÉ and **ROSÉ** Grenache Blanc, Grenache Gris, Grenache Noir, Macabeo, Tourbat (also called Malvoisie du Roussillon), Muscat à Petits Grains, and Muscat d'Alexandrie (a maximum of 20 per cent). Ambré is aged for at least 30 months. Rosé is bottled no later than 31 December of the year following the harvest. Because much of the red version can be lightened after lengthy maturation in wood, all Rivesaltes wines eventually merge into one tawny style with time. The whites do not, of course, have any tannic astringency and are more oxidative and raisiny, with a resinous, candied-peel character.

Hors d'Age reserved for *Tuilé* and *Ambré* with a minimum of five years ageing. *Rancio* reserved for *Tuilé* and *Ambré*, denoting the particular aged style of wine.

🍷 *10–20 years*

✓ *Des Chênes • Cazes Frères* ❸ *• Gardies • Château de Nouvelles (Tuilé) • Sarda-Malet (La Carbasse) • Château de Sau (Ambré Hors d'Age) • Terrassous (Ambré Vinifié en Fûts de Chêne) • Arnaud de Villeneuve (Ambré Hors d'Age) • Domaine de Rancy • Domaine Fontanel*

ROQUEBRUN ST-CHINIAN AOC

See St-Chinian (Villages) AOC

ST-CHINIAN AOC

One of the earlier appellations of the Languedoc, St-Chinian AOC was created in 1982 for red and rosé wine, with white wine following in 2004. In simple terms, the soil here is schist to the north and west of the river Vernazobre

TOWERING MOUNTAINS FRAME VINEYARDS IN HÉRAULT
This region is part of the rugged Massif Central, which includes the Cévenne mountain range. Vineyards nestle at the foot of the mountains, as well as on the plateau of Espinouse.

MORE GREEN WINES

In addition to the producers recommended in this directory that are either biodynamic or organic, there are also the following. No negative inference of quality should be taken from the fact that they are not featured among the other recommended producers. There are a number that have been recommended in other editions and still make some fine wines but have been culled to make room for others.

De Bila-Haut (Latour de France), De Fontedicto (Caux), Le Petit Domaine de Gimios (St Jean de Minervois), De Camplazens (Combaillaux), Château de Caraguilhes (St Laurent de la Cabrerisse), Costeplane (Cannes et Clairan), Mas Coutelou (Puimisson), Mas Jullien (Jonquières), Loupia (Pennautier), Château Maris (La Livinière), De Montahuc (St Jean de Minervois), Château Ste Cécile du Parc (Pezenas), Château Pech-Latt (Lagrasse), De Petit Roubié (Pinet), Château des Auzines (Auzines), Bassac (Puissalicon), Peyre-Rose (St Pargoire), Château Roubia (Roubia), Château St Auriol (Lézignan-Corbières), La Triballe (Guzargues), Zumbaum-Tomasi (Claret), De Brau Ilemoustaussou),, Joliette (Espira de l'Agly), Franck Jorel (St Paul de Fenouillet), Château de L'Ou (Montescot), Laguerre (St Martin de Fenouillet), Le Casot de Mailloles (Banyuls-sur-Mer),Olivier Pithon (Calce), Du Soula (St Martin de Fenouillet), Domaine du Clos Roca, Roc des Anges, Mas Amiel, Domaine St Antonin, Aubai Mema, Domaine de Bon Augure, Clos des Augustins, Domaine d'Aupilhac, Mas Baux, Château Beaubois, Château Beauregard Mirouze, Prieuré de St Jean de Bébian, Mas Becha, Domaine Benezech-Boudal, Domaine Bertrand-Bergé, Domaine de Borde-Rouge, Domaine Boucabeille, Domaine Bourdic, Mas Bruguière, Mas Cal Demoura, Domaine Canet-Valette, Domaine Cazaban, Château de Cazeneuve, Domaine de Cébène, Domaine de la Cendrillon, Domaine Alain Chabanon, Mas des Chimères, Domaine Costes-Cirgues, Domaine Pierre Clavel, Les Clos Perdus, Domaine la Colombette, Mas Conscience, Château Coupe-Roses, Mas de Cynanque, Villas Dondona, Château la Dournie, Mas de l'Ecriture, Domaine les Eminades, Domaine Pas de l'Escalette, Domaine Borie de Maurel, Mas d'Espanet, Château des Estanilles, Famille Fabre, Les Fusionels, Mas Gabriel, Mas de la Rime, Les Chemins de Carabote, Château de Gaure, Mas des Quernes, la Grange des Quatre Sous, Clos du Gravillas, Clos de l'Anhel, Domaine Henry, Domaine de l'Horizon, Domaine de l'Horte, La Jasse Castel, Domaine du Joncas, Château de Jonquières, Domaine des Jougla, Domaine Lacroix-Vanel, Domaine Laguerre, Domaine Lanye-Barrac, Château de Lascaux, Domaine Près Lasses, Domaine du Météore, Château la Baronne, Château la Liquière, Domaine la Louvière, Domaine de la Marfée, Mas de l'Oncle, Domaine de l'Ancienne Mercerie, Domaine Modat, Domaine Monplézy, Domaine Montcalmès, Domaine de Mortiès, Château Mourgues du Grès, Clos des Nines, Mas du Novi, Domaine Ollier-Taillefer, Mas Onésime, Domaine de la Font des Ormes, Domaine du Clot de l'Oum, Domaine de Rancy, Domaine de Ravenès Château Pech Redon, Domaine Frédéric Brouca, Domaine la Tour Penedesses, Château Coujan, Domaine Pouderoux, Domaine de Rancy, Domaine de Ravanes, Domaine de la Reserve d'O, Domaine Ribiera, Clos Riveral, Domaine de Rolland, Domaine de Roquemale, Domaine la Rouviole, Clos Centeilles, Domaine de Saumarez, Domaine les Chemins de Bassac, Domaine des Schistes, Mas de la Séranne, Domaine la Clos du Serres, Domaine des Soulanes, Domaine Stella Nova, Château de Stony, Villa Symposia, Domaine Villa Tempora, Domaine la Tour Boisée, Abbaye de Valmagne, Domaine Allegria, Domaine Virgile Joly, Borie la Vitarèle, Domaine Verena Wyss

---

and clay and limestone to the south. The two villages of Berlou and Roquebrun give their names to the two *crus* of the appellation.

RED Medium colour with ripe spicy flavours and weight.

Grenache Noir, Lledoner Pelut, Mourvèdre, and Syrah, Carignan, and Cinsaut

2-6 years

WHITE Fresh, grassy fruit with good acidity.

Grenache Blanc, Marsanne, Roussanne, and Vermentino (main varieties), with Bourboulenc, Carignan Blanc, Clairette, Viognier, Macabeo, and Piquepoul Blanc also allowed

1-2 years

ROSÉ Dry and delicately fruity rosés with an attractive, fragrant bouquet and flavour.

Same as red

1-3 years

✓ Clos Bagatelle • Mas Champart • Château La Dournie • Des Jougla • Thierry Navarre • Domaine la Madura • Borie la Vitarèle • Domaine des Païssels .

ST-CHINIAN BERLOU AOC

The emphasis her is on Carignan – which is not surprising, because there are some interesting old Carignan vineyards in the area – but the wines may also include Grenache Noir, Syrah, Mourvèdre, and Cinsaut.

✓ CV de Berlou • Domaine Rimbert • Domaine la Grange Léon

ST-CHINIAN ROQUEBRUN AOC

Schist is the dominant soil in this appellation, and Syrah is the main grape variety, with the village cooperative working well for the *cru*; however most of the independent growers in the village prefer to work outside the *cru*.

✓ CV de Roquebrun

ST-CHRISTOL AOC

Just north of Lunel, the clay and limestone soil of St-Christol produces ripe, spicy, well-balanced red and rosé wines. The local *coopérative* dominates production.

✓ Domaine la Coste Moynier • Frères Guinand • CV de St-Christol

ST-DRÉZÉRY AOC

Only red wine is allowed under this village appellation, which lies north of Montpellier, with a soil of galets roulées. One producer dominates the appellation. Any white and rosé is Languedoc AOC without the mention of the village.

✓ Château Puech Haut • Mas d'Arcaÿ

ST-GEORGES-D'ORQUES AOC

Just west of Montpellier, the St-Georges-d'Orques appellation extends over five communes and produces mostly red and some rosé wines, of good colour, with plenty of fruit, and some character

✓ Château de l'Engarran (Quetton Saint-Georges) • Domaine Belles Pierres • Domaine de la Marfée . Domaine Henry • Domaine la Prose

ST-SATURNIN AOC

Named after a Roman notable who settled here. Neighbour of Montpeyroux but very much less dynamic as production has been dominated by the village cooperative. St Saturnin has a reputation for rosé and also produces Vin d'Une Nuiit and en *primeur* wines from the middle of October.

✓ Domaine Virgile Joly

SOMMIÈRES AOC

Recognised in 2011 as part of the Languedoc appellation, for red wines only, from vineyards around the attractive town of Sommières. Syrah and Grenache Noir account for 50 per cent of the blend, and the wines are aged for 15 months before sale.

✓ Domaine de Coursac, Mas Granier, Mas des Cabres

TERRASSES DE BÉZIERS AOC

Although the syndicat representing this appellation has existed since 1990, the wines have yet to appear on the shelf. Fonséranes is another suggestion for the area, but again nothing concrete.

TERRASSES DU LARZAC AOC

Terrasses du Larzac is a new appellation created in 2015 for red wines only; it has quickly established a reputation for wines coming from some of the coolest and highest vineyards of the Languedoc, attracting many new producers creating highly individual estates. It covers a broad area from Octon in the west to Aniane in the east. Think of it as the Upper Gallery of the amphitheatre of the Languedoc. A minimum of three varieties, but no percentage requirements. A minimum of twelve months ageing before sale.

✓ Mas Cal Demoura • Mas de l'Ecriture • Mas Jullien • Mas des Chimères. Domaine du Pas de l'Escalette • Mas Haut Buis • Clos Maia • Clos du Serres • La Traversée • Château de Jonquières • Mas Conscience • Domaine du Causse d'Arboras • Domaine de Montcalmès

Provence and Corsica

The range of varieties lends a fascinating Italian twist to these wines, which makes both Provence and Corsica unique among the wine regions of France. When the Phocaeans founded Marseille 2,600 years ago, they introduced the grapevine into France, historically making Provence France's very first wine region. The dramatic landscape of Corsica provides an ample palette of terroirs which, combined with a host of indigenous grapes, can be a delightful discovery for wine lovers.

Take off those rose-tinted spectacles! It seems that in recent times whilst Corsica has continued doing what it does best, Provence has entered a space race of sorts for bottle design. Here, for some, the quality of the wine is secondary to how pretty the bottle looks – all in clear glass to show off the colour of the wine but tempting light-strike to tamper with their creations. In amongst textured and ingeniously shaped bottles, a whole raft of different punt designs and changing closure types, let's hope that the next coup for Provence is to convince their fan base that the most important aspect is that their wine is protected, either in coloured glass or tinted wrappers. In this way the wine may remain stable and speak for itself, still satisfying that rosé-tippling, cool, and refreshing schtick that Provence pink does best, but with a higher hit rate of being a sound bottle.

They have been making wine here for more than 2,600 years – and by the time the Romans arrived, in 125 BC, it was already so good that they immediately exported it back to Rome – but ask yourself this question: do you think that wine was a rosé? Of course not: Provence rosé is a relatively recent phenomenon and, some would cynically say, one that was deliberately designed to swindle the gold-draped *nouveau riche* who flock here.

Today, however, the majority of Provence production is rosé; in 2017 it accounted for 89 per cent of the production, some 1,165,389 hectolitres. The export side is strong, with 30 per cent sold outside of France and the United States happily accounting for half of that.

Perhaps the property that expresses Provence rosé best is Château d'Esclans, purchased by Sacha Lichine (son of Alexis Lichine of Bordeaux fame) in 2006. With the help of Michel Rolland, Sacha set about perfecting that most elusive of Provençal wines: a truly fine rosé. The result is a range of beautifully refined, ultra-premium rosés, from the hysterically popular Whispering Angel as an entry-level wine to the top-of-the-range Garrus. LVMH have purchased a majority stake in the winery in 2019, and the plan seems to be to extend it by some 60 hectares (148 acres) and continue with the recipe that has proven to work so well.

THE TRUE CLASSIC WINES OF PROVENCE

For most people, Provence evokes the beaches of St-Tropez or the rich, bouillabaisse-laden aromas of backstreet Marseille, but there are other experiences to be had in this sun-blessed corner of southern France. The wines of Provence may not have the classic status of Burgundy or Bordeaux, but the reds, such as the magnificent Bandol, the darkly promising Bellet, and the aptly named Cassis, have an abundance of spice-charged flavours that show more than a touch of class. Silly-shaped bottles are being discarded by the more serious winemakers, who find the classic, understated lines of Bordeaux- and Burgundy-style bottles better reflect the quality of their wines.

The top estates of the Riviera produce red wines of a far more serious calibre than might be expected from an area better known for its play than its work. Maximum yields are low, however, and the potential for quality can be high. The all-embracing AOCs of Côtes de Provence and Coteaux d'Aix-en-Provence are absolutely full of fine vineyards capable of producing exciting red wine that could surely be Provence's future.

CORSICA – THE ISLE OF BEAUTY

In Corsica, the advent of France's *vin de pays* system meant that about a third of the island's vineyards were uprooted and put to better use. If the *vin de pays* system was intended to encourage the production of superior-quality wines from the bottom of the market upwards, then here at least it has been successful, for this island is no longer the generous contributor to Europe's "wine-lake" it used to be.

The days when Corsica's wines were fit only for turning into industrial spirit are long gone, but nor is this a potentially fine-wine region. Just 15 per cent of its vineyards are AOC, and these yield barely more than 5 per

---

FACTORS AFFECTING TASTE AND QUALITY

LOCATION
Provence is situated in the southeast of France, between the Rhône delta and the Italian border. A farther 110 kilometres (68 miles) southeast lies Corsica, surrounded by the Mediterranean Sea.

CLIMATE
Winters are mild, as are springs, which can also be humid. Summers are hot and stretch into long, sunny autumns. A vine requires 1,300 hours of sunshine in one growing season – 1,500 hours is preferable, but in Provence, it luxuriates in an average of 3,000 hours. The close proximity of the Mediterranean, however, is also capable of inducing sharp fluctuations in the weather. Rain is spread over a limited number of days in autumn and winter.

ASPECT
The vineyards of both Provence and Corsica run down hillsides and on to the plains.

SOIL
The geology of Provence is complex. Many ancient soils have undergone chemical changes and numerous new soils have been created. Sand, red sandstone, and granite are, however, the most regular common denominators, with limestone outcrops that often determine the extent of superior *terroirs*: the Var *département* has mica-schist, chalky scree, and chalky tufa as well as granite hillsides; there are excellent flinty-limestone soils at Bandol; and pudding stones (conglomerate pebbles) that are rich in flint at Bellet. The south of Corsica is mostly granite, while the north is schistous, with a few limestone outcrops and deposits of sandy and alluvial soils in between.

VITICULTURE & VINIFICATION
All the vines in these regions used to be planted in gobelet fashion, but most are now trained on wires. The recent trend towards Cabernet Sauvignon has stopped, although many excellent wines are still made from this grape. The current vogue is to re-establish a true Provençal identity by relying exclusively (where possible) on local varieties, and the laws have been changed to encourage this particular evolution. Much of the rosé has been improved by modern cool-vinification techniques, although most remains tired and flabby.

GRAPE VARIETIES
Aléatico, Aragnan, Aramon, Aramon Gris, Barbarossa, Barbaroux, Barbaroux Rosé, Biancu Gentile, Braquet, Brun-Fourcat, Cabernet Sauvignon, Codivarta, Carignan, Castets, Chardonnay, Clairette, Clairette à Gros Grains, Clairette à Petits Grains, Clairette de Trans, Colombard, Counoise, Doucillon (Bourboulenc), Durif, Fuella, Genovèse, Grenache, Grenache Blanc, Manosquin (Téoulier), Marsanne, Morrastel, Mayorquin, Muscat Blanc à Petits Grains (see individual appellations for local synonyms), Muscat Rosé à Petits Grains, Nielluccio, Pascal Blanc, Pecoui-Tour (Calitor), Petit-Brun, Picardan, Picpoul, Pignerol, Plant d'Arles (Cinsault), Rolle (Vermentino), Rossola Nera (Mourvèdre), Sauvignon Blanc, Sémillon, Sciacarello, Syrah, Terret Blanc, Terret-Bourret (Terret Gris), Terret Noir, Terret Ramenée, Tibourenc (Tibouren), Ugni Blanc, Ugni Rosé

cent of Corsica's total wine production. Although this is a fair and accurate reflection of its true potential, most critics agree that even within the modest limits of *vin de pays* production there should be many more individually expressive domaines. What has restricted their numbers is, however, something few are willing to talk about – intimidation and extortion by the local Mafia, which has caused more than one brave new Corsican venture to founder.

WINE REGIONS OF PROVENCE AND CORSICA
(see also p131)

One part of this wine-producing area is an island and the other part lies on the French mainland, but both these sun-soaked regions are subject to the same capricious Mediterranean weather conditions, although frost is rare in both.

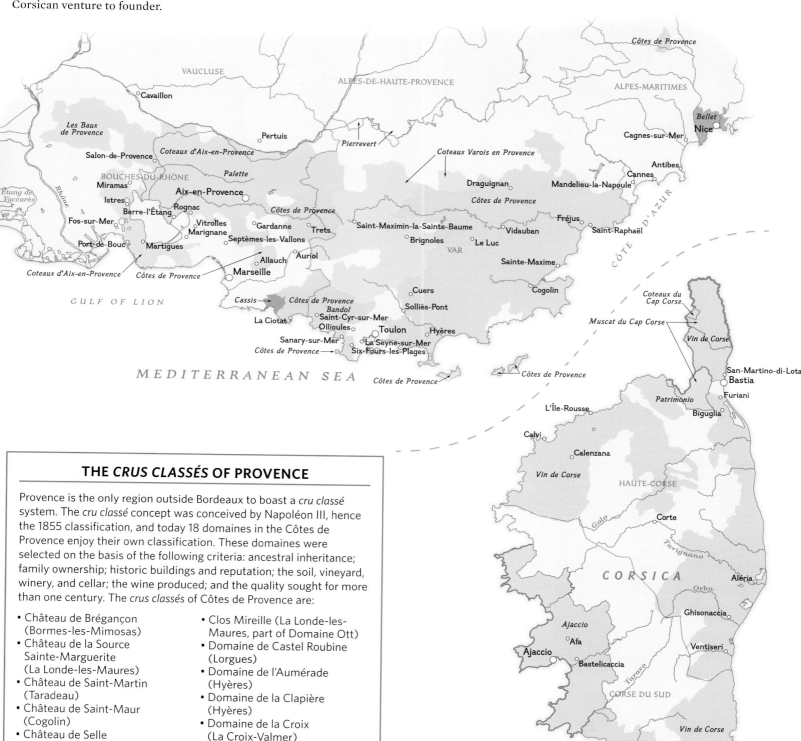

THE *CRUS CLASSÉS* OF PROVENCE

Provence is the only region outside Bordeaux to boast a *cru classé* system. The *cru classé* concept was conceived by Napoléon III, hence the 1855 classification, and today 18 domaines in the Côtes de Provence enjoy their own classification. These domaines were selected on the basis of the following criteria: ancestral inheritance; family ownership; historic buildings and reputation; the soil, vineyard, winery, and cellar; the wine produced; and the quality sought for more than one century. The *crus classés* of Côtes de Provence are:

- Château de Brégançon (Bormes-les-Mimosas)
- Château de la Source Sainte-Marguerite (La Londe-les-Maures)
- Château de Saint-Martin (Taradeau)
- Château de Saint-Maur (Cogolin)
- Château de Selle (Taradeau, part of Domaine Ott)
- Château du Galoupet (La Londe-les-Maures)
- Château Minuty (Gassin)
- Château Sainte-Roseline (Les Arcs-sur-Argens)
- Clos Cibonne (Pradet)
- Clos Mireille (La Londe-les-Maures, part of Domaine Ott)
- Domaine de Castel Roubine (Lorgues)
- Domaine de l'Aumérade (Hyères)
- Domaine de la Clapière (Hyères)
- Domaine de la Croix (La Croix-Valmer)
- Domaine de Mauvanne (Salins d'Hyères)
- Domaine de Rimaurescq (Pignans)
- Domaine du Jas d'Esclans (La Motte)
- Domaine du Noyer (Bormes-les-Mimosas)

ROWS OF VINES GROW SURROUNDED BY OLIVE TREES IN THE REGINO VALLEY OF CORSICA'S BALAGNE REGION
Closer to Italy than mainland France, this island shows a blend of wine influences.

RECENT PROVENÇAL AND CORSICAN VINTAGES

2018 As with many other regions 2018 brought extended periods of sun and ripening; fruity with decent yields across the board.

2017 In Provence, despite some particularly arid areas suffering from water-stress, the harvest was well-balanced in terms of maturity with yields generally down.

2016 The exceptionally mild winter and early spring favoured early budding and flowering for the Corsican vintage leading to an excellent year. For Provence a short frost at the end of April affected mostly the IGP areas but otherwise decent.

2015 Exceptional Provence vintage; hot and dry leading to an early harvest, good red structure with Bandol especially being outstanding, rosé quality good. Good vintage for Corsica across the board.

2014 The cooler, wetter conditions yielded elegant, fresh wines in Corsica but slightly dilute ones in Provence.

THE APPELLATIONS OF
PROVENCE AND CORSICA

Note: An asterisk (*) denotes producers who are particularly recommended for rosé.

AJACCIO AOC

This is a predominantly red-wine appellation on the west coast of Corsica.

RED When successful, this will inevitably be a medium-bodied Sciacarello wine with a good bouquet.

🍷 *A minimum of 60% (in total) of Barbarossa, Nielluccio, Vermentino, and Sciacarello (the last of which must account for a minimum of 40%), plus Grenache, Cinsault, and Carignan (which must not exceed 15%). Aleatico, Carcajolo, Morrastel are optional but collectively must not exceed 10%.*

WHITE Decently dry and fruity; the best have a good edge of Ugni Blanc acidity.

🍷 *A minimum of 80% Vermentino, plus Biancu Gentile, Codivarta, Genovese, and Ugni Blanc (Biancu Gentile, Codivarta, and Genovese combined must not exceed 10%)*

ROSÉ Average to good-quality dry rosé with a typical southern roundness.

🍷 *Same as for red*

🍼 *All wines: 1–3 years*

✓ *Comte Abbatucci (Faustine Abbatucci) • Comte Peraldi*

BANDOL or VIN DE BANDOL AOC

The red wine of this Provence appellation is a true *vin de garde* and deserves recognition.

RED The best of these wines have a dark purple-black colour, a deep and dense bouquet, masses of spicy-plummy Mourvèdre fruit, and complex after-aromas that include vanilla, cassis, cinnamon, violets, and sweet herbs.

🍷 *Between 50% and 95% Mourvèdre, plus Grenache, Cinsault, Carignan, and Syrah (Carignan and Syrah together cannot be more than 10%)*

🍼 *3–12 years*

WHITE These dry wines are now fresher and more fragrant than they were, but they are nothing special compared with the reds.

🍷 *Between 50% and 95% Clairette, then Bourboulenc, Ugni Blanc, Marsanne, Sauvignon Blanc, Semillon, and Vermentino (out of the last four listed, no one varietal can be represent more than 10% of the blend, and collectively they cannot be more than 20% of the blend)*

🍼 *Within 1 year*

ROSÉ Well-made, attractive dry rosés, with body, structure, and a fine individual character.

🍷 *Between 20% and 95% Mourvèdre, plus Grenache, Cinsault, Carignan, Syrah, Clairette, Bourboulenc, and Ugni Blanc (out of the last five listed, no one varietal can be represent more than 10% of the blend, and collectively they cannot be more than 20% of the blend)*

🍼 *1–2 years*

✓ *Château des Baumelles • Bunan (Moulin des Costes Charriage, Château la Rouvière, Mas de la Rouvière) • Dupéré-Barrera (India) • Du Gros'Noré • Lafran-Veyrolles • Château de Pibarnon • Château Pradeaux • Château Ste-Anne ❶ • Des Salettes • La Suffrene (Les Lauves) • Tempier • Terrebrune • Vannières.*

LES BAUX DE PROVENCE AOC

Excellent red and rosé. This appellation also now includes whites, but these represent only 10 per cent of the production.

RED Deep, dark, and rich, with creamy-spicy cherry and plum flavours.

🍷 *Carignan, Cinsault, and Counoise (no single varietal of these three being more than 30%), Grenache, Syrah, Mourvèdre, and (a maximum of 20%) Cabernet Sauvignon*

🍼 *4–10 years*

WHITE Fresh, aromatic, expressing stone fruit, anise, and rosemary.

🍷 *Clairette, Grenache, and Vermintino, (between 10% and 30%) Rousanne, Bourboulenc, Marsanne, and Ugni Blanc (the last three collectively cannot be more than the amount of Rousanne present)*

ROSÉ De la Vallongue is rich and ripe.

🍷 *Carignan, Mourvèdre, and Counoise (no single varietal of these being more than 30%), Grenache, Syrah, Cinsault and (a maximum of 20%) Cabernet Sauvignon. It is also permissible to include the white varietals listed above but collectively no more than 10%.*

🍼 *1–2 years*

✓ *Hauvette ❶ • Mas Sainte-Berthe (Louis David) • De la Vallongue* ❶

BELLET or VIN DE BELLET AOC

This tiny Provence appellation is cooled by Alpine winds and produces exceptionally fragrant wines for such a southerly location.

ROSÉ Fine, floral, dry rosés that are exceptionally fresh and easy to drink.

🍷 *A minimum of 60% Braquet and Fuella, with Cinsault, Clairette, and Vermentino together being no more than 15%, Grenache, plus Bourboulenc, Blanqueiron, Mayorquin, and Ugni Blanc, which collectively cannot exceed 5%*

🍼 *1–2 years*

RED These wines have a good colour and structure, with a well-perfumed bouquet.

🍷 *A minimum of 60% of Braquet and Fuella, (less than 15%) Cinsault, plus Grenache*

🍼 *4–10 years*

WHITE Fine, firm yet fragrant, and highly aromatic, dry white wines of unbelievable class and finesse.

🍷 *A minimum of 60% Vermentino, Clairette, and Chardonnay, plus Bourboulenc, Blanqueiron, Ugni Blanc, Mayorquin, and Muscat à Petits Grains, which collectively cannot exceed 5%*

🍼 *3–7 years*

✓ *Château de Bellet • Clos Saint-Vincent (Clos)*

CASSIS AOC

This is a decent but overpriced Provence appellation located around a beautiful rocky bay, a few kilometres east of Marseille. In all but a few enterprising estates, these vineyards are on the decline.

RED These solid, well-coloured red wines can age, but most do not improve.

🍷 *A minimum of 70% Grenache, Mourvèdre, and Cinsault collectively, plus Carignan, Barbaroux, and maximum of 5% Terret Noir*

WHITE These dry wines have an interesting bouquet of herby aromas of gorse and bracken but are sometimes flabby and unbalanced on the palate. Bagnol and St-Magdeleine produce excellent, racy whites for such southerly vineyards, however, and Ferme Blanche can be almost as good.

🍷 *A minimum of 60% Clairette and Marsanne (where Marsanne must be between 30% and 80%), Pascal Blanc, Bourboulenc, Sauvignon Blanc, Ugni Blanc, and a maximum of 5% Terret Blanc*

ROSÉ Pleasantly fresh, dry rosés of moderately interesting quality.

🍷 *A minimum of 70% Grenache, Mourvèdre, and Cinsault collectively, plus Bourboulenc, Clairette, Marsanne, Pascal Blanc, Sauvignon Blanc, and Ugni Blanc, which together cannot exceed 20%, and a maximum of 5% Terret Noir*

🍼 *All wines: 1–3 years*

✓ *Du Bagnol • Château Barbanau (Clos Val Bruyère)*

COTEAUX D'AIX-EN-PROVENCE AOC

This large appellation has many fine estates, several of which have been replanted and re-equipped.

RED The best are deeply coloured vins de garde with lots of creamy-cassis, spicy-vanilla, and cherry flavours of some complexity.

🍇 *A minimum of two varieties from Cinsault, Counoise, (minimum 20%) Grenache, Mourvèdre, and Syrah, plus Cabernet Sauvignon, Carignan, and (maximum 10%) Caladoc, which collectively cannot exceed 30%*

🍷 *3–12 years*

WHITE Dry and fruity white wines that are of moderate quality and are certainly improving.

🍇 *A maximum of 50% of Vermentino, a minimum of 30% Clairette, Grenache Blanc, Ugni Blanc, and Sauvignon Blanc collectively, plus Bourboulenc and Sémillon*

🍷 *Upon purchase*

ROSÉ Fine-quality dry rosés that are light in body but bursting with deliciously fresh and ripe fruit.

🍇 *As for red, plus up to 20% (in total) of the white varietals*

🍷 *1–2 years*

✓ Des Béates 🅱 (Terra d'Or) • Château de Beaupré • Château de Calissanne • De Camaissette • Jean-Luc Colombo (Pin Couchés) • Château la Coste* • Hauvette 🅾 • Château Pigoudet (La Chapelle) • Château St-Jean*

COTEAUX VAROIS EN PROVENCE AOC

Upgraded to AOC in 1993, this appellation covers an area of pleasant country wines in the centre of Provence.

RED The best have good colour, a deep fruity flavour, and some finesse.

🍇 *At least two of the following must represent a minimum of 80% of the entire blend: Cinsault, Grenache, Mourvèdre, and Syrah, plus Cabernet Sauvignon, Carignan, and Tibouren. No single variety may exceed 90%.*

WHITE Soft and fresh at best.

🍇 *A minimum of 30% Vermentino, plus Clairette, Grenache Blanc, a maximum of 30% Sémillon, and a maximum of 25% Ugni Blanc*

ROSÉ These attractive, easy-to-drink, dry rosés offer better value than some of the more famous, pretentious wines of Provence.

🍇 *As for red, with the inclusion of the white varietals up to a maximum of 20%*

🍷 *All wines: 1–3 years*

✓ Des Chaberts (Prestige) • De Garbelle (Les Barriques de Barbelle) • Château Lafoux • De Ramatuelle • Château Thuerry (Les Abeillons) • Château Trians

CÔTES DE PROVENCE AOC

While this AOC is famous for its rosés, it is the red wines of Côtes de Provence that have real potential, and they seem blessed with good vintages, fine estates, and talented winemakers. Inferior wines are made, but drink the best and you will rarely be disappointed. The major drawback of this all-embracing appellation is its very size. There are, however, several areas with peculiarities of soil and specific microclimates that have long made it obvious that such a large and wide-ranging AOC could support more defined internal zones, therefore four sub-appellations (Sainte-Victoire, Fréjus, Pierrefeu, and Notre-Dame des Anges) were introduced for red and rosé wines only, and La Londe for white, red, and rosé.

FRÉJUS

At the opposite end of the Côtes de Provence to Sainte-Victoire, Fréjus is as sunny as anywhere else in the region, but its maritime influence provides the highest amount of beneficial rainfall, which drains through its sandy-clay soils from west to east into the Argens River.

LA LONDE

This sub-appellation is at the very centre-south of the Côtes de Provence, where the foothills of the Massif des Maures meet those of La Londe les Maures. The schist and quartz soils provide a superior distinctive *terroir*.

NOTRE-DAME DES ANGES

Located at the summit of the Massif des Maures, which overlooks the central basin of the Permian depression, this sub-appellation was instantiated in 2019.

PIERREFEU

Added in 2012, this is a sub-appellation whose member communes could unromantically be described as following the southern half of the A57 road. It corresponds to the southwest end of the Permian depression, consisting of Permian sandstone often covered with calcareous deposits to the north and shale debris to the south.

SAINTE-VICTOIRE

This is on the deep, dry, stony lower slopes of Mont Sainte-Victoire, on the western periphery of the Côtes de Provence.

RED There are too many exciting styles to generalize, but the best have a deep colour and many show an exuberance of silky Syrah fruit and plummy Mourvèdre. Some have great finesse; others are more tannic and chewy. The southern spicy-cassis of Cabernet Sauvignon is often present and the Cinsault, Grenache, and Tibouren grapes also play important roles. The make-up of the blends do differ slightly across the sub-appellations but are largely similar, the below pertains to the main appellation.

🍇 *A minimum of 70% Cinsault, Grenache, Mourvèdre, Syrah, and Tibouren (with no single variety accounting for more than 90% of the entire blend), Clairette, Semillon, and Ugni Blanc collectively not exceeding 20%, but with no single variety being more than 10%. Cabernet Sauvignon and Carignan are allowed, as are Barbaroux and Calitor, but only where the last two were planted before 31 July 1994.*

🍷 *3–10 years*

WHITE Moderate but improving; soft, dry, fragrant, and aromatic wines from La Londe.

🍇 *A minimum 50% Vermentino, with Clairette, Sémillon, and Ugni Blanc*

🍷 *1–2 years*

ROSÉ Mediterranean sun is integral to the enjoyment of these wines, even the best of which fail to perform against other rosés under blind conditions; their low acidity makes them seem flat and dull. Château d'Esclans is, so far, the only exception. Its purity of fruit, so beautifully integrated with oak in three of four rosés (the "bottom" of the range is exclusively fruit-driven), demonstrates to other producers the heights to which this previously boring wine can aspire. The acidity is so perfectly balanced that the rosés of Château d'Esclans never lose site of their expression of *terroir*.

🍇 *Same as for red*

🍷 *1–2 years*

✓ Clos d'Alari (Manon) • Château des Anglades • Château de l'Aumerade (Louis Fabre) • Château Barbeyrolles • De la Bastide Neuve • Ludovic de Beauséjour • De la Courtade • Château Coussin Sainte-Victoire (César) • Château d'Esclans* • Dupéré-Barrera (En Caractère) • Domaines Gavoty • Du Grand Cros (Nectar) • De Jale (La Bouïsse) •

L'AMPHORE DE PROVENCE, A CÔTES DE PROVENCE AOC WINE

Its amphora shape illustrates the propensity of Provençal winemakers to bottle their wares in clear glass of unusual shape. This may appeal to less-savvy customers, but the clear glass can actually adversely affect the wine's quality.

De Lauzade • De la Malherbe* • Château Maravenne • Château Minuty (Prestige) • Commanderie de Peyrassol* • De Rimauresq • Saint-André de Figuière • Château St-Baillon • Château Sainte-Roseline • Château Sarrins* • Domaine Ott Château de Selles* • Château de la Tour l'Evéque* • Vignerons Presqu'Ile St-Tropez (Château de Pampelonne, Carte Noire) • Vannières

MUSCAT DU CAP CORSE AOC

The Muscat du Cap Corse appellation overlaps Vin de Corse Coteaux du Cap Corse and five of the seven communes that comprise the Patrimonio AOC.

WHITE The wines of Clos Nicrosi (in particular) demonstrate that Corsica has the ability to produce one of the most fabulous Muscats in the world. These wines are so pure and succulent, with wonderful fresh aromas, that they have very little to gain through age.

🍷 Muscat Blanc à Petits Grains

🍾 Upon purchase

✓ Clos Nicrosi • Leccia • Orenga de Gaffory

MORE GREEN WINES

In addition to the producers recommended in this directory that are either biodynamic or organic, there are also the following. No negative inference of quality should be taken from the fact that they are not featured among the other recommended producers. There are a number that have been recommended in other editions and still make some fine wines but have been removed to make room for others.

BIODYNAMIC

Château la Canorgue (Bonnieux), Les Fouques (Hyères), Château Romanin (St Rémy de Provence)

ORGANIC

Des Terres Blanches (St Rémy-de-Provence), Du Thouar (Le Muy), De l'Adret des Salettes (Lorgues), Des Alysses (Pontèves), Des Annibals (Brignoles), L'Attilon (Mas Thibert), Les Bastides (Le Puy Ste Réparade), De Beaujeu (Arles), Château de Beaulieu (Rognes), Château La Calisse (Pontèves), CV des Vignerons de Correns et du Val (Correns), De Costebonne (Eygalières), Mas de la Dame (Les Baux de Provence), Château du Duvivier (Pontevès), Château d'Esclans (La Motte), Château Les Eydins (Bonnieux), Mas de Gourgonnier (Mouriès), Hauvette (St Rémy de Provence), De l'Isle des Sables (Fourques), Du Jas d'Esclans (La Motte), La Bastide du Puy (St Saturnin Lès Apt), De Landue (Solliés-Pont), Mas de Longchamp (St Rémy de Provence), Château Miraval (Correns), De Pierrascas (La Garde), De Pinchinat (Pourrières), Rabiega (Draguignan), De Révaou (La Londes les Maures), Richeaume (Puyloubier), Robert (Rognes), De la Sanglière (Bormes Les Mimosas), SCIEV (Molleges), De Séoule (St Saturnin Lès Apt), St André de Figuière (La Londe des Maures), De St-Jean-de-Villecroze (Villecroze)

PALETTE AOC

It does include more varietals than many appellations but nevertheless, Palette is one of the best in Provence. It can be considered the equivalent of a *grand cru* of Coteaux d'Aix-en-Provence, standing out from surrounding vineyards by virtue of its calcareous soil. Just one property occupies three-quarters of this appellation – Château Simone.

RED Though not in the blockbusting style, this is a high-quality wine with good colour and firm structure and can achieve finesse.

🍷 A minimum of 50% (in total) of Mourvèdre, Grenache, and Cinsault, however Mourvèdre must be present at a minimum of 10% and no single varietal over 80%, plus Téoulier, Durif, and Muscat (à Petits Grains, de Hambourg), Carignan, Syrah, Terret Gris, Tibouren, and Cabernet Sauvignon

🍾 7–20 years

WHITE Firm but nervy dry wine with a pleasantly curious aromatic character.

🍷 A minimum of 55% (in total) of Clairette (any variety planted), Araignan, and Bourboulenc, plus Ugni Blanc, Grenache Blanc, Muscat (any variety planted), Picpoul, Pascal Blanc, Colombard, and a maximum of 20% Terret-Gris

🍾 Upon purchase

ROSÉ A well-made but unexceptional wine that is perhaps made too seriously for its level of quality.

🍷 As for red with the inclusion of the white varietals to a maximum of 15%

🍾 1–3 years

✓ Château Crémade • Château Simone

PATRIMONIO AOC

This is a small appellation situated west of Bastia in the north of Corsica.

RED Some fine-quality red wines of good colour, body, and fruit are made.

🍷 A minimum of 90% Nielluccio, plus Grenache, Sciacarello, and Vermentino

WHITE Light and dry wines of a remarkably fragrant and floral character for Corsica.

🍷 Vermentino

ROSÉ Good-value dry rosés that have a coral-pink colour and an elegant flavour.

🍷 A minimum of 75% Nielluccio, plus Grenache, Sciacarello, and Vermentino

🍾 All wines: 1–3 years

✓ Aliso-Rossi • Clos de Bernardi • De Catarelli • Gentile Giacometti (Cru des Agriate) • Leccia • Orenga de Gaffory • Pastricciola • Clos Teddi

PIERREVERT AOC

An appellation in the Alpes-de-Haute-Provence department where the red and rosé wines represent 85 per cent of the volume produced.

RED Fruity, often with scents of spices and *garrigue*.

🍷 A minimum 50% of Grenache and Syrah together (where Grenache is a minimum of 15% and Syrah is a minimum of 30%), Mourvèdre and Téoulier not exceeding 10% collectively, Carignan, Cinsault, plus Clairette, Grenache Blanc, Marsanne, Picpoul, Rousanne, Ugni Blanc, Viognier, and Vermentino, in which these white varietals do not exceed 10% of the blend

WHITE Dry with floral and fruity notes.

🍷 A minimum 50% of Vermentino and Grenache together, a maximum 10% Marsanne, Picpoul, and Viognier collectively, plus Clairette, Rousanne, and Ugni Blanc

ROSÉ Fresh and fruity.

🍷 A minimum 70% of Grenache, Syrah, and Cinsault collectively, Mourvèdre and Téoulier together not exceeding 10%, Carignan, plus Clairette, Grenache Blanc, Marsanne, Picpoul, Rousanne, Ugni Blanc, Viognier, and Vermentino, where these white varietals do not exceed 20% of the blend

---

and the individual proportion of Marsanne, Picpoul and Viognier do not exceed 10% each

🍾 All wines: 1–3 years

VIN DE CORSE OR CORSE AOC

This is a generic appellation covering the entire island. There are five sub-appellations which are covered below.

RED These honest wines are full of fruit, round, and clean, with rustic charm.

🍷 A minimum of 50% of Grenache, Nielluccio, and Sciacarello collectively (Nielluccio and Sciacarello must together account for at least a third of the blend), plus Aléatico, Barbarossa, Caracjolo, Carignan, Cinsault, Morrastel, Mourvèdre, Syrah, and Vermentinom, with Carignan and Vermentino each not exceeding 20% of the entire blend and Aléatico, Barbarossa, and Carajolo together not exceeding 10% of the blend

WHITE The best are well made, clean, and fresh, but not of true AOC quality.

🍷 A minimum of 75% Vermentino, plus a maximum of 25% Ugni Blanc and Biancu Gentile, Codivarta and Genovèse (collectively no more than 10%)

ROSÉ Attractive, dry, fruity, easy-to-drink wines.

🍷 Same as for red

🍾 All wines: 1–3 years

✓ Andriella • Casabianca • Clos Colombu • CV de L'Ile de Beauté (Réserve du Président) • Maestracci • Clos de L'Orlea • Domaine Terra Vecchia

VIN DE CORSE CALVI AOC

Lying north of Ajaccio, this sub-appellation of Vin de Corse AOC requires the same grape varieties and meets the same technical level.

✓ Colombu • Maestracci • Renucci

VIN DE CORSE COTEAUX DU CAP CORSE AOC

A sub-appellation of Vin de Corse.

RED These honest wines are full of fruit, round, and clean, with rustic charm.

🍷 A minimum of 60% Grenache, Nielluccio, and Sciacarello collectively (Nielluccio and Sciacarello must together account for at least a third of the blend), plus Aléatico, Barbarossa, and Carajolo, Carignan (maximum 20%), Cinsault, Morrastel, Mourvèdre, Syrah, and Vermentino (maximum 20%), with Aléatico and Carajolo together not exceeding 10% of the blend.

WHITE The best are well made, clean, and fresh, but not of true AOC quality.

🍷 A minimum of 80% Vermentino, plus a maximum of 20% Ugni Blanc and Codivarta together, plus a maximum of 10% Biancu Gentile and Genovèse together

ROSÉ Attractive, dry, fruity, easy-to-drink wines.

🍷 Same as for red

✓ Clos Nicrosi • Pieretti

VIN DE CORSE FIGARI AOC

Situated between Sartène and Porto Vecchio, Vin de Corse Figari is a sub-appellation of Vin de Corse AOC and requires the same grape varieties and technical level.

✓ De Petra Bianca (Vinti Legna)

VIN DE CORSE PORTO VECCHIO AOC

The southeastern edge of Corsica, around Porto Vecchio, is a sub-appellation of Vin de Corse AOC and requires the same grape varieties and technical level.

✓ De Torraccia

VIN DE CORSE SARTENE AOC

South of Ajaccio, this sub-appellation of Vin de Corse AOC requires the same grape varieties and conforms to the same technical level.

✓ Clos d'Alzeto • Saparale

Vins de France and Vins de Pays

These wines may be at the lower end of the appellation pyramid, but benefitting from looser regulations does not necessarily mean lower-quality wines.

WHAT IS IGP?

In 1992 the European Union launched an initiative to protect their agricultural products worldwide. The PGI (Protected Geographical Indication) system – *Indication Géographique Protégée* (IGP) in French – has been taking over wine labels from the Vin de Pays designation. Legally, however, *vins de pays* have not disappeared or been replaced by IGP as some reports suggest. All *vins de pays* will carry the EU's IGP certification, just as all AOCs carry its AOP certification. Many *vins de pays* are switching entirely to IGP, but technically there is no difference in their status. The two are synonyms in the French system and are used interchangeably in this book's text.

This category of wine includes some of the most innovative and exciting wines being produced in the world today, yet most of the 177 *Indication Géographique Protégée* (IGP) denominations are superfluous and confusing. The success of the *vin de pays* system lies not in creating more appellations, but in freeing producers from them, which allows the most talented individuals to carve out their own reputations.

Vins de pays, or "country wines", are supposed to be unpretentious, but many are better than average-quality AOC wines, and the best rank as some of the finest wines that France can produce. This was never the intention: a *vin de pays* was merely meant to be a quaffing wine that should display, in a very rudimentary sense, the broadest characteristics of its region's greatest wines. Yet, because of this modest aim, *vin de pays* had fewer restrictions than higher appellations. This encouraged the more creative winemakers to produce wines that best expressed their *terroir* without being hampered by an over-regulated AOC system, and in so doing, they managed to equal and occasionally surpass the quality of the more famous local appellations. As news of these exciting vins de pays hit the headlines of the international wine press, so the thought of shedding the shackles of AOC restrictions attracted a new generation of winemakers, including a number of Australians. The combination of French and foreign winemakers opened up the *vin de pays* system, turning it into something its creators had never imagined.

COUNTRY ORIGINS

The expression "Vin de Pays" first appeared in the statute books in a decree dated 8 February 1930. The law in question merely allowed wines to refer to their canton of origin provided they attained a certain alcoholic degree, for example, "Vin de Pays de Canton X". These cantons were not controlled appellations as such: there was no way of enforcing a minimum standard of quality, and the relatively small amounts of these so-called vins de pays were often the product of inferior hybrid grapes. It was not until 1973 that the concept was officially born of vins de pays that were a superior breed of *vins de table,* originating from a defined area and subject to strict controls. By 1976 a total of 75 vins de pays had been established, but all the formalities were not worked out until 1979, and between 1981 and 1982 every single existing *vin de pays* was redefined and another 20 created. Currently there are 177, although the number fluctuates as vins de pays are upgraded to AOC, when new IGPs are created, or, as has happened recently, when some *départements* are denied the right to produce a *vin de pays* (see Vins de Pays at a Glance, p333).

Officially, a *vin de pays* was originally a *vin de table* from a specified area that conforms to quality-control laws that were and are very similar to those regulating AOC wines, although obviously not quite as strict. *Vin de pays* came of age in 1989, when Vin de Table Français was dropped from the label.

VIN DE PAYS, MAP A
(see also p131 and Map B on next page)
Languedoc-Roussillon encompasses the greatest concentration of zonal *vins de pays* and is dominated by the regional IGP Pays d'Oc, by far the most successful IGP.

VIN DE PAYS, MAP B
(*see also* Map A on previous page)
This map shows all the regional and
départementale vins de pays, plus those zonal
appellations that are situated beyond the
boundaries of the Languedoc-Roussillon region.

Pays d'Oc	Vin de Pays-producing region
Dordogne	Vin de Pays-producing département
Périgord	Vin de Pays-producing zone
Île de Ré	Vin de Pays-producing sub-zone
Marne	Non-Vin de Pays-producing département
•	Small appellation

0 mi   50
0 km   50

VARIETY IS THE SPICE OF LIFE FOR *VIN DE PAYS*

The universally recognized great wines produced in the most famous
regions of France remain, but they are very much in a cosy market of their
own: their volume in global terms is minute, and the average-quality AOC
wines have been losing the export battle against the New World, particularly "Brand Australia". It was left to the up-and-coming *vins de pays* to
fight a rear-guard action, while the AOC regions sorted out their problems
of perennial mediocrity (if indeed they ever can); yet, even with fewer
restrictions than AOCs, *vins de pays* have had to fight with one hand tied
behind their back. The first and most important disadvantage was that
vins de pays were marketing wines under names such as Haute-Vallée de
l'Aude or Comté Tolosan, while the opposition was simply selling
Chardonnay or Cabernet Sauvignon. The varietal concept might cause a
"sense of place" crisis for the New World's finer wines when they try to
ride the export surf into Europe, but it matters
not a jot in the United States, and the *vins de pays*
are not up against such wines. At the price point
of most *vins de pays* and their New World oppo-
nents, the appellation per se is irrelevant. All
readers of this encyclopedia will drink such
wines and delight in searching out those that are
the best and most interesting, but the vast major-
ity of consumers neither know nor care about appellations. They see
Chardonnay on two or three bottles and grab one. Why? Probably because
they have had it before, and it was okay, or it has a more interesting label
than another – but the reason is immaterial. What is important is that in
making their choice, they – that is, the vast majority – have gone straight
past the Haute-Vallée de l'Aude or other *vin de pays* offerings.

This is the old story of customers proclaiming, "I don't want Chablis, I want Chardonnay", and the French authorities have agonized about how to get around the problem. The French being French could not bring themselves to allow *vins de pays* to be sold as pure varietal wines. Instead, in 1996, they tried to divert consumer attention from single-varietal names with the twin-varietal concept, enabling producers to market wines as Cabernet-Merlot, Chardonnay-Sémillon, and so on. Some very good wines were produced, but the idea never really took off. It is fine in a New World country, where the varietal-wine concept is all, and a twin-varietal is simply seen as a logical move, but for the *vins de pays*, it was like learning to run before they could walk.

The French realized this, and in 2000 issued wide-sweeping decrees authorizing varietal wines for almost all *vins de pays*. It was still an unfair fight because the regulations insisted that such wines be made from 100 per cent of the variety indicated, whereas New World imports merely had to comply with the EU minimum of 85 per cent. At the price point of most *vins de pays* and New World varietals, the flexibility to strengthen or soften a variety to one degree or another in a given year is a major advantage. Another advantage the opposition had was to use oak chips. It is one thing to expect a *cru classé* château to use casks rather than chips, but it is impossible for *vins de pays* to compete at the same price as the New World wines if they have to use oak casks or nothing. With casks, they cannot even get the wine on the shelf at the same time, let alone the same price. Fortunately, in July 2004 producers were authorized to use oak chips and to blend up to 15 per cent of another variety and vintage into a single-varietal *vin de pays*.

SCARLET AND PINK FLOWERS DECORATE AN OLD WAGON OF WINE BARRELS IN LANGUEDOC-ROUSSILLON, FRANCE
Pays d'Oc is the best known of the *vins de pays and is* arguably the most important, producing the majority of France's IGP wines. Its catchment area roughly corresponds to the Languedoc-Roussillon wine region and covers a wide range of *terroirs*.

VINS DE PAYS AT A GLANCE

177 VIN DE PAYS DENOMINATIONS IN TOTAL
From virtually zero production in 1973 to an annual average of over 12 million hectolitres (133 million cases), *vins de pays* now represent over 30 per cent of the total French wine production, and this expansion is continuing.

TYPES OF VINS DE PAYS
There are three basic categories of *vins de pays*: regional, *départementale,* and zonal. Although no official quality differences exist between these, it is not an unreasonable assumption that the zonal *vins de pays* may sometimes show more individual character than the much larger *vins de pays départementaux* or all-encompassing regional *vins de pays*, but this is not always so.

A grower within a specific zone may find it easier to sell wine under a more generic *vin de pays*, which explains why so many individual wines are to be found under the vast IGP Pays d'Oc denomination. Also, the geographical size of a *vin de pays* can be deceptive, with many *départements* producing relatively little wine compared with some more prolific, relatively minute, zonal *vins de pays*. In the Loire, few producers bother with the *départementale* denominations.

REGIONAL VINS DE PAYS (49 per cent of total vin de pays output)
There are now six of these wide-ranging appellations: Atlantique, Comté Tolosan, Comtés Rhodaniens, Val de Loire (which includes a further two sub-regional denominations), Oc, and Méditerranée. Each encompasses two or more *départements*. They represented only 12 per cent of *vin de pays* production in 1990, yet IGP Pays d'Oc alone accounts for 33 per cent today.

DÉPARTEMENTALE VINS DE PAYS (24 per cent of total vin de pays output)
These cover entire *départements,* and though 55 are officially in use, some are effectively redundant as producers have opted for one of the more widely supported regional appellations, which are often easier to promote. In theory every *département* in France can claim a vin de pays under its own *départementale* denomination, but to avoid confusion with AOC wines the following were in 1995 expressly forbidden from exercising this right: the Marne and Aube (in Champagne), the Bas-Rhin and Haut-Rhin (in Alsace), the Côte-d'Or (in Burgundy), and the Rhône (which encompasses Beaujolais but is likely to cause confusion with the Côtes du Rhône AOC). Jura, Savoie, and Haut-Savoie have also been disenfranchised.

ZONAL VINS DE PAYS (27 per cent of total vin de pays production)
There are 93 zonal *vins de pays,* plus 14 subzonal denominations, although many of the former and some of the latter are seldom encountered.

VINS DE CÉPAGE
After a widespread authorization of pure varietal *vins de pays* in 2000, regulations for which lagged well behind the actual production, this category has taken off to such an extent that it now accounts for half of all the *vins de pays* produced each year, with 80 per cent sold under one of six regional *vins de pays*, of which 70 per cent is IGP Pays d'Oc.

47% red

28% white

25% rosé

PERCENTAGES OF RED, WHITE, AND ROSÉ VINS DE PAYS

VINS DE PAYS PRIMEUR
Since 1990 all *vin de pays* denominations have been permitted to produce these wines, which may be marketed from the third Thursday of October following the harvest (much earlier than Beaujolais Nouveau). The regulations allow for red and white wines to be made as *vins primeurs,* but strangely not rosé, although by the very nature of their production, many reds are lighter than some rosés, so it is a somewhat moot point. White wines are vinified by cool-fermentation techniques and red wines by one of three methods: carbonic maceration (*see* Micropedia), part-carbonic maceration (some of the grapes are crushed and mixed with the whole bunches), and "short-classic" (traditional vinification with minimum skin contact). Cool-fermented whites and carbonic maceration reds are dominated by amylic aromas (peardrop, banana, nail varnish). "Short-classic" rarely provides enough fruit, depth, or colour, whereas part-carbonic maceration can be very successful in enhancing the fruitiness of a wine without drowning it in amylic aromas, but it has to be expertly applied.

AGENAIS

Zonal *vin de pays* Map B

These red, white, rosé, and late harvest white wines are produced from a combination of classic *bordelais* grapes and some rustic regional varieties, including the Tannat and Fer. Although more than three-quarters of the output is red, it is the rosé made primarily from Abouriou that is best known.

ALPES DE HAUTE PROVENCE

Départementale vin de pays Map B

Most of these wines come from the Durance Valley in the east of the *département*. Production is mostly red, made from Carignan, Grenache, Cinsault, Cabernet Sauvignon, Merlot, and Syrah. Rosé accounts for 15 per cent and white made from Ugni Blanc, Clairette, Chardonnay, and Muscat for just 7 per cent. Pure varietal wines have been allowed since 2000.

✔ *La Madeleine (Cabernet Sauvignon)* • *CV de Pierrevert (Saint Patrice Blanc)*

ALPES MARITIMES

Départementale vin de pays Map B

Some 70 per cent of production is red and 30 per cent rosé, made from Carignan, Cinsault, Grenache, Ugni Blanc, and Vermentino grapes, mostly from the communes of Carros, Mandelieu, and Mougins. White wines may also be produced.

✔ *de Toasc (Lou Vin d'Aqui)*

ALPILLES

Zonal *vin de pays* Map B

Geographically, this is approximately the *vin de pays* equivalent of Les Baux de Provence. Mostly red is produced, but rosé and white are also permitted.

ARDÈCHE

Départementale vin de pays Map B

Red and white wines, from a wide range of grapes. Coteaux de l'Ardèche is a sub-IGP of this appellation.

✔ *Vignerons Ardéchois (Cuvée Orélie)*

ARDÈCHE COTEAUX DE L'ARDÈCHE

Zonal *vin de pays* Map B

This large denomination has a very significant output and is particularly well known for its dark, spicy red wine, which accounts for 80 per cent of the total production. Often Syrah-dominated, this wine may also include Cabernet Sauvignon, Carignan, Cinsault, Grenache, Gamay, and Merlot. Louis Latour's Chardonnay put this *vin de pays* on the map during the 1990s, and the same firm's stunning Pinot Noir is proving to be an even greater success. Just over 10 per cent of this denomination's total output is rosé, and just under 10 per cent is white, the latter made from Bourboulenc, Marsanne, Viognier, Roussanne, Marsanne, Sauvignon Blanc, and Ugni Blanc. About 30 per cent of all wines produced are sold as pure varietal.

✔ *CV Ardéchois* • *de Bournet (Chris)* • *Mas d'Intras* • *Louis Latour* • *des Louanes (L'Encre de Sy)*

ARIÈGE

Départementale vin de pays Map B

A small production of red, white, and rosé, and late-harvest white, Ariège is in the foot of the Pyrénées, overlapping the notional borders of Southwest France and Languedoc-Roussillon. Varietal wines and blends from Syrah, Merlot, Cabernet Sauvignon, Cabernet Franc, Tannat, Pinot Noir, Chardonnay, Chenin Blanc, Sémillon, Sauvignon, and Petit Manseng allowed. Two sub-appellations can attach their names to Ariège: Coteaux de la Lèze and Coteaux du Plantaurel.

ARIÈGE COTEAUX DE LA LÈZE

Zonal *vin de pays* Map A

Very small appellation around Fossat and Mas-d'Azil within the Ariége IGP.

ARIÈGE COTEAUX DU PLANTAUREL

Zonal *vin de pays* Map A

Small appellation around the villages of Lavelanet, La Bastide-de-Sérou, Foix, Foix-Rural, Mirepoix, Pamiers-Ouest, and Varilhes within the Ariége IGP.

ATLANTIQUE

Regional *vin de pays* Map B

Encompassing the *départements* of Charente, Charente-Maritime, Dordogne, Gironde, and the northwestern corner of Lot-et-Garonne. Red, white, and rosé wines made mainly from Bordeaux varieties, although a large pool of grapes permitted. Not to be confused with IGPs Val de Loire-Atlantique and Comte Tolosan Pyrénées Atlantiques.

AUDE

Départementale vin de pays Map B

Although still the second- or third-largest *vin de pays*, this used to boast almost exactly the same output as IGP Pays d'Oc in 1993, but while the production of the latter appellation has more than trebled, Vin de Pays de L'Aude has virtually stood still. Nevertheless, a lot of fresh and fruity wine is made here, most of which is red, with just 25 per cent of rosé and 5 per cent white. Mediterranean grape varieties dominate. The following zonal IGPs are included in the Aude: Coteaux de la Cabrerisse, Coteaux de Miramont, Côtes de Lastours, Côtes de Prouilhe, Hauterive, La Côte Rêvée, Pays de Cucugnan, Val de Cesse, and Val de Dagne

✔ • *Domaine de La Bouysse (Syrah)* • *Château de Fontenelles (Le Poête)*

VIN DE FRANCE

Vin de France is a revamped *vin de table* that permits, for the first time, the vintage and grape variety to be shown on the label. Rather than a bottomless pit of anonymous wines, it is now possible to find a *vin de France* labelled with a vintage and one or more grape varieties. There is no geographic origin for these wines other than France itself, the idea being to allow vintaged varietal wines to be blended across two or more regions in the hope of providing a consistency of quality at a good-value price. It is not obligatory for a *vin de France* to be a blend of two or more regions, but that has been the prime locomotive for change in this category of wine: to enable France to compete with entry-level vintaged varietal wines from the New World.

✔ *Auriol (So Light Sauvignon Blanc)* • *Brise de France (Cinsault Rosé, Merlot)* • *Lacheteau (Kiwi Cuvée Sauvignon Blanc)* • *French Connection (Reserve Chardonnay, Reserve Merlot, Grande Reserve Shiraz)* • *LGI (La Campagne Sauvignon Blanc)* • *Simonnet-Febvre (Chardonnay des Lyres)* • *Guy Saget (La Petite Perrière Pinot Noir)*

AUDE COTEAUX DE LA CABRERISSE

Zonal *vin de pays* Map A

This zonal IGP is for mostly red wine, as might be expected from what is after all the heart of Corbières, with 10 per cent rosé wines and just 3 per cent white, including some *vins primeurs*. There is a wide range of grape varieties permitted, particularly for the reds. Regulation for maximum yield is unusually strict, with just 85 hectolitres per hectare.

AUDE COTEAUX DE MIRAMONT

Zonal *vin de pays* Map ???

Red wines from a typically Mediterranean range of *bordelais* and southern Rhône grapes dominate this denomination, which bridges Malepère and Corbières. Rosé and white together account for less than 5 per cent of output, including some *vins primeurs*.

AUDE CÔTES DE LASTOURS

Zonal *vin de pays* Map A

This area roughly corresponds to the Cabardès AOC and produces mostly red wines from a similar range of grapes. Just 4 per cent rosé and 3 per cent white (Mauzac, Chenin, Chardonnay, and Ugni Blanc) is made, with *vins primeurs* a local speciality.

AUDE CÔTES DE PROUILHE

Zonal *vin de pays* Map A

These red, white, and rosé wines come from the Aude *département*. The area benefits from its protected position and has a mild maritime climate.

AUDE LA CÔTE RÊVÉE

Zonal *vin de pays* Map A

Zonal IGP for the coastal area between Perpignan and Narbonne. Mainly reds from southern French varieties, but rosés and whites are also permitted.

AUDE HAUTERIVE

Zonal *vin de pays* Map A

Formerly called Vin de Pays d'Hauterive en Pays d'Aude, this appellation was renamed in 2004, when it was merged with the former *vins de pays* of Coteaux de Termenès, Côtes de Lézignan, and Val d'Orbieu. Production is primarily red, made from a range of typical Languedoc and southwestern grapes in the Corbières. A little *Vin primeur* is made, as well as small amounts of white and rosé.

✔ *Domaine du Cerbier (Pierre de Lune)*

AUDE PAYS DE CUCUGNAN

Zonal *vin de pays* Map A

A tiny denomination of essentially red wines in the midst of the Côtes du Roussillon Villages district, produced from Rhône varieties boosted by Cabernet Sauvignon and Merlot. Rosé may be made, but it seldom is, although there is an average of 1 per cent dry white from Mauzac, Chenin Blanc, Chardonnay, and Ugni Blanc.

AUDE VAL DE CESSE

Zonal *vin de pays* Map A

These are mostly red wines, plus 10 per cent rosé and 5 per cent white from local grape varieties in the Minervois area.

AUDE VAL DE DAGNE

Zonal *vin de pays* Map A

This area overlaps part of Malepère and produces almost entirely red wines from Carignan, Grenache, Terret Noir, Merlot, Cabernet Sauvignon, Cabernet Franc, and the Teinturier Alicante Bouschet. Very little rosé and white is made.

AVEYRON
Départementale vin de pays Map B

Most of the wines produced in this *département,* which is situated between Cahors and Hérault, used to claim the Vin de Pays des Gorges et Côtes de Millau (now Millau AOC) appellation. Today most *vin de pays* producers prefer to use the wider Comté Tolosan appellation.

CALVADOS
Départementale vin de pays Map B

Growers are legally entitled to produce a *vin de pays* named after the *département* in which the vines are located, but after all the fuss about *départementale* names that could be construed to come from a classic wine region, it is surprising to see Calvados approved. Although a *départementale vin de pays,* the area of vines is very small, mostly located at Grisy (*see* separate IGP below).

CALVADOS GRISY
Zonal vin de pays Map B

The only relevant winemaking area in the Calvados IGP, between St Pierre-sur-Dives and Vendeuvre. Varietal red, white and rosé made from Bordeaux and Alsace varieties.

CÉVENNES
Zonal vin de pays Map A

This is an amalgamation of four former *vins de pays,* now non-official subzones: Coteaux Cévenols, Coteaux du Salavès , Côtes du Libac (originally Serre du Coiran), Mont Bouquet , and Uzège . Most of the wines are red, and 15 per cent are rosé, with *saignée* rosé a speciality. Very small quantities of white *Vin primeur* are also produced, and white, rosé, and red late-harvest styles are also permitted. Production is mostly of an honest, fruity Languedoc style.

✓ *Les Terrasses (Roche Fourcade)*

CHARENTAIS
Zonal vin de pays Map B

The dry white wines are really good, even though 50 per cent of the grapes used are Ugni Blanc. It seems that this lowly variety and, indeed, the Colombard make light but very fresh, crisp, and tangy wine in the Charente, which is not only well suited to distilling Cognac, but makes a cheap, cheerful, quaffable wine. Some red and rosé wines from Gamay and *bordelais* grapes are also made. The following subzonal areas are allowed to add their names: Île d'Oléron, Île de Ré, Charente, Charente-Maritime, and Saint Sornin.

✓ *De La Chauvillière (Chardonnay)* • *Gardrat (Colombard)*

CHARENTAIS: CHARENTE
Départementale vin de pays Map B

Hardly used *départementale vin de pays,* most producers preferring the wider-known Vin de Pays Charentais.

CHARENTAIS: CHARENTE-MARITIMES
Départementale vin de pays Map B

Hardly used *départementale vin de pays,* most producers preferring the wider-known Vin de Pays Charentais.

CHARENTAIS ÎLE DE RÉ
Subzonal vin de pays Map B

Covers the districts of d'Ars en Ré and Saint Martin de Ré within the Charentais IGP.

CHARENTAIS ÎLE D'OLÉRON
Subzonal vin de pays Map B

Covers the villages near Île d'Oléron within the Charentais IGP.

CHARENTAIS SAINT SORNIN
Subzonal vin de pays Map B

Covers the districts of Montbron and La Rochefoucauld within the Charentais IGP.

CILAOS
Zonal vin de pays Map B

This IGP is still pending. *Cilaos* means "the place you never leave". The most remote of all French wines, it was created in January 2004 for vines growing on the island of Réunion, a French protectorate in the Indian Ocean, 680 kilometres (420 miles) east of Madagascar. Grape varieties already grown include Chenin Blanc, Pinot Noir, and Malbec. The styles produced are red, white, rosé and, because the tendency is for grapes to overripen rather too easily, *moelleux.*

CITÉ DE CARCASSONNE
Zonal vin de pays Map A

Red wines account for over 50 per cent of production, rosé 40 per cent, and white wine barely 10 per cent. The wines come from 11 communes around Carcassonne. Red and rosé are made from Cabernet Franc, Cabernet Sauvignon, Malbec, Merlot, Pinot Noir, and Syrah, Alicante Bouschet, Arinarnoa, Caladoc, Carignan, Chenanson, Cinsault, Eiodola, Grenache, Marselan, and Portan. White wine may be made from any of the following, including blends: Bourboulenc, Carignan Blanc, Chardonnay, Chasan, Chenin Blanc, Clairette, Colombard, Grenache Blanc, Macabéo, Marsanne, Mauzac, Muscat d'Alexandrie, Muscat à Petits Grains Blanc, Picpoul, Roussanne, Sauvignon Blanc, Sémillon, Terret, Turbat, Ugni Blanc, and Viognier.

✓ *Auzias*

COLLINES RHODANIENNES
Zonal vin de pays Map B

A large denomination at the centre of Comtés Rhodaniens, this *vin de pays* straddles five *départements* (Rhône, Isère, Drôme, Ardèche, and Loire) and produces primarily red wines (65 per cent) from Gamay, Syrah, Merlot, and Pinot Noir. Some smaller quantities are made of rosé (10 per cent) and white (25 per cent), the latter coming from mainly Marsanne, Roussanne, and Viognier. Sparkling wines are also permitted. With the range of body, fatness, aromatic complexity, and acidity that a successful blend of these varieties could bring, there is ample scope for Collines Rhodaniennes to develop a first-rate reputation for white wines.

✓ *Cuilleron-Gaillard-Villard (Sotanum)* • *François Villard (Syrah)*

COMTÉ TOLOSAN
Regional vin de pays Map B

About half of the production is white (of which Ribonnet stand out), the rest is split evenly between rosé and red. The wines should be made from the grapes of the Southwest and widespread international varieties. Certainly the white wines are the most successful. The following zonal IGPs are inside Comté Tolosan: Bigorre, Coteaux et Terrasses de Montauban, Pyrénées Atlantiques, Tarn et Garonne, Haute-Garonne, and Cantal.

✓ *Claude Nicolas (Les Castellanes Blanc)* • *De Ribonnet*
• *Tarani (Malbec, Sauvignon)* • *Terréo (Negrette Rosé)*
• *Clos Triguedina (Vin de Lune le Molleux du Clos)*

COMTÉ TOLOSAN BIGORRE
Zonal vin de pays Map B

Mostly full, rich, Madiran-type red wine is made here, plus a little good crisp, dry white. Rosé and sparkling wine may also be produced.

COMTÉ TOLOSAN CANTAL
Départementale vin de pays Map B

Better known for its excellent mountain cheese, Cantal has few vines, mostly in the southwestern corner.

COMTÉ TOLOSAN COTEAUX ET TERRASSES DE MONTAUBAN
Zonal vin de pays Map B

Still red, white and rosé still and sparkling, plus late harvest white wines from the area around Montauban.

✓ *De Montels (Louise)*

COMTÉ TOLOSAN HAUTE GARONNE
Départementale vin de pays Map B

Robust red and rosé wines made from Jurançon Noir, Négrette, and Tannat grapes grown in old vineyards south of Toulouse, with a little Merlot, Cabernet, and Syrah. A minuscule amount of dry white is made, and total production of all styles is very small, but De Ribonnet has put this denomination on the viticultural map and is in a class apart from all the other producers.

✓ *De Ribonnet*

COMTÉ TOLOSAN PYRÉNÉES ATLANTIQUES
Départementale vin de pays Map B

Two-thirds red and one-third white from traditional southwestern grape varieties is produced under this appellation.

COMTÉ TOLOSAN TARN ET GARONNE
Départementale vin de pays Map B

Red wine accounts for 90 per cent of the tiny production of this appellation, although some rosé and a minuscule amount of white is also made. Most of these vines are west of Montauban. The grape varieties are similar to those for Lavilledieu, except for the fact that they can also include Merlot, Cabernet, and Tannat.

COMTÉS RHODANIENS
Regional vin de pays Map B

Created in 1989, Comtés Rhodaniens encompasses eight *départements,* and its name can be given only to wines that have already qualified as a zonal vin de pays. It is the smallest of the five regional *vins de pays,* with just a tenth of the production of the second-smallest regional *vin de pays* (Comté Tolosan), and a mere one-thousandth of the largest (Oc).

✓ *CV Ardechois*

CÔTE VERMEILLE
Zonal vin de pays Map A

Dry, late-harvest, and *rancio* style wines of all three colours are permitted in this IGP, which covers the area around Banyuls and Collioure.

COTEAUX DE L'AIN
Départementale vin de pays Map B

This *départementale vin de pays* encompasses unclassified vineyards in southern Burgundy and has relatively recently come into usage, but the level of production remains insignificant. Still red, white and rosé, plus sparkling white, and rosé are made. Four distinct areas may attach their names to the appellation on the label: Pays de Gex, Revermont, Val de Saône, and Valromey.

COTEAUX DE L'AIN PAYS DE GEX
Zonal vin de pays Map B

Wedged between the Jura mountains and Lac Léman, these vineyards are huddling the town of Challex. Winemaking is similar to neighbouring Switzerland: reds made of Gamay and Pinot Noir and whites of Chardonnay and Chasselas.

COTEAUX DE L'AIN REVERMONT
Zonal vin de pays Map B

The southernmost part of the Jura, this area is struggling to revive its grape-growing heritage after the devastation of phylloxera and the negative effects of fruit shipped there from the south.

COTEAUX DE L'AIN VAL DE SAÔNE
Zonal vin de pays Map B

Only nine villages (Amareins, Beauregard, Chaleins, Fareins, Francheleins, Guéreins, Lurcy, Messimy-sur-Saône, and Montmerle-sur-Saône) can use this appellation, with vineyard acreage dwindling.

COTEAUX DE L'AIN VALROMEY
Zonal vin de pays Map B

This zonal IGP is situated in the Alps foothills around the Bugey area. After the phylloxera crisis vineyards are limited to the vicinity of Belmont.

COTEAUX DE L'AUXOIS
Zonal vin de pays Map B

This zonal *vin de pays* has replaced the *départementale* denomination Côte d'Or (which received a new AOC designation in 2017) and may produce red and rosé wines from one or more of Gamay, Gamaret, Merlot, Pinot Gris, and Pinot Noir. White wines are made from one or more of Aligoté, Auxerrois, Chardonnay, Chasselas, Pinot Gris, Melon de Bourgogne, Viognier, Sauvignon Blanc, and Sauvignon Gris. Late-harvest wines of all three colours are also permitted.

✓ *CV de Flavigny (Chardonnay Fûts de Chêne)*

COTEAUX DES BARONNIES
Zonal vin de pays Map B

The boundaries of this IGP are clearly defined by the Baronnies mountains, which runs parallel to the Rhône Valley. Around 55 per cent of Coteaux des Baronnies wines are red, almost 10 per cent rosé made from Cabernet Sauvignon, Cinsault, Syrah, Grenache, Pinot Noir, Gamay, and Merlot, and 35 per cent is white, with varietal wines generally increasing and sparkling production is permitted.

✓ *Du Rieu Frais (Cabernet Sauvignon)* • *La Rosière (Merlot)*

COTEAUX DE BÉZIERS
Zonal vin de pays Map B

Situated around Béziers itself, this IGP produces mainly red wines, made from Carignan, Grenache, Cinsault, Cabernet Sauvignon, Merlot, and Syrah. Some 20 per cent of the output is rosé and less than 15 per cent white, the latter being made mainly from Ugni Blanc, with a touch of Terret and Clairette.

✓ *Terroirs en Garrigues (Fou de Bassan Rouge)*

COTEAUX DU CHER ET DE L'ARNON
Zonal vin de pays Map B

The reds and *vin gris* are made from Gamay, Pinot Noir and Pinot Gris and dry white from Sauvignon Blanc, Chardonnay, and Pinot Blanc.

COTEAUX DE COIFFY
Zonal vin de pays Map B

A relatively new and as yet untested *vin de pays* from way up north, between Champagne and Alsace in the southeastern corner of the Haute-Marne *département*. Still and traditional-method sparkling wines of all three colours are permitted. Red and rosé mainly from Gamay and Pinot Noir (although Merlot and Syrah are also allowed); whites from Aligoté, Auxerrois, Chardonnay, Muscat, Pinot Blanc, Pinot Gris, and Petit Meslier.

✓ *CV les Coteaux de Coiffy (Auxerrois)*

COTEAUX D'ENSÉRUNE
Zonal vin de pays Map A

This *vin de pays* west of Béziers produces two-thirds red and one-third rosé, made from a typical range of Languedoc grapes, and a tiny production of gently aromatic, fresh whites.

✓ *Foncalieu (Enseduna)*

COTEAUX DE GLANES
Zonal vin de pays Map B

This small appellation produces mainly red wines, which are Gamay or Merlot dominated, plus a little rosé and whites of Chenin Blanc and Chardonnay.

COTEAUX DE NARBONNE
Zonal vin de pays Map A

Mostly soft red wines and a little white and rosé come from this coastal edge of Corbières, where the vines overlap with those of Coteaux du Languedoc.

COTEAUX DE PEYRIAC
Zonal vin de pays Map A

Full and rustic red wines, made from local grape varieties augmented by the Syrah, Cabernet, and Merlot, account for more than 80 per cent of the production of this *vin de pays* in the heart of Minervois. The remainder is mostly rosé, with less than 1 per cent white.

✓ *La Bouscade (Old Vine Carignan)*

COTEAUX DE PEYRIAC HAUT DE BADENS
Zonal vin de pays Map A

Red and rosé wines produced in a rustic Minervois style.

COTEAUX DU PONT DU GARD
Zonal vin de pays Map A

The Pont du Gard is a stunningly beautiful, three-tier Roman aqueduct from the first century BC. Its bottom tier is still strong enough to support modern traffic 2,000 years later, but the wines of Coteaux du Pont du Gard, though mostly rich and powerful reds, are built for a considerably shorter life span. A small amount of white and rosé is also produced, as well as late-harvest wines, and *vins primeurs* are a local speciality.

COTEAUX DE TANNAY
Zonal vin de pays Map B

White wines are made from Chardonnay or Melon de Bourgogne and represent 90 per cent of the production. Red and rosé come from Gamay and Pinot Noir, plus Gamay Teinturier de Bouze and Gamay de Chaudenay. Pure varietal wines are allowed, as well as white, rosé, and red traditional-method sparkling wines

CÔTES CATALANES
Zonal vin de pays Map A

Much enlarged following the absorption of and merger with the former *vins de pays* of Catalan, Coteaux de Fenouillèdes, and Vals d'Agly. This nonsensical exercise of bureaucratic expediency has angered many, not least those in the Coteaux de Fenouillèdes, who were just beginning to make a name for themselves when their *vin de pays* was suppressed in September 2003. Furthermore, the inhabitants of the Fenouillèdes are not Catalan, nor do they speak Catalan. Red (55 per cent), rosé (30 per cent), and white (15 per cent) are produced, and *rancio* style is also allowed.

✓ *Boudau (Le Petit Clos Rosé, Muscat Sec)* • *Calvet-Thunevin (Cuvée Constance)* • *Caves de Baixas (Rozy)* • *Cazes Frères* **ⓑ** • *La Différence (Carignan, Grenache Noir, Viognier Muscat)* • *Galhaud (Oenoalliance)* • *Gauby* **ⓑ** *(La Soula)* • *Lauriga* • *Mas Baux (Velours Rouge)* • *Mas Karolina (Blanc Sec)* • *Mas de Lavail (Ballade)* • *Matassa* **ⓑ** • *de Pézilla* • *Salvat (Fenouil Rouge)* • *Arnaud de Villeneuve (Muscat Moelleux, Prieuré de la Garrigue Chardonnay)*

CÔTES CATALANES PYRÉNÉES ORIENTALES
Départementale vin de pays Map B

From the most southerly French *département*, bordering Spain's Andorra region, much of the *vin de pays* in the Pyrénées-Orientales comes from the same area as the AOC Côtes-du-Roussillon. About half of the production consists of full, fruity reds, with 30 per cent rosé and 15 per cent white.

CÔTES DE LA CHARITÉ
Zonal vin de pays Map B

Formerly known as Vins de Pays Coteaux Charitois. Basic white wines from Chardonnay, Sauvignon Blanc, Pinot Blanc, and Pinot Gris and light reds and rosés mainly from Pinot Noir and Gamay, plus sparkling white and rosé from the Loire Valley.

CÔTES DE GASCOGNE
Zonal vin de pays Map B

These deliciously tangy, dry white wines are the undistilled produce of Armagnac. Those made from the Colombard grape are the lightest, while those from the Ugni Blanc are fatter and more interesting. Manseng and Sauvignon Blanc are also used to good aromatic effect in some blends. Less than 20 per cent of production is red and barely 1 per cent rosé.

✓ *Les Acacias (Petit Manseng)* • *Alain Brumont (Les Menhirs)* • *des Cassagnoles (Gros Manseng Sélection)* • *Chiroulet (Soleil Automne, Terres Blanche)* • *Famille Laplace (Aramis)* • *de Joy (Sauvignon Gros Manseng)* • *de Lartigue* • *De Pellehaut (Ampelomeryx)* • *Plaimont* • *Répertoire (Hidden Treasures)* • *SDU (Domus Gascogne Rosé)* • *Sovino Cox* • *du Tariquet* • *UBY (Sauvignon-Chardonnay-Muscadelle)* • *Vignoble de Gascogne (Face à Face)*

ABBAYE SAINT-MICHEL DE CUXA OFFERS A RED WINE FROM THE CÔTES CATALANES IGP
Côtes Catalanes, a *vin de pays* in the Languedoc-Roussillon region of southern France, covers a wide range of *terroir* suited to classic Mediterranean grape varieties, such as Grenache, Mourvèdre, Cinsaut and Carignan.

CÔTES DE GASCOGNE CONDOMOIS
Zonal *vin de pays* Map B

Production in this zonal *vin de pays* includes 60 per cent red wine, dominated by the Tannat grape, and 40 per cent white wine, from the Colombard or Ugni Blanc. A little rosé is also produced.

CÔTES DU LOT
Départementale vin de pays Map B

Most wines are red or rosé (60 per cent and 35 per cent respectively) and claim either the regional Comté Tolosan denomination or IGP Coteaux de Glanes.

CÔTES DU LOT ROCAMADOUR
Zonal *vin de pays* Map B

Single-commune IGP within the Lot *département*. A small but dynamic group of growers are aiming to replant some of the 200 hectares (495 acres) of vineyards that were destroyed by phylloxera.

CÔTES DE MEUSE
Zonal *vin de pays* Map B

This small IGP covers 40 hectares (100 acres) of vineyards in 15 communes in the Meuse *département*, halfway between Champagne and Alsace, in an area that fell into neglect following phylloxera at the end of the 19th century. Today it produces red and rosé wines, including a pale *vin gris*, from Pinot Noir and Gamay. Some white wine from Chardonnay, Aligoté, and Auxerrois is also made.

✔ *De Coustille (Auxerrois)*

CÔTES DU TARN
Zonal *vin de pays* Map B

Côtes du Tarn is the *vin de pays* equivalent of Gaillac: some 55 per cent is red, 25 per cent white, and the rest rosé. These wines are made from *bordelais* and southwestern grape varieties, plus, uniquely for France, the Portugais Bleu. Gamay is often used for Côtes du Tarn Primeur, while together with Syrah it makes an appealing, delicate *saignée* rosé. The areas of Cabanés and Cunac may attach their names to the IGP on the label.

✔ *Guy Fontaine (Les Vignes des Garbasses Syrah)*

CÔTES DU TARN CABANÉS
Zonal *vin de pays* Map B

This small sub-IGP if the Côtes du Tarn encompasses the villages of Cabanes, Briatexte and Graulhet.

CÔTES DU TARN CUNAC
Zonal *vin de pays* Map B

This IGP within the Côtes du Tarn covers the village of Cunac and nine neighbouring communes.

CÔTES DE THAU
Zonal *vin de pays* Map A

A small denomination along the shore of Etang de Thau lagoon, producing almost equal quantities of red, white, and rosé from all the usual Languedoc varieties. These wines used to be used for vermouth until the upswing in vin de pays. Sparkling wines are also allowed in all three colours.

✔ *Caves Richemer (Syrah)* • *Hugues de Beauvignac (Syrah)*

CÔTES DE THAU CAP D'AGDE
Zonal *vin de pays* Map A

This is a two-commune IGP, encompassing the volcanic hill around Agde and Marseillan

CÔTES DE THONGUE
Zonal *vin de pays* Map A

Covers the triangle between Faugères, Pézenas and Béziers, producing mostly red wine, from grapes including Grenache, Cinsault, Cabernet Sauvignon, Merlot, Syrah, and Carignan, with *vins primeurs* of the last three varieties a speciality. Emphasis is placed on pure varietal wines of all styles. Some 15 per cent rosé and 10 per cent white.

✔ *Chemins de Bessac* • *de Brescou (Syrah)* • *Coste Rousse* • *CV d'Alignan du Vent (Icare, Montarels)* • *Clos de l'Arjolle* • *de la Croix Belle (Cascaïllou, No.7 Rouge)* • *Deshenrys (Lissac)* • *Les Filles de Septembre* • *Mont d'Hortes (Sauvignon)* • *de Montmarin* • *Montplézy (Félicité)* • *Saint Rose (Roussanne)* • *Les 3 Poules (La Coquine)* • *Vignerons de l'Occitane (Clamery Réserve Rouge)*

DRÔME
Départementale vin de pays Map B

Red, white, and rosé wines are produced from typical Rhône varieties augmented by Gamay, Cabernet Sauvignon, and Merlot. More than 90 per cent of the output is red and similar in style to Grignan-Les-Adhémar AOC. A small amount of decent rosé is made. Two sub-IGPs are allowed to attach their names to Drôme: Comté de Grignan and Coteaux de Montélimar.

DRÔME COMTÉ DE GRIGNAN
Zonal *vin de pays* Map B

This zonal IGP covers the same area as Mediterranée Comté de Grignan, *see p338*.

DRÔME COTEAUX DE MONTÉLIMAR
Zonal *vin de pays* Map B

This zonal IGP covers the same area as Mediterranée Coteaux de Montélimar, *see p338*.

FRANCHE - COMTÉ
Zonal *vin de pays* Map B

A vast area overlapping the Jura and Savoie, Franche Comté produces fresh, clean, crisp, but otherwise unremarkable red and rosé wines. Only the white, made from Chardonnay, Pinot Gris, and Pinot Blanc, is of any note. Doubs and Haute-Saône *départements* may attach their names to the region's name. Wines produced in the villages of Champlitte, Vuillafans, Buffard, Offlanges and Motey-Besuche also have the right to add their subzonal name to the Franche Comté IGP, as well as the areas around Hugier and Gy.

✔ *Vignoble Guillaume (Pinot Noir Collection Réservée, Chardonnay Collection Réservée)*

FRANCHE - COMTÉ BUFFARD
Zonal *vin de pays* Map B

Small communal appellation for sparkling and still white, red, and rosé wines.

FRANCHE - COMTÉ COTEAUX DE CHAMPLITTE
Zonal *vin de pays* Map B

Small communal appellation for sparkling and still white, red, and rosé wines.

FRANCHE COMTÉ DOUBS
Départementale vin de pays Map B

Recent *départementale vin de pays* with minute production, Doubs is in the very east of France, between the Jura and Alsace.

FRANCHE - COMTÉ GY
Zonal *vin de pays* Map B

Small zonal IGP covering the villages of Bucey-les-Gy, Gy, Charcenne, Choye, Virey, Avrigney, and Autoreille in the Haute-Saône *département*.

FRANCHE COMTÉ HAUTE SAÔNE
Départementale vin de pays Map B

This IGP covers the entire Haute Saône *département* for red, white, and rosé wines from unclassified vineyards in the south of Burgundy.

FRANCHE - COMTÉ HUGIER
Zonal *vin de pays* Map B

Small zonal IGP covering the villages Hugier and Tromaray in the Haute-Saône *département*.

FRANCHE - COMTÉ MOTEY-BESUCHE
Zonal *vin de pays* Map B

Small communal appellation for sparkling and still white, red, and rosé wines.

FRANCHE - COMTÉ OFFLANGES
Zonal *vin de pays* Map B

Small communal appellation for sparkling and still white, red, and rosé wines.

FRANCHE - COMTÉ VUILLAFANS
Zonal *vin de pays* Map B

Small communal appellation for sparkling and still white, red, and rosé wines.

GARD
Départementale vin de pays Map B

A large production of mostly (40 per cent) rosé wines made from Carignan, Grenache, Cinsault, Cabernet Sauvignon, Merlot, and Syrah. Some clean, straight-forward and mainly Ugni Blanc-based white wines (with Viognier a growing niche within this small percentage).

✔ *De Guiot*

GERS
Départementale vin de pays Map B

Mostly white wines, similar in style to Côtes de Gascogne, but red and rosé in dry and late-harvest style are also permitted.

✔ *Château Bouscassé* • *Plaimont (Rive Haute)*

HAUTE MARNE
Départementale vin de pays Map B

Located adjacent to Champagne's Aube and Burgundy's Côte d'Or, it is not surprising that Le Muid Montsaugeonnais has made some pretty good Pinot Noir. Still and sparkling reds, whites, and rosés are permitted.

✔ *Le Muid Montsaugeonnais (Pinot Noir Elevé en Fûts de Chêne)*

HAUTES-ALPES
Départementale vin de pays Map B

Very little wine is produced under this denomination, which is the *département* just south of the Savoie. Still and sparkling production is allowed in all three colours, of which 50 per cent red, 30 per cent rosé, and 20 per cent white.

✔ *Domaine Allemand (Blanc)* • *de Trésbaudon (Rouge)*

HAUTE-VALLÉE DE L'AUDE
Zonal *vin de pays* Map A

Mostly dry whites, from Chardonnay, Chenin, Viognier, and Pinot Blanc/Gris grapes grown in the Limoux district. A third of the output is red, from *bordelais* grapes, with just 5 per cent rosé.

✔ *De l'Aigle (Pinot Noir Aigle Royal)*

HAUTE VALLÉE DE L'ORB
Zonal *vin de pays* Map B

A limited amount of red, white, and rosé, primarily from Pinot Noir, Chardonnay, and Viognier. Sparkling and dry or late-harvest still wines are allowed in all three colours.

HAUTE VIENNE
Départementale vin de pays Map B

Rarely used, minor *départementale vin de pays* for red, white, and rosé wines east of the Cognac region.

ÎLE DE BEAUTÉ
Zonal *vin de pays* Map B

With AOC wines accounting for just 40 per cent of Corsica's wine output, its one all-encompassing (though technically not *départementale*) *vin de pays* has a huge impact on the perceived quality of this island's wine. Almost 60 per cent of this *vin de pays* is red and 25 per cent rosé, both made from a wide range of grapes, including many indigenous Corsican varieties, some of which may be related to certain Italian grapes. Grape varieties used include: Aleatico (red Muscat-like variety), Cabernet Sauvignon, Carignan, Cinsault, Grenache, Merlot, Syrah, Barbarossa, Nielluccio (the widest-planted on the island and said to be related to Sangiovese), Sciacarello (echoes of Provence's Tibouren), and Pinot Noir. White wines from Chardonnay, Ugni Blanc, Muscat, and Vermentino are usually more demonstrative of clean, anaerobic handling and very cool fermentation than they are of any expression of *terroir*.

✓ *Du Mont Saint-Jean (Aleatico, Pinot Noir)* • *De Patrapiana (Nielluciu)* • *De Saline* • *Skalli (Terra Vecchia Rouge)* • *Union des Vignerons (Marestagno, Réserve du Président Muscat)*

ÎLE-DE-FRANCE
Zonal *vin de pays* Map B

Not yet approved at the time of writing (mid-2020), but in the final stages of legislation after 21 years of petitioning. Once one of the most important vine-growing areas, the Île-de-France used to cultivate grapes on 42,000 hectares (103,785 acres) – more than the size of Burgundy today –– in the 18th century. Today the total vineyard area is just over 10 hectares (25 acres), but this new recognition might breathe life into the region once again. Red, rosé, and white are permitted from a number of local and international grapes. The Île-de-France IGP has five subzones that may add their names on the label: Coteaux de Suresnes-Mont-Valérien, Coteaux de Blunay, Coteaux de Provins, Guérard, and Paris.

ÎLE-DE-FRANCE
COTEAUX DE SURESNES-MONT-VALÉRIEN
Zonal *vin de pays* Map B

This single-commune IGP covers the town of Suresnes where 4,500 bottles are produced annually at the 1-hectare (2.5-acres) Clos du Pas Saint-Maurice.

ÎLE-DE-FRANCE
COTEAUX DE BLUNAY
Zonal *vin de pays* Map B

Single-commune IGP covering the town of Melz-sur-Seine, where about 10,000 bottles of Chardonnay and Pinot Noir are produced on 2 hectares (5 acres).

ÎLE-DE-FRANCE:
COTEAUX DE PROVINS
Zonal *vin de pays* Map B

Single-commune IGP covering the town of Provins. Local grower Patrice Bersac uprooted the old vineyard and planted 500 vines of the resistant grape variety Cabernet Cortis, but also Solaris and Muscaris.

ÎLE-DE-FRANCE GUÉRARD
Zonal *vin de pays* Map B

A small Subzonal IGP covering the towns of Guérard, Tigeaux, and Crécy-la-Chapelle. Currently the only grape-growing area is Daniel Kiszel's vineyard (Chardonnay and Pinot Noir) covering 1 hectare (2.5 acres) on a historic site, with plans to extend up to 3 hectares (7.5 acres).

ÎLE-DE-FRANCE PARIS
Zonal *vin de pays* Map B

Single-commune IGP covering Paris. Currently there are insignificant plantings: 200 vines at the Clorivière institute and 38 at the Lycée Albert de Mun.

ISÈRE
Départementale vin de pays Map B

Rarely used *départementale vin de pays*: most producers prefer one of a number of zonal IGPS for the unclassified vineyards in this part of the northern Rhône. Mostly white wine (65 per cent); the rest is split between red and rosé.

ISÈRE BALMES DAUPHINOISES
Zonal *vin de pays* Map B

In the denomination of Balmes Dauphinoises, dry white, made from mainly the Jacquère and Chardonnay varieties, accounts for 60 per cent of production, and red, made from Gamay and Pinot Noir, accounts for 40 per cent. Rosé may also be produced.

ISÈRE COTEAUX DU GRÉSIVAUDAN
Zonal *vin de pays* Map B

Red and rosé Savoie-style wines made from Gamay, Pinot, and Etraire de la Dui and Jacquère-based dry whites.

LANDES
Départementale vin de pays Map B

Red, white, and rosé wines are produced, with approximately 80 per cent red, and the grapes used are traditional southwestern varieties. Reds and rosés are made from Tannat supported by *bordelais* varieties; whites are primarily Ugni Blanc plus Colombard, Gros Manseng, and Baroque. The following subzonal areas are allowed to attach their names to Landes on the labels: Coteaux de Chalosse, Côtes de l'Adour, Sables Fauves, and Sables de l'Océan.

✓ *Michel Guérard (Rouge de Bachen)*

LANDES COTEAUX DE CHALOSSE
Zonal *vin de pays* Map B

The largest area in the south of the Landes *département*, where vines grow in and around Dax and Mugron.

LANDES CÔTES DE L'ADOUR
Zonal *vin de pays* Map B

Small subzone of vines around Aire-sur-Adour and Geaune.

LANDES SABLES DE L'OCÉAN
Zonal *vin de pays* Map B

The romantic-sounding "Sands of the Ocean" refers to vines growing in the sand dunes around Messanges. Still red, white, and rosé wines, traditional-method white and rosé, plus late-harvest whites are allowed.

LANDES SABLES FAUVES
Zonal *vin de pays* Map B

The "Wild Sands" is a tiny enclave of vines west of Eauze.

LAVILLEDIEU
Zonal *vin de pays* Map B

Small zonal IGP covering 13 communes west of Montauban in the Tarn-et-Garonne *département*. A small production of red and rosé wines that must be blends of a minimum 10 per cent of Cabernet Franc, Gamay, Négrette, Syrah, and Tannat, plus up to 10 per cent Fer and Milgranet.

MAURES
Zonal *vin de pays* Map B

This denomination has a large production (60,000 hectolitres) and covers most of the southern part of the Côtes de Provence. Rosé of a similar style and quality to those of the IGP Var account 68 per cent of the wines. The rest is red, which is rich, warm, and spicy, while white wine accounts for just 7 per cent of total output.

MEDITERRANÉE
Regional *vin de pays* Map B

This vast regional IGP encompasses the *départements* of Drôme, Ardèche, Alpes de Haute Provence, Hautes-Alpes, Alpes-Maritimes, Bouches-du-Rhône, Var, Vaucluse, Corse-du-Sud, and Haute-Corse entirely and parts of Isère, Loire, and Rhône. Production sits around 56.5 million bottles, at least two-thirds of which is rosé. Sparkling wine production is also significant, with about 1.3 million bottles of mostly tank-method rosé fizz made each year.

MEDITERRANÉE
COMTÉ DE GRIGNAN
Zonal *vin de pays* Map B

Most Comté de Grignan is red, and most of that Grenache-dominated, although Syrah, Cinsault, Gamay, Carignan, Merlot, and Cabernet Sauvignon are also allowed. Whites are Grenache Blanc based, with Viognier, Chardonnay, Muscat and Vermentino permitted, amongst others. Sparkling wines are made by traditional or tank method of all three colours, too.

MEDITERRANÉE
COTEAUX DE MONTÉLIMAR
Zonal *vin de pays* Map B

This *vin de pays* encompasses vineyards between Montélimar and Dieulefit in the Drôme *département*, just north of the Grignan-Les-Adhémar AOC. Red and rosé must be made from one or more of the following: Cabernet Sauvignon, Carignan, Cinsault, Gamay, Grenache, Marselan, Merlot, Pinot Noir, and Syrah. White wines from one or more of Chardonnay, Clairette, Grenache Blanc, Marsanne, Muscat à Petits Grains, Roussanne, Sauvignon Blanc, and Viognier.

MONT CAUME
Zonal *vin de pays* Map B

This is effectively the *vin de pays* of the commune of Bandol, which accounts for about 220 hectares (545 acres) of the 2,400 hectares (5,930 acres) total vine area in the region. It is indeed the red wines that are the best here. The main varieties for most red and rosé wines (which account for over 75 per cent of the total production) are Carignan, Grenache, Cinsault, Syrah, and Mourvèdre. A little white is produced as well as sparkling wines of all colours, but they are of little interest.

PAYS DES BOUCHES-DU-RHÔNE
Départementale vin de pays Map B

This denomination has one of the largest *vin de pays* productions in the country, some 50 per cent of which is rosé and most of that comes from the Coteaux d'Aix-en-Provence area. A little white is also made from Ugni Blanc, Clairette, Bourboulenc, Vermentino, Chardonnay, and Chasan (a slightly aromatic Listan and Chardonnay cross). Domaine de Trévallon is not only the finest wine of the appellation, but one of the greatest *vins de pays*, producing a quality easily equal to some of the best crus classés of Bordeaux. There is one zonal IGP within the appellation: Terre de Camargue

✓ *Jean-Paul Luc (Minna Vineyard Rouge)* • *Domaine de Trévallon*

PAYS DES BOUCHES-DU-RHÔNE
TERRE DE CAMARGUE
Zonal *vin de pays* Map A

Varietal wines take a lead in this small appellation, reds, rosés and whites made from Cabernet Sauvignon, Merlot, Chardonnay, Sauvignon Blanc and Viognier.

PAYS DE BRIVE
Zonal *vin de pays* Map B

A recent (2017) zonal IGP in the *département* of Corrèze, permitting dry and late-harvest reds and whites and dry rosés. The whites are produced in the north of the appellation and made from Chardonnay, Chenin, and Sauvignon Blanc. Reds and rosés come from the south of the IGP and are also made of Loire varieties.

LE PAYS CATHARE

Zonal vin de pays Maps A and B

This area overlaps the Aude and Ariège *départements*. Red wines are made from the following: Merlot, Cabernet Franc, Cabernet Sauvignon, and Merlot, Arinarnoa, Caladoc, Carignan, Cinsaut, Chenanson, Eiodola, Grenache, Malbec, Marselan, Pinot Noir, Pinot Gris, Pinot Blanc, Portan, and Syrah. For white wine, Chardonnay and/or Sauvignon Blanc, Bourboulenc, Chasan, Chenin Blanc, Grenache, Grenache Blanc, Macabéo, Marsanne, Mauzac, Roussanne, Sémillon, Ugni Blanc, Vermentino, and Viognier are used. Rosé is made from Merlot, Cabernet Franc, Cabernet Sauvignon, and Syrah, Caladoc, Cinsaut, Chanson, Grenache, and Portan.

PAYS D'HÉRAULT

Départementale vin de pays Map B

Red wine accounts for 70 per cent of production and is increasing. The choice of grapes is quite wide, but most blends are based on Carignan, Cinsault, Grenache, and Syrah, boosted by a small proportion of *bordelais* varieties. Whites are typically made from Clairette, Macabéo, Grenache Blanc, and Ugni Blanc. Production is vast, 10 million cases, making it the second- or third-largest *vin de pays*, together with Vin de Pays de l'Aude. Improved viticultural techniques have transformed what was once the heart of Midi mediocrity into delicious, brilliant-value wine in some cases. The following zonal IGPs may use their name on the label: Bérange, Bénovie, Pays de Bessan, Cassan, Pays de Caux, Cessenon, Collines de la Moure, Coteaux de Bessilles, Coteaux de Fontcaude, Coteaux de Laurens, Coteaux de Murviel, Coteaux du Salagou, Côtes du Brian, Côtes du Ceressou, Mont Baudile, and Monts de la Grage. The most famous wine bearing this modest *vin de pays* denomination is, of course, Mas de Daumas Gassac. Situated northwest of Montpellier at Aniane, this property has been dubbed "the Lafite of the Languedoc", but when Aimé and Véronique Guibert purchased it, wine was the last thing on their minds. That was, however, before the visit of Professor Enjalbert, the Bordeaux geologist and author who discovered Daumas Gassac had a rare, fine, powdery volcanic soil that is an incredible 20 metres (65 feet) deep. Enjalbert predicted that it would yield a world-class wine if cultivated as a grand cru, and that is exactly what Aimé Guibert set out to do. He passed away in 2016, but four of his children run the estate today.

✓ *Mas de Daumas Gassac (Rouge)* • *La Grange des Pères* • *Mas de Janiny (Cabernet Sauvignon)* • *Moulin de Gassac (Guilhem, Villeveyrac)* • *Les Quatre Pilas (Mouchère)* • *Valjulius*

PAYS D'HÈRAULT BÉNOVIE

Zonal vin de pays Map A

This denomination is located at the eastern extremity of the Languedoc and overlaps part of the Muscat-de-Lunel area. Almost 80 per cent of the production is a light, fruity red, made predominantly from Carignan, Grenache, Cinsault, and Syrah, although Merlot and Cabernet Sauvignon may also be used. A fair amount of attractive rosé is made, mostly by the *saignée* method, plus a small quantity of Ugni Blanc–based dry white.

PAYS D'HÈRAULT BÉRANGE

Zonal vin de pays Map A

Production in Bérange is 75 per cent red, 20 per cent rosé, and 5 per cent white, made from a range of grape varieties very similar to those used in the Languedoc and neighbouring Bénovie.

PAYS D'HÈRAULT BESSAN

Zonal vin de pays Map A

This tiny, single-village denomination just east of Béziers is best known for its dry, aromatic rosé, which now accounts for 65 per cent of production. Just 10 per cent red is made, with simple Ugni Blanc–based white accounting for the balance.

✓ *Le Rose de Bessan (Cuvée Spéciale)*

PAYS D'HÉRAULT CASSAN

Zonal vin de pays Map A

This zonal *vin de pays* is for an area in the Coteaux du Languedoc north of Béziers and overlaps part of Faugères. Almost three-quarters of the wine from this IGP is red and full bodied, from Carignan, Cinsault, Grenache, Cabernet Sauvignon, Merlot, and Syrah. The balance is split between a well-flavoured rosé and a crisp, dry white made primarily from Ugni Blanc, although Clairette and Terret may also be used. Since 2000, however, it has been permissible to market wines as pure varietals with a very wide range of grapes to choose from.

PAYS D'HÉRAULT CESSENON

Zonal vin de pays Map A

This appellation produces red wines of a rustic St-Chinian style, plus a little rosé.

PAYS D'HÉRAULT COLLINES DE LA MOURE

Zonal vin de pays Map A

More than 80 per cent of production is basic red made from a choice of Carignan, Cinsault, Grenache, Syrah, Cabernet Sauvignon, and Merlot, the last two of which are often vinified separately. A light, dry rosé accounts for about 15 per cent of production, but very little white is made, and it is mostly from Ugni Blanc.

✓ *Montlobre (La Chapelle)* • *de Mujolan (Mas de Mante Vertige)*

PAYS D'HÉRAULT COTEAUX DE BESSILLES

Zonal vin de pays Map A

Fresh, light, rustic-style wines from the Coteaux du Languedoc immediately north of Pinet. Production is two-thirds red, one-quarter rosé, and the balance white.

PAYS D'HÉRAULT COTEAUX DE FONTCAUDE

Zonal vin de pays Map A

This area overlaps much of St-Chinian in the eastern section of the Coteaux du Languedoc. Almost 80 per cent of the wines are red, and these are made in a light, fresh but interesting style from Carignan, Cinsault, Grenache, Cabernet Sauvignon, and Syrah. Rosé and a little white are also to be found.

PAYS D'HÉRAULT COTEAUX DE LAURENS

Zonal vin de pays Map A

Some 85 per cent of these wines are a red *vin de pays* equivalent of Faugères, made mostly from Carignan, Cinsault, Syrah, and Grenache, but some of the better wines also include Cabernet Sauvignon and Merlot. A small amount of rosé and even less white is made.

PAYS D'HÉRAULT COTEAUX DE MURVIEL

Zonal vin de pays Map A

Just next to the St-Chinian area, this *vin de pays* produces some very good reds, which account for 85 per cent of the output, made from Carignan, Cinsault, Grenache, Cabernet Sauvignon, Merlot, and Syrah. There are a number of rosés, but very few whites.

PAYS D'HÉRAULT COTEAUX DU SALAGOU

Zonal vin de pays Map A

Production is 80 per cent red and 20 per cent rosé, from a range of typical Languedoc grapes.

PAYS D'HÉRAULT CÔTES DU BRIAN

Zonal vin de pays Map A

An area in Minervois, Côtes du Brian produces 90 per cent red, from Carignan for the most part, although Grenache, Cinsault, Cabernet Sauvignon, and Syrah may also be used. The balance of the production is principally rosé, with very few white wines.

PAYS D'HÉRAULT CÔTES DU CÉRESSOU

Zonal vin de pays Map A, No.53

Typically light and fruity Languedoc wines are produced in fairly large amounts; 60 per cent is red, 15 per cent white, and 25 per cent rosé.

PAYS D'HÉRAULT MONT BAUDILE

Zonal vin de pays Map A

Production is mostly red wine, from a typical Languedoc range of grapes grown in the foothills of the Causses de Larzac. Rosé and white account for less than 15 per cent of the total output, although crisp, dry whites are on the increase.

✓ *D'Aupilhac*

PAYS D'HÉRAULT MONTS DE LA GRAGE

Zonal vin de pays Map A

These red and rosé wines are produced in a basic Languedoc style, often beefed up with Syrah grapes. White wines must be truly dry (less than 2.5 grams per litre residual sugar).

PAYS D'HÉRAULT PAYS DE CAUX

Zonal vin de pays Map A

This *vin de pays* produces a good, typical dry and fruity rosé in a Languedoc style from vines growing north of Béziers. Mainly red, a little white, and rosé are also made. Pure varietal wines have been allowed since 2000.

✓ *Caves Molière (Barbier Gely Rouge, Cuvée des Comédiens Rouge)* • *Pech Rome (Tempranillo)*

PAYS D'OC

Regional vin de pays Map B

This vast appellation is the most successful of all IGPs, commercially and qualitatively. It sells well because of the simplicity of its name, which even Anglo-Saxons have no difficulty pronouncing, and it encompasses most of the best *vins de pays* produced in France. The "Oc" of this regional *vin de pays* is southern dialect for "yes", it being "*oui*" elsewhere in France, thus Languedoc, the "tongue of Oc". Some 62 per cent of production is red, with the balance split equally between white and rosé. Even though most of the finest *vins de pays* are produced here, the volume of output is so vast that the quality will inevitably be variable. What has enabled so many truly fine wines to emerge in this denomination is the role of the so-called *cépages améliorateurs*. These grapes, such as Cabernet Sauvignon, Merlot, Syrah, Chardonnay, and Sauvignon Blanc, are not traditional in the region, but add greatly to the quality and finesse of more rustic local varieties. They also make excellent varietal wines. Viognier is on the increase as a varietal wine and makes a good-value, juicy-fruity, more exotic alternative to Chardonnay. Marselan (Cabernet Sauvignon x Grenache) has been authorized since December 2005.

✓ *Des Aires Hautes)* • *Saint Auriol* ◉ *(Pinot Noir, Chardonnay)* • *Barton & Guestier (Cabernet Sauvignon)* *Gérard Bertrand* • *Jean-Marc Boillot (De la Truffière)* • *Des Bons Auspices (Emparitz Cabernet Sauvignon)* • *Du Bosc (Petit Verdot)* • *Camplazens (Syrah)* • *CV de Cers Portiragnes (Cabernet Sauvignon Sensation)* • *Chantovent (Merlot Bistro Marquisat)* • *Jean-Louis Denois* • *La Différence* • *Le Fort (Blanche Chardonnay)* • *Foncalieu)* • *La Grange des Quatre Sous* • *de Granoupiac (Merlot)* • *Jeanjean (L'Incompris)* • *Lalande (Pinot Noir)* • *Michel Laroche (La Croix Chevalière)* • *Lorgeril (Chardonnay de Pennautier)* • *Paul Mas* • *Mas Carlot (Syrah-Grenache)* • *Gabriel Meffre* • *Laurent Miquel* • *Montlobre (Tête de Cuvée Blanc)* • *Château d'Or et de Guelles* • *L'Ormarine (Haut de Sénaux Syrah)* • *de l'Orviel (Chardonnay)* • *Peirière (Elevé en Fûts de Chêne: Cabernet Sauvignon, Merlot, Syrah)* • *Michel Picard (Chardonnay)* • *Preignes le Vieux (Petit Verdot)* • *Antonin Rodet (Syrah)* • *Saint Marc (Sauvignon)* • *St-Hilaire* • *de Sérame (Muscat Réserve))* • *Vedeau (Chardonnay Terre d'Amandiers)* • *Arnaud de Villeneuve (Grenache, Syrah)* • *Vindivin (Grenache Blanc)* • *Les Yeuses*

PÉRIGORD
Zonal *vin de pays* Map B

This IGP encompasses the Dordogne *département* and one commune (Salviac) in the Lot. Most vineyards are growing on the *causse*, or limestone region, of Périgord, where grape-growing flourished in the 19th century. The sub-zonal Vin de Domme and the *départementale* Dordogne are allowed to add their name to this IGP.

✓ *Grand Gaillard (Sauvignon Blanc)*

PÉRIGORD DORDOGNE
Départementale *vin de pays* Map B

The *département* is named after one of the two great rivers of Bordeaux and includes AOCs such as Bergerac and Monbazillac, thus should have plenty of potential for aspiring *vin de pays* producers. Grape varieties include Cabernet Franc, Cabernet Sauvignon, Gamay, Malbec, Merlot, and Pinot Noir for reds or rosés, and Chardonnay, Chenin Blanc, Colombard, Folle Blanche, Sémillon, and Ugni Blanc for whites.

PÉRIGORD VIN DE DOMME
Zonal *vin de pays* Map B

Small zonal IGP covering 16 communes north of Cahors.

PUY-DE-DÔME
Départementale *vin de pays* Map B

Red, white, and rosé wines of simple, rustic quality are made under this appellation. A *puy* is a volcano stack, a number of which are found in the hilly terrain of the Puy-de-Dôme *département*, which has been shaped by volcanic activity over time.

SABLE DE CAMARGUE
Zonal *vin de pays* Map A

Formerly called Vin de Pays Sables du Golfe du Lion, this is a very go-ahead IGP situated along the Mediterranean coast of Gard, Hérault, and Bouches-du-Rhône, where the vines are mostly ungrafted and grown on an amazing sandbar with seawater on both sides. Production is, unusually, two-thirds rosé, including a large proportion in pale *vin gris* style, plus 30 per cent red and a small amount of white wine.

SAINT-GUILHEM-LE-DÉSERT
Zonal *vin de pays* Map A

Formerly called Vin de Pays des Gorges de L'Hérault, this denomination is in Coteaux du Languedoc and overlaps most of the Clairette du Languedoc appellation, yet less than 10 per cent of its wines are white, with Ugni Blanc and Terret supporting Clairette. They are, however, better known than the reds, which account for over 80 per cent of output. Some rosés and late harvest wines of all colours are also made.

✓ *D'Aupilhac (Plos des Baumes)*

SAINT-GUILHEM-LE-DÉSERT CITÉ D'ANIANE
Zonal *vin de pays* Map A

A tiny single-commune IGP in the Hérault *département*.

SAINT-GUILHEM-LE-DÉSERT VAL DE MONTFERRAND
Zonal *vin de pays* Map A

Mostly red wines, but rosé accounts for 25 per cent of production and white for 15 per cent in this denomination, which encompasses the *garrigues* below the Cévennes. The usual Languedoc grapes are used. *Vin primeur* and *vin d'une nuit* are local specialities produced in relatively large quantities.

✓ *L'Hortus • Mas de Martin (Roi Patriote)*

SAINTE-MARIE-LA-BLANCHE
Zonal *vin de pays* Map B

This *vin de pays* area straddles the border between Côte d'Or and Saône-et-Loire. Those in the Côte d'Or are just a few kilometres east of Beaune and used to have the right to the Vin de Pays de la Côte d'Or until it was suppressed in 1998. Red wines are made from one or more of Gamay, Pinot Gris, and Pinot Noir, while white wines are made from one of the following: Aligoté, Auxerrois, Chardonnay, Melon de Bourgogne, Pinot Blanc, and Pinot Gris.

SAÔNE-ET-LOIRE
Départementale *vin de pays* Map B

This is a rarely exported *départementale vin de pays* for mostly red and white wines from unclassified vineyards in the Burgundy region.

TERRES DU MIDI
Zonal *vin de pays* Map B

This large area covers the Aude, Hérault, Gard, and Pyrénées-Orientales dpartements, and the Southern part of Lozère, producing around 1.5 million hectolitres (200 million bottles) of red, rosé, and white still wines.

THÉZAC-PERRICARD
Zonal *vin de pays* Map B

This is a small IGP adjoining the western extremity of Coteaux du Quercy, producing Cahors-like reds from Cabernet Franc, Cabernet Sauvignon, Gamay, Malbec, Merlot, and Tannat.

URFÉ
Zonal *vin de pays* Map B

This regional *vin de pays* produces red and rosé from Gamay, Pinot Noir, and Syrah are made in modest quantities. White may also be produced from Chardonnay, Viognier, Gewürztraminer, Roussanne, Pinot Gris, and Aligoté. The villages Ambierle and Trelins may attach their names to the IGP for rosé only.

VAL DE LOIRE
Regional *vin de pays* Map B

This regional *vin de pays* produces red, white, and rosé wines from 14 *départements* covering most of the Loire Valley. It would seem, from the wines most commonly encountered, that the majority of this vast output must be dry white, dominated by either the Chenin Blanc or Sauvignon Blanc varieties, but, according to the statistics, 60 per cent is red. Val de Loire is a generally disappointing appellation, with the thin, acid Chardonnays being the most depressing, but the Cabernet Sauvignon from Pierre & Paul Freuchet is surprisingly soft and perfumed for this northerly denomination.

✓ *Ampelidae (Marigny-Neuf) • L'Aujardière (Fié Gris) • Paul Boutinot (Signature Chardonnay) • De la Coche • Adéa Consules (Ligeria) • Bruno Cormerais (Elevé en Fûts de Chêne) • De la Couperie (Clyan Elevé en Fûts de Chêne) • Marquis de Goulaine (Chardonnay) • De la Houssais • Levin (Sauvignon Blanc) • Manoir La Perrière (Chardonnay) • Henry Marionnet (Provinage) • Du Petit Château (Chardonnay) • Petiteau-Gaubert (Domaine de la Tourlaudière Cabernet Sauvignon Rosé) • Château de La Ragotière • De la Saulze (Gamay Rosé)*

VAL DE LOIRE ALLIER
Départementale *vin de pays* Map B

This is a relatively recent *départementale vin de pays* within the IGP Val de Loire. It produces red and rosé wines primarily from Cabernet Franc, Cabernet Sauvignon, Gamay, and Grolleau, but Abouriou, Pinot d'Aunis, and Pinot Noir are also allowed. Whites are made primarily from Chardonnay, Chenin Blanc, and Sauvignon Blanc, but Orbois, Folle Blanche, Grolleau Gris, Melon de Bourgogne, and Pinot Blanc are also allowed.

VAL DE LOIR CHER
Départementale *vin de pays* Map B

Mostly Touraine-like, Gamay-based red wine, plus a small amount of dry rosé in a light *vin gris* style.

VAL DE LOIRE INDRE
Départementale *vin de pays* Map B

Red, white, and rosé wines, including a pale *vin gris* style, from traditional Loire grape varieties are produced under this appellation.

VAL DE LOIRE INDRE ET LOIRE
Départementale *vin de pays* Map B

From vineyards east of Tours, this *vin de pays* is 80 per cent red, made primarily from Gamay, although Cabernet Franc and Grolleau are also used. A little white, but even less rosé, is made.

VAL DE LOIRE LOIRE-ATLANTIQUE
Départementale *vin de pays* Map B

Red and rosé from Gamay and Grolleau and light whites from Melon and Folle Blanche are made. Some interesting Chardonnays also started to emerge.

VAL DE LOIRE LOIRET
Départementale *vin de pays* Map B

A small amount of red and rosé, including a pale *vin gris* style, made from Gamay. A Sauvignon Blanc white is also made.

VAL DE LOIRE LOIR-ET-CHER
Départementale *vin de pays* Map B

Red, white, and rosé wines, including a pale *vin gris*, are made from traditional Loire grape varieties, mostly from around Blois in the Cheverny area. Almost half the production is red and the rest white, but a touch more of the former is made, with very little rosé.

VAL DE LOIRE MAINE ET LOIRE
Départementale *vin de pays* Map B

Production is 85 per cent red, and white and rosé account for the rest, made from traditional Loire grape varieties in the Anjou-Saumur district. Substantial quantities are produced, but most are sold from the back door.

VAL DE LOIRE MARCHES DE BRETAGNE
Subregional *vin de pays* Map B

Wines from the Marches de Bretagne subregion have the right to add this name to the IGP Val de Loire. Production includes red and rosé wines, made predominantly from Abouriou, Gamay, and Cabernet Franc, plus a little white from Muscadet and Gros Plant.

VAL DE LOIRE NIÈVRE
Départementale *vin de pays* Map B

Red, white, and rosé wines, production of which is mostly confined to the areas of Charité-sur-Loire, La Celle-sur-Nièvre, and Tannay.

VAL DE LOIRE PAYS DE RETZ
Zonal *vin de pays* Map B

This denomination may be added to the IGP Val de Loire. Best known for its rosé made from the Grolleau Gris, although red wines are also made from Grolleau, Gamay, and Cabernet Franc and account for as much as 70 per cent of production. A tiny amount of white from Grolleau Gris can be found.

VAL DE LOIRE SARTHE
Départementale *vin de pays* Map B

Red, white, and rosé wines may be made, although production of this *départementale vin de pays* is minuscule and, as far as I am aware, limited to just one grower in Marçon.

VAL DE LOIRE VENDÉE
Départementale *vin de pays* Map B

Red, white, and rosé wines are produced in small quantities in this *départementale vin de pays*, in a similar style to Fiefs Vendéens (which was a *vin de pays* until it was promoted to VDQS status in December 1984 and then to AOC in 2011).

VAL DE LOIRE VIENNE

Départementale vin de pays Map B

This *départementale vin de pays* produces red, white, and rosé wines, most of which are grown in and around the Haut-Poitou district.

✓ *Ampelidae*

VALLÉE DU PARADIS

Zonal *vin de pays* Map A

Mostly red wines from an area that overlaps Corbières and Fitou, made from Carignan, Syrah, Grenache, Cinsault, Cabernet Sauvignon, and Merlot. Very little white wine is made (usually as *vin primeur*) and even less rosé. A heavenly name, even to Anglo-Saxon ears, which no doubt helps it on export markets, which account for almost half its sales.

✓ *De Garrigotes (Rouge)* • *Mont Tauch*

VALLÉE DE TORGAN

Zonal *vin de pays* Map A

As this is geographically the *vin de pays* equivalent of Fitou AOC, it is no surprise to discover that 98 per cent of its wines are red. They are in fact made from a very similar range of grapes as that used to make Fitou, but certain other grapes are also permitted, including *bordelais* varieties and the Teinturier Alicante Bouschet. Rosé may be made, but seldom is, and the minuscule amount of aromatic dry white may be made from Clairette, Macabéo, and Marsanne. This IGP used to be known as Coteaux Cathares.

✓ *Bertrand-Bergé (Le Méconnu Blanc)* • *Mont Tauch (Le Garrigues)*

VAR

Départementale vin de pays Map B

This denomination is one of the largest producers of *vins de pays*. Covering the vast majority of the Côtes de Provence AOC, it not surprisingly produces a large quantity of rosé: no less than 70 per cent, in fact. The wines are usually made by the *saignée* method and range from a very pale *vin gris* style, through orange to almost cherry red. The quality is equally as variable. The big, rich, spicy reds (20 per cent of total production) are much better and of a more consistent quality, often using Syrah, Mourvèdre, and Cabernet Sauvignon to boost the Grenache, Cinsault, Carignan, and Roussanne du Var, on which the rosés mainly rely. Some white wines are also made but represent just 10 per cent of production. They are not usually very exciting and are mostly made from Chardonnay, Viognier, and Vermentino. Three zonal IGPs are included in the Var appellation: Argens, Coteaux du Verdon, and Sainte-Baume.

✓ *Les Caves du Commandeur (Rolle, Syrah rosé)* • *Domaine de Garbelle (Vermentino)* • *Le Saint André* • *Château Sarrins (Blanc de Rolle)* • *Château Thuerry (L'Exception)*

VAR ARGENS

Zonal *vin de pays* Map B

Almost 50-50 rosé and red are made is this zonal *vin de pays* from Carignan, Cinsault, Syrah, Roussanne du Var, Mourvèdre, and Cabernet Sauvignon. A little white wine is also produced.

VAR COTEAUX DU VERDON

Zonal *vin de pays* Map B

This *vin de pays* in the northern hinterland of Côtes de Provence and Coteaux Varois was created in 1992.

VAR SAINTE BAUME

Zonal *vin de pays* Map B

This *vin de pays* encompasses vineyards in the villages of Bras, Brignoles, La Celle, Mazaugues, Méounes-lès-Montrieux, Nans-les-Pins, Néoules, Ollières, Plan-d'Aups, Riboux, La Roquebrussanne, Rougiers, Saint-Maximin, Saint-Zacharie, Signes, and Tourves in the Var *département*. A large number of varieties permitted, and pure varietal wines are allowed.

VAUCLUSE

Départementale vin de pays Map B

Three-quarters of output of this *départementale vin de pays* is red, similar to basic Côtes-du-Rhône. Rosé can be fresh, and accounts for just 10 per cent of output; white is generally the least interesting. Pure varietal wines are allowed. There are two zonal IGPs within Vaucluse: Aigues and Principauté d'Orange.

✓ *Domaine de la Citadelle (Court Metrage Cabernet Sauvignon)* • *Fontaine du Clos (Aura Chardonnay blend)*

VAUCLUSE AIGUES

Zonal *vin de pays* Map B

Created in 1993 within IGP Vaucluse, this *vin de pays* roughly corresponds to the Coteaux du Lubéron and has a similar range of grape varieties.

VAUCLUSE PRINCIPAUTÉ D'ORANGE

Zonal *vin de pays* Map B

This is a full red wine, made predominantly from Rhône grape varieties grown around Orange, north of Avignon within the IGP Vaucluse, in an area that bridges Châteauneuf-du-Pape and Côtes-du-Rhône Villages. Just 4 per cent rosé and barely 1 per cent white is made.

✓ *De l'Ameillaud* • *Font Simian* • *Les Pialons (rosé)*

VICOMTÉ D'AUMELAS

Zonal *vin de pays* Map A

Overlapping part of Cabrières, this *vin de pays* produces mostly (50 per cent) red wines of a simple, fresh, fruity style, from traditional southwestern grape varieties. About a quarter of the production is rosé and the same amount of crisp, dry white is produced.

✓ *D'Aubaret (Sauvignon)*

VICOMTÉ D'AUMELAS VALLÉE DORÉE

Zonal *vin de pays* Map A

A small subzonal appellation covering 13 villages along the Hérault river.

VIN DES ALLOBROGES

Zonal *vin de pays* Map B

This is the *vin de pays* equivalent of the Vin de Savoie, although it extends beyond the borders of that AOC into cattle country where the existence of vines is very sporadic. The wines are similar to Vin de Savoie, if somewhat lighter and more rustic in style. Dry, sweet, and sparkling whites and dry reds and rosé are allowed. Almost 90 per cent of production is white, made primarily from Jacquère and Chardonnay. Reds may be made from Gamay, Mondeuse, and Pinot Noir, amongst many other varieties, with rosé accounting for less than 1 per cent.

✓ *Demeure-Pinet (Jacquère)*

YONNE

Départementale vin de pays Map B

The Yonne is a minor *départementale* appellation, and its production of wines is quite small. Reds from Pinot Noir, Gamay, César and Tressot, pure Pinot Noir rosés, and whites from Chardonnay, Aligoté, Pinot Blanc Pinot Gris, and Sauvignon Blanc. White and rosé sparkling are made of Chardonnay and Pinots.

A VINEYARD WORKERS CHECKS THE RIPENESS OF BLACK GRAPES BEFORE HARVESTING
A long list of black grapes, such as Carignan, Cinsault, Grenache, Gamay, Cabernet Sauvignon, Pinot Noir, and Syrah, make up the allowed varieties in many of the *vin de pays*, with lesser-known varieties like Roussanne du Var, Abouriou, Pinot d'Aunis, Mourvèdre, and Jacquère added to the mix.

The Wines of Italy

WINEMAKING HAS ALWAYS BEEN AN INTEGRAL PART of Italian culture. The Etruscans and the Romans largely developed and diffused this culture throughout Europe and initiated a vast latitude of styles and grape varieties grown in every part of the country. Vineyards are planted in mountains, hills, coastal regions, and also in volcanic areas. Italy now counts more than 650,000 hectares (1,606,185 acres) of surface under vineyard, making it the world's biggest wine producer. With a climate very suitable for winegrowing, Italy is the country with the highest surface under organic regime. In France wine styles are more relative to the regions, but in Italy there is a more "provincial" approach that reflects the massive amount of indigenous varieties, and this has produced a staggering number of appellations. Italy has evolved in the last decades from thin and dilute styles of wines, predominantly consumed domestically, to distinctive world-class wines that make their most significant impact in key export markets. This achievement is also boosted and driven by the high diffusion of Italian cuisine outlets around the world; the impressive diversity and variability of wines strongly reflects and complements the equally various and creative dishes throughout the country. Alongside wines that have now become classics, this immense variability of styles, cultivars, and appellations results in a somewhat confusing image that has to be managed to achieve better results.

VINEYARDS SPILL ALONG THE HILLSIDES BENEATH THE COMMUNE OF CASTIGLIONE FALETTO IN THE BAROLO WINE REGION
Sometimes called "the king of wines", Barolo's rich taste makes it one of Italy's most-sought-after wines. It is made from the Nebbiolo grape grown on vineyard slopes surrounded by the hills of the Langhe area of Piedmont.

Italy

Millennia of wine heritage, endemic fragmentation, and substantial localism are the recipe for the most creative, captivating, and bewildering wine country. Italy has the broadest palette of indigenous varieties in cultivation and a tremendous amount of geographical indications, making its discovery both engaging and daunting.

Named Enotria ("land of wine") by the Greeks, Italy boasts a 4,000-year wine history. In fact, the Etruscans, whose civilization predates Rome, were already well known as producers of quality wines. Sited in the middle of the Mediterranean Sea, the Italian peninsula has always been a crossroads for populations. This actively influenced the heterogeneity among cultivars, techniques, styles, and winemaking approaches. Although during the Middle Ages wine was primarily produced for sacramental purposes, it was during the Renaissance that the upper classes became more engaged with the pleasures of wine. The devastating phylloxera epidemic of the late 1800, however, forced the industry to focus on quantity over quality, resulting in the deteriorated reputation of Italian wine.

EXCEPTIONAL BIODIVERSITY

With its peninsular boot shape and north-to-south extension, Italy enjoys a variety of climates that range from hotter Mediterranean in Puglia and Sicily to a cooler continental climate in Val d'Aosta and Alto Adige. The number of interesting indigenous grape varieties is greater than can be found in any other major winemaking country of the world, including France. The great orography variability, with 35 per cent mountainous terrain and 40 per cent hillside slopes, along with a number of coastal plantings, also make notable difference in styles and varieties grown. A staggering 1,000 different varieties are distributed along the country's territory, to the point that the most widely planted red grape variety, Sangiovese, only accounts for a mere 8 per cent of plantings. To a certain extent it is possible

CHIANTI IN THE TRADITIONAL *FIASCO* BOTTLES FILL A BASKET IN A TOURIST SHOP
The *fiasco* style of bottle, with its round bottom covered with a close-fitting straw basket allows the bottles to be packed with the necks inverted for safe transport. They are closely associated with the wines of Italy, especially Chianti, although they are rarely used today. These days, *fiaschi* are more likely to be sold to tourists as souvenirs.

to generalise about Italian varieties, saying that these are often producing higher acidity and lower alcohol levels, making them very suitable in a climate change scenario. In the last few years these varieties also started to be planted in New World countries, where historically French cultivars had been used for decades.

UNIQUE HERITAGE AND TRADITION

Italian wine should be regarded not only as a mandatory pairing to traditional dishes but also as a complementary element to Italy's varied territories. It is viticulture that strongly influences the landscape, but it is also the peculiar characteristics of an area that determines the type of vine growing. Ranging from the steep coastal slopes of Costa d'Amalfi and Cinque Terre to the verdant flatlands of Emilia-Romagna and from the harsh terraces of Valtellina to the rolling hills of Montalcino and from the black volcanic soils of Etna to the pure bright marl of Barolo, Italy has just about all terrains. As opposed to France, where a small number of highly identifiable regions have their style and reputation, in Italy wine comes from everywhere and, often, in a very blundering way. It is this primordial disorder that makes Italy one of the most dynamic wine countries of today.

THE ITALIAN WINE RENAISSANCE

Italy has always been consuming its own wine, while exporting only a small proportion of it. It used to be thin and dilute inexpensive everyday wine often lacking ripeness and proper alcohol level, because the industry was more focused on quantity rather than quality. In 1986 the so-called methanol scandal (when methanol was used to raise the alcohol level of basic Piedmontese wines) represented a real turning point in Italy's wine history. Since then, investments on techniques, and traceability have been

WINE REGIONS OF ITALY
As might be expected of a nation so geographically and culturally diverse, Italy produces a vast array of different types of wine. Away from the mainland, the islands of Sicily and Sardinia both have thriving wine industries.

made to raise the status of the industry. A new awareness of the potential of each *terroir* and variety could have on the market led to a real Italian wine renaissance. Limits on yields were imposed and vineyard management techniques were enhanced to produce a riper and more age-worthy style of wines. Technology in the cellars was implemented to master cleanliness, precise extraction, and purity, taking control of wine quality.

THE QUALITY PYRAMID

In 1963 Italy introduced *Denominazione di Origine controllata* (DOC), appellations to protect specific areas of production and to regulate their style and production methods as done in France with their AOC designations. In 1980 the first DOC wines were promoted to DOCG, a more restrictive type of appellation that imposes a tasting commission to approve and entitle bottle for a quality seal. In the last several years, even the DOC wines must be approved by a panel of technical tasters to ensure that the style,

typicity, and overall quality of the samples are in line with the appellation regulations. A new and broader type of appellation has also been introduced as per EU implementation: *Indicazione geografica Tipica/Indicazione Geografica Protetta*. These types of appellations are in between the DOC and the table wine (now called *vino d'Italia* or just *vino*) and regulate lesser capped yields, higher stylistic latitude, and broader appellation borders. At present, there are 75 DOCGs, 331 DOCs, and 118 IGT/IGPs, making Italian wine education rather difficult follow. Many of these appellations cover a very limited surface, however, and an equally minute, if any, production is claimed. Further, a good portion of these appellations are unknown to export markets, because their wines are mostly consumed locally.

THE MEDIEVAL HILL TOWN OF SAN GIMIGNANO LOOMS OVER THE VINEYARDS SCATTERED OVER THE TUSCAN COUNTRYSIDE
Known as the Town of Fine Towers, San Gimignano, gave its name to Italy's first DOC, which was designated in 1966 before being elevated to DOCG status in 1993. Vernaccia is the white grape variety from which its wines are made.

RECENT ITALIAN VINTAGES

2019 An old-fashioned style of vintage, with grape harvest often delayed to the beginning of October. This produced wines of classic and subdued finesse in northern and central Italy. In the south and in the islands, yields were significantly reduced, resulting in denser and weightier reds.

2018 A more classic and austere harvest year that produced refreshing and upright wines yet beefed up by a notable degree of concentration. Late rainfall during harvest was responsible for more moderate alcohol levels throughout the country.

2017 Overall a very hot vintage, with early bud-burst compromised by some spring frost, especially in the northern and central regions of the country. Yields were incredibly low, also due to the extremely dry season that paired with high temperatures and low diurnal variation that resulted in very dense and concentrated wines that are very ready to drink in their early stages.

2016 Very similar to 2015, it featured overall cooler temperatures, resulting in a more vibrant and upfront style of wine that is more redolent of a classic Italian taste. Surely less approachable than 2015 in its earlier stages, one hopes it will show its age-worthiness.

2015 Regarded as one of the best and most consistent harvest years throughout Italy. It featured warmer, if not hot, weather during the growing season, with always providential and never extreme rainfall events that resulted in a very easy vineyard management. Quality, ripeness, and depth, along with significant yield, are the characteristics of this vintage.

2014 Generally a very difficult year due to high rainfall and cooler temperatures throughout the country. It led many appellations to strongly reduce production to maintain quality levels. Wines from the north and central Italy have austere character with higher acidity and slender body. In Sicily, however, this is a vintage that was very late ripening with more mild rain, yielding elegant and refined reds and aromatically lifted whites.

2013 A more classic and fresher style of vintage with overall more floral character due to the longer and temperate growing season. The southern regions can be defined as cooler and never dry, yielding very balanced and delicate reds and aromatic, crisp whites.

2012 Quite dry and hot with higher alcohol levels and, for the reds, firm tannins. Surely age-worthy, yet bold and powerful. Some problems of drought reduced the yields in central and southern regions.

2011 Rather warm style of vintage, with riper examples especially from the southern regions. More temperate climate until august in northern and central Italy produced very good Valpolicellas and exceptional Barolos. Tuscan wines enjoyed an earlier maturity and drinkability. Overall a good year.

2010 This was a difficult and inconsistent year in the northwest, where only the most advantageously located vineyards operating on best practices stood out. The northeast was probably the most consistent of all, producing fresh, crisp whites with vivid fruit of excellent quality, while central Italy was possibly the most inconsistent, with the quality ranging from dire to truly great. Some excellent wines were also produced in the south.

ITALIAN LABEL LANGUAGE

Abboccato Slightly sweet

Amabile Sweeter than *abboccato*

Amaro Bitter or very dry

Annata "Year"; often precedes or follows vintage date. At least 85 per cent of the wine must be from the vintage indicated. *See also Vendemmia.*

Appassimento Includes the partial use of semi-dried grapes

Asciutto Bone dry

Auslese German term used in the Alto Adige for wines from selected grapes

Azienda, Azienda agricola, Azienda agraria, or **Azienda vitivinicola** Estate winery

Bianco White

Cantina Winery

Cantina sociale or **Cooperativa** A cooperative winery

Cascina Northern Italian term for a farm or estate

Casa vinicola A commercial winery

Cerasuolo Cherry red, used for vividly coloured *rosato* wines

Chiaretto Wines falling between very light red and genuine *rosato*

Classico The best, oldest, or most famous part of a DOC zone

Consorzio A group of producers that controls and promotes wine, usually insisting on higher standards than DOC regulations enforce

Denominazione di Origine Controllata (DOC) There are 331 DOCs, but some are multiple-varietal appellations covering as many as 12 different wines, resulting in more than 700 DOC names. This is similar to the French AOC system.

Denominazione di Origine Controllata e Garantita (DOCG) The highest-quality official denomination, based on reduced yields, restrictive production guidelines, and a tasting panel to guarantee, not only the quality level, but also and more important, the typicity and compliance to the regulations. Created in 1980, at present there are 75 DOCG encompassing wider areas as well as isolated pockets of truly fine wine.

Dolce Sweet (minimum 50 grams of residual sugar per litre)

Fattoria Wine estate

Fermentazione naturale Method of producing sparkling wine by natural refermentation in a tank or bottle

Fiore Term meaning "flower". Often part of a name, it indicates quality, implying that the first grape pressing has been used: the "free-run".

Frizzante Semi-sparkling, the equivalent of *pétillant*

Frizzantino Very lightly sparkling

Imbottigliato all'origine Estate-bottled

Incrocio Literally meaning a cross (as in cross-bred grape); the name following *incrocio* is the name of that cross.

Indicazione Geografica Tipica (IGT) The Italian equivalent of a French vin de pays. Since 2009, both in France and Italy, this category has been unified by the new EU law as IGP (*Indicazione Geografica Protetta,* or "Protected designation of origin". These appellation sits between the basic *"vino"* (vin de table in France) and the more restrictive DOC.

Liquoroso Usually fortified and sweet, but may also be dry wine that is simply high in alcohol

Località, Ronco, or **Vigneto** Indicates a single-vineyard wine

Messo in bottiglia nell'origine Estate-bottled

Metodo classico or **Metodo tradizionale** The Italian for "traditional method"

Occhio di pernice Literally "partridge's eye", a term traditionally applied to a sweet red *vin santo*

Pas dosé Used for sparkling wine that has not received any dosage: equivalent to *brut nature* in Champagne

Passito Strong, often sweet wine made from semi-dried grapes

Passito annoso Aged *passito*

Pastoso Medium-sweet

Ramato Copper-coloured wine made from Pinot Grigio grapes that are briefly macerated on their skins

Recioto Strong, sweet wine made from semi-dried grapes

Ripassa, Ripassato, or **Ripasso** Wine refermented on the lees of a Recioto wine

Riserva DOC or **Riserva DOCG** Wines that have been matured for at least two years for reds and one year for white wines

Ronco Hillside vineyard or estate

Rosato Rosé

Rosso Red

Secco Dry

Semi-secco Medium-sweet

Spumante Fully sparkling

Stravecchio Very old wines aged according to DOC or DOCG rules

Superiore When part of the appellation, usually refers to a higher minimum alcohol content

Talento A registered trademark signifying a sparkling wine made by the traditional method

Tenuta Estate

Uva Grape variety

Uvaggio Wine blended from various grape varieties

Vecchio Old

Vendemmia This means "harvest" and often precedes or follows the vintage date. At least 85 per cent of the wine must be from the vintage indicated. *See also Annata.*

Vendemmia tardiva Late-harvest sweet wine

Vigna or **Vigneto** Vineyard

Vin santo or **vino santo** Traditionally sweet, occasionally dry, white wine made from *passito* grapes stored in sealed casks and not topped up for several years

Vino Wine

Vino novello Italian equivalent of v*in de primeur* or *vin nouveau*

Vino da pasto Ordinary wine

Vino/Vino d'Italia Once known as "*Vino da Tavola*", literally "table wine", this is Italy's most basic category. It can encompass both very ordinary up to fine or even iconic wines, strongly depending on producer's name.

ITALY TODAY

Italy it is not only the country of Prosecco – the biggest commercial wine success of the last decade – it is capable of keeping a positive trend in the volume of export for bottled wine production. Recent analyses are showing a promising trajectory in which Italy is still growing in the export market, with an increase in both volume and value supported by high-quality productions from three macro-regions and their major appellations. Brunello di Montalcino, Chianti Classico, and Bolgheri from Toscana; Barolo and Barbaresco in Piedmont; and Ripasso and Amarone della Valpolicella from Veneto are consolidating their image of fine wines amongst the highly competitive global wine scene. Together with these established appellations, new Italian wine trends are echoing in the international context and deserve attention. Valpolicella Ripasso is dominating the Scandinavian markets despite its recent history, and luscious and smooth reds from Primitivo in Puglia are following suit, while a consistent increase in interest is registered towards traditional-method sparkling wines. Franciacorta DOCG and Trento DOC guide this trend on value, yet more emerging areas such as Alta Langa in Piedmont and Friuli for Ribolla have demonstrated a wide-spread Italian culture on sparkling wines, figured not only on the method (Charmat versus traditional) but also on varietals (Ribolla, Prosecco, Pignoletto). In addition, Etna DOC has firmly resurged in interest for its high quality; it is capable of translating a volcanic *terroir* into wines with a highly compelling character.

Italy was once often depicted as the land of autochthonous grape varieties, producing muted and phenolic yet fresh white wines and long-aged tannic reds. This is a legacy of the past: the progress shown in the viticulture of the last decades and the substantial investments in technology and innovation in the winery have dramatically changed the course and shape of Italian wine industry. Today, thanks to these elements, coupled with the unequivocal climate change, Italy could claim an incredible range of white wine excellences ranging from aromatic varietals (Moscato from Piedmont and Gewürztraminer and Sauvignon Blanc from Friuli), semi-aromatic (Grillo in Sicily, Pinot Grigio from Veneto-Friuli-Trentino, Verdicchio from Marche, and Vermentino from Liguria and Toscana), and neutral grapes (Catarratto in Sicily, Fiano in Campania-Puglia, Arneis and Cortese from Piedmont, and Trebbiano from Abruzzo). Autochthonous whites are performing with more distinctive profiles and less phenolic character and with a remarkable natural freshness that adds value when compared to international varieties, often less adapted to most microclimates and soils of Italian wine regions.

Climate change of the last two decades has boosted the phenolic ripeness and increased quality for classic reds too, helped by recent innovations in viticulture and winemaking. Also here, the natural outcome has moved towards a higher recognizability but also to a distinct Italian red wine style made of a highly tannic frame and fresh acidic backbone with dominant floral and red berries aromas. Italy has developed wine styles with different declinations attributed to each variety and origin, but with a common traditional winemaking approach. The Italian method for oak ageing, as opposed to the French and Spanish ones, interact with the use of big casks and long ageing periods to respect the vibrancy and typicality of the wine and make it ultimately distinguishable from its European cousins.

Northwest Italy

This area includes the great wine region of Piedmont, as well as the regions of Liguria, Lombardy, and Valle d'Aosta.

Few areas encompass such contrasting topographies as Northwest Italy, from the alpine pistes of the Valle d'Aosta and the Apennines of Liguria to the alluvial plains of the River Po. Contrast is also evident in the character of its most famous wines, the monumental, firmly tannic Barolo and the light, water-white, effervescent, Asti.

PIEDMONT (PIEMONTE)

Piedmont is dominated by two black grapes (Nebbiolo and Barbera) and one white (Moscato). Nebbiolo makes the magnificently rich and long-lived Barolo and the elegant, more feminine, yet sometimes just as powerful, Barbaresco. Nebbiolo is also present in smaller northern Piedmont appellations as Gattinara, Ghemme, Lessona, and Boca. Barbera can potentially be as fine as Nebbiolo, while featuring almost opposite tasting qualities. It is softer in tannin, at least as high in acidity, yet with darker, sometimes brooding, fruit character and excels around Alba and Asti. Dolcetto is another autochthonous black variety, often vinified to be enjoyed early, equipped with crunchy tannins and lively character. The most popular white variety is the Moscato (Muscat Blanc à Petits Grains), produced as still, *frizzante,* or *spumante.* Amongst dry still whites, an important role is played by the delicate Cortese variety, which yields Gavi di Gavi DOCG. Rapidly growing in interest is Arneis, a rather neutral and vibrantly refreshing variety, and Timorasso, a semi-aromatic and age-worthy cultivar that boasts very sharp acidity, powerful character, and petrol-like aromas with age. Piedmont has also been experimenting with long-lees-aged bottle-fermented Chardonnay and Pinot Noir from the cooler part of the Langhe area, known since 2011 as Alta Langa DOCG.

LOMBARDY (LOMBARDIA)

Northeast of Piedmont, Lombardy stretches from the flat plains of the Po Valley to snow-clad Alpine peaks. The region's finest wines areas include the reds from Valtellina terraces, Franciacorta's bottle-fermented sparkling wines, and the more easy-drinking and varied Oltrepò Pavese, ranging from full reds from indigenous varieties (Barbera, Uva Rara, Vespolina, and Croatina) and tank-method Pinot Noir. As for the whites, the Lugana appellation – shared with Veneto – produces delicate and perfumed wines with a distinctive almondy finish that are rapidly gaining traction in key export markets.

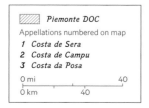

Piemonte DOC

Appellations numbered on map
1 Costa de Sera
2 Costa de Campu
3 Costa da Posa

0 mi   40
0 km   40

WINE REGIONS OF ASTI AND ALBA
(see also main Northwest Italy map, below*)*
The complexity of appellation upon appellation in the Asti-Alba area is such that they have to be divided between the main map and this additional map.

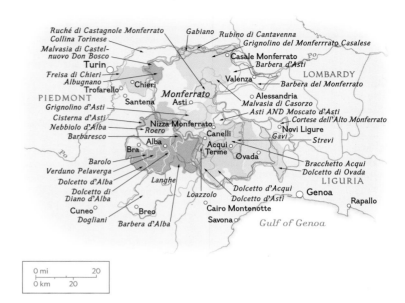

0 mi   20
0 km   20

WINE REGIONS OF NORTHWEST ITALY
(see also p345*)*
The presence of the Alps gives this largely hilly region a hot growing season and a long autumn. The finest wines come from the foothills of Piedmont, which provide ideal growing conditions for the late-ripening Nebbiolo grape.

LIGURIA

This is a small, crescent-shaped region, nestled between the Maritime Alps and the coast, which enables a definite temperature shift. This peculiar situation makes Liguria an excellent place for the production of briny, mineral, and lively aromatic whites from Vermentino, locally known as Pigato. Cinque Terre, which is the best-known Ligurian wine, is named after the Cinque Terre, or "five villages", which are perched along the Ligurian coast, above which the steep, intricately terraced vineyards make grape growing heroic here. The western part of Liguria, bordering France, is the home for the local variety, Rossese, that produces a light and juicy red wine of lively energy.

VALLE D'AOSTA

Italy's smallest and most mountainous wine region, Valle d'Aosta has picturesque vineyards that produce some enjoyable wines with a focus on red varieties. High in the Alps, overlooked by Mont Blanc and the Matterhorn, the Valle d'Aosta looks at first as if it could be a part of France or Switzerland, but the only easy, natural access is from Piedmont. The region has only one DOC, which has five subzones and many different styles from a significant number of autochthonous grape varieties. Some of the wines from this region are easy-drinking and unpretentious, but the better examples are extremely distinctive and full of character.

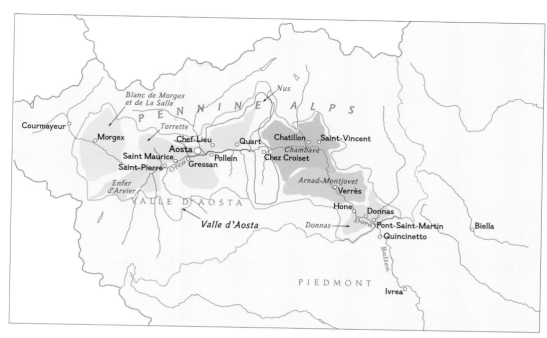

| Valle d'Aosta | Region |
| Donnas | Sub-region |

THE VALLE D'AOSTA WINE REGION
(see also main Northwest Italy map, opposite) The winters here are cold and snowy, but summers in the valley can be very hot with contrastingly cold nights, which should make for some exciting wines.

BOTTLES OF MONDORO ASTI CHILL ON ICE
Mondoro is a highly esteemed producer of Asti, one of Italy's iconic wines. Asti Spumante exploded in international markets when soldiers returning home from Italy after World War II brought the sweet sparkler with them. The rise in popularity went along with a lowering of quality, but things improved after Asti was promoted to DOCG status in 1993. Producers began making dryer versions and dropped the "Spumante" from the name.

ALBA DOC
Piedmont

Red-wine-only tiny appellation producing Nebbiolo-Barbera blends and covering all of Roero, Barolo, Barbaresco, and Dogliani communes.

ACQUI DOCG

See Brachetto d'Acqui DOCG

ALBUGNANO DOC
Piedmont

Nebbiolo-based red and *rosato* wines produced at Albugnano and four surrounding villages in a small, hilly area to the north of Asti, where the vineyards climb to 520 metres (1,700 feet) in altitude.

ALPI RETICHE IGP
Lombardy

This IGP covers numerous communes in the province of Sondrio and produces white, *rosato*, and red wines, either blended or varietal, from any grape varieties approved for the region of Lombardy. Traditional-method sparkling wines (white or *rosato*) are also made from any proportions of Chardonnay, Nebbiolo, Pignola, Pinot Bianco, and/or Rossola or as varietals. A white or red *vendemmia tardiva,*and *passito* are also produced.

✓ *Nino Negri*

ALTA LANGA DOCG
Piedmont

This DOCG is dedicated to the production of traditional-method sparkling wines (either white or *rosato*) from minimum 90 per cent Chardonnay and/or Pinot Nero in the southern half of the Langa district. A *riserva* version is also produced (minimum 36 months lees ageing). The wines are of very good to outstanding quality.

✓ *Enrico Serafino • Ettore Germano*

ALTO MINCIO IGP
Lombardy

Geographically, this IGP is roughly equivalent to the Colli Mantovani DOC. Anything goes, but most often utilized for Pinot Grigio at the moment.

✓ *Prendina*

ARNAD-MONTJOVET
Valle d'Aosta

A red-only subzone of the regional Valle d'Aosta DOC producing Nebbiolo-based red wines, with up to 30 per cent Dolcetto, Freisa, Neyret, Pinot Noir, and Vien de Nus. *See* Valle d'Aosta DOC.

ASTI DOCG
Piedmont

One of the most famous sparkling wines in the world, Asti is made by *cuve close,* which is far superior to the traditional method when producing an aromatic, sweet sparkling wine. The grapes used are grown in 52 communes throughout the provinces of Asti, Cuneo, and Alessandria. The best Asti has a fine *mousse* of tiny bubbles, a fresh and grapey aroma, a luscious sweetness, and a light, delicately rich flowery-fruitiness that hints at peaches, and it is best when consumed as young as possible. Since 2017 a new version of extra-dry Moscato Spumante "Asti *secco*" is now part of the production regulation. The Asti DOCG also includes the production of Moscato d'Asti, a wine that is similar in flavour to Asti but with a minimum pressure of three atmospheres as opposed to five. Still, slightly *frizzantino,* and positively *frizzante* examples of Moscato d'Asti exist. Those with some

degree of effervescence are bottled without any dosage, while the first fermentation is still underway, as opposed to Asti plain and simple, which is made by the *cuve close* method. Compared with Asti, Moscato d'Asti is generally less alcoholic (5.5 to 8 per cent instead of 7.5 to 9 per cent) and almost always much sweeter. This luscious, succulently sweet wine exudes mesmerising grapey aromas. A *vendemmia tardiva,* (late harvest) Moscato Bianco is also allowed. The Asti DOCG includes three subzones: Cannelli (Moscato d'Asti only), Santa Vittoria d'Alba (Moscato d'Asti and *vendemmia tardiva,* only), and Strevi (Moscato d'Asti only).

🍷 *Upon purchase*

✓ *Moscato d'Asti: Araldica • Banfi Piemonte • Alfiero Boffa • Borgo Maragliano • Redento Dogliotti • Tenuta il Falchetto • Gatti • Marchesi di Gresy • Villa Lanata • Luciana Rivella • La Spinetta-Rivetti • Scagliola • Bosio; Asti: Barbero (Conte di Cavour) • Walter Barbero • Batasiolo • Bersano • Capetta • Villa Carlotta • Cerutti (Cesare) • Conte di Cavour • Giuseppe Contratto • Cuvage • Romano Dogliotti • Fontanafredda • Marenco ❶ • De Miranda • Mondoro • Perlino • Sperone • Tosti • Cantina Sociale Vallebelbo • Bosio*

BARBARESCO DOCG
Piedmont

Generally more feminine and elegant than Barolo, Barbaresco has a greater suppleness, softer fruit, and a more obvious charm, although some producers overlap the weightier Barolo style. Produced from Italy's greatest indigenous grape variety, Nebbiolo, these wines must be aged for a minimum of 26 months, 9 of which must be in oak or chestnut casks. Since 2015, 66 "additional geographical definitions" or MeGAs, may be displayed on the label.

🍷 *5–20 years*

✓ *Adriano Marco & Vittorio • Brema • Ca' del Baio • Burlotto • Piero Busso • Cascina Luisin • Ceretto • Pio Cesare • Fontanabianca • Fratelli Cigliuti • Giuseppe Cortese • Gaja • Bruno Giacosa • Piero Busso • Cantina del Glicine • Ugo Lequio • Marchesi di Gresy • Moccagatta • Fiorenzo Nada • Castello di Neive • Paitin • Piazzo Armando • Pelissero • Produttori di Barbaresco • Alfredo Prunotto • Albino Rocca • Bruno Rocca • Scarpa • Sottimano • La Spinetta • Roagna*

BARBERA D'ALBA DOC
Piedmont

The most prolific Piedmont vine, Barbera has suffered unfairly from a somewhat lowly image. In fact, it is one of Italy's great grapes, and the best Barbera from Alba are magnificently rich and full of flavour. The production of Barbera d'Alba is very small, in fact often minute, in comparison with that of Barbera d'Asti.

🍷 *5–12 years*

✓ *Marziano Abbona • Pio Cesare • Fratelli Cigliuti • Clerico • Aldo Conterno • Giacomo Conterno • Conterno Fantino • Damilano • Damonte • Franco Fiorina • Gepin • Elio Grasso • Manzone • Giuseppe Mascarello • Prunotto • Renato Ratti • Bruno Rocca • Vajra • Vietti • Roberto Voerzio • Olivero Mario*

BARBERA D'ASTI DOCG
Piedmont

This DOCG is similar in character to Barbera d'Alba, but fuller bodied, softer, and suppler, with simpler generic wines and equally profound single-vineyard wines. The appellation includes two subzones: Colli Astiani (or Astiano) and Tinella, where only the *superiore* version is produced (13 per cent ABV required).

🍷 *3–8 years*

✓ *Marchesi Alfieri • Paolo Avezza • Cascina la Barbatella • Bava (Stradivario) • Alfiero Boffa • Braida (Bricco della Bigotta) • Brema • Cascina Castelet (Passum) • Chiarlo (Valle del Sole) • Coppo • Cossetti (Cascina Salomone) • Franco Martinetti • Neirano (Le Croci) • Olim Bauda • Scarpa • Scrimaglio • Luigi Spertino (La Mandorla) • Tenuta Santa Caterina*

BARBERA DEL MONFERRATO DOC
Piedmont

Most of the wines produced by Barbera Del Monferrato are lesser versions of Barbera d'Asti. Semi-sweet or *frizzante* styles may also be made.

✓ *Accornero & Figli • CS di Castagnole Monferrato (Barbera Vivace) • Montaldo • Pico Gonzaga*

BARBERA DEL MONFERRATO SUPERIORE DOCG
Piedmont

Compared to Barbera del Monferrato the superiore wines must reach a minimum 13 per cent ABV and must be aged for a minimum of 14 months before release. Those wines display more depth and concentration of fruit with hints of spice.

BAROLO DOCG
Piedmont

Barolo is unquestionably Italy's greatest wine appellation, and its finest wines are the ultimate expression of the Nebbiolo grape. All the best vineyards of Barolo are located on a small, raised area of mostly gentle, but occasionally steep, slopes surrounded by the hills of the Langhe. The soil is essentially calcareous marl, the northwestern half high in magnesium and manganese, while the southeastern half more iron-rich. This small difference is deemed to be the reason why Barolo wines from the northwest have more elegance, while those from the southeast are fuller in body. The

NORTHWEST ITALY WINE REGIONS AT A GLANCE

REGION	TOTAL AREA UNDER VINE Hectares (Acres)	IGP/DOC/DOCG AREA UNDER VINE Hectares (Acres)	IGP/DOC/DOCG PRODUCTION Hectolitres	TOTAL PRODUCTION Hectolitres
Piedmont	45,979 (113,600)	40,451 (89,000)	2 million (50-50 red/white)	2.5 million (27.7 million cases)
Lombardy	23,919 59,100)	23,156 (57,200)	983,000 (51-49 red/white)	1.4 million (15.5 million cases)
Liguria	1,591 (3,900)	800 (1,980)	36,190 (25-75 red/white)	79,120 (0.9 million cases)
Valle d'Aosta	450 (1,100)	319 (790)	13,520 (67-33 red/white)	14,470 (160,500 cases)

Source: Osservatorio del Vino UIV - ISMEA

FACTORS AFFECTING TASTE AND QUALITY

LOCATION
Flanked to the north and west by the Alps and by the Ligurian Sea to the south, Northwest Italy contains the regions of Piedmont, Lombardy, Liguria, and Valle d'Aosta.

CLIMATE
Different climatic areas are identifiable, from the warm Mediterranean Ligurian area to the cool continental climate in Piedmont and Valtellina to the dry and temperate Oltrepò and the alpine climate of Val d'Aosta. Long autumns enable the late-ripening Nebbiolo grape to be grown very successfully.

ASPECT
This area covers mountains, foothills (the wotd *piedmont* means "foothill"), and the valley of Italy's longest river, the Po. Grapes are grown on hillsides that provide good drainage and exposure to the sun. In classic areas such as Barolo, every south-facing hillside is covered with vines, while in Lombardy many vineyards extend down to the rich, alluvial plains of the Po Valley. Both in Valtellina and in most parts of

Val d'Aosta viticulture is heroic because vines are planted in steep and narrow terraces.

SOIL
There is a wide range of soils with many local variations; the predominant type is calcareous marl, which may be interlayered or intermingled with sand and clay. Piemonte hosts calcareous marl in the Langhe area, while in the northern part of the region and in Lombardy's Valtellina acidic sand is found. Oltrepò Pavese sits on sandy soils with a high clay proportion, while Franciacorta soils have morainic origin and are rich in sand and poor in clay.

VITICULTURE & VINIFICATION
Both viticultural and vinification aspects are extremely heterogeneous in this area. Piemonte displays the effect of the French influence with contour plantings and Guyot pruning. Valtellina has very densely planted terraces and double-arch pruning. Sylvoz training system is the style in Oltrepò Pavese. Long post-fermentation maceration and cask maturation are typical of Nebbiolo and *appassimento* for the bunches in

Valtellina for Sforzato production. Franciacorta has a distinctive bottle-fermented category, "Satèn", with lower pressure. Tank method, locally known as Martinotti, is used for Moscato d'Asti, Asti, and Brachetto.

GRAPE VARIETIES
Primary varieties: Barbera, Chiavennasca (Nebbiolo), Moscato (Muscat Blanc à Petits Grains)
Secondary varieties: Arneis, Blanc de Morgex, Bonarda, Croatina, Brachetto, Brugnola, Buzetto (Trebbiano), Cabernet Franc, Cabernet Sauvignon, Chardonnay, Cortese, Erbaluce, Favorita, Freisa, Fumin, Gamay, Grenache, Grignolino, Gropello, Lambrusco, Malvasia, Marzemino, Mayolet, Merlot, Neyret, Oriou (Petit Rouge), Ormeasco in Liguria (Dolcetto), Petite Arvine, Pigato, Pignola Valtellina, Pinot Bianco (Pinot Blanc), Pinot Grigio (Pinot Gris), Pinot Nero (Pinot Noir), Premetta, Riesling Italico (Welschriesling), Riesling Renano (Riesling), Rossese, Rossola, Ruché, Schiava Gentile, Syrah, Timorasso, Uva Rara, Vermentino, Vespolina, Vien de Nus

---

biggest factor in the quality of Barolo has, however, been due more to the human than to soil, especially with the trend over the past 10 years to isolate the very best Barolo and market them under single-vineyard names. A vineyard name will not necessarily guarantee excellence, but hardly any of the very top Barolo are blended these days, and quality is steadily rising since MeGaS ("additional geographic definitions") have significantly lower yields compared to standard blended Barolo. Back in the 1980s, the appellation was invested by a movement called "Barolo Boys", setting forth the birth of modern, *barrique*-aged Barolo as opposed to the traditional way of using old big *"botti"*. Today both the modern and traditional schools of producers make stunning Barolo. Modern styles are riper and more creamy than traditional ones, supported by the vanilla of new oak, whereas the traditional styles are arguably more complex, with tobacco, tar, and smoky aromas replacing the clean-cut vanilla. The best of both, however, share lashings of fruit and a distinctive violet flower note. All Barolo should be powerfully built, even the more elegant, earlier-drinking styles, and the best have surprising finesse for such weighty wines.

Barolo now has official subzoning (MeGAs) that can identify about 181 specific *crus* throughout the appellation, including 11 commune names. The most important and recognised *crus* are Bussia (one of the biggest), Cannubi, Brunate, Lazzarito, and Asili.

Barolo Chinato is a sweet red aromatized wine made from a base of Barolo DOCG, flavoured with herbs and flavouring ingredients—notably *chinino* (quinine). Barolo Chinato is not itself a DOCG product.

🍷 *8–25 years*

✓ Marziano Abbona • Accomasso • Elio Altare • Abbazia dell'Annunziata • Ascheri • Azelia • Fratelli Barale • Luigi Baudana • Enzo Boglietti • Giacomo Borgogno • Gianfranco Bovio • Brezza • Bricco Asili • Bricco Rocche • Brovia • Burlotto • Cappellano • Castello di Verduno • Cavalotto • Ceretto • Michele Chiarlo • Domenico Clerico • Elvio Cogno • Aldo Conterno • Giacomo Conterno • Paolo Conterno • Conterno-Fantino • Corino • Franco-Fiorina • Fratelli Alessandria • Gagliasso • Bruno Giacosa • Elio Grasso • Icardi • Manzone • Marcarini • Marchesi di Barolo • Mario Marengo • Bartolo Mascarello • Giuseppe Mascarello • Massolino • Oddero • Pelassa • Pianpolvere Soprano • Piazzo Armando • Pio Cesare • Luigi Pira • Alfredo Prunotto • Renato Ratti • Michele Reverdito • Rinaldi Giuseppe • Roagna • Giovanni Rosso • Luciano Sandrone • Scarpa • Paolo Scavino • Schiavenza • Silvio Grasso • Edoardo Sobrino • Alessandro Veglio • Vietti • Roberto Voerzio

BENACO BRESCIANO IGP
Lombardy

Red and white wines produced in the foothills west of Lake Garda in the province of Brescia, where the best so far appear to be red wine blends of three or four varieties, including Cabernet Franc, Cabernet Sauvignon, Marzemino, Massimo, Merlot, and Sangiovese.

✓ *Terra Lunari • Cà dei Frati (Ronchedone)*

BERGAMASCA IGP
Lombardy

These are wines from the hillside vineyards just outside of Bergamo, where all styles are permitted, and the most successful grape varieties include Manzoni Bianco, Moscato, and Pinot Bianco for whites and Cabernet Sauvignon, Merlot, and Schiava for red and *rosato* wines. Mostly produced by the local cooperative.

BOCA DOC
Piedmont

These are medium-bodied to full-bodied, spicy red Nebbiolo wines from the Novara hills in the north of Piedmont. They are hard to find, but wine has been made here since Roman times and when found can be a good value.

🍷 *3–6 years*

✓ *Antonio Vallana • Le Piane • Podere ai Valloni*

BOTTICINO DOC
Lombardy

Full-bodied, Barbera/Marzemino-based red wines with a good level of alcohol and a light tannic structure.

🍷 *3–5 years*

✓ *Miro Bonetti • Benedetto Tognazzi*

BONARDA DELL'OLTREPÒ PAVESE DOC
Lombardy

Formerly part of the Oltrepò Pavese DOC. Established as a separate DOC in 2010, it produces a varietal Croatina (locally Bonarda) in a either still, *frizzante*, or sweet style. Particularly fruity wines for relatively early consumption.

BRACHETTO D'ACQUI/ACQUI DOCG
Piedmont

A sort of red Asti, Brachetto d'Acqui is as sweet as Australian sparkling Shiraz but without the weight, tannin, or oak. Totally fruit-driven fizz, the best of which have lovely, fresh, aromatic grapey fruit with a lush, soft *mousse* and impressive raspberry acidity underpinning the sweetness. Must be 97 per cent Brachetto grape and may be sold simply as Acqui.

✓ *Batasiolo • Contero • Duchessa Lia • Banfi (Rosa Regale)*

BRAMATERRA DOC
Piedmont

Good value, full-bodied red wines produced primarily from Nebbiolo, but which may include Bonarda, Croatina, and Vespolina grapes. A *riserva* can also be produced (minimum 34 months ageing requirement)

🍷 *3–6 years*

✓ *Luigi Perazzi • Fabrizio Sella*

BUTTAFUOCO DELL'OLTREPÒ PAVESE/BUTTAFUOCO DOC
Lombardy

This DOC produces a deep-coloured, fruity, dry or semi-sweet *frizzante* red made from 25 to 65 per cent Barbera, 25 to 65 per cent Croatina, and a maximum of 45 per cent Uva Rara and/or Vespolina.

CALOSSO DOC
Piedmont

This is a red wine–only tiny appellation producing Gamba Rossa wines. They generally are medium-bodied wines with some grip, mainly fruit driven with some spicy and balsamic notes.

CANAVESE DOC
Piedmont

The best-known wine produced from these vineyards in the hills surrounding Lake Viverone is Erbaluce, but reds and *rosato* (still or sparkling) wines are also made with Barbera, Bonarda, Freisa, Nebbiolo, and Uva Rara. The reds are best. A white Erbaluce sparkling can also be made.

✓ *CV del Canavese (Rosso) • Ferrando (Rosso)*

CAPRIANO DEL COLLE DOC
Lombardy

Rarely encountered red wines made from Marzemino and Merlot with the addition of up to 10 per cent Sangiovese and a Trebbiano-based white in either *frizzante* or still style. A varietal Trebbiano (still or *frizzante*) and a varietal Marzemino are also allowed.

CAREMA DOC
Piedmont

Soft, medium-bodied Nebbiolo wines that are grown on the mountainous slopes close to the border with the Valle d'Aosta. They are good and reliable wines.

🍷 *2–5 years*

✓ *Luigi Ferrando*

CASTEGGIO DOC
Lombardy

This small appellation produces a Barbera-based blend in the province of Pavia.

CELLATICA DOC
Lombardy

This aromatic and flavoursome red wine has been made in the hills overlooking Brescia for at least 400 years. Permitted grapes are Barbera, Marzemino, Schiava Gentile, and Incrocio Terzi.

🍷 *2–6 years*

CHAMBAVE
Valle d'Aosta

A subzone of the regional Valle d'Aosta DOC, Chambave produces attractively scented, crisp red wines primarily from Petit Rouge grapes, plus up to 40 per cent Dolcetto, Gamay, and Pinot Noir. Two white wines are also permitted: one sweet, long-lived, and *passito* in style, the other a highly perfumed, early-drinking, dry- to off-dry white, both from the Moscato grape. *See* Valle d'Aosta DOC.

🍷 *2–3 years (red and* passito*); Upon purchase (white)*

✓ *La Crotta di Vegneron • Ezio Voyat*

CINQUE TERRE/ CINQUE TERRE SCIACCHETRÀ DOC
Liguria

This DOC of scenic coastal vineyards produces good, though not exactly spectacular, delicately fruity, dry white wines, primarily from Bosco, Albarola, and Vermentino grapes (plus a maximum of 20 per cent optional other local varieties) and the more exciting Cinque Terre Sciacchetrà, a medium-sweet *passito* wine. The permitted sub-appellations of Costa da Posa, Costa de Campu, and Costa de Sera are all named sites that overlook the pretty coastal town of Riomaggiore.

🍷 *1–3 years*

✓ *Walter de Batté • Buranco • Forlini Cappellini*

CISTERNA D'ASTI DOC
Piedmont

This DOC is named after the Cisterna fortress, built where Saint Paul supposedly stopped while attempting to convert the local population. Its red wines are made from at least 80 per cent Croatina.

COLLI DI LUNI DOC
Liguria

Sangiovese-based reds and Vermentino Albarola whites from the eastern extremity of Liguria, bordering and even overlapping part of Tuscany.

🍷 *2–5 years (red) and 1–2 years (white)*

✓ *Bosoni • La Colombiera • Lambruschi • Terenzuola*

COLLI TORTONESI DOC
Piedmont

Most are robust and rather rustic, full-bodied Barbera reds and crisp, dry, rather lightweight, still, *frizzante,* or sparkling Cortese whites. In recent years, however, a few committed producers have started to craft some beautiful reds that walk the tightrope between fine minerality and sensational softness. Three other varietal whites can be made (the aromatic Favorita, as well as Moscato and Timorasso) and three reds (Croatina, Dolcetto, and Freisa). A *rosato* wine called *chiaretto* is produced from any proportions of Aleatico, Barbera, Bonarda, Cabernet Franc, Cabernet Sauvignon, Croatina, Dolcetto, Freisa, Grignolino, Lambrusca di Alessandria, Merlot, Nebbiolo, Pinot Nero, and/or Sangiovese. The DOC includes two subzones: Monleale (Barbera only) and Terre di Libarna (*bianco*, Timorasso, *rosso,* and *spumante* only).

✓ *Luigi Boveri • La Colombera • Claudio Mariotto • Vigneti Massa*

COLLINA DEL MILANESE IGP
Lombardy

The geographical equivalent of San Colombano DOC, this IGP allows all styles of wine from many different grape varieties, but much of the production is in the form of a rustic *frizzante* from one or more of the following varieties: Bonarda, Cabernet Sauvignon, Chardonnay, Cortese, Malvasia, Merlot, Trebbiano, and Verdea.

✓ *Poderi San Pietro*

COLLINA TORINESE DOC
Piedmont

From vineyards just outside of Turin, on foothills overlooking the city, these red wines are made primarily from the Barbera or Freisa varieties, although pure Barbera, Bonarda, Pelaverga (locally called Cari and made in a sweet style), and Malvasia di Schierano (sweet) varietal red wines are also produced.

COLLINE DEL GENOVESATO IGP
Liguria

This small IGP covers a portion of the province of Genova included in the Golfo del Tigullio, Riviera di Ponente, and Val Polcèvera DOCs and produces white, *rosato,* and red wines from any proportions of grape varieties approved for the region of Liguria. A varietal Pigato and a varietal red called Granaccia – aka Alicante – (minimum 85 per cent) are also made, as well as a red *passito* from any proportions of red grape varieties approved for the region of Liguria.

COLLINE DI LEVANTO DOC
Liguria

This DOC consists of vineyards clinging precariously to the cliffs and hills in the eastern extremity of the spectacular Cinque Terre vineyards; it thus enjoys the same combination of soil, climate, and grape varieties, in addition to which red wines are also produced (primarily from Sangiovese and Ciliegiolo grapes).

COLLINE NOVARESI DOC
Piedmont

This DOC produces red, *rosato*, and white wines from the Novarese province northwest of Milan; the white is dry and made exclusively from the Erbaluce grape, whereas the red is a blend of at least 50 per cent Nebbiolo. Five varietal reds are produced: Barbera, Croatina, Nebbiolo, Uva Rara, and Vespolina.

COLLINE SALUZZESI DOC
Piedmont

This DOC produces mainly red wines which may be either blended from Pelaverga, Nebbiolo, Chatus, and Barbera or pure varietals. A *rosato* Pelaverga can be made. A still or sparkling red wine varietal is also allowed for Quagliano, an ancient local variety that might have become extinct

TERRACED VINEYARDS LINE THE HILLSIDES BELOW AND ACROSS FROM A CLUSTER OF SHERBET-COLOURED BUILDINGS IN VOLASTRA, ONE OF LIGURIA'S CINQUE TERRE VILLAGES
One of a string of five fishing villages known as the Cinque Terre ("Five Lands"), Volastra tumbles along rocky promontories along the coastline of the Italian Riviera. Until recently connected by just mule tracks and accessible only by rail or water, the villages are now part of the Cinque Terre National Park, a UNESCO World Heritage Site. The Cinque Terre DOC produces fruity, dry white wines.

had the growers in this DOC not planted well-known varieties such as Nebbiolo and Barbera, which enabled their businesses to survive.

COLLINE SAVONESI IGP
Liguria

Freed of any DOC restrictions, the hills overlooking the Ligurian Riviera make a fascinating source of long-forgotten grape varieties, including Lumassina (locally Mataòssu), Pigato, and even the local Granaccia, which is an assertively independent localized clone of the better-known Alicante.

✓ *Bruna (Pulin)* • *Punta Crena*

CORTESE DELL'ALTO MONFERRATO DOC
Piedmont

Dry, still, *frizzante,* and fully sparkling white wines made from the Cortese grape in the hills of Monferrato. The wines are typically very clean and fresh and relatively light-bodied with a crisp acidity.

COSTE DELLA SESIA DOC
Piedmont

Red wines produced from Vespolina, Croatina, and Nebbiolo grapes growing in a hilly area overlapping Lessona, Bramaterra, and Gattinara. A white wine from Erbaluce and a Nebbiolo-based *rosato* are also made.

✓ *Antoniolo*

CURTEFRANCA DOC
Lombardy

Part of the original Franciacorta appellation since 1967 and once known as Terre di Franciacorta, it represents the still wine appellation for wines produced northeast of Milan, on hilly slopes near Lake Iseo. The red wines are typically well coloured, medium to full bodied, made from Cabernet Franc and Carmènere, with the possible addition of Merlot and Cabernet Sauvignon. Many are richly flavoured and show some finesse. The dry whites, primarily from Chardonnay with a little Pinot Blanc allowed, have demonstrated that they will age beautifully.

🍷 *3–8 years (red), 1–3 years (white)*

✓ *Ca'del Bosco* • *Enrico Gatti* • *Ragnoli*

DOGLIANI DOCG
Piedmont

This DOCG produces red wines only from 100 per cent Dolcetto grapes. It comprises 76 *menzioni geografiche aggiuntive* ("additional geographical definitions") and a superiore version is also made. The generic red is usually made in a easy-to-drink style, fruit driven and characterised by low acidity and low soft tannin. The superiore versions are generally more complex and perfumed with hints of liquorice and almonds.

✓ *Abbona* • *Chionetti* • *Luigi Einaudi* • *Cascina Corta* • *Pecchenino*

DOLCEACQUA/ ROSSESE DI DOLCEACQUA DOC
Liguria

This appellation is located at the extreme western edge of the Ligurian Riviera, bordering Provence in France, where red wine is produced from the so-called Rossese di Ventimiglia variety, which is probably the Tibouren grape. The best wines are light, easy-drinking reds that are capable of rich, lush fruit, a soft texture, and a spicy aromatic aftertaste. If qualified by Superiore, this wine must be aged for a minimum of 12 months prior to release. The DOC includes 38 *menzioni geografiche aggiuntive* ("additional geographical definitions") plus seven communes: Camporosso, Dolceacqua, Perinaldo, San Biagio della Cima, Soldano, Vallecrosia, and Ventimiglia.

🍷 *1–4 years*

✓ *Terre Bianche (Bricco Arcagna)* • *Maccario Dringenberg* • *Giobatta Cane* • *Lupi* • *Antonio Perrino*

DOLCETTO D'ACQUI DOC
Piedmont

Dolcetto is a plump grape with a low acid content that is traditionally used to make cheerful, Beaujolais-type wines that are deep purple in colour and best enjoyed young.

🍷 *1–3 years*

✓ *Viticoltori dell'Acquese*

DOLCETTO D'ALBA DOC
Piedmont

The Dolcetto d'Alba DOC produces some soft, smooth, juicy wines that should be drunk while they are young, fresh, and fruity.

🍷 *1–3 years*

✓ *Elio Altare* • *Azelia* • *Batasiolo* • *Brovia* • *Ca' Viola* • *Ceretto* • *Pio Cesare* • *Fratelli Cigliuti* • *Aldo Conterno* • *Cascina Drago* • *Franco Fiorina* • *Elio Grasso* • *Bartolo Mascarello* • *Pira* • *Roberto Voerzio*

DOLCETTO D'ASTI DOC
Piedmont

This red-only DOC, based on the Dolcetto grape, produces wines that are lighter than Dolcetto d'Alba.

DOLCETTO DI DIANO D'ALBA or DIANO D'ALBA DOCG
Piedmont

Diano is a hilltop village just south of Alba, which produces wines that are fuller and grapier than most other Dolcetto wines. This DOCG includes 75 *menzioni geografiche aggiuntive* ("additional geographical definitions").

🍷 *3–5 years*

✓ *Alario* • *Casavecchia* • *Colué* • *Fontanafredda* • *Giuseppe Mascarello* • *Cantina della Porta Rossa* • *Mario Savigliano* • *Veglio & Figlio*

DOLCETTO DI OVADA DOC
Piedmont

These are the fullest and firmest wines of the Dolcettos.

🍷 *3–6 years (up to 10 for the very best)*

✓ *Cascina Scarsi Olivi*

DOLCETTO DI OVADA SUPERIORE/OVADA DOCG
Piedmont

This DOCG produces Dolcetto di Ovada Superiore only. A minimum 12 months ageing is required and a minimum of 20 months for *superiore* with a named vineyard. The wines of this appellation have more concentration and depth than the simple Ovada ones.

DONNAS
See Valle d'Aosta DOC

ENFER D'ARVIER
See Valle d'Aosta DOC

ERBALUCE DI CALUSO/ CALUSO DOCG
Piedmont

These are fresh, dry, light-bodied white wines made from the Erbaluce grape variety. Erbaluce's most famous – even if rare – manifestation is the golden sweet fragrant, yet full-bodied *passito.*

🍷 *1–3 years*

✓ *Luigi Ferrando* • *Orsolani* • *Vittorio Boratto*

FARA DOC
Piedmont

In the hands of someone like Dessilani, Fara DOC proves to be an underrated, enjoyable, Nebbiolo-based wine with lots of fruit combined with a spicy-scented character. A *riserva* version can also be made.

✓ *Luigi Dessilani (Caramino)*

FRANCIACORTA DOCG
Lombardy

The sparkling white and *rosato* wines made by Franciacorta were promoted to DOCG status in September 1995. Produced by the traditional method with 18 months of ageing on the lees (or 60 if *riserva*), from noble varieties Chardonnay and Pinot Noir, with a smaller influence of Pinot Blanc. Franciacorta had already shown its potential for producing fine, biscuity *brut,* and brightly rich *rosato* sparkling wines, though its distinctive Satèn style makes it unique amongst the best-known sparkling wine regions. Satèn can be only be a *blanc de blanc* with pressure not exceeding 5 atmospheres, which translates into a creamy finesse of persistent *perlage,* refined nutty aromas, and velvety minerality. Franciacorta was the first Italian wine to insist on bottle fermentation, and despite its recent history, it is together with Trento DOC, a sparkling wine appellation that can demand respect from the rest of the world. It is, however, prevalently consumed within Italy's borders, leaving less than 10 per cent of volume produced for the export market. *See* Curtefranca DOC for still wines.

🍷 *2–5 years*

✓ *Antica Fratta* • *Bellavista* • *Biondelli* • *Bonfadini* • *Ca'del Bosco* • *Ca' d'Or* • *Fratelli Berlucchi* • *Guido Berlucchi* • *Contadi Castaldi* • *Ricci Curbastro* • *Lantieri De Paratico* • *Majolini* • *La Montina* • *Monogram* • *Mosnel* • *Barone Pizzini* • *Le Quattro Terre* • *Lo Sparviere* • *Uberti*

FREISA D'ASTI DOC
Piedmont

Known for centuries, and a favourite of King Victor Emmanuel, this is the original and most famous Freisa. These fruity red wines may be made fully sparkling (sweet only) or still (dry or sweet) from 100 per cent Freisa.

🍷 *3–6 years (up to 10 for the very best)*

✓ *Gilli*

FREISA DI CHIERI DOC
Piedmont

Freisa di Chieri produces fully sparkling, *frizzante,* or still Freisa in both dry and sweet styles from just outside Turin. As for Freisa d'Asti, all the sparkling wines are made using the Charmat method.

GABIANO DOC
Piedmont

Full-bodied red Barbera wines made in the town of Gabiano in Monferrato. A small amount of Freisa and/or Grignolino can be added to the blend.

✓ *Castello di Gabiano*

GARDA DOC
Lombardy, Veneto

Garda is a wide-ranging appellation encompassing 18 grape varieties and innumerable other local non-aromatic varieties grown over three separate areas of hillsides around Lake Garda, the largest of which is in neighbouring Lombardy. In addition to blended red, *rosato,* white, *passito,* and *frizzante* wines, there are six varietal red wines (Cabernet Franc, Cabernet Sauvignon, Corvina, Marzemino, Merlot, and Pinot Nero) and seven varietal white wines: Chardonnay (still and *frizzante*), Cortese, Garganega (still or dry/sweet *frizzante*), Pinot Bianco, Pinot Grigio (still or *frizzante,* dry or sweet), Riesling and/or Riesling Italico, and Sauvignon Blanc. Sparkling wines are also and can be either Chardonnay-based, sparkling Pinot Grigio, or *rosato spumante* (minimum 50 per cent Cabernet Franc, Cabernet Sauvignon, Carmenere, Corvina, Merlot, and/or Pinot Nero).

✓ Cavalchina • Costaripa • Delai (Vigna Nobile) • La Guarda (Gropello, Rosso Sabbioso) • Monte Cicogna (Don Lisander, Rubinere)

GARDA COLLI MANTOVANI DOC
Lombardy

Although vines have been cultivated since ancient times in Lake Garda, today viticulture is one of the least important crops. The wines are dry, light-bodied, red, white, and *rosato*, produced from a wide variety of grapes.

GATTINARA DOCG
Piedmont

From Nebbiolo and up to 10 per cent Uva Rara and a maximum of 4 per cent Vespolina grown on the right bank of the Sesia River in northern Piedmont, Gattinara can be a fine wine, especially now that yields have been reduced since Gattinara attained DOCG status. When young, the fruit can be chunky and rustic when compared with Barbaresco or Barolo, but the best Gattinara wines develop a fine, silky-textured flavour and a graceful, violet-perfumed finesse when mature. Gattinara earned its DOC and then full DOCG status through the almost single-handed efforts of Mario Antoniolo, whose wines have always been among the greatest produced in the area.

🍷 6–15 years

✓ Mario Antoniolo • Le Colline • Nervi • Travaglini

GAVI or CORTESE DI GAVI DOCG
Piedmont

This is the most interesting expression of the Cortese grape in Piedmont. At best, the wine is soft-textured, sometimes with a slight *frizzantino* when young, and can develop a honey-rich flavour after a couple of years in bottle. A traditional-method sparkling Cortese is also made as well as a *riserva* version.

🍷 2–3 years

✓ Nicola Bergaglio • Gian Piero Broglia • Chiarlo • Carlo Deltetto • La Scolca • Villa Sparina

GHEMME DOCG
Piedmont

A Nebbiolo-based wine produced on the bank opposite Gattinara, Ghemme is generally very consistent and good value. Most Ghemme has as much colour, body, and flavour as Gattinara in general and starts off with a fine bouquet and very elegant fruit. In addition to Nebbiolo, Vespolina and Uva Rara may be used.

🍷 4–15 years

✓ Antichi Vigneti di Cantelupo • Le Colline • Luigi Dessilani • Torraccia del Piantavigna

GOLFO DEL TIGULLIO-PORTOFINO/ PORTOFINO DOC
Liguria

This DOC covers a number of different wines, primarily varietals from Vermentino, Bianchetta Genovese, and Moscato for whites and Ciliegiolo (including *novello*) for reds. Except for Moscato, which must be sweet and may be *passito*, all of these grapes may be made into either dry still wine or *frizzante*. Still and *frizzante* blends are also permitted in both red and white. A *passito* from a minimum of 60 per cent Bianchetta Genovese and/or Vermentino can be made. A dry white blended *spumante* is also allowed.

GRIGNOLINO D'ASTI DOC
Piedmont

Grignolino d'Asti DOC produces relatively low alcohol, slightly grippy, fruity red wines in which a small amount of Freisa may be used (up to 10 per cent).

GRIGNOLINO DEL MONFERRATO CASALESE DOC
Piedmont

These are light, crisp, fresh red wines made from Grignolino around the Casalese Monferrato.

GRUMELLO DOCG
See Valtellina Superiore DOCG

INFERNO DOC
See Valtellina Superiore DOCG

LAMBRUSCO MANTOVANO DOC
Lombardy

Not all Lambrusco wines comes from Emilia-Romagna. This one is produced just across the border on the plains of Mantua (Mantova). Formerly a *vino da tavola*, Mantovano is red and *frizzante* with a pink foam, and it may be dry or sweet. A *rosato* version is also produced. Two subzones of the DOC can be labelled as Oltre Po Mantovano and Viadanese Sabbionetano.

🍷 Upon purchase

✓ CS di Quistello

LANGHE DOC
Piedmont

The Langhe is an historical area for vine-growing and many of the wines made here use traditional, well-established grape varieties such as Arneis and Favorita (Vermentino) for whites and Nebbiolo, Dolcetto, and Freisa for reds. The DOC covers a very wide area that also includes the Barolo, Barbaresco, Asti, and Dogliani communes. For wines that do not conform to the production criteria (production area, grape varieties, or winemaking techniques) associated with these prestigious names, the Langhe DOC, which has more relaxed production restrictions, allows winemakers to experiment with varieties and techniques. Since its introduction in 1994, the Langhe DOC has gained considerable reputation for its innovative viticulture and use of international varieties (Cabernet Sauvignon and Sauvignon Blanc especially) and, above all, for the outstanding quality of some of the wines produced under this denomination. Gaja, a man whose wines have become more iconic than the famous appellations from which they come, turned Langhe and Barolo on their heads by (for a long time) bottling his very best Barolo vineyards – Sperss, Conteisa, and Dagromis – under the Langhe DOC.

The Langhe DOC also allows the production of *passito* wines (either blends or varietal, white and red). The DOC includes the subzone Nascetta del Comune di Novello, where a Nascetta wine or Passito Nascetta are only made.

✓ Elio Altare • Ca' Viola • Chionetti • Aldo Conterno • Gaja • Ettore Germano • Giuseppe Mascarello • Olivero Mario • Icardi • Bosio

LESSONA DOC
Piedmont

These are Nebbiolo-based red wines that can be delightfully scented, with rich fruit and great finesse.

✓ Sella • Proprietà Sperino

LIGURIA DI LEVANTE IGP
Liguria

This IGP covers the entire province of La Spezia and produces a wide range of styles: whites, reds, and *rosati* from any proportions of grape varieties approved for the region of Liguria and varietal whites (Malvasia Bianca Lunga or Trebbiano Toscano) and reds (Canaiolo Nero, Ciliegiolo, Merlot, Pollera Nera, Sangiovese, Syrah, or Vermentino Nero). A white or red *passito* can also be made.

LOAZZOLO DOC
Piedmont

From the village of Loazzolo, south of Canelli, this sweet Moscato or *vendemmia tardiva*, is aged for two years,

including six months in *barriques*, to produce a golden-hued, richly flavoured, lusciously textured, exotically sweet wine. A wine that can age, but which is best drunk young.

🍷 2–4 years

✓ Borgo Maragliano • Borgo Sambui • Bricchi Mej • Giancarlo Scaglione

LUGANA DOC
Lombardy

Lugana is a small DOC that straddles the border between Lombardy and Veneto. These are soft, smooth, dry white wines made from the Trebbiano di Lugana, the same as Trebbiano di Soave (Verdicchio), grown on the shores of Lake Garda, slightly overlapping Veneto in Northeast Italy. It thrives in the zone's calcareous clay soils, which are rich in mineral salts and help the fruit to reach high levels of ripeness. All Lugana wines must contain a minimum of 90 per cent of this variety and tend to be fragrant and concentrated with floral aromas and hints of spice, being sometimes similar in character to the wines of Soave Classico. A sparkling version (Charmat or traditional method) and *vendemmia tardiva* (late harvest) are also allowed.

🍷 1–2 years

✓ Ca' dei Frati • Provenza • Visconti • Montonale

MALVASIA DI CASORZO D'ASTI/ MALVASIA DI CASORZO/ CASORZO DOC
Piedmont

These are lightly aromatic, sweet, red, and *rosato* wines that may also be sparkling or *passito*.

🍷 1–2 years

✓ Bricco Mondalino

MALVASIA DI CASTELNUOVO DON BOSCO DOC
Piedmont

This DOC produces attractive, lightly aromatic, sweet, red, still, and sparkling wines.

🍷 1–2 years

✓ Bava

MAROGGIA DOCG
See Valtellina Superiore DOCG

MONFERRATO DOC
Piedmont

This DOC, overlapping the Asti region, was introduced in 1995 and produces a generic white, a generic red, and claret or *chiaretto* (rosé). Four varietals wines are made here: a white Cortese from the subzone Casalese and the red Dolcetto, Freisa, and Nebbiolo. The *rosato* or *chiaretto* must be made from a minimum of 85 per cent Barbera, Bonarda, Cabernet Franc, Cabernet Sauvignon, Dolcetto, Freisa, Grignolino, Nebbiolo, and/or Pinot Nero.

✓ La Barbatella • Martinetti • Villa Sparina • La Spinetta • Tenuta Santa Caterina

MONTENETTO DI BRESCIA IGP
Lombardy

These wines are made from vineyards in Brescia and half a dozen surrounding villages from Barbera, Cabernet Franc, Cabernet Sauvignon, Marzemino, Merlot, or Sangiovese for reds and Chardonnay, Pinot Bianco, or Trebbiano for whites. *Novello*-style reds are allowed, provided that one or more of the following grapes represents at least 70 per cent: Marzemino, Merlot, and Sangiovese.

MORGEX ET LA SALLE
Valle d'Aosta

A subzone of the regional Valle d'Aosta DOC, the vineyards of these two communes reach as high as 1,300 metres (4,265 feet), which makes it one of the highest wine-growing areas in Europe. Most vines in these two villages are grown between 900 and 1,000 metres (2,952 and 3,280 feet), however, although even this is remarkable as, in theory, grapes do not ripen above 800 metres (2,625 feet) in the Valle d'Aosta. Yet in practice they ripen and produce a fine, dry, *frizzantino* white. *Brut, extra brut,* and *demi-sec spumante* versions are also made. *See also* Valle d'Aosta DOC.

🍷 *1–3 years*

✓ *Alberto Vevey*

MOSCATO DI SCANZO/ SCANZO DOCG
Lombardy

This highly aromatic red Moscato wine is a real speciality from Scanzorosciate in the province of Bergamo. It is made from the Moscato di Scanzo grape and can only be made in a sweet style. Grapes must be dried for at least 21 days to achieve the minimum sugar level of 280 grams per litre (28 per cent). The finer examples display aromas of dried sage, rose, maraschino cherries, acacia honey, and even a hint of sweet spice leaning towards cinnamon, clove, and licorice.

✓ *Monzio Compagnoni • Il Cipresso (Serafino)*

NEBBIOLO D'ALBA DOC
Piedmont

Pure Nebbiolo wines come from between Barolo and Barbaresco. Most are fine, full, rich, and fruity. Sweet and sparkling versions are allowed.

🍷 *4–10 years*

✓ *Ascheri • Fabrizio Battaglino • Tenuta Carretta • Ceretto • Pio Cesare • Aldo Conterno • Giacomo Conterno • Franco Fiorina • Bruno Giacosa • Hilberg-Pasquero • Giuseppe Mascarello • Val del Prete • Vietti*

NIZZA DOCG
Piedmont

Formerly part of Barbera D'Asti DOCG, the Nizza subzone was approved as a separate DOCG as of the 2014 harvest. It covers 18 communes and produces wines from 100 per cent Barbera, which must be aged for a minimum of 18 months for the *rosso* and minimum 30 months for the *riserva* before release. The Nizza wines range from very good to outstanding quality, and they are full-bodied wines displaying ripe red and black fruit with hints of spice and licorice, always lifted by refreshing, crisp acidity. The fullest-bodied, some Barbera d'Asti Superiore wines from the Nizza subzone will still be available on the market for a few more years.

✓ *Prunotto*

NUS
Valle d'Aosta

A subzone of the Valle d'Aosta DOC, Nus makes an interesting red from Vien de Nus, Petit Rouge, and Pinot Noir grapes. Dry and *passito* styles of white wine are made from Pinot Gris or Malvoisie de Nus. *See also* Valle d'Aosta DOC.

🍷 *2–4 years*

✓ *La Crotta di Vegneron*

OLTREPÒ PAVESE DOC
Lombardy

The Oltrepò Pavese covers 42 communes south of the Po, although much of the production is not marketed under this DOC but sold to specialist wineries in Piedmont. A wide range of styles are produced: a white wine made from a minimum of 60 per cent Riesling and/or Welschriesling (aka Riesling Italico) and a maximum of 40 per cent Pinot Nero; a red or *rosato* made from 25 to 65 per cent Barbera; a red or *rosato* made from 25 to 65 per cent Croatina; and a red or *rosato* made from a maximum of 40 per cent Pinot Nero, Uva Rara, and/or Vespolina (locally Ughetta). There are 10 varietals permitted: Chardonnay, Cortese, Malvasia,

Moscato, Pinot Nero Bianco, Riesling, and Sauvignon Blanc for whites; Pinot Nero for *rosato,* and Barbera and Cabernet Sauvignon for reds. Varietal sparkling wines are made using the Charmat method only from the following grapes: Chardonnay, Cortese, Pinot Nero (vinified as a white wine), Riesling, or Sauvignon. A Pinot Nero *rosato spumante,* a Malvasia *spumante,* and a Moscato-based *passito* or fortified wine are also produced.

🍷 *1–3 years (white, rosato, and sparkling); 2–5 years (red)*

✓ *Giacomo Agnes • Angelo Ballabio • Bianchina Alberici • Castello di Cigognola • Doria (Pinot Nero) • Frecciarossa • Lino Maga (see also Barbacarlo) • Tenuta Mazzolino • Monsupello • Piccolo Bacco dei Quaroni • Tronconero (Bonarda) • Terre Bentivoglio*

OLTREPÒ PAVESE METODO CLASSICO DOCG
Lombardy

This DOCG is for traditional-method wines made in the province of Pavia. For standard Oltrepò Pavese Metodo Classico wines, Pinot Nero must make up 70 percent or more of the final blend, and a maximum 30 per cent of Chardonnay, Pinot Grigio, and/or Pinot Bianco is allowed. In terms of ageing, a minimum 15 months on the lees is required, whilst for *millesimato* (vintage) the minimum must be 24 months. Significant amounts of bulk wine have always been sold in nearby Milan, and this, together with the significant role played by co-operatives, has led to a trend for making quantity rather than quality. Even if the vast majority of the Oltrepò's production is not particularly interesting, there is no doubt that good, and occasionally very good, wine is made under this DOCG.

OLTREPÒ PAVESE PINOT GRIGIO DOC
Lombardy

This DOC applies to still and *frizzante* wines containing at least 85 per cent Pinot Grigio. It was formerly part of the Oltrepò Pavese DOC and was established as a separate DOC in 2010.

PIEMONTE DOC
Piedmont

This DOC embraces all the other DOC areas in Piedmont and produces all styles of wine from a very wide range of grapes. The principal red grape varieties are Albarossa, Barbera, Bonarda, Brachetto, Cabernet Franc, Cabernet Sauvignon, Carménère, Croatina, Dolcetto, Freisa, Grignolino, Merlot, Nebbiolo, Pinot Nero, and Syrah, while the principal white varieties used are Bussanello, Chardonnay, Cortese, Erbaluce, Favorita, Moscato, Pinot Bianco, Pinot Grigio, Riesling, Sauvignon Blanc, Viognier, and Welschriesling (Riesling Italico). The DOC includes the subzone Marengo Storico (*frizzante* or sparkling Cortese) and Vigneti di Montagna, for those vineyards with a minimum average vineyard elevation of 500 metres (1,640 feet) and a slope of at least 30 per cent.

✓ *Coppo • Paolo Saracco (Moscato)*

PINEROLESE DOC
Piedmont

This appellation includes varietal wines (Barbera, Bonarda, Dolcetto, the rare Doux d'Henry, and Freisa), a blended red called Ramie (produced from Avanà, Avarengo, and Neretto, plus a maximum of 40 per cent other local non-aromatic red varieties), and unnamed red and *rosato* blends (from Barbera, Bonarda, Nebbiolo, and Chatus, plus a maximum of 50 per cent other local non-aromatic red varieties).

✓ *Le Marie (Barbera Colombè, Debàrges)*

PINOT NERO DELL'OLTREPÒ PAVESE DOC
Lombardy

Formerly part of the Oltrepò Pavese DOC, it was established as a separate DOC in 2010. Whilst Pinot Nero was

traditionally used in the making of Oltrepò Pavese's sparkling wines, this DOC is dedicated to the production of a still Pinot Nero. A *riserva* is also allowed, with a minimum of two years ageing required before release.

PORNASSIO/ ORMEASCO DI PORNASSIO DOC
Liguria

A dry red or sweet *passito* wine made from the Ormeasco grape, a localized clone of Dolcetto that has been growing in the Riviera di Ponente area since the 14th century. The dry red wine may be labelled Superiore if aged 12 months prior to release. Sciac-trà, or Sciacchetrà, is a dry *rosato* wine made from Dolcetto. A red *passito,* either sweet or *liquoroso,* from Dolcetto is also allowed.

✓ *Fontanacota • Nirasca (Superiore) • Lorenzo Ramò (Rosso)*

PROVINCIA DI MANTOVA IGP
Lombardy

All grape varieties and wine styles are allowed for this IGP covering the southeastern corner of Lombardy close to Emilia-Romagna. Of its wines, Lambrusco is by far the best known, and most is sold under the Lambrusco Mantovano DOC, although this IGP does allow *frizzante* Lambrusco, and some is indeed produced.

PROVINCIA DI PAVIA IGP
Lombardy

The IGP equivalent of the Oltrepò Pavese DOC. It is located in foothills that rise south of the Po Valley, where Pinot Grigio is most commonly used for this IGP, although any locally grown variety is permitted, and Cortese, Pinot Nero, and Riesling Italico are also often used.

✓ *Picchioni (Arfena Pinot Nero)*

QUISTELLO IGP
Lombardy

Mostly produced by the Quistello cooperative, whose wines include both still and *frizzante* red, *rosato,* and whitefrom grapes grown in the same general area as the Mantova IGP in the southeastern corner of Lombardy, close to Emilia-Romagna.

RIVIERA DEL GARDA CLASSICO DOC
Lombardy

This area makes light, fruity, slightly bitter red wines that are no more interesting than the bulk of Valpolicella produced on the opposite bank of Lake Garda. Most are blended from Gropello, Sangiovese, Marzemino, and Barbera, but some are pure varietal wines from the Gropello grape, and these tend to be superior. A white wine only is made and must contain a minimum of 50 per cent Riesling Italico. The most successful wine here has been the *rosato,* or *chiaretto,* which, although it is made from exactly the same varieties as the *rosso,* can be much softer and easier to drink. A sparkling rosé made from a blend of Gropello, Barbera, Marzemino, and Sangiovese is also allowed. A subzone can be labelled as such: Valtènesi (*chiaretto, rosso,* and *rosso riserva* only). Within Valtènesi, the following communes and villages can also be mentioned: Manerba, Mocasina, Moniga, Padenghe, Picedo, Polpenazze, Portese, Puegnago, Raffa, and San Felice.

🍷 *Upon purchase*

✓ *Costaripa*

RIVIERA LIGURE DI PONENTE DOC
Liguria

This appellation covers the western Riviera of Liguria and produces three varietal whites (an aromatic Moscato, a characterful Pigato, and a rich, full-bodied Vermentino), two varietal reds (Granaccia or Alicante and Rossese), and three sweet wines (a late-harvest wine from Moscato Bianco, a varietal red or white *passito,* and a *passito* liquoroso). The DOC includes five subzones: Albenganese (Pigato, Vermentino, and Rossese only), Finalese (Pigato,

Vermentino, and Rossese only), Quiliano (Granaccia only), Riviera dei Fiori (Pigato, Vermentino, and Rossese only), and Taggia (Moscatello still or *frizzante, vendemmia tardiva,* and *passito* only).

✓ *Maria Donata Bianchi • Bruna • Lupi (Le Serre) • Poggio dei Gorleri*

ROERO DOCG
Piedmont

This DOCG covers the sandy hills on the left bank of the Tanaro. It is an increasingly important vineyard area that deserves to be better known given the quality and the charm of the Nebbiolos produced under this title. The wines are bold and fragrant and tend to be softer and earlier maturing than those from Barbaresco and Barolo. The area is also particularly known for its refreshing white wines made from Arneis. The classic Roero Arneis is dry and crisp with blossom-like aromas and flavours of fresh pear, apricot, and hints of hazelnut and generally unoaked. Alongside Gavi it is one of Piedmont's most interesting and highly regarded white wines, and its popularity continues to grow. A *spumante* version, either dry or sweet, can also be made. The DOCG includes 153 *menzioni geografiche aggiuntive* ("additional geographical definitions" or MeGAs), including 19 commune names.

🍷 *1–2 years*

✓ *Cascina Ca' Rossa • Carlo Deltetto • Giovanni Almondo • Matteo Correggia • Monchiero Carbone • Filippo Gallino • Malvirà • Ceretto • Bruno Giacosa • Castello di Neive • Vietti*

RONCHI DI BRESCIA IGP
Lombardy

This IGP encompasses red (which may be *novello*) and white (still, *frizzante,* or *passito*) wines from the villages of Bovezzo, Caino, Cellatica, Concesio, Collebeato, Naivo, and Villa Carcina to the north of the city of Brescia and Botticino, Nuvolento, Nuvolera, and Rezzato to the east of the city. They must be made exclusively from Chardonnay, Invernenga, Pinot Bianco, Trebbiano di Soave, and Trebbiano Toscano for whites and Barbera, Cabernet Franc, Cabernet Sauvignon, Incrocio Terzi No.1, Marzemino, Merlot, Sangiovese, and Schiava for reds.

RONCHI VARESINI IGP
Lombardy

White wines from any proportion of white non-aromatic grape varieties approved for the region of Lombardy and *rosato* and red wines from a minimum of 60 per cent Barbera, Croatina, Merlot, and Nebbiolo.

ROSSO DI VALTELLINA DOC
See Valtellina Rosso DOC

RUBINO DI CANTAVENNA DOC
Piedmont

This DOC produces a full-bodied red wine blend of mostly Barbera, Grignolino, and sometimes Freisa. It is made by a *cooperativa* called Rubino, which was responsible for creating the appellation.

RUCHÉ DI CASTAGNOLE MONFERRATO DOCG
Piedmont

When grown on the vineyards overlooking Castagnole Monferrato, the Ruché grape produces an aromatic red wine that ages like Nebbiolo. Up to 10 per cent Barbera and Brachetto may be used in production.

🍷 *3–5 years*

✓ *Piero Bruno • Ruché del Parroco*

SABBIONETA IGP
Lombardy

In this IGP, white, red, *rosato,* and *frizzante* styles are allowed from any grapes grown in this area of lush, fertile, sandy meadows on the north bank of the River Po, but Lambrusco dominates.

SAN COLOMBANO AL LAMBRO or SAN COLOMBANO DOC
Lombardy

Rich, robust, if somewhat rustic red wines made from Croatina, Barbera, and Uva Rara are produced by this DOC. A Chardonnay-based white is also allowed. San Colombano is the only DOC in the Milan province.

🍷 *2–5 years*

✓ *Carlo Pietrasanta*

SAN MARTINO DELLA BATTAGLIA DOC
Lombardy

Made from the Friulano (aka Sauvignonasse) grape variety grown in one of three separate areas overlooking the southern shores of Lake Garda – one of which is in Veneto – this is a dry, full-flavoured white wine with a flowery aroma and a slightly bitter aftertaste. A sweet *liquoroso* version is also produced.

SANGUE DI GIUDA DELL'OLTREPÒ PAVESE/SANGUE DI GIUDA DOC
Lombardy

Soft, sweet red *spumante* made from Barbera and Croatina and smaller proportions of Pinot Nero, Uva Rara, and/or Vespolina (locally Ughetta). A sweet wine either still or *frizzante* is also allowed from the same grapes.

SASSELLA DOCG
See Valtellina Superiore DOCG

SEBINO IGP
Lombardy

Effectively the IGP equivalent of Franciacorta, where, freed of the DOC restrictions of Terre di Franciacorta, several top-quality Chardonnay, Pinot Nero, and Bordeaux blends are produced.

✓ *Maurizo Zanella*

SFORZATO DI VALTELLINA or SFURSAT DI VALTELLINA DOCG
Lombardy

Sforzato, or *sfursat,* literally means "strained", and this wine is made from grapes that have been left to shrivel on the vine and have been vinified dry, with a minimum of 14.5 per cent alcohol, making this highly concentrated red wine the equivalent of an Amarone. All wines must be aged for a minimum of 18 months prior to release. *See also* Valtellina Superiore DOCG.

✓ *Nino Negri • Mamete Prevostini • Dirupi*

SIZZANO DOC
Piedmont

These are good, full-bodied red wines that are produced from a Gattinara-like blend on a bank of the river Sesia, just south of Ghemme.

STÄGAFÄSSLI DOCG
See Valtellina Superiore DOCG

STREVI DOC
Piedmont

A sweet *passito* wine made from Moscato grapes grown in the vineyards surrounding Acqui Terme.

TERRAZZE DELL'IMPERIESE IGP
LIGURIA

A relatively new, tiny IGP covering the entire province of Imperia and producing aromatic white wines from Pigato and/or Vermentino-based and perfumed fruit-driven *rosato* and red wines from a minimum of 40 per cent Dolcetto and/or Rossese.

TERRE ALFIERI DOC
Piedmont

A small appellation covering an area between Asti and Cuneo producing a varietal Arneis and a varietal Nebbiolo.

TERRE DEL COLLEONI/ COLLEONI DOC
Lombardy

The official Terre del Colleoni viticultural area spans the middle of the Bergamo province from east to west on either side of Bergamo town. This is almost exactly the same area covered by the Valcalepio DOC, but it produces different wine styles. There are five varietal whites (Chardonnay, Incrocio Manzoni, Moscato Giallo, Pinot Bianco, and Pinot Grigio), one Shiava-base *rosato,* and six red varietals (Cabernet Sauvignon, Franconia, Incrocio Terzi, Marzemino, and Merlot). Sparkling wines, either blended or varietals, are also made (using Charmat or traditional method) from any proportions of Chardonnay, Manzoni Bianco, Pinot Bianco, Pinot Grigio, and/or Pinot Nero. A Moscato Giallo *passito* is also allowed.

TERRE LARIANE IGP
Lombardy

This IGP produces white, *rosato,* and red wines in either dry or sweet style in the hilly and mountainous parts of the provinces of Como and Lecco. *Passito* wines are also allowed.

TORRETTE
Valle d'Aosta

A red-only subzone of the regional Valle d'Aosta DOC, Torrette produces deep-coloured wines of good bouquet and body from the Petit Rouge, plus up to 30 per cent Dolcetto, Fumin, Gamay, Mayolet, Pinot Noir, Premetta, and Vien de Nus. *See also* Valle d'Aosta DOC.

✓ *Elio Cassol • Grosjean*

VALCALEPIO DOC
Lombardy

An up-and-coming appellation for well-coloured, deeply flavoured red wines made from a blend of Merlot and Cabernet Sauvignon and for light, delicately dry white wines made from Pinot Blanc and Pinot Gris. A red Moscato *passito* from 100 per cent Moscato di Scanzo or Moscato is also allowed.

🍷 *1–3 years (white), 3–7 years (red)*

✓ *Tenuta Castello*

VALCAMONICA IGP
Lombardy

The Camonica Valley is best known as a UNESCO site, boasting the greatest complex of ancient rock drawings in Europe, but the area also contains some of the world's most spectacular, stunningly steep terraced vineyards. The IGP covers numerous communes in the province of Bresca and produces a white wine (minimum 60 per cent Manzoni Bianco, Müller-Thurgau, and/or Riesling) a Marzemino/Merlot-based red and wo varietals (Marzemino and Merlot). A white *passito* from a minimum of 60 per cent Manzoni Bianco, Müller-Thurgau, and/or Riesling can also be made.

VALLE D'AOSTA or VALLÉE D'AOSTE DOC
Valle d'Aosta

A regional DOC encompassing 20 different styles of wine, Valle d'Aosta took under its wing the only two DOCs formerly recognized in the region: Donnas and Enfer d'Arvier. These are now two of the seven subzones within this DOC, the other five being Arnad-Montjovat, Chambave, Morgex et La Salle, Nus, and Torrette (*see* individual entries). The Valle d'Aosta DOC has revitalized this region's tiny production of somewhat low-key wines. Apart from blended red, white, and *rosato*, the following varietal wines are permitted: Fumin, Gamay, Nebbiolo, Petit Rouge, Pinot Noir, and Premetta for reds and Chardonnay, Müller-Thurgau, Petit Arvine, and Pinot Gris for whites. There is also a Bianco di Pinot Nero, which you might see labelled as a Blanc de Noir de Pinot Noir, because all Valle d'Aosta wines have official alternative French designations.

🍷 *1–3 years*

✔ *Anselmet • Brégy & Gillioz • Les Crêtes • La Crotta di Vegneron • CV de Donnas • Les Granges • Grosjean • Lo Triolet*

VALLI OSSOLANE DOC
Piedmont

This very tiny appellation covering 8 hectares (20 acres) in the province of Verbano-Cusio-Ossola produces a white wine based on Chardonnay and two reds: an unnamed *rosso* made from a minimum of 60 per cent Croatina, Merlot and/or Nebbiolo, and a varietal Nebbiolo.

VAL POLCÈVERA DOC
Liguria

Genoa's very own wine, Val Polcèvera, encompasses a wide range of wine styles: white (which can be still, *frizzante*, *spumante*, or *passito* and must be at least 60 per cent Vermentino, Bianchetta Genovese, or Albarola, plus optional Pigato, Rolle, and Bosco), red or *rosato* (which can be still or *frizzante* and must contain a minimum of 60 per cent Ciliegiolo, Dolcetto, or Sangiovese, plus optional Barbera). Vermentino is the only pure varietal wine allowed. The DOC includes the subzone Coronata, which can be shown on the label for white wines coming from this ancient village perched high above the sea at the southernmost tip of Genoa, where very few vines survive today.

VALSUSA DOC
Piedmont

Situated just south of the Valle d'Aosta and close to the French border, the Valle di Susa is where this red wine is blended from at least 60 per cent Avanà, Barbera, or Dolcetto, plus optional Beretta Cuneese and other local non-aromatic varieties. Two varietal reds, Avanà and Becuet, are also produced.

✔ *Carlotta*

VALTELLINA ROSSO/ ROSSO DI VALTELLINA DOC
Lombardy

Based on Nebbiolo (locally known as Chiavennasca), this appellation encompasses 19 communes of the province of Sondrio. Most are light-scented, medium-bodied, and crunchy red wines of simple but pleasing character. Fine wines are virtually all classed as Valtellina Superiore. *See also* Valtellina Superiore DOC.

VALTELLINA SUPERIORE DOCG
Lombardy

The best wines of Valtellina are produced from a narrow strip of vineyards on the north bank of the river Adda near the Swiss border and must contain a minimum of 12 percent alcohol, as opposed to the 11 per cent required for Valtellina DOC. Most of the wines come from five subdistricts: Grumello (the lightest), Inferno (supposedly the hottest, rockiest part of the valley), Maroggia (the smallest subregion at just 24 hectares, or 60 acres), Sassella (the best), and Valgella (the most productive but least interesting). Stägafässli (reserved for Valtellina Superiore bottled over the border in Switzerland) may also appear on the label, despite not being a subzone. Essentially produced from Chiavennasca, the local synonym for Nebbiolo, all wines must be aged for a minimum of 24 months prior to release or 36 months if the term *riserva* is used. The richness of the best of these wines is confirmed by their elegance. They have good colour and are capable of developing exquisite finesse after several years in bottle. *See also* Sforzato di Valtellina DOCG.

🍷 *5–15 years*

✔ *Enologica Valtellinese • Sandro Fay • Fondazione Fojanini • Nino Negri • Conti Sertoli-Salis • Fratelli Triacca • Mamete Prevostini • Aldo Rainoldi • Dirupi*

VERDUNO or VERDUNO PELAVERGA DOC
Piedmont

The Pelaverga of Verduna, known locally as the Pelaverga Piccolo, is a distinctly different variety from the Pelaverga grown in the Colline Saluzzesi and elsewhere. The red wine it produces is fragrant, verging on spicy, with good acidity and minerality.

✔ *Fratelli Alessandria • Ascheri • Burlotto • Michele Reverdito • Castello di Verduno*

TERRACED ROWS OF PERGOLA-TRAINED VINES CLIMB THE PRECIPITOUS SLOPES OF THE VALLE D'AOSTA WINE REGION
Grape-growing in the Valle d'Aosta is concentrated primarily on south-facing slopes of a single Alpine valley, where the topography is nearly vertical, rising to 4,575 metres (15,000 feet). Here, the steep hillsides call for a terraced system, with vines trained onto stone pergolas, which help regulate the ground heat on cooler nights.

Northeast Italy

Freshness, crisp acidity, and purity of varietal character exemplify the wines of the Trentino-Alto Adige, Veneto, and Friuli-Venezia Giulia regions of Northeast Italy. This is the biggest production area of the country, the one that is most populated with cooperative cellars, and it is the home of interregional Prosecco and Pinot Grigio appellations. Yet top-end wineries and artisanally crafted world-class wines are also common, especially in Collio, Alto-Adige, and Valpolicella.

N ortheast Italy is a more mountainous area than the Northwest (with the exception of that region's Valle d'Aosta): just over half the land is occupied by the Dolomites and their precipitous foothills. Some of the finest wines are grown in the lush, verdant vineyards of the South Tyrol (Alto Adige), just over the border from Austria. The broad area encompassing the provinces of Treviso, Venice, and Pordenone, once devoted to production of Bordelais varieties, has been completely converted to Pinot Grigio and Italy's best-known sparkling wine: Prosecco. As a result, a great deal of wine is exported, including the mineral and refreshing whites of Soave and the cherry fruit–driven reds from Valpolicella. The latter, in the last decades, experienced an impressive growth, thanks to its soft and supple Ripasso and its flagship top-end red Amarone. Peculiar styles, such as orange wines from Ribolla in the Collio area, along with upcoming areas such as Lugana, with its refreshing and creamy unoaked dry style, and the bottle-fermented Trento DOC are now synonyms of more artisanal approach to production.

REGIONS AT A GLANCE

REGION	TOTAL AREA UNDER VINE Hectares (Acres)	IGP/DOC/ DOCG AREA UNDER VINE Hectares (Acres)	IGP/DOC/ DOCG PRODUCTION Hectolitres	TOTAL PRODUCTION Hectolitres
Veneto	94,291 (233,000)	89,190 (220,400)	7.1 million (45-55 red/white)	9.7 million (108 million cases)
Trentino-Alto Adige	15,690 (38,800)	15,153 (37,450)	610,000 (70-30 red/white)	1.2 million (13.3 million cases)
Friuli-Venezia Giulia	26,298 (65,000)	15,214 (37,600)	1.3 million (40-60 red/white)	1.9 million (21 million cases)

Source: Osservatorio del Vino UIV - ISMEA

TRENTINO-ALTO ADIGE

This is the most westerly and spectacular of Northeast Italy's three regions, and more than 90 per cent of its area is covered by mountainous countryside. It is made up of two autonomous provinces: the Italian-speaking Trento in the south and the German-speaking Bolzano, or South Tyrol, in the north. The Italian Tyrol is characterized by high-quality co-ops that are focussing on single-varietal whites. In the past, this area was strongly focussed on Pinot Grigio and Gewürztraminer, while nowadays Kerner, Sylvaner, and Pinot Blanc are gaining traction. Interesting things are also coming from Müller-Thurgau, Riesling, and the revitalised red Schiava. Italian Tyrol represents the most important area of Italy for Pinot Noir, and, given its extreme parcelisation, the region is actively interested in developing vineyard selections (*cru*). The southern area, Trentino, is characterised by a different orographic conformation, with an extensive central valley and several smaller satellite valleys. Here, there are two bigger commercial cooperative cellars, incorporating a plethora of smaller co-ops more involved more

WINE REGIONS OF NORTHEAST ITALY
(*see also* p345)
The variety of sites offered by the mountains and the hills of this area enables many grape varieties to be grown in addition to local ones. The most exciting wines come from the high vineyards of the South Tyrol, the hills of Friuli, and around Vicenza in the Veneto.

localised markets and in the production of autochthonous varieties: the white Nosiola in Toblino and Müller-Thurgau in Cembra Valley and the red Teroldego in the Piana Rotaliana and Marzemino in Isera. In the last decades, leading wineries have succeeded in establishing Trentino's bottle-fermented sparkling wine: Trento DOC, based on Chardonnay and Pinot Noir.

VENETO

Veneto stretches from the river Po to the sheltering Alps on the Austrian border, between Lake Garda to the west and Friuli-Venezia Giulia to the east. It is the first Italian region for wine produced because it is home to Pinot Grigio and Prosecco. These two interregional appellations, widely planted in flat alluvial lands, are increasingly successful and supported the growth of Italy as a major wine exporting country. On the other hand, beautiful and dramatic hillside vineyards with a cooler climate provide the crisp Cartizze, the mineral whites of Soave and Durello, the fruity Bardolino, and the salty Valpolicella. Valpolicella, known for wine production since Roman time, has been a trailblazer in the development of a hugely appreciated style of wine from dried grapes (Amarone) and second fermentation (Ripasso). Veneto is also a vital basin for indigenous grape varieties such as the white Glera, Garganega, Trebbiano di Soave, Lugana, and Durella and red Corvina, Corvinone, Rondinella, and Raboso.

FRIULI-VENEZIA GIULIA

Friuli is situated in the northeastern corner of Italy, between the Venice Laguna, the ex-Yugoslav border, and the Alps. This predominantly mountainous region has grown many non-Italian varieties since phylloxera wiped out its vineyards in the late 19th century. Friuli today has two primary areas, Grave and Collio, boasting entirely different aims. The Grave is a highly productive flatland that hosts mostly Pinot Grigio and Prosecco but with also a good potential (thanks to gravelly soils) for Sauvignon Blanc. Collio is a gorgeous hillside area, once more oriented to oaky Chardonnay production. Today Collio has a significant interest in indigenous grapes as Ribolla (Rebula) and Friulano.

The region combines a volumetric approach with more artisanal methods and has seen a steady development of wines in amphora, also thanks to the Slovenian community.

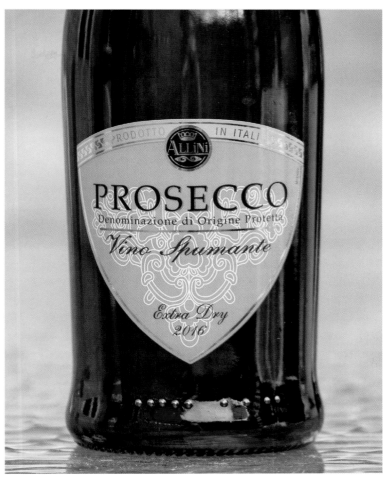

ALLINI PROSECCO VINO SPUMANTE 2016
Aided by its comparatively low price and higher standards, Prosecco saw a sharp rise in popularity in recent decades. This sparking wine, whether DOC or DOCG, is made from Glera grapes in a fully sparkling, *spumante*, or semi-sparkling, *frizzante*, style.

FACTORS AFFECTING TASTE AND QUALITY

LOCATION
Northeast Italy is bordered by the Dolomites to the north and the Adriatic Sea to the south.

CLIMATE
Similar to the northwest in that summers are hot and winters cold and harsh, but fog is less of a problem and hail more frequent. There are unpredictable variations in the weather from year to year, so vintages are important, particularly for red wines.

ASPECT
Vineyards are found on a variety of sites ranging from the steep, mountainous slopes of Alto Adige to the flat, alluvial plains of the Veneto and Friuli-Venezia Giulia. The best vineyards are always sited in hilly countryside, as on the terraces of Valpolicella.

SOIL
Soils of valley floor vineyards are on glacial moraine – a gritty mixture of sand, gravel, and sediment deposited during the Ice Age. Hillside vineyards of Friuli have more clay or sandy clay, with top vineyards sited in galestro schist (*ponca*). Soils in Alto Adige change from volcanic porphyry around Bolzano, weathered primitive rock soils with quartz, slate, and mica in the Isarco Valley and the Val Venosta, up to calcareous and dolomite rock in the southern part of the province. The best appellations of Verona's vineyards are hosted in the limestone and reddish irony clay soils of Valpolicella, the volcanic soils of Soave, the morenic soil of Bardolino, and the clayish soil of Lugana.

VITICULTURE & VINIFICATION
In this region there is a unique combination of the traditional pergola trellising system, especially on the hillsides of Trento and Verona and the diamond-shaped Bellussi of Prosecco, together with the modern and mechanisable VSP of the high-yielding flat areas, with some local interpretation such as *"doppio capovolto"*. Co-ops are very well equipped and organised, with grape sorting, temperature control, and gas blanketing resulting in very clean and pure styles of wine. Charmat method is common practice for sparkling Prosecco, while Trento DOC requires second fermentation to occur in the bottle. Wines from dried grapes are part of the culture of the region, with state-of-the-art innovative drying warehouses with ventilation and humidity control. Classic producers are still investing in a natural drying process and in the second fermentation on a partially sweet pomace of Recioto and Amarone (Ripasso technique).

GRAPE VARIETIES
Bianca Fernanda (Cortese), Cabernet Franc, Cabernet Sauvignon, Carménère, Chardonnay, Corvinone, Corvina, Durello, Friulano, Garganega, Gewürztraminer, Incrocio Manzoni 215 (Glera x Cabernet Sauvignon), Incrocio Manzoni 6013 (Riesling x Pinot Blanc), Kerner, Lagrein, Lambrusco, Limberger (Franconia), Malbec, Malvasia Istriana, Marzemino, Merlot, Moscato (Muscat Blanc à Petit Grains), Müller-Thurgau, Nosiola, Oseleta, Petit Verdot, Picolit, Pinot Bianco (Pinot Blanc), Pinot Grigio (Pinot Gris), Pinot Nero (Pinot Noir), Raboso, Rebo (Teroldego x Merlot), Refosco, Ribolla, Riesling, Rondinella, Molinara, Sauvignon Blanc, Sauvignonasse (Friulano), Schioppettino, Sylvaner, Glera, Tazzelenghe, Teroldego, Terrano (Mondeuse), Trebbiano (Ugni Blanc), Trebbiano di Soave, Trebbiano di Lugana (Torbiana), Veltliner (Rossola), Verduzzo, Vernatsch (Schiava), Vespaiola

THE APPELLATIONS OF
NORTHEAST ITALY

ALTO ADIGE or SÜDTIROL DOC
Trentino-Alto Adige

This generic appellation covers the entire Alto Adige, or South Tyrol, which is the northern half of the Valdadige DOC. There are six red wine varietals: Cabernet covers either Cabernet Sauvignon or Cabernet Franc, or both, and ranges from simple, but delightful, everyday wines to deeper-coloured, fuller-bodied, richer wines that become warm, mellow, and spicy after 5 to 10 years; Lagrein is made from an underrated, indigenous grape and can have a fine, distinctive character and good colour; Malvasia Nera, which is produced here as a red wine, although it gives best results when vinified as *rosato;* Merlot can be simply light and fruity or can attain greater heights, with a good spicy, sweet-pepper aroma and a fine, silky texture; Pinot Noir, which is a difficult varietal to perfect, but those from Mazzon are good and a speciality of this region (they will have Mazzoner on the label); and Schiava, which accounts for one in five bottles produced under this appellation and is the most popular tavern wine. Dual varietals are also allowed as follows: Cabernet Sauvignon or Franc, Lagrein, or Merlot: 50 to 84 per cent of first-named variety and 16 to 50 per cent of other variety.

There are 11 permitted white varietals: Chardonnay can range from light, delicately fruity to fuller versions with recognizable varietal characteristics; Moscato Giallo made either in a dry or sweet style;

Pinot Bianco, which is the most widely planted of all the white grape varieties and produces most of Alto Adige's finest white wines; Pinot Grigio, potentially as successful as Pinot Blanc but does not always occupy the best sites; Riesling Italico, which is produced in very small amounts; Riesling Renano, which is fine, delicate, and attractive at the lowest level but can be extraordinarily good in exceptional vintages;

Müller-Thurgau is relatively rare, which is a pity because it achieves a lively spiciness in this region;

Sauvignon was very scarce in the early 1980s, but as the vogue for this grape spread so its cultivation increased and the crisp, dry, varietally pure Sauvignon is now one of the most successful white wines of the Alto Adige; Sylvaner is made mostly to be drunk young and fresh; Traminer Aromatico, or Gewürztraminer, is far more restrained than the classic Alsace version, even if, as some claim, the variety originated in the village of Tramin or Termeno, between Bolzano and Trento – the delicate aroma and flavour of this wine can have a certain charm of its own; Kerner, which produces Riesling-like wines, displaying very similar flavours and a slightly leafy character.

Only four dry varietal *rosato* wines are allowed: Lagrein Rosato soft and fruity; Merlot Rosato, which is a relatively recent addition and makes a curiously grassy or bell-pepper *rosato;* Pinot Nero Rosato, or Blauburgunder Kretzer, is more successful as a *rosato* than as a red; and the rarer Moscato Rosa, which are very flamboyant, deep-pink to scarlet-red, semi-sweet to sweet wines with an unusually high natural acidity, intense floral aromas, and notes of rose, orange peel, and berry fruit on the palate.

There is one sparkling-wine category for pure or blended Pinot Blanc, Chardonnay, Pinot Bianco, Pinot Grigio, or Pinot Noir. All sparkling wines must undergo second fermentation in bottle.

The DOC Alto Adige includes six subzones. Colli di Bolzano, or Bozner Leiten, is for soft, fruity, early-drinking, Schiava-based red wines from the left bank of the river Adige and both banks of the Isarco near Bolzano. Meranese, or Meranese di Collina or Meraner Hügel, is from the hills around Merano, north of Bolzano, where the Schiava grape produces a light, delicately scented red wine. Santa Maddalena, or Sankt Magdalener, is in the heart of the Colli di Bolzano (*see* above), where the Schiava-based red wines are just as soft and fruity, but fuller in body and smoother in texture. Terlano, or Terlaner, which overlaps most of Caldaro DOC on the right bank of the Adige, running some 16 kilometres (10 miles) northwest of Bolzano to 24 kilometres (15 miles) south, and covers one blended and the following eight varietal white wines: Chardonnay, Müller-Thurgau, Pinot Bianco (Weissburgunder), Pinot Grigio, Riesling Italico (Welschriesling), Riesling Renano (Rheinriesling), Sauvignon Blanc, and Sylvaner.

Valle Isarco, or Eisacktaler, takes in both banks of the river Eisack (Isarco) from a few miles north of Bolzano to just north of Bressanone, where the vineyards are high in altitude and the vines trained low to absorb ground heat. Six white wine varietals are produced: Kerner, Müller-Thurgau, Pinot Grigio (or Ruländer), Sylvaner, Traminer Aromatico (or Gewürztraminer), and Veltliner. Valle Isarco also includes a geographical blend called Klausner Leitacher for red wines from the villages of Chiusa (also known as Klausen), Barbiano, Velturno, and Villandro, which must be made from at least 60 per cent Schiava, plus up to 40 per cent Portoghese and Lagrein. Furthermore, any Valle Isarco wine produced in Bressanone or Varna can be labelled "di Bressanone" or "Brixner". The last subzone is Valle Venosta, which covers all white non-sparkling types (including *vendemmia tardiva, passito,* and *riserva*) made from Chardonnay, Gewürztraminer, Kerner, Müller-Thurgau, Pinot Bianco, Pinot Grigio, Riesling, and Sauvignon Blanc, as well as varietal Pinot Nero, Pinot Nero Riserva, and Schiava.

🍷 *2–10 years (red and* rosato*), 2–5 years (white), 1–3 years (sparkling)*

✓ *Abbazia di Novacella • Cantina Caldaro • Cantina Gries-Bolzano • Cantina Terlano • Castel Juval • Colterenzio • Cortaccia • Anton Gojer • Giorgio Grai • Franz Haas • Haderburg • Hofstätter • Kettmeir • Günther Kershbaumer • Klosterkellerei Eisacktaler • Klosterkellerei Schreckbichl • Alois Lageder • Josephus Mayr • Foradori • Muri-Gries • Nalles Niclara Magre • Nals Margreid • Stiftskellerei Neustift • Niedermayr • Manfred Nössing • Peter Pliger • Plattner-Waldgries • Gumphof Markus Prackweisser • Franz Pratzner • Prima & Nuova • Heinrich Rottensteiner • San Michele (Sanct Valentin) • Santa Margherita • Schloss Rametz • Schloss Sallegg • Schloss Schwanburg • Termeno • Tiefenbrunner • Elena Walch • Baron Georg von Widmann*

ALTO LIVENZA IGP
Friuli-Venezia Giulia and Veneto

All grapes and styles (white, red, and *rosato* made either still or *frizzante*) are possible from this IGP, which straddles the Friuli-Veneto border between the Colli di Conegliano and the Friuli-Grave DOCs, including 15 communes in the province of Pordenone and Treviso.

AMARONE DELLA VALPOLICELLA DOCG
Veneto

Amarone is a derivative of Recioto (*see* Recioto della Valpolicella DOCG), with a similar deep colour but in a dry or off-dry style, which seems to make the flavours more powerful and chocolaty-spicy with a distinctly bitter finish. There is also something very specific that marks the fruit in Amarone wines, which wine writer Oz Clarke once described as a "bruised sourness". This is an oxidative character, coming from the fact that the wines are made by drying grapes for at least two months, long maceration during the wintertime, and long cask ageing. Bertani and Quintarelli are the most outstanding producers of a more traditional and classic dry style of Amarone. More powerful, richer, and sweeter styles are produced by Romano dal Forno and Allegrini. Formerly part of Valpolicella DOC, Amarone della Valpolicella was established as a separate DOCG in 2010, and it is now among the most sought-after high-end appellations of Italy, given its notable complexity and the capability of such wines to age wonderfully. *See also* Valpolicella DOC.

✓ *Allegrini • Begali (Monte Ca' Bianca) • Cecilia Beretta • Bertani • Brigaldara • Tommaso Bussola • Giuseppe Campagnola • Garbole • Marion • Masi • Pasqua •*

Quintarelli • Rizzardi (Calcarole) • Romano Dal Forno (Vigneto di Monte Lodoletta) • Sant'Alda • Sant'Antonio (Campo dei Gigli) • Santi (Proemio) • Speri (Vigneto Monte Sant'Urbano) • Tedeschi (Capitel Monte Olmi) • Tommasi • Trabucchi d'Illasi • Venturini (Campo Masua) • Viviani (Casa dei Bepi) • Zenato (Sergio Zenato)

ARCOLE DOC
Veneto

The Arcole DOC abuts the southeastern corner of Valpolicella, and it covers a wide range of wine styles: white (which can be still, *frizzante, spumante,* or *passito* and must be at least 50 per cent Garganega), red (only still, either *rosso* or *riserva* from a minimum of 50 per cent Merlot) or *rosato* (which can be *frizzante* or still, dry, or sweet and must contain a minimum of 50 per cent Merlot), and the following varietal wines: Chardonnay (can be *frizzante*), Pinot Grigio, and Merlot.

✓ *Sartori*

ASOLO PROSECCO DOCG
Veneto

This DOC produces still or *frizzante* white wines, a Spumante Superiore (made by tank method) or "Sui Lieviti" (made from a single harvest year and fermented in bottle with minimum 90 days on the lees), which must account a minimum 85 per cent Glera.

✓ *Montelliana*

BAGNOLI or BAGNOLI DI SOPRA DOC
Veneto

This typically Italian all-purpose appellation covers still, *passito, frizzante,* and *spumante* wines from a wide range of grapes grown southwest of Venice in a large area sandwiched between the Colli Euganei and Corti Benedettine del Padovano. The most important grape varieties are Cabernet Franc, Cabernet Sauvignon, Carménère, Merlot, Raboso Piave, and Raboso Veronese for reds and Chardonnay, Sauvignon, and Sauvignon Blanc for whites, while the *rosato* must be at least 50 per cent Raboso del Piave or Raboso Veronese. Varietal wines are Marzemina Bianca, Cabernet (minimum 85 per cent Cabernet Franc, Cabernet Sauvignon, and/or Carménère), Cabernet Franc, Cabernet Sauvignon, Carménère, Cavrara, Corbina, Merlot, Refosco, and Turchetta. A sparkling white or *rosato* and a red fortified wine from Raboso Piave are also allowed. The DOC includes a *classico* subzone that covers the commune of Bagnoli di Sopra only.

BAGNOLI FRIULARO/ FRIULARO DI BAGNOLI DOCG
Veneto

Formerly part of the Bagnoli di Sopra DOC, it was established as a separate DOCG in 2011. This is a red wine–only DOCG producing dry full-bodied reds with high acidity and high tannin from Raboso del Piave. *Vendemmia tardiva,* and *passito* versions are also allowed.

BARDOLINO DOC
Veneto

From the hills surrounding the eastern shore of Lake Garda, the wines of the Bardolino DOC are made from 35 to 80 per cent Corvina and 10 to 40 per cent Rondinella, plus a maximum of 20 per cent optional authorised varieties, with no single variety exceeding 10 per cent of the total blend. They may be dry, still, *spumante,* red, *chiaretto,* or *novello.* The wines are generally light, red fruit––driven, and attractive, yet overall modest. The best examples fall under Bardolino Superiore DOCG.

✓ *Corte Gardoni • Le Fragh • Guerrieri-Rizzardi*

BARDOLINO SUPERIORE DOCG
Veneto

Bardolino Superiore is the higher-quality DOCG variant (higher vine density and lower yields are required) of the light red Bardolino wine made on the eastern shores of Lake Garda. Compared with Bardolino DOC wines, to gain the DOCG status the wines must reach a minimum of 12 percent ABV (against 10.5 percent) and must be aged for at least one year before release; the wines therefore show more concentration and character. Bardolino is made from a blend of Corvina, Corvinone, and Rondinella grapes, complemented by up to 15 percent Molinara. In the past decade or so the traditional blend has been augmented with additions of such grapes as Barbera, Sangiovese, Marzemino, Merlot, and Cabernet Sauvignon, which are permitted up to 20 percent in total under the DOCG's production laws. Bardolino DOCG only allows the production of dry red wine and includes the *classico* subzone that covers the communes of Affi, Bardolino, Cavaion, Costermano, Garda, and Lazise.

✓ *La Prendina (San Lucia)* • *Corte Gardoni (Le Fontane)*

BIANCO DI CUSTOZA/CUSTOZA DOC
Veneto

Effectively the white wines of Bardolino, the wines of this DOC can be scented with a smooth aftertaste and may sometimes be sparkling. Some critics compare Bianco di Custoza to Soave. The grapes used are Cortese, Friulano, Garganega, and Trebbiano with lower proportions of Chardonnay, Malvasia, Manzoni Bianco, Pinot Bianco, Riesling, and/or Welschriesling. Substantial research in soil composition and meso-climates was executed in 2010, resulting in either innovation or an increase in quality also due to the increased proportion of the Cortese grape in the blends.

✓ *Corte Gardoni (Mael)*

BOZNER LEITEN or COLLI DI BOLZANO DOC
See Alto Adige DOC

BREGANZE DOC
Veneto

This is one of Italy's unsung DOC heroes, largely due to the excellence of Maculan. As a generic red, Breganze is an excellent Merlot-based wine with the possible addition of Pinot Noir, Freisa, Marzemino, Gropello Gentile, Cabernet Franc, and Cabernet Sauvignon. As a white, it is fresh, zesty, and dry and made from Friulano, with the possible addition of Pinot Blanc, Pinot Gris, Welschriesling, and Vespaiola. But it is mostly sold as a varietal, of which there are four reds (Cabernet, Cabernet Sauvignon, Marzemino, and Pinot Nero) and six white (Chardonnay, Pinot Bianco, Pinot Grigio, Sauvignon Blanc, Tai Rosso, and Vespaiola). A *spumante* and a sweet wine from Vespaiola grape are also allowed.

🍷 *2–5 years (red), 1–2 years (white)*

✓ *Bartolomeo da Breganze* • *Maculan* • *Vigneto Due Santi*

CALDARO/LAGO DI CALDARO/ KALTERERSEE/KALTERER DOC
Trentino-Alto Adige

This overlaps the heart of the Terlano area, so Caldaro might as well be yet another sub-appellation of the Alto Adige DOC, along with Terlano and the other former DOCs that have now been absorbed. These Schiava-based red wines have the possibility of adding up to 15 per cent in total of Pinot Nero or Lagrein and are mostly soft and fruity, hinting of almond and easy to drink. Later-harvested wines may be labelled Auslese, Klassisch Auslese, or Scelto. The Caldaro, or Lago di Caldaro DOC, may also be labelled in German as Kalterer, or Kaltereresee DOC.

✓ *Cantina di Caldaro* • *Erste & Neue* • *Manincor* • *Thomas Pichler (Alte Reben)*

CARSO/CARSO-KRAS DOC
Friuli-Venezia Giulia

Carso Terrano must consist of at least 85 per cent Terrano (also known as Cagnina), which makes a deep, dark, and full red wine that can be fat and juicy. Carso Malvasia is a rich and gently spicy, dry white wine. Several other varietal wines can be produced: Chardonnay, Traminer, Glera, Pinot Grigio, Sauvignon Blanc, and Vitovska for whites; Cabernet Franc, Cabernet Sauvignon, Merlot, and Refosco dal Peduncolo Rosso for reds. The DOC also includes a *classico* subzone (from the communes of Duino-Aurisina, Monrupino, Sgonico, and Trieste only).

🍷 *1–3 years*

✓ *Edi Kante* • *Skerk* • *Zidarich*

CASTELLER DOC
Trentino-Alto Adige

The thin strip of hilly vineyards that produces these wines is virtually identical to the Trentino DOC. Dry red wines for everyday drinking from Schiava, Merlot, and Lambrusco grapes.

COLLI BERICI DOC
Veneto

This DOC includes a sparkling Garganega-based white wine, a traditional-method Chardonnay-based sparkling white, a Tai Rosso *spumante*, a generic white Garganega-based, a generic red Merlot-based, and several varietals wines: Cabernet, Cabernet Franc, Cabernet Sauvignon, Carménère, Merlot, Pinot Nero, and Tai Rosso for reds; Chardonnay, Garganega, Manzoni Bianco, Pinot Bianco, Pinot Grigio, Sauvignon Blanc and Tai for whites. A *passito* Garganega-based wine can also be made. The DOC includes the subzone Babarano, which produces the Merlot-based *rosso* and Tai Rosso *spumante* only. The Cabernet can be rich and grassy with chocolatey fruit, and the Merlot, in the right hands, is plump and juicy The Tai Rosso "i Rosso" is unusual and interesting, and the Chardonnay-based sparkling white is only made by traditional method with a minimum of 15 months on the lees.

🍷 *2–5 years (red), 1–2 years (white)*

✓ *Castello di Belvedere* • *Portogodi (Pozzare)* • *Villa dal Ferro*

COLLI DI CONEGLIANO DOCG
Veneto

The DOCG's default wine is a dry white based on Manzoni Bianco with either Pinot Bianco or Chardonnay (or both). A hint of Sauvignon Blanc and Riesling is also permitted (limited to a combined maximum of 10 per cent). The red wines (much less common) are based on the Bordeaux varieties Cabernet Franc, Cabernet Sauvignon, and Merlot. The most interesting wines made under this title are, however, the sweet wines that are produced in defined areas within the denomination. First of these is sweet, red Colli di Conegliano Refrontolo Passito, a dried-grape wine based almost exclusively on Marzemino. This is complemented by white Torchiato di Fregona, which is made from air-dried Glera, Verdiso, and Boschera grapes.

✓ *Sorelle Bronca (Ser Bele)*

COLLI EUGANEI DOC
Veneto

Bordering Colli Berici to the southeast, this area produces soft, full-bodied red-wine blends (from Merlot, Cabernet Franc, Cabernet Sauvignon, and Raboso) and six red varietals (Cabernet, Cabernet Franc, Cabernet Sauvignon, Carménère, Merlot, and Raboso). A generic Garganega and Glera-based white wine is made, together with varietal dry whites (Chardonnay, Garganega, Manzoni Bianco, Pinot Bianco, Sauvignon Blanc, and Tai) and a varietal Moscato (dry or sweet), a dry or sweet Pinello *frizzante* and a dry or sweet Serprino *frizzante* (minimum 85 per cent Glera). A fully sparkling Pinello or Serprino and a sweet sparkling Moscato are also allowed.

🍷 *2–5 years (red), 1–2 years (white)*

✓ *Cavalchina (Amedeo)* • *Monte de Frà (Ca' del Magro)* • *Villa Sceriman* 🅓 • *Vignalta*

COLLI EUGANEI FIOR D'ARANCIO/FIOR D'ARANCIO COLLI EUGANEI DOCG
Veneto

This DOCG is dedicated to the production of high-quality, highly aromatic white wines from Moscato Gialo: a generic white (dry or sweet), a sweet *spumante*, and a *passito* version with a fresh, delicate, blossom-like aroma are allowed.

COLLI ORIENTALI DEL FRIULI PICOLIT DOCG
Friuli-Venezia Giulia

This sweet white wine is made from Picolit, a grape variety that derives its name from the small, or *piccolo*, quantity of grapes it produces, thanks to its exceptionally poor pollination rate in the vineyard.

Some better estates still make the wine much as it was in the past, with the bunches harvested late in mid-October and then left to dry and raisin on mats before pressing. Other producers have opted for a late-harvest style, with the grapes left even longer in the vineyard but not raisined after picking. The wines are generally matured in barrels. Although Picolit is generally considered a dessert wine, it is not luscious and is best considered a *vino da meditazione*, a wine to be sipped alone in order to appreciate its delicate floral aromas and its light sweetness that suggests peaches and apricots. These wines must be aged for at least 12 months prior to release, unless made in the subzone "Cialla", in which case it must be aged for at least 24 months.

✓ *Ronco delle Betulle*

COLLI TREVIGIANI IGP
Veneto

This IGP covers numerous communes in the province of Treviso and produces a wide range of wines and styles. The generic white *rosato* and red can be made with any proportions of grape varieties approved for the province of Treviso in a dry, sweet, still, or *frizzante* style. Varietals from the same grapes are equally allowed.

COLLIO OR COLLIO GORIZIANO DOC
Friuli-Venezia Giulia

A large range of mostly white wines from a hilly area close to the Slovenian border, Collio now encompasses 19 different wines. With some top producers making very high-quality wines, this appellation is clearly in line for full DOCG status in the near future. As well as a blended dry white (primarily from Ribolla, Malvasia, and Friulano), which can be slightly *frizzantino*, and blended red (Merlot, Cabernet Sauvignon, and Cabernet Franc, plus Pinot Noir and possibly one or two others), there are 5 red varietals (Cabernet, Cabernet Franc, Cabernet Sauvignon, Merlot, and Pinot Nero) and 12 white (Chardonnay, Malvasia, Müller-Thurgau, Picolit, Pinot Bianco, Pinot Grigio, Riesling Italico, Riesling Renano, Friulano, Ribolla, Sauvignon, and Traminer Aromatico).

🍷 *1–4 years (red), 1–3 years (white)*

✓ *Borgo del Tiglio* • *Branko* • *Castello di Spessa* • *Eugenio Collavini* • *Colle Duga* • *Josko Gravner* • *Jermann* • *Edi Keber* • *Dorino Livon* • *Isidoro Polencic* • *Doro Princic* • *Dario Raccaro* • *Ronco dei Tassi* • *Russiz Superiore* • *Schiopetto* • *Franco Toros* • *Venicia & Venicia* • *Villa Russiz* • *Borgo Savaian*

CONEGLIANO VALDOBBIADENE PROSECCO DOCG
Veneto

The secret behind Italy's most-appreciated sparkling wine is all about maintaining freshness and preserving as many primary aromas possible. After a huge success in the mass market, the appellation started to develop a vertical segmentation because the best Prosecco invariably come from the hilly Conegliano-Valdobbiadene region, which was upgraded to DOCG in 2009. Higher in altitude and therefore cooler in climate compared to the broader DOC area, Valdobbiadene can produce inimitable floral, crunchy, and savory sparklings with more refined bubbles. Most Prosecco is made in the extra-dry style, but with an average of just 16 grams of residual sugar, it is not one of the sweeter extra-dry sparkling wines on the market. The strategy is to enhance the primary aromas and grapiness, rather than project any sweetness as such. One Prosecco that definitely has a sweetness about it is Cartizze, the so-called *grand cru* of Prosecco. A recent addition is Rive, which in the local dialect refers to high hillsides, where the slopes are so steep that they can be very difficult to work. A Prosecco with a Rive denomination will come from a specific locality within the Conegliano-Valdobbiadene region. Rive is therefore an extension of the Cartizze concept but for a drier style. Labeling options abound for this denomination and can be confusing. First, all sparkling wines from this denomination – which is to say, nearly all its wines – qualify for the "Superiore" adjective; only the still and *frizzante* versions do not. The use of Superiore is optional, however, as is the word Prosecco itself. In addition, wines from the commune of Valdobbiadene can drop "Conegliano", and those from the commune of Conegliano don't need to include "Valdobbiadene" on the label. Sparkling wines from the subzone of Cartizze must use the phrase "Superiore di Cartizze." Most Prosecco is not vintaged, but for a wine that you should consume as fresh as possible, knowing the year of production is vital. *See also* Prosecco DOC.

🍷 *Upon purchase*

✔ Andreola • Astoria (Cuvée Tenuta Val de Brun) • Bisol, Bonfadini (Opera) • Carpenè Malvolti • Fratelli Bortolin (Cartizze) • Le Colture, Masottina • CS Montelliana • Nino Franco • Col Vetoraz (Cartizze) • Sanfeletto • La Tordera • Tenuta degli Ultimi • Col Vetoraz (Cartizze)

CONSELVANO IGP
Veneto

This large IGP west of the Venetian lagoon encompasses a defined area in the province of Padova, including most of the DOCs of Colli Euganei, Bagnoli di Sopra, Corti Benedettine del Padovano, and the Riviera del Brenta. Consequently, it offers lots of potential but is seldom used.

CORTI BENEDETTINE DEL PADOVANO DOC
Veneto

This large DOC covers the southeastern area of the Padova province and produces various wine styles: white, *rosato*, red, sweet sparkling, and sweet. The basic red or *rosato* must be a blend of 60 to 70 per cent Merlot, 10 per cent Raboso, and up to 30 per cent other local or Bordeaux red varieties. The basic white must be a blend of at least 50 per cent Friulano and a maximum of 50 per cent Chardonnay and/or Pinot Bianco and/or Pinot Grigio and/or Sauvignon Blanc. A *passito* style is permitted, as long as it consists of at least 70 per cent Moscato Giallo (optional varieties are Chardonnay, Pinot Bianco, Pinot Grigio, and Sauvignon Blanc). A blend of Cabernet Franc and Cabernet Sauvignon may be sold as Cabernet, and Cabernet Sauvignon may be produced as a pure varietal wine as well as Merlot, Raboso, and Refosco dal Peduncolo Rosso. A sparkling Chardonnay-based and a sweet Moscato *spumante* are also produced.

DELLE VENEZIE DOC
Friuli-Venezia Giulia, Trentino-Alto Adige, and Veneto

Considering this vast appellation covers all grape varieties and styles of wine produced in Friuli-Venezia Giulia, Trentino-Alto Adige, and Veneto, it has a remarkably reliable reputation. This DOC was established in 2017, elevating Pinot Grigio from IGP delle Venezie, and indeed most wine produced today is the varietal Pinot Grigio that can be made either in a still or sparkling style (tank method). The list of grape varieties sanctioned for use in delle Venezie wines is long. These include Chardonnay, Friulano, Müller-Thurgau, Pinot Bianco, Verduzzo, and Garganega. Until 2027, other nonaromatic white grapes may be used in addition to those listed.

✔ Albino Armani • Tommasi

FRIULI ANNIA DOC
Friuli-Venezia Giulia

Named after the Via Annia, an ancient Roman road built by Titus Annius Rufus in 131 BC, this DOC has a relatively small production that is focused on the Friulano (formerly Tocai Friulano). The combination of wines allowed is typically large, however, with blended red, white, and *rosato* wines, *frizzante* and *spumante,* and plenty of varietal wines (Cabernet Franc, Cabernet Sauvignon, Chardonnay, Friulano, Malvasia, Merlot, Pinot Bianco, Pinot Grigio, Refosco dal Peduncolo Rosso, Sauvignon, and Verduzzo Friulano).

✔ Emiro Bortolusso

FRIULI AQUILEA DOC
Friuli-Venezia Giulia

Formerly known simply as Aquileia, this is a wide-ranging appellation that covers 5 varietal red wines (Merlot, Cabernet, Cabernet Franc, Cabernet Sauvignon, and Refosco); 10 white varietals (Chardonnay, Friulano, Malvasia Istriana, Müller-Thurgau, Pinot Bianco, Pinot Grigio, Riesling, Sauvignon, Traminer, and Verduzzo Friulano); and a generic *rosato* of least 70 percent Merlot plus Cabernet Franc, Cabernet Sauvignon, and Refosco. There is also a Chardonnay *spumante*. All are generally light, crisply balanced wines, although some producers excel in better years.

🍷 1–4 years (red), 1–3 years (white, rosato, and sparkling)

✔ Ca' Bolani

FRIULI COLLI ORIENTALI DOC
Friuli-Venezia Giulia

The Colli Orientali hills are home to some of Friuli's most prestigious vineyards. The *terroir* is similar to that of the hills around Gorizia (*see* Collio Goriziano) and the two areas use many of the same grape varieties. Friulano is the leading grape; other important Italian varieties include Ribolla Gialla, Verduzzo, and Picolit. International grapes are also grown widely and include Sauvignon Blanc, Chardonnay, Riesling, Pinot Grigio, and Pinot Bianco. Most of the Friuli-Venezia Giulia region is renowned for its white wine production; however, within this DOC reds of outstanding quality are also made. The varietal gems are led by Refosco, which delivers wines with incredible blackberry fruit intensity that is underpinned by a strong line of minerality. This is followed closely by Pignolo and Merlot. The rest of the reds are made from Bordeaux varieties and local Schioppettino and Tazzelenghe.

🍷 3–8 years (red), 1–3 years (white)

✔ Bastianich • Borgo del Tiglio • Girolamo Dorigo • Le Due Terre • Livio Felluga • Dorino Livon • Miani • Ronc di Vico • Ronchi di Manzano • Scubla • Giordano Sirch • La Tunella • Le Vigne di Zamò • Volpe Pasini

FRIULI GRAVE DOC
Friuli-Venezia Giulia

Formerly called Grave del Friuli, this massive appellation spreads out on either side of the river Tagliamento between Sacile in the west and Cividale di Friuli in the east and accounts for over half of the region's total production. It is a huge and complicated, but rapidly improving, multi-varietal DOC in which several winemakers regularly produce fine wines. In addition to generic red and white blends, there are six red varietals (Cabernet, Cabernet Franc, Cabernet Sauvignon, Merlot, Refosco, and Pinot Noir) and eight white varietals (Chardonnay, Pinot Bianco, Pinot Gris, Riesling Renano, Sauvignon, Friulano, Traminer Aromatico, and Verduzzo Friulano). There is a Chardonnay *frizzante*, Verduzzo *frizzante*, Chardonnay *spumante*, and a blended generic *spumante*.

🍷 1–4 years (red), 1–3 years (white, rosato, and sparkling)

✔ Borgo Magredo • Pighin • Vigneti le Monde • Vigneti Pittaro

FRIULI ISONZO or ISONZO DEL FRIULI DOC
Friuli-Venezia Giulia

Just south of the Collio, nestling close to the border with Slovenia, Friuli Isonzo encompasses red, white, *rosato*, *frizzante*, *spumante,* and *vendemmia tardiva*, the inevitable Cabernet for a blend of Cabernet Franc and Cabernet Sauvignon, Pinot Spumante for a blend of Pinot Bianco, Pinot Grigio, Pinot Nero, and a whole raft of varietal wines (Cabernet Franc, Cabernet Sauvignon, Chardonnay, Franconia, Friulano, Malvasia, Merlot, Moscato Giallo, Moscato Rosa, Pinot Bianco, Pinot Grigio, Pinot Nero, Refosco dal

VERDANT ROWS OF GRAPEVINES CURVE ALONG THE HILLS IN FRIULI VENEZIA-GIULIA
This large vine-growing region in the far northeast of Italy is bordered by the Veneto region of Italy, Austria, Slovenia, and the Gulf of Venice. A wide range of grapes are found here, including local and international varieties.

Peduncolo Rosso, Riesling, Riesling Italico, Sauvignon, Schioppettino, Traminer Aromatico, and Verduzzo Friulano), often of good quality and value.

🍷 *1–4 years (red), 1–3 years (white)*

✓ *Borgo Conventi • Borgo San Daniele (Friulano) • Mauro Drius (Pinot Bianco) • Lis Neris-Pecorari • Ronco del Gelso • Vie di Romans (Sauvignon) • Tenuta Villanova (Malvasia Saccoline) • Borgo Savaian*

FRIULI LATISANA DOC
Friuli-Venezia Giulia

This area stretches from the central section of the Friuli Grave to the Adriatic coast at Lignano Sabbiadoro. In addition to a generic, blended *rosato* (Bordeaux varieties) and *spumante* (Chardonnay and Pinot varieties) there are eight red varietal wines (Merlot, Cabernet Franc, Cabernet Sauvignon, Carménère, Franconia, Pinot Nero, and Refosco); and nine white varietals (Chardonnay, Malvasia, Pinot Bianco, Pinot Grigio, Riesling Renano, Sauvignon, Friulano, Traminer Aromatico, and Verduzzo Friulano). Chardonnay, Malvasia, Pinot Bianco, and Verduzzo Friulano may also be produced in a *frizzante* style.

FRIULI-VENEZIA GIULIA DOC
Friuli-Venezia Giulia

This large DOC produces a wide range of styles. A generic white made from any proportion of Chardonnay, Friulano, Gewürztraminer, Malvasia Istriana, Pinot Bianco, Pinot Grigio, Riesling, Ribolla Gialla, Sauvignon Blanc, and/or Verduzzo Friulano; a generic red made from any proportions of Cabernet Franc, Cabernet Sauvignon, Carménère, Merlot, Pinot Nero, and/or Refosco dal Peduncolo Rosso; nine varietal whites (Chardonnay, Friulano, Malvasia, Pinot Bianco or Pinot Blanc, Pinot Grigio or Pinot Gris, Riesling, Sauvignon, Traminer Aromatico, and Verduzzo Friulano); and five varietal reds (Cabernet Franc, Cabernet Sauvignon, Merlot, Pinot Nero, and Refosco dal Peduncolo Rosso); and a generic sparkling wine and a *spumante metodo classico*, both made from any proportions of Chardonnay, Pinot Bianco, Pinot Grigio, and/or Pinot Nero. A sparkling Ribolla Gialla (made either using tank method or traditional method) is also allowed.

GAMBELLARA DOC
Veneto

Produces crisp and refreshing Garganega-based wines, with notes of lemon sherbet, almond, and a hint of sweet, fragrant spice, which can be either made in a still or sparkling style. The DOC includes seven subzones: Classico (where only the generic *bianco* and a *vin santo*, as well as Garganega are allowed), Creari, Monti di Mezzo, Faldeo, San Marco, Selva, and Taibane.

🍷 *2–5 years*

✓ *La Biancara • Luigino del Masso (Riva del Molino)*

GARDA DOC
Lombardy, Veneto

A wide-ranging appellation encompassing 18 grape varieties and innumerable other local non-aromatic varieties grown over three separate areas of hillsides around Lake Garda, the largest of which is in neighbouring Lombardy. In addition to blended red, *rosato*, white, *passito*, and *frizzante* wines, there are six varietal red wines (Cabernet Franc, Cabernet Sauvignon, Corvina, Marzemino, Merlot, and Pinot Nero) and seven varietal white wines: Chardonnay (still and *frizzante*), Cortese, Garganega (still or dry/sweet *frizzante*), Pinot Bianco, Pinot Grigio (still or *frizzante*, dry or sweet), Riesling and/or Riesling Italico, and Sauvignon Blanc. Sparkling wines are also allowed and can be either Chardonnay-based, sparkling Pinot Grigio, or *rosato spumante* (minimum 50 per cent Cabernet Franc, Cabernet Sauvignon, Carmenere, Corvina, Merlot, and/or Pinot Nero).

✓ *Cavalchina • Costaripa • Delai (Vigna Nobile) • La Guarda (Gropello, Rosso Sabbioso) • Monte Cicogna (Don Lisander, Rubinere)*

LESSINI DURELLO/ DURELLO LESSINI DOC
Veneto

This appellation is found just northeast of Soave, in the high hills between the provinces of Verona and Vicenza and produces floral, citrusy sparkling wines (made by either the Charmat or traditional method) from the Durella grape. The *riserva* version must be made using the traditional method only.

✓ *Marcato*

LISON DOCG
Friuli Venezia Giulia, Veneto

This small DOCG covering an area of eastern Veneto and western Friuli Venezia Giulia produces a dry white wine made from Friulano (locally Lison). The wines are generally light bodied with floral aromas and pronounced green almond notes. A *classico* subzone is also included within the appellation.

LISON-PRAMAGGIORE DOC
Veneto and Friuli-Venezia Giulia

This DOC, which is in the very east of the Veneto and overlaps a small part of Friuli, originally combined three former DOCs (Cabernet di Pramaggiore, Merlot di Pramaggiore, and Tocai di Lison) into one. It has since been expanded to encompass a total of four white varietal wines (Chardonnay, Pinot Grigio, Sauvignon Blanc, and Verduzzo) and seven reds (Cabernet, Cabernet Franc, Cabernet Sauvignon, Carménère, Malbec, Merlot, and Refosco dal Peduncolo Rosso). The reds are fresh and pleasurable, if not memorable, with Cabernet Franc regularly giving the most interesting results. Some of the best examples of Cabernet and Merlot are still made in a rich, delicious, chocolatey style.

🍷 *3–8 years (red), 1–3 years (white)*

✓ *Santa Margherita • Russola • Tenuta Sant'Anna • Villa Castalda (Cabernet Franc)*

LUGANA DOC
Veneto, Lombardy

This DOC is split into two sections, both at the southern end of Lake Garda, with one part in Veneto and the other in Lombardy. Veneto Lugana is based on Trebbiano di Lugana (Verdicchio) grown around Lake Garda near the town of Desenzano del Garda. Wines have delicate nose of stone fruit and yellow flowers and a lean body with gentle – if any – oak character and a typical almondy finish. Some nutty character from lees work characterise the best examples. *See* Lugana DOC (Lombardy) under Northwest Italy.

✓ *Ottella (Molceo) • Santa Cristina • Cà Maiol • Cà dei Frati • Le Morette*

MARCA TREVIGIANA IGP
Veneto

This IGP covers the entire province of Treviso, and the wines may be made in any one of various styles: red, white, or rosé; blend or varietal; and still, *frizzante*, or sweet. The *frizzante* styles are less common, as most of the wines are dry, still white wines made from any proportions of grape varieties approved for the province of Treviso. French white wine varieties Sauvignon Blanc and Chardonnay are also increasingly used. Red wines sold under the appellation are typically light and made from Merlot, Cabernet Sauvignon, Pinot Noir, and/or local varieties (Marzemino, Raboso, Refosco dal Peduncolo Rosso, and others).

✓ *Sacchetto • Villa Sandi*

MERLARA DOC
Veneto

Established as recently as 2000, this DOC is split in two, with one part in the province of Verona, the other in the province of Padua. The wines include a Friulano-based generic white that may be *frizzante* and a Merlot-based

generic red that may be *novello*. There is a blended Cabernet and various varietal wines: Cabernet Sauvignon, Marzemino (which must be *frizzante*), Merlot, Raboso, and Refosco dal Peduncolo Rosso for reds and Chardonnay, Malvasia, Pinot Bianco, Pinot Grigio, Riesling, and Tai for whites.

MITTERBERG IGP
Trentino-Alto Adige

This IGP is centred on Merano at the northern end of the Valdadige and covers the entire province of Bolzano. Although all grape varieties and types of wine are allowed, Schiava and Lagrein seem to make the most expressive reds, while Pinot Bianco is probably the most consistent for white wines.

✓ *Cantina Vini Merano*

MONTELLO-COLLI ASOLANI DOC
Veneto

From vineyards at the foot of the aptly named Monte Grappa, most of the wines are varietals, comprising six reds (Cabernet, Cabernet Franc, Cabernet Sauvignon, Carménère, Merlot, and Recantina), five whites (Bianchetta, Chardonnay, Manzoni Bianco, Pinot Bianco, and Pinot Grigio), and two *spumante* (Chardonnay or Pinot Bianco). Any red wines that do not carry a varietal name will be primarily Cabernet-Merlot blends. The DOC also includes the subzone Venegazzù, which produces red wines only.

🍷 *1–3 years*

✓ *Fernando Berta • Abbazia di Nervesa*

MONTELLO ROSSO/ MONTELLO DOCG
Veneto

This is a relatively new DOCG that covers the top-quality, red Bordeaux blend-styled wines.

MONTI LESSINI DOC
Veneto

Monti Lessini DOC covers a greater area than just Lessinia (the official name for the area): it stretches eastwards far beyond the bounds of the Lessinia National Park (Parco Regionale della Lessinia). Rather than stopping in line with the Valpolicella DOC's eastern edge, Monti Lessini continues to cover the area immediately north of both Soave and Gambellara. This DOC offers a range of styles from various varieties. Monti Lessini Durello is a traditional white with plenty of body and is suitable for long ageing. Grapes are dried to make honeyed Monti Lessini Durello Passito. Both of these wines must contain 85 per cent Durella. Other grapes that may be included are Garganega, the Pinots Nero, Bianco and Grigio, Sauvignon Blanc, and Chardonnay. A red wine made from Pinot Noir is also allowed.

PIAVE DOC
Veneto

A large area to the west of Lison-Pramaggiore produces four varietal reds (Merlot, Cabernet, Carménère, and Raboso) and four whites (Chardonnay, Manzoni Bianco, Friulano (locally Tai), and Verduzzo). The Cabernet and Raboso, generally fruity and fresh, can be particularly good. A white Verduzzo *passito* and a red Raboso *passito* are also made.

🍷 *1–3 years*

✓ *Rechsteiner*

PIAVE MALANOTTE/ MALANOTTE DEL PIAVE DOCG
Veneto

A small DOCG formerly part of the Piave DOC dedicated to the production of the red Raboso Piave only. Between 15 and 30 per cent of the grapes must be dried, arguably to counteract the grape's tannic, high-acid character. But in the right hands this variety can produce high-quality, long-lived, complex red wines.

PROSECCO DOC
Friuli-Venezia Giulia, Veneto

Prosecco is the wine that created a new way of consumption as a cheaper and more approachable alternative to bottle-fermented sparklings for popular and mainly uninvolved consumers. Produced especially in the extra-dry style, it exudes perfumed pear-drop notes, along with a delicate appley palate. This DOC covers both *spumante* and *frizzante* styles of Prosecco (minimum 85 per cent Glera) produced and bottled in the provinces of Treviso, Vicenza, Venice, Padua, and Belluno in Veneto and Pordenone, Udine, Gorizia, and Trieste in Friuli-Venezia Giulia. This DOC was formed when Prosecco di Conegliano-Valdobbiadene was promoted to DOCG status, and Prosecco IGT was banned (last vintage permissible being 2008). There are two subzones within the DOC: Provincia di Treviso and Provincia di Trieste. The Consorzio of Prosecco producers has, since 2019, authorised a rosé version of Prosecco that can contain up to 15 percent of Pinot Nero in addition to a majority of Glera. Prosecco Rosé can be produced as *millesimato*, therefore stating the harvest year on the label.

✓ *Bosco del Merlo • Paladin*

RAMANDOLO DOCG
Friuli-Venezia Giulia

A sweet late-picked wine that is almost, but not quite, a *passito* from a local clone of Verduzzo Friulano called Ramandolo. The wine, golden in colour, is usually dense with honey-eyed aromas. This particular variety is grown in the foothills of the Alps above Nimis, where the vineyards are so steep that the vines are trained, pruned, and harvested by the same methods as those used at the time of Pope Gregory XII, when Ramandolo was served at the Council of 1409.

✓ *Anna Berra • Dario Coos*

RECIOTO DELLA VALPOLICELLA DOCG
Veneto

Intensely flavored, sweet red wine made from dried (*passito*) grapes in Valpolicella. The grapes allowed in the blend are 45 to 95 per cent Corvina and/or Corvinone, 5 to 30 per cent Rondinella, and a maximum of 25 per cent of other grapes approved by the Valpolicella region (of which no single variety can exceed 10 per cent, and all aromatic varieties combined cannot exceed 10 per cent). A *spumante* version can also be made. The DOCG includes two subzones that can be labelled as such: Classico, comprising the communes of Fumane, Marano, Negrar, San Pietro in Cariano, and Sant'Ambrogio (Recioto), and Valpantena (Recioto and *spumante*).

✓ *Stefano Accordini (Acinatico) • Allegrini • Lorenzo Begali • Tommaso Bussola (TB) • Masi • Quintarelli • Serègo Alighieri (Casel dei Ronchi) • Tedeschi • Trabucchi d'Illasi*

RECIOTO DI GAMBELLARA
Veneto

White sweet wine made from dried (*passito*) grapes in Gambellara displaying notes of ripe fruit and a hint of spice. On the palate they remain lively, fresh, and mineral. Two versions both from 100 per cent Garganega can be made: *spumante* and the so called *classico* (which refers to the style rather than a subzone).

RECIOTO DI SOAVE DOCG
Veneto

Naturally sweet Soave made from *passito* grapes, yielding wines that are either sweet or *spumante* from a minimum of 70 per cent Garganega. The DOCG includes a *classico* subzone, essentially comprising the communes of Soave and Monteforte d'Alpone (Recioto only).

✓ *Pieropan (Le Colombare Vendemmia Tardiva)*

RIVIERA DEL BRENTA DOC
Veneto

From a large appellation inland from the Venetian lagoon, this DOC centres primarily on vineyards along the banks of the river Brenta, where the wines include a Friulano-based generic white that may be *frizzante* and a Merlot-based generic red and *passito* that may be *novello*. There is a blended Cabernet and various varietal wines: Merlot, Raboso, and Refosco dal Peduncolo Rosso, with white varietals of Chardonnay (which must be *frizzante* or *spumante*), Pinot Bianco (which must be *frizzante* or *spumante*), Pinot Grigio, and Tai (Friulano).

ROSAZZO DOCG
Friuli-Venezia Giulia

This small DOCG is dedicated to the production of a white wine made from mainly Friulano and including Pinot Bianco, Sauvignon Blanc, Chardonnay, and a small amount of Ribolla Gialla. The wines, usually of outstanding quality, are very complex and display flavours of woody herbs, spices, and dried fruit, with a characteristic underpinning of minerality.

SANTA MADDALENA or SANKT MAGDALENER DOC
See Alto Adige DOC

SAN MARTINO DELLA BATTAGLIA DOC
Veneto, Lombardia

This DOC produces wines from Friulano grapes grown in one of three separate areas overlooking the southern shores of Lake Garda, two of which are located in Lombardy. This is a dry, full-flavoured white wine with a flowery aroma and a slightly bitter aftertaste, typical of the variety. A sweet *liquoroso* version is also produced.

SOAVE DOC
Veneto

Most Soave is made in an easy-to-drink style, light-bodied and rather neutral; however, an increasing number of producers seek a more ambitious and *terroir*-driven style. Soave DOC may be produced and sold as *spumante*, although a classic Soave is definitely a dry still wine. As with all Soave appellations, the wine must be at least 70 per cent Garganega, to which Trebbiano, Chardonnay, and up to 5 per cent other local varieties may be added. The appellation is divided into two subzones: Soave Colli Scaligeri and Soave Classico. Soave Classico wines are restricted to the central, hilly classic area in the communes of Monteforte and Soave itself, just two of the 13 communes that make up the entire Soave DOC area. Soave Classico is officially superior to Soave DOC but inferior to Soave Superiore DOCG. Of most of the top-quality Soave produced, however, Soave Classico DOC, not Soave Superiore DOCG, usually shines. In addition to the Soave appellations below, *see* Recioto di Soave DOC.

Consortium of Soave is very active in developing a *"cru"* culture. There are now 33 different *crus* from the *classico* part of the appellation that can be mentioned on the label: Castelcerino, Colombara, Froscà, Fittà, Foscarino, Volpare, Tremenalto, Carbonare, Tenda, Corte Durlo, Rugate, Croce, Costalunga, Coste, Zoppega, Menini, Monte Grande, Ca 'del Vento, Castellaro, Pressoni, Broia, Brognoligo, Costalta, Paradiso, Costeggiola, Casarsa, Monte di Colognola, Campagnola, Pigno, Duello, Sengialta, Ponsarà, and Roncà–Monte Calvarina.

🍷 *1–3 years*

✓ *Ca' Rugate (Monte Alto, Monte Fiorentine) • Campi (Campo Vulcano) • Inama (Vigneto du Lot) • Pieropan (Vigneto Calvarino, Vigneto la Rocca) • Prà (Staforte) • Suavia (Le Rive, Monte Carbonare) • Tedeschi • Bertani • Gini • Filippi (Castelcerino)*

SOAVE SUPERIORE DOCG
Veneto

Geographically, this DOC is second in size to the basic Soave DOC only, encompassing 11 of that appellation's 13 communes (Buon Albergo, Caldiero, Cazzano di Tramigna, Colognola ai Colli, Illasi, Lavagno, Monteforte d'Alpone, Roncà, San Martino Buon Albergo, San Giovanni Ilarione, and last but not least, Soave). Soave Superiore DOCG must be dry and aged for a minimum of 10 months (24 months if sold as a *riserva*).

🍷 *1–5 years*

TEROLDEGO ROTALIANO DOC
Trentino-Alto Adige

This DOC is for full-bodied red wines made from the Teroldego in the Rotaliano area, where the grape is said to have originated. There is also a fuller Superiore version and an attractive *rosato*.

🍷 *1–4 years*

✓ *Bolognani • Cavit (Maso Cervara) • Foradori (Vigneto Morei) • Conti Martini • Mezzacorona • Zeni*

TRENTINO DOC
Trentino-Alto Adige

This appellation represents the southern half of the Valdadige DOC, and its wines are generally softer and less racy than those from Alto Adige to the north, although there is an equally bewildering number of varietal wines. If no variety is shown, white wines will be Chardonnay–Pinot Blanc blends and reds Cabernet-Merlot. There are 11 red wine varietals (Cabernet, Cabernet Franc, Cabernet Sauvignon, Lagrein, Marzemino, Merlot, Moscato Rosa, Rebo, Schiava, Schiava Gentile, and Pinot Nero); and 12 white (Chardonnay, Kerner, Manzoni Bianco, Moscato Giallo, Müller-Thurgau, Nosiola, Pinot Bianco, Pinot Grigio, Riesling Italico, Riesling Renano, Sauvignon Blanc, and Traminer Aromatico). Furthermore, the Nosiola can be made in *vin santo* style, while there is also a bright scarlet, lusciously sweet Moscato Rosa, and both Moscato Rosa and Moscato Giallo may be made in *liquoroso* style.

Sorni is a sub-appellation for wines just south of Mezzolombardo at the confluence of the rivers Avisio and Adige. Here we find soft Schiava-based reds are often improved by the addition of Teroldego and Lagrein. Light, fresh, delicate Nosiola-based white wines are usually charged with a dollop of Müller-Thurgau, Silvaner, and Pinot Blanc. Other four subzones can be labelled Castel Beseno or Beseno (Moscato Giallo Superiore, Passito, and *vendemmia tardiva* only), Isera (Marzemino Superiore only), Valle di Cembra or Cembra (Müller-Thurgau Superiore, Riesling Renano Superiore, Pinot Nero Superiore, Schiava Superiore, and Schiava Gentile Superiore only) and Ziresi (Marzemino Superiore only).

✓ *Barone de Cles • Càvit • Cesconi • LaVis • Longariva • Madonna del Vittoria • Conti Martini • Maso Poli • Pojer & Sandri • Giovanni Poli • Tenuta San Leonardo • Armando Simoncelli • Sorini • de Tarczal • Vallarom • Zeni*

TRENTO DOC
Trentino-Alto Adige

The only traditional-method sparkling wine appellation to be created since Franciacorta DOCG, Trento must be made solely with Chardonnay, Pinot Bianco, Pinot Meunier, or Pinot Nero grapes, with at least 15 months on lees for non-vintage, 24 months for vintage, and 36 months for riserva. Compared to Franciacorta, Trento DOC wines have more upright character and a more assertive *perlage*.

✓ *Altemasi • Bellaver • Cantina Aldeno • Cantina d'Isera • Cantina Toblino • Cavit • Ferrari • Letrari • Maso Martis • Pisoni • Rotaliana • Rotari*

TREVENEZIE IGP
Friuli-Venezia Giulia, Trentino, Veneto

This geographic indication was once called IGP Delle Venezie and was renamed in 2017, when a new denomination primarily for Pinot Grigio was created as Delle Venezie DOC. Wines are made in a variety of styles from the entire regions of Friuli-Venezia Giulia and Veneto, plus the entire province of Trento in Trentino. Still or *frizzante* white, *rosato*, and red are produced from any proportions of white grape varieties approved for the production area. Varietals are equally allowed.

✓ *Pirovano • Dario Coos*

VALDADIGE or ETSCHTALER DOC
Trentino-Alto Adige and Veneto

This large denomination encompasses both the Alto Adige and Trentino DOCs and extends well into the Veneto. Both blended wines and varietals are made under this appellation, as well as semi-sparkling *frizzante* wines made from either Pinot Bianco or Chardonnay. The varietals are also made from Pinot Grigio and the flagship red grape, Schiava, which delivers light and soft, fruit-driven wines. The generic *bianco* is made from a minimum of 20 per cent Chardonnay, Müller-Thurgau, Pinot Bianco, Pinot Grigio, and/or Welschriesling and a maximum of 80 per cent Garganega, Nosiola, Sauvignon Blanc, and/or Trebbiano Toscano. The generic *rosso* must account for a minimum of 50 per cent Lambrusco a Foglia Frastagliata and/or Schiava. A Lambrusco a Foglia Frastagliata-based *rosato* wine is also produced.

VALDADIGE TERRADEIFORTI/ TERRADEIFORTI DOC
Trentino-Alto Adige and Veneto

Three varietal wines from vines growing on the southeastern corner of the Valdadige, where it overlaps the Veneto region are allowed: Pinot Grigio, Enantio (a local synonym for Lambrusco, a Foglia Frastagliata, with fuller, softer, spicier fruit than might otherwise be expected from any Lambrusco), and Casetta (aka Foja Tonda, an obscure black-skinned grape rescued from possible extinction by Albino Armani – an interesting variety with a tannic bite).

✓ *Albino Armani (Foja Tonda) • Roeno (Enantio)*

VALLAGARINA IGP
Trentino-Alto Adige and Veneto

This IGP covers the extreme southern tip of the Valdadige, including a defined area in the province of Trento and Verona. For Vallagarina *bianco, rosso,* and *rosato* blends, winemakers can use any of the varieties authorized in their respective province. For varietal and blended wines, Trento producers have similar flexibility. Their neighbors in Verona, however, have "only" around 20 grape varieties to work with. Varietal *spumante* wines may be made from 85 percent or more of Chardonnay, Müller-Thurgau, Pinot Bianco, or Pinot Nero. These varieties plus Pinot Grigio may be combined in any proportions in nonvarietal sparkling wines.

✓ *Tenuta San Leonardo*

VALPOLICELLA DOC
Veneto

Valpolicella wines are always blends from 45 to 95 per cent Corvina and/or Corvinone, with the possible addition of 5 to 30 per cent of Rondinella and 25 of other authorised grape varieties. There are an increasing number of wines that are full of juicy cherry-fruit flavours and have light body and supple tannins balanced by a refreshing acidity. Two subzones are recognised by the DOC: *classico*, which covers the historical vineyards, and Valpantena for the vineyards around Quinto in the centre of the Valpolicella district. A Superiore version is also made with minimum one-year ageing requirement.

🍷 *Upon purchase for most, 2–5 years for recommendations*

✓ *Allegrini (La Grola) • Bertani • Bolla (Vigneti di Jago) • Corte Sant'Alda (Mithas) • Grassi (Superiore) • Guerrieri-Rizzardi (Villa Rizzardi Poiega) • Marion (Superiore) • Quintarelli • Le Ragose • Romano dal Forno (Lodoletta) • Serègo Alighieri • Tedeschi • Tommasi (Vigneto del Campo Rafael) • Viviani (Campo Morar) • Monte dall'Ora*

VALPOLICELLA RIPASSO DOC
Veneto

Ripasso ("re-passed") wine has long been traditional in the Valpolicella. The best young Valpolicella is put into tanks that still contain the lees of Recioto or Amarone for which they were previously used. As a second fermentation occurs, the result is a wine that combines the freshness and the brightness of a dry Valpolicella with riper sour cherry–

GRAPES DRY ON WOOD RACKS IN A VENETO DRYING HOUSE
In a process call *appassimento,* harvested grapes are dried prior to fermentation to concentrate sugars and influence their flavour. The *appassimento* method is employed to make Amarone, Recioto, Valpolicella Ripasso, and Sforzato.

fruit profile and soft, Recioto-like tannins. For more than a decade, Ripasso has represented an important category in several key export markets, such as Scandinavian monopolies and Canada. As for Valpolicella DOC, two subzones can be mentioned on the label (Classico and Valpantena), and the same grapes in the same proportions are used.

🍷 *6–15 years*

✓ *Bertani (Catullo) • Boscaini (Le Cane) • Tedeschi (Capitel San Rocco) • Zenato (Ripassa)*

VENETO IGP
Veneto

This area encompasses all of Veneto, covering a broad range of wine styles thst can be made from any proportions of grape varieties approved for the region of Veneto or as varietals. The wines are most often based on such varieties as Pinot Grigio, Garganega, and Corvina – the traditional and most widely used grape varieties of the Veneto region. The relatively liberal rules that govern the production of IGP wines mean that a number of non-native grapes are also used, most notably the Bordeaux varieties Cabernet Sauvignon, Cabernet Franc, Carménère, and Merlot.

✓ *Anselme • Bortolomiol • Bosco del Merlo • Coffele • Costadoro • Inama • Maculan • Sacchetto • Villa Sandi • Zenato*

VENETO ORIENTALE IGP
Veneto

This IGP covers the eastern half of the province of Venezia, plus the communes of Meduna di Livenza and Motta di Livenza in the province of Treviso, and produces a variety of styles made from any grape varieties approved.

✓ *Santa Margherita*

VENEZIA DOC
Veneto

This large DOC produces a wide range of styles. A generic white Friulano-based, a Raboso-based *rosato*, and a Merlot-based red. Several varietal wines are produced: Chardonnay (still or *frizzante*), Manzoni Bianco, Pinot (still or *frizzante*, a minimum of 85 per cent Pinot Bianco, Pinot Grigio, and/or Pinot Nero, Pinot Bianco, Pinot Grigio (still or *frizzante*), Sauvignon Blanc, Tai (Friulano), Traminer. and Verduzzo for whites and Cabernet, Cabernet Franc, Cabernet Sauvignon, Carménère, Malbec, Merlot, Pinot Nero, and Refosco dal Peduncolo Rosso for reds. A *riserva* version is also made (can contain up to 30 per cent dried grapes) with minimum 24-month ageing requirement. Dual varietals are also allowed. A wide range of sparkling

wines is also produced (both white and *rosato*), as well as a white *passito* blend (Friulano, Glera, and/or Verduzzo) and a varietal Verduzzo *passito*.

VENEZIA GIULIA IGP
Friuli-Venezia Giulia

All locally grown grape varieties and styles apply. Exceptionally fresh and vividly fruity styles from the best producers.

✓ *Jermann • San Simon • Schiopetto • Zidarich*

VERONA/PROVINCIA DI VERONA/VERONESE IGP
Veneto

All varieties and styles abound, but some of the best varietals include Cabernet Sauvignon, Corvina, Garganega, Merlot, and Rondinella.

✓ *Allegrini • Musella • Sartori*

VICENZA DOC
Veneto

The wines include a Garganega-based generic white that may be still, *frizzante*, or *spumante* and dry or sweet and a Merlot-based generic red or *rosato*. Varietal wines are also produced, including seven whites (Chardonnay, Garganego, Manzoni Bianco, Pinot Grigio, Riesling, and Sauvignon Blanc) and five reds (Cabernet, Cabernet Sauvignon, Merlot, Pinot Nero, and Raboso), which can also be labelled as *riserva* if aged for a minimum of two years).

✓ *La Berolà*

VIGNETI DELLA SERENISSIMA/ SERENISSIMA DOC
Veneto

A small appellation dedicated to the production of traditional-method sparkling wines (white or *rosato*) made from any proportions of Chardonnay, Pinot Bianco, and/or Pinot Nero. A *millesimato* or vintage (with minimum 24 months on lees) and a *riserva* (minimum 36 months on lees) are also produced.

VIGNETI DELLE DOLOMITI IGP
Trentino-Alto Adige and Veneto

This IGP covers the foothills of the Dolomites, either side of the border between Trentino–Alto Adige and Veneto and produces a wide range of styles by any grape varieties approved for the relevant province.

✓ *Endrizzi • Foradori • MezzaCorona • Pojer & Sandri • Tenuta San Leonardo*

West Central Italy

The heart of Italy is also the centre of the country's most important quality-wine exports, which are dominated by famous red Sangiovese wines from the rolling Tuscan hills.

Together with classic areas of Chianti and Brunello di Montalcino, Tuscany gave birth to a crop of rising stars from Bordeaux blends, called Super Tuscans, nowadays among the most sought-after wines of the world. Umbria's dramatic and multicoloured landscape is capable of variegated vinous expressions: lively Sangiovese, structured Sagrantino, and luscious botrytised dessert wines. Latium, with its volcanic soils has a focus on light and lifted white wines, mainly consumed in-situ.

TUSCANY (TOSCANA)

The home of traditional winemaking, Tuscany has also been the main focus of experimentation. Its powerful reds, Brunello di Montalcino and Vino Nobile di Montepulciano, were Italy's first DOCGs, in 1980, followed by Chianti, Carmignano, and Vernaccia di San Gimignano. But not all of its finest wines bear these famous appellations, a fact recognized by the Tuscan producers themselves, who, on the one hand, sought the ideal DOCG solution for Chianti, while on the other began to invest in premium wines that were not restricted by the DOC. It was the uncompromising quality of their Super Tuscan wines that encouraged premium *vini da tavola* throughout the rest of Italy. Tuscany is mainly a red wine region, with most of its top-end appellations based on the Sangiovese variety, the most extensively cultivated variety in Italy, that here gives its best. Chianti, Brunello di Montalcino, Nobile di Montepulciano, Carmignano, and many

REGIONS AT A GLANCE

REGION	TOTAL AREA UNDER VINE Hectares (Acres)	IGP/DOC/DOCG AREA UNDER VINE Hectares (Acres)	IGP/DOC/DOCG PRODUCTION Hectolitres	TOTAL PRODUCTION Hectolitres
Tuscany	60,513 (150,000)	57,650 (142,450)	2.2. million (86-14 red/white)	2.8 million (31 million cases)
Umbria	12,495 (30,900)	9,729 (24,040)	260,680 (20-80 red/white)	765,000 (0.85 million cases)
Latium	18,200 (45,000)	11,772 (29,100)	512,000 (5-95 red/white)	1,7 million (18.8 million cases)

Source: Osservatorio del Vino UIV - ISMEA

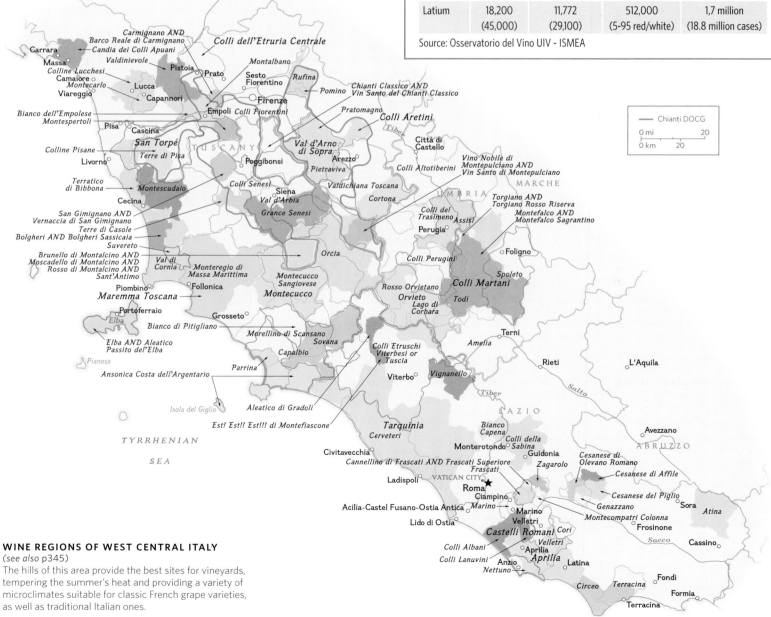

WINE REGIONS OF WEST CENTRAL ITALY
(see also p345)

The hills of this area provide the best sites for vineyards, tempering the summer's heat and providing a variety of microclimates suitable for classic French grape varieties, as well as traditional Italian ones.

VINEYARDS OF THE VINO NOBILE DI MONTEPULCIANO DOCG SPREAD OUT BELOW THE TOWN OF MONTEPULCIANO
Known worldwide for the wines produced there, the lands surrounding the medieval hill town of Montepulciano in the province of Siena in southern Tuscany combine location, climate, aspect, and soil to provide fertile ground for wine grapes. The town also gives its name to a red wine grape cultivated throughout Central and Southern Italy.

other DOCs are all different expressions of Sangiovese, exploiting the regions' dramatic differences in soil, climate, and orography. The experimentation with French varieties, however, led also to the creation of the rising star of the region coastal area, Bolgheri DOC. Here, the mean temperature during the growing season in slightly higher compared to Bordeaux, yielding powerful yet elegantly done Cabernets and Merlot blends. Currently, Tuscany has no white-wine sister for its red Sangiovese grape. Together with Chardonnay, grown in ambitious style throughout the region, Vermentino is a variety that better adapts to the Tuscan climate, especially when planted on the sandier soils of the Tyrrhenian Coast. Yet at the moment, there is only one DOCG for white grapes, Vernaccia di San Gimignano, that produces savoury textured dry whites, often referred to as disguised reds.

Birth of the Super Tuscans

Most of the exceptional Tuscan wines 40 years ago were the then relatively new *barrique*-aged Super Tuscans. Their story began in 1948 when the now-famous Sassicaia wine was made for the first time by Incisa della Rochetta using Cabernet Sauvignon vines reputedly from Château Lafite-Rothschild. This was an unashamed attempt to produce a top-quality Italian wine from Bordeaux's greatest grape variety, decades before the idea became old hat in the wine world. It became so successful that in the wake of the 1971 vintage, a new red called Tignanello was introduced by Piero Antinori with a Sangiovese base and 20 per cent Cabernet Sauvignon, as a compromise between Tuscany and Bordeaux. Although Frescobaldi had used Cabernet Sauvignon in its Nipozzano Chianti for over a century, and it grew in the Carmignano area in the 18th century, nobody had truly appreciated the harmony that could be achieved between the two grapes until Tignanello appeared. The blend was akin to the natural balance of Cabernet and Merlot, only the Cabernet added weight to the Sangiovese and provided balance through a more satisfying flavour. Tignanello thus sparked off a new wave of Super Tuscan *vini da tavola*. Yet as the numbers

grew, so they became an embarrassment, as observers realized that very few of the region's most celebrated wines qualified for DOC. At the same time, however, many of the winemakers responsible for these French-influenced Super Tuscans were also working hard to make the Sangiovese stand alone. After extensive clonal and site selection, reduced yields, and improved viticultural practices and vinification techniques, a new breed of Super Tuscan emerged, first as Sangiovese-dominated blends, such as Tignanello, and then as pure Sangiovese wines.

UMBRIA

Once known almost solely for its Orvieto, which yielded golden and gently sweet whites, Umbria is a region that is capable of producing a vast array of high-quality wines. Its signature grape, Sagrantino, provides oak-aged and concentrated, impressively tannic reds that can be also produced as *passito* and labelled under Montefalco Sagrantino DOCG. Sangiovese also plays a paramount role in Umbria, especially in the Torgiano area, where it gained the DOC status in 1968. The area surrounding Lago di Corbara (Lake Corbara) on the border with Latium, is gaining traction due to reputable wineries producing dry white Grechetto, red French varietals, and luscious botrytised wine from Grechetto and Sauvignon Blanc.

LATIUM (LAZIO)

One of Italy's largest regions, with a staggering number of appellations, Latium has a high incidence of volcanic soils that make it particularly suited to vine-growing. Historically, its wine production was reserved for Italy's capital city, Rome, and consequently was crafted for the taste of Romans. Malvasia, with its soft and gently aromatic taste, is the base for Latium's most famous wine, Frascati. The region is also capable of producing red autochthonous varieties, such as the savoury and spicy, red-fruited Cesanese in the inland foothills. Latium can boast of several examples of high-end *vino da tavola* produced from Cabernet and Merlot with a reputation for age-worthiness and deliciousness, too.

ALEATICO DI GRADOLI DOC
Latium

Sweet, aromatic red from Aleatico grapes growing on the northern shore of the crater lake of Bolsena. *Vino liquoroso* (fortified version) is also allowed with up to 15 per cent ABV.

ALLERONA IGP
Umbria

Mostly Grechetto from an area encompassing Orvieto. Wines made from any proportion of grape varieties approved for the region of Umbria can be made in both dry and sweet styles.

ALTA VALLE DELLA GREVE IGP
Tuscany

This IGP is centred on the vineyard's around Greve in the Chianti district, and Poggio Scalette's dark, dense, and imposing *barrique*-aged Sangiovese called Il Carbonaione is a wine that some rank as high as the very best Chianti itself.

✓ *Poggio Scalette*

AMELIA DOC
Umbria

This appellation is found in the southwestern corner of Umbria, in a hilly area far more famous for olive oil than for wine. The wines include a generic Trebbiano-based white, two varietal whites (Grechetto and Malvasia), a generic Sangiovese-based red, and three varietal reds (Ciliegiolo, Merlot, and Sangiovese). A *riserva* must be red and aged for a minimum of 24 months prior to release. A *novello* red is allowed. The DOC also includes a *vin santo* (which may be dry or sweet).

✓ *Cantina dei Colli Amerini (Ameroe, Carbio)*

ANAGNI IGP
Latium

This is a 1-hectare (2.5-acre) appellation covers the township of Anagni in the Frosinone province and produces two wines: a white made from a minimum of 50 per cent Malvasia del Lazio (locally Malvasia Puntinata) and/or Passerina and a maximum of 35 per cent Bellone, Chardonnay, Grechetto, and/or Manzoni Bianco and a red wine made from a minimum of 50 per cent Cabernet Franc and/or Cabernet Sauvignon, a maximum of 25 per cent Merlot, and 10 to 20 per cent Cesanese di Affile.

ANSONICA COSTA DELL'ARGENTARIO DOC
Tuscany

Theoretically the same variety as Sicilian Inzolia. In southern Tuscany, bordering Latium, the localized version of this grape produces a distinctive, savoury, and characterful wine, albeit with little finesse.

APRILIA DOC
Latium

This DOC produces a generic Sangiovese-based red and one red varietal, Merlot, and one white, Trebbiano, which is uninspiring.

ASSISI DOC
Umbria

In the hills surrounding Assisi, birthplace of St Francis of Assisi, this DOC encompasses a Trebbiano-based generic white, a varietal white Grechetto, a Sangiovese-based generic red, and three varietal reds (Cabernet Sauvignon, Merlot, and Pinot Nero). The *riserva* is a blend of any proportion of Cabernet Sauvignon, Merlot, and Pinot Nero, with a minimum 24 months ageing required before release.

✓ *Sportoletti*

ATINA DOC
Latium

A mainly red wine DOC located in the hills between Sora (which famously fought off Hannibal) and Cassino (where the Germans fiercely defended the Monte Cassino during World War II). Just two wines are allowed: a generic red (a minimum of 50 per cent Cabernet Sauvignon, 10 per cent Cabernet Franc, 10 per cent Merlot, 10 per cent Syrah, and a maximum of 20 per cent other local non-aromatic varieties) and a blended Cabernet that must contain at least 85 per cent Cabernet Sauvignon and/or Cabernet Franc. For whites only a varietal Semillon is allowed.

BARCO REALE DI CARMIGNANO DOC
Tuscany

After Carmignano became a DOCG, this appellation was adopted for easy-drinking, Sangiovese-dominated (with a touch of Cabernet) red wines in order to retain simple DOC status. This acted as a selection instrument to increase and maintain quality for the superior denomination. Since 2013 the DOC includes both red and *rosato* styles, while the sweet wines come under the Vin Santo di Carmignano DOC. Comparing Barco Reale and Rosato with Carmignano DOCG, the blend for all is the same: at least 50 per cent Sangiovese, a mandatory dose of Cabernet Franc and/or Cabernet Sauvignon (between 10 and 20 percent), and potentially small amounts of other grapes. Barco Reale and Rosato, however, require only 11 per cent alcohol and have no minimum ageing requirements. *See also* Carmignano DOCG.

✓ *Capezzana*

BETTONA IGP
Umbria

This is a seldom-used IGP in the Orvieto region.

BIANCO CAPENA DOC
Latium

This blended white wine is produced between Monterotondo and Tivoli, northeast of Rome, from at least 55 per cent Malvasia, 25 per cent Trebbiano, Romagnolo, or Giallo, and up to 20 per cent Bellone and Bombino.

BIANCO DELL'EMPOLESE DOC
Tuscany

This is a rather neutral dry white Trebbiano-based wine from the Empoli hills, west of Florence. A *passito* may be sold as *vin santo* under the same denomination. For both styles a minimum of 60 per cent Trebbiano is required.

BIANCO DI PITIGLIANO DOC
Tuscany

Delicate, refreshing, dry, and easy-drinking Trebbiano-based (minimum 40 per cent) white wines, which are usually blended with Ansonica (maximum 60 per cent) and improved by the possible inclusion of (for a maximum of 15 per cent of the blend) Malvasia, Grechetto, Verdello, Chardonnay, Sauvignon Blanc, Pinot Blanc, Viognier, and Welschriesling. A *spumante* version and *vin santo* are also allowed.

🍷 *1-2 years*

✓ *La Stellata*

BOLGHERI DOC
Tuscany

This was once a pleasant but hardly exciting DOC established in 1983, producing delicate, dry whites and dry, slightly scented, Sangiovese *rosato*. The red wines of the area, including Sassicaia, were not recognized by the denomination, and they continued to qualify for only *vino da tavola* status until 1994 when the DOC rules were finally revised to encompass Cabernet-based red wines, and Sassicaia was given special status and later established as a DOC in its own right (*see* Bolgheri Sassicaia DOC). The Bolgheri DOC today includes the following styles: white, which can be blended Bianco or varietal Sauvignon Blanc or Vermentino. The *rosato* and red wines can contain up to 100 per cent Cabernet Franc, Cabernet Sauvignon, or Merlot, alone or in combination, along with up to 50 per cent Sangiovese and/or Syrah. The reds must be aged for about a year, and if they are aged for two years with at least a year in oak, they can be labelled Superiore. Some of Italy's most renowned vineyards lie within this DOC, making Bolgheri a mecca for wine lovers.

✓ *Antinori • Castello di Bolgheri • Gaja (Ca' Marcanda) • Le Macchiole • Grattamacco • Podere Sapaio • Enrico Santini • Michele Satta • Tenuta dell'Ornellaia • Tenuta San Guido*

BOLGHERI SASSICAIA DOC
Tuscany

The modern history of Bolgheri began at the end of World War II when Marchese Mario Incisa got the idea to plant Cabernet Sauvignon grapes on his estate, in an area otherwise known only for rustic Sangiovese and Trebbiano wines. Liking the results, he went on to plant his Sassicaia vineyard with Cabernet in the early 1960s. Teaming up with his brother-in-law Nicolò Antinori and winemaker Giacomo Tachis, Incisa improved his Sassicaia wines during the 1970s, and they gained international fame, putting Bolgheri on the map. Sassicaia's reputation attracted other winemakers who were interested in exploring the possibilities Incisa had demonstrated and making great wines primarily with French grapes. in 1994 Sassicaia was given special status as a subzone of the Bolgheri DOC (*See* Bolgheri DOC) and in 2013 was established as a separate DOC.

✓ *Tenuta San Guido*

BRUNELLO DI MONTALCINO DOCG
Tuscany

One of Italy's most prestigious wines, made from "Sangiovese Grosso", a localized clone of Sangiovese. Established as a DOC in 1966, it became a DOCG in 1980. Brunello is Montalcino's statement to making wine without the help of foreign varieties. The Montalcino area is a hill with a patchwork of ancient soils and offers a riper and more powerful style compared to Chianti Classico. Naturally firm tannins of Sangiovese are resolved through at least 24 months in oak, which can be French or Slavonian. Wines are released the fifth year after the harvest and are reputed to be extremely age-worthy, while having become significantly approachable in their earlier years in the last few decades. As a top-end appellation, Montalcino is extremely parcelised, counting more than 220 producers for 1,500 hectares (3,700 acres) of vineyards surface. The notable variability of soils, aspect, and altitude, ranging from 200 up to 600 metres (655 to 1,970 feet), results in a wide range of expressions for Brunello wines. Wines from the northern area usually have a more vibrant and upright character and may be quite austere in cooler vintages. Wines produced in the southern slope are naturally riper and fuller, yet they may suffer from excessive heat. For that reason, many producers have vineyards all around the hill of Montalcino, enabling them to produce a solid and consistent style. In the last few years, a strong focus on single varietal wines – that have more restricted yields – has evidenced the difference among the *terroirs* of the appellation. Brunello di Montalcino Riserva must be released the sixth year following the harvest, and it is prevalently produced in exceptional years. Brunello has a garnet colour and complex nose of spices and red berries, coupled with a notable tension between the fruit weight, lively acidity, and firm tannins.

✓ *Altesino • Argiano • Castello Banfi • Fattoria dei Barbi • Biondi Santi • Campogiovanni • Canalicchio di Sopra*

FACTORS AFFECTING TASTE AND QUALITY

LOCATION
Located between the Apennines to the north and east and the Tyrrhenian Sea to the west.

CLIMATE
Summers are long and fairly dry, and winters are less severe than in northern Italy. Heat and lack of rain can be a problem throughout the area during the growing season.

ASPECT
Vineyards are usually sited on hillsides for good drainage and exposure to the sun. Deliberate use is made of altitude to offset the heat, with grapes grown at up to 700 metres (2,275 feet), as well as on flatlands at sea level in coastal areas.

SOIL
These are very complex soils with gravel, limestone, and clay outcrops predominating. In Tuscany a rocky, schistose soil, known in some localities as alberese, and schist in galestro structure covers most of the best vineyards of Chianti Classico and Montalcino. These poor soils infuse Sangiovese top-end wines with red cherry, blood orange aromas and gamey complexity, along with thick angular tannic structure, giving great age-worthiness. More alluvial soils and a mix of clay, sand, and loam, with high variability of composition, are typical of coastal areas such as Bolgheri, aspects that help complete phenolic maturation of Bordelaise grape varieties. The tuffaceous soil of San Gimignano and Pitigliano results in light and fruity white wines with a savoury palate. Umbria is known for high-clay soils that characterise the firm tannic level of Sagrantino and the more phenolic profile in its white Orvieto. Latium has a large array of soils, with more sand and loam in the coastal part, more clayish inland, and with volcanic origin in the Frascati area.

VITICULTURE & VINIFICATION
After much experimentation with classic French grapes, particularly in Tuscany, the trend recently has been to develop the full potential of native varieties. Many stunning Cabernet-influenced Super Tuscan wines exist, but top-performing producers are seeking clones, terroirs, and techniques to maximize the fruit and accessibility of their own noble grapes. A traditional speciality is the sweet, white vin santo, which is made from passito grapes dried on straw mats in attics. It is aged for up to six years, often in a type of solera system. Moreover, botrytis-affected white grapes grown in proximity to bodies of water are used to produce classic luscious dessert wines.

GRAPE VARIETIES
Primary varieties: Brunello/Morellino/Prugnolo/Sangioveto/Tignolo/Uva Canina (Sangiovese) Malvasia, Procanico (Ugni Blanc)
Secondary varieties: Abbuoto, Aglianico, Albarola, Aleatico, Ansonica (Inzolia), Barbera, Bombino, Cabernet Franc, Cabernet Sauvignon, Cesanese, Chardonnay, Ciliegiolo, Colorino, Drupeggio in Umbria (Canaiolo), Gamay, Greco (Grechetto), Albana, Mammolo, Merlot, Montepulciano, Moscadello/Moscato (Muscat Blanc à Petits Grains), Pinot Bianco (Pinot Blanc), Pinot Grigio (Pinot Gris), Pulciano (Grechetto), Roussanne, Sagrantino, Sauvignon Blanc, Sémillon, Syrah, Uva di Spagna (Carignan), Verdello, Vermentino, Vernaccia

---

• Capanna • Caparzo • Caprili • Casanova di Neri • Case Basse-Soldera • Castelgiocondo • Sesti • Castello Romitorio • Cerbaiona • Col d'Orcia • Costanti • Fuligni • Cupano • Lisini • Mastrojanni • Ciacci Piccolomini d'Aragona • Baricci • Salicutti • Poggio Antico • Salvioni • Il Poggione • Tenuta Silvio Nardi • Siro Pacenti • Talenti • Uccelliera • Val di Suga • Castiglion del Bosco • Cava d'Onice • Le Ragnaie • San Polo • Valdicava • Cortonesi

CANDIA DEI COLLI APUANI DOC
Tuscany

Delicate, slightly aromatic, dry or semi-sweet whites from Vermentino and Albarola grapes. The DOC also includes rosato wines made from Vermentino Nero or are Sangiovese based, and red wines are made from Barsaglina, Sangiovese, and Vermentino Nero.

CANNARA IGP
Umbria

This IGP covers the floodplain of Perugia, celebrated more for its onions than wines.

CANNELLINO DI FRASCATI DOCG
Latium

Formerly part of the Frascati DOC, it was established as a separate DOCG in 2011. It produces sweet wines (late harvest) only from a minimum of 70 per cent Malvasia Bianca di Candia or Malvasia del Lazio (locally Malvasia Puntinata) and a maximum of 30 per cent Bellone, Bombino Bianco, Greco, and/or Trebbiano.

CAPALBIO DOC
Tuscany

This appellation covers the rolling hillsides of southern Grosseto, encompassing Parrina and large parts of four other DOCs (Ansonica Costa dell'Argentario, Bianco di Pitigliano, Morellino di Scansano, and Sovana). The wines include a generic Trebbiano-based white, a generic Sangiovese-based red and passito, a vin santo (which may be dry or sweet), one varietal white (Vermentino), and two varietal reds (Cabernet Sauvignon and Sangiovese). A riserva must be red and aged for a minimum of 30 months prior to release.

CARMIGNANO DOCG
Tuscany

Historic red wine made from 16 kilometres (10 miles) northwest of Florence in a zone noted as one of Tuscany's finest for red wine production since the Middle Ages. Carmignano was amongst the earliest protected denominations in the world, being named by Grand Duke Cosimo III de' Medici – along with Chianti, Pomino, and Valdarno di Sopra – as a quality wine region. In modern times, Carmignano was at first incorporated under the Chianti banner in the Montalbano subzone, but in large measure due to the efforts of Count Ugo Contini Bonacossi of Tenuta di Capezzana, it received separate DOC status in 1975 and was elevated to DOCG in 1991 (from the 1988 vintage). The wines must be a minimum of 50 per cent Sangiovese blended with up to 20 per cent Cabernet Franc or Cabernet Sauvignon. Other Tuscan indigenous varieties are allowed in the blend (up to 20 per cent Canaiolo Nero and up to 10 per cent white varieties also allowed). The wines must have a minimum alcohol level of 12.5 per cent and at least a year and a half of ageing (including 8 months in barrel) – or, for riserva, 3 years (including 12 months in barrel). The result is wines of similar structure to Chianti, with less acidity and a chocolaty finesse to the fruit given by the Cabernet content.

🍷 4-10 years

✓ Fattoria di Ambra • Tenuta di Artimino • Fattoria di Bacchereto • Contini Bonacossi (Villa di Capezzana, Villa di Trefiano) • Fattoria Il Poggiolo • Piaggia • Fattoria di Bacchereto (Terre a Mano)

CASTELLI ROMANI DOC
Latium

Covering a vast area of volcanic hillsides south and southeast of Rome, Castelli Romani encompasses no fewer than nine other DOCs, including Frascati. One generic white is produced from a minimum of 70 per cent Malvasia Bianca di Candia, Malvasia del Lazio (locally Malvasia Puntinata), and/or Trebbiano (di Soave, Giallo, or Toscano); one rosato from any proportions of Cesanese, Malvasia, Merlot, Montepulciano, Nero Buono, Sangiovese, and/or Trebbiano; and one red from a minimum of 85 per cent Cesanese, Merlot, Montepulciano, Nero Buono, and/or Sangiovese.

CERVETERI DOC
Latium

Rustic Sangiovese/Montepulciano-based reds and Trebbiano-Malvasia whites of decent, everyday quality.

CESANESE DEL PIGLIO or PIGLIO DOCG
Latium

This is a red-only DOCG producing one red wine from a minimum of 90 per cent Cesanese. A Superiore version (minimum 13 per cent ABV and 18 months ageing) and Superiore Riserva (minimum 20 months ageing period) version are also made.

CESANESE DI AFFILE or AFFILE DOC
Latium

This DOC produces the same styles as the Cesanese del Piglio DOC from a neighbouring area with the addition of a red sweet version.

CESANESE DI OLEVANO ROMANO or OLEVANO ROMANO DOC
Latium

This is a much smaller appellation than the previous two nearby Cesanese DOC and DOCG, but it covers the same styles (both dry and sweet).

CHIANTI DOCG
Tuscany

Arguably Italy's most famous red wine, based on the Sangiovese grape. Chianti was established as a DOC in 1967 and became a DOCG in 1984; however, its guidelines for production date back to 1716, making it the first ever appellation in history. The best basic Chianti wines are full of juicy cherry, raspberry, and plummy fruit flavours, which makes an enjoyable quaffer, also given their likely lower oak impact, compared to Chianti Classico. Chianti DOCG comprises seven sub-appellations, covering the peripheral areas surrounding Chianti Classico: Colli Aretini, Colli Fiorentini, Colline Pisane, Colli Senesi, Montalbano, Montespertoli, and Rufina. Rufina (from a high-elevation area northeast of Florence, which should not be confused with Ruffino, the brand name) and Colli Fiorentini (which bridges the Classico and Rufina areas) are capable of producing *classico*-like quality. Colli Senesi is the largest and most varied, with very few wines claiming this provenance, two portions of which are better known for Brunello di Montalcino and Vino Nobile di Montepulciano. Colline Pisani is lightest of all Chianti, and Colli Aretini is a young, lively one. Montalbano is in effect the second wine of Carmignano. The traditional governo process of enriching the wine by allowing it to finish fermenting in contact with unfermented dried grapes is still permitted. All Chianti must contain at least 70 per cent Sangiovese, plus the possibility of a maximum of 15 per cent Cabernet, in addition to traditional Canaiolo Nero, Trebbiano, or Malvasia or any other specified black grape varieties.

🍷 *3–5 years (inexpensive, everyday drinking), 4–8 years (more serious Chianti)*

✔ *Poggio Bonelli • Bindi Sergardi • Cecchi • Peraccio • Ormanni • Barone Ricasoli • Antinori • Frescobaldi • Selvapiana • Colognole • Pietro Beconcini • CS Viticoltori Senesi-Aretini • Fattoria di Poggiopiano*

CHIANTI CLASSICO DOCG
Tuscany

Established as a subzone of the Chianti DOC in 1967, Chianti Classico became a separate DOCG in 1996. Compared to Chianti DOCG, Chianti Classico accounts for a mere fraction of vineyard area and represents the original classic area of production. Historically Chianti was required to have a small proportion of white grapes in its blend, usually from Malvasia or Trebbiano. Nowadays a minimum of 80 per cent Sangiovese is required, the rest can be made of other grape varieties authorised in Tuscany. Starting from vintage 2010, the "Gran Selezione" category was created for Chianti Classico. Gran Selezione is a sort of enhanced *riserva* status that must be produced uniquely from estate-grown grapes and aged for at least 30 months. Gran Selezione was initially designed to get the same reputation of the Super Tuscans and today accounts for a minute proportion of Chianti Classico sales. Tastewise, Chianti is ruby red, tending to garnet and often exudes violet flower and crunchy red-berried fruit. Mouthfeel is agile, thanks to

crisp and firm and seldom-austere tannins, letting the wine linger for long in the aftertaste.

🍷 *6–30 years*

✔ *Antinori • Badia a Coltibuono • Barone Ricasoli (Castello di Brolio) • Carobbio (Riserva) • Caparsa • Castellare di Castellina • Castell'in Villa • Castello di Ama • Castello di Cacchiano • Castello di Fonterutoli • Castello Querceto • Castello di Radda • Castello di Rampolla • Castello di San Polo in Rosso • Castello di Volpaia • Corzano e Paterno • Felsina • Fontodi • Isole e Olena • La Massa (Giorgio Primo) • Monsanto • Podere Il Palazzino • Poggerino • Poggio al Sole (Casasilia) • Querciabella • Riecine (Riserva) • Rocca di Castagnoli • Rocca della Macie (Roccato) • Rocca di Montegrossi • San Fabiano Calcinaia • Ormanni • San Giusto (Gaio) • San Vicente (Riserva) • Terrabianca • Uggiano • Vecchie Terre di Montefili • Villa Vignamaggio • Fietri*

CIRCEO DOC
Latium

The most southerly DOC in Latium, Circeo is where the sorceress Circe once lived, according to Homer's *Odyssey*. Dry or *frizzante* whites may be produced from at least 55 per cent Trebbiano, up to 30 per cent Malvasia, up to 30 per cent Chardonnay, and up to 15 per cent other local varieties. Dry or *frizzante* red may be produced from at least 55 per cent Merlot, a maximum of 30 per cent Cabernet Sauvignon, and a maximum of 30 per cent Sangiovese, and two varietals are allowed – Sangiovese for still reds and Trebbiano for still whites. Trebbiano-base sparkling wines are also allowed.

✔ *Sant'Andrea (Il Sogno)*

CIVITELLA D'AGLIANO IGP
Latium

This IGP covers the southern half of the Orvieto region and is starting to produce some exciting wines, particularly for the white and red Grechetto but also for Merlot and Cabernet Sauvignon.

✔ *Isabella Mottura • Sergio Mottura*

COLLI ALBANI DOC
Latium

A white-only appellation, Colli Albani DOC is southeast of Rome and must contain 5 to 45 per cent Malvasia del Lazio, 25 to 50 per cent Trebbiano, and up to 60 per cent Malvasia di Candia, plus a maximum of 10 per cent other local non-aromatic varieties. The wine must be dry or sweet, still or *spumante*. *Novello* is permitted for dry or sweet styles, but they must be still, not *spumante*.

COLLI ALTOTIBERINI DOC
Umbria

An interesting DOC in the hilly upper Tiber Valley area, which produces dry white wines from Trebbiano and Malvasia, and firm, fruity reds from Sangiovese and Merlot. It is the crisp, fragrant *rosato* wines that most people prefer, though. Sparkling wines are also allowed and must be made with a minimum 50 per cent Grechetto, Pinot Bianco, Pinot Grigio, and/or Pinot Nero.

COLLI CIMINI IGP
Latium

This lies outhwest of Vignanello DOC. The most widely used grapes are Ciliegiolo and Sangiovese for reds and Trebbiano and Malvasia for whites. The reds are often produced by carbonic maceration.

COLLI DELL'ETRURIA CENTRALE DOC
Tuscany

This is one of the largest DOCs in Tuscany, covering Chianti and all its subzones. It was created in 1990 to allow greater creative freedom amongst the region's winemakers and to distinguish these wines from those made as Chianti. The DOC applies to a wide range of wines. A *bianco* must feature at least 50 per cent Trebbiano Toscano, while the *rosso* (which can also be sold as *novello*) and *rosato* are Sangiovese based. White *vin santo* may be made from at least 70 per cent Malvasia Bianca Lunga and/or Trebbiano. A pink or red *vin santo* Occhio di Pernice must feature at least 50 per cent Sangiovese. Both may be labelled as *riserva* if they are aged for four years in caratelli (small barrels) rather than the standard three years.

✔ *Bindella (Dolce Sinfonia Vin Santo Occhio di Pernice)*

COLLI ETRUSCHI VITERBESI/ TUSCIA DOC
Latium

This massive appellation encompasses a large swathe of northern Latium and a wide variety of wines. The wines include a generic Trebbiano/Malvasia-based white, a generic Sangiovese-Montepulciano-based red, a white Moscato-based *passito*, four varietal white wines (Grechetto, Moscatello, Procanico, and Rossetto), and five varietal reds (Canaiolo, Greghetto, Merlot, Sangiovese, and Violone). The generic white may be dry or sweet, still or *frizzante*, the generic red, dry or sweet, still or *frizzante* and may be sold as *novello*, while the *passito* can only be sweet. For the varietal whites, Grechetto may be still or *frizzante* but must be dry; Moscatello may be dry or sweet, still or *frizzante*; and Procanico and Rossetto are both distinctively different localized clones of Trebbiano. (The Grechetto must be dry and may be still or *frizzante*, while Rossetto may be dry or sweet but can only be produced in a normal still-wine style). The varietal reds Greghetto, Merlot, and Violone must be produced in a classic dry still-wine style (Greghetto is in fact a black-skinned variant of Grechetto, and Violone is none other than the Montepulciano) whilst the varietal Canaiolo and Sangiovese can be made in either dry or sweet style. For the Sangiovese a *frizzante* version is also allowed.

COLLI LANUVINI DOC
Latium

Smooth, white wines, either dry or semi-sweet. Merlot-based reds and sparkling wines made from Trebbiano and Malvasia can also be made.

A GIANT CORK EMBLAZONED WITH THE GALLO NEGRO STANDS NEAR THE ENTRANCE TO THE CELLARS OF THE CASTELLO DI VERRAZZANO WINERY NEAR GREVE IN CHIANTI
The Gallo Nero, or Black Rooster, is the historic symbol of Chianti wines and the trademark for Chianti Classico.

COLLI DI LUNI DOC

Tuscany

See Colli di Luni DOC (Liguria, Northwest Italy)

COLLI MARTANI DOC

Umbria

This covers a few varietal wines from a large but promising area encompassing the Montefalco DOC: Sangiovese, Cabernet Sauvignon, Merlot, and Vernaccia Nera for reds; Trebbiano, Grechetto, Chardonnay, Riesling, and Sauvignon Blanc for whites. The subzone Cannara produces sweet red Vernaccia, and in the Todi subzone only wines made from Grechetto are allowed.

COLLI PERUGINI DOC

Umbria

Dry, slightly fruity, Trebbiano-based white wines, full-bodied red wines, and dry, fresh *rosato* wines, primarily from Sangiovese grapes. Produced in a large area between Colli del Trasimeno and the Tiber, covering six communes in the province of Perugia and one in the province of Terni.

COLLI DELLA SABINA DOC

Latium

Named after the Sabinium, an ancient neighbour and enemy of Rome, this appellation is located along the banks of the Tiber to the northeast of the city. The wines include a generic Malvasia del Lazio–based white and a generic Sangiovese-based red.

COLLI DELLA TOSCANA CENTRALE IGP

Tuscany

A superstar IGP for Super Tuscan wines, this appellation comes from hilly areas of central Tuscany and, consequently, has the potential to produce some of the finest wines in Italy. Expect Sangiovese with or without backing from Bordeaux varieties for reds and Chardonnay for the much rarer whites, as well as some quirky stuff such as Pinot Nero.

✓ *Fontodi* • *Gagliole* • *Castello di Querceto* • *Piazzano* • *Torraccia di Presura* • *Vecchie Terre di Montefili*

COLLI DEL TRASIMENO DOC

Umbria

A very large DOC on the Tuscan border. The dry and off-dry whites are ordinary, but the reds, in which the bitter edge of Sangiovese is softened with Gamay, Ciliegiolo, Malvasia, and Trebbiano, are more interesting. Traditional-method sparkling wines are also produced with a minimum of 70 per cent Chardonnay, Grechetto, and/or Pinot Nero for the white version and a minimum 50 per cent Pinot Nero for the *rosato*. *Vin santo* (dry or sweet) is also allowed.

🍷 *2–5 years*

COLLINE LUCCHESI DOC

Tuscany

This DOC produces light, soft, Chianti-like reds and dry, rather neutral Trebbiano-based whites. Varietal wines made from Sauvignon Blanc and Vermentino can be produced too. *Vin santo* (from white grapes) and *vin santo* Occhio di Pernice (from red grapes) are also made.

CORI DOC

Latium

Little-seen and rarely exciting dry white wines made from the indigenous Bellone grape and dry reds from the indigenous Nero Buono.

CORTONA DOC

Tuscany

Located on the southern border close to Umbria, this is primarily an all-varietal DOC, with four red wine varietals (Cabernet Sauvignon, Merlot, Sangiovese, and Syrah), all made in a classic dry still-wine style, and three white wine varietals (Chardonnay, Grechetto, and Sauvignon Blanc), all made in a classic dry still-wine style. There is also a *vin santo*, which must be at least 70 per cent Trebbiano, Grechetto, or Malvasia Bianca Lunga, and a sweet red or *rosato vin santo* Occhio di Pernice, which can be made from any proportion of Malvasia Nera and/or Sangiovese. Nonetheless almost 80 per cent of the wines labelled as Cortona DOC are based on Syrah. There is a long history of Syrah cultivation in this area, driven by a climate resemblance with the Rhône Valley in France. Syrah from Cortona is dark-berried and spicy on the nose, and the palate follows through with voluptuousness given by the sustained acidity and the succulent tannic profile.

✓ *Tenimenti d'Alessandro* • *Stefano Amerighi* • *Fabrizio Dionisio* • *Capoverso*

COSTA ETRUSCO ROMANA IGP

Latium

Small IGP producing wines from the communes of Cerveteri, Fiumicino, Ladispoli, Santa Marinella, and Tolfa in the province of Rome. One white wine with a minimum of 60 per cent Malvasia del Lazio (locally Malvasia Puntinata) and Vermentino combined (a minimum of 25 per cent of each) and a maximum of 25 per cent Chardonnay and/or Fiano is made. White varietals from Chardonnay, Fiano, Malvasia, or Vermentino are allowed in either dry or sweet styles. One generic red is produced from a minimum of 60 per cent Montepulciano and Sangiovese combined (minimum 25 per cent of each) and a maximum 25 per cent Merlot; varietals red produced (dry and sweet) are Cabernet Sauvignon, Merlot, Sangiovese, or Syrah.

COSTA TOSCANA IGP

Tuscany

A wide range of styles can be made from several varieties (both indigenous and others), including white, red, *rosato*, and dessert wines. Grapes must be sourced from designated communes in the coastal provinces of Grosseto, Livorno, Lucca, Massa Carrara, and Pisa.

✓ *Duemani*

ELBA DOC

Tuscany

The range of wines from the holiday isle of Elba have been expanded to include 10 types: Trebbiano-based dry white, Sangiovese-based red and *riserva* red, *rosato*, Ansonica dell'Elba (dry white from the Ansonica, better known as the Inzolia grape of Sicily), Ansonica Passito dell'Elba, Aleatico dell'Elba, *vin santo* dell'Elba, *vin santo* dell'Elba Occhio di Pernice, and a white *spumante*. These wines are mostly made for tourists.

🍷 *Tastes best on site*

✓ *Acquabona*

ELBA ALEATICO PASSITO/ALEATICO PASSITO DELL'ELBA DOCG

Tuscany

This sweet red wine style is made from dried (*passito*) Aleatico grapes grown on the small island of Elba, just off the Tuscan coast. The wines are fairly high in alcohol but maintain a good level of acidity. They have a distinctive mulberry bouquet, and they are particularly age-worthy wines.

✓ *Tenuta delle Ripalte*

EST! EST!! EST!!! DI MONTEFIASCONE DOC

Latium

The name is the most memorable thing about these dry or semi-sweet white wines made from Trebbiano and Malvasia grapes grown around Lake Bolsena, adjacent to the Orvieto district. Traditionally, the name dates to the 12th century, when a fat German bishop called Johann Fugger had to go to Rome for the coronation of Henry V. In order to drink well on his journey, he sent his majordomo ahead to visit the inns along the route and mark those with the best wine with the word Est, short for "Vinum est bonum". When he arrived at Montefiascone, the majordomo so liked the local wine that he chalked "Est! Est!! Est!!!". Fugger must have agreed with him, because once he had tasted the wine, he cancelled his trip and stayed in Montefiascone until his death. The truth of the story is uncertain, for, although a tomb in the village church bears Fugger's name, whether it contains his 800-year-old body or not is unknown. Trebbiano-based sparkling wines are also allowed. Within the DOC there is a *classico* zone that includes the historic production area of the communes of Bolsena and Montefiascone (*bianco* only).

✓ *Falesco* • *Cantina Stefanoni*

FRASCATI DOC

Latium

Most Frascati used to be flabby or oxidized, but with recent improvements in vinification techniques they are now invariably fresh and clean, although many still have a bland, pear-drop aroma and taste. The few exceptions come from virtually the same group of top-performing producers as they did a decade or so ago and these wines stand out for their noticeably full flavour, albeit in a fresh, zippy-zingy style. Frascati is made from mainly Malvasia grapes (minimum 70 per cent) and Bellone, Bombino Bianco, Greco, and/or Trebbiano grapes (maximum 30 per cent), primarily dry, but semi-sweet, sweet, and *spumante* styles are also made.

🍷 *1–2 years*

✓ *Colli di Catone* • *Fontana Candida* • *Poggio Le Volpi* • *Villa Simone* • *Principe Pallavicini*

FRASCATI SUPERIORE DOCG

Latium

Formerly part of the Frascati DOC, it was established as a separate DOCG in 2011. It produces white wines only from mainly Malvasia Bianca di Candia, Malvasia del Lazio (locally Malvasia Puntinata), and a maximum of 30 per cent Bellone, Bombino Bianco, Greco, and/or Trebbiano. This DOCG produces higher quality and more concentrated wines compared to Frascati DOC with a minimum ABV content of 12 per cent for Superiore and 13 per cent ABV for the Superiore Riserva. See also Frascati DOC

✓ *Principe Pallavicini*

FRUSINATE/DEL FRUSINATE IGP

Latium

This appellation is located in the rugged hillside country of Latium's Piglio area and is thus the IGP equivalent of Cesanese del Piglio DOC. Indeed, some of the red and *rosato* wines produced under this IGP are from the Cesanese grape, but there are also wines from Sangiovese, Cabernet, Merlot, Olivella (aka Sciascinoso), and Syrah, plus whites from Albana (known locally as the Passerina), Bellone, Bombino, Malvasia, Moscato, and Pinot Bianco.

✓ *Giovanni Terenzi*

GENAZZANO DOC

Latium

Located in the rolling foothills of Monti Prenestini to the east of Rome, where just two wines are produced. The red is from a minimum of 85 per cent Ciliegiolo and an optional maximum of 15 per cent other local black varieties; the white is from a minimum of 85 per cent Malvasia di Candia and an optional maximum of 15 per cent other local white varieties.

GRANCE SENESI DOC

Tuscany

Very tiny appellation located in the province of Siena. A variety of styles are produced: a white blend of Trebbiano and Malvasia Bianca Lunga and a red Sangiovese-based or Cabernet Sauvignon-Merlot blend. *Vendemmia tardiva*, (late harvest) and *passito* are also made.

LAGO DI CORBARA DOC
Umbria

This DOC is for red wines from the Orvieto region but without being lumbered by the Orvieto name (see Orvietano Rosso), which most people associate with white wines. The wines from this DOC include a generic blend from at least 70 per cent Cabernet Sauvignon, Merlot, Pinot Nero, or Sangiovese and up to 30 per cent from a choice of Aleatico, Barbera, Cabernet Franc, Canaiolo, Cesanese, Ciliegiolo, Colorino, Dolcetto, and Montepulciano. White wines can also be made from Chardonnay, Grechetto, Sauvignon Blanc, and Vermentino.

✓ Castello di Corbara • Barberani • Tenuta di Salviano

LAZIO IGP
Latium

A regional IGP claimed by several top producers turns out some really exciting red and white wines.

✓ Falesco • Casale del Giglio • Sant'Isidoro (Soremidio)

MAREMMA TOSCANA DOC
Tuscany

This DOC covers the coastal Maremma west of Grosseto – an area that has traditionally been famous for its long-horned Maremma cattle but, over the past 20 years or so, has also built up an extraordinary reputation for its typically Super Tuscan wines. The DOC also includes white wines made from Ansonica, Chardonnay, Malvasia, Sauvignon Blanc, Trebbiano, Vermentino, Viognier, and rosato in which a minimum of 40 per cent Cliegiolo or Sangiovese is required.

✓ Ampeleia • Belguardo • Brancaia • Elisabetta Geppetti • Loacker • Magliano • Poggio Argentiera • Castello del Terriccio • Verrano • Le Mortelle

MARINO DOC
Latium

A typically light Malvasia blend that may be dry, semi-sweet, sweet (late harvest or passito), or spumante. Paola di Mauro's deliciously rich and caramelized Colle Picchioni Oro stands out due to its relatively high proportion of Malvasia grapes and the fact that it receives a pre-fermentation maceration on its skins and is matured in barriques. Also varietal whites from indigenous Bellone, Bombino, Greco, Malvasia del Lazio, and Trebbiano Verde are allowed. The DOC includes a classico subzone with the oldest and best located vineyards.

🍷 1–4 years

✓ Colle Picchioni (Oro)

MONTECARLO DOC
Tuscany

Some interesting dry white wines are starting to appear in this area situated between Carmignano and the coast. The wines were traditionally Trebbiano based, but other varieties (Roussanne, Sémillon, Pinot Grigio, Pinot Bianco, Sauvignon Blanc, and Vermentino) may account for up to 70 per cent of the blend, allowing growers to express individual styles, from light and delicate to full and rich, either with or without barrique-ageing. Red wines may be made from Sangiovese, Canaiolo, Ciliegiolo, Colorino, Syrah, Malvasia, Cabernet Franc, Cabernet Sauvignon, and Merlot. A white vin santo and a pink Occhio di Pernice vin santo are also allowed.

🍷 4–10 years

✓ Tenuta del Buonamico • Carmignani • Fattoria Michi • Vigna del Greppo

MONTECASTELLI IGP
Tuscany

Red and white wines exclusively made in the communes of Castelnuovo Val di Cecina Volterra and Pomarance in the province of Pisa. Several varieties are allowed.

MONTECOMPATRI-COLONNA/ MONTECOMPATRI/COLONNA DOC
Latium

These dry or semi-sweet Malvasia-based white wines may bear the name of one or both of the above towns on the label. Sweet frizzante versions can also be made.

MONTECUCCO DOC
Tuscany

Abutting Brunello di Montalcino to the southwest, this DOC consists of Sangiovese-based reds, a Trebbiano-based white, and one varietal (Vermentino). Whilst in the past rather generic wines were made in the area, this is now recognised as a source of quality wines and some inspiring examples (see Montecucco Sangiovese DOCG). Vin canto and vin santo Occhio di Pernice are also made.

✓ Colle Massari • Salustri (Grotte Rosse)

MONTECUCCO SANGIOVESE DOCG
Tuscany

Established as part of the Montecucco DOC in 1998, Montecucco Sangiovese became a separate DOCG in 2011 and represents a source of high-quality wines, albeit often consumed locally. Only red Sangiovese can be made (90 per cent minimum) in two versions: the rosso must be aged for 17 months and the riserva for a minimum of 34 months ageing before release. Seven subzones are also recognised, taking their names from the related communes (names can be shown on the label). The wines show great character and complexity, with earthy notes and berry flavours. Best examples, especially riserva, benefit with bottle ageing.

✓ Colle Massari

MONTEFALCO DOC
Umbria

There is a significant difference in quality between these basic reds and whites from the Monetefalco DOC and the more interesting, characterful Montefalco Sagrantino DOCG (see below).

MONTEFALCO SAGRANTINO DOCG
Umbria

These distinctive red wines in dry and sweet passito styles are made exclusively from Sagrantino, which has the advantage of being grown on the best-exposed hillside vineyards southwest of Perugia. The passito wines are the most authentic in style, dating back to the 1400s, but the dry table-wine style, hinting of ripe, fresh-picked blackberries, is the best and most consistent.

🍷 3–12 years

✓ Adanti • Antonelli • Villa Antico • Arnaldo Caprai • Colpetrone • Villa Mongalli • Perticaia • Tabarrini

MONTEREGIO DI MASSA MARITTIMA DOC
Tuscany

These are red, white, rosato, novello, and vin santo from the northern part of the province of Grosseto. A dry white "Vermentino varietal wine and a vin santo called Occhio di Pernice are also included within the DOC.

MONTESCUDAIO DOC
Tuscany

A range of wines from the Cecina Valley: Trebbiano-based dry white wine and three monovarietal whites (Chardonnay, Sauvignon Blanc, and Vermentino); reds tend to be soft, slightly fruity, Sangiovese-based but also monovarietals are made (Cabernet Franc, Cabernet Sauvignon, Merlot, and Sangiovese). Vin santo is also produced.

🍷 1–3 years

✓ Poggio Gagliardo • Sorbaiano

MORELLINO DI SCANSANO DOCG
Tuscany

This DOCG produces wines from Sangiovese (minimum 85 per cent), which can be made in an either fruity easy-to-drink style or as Brunello-like wines that show complexity and with higher but still soft ripe tann-in and spicy notes coming from barrel ageing. Some of the best examples are also age-worthy. A riserva version is made with a minimum of two years ageing required before release.

🍷 4–8 years

✓ Erik Banti • Poggio Argentiera (Capa Tosta) • Motta • Fattoria Le Pupille • Podere 414 • Roccapesta

MOSCADELLO DI MONTALCINO DOC
Tuscany

An ancient style of aromatic, sweet Muscat that was famous long before Brunello. Late-harvest versions are also possible.

🍷 Upon purchase

✓ Banfi (Florus) • Col d'Orcia • Il Poggione • Caprili • Capanna

NARNI IGP
Umbria

This is the IGP equivalent of Colli Amerini DOC. The best producers here manage to tease a bit more Sangiovese out of the heritage of the Ciliegiolo grape than anywhere else. (See Colli Amerini DOC.)

✓ Cantina dei Colli Amerini (Ciliegiolo, Ciliegiolo Ani)

NETTUNO DOC
Latium

Abutting Circeo on the Tyrrhenian coast south of Rome, farther inland, this DOC also overlaps the southern tip of Aprilia. Nettuno includes a generic Merlot-Sangiovese red, and a generic Bellone-Trebbiano dry white, plus a varietal Bellone, which may be labelled Cacchione and is thought to be indigenous to the area. The red may be sold as novello, and the white and rosé may be frizzante.

ORCIA DOC
Tuscany

Mostly Sangiovese-based red from a large area overlapping much of Montalcino and Montepulciano, encompassing the Val d'Orcia. This appellation covers a vast latitude of styles: from the crunchy and lean unoaked Sangiovese to more robust and powerfully oaked examples blended with international varieties.

✓ Fattoria del Colle • Capitoni • Le Buche • Poggio Grande • SassodiSole

ORVIETANO ROSSO or ROSSO ORVIETANO DOC
Umbria

This DOC allows a Cabernet blend, as well as a generic red blend, at least 70 per cent of which must consist of two or more of the following: Aleatico, Cabernet Franc, Cabernet Sauvignon, Canaiolo, Ciliegiolo, Merlot, Montepulciano, Pinot Nero, and Sangiovese, plus an optional choice of Aleatico, Barbera, Cesanese, Colorino, or Dolcetto. Orvietano is nonetheless primarily a varietal appellation for Aleatico, Cabernet Franc, Cabernet Sauvignon, Canaiolo, Ciliegiolo, Merlot, Pinot Nero, and Sangiovese, all of which must be produced in a classic, dry still style.

ORVIETO DOC
Umbria and Latium

The vineyards for this popular, widely exported, dry or semi-sweet (more rarely sweet) white wine based on Grechetto and/or Trebbiano (locally Procanico) are primarily located in Umbria. Within the extensive DOC is a historic classico zone where some of the best examples from a bigger portion of the Grechetto grape are made. Due to the proximity of the lakes Corbara and Bolsena and frequent fogs, Orvieto is one of the very few places in Italy regularly affected by noble rot. Those fully botrytised, or muffato, Orvietos are well worth tracking down because they offer a fabulous combination of elegance, concentration, and youthful succulence. Although Orvieto was associated in the past with rather uninteresting Procanico-based wines, recent changes to the production rules have allowed producers to use higher percentages of Grechetto and also to make varietal Grechetto, showing the grape's potential. Some producers also make single-vineyard Grechetto, leading the trend towards quality rather than quantity.

🍷 *Upon purchase*

✓ *Antinori (Campogrande, San Giovanni della Sala)*
• *Barberani* • *Bigi* • *Decugnano dei Barbi* • *Lungarotti*
• *Palazzone* • *Argillae*

PARRINA DOC
Tuscany

This is the most southerly of Tuscany's DOCs, where Sangiovese-based reds are made together with monovarietal Cabernet Sauvignon and Merlot. Wines used to be soft, light, and attractive, but of late much darker, fuller, and richer wines have been made. Some examples are reminiscent of Chianti, with just a touch of oak on the finish and even an occasional wisp of mint in the aftertaste.

🍷 *3–7 years*

✓ *Franca Spinola*

POMINO DOC
Tuscany

This wine dates back to 1716, and a Pomino was marketed as a single-vineyard Chianti by Marchesi de' Frescobaldi long before it was resurrected as its own DOC in 1983. The white is a blend of Chardonnay (minimum 70 per cent), Pinot Blanc, and/or Pinot Grigio, although Frescobaldi's Il Benefizio is pure Chardonnay. The red is a blend of Sangiovese, Merlot, and Pinot Nero. A *vin santo* and late-harvest wines are also made in both red and white styles.

🍷 *1–3 years (blended white), 3–7 years (red and Il Benefizio)*

✓ *Frescobaldi* • *Fattoria Petrognano*

ROSSO DI MONTALCINO DOC
Tuscany

This appellation is for lesser or declassified wines of Brunello di Montalcino or wines made from younger vines. Although there has been a tendency in recent years to produce more powerful and succulent wines, as a rule, the finest Rosso di Montalcino wines are much more accessible in their youth than Brunello, which some readers may prefer.

🍷 *3–15 years*

✓ *Altesino* • *Castelgiocondo* • *Salvioni* • *Costanti* • *Lisini* • *Poggio di Sotto* • *Il Poggione* • *Val di Suga* • *Baricci* • *Gorelli* • *Podere Le Ripi* • *Mastrojanni* • *Canalicchio di Sopra* • *Ridolfi* • *Tenuta Buon Tempo* • *Ciacci Piccolomini d'Aragona* • *Capanna*

ROSSO DI MONTEPULCIANO DOC
Tuscany

This DOC is for the "second wines" of Vino Nobile di Montepulciano DOCG and, like Rosso di Montalcino DOC, its wines are softer and more approachable when young. In this area, the high percentage of sand in the soil promotes aromatic expression and tannin refinement, yielding a crunchy red fruit bouquet and delicate mouthfeel.

🍷 *3–15 years*

✓ *Avignonesi* • *Bindella* • *Boscarelli* • *Le Casalte* • *Contucci* • *Gracciano della Seta* • *Poliziano* • *Tenuta Trerose*

ROMA DOC
Latium

This DOC covers Rome and more than 60 communes falling in the Rome province. A generic Malvasia-based white (still or sparkling) and two white varietals (Bellone and Malvasia Puntinata) are made. A generic red and *rosato*

Montepulciano-based are also allowed. The *classico* subzone includes the township of Rome only.

✓ *Principe Pallavicini*

ROSSO ORVIETANO DOC
See Orvietano Rosso DOC

SANT'ANTIMO DOC
Tuscany

This DOC in Montalcino includes a generic red and white from 100 per cent unnamed local varieties, three red varietals (Cabernet Sauvignon, Merlot, and Pinot Nero), three white varietals (Chardonnay, Pinot Grigio, and Sauvignon Blanc), a dry or sweet Trebbiano-Malvasia-based *vin santo*, and a sweet red *vin santo* Occhio di Pernice with a base of Sangiovese and Malvasia Nero.

✓ *Castello Banfi* • *Casanova di Neri* • *Ciacci Piccolomini d'Aragona* • *Fanti*

SAN GIMIGNANO DOC
Tuscany

When Vernaccia was promoted to DOCG status, all other white varieties remained DOC only, together with red and *rosato* wines.

SAN TORPE' DOC
Tuscany

This very small appellation – just 2 hectares (5 acres) – produces Trebbiano-based white wines, mono-varietals (Chardonnay, Sauvignon Blanc, Trebbiano, and Vermentino), and *rosato* (minimum 50 per cent Sangiovese). *Vin santo* and *vin santo riserva* are also allowed.

SOVANA DOC
TUSCANY

This DOC in the south of Tuscany includes a generic Sangiovese-based red and a dry *passito*, as well as four red varietals: Aleatico, Cabernet Sauvignon, Merlot, and Sangiovese. *Rosato* wines with a minimum of 5 percent Sangiovese can also be made.

✓ *San Lorenzo*

SPELLO IGP
Umbria

This very tiny appellation of just 3 hectares (7.5 acres) covers the township of Spello. From vines growing around Spello and Assisi to the southeast of Perugia, Sportoletti produces delicious wines from Merlot, Cabernet Franc, and Cabernet Sauvignon for reds and from Grechetto and Chardonnay for whites.

✓ *Sportoletti*

SPOLETO DOC
Umbria

Established in 2011, this DOC produces mainly Trebbiano-based white wine and varietal Trebbiano Spoletino. Also sparkling and *passito* can be made from the same grape.

SUVERETO DOCG
Tuscany

Formerly a subzone of the Val di Cornia DOC, it was established as a separate DOCG in 2011. It includes Merlot- or Cabernet Sauvignon–based reds and three monovarietal reds (Sangiovese, Cabernet Sauvignon, and Merlot).

TARQUINIA DOC
Latium

This vast appellation produces a generic Montepulciano/Sangiovese-based red and *passito* and a generic Chardonnay/Malvasia-based white (both dry and sweet).

✓ *Sant'Isidoro*

WORKERS HARVEST GRAPES IN MONTALCINO
This hill town is best known for its Brunello wine, but along with the Brunello di Montalcino DOCG, Montalcino winemakers can produce wines under the Rosso di Montalcino, Sant'Antimo, and Moscadello di Montalcino appellations.

TERRATICO DI BIBBONA DOC
Tuscany

East and south of Livorno, this large DOC allows a generic Sangiovese/Merlot-based red, a Vermentino-based generic dry white, four red varietal wines (Cabernet Sauvignon, Merlot, Sangiovese, and Syrah), and two generic dry white varietal wines (Trebbiano and Vermentino).

TERRE DI CASOLE DOC
Tuscany

This small appellation north of Siena includes a dry Chardonnay-based white, a Sangiovese-based red, and a *passito* made mainly from Chardonnay grapes.

TERRE DI PISA DOC
Tuscany

This Tuscan appellation produces a Cabernet Sauvignon/Merlot/Sangiovese-based or Syrah-based red and a mono-varietal Sangiovese.

TERRACINA/MOSCATO DI TERRACINA DOC
Latium

The DOC covers the province of Latina and the communes of Monte San Biagio, Terracina, and Sonnino and produces the aromatic Moscato di Terracina in a dry, sweet, and sparkling style. A *passito* version is also allowed.

TODI DOC
Umbria

This DOC allows the production of white Grechetto-based wines and varietal Grechetto. Red wines can be Sangiovese based or varietal (Merlot or Sangiovese). A Grechetto *passito* is also produced. The Colli Martani DOC encompassing the Todi DOC also has a subzone Todi for the production of Grechetto only (*see* Colli Martani DOC).

TORGIANO DOC
Umbria

This DOC was built on the back of the reputation of one producer, Lungarotti (*see also* Torgiano Riserva DOCG). As before, Torgiano DOC covers generic Sangiovese blends for red and *rosato* and white (Trebbiano, Grechetto, Malvasia, and Verdello), as well as sparkling (Pinot Noir and Chardonnay) and six varietals (Chardonnay, Pinot Grigio, Riesling Italico, Cabernet Sauvignon, Pinot Noir, and Merlot).

🍷 *3–8 years (red), 1–5 years (white and rosato)*

✓ *Lungarotti*

TORGIANO ROSSO RISERVA DOCG
Umbria

Lungarotti's best *rosso*, the *riserva* is a model of how all DOCG denominations should work and has deservedly been upgraded from DOC. Wines must be a minimum of 70 per cent Sangiovese and aged for a minimum of three years (including six months in bottle) before release.

🍷 *4–20 years*
✓ *Lungarotti*

TOSCANO/TOSCANA IGP
Tuscany

Anything goes for this region-wide IGP, and the freedom it allows has encouraged the best producers to let rip with the entire gamut of local and international grape varieties. There are ways to include almost any variety grape in Tuscany, but the specifically authorized varieties are Chardonnay, Malvasia, Pinot Grigio, Sauvignon, Traminer, Trebbiano, Verdello, and Vermentino for whites and Aleatico, Alicante, Cabernet Franc, Cabernet Sauvignon, Canaiolo, Ciliegiolo, Merlot, Pinot Nero, Sangiovese, and Syrah for reds.

✓ *Avignonesi & Capannelle (50&50) • Badia di Morrona • Barone Ricasoli • Bindella • Poggio Bonelli • Boscarelli • Tenuta Prima Pietra • Tenuta di Capezzana • Castello di Ama • Caiarossa • Castellare di Castellina • Tenuta Masseto • Dievole • Felsina (Fontalloro) • Folonari • Tenuta di Ghizzano • Isole e Olena • Lornano (Commendator Enrico) • Marchesi Antinori • Le Macchiole • Montevertine • Podere Orma • Querciabella • Rocca di Castagnoli • Ruffino (Modus, Romitorio di Santedame) • San Fabiano Calcinaia • San Felice • Tua Rita*

UMBRIA IGP
Umbria

This is a region-wide IGP that allows all grape varieties and every style of wine, including one of the world's very few examples of a dry Aleatico (Alea Viva, produced by Andrea Occhipinti).

✓ *Antinori (Cervaro della Sala) • Occhipinti • Argillae • Barberani*

VAL D'ARBIA DOC
Tuscany

A large area south of the Chianti Classico district, it produces a dry, fruity, Malvasia/Trebbiano-based, wine that may be dried prior to fermentation for dry, semi-sweet, or sweet *vin santo*. Six mono-varietals can also be made (Chardonnay, Grechetto, Pinot Bianco, Sauvignon Blanc, Trebbiano, and Vermentino). A Sangiovese-based *rosato* version is also allowed.

VAL D'ARNO DI SOPRA/ VALDARNO DI SOPRA DOC
Tuscany

A relatively recent appellation that includes 11 communes in the province of Arezzo. A wide range of styles (white, *rosato*, red, sparkling, and sweet) can be made under this DOC, where especially in the last 15 years wines of increasing quality and some outstanding examples are produced. White wines include a Chardonnay-based wine boosted by Malvasia Bianca and two monovarietals (Sauvignon Blanc and Chardonnay). A white Sauvignon Blanc–based and monovarietal Malvasia Bianca are allowed in the Pietraviva and Pratomagno subzone. Reds can be made from either Tuscan (Pugnitello, Malvasia Nera, or Canaiolo) or international varieties, even if the best examples tend to be Sangiovese based. This area is also home to some outstanding Super Tuscans, some of which switched from Toscana IGP to DOC status.

✓ *Il Borro • La Salceta • Tenuta Sette Ponti • Il Carnasciale • Petrolo*

VALDICHIANA TOSCANA DOC
Tuscany

This hugely fertile plain was once swampland but has long since been turned to agricultural use. A wide range of styles are made (white, red, sparkling, *rosato*, and sweet). It is better known, however, for Chiana, one of the oldest breeds of cattle, than for wine.

VAL DI CORNIA DOC
Tuscany

A large area of scattered vineyards on the hills east of Piombino and south of Bolgheri produce Vermentino-based dry whites, monovarietal Ansonica and Vermentino, and Sangiovese reds and *rosato*. The DOC also includes monovarietal Cabernet Sauvignon and Merlot and sweet Ansonica Passito and Aleatico Passito.

VAL DI CORNIA ROSSO/ ROSSO DELLA VAL DI CORNIA DOCG
Tuscany

Formerly part of the Val di Cornia DOC, it was established as a separate DOCG in 2011. It includes only red wines made from Sangiovese (minimum 40 per cent) blended with Cabernet Sauvignon and/or Merlot. Both *rosso* and *riserva* (minimum 26 months ageing) are made.

VAL DI MAGRA IGP
Tuscany

A vast appellation covering the province of Massa e Carrara at the very northern tip of Tuscany, where the following varieties are permitted: Ciliegiolo, Groppello, Merlot, and Pollera for red wines and Albarola, Durella, Trebbiano, Verdello, and Vermentino for whites.

VALDINIEVOLE DOC
Tuscany

This small appellation located southwest of Pistoia in the Nievole Valley produces Trebbiano-based white wines and Sangiovese-based reds. *Vin santo* is also produced.

VELLETRI DOC
Latium

Rather uninspiring, dry or semi-sweet white wines and reds from the Castelli Romani area.

VERNACCIA DI SAN GIMIGNANO DOCG
Tuscany

This dry white wine was Italy's first-ever DOC in 1966 and became a DOCG in 1993. The best Vernaccias have always been deliciously crisp and full of vibrant fruit, which makes them well worth seeking out. Its signature austere texture has resulted in Vernaccia di San Gimignano often being referred to as a disguised red wine. *Riserva* wines are also made (minimum 11 months ageing before release), the best examples of which show a minerally, slightly savoury character and develop flinty notes in the bottle.

🍷 *1–3 years*
✓ *Falchini • Panizzi • Teruzzi & Puthod • Montenidoli*

VIGNANELLO DOC
Latium

This DOC covers the township of Vignanello and six other communes in the Viterbo province. A generic Trebbiano-based white and one white varietal from Greco are made, as well as a generic Sangiovese-based red and a sparkling Greco. Two sweet wines are also made: a Trebbiano-based late harvest and a late-harvest Greco (minimum 85 per cent).

VINO NOBILE DI MONTEPULCIANO DOCG
Tuscany

Established as a DOC in 1966, it became a became a DOCG in 1980. Made largely from the Sangiovese clone Prugnolo Gentile, plus Canaiolo and other local grapes, including white varieties, these wines are exclusively made in the township of Montepulciano 120 kilometres (75 miles) southeast of Florence. Stylistically Vino Nobile sits between Chianti Classico and Brunello di Montalcino, generally displaying a ruby to garnet colour and generous ripe-fruit flavours hinting of cherry and wild berries. In these sandy soils Sangiovese tends to develop lighter, sweeter, and more approachable wine. Some of the best examples and especially the *riservas* can be more austere and muscular wines that demand bottle ageing.

🍷 *6–25 years*
✓ *Avignonesi • Bindella • Poderi Boscarelli • Le Casalte • Poliziano • Tenuta Trerose • Gracciano della Seta • Dei*

VIN SANTO DEL CHIANTI DOC
Tuscany

This appellation includes *vin santo* made from a minimum of 70 per cent Malvasia Bianca and/or Trebbiano and a pink *vin santo* Occhio di Pernice made from a minimum of 50 per cent Sangiovese. Seven different subzones are recognised: Colli Aretini, Colli Fiorentini, Colline Pisane, Colli Senesi, Montalbano, Montespertoli, and Rufina. When it comes to *vin santo*, there is no evidence in the bottle that Chianti is inferior to Chianti Classico but is just the expression of a different *terroir*.

✓ *Cantagallo • Frascole • Grignano • Lanciola • Castello di Monastero (Lunanuova) • Castello della Paneretta • Torre a Cona (Merlaia) • Travignoli (Rufina) • Villa Petriolo • Villa Pillo • Villa Vignamaggio*

VIN SANTO DEL CHIANTI CLASSICO DOC
Tuscany

Inexplicably the *vin santo* remained a DOC when Chianti Classico was elevated to DOCG. The appellation includes sweet *vin santo* made from a minimum of 60 percent Malvasia Bianca and/or Trebbiano and a pink *vin santo* Occhio di Pernice made from a minimum of 80 per cent Sangiovese. Both styles have a minimum three years ageing requirement with a minimum of 24 months in traditional caratelli (small barrels usually between 55 to 110 litres, to 15 to 30 gallons). Grapes must come from the delimited Chianti Classico zone.

✓ *Avignonesi • Folonari • Isole e Olena • Rocca di Montegrossi • La Ripa • La Sala • San Felice • Borgo Scopeto • Villa Vignamaggio • Vistarenni*

VIN SANTO DI CARMIGNANO DOC
Tuscany

Established as part of the Barco Reale di Carmignano DOC in 1994, it became a separate DOC in 2013. The DOC includes Malvasia- and/or Trebbiano-based *vin santo* and a pink Sangiovese-based *vin santo* Occhio di Pernice. All the grapes must come from the delimited Carmignano zone northwest of Florence.

✓ *Capezzana • Tenuta di Artimino*

VIN SANTO DI MONTEPULCIANO DOC
Tuscany

This small appellation in the province of Siena produces Grechetto/Malvasia/Trebbiano-based *vin santo* and pink Sangiovese-based *vin santo* Occhio di Pernice.

✓ *Poliziano • Tenuta TreRose • Avignonesi*

ZAGAROLO DOC
Latium

This DOC has a tiny production of dry or semi-sweet white from Malvasia and Trebbiano grapes, which are grown east of Frascati in an area more famed for its wines half a millennium ago than it is now.

East Central Italy

This area comprises the regions of Emilia-Romagna, Marche, Abruzzo, and Molise. The best-quality wines come from Marche and the Abruzzo, but the best-known is Emilia-Romagna's juicy red sparkling Lambrusco, a wine exported in vast quantities.

The East Central region, which extends across almost the entire width of northern Italy into Piedmont, appears geographically to wander off its central-east designation. Yet it certainly does not do so topographically, for every hectare of it lies east of the Apennines on the initially hilly ground that flattens out into alluvial plains stretching towards the Adriatic.

EMILIA-ROMAGNA

Emilia-Romagna is protected on its western flank by the Apennines, the source of seven major, and many minor, rivers. The fertile soil results in abundant grape production, the most prolific varieties being Lambrusco, Trebbiano, and Albana, which deliver savoury white wines that have been given Italy's first DOCG for a white wine. Among white varieties, Pignoletto is capable of producing refreshingly aromatic dry still and Charmat-method sparkling wines. Sangiovese di Romagna is nowadays resurging in interest; it is able to provide delicious, wild-berried reds that are easier to drink compared to their Tuscan counterparts.

ABRUZZO (THE ABRUZZI)

Given its notable variety of soils and microclimates, ranging from coastal to alpine areas, Abruzzo is capable of producing both robust, tannic wines and more ordinary, light, and fruit-driven Montepulciano-based reds. Trebbiano d'Abruzzo provides a lean and neutral refreshing white, produced in large quantities. Pecorino and Passerina varieties are produced close to the northern border with the Marche region and account for lower volumes, although they are rapidly growing.

MARCHE (THE MARCHES)

The Marche region has a beautiful coastline complemented by hilly areas, and it is homeland to one of Italy's most noble white grapes: the Verdicchio. This variety is produced in a vast latitude of different styles; however, it can properly show its greatness in aged dry examples. Complex and muscular reds based on Montepulciano and Sangiovese are produced in the limestone peninsula of Conero. The rediscovered Pecorino variety yields sage-perfumed and vibrantly exotic-flavoured dry whites. Passerina has lower acidity and provides more ordinary easy-drinking still and sparkling whites.

MOLISE

Once combined with Abruzzo, Molise is a relative newcomer to the Italian wine scene, having gained its first DOC only in 1983. Montepulciano and Trebbiano represent the largest vineyard area, being the base for the Biferno DOC and Molise DOC appellations. Tintilia del Molise DOC, based on the homonymous grape variety, yields a light and perfumed *rosato* and bright, spicy, enjoyable reds.

WINE REGIONS OF EAST CENTRAL ITALY
(see also p345)
With the Apennines forming the region's western border, the eastern part of Central Italy is dominated by their foothills and the plains.

REGIONS AT A GLANCE

REGION	TOTAL AREA UNDER VINE Hectares (Acres)	IGP/DOC/ DOCG AREA UNDER VINE Hectares (Acres)	IGP/DOC/ DOCG PRODUCTION Hectolitres	TOTAL PRODUCTION Hectolitres
Emilia-Romagna	50,846 (125,643)	50,841 (125,630 acres)	2.3 million (75-25 red/white)	7.3 million (81 million cases)
Abruzzo	33,078 (81,740)	20,406 (50,400)	1,26 million (90-10 red/white)	3 million (33,2 million cases)
Marche	17,332 (42,830)	13,9640 (34,500)	388,8000 (25-75 red/white)	958,500 (10.6 million cases)
Molise	5,361 (13,247)	1,400 (3,460)	38,380 (80-20 red/white)	231,860 (2.6 million cases)

Source: Osservatorio del Vino UIV - ISMEA

VINEYARDS AND OLIVE GROVES SIT SIDE BY SIDE IN ABRUZZO
The rugged terrain of this mountainous region does not deter viticulture in Abruzzo. The DOC produces millions of cases of wine annually, with the red Montepulciano and the white Trebbiano d'Abruzzo the predominant grape varieties.

FACTORS AFFECTING TASTE AND QUALITY

LOCATION
This area stretches along the Adriatic coast, from Molise in the south right up to Emilia-Romagna in the north.

CLIMATE
In this region, the influence of the Mediterranean Sea provides generally hot and dry summers, which become progressively hotter as one travels south, and cool winters. In the hilly regions, microclimates are created by the effects of altitude and aspect.

ASPECT
The best vineyards are invariably to be found on well-drained, foothill sites, but viticulture is spread across every imaginable type of terrain, with a heavy concentration on flat plains, particularly along the Po Valley in Emilia-Romagna, where grapes are produced in abundance

SOIL
The soil is mostly alluvial and clayish. Some limestone outcrops are found in proximity to the Apennines mountain range and the Conero peninsula. Both in Marche and Abruzzo, there is a gradient of soil types, ranging from sandier in the coastal areas to richer in clay farther inland.

VITICULTURE & VINIFICATION
A wide variety of viticultural practices and vinification techniques are used here. Much bulk-blended wine originates here (especially in Emilia-Romagna and Abruzzo), but some producers retain the most worthwhile traditions and augment them with modern methods.

GRAPE VARIETIES
Aglianico, Albanella/Campolese/ Trebbiano (Ugni Blanc), Altra Uva (Ortrugo), Ancellotta, Barbarossa, Barbera, Beverdino, Biancame/Bianchello/Greco di Ancona (Albana), Bianchino (Montù), Croatina, Cabernet Franc, Cabernet Sauvignon, Cagnina (Mondeuse), Chardonnay, Ciliegiolo, Fruttana (Fortana), Gallioppa/Galloppo (Lacrima), Incrocio Bruni 54 (Verdicchio x Sauvignon), Lambrusco, Maceratino, Malvasia, Merlot, Montepulciano, Montù (Montuni), Moscato (Muscat Blanc à Petits Grains), Pagadebit (Bombino Bianco), Passerina, Pecorino, Pignoletto, Pinot Bianco (Pinot Blanc), Pinot Grigio (Pinot Gris), Pinot Nero (Pinot Noir), Refosco, Riesling Italico (Welschriesling), Sangiovese, Spergola (Sauvignon Blanc), Terrano (Refosco), Uva d'Oro (Fortana), Verdicchio, Vernaccia Nera

THE APPELLATIONS OF
EAST CENTRAL ITALY

ABRUZZO DOC
Abruzzo

The Abruzzo DOC, a large appellation covering the provinces of Chieti, L'Aquila, Pescara, and Teramo, produces a wide range of styles: a generic Trebbiano-based white, a generic Montepulciano-based red, and five white varietals (Cococciola, Malvasia, Montonico, Passerina, and Pecorino). Sparkling wines are made either by tank method or by traditional method (Chardonnay-based for the white version and Montepulciano/Pinot Nero-based for the rosé). The traditional-method sparklers are also made as vintage. *Passito* can be either white (made from a minimum of 60 per cent Gewürztraminer, Malvasia, Moscato, Passerina, Pecorino, Riesling, and/or Sauvignon Blanc) or red (minimum 60 per cent Montepulciano).

BIANCHELLO DEL METAURO DOC
Montù

Dry, delicate white wines are made from the Bianchello grape (possibly with some Malvasia), which is grown in the lower Metauro Valley. Once a pretty obscure appellation, this has seen an increase in awareness due to lack of basic Chablis in difficult vintages.

✓ *Umani Ronchi*

BIANCO DEL SILLARO/SILLARO IGP
Emilia-Romagna

This a vast IGP covering generic Albana-based white wines from the provinces of Bologna, Ravenna, Forlì-Cesena, and Rimini.

BIFERNO DOC
Molise

Smooth, slightly tannic, red and fruity *rosato* wines from Montepulciano and Aglianico grapes and dry, lightly aromatic white wines from Trebbiano, Bombino, and Malvasia.

BOSCO ELICEO DOC
Emilia-Romagna

This is a large coastal region separated from the major Emilia-Romagna viticultural areas in the northeast. There is a quite rustic red varietal from the mysterious Fortana grape, also called the Uva d'Oro. Bosco Eliceo Fortana can be dry or sweet, has a slightly bitter tannic bite and is *frizzante*; it is not dissimilar to Lambrusco. There is also a generic blended white (Trebbiano, Sauvignon Blanc, and Malvasia and may be *frizzante*), two pure varietals, a Merlot (sometimes *frizzantino*), and a Sauvignon Blanc (may also be *frizzantino*).

CASTELFRANCO EMILIA IGP
Emilia-Romagna

Until 2013 this IGP was called Bianco di Castelfranco Emilia and allowed only the Montù-based (aka Montini) *bianco*. This IGP now allows Moscato and Trebbiano from the provinces of Bologna and Modena.

CASTELLI DI JESI VERDICCHIO RISERVA DOCG
Montù

Shares the same area of Verdicchio dei Castelli di Jesi DOC. The wine here must be still and dry and aged for at least 18 months, including 6 months in bottle. Can be surprisingly age-worthy. There are 99 *menzioni geografiche aggiuntive* ("additional geographical definitions") and 18 communes entitled to be mentioned on the label.

🍷 *2–20 years*

✓ *Bucci • Tenuta San Sisto • Garofoli*

CERASUOLO D'ABRUZZO DOC
Abruzzo

These are light and vibrant rosé wines produced from the Montepulciano grape including a superiore version (12.5

per cent minimum ABV required). They show a bright, fruity, and floral character that is achieved by leaving the fermenting juice in contact with the skins for a very short time, resulting in an attractive cherry-like color that is behind the name Cerasuolo ("cherry" in Italian) and very low tannins.

COLLI APRUTINI IGP
Abruzzo

An IGP for white (still, *frizzante*, or *passito*), red (still, *novello*, *frizzante*, or *passito*), and *passito* (still, *novello*, or *frizzante*) from a vast range of grapes grown in the province of Teramo.

✓ *Cvetic (Iskra)*

COLLI BOLOGNESI DOC
Emilia-Romagna

This appellation covers four red varietals (Barbera, Cabernet Sauvignon, Pinot Nero, and Merlot) and four whites (Riesling Italico, Sauvignon Blanc, which may be *frizzantino*, Pinot Bianco, and Chardonnay). *Spumante* can come from the subzone Bologna only, and it is usually based on Pinot Bianco or Chardonnay

🍷 *1-3 years*

✓ *Terre Rosse*

COLLI BOLOGNESI PIGNOLETTO DOCG
Emilia-Romagna

Formerly part of the Colli Bolognesi DOC, it was established as a separate DOCG in 2010, with a name change and area expansion in 2014. This wine is, however, restricted to the Grechetto grape (known locally as Pignoletto) produced in a classic dry white style or in a *spumante* version.

COLLI DI FAENZA DOC
Emilia-Romagna

Extremely minute appellation that prodices a Chardonnay-based dry generic white, a Cabernet Sauvignon-based generic red, two varietal whites (Pinot Bianco and Trebbiano), and one varietal red (Sangiovese).

✓ *Il Pratello*

COLLI D'IMOLA DOC
Emilia-Romagna

This DOC includes a varietally non-specific generic white, which must be dry and may be *frizzante*; a varietally non-specific generic red, which may be *frizzante*; two varietal whites (Chardonnay and Trebbiano); and three varietal reds (Barbera, Cabernet Sauvignon, and Sangiovese).

COLLI MACERATESI DOC
Montù

This huge area, while having fewer than 100 hectares (250 acres) of vineyard, produces a dry or sweet from a late-harvest white wine blend that used to be Trebbiano based but now must be based on Maceratino, Verdicchio, Malvasia, and Chardonnay. Reds must be based on Sangiovese blended with Cabernet Franc, Cabernet Sauvignon, Ciliegiolo, Lacrima, Merlot, Montepulciano, and/or Vernaccia Nera.

COLLI DI PARMA DOC
Emilia-Romagna

Solid, slightly *frizzantino* red wines from local varieties and two principal white varietals, Malvasia and Sauvignon Blanc, in dry, sweet, still, *frizzante*, or *spumante* styles, the latter based on Chardonnay, Malvasia, and Pinot Bianco.

COLLI PESARESI DOC
Montù

Originally a Sangiovese DOC, Colli Pesaresi now includes many varieties and styles, although the deeply flavoured Sangiovese-based red still stands out, with the best showing real class. Subzone Focara may contain up to 15 percent Pinot Noir but displays Sangiovese-like characteristics.

There is also a Biancame or Trebbiano-based dry white and a very similar wine from the subzone Roncaglia. Parco Naturale Monte San Bartolo is another subzone, which produces Cabernet Sauvignon and Sangiovese wines only.

🍷 *3-8 years*

✓ *Tattà • Vallone • Villa Pigna*

COLLI PIACENTINI DOC
Emilia-Romagna

This large DOC includes a wide range of styles, including a generic Chardonnay-based white (still or *frizzante*) and three white varietals (Malvasia, Pinot Grigio, and Sauvignon Blanc) that can be either still or *frizzante*. There are also three white blends from three specific subzones: an aromatic wine called Monterosso Val d'Arda (Malvasia, Moscato, Trebbiano, and Ortrugo, possibly with Beverdino and Sauvignon) in dry, sweet, still, *frizzantino*, or *frizzante* styles; a wine called Valnure (Malvasia, Moscato, Trebbiano, and Ortrugo) that may be dry or sweet, still, *frizzante*, or *spumante*; and a wine called Trebbianino from Val Trebbia (Ortrugo, Malvasia, Moscato Bianco, Trebbiano Romagnolo, and Sauvignon Blanc) that can be still or *frizzante*. Four red varietals reds are made: Barbera and Pinot Nero (either still or *frizzante*), Bonarda (still, *frizzante*, and sparkling, either dry or sweet), and a dry still Cabernet Sauvignon. A *passito* wine from Malvasia and a dry or sweet Vinsanto from Ortrugo, Malvasia, Marsanne, Sauvignon Blanc, and/or Trebbiano are also allowed. The sparklers can be made either by Charmat or traditional method. The DOC includes four subzones: Val d'Arda, Valnure, Val Trebbia, and Vigoleno.

COLLI DI RIMINI DOC
Emilia-Romagna

This DOC allows a generic Trebbiano-based dry white, a generic Sangiovese-based red, two varietal whites (Biancame – aka Albana – and Rebola), and two varietal reds (Cabernet Sauvignon and Sangiovese).

COLLI ROMAGNA CENTRALE DOC
Emilia-Romagna

This DOC allows a generic Chardonnay-based dry white, a generic Cabernet Sauvignon–based red, two varietal whites (Chardonnay and Trebbiano), and two varietal reds (Cabernet Sauvignon and Sangiovese).

COLLI DEL SANGRO IGP
Abruzzo

This IGP allows white (still, *frizzante*, or *passito*), red (still, *novello, frizzante*, or *passito*), and *passito* (white or red) from a wide range of grapes growing in Torino di Sangro and surrounding villages in the province of Chieti.

COLLI DI SCANDIANO E CANOSSA DOC
Emilia-Romagna

Formerly Bianco di Scandiano, this DOC has been considerably widened both geographically and in the scope of wines that may be produced, which now includes a white made from a minimum of 85 per cent indigenous Spergola, five white varietals (Chardonnay, Malvasia, Pinot, Sauvignon Blanc, and Spergola), a generic Marzemino-based red, and six varietal reds (Cabernet Sauvignon, Lambrusco, Lambrusco Grasparossa, Lambrusco Montericco, Malbo Gentile, and Marzemino). *Passito* wines, either white (Malvasia, Sauvignon, or Spergola) or red (Malbo Gentile or Marzemino), can also be made. All wines except Cabernet Sauvignon can be either still or *frizzante*. All whites may also be made in a full *spumante* style, with the exception of Sauvignon Blanc. The DOC also includes a *classico* subzone source for the best grapes.

✓ *Casali Viticultori*

COLLINE FRENTANE IGP
Abruzzo

This appellation allows white (still, *frizzante*, or *passito*), red (still, *novello, frizzante*, or *passito*), and *passito* (white

---

or red) from a wide range of grapes growing in Castel Frentano and surrounding villages in the province of Chieti.

COLLINE PESCARESI IGP
Abruzzo

This IGP allows white (still, *frizzante*, or rosato), red (still, *novello, frizzante*, or *passito*), and *passito* (white or red) from a vast range of grapes growing in the province of Pescara.

✓ *Pasetti*

COLLINE TEATINE IGP
Abruzzo

This IGP allows white (still, *frizzante*, or *passito*), red (still, *novello, frizzante*, or *passito*), and *passito* (white or red) from a wide range of grapes growing in Ripa Teatina, San Giovanni Teatino, and surrounding villages in the province of Chieti.

✓ *Masciarelli*

COLLINE TERAMANE MONTEPULCIANO D'ABRUZZO DOCG
Abruzzo

A gem within the Abruzzo region. This DOCG is for red wine only and produces almost straight Montepulciano (minimum 90 per cent required) with 10 per cent Sangiovese on the hills of the province of Teramo. A *riserva* version can be made with minimum three years ageing requirement of which one year must be in barrel. This territory in the province of Teramo has a special microclimate and good soil structure, located between the high peaks of the Gran Sasso National Park and the Adriatic coast. Passionate local wine growers aim for reasonably low yields to get first-class fruit. The wines are very expressive with an intense purity of black fruit with hints of spice and smoke due to the oak ageing. Some can be muscular and earthy but at the same time offer an elegant and velvet-smooth texture.

CONERO DOCG
Montù

Starting with the 2004 vintage, the *riserva* category of Rosso Conero DOC was renamed simply as Conero and elevated to DOCG status. Conero DOCG comes from the same area of production as Rosso Conero DOC, but it is restricted to lower yields and fractionally riper grapes, and the wines must be aged at least 24 months prior to release. *See also* Rosso Conero DOC.

🍷 *6-25 years*

✓ *Moroder (Dorico) • Piantate Lunghe (Rossini) • Silvano Strologo • Umani Ronchi (Cùmaro)*

CONTROGUERRA DOC
Abruzzo

This DOC produces a Trebbiano-based generic white that may be dry or sweet (*passito* or *passito annoso*) or *spumante*, a Montepulciano-based generic red that may be dry or sweet (*passito* or *passito annoso*) and also sold as *novella*, three varietal whites (Chardonnay, Passerina, and Pecorino), and two red varietals (Cabernet and Merlot).

EMILIA IGP
Emilia-Romagna

This IGP allows white (still, *frizzante*, or *passito*), red (still, *novello, frizzante*, or *passito*), and *passito* (still, *novello*, or *frizzante*) from a vast range of grapes growing in the provinces of Ferrara, Modena, Parma, Piacenza, Reggio Emilia, and part of Bologna.

✓ *Bassi • La Stoppa • Ceci*

ESINO DOC
Montù

This very large DOC encompasses four other DOCs, including Verdicchio dei Castelli Jesi and Verdicchio di Matelica, yet Esino's generic dry white is also Verdicchio based. Go figure. Just two wines are made under this DOC, the other being a generic Montepulciano/Sangiovese-based red.

FALERIO DOC
Montù

This DOC produces dry, lightly scented white wines made from a blend of Trebbiano, Passerina, Verdicchio, Malvasia, Pinot Blanc, and Pecorino.

🍷 *1–3 years*

✓ *Cocci Grifoni (Tarà)*

FORLÌ IGP
Emilia-Romagna

White (still, *frizzante*, or *passito*), red (still, *novello*, *frizzante*, or *passito*), and *passito* (still, *novello*, or *frizzante*) from a large range of grapes growing in the province of Forlì-Cesena.

✓ *Castelluccio*

FORTANA DEL TARO IGP
Emilia-Romagna

This varietal IGP is restricted to Brugnola (known locally as Fortana) grown in the province of Parma. It can be made still or *frizzante* and may be sold as *novello*.

GUTTURNIO DOC
Emilia-Romagna

Produced in the Piacenza hills, where Julius Caesar's father-in-law made a wine that was traditionally drunk from a vessel called a *gutturnium*.

HISTONIUM OR VASTESE IGP
Abruzzo

This IGP allows white (still or *frizzante*), red (still, *novello*, or *frizzante*), and *passito* (white or red) from a wide range of grapes growing in Vasto (Histonium in Latin) and surrounding villages in the province of Chieti.

I TERRENI DI SANSEVERINO DOC
Montù

This tiny denomination of just 2 hectares (5 acres) covering the commune of Sanseverino in the province of Macerata produces red wine only: a Vernaccia Nera–based (dry and sweet) and the so-called Moro, which must be a minimum of 60 per cent Montepulciano.

LACRIMA DI MORRO/ LACRIMA DI MORRO D'ALBA DOC
Montù

Nothing to do with Lacryma Christi or the town of Alba in Piedmont, this DOC is for a soft, medium-bodied red (dry and *passito*) from the mysterious Lacrima grape grown in and around Morro d'Alba in Ancona.

✓ *Umani Ronchi*

LAMBRUSCO GRASPAROSSA DI CASTELVETRO DOC
Emilia-Romagna

Dry or semi-sweet, vinous, *frizzantino*, or sparkling reds and *rosati* are made from Lambrusco Grasparossa grapes grown around the town of Castelvetro di Modena. The red sparkling is deep purple in colour, and it has pronounced aromas of violets, strawberries, fresh plums, and black cherries. It is fuller bodied and higher in alcohol than other Lambrusco wines, and it also has the firmest tannin structure.

✓ *Cantina Settecani • Corte Manzini*

LAMBRUSCO SALAMINO DI SANTA CROCE DOC
Emilia-Romagna

These dry or semi-sweet, vinous, semi-*spumante* red and *rosato* wines are the most aromatic and lighter-bodied of the Lambruscos, displaying a delicate bouquet and refreshing fruitiness.

✓ *Cantina di S Croce*

LAMBRUSCO DI SORBARA DOC
Emilia-Romagna

Mostly dry, although sometimes semi-sweet, these are medium-bodied, *frizzantino* or sparkling reds or *rosati* with more body and depth of flavour than most.

✓ *Cleto Chiarli*

MARCHE IGP
Montù

This regional IGP allows white (still, *frizzante*, or *passito*), red (still, *novello*, *frizzante*, or *passito*), and *passito* (white, *rosato*, and red) from Barbera, Cabernet Franc, Cabernet Sauvignon, Chardonnay, Grechetto, Merlot, Passerina, Pinot Bianco, Pinot Grigio, Pinot Nero, Sangiovese, Sauvignon Blanc, and Trebbiano growing anywhere in the Marche region.

✓ *Il Pollenza • Valturio*

MODENA DOC
Emilia-Romagna

The wines produced here are all almost *frizzante* or sparkling in a variety of styles – red, rose, white, and varietal. The only dry white (still or *frizzante*) made must be a minimum of 85 per cent Grechetto Gentile, Montù, or Trebbiano. The other wines produced are a generic *rosato* blend and a Lambrusco Rosato; a generic Lambrusco-based red and a varietal Lambrusco, both also sold as novella; four versions of red sparkling Lambrusco-based/Lambrusco; and one sparkling white made from a minimum of 85 per cent Grechetto Gentile, Montù, and/or Trebbiano.

✓ *Opera 02*

MOLISE DOC
Molise

This DOC covers a huge portion of Molise. The wines allowed include a generic Montepulciano red (still or sparkling), a generic Chardonnay-based *spumante*, 5 varietal reds (Aglianico, Cabernet Sauvignon, Sangiovese, Merlot, and Pinot Nero), and 10 varietal whites (Chardonnay, Falanghina, Fiano, Greco Bianco, Malvasia, Moscato Bianco, Pinot Bianco, Pinot Grigio, Sauvignon, and Trebbiano). All wines may be produced as either still or *frizzante*. Chardonnay, Falanghina, Fiano, Moscato, Pinot Grigio, and Pinot Bianco may also be produced as a *spumante*. *Passito* versions of Falanghina and Moscato are also allowed.

✓ *Di Majo Norante • Colle Sereno*

MONTEPULCIANO D'ABRUZZO DOC
Abruzzo

Produced in the provinces of Chieti, L'Aquila, Pescara, and Teramo, two distinct styles are made from Montepulciano with up to 15 per cent Sangiovese. Both are very deep in colour, but one is full of soft, luscious fruit; the other is firmer and more tannic. The subzone Casauria must be 100 per cent Montepulciano. The other four subzones can be labelled as such: Alto Tirino and Terre dei Peligni (minimum 95 per cent Montepulciano) and Teate and Terre dei Vestini (minimum 90 per cent Montepulciano). A *riserva* version can also be made, with minimum 24-month ageing requirement.

🍷 *4–8 or 8–20 years (red)*

✓ *Agriverde • Barba (Vigna Franca) • Cataldi Madonna • Illuminati (Ilico) • Masciarelli • Emidio Pepe ⓞ • Tenuta del Priore • Torre dei Beati • La Valentina (Bellovedere) • Valentini • Valle Reale (San Calisto) • Villa Medoro • Cantina Tollo*

MONTEPULCIANO D'ABRUZZO COLLINE TERAMANE DOCG
Abruzzo

Formerly part of the Montepulciano d'Abruzzo DOC, it was established as a separate DOCG in 2003 and the name changed to Montepulciano d'Abruzzo Colline Teramane in 2016. The hilly terrain of the province of Teramo has always been considered superior, which is why it was originally declared a subzone of Montepulciano d'Abruzzo that could be indicated with the appellation on the label. The

wines from these hillsides have now been upgraded to DOCG and must be at least 90 per cent Montepulciano (as opposed to 85 per cent for the DOC, although the DOC's two subzones are strictly 100 per cent), with lower yields and aged a minimum of 24 months prior to release. Wines claiming *riserva* status must be aged at least 36 months.

✓ *Illuminati (Pieluni, Zanna) • Valentini • Villa Medoro*

OFFIDA DOCG
Montù

At the southern end of the Marche region, just to the north of Montepulciano d'Abruzzo, this DOCG includes a generic Montepulciano red, a Pecorino varietal dry white, and a varietal Passerina white.

✓ *Cocci Grifoni • Tenuta Spinelli*

ORTONA DOC
Abruzzo

A Trebbiano-based white and a Montepulciano-based red from grapes growing in the township of Ortona only.

ORTRUGO DEI COLLI PIACENTINI
Emilia-Romagna

This DOC produces a white wine from a minimum of 90 per cent Ortrugo, either still or sparkling. The sparkling version can be made either by Charmat or traditional method.

PENTRO OR PENTRO DI ISERNIA DOC
Molise

This DOC produces smooth, slightly tannic reds and dry, fruity rosati from Montepulciano and Tintilia and dry, fresh whites from Falanghina and Trebbiano.

PERGOLA DOC
Montù

Pergola is a small DOC on the Adriatic coast of the Marche region. It covers a variety of wine styles (red and *rosato* still, sparkling, and *passito*) made predominantly from the Muscat-scented Aleatico grape.

PIGNOLETTO DOC
Emilia-Romagna

Appellation for white wines made from Grechetto Gentile, which may be still, sparkling, late harvest, or *passito*. The DOC includes numerous communes in the provinces of Bologna and Modena, plus four communes in the province of Ravenna. Three subzones are recognised and can be mentioned on the label: Colli d'Imola, Modena, and Reno.

✓ *Poderi Morini*

RAVENNA IGP
Emilia-Romagna

This IGP allows white (still, *frizzante*, or *passito*), red (still, *novello*, *frizzante*, or *passito*), sparkling (dry or sweet) and *passito* (still, *novello*, or *frizzante*) from a large range of grapes growing in the province of Ravenna.

✓ *Zerbina • Poderi Morini*

REGGIANO DOC
Emilia-Romagna

Although formerly known as Lambrusco Reggiano, this remains an exclusively Lambrusco appellation, which may be red or *rosato*, dry or sweet, and still or *frizzante*, with *novello* allowed for all reds.

✓ *Albinea Canali • Cantine Lombardini • Ermete Medici & Figli (Secco Concerto) • Casali Viticultori*

RENO DOC
Emilia-Romagna

This white-only DOC includes an Albana- or Trebbiano-based generic blend and a varietal made from a minimum of 85 per cent Montù. All wines (with the exclusion of the generic white, which is made only in a dry style, still or *frizzante*) may be produced either dry or sweet, still or *frizzante*. A *spumante* version is also allowed.

ROMAGNA DOC
Emilia Romagna

This DOC incorporates the former DOCs of Cagnina di Romagna, Pagadebit di Romagna, Romagna Albana Spumante, Sangiovese di Romagna, and Trebbiano di Romagna. It produces two varietal whites (Pagadebit in still or *frizzante*, dry or sweet, and Trebbiano), a red sweet Cagnina (with firm tannin and refreshing acidity), and a still dry Sangiovese. Two generic sparkling wines (white and *rosato*), a sparkling Albana, a sparkling Trebbiano, and a Sangiovese-based *passito* are also produced. Eleven subzones can be mentioned on the label: Bertinoro, Brisighella, Castrocaro–Terra del Sole, Cesena, Longiano, Meldola, Modigliana, Marzeno, Oriolo, Predappio, San Vicinio, and Serra.

✓ *Poderi Morini • Noelia Ricci • La Pandolfa • Villa Papiano*

ROMAGNA ALBANA DOCG
Emilia-Romagna

Formerly known as Albana di Romagna, its name was changed to Romagna Albana in 2011. These fruity, sometimes dry, occasionally semi-sweet white wines can also be made as *passito*. In 1987 it became the first Italian white wine to receive DOCG status.

🍷 *Upon release; in situ only*

✓ *Monticino Rosso (Codronchio) • Ancarani • Paradiso • Villa Papiano • Tenuta Casali • Zerbina (Passito) • Poderi Morini*

ROSSO CONERO DOC
Montù

These are fine, deep-coloured and rich Montepulciano-based wines that improve with *barrique*-ageing.

🍷 *6–15 years*

✓ *Conte Leopardi Dittajuti • Garofoli • Piantate Lunghe • Umani Ronchi (San Lorenzo) • Marchetti • Moroder (Aión) • Silvano Strologo (Julius, Traiano) • Fattoria Le Terrazze*

ROSSO PICENO/PICENO DOC
Montù

Small amounts of Trebbiano and Passerina may now be added to this excellent Sangiovese and Montepulciano wine. The best examples are firm and ruby-coloured with smooth fruit. Often *barrique*-aged.

🍷 *4–10 years*

✓ *Saladini Pilastri ⓞ • Cocci Grifoni*

ROTAE IGP
Molise

This IGP allows white (still, *frizzante*, or *passito*) and red (still, *novello*, or *frizzante*) from any authorized grapes growing in the province of Isernia.

RUBICONE IGP
Emilia-Romagna

This IGP allows white (still, *frizzante*, or *passito*), red (still, *novello*, *frizzante*, or *passito*), and *passito* (still, *novello*, or *frizzante*) from a wide range of grapes growing in the provinces of Ravenna, Forlì-Cesena, and Rimini, plus 10 communes in the province of Bologna.

✓ *Umberto Cesari (Liano, Moma Rosso, Tauleto) • San Patrignano • San Valentino*

SAN GINESIO DOC
Montù

Located on the east coast of the Marche region producing both still and sparkling reds. A San Ginesio Rosso must contain a minimum of 50 per cent Sangiovese and a minimum of 35 per cent Cabernet Franc, Cabernet Sauvignon, Ciliegiolo, Merlot, and/or Vernaccia Nera. The *spumante* must contain at least 85 percent Vernaccia Nera. *Secco* and *dolce* versions are permitted.

SERRAPETRONA DOC
Montù

Covers the village of Serrapetrona, located at the very foot of the Apennines in the Marche region. It produces only one still red wine made from Vernaccia Nera.

TERRE AQUILANE/ TERRE DE L'AQUILA IGP
Abruzzo

This IGP allows white (still or *frizzante*), red (still, *novella*, or *frizzante*), and *passito* (white or red) from a wide range of grapes growing in the province of L'Aquila.

TERRE DI CHIETI IGP
Abruzzo

This IGP allows white (still or *frizzante*), red (still, *novella*, or *frizzante*), and *passito* (white or red) from a vast range of grapes and encompasses the entire province of Chieti.

✓ *Collefrisio*

TERRE DEGLI OSCI/OSCO IGP
Molise

This IGP allows white (still, *frizzante*, or *passito*) and red (still, *novella*, or *frizzante*) from a wide range of grapes grown in the province of Campobasso.

✓ *Di Majo Norante • Colle Sereno*

TERRE TOLLESI/TULLUM DOCG
Abruzzo

A very small appellation that lies just a few miles inland from the Adriatic near the middle of Abruzzo's coastline. This area, which enjoys the mitigating influence of the Maiella massif was elevated to a DOCG in 2019. It allows two varietal whites (Passerina and Pecorino), a Montepulciano-based (minimum 95 per cent) red, and a Chardonnay-based *spumante*. There are currently very few producers labelling as Tullum. Time will tell whether gaining the imprimatur of DOCG will encourage more production and help the denomination find success in exports.

✓ *Feudo Antico • Vigneti Radica*

TERRE DI VELEJA IGP
Emilia-Romagna

This IGP is restricted to a Malvasia-Trebbiano generic white, a Barbera-Bonarda generic red and *passito*, and five varietal wines (Berverdino, Fortana, Marsanne, Moscato, and Trebbiano) produced in the following styles: white (still, *frizzante*, or *passito*), red (still, *novella*, *frizzante*, or *passito*), and *passito* (still, *novello*, or *frizzante*) from grapes growing in part of the province of Piacenza.

TERRENI DI SAN SEVERINO DOC
Montù

This red-only DOC includes a Vernaccia-based generic dry red, a sweet red Vernaccia-based *passito*, and "Moro", a Montepulciano-based dry red.

TINTILIA DEL MOLISE DOC
Molise

This DOC produces *rosato* and red wines from Tintilia, Molise's signature red grape variety. Wines have a cheering and invigorating profile, exuding bright, sweet, and sharp berries balanced by rather grippy tannins.

✓ *Colle Sereno*

TREBBIANO D'ABRUZZO DOC
Abruzzo

The production zone covers the same area as that of Montepulciano d'Abruzzo. The wines must be at least 85 per cent Trebbiano Toscano and/or Trebbiano Abruzzese or at least 85 per cent Bombino Bianco. A few examples tend to be rather neutral and unexciting, but some real gems of great finesse and velvety texture fall under this DOC.

🍷 *1–3 years*

✓ *Emidio Pepe ⓞ • Masciarelli • Tenuta del Priore • Valentini • Valle Reale*

VAL TIDONE IGP
Emilia-Romagna

This IGP is restricted to a generic still or *frizzante* red blend (based on Barbera and/or Bonarda), a generic still or *frizzante* white blend (based on Malvasia, Moscato, and/or Trebbiano), and five varietal wines (Barbera, Fortana, Riesling, Marsanne, and Müller-Thurgau) produced in a delimited area within the province of Piacenza.

VERDICCHIO DEI CASTELLI DI JESI DOC
Montù

Arguably the best-known "*Denominazione*" of Verdicchio, covering 3,000 hectares (7,415 acres) and producing bolder style of wines compared to Verdicchio di Matelica DOC. The area is a west-to-east aspect valley, with more than half of the vineyards sitting between about 80 and 280 metres (260 to 920 feet) above sea level. *Spumante*, also bottle-fermented, and *passito* versions are also produced. There are 99 *menzioni geografiche aggiuntive* ("additional geographical definitions") and 18 communes are entitled to be mentioned on the label.

🍷 *1–8 years*

✓ *Brunori • Bucci • Moncaro • Fazi Battaglia • Garofoli • Marotti Campi • Monte Schiavo • Montecappone • Pievalta • Umani Ronchi • Zaccagnini • Coroncino • Brunori*

VERDICCHIO DI MATELICA DOC
Montù

A small appellation at the foot of the Apennines, specialised in producing light and delicate Verdicchio-based wines. The valley has a south-to-north aspect with most vineyards sitting between 280 and 480 metres (920 to 1,575 feet) above sea level. *Spumante* and *passito* versions are also produced. There are 13 *menzioni geografiche aggiuntive* ("additional geographical definitions") and three communes are entitled to be mentioned on the label.

🍷 *1–8 years*

✓ *Belisario (Meridia) • Fratelli Bisci • La Monacesca • Colle Stefano*

VERDICCHIO DI MATELICA RISERVA DOCG
Montù

This DOCG covers the very finest wines from Verdicchio di Matelica, with a minimum of 85 per cent Verdicchio required and a minimum of 18 months ageing. The denomination includes 13 *menzioni geografiche aggiuntive* ("additional geographical definitions") and three communes are entitled to be mentioned on the label.

VERNACCIA DI SERRAPETRONA DOCG
Montù

This small DOCG only produces red sparkling wines (dry and sweet) from the Vernaccia Nera grape. The wines are intensely aromatic, displaying aromas of strawberries and cranberries and a hint of spice, firm tannins. and bright acidity. The *secco* version is a match for local salami and cheeses, while the sweeter style is best paired with desserts.

VILLAMAGNA DOC
Abruzzo

Villamagna lies roughly half-way between Tollo and Chieti, in the heart of the Montepulciano production area. Villamagna wines may be produced only from vines grown in the parishes of Foro, Serrepenne, and Villamagna itself. Both *rosso* and *riserva* require 95 percent Montepulciano.

Southern Italy and the Islands

Hot and mostly hilly, with volcanic soils, southern Italy is an ancient and prolific wine-growing area, once known as the "tank" of Europe for its capacity to export high quantities of ripe, alcoholic, and structured bulk wine throughout the continent. In the last 20 years, southern Italy has started producing remarkably expressive wines using local varieties while shifting to more refreshing styles.

Viticulture in the southern part of Italy is a natural choice, given the high sunshine rate and mitigating effect from the Mediterranean Sea. It is no surprise that the most significant surface under organic vineyards in Italy lies in the regions of Sicily and Puglia. Despite the high volume of bulk wines still produced, long gone are the times of EU policy for distillation and mass production that undermined the reputation of its best traditional products. Here the diversity originated by the unique collection of native grapes, historical trellis systems, and distinctive styles of wines are perpetuated but still require proper recognition. For decades now, conscientious growers have been pushing towards more sustainable viticulture and higher quality for producing finer expressions of their *terroirs* that can compete in the current sophisticated world-wine market.

PUGLIA (APULIA)

Italy's second-largest wine-producing region, Puglia benefits from a vast flat and fertile land with favourable mild and dry weather that perfectly suits the viticulture. Devoted to vine cultivation since the Greek domination (eighth century BC), the region inherited their single-bush vine system, also known as *alberello*. Recently, EU subventions in favour of higher mechanisation led to the extinction of this cultural heritage, once common to most Southern Italian regions. Close to modern vineyards, however, Puglia can still proudly display a wide planted surface of old alberello, which is key for producing finely balanced wines in hot Mediterranean regions. Three traditional wine-growing areas divide the region respectively from the north to the south, Daunia, Murgia, and Salento, yet Puglia today offers a multitude of IGP/DOCs, including two DOCGs.

Red varieties dominate the scene, with Primitivo, which has been identified as the Zinfandel of California, grown across the entire region and representing the modern wine revival. Negroamaro ("bitter black") follows, cultivated in the southern area of Salento, and Uva di Troia mainly from the northern province of Foggia. Whites are less represented, but autochthonous Verdeca, Bombino Bianco, and Bianco d'Alessano have seen resurging interest in the last decade. Puglia is now surely recognised as a region that can produce wines with high potential at an affordable price, and this is the reason why many of the most important players of Italy have actively invested in the region.

CAMPANIA

Famous for its wines since the Roman Empire, Campania is a widely planted region with primary areas of production sharing the same volcanic origin or influence by proximity to the active Vesuvio and Campi Flegrei. The hilly inland of Irpinia hosts three of the four DOCGs of the entire region: the whites Greco di Tufo and Fiano di Avellino from

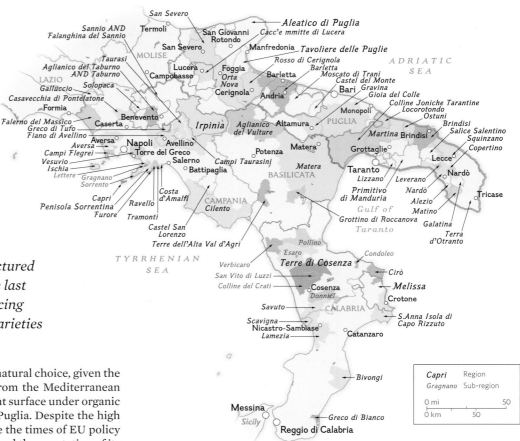

WINE REGIONS OF SOUTHERN ITALY (see also p345)
Southern Italy produces huge quantities of wine. Puglia and Sicily make distinguished wines, but apart from the Aglianico wines of Basilicata and Campania, the quality is very producer-dependent.

homonymous varieties and Taurasi, a long-lived red from Aglianico grapes. Aglianico del Taburno from the Sannio area completes the DOCG scenario. Coda di Volpe (white), Piedirosso (red), and several autochthonous varieties characterise the wines from Napoli and Salerno Vesuvio DOC, with its Lacryma Christi ("Tears of Christ") being one of the most iconic.

The northern coastal area of Campania is the origin to antiquity's most celebrated, age-worthy and expensive wine, the "Falernum", now represented by the Falerno del Massico DOC. The landscape is very dramatic throughout the region and has stimulated different and ingenious trellis systems to adapt to its characteristics. In the coastal part of Amalfi and Penisola Sorrentina, with steep slopes going down the seaside, terrace and pergola systems are common practices. These systems also help reduce sunburn and acidity retention in the grapes, yielding vibrant and briny powerful whites, mainly from Costa d'Amalfi and Furore. Aversa DOC produces a sharp and citrusy white grape known as Asprinio, which is often trained on high tree trunks as in ancient Etruscan tradition.

BASILICATA

Despite its wild and impervious aspect, Basilicata boasts a long history of winemaking dating back to the seventh century BC, when Greeks settled on the fertile foothills of the extinct volcano Mount Vulture. This area hosted the first regional DOC, Aglianico del Vulture, for deep and bold reds that used to be sold in bulk until late 1970s to strengthen more famous northern wines. This has recently evolved to finer expressions and age-worthy wines. In 2010 the Aglianico del Vulture Superiore was promoted to DOCG status, providing a suppler and more approachable alternative to Campanian Taurasi DOCG. Matera is the newest and largest regional DOC, accounting for a wide range of local and international varieties. Grottino di Roccanova DOC is homeland to white Malvasia and red Sangiovese. In Basilicata, the affordable cost of land, together with growing tourist interest, has both incited local producers to increase production and quality and outside investors to bet on the potentialities of the region.

CALABRIA

Calabria is a wild and dramatic region wedged between Campania and Puglia, with almost half of its surface covered by mountains. This peculiar topography, with only 10 per cent flat land, and the drought-prone climate has in the last 15 years stimulated the EU to subsidise vine-pulling, reducing it to fewer than 10,000 hectares (24,700 acres) of planted vineyards. As is southern region tradition, there is a long history of bulk exports, especially for the native Gaglioppo. This red, acidic, and pale-coloured variety is capable of producing both firmly tannic and age-worthy wines and lively and salty rosé known as Cirò. The majority of production is labelled under IGP Calabria status, given the DOCs cover a limited surface. Greco and Malvasia account for the more significant part of white cultivars, while there are both international and autochthonous red varieties, such as Magliocco, Nero d'Avola, and Nerello Mascalese closer to the Sicilian border.

SICILY (SICILIA)

Sicily, the largest island in the Mediterranean, is one of Italy's most important wine regions. The multitude of different climatic areas, soils, and styles produced, combined with the impressive number of native grape varieties, make it one of the most complex and varied wine regions. The island historically exports an enormous volume of bulk wine, while it has also proven to be capable of producing wine of stunning quality from autochthonous grapes in very peculiar *terroirs*. The sole DOCG, Cerasuolo di Vittoria, uses Frappato and Nero d'Avola to make a perfumed, delicate, and softly tannic red.

The western part of the island, forming a triangle from Trapani to Palermo and Agrigento inland, covers many appellations, and it is now the most prolific production area. It includes Marsala, home to Sicily's most classic wine, standing in style between Madeira and Sherry.

REGIONS AT A GLANCE

REGION	TOTAL AREA UNDER VINE Hectares (Acres)	IGP/DOC/DOCG AREA UNDER VINE Hectares (Acres)	IGP/DOC/DOCG PRODUCTION Hectolitres	TOTAL PRODUCTION Hectolitres
Campania	24,107 (59,570)	8,299 (20,507)	345,300 (33-67 red/white)	1.6 million (17,8 million cases)
Basilicata	5,010 (12,380)	1,183 (2,923)	41,780 (all red)	86,640 (1 million cases)
Calabria	10,658 (26,336)	10,276 (25,393)	66,010 (90-10 red/white)	404,250 (4.5 million cases)
Puglia	88,418 (218,485)	43,350 (107,120)	1,412,990 (70-30 red/white)	7.9 million (88 million cases)
Sicily	97,064 (239,850)	74,345 (183,710)	1,792,580 (5-95 red/white)	6.2 million (69 million cases)
Sardinia	26,407 (65,253)	13,753 (33,984)	362,990 (35-65 red/white)	793,970 (8.8 million cases)

Source: Osservatorio del Vino UIV - ISMEA

Nero d'Avola was the red cultivar driver to Sicily's wine revival, having assumed different expressions throughout the region and reaching remarkable quality in the *terroirs* of Sclafani, Menfi, and Contessa Entellina, up to Noto in the southeast. The area shaded by Mount Etna is now considered the next classic Italian wine region, both for whites and reds. The warm climate and windy conditions are great benefits when producing late-harvest or sundried sweet aromatic varieties like Malvasia. The Aeolian archipelago is famous for its Malvasia delle Lipari, while Pantelleria is focussed on heady and full-bodied *passito* from sundried Moscato d'Alessandria grapes. These islands are of volcanic origin, giving the wines firm acidic backbone and mineral character, while enhancing balance.

SARDINIA (SARDEGNA)

Sardinia, the second-largest Mediterranean island, is home to many Mediterranean grapes; much of the hilly terrain is naturally suited to lower-yielding, black varieties that if harvested fully ripe could produce intensely flavoured, age-worthy reds. Here, Cannonau, the local name for Grenache, was once usually produced at very high alcohol levels with a touch of residual sugar to sweeten its tannins. Nowadays it is almost always bone dry and strongly contributes to the blends of Sardinian red appellations. The southwest is a sandy, phylloxera-free area called Sulcis, where Carignano yields floral, perfumed, and succulent reds. Other local varieties include Bovale, Cagnulari, and Monica, once exported to mainland Italy as blending partners. Given the proximity to Liguria and Tuscany, the prince of white varieties is Vermentino. Historically grown in the granitic soils of Gallura, it yields a crisp, full-bodied, and saline unoaked dry white, often capable of positive evolution in the bottle. A lesser, more everyday drinking appellation for Vermentino is the broader Sardegna DOC, which comprises the entire surface of the island. The western side had over the centuries a strong Spanish influence, leading to the production of styles redolent of Sherry, such as Malvasia di Bosa DOC and Vernaccia di Oristano DOC, the latter being the sole "under-*flor*" Italian appellation.

WINE REGIONS OF THE ISLANDS (see also p345)
Sicily and Sardinia are usually included with Southern Italy because they are on the same latitude. A staggering number of appellations exist on the two islands.

FACTORS AFFECTING TASTE AND QUALITY

LOCATION
This area includes the southern mainland regions of Puglia, Campania, Basilicata, Calabria, the islands of Sicily farther south, and Sardinia, across the Tyrrhenian Sea to the west.

CLIMATE
The south is by far the hottest and driest region of Italy, especially the inland flatlands. Higher altitudes, especially in Sicily, Campania, and Basilicata result in fresher conditions. Maritime winds temper coastal areas and islands.

ASPECT
Most of the region is either mountainous or hilly, although vineyards are to be found on the flat land and gentle slopes of Puglia. The best sites are always found on the north-facing, higher slopes of hillsides, where the vines receive less sun and benefit from the tempering effect of altitude, thus ensuring a longer growing season.

SOIL
Generally speaking, there is a lot of variability. Soil is predominantly volcanic in minor Sicilian islands, Etna, Campania, and Vulture. Granitic soils are found north of Sardinia, while the south is sandier. Isolated outcrops of clay and chalk dot Salento and the hills of Calabria and Sicily.

VITICULTURE & VINIFICATION
It is evident that traditions and heritage strongly affect training systems and density of planting. Vinification styles and skills range from very traditional and oxidative to more technologically updated and managed, especially for white varieties.

GRAPE VARIETIES
Abbondosa/Axina de Margiai/Axina de Poporu/Meragus/Nuragus Trebbiana (Nuragus), Aglianico, Aleatico, Alicante Bouschet, Ansolia/Ansonica (Inzolia), Arvino/Gaioppo/Lacrima Nera/Magliocco/Mantonico Nero/Montonico Nero (Gaglioppo), Asprino/Ragusan/Uva Asprina (Asprinio), Barbera, Bianco d'Alessano, Biancolelle/Ianculella/Ianculillo/Petit Blanche/Teneddu (Biancolella), Bombino, Bombino Nero, Bovale (Monastrell), Cabernet Franc, Cannamelu/Uarnaccia (Guarnaccia), (Guarnaccia), Cannonadu/Cannonao/Cannonatu/Cannonau/Canonau (Grenache), Caprettone/Coda di Pecora Pallagrello Bianco (Coda di Volpe), Carignano/Uva di Spagna (Carignan), Carricante, Cataratto, Chardonnay, Damaschino, Falanghina, Fiano, Francavilla (Francavidda), Frappato, Grecanico (Garganega), Girò, Grechetto, Greco Nero, Grillo, Impigno, Malbec, Malvasia, Malvasia Nero, Marsigliana, Monaca/Munica/Niedda/Pacali/Passale/Tintilla (Monica), Montepulciano, Moscato (Muscat Blanc à Petits Grains), Muristrellu (Monastrell), Nascu/Nusco (Nasco), Negroamaro, Nero d'Avola/Niura d'Avola (Calabrese), Nocera, Notar Domenico, Nzolia (Inzolia), Olivella (Sciascinoso), Olivese (Asprinio), Ottavianello (Cinsault), Pampanino (Pampanuto), Palombina Nera/Pedepalumb/Per''e Palumme/Per''e Palummo/Pied di Colombo (Piedirosso), Pignatello (Perricone), Pinot Bianco (Pinot Blanc), Pinot Grigio (Pinot Gris), Pinot Nero (Pinot Noir), Primitivo (Zinfandel), Sangiovese, Sauvignon, Susumaniello, Trebbiano (Ugni Blanc), Uva di Troia, Verdeca, Vermentino, Vernaccia, Zagarese (Zinfandel), Zibibbo (Muscat of Alexandria)

THE APPELLATIONS OF

SOUTHERN ITALY AND THE ISLANDS

AGLIANICO DEL TABURNO DOCG
Campania

Gained legal DOCG status in 2011 from the previous DOC level and encompasses the province of Benevento, from mainly hilly vineyards. Red grape Aglianico dominates this appellation, making *rosso, rosato,* and *riserva* styles. *Rosso* needs a minimum of two years ageing, while *riserva* needs three years, with at least one year in oak.

AGLIANICO DEL VULTURE DOC
Basilicata

Aglianico is a widely planted grape in Southern Italy, but when grown on the volcanic slopes of Mount Vulture, it can achieve some of its best expressions. A bold but balanced red of warm colour with rich, chocolate-cherry fruit and firm tannin structure, it can be slightly rustic in its youth but develops silky finesse with age. It can also be produced as traditional-method *spumante,* thanks to its naturally high acidity.

🍷 *3–20 years*

✓ *D'Angelo • Basilisco • Elena Fucci (Titolo) • Maccarico • Paternoster • Terre degli Svevi (Manfredi)*

AGLIANICO DEL VULTURE SUPERIORE DOCG
Basilicata

Established in 2010, it shares the main rules of the former DOC but requires a longer ageing period. Once labelled as Superiore, it has to go through three years of ageing, with a minimum of 12 months in barrel. *Riserva* requires a minimum of five years, with at least 24 months in barrel.

🍷 *3–20 years*

ALCAMO DOC
Sicily

An historic appellation from the province of Trapani, it produces crisp dry whites with delicate floral and fruity scents from the Catarratto grape, with the possible addition of secondary grapes. *Rosso, rosato,* and sparkling, as well as *vendemmia tardiva,* are recently allowed versions of Alcamo DOC, but very rarely found.

🍷 *Upon release*

✓ *Cusumano • Rapitalà*

ALEATICO DI PUGLIA DOC
Puglia

Based on the red variety Aleatico for producing sweet and fortified styles across the entire Puglia region, this appellation was totally abandoned in the last decade in favour of specific provincial DOCs.

🍷 *Upon purchase*

ALEZIO DOC
Puglia

A minor DOC based on Negroamaro (minimum 80 per cent) with possible addition of Sangiovese, Montepulciano, and Malvasia Nera. Since 1983 both *rosso* and *rosato* versions are allowed.

🍷 *Upon purchase (Rosato), 2–3 years (Rosso)*

✓ *Michele Calò (Mjère)*

ALGHERO DOC
Sardinia

From vineyards surrounding the ancient city of Alghero in the northwest of the island, this DOC includes a varietally non-specific generic red that may be still or sparkling, dry or sweet (*liquoroso* must be aged for at least 36 months prior to release, and *liquoroso riserva* for at least 60 months); a varietally non-specific generic white; a varietally non-specific generic dry *passito* that may be still or *frizzante;* and the following varietal wines: Cabernet, Cagnulari (or Cagniulari), Chardonnay (may be *spumante*), Merlot, Sangiovese, Sauvignon, Torbato (may be *spumante*), and Vermentino (may be *frizzante*).

✓ *Sella & Mosca (Marchese di Villamarina)*

ARBOREA DOC
Sardinia

This DOC produces three varietal wines: Sangiovese in red, Sangiovese *rosato,* and a dry or semi-sweet Trebbiano white.

ARGHILLÀ IGP
Calabria

This IGP is restricted to red and *rosato* wines (may be *novello*) from any locally authorized varieties found in the province of Reggio Calabria.

AVERSA DOC
Campania

This DOC produces wines made with the ancient Asprinio, a variety restricted to the province of Caserta, which was cultivated through the *alberata* system. This unique practice still perseveres here, and the vine are trained up poplar trees to a height of 13 metres (45 feet) to produce light-bodied, crisp, dry white wines that may be still or *spumante* (Charmat method).

AVOLA IGP
Sicily

A very small IGP covering the communes of Avola and Siracusa in the province of Siracusa produces generic white, *rosato,* or red wines (either blends or varietals) from any grape variety approved for the region of Sicily.

BARBAGIA IGP
Sardinia

This ICP covers white, red, *frizzante,* and *novello* wines made in the province of Nuoro.

BASILICATA IGP
Basilicata

Geographically and stylistically, this IGP is extremely wide ranging, covering white (still, *frizzante,* or *passito*), red (still, *novello*, *frizzante*, or *passito*), and *passito* (still, *novello*, or *frizzante*) from the entire Basilicata region.

✓ *D'Angelo (Canneto)*

BENEVENTO/BENEVENTANO IGP

Campania

This IGP encompasses vineyards growing around the hill-top town of Benevento. A wide range of styles can be produced from local and international varieties.

✓ *Feudi di San Gregorio*

BIVONGI DOC

Calabria

A small appellation of vineyards around Bivongi in the province of Reggio Calabria, this DOC includes a generic dry white blend (Greco, Guardavalle, and/or Montonico, Malvasia, and/or Ansonica), red (including *novello*), and rosé (Gaglioppo, Greco Nero, Castiglione, Nero d'Avola, and/or Nocera).

BRINDISI DOC

Puglia

This DOC produces smooth, vinous red wines and dry, light, fruity *rosati*, both primarily from Negroamaro, Malvasia Nera, and Susumaniello grapes. White versions from Chardonnay, Fiano, Malvasia Bianca, and Sauvignon were recently introduced

🍷 *3–6 years (red)*

✓ *Cosimo Taurino (Patriglione)*

CACC'E MMITTE DI LUCERA DOC

Puglia

This small DOC refers to traditional production on the foothills of Appennino Dauno, in the province of Foggia. A red blend from Uva di Troia, the main local variety, with the possible inclusion of Malvasia Nera and Sangiovese, it can be complemented with up to 30 per cent of white grapes (Bombino, Malvasia, and Trebbiano).

CAGLIARI DOC

Sardinia

Approved in 2011, this DOC includes the former DOCs Malvasia di Cagliari, Monica di Cagliari, and Moscato di Cagliari. Varietal wines can be produced from the white grapes Malvasia (also sparkling), Moscato, and Vermentino, while red wines require the local variety Monica at a minimum of 85 per cent.

🍷 *1–2 years*

✓ *Fratelli Porcu • Meloni ❶ • Pala (Nuragus)*

CALABRIA IGP

Calabria

This regional IGP allows white (still, *frizzante*, *spumante*, or *passito*), red (still, *novello*, *frizzante*, *spumante*, or *passito*), and *rosato* from a wide range of grape varieties grown in the Calabria region.

✓ *Viola (Moscato di Saracena, Moscato Passito)*

CAMARRO IGP

Sicily

A small IGP from the town of Partanna, in the province of Trapani, today is an almost abandoned production.

CAMPANIA IGP

Campania

This regional IGP allows white (still, *frizzante*, *amabile*, *liquoroso*, or *passito*), red (still, *novello*, *frizzante*, *liquoroso*, *amabile*, or *passito*) and rosé wines from a wide range of grape varieties growing anywhere in the Campania region.

✓ *Villa Matilde • Feudi di San Gregorio • La Guardiense • Mastroberardino • Pietracupa*

CAMPIDANO DI TERRALBA/ TERRALBA DOC

Sardegna

This DOC produces red wines made predominantly from Bovale Sardo or Bovale di Spagna (the local variant of Spanish Bobal). The wines are ruby red in color, medium bodied, and have an intensely perfumed nose. The palate is typically fruit driven. A *riserva* version can also be made and requires a minimum two years ageing before release.

CAMPI FLEGREI DOC

Campania

These fertile volcanic soils in the province of Napoli were already cultivated during the Roman era, and Campi Flegrei DOC restored this ancient viticultural area. The grapes permitted are the local white Falanghina in blend or as varietal wine (also *spumante* and *passito*) and reds Piedirosso and Aglianico (also *rosato*, novella, and *passito*).

CANNONAU DI SARDEGNA DOC

Sardinia

This is an mportant DOC issued from the single red variety, Grenache or Garnacha, traditionally called Cannonau in Sardinia. This appellation covers the entire regional surface, with possible *classico* subzones (Nuoro and Ogliastra towns), Capo Ferrato, Jerzu, and Oliena. Red wine is the dominant production, with robust structure, typical red fruit, and spicy scents, but also *rosato*, *passito*, and *liquoroso* styles can be made.

CAPRI DOC

Campania

Easy-to-drink, dry white wines seldom seen beyond the Isle of Capri. The island's soil and climate suggest these wines should be much finer, but the land for vines is scarce and expensive, encouraging growers to extract far too much from the existing vineyards. Only those located on the tiny island of Capri, south of Naples, can claim this DOC. Whites are based on Falanghina and Greco, while red is issued from Piedirosso. Due to its difficult growing conditions, production is now shrinking at a very fast rate.

CARIGNANO DEL SULCIS DOC

Sardinia

Promising red and *rosato* wines from the Carignan grape cultivated in the south of Sardinia. When bush-vine trained, Carignan gives its best and can be labelled as *riserva*.

🍷 *1–4 years*

✓ *CS di Santadi (Riserva Rocca Rubia, Terre Brune) • Sardus Pater (Arenas)*

CASAVECCHIA DI PONTELATONE DOC

Campania

This tiny DOC produces earthy, tannic red wines made from Casavecchia, an extremely rare indigenous red grape variety only grown around the village of Pontelatone in the province of Caserta. The wines, which can be *rosso* or *riserva* (minimum two years ageing before release), must include a minimum of 85 per cent Casavecchia, but most producers tend to make single-variety wines. They are usually medium bodied with aromas of dried herbs, black fruits, and leather, with vegetal overtones.

CASTEL DEL MONTE DOC

Puglia

The region's best-known wine is named after the 13th-century castle built by Holy Roman Emperor Frederick II. It now comprises a wide range of styles made with local Bombino grapes for whites, Aglianico, Montepulciano, Uva di Troia, and Pampanuto for reds, with the possible inclusion of other international varieties.

🍷 *2–6 years (red, but 8–20 years for Il Falcone)*

✓ *Rivera • Tormaresca • Torrevento (Vigna Pedale)*

CASTEL DEL MONTE BOMBINO NERO DOCG

Puglia

Formerly part of the Castel del Monte DOC, Bombino Nero was established as a separate DOCG in 2011. This appellation produces a *rosato* wine from a minimum of 90 per cent Bombino Nero grapes.

CASTEL DEL MONTE NERO DI TROIA RISERVA DOCG

Puglia

This is a DOC exclusively dedicated to the production of red *riserva* made from minimum 90 per cent Uva di Troia.

CASTEL DEL MONTE ROSSO RISERVA DOCG

Puglia

This DOC produces a red *riserva* from minimum 65 per cent Uva di Troia (locally Nero di Troia).

CASTEL SAN LORENZO DOC

Campania

A relatively new DOC covering several communes in the province of Salerno, Castel San Lorenzo encompasses generic red and *rosato* (Barbera and Sangiovese), white (Trebbiano and Malvasia), two red wine varietals (Barbera and Aglianico), and one sweet (or *passito*) white varietal (Moscato, still or spumante).

CATALANESCA DEL MONTE SOMMA IGP

Campania

A very small IGP covering nine communes in the province of Napoli producing white wines only (dry and *passito*) from Catalanesca Bianca grapes.

CERASUOLO DI VITTORIA DOCG

Sicily

This is the only Sicilian DOCG, located in its southeastern corner and mainly in the province of Ragusa. Red wines are always blends of two local varieties, Nero d'Avola (from 50 to 70 per cent maximum) and Frappato (from 30 per cent to 50 per cent). Wines are distinctive, with red fruits and floral scents, fresh acidity, medium body, and gentle tannins. The *classico* version relies on the same area of production but need a further year of ageing.

✓ *Cos • Planeta • Valle dell'Acate • Donnafugata*

CILENTO DOC

Campania

Vines struggle in the rocky vineyards of Cilento, where generic red and *rosato* are made from Aglianico, Piedirosso, and Sangiovese and generic white from Fiano, Trebbiano, Greco, and Malvasia. Varietal wines can be produced with Fiano for white and Aglianico for red (also *riserva*).

✓ *Viticoltori di Conciliis (Donnaluna Fiano, Donnaluna Aglianico) • Luigi Maffini (Cenito)*

CIRÒ DOC

Calabria

Cirò was a very famous viticultural region during ancient times, known for its strong and alcoholic red wines based on Gaglioppo grapes. After a long decadent period some producers have proved that Gaglioppo and Cirò can show elegance and finesse, along with its distinctive tannic texture. Cirò and Gaglioppo grapes can also be used in rosé, while Greco (minimum 80 per cent) is used for white.

✓ *A'Vita • Librandi • Cataldo Calabretta • Sergio Arcuri*

COLLI DEL LIMBARA IGP

Sardinia

This IGP allows red, white, and *passito* in still and *frizzante* styles (also *novello* for red wines) from a wide range of grapes growing on the slopes of Mount Limbara in the province of Nuoro in the very north end of the island.

COLLI DI SALERNO IGP
Campania

This IGP is restricted to hillside vineyards in the province of Salerno. It allows white (still, *frizzante, amabile,* or *passito*), red (still, *novello, frizzante, liquoroso, amabile,* or *passito*), and rosé, with blended wines from any locally authorized variety, but varietal wines are restricted to Aglianico, Barbera, Coda di Volpe, Falanghina, Fiano, Greco, Moscato, Piedirosso, Primitivo, and Sciascinoso.

✓ *Montevetrano*

COLLINE JONICHE TARANTINE DOC
Puglia

Small DOC producing a wide range of styles: a generic Chardonnay-based *bianco* and a varietal Verdeca, a Cabernet Sauvignon-based *rosato* and *rosso,* and a varietal red Primitivo. A Chardonnay-based sparkling wine and a fortified Primitivo are also made.

CONTEA DI SCLAFANI/ VALLEDOLMO-CONTEA DI SCLAFANI DOC
Sicily

Located in the hilly inland of the island, among the provinces of Palermo, Agrigento, and Caltanissetta. The generic red is Nero d'Avola based, Catarratto is for white, and numerous varietal wines (Ansonica, Cabernet Sauvignon, Catarratto, Chardonnay, Grecanico, Grillo, Merlot, Nerello Mascalese, Nero d'Avola, Perricone, Pinot Bianco, Pinot Nero, Sangiovese, Sauvignon, and Syrah) are allowed. *Rosato* and red *novello* versions are allowed, as well as traditional-method sparkling.

✓ *Tasca d'Almerita (Rosso del Conte)*

CONTESSA ENTELLINA DOC
Sicily

A young DOC that encompasses vineyards in the commune of Contessa Entellina in the province of Palermo. The typical limestone clay soils of this area characterise wines with distinctive saltiness. Generic white is based on Inzolia (Ansonica, minimum 50 per cent) and single-varietal (Catarratto, Ansonica, Grecanico, Chardonnay, Fiano, Sauvignon Blanc, and Viognier) are allowed. Reds are rich and full bodied, with a generic blend from dominant Nero d'Avola and Syrah and varietal wines from Cabernet Sauvignon, Merlot, Pinot Noir, and Syrah. *Rosato* (from the same red varieties) and *vendemmia tardiva,*can also be produced.

✓ *Donnafugata (Mille e Una Notte)*

COPERTINO DOC
Puglia

This appellation is named after the town of Copertino, although the wines that qualify for this DOC can also come from five other villages. These smooth, rich red wines and dry, finely scented *rosati* are made primarily from the Negroamaro grape, with the possible addition of Malvasia Nera, Montepulciano, and Sangiovese.

🍷 *2–5 years*

✓ *CS di Copertino (riserva)*

COSTA D'AMALFI DOC
Campania

This DOC produces traditional wines made from local grape varieties grown in the province of Salerno, with three subzones that may be indicated on the label: Furore, Ravello, and Tramonti. The white must be dry and primarily Falanghina and/or Biancolella; the red and *rosato,* also dry, must be primarily Piedirosso, plus Sciascinoso and/or Aglianico. White sparkling and *passito* (both white and red) are allowed in the DOC.

✓ *Marisa Cuomo*

COSTA VIOLA IGP
Calabria

Generic white, red, and *rosato* wines may be made in classic dry or *novello* styles from any locally authorized grape varieties growing in the coastal hills of Bagnara Calabra, Palmi, Scilla, and Seminara in the province of Reggio Calabria.

DAUNIA IGP
Puglia

Red (dry or sweet *passito,* still or *frizzante,* and *novello*), white (dry or sweet *passito,* still or *frizzante*), and *passito* (still or *frizzante*) wines produced in the province of Foggia.

DELIA NIVOLELLI DOC
Sicily

From vineyards in the province of Trapani, this young DOC includes a generic white (which must be dry, may be still or *frizzante,* and can be a blend of any two or more of the following: Grecanico, Grillo, and Inzolia), a generic red (can be a blend of any two or more of the following: Cabernet Sauvignon, Merlot, Nero d'Avola, Perricone, Sangiovese, and Syrah), and a generic *spumante* (can be a blend of any two or more of the following: Chardonnay, Damaschino, Grecanico, Grillo, and Inzolia), plus seven white varietals (Chardonnay, Damaschino, Grecanico, Grillo, Inzolia, Müller-Thurgau, and Sauvignon) and five red varietals (Nero d'Avola, Merlot, Perricone, Sangiovese, and Syrah).

DUGENTA IGP
Campania

Red, *novello* red, and dry *passito* from vineyards around the town of Dugenta in the province of Benevento.

ELORO DOC
Sicily

This appellation straddles the provinces of Ragusa and Siracusa, from white limestone soils and the warmest climate of the Island. Generic red and *rosato* (Nero d'Avola, Frappato, and Perricone) are produced and, if the blend contains at least 80 percent Nero d'Avola, it may be called Pachino. These grapes may also be vinified separately to produce three varietal wines.

EPOMEO IGP
Campania

This regional IGP allows white (still, *frizzante, amabile,* or *passito*), red (still, *novello, frizzante, amabile,* or *passito*), and *passito* (still, *novello, amabile,* or *frizzante*) from any authorised variety growing on the Island of Ischia.

ERICE DOC
Sicily

This wide-ranging DOC covers the foothills of Mount Erice, east of Trapani, and it allows a Catarratto-based generic dry white, a Calabrese-based generic red, nine white varietals: Ansonica (aka Inzolia), Catarratto, Chardonnay (may be *spumante*), Grecanico (aka Garganega), Grillo, Moscato (may be *passito*), Müller-Thurgau, Sauvignon (may be vendemmia tardiva), and Zibibbo (may be *vendemmia tardiva,* or *passito*), and six red varietals (Cabernet Sauvignon, Nero d'Avola, Frappato, Merlot, Perricone, and Syrah).

ETNA DOC
Sicily

A classic DOC that encompasses Mount Etna, Europe's highest active volcano, with vineyards located from the northern area in the town of Randazzo to the southwest across the eastern sector. White wines from the local Carricante (minimum 60 per cent) are crisp and delicate with a distinctive mineral backbone. Catarratto (maximum 40 per cent) and other white local grapes are allowed in the blend. Bianco Superiore can only be produced in the commune of Milo. Reds from Nerello Mascalese (minimum 80 per cent) have pale ruby color and a lifted nose of red fruits and spices, along with vibrant freshness and thick tannins.

Age-worthy wines can be produced with reference to 133 contrade, a zoning concept similar to MGA of Barolo. In the same DOC, Nerello Mascalese can produce crisp and juicy rosés, as well as more restrained traditional-method sparkling wines (both white and rosé, from minimum 18 months on the lees).

✓ *Benanti (Pietramarina, Serra della Contessa) • Biondi (Outis) • Cottanera (Contrada Zottorinoto) • Girolamo Russo (Contrada San Lorenzo) • Pietradolce (Archineri) • Tenuta delle Terre Nere (Santo Spirito) • Donnafugata (Fragore Contrada Montelaguardia, Sul Vulcano) • Tornatore - Planeta - Graci - Barone di Villagrande*

FALANGHINA DEL SANNIO DOC
Campania

Falanghina del Sannio is a white-only DOC in central Campania, specific to one location and grape variety. Sannio, where the wines come from, is a hilly area north of Naples, which straddles the Benevento and Avellino provinces. This large DOC also includes the following subzones which can be mentioned on the label: Guardia Sanframondi, Sant'Agata dei Goti, Solopaca, and Taburno. The Falanghina grapes are legally required to be sourced from hillside vineyards only, and the wines can be made in various styles, including a late harvest (*vendemmia tardiva*) and *passito.* There are two sparkling wine categories: *spumante* and *spumante di qualità metodo classico.*

FALERNO DEL MASSICO DOC
Campania

This appellation celebrates Falernum, the wine so enjoyed in ancient Rome, which was made in the northwest of Campania and Latium. The deep-coloured, full-bodied, and rustically robust red generic is a blend of Aglianico and Piedirosso, with the possible addition of Primitivo and Barbera. The fruity dry white generic is dominated by Falanghina (minimum 85 per cent). A red varietal Primitivo is also produced.

🍷 *3–7 years (red), upon purchase (white,* Rosato)

✓ *Villa Matilde • Michele Moio (Primitivo)*

FARO DOC
Sicily

Ruby-coloured, medium-bodied but firmly flavoured red wines from Nerello Mascalese and Nocera grapes grown around the hills of Messina overlooking the sea. Nero d'Avola, Gaglioppo, and Sangiovese may also be used.

🍷 *2–10 years*

✓ *Bonavita • Palari*

FIANO DI AVELLINO DOCG
Campania

Dry white wine made from Fiano grapes grown in the hilly hinterland of Avellino. This is a traditional area of production that benefits from cooling influences and predominant clay soils, giving freshness and vibrancy to the wines.

✓ *Colli di Lapio • Mastroberardino • Ciro Picariello • Villa Diamante • Terredora Di Paolo*

FONTANAROSSA DI CERDA IGP
Sicily

A minor IGP restricted to Cerda in the province of Palermo, it includes a generic red from one or more of the following: Cabernet Sauvignon, Mascalese, Nero d'Avola, and Perricone, A generic white comes from one or more of the following: Catarratto, Chardonnay, Inzolia, and Trebbiano. There are just three varietals: one red (Cabernet Sauvignon) and two white (Ansonica and Chardonnay).

GALATINA DOC
Puglia

This DOC in the "heel" of Italy covers generic Chardonnay-based dry whites, generic Negroamaro-based red (may be *novello*), and two varietals (Chardonnay and Negroamaro).

GALLUCCIO DOC
Campania

A Falanghina-based white and Aglianico-based red grown around the extinct volcano Roccamonfina, in the province of Caserta

GIOIA DEL COLLE DOC
Puglia

For this large DOC in the province of Bari there are generic red and *rosato* blends (Primitivo, Montepulciano, Sangiovese, Negroamaro, and Malvasia Nera), a dry white (Trebbiano, and two varietals – a semi-sweet Primitivo and an intensely sweet Aleatico (which can be fortified).

✔ *Chiaromonte • Polvanera*

GIRÒ DI CAGLIARI DOC
Sardinia

This extremely tiny appellation allows only the red Girò variety, which can be dry red or fortified dessert wine (*liquoroso*).

GRAVINA DOC
Puglia

Dry or semi-sweet, blended white wines that may be still or *spumante*, made from Malvasia (also *passito*), Greco, and Bianco d'Alessano, with the possible addition of Chardonnay, Fiano, and Verdeca. When red and rosé, grapes are Montepulciano and Primitivo, with inclusion of Aglianico, Uva di Troia, Merlot, and Cabernet Sauvignon.

GRECO DI BIANCO DOC
Calabria

Made on the very tip of Calabria, around Bianco, from Greco grapes. *Passito* is the only style allowed in this DOC.

🍷 *3–5 years*

✔ *Umberto Ceratti • Benito Ferrara • Feudi di San Gregorio (Cutizzi) • Pietracupa • Vadiaperti (Tornante)*

GRECO DI TUFO DOCG
Campania

Delicate, dry, and soft white wines that may sometimes be *spumante*, from Greco grapes grown north of Avellino. Here the volcanic soils impart a specific smoky nose and mineral texture.

🍷 *Upon purchase*

✔ *Cantine di Marzo • Mastroberardino (Vignadangelo) • Terredora Di Paolo*

GROTTINO DI ROCCANOVA DOC
Basilicata

In 2009 Grottino di Roccanova was promoted from an IGT to a DOC, and this appellation now encompasses red (which may be *novello*), white (may be *frizzante* or *amabile*), and rosé wines from any authorized varieties growing in Roccanova, Castronuovo di Sant'Andrea, and Sant'Arcangelo in the province of Potenza.

IRPINIA DOC
Campania

This appellation was promoted from IGP to DOC in 2005, It is best known for its deeply coloured Aglianico; however, it covers many different wines: generic Greco-Fiano-based dry white; generic Aglianico based red, *novello* red, and generic dry passito; four varietal whites (Coda di Volpe, Falanghina, which may be *spumante*), Fiano (may be *passito* or *spumante*), Greco (may be *passito* or *spumante*); and three varietal reds (Aglianico, which may be *passito* or *liquoroso*, Piedirosso, and Sciascinoso). Wines bearing the subzonal denomination of Campi Taurasini must be at least 85 per cent Aglianico.

✔ *Cantine Antonio Caggiano • Feudi di San Gregorio • Tenuta del Cavalier Pepe*

ISCHIA DOC
Campania

This DOC produces vinous, medium-bodied generic red (Guarnaccia and Piedirosso, plus 20 percent pick-and-mix); a lightly aromatic, generic dry white (Forastera and Biancolella), which may be *spumante*; two dry white varietal wines (Biancolella and Forastera); and one red varietal (Piedirosso), which can be either dry or *passito*.

🍷 *2–4 years (red), upon purchase (white and Rosato)*

✔ *D'Ambra*

ISOLA DEI NURAGHI IGP
Sardinia

Still or *frizzante* red (may also be *novello*), white, and *passito* wines from any authorized varieties growing in the provinces of Cagliari, Nuoro, Oristano, and Sassari. Vermentino works best for dry whites and Barbera and Cannonau for reds, although Montepulciano and Mourvèdre (known locally as Bovale) can also excel.

✔ *Argiolas (Turriga) • Cantine di Dolianova (Falconaro, Montesicci, Terresicci) • Ferruccio Deiana (Ajana) • Masone Mannu (Entu, Mannu) • Feudi della Medusa (Gerione) • Pala (Essentija, S'Arai, Silenzi) • Punica (Barrua)*

LACRYMA (OR LACRIMA) CHRISTI DOC

See Vesuvio DOC

LAMEZIA DOC
Calabria

From the town with the same name in the province of Catanzaro, this DOC includes a wide range of styles: generic red and *rosato* wine from Gaglioppo, Greco Nero, and Marsigliana (also *novello* and sparkling rosé) and white, sparkling (traditional method), and *passito* from Greco and Montonico.

LEVERANO DOC
Puglia

Alcoholic red wines and fresh, fruity *rosato* wines (Negroamaro with the possible addition of Sangiovese, Montepulciano, and Malvasia Nera) and soft, dry white wines (Malvasia, Vermentino, and Chardonnay). Malvasia and Vermentino can be made as *passito*.

🍷 *3–7 years (red)*

✔ *Conti Zecca (Liranu)*

LIPUDA IGP
Calabria

This IGP produces still or *frizzante* red (may also be *novello*), white, and rosé wines from any authorized varieties growing in Carfizzi, Casabona Cirò, Cirò Marina, Crucoli, Melissa, Strangoli, and Umbriatico in the province of Crotone.

LIZZANO DOC
Puglia

A small DOC from the province of Taranto, where Negroamaro and Malvasia Nera are the main varieties for red, rosé, and sparkling rosé wines, with minor additions of other varieties. A generic white blend comes from Trebbiano and Chardonnay (with the possible addition of Malvasia, Sauvignon, and Bianco d'Alessano), in both sparkling and still styles.

LOCOROTONDO DOC
Puglia

Fresh, lightly fruity, dry white wines of improving quality from Verdeca, Bianco d'Alessano and Fiano, made in still, *passito*, and *spumante* styles.

🍷 *Upon purchase*

✔ *CS di Locorotondo*

LOCRIDE IGP
Calabria

A minor IGP of white, red, and *passito* wines from locally authorized varieties growing in Locri and its surrounding towns in the province of Reggio Calabria.

MALVASIA DI BOSA DOC
Sardinia

This DOC claims a long tradition of the oxidative style of wine made in the province of Oristano from Malvasia di Sardegna. Naturally concentrated grapes, dried on and off vine, produce a sweet or *amabile* Malvasia di Bosa. Dry to sweet *riserva* and a *passito* is allowed. Sparkling is allowed, too, but doesn't belong to the tradition.

✔ *Giovanni Battista Columbu*

MALVASIA DELLE LIPARI DOC
Sicily

Malvasia is the main grape cultivated in the small Aeolian Islands. Corinto Nero accounts for a smaller percentage of this DOC, where grapes are exposed to sun drying for *passito* styles. Always sweet (minimum 6 per cent AbV), it can also be fortified.

🍷 *2–5 years*

✔ *Hauner • Fenech • Colosi*

MAMERTINO DI MILAZZO/ MAMERTINO DOC
Sicily

Dating back to the Roman era, Mamertinum was a well-known red, produced in the actual province of Messina. A red blend of Nero d'Avola and Nocera is allowed, along with a varietal (Nero d'Avola); whites come from Ansonica, Grillo, and Catarratto.

MANDROLISAI DOC
Sardinia

From bush-trained old vines cultivated on the hilly countryside at the center of Sardinia, among the provinces of Nuoro and Oristano. This DOC is a blend of Bovale (Monastrell), Cannonau, and Monica, giving rich and lifted reds. Rosé can be produced from the same grapes.

MARMILLA IGP
Sardinia

A wide range of styles is produced in this IGP from the provinces of Cagliari and Oristano. Any variety allowed in Sardinia is allowed here.

MARSALA DOC
Sicily

Marsala DOC was among the first Italian appellations recognised in 1969 to protect a classic fortified and oxidative style produced in the homonymous town, in the western corner of Sicily. The long tradition of producing naturally high alcoholic dry wine, locally called *perpetuum*, was exploited by English merchant John Woodhouse in 1773. He shipped the *perpetuum* wine to London, adding neutral grape spirit to preserve it from transport spoilage, thus inventing the fortified Marsala. The area of production covers the majority of the province of Trapani. Best-located vineyards surround the coastal area of Marsala and Petrosino, where vines struggle on the arid calcareous soils and give naturally concentrated grapes. The grapes used include white Grillo, Catarratto, Inzolia, and Damaschino for oro and ambra styles and red Nero d'Avola, Perricone, and Nerello Mascalese for rubino. Marsala can also be categorised by its sweetness, from dry (less than 40 grams of residual sugar per litre), demi-sweet (40 to 100 g/l), and sweet (over 100 g/l), which is achieved by the addition of mistella (a mix of must and grape spirit) or cooked must. The ageing requirements determine the quality levels of this appellation starting from a mere one year (Fine), two years (Superiore), four years (Superiore Riserva) to at least five for the driest version called Vergine (no sweetness allowed) to be spent in oak casks. Despite the now-banned flavoured and sweetish cooking versions that have undermined the world-class reputation of this unique fortified style, the best examples of Marsala are capable of showing the maritime influence of the area with iodine scents and candied orange and dried fruits, along with a salty texture and mouthwatering length.

🍷 *2 years up to decades for Vergine*

✔ *Marco de Bartoli • Florio • Carlo Pellegrino - Buffa - Curatolo Arini - Martinez*

MARTINA OR MARTINA FRANCA DOC
Puglia

These dry white wines in still or *spumante* styles are very similar to those of Locorotondo.

MATERA DOC
Basilicata

From the town famous for its Sassi (prehistoric cave dwellings) this DOC allows styles that include a Malvasia-based generic dry white (which may be *spumante*), a Sangiovese-Primitivo-based generic red, and two red varietals (Moro and Primitivo).

MATINO DOC
Puglia

Dry, robust red and slightly vinous *rosato* wines made from Negroamaro with the possible addition of Malvasia Nera and Sangiovese, in the province of Lecce.

MELISSA DOC
Calabria

An appellation of full-bodied reds (Gaglioppo, Greco Nero, Grechetto, Trebbiano, and Malvasia) and crisp, dry whites (Grechetto, Trebbiano, and Malvasia).

MENFI DOC
Sicily

Made in and around Menfi at the western end of the south coast, these wines include a generic red (at least 70 per cent from one or more of the following varieties: Cabernet Sauvignon, Merlot, Nero d'Avola, Sangiovese, and Syrah), a generic white (at least 75 per cent from one or more of the following: Catarratto, Chardonnay, Inzolia, and Grecanico), and a generic *vendemmia tardiva*, (from one or more of the following: Catarratto, Chardonnay, Inzolia, and Sauvignon), There are two subzones, Bonera (at least 85 per cent from one or more of the following: Cabernet Sauvignon, Merlot, Nero d'Avola, Sangiovese, and Syrah) and Feudo dei Fiori {at least 85 per cent Chardonnay and/or Inzolia). Five red varietals are allowed (Cabernet Sauvignon, Merlot, Nero d'Avola, Sangiovese, and Syrah) and three white varietals (Chardonnay, Grecanico, and Inzolia).

MONICA DI SARDEGNA DOC
Sardinia

This a red-only DOC producing a wine made from minimum 85 per cent Monica grape, which can be either still (dry or sweet) or *frizzante* (dry or sweet).

MONREALE DOC
Sicily

A wide-ranging DOC, Monreale includes a Catarratto-and/or Inzolia-based generic white (which may be dry or, if *vendemmia tardiva,* sweet), and a Nero d'Avola-Perricone-based generic red dry. There are five white varietals: Ansonica (aka Inzolia), Catarratto, Chardonnay, Grillo, and Pinot Bianco and six red varietals: Cabernet Sauvignon, Nero d'Avola, Merlot, Perricone (aka Pignatello), Sangiovese, and Syrah.

✔ *Sallier de la Tour (La Monaca)*

MOSCATO DI SARDEGNA DOC
Sardinia

This DOC produces sweet Moscato Bianco–based wines, which can be still, sparkling, late harvest, or *passito*.

MOSCATO DI SORSO-SENNORI DOC
Sardinia

This DOC comes from the homonymous communes in the province of Sassari with Moscato grapes. It can produce either sweet and aromatic white Moscato, sparkling and fortified (both dry and sweet).

MOSCATO DI TRANI DOC
Puglia

Tiny production of luscious, smooth, and sweet white Moscato wines from the *appassimento* method. It can also be fortified.

🍷 *Upon purchase*

✔ *Fratelli Nugnes*

MURGIA IGP
Puglia

A recently added IGP with a wide range of styles from any variety permitted in the province of Bari.

NARDÒ DOC
Puglia

Tiny DOC encompassing the communes of Nardò and Porto Cesareo, in the province of Lecce. Reds and rosés from Negroamaro, with smaller percentages of Malvasia and Montepulciano (maximum 20 per cent).

NASCO DI CAGLIARI DOC
Sardinia

Small DOC producing a white wine (dry or sweet) made from a minimum of 95 per cent Nasco. A fortified version is also made.

NEGROAMARO DI TERRA D'OTRANTO DOC
Puglia

This DOC produces Negroamaro-based wines (minimum 90 per cent) in the provinces of Brindisi, Lecce, and Taranto in a variety of styles: *rosato* (still or *frizzante*), red (still *rosso*, or *riserva*), and sparkling *rosato*.

NOTO DOC
Sicily

Traditional area of production in the province of Siracusa, already famous during ancient times for its *passito* style from Moscato. This variety can produce Moscato (also dry) and Passito di Noto together with *spumante* and a fortified sweet. Reds from this area can be rich and luscious, based mainly on Nero d'Avola (as a varietal wine and a generic red for a minimum of 65 per cent).

✔ *Marabino - Planeta*

NURAGUS DI CAGLIARI DOC
Sardinia

Important DOC based on the traditional white variety Nuragus. Despite its name, it can be produced not only in Cagliari but also in the provinces of the south of Sardinia, Oristano, and Medio Campidano. A small percentage of any permitted variety can be used for both still and *frizzante* styles.

NURRA IGP
Sardinia

This IGP produces wines from authorized varieties growing in and around the province of Alghero. Still white, red, and *rosato* styles are allowed, but no production is reported at the moment

OGLIASTRA IGP
Sardinia

This IGP produces still and *frizzante* white; still, *frizzante*, and *novello* red and *rosato;* and still and *frizzante passito* from authorized varieties growing in numerous communes in the province of Ogliastra, plus the communes of San Vito and Villaputzu in the province of Cagliari

✔ *Jerzu (Akratos, Radames)*

ORTA NOVA DOC
Puglia

Minuscule DOC from the villages of Orta Nova and Ordona in the province of Foggia, producing red and rosé mainly from Sangiovese grapes.

OSTUNI DOC
Puglia

This DOC produces a delicate, dry white made from Impigno and Francavidda and a light-bodied red from the Cinsault grape, known locally as Ottavianello, with the possible addition of Negroamaro, Notardomenico, Malvasia Nera, and Susumaniello.

PAESTUM IGP
Campania

This IGP extends throughout most of the province of Salerno, encompassing white (still, *frizzante, amabile*, or *passito*), red (still, *novello, frizzante, amabile*, or *passito*), and *passito* (still, *amabile*, or *frizzante*) wines blended from any locally authorized variety, but varietal wines are restricted to the following: Aglianico, Barbera, Coda di Volpe, Fiano, Greco, Moscato, Piedirosso, Primitivo, and Sciascinoso.

✔ *De Conciliis (Naima) • Luigi Maffini (Pietraincatenata) • Montevetrano*

PALIZZI IGP
Calabria

This IGP produces red (also *novello*) and rosé wines from locally authorized varieties growing in Palizzi, Staiti, and the neighboring towns in the province of Reggio Calabria.

PANTELLERIA DOC
Sicily

Formerly known as Moscato di Pantelleria DOC, the appellation was renamed in 2013. The volcanic island of Pantelleria is located in the Sicilian Channel, closer to Tunisia than to the Sicilian coast, and it is dominated by winds and a wild, hilly landscape. Viticulture here is only possible through terraces and an autochthonous bush vine-training system called *alberello pantesco*, now protected by UNESCO as World Heritage Site. Zibibbo, the local name for white aromatic Muscat of Alexandria, is the only permitted variety in the entire DOC. It can make still (dry and sweet, also *frizzante*) and sweet sparkling, but it is best known for its dessert wines. Moscato di Pantelleria is sweet and produced from fresh grapes, while the more famous Passito di Pantelleria requires a sun-drying method to naturally concentrate not only the sugars but also the aromas and flavours. Apricot and peach jam scents, dried figs and dates, along with a thick palate of finely balanced sweetness and vibrant acidity put Passito di Pantelleria amongst the best sweet wines of the world, and it also capable of long bottle ageing. A fortified *passito* complete the possible styles of the DOC.

🍷 *1-10 years*

✔ *Donnafugata (Ben Ryé, Kabir) • Salvatore Ferrandes • Marco de Bartoli (Bukkuram) • Coste di Ghirlanda • Pellegrino*

PARTEOLLA IGP
Sardinia

This IGP produces still and *frizzante* red and white wine from locally authorized varieties growing in Dolianova,

Donori, Monastir, Serdiana, Soleminis, and Ussana, in the province of Cagliari.

PELLARO IGP
Calabria

This IGP produces red, *novello* red, and dry *rosato* wines from locally authorized varieties growing in Motta San Giovanni and its surrounding towns in the province of Reggio Calabria.

PENISOLA SORRENTINA DOC
Campania

This DOC from the Sorrento peninsula includes a generic red from Aglianico (minimum 60 per cent), with the possible addition of Piedirosso and Sciascinoso. Biancolella, Falanghina, and Greco take part in the generic dry white. Three specific subzones with higher quality standards are allowed, including Sorrento (for white and still red), Gragnano, and Lettere (only red *frizzante*).

PLANARGIA IGP
Sardinia

This is a minuscule appellation for still and *frizzante* red, white, and *rosato* wine from locally authorized varieties growing in Bosa, Flussio, Magomadas, Modolo, Sagama, Suni, and Tinnura in the province of Nuoro and Tresnuraghes in the province of Oristano.

POMPEIANO IGP
Campania

Obviously alluding to the ancient city of Pompeii, this IGP allows white (still, *frizzante, amabile,* or *passito*), red (still, *novello, frizzante, amabile,* or *passito*), and *passito* (white and red) from a wide range of grape varieties growing anywhere on the mainland in the province of Naples, but varietal wines are restricted to the following: Aglianico, Coda di Volpe, Falanghina, Piedirosso, and Sciascinoso.

✓ *Mastroberardino*

PRIMITIVO DI MANDURIA DOC
Puglia

This DOC produces red wines made from Primitivo grapes (minimum 85 per cent) cultivated in the provinces of Taranto and Brindisi. The *riserva* requires a minimum of 24 months of ageing. Voluptuous wines are produced from the Primitivo variety, especially from those grown in the western part of Salento's peninsula.

🍷 *3–10 years*

✓ *Cantolio Manduria • Fino (ES) • Gianfranco • Giordano • Pervino (Archidamo) • Vinicola Savese • Soloperto (CentoFuochi) • Vigne & Vini (Moi)*

PRIMITIVO DI MANDURIA DOLCE NATURALE DOCG
Puglia

Formerly part of Primitivo di Manduria DOC, it was established as a separate DOCG in 2011. A red dessert wine made of 100 per cent Primitivo grapes, both from late-harvest or further air-dried bunches, is allowed.

PROVINCIA DI NUORO IGP
Sardinia

This regional IGP allows still or *frizzante* red (may also be *novello*), *rosato*, and white wines from any authorized varieties growing in the Nuoro province. No production is reported at the moment.

PUGLIA IGP
Puglia

This region-wide IGP covers still and *frizzante* red (may also be *novello* or *passito*), white (may also be *passito*), sparkling (white and rosé), still *rosato*, and *passito* (white and red) wines from locally authorized varieties growing anywhere in the Puglia region, plus many specified varietal wines: Aglianico, Aleatico, Bianco d'Alessano, Bombino, Bombino Nero, Cabernet Franc, Cabernet Sauvignon, Chardonnay, Falanghina, Fiano, Greco, Lambrusco, Malvasia, Malvasia Nera, Moscatello Selvatico, Moscato, Negroamaro, Pampanuto, Pinot Bianco, Pinot Nero, Primitivo, Riesling, Sangiovese, Sauvignon, Trebbiano, Uva di Troia, and Verdeca.

✓ *Rasciatano • TorreVento • Varvaglione*

RIESI DOC
Sicily

One of the historical areas of production for Nero d'Avola – also known as Calabrese here – yielding very dense and inky dark-fruited reds. Varietally labelled wines are allowed from Cabernet Sauvignon, Merlot, and Syrah. Ansonica- and/or Chardonnay-based white, which may be dry and still, *spumante,* or sweet *vendemmia tardiva,* and a Calabrese-based dry *passito* are allowed, as well as *rosato* from Nero d'Avola, Nerello Mascalese, and/or Cabernet Sauvignon. White versions based on Inzolia and Chardonnay grapes can be still, sparkling, and late harvest.

ROCCAMONFINA IGP
Campania

This IGP includes white (still, *frizzante, amabile,* or *passito*), red (still, *novello, frizzante, amabile,* or *passito*), and *rosato* (still, *amabile,* or *frizzante*) from a wide range of varieties growing in Roccaromana and surrounding villages in the province of Caserta, but varietal wines are restricted to Aglianico, Coda di Volpe, Falanghina, Fiano, Greco, Piedirosso, Primitivo, and Sciascinoso.

✓ *Villa Matilde (Cecubo, Eleusi)*

ROMANGIA IGP
Sardinia

This IGP includes still, *frizzante,* and *passito* white; still, *frizzante,* and *novello* red; and still and *frizzante rosato* from authorized varieties growing in Castelsardo, Osilo, Sennori, Sorso, and Valledoria in the province of Sassari, with the following specific varietal exceptions: Cannonau, Carignano, Girò, Malvasia, Monica, Moscato, Nasco, Nuragus, Semidano, Vermentino, and Vernaccia.

✓ *Tenute Dettori (Bianco, Rosso, Tenores)*

BARLETTA DOC
Puglia

Medium-bodied, ruby-coloured, everyday-quality red wines based on Uva di Troia variety are made here. Most Barletta red is consumed locally when very young. White versions are made with Malvasia Bianca.

ROSSO DI CERIGNOLA DOC
Puglia

This DOC prouduces rare, rustic reds based on Uva di Troia with the possible addition of Negroamaro, Sangiovese, Barbera, Montepulciano, Malbec, and Trebbiano.

SALAPARUTA DOC
Sicily

This DOC on the eastern fringe of Trapani allows a Catarratto-based generic dry white, a Nero d'Avola–based generic red, a Nero d'Avola/Merlot-based generic dry *passito,* four white varietals (Catarratto, Chardonnay, Grillo, and Inzolia), and four red varietals (Cabernet Sauvignon, Merlot, Nero d'Avola, and Syrah).

STONEWORK AND TERRACING ARE HALLMARKS OF VINEYARDS IN PANTELLERIA, SICILY
As does the larger main island of Sicily, the smaller satellite island of Pantelleria, off the coast of Tunisia in North Africa, has mineral-rich lava soil. On this volcanic island, the Zibibbo grape (a Muscat variety) thrives and is used to make a range of white wines, but it is best known for the sweet Passito di Pantelleria.

GRAPES DRYING FOR *PASSITO* WINE
To make *passito* wines, sometimes known as sweet raisin wines, grapes are allowed to rasinate (dry up and shrivel) to concentrate their juice.

SALEMI IGP
Sicily

This IGP is restricted to the commune of Salemi, in the hilly countryside of the province of Trapani. Wines can be still and *frizzante rosato*; still, *frizzante,* and *novello* red and still and *frizzante* white from Ansonica, Catarratto, Grecanico, and Damaschino grapes only. The wines may be blended or pure varietal.

SALENTO IGP
Puglia

This IGP includes white (still, *frizzante,* or sparkling), red (still, *novello,* and *frizzante*), rosato (still, *frizzante,* and sparkling), and *passito* (white and red) from a wide range of grape varieties growing in the provinces of Brindisi, Lecce, and Taranto, but varietal wines are restricted mainly to the following: Aglianico, Cabernet Franc, Cabernet Sauvignon, Lambrusco, Negroamaro and Primitivo for reds and Bombino, Chardonnay, Fiano, Garganega, Greco, Malvasia, Moscato, Pinot Bianco, Sauvignon, Trebbiano, Verdeca, and Vermentino for whites.

✓ *Conti Zecca • Vinicola Mediterranea • Morella • Masseria Li Veli • Tenute Rubino • Tormaresca • Vallone • Vetrere*

SALICE SALENTINO DOC
Puglia

This DOC produces full-bodied reds and smooth, alcoholic rosati from Negroamaro, with the possible addition of Malvasia Nera, and more recently two dry whites: Chardonnay, which may be *frizzante,* and Pinot Bianco, which may be spumante; plus a sweet red Aleatico, which can be fortified. The Negroamaro-based reds are still the best wines in this DOC.

🍷 *3–7 years*

✓ *Francesco Candido • Due Palme • Leone de Castris • Vinicola Mediterranea (Granduca, Sirio) • Cosimo Taurino • Vallone • Salice Salentino*

SALINA IGP
Sicily

This IGP includes white (still, *frizzante,* or *passito*), red (still, *novello,* or *frizzante*), and *passito* (still or *frizzante*) from the island of Lipari. All grapes authorized for the province of Messina are allowed, but local resources are limited, and the most likely varieties in any blend will probably be Alicante, Calabrese, Corinto, Nerello, Nero d'Avola, Nocera, or Sangiovese for reds and Catarratto, Grecanico, Grillo, or, mostly, Malvasia for whites.

✓ *Hauner (Hierà, Malvasia Passito)*

SAMBUCA DI SICILIA DOC
Sicily

This DOC produces an Ansonica-based white that may be either dry or *passito* and a Nero d'Avola dry red and *passito,* plus five red varietals (Cabernet Sauvignon, Merlot, Nero d'Avola, Sangiovese, and Syrah) and three white varieties (Ansonica, Chardonnay, and Grecanico).

SANNIO DOC
Campania

This is the modern equivalent of Samnium – a famous ancient Roman wine described by Pliny, Columella, Cato, and Horace – but rather than focussing on one style to resurrect that reputation, this is yet another typically wide-ranging DOC. It allows Trebbiano-based generic white, which may be still and dry, or *frizzante*; a Sangiovese-based red and *passito,* which must be dry and may be still but could also be *frizzante* or *novello*; a generic *spumante* that must contain Aglianico, Greco, or Falanghina to the exclusion of all other varieties and must be aged at least 14 months before release; five red varietals (Aglianico, Barbera, Piedirosso, Primitivo, and Sciascinoso); and five white varietals (Coda di Volpe, Falanghina, Fiano, Greco, and Moscato).

✓ *Mastroberardino • Feudi di San Gregorio (Falanghina)*

SANT'AGATA DEI GOTI DOC
Campania

A relatively new DOC for generic red, dry white, and *rosato* from Aglianico and Piedirosso (thus the white is a blanc de noirs); two red wine varietals, Aglianico and Piedirosso; and two white varietals, Greco and Falanghina.

SANT'ANNA DI ISOLA CAPO RIZZUTO DOC
Calabria

This DOC produces vinous reds and rosati from Gaglioppo, Greco Nero, and Guarnaccia grapes from the Ionian hills.

SANTA MARGHERITA DI BELICE DOC
Sicily

This DOC produces an Asonica-, Grecanico-, or Catarratto-based white and a Nero d'Avola-, Sangiovese-, or Cabernet Sauvignon-based red; plus two red varietals (Nero d'Avola and Sangiovese) and three white varietals (Ansonica, Catarratto, and Grecanico).

SAN SEVERO DOC
Puglia

This DOC produces dry, vinous red and *rosato* wines made from Montepulciano and Sangiovese and dry, fresh whites from Bombino and Trebbiano, with the possible addition of Malvasia and Verdeca.

SARDEGNA SEMIDANO DOC
Sardinia

This DOC is for wines made from grapes grown throughout the island, although in practice they are generally restricted to traditional areas. For Moscato, vines must not be grown above 450 metres (1,476 feet). Moscato di Sardegna is sweet and surprisingly delicate, usually in a natural still-wine style, although *spumante* is produced, and the sub-appellations Tempio Pausania, Tempio, and Gallura are reserved for sparkling wines in the Gallura area. Monica di Sardegna is a fragrant red wine that may be dry or sweet, still or *frizzante,* but unlike Monica di Cagliari it is never fortified. Vermentino di Sardegna is a light-bodied, soft, clean, unexciting dry white wine that all too often has its flavours washed out by cool-fermentation techniques. Cannonau (aka Grenache), is this DOC's most successful varietal, especially as a dry red, although it may also be produced as semi-sweet or sweet red, *rosato,* and *liquoroso.*

✓ *Antonio Argiolas • Carpante (Vermentino Frinas, Vermentino Longhera) • Pala (Monica, Vermentino) • Pedres (Moscato) • CS di Santadi (Vermentino Cala Silente)*

Cannonau: *Carpante • Contini (Inu Reserva, Tonaghe) • Ferruccio Deiana (Sìleno) • Cantine Dolianova (Azenas) • Dorgali • Feudi della Medusa • Giuseppe Gabbas • Giorgantinu • Jerzu • Alberto Loi • Piero Mancini • Giuseppe Sedilesu (Mamuthone) • Tenute Soletta*

SAVUTO DOC
Calabria

This DOC produces fresh, fruity red or *rosato* wines from Gaglioppo, Greco Nero, Nerello, Sangiovese, Malvasia, and Pecorino grapes grown in the province of Catanzaro.

🍷 *1–3 years*

✓ *Odoardi (Vigna Vecchia)*

SCAVIGNA DOC
Calabria

This DOC produces generic red, *rosato* (Gaglioppo, Nerello, and Aglianico), and a fruity dry white (Trebbiano, Chardonnay, Grechetto, and Malvasia) from the communes of Nocera Terinese and Falerna.

SCIACCA DOC
Sicily

This DOC produces an Inzolia-, Grecanico-, or Chardonnay-based white and a Merlot-, Nero d'Avola–, Sangiovese-, or Cabernet Sauvignon–based red and dry *passito,* plus four red varietals (Cabernet Sauvignon, Merlot, Nero d'Avola, and Sangiovese) and three white varietals (Ansonica, Catarratto, and Grecanico). Riserva Rayna is a subzonal denomination for a Catarratto- and/or Inzolia-based dry white that must be aged for at least 24 months.

SCILLA IGT
Calabria

This IGT produces red, novella, and dry *passito* from any locally authorized variety growing in Scilla in the province of Reggio Calabria.

SIBIOLA IGP
Sardinia

This IGP produces still or *frizzante* red, white, and *passito* from Serdiana and Soleminis in the province of Sassari from any locally authorized grape variety, with the specific exclusion of Cannonau, Carignano, Girò, Malvasia, Monica, Moscato, Nasco, Nuragus, Semidano, Vermentino and Vernaccia. Reds may also be sold as *novello.*

SICILIA DOC
Sicily

Officially established in 2011 from previous IGP status, the production area encompasses the entire administrative region of Sicily, including its smaller islands. Given the incredible diversity of *terroirs,* this DOC allows a plethora of styles of wines to be produced from all cultivable varieties but with stricter rules than the alternative Terre Siciliane IGP. Since 2017 Grillo and Nero d'Avola, the most internationally revered white and red grape varieties of the region, could be used only under DOC rules. Today fine and age-worthy wines, as well as less intriguing yet approachable fruit-driven wines, can be found on the market under the same appellation, so attention should be paid to the producer name.

✓ *Benanti • Cottanera • Cusumano • Donnafugata • Feudo Principi di Butera • Feudi del Pisciotto • Feudo Maccari • Firriato • Gulfi • Morgante • Planeta • Gorghi Tondi • Tasca d'Almerita • Tenute Rapitalà • Valle dell'Acate • Alessandro di Camporeale*

SIRACUSA DOC
Sicily

Established as the Moscato di Siracusa DOC in 1973, it was expanded as Siracusa DOC in 2011. This DOC produces the traditional sweet, smooth white wines made from semi-*passito* Moscato Bianco grapes, but also dry white wines (a generic Moscato Bianco–based white and a varietal Moscato Bianco) and dry reds (a generic Nero d'Avola–based red and the varietal Nero d'Avola and Syrah). Sparkling wines can also be produced from a minimum 85 per cent Moscato Bianco.

✓ *Pupillo*

SQUINZANO DOC
Puglia

This DOC produces full-bodied red wines and lightly scented *rosato* wines from Negroamaro, with the possible addition of Sangiovese and Malvasia Nera, grown in Squinzano and the surrounding communes.

🍷 *2–4 years (red), upon purchase (Rosato)*

✓ *Villa Valletta*

TARANTINO IGP
Puglia

This IGP produces still and *frizzante* red (may also be *novello* or *passito*), white (may also be *passito*), and *passito* wine from locally authorized varieties (with the specific exception of either Montepulciano or Ottavianello) growing in the province of Taranto. Only two varietal wines are allowed: Negroamaro and Malvasia Nera.

✓ *Varvaglione*

TAURASI DOCG
Campania

Red wines made primarily from Aglianico grapes (up to 15 percent in total of Barbera, Piedirosso, and Sangiovese may be added) that are grown in Taurasi and 16 nearby villages. Along with Basilicata's thicker-tasting Aglianico del Vulture, it is one of the country's greatest wines from this underrated grape variety, in an equally age-worthy style. Mastroberardino's *riserva* excels.

🍷 *5–10 years (some may last for as long as 20)*

✓ *Antonio Caggiano • Di Prisco • Mastroberardino • Perillo • Giovanni Struzziero • Urciuolo*

TAVOLIERE DELLE PUGLIE/ TAVOLIERE DOC
Puglia

This DOC produces a Uva di Troia-based *rosato* and red wine and a varietal Uva di Troia.

TERRE DELL'ALTA VAL D'AGRI DOC
Basilicata

This is a relatively small DOC in the hills of Viggiano, Moliterno, and Grumento Nova, on the southeastern border of Campania, where a red and dry *passito* Merlot-Cabernet can be made.

TERRE DI COSENZA DOC
Calabria

This DOC covers the entire province of Cosenza and produces a wide range of styles: generic white wines from Greco, Guarnaccia, Montonico Bianco, and/or Pecorello grapes, six varietal whites (Chardonnay, Greco Bianco, Guarnaccia Bianca, Malvasia Bianca, Mantonico, Pecorello), and red or *rosato* wines mainly from Magliocco Canino but several other varieties are allowed (both indigenous and international). Sparkling wines and sweet wines (*passito* or late harvest) are also made. Seven subzones can be mentioned on the label: Colline del Crati, Condoleo, Donnici, Esaro, Pollino, San Vito di Luzzi, Verbicaro.

TERRA D'OTRANTO DOC
Puglia

This is a small DOC producing white wines from Chardonnay, Fiano, Malvasia Bianca, and Verdeca; *rosato* wines from Malvasia Nera and *rosato frizzante* from Negroamaro; and red wines from Aleatico, Malvasia Nera, and Primitivo. Sparkling wines from Chardonnay and a *rosato spumante* from Negroamaro are also made.

TERRE SICILIANE IGP
Sicily

Renamed in 2011 from the previous Sicilia IGP, it covers the entire administrative region of Sicily and allows wines from all the possible varieties cultivated in Sicily. A wide range of styles, both varietal and blended, can be found under this IGP, though quality is very producer-dependent. In 2017 rules were passed stipulating that all wines made from either Grillo or Nero d'Avola had to be classified as Sicilia DOC. *See* Sicilia DOC

✓ *Cusumano • Passopisciaro • Firriato • Gulfi • Mandrarossa • Duca di Salaparuta • Principi di Spadafora • Tenute Rapitalà • Feudi Spitaleri*

TERRE DEL VOLTURNO IGP
Campania

This IGP covers most of the province of Caserta, plus the towns of Giugliano, Qualiano, and Sant'Antimo in the province of Naples. The wines include a white (still, *frizzante*, *amabile*, or *passito*), red (still, *novello*, *frizzante*, *amabile*, or *passito*), and *passito* (still, *amabile*, or *frizzante*) from a wide range of grape varieties growing in Roccaromana and surrounding villages in the province of Caserta, but varietal wines are restricted to the following: Aglianico, Asprinio, Casavecchia, Coda di Volpe, Falanghina, Fiano, Greco, Pallagrello Nero, Primitivo, and Sciascinoso.

✓ *Vestini (Cavavecchia, Kaja Nero) • Terre del Principe (Centomoggio)*

THARROS IGP
Sardinia

This IGP covers still or *frizzante* red, white, and *passito* produced in the province of Oristano from any locally authorized grape variety, with the specific exclusion of Cannonau, Carignano, Girò, Malvasia, Monica, Moscato, Nasco, Nuragus, Semidano, Vermentino, and Vernaccia. Reds may also be sold as *novello*.

TREXENTA IGP
Sardinia

This IGP covers still or *frizzante* red, white, and *passito* produced in part of the province of Cagliari from any locally authorized grape variety, with the specific exclusion of Cannonau, Carignano, Girò, Malvasia, Monica, Moscato, Nasco, Nuragus, Semidano, Vermentino, and Vernaccia. Reds may also be sold as *novello*.

✓ *Cantina Trexenta*

VALLE BELICE IGP
Sicily

This IGP includes white (still, *frizzante*, or *passito*), red (still, *novello*, or *frizzante*), and *passito* (still or *frizzante*) from locally authorized grape varieties growing in the Belice Valley as it passes through Menfi, Montevago, and Santa Margherita Belice in the province of Agrigento and in Contessa Entellina in the province of Palermo.

VALLE D'ITRIA IGP
Puglia

This IGP includes white (still, *frizzante*, or *passito*), red (still, *passito*, *novello*, or *frizzante*), and *passito* (still or *frizzante*) from locally authorized grape varieties growing in the Itria Valley throughout the provinces of Bari, Brindisi, and Taranto. Varietal wines are restricted to the following: Aleatico, Cabernet Franc, Cabernet Sauvignon, Malvasia Nera, Negroamaro, Pinot Nero, Primitivo, and Sangiovese for red and *passito* wines and Bianco d'Alessano, Bombino, Chardonnay, Fiano, Impigno, Malvasia, Moscatello Selvatico, Moscato, Pinot Bianco, Sauvignon, Trebbiano, and Verdeca for whites.

VAL DI NETO IGP
Calabria

This IGP includes white (still, *frizzante*, or *passito*), red (still, *novello*, or *frizzante*), and *passito* (still or *frizzante*)

from locally authorized grape varieties in the Neto Valley in the province of Crotone.

✓ *Librandi (Gravello, Magno Megonio, Le Passule)*

VALLI DI PORTO PINO IGP
Sardinia

This IGP covers still or *frizzante* red, white, and *passito* produced in Giba, Masainas Narcao, Nuxis, Perdaxius, Piscinas, Santadi, Sant'Anna Arresi, Teulada, Tratalias, and Villaperuccio in the province of Cagliari from any locally authorized grape variety, with the specific exclusion of Cannonau, Carignano, Girò, Malvasia, Monica, Moscato, Nasco, Nuragus, Semidano, Vermentino and Vernaccia. Reds may also be sold as *novello*.

VALLE DEL TIRSO IGP
Sardinia

This IGP covers still or *frizzante* red, white, and *passito* produced in the Tirso Valley in the province of Oristano from any locally authorized grape variety, with the specific exclusion of Cannonau, Carignano, Girò, Malvasia, Monica, Moscato, Nasco, Nuragus, Semidano, and Vermentino. Reds may also be sold as *novello*.

VALDAMATO IGP
Calabria

This IGP includes white (still, *frizzante*, or *passito*), red (still, *novello*, or *frizzante*), and *passito* (still or *frizzante*) from locally authorized grape varieties growing in Curinga, Feroleto, Gizzeria, Lamezia Terme, Maida, and Maida Pianopoli San Pietro in the province of Catanzaro.

VERMENTINO DI GALLURA DOCG
Sardinia

This is the best Vermentino Sardinia offers, with the hilly *terroir* of Gallura producing intense flavours.

🍷 *1–2 years*

✓ *Capichera • CS di Gallura • Giorgantinu • Piero Mancini • Masone Mannu • Pedres • CS del Vermentino*

VERMENTINO DI SARDEGNA DOC
Sardinia

This is large DOC producing a Vermentino-based (minimum 85 per cent) white wine, which can be still or *frizzante*, dry or sweet. The sparkling version is also made from Vermentino grapes.

VERNACCIA DI ORISTANO DOC
Sardinia

The DOC of Vernaccia di Oristano makes dry and lightly bitter fortified white wines, which are Sherry-like in style. *Liquoroso* sweet and dry wines are also produced.

🍷 *Upon purchase*

✓ *Contini • Fratelli Serra*

VESUVIO DOC
Campania

Restricted to vines on the volcanic slopes of Mount Vesuvius, a generic dry white (Coda di Volpe and Verdeca grapes, with the possible addition of Falanghina and Greco), and a generic *rosato* (Piedirosso and Sciascinoso, with the possible addition of Aglianico) are produced by Vesuvio DOC. Those that are labelled Lacryma (or Lacrima) Christi del Vesuvio are usually dry white wines, but less commonly they may also be *spumante* or sweet and fortified.

VITTORIA DOC
Sicily

When Cerasuolo di Vittoria gained its DOCG status in 2005, this DOC was created for red wines that are also produced from Nero d'Avola (minimum 50 per cent) and Frappato grapes. *Novello* is permitted, as well as three varietal wines: Ansonica (aka Inzolia), Nero d'Avola, and Frappato.

The Wines *of* Spain

SPAIN IS DEFINITELY OUT OF LAST CENTURY'S DOLDRUMS. Affluence, globalisation, and impressively higher production values are behind the resurgence of self-confidence among Spanish wine producers. Spain is now in a position to exploit its huge fine wine potential, the result of a combination of a diversity of *terroirs,* a nice genetic heritage of many autochthonous grape varieties, and a long history. All of Spain is suitable for viticulture. Besides, it is one of the most mountainous countries in Europe, with a great diversity of soils, and a very long coastline. Finally, it is the only country in Europe with four distinctive climates (including the Canary Islands). These provide the perfect ingredients for many different top wines but, so far, only a few Spanish wines have reached iconic status at world level. For the immediate future, while some jewels in Rioja, Ribera del Duero, Cataluña, and Jerez will continue to shine, one can expect a good number of top-notch wines coming from other regions. Do not expect a characteristic Spanish style, but rather a large number of different wine styles coming from the country, styles as varied as its geography. Never has an established, traditional winemaking nation got its flagging act together as quickly or as thoroughly as Spain.

**BODEGA DE LOS HEREDEROS
DEL MARQUÉS DE RISCAL, RIOJA ALAVESA**

The vineyards of Marqués de Riscal had it first harvest in the 1860s, but the hotel that forms the centrepiece of the property in the village of Elciego is thoroughly modern with its flowing, coloured curves. Marqués de Riscal now produces more than six million bottles a year.

Spain

For millennia, Spain supplied top wines to nations in power: Alicante, Tarragona, and Sherry to the Romans; Ribadavia, Canary, and again Sherry to the British; Rioja to the French and the Americans; Málaga, Valdepeñas, and Cariñena to many others. Yet, by the 1970s, Spanish wines had gained a bad reputation, the result of years of standardisation and high productivity policies under Franco's dictatorship. Democracy and the nation's entry into the European Union brought swift change, however.

That Spain could achieve such a rapid turn-around in the quality of its wines can be attributed to a country-wide effort to get into the fine wine markets, realised through a number of different avenues, but basically based upon increased self-confidence, higher skills, and strong public support. Yet even now there are two Spains: the fine wine country . . . and the bulk wine behemoth.

Spain is the number-one wine exporter in the world in terms of quantity, but rank only third in terms of value. Bulk wine exports account for more than half the volume, but they account for only one-fifth of the value. Spain is also the world leader in terms of delivering quality wine at affordable prices, but it is still far behind France, Italy, and even the United States in terms of the number of iconic wines. At premium levels, Rioja, Cava, and Sherry may be relied upon to produce high quantities of decent-quality wines. Reliability and good value are the key words for Spain in wide markets, while diversity and little-known jewels are the secret wild cards for connoisseurs.

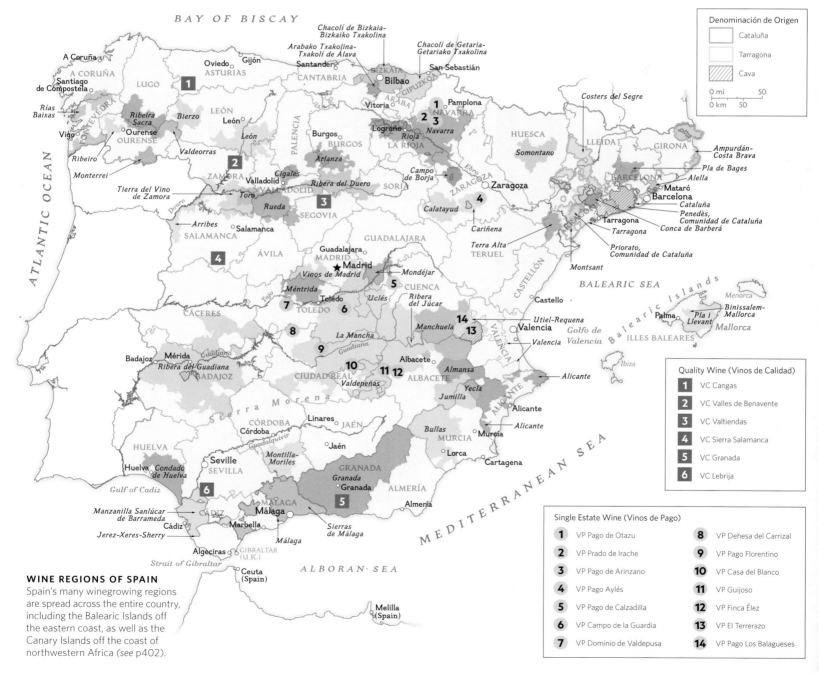

WINE REGIONS OF SPAIN
Spain's many winegrowing regions are spread across the entire country, including the Balearic Islands off the eastern coast, as well as the Canary Islands off the coast of northwestern Africa (*see p402*).

Denominación de Origen
- Cataluña
- Tarragona
- Cava

Quality Wine (Vinos de Calidad)
- **1** VC Cangas
- **2** VC Valles de Benavente
- **3** VC Valtiendas
- **4** VC Sierra Salamanca
- **5** VC Granada
- **6** VC Lebrija

Single Estate Wine (Vinos de Pago)
- **1** VP Pago de Otazu
- **2** VP Prado de Irache
- **3** VP Pago de Arinzano
- **4** VP Pago Aylés
- **5** VP Pago de Calzadilla
- **6** VP Campo de la Guardia
- **7** VP Dominio de Valdepusa
- **8** VP Dehesa del Carrizal
- **9** VP Pago Florentino
- **10** VP Casa del Blanco
- **11** VP Guijoso
- **12** VP Finca Élez
- **13** VP El Terrerazo
- **14** VP Pago Los Balagueses

ANCIENT ORIGINS

Vines were first planted in Spain around Cádiz circa 1100 BC by, it is believed, the Phoenicians. The Greeks came 300 years later and planted vines in selected spots on the Mediterranean coast. The earliest wines were by all accounts tradable, so one can assume that they were rich and sweet to resist travel vagaries. The Romans moved farther inland, expanding viticulture to most of Hispania. From the start of the 8th century until the end of the 15th, however, southern Spain was under the rule of the Moors and, being Mohammedans, wine production was not a priority for them. The Moors were defeated at Covadonga and Poitiers, and the Reconquista ("Reconquest") ensued for 800 years, re-creating a fine wine map of Spain during a time when international trade was not a priority. Top-quality vineyards in the northern half of Spain were patiently planted during the first part of the Reconquista. Viticulture in the southern half of Spain and the islands, with the exceptions of Málaga and Sherry, is more the result of the Moors debacle after the Battle of Las Navas de Tolosa and the establishment of colonial exploitation of the land. When Spain became an empire, much wine was sold in the American colonies, but after this the wine industry stagnated – most of Spain was isolated from trade routes, and the country was poor.

FREIXENET HEADQUARTERS IN SANT SADURNI
Freixenet is one of Spain's leading makers of the sparkling wine Cava. Cava, Sherry, and Rioja are Spanish wine styles well-known throughout the world.

SPANISH LABEL LANGUAGE

Abocado With some residual sugar

Añada Vintage

Añejo Wines aged for a minimum period of 24 months in total, either in oak barrels of 600 litres maximum capacity or in bottle

Blanco White

Bodega Winery

Clarete Mid-way between light red and dark rosado

Cosecha Vintage

Criado y embotellado por Blended and bottled by

Crianza Wines other than sparkling, semi-sparkling and liqueur wines, that fulfil the following conditions:

- Red wines must have a minimum period of ageing for 24 months, of which they must remain at least 6 months in oak barrels of 330 litres maximum capacity.

- White and rosé wines must have a minimum period of ageing for 18 months, of which they must remain at least 6 months in oak barrels of 330 litres maximum capacity.

Crianza corta Wines that have less than the legal minimum cask-age for any cask-age designation. Synonymous with *sin crianza*

Denominación de Origen (DO) Appellation of Origin, according to EU legislation

Denominación de Origen Calificada (DOCa) Appellation of origin with extra requirements in terms of notoriety, bottling, and quality control, above *Denominación de Origen*

Doble pasta This term refers to red wines that have been macerated with double the normal proportion of grape skins to juice during fermentation. Such wines are opaque, with an intense colour, and may be sold in the bottle or for blending.

Dulce Sweet

Embotellado por Bottled by

Espumoso A sparkling wine made by any method

Generoso A fortified wine, except in Montilla-Moriles, where it is a minimum 15% ABV non-fortified wine

Gran Reserva Wines other than sparkling, semi-sparkling, and liqueur that fulfil the following conditions:

- Red wines must have a minimum period of ageing of 60 months, of which they shall remain at least 18 months in oak barrels of 330 litres maximum capacity, and in bottle the rest of this period.

- White and rosé wines must have a minimum period of 48 months ageing, of which they shall remain at least 6 months in oak barrels of the same maximum capacity and in bottle the rest of this period.

Indicación Geográfica Protegida (IGP) Protected Geographical Indications

Noble Aged during a minimum period of 18 months in total, in oak barrels of 600 litres maximum capacity or in bottle

Nuevo A fresh, fruity, "new" or nouveau-style wine, synonymous with *vino joven*

Pago *Finca,* plot, property. The term was privatised by the Spanish law (*see Vinos de Pago*).

Rancio Wines that have followed a process of ageing, noticeably rusted, with abrupt changes of temperature in presence of air, or in wood packages or crystal packages

Reserva Wines other than sparkling, semi-sparkling and liqueur wines that fulfil the following conditions:

- Red wines must have a minimum period of ageing of 36 months, of which they shall remain at least 12 months in oak barrels of 330 litres maximum capacity and in bottle for the rest of this period.

- White and rosé wines must have a minimum period of ageing of 24 months, of which they shall remain at least 6 months in oak barrels of the same maximum capacity and in bottle for the rest of this period.

Rosado Rosé

Seco Dry

Semidulce Medium-sweet

Semiseco Medium-dry (which is as "dry" as *demi-sec* in France)

Tinto Red

Viejo Wine aged at least 36 months, with a rusted character noticeably due to the action of the light, oxygen, heat, or any or all of these factors

Viña or **viñedo** Literally "vineyard"; often used as part of a brand name

Vino de tea Wine of the north subzone of the La Palma PDO aged in pinewood vessels of *Pinus canariensis* ("tea") during a maximum of 6 months

Vino añejo A wine that has been aged in oak or in bottle for at least 24 months

Vino de aguja A semi-sparkling, or *pétillant,* wine

Vino de Calidad con Indicación Geográfica (VC) An overarching term for *Vino de la tierra* and *Denominación de Origen*

Vino de la Tierra Legal term for wines with geographical indication but not appellation of origin; a kind of broader, less-stringent protection

Vino de licor A sweet, possibly fortified wine

Vino de mesa "Table wine"; wine with no indication of origin; likely seen on labels of inexpensive, ordinary wines, but some producers follow the Tuscan example and label their top wines as *vino de mesa.*

Vino de pasto An ordinary, inexpensive style of wine; it is often bulk wine

Vino de Pago Single-vineyard appellation of origin that is theoretically at the peak of the quality pyramid; in practice the term is used only within the less prestigious appellations, as a way to get out of them.

Vino joven Wine with no ageing, normally to be drunk within the year

Vino noble A wine that has been aged in oak or in bottle for at least 18 months

Vino viejo A wine that has been aged in oak or in bottle for at least 36 months

As a consequence, quality wine disappeared for several centuries, with the exception of those wines that were successfully exported: Canary, Sherry, Málaga, Alicante, Ribadavia (now Ribeiro), and a few others. Rioja gained recognition as a top-quality region only in the 1860s; Ribera del Duero and Cava in the 20th century; Rías Baixas, Priorat (Priorato, Comunidad de Cataluña), and Bierzo not until the 21st. The rescue of old autochthonous varieties and remote wine regions is an endeavour of the last 15 years. So, though Spain can claim a 3,000-year wine history, one could almost consider it a New World country, which gives a mouth-watering hint of its future potential.

SPAIN'S BEST WINES AND ITS APPELLATION SYSTEM

Most of the best wines of Spain are concentrated in two appellations of origin: Rioja and Sherry. There are several top wines from other regions, which are at the basis of newer appellations of origin (*Denominación de Origen*, or DO, in Spanish). The most distinguished one is Vega Sicilia in Ribera del Duero. Torres is probably the benchmark for Penedès, Riscal for Rueda, and Codorniù and Freixenet for Cava.

Until this century, top Spanish wines used to be brands; they could be regional blends or single-vineyard, but the relevant value was the brand. Best examples are Viña Tondonia, GR 890, Ygay, Pérez Pascuas GR, Corullón, Sacristía, or Cirsión. Nowadays a new generation of single-vineyard wines are also at the summit. This is creating a lot of pressure for classifying *terroirs* in most DOs, and also for allowing an increasing number of ambitious producers to move out of their traditional DO. The Spanish appellation system, approved in 2003, is designed to favour this kind of movement, having set at the top of the quality scale the *"Vino de Pago"* kind of single-vineyard appellations. *Vinos de Pago* are by no means the best wines of Spain, but they set a trend against DO. Indeed, the most advanced DO, such as Rioja, Bierzo, or Priorat have already set *terroir* classifications systems, but many Spanish DO are ridiculously huge regions, often managed by high-volume, mediocre-quality producers and conservative cooperatives, which are against quality heroes.

RECENT SPANISH VINTAGES

2017 A vintage to forget in many parts of Spain, with a terrible April frost that decimated yields and then resulted in late ripening and average wines.

2016 For once, an underrated vintage in Rioja, with many wines of excellent quality and a good quantity overall. In the Mediterranean regions, drought greatly reduced yields, often resulting in unbalanced wines.

2015 Great vintage in Priorat and Ribera del Duero, a bit too hot in Rioja. Again a warm year, in some regions too hot.

2014 A relatively wet year, with problems of rot in several regions. Great in Bierzo and Ribeiro.

2013 Irregular vintage, excellent for Cava, quite uneven for Rioja.

2012 Very dry year in most of Spain, giving low yields and great concentrations. Old vines were the best.

2011 In general a very hot year, resulting in powerful red wines of good quality. In Galicia, is was a top year for Godello.

2010 The cool but dry summer and autumn allowed the crop to ripen slowly, preserving better acidity levels than usual and resulting in a potentially excellent vintage the length and breadth of Spain.

GREAT WINE PRODUCERS OF
SPAIN

Note: This list does not include sparkling wines and sherry.

AALTO
Ribera del Duero
★

An ambitious project from Mariano García, former chief winemaker at Vega Sicilia, and Javier Zaccagnini, former director of the Ribera del duero DO. They started by renting a large number of small plots in the best Ribera areas and have recently planted vines around the winery. A benchmark for one of the Ribera styles, powerful, oaky, concentrated, needing some years in bottle to soften up.
✓ *Aalto PS*

ABADÍA RETUERTA
★★

A 12th-century abbey lent its name to this winery and its vineyards. The vineyard was planted with the particular style of wine already decided, and now, after 25 years, the results are reaching the expected plateau of personality, quality, and complexity. A new *Vino de Pago* very likely looms in the near future.
✓ *Pago Valdebellón Cabernet Sauvignon • Pago Negralada Tempranillo*

ABEL MENDOZA
Rioja
★

An artisan of the vine, deftly made whites and serious reds.
✓ *5V • Graciano Grano a Grano*

ALEJANDRO FERNÁNDEZ
GRUPO PESQUERA
★

Vega Sicilia may have originally won the accolade for Ribera as a fine wine region, but it is the merit of Alejandro Fernández to have put Ribera in the mouths and memories of consumers all over the world. His Tinto Pesquera was for some time the best ambassador for the region, winning praise in many countries. He also has wineries in Zamora and La Mancha. His wines are no longer at the top, having been usurped by one or two others, but they are still a good reference.
✓ *Tinto Pesquera Reserva • El Vínculo Paraje La Golosa GR (La Mancha)*

ALFREDO MAESTRO
★

The apostle of pure wines is an explorer of vineyards in and around Ribera del Duero, Cigales, Navalcarnero (Madrid), and Gredos. Alfredo's creativity has resulted in at least 15 different wines with little commercial sense but with lots of sensual appeal. If only he could produce more.
Lovamor • 46 Cepas Merlot • Viña Almate La Asperilla (no DO)

ALTA ALELLA
★

This winery is a great representative of Alella and its unique *sauló* soils. Also a Cava producer.
✓ *AA GX • AA Cau d'en Genìs*

ALVEAR
★★★

Founded in 1729, this is one of the oldest wineries in Spain, with a most impressive stock of old wines. It is a benchmark for Montilla-Moriles that is still pursuing innovation. Its joint project with Envinate, Tres Miradas, is an amazing new expression of the albariza soil in their region. Alvear also invested in Extremadura, creating the Palacio Quemado winery, with probably the best wine of the Badajoz province.
✓ *Fino Capataz • Criadera A • Solera Fundación Amontillado (Montilla-Moriles) • Palacio Quemado Los Acilates (Ribera del Guadiana)*

ARTADI
★★

Juan Carlos López de Lacalle created this group, now also present in Alicante and Navarra. It first became famous for the amazing quality of their single-vineyard Riojas and then because they abandoned the DOCa Rioja to show their dissatisfaction with the institutional approach.
✓ *Viña El Pisón • La Poza de Ballesteros (no DO in Rioja) • Santa Cruz de Artazu (Navarra)*

ARTUKE
Rioja
★

Some of the best single-vineyard wines in Rioja. Member of Rioja 'n' Roll, a group for wines with individuality.
✓ *La Condenada*

AVANTESELECTA
★★

This group owns Dominio de Atauta, sited in Ribera's most remote area, with pre-phylloxera vines that survive in sandy soils in the middle of forests. Look for wines with lots of personality. The group owns seven more wineries all over Spain, the most relevant being Cénit in Zamora, Alvaro Domecq in Jerez, Naia in Rueda, and Pazos del Rey in Monterrei.
✓ *Atauta La Mala (Ribera del Duero) • Cénit (Zamora) • Naia (Rueda)*

BAIGORRI
Rioja
★

This flagship of Rioja Alavesa offers one of the most beautiful wineries and an excellent infrastructure for tourism.
✓ *B70 • Garnacha*

BAJA MONTAÑA
Navarra
★

This is Spain's only winery producing solely rosé wines from Navarra. It intends to demonstrate that rosé is by no means a lesser wine.
✓ *Arbayun*

BARÓN DE LEY – MUSEUM
★

A large group based in Navarran Rioja, with wineries in Rioja, La Mancha,

Cigales. It owns the largest vineyard in Rioja. A reliable producer for Rioja Oriental and Cigales.

✓ *7 Viñas Reserva (Rioja)* • *Museum Reserva (Cigales)*

BASILIO IZQUIERDO
Rioja
★

Master winemaker Izquierdo's personal project is small but very attractive. His rosé is complex and deep; his whites finely balanced, made for the long run.

✓ *B de Basilio (red, white, and rosé)*

BELONDRADE
★

A deft use of oak on Verdejo wines coming from selected vineyards put this winery quickly at the summit of prestige in Rueda.

✓ *Belondrade y Lurton (Rueda)* • *Quinta Apolonia (Castilla y León IGP)*

BHILAR
★

David Sampedro's winery, a homage to Rioja's old vineyards and biodynamics, offers highly distinctive new wave top wines. He also produces wine in Salamanca with the Rufete grape variety.

✓ *Phincas* • *Thousand Mils (Rioja)* • *Phinca Encanto (Salamanca)*

BODEGA MUSTIGUILLO
★★

The only *Vino de Pago* with a real sense of origin, may be the best expressions of the red Bobal and white Merseguera grapes.

✓ *Quincha Corral* • *Finca Terrerazo*

BODEGAS PALACIO 1894
★

This former classic, now a multiregion group, has wineries in Rioja, Ribera, Rueda, and Toro, mostly with great price-quality ratios and a few big wines.

✓ *El Secreto (Ribera)* • *Glorioso GR* • *Cosme Palacio 1894 Blanco (Rioja)*

BORSAO
Campo de Borja
★

This is one of the most prosperous cooperatives in Spain with a well-known brand and remarkable export success.

✓ *Tres Picos*

LA CALANDRIA

This boutique *bodega* was established by three friends, on a mission to save some of the oldest Garnacha vineyards in Navarra and Aragón. Their wines are among the most exciting pure Garnacha. A *calandria* is a type of lark, and the ambition of this winery is to nest in as many old Garnacha vineyards as possible.

✓ *Tierga* • *Cientruenos*

CAN RAFOLS DEL CAUS
Penedès
★

Founder Carlos Esteva represents one of the best examples of Mediteranean ingenuity. He planted 28 varieties in this hilly area of Garraf and built a state-of-the-art winery allowing for micro vinifications. With so many factors involved, it is difficult to see a house style, but some of the wines possess a distinctive personality. A trend-setter, one to follow.

✓ *Gran Caus Rosado* • *Sumoll*

CASTILLO DE CUZCURRITA
Rioja
★

A domaine concept in a picturesque area of Rioja, great for a winery visit.

✓ *Cerrado del Castillo de Cuzcurrita*

CONTADOR
Rioja
★★

Recent winery in the stratosphere from the start, with amazing single-vineyard wines.

✓ *La Viña de Andrés Romeo*

CUATRO RAYAS
★

A cooperative with the most important stock of very old Verdejo bush vines, their best wines are much better than 98 per cent of Rueda.

✓ *Viñedos Centenarios* • *40 Vendimias*

CVNE/VIÑA REAL/CONTINO
Rioja
★★

This is one of the most classic wineries, with an impressive portfolio and sound mid-range wines.

✓ *Imperial* • *Viña Real GR* • *Contino Viña del Olivo*

DESCENDIENTES DE J. PALACIOS
Bierzo
★★★

Although this winery should go under the Alvaro Palacio heading, the key role played by Alvaro's nephew, Ricardo, in this property deserves a separate recognition. Alvaro and Ricardo brought Bierzo back to prestige with the Corullón wines and, later, the vineyard selections. They have also brought the most authentic *terroir* classification to the region, along with a productive approach in which solidarity and respect of the environment are key. A great example, with some of the greatest Spanish wines.

✓ *La Faraona* • *Moncerbal*

DOMINIO DE TARES
Bierzo
★

Founded with the new century, they have become one of Bierzo's leading wineries. They manage a small plot of old Godello and Mencía vines.

✓ *Cepas Viejas*

DOMINIO DEL AGUILA
Ribera del Duero
★★

Jorge Monzón is probably the only professional fine vine grower at Ribera, selling to famous wineries the best quality grape from his vineyards. When he decided to produce some wine from his oldest vines, he got two benchmarks for the future of the whole region, an impressively subtle red, the Peñas Aladas, and a complex and open rosé, the Pícaro.

DOMINIO DEL BENDITO
Toro
★

Frenchman Antony Terrin brought to the Toro *terroir* a mix of originality and loyalty, resulting in some of the most interesting wines in this region.

✓ *El Titán del Bendito (Toro)*

EMILIO MORO
Ribera del Duero
★★

Probably the most important winery now at Pesquera, they obtained fame thanks to their reliable basic styles and their two top wines, benchmarks for the area.

✓ *Malleolus de Valderramiro* • *Malleolus de Sanchomartín*

ENRIQUE MENDOZA
★

For many years, the leading producer of international-style wines in Alicante, this winery has come under the lead of Pepe Mendoza, the champion of the high-altitude delicate Monastrell wines from Villena. These wines must be tasted to understand Alicante's potential.

✓ *Santa Rosa* • *Las Quebradas* • *Estrecho (Alicante)*

ENVÍNATE
★★

A group of four winemakers that are revolutionising Spain's wine scene with their well-focused search for identity, freshness, and refinement in an increasing number of Spanish regions. One to follow closely. Their main projects are in Galicia (Lousas in Ribeira Sacra), Tenerife (Taganan and Benje), Almansa (Albahra), and Extremadura (T. Amarela Parcela Valdemedel).

EL ESCOCÉS VOLANTE
★

Norrel Robertson's adventure in Spain appears crazy: a Scot moving to deep Spain, the hot Calatayud, and rescuing old Garnacha vines from oblivion or extinction. Yet Norrel has become a Garnacha champion and a prestigious winemaker in other appellations.

✓ *El Puño (Calatayud)* • *En Sus Trece (No DO)*

FAUSTINO
★

With 150 years of history and thousands of barrels and millions of bottles in stock, Faustino leads a large group of wineries in Rioja, La Mancha, Ribera del Duero, and Navarra. Everything about Faustino is big, but for some years price seemed to be more important than quality in the winery. After the glorious 2001 vintage, however, the company seemed to back on the right track. The emblematic Faustino I Gran Reserva continues to be one of the cheapest great wines in the world.

FERRER BOBET
Priorat
★★

Sergi Ferrer and Raúl Bobet are the perfect combination of passion and science, producing wines that reflect their origin and their makers. The Ferrer Bobet Vinyes Velles is a Priorat benchmark.

FINCA ALLENDE
Rioja
★★

The exaltation of *terroir* by Miguel Angel de Gregorio, who selected possibly the best vineyards in Briones to produce his wines. Balance and freshness come to mind when tasting them.

✓ *Calvario* • *Allende (white)*

FINCA VILLACRECES
★★

A quite different style that its neighbour, Vega Sicilia, it offers fruitier wine that is more accessible when young and with the clear influence of Cabernet Sauvignon in the blend. The winery is a key reference in Ribera's Golden Mile and is part of the Artevino group, with wineries in Rioja (Izadi and Orben), and Toro (Vetus).

✓ *Finca Villacreces (Ribera del Duero)* • *Celsus (Toro)*

FRONTONIO
★

The personal project of Fernando Mora, this new star started from almost unknown Valdejalón IGP, with fruity Garnacha. Mora is now making wine in Campo de Borja. A brilliant career has just begun.

GÓMEZ CRUZADO
★

A classic winery too long in the doldrums. Juan Antonio Leza and David González are bringing it back to shape. They also produce excellent wine in Arlanza.

✓ *Pancrudo* • *Montes Obarenes (Rioja)* • *Sabinares (Arlanza)*

GONZÁLEZ BYASS GROUP
(except Tío Pepe)
★★

A group with wineries in several regions of Spain and in Chile. Highlights include: Blecua, Viñas de Vero's boutique winery in Somontano, with just one wine of great complexity; Beronia, developed under the lead of Matías Calleja into a top Rioja brand with a broad portfolio of reliable wines (his top wines win accolades everywhere); Finca Moncloa in Cádiz and Finca Constancia in La Mancha boast a portfolio of more international wines; and the lovely Tintilla de Rota in Cádiz.

✓ *Blecua (Somontano)* • *Beronia III A.C.* • *Crianza Edición Limitada (Rioja)* • *Finca Moncloa Tintilla de Rota (IGP Cádiz)*

GRANDES VINOS Y VIÑEDOS
Cariñena
★

The largest co-op in Cariñena, it one of the leaders of an effort to position Cariñena as a top Garnacha and Carignan region. They have access to many soil types and very old vines, with an excellent winemaker, the Chilean Marcelo Morales.

✓ *Anayón Terracota* • *Anayón Cariñena*

GRUPO CODORNÍU
★★

Bought by the Carlyle Fund, the group will likely undergo many changes. Codorníu includes 10 major wineries. Recent strategy focuses on quality over cheap wine. With Bilbaínas in Rioja, they enhanced the value of Viña Pomal, a sacred Rioja brand. In Abadía de Poblet they produce one of the best Cistercian-tradition Pinot Noirs. Legaris in Ribera del Duero is improving after a difficult start. Raimat in Costers del Segre aims to be the most eco-friendly winery. Scala Dei in Priorat is now releasing wines with a distinctive personality, an alternative to the classic *liquorella* wines of Priorat.

✓ *Viña Pomal Gran Reserva* • *La Vicalanda Reserva (Rioja)* • *Legaris Calmo (Ribera del Duero)* • *Raimat Natura Negre (Costers del Segre)* • *Scala dei Cartoixa (Priorat)*

GUELBENZU
★

One of the first "escapists" from DOs, Ricardo Guelbenzu abandoned Navarra in 2000 to follow his own way. He became the lead of the new Ribera del Queiles IGP. His wines are benchmark NW style.

✓ *Lautus*

HERMANOS PÉREZ PASCUAS
Ribera del Duero
★★★

They are the apostles of long ageing for top wines in the Roa part of Ribera del Duero. Their wines become as complex as Vega Sicilia after 20 years in bottle. An understated winery.

✓ *Viña Pedrosa Gran Reserva* • *Pérez Pascuas Gran Selección*

J. CHIVITE WINE ESTATES
★

This iconic winery in the south produces one of the best Chardonnays in Spain and lovely *rosados*. It has been bought by Castillo Perelada, ideally for the good.

✓ *Colección 125 (red and white)*

JOSÉ PARIENTE
Rueda
★

This is probably the cleanest expression of modern aromatic Verdejo from Rueda, far from the influence of selected yeasts as many other wines.

✓ *Cuvée Especial*

JUAN CARLOS SANCHA
Rioja
★★

Serious, natural wines that are the result of scientific research rather than ideologies. Greatest Garnacha in Rioja Alta.

LAGRAVERA
Costers del Segre
★

Located in an old gravel quarry converted to a vineyard over the rocks, this winery is a leading example of environmental con-

sciousness, following biodynamics principles and specialising in natural wines. Their wines have a message and are good.

✓ *Onra Moltahonra (red and white)*

LUIS CAÑAS
★

This winery shifted towards higher quality wines during the great development period of the 1970s. They offer great value in their range, with a unique top wine.

✓ *Hiru 3 Racimos* • *Selección Familia*

LUPIER
Navarra
★

A young couple who have recovered some of the best old Garnacha vines to produce delicate wines; an icon in the making.

✓ *La Dama* • *El Terroir*

MANZANOS WINES
★

Two brothers created this mid-size group with wineries in Rioja and Navarra and several hundred different wines from an important bulk wine and vineyard business managed by their family for four generations.

✓ *Manzanos 125 Aniversario* • *Voché Selección Graciano (Rioja)*

MARQUÉS DE CÁCERES
★

In 1970 Enrique Forner founded a winery that was revolutionary for its time. It was a great success, but today it struggles to keep in the same league as the very best.

✓ *Gaudium MC*

MARQUÉS DE MURRIETA
★★★

One of the two founders of new Rioja, 150 years ago. They passed through a difficult period at the end of last century, now are back to full splendour. Their wines from Galicia are also among the best.

✓ *Castillo de Ygay Gran Reserva Especial (red and white)* • *Dalmau (Rioja)* • *La Comtesse Pazo de Barrantes (Rías Baixas)*

MARQUÉS DE RISCAL
★★★

One of the founders of Rioja, they produce the most successful wine in international markets in terms of volume by price, as well as several rare jewels.

✓ *Reserva* • *150 Aniversario (Rioja)* • *Barón de Chirel (red from Rioja and white from Rueda region with no DO)*

MARQUÉS DE VARGAS
★

This multi-regional family project gives good quality overall and is not yet at the peak of its potential.

✓ *Marqués de Vargas Selección Privada (Rioja)* • *Pazo de San Mauro (Rías Baixas)*

MARTÍN CODAX
Rias Baixas
★

A top co-op in Rias Baixas, it has become a leading figure for innovation and visi-

bility within the whole appellation. It now makes wine in other regions as well.

✓ *Lías* • *Gallaecia*

MARTÍNEZ LACUESTA
Rioja
★

This winery classic from Haro has more than a century of history. Not very well known in export markets, it is one of the Madrilenos' favourite brands. They also produce a stupendous vermouth.

✓ *Campeador Reserva* • *Selección de Añada Reserva*

MAS DOIX
Priorat
★

Valentí Llagostera is one of the champions of the second wave of Priorat heroes. The first wave brought the region to the attention of the fine wine world by exploiting the energy in its *terroir*; the second returned to the original Cariñena and Garnacha vineyards to find the purity. There is no better Cariñena than Valentí's.

✓ *Mas Doix* • *1902*

MAURO
★

Founded in 1978, before the creation of the Ribera del Duero DO, this great winery was oddly left outside the DO, even though their key vineyards are well inside it. Mauro was created by Mariano García, a famous restaurateur. The wines reflect his style: powerful, rich, and rounded, needing long bottle ageing to soften up. Recently they have launched a white wine in Bierzo, with the autochthonous grape Godello.

✓ *Mauro VS* • *Terreus (no DO)* • *Mauro Godello (Bierzo)*

MUGA
Rioja
★★★

One of the most amazing progressions in the region, from bulk wine traders in the 1960s to icon Rioja wine producers now.

✓ *Aro* • *Torre Muga* • *Prado Enea*

NUMANTHIA
Toro
★

The property of LVMH, Numanthia is the archetype of powerful, densely concentrated Toro wines. They bought the property from Marcos Eguren, including many 100-year-old vines. Quite demonstrative but top quality wines.

✓ *Numanthia* • *Termanthia (Toro)*

OCHOA
Navarra
★

This family firm is known for lots of creativity and high quality, a benchmark for innovation and reliability in Navarra.

✓ *8A Moscato de Ochoa* • *8A Mil Gracias*

OJUEL
Rioja
★

This small family company offers decent wines and a great old-time speciality, the

supurao, a delicious sweet wine made in the Amarone style with dessicated grapes.

✓ *Supurao Tinto*

OLIVIER RIVIÈRE
★

A leader of the new wave of producers in Rioja, this winery produces stupendous Arlanza. One to follow closely.

✓ *El Cadastro (Arlanza)* • *Mirando al Sur (Rioja)*

OXER WINES
★

A leader in mastering notable new wines from Rioja Alavesa and Biscay Txakoli. Great design and excellent quality.

✓ *Suzanne* • *Kalamity (Rioja)* • *Marko Late Harvest (Bizkaiko Txakolina)*

PAGO DE LOS CAPELLANES
Ribera del Duero
★

The "estate of the chaplains" refers to the monastic origins of this property, now owned by the Rodero family. Their top *cuvées* are now the references for the DO.

✓ *Finca el Picón*

PAGO DE CARRAOVEJAS + OSSIAN
★

A very successful winery in Spain, it produces a brand that sells even its basic wines at a premium. Owned by José María Ruiz, a famous restaurateur, together with Ossian from Rueda (but not declaring its wines as such).

✓ *Pago de Carraovejas Cuesta de las Liebre (Ribera del Duero)* • *Ossian (no DO, in Rueda)*

PALACIOS REMONDO + ALVARO PALACIOS
★★★

An *enfant terrible* of the Spanish wine scene is now a top senior producer, year after year honing the finesse of his wines.

✓ *Finca Propiedad (Rioja)* • *L'Ermita (Priorat)*

PAZO DE SEÑORANS
Rias Baixas
★★

Instrumental in designing modern Rias Baixas through the work of Marisol Bueno, the owner and president of the appellation for 21 years. They have been wine pioneers and the discoverers of a new style, Albariño, with long ageing on the lees.

✓ *Selección de Añada*

PEÑA EL GATO
La Rioja Alta
★★★

One of the greatest classics, always getting better. Their wineries in Rioja Alavesa, Ribera del Duero, and Rías Baixas also deserve praise. A must try.

✓ *Gran Reserva 890* • *GR 904* • *Viña Ardanza (Rioja)* • *Aster (Ribera)* • *Martelo (Rioja Alavesa)*

PERELADA
Empordà
★

The largest winery in Empordà has a broad portfolio of wines and great success in international markets. Their newest development, single-vineyard wines, is a step up in quality. They have bought Chivite from Navarra and Viña Salceda in Rioja.

✓ *Aires de Garbet • Finca Garbet • Gran Claustro*

PÉREZ BARQUERO
Montilla-Moriles
★

A great classic: they manage some of the best Albariza PX vineyards both in Montilla and in Moriles.

✓ *Gran Barquero • Solera 1905 Amontillado*

PERNOD RICARD
(CAMPO VIEJO + YSIOS)
★

Elena Adell, one of Spain's best winemakers, brilliantly oversees Pernod Ricard's Spanish subsidiary. Even the basic Campo Viejo is decent and the Dominio is the cheapest luxury wine in Rioja. Ysios is the group's boutique winery with a building designed by renowned Spanish architect Calatrava, and the wines are excellent.

✓ *Dominio de Campo Viejo • Ysios Edición Limitada*

PINGUS + HACIENDA MONASTERIO
Ribera del Duero
★★★

Pingus is the Spanish Screaming Eagle, the wine that reached the pinnacle in record time. Peter Sisseck, the man behind such success, is an excellent winemaker. He looked for more delicacy and less oak and obvious fruit, so the wine is now even better than when it first became an icon. Peter is also the winemaker at Hacienda Monasterio, a less glossy property that put out a high-quality Ribera, with a great capacity to improve with age.

✓ *Pingus • Hacienda Monasterio Reserva Especial*

PROTOS
Ribera del Duero
★

Formerly called Bodega Cooperativa Ribera del Duero, they gave their name to the whole appellation. At one time they were a reference, together with Alejandro Fernández, for great Ribera. Their winery is impressive, and they have access to a lot of vineyards. It should be only a matter of time until they get back to the league of the best.

✓ *Finca El Grajo Viejo*

PUJANZA
Rioja
★★

This young winery has experienced a swift career up the ladder thanks to their exceptional Norte and Cisma wines. Now they also produce a white wine, Añadas Frías, at top level, a kind of *grand cru* Rioja.

R. LÓPEZ DE HEREDIA VIÑA TONDONIA
Rioja
★★★

The most classic Rioja producer; they have never changed their production process; their best wines are wonderful and can keep for 80 years at least.

✓ *Viña Tondonia Gran Reserva (red, white and rosé)*

RAFAEL PALACIOS
Valdeorras
★★

Rafael decided not to compete with his brother Alvaro on the red wine turf, but to be the maker of the best white wine in Spain. There he is.

✓ *O Soro • As Sortes*

RAMÓN BILBAO
★

This large group with an ambitious revamp project provides reliable mid-range wines while positioning others in the upper-market echelons.

✓ *Mirto • Edición Limitada*

RAÚL PÉREZ
★★★

Probably the best winemaker in Spain and one of the most distinguished in the world, he is capable of finding finesse and delicacy in many different regions and has become the godfather of a new style of wine built upon subtle complexity rather than fruit or concentration. He makes an impossible number of wines and consults for several wineries, with many jewels in his portfolio.

✓ *Ultreia (Bierzo) • Los Arrotos del Pendón (Tierra de León) • Sketch (Rías Baixas) • El Pecado (Ribeira Sacra) • La Patena (no DO)*

REMELLURI – TELMO RODRÍGUEZ

Advocates of *terroir* and single-vineyard wines, they produce excellent wines in Rioja and other regions.

✓ *Granja de Remelluri GR • Las Beatas (Rioja) • Gaba do Xil (Valdeorrras) • Pegasus (Gredos, with no DO)*

REMIREZ DE GANUZA
Rioja
★★

Perfectionist Fernando Remirez de Ganuza akes only the shoulders of the best bunches for his top wines and is in the highest Rioja league. His newer white wines are also outstanding.

✓ *Trasnocho • Remirez de Ganuza Blanco Reserva*

LA RIOJA ALTA
★★★

One of the greatest classics, always getting better. Their wineries in Rioja Alavesa, Ribera del Duero and Rías Baixas also deserves praise. A must.

✓ *Gran Reserva 890 • GR 904 • Viña Ardanza (Rioja) • Aster (Ribera) • Martelo (Rioja Alavesa)*

RODA
★★

This relatively modern winery introduced the concept of the optimum classic Rioja blends from the best *terroirs*. Their finest wines are top class.

✓ *Cirsión, Roda I (Rioja) • Corimbo I (Ribera del Duero)*

RODRIGUEZ SANZO
★

A flying wine company, their speciality is Toro, although their top wines from Rioja and Bierzo are also worth mentioning.

✓ *Las Tierras de Javier Rodriguez El Teso (Toro) • La Senoba (Rioja) • Vitis Extrema (Bierzo)*

SIERRA CANTABRIA + VIÑEDOS DE PÁGANOS + TESO LA MONJA
★★★

The magic of Marcos Eguren is behind some of the top wines in Rioja and Toro. La Nieta is the paradigm of the finest complexity for a Rioja single-vineyard.

✓ *La Nieta • Amancio (Rioja) • Victorino (Toro)*

SPECTACLE + CLOS MOGADOR
★★

René Barbieris one of the rare Priorat pioneers who kept his original formula. Clos Mogador is still a blend of international and Spanish varieties; Clos Manyetes is pure Cariñena. He repeated this formula in a new area with Spectacle, one of the most spectacular Garnacha wines in Spain.

✓ *Clos Mogador • Clos Manyetes (Priorat) • Spectacle (Montsant)*

TORRES + JEAN LEON
★★★

The world of wine would be very different without Torres. Generation after generation, they set trends, bringing in foreign grape varieties last century and now recovering near-extinct autochthonous varieties. They broought industrial efficiency into wineries and are now pioneers in climate-conscious growing and winemaking; they fostered the Cataluña DO to have more flexibility sourcing grapes and are also behind the first *Vi de Finca* ("*Vino de Pago*") in Cataluña, those of Jean Leon.

✓ *Reserva Real (Penedès) • Grans Murailles (Conca de Barberà) • Mas La Plana (Penedès) • Perpetual (Priorat) • Jean Leon Vinya Gigi (Vi de Finca Penedès)*

LA UNIVERSAL + MAS MARTINET
★★

Children of two of the pioneers of Priorat, Sara Pérez and René Barbier IV have taken over their respective wineries, Mas Martinet (Sara) and Clos Mogador (René), while launching this joint project in Montsant. Sara takes after Raúl Pérez: both have the magic of delicacy in their wines.

✓ *Venus la Universal (Montsant) • Clos Martinet (Priorat)*

URBINA
Rioja
★

A classic *bodega* selling its wines only after a long ageing period, normally longer than 10 years, at very interesting prices. Their blog is among the best in the sector. Traditional Rioja at its best.

✓ *Reserva Especial • Gran Reserva*

VALDUERO
Ribera del Duero
★★

The García Viadero sisters manage a project with lots of ambition and the means to realise it. This is a great *bodega*, excavated in the hill, with lots of good bush vineyards (they do not use VSP), and the masterful ageing, for as long as necessary, of their top wines. One of the best.

✓ *Una Cepa • 6 Años • 12 Años*

VALENCISO
Rioja
★

One winery, one wine. A theoretically simple concept that results in one of the most reliable Rioja wines.

VEGA SICILIA
★★★

The most iconic winery in Spain, an accolade that Pablo Alvarez keeps alive with smart management and very particular wines. Vega Sicilia has become an international group, with wineries in Hungary and in Toro, Rioja, on top of Aliön, also in Ribera del Duero. Vega Sicilia wines take a very long time to produce and are made to last many years. They are as much an object of desire as they are a pleasure to drink. Unique.

✓ *VS Unico • VS Reserva Especial • Valbuena (Ribera del Duero) • Macän (Rioja)*

VIÑA MAGAÑA
Navarra
★★

The best Merlot in Spain. A forgotten Navarra hero, with wines that improve over many years.

✓ *Viña Magaña Gran Reserva*

VINICOLA REAL
Rioja
★

When launched, this winery's brand, 200 Monges, was an innovative concept that put together tourism and fine wine. Its a young winery, but the wines are benchmark classics to be kept for many years.

✓ *200 Monges Gran Reserva*

VINYES DOMENECH
Montsant
★

Joan Ignasi Domenech, a man from the sustainable energy sector, decided to invest his life in a project for and with his family, buying an isolated vineyard in the wildest part of Montsant and producing wonderful Garnacha wines. One of those wines that help prove that Montsant is not inferior to Priorat.

✓ *Teixar*

VIVANCO
★

Visiting Vivanco is a must; they have the best wine museum in the world, with no exaggeration. The art, books, and artifacts collection is amazing. Their wines are also worth a try.

✓ *4 Varietales • Parcelas de Graciano*

Rioja

The first wine to receive DOCa status, Spain's highest classification, Rioja still produces more fine wine than any other region in this country. Rioja is best known for its red wines, the historic gran reservas, *the classic* reservas *and the affordable* grianzas, *but it excels as well with white wines – some of them amazingly unique – its rosés, its sparkling* méthode traditionelle, *and even some rare sweet wines (the* supurao, *a kind of Recioto della Valpolicella).*

Despite the recent resurgence of many high-quality wine regions in Spain, one can still say that Rioja is the top name for fine wine in Spain. So important is Rioja that politicians named a whole political region after the wine region (instead of calling it Logroño, as the province was called), turning a blind eye to the fact that parts of La Rioja are not in Rioja DOCa, and parts of Rioja DOCa are in Navarra, Basque Country, and Castilla y León .

Modern Rioja was created by two emigrants returning home: Marqués de Riscal and Marqués de Murrieta. Independently from each other, they brought Bordeaux expertise to Rioja and quickly changed the paradigm in the whole region. They succeeded because of the natural conditions in the region and because at that time (1860s), Rioja became connected to the rest of the world, thanks to the train. Huge market success was supported by relevant foreign investments during the period phylloxera had wreaked havoc on French vineyards but had not yet reached Rioja. This went hand in hand with a qualitative approach, which even the misfortunes of the wars did not alter. Only in the 1970s did Rioja relax in quality, with heavy consequences; it took more than 20 years to recover former quality levels.

There are two not-so-evident features in Rioja that help one to understand the region and its wines. First, in the late 19th century, fine Rioja was red wine fermented and aged in oak vats. At the beginning of the 20th century, some wineries were producing amazing white wines, also fermented and aged in *barriques,* with the capacity to improve in bottle for decades. Some red Rioja wines from the 1870s and whites from the 1920s are still in pristine condition today, complex and fresh. This is Rioja's major asset: the unique capacity to age for a very long time, first in barrel and then in bottle. For more than a century, ageing time was, together with the brand, the best quality indicator. This is the reason for Rioja's unique quality

classification based upon time of ageing in oak. Ageing is called *crianza* in Spanish, meaning "breeding", probably a much more appropriate term: the wines are not aged but brought up to excellence. That unique potential was all but destroyed during the 1960 and 1970s, a period in which quantity was more relevant than quality. As a result, most *gran reserva* from the 1970s and 1980s are thin, dry wines of little interest. The lesser qualities, *reserva* and *crianza,* are no better. High international criticism ensued, but it was focused on the decadent barrel stock and the very classification system, instead of on the real causes of the problem: lack of financing, inadequate market positioning, and poor viticulture.

FACTORS AFFECTING TASTE AND QUALITY

LOCATION
Most people know the three subregions in Rioja, but some connoisseurs maintain that the real zonification of the region should be based upon its seven valleys, as many as the tributaries flowing from the Sistema Ibérico into the Ebro River, which together form the Ebro basin. The seven river valleys are the Oja, Najerilla, Iregua, Leza, Jubera, Cidacos, and Alhama.

CLIMATE
Climate in Rioja is a key factor for uniqueness and quality, because it integrates three different influences. In the north, the Sierra Cantabria, a mountain range that is modest in elevation yet impressive in structure, provides a major key to the quality of Rioja, somewhat protecting the region from the Atlantic wet winds. The region is bisected by the Ebro river, opening it up to Mediterranean influences. Finally, to the south are two wide mountain ranges: the Sierra de la Demanda and the Sierra de Cameros, which separate it from La Meseta, keeping some continental influence for the region. The three Rioja subzones, Alta, Alavesa, and Oriental, present relevant climatic differences upon this pattern.

ASPECT
Vineyards are variously located and can be exposed to any orientation, although a southern exposure is favoured. Land is more hilly in Alavesa and Alta than in Oriental, where too much exposure to the sun is not good for top Tempranillo, but better for Garnacha.

SOIL
Although soils do vary a lot, the most relevant type of soil for fine wines is limestone. Limestone with either sandstone or calcareous clay and slatey deposits dominate the Alavesa and are quite common in Alta, while a ferruginous-clay and a silty-loam of alluvial origin cover a limestone base in many parts of Oriental. The most common type of soil is alluvial, providing excellence only in a few cases.

VITICULTURE & VINIFICATION
Most red wines used to be a blend of at least three grapes from different areas within a single appellation. An increasing number of wines are pure varietal (mostly Tempranillo); there is also a growing trend for single-estate wines. Viura is the most important white variety, although Garnacha Blanca, and Malvasía used to play a relevant role in the past. Recent normative modification allowed for the introduction of Verdejo, Sauvignon, and Chardonnay, something difficult to understand from the viewpoint of typicity. There were two traditional vinification processes. The first, the one used before the arrival of the Bordelais with their winemaking know-how, consisted of semi-carbonic maceration carried out in open vats: the grapes were trodden after the first few days of inter-cellular fermentation. The second, and most usual process, was based upon extended maceration after alcoholic fermentation and ageing in oak barrels. Although recent trends favour shorter oak-ageing and longer bottle-maturation, the unique character of Rioja still relies heavily on oak, and it is essential for its future that it should remain so.

GRAPE VARIETIES
Tempranillo, Garnacha, Graciano, Mazuelo (Cariñena), Maturana Tinta, Viura (Macabéo), Malvasia, Garnacha Blanca, Tempranillo Blanco, Maturana Blanca, Turruntés, Chardonnay, Sauvignon Blanc, Verdejo

BODEGAS YSIOS WINERY, LAGUARDIA
This truly avant-garde building lies at the foothills of the Cordillera Cantábrica. The pixilated-looking landmark winery was designed by architect Santiago Calatrava.

WIRE-TRAINED VINES FORM NEAT ROWS OUTSIDE HARO IN THE EBRO VALLEY, RIOJA ALTA DO
This region favours the planting of Tempranillo grapes, a variety that produces the classic Rioja-style red wines. Other important varieties include Garnacha and Graciano.

Second, Rioja was not "created" by vine-growers or domaine owners, but by wineries that used to buy grapes. Until 50 years ago, most wineries did not own relevant vineyards. This, together with the convenience of blending grape varieties of different origins as an insurance against vintage variation, is the reason why classic Riojas are geographical and varietal blends. Even today, some of the best Rioja wines are intra-regional blends, with proportions of Garnacha from Rioja Oriental blended with Tempranillo and Mazuelo from Alta, for instance.

At the end of the 20th century some producers took new-wave Bordeaux and New World fine wines as their model, placing more emphasis on fruitiness, colour intensity, the use of French oak instead of American oak, and shorter oak ageing times. As a crucial advantage, they had access to EU subsidies and could afford such a change. Shortly afterwards Rioja joined the trend for single-vineyard wines as an expression of top class. Many of those producers decided to abandon the traditional quality terms in order to keep flexibility in deciding the ageing period for each vintage, or simply as a way of stating their modern status, compared to the recent past of thin, ordinary wines. Seemingly, most of them claimed (quite correctly) the right to identify and provide a guarantee of the origin of the grapes, as closely as possible. The DO authorities gave in and conceded a recent new classification, based upon origin, in four categories: generic Rioja, the sub-zones (Alavesa, Alta, and Oriental – ex-Baja), *vinos de municipio* (similar to Burgundy villages, but referring to wineries rather than vineyards, a nod to Rioja's particular past), and, at the top, single-vineyard wines. The newly classified wines were released in 2019. Both quality classifications, the former one based upon ageing time and the new one based upon origin, now co-exist.

The improvements in vineyard and winery management is not confined only to modern producers, but to the whole region. As a consequence, present wines with a long oak ageing (*gran reserva*) are again normally high quality. *Crianza* is a mid-market category, useful to identify decent wines with evidence of oak ageing. *Reserva* is an in-between and also includes some excellent wines.

ORIGINS

Wine has been made in Rioja since at least the 11th century BC, when the Romans conquered the area. Rioja was well respected enough by 1560 that its producers forbade the use of grapes from outside the region and guaranteed the authenticity of their wines with a brand on the *pellejos* (goat skins) in which they were transported. Wooden barrels came into use in the 18th century but were five times the size of casks today. It was not until the 1850s that Marqués de Riscal and Marqués de Murrieta brought Bordeaux winemaking know-how and access to national and international markets. Indeed, Manuel Quintano was the first to bring the new techniques, but he did not have the marketing capacity of the Marquises. Camilo Hurtado de Amezaga, the Marqués de Riscal, persuaded the Alava government to hire a distinguished Bordeaux winemaker, Jean Pineau, to teach French methods to local growers. Pineau was keen to accept the job because it seems that he was not happy with his boss, the owner of Château Lalanne, who did not apply at the 1855 classification competition. Upon completion of his contract, Pineau was employed by Riscal and started a story that was swiftly followed by the wineries built around the then-new Haro Train Station, the most relevant engine for ensuring the access of Rioja wines to fine wine markets. Phylloxera arrived in Rioja in 1899, but a few years of hard work in the vineyards meant that by World War I, Rioja was again on the road to success. The Spanish Civil War (1936–39) did not directly affect Rioja, because the region was already within Franco's territory. But the 1960s brought a period of decadence that continued until several years after the entry of Spain into the European Union (1986).

RIOJAN AGEING CRITERIA

Genérico No indication of ageing. Valid for lots of young wines and for many fine wines whose makers do not want to commit to minimum ageing periods, or simply do not want to use the classification.

Crianza Rioja ageing terms have become the benchmark for most other Spanish DOs, but the specific periods can be different. In Rioja, a red *crianza* is a wine that is released in its third year after the harvest, having been aged for at least one year in oak barrels. White and rosé *crianza* wines should have at least six months in oak barrels.

Reserva Wines that have been aged for at least three years, at least one of them in oak *barriques,* and at least 6 months in bottle. For white and rosé wines, two years aging with a minimum of six months in *barrique.*

Gran Reserva Red wines aged for at least five years, at least two of them in oak barrels and three in bottle (the greatest tend to age for much longer). For white and the only rosé *gran reserva* wine (*Viña Tondonia GR Rosado*), minimum ageing is four years, with at least six months in barrel.

RIOJA

Rioja's vineyards are located along the Ebro Valley, between Haro and Alfaro, and throughout its hinterland, with vines clustered around many of the Ebro's tributaries, one of which, the Oja River, has given its name to the region. Much Rioja is red and blended from wines or grapes (primarily Tempranillo and Garnacha) originating from one or several of the region's three districts (Rioja Alta, Rioja Alavesa, and Rioja Baja), although many of the best quality wines are single-district wines, and a handful of single-estate Riojas have also emerged in recent years.

RIOJA ALTA

Logroño and Haro, the principal towns of Rioja, are both in the Rioja Alta. Rioja Alta's vineyards lie primarily to the south of the Ebro River in river valleys and on terraces built into the hillsides of the Sierra de la Demanda. Tempranillo is now the predominant variety (60 years ago it used to be Garnacha), but most wines also include Mazuelo and/or Graciano in the blend. The area's wine is Rioja's fullest in terms of fruit and concentration and can be velvety smooth. Classic oaked white wines can also be great, if properly made. The neutral Viura grape makes wines that should never be drunk young, but which become beautiful with age. The Rioja Alta also has good potential for *méthode champenoise* sparkling wines made, mostly, with Viura and Garnacha.

RIOJA ALAVESA

There are no large towns in the Alavesa, a district that is slightly cooler than Alta, although the real difference between Alta and Alavesa is more political (Alavesa is Basque Country, Alta is La Rioja) and historical (the vine-growers have a different social structure in Alavesa than natural. Tempranillo is practically the only variety here, planted together with Viura and Malvasía in old vineyards. Wines produced here tend to have greater acidity.

RIOJA ORIENTAL

Formerly Baja, Rioja Oriental is a bizarre new name for this subzone. Until recently it was barely used by wineries, which prefer to use just Rioja. Baja is a semi-arid area influenced by the Mediterranean and is hotter, sunnier, and drier than the Alta and the Alavesa, with rainfall averaging between 38 and 43 centimetres (15 and 17 inches) per year, but falling to as low as 25 centimetres (10 inches) at Alfaro in the south. Although much Tempranillo is planted, Garnacha from this area can in some cases can be magnificent. One of the most expensive wines in Spain, the pure Garnacha Quiñón de Valmira by Alaro Palacios, is produced here. But far too many wineries lag behind, trying to use poorly adapted Tempranillo.

VINO DE MUNICIPIO

A new idea based on the Burgundy villages model, with the added (rather strange) obligation of bottling at least 85 per cent of the wine in the same village, as if the bottling venue were a crucial *terroir* component. A second, very relevant, difference to the Burgundy model is that 142 municipalities in the Rioja DOCa are eligible, no matter how painful their quality track record is. These wines were released in 2019. They are likely to be mid-market wines, labelled as such by wineries with single-vineyard wines in their portfolio.

VIÑEDO SINGULAR

At first impression, this new term looks quite similar to Burgundy Grand Cru. But it is not. This term should be translated into English as "Special Vineyard" rather than "Single Vineyard". Indeed, a wine from a *Viñedo Singular* could come from several single vineyards. The key differentiation factor for a *Viñedo Singular* wine is that the winery should have full ownership of vineyards or effective monopoly through long-term contracts to buy grapes from the protected plots of land. No historic track record of quality is required, nor a probation period. Nothing to do with Clos de Vougeot. The initial group of wines were released in 2019.

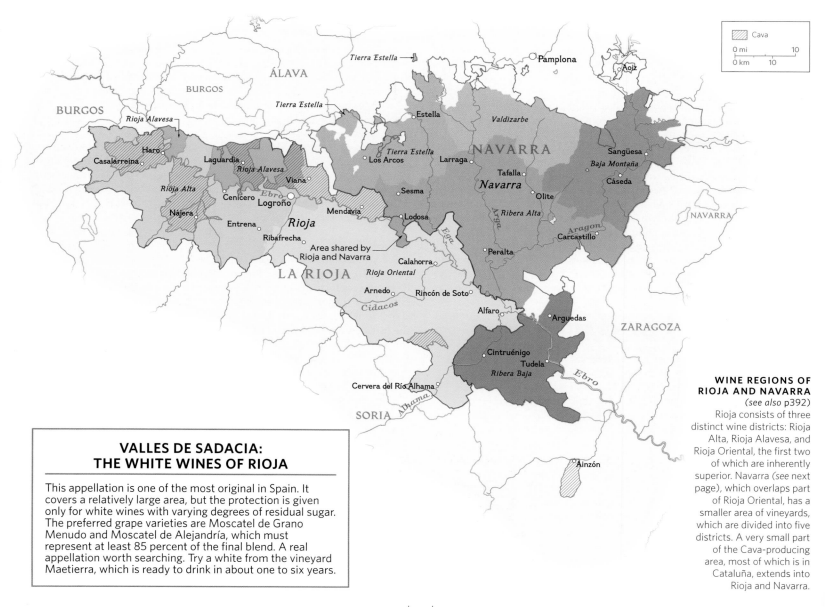

VALLES DE SADACIA: THE WHITE WINES OF RIOJA

This appellation is one of the most original in Spain. It covers a relatively large area, but the protection is given only for white wines with varying degrees of residual sugar. The preferred grape varieties are Moscatel de Grano Menudo and Moscatel de Alejandría, which must represent at least 85 percent of the final blend. A real appellation worth searching. Try a white from the vineyard Maetierra, which is ready to drink in about one to six years.

WINE REGIONS OF RIOJA AND NAVARRA
(*see also* p392)
Rioja consists of three distinct wine districts: Rioja Alta, Rioja Alavesa, and Rioja Oriental, the first two of which are inherently superior. Navarra (*see next page*), which overlaps part of Rioja Oriental, has a smaller area of vineyards, which are divided into five districts. A very small part of the Cava-producing area, most of which is in Cataluña, extends into Rioja and Navarra.

Navarra and Aragón

Probably the most classic wine regions of Spain, Navarra and Aragón are in all likelihood the cradle of Garnacha (Grenache) and Cariñena (Carignan) grapes.

Navarra and Aragón were highlighted in the late Middle Ages as quality regions, but have since faded in comparison to their successful neighbour Rioja. Navarra reacted by grubbing up old Garnacha vines and fostering the planting of international varieties and Tempranillo, which has proved useful for commercial wines but counterproductive for fine wines. In Aragón, Somontano did the same, becoming a DO with little character. In the rest of Aragón, although the same fashion was followed for planting a lot of Tempranillo and French varieties, growers were less enthusiastic and kept many old-vine Garnacha; that stock is now the source of an spectacular renaissance of the region.

The Pyrenees form the northern boundary of Aragón, which is a landlocked region comprising the provinces of Huesca, Zaragoza, and Teruel. There are four DOs in the region (*see* below).

PROPIEDAD DE ARÍNZANO, NAVARRA
Situated in the northeast between Rioja and Bordeaux, Propiedad de Arínzano practices environmentally sustainable vine-growing. Arínzano is one of the few estates in all of Spain to be recognized with Pago status and was the first in the north of Spain.

THE APPELLATIONS OF
NAVARRA AND ARAGÓN

NAVARRA

A large wine region, with several distinctive mesoclimates, some of them, close to Rioja Oriental, are too hot for good Tempranillo, others too cold. Navarra was once virtually synonymous with *rosado*, and this style still accounts for a third of total production. Until recently, *rosado* was considered a lesser wine. Navarra wanted to emulate Rioja's success and endeavoured to produce higher value wines by fostering both Tempranillo and international varieties from Rioja, which sometimes give good quality, but never excel. There is now no distinctive Navarra style: upon the same name one can drink international-style Cabernet Sauvignon or classic Garnacha Rosado or Muscatel or Chardonnay. The region is, however, capable of producing fine wines of exceptional value, even some great *rosados*. Similar to Australia's decision to rip out its oldest Riesling vineyards just before its top-quality Riesling wines started to receive the recognition they long deserved, so Navarra pulled up some of its oldest Garnacha vineyards just as wine lovers the world over began to appreciate just how stunning Garnacha could be from low-yielding old vines.

🍇 Tempranillo, Viura (Macabéo), Cabernet Sauvignon, Chardonnay, Cinsault, Garnacha (Grenache), Garnacha Blanca (Grenache Blanc), Graciano, Malvasia, Mazuelo (Cariñena), Monastrell (Mourvèdre), Moscatel Grano Menudo (Muscat Blanc à Petits Grains)

🍷 1–10 years (red), 1–4 years (white) on purchase (rosé)

✓ Castillo de Monjardín • J. Chivite • Bodegas Ochoa • Viña Magaña • Domaines Lupier • Bodegas Inurrieta • Artazu • Viña Zorzal • Pago de Larrainzar • Señorío de Sarria.

VINOS DE PAGO OF NAVARRA

Partly as a consequence of the lack of prestige for the Navarra DO, several *Vino de Pago* appellations have been created in the region, all of them blends of Rioja and international varieties. They are:

- La Finca Bolandín
- Pago de Arínzano
- Pago de Otazu
- Prado de Irache

ARAGÓN

CALATAYUD

Rescued from oblivion by Norrel Robertson and other pioneers, its best asset is a stock of very old vines and an extremely dry continental climate that protects the vines from major diseases. Many decent simple wines made by co-ops.

🍇 Garnacha, Mazuela, Cabernet Sauvignon, Syrah, Tempranillo, Merlot, Monastrell, Garnacha Blanca, Malvasía, Macabeo, Chardonnay, Moscatel

🍷 Upon purchase, except for the best 3–8 years

✓ Augusta Bilbilis • El Escocés Volante • Bodegas y Viñedos del Jalón • Breca

CAMPO DE BORJA

The notorious Borgia family took their name from this region. A big success in US and UK markets with their Garnacha-based *crianza* wines, they are refocusing on this variety at the expense of experiments with Tempranillo, Syrah, and others. Whites are fresh, but somewhat neutral.

🍇 Garnacha, Mazuela, Tempranillo, Garnacha Tintorera, Syrah, Cabernet Sauvignon, Merlot, Garnacha Blanca, Viura, Chardonnay, Verdejo, Moscatel, Sauvignon Blanc

🍷 3–8 years (red), 9–15 months (white and rosado)

✓ Bordeje • Borsao • Alto Moncayo • Aragonesas • Cuevas de Arom Coop Agraria del Santo Cristo • CA del Campo San Juan Bautista

CARIÑENA

The most classic region in Aragón has the most potential for quality. Amazing soil diversity, good stocks of very old Garnacha vines, but only a bit of their emblematic grape, Cariñena (Carignan). Production is dominated by co-ops, which are well organised and capable of selecting the best plots to make outstanding wines. Some deliberately maderised *rancio* wines are still produced. This is a wine area to watch.

🍇 Cabernet Sauvignon, Chardonnay, Garnacha blanca, Garnacha Tinta, Juan Ibañez, Macabeo, Cariñena, Merlot, Monastrell, Moscatel de Alejandría, Parellada, Syrah, Tempranillo, Vidadillo

🍷 3–10 years (red), 9–18 months (white and rosado)

✓ San Valero • Grandes Vinos y Viñedos • Solar de Urbezo • Paniza • Quinta Mazuela

SOMONTANO

Set at the foot of the Pyrenees, between Penedès and Navarra, the Somontano region received universal plaudits when Bodegas Lalanne was created in the 19th century. The region was then almost abandoned until the 1980s when huge investments in Chardonnay, Cabernet, and Tempranillo vineyards meant it once again became one of Spain's greatest wine regions. Now, it struggles to get recognition. some of the region's best wines are produced with autochthonous varieties and the potential for natural quality is still there.

🍇 Parraleta, Pinot Noir, Moristel, Garnacha Tinta, Syrah, Chardonnay, Gewürtraminer, Macabeo, Garnacha blanca, Alcañón, Riesling, Sauvignon Blanc.

🍷 2–15 years (red), 1–3 years (white)

✓ Blecua • Otto Bestué • Viñas del Vero • Enate • Lalanne • Pirineos • El Grillo y La Luna

VALDEJALON IGP

The region was all but forgotten until it was made famous by Fernando Mora. Lovely Garnacha wines from old vineyards, punchy and expressive.

🍇 Cabernet Sauvignon, Garnacha Tinta, Graciano, Mazuela, Merlot, Monastrell, Syrah, Tempranillo, Chardonnay, Garnacha Blanca, Macabeo, Moscatel

🍷 1–5 years (red)

✓ Frontonio

REMOTE IGPs OF ARAGÓN

These are small IGPs in relatively remote areas, with a history of producing wine. Very few wineries and no memorable wines yet.

- Bajo Aragon
- Ribera del Gallego – Cinco Villas
- Ribera del Jiloca
- Ribera del Quieles
- Valle del Cinca

The Spanish Islands

Romans first planted vines in the Balearics around 121 BC, although the wine industry there is still growing. The first of the Canary Island appellations was established in 1992, but "Canary sack" was famous in Shakespeare's time.

The two Spanish archipelagos have only one thing in common: they both attract millions of tourists each year. All other basics are different: The climate in the Balearics is textbook Mediterranean, while the Canary Islands (Islas Canarias) enjoy a unique combination of a subtropical and Atlantic climate. The soils in the Balearics are sedimentary and limestone; in the Canaries all soils are volcanic. Viticulture in Mallorca was influenced by the Kingdom of Aragón; in the Canaries the influences are Portuguese and Castilian.

The Balearics exported a lot of ordinary wine to France during the French phylloxera crisis, until the disease wreaked havoc in their own vineyard. It never reached the Canary Islands, however, and "Canary" was one of the most expensive and appreciated wines in Great Britain and America in the 17th century; they were then forgotten for centuries, until now.

WINE REGIONS OF THE CANARY ISLANDS

Along with the Islas Canarias PDO that covers the entire archipelago, the Canary Island appellations are spread over these volcanic islands: Tenerife alone has five wine-producing zones. The Balearic Islands (*see* map p392) have two official DO wine titles, both of which are on the island of Mallorca, but a number of *Vino de la Tierra* zones are spread throughout the other islands.

ROWS OF VINES GROW IN CONE-SHAPED GERIA IN THE VOLCANIC LANDSCAPE OF LANZAROTE
The challenging landscape of the Canary Islands calls for an ingenious system of planting vines in geria: cone-shaped hollows excavated several metres deep in the volcanic gravel.

THE APPELLATIONS OF
THE SPANISH ISLANDS

CANARY ISLANDS

ABONA
Tenerife

Claiming to protect the vineyards at the highest altitude in Europe, this small appellation was created out of the effort of one cooperative. It is not very clear how their wines are different from those of other appellations in Tenerife, especially when considering that they used to be a white wine region with one main variety, Listán Blanco, and now accept practically all varieties cultivated on the island.

🍷 *Listán Blanco, Bual, Verdello, Bermejuelo, Malvasía, Sabro, Listán Negro, Negramoll, Cabernet, Tempranillo, Merlot, Syrah, Ruby Cabernet, Castellana, Vijariego Negro, Baboso, Tintilla.*

🍾 *1–3 years (red and white)*

✓ *Cumbres de Abona • Altos de Trevejos*

TACORONTE-ACENTEJO
Tenerife

The first appellation on the island, it launched the recovery effort. The best wineries are probably not here, but Tacoronte produces relatively high volumes of very decent wine, and includes some of the most interesting visits in the island. There are 32 authorised grape varieties: most of the autochthonous ones, plus a number of international varieties, the result of early institutional efforts to push for a quality appellation with all possible solutions.

🍾 *1–3 years (red and white)*

🍾 *Bodegas Insulares Tenerife • Cráter*

VALLE DE GÜIMAR
Tenerife

Very similar to Abona, in all respects, including authorised grape varieties and wine styles.

🍷 *1–3 years (red and white)*

🍾 *Valle de Güímar, El Borujo*

VALLE DE LA OROTAVA
Tenerife

One of the most exciting appellations in the island, with viticultural practices that are different to the rest of Tenerife, producing some of the most distinguished wines. Their trellising system, *cordón trenzado*, is unique in the world and spectacular, even more so when considering that most vines are 100 years old. There are two different areas in this small region: Los Realejos, specialising in white wines, and La Orotava, specialing in red wines.

🍷 *Listan Blanco, Listan Negro, Tintilla, Moscatel, Albillo, Vijariego, Marmajuelo, Malvasía*

🍾 *1–7 years (red and white)*

✓ *Suertes del Marqués • Tajinaste • Valleoro • El Penitente*

YCODEN-DAUTE-ISORA
Tenerife

The resurgence of Canary autochthonous varieties started in this appellation, thanks to the initiative of Juan Jesús Méndez at Bodegas Viñátigo. He is the man who showed the value of Baboso, Marmajuelo, Gual, Vijariego, and other grape varieties. Now other wineries are following his path with success. Together with Orotava, this is the home of most of the mythic Canary wine of the 17th century

🍷 *Tintilla, Listán Negro, Malvasía Rosada, Negramoll, Castellana, Baboso Negro, Bastardo Negro, Moscatel Negro, Vijariego Negra, Marmajuelo, Gual, Malvasía, Moscatel, Pedro Ximénez, Verdello, Vijariego, Albillo, Baboso Blanco, Bastardo Blanco, Forastera Blanca, Listán Blanca, Sabro, Torrontés*

🍾 *1–7 years (red and white)*

✓ *Viñátigo • Envinate • Ignios Orígenes • La Guancha*

BODEGAS JOSÉ L FERRER, BINISSALEM DO
This Mallorca winery displays some of its barrels. Wine barrels have become a symbol of the island.

EL HIERRO

This small island is a separate ecosystem, with unique species and a very particular viticulture. Some of the indigenous grape varieties that have been recovered come from here. The main player in the island is the cooperative, very professional and innovative. Vines are normally not trellised; they creep on the lava soil.

🍷 *Listán Negro, Verijadiego Negro, Baboso Negro, Listán Blanco, Verijadiego Blanco*

🍾 *1–4 years (red and white)*

✓ *Cooperativa Frontera*

GRAN CANARIA

Recent appellation for the second-largest island in the archipelago, definitely with a good quality potential, but until now short of the personality of the other islands, each one of them with their flagship varieties and wines styles. Progressing quickly though.

🍷 *Listán Negro, Negramoll, Tintilla, Malvasía Rosada, Malvasía Volcánica, Vijariego Blanco, Marmajuelo, Gual, Albillo, Moscatel*

🍾 *1–4 years (red and white)*

✓ *Bentayga • Higuera Mayor • Lava • Las Tirajanas • Los Berrazales • Volcán*

LA GOMERA

Small island with a speciality, the Forastera Blanca (*forastero* means "foreigner", a curious name for an autochthonous variety), cultivated in creeping vines in steep slopes. There are also many field blends; this is the justification given for having as many as 31 authorised varieties in the appellation

🍷 *Forastera and 30 varieties*

🍾 *1–2 years (red and white)*

✓ *Cumbres de Garajonay*

LA PALMA

An island with spectacular vineyards in three completely different zones: the green north, the high-altitude central, and the warm semi-desert south. The best grape varieties in La Palma are Malvasía and Albillo, although they produce good wines with other varieties. They have a speciality, the *vino de tea*, a kind of Canary retsina, because the wine is fermented and aged in *Pinus canariensis* (Canary Pine) barrels; very original. There are some great classic Canary dessert wines as well.

🍷 *Malvasía, Sabro, Bujariego, Gual, Almuñeco, Verdello, Albillo, Negramoll, Listán Prieto, Listán Blanco*

🍾 *1–3 years (red and white), 1–10 years (sweet)*

✓ *Llanovid • El Níspero • Matías i Torres • Vega Norte*

LANZAROTE

This island's vineyards are worth a visit, even for those who do not like wine. At La Geria, vines are planted in holes excavated in a lunar-like landscape and protected with small walls, and camels are used for work in the vineyards. Such a beautiful vineyard has started to produce good original Malvasía Volcánica wines. There is no other place with this grape; the variety's parentage and origin are still a mystery.

🍷 *Malvasía Volcánica, Moscatel, Listán Negro, Listán Blanco, Diego, Syrah*

🍾 *1–5 years (white), 1–10 years (sweet)*

✓ *El Grigo • La Geria • Los Bermejos • Stratus, Rubicon*

ISLAS CANARIAS

What should have been an umbrella appellation, covering all seven islands in the archipelago with the purpose of ensuring supply, has become a sad competitor of the other island appellations. They have registered the name "Canary" within their appellation, which, even though legal, is a fake claim. *All* wines from Canary Islands *should* be Canary. Seemingly, consumers are not allowed to read on the label that, for instance, Abona is in the Canary Islands. Political nonsense, and quite harmful for Canary wine. Some good producers nonetheless still sell wines under this appellation.

🍷 *Baboso, Vijariego Negro, Vijariego Blanco, Titilla, Negramoll, Verdello, Moscatel, Marmajuelo, Malvasía Aromática, Gual, Listán Prieto, Listán Blanco*

🍾 *1–6 years (red and white)*

✓ *Viñátigo • El Lomo • Monje*

BALEARIC ISLANDS

BINISSALEM

This appellation owes its existence primarily to José Ferrer, a man convinced of the value of the main grape variety, Manto Negro. Other producers are making interesting wines now, some of them worth ageing.

🍷 *Manto Negro , Callet, Tempranillo, Monastrell, Cabernet Sauvignon, Syrah, Merlot, Moll or Prensal Blanc, Giró Ros, Macabeo, Parellada, Chardonnay, Moscatel.*

🍾 *1–3 years (red and white)*

✓ *José Luis Ferrer • Antonio Nadal*

PLA I LLEVANT

This appellation is structured around the fame of Anima Negra and their Callet grape variety, although it is quite open to many innovations.

🍷 *Manto Negro, Callet, Prensal Blanc, Fogoneu, and a long list of international varieties*

🍾 *1–10 years (red)*

✓ *Anima Negra • Miquel Oliver • Gelabert*

OTHER IGPs OF THE BALEARICS

These six IGPs were created to provide a protection for the names of wines produced in the respective islands or areas, outside the DO framework, but they do not so far have any organoleptic or cultural identity.

- Ibiza
- Formentera
- Illes Baleares
- Menorca
- Mallorca
- Serra de Tramuntana – Costa Nord

Cava and Sparkling Wine Country

Cava is, together with Sherry and Rioja, the most popular Spanish wine name in the world. It was born as a Catalan version of champagne and has developed into a very Spanish sparkling wine with a broad range of quality levels, from sublime to mediocre. For legal and political reasons, it is indeed an appellation of many origins.

Cava can be made in seven Spanish regions, yet 95 per cent is made in Penedès, Cataluña. Because of the growing market for quality sparkling wine, and of the many dissentions inside the Cava DO, a number of producers make high-quality sparkling wine outside the Cava DO.

Prior to phylloxera, which struck Penedès in 1876, more than 80 per cent of the vineyards here were planted with black grapes. When the vines were grafted on to American rootstock white varieties were given priority due to the growing popularity of sparkling white wines. It is easy to recognise the classic imported varieties in the vineyards because they are trained along wires, whereas traditional Spanish vines grow in little bushes.

Cava was born as *xampany* ("Champagne"), something that the French did not appreciate. Subsequently the Cava name was adopted in 1972. It was created as a term defining the winemaking method, rather than as an indication of origin, so that wineries in places as far away as Rioja, Extremadura, or Valencia were entitled to produce Cava. When Spain joined the EU, Cava had to be translated into a DO. Because of that it is split in many areas throughout Spain.

The first Spanish sparkling wine was made by Antoni Gali Comas some time prior to 1851, when he entered it in a competition in Madrid. He did not persevere, and the next milestone was with Luis Justo y Villanueva, the laboratory director at the Agricultural Institute of Sant Isidre in Cataluña. It was under Villanueva that all the earliest commercial producers of Spanish sparkling wine learned the Champagne process. In 1872 three of his former students, Domènec Soberano, Francesc Gil, and Agustí Vilaret, entered their sparkling wines in a Barcelona competition. All used classic Champagne grapes grown in Cataluña. Soberano and Gil were awarded gold medals, while Vilaret, who used raspberry liqueur in the dosage, received a bronze. Vilaret's firm, Caves Mont-Ferrant, is the only one of these enterprises to have survived.

The above facts are fully documented and thus contradict Codorníu's claim that it was the first to produce Spanish sparkling wine in 1872. Codorníu did not, in fact, sell its first sparkling wines until 1879, and it would not be until 1893 that its production hit 10,000 bottles, a level that the three original firms achieved in the 1870s. What everyone seems to agree on, however, is that José Raventós i Fatjó of Codorníu was the first to make bottle-fermented sparkling wine out of Parellada, Macabéo, and Xarel-lo and that these grapes came to form the basis of the entire Cava industry.

THE PROBLEMS OF CAVA

Cava is probably Spain's most successful wine in world markets. In 2017 Cava exports amounted to 162 million bottles (for a total production of 252 million bottles). Yet such a success has its problems. Most of Cava is cheap and cheerful, but Cava vineyards (mostly but not only in Penedès) have the

Denominación de Origen

Tarragona

Cava

0 mi 10
0 km 10

CAVA-PRODUCING REGIONS OF PENEDÈS
(see also p392)
Spreading out behind the coastal cities of Barcelona and Tarragona, the Penedès DO in Cataluña has a flourishing Cava industry.

potential for much higher quality. The two Cava behemoths, Codorníu and Freixenet, fought for years for the base market, spiralling down the perceived quality. This gave entry to low-cost specialists such as García Carrión, who first accelerated the trend and then almost destroyed both competitors (both Freixenet and Codorníu were finally sold to foreign investors).

Meanwhile, a number of Cataluña producers focused on quality and paid much attention to their vineyards. They became the force behind the best Cava wines, which have a precisely defined expression of their respective *terroirs*.

The Cava board reacted against the negative image of Cava first by fostering premium Cava, wines with a longer ageing on the lees, and then by creating the Cava de Paraje Calificado (CPC) category, indicating a top-quality single-vineyard Cava. Meanwhile, a relevant group of wineries abandoned the appellation and were welcomed by the Penedès DO, which created a category, Clàssic Penedès, for them. Others, like Raventòs and Torres, decided to go out on their own. Finally, another group, Corpinat, could as well decide to abandon Cava. Things get more complicated, as usual, thanks to politicians. Catalan Cava is boycotted in several Spanish regions because of the separatism crisis. Cava producers in other regions benefit from that, with the shocking support of their own politicians. Besides, Cava is, together with Rioja and Jumilla, managed by the central government, because it covers several regions, while all other DO are managed by regional government. Some Cataluña politicians would like to see Cava destroyed and a purely Catalan appellation take over. This is a mess that has nothing to do with wine.

CAVA AGEING CRITERIA

Cava Aged on the lees in bottle for at least 9 months.

Cava Reserva Aged on the lees in bottle for at least 15 months.

Cava Gran Reserva Aged on the lees in bottle for at least 15 months.

Cava de Paraje Calificado (CPC) Made from a wine using grapes from a single parcel of with specific *terroir* and microclimatic conditions, together with quality criteria of its production and ageing. Minimum ageing 36 months.

FACTORS AFFECTING TASTE AND QUALITY

LOCATION
Lying in the northeast corner of Spain, where Rioja's River Ebro enters the Mediterranean, Penedès is part of Cataluña. The 5 per cent of cava produced outside of Penedès comes from several regions: Aragón, Castilla y León, Extremadura, La Rioja, the Basque Country, Navarra, and Valencia.

CLIMATE
A mild Mediterranean climate prevails in Penedès, becoming more continental (hotter summers and colder winters) moving westwards and inland towards Terra Alta. In the same way, problems with fog in the northeast are gradually replaced by the hazard of frost towards the southwestern inland areas. In the high vineyards of Alta Penedès, white and aromatic grape varieties are cultivated at greater altitudes than traditional ones, because they benefit from cooler temperatures.

ASPECT
Vines are grown on all types of land, ranging from the flat plains of the Campo de Tarragona, through the 400-metre- (1,300-foot) high plateaux of Terra Alta, to the highest vineyards in the Alta Penedès, which reach an altitude of 800 metres (2,620 feet). For every 100-metre (330-foot) rise in altitude, the temperature drops 1°C (1.8°F).

SOIL
There is a wide variety of soils, ranging from granite in Alella, through limestone-dominated clay, chalk, and sand in Penedès to a mixture of mainly limestone and chalk with granite and alluvial deposits in Tarragona.

VITICULTURE & VINIFICATION
Cataluña is a hotbed of experimentation. Ultra-modern winemaking techniques have been pioneered by Cava companies such as Codorníu and fine-wine specialists like Torres.

GRAPE VARIETIES
Macabeo (Viura), Xarel-lo, Parellada, Malvasía (Subirat Parent), Chardonnay, Garnacha tinta, Monastrell, Pinot Noir, Trepat

TOP SPARKLING WINE PRODUCERS OF

SPAIN

CAVA

ALTA ALELLA
The only Cava producer in Alella with lots of personality in their Mirgin.

FREIXENET/CASA SALA
One of the two leaders, along with Codorníu, it is now focussing on quality. Casa Sala Gran Reserva is a jewel, pressed in a 100-year old champagne *pressoir*, aged for a long time, the best *parellada*-based Cava.

✓ *Cuvée DS (Freixenet)*

GRUPO CODORNÍU
Freixenet's competitor and alter ego. Before being sold to the Carlyle Fund, Codorníu was focusing on quality and abandoning low price segments.

✓ *Codorniu Ars Collecta series*

GRAMONA
The heroes of long ageing are now also heroes of biodynamics. Their top Cavas are simply unique, sometimes even shocking because of their complexity.

✓ *Enoteca • Celler Batlle*

JUVÉ Y CAMPS
This family-owned winery located in San Sadurní d'Anoia, Juve y Camps are Cava ambassadors all over the world, with trending very dry styles and some special old-ageing wines.

✓ *Reserva de la Familia • Gran Juvé y Camps, La Capella*

MESTRES
The inventors of Cava brut nature (no dosage), and the first ones to age Cava for a very long time, first in barrel and then in bottle. Unique wines, not sufficiently well known outside Cataluña.

✓ *Mas Via • Vintage Clos Damiana*

RECAREDO
This pioneer established Penedès *terroir* as the most important condition for top Cava, implemented biodynamics, and realised that the *terroir* would become evident only after long ageing. Probably the first *grand cru* in Penedès, with their Turò d'en Mota. Also Subtil and Serral del Vell.

SABATE I COCA CASTELLROIG
Small family winery, focused on terroir.

✓ *Reserva Familiar CPC*

TORELLÓ
A domaine style with remarkable quality. Gran Torelló Brut Nature

VINS EL CEP
Original styles, with biodynamic viticulture, non-interventionist winemaking and lots of vinosity in their Cavas.

✓ *Claror Cava Paraje Calificado*

CLASSIC PENEDÈS

COLET
Compromise for complexity and old ageing. Colet produces some amazing wines with sherry dosage.

✓ *Assemblage • Gran Cuvée • Colet-Navazos*

LOXAREL
An original creative producer, Loxarel is the only winery daring enough to sell a non-disgorged sparkling after nine years ageing, and it is delicious.

✓ *109 • Gran Elisenda*

NO APPELLATION

RAVENTÓS I BLANC
Josep Maria, black sheep of the Raventós family, left Codorníu to set up his own Cava house, producing lively *cuvées* of excellent quality.

✓ *Manuel Raventòs Negra • Mas del Serral • De Nit*

TORRES
Torres's newest development has been an elegant and complex Chardonnay, Pinot noir, and Xarel-lo sparkling wine, Vardon Kennet. It has not yet been declared as Cava because climate-conscious Miguel Torres wanted the freedom to plant in new mountain areas.

Sherry Country

After many years in the doldrums, Sherry seems to have had a promising rebirth, with much higher quality and lots of research and innovation on its own wines. Traditional wines are amazing again, and new styles show up. Yet the market does not pay attention, and sherry continues to keep a low profile. Little else can be done with respect to achieving top quality with traditional old wines, yet present challenges lie in the vineyard and probably in producing lower-alcohol wines.

The vinous roots of Sherry penetrate a thousand years of history, back to the Phoenicians, who founded Gadir (today called Cádiz) in 1100 BC. They quickly deserted Gadir because of the hot, howling *levante* wind that is said to drive men mad, and they established a town farther inland called Xera, which some historians believe may be the Xérès, or Jerez, of today. It was probably the Phoenicians who introduced viticulture to the region. If they did not, then the Greeks certainly did, and it was the Greeks who brought with them their *hepsema,* the precursor of the *arropes* and *vinos de color* that add sweetness, substance, and colour to modern-day sweet or cream sherries. In the Middle Ages, the Moors introduced to Spain the *alembic,* a simple pot-still with which the people of Jerez were able to turn their excess wine production into grape spirit, which they added, along with *arrope* and *vino de color,* to their new wines each year to produce the first crude but true Sherry. The repute of these wines gradually spread throughout the Western world, helped by the English merchants who had established wine-shipping businesses in Andalucía at the end of the 13th century.

PEDRO XIMÉNEZ GRAPES GROWING IN *ALBARIZA* SOIL
Pedro Ximénez, or PX, thrives in the brilliant white *albariza* soil of the region. This grape variety produces the intensely sweet, dark Sherry wine named after it.

FACTORS AFFECTING TASTE AND QUALITY

LOCATION
This winemaking region is situated in the province of Cádiz, around Jerez de la Frontera in the southwest of Spain.

CLIMATE
Generally, the climate is Mediterranean with a strong Atlantic influence, accentuated with the predominant winds. Mild winters and dry very hot summers.

ASPECT
Vines are grown on all types of land, from the virtually flat coastal plains producing manzanilla, through the slightly hillier Sherry vineyards rising to 100 metres (330 feet).

SOIL
The predominant soil in Jerez is a deep lime-rich variety known as *albariza,* which soaks up and retains moisture. Its brilliant white colour also reflects sun on to the lower parts of the vines. Sand and clay soils also occur but, although suitable for vine-growing, they produce second-rate sherries. The equally bright soil to the east of Jerez is not *albariza,* but a schisto-calcareous clay.

VITICULTURE & VINIFICATION
There is a typical pruning system for Palomino vines, *pulgar y vara,* similar to guyot (cane pruning). Vinification is the key to the production of the great fortified wines for which this area is justly famous. Development of a *flor* yeast and oxidation by deliberately underfilling casks are vital components of the vinification, as, of course, is the *solera* system that ensures a consistent product over the years. The larger the *solera,* the more efficient it is, because there are more butts.

GRAPE VARIETIES
Moscatel Gordo Blanco (Muscat d'Alexandrie), Palomino, Pedro Ximénez

After Henry VIII broke with Rome, Englishmen in Spain were under constant threat from the Inquisition. The English merchants were rugged individualists and they survived, as they also did, remarkably, when Francis Drake set fire to the Spanish fleet in the Bay of Cádiz in 1587. Described as the day he singed the King of Spain's beard, it was the most outrageous of all Drake's raids, and when he returned home, he took with him a booty of 2,900 casks of Sherry. The exact size of these casks is unknown, but the total volume is estimated to be in excess of 150,000 cases, which makes it a vast shipment of a single wine for that period in history. It was, however, eagerly consumed by a relatively small population that had been denied its normal quota of Spanish wines during the war. England has been by far the largest market for Sherry ever since.

THE UNIQUENESS OF JEREZ SHERRY

It is the combination of Jerez de la Frontera's soil and climate that makes this region uniquely equipped to produce Sherry, a style of wine attempted in many countries around the world but never truly accomplished outside Spain. Sherry has much in common with Champagne, as both regions are inherently superior to all others in their potential to produce a specific style of wine. The parallel can be taken further: both Sherry and Champagne are made from neutral, unbalanced base wines that are uninspiring to drink before they undergo the elaborate process that turns them into high-quality, perfectly balanced finished products.

The famous *albariza* soil

Jerez's *albariza* soil, which derives its name from its brilliant white surface, is not chalk but rather a soft marl of organic origin formed by the sedimentation of diatom algae during the Triassic period. The *albariza* begins to turn yellow at a depth of about 1 metre (3 feet) and turns bluish after 5 metres (16 feet). It crumbles and is super absorbent when wet, but extremely hard when dry. This is the key to the exceptional success of *albariza* as a vine-growing soil. Jerez is a region of baking heat and drought; there are about 70 days of rain each year, with a total precipitation of some 50 centimetres (20 inches). The *albariza* soaks up the rain like a sponge,

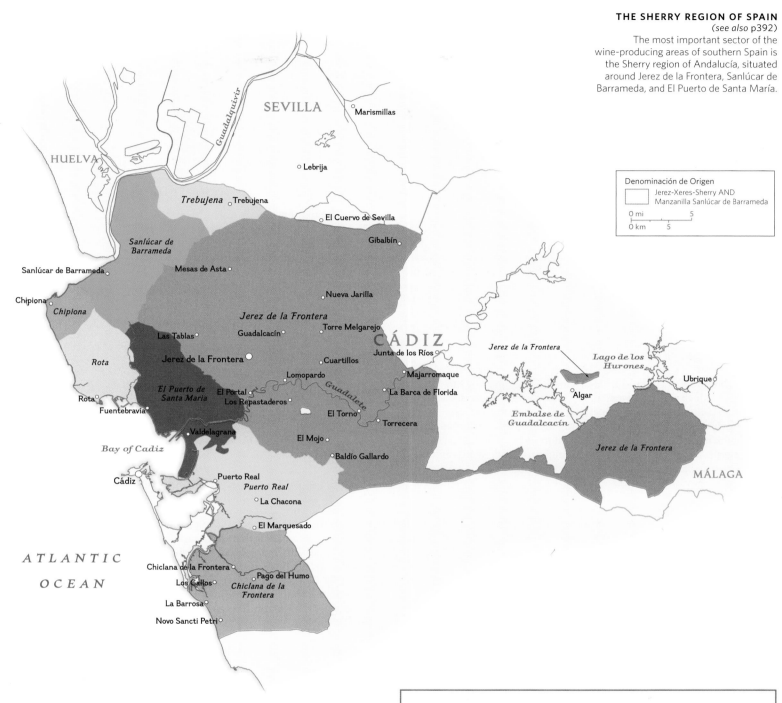

THE SHERRY REGION OF SPAIN
(*see also* p392)
The most important sector of the
wine-producing areas of southern Spain is
the Sherry region of Andalucía, situated
around Jerez de la Frontera, Sanlúcar de
Barrameda, and El Puerto de Santa María.

Denominación de Origen
Jerez-Xeres-Sherry AND
Manzanilla Sanlúcar de Barrameda
0 mi — 5
0 km — 5

and, with the return of the drought, the soil surface is smoothed and hardened into a shell that is impermeable to evaporation. The winter and spring rains are imprisoned under this protective cap, and remain at the disposal of the vines, the roots of which penetrate some 4 metres (13 feet) beneath the surface. The *albariza* supplies just enough moisture to the vines, without making them too lazy or over-productive. Its high active-lime content encourages the ripening of grapes with a higher acidity level than would otherwise be the norm for such a hot climate. This acidity safeguards against unwanted oxidation prior to fortification.

The *levante* and *poniente* winds

The hot, dry *levante* is one of Jerez de la Frontera's two alternating prevailing winds. This easterly wind blow dries and vacuum-cooks the grapes on their stalks during the critical ripening stage. This results in a dramatically different metabolisation of fruit sugars, acids, and aldehydes, which

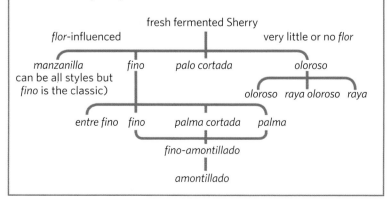

THE EVOLUTION OF SHERRY STYLES

The tree shows the course taken by each Sherry to become one of the well-known styles by which it is sold.

fresh fermented Sherry

flor-influenced — very little or no *flor*

manzanilla — *fino* — *palo cortada* — *oloroso*
(can be all styles but *fino* is the classic)

oloroso raya oloroso raya

entre fino fino — *palma cortada* — *palma*

fino-amontillado

amontillado

produces a wine with an unusual balance peculiar to Jerez. Alternating with the *levante* is the wet Atlantic *poniente* wind. This is of fundamental importance, because it allows the growth of several *Saccharomyces* strains in the microflora of the Palomino grape. This is the poetically named Sherry *flor* (*see* below), without which there would be no *fino* in Jerez.

SHERRY'S CLASSIC GRAPE VARIETIES

British Sherry expert Julian Jeffs believes that as many as 100 different grape varieties were once traditionally used to make Sherry and, in 1868, Diego Parada y Barreto listed 42 then in use. Today only three varieties are authorised: Palomino, Pedro Ximénez, and Moscatel Fino. The Palomino is considered the classic Sherry grape, and most Sherries are, in fact, 100 per cent Palomino, though they may be sweetened with Pedro Ximénez for export markets.

HOW GREAT SHERRY IS MADE

The harvest

Twenty or more years ago, it was traditional to begin the grape harvest in the first week of September. Today Palomino grapes from the inland vineyards are picked as early as the first days of August. After picking, Palomino grapes were left in the sun for 12 to 24 hours; Pedro Ximénez and Moscatel, for 10 to 21 days. At night, the grapes were covered with *esparto* grass mats as a protection against dew. This sunning is called the *soleo,* and its primary purpose is to increase sugar content. Palomino grapes are not left in the sun any longer.

The pressing

Traditionally, four labourers called *pisadores* were placed in each lagar (open receptacle) to tread the grapes, not barefoot but wearing *zapatos de pisar,* heavily nailed cow-hide boots, to trap the pips and stalks undamaged between the nails. Each man tramped 58 kilometres (36 miles) on the spot during a typical session lasting from midnight to noon. Today, automatic horizontal, usually pneumatic, presses are in common use.

Fermentation

Some Sherry houses still ferment their wine in small oak casks purposely filled to only 90 per cent capacity. After 12 hours, the fermentation starts and continues for between 36 and 50 hours at 25° to 30°C (77° to 86°F), by which time as much as 99 per cent of available sugar is converted to alcohol. Current methods often use stainless steel fermentation vats and yield wines that are approximately 1 per cent higher in alcohol than those fermented in casks; this is due to an absence of absorption and evaporation.

THE MAGICAL *FLOR*

For the majority of Sherry drinkers, fino is the quintessential Sherry style. It is a natural phenomenon called *flor* that determines whether or not a Sherry will become a fino. *Flor* is a strain of *Saccharomyces* yeast that appears as a

---

GRAPE-BASED SWEETENING AND COLOURING AGENTS

The most traditional and most important sweetening agent in the production of Sherry is that made from pure, overripe, sun-dried Pedro Ximénez grapes, also known as PX. After the *soleo,* or sunning, of the grapes (*see* above), the sugar content of the PX increases from around 23 per cent to between 43 and 54 per cent. The PX is pressed and run into casks containing pure grape spirit. This process, known as muting, produces a mixture with an alcohol level of about 9 per cent and some 430 grams of sugar per litre. This mixture is tightly bunged and left for four months, during which time it undergoes a slight fermentation, increasing the alcohol by about one degree and reducing the sugar by some 18 grams per litre. Finally, the wine undergoes a second muting, raising the alcoholic strength to a final 13 per cent, but reducing the sugar content to about 380 grams per litre.

HOW THE FERMENTED SHERRY DEVELOPS

Some *bodegas* like to make something of a mystery of the *flor,* declaring that they have no idea whether or not it will develop in a specific cask. There is some historic justification for this – one cask may have a fabulous froth of *flor* (looking like dirty soapsuds), while the cask next to it may have none. Nowadays the best *bodegas* control the winemaking. First pressing must is destined to *fino* and *flor* ageing, while second pressing goes to oxidative ageing (*oloroso).* One exception is *palo cortado,* a *fino* quality wine that pursues oxidative ageing.

---

grey-white film floating on a wine's surface, and it occurs naturally in the microflora of the Palomino grape grown in the Jerez district. It is found to one degree or another in every butt or vat of Sherry and *manzanilla,* but whether or not it can dominate the wine and develop as a *flor* depends upon the strength of the *Saccharomyces* and the biochemical conditions. The effect of *flor* on Sherry is to absorb remaining traces of sugar, diminish glycerine and volatile acids, and greatly increase esters and aldehydes.

To flourish, *flor* requires:
- An alcoholic strength of between 13.5 and 17.5 per cent. The optimum strength is 15.3 per cent, the level at which vinegar-producing acetobacter is killed.
- A temperature of between 15° and 30°C (59° and 86°F).
- A sulphur dioxide content of less than 0.018 per cent.
- A tannin content of less than 0.01 per cent.
- A virtual absence of fermentable sugars.

CASK CLASSIFICATION AND FORTIFICATION

The cellarmaster's job is to sniff all the casks of Sherry and then mark in chalk on each one how he or she believes it is developing according to a recognised cask-classification system. At this stage, lower-grade wines (those with little or no *flor*) are fortified to 18 per cent to kill any *flor,* thus determining their character once and for all and thereafter protecting the wine from the dangers of acetification. The *flor* itself is a protection against the acetobacter that threaten to turn the wine into vinegar, but it is by no means invincible, and it will be at great risk until it is fortified to 15.3 per cent or above, the norm for *fino.* It is not truly safe until it is bottled. A 50-50 mixture known as *mitad y mitad, miteado,* or *combinado* (half pure alcohol, half grape juice) is usually used for fortification. Some producers, however, prefer to use mature Sherry for fortification instead of grape juice.

Further cask classification

The wines are often racked prior to fortification, and always after. A fortnight later, they undergo a second, more precise classification, but no further fortification, or other action, will take place until nine months have elapsed, after which they will be classified regularly over a period of two years to determine their final style.

The *solera* blending system

Once the style of the Sherry has been established, the wines are fed into fractional-blending systems called *soleras.* A *solera* consists of a stock of wine in cask, split into units of equal volume but different maturation. The oldest stage is called the *solera;* each of the younger stages that feed it is a *criadera,* or "nursery". There can be more than seven *criaderas* in a Sherry *solera.* Up to one-third (the legal maximum) of the *solera* may be drawn off for blending and bottling, although some *bodegas* may restrict their very high-quality, old *soleras* to one-fifth. The amount drawn off from the mature *solera* is replaced by an equal volume from the first *criadera,* which is topped up by the second *criadera,* and so on. When the last *criadera* is emptied of its one-third, it is refreshed by an identical quantity of *añada,* or "new wine". This comprises like-classified Sherries from the current year's production, aged up to 36 months, depending on the style and exactly when they are finally classified.

THE STYLES OF
SHERRY

ALMACENISTA
or BODEGAS DE ALMACENADO

This is not a style, but a category of increasing interest among Sherry enthusiasts. An *almacenista* is a private stockholder whose pure, unblended Sherries are held as an investment for 30 years or more, after which they are in great demand by large *bodegas,* who use them to improve their commercial blends. Lustau, itself an *almacenista* until the 1950s (and now part of the Caballero group), was the first firm to market the concept, making these purest of sherries available to consumers (and they registered "Almacenista" as a trademark in the process). All styles of Sherry and *manzanilla* exist in *almacenista* form and are, almost by definition, guaranteed to be of extraordinary quality. Fractions on the label such as ⅛, ⅐, or ¹⁄₄₀ indicate the number of barrels in the *solera* from which it was drawn, therefore the lower the denominator (the number below the line), the greater the rarity of the wine and consequently the more expensive it will be.

✓ *Lustau*

AMONTILLADO

With age, the *flor* get weaker and *fino* develops an amber colour in cask and becomes a *fino-amontillado,* then, after at least eight years, a full *amontillado,* when it takes on a nutty character and acquires more body. A true *amontillado* is completely dry, with between 16 and 18 per cent alcohol, but can sometimes be sweetened to a medium style for export markets. The term *amontillado* means "Montilla style"; it was originally used to distinguish a Sherry with characteristics similar to those of Montilla. Ironically, it was illegal for producers of Montilla to use the term *amontillado* under the Spanish republic, thus Sherry could be made in a Montilla-style, but not Montilla. Under the EU, however, this has changed and once again Montilla houses are shipping Montilla *amontillado.*

✓ *Domecq (1730)* • *Maestro Sierra* • *Gonzalez Byass (Cuatro Palmas)* • *Osborne (51 1ª)* • *Delgado Zuleta (Viejo)* • *Valdespino (Coliseo, Don Tomás)* • *Wisdom & Warter (Very Rare Solera Muy Viejo)*

CREAM OR DARK CREAM

An *oloroso* style that is usually sweetened with Pedro Ximénez, the quality of which can range from the commercial to extremely good.

✓ *Diego Romero (Jerezana)* • *Gutierrez Colosia*

EAST INDIA

Some sources believe that this rich, sweet, Madeira-like style of Sherry dates back to as early as 1617, but the practice of shipping Sherry to the Far East and back gradually disappeared during the 19th century with the advent of steam-driven ships. It was revived in 1958 by the owners of the Ben Line and Alastair Campbell, an Edinburgh wine merchant, when they sent a hogshead of Valdespino Oloroso on a 32,000-kilometre (20,000-mile) round trip to the Far East, but although the style survives, the effects of the sea voyage, as with Madeira, are now replicated in the cellar.

✓ *Lustau (Old East India)* • *Osborne (India Rare Solera Oloroso)*

FINO

A *palma* is the highest quality of *fino* Sherry and may be graded in a rising scale of quality: *dos palmas, tres palmas, cuatro palmas.* A *palma cortada* is a *fino* that has developed more body, has a very dry, but smooth, almondy flavour and is veering towards *amontillado.* An *entre fino* has little merit. Few *finos* remain *fino* with age in cask, which is why genuine Old Fino Sherry is rare. A *fino* is light, dry, and delicate; its flor nose should overpower any acetaldehyde.

This style is best appreciated straight from the cask, when it is crisp and vital; it quickly tires once bottled and further declines rapidly if filtered. Until producers are required to declare the bottling date on the label (fat chance), the only sensible advice is either to buy *fino en rama* (unfiltered) or to buy it when you want to drink it and, once opened, consume the entire contents. The wines are invariably 100 per cent Palomino with an alcoholic strength of between 15.5 and 17 per cent.

✓ *Tomás Abad* • *Lustau (La Ina)* • *Gonzalez Byass (Una Palma, Tio Pepe en Rama)* • *Osborne (Fino Quinta)* • *Valdespino (Ynocente)* • *Williams & Humbert (Fino de añada)* • *Emilio Hidalgo (La Panesa)*

MANZANILLA

Sea winds in the Sanlúcar de Barrameda area create a more even temperature and higher humidity than those found in Jerez itself, which with the tradition of allowing more ullage (empty volume) in *manzanilla* casks, encourages a thicker, whiter, and more vigorous growth of *flor*. A true *manzanilla* is pale, light-bodied, dry, and delicate with a definite *flor* nose, a touch of bitterness on the palate, and sometimes even a slightly saline tang (*see* general *Fino* style above about freshness and when to consume). These wines are usually 100 per cent Palomino with an alcoholic strength of 15.5 to 17 per cent.

✓ *Barbadillo (Pastora, Solear en rama)* • *La Cigarrera* • *Delgado Zuleta (Barbiana en Rama)* • *Duff Gordon (Cabrera)*

MANZANILLA AMONTILLADA

Fuller than a *pasada,* but lighter and more fragrant than Jerez *amontillado,* this is less common than the previous two, but can be excellent.

✓ *Barbadillo (Principe)* • *Lustau (Manuel Cuevas Jurado)*

MANZANILLA PASADA

When a *manzanilla* begins to age, it loses its *flor,* gains alcoholic strength, and becomes the equivalent of a *fino-amontillado,* known in Sanlúcar de Barrameda as a *pasada.* These wines are invariably 100 per cent Palomino with an alcoholic strength of up to 20.5 per cent.

✓ *Barbadillo (Solear)* • *Delgado Zuleta (Goya XL)* • *Hidalgo (Pastrana)*

MOSCATEL

Moscatel is a naturally sweet wine. Releases of this wine can occasionally be rich, raisiny delights.

✓ *Valdespino (Toneles)* • *Gonzalez Byass (Pio X)*

OLOROSO

Oloroso means "fragrant", and when it is genuinely dry, rich, and complex from age, Sherry lovers find it a most rewarding wine. Much of its character is due to the relatively high fortification it receives and the generous glycerine content that develops without the aid of *flor.* The alcoholic strength of these wines usually ranges between 18 and 20 per cent. Some high-quality, sweeter, dessert-style *oloroso* wines are also produced.

✓ *Dry Maestro Sierra (Extra Viejo)* • *Tradición (VORS)* • *Alvaro Domecq (1730)* • *Barbadillo (Cuco VORS)* • *Diez-Mérito (Victoria Regina)* • *Gonzalez Byass (Alfonso, Apostoles)* • *Hidalgo (Oloroso Seco)* • *Lustau (Don Nuno, Emperatríz Eugenia, Principe Rio, Tonel)* • *Osborne (Bailén, Alonso el Sabio)* • *Diego Romero (Jerezana)* • *Sandeman (Dry Old Oloroso)* • *Valdespino (Don Gonzalo)*

Dessert style *Gonzalez Byass (Matúsalem)* • *Sandeman (Royal Corregedor)* • *Valdespino (Solera 1842)*

PALO CORTADO

Palo cortado is presented as a mystery wine, as something that cannot be deliberately made. In reality, it is a very refined *oloroso,* coming from the best must (must for *oloroso* is normally second pressings and rougher). A law unto itself, it is a naturally dry wine with a style somewhere between *amontillado* (on the nose) and *oloroso* (on the pal-ate), but this description does not by any means convey the stunning richness, nutty complexity, and fabulous finesse, which really must be experienced to be believed.

It should be totally dry, but some sweeter dessert-style *palo cortado* wines are produced and can be wonderful. Like *palma, palo cortado* may be graded: *dos cortados, tres cortados, cuatro cortados.*

✓ *Fernando de Castilla (Antique)* • *Alvaro Domecq (Sibarita)* • *Hidalgo (Wellington VORS)* • *Lustau (Peninsula)* • *Rosario Fantante (Dos Cortados)* • *Valdespino (Cardenal)* • *Williams & Humbert (Dos Cortados)* • *Barbadillo (Obispo Gascón)* • *Gonzalez Byass (Añada 1987)* • *Osborne (Capuchino)*

Dessert style *Osborne (Abocado Solera)* • *Sandeman (Royal Ambrosante)* • *Wisdom & Warter (Tizón)*

PEDRO XIMÉNEZ

All wineries in the Sherry area except two, Gonzalez Byass and Ximénez Spinola, get their PX wines from Montilla, and age them in the Sherry area. Although it is primarily produced as a sweetening agent, Pedro Ximénez is occasionally released in limited bottlings that are invariably very old and utterly stunning. These are huge, dark, deep, powerfully rich wines piled high with complex yet succulent, raisiny, Muscovado flavours. These bottlings of Pedro Ximénez can be compared in quality, weight, and intensity – though not in character – with only some of the oldest and rarest Australian liqueur Muscats.

✓ *Gonzalez Byass (Noe)* • *Lustau (Murillo, San Emilio)* • *Sanchez Romate (Superior)* • *Valdespino (Solera Superior)* • *Williams & Humbert (20 Years)* • *Wisdom & Warter (Viale Viejisimo)* • *Ximenez Spinola (Muy Viejo)*

AN OSBORNE BULL IN JEREZ DE LA FRONTERA
On hilltops along roadsides in many parts of Spain stand these black silhouettes of a bull. They began in 1956 as advertisements for a brandy from Osborne, who also produce a range of Sherries. The bulls have since become officially protected cultural symbols.

Other Wine Regions of Spain

The classics of Rioja, Sherry, and traditional method sparkling wines enjoy continued popularity around the world, but the dazzling array of Spanish wine styles offer variety and potential for quality every step of the way.

Other, less well-known, Spanish wine regions cultivate obscure indigenous grapes side-by-side with internationally recognised varieties to tempt the palates of a wide range of wine lovers. There are wines to pique the gastronomic interest of the most discerning connoisseur scouting new horizons and wines that offer excellent value for the bargain hunter eager to be the first to discover something exciting. The ancient and recent appellations form a mosaic of well-established and "work in progress" origins, which are equally worth seeking out and following.

The appellations appearing in the following list are grouped geographically and can be found under the Cataluña, Castilla y León, Central and Southern Spain, and Northwest Spain headings.

VINEYARDS AT THE ROYAL ABBEY OF SANTA MARÍA DE POBLET
Located in the Conca de Barberá DO in the south of Cataluña, this Cistercian monastery is a UNESCO World Heritage site. The basement of the building still contains evidence of the monks' original wine production.

OTHER APPELLATIONS OF
SPAIN

CATALUÑA

Cataluña is a kind of Spanish Piemonte, with a hilly landscape from the Mediterranean coast to the Pyrenees, a very long history, and – most of all – a culture open to innovation. All types of wine can be found in the region, as well as a large quality range, from the mediocre to the sublime, from the openly international to the extremely local, and from extremely efficient automated winemaking to making wines with natural methods.

DENOMINACIÓNES DE ORIGEN

ALELLA

This is a tiny, predominantly white wine appellation just north of Barcelona, where grapes are traditionally grown on a unique type of soil, the *saulo*, a kind of sandy granite. Due to urban development, the DO was extended in 1989 into the colder, limestone valleys of the Cordillera Cataluña. The best red wines of Alella are soft Garnacha, and the flagship white wines are Pansa Blanca, a local Xarel-lo biotype. Alella is home to Alta Alella, a top Cava producer.

🍷 *Garnacha Blanca, Moscatel, Pansa Blanca (Xarel-lo) Chardonay, Chenin, Macabeo, Malvasía, Moscatel de Grano Menudo, Parellada, Picapoll Blanco, Sauvignon Blanc, Garnacha, Cabernet Sauvignon, Garnacha Peluda, Merlot, Monastrell, Pinot Noir, Samsó (Mazuela), Sumoll Negre, Syrah, Ull de llebre (Tempranillo)*

🍸 *1–5 years (red), 1–2 years (white and rosado), 1–4 years (sweet)*

✓ *Alella Vinícola • Alta Alella • Can Roda • Roura • Quim Batlle*

CATALUNYA / CATALUÑA

An appellation of origin with little if any sense of origin; it was invented to facilitate buying grapes in the whole region and, theoretically, to offer more flexibility to innovative producers. It would make sense as an IGP. There are 36 authorised grape varieties, which makes it a nonsense to list them. Most top Cataluña producers are registered in this DOP, in case they decide to trade a wine outside real DOs. Torres is the main supporter of this DO, for mid-range wines.

CONCA DE BARBERÁ

This once little-known appellation for red, white, and *rosado* wines from the hilly hinterland of Penedès is home to Poblet Monastery and the oldest cooperative in Spain. It is Cataluña's heartland and can now claim some of the best Cataluña wines, thanks mostly to Torres (Grandes Murailles). Their authoctonous grape variety, Trepat, is used for sparkling and rosé wines. Most of the area is still dedicated to providing grapes for Cava producers, but things are likely to change in the coming years. Conca de Barberà is one of Spain's most exciting areas.

🍷 *Garnacha Blanca, Xarel-lo, Chardonay, Chenin, Macabeo, Malvasía, Moscatel de Grano Menudo, Parellada, Picapoll Blanco, Sauvignon Blanc, Garnacha, Cabernet Sauvignon, Garnacha Peluda, Merlot, Monastrell, Pinot Noir, Samsó, Mazuela, Sumoll Tinto, Syrah, Tempranillo*

🍸 *1–5 years*

✓ *Torres • Abadía de Poblet • Celler Escoda-Sanahuja*

COSTERS DEL SEGRE

This is an appellation that was created as the flagship of the new Spain – for high-tech vineyards and wineries with imported international varieties. Raimat was the catalyst for this appellation. The DO is indeed split in four different areas, each with little in common with the others (a typical Spanish approach, where politics prevail over nature). The DO hosts a good number of excellent wineries and shows the promise of great developments in the future. The amazing number of 24 grape varieties are authorised in this DOP. Practically anything that can grow here.

🍸 *2–6 years (red), 1–4 years (white)*

✓ *Castell d'Encus • Castell del Remei • Mas de Mora • Cérvoles • L'Olivera • Lagravera • Purgatori • Cusiné • Raïmat*

EMPORDÀ

Part of the appellation (Alt Empordà) lies at the foot of the narrowest section of the Pyrenees; this is the closest Spanish wine appellation to France. The other part (Baix Empordà) includes the heart of Costa Brava. The best wines are made of Cariñena and Garnacha (known locally as Samsó and Lledoner, respectively). The region is a traditional rosé producer, and rosé is very popular among tourists. Whites are fruity and may be off-dry or slightly sweet, with a pale, often greenish-tinged colour, and are sometimes *pétillant* (lightly sparkling).

🍷 *Garnacha Blanca, Garnacha Roja, Macabeo, Moscatel de Alejandría, Cariñena, Garnacha, Chardonnay, Gewurztraminer, Malvasía, Moscatel de Grano Pequeño, Picapoll Blanco, Sauvignon Blanc, Xarel-lo, Cabernet Sauvignon, Cabernet Franc, Merlot, Monastrell, Tempranillo, Syrah*

🍸 *2–5 years (red), 9–18 months (white and rosado)*

✓ *La Vinyeta • Clos de'Agon • Mas Anglada • Espelt • Perelada • Marti Fabra • Oliveda*

MONTSANT

A recent appellation (split off from Tarragona) that practically encircles Priorat, Montsant was born as a lesser DO, but nowadays it is at least as good as the mother, thanks to a outsiders investing in the region and, partly, to climate change, which benefits this slightly cooler area over Priorat. Most wines are red. Top quality is to be found with Garnacha and Cariñena, the other wines being less exciting.

🍷 *Cabernet Sauvignon, Cariñena (Mazuelo), Garnacha Blanca, Garnacha Gris, Garnacha Peluda, Garnacha Roja, Garnacha Tinta, Macabeo, Moscatel de Grano Menudo, Picapoll Negro, Tempranillo, Chardonnay, Merlot, Monastrell, Pansal, Syrah*

🍸 *2–10 years (red)*

✓ *Acustic • Falset-Marçà • Clos Pissarra • Celler de Capçanes • Cellers el Masroig • Can Blau • Capafons-Ossò • Joan d'Anguera • Spectacle • Venus l'Universal • Vinyes Domenech, Torres*

PENEDÈS

Penedès sells two-thirds of its production inside Cataluña. It is the heartland of Cava territory, but it prevents Cava from claiming its real origin. A new sparkling wine brand, Penedès Classic, has been created within the appellation. This brand is intended to give shelter to wineries abandoning Cava. Again, in Spain politics prevail over common sense. The DO is open to all types of wine, stimulating innovation, and one requirement for wines given the

Penedès Classic label is that they are produced via organic viticulture. It houses the most important Cataluña wineries, although most of them use other DO for some or all their wines. It is one of the most advanced Spanish DO in terms of soil classification. Politics apart, it is an exemplary DO. Penedès also houses a unique and delicious type of sweet wine, Malvasía de Sitges.

🍇 *Macabeo, Xarel-O, Parellada, Subirat Parent, Moscatel de Alejandría, Moscatel de Grano Menudo, Malvasía de Sitges, Chardonnay, Sauvignon Blanc, Riesling, Gewurztraminer, Chenin, Garnacha Tinta, Cab. Sauvignon, Merlot, Monastrell, Pinot Noir, Samsó, Sumoll Tinto, Syrah, Tempranillo*

🍷 *2-15 years (red), 1-8 years (white and rosado)*

✓ *Agustí Torellò Mata • Albet i Noya • Can Feixes • Can Rafols del Caus • Castellroig • Clos Lentiscus • Girò Ribot • Gramona • Heretat Mestres • Janè Ventura • Jean Leon • Loxarel • Mas Comtal • Mascarò • Torres • Mont Marsal • Nadal • Parés Baltà • René Barbier • Sumarroca*

PLA DE BAGES

Winemaking is an old tradition in Bages, and yet this is an up-and-coming small DO, with many newly planted vineyards and a small but growing band of modern *bodegas*. Traditional Picapoll and Macabeo are being complemented with international varieties but the appellation's unique personality is still based on Picapoll wines.

🍇 *Macabeo, Parellada, Picapoll, Chardonnay, Gerwürztraminer, Garnacha, Tempranillo, Merlot, Cabernet Sauvignon, Cabernet Franc, Syrah, Sumoll*

🍷 *2-15 years (red), 1-8 years (white)*

✓ *Abadal • Massies d'Avinyó*

PRIORAT DOCA (DOQ)

Priorat means "priory". This area was identified in the 1970s as one of the top wine *terroirs* in the world. A group of (then) young winemakers, including Alvaro Palacios, René Barbier, Carles Pastrana, and others, came in and achieved impressive worldwide success with their wines. Priorat has a dry climate and poor soil in which the vines' roots spread everywhere in search of moisture, the local saying being that Priorat vines can suck water out of stone. There is a type of slate soil, the *licorella*, which is considered as the most important quality feature in the region, although some vines planted in clay soils also give good quality. The wines were at first distinguished by their power; a second generation of winemakers brought to the region a more delicate style, giving more relevance to old-vine Cariñena (Carignan) and less to international varieties. Garnacha is the queen at Priorat. Priorat is one of the Spanish wine industry's superstars. The use of the spelling "Priorato" has been dropped on bottles in favour of the Catalan version "Priorat", and DOQ, the Catalan equivalent of DOCa, is now prevalent.

🍇 *Cariñena, Garnacha, Garnacha Peluda, Tempranillo, Picapoll Negre, Cabernet Sauvignon, Cabernet Franc, Pinot Noir, Merlot, Syrah, Garnacha Blanca, Macabeo, Pedro Ximénez, Chenin Blanc, Moscatel de Alejandría, Moscatel de Grano Menudo, Blanquilla, Picapoll Blanc, Viognier*

🍷 *5-25 years (red), 2-10 years (whote)*

✓ *Alvarez Durán • Alvaro Palacios • Mas Alta • Buil & Giné • Familia Nin Ortiz • Mas Doix • Vall Llach • Capafons-Ossò • Cellers de Scala Dei • Cims de Porrerà • Clos Mogador • Clos Erasmus • Costers del Siurana (Clos de l'Obac • Miserere) • De Muller • Ferrer Bobet • Gratavinum • Mas Igneus • Mas Martinet • Terroir al Limit • Torres • Mas d'En Gil • Pasanau (Ceps Nous) • Joan Sangenís • Vinicola del Priorato*

TARRAGONA

Tarragona has been renowned for its sweet red wines since the times of the ancient Romans, when it was exported to Rome. The area under vines expanded constantly over the centuries to the point that Tarragona became a common name for a type of wine, and in 1900 over 50 per cent of the province was covered by vineyards. It was badly affected by the outbreak of the phylloxera plague at that period and unfortunately the DO never recovered. Sweet wines are not fashionable, and the zones within Tarragona that were more suitable for fine wine and where more independent producers concentrated their efforts got an independent DO status (Terra Alta, Montsant, Priorat). Most of the production now goes into Cava, but a reduced number of winemakers are doing their best to recover some of the region's former glories, including 13 co-ops with a renewed approach. Ironically, Tarragona hosts one of the best universities dedicated to winemaking in Spain. The local fortified wine, which is sold as Tarragona Classico, is worth looking out for.

🍇 *Macabeu, Cartoixà (Xarel-lo), Parellada, Moscatell Frontignan, Moscatell d'Alexandria, Garnatxa Blanca, Malvasia de Sitges, Subirat Parent, Xarel-lo Vermell, Sumoll Blanc, Chardonnay, Sauvignon Blanc, Vinyater, Ull de Llebre / Tempranillo, Sumoll, Cariñena, Garnacha Negra, Merlot, Syrah, Pinot Noir, Cabernet Sauvignon*

🍷 *1-5 years (red), 1-2 years (white and rosado)*

✓ *Fustel • Vins Suñer • De Müller • Pedro Rovira*

TERRA ALTA

Closely related to neighbouring Tarragona, this appellation improved greatly from the increased professionalisation of co-ops and the leading efforts of Barbara Forès. Unfortunately, there are still many vineyards planted with less suitable varieties as a result of early revamping efforts last century, but the situation is improving. If and when the co-ops achieve the quality they are aiming for, the region will get recognition.

🍇 *Cabernet Sauvignon, Cariñena, Garnacha, Garnacha Blanca, Merlot, Moscatel, Parellada, Tempranillo, Viura*

🍷 *1-6 years*

✓ *Pedro Masana • Barbara Forès • Piñol • Edetària • Torres • Les Vinyes d'Andreu • Vinicola de Gandesa • Pedro Rovira (Alta Mar, Viña d'Irto) • Vinalba dels Arcs (Vall de Berrús)*

GOBELET-TRAINED VINES IN BURGOS
Located in the Ribera del Duero DO, Burgos was the historic capitol of Castilla.

CASTILLA Y LEÓN

The Castilla y León region currently occupies the northern part of the Spanish Meseta Central (plateau). It is one of the largest regions in Europe, and the one with the richest heritage in Spain. The region is surrounded by high mountains, which determines its continental dry climate. There are many different soil types and relevant mesoclimate variations, resulting in many different *terroirs*. The plateau consists mostly of the Duero River Basin, although the influence of the river itself is not very relevant. This is a region with a potential – not yet realised – to produce high-quality Tempranillo wines to rival those of Ribera del Duero and other appellations.

DENOMINACIÓNES DE ORIGEN

ARLANZA

A recent DOP with a very long history of glory in the Middle Ages before it faded into oblivion. Thanks to climate change and some pioneers, its high-altitude Tempranillo and Albillo vineyards are showing their great potential, which is being realised little by little. Maybe this appellation is waiting for its own Alvaro Palacios? Anyway, Arlanza is a name to follow.

🍇 *Albillo, Viura, Tinta del País (Tempranillo), Garnacha Tinta, Mencía Cabernet Sauvignon, Merlot, Petit Verdot*

🍷 *1-7 years (red), 1-4 years (white)*

✓ *Olivier Rivière • Sabinares, Ferozia*

ARRIBES

This is another DO finalised as recently as 2008, and the vineyards of Arribes currently cover just 750 hectares (1,850 acres) but used to be prolific, stretching from Zamora south through Salamanca, along the border with Portugal. Indeed this DO could be Iberian rather than Spanish, so related is it with Portugal. This is the region of a relatively unknown grape variety, Juan García. It is often blended with Tempranillo (not a great idea), with Rufete (known in Portugal as Tinta Pinheira), and, increasingly so, with Bruñal (the Spanish name for Alfrocheiro), resulting in wines with great personality.

🍇 *Juan García, Rufete, Bruñal, Doña Blanca, Albillo Mayor, Albillo Real, Verdejo, Tempranillo, Garnacha Tinta, Mencía,*

🍷 *1-7 years (red), 1-2 years (white)*

✓ *Arribes de Duero • Alma Roja • Las Gavias*

BIERZO

Probably the most important rediscovery in Spain in the 21st century, Bierzo is a region with an amazing viticultural history (the Romans brought the vines to quench the thirst of their gold miners), a unique climate and landscape, and an almost exclusive grape variety, Mencía. Two outsiders and a man from the land, the Palacios cousins and Raül Pérez, are the heroes of Bierzo, which in 1999 was almost unknown and is now a top-quality wine region on its way to becoming an icon. Next big thing in this region: the Godello-based white wines.

🍇 *Mencía, Godello, Doña Blanca, Garnacha, Palomino, Malvasía, Garnacha Tintorera*

🍷 *1-15 years (red), 1-7 years (white), 1 year (rosé)*

✓ *Raúl Pérez • Peique • Mengoba • Mauro • Descendientes de J Palacios • Luna Beberide • Dominio de Tares • Castroventosa • Demencia de Autor • Losada • Pittacum • Valtuille • Pérez Caramés*

CEBREROS

Cebreros is the most recent appellation in Spain (2017), it is still not confirmed. It is in the Gredos area, occupying the territory within Castilla y León. The name corresponds to a renowned village in the region. It is too early to speak about producers.

🍷 *Garnacha, Albillo Real*

CIGALES

Until last century Cigales was a classic region of national fame thanks to its clairet wines, which fall between rosé and red. Those wines fell out of favour, and the leading producers tried hard with Tempranillo-based reds very similar to those in neighbouring Ribera del Duero; some of them, like Museum, are very good, but most are rather indistinct. There is now a trend to try again with the light wines of the past, and some examples are outstanding.

🍇 *Albillo, Garnacha, Palomino, Tempranillo, Verdejo, Viura*

🍷 *Rodriguez Sanz • Finca Museum • Traslanzas • Cesar Principe • La Legua • Sinforiano • Lezcano-Lacalle*

RIBERA DEL DUERO

Awkward as it seems, fewer than 50 years ago Ribera del Duero was just the name of a cooperative winery. There was a very long tradition of winemaking throughout the region, particularly around Roa, and there was the unique and isolated Vega Sicilia, but there was no real fine wine industry. Alejandro Fernández of Pesquera fame was the real hero of the region, the one who conquered the world with wines that did not try to be similar to Vega Sicilia but something great on their own. His success attracted many investors, in such a way that Ribera started to boom. The 2008 financial crisis brought the bust, eliminating many underperformers. Unfortunately, the region still has a lot of work to do, first of all by declassifying millions of mediocre wines labelled as "*roble*" (oak), second by allowing zone identification on the labels, and finally by preventing the planting of vineyards in cereal fields. The quality potential is huge, Ribera could and should rival Burgundy, but the (bad) governance and cultural traditions are preventing its realisation. Despite this, producers like Dominio del Aguila, Perez Pascuas, Pingus, Pago de los Capellanes, Emilio Moro, Vega Sicilia, Atauta, and some others are producing superb wines. They justify and keep alive the whole region. Indeed, the best Ribera wines are not heavy and overoaked as common wines are, but are delicate, finely grained, and complex. Ribera del Duero can also deliver amazingly deep and original rosé wines.

🍷 *Tinta Fina (Tempranillo), Cabernet Sauvignon, Garnacha, Malbec, Merlot, Albillo Pardina, Tinto del País*

🍷 *3–75 years (red), 1–8 years (rosado)*

✓ *Aalto • Pesquera • Alión • Antidoto • ArzuagavBaden NumenvBohorquez • Carramimbre • Dominio de Atauta • Emilio Moro • Félix Callejo • Hacienda Monasterio • Pérez Pascuas • Hermanos Sastre • Ismael Arroyo • La Horra • Teofilo Reyes • Rodero • Tionio • Vega Sicilia • Vinum Vitae • Alión • Ortega Fournier • Tábula • Valderiz • Valtravieso • Cillar de Silos • Dominio de Pingus • Dominio del Aguila • Finca Villacreces • Garmon • Grupo Bodegas Palacios 1894 • Hornillos Ballesteros • La Loba • Legaris • Nexus • Pago de Carraovejas • Pago de los Capellanes • Protos • Real Sitio de Ventosilla • Sei Solo • Torres • Alonso del Yerro • Aster*

RUEDA

This wine region was much appreciated by the Spanish Court in the Middle Ages but fell into oblivion until it was rediscovered by Marqués de Riscal in the 1970s. Now it is at a similar stage of development as Marlborough, New Zealand, was in the 2000s: lots of plantings of the autochthonous variety, in Rueda's case Verdejo, to supply formulaic aromatic varietal wines. Differentl from Marlborough, there are still some vineyards with very old-bush Verdejo vines that produce wines with lots of personality and depth. The cooperative and some small producers have real value selections, but, unfortunately, two of the most distinguished producers, Riscal and Ossián, have decided not to label those wines as Rueda. It is still possible to find traditional oxidative style *pálido* (kind of rough *fino*) and the lovely *dorado*, a nice speciality. A *méthode traditionnelle* Rueda Espumoso is also allowed.

🍷 *Verdejo, Palomino, Sauvignon Blanc, Viura (Macabéo), Tempranillo, Cabernet Sauvignon, Merlot, Garnacha*

🍷 *1–5 years*

✓ *Marqués de Riscal • Félix Sanz • Menade • Ramón Bilbao • Finca Montepedroso • Vinos Sanz • Belondrade • José Pariente • Naia • Protos • Javier Sanz • Cuatro Rayas Agricola Castellana • Viñedos de Nieva*

SIERRA DE SALAMANCA

Recent (2010) and very small DO (just six wineries), with one attractive speciality, the Rufete grape variety. Several biotypes specific to the region have been identified, giving wines with lots of personality. Worth seeking.

🍷 *Rufete, Garnacha, Tempranillo*

🍷 *1–5 years*

✓ *Viñedos del Cámbrico • Compañía de Vinos La Zorra*

TIERRA DE LEÓN

It is no coincidence that after endless fields of wheat and grain, this outcrop of vineyards is located where two pilgrim roads, the *Via de la Plata* and *Camino de Santiago*, cross. Traditionally, the wines here were *rosado* or *clarete* and produced from a blend of black and white grape varieties; however, although all styles – red, white, and *rosado* – can be produced under this DO, the current focus is firmly fixed on realising the potential of the indigenous grape variety, Prieto Picudo. Although some good full-bod-

ied reds are made, claret styles seem to be more successful. A new style of *rosado* is being developed in this region.

🍷 *Prieto Picudo, Albarín Blanco, Mencía, Verdejo, Godello, Tempranillo, Garnacha, Malvasía, Palomino*

🍷 *1–5 years*

✓ *Raúl Pérez • Andrés Marco Tampesta • Coop Los Oteros • Fuentes del Silencio • Leyenda del Páramo • Pardevalles*

TIERRA DEL VINO DE ZAMORA

Abutting the western fringes of Rueda and Toro, this recent DO intends to re-create something of its past glory, through modern wineries and vineyards. Mostly Tempranillo, although other varieties include Garnacha and Cabernet Sauvignon for red and rosado wines. For white wines, a local biotype of Malvasía is the most planted variety.

✓ *Viñas del Cénit*

TORO

The exciting quality of Toro wines today is a very recent phenomenon, built on a sudden influx of high-profile names, which has multiplied the number of *bodegas* since 2000. When this DO was established in 1987 there were just four *bodegas*, the decrepit local cooperative dominated production, and the quality was really quite dire. This was quite unfair to the real value and historic background of the region. It has the most continental and driest climate in Castilla y León. Its soils are quite diverse, including sandy soils that are phylloxera-free and provide balance to the vines. But the recent success of the region has some drawbacks. It has been built upon Tempranillo (often with Rioja clones), which is not always the best-suited variety for such a climate, given its low acidity. Toro is already great, but it could be much greater with more attention to its *terroir* and less attention to market studies. Garnacha from Toro and old autochthonous biotypes of Tinta de Toro (or Tempranillo), can produce excellent wines.

🍷 *Tinta de Toro (Tempranillo), Garnacha, Malvasía, Verdejo*

🍷 *2–10 years (red), 1–3 years (white)*

✓ *Pintia • Domaine Magrez España • Dominio del Bendito • Elías Mora • Estancia Piedra • Fariña • Liberalia • Numanthia • Quinta de la Quietud • Rodríguez Sanzo • San Román • Teso La Monja • Vetus • Campo Eliseo • Covitoro*

INDICACIÓN GEOGRÁFICA PROTEGIDA

CASTILLA Y LEÓN IGP

An overarching IGP, allowing for 38 grape varieties and any type of wine. Used by producers intending to stay outside the DO framework or byt those unable to be inside. The most relevant ones, in qualitative terms, are Abadía Retuerta, Mauro, El Lagar de Isilla, Ermita del Conde, Leda, Barcolobo, Fuentes del Silencio, La Mejorada, El Jorco, Dehesa La Granja, Tridente, Hacienda Zorita, and Ossian. The regional government has recently launched an initiative aimed to give those wineries a *Vino de Pago* status – something quite worrying and likely to create confusion rather than clarify quality levels. Some of those wineries are based in areas with a long wine tradition and are worth seeking out for that reason, rather than because they have a single-vineyard designation that really means very little in terms of quality.

VINOS DE CALIDAD CON INDICACIÓN GEOGRÁFICA

VALLES DE BENAVENTE VC

Immediately south of the Tierra de Léon DO, across the provincial border in Zamora, the vineyards of Benavente were once famous for producing inexpensive *pétillant rosado* from the Prieto Picudo variety, on its own or blended with Tempranillo.

🍷 *Bodegas Otero*

VALTIENDAS VC

Not yet an appellation of origin, Valtiendas has many of the elements required to produce great wines. It neighbours

Ribera del Duero, but Valtiendas has been deemed less interesting for winemaking because of its higher altitude, 900 metres (2,953 feet) above sea level, the fact that many vineyards face North, the predominance of the white Albillo grape variety, and the poor *cascajo* (calcareous with stones) soils. Ironically, as climate change occurs and mean temperatures increase, those are the very elements deemed crucial for quality. The white wines are already very interesting, and the reds have a lovely surplus of freshness.

🍷 *Albillo, Tempranillo, Garnacha, Cabernet Sauvignon, Merlot, Syrah*

🍷 *2–5 years (red), 1–5 years (white)*

✓ *Alfredo Maestro (in the area but not in the IGP) • Tinto Redreja • Vagal • Navaltallar*

BARTENDER POURS PAJARETE IN BODEGA ANTIGUA CASA DE GUARDIA, MÁLAGA
The oldest bar in Málaga, this small, charming tavern dates to 1840. It pours its wide variety of wines, including pajarete, a Málaga speciality, from old barrels.

CENTRAL AND SOUTHERN SPAIN

This large area represents most of the Spanish wine production in volume, but it is less relevant for the fine wine markets, although there are remarkable exceptions. Six political regions are included in this section: Andalucía (except sherry), Valencia, Murcia, Castilla–La Mancha, Madrid, and Extremadura.

ANDALUCÍA
DENOMINACIÓNES DE ORIGEN

CONDADO DE HUELVA

A very old appellation with a glorious history: its wines from the municipalities of Bollullos and Manzanilla were illustrious in the 16th century, and large quantities were exported to America. But all of this finished, and now the appellation struggles to get recognition out of its main grape, Zalema, and experiments with international varieties.

🍷 *Zalema, Palomino Fino, Listán de Huelva, Garrido Fino, Moscatel de Alejandría, Pedro Ximénez, Colombard, Sauvignon Blanc, Chardonnay, Syrah, Tempranillo, Merlot, Cabernet Sauvignon, Cabernet Franc.*

🍷 *Upon purchase (white), 1–10 years (fortified)*

✓ *Andrade*

GRANADA

This is a recent appellation still trying hard to find an identity. Granada has a lovely landscape and an indigenous grape variety, Vijiriego. Many other varieties – too numerous to list – are authorised here.

🍇 *Vijiriego, and a long series of international varieties*

🍷 *1–2 years (red and white)*

✔ *Señorío de Nevada*

MÁLAGA AND SIERRAS DE MÁLAGA

These two appellations share the same territory. Málaga is known for its sweet and fortified wines, and Sierras de Málaga is known for it dry still wines. Málaga was the most important wine province of Spain in the early 19th century. The Muscat vineyards produced wines popular in Europe. Unfortunately, phylloxera destroyed the whole vineyard, which never fully recovered. Some very promising wines are now being produced here, noticeably some reds made with the Romé grape and whites made with Muscat. This region should be followed closely.

🍇 **Málaga** *Pedro Ximénez, Moscatel*
Sierras de Málaga *Pedro Ximénez, Moscatel de Alejandría, Moscatel Morisco, Chardonnay, Macabeo, Colombard, Sauvignon Blanc, Lairen, Doradilla, Gewürztraminer, Riesling, Verdejo, Viognier, Romé, Cabernet Sauvignon, Merlot, Syrah, Tempranillo, Garnacha, Cabernet Franc, Pinot Noir, Petit Verdot, Graciano, Malbec, Monastrell, Tintilla*

🍷 *1–10 years (Málaga), 1–3 years (Sierras de Málaga)*

✔ *F Schatz • Bentomiz • Jorge Ordoñez • Málaga Virgen • Quitapenas • Telmo Rodríguez • Descalzos Viejos • Los Aguilares • Sedella • Victoria Ordoñez*

MONTILLA-MORILES

This very old appellation is behind one of the most surprising exceptions to the appellation of origin laws. Most of the Pedro Ximénez wine aged in Jerez is produced here in Montilla –Moriles, which is the consequence of the secular advantage of Sherry producers to access markets. Pedro Ximénez, or PX, is the main grape variety in the region, which produces all Sherry categories with PX. *Fino* is not fortified and can be really complex. The best soils of this DO are *albariza*, as in Sherry area. The region is likely to get into the top league, thanks to Alvear and Envínate efforts to highlight *terroir* over other commercial considerations.

🍇 *Pedro Ximénez, Airén, Baladí, Verdejo, Moscatel de Grano Menudo, Moscatel de Alejandría, Torrontés, Chardonnay, Sauvignon Blanc, Macabeo*

🍷 *1–100 years (PX, oloroso, palo cortado, amontillado), 1–2 years (fino)*

OTHER IGPs OF ANDALUCÍA

These 16 IGP were created in order to provide a protection to the names of wines produced in the respective islands or areas, outside the DO framework, but they do not have any organoleptic or cultural identity so far.

- Altiplano de Sierra Nevada
- Bailén
- Cádiz
- Córdoba
- Cumbres Del Guadalfeo
- Desierto De Almería
- Laderas Del Genil
- Laudar -Alpujarra
- Lebrija
- Los Palacios
- Norte de Almería
- Ribera del Andarax
- Sierra Norte de Sevilla
- Sierra Sur de Jaén
- Torreperogil
- Villaviciosa se Córdoba

COMUNIDAD VALENCIANA
VINO DE PAGO

EL TERRERAZO VINO DE PAGO

An exception to the national rule for *Vinos de Pago*, El Terrerazo concentrates in the two autochthonous grape varieties, Bobal for red wines and Merseguera for whites, with superb results. Winemaker Toni Sarrión showed the world how to produce top quality wines from those grapes at his Bodega Mustiguillo. Now it is one of the best wineries in Spain.

🍇 *Bobal, Merseguera*

🍷 *2–20 years (red), 1–5 years (white)*

DENOMINACIÓNES DE ORIGEN

ALICANTE

A very old wine name, in the past Alicante was a type of wine as famous as port, burgundy, or claret. The appellation is divided into eight districts within two different regions: one inland specialising in high-altitude Monastrell wines – some of them excellent – and the other along the coast specialising in Muscatel, as well as one of the rarest and most amazing wines in Spain, *fondillón*. This wine is made with Monastrell grapes allowed to over-ripen on the vine, which results in a wine, after natural fermentation, with high residual sugar and high alcohol content. It is aged for at least 10 years, often much longer, in old oak vats, through *solera* or single-vintage systems. There is no fortification.

🍇 *Moscatel, Garnacha, Monastrell, Airén, Subirat Parent (Malvasia), Chardonnay, Macabeo, Merseguera, Planta Fina De Pedralba, Sauvignon Blanc, Verdil, Garnacha Tintorera, Bobal, Cabernet Sauvignon, Merlot, Pinot Noir, Petit Verdot, Syrah, Tempranillo*

🍷 *1–15 years (red), upon purchase (white), 1–100 years (fondillón)*

✔ *Enrique Mendoza • Monovar • El Sequé • Primitivo Quiles • MG Wines • Volver*

UTIEL-REQUENA

This region, dedicated for many years to producing bulk wines, found its fine wine market niche with the development of the native Bobal grape, which represents 80 per cent of the production. The best wines are balanced and open, very suave and often complex. They age relatively well.

🍇 *Bobal, Tempranillo, Garnacha Tinta, Garnacha Tintorera, Cabernet Sauvignon, Merlot, Syrah, Pinot Noir, Petit Verdot, Cabernet Franc, Tardana, Macabeo, Merseguera, Chardonnay, Sauvignon Blanc, Parellada, Verdejo, Moscatel de Grano Menudo*

🍷 *1–5 years (red)*

✔ *Ecovitis • Dominio de la Vega • Hispano-Suizas • Chozas Carrascal • Pago de Tharsys • Cerro Gallina • Noemi Wines • Murviedro • Cherubino Valsangiacomo • Vicente Gandïa Plá*

VALENCIA

This is one of those Spanish anti-appellations, because the area covered is the whole Valencia region, with no specific identity to be found. It should be an IGP and allow some smaller areas within the appellation, Alto Turia, Valentino and Clariano, getting their own DO status. Here, 34 grape varieties are authorised. Valencia can produce anything from mediocre to excellent wines.

🍷 *1–5 years (red), 1–2 years (white)*

✔ *Celler del Roure • Mustiguillo • Mitos • Hispano-Suizas • Iranzo • Murviedro • Cherubino Valsangiacomo • Vicente Gandía Plá • Vinya Alforí*

✔ *Mustiguillo*

INDICACIÓN GEOGRÁFICA PROTEGIDA

CASTELLÓ IGP

This IGP should set the basis to recover quality viticulture in some areas of this province, which 200 years ago was quite important.

MURCIA
DENOMINACIÓNES DE ORIGEN

BULLAS

This small appellation close to Jumilla is trying hard to get out of the bulk wine market.

🍇 *Monastrell, Syrah, Tempranillo, Merlot, Petit Verdot, Cabernet Sauvignon, Garnacha, Airén, Macabeo, Chardonnay, Sauvignon, Verdejo, Moscatel, Malvasía*

🍷 *1–4 years (red)*

✔ *Balcona*

JUMILLA

The heartland of Monastrell (Mourvèdre), with a uniquely harsh climate, semi-desert and continental, produces some great wines with lots of personality and power, often from ungrafted old vines. Even phylloxera has problems surviving here. Jumilla used to be synonymous with heavy *doble pasta* alcoholic wines, but a more modern style of winemaking now dominates, and the resulting wines compete, often advantageously, with the best Australian Shiraz. Monastrell accounts for 80 per cent of production.

🍇 *Monastrell, Cencibel, Garnacha Tintorera, Garnacha, Cabernet Sauvignon, Merlot, Syrah, Petit Verdot, Airen, Macabeo, Pedro Ximénez, Malvasía, Chardonnay, Sauvignon Blanc, Moscatel de Grano Menudo, Verdejo*

🍷 *1–10 years (red)*

✔ *Casa Castillo • Casa de la Ermita • Carchelo • Juan Gil • El Nido • Xenysel*

YECLA

This is a small enclave between Jumilla, Almansa, and Alicante, dedicated to Monastrell grapes, following the lead of Jumilla.

🍇 *Monastrell, Garnacha, Merseguera, Verdil*

🍷 *1–5 years (red)*

✔ *Castaño*

INDICACIÓNES GEOGRÁFICA PROTEGIDA

CAMPO DE CARTAGENA IGP

This is a small area with only one winery.

MURCIA IGP

A typical IGP created to provide protection to wines that for one reason or another are outside the DO.

CASTILLA–LA MANCHA
DENOMINACIÓNES DE ORIGEN

ALMANSA

The Almansa region found its way up the quality ladder, from bulk wine production to a clear identity, with an unexpected variety, Alicante Bouschet (locally called Garnacha Tintorera), a cross developed in France in the 1860s. Situated next to Alicante, some areas produce good Monastrell (Mourvèdre).

🍇 *Garnacha Tintorera, Monastrell, Merlot, Syrah, Cabernet Sauvignon, Petit Verdot, Tempranillo, Garnacha, Chardonnay, Sauvignon Blanc, Verdejo, Moscatel*

🍷 *1–5 years (red)*

✔ *Piqueras, Dehesa El Carrascal*

LA MANCHA

One of the largest appellations in the world: this is a debatable honour, however, still difficult to understand in wine terms, probably because the DO is just a political creation. Most of the wine produced in the region is sold anonymously. Most of the best producers escape the appellation thanks to the lenience of the regional government for approving *Vinos de Pago*. Lots of wineries and wines, mostly produced by cooperatives.

🍷 *Airén, Macabeo, Chardonnay, Sauvignon Blanc, Verdejo, Moscatel de grano menudo, Pedro Ximénez, Parellada, Torrontés, Gewürztraminer, Riesling, Viognier, Tempranillo, Garnacha, Moravia, Cabernet Sauvignon, Syrah, Merlot, Petit Verdot, Graciano, Malbec, Cabernet Franc, Pinot Noir*

🍷 *1–3 years (red and white)*

✓ *Finca Antigua • Verum • Ayuso*

MANCHUELA

Bordering Valencia, with relevant stocks of old-vine Bobal, Garnacha, and other varieties, such as Rojal and Moravia Agria, this area is promising. Unhelpfully, they accept an impossible-to-list number of other grape varieties.

🍷 *1–10 years (red)*

✓ *Alto Landón, Iniesta, Finca Sandoval*

MÉNTRIDA

As in many other cases, this appellation is more the result of national politics than of natural conditions. It should be part of an inter-regional appellation for the Gredos mountains, which has a clear historical and natural identity. But such an appellation would need three regions to come together in agreement, something politically very difficult in this country. A pity. The main grape is Garnacha, which, at high altitude, can produce delicious wines. The appellation authorises many other varieties, mainly international, with little interest.

🍷 *1–7 years (red)*

✓ *Jimenez-Landi • Finca Constancia • Canopy*

MÓNDEJAR

This appellation of the classic Alcarria región is still to release something great.

🍷 *Tempranillo, Cabernet Sauvignon, Syrah, Garnacha, Malvar, Macabeo, Sauvignon, Torrontés*

🍷 *Upon purchase*

RIBERA DEL JÚCAR

Indistinct recent appellation, created by a group of co-ops. No wines worth international recognition. Bobal is the most promising variety

🍷 *Tempranillo, Cabernet Sauvignon, Merlot, Syrah, Bobal, Petit Verdot, Cabernet Franc, Moscatel de Grano Menudo, Sauvignon Blanc*

🍷 *Upon purchase*

UCLÉS

This is a recent and very dynamic appellation that focusses on modern wines, environmental concerns, and wine values. A real game changer in Spain.

🍷 *Verdejo, Chardonnay, Macabeo, Sauvignon, Moscatel, Tempranillo, Cabernet Sauvignon, Merlot, Syrah, Garnacha*

🍷 *1–5 years (red)*

✓ *Fontana • Finca La Estacada*

VALDEPEÑAS

Valdepeñas was known in the 19th century as the origin of the best wines drunk in Madrid. It then supplied cheap wine for the new industrial classes in Spain. One of the most efficient wine companies in Spain, Felix Solis, is based here. Their Viña Albali can be found almost anywhere in the world.

🍷 *Airén, Macabeo, Chardonnay, Sauvignon, Verdejo, Moscatel, Tempranillo, Garnacha, Cabernet, Merlot, Syrah, Petit Verdot*

🍷 *1–5 years (red)*

✓ *Félix Solís • Vinartis*

OTHER APPELLATIONS OF SOUTHERN AND CENTRAL SPAIN
DENOMINACIÓNES DE ORIGEN

VINOS DE MADRID

One more appellation in which pseudo-political local interests prevail over natural, social, or wine conditions. Madrid has never been a region, but the province around the capital city. Within its boundaries are three wine areas, each with different characteristics and history. San Martín should be part of the Gredos area. Arganda and Navalcarnero are typical Castilla–La Mancha climates. Yet Madrid is an effective selling name. Vienna, Austria is the only other capital city with a DO, although in Vienna the DO is real. Here it is more misleading, with vineyards situated far away from the city. Still, some very good wines are produced here.

🍷 *Albillo, Malvar, Garnacha, Tempranillo, Airen, Macabeo, Parellada, Alarije, Moscatel, Cabernet Sauvignon, Merlot, Petit Verdot, Syrah, Graciano*

🍷 *1–10 years (red), on purchase (white)*

✓ *El Regajal • Jesús Díaz • Finca El Rincón • 4 Monos • Comando G • Marañones • Bernabeleva • Las Moradas de San Martín • Valquejigoso (not in the DO)*

RIBERA DEL GUADIANA

This DO covers most of the Extremadura region, bordering Portugal. It is anything but a real DO, but instead a huge area with no real identity. Mountain areas like Montánchez and Cañamero co-exist with fertile valleys like Tierra de Barros only for administrative reasons. To make things worse, the government encouraged growers to plant Tempranillo and international varieties, poorly adapted to those climates, while allowing for the grubbing up of bush Garnacha vines and even forbidding Portuguese varieties in Tierra de Barros, the ones really adapted to the climate. There are 30 authorised varieties. Let's hope that one day they will understand.

🍷 *1–5 years (red), upon purchase (white)*

✓ *Pago Los Balancines • Palacio Quemado*

INDICACIÓN GEOGRÁFICA PROTEGIDA

EXTREMADURA IGP

Just to provide coverage for those wineries not using the Ribera del Guadiana DO.

TERRACED VINEYARDS LINE THE STEEP SLOPES LEADING DOWN TO THE SIL RIVER
Ribeira Sacra, a red wine DO in Northwest Spain, shows why this wine region is also called "Green Spain". This DO encompasses the area of inland Galicia where the rivers Sil and Miño meet.

NORTHWEST SPAIN

The continental Atlantic coast in Spain is exposed to predominant rainy winds and the temperate effect of the Gulf Stream. Consequently the climate is humid and moderate, which results in stunning verdant landscapes (the reason this region is also called "Green Spain". The region is a pleasure to the eyes but also provides very fertile soils for farming: the food of the region is delicious. Vine cultivation is more difficult with so much water and so little insulation. Great white wines, fresh and tense, come from this region, thanks to genetic adaptation, high trellises, and cultivation on steep slopes. Market demand and science are also bringing to the consumer attractive Atlantic red wines from the area. This is Spain's cool climate hearth.

Northwest Spain consists of four regions: Galicia, Asturias, Cantabria, and Basque Country.

GALICIA
DENOMINACIÓNES DE ORIGEN

MONTERREI

Most vineyards in this DO are south-facing, basking in the sun. A small region with a long history, it has only recently rejoined the fine wine scene thanks to the efforts of José Luis Mateo, a great pioneer in recovering old varieties and experimenting winemaking approaches.

🍷 *Godello, Dona Branca, Treixadura, Albariño, Caíño Blanco, Loureira, Blanca de Monterrei, Mencía, Merenzao, Araúxa (Tempranillo), Caíño Tinto, Sousón*

🍷 *1–5 years (red and white)*

✓ *Quinta da Muradella • Crego e Monaguillo • Pazo de Valdeconde*

RIAS BAIXAS

The best-known Galician appellation, mostly because of its Albariño grape, giving aromatic, fresh well-structured wines, some of them magnificent. But there is more in Rías Baixas than top Albariño. The DO is divided into five separate areas: Val de Salnés, Condado de Tea, O Rosal, Soutomaior, and Ribeira del Ulla. Single-variety Albariño wines are typical only of Salnés, the main region in terms of volume (more than half of the total production), and Soutomaior. In Condado de Tea and O Rosal, blends with Loureira, Treixadura, and Caíño are more typical. Red wines are becoming more important, thanks to their lovely freshness and uniqueness. Sparkling wines also have a market.

DOC or DOP *Albariño, Loureira/Loureiro Blanco, Treixadura, Caíño Blanco, Torrontés, Godello, Caíño Tinto, Espadeiro, Loureira Tinta, Sousón, Mencía, Pedral, Brancellao*

🍷 *1–7 years (red), 1–10 years (white)*

VINOS DE PAGO OF CENTRAL AND SOUTHERN SPAIN

An original legislative development in Spain, allowing individual landowners to create a publicly endorsed monopoly on the basis of a self-declared prestige and uniqueness. *Vinos de Pago*, officially the summit of the Spanish wine quality pyramid, have been approved only in those regions where the beneficiary landowners do not want to tell the customers where they come from, lest it is bad for their prestige. Castilla-La Mancha is the main turf for *Vinos de Pago*, although Valencia and Navarra are also active. Most *Vinos de Pago* are made from high-tech irrigated vineyards of Tempranillo and international red and white varieties. Dominio de Valdepusa was the first one, and it is still one of the most remarkable.

- Calzadilla
- Campo la Guardia
- Casa del Blanco
- Dehesa del Carrizal
- Dominio de Valdepusa
- Finca Elez
- Guijoso
- Pago Florentino
- Pago de Vallegarcia
- Pago de la Jaraba
- Pago los Cerrilo

✓ Condes de Albarei • Morgadío • Attis • Palacio de Fefiñanes • Fillaboa • Forjas del Salnés • Gerardo Méndez Do Ferreiro • La Val • Martín Codax • Santiago Ruiz • Terras Gauda • Trico • Eulogio Pomares • Mar de Frades • Pazo de Barrantes • Pazo de Señorans • Señorío de Rubiós • Valdamor

RIBEIRA SACRA

One of the most beautiful vineyard regions in the world, with very steep slopes finishing in the Sil river, impossible terraces, a green scenery all around, and evidence of a long and intense history. For many years the wines did not reflect the beauty of the region, but Raúl Pérez and a few other winemaking heroes are changing this. Basically a red wine area, although some white wines are interesting. Mencía is the main variety, but not necessarily the one with the highest quality potential. Merenzao, Brancellao, and other varieties are likely to render very nice wines in the near future.

🍷 Albariño, Loureira, Treixadura, Godello, Doña Blanca, Torrontés, Mencía, Brancellao, Merenzao, Tempranillo, Sousón, Caíño Tinto, Garnacha Tintorera, Mouratón

🍷 1–8 years (red), 1–3 years (white and rosado)

✓ Algueira • Raúl Pérez • Moure • Rectoral de Amandi

RIBEIRO

The most classic and historic Galician appellation, heir to 16th-century Ribadavia, a wine that was most appreciated in the English Court. Ribeiro is the land of a thousand *terroirs*, the top expression of Galician soils and microclimates, a region with extraordinary complexity and top potential, both for white and red wines, not always sensibly exploited by its vine-growers. There is a unique feature in Ribeiro, the *colleiteiro* wines, made by very small wineries. The region suffered greatly in terms of viticultural practise during Franco's regime, and it is only now beginning to recover from that past. In the future, Ribeiro should become the top white wine appellation in Spain. Ribeiro wines used to be blends of varieties.

🍷 Treixadura, Torrontés, Godello, Albariño, Loureira, Lado, Caíño blanco, Palomino, Albillo, Caíño longo, Caíño bravo, Caíño tinto, Ferrón, Sousón, Mencía, Brancellao, Garnacha Tintorera, Tempranillo

✓ Emilio Rojo • Coto de Gomariz • Dominio do Bibei • El Paraguas • Señorío de Beade • Viña Mein • Adega do Moucho • Manuel Rojo

VALDEORRAS

The cradle of Godello, one of the most fashionable grape varieties in Spain. It owes its fame to two people. First, Horacio Fernández, a public servant who persuaded the government to initiate a programme to recover the variety, which was almost extinct at the time. Then, Rafael Palacios, Alavro's younger brother, who put Valdeorras on the international wine scene with his As Sortes. The region is very close to Castilla y León and has a quasi-continental climate, the warmest in Galicia. The best wines are delicious, but there is still a great quality gap among the best and the others.

🍷 Godello, Dona Branca, Doña Blanca, Loureira, Treixadura, Albariño, Torrontés, Lado, Palomino, Mencía, Tempranillo, Brancellao, Sousón, Caíño Tinto, Espadeiro, Ferrón, Merenzao, Gran Negro, Garnacha Tintorera, Mouratón

🍷 1–3 years (red), 1–10 years (white)

✓ A Coroa • Joaquín Rebolledo • Rafael Palacios • Godeval • Avancia • Guitián • Valdesil

INDICACIÓNES GEOGRÁFICA PROTEGIDA

BARBANZA E IRIA IGP

A very small district producing red and white wines from Galician autochthonous varieties, similar to Salnés from Rias Baixas in natural conditions.

BETANZOS IGP

Until recently a small district for green acidic wines to be drunk locally. Climate change is increasing the potential of this region and is bound to become a fully-fledged DO in a few years. Look for elegant white wines with much tension. A region to follow closely.

RIBEIRAS DO MORRAZO IGP

The Ribeiras do Morrazo is a new IGP, approved in 2018. There are just four wineries here, located in the province of Pontevedra, dedicated to Albariño and other autochthonous varietal wines.

VAL DO MIÑO - OURENSE IGP

The oldest Galician IGP, it is a small area around the city of Ourense. Only four wineries

CANTABRIA
DENOMINACIÓNES DE ORIGEN

CANGAS

This appellation has everything required to become a most renowned wine name. The landscape of Cangas is impressive, with steep slopes in green mountains facing primitive forests. Viticulture here is for heroes. Grown here are several autochthonous grape varieties with great quality potential: Carrasquín, Verdejo Negro (nothing to do with Verdejo), Albarín Negro, and Albarín Blanco. The red wines are particularly suave, balanced, and elegant. The white wines are fresh and fruity. For the time being, Cangas is still a hidden jewel. Social issues (most of the people in the area are retired miners, reluctant to sell their vineyards to investors) and the small size of the appellation, 70 hectares (173 acres) in all (half of them planted with Mencía, which is less interesting in this area), are keeping the area in the shadow.

🍷 Carrasquín, Verdejo Negro, Albarín Negro, Mencía, Albarín Blanco, Albillo, Moscatel

🍷 1–6 years (red), 1–3 years (white)

✓ Vidas • Monasterio de Corias • Chacón Buelta

INDICACIÓN GEOGRÁFICA PROTEGIDA

LIÉBANA IGP

The Liébana IGP is an inland area in Cantabria, close to Asturias and Castilla. It is working on some interesting experiments with foreign cool climate grape varieties.

COSTA DE CANTABRIA IGP

The Costa de Cantabria IGO runs inland from the coast up to an altitude of 600 metres (1,970 feet). Some investors are trying to define this area, which is potentially good for aromatic white wines.

BASQUE COUNTRY

Most of the production of wine in the Basque Country takes place in Rioja Alavesa and is labelled as such. This region is included within Rioja

ARABAKO TXAKOLINA-TXAKOLI DE ALAVA

This is the most recent *chacolí* appellation, completely inland, giving wines that are quite different from the other *chacolis* (sometimes written *Txakoli*). These are slightly sparkling, very dry white wines that are rounder and firmer. Great potential, some excellent wines.

🍷 Hondarribi Zuri, Gros Manseng, Petit Manseng, Petit Courbu

🍷 1–3 years (white)

✓ Astobiza • Goiaena

CHACOLI DE BIZKALA-BIZKALKO TXAKOLINA

This appellation is one of the most interesting discoveries in Spain. Originally a small area producing thin wines for local consumption, it is now a reliable source of good quality and original white wines. The work of Ana Martín and Pepe Hidalgo, renowned Spanish consultants, and the leading role of the Itsasmendi and Gorka Izagirre wineries are behind such a change. The appellation is small, but consists of three subzones giving distinctive styles. Some producers are trying red wines as well, with promising results.

🍷 Hondarribi Zuri, Hondarribi Zuri Zerratia (Petit Courbu), Hondarribi Beltza, Folle Blanche, Petit Manseng, Gros Manseng, Sauvignon Blanc, Riesling, Chardonnay

🍷 1–5 years (white), 1–3 years (red)

✓ Gorka Izagirre • Itsasmendi • Gorrondona • La Antigua • Bikandi

CHACOLI DE GETARIA-GETARIAKO TXAKOLINA

The most classic *txakoli*, with very high acidity and lots of personality, thanks to the strong maritime climate. Many vines face the Cantabric sea. The appellation is changing swiftly, because of the success of *txakoli* from Biscay, which opened new markets for all *txakolis*, and convinced Basque consumers of its quality. Local restaurants, many of them at the summit of world gastronomy, often support Getaria *txakoli*. Some changes are good: recent improvements in viticulture and oenology have resulted in much more palatable wines, which can be enjoyed with food. Until recently, *txakoli* used to be served (and considered) as cider. The second change, effected in 2007, was not so positive. The appellation, originally restricted to a small well-defined area, now covers the whole province. This could result in loss of typicity.

🍷 Hondarribi Zuri, Hondarribi Beltza, Ondarrabi Zuri Zerratia (Petit Courbu), Izkiriota (Gros Manseng), Riesling, Chardonnay

🍷 1–5 years (white)

✓ Txomin Etxaniz • Elkano • Etxetxo • Gaintza

GETARIA VINEYARDS GROWING GRAPES FOR TXAKOLI WINE

Txakoli (pronounced "cha-koh-lee") is a slightly sparkling, spritzy white wine produced in Spain's Basque Country. Getaria is a coastal town located in the province of Gipuzkoa in the north of Spain.

The Wines *of* Portugal

FROM THE FRESH AND BRIGHT VINHO VERDE
to richly WARMING Port and Madeira, the wines
of Portugal have a long and venerable history.
With an ancient winemaking heritage generally
associated with the classic Ports and Madeiras,
Portugal has managed to honour its past as it
nonetheless looks to the future. An injection of
capital after formally joining the European Union in
1986 allowed Portugal the chance to redefine and
expand the horizons of its wine scene to meet the
demands of current trends. Impressive, inexpensive,
and even some world-class wines started appearing.
Long gone were the days of oxidized whites and
tired reds. Wineries large and small began using
modern technology to craft reds of previously
unimaginable elegance and finesse, and they
have also looked to the *terroir* for an expression
of minerality in some of Portugal's finest white
wines. Thankfully, Portuguese producers are
still using mostly, and above anything else, native
grapes, displaying tradition and a passion
for their own culture to the world.

**CURVING ROWS OF TERRACED VINEYARDS
LINE THE BANKS OF THE DOURO RIVER**
The colours of autumn bring out the best of the Douro
Valley wine region, which is famed for its enchanting
landscape as much as it is for its world-famous Port wine.
The hilly topography of this region means that growers must
carefully plan the best aspects and elevations on which to
plant their grapevines in order to get the best results.

Portugal

Portuguese wines have left behind their former reputation for indifferent quality. Quite the opposite, in fact: They are now energetic, vibrant, and, when matched with the right food, so pleasing to the palate you will want to try them again and again.

It's been a steep learning curve from grape to glass across the entire country, but thanks to the heavy investment of EU funds in the 1990s, the infrastructure of the Portuguese wine industry has been transformed. So-called international varieties have played an intermediate role in boosting the quality of Portugal's indigenous varieties. Yet it has been a much smaller role than in Spain, and there are now renewed efforts to explore the native varieties, some of which are turning out to be Spanish grapes under a different name . . . and vice versa.

The Douro region leads the way – not only for Port, but also for unfortified red and white wines. Improvements usually commence at the top end of the wine-quality pyramid, where price can pay for the investments necessary to achieve the desired aims and then take time to filter down to cheaper, larger-volume products. But Portugal's makeover started at the bottom and quickly worked its way up. It is even possible to pinpoint where and when this revolution began to materialize – in 1995, when deliciously fruity upfront reds and fresh, crisp whites came on-stream with wine brands such as Alta Mesa and Ramada from (the then) Estremadura.

There were many top-quality reds produced in the late 1990s, but the defining moment was the release of Charme Douro by Niepoort. If a producer in the sweltering hot Douro can produce a wine of such beautifully soft and silky fruit with velvety tannins, then Portugal must surely be one of the most potentially diverse winemaking countries in the world.

Some of the local varieties have proven to be so successful that, in an effort to tackle the effects of climate change, Bordeaux has recently allowed the use of native Portuguese grapes such as Touriga Nacional and Alvarinho in small amounts in their wines.

Regions like Douro, Dão, Bairrada, and Alentejo are constantly increasing in quality, while lesser-known regions like Beira Interior or Algarve are working their way through. High-quality Madeira can be a unique experience but it is rare, representing only about 3 per cent of total production. Port, however, is gradually being overshadowed by high-quality table wines from the Douro Valley.

Wine production and export are on the increase, and a series of excellent vintages guarantees consistently good quality for the markets.

BASALTIC ROCKS FORM A COMPLEX NETWORK OF DRY-STONE WALLS THAT PROVIDE SHELTER FOR VINEYARDS ON PICO ISLAND IN THE AZORES
The walls (called *paredes* or *murinhos*) delineate vineyard plots (*currais*) that run inland from, and parallel to, the rocky shore, with the Pico volcano rising beyond them. Pico Island has its own VR and is also part of the greater Açores appellation that covers the entire Azores archipelago. Pico produces white, sparkling, and fortified wines.

WINE REGIONS OF PORTUGAL

Its wine regions span the length and breadth of the country, but it is in the north of Portugal that the most famous wines – Port and Vinho Verde – are produced, along with the most upwardly mobile, those from Bairrada and Dão.

Vinho Verde DOC (DOP)

Miño

1a

Lima

2a

Chaves

Trás-os-Montes DOC (DOP)

Cavado

Minho Vinho Regional (IGP/IG)

○ Braga

Transmontano Vinho Regional (IGP/IG)

Basto

Tâmega

Valpaços

Ave

○ Guimarães

Planalto Mirandês

Sousa

Amarante

Beira Atlântico Vinho Regional (IGP/IG)

○ Porto

1b

3a *Douro*

Baião

Cima Corgo

Douro Superior

Duriense Vinho Regional (IGP/IG)

Douro e Porto DOC (DOP)

Távora-Varossa DOC (DOP)

5a Pinhel ○

Terras de Cister Vinho Regional (IGP/I...)

ATLANTIC OCEAN

Lafões DOC (DOP)

○ Aveiro

Dão DOC (DOP)

Pinhel

4c 4a

Beira Interior DOC (DOP)

Bairrada DOC (DOP)

Besteiros

4d

Terras do Dão Vinho Regional (IGP/IG)

4e 4b

Alva

○ Coimbra

Terras da Beira Vinho Regional (IGP/IG)

Mondego

Cova da Beira

Encostas d'Aire DOC (DOP)

Zêzere

Lisboa Vinho Regional (IGP/IG)

Ourém

Tomar ○

Tomar

Obidos DOC (DOP)

Alcobaça ○

Alcobaça

Tejo Vinho Regional (IGP/IG)

Chamusca

Portalegre ○

Portalegre

Lourinhã DOC (DOP)

Alenquer DOC (DOP)

Santarém ○ *Santarém*

Almeirim

Tejo

Do Tejo DOC (DOP)

Alentejo DOC (DOP)

Torres Vedras DOC (DOP)

Torres Vedras ○

Cartaxo

Arruda DOC (DOP)

Coruche ○ Coruche

Borba

Bucelas DOC (DOP)

Portalegre

Colares DOC (DOP)

Sorraia

Amadora ○ ★ Lisbon

Carcavelos DOC (DOP)

Redondo

Palmela DOC (DOP)

Évora ○

Setúbal DOC (DOP)

Setúbal ○

Évora

8b

Peninsula de Setúbal Vinho Regional (IGP/IG)

Sado

Ardila

Vidigueira

8a

Algarve Vinho Regional (IGP/IG)

Beja ○

Moura

Alentejo Vinho Regional (IGP/IG)

Guadiana

Chanza

Mira

Lagos DOC (DOP)

Lagos ○ ○ Portimão

Tavira DOC (DOP)

○ Faro

Portimão DOC (DOP)   *Lagoa DOC (DOP)*

1a	Monção e Melgaço
1b	Paiva
2a	Chaves
3a	Baixo Corgo
4a	Castendo
4b	Serra da Estrela
4c	Silgueiros
4d	Terras de Azurara
4e	Terras de Senhorim
5a	Castelo Rodrigo
8a	Granja-Amareleja
8b	Reguengos

0 mi — 40
0 km — 40

A RED VINHO VERDE SITS IN ICE

Vinho Verde originated in the Minho region in the far north of Portugal. The name literally translates to "green wine" but refers to a "young wine" that is released three to six months after the grapes are harvested. A Vinho Verde can be red, white, or rosé and even sparkling or a brandy. It shoud be consumed soon after bottling.

THE APPELLATIONS OF
PORTUGAL

Note: DOP or DOC stands for *Denominação de Origem Protegida/Controlada*, which is Portugal's equivalent of France's AOC, Spain's DO etc. IGP stands for *Indicação Geográfica Protegida*. VR stands for *Vinho Regional*, a sort of large-sized *vin de pays*.

AÇORES VR or IGP

The Açores appellation is located in the Azores, an archipelago composed of nine islands in the middle of the Atlantic Ocean, which provides a maritime climate 1,600 kilometres (1,000 miles) from the Portuguese coast. Three of these islands produce wine, with vines grown in volcanic soil, and are now enjoying full DOP status. The regional appellation covers the entire Azores. These islands are green, with plenty of pasture, as well as rain and wind from

the ocean. To protect the vines, but also to radiate heat at night, traditional volcanic stone walls that form *currais* of 2 to 3 square metres (21.5 to 33 square feet) were built. The most impressive vineyards are in Pico and, with their spectacular views and landscapes, are a UNESCO World Heritage site. Wine styles under the Açores VR range from fortified to unfortified whites, reds, and rosés. *See also* Biscoitos DOP, Graciosa DOP, and Pico DOP.

🍷 *Arinto (Pedernã), Cabernet Franc, Cabernet Sauvignon, Merlot, Saborinho, Syrah, Terrantez, Verdelho*

ALENQUER DOC or DOP

Red wines from the Alenquer appellation are vinous, lively, and with shiny tones while young. They can be intense aro-

matically when well matured and aged. White wines are of good quality, generally coming from vines grown on middle slopes with exposures to the southwest. Alenquer wines possess long persistence and aromatics, and they are full of flavour.

🍷 *Alicante Bouchet, Alicante Branco, Alvarinho, Amostrinha, Aragonez (Tinta Roriz), Arinto (Pedernã), Baga, Cabernet Sauvignon, Caladoc, Castelão (Periquita), Chardonnay, Fernão Pires (Maria Gomes), Jampal, Malvasia Rei, Rabo de Ovelha, Ratinho, Seara Nova, Tinta Miúda, Touriga Nacional e Trincadeira (Tinta Amarela), Vital, Viosinho*

🍷 *1–3 years (new-wave, fruity style), 2–5 years (others)*

✓ *Quinta das Setencostas • Quinta de Abrigada • Quinta de Plantos • Quinta do Carneiro*

PORTUGUESE LABEL LANGUAGE

Adamado Sweet

Adega Literally "cellar"; commonly used as part of the name of a company or cooperative, similar to the Spanish term *bodega*

Aperitivo Apéritif

Branco White

Bruto Portuguese adaptation of the French *brut*; used to describe a dry sparkling wine

Carvalho Oak

Casa Refers to the property or estate and may also indicate a single-vineyard wine

Casta Grape variety

Casta predominante Refers to the major grape variety used in a wine

Clarete Bordeaux-style or deep rosé

Claro New, or *"nouveau"*, wine

Colheita States vintage and is followed by the year of harvest.

Denominação de origem controlada/protegida (doc/dop) Roughly equivalent to a French AOC

Doce Sweet

Engarrafado na origem Estate-bottled

Engarrafado por Bottled by

Escolha Choice or selection

Espumante A sparkling wine that may be made by any method unless qualified by another term

Garrafa Bottle

Garrafeira This term may be used only for a vintage-dated wine that possesses an extra 0.5 per cent of alcohol above the minimum requirement. Red wines must have a minimum of three years' maturation, including one year in bottle; white wines must have one year, with six months in bottle. The wine may come from a demarcated region or be blended from various areas.

Generoso An apéritif or dessert wine rich in alcohol and usually sweet

Indicação de Proveniência Regulamentada) (IPR) Roughly equivalent to a French VDQS

Licoroso A fortified wine

Maduro "Matured"; often refers to a wine that has been aged in a vat

Palácio Refers to the property or estate and may also indicate a single-vineyard wine

Produzido e engarrafado por Produced and bottled at or by

Quinado Tonic wine

Quinta Farm or estate

Regulamentada (IPR) Roughly equivalent to a French VDQS

Reserva A term that can be used only to qualify a vintage year of "outstanding quality", in which the wine has an alcoholic strength of at least 0.5 per cent above the minimum

requirement. The wine may come from a demarcated region, but it could be blended from different areas.

Rosado Rosé

Seco Dry

Solar Refers to the property or estate; may also indicate a single-vineyard wine

Tinto Red

Velho Literally means "old" and, in the past, this term had no legal definition. Now it can only be applied to wines with a strict minimum age: three years for red wines and two years for whites.

Vinha Vineyard

Vinho de mesa Table wine equivalent of French *vin de table*. If the label does not state a specific DOC or Garrafeira, the wine will be a cheap blend.

Vinho Regional (VR) Similar to large-sized *vin de pays* and may also be labelled IGP

ALENTEJANO VR or IGP

This is the biggest region in the country but not the one that would necessarily produce the most wine. Here olives grow in long plains, as well as oak trees and many other crops. The Alentejo can be extremely hot in the summer, hence locals must sleep their *sesta* in the afternoon when temperatures rise to over 40°C (104°F). On the other hand, winter is freezing. The region has few hills and plains, increasing to a higher altitude at the Serra de São Mamede in the subregion of Portalegre in the northeast. Here, altitude influences the wines, giving them complexity and freshness. Soils vary from schist, pink marble, granite, limestone, and even clay. There are eight subregions; seven of them are fairly clustered together. Granja-Amarela borders with Spain and is the easternmost. Moura is the southernmost and the smallest. Wines of good quality are made all around the Alentejo region, especially the powerful reds.

🍇 *Alfrocheiro, Alicate-Bouschet, Alicante-Branco, Alvarinho, Antão Vaz, Aragonez (Tinta Roriz), Arinto (Pedernã), Baga, Bical, Cabernet Sauvignon, Caladoc, Carignan, Castelão (Periquita), Cinsault, Chardonnay, Chasselas, Corropio, Diagalves, Encruzado, Fernão-Pires (Maria Gomes), Gouveio, Grand Noir, Grenache, Larião, Manteúdo Preto, Malvasia-Fina, Malvasia-Rei, Manteúdo, Merlot, Moreto, Moscatel-Graúdo, Mourisco-Branco, Petit Verdot, Perrum, Pinot Noir, Rabo-de-Ovelha, Syrah, Tannat, Tinta Barroca, Tinta Caiada, Tinta Carvalha*

🍾 *1–3 years (new-wave, fruity style), 2–5 years (others)*

✓ Cortes de Cima • Esporão • Fitapreta • Herdade de Mouchão • J B (Júlio B Bastos Alicante Bouschet, Dona Maria Reserva) • José de Sousa • Paolo Laureano (Vinea Julieta Talhão 24) • A C de Reguengos • Quinta do Mouro (Rótulo Dourado) • Tapada do Chaves

ALENTEJO DOC or DOP

This DOP divides itself into eight smaller DOPs (Borba, Évora, Granja-Amarela, Moura, Portalegre, Redondo, Reguengos, and Vidigueira). A natural feature that influences the region is the Serra de São Mamede, a mountain range rising to 1,025 metres (3,363 feet) in altitude. A mixture of Mediterranean and continental climate marks the region.

🍇 *Alfrocheiro, Alicante-Bouschet, Antão Vaz, Aragonez (Tinta Roriz), Arinto (Pedernã), Cabernet Sauvignon, Castelão*

(Periquita), Fernão-Pires (Maria Gomes), Manteúdo, Perrum, Rabo-de-Ovelha, Síria (Roupeiro), Syrah, Tamarez, Touriga Nacional, Trincadeira (Tinta Amarela), Trincadeira das Pratas

🍾 *1–3 years (fruity style), 2–5 years (others)*

✓ Herdade do Rocim • Tapada do Chaves • Esporão • Herdade da Malhadinha (Matilde) • Herdade do Menir • Pêra Manca • Quinta do Centro • A C de Reguengos • Sogrape (Vinha do Monte) • António Saramago (Dúvida) • José de Sousa

ALENTEJO BORBA DOC or DOP

This was the first sub-appellation of the Alentejo region to gain recognition outside of Portugal itself, especially for its inexpensive, well-made red wines.

🍇 *Alfrocheiro, Alicante Bouschet, Alicante Branco, Antão Vaz, Aragonez (Tinta Roriz), Arinto (Pedernã), Cabernet Sauvignon, Carignan, Castelão (Periquita), Grand Noir, Moreto, Perrum, Rabo de Ovelha, Síria (Roupeiro), Tinta Caiada, Trincadeira (Tinta Amarela), Trincadeira das Pratas,*

🍾 *1–3 years (new-wave, fruity style), 2–5 years (others)*

✓ AC de Borba (Reserva)

ALENTEJO ÉVORA DOC or DOP

Located to the west of Redondo and Reguengos, this region can make smooth, harmonious, and easy-drinking reds. Évora is the capital of the Alentejo.

🍇 *Alfrocheiro, Alicante Bouschet, Antão Vaz, Aragonez (Tinta Roriz), Arinto (Pedernã), Cabernet Sauvignon, Castelão (Periquita), Diagalves, Fernão Pires (Maria Gomes), Grand Noir, Malvasia Rei, Manteúdo, Moreto, Perrum, Síria (Roupeiro), Rabo de Ovelha, Tinta Caiada, Trincadeira (Tinta Amarela), Trincadeira das Pratas*

🍾 *2–5 years*

✓ Herdade da Calada • João Gonçalves Gomes (Monte das Serras) • Herdade da Cartuxa • Pêra Manca

ALENTEJO GRANJA AMARELEJA DOC or DOP

This region surrounds the town of Mourão, bordering Spain in the east. Challenging conditions include soils of poor clay and schist and a climate that is very hot and dry. Extreme conditions produce wines with a distinctive

character – warm, smooth wines with high alcohol. The Moreto grape is well suited to the harshness of the region and considered one of the typical varieties of the area.

🍇 *Alfrocheiro, Antão Vaz, Aragonês (Tinta Roriz), Carignan, Castelão (Periquita), Diagalves, Manteúdo, Moreto, Perrum, Rabo de Ovelha, Síria (Roupeiro), Tinta Caiada, Trincadeira (Tinta Amarela), Trincadeira das Pratas*

ALENTEJO MOURA DOC or DOP

This locality is known for very poor clay soils and limestone. The Castelão grape dominates in the subregion. Wines are typically warm and soft, with corresponding high alcohol levels. Moura has a strong continental climate defined by cold, harsh winters and long, dry summers.

🍇 *Alfrocheiro, Alicante Bouschet, Antão Vaz, Aragonez (Tinta Roriz), Arinto (Pedernã), Cabernet Sauvignon, Castelão (Periquita), Fernão Pires (Maria Gomes), Moreto, Rabo de Ovelha, Síria (Roupeiro), Branco, Bical, Chardonnay, Moscatel Graúdo, Perrum, Tinta Carvalha Trincadeira (Tinta Amarela), Trincadeira das Pratas*

ALENTEJO PORTALEGRE DOC or DOP

This is the northernmost subregion of the Alentejo. Vines are grown on the foothills of the Serra de São Mamede, and some are also planted on steep slopes reaching 1,000 metres (3,280 feet). Higher altitudes mean lower temperatures, more elegant wines. Soils are composed of granite interspersed with patches of schist. Here, some of the vineyards may be over 70 years old. Grand Noir is one of the original, still- important grapes planted in Portalegre.

🍇 *Alicante Bouschet, Alicante Branco, Aragonez (Tinta Roriz), Arinto (Pedernã), Castelão (Periquita), Cinsaut, Diagalves, Fernão Pires (Maria Gomes), Grand Noir, Malvasia Rei, Manteúdo, Moreto, Síria (Roupeiro), Trincadeira (Tinta Amarela), Trincadeira das Pratas*

✓ Adega de Portalegre

ALENTEJO REDONDO DOC or DOP

The hills of Serra d'Ossa demarcate the subregion of Redondo, which is one of the highest hills of the Alentejo, with peaks reaching 600 metres (1,970 feet). These hills protect the vineyards from northerly and easterly winds, contributing to consistent cold, dry winters with the balance of hot, sunny summers. Schist and granite again mark this subregion.

🍷 *Alfrocheiro, Alicante Bouschet, Antão Vaz, Arinto (Pedernã), Aragonez (Tinta Roriz), Cabernet Sauvignon, Carignan, Castelão (Periquita), Diagalves, Fernão Pires (Maria Gomes), Grand Noir, Manteúdo, Moreto, Rabo de Ovelha, Síria (Roupeiro), Tinta Caiada, Trincadeira (Tinta Amarela), Trincadeira das Pratas*

🍽 *1–3 years*

✓ *A C de Redondo*

ALENTEJO REGUENGOS DOC or DOP

The largest subregions of the Alentejo, Reguengosis is marked by a continental climate, with cold winters and hot summers. It is home to some of the oldest vineyards in the Alentejo and produces powerful, full-bodied, and age-worthy wines. Soils are poor and stony, composed of schist.

🍷 *Alfrocheiro, Alicante Bouschet, Antão Vaz, Aragonez (Tinta Roriz), Arinto (Pedernã), Cabernet Sauvignon, Carignan, Castelão (Periquita), Corropio, Diagalves, Fernão Pires (Maria Gomes), Grand Noir, Manteúdo, Moreto Perrum, Rabo de Ovelha, Síria (Roupeiro), Tinta Caiada, Trincadeira (Tinta Amarela), Trincadeira das Pratas*

🍽 *1–3 years (new-wave, fruity style), 2–5 years (others)*

✓ *Herdade do Esporão • Conde d'Ervideira • A C de Reguengos (tinto) • J P Vinhos (Quinta da Anfora)*

ALENTEJO VIDIGUEIRA DOC or DOP

Separating the Alto (upper) Alentejo from the Baixo (lower) Alentejo, Vidigueira has one of the mildest climates of the region. The soils are quite infertile, composed of granite and schist like most of the region. The DOC is home to the Tinta Grossa grape made famous by Paulo Laureano.

🍷 *Alfrocheiro, Alicante Bouschet, Alicante Branco, Antão Vaz, Aragonez (Tinta Roriz), Arinto (Pedernã), Cabernet Sauvignon, Castelão (Periquita), Diagalves, Fernão Pires (Maria Gomes), Grossa, Larião, Manteúdo, Moreto, Mourisco Branco, Perrum, Rabo de Ovelha, Síria (Roupeiro), Tinta Caiada, Trincadeira (Tinta Amarela), Trincadeira das Pratas*

✓ *Ribafreixo Wines • Herdade do Sobroso*

ALGARVE VR or IGP

Marked by a strong Mediterranean climate, the Algarve is protected from cold winds from the north by a barrier made of mountains and its southern exposure, allowing the region to experience an average of 3,000 hours of sunshine a year. It encompasses four DOPs: Lagos, Lagoa, Portimão, and Tavira.

🍷 *Arinto, Bastardo, Diagalves, Moreto, Negra Mole, Periquita, Perrum, Rabo de Ovelha, Tamarêz d'Algarve*

✓ *Adega do Cantor*

ARRUDA DOC or DOP

A small DOP directly south of Alenquer and Torres Vedras. From 2002 on, grape restrictions were relaxed to allow the use of new indigenous and international varietals such as Cabernet Sauvignon, Touriga Franca, Syrah, Sauvignon Blanc, and Chardonnay.

🍷 *Alicante Branco, Aragonez (Tinta Roriz), Arinto (Pedernã), Bouschet, Cabernet Sauvignon, Caladoc, Camarate, Castelão (Periquita), Chardonnay, Fernão Pires (Maria Gomes), Jaen, Jampal, Malvasia Rei, Rabo de Ovelha, Seara Nova, Syrah, Tinta Barroca, Tinta Miúda, Touriga Franca, Touriga Nacional, Trincadeira (Tinta Amarela), Viosinho, Vital*

🍽 *1–3 years (new-wave, fruity style), 2–5 years (others)*

✓ *AC de Arruda*

BAIRRADA DOC or DOP

This region has a maritime climate on the Atlantic coast with the Dão region to the east. The soil structure changes here from the northern regions to limestone and sand. For a sense of Bairrada, imagine you are in a restaurant and you have the smell of wood-fired ovens roasting suckling pigs that will be served with potato crisps and a fresh salad. Outside it is 40°C (104°F), so all you want is a fresh, clean, and sharp glass of sparkling Bairrada, with its naturally high acidity.

🍷 *Alfrocheiro, Aragonez (Tinta Roriz), Arinto (Pedernã), Baga, Bastardo, Bical, Cabernet Sauvignon, Camarate, Castelão (Periquita), Cercial, Chardonnay, Fernão Pires (Maria Gomes), Jaen, Merlot, Pinot Blanc, Pinot Noir, Rabo de Ovelha, Rufete, Sauvignon, Sercealinho, Syrah, Tinta Barroca, Tinto Cão, Touriga Franca, Touriga Nacional, Verdelho*

🍽 *3–12 years (reds), 1–3 years (rosés and whites)*

✓ *Casa do Canto • Encontro • Marquês de Marialva • Quinta dos Abibes Sublime • Quinta da Lagoa Velha • Adega Rama • Caves Aliança • Gonçalves Faria (Reserva) • Luis Pato • Filipa Pato • Quinta de Pedralvites • Casa de Saima • Caves São João • Terra Franca (tinto)*

BEIRA INTERIOR DOC or DOP

This region's continental climate offers hot and dry summers but very cold and long winters. Beira Interior has hosted three subregions since 2005: Cova da Beira, in which ripening is easier; the high-altitude Castelo Rodrigo; and Interior Pinhel.

🍽 *1–3 years (new-wave, fruity style), 2–5 years (others)*

✓ *Quinta do Cardo • Almeida Garrett • Quinta dos Currais (Colheita Selecionada) • A C do Fundão (Fundanus Prestige)*

BEIRA INTERIOR CASTELO RODRIGO DOC or DOP

Known for its high altitude, poor soils, and high acidity, this subregion borders Spain.

🍷 *Arinto, Bastardo, Fonte Cal, Marufo, Rufete, Touriga Nacional*

🍽 *1–3 years (new-wave, fruity style), 2–5 years (others)*

✓ *Quinta do Cardo*

BEIRA INTERIOR COVA DE BEIRA DOC or DOP

A subregion of the Beira Interior DOP since 2005, Cova da Beira is located between Dão and Vinho Verde DOCs.

🍷 *Aragonez, Malvasia Fina, Síria (Roupeiro), Touriga Franca, Touriga Nacional*

BEIRA INTERIOR PINHEL DOC or DOP

One of the more mountainous subregions of the Beira Interior, it produces sparkling wines of good quality and also some fine rosés.

🍷 *Aragonez, Alfrocheiro, Arinto, Bastardo, Fernão Pires, Fonte Cal, Malvasia, Marufo, Rufete, Síria, Touriga Franca, Touriga Nacional, Trincadeira*

BEIRAS VR or IGP

A wide region, stretching from the Atlantic coast into the border with Spain, Beiras encompasses DOPs like Bairrada and its limestone and clay soils; Beira Interior and its three subregions; and the Dão, with its alluvial soils. Although heavily influenced by the Atlantic Ocean, reflecting a maritime climate, farther inland it is also continental, with hot and dry summers.

🍷 *Aragonez, Malvasia Fina, Síria, Touriga Franca, Touriga Nacional*

🍽 *1–3 years (new-wave, fruity style) 2–5 years (others)*

✓ *Buçaco • Conde de Santar (Reserva) • Entre Serras • Filipa Pato • João Pato • Quinta dos Cozinheiros • Quinta de Foz de Arouce*

BISCOITOS DOC or DOP

Biscoitos produces solely fortified wine from white varietals on the island of Terceira in the Azores. These wines must be matured for a minimum 36 months in wood. Dark and stony soils characterize the region.

🍷 *Arinto (Pedernã), Terrantez, Verdelho*

BUCELAS DOC or DOP

This region of loamy soils produces white wines and sparkling wines only. The wines are dry, fresh, and crisp and made with a minimum of 75 per cent Arinto plus

Rabo de Ovelha and Sercial. Good-quality sparkling wines can be found.

🍷 *Arinto (Pedernã), Rabo de Ovelha, Sercial (Esgana Cão)*

🍽 *Upon purchase*

✓ *Quinta da Romeira (Prova Régia) • Quinta do Boição Grande*

CARCAVELOS DOC or DOP

Carcavelos lies to west of Lisbon. The region makes tiny quantities of fortified wines that are most often sweet. These wines can be vintage and non-vintage, and white or red grapes can be used.

🍷 *Arinto (Pedernã), Castelão (Periquita), Galego Dourado, Preto Martinho, Ratinho*

🍽 *5–20 years*

✓ *Quinta dos Pesos • Villa Oeiras*

COLARES DOC or DOP

Famous for the surfing beach of Guincho and a fantastic Michelin-star restaurant, Fortaleza do Guincho, Colares is also known for the ungrafted Ramisco vines. These are planted in sandy soils, scattered amongst the dunes that protect them from phylloxera and also from strong winds from the Atlantic Ocean. Two types of soil exist here: *Chão de areia*, or sandy soils; this is where the DOP wines come from, composed of Ramisco for reds or Malvasia for whites. They can also come from *Chão rijo* (clay soils), however, provided only a maximum of 10 per cent is used. Here, grapes such as Castelão (Periquita) and Malvasia *(branca)* must represent a minimum of 80 per cent.

🍷 *Malvasia, Ramisco*

🍽 *15–30 years (red)*

✓ *Antonio Bernardino Paulo da Silva (Chitas) • Adega Regional de Colares • Adega Viúva Gomes*

DÃO DOC or DOP

Located in the mountainous central north of Portugal, where altitude may reach 400 to 800 metres (1,310 to 2,625 feet), Dão is surrounded by the Caramulo, Buçaco, Nave, and Serra da Estrela Mountains. These act as a barrier from humidity and strong continental winds. Soils are very granitic, with some schist in the mix as well. The climate is moderate, with most rain occurring during winter but dry and hot in summer. The red wines are firm, but the best have a good balance of fruit; the whites clean and fresh.

🍷 *Alfrocheiro, Encruzado, Baga, Bastardo, Bical, Cercial, Jaen, Malvasia Fina, Rabo de Ovelha, Tinta Pinheira, Tinta Roriz (Aragonez), Touriga Nacional, Verdelho*

🍽 *3–8 years (reds), 1–3 years (whites)*

✓ *Campos da Silva Oliveira • Casa da Passarella • Quinta da Ponte Pedrinha • Quinta da Gandara • Duque de Viseu • José Maria da Fonseca (Garrafeira P) • Paço das Cunhas • Quinta de Cabriz (Four C) • Quinta dos Carvalhais (Encruzado) • Quinta da Falorca (Garafeira) • Quinta dos Roques (Reserva) • Quinta de Saes (Reserva Branco)*

DÃO ALVA DOC or DOP

Named after the Alva River, which forms the southern border of the subregion in the Coimbra district.

DÃO BESTEIROS DOC or DOP

With a warmer, more Mediterranean climate than the rest of Dão, Besteiros is located southwest of Viseu but still in the Viseu district, where the soils are composed of a more fertile, slightly acid, granitic-based loam.

DÃO CASTENDO DOC or DOP

This subregional DOC encompasses three valleys – the Dão, Vouga, and Caja – towards the northeastern corner of the region.

DÃO SERRA DA ESTRELA DOC or DOP

Located in the Guarda district in the higher-altitude foothills of the Serra da Estrela, this DOC features vineyards planted at

700 to 750 metres (2,300 to 2,460 feet) above sea level. The region is well known for the Serra da Estrela cheese, a creamy, strong ewe's milk cheese with a spicy finish. Due to the altitude, the vineyards are more exposed to the maritime effects of the Atlantic than elsewhere in the region, much of which is protected by the Serra do Caramulo to the west.

✓ *Quinta das Maias*

DÃO SILGUEIROS DOC or DOP

This DOC is centred around Viseu, the capital of the Dão, in the northwestern corner of the region.

✓ *Quinta da Falorca • Quinta do Perdigão*

DÃO TERRAS DE AZURARA DOC or DOP

As with both Terras subregions, these vineyards are located between the two rivers, the Dão and Modego, at the very heart of the Dão region. The Terras de Azurara is farther upstream, where the vines grow mostly on red granitic sand, with some schist and clay, at heights of up to 450 metres (1,475 feet) above sea level.

DÃO TERRAS DE SENHORIM DOC or DOP

Terras de Senhorim is farther upstream than Terras de Azurara, but as with both Terras subregions, these vineyards are located between the two rivers, the Dão and Modego, at the very heart of the Dão region.

TEJO DOC or DOP

Formerly known as Ribatejo, Tejo hosts six subregions, which have some of the most affordable wines in the country. Tejo is characterized by three different *terroirs*, defined by the river: Bairro, on the north of the river, has hills composed of rich limestone and clay; the Charneca, south of the Tejo, where the landscape is dry and flat, has sandy soils forcing vines to fight for nutrients and water, producing richly flavoured grapes. This is also the warmest area of the three. Last, there is the Campo, right by the edge of the river, with a more maritime climate and more moderate temperatures.

🍷 *Alicante Branco, Alicante Bouchet, Aragonez, Arinto, Cabernet Sauvignon, Caladoc, Camarate, Castelão, João de Santarém, Fernão Pires (Maria Gomes), Pedernã Periquita, Syrah, Tempranillo, Tinta Amarela, Trincadeira, Trincadeira Preta, Tinta Roriz, Touriga Nacional*

🍷 1–5 years (reds), 1–3 years (whites)

✓ *Bridão • Falcoaria • Quinta do Casal Monteiro*

TEJO ALMEIRIM DOC or DOP

This is an up-and-coming area in the Tejo region. Producers such as Casal da Coelheira Sociedade Agrícula are doing well in the region.

🍷 *Arinto, Castelão Nacional, Fernão Pires, Periquita, Poeirinha, Rabo de Ovelha, Tália, Tamarêz d'Azeitão, Trincadeira Preta, Vital*

🍷 6–18 months

✓ *Casal da Coelheira Sociedade Agrícula*

TEJO CARTAXO DOC or DOP

The flat, fertile area of Tejp Cartaxo produces good, fruity, value-for-money reds and whites.

🍷 *Arinto, Castelão Nacional, Fernão Pires, Periquita, Preto Martinho, Tália, Tamarêz d'Azeitão, Trincadeira Preta, Vital*

🍷 1–3 years

✓ *Adega do Cartaxo*

TEJO CHAMUSCA DOC or DOP

This sub-appellation of the Tejo region is adjacent to Almeirim and produces similar wines, but not quite of the same potential. It is well known for the Lusitano horses from Golegã.

🍷 *Arinto, Castelão Nacional, Fernão Pires, Periquita, Tália, Tamarêz d'Azeitão, Trincadeira Preta, Vital*

TEJO CORUCHE DOP or DOP

This appellation covers wines made from sandy, well-irrigated plains covering the southern half of the Ribatejo region, but Coruche is seldom encountered and has yet to make its mark.

🍷 *Fernão Pires, Periquita, Preto Martinho, Tália, Tamarêz d'Azeitão, Trincadeira Preta, Vital*

TEJO SANTARÉM DOP or DOP

This DOP encompasses the districts of Rio Maior and the capital of the Tejo region, Santarém.

🍷 *Arinto, Castelão Nacional, Fernão Pires, Periquita, Preto Martinho, Rabo de Ovelha, Tália, Tamarêz d'Azeitão, Trincadeira Preta, Vital*

TEJO TOMAR DOP or DOP

Red and white wines are grown on limestone slopes on the right bank of the River Tagus in the Tejo region.

🍷 *Arinto, Baga, Castelão Nacional, Fernão Pires, Malvasia, Periquita, Rabo de Ovelha, Tália*

DOURO DOC or DOP

Any picture taken of the region at the end of summer, with a sunset exposed in between mountain, shows its breathtaking golden colours. This is a region known mostly for the fortified wines of Porto DOP, along with the intense reds from Touriga Nacional, Touriga Franca, and many others. It is also known for the fresh, clean, and correct white wines that are made here today. Some important factors of the local vineyards include the exposure, altitude, and the schistous soils. The region is also characterized by three distinct types of vineyard plantings: Vinhas ao Alto, where rows run vertically downhill and soils are prone to erosion; Patamares are the traditionally cultivated terraced slopes; and, finally, the Socalcos are newer machine-workable terraces. In fact, the diversity and beauty of the region caused it to be chosen as a UNESCO World Heritage site. Dozens of grapes are planted in the vineyards of the Douro – some old vineyards are planted randomly, as they traditionally were, Quinta do Crasto "Vinha Maria Teresa" being one of the greatest examples. *See also pp 425–431.*

🍷 *Aragonez (Tinta Roriz), Côdega de Larinho, Dona Branca, Donzelinho Branco, Folgasão, Gouveio, Malvasia Fina, Moscatel Galego Branco, Rabigato, Tinta Barroca, Tinta Francisca, Tinto Cão, Touriga Franca, Touriga Nacional, Trincadeira (Tinta Amarela), Viosinho*

🍷 2–10 years (red, but up to 25 for Barca Velha), 1–4 years (white)

✓ *Apegadas (Quinta Velha Grande Reserva) • CARM (Reserva, Quinta do Côa, Quinta do Côa Reserva) • Churchill (Touriga Nacional, Quinta da Gricha) • Domingos Alves de Souza (Abandonado) • Duorum (Reserva Vinhas Velhas) • Encostas do Douro (Palestra) • Ferreirinha (Barca Velha, Reserva Especial, Callabriga, Vinha Grande) • J & F Lurton (Quinta do Malhô) • Niepoort (Redoma and especially Charme Douro) • Jorge Nobre Moreira (Poeira) • Prats & Symington (Chryseia, Post Scriptum de Chryseia) • Quanta Terra • Quinta do Côtto (Grande Escholha) • Quinta do Crasto • Quinta da Foz • Quinta do Noval • Quinta da Padrela (Grande Reserva) • Quinta de Passadouro • Quinta de Pellada • Quinta do Portal (Auru) • Quinta da Romaneira • Quinta de la Rosa • Quinta do Vale Meão • Quinta do Vale Dona Maria • Quinta do Vale da Raposa • Quinta de Vallado • Quinta de Ventozelo (Touriga Nacional) • Quinta do Vesuvio • Ramos-Pinto (Duas Quintas) • Secret Spot (Moscatel) • Sogrape (Reserva) • Vale do Bomfim (Reserva) • Vinhas de Ciderma (Reserva) • Vinilourenço (D Graça Grande Reserva, D Graça Reserva Especial) • Wine & Soul (Pintas Character)*

DOURO BAIXO CORGO DOC or DOP

Encompassing the districts of Mesão Frio and Peso da Régua, the Baixo (or Lower) Corgo is the smallest of the three subregions and the westernmost. It has a continental climate with rainfall occurring mostly during winter; summers are usually very hot and dry.

DOURO CIMA CORGO DOC or DOP

The most important of the three subregions, its main town is Pinhão, where some of the best *quintas* are located, including the famous Quinta do Noval.

DOURO SUPERIOR DOC or DOP

The largest of the three subregions, it is also the driest. Viticulturally the least important, but nowadays making very good wines such as Quinta do Orgal from Quinta do Vallado, which is made with organic grapes.

DURIENSE VR or IGP

White wines, rosés, and reds are allowed in a VR of the same boundaries of the Douro DOP, but with a greater regulatory flexibility.

🍷 *Alvarelhão, Arinto, Avesso, Bastardo, Boal, Branco Sem Nome, Cabernet Franc, Cabernet Sauvignon, Casculho, Castela, Cercial, Chardonnay, Côdega, Cornifesto, Coucieira, Donzelinho (Branco and Tinto), Esganação, Fernão Pires, Folgosão, Gewürztraminer, Gouveio, Malvasia (Corada, Fina, Parda, Preta, and Rei), Merlot, Moreto, Moscatel Galego, Mourisco (aka Palomino), Mourisco (de Semente and Tinto), Pederna, Periquita, Pinot Noir, Praça, Rabigato, Rufete, Samarrinho, Sauvignon Blanc, Sémillon, Sousão, Syrah, Tinta Amarela, Tinta Bairrada, Tinta Barroca, Tinta Cão, Tinta Carvalha, Tinta da Barca, Tinta Francisca, Tinta Roriz, Tinto Martins, Touriga Branca, Touriga Brasileira, Touriga Francesa, Touriga Naçional, Viosinho*

🍷 1–5 years - up to 10 years for some ultra-serious quality red), 1–3 years (white, with 1–2 years for Sauvignon Blanc)

✓ *Carla Ferreira (Unipessoal Bastardo Duriense) • Conceito Bastardo • Casa de Vila Verde (Alvarinho) • Quinta do Noval (Cedro, Labrador Syrah)*

ENCOSTAS D'AIRE DOC or DOP

Situated on the western slopes and hills of the Candeiros and Aire mountains, this is the northernmost DOP of the VR Lisbon region. It is also the largest, with soils based on limestone. Good, rich reds and modern whites are being made from traditional grapes, low in alcohol and with fresh acidity.

🍷 *Alfrocheiro, Alicante Bouchet, Amostrinha, Aragonez (Tinta Roriz), Arinto (Pedernã), Baga, Bastardo, Bical, Boal Branco, Cabernet Sauvignon, Caladoc, Castelão (Periquita), Cercial, Chardonnay, Diagalves, Fernão Pires (Maria Gomes), Grand Noir e Syrah, Jampal, Malvasia Fina, Rabo de Ovelha, Ratinho, Seara Nova, Rufete, Tamarez, Alicante Branco, Tinta Miúda, Touriga Franca, Touriga Nacional, Trincadeira (Tinta Amarela), Trincadeira Branca*

ENCOSTAS D'AIRE ALCOBAÇA DOC or DOP

A beautiful Roman Catholic monastery is located here, which was built by first Portuguese king, Afonso Henriques. The region produces mainly white wines, but reds and rosés are also found.

🍷 *Alicante Bouchet, Alicante Branco, Amostrinha, Aragonez (Tinta Roriz), Arinto (Pedernã), Baga, Castelão (Periquita), Bical, Boal Branco, Cercial, Chardonnay, Diagalves, Fernão Pires (Maria Gomes), Jampal, Malvasia Fina, Rabo de Ovelha, Ratinho, Rufete, Seara Nova, Syrah, Tamarez, Tinta Miúda, Touriga Franca, Touriga Nacional, Trincadeira Branca, Vital*

ENCOSTAS D'AIRE OURÉM DOC or DOP

This region is located the northernmost part of the Encostas D'Aire; alas, no producers of note are yet found here. Two grapes are allowed and qualify for the DOP: Trincadeira (Tinta Amarela) for the reds and Fernão Pires (Maria Gomes) for the whites.

GRACIOSA DOC or DOP

Now a full DOP of Azores, it only produces white wines. These wines are of good quality: light, fresh, and fruity.

🍷 *Arinto, Fernão Pires, Malvasia Fina, Terrantez, Verdelho*

LAFÕES DOC or DOP

Heading north from Dão lies Lafões. Soils are of granitic origin mixed with schist. This area is similar to the Vinho Verde region in the way vineyards are planted but also in the characteristics of the wine itself – low alcohol, higher acidity, and very fruity.

🍇 *Amaral Arinto, Cerceal, Dona Branca, Esgana Cão, Jaen, Rabo de Ovelha*

LAGOA DOC or DOP

This is the largest and most central DOP of the Algarve.

🍇 *Alicante Bouschet, Aragonez (Tinta Roriz), Arinto (Pedernã) e Síria (Roupeiro), Cabernet Sauvignon, Castelão, (Periquita), Manteúdo, Monvedro, Moreto, Moscatel Graúdo, Negra Mole e Trincadeira (Tinta Amarela), Perrum, Rabo de Ovelha e Sauvignon, Syrah, Touriga Franca, Touriga Nacional*

LAGOS DOC or DOP

The red wines produced here are velvety and light in body, with fruity aromas and low tannins. The whites are delicate and smooth, with tropical notes characteristic of a warmer region.

🍇 *Alicante Bouschet, Aragonez (Tinta Roriz), Arinto (Pedernã), Bastardo, Cabernet Sauvignon, Castelão (Periquita), Malvasia Fina e Síria (Roupeiro), Manteúdo, Moscatel Graúdo e Perrum, Negra Mole e Trincadeira (Tinta Amarela), Monvedro e Touriga Nacional*

LISBOA VR or IGP

A very large wine region in terms of production, Lisboa extends from just south of Bairrada all the way down to its border on the Tagus River. There are great producers in the region that make some very interesting wines; Quinta do Monte d'Oiro, a member of the Lisbon Family Vineyards, is a good example. Here whites, rosés, reds, and fortified wines are produced.

🍇 *Alicante Bouschet, Caladoc, Castelão (Periquita) Fernão Pires, Syrah*

🍷 *2–4 years (reds, 4–8 years for better wines), 1–3 years (whites)*

✓ *A C do Arruda (selected cuvées) • A C de São Mamede da Ventosa • A C do Torres Vedras (selected cuvées) • Baron Bruemmer (Senhor d'Adraga) • Espiga • Palha Canas • Quinta do Chocapalha (Arinto) • Quinta da Folgorosa • Quinta do Monte d'Oiro (Ex Aequo) • Quinta de Pancas*

LOURINHÃ DOC or DOP

This cool and windy DOP is located between Óbidos and the Atlantic Ocean, where the grapes struggle to ripen. Production is restricted to distillation, and *aguardente* (brandy) is made.

🍇 *Alicante Branco, Boal, Cabinda, Malvasia Rei, Marquinhas*

MADEIRA DOC or DOP

See Madeira, pp 432–433

MINHO VR or IGP

The region produces still wines from the classical Vinho Verde varietals, but some international varietals are also allowed. Look for generic, medium-bodied red wines offering aromas of ripe red fruits, which are better drunk in their youth. Whites have green tones and are aromatically delicate with floral and citric notes. The region is subdivided into nine smaller regions.

🍇 *Alfrocheiro, Alicante Bouschet, Alvarelhão, Alvarinho, Amaral, Aragonez (Tinta Roriz), Arinto (Pedernã), Avesso, Azal, Baga, Batoca, Borraçal, Cabernet Franc, Cabernet Sauvignon, Caínho, Cascal, Castelão (Periquita), Chardonnay, Chenin, Colombard, Diagalves, Doçal, Doce, Esganinho, Esganoso, Espadeiro Fernão Pires (Maria Gomes), Folgasão, Godelho, Lameiro, Loureiro, Malvasia Fina*

✓ *Covela Escolha • Casa de Santa Eulalia • Casa de Vila Verde • Quinta do Gomariz*

ÓBIDOS DOP or DOP

The picturesque medieval village of Óbidos surrounds a 12th-century castle. Climatically, this is a windy and cold region due to the proximity of the coast. Some good sparkling wines are made, but the reds are growing in importance, Quinta das Cerejeiras Reserva being a fantastic example.

🍇 *Alicante Bouchet, Alicante Branco, Alvarinho, Amostrinha, Antão Vaz, Aragonez (Tinta Roriz), Arinto (Pedernã), Baga, Cabernet Sauvignon, Caladoc, Camarate, Carignan, Castelão, Chardonnay, Encruzado, Fernão Pires (Maria Gomes), Jaen, Jampal, Loureiro, Malvasia Rei, Merlot, Moscatel Graúdo, Pinot Noir, Preto Martinho, Syrah, Tinta Barroca, Tinta Miúda, Touriga Franca, Touriga Nacional, Trincadeira (Tinta Amarela), Vital*

✓ *Quinta das Cerejeiras*

PALMELA DOC or DOP

Palmela is known for its Castelão-based red, grown in a warm maritime climate and sandy soils. These wines can be full bodied, with intense colour and notes of dried fruits and spices. Ageing brings smoothness and finesse.

🍇 *Alicante-Bouschet, Alvarinho, Antão-Vaz, Aragonez (Tinta-Roriz), Arinto (Pedernã), Bastardo, Cabernet-Sauvignon, Castelão (Periquita), Chardonnay, Fernão-Pires (Maria-Gomes), Loureiro, Malvasia-Fina, Merlot, Moscatel-Galego-Branco, Moscatel-Galego-Roxo (Moscatel-Roxo), Moscatel-Graúdo (Moscatel-de-Setúbal), Petit-Verdot, Syrah, Tannat, Tinta-Miúda,*

✓ *Quinta do Monte Alegre*

PENÍNSULA DE SETÚBAL VR or IGP

This region covers a large area fanning out from the Sado Estuary far beyond the Setúbal Peninsula, where many innovative wines originated, but their future has been threatened by the urban sprawl south of Lisbon. Adega de Pegões, Casa Ermelinda Freitas, and Quinta da Bacalhôa are good reliable growers from the region. A variety of styles are produced, from reds to whites, rosés to sparkling and fortified wines.

🍷 *1–3 years (new-wave style), 2–5 years (others)*

✓ *J M Fonseca (João Pires, Periquita, Quinta de Camarate Tinto) • J P Vinhos (Quinta da Bacalhôa, Cova da Ursa)*

PICO DOC or DOP

One of the Acores DOPs for white, sparkling, and fortified wines. The volcanic peak that gives its name to the island is also the highest mountain in Portugal. Vines are planted in volcanic and hard rock soils. The fortified wines can be complex, full bodied, and well structured. In 2004 Pico was declared a UNESCO World Heritage site.

🍇 *Arinto (Pedernã), Terrantez, Verdelho*

PORT OR PORTO DOC or DOP

Breathtaking views from Vila Nova de Gaia to Porto over the city lights makes visitors appreciate the people, culture, food, and wines.

PORTIMÃO DOC or DOP

The fruity reds, are light in acid but with accentuated alcohol. The whites are smooth and delicate.

🍇 *Alicante, Arinto (Pedernã) e Síria (Roupeiro), Aragonez (Tinta Roriz), Bouschet, Cabernet Sauvignon, Castelão (Periquita), Manteúdo, Monvedro, Moscatel Graúdo, Negra Mole e Trincadeira (Tinta Amarela), Perrum e Rabo de Ovelha, Syrah e Touriga Nacional*

PORTO BAIXO CORGO DOC or DOP

See Douro Baixo Corgo DOP

PORTO CIMA CORGO DOC or DOP

See Douro Cima Corgo DOP

PORTO DOURO SUPERIOR DOC or DOP

See Douro Superior DOC

SETÚBAL DOC or DOP

Directly south of Lisbon, across the river Tejo, lies Setúbal and its mountain, Serra da Arrábida, where soils are composed of limestone or clay-limestone. Moscatel de Setúbal grapes are grown on these cooler, elevated slopes. Setúbal has a Mediterranean climate where summers can be hot and dry. Winters are rainy and mild. Moscatel Roxo (red Muscatel) is a rarity as the vine was nearly extinct but rescued by local producer José Maria da Fonseca. Red wine production is tiny.

🍇 *Antão-Vaz, Aragonez (Tinta Roriz), Arinto (Pedernã), Bastardo, Castelão (Periquita), Fernão-Pires (Maria-Gomes), Malvasia-Fina, Moscatel-Galego-Branco, Moscatel-Galego-Roxo (Moscatel-Roxo), Moscatel-Graúdo (Moscatel-de-Setúbal), Rabo-de-Ovelha, Roupeiro-Branco, Touriga-Franca, Touriga-Nacional, Trincadeira (Tinta-Amarela), Verdelho, Viosinho*

🍷 *Upon purchase (but will last many years)*

✓ *J M Fonseca*

A TRAIN GLIDES PAST THE CROFT WINERY ON THE BANKS OF THE DOURO RIVER
Located in Vila Nova de Gaia near Pinhão, Croft is the oldest firm still active today as a Port wine producer.

TAVIRA DOC or DOP

The easternmost DOP of the Algarve features a warm Mediterranean climate. Soils are sand, clay, and limestone.

🍷 *Alicante Bouschet, Arinto (Pedernã) e Síria (Roupeiro), Aragonez (Tinta Roriz), Cabernet Sauvignon, Castelão (Periquita), Diagalves, Manteúdo, Moscatel Graúdo e Tamarez, Negra Mole e Trincadeira (Tinta Amarela), Syrah e Touriga Nacional*

TÁVORA-VAROSA DOC or DOP

This DOC is encapsulated between Douro in the north and VR Beiras to the south. The region shares the same soil structure as Douro, schist and granitic soils, with an altitude between 500 to 800 meters (1,640 to 2,625 feet). Távora-Varosa is mostly known for its high-acid sparkling wines; its DOC was created in 1989.

🍷 *Alvarelhão, Aragonez, Bastardo, Malvasia Fina, Malvasia Rei, Malvasia Preta*

TEJO VR or IGP

This large region located between Lisboa and Alentejo boasts a moderate climate, while the rich alluvial plains of the river Tagus encourage high yields. Some very good wines – at extremely reasonable prices – are made here by those who restrict yields and use modern techniques to vinify soft, fruity reds.

🍷 *Arinto (Pedernã), Camarate, Carignan, Castelão (Periquita), Chardonnay, Esgana Cão, Fernão Pires, Jampal, Malvasia Fina, Malvasia Rei, Rabo de Ovelha, Merlot, Periquita, Pinot Noir, Sauvignon Blanc, Syrah, Tália, Tamarêz, Tamarêz d'Azeitão, Tinta Muida, Touriga Nacional, Trincadeira Preta, Vital*

🍷 *1–5 years (reds), 1–3 years (whites)*

✓ *Quinta da Alorna (Marquesa de Alorna Reserva) • Bright Brothers • Companhia das Lezírias (Reserva) • Falua • Fuiza • Rui Reguinga (Tributo) • Terra de Lobos • Vale D'Algares (D)*

TERRAS DÃO VR or IGP

Found in the central north of Portugal, it covers the Dão and Lafões DOPs.

TORRES VEDRAS DOC or DOP

Originally called simply "Torres" until Miguel Torres objected, these high-yielding vineyards in the Lisboa region of Portugal have traditionally supplied the largest producers with bulk wines for their high-volume branded *vinho de mesa*.

🍷 *Arinto, Camarate, Fernão Pires, Jampal, Mortágua, Periquita, Rabo de Ovelha, Seara Nova, Tinta Miuda, Vital*

TRANSMONTANO VR or IGP

A large area north of the Douro and west of Rios do Minho, Transmontano encompasses the Trás-os-Montes DOC and its three subzones.

TRÁS-OS-MONTES DOC or DOP

Situated in the most northeastern corner of Portugal, it is subdivided into three smaller regions: Chaves, which is the northernmost and right on the border with Spain; Valpaços, which is directly southeast of the first; and, finally, Planalto Mirandês, which sits on the Serra do Mogadouro in the southeast, bordering once again with Spain. This region is marked by a more continental climate and is protected by mountains, making it extremely hot in the summer and very cold in the winter.

🍷 *Amarela, Bastardo, Cabernet Franc, Cabernet Sauvignon, Chardonnay, Donzelinho, Gewürztraminer, Gouveio, Malvasia Fina, Merlot, Mourisco Tinto, Pinot Noir, Rabigato, Sauvignon Blanc, Sémillon, Tinta Tinta Barroca, Tinta Cão, Tinta Roriz, Touriga Francesa, Touriga Nacional, Viosinho*

🍷 *1–3 years (new-wave style), 2–5 years (others)*

✓ *Casal de Valle Pradinhos • Quintas dos Bons Ares*

TRÁS-OS-MONTES CHAVES DOC or DOP

Similar in style to that of the Douro wines, but the whites, reds, and *rosados* of this region are lighter than in the Douro. The region also produces *aguardente bagaceira* ("burning water" or "brandy-to-be").

🍷 *Bastardo, Boal, Codega, Gouveio, Tinta Carvalha, Tinta Amarela*

TRÁS-OS-MONTES PLANALTO MIRANDÊS DOC or DOP

Located in the Trás-os-Montes region, in the very northeastern corner of Portugal, Planalto Mirandês borders Spain and produces full-bodied reds and heavy whites.

🍷 *Bastardo, Gouveio, Malvasia Fina, Mourisco Tinto, Rabigato, Tinta Amarela, Touriga Francesa, Touriga Nacional, Viosinho*

TRÁS-OS-MONTES VALPAÇOS DOC or DOP

Firm reds and slightly *pétillant* rosés are produced in the upper reaches of the Tua, a tributary of the Douro, in the Trás-os-Montes region.

🍷 *Bastardo, Boal, Cornifesto, Codega, Fernão Pires, Gouveio, Malvasia Fina, Mourisco Tinto, Rabigato, Tinta Amarela, Tinta Carvalha, Tinta Roriz, Touriga Francesa, Touriga Nacional*

VINHO VERDE DOC or DOP

Verde translates to "green", representative of the green part of the country with its plentiful vegetation and also of the wines themselves, with their green, unripe, and crisp flavours. This is the largest DOC in Portugal. The vines are grown in granitic soils, and the Minho River flows all the way from the Portuguese coast, marking the frontier with "*Nuestros Irmanos*", the Spaniards, up into the northern part of the country. The IGP or Vinho Regional Minho and the DOC Vinho Verde share the same delimitations in this cool, wet region, which is prone to diseases, making winemaking challenging. It is the reason pergolas, locally called *enforcado*, exist. These raised frames let the wind blow underneath the vines while also allowing other crops to be planted on the ground, traditionally using all available space. This is changing in the region, however, which is now peppered with low-trained vineyards. Monção and Melgaço are still the best-recognized subregions of the DOC Vinho Verde, where the Alvarinho grape really shows its ripe citrus character, ranging from lemon, tangerine, and orange to stone fruit and often floral. These days local restaurants will serve their deep purple Vinho Verde Tinto in white clay mugs so one can clearly see the staining of the inky color of the youthful and freshly sparkling Vinhão. This local variety is one of the few *teinturier* (dark-fleshed) grapes in the world.

🍷 *Alvarelhão, Alvarinho, Avesso, Azal, Azal Tinto, Batoca, Borraçal, Espadeiro, Loureiro Trajadura, Padeiro, Pedral, Rabo de Anho, Vinhão*

🍷 *Upon purchase (9–18 months maximum)*

✓ **Red single-*quintas*:** Afros • Casa do Valle • Ponte de Lima
White single-*quintas*: Casa de Sezim • Anselmo Mendes • Morgadio de Torre • Palácio da Brejoeira • Ponte de Lima • Quinta de Azevedo • Quinta do Convento da Franqueira • Quinta de Gomariz (Grande Escolha) • Quinta da Tamariz • Soalheiro Primeiras Vinhas • Solar de Bouças
White commercial blends: Cepa Velha (Alvarinho) • Chello • Gazela • Grinalda

VINHO VERDE AMARANTE DOC or DOP

As one of the interior subregions, Amarante is protected from much of the Atlantic's maritime influence. Consequently, it favours later-ripening varieties and tends to produce wines with a higher alcoholic degree than the regional norm. The red Vinho Verde is particularly deep coloured. Azal is one of the dominant varieties in interior subregions such as Amarante, where it has a citrus and green-apple aroma. Its black-skinned sibling, the Azal Tinto, is grown here for red Vinho Verde.

VINHO VERDE AVE DOC or DOP

Vineyards are located on an irregular topography around the Ave River basin, where the vines are exposed to the Atlantic influence. As a result, this subregion favours earlier-ripening varieties and tends to produce crisp wines with lower alcohol. Loureiro tends to grow in the coastal area to the west of this subregion.

VINHO VERDE BAIÃO DOC or DOP

As one of the interior subregions, Baião is protected from much of the Atlantic's maritime influence. Consequently, it favours later-ripening varieties and tends to produce wines with a higher alcoholic degree than the regional norm. Azal is one of the dominant varieties in interior subregions, such as Baião, where it has a citrus and green-apple aroma. Its black-skinned sibling, the Azal Tinto, is grown here for red Vinho Verde. Baião is renowned for a lingering floral-almond aroma and a lively acidity of wines from the Avesso grape.

VINHO VERDE BASTO DOC or DOP

As one of the most interior subregions, Basto is particularly protected from the Atlantic, although with the Lima subregion, it has one of the highest levels of rain (mostly in the winter). Azal is one of the dominant varieties in interior subregions such as Basto, where it has a citrus and green-apple aroma. Batoca is another important grape, and it brings smoothness to the mid-palate. Basto produces a high proportion of red Vinho Verde, and the black-skinned Azal Tinto, Padeiro, and Rabo de Anho grapes all grow well here.

✓ *Casa do Valle*

VINHO VERDE CÁVADO DOC or DOP

Vineyards are located on an irregular topography around the Cávado River basin, where the vines are low in altitude and exposed to the Atlantic influence. Consequently, this subregion favours earlier-ripening varieties, but the wines are not as crisp as in other coastal subregions. Loureiro tends to grow in the coastal area to the west, but this subregion is best known for its red Vinhos Verdes made from Vinhão and Borraçal grapes.

VINHO VERDE LIMA DOC or DOP

The vineyards here are not as exposed to the effects of the Atlantic as other coastal subregions, despite its combination of proximity and northerly location. Loureiro tends to grow in the coastal area to the west of this subregion, and its wines are locally considered the best, although Lima also has a reputation for red Vinhos Verdes made from Vinhão and Borraçal grapes.

✓ *Afros • Quinta do Ameal • Ponte de Lima (Loureiro Seleccionada)*

VINHO VERDE MONÇÃO DOC or DOP

This is the "*grand cru*" of Vinho Verde, and its success is built on a series of exceptions. The only white variety growing in Monção is the Alvarinho, which is the best Vinho Verde grape, and relatively little Alvarinho is grown elsewhere in the Minho. Monção is also the only Vinho Verde subregion with vineyards that are actually on the banks of the Minho River. Even red Vinho Verde is special: the black-skinned Alvarelhão and Pedral grapes are the only other varieties grown here, yet they are rarely encountered elsewhere in the region.

✓ *Palácio da Brejoeira • Anselmo Mendes (Contacto Alvarinho) • Palácio da Brejoeira*

VINHO VERDE PAIVA DOC or DOP

A small, high-altitude area close to the Douro, Paiva is a semi-interior subregion of the Minho and can be scorching hot, which makes its vineyards better suited to black-skinned varieties. Azal Tinto is grown here, but Paiva is most famous for red Vinhos Verdes from Amaral and, particularly, Vinhão.

VINHO VERDE SOUSA DOC or DOP

This is another semi-interior subregion of the Minho, but with a far more temperate climate than neighbouring Paiva. Azal is one of the dominant varieties in Sousa, where it has a citrus and green-apple aroma. Its black-skinned sibling, the Azal Tinto, is grown here for red Vinho Verde, although Borraçal and Vinhão are more common.

Port: The Douro Valley

The translation for the name Douro is "golden" in Portuguese. Imagine a picture taken of the region at the end of summer with a setting sun flashing in between valleys, the breathtaking golden colours shining through the vineyards and the river. These vineyards are dotted with dozens of native grape varieties, while different exposures and aspects define this region, known as being one of the most beautiful in the world.

It is hard to imagine how such a wonderful winter-warming drink as Port could ever have been conceived in such a hot and sunny country as Portugal. Popular belief has it that that it was not the Portuguese but the British who were responsible for Port; however, this is not entirely accurate. We can thank the Portuguese for dreaming up this most classic of fortified wines; the British merely capitalised on their original idea.

In 1678 a Liverpool wine merchant sent two Englishmen to Viana do Castello, north of Oporto, to learn the wine trade. Holidaying up the River Douro, they were regally entertained by the Abbot of Lamego. Finding his wine "very agreeable, sweetish, and extremely smooth", they asked what made it exceptional amongst all others tasted on their journey. The Abbot confessed to doctoring the wine with brandy, but the Englishmen were so pleased with the result that they purchased the entire stock and shipped it home.

THE ORIGIN OF THE PORT TRADE

The ancient house of C N Kopke & Co had been trading in Douro wines for nearly 40 years by the time the above encounter took place. Eight years before they stumbled upon the Abbot of Lamego, another Englishman named John Clark was busy building up a business that would become Warre & Co. In 1678, the same year as the encounter, Croft & Co was established, and this was followed by Quarles Harris in 1680 and Taylor's in 1692. By the time the Methuen Treaty of 1703 gave Portuguese wines preferential rates of duty in Britain, many British firms had set up trade in Oporto. Other nationalities followed, but it was the British shippers who virtually monopolized the trade, frequently abusing their power. In 1755 the Marquis of Pombal, who had assumed almost dictatorial powers over Portugal, put a curb on their activities through the Board of Trade. The privileges enjoyed by the British under two 100-year-old treaties were restricted. He also established the Oporto Wine Company, endowing it with the sort of powers to which the British had been accustomed.

This infuriated the British, but their protests were to no avail, and Pombal went on to instigate many worthy, if unpopular, reforms, including limiting the Douro's production area to the finest vineyards, the banning of elderberries for colouring the wine, and the outlawing of manure, which reduced yields but greatly improved quality.

FACTORS AFFECTING TASTE AND QUALITY

LOCATION
Port is made in the Cima Corgo, Baixo Corgo, and Douro Superior districts in the north of Portugal.

CLIMATE
The summers are dry and hot, the winters mild and wet, becoming more continental in the upper Douro Valley, where the summers are extremely hot, rainfall is high – 52 centimetres (20.5 inches) – and winters can be very cold.

ASPECT
Vines are planted on generally hilly land that becomes very steep in some parts. The region is characterized by three distinct types of plantings: *Vinhas ao Alto*, where rows run vertically and are prone to erosion; *Patamares* are the traditionally cultivated terraced slopes, and the *Socalcos* are newer, machine-workable terraces.

SOIL
The Douro is a patchwork of hard, sun-baked, granite, and schist soils, with the finest Ports originating from the schist vineyards that dominate upriver. Because of the importance of schist to Port production, Douro table wines are relegated to granite soils.

VITICULTURE & VINIFICATION
Terracing in the Douro Valley is widespread to maximize the use of the hilly land, although the current trend is to make wider terraces, thus enabling mechanization. Some of the less precipitous slopes are now cultivated in vertical rows. Steep terraces mean labour-intensive viticulture, and the hard Douro soil often requires blasting to enable planting. Port wines are made and fortified in the Douro, but most are blended and bottled at lodges in Vila Nova de Gaia.

GRAPE VARIETIES
See Port Grape Varieties, p427.

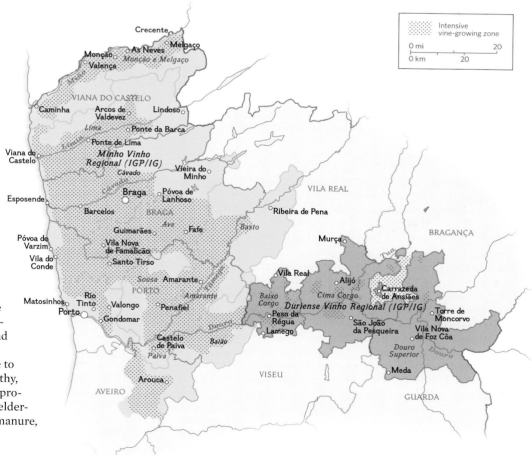

THE DOURO AND THE MINHO *(see also p419)*
These two northern areas produce the celebrated Port and Vinho Verde. The River Douro has long been crucial to the Port trade.

A VINEYARD WORKER HAULS A BARREL OF HAND-PICKED GRAPES DURING THE SEPTEMBER HARVEST IN PESO DA RÉGUA NEAR THE DOURO RIVER
Although much of the Port winemaking process is now often mechanised, grapes are still picked by hand. The ancient terraces of the Douro are too narrow for tractors.

The production of Port had not been perfected at this time. Fifty years after the encounter with the Abbot of Lamego, the trade had widely accepted the practice of brandying, but the importance of when, and in what quantity, to administer the brandy had not been recognized. Ironically, the Abbot's wine was superior because he had added the brandy during, and not after, the fermentation, thus interrupting, or "muting", the process with the natural sweetness that had so attracted the two Englishmen. Even after several centuries, the Port trade is still dominated by the British. It is even possible to generalize about the stylistic difference between the wines of British- and Portuguese-owned Port houses. The British tend to go for bigger, darker, sweeter, fruit-driven wines and have made vintage Port their particular niche, while the Portuguese opt for lighter, elegant, more mellow styles, the best of which are exquisitely aged tawnies. At one time, some Portuguese houses rarely even declared a vintage, as they majored on tawny. When they did, the wines were so much lighter in style that export countries often dismissed them as inferior. Although a few were indeed inferior, many were just different. Now, of course, world markets demand that if they are to survive, the Portuguese houses must declare vintages as often as British-owned houses.

When considering stylistic differences between vintage Port and the lighter, tawny styles, it is probably more accurate to categorize the fuller, fatter vintage Port as North European, not simply British, because this style is also preferred by Dutch, German, and French Port houses.

HOW PORT IS MADE

If any wine is perceived as having been "trodden", then it is Port. This is perhaps because the pressing and winemaking traditionally takes place in the vineyards where, until relatively recently, affairs were conducted on farms in a rather rustic style. Nowadays, few Ports are trodden, although several houses have showpiece *lagares* – troughs used for crushing grapes – for tourists. Many houses have "autovinificators" – rather antiquated devices that rely on the build-up of carbonic gas pressure given off during fermentation to force the juice up and over the *manta* (cap) of grape skins. The object is to extract the maximum amount of colouring matter from the skins, because so much is lost by fortification. It is possible also to achieve this using special vats and several have been installed throughout the Port industry.

FERMENTATION AND FORTIFICATION

The initial fermentation phase of Port differs little from that in the rest of the world, except that vinification temperatures are often as high as 32°C (90°F). This has no detrimental effect on Port, and in fact probably accounts for its chocolaty, high pH complexity. When a level of about 6 to 8 per cent alcohol has been achieved, the wine is fortified, unlike Sherry, for which the fermentation process is allowed to complete its natural course. Port derives its sweetness from unfermented sugars, whereas sweet Sherries are totally dry wines to which a syrupy concentrate is added. The timing of the addition of brandy is dependent upon the sugar reading, not the alcohol level. When the sweetness of the fermenting juice has dropped to approximately 90 grams per litre of sugar, the alcoholic strength will normally be between 6 and 8 per cent, but this varies according to the richness of the juice, which in turn is dependent upon the grape variety, where it is grown, and the year.

The use of the word "brandy" is somewhat misleading. It is not, in fact, brandy in the true sense, but a clear, flavourless, grape-distilled spirit of 77 per cent alcohol, known in Portugal as *aguardente*. It adds alcoholic strength to a Port, but not aroma or flavour. *Aguardente* is produced either from wines made in southern Portugal or from excess production in the Douro itself. Its price and distribution to each shipper are strictly rationed. On average, 110 litres (24 gallons) of *aguardente* is added for every 440 litres (97 gallons) of wine, making a total of 550 litres (121 gallons) – which is the capacity of a Douro pipe, a specific size of cask used for shipping wine from the valley to the lodges at Vila Nova de Gaia. A drier Port has a slightly longer fermentation and requires less than 100 litres (22 gallons) of *aguardente*, while a particularly sweet (or *geropiga*) Port is muted with as much as 135 litres (30 gallons).

If gauged correctly, the brandy that has been added to arrest the fermentation will eventually harmonize with the fruit and the natural sweetness of the wine. Our conception of balance between fruit, alcohol, and sweetness is, of course, greatly affected by what we are used to drinking. In deepest Douro a local farmer is likely to use a far higher proportion of alcohol for Port he intends to drink himself than for the Port he makes for export. Generally, British shippers prefer more fruit and less brandy, but all commercial shippers, British or Portuguese, would consider the domestic Port of a Douro farmer to lack sufficient body to match the brandy.

MATURATION AND BLENDING

Until 1986, by law, all Port had to be matured and bottled at Vila Nova de Gaia on the left bank of the Douro estuary opposite the city of Oporto. At some 75 kilometres (47 miles) from the region of production, this was like insisting that all Champagne be blended, bottled, and disgorged at Le Havre in Normany. This law was made in 1756, and it was ostensibly created by and for the big shippers, because it effectively prevented small growers from exporting their wine, because they could not possibly afford the expense of a lodge at Vila Nova de Gaia. By the late 18th century, most of the famous Port names were already established, and this restrictive law enabled them to maintain the status quo, especially on the lucrative international markets. All this has changed, and though most Ports still come from the big shippers' lodges, many new Ports now find their way on to the market direct from privately owned Douro *quintas*.

WINE FERMENTATION TANKS IN A DOURO FACILITY
Rather than relying on traditional foot-treading to crush grapes, most Port producers now use more modern methods. The wine grapes are placed in *lagares*, wide, open-top fermenting tanks made from stone or neutral concrete, and then crushed by machine.

PORT GRAPE VARIETIES

There are 48 grape varieties permitted in the production of Port. This simple fact goes a long way to explaining the great variation in quality and character of Ports within the same basic style. The official classification of the grapes is as follows:

Very good black grapes
Bastardo, Donzelinho Tinto, Mourisco, Tinta Cão, Touriga Franca (Touriga Francesca), Tinta Francisca, Tinta Roriz (Tempranillo), Touriga Nacional

Good black grapes
Cornifesto, Malvasia Preta, Mourisco de Semente, Periquita, Rufete, Samarrinho, Sousão, Tinta Amarela, Tinta da Barca, Tinta Barroca, Tinta Carvalha, Touriga Brasileira

Average black grapes
Alvarelhão, Avesso, Casculho, Castela, Coucieira, Moreto, Tinta Bairrada, Tinto Martins

Very good white grapes
Boal (Malvasia Fina), Donzelinho, Esgana Cão (Sercial), Folgosão, Gouveio (Verdelho), Malvaisia Rei, Rabigato, Viosinho

Good white grapes
Arinto, Boal (Malvasia Fina), Cercial, Códega, Malvasia Corada, Moscatel Galego

Average white grapes
Branco sem Nome, Fernão Pires, Malvasia Parda, Perdernã, Praça, Touriga Branca

THE SIX BEST PORT GRAPE VARIETIES

The six Port grape varieties currently considered by growers and winemakers to be the best.

Touriga Nacional
Almost universally agreed to be the very best of all Port grapes, it has tiny berries and produces pitch-black wine with intense aromatic properties, great extract, and high tannin content. It prefers hotter climates but is a poor pollinator and does not yield a large crop. Cloning is underway to increase production by 15 per cent and sugar content by 10 per cent; so far the most successful clone is R110.

Tinta Cão
This variety can add finesse and complexity to a blend, but it enjoys a cooler environment and requires training on wires to produce a decent crop. As traditional cultivation methods make it a poor producer, growers are not overly eager to cultivate it; the grape's survival will depend upon the willingness of Port shippers to maintain the variety on their large, managed estates.

Tinta Roriz
Not really a member of the Tinta family, this grape is sometimes simply called the Roriz. It is, in fact, the well-known Spanish variety, Tempranillo, the mainstay of Rioja. In the Douro it likes heat and fares best on the sunniest front rows of terraces cut into south- or west-facing slopes. Its dark grapes have thick skins, high sugar content, and low acidity, providing great colour, tannin, and succulence in a blend. Some believe this to be better than the Touriga Nacional.

Tinta Barroca
The wine produced by this grape is quite precocious and is therefore useful in younger-drinking Ports or to dilute wines that are too tannic and distinctive. Like Tinta Cão, this variety prefers cooler situations and particularly enjoys north-facing slopes.

Touriga Francesa
A member of the Touriga family, this grape has no connection with the similarly named Tinta Francisca of the Tinta family. According to the late Bruce Guimaraens of Fonseca and Taylor's, this high-quality variety is useful for "filling in the gaps" in vineyards between those areas occupied by vines that like either hot or cool situations. It gives fruit and aroma to a blend.

Tinta Amarela
This grape variety is dark-coloured and very productive. Its importance has been on the increase in recent years. As this high-quality vine is susceptible to rot, it performs best in the hottest, driest areas.

THE *QUINTA* CLASSIFICATION

A *quinta* is a wine-producing estate or vineyard. The Douro Valley covers 247,420 hectares (611,000 acres), of which 44,000 hectares (108,725 acres) are cultivated. Within this area, there are approximately 80,000 individual vineyards or quintas owned by 22,000 growers. Each vineyard is classified according to a points system allocated for the categories listed below. The better the classification, the higher official price a vineyard receives for its grapes and the greater its permitted production.

CATEGORY	MINIMUM	MAXIMUM
Location	-50	+600
Aspect	1,000	+250
Altitude; lowest is best	(-900)	(+150)
Gradient; steepest is best	(-100)	(+100)
Soil	(-350)	(+100)
Schist	(N/A)	(+100)
Granite	(-350)	(N/A)
Mixture	(-150)	(N/A)
Microclimate; sheltered is best	0	60
Vine varieties; official classification	-300	+150
Age of vines; oldest is best	0	+60
Vine density; lowest is best	-50	+50
Productivity; lowest is best	-900	+120
Vineyard maintenance	-500	+100
Total	**-3,150**	**+1,490**

Vineyards are classified from A, for best, to F, for worst, as follows: **Class A** (1,200 points or more); **Class B** (1,001–1,199 points); **Class C** (801–1,000 points); **Class D** (601–800 points); **Class E** (400–600 points); **Class F** (400 points or below).

CLASS A
Aciprestes (Royal Oporto), Atayde (Cockburn), Bomfin (Dow), Bom-Retiro (Ramos-Pinto), Carvalhas (Royal Oporto), Carvalheira (Calem & Filho), Boa vista (Offley Forrester), Corte (privately owned, managed by Delaforce), Corval (Royal Oporto), Cruzeiro St. Antonio (Guimaraens), Cavadinha (Warre), Eira Velha (Taylor's), Ervamoira (Ramos-Pinto), Fontela (Cockburn), Fonte Santa (Kopke), Foz (Calem & Filho), La Rosa (privately owned), Lobata (Barros, Almeida), Madalena (Warre), Malvedos (Graham), Mesquita (Barroa, Almeida), Monte Bravo (privately owned, managed by Dow), Nova (privately owned, managed by Warre), Panascal (Guimaraens), Passa Douro (Sandeman), Sagrado (Càlem & Filho), Santo Antonio (Càlem & Filho), Sibio (Royal Oporto), St Luiz (Kopke), Terra Feita (Taylor's), Tua (Cockburn), Vale de Mendiz (Sandeman), Vale Dona Maria (Smith Woodhouse), Vargellas (Taylor's), Vedial (Càlem & Filho), Zimbro (privately owned, managed by Dow)

CLASS A–B
Aradas (Noval), Avidagos (Da Silva), Casa Nova (Borges & Irmao), Ferra dosa (Borges & Irmao), Hortos (Borges & Irmao), Junco (Borges & Irmao), Leda (Ferreira), Marco (Noval), Meao (Ferreira family), Noval (Noval), Porto (Ferreira), Roeda (Croft), Seixo (Ferreira), Silho (Borges & Irmao), Silval (Noval), Soalheira (Borges & Irmao), Urqueiras (Noval), Velho Roncao (Pocas), Vezuvio (Symington family)

CLASS B
Carvoeira (Barros, Almeida), Dona Matilde (Barros, Almeida), Laranjeira (Sandeman), San Domingos (Ramos-Pinto), Urtiga (Ramos-Pinto)

CLASS B–C
Sta Barbara (Pocas)

CLASS C
Porrais (Ferreira family), Quartas (Pocas), Valado (Ferreira family)

CLASS C AND D
Sidro (Royal Oporto)

CLASS C, D, AND E
Granja (Royal Oporto)

CLASS D
Casal (Sandeman), Confradeiro (Sandeman)

CLASS NOT DISCLOSED
Agua Alta (Churchill), Alegria (Santos), Cachão (Messias), Côtto (Champalimaud), Crasto (privately owned), Fojo (privately owned), Forte (Delaforce), Infantado (privately owned), Rosa (privately owned), Val de Figueria (Càlem & Filho), Vau (Sandeman)

THE STYLES OF
PORT or PORTO

Note: With the exception of white Port, there are just two basic styles from which all variants stem: ruby and tawny. What distinguishes these two styles is bottle-ageing for rubies and cask-ageing for tawnies.

AGED TAWNY PORT
These 10-, 20-, 30-Year-Old, and Over-40-Years-Old Ports are traditionally known as "Fine Old Tawnies", but like fine old rubies, there are no legal requirements. By constant racking over a period of 10, 20, or more years, Port fades to a tawny colour, falls bright, and will throw no further deposit. It assumes a smooth, silky texture, a voluptuous, mellow, nutty flavour, and complex after-aromas that can include freshly ground coffee, caramel, chocolate, raisins, nutmeg, and cinnamon. The years are merely an indication of age; in theory, a 20-Year-Old tawny could be just a year old, if it could fool the Port Wine Institute. In practice, most blends probably have an age close to that indicated.

Most tawny Port experts find 20-Year-Old the ideal tawny, but 30-Year-Old and Over-40-Years-Old tawnies are not necessarily past it. The only relatively negative aspect of a good 30-Year-Old or Over-40-Years-Old is that it will be more like a liqueur than a Port.

🍷 *Upon purchase*

✓ **10-Year-Old:** *Càlem • Churchill's • Cockburn's • Croft • Dow's • Ferreira (Quinta do Porto) • Fonseca • Graham's • Niepoort • Offley (Baron Forrester) • Ramos-Pinto • Robertson's (Pyramid) • Smith Woodhouse • Taylor's • Warre's (Optima, Sir William);* **20-Year-Old:** *Barros Almeida • Burmester • Càlem • Cockburn's • Croft (Director's Reserve) • Dow's • Ferreira (Duque de Bragança) • Fonseca • Graham's • Niepoort • Noval • Offley (Baron Forrester) • Ramos-Pinto (Quinta do Bom) • Robertson's (Private Reserve) • Sandeman (Imperial) • Taylor's;* **30-Year-Old:** *Càlem • Croft • Dow's • Fonseca • Niepoort • Ramos-Pinto*

Over 40-Years-Old: *Càlem • Fonseca • Graham's • Noval • Sandeman • Taylor's*

CRUSTED PORT or CRUSTING PORT
The greatest non-vintage ruby Port, this is a blend of very high-quality wines from two or more years, aged up to four years in cask and, ideally, at least three years in bottle. Like vintage, it throws a deposit in bottle, hence the name. Ready to drink when purchased, providing it is carefully decanted.

🍷 *1–10 years*

✓ *Churchill's • Graham's • Martinez • Smith Woodhouse*

LATE-BOTTLED VINTAGE PORT (LBV)
A late-bottled vintage is a pure vintage Port from a good but not necessarily great year. LBVs are usually made in

lighter, generally undeclared years. Lighter, more precocious vintages are usually chosen for this less expensive category, and the wines are matured between four and six years in cask to bring them on even more quickly. LBVs are thus ready for drinking when sold but will continue to improve in bottle for another five or six years.

🍷 5–10 years

✔ *Burmester • Churchill's • Graham's • Quinta de la Rosa • Ramos-Pinto • Smith Woodhouse • Warre's*

PINK PORT

Croft created and launched the first Pink Port in 2008 to entice women and generally younger customers who might otherwise think of Port as old-fashioned. The production method is similar to a non-oxidative white port, which is then back-blended with pressed skins to tease out a little colour. It's a very fresh, in-your-face, fruit-driven style that is meant to be drunk young and, because of its fortified structure, can be served on the rocks. I was initially doubtful, but it works, certainly as a bar drink.

🍷 Upon purchase

✔ *Croft • Porto Cruz • Quinta & Vineyard*

RUBY PORT

The cheapest red Ports are rubies with less than a year in cask and often no time at all. Sold soon after bottling, they do not improve if kept. Inexpensive rubies have a basic, pepper-grapey flavour that can be quite fiery. Superior-quality wines are blended from various vintages with up to four years in cask, giving a more homogenous taste, though they should still have the fruity-spice and warmth of a young ruby.

🍷 Upon purchase

✔ *Cálem • Churchill's • Cockburn's (Special Reserve) • Ferreira • Fonseca (Bin 27, Terra Prima Organic Reserve) • Graham's (Six Grapes) • Quinta de la Rosa (Finest Reserve) • Sandeman (Signature) • Warre's (Warrior)*

SINGLE-QUINTA PORT

A wine from a single vineyard, this may be a classic vintage Port from an established house or a special release from an undeclared vintage. With interest in these Ports increasing, the change in the law allowing Port to be matured at the *quintas* and the fact that individual domaine wines are seen as more prestigious, this category has become as esteemed as vintage Port itself.

🍷 8–25 years

✔ *Quinta de Agua Alta (Churchill's) • Quinta do Bom • Quinta da Cavadinha (Warre's) • Quinta do Côtto (Champalimaud) • Quinta de Foz (Cálem) • Quinta Nova • Quinta do Noval • Quinta do Panascal (Fonseca) • Quinta Passadouro • Quinta do Passadouro (Niepoort) • Quinta Portal • Quinta Prelada • Quinta da Roeda (Croft) • Quinta de Roriz • Quinta de la Rosa • Quinta do Sagrado • Quinta do Seixo (Ferreira) • Quinta Senhora da Ribeira (Dow's) • Quinta do Tedo • Quinta da Urtiga (Ramos-Pinto) • Quinta Vale Dona Maria • Quinta do Vallado • Quinta de Vargellas (Taylor's) • Quinta Ventozelo • Quinta de Vesuvio*

SINGLE-QUINTA TAWNY PORT

Although most single-vineyard Ports are vintage and ruby in style, some are made as tawny, either with an indicated age or vintage-dated.

🍷 Upon purchase

✔ **10-Year-Old:** *Quinta do Sagrado • Ramos-Pinto (Quinta da Ervamoira);* **20-Year-Old:** *Ramos-Pinto (Quinta do Bom-Retiro);* **Vintage-dated:** *Borges' Quinta do Junco*

TAWNY PORT

These are often a blend of red and white. Some skilful blends can be very good and have even the most experienced Port-tasters guessing whether they are tawny by definition or blending. It is wise to pay more to ensure you are buying an authentically aged product. The best tawnies are usually eight years old, although this is seldom indicated.

🍷 Upon purchase

✔ *Delaforce (His Eminence's Choice) • Dow's (Boardroom) • Warre's (Nimrod)*

VINTAGE CHARACTER PORT

This misleading term was banned in 2002, although it took some time for stocks to clear the global distribution systems. Most producers have rebranded these wines as Reserve.

RUBY RESERVE or PREMIUM RUBY

This term replaced "Vintage Character" and has more complexity and character than a basic Ruby Port. Vintage Character was a misleading term and banned in 2002.

VINTAGE-DATED TAWNY or COLHEITA PORT

These excellent-value, often sublime, cask-aged wines are from a single vintage, and may have 20 or 50 years in cask. They must not be confused with the plumper, fruitier vintage Ports. There should be an indication of when the wine was bottled or a term such as "Matured in Wood". Some vintage Ports are simply labelled "vintage" and tawnies "Colheita". Other clues are "Reserve", "Reserva", or "Bottled in" dates.

🍷 Upon purchase

✔ *Barros Almeida • Burmester • Cálem • Delaforce • Offley (Baron Forrester) • Niepoort • Noval*

VINTAGE PORT

By law, a vintage Port must be bottled within two years. Maturation in bottle is more reductive than cask-ageing, and the wine that results has a certain fruitiness that will not be found in any Fine Old Tawny. When mature, a fine vintage Port is a unique taste experience with a heady bouquet and a sultry flavour. A warming feeling seems to follow the wine down the throat – more an "afterglow" than an aftertaste. The grape and spirit are totally integrated and the palate is near to bursting with warm, spicy-fruit flavours.

🍷 12–30 years (but see *individual producers*)

✔ *Churchill's • Champalimaud (Quinta do Côtto) • Dow's • Ferreira • Fonseca • Fonseca Guimaraens • Gould Campbell • Graham's • Niepoort • Pintas • Quinta do Noval (especially Nacional) • Smith Woodhouse • Taylor's • Warre's*

WHITE PORT

Most dry white Ports taste like flabby Sherry, but there are some interesting sweet Ports such as Ferreira's Superior White, which is creamy-soft and delicious. Truly sweet white Port is often labelled Lagrima. Niepoort pure Moscatel can make an nice alternative to Moscatel de Setúbal.

🍷 Upon purchase

✔ *Andresen 10-Year-Old Dry White • Ferreira's Superior White*

CERAMIC TILES ON A FOUNTAIN IN THE DOURO VALLEY DEPICT THE WINEMAKING PROCESS
The fountain is located by the side of N22, a national highway crossing in the neighbourhood of Quinta da Gricha, home of Churchill's Port. New by Port house standards, Churchill's was established in 1981, the first in 50 years.

THE DOURO VALLEY

Note: The age range for 🍷 (when to drink) refers to the optimal enjoyment of vintage Ports and appears only when this style is recommended. When the entire range is recommended, this includes all wines with the exception of the most basic ruby and tawny or any white Port, unless otherwise stated.

BARROS ALMEIDA
☆ Ⓥ

Manuel de Barros started work as an office boy at Almeida & Co, ending up not only owning the company, but also building a Port empire, as he and his descendants took over Douro Wine Shippers and Growers, Kopke, Feist, Feuerheerd, Hutcheson, Santos Junior, and Viera de Sousa. Only Kopke and Barros Almeida itself retain any form of independence. In 2006 Barros Almeida was taken over by Sogevinus, the drinks arm of the Spanish Caixanova bank and owners of Port houses Burmester, Cálem, and Gilberts.

✓ *20-Year-Old Tawny • Vintage-Dated Tawny • Colheitas*

BORGES & IRMÃO
Ⓧ

Established in 1884, the Portuguese-owned house of Borges & Irmão is best known for its Soalheira 10-Year-Old and Roncão 20-Year-Old tawnies, but the quality of their Port has suffered under a lack of direction from Portuguese government ownership.

BURMESTER
★★ Ⓥ

Of Anglo-German origin, Burmester is an underrated house that is best for mature tawny styles, but also makes good vintage. *See* Cálem.

✓ *Late-Bottled • Vintage • 20-Year-Old Tawny • Colheitas • Vintage*

CÁLEM
★

Cálem was established in 1859 and today is owned by Sogevinus. Consistently elegant tawny, single-*quinta,* and vintage Port. Also owns Barros, Burmester, and Gilberts.

🍷 *15–25 years*

✓ *0-, 20-, 30-, and 40-Year-Old Tawny • Vintage Character • Quinta de Foz • Vintage*

CHAMPALIMAUD
Quinta do Côtto
★☆

Miguel Montez Champalimaud can trace his family's viticultural roots in the Douro Valley back to the 13th century. He did not, however, produce his first Port until 1982, when it became the world's first estate-bottled single-*quinta* Port, making him the fastest-rising star among the new wave of privately owned grower Ports.

🍷 *12–25 years*

✓ *Quinta do Côtto*

CHURCHILL'S
★★☆

Established in 1981 by Johnny Graham, a member of the Graham's Port family who was married to Caroline (née Churchill), this was the first Port shipper to be established in recent times, and it has quickly risen to the top.

🍷 *12–25 years*

✓ *Entire range*

COCKBURN'S
★★

Founded in 1815 by Robert Cockburn, a Scot who married Mary Duff, who was much admired by Lord Byron in his childhood. The Cockburn's brand was formerly owned by Jim Beam Global, but it was sold to the Symington Group in 2010. Cockburn's best-selling Special Reserve and Fine Old Ruby are surprisingly good despite being produced in vast quantities. The vintage Ports are consistently excellent in quality.

🍷 *15–30 years*

✓ *Entire range*

CROFT
☆

To many people, Croft is best known for its Sherries, yet it is one of the oldest Port houses in the country, having started production in 1678. Croft's elegant style best suits Fine Old Tawnies. In all but a few years, its vintage Ports are among the lightest. Now part of Taylor's.

✓ *10-, 20-, and 30-Year-Old Tawny*

CRUZ
Ⓧ

The lacklustre quality of Porto Cruz says a lot about the average Frenchman's appreciation of the world's greatest fortified wine. France takes nearly half of all the Port currently exported, and this is the best-selling brand. The French could be forgiven if they cooked with it, but they actually drink the stuff. Cruz is now under the same ownership as C da Silva. Pink Port seems to be the limit of Porto Cruz's capabilities.

DELAFORCE
★

George Henry Delaforce founded the company in 1858 as the House of Delaforce. The style of Delaforce is very much on the lighter-bodied side, which particularly suits His Eminence's Choice, an exquisite old tawny with a succulent balance as well as a lingering fragrance.

✓ *His Eminence's Choice Superb Old Tawny*

DIEZ HERMANOS
❶

Interesting recent vintages make this low-key producer one to watch.

DOURO WINE SHIPPERS AND GROWERS
See Barros Almeida

DOW'S
Silva & Cosens
★★

Although the brand is Dow's, this firm is actually called Silva & Cosens. It was established in 1862, and James Ramsay Dow was made a partner in 1877, when his firm, Dow & Co, which had been shipping Port since 1798, was merged with Silva & Cosens. Dow's is now one of the many brands belonging to the Symington family, but consistently makes one of the very greatest vintage Ports. *See also* Cockburn's, Gould Campbell, Graham's, Quarles Harris, Smith Woodhouse, and Warre's.

🍷 *18–35 years*

✓ *Entire range*

FEIST
See Barros Almeida

FERREIRA
★★

This house was established in 1761 and was family-owned until 1988, when it was taken over by Sogrape, the largest wine shipper in Portugal. The brand leader in Portugal itself, Ferreira did not establish its present reputation until the mid-19th century, when it was under the control of Dona Antonia Adelaide Ferreira, who was so successful that when she died she left an estate of £3.4 million, valued at 1896 rates. Ever mindful of its Portuguese roots, it is not surprising that elegantly rich tawny Port is Ferreira's forte. Equally consistent is the style of its vintage Port. While this wine is typically light, smooth, and mellow, it does age beautifully.

🍷 *12–30 years*

✓ *Entire range*

FEUERHEERD
See Barros Almeida

FONSECA
Fonseca Guimaraens
★★☆

This Port house was originally called Fonseca, Monteiro & Co, but it changed its name when it was purchased by Manuel Pedro Guimaraens in 1822. Although it has been part of Taylor, Fladgate & Yeatman since 1948, it operates in a totally independent fashion, utilizing its own *quintas* and contracted suppliers. Fonseca has always been the top-quality brand, but even its second wine, Fonseca Guimaraens, puts many Port houses to shame.

🍷 *15–30 years*

✓ *Entire range*

GOULD CAMPBELL
★

Although at the cheaper end of the Symington Group, Gould Campbell, which was established in about 1797, produces fine, sometimes stunningly fine, vintage Ports. *See also* Cockburn's, Dow's, Graham's, Smith Woodhouse, Quarles Harris, and Warre's.

🍷 *12–25 years*

✓ *vintage*

GRAHAM'S
★★☆

W&J Graham & Co was founded as a textile business and only entered the Port trade when, in 1826, its Oporto office accepted wine in payment of a debt. Along the way, Graham's took over the firm of Smith Woodhouse, and both of them became part of the Symington Port empire in 1970. It is the vintage Port that earned Graham's its reputation, but the entire range is of an equally high quality, category for category. *See also* Cockburn's, Dow's, Gould Campbell, Quarles Harris, Smith Woodhouse, and Warre's.

🍷 *18–40 years*

✓ *Entire range, including basic Six Grapes Ruby*

GUIMARAENS
See Fonseca

HUTCHESON
See Barros Almeida

KOPKE
See Barros Almeida

MARTINEZ GASSIOT
★☆ Ⓥ

Martinez Gassiot was founded in 1797 by Sebastian González Martinez, a Spaniard, but it became affiliated with Cockburn's in the early 1960s. Rather underrated, Martinez produces small quantities of high-quality vintage and great-value crusted and is also a source of extraordinarily good own-label Port. *See also* Cockburn's.

🍷 *12–20 years*

✓ *Entire range*

MORGAN
☆ Ⓥ

One of the oldest Port houses, Morgan was acquired by Croft in 1952, but Croft itself was sold by Diageo to Taylor's in 2001. The Morgan brand was retained by Diageo (owners of Captain Morgan Rum), but its production has stopped and its future remains uncertain. A number of excellent own-label Ports have filtered on to the market bearing the Morgan Brothers name in small type, and Taylor's sold off a large cache of excellent Morgan vintages at auction in 2007.

NIEPOORT
★★

This small Dutch-owned firm is best known for its elegantly rich tawnies, especially the *colheitas,* but also makes fine, underrated vintage Port.

🍷 *12–25 years*

✓ *Entire range*

NOVAL
★☆

Originally a Portuguese house, having been founded in 1813 by Antonio José da Silva, Quinta do Noval was controlled by the Dutch van Zeller family for four generations until it was purchased by the French AXA group. Although officially known as Quinta do Noval, it carefully markets most products under the Noval brand, reserving the *quinta* name for those Ports made exclusively from the property itself. Their wines are generally good, but their Nacional rates three stars.

🍷 15–35 years (25–70 years for Nacional)
✓ Vintage • Nacional • 20- and 40-Year-Old Tawny

OFFLEY
Offley Forrester
★☆

Established in 1737, this firm once belonged to the famous Baron Forrester, but it is now owned by Sogrape. Offley makes excellent quality *colheitas*, elegant vintage Ports, and a deeper-coloured single-*quinta* Boa Vista.

🍷 12–25 years
✓ Vintage • Boa Vista • Colheita • 10- and 20-Year-Old Tawny

POÇAS JUNIOR

Established in 1918, like many Portuguese-owned firms, Poças Junior started declaring vintage Ports relatively recently (in the 1960s) and is primarily known for its tawnies and *colheitas*. Second labels include Lopes, Pousada, and Seguro.

QUARLES HARRIS

Established in 1680 and the second-largest Port house in the 18th century, when it was taken over by Warre's, Quarles Harris is now one of the smallest Port brands under the control of the progressive Symington Group. *See also* Cockburn's, Dow's, Gould Campbell, Graham's, Quarles Harris, Smith Woodhouse, and Warre's.

QUINTA DE LA ROSA
★

This house produces a promising new estate-bottled Single-Quinta Port. Look out for the new 10-Year-Old tawny.

🍷 8–25 years
✓ Finest Reserve • Late Bottled Vintage • Vintage

QUINTA DO CÔTTO
See Champalimaud

QUINTA DO INFANTADO
★Ⓥ

The vintage Port from this estate-bottled single-*quinta* have been rather lightweight, but the 1991 late-bottled vintage was a real gem, so it might be worth keeping an eye on future releases from this independent producer.

QUINTA DO NOVAL
See Noval

QUINTA DO VESUVIO

This massive, 400-hectare (990-acre), former Ferreira property was considered to be the greatest *quinta* in the Douro when planted in the 19th century. Owned since 1989 by the Symington group, which has already established its vintage Port as exceptional, but how exceptional remains to be seen.

🍷 12–30 years
✓ Vintage

RAMOS-PINTO
★☆

Established in 1880 by Adriano Ramos-Pinto when he just 20, this firm remained family-owned until 1990, when it was purchased by Champagne Louis Roederer. Ramos-Pinto has always produced excellent tawnies, particularly the single-*quintas* 10-Year-Old and 20-Year-Old and its *colheitas*, but if the 1994 is anything to go by, vintage Port could be its future forte.

🍷 12–25 years
✓ Late-Bottled Vintage • Single-Quinta (Quinta da Ervamoira 10-Year-Old, Quinta do Bom-Retiro 20-Year-Old) • Colheitas • Vintage (from 1994)

REBELLO VALENTE
See Robertson Brothers and Co

ROBERTSON BROTHERS AND CO
☆

Established in 1881, this small Port house is now a subsidiary of Sandeman and thus owned by Sogrape. Best known for its tawnies and, under the famous Rebello Valente label, very traditional vintage Ports.

🍷 12–25 years
✓ 10- and 20-Year-Old Tawnies • Vintage

ROYAL OPORTO
Real Companhia Velha
☆Ⓥ

Founded in 1756 by the Marquis de Pombal to regulate the Port trade. Its reputation as a good source of inexpensive tawny and vintage character Ports has ensured its survival, but it is focussing its greatest efforts on non-fortified Douro wines.

✓ Vintage Character (The Navigator's)

ROZES
☆

Part of the giant Moët-Hennessy group, Rozes has yet to establish a reputation beyond France.

SANDEMAN
☆

Also known for its Sherry and Madeira, Sandeman was established in 1790 and took over Robertson's and Rebello Valente in 1881, after which it acquired Diez and Offley, and is itself owned by Sogrape. The quality of Sandeman's vintage Port is variable and seldom blockbusting, but its tawnies are reliable, especially Imperial 20-Year-Old.

✓ 20-Year-Old Tawny (Imperial) • Vintage Character (Signature)

BARRELS OF COCKBURN'S SIT IN A RABELOS BOAT ON THE DOURO RIVER
Before roads or railways, these flat, wooden cargo boats were the most efficient way to carry Porto wine from the Douro Valley vineyards to the cellars in Vila Nova de Gaia. Producers now line the river in Porto with these vessels to advertise their brands.

SANTOS JUNIOR
See Barros Almeida

SILVA & COSENS
See Dow's

SILVA, C DA

Now under the same ownership as Cruz.

SMITH WOODHOUSE
★☆Ⓥ

The products of this famous British Port house are not as well known in Portugal as they are on export markets, where they are much appreciated for tremendous value across the entire range, and also for some of the very greatest vintage Ports ever made. *See also* Cockburn's, Dow's, Gould Campbell, Graham's, Smith Woodhouse, Warre's.

🍷 12–25 years
✓ Entire range

SYMINGTON GROUP

The Symingtons are one of the most well-known and successful of all Port families, owning seven brands, yet not one of these boasts the family name. *See also* Cockburn's, Dow's, Gould Campbell, Graham's, Quarles Harris, Smith Woodhouse, and Warre's.

TAYLOR, FLADGATE & YEATMAN
★★★

Founded in 1692 by Job Bearsley, this house underwent no fewer than 21 changes of title before adopting its present one, which derives from the names of various partners: Joseph Taylor in 1816, John Fladgate in 1837, and Morgan Yeatman in 1844. At one time the firm was known as Webb, Campbell, Gray & Cano (Joseph Cano was, as it happens, the only American ever to be admitted into the partnership of a Port house). In 1744 Taylor's became the first shipper to buy a property in the Douro, but its most famous acquisition was Quinta de Vargellas in 1890, a prestigious property that provides the heart and soul of every Taylor's vintage Port.

🍷 20–40 years
✓ Entire range

VIERA DE SOUSA
See Barros Almeida

WARRE'S
★★☆

Only the German-founded house of Kopke can claim to be older than Warre's, which was established in 1670, although it only assumed its present name in 1729 when William Warre entered the business. This house, which now belongs to the entrepreneurial Symington family, normally vies with Graham's, another Symington brand, as the darkest and most concentrated vintage Port after that of Taylor's. *See also* Cockburn's, Dow's, Gould Campbell, Graham's, Quarles Harris, and Smith Woodhouse.

🍷 18–35 years
✓ Entire range

WEISE & KROHN

Weise & Krohn Port house was originally founded in 1865 by two Norwegians of German extraction, Theodore Weise and Dankert Krohn. It has been owned by the Portuguese Carneiro family since the 1930s. It makes rather insubstantial vintage Port, and is best known for its tawnies.

✓ Colheitas

Madeira

The island of Madeira gives its name to the only wine in the world that must be baked in an oven! This fortified wine is deliberately heated to replicate the voyages of old during which the wine accidentally underwent maderisation at equatorial temperatures.

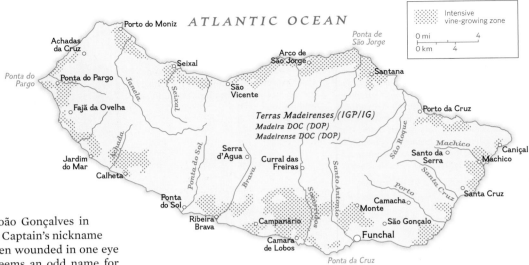

WINE-PRODUCING REGIONS OF MADEIRA
The island of Madeira, in the Atlantic Ocean west of Morocco, is famed for its fortified wine. Funchal is the capital of the island, where many wine lodges are found.

Prince Henry the Navigator sent Captain João Gonçalves in search of new lands in the 15th century. The Captain's nickname was Zarco, or "squinter", becaise he had been wounded in one eye while fighting the Moors. Although "squinter" seems an odd name for someone whose job it is to keep his eyes peeled for events on the horizon, particularly at that time, when every sailor was worried about falling off the edge of the world, this nickname is fact not fiction. Indeed, he was so proud of being called the "squinter" that he adopted it as a surname, becoming João Gonçalves Zarco. Despite this impediment to his sight, Zarco not only navigated his ship safely but actually managed to discover places. In 1418, for example, he found Madeira, although it must be admitted that he was searching for Guinea at the time. Zarco actually thought it was a cloud, but the more he squinted, the more suspicious he became: it was always in the same position. One day he chased the cloud, sailed through it and bumped into an island. It was entirely covered in dense forest and the cloud was merely the mist of transpiration given off in the morning sun. The forest was impenetrable (Madeira means "island of woods"), so Zarco lit fires to denude areas of the island, sat back, and waited. By all accounts he had a long wait, as the fires raged for seven years and consumed every bit of vegetation, infusing the permeable volcanic soil with potash, which by chance rendered it particularly suitable for vine-growing.

THE ORIGIN OF MADEIRA'S DISTINCTIVE WINE
As a source of fresh food and water, the island soon became a regular port of call for eastbound ships, which would often transport barrels of Madeira wine for sale in the Far East or Australia. As the ships journeyed through the tropics, the wine was heated to a maximum of 45°C (113°F) and cooled again during the six-month voyage, giving it a very distinctive character. The winemakers of Madeira were totally unaware of this until one unsold shipment returned to the island. Since then, special ovens, called *estufas*, have evolved in order that this heating and cooling can be replicated in the *estufagem* process (cheaper wines do this in large concrete vats, while the casks of better-quality wines experience a much longer, gentler process in warm rooms). All Madeiras undergo a normal fermentation prior to the *estufagem* process. Drier wines are fortified prior to *estufagem*, the sweeter styles afterwards.

MADEIRA FACTS
Since 1997 Madeira has seen a consistent average production of about 3.3 million litres on its mere 500 hectares (1,236 acres) of vine-growing land. Vineyards are irrigated through canals called *levadas*, where water is captured from springs on their way down the hills. The traditional training system is called *latada* (a training style similar to the pergola system): Vines are laid out horizontally, but newer trainings are now also being applied. France is the biggest importer of Madeira wine, followed by Japan and Germany, the United Kingdom, and the United States. Belgium also makes an impact on wine exports. Unfortunately, most Madeira sold is 3 years old and, to a lesser extent, 5 and 10 years old. Most exported wines are the sweeter styles, Malvasia and Boal being the highest proportion. Tinta Negra, previously called Tinta Negra Mole, represents 85 per cent of the production.

THE STYLES OF
MADEIRA

There have traditionally been four basic styles of Madeira, named after the following grape varieties: Sercial, Bual, Verdelho, and Malmsey. During the 20th century, however, the majority of Madeira produced was blended with a high proportion of Tinta Negra, Complexa (a red-fleshed cross), or American hybrids. Since 1990, however, no hybrids have been allowed, and since 1993, those wines made from Tinta Negra may use only generic terms of sweetness (and not the stylistic terms listed below), such as *seco* (dry), *meio seco* (medium dry), *meio doce* (medium sweet), and *doce* (sweet), accompanied by descriptors such as pale, dark, full, or rich, while at least 85 per cent of a varietal wine must be made from the grape indicated.

Note: Madeira should be ready to drink when sold and good Madeira will keep indefinitely, so no optimal drinking periods are given below.

BASTARDO
Another rarity from Madeira's once-glittering past, this black grape variety is related to the Douro variety of the same name (which is called Trousseau in France and elsewhere). There is no doubting Bastardo's great potential quality, making it yet another golden oldie that is long overdue a revival.

✔ *D'Oliveira's (1927)*

BOAL or BUAL
Definitely sweeter and darker than the Sercial and Verdelho styles, Bual can often be recognised under blind conditions by its khaki-coloured meniscus. This style has soft, gentle fruit with noticeable fatness and ripeness, underscored by a baked, smoky complexity. The legal range for residual sugar is 78 to 96 grams per litre.

✔ *Blandy's (1977, 1968, 1920)* • *H M Borges (1977)* • *Cossart*

(Colheita 1995) • *D'Oliveira's (1933, 1982, 1987)* • *Henriques & Henriques (Grand Old Boal, 15-Year-Old Boal)*

COLHEITA
Relatively rare, cask-aged, vintage Madeira; it may be pure varietal but does not have to be.

✔ *Barbeito (Cask 8a & d)* • *Blandy's (Malmsey: 2000, 1992, 1990)* • *Cossart (Bual 1995, Sercial 1988)*

EXTRA RESERVE or OVER-15-YEAR-OLD MADEIRA
This style is rarely encountered but is always significantly richer and more complex than 10-Year-Old Madeira. The "15-years" reference is excluded from some wines that are much older.

✔ *Barbeito (25-Year-Old Bual)* • *Cossart (Malmsey 15-Year-Old Malmsey, Very Old Duo Centenary Celebration Bual,*

Very Old Duo Centenary Celebration Sercial) • Henriques & Henriques (Verdelho, Boal)

FINEST or 3-YEAR-OLD MADEIRA

This, the lowest level of Madeira, will be the style of any Madeira whose age is not given. It consists primarily of Tinta Negra, and some Moscatel. Three years is too short a time for any decent wine to evolve into a true Madeira, and much is sold in bulk for cooking and cannot be recommended.

FRASQUEIRA or GARRAFEIRA

All Madeira wines at one time bore a vintage, but this practice is unusual today, because most are blends or products of *solera* systems. Current regulations stipulate that vintage Madeira must spend at least 20 years in cask, but if Madeira is to prosper, this outdated approach must be discarded, and a system similar to that used for vintage port must be adopted. This is not to suggest that vintage Madeira should not spend 20 years in cask, just that this rather lengthy process ought not to be mandatory. Consumers worldwide perceive vintage wines to be superior to non-vintage ones, and it does not seem to matter that the opposite is sometimes true. If Madeira is to reclaim the world stage and capture international imaginations, producers simply must put the spotlight on vintage Madeira, and the world is unlikely to wait 20 years for this to happen. Madeira must therefore release some of its vintages when they are much younger and put the onus on consumers to age the wines themselves. This would have the added advantage of feeding the auction circuit with new investment and, as the different vintages are successively tasted and retasted and reported and discussed by critics, this would raise consumer awareness of these wines, thus very usefully enhancing the reputation of Madeira in general.

MALMSEY

Made from both white and black varieties of Malvasia grown on the island's lowest and warmest regions, Malmsey is the ultimate Madeira and my own favourite. It is the most luscious, the sweetest, and most honeyed of all Madeira styles. Potentially the most complex and long-lived Madeira, Malmsey matures to an ultra-smooth, coffee-caramel succulence, which lasts forever in the mouth. The legal limit for residual sugar is 96 to 135 grams per litre.

✔ *Blandy's (1985, Colheitas: 2000, 1992, 1990) • Justino's (1933) • Quinta & Vineyard (2001)*

MOSCATEL

Madeira has made some interesting Moscatel, but only a few remain available for tasting today. Yet with so many fortified Muscats of superb quality made throughout the world, Madeira's Moscatel should perhaps be reserved for blending purposes only.

RAINWATER

This is effectively a paler, softer version of a medium-dry Verdelho but, because varietal names are never used with Madeira, making Rainwater with Verdelho would be a waste of this good-quality classic grape. Rainwater is more likely to be a paler, softer version of Tinta Negra trying to emulate Verdelho. There are two theories about the origin of this curiously named Madeira style. One is that it came from the vines, which were grown on hillsides where it was impossible to irrigate, thus growers had to rely on rainwater. The other theory concerns a shipment that was bound for Boston in the United States: even though the wine was accidentally diluted by rain en route, the Madeira house in question had hoped to get away with it, and it was shocked when the Americans loved it so much they wanted more.

RESERVE or 5-YEAR-OLD MADEIRA

This is the youngest age at which the noble varieties (Sercial, Verdelho, Bual, Malmsey, Bastardo, and Terrantez) may be used, so if no single variety is claimed, you can be sure that it is mostly, if not entirely, Tinta Negra. Five years was once far too young for a classic Madeira from noble grapes, but they are much better than they used to be and some bargain five-year-old varietals can be found.

✔ *Barbeito (Malmsey) • Blandy's (Sercial, Bual, Malmsey) • Cossart (Sercial)*

SERCIAL

This grape is known on the Portuguese mainland as Esgana Cão, or "dog strangler", and it is grown in the island's coolest vineyards. Sercial is the palest, lightest, and driest Madeira, but matures to a rich, yet crisp, savoury flavour with the sharp tang of acidic spices and citrus fruits. The legal limit for residual sugar is 18 to 65 grams per litre (the bottom end of this range tastes bone dry when balanced against an alcohol level of 17 per cent in such an intensely flavoured wine, and even 40 grams per litre can seem dry).

✔ *Cossart (Colheita 1988) • Barbeito (1956) • D'Oliveira's (1999)*

SOLERA MADEIRA

An old *solera* will contain barely a few molecules from the year on which it was based and which it boasts on the label, but the best examples of authentic Madeira *soleras* (there have been several frauds) are soft, sensuous, and delicious.

✔ *Blandy's Solera 1863 Malmsey • Blandy's Solera 1880 Verdelho • Cossart Solera 1845 Bual*

SPECIAL RESERVE or 10-YEAR-OLD MADEIRA

This is where serious Madeira begins, and even non-varietals may be worthy of consideration, because, although some wines in this style may be made with Tinta Negra, producers will risk only superior wines from this grape for such extended ageing.

✔ *Blandy's (Sercial, Malmsey) • Cossart (Verdelho, Malmsey) • Henriques & Henriques (Sercial, Malmsey) • Power Drury (Malmsey) • Rutherford & Miles (Bual, Malmsey)*

TERRANTEZ

This white grape variety is virtually extinct on the island, but the highly perfumed, rich, powerfully flavoured, and tangy-sweet Madeira wine it produced is so highly regarded that old vintages can still attract top prices at auction. If and when the Young Turks ever reach Madeira, this is one variety they will be replanting in earnest.

✔ *Blandy's (1976) • D'Oliveira's (1977, 1988)*

20-YEAR-OLD

Although 20-year-old Madeira is not specified in the regulations in the same way as 3-Year-Old (Finest), 5-Year-Old (Reserve), 10-Year-Old (Special Reserve), or 15-Year-Old (Extra Reserve) Madeiras, there is nothing in the regulations that prohibits its use.

✔ *Henriques & Henriques (Malmsey)*

VERDELHO

Some modern renditions of this style are almost as pale as Sercial, but Verdelho, both white and black varieties of which are grown on the island, traditionally produces a golden Madeira whose colour deepens with age. It is, however, always made in a medium-dry to medium-sweet style, which gives it somewhat more body than Sercial. This can make it seem softer and riper than it actually is, although its true astringency is revealed on the finish. The legal limit for residual sugar in these wines is 49 to 78 grams per litre.

✔ *Blandy's (1968, Colheita 2000) • Henriques & Henriques (15-Year-Old Verdelho) • D'Oliveira's (1985, 1988, 2000)*

FESTIVAL-GOERS DON TRADITIONAL COSTUMES TO PICK GRAPES AT THE MADEIRA WINE HARVEST FESTIVAL IN THE ESTREITO DE CÂMARA DE LOBOS PARISH OF FUNCHAL

Every September locals and visitors flock to this city to celebrate the island's traditional winemaking, including picking and treading grapes, as well as enjoying food, shows, and parades. The wine festival is one of Madeira's main tourist attractions that benefits the island's economy and brings Madeira's unique wine to the attention of consumers.

The Wines *of* Germany, Austria, *and* Switzerland

CLASSIC GERMAN RIESLING IS INCOMPARABLE. No other wine can offer in a single sip as much finesse, purity of fruit, intensity of flavour, and thrilling acidity as a fine Riesling, yet it remains underrated. German Pinot Noir (Spätburgunder), hardly known across the country's borders for many years, is also beginning to establish itself on the export markets. Almost 40 per cent of Germany's vineyards are planted with black grapes and, of that, one-third is Pinot Noir. Top Spätburgunder from the Ahr, Baden, and the Pfalz can be compared in quality with some of the best *premiers crus* in Burgundy. The only question is whether such wines can be produced at prices that do not make good Burgundy seem cheap by comparison! Austria also makes beautiful, under-appreciated Riesling and Pinot Noir, in addition to which, top-quality examples of one of its indigenous varieties, the highly distinctive Grüner Veltliner, are at last being taken seriously. And if German red wine is a surprise, nearly half of all Swiss wine is red, with smooth Pinot Noir and dense-yet-velvety Merlot the up-and-coming stars. These three countries will offer consumers some of the most exciting and diverse wine experiences of the new millennium.

SCHLOSS SCHWANDEGG IN CANTON ZÜRICH, SWITZERLAND
A medieval castle turned hotel and restaurant is surrounded by hillsides planted in grapevines. Canton Zürich is one of the 19 cantons in German-speaking Switzerland, the third-largest of the country's winemaking regions. The region grows mostly Pinot Noir grapes.

Germany

Qualitatively, Germany has turned itself around literally at root level, with Riesling and Spätburgunder (Pinot Noir) now representing one-third of the total area under vine, compared to 1995, when Müller-Thurgau was still the widest-cultivated variety in the country and Pinot Noir accounted for less than 7 per cent, even though that figure was double what it had been another 15 years earlier.

In terms of area under vine and volume of production, Germany may not be in the league of France, Italy, or Spain, but the best of its wines are world-renowned and, from the 18th century to the turn of the 20th century, the best of them fetched even higher prices than the famous *crus* of Bordeaux or Burgundy. "Iron Chancellor" Otto von Bismarck used to refer to German wine as his best ambassador. Two world wars and troubled times in between did nothing to preserve that reputation, and when the country began to rebuild its wine industry after 1945, its face was changed by large commercial wineries and cooperatives, which put vast efforts into mass production, blending wines from many different vineyards and varieties. In the 1970s and 1980s Riesling, which the international reputation of German wine had been built on over centuries, had to concede top rank in the country's vineyards to Müller-Thurgau, an earlier-ripening variety capable of much larger crops and therefore far better suited to serve the demand for inexpensive, mild-tasting plonk. Even at very high yields Müller-Thurgau still managed to exhibit a fairly perfumed fragrance, as did its then most-popular partners in commercial *cuvées*, Kerner and Morio-Muskat. In the post-war decades the average wine drinkers' palates also hankered for a sweeter style of wine, and with the technological advances in the sterilisation of grape must this could be achieved at low cost by adding the required amount of Süßreserve (sweet reserve) to the wine (*see Süßreserve, p442*). The most successful beneficiary of this practice became a brand with the phantasy name

"Liebfraumilch", which in the 1980s accounted for virtually 60 per cent of the total of German wine exports. With less than a quarter of the annual 13 million cases sold in its heyday, today Liebfraumilch is no longer as significant as it used to be, but of all German wines designated as *Qualitätswein* (quality wine) it is still the only one that can be a blend of constituents from more than one region. This exemption was granted by the wine law of 1971, and this may therefore be the right place to expose some other flaws of this ill-conceived piece of legislation.

THE 1971 GERMAN WINE LAW

In 1971, a new wine law was established to take account of EEC legislation regulating wine production. The declared objective of this law was to protect the consumer by offering greater clarity of information regarding quality and origin and to reduce the risk of confusion or fraud.

Considered the most important aspect was the classification into the three quality categories of *Tafelwein* (table wine), *Qualitätswein* ('quality wine"), and *Qualitätswein mit Prädikat* ("quality wine with special attribute)". But whereas in countries like France and Italy geographical, geological, and climatic conditions of a site or region tend to determine which category of quality a wine falls into, the German wine law of 1971 made the sugar content of the grape must, measured in degrees of *Oechsle*, the sole deciding factor. There are too many flaws in this system to mention them all, but suffice it to say that some common or garden varieties, more often than not "bred" for that particular purpose, tend to accumulate grape sugar faster than the late-ripening noble Riesling and would therefore easily qualify for the top tier of the classification, after the simplistic motto of "the sweeter the grapes the better the wine".

The requirements for the base tier of *Tafelwein* were set very low, with, for example, a minimum natural alcohol content between 5 to 6 per cent, which would then have to be chaptalised to a minimum actual alcohol content of 8.5 per cent, with 15 per cent the upper limit. Naturally very few wines fall into this category, between 0.4 per cent and 6 per cent in the last 20 years. A distinction was made between wines made from grapes harvested in Germany (*Deutscher Tafelwein*) and permitted blends from various countries of the EEC.

In most vintages the largest amount of German wine production falls into the middle tier called *Qualitätswein*. One could be slightly sceptical that over 90 per cent of German wine should qualify as at least quality wine, particularly when one takes into account that even the cheapest and most ignominious bottles of Liebfraumilch carry this distinction. Consumers abroad could not be blamed for thinking that Germany had set its sights very low in the definition of what is quality. What makes the issue even more confusing is that a number of years ago producers of some of the greatest dry wines of Germany called *Große Gewächse*, decided, (for reasons explained in VDP Classification, *see p439*) to market these as *Qualitätsweine* too.

The wines in the top tier of the classification qualified for the designation *Qualitätswein mit Prädikat* ("quality wine with predicate/special attribute"). It is the only category for which chaptalisation was not allowed. The minimum *Oechsle* (Oe) requirements for the individual *Prädikate* of (in ascending order) *Kabinett* (73°Oe), *Spätlese* (85°Oe), *Auslese* (95°Oe), *Eiswein* (125°Oe), *Beerenauslese* (125°Oe), and *Trockenbeerenauslese* (150°Oe) could vary marginally depending on region and grape variety, the values given in brackets applying to Riesling in the Rheingau (*see also* Quality Structure Overview, *p441*).

RECENT GERMAN VINTAGES

2018 The proverbial vintage of the century. A dry and hot summer produced very healthy grapes in large quantities. The harvest began as early as late August and by the beginning of October most of the crop had been picked. Growers were pleasantly surprised by how well the grapes had retained their acidity, the one difference to the otherwise very similar vintage of 2003.

2017 This was a challenging vintage for German growers. Early budding of the vines was punished by severe late spring frost, which led to a serious reduction in quantity. A warm and dry summer accelerated the ripening of the grapes. Late, in some regions prolific, summer rain necessitated a speedy harvest to avoid fungal disease. Relatively high acidity gives the best wines great aging potential.

2016 Capricious weather in the early part of the growing season, including late spring frost, hail, high humidity, and pressure from *Peronospora*, made growers work hard for their living. Those who managed to keep their fruit healthy throughout these adverse conditions were rewarded with a sunny and dry autumn.

2015 A long dry summer saw some vines suffer from the heat and water stress. September rains prevented a crisis. Ripe grapes were blessed with high sugar levels, but sometimes at the expense of acidity. Vineyards in cooler locations and planted with older vines produced the best results.

2014 This year had some serious challenges in store for growers. Prolific precipitation in the summer, subsequent mildew, and the attack of the cherry vinegar fly caused some severe losses for early-ripening varieties. Healthy grapes with a decent ripeness could only be achieved with maximum selectivity in the vineyards, but a sunny dry autumn yielded some good Riesling and Spätburgunder wines.

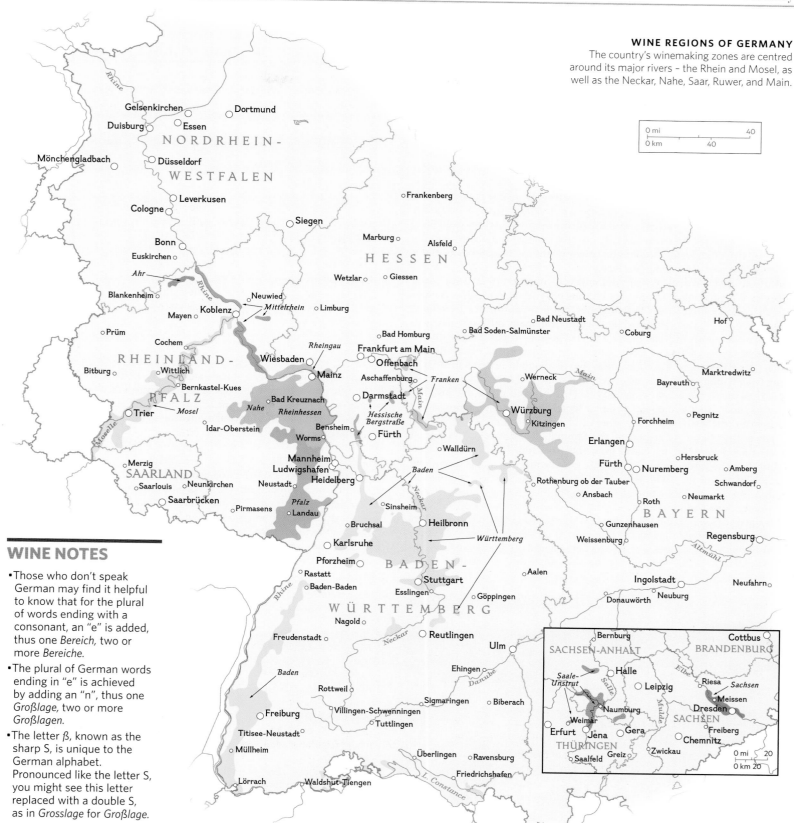

WINE NOTES

- Those who don't speak German may find it helpful to know that for the plural of words ending with a consonant, an "e" is added, thus one *Bereich,* two or more *Bereiche.*

- The plural of German words ending in "e" is achieved by adding an "n", thus one *Großlage,* two or more *Großlagen.*

- The letter ß, known as the sharp S, is unique to the German alphabet. Pronounced like the letter S, you might see this letter replaced with a double S, as in *Grosslage* for *Großlage.*

A fourth category was added in 1982 with *Landwein,* positioned between *Tafelwein* and *Qualitätswein,* trying to emulate the French designation of *vin de pays,* but it never became popular with growers and only negligible amounts are marketed in this category today.

THE GREAT *GROßLAGE* SWINDLE

With regard to specification of origin, the wine law of 1971 also regulated which geographical designations producers would be permitted to use for wines in the three different categories of the classification.

For the basic category of *Deutscher Tafelwein* they were not allowed to give any more detailed information about origin than the region the wine came from. This would later also apply to *Landwein,* although the regions for this category were defined a little more narrowly (*see also* The Landwein Regions of Germany, p445). Only *Qualitätsweine* or *Qualitätsweine mit Prädikat* were permitted to specify the smallest geographical unit of an individual site as their source of origin. The name of the vineyard would always be proceeded by the name of the commune within the boundaries of which it was located.

Allegedly in order to spare the consumer getting confused, the wine law of 1971 reduced the number of some 25,000 *Einzellagen* (individually named sites) to 2,600. In historical understanding, an *Einzellage* would be a delimited piece of land with particular geographical, geological, and climatic conditions, which would set it apart from other sites. In 1971 the law giver took the random decision that this piece of land would have to be at least 5 hectares (12.5 acres) in size, which meant that many until-then-existing smaller *Einzellagen* were wiped out as appellations of origin, because they were merged into one larger one. It was also deemed that the new *Einzellagen* could be as big as 100 hectares (250 acres) or more, "if growing conditions were similar". I am not sure how this similarity was defined, but when you visit the Würzburger Stein in the region of Franken – with 85 hectares (210 acres) the country's largest *Einzellage* – you will find great variations in soil, exposition, and gradient. Shame to who thinks evil of it, but the new larger sites were usually given the name of the most famous and popular of its former individual constituents. So, to avoid consumers getting confused, they were deceived.

But the powers that be had another even more artful trick up their sleeves. Winemaking is an industry like any other, where small individual producers have no political clout or lobby, but mass producers do. Whose interest could it have been in to blend wines from many different places to make vast amounts of basic plonk, but still be able to put a name on the label that made it sound as if the wine came from a special place? Best of all, if part of that name was the same as that of a commune that had already long been famous as the source of origin of some truly outstanding wines. And thus the *Großlage* was born in 1971. This was to become the designation for many individual sites lumped together under one name. Enter Piesporter Michelsberg (among others)!

The pretty village of Piesport on the Mosel had over many years earned international fame for its wines from the Goldtröpfchen, Domherr, and Treppchen vineyards, sites of limited size gifted with particularly favourable microclimates, planted with the finest German grape Riesling, and blessed with a few winemakers who knew how to get the best out of both. Would consumers know or notice that Piesporter Michelsberg was not one of those privileged sites, but a conglomerate of 1,375 hectares (3,400 acres) of different vineyards? Would they realise that most of this wine did not even come from Piesport, but villages very few had ever heard of? And would they be aware that most of this wine did not contain a single drop of Riesling?

COMPARISON OF GERMAN HARVESTS 2015–2018

Production varies enormously in Germany – from as much as 15.4 million hectolitres (171 million cases) in 1982 to as little as 5.2 million hectolitres (58 million cases) in 1985 – as does the spread of wines across the range of quality categories, shown by the table below.

	2015	2016	2017	2018
QUALITY CATEGORY				
Tafelwein to Landwein	3.5%	4%	2.2%	4%
QbA	49.9%	58.4%	59.4%	59%
QmP	46.6%	37.6%	38.2%	37%
SIZE OF HARVEST				
Millions of hectolitres	8.9	9.1	7.5	10.5
Millions of cases	98.9	100.8	83.3	116.6
Hectares under vine	99,906	99,702	100,039	102,873
Hectolitres per hectare	88.8	90.7	75	103

What followed this piece of legislation in the 1970s and 1980s were the dark ages of German wine. One of the biggest problems was overproduction. A reform of the wine law in 1994 authorised regional governments to introduce limits for yields, but otherwise did very little to address any of the other contentious issues.

CHANGING TASTES AND ATTITUDES
If until the 1980s the export markets could be relied on as loyal consumers of the bulk of cheap and mild mass-produced wines such as Niersteiner Gutes Domtal, Liebfraumilch, and Piesporter Michelsberg, this situation changed dramatically when countries like Australia, New Zealand, and Italy expanded their production and entered international competition with the trendy varieties Chardonnay, Sauvignon Blanc, and Pinot Grigio, meeting the increasing global taste for fresher and drier wines.

PIESPORT ON THE MOSEL
On the right bank of the river, Piespont has long produced fine Rieslings, such as Piesporter Goldtröpfchen. The municipality's name was co-opted by winemakers for Piesporter Michelsberg wines, which are made from grapes grown in the area and not in the first-class vineyards of Piesport itself.

But it was not only on the export markets, where consumers' preference had changed towards drier wines, but on the home market too. In 1975 only 2 per cent of the total of German quality wine was produced as dry wine, with 9 grams per litre of residual sugar the legally defined upper limit for the *trocken* designation. With the country's economy doing well, eating out and drinking wine with dinner became more commonplace among Germans, and local growers did not want to miss out to foreign competition in this important sector of the market. Within 25 years the share of *trocken* rose to one-third of the country's total production, although for many producers the initial switch from sweeter to drier wines was not without its problems. Georg Breuer, a grower from Rüdesheim in the Rheingau, was one of the first to recognise that it would be necessary to implement some strict quality criteria for the production of top-quality dry wines. He became one of the founders of Charta, a private growers' association, which in 1987 drew up a register of the best vineyards of the Rheingau region suited to producing premium dry wines from Riesling and Spätburgunder grapes. These wines were identified by the title *Erstes Gewächs (premier cru)* on the label.

In 2000 the Federal German wine law introduced the designations of "Classic" for dry wines of superior quality and typical of their region and "Selection" for the best dry wines of the vintage, but whereas these were well received by cooperatives and larger merchant-bottlers, their acceptance among individual growers was not enthusiastic. Today neither designation is of any real relevance.

VDP CLASSIFICATION

Taking their inspiration from the Rheingau Charta, members of the *Vereinigung Deutscher Prädikatsweingüter* (VDP), a private association of German premium wine growers, originally founded in 1910 to sell their best wines together at auction, in 2001 passed a resolution for a three-tier classification of the wines of its member estates with the three categories of *Gutswein* ("estate wines of regional origin"), *Ortswein* ("cru villages") and *Erste Lage* ("first growth"). A set of production rules was established that determined the quality criteria for each category, with requirements getting more stringent according to rise in status. Some of the most important matters addressed by these regulations concerned permitted grape varieties, yield restrictions, and viticultural and oenological practices. It was left to the individual regional chapters of the VDP to determine which of their members' vineyards would be included in the roll of *Erste Lagen*. Dry wines from *Erste Lagen* were to be marketed under the designation *Großes Gewächs (grand cru)*. No provision was made at that time for the status of fruity or noble sweet wines from the best vineyards, an issue addressed later in 2006, when they were also granted the right to show the designation of *Erste Lage* on the label in conjunction with their respective predicate, such as *Spätlese, Auslese* etc.

In 2012 VDP members unanimously passed a resolution to further improve their classification system, bringing it even more in line with the Burgundy model. It allowed the regional VDP associations to split the top

QUALITY REQUIREMENTS AND HARVEST PERCENTAGES

Before climate change Germany used to be notorious for the variation of its vintages, and therefore, as a form of insurance, much of the crop was harvested early. This secured a minimum income from *Deutscher Tafelwein* and QbA wines, with those grapes that remained on the vine providing a possibility of higher-quality wines from late harvests. The degree of ripeness attained is expressed in degrees *Oechsle* and the minimum level for each QmP can vary according to the region and grape variety in question. Each degree of *Oechsle* is the equivalent of 2 to 2.5 grams per litre of sugar. The exact amount of sugar varies according to the level of unfermentable extract also present. (For a survey of each region, see the individual Quality Requirements and Quality Wine Production charts in the regional sections.)

DECODING THE AP NUMBER

All *Qualitätsweine*, including *Deutscher Sekt,* must carry an AP number (*Amtliche Prüfnummer*). This proves that a wine has undergone and passed various tasting and analytical tests and its origin has been established to the board's satisfaction. Each time a producer applies for an AP number, specially sealed samples of the approved wine are kept by both the board and the producer. Should there be a fault or a fraud enquiry, these samples are analysed and checked against a sample of the product on the market that has given rise to complaint or investigation. This sounds foolproof, and it is indeed more stringent than the EU's Lot Number, which has been in operation since 1989. But like all systems it can be abused, which in this case would simply mean printing fictitious AP numbers on the label, although if noticed the penalty for this would be imprisonment, plus a ban from operating for the company, and a large fine. Understanding the AP number can be very useful. If the wine is a non-vintage QbA, it can give you some idea of how long it has been in bottle.

2016 DÖNNHOFF RIESLING *TROCKEN QUALITÄTSWEIN*

7 = Examination Board number

753 = number of the commune where the wine was bottled

010 = the bottler's registered number

04 = the bottler's application number

17 = the year in which the bottler made the application

AP Nr 7 753 010 04 17

The communes' and bottlers' numbers run into thousands and are of the least significance to the consumer. It is the last two sets of numbers that provide the most useful information. In this instance, the 04 and 17 reveal that this is Dönnhoff's 4th application made in 2017. The application numbers are sequential, but do not necessarily apply to the same wine, ie the previous application made by Dönnhoff in 2016 need not necessarily apply to its 2016 Riesling *trocken.* You will find that some references state that the application number actually refers to a specific cask the wine came from (which in this case would be *fuder* 04), but although this is so in some cases (Fritz Haag or Von Schubert's Maximin Grünhaus, for example), this is only because the producer in question will number his cask to coincide with the application number.

What the last two sets of digits tell us is that if they are the same, then the wine is *exactly* the same. If you find two bottles of 2016 Dönnhoff Riesling *trocken* both bearing the AP number 7 753 010 04 17, then you know beyond any doubt that the contents of both bottles swam with each other before being bottled together. On the other hand, if you find two bottles of 2016 Dönnhoff Riesling *trocken* bearing the AP number in which either or both of the last two sets of digits are different, then you know that the contents are, at the very least, subtly different. They might simply be bottled on a different date or from a different cask, but when the designation is as broad as Riesling *trocken Qualitätswein*, the grapes could have been grown miles apart, picked on different days, vinified in different tanks, with radically different sugar, alcohol, and acidity readings.

HAND-HARVESTING GRAPES
A vineyard worker unloads his latest haul. The VDP mandates that grapes for certain German wines, such as *Auslese* and *Sekt,* must be hand-picked.

tier of premium vineyards into *Große Lagen* (*grands crus*) and *Erste Lagen* (*premiers crus*), factually creating a four-tier quality ranking. Only dry wines from *Große Lagen* would now be entitled to use the designation *Großes Gewächs,* showing on the label just the name of the vineyard, but not that of the commune anymore. Dry wines from *Erste Lagen* would give the name of the vineyard, the commune it was located in, and the attribute of *Qualitätswein trocken.*

In 2018 this four-tier system was approved for *Sekt* too: all VDP *Sekt* must be hand-picked from VDP sites and bottle fermented, as well as 15-months minimum lees ageing for *Gutssekt* and *Ortssekt* and 36 months for *Erste Lage* and *Große Lage.*

It will probably take the export markets some time to come to terms with this system, but the intentions were laudable. It also has to be remembered that this is an internal classification of a private growers' association, although it is benignly tolerated by the official authorities and has been imitated by quite a few other producers' alliances. To identify the original, one always needs to look for the VDP emblem on the capsule or label.

THE UNSTOPPABLE RISE OF *TROCKEN*, RED AND WHITE

If in 1975 the proportion of wines legally defined as *trocken* was only 2 per cent of the country's overall production; by 2018 this share had risen to a staggering 48 per cent. Another 22 per cent fell into the category of halb-*trocken,* taking the output of dry and semi-dry wines from German vineyards for the 2018 vintage to 70 per cent. This dramatic change of style over the last 40 years is not down to one single cause, but the result of a several factors. In the 1970s climate change had not taken any real effect yet, meaning that grapes in Germany rarely ripened to a degree that would have allowed a regular production of dry quality wine. Acidity levels

GERMAN LABEL LANGUAGE

Abfüllung Bottling

Auslese A category of German QmP wine that is very sweet, made from late-picked grapes, and which may also contain some botrytised grapes, *Auslese* is one step of sweetness above *Spätlese* but one below *Beerenauslese,* and most of the best lean upwards rather than downwards

Badisch Rotgold Sometimes also called *Badischer Rotling,* this is a speciality rosé wine made from a blend of Grauburgunder and Spätburgunder; *Badischer Rotgold* must be of at least *Qualitätswein* level, and can be produced only in Baden.

Barriquewein Fermented or matured in new oak casks

Beerenauslese A category of German QmP wine that comes above *Auslese* in sweetness, but beneath *Trockenbeerenauslese* and is made from botrytised grapes. It has more finesse and elegance than any other intensely sweet wine, with the possible exception of *Eiswein.*

Bereich The 13 German wine regions are divided into 41 *Bereiche* (subregions). If the wine carries the appellation of a *Bereich,* it will simply state this – "*Bereich* Burg Cochem", for example.

Deutscher German

Deutscher Qualitäts-schaumwein or **Deutscher Sekt bestimmter Anbaugebiete** A sparkling wine made by any method (but probably *cuve close*) from 100 per cent German

grapes grown in one specified region, although it may indicate an even smaller area of origin if at least 75 per cent of the grapes come from that named area.

Deutscher Sekt or **Deutscher Qualitätsschaumwein** A sparkling wine made by any method (but probably *cuve close*) from 100 per cent German grapes. It may indicate a maximum of two grape names and should be at least 10 months old when sold.

Deutscher Wein Lowest tier of the legal classification of German wines without appellation of origin, which replaces the former designation of *Deutscher Tafelwein.* Grapes must come from Germany.

Deutsches Weinsiegel An official seal of quality awarded to wines that receive a higher score than the minimum required for an AP number. *Deutsche Weinsiegels* use a colour-coding system: *trocken* wines bear a bright yellow seal; *halbtrocken* a lime green seal, and sweeter wines a red seal.

Edelfäule The German term for "noble rot"

Einzellage A single-vineyard wine area; the smallest geographical unit allowed under German wine law

Eiswein A rare wine resulting from the tradition of leaving grapes on the vine until temperatures drop below -7°C (19.4°F). The grapes are frozen by frost or snow, and

then harvested and pressed while frozen. This is done because only the ice freezes and, as it rises to the top of the vat, it can be separated from the concentrated juice that produces a wine with a unique balance of sweetness, acidity, and extract.

Erstes Gewächs Designation introduced in the Rheingau in 1999 for premium category of dry wines from Riesling or Spätburgunder. Today only used by growers of the region, who are not members of the VDP.

Erste Lage Second tier of the VDP classification, equivalent to Burgundy Premier Cru

Erzeugerabfüllung At one time this was considered to mean estate-bottled, but so many cooperatives use it (because, they claim, their members' estates are the cooperative's estates) that it now means no more than producer-bottled, thus some estates now use *Gutsabfüllung.*

Flaschengärung A bottle-fermented *Sekt* that is not necessarily made by the traditional method

Flaschengärung, klassische or **traditionelle** A *Sekt* made by the traditional method

Feinherb This is a traditional term for off-dry wines, which is generally used for wines with a little more residual sugar than *halbtrocken,* although there is no legally prescribed limit for residual sugar.

Fuder A large oval cask with a capacity of 1,000 litres, more prevalent in the Mosel than the Rhine

Große Lage Top tier of the VDP classification and reserved for wines from the most highly rated vineyards with the traditional *Prädikate* of *Kabinett, Spätlese, Auslese, Beerenauslese, Trockenbeerenauslese,* or *Eiswein.* All wines with the designation *Große* or *Erste Lage* must be hand-harvested to a maximum yield of 50hl/ha.

Großes Gewächs The designation for a dry wine from a *Große Lage,* which falls into the legally delimited *trocken* category. The term was coined by the VDP but was not registered and is therefore also used by other growers. All VDP *Große Gewächse* are marketed as *Qualitätswein trocken,* because the association wants to reserve *Prädikate* like *Spätlese* or *Auslese* for the sweet wine styles they were traditionally associated with. *Große Gewächse* must have a minimum must weight corresponding to that required for *Spätlese,* be subjected to a VDP taste test, and cannot be released until September of the year following that of the harvest (two years for red wines).

Großlage A "collective site", or a large area under which name cheap, bulk-blended wines are sold.

Gutsabfüllung This term literally means "bottled on the property" and has been taken up by those who feel that *Erzeugerabfüllung* has been so

tended to be too high, particularly for Riesling, and many of the early attempts to meet the rising demand for dry wines were quite sharp, thin, and sour. But not only did summers begin to become noticeably hotter and drier towards the late 1980s, but growers had also learned from their early mistakes, using only fully ripe and healthy grapes without botrytis for their *trocken* production. This resulted in wines with more body, greater extract, and less acidity – more harmonious altogether.

Another significant reason for the rise of *trocken* was the ever-increasing demand for red wines. In 1980 the share of dark grapes in German vineyards was just over 11 per cent, yet by 1990 it had risen to 19 per cent and five years later it reached 26 per cent. By the turn of the century almost 37 per cent of the country's total area under vine was planted with red varieties. On the one hand this was due to the increase in popularity of dry wines from the noble Spätburgunder grape (Pinot Noir), taking its share from 4.5 per cent in 1985 to 11.5 per cent within 20 years, but on the other hand it was due to the rise of the not-quite-so-noble but useful newcomer Dornfelder from 0 per cent in 1985 to 8 per cent in 2005.

But while within minor fluctuations Riesling's share of German vineyards has remained steady at around 23 per cent for the last 50 years, albeit with a major change of direction from sweet to dry, the real and most recent winners of the *trocken* boom turned out to be the Pinot varieties, not just Spätburgunder, but also the two white protagonists Weißburgunder (Pinot Blanc), from 1 per cent in 1990 to 5.2 per cent in 2017, and, for the same period of time, Grauburgunder (Pinot Gris), up from 2.4 to 6.2 per cent. Where there are winners, there have to be losers, and these are Silvaner, Müller-Thurgau, and Kerner for white and Portugieser for red.

Although global warming, the switch from mild to dry wines, and the growing taste for red have all had an influence on the transformation of

QUALITY STRUCTURE OVERVIEW

This is a simplistic overview because each category varies according to the grape variety and its area of origin. More detailed analyses are given under each region (*see* Quality Requirements and Harvest Percentages, p439).

QUALITY CATEGORY	MINIMUM OECHSLE	MINIMUM ALCOHOL / ORIGIN
* Deutscher Wein	44-50°	8.5% / 100% German grapes)
* Landwein	47-55°	8.5% / 85% grapes from named Landwein region
* Qualitätswein	55-72°	7% / 100% grapes from named region

QUALITÄTSWEIN MIT PRÄDIKAT (QMP)

QUALITY CATEGORY	MINIMUM OECHSLE	MINIMUM ALCOHOL
Kabinett	67-82°	7%
Spätlese	76-90°	7%
Auslese	83-100°	7%
Beerenauslese	110-128°	5.5%
Eiswein	110-128°	5.5%
Trockenbeerenauslese (TBA)	150-154°	5.5%

* Chaptalisation is allowed and will be necessary if the wine has a potential alcoholic strength of less than 8.5 per cent.

---

debased by the cooperatives that it no longer stands for a wine that is truly estate-bottled. *Gutsabfüllung* can only be used by a winemaker who holds a diploma in oenology. This means any naturally gifted winemaker who has not sat the examination, but owns an estate and makes brilliant wines, has no legal way of indicating his or her wine is an authentic product that has been grown on, fermented at, and bottled on an individual wine estate.

Gutswein and **Ortswein** These terms ("estate wine" and "commune wine") are applicable to wines observing the VDP's general standards, which include viticultural methods adhering to strict controls, regular inspection of vineyards, higher must weights than those prescribed by law, and a maximum yield of 75hl/ha.

Halbtrocken Literally "half-dry", this wine does not contain more than 18 grams per litre of residual sugar and 10 grams per litre of acid

Jahrgang Vintage year.

Klassifizierter Lagenwein This term ("wine from a classified site") is restricted to wines from classified vineyards that impart site-specific traits, with a maximum yield of 65hl/ha.

Kabinett The predicate at the end of this wine's name (it can come either before or after the grape variety) reveals that it is the lightest of *Qualitätsweine*, traditionally moderately sweet, but can also be vinified as *trocken, halbtrocken,* or *feinherb*.

Landwein This Teutonic version of the French *vin de pays* is not only a failure in commercial terms (unlike its Gallic compatriot), but it is also rather strange in that it may, with the exception of Landwein Rhein, be produced in only *trocken* and *halbtrocken* styles.

Lieblich Technically medium-sweet, although nearer to the French moelleux, this wine may have up to 45 grams per litre of residual sugar.

Ortswein *See* Gutswein.

Qualitätswein mit Prädikat (QmP) The wine comes from the highest quality category of German wine and will carry one of the predicates that range from *Kabinett* up to *Trockenbeerenauslese*.

Perlwein Cheap, semi-sparkling wines made by carbonating a still wine; they are mostly white, but may be red or rosé.

QbA (Qualitätswein bestimmter Anbaugebiete) This is the theoretical equivalent of the French AOP.

QmP (Qualitätswein mit Prädikat) This literally means a "quality wine with predication", and is a term used for any German wine above QbA, from *Kabinett* upwards. The predication carried by a QmP wine depends upon the level of ripeness of the grapes used in the wine.

Qualitätsschaumwein A "quality sparkling wine" can be produced by any member state of the EU, but the term should be qualified by the

country of origin (of the wine). Only *Deutscher Qualitätsschaumwein* will necessarily be from Germany.

Rebe Grape variety

Restsüsse Residual sugar

Roséwein or **Roseeewein** Rosé

Rotling A rosé wine that may be made from black grapes or a mixture of black and white, which must be fermented together. This designation must be indicated on the label of any category of Tafelwein up to and including Landwein and must also be featured on the label of QbA wine, although it is optional for QmP.

Rotwein Red wine. This designation must be indicated on the label of any category of *Tafelwein* up to and including *Landwein* and must also be featured on the label of QbA wine, although it is optional for QmP.

Schaumwein If there is no further qualification referred to, such as *Qualitätsschaumwein*, this term indicates the cheapest form of sparkling wine, probably a carbonated blend of wines from various EU countries.

Schillerwein A style of wine produced in Württemberg that is the same as a Rotling

Sekt Sparkling wine

Spätlese A QmP wine that is traditionally one level of sweetness above *Kabinett*, but one below Auslese. Literally translated, it means "late harvest", but now the grapes for this designation can be picked at

any time, as long as they have the legally prescribed sugar content. The classic style is sweet, but with most or all of the sugar fermented, *Spätlesen* can also be *halbtrocken* or *trocken*.

Süß A sweet wine with in excess of 45 grams per litre of residual sugar

Tafelwein A lowly table wine or *vin de table* used for blends from countries of the EU

Trocken Literally "dry", a *trocken* wine must not contain more than 4 grams per litre of residual sugar, although up to 9 grams per litre is permitted if the acidity is at least 2 grams higher per litre than the residual sugar.

Trockenbeerenauslese A wine produced from individually picked, botrytised grapes that have been left on the vine to shrivel. It can be anything from deep golden to amber in colour, intensely sweet, viscous, and very complex.

VDP (Verband Deutscher Prädikats- und Qualitätsweingüter) A private association of top wine estates

Weißherbs A single-variety rosé produced from black grapes only, the variety of which must be indicated

Weißwein White wine. This designation must be indicated on the label of any category of *Tafelwein* up to and including *Landwein*. It may also be featured on the label of QbA and higher-quality wines, but this is not mandatory.

SÜßRESERVE

Süßreserve is sterilised grape juice that may be added to a wine after fermentation to increase its level of sweetness and improve the balance of a wine with high acidity or alcohol. The alcohol content of *Süßreserve* cannot exceed 8 grams per litre, which is approximately 1 per cent alcohol by volume. Its use in Germany is allowed for all categories from *Deutscher Wein* to *Prädikatsweine*. The origin of Süßreserve must, in essence, be the same as the wine to which it is added. Its quality, or degree of ripeness, should by law be at least the equivalent of the wine itself. Under German law, no more than 15 per cent of the wine's final volume may be *Süßreserve*. The practice of sweetening wines with *Süßreserve* has decreased significantly. Most producers who wish to adjust the residual sugar content of a wine prefer to arrest fermentation before all the sugars have been converted to alcohol. Adding Süßreserve also changes the composition of sugars. During fermentation glucose is fermented more quickly than fructose. A halted fermentation will thus produce a wine with residual sugar composed mainly of fructose, while the addition of Süßreserve will yield a wine with a residual sweetness composed of both glucose and fructose.

German wine from bland and stuffy to sexy and exciting, the arguably most important contribution has come from a new generation of well-educated and widely travelled growers. It was not continuation of tradition that fuelled their interest in wine, but the realisation that their vineyards had the potential to craft wines of outstanding quality and unique character. Chardonnay is produced all over the world, but Riesling, just like Pinot Noir, is pickier, and – despite climate change – the relatively moderate climate of the German regions offers ideal conditions to make a world class wine from this variety that cannot be simulated elsewhere. Whether it be in established associations like the VDP and the Bernkasteler Ring or in newly forged alliances such as *Fünf Freunde,* Generation Riesling, and *Maxime Rheinhessen,* it was the discussion of new ideas and exchange of personal experiences among enthusiastic friends that has been more responsible than any other factor for the revival of German wine.

ROTKÄPPCHEN *SEKT*
Bottles of *Halbtrocken Sekt* for sale from famous East German winery Rotkäppchen ("Little Red Riding Hood"). The stiff competition amongst the producers of *Sekt* keeps the prices of these sparkling wines at rock-bottom levels.

A NEW ORDER IN 2020

As the EU wine reforms of 2009 are beginning to make an impact on the legislation of the individual member states, Germany is looking to adopt the intended division into the just two principal categories of wines with (*Landwein, Qualitätswein*) and without (*Deutscher Wein, Europäischer Gemeinschaftswein*) appellation of origin, while at the same integrating its traditional model of designations including *Prädikate* into the new system. This goal is to be achieved by a new German wine law scheduled to be passed in 2020. Contrary to other European countries, wines without appellation of origin will not be allowed to state grape varieties. Producers of wines with an appellation of origin may also once again indicate the provenance from physio-geographically privileged parcels with a historical track record, something the law of 1971 had prohibited.

The greatest challenge for the ministry responsible for the new legislation will be to accommodate the differing interests of mass producers, cooperatives, and individual quality winegrowers. For wines with an appellation of origin there appears to be agreement about a quality pyramid similar to that of the VDP with a three-tier hierarchy of region, commune, and individual sites. A consensus exists that requirements regarding permitted grape varieties, yields, and ripeness of grapes should be more stringently defined in line with rising status, but the issue of *Großlage* remains a contentious one. Although an alliance of quality-orientated growers insists that this designation makes false pretences of origin, the Federal association of wine merchants and traders insists that that they have to be retained to enable them to market these wines with a declaration of regional origin. The truth of Müller-Thurgau Mosel just wouldn't sell as well as the phantasy of Piesporter Michelsberg.

SEKT: GERMANY'S SPARKLING WINE

Germany's *Sekt,* or sparkling wine, industry is very important, producing some 260 million litres in 2018, more than the total output of Champagne. This is almost entirely made by *cuve close,* and over 85 per cent is not *Deutscher Sekt* at all, but *Sekt* plain and simple, made from imported base wines. Most of it is consumed domestically, and with 4 litres per head Germans hold the world lead in the consumption of sparkling wine, with very little left for export. With Henckel-Freixenet, Schloß Wachenheim, and Rotkäppchen-Mumm, Germany is home of some of the largest sparkling wine operations in the world, and fierce competition in the bargain basement section has seen prices drop as low as three euros per bottle, one of which goes to the taxman. *Deutscher Sekt,* made by German growers from German grapes, accounts for less than 2 per cent of production. The best of those made by the traditional method with second fermentation in bottle are worth seeking out, as top growers are increasingly using Chardonnay and the Pinot varieties in addition to Riesling. At the opposite end of the market the gigantic Rotkäppchen-Mumm Winery in Freyburg produces more than 120 million bottles a year.

LEVELS OF *SEKT* SWEETNESS

Sekt can be made with varying levels of residual sweetness. These designations of styles are defined by German wine law.

NATURHERB or BRUT NATURE	0 to 3 grams per litre
EXTRA HERB or EXTRA BRUT	0 to 6 grams per litre
HERB OR BRUT	0 to 12 grams per litre
EXTRA TROCKEN or EXTRA DRY	12 to 17 grams per litre
TROCKEN*	17 to 32 grams per litre
HALBTROCKEN*	32 to 50 grams per litre
MILD, SÜSS, DRUX, DOUX, or SWEET	more than 50 grams per litre

* Not to be confused with still-wine limits, which are no more than 9 grams per litre for trocken and 18 grams per litre for *halbtrocken*.

HOW TO USE THE "APPELLATIONS OF . . ." SECTIONS

The entries that follow for each of Germany's wine regions are set out in a logical order. Regions, or *Anbaugebiete*, are subdivided into Bereiche, which were created to be able to market larger volumes of wine under a unitary name. This is followed by the *Großlagen* within that particular Bereich. For our purposes we use the *Großlagen* solely as superordinate geographical districts within which the best villages are listed where applicable and, within each of these their finest *Einzellagen*, or vineyards. In Germany, vineyards are rarely owned only by one estate – as they are in Bordeaux, for example. It is therefore important to be aware of not only which are the top vineyards, but also who makes the best wines there, which is why we recommend growers and estates within each of the finest vineyards. The term "village" covers both *Gemeinden* – communes, which may be anything from a tiny village to a large, bustling town – and *Ortsteile*, which are villages absorbed within the suburbs of a larger Gemeinde. The term *"Großlagenfrei"* refers to villages and *Einzellagen* within one *Bereich* that are not part of any of its *Großlagen*. Under the term "Non-*Einzellagen*" we list outstanding wines, which are *cuvées* from a number of *Einzellagen*.

All *Großlagen* obviously comprise a varying number of individual sites, but if none are named under a specific *Großlage* heading, this means that they may produce perfectly drinkable wines, but not of outstanding quality in our consideration.

All styles and descriptions of wines refer to Riesling, unless otherwise stated. The following sample entry is given from the Franken region.

BEREICH	BEREICH STEIGERWALD
GROßLAGE	GROßLAGE SCHLOßBERG
BEST VILLAGE	RÖDELSEE
BEST VINEYARD AND/OR BEST GROWERS	**Vineyard:** ✔ *Küchenmeister* • **Grower:** *Paul Weltner Würzburg* **Vineyard:** *Schwanleite* • **Grower:** *Johann Ruck*

THE WINE STYLES OF

GERMANY

Note: From *Wein* to *Trockenbeerenauslese*, the categories below are in ascending order of quality, and not in alphabetical order, after which different styles are listed. Only the broadest character can be given for each category, since so much depends upon the grape variety. The grape variety can even determine the colour of a wine, although some 90 per cent of the production is white. Most grape varieties likely to be encountered, including all the most important crosses, are included in the ABC of Grape Varieties (*see* pp 39–59).

The area of origin also has a strong influence on the type of wine a particular grape variety will produce. A winemaker can either make the most of what nature brings, or fail. For full details on these two aspects, see the regional chapters in which styles are described and growers recommended.

WEIN AUS DER EUROPÄISCHEN GEMEINSCHAFT

This may be a blend from different EU countries or wine made in one member country from grapes harvested in another. Known as "Euroblends", these products are sometimes dressed up to look like German wines, but the wine beneath the Gothic writing and German language usually turns out to be an Italian or multi-state blend.

TAFELWEIN

This designation was abolished in 2009.

DEUTSCHER WEIN

In 2009 the designation of *Deutscher Tafelwein* was abolished and replaced by *Deutscher Wein*. Wines marketed under this name have to come exclusively from German vineyards and can only be made from officially permitted grape varieties. Vintage and variety may be stated on the label. Alcohol content must be a minimum of 8.5 per cent, with a maximum of 15 per cent. As *Deutscher Tafelwein* falls into the European category of wines without declaration of origin, the former specification of particular *Tafelwein* regions has been abolished. This is the lowest grade of pure German wine, which, depending on the vintage accounts for 1 to 4 per cent of the country's production.

🍷 *Upon purchase*

LANDWEIN

Landwein is a *Deutscher Wein* from a specific region and the label must contain both terms. The major difference is that the former must be made either as *trocken*, with up to 9 grams per litre residual sugar, or as *halbtrocken*, with a maximum of 18 grams per litre, except for the region Landwein Rhein, for which the production of sweet wines is also permitted. *Landwein* was introduced in 1982 to parallel the French *vin de pays* system, although there are significant differences. The 153 or so *vins de pays* represent a group of wines that aspire to VDQS and, theoretically, AOC status, and it has become one of the most exciting categories of wine in the world (*see* pp 331–341). The German category is not a transitional one, it consists of 26 fixed areas that often overlap with quality wine regions, but with less stringent requirements than those for the production of *Qualitätsweine*.

It would be fair to say that the original concept behind the introduction of the *Landwein* category was to create something similar in quality and status to the French *vins de pays*, but the idea was ill-conceived, as the quality wine structure of Germany is totally different and with the country's total output about a fifth of French production the necessary quantities to make this category a significant one would never be available.

On the other hand, a growing number of growers in the Baden region are using the category to their advantage in a similar way that ambitious and innovative producers in Italy once used the lowly denomination of *vino da tavola* to create exciting and unconventional wines, which the restrictions and requirements of their regional DOC regulations would have not permitted them to make. Having once been refused an official *Qualitätswein* approval for one of his wines because it was judged as untypical for the category it was entered for, Baden grower Hanspeter Ziereisen of Efringen-Kirchen decided to declare all his wines as Badischer Landwein. He has since achieved cult status and his wines are widely sought after on the international wine scene.

🍷 *The best will age well.*

QUALITÄTSWEIN BESTIMMTER ANBAUGEBIETE or QBA

A QbA is literally a quality wine from one of the 13 specified regions. These wines may be (and often are) chaptalised to increase the alcohol content and can be sweetened with *Süßreserve*, although the majority of them are nowadays produced as *trocken* or *halbtrocken*. The legal minimum potential alcoholic strength of a QbA is merely 7.0 per cent, so it can be understood why the alcohol level is not increased for its own sake, but to give the wine a reasonable shelf-life. Traditionally this category has always included generic sweet and semi-sweet wines such as Liebfraumilch and the vast majority of Niersteiner Gutes Domtal and Piesporter Michelsberg. But today it is also used by premium producers at the top end of the quality spectrum of dry wines from individual vineyards, as they dispense with designations such as *Spätlese* or *Auslese* for for their *trocken* wines. These *Prädikate*, they feel, should be used to describe the classic fruity and noble sweet styles in their traditional expression.

🍷 *1–3 years (up to 20 years for exceptional dry wines)*

QUALITÄTSWEIN MIT PRÄDIKAT or QMP

This means "a quality wine affirmed (predicated) by ripeness" and covers the categories *Kabinett, Spätlese, Auslese, Beerenauslese, Eiswein,* and *Trockenbeerenauslese*. The grower must give the authorities prior notice of his intention to harvest for any QmP wine and, whereas a QbA can be blended from grapes gathered from all four corners of an *Anbaugebiet*, providing it carries only the name of that region, the origin of a QmP must be a geographical unit of no greater size than a *Bereich*. A QbA can be chaptalised, but a QmP cannot. It is, however, permissible to add *Süßreserve*, although many growers maintain that it is not the traditional method and claim that sweetness is best achieved by stopping the fermentation before all the natural grape sugars have been converted to alcohol. (*See* the various QmP entries.)

KABINETT QMP

The term *"Kabinett"* once referred to wines that were stored for their rare and exceptional qualities, much as "Reserve" is commonly used today. In this context it originated at Kloster Eberbach in the Rheingau in the early 18th century. The first documented existence of this word appeared as *"Cabernedt"* on an invoice from the Elville master cooper Ferdinand Ritte to the Abbot of Eberbach in 1730. Just six years later, another bill in Ritter's hand refers to the *"Cabinet-Keller"*. (The meaning of *cabinet* as a treasured store found its way into the French language as early as 1547 and is found in German literature in 1677.) *Kabinett* is the first of the predicates in the *Oechsle* scale. Its grapes must reach an *Oechsle* level of 67° to 82°, the exact minimum varying according to the grape variety concerned and its geographical origin. With no chaptalisation, this means that the wine has a minimum potential alcoholic strength of between 8.6 and 11.4 per cent by volume. Because many producers resolutely refuse to bolster the wine with *Süßreserve*, this makes their *Kabinett* the lightest and, for some, the purest of German wine styles.

🍷 *2–5 years (up to 10 years in exceptional cases)*

♛ **Riesling: Mittelrhein:** *Toni Jost* • **Rheingau:** *Leitz, Künstler* • **Mosel:** *Falkenstein, A.J.Adam, Fritz Haag* • **Nahe:** *Diel, Dönnhoff, Kruger-Rumpf* • **Rheinhessen:** *Groebe, Wagner-Stempel*

SPÄTLESE QMP

Spätlese is a quality designation in the top tier of German wine classification *Qualitätswein mit Prädikat* which literally means "late harvest", but practically describes a wine from grapes picked at a ripeness of around 85° *Oechsle*, exact minimum requirements depending on the respective region. The traditional light racy sweet style of *Spätlese* is produced by cooling down the fermenting must and filtering out the remaining active yeasts when the grower feels that the right balance between alcohol and residual sugar has been achieved. As the demand for dry wines began to rise during the 1990s, an increasing number of producers let the yeasts convert most or all of the grape sugar to make wines that they could market as *Spätlese trocken*. There is no doubt that this led to some confusion among consumers, as wines declared as *Spätlese* could now be sweet, medium dry or dry. The German association of premium growers VDP has decreed that its members should use the designation only for the classic sweet style and not for *trocken* or *halbtrocken* wines.

🍷 *3–8 years (up to 25 in exceptional cases)*

♛ **Mittelrhein:** *Toni Jost, Weingart* • **Mosel:** *J J Prüm, Fritz Haag, Grans-Fassian, Molitor* • **Nahe:** *Schäfer-Fröhlich, Dönnhoff, Emrich-Schönleber* • **Rheingau:** *Weil, Spreitzer* • **Rheinhessen:** *Gunderloch, Keller, Groebe* • **Pfalz:** *A Christmann, Müller-Catoir* • **Baden:** *Andreas Laible* • *Württemberg: Karl Haidle*

BLANKETED GRAPES AWAIT HARVESTING FOR *EISWEIN*
Healthy grapes are allowed to freeze and then rushed to a winery after harvesting (ideally by hand) for the production of *Eiswein*, a sweet dessert wine. In Germany, Riesling grapes are most often used to make the finest *Eisweins*. In Canada, where it is called icewine or ice wine, white Vidals are often used, as well as black Cabernet Franc, amongst others. White varieties yield a pale yellow or light gold wine that deepens to richer golden tones as it ages. Red and black grapes yield a pink wine.

AUSLESE QMP

This predicated wine is made from bunches of grapes left on the vines after the *Spätlese* harvest and as such is truly late-harvested. The regulations state that bunches of fully ripe to very ripe grapes free from disease or damage must be selected for this wine – how this can be achieved by machine-harvesting, which has been permitted for *Auslese* since the new German wine laws of 1994, is beyond the imagination of every quality-conscious German winemaker I have spoken to.

Auslese must also possess an *Oechsle* reading of 83° to 100°, the exact minimum varying according to the grape variety concerned and its geographical origin. With no chaptalisation, this means that the wine has a minimum potential strength of between 11.1 and 14.5 per cent by volume.

Traditionally this rich and sweet wine is made in exceptional vintages only. There may be some hint of *Edelfäule* (botrytis), especially if the wine comes from a top estate that has a policy of under-declaring its wines, and thus an *Auslese* might be borderline *Beerenauslese*, but even without *Edelfäule*, an *Auslese* is capable of some considerable complexity. It is possible to find totally dry *Auslese* wine and this may or may not be labelled *Auslese trocken*, depending on the whim of the winemaker. Ideally a different designation should be found for dry wines of *Auslese* ripeness to distinguish them from their sweet counterparts.

🍷 *5–20 years*

♛ **Sweet style: Franken:** *Horst Sauer* • **Rheingau:** *Peter Jakob Kühn, Wegeler, Weil* • **Mittelrhein:** *Weingart, Matthias Müller* • **Mosel:** *Thanisch-Erben Thanisch, Nik Weis, Fritz Haag, J J Prüm, Egon Müller, Molitor, Selbach-Oster, Willi Schaefer, Zilliken Nahe: Dönnhoff, Emrich-Schönleber, Gut Hermannsberg* • **Pfalz:** *Neiss, PhilippKuhn, Müller-Catoir* • **Rheinhessen:** *Gunderloch, Keller, Wagner-Stempel, Wittmann*

BEERENAUSLESE QMP

A very rare wine made only in truly exceptional circumstances from overripe grapes that have been affected by *Edelfäule*. According to the regulations, each berry should be shrivelled and be individually selected on a grape-by-grape basis. They must achieve an *Oechsle* level of 110 to 128°, the exact minimum varying according to the grape variety concerned and its geographical origin. With no chaptalisation, this means that the wine has a minimum potential strength of between 15.3 and 18.1 per cent by volume, but only 5.5 per cent need actually be alcohol, with residual sugar accounting for the rest.

These intensely sweet, full-bodied wines are remarkably complex and elegant. I actually prefer *Beerenauslese* to the technically superior *Trockenbeerenauslese* – it is much easier to drink and be delighted by the former, whereas the *Trockenbeerenauslese* requires concentration and appraisal.

🍷 *10–35 years (up to 50 years in exceptional cases)*

♛ **Baden:** *Andreas Laible* • **Mosel:** *Molitor, Fritz Haag, J J Prüm, Max.Ferd, Richter, Schloß Lieser, Zilliken* • **Nahe:** *Dönnhoff, Emrich-Schönleber* • **Rheingau:** *Peter Jakob Kühn, Wegeler, Weil* • **Rheinhessen:** *Keller* • *Pfalz: Müller-Catoir*

EISWEIN QMP

Until 1982 this was a qualification used in conjunction with one of the other predicates. It was previously possible to obtain *Spätlese Eiswein, Auslese Eiswein*, and so on; but *Eiswein* is now a predicate in its own right, with a minimum *Oechsle* level for its grapes equivalent to *Beerenauslese*. An *Eiswein* occurs through extremely unusual circumstances whereby healthy grapes left on the vine are frozen by frost or snow. They are harvested, rushed to the winery, and pressed in their frozen state. Only the water inside the grape actually freezes: this either remains caked inside the press or rises to the top of the vat in the form of ice, which is then skimmed off. What remains available for fermentation is a freeze-condensed juice that is as rich and concentrated in extract as a *Beerenauslese* or *Trockenbeerenauslese*.

Because the temperature has to drop to at least -7°C (19.4°F), the harvest rarely occurs before November, but can be as late as January the following year. The finest *Eisweins* are always made from healthy grapes and therefore taste quite different to *Beerenauslesen* or *Trockenbeerenauslesen*, which are produced from grapes affected by noble rot. *Eiswein* tends to taste fresher, fruitier, and racier, because the frost does not only condense the grape sugars but also the acidity. Ideally *Eiswein* should only be made from hand-picked grapes, but the lobby of industrial bottlers fended this off, making it legal for it to be machine-harvested, although it is impossible to collect authentic *Eiswein* grapes this way.

🍷 0–50 years

👑 **Mosel:** *Dr Loosen, Molitor, Max Ferd Richter, Karthäuserhof (Ruwer), Egon Müller (Saar)* • **Nahe:** *Schloßgut Diel, Dönnhoff, Emrich-Schönleber, Schäfer-Fröhlich* • **Rheingau:** *Weil* • **Rheinhessen:** *Keller* • **Franken:** *Fürstlich Castellsches Domäneamt, Max Müller, Horst Sauer*

TROCKENBEERENAUSLESE QMP or TBA

Germany's legendary TBA is produced from heavily botrytised grapes, left on the vine to shrivel into raisin-like berries that must be individually picked. These grapes must reach an *Oechsle* level of 150 to 154°. With no chaptalisation, the wine will have a minimum potential strength of between 21.5 and 22.1 per cent by volume, although only 5.5 per cent need be alcohol, the rest being residual sugar.

Consulting charts, however, does little to highlight the difference in style that exists between a *Beerenauslese* and a TBA. The first noticeable difference is often the colour. Going from *Auslese* to *Beerenauslese* is merely to progress from a light to a rich gold or buttercup yellow. The TBA colour range extends from a deep gold to amber, with some distinctly odd orange-tawny hues in between. The texture is very viscous and its liqueur-like consistency is just one of many reasons why it is impossible to drink TBA. Taking a good mouthful of TBA is about as easy as swigging cough mixture – one can merely sip it. Its intensity and complexity and the profundity of its aromas and flavours really must be experienced. A deliciously luscious liquid to be revered by the thimble-full!.

🍷 12–50 years

👑 **Mosel:** *Fritz Haag, Schloß Lieser, Egon Müller (Saar), Molitor* • **Nahe:** *Dönnhoff, Emrich-Schönleber, Schäfer-Fröhlich* • **Rheingau:** *Wegeler, Weil* • **Franken:** *Horst Sauer, Bürgerspital* • **Rheinhessen:** *Gunderloch, Keller* • **Pfalz:** *Neiss, Müller-Catoir*

RED WINE

Outside of Germany it is still a little-known fact that almost 37 per cent of the country's vineyards are planted with black grape varieties. The most commonly encountered German red wines are Spätburgunder (Pinot Noir), Dornfelder, and Regent, and many of the best are sold simply as *Qualitätswein* or even *Tafelwein* to allow producers a greater freedom of expression. Dornfelder is grown mostly in the Pfalz, Rheinhessen, and Württemberg, and producers have developed three basic styles: *barrique*-aged for serious drinking; early-drinking Beaujolais-style; and an off-dry grapey style. Regent is a relatively new hybrid – (Silvaner x Müller-Thurgau) x Chambourcin – grown for its resistance to rot and mildew, and though mainly produced in a quaffing style, it can also yield wines with reasonable quality aspirations. Spätburgunder (Pinot Noir) is the most widely planted black grape variety in the country, and over the past 30 years a host of talented winemakers have produced outstanding red wines from it, which can challenge Burgundy at least at *premier cru* level. Frühburgunder, an early-ripening mutation of Pinot Noir, is a speciality of the Ahr region, but has its true master in Franken where the estate of Rudolf Fürst in Bürgstadt manages to coax some exquisitely elegant wines from it by lowering yields. Even the international markets have by now realised that Germany is a world-class producer of quality Pinot Noir. Much more of an insider tip are the red wines from the Lemberger grape, better known under the name Blaufränkisch in Austria. It is only in the last 15 years that some growers in Württemberg, the only region where this variety is grown, have woken up to the potential of Lemberger as a wine that at its best can challenge Spätburgunder in terms of quality.

🍷 2–8 years

👑 **Spätburgunder: Ahr:** *Meyer-Näkel, Burggarten,* Deutzerhof, Stodden • **Pfalz:** *Friedrich Becker, Bernhart, A Christmann, Philipp Kuhn, Knipser)* • **Baden:** *Huber, R & C Schneider, Salwey, Fritz Wassmer, Martin Wassmer, Ziereisen* • **Franken:** *Fürst, Benedikt Baltes* • **Rheingau:** *August Kesseler, Chat Sauvage* • **Rheinhessen:** *Keller* • **Mosel:** *Molitor, Daniel Twardowski*

Frühburgunder: Franken: *Fürst* • **Pfalz:** *Neiss* • **Ahr:** *Kreuzberg*

Lemberger: Württemberg: *Aldinger, Dautel, Jürgen Ellwanger, Haidle, Graf Neipperg, Schnaitmann, Wachtstetter*

SEKT or SPARKLING WINE

The common method of production for this anonymous sparkling wine is *cuve close*, although when the industry was born in the 1820s, all *Sekt* was bottle-fermented. Today, not only is it an industrialized product, but most is made from grapes grown outside Germany, usually from Italy or the Loire Valley in France. Furthermore, until 1986, this Euro-fizz could be called *Deutscher Sekt*. This was because *Sekt* was considered to be a method of production and, so it was argued, if production took place in Germany, it must logically be German or *Deutscher Sekt*. As the vast majority of Germany's huge *Sekt* industry used imported grapes, juice, or wine (and still does), this was more a matter of lobbying by those with vested interests, than logic. Much of the effort to dispose of this false logic, however, came from honourable sectors within the German wine industry itself. Common sense finally won through on 1 September 1986, when an EU directive brought this appellation into line with the general philosophy of the EU's wine regime. There has been a noticeable upsurge in the numbers of what are now genuine German *Deutscher Sekt*, although *Sekt* plain and simple still dominates the market.

DEUTSCHER SEKT

Deutscher Sekt must be made from 100 per cent German grapes. For a long time the best were Riesling-based, but this is no longer the case, and whereas once autolysis (the enzymatic breakdown of yeast) was thought to interfere with the pure varietal aroma and flavour of the wine, this is no longer the case. Most of the top producers are now also using Chardonnay, Pinot Noir, Pinot Meunier, and Pinot Blanc to make wines more in the image of Champagne than German sparkling wine of the past.

🍷 3–15 years

👑 **Rheinhessen:** *Raumland* • **Heßische Bergstraße:** *Griesel & Compagnie* • **Württemberg:** *Aldinger* • **Pfalz:** *von Buhl, Frank John* • **Rheingau:** *Barth, F.B.Schönleber, Solter*

DEUTSCHER SEKT BESTIMMTER ANBAUGEBIETE or DEUTSCHER SEKT BA

See *Deutscher Qualitätsschaumwein bestimmter Anbaugebiete*

DEUTSCHER QUALITÄTSSCHAUMWEIN BESTIMMTER ANBAUGEBIETE or DEUTSCHER QUALITÄTSSCHAUMWEIN BA

This *Deutscher Sekt* must be made entirely from grapes grown within one specified wine region and may come from a smaller geographical unit, such as a *Bereich*, *Großlage*, or *Einzellage*, providing that at least 85 per cent of the grapes used come from the area indicated. An alternative appellation is *Deutscher Sekt bestimmter Anbaugebiete*.

🍷 3–15 years

TROCKEN/HALBTROCKEN/ FEINHERB WINES

Wines of the *trocken* category must not contain more than 9 grams per litre of residual sugar or 18 grams for the off-dry *halbtrocken* category. Both these designations are defined by the German law, but *feinherb* is not. A traditional term for off-dry wines, it was abolished by the German wine law of 1971, but had to be re-admitted in 2003 after a legal battle, which the late Annegret Reh-Garner of the Ruwer estate Reichsgraf von Kesselstatt won against the authorities. It

THE *LANDWEIN* REGIONS OF GERMANY

This list shows the 26 *Landwein* regions and the *Qualitätswein* regions they fall in.

REGION	QUALITÄTSWEIN REGION
1 Ahrtaler Landwein	Ahr
2 Badischer Landwein	Baden
3 Bayerischer Bodensee-Landwein	Württemberg
4 Brandenburger Landwein	unattached
5 Landwein Main	Franken
6 Landwein der Mosel	Mosel
7 Landwein Neckar	Württemberg
8 Landwein Oberrhein	Baden
9 Landwein Rhein	Ahr, Hessische Bergstraße, Mittelrhein, Mosel, Nahe, Pfalz, Rheingau, Rheinhessen
10 Landwein Rhein-Neckar	Baden, Württemberg
11 Landwein der Ruwer	Mosel
12 Landwein der Saar	Mosel
13 Mecklenburger Landwein	unattached
14 Mitteldeutscher Landwein	Saale-Unstrut
15 Nahegauer Landwein	Nahe
16 Pfälzer Landwein	Pfalz
17 Regensburger Landwein	Untere Donau
18 Rheinburgen-Landwein	Mittelrhein
19 Rheingauer Landwein	Rheingau
20 Rheinischer Landwein	Rheinhessen
21 Saarländischer Landwein	Mosel
22 Sächsischer Landwein	Sachsen
23 Schleswig-Holsteinischer Landwein	unattached
24 Schwäbischer Landwein	Württemberg
25 Starkenburger Landwein	Hessische Bergstraße
26 Taubertäler Landwein	Baden

describes wines which usually lie between 18 and 45 grams per litre of residual sugar, although I have come across drier and sweeter ones. Literally translated the expression *feinherb* means "subtly dry". While previous editions of this encyclopaedia may have been justified in complaining that many dry wines from Germany were too thin and acidic, climate change and the growing expertise of producers have managed to add the body, structure, and weight that many of the early contenders seemed to be lacking.

🍷 2–20 years

👑 **Riesling Trocken: Nahe:** *Emrich-Schönleber, Dönnhoff, Schäfer-Fröhlich* • **Pfalz:** *Bassermann-Jordan, Bürklin-Wolf, A Christmann, Knipser, Philipp Kuhn, Öknomierat Rebholz* • **Rheingau:** *Breuer, Weil, Künstler, Peter Jakob Kühn* • **Rheinhessen:** *Battenfeld-Spanier, Groebe, Gunderloch, Keller, Kühling-Gillot, Wagner-Stempel Wittmann* • **Mosel:** *Schloß Lieser, Clemens Busch, Peter Lauer, Carl Loewen, van Volxem*

Riesling Feinherb: Mosel: *Falkenstein, Lubentiushof, Peter Lauer, von Kesselstatt*

Grau- and Weißburgunder Trocken: Baden: *Bercher, Heger, Huber, Franz Keller, R u C Schneider* • **Pfalz:** *Kranz, Dr Wehrheim, Ökonomierat Rebholz, Knipser, Bernhard Koch*

Chardonnay Trocken: Pfalz: *Ökonomierat Rebholz, Knewitz* • **Baden:** *Huber, Martin Wassmer* • **Franken:** *Fürst*

Silvaner Trocken: Franken: *Bürgerspital, Zehnthof-Luckert, Rainer Sauer, Weltner, Rudolf May*

The Ahr

The well-preserved remains of a Roman villa at Walporzheim prove that the Romans definitely lived here, but it is debatable whether or not they introduced vines to the valley. Nevertheless, the first documented evidence of grapes grown in the valley dates back as far as AD 770. It was also at the village of Mayschoss that the first German growers' cooperative was founded in 1868. With cultivation reflecting more than 80 per cent black grapes, the Ahr is red wine country, and Spätburgunder (Pinot Noir) accounts for more than 60 per cent of the region's vines. Its best growers consistently produce some of Germany's greatest reds.

Spätburgunder can grow in this northerly location because the vineyards along the small river Ahr lie on the steep slopes of a deep and narrow valley. Protected by the Hohe Eifel hills, it captures and stores the sunlight, with temperatures often reaching Mediterranean levels even as late as October. An exceptionally long vegetation period of between 120 and 130 days is one of the reasons why the Pinot Noir yields such excellent wines here; another is the slaty, rocky slopes of the upper valley and the partly volcanic, partly loess-dominated soils of the lower. If from the 1950s to the 1970s the region was still more infamous than famous for its mild, dilute-red plonk, popular with social clubs on weekend outings, this started to change in the mid-1980s, when the visionary Werner Näkel began experimenting with limited yields of riper grapes and the use of *barriques*. A few others followed suit, and the legendary 1990 vintage convinced even dedicated Pinot Noir disciples that the region was capable of red wines of international class. Quantities will always be restricted due to a vineyard area of no more than 560 hectares (1,383 acres), and this scarcity combined with outstanding quality has seen prices of the best rise to Burgundy *premier pru* level.

It is hardly surprising that most of the wines are sold directly from the cellars of the estates to the end consumer. Many of the visitors come the valley not just for its wines but also its beautiful landscape, which can be fully enjoyed by hiking along the *Rotweinwanderweg*, a spectacular walking trail along a 30-kilometre (19-mile) stretch of vine-clad slopes from Altenahr in the west to Bad Bodendorf in the east.

Frühburgunder may only account for just over 6 per cent of the region's vineyards, but it is a true speciality of the valley and so highly rated that it has been accepted in the *Verband Deutscher Prädikatsweingüter* (VDP) list of *Große Gewächse*. It's an early-ripening mutation of Pinot Noir (registered in France as Pinot Madeleine and Pinot Noir Précoce), slightly more vegetal and crunchier in expression than Spätburgunder, with fresh herbal and peppery notes giving it its very own character.

QUALITY REQUIREMENTS AND QUALITY WINE PRODUCTION 2017

AHR'S MINIMUM *OECHSLE*	QUALITY CATEGORY	HARVEST BREAKDOWN
44°	Deutscher Wein	0.5%
47°	Landwein	0.5%
50-60°	* Qualitätswein	98%
67°+	* Prädikatswein	1%

* Minimum *Oechsle* levels vary according to grape variety; those that have a naturally lower sugar content may qualify at a correspondingly lower level.

INTENSIVE VINE-GROWING ZONES OF THE AHR
(*see also* p437)
The most northerly of Germany's wine-producing regions prior to reunification, the Ahr is made up of districts close to the Ahr, a tributary of the Rhine.

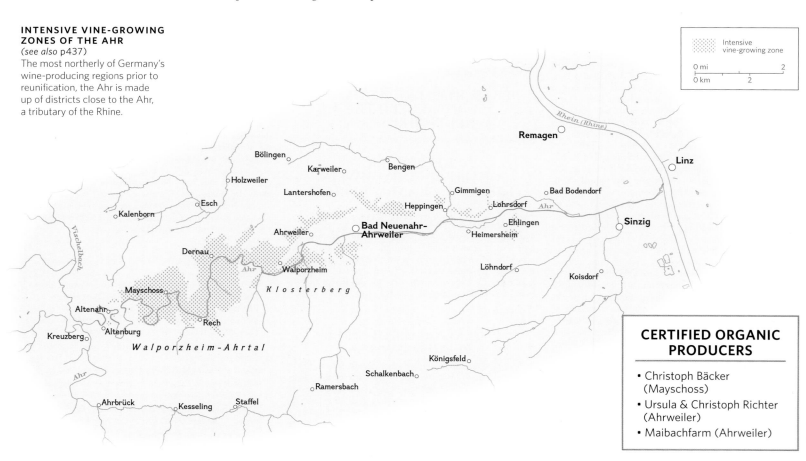

CERTIFIED ORGANIC PRODUCERS

• Christoph Bäcker (Mayschoss)

• Ursula & Christoph Richter (Ahrweiler)

• Maibachfarm (Ahrweiler)

AUTUMN VINEYARDS IN BAD NEUENAHR-AHRWEILER
Vines are grown on terraced plots on the on rocky slopes of the Ahr Valley.

FACTORS AFFECTING TASTE AND QUALITY

LOCATION
The Ahr is one of Germany's northernmost wine regions, with vineyards extending only 24 kilometres (15 miles) along the lower reaches of the River Ahr, 30 kilometres (19 miles) south of Bonn.

CLIMATE
Despite its northerly position, the deep Ahr Valley is sheltered by the surrounding Hohe Eifel hills and maintains temperatures that are favourable for viticulture, with a greenhouse-like climate on some of the steeper sites.

ASPECT
The best vineyards of the Ahr are sited on steeply terraced rocky valley sides, with higher-yielding vineyards on gentler slopes in the broader, eastern end of the valley.

SOIL
Deep, rich loess soils and volcanic stone in the lower Ahr Valley (eastern); slate and rocky soils in the middle Ahr; and basalt and slaty stone soils with some tufa in the upper Ahr Valley (western).

VITICULTURE & VINIFICATION
Three-quarters of vineyards are worked by part-time farmers, whose wines are sold by cooperatives. Most vines are cultivated under labour-intensive conditions, hence the wines are expensive to produce. Nearly all of the region's wine is consumed locally or sold to tourists. Pure varietal Weißherbst, usually Spätburgunder, is a speciality, but in terms of pure quality, this easy-drinking rosé has given way to fully red styles. Spätburgunder has always been the classic Ahr variety.

GRAPE VARIETIES
Primary varieties: Spätburgunder
Secondary varieties: Dornfelder, Frühburgunder, Kerner, Müller-Thurgau, Portugieser, Riesling

---

THE APPELLATIONS OF

THE AHR

BEREICH WALPORZHEIM-AHRTAL

This is the only *Bereich* in the Ahr region. Very few wines are actually sold under this designation, with the varieties Portugieser Dornfelder, Regent, Kerner, and Müller--Thurgau the most likely candidates.

At one time the wines used to be fairly sweet, but the trend nowadays is for drier wines. Try the Rotwein with *Rauchfleisch* (smoked meats), the Weißherbst with *Schinken* (ham), and the Riesling with trout or with the salmon that has recently been reintroduced and is now thriving in the river.

AHRWEILER
Vineyard: ✓ *Silberberg* • **Growers:** *Kreuzberg (Spätburgunder), Meyer-Näkel (Spätburgunder)*
Vineyard: *Rosenthal* • **Growers:** *J J Adeneuer (Spätburgunder), Julia Bertram (Spätburgunder), Jean Stodden (Spätburgunder)*
Vineyard: *Forstberg* • **Grower:** *Julia Bertram (Spätburgunder)*

ALTENAHR
Vineyard: ✓ *Eck* • **Growers:** *Deutzerhof (Spätburgunder), Sermann (Riesling, Spätburgunder)*

DERNAU
Vineyard: ✓ *Pfarrwingert* • **Grower:** *Meyer-Näkel (Spätburgunder)*
Vineyard: *Hardtberg* • **Growers:** *Kreuzberg (Frühburgunder), Sermann (Frühburgunder)*

HEIMERSHEIM
Vineyard: ✓ *Landskrone* • **Growers:** *Deutzerhof (Spätburgunder), Nelles (Spätburgunder)*
Vineyard: *Burggarten* • **Growers:** *Weingut Burggarten (Spätburgunder), Nelles (Spätburgunder)*

MARIENTHAL
Vineyard: *Trotzenberg* • **Grower:** *Paul Schuhmacher (Spätburgunder)*

MAYSCHOSS
Vineyard: ✓ *Moenchberg* • **Growers:** *Deutzerhof (Spätburgunder), Josten & Klein (Spätburgunder), Stodden (Spätburgunder)*

NEUENAHR
Vineyard: ✓ *Schieferlay* • **Grower:** *Kreuzberg (Spätburgunder)*
Vineyard: *Sonnenberg* • **Growers:** *J J Adeneuer (Spätburgunder, Frühburgunder), Burggarten (Frühburgunder), Kreuzberg (Spätburgunder), Meyer-Näkel (Spätburgunder), Jean Stodden (Spätburgunder)*
Vineyard: *Kirchtürmchen* • **Grower:** *Deutzerhof (Spätburgunder)*

RECH
Vineyard: ✓ *Herrenberg* • **Grower:** *Jean Stodden (Spätburgunder, Frühburgunder)*

WALPORZHEIM
Vineyard: ✓ *Gärkammer* • **Grower:** *J J Adeneuer (Spätburgunder)*
Vineyard: *Kräuterberg* • **Growers:** *Meyer-Näkel (Spätburgunder), Burggarten (Spätburgunder), Peter Kriechel (Spätburgunder, Portugieser)*
Vineyard: *Alte Lay* • **Grower:** *Brogsitter (Spätburgunder Hommage, Riesling Hommage)*
Premium Red Wine Blends: Growers: *J J Adeneuer (Spätburgunder No.1), Deutzerhof (Spätburgunder Caspar C, Grand Duc), H J Kreuzberg (Spätburgunder Devonschiefer), Peter Kriechel (Frühburgunder Jubilus Goldkapsel), Meyer-Näkel (Spätburgunder Blauschiefer, S, Goldkapsel), Nelles (Spätburgunder B48, B52, B59), Paul Schuhmacher (Spätburgunder Magna Essentia), Jean Stodden (Spätburgunder Alte Reben, Goldkapsel), Winzergenossenschaft Mayschoss-Altenahr (Spätburgunder Ponsart, Pinot Noir "R")*

THE REGION AT A GLANCE

AREA UNDER VINE
561ha, or 1,386 acres (increasing)

AVERAGE YIELD (2008-2017)
82hl/ha

RED WINE
83%

WHITE WINE
17% (decreasing)

MOST IMPORTANT GRAPE VARIETIES
65% Spätburgunder, 6% Portugieser, 8% Riesling, 21% others

INFRASTRUCTURE
Bereiche 1, *Großlagen* 1, *Einzellagen* 40

Note: The vineyards of the Ahr straddle 11 *Gemeinden* (communes), the names of which may appear on the label.

The Mittelrhein

Possessing precariously perched vineyards that have declined by almost 50 per cent since 1965, the Mittelrhein is a region that is all too often overlooked by serious wine drinkers. Yet it offers some of Germany's finest and most underrated wines.

It was from the Mittelrhein that the Celts spread out across Europe. With such ancient roots, it is not surprising that this region is so steeped in Germany's mythical history. It was at the Drachenfels in Königswinter, for instance, that Siegfried slew the dragon, and, in fact, the vineyards in this area produce a red Spätburgunder wine known as *Drachenblut,* or "dragon's blood". The Rhine River, associated with many myths and fables, flows past numerous medieval castles and towers and rushes through the Rhine Gorge, past the famous "Loreley" rock on to which the siren lured many ships to their final and fatal destination.

SHRINKING VINEYARDS, GROWING TOURISM

The difficulty of working the steepest of the Mittelrhein's vineyard slopes has encouraged many of the workforce to forsake them and seek higher wages for easier work in Germany's industrial cities. This has led to a decline in the number of vineyards from 1,000 hectares (2,470 acres) in 1970 to today's 469 hectares (1,160 acres), with the positive consequence that only the best sites were retained. But as an area of outstanding natural beauty, with its dramatic scenery and almost countless castles, the Mittelrhein will always be a tourist hotspot, as announcements even in Japanese on the riverboats cruising from Cologne to Rüdesheim demonstrate. In this region, where many tiny tributaries provide valley vineyards of a superior natural aspect for viticulture than most others on the Rhine, there is much potential for producing high-quality Riesling on its slaty soil. There are a few excellent estates making exciting wines that display a vigorous varietal character, intense flavour, and splendid acidity. The rocky slate and greywacke slopes exponentiate the effects of long sunshine hours, so Riesling quite regularly manages to rise to the heights of delicious Spätlesen and Auslesen.

QUALITY REQUIREMENTS AND QUALITY WINE PRODUCTION 2017

MITTELRHEIN'S MINIMUM *OECHSLE*	QUALITY CATEGORY	HARVEST BREAKDOWN
44°	Deutscher Wein	0.3%
47°	Landwein	0.3%
50-60°	* Qualitätswein	75%
67°+	* Prädikatswein	24%

* Minimum *Oechsle* levels vary according to grape variety; those that have a naturally lower sugar content may qualify at a correspondingly lower level.

INTENSIVE VINE-GROWING ZONES OF THE MITTELRHEIN
(*see also* p437)
North and south of the city of Koblenz, the Mittelrhein's vineyards run up towards the steep, rocky escarpments that closely border this stretch of the Rhein.

A STATUE OF LORELEY SITS ATOP THE LORELEY ROCK
The statue overlooks the bend of the Rhein at Sankt Goarshausen and the vine-planted slopes beyond.

FACTORS AFFECTING TASTE AND QUALITY

LOCATION
A 140-kilometre (87-mile) stretch of the Rhine Valley between Bonn and Bingen.

CLIMATE
The benefits of the sun are maximized by the steep valley sides that afford protection from cold winds. The river acts as a heat-reservoir, tempering the low night and winter temperatures.

ASPECT
Vines on the steep valley sides benefit from any available sunshine. North of Koblenz the vineyards are on the east bank, while to the south most are on the west bank.

SOIL
Primarily clayish slaty soil and greywacke. There are also small, scattered deposits of loess and, towards the north, and some vineyards are of volcanic origin.

VITICULTURE & VINIFICATION
Virtually all remaining vineyards have been *flurbereinigt* (consolidated by exchanging plots between growers) and many slopes are a patchwork of steep vineyards. With a high proportion of Riesling giving an average yield that is very low by German standards, quality is generally high. About 80 per cent of growers are part time, and a quarter of the harvest is processed by the coops using normal white-wine techniques.

GRAPE VARIETIES
Primary varieties: Riesling
Secondary varieties: Dornfelder, Müller-Thurgau, Spätburgunder

THE REGION AT A GLANCE

AREA UNDER VINE
469ha or 1,158ac (decreasing)

AVERAGE YIELD (2008–2017)
64hl/ha

RED WINE
9%

WHITE WINE
82% (decreasing)

ROTLING
9%

MOST IMPORTANT GRAPE VARIETIES
67% Riesling (decreasing), 10% Spätburgunder, 4.5% Müller-Thurgau (decreasing), 18% others

INFRASTRUCTURE
Bereiche 2, *Großlagen* 11, *Einzellagen* 111

Note: The vineyards of the Mittelrhein straddle 59 *Gemeinden* (communes), the names of which may appear on the label.

MEDIEVAL TOWN OF BACHARACH
Nestled in the wine country of the UNESCO Upper Middle Rhine Valley site, Bacharach thrives on tourism and wine. It is located in an ideal position along the river, so that in the Middle Ages, the town served as a shipping station for wine exportation.

THE APPELLATIONS OF

THE MITTELRHEIN

BEREICH LORELEY

LEUTESDORF

Vineyard: *Gartenlay* • **Growers:** *Josten & Klein (Riesling: trocken, Sauvignon Blanc, Grauburgunder); Scheidgen (Riesling: trocken, Spätburgunder); Sturm (Riesling: trocken, Prädikate, Pinot Noir)*

BOPPARD HAMM

Vineyard: ✓ *Engelstein* • **Growers:** *Matthias Müller (Riesling: GG, trocken, Prädikate), Weingart (Riesling: trocken, Prädikate)*

Vineyard: *Feuerlay* • **Growers:** *Heilig Grab (Riesling: trocken, Prädikate), Matthias Müller (Riesling: trocken, Prädikate),*

Vineyard: *Mandelstein* • **Growers:** *Matthias Müller (Riesling: trocken, Prädikate)*

Vineyard: *Ohlenberg* • **Growers:** *Weingart (Spay) (Riesling: trocken, Prädikate), Matthias Müller (Riesling: trocken)*

OBERWESEL

Vineyard: *Ölsberg* • **Grower:** *Dr Kauer (Riesling trocken, feinherb)*

BACHARACH

Vineyard: ✓ *Hahn* • **Grower:** *Toni Jost (Riesling trocken, GG, Prädikate)*

Vineyard: *Kloster Fürstental* • **Grower:** *Dr Kauer (Riesling trocken, Prädikate)*

Vineyard: *Wolfshöhle* • **Grower:** *Ratzenberger (Riesling GG, Prädikate); Bastian (Riesling feinherb)*

OBERDIEBACH

Vineyard: *Schloß Fürstenberg* • **Grower:** *Dr Kauer*

(Riesling Prädikate); Ratzenberger (Riesling trocken)

STEEG

Vineyard: *St. Jost* • **Grower:** *Ratzenberger (Riesling trocken, GG)*

BEREICH SIEBENGEBIRGE

A single *Großlage Bereich* covering the same area as *Großlage* Petersberg.

NIEDERDOLLENDORF

Vineyard: *Heisterberg* • **Grower:** *Kay Weine (Riesling trocken)*

KÖNIGSWINTER

Vineyard: *Drachenfels* • **Grower:** *Pieper (Riesling trocken, Prädikate)*

Mosel

The greatest Rieslings grown along the Mosel River have a legendary acidity that can only be relieved through a knife-edge balance of sweetness. Unlike in the warmer Rhine regions, the Riesling grape is at its best here in hot vintages.

If any grape is intrinsically racy, it is the vigorous Riesling; and if any region can be singled out for emphasizing this raciness, it must be Mosel. Grown on the steepest, slaty slopes, Riesling combines a relatively high acidity with an irrefutable suggestion of lightness and elegance. But a fine Mosel is never thin, because these wines have surprisingly high extract levels that, together with the acidity, intensify the characteristics of flavour.

When the trend towards dry white wines began to reach German palates in the late 1980s and early 1990s, the Mosel found it difficult to change track, and many of its early Riesling *trocken* renditions lacked body and extract, bordering onto the thin and sour. The main problem was that yields perfect for *Prädikatsweine* were too high for dry wines. It took the majority of growers a long time to adjust, but in the last 10 years they have certainly managed to close the quality gap on their rivals from the other regions.

The noble sweet *Auslesen* and *Beerenauslesen* manage to remain racy in even the hottest vintages, while those from the other regions can appear fat and overblown by contrast. Even the most modest wines retain a freshness and vitality in sun-blessed years that will be lacking in those from the warmer regions. But whereas until as recently as 30 to 40 years ago growers' most frequent nightmare would have been that grapes might not always fully ripen even when left on the vines until late October, today they have to concern themselves with picking fruit almost a month earlier to avoid loss of acidity and freshness.

THE GOOD DOCTOR

There are many great vineyards in this region, but none so famous as the legendary Bernkasteler Doctor, which produces Germany's most expensive wine. The story is that Boemund II, the Archbishop of Trier in the 14th century, was so ill that his doctors could do nothing for him. A winegrower from Bernkastel recommended the restorative powers of the wine produced from his vineyard. Boemund drank some, made a miraculous recovery, and declared "The best doctor grows in this vineyard in Bernkastel". More recently, the Doctor vineyard has been the subject of a lengthy court case. The original vineyard comprised 1.35 hectares (3.33 acres), but in 1971 the new German wine law proscribed a ban on all vineyards of less than 5 hectares (12.33 acres). The authorities planned to expand the Doctor almost equally to the west (into the *Einzellage* Graben) and to the east (into an area classified merely as *Großlage* Badstube). This enabled 13 different producers to make and sell Bernkasteler Doctor, whereas only three owners of the true Doctor vineyards had existed before. It is not surprising that the owners of the original Doctor vineyard felt strong enough to take their objections to court. The case continued until finally, in 1984, after an exhaustive study had been made of the vineyard's *terroir*, the court decided that the Doctor vineyard could legitimately be stretched to include all the Graben element and a small portion of the Badstube, making a total of 3.26 hectares (8 acres). The main reason why the Doctor vineyard was primarily expanded westwards, rather than eastwards, is that these westerly exposures benefit from longer hours of sunshine. Yet the 10 owners of the Badstube section excluded from the Doctor vineyard found themselves in possession of vineyards that, having for 13 years been accorded the status of Germany's most prestigious wine, were now nameless. They proposed that they should be allowed to use the *Einzellage* name of Alte Badstube am Doctorberg ("Old Badstube on the Doctor's Hill") and, despite protests from the original owners, this was accepted by officials.

QUALITY REQUIREMENTS AND QUALITY WINE PRODUCTION 2017

MOSEL'S MINIMUM *OECHSLE*	QUALITY CATEGORY	HARVEST BREAKDOWN
44°	Deutscher Wein	1%
47°	Landwein	1%
50–60°	* Qualitätswein	79%
67°+	* Prädikatswein	19%

* Minimum *Oechsle* levels vary according to grape variety; those that have a naturally lower sugar content may qualify at a correspondingly lower level.

FACTORS AFFECTING TASTE AND QUALITY

LOCATION
This region follows the Mosel river, as it meanders for 250 kilometres (150 miles) from the border triangle of France, Luxemburg, and Germany at Schengen north to the confluence of the Mosel and Rhine at Koblenz. It includes the vineyards of two major tributaries, the Saar and the Ruwer, which flow into the Mosel from the south.

CLIMATE
The Mosel Valley is protected by the mountain ranges of the Eifel in the west and Hundsrück in the east, so precipitation in the area is relatively moderate. The generous layers of slate on the steep, mostly south-facing slopes absorb the heat during the day and give off warmth at night to guarantee full ripening of the grapes, particularly in these days of climate change.

ASPECT
The Mosel has more loops and bends than any other German river, and most of the vines grow at an altitude of between 100 and 350 metres (330 and 1,150 feet). This valley provides slopes of every aspect, with many that are spectacularly steep, boasting gradients as high as 70 degrees.

SOIL
Soils in this region vary from sandstone, shell-limestone, and red marl in the upper Mosel to Devon slate in the middle Mosel, Saar, and Ruwer and clay slate and grey stony soil in the lower Mosel. Alluvial sand and gravel soils are also found in lower sites. Classic Riesling sites are slaty; the Elbling prefers limestone.

VITICULTURE & VINIFICATION
The vineyards with their approximately 60 million vines are cultivated by some 3,200 growers. The average size estate owns about 2.7 hectares (6.7 acres) of vineyard. While in its hay-days the region's area under vine comprised more than 12,000 hectares (29,650 acres), due to structural changes and lack of interest by the next generation, many small estates gave up in the nineties and almost 30% of vineyard were abandoned. The last decade has seen a consolidation at 8,870 hectares (21,920 acres). With 3,530 hectares (8,723 acres) of its sites on slopes with a gradient of 30 per cent or higher the Mosel is the largest steep slope winegrowing area in the world. In these notoriously difficult sites, each vine has to be tied to its own 2.4-metre (8-foot) wooden stake, and the slate soil that is washed down to the bottom of the slope by the winter rains has to be carried back up every year. Tending the vines is thus unavoidably labour-intensive, and this, combined with a longer winter than experienced elsewhere in Germany, accounts for the higher prices asked for fine Mosel wines. The early onset of winter causes fermentation to take place at cool temperatures, and when the wines are bottled early they retain more carbonic gas, which emphasizes the crisp, steely character of the Riesling grape.

GRAPE VARIETIES
Primary varieties: Riesling
Secondary varieties: Elbling, Müller-Thurgau, Spätburgunder

THE REGION AT A GLANCE

AREA UNDER VINE
8,870ha (21,91ac)

AVERAGE YIELD (2008–2017)
85hl/ha

RED WINE
9%

WHITE WINE
90% (decreasing)

ROSÉ
1%

**MOST IMPORTANT
GRAPE VARIETIES**
61.5% Riesling, 11.2% Müller-Thurgau,
5.6% Elbling, 4.6% Spätburgunder, 3.9%
Weißburgunder, 13.2% others

INFRASTRUCTURE
Bereiche 6, *Großlagen* 19, *Einzellagen* 520

Note: The vineyards of Mosel straddle
192 *Gemeinden* (communes), the names
of which may appear on the label.

INTENSIVE VINE-GROWING ZONES OF MOSEL
(see also p437)
Formerly known as the Mosel-Saar-
Ruwer, this region underwent a name
change, being simplified to Mosel from
the 2007 vintage, according to the
latest wine-legislation update.

Intensive
vine-growing zone

Bereich boundary

Großlage boundary

THE BERNKASTELER DOCTOR, BERNKASTEL-KUES
This tiny vineyard, sited on a steep slope, is known the world over for
producing the finest of Rieslings. Barbara Rundquist-Müller owns the entire
Dr H Thanisch Müller-Burggraef estate, which the Doctor vineyard is a part of.

THE TOP APPELLATIONS OF

MOSEL

BEREICH BERNKASTEL

This *Bereich* covers the entire Mittelmosel, encompassing all of the river's most famous villages and towns, and most of its best vineyards, as well. But it also comprises the three vast *Großlagen*: Badstube, Kurfürstlay, and Piesporter Michelsberg, under the names of which much unexciting wine is produced for cheap discounters and supermarket chains.

GROßLAGE BADSTUBE
BERNKASTEL

Vineyard: ✓ *Doctor* • **Growers:** *Wwe Dr H Thanisch-Erben Thanisch (Riesling GG, Prädikate), Wwe Dr H Thanisch-Erben Müller-Burgraef (Riesling Prädikate), Wegeler-Gutshaus Mosel (Riesling GG, Prädikate), Markus Molitor (Riesling Prädikate), Schloß Lieser (Riesling Prädikate)*

Vineyard: *Alte Badstube am Doctorberg* • **Grower:** *Dr Pauly-Bergweiler (Riesling Prädikate)*

Vineyard: *Großlage Badstube* • **Grower:** *Wwe Dr H Thanisch-Erben Thanisch (Riesling Prädikate)*

Vineyard: *Lay* • **Grower:** *Wwe Dr H Thanisch-Erben Müller-Burgraef (Riesling Prädikate)*

GROßLAGE KURFÜRSTLAY
BRAUNEBERG

Vineyard: ✓ *Juffer* • **Growers:** *Fritz Haag-Dusemonder Hof (Riesling GG, Prädikate), Paulinshof (Riesling Prädikate), Max Ferd Richter (Riesling trocken, Prädikate), Günther Steinmetz (Riesling Prädikate)*

Vineyard: *Juffer Sonnenuhr* • **Growers:** *Fritz Haag-Dusemonder Hof (Riesling GG, Prädikate), Reichsgraf von Kesselstatt, Schloß Lieser (Riesling Prädikate), Max Ferd Richter (Riesling trocken, Prädikate), Paulinshof (Riesling Prädikate), Günther Steinmetz (Riesling Prädikate)*

Vineyard: *Kammer* • **Grower:** *Paulinshof (Riesling GG, feinherb, Prädikate)*

Vineyard: *Klostergarten* • **Grower:** *Markus Molitor (Pinot Noir)*
Vineyard: *Mandelgraben* • **Grower:** *Markus Molitor (Pinot Noir)*

KESTEN

Vineyard: *Paulinshofberg* • **Growers:** *Bastgen (Riesling Prädikate), Meierer (Riesling Prädikate), Paulinshof (Riesling GG, feinherb), Günther Steinmetz (Riesling GG)*

Vineyard: *Paulinsberg* • **Growers:** *Meierer (Riesling trocken, feinherb, Prädikate), Günther Steinmetz (Riesling Prädikate)*

LIESER

Vineyard: ✓ *Niederberg Helden* • **Growers:** *Sybille Kuntz, Schloß Lieser (Riesling GG, Prädikate), Axel Pauly (Riesling trocken, Prädikate)*

LONGUICH

Vineyard: *Maximin Herrenberg* • **Grower:** *Carl Loewen (Riesling trocken, GG, Prädikate)*

MARING

Vineyard: *Sonnenuhr* • **Grower:** *Steffen-Prüm (Riesling Prädikate)*

MÜLHEIM

Vineyard: *Sonnenlay* • **Grower:** *Günther Steinmetz (Pinot Noir)*

WINTRICH

Vineyard: *Ohligsberg* • **Grower:** *Julian Haart (Riesling trocken, Prädikate), Haart (Riesling GG), Günther Steinmetz (Riesling GG)*

Vineyard: *Geierslay* • **Grower:** *Günther Steinmetz (Riesling GG)*

GROßLAGE MICHELSBERG
DHRON

Vineyard: *Hofberg* • **Growers:** *A J Adam (Riesling trocken, Prädikate), Grans-Fassian (Riesling GG), Günther Steinmetz (Riesling GG), Daniel Twardowski (Pinot Noir)*

Vineyard: *Häschen* • **Grower:** *A J Adam (Riesling trocken, Prädikate)*

NEUMAGEN

Vineyard: *Rosengärtchen* • **Grower:** *Günther Steinmetz (Riesling trocken)*

PIESPORT

Vineyard: ✓ *Domherr* • **Growers:** *Kurt Hain (Riesling trocken, Prädikate), Später-Veit (Riesling trocken, Prädikate)*

Vineyard: *Goldtröpfchen* • **Growers:** *Julian Haart (Riesling trocken, Prädikate), Haart (Riesling GG, Prädikate), Kurt Hain (Riesling trocken, Prädikate), Lothar Kettern (Riesling Prädikate), Lehnert-Veit (Riesling GG, Prädikate; Spätburgunder), Schloß Lieser (Riesling GG, Prädikate), Molitor Rosenkreuz (Riesling GG, Prädikate), Später-Veit (Riesling feinherb, Prädikate), Günther Steinmetz (Riesling Prädikate), Nik Weis-St Urbans-Hof (Leiwen) (Riesling Prädikate), Reichsgraf von Kesselstatt (Riesling Prädikate)*

TRITTENHEIM

Vineyard: ✓ *Apothek* • **Growers:** *Ernst Clüsserath, Ansgar Clüsserath (Riesling trocken, feinherb, Prädikate), Clüsserath-Eifel (Riesling feinherb, Prädikate), Grans-Fassian (Riesling GG, Prädikate), Loersch (Riesling trocken, feinherb, Prädikate), Josef Rosch (Riesling trocken, Prädikate)*

Vineyard: *Altärchen* • **Growers:** *Ernst Clüsserath (Riesling trocken, feinherb, Prädikate), Bernhart Eifel (Riesling trocken, feinherb, Prädikate), Franz-Josef Eifel (Riesling trocken, feinherb, Prädikate)*

GROßLAGE MÜNZLAY
GRAACH

Vineyard: ✓ *Domprobst* • **Growers:** *Kees-Kieren (Riesling trocken, feinherb, Prädikate), Kerpen (Riesling GG, Prädikate), Markus Molitor (Riesling trocken, Prädikate), S A Prüm (Riesling GG, Prädikate), Willi Schaefer (Riesling Prädikate)*

Vineyard: *Himmelreich* • **Growers:** *Kees-Kieren (Riesling trocken, feinherb, Prädikate), Markus Molitor (Riesling Prädikate, Pinot Noir), J J Prüm (Riesling Prädikate), Max Ferd Richter (Riesling Prädikate), Willi Schaefer (Riesling Prädikate)*

Vineyard: *Josephshof* • **Grower:** *Reichsgraf von Kesselstatt (Riesling feinherb, Prädikate)*

WEHLEN

Vineyard: ✓ *Sonnenuhr* • **Growers:** *Kerpen (Riesling GG, Prädikate), Dr Loosen (Riesling GG, Prädikate), Markus Molitor (Riesling Prädikate), S A Prüm (Riesling Prädikate), J J Prüm (Riesling Prädikate), Willi Schaefer (Riesling Prädikate), Schloß Lieser (Riesling GG, Prädikate), Wwe Dr H Thanisch-Erben Müller-Burgraef (Riesling Prädikate), Wegeler-Gutshaus Mosel (Riesling GG, Prädikate)*

ZELTINGEN

Vineyard: ✓ *Himmelreich* • **Grower:** *Selbach-Oster (Riesling Prädikate)*

Vineyard: ✓ *Schloßberg* • **Grower:** *Selbach-Oster (Riesling Prädikate)*

Vineyard: *Sonnenuhr* • **Growers:** *Gessinger (Riesling trocken, Prädikate), Markus Molitor (Riesling trocken, Prädikate), Selbach-Oster (Riesling Prädikate)*

GROßLAGE NACKTARSCH
KRÖV

Vineyard: *Steffensberg* • **Growers:** *Christian Klein (Riesling Prädikate), Martin Müllen (Riesling trocken), Vollenweider (Riesling trocken, Prädikate)*

Vineyard: *Paradies* • **Grower:** *Martin Müllen (Riesling trocken, Prädikate)*

Vineyard: *Letterlay* • **Grower:** *Martin Müllen (Riesling trocken, Prädikate)*

GROßLAGE PROBSTBERG
SCHWEICH

Vineyard: *Annaberg* • **Grower:** *Bernhart Eifel (Riesling trocken, feinherb, Prädikate)*

GROßLAGE RÖMERLAY

GROßLAGE SANKT MICHAEL
KLÜSSERATH

Vineyard: *Bruderschaft* • **Grower:** *Regnery (Riesling trocken, Prädikate, Spätburgunder, Syrah)*

LEIWEN

Vineyard: ✓ *Laurentiuslay* • **Growers:** *Clüsserath-Eifel (Riesling Prädikate), Carl Loewen (Riesling trocken, Prädikate), Grans-Fassian (Riesling GG), Nikolaus Köwerich (Riesling Prädikate), Nik Weis-St Urbans-Hof (Riesling GG, Prädikate)*

MEHRING

Vineyard: *Layet* • **Grower:** *Nik Weis-St Urbans-Hof (Leiwen) (Riesling GG, Prädikate)*

THÖRNISCH

Vineyard: *Ritsch* • **Grower:** *Carl Loewen (Riesling GG, Prädikate)*

GROßLAGE SCHWARZLAY
ENKIRCH

Vineyard: *Batterieberg* • **Grower:** *Immich Batterieberg (Riesling feinherb)*

Vineyard: *Ellergrub* • **Growers:** *Weiser-Künstler (Riesling trocken, Prädikate), Immich Batterieberg (Riesling trocken)*

Vineyard: *Zeppwingert* • **Grower:** *Immich Batterieberg (Riesling feinherb)*

Vineyard: *Steffensberg* • **Growers:** *Immich Batterieberg (Riesling trocken), Weiser-Künstler (Riesling trocken)*

ERDEN

Vineyard: ✓ *Treppchen* • **Growers:** *Dr Hermann (Riesling Prädikate), Dr Loosen (Riesling Prädikate), Lotz (Riesling Prädikate), Meulenhof (Riesling Prädikate), Mönchhof-Robert Eymael (Riesling Prädikate), Kees-Kieren (Riesling Prädikate), Andreas Schmitges (Riesling GG, Prädikate), Steffen-Prüm (Riesling Prädikate)*

Vineyard: *Prälat* • **Growers:** *Dr Loosen (Riesling GG, Prädikate), Mönchhof-Robert Eymael (Riesling Prädikate), Andreas Schmitges (Riesling GG, Prädikate)*

KINHEIM

Vineyard: *Rosenberg* • **Grower:** *Jakoby-Mathy (Riesling trocken, Prädikate)*

Vineyard: *Hubertuslay* • **Grower:** *Markus Molitor (Riesling Prädikate)*

TRABEN

Vineyard: *Gaispfad* • **Grower:** *Weiser-Künstler (Riesling trocken)*

TRARBACH

Vineyard: ✓ *Schloßberg* • **Grower:** *Markus Molitor (Pinot Noir)*

Vineyard: *Hühnerberg* • **Grower:** *Martin Müllen (Riesling trocken, Prädikate)*

ÜRZIG

Vineyard: ✓ *Würzgarten* • **Growers:** *Dr Loosen (Riesling Prädikate), Karl Erbes (Riesling feinherb, Prädikate), Dr Hermann (Riesling Prädikate), Markus Molitor (Riesling Prädikate), Rebenhof (Riesling GG, Prädikate)*

WOLF

Vineyard: ✓ *Goldgrube* • **Grower:** *Vollenweider (Riesling GG, Prädikate)*

Vineyard: *Sonnenlay* • **Grower:** *Weiser-Künstler (Riesling Prädikate)*

GROßLAGE VOM HEISSEN STEIN
PÜNDERICH

Vineyard: ✓ *Marienburg* • **Growers:** *Clemens Busch (Riesling GG, Prädikate), Frank Brohl (Riesling trocken, Prädikate)*

REIL

Vineyard: ✓ *Goldlay* • **Grower:** *Melsheimer (Riesling Prädikate)*

Vineyard: *Mullay-Hofberg* • **Grower:** *Melsheimer (Riesling trocken, Prädikate)*

CERTIFIED ORGANIC AND BIODYNAMIC PRODUCERS

BIODYNAMIC
- Öko-Rita & Rudolf Trossen (Kinheim-Kindel)
- Ulrich Treitz (Traben-Trarbach)

ORGANIC
- Ferienhof Josef Luy (Konz)
- Alfred Cuy (Zell-Merl)
- Alter Weinhof (Erden)
- Arns und Sohn (Reil)
- Artur Mentges (Kröv)
- Caspari-Kappel (Enkirch)
- Christoph Rimmele (Zell/Kaimt)
- Christopher Koenen (Minheim)
- Clemens Busch (Pünderich)

- Dr Renate Wilkomm (Bernkastel-Kues)
- Frank Brohl (Pünderich)
- Franz_Josef Eifel (Trittenheim)
- Gerhard Lönnartz (Ernst)
- Günther (Wincheringen)
- Hoffmann-Simon (Piesport)
- Hugo Schorn (Burg)
- Joachim Deis (Senheim)
- Johannes Schneider (Maring-Noviand)
- Kirsten (Klüsserath)
- Klaus Schweisel (Osann-Monzel)
- Klaus Stülb (Zell/Kaimt)
- Laurentiushof (Bremm)
- Louis Klein (Traben-Trabach)

- Paul Schwarz (Neumagen)
- Peter IG Frommer (Traben-Trabach)
- Melsheimer (Reil)
- Dr Melsheimer (Traben-Trabach)
- Peter Mentges (Bullay)
- Rudolf Holbach (Wincheringen)
- Sekt und Karl Weber (Wincheringen)
- Schömann (Zeltingen)
- Steffens-Keß (Steffens-Keß (Reil)
- Udo Wick (Kröv)-
- Volker Manz (Konz Kommlingen)
- Weingut-Sektkellerei Uwe Kreuter (Alf)
- zur Römerkelter (Maring-Noviand)

MOSEL SINUOSITY AT TRITTENHEIM
The third-largest of Germany's wine regions, Mosel takes its name from the Mosel River, which twists and turns its way between Trier and Koblenz, where it then discharges into the Rhine. Since the Middle Ages many "wine villages" sprouted along the river, and scores of vineyards line its banks.

The Nahe

In the Nahe region, a sunny microclimate and varied soils combine to produce wines that have the elegance of a Rheingau, the body of a light Rheinhessen, and the acidity of a Mosel. This diversity of attributes is owed to the fact that despite its limited size the region can boast more than 180 different soil formations. Nevertheless, throughout its existence as a winegrowing area, the Nahe has always been overshadowed by its far more prominent neighbour, the Mosel. This only changed over the last three decades, with the emergence of a handful of truly outstanding winemakers.

Despite almost 2,000 years of viticulture the Nahe remained a relative backwater of wine production. There are a number of reasons for that, one of the most plausible being that the river that gives its name to the region was not wide and deep enough to be able to ferry wine or any other products by boat. Most growers who marketed their own wine adorned their labels with pictures of good old father Rhine, foregoing the opportunity to create their own regional identity. Another factor that hampered progress was the fragmentation of the area under vine into small, scattered vineyards, because an underdeveloped infrastructure hindered not only transport but also cooperation between estates. Even the setting up of a college for viniculture in Bad Kreuznach and the founding of a state domain at Niederhausen over 100 years ago did little to promote a regional identity. But since the early 1990s the area has been propelled into the forefront of German quality wine production by some truly stunning wines from masters such as Helmut Dönnhoff, Frank Schönleber, and Armin Diel.

THE REGION AT A GLANCE

AREA UNDER VINE
4,225ha (10,440ac)

AVERAGE YIELD (2008–2017)
75hl/ha

RED WINE
25%

WHITE WINE
75.8%

ROSÉ
4%

MOST IMPORTANT GRAPE VARIETIES
28.7% Riesling (increasing), 12.6% Müller-Thurgau (decreasing), 10.1% Dornfelder, 7.5% Grauburgunder, 6.8% Weißburgunder, 6.7% Spätburgunder, 27.6% others

INFRASTRUCTURE
Bereiche 1, *Großlagen* 7, *Einzellagen* 310

Note: The vineyards of the Nahe straddle 80 *Gemeinden* (communes), the names of which may appear on the label.

QUALITY REQUIREMENTS AND QUALITY WINE PRODUCTION 2017

THE NAHE'S MINIMUM *OECHSLE*	QUALITY CATEGORY	HARVEST BREAKDOWN
44°	Deutscher Tafelwein	1%
47°	Landwein	1%
50–60°	* QbA	76%
67–73°	* Kabinett	17%
76–85°	* Spätlese	1%
83–88°	* Auslese	1%
110°	Beerenauslese	1%
110°	Eiswein	1%
150°	Trockenbeerenauslese	1%

* Minimum *Oechsle* levels vary according to grape variety; those that have a naturally lower sugar content may qualify at a correspondingly lower level.

INTENSIVE VINE-GROWING ZONES OF THE NAHE
(*see also* p437)
Between Rheinhessen and the Mittelrhein nestles the self-contained wine-producing region of the Nahe. Its namesake river has many tributaries running between spectacular overhanging cliffs.

Intensive vine-growing zone

Großlage boundary

0 mi — 4
0 km — 4

FACTORS AFFECTING TASTE AND QUALITY

LOCATION
The region balloons out from between Rheinhessen and Mittelrhein around the Nahe River, which runs parallel to and 40 kilometres (25 miles) southeast of the Mosel.

CLIMATE
A temperate, sunny climate with adequate rainfall and no frosts. Local conditions are influenced by the Soonwald forest to the northeast and heat-retaining, rocky hills to the east. Protected, south-facing vineyards enjoy microclimates that are almost Mediterranean.

ASPECT
Vineyards are found on both the gentle and steep slopes of the Nahe and its hinterland of many small tributary river valleys. Vines grow at altitudes of between 100 and 300 metres (330 and 985 feet).

SOIL
Quartzite and slate are the predominant soil formations along the lower reaches of the region. Porphyry (hard rock poor in lime), melaphyry (hard rock rich in lime), and coloured sandstone are primarily found in the middle and upper reaches. Near Bad Kreuznach weathered clay, sandstone, limestone, loess, and loam soil may also be found. The greatest Riesling wines grow on sandstone.

VITICULTURE AND VINIFICATION
Since the mid-1960s cultivation of Riesling and Silvaner has declined by 20 and 15 per cent respectively, but Riesling has managed to buck this trend, remaining the most widely planted variety with 28 per cent of the Nahe's vineyards. A passing fad saw the temporary emergence of new-fangled crosses, such as Kerner, Scheurebe, and Bacchus, but these aromatic varieties have had to make way in recent years for the Pinots and their far greater potential for making superior dry wines. As much as 40 per cent of all Nahe wine is processed by small growers who often sell direct to passing customers, with the remaining 40 per cent belonging to the traditional trade and export houses.

GRAPE VARIETIES
Primary varieties: Riesling
Secondary varieties: Müller-Thurgau, Grauer Burgunder, Weißburgunder, Spätburgunder, Dornfelder

---

THE APPELLATIONS OF

THE NAHE

BEREICH NAHETAL
This region-wide *Bereich* replaces the former Bereich of Kreuznach, which covered the area once called *Untere Nahe*, or "Lower Nahe", and Schloßböckelheim, which is the most famous.

GROßLAGE BURGWEG
ALSENZ

Vineyard: Elkersberg • **Grower:** Hahnmühle (Riesling trocken)
ALTENBAMBERG

Vineyard: Rotenberg • **Grower:** Gut Hermannsberg (Riesling trocken, Prädikate)
AUEN

Vineyard: Elkersberg • **Grower:** Hahnmühle (Riesling trocken)
Vineyard: Römerstich • **Grower:** Hees (Riesling trocken, Prädikate)
EBERNBURG

Vineyard: Schlossberg • **Grower:** Hahnmühle (Riesling trocken)
NIEDERHAUSEN

Vineyard: ✔ Felsensteyer • **Grower:** Dr Crusius (Riesling Prädikate)
Vineyard: Hermannsberg • **Grower:** Gut Hermannsberg (Riesling GG)
Vineyard: ✔ Hermannshöhle • **Growers:** Hermann Dönnhoff (Riesling GG, Prädikate), Jakob Schneider (Riesling trocken, Prädikate)
Vineyard: Steinberg • **Grower:** Gut Hermannsberg (Riesling GG, Prädikate)
NORHEIM

Vineyard: ✔ Kirschheckg • **Grower:** Hermann Dönnhoff (Riesling Prädikate)
Vineyard: Dellchen • **Grower:** Hermann Dönnhoff (Riesling GG)
OBERHAUSEN

Vineyard: ✔ Brücke • **Grower:** Hermann Dönnhof (Riesling Prädikate)
Vineyard: Leistenberg • **Grower:** Hermann Dönnhof (Riesling Prädikate)
OBERNDORF

Vineyard: Aspenbergg • **Grower:** Hahnmühle (Riesling trocken)
ODERNHEIM

Vineyard: Montfort • **Grower:** Klostermühle Odernheim (Chardonnay, Spätburgunder)

SCHLOßBÖCKELHEIM

Vineyard: ✔ Kupfergrube • **Growers:** Dr Crusius (Riesling GG), Gut Hermannsberg (Riesling GG), Schäfer-Fröhlich (Riesling GG)
Vineyard: Felsenberg • **Growers:** Dr Crusius (Riesling GG, Prädikate), Hermann Dönnhoff (Riesling GG), Gut Hermannsberg (Riesling GG), Schäfer-Fröhlich (Riesling GG)
TRAISEN

Vineyard: ✔ Bastei • **Growers:** Dr Crusius (Riesling GG, Prädikate), Gut Hermannsberg (Riesling GG)
Vineyard: Rotenfels • **Grower:** Dr Crusius (Riesling trocken)

GROßLAGE KRONENBERG
BAD KREUZNACH

Vineyard: ✔ Paradies • **Grower:** Korrell-Johanneshof (Riesling trocken, Prädikate)

GROßLAGE PARADIESGARTEN
MEDDERSHEIM

Vineyard: ✔ Rheingrafenberg • **Growers:** Bamberger (Riesling feinherb, Prädikate; Grauburgunder, Weißburgunder), Hexamer (Riesling trocken, Prädikate)
MONZINGEN

Vineyard: Auf der Ley • **Grower:** Emrich-Schönleber (Riesling GG)
Vineyard: ✔ Frühlingsplätzchen • **Grower:** Emrich-Schönleber (Riesling GG, Prädikate)
Vineyard: Halenberg • **Growers:** Emrich-Schönleber (Riesling GG, Prädikate), Schäfer-Fröhlich (Riesling GG)
SOBERNHEIM

Vineyard: ✔ Marbach • **Grower:** Hexamer

GROßLAGE IM PFARRGARTEN

Vineyard: ✔ Felseneck • **Grower:** Prinz Salm (Riesling GG, Prädikate)

GROßLAGE ROSENGARTEN
BOCKENAU

Vineyard: ✔ Felseneck • **Grower:** Schäfer-Fröhlich (Riesling GG, Prädikate)
Vineyard: Stromberg • **Grower:** Schäfer-Fröhlich (Riesling GG)
Non-*Einzellagen* **Grower:** Schäfer-Fröhlich (Weißburgunder)

GROßLAGE SCHLOßKAPELLE
BURG LAYEN

Vineyard: ✔ Schloßberg • **Growers:** Schloßgut Diel (Riesling trocken, Prädikate), J B Schäfer (Riesling GG, Prädikate)
DORSHEIM

Vineyard: ✔ Burgberg • **Growers:** Schloßgut Diel (Riesling GG), Kruger-Rumpf (Riesling GG, Prädikate)
Vineyard: Goldloch • **Growers:** Schloßgut Diel (Riesling GG, Prädikate), J B Schäfer (Riesling GG, Prädikate)
Vineyard: Pittermännchen • **Growers:** Schloßgut Diel (Riesling GG, Prädikate), J B Schäfer (Riesling GG, Prädikate)
LAUBENHEIM

Vineyard: ✔ Karthäuser • **Grower:** Tesch
Vineyard: Krone • **Grower:** Montigny (Riesling trocken, St Laurent)
Vineyard: St Remigiusberg • **Grower:** Tesch
Non-*Einzellagen* **Grower:** ✔ Poss (Chardonnay, Grauburgunder, Weißburgunder)
MÜNSTER (-SARMSHEIM)

Vineyard: ✔ Dautenpflänzer • **Growers:** Göttelmann (Riesling trocken), Kruger-Rumpf (Riesling GG)
Vineyard: Rheinberg • **Grower:** Göttelmann (Riesling Prädikate)
Vineyard: Pittersberg • **Grower:** Kruger-Rumpf
Vineyard: Kapellenberg • **Grower:** Göttelmann (Riesling Prädikate)
Vineyard: Im Pittersberg • **Grower:** Kruger-Rumpf (Riesling GG, feinherb, Prädikate)
Non-*Einzellagen* **Wine:** Kruger-Rumpf (red)
WINDESHEIM

Vineyard: Römerberg • **Grower:** Gebrüder Kauer (Riesling trocken, Prädikate, Weißburgunder)
Vineyard: Rosenberg • **Grower:** Lindenhof (Frühburgunder, Spätburgunder)
Vineyard: Sonnemorgen • **Grower:** Lindenhof (Chardonnay, Weißburgunder)
Non-*Einzellagen* **Grower:** ✔ Lindenhof (red)

GROßLAGE SONNENBORN
LANGENLONSHEIM

Vineyards: ✔ Löhrer Berg, Karthäuser, St Remigiusberg • **Grower:** Tesch (Riesling trocken)

The Rheingau

Rheingau Riesling owes its quality and fame to a whim of nature. Between the towns of Mainz and Wiesbaden the river Rhine changes the direction of its flow from northwards to westwards, giving all the vineyards on its right bank between Wiesbaden and Rüdesheim perfect southern exposure, and thus most favourable conditions for the ripening of grapes.

To the east of the bend of this majestic stream, along the north bank of the river Main and just before its confluence with the Rhine, lies the vinous exclave of Hochheim with its Königin Victoriaberg vineyard, named in honour of England's Queen Victoria because this is where her favourite tipple came from. Hochheim proved just one step too far for British royalty to pronounce and for simplicity's sake it was therefore referred to as "Hock". Under this synonym Riesling from the Rheingau, and soon from anywhere else along the Rhine, began to conquer the world.

HOME OF SPÄTLESE

In the year 1100 the monks of the Benedictine abbey of Mainz built a new basilica at a location in the hills of a Rheingau that would become known as Sankt Johannisberg in the middle of the 12th century. When this became the property of the prince-bishop of Fulda in 1716, he ordered the monks to plant its vineyards with 260,000 Riesling vines, making it the first vineyard in history to be entirely given over to this variety. This is also the reason why Riesling subsequently became known under the synonym of Johannisberg Riesling in many parts of the world. In 1775 a courier who had been sent from the estate to Fulda to obtain official permission for the harvest was held up for several weeks on his journey back, and by the time he finally returned, the grapes had already shrivelled and been beset by an unknown fungus. The desperate cellar master still decided to make wine from the unsightly fruit, and to everybody's amazement the result was an utterly delicious sweet wine. The fungus responsible for this unique taste entered the annals of wine history as "noble rot".

QUALITY REQUIREMENTS AND QUALITY WINE PRODUCTION 2017

RHEINGAU'S MINIMUM *OECHSLE*	QUALITY CATEGORY	HARVEST BREAKDOWN
44°	Deutscher Wein	0.3%
53°	Landwein	0.3%
57–60°	* Qualitätswein	57.4%
73–80°	* Prädikatswein	42%

* Minimum *Oechsle* levels vary according to grape variety; those that have a naturally lower sugar content may qualify at a correspondingly lower level.

THE RHEINGAU'S INITIAL CHARTA FOR QUALITY

When towards the end of the 1980s it became obvious that the growing taste for dry wines was not just a passing trend, a number of Rheingau estates became concerned that most of the wines exported were bulk blended and of low commercial quality. Because these properties had traditionally produced naturally drier wines, they believed that a continuance of poor *trocken* wines could damage their own image, so they banded together to protect it. In 1983, the Association of Charta (pronounced "karta") Estates was launched to "further the classic Rheingau Riesling style, to upgrade the quality of Rheingau wines, and to make them unique among wines from other growing areas". Although this could not have succeeded without the active support of almost all the best Rheingau producers, the engine that relentlessly drove the Charta organization was, until his untimely death, Georg Breuer. Under his direction, Charta became the antithesis of any official quality-control system, as it aspired to the highest possible quality, rather than succumbing to the lowest common denominator. When first drawn up, the Charta's rules were uncommonly stringent, yet every few years, Breuer would tighten the ratchet a bit more by instigating even tougher rules.

INTENSIVE VINE-GROWING ZONES OF THE RHEINGAU
(*see also* p437)

This region's vineyards cling to the northern banks of the Rhein and Main rivers, in the area that curves gently between the towns of Mainz and Wiesbaden.

Legend:
- Intensive vine-growing zone
- Großlage boundary

0 mi — 4
0 km — 4

TROCKEN WINES

At that time one in every five bottles of German wine was *trocken*, or dry, but many of these were thin and anaemic. The intrinsically riper grapes grown by the Rheingau's Charta estates produced a fatter, fuller style of wine that adapted naturally to the trocken style, especially when drunk with food. In the wake of this success other German regions followed the Rheingau's lead, and today the well-balanced dry Riesling can be found anywhere, including regions traditionally associated with fruity and noble sweet wines, like the Mosel or Nahe.

Many of the former Charta producers are now members of the VDP, and since they adopted that association's quality classification in 2012, their dry premium wines under the *Großes Gewächs* designation have become drier than they used to be as Charta Erste Gewächse.

WINE CELLAR AT KLOSTER EBERBACH
The Abbey is a former Cistercian monastery near Eltville. From the 12th to 16th century it flourished, principally from the cultivation of vineyards and the production of wine.

FACTORS AFFECTING TASTE AND QUALITY

LOCATION
The Rheingau is a compact region only 36 kilometres (22 miles) long, situated on the northern banks of the Rhine and Main rivers between Bingen and Mainz, from Hochheim in the east to Lorch in the west.

CLIMATE
Dry and steep southerly slopes have proven the ideal environment for the production of fine Riesling and Spätburgunder wines. Mild winters and warm summers with an annual mean temperature of 10.6°C (51°F) provide most beneficial ripening conditions and are supported in their good work by the heat-reflecting water surface of the Rhine.

ASPECT
The vines grow at an altitude of 100 to 300 metres (330 to 985 feet) on a superb, fully south-facing slope.

SOIL
Quartzite and weathered slate-stone in the higher situated sites produce the greatest Riesling, while loam, loess, various types of clay, and sandy gravel soils of the lower vineyards give a fuller, more robust style. The blue phyllite-slate around Assmannshausen is traditionally thought to favour the Spätburgunder.

VITICULTURE & VINIFICATION
In contrast to other German regions, the Rheingau consists of a large number of independent wine estates (500-odd) in a relatively compact area. There are approximately 2,600 private growers, many of whom make and market their own wines, while others supply the region's 10 cooperatives. The Riesling grape variety represents almost 80 per cent of the vines cultivated and it is traditional in this region to vinify all but the sweetest wines in a drier style than would be used in other regions. Assmannshausen is famous for its red wine, one of Germany's best.

GRAPE VARIETIES
Primary varieties: Riesling, Spätburgunder
Secondary varieties: Müller-Thurgau, Silvaner

THE APPELLATIONS OF

THE RHEINGAU

BEREICH JOHANNISBERG

This *Bereich* covers the entire Rheingau. Trading off the famous village of Johannisberg, situated behind Schloß Johannisberg, lesser wines sell well under it. Some producers are using *Großlage* names on labels, and there are some decent wines to be had under the *Großlage* appellations of Daubhaus and Erntebringer.

GROßLAGE BURGWEG

GEISENHEIM

Vineyard: ✓ *Rothenberg* • **Grower:** *Wegeler-Gutshaus Rheingau (Riesling GG, Prädikate)*

Vineyard: *Kläuserweg* • **Grower:** *Goldatzel (Spätburgunder)*

LORCH

Vineyard: *Bodental-Steinberg* • **Grower:** *Paul Laquai (Spätburgunder)*

Vineyard: *Pfaffenwies* • **Growers:** *Altenkirch (Riesling 1G, Spätburgunder 1G), von Kanitz (Riesling GG)*

Vineyard: *Kronenvon Kanitz (Riesling Prädikate)*

Vineyard: *Schlroßsberg* • **Growers:** *Chat Sauvage (Pinot Noir), Eva Fricke (Riesling feinherb, Prädikate), von Kanitz (Spätburgunder), August Kesseler (Riesling feinherb), Paul Laquai (Riesling Prädikate)*

Vineyard: *Kapellenberg* • **Growers:** *Chat Sazuvage (Pinot Noir), Carl Erhard (Riesling trocken, Prädikate), von Kanitz (Riesling GG)*

LORCHHAUSEN

Vineyard: *Seligmacher* • **Growers:** *Eva Fricke (Riesling feinherb), August Kesseler (Riesling GG)*

RÜDESHEIM

Vineyard: *Bischofsberg* • **Grower:** *Carl Erhard (Riesling trocken)*

Vineyard: ✓ *Berg Roseneck* • **Growers:** *Allendorf (Riesling GG); Georg Breuer (Riesling trocken), Dr Corvers-Kauter (Riesling trocken, Prädikate), Carl Erhard (Riesling trocken, Prädikate), Leitz (Riesling GG, Prädikate), Solter (Riesling Sekt)*

Vineyard: *Berg Rottland* • **Growers:** *Allendorf (Riesling Prädikate), Asbach-Kretschmar (Riesling trocken), Georg Breuer (Riesling trocken), Hessische Staatsweingüter Kloster Eberbach (Riesling GG), Carl Erhard (Riesling trocken, Prädikate), Johannishof (Riesling GG, Prädikate), Künstler (Riesling GG), Leitz(Riesling GG), Balthasar Ress (Riesling GG), Wegeler-Gutshaus Rheingau (Riesling Prädikate)*

Vineyard: *Berg Schlroßberg* • **Growers:** *Georg Breuer (Riesling trocken), Bibo Runge (Riesling Prädikate), Dr Corvers-Kauter (Riesling trocken), August Kesseler (Pinot Noir GG), Jörnwein (Riesling trocken), Josef Leitz (Riesling GG), Hessische Staatsweingüter Domäne Assmannshause (Spätburgunder), Hessische Staatsweingüter Kloster Eberbach (Riesling Prädikate), Künstler (Riesling GG), Balthasar Ress (Riesling GG), Schloss Schönborn (Riesling 1G), Wegeler-Gutshaus Rheingau (Riesling GG)*

Vineyard: *Drachenstein* • **Growers:** *Chat Sazuvage (Pinot Noir), Dr Corvers-Kauter (Pinot Noir)*

Vineyards; *Kaisersteinfels, Rosengarten* • **Grower:** *Leitz (Riesling GG)*

Vineyard: *Klosterlay* • **Grower:** *Asbach-Kretschmar (Spätburgunder)*

Vineyard: *Pfaffenberg* • **Grower:** *Schloss Schönborn (Riesling Prädikate)*

Non-*Einzellagen* Grower:s *Bischöfliches Weingut Rüdesheim (Pinot Noir), Leitz (Riesling Reserve)*

Premium Red Wine Blends Growers: *Allendorf (Spätburgunder "Quercus"), Breuer (Spätburgunder "S"), Kesseler (Cuveé Max)*

GROßLAGE DAUBHAUS

HOCHHEIM

Vineyard: ✓ *Domdechaney* • **Grower:** *Domdechant Werner'sches Weingut (Riesling GG)*

Vineyard: *Kirchenstück* • **Growers:** *Domdechant Werner'sches Weingut (Riesling GG), Franz Künstler (Riesling GG, Prädikate)*

Vineyard: *Hölle* • **Grower:** *Domdechant Werner'sches Weingut (Riesling Prädikate), Franz Künstler (Riesling GG)*

Vineyard: *Königin Victoriaberg* • **Grower:** *Joachim Flick (Riesling trocken, Prädikate)*

Vineyard: *Reichestal* • **Grower:** *Franz Künstler (Spätburgunder)*

Vineyard: *Stein* • **Grower:** *Himmel (Spätburgunder)*

Vineyard: *Weiß Erd* • **Grower:** *Franz Künstler(Riesling GG)*

WICKER

Vineyard: ✓ *Nonnberg* • **Grower:** *Joachim Flick (Riesling trocken; Spätburgunder)*

GROßLAGE DEUTELSBERG

HATTENHEIM

Vineyard: *Nussbrunnen* • **Grower:** *Balthasar Ress (Riesling GG)*

Vineyard: ✓ *Pfaffenberg* • **Growers:** *Balthasar Ress (Riesling GG), Domäne Schloss Schönborn (Riesling Prädikate)*

**PREPARING AUTUMN VINES
FOR EISWEIN PRODUCTION**
The plastic covering protects the grapes in this Rheingau vineyard from birds and rain until harvest. Most German winegrowers leave a portion of their grapes on the vine, hoping temperatures fall to the required -7°C (19.4°F).

KIEDRICH

Vineyard: ✓ *Gräfenberg* • **Grower:** *Robert Weil (Riesling GG, Prädikate)*

Vineyard: *Klosterberg* • **Grower:** *Robert Weil (Riesling Prädikate)*

Vineyard: *Turmberg* • **Grower:** *Robert Weil (Riesling Prädikate)*

GROßLAGE HONIGBERG
ERBACH

Vineyard: *Hohenrain* • **Grower:** *Jakob Jung (Riesling GG), Achim von Oetinger (Riesling GG)*

Vineyard: ✓ *Marcobrunn* • **Grower:** *Hessische Staatsweingüter Kloster Eberbach (Riesling GG, Prädikate), Achim von Oetinger (Riesling GG), Schloß Vaux (Riesling Sekt)*

Vineyard: *Michelmark* • **Grower:** *Crass (Merlot)*

Vineyard: *Siegelsberg* • **Grower:** *Crass (Riesling, Spätburgunder), Jakob Jung (Riesling GG), Achim von Oetinger (Riesling GG)*

Vineyard: *Steinmorgen* • **Grower:** *Crass (Frühburgunder)*

Vineyard: *Schloßberg* • **Grower:** *Schloß Reinhartshausen (Riesling GG, Prädikate)*

Non-*Einzellagen* Growers: *Sektmanufaktur Bardong (Sekt Chardonnay, Riesling, Spätburgunder Blanc de Noirs), Jakob Jung (Spätburgunder)*

GROßLAGE MEHRHÖLZCHEN
HALLGARTEN

Vineyard: ✓ *Jungfer* • **Growers:** *Bibo Runge (Riesling Prädikate), Prinz (Riesling GG, Prädikate)*

Vineyard: *Schönhell* • **Growers:** *Asbach-Kretschmar (Riesling Prädikate), Prinz (Riesling GG)*

Vineyard: *Hendelberg* • **Growers:** *Peter-Jakob Kühn (Riesling trocken), Prinz (Spätburgunder GG)*

Vineyard: *Würzgarten* • **Growers:** *Spreitzer (Riesling Prädikate)*

GROßLAGE STEIL
ASSMANNSHAUSEN

Vineyard: ✓ *Höllenberg* • **Growers:** *Allendorf (Spätburgunder), Dr Corvers-Kauter (Pinot Noir), August Kesseler (Pinot Noir GG), Hessische Staatsweingüter Domaine Assmannshausen (Spätburgunder), Weingut Krone (Spätburgunder), Künstler (Spätburgunder GG)*

Vineyard: *Frankenthal* • **Grower:** *Robert König (Spätburgunder)*

Non-*Einzellagen* Grower: *Bischöfliches Weingut Rüdesheim (Pinot Noir)*

GROßLAGE STEINMÄCHER
ELTVILLE

Vineyard: ✓ *Sonnenberg* • **Grower:** *J Koegler (red and white)*

Vineyard: *Rheinberg* • **Grower:** *J B Becker (Spätburgunder)*

MARTINSTHAL

Vineyard: *Langenberg* • **Grower:** *Diefenhardt (Riesling GG)*

Vineyard: *Rödchen* • **Grower:** *J B Becker (Riesling Prädikate)*

Vineyard: *Schlenzenberg* • **Grower:** *Diefenhardt /Spätburgunder)*

Vineyard: *Wildsau* • **Grower:** *Diefenhardt (Riesling feinherb, Spätburgunder)*

RAUENTHAL

Vineyard: ✓ *Baiken* • **Grower:** *Hessische Staatsweingüter Kloster Eberbach (Riesling Prädikate)*

Vineyard: *Nonnenberg* • **Grower:** *Georg Breuer (Riesling trocken)*

WALLUF

Vineyard: ✓ *Walkenberg* • **Grower:** *J B Becker (Riesling trocken, Prädikate, Spätburgunder)*

Vineyard: *Steinberg* • **Grower:** *Hessische Staatsweingüter Kloster Eberbach (Riesling Prädikate)*

Vineyard: *Wisselbrunnen* • **Growers:** *Barth (Riesling GG), Kaufmann (Riesling GG), Georg Müller Stiftung (Riesling Prädikate), Balthasar Ress (Riesling GG), Josef Spreitzer (Riesling GG)*

Vineyard: *Hassel* • **Growers:** *Barth (Riesling GG, Sekt Brut Nature), Kaufmann (Pinot Noir), Georg Müller Stiftung (Riesling GG)*

Vineyard: *Schützenhaus* • **Grower:** *Barth (Riesling GG, Sekt Brut Nature)*

Non-*Einzellagen* Growers: *Georg Müller Stiftung (Spätburgunder Hommage a Georg), Balthasar Ress (Pinot Noir Caviar)*

GROßLAGE ERNTEBRINGER
GEISENHEIM

Vineyard: *Kläuserweg* • **Grower:** *Sohns (Riesling feinherb, 1G, Prädikate)*

Vineyard: ✓ *Rothenberg* • **Growers:** *Schumann-Nägler (Riesling 1G, Prädikate), Wegeler Gutshaus Rheingau (Riesling GG, Prädikate)*

Non-*Einzellagen* Grower: *Solveigs (Pinot Noir)*

JOHANNISBERG

Vineyard: ✓ *Hölle* • **Growers:** *Chat Sauvage (Pinot Noir), Johannishof (Riesling GG)*

Vineyard: *Schloß Johannisberg* • **Growers:** *Domäne Schloß Johannisberg (Riesling GG, Prädikate), Trenz (Riesling Prädikate)*

Non-*Einzellagen* Grower: *Chat Sauvage (Chardonnay, Pinot Noir)*

MITTELHEIM

Vineyard: ✓ *St Nikolau* • **Growers:** *Peter Jakob Kühn (Riesling GG), F B Schönleber (Riesling GG), Spreitzer (Riesling GG)*

Non-*Einzellagen* Grower: *Perter Jakob Kühn (Riesling Schlehdorn, "R")*

WINKEL

Vineyard: *Dachsberg* • **Grower:** *Prinz von Hessen (Riesling GG)*

Vineyard: *Hasensprung* • **Growers:** *Bibo Runge (Riesling Prädikate), Prinz von Hessen (Riesling GG, Prädikate)*

Vineyard: *Jesuitengarten* • **Growers:** *Allendorf (Riesling GG), F B Schönleber (Riesling GG), Wegeler Gutshaus Rheingau (Riesling GG)*

Vineyard: ✓ *Schloß Vollrads* • **Grower:** *Schloß Vollrads (Riesling GG, Prädikate)*

GROßLAGE GOTTESTHAL
OESTRICH

Vineyard: ✓ *Lenchen* • **Growers:** *Ferdinand Abel (Riesling 1G), Peter Jakob Kühn (Riesling GG), Josef Spreitzer (Riesling Prädikate), Querbach (Riesling Prädikate)*

Vineyard: *Doosberg* • **Growers:** *Josef Spreitzer (Riesling trocken), Querbach (Riesling Prädikate)*

Vineyard: *Klosterberg* • **Grower:** *Peter Jakob Kühn (Riesling trocken)*

Non-*Einzellagen* Grower: *Ferdinand Abel (Spätburgunder)*

GROßLAGE HEILIGENSTOCK

The best product of this small *Großlage* is the fabulous, peach-and-honey Riesling of Kiedrich, often culminating in some of the world's most noble sweet wines. The *Großlage* wine can also be good.

THE REGION AT A GLANCE

AREA UNDER VINE
3,191h (885 ac)

AVERAGE YIELD (2008–2017)
67hl/ha (increasing)

RED WINE
7%

WHITE WINE
86%

ROSÉ
8%

MOST IMPORTANT GRAPE VARIETIES
79% Riesling, 12% Spätburgunder, 9% others

INFRASTRUCTURE
Bereiche 1, *Großlagen* 10, *Einzellagen* 129

Note: The vineyards of Rheingau straddle 28 *Gemeinden* (communes), the names of which may appear on the label.

Rheinhessen

Rheinhessen is the largest of all Germany's
Qualitätswein regions and produces a
quarter of the country's wine.

The name Rheinhessen is slightly misleading because this region does not lie in
the state of Rheinhessen, but in
Rheinland-Pfalz. The diversity of its soils
and grape varieties makes it impossible to
convey a uniform impression of its wines.
Up to the 1980s Rheinhessen was invariably associated with mass production, but
since then a quantum leap in quality and
well-concerted marketing efforts by various
growers' alliances have led to a total change of
image. Müller-Thurgau, stalwart of the
cheap and not always cheerful, is on the
retreat, the classics Riesling and Silvaner
are regaining territory, and the Pinot varieties Weißburgunder, Grauburgunder, and
Spätburgunder are the new contenders.

Rheinhessen is indisputably linked with
Liebfraumilch – in the 18th century still a
noble wine from the famous Liebfraumilch-
Kirchenstück vineyard in Worms – which lost its
good reputation when wines that were even surplus
to *Großlagen* requirements, found their final resting
place under this ubiquitous blend. Brands like Blue Nun
or Black Tower were cleverly marketed as the *grand crus* of
Liebfraumilch, and were probably quite superior to those that
simply sold on price. Nierstein carries a similar stigma to that of
Liebfraumilch in the minds of the older generation of drinkers, due
to the huge quantities of the cheap Bereich Nierstein and Niersteiner
Gutes Domtal that flooded the market and, in their wake, did so much
damage to the reputation of the high-quality Niersteiner *Einzellagen*.

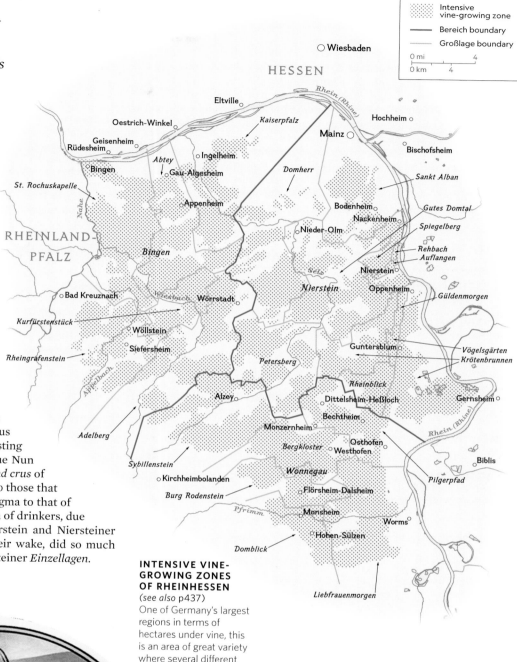

**INTENSIVE VINE-
GROWING ZONES
OF RHEINHESSEN**
(*see also* p437)
One of Germany's largest
regions in terms of
hectares under vine, this
is an area of great variety
where several different
grapes are grown.

VINTAGE BLUE NUN WINE LABEL
Blue Nun Liebfraumilch introduced German wines to the World War II generation. By
the 1970s it was heavily advertised in American markets and sales rose further. By the
1990s semi-sweet wines were losing popularity, and Blue Nun was considered dated.

QUALITY REQUIREMENTS AND QUALITY WINE PRODUCTION 2017

RHEINHESSEN'S MINIMUM *OECHSLE*	QUALITY CATEGORY	HARVEST BREAKDOWN
44°-53°	Deutscher Wein / Landwein	7.7%
53°	Landwein	0.5%
60-62°	* Qualitätswein	77.8%
73-76°	* Prädikatswein	14.5%

* Minimum *Oechsle* levels vary according to grape variety; those that
have a naturally lower sugar content may qualify at a correspondingly
lower level.

ALL CHANGE: THE NEW RHEINHESSEN

If for many decades the fame of the region relied solely on the so-called Rheinfront between Nackenheim and Schwabsburg, with the legendary single sites of Nierstein in its epicenter, in the last 20 years the wines from the *Bereich* Nierstein were overshadowed for quite a while by those of the *Bereich* Wonnegau to the south of it. Unspectacular villages, which can't even call any steep slopes their own with which to impress wine tourists, were pushed into the limelight by a new generation of growers, who came (home from wine college), saw (the potential of their *terroir*), and conquered (with dry Riesling such as the wine world had never seen). And whereas the old guard might have still viewed even their closest neighbours as rivals, their successors work closely together in alliances such as Maxime Herkunft Rheinhessen, Message in a Bottle, or Selection Rheinhessen, not just to support one another with advice and exchange of information, but also carry the message of the new, quality-oriented Rheinhessen to a worldwide audience.

THE REGION AT A GLANCE

AREA UNDER VINE
26,617ha, or 65,77ac (increasing)

AVERAGE YIELD (2008–2017)
93hl/ha

RED WINE
64%

WHITE WINE
86%

ROSÉ
9.3%

MOST IMPORTANT GRAPE VARIETIES
17.4% Riesling (increasing), 16.1%

Müller-Thurgau (decreasing), 12.6% Dornfelder (decreasing), 8.4% Silvaner (decreasing), 4.4% Portugieser (decreasing), 3% Kerner (decreasing), 5.5% Spätburgunder, 6.5% Grauburgunder, 4.9% Weißburgunder, 24% others

INFRASTRUCTURE
Bereiche 3, *Großlagen* 24, *Einzellagen* 414

Note: The vineyards of Rheinhessen straddle 167 *Gemeinden* (communes), the names of which may appear on the label.

FACTORS AFFECTING TASTE AND QUALITY

LOCATION
To the north and east the Rhine represents the boundaries of the region. To the west it borders onto the Nahe region, to the south onto the Pfalz.

CLIMATE
With annual averages of 1,600 hours of sunshine, 10°C (50°F) temperature, and 50 centimetres (20 inches) of rain, Rheinhessen enjoys a fairly warm and dry climate. The Donnersberg mountain range offers protection against rains coming in from the west, while the hills of the Taunus and Odenwald keep out most of the cold northerly winds.

ASPECT
Steep slopes are only to be found in the area around Nierstein, Nackenheim, and Bingen, while the landscape of the hinterland is shaped by gently undulating hills.

SOIL
Mainly loess deposited during inter-glacial sandstorms, but also limestone, sandy-marl, quartzite, porphyry-sand, and silty-clay. Riesling growers favour heavier marl soil, except near Bingen, where there is an outcropping of quartzite-slate. A red, slaty-sandy clay soil known as *rotliegendes* is found on the best, steep riverfront vineyards of Nackenheim and Nierstein. KP Keller, Spanier, and Gunderloch have given these sites a new lease of life.

VITICULTURE & VINIFICATION
There are still many part-time growers, who supply the large bottling merchants and cooperatives, but the region has seen a true revival of individual estates, propelled by an innovative, enterprising band of winemakers. Klaus Peter Keller, Philipp Wittmann, and Oliver Spanier are among the prophets of a new quality-driven ideology. Flörsheim-Dalsheim and Westhofen represent the new Mecca and Medina, Hubacker and Morstein the new temples.

GRAPE VARIETIES
Primary varieties: Riesling, Silvaner, Müller-Thurgau, Dornfelder
Secondary varieties: Spätburgunder, Weißburgunder, Grauburgunder, Bacchus, Faberrebe, Huxelrebe, Kerner, Morio-Muskat; Portugieser and Scheurebe are rapidly losing ground.

THE APPELLATIONS OF

RHEINHESSEN

BEREICH BINGEN

Abutting the Nahe region to the west and separated by the Rhine from the Rheingau to the north, *Bereich* Bingen is the smallest of Rheinhessen's three *Bereiche* and the least important in terms of both the quantity and quality of the wine produced.

GROSSLAGE ABTEY
APPENHEIM

Vineyard: *Hundertgulden* • **Growers:** *Hofmann (Riesling trocken), Knewitz (Riesling trocken)*

Vineyard: *Eselspfad* • **Grower:** *Knewitz (Weissburgunder trocken)*
GAU-ALGESHEIM

Non-*Einzellagen* **Grower:** *Teschke (Silvaner trocken, Spätburgunder trocken)*
NIEDER-HILBERSHEIM

Vineyard: *Steinacker* • **Grower:** *Knewitz (Riesling trocken, Prädikate)*

Non-*Einzellagen* **Grower:** *Knewitz (Chardonnay)*

GROSSLAGE ADELBERG
WENDELSHEIM

Non-*Einzellagen* **Grower:** *Bäder (Riesling trocken, Grauburgunder)*

GROSSLAGE KAISERPFALZ
INGELHEIM

Vineyard: *Horn* • **Grower:** *J Neus (Spätburgunder GG)*
Vineyard: *Pares* • **Grower:** *J Neus (Spätburgunder GG)*
Vineyard: ✓ *Sonnehang* • **Grower:** *Arndt F Werner (Portugieser)*

GROSSLAGE KURFÜRSTENSTÜCK

GROSSLAGE RHEINGRAFENSTEIN
SIEFERSHEIM

Vineyard: ✓ *Heerkretz* • **Growers:** *Bischel (Riesling GG, Prädikate), Wagner-Stempel (Riesling GG, Prädikate)*
Vineyard: *Höllberg* • **Grower:** *Wagner-Stempel (Riesling GG, Prädikate)*

GROSSLAGE SANKT ROCHUSKAPELLE
BINGEN

Vineyard: *Scharlachberg* • **Grower:** *Bischel (Riesling Prädikate)*

BEREICH NIERSTEIN

Although Nierstein is a famous *Bereich* with many superb sites and great growers, most wines sold under its label include some of the most dull, characterless, and lacklustre in all Germany. Wine enthusiasts will find it best to choose the great *Einzellagen* and leave these *Großlage* products alone.

GROSSLAGE AUFLANGEN
NIERSTEIN

Vineyard: *Schloss Schwabsburg* • **Grower:** *Georg Gustav Huff (Riesling Prädikate)*

Vineyard: *Oelberg* • **Growers:** *Fritz Ekkehard Huff (Riesling Prädikate), Kühling-Gillot (Riesling GG), St Antony (Blaufränkisch Lange Berg)*

GROSSLAGE DOMHERR
SAULHEIM

Vineyard: *Hölle* • **Grower:** *Thörle (Riesling trocken, Prädikate, Spätburgunder)*

Vineyard: *Probstey* • **Grower:** *Thörle (Spötburgunder)*

GROSSLAGE GÜLDENMORGEN
OPPENHEIM

Vineyard: ✓ *Sackträger* • **Grower:** *Manz*

Non-*Einzellagen* **Grower:** *Kühling-Gillot (Chardonnay)*

VINEYARDS ABUT A HOUSING PROJECT IN WORMS
Since Roman times, Worms has been linked with the cultivation of grapes and the production of wine, and even the modern parts of the city are surrounded by vineyards.

GROßLAGE GUTES DOMTAL

This district covers a vast area of Rhine hinterland behind the better *Großlagen* of Nierstein. Although this *Großlage* encompasses 15 villages, most wine is sold under the ubiquitous Niersteiner Gutes Domtal (sometimes Domthal) name. Much is decidedly inferior and cheapens the reputation of Nierstein's truly great wines.

WEINOLSHEIM

Vineyard: ✓ *Kehr* • **Grower:** *Manz (Riesling Prädikate)*
Non-*Einzellagen* **Grower:** *Manz (Red cuvée M)*

GROßLAGE KRÖTENBRUNNEN

GROßLAGE PETERSBERG

GROßLAGE REHBACH

NIERSTEIN

Vineyard: ✓ *Hipping* • **Growers:** *Georg Albrecht Schneider, St Antony (Riesling GG), Gunderloch (Riesling GG), Keller (Riesling GG, Prädikate), Kühling-Gillot (Riesling GG, Prädikate), Schätzel (Riesling GG, Prädikate)*
Vineyard: ✓ *Pettenthal* • **Growers:** *St Antony (Riesling GG), Gunderloch (Riesling GG), Keller (Riesling GG, Prädikate), Kühling-Gillot (Riesling GG, Prädikate), St Antony (Blaufränkisch Rothe Bach), Schätzel (Riesling GG, Prädikate)*

GROßLAGE RHEINBLICK

GROßLAGE ST ALBAN

GROßLAGE SPIEGELBERG

NACKENHEIM

Vineyard: ✓ *Rothenberg* • **Growers:** *Gunderloch (Riesling GG, Prädikate), Kühling-Gillot (Riesling GG, Prädikate)*

GROßLAGE VOGELSGARTEN

BEREICH WONNEGAU

The least-known of Rheinhessen's three *Bereiche*, Wonnegau contains the world-famous (although not world-class) Liebfrauenstift *Einzellage*, which had the rather dubious honour of giving birth to Liebfraumilch. *Wonnegau* means "province of great joy".

GROßLAGE BERGKLOSTER

WESTHOFEN

Vineyard: ✓ *Abtserde* • **Grower:** *Keller (Riesling GG, Prädikate)*
Vineyard: *Aulerde* • **Growers:** *Wittmann (Riesling GG), K F Groebe (Riesling GG, Prädikate)*
Vineyard: *Brunnenhäuschen* • **Growers:** *Gutzler (Spätburgunder GG), Wittmann (Riesling GG)*
Vineyard: *Morstein* • **Growers:** *Wittmann (Riesling GG, Prädikate), Gutzler (Spätburgunder GG), Keller (Riesling GG, Prädikate, Spätburgunder GG), Seehof-Ernst Fauth (Riesling trocken, Prädikate, Scheurebe Prädikate)*
Vineyard: *Kirchspiel* • **Growers:** *Wittmann (Riesling GG), K F Groebe (Riesling GG, Prädikate), Keller (Riesling GG)*

GROßLAGE BURG RODENSTEIN

DAHLSHEIM
(an Ortsteil of Flörsheim-Dalsheim)

Vineyard: ✓ *Hubacker* • **Grower:** *Keller (Riesling GG, Prädikate)*
Vineyard: *Bürgel (red)* • **Grower:** *Geils Sekt- und Weingut (Spätburgunder), Keller (Spätburgunder GG)*
Non-*Einzellagen* **Growers:** *Keller (Riesling G-Max), Raumland (Sekt Chardonnay, Pinot Noir, Riesling)*

NIEDERFLÖRSHEIM
(an Ortsteil of Flörsheim-Dalsheim)

Vineyard: ✓ *Frauenberg* • **Growers:** *Battenfeld-Spanier (Riesling GG), Keller (Spätburgunder GG)*

GROßLAGE DOMBLICK

HOHEN-SÜLZEN

Vineyard: ✓ *Kirchenstück* • **Grower:** *Battenfeld-Spanier (Riesling GG)*

MÖLSHEIM

Vineyard: *Zellerweg am schwarzen Herrgott* • **Grower:** *Battenfeld-Spanier (Riesling GG, Prädikate)*

MÖRSTADT

Vineyard: *Im Nonnegarten* • **Grower:** *Milch (Chardonnay)*
Vineyard: *Im Wasserland* • **Grower:** *Milch (Chardonnay)*

MONSHEIM

Vineyard: *Im Blauarsch* • **Grower:** *(Chardonnay, Grauer Burgunder trocken)*
Vineyard: ✓ *Silberberg* • **Grower:** *Keller (Riesling Prädikate)*

GROßLAGE GOTTESHILFE

BECHTHEIM

Vineyard: ✓ *Geyersberg* • **Growers:** *Dreißigacker (Riesling trocken, Chardonnay), Joahnn Geil Erben (Riesling Prädikate)*

GROßLAGE LIEBFRAUENMORGEN

This familiar sounding *Großlage* includes the famous Liebfrauenstift-Kirchenstück vineyard in Worms, the birthplace of Liebfraumilch.

WORMS

Vineyard: ✓ *Liebfrauenstift-Kirchenstück* • **Grower:** *Gutzler (Riesling GG)*

GROßLAGE PILGERPFAD

DITTELSHEIM

Vineyards: ✓ *Geyersberg, Leckerberg, Kloppberg* • **Grower:** *Winter (Riesling GG)*

MONZERNHEIM

Non-*Einzellagen* **Grower:** *Weedenborn (Sauvignon Blanc)*

GROSSLAGE SYBILLENSTEIN

BEST ORGANIC AND BIODYNAMIC PRODUCERS

- Alte Schmiede (Siefersheim)
- Barth (Alzey)
- Battenfeld-Spanier (Hohen-Sülzen)
- Brüder Dr Becker (Ludwigshöhe)
- Dreissigacker (Bechtheim)
- Feth-Wehrhof (Flörsheim-Dalsheim)
- Frey (Ober-Flörsheim)
- A Gänz (Hackenheim)
- Götz (Uelversheim)
- Goldschmidt (Worms-Pfeddersheim)
- Gysler (Alzey-Weinheim)
- Hemer (Worms-Abenheim)
- Hirschhof (Westhofen)
- Hothum (Aspisheim)
- Huster (Ingelheim)
- Julius (Gundheim)
- Kampf (Flonheim)
- Knoböoch (Ober-Flörsheim)
- Kronenhof (Gau-Algesheim)
- Kühling (Gundheim)
- Kühling-Gillot (Bodenheim)
- Landgraf (Saulheim)
- Lorenz (Friesenheim)
- Karl May (Osthofen)
- Meyerhof Flonheim)
- Neverland (Vendersheim)
- Raumland (Flörsheim-Dalsheim)
- Riffel (Bingen)
- Sander (Mettenheim)
- St Antony (Nierstein)
- Eugen Schönhals (Biebelnheim)
- Strauch Sektmanufaktur (Osthofen)
- Dr Eva Vollmer (Mainz-Ebersheim)
- Wagner-Stempel (Siefersheim)
- Wedekind (Nierstein)
- Weinreich (Bechtheim)
- Weitzel (Ingelheim-Grosswinternheim)
- Arndt F Werner (Ingelheim)
- Wittmann (Westhofen)

The Pfalz

The Pfalz is Germany's second-largest wine region, bordering onto Rheinhessen in the north and the French wine region of Alsace to the south. With protection from westerly winds and rains by the Haardt mountain range and the country's largest forest, the Pfälzer Wald, it enjoys a warm and fairly dry climate. On a stretch of 80 kilometres (50 miles) to Pfeddersheim, it features the famous Deutsche Weinstraße, established in 1935 and thus the first official wine route of the world. Divided into the two Bereiche of Mittelhaardt and Südliche Weinstraße, the region can probably also boast more annual wine festivals than any other wine district in the world, with the Dürkheimer Wurstmarkt on the first two weekends of September attracting more than 600,000 visitors every year.

In times gone by the Mittelhaardt – with such solubrious flagship estates as Bassermann- Jordan, von Buhl, Bürklin-Wolf, and von Winning – may have got all the plaudits, but there is certainly healthy competition today from the north and the south of the region. The rather unspectacular villages of Laumersheim and Kindenheim in the lower Haardt, as well as Schweigen, Birkweiler and Siebeldingen in the upper Haardt, might not be able to rival the charm and beauty of places such as Deidesheim, Forst, and Wachenheim in the Mittelhaardt, but they can still certainly match them for the best of dry Riesling, and they are in a class of their own when it comes to top-quality red wines made from the Spätburgunder grape.

FROM ROMAN TO FRENCH EMPERORS

Celtic graves from 550 BC show evidence of winemaking in the Pfalz, but as with so many regions in Europe, it was the Romans who truly brought the culture of wine here, for which there is even the physical evidence of a 1,600 glass amphorae, filled with a syrupy golden liquid, at the wine museum of Speyer. In the Middle Ages the Palatinate was referred to as the "wine cellar of the Holy Roman Empire", and the bishop of Speyer owned the best vineyards of the region until they were usurped by Napoleon.

After the great Corsican emperor somewhat reluctantly left the region, the socio-economic composition of the Pfalz changed dramatically and irrevocably. With the restructuring came a considerably less monopolistic form of land ownership.

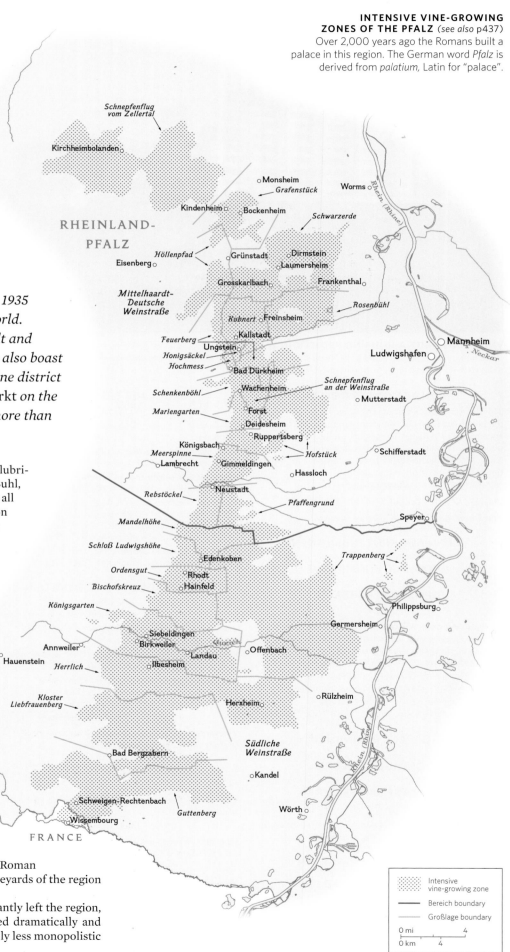

INTENSIVE VINE-GROWING ZONES OF THE PFALZ *(see also p437)* Over 2,000 years ago the Romans built a palace in this region. The German word *Pfalz* is derived from *palatium*, Latin for "palace".

Legend:
- Intensive vine-growing zone
- Bereich boundary
- Großlage boundary

0 mi 4
0 km 4

FACTORS AFFECTING TASTE AND QUALITY

LOCATION
The second-largest German wine region, it stretches 80 kilometres (50 miles) from Rheinhessen to Alsace, bounded by the Rhine on the east and the Haardt Mountains on the west.

CLIMATE
The Pfalz is the sunniest and driest wine-producing region in Germany, sheltered by the Haardt Mountains and Donnersberg hills.

ASPECT
Vineyards are sited mainly on flat land or gentle slopes, at an altitude of between 100 and 250 metres (330 and 820 feet).

SOIL
Loam is prevalent, often interspersed with other soils, from loess, sand, and weathered sandstone to widely dispersed "islands" of limestone, granite, porphyry, and clayish slate.

VITICULTURE & VINIFICATION
The Pfalz vies with Rheinhessen as the largest wine producer among Germany's wine regions. The leading estates have a relatively high proportion of Riesling in their vineyards, and produce wines of the very highest quality, although speciality Gewürztraminers and Muskatellers can be extraordinarily good when vinified dry. One-third of Pfalz wine is sold directly to consumers and half is marketed through large commercial wineries and two dozen or so cooperatives.

GRAPE VARIETIES
Primary varieties: Riesling, Dornfelder
Secondary varieties: Müller-Thurgau, Gewürztraminer, Kerner, Portugieser, Spätburgunder, Grauburgunder, Weißburgunder

QUALITY REQUIREMENTS AND QUALITY WINE PRODUCTION 2017

THE PFALZ'S MINIMUM OECHSLE	QUALITY CATEGORY	HARVEST BREAKDOWN
44°–50°	Deutscher Wein / Landwein	4.5%
60–62°	* Qualitätswein	76%
73–76°	* Prädikatswein	19.5%

* Minimum *Oechsle* levels vary according to grape variety; those that have a naturally lower sugar content may qualify at a correspondingly lower level.

FOOTPATH MARKER ON THE *DEUTSCHE WEINSTRAßE*
The German Wine Route begins in Schweigen-Rechtenbach on the French border and continues 85 kilometres (53 miles) northward through Pfalz until it reached Bockenheim.

BEST ORGANIC AND BIODYNAMIC PRODUCERS

- Ackermann (Ilbesheim)
- Andres (Ruppersberg)
- Andres&Mugler (Ruppertsberg)
- Bassermann-Jordan (Deidesheim)
- Benzinger (Kirchheim)
- Bergdolt-St Lamprecht (Neustadt-Duttweiler)
- Bernhart (Schweigen-Rechtenbach)
- Brand (Dittelsheim-Hessloch)
- Von Buhl (Deidesheim)
- Christmann (Neustadt-Gimmeldingen)
- Ehrhart (Eschbach)
- Eymann (Gönnheim)
- Fitz-Ritter (Bad Dürkheim)
- Fusser (Niederkirchen)
- Gabel Herxheim am Berg)
- Hahn-Pahlke (Battenberg)
- Frank John (Neustadt)
- Andreas Kopf (Landau)
- Kranz (Ilbesheim)
- Lebenshilfe (Bad Dürkheim)
- Leiner (Ilbesheim)
- Lucashof (Forst)
- Manderschied (Kapellen-Drusweiler)
- Mehling (Deidesheim)
- Theo Minges

- (Flemlingen)
- Müller-Catoir (Neustadt-Haardt)
- Nauerth-Gnägy (Schweigen-Rechtenbach)
- Neuspergerhof (Rohrbach)
- Pfirmann (Landau-Wollmesheim)
- Pflüger (Bad Dürkheim)
- Porzelt (Klingenmünster)
- Ökonomierat Rebholz (Siebeldingen)
- Rings (Freinsheim)
- St Annaberg (Burrweiler)
- Heiner Sauer (Böchingen)
- Schädler (Maikammer)
- Karl Schaefer (Bad Dürkheim)
- Schuhmacher (Herxheim am Berg)
- Schwedhelm (Zellertal)
- Georg Siben (Deidesheim)
- Heinrich Spindler (Forst)
- Stortz-Nicolaus (Neustadt-Diedesfeld)
- Dr Wehrheim (Birkweiler)
- Winterling (Niederkirchen)
- Wöhrle (Bockenheim)

THE APPELLATIONS OF

THE PFALZ

Note: In the Pfalz wine region, each *Großlage* can be pre-fixed only by a specified village name. The village name has been listed here under the name of the *Großlage*.

BEREICH MITTELHAARDT-DEUTSCHE WEINSTRAßE

GROßLAGE FEUERBERG
Bad Dürkheim

GROßLAGE GRAFENSTÜCK
BOCKENHEIM

Vineyard: *Vogelsang* • **Grower:** *Ludi Neiss (Riesling* Prädikate, *Frühburgunder)*

Vineyard: *Schloßberg* • **Grower:** *Ludi Neiss (Chardonnay)*

KINDENHEIM

Vineyard: *Burgweg* • **Grower:** *Ludi Neiss (Riesling trocken,* Prädikate, *Rieslaner* Prädikate, *Weißburgunder)*

Non-*Einzellagen* Grower: *Ludi Neiss (Cabernet Franc, Syrah)*

GROßLAGE HOCHMESS
Bad Dürkheim

BAD DÜRKHEIM

Vineyard: ✓ *Michelsberg* • **Grower:** *Karl Schaefer (Riesling GG)*

Vineyard: *Spielberg* • **Growers:** *Karl Schaefer (Riesling trock-en), Egon Schmitt (Riesling trocken)*

Non-*Einzellagen* Grower: *SOPS Im Dambach (Chardonnay, Spätburgunder, Syrah)*

GGROßLAGE HOFSTÜCK
Deidesheim

RUPPERTSBERG

Vineyard: ✓ *Gaisböhl* • **Grower:** *Bürklin-Wolf (Riesling GG)*

Vineyard: *Reiterpfad* • **Growers:** *Bergdolt-St Lamprecht (Riesling GG), Bürklin-Wolf (Riesling GG), von Buhl (Riesling GG), A Christmann (Riesling GG), von Winning (Riesling GG)*

GROßLAGE HÖLLENPFAD
Grünstadt

ASSELHEIM

Vineyard: *Schloß* • **Grower:** *Matthias Gaul (Pinot Noir)*

Vineyard: *St Stephan* • **Grower:** *Matthias Gaul (Chardonnay), Metzger (Chardonnay)*

Vineyard: *Steinrassel* • **Grower:** *Matthias Gaul (Pinot Noir)*

Non-*Einzellagen* Grower: *Matthias Gaul (Cabernet Franc), Metzger (Pinot Noir)*

GROßLAGE HONIGSÄCKEL
Ungstein

UNGSTEIN

Vineyard: ✓ *Herrenberg* • **Growers:** *Pfeffingen (Riesling GG), Egon Schmitt (Riesling trocken)*

Vineyard: *Weilberg* • **Growers:** *Pfeffingen (Riesling GG), Rings (Riesling GG, Weißburgunder GG), Karl Schaefer (Riesling GG)*

Non-*Einzellagen* Grower: *Pfeffingen (Chardonnay, Scheurebe trocken)*

GROßLAGE KOBNERT
Kallstadt

FREINSHEIM

Vineyard: *Musikantenbuckel* • **Grower:** *Krebs (Viognier)*

Non-*Einzellagen* Growers: *Krebs (Portugieser), Rings (red cuvées)*

HERXHEIM AM BERG

Vineyard: ✓ *Berg Honigsack* • **Growers:** *Gabel (Spätburgunder), Krebs (Riesling trocken)*

KALLSTADT

Vineyard: ✓ *Saumagen* • **Growers:** *Benderhof (Riesling trocken), Koehler-Ruprecht (Riesling trocken, Prädikate), Philipp Kuhn (Riesling GG), Am Nil (Riesling trocken), Rings (Riesling GG)*

Vineyard: *Steinacke* • **Growers:** *Benderhof (Spätburgunder), Rings (Spätburgunder trocken)*

Non-*Einzellagen* Grower: *Rings (Spätburgunder, Syrah, red cuvées)*

GROßLAGE MARIENGARTEN
Forst an der Weinstraße

DEIDESHEIM

Vineyard: ✓ *Kalkofen* • **Grower:** *Bassermann-Jordan (Riesling GG)*

Vineyard: *Kieselberg* • **Growers:** *Georg Mosbacher (Riesling GG), Stern (Riesling GG)*

Vineyard: *Herrgottsacker* • **Grower:** *von Buhl (Riesling GG)*

Vineyard: *Hohenmorgen* • **Grower:** *Bassermann-Jordan (Riesling GG)*

Vineyard: *Langenmorgen* • **Growers:** *Bürklin-Wolf (Riesling GG), A Christmann (Riesling GG)*

Vineyard: *Paradiesgarten* • **Grower:** *von Buhl (Riesling GG)*

Non-*Einzellagen* Grower: *von Buhl (Sekt), von Winning (Riesling, Sauvignon Blanc, Scheurebe* Prädikate)

FORST

Vineyard: ✓ *Freundstück* • **Growers:** *von Buhl (Riesling GG), Georg Mosbacher (Riesling GG)*

Vineyard: *Jesuitengarten* • **Growers:** *Acham-Magin (Riesling GG), Bassermann-Jordan (Riesling GG), von Buhl (Riesling GG), Bürklin-Wolf (Riesling GG), Georg Mosbacher (Riesling GG), Heinrich Spindler (Riesling GG)*

Vineyard: *Kirchenstück* • **Growers:** *Acham-Magin (Riesling GG), Bassermann-Jordan (Riesling GG), Bürklin-Wolf (Riesling GG), Eugen Müller (Riesling trocken)*

THE REGION AT A GLANCE

AREA UNDER VINE
23,652ha (57,800ac)

AVERAGE YIELD (2008–2017)
93hl/ha

RED WINE
30%

WHITE WINE
57%

ROSÉ
13%

MOST IMPORTANT GRAPE VARIETIES
24.8% Riesling, 12.5% Dornfelder, 8.3% Müller-Thurgau, 7.1% Spätburgunder, 6.9% Grauburgunder, 6% Portugieser, 5.4% Weißburgunder, 3.4% Kerner, 25.6% others

INFRASTRUCTURE
Bereiche 2, Großlagen 25, Einzellagen 323

Note: The vineyards of Pfalz straddle 170 *Gemeinden* (communes), the names of which may appear on the label.

Vineyard: *Pechstein* • **Grower:s** *Acham-Magin (Riesling GG), Bassermann-Jordan (Riesling GG), Bürklin-Wolf (Riesling GG), Georg Mosbacher (Riesling GG), Heinrich Spindler (Riesling GG)*

Vineyard: *Ungeheuer* • **Growers:** *Acham-Magin (Riesling GG), Bassermann-Jordan (Riesling GG), Bürklin-Wolf (Riesling GG), Von Buhl (Riesling GG), Georg Mosbacher (Riesling GG,* Prädikate), *Heinrich Spindler (Riesling GG), Stern (Riesling GG)*

WACHENHEIM

Vineyard: *Altenburg* • **Growers:** *Bürklin-Wolf (Riesling trock-en), Zimmermann (Riesling trocken)*

Vineyard: ✓ *Böhlig* • **Grower:** *Bürklin-Wolf (Riesling trocken)*

Vineyard: *Goldbächel* • **Grower:** *Bürklin-Wolf (Riesling trocken)*

Vineyard: *Gerümpel* • **Grower:** *Bürklin-Wolf (Riesling trocken)*

GROßLAGE MEERSPINNE
Neustadt-Gimmeldingen

GIMMELDINGEN

Vineyard: *Kapellenberg* • **Grower:** *A Christmann (Riesling trocken)*

Vineyard: ✓ *Mandelgarten* • **Growers:** *A Christmann (Riesling GG), Müller-Catoir (Riesling GG,* Prädikate)

HAARDT

Vineyard: *Herzog* • **Grower:** *Müller-Catoir (Rieslaner* Prädikate)

Vineyard: ✓ *Bürgergarten* • **Grower:** *Müller-Catoir (Riesling GG)*

Vineyard: *Herrenletten* • **Grower:** *Müller-Catoir (Riesling trocken)*

Vineyard: *Mandelring* • **Grower:** *Müller-Catoir (Scheurebe trocken,* Prädikate)

KÖNIGSBACH

Vineyard: ✓ *Idig* • **Grower:** *A Christmann (Riesling GG, Spätburgunder)*

Vineyard: *Ölberg-Hart Kapelle* • **Grower:** *A Christmann (Riesling GG)*

Non-*Einzellagen* Grower: *Frank John (Riesling trocken, Sekt, Pinot Noir)*

GROßLAGE PFAFFENGRUND
Neustadt-Diedesfeld

GROßLAGE REBSTÖCKEL
Neustadt-Diedesfeld

GROßLAGE ROSENBÜHL
Freinsheim

GROßLAGE SCHENKENBÖHL
Wachenheim

DÜRKHEIM

Vineyard: *Fronhof* • **Grower:** *Pflügler (Spätburgunder).*

GROßLAGE SCHNEPFENFLUG AN DER WEINSTRAßE
Forst an der Weinstraße

GROßLAGE SCHNEPFENFLUG VOM ZELLERTAL
Zell

NIEFERNHEIM

Vineyard: *Apotheker* • **Grower:** *Bremer (Pinot Noir GG)*

ZELL

Vineyard: *Schwarzer Herrgott* • **Grower:** *Philipp Kuhn (Riesling GG)*

Non-*Einzellagen* Grower: *Bremer (Pinot Noir Reserve)*

GROßLAGE SCHWARZERDE
Kirchheim

BISSERSHEIM

Vineyard: *Goldberg* • **Grower:** *Wageck-Pfaffmann (Spätburgunder)*

Non-*Einzellagen* **Grower:** *Wageck-Pfaffmann (Chardonnay, Sauvignon Blanc, Portugieser)*

DIRMSTEIN

Vineyard: ✓ *Mandelpfad* • **Grower:** *Knipser (Riesling GG, Spätburgunder GG)*

GROßKARLBACH

Vineyard: ✓ *Burgweg* • **Growers:** *Knipser, (Spätburgunder GG), Philipp Kuhn (Riesling GG), Wageck-Pfaffmann (Spätburgunder)*

LAUMERSHEIM

Vineyard: ✓ *Kirschgarten* • **Growers:** *Knipser (Weißburgunder GG, Spätburgunder GG), Philipp Kuhn (Pinot Blanc GG, Riesling GG, Prädikate, Pinot Noir GG), Zelt (Riesling trocken, Chardonnay, Weißburgunder)*

Vineyard: *Steinbuckel* • **Growers:** *Knipser (Riesling GG), Philipp Kuhn (Riesling GG, Spätburgunder GG), Zelt (Riesling trocken)*

Non-*Einzellagen* **Growers:** *Knipser (Chardonnay, Spätburgunder, Syrah, red cuvées), Philipp Kuhn (Rieslaner Prädikate, Cabernet Franc)*

BEREICH SÜDLICHE WEINSTRAßE

GROßLAGE BISCHOFSKREUZ

Walsheim

BÖCHINGEN

Vineyard: *Rosenkranz* • **Grower:** *Theo Minges (Weißburgunder GG)*

BURRWEILER

Vineyard: *Auf der Hohl* • **Grower:** *Herbert Messmer (Spätburgunder GG)*

Vineyard: *Schäwer* • **Growers:** *Herbert Messmer (Riesling GG), Theo Minges (Riesling GG)*

Vineyard: ✓ *Schloßgarten* • **Grower:** *Herbert Messmer (Grauburgunder trocken, Prädikate)*

Non-*Einzellagen* **Grower:** *Theo Minges (Chardonnay, Grauer Burgunder)*

FLEMLINGEN

Vineyard: *Vogelsprung* • **Grower:** *Bernhard Koch (Scheurebe Prädikate)*

Vineyard: *Herrenbucke* • **Grower:** *Theo Minges (Gewürztraminer Prädikate)*

GLEISWEILER

Vineyard: ✓ *Hölle* • **Grower:** *Theo Minges (Riesling GG)*

WALSHEIM

Vineyard: ✓ *Silberberg* • **Grower:** *Benhard Koch (Weißburgunder), Karl Pfaffmann (Grauer Burgunder)*

GROßLAGE GUTTENBERG

Schweigen

When is a German wine not German? When it's grown in France, of course, where 130 hectares (321 acres) of Schweigen's "sovereign" vineyards are located. Every year the growers pick their grapes in France and trundle across the border to vinify them. Weird as it might sound, the grapes are German if pressed in Germany, but French if pressed in France! The *terroir* of this *Großlage* is most suited to the Burgunder varietals and Gewürztraminer, reflecting the area's northern extension of the Alsace vineyards.

SCHWEIGEN

Vineyard: ✓ *Sankt Paul* • **Grower:** *Friedrich Becker (Spätburgunder GG)*

Vineyard: *Sonnenberg "Rädling"* • **Growers:** *Bernhart (Weißburgunder GG, Spätburgunder GG, St Laurent), Jülg (Riesling trocken, Weißburgunder, Spätburgunder), Scheu (Riesling trocken, Weißburgunder)*

Vineyard: *Strohlenberg* • **Grower:** *Scheu (Weißburgunder)*

Vineyard: *Kammerberg* • **Grower:** *Friedrich Becker (Spätburgunder GG)*

Vineyard: *Heydenreich* • **Grower:** *Friedrich Becker (Spätburgunder GG)*

Vineyard: *Steinwingert* • **Grower:** *Friedrich Becker (Pinot Noir GG)*

Non-*Einzellagen* **Growers:** *Friedrich Becker (Chardonnay), Jülg (Sauvignon Blanc, Weißburgunder, Pinot Noir), Scheu (Gewürztraminer Prädikate)*

GROßLAGE HERRLICH

Eschbach

ILBESHEIM

Vineyard: ✓ *Kalmit* • **Growers:** *Gies-Düppel (Spätburgunder), Klein (Riesling trocken), Kranz (Riesling GG, Weißburgunder GG, Spätburgunder GG), Leiner (Riesling trocken)*

LEINSWEILER

Vineyard: *Kalmitv* • **Grower:** *Pfirmann (Weißburgunder)*

Vineyard: ✓ *Sonnenberg* • **Grower:** *Siegrist (Riesling GG)*

Non-*Einzellagen* **Grower:** *Siegrist (Chardonnay, Pinot Noir)*

WOLMERSHEIM

Vineyard: *Mütterle* • **Grower:** *Pfirmann (Weißburgunder)*

GROßLAGE KLOSTER LIEBFRAUENBERG

Bad Bergzabern

KLINGENMÜNSTER

Vineyard: *Kirchberg* • **Grower:** *Porzelt (Chardonnay, Silvaner, Weißburgunder)*

Vineyard: *Maria Magdalena* • **Grower:** *Porzelt (Riesling trocken, Portugieser)*

GROßLAGE KÖNIGSGARTEN

Godramstein

ARZHEIM

Vineyard: *Am Fürstenweg* • **Grower:** *Kranz (Chardonnay)*

ALBERSWEILER

Vineyard: *Latt* • **Grower:** *Gies-Düppel (Riesling trocken, Prädikate):*

BIRKWEILER

Vineyard: *Am Dachsberg* • **Growers:** *Gies-Düppel (Riesling trocken), Dr Wehrheim (Riesling trocken)*

Vineyard: ✓ *Kastanienbusch* • **Growers:** *Ökonomierat Rebholz (Riesling GG), Gies-Düppel (Riesling trocken), Siener (Riesling trocken, Prädikate), Kleinmann (Chardonnay), Dr Wehrheim (Riesling GG, Spätburgunder GG), Wolf (Riesling trocken, Weißburgunder)*

Vineyard: *Mandelberg* • **Growers:** *Ökonomierat Rebholz (Weißburgunder GG), Dr Wehrheim (Weißburgunder GG), Gies-Düppel (Weißburgunder)*

Vineyard: *Rosenberg* • **Grower:** *Dr Wehrheim (Chardonnay, Weißburgunder)*

Non-*Einzellagen* **Growers:** *Kleinmann (Syrah), Ökonomierat Rebholz (Chardonnay, Spätburgunder)*

GODRAMSTEIN

Vineyard: ✓ *Münzberg "Schlangenpfiff"* • **Grower:** *Münzberg (Spätburgunder GG)*

SIEBELDINGEN

Vineyard: ✓ *Im Sonnenschein* • **Grower:** *Ökonomierat Rebholz (Riesling GG, Weißburgunder GG, Spätburgunder GG)*

Non-*Einzellagen* **Grower:** *Wilhelmshof (Spätburgunder Sekt)*

GROßLAGE MANDELHÖHE

Maikammer

KIRRWEILER

Vineyard: ✓ *Heiligenberg* • **Grower:** *Faubel (Gewürztraminer trocken)*

Vineyard: *Mandelberg* • **Grower:** *Bergdolt-St Lamprecht (Weißburgunder GG, Riesling Prädikate)*

MAIKAMMER

Vineyard: *Heiligenberg* • **Grower:** *Faubel (Gewürztraminer trocken)*

Vineyard: ✓ *Schlangengässel* • **Grower:** *Faubel (Riesling trocken)*

GROßLAGE ORDENSGUT

Edesheim

EDESHEIM

Vineyard: *Rosengarten* • **Grower:** *Borell-Diehl (Riesling trocken, Spätburgunder)*

HAINFELD

Vineyard: *Kapelle* • **Grower:** *Borell-Diehl (Rieslaner Prädikate)*

Vineyard: *Kirchenstück* • **Grower:** *Klein (Frühburgunder)*

Vineyard: *Letten* • **Growers:** *Borell-Diehl (Spätburgunder), Bernhard Koch (Chardonnay, Piniot Noir)*

Non-*Einzellagen* **Growers:** *Borell-Diehl (Chardonnay, Pinot Noir), Klein (Syrah), Bernhard Koch (Pinot Noir)*

WEYHER

Vineyard: *Michelsberg* • **Grower:** *Meier (Riesling trocken)*

GROßLAGE SCHLOß LUDWIGSHÖHE

Edenkoben

ST MARTIN

Vineyard: *Am Gukuckberg* • **Grower:** *Alois Kiefer (Riesling trocken, Pinot Noir)*

GROßLAGE TRAPPENBERG

Hochstadt

ESSINGEN

Non-*Einzellagen* **Grower:** *Frey (Noble sweet: Chardonnay, Sauvignon Blanc, St Laurent, Merlot, Spätburgunder)*

HOCHSTADT

Non-*Einzellagen* **Grower:** *Stern (Pinot Noir)*

The Hessische Bergstraße

Situated between the Rheinhessen and Franken, the northern tip of Baden's vineyards is the Hessische Bergstraße. This is the smallest and least known of Germany's Qualitätswein regions, and its wines are marked by plenty of fruit and extract, with two-thirds made in a dry style.

This region corresponds to the northern section of the old Roman *strata montana*, or "mountain road", hence Bergstraße. This ancient trade route ran from Darmstadt to Wiesloch, which is south of Heidelberg in what is now the Baden region. The Romans brought viticulture to this area, but without the monasteries, which developed, spread, and maintained the vineyards throughout the medieval period, the tradition of winemaking would have ceased long ago. Earliest documented evidence of winegrowing in the area can be found in the *Codex Laureshamensis* written in the latter part of the 12th century.

It was only in 1971 that the Hessische Bergstraße, formerly part of the Badische Weinstraße, became an autonomous wine region, when, because of its location in the state of Hessen, its wines were no longer recognised as Baden wines.

The vineyards of the Hessische Bergstraße are dotted around picturesque villages and towns in a strip of foothills along the western edge of Odenwald. Protected by the Odenwald, in springtime these hills are redolent with the fragrance of fruit trees in full bloom. Indeed, Odenwald's forested mountains offer such effective protection that the sun-trap vineyards of Bensheim boast the highest annual mean temperatures in Germany. This exceptional heat is, of course, relative to Germany's cool northern climate, and the white wines produced here are rich in extract, with a sound level of acidity. In recent years red wines from the Spätburgunder grape have also begun to make an impact.

The vineyards are farmed by more than 1,000 individual growers with the average-sized plot being barely more than one-third of a hectare (four-fifths of an acre). Most of these growers are part-timers who tend their plots at weekends. Most of the region's wines are consumed locally by a thriving tourist community.

QUALITY REQUIREMENTS AND QUALITY WINE PRODUCTION 2017

HESSISCHE BERGSTRAßE MINIMUM *OECHSLE*	QUALITY CATEGORY	HARVEST BREAKDOWN
44°	Deutscher Wein	-
47°	Landwein	-
50–60°	* Qualitätswein	73%
67–73°	* Prädikatswein	27%

* Minimum *Oechsle* levels vary according to grape variety; those that have a naturally lower sugar content may qualify at a correspondingly lower level.

THE REGION AT A GLANCE

AREA UNDER VINE
462ha, or 1,142ac (decreasing)

AVERAGE YIELD (2008–2017)
65hl/ha (increasing)

RED WINE
19%

WHITE WINE
76%

ROTLING
5%

MOST IMPORTANT GRAPE VARIETIES
45.6% Riesling (decreasing), 10.4% Spätburgunder, 10% Grauburgunder, 34% others

INFRASTRUCTURE
Bereiche 2, *Großlagen* 3, *Einzellagen* 23

Note: The vineyards of the Hessische Bergstraße straddle 10 *Gemeinden* (communes), the names of which may appear on the label.

INTENSIVE VINE-GROWING ZONES OF HESSISCHE BERGSTRAßE
(see also p437)
This area is called the "spring garden", because of the early flowering of its fruit and almond orchards, between which vineyards are planted.

THE HISTORIC OLD TOWN OF HEPPENHEIM
The Heppenheim market square boasts beautiful half-timbered houses. For more then 50 years, this Hessian town has hosted the Bergsträßer Weinmarkt, a major wine festival that lures 80,000 visitors each year. The festival features not only Bergstraße wines and Secco and *Sekt* sparkling wine varieties, but also other regional specialities.

FACTORS AFFECTING TASTE AND QUALITY

LOCATION
The main, 7-kilometre (4.3-mile) short stretch of vineyards lies between Pfungstadt in the north and Heppenheim in the south, to the west of the Odenwald Mountains, with the Rhine to the west and the Main to the north. Three tiny exclaves are dotted around Gross-Umstadt, Ober-Ramstadt, and Neu-Isenburg a little farther north.

CLIMATE
Protected by the mountain range of the Odenwald from easterly air streams, the Hessische Bergstraße enjoys one of the mildest climates of all German regions, with cherry trees, almond trees, and magnolias already in full blossom when the rest of the country is still struggling with the aftermath of winter. The central town of Bensheim is supposed to be one of the warmest places in Germany.

ASPECT
Most of the steeper slopes are located around Heppenheim, Bensheim, and Auerbach. The general terrain is more one of hilly hinterland than steep valleys.

SOIL
Various types of soil include weathered granite, basalt, loess, sandstone, and limestone.

VITICULTURE & VINIFICATION
The vineyards are not contiguous but are often planted on old established terraces among orchards. A great many individuals grow grapes, but much of the crop is processed by local cooperatives. The most important varieties are Riesling, Grauburgunder, and Spätburgunder. Two-thirds of the total wines produced in this region are technically dry (*trocken*). The State Wine Domain in Bensheim is the region's largest estate owner, with a share of 8 per cent of the total area under vine.

GRAPE VARIETIES
Primary varieties: Riesling
Secondary varieties: Müller-Thurgau, Grauburgunder, Spätburgunder, Silvaner, Kerner, Weißburgunder, Dornfelder

THE APPELLATIONS OF

THE HESSISCHE BERGSTRAßE

BEREICH STARKENBURG

This is the larger of this region's *Bereiche* and the best in terms of quality. Riesling is planted in most of its vineyards.

GROßLAGE ROTT
AUERBACH

Vineyard: *Fürstenlager* • **Growers:** Griesel & Compagnie (Sekt Riesling, Spätburgunder, Chardonnay), Schönberg (Grauburgunder, Weißburgunder, Spätburgunder)
Vineyard: ✔ *Höllberg* • **Grower:** Simon-Bürkle (Riesling, Spätburgunder)

GROßLAGE SCHLOßBERG
HEPPENHEIM

Vineyard: ✔ *Centgericht* • **Grower:** Hessische Staatsweingüter Domaine Bergstraße (Riesling, Grauburgunder)
Vineyards: *Eckweg, Maiberg, Schloßberg* • **Grower:** Bergsträsser Winzer (Riesling trocken, Prädikate, Goldmuskateller)

GROßLAGE WOLFSMAGEN
This includes the southern section of Bensheim with the two Ortsteile of Zell and Gronau.

BENSHEIM

Non-*Einzellagen* Grower: Weingut der Stadt Bensheim (Riesling. Goldmuskateller)

BEREICH UMSTADT

Its six *Einzellagen* are *Großlagenfrei*. The grape varieties Müller-Thurgau, Ruländer, and Silvaner dominate.

ROSSDORF

Vineyard: *Rossberg* • **Grower:** Edling (Grauburgunder, Gelber Muskateller trocken, Merlot, Spätburgunder)

Franken

In many parts of the international wine world Franken is still an unknown entity. The traditional bottle shape called Bocksbeutel (literally translated "goat's scrotum") and the Silvaner variety represent centuries of winegrowing history. Yet the present generation of growers is not averse to innovation, as is demonstrated by experiments with new methods of vinification: prolonged maceration of white wines on their skins, the use of amphorae and concrete eggs, and the production of unfiltered PetNat bubbly.

Whereas in all other regions of Germany there's been a serious decline in the cultivation of the classic Silvaner variety, in its homeland of Franken, where it was first planted in 1659 in the vineyards of the Domaine Castell, it has seen a true revival. There can be no doubt that this trend is benefiting from wine drinkers' still-growing preference for *trocken* wines, in which it can demonstrate its natural affinity for minerality. This has always been its strength, and at its finest and freshest it has much in common with the best of dry and steely wines from Chablis. Even Germany's greatest poet laureate Johann Wolfgang von Goethe would get quite miserable when he couldn't lay his hands on a bottle of his favourite Würzburger Stein.

Unfortunately efforts to pep up the somewhat stuffy image of the *Bocksbeutel* with a modernised design was not successful, as no agreement could be reached on what particular style or quality of wine it should be used for. For the cooperatives it will remain a strong sales point, but many individual growers are beginning to switch to flute- and Burgundy-shaped bottles.

Although most of Franken's vines grow on or near the slopes along the river Main, the regions' vineyards are very fragmented and are best categorised by three distinctive soil types. At the eastern end in the *Bereich* Steigerwald the soil is composed mainly of keuper, which tends to yield sturdy, earthy, powerful wines, while the shell limestone-dominated strata around Würzburg is better known for wines with herbal piquancy and a slightly rounder style. To the west, in the *Bereich* Mainviereck with the vinous hotspots of Bürgstadt and Klingenberg, the red sandstone soils favour the Pinot varieties, with mineral and elegant Spätburgunder wines that have caught the attention of the international wine scene.

Rieslaner (Riesling x Silvaner), Bacchus, and Kerner are all successful, particularly as QmP, although it is rare to find a wine above *Auslese* level

QUALITY REQUIREMENTS AND QUALITY WINE PRODUCTION 2017

FRANKEN'S MINIMUM *OECHSLE*	QUALITY CATEGORY	HARVEST BREAKDOWN
44°	Deutscher Wein	0.4%
50°	Landwein	-
60°	* Qualitätswein	52.6%
76–80°	* Prädikatswein	47%

* Minimum *Oechsle* levels vary according to grape variety; those that have a naturally lower sugar content may qualify at a correspondingly lower level.

INTENSIVE VINE-GROWING ZONES OF FRANKEN *(see also p437)*
This region is in the very heart of Germany. At the centre of Franken lies Würzburg, which is actually famous for its beer. Yet the Würzburger Stein, one of Germany's largest individual plots, has produced Steinwein here since at least the eighth century.

Legend:
- Intensive vine-growing zone
- Bereich boundary
- Großlage boundary

0 mi — 10
0 km — 10

STAATLICHER HOFKELLER WÜRZBURG SILVANER TROCKEN
Franconian wines are traditionally bottled in the distinctively shaped *Bocksbeutel,* which take the form of a flattened ellipsoid.

FACTORS AFFECTING TASTE AND QUALITY

LOCATION
Situated in Bavaria, Franken was the most northerly of Germany's wine regions until reunification.

CLIMATE
This area offers the most continental climate of Germany's wine regions, with dry warm summers and cold winters. Severe frosts may affect yields.

ASPECT
Many vineyards face south and are located on the slopes of the valleys of the Main and its tributaries, as well as on sheltered sites of the Steigerwald.

SOIL
Franken's three *Bereiche* have different soil structures: Mainviereck has predominantly weathered coloured sandstone; Maindreieck has limestone with clay and loess; and Steigerwald is keuper (weathered red marl).

VITICULTURE & VINIFICATION
The classic Franconian vine, the Silvaner, in decline for a while, has seen a revival that puts in on a par with Müller-Thurgau, which makes more dry wine here than in any other region. Rieslaner is something of a local speciality. Bacchus is appreciated for its aromatic Sauvignon-like qualities. Franken wines generally are usually drier than most in Germany and accompany food well. There are 6,000 growers, allowing for a great range of styles, although 40 per cent of the wines are processed by the regional cooperative in Kitzingen and smaller cooperatives. Exports are insignificant, with four out of every five bottles of Franken wine consumed within a 50-kilometre (155-mile) radius of where they were produced.

GRAPE VARIETIES
Primary varieties: Müller-Thurgau, Silvaner
Secondary varieties: Bacchus, Kerner, Rieslaner, Riesling, Scheurebe, Spätburgunder, Traminer (Savagnin Blanc)

THE APPELLATIONS OF
FRANKEN

Note: In Franken, certain *Großlagen* can be prefixed only by a specified village name, which is listed immediately beneath the relevant *Großlage* name.

BEREICH MAINDREIECK
Most of the vineyards in this *Bereich* are in the vicinity of Würzburg. Grapes grown on the limestone soils can produce exceptional wines.

GROßLAGE BURG
Hammelburg

GROßLAGE ENGELSBERG
Vineyard: *Katzenkopf* • **Grower:** *Glaser-Himmelstoss*

GROßLAGE EWIG LEBEN
RANDERSACKER

Vineyard: ✔ *Pfülben* • **Growers:** *Reiss (Riesling* Prädikate), *Schmitt's Kinder (Riesling GG, Silvaner GG)*
Vineyard: *Marsberg* • **Growers:** *Stefan Barsdorf (Riesling* Prädikate, *Silvaner trocken, Prädikate), Schmitt's Kinder (Riesling trocken, Prädikate, Silvaner)*
Vineyard: *Sonnenstuhl* • **Growers:** *Schmitt's Kinder (Riesling* Prädikate, *Spätburgunder GG), Stahl (Silvaner), Am Stein – Ludwig Knoll (Rieslaner* Prädikate)*, Störrlein Krenig (Riesling GG, Silvaner GG, Spätburgunder GG)*
Vineyard: *Teufelskeller* • **Grower:** *Störrlein Krenig (Weißburgunder, Spätburgunder GG)*

Non-*Einzellagen* Grower: *Stahl (Chardonnay, Silvaner)*

GROßLAGE HOFRAT
Kitzingen

SULZFELD

Vineyard: ✔ *Sonnenberg* • **Grower:** *Zehnthof Theo Luckert (Chardonnay, Riesling trocken, Silvaner, Frühburgunder)*
Vineyard: *Maustal* • **Growers:** *Brennfleck (Silvaner trocken), Zehnthof Theo Luckert (Riesling GG, Silvaner GG)*

GROßLAGE HONIGBERG

GROßLAGE KIRCHBERG
Volkach

ASTHEIM

Vineyard: *Karthäuser* • **Grower:** *Rudolf Fürst (Chardonnay)*
ESCHERNDORF

Vineyard: ✔ *Lump* • **Growers:** *Michael Fröhlich (Riesling trocken, Rieslaner Prädikate), Max Müller I (Gemischter Satz, Riesling trocken, Silvaner), Horst Sauer (Riesling Prädikate), Rainer Sauer (Riesling GG)*
Vineyard: *Am Lumpen 1655* • **Growers:** *Michael Fröhlich (Riesling GG, Silvaner GG), Horst Sauer (Riesling GG, Silvaner GG), Rainer Sauer (Riesling Prädikate, Silvaner GG), Egon Schäffer (Riesling GG, Silvaner GG)*
SOMMERACH

Vineyard: ✔ *Katzenkopf* • **Growers:** *Max Müller I (Silvaner), Richard Östreicher (Chardonnay, Weißburgunder, Silvaner)*
VOLKACH

Vineyard: ✔ *Karthäuser* • **Grower:** *Juliusspital (Weißburgunder GG)*

Vineyard: *Ratsherr* • **Grower:** *Max Müller I (Riesling Prädikate, Silvaner trocken)*

GROßLAGE MARIENBERG
WÜRZBURG

Vineyard: ✔ *Stein* • **Growers:** *Bürgerspital zum Heiligen Geist (Chardonnay, Gewürztztraminer, Silvaner GG, Blaufränkisch, Spätburgunder), Juliusspital Würzburg (Riesling GG, Prädikate, Silvaner GG)*

THE REGION AT A GLANCE

AREA UNDER VINE 6,139ha, or 170ac (increasing)	**MOST IMPORTANT GRAPE VARIETIES** 26% Müller-Thurgau, 24% Silvaner, 12% Bacchus (static), 38% others
AVERAGE YIELD (2008–2017) 72 hl/ha	
RED WINE 14%	**INFRASTRUCTURE** *Bereiche* 3, *Großlagen* 22, *Einzellagen* 216
WHITE WINE 78% (decreasing)	Note: The vineyards of Franken straddle 125 *Gemeinden* (communes), the names of which may appear on the label.
ROSÉ 8%	

Vineyard: *Stein Hagemann* • **Grower:** *Bürgerspital zum Heiligen Geist (Riesling GG)*

Vineyard: *Stein Harfe* • **Grower:** *Bürgerspital zum Heiligen Geist (Riesling GG, Silvaner GG)*

Vineyard: *Pfaffenberg* • **Grower:** *Bürgerspital zum Heiligen Geist (Silvaner Prädikate)*

GROßLAGE MARKGRAF BABENBERG
FRICKENHAUSEN

Vineyard: ✓ *Kapellenberg* • **Grower:** *Bickel-Stumpf (Silvaner)*

Vineyard: *Mönchshof* • **Grower:** *Bickel-Stumpf (Silvaner)*

GROßLAGE OELSPIEL

GROßLAGE RAVENSBURG
Thüngersheim
RETZBACH

Vineyard: *Benediktusberg* • **Grower:** *Rudolf May (Spätburgunder)*

Vineyard: *Scharlachberg* • **Grower:** *Reiss (Rieslaner Prädikate)*

THÜNGERSHEIM

Vineyard: *Rothlauf* • **Grower:** *Bickel-Stumpf (Silvaner GG)*

GROßLAGE ROSSTAL
Karlstadt
GAMBACH

Grower: *Giegerich (Chardonnay, Spätburgunder)*

Vineyard: *Kalbenstein* • **Grower:** *Höfling (Spätburgunder GG)*

Non-*Einzellagen* Grower: *Stefan Vetter*

CERTIFIED ORGANIC AND BIODYNAMIC PRODUCERS

- Anton Hell (Wiesenbronn)
- Bausewein (Iphofen)
- Benedikt Baltes (Klingenberg)
- Erwin Christ (Nordheim)
- Deppisch (Theilheim)
- 3 Zeilen (Rödelsee)
- Franziska Schömig (Rimpar)
- Fred Ruppert (Prichstenstadt-Kirch)
- Geier (Königheim)
- Gerhard Roth (Wiesenbronn)
- Hamdorf (Klingenberg)
- Helmut Christ (Nordheim)
- Hench (Bürgstadt)
- Hermann Kramer (Simmershofen)
- Hofmann (Ergersheim)
- Kraemer (Auernhofen)
- Lauerbach (TBB-Impfingen)
- Mainstockheim (Mainstockheim)
- Manfred Rothe (Nordheim)
- Manfred Schwab (Iphofen)
- Rainer Zang (Nordheim am Main)
- Roland Hemberger (Rödelsee)
- Rudolf May (Retzstadt)
- Schloß Saaleck (Hammelburg)
- Schmachtenberger (Randersacker)
- Stritzinger (Klingenberg)
- Stefan Vetter (Gambach)
- Wille Gebrüder (Esselbach)
- Willert-Eckert (Müdersheim)
- Zehnthof-Theo Luckert (Sulzfeld)
- Zehntkeller (Iphofen)

WÜRZBURGER STEIN
Sited high on the hilly slopes overlooking the river Main outside the city of Würzburg, the Würzburger Stein, one of Germany's largest vineyard, has produced *steinwein* in this region since at least the eighth century.

RETZSTADT

Vineyard: *Benediktusberg* • **Grower:** *Rudolf May (Silvaner trocken)*

Vineyard: *Himmelspfad* • **Grower:** *Rudolf May (Silvaner GG)*

Vineyard: *Langenberg* • **Grower:** *Rudolf May (Riesling trocken, Silvaner Prädikate, Weißburgunder)*

Vineyard: *Der Schäfer* • **Grower:** *Rudolf May (Silvaner trocken)*

STETTEN

Vineyard: ✓ *Stein* • **Growers:** *Höfling (Silvaner GG), Am Stein-Ludwig Knoll (Riesling GG, Silvaner GG)*

GROßLAGE TEUFELSTOR
EIBELSTADT

Vineyard: ✓ *Kapellenberg* • **Grower:** *Schloß Sommerhausen*

GROßLAGENFREI

Many vineyards in this *Bereich* are *Großlagenfrei* (composed of individual *Einzellagen* that are not grouped under any *Großlagen*).

HALLBURG

Vineyard: ✓ *Schloßberg* • **Grower:** *Graf von Schönborn Schloß Hallburg (Riesling trocken, Silvaner)*

HOMBURG

Vineyard: ✓ *Kallmuth* • **Grower:** *Fürst Löwenstein estate at Kleinheubach (Riesling GG, Silvaner GG)*

RÖTTINGEN

Vineyard ✓ *Feuerstein* • **Grower:** *Krämer (Silvaner trocken)*

BEREICH MAINVIERECK

Mainviereck is the smallest of the *Bereiche*, as well as being the most westerly, and the wines it produces are modest.

GROßLAGE HEILIGENTHAL

GROßLAGE REUSCHBERG

GROßLAGENFREI

Most of *Bereich* Mainviereck's vineyards are *Großlagenfrei* (individual *Einzellagen* not grouped under any *Großlagen*).

GROßWALLSTADT

Vineyard: *Lützeltalerberg* • **Grower:** *Giegerich (Chardonnay, Spätburgunder)*

WÖRTH AM MAIN

Vineyard: *Campestres* • **Grower:** *Giegerich (Frühburgunder)*

BÜRGSTADT

Vineyard: ✓ *Centgrafenberg* • **Grower:** *Rudolf Fürst (Riesling GG, Spätburgunder GG, Frühburgunder R)*

Vineyard: *Hundsrück* • **Growers:** *Benedikt Baltes (Spätburgunder GG), Rudolf Fürst (Spätburgunder GG), Josef Walter (Spätburgunder)*

Non-*Einzellagen* Grower: *Rudolf Fürst (Chardonnay, Weißburgunder, Frühburgunder)*

GROßHEUBACH

Vineyard: *Bischofsberg* • **Grower:** *Benedikt Baltes (Spätburgunder)*

KLINGENBERG

Vineyard: ✓ *Schloßberg* • **Growers:** *Benedikt Baltes (Spätburgunder), Rudolf Fürst (Spätburgunder GG)*

BEREICH STEIGERWALD

The distinctive, earthy character of Franconian wine is evident in *Bereich* Steigerwald, where heavier soils result in fuller-bodied wines.

GROßLAGE BURGBERG

GROßLAGE BURGWEG
IPHOFEN

Vineyard: ✓ *Julius-Echter Berg* • **Growers:** *Johann Ruck (Riesling trocken, Silvaner GG), Juliusspital Würzburg (Silvaner GG), Ernst Popp (Riesling Prädikate, Silvaner), Hans Wirsching (Riesling GG, Prädikate, Silvaner GG)*

Vineyard: *Kronsberg* • **Grower:** *Hans Wirsching (Riesling GG, Silvaner GG)*

GROßLAGE HERRENBERG
CASTELL

Vineyard: *Schloßberg* • **Grower:** *Castellsches Domänenamt (Silvaner GG)*

GROßLAGE KAPELLENBERG

GROßLAGE SCHILD

GROßLAGE SCHLOßBERG
RÖDELSEE

Vineyard: ✓ *Küchenmeister* • **Growers:** *Drei Zeilen (Silvaner GG, Sauvignon Blanc, Spätburgunder), Roth (Rieslaner Prädikate), Weltner (Riesling trocken, Silvaner trocken, GG)*

Vineyard: *Hoheleite* • **Grower:** *Weltner (Riesling GG, Silvaner GG)*

GROßLAGE SCHLOßTÜCK

GROßLAGE STEIGE

GROßLAGE ZABELSTEIN

Württemberg

Württemberg's wines have not yet made their mark outside the region, but this is about to change. The locals' tradition of a glass or two of the light and mild Trollinger or a rosé Schillerwein at the end of a day's hard work is wavering, while the reputation for great reds from the Lemberger grape is spreading.

With a share of some 70 per cent of its total annual output, Württemberg produces more red wine than any other region in Germany. Around 80 per cent of the harvest is marketed by more than 50 cooperatives. The region is divided into six *Bereiche*: Bayrischer Bodensee, Kocher-Jagst-Tauber, Württembergisch Unterland, Oberer Neckar, Remstal-Stuttgart, and Württembergischer Bodensee. It stretches from Taubergrund in the north to the Bodensee, although most vineyards around the lake belong to the Baden region. The centre of Württemberg's wine industry lies in a patchwork of vineyards between Ludwigsburg, Heilbronn, and the eastern fringe of Stuttgart.

Most of the vineyards are located in the valleys of the river Neckar and its tributaries Enz, Rems, and Tauber. Protection from cold winds offered by the mountains and hillsides of the Black Forest, Odenwald, and Schwäbische Alb guarantees a mild climate in sheltered sites. The mountains rise between 200 and 400 metres (656 to 1,312 feet) high, however, so temperatures there are a little lower than in Baden to the west. Compared to other regions, the area under vine is rather fragmented into many small pockets of vineyard, mostly on southerly slopes with favourable exposure towards the sun. Specialities of Württemberg are the red varieties: Trollinger, Lemberger, and Schwarzriesling (Pinot Meunier), which are rarely found in any other region of Germany.

Yet red is not considered the be all and end all by the region's growers, as a new trend towards light and fresh Riesling *Kabinett* demonstrates. Even a tentative return to a formerly thriving sparkling wine industry can be observed. Kerner on the other hand, a cross of Riesling and Trollinger created at the Württemberg viticultural institute in Heilbronn and quite popular right across Germany in the 1970s and 1980s, has gone into a rapid decline because its mild and sweetish white wines have fallen out of favour.

INTENSIVE VINE-GROWING ZONES OF WÜRTTEMBERG (*see also* p437)
Most of the region's vineyards, in fertile districts near the Neckar River, are interspersed with farmland. In the Middle Ages, Württemberg, together with Franken, had more vines planted than any other part of Germany, four times as many as today.

Map legend:
- Intensive vine-growing zone
- Bereich boundary
- Großlage boundary
- 0 mi — 10
- 0 km — 10

QUALITY REQUIREMENTS AND QUALITY WINE PRODUCTION 2017

WÜRTTEMBERG'S MINIMUM *OECHSLE*	QUALITY CATEGORY	HARVEST BREAKDOWN
40°	Deutscher Wein / Landwein	1%
57–63°	* Qualitätswein	86%
72–78°	* Prädikatswein	13%

* Minimum *Oechsle* levels vary according to grape variety; those that have a naturally lower sugar content may qualify at a correspondingly lower level.

FACTORS AFFECTING TASTE AND QUALITY

LOCATION
The main winegrowing zone lies between the major towns of Stuttgart and Heilbronn, with a small southern outpost on the northern shore of Bodensee (Lake Constance).

CLIMATE
Protected by the mountain ranges of the Black Forest to the west and the Swabian Alb to the southeast, the vineyards of Württemberg enjoy a mild climate with an annual average temperature of 10°C (50°F) and around 1,660 hours of sunshine yearly.

ASPECT
Most vineyards lie on south- and southwest-facing slopes of the river Neckar and several tributaries at an altitude between 200 and 400 metres (656 and 1,312 feet).

SOIL
The vineyards of Württemberg are very fragmented, so there are many different soil types, but various keuper formations are the most prevalent, with pockets of shell limestone in the central Neckar region.

VITICULTURE & VINIFICATION
Unusually for Germany, the proportion of wines vinified mild and sweet is quite high, between 40 and 45 per cent, with truly dry (*trocken*) accounting for less than 30 per cent, and the rest falling into the off-dry (*halbtrocken*) sector.

GRAPE VARIETIES
Primary varieties: Trollinger (Schiava), Lemberger, Schwartzriesling (Pinot Meunier)

THE APPELLATIONS OF
WÜRTTEMBERG

BEREICH BAYERISCHER BODENSEE

GROßLAGE LINDAUER SEEGARTEN

BEREICH KOCHER-JAGST-TAUBER

GROßLAGE KOCHERBERG

GROßLAGE TAUBERBERG

BEREICH OBERER NECKAR

BEREICH REMSTAL-STUTTGART

GROßLAGE HOHENNEUFFEN

GROßLAGE KOPF
KORB

Vineyard: *Steingrüble* • **Grower:** *Escher (Grauburgunder)*
Non-*Einzellagen* **Grower:** *Albrecht Schwegler (red cuvées); Zimmerle (Chardonnay, Lemberger, Zweigelt, red cuvées)*

GROßLAGE SONNENBÜHL

GROßLAGE WARTBÜHL

Non-*Einzellagen* **Grower:** *Escher (Sauvignon Blanc, Cabernet Franc, Merlot, Lemberger, Spätburgunder, red cuvées)*
BEUTELSBACH

Vineyard: *Sonneberg* • **Grower:** *Knauss (Chardonnay)*
HEBSACH

Vineyard: *Berg* • **Grower:** *Jürgen Ellwanger (Lemberger GG)*
Vineyard: *Linnebrunnen* • **Grower:** *Jürgen Ellwanger (Spätburgunder GG)*
Non-*Einzellagen* **Grower:** *Jürgen Ellwanger (Zweigelt,"Nicodemus")*
GROßHEPPACH

Vineyard: *Steingrüble* • **Grower:** *Wolfgang Klopfer (Weißburgunder, Spätburgunder)*
Non-*Einzellagen* **Grower:** *Wolfgang Klopfer (red cuvées)*
SCHNAIT

Vineyard: *Altenberg* • **Grower:** *Knauss (Riesling trocken, Lemberger)*
STETTEN

Vineyard: *Mönchberg-Berge* • **Grower:** *Karl Haidle (Lemberger GG)*
Vineyard: *Mönchberg-Gehrnhalde* • **Grower:** *Karl Haidle (Lemberger GG)*

Vineyard: ✓ *Pulvermächer* • **Growers:** *Karl Haidle (Riesling Prädikate), Beurer (Grauburgunder GG, Riesling GG)*
STRÜMPFELBACH

Vineyard: *Nonnenberg* • **Grower:** *Knauss (Spätburgunder)*

GROßLAGE WEINSTEIGE
BAD CANSTATT

Vineyard: *Zuckerle* • **Grower:** *Klopfer (red cuvée)*
ESSLINGEN

Non-*Einzellagen* **Grower:** *Kusterer (Grauburgunder, Spätburgunder)*
FELLBACH

Vineyard: ✓ *Lämmler* • **Growers:** *Aldinger (Riesling GG, Lemberger GG, Spätburgunder GG), Heid (Lemberger GG, Spätburgunder GG), Schnaitmann (Grauburgunder GG; Riesling GG, Lemberger GG, Spätburgunder GG)*
Non-*Einzellagen* **Grower:** *Schnaitman (Cabernet Franc, Frühburgunder)*
UNTERTÜRKHEIM

Vineyard: *Gips* • **Grower:** *Aldinger (Weißburgunder GG)*
Vineyard: *Gips-Marienglas* • **Grower:** *Aldinger (Riesling GG, Spätburgunder GG)*
Vineyard: *Herzogenberg* • **Grower:** *Wöhrwag (Grauburgunder GG, Riesling GG, Lemberger GG, Pinot Noir GG)*
Non-*Einzellagen* **Grower:** *Aldinger (Cabernet Sauvignon, Sauvignon Blanc; Sekt), Weinmanufaktur Untertürkheim (Lemberger)*

BEREICH WÜRTTEMBERGISCH BODENSEE

A one-*Einzellage Bereich* on Lake Constance.

BEREICH WÜRTTEMBERGISCH UNTERLAND

GROßLAGE HEUCHELBERG
CLEEBRONN

Vineyard: *Michaelsberg* • **Grower:** *Dautel (Lemberger GG)*
Non-*Einzellagen* **Grower:** *Weingärtner Cleebronn-Güglingen (Merlot, Lemberger)*
NEIPPERG

Vineyard: ✓ *Schloßberg* • **Grower:** *Graf zu Neipperg (Weißburgunder GG, Lemberger GG, Spätburgunder GG)*
PFAFFENHOFEN

Vineyard: ✓ *Hohenberg* • **Grower:** *Wachtstetter (Riesling GG, Lemberger GG, Spätburgunder GG)*
SCHWAIGERN

Vineyard: ✓ *Ruthe* • **Grower:** *Graf zu Neipperg (Lemberger GG)*

Non-*Einzellagen* **Grower:** *Graf zu Neipperg (Cabernet Sauvignon, Merlot)*
VAIHINGEN

Non-*Einzellagen* **Grower:** *Sonnenhof (Lemberger)*

GROßLAGE KIRCHENWEINBERG

GROßLAGE LINDELBERG
VERRENBERG

Vineyard: ✓ *Verrenberg* • **Grower:** *Fürst zu Hohenlohe Öhringen (Riesling GG, Lemberger GG, Spätburgunder GG)*

GROßLAGE SALZBERG

GROßLAGE SCHALKSTEIN

GROßLAGE SCHOZACHTAL

GROßLAGE STAUFENBERG
HEILBRONN

Vineyard: *Vorderer Hundsberg* • **Grower:** *G A Heinrich (Lemberger)*
Vineyard: *Stiftsberg* • **Grower:** *Kistenmacher-Hengerer (Spätburgunder GG)*
Vineyard: *Wartberg* • **Grower:** *Kistenmacher-Hengerer (Lemberger GG)*
Non-*Einzellagen* **Grower:** *Drautz-Able ("Jodokus")*
WEINSBERG

Vineyard: *Schemelsberg* • **Grower:** *Staatsweingut Weinsberg (Lemberger GG)*
Non-*Einzellagen* **Grower:** *Staatsweingut Weinsberg (Sauvignon Blanc, Syrah)*

GROßLAGE STROMBERG
BÖNNIGHEIM

Vineyard: ✓ *Schupen* • **Grower:** *Ernst Dautel (Spätburgunder GG)*
Vineyard: *Steingrüben* • **Grower:** *Ernst Dautel (Riesling GG)*
Non-*Einzellagen* **Grower:** *Ernst Dautel (Kreation S, Chardonnay, Weißburgunder)*

GROßLAGE WUNNENSTEIN
KLEINBOTTWAR

Vineyard: ✓ *Süssmund* • **Grower:** *Graf Adelmann (Riesling GG)*
Vineyard: *Oberer Berg* • **Grower:** *Graf Adelmann (Lemberger GG)*

THE REGION AT A GLANCE

AREA UNDER VINE 11,360ha (28,070ac)	Spätburgunder, 22% Schwarzriesling, 15% Riesling, 12% Lemberger, 13% others
AVERAGE YIELD (2008–2017) 93hl/ha (static)	
RED WINE 60%	**INFRASTRUCTURE** *Bereiche* 6, *Großlagen* 17, *Einzellagen* 210
WHITE WINE 24%	
ROSÉ 16%	Note: The vineyards of Württemberg straddle 230 communes *Gemeinden* (communes), the names of which may appear on labels.
MOST IMPORTANT GRAPE VARIETIES 19% Trollinger, 19%	

BURGRUINE WEIBERTREU, WEINBERG
Surrounded by terraced vineyards, the ruins of this circa 11th-century castle look down over the town.

Baden

Often described as Germany's southernmost wine-producing region, Baden is not so much one region as a hotchpotch of politically diverse districts that once produced wine in the now-defunct Grand Duchy of Baden.

Baden is the most diverse of all German wine regions. It comprises nine *Bereiche,* which are spread over 400 kilometres (248 miles) from Tauberfranken in the north to the Kraichgau and Badische Bergstraße on the right bank of the Rhine facing Alsace on the other side of the river, right down to Bodensee (Lake Constance) in the south, separating Germany from Switzerland. This means significant differences in terms of landscape and climate. Soils vary from gravel, marl, and clay to loam and loess, as well as shell limestone and keuper. Subsequently there is a rich spectrum of different wines. With all other German wine regions assigned to Europe's coolest winegrowing designation, zone A, Baden, because of its warmer climate, is the only one allocated to zone B, the same as the Loire, Champagne, and all of Austria's wine districts. The southern slopes of the Kaiserstuhl regularly register the highest temperatures in the whole country, ideal for cultivation of the vine and making Baden a hub of viticulture as early as the 16th century. In the present times of climate change, growers in the hottest areas face a different challenge, needing to protect their grapes from too much sunshine and subsequent high alcohol levels of their wines.

Today Baden is the country's third-largest wine region with almost 16,000 hectares (3,953 acres), but that is only half of the area under vine 200 years ago, when it could claim to be the country's biggest grower.

Ironically, it was when Germany acquired a wealthy and viticulturally prolific Alsace in 1871 as one of the spoils of the Franco-Prussian War, that Baden's vineyards began to decline. The downward trend continued, despite the formation of the Baden Wine Growers' Association founded by priest and winemaker Heinrich Hansjakob at Hagnau in 1881. Even after the return of Alsace to French sovereignty in 1918, Baden's wine production continued to decline, primarily through lack of investment. In the 1920s, wine production was adversely affected by inheritance laws that split Baden's vineyards into smaller and smaller units. By 1950, with barely 6,000 hectares (14,800 acres) of vines, Baden's wine industry was at its lowest ebb.

The eventual resurgence of Baden's wine industry began in 1952 with the formation of the *Zentralkellerei Kaiserstuhler Winzergenossenschaften* (Central Winery of Kaiserstuhl), which two years later expanded into the *Zentralkellerei Badischer Winzergenossenschaften* (Central Winery of Baden), or ZBW for short. ZBW built a 25-million-pound vinification and storage plant at Breisach, producing vast amounts of wine that it promoted with an aggressive marketing policy on the domestic scene. Even up to today the majority of Baden's wines is still marketed by cooperatives.

Legend:
Intensive vine-growing zone
—— Bereich boundary
—— Großlage boundary

0 mi   20
0 km   20

INTENSIVE VINE-GROWING ZONES OF BADEN
(see also p437)
The vineyards of this huge wine-producing region are mostly spread along a strip extending beside the western boundary of the Black Forest, between it and the border with France.

QUALITY REQUIREMENTS AND QUALITY WINE PRODUCTION 2017

BADEN'S MINIMUM *OECHSLE*	QUALITY CATEGORY	HARVEST BREAKDOWN
50–55°	Deutscher Wein / Landwein	1%
60–72°	* Qualitätswein	81%
76–85°	Prädikatswein	19%

* Minimum *Oechsle* levels vary according to grape variety; those that have a naturally lower sugar content may qualify at a correspondingly lower level.

In recent years a massive program of restructuring and rationalisation of Baden's vineyards has seen a considerable growth of the area under vine, although ecological aspects did not always get the attention they might have done. Of the many different varieties, it's the so-called Burgundian grapes – Spätburgunder, Chardonnay, Weißburgunder, Grauburgunder, and Auxerrois – which feature most prominently. They are almost always produced as dry wines, and therefore they are particularly appreciated for their eminent suitability to accompany the excellent regional cuisine. Gutedel (the Swiss Chasselas) and Müller-Thurgau meet the demand for lighter, fresher, and thirst-quenching wines. That is also the market for the local speciality Badisch Rotgold, for which Spätburgunder and Grauburgunder grapes are vinified together. That Baden is not standing still is best exemplified by the project known as Generation Pinot, an alliance of over 50 well-educated and quality-oriented young growers, who work closely together to increase awareness of the region's wine excellence well beyond national borders. Led by the charismatic Hanspeter Ziereisen there is also a small but flourishing *Landwein* community.

FACTORS AFFECTING TASTE AND QUALITY

LOCATION
The longest Anbaugebiete, Baden stretches for approximately 400 kilometres (250 miles), from Franken in the north, past Württemberg and the Badische Bergstraße to Bodensee, home to some of Germany's most southerly vineyards.

CLIMATE
Compared with the rest of Germany, the bulk of Baden's vineyards have a sunny and warm climate, due in part to the shelter afforded by the Black Forest and the Odenwald Mountains.

ASPECT
Most vineyards are on level or gently sloping ground. Some, however, are to be found higher up the hillsides, and these avoid the frosts of the valley floors.

SOIL
Baden's soils are rich and fertile, varying from heat-retaining gravel near Bodensee, through limestone, clay, marl, loam, granite, and loess deposits to limestone and keuper, a sandy-marl, in the Kraichgau and Taubergrund. Volcanic bedrock forms the main subsoil structure throughout most of the region.

VITICULTURE & VINIFICATION
The relatively flat and fertile vineyards of this region are easily mechanised. Although the geographical spread and variety of soils has led to a large number of different, traditional styles of wine, they are over-shadowed by the mild and neutrally fruity, bulk-produced Baden QbA marketed by the Badischer Winzerkeller. Approximately 70 per cent of the winemaking in Baden is conducted by its many cooperatives, but there are also over 800 independent estates among the region's 20,000 grape growers. Spätburgunder is by far Baden's most important grape, and more than half the wines produced from all varieties in this region are made in a dry, or *trocken*, style. Grauburgunder has undergone a total transformation, from the time when it was produced as a mostly sweet and heavy wine under the synonym of Ruländer, to today's sometimes full and dry renditions, frequently with a touch of *barrique,* but also in a fresher, lighter, non-oaked style. A speciality that is unique to Baden is Badisch Rotgold, a soft, delicately flavoured rosé made from pressing together Grauburgunder and Spätburgunder grapes.

GRAPE VARIETIES
Primary varieties: Trollinger (Schaiva), Lemberg Spätburgunder
Secondary varieties: Bacchus, Kerner, Rieslaner, Riesling, Scheurebe, Spätburgunder, Traminer (Savagnin Blanc)

ROWS OF GRAPEVINES COVER THE LANDSCAPE OF OBERBERGEN
The volcanic soils of Oberbergen are covered with fertile loess, which provides excellent ground for growing vines.

THE REGION AT A GLANCE

AREA UNDER VINE
15,906ha (39,304ac)

AVERAGE YIELD (2008–2017)
79hl/ha

RED WINE
25%

WHITE WINE
60%

ROSÉ
15%

MOST IMPORTANT GRAPE VARIETIES
34% Spätburgunder, 15% Müller-Thurgau, 13% Grauburgunder, 10% Weißburgunder, 7% Riesling, 7% Gutedel, 14% others

INFRASTRUCTURE
Bereiche 9, *Großlagen* 16, *Einzellagen* 315

Note: The vineyards of Baden straddle 315 communes *Gemeinden* (communes), the names of which may appear on labels.

THE APPELLATIONS OF
BADEN

BEREICHE BADISCHE BERGSTRAßE/ KRAICHGAU

This two *Bereiche* have four *Großlagen*, two in the Badische Bergstraße, north and south of Heidelberg, and two in the Kraichgau, a larger but sparsely cultivated area farther south.

GROßLAGE HOHENBERG

GROßLAGE MANNABERG
LEIMEN

Vineyard: ✓ *Herrenberg* • **Grower:** *Seeger (Auxerrois, Chardonnay, Grauburgunder GG, Weißburgunder GG, Sauvignon Blanc, Blaufränkisch GG, Spätburgunder GG, Frühburgunder)*

UNTERÖWISHEIM

Vineyard: ✓ *Kirchberg* • **Grower:** *Klumpp (Chardonnay, Grauburgunder, Weißburgunder)*

GROßLAGE RITTERSBERG
SCHRIESHEIM

Vineyard: *Madonnenberg* • **Grower:** *Rainer Baumann (Sauvignon Blanc, Spätburgunder)*

GROßLAGE STIFTSBERG
HILLSBACH

Vineyard: *Eichelberg* • **Grower:** *Heitlinger (Pinot Blanc GG)*
TIEFENBACH

Vineyard: *Heinberg* • **Grower:** *Heitlinger (Chardonnay GG)*
Vineyard: *Schellenbrunnen* • **Grower:** *Heitlinger (Riesling GG)*
Vineyard: *Wormsberg* • **Grower:** *Heitlinger (Pinot Noir GG)*
SULZFELD

Vineyard: *Löchle* • **Grower:** *Burg Ravensburg (Grauburgunder GG, Weißburgunder GG, Pinot Noir GG)*

Non-*Einzellagen* Grower: ✓ *Burg Ravensburg (Blaufränkich)*

BEREICH BODENSEE

GROßLAGE SONNENUFER
MEERSBURG

Vineyard: *Kriesemann* • **Grower:** *Aufricht (Spätburgunder)*
Vineyard: *Mocken* • **Grower:** *Aufricht (Spätburgunder)*
Vineyard: ✓ *Sängerhalde* • **Grower:** *Aufricht (Grauburgunder, Riesling trocken, Sauvignon Blanc, Spätburgunder)*
Vineyard: *Trielberg* • **Grower:** *Aufricht (Spätburgunder)*

BEREICH BREISGAU

GROßLAGE BURG LICHTENECK
BOMBACH

Vineyard: ✓ *Sommerhalde* • **Grower:** *Bernhard Huber (Grauburgunder, Spätburgunder GG)*
HECKLINGEN

Vineyard: ✓ *Schloßberg* • **Grower:** *Bernhard Huber (Spätburgunder GG)*
HERBOLZHEIM

Vineyard: *Kaiserberg* • **Grower:** *Fritz Wassmer (Cabernet Franc, Spätburgunder)*
MALTERDINGEN

Vineyard: ✓ *Bienenberg* • **Grower:** *Bernhard Huber (Chardonnay GG, Spätburgunder GG)*
Vineyard: *Bienenberg-Wildenstein* • **Grower:** *Bernhard Huber (Spätburgunder GG)*

KENZINGEN

Vineyard: *Roter Berg* • **Grower:** *Fritz Wassmer (Chardonnay, Merlot, Spätburgunder)*

Non-*Einzellagen* Grower: *Shelter Winery (Chardonnay, Pinot Noir)*

GROßLAGE BURG ZÄHRINGEN
DENZLINGEN

Vineyard: *Sonnhalde-Steinhalde* • **Grower:** *Otto und Martin Frey (Grauburgunder)*

FREIBURG

Vineyard: *Schloßberg* • **Growers:** *Staatsweingut Freiburg (Chardonnay), Stigler (Spätburgunder)*
GLOTTERTAL

Vineyard: *Eichberg* • **Grower:** *Otto und Martin Frey (Chardonnay, Grauburgunder, Spätburgunder)*

GROßLAGE SCHUTTER-LINDENBERG
LAHR

Vineyard: *Gottesacker* • **Grower:** *Wöhrle (Chardonnay GG)*
Vineyard: *Herrentisch* • **Grower:** *Wöhrle (Weißburgunder GG)*
Vineyard: *Kirchgass* • **Grower:** *Wöhrle (Grauburgunder GG, Spätburgunder GG)*
Vineyard: ✓ *Kronenbühl* • **Grower:** *Wöhrle (Riesling trocken, Grauburgunder, Weißburgunder, Spätburgunder)*
MÜNCHWEIER

Non-*Einzellagen* Grower: *Enderle-Moll (Pinot Noir)*

BEREICH KAISERSTUHL

GROßLAGE VULKANFELSEN
ACHKARREN

Vineyard: ✓ *Schloßberg* • **Growers:** *Dr Heger Spätburgunder GG), Franz Keller (Grauburgunder GG, Spätburgunder GG), Michel /Chardonnay, Grauburgunder, Weißburgunder, Spätburgunder), Fritz Wassmer (Grauburgunder, Syrah)*
BICKENSOHL

Vineyard: ✓ *Herrenstück* • **Grower:** *Holger Koch (Grauburgunder, Pinot Noir)*

Non-*Einzellagen* Grower: *Holger Koch (Grauburgunder)*
BISCHOFINGEN

Vineyard: *Enselberg* • **Grower:** *Abril (Chardonnay, Gewürztraminer Prädikate, Spätburgunder), Johner (Gewürztraminer Prädikate, Pinot Noir)*
BLANKENHORNSBERG

Vineyard: ✓ *Doktorgarten* • **Grower:** *Staatsweingut Freiburg (Grauburgunder GG, Spätburgunder GG)*
BURKHEIM

Vineyard: ✓ *Feuerberg* • **Grower:** *Bercher (Spätburgunder)*
Vineyard: ✓ *Feuerberg Haslen* • **Grower:** *Bercher Grauburgunder GG, Weißburgunder GG)*

Non-*Einzellagen* Grower: *Bercher (Chardonnay, Scheurebe&Chenin Blanc trocken)*
ENDINGEN

Vineyard: *Diel* • **Grower:** *Reinhold und Cornelia Schneider (Spätburgunder)*
Vineyard: ✓ *Engelsberg* • **Growers:** *Knab (Grauburgunder, Weißburgunder, Spätburgunder), Reinhold und Cornelia Schneider (Spätburgunder)*
Vineyard: *Schönenberg* • **Grower:** *Reinhold und Cornelia Schneider (Spätburgunder)*

Non-*Einzellagen* Grower: *Reinhold und Cornelia Schneider (Chardonnay, Ruländer trocken, Weißburgunder)*
EICHSTETTEN

Non-*Einzellagen* Growers: *Hiss (Chardonnay, Grauburgunder,*

Gewürztraminer Prädikate, Spätburgunder), Höfflin (Grauburgunder, Spätburgunder), Arndt Köbelin (Grauburgunder, Weißburgunder, Spätburgunder)
IHRINGEN

Vineyard: ✓ *Winklerberg* • **Growers:** *Freiherr von Gleichenstein (Spätburgunder), Dr Heger (Muskateller Prädikate, Spätburgunder), Konstanzer (Weißburgunder, Spätburgunder), Stigler (Grauburgunder GG)*
Vineyard: ✓ *Vorderer Winklerberg* • **Growers:** *Dr Heger (Spätburgunder GG), Stigler (Spätburgunder GG)*
Vineyard: *Winklerberg-Wanne* • **Grower:** *Dr Heger (Spätburgunder GG)*
Vineyard: ✓ *Winklerberg-Winklen* • **Growers:** *Dr Heger (Spätburgunder GG), Konstanzer (Grauburgunder, Muskateller trocken)*
Vineyard: *Winklerberg-Hinter Winklen* • **Grower:** *Dr Heger (Chardonnay GG, Weißburgunder GG)*
JECHTLINGEN

Vineyard: *Enselberg* • **Grower:** *Franz Keller (Spätburgunder GG)*
OBERBERGEN

Vineyard: ✓ *Bassgeige* • **Grower:** *Franz Keller (Grauburgunder, Weißburgunder, Spätburgunder)*
OBERROTTWEIL

Vineyard: ✓ *Eichberg* • **Growers:** *Freiherr von Gleichenstein (Grauburgunder, Spätburgunder), Franz Keller (Spätburgunder GG), Landerer (Spätburgunder), Salwey (Grauburgunder GG)*
Vineyard: *Henkenberg* • **Growers:** *Landerer (Chardonnay, Grauburgunder), Salwey (Grauburgunder GG, Weißburgunder GG; Spätburgunder GG)*
Vineyard: *Kirchberg* • **Growers:** *Franz Keller (Chardonnay GG, Spätburgunder GG), Salwey (Weißburgunder GG); Spätburgunder GG)*
SASBACH

Vineyard: ✓ *Limburg* • **Grower:** *Bercher (Weißburgunder, Spätburgunder)*
SCHELINGEN

Vineyard: ✓ *Kirchberg* • **Grower:** *Schätzle (Chardonnay, Grauburgunder, Spätburgunder)*

Non-*Einzellagen* Grower: *Hermann (Chardonnay, Grauburgunder, Pinot Noir)*

BEREICH MARKGRÄFLERLAND

GROßLAGE ATTILAFELSEN

GROßLAGE BURG NEUENFELS
BADENWEILER

Vineyard: ✓ *Römerberg* • **Grower:** *Dörflinger (Grauburgunder)*
BRITZINGEN

Vineyard: *Muggardter Berg* • **Grower:** *Harmut Schlumberger (Chardonnay)*
DOTTINGEN

Vineyard: *Castellberg* • **Grower:** *Martin Wassmer (Chardonnay, Grauburgunder, Weißburgunder, Syrah, Pinot Noir)*
LAUFEN

Vineyard: *Wingerte* • **Grower:** *Hartmut Schlumberger (Grauburgunder GG, Weißburgunder GG, Pinot Noir GG)*
MÜLLHEIM

Vineyard: *Pfaffenstück* • **Grower:** *Dörflinger (Spätburgunder)*
SCHLIENGEN

Vineyard: *Sonnenstück* • **Grower:** *Fritz Blankenhorn (Chardonnay GG, Grauburgunder GG)*

GROßLAGE LORETTOBERG
EHRENSTETTEN

Vineyard: *Ölberg* • **Grower:** *Martin Wassmer*
(Spätburgunder)

SCHERZINGEN

Vineyard: ✓ *Batzenberg* • **Grower:** *Heinemann (Chardonnay, Muskateller trocken, Spätburgunder)*

SCHLATT

Vineyard: *Maltesergarten* • **Grower:** *Martin Wassmer (Pinot Noir)*

STAUFEN

Vineyard: *Schloßberg* • **Growers:** *Fritz Wassmer (Chardonnay, Weißburgunder, Cabernet Sauvignon), Martin Wassmer (Weißburgunder)*

Non-*Einzellagen* Growers: *Achim Jähnisch/Chardonnay, Grauburgunder, Riesling trocken, Spätburgunder, Pinot Noir), Fritz Wassmer (Spätburgunder)*

GROßLAGE VOGTEI RÖTTELN
EFRINGEN-KIRCHEN

Non-*Einzellagen* Grower: ✓ *Ziereisen (Chardonnay, Gutedel, Grauburgunder, Weißburgunder, Pinot Noir, Spätburgunder, Syrah)*

WEIL

Vineyard: ✓ *Schlipf* • **Grower:** *Claus Schneider (Chardonnay, Pinot Blanc, Pinot Gris, Pinot Noir, Spätburgunder)*

BEREICH ORTENAU

GROßLAGE FÜRSTENECK
DURBACH

Vineyard: *Bienengarten* • **Grower:** *Andreas Männle (Gewürztraminer)*

Vineyard: ✓ *Kochberg* • **Grower:** *Heinrich Männle (Weißburgunder feinherb, Spätburgunder)*

Vineyard: *Plauelrain* • **Growers:** *Winzergenossenschaft Durbach (Gewürztraminer Prädikate, Scheurebe Prädikate), Andreas Laible (Riesling trocken, GG, Prädikate, Grauburgunder GG. Gewürztraminer. Traminer Prädikate)*

Vineyard: *Schloßberg* • **Growers:** *Schloß Staufenberg-Markgraf von Baden (Spätburgunder), Gräflich Wolff Metternich'sches Weingut (Riesling trocken)*

Non-*Einzellagen* Grower: *Alexander Laible (Riesling trocken, Blaufränkisch)*

ZELL-WEIERBACH

Vineyard: *Abtsberg* • **Grower:** *Freherr von und zu Franckenstein (Grauburgunder GG)*

Vineyard: *Neugesetz* • **Grower:** *Freherr von und zu Franckenstein (Riesling GG)*

GROßLAGE SCHLOß RODECK
BADEN-BADEN

Non-*Einzellagen* Grower: *Sven Nieger (Riesling trocken)*

NEUWEIER

Vineyard: *Goldenes Loch* • **Grower:** *Schloß Neuweier (Riesling GG)*

Vineyard: *Mauerberg* • **Growers:** *Holger Dütsch (Chardonnay, Riesling trocken), Schloß Neuweier (Riesling trocken, Prädikate)*

Vineyard: *Schloßberg* • **Grower:** *Schloß Neuweier (Riesling trocken)*

Vineyard: *Heiligenstein* • **Grower:** *Schloß Neuweier (Pinot Noir)*

VARNHALT

Vineyard: *Sonnenberg* • **Grower:** *Kopp (Spätburgunder)*

Vineyard: *Sonnenhalde* • **Grower:** *Kopp (Spätburgunder)*

BEREICH TAUBERFRANKEN

The most northerly of Baden's vineyards, *Bereich* Tauberfranken bridges Franken and Württemberg.

GROßLAGE TAUBERKLINGE
REICHHOLZHEIM

Vineyard: ✓ *First* • **Grower:** *Konrad Schlör (Riesling trocken, Schwarzriesling)*

Vineyard: *Oberer First* • **Grower:** *Konrad Schlör (Weißburgunder GG, Spätburgunder GG)*

Non-*Einzellagen* Grower: *Konrad Schlör (Schwarzriesling)*

BEREICH TUNIBERG

Non-*Einzellagen* Grower: *von der Mark (Grauburgunder, Weißburgunder, Sauvignon Blanc, Spätburgunder)*

CERTIFIED ORGANIC AND BIODYNAMIC PRODUCERS

BIODYNAMIC
- Abril (Vogtsburg-Bischoffingen)
- Alfons Schüber (Sasbach-Jechtingen)
- Baumann (Gerlachsheim)
- Breitehof Hubert Schies (Vogtsburg-Burkheim a Kaiserstuhl)
- Burg Ravensburg (Sulzfeld)
- Christoph Brenneisen (Sulburg-Laufen)
- Daniel Bach (Kenzingen-Hecklingen)
- Erhard Heitlinger (Östringen-Tiefenbach)
- Guido Friderich (Sasbach)
- Josef u Klara Vögtle (Sasbach-Jechtingen a Kaiserstuhl)
- Josef Wörner (Durbach)
- Markus Bürgin (Fischingen)
- Max und Alice Schneider (Sasbach)

ORGANIC
- Andreas Wöhrle (Bockenheim)
- Andrea und Heiner Renn (Hagnau)
- Andreas Fritz (Bühlertal)
- Bettina Beck (Bahlingen)
- Daniel Feuerstein (Heitersheim)
- David Klenert (Kraichtal)
- Edmund Eisele/Schelb (Ehrenkirchen)
- Eduard Mannsperger (Sinsheim-Dühren)
- Edwin Menges (Rauenberg)
- Erwin Mick (Nimburg)
- Freiburg (Heiliggeist Freiburg im Breisgau)
- Friedhelm Rinklin (Eichstetten)
- Friedrich & Bärbel Ruesch (Buggingen)
- Fritz Lampp (Heitersheim)
- Gallushof/Hügle (Teningen-Heimbach)
- Geier (Königheim)
- Gerd Köpfer (Staufen-Grunern)

- Gerhard Aenis (Binzen)
- Gretzmeier (Merdingen)
- Gudrun Lauerbach (Tauber-bischofsheim-Impfingen)
- Günther Kaufmann (Efringen-Kirchen)
- H & J Sprich (Weil-Haltingen)
- Hans Hardt (Herbolzheim)
- Harald Süssle (Merdingen)
- Harteneck (Schliengen)
- Hermann Helde & Sohne (Jechtingen)
- Hermann Neuner-Jehle (Immenstaad)
- HP & Helmut Grether (Laufen-Sulzburg)
- Hubert Lay (Ihringen)
- Hummel (Malsch)
- Jette Krumm (Auggen)
- Joachim Netzhammer (Klettgau-Erzingen)
- Johannes Haug (Nonnenhorn)
- Jürgen Landmann (Freiburg-Waltershofen)
- Kirchberghof (Kenzingen-Bombach)
- Klaus Benz (Ballrechten-Dottingen)
- Klaus Bischoff (Kelltern)
- Klaus Vorgrimmler (Freiburg-Munzingen)
- Koehly-Harteneck (Bad Bellingen)
- Kuckuckshof (Karlsbad-Ittersbach)
- Kurt Breisacher (Riegel)
- Langwerth Öko-Wein (Esselbach)
- Ludwig Missbach (Ebringen)
- Manfred & Eva Maria Schmidt (Vogtsburg-Bischoffingen)
- Manfred Dannmeyer (Bamlach)
- Markgräfler Wyhus (Efringen-Kirchen)
- Martin Hämmerlin (Buggingen)
- Martin Küchlin (Buggingen-Betberg)

- Martina Heitzmann (Bahlingen)
- Matthias Höfflin/Schambachhof (Bötzingen)
- Matthias Seywald (Ballrechten-Dottingen)
- Matthias Wolff (Riedlingen)
- Peter Kaiser (Bahnbrücken)
- Peter Landmann (Staufen)
- Philip Isele (Achkarren)
- Philipp Rieger (Buggingen-Retberg)
- Pix (Ihringen A K)
- Rabenhof (Vogtsburg-Bischofingen)
- Reblandhof (Sulzfeld)
- Reinhard Burs (March)
- Richard G Schmidt (Eichstetten)
- Scherer&Zimmer (Bad Krozingen)
- Schlumberger (Sulzburg-Laufen)
- Schneider-Pfefferle (Heitersheim)
- Siegfried Frei (Wasenweiler)
- Sonnenbrunnen (Freiburg-Opfingen)
- Stadt Lahr (Lahr/Schwarzwald)
- Stränglehof (Leiselheim)
- Thomas Schaffner (Bötzingen am Kaiserstuhl)
- Thomas Selinger (Merdingen)
- Tobias Kehnel (Broggingen)
- Trautwein (Bahlingen)
- Ulrich Klumpp (Bruchsal)
- Volker Maier (Baden-Baden-Haunebersteim)
- Walter J Schür (Vogtsburg-Oberrotweil)
- Wendelin Brugger (Laufen)
- Wilhelm Zähringer (Heitersheim)
- Willi Frey (Freiburg-Tiengen)
- Winzerhof Leber (Vogtsburg-Schelingen)
- Wolfgang Ibert (Wallburg)

Saale-Unstrut

Even some of the vineyards of England and Poland lie on a slightly more southern latitude than those in the valleys of the rivers Saale and Unstrut, but it's a mix of continental climate and geological structure that favours viticulture in this borderline region. Terraced vineyards with southern exposure afford a vegetation period long enough for grapes to ripen.

There are records that prove the existence of viticulture in this area as long ago as AD 998, and in the 16th century the area under vine extended to an unbelievable 10,000 hectares (24,710 acres). This thriving wine industry was mainly due to the dedication of the Cistercian monks from the Abbey of Sancta Maria Schulpforta, which had been founded in 1137. Wine was not considered a luxury in those days but a healthier alternative to drinking water, which was polluted more often than not. After World War II winegrowing barely survived, thanks to the dedication of a few single-minded enthusiasts and, subsequently, the nationalisation of the industry. During the winter of 1986–1987 temperatures fell to -30°C (-22°F) and nearly wiped out what was left of the vineyards. It is quite probable that the following revival of the wine industry had much to do with Germany's reunification in 1990, giving growers a chance to fulfil the long-suppressed dream of growing their own grapes. Only a fifth of them, though, make their own wine, with the rest supplying larger estates or cooperatives, such as the Winzervereinigung Freyburg-Unstrut and the state domaine of Kloster Pforta. The wines are mainly vinified dry, as the addition of *Süßreserve* never had any tradition here. The danger of early autumn frosts means that most of the varieties planted are early-ripening, with Müller-Thurgau a natural contender. The red

Zweigelt may seem an odd choice, but it is explained by the fact that during the communist regime it was not a question of what one would have liked to plant, but rather what one could get hold of. Most likely its proliferation at the time was down to a trade deal with Czechoslovakia. There are not many things global warming is good for – viticulture in the Saale-Unstrut region probably one of the few. Top producers of the region seem to be banking on it, as their increased output in the premium Riesling and Pinot varieties would indicate.

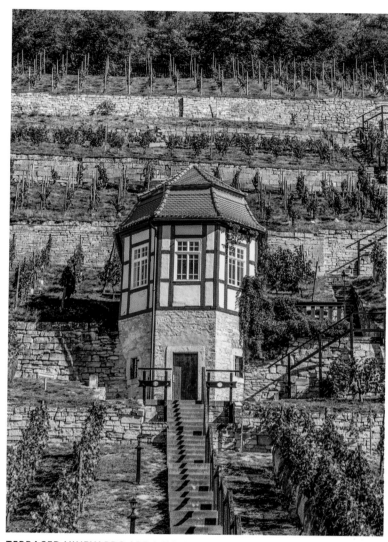

TERRACED VINEYARDS ARE A HALLMARK OF FREYBURG WINE COUNTRY
Stone retaining walls protect the south-facing terraces of grapevines planted along the steep and rocky slopes of the Saale-Unstrut wine region.

QUALITY REQUIREMENTS AND QUALITY WINE PRODUCTION 2017

SAALE-UNSTRUT MINIMUM *OECHSLE*	QUALITY CATEGORY	HARVEST BREAKDOWN
44/47°	Deutscher Wein / Landwein	0.2%
50°	* Qualitätswein	76.8%
75°	Prädikatswein	13%

* Minimum *Oechsle* levels vary according to grape variety; those that have a naturally lower sugar content may qualify at a correspondingly lower level.

FACTORS AFFECTING TASTE AND QUALITY

LOCATION
Just beyond 51 degrees of northern latitude, the two main areas under vine are located in the Saale Valley between Weißenfels and Bad Kösen and the Unstrut Valley between Stigra and Freyburg. There's also a small exclave of vines just south of Halle.

CLIMATE
Protected by the Harz Mountains, the region is one of Germany's driest, with precipitation at an annual average of 50 centimetres (20 inches). Winters can be very cold. Early

frosts present the greatest danger to the crop in the autumn and can cut the vegetation period to as little as 155 days.

ASPECT
Vines are grown mainly on terraced slopes facing south, often protected by sometimes centuries-old dry-stone walls.

SOIL
In the area of Freyburg shell limestone dominates the vineyard soil; around Naumburg sandstone is more prevalent.

VITICULTURE & VINIFICATION
Around 30 different varieties of grapes are grown in the Saale-Unstrut region with Müller-Thurgau and Weißburgunder the most popular varieties. A quarter of the vineyard hectarage is given over to red grapes, with white grape varieties making up about 74 per cent of Saale-Unstrut's vineyards.

GRAPE VARIETIES
Müller-Thurgau, Weißburgunder, Spätburgunder, Dornfelder, Portugieser, Spätburgunder, Zweigelt

INTENSIVE VINE-GROWING ZONES OF SAALE-UNSTRUT AND SACHSEN

(see also p437)

Almost 800 hectares (1,700 acres) of vines grow in these areas, which were formerly part of East Germany. Production is centred at Bad Kösen, Freyburg, and near the confluence of the Saale and Unstrut rivers at Naumburg.

THE APPELLATIONS OF

SAALE-UNSTRUT

BEREICH SCHLOß NEUENBURG

GROßLAGE BLÜTENGRUND

GROßLAGE GÖTTERSITZ-
NAUMBURG

Vineyard: ✓ *Steinmeister* • **Growers:** *Fröhlich-Hake (Riesling trocken), Gussek (Riesling trocken, Prädikate), Hey (Riesling trocken, Prädikate, Weißburgunder)*
PFORTA

Vineyard: *Köppelberg* • **Growers:** *Kloster Pforta (Bacchus, Weißer Heunisch, Blauer Silvaner)*

GROßLAGE KELTERBERG

GROßLAGE SCHWEIGENBERG
FREYBURG

Vineyard: *Schweigenberg (Großlage-designation)* • **Grower:** *Böhme & Töchter (Chardonnay, Riesling trocken, Traminer, Weißburgunder)*
Vineyard: ✓ *Edelacker* • **Grower:** *Pawis (Riesling GG, Grauburgunder GG, Weißburgunder GG)*
KARSDORF

Vineyard: *Hohe Grät* • **Grower:** *Lützkendorf (Riesling GG, Traminer GG, Prädikate, Weißburgunder GG)*

WEISCHÜTZ

Vineyard: *Nüssenberg* • **Grower:** *Lüttmer (Frühburgunder)*
ZSCHEIPLITZ

Vineyard: *Himmelreich* • **Grower:** *Böhme & Töchter (Spätburgunder)*

GROßLAGEFREI
(not belonging to a Großlage)
DORNDORF

Vineyard: *Rappental* • **Grower:** *Klaus Böhme (Riesling trocken, Prädikate, Weißburgunder, Traminer feinherb, Frühburgunder)*

BEREICH THÜRINGEN

GROßLAGEFREI
(not belonging to a Großlage)
KAATSCHEN

Vineyard: *Dachsberg* • **Grower:** *Gussek (Riesling trocken, Grauburgunder, Blauer Zweigelt)*

BEREICH MANSFELD SEEN

BEREICH WERDER (HAVEL)

THE REGION AT A GLANCE

AREA UNDER VINE
772hc, or 1,907ac (increasing)

AVERAGE YIELD (2008–2017)
74hl/ha (increasing)

RED WINE
21%

WHITE WINE
72% (decreasing)

ROSÉ
7%

MOST IMPORTANT GRAPE VARIETIES
14.8% Müller-

Thurgau, 14% Weißburgunder, 9% Riesling (increasing), 7% Dornfelder, 6.5% Silvaner, 6% Grauburgunder, 42.7% others

INFRASTRUCTURE
Bereiche 4, Großlagen 4, Einzellagen 34

Note: The vineyards of Saale-Unstrut straddle 38 communes *Gemeinden* (communes), the names of which may appear on labels.

Sachsen

Although Sachsen lies on the northeastern periphery of European winegrowing, it still has an annual average of 1,600 sunshine hours. The early-ripening Müller-Thurgau has always been the grape variety of choice, but the best wines are made from Riesling and the white Pinots. About 500 hectares (1,235 acres) under vine make Sachsen the smallest winegrowing region of Germany, which is also the reason its wines are rarely ever seen outside the area.

Wine has been made in the valley of the river Elbe for more than 800 years, as a document from the year 1161 testifies. For the resident dukes and kings it was always a matter of course to have their own vineyards. For many centuries the nobility and the church were the prime protagonists of the region's wine industry. In the 17th century the area under vine was 10 times as large as it is today. The years between World War II and reunification were the dark days of viticulture in Sachsen, because winegrowing was not an important matter on the agenda of the communist regime. This changed in the 1990s, with the influence of the castle domaines of Schloß Wackerbarth and Schloß Proschwitz, two of the major driving forces behind the revival of winegrowing.

Today, in a time of climate change, producers seize the opportunities their relatively cool climate offers them to craft fresh wines with plenty of vibrancy. Traminer is their legacy, the Pinots their future. Schloß Wackerbarth is giving the manufacture of fine *Sekt* a new lease of life.

THE REGION AT A GLANCE

AREA UNDER VINE
497ha, or 1,215ac
(increasing)

**AVERAGE YIELD
(2008–2017)**
62hl/ha (increasing)

RED WINE
10.5%

WHITE WINE
84%

ROSÉ
5.5%

**MOST IMPORTANT
GRAPE VARIETIES**
14.5% Müller-Thurgau, 14.3% Riesling (increasing), 11.9% Weißburgunder, 9.3% Grauburgunder, 8.2% Spätburgunder, 4.5% Dornfelder, 37.3% others

INFRASTRUCTURE
Bereiche 2, *Großlagen* 4, *Einzellagen* 17

Note: The vineyards of Sachsen straddle 14 communes *Gemeinden* (communes), the names of which may appear on labels.

THE BELVEDERE SUMMER HOUSE AT SCHLOSS WACKERBARTH
The summer house near the top of the sloping terraces around Schloß Wackerbarth is the centrepeice of the castle's Baroque gardens. Along with the castle, gardens, and vineyards there is a modern factory on the grounds for producing wine. Wackerbath is the second-oldest sparkling wine producer in Germany.

FACTORS AFFECTING TASTE AND QUALITY

LOCATION

With approximately 500 hectares (1,235 acres) of area under vine, Sachsen is the smallest winegrowing region of Germany. Its vineyards lie on a 55 kilometres (34 mile) stretch along the river Elbe from Pirna near the Czech border in the east to the sleepy village of Diesbar-Seusslitz in the west. The town of Meissen, best known for its porcelain manufacturing history, handles most of the wine production of the region.

CLIMATE

The climate is relatively dry and continental, with the widest temperature variations of all regions between summer and winter. Late spring and early autumn frosts are the biggest enemy of the growers and can lead to considerable differences of quality and quantity from year to year.

ASPECT

The best sites lie on south and southeast-facing valley slopes of the river Elbe. Many of the vineyards are located on terraces supported by dry-stone walls.

SOIL

Soils are very diverse and range from weathered rock in the area of Pillnitz to granite with a top layer of loess on the slopes of Meissen and Radebeul. Variegated sandstone prevails at Zadel, and pockets of colluvial and alluvial deposits can also be found.

VITICULTURE & VINIFICATION

More than 90 per cent of some 2,100 grape growers supply cooperatives or larger estates, and there were only 37 independent producers in 2018. The cooperative of Meissen is the largest producer of the region, processing approximately two-thirds of the crop. The annual average production of the region amounts to 26,000 hectolitres, and 86% of all wines are vinified dry. The main variety is Müller-Thurgau, followed by Riesling, Weißburgunder (Pinot Blanc), and Grauburgunder (Pinot Gris). Goldriesling is a Saxonian speciality not to be found anywhere else in Germany, with Alsace its presumed origin.

GRAPE VARIETIES

Primary varieties: Müller-Thurgau, Riesling, Weißburgunder, Grauburgunder
Secondary varieties: Dornfelder, Goldriesling, Kerner, Regent, Traminer, Spätburgunder, Scheurebe, Zweigelt

MÜLLER-THURGAU GRAPES
The Müller-Thurgau is the most widely planted grape in this region. Wines made with this white variety may be drunk while fairly young and are rarely considered to improve with age.

QUALITY REQUIREMENTS AND QUALITY WINE PRODUCTION 2017

SACHSEN MINIMUM *OECHSLE*	QUALITY CATEGORY	HARVEST BREAKDOWN
44/47°	Deutscher Wein / Landwein	3.6%
50°	* Qualitätswein	51.6%
75°	Prädikatswein	44.8%

* Minimum *Oechsle* levels vary according to grape variety; those that have a naturally lower sugar content may qualify at a correspondingly lower level.

THE APPELLATIONS OF
SACHSEN

BEREICH ELSTERTAL

This is a *Bereich* consisting of three villages at the southern end of Sachsen. There are relatively few vines and, so far, no village has produced any outstanding wines.

BEREICH MEISSEN

GROßLAGE ELBHÄNGE
PILLNITZ

Vineyard: *Königlicher Weinberg* • **Grower:** *Klaus Zimmerling* (Riesling, Gewürtztraminer, Weißburgunder GG, Grauburgunder)

GROßLAGE LÖSSNITZ
MEISSEN

Vineyard: *Kapitelberg* • **Growers:** *Martin Schwarz* (Riesling trocken), *Schuh* (Riesling trocken)

Vineyard: *Klausenberg* • **Grower:** *Schuh* (Weißburgunder)
RADEBEUL

Vineyard: *Friedstein* • **Growers:** *Martin Schwarz* (Chardonnay, Riesling trocken, Traminer trocken, Pinot Noir)

Vineyard: *Goldener Wagen* • **Growers:** *Schloß Wackerbarth* (Riesling trocken, Prädikate, Traminer Prädikate), *Kastler Friedland* (Bacchus, Riesling trocken, Grauburgunder, Silvaner)

Vineyard: *Wackerbarthberg* • **Grower:** *Schloß Wackerbarth* (Riesling Prädikate)

Non-*Einzellagen* Grower: *Karl Friedrich Aust* (Riesling feinherb, Traminer feinherb)

GROßLAGE SCHLOß-WEINBERG

GROßLAGE SPAARGEBIRGE
PROSCHWITZ

Vineyard: ✓ *Schloß Proschwitz* • **Grower:** *Schloß Proschwitz* Grauburgunder GG, Weißburgunder GG, Spätburgunder GG)

Austria

The identity of Austrian wines is stronger than ever, and with the gradual shift to the "Roman system" the DAC appellations have increased over the last decades. It's more about the place of origin than purely the variety. Austrian wines are now role models for many other Central and Eastern European wine markets due to their excellent communication, precise winemaking, and innovative style.

In the late 1990s Austria had just gone through some turmoil, but it emerged stronger than ever by establishing the Austrian Wine Marketing Board and later founding the Austrian Wine Academy (*Weinakademie Österreich*). The country not only has some of the most strictly controlled and safest wine industries in the world, but with the introduction of the Districtus Austriae Controllatus (DAC) it also made further huge steps to define typicity of specified growing regions. Today almost all regions are part of this system, the very foundation of which lies in the concept of appellation systems.

THE MODERN WINE TRADE

Austria has 45,500 hectares (112,432 acres) under vineyard, producing roughly 2.6 million hectolitres of wine per year (depending on the vintage). Two-thirds of the vineyards are planted with white varieties, and Grüner Veltliner is still the undisputed leader amongst the varieties, followed by the red Zweigelt. Its wine export quantity seems to remain steady over the years; the average price is increasing, however, and it recently experienced a huge jump. The most important market is by far Germany.

Austria delivers consistent wines with a clear and clean fruit profile, as well as personality and diversity, from fizzy styles to still and sweet wines, in all quality segments.

RECENT SWISS VINTAGES

2019 A rather uneventful yet excellent vintage with no extremes and fine fruit with crisp, fresh acidity. Winter was mild and dry, and some early budding took place. The summer was warm to hot with moderate rainfall and good picking conditions delivered high quality.

2018 A hot and early vintage with average to moderate rainfall resulted in very good quality and quantity throughout of Austria. Wines are usually of higher alcohol and slightly softer acidity. It all started with a warm spring followed almost immediately by a hot summer with lower rainfall. Very promising vintage with good volumes.

2017 This vintage delivered good quality and rich wines across Austria. The year started with a cold January, yet some of the warmest spring temperatures were measured along with moderate rainfall. A vintage with some extremes.

2016 A challenging vintage: Burgenland and Steiermark had huge losses due to frost and hail. The sunny and dry September, however, saved the vintage and helped to achieve good results. Fine, aromatic, fruity, fresh wines with moderate alcohol were the result.

2015 A beautiful flowering and a hot, dry summer with some rainfall in August, sunny autumn days, and cool nights all the way to November helped the grapes to achieve perfect maturity. The result: expressive, fruit-driven wines with moderate to high alcohol, depth, and richness. In Burgenland some fine examples of sweet botrytis wines can be found too.

2014 A very difficult year with plenty of rainfall and mostly cool conditions. Some fine Blaufränkisch wines were produced, and growers who fought hard against the challenging conditions could expect good results. Short window for harvests due to rain, very variable vintage.

THE AUSTRIAN WINE LAW

As a member of the European Union (EU), Austria's wine laws are integrated into the EU wine laws as well. *Prädikatsweine* is produced within the Quality wine category according to specific ripeness level of the berry, somewhat similar to Germany. In Austria the measurement scale is called Klosterneuburg Must Weight (KMW), and there are usually higher must degrees required for identical styles compared to Germany. With the increasing importance of the appellation system, however, if a specified wine-growing region (eg Weinviertel) in which DAC allows only the production of dry wine, for instance (in the Weinviertel DAC only Grüner Veltliner is permitted), the *Prädikatswines* have to be referred to as generic regions (eg Niederösterreich) on the label. Hence, if you make a sweet or a red wine from the Weinviertel area you may only label it as Niederösterreich and not Weinviertel DAC.

QUALITY CATEGORY	MINIMUM *OECHSLE*	
	AUSTRIA	GERMANY
Wine or Wine without GI	51°	44–50°
Landwein	70°	47–55°
Qualitätswein (Steinfeder*)	75°	50–72°
Kabinett (Federspiel*)	85°	67–85°
Spätlese	94°	76–95°
Auslese (Smaragd*)	105°	83–105°
Beerenauslese	127°	110–128°
Eiswein	127°	110–128°
Ausbruch	156°	N/A
Trockenbeerenauslese	156°	150–154°

*Special terms in the Wachau

A STORAGE CELLAR BURROWS ALONGSIDE A BURGENLAND VINEYARD
At the edge of Central Europe's Pannonian plain, Burgenland lies on Austria's eastern border. This region produces standard whites, rich reds, and sweet botrytised wines.

RUSTER AUSBRUCH: A RELIC OF THE AUSTRO-HUNGARIAN EMPIRE

Basically a Trockenbeerenauslese (minimum 30 KMW/156°Oe), Ruster Ausbruch is produced only in the Free City of Rust in Burgenland at the Neusiedler Lake. The name *Ausbruch* translates literally as "break out", referring to "noble rot" and its shrivelling effect on the botrytised berries. Rust gained the right to become a Free City in 1681 by paying with the local currency and "500 Eimer" – about 28,000 litres of sweet wine. The lake plays an important role in the development of noble rot and only hand-harvesting is permitted. The choice of grape varieties has shifted over the last decade: Furmint and Muskateller were once the popular choices, but today producers are not shy about using Welschriesling, as well as Pinot varieties (Pinot Blanc, Pinot Gris, Chardonnay) amongst many others. The Cercle Ruster Ausbruch is a group of winemakers who stand for high-quality production. The wines show sweetness, freshness, great balance, and plenty of rich, bold fruit character.

| *Burgenland* | Wine region |
| *Eisenberg* | Wine sub-region |

0 mi   10
0 km   10

WINE REGIONS OF AUSTRIA
The federal states of Burgenland, Niederösterreic, and Steiermark (Styria) are defined as distinct wine regions, There are also 17 smaller wine regions, including the Wien DAC, which surrounds the capital city of Vienna.

FACTORS AFFECTING TASTE AND QUALITY

LOCATION
The vineyards of Austria lie in the east of the country, along the Danube, as well as north and south of Vienna, bordering the Czechia, Hungary, and Slovenia.

CLIMATE
The Austrian climate is warm, dry, and continental, with annual rainfall between 50 and 70 centimetres (20 to 28 inches). The warmest and driest area is Burgenland, where during the warm autumns mists rise from the Neusiedler Lake, which help promote *Botrytis cinerea* (the "noble rot").

ASPECT
Vines are grown on all types of land, from the plains of the Danube to the sides of its valleys – which are often very steep and terraced – and from the hilly Burgenland to the slopes of the mountainous Steiermark (Styria).

SOIL
Soils vary from generally stony schist, limestone, and gravel (though occasionally loamy) in the north, through predominantly sandy soils on the shores of the Neusiedler Lake in Burgenland, to mainly clay with some volcanic soils in Steiermark.

VITICULTURE & VINIFICATION
Not surprisingly, Austria's methods are similar to those of Germany's, but, although modern techniques have been introduced here, far more Austrian wine is produced by traditional methods than is the case in Germany. More than 85 per cent of the country's vines are cultivated by the Lenz Moser system. This is a method of training vines to twice their normal height, thereby achieving a comparatively higher ratio of quality to cost by enabling the use of mechanized harvesting.

This system brought fame to its Austrian inventor, Lenz Moser, and has been adopted by at least one grower in most winemaking countries. Harvesting is traditionally accomplished in tries, particularly on steeper slopes and for sweeter styles.

GRAPE VARIETIES
Blauer Portugieser, Blauer Wildbacher, Blaufränkisch (Lemberger), Bouviertraube (Bouvier), Cabernet Franc, Cabernet Sauvignon, Chardonnay, Frühroter Veltliner, Furmint, Gewürztraminer, Goldburger, Grauer Burgunder (Pinot Gris), Grüner Veltliner, Merlot, Müller-Thurgau, Muscat Ottonel, Muskateller (Muscat Lunel), Neuburger, Pinot Blanc, Pinot Noir, Riesling, Roter Veltliner, Rotgipfler, St-Laurent, Sauvignon Blanc, Scheurebe, Silvaner, Trollinger (Schiava), Welschriesling, Zierfandler, Zweigelt

THE WINE STYLES OF
AUSTRIA

BLAUBURGER

RED A Blauer Portugieser/Blaufränkisch cross that produces a well-coloured wine of little distinction, often used as a blending partner in *cuvées*.

🍷 *1–3 years*

👑 *Hoffmann • Schuckert • Sepp*

BLAUER PORTUGIESER

RED Introduced to Austria in 1770, it was once the most widely planted black grape variety and is still commonly found in the Weinviertel and the Thermenregion. It makes a light-bodied but well-coloured red wine with mild flavour, hinting of violets.

🍷 *1–2 years*

👑 *Johann Gipsberg • Weinfried-Harald Schachl*

BLAUER WILDBACHER

RED/ROSÉ This late-ripening variety, the traditional grape for Schilcher (dark rosé) wine, is produced in the Weststeiermark region. Very crisp, tart, dry, and fruity rosés are mostly produced from the variety.

🍷 *1–2 years*

👑 *Jöbstl • Langmann • Strohmeier*

BLAUFRÄNKISCH

RED Called Lemberger in Germany and Kékfrankos in Hungary, this variety is popular all across Burgenland, where it produces tart, fruity wines with good tannin and an underlying hint of cherries and spice. The best Blaufränkisch wines express their origin.

🍷 *2–4 years*

👑 *Gesellmann • Hans Igler • Kerschbaum • Kirnbauer • Kiss • Kollwentz • Krutzler • Nittnaus • Paul Achs • Prieler • Tesch • Wachter-Wiesler • Weninger • Wolhmuth*

BOUVIER

WHITE This early-ripening table grape has a low natural acidity, provides some aromatic notes, and is often used for high-quality *Prädikatswein*, especially around Burgenland.

🍷 *1–3 year (Qualitätswein), 2–8 years (Prädikatswein)*

👑 *Willi Opitz*

CABERNET SAUVIGNON

RED Very little of this variety used to be grown in Austria. Until 1982 the only commercial cultivation of Cabernet Sauvignon was by Schlumberger at Bad Vöslau. Then Lenz Moser received special permission to cultivate 2.5 hectares (6 acres) at Mailberg, and the first Cabernet Sauvignon was grown in 1986 in Lower Austria. Nowadays, everyone seems to be growing Cabernet Sauvignon, especially in Burgenland.

🍷 *5–6 years (Qualitätswein), 10–15 years (Prädikatswein)*

👑 *Gesellmann • Hahn • Igler • Juris • Leberl • Scheiblhofer • Wieder*

CHARDONNAY

WHITE This grape variety has been grown for decades in Steiermark, where it is known as Morillon. Cultivation of this grape has risen rapidly since the early 1990s, producing some fat, rich wines, but occasionally even sweet wine is made of variety at the Neusiedler Lake.

🍷 *2–5 years*

👑 *Bründlmayer • Kollwentz • Juris-Stiegelmar • Andreas Schafler • Gottfried Schellmann • Tement • Velich • Wieninger*

FRÜHROTER VELTLINER

WHITE The importance of the Frühroter Veltliner variety is decreasing, and it is mainly found in the Thermen region, Wagram, and Weinviertel. It often expresses soft acidity with a touch of floral and almond character.

🍷 *1–2 years*

👑 *Leth*

FURMINT

WHITE This is a rarely encountered Hungarian dry, medium- to full-bodied, rich, and fruity varietal that does well at Rust in Burgenland.

🍷 *3–5 years*

👑 *Tremmel • Seiler • Robert Wenzel*

GEWÜRZTRAMINER

WHITE Usually labelled Traminer or Roter Traminer, the wines from this variety range from light and floral to intensely aromatic. They may be dry or have various shades of sweetness. They are the fullest, richest, and most pungent of *Trockenbeerenauslesen*.

🍷 *3–6 years*

👑 *Neumeister • Polz • Tement • Wohlmuth*

GOLDBURGER

WHITE This Welschriesling/Orangetraube cross produces white wines that are used for blending or as a table grape.

🍷 *1–3 years*

GRÜNER VELTLINER

WHITE This grape has reinvented itself following the much-publicized thrashing it gave Burgundy in the tasting organized by Jancis Robinson MW and Tim Atkin MW in 2002. The new-found respect for Grüner Veltliner could be just as quickly lost, however, if the current trend for bigger, fatter, and less varietally typical wines continues. It won the contest not because the Burgundies chosen were inferior or too young – they included *grands crus* with good bottle-age from top growers, not to mention a number of top-quality international Chardonnay wines. (Burgundy should be thankful that Grüner Veltliner stole the limelight; being trounced by Chardonnays from Australia, Austria, California, Italy, and South Africa might have had an even greater negative effect.) No, it won because of its intrinsic minerality – the distinctive hint of fiery, freshly ground pepper that comes through on the finish. Yet many producers have forgotten these roots and pushed ripeness levels so high that some of the wines they make have become parodies of the Chardonnay they so roundly defeated. Instead of mimicking Chardonnay, growers of Grüner Veltliner should concentrate on exploiting their *terroir* to express the truly diverse range of varietal styles that can be achieved – from the rich but distinctive *barrique*- fermented wines that can beat the greatest Chardonnay without being bigger or heavier, to the leanest and raciest Veltliners, which can have such fantastically fine minerality that they may be compared to the best Rieslings.

🍷 *1–4 years (Qualitätswein), 3–10 years (Prädikatswein)*

👑 *Angerer • Anton Bauer • Bründlmayer • Ebner Ebenauer • Brigit Eichinger • Domäne Wachau • Geyerhof • Graf Hardegg • Gritsch Mauritiushof • Heidler • Hirtzberger (Spitzer) • Huber • Jurtschitsch-Sonnof • Knoll • Laurenz V • Loimer • Metternich-Sandor • Nikolaihof • F X Pichler • Franz Prager • Rabl • Schloss Gobelsburg • Steininger • Salomon Undhof • Petra Unger • Wieninger*

JUBILÄUMSREBE

WHITE A Grauer Portugieser/Frühroter Veltliner cross presented at the 100-year anniversary of the Klosterneuburger Viticulture School, virtually only producing *Prädikatswein*, yet it never gained a widespread level of acceptance.

🍷 *3–7 years*

MERLOT

RED Officially permitted since 1986, small amounts of Merlot are grown in Krems, Mailberg, and Furth in Lower Austria, and this variety has gained importance over the last years in Burgenland. Merlot has potential in Austria, as it can produce richly coloured red wines with soft, spicy-juicy fruit.

🍷 *1–3 years (Qualitätswein), 3–5 years (Prädikatswein)*

👑 *Netzl • Niki Windisch • Prieler • Reeh • Scheiblhofer • Schwertführer*

MÜLLER-THURGAU

WHITE Sometimes also labelled as Rivaner. The key attributes of Müller-Thurgau are its usually soft acidity and lower alcohol content. Often used as a blending partner in young wines, such as the Steirischer Junker, it has decreased in importance over the last decade.

🍷 *1–2 years*

👑 *Arndorfer • Hirtzberger • Schützenhof Fam Korper*

MUSKATELLER

WHITE With its pronounced, intense aromatic character, the Muskatller variety produces some fine dry wines of crisp acidity in Steiermark as well as elegant, dry, and sweet wines in Burgenland.

🍷 *1–3 years (Qualitätswein), 2–10 years (Prädikatswein)*

👑 *Erwin Sabathi • Leth • Tement • Tremmel • Wohlmuth*

MUSKAT-OTTONEL

WHITE Less weighty than the Muskateller grape, the Muskat-Ottonel variety has more immediate aromatic appeal and is best drunk when young.

🍷 *2–4 years*

👑 *Kracher • Nittnaus • Velich • Willi Opitz*

A PLOT OF GRÜNER VELTLINER IS MARKED OUT IN DOMÄNE WACHAU'S VINEYARDS
Like many wineries in the Wachau region, Domäne Wachau cultivates Grüner Veltliner for many of its wines, including a range of single-vineyard whites.

NEUBURGER

WHITE This Austrian variety excels on chalky soil and in drier climates, making full-bodied wines with a typically nutty flavour in all categories of sweetness. Fine examples can be found in the Thermenregion and along the Danube in Lower Austria as well.

🍷 *2–4 years (Qualitätswein), 3–8 years (Prädikatswein)*

👑 *Dipl Ing Karl Alphart • Domäne Wachau • Hirtzberger • Hofstätter*

PINOT BLANC

WHITE Often called Klevner or Weißburgunder, this produces fresh, light-bodied, easy-to-drink wines, usually softer and lighter compared to Chardonnay. In exceptional years, it can develop a fine spicy-richness in the upper Prädikatswein categories.

🍷 *2–4 years (Qualitätswein), 3–8 years (Prädikatswein)*

👑 *Erwin Sabathi • Lackner - Tinnacher • Neumeister • Rudi Pichler*

PINOT GRIS

WHITE Sold in Austria as Grauburgunder, this is a fuller, spicier version of the Pinot varieties, but it has a typical nutty richness.

🍷 *2–4 years (Qualitätswein), 3–8 years (Prädikatswein)*

👑 *Edlmoser • Erwin Sabathi • Neumeister • Tement*

PINOT NOIR

RED Labelled Blauer Burgunder, Blauer Spätburgunder, or Blauburgunder in different parts of Austria, it has a rather minor quantity, yet increasing importance in Austria; the best examples can be found in the Thermenregion and Burgenland.

🍷 *1–3 years*

👑 *Aumann • Bründlmayer • Johann Gipsberg • Juris-Stiegelmar • Jurtschitsch-Sonnof • Johanneshof Reinisch • Kollwentz • Scheiblhofer • Wieninger • Wohlmuth*

RIESLING

WHITE Not related to the Welschriesling, it is a superior variety and has special importance in Lower Austria, where fine DAC wines are produced with typicity and high quality. Offering gentle aromatic notes with high acidity and fruit-driven style, the finest examples are to be found in Wachau and Kremstal.

🍷 *2–6 years (Qualitätswein), 4–12 years (Prädikatswein)*

👑 *Bründlmayer • Domäne Wachau • Hirtzberger • F.X Pichler • Knoll • Mayer am Pfarrplatz • Nigl • Prager*

ROTER VELTLINER

WHITE To achieve good results strict yield control must be applied. Fine, spicy aromatics with some honey are not unusual; the best examples can be found in Wagram.

🍷 *1–3 years*

ROTGIPFLER

WHITE This makes a robust, full-bodied, spicy wine of not dissimilar character to the Zierfandler, which is often made in a dry style, although the semi-sweet Rotgipfler of Gumpoldskirchen may be the most famous rendition of this grape.

🍷 *3–7 years*

👑 *Auer • Aumann • Biegler • Johanneshof Reinisch • Stadlmann*

ST-LAURENT

RED This typically produces a light, mild-flavoured wine of quaffing character. St-Laurent is thought to be related to Pinot Noir, and it has to be said that some producers can craft lovely, velvety wines of Pinot-like quality and style.

🍷 *1–3 years*

👑 *Johannes Trapl • Juris • Philipp Grassl • Schneider • Johanneshof Reinisch • Pittnauer • Schloss Halbturn • Umathum*

SAUVIGNON BLANC

WHITE Called Muskat-Silvaner in the 19th century, this variety – grown in Steiermark – is a real shooting star.

🍷 *2–4 years (Qualitätswein), 4–10 years (Pradikätswein)*

👑 *Erwin Sabathi • Neumeister • Polz • Sattlerhof • Tement*

SCHEUREBE

WHITE This variety is not very pleasant at *Qualitätswein* level but develops a beautiful aromatic character at higher levels of *Prädikatswein*.

🍷 *2–4 years*

👑 *Gessellmann • Kollwentz • Leberl • Tschida • Willi Opitz*

SILVANER

WHITE Sometimes spelled Sylvaner, the simple, earthy, restrained aromas are the most common characters.

🍷 *1–2 years*

WELSCHRIESLING

WHITE Austria's third-most-prolific variety, making very ordinary dry wines, but they can be rich and stylish at upper levels of *Prädikatswein*.

🍷 *1–3 years (Qualitätswein), 2–8 years (Pradikätswein)*

👑 *Bründlmayer • Lackner Tinnacher • Gross • Neumeister • Willi Opitz • Weinhof Platzer • Tement*

ZIERFANDLER

WHITE Also known as Spätrot, this variety makes full-bodied and well-flavoured dry wine, especially in the Thermenregion.

🍷 *2–6 years*

👑 *Biegler • Johanneshof Reinisch • Stadlmann • Gottfried Schellmann*

ZWEIGELT

RED Another mild red wine variety, Zweigelt is sometimes given the name Blauer Zweigelt or Rotburger. The best examples have a structure that is suited to food, with big, peppery fruit, but the norm is rather light and lacklustre.

🍷 *1–3 years*

👑 *Grassl • Iro Markus • Kiss • Netzl • Pfaffl • Umathum • Johann Gipsberg • K+K Kirnbauer • Leth*

BLENDED RED

The finest of these blended reds are usually oak-aged in new *barriques* and Cabernet-Sauvignon-based for backbone, with Blaufränkisch, St-Laurent, and sometimes Zweigelt added for fruit and softness.

🍷 *2–8 years*

👑 *Gessellmann (Opus Eximium, Bela Rex) • Igler (Cuvée Vulcano) • Josef Pöckl (Admiral) • Ernst Triebaumer (Blaufränkisch-Cabernet) • Umathum (Ried Hallebühl Cuvée Rot) • Grassl (Bärnreiser) • Nittnaus (Comondor) • Gager (Gablot) • Pöckl (Admiral) • Heinrich (Gabarinza) • Markowitsch (Rosenberg), • Heinrich (terra o.)*

BLENDED WHITE

Varietal wines are more common, but fresh, young, *primeur*-like white blends are also made in Steiermark; called "Junker" (blend of Welschriesling, Müller Thurgau, and other early varieties), top sweet wines as well as some mineral-driven *cuvées* from specific *terroirs* and field blends, such as "Gemischter Satz" are still popular, the latter especially in Vienna. While around the Leithaberg the Chardonnay and Pinot Blanc as well as Grüner Veltliner dominate, when it comes to the mixed set (Gemischter Satz) virtually anything goes . . . from Riesling, Chardonnay, and Grüner Veltliner as a field blend.

👑 *Altenburger • Mayer am Pfarrplatz • Ott • Schwarz • Wieninger*

BOTTLE-FERMENTED SPARKLING

Austrian *Sekt* received new regulations in 2015, a three-tier pyramid of Klassik, Reserve, and Grosse Reserve (Grand Reserve) categories, the major difference being the minimum time spent on lees. For Sekt Klassik, minimum lees ageing is 9 months; tank and bottle fermentation are both allowed, while for Reserve whole cluster pressing, 18 months lees and traditional method is required. With Grosse Reserve (Grand Reserve) Sekt the lees ageing is increased to 30 months; it must be of *brut* style, and single-vineyard (Ried in Austrian) designation is permitted on the label. The classic varieties for sparkling wine production are favoured (eg Pinot noir, Chardonnay), yet you will sometimes find Welschriesling as well as the local Grüner Veltliner. Austria's best-known sparkling wines are the bottle-fermented Schlumberger, produced in Vienna and Bründlmayer from Kamptal.

👑 *Bründlmayer • Schlumberger • Szigeti*

THE APPELLATIONS OF

AUSTRIA

Note: The designation DAC (*Districtus Austriae Controllatus*) was introduced in 2001, with Weinviertel the first to receive this status. Its aim is to develop region typical profiles with special focus on the permitted grape varieties, as well as production method. Not all the wine districts have a designated DAC; due to the permitted grape varieties and allowed methods of production the styles have a wider range.

BURGENLAND

Austria's easternmost region is also its warmest, producing a wide range of styles from generic whites to medium- and full-bodied, rich red wines, as well as some of the country's most fascinating sweet wines around the Neusiedler Lake.

EISENBERG DAC (FORMERLY SÜDBURGENLAND)

Most of the wines are sold locally: crispy whites and juicy reds, as well as the so-called *Uhudler* (hybrid) wines. The DAC is reserved for 100 per cent Blaufränkisch from the iron-rich soil of the "Iron mountain", delivering spicy, tight, yet mineral-driven reds.

✔ **Villages:** *Deutsch Schützen, Eisenberg, Rechnitz, Weinberg, Welgersdorf* • **Growers:** *Jalits, Krutzler, Schiefer, Wachter*

KÄRNTEN

The wines of Kärnten are confined to a small 170-hectare (420-acre) area, mainly around Lake Läng. Predominantly white wines are produced and consumed through the thriving tourism in the region. Some promising results have been achieved in the recent years, however.

LEITHABERG DAC (FORMERLY NEUSIEDLERSEE-HÜGELLAND)

The western side of Neusiedler Lake is influenced strongly by the Leitha mountain range, with the limestone, gneiss, and schist soils delivering mineral-driven, spicy, fresh wines. This DAC produces mainly Blaufränkisch in red and Grüner Veltliner and Burgundy varieties in white. The city of Rust is the home of the luscious, rich, yet vibrant Ruster Ausbruch wines.

✓ **Villages:** *Donnerskirchen, Großhöflein, Rust* • **Growers:** *Altenburger, Feiler-Artinger, Kollwentz, Prieler, Schandl, Schröck, Tinhof, Triebaumer, Wenzel*

MITTELBURGENLAND DAC

Mittelburgenland (often referred to as "Blaufränkischland") is where more than 90 per cent of vines grown is red grape varieties, from which Blaufränkisch is the permitted variety to produce regional-styled wines in either fruit-forward style, matured in oak, or from a specific single-vineyard site. Deep loam and clay soil with limestone and pebbles are the main characteristics here in the home of some of the best Austrian reds.

✓ **Villages:** *Deutschkreuz, Horitschon, Neckenmarkt* • **Growers:** *Johann Heinrich, K+K Kirnbauer, Gesellmann, Igler, Moric, Weninger, Wohlmuth*

NEUSIEDLERSEE DAC

Along the eastern side of the lake, vineyards are protected from the northerly wind; silt-sand, gravel, loam and sandier, quartz-rich soils are to be found on the southern edge. A wide range of styles are produced from dry to noble sweet, yet the DAC status can only be applied to Zweigelt or in Reserve wines Zweigelt and blend of local varieties.

✓ **Villages:** *Apetlon, Frauenkirchen, Gols, Halbturn, Illmitz, Podersdorf, Schützen* • **Growers:** *Juris, Heinrich, Alois Kracher, Willi Opitz, Pittnauer, Prieler, Pöckl, Schloss Halbturn, Umathum, Velich*

NIEDERÖSTERREICH

This is Austria's premier dry white wine region. It is famous for its peppery Grüner Veltliners and the elegance of its Rieslings. Top Rieslings can be light and airy in their youth, but they attain a certain richness after a few years in bottle, and they can achieve a fine, racy balance comparable to that of some of the best German Rieslings.

CARNUNTUM DAC

One of Austria's smallest regions, the ancient Roman settlement Carnuntum, east of Vienna, lies along the right side of the Danube. Sand, loam, and loess are the dominant soil types. With the 2019 vintage the regional wines have to be dominated by Blaufränkisch and Zweigelt in the reds and Grüner Veltliner, Chardonnay, and Pinot Blanc in the whites.

✓ **Villages:** *Göttlesbrunn, Rohrau* • **Growers:** *Glatzer, Grassl Markowitsch, Pittnauer, Artner, Netzl*

KAMPTAL DAC

North of Kremstal with Austria's largest wine-producing town in the centre, it produces some of Austria's finest Rieslings and Grüner Veltliners and is home to some of the country's most gifted, innovative, and expressive winemakers.

✓ **Villages:** *Gobelsburg, Langenlois, Strass* • **Growers:** *Allram, Angerer, Brandl, Bründlmayer, Brigit Eichinger, Heidler, Hirsch, Jurtschitsch Sonnhof, Laurenz V, Rabl, Schloss Gobelsburg, Steininger*

KREMSTAL DAC

Varied soil types and climate make this a very diverse and exciting region. Only Grüner Veltliner and Riesling are permitted as DAC wines. Some of the most exciting single vineyards can be found here throughout Austria.

✓ **Villages:** *Krems, Stein* • **Growers:** *Buchegger, Dockner, Geyerhof, Göttweig, Mantler, Nigl, Salomon Undhof, Petra Unger, Stadt Krems*

THERMENREGION

Named after the hot springs, Thermenregion is also one of Austria's warmest regions. Both red and white, with the Blauer Portugieser, Pinot Noir, and Neuburger grape varieties dominating. Zierfandler and Rotgipfler are specialities of Gumpoldskirchen.

✓ **Villages:** *Bad Vöslau, Gumpoldskirchen* • **Growers:** *Alphart, Auer, Hartl, Johanneshof Reinisch, Andreas Schafler, Gottfried Schellmann, Stadlmann*

TRAISENTAL DAC

Formerly the southwestern half of Donauland, Traisental achieved independence as a regional appellation in 1994.

✓ **Villages:** *Herzogenburg, Reichersdorf* • **Growers:** *Huber, Neumayer*

WACHAU

Lower Austria's top-performing district, Wachau (a DAC from 2020 on), makes many fine varietals, including Grüner Veltliner and Riesling, definitely the best. Some of the most sought-after top vineyards and the full-bodied Smaragd wines can be found in this exciting wine region.

✓ **Villages:** *Durnstein. Loiben, Spitz* • **Growers:** *Alzinger, Donabaum, Gritsch-Mauritiushof, Hirtzberger, Höllmüller, Knoll, Jamek, Lagler, Nikolaihof, F X Pichler, Prager, Wess*

WAGRAM

Formerly Donauland, or land of the Danube, the Wagram region houses the world's oldest viticultural college at Klosterneuburg. Loess and limestone are the distinct soil types.

✓ **Villages:** *Kirchberg, Klosterneuburg, Wagram* • **Growers:** *Anton Bauer, Josef Bauer, Fritsch, Leth, Ott, Söllner*

WEINVIERTEL DAC

A large wine region with a wide range of soil types, it has the coldest average temperatures of the wine-producing areas. Expect well-made, easy-going whites and reds with distinct freshness. The DAC wines must be Grüner Veltliner, made dry with a definite peppery character.

✓ **Villages:** *Ebenthal, Falkenstein, Mailberg, Poysdorf, Retz* • **Growers:** *Ebner-Ebenauer, Graf Hardegg, Jassek, Pfaffl, Malteser Ritterorden, Setzer, Herbert Zillinger, Zull*

OBERÖSTERREICH

A large area to the west of Wachau with just 45 hectares (110 acres) of vineyards. The focus is on white and Zweigelt.

STEIERMARK or STYRIA

Situated in the southeast corner of Austria, this region has a high level of rainfall, interspersed with exceptional levels of sunshine and warmth. Sandy, marl, loam, limestone, and even volcanic soils exist in the wide region. Roughly two-thirds of the production is white, usually varietal led, fresh-fruit driven. There are three wine districts in this region.

SÜDSTEIERMARK DAC

The very best wines of this area have Steiermark's naturally high acidity, but this is combined with delicate and pure fruit flavours, making for exceptional finesse, particularly in varieties such as Gewürztraminer, Chardonnay (known locally as Morillon), and Sauvignon Blanc. Some of the area's finest single vineyards can be found here.

✓ **Villages:** *Gamlitz, Leibnitz, Leutschach* • **Growers:** *Gross, Maitz, Polz, Sattler, Skoff, Tement, Erwin Sabathi, Sattlerhof, Wohlmuth*

VULKANLAND STEIERMARK DAC (FORMERLY SÜDOSTSTEIERMARK)

This DAC id for white varieties only. Klöch is famous for the Gewürztraminer on volcanic soil, and the Welschriesling and the Pinot family dominate in the other areas.

✓ **Village:** *Kloch* • **Growers:** *Frühwirth, Neumeister, Platzer, Winkler-Hermaden*

WESTSTEIERMARK DAC

The leading variety undoubtedly is the Blauer Wildbacher, from which Schilcher (dark rosé) is produced on crystalline rock. Tart, utterly crisp (usually) rosé wines are produced here with distinct strawberry fruitiness.

✓ **Village:** *Stainz* • **Growers:** *Friedrich. Jöbstl. Langmann Müller*

TIROL

The majority of producers are based in the villages of Silz, Tarrenz, and Haiming, and vineyard surface is barely 5 hectares (12 acres). Producers mainly focus on Burgundian varieties.

VIENNA or WIEN DAC

Vienna is not the only capital city that features wine production within its boundaries, but it's perhaps the most famous. Vineyards are mainly to be found on the northern edge. Most of the wines are sold by the pitcher as Wiener Heuriger in the city's many bars called Heurigen (also known as Buschenschanken). The Gemischter Satz DAC is the typical quality wine, in which at least three different grape varieties have to be harvested and processed together.

✓ **Growers:** *Franz Mayer, Mayer am Pfarrplatz, Wieninger*

VORARLBERG

Vorarlberg is the westernmost region of Austria. Once home to 500 hectares (1,235 acres), today the community of viticulturists have roughly 60 members on 10 hectares (24 acres).

WHITE WINE GRAPES AWAIT HARVEST IN SOUTHERN STEIERMARK
The various subregions of the Steiermark (Styria) are mainly devoted to cultivating white varieties, including Grüner Veltliner, Riesling, Gewürztraminer, Chardonnay (known locally as Morillon), and Sauvignon Blanc.

Switzerland

*Environmentally sound vineyards, a huge diversity of grape varieties, and alpine terroir
are the base of the Swiss wine culture, which dates back to Roman times.
Exports are still low, so to discover the wines, you must travel to the country itself.*

The alpine country's diverse wine landscape is defined by its four national languages (German, Italian, French, and Romansch) and their distinct cultural heritages. Switzerland's nearly 15,000 hectares (37,065 acres) of vineyards are divided into six official wine regions: Valais (4,976 hectares/12,295 acres), Vaud (3,784 hectares/9,350 acres), German-speaking Switzerland (2,852 hectares/7,047 acres), Genève (1,435 hectares/3,546 acres), Ticino (1,105 hectares/2,730 acres), and the region of the Three Lakes (975 hectares/2,409 acres).

Almost 60 per cent of Swiss vineyards are planted with red grape varieties. Number one is Pinot Noir, followed by Gamay and Merlot. For the whites, Chasselas is comes in first. Pinot Noir and Chasselas account for more than 55 per cent of the total production.

Chasselas is the symbol of Swiss white wine and one of the most widespread white grape varieties in the world, originating from Canton Vaud around Lac Léman (Lake Geneva). A good Chasselas is dry, delicate, and very refreshing with lots of minerality and a slight floral expression. It is the perfect wine to drink when you want to give your palate a rest or when you are thirsty – the local name for it is *vin de soif,* or "wine of thirst".

It would be wrong to think that Swiss wines are only light bodied and easy drinking. In the last 15 years a new generation of wine producers all over the country have improved and invested in their domains, creating a wide range of high-quality and unique wines. Of special note are the Pinot Noirs from the Grisons and Neuchâtel, the Merlots from Ticino, and the indigenous vines (dry and sweet) from Valais, as well as the Syrahs and of course the unique Chasselas from Lavaux.

Compared to other wine regions Switzerland so far is blessed with a healthy internal market that sells most of the production and still imports a lot of wine – even though consumption continuously goes down. Nevertheless, at the moment, the best way to discover and taste Swiss wine is still by visiting the country and exploring this huge diversity that celebrates itself on such a small surface.

RECENT SWISS VINTAGES

2018 Difficult start, but then a very good vintage for quality and quantity.

2017 Very difficult vintage with lots of frost. Some regions lost up to 80 per cent of production. What could be harvested is of very good quality with a vibrant acidity.

2016 Less complex vintage than 2015 but considered also a very good one. Second part of summer was hot and dry.

2015 A great vintage for both reds and whites. Outstanding in Valais where the harvest began early September.

2014 Winter in 2014 was the warmest in the last 150 years. Spring was very dry and hot and bud break occurred 10 days earlier than the average vintage. July and August were very wet, and this humidity attracted a new insect, *Drosophila suzukii,* which did a lot of damage. The result is a low-quantity vintage. To achieve good quality, a crucial selection at harvest was necessary.

2013 Not an easy start, but a great vintage with good acidity and high concentration. Outstanding year for Ticino.

2012 Difficult year that needed a lot of work in the vineyard. The result in the end was good, but the quantity was significantly less, especially for white wines.

2011 Very good year with balanced and complex wines.

WINE REGIONS OF SWITZERLAND
The language spoken in an area divides Switzerland into three parts: French-speaking, German-speaking, and Italian-speaking. Within the six wine regions, some of the wine-producing cantons cross these "international" boundaries and possess alternative names, with most vineyards concentrated around lakes and rivers.

	Valais	Wine-growing region
	Valais	Appellations of origin (AOC)

AOC numbered on map

1	Fully Grand Cru
2	Saillon Grand Cru
3	Leytron Grand Cru
4	Chamoson Grand Cru
5	Vétroz Grand Cru
6	Conthey Grand Cru
7	Ville de Sion Grand Cru
8	Saint-Léonard Grand Cru
9	Sierre Grand Cru
10	Salquenen Grand Cru

0 mi ——— 50
0 km ——— 50

FACTORS AFFECTING TASTE AND QUALITY

LOCATION
Situated between the south of Germany, north of Italy, and the central-east French border.

CLIMATE
Continental Alpine conditions prevail, with local variations due to altitude, the tempering influence of lakes, and the sheltering effects of the various mountain ranges. An Alpine wind called the Foehn raises, rather than lowers, temperatures in some valleys. Rainfall is relatively low, and a number of areas, such as the Valais in the south, are very dry indeed. Spring frosts are a perennial threat in most areas.

ASPECT
Vines are grown in areas ranging from valley floors and lake shores to the steep Alpine foothill sites, which start at an altitude of 270 metres (885 feet) above sea level with the highest at nearly 1,200 metres (2,460 feet). The best sites are found on south-facing slopes, which maximise exposure to sun, and where the incline is too steep to encourage high yields.

SOIL
Mostly a glacial moraine (scree) of decomposed slate and schist, often with limestone over sedimented bedrock of limestone, clay, and sand. In the Vaud, Dézaley is famous for its "pudding stones", and the Rhône Valley encompasses a diversity of soil types, three of which are most notable: black slate, alluvial gravel, and silt.

VITICULTURE & VINIFICATION
Terracing is required on the steeper sites and so too is irrigation (with mountain water) in dry areas such as the Valais. Most vineyard work is labour-intensive, except in a few more gently sloping vineyards such as those around Lac Léman (Lake Geneva), where mechanical harvesting is practical. Careful viticulture produces remarkably high yields. The cultivation of red wine is slightly dominant while environmentally friendly production is common with a strong movement toward organic production.

GRAPE VARIETIES
More than 240 grapes are cultivated. Among them: Aligoté, Amigne, Ancellota, Arvine (Petite Arvine), Blauburgunder, Bonarda, Bondola, Cabernet Dorsa, Cabernet Franc, Cabernet Sauvignon, Chardonnay, Chasselas (Fendant), Completer, Cornalin, Diolinoir, Divico, Doral, Dornfelder, Ermitage (Marsanne), Freisa, Galotta, Gamaret, Gamay, Garanoir, Gewürztraminer, Heida (Savagnin Blanc), Humagne Blanc, Humagne Rouge, Johanniter, Kerner, Malbec, Malvoisie, Merlot, Muscat, Paien (Savagnin Blanc), Pinot Blanc, Pinot Gris, Räuschling (Elbling), Regent, Riesling x Sylvaner (Müller-Thurgau), Sauvignon Blanc, Sémillon, Seyval Blanc, Solaris, Sylvaner, Syrah, Traminer, Viognier, Zweigelt

THE WINE STYLES OF

SWITZERLAND

AMIGNE
WHITE Old Valais variety mostly grown at Vétroz, the Amigne usually makes a full, rustic, smooth, dry white, but Germanier-Balavaud makes a sweet, luscious, late-harvest style.

🍷 *1–4 years*

👑 *André Fontannaz • Germanier-Balavaud (Mitis)*

ARVINE or PETIT (PETITE) ARVINE
WHITE Another old Valais variety that makes an even richer dry white, with a distinctive grapefruit character and good acidity. The Petite Arvine (to distinguish it from the lesser-quality Grosse Arvine) adapts well to late-harvest styles and is highly regarded by Swiss wine lovers in both dry and sweet styles.

🍷 *1–3 years*

👑 *Valentina Andrei • Gérald Besse • Marie-Thérèse Chappaz (especially Grain Noble) • Benoît Dorsaz • René Favre & Fils • Simon Maye • Maître de Chais Provins*

BONDOLA
RED An indigenous grape producing rustic reds on its own, Bondola is at its best in a wine called Nostrano, in which it is blended with Bonarda, Freisa, and other local varieties.

BONVILLARS
WHITE Restricted to Chasselas grown around Lac de Neuchâtel in the villages of Côtes de l'Orbe and Vully, Bonvillars is lighter and more delicate and lively than Vaud wines of this variety grown farther south.

👑 *CV Bonvillars (Vin des Croisés)*

CHARDONNAY
WHITE Not so popular as in other countries but some beautiful wines can be found.

🍷 *1–4 years*

👑 *Château d'Auvernier • Bovel • Weingut Donatsch • Gantenbein • Tom Litwan*

CHASSELAS
WHITE Called Fendant in the Valais, this is the most important Swiss grape variety and can be nicely flavoured, often having a charming *pétillance*. When grown on light, sandy soils it may have a lime-tree blossom aroma. The grape is sensitive to soil, with different characteristics depending on whether the soil is limestone, flint, gypsum, marl, schist, or any other. Although it is possible to find relatively full, potentially long-lived wines from this variety, it is a wine best enjoyed young. Exceptions are the wines from Dézaley or Calamin – they can age for decades.

🍷 *1–3 years*

👑 *Domaine Louis Bovard • Clos du Boux • Domaine La Colombe • Domaine Henri Cruchon • Blaise Duboux • Château Vinzel*

COMPLETER
WHITE The Completer is rare grape variety of ancient origin that makes a fascinating, rich, *Auslese*-style of wine from scattered plots found in Graubünden. Can age for decades.

🍷 *3–7 years*

👑 *Adolf Boner • Weingut Donatsch • Peter & Rosi Hermann • Malans*

CORNALIN
RED Very popular indigenous Swiss variety from the Valais that produces dark, powerful, and complex wines that are rich and concentrated, with a full, spicy complexity.

🍷 *3–7 years*

👑 *Histoire d'Enfer • Jean-René Germanier • Denis Mercier • Domaines Rouvinez • Maurice Zufferey*

LA CÔTE
RED AND WHITE Grown exclusively at La Côte on the shore of Lac Léman (which encompasses the villages of Aubonne, Begnins, Bursinel, Coteau de Vincy, Féchy, Luins, Mont-sur-Rolle, Morges, Nyon, Perroy, and Vinze, as well as La Côte itself), where the Chasselas grape is at its most floral and aromatic, and fruity reds are often blended from two or more varieties, including Gamay, Gamaret, and Garanoir.

🍷 *1–2 years*

👑 *Cave de la Côte • Château de Châtagnereâz • Domaine de Autecour • Domaine de Sarraux • Château de Vinzel*

DÔLE
RED No longer just a light red since legislation changed for its production. A Dôle wine is a Valais speciality that must predominantly consist of Pinot Noir and Gamay (minimum 85 per cent of the blend). The rest can consist of Gamaret, Garanoir, Carminoir, Ancellotta, Diolinoir, Merlot, or Syrah.

🍷 *1–3 years*

👑 *Marie-Thérèse Chappaz • Simon Maye • Provins • Robert Gilliard*

DÔLE BLANCHE
WHITE The *blanc de noirs* version is popular, but does not carry its own appellation.

ERMITAGE
WHITE A synonym for the Marsanne used in the Valais, where it makes a delicately rich, dry white wine that is often aged too long in bottle by its aficionados.

🍷 *1–4 years*

👑 *Marie-Thérèse Chappaz • Dom Cornulus • Adrian Mathier • Philippoz Frères*

FENDANT
WHITE This Chasselas synonym is used in the Valais, where it produces a fleshy-styled, dry white wine, the best of which can have a flinty, lime-tree-blossom aroma and a crisp, minerality of fruit. Ideal with cheese fondue, raclette, or sushi. In the past all Chasselas grown in Switzerland was called Fendant.

🍷 *1–3 years*

👑 *Cave Mabillard-Fuchs • Adrian & Diego Mathier • Domaine Jean-Louis Mathieu • Domaine des Muses • Simon Maye • Provins*

GAMAY
RED Until recently, this Beaujolais grape rarely made red wines of distinction in its pure varietal form and was much more successful when blended with Pinot Noir (see Dôle).

🍷 *1–3 years*

👑 *Gérald Besse • Jean-René Germanier • Les Hutins • Madeleine et Jean-Yves Mabillard-Fuchs*

GEWÜRZTRAMINER

WHITE Rarely encountered.

🍷 1–5 years

🍾 *Dominique Passaquay (Passerillé)* • *Maurice Dupraz*

HEIDA

WHITE Fresh, lightly aromatic, dry, and off-dry wine made from the Savagnin (Traminer) grape. Very popular in Valais. The most famous Heidas are cultivated in Visperterminen on 1,150 metres (3,773 feet) above sea level. In German, the name Heida is related to the word *Heide*, or "heathen", and in the Upper Valais dialect, the term also meant "old", "ethnic", "pre-Christian", or "in heathen times". It is also found in the name Heidenhäuser ("old houses"), or Heido, for the oldest water conduct of Visperterminen. Since 1812 the French-speaking population of the Valais has also called the Heida "Païen".

🍷 1–6 years

🍾 *Chanton Weine* • *Simon Maye (Païen)* • *St Jodernkellerei*

HUMAGNE BLANCHE

WHITE Once perceived as a wine of the nobility and bishops; it is also known as the "wine for women in childbed" because it was credited with having a high iron content, thus helping women regain their strength after childbirth. The Humagne Blanche is produced as a dry wine and normally has a fairly low alcohol content. When it is young, the yellow-gold wine is appreciated for its fine acidity and delicate fruitiness with soft flowery notes. Its slightly resinous aromas and notes of acacia honey develop after three to four years of storage, giving it its own distinctive character. On the palate, it impresses with beautiful fullness and taste.

🍷 1–10 years

🍾 *Bonvin* • *Histoire d'Enfer* • *Provins*

HUMAGNE ROUGE

RED Not related to Humagne Blanche; however, it is identical with the one called Aosta Valley Cornalin. Humagne Rouge has a long tradition in the Valais and is characterised by a slightly rustic and unruly note. The wines are complex and go well with feathered game.

🍷 1–4 years

🍾 *Histoire d'Enfer* • *Jean-René Germanier*

JOHANNISBERG

WHITE This is not the Johannisberg Riesling, but the official synonym for Sylvaner in the Valais, where it is supposedly musky, although the wines are usually sappy and savoury with some sweetness.

🍷 Upon purchase

🍾 *Adrian Mathier* • *Albert Mathier & Söhne* • *Claudy Clavien* • *Domaine du Mont d'Or* • *Cave du Rhodan*

LAFNETSCHA

WHITE This Completer x Humagne Blanche cross has long grown in the Haut Valais, where it produces dry whites with a leafy aroma.

🍾 *Chanton Weine*

LAVAUX

WHITE Grown exclusively at Lavaux (which encompasses the villages of Calamin, Chardonne, Dézaley, Epesses, Lutry, Saint-Saphorin, Vevey-Montreux, and Villette, as well as Lavaux itself), the Chasselas is at its softest yet fullest. This terraced landscape of the Lavaux was created in the 12th century by Cistercian monks and then maintained, expanded, and improved by generations of wine growers and is today part of the UNESCO world heritage site. Part of the Lavaux are the two famous *terroirs* Dézaley and Calamine.

🍷 1–6 years

🍾 *Domaine Louis Bovard* • *Clos du Boux* • *Henri & Vincent Chollt* • *Blaise Duboux* • *Pierre-Luc Leyvraz* • *Pierre & Basile Monachon*

MERLOT DEL TICINO

RED Merlot grape accounts for more than 80 per cent of the vines planted in Italian-speaking Switzerland, where it produces red wines that range from young and light-bodied to fuller, better-coloured wines that have a nice varietal perfume and are often aged in *barriques*. The VITI classification on a bottle of Merlot del Ticino used to be seen as a guarantee of superior quality, but now merely indicates the wine has had one year's ageing before being bottled.

🍷 2–4 years (lighter styles), 3–10 years (barrique wines)

🍾 *Fratelli Corti* • *Adriano Kaufmann* • *Werner Stucky* • *Enrico Trapletti* • *Huber Vini* • *Cantina Kopp von der Crone Visini* • *Luigi Zanini* • *Christian Zündel* • *Domaine Grand'Cour*

MÜLLER-THURGAU

WHITE Müller-Thurgau is the most important white grape variety in German-speaking Switzerland. The name repeatedly causes confusion, as it is partly known under its actually false name "Riesling x Sylvaner" (or Riesling-Sylvaner). This grape was created in 1882 by the Thurgau Professor Hermann Müller (1850–1927) in Germany from a cross between Riesling and Chasselas – and not Sylvaner, as he originally thought. However, the resulting mild, grapey wine is very aromatic and refreshing in style.

🍷 1–2 years

🍾 *Schloßgut Bachtobel* • *Baumann* • *Daniel Marugg* • *Weingut Pircher* • *Obrecht - Weingut zur Sonne* • *Weingut Familie Zahner*

MUSCAT

WHITE This variety is traditionally grown in small quantities in the Valais. It makes wines of very light body, with the grape's unmistakeable varietal flowery aroma. Some interesting late-harvest styles are beginning to emerge.

🍷 2–4 years

🍾 *Cave Emery (Grains Nobles de la Saint Nicolas – Séduction)* • *Domaine des Muses* • *Stéphane Clavien (Coteaux de Sierre)*

NON-FILTRE

WHITE Meaning simply unfiltered, Non-Filtré is a Neuchâtel speciality dating back to 1974. At that time a small harvest resulted in a lack of white wine in the canton. At the urging of "thirsty" customers, a cellar bottled a portion of the new wine unfiltered. This cloudy wine was so well received that the Non-Filtré now accounts for 10 per cent of total production. It is presented every year on the third Wednesday in January. A Neuchâtel Non-Filtré always consists of 100 per cent Chasselas. The yeasts are left in the wine after fermentation and give it a very typical character.

🍷 1–8 years

🍾 *Château d'Auvernier* • *La Maison Carrée*

NOSTRANO

RED Once an inexpensive blend of *vin ordinaire* quality, produced from grapes that failed to reach appellation standard, it now has its own official classification, and quality has certainly improved. Nostrano means "our" and refers to the local varieties used, as opposed to American hybrids.

OEIL DE PERDRIX

ROSÉ *Oeil de Perdrix*, which literally means "partridge eye", is a French term for pale, dry rosé wines made from free-run Pinot Noir. Best wines are from the Neuchâtel area.

🍷 Upon purchase

🍾 *Château d'Auvernier*

PAÏEN

The local synonym for the Heida, aka Savagnin. *See* Heida.

PINOT BLANC

WHITE Recently planted, fashionable variety, but few wines stand out.

🍷 1–3 years

🍾 *Bad Osterfingen* • *Domaine des Abeilles d'Or* • *Domaine Saint-Raphaël*

PINOT GRIS

WHITE This famous Alsace grape is becoming quite fashionable in Switzerland, where it makes more of a Pinot Grigio style, with some very successful *vendange tardive* emerging.

🍷 1–3 years

🍾 *Cave Saint-Pierre* • *Caves du Château d'Auvernier (Vendange Tardive)* • *La Colombe* • *Domaine E de Montmollin Fils (Vendange Tardive)* • *Schloßgut Bachtobel/Hans Ulrich Kesselring* • *Urs Pircher Stadtberg (Egfisauer Stadtberger)*

PINOT NOIR

RED Main red varietal. In a country like Switzerland, which adores burgundy, it is not surprising that top producers make some of Switzerland's most serious wines with this grape variety – well coloured, velvety, with soft, cherry fruit and a touch of oak. Pinot Noir or Blauburgunder can be found all over the country, while the wines from the Grisons and Neuchatel are highly praised.

🍷 3–6 years (up to 10 years in exceptional cases)

🍾 *Bovel* • *Georg Fromm* • *Daniel & Martha Gantenbein* • *Christian Hermann* • *Schloßgut Bachtobel* • *Peter Wegelin* • *Tom Litwan* • *Annatina Pelizzatti* • *Markus Stäger* • *La Maison Carrée* • *Weingut Donatsch* • *Jacques Tatasciore* • *Weingut Eichholz* • *Histoire d'Enfer*

RÄUSCHLING

WHITE This rare variety is in decline and is now found only along the shore of Lake Zurich, where it is prized for its very fresh scents and fine acidity.

🍷 Upon purchase

🍾 *Weingut Pircher* • *Schwarzenbach*

RIESLING

WHITE Apart from exceptional botrytised wines, Swiss Riesling does not have the quality or varietal intensity of Germany or Austria.

🍾 *Schloßgut Bachtobel*

SALVAGNIN

WHITE A light, supple red-wine blend of Pinot Noir and Gamay, this is the Vaud's equivalent of the more famous Dôle.

🍷 1–3 years

🍾 *Henri Cruchon*

SAUVIGNON BLANC

WHITE This grape represents less than 1 per cent of Switzerland's vineyards, but a few interesting interpretations are beginning to emerge, especially from Genève *terroir*.

🍾 *Domaine Les Faunes Dardagny* • *Pierre Dupraz et Fils*

SÜSSDRUCK

ROSÉ A soft, fresh, dry rosé produced in eastern Switzerland from free-run Pinot Noir.

SYRAH

RED This grape is at home in the heat and exposure of the northern Côtes du Rhône, and in Valais the Rhône River is as northerly as it gets. Peppery black fruits distinguish these fast-improving, increasingly fashionable wines.

🍷 3–10 years

🍾 *Gérald Besse* • *Jean-René Germanier* • *Jean-Michel Novelle* • *Simon Maye* • *Romain Papilloud* • *Henri Valloton* • *Zufferey*

VIN DU GLACIER

Although *vin de glacier* wine, Switzerland's equivalent of Germany's *Eiswein*, tends to be oxidative, this rare product does have its enthusiastic followers.

THE WINE REGIONS OF
SWITZERLAND

The country is divided into six official wine regions. Each region and subregion has an identity defined by its landscape, its geology, and its specific climate.

Note: Under the federal appellation system, all cantons have the right to their own appellation, and each winemaking village within each canton can register its own appellation. Genève was the first canton to comply in 1988, followed by Valais in 1991, Neuchâtel in 1993, and Vaud in 1995. Along with these village appellations, each canton may also designate generic district and stylistic appellations, the number of which is increasing all the time.

VALAIS

A characteristic of Swiss wine is its extraordinary diversity of grape varieties and a large number of indigenous grapes, which are rarely found in other countries. Most of these varieties are found in the largest wine-growing region of Switzerland, the Valais. This area is a region of contrasts. It has glaciers and palm trees, saffron and cheese, Chasselas and Heida. With almost 5,000 hectares (12,355 acres) of vineyards, the Valais produces about a third of Switzerland's total production. Mainly Pinot Noir and Chasselas, that here is called Fendant, and then more than 50 different local and indigenous grapes such as Petite Arvine, Heida, Humagne Rouge, Humagne Blanche, or Cornalin. The Valais also has a language barrier between the German-speaking Upper Valais and the French-speaking Lower Valais. The Rhône flows through this area, since the river originates from here and one often hears wines described as being from the Swiss Rhône valley.

Growers: *Adrian & Diego Mathier • Albert Mathier & Söhne • Domaine Jean-Louis Mathieu • André Fontannaz • Caves Fernand Cina • Chanton Weine • Charles Bonvin • Les Fils de Charles Favre • Château Lichten • Denis Mercier • Didier Joris • Domaine des Muses • Domaine du Mont d'Or • Cave Dubuis et Rudaz • Frédéric Dumoulin • Gérald Besse • Jean-René Germanier • Madeleine et Jean-Yves Mabillard-Fuchs • Marie-Thérèse Chappaz • Maurice Gay • Maurice Zufferey • René Favre & Fils • Romain Papilloud • Rouvinez • Clos de Tsampéhro • Provins • Valais Mundi • Valentina Andrei • Simone Maye & Fils • Histoire d'Enfer • Domaines Chevaliers • St. Jodernkellerei • Domaine Cornulus • Benoît Dorsaz • Domaine Cornulus • Cave Fin Bec • Maison Gilliard • Domaine La Rodeline • Cave du Rhodan*

VAUD

In the neighbouring canton Vaud (the second-largest wine-growing region) the scenery looks completely different. While the Valais is marked by alpine landscape and grape diversity, here everything turns around the Chasselas grape – 60 percent (2,268 hectares/5,604 acres) are planted with it. DNA analysis by Dr. José Vouillamoz proved in 2009 that the origins of the Chasselas are in the canton Vaud, and not Egypt or Turkey as previously thought. The canton of Vaud is composed of the following wine regions: La Côte, Lavaux (a protected UNESCO world heritage site), Chablais, Côtes de l'Orbe, Vully, and Bonvillars. Typical of Vaud Chasselas wines is the reference to *terroir*. Most wines are not labelled as Chasselas, but with the name of the village, the home community, or the vineyard they come from (as in Burgundy). For example, St Saphorin, Yvorne, Aigle, Féchy, Dézaley, or Epesses. The best wines have their origin in terraced vineyards along Lac Léman. Chasselas is an aromatic delicate and rather neutral (ideal for Switzerland) grape variety that responds very quickly to the different soils or climatic conditions.

Growers: *Château d'Allaman • Château de Vinzel • Cave de la Côte • Château Maison Blanche • Clos du Châtelard • Clos de la George • Domaine du Martheray • Obrist • Hammel • Henri Badoux • Domaine Louis Bovard • Domaine La Colombe • Domaine Henri Cruchon • Blaise Duboux • Patrick Fonjallaz • Pierre-Luc Leyvraz • Domaine Mermetus • Bernard Cavé • Cave de Bonvillars • Les Frères Dubois • Bolle & Cie. • Domaine de*

Autecour • Domaine Blondel • Les Dames de Hautecour • Basil und Pierre Monachon • Hammel • Terroir du Crosex Grillé • Anne Muller • Domaine de l'Ovaille • Château de Châtagneréaz • Domaines de la Ville de Lausanne • Christophe Chappuis • Domaine de Penloup • J&M Dizerens • Château La Bâtie • Domaine du Burignon • Domaine La Capitaine • Domaine Es Cordelières • Domaine Bovy • Château Vufflens • Le Satyre

GERMAN-SPEAKING SWITZERLAND

The third-largest wine region is German-speaking Switzerland. The most important wine cantons are Zurich, Schaffhausen, Graubünden, and Aargau. But vines are also cultivated, for example, in Appenzell, Zug, and Glarus. Nineteen cantons are united in this wine region. The main variety here is the Pinot Noir. Müller-Thurgau (*Riesling x Sylvaner*) plays an important role for the white wines. The best Pinot Noirs come from Graubünden, where 80 per cent of the vineyards are planted with Pinot, giving the canton its nickname "Burgundy of Switzerland".

By far the largest of the three cultural divisions, German-speaking Switzerland (eastern Switzerland) covers almost two-thirds of the country, yet it encompasses only one-sixth of its total viticultural area. Aargau is best known for its red wines, but also produces off-dry, light, fragrant white wines of low alcohol content. Basel is a producer of white and red, but in small quantities. Saint-Gallen also encompasses the upper reaches of the Rhine and produces mostly red wine. The vines of Schaffhausen grow within sight of the famous falls of the Rhine and enjoy enormous popularity. Thurgau is, of course, home of the famous Dr Müller, who left his mark on world viticulture with his prolific Müller-Thurgau cross, and inevitably the grape grows there, although more than 50 per cent of the vines are Pinot Noir, and the fruity wine that grape produces is far superior. Zürich is the most important canton in German-speaking Switzerland, but it produces some of the most average wines in the entire country, although Pinot Noir grown on sheltered slopes with a good exposure to the sun can produce attractive fruity wines.

LAVAUX WINE REGION, CANTON VAUD
The famous Lavaux wine region overlooking the northern shores of Lac Léman (Lake Geneva) has been a UNESCO World Heritage Site since 2007.

Growers: *Adolf Boner • Daniel & Martha Gantenbein • Georg Fromm • Christian Hermann • Schlossgut Bachtobel • Peter & Rosi Hermann • Peter Wegelin • Urs Pircher • Erich Meier • Tom Litwan • Weingut zur Sonne • Weingut Martin Donarsch • Weinbau von Tscharner • Weingut Sprecher von Bernegg • Winzerkeller Strasser • Weingut Schwarzenbach • Markus Ruch • Annatina Pelizzatti • Möhr-Niggli • Christian Hermann • Weingut Bad Osterfingen • Michael Broger • Markus Hedinger • Weingut Baumann • Weingut Eichholz • Saxer • Urs Pircher • Aagne • Hansruedi Adank • Weingut Zahner • Weimgut zum Sternen • Bovel • Schmid Wetli • Jauslin • Zweifel*

GENÈVE,

The fourth-largest wine growing region, Genève, (western Switzerland) is again marked by diversity. The hills around the western end of Lac Léman favour a very diverse production and allow the use of the most modern wine-producing techniques. Here 56 per cent of the vineyards are red: Gamay, Pinot Noir, Gamaret, Garanoir, Cabernet Franc, Merlot, and Syrah are dominant. For the whites, there are Chasselas, Chardonnay, Sauvignon Blanc, and Aligote. In Genève, the majority of vines grow on gentle south-facing slopes that are protected from the Jura mountains.

Growers: *Domaine du Paradies • Domaine Les Hutins • La cave de Genève • Domaine des Charmes • Domaine Grand Cour • Domaine Dugerdil • Domaine du Centaure • Jean-Michel Novelle • Roger Burgdorfer • Domaine Les Perrières • Domaine des Bonnettes • Domaine des Bossons • Cave Les Baillets • Domaine des Curiades • Château du Crest • Domaine Villard & Fils • Domaine des Graves*

ITALIAN-SPEAKING SWITZERLAND

Italian-speaking Switzerland consists of the cantons of Ticino (also known as Tessin) and the southern fringes of Graubünden (known in Italian as Grigioni) and more than 80 per cent of the vineyards are planted with Merlot. Merlot del Ticino is consistently one of the top wines of Switzerland. Labour costs are nowhere near as high as they are in the rest of the country, which also makes Merlot del Ticino the least expensive of Switzerland's finest wines. The Merlot Bianco, an elegant white wine that is gaining in popularity, also comes from this same grape variety. Ticino is divided by the Monte Ceneri in two major wine regions. The northern Sopraceneri and the southern Sottoceneri.

Growers: *Adriano Kaufmann • Agriloro • Christian Zündel • Huber Vini • Luigi Zanini • Tamborini Vini • Werner Stucky • Fratelli Corti • Enrico Trapletti • Cantina Kopp von der Crone Visini • Gialdi Vini • Brivio Vini • Delea Vini • Cantina Monti • Chiodi • Klausener*

THREE LAKES

The sixth and smallest wine-growing region is that of the Three Lakes. Actually it should be called "three lakes, two languages, and four cantons", because the region comprises Neuchâtel, Biel, Vully, and Jura. The spoken language varies from German to French, and Pinot Noir and Chasselas are mainly cultivated. Very typical is a rosé wine made from Pinot Noir Rosé known as Oeil de Perdrix. The vineyards located on the banks of Lac de Neuchâtel account for the largest part of the production in this region. The vineyards stretch along the lake from Vaumarcus in the south to Cressier in the north. The other wines from the region (issued for the most part from the Chasselas and the Pinot Noir grapes) come from the vineyards bordering the Lakes of Bienne (canton of Bern) and of Morat (canton of Fribourg and Vaud).

Growers: *Château d'Auvernier • Domain Chambleau • Domaine Souaillon • Château de Praz • Jacques Tatasciore • La Maison Carrée • Cru de l'Hôpitel • Domaine Bouvet-Jabloir • Domaine de Vaudijon • Domaine Montmollin • Mauler • Grillette • Domaine Chevret • Steiner Schernelz Village*

The Wines *of* Northwestern Europe

THE MAJORITY OF THESE COUNTRIES LIE at the very edge of Europe, jutting out into the Atlantic Ocean and North Sea – not exactly the mellow, sun-warmed hills of Burgundy for producing wine. Unlike Luxembourg with its milder Germanic climate and *terroir*, England is located on the same harsh latitude as Labrador yet manages to have a world-class wine industry, specifically when it comes to sparkling wines. And if that seems surprising, consider that there are vineyards in even more northerly reaches of Europe, in Belgium, the Netherlands, and even on an island north of Poland. It's never an easy haul for these winegrowers: the uncertain, wet, windy maritime climate, in spite of the balmy presence of the North Atlantic Current, limits their scope. The fewer hours of daylight and the greater angle at which the sun is spread causes their grapes to ripen slowly. Still, the output of these hardy vintners, though sometimes variable, can be surprisingly fine.

GRAPEVINES GROW ON THE CITY WALLS NEAR THE CHURCH OF SAINT JEAN DU GRUND IN LUXEMBOURG CITY
The tiny Grand Duchy of Luxembourg produces some wonderful wine but most of never reaches markets farther away than Belgium or Germany. Most of its vineyards line the Moselle River, which gives its single appellation its name: Moselle Luxembourgeoise.

The British Isles

The UK climate might be evermore uncertain, but the success of English sparkling wine, despite its relatively high price, has made the British Isles a hot topic throughout the global fizz community. And the fact that sales have continued to increase over the last 10 years suggests that its future is very certain indeed.

The quality of Welsh wine has yet to make an impact, but wherever wine is produced on these islands, there is only one world-class wine style: sparkling. The Bacchus grape can produce some wonderfully fresh, crisp, and dry herbaceous-aromatic white wines. Occasionally a gorgeous red wine like Gusbourne's Pinot Noir (Boot Hill) will also grab headlines, particularly in warmer vintages, but Bacchus will never earn global fine wine respect, and nobody should give up sparkling to focus on red wine.

ANCIENT ORIGINS

We tend to think of English and Welsh wine as a relatively recent phenomenon, and although *world-class* English *sparkling* wine certainly is, the vineyards in this country date back to the first century AD, and every important Roman villa had a garden of vines. It was not until 1995, however, when an 8-hectare (20 acres) Roman vineyard was excavated at Wollaston in Northamptonshire, that the true scale of this early viticulture was realized.

Evidence of at least 20 Roman vineyards exists, primarily in the southeast, although Wollaston remains the largest and most extensively excavated site. In 1087 the Normans had recorded 46 English vineyards in the Domesday Book, many of which were owned by Norman nobles and had either been recently planted or extended. Only one-quarter of the vineyards belonged to monasteries, mostly Benedictine, but in the wake of William's pope-blessed invasion of 1066 many more monasteries would be planted throughout Kent, Sussex, and Hampshire in the southeast and Somerset, Gloucestershire, Herefordshire, and Worcestershire in the southwest as various orders followed. In *De Gestis Pontificum* (written in 1125, but published much later), the historian William of Malmesbury was particularly impressed by the Vale of Gloucester, stating that "No county has more or richer vineyards, or which yield greater plenty of grapes, or of a more agreeable flavour. The wine has a not disagreeable sharpness to the taste, as it is little inferior to that of France in sweetness."

Medieval English vineyards probably peaked between 1150 and 1348, the latter date coinciding with the first and most rapid demise of the population due to the Black Death. This would mark the beginning of a 600-year decline of English viticulture, assisted by numerous recurrences of the plague (the last being 1665), the dissolution of the monasteries (1536–41), and an increasingly colder climate (the so-called Little Ice Age from the 16th to 19th century).

By 1509, when Henry VIII (whose impressive wine cellar is beautifully preserved beneath Ministry of Defence buildings in London today) was crowned, the number of vineyards had dropped to 139, according to Hugh Barty-King, who tells us that 11 were owned by the Crown, 67 by noble families, and 52 by the church. Either Barty-King could not count or 7 were owned by non-nobles.

CASTELL COCH, THE SITE OF THE BRITISH ISLES' FIRST COMMERCIAL VINEYARD, SITS ABOVE THE VILLAGE OF TONGWYNLAIS IN SOUTH WALES
The vineyards here were established in 1875 by the third Marquess of Bute, who was bluntly honest about the sub-par quality of his early efforts. Yet despite reviews such as one in *Punch* magazine that noted Castell Coch wine would be so awful that "it would take four men to drink it – two to hold the victim and one to pour the wine down his throat", Bute continued on and did achieve success and critical acclaim. Sugar shortages during World War I interrupted production, and the vineyards were uprooted in 1920.

WHO INVENTED CHAMPAGNE?

In 1662, six years before Dom Pérignon set foot in Champagne, Christopher Merret described the process English vintners used to intentionally turn "all sorts of wines" into fully sparkling by the use of a repeatable second fermentation ("stum perform"). Merret did not specify Champagne, but as the term "sparkling Champaign [sic]" was first used by Sir George Etherege just 14 years later – and that was 40 years before the French retrospectively claimed Dom Pérignon had invented sparkling Champagne – it is reasonable to assume that still Champagne shipped in barrels would not only have been one of those "sorts of wines", but it would also turn out to be far the most suitable. Merret also did not mention bottling the wine, but Etherege talks about going to the park with "champaign", and he was hardly referring to people lugging around barrels of the stuff. Furthermore, in March 1679, just four years after Etherege, Pepys, who knew Merret, mentioned him coming to dine at his house. Pepys noted that he went to Hyde Park in his carriage for "the first time this year, taking two bottles of champagne" in a spookily similar quote by Etherege.

In *The Compleat Vineyard* (William Hughes, London, 1665), written just after Merret and almost 10 years before Etherege, the author discusses wine-producing vineyards in Kent, Essex, and the West of England and asks rhetorically "if the wine be not brisk, how shall we make it without the addition of Sugar, Vinegar, Vitriol to sparkle or rather bubble in the Glass?" raising the intriguing possibility of sparkling wine made from English-grown grapes in the 17th century.

By the 18th century there were barely 20 English vineyards left. Most historians have concluded that the decline of English viticulture at this juncture was the inevitable consequence of the worsening climate, but if this were so it would be reasonable to assume that any surviving vines would be confined to the warmest climes of the southeast and southwest of England. The problem with the climate theory is that quite a few vines were growing as far north as Yorkshire. As the famed 18th-century gardener William Speechly wrote in his *Treatise on the Culture of the Vine*, "At Northallerton, in Yorkshire, there is a Vine now growing that once covered a space containing 137 square yards; and it is judged, that if it had been permitted, when in its greatest vigour to extend itself, it might have covered three or four times that area. The circumference of the trunk or stem a little above the surface of the ground is three feet eleven inches. It is supposed to have been planted 150 years ago but from its great age and from an injudicious management, it is now, and has long been, in a very declining state. There are many other Vines growing at Northallerton, which are remarkable for their size and vigour." It is now believed to have been planted by the Carmelite friars prior to the dissolution, which would have made it 250 years old in Speechly's time (1789), and contrary to the great gardener's gloomy prediction of its imminent demise, the Great Vine was still going strong at least 140 years later. Apparently it all but disappeared in the 1970s, and although shoots are said to have appeared shortly after, it is probably more valid to believe that Northallerton's Great Vine was in the 1970s at an age of some 430 years. Not bad for a country widely considered too cold for viticulture.

The last truly commercial vineyards planted in this country prior to the post-WWII revival were established in the late 19th century by John Patrick Crichton-Stuart, the third Marquess of Bute and one of the richest men in the world. He planted his first vineyard at Castell Coch in 1875 and sold the first vintage (1878) through the Angel Hotel in Cardiff, with later vintages becoming available from London wine merchants. In 1885 the Marquess planted a second, larger vineyard at Swanbridge overlooking the Bristol Channel and shortly afterwards a third at St Quentin near Cowbridge. Although pleased with his success, the Marquess was refreshingly honest about his own wine and would often say "You wouldn't want to trade hock for Coch".

On 6 February 1897 *The Spectator* reported "Lord Bute's vineyards at Castle [sic] Coch, near Cardiff, have produced another great crop. The incessant rain of September made it difficult to gather the grapes dry. But the quantity made from two vineyards – a second is now in bearing at Swanbridge, seven miles from Cardiff – was forty hogsheads, the same yield as that of 1893. The kind of vine grown is the Gamay Noir, used in the vineyards near Paris and in the colder parts of France, and it is planted in rows

three feet apart, and trained to stakes four feet high. At the end of the season the vines are pruned close back, leaving only two buds of last year's growth."

The most fascinating aspect of this report in *The Spectator* was its uncanny prediction of the future of UK vineyards: "The quality of the native juice of the Castle Coch grapes must be very high for the process used in their vinification is very simple and the only ingredient added is some sugar, with probably some spirits to prevent decay. But the result, as we have said before, is a still champagne. This seems to show the direction in which future owners of vineyards in this country may look for profit. *They must learn to make, not still champagne, but sparkling wines.*" I have italicised the last sentence, not *The Spectator,* to draw attention to such a prophetic statement written more than 120 years ago!

The Marquess died in 1900 and was succeeded his 19-year-old son, John Crichton-Stuart, the fourth Marquess of Bute. All seemed fine until August 1914, when war was declared, and it became impossible to obtain the sugar for chaptalisation. When peace returned after four years of slaughter, which claimed 700,000 English lives, including the Marquess's brother Edward, production never resumed.

THE MODERN ERA

George Ordish planted a few vines in 1939 in his garden at Yalding in Kent, renowned gardener Edward Hyams planted a vineyard nearby in 1946 in preparation for his book *The Grape Vine in England* (Bodley Head, 1949), and

---

FACTORS AFFECTING TASTE AND QUALITY

LOCATION
Most of the vineyards are located in England and Wales, south of a line drawn through Birmingham and the Wash.

CLIMATE
Great Britain is at the northerly extreme of climates suitable for the vine, but the warm Gulf Stream tempers the weather sufficiently to make viticulture possible. Rainfall is relatively high, and conditions vary greatly from year to year, making yields highly volatile. High winds can be a problem, especially near the coast and above 120 metres (395 feet) altitude, but winter frosts are less troublesome than in many wine regions. Kent has an average annual temperature that is 1.5°C (34.7°F) higher than Champagne, but it is subject to the full force of the Atlantic, whereas Champagne is moderated by its location within the Oceanic-Continental Corridor, thus it is climatically far more uncertain.

ASPECT
Vines are planted on all types of land, but the best sites are usually sheltered, south-facing slopes with the consequent microclimate advantages that can be crucial for wine production in this marginal viticultural region.

SOIL
Chalk is a fabulous soil to grow vines on, particularly for sparkling wine, but it is not the only soil: vines thrive in this country on a wide range of soils, from granite through limestone, chalk, greensand, gravel, and clay.

VITICULTURE & VINIFICATION
High-trained vine systems indicate that vineyards are poorly sited in frost-prone areas. Vigour can be a problem, with sap passing the fruit to produce an excessively luxuriant canopy of leaves, but this has become less of a problem in the last 10 years. Some vineyards have introduced sheep to manage grass and weeds, with Texel and Southdown the most preferred choice (the Baby Dolls used in New Zealand are related to Southdowns). The focus of production is on bottle-fermented sparkling wine, and the ideal is to have all operations in-house, with minimal movement between winemaking and ageing, and ageing and disgorgement. Being such a young industry, there is much less clear-glass in the bottling of English and Welsh sparkling wine than there is anywhere else. Jetting and MytiK are becoming more prevalent.

GRAPE VARIETIES
Primary varieties: Bacchus, Chardonnay, Meunier, Pinot Noir, Seyval Blanc
Secondary varieties: Madeleine Angevine, Müller-Thurgau, Ortega, Phoenix, Pinot Blanc, Pinot Gris, Pinot Noir Précoce, Regent, Reichensteiner, Rondo, Schönburger, Solaris

---

A WINE SHOP DISPLAYS CAMEL VALLEY SPARKLING WINES
Located in Nanstallon, this winery puts out a range of fizzy sparklers, including its Cornwall Brut and Pinot Noir Rosé Brut, as well as still wines from Bacchus and other varieties.

Ray Barrington Brock planted a vineyard of experimental varieties in 1946 at what would become the Oxted Viticultural Research Station, but the first truly commercial vineyard since the Marquess of Bute's 19th-century venture was planted in 1952 by Sir Guy Salisbury-Jones at Hambledon in Hampshire.

As Stephen Skelton MW points out in his *UK Vineyards Guide 2010,* "the revival of commercial viticulture in the early 1950s was based upon the discovery by Brock and Hyams that Müller-Thurgau and Seyval Blanc were varieties that would both fruit and ripen in our climate", and in the 1970s, the very thought of English wine was a joke. As the late, great Peter Ustinov put it: "I imagine hell like this: Italian punctuality, German humour and English wine".

It needed people like Barrington Brock and Hyams to re-establish UK viticulture in the 1950s, but the broad base of consumers had evolved in the early 1970s and could not understand what was so English about wines made from grapes with names like Müller-Thurgau and Seyval Blanc, let alone Reichensteiner, Huxelrebe, or Siegerrebe. Cheap holidays abroad had expanded horizons of ordinary people who, after devouring Hugh Johnson's *World Atlas of Wine* (1971), at least knew that fine wines were made from classic varieties, not grapes with funny names.

When interested consumers asked why the English were growing German grapes that were shunned in Germany and French hybrids that were illegal in France, vineyard owners pointed out that nothing else could survive or ripen in the UK's inhospitable climate. So, at that juncture, British viticulture was not about the quality of wine, but the survivability of the crop. They might have been better off growing rape seed and, eventually, many were.

This was the state of play until Nyetimber's original American owners Stuart and Sandy Moss arrived on the scene in the 1980s. They had this crazy notion of wanting to make a top-quality English fizz from classic Champagne varieties. In preparation for planting their vineyard, the Mosses sought expert advice locally, only to be told that the Chardonnay would not ripen and the Pinot Noir would just rot. They were advised to plant all sorts of crosses and hybrids, but they were not interested. Knowing they would not be able to build an international reputation on such varieties, the Mosses did the damnedest thing: they asked the *champenois* for help. Who would have thought it? Certainly not any English vineyard owner at that time. Most had trained at Geisenheim, the home of

Müller-Thurgau, and asking the French anything would have been the last thing on their "To Do" list. I had a lot of respect for the late Professor Helmut Becker, who headed up Geisenheim, but he did run the funniest of all funny farms when it came to creating a multitude of boring crosses, and this had had an unfortunate effect on how the English wine industry viewed itself at the time.

The vines went in the ground at Nyetimber in 1988, and the first commercial wine was produced in 1992. I was completely unaware of this until 1995 when winemaker Kit Lindlar telephoned to ask whether he could bring around a sparkling wine he had made. All he wanted was an honest opinion about whether or not it was ready to be disgorged, and I said fine, as long as he understood that I did belong to "the longer the better" school of thought for yeast-ageing. I told him that IMHO some of the finest Champagnes could and should be disgorged earlier rather than later, so the odds were (back then) that I would not recommend extended lees contact for any aspiring English sparkling wine. As soon as I tasted the Nyetimber (the identity of which he revealed only after we had discussed the wine), however, its potential for further development was so obvious that I felt compelled to recommend at least another 12 to 18 months on yeast. At the time I wondered why Kit did not look too happy when I congratulated him on producing England's first world-class wine, but later I discovered his contract with the Mosses was to receive payment in three equal amounts: the first on signing, the second on bottling, and the third on disgorging. I had just delayed his paycheck more than a year!

Nyetimber 1992 was received with such acclaim when it was launched that it was selected for Queen Elizabeth II's Golden Anniversary Lunch in 1997. This was such an honour. After all, despite the publicity, no one had really heard of Nyetimber beyond an insignificant few in the wine trade. It was a completely new brand and not even English-owned. Yet the legendary reputation of Nyetimber was firmly established, and we can trace the emergence of world-class English sparkling wine directly from these events.

Stuart and Sandy Moss were not the first to cultivate Chardonnay or Pinot Noir in an English vineyard. Nor were they the first to make a traditional-method sparkling wine from these classic varieties. Piers Greenwood successfully grew a tiny amount of Chardonnay and Pinot Noir at New Hall vineyard at Purleigh, Sussex, in 1983, and Kenneth McAlpine used those grapes as part of his sparkling wine blend at Lamberhurst Vineyards. There were other pioneers of these Champagne grapes between Greenwood and the Mosses, but their plots of vines were very small indeed, often poorly sited, experimental in nature, and planted with inappropriate clones that frequently failed to ripen.

The difference at Nyetimber was that the Mosses sought the advice of *champenois* Jean-Manuel Jacquinot. He might not have been an outstanding Champagne producer, but Jacquinot was one of the most important nurseryman in the Champagne and knew his stuff. He selected early-ripening clones (such as Dijon Chardonnay clone 95 & 96). The Mosses were no mugs. They also hired Jacquinot to supervise the preparation of the ground, the planting, and the training of the vines. If the vineyard did not take, Jacquinot would not have been able to blame the Mosses, and not employing someone else, the Mosses avoided a planter blaming Jacquinot's cuttings. The first time I visited Nyetimber I was shocked to see how low to the ground the vines were trained because although this might be the norm in Champagne, most vineyards in the UK at that time were trained high to avoid ground frost. At Nyetimber, however, there was a natural air-drainage that swept through the vines and was drawn down to the open golf course below.

The first edition of *Christie's World Encyclopedia of Champagne & Sparkling Wine* (Absolute Press, 1998) included an entry on Nyetimber, but England's viticultural reputation was, in general, still one of "farmers-playing-at-winemakers" growing unfashionable grapes because nothing else would grow. Yet, by the second edition (2003), English sparkling wine had "a world class potential" and this sea change took place within just five years. Such was the Nyetimber effect and the fact that Ridgeview was already snapping at its heels. Few people know that Mike Roberts had taken part in one of Nyetimber's earliest harvests, and that experience had convinced him to plant what was in 2003 the second-largest vineyard of Chardonnay, Pinot Noir, and Meunier vines. Ridgeview chased Nyetimber in those early years, often equalling the UK's premier sparkling wine producer and sometimes (when Nyetimber went through a patch of dodgy disgorging that left beautiful wines with mercaptan aromas) even trouncing it.

The English wine industry underwent a total revolution at the turn of the new millennium. The difficulty of marketing amateurish wines made from German crosses and French hybrids had resulted in a drop of English vineyards from a high of 1,065 hectares (2,632 acres) in 1993 to a low of 697 hectares (1,722 acres) in 2007. However, the total area planted was misleading, whereas the increase and decrease of individual varieties told a different story.

In 1995 Mike Roberts had planted Chardonnay, Pinot Noir, and Meunier at Ridgeview, and the international acclaim received initially by Nyetimber and quickly followed by Ridgeview attracted the attention of other entrepreneurs and investors, who began buying land and planting it exclusively with these classic Champagne grapes. In 1999 there were just 18.5 hectares (46 acres) of Chardonnay in the entire country, and most of those were growing at Nyetimber and Ridgeview. By 2006 the total area under vine was still declining, and yet there were in excess of 91 hectares (225 acres) of Chardonnay, a five-fold growth. This illustrates that although the UK wine industry appeared to be in decline to outsiders, it was in fact gearing up for a fundamental change in direction that would determine its future, steering away from still country wines made by farmers to world-class sparkling wines produced by professionals, both in the winery and running the business. There are now 1,063 hectares (2,627 acres) of Pinot Noir (including Pinot Précoce), 1,034 hectares (2,555 acres) of Chardonnay, and 394 hectares (974 acres) of Meunier, representing 70 per cent of all vines planted. Curiously, even today Seyval Blanc and Bacchus remain in the top five most widely planted varieties.

OUT WITH THE OLD, IN WITH THE NEW

The UK wine industry as a whole is currently doubling every four or five years, which is impressive enough, but the sparkling wine component of that industry is moving a lot faster than that. To establish English (and hopefully Welsh) wine as a strong international brand, exports must increase beyond their present 8 per cent (up from 4 per cent in 2017) to 50 per cent. As global interest has already been piqued, it is evidently an achievable goal, but producers should not forget what has grabbed attention in the first place: a classic, *brut*-style, bottle-fermented sparkling wine that was launched at a relatively expensive price and has managed to hold its aspirational price for at least 10 years. Any deviation from this would be reckless, thus still wines should be kept below the counter or as a "sommelier's secret" on the domestic market, while sparkling wine remains the focus of attention for exports.

English sparkling wine's reputation has been built upon a relatively small number of award-winning producers. Many more have clung on to their coattails and benefitted from unrealistically high prices for the short term, but unlike the award-winning wines, their products of crosses and hybrids have not been worth it. Consequently a number of the Old School have fallen to the wayside as the marketplace, not the producer, decides how much a wine is worth. This drop-off continues, but as we have seen, particularly over the last five years, there are many more exciting new ventures in the pipeline ready to take their place. Almost all the most successful English sparkling wine producers have started up after the

RECENT UK VINTAGES

2019 A bumper crop (12 million bottles) of fair to good quality, where sorting in the vineyard was strictly applied. After a promising start, August was wet and warm, creating disease pressure. September and October were also wet, dashing hopes of another 2018.

2018 A bumper crop (13.2 million bottles) of exceptional quality.

2017 Despite 30 to 40 per cent losses due to spring frost and a wet August, the wines turned out to be better quality than many feared during the harvest. Chardonnay yields were high and the quality much higher than for Pinot Noir, which suffered some rot.

2016 A medium- to good-sized harvest of 4.15 million bottles, with vineyards in the east and southeast generally faring better than those in the southwest in terms of both quality and quantity.

2015 A late start to the growing season resulted in a harvest that lasted well into November, but this had no adverse on the volume (5.06 million bottles).

2014 At the time, this was the largest crop on record (6.3 million bottles) and the greatest quality vintage in memory. This is a year that has been described as "the vintage of dreams" (Camel Valley).

2013 After a late bud-break and a long, warm summer, the harvest was due to be late anyway, but was further delayed by a cool autumn and rain. A good-sized crop (4.45 million bottles) with surprisingly little rot and generally good quality and a few examples of really exceptional Chardonnay.

2012 A tiny crop (1.03 million bottles) of poor quality, although an odd few examples bucked the trend and managed to produce something pretty good.

2011 Believe it or not, 3.02 million bottles was not a bad-sized crop at the time. Berry sizes were small and uneven. Curiously, acids started dropping early and without sugar levels rising as much as needed, but an Indian Summer saved the crop, particularly the Chardonnay.

Nyetimber phenomenon – either as completely new ventures or old established vineyards, like Hambledon, who have ripped up their hybrids and crosses to start afresh under a new regime.

The biggest obstacle for any producer in the United Kingdon hoping to build and maintain sales, whether for home or export, has to be the UK's fluctuating yields. They represent the price that Champagne has to pay for having the requisite knife-edge climate for a long, drawn-out ripening process that is primarily responsible for the superior quality that has allowed its sparkling wines to conquer the world's markets. Fluctuating yields are also the price that UK sparkling wine producers have to pay, and their vineyards are for the moment on the wrong side of that knife edge, which is why the volatility of English and Welsh crops is considerably greater. It took the *champenois* of France 80 years to perfect an emergency reserve system to counter their own volatility (initially through regional *blocage* and *déblocage* in 1938, then *réserve qualitative individuelle* since 2008). If UK producers steal this concept, it should be much easier and quicker to implement, without having to jump through any hoops of an appellation system.

Wine tourism will also play an increasingly important role. Of the current 750 vineyards, only 150 are set up to receive visitors. Although that is a good thing, that number is evidently insufficient, as most of these premises either do not have a winery or have a winery that lacks the full range of facilities to show visitors and to explain the process from grape to glass, including vines, grape reception, ageing, *remuage*, disgorgement, *dosage*, and corking. The more the public is educated, the quicker they will see what is missing on a visit, but the potential gain from wine tourism is enormous: once anyone has been to a winery, walked the vineyards, seen how the wine is processed, and enjoyed a well-organised tasting, that person will be a customer for life, opening bottles at the table and doing the producer's selling for him. WineGB should set up a wine tourism consultancy.

THE WINE REGIONS OF
THE BRITISH ISLES

Regional characteristics have not yet been established in the wine regions of the British Isles, but as the French have taught us and the Italians have failed to listen to, it is easier to build-up a spatial awareness of where vineyards are located if the country is first broken down into a small number of workable areas. As from this edition, the regions have moved from the old (and still current) vineyard associations to the regional breakdown used by Wines of Great Britain (WineGB), which is the national association for the English and Welsh wine industry, for all of its data.

EAST ANGLIA
4% of total UK area under vines
(92 vineyards covering 143 hectares, or 353 acres)

Bedfordshire, Cambridgeshire, Essex, Hertfordshire, Norfolk, Suffolk

The flattest and most exposed of all the regions, East Anglia is subject to bitterly cold easterly and northeasterly winds but has fertile soils, which can encourage significantly higher yields than in some areas of England. Quite often this has the effect of trading off quality for an attempt to preserve consistency.

MERCIA AND REST OF THE UNITED KINGLDOM
6% of total UK area under vines
(145 vineyards covering 215 hectares, or 531 acres)

Mercia: Cheshire, Derbyshire, Greater Manchester, Lancashire, Leicestershire, Lincolnshire, Northamptonshire, Nottinghamshire, Rutland, Shropshire, Staffordshire, Warwickshire, West Midlands, Yorkshire

(East Riding, North, South, and West); Rest of UK: Channel Islands, Greater London, Isle of Wight

This is a very large and diverse region, with vineyards planted both on flat ground and slopes, with soils ranging from light sand to heavy clay. About the only common factor (for Mercia) is Madeleine Angevine, which seems to be the most reliable variety the farther north vineyards progress. La Mare vineyard and winery on Jersey still exists, but the vines are on the wrong (unprotected) side of the island, and its barrique-aged apple brandy is far more successful. UK billionaire identical twins the Barclay brothers had invested millions in an ambitious 60-hectare vineyard on Sark, also part of the Channel Islands, but it barely lasted six years. Apparently, it was taxed out of existence by the fiscally autonomous island government, which had only just given up feudalism in 2008 and evidently did not want a winery on its doorstep or the attention it might bring.

SOUTH EAST
76% of total UK area under vines
(341 vineyards covering 2,720 hectares, or 6,721 acres)

Berkshire, Buckinghamshire, Buckinghamshire, Hampshire, Kent, Oxfordshire, Surrey, Sussex (East and West)

This region covers the "Garden of England", which probably says it all. It is not as wet as the West Country and is milder than East Anglia, but not as warm as Thames and Chiltern. Soils range from clay to chalk. The latter forms the same basin that creates the chalk cliffs at Dover and goes under the Channel to rise in Champagne. Pommery (Hampshire) and Taittinger (Kent) have both planted vineyards in this region.

SOUTH WEST
13% of total UK area under vines
(185 vineyards covering 465 hectares, or 1,150 acres)

Bristol, Cornwall, Devon, Dorset, Gloucestershire, Herefordshire, Somerset, Wiltshire, Worcestershire

Geographically, the South West is the largest wine region of the British Isles, with vineyards as far apart as Worcestershire south to Cornwall, where the Gulf Stream has its greatest effect. Although prevailing westerly winds make conditions difficult in the west of Cornwall, the region is generally milder than the east coast, with bud-break and flowering significantly earlier than elsewhere in the UK. Soils vary tremendously from limestone through shale, clay, and peat to granite.

WALES
1% of total UK area under vines
(31 vineyards covering 36 hectares, or 89 acres)

Clwyd, Dyfed, Glamorgan (Mid, South, and West), Gwent, Gwynedd, Powys

Although there is a vineyard as far north as Anglesey, it currently survives blending with Andalusian Tempranillo from Spain. Most vineyards are confined to the southern strip of Wales, where the topography of the region ranges from almost mountainous to flat, and where there is high annual rainfall. Only a handful of Welsh wines are commercially available, and those are primarily from Glamorgan (particularly Monmouthshire) and Powys. The bravely biodynamic Ancre Hill Estate (not profiled below) is the leading producer.

THE WINE PRODUCERS OF
THE BRITISH ISLES

A'BECKETT'S VINEYARD
Devizes, Wiltshire
☆

Hard-to-source wines from Paul and Lynn Langham's 4.5-hectare (11-acre) vineyard on chalk soil in the village of Littleton Panell. It is nonetheless well worth a visit to taste and buy.

✔ *Still (Penruddocke's Pinot Noir)*

ALBOURNE ESTATE
Albourne, West Sussex

This family winery is run by Plumpton College graduate Alison Nightingale. It offers still and sparkling wines and a vermouth in stylish packaging.

✔ *Sparkling (Blanc de Blancs, Blanc de Noirs)*

AMBRIEL
Pulborough, West Sussex
☆

These wines are sourced from Redfold Vineyards, close to Nyetimber, by banker Charles and Wendy Outhwaite. The quality is more potential than actual at the moment, with oxidative aromas in some of the sparkling wines, although the 2014 Rosé was delightful.

✔ *Sparkling (Rosé)*

ASTLEY
Stourport on Severn, Worcestershire
❶

Due to it being under new ownership since 2017, I am unable to assess this winery, which had been an award winner underits former owners, whose Late Harvest is still something of a legend.

✔ *Still (Late Harvest)*

BALFOUR
See Hush Heath

BLACK CHALK
Cottonworth, Hampshire
☆

Launched in 2018 by Jacob Leadley and family, this venture is based on the négociant-manipulant model in Champagne, producing grapes from growers they work closely with.

✔ *Sparkling (Wild Rosé)*

BLACKBOOK
Battersea, London
☆

The English wine brand of an urban contract winemaking services.

✔ *Still (Painter of Light Chardonnay)*

BLACK DOG HILL VINEYARD
Ditchling, East Sussex
☆

Jim and Anja Nolan have a good location, and Wiston's Dermot Sugrue is an excellent choice for winemaker, which makes the oxidative tendency of these sparkling wines a bit puzzling, but the potential is definitely there.

BLUEBELL VINEYARD ESTATES
Furners Green, East Sussex
☆

Kevin Sutherland, who is to be congratulated for building a winery and planting 9 clones of Chardonnay, 14 of Pinot Noir, and 4 of Meunier, established Bluebell in 2004. Yet for such a brave new future, it's odd to find these classic Champagne varieties growing alongside a curious mix of old-style Seyval Blanc and German crosses.

✔ *Sparkling (Hindleap Blanc de Blancs)*

BOLNEY WINE ESTATE
Bolney, Sussex
★☆

Sam Linter makes some lovely sparkling wines when she focuses on Champagne varieties, and especially when they are offered in magnum.

✔ *Sparkling (Blanc de Blancs, Rosé)*

BREAKY BOTTOM
Rodmell, Lewes, East Sussex
☆

If and when the time comes that there is only one old-style English wine producer left standing, let it be Peter Hall. Yet if you do not know why, you will just have to visit him to find out, and don't be surprised if you end up buying a case of something – occasionally he comes up with a stunner.

BRIDE VALLEY
Litton Cheyney, Dorset
★

Owners Steven and Bella Spurrier planted their 10 hectares (25 acres) exclusively with various clones of Chardonnay, Pinot Noir, and Meunier. Steven Spurrier is, among many things, famous for being the architect of "The Judgment of Paris", the events of which were immortalized and, Spurrier claims, fictionalized in the film *Bottle Shock*, starring Alan Rickman as Spurrier himself. The wines are made by Ian Edwards of Furleigh Estate.

✔ *Entire range*

ENGLAND'S ICONIC WHITE CLIFFS OF DOVER

The White Cliffs of Dover, part of the North Downs formation, stand over the Strait of Dover, facing France. These chalk cliffs are part of the same geological system as the Côte d'Albâtre (Alabaster Coast) of Normandy. Chalk is often mentioned in hushed tones when discussing sparkling wine, but it is not the primary factor determining quality. Although great English sparkling wines are produced on other soils, there is no doubt that a chalk-based *terroir* does produce sparkling wine of a certain typicity. It is interesting, therefore, to see where the chalk actually extends throughout the country.

WINE NOTES

• The term *British wine* refers to a drink that is neither British nor wine made from fresh grapes, let alone fresh grapes grown in British vineyards. It is a (legal) misnomer for wines made from reconstituted grape concentrate sourced from anywhere in the world.

MAJOR VINEYARDS AND WINERIES OF GREAT BRITAIN

There are 794 vineyards and 154 wineries in Great Britain and Wales, with most of them located south of 53°N latitude The best wines of the British Isles are sparkling, often from vines growing in chalk (dotted white on the map).

BRIGHTWELL VINEYARD

Wallingford, Oxfordshire

☆

The Sparkling Chardonnay from this Oxfordshire producer has come of age, and it is a thousand times better than its old-fashioned presentation might have you think. Depending on the vintage, the Pinot Noir can be convincing too.

✓ *Sparkling Chardonnay, Pinot Noir (Still red)*

CAMEL VALLEY

Nanstallon, Cornwall

★★

All the wines are worth drinking here, but Bob and Sam Lindo's Pinot Noir Rosé is, without doubt, one of the greatest and most consistent sparkling wines that has ever been produced in the UK.

✓ *Entire range*

CASTLE BROOK

★

In 2004 Waitrose's premier asparagus farmers, the Chinn Family, planted 21 clones of three major Champagne varieties, with the vines growing on the site of an ancient Roman vineyard, a steep south-facing slope in the rain-shadow of the Black Mountains. Made at Ridgeview, the wines are soft and stylish.

✓ *Entire range*

CHAPEL DOWN

Tenterden, Kent

★

As this group grew and evolved, so the ups and downs in the range of its output increased. Some 400,000 bottles of wines are produced from 100 hectares (250 acres) owned and another 120 hectares (300 acres) bought-in, and they are spread across 20-odd labels. Chapel Down even has plans for a future target of 2.5 million bottles. Then there are the beers, ciders, gin, vodka, brandy, and restaurants (very good informal dining ones too). Potentially there is a brilliant business at the core here. There are always good reasons to diversify, but they have just made a million-pound loss. I think it is time to focus. Time to let the quality shine through. Kit's Coty is supposed to be the ultimate here, but the best wine in that range is the still Estate Chardonnay. The greatest sparkling wine produced here in recent years has been the 2009 Blanc de Noirs marketed in 2015.

✓ *Sparkling (Kit's Coty, Three Graces, Blanc de Noirs), Chardonnay (Kit's Coty), Bacchus*

COATES & SEELY

Whitchurch, Hampshire

★☆

Except for the occasional blip, this joint venture between financial wizard Nicholas Coates and Christian Seely (who runs Château Pichon-Longueville-Baron in Pauillac and Quinta do Noval in Portugal, among other wine estates, for AXA Millésimes) continues to produce some of the most exciting, fresh, and vibrant sparkling wines in the UK.

✓ *Entire range*

COTTONWORTH

Cottonworth, Andover, Hampshire

★

Hugh Liddell fell in love with working a vineyard in Burgundy, and his 12 hectares (30 acres) of exclusively sparkling wine vines in the Test Valley are just beginning to show potential.

✓ *Classic Cuvée, Sparkling Rosé*

DAVENPORT

Rotherfield, East Sussex, and Horsmonden, Kent

☆

Roseworthy graduate Will Davenport was one of the first and bravest vineyard owners to go organic in the United Kingdom, but it is difficult playing the organic game fairly in this country, and I fear the quality of his wines has suffered as a consequence. I admire the man for sticking to his principles, and when his wines are fresh, crisp, and correct, they make great drinking.

✓ *Sparkling (Blanc de Blancs)*

DENBIES WINE ESTATE

Dorking, Surrey

★

When John Worontschak of Litmus Wines took over in 2008, there must have been a lot of heavy lifting to get Denbies out of the mire. It's always been successful as a tourist attraction, albeit its Tesco-looking façade, but the quality has been extremely inconsistent at best. Such is the lead time required for sparkling wine that 12 years on and we are just beginning to see the changes in store for fizz lovers, particularly the Cubitt Blanc de Noirs.

✓ *Cubitt Blanc de Noirs, Greenfields*

DIGBY FINE ENGLISH

Arundel, West Sussex

★☆

Named after the swashbuckling glass-maker Sir Kenelm Digby, these award-winning wines are produced from grapes supplied by growers that owners Trevor Clough and Jason Humphries work closely with in Kent, Sussex, and Hampshire. The 2009 Reserve (made by Dermot Sugrue of Wiston) was winner of Best English Sparkling Wine at the inaugural Champagne & Sparkling Wine World Championships (CSWWC).

✓ *Entire range*

DOMAINE ÉVREMOND

Chilham, Kent

Of the 69 hectares (170 acres) purchased by Champagne Taittinger for this project, 20 hectares (50 acres) were planted in 2018, and a further 20 will follow shortly, all on chalk. Great things are expected here, although the name is an amusing choice. Charles de Saint-Évremond (1614–1703) was the first true "ambassador" of Champagne in England, but it was at a time when Champagne was a still wine in his native France. When he arrived in London, he was horrified to discover that Champagne turned fizzy after bottling by London merchants and appalled that England's fashionable set absolutely adored it. He was a founding member of *L'Ordre des Coteaux de Champagne*, and they all hated bubbles!

DOMAINE OF THE BEE

☆

Currently just one *cuvée*, which is made from grapes grown by the Chinn family (*see* Castle Brook) at the Ridgeview winery. The presentation is old-fashioned and has nothing in common with the clean and elegantly understated labelling Domaine of the Bee use for the still wines from other regions in its range.

✓ *Sparkling (Hart of Gold)*

EGLANTINE

Costock, Nottinghamshire

☆

Tony Skuriat is still self-freezing his grapes to produce his award-winning *Eiswein*-style North Star.

✓ *North Star*

EXTON PARK

Exton, Hampshire

☆

Owned by Malcolm Isaac, retired king of Hampshire watercress, who purchased Exton Park ready-planted in 2009. After initially supplying Coates & Seely with grapes, Isaac constructed a winery in 2011 and hired former Coates & Seely winemaker Corinne Seely. Lovely rosé, despite some light-stuck issues.

✓ *Sparkling (Rosé, Blanc de Noirs)*

FORTY HALL VINEYARD

Enfield, London

The first release was unimpressive, but so was the first release of Roederer Estate in the Anderson Valley, and that venture became the first world-class sparkling wine outside of Champagne, so we shall see.

FOX & FOX

Mayfield, East Sussex

★

Owners Jonica and Gerard Fox planted two delightful sites totalling 12.5 hectares (31 acres) with Champagne varieties, originally selling under both Mayfield and Fox & Fox labels, but now only the latter. The wines are made by Will Davenport (*see* Davenport) and have steadily improved in recent years.

✓ *Sparkling (Entire range)*

FURLEIGH

Bridport, Dorset

This small vineyard and winery is owned by Ian and Rebeca Edwards. Ian also produces Steven and Bella Spurrier's Bride Valley sparkling wines.

✓ *Sparkling (Entire range), Still (Sea Pink, Bacchus Fumé)*

GREYFRIAR'S VINEYARD

Puttenham, Surrey

Nice site, tidy vines, which is why I shall continue to keep an eye on this 30-year-old venture, but it has yet to achieve anywhere near its potential.

GUSBOURNE ESTATE

Appledore, Ashford, Kent

★★☆

Winemaker Charlie Holland is at the top of his game and since 2013, when all wine operations have been brought under one roof at Appledore, Gusbourne has achieved a remarkable consistency. The Blanc de Blancs is Charlie's signature wine, but his Boot Hill Pinot Noir is ridiculously good for a UK red wine.

✓ *Sparkling and Still (Entire range)*

HAMBLEDON VINEYARD

Hambledon, Hampshire

★

Initially planted in 1952 by Major-General Sir Guy Salisbury-Jones, Hambledon was the UK's first commercial vineyard and caught in the economic shadow of the post-Nyetimber boom, it would have remained little more than a viticultural relic of the past, had it not been for Sir Ian Kellett, who purchased the rundown property in 2004. He grubbed-up all the hybrids and crosses and replanted the vineyards exclusively with Champagne varietals in 2005, hired *champenois* Hervé Jestin, a leading bio-dynamic consultant, and built a new gravity-fed winery equipped with Coquard PAI presses. With 30 hectares (75 acres) in production, and another 60 hectares (150 acres) planted in 2018, this winery will shortly be capable of producing 500,000 bottles per year and ageing them on-site in 1.8-million capacity underground cellars, recently excavated from the chalk subsoil. The site is ideal, and all the infrastructure is perfect, yet, after superbly promising initial releases, some (at least) of these wines have shown an oxidative tendency. Hervé is one of the sweetest guys in the business, but his philosophical stance often results in a lack of sufficient SO2 at disgorgement. I'm sure that Kellett knows. He is, after all, a trained biochemist himself, and it is his prerogative to allow his consultant whatever leeway he wishes. My recommendation is to taste these wines the minute they are released, and when you find a stunner, as you most assuredly will, buy it, preferably in magnum or jeroboam, and drink it: do not keep it.

✓ *Entire range*

HARROW & HOPE
Marlow, Buckinghamshire
★

The family venture of Henry Laithwaite and his wife, Kaye, who named this 6.2-hectare (15.3-acre) vineyard when it was made ready for planting, because it was so full of large boulders that they lost a few ploughs and just had to hope they would not lose another. A winery was built in 2012, with professional advice from Ridgeview and the late Dr Tony Jordan. The delicate *blanc de blancs* and rosé are the signature *cuvées* here.

✓ *Entire range*

HATTINGLEY VALLEY WINES
Lower Wield, Hampshire
★★☆

It is hard to imagine that the first Hattingley wines hit the shelves as recently as August 2013, and yet it is now so far ahead of the field that it is neck and neck with Gusbourne chasing for the number-two spot behind Nyetimber. Owner Simon Robinson is the brains behind the operation and there is a well-oiled machine of people who all do their bit, but the secret of Hattingley's success is, without doubt, the humble and unassuming, but amazingly competent and talented winemaker, Emma Rice. She's a human hoover, taking in every opportunity to learn and benefit from other people's experiences, such as the late Dr Tony Jordan, and Pommery's Thierry Gasco. I hope Simon has Emma on a golden handcuff contract. He has already made her a director, but maybe he should take a leaf out of the Roederer's book and make her vice-chairman? All these wines are beautifully made, but the best has to be the rosé, closely followed by the *blanc de blancs*.

✓ *Entire range*

HENNERS
Herstmonceux, East Sussex
★

When Lawrence Henners established Henners in 2007, his intention was to reach the same level as success as he had enjoyed first as a Formula One race engineer with Stewart Grand Prix, and then with his automotive engineering consultancy. This he achieved from 2.8 hectares (7 acres) of vines and his own first-stage winery, with Dermot Sugrue

consultancy. His 2009 and 2010 vintages were outstanding, but the 2011 and 2014 less convincing. Also some disgorgements were clearly superior to others. Henners got out on top, however, selling the business to his distributor, Boutinot, who have their own winemaking team under Eric Monnin. Although Monin might be more at home in the south of France, that has not stopped others mastering the art of bubbles. Initially I thought that Henners could be the Charles Heidsieck of English sparkling wine, jumping from my initial ★☆ to, hopefully, ★★. Uncertainty and inconsistency have instead dropped a half-star instead of gaining it, but the entire range is worth following, and I look to Monin and technically astute Boutinot to reverse the star-rating trend in the next edition.

✓ *Entire range*

HERBERT HALL
Marden, Tonbridge, Kent
★

Named after owner Nick Hall's grandfather, Herbert, who grew hops, apples, pears, and plums and swore it was the best plot of land he had ever farmed. With fellow winemaker Kirsty Smith, Nick produces some of the most exciting barrel-fermented sparkling wines of any organic, boutique venture in the country.

✓ *Entire range*

HIGH CLANDON VINEYARD
High Clandon, Surrey
★

From here comes absolutely tiny volumes of ultra-boutique sparkling wine.

✓ *Sparkling (Elysium Cuvée, The Aurora Cuvée)*

HINDLEAP
See Bluebell Vineyard Estates

HOFFMANN & RATHBONE
Mountfield, East Sussex
★

Ulrich Hoffmann cut his winemaking teeth in various wineries, from Baden-Württemberg to Bordeaux, Navarra, and the Napa Valley, not to mention Bolney Estate and Gusbourne in the UK before he set up his own venture with his wife, Birgit Rathbone.

✓ *Entire range*

HUSH HEATH VINEYARD
Cranbrook, Kent
★☆

A beautiful Tudor timber-framed manor set in impeccably maintained Italianate gardens, Hush Heath is owned by Richard and Leslie Balfour-Lynn, hence the name of its Balfour *cuvées*. Prior to the on-site winery, built in 2010, these wines were produced at Chapel Down. Victoria Ash is the winemaker, with former Chapel Down winemaker Elias Owen consulting, but the man pulling all the strings is and always will be Richard Balfour-Lynn.

✓ *Sparkling (Entire range)*

JENKYN PLACE
Bentley, Hampshire
★

Simon and Rebecca Bladon's 6-hectare (15-acre) vineyard has started to show its potential in recent years. Winemaking carried out by Dermot Sugrue at Wiston.

✓ *Sparkling (Brut, Blanc de Blancs)*

LANGHAM WINE ESTATE
Crawthorne, Dorset
★

Owned by John Langham, this 12-hectare (30-acre) vineyard at the heart of the historic 1,000-hectare (2,470-acre) Bigham's Melcombe Manor estate, the Classic Cuvée Reserve Brut and, in magnum, the Blanc de Blancs are the standouts so far, but it is early days and both the Rosé and Blanc de Blancs show excellent promise.

✓ *Sparkling (Entire range)*

LAVERSTOKE PARK
Laverstoke, Hampshire

Great things were expected from owner Jody Scheckter, the former Formula 1 world-champion racing driver, but his extreme application of biodynamic principles wreaked havoc in 2014, when he lost almost all of his crop, while the rest of the country enjoyed the greatest English sparkling wine vintage in history. This disaster did at least help the late Tony Jordan persuade Jody to allow some limited spraying, if and when required, from 2015 on; however, the wines have been dire, despite being made at Hattingley under the guidance of Jordan. The intrinsic problem here is the siting of the first (and only vineyard) and the low volumes of less-than-perfect fruit it has produced. This project has now reached an impasse in which there are only two options: to take it more seriously or to pack it in. Taking it more seriously would entail trebling the current acreage by planting 20 hectares (50 acres) on a more low-lying, protected site to balance the current higher, more-exposed site to provide a better-balanced selection of fruit in both good and bad years; building a press-house, winery, and cellar; and hiring a truly world-class sparkling wine consultant (such as Claude Thibaut).

LECKFORD ESTATE
Leckford, Hampshire
☆

This 5-hectare (12-acre) vineyard belongs to Waitrose, an upmarket UK supermarket group and major wine retailer, and is located on its 1,600-hectare (3,955-acre) farm. The style is successfully fruit-driven.

✓ *Entire range*

LYME BAY WINERY
Axminster, Devon
☆

Not impressed by the Seyval Blanc Brut Reserve, but I love the Sparkling Rosé, particularly if kept for an additional 6 to 12 months.

✓ *Sparkling (Sparkling Rosé)*

NUTBOURNE VINEYARD
Pulborough, West Sussex
☆

Having dropped Reichensteiner from the Nutty's blend, the quality has improved, but there are still far too many wines produced here. With a vineyard across the lane from Nyetimber, the strategy should be to chip-bud all existing hybrid and German crosses to Champagne varieties and to focus on a small number of top-class sparkling wines selling at twice the price.

✓ *Sparkling (Nutty's)*

NYETIMBER
Pulborough, Sussex
★★★

If not for Nyetimber, there would be no such thing as an English sparkling wine industry today, only a few crude fizzy wines made from hybrid grapes and German crosses, yet Nyetimber's first vintage, a 1992 *blanc de blancs*, was not released until the end of 1996. That is how young the concept of world-class English sparkling wine is, and it only exists thanks to the stubbornness of Nyetimber's two American founders, Stuart and Sandy Moss, along with, of course, the winemaker of the legendary 1992 *blanc de blancs*, Kit Lindlar. The arrival of Cherie Spriggs and Brad Greatrix in 2007 marked another milestone, one of professionalism, and that could not have happened had it not been for Eric Heerema, the Dutch financier who purchased Nyetimber the year before. Cherie and Brad, a winemaking couple from Canada, might not make the greatest sparkling wines in the world, but they do make the greatest sparkling wines in the UK. Like all so-called "greatest" producers, someone else can make something better in any given year, but for consistency across the range and from year to year, no one can touch Cherie and Brad. From how they plan a winery, press house or an event, to considering what problems the future might bring and how to tackle them, they are in a class of their own. For example, they switched to almost black bottles (very dark amber) for the entire range in 2009 to avoid light-struck faults and, yet, more than 10 years later the world is full of idiots who still think they should risk the quality of their wine and the reputation of their brand by bottling in clear glass. Champagne is the worst culprit as far as the proliferation of clear-glass bottles is concerned, whereas it is much less of a problem in the UK, thanks to the lead and professionalism of Nyetimber. With 258 hectares (638 acres) planted over three counties (West Sussex, Hampshire, and Kent), Nyetimber is destined to become the most important, quality-focused sparkling wine operation in the UK.

✓ *Entire range*

OASTBROOK ESTATE
Robertsbridge, East Sussex

Plumpton College alumni America Brewer only planted her first vineyard in 2018, but she launched her sparkling rosé from purchased wine (with help from Dermot Sugrue of Wiston fame) the same year. Since then a lovely Pinot Gris

NEWLY PLANTED VINES ARE SUPPORTED BY PROTECTIVE TUBING IN A SUSSEX VINEYARD
Sussex is part of Great Britain's South East wine region, where more area is under vine than any other region in the country.

cuvées as a hedge against extreme volatility in yields. A new, eco-friendly winery and underground cellarage for ageing has also been constructed. Ridgeview 2009 Limited Release Blanc de Noirs in magnum is still one of the greatest English sparkling wines ever produced.

✓ *Entire range*

RYEDALE VINEYARDS
Westow, North Yorkshire
☆

Planted in 2006, Ryedale is still the country's most northerly commercial vineyard, with 16 varieties growing on two sites north of York.

✓ *Still (Wold's View)*

SQUERRYES
Westerham, Kent
★☆

Recipient of Best English Sparkling Wine at the CSWWC in 2016, this venture grew out of interest expressed by Duval-Leroy, who were considering planting a vineyard at the time. The Champagne firm did not prefer another location, but for reasons of its own, decided not to invest in English sparkling wine, so in 2006 the Warde family planted 15 hectares (37 acres) on their 1,000-hectare (3,953-acre) estate and have not looked back since.

✓ *Sparkling (The Trouble with Dreams)*

SUGRUE
Washington, West Sussex
★

The own label of Wiston's winemaker, Dermot Sugrue.

✓ *Sparkling (The Trouble with Dreams)*

THREE CHOIRS
Newent, Gloucestershire

This winery really needs to bring its vineyard varietals and wines into the 20th century at least. The 21st, preferably. One of the country's largest wine producers, Three Choirs is an important name when it comes to disgorging and dosaging sparkling wines made by others, but frankly I wonder how, when its own singular, Seyval-based sparkling wine blend fails to inspire. Come on guys, grub-up or chip-bud all existing hybrid and German crosses to Champagne varieties, produce two or three top-class cuvées and more upmarket still wines, and get a higher price and reputation.

WISTON
Washington, West Sussex
★☆

Owned by Harry and Pip Goring, these wines are made by Dermot Sugrue, one of the most engaging characters in the entire English wine industry. You might see Wiston in a clear-glass bottle here and there, but if so, the wines will be from the 2010 vintage or earlier, as the complete range moved to very dark-amber bottles as from the 2011 harvest.

✓ *Entire range*

---

and another vintage of sparkling rosé has been released. One to watch for when the vineyard starts producing.

✓ *Sparkling (Brut Rosé), Still (Pinot Gris)*

PLUMPTON ESTATE
Plumpton, East Sussex
★☆

The only educational facility in Europe offering a BSc in viticulture and oenology in the English language, Plumpton College also possesses two vineyards totalling almost 9 hectares (22 acres) along with a well-equipped winery. Few educational establishments around the world produce commercial wines that truly excel, yet Plumpton's sparkling wines regularly win gold medals and Best in Class awards at the CSWWC. As a Centre of Excellence that evidently practises what it preaches, perhaps it can offer an MSc specialising in Sparkling Wine Viticulture, Technology, and Production?

✓ *Sparkling (Brut Classic, Brut Rosé – both formerly "The Dean")*

LOUIS POMMERY
Old Alresford, Hampshire
★

Pommery started planting its 40-hectare (100-acre) chalkland vineyard in 2017, and the fruit structure of the sparkling wine produced is more like Champagne than English sparkling wine, although it does not resemble Champagne Pommery in the slightest.

✓ *Sparkling (Brut)*

RATHFINNY ESTATE
★☆

With 92 hectares (222 acres) planted at the time of writing, a target of 160 hectares (395 acres) soon to be achieved, and a 1-million-bottle capacity winery already built, Rathfinny is due to become the largest single-estate producer of

English sparkling wine. The speed of development is almost as breath-taking as its location, which is set in a gorgeous, if somewhat windswept, escarpment depression of the South Downs, barely a stone's throw from the sea, and acquired by Sarah and Mark Driver as recently as 2010. The first vintage was 2014, released in 2018, and French winemaker Jonathan Médard seems easily able to take future vintages of these wines to the very top.

✓ *Entire range*

RIDGEVIEW ESTATE
Ditchling, Sussex
★☆

Founded by the late Mike Roberts, this was the first commercial producer to follow in the footsteps of Nyetimber. Mike's son Simon is head winemaker, the presentation of the Ridgeview range has been updated, and some of the wines have been converted into non-vintage

Luxembourg

A small but varietally diverse wine-producing country on the northern cusp of commercially viable viticulture, Luxembourg can boast that its best vineyards consistently produce wonderfully fresh and pure dry white wines that are delicious to drink with food or without. They also fit conveniently into the niche vacated by so many Alsace producers as that great French wine region makes higher-alcohol wines of increasing residual sugar.

Luxembourg derives its name from *Lucilinburhuc*, or "Little Fortress", which is the Saxon name for a defensive position originally built by the Romans on a rocky promontory around which the city of Luxembourg was destined to grow. Under the Romans, this area held no geopolitical identity, but was part of the Trier region, where vines were grown 2,000 years ago.

During the Middle Ages, the monasteries spread viticulture beyond the Moselle, along the banks of the rivers Sûre and Our into Ösling, to the very north of (modern-day) Luxembourg, but those vineyards were destroyed by the Great Frost of 1709 and, gradually many of the surviving Moselle vineyards began to diminish.

Demand from the *Sekt* industry prompted a revival of Luxembourg's vineyards by 1880, when almost 90 per cent of the vines grown were Elbling, the favoured variety of many German fizz producers. The first Luxembourg sparkling wine was not produced until 1921, the same year that Vinsmoselle, the country's first wine cooperative was established. In 1935 the Marque Nationale points-based appellation system was created. By 1959 this was extended to include the designations *Vin Classé, Premier Cru,* and *Grand Premier Cru* and by 1988 also incorporated sparkling wines. In 1991 Luxembourg's sparkling wines became known as Crémant de Luxembourg and, rather bizarrely, competed with the *crémant* appellations of France at the French Concours National des Crémants. In 2000 the designations of *Vin de Glace, Vendages Tardives,* and *Vin de Paille* were added to the Marque Nationale appellation system.

The Marque Nationale system was replaced by AOP in 2014. A points-based system achieved through blind tasting is fine in theory, but not so much in practice and, as this encyclopedia has pointed out since 2005, classifying a wine as a *cru* (growth) of any sort on the basis of tasting, whether or not the wine comes from a single site, is irrational and contrary to the spirit of the European Union's wine regime. It was inevitable, therefore, that this would change.

THE VINEYARDS

Today, the vineyards are scattered around and between 28 towns and villages, stretching from Wasserbillig in the north to Schengen in the south. This viticultural area is the mirror image of the German *Bereiche* of Obermosel and Moseltor across the Moselle River, which are situated on the right bank and generally produce less-exciting wines than those of Luxembourg's vineyards opposite. It is from the German side of the river that the topography of Luxembourg's vineyards can best be observed, especially when the river bends north and south. This strip of the Moselle's left bank is officially designated the *perimeter viticole* for the Moselle Luxembourgeoise AOP. It is about 1,300 hectares (3,210 acres), and vines are not allowed to be planted anywhere else in Luxembourg.

These vineyards are at their narrowest in the northern half of Luxembourg's Moselle Valley, where the hard dolomite Muschelkalk rock has resisted erosion, leaving a narrower valley with steeper slopes.

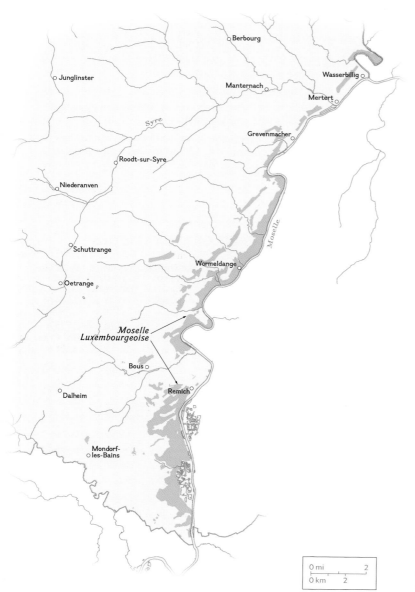

> ### THE NEW MOSELLE-LUXEMBOURGEOISE AOP
>
> Maximum yields have dropped from 140 hectolitres per hectare for Elbling, 115 for Rivaner, and from 120 to 100 for all other varieties. White wines are still allowed to use the terms *Vin Classé, Premier Cru,* and *Grand Premier Cru,* but until there is a geographical-based classification of Luxembourg's vineyards, such terms remain meaningless. Although no legal definition yet exists: the use of "Côtes de" is reserved for entry-level wines and is usually canton-based, the use of "Coteaux de" is considered to be premium quality and usually municipal-based, and *lieu-dit* wines offer true interest for *terroiriste* consumers.

THE MOSELLE-LUXEMBOURGEOISE WINE REGION
The appellation follows the Moselle River in southeastern Luxembourg, along the river's border with Germany. Luxembourg is a country united by other tongues: French is used by its government and courts, with most newspapers published in German, leaving the national dialect with no official role.

This pinkish brown bedrock can be seen overhanging the old Caves St Martin – now owned by the firm of Gales – which were carved out in 1919 to 1921. Going south from Remich, much softer marly and clayey soils dominate; consequently, the area has been prone to erosion, which has widened the valley, leaving a softer landscape of gently rolling hills. As a rule of thumb, the wines in the south are softer, fuller, and best drunk young, whereas those in the north are crisper with more minerality, requiring time in bottle.

Owners of Luxembourg's vineyards have dropped from 831 in 1992 to 340 today, as younger generations have moved away from viticulture. Yet when sold, the vineyards do not disappear, they are merely absorbed by other growers.

WINE PRODUCERS

There is a closed-shop system of 60 wine producers, including six commercial shippers and six cooperatives. The Vinsmoselle cooperative dominates production through its membership, which owns two-thirds of all of Luxembourg's vineyards and thus produces two-thirds of all of its wines. Do not be put off by the size of Vinsmoselle: there are very few poor wines produced by anyone in Luxembourg, particularly amongst the white wines and Crémant de Luxembourg, and the average quality for Vinsmoselle is undeniably good, with some wines that are regularly stunning. Because Vinsmoselle benefits from various subsidies and tax benefits, like cooperatives everywhere, a number of private producers formed Charta groups declaring lower yields and other quality criteria. The first was Domaine et Tradition in 1985, which has been followed by Charta Luxembourg (aka Charta Privatwënzer) and Charta Schengen Prestige. Producers are not compelled to produce all their wines under the Charta label to which they belong, but if they do claim the label, it is usually regarded as a premium wine.

RIVANER GRAPES GROW IN A MOSELLE VINEYARD
These white grapes, also known as Müller-Thurgau, are the most widely planted variety in Luxembourg – even though they produce rather uninspired wine.

RECENT LUXEMBOURG VINTAGES

2019 A warm spring gave way to a damaging late frost at the beginning of May. With summer temperatures over 40°C (104°F), exposed grapes in a number of vineyards suffered sunburn. This resulted in the smallest crop of the decade (well below 2012), but the grapes were healthy and the wines comparable to 2018.

2018 A mild winter and sun-drenched summer ensured strong, aromatic wines of a quality and yield that made everyone smile.

2017 The beginning and end of this year were particularly painful, with a late spring frost of a magnitude rarely seen and heavy rains at harvest time. Under the circumstances, the quality was just about satisfactory, although the yield was even lower than in 2012.

2016 The year started unusually mild, with no winter snow or frost and was followed by a cold and wet spring, with frost damage in late April. A dry, hot, and sunny August brought on ripeness, however, and exceptional weather in September resulted in finely balanced wines of exceptional complexity and harmony.

2015 The quality was high considering the periods of drought induced by heatwaves. A slightly smaller harvest than normal, but the wines showed well, thanks to high sugar levels and well-balanced acidity.

2014 An average yield was reduced by some 8 per cent due to rot caused by the wet weather during the latter part of the harvest. Thanks to earlier exceptional weather, the quality was, however, high, with optimal ripeness and well-balanced acidity.

2013 After an unusually mild winter, growers expected late frosts but were relieved when these fears did not materialise. Due to rain and cold weather during bud break, flowering, and maturation phase, however, the crop was smaller than normal (but larger than 2012), and the quality was just about satisfactory.

2012 Temperatures in September and October were slightly lower than normal, but conditions were dry, enabling ripe and clean grapes to be harvested. The quality was very satisfactory, but the volume was small.

2011 Due to the dry and sufficiently sunny climatic conditions, the grapes ripened without problem and the quality was high, particularly Vendanges Tardives. A good year for white wine, producing Burgundian varieties Auxerrois, Pinot Blanc, and Pinot Gris.

LUXEMBOURG'S GRAPE VARIETIES

VARIETY	HECTARES 2018	CURRENT % OF VINES 2018	MAXIMUM YIELD HL/HA ALLOWED	AVERAGE YIELD HL/HA 2009–2018	AVERAGE % OF WINES 2009–2018
Rivaner	288.77	23%	120	108.1	29.3%
Pinot Gris	194.11	15.6%	100	72.5	13.2%
Auxerrois	185.34	14.9%	100	92.1	16.0%
Riesling	157.54	12.6%	100	74.2	11.0%
Pinot Blanc	157.47	12.6%	100	91.4	13.5%
Pinot Noir	123.07	9.9%	100	65.2	7.5%
Elbling	72.17	5.8%	120	115.5	7.8%
Chardonnay	4.78	2.8%	100	62.9	0.3%
Gewürztraminer	20.39	1.6%	100	45.2	0.9%
Others (Gamay, St Laurent, Muscat)	14.82	1.2%	100	43.2	0.6%
TOTAL	**1,248.46**	**100%**		**88.6**	**100%**

FACTORS AFFECTING TASTE AND QUALITY

LOCATION
The vineyards are found on a 42-kilometre (260mile) strip of land along the country's southeastern flank, where 1,300 hectares (3,200 acres) of vines grow on the left bank of the Moselle River, overlooking the southern-most section of Germany's Mosel region.

CLIMATE
As part of the modified continental climate with mild winters and cool summers, the rainiest months coincide with ripening and harvest. Frost can be dangerous, reducing yields by 80 per cent on rare occasions, but rain and cold temperatures at flowering more commonly reduce yields. Average annual rainfall is 70 to 72.5 centimetres (27.5 to 28.5 inches).

ASPECT
Going south from Remich, the valley widens, facilitating the growth of vines. Going north from Remich, it narrows, and the slopes become steeper. More than 290 hectares (716 acres) of vines are grown on slopes of above 30 per cent incline, 32 hectares (79 acres) are on slopes in excess of 45 per cent incline, and 8 hectares (20 acres) are still terraced. The best sites are either southeast or southwest facing at an altitude of 150 to 200 meters (492 to 646 feet).

SOIL
This is a tale of two soils, dividing the vineyards almost exactly in half, with soft, marly-clayey Keuper to the south and Muschelkalk – a hard, pinkish brown, fissured Dolomite limestone – to the north.

VITICULTURE & VINIFICATION
Luxembourg's vineyards have undergone a restructuring, replacing traditional terracing with wider-spaced, vertically sloping rows. Called *remembrement,* this practice started in Wormeldange in 1970 and is nearly complete. Most vines are trained using a single Guyot system, though some are trained using the Trierer Rebenrad. Yields are still much too high, but a 10-year average of 89hl/ha is a huge improvement compared to 103 hl/ha 10 years ago. Some *barriques* are used, and a number of traditional producers have large old oak *fûdres,* but stainless-steel predominates.

GRAPE VARIETIES
Primary varieties: Auxerrois, Pinot Blanc, Pinot Gris, Pinot Noir, Riesling, Rivaner (Müller-Thurgau)
Secondary varieties: Chardonnay, Elbling, Gamay, Gewürztraminer, Muscat, St Laurent

THE WINE STYLES OF
LUXEMBOURG

ÄISWÄIN
See Vin de Glace

AUXERROIS
One of Luxembourg's most successful grape varieties, which is amazing considering that its yield is second only to the rampant Rivaner. Yet Auxerrois ranks with the best you can find in the Haut-Rhin. Some producers leave a little residual sugar in these wines, but they are essentially light, dry, and delicious to drink.

1-2 years

✔ Aly Duhr (Wormeldange Koeppchen) • Max-Lahr et Fils (Wormeldange Heiligenhäuschen) • Pundel-Hoffeld (Machtum Alwengert) • R Kohll-Leuck (Ehnen Ehnerberg) • Vinsmoselle Caves de Stadtbredimus (Stadtbredimus Primerberg)

CHARDONNAY
The clones planted are 95 and 96, which are good for both still and sparkling wines, although 96 fares better at the latter, and this far north growers should really be trying 75 and 121. Everyone who had this variety told me that their Chardonnay had been planted for sparkling wine, yet they all just happened to be making a still wine from it, none of which was any better than a Belgian Chardonnay (Clermont, Genoels-Elderen, and others). A certain warmth is required to make premium-quality Chardonnay, and Luxembourg just does not have this on any regular basis. If decent-quality, still Chardonnay is extremely rare in Champagne, what chance has Luxembourg?

1-2 years

✔ Gales (Coteaux de Stadtbredimus)

ELBLING
Traditionally grown as low-quality, low-tax fodder for Germany's *Sekt* industry, Elbling typically provides a neutral-flavoured, high-acid wine. I once thought that it might be interesting to look out for the odd wine made from low-yield Elbling but gave up when it became patently clear that no such thing exists.

GEWÜRZTRAMINER
This grape is not as abundant as it is in Alsace, and it lacks the gravitas and broader notes found in the latter region's truly classic examples. The significantly lower level of ripeness reduces the availability of spice-laden terpenes, while its lower alcohol content (3 to 4 per cent lower) gives an entirely different, much lighter, less viscous mouthfeel. If anyone in Luxembourg is really serious about this grape, it must be grown on the very best, south-facing sites that have been traditionally reserved for Riesling, with yields halved and harvested as late as possible. With so few consumers who understand or appreciate Gewurztraminer, I would not dream of sacrificing my best Riesling sites if I were a grower in Luxembourg, but if I were a Gewürztraminer fanatic of a consumer, I would love to see the results. For those winemakers who are equally fanatical, I suggest washing the grapes, a longer pre-fermentation cold-soak, less RS, not worrying about low acidity, and cellaring the wines an extra 18 months before release.

1-2 years

✔ A Gloden (Schengen Markusberg) • Charles Decker (Remerschen Kreitzberg) • Bernard-Massard • Domaine Mathis Bastian Y (Domaine et Tradition) • Schmit-Fohl (Ahn Goellebour)

PINOT BLANC
This is the crystal-clear, refreshing, light, dry, and fruity entry-level wine for any half-serious Luxembourg wine drinker. The sooner that plantings of Rivaner and Elbling are reduced to a statistical insignificance, the sooner that Auxerrois will be the entry-level wine for all consumers, with Pinot Blanc the flagship, and the better Luxembourg's wine reputation will be.

1-2 years

✔ Cep d'Or (Stadtbredimus Goldberg) • Domaine Gales (Domaine et Tradition) • Domaine Gales (Bech-Macher Rëtschelt) • Häremillen (Wormeldange Weinbour) • Mathes (Wormeldange Mohrberg) • Mathis Bastian (Coteau de Remich) • Pundel-Hoffeld (Machtum Gëllebour) • R Kohll-Leuck (Ehnen Rousemen) • Schram & Fils (Bech-Kleinmacher Falkenberg) • Schumacher-Lethal (Wormeldange Heiligenhäuschen)

PINOT GRIS
Luxembourg's Pinot Gris has none of the spice found in Alsace, and is simply a slightly richer, equally delicious version of Pinot Blanc – everything Italy's Pinot Grigio should be, yet seldom is. It is probably impossible for Pinot Gris to ripen sufficiently in Luxembourg's climate to create enough terpenes in the grape's skin for these to develop into spicy bottle-aromas.

1-4 years

✔ A Gloden & Fils (Wellenstein Foulschette) • Cep d'Or (Stadtbredimus Primerberg Signature Terroir et Cépage) • Häremillen (Mertert Herrenberg) • Jean Schlink-Hoffeld (Machtum Ongkâf "Arômes et Couleurs") • Krier Welbes (Bech-Kleinmacher Naumberg) • Pundel-Hoffeld (Machtum Widdem) • Mathes (Wormeldange Woûsselt) • R Kohll-Leuck (Ehnen Kelterberg) • Schmit-Fohl (Ahn Goellebour)

PINOT NOIR
The clones planted are 115 and 114, which are good, but also 375, which is not (because it is an over-cropper and produces low-quality wine with poor varietal character), and this far north growers should be trying 779 and 927. The wines have a certain sour-cherry varietal character but, in general, lack fruit, structure, finesse, and complexity. I have tasted a few delights, but unless growers are relying on global warming, Pinot Noir is fighting the climate in Luxembourg. I don't want to say no because even in the UK's more climatically challenged climate, Gusbourne can produce wonders. For the fanatic (and it is many winemakers' ultimate goal), maybe choose better clones, ring-fence the vines, and only produce in years like 2018. As an aesthetic and marketing point, too many Pinot Noir red wine producers need to see a psychologist when it comes to bottle-shape. At a recent Luxembourg Pinot Noir tasting, no fewer than 25 per cent of the bottles were high-shouldered and straight-sided, which most knowledgeable consumers (and, let's face it, Pinot Noir usually appeals to knowledgeable consumers) associate with Bordeaux varieties. Two were even in Riesling-shaped bottles.

✔ Caves Burna • Cep d'Or (Stadtbredimus Goldberg)

RIESLING
This is Luxembourg's finest potential wine, but most Riesling growers are not realizing this potential. In such a northern, Atlantic-influenced area, it was madness to average almost 100 hectolitres per hectare 10 years ago. This has dropped to 74.2 hl/ha, which is admirable; however, more discipline is required to succeed on the world stage. It is widely understood that Riesling can be successfully cultivated only on fully south-facing sites in Luxembourg (practicalities extend that to south-by-southeast and south-by-southwest), yet the growers throw away the potential of the country's greatest sites by over-yielding. They should be aiming for 12 to 13 per cent alcohol without chaptalisation, but they are averaging barely 9 per cent. Only in years like 2018 can Luxembourg's Riesling achieve a country-wide average of 91 Oechsle at 84 hl/ha. That is a

VINEYARDS OVERLOOK THE BANKS OF THE MOSELLE NEAR WORMELDANGE
Wormeldange is home to some of Luxembourg's top vineyards, including the outstanding Wormeldange Koeppchen of Domaine Alice Hartmann. The aromatic Reisling is the favoured variety of grape in the region.

natural ABV of just over 12 per cent, which should be more than enough to produce classic dry Riesling with great freshness, linearity and minerality, but it is impossible to find an authentically dry Riesling of 5 grams per litre of residual sugar or less in Luxembourg. If you bring this up with sommeliers in Luxembourg, you'll likely find they have been brainwashed. They start defending the sweetness by saying you cannot detect the sugar because of the acidity. They truly believe they taste dry. God forbid, I went through this in Germany 20 years ago! Luxembourg has a great future with Riesling, if winemakers do not leave some sweetness, and although I do not condemn chaptalisation when it is necessary, it should be illegal to chaptalise *and* leave residual sweetness. One further constructive criticism: I have noticed some Riesling producers using blue or blue-green glass bottles. Please stop: blue glass is only marginally less dangerous than clear glass in terms of leaving wine prone to light-strike faults.

🍷 *1–8 years*

✓ *Aly Duhr (Domaine et Tradition) • Cep d'Or Riesling (Stadtbredimus Fels "Signature Terroir et Cépage") • Domaine Alice Hartmann (Wormeldange Koeppchen and Les Terrasses de la Koeppchen la Chapelle) • Domaine Gales (Wellenstein Kurschells) • Pundel-Hoffeld Riesling (Wormeldange Elterberg "Cuvée Spéciale") • Clos du Rochers • Domaine Henri Ruppert • Schumacher-Knepper (Wintrage Felsberg and Ancienne Propriété Constant Knepper) • Steinmetz-Jungers (Grevenmacher Fels) • Vinsmoselle (Coteaux de Schengen) • Vinsmoselle Caves de Grevenmacher (Mertert Herrenberg)*

RIVANER

Also known as Müller-Thurgau. Although this ubiquitous grape accounts for almost one-third of all the vines growing in Luxembourg today, it is definitely on the decline. Twenty-five years ago every other bottle of wine produced in the Grand Duchy was made from Rivaner. This grape is grown at very high yields and turned into a sweetish, Liebfraumilch-type wine, making it Luxembourg's least interesting variety, but one that enjoys high-volume sales. Rivaner prices are low, and the popularity of this style of wine extends beyond the domestic market to Belgium, where it accounts for more than half of global exports of Vins Moselle Luxembourgoise. As long as the cultivated area of this variety continues to decline, there seems little point in getting high-minded about its yield. It will eventually bottom out, and lowering yield will hardly produce a noticeably finer quality of wine.

🍷 *Upon purchase*

✓ *Aly Duhr (Sélection Ahn Hohfels) • Cep d'Or (Stadtbredimus Goldberg) • Domaine Viticole Kox (Remich Primerberg) • Krier-Welbes (Bech-Kleinmacher Naumberg)*

OTHER GRAPE VARIETIES

Gamay did not impress me, although the three Muscats demonstrated that it might be worth persevering with this variety, especially Charles Decker's Muscat Ottonel (which has been deliciously creamy or evocatively strawberry).

🍷 *1–4 years*

✓ *Charles Decker (Muscat Ottonel, Remerschen Kreitzberg)*

OTHER STYLES
CRÉMANT DE LUXEMBOURG

Champagne – real Champagne – was in fact produced in Luxembourg by Mercier from 1886 up to almost the beginning of World War II. During this time, Mercier was known as "Champagne Mercier, Epernay-Luxembourg", and both establishments were famous for being completely powered by 19th-century electricity. Today, of course, Champagne is a prohibited term, and the Crémant de Luxembourg appellation was created as part of the deal with Champagne to not use the term *méthode champenoise*. Curiously, it gets lumped in with French Crémant appellations (Alsace, Bordeaux, Bourgogne, Die, Jura, Limoux, Loire, and Luxembourg) at the annual Concours des Crémants. Most Crémant de Luxembourg encountered outside the country is boring, but the general standard is much higher, primarily because it is much fresher, and these sparkling wines are a delight to drink young. The style is fresh, soft, and elegant, with good acids providing a crisp, fruity finish. The only exception is Domaine Alice Hartmann Brut, which is chock-a-block full of flavour and character, perhaps at the expense of finesse, and could be criticized for being too big and too oaky, but has to be admired, and is currently Luxembourg's greatest sparkling wine. It's a style that I would not like to see more than two or three other producers emulate, however, as Luxembourg is more naturally attuned to the Gales style of freshness and finesse. Whereas Crémant de Luxembourg must be made exclusively from grapes grown within the appellation Moselle Luxembourgoise, readers should be aware that any Luxembourg wines labelled *Vin Mousseux, Vin Pétillant, Perlwein,* or sold as *"boissons effervescents à base de vin"* may be blended from imported grapes, juice, or wine.

🍷 *1–2 years*

✓ *Aly Durhr • Domaine Alice Hartmann (Brut) • Gales (Heritage) • Gales (Jubilée) • Gales (Cuvée Premier Rosé) • Domaine Henri Ruppert (Gelle Fra Rosé) • Stephane Singery (Clos Jangli Vendanges Traditionelles) • Vinsmoselle (Poll-Fabaire Cuvée Art Collection)*

VENDANGE TARDIVE

Only Auxerrois, Pinot Blanc, Pinot Gris, Riesling, and Gewürztraminer are authorized. The minimum ripeness level of 243 grams per litres (220 for Riesling) of natural sugar content is as tough as the vendange tardive law introduced in Alsace in 1984, and raising ripeness levels is more difficult in Luxembourg's colder, more northerly climate. Alsace raised the bar in 2001 (to 235 to 257 grams per litre), and Luxembourg should do likewise, confident in the knowledge that its vineyards are nothing like the suntraps in Alsace, thus the Grand Duchy will never be in danger of losing its primary crop of dry wine grapes. Vendange Tardive is a new, emerging category of wine, and too many of the wines are noticeably high VA, with a bitter finish. This will change as producers get used to the style – some, like Domaine Thill Frères, already are.

🍷 *1–8 years*

✓ *Domaine Thill (Château de Schengen Riesling)*

VIN DE GLACE OR ÄISWÄIN

Only Pinot Blanc, Pinot Gris, and Riesling are authorized. The minimum ripeness level is 292 grams of natural sugar per litre. Luxembourg seems to be more climatically adept at *vin de glace* than *vendange tardive,* but it is a style that generally attracts a high VA level, and the Grand Duchy is no exception.

🍷 *1–8 years*

✓ *Charles Decker • Domaine Gales*

VIN DE PAILLE

Only Auxerrois, Pinot Blanc, Pinot Gris, and Gewürztraminer are authorized. The minimum ripeness level is 317 grams per litre of natural sugar. Only one outstanding example so far.

🍷 *1–8 years*

✓ *Caves Sunnen-Hoffmann (Auxerrois)*

THE APPELLATIONS OF
LUXEMBOURG

Note: Due to the varying usage of three languages in the country – Luxembourgish, German, and French – and sometimes Dutch/Low German, spellings of some villages and *lieux-dits* can vary. The most common variations of the *lieux-dits* names are shown below the appellation name. To list all the villages and all the *lieux-dits* in use today in all languages, not to mention the odd variant in each language, would confuse more than it would inform. It is obvious that Luxembourg must clean up its labelling act, obliging everyone to use the same nomenclature, and most observers would agree that Luxembourgish should be the dialect of choice.

AHN
Palmberg, Vogelsang, Elterberg, Hohfels/Houfels, Göllebour, Pietert, Nussbaum

Warmer vineyards enable Gewürztraminer to succeed in exceptionally hot years. It is said that Luxembourg's first Traminer and Muscat vines were grown here at the beginning of the 20th century. Also good for Pinot Gris, Auxerrois, Riesling, and occasionally Crémant de Luxembourg. Domaine Viticole Mme Aly Duhr & Fils is the most famous producer here, but I have been at least as pleased by Schmit-Fohl, followed by Steinmetz-Duhr. Other Ahn wines that have impressed me on occasions have been Caves Berna (Pinot Noir), Clos du Rochers (Bernard-Massard), and Jean Ley-Schartz.

BECH-KLEINMACHER
Enschberg, Jongeberg, Falkenberg, Naumberg, Fuusslach, Gaalgeberg, Brauneberg, Gottesgôf, Kurschels, Rëtschelt

Two villages, Bech and Macher, merged 100 years or so ago, when they built a communal church. There are just two producers resident in the town (Krier-Bisenius and Schram et Fils), but growers sell most grapes to the cooperative at Wellenstein and Caves Gales for their sparkling wine (they also grow their own grapes here). Naumberg has been the most successful *lieu-dit* in my tasting, with fine Pinot Gris from Krier Welbes and the cooperative at Wellenstein. Schram makes a very good Pinot Blanc on both Falkenberg and Kurschells.

BOUS
Johannisberg, Fels

There is just one local producer (Caves Beissel) in Bous and nothing outstanding so far.

EHNEN
Kelterberg, Mesteschberg, Ehnerberg/Einerberg, Rousemen, Bromelt, Wousselt

Ehnen is one of the prettiest villages on the Luxembourg wine route. I have enjoyed wines from Jean Linden-Heinisch and M Kohll-Reuland. Pinot Gris and Auxerrois fare well, with Mesteschberg and Kelterberg among the most successful *lieux-dits*.

ELLINGEN/ELLANGE
None

Two very good producers are located here, Caves Gales and Krier Welbes, but no *mono-cru* wines are produced. Growers either deliver their grapes to the cooperative or sell them to one of the six commercial wineries.

ELVANGE/ELVINGEN
NONE

No producers are located here. Although there is one grower who owns three plots within this village that are delimited for the production of *Appellation Moselle Luxembourgeoise Contrôlée* and have been exploited in the recent past, there are no vines growing here at the moment.

ERPELDINGEN
None

The best producer here is Stephane Singery (Clos Jangli). Growers either deliver to the cooperative or sell to one of the six commercial wineries.

GOSTINGEN
Bocksberg, Häreberg

There is just one local producer (Caves Fernand Rhein) in Gostingen and nothing outstanding so far.

GREIVELDANGE/ GREIWELDIGEN
Hutte/Hëtt, Dieffert, Herrenberg/Häreberg, Hëttermillen

The village is 2 kilometres (1.25 miles) from the Moselle, but the vineyards extend down to the river's banks. Of the two local producers, Beck-Franck and Stronck-Pinnel, the latter makes one of Greiveldange's best Rieslings from the Hëtt *lieu-dit*, while the former makes better wines from other villages. Most of the growers belong to the village cooperative, however, and Vinsmoselle makes most of the best wines here, especially Dieffert Pinot Gris and Herrenberg Auxerrois.

GREVENMACHER
Fels, Rosenberg/Rouseberg, Leitschberg/Leiteschberg, Pietert, Groaerd/Groärd

The best-known producer in Grevenmacher is Bernard-Massard, one of Luxembourg's truly international exporters. The Fels *lieu-dit* is for Riesling and Leitschberg and Pietert for Pinot Blanc. Steinmetz-Jungers, Vinsmoselle's Caves de Grevenmacher, and Fédération Viticole make the finest local wines.

LENNINGEN
Häreberg

There is just one local producer (Caves Fernand Rhein) in Lenningen and nothing outstanding so far.

MACHTUM
Alwengert, Hohenfels, Ongkâf, Goellebour/Göllebour, Widdem

Matchum is supposedly known for its Ongkâf Riesling, but it produces better Pinot Gris, and to be truthful, Göllebour vies with Ongkâf as the top *lieu-dit* in the village. Matchum also makes good Pinot Blanc. The Riesling is good, but no better than Auxerrois and, in hotter years, Gewürztraminer. The two local producers, Jean Schlink-Hoffeld and Pundel-Hoffeld, are both very good, with Vinsmoselle and Cep d'Or the best producers from outside the village.

MERTERT
Bocksberg, Herrenberg/Häreberg

The best *lieu-dit* in this ugly, industrial port is Herrenberg, which thankfully is closer to Wasserbillig. Pinot Gris is Herrenberg's most successful variety, with Häremillen and the local Vinsmoselle Caves de Grevenmacher the top-performing producers.

MONDORF
None

No producers are located in Mondorf. Growers either deliver their grapes to the cooperative or sell to one of the six commercial wineries.

NIEDERDONVEN
Fels, Gölleberg, Diedenacker

No producers are located in Niederdonven. Growers either deliver their grapes to the cooperative or sell to one of the six commercial wineries.

OBER-WORMELDINGEN
None

No producers are located in Ober-Wormeldingen. Growers either deliver their grapes to the cooperative or sell to one of the six commercial wineries.

OBERDONVEN
None

No producers are located in Oberdonven. Growers either deliver their grapes to the cooperative or sell to one of the six commercial wineries.

REMERSCHEN
Jongeberg, Kräitchen, Kreitzberg, Rodernberg/Roudeberg, Reidt

Charles Decker is the award-winning producer here; many of his medals are for wines that were denied the most-lowly appellation, having been rejected by the IVV. In the best years, Decker can produce good Muscat Ottonel and Gewürztraminer on the Kreitzberg *lieu-dit*, while Krier Frères has made good Riesling on Jongeberg.

REMICH
Fels, Goldberg, Gaalgeberg, Hélwéngert, Hôpertsbour, Primerberg, Scheuerberg

Primerberg is the most important *lieu-dit* here and Riesling its best variety, with Krier Frères, Stephane Singery, and Mathis Bastian the top producers, but this vineyard also makes fine Pinot Blanc and Auxerrois (St Remy Desom) and Pinot Gris (Gales). Overall Remich is most successful for Pinot Blanc, with Mathis Bastian (Coteau de Remich) and Krier Frères (Hôpertsbour) both vying with St Remy Desom.

BERNARD-MASSARD CRÉMANT DE LUXEMBOURG
Located in Grevenmacher, Bernard-Massard produces a large range of wines, including *crémant* and still wine.

ROLLING
None

No producers are located in Rolling. Growers either deliver their grapes to the cooperative or sell to one of the six commercial wineries.

ROSPORT
None

Everyone in the Luxembourg wine trade will tell you that Wasserbillig is the very northern limit of *Appellation Moselle Luxembourgoise Controlée*, yet I found these vines 14 kilometres (8.5 miles) farther north, and they are officially within the delimited region, even though they are not on the Moselle, or even in its hinterland, but located on a tributary called the Sauer.

SCHENGEN
Fels, Markusberg

Although Fels Riesling is supposed to be best known, this most southerly of Luxembourg's wine villages is definitely Pinot country (Blanc and Gris), with Vinsmoselle dominating, though A Gloden & Fils and Domaine Henri Ruppert also makes some fine wines.

SCHWEBSINGEN/SCHWEBSANGE
Letscheberg, Steilberg, Hehberg, Kolteschberg

Supposedly good for Gewürztraminer, but I have yet to confirm that. The best *lieu-dit* is Kolteschberg, where

Vinsmoselle Caves de Wellenstein and Laurent & Benoît Kox produce some good Pinot Gris and Riesling.

STADTBREDIMUS
Primerberg, Goldberg, Dieffert, Fels

Only Wormeldingen produces more top-quality Luxembourg wines than Stadtbredimus. Most of the best come from Vinsmoselle Caves de Stadtbredimus and Cep d'Or, but some also comes from Gales and Laurent & Benoît Kox. The largest *lieu-dit* is the Primerberg, but Dieffert and Fels are the best, producing some of the finest Rieslings in the country. Pinot Gris is the second-best variety in Stadtbredimus, and Primerberg, Goldberg, and Fels the most successful *lieux-dits* for this grape.

WASSERBILLIG
Bocksberg, Häreberg

Supposedly the most northerly of Luxembourg's main drag of vineyards, but not strictly so (*see* Rosport).

WELLENSTEIN
Foulschette, Kurschels, Veilchenberg, Roetchelt, Brauneberg, Knipp, St Annaberg

Kurschels stands out as the best *lieu-dit* and the best *lieu-dit* for Riesling. Foulschette is the second-best *lieu-dit* and the best *lieu-dit* for Pinot Gris. Krier Welbes and Gales are the top producers in this region, followed closely by Mathis Bastian and Vinsmoselle (Série Limitée Art).

WINTRANGE/WINTRINGEN
Hommelsberg, Felsberg

Felsberg is the best-known *lieu-dit*, with its statue of St Donatus crowning the hill, where Pinot Gris and Riesling are supposed to be among the best in Luxembourg, although I have tasted only one wine that matches up to that sort of reputation (and that is Vinsmoselle Caves de Remerschen Riesling Grand Premier Cru Wintrange Felsberg).

WORMELDANGE/WORMELDINGEN
Elterberg, Heiligenhäuschen, Koeppchen/Köppchen, Moorberg, Nidert, Nussbaum, Péiteschwengert, Pietert, Weinbour, Wousselt

Wormeldingen is the superstar of Luxembourg's wine villages, with Koeppchen its greatest *lieu-dit*, Domaine Alice Hartmann its best producer and Riesling its king of grapes. Domaine Alice Hartmann is by no means the only top producer making wines from this village. I have also been impressed by Wormeldingen wines from Aly Duhr, Fédération Viticole, Häremillen, Jean Schlink-Hoffeld, Mathes, Pundel-Hoffeld, Schumacher-Lethal, St Remy Desom, Steinmetz-Jungers, and Vinsmoselle Caves de Wormeldange. I have tasted some good Pinot Gris, Pinot Blanc, and Auxerrois from these producers, but with one exception (Aly Duhr), they have all been from other *lieux-dits*. If I have learned anything about the wines of Luxembourg, it is that the best parts of Koeppchen produce the country's finest Riesling, and it would be a great shame to plant this *terroir* with any other variety.

VIVID FLOWERS BRIGHTEN A ZIELONA GÓRA VINEYARD IN POLAND
The first vineyards are said to have been first planted here in about 1150, but by the 20th century few vineyards had survived World War II. Since the turn of the millennium, viticulture in the region has begun to recover and regain its focus on winemaking. Tourists to the region can visit the Wine Museum there and celebrate at Zielona Góra Wine Fest.

Other Winemaking Countries of Northwestern Europe

In the Northern Hemisphere, the farther north a vine grows, the fewer hours of daylight it receives and the greater the angle at which the sunlight is spread out, thus the longer it takes to ripen grapes. Factor in the wet, windy, and uncertain climes of the Atlantic, and it may seem surprising that these countries produce any wine.

BELGIUM

In Belgium, winegrowing dates back to Roman times. Today, more than 185 hectares (457 acres) of vineyards are cultivated by more than 100 growers. Most of these growers are part-time or hobbyists, however, although there are around 30 who could be described as commercial to one extent or another.

✓ *Genoels-Elderen (Chardonnay Goud, Zwarte Parel, Maastricht Riesling) • Peter Colemont (Chardonnay Clos d Opleeuw) • Domaine Viticole du Chenoy • Chateau de Bioul • Domaine du Ry d'Argent*

DENMARK

There are 200 hectares (more than 490 acres) of vines in this country. Even more amazingly, there are 90 commercial producers, the oldest of which is Domain Aalsgaard, which was established in 1975. Sparkling wines from the Dons area received PDO status in 2018. Vingården Lille Gadegård produced its first vintage in 2003 on Bornholm, a Danish island in Eastsee, to the north of Poland.

✓ *Skaersoegaard • Dyrehoj Vineyard*

FINLAND

This Scandinavian nation is not currently listed as an official wine-producing country by the EU. There are only a handful of grape growers, and none of them are commercial. Sundom Winery is a pioneer in western Finland, growing Riesling, Merlot, and Chardonnay on 0.3 hectares (0.74 acres) near the town of Vaasa at 63°N latitude, which makes it a good contender for the most northerly vineyard in the world.

IRELAND

Better suited to stout and whiskey, Ireland has no winemaking history, however, the country is now officially listed by the EU as a wine-producing country. The story started when, in 1972, Michael O'Callaghan started planting Müller-Thurgau vines at Mallow in County Cork. In its heyday O'Callaghan had just over 1.5 hectares (4 acres) of exclusively Reichensteiner vines, but the vineyard was replanted with orchard fruit after his death in 2010. Just 5 kilometres (3 miles) away from Mallow was another Irish vineyard experiment, called Blackwater Valley. Here 5 hectares (12 acres) of Reichensteiner, Madeleine Angevine, and Seyval Blanc were grown until 2006. Another venture, West Waterford Vineyards at Cappoquin, has 2,000 vines, promising 500 bottles of wine in a successful year. Owner David Dennison, who runs a wine import business, has been planting grapes since 2011 in Waterford and today has just under 1 hectare (2 acres) of vines, mainly Pinot Noir, Rondo, and several white varieties.

One of Ireland's oldest wine producers, cultivating vine here since the mid 1980s, is Thomas Walk Vineyard near Kinsale in the south of the country. Here, he is growing up to 1.2 hectares (3 acres), also using the Rondo variety, following organic practices.

One of the only commercial winemakers in this country is David Llewellyn (who also grows apples) is a trained horticulturist and grows mostly Cabernet Sauvignon, Merlot, and Rondo on half an acre of vineyard north of Dublin.

LATVIA

Jukka Huttunen tends 160 wine-producing vines planted in 2001 on ground warmed by buried coolant pipes from Olkiluoto nuclear power station. At 61° 13'N latitude, this is the most northerly vineyard in the world. Sabile Wine Hill used to claim this title too. There are approximately 30 different varieties planted on 1.5 hectares (3.7 acres). The variety Zilga is the most popular.

LIECHTENSTEIN

Situated between Germany and Switzerland, this tiny principality grows a tiny amount of vines, but it has its own AOC and can indicate the following vineyard names on the label: Balzers, Bendern, Eschen, Eschnerberg, Gamprin, Mauren, Ruggell, Schaan, Schellenberg, Triesen, and Vaduz. The vines are primarily Pinot Noir and Chardonnay and are mostly owned by the Crown Prince through his private winery, the Hofkellerei Fürsten von Liechtenstein.

NETHERLANDS

The Netherlands is north of Belgium and would therefore seem an even more unlikely location for grape-growing, but almost all its vines grow east of Maastricht, so they are actually situated farther south than the vineyards in the Belgian part of Brabant. There are over 150 commercial vineyards – mostly in the south in Gelderland and Limburg – growing mainly Riesling, Müller-Thurgau, Auxerrois, and the inevitable German crosses.

✓ *Apostelhoeve (Auxerrois, Riesling)*

NORWAY

In 1992 the first wine-producing plot in Hallingstad vineyard was planted close to Harten, which sits at a latitude of 59° 24'N latitude, by Sveier Hansen. Today there are almost 40 commercial vineyards. Some of the better-known Norwegian producers are Lerkeasa Vingaard, Kvelland Gaard, and Egge Gaard.

POLAND

Vines have been grown in Poland since at least the 14th century but ceased production during the Communist era. Eventually, a few isolated areas were replanted a few years before Solidarity changed the face of Europe forever in 1989. These vineyards were confined to the Carpathian Mountains in the very south of the country, but when Poland became a member of the EU in 2003, the Polish Wine Institute was established, and hundreds of vineyards were created in six Polish wine regions (Zielona Góra-Wielkopolskie, Central and Northern Poland, Lower Silesia, Malopolska-Vistula-Lubelskie, Malopolska-Vistula-Świętokrzyskie, and Carpathia), where today no fewer than 150 vineyards officially exist sharing 200 hectares (495 acres) amongst them, although most of these are not commercial enterprises by international standards. Four of the more important wineries are Winnica Golesz, Winnica Jasiel, Winnice Jaworek, and Winnica Pałac Mierzęcin, Winnica Milosz, Winnica Hople in Paczkow, and Winnica Plochockich.

SWEDEN

Although Åkesson has produced sparkling wine since 1985, the grapes used were not grown in Sweden. The first three vineyards to be established in this country are Blaxta Vingård (planted in 2000, first wines in 2002) southwest of Stockholm near Flen (at a latitude of 59° 03'N), and two in the Skåne region farther south: Kullabygden Vingård (planted in 2000, first wines in 2002) near Helsingborg and the Nangijala Vingård (planted in 2001, first wines in 2003) near Malmö. Today there are around 30 commercial vineyards, cultivating 100 hectares (247 acres).

The Wines *of* Southeastern Europe

SOUTHEASTERN EUROPE HAS BEEN AN ALMOST FORGOTTEN corner of the wine world until recently, at least as far as quality is concerned. These countries were probably better known in the 18th and 19th centuries than today, supplying royal courts across Europe with renowned wines like Tokaji, Cotnari, Cadarcă Aszú, and others. Decades of communism almost completely broke the links between land, people, and wine quality, but since independence there has been a complete revolution in winemaking as each country, and its winemakers, has worked hard to rediscover and reinvent their wines, with a modern interpretation. Wine is a great lens to see how these countries have emerged from the past, reflecting as it does both culture and *terroir*. Today this is arguably the last undiscovered part of the wine world, but with so much exciting and authentic wine on offer, it is already rewarding to explore. There will be more to come as winemakers truly get to grips with their *terroirs* and the exciting potential of their best indigenous grapes.

A SLEEK TRACTOR GARAGE SITS AMID FURMINT VINEYARDS AT THE DISZNÓKŐ TOKAJI ESTATE IN HUNGARY
Designed by prize-winning architect Dezső Ekler, the garage is a prime example of Hungary's architectural style known as "organic". The vineyard is located in the Tokaj-Hegyalja OEM, a region that produces Hungary's most recognisable wines. The grapes from these Furmint vines are destined for Disznókő's best Aszú wines.

Bulgaria

*Bulgarian wines, particularly Cabernet Sauvignon, were affordably fashionable in the West
between the early 1980s and mid-1990s, but the wine industry lost its way after
the collapse of communism and has taken a long time to find its feet in a free-market economy.*

From the mid-1950s to the late 1970s Bulgarian wine exports were dominated by sales to the Soviet Union and Comecon countries. From the 1950s the planned economy had led to widespread plantings of international varieties, most notably Cabernet Sauvignon, which became the key to successful exports to the West from the 1980s onwards. Cabernet Sauvignon was the most popular wine grape, and Bordeaux reigned supreme, but wine drinkers on the lookout for cheaper alternatives found that Bulgarian versions, most famously from Bulgaria's Suhindol region, were not just cheap, but also a darn sight better than a lot of unclassified claret at twice the price. Bulgaria had put a lot of research effort into its vineyards and winery equipment, but undoubtedly the arrival of PepsiCo influenced winemaking styles that were attractive to Western markets. PepsiCo wanted to sell cola behind the iron curtain, so, in a barter deal, it provided the country's winemakers with expertise from UC Davis, leading them to produce fruit-driven varietal wines that could be sold for hard currency and thus open the door for sales of Pepsi. In 1980 the then state monopoly Vinimpex set up a subsidiary in London, and Bulgaria grew rapidly to become the fourth-largest red wine supplier to the UK by 1990.

In 1990 the state monopoly was suddenly disbanded and free-market reforms were introduced. Unfortunately, it took a long time to resolve issues of privatisation of vineyards and winery ownership, which meant the consistent quality of the previous era could not be maintained. Domaine Boyar was the first private wine company to be formed in 1991. This put it one step ahead in terms of sourcing by enabling the company to forge links with the best vineyards, as well constructing the new-world-influenced greenfield Blue Ridge winery, though unlike most wineries, it did not invest in vineyards. In spite of its sometimes-difficult history, Domaine Boyar today is Bulgaria's third-largest producer, and arguably most visible exporter in Western markets. It has a separate boutique cellar at Korten, producing some very high-quality wines under Solitaire and Korten labels. Today's industry has 263 registered wineries, producing around 1.2 million hectolitres of wine from just under 37,000 hectares (91,420 acres) in 2016. Exports remain significant – to Poland, Sweden, the UK, and Czechia and the fickle Russian market. At the same time, domestic sales of bottled wine have grown and are especially important for more premium wines.

The country has reached a turning point recently: wineries are more settled in their vineyard ownership and fruit-sourcing and are therefore better at understanding the potential of their vines. This means winemaking is starting to rein back on sheer power and lashings of new oak in favour of more restraint and elegance. Producers are also starting to explore their local grape varieties in search of a flagship. Dark-skinned Mavrud is the most widely planted of the local varieties and, at its best, can make exciting wines in its own right or in blends. Not everyone agrees, and some producers argue in favour of Rubin (an inky dark cross of Nebbiolo and Syrah created in 1944), early Melnik (also known as Melnik 55, the earlier-ripening, more supple offspring of the traditional Shiroka Melnishka Loza, but only grown with any success in the Struma Valley) or even Gamza (a pale-skinned red grown in the northwest, aka Kadarka). At the same time, the local market has strong preferences for international grapes rather than the local varieties that their grandparents used to drink. Organic, biodynamic, pet-nat, and skin-contact whites are beginning to appear but remain niche, yet it is encouraging to see producers

---

BULGARIAN LABEL LANGUAGE

Note: Most wines with export or quality aspirations will be labelled in English as well as Cyrillic, though sometimes Cyrillic only. Label law complies with EU standards.

Protected Geographical Indication (PGI) Either Danubian Plain or Thracian Lowlands

Protected Denomination of Origin (PDO) In law, 52 exist but rarely in use commercially

Reserve/Reserva Single-grape variety, aged for one year after harvest, often with oak ageing

Premium Single-grape variety, maximum 10 per cent of the volume of a wine

Premium Reserve Single-grape variety, best lot, released later than premium

Special Reserve Single-grape variety or blend, PGI or PDO, aged one year longer than relevant minimum for high quality wine

---

FACTORS AFFECTING TASTE AND QUALITY

LOCATION
Situated in the centre of the Balkan Peninsula, Bulgaria's eastern border is formed by the Black Sea, while its neighbours to the north, west, and south are respectively Romania, Serbia/Macedonia, and Greece/Turkey.

CLIMATE
The climate is warm continental, although it is cooler in the north and more temperate towards the east due to the influence of the Black Sea. Annual rainfall averages between 47 and 85 centimetres (19–33 inches).

ASPECT
Vines are grown largely on the flat valley floors and coastal plains, though in places vineyards are returning to hilly locations.

SOIL
Predominantly fertile alluvium in the valleys, with chernozem (fertile humus-rich black soils) over limestone in the northern region and brown forest and alluvial soils in the southern region. The eastern region has both black and brown forest soils, while there are areas of sand in the vineyards of the southwest.

VITICULTURE & VINIFICATION
Most vineyards are concentrated in the valleys of the rivers Danube, Struma, and Maritsa.

The lengthy and difficult process of privatisation was not a positive factor for Bulgaria, leaving it for a long time with fragmented and poorly maintained vineyards. Today most wineries own or closely control their vineyards, so fruit quality is much improved, though the area in commercial production has fallen to just under 37,000 hectares (91,429 acres) from the officially declared 135,760 hectares (335,470 acres) just before Bulgaria joined the EU in 2006. The wineries are well equipped, often thanks to EU subsidy, with modern vinification as standard.

GRAPE VARIETIES
Primary varieties: Merlot, Cabernet Sauvignon, Pamid, Rkatsiteli, Red Misket, Muscat Ottonel, Dimiat, Mavrud, Syrah, Shiroka Melnishka Loza, Sauvignon Blanc, Gamza (Kadarka), Traminer
Secondary varieties: Alicante Bouschet, Bouquet, Cabernet Franc, Chardonnay, Dornfelder, Evmolpia, Gamay, Gergana, Grenache, Keratsuda, Malbec, Marselan, Melnik 55, Melnik 1300, Mourvedre, Petit Verdot, Pinot Gris, Pinot Noir, Riesling, Rubin, Ruen, Sandanski Misket, Tamianka (Muscat Blanc à Petits Grains), Varnenski Misket, Vermentino, Viognier, Vrachanski Misket

WINE REGIONS OF BULGARIA
Situated in the centre of the Balkan Peninsula, Bulgaria has an extremely varied landscape, many different soil types, and an attractive climate.

with the confidence to experiment. It may have been single-varietal wines that originally put Bulgaria on the international wine map, but today many of the most exciting wines are blends. Most producers do not use the restrictive PDO (Protected Denomination of Origin) category, so they have a lot of freedom to experiment with less common blends. The widely planted Bordeaux varieties are usually the core but supported by grapes like Syrah, Malbec, Petit Verdot, Cabernet Franc, Marselan, Nebbiolo, Regent, or something local like Mavrud or Rubin to add a sense of real place. Cabernet Franc is also looking increasingly thrilling in its own right. The white picture is less exciting; the local white varieties at best offer appealing uncomplicated drinking, though some producers are experimenting with skin contact, oak, and amphora to add complexity, especially to local Dimiat and Red Misket. In terms of sheer white quality, Chardonnay leads the way with some increasingly convincing Viogniers appearing too.

At the top end Bulgarian wine is better than ever and still pushing forward for improvement. The tricky bit is how to communicate all this to a wider world that still links Bulgaria with cheap, cheerful Cabernet.

RECENT BULGARIAN VINTAGES

2018 Promising fresh aromatic whites. Good for late reds, but poor summer weather hit early reds.

2017 Very good for reds and whites with balanced acidity.

2016 Small harvest but excellent fruit expression and phenolic ripeness, balanced acids.

2015 Rich ripe year, full-bodied wines but balanced.

2014 Challenging year, especially in north, so careful selection required. Cooler more elegant reds, long-lived Chardonnay good for sparkling too.

2013 Good to excellent.

2012 Warm low-yielding year, but good very ripe full-bodied reds.

RKATSITELI GRAPES DISPLAY THE RED STEMS THAT INSPIRED THEIR NAME
The name of this ancient pale-skinned grape literally translates to "red stem". Rkatsiteli is one of the primary varieties grown in Bulgaria.

THE APPELLATIONS OF

BULGARIA

Bulgaria's quality wines are divided into two categories: wines with Protected Geographical Indication (PGI) and wines with Protected Denomination of Origin (PDO), of which there are 52 in law but very few in commercial use (less than 1 per cent of wine produced). The division of the country into just two PGIs is widely held to have been a political decision due to lobbying from large producers before vineyard ownership by wineries was the norm. The four regions (plus one significant subregion, the Valley of the Roses) of the previous era are still widely referred to because they more closely reflect regional differences and were based on a solid foundation of detailed research in the 1950s and 1960s. The National Vine and Wine Chamber itself now refers to four viticultural regions, though this is not formalised in law as yet. In practice, producer reputation is a more helpful guide to choosing wine than any regional classification. Expect this to change in the next few years as smaller regions such as Struma Valley and South Sakar lobby for new PDOs. A number of the country's largest producers still source across multiple regions though most also have vines of their own. These wineries include Black Sea Gold, Domaine Boyar, Katarzyna Estate, Lovico Suhindol, Domain Menada, Power Brands (including Vinprom Peshtera and the competent and good value Villa Yambol), Vinex Preslav (also owns Khan Krum), Vinex Slavyantsi, and SIS Industries, owners of Karnobat/Minkov Brothers.

DANUBIAN PLAIN

The cooler northern half of the country produces slightly more white wine than red, though it is the northeastern corner where whites predominate. Vineyards in commercial production cover just under 8,000 hectares (19,770 acres). In general the northern zone is better regarded for refreshing whites with good aromatic expression, though there are also good lighter reds from Pinot Noir, and several producers are now taking a serious look at Gamza again. There is also potential for elegant reds from Bordeaux varieties, thanks to long sunshine hours, but site selection is critical to avoid greenness. Wineries range from the tiny Borovitza nestled among the rocks of the Belogradchik national park, producing handcrafted wines from individual plots, to *garagiste* Santa Sarah, revitalised former vinproms like Svishtov and Magura, and newer projects like Burgozone, Bononia, Maryan, Salla, Tohun, and Varna winery.

🍇 *Cabernet Sauvignon, Merlot, Chardonnay, Gamza, Muskat Ottonel, Pinot Noir, Red Misket, Dimiat Rkatsiteli, Sauvignon Blanc, Traminer, Vrachanski Misket, Varnenski Misket*

✓ *Borovitza • Bononia • Burgozone • Magura • Maryan • Rousse Wine House • Salla Estate • Santa Sarah • Svishtov • Tohun • Varna Winery*

THRACIAN LOWLANDS

The southern half of Bulgaria accounts for around 29,000 hectares (71,660 acres) of the vineyards that are in commercial production. Production is close to 50:50 red to white, though whites become more important close to the moderating effect of the Black Sea. The far southwestern zone – the Struma Valley – is quite distinct, by far the warmest region and with a Mediterranean influence where it opens towards Greece and the Aegean. Grapes grown here are distinctive too, dominated by the traditional ancient vine Shiroka Melnishka Loza (the broad-leafed vine of Melnik). This was famously enjoyed by Winston Churchill but is tricky to make into consumer-friendly reds – it's rather like a tougher version of Nebbiolo. It's proving more useful for rosé and some good sparkling wines (such as Logodaj's Satin), though it appears in some successful blends. This vine is also parent of several promising varieties, such as the increasingly popular early Melnik (aka Melnik 55), Melnik 1300, Ruen, and the grapey Sandanski Misket. The area around Plovdiv is particularly notable for some serious reds, as it has a long growing season. This particularly suits the ancient local red grape Mavrud, a particular specialism for Rumelia at Panagyurishte and Zagreus (whose top version is Vinica, made from partially dried grapes). Pazardjik is the home of one of Bulgaria's first real estates – Bessa Valley (excellent Bordeaux-based reds and Syrah), an investment by Count von Niepperg (of Château Canon-la-Gaffelière) and his German partner. The region of South Sakar around Harmanli is notable for its concentration of good producers such as Eolis, Bratanov, Terra Tangra, and Castra Rubra, while Katarzyna lies even farther south in the border zone. The area around Stara Zagora is also notable, home to boutique producers like Better Half, Alexandra Estate, and the larger Italian investment of Edoardo Miroglio.

🍇 *Merlot, Cabernet Sauvignon, Chardonnay, Dimiat, Gewürztraminer (Traminer), Pamid, Rkatsiteli, Red Misket, Riesling (Rheinriesling), Sauvignon Blanc, Tamianka, Varnenski Misket*

✓ *Alexandra Estate • Angel's Estate • Bessa Valley • Better Half • Black Sea Gold • Bratanov • Domaine Boyar • Dragomir • Edoardo Miroglio • Eolis • Midalidare • Rossidi • Rumelia • Vinex Slavyantsi • Terra Tangra • Villa Yustina • Via Vinera • Yamantiev's • Zagreus • Zelanos*

In Struma Valley: *Abdyika • Damianitza • Logodaj • Medi Valley • Orbelia, Orbelus • Rupe • Villa Melnik • Zlaten Rozhen*

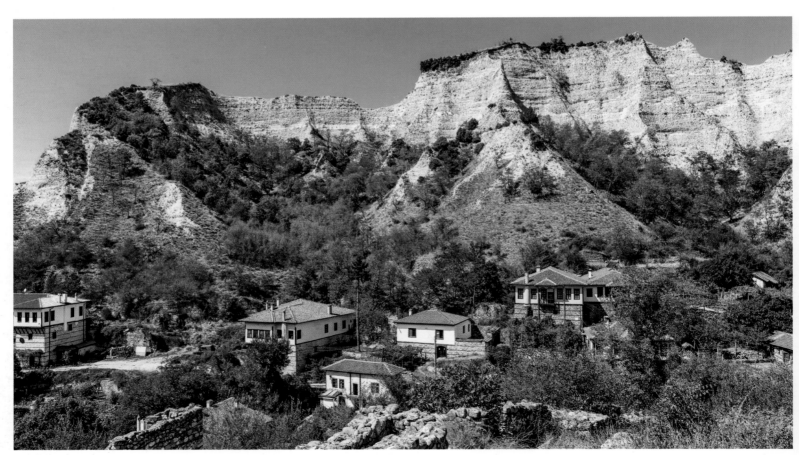

THE TINY TOWN OF MELNIK IN BULGARIA HAS A STORIED AND ROBUST WINEMAKING TRADITION
Nestled in the foothills of the Pirin mountain range of southwestern Bulgaria, Melnik is the smallest town in the nation, with original mediaeval houses, the preserved ruins of a fortress, and a nearby monastery. The town is famous for its wine cellars producing strong red wine since at least the 14th century.

Hungary

Hungary is quietly re-establishing itself as one of Central Europe's leading wine lights. It has one significant advantage over its neighbours in the global recognition of Tokaji as one of the world's great wines. Yet wonderful as the luscious golden Tokaji Aszú wines are, sweet wines of this quality and price will always be expensive and difficult to sell. The region's response to this has been to develop a new generation of high-quality dry wines. The main grape of Tokaj is Furmint and has the potential to be the next big thing in wine grapes. It's easy to pronounce and incredibly versatile, producing wines that range from bone dry and steely to complex and Burgundian to sparkling and amazing sweet wines. It's a grape that can truly reflect its terroir as well.

There's more to Hungary than just Tokaj though. It has 65,477 hectares (161,800 acres) under vine, with 22 wine regions across six PGI areas and nearly 160 grape varieties planted, two thirds of which are white. The seven-year average harvest is around 2.5 million hectolitres, putting it a little behind Romania. There's a significant presence for international grapes, not least Pinot Grigio, which is widely sold as a superior Italian look-alike at supermarket prices.

At the same time there's a huge diversity of local grapes led by Kékfrankos (aka Blaufränkisch) which almost certainly originated in the region of Lower Styria in the Middle Ages (today's Slovenia, but old Hungary then) so is arguably a Hungarian grape after all. It's by far the most planted grape in this country at 7,825 hectares (19,335 acres), though was once treated as a volume workhorse – easier to grow than Kadarka, which had previously held sway. Today the grape is treated with much more respect, and winemakers have understood that it is important to treat it gently, more like Pinot Noir without trying too hard to extract deep colour and structure. It

A CLUSTER OF FURMINT GRAPES DISPLAYS THE EFFECTS OF BOTRYTIS
To make Tokaji Aszú wines, these grapes must be harvested by hand. Keep in mind, Tokaj is the place and Tokaji the wine.

FACTORS AFFECTING TASTE AND QUALITY

LOCATION
Hungary is sandwiched between Austria and Romania, north of Croatia, in the Carpathian basin south of Slovakia.

CLIMATE
Generally a continental climate, with some Mediterranean influence in the southern wine districts. The annual mean temperature is 10.5°C (51°F), with an average rainfall of 60 centimetres (23 inches). In the Carpathian foothills of Tokaj, in northeastern Hungary, long misty autumns encourage botrytis.

ASPECT
Over a third of the country's vines are grown on the Great Plain, where the land is flat or gently undulating river plains. The rest of the country is hilly.

SOIL
Much of Hungary is underlaid by volcanic bedrock so hot springs are common. Topsoils may be slate, basalt, clay, and loess in the west; sandy soils in the Great Plain; clay, loess, and volcanic rock in the northeast; and limestone in the south.

VITICULTURE & VINIFICATION
Tokaji Aszú requires hand-harvesting of individual botrytized grapes. Otherwise, all Hungarian red and white wines are produced using standard vinification techniques and most wineries are fully equipped with modern technology. There is also a trend towards more "natural" winemaking and some producers experimenting with amphorae. Hungarian oak is widespread – there are extensive forests of *Quercus petraea* here, which produce high-quality barrels.

GRAPE VARIETIES
Cabernet Franc, Cabernet Sauvignon, Chardonnay, Cserszegi Fűszeres, Ezerjó, Furmint, Hárslevelű, Irsai Olivér, Juhfark, Kadarka, Kékfrankos (Blaufränkisch), Portugieser, Királyleányka (Fetească Regală) Leányka (Fetească Albă), Olaszrizling (Welschriesling), Pinot Blanc, Szürkebarát (Pinot Gris), Rajnai rizling (Riesling), Rizlingszilváni (Muller-Thurgau), Sauvignon Blanc, Tramini (Gewürztraminer), Zweigelt, Zöld veltelini (Gruner Veltliner)

has naturally high acidity and rarely gets too alcoholic, so it suits today's drinking trends with its refreshing elegance and ability to match rather than dominate food.

Kékfrankos is also the backbone of the new generation of Bikavér (literally meaning "bull's blood") that has been reinvented as a flagship red blend based on Carpathian varieties, though it is only permitted in the two regions of Eger and Szekszárd. Today's best versions are a world away from the rustic cheap blends of the past. Kadarka fell out of favour because it was tricky to grow and vinify, and its light body and colour was not appreciated by the local market who like their reds with body, power, and lots of oak. It is also gaining renewed focus, however, adding a spicy note in blends or in lighter, elegant reds.

Another red strong point in Hungary worth highlighting is the success of Cabernet Franc in southern Hungary, especially Villány, where it does better than Cabernet Sauvignon and is well worth looking out for. Hungary is an exciting place to explore – offering an incredible diversity of styles from the vivid elegant reds of Sopron to the mineral rich whites of Somló, aromatic Sauvignon and elegant Pinot Noir from Etyek-Buda or exotic Irsai Olivér from Balaton. Eger, Villány, and Szekszárd vie for the red crown – Eger producing cooler more Burgundian reds, while Villány is best regarded for its Bordeaux varieties, especially Cabernet Franc, and Szekszárd for its rich plush reds. And there's so much more to discover.

When democracy arrived in 1989, Hungary had a clear advantage over other former Eastern Bloc countries because even under Communist rule it had long been dabbling with a mixed economy. Exploiting this to the full, the transition from a centrally planned economy to a market-driven one has been relatively smooth, and because producers could keep small plots of their own from 1950s onwards the link between land and winemaking was less broken than farther east. Foreign investment was major factor in the early development of Tokaji in particular but has been less significant in the rest of the country. Here it is largely passionate Hungarians who have realised the potential for quality wine, fortunately with a loyal domestic market prepared to buy them. It's encouraging to see a second generation of winemakers appearing now in every region – enthusiastic, talented, and open-minded young people who travel and taste together and make better wines for it.

RECENT HUNGARIAN VINTAGES

2018 A late spring frost reduced yields in some areas, and summer weather was extreme at times. An early harvest has produced good results though.

2017 A mild year without extremes and with adequate rainfall. Very good aromatic expression in reds and whites. Balanced wines.

2016 Difficult rainy vintage in Tokaj, lots of botrytis needing careful selection. In the south producers are very optimistic reporting excellent sugar: acid balance and very good flavour expression.

2015 Very even healthy vintage. Good for dry whites and reds. Not a widespread Aszú year.

2014 The worst vintage for decades. Very wet and cold. A few carefully selected Tokaji Aszú are decent.

2013 Lighter but elegant reds, fresh fine whites. Superb quality and quantity of Tokaji Aszú.

2012 Very dry warm vintage with low yields. Ripe reds, good whites provided drought stress was avoided wines, not an Aszú year.

2011 Warm dry vintage but not stressed due to good soil moisture after 2010. Ripe reds, full-bodied dry wines, not an Aszú year.

2010 A very wet difficult year.

WINE NOTES

• Hungarian Zsigmond Teleki played a major role in developing the rootstocks that saved the European wine industry post phylloxera.

INVESTMENT IN TOKAJ

The first area that received investment was Tokaj. This is understandable, as Tokaji is the only classic wine of authentic historical reputation in all Eastern Europe. Tokaj was the first region in the world to classify vineyards in 1720s and established the first controlled appellation too. In 1737 a royal decree defined which villages were allowed to use the Tokaji wine name.

HUNGARIAN LABEL LANGUAGE

Aszú Botrytis-affected and shrivelled berries

Bor Wine

Borvidék Wine region

DHC (Districtus Hungaricus Controllatus) Certain PDO wines, notably from Villány

Dűlő Vineyard

Édes Sweet

Fehér White

Fordítás A style of sweet Tokaji produced from a second use of the aszú pomace. Less sweet and more tannic than Aszú

Hordó Cask or barrel

Jégbor Ice wine, or Eiswein

Késöi Szüretelésű Late harvest

Minőségi Quality

OEM (Oltalom Alatt Álló Eredetmegjelölések) Term for PDO wines. Traditional term, Védett Eredetű, also in use for PDO wines

OFJ (Oltalom Alatt Álló Földrajzi Jelzések) Term for PGI wines

Palack Bottle

Pezsgő Sparkling

Pince Cellar

Siller A dark rosé, almost a light red

Szamorodni Wine made from whole bunches with varying amounts of noble rot and with minimum oak ageing of six months. The édes (sweet) style has a minimum of 45 g/l residual sugar. There's also a rare but intriguing flor-aged dry (száraz) category of Szamorodni still made by Samuel Tinon and Dereszla.

Száraz Dry

Szőlő Grapes

Szőlőbirtok Wine estate

Szőlőskert Vineyard

Termelte és palackozta Produced and bottled by

Válogatás selection

Vörös Red

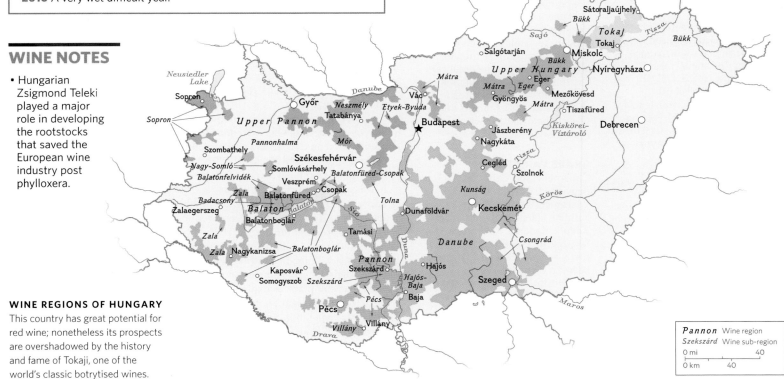

WINE REGIONS OF HUNGARY

This country has great potential for red wine; nonetheless its prospects are overshadowed by the history and fame of Tokaji, one of the world's classic botrytised wines.

| *Pannon* Wine region |
| *Szekszárd* Wine sub-region |

0 mi — 40
0 km — 40

Records of Aszú grapes now go back to 1527 after a new document was recently found. A number of foreign investors arrived in the early 1990s, led by Hugh Johnson and others at Royal Tokaji and followed by the likes of Vega Sicilia at Oremus, AXA at Disznókő, and Jean-Louis Laborde of Château Clinet (Château Pajzos & Château Megyer), and more recently Michel Reybier (Cos D'Estournel) at Hétszőlő and Anthony Hwang at Királyudvar. But much of Tokaji's revolution more recently has come from Hungarians and the large number of small producers has made this a very dynamic region. Unfortunately around 50 per cent of Tokaji wine is still produced by the state-owned Grand Tokaj. It has recently seen considerable investment with a new winemaker and viticulturalist, and the best wines are now very good, but there's still a lot of volume of non-descript stuff sold cheaply to domestic supermarkets and export customers like Poland.

The early controversy about new investors changing the style of Tokaji has now disappeared – the oxidation of the communist period is no longer found. Indeed this was almost certainly never part of Tokaji winemaking pre-communism when barrels were kept topped up and wines often exported in those barrels so new wood would have been the norm. The flagship style, Tokaji Aszú, is made from individual shrivelled and noble-rotted berries, typically a mixture of Furmint, Hárslevelű, and yellow Muscat (Zéta, Kabar, and Kövérszőlő are also permitted). These are trodden or mashed into a paste and soaked in new or fermenting must or occasionally finished wine, for 12 to 60 hours before ageing in oak for at least 18 months. Since 2013, 5 and 6 Puttonyos are the only permitted Aszú styles, based on the measured residual sugar (at least 120g/l for 5 Puttonyos), plus Eszencia, which is the thick elixir that trickles from the Aszú berries but rarely ferments to more than 3 per cent alcohol. Other sweet styles include late harvest – usually made from whole bunches with less botrytis and short, if any, oak ageing – while the more authentic Tokaji category of Szamorodni is also enjoying a renaissance. It's a Polish word meaning "as it comes" for wine made from whole bunches with varying amounts of noble rot and with minimum oak ageing of six months. There's also a rare but intriguing *flor*-aged dry (*száraz*) category of Szamorodni still made by Samuel Tinon and Dereszla. Perhaps the most exciting development in the last couple of decades has been the development of serious dry wines – due to the combination of Tokaj's volcanic soils and superb grapes in the shape of Furmint, and the less well known Hárslevelű. The legend in his own lifetime István Szepsy pioneered this category in 2000 but has been joined by a multitude of exciting producers (and remarkably, in this sometimes-macho culture, many of the best are women).

THE REGIONS AND PROTECTED DENOMINATIONS OF
HUNGARY

Hungary has five Protected Geographical indications, or OFJ, which are Balatonmelléki, Duna-Tisza-közi, Dunántúli, Felső-Magyarországi, and Zempléni, and 32 OEM, most of which coincide with the name for a wine region, though some apply to a specified wine such as Debrői Hárslevelű. The country is divided into six wine zones in which the districts share characteristics of climate and soils

UPPER (OR NORTHERN) PANNONIA

Dunántúl is the OFJ for this region. It comprises five smaller wine districts with a continental climate, brown forest, and alluvial soils with areas of clay, slate, and loess.

ETYEK-BYUDA OEM

Lies to the west of Budapest itself and is the closest wine district. Rolling limestone hills overlaid with black soils and a cool climate mean this area is particular good for crisp whites, especially aromatic Sauvignon and Irsai Olivér, fine Chardonnay, and elegant Pinot Noir. It's also a good source of base wine for leading sparkling wine brands from giant Törley.

🍇 Chardonnay, Irsai Olivér, Királyleányka, Pinot Blanc, Pinot Gris, Pinot Noir, Sauvignon Blanc, Zenit
✓ Etyeki-Kuria • Haraszthy • Kertész • Nyakas • Rókusfalvy • Törley

NESZMÉLY OEM

Winemaking traditions date back to the Middle Ages. Vineyards are on hills overlooking the Danube and most soils are loess-based. The climate is milder than the surrounding Great Plains area, with less risk of spring and autumn frost. Best for fresh, fragrant, lively white wines. The modern Hilltop winery is by far the most significant producer here.

🍇 Chardonnay, Cserszegi Fűszeres, Királyleányka, Irsai Olivér, Olaszrizling, Sauvignon Blanc, Szürkebarát (Pinot Gris)
✓ Hilltop

PANNONHALMA OEM

An eight-hundred-year wine history was revived in the early 2000s by the dramatic hilltop Benedictine abbey of Pannonhalma, which built a gravity-fed winery and replanted 52 hectares (128 acres) with mainly Alsace varieties plus Olaszrizling.
✓ Pannonhalma Abbey Winery

MÓR OEM

A small, cool region of limestone-rich vineyards, home to the fiery Ezerjó grape and crisp Chardonnay

🍇 Ezerjó, Olaszrizling, Chardonnay
✓ Csetvei

SOPRON OEM

One of the most ancient wine regions of Hungary, the Sopron lies on the Austrian border in the northwest of the country where Lake Fertő moderates the climate, enabling red grapes to ripen well. Until the 18th century Sopron was reputedly the largest wine-trading centre in central Europe. Sopron is establishing itself as a fine wine region, where the Kékfrankos grape is most famous, particularly thanks to the efforts of the biodynamic Franz Weninger Jr, whose best vineyard, Spern Steiner, is on degraded slate. Although essentially a red wine region, Sopron does also produce some interesting whites particularly from Zöld Veltelini.

🍇 Cabernet Franc, Kékfrankos, Leányka, Sauvignon Blanc, Syrah, Zöld Veltelini, Zweigelt
✓ Franz Weninger • Ráspi • Winelife • Luka • Pfneiszl

BALATON

The Balaton wine region comprises six districts close to the largest freshwater lake in central Europe which modifies the climate and makes it less extremely continental. Soils are very varied; from basalt, sandstone, and limestone bedrocks with soils of loess, marl, sand, and clay. The OFJs here are Dunántúl and Balatonmellék, plus there are several smaller OEMs including Káli, Tihany, and Zala that have minor commercial significance.

BADACSONY OEM

With its south-facing volcanic slopes on the northerly shores of Balaton, Badacsony is noted for its own indigenous grape, called Kéknyelű. This rare, but being replanted, white variety has excellent potential for full-bodied and long-lived white wines. Riesling also does notably well here and long-lived, intense Szürkebarát.

🍇 Kékfrankos, Kéknyelű, Olaszrizling, Riesling, Szürkebarát
✓ Laposa • Szeremley • Sandahl • Szászi • Válibor • Villa Tolnay

BALATONBOGLÁR OEM

This district to the south of the lake shares its name with Törley's biggest winery. In the past, it was dismissed as only suitable for inexpensive commercial quality wines, and largely white. Today the region still produces sizeable volumes of wines for big brands like Chapel Hill and I Heart but is also recognised for high-quality reds thanks to Konyári and Ikon.

🍇 Cabernet Franc, Chardonnay, Cserszegi Fűszeres, Kékfrankos, Királyleányka, Merlot, Olaszrizling, Szürkebarát (Pinot Gris), Sauvignon Blanc
✓ Bujdosó • Garamvári • Ikon • Konyári • Légli

BALATONFELVIDÉK OEM

The Balaton uplands to the north of the lake are a popular tourist destination. Cooler air rolls in from mountains to the north, while summer heat is tempered by the lake. The best wines produced here are full-bodied, with a fine fragrance and crisp acidity.

🍇 Chardonnay, Olaszrizling, Szürkebarát

BALATONFÜRED-CSOPAK OEM

On the northern shore of Lake Balaton, with very varied geology including red soils around Csopak itself and basalt on the Tihany peninsula. Producers here have united to promote Olaszrizling under a specific "Kodex" that guarantees a set standard, and there are a number of single-vineyard selections available. The wines are full-bodied yet fragrant, with fine acidity.

🍇 Cabernet Franc, Cabernet Sauvignon, Chardonnay, Kékfrankos, Merlot, Olaszrizling, Rizlingszilváni, Tramini
✓ Figula • Homola • Jásdi • Szent Donát

NAGY-SOMLÓ OEM

Often just written Somló, this is the second-smallest wine region of Hungary. Its wines are fiery, with a firm mineral character, often with high acid and alcohol content. The vineyards are on the lower slopes of an extinct volcano that is said to give the wines special power to guarantee an heir. Traditionally the wines were long aged in barrel, and wines like these can still be found, though the most consistent wines now come from bigger producers such as Kreinbacher (producer of Hungary's most exciting sparkling, based on Furmint) and Tornai. This is predominantly a white region and specially noted for the firmly structured Juhfark ("sheep's tail").

🍇 Furmint, Hárslevelű, Juhfark, Olaszrizling
✓ Kreinbacher • Kolonics • Tornai • Zsirai

PANNON REGION

This covers the landscape to the south and southwest of the region, across four wine districts. The continental climate is moderated by Mediterranean influences from the south. Tolna and Szekszárd lie on clayey sedimentary bedrock, while Villány has limestone beneath its roots. The OFJ for this region is Dunántúl.

PÉCS OEM

Vineyards here scattered over more than 80 kilometres (50 miles) of warm slopes around the historic city of Pécs, which has long been noted for Olaszrizling, Cirfandli and Pinot Noir.

🍇 *Cabernet Franc, Cabernet Sauvignon, Chardonnay, Cirfandli, Merlot, Olaszrizling, Pinot Noir, Riesling, Rizlingszilváni, Sauvignon Blanc*

✓ *Ebner*

SZEKSZÁRD OEM

The Szekszárd vineyards have been noted for their wines since Roman times and for red wines since the 15th century. In this sunny wine region, Hungary's warmest, the vineyards grow on low hills up to around 200 metres (656 feet). It's most noted for its ripe velvety red wines that are often slightly overshadowed in reputation, if not drinkability, by its southern neighbour Villány. This is one of two regions allowed to make Bikavér, and the rules in Szekszárd insist on a minimum 45 per cent Kékfrankos and at least 5 per cent Kadarka, with at least two other grapes, plus a year in oak. The indigenous Kadarka variety is also produced by itself, and at its best gives elegant spicy light reds.

🍇 *Cabernet Sauvignon, Cserszegi Fűszeres, Kadarka, Kékfrankos, Merlot, Syrah, Viognier*

✓ *Dúzsi Tamás • Eszterbauer • Heimann • Márkvárt • Pósta • Sebestyén, • Szent Gaál • Takler Pince • Vesztergombi • Vida*

TOLNA OEM

Although Tolna became a wine region in its own right in 1998, winemaking traditions go back centuries, and were strengthened in the 16th century, when the area was settled by Germans. Today the most significant producer is the Antinori-owned Tűzkő

🍇 *Chardonnay, Merlot, Kékfrankos, Tramini, Zöld Veltelini*
✓ *Tűzkő*

VILLÁNY DHC

In Villány, the more Mediterranean climate produces some of Hungary's best red wines, including impressive Bordeaux varietals and especially Cabernet Franc, which is recognised through the Villány Franc classification at different quality levels. The red wine potential of Villány is probably the greatest of all of the 22 Hungarian wine regions. It also includes the slightly cooler vineyards of Siklós, once known for whites but now for more elegant reds.

🍇 *Cabernet Franc, Cabernet Sauvignon, Hárslevelű, Kékfrankos, Portugieser, Olaszrizling, Pinot Noir, Syrah*

✓ *Bock • Gere Attila • Gere Tamás & Zsolt Pincészet • Jackfall Heumann • Lelovitz • Malatinszky • Polgár • Sauska • Riczu • Ruppert • Stier • Vylyan Pincészet*

DUNA

The great plain of Hungary covers nearly 24,000 hectares (59,300 acres) but is generally the source of much low-quality wine. The climate is very continental, and winter cold can be a problem. Soils are sandy with occasional deposits of alluvial soils and chernozems. There are OEMs for Duna, Csongrád, Monor, and Izsáki Arány Sárfehér (for this rather non-descript grape grown in a single commune). Duna-Tisza-közi is the OFJ applicable here.

HAJÓS-BAJA OEM

An area of loess and sand producing medium-bodied reds and whites at good value prices.

🍇 *Cabernet Franc, Cabernet Sauvignon, Kadarka, Kékfrankos, Merlot, Olaszrizling, Riesling*
✓ *Koch*

TOKAJI WINE IN BUDAPEST'S CENTRAL MARKET
At right is a 2009 vintage Oremus Tokaji Aszú 5 Puttonyos. The unit *puttonyos* denotes sugar level and, hence, the sweetness of a Hungarian dessert wine.

KUNSÁG OEM

This Great Plain wine region is the largest wine region in Hungary. The wines of this region can be rich in fragrance and flavours with decent acidity, though are by no means Hungary's most complex.

🍇 *Ezerjó, Cserszegi Fűszeres, Generosa, Irsai Olivér, Kékfrankos, Kövidinka, Olaszrizling*
✓ *Frittmann • Font • Gál*

BÜKK OEM

Grown on the southern and southwestern slopes of the Bükk Mountains, where Kékfrankos and Leányka are the major varieties, but produce wines that are lighter than those of neighbouring Eger. As in Eger and Tokaj farther east, cellars have long been hewn out of rhyolite-tuff rock.

🍇 *Cserszegi Fűszeres, Kékfrankos, Olaszrizling, Leányka*

UPPER HUNGARY (FELSŐ-MAGYARORSZÁG)

The vineyards of Upper Hungary lie on the foothills and slopes of the Mátra and Bükk mountains, so the peaks provide shelter from winds from the north. Felső-Magyarország is the OFJ for this region. Soils are typically loess and forest soils over rhyolite tuff with occasional outcrops of limestone such as Eger's famous Nagy-Eged hill.

DEBRŐI HÁRSLEVELŰ OEM

A single-commune, pure-varietal appellation for the famed Hárslevelű growing in Debrői and surrounding villages in the Eger region.

EGER OEM

Located halfway between Budapest and Tokaj is Eger, a region famous for the legend of Egri Bikavér, or "Bull's Blood of Eger". The story dates from 1552, when the fortress of Eger, fiercely defended by István Dobó and his Magyars, was besieged by the numerically superior force of the Turkish army, led by Ali Pasha. It is said that, throughout the battle, the Magyars drank copious quantities of the local wine and that when the Turks saw the beards of their ferocious enemies stained red with wine, they ran in terror, thinking that all Magyars gained their strength by drinking the blood of bulls. Hence the name of this wine was born. Unfortunately the fact that there was no red wine in the region at the time rather undermines this colourful story. Egri Bikavér must be a blend of at least three grape varieties, based on Kékfrankos. It's a complex region on a similar latitude to the bottom of Burgundy and the top of the Rhône so Chardonnay, Pinot Noir and Syrah can all do well here as well a raft of local grapes.

🍇 *Blauburger, Cabernet Franc, Cabernet Sauvignon, Furmint, Kadarka, Kékfrankos, Hárslevelű, Leányka, Merlot, Olaszrizling, Pinot Noir, Syrah*

✓ *Bolyki • Gál Tibor • Kovács Nimród • Ostoros • Pók • St Andrea • Thummerer*

MÁTRA OEM

At Gyöngyös the first truly excellent Hungarian dry white wine was produced in the early 1980s, and no prizes for guessing that it was made from Chardonnay. But it was essential to demonstrate what could be done with grapes of a certain quality and the right technology. Today the region is still a reliable producer of commercial whites and rosé, for instance under the Nagyréde label, but there are also more ambitious producers aiming for higher quality.

🍇 *Chardonnay, Hárslevelű, Kékfrankos, Müller-Thurgau, Muscat Ottonel, Muskotály, Olaszrizling, Tramini, Rizlingszilváni, Sauvignon Blanc, Zweigelt*

✓ *Benedek • NAG • Gábor Karner*

TOKAJ

TOKAJ-HEGYALJA OEM

As with all great sweet wines, Tokaji owes its quality and character to over-ripe, semi-dried grapes (Furmint and Hárslevelű predominantly), which have been affected by *Botrytis cinerea*, or noble rot. The foggy autumn mornings encourage noble rot, while sunny and breezy afternoons on this hilly volcanic landscape result in very shrivelled berries (Aszú) that cannot simply be pressed for juice. A complex winemaking method evolved in which crushed Aszú berries measured in a kind of wooden hod called a *puttony* were added to young wine to soak and ferment. The sweetness of the Tokaji depended on how many *puttonyos* were added to the dry base wine. Today the process has been modernised with no oxidation and much shorter oak ageing – allowing the quality of the fruit flavours and botrytis to show. And the *puttonyos* statement is now optional though if used, the residual sugar level must be 120g/l minimum (5 *puttonyos*). Fully sweet wines account for around 21 per cent of the region's production, the rest being serious dry wines, or lower quality semi-dry wines for supermarkets and certain export customers. The rainy cold 2010 vintage proved to be a turning point for sparkling wines – Furmint's high natural acidity makes it a good base wine for bottle fermentation. The PGI in this region is Zemplén and accounts for less than 15 per cent of the region's production, including small volumes of international grapes such as Sauvignon Blanc, Chardonnay, and Pinot Noir.

🍇 *Furmint, Hárslevelű, Kabar, Kövérszőlő, Sárga Muskotály (Yellow Muscat), Zéta*

✓ *Árvay • Barta • Balassa • Bardon • Basilicus • Béres • Bodrog Borműhely • Bott • Carpinus • Demeter Zoltán • Demetervin • Dereszla • Disznókő • Dobogó • Erzsebet • Füleky • Gizella • Grand Tokaj • Hétszőlő • Holdvölgy • Kikelet Királyudvar • Nobilis Oremus • Pajzos • Patricius • Royal Tokaji • Sauska • Samuel Tinon • Szent Tamás Winery • Szepsy István*

Romania

Romania has at least as much potential as any other Eastern European winemaking country. It has been growing French varieties since its recovery from phylloxera and has an increasing international reputation for Pinot Noir. Unlike its neighbours in the region, it has also retained considerable areas of local grape varieties, which are proving capable of exciting quality. EU subsidies and foreign investment have had a considerable impact on the wine industry here and have encouraged the rise of numerous small premium estates. Overall it's a mixed picture, with a strong and often unfussy domestic market dominated by a small number of large wineries but a very dynamic and competitive scene among the smaller producers. Fortunately there's a rapidly improving food and wine culture and growing interest amongst young drinkers in better quality wine.

The biggest problem and opportunity in Romania has been its local market. The domestic market still has a preference for white, semi-dry and semi-sweet wines and grey market, home-made wine from hybrids sold by the roadside may still account for as much as 50 per cent of wine consumption. The appearance of so many small, quality-driven estates in the last few years, however, is helping to change perceptions and drive quality improvements at the large former state dinosaurs too. Romania has been net importer of wine since 2006, either cheap stuff from Spain that undercuts anything local producers can match or premium wines from Italy, France and certain new-world styles that Romania can't produce. Foreign and non-industry investment is closely associated with the most dynamic wine producers – often Italian due to language similarities, but also Austrian, Hungarian, British, and German. Modern equipment and vineyard ownership is now widespread, and with Romania's huge diversity of vineyard landscapes, the scene is set for an exciting future – the question is whether the market is ready yet.

MORE IMPORTANT THAN HUNGARY

Viticulture and winemaking is an old tradition in Romania, dating back at least 4,000 years and possibly a couple of millennia more. In Europe, it is the fifth-biggest country by area under vine in Europe with 182,000 hectares (449,730 acres) planted, though worth noting that these data include at least 83,000 hectares (205,100 acres) of interspecific hybrids, while one industry source believes that only 68,000 hectares (168,030 acres) are realistically in commercial production. Romania is just ahead of Hungary in winemaking terms, producing just under 2.8 million hectolitres from noble grapes in 2017 (compared to Hungary's seven-year average of around 2.5 million hectolitres). As in Bulgaria, there was a massive planting programme in the 1960s, switching to grapes capable of high yield and consistent cropping, and allowing mechanisation with large Soviet tractors. Some exports of sweet reds were destined for the Soviet Union, but mostly white wines were produced to supply the large domestic market. The arrival of dictator Ceaușescu had an impact on exports, as he sought to raise money for his monumental building schemes, though wine exports only accounted for 7 per cent of production in the period 1985 to 1989. As in Bulgaria, PepsiCo was involved in early exports in the 1970s and then

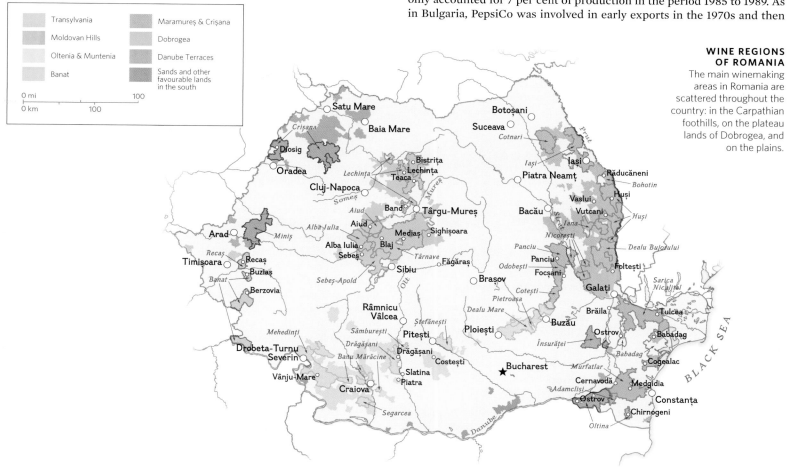

WINE REGIONS OF ROMANIA
The main winemaking areas in Romania are scattered throughout the country: in the Carpathian foothills, on the plateau lands of Dobrogea, and on the plains.

Map legend:
- Transylvania
- Moldovan Hills
- Oltenia & Muntenia
- Banat
- Maramureș & Crișana
- Dobrogea
- Danube Terraces
- Sands and other favourable lands in the south

the mid-1980s saw the first significant exports of Pinot Noir to the UK by John Halewood, who was seeking an alternative source of wine once Bulgaria set up its own UK subsidiary. Romania never quite reached the export heights of Bulgaria, but at its peak (1996) sold a million cases per year in the UK. By the fall of communism in 1989, vineyard area had increased to around 275,000 hectares (679,540 acres) but the process of privatization was slow and difficult and is still subject to legal disputes today. Many vineyards fell into disrepair during this period, and fragmented tiny land-holdings remain an issue – there are over 800,000 growers with fewer than 0.5 hectares (1.24 acres) – many of these being hybrids that are easier to grow on small plots. EU accession undoubtedly drove considerable changes, allowing consolidation of plots, replanting with better varieties, buying equipment, and so on. Indeed the wine industry received 210 million euros between 2007 and 2013 and has been supported by a further 47.7 million euros per year up to 2018. This has included more than 25,575 hectares (63,200 acres) of new plantings or conversions from poor varieties to better ones. This means that today most wineries own vineyards and have modern equipment.

Romanian wine law is fully compliant with EU regulations. There are eight wine regions, 12 PGIs (*Vin cu Indicație Geografică*), and 33 PDOs (38 are listed by EU because some regions have multiple registrations for different wine styles). Romania's term for PDO is DOC (*Denumire de Origine Controlată*), meaning high-quality wines from a delimited area, produced within that area, and subject to certain other controls such as yield and grape variety. The basic DOC category is supplemented by further classifications according to maturity of grapes: DOC-CMD (*Cules la maturitate deplină*) for wines harvested at full maturity; DOC-CT (*Cules târziu*) for late harvest wines; DOC-CIB (*Cules la înnobilarea boabelor*) for noble late harvest wines. DOC, PGI, and varietal wines accounted for just over 1.2 million hectolitres in 2017, with the significant remainder being table wines.

THE FUTURE

Accession to the EU in 2007 and ongoing availability of EU subsidies continue to make a dramatic difference to Romanian wine; however, some of this investment has brought problems of their own. Some of the new vineyards were either badly planted simply to claim the subsidy with no intention of producing grapes or did not build a winery to process fruit, so there is an excess on the market. The biggest seven wineries held 65 per cent market share in 2016, dominating the Romanian market (though the largest, Murfatlar, collapsed in 2017 due to alleged financial fraud) and leaving more than 200 wineries fighting for a share of the rest. The problem of black market wines is still huge – people drinking these wines are not buying commercial wine, nor will they have taste for it. However, dealing with

---

FACTORS AFFECTING TASTE AND QUALITY

LOCATION
The main wine-producing areas in Romania spread from Iaşi in the northeast of the country down through the Carpathian foothills to the Danube River.

CLIMATE
A continental climate, with hot summers and cold winters, which are tempered in the southeast by the Black Sea and by altitude in the more mountainous regions.

ASPECT
Romanian vines are grown on all types of land – from the plains through the plateau land of Dobrogea to the slopes of the Carpathian foothills.

SOIL
A wide variety of soil types, including the generally sandy-alluvial plains and the stony, hillside soils of Banat; the limestone of Dobrogea; the oolitic limestone of Muntenia's Pietroasele vineyards; and the stony fringes of the Carpathian Mountains.

VITICULTURE & VINIFICATION
Thanks to EU subsidies and outside investment, most wineries now own vineyards and have modern equipment. The best wines are now world class, but many poor-quality table wines and grey-market hybrid wines are still an issue.

GRAPE VARIETIES
Fetească Regală, Fetească Albă, Merlot, Welschriesling, Sauvignon Blanc, Aligoté, Cabernet Sauvignon, Muscat Ottonel, Fetească Neagră, Rosioara, Băbească Neagră, Pinot Noir, Chardonnay, Tămâioasă Romanească, Pinot Gris, Blaufränkisch, Grasă de Cotnari, Traminer, Syrah, Crâmpoșie Selecționată, Busuioacă de Bohotin, Galbenă de Odobești, Frâncușă

---

hybrids and large numbers of rural voters is too big a political problem. Exports are developing strongly, but largely thanks to a small number of producers, led by Cramele Recaş. Bristol-born Philip Cox, with his Romanian wife and partners, bought this former state winery. Winemaking know-how comes from Australian Hartley Smithers and his Spanish wife, Nora Iriate, delivering large volumes of good-value, reliable, commercial wines that over-deliver at their price points. They also have a successful range of premium wines. The collapse of Murfatlar has helped Recaş and other wineries pick up domestic share too and accelerate the change of taste towards modern, dry wines. This strong domestic market is a double-edged sword – consumers still enjoy semi-dry and semi-sweet whites with sugar that often covers up sloppy winemaking standards and at prices that are not competitive for export. The preference for reds is dark, alcoholic, and oaky – Pinot Noir is notably the most difficult international variety to sell in Romania. At the same time, there is a burgeoning culture of good wine, wine bars, and specialist shops – in Bucharest and richer cities like Timisoara and Cluj-Napoca. And this market is used to paying quite high prices, which is helping to support large numbers of the newer winery projects, but then also makes discounting for export unattractive. Another mixed blessing for Romanian wine is those large plantings of local grapes: especially Fetească Albă (12,383 hectares, or 30,600 acres), its Transylvanian offspring Fetească Regală (12,661 hectares, or 31,290 acres), and the unrelated "black maiden grape" Fetească Neagră (2,950 hectares, or 7,290 acres). All three have the potential to make high-quality wines, alone or in blends, and can thus act as local flagships. There are now plantings of Fetească Neagră in just about every region today as producers explore its potential and with selected clones available commercially and better understanding of how to manage its vigour. It also works well in blends – for instance in iconic wines like SERVE's Cuvee Charlotte and Davino's Domaine Ceptura and Flamboyant. Yet selling grapes no one has ever heard of from a country with little global wine reputation to export markets can be tough, and local drinkers still prefer the glamour of international varieties – Merlot and Sauvignon Blanc are top sellers. Romania has a multitude of interesting smaller local varieties that also have promise – Crâmpoșie Selecționată, Negru de Drăgăşani, Novac, Grasă de Cotnari, Busuioacă de Bohotin, Frâncușă, Galbenă, Grasă de Cotnari, Mustoasă de

---

ROMANIAN LABEL LANGUAGE

Crama/Cramele Cellar/cellars

DOC (*Denumire de Origine Controlată*) PDO wines

DOC-CIB (*Cules la înnobilarea boabelor*) For noble late-harvest wines

DOC-CMD (*Cules la maturitate deplină*) For wines harvested at full maturity

DOC-CT (*Cules târziu*) For late-harvest wines

Domeniile Domaine

Dulce Sweet

Recolta Vintage

Rezervă For wine matured for at least six months in oak and six months in bottle

Sec Dry

Spumante Sparkling

Strugure Grape

Vie Vine

Viile Vineyard

Vin alb White wine

Vin cu Indicație Geografică PGI wines

Vin de Vinotecă Wine matured for at least one year in oak and four in bottle

Vin roșu Red wine

Vin roze Rosé wine

Vin Table wine

Vin tânăr (young wine) Wine sold before the end of the year of production

Mǎderat, and Şarbǎ, while the oft claimed to be local Tǎmâioasǎ Românescǎ is actually Muscat à Petit Grains. There are small-scale efforts to research pre-phylloxera varieties too. It's worth a brief look at Cotnari too, as its sweet wines once rivalled Tokaji across Europe. Both regions went through a communist regime that destroyed quality, but Tokaji has emerged to be flagship for Hungary thanks to foreign investment and the development of many small wineries. Cotnari took a different route with the state factory privatised as a whole and no small wineries to provide competition. SC Cotnari is incredibly commercially successful with its predominantly semi-dry/semi-sweet wines so has no reason to change its model, and while the children of the owners have started Casa de Vinuri Cotnari to focus more on dry wines, this is not enough to make Cotnari an international marketing flagship as Tokaji has become. The biggest challenge of all facing Romania on the global stage is communication – it needs to offer more than simply competing on price. It's coming from behind because the country has not always had the most positive reputation, but the best of its wines, combined with its stunning wild landscapes, could put Romania back on the world wine map.

RECENT ROMANIAN VINTAGES

2018 Late reds are intense and aromatic; early reds and whites needed careful selection due to rainy summer. Not as good as 2017.

2017 Spring frost reduced the crop in some areas, but then even ripening with good water availability resulted in healthy, ripe, and balanced wines with good aromatic expression and fresh acidity in whites.

2016 A mild summer and long autumn gave very good results in both reds and whites. Excellent aromas, phenolic maturity and good acidity.

2015 Another very good, warm year – with good quality and quantity.

2014 Rainy and challenging but concentrated year – some very good, firm reds with careful selection.

2013 Very good year with ripe but balanced wines and fine acid balance.

2012 Cold and wet winter and spring reduced yields, and then a hot dry summer gave an early harvest of very concentrated wines.

THE APPELLATIONS OF
ROMANIA

BANAT

Although Banat is one of the better-known wine regions in Romania, it is the smallest by a large margin, with 2,845 hectares (7,030 acres) under vine. It lies towards the west of the country and has a mild Mediterranean climate with Adriatic influences. There are two DOC regions: Banat (subregions: Moldova Nouǎ, Dealurile Tirolului, and Silagiu) and Recaş (the latter is more commercially notable). Vineyards have been planted in the Recaş region since Roman times, and the arrival of Saxons from the middle and lower Rhine in the early Middle Ages may have shaped wine growing here too.

🍇 *Blaufränkisch (known as Burgund Mare), Cabernet Sauvignon, Cadarcǎ, Chardonnay, Feteascǎ Neagrǎ Feteascǎ Regǎlǎ, Merlot, Pinot Gris, Pinot Noir, Riesling, Sauvignon Blanc, Syrah*
✓ *Cramele Recaş • Petro Vaselo • Thesaurus*

CRIŞANA AND MARAMUREŞ

The most northerly region in Romania, lying at 48°N, though a little warmer than Transylvania. Vineyards may be as high as 500 metres (1,640 feet). There are two DOCs within this region: Miniş in Arad county, influenced by Lake Arad, and Crişana (with sub-denominations of Diosig, Biharia, and Şimleu Silvaniei). Total vineyard area was around 9,693 hectares (23,950 acres) in 2017. Long, warm autumns means reds ripen well, and the area was historically famous for a sweet red Aszú wine made from Cadarcǎ. There's a Hungarian influence to grapes grown here, including Cadarcǎ and Furmint, as well as good Cabernet Franc, Feteascǎ Neagrǎ, and local specialities like the fresh crisp and light Mustoasa de Mǎderat in still and sparkling versions.

🍇 *Blaufränkisch (known as Burgund Mare), Cabernet Sauvignon, Cadarcǎ, Cabernet Franc, Feteascǎ Regǎlǎ, Feteascǎ Neagrǎ, Merlot, Mustoasǎ de Mǎderat, Pinot Noir, Furmint*
✓ *Balla Géza • Carastelec • Nachbil*

DOBROGEA

This region lies in the southwest, between the Danube and the Black Sea. Altitude is low, averaging just 71 metres (230 feet), and it can have as many as 300 days of sunshine each year, which can cause problems with drought. Winters are mild due to the close proximity to the Black Sea, but the climate is distinctly warm. There were 16,948 hectares (41,880 acres) of vines across four DOCS: Murfatlar, Babadag, Adamclisi, and Sarica-Niculiţel (which includes the subregion Tulcea). Murfatlar, with its two subregions of Medgidia and Cernavodǎ, is the most important wine region here. In the past, the zone was best known for sweet whites, especially from Chardonnay, but today it is better known for reds. The dry climate is allowing some producers to farm organically.

🍇 *Cabernet Sauvignon, Chardonnay, Muscat Ottonel, Pinot Gris, Pinot Noir, Riesling, Feteasca Neagrǎ, Feteasca Regalǎ, Aligoté, Bǎbeascǎ Neagrǎ*
✓ *Domeniul Bogdan • Via Viticola Sarica Niculitel • Clos des Colombes*

MOLDOVAN HILLS

This is the eastern part of the old Romanian principality of Moldavia and is much the biggest wine region in Romania in terms of grapevine area, with 69,154 hectares (170,880 acres) in 2017. It stretches for several hundred kilometres southwards from the hills of Cotnari at 47°N, so climate and growing conditions vary considerably. Vineyards are typically sited on the slopes of south- and southwest-facing amphitheatres that protect the vines from the harsh north winds, and altitudes can vary from 200 to almost 500 metres (655 to 1,640 feet). The DOC regions include Bohotin, Coteşti, Cotnari, Dealu Bujorului, Huşi, Iaşi, Iana, Nicoreşti, Odobeşti, and Panciu. The most famous of these is Cotnari, once renowned for its rich sweet wines to rival Tokaji, but today largely a region of semi-dry to semi-sweet wines, dominated by the privatized former state cellar.

🍇 *Cabernet Sauvignon, Feteascǎ Albǎ, Feteascǎ Neagrǎ, Frâncuşǎ, Grasǎ de Cotnari, Merlot, Pinot Noir, Tǎmâioasǎ Românescǎ, Plǎvaie, Şarbǎ, Chardonnay, Aligoté, Galbenǎ de Odobeşti, Zghiharǎ*
✓ *Casa de Vinuri Cotnari • Gramma Winery • Crama Gîrboiu • Crama Avereşti • Domeniile Panciu • SC Cotnari, Crama Hermeziu*

OLTENIA AND MUNTENIA

This large area in the Carpathian foothills, north of the capital Bucharest, covered 53,601 hectares (132,450 acres) in 2017. DOC areas include Dealu Mare, Drǎgǎşani, Ştefǎneşti, Sâmbureşti, Banu Maracine, Mehedinţi, Cernǎteşti-Podgoria, Pietroasa, and Segarcea. Most famous is Dealu Mare, which lies on the 45th parallel, along with Bordeaux and Tuscany. It has a temperate continental climate, with hot summers and mild, dry autumns, and is noted for some of the country's best reds. Another notable DOC is Pietroasa, with its outcrop of calcareous soil at 300 metres (985 feet), making the district particularly famous for its lusciously sweet late-harvest wines, especially Tǎmâioasǎ and Busuioacǎ, and in some years noble rot occurs. Heading southwest lies the hilly area of Drǎgǎşani, where a group of dynamic small wineries focus on local grapes such as Crâmpoşie Selecţionatǎ, Negru de Drǎgǎşani, and Novac. Even farther southwest, the climate becomes temperate continental with DOCs of Segarcea and Mehedinţi.

🍇 *Cabernet Sauvignon, Feteascǎ Neagrǎ, Feteascǎ Regǎlǎ, Crâmpoşie Selecţionatǎ, Merlot, Negru de Drǎgǎşani, Novac, Pinot Gris, Pinot Noir, Riesling, Sauvignon Blanc, Syrah, Tǎmâioasǎ Românescǎ, Tǎmâioasǎ Roze*
✓ *Apogeum, Avincis • Crama Bauer • Budureasca, Basilescu • Corcova Roy & Damboviceanu • Domeniul Catleya • Dagon Clan • Davino, Domeniile Franco-Române • Halewood • Lacerta • Licorna Winehouse • SERVE (Cuvée Charlotte, Terra Romana) • Vinarte (Zoresti, Bolovanu, Starmina) • Prince Ştirbey • Aurelia Vişinescu • Viile Metamorfosis • Domeniul Coroanei Segarcea • Crama Oprişor*

TRANSYLVANIAN PLATEAU

This high central region has some of Romania's coolest vineyards and produces mainly white wines, which can have fresh acidity and good aromatic expression when well made. Altitudes range up to 600 metres (1,970 feet) while lower lying vines are often earthed-up to protect them from the winter cold, which can fall as low as -30°C (-22°F). DOCs within this region are: Târnave (with its subregions of Jidvei, Blaj, and Mediaş), Alba Iulia, Aiud, Sebeş-Apold, and Lechinţa (recently revived by an Austrian project called Liliac). Evidence of the medieval immigration of Saxon settlers from the Mosel valley is still seen in the architecture and in the wine styles. One of the most important DOCs in production is Jidvei, which is mentioned in documents dating from as early as 1309. Jidvei is also the name of the privatized former state winery (also Romania's largest single vineyard at over 2,460 hectares, or 6,080 acres).

🍇 *Feteascǎ Albǎ, Feteascǎ Regǎlǎ, Muscat Ottonel, Pinot Gris, Riesling, Sauvignon Blanc, Traminer, Pinot Noir*
✓ *Liliac • Lechburg • Villa Vinèa • Jidvei (Owner's Choice) • La Salina*

DANUBE TERRACES

This region stretches along the lower banks of the Danube, opposite the Bulgarian border. There were 11,210 hectares (27,700 acres) of vines in 2017. Table grapes are more typical here, and the only DOCs are Oltina and Însurǎţei. Vineyards are planted facing northwards, on terraces close to the Danube.

🍇 *Cabernet Sauvignon, Feteascǎ Neagrǎ, Merlot, Sauvignon Blanc*
✓ *Alira*

SANDS AND OTHER FAVOURABLE LANDS IN THE SOUTH

This is an area of deep, sandy soils and a warm climate similar to the Danube Terraces. There were 12,943 hectares (31,980 acres) in 2017, and there are no DOCs or significant commercial producers here.

Slovakia and Czechia

In 1993 in what is known as the "Velvet Divorce", Czechoslovakia split into two independent states, Slovakia and the Czech Republic (now called Czechia). The two countries share not just a political history, but a viticultural one as well. Vines have been planted in this region since as early as the Roman times in the third century AD, and possibly much earlier.

At times, the wines of this region (once Upper Hungary) enjoyed fame in royal courts but as with the rest of the former Eastern Bloc wine-producing countries, the period of centralised processing under communist rule focused on quantity and destroyed any reputation for quality. The last three decades have seen producers in both countries reinvent themselves, but thirsty home markets and wine industries that can only supply part of this demand means there are few exports.

SLOVAKIA

Slovakia as a country has existed for fewer than 30 years, splitting from Czechoslovakia in 1993. This region, where the Slovaks have lived for a thousand years, has a long and noble history of wine production, supplying the royal courts of Upper Hungary and the Austro-Hungarian monarchy. Vineyard area was 30,000 hectares (74,130 acres) in 1989, but has decreased since independence and by 2017 had fallen to around 15,000 hectares (37,065 acres) registered and just 8,500 hectares (21,000 acres) actually being harvested, yielding 46,000 tonnes that year and close to 300,000 hectolitres of wine. Domestic production can only supply around 20 per cent of all wine consumed in Slovakia, which shows why so little Slovak wine is seen in any export markets. Wine quality was highly regarded in the past (such as Ausbruch wines from near Bratislava and reds from Rača and Cachtiče, and more), but then the communist era brought very mediocre mass production. There are two strong trends today. One is the development of classic high-quality wine, notably led by Egon Müller's involvement at Chateau Belá, initially showing that Slovakia is capable of stunning Riesling (in a richer, more full-bodied style than Müller's fine Saar wines) but then developing a reputation for reds and sweet wines too. Another notable producer in southern Slovakia is Bott Frigyes, which identifies as Hungarian and actually sells most of its excellent wine there. Karpatská Perla and Pavelka are also trend-setters, being significant vineyard owners and producers of very good, value-for-money varietal wines. The other trend is a reaction against the perception of poor quality in Slovakia's mass market wines that has led to a dynamic group of small family producers making more "natural" low-intervention and organic wines, including skin-contact and amphora styles. Some of these have been taken up with enthusiasm by sommeliers all over the world, though in tiny volumes. For most producers, exports are only on the agenda through hand-selling to specialists looking for niche, quality wines.

Slovakia joined the EU in 2004, updating its wine law in 2009. There is one PGI for the whole nation and six PDOs that apply to its six wine regions. These are Malokarpatská, Južnoslovenska, Nitrianska, Stredoslovenská, Vychodoslovenská, and Vinohradn výber cka oblast' Tokaj. Two wines, Karpatská Perla and Skalický Rubín, have their own PDO. There are also categories of monovarietal wines for attributes based on sugar levels, botrytis etc (similar to German law). In general though it is still better to rely on producer name rather than regulations. Perhaps the most controversial of the regions is Tokaj, which Hungary has historically objected to, though this has been resolved. Tokaj is a single wine region that happens to extend north eastwards over a national border and indeed the part that is in Slovakia today was included in the Tokaj vineyard classification of 1737. Slovak Tokaj is smaller than Hungary's part of the region, with up to 929 hectares (2,295 acres) permitted for winemaking, and growing exactly the same grapes. The style of Slovak Tokaj tends to be a little bit more oxidative and structured than in Hungary, and the region has not had anything like the external investment and buzz that has benefitted the

Hungarian part. Slovakia's wine law still allows *putňový* (*puttonyos* in Hungarian) for its *výber* (equivalent of Aszú) wines, covering categories from three to six (unlike Hungary which has now banned styles less sweet than five *puttonyos*) while in Slovakia, Výber Tokaj must include all three varieties of Furmint, Lipovina (Hárslevelű), and Muškát žltý (Yellow Muscat) in specified proportions.

Overall Slovakia is close to the northernmost limit of where grape growing is feasible, lying between 47.5° to 48.5° N and it has a cool continental climate. It's a mountainous landscape with very varied bedrock and soils, including some volcanic areas but also granite, schist, carbonate clays, sandstone, loess, eolian, and alluvial deposits. Most of the vineyards lie on gentle slopes in the foothills, largely south-facing. Winter cold and spring frost can be an issue, as can ripening of grapes. As a result, a number of local grape varieties designed to cope with the climate have been developed such as Dunaj, Nitria, Hron, and Devín. Overall grape variety plantings are led by Grüner Veltliner (Veltlínské zelené), Welschriesling (Rizling Vlasšky), Blaufränkisch (Frankovka Modra), Müller-Thurgau, Svätovavrinecké (St Laurent), Riesling (Rizling Rýnsky), Pinot Blanc (Rulandské Biele), Cabernet Sauvignon, Alibernet, Pinot Noir, and André. Vineyards are around 75 per cent white grapes and the most typical style is fairly light-bodied dry to semi-dry whites with racy acidity. Južnoslovenska especially around Mužla and Strekov is best known for reds. As in much of southeastern Europe, the wine scene is fascinating and evolving rapidly, but anyone who wants to know more will have to visit the country.

✔ *Chateau Belá • Bott Frigyes • Tokaj Macik • Tokaj & Co • Ostrožovič, • Martin Pomfy – Mavín • Karpatská Perla • VPS (Vinohradníctvo Pavelka a Syn) • Berta, Slobodné Vinárstvo • Mátyás • Domin & Kušický • Strekov 1075*

AN OBSERVATION TOWER AFFORDS A VIEW OF A MALÁ TŘŇA VINEYARD IN THE FAMED TOKAJ WINE REGION OF SLOVAKIA
Located in southeastern Slovakia and northeastern Hungary, the Tokaj Wine Region Historic Cultural Landscape was inscribed on the World Heritage List in 2002.

0 mi — 60
0 km — 60

A COLOURFUL BUILDING BRIGHTENS A VINEYARD IN MODRA, SLOVAKIA
Modra is one of the cities along the Small Carpathian Wine Route that runs below the mountain range of Malé Karpaty. As well, as Modra, the Wine Road of Malé Karpaty leads through the former royal towns Bratislava, Svätý Jur, Pezinok, and other contiguous villages, and then ends in Trnava.

WINE REGIONS OF SLOVAKIA AND CZECHIA
Most of this region's vineyards are found in Slovakia, along the border with Austria and Hungary.

CZECHIA

Czechia (Czech Republic) is better known for its beer than its wine, but it has 17,700 hectares (43,740 acres) of vineyards in cultivation and a history dating back to the time of the Roman Emperor Probus (276–282) in Morava. Viticulture developed particularly strongly in the Middle Ages, and there are claims that King Wenceslas himself founded the vineyards around Prague castle in the 10th century. Holy Roman Emperor Charles IV introduced red grapes by the 4th century, and he also drew up laws protecting the vineyards. The golden age of wine here continued over the following centuries, but in time it was damaged by the Thirty Years' War, politics and phylloxera. As with other Central and Eastern European countries, quality winemaking was disrupted during the communist period of Czechoslovakia between World War II and 1989. During this period grape-growing and winemaking were collectivised into cooperatives and state enterprises with volume as the aim, imposing heavy mechanisation and increased plantings of the most productive varieties, especially Welschriesling, Gruner Veltliner, and Muller-Thurgau. In 1993, after a peaceful separation, Czechia and Slovakia emerged from Czechoslovakia (which had been created in 1918) and both joined the EU in 2004.

Today's wine industry averages 657,000 hectolitres of wine, two-thirds of which is white, though this is only enough to supply a quarter to a third of home-market wine consumption. Imports are significant and often undercut the local market. Other issues here include fragmentation: there are more than 18,000 growers and an estimated 1,200 wine producers, though only 195 have more than 5 hectares (12 acres). There is a tendency for producers, even small ones, to make small volumes of very many different wines, rather than focusing on their particular strengths. This means volumes available for export or even domestically for bigger retail customers are very limited, and direct sales from the winery are often an important channel. Semi-dry and semi-sweet wines also remain popular, a category that is often disheartening, using sugar to compensate for old-fashioned winemaking and failing to show off the best of the *terroir*. Perhaps the country's most famous name is the sparkling wine Bohemia Sekt that sells nearly 10 million bottles a year and is wholly owned by Germany's Henkell group. It has 70 years of history and a range named after the Champenois native Louis Girardot, who brought yeast and know-how to the cellar, though today most of the brand is made by the *charmat* method.

The positive news though is that good-quality wines from Czechia are becoming more widely recognised through competition results, tourism, and specialist importers in Western markets, along with promotional efforts by the Czech Winemakers' Union, Czech Wine Fund, The National Wine Centre, and the brand campaign *Vína z Moravy, vína z ᐸech* ("Wines from Moravia, Wines from Czechia"). And at their best, both dry and sweet styles can be very good indeed, with crispness and racy acidity that reflect this marginal northerly climate. A significant number of local winemakers are moving into a new phase of maturity and starting to respect *terroir* and origin, as well as winemaking traditions. This trend is supported by the VOC (Vína Originální Certifikace) system introduced in 2009. It is based on specific origins for wine equivalent to Italy's DOC or Austria's DAC. It is limited to specific locations and specified grape varieties with rules on winemaking and style, and by 2017 there were 10 VOC associations registered. This is a parallel system running alongside the Germanic 2004 classification based on sugar level in grape must at harvest and with specific categories for certain styles such as Eiswein and straw wine. As with Slovakia, it continues to be more helpful to select wines by producer name rather than by formal classification, as there is still a considerable amount of indifferent winemaking. The good news is that the best of today's winemakers are often young and fiercely independent and really focusing on quality, along with organic methods, local varieties, natural fermentations, and techniques such as the trendy pet-nat, skin maceration, amphorae, and acacia barrels.

Czechia is a landlocked nation around 49° to 50°N, and its climate is notably continental, though milder in the south. There are two wine-growing regions and most vineyards (96 per cent) are in the slightly warmer region of Morava close to Austria and Slovakia, where the slopes around the river valleys are often planted. Morava is a natural continuation of Austria's Weinviertel and has four subregions (Znojmo, Velké Pavlovice, Mikulov, and Slovácko). Of these, Znojmo is the warmest, in the rain shadow of the Czech-Moravian highlands and largely stony soils. Cool air from the mountains slows down ripening and allows for good aromatic expression. The most notable feature of the Mikulov subregion is the massif of the Pálava hills, which is also a UNESCO biosphere reserve. This limestone outcrop provides some great locations on its lower slopes where limestone-clay, sandy soils, and loess deposits can be found. Velké Pavlovice is the heart of red wine production in Morava. The northern parts are sandy, while the southern vineyards grow on limestone, clay, marl, and grit, with deep loess deposits. The Slovácko region is very varied, with vineyards scattered along the Morava river valley where northwesterly breezes bring a cooling influence. The eastern part of the region lies close to the Carpathian foothills where warm breezes help grapes ripen and give full-flavoured wines. Bohemia's 700 hectares (1,730 acres) of vineyards are the remnants of what was once over 3,000 hectares (7,415 acres) and are relatively cool. There are two subregions: Litoměřice and Mělník) close to Prague. The Mělník subregion typically has limestone subsoil or sandy gravel, covered with an alluvial sandy topsoil. The subsoil of Litoměřice is often basalt with areas of limestone (especially opuka soils) on the lower slopes.

Czechia has as many as 50 grape varieties in cultivation. Among whites, Gruner Veltliner (Veltlínské Zelené) leads the way, followed by Müller-Thurgau, Riesling (Ryzlink Rýnský), Welschriesling (Ryzlink Vlašsky), Pinot Gris (Rulandské Bílé), Sauvignon, and Chardonnay. There are also fairly significant plantings of local selections such as Pálava, Muškát Moravský, and Aurelius. Reds are led by St Laurent (Svatovavřinecké), Blaufränkisch (Frankovka), Zweigeltrebe, Pinot Noir (Rulandské modré), and Portugieser (Modrý Portugal), though there are a number of red local grapes too, including André, Cabernet Moravia, and Neronet.

✓ *Bohemia Sekt • Dobra Vinice • Dufek Josef • Lobkowicz • Novàk • Osička • Zámecké Vinařství Bzenec • Sonberk • Lahofer • Nové Vinařství • Gotberg • Volarik • Vinařství Čech • Stávek • Spielberg • Mikrosvin Mikulov • Vinselekt Michlovský • Znovin Znojmo • Chateau Valtice • Stapleton & Springer • Krásná Hora*

THE RED ROOFS OF THE TOWN OF VELKÉ PAVLOVICE PROVIDE A CONTRAST TO THE BRIGHT GREEN OF THE SURROUNDING VINEYARDS
An aerial view of Velké Pavlovice near the Nové Mlýny reservoirs in southern Moravia shows the village surrounded by hillside vineyards.

The Western Balkans

Few regions in the world have undergone so much political turmoil over such a relatively short period as the Western Balkans. Until the early 1990s most of the region was the Socialist Federal Republic of Yugoslavia, plus the People's Socialist Republic of Albania. At its peak, vineyards covered around 220,000 hectares (543,630 acres), making it one of the world's top 10 volume wine producers. Today each of its individual countries has gone its own way and developed a distinctive identity.

Two themes are emerging across the region. One is the use of international varieties – most commonly Merlot, Cabernet, and Pinot Noir. In part, this is because these are glamorous and exciting to local consumers who want to drink wines different from their parents and grandparents, but also because producers want to show that they too have the *terroir* to make world-class wines that people anywhere can understand. The other, more recent theme, is the rescue of workhorse local varieties. Such vines were once pushed to their limits for yield with no thought of quality, but with the right attention, several are showing real potential for exciting wines with a sense of place.

ALBANIA

Although vineyards have been cultivated in Albania since pre-Roman times, the influence of Ottoman rule meant wine production was not of much importance in recent historical terms. The late arrival of phylloxera in 1933 and the rise of communism meant vineyard area had fallen to just 2,430 hectares (6,000 acres) by 1950. Today the country has seen quite a renaissance which, coupled with some substantial Italian support in the wine industry, has resulted in an estimated 10,178 hectares (25,150 acres) under vine by 2015. There's a wine road, and several wineries are open to tourists. There are four regions: the coastal plain near the capital Tirana; the central hilly region at 300 to 600 metres (985 to 1,970 feet) above sea level, the eastern mountainous region at 600 to 800 metres (1,970 to 2,625 feet) above sea level, and the highlands where vineyards may reach 1,000 metres (330 feet) above sea level. Plantings include international varieties and grapes like Vranac, Pamid, and Mavrud from Balkan neighbours, but most significant is Kallmet, which is genetically identical to Kadarka. There are also several unique indigenous varieties such as red Shesh i Zi, and white Shesh i Bardhe, which apparently account for around one-third of the harvest, plus Pulez, Debine e bardhe, Debine e zeze, Vlosh, and Serine. Producers that have been recommended include Arbëri, Bardha, Çobo, and Kokomani.

MONTENEGRO

This dramatic mountainous country is dominated by one major producer: 13 Jul Plantaže, which owns one of Europe's largest vineyards covering 2,310 hectares (5,710 acres) – from the country's total of around 4,500 hectares (11,120 acres) – but punches above its weight in wine-quality terms. It lies close to the dramatic Lake Skadar, possibly the original home of Kadarka and Kratošija (aka Zinfandel/Primitivo). The inky-dark Vranac dominates in the limestone-based vineyards here, enjoying the long sunshine hours and warm climate. Other varieties grown include Cabernet Sauvignon and Merlot. Whites are only 20 per cent of plantings, including indigenous varieties Krstač and Žižak, with Chardonnay, Rkatsiteli, Sauvignon, and Pinot Blanc present as well. A few small family producers are also starting to appear. Important wine regions are Podgorica and Crmnica.

✔ *13 Jul Plantaže • Lipovac • Sjekloča*

BOSNIA-HERZEGOVINA

A small country on the western Balkan peninsula with just 12 kilometres (7.5 miles) of Adriatic coastline and high central mountains, it is divided into two entities: the Federation of Bosnia and Herzegovina and Republica Srpska, plus the district of Brčko. A vineyard register is still a work in progress, but estimates put vineyard area at around 3,500 hectares (8,650 acres) with over 40 commercial wineries, which own 1,583 hectares (3,910 acres) and

produce just over 4.5 million litres (1.19 million gallons) of wine. The remainder belong to around 11,000 small households producing for the grey market or home consumption. The wine industry was badly scarred by civil war but is certainly recovering in terms of quality, if not quantity. Ninety per cent of the vineyards are in the Herzegovina area, especially around Mostar with its dramatic, rebuilt Ottoman bridge. The climate here is Mediterranean, with limestone/marl soils. Production is dominated by two indigenous varieties: the peachy, full-bodied Žilavka for white wine and Blatina that offers genuinely enjoyable reds, full of fruity, crushed berry flavours and velvety tannins, though it needs a cross-pollinator, which is typically Kambuša (Alicante Bouschet), Trnjak, or Merlot. There are also good examples of Vranac, especially around Trebinje. Other local whites include Krkošija, Bena, and Smederevka (aka Bulgaria's Dimiat).

✔ *Hercegovina Produkt • Tvrdos Monastery • Vilinka • Crjnac & Zadro • Čitluk, Škegro • Carski • Keža • Nuić, • Andrija*

KOSOVO

Kosovo declared its independence in 2008 and is now recognised by 103 nations (as of 2019), though not Serbia. Historically, wine has been produced in this region for around 2,000 years. In the 1950s the Yugoslav regime developed four large-scale state wineries largely shipping wine to Belgrade for blending, with the notable exception of Amselfelder, a light medium-sweet red sold in Germany. Shipments of Amselfelder reached 32 million litres (8.5 million gallons) in the 1980s, and vineyards at this time covered around 9,000 hectares (22,240 acres). Today, vineyards cover around 3,220 hectares (7,960 acres) with plantings led by Smederevka, Welschriesling, Vranac, Prokupac, Gamay, Chardonnay, and Pinot Noir. Wine quality lags behind its neighbours in spite of privatisation. There are around 15 wineries ranging from tiny family operations to the 600-hectare (1,480-acre) former state winery. Wine law still falls between Kosovo's own, recognising one region of Dukagjini, and Serbia's which defines two: South Metohija and North Metohija.

REPUBLIC OF MACEDONIA

This small landlocked country has been at the centre of trade routes across Europe for millennia. It has been caught up in an identity crisis but has recently agreed to rename itself the Republic of Northern Macedonia to resolve disagreements with Greece (and its region of Macedonia), though that is still subject to final confirmation. Once the world's most powerful nation (Philip of Macedon and Alexander the Great enjoyed wines from the area), this country is today one of the most dependent on wine in the world – its second-most important export after tobacco. Macedonia lies between the latitudes of 40° to 43°N. It is landlocked and geographically defined by a central valley formed by the Vardar River. The country is seismically active, and the climate is transitional from Mediterranean to continental. Today wine vineyards cover around 25,000 hectares (61,775 acres), and there are 74 registered wineries, though only 10 or so are serious producers of bottled premium wines (bulk exports especially to Germany remain big business). Wine production ranges from 95 to 120 million litres (25 to 32 million gallons). There is now a single PGI for the whole country, which has three wine regions: the eastern region (Pcinja-Osogovo), the central region (Vardar River Valley), and the western region (Pelagonija-Polog). Of these, Vardar River Valley accounts for 85 per cent of production, and within that the Tikveš district is most important, with close to 50 per cent of the country's vines. Low summer rainfall also means that vines need very little intervention or spraying so the vineyards are full of wildlife – even tortoises roam among the vines. Vranec (whose name means "black stallion") is the flagship red grape, now grown on 9,000 hectares (22,240 acres). Kratošija (Zinfandel) is also important, and there are plantings of Cabernet Sauvignon, Merlot, Pinot Noir, and Syrah, plus local light red Stanušina. White varieties include Smederevka, Rkatsiteli, Temjanika, Žilavka, Župljanka, Chardonnay, Grenache Blanc, and Sauvignon. Unusually it's the large producers that are the most dynamic in switching to quality away from quantity, led by giant Tikveš with its young Serbian-born but French-trained winemaker.

✓ Bovin • Chateau Kamnik • Dalvina • Ezimit • Lazar • Popov • Popova Kula • Stobi • Tikveš

SERBIA

Wine production in Serbia (Srbija) has a chequered history. The Turks did their best to rout the vine; the Habsburgs positively encouraged it. Serbia was a late starter in reinventing its wines in the new, post-Yugoslavia era but, as the old industrial giants fade, a new generation of often small and incredibly dynamic wineries has appeared, taking the country's total up to around 400. Serbia's vineyards have amazing potential. Hampered until recently by the usual post-communist issues of ownership and fragmentation (there are still around 120,000 growers for the country's 25,000 hectares (61,775 acres) of wine grapes) but rapidly catching up with its neighbours. Good Cabernet Sauvignon and Merlot-based wines and even Pinot Noir have been established for a while, but perhaps most exciting is the rediscovery of Prokupac. Once popular for it vigour, now low yields can produce elegant, Pinot-esque wines. Other local varieties like Morava, Neoplanta, Probus, Seduša, and Začinak are also gaining more

attention. Very few Serbian wines currently make it to onto export markets – the local market seems to be thirsty enough for these new quality wines.

The northern autonomous province of Vojvodina shares the climatic extremes of the Pannonian Plain with Hungary to the north. Welschriesling is common here and is increasingly labelled as Grašac (to overcome the grape's poor reputation with Serbian consumers), while Pinots of all three colours currently offer promise. Today there are 20 wine regions, the most notable including Šumadija, Tre Morave, Srem, Leskovac, Toplica, Negotinska-Krajna, Nis, and Banat. The Fruška Gora region on the hills that relieve the flatness of Vojvodina along the Danube north of Belgrade is now the most dynamic wine region. Here numerous young winemakers are experimenting with natural, biodynamic, organic, and amphora wines. The sandy soils of Subotica are home to some of the oldest Kadarka vines in the world, dating back to 1880. The town of Smederevo south of Belgrade gives its name to the white Smederevka grape (aka Bulgaria's Dimiat). The hilly Šumadija region is producing good results with Riesling, Sauvignon, Chardonnay, and Pinot Noir. Wine laws harmonised with EU regulations.

✓ Aleksandrović • Radovanović • Kovačević • Deurić • Janko • Despotika • Čokot • Botunjac • Doja • Ivanović • Temet • Matalj • Cilić • Virtus • Budimir • Zonko Bogdan

WINE REGIONS OF THE WESTERN BALKANS
This region possesses an impressive combination of indigenous grape varieties and *terroirs* in which they can be exploited.

A RESTAURANT IN ULCINJ, MONTENEGRO, DISPLAYS BOTTLES OF PLANTAŽE WINE
Plantaže, a Montenegrin producer. owns one of the largest vineyards in Europe.

Croatia

Croatia's dramatic coastline, sunny climate, and stunning coastal towns attracted more than 17 million tourists in 2017, consuming a lot of wine in Croatia and meaning that exports are relatively insignificant.

Viticulture dates back to the ancient Greeks and possibly earlier with the Illyrians. Grape-growing became more organised under Roman occupation, though later the Ottoman empire subdued winemaking. Under the Habsburg empire, grape-growing flourished again until the arrival of phylloxera. In the Yugoslavian era, the industry was collectivized, and viticulture suffered during the brutal wars of independence of the early 1990s.

MODERN DAY

Croatia is a very diverse country with 120 indigenous grapes, a fascinating range of climatic influences, and 19,585 hectares (48,395 acres) of vineyards in 2018, with 1,575 producers making 640,000 hectolitres of wine. The country has four large wine regions and 16 PDOs. Wine categories include premium quality wines (Vrhunsko Vino), quality wines (Kvalitetno Vino), and table wines (Stolno Vino).

ISTRIA AND KVARNER

The Istria and Kvarner region share the mild climate and cultural influences of Italy and Slovenia. White wines are most important, particularly the characterful local Malvazija Istarska, often made without oak in a fresh, fruity style. It can also be vinified as full-bodied, complex, food-friendly wine, often with oak or extended skin contact. Occasionally blended with Merlot, the local Teran, with its typically firm acidity and red fruit flavours, is the most significant red grape. International whites and various Muscats are also produced as semi-sweet or sweet styles. The island of Krk's speciality is the delicate white Žlahtina.

✓ *Arman Franc • Benvenuti • Boškinac • Cattunar • Coronica • Damjanić • Degrassi • Fakin • Gerzinić • Katunar • Kabola • Kozlović • Matošević • Meneghetti • Pilato • PZ Vrbnik • Radovan • Roxanich • Saints Hills • Trapan • Tomaz • Vina Laguna*

DALMATIA

This coastal region of Dalmatia stretches south from Zadar. Its sun-drenched, often steep, rocky vineyards are a treasure trove of native varieties and produce some of the country's most powerful and full-bodied reds. Plavac Mali is the grape behind the designations of Dingač and Postup from the seaside terraces of the Pelješac peninsula and Ivan Dolac from Hvar. Crljenak Kaštelanski, rescued from just nine lingering vines at Kastela near Split, is identical to Zinfandel (aka Tribidrag) and is also a parent of Plavac Mali (probably originating in Montenegro). Babić is another potentially good-quality red grape grown near Šibenik and Primošten. Pošip is the most important white grape and is producing increasingly exciting wines, especially from Korčula, Hvar, and Brač. Other interesting white varieties include Grk, Debit, Vugava, Bogdanuša, Gegic, and Maraština.

✓ *Ahearne • Badel 1862 • BibicH • Bura-Mrgudić • Gracin • Grgić • Korta Katarina • Krajančić • Madirazza • Miloš • Nerica • Saints Hills • Skaramuča, Tomić • Stina • Zlatan Otok*

SLAVONIA AND CROATIAN DANUBE

The inland zone of Slavonia and Croatian Danube, a relatively flat region with a few low hills and a continental climate, is the home of Croatia's most important white grape, Graševina (aka Welschriesling). The long, warm autumns allow some impressive sweet late-harvest styles to be produced. Traminac (Gewurztraminer), especially around Ilok, as well as Chardonnay and Sauvignon, are proving successful. Reds are led by Frankovka (aka Blaufränkisch), with some recent promising results from Pinot Crni (Pinot Noir), Merlot, and Cabernet Sauvignon. Slavonia is also famous for its oak, beloved of Italian producers for their large oak *botti*.

✓ *Adzič • Antunović • Bartolović • Belje • Enjingi • Feravino • Galić • Krauthaker • Kutjevo • Iločki Podrumi • Orahovica*

THE CROATIAN UPLAND

The Croatian Upland region surrounding Zagreb is the coldest zone, with small family vineyards dotted across green hillsides. International white grapes (Sauvignons, Rieslings, and recently Pinot Gris and Pinot Noir) are doing well, especially in Plešivica and Zagorje. There are also a few impressive ice wines. Furmint (aka Moslavac or Pušipel) and the native variety Škrlet, found in the Moslavina district, makes fresh, fruity but quite simple wines.

✓ *Bodren • Bolfan • Korak • Tomac*

WINE REGIONS OF CROATIA
This country not only has its own viticultural heritage but also established the New Zealand wine industry via the likes of Babich, Brajkovich, Delegat Fistonich, Nobilo, Selaks, and others.

Dalmatia	Wine region
Slavonija	Wine sub-region

0 mi — 40
0 km — 40

Slovenia

Slovenia, with its stunning green hillsides, is increasingly recognised as one of Central Europe's most exciting wine countries. Two-thirds of the vineyards are planted to whites, but red quality is catching up fast. The winemaking scene is incredibly dynamic and has also become quite a hotspot for "orange" wines too.

Slovenia gained its independence from Yugoslavia early and relatively peacefully compared to some of its neighbours. It quickly progressed as a modern European nation and joined the EU in 2004. Today Slovenia has just under 16,000 hectares (39,537 acres) of registered vineyards and an estimated 5,000 hectares (12,355 acres) further in unregistered tiny plots owned by nearly 31,000 growers. This shows quite how important wine and grapes are here, with Slovenians usually in the top 5 per capita wine consumers in the world. Wine here dates back to the Celts in 400 BC and prospered under the Romans and later the Church. The post WWII era saw winemaking concentrated into cooperatives producing volume (though some archive wines from this era are still drinking beautifully), but in the 1970s the first private producers started to emerge. This accelerated with independence, and a new generation of producers started to make wine with a fresh mindset and a focus on quality, rather than volume. An understanding of *terroir* has emerged along with much better, more subtle, use of oak. Organic wine production is also growing with 320 certified grape growers.

The vineyards of Slovenia are divided into one coastal region, Primorje, and two inland regions, Posavje and Podravje.

WINE REGIONS OF SLOVENIA
Slovenia is a jewel in the crown of Central European winemaking. Its stunning hilly vineyards are producing some of the region's most exciting and consistently high-quality wines.

PRIMORJE

The southwestern Primorje region enjoys a mild Mediterranean climate with its proximity to the sea and cooling influences from the Italian Alps. There are 6,534 hectares (16,145 acres) divided into four districts: Goriška Brda, Vipavska Dolina, Kras, and Slovenska Istra. Goriška Brda is continuation of Italy's famous Collio region, though arguably Slovenia got the best of the hilly vineyard sites – indeed a number of vineyards straddle the border. Soils here are called Opoka: clay, marl, and limestone that provide good drainage in this relatively rainy area. This region pioneered quality in the independent era and is home to many of Slovenia's wine superstars. There is a mixture of international grapes (Merlot, Sauvignon, and Chardonnay can be impressive), plus a few local ones, notably Rebula (aka Ribolla Gialla), which is incredibly versatile making wines ranging from fresh, steel-fermented styles to complex, rich skin-contact dry whites and sweet passito sweet wines too. The dramatic Vipava valley, or Vipavska Dolina, is noted for incredibly strong winds called Burja that blow in the late summer and help keep grapes healthy. Its dynamic region of smaller producers with several working organically and a few biodynamic names. Local grapes here include Zelen and Pinella. Elegant whites and lighter reds, especially Pinot Noir, are the valley's strength. Kras is particularly noted for its red soils over limestone karst, especially Kraški Teran made from Refošk (and cause of quite a political battle with Croatia over the name *Teran)*, while Slovenska Istra lies close to Slovenia's few kilometres of coastline, continuing into Croatia and featuring similar grapes, in particular Malvazija Istarska and Refošk.

POSAVJE AND PODRAVJE

These two inland regions are surrounded by Austria, Hungary, and Croatia. Podravje is the largest and most northerly, comprising Štajerska Slovenija and Prekmurje. Štajerska Slovenija is a continuation of Austria's Styria and has 6,585 hectares (16,272 acres) of vineyards. The climate is more continental with hot summers and cold winters but all-important cool nights in September allowing for crisp, vibrant whites and some excellent sweet wines. Laški Rizling (aka Welschriesling) is the most important grape, but there are also impressive Sauvignons, Riesling, Muscat, and Furmint (called Šipon here). Good lighter reds have also started to emerge recently, notably Pinot Noir and Blaufränkisch (here called Modra Frankinja – a grape which probably arose in this region and not Austria after all). Posavje (three regions: Bela Krajina, Dolenjska, and Bizeljsko-Sremić) has long been overlooked with its many micro producers and reliance on traditional styles like the tart light red Cviček. Recent developments are showing that the region has strong potential for lively sparkling wines (including local Rumeni Plavec and Žametovka), fresh dry whites, and some very promising Modra Frankinja reds. Sweet wines can be stunning.

✓ *Bjana • Blažič • Burja • Čotar • Dolfo • Edi Simčič • Guerila • Jakončič • Kabaj • Klet Brda • Klinec • Kristančič • Marjan Simčič • Medot • Movia • Pasji Rep • Santomas • Ščurek • Sutor • Tilia • Vinakoper • Vina Kras*

✓ *Dveri-Pax • Domaine Slapšak • Doppler • Gaube • Kogl • Kupljen • Gross • Joannes • Istenič • Kobal Wines • Kobal Family Winery • Marof • PRA-vinO • Ptujska Klet • Prus • Puklavec Family Wines • Steyer • Šturm • Valdhuber • Verus • Žaren*

Black and Caspian Seas

Classified as a lake, the Caspian Sea is the world's largest inland body of water and is located between Southern Europe and western Asia, east of the Caucasus Mountains. Five countries – Iran, Russia, Azerbaijan, Turkmenistan, and Kazakhstan – border the Caspian. About 500 kilometres (310 miles) to the west lies the Black Sea, bordered by Ukraine, Romania, Bulgaria, Turkey, Georgia, and Russia. This region in general, and Georgia in particular, is thought to have provided a safe haven for the vine during the last ice age.

Although Russia is the largest wine-producing country in this region, the wines of Georgia are the most famous, building on the country's reputation for its 8,000-year wine history and UNESCO listing of its qvevri winemaking as an intangible asset of humanity. Moldova relied heavily on exports to Russia until total bans in 2006 and 2013 have persuaded it to refocus its efforts on the West. It is now making the sort of clean and bright wines that such countries want. Armenia has attracted renowned winemaking consultants like Alberto Antonini and Michel Rolland, which is a sign of real potential in these ancient vineyards. And although Ukraine may have an interesting history, its wine industry has had to face the annexation of its most important wine region of Crimea by Russia after a referendum.

ARMENIA

Armenia hit the headlines in 2011 with publication of the discovery of the world's oldest-known winery in the Areni-1 cave in the province of Vayots Dzor. There was a rudimentary wine press, clay jars, and remnants of grapes, all dated to around 6,100 years old. Winemaking in the region appears to have hung on ever since, despite periods of Muslim rule. Today's Republic of Armenia is only 100 years old, though its membership in the USSR from 1922 had a significant effect on wine production. As part of the Soviet planned economy, the country was designated a brandy producer, supplying 25 per cent of all the Soviet Union's brandy, and still today the vast majority of wine is distilled (only 10 to 15 per cent of the harvest remains as wine, though this sector is growing). The revival of the modern wine industry and the switch towards quality didn't

begin until the turn of the 20th century. Zorah wines was a pioneer in Vayots Dzor, rescuing the native Areni Noir vine from obscurity and reviving the use of traditional *karas* (large clay jars) under the guidance of the renowned Alberto Antonini. Other Armenian diaspora have also returned to their roots to establish wineries, and the involvement of other high-profile international wine figures like Paul Hobbs from California and Michel Rolland suggests further potential. One of Armenia's advantages is altitude, with vineyards going up to 1,600 metres (5,250 feet) above sea level, and in places untouched by phylloxera. The country has 15,840 hectares (39,140 acres) of vines, 40 to 50 wineries, and claims to have more than 400 native grapes, though only 55 are actually grown and of these 31 are used for wine and distillation. Of the wine grapes, Areni Noir, Voskeat, Garandmak, and Akhtanak are the most highly regarded for winemaking so far with others under investigation.

✔ *Zorah* • *Trinity Canyon* • *Van Ardi* • *Kataro* • *Voskevaz*

WINE NOTES

• The Black and Caspian Seas region in general, and Georgia in particular, is thought to have provided a safe haven for the vine during the last ice age.

WINE REGIONS OF THE BLACK AND CASPIAN SEAS
The wine-producing nations of this region include Armenia, Azerbaijan, Georgia, Kazakhstan, Moldova, Russia, and Ukraine.

AZERBAIJAN

Azerbaijan, along with its Caucasus neighbours, has a strong claim as a very ancient wine region, with grape remains found at Shomu Tepe dated to the fifth or fourth millennium BC. It lies south of Georgia and east of Armenia (where the conflict over Nagorno-Karabakh is still unresolved). Wine continued to play an important role here even after the adoption of Islam. German immigrants were significant in developing the modern wine industry from the 1820s in Tsarist Russia and later, as a Soviet Republic, Azerbaijan's vineyard area rose to 284,000 hectares (70,1780 acres) by 1984, accounting for 40 per cent of the Republic's budget. Gorbachev's anti-alcohol edict hit the country hard, destroying an estimated 160,000 hectares (395,370 acres). Vineyard area continued to decrease, only starting to rebuild in the 2000s. By 2017 there were just over 16,000 hectares (39,540 acres) and a crop of 150,000 tonnes, producing 100,000 hectolitres of wine with a mix of international and Soviet grapes as well as some local ones, notably Matrasa. There are around 17 wineries, including several significant large-scale and high-profile investments, though sales are largely to Russia, China, and the home market.

GEORGIA

Georgia has one of the oldest winemaking traditions in the world, with evidence of wine made from grapes found in excavations south of Tbilisi dating back around 8,000 years. And for millennia wine has been an important part of Georgian culture: it has been claimed that the Georgian word for wine, *ghvino,* may predate the Latin word *vino.* Tradition says that Saint Nino, who brought Christianity to Georgia in the fourth century ad, carried a cross of vine branches, and wine still plays a central role in feasts hosted by a *tamada* (toastmaster). In the Soviet era, winemaking was collectivised under the state monopoly Samtrest. This was an era of quantity over quality, with vineyards covering 150,000 hectares (370,660 acres). Georgia developed a reputation for producing some of the best wines of all the Soviet states, though after the collapse of the USSR, it suffered from considerable fraud and imitation. Transition to a free market economy hit Georgian wine hard, and by 2006 vineyard area had fallen to 35,000 hectares (86,490 acres). Georgia was selling an estimated 80 to 90 per cent of its production to Russia by then, but (along with Moldova) Russia imposed a total ban in 2006. The embargo was devastating at the time, but it was also a turning point that forced producers to think about quality instead of quantity. Today, Georgia is particularly famous for its *qvevri* winemaking – wine vinified in clay amphorae, along with skins for several months, producing (in the case of white grapes) amber or orange wines. This method has been recognised by UNESCO as a cultural asset of humanity. But in reality, only around 1 to 3 per cent of Georgia's wines are made this way. Also of interest is Georgia's claim to 525 local grapes, though 2 grapes dominate – the red Saperavi (40 per cent of plantings) and

white Rkatsiteli (50 per cent of vines). Other grapes in commercial production include Mtsvane, Kisi, Goruli Mtsvane, Krakhuna, Tsitska, Tsolikouri, Mujuretuli, Otskhanuri Sapere, Aleksandrouli, Ojaleshi, and others. Today Georgia grows about 55,000 hectares (135,900 acres), and the country is divided into several wine regions, including the Black Sea coastal zone (in mythology where Jason and the Argonauts found the Golden Fleece), Meshketi, Racha, Kartli, Imereti, and Kakheti. Of these, Kakheti is by far the most important, accounting for around 90 per cent of all vineyards, which are planted around the Alazani river valley at 400 to 700 metres (1,310 to 2,300 feet) above sea level on black and alluvial soils over limestone. There are also 18 PDOs, which can be a useful guide to wine style. For instance, Mukuzani is dry red made from Saperavi; Khvanchkara and Kindzmaruli are semi-sweet red; and Tsinandali is a semi-sweet white. Winemaking has improved considerably in the last few years, and a number of small boutique producers have appeared too.

✓ *Chandrebi Estate • Badagoni • Chateau Mukhrani • Khareba • GWS (Tamada) • Pheasant's Tears • Telavi Wine Cellar*
Teliani Valley: *Tbilvino • Orgo • Vita Vinea • Shumi • Schuchmann • Kindzmarauli Marani*

KAZAKHSTAN

Of the former Soviet republics, Kazakhstan is geographically the largest and has substantial oil and mineral reserves. Grapes were first mentioned in a Chinese document from around the seventh century, but wine culture largely disappeared until the Soviets planted around 60,000 hectares (148,260 acres) here, though in the 1980s Gorbachev's anti-alcohol programme destroyed huge areas of vines, and then privatisation only increased the area that was abandoned. Estimates today range from 6,000 to 13,000 hectares (14,825 to 32,125 acres) producing 172,000 hectolitres in 2017, though not generally regarded as high quality, as sweet reds remain popular.

The most high-profile investment of the 21st century is Arba Wine, which, with Italian consultancy, has revived forgotten Soviet vineyards at 1,000 metres (3,280 feet) above sea level and is producing good wines as a result. Wine consumption is only 2.5 litres (0.66 gallons) per capita, but still Kazakhstan's own production supplies just 20 per cent of local demand, so exports are almost non-existent.

MOLDOVA

Moldova has more vines per person than any other country on earth, and its economy has long depended on wine. Today's Republic of Moldova was once the eastern part of the principality of Moldavia, so it shares wine history with the Moldovan Hills region of Romania and also many of its grape varieties. It became part of the Russian empire in 1812 and then a Soviet Republic, so its wine history has been very different to Romania's. The 19th century saw the founding of some of its most famous and historic wineries, including Vinaria Purcari in 1827, Romaneşti (apparently by the Romanovs) in 1850, and Castel Mimi in 1893 by the last governor of Bessarabia. The Soviet era brought collectivisation of the industry into state and collective farms, and under Stalin there was a drive to increase wine production as Moldova became one of the USSR's great agricultural centres. Moldova became a huge wine supplier to the Soviet Union, at its peak supplying 9.6 million hectolitres. As with its neighbours, land reform was far from smooth, and many vineyards fell into decline. From the early 1990s onwards, various Western-led projects attempted to make and export wine, but economic difficulties meant these failed. By the early 2000s Russia became the focus again – an unfussy customer for cheap semi-sweet wines, never mind the quality. But a total Russian ban in 2006 kick-started industry reforms, helped by the US government and other aid projects.

Today the industry has developed three PGI regions (Codru, Ştefan Vodă, and Valul lui Trajan), and legislation changes have allowed the emergence of a passionate group of small wineries. At the same time larger wineries have also reformed and invested in vineyards. Modern equipment and winemaking know-how has transformed wine quality over the last decade. By 2017 there were 81,000 hectares (200,155 acres) of commercial vineyards plus a further 9,600 hectares (23,720 acres) in back gardens. The harvest reached 311,000 tonnes giving 1.8 million hectolitres of

WINE BOTTLES AND GLASSES FORM AN IMAGINATIVE WATER FOUNTAIN AT THE MILEŞTII MICI WINERY IN MOLDOVA
This Moldovan wine producer boasts vast cellars extending for 200 kilometres (120 miles), of which only 55 kilometres (34 miles) are now in use. It also holds the Guinness World Record for biggest wine collection in the world, holding nearly 2 million bottles.

wine, with 106 wineries making wine that year. Varieties are mostly international ones, though there's growing interest in local varieties such as Viorica, Alb de Onițcani, and Rară Neagră (aka Băbească Neagră). Moldova is poor though, so it will always be heavily reliant on exports – largely to Poland, China, Romania, and former Soviet states. Western markets have recently picked up on the fact that Moldova can offer exactly the sort of wines that Western drinkers want – bright, modern varietal wines at value-for-money pricing. At its best Moldovan wine has more to offer though, especially from the small producers but also from larger wineries like Purcari, which has revived the historic Negru de Purcari blend (famously the only wine exported with an English label in the Soviet era – to Queen Elizabeth II). Moldova is also particularly proud of its two huge cellars at Cricova and Milештii Mici. Both are former limestone mines with miles of tunnels that can be driven through. Cricova has a focus on sparkling wine, made here since the 1950s, and Milештii Mici claims the Guinness world record for the largest collection of bottled wines in the world.

✓ *Asconi • Atú • Bostavan • Carpe Diem • Castel Mimi • Chateau Vartely • Doina-Vin • Et Cetera • Equinox • Fautor • Gitana • Gogu • Minos Terrios • Novak • Vinaria Purcari • Sălcuța • Vinaria Din Vale • Vinaria Nobilă • Pelican Negru*

RUSSIA

With 91,700 hectares (226,595 acres) under vine, Russia is now the biggest vine grower in this region. Climate is something of an obstacle to viticulture in much of the country, because winters are generally very cold indeed, often -30°C (-22°F), and the vines in many regions, such as the Don Valley, have to be earthed up to survive. Summers are very hot and dry but are modified by the effect of the Black, Caspian, and Azov seas, especially around the Kuban peninsula. Wine probably came with the Greeks around 2,500 years ago, but the modern era developed in the 19th century under Tsar Alexander II, who invited French winemakers to the country. Crimea's famous Massandra winery was built in 1894 by Prince Golitsyn, who was a significant figure in Russian wine. Massandra housed wine collections for the nobility and is especially proud of its long-lived fortified wines. The oldest is Jerez de la Frontera from 1775: scandalously, Vladimir Putin and Silvio Berlusconi drank a bottle during a cellar tour in 2015. The major wine-growing districts (with PGI status) in Russia are: Kuban (Krasnodar region) 25,000 hectares (61,775 acres) Dagestan 24,800 hectares (61,280 acres); Don Valley (Rostov region) 4,300 hectares (13,100 acres); Terek Valley (Kabardino-Balkarian Republic) 4,300 hectares (13,100 acres); Stavropol Territory 5,900 hectares (14,580 acres); Lower Volga (Volgograd Region) 600 hectares (1,480 acres). As of 2014, after a referendum, Crimea (Autonomous Republic of Crimea and Sevastopol) has come under Russian administration. Crimea has 18,700 hectares (46,200 acres) under vine and Sevastopol 5,900 hectares (14,580 acres).

There are 59 licensed small-to-medium wineries in Russia, with 25 large producers, and in 2017 the vineyards yielded 580,000 tonnes. Government statistics show domestic wine production of 6.71 million hectolitres the same year, and this illustrates one of Russian wine's biggest problems – more wine is sold as "Russian" than can be produced from Russian-grown grapes. The law here allows wine to be declared Russian if it is finished and bottled in Russia, and a lot of low-quality, cheap bulk is imported to supply this sector, undercutting and undermining domestic production. There are also considerable volumes of bottled wine imports led by Spain, Italy, and France. The modern wine industry is still relatively young but has seen the number of wineries growing and, since 2010, with a clearer vision for quality. There is still a need to tighten up wine law and clarify which wines are made from Russian grapes as well as developing wine education, though there is work in progress on this. There is also a plan to increase area under vine to an ambitious 140,000 hectares (345,950 acres) by 2020. Grape varieties planted include Aligoté, Cabernet Sauvignon, Merlot, Muscat Blanc à Petits Grains, Muscat Noir, Pinot Blanc, Pinot Noir, Riesling, Semillon, Silvaner, Sauvignon Blanc, Chardonnay, Rkatsiteli, and Saperavi. There is also increasing interest in local varieties, such as Bastardo Magarachskiy, Golubok, Dostoynyy, Kefesiya, Kokur Belyy, Krasnostop Zolotovskiy, and Tsimlyanskiy Chernyy. A number of these were bred at the Magarach Research Institute near Yalta during the Soviet

ABRAU DURSO SPARKLING WINE
Abrau-Durso, a Russian wine company established in 1870 by decree of Tsar Alexander II, produces sparkling wines and also operates a hotel, spa, and restaurant.

era. In terms of customer preference, semi-sweet wines and high-alcohol reds remain popular, but there is trend towards dry aromatic whites and brut sparkling styles, helped by the development of gastronomy.

✓ *Krasnodar: Abrau Durso • Château le Grand Vostock • Fanagoria • Kuban Vino (Ch Tamagne) • Sikory • Villa Victoria • Lefkadia • Myskhako • Usadba Divnomorskoye*
Rostov: *Vedernikov*
Republic of Crimea and Sevastopol: *Satera • Alma Valley • Sun Valley • Uppa Valley*

UKRAINE

Viticulture has long been an important industry in what is Ukraine today, dating back around 2,500 years. Famously a Swiss colony was established at Chabag in Bessarabia (Shabo today) in 1822 to bring winemaking knowledge in the Tsarist era. Wine production developed strongly during the Soviet period – reaching 830,000 tonnes from 246,000 hectares (607,880 acres) at its peak in the 1970s. Gorbachev's anti-alcohol edict in 1980s hit the industry hard, and then the transition to a free market economy brought further woes. There's been a steady downward trend from 175,000 hectares (432,435 acres) in 1990 to today's 43,500 hectares (107,490 acres) that produced around 1.2 million hectolitres of wine from 151,000 tonnes in 2017 (less than Moldova). Ukraine lost its most historically renowned wine region of Crimea when it was annexed by Russia in 2014. Ukraine today has 121 entities with a licence to produce alcohol in general, and 100 producing wine only. The main wine region is Odessa with 64 per cent of vineyards, and the rest in Mykolaiv (aka Nikolaev) with 13 per cent, Kherson with 11 per cent) and Zakarpattia with 8 per cent. Varieties are led by Aligoté, Rkatsiteli, Cabernet Sauvignon, Muscat, Chardonnay, Sauvignon, and Riesling, though the country has some indigenous vines such as Telti-Kuruk, Odessa Black, and Sukholimansky White. Wine grapes are second only to fruit-growing in profitability terms so are an important contributor to the economy. Home-grown grapes can only supply a small percentage of market demand, so there's considerable trade in winemaking raw materials and wine imports (4.3 million hectolitres). Sparkling wines are a strong feature of the market here with only about half the crop going to still table wines and a third going for sparkling products. The untapped potential of Ukraine's best vineyards is arguably shown by some of the private wineries that have been developed recently, largely focussing on tourism and the local market. Shabo, for instance, has attracted Stéphane Derenoncourt as its consultant. It is the first winery in Ukraine to produce wine with protected origin status, and it is proud of its cultural centre and social responsibility. It is one of the few to be found in export markets.

✓ *Shabo*

The Wines *of the* Eastern Mediterranean

FEW REGIONS PROVIDE THE COMBINATION of mild climate and hilly terrain that wine grapes require to thrive as does the Mediterranean. Greece's wine heritage goes back thousands of years – it was the first country to truly develop viniculture and winemaking. The early Israelites also produced wine, both for daily consumption and for religious observances. Wine was a mainstay in the Levantine countries of Lebanon, Turkey, Cyprus, Egypt, Syria, and Jordan, where the most ancient remnants of winemaking and consumption have been found. In modern times, however, this corner of the Mediterranean has been overlooked by the wine lover. And with good reason: for too long Greek wines were worse than indifferent, and only a handful of vineyards were producing commercial wine. Meanwhile, the Levantine countries had little to offer in the way of quality or consistency. But that all began to change in the post-millennial world. Greece upped its game considerably and is now one of the most outstanding wine producers in Europe. Israel and Lebanon are now home to a record number of boutique wineries, Turkey is now showcasing their native wines, and Cypress is proving it can produce more than delicious dessert wine. Even Egypt has made inroads on improving the quality of their notoriously poor wines.

THE VINEYARDS OF METEORA LIE NEAR THE GREEK VILLAGES OF KALABAKA AND KASTRAKI
Meteora, an imposing rock formation in central Greece, is the site of an impressively large complex of Eastern Orthodox monasteries that perch amongst its boulders. The area, especially Kastraki, is known for its red wines, and its viniculture has received a boost with the designation of the PGI Meteora.

Greece

Greece is now it is one of Europe's most exciting wine-producing countries and, with thousands of unidentified ancient grape varieties preserved in its nurseries, it should be able to delight and surprise wine lovers for centuries to come.

Greece is far and away one of the most exciting wine-producing countries in the modern world of viticulture. Producing less than half of the amount of wine made in Bordeaux in an average year, while consuming more than 80 per cent locally, Greece is the home of some of the rarest wines around. The fantastic interplay of factors such as several hundred autochthonous grape varieties, many producers with enquiring minds, and infertile soils on an essentially mountainous terrain, yet surrounded by sea, is reflected in an array of distinctive wine styles that remain superb value for money. Greek wines used to be too esoteric, too complex, and just too different for most wine drinkers of past decades. Now these qualities are what many people are seeking. Their growing popularity and increasing exports are a magnificent testament to that fact and the only way is up.

There are very few regions in Greece that do not have local wine production, but the main areas, in terms of volume, are Peloponnese, Central Greece, Crete, and Macedonia. The islands of both the Aegean and Ionian seas are producing small quantities, but with their high quality, their wines are frequently in the spotlight. There are more than a thousand Greek wineries, most of which are very small, family owned enterprises, selling their wines locally and possibly bottling only a small proportion of their annual production. The number of good-quality producers, with a broad distribution across the country and also doing some exports, sits around the 400 mark. Still, among this group, the average output is rarely above 100,000 cases per year.

Greek wines, having a very strong consumer base at home, remained largely unaffected by the increasing globalisation and harmonisation of taste that can be found nowadays in a significant number of wine-producing regions and countries around the world. A Greek wine must be food-friendly above everything else, because Greeks will never drink wine without at least some nibbles on the side. This symbiotic relationship results in a style that is based on a moderate level of alcohol, crisp acidity, medium body, firm tannins, and palate structures that are dense but never saturating. The ultimate quality criterion for wine in Greece is a bottled that has been emptied.

RECENT GREEK VINTAGES

2018 Perfect weather condition during harvest. Very low yields in some places, such as Santorini.

2017 More drought and heat stress. Greek varieties fared best. Nemea, Naousa, and Mantinia the most successful.

2016 Drought and early harvest led to reduced crop. Red varieties showed perfect phenolic maturation.

2015 A moderate summer and high rainfall during winter helped grapes, especially white, to accumulate flavour complexity and sugar levels. Reds were deep coloured with medium alcohol.

2014 A challenging year with disrupted fruit set. Despite lower temperatures than usual and rain in September quality looked good with low-to-moderate alcohol levels and good flavour concentration.

☐ Protected Designation of Origin (PDO)

Protected Geographic Indication (PGI)

▨ District

● Area (local)

1 Anavyssos
2 Ilion
3 Koropi AND Retsina of Koropi
4 Lilantio Pedio
5 Mantzavinata
6 Markopoulo AND Retsina of Markopoulo
7 Martino
8 Metaxata
9 Opountia Lokris
10 Paiania AND Retsina of Paiania
11 Pallini AND Retsina of Pallini
12 Retsina of Chalkida
13 Retsina of Mesogia
14 Retsina of Pikermi
15 Retsina of Thiva
16 Ritsona
17 Slopes of Aigialeia
18 Slopes of Ainos
19 Slopes of Kitherona
20 Slopes of Parnitha
21 Slopes of Penteliko
22 Slopes of Petroto
23 Spata AND Retsina of Spata
24 Thiva
25 Valley of Atalanti
26 Verdea of Zakynthos

0 mi — 50
0 km — 50

VINEYARDS SURROUND THE CHURCH OF AGIOS NEKTARIOS AT KISSAMOS, WESTERN CRETE
This largest of the Greek isles is one of the country's major wine-producing regions and is host to seven PDOs and six PGIs (which are a mix of Regional, District, and Area PGIs).

GREEK LABEL LANGUAGE

ΑΦΡΩΔΗΣ ΟΙΝΟΣ, ΗΜΙΑΦΡΩΔΗΣ;
ΑφρώδηςΟίνος, Ημιαφρώδης Sparkling wine, semi-sparkling

ΕΛΛΗΝΙΚΟ ΠΡΟΙΟΝ; ΕλληνικόΠροϊόν Product of Greece

ΠΟΙΚΙΛΙΑΚΟΣ ΟΙΝΟΣ; ΠοικιλιακόςΟίνος Varietal wine

ΕΡΥΘΡΟΣ; Ερυθρός Red

ΓΛΥΚΟΣ; Γλυκός Sweet

ΗΜΙΓΛΥΚΟΣ; Ημίγλυκος Semi-sweet

ΗΜΙΞΗΡΟΣ; Ημίξηρος Semi-dry

ΛΕΥΚΟΣ; Λευκός White

ΞΗΡΟΣ; Ξηρός Dry

ΟΙΝΟΣ; Οίνος or **ΚΡΑΣΙ; Κρασί** Wine

ΟΙΝΟΠΟΙΕΙΟΝ; Οινοποιείον Winery

ΟΝΟΜΑΣΙΑ ΚΑΤΑ ΠΑΡΑΔΟΣΗ; Ονομασία κατα παράδοση Traditional Designation (reserved for Retsina and Verdea of Zakynthos)

ΠΑΡΑΓΩΓΗ ΚΑΙ ΕΜΦΙΑΛΩΣΗ; Παραγωγή και Εμφιάλωση Produced and bottled by

ΠΡΟΣΤΑΤΕΥΟΜΕΝΗ ΟΝΟΜΑΣΙΑ ΠΡΟΕΛΕΥΣΗΣ;
Προστατευόμενη Ονομασία Προέλευσης (ΠΟΠ) Protected Designation of Origin (PDO)

ΠΡΟΣΤΑΤΕΥΟΜΕΝΗ ΓΕΩΓΡΑΦΙΚΗ ΕΝΔΕΙΞΗ;
Προστατευόμενη Γεωγραφική Ένδειξη (ΠΓΕ) Protected Geographical Indication (PGI)

ΡΕΤΣΙΝΑ; Ρετσίνα Retsina

ΡΟΖΕ; Ροζέ Rosé

Vintage Greece belongs to the EU, so the term vintage on the label should mean that at least 85 per cent of the wine comes from the year indicated.

WINE REGIONS OF GREECE

The mainland areas and islands that form modern-day Greece provide a plethora of wine regions. Cooler northern regions produce the best wines, while the island of Samos produces great sweet wines. See inset map below for the major Region Protected Geographic Indications and opposite page for the numbered key to Area designations.

RETSINA ACCOMPANIES A MEAL OF YOGHURT, TOMATOES, AND OLIVES
One of Greece's best-known and best-loved wines is Retsina, a white or rosé that is made from Savatiano and Rhoditis grapes infused with resin from the Aleppo pine tree.

FACTORS AFFECTING TASTE AND QUALITY

LOCATION
Greece is located halfway between southern Italy and Turkey. Vines grow throughout the mainland and on its many islands, but most are in Macedonia and Peloponnese.

CLIMATE
Mild winters are followed by sub-tropical summers, during which the heat in many vineyards is tempered by sea breezes called *meltemi* while they are bathed in around 3,000 hours of sunshine. Extremes are found in mountain vineyards in Macedonia, where grapes sometimes fail to ripen, and on Crete, where the summer brings five months of intense heat and drought.

ASPECT
Vines grow on all types of land, from flat coastal sites on the mainland, through rolling hills and river valleys, to mountain slopes. The best sites are north-facing (to avoid the hot sun) slopes up to 610 metres (2,000 feet) above sea level.

SOIL
Mainland soils are mostly limestone, while those on the islands are rocky and volcanic.

VITICULTURE & VINIFICATION
The Greek cooperatives have long been in possession of thermostatically controlled stainless-steel vats as well as other modern equipment, but they have only started to take advantage of this technology since the mid-1990s. The majority of the best Greek wines on the market today are made by small, high-tech, boutique wineries.

GRAPE VARIETIES
Primary varieties: Assyrtiko, Muscat, Savatiano, Robola, Roditis, Moschofilero, Agiorgitiko, Xinomavro, Liatiko, Limnio, Mandilaria, Mavrodaphne
Secondary varieties: Cabernet Franc, Cabernet Sauvignon, Chardonnay, Syrah

THE APPELLATIONS OF

GREECE

Note: The spellings of the following appellations are from the latest official list but may differ slightly from what you see on the labels of some wines because the transliteration of the Greek alphabet is not a precise science, and individual producers have different ideas. With time, one hopes the official Latinized names will be adopted by all producers.

The appellations are Traditional Designation, PDO (Protected Designation of Origin), and PGI (Protected Geographical Indication). Only existing primary grape varieties grown are provided for PGI wines because the inherent flexibility of PGI regulations permits an almost endless list, unless restrictions are imposed locally. This is the first attempt by any reference work to list all the PGIs, most of which are in their earliest formative years, and many have not yet been produced, thus it is possible only to provide the briefest of styles permitted.

ACHAIA PGI
Peloponnese

This PGI District is for white (dry, semi-dry, semi-sweet, or sweet), rosé (dry or semi-dry), and red (dry, semi-dry, semi-sweet, or sweet), and semi-sparkling white and rosé from Achaia.

🍷 *Assyrtiko, Roditis, Malagousia, Muscat Blanc, Robola, Athiri, Lagorthi, Sauvignon Blanc, Chardonnay, Riesling, Ugni Blanc, Sideritis, Mavrodaphne, Mavro Kalavritino, Cabernet Sauvignon, Cabernet Franc, Grenache Rouge, Merlot, Syrah*
✔ *Mega Spileo*

ADRIANI PGI
Macedonia

This PGI Area is for dry white, rosé, and red from the vineyards in the southern and central part of the district of Drama.

🍷 *Viognier, Chardonnay, Sémillon, Trebbiano, Cabernet Sauvignon, Cabernet Franc, Merlot, Syrah*

AEGEAN SEA PGI
Aegean Sea

This Regional PGI is for white (dry, semi-dry, semi-sweet, or sweet), rosé (dry, semi-dry, semi-sweet, or sweet), and red (dry, semi-dry, semi-sweet, or sweet), and white and rosé sparkling (dry, semi-dry, or sweet) and white and rosé semi-sparkling (dry, semi-dry, or semi-sweet). There are 33 varieties are allowed from the North and South Aegean regions, including grapes authorised for any of the PDOs in the region indicated.

AGION OROS (MOUNT ATHOS) PGI
Macedonia

This PGI Area is for white (dry, semi-dry, or sweet), rosé (dry or semi-dry), and red (dry, semi-dry, or sweet) from the Agion Oros (Mount Athos) region.

🍷 *Assyrtiko, Roditis, Athiri, Sauvignon Blanc, Chardonnay, Syrah, Xinomavro, Limnio, Cabernet Sauvignon, Grenache Rouge*
✔ *Tsantalis • Ierokeli Agiou Efstathiou "Milopotamos"*

AGORA PGI
Macedonia

This PGI Area is for white (dry), rosé (dry), and red (dry) from the Drama region.

🍷 *Roditis, Muscat of Alexandria, Sauvignon Blanc, Chardonnay, Ugni Blanc, Cabernet Sauvignon, Cabernet Franc, Merlot, Grenache Rouge*
✔ *Nico Lazaridi*

AMYNTAIO PDO
Macedonia

The Amyntaio appellation is 100 per cent devoted to the Xinomavro variety, producing wines that express the grape in far more ways than other Xinomavro *terroirs*. Amyntaio is the coolest viticultural region in Greece and is therefore responsible, apart from the reds, for the production of significant quantities of still rosé and sparkling wines, as well as of some excellent *blancs de noirs*.

🍷 *Xinomavro*
✔ *Alpha Estate • Kir-Yianni*

ANAVYSSOS PGI
Central Greece

This PGI Area is for dry white wines from Attiki (Attica).

🍷 *Assyrtiko, Roditis, Savatiano, Ugni Blanc*

ANCHIALOS PDO
Thessaly

This PDO refers to wines (Retsina) with a minimum of 75 per cent Roditis and Savatiano from the low hilly terrain west and northwest of the Pagasitikos Gulf in the district of Magnisia.

ARCHANES PDO
Crete

This is a dry red wine appellation, producing wines with firm structure and a slightly rustic character.

🍷 *Kotsifali, Mandilaria*

ARGOLIDA PGI
Peloponnese

This PGI District is for white (dry, semi-dry, or sweet), rosé (dry, semi-dry, or sweet), and red (dry, semi-dry, or sweet) from the Argolida region.

🍷 *Assyrtiko, Savatiano, Asproudes, Malagousia, Rokaniaris, Chardonnay, Viognier, Roditis, Moschofilero, Agiorgitiko,*

Voidomatis, Mavroudi, Cabernet Sauvignon, Cabernet Franc, Merlot, Syrah
✓ *Skouras*

ARKADIA PGI
Peloponnese

This PGI District is for white (dry), rosé (dry), and red (dry), and white semi-sparkling (dry or semi-dry) and rosé semi-sparkling (dry or semi-dry) from Arkadia region.

Assyrtiko, Savatiano, Asproudes, Robola, Riesling, Syrah, Chardonnay, Roditis, Moschofilero, Agiorgitiko, Koliniatiko, Mavroudia, Skylopnichtis, Cabernet Sauvignon, Cabernet Franc, Merlot
✓ *Tselepos • Semeli Estate*

ATTIKI PGI
Central Greece

This PGI District is for white (dry, semi-dry, semi-sweet, or sweet), rosé (dry, semi-dry, semi-sweet, or sweet), and red (dry, semi-dry, semi-sweet, or sweet) from the Attiki (Attica) region.

Assyrtiko, Savatiano, Aidani, Athiri, Malagouzia, Ugni Blanc, Sauvignon Blanc, Chardonnay, Roditis, Agiorgitiko, Limniona, Mouchtaro, Mandilaria, Cabernet Sauvignon, Merlot, Syrah, Grenache Rouge
✓ *Mylonas • Matsa Estate*

AVDIRA PGI
Thrace

This PGI Area is for white (dry, semi-dry, or semi-sweet), rosé (dry, semi-dry, or semi-sweet), and red (dry, semi-dry, or semi-sweet) from Avdira.

Assyrtiko, Roditis, Zoumiatiko, Muscat of Alexandria, Athiri, Malagouzia, Sauvignon Blanc, Chardonnay, Ugni Blanc, Mavroudi, Limnio, Pamidi, Cabernet Sauvignon, Merlot, Syrah, Grenache Rouge
✓ *Vourvoukelis • Anatolikos Vineyards*

CANDIA PDO
Crete

Also called Chandakas-Candia or Handakas-Candia, this PDO is for dry white wines that must be made from at least 85 per cent Vilana grapes, with other varieties, such as Assyrtiko, Vidiano, Athiri, and Thrapsathiri. Dry red wines are made from at least 70 per cent Kotsifali and Mandilaria.

CHALIKOUNA PGI
Ionian Islands

This PGI Area is for dry white wine from Kerkyra (Corfu).
Kakotrygis

CHALKIDIKI PGI
Macedonia

This PGI District is for white (dry, semi-dry, or semi-sweet), rosé (dry, semi-dry, or semi-sweet), and red (dry, semi-dry, or semi-sweet) from Chalkidiki.

Assyrtiko, Malagousia, Athiri, Muscat of Alexandria, Sauvignon Blanc, Ugni Blanc, Roditis, Limnio, Xinomavro, Grenache Rouge, Cabernet Sauvignon, Merlot, Syrah
✓ *Claudia Papayianni*

CHANIA PGI
Crete

This PGI District is for white (dry, semi-dry, semi-sweet, or sweet), rosé (dry, semi-dry, semi-sweet, or sweet), and red (dry, semi-dry, semi-sweet, or sweet) from Chania. on the island of Crete. There are 15 varieties allowed, including the Vilana, Thrapsathiri, Muscat Blanc, Romeiko, Kotsifali, Mandilaria, Fokiano, Cabernet Sauvignon, Carignan, and Syrah.

CHIOS PGI
North Aegean

This PGI District is for white (dry, semi-dry, or semi-sweet) and red (dry, semi-dry, semi-sweet, or sweet) from Chios.

Assyrtiko, Muscat Blanc, Athiri, Saviatiano, Agianiotiko, Chiotiko Krassero, Mandilaria
✓ *Ariousios*

CENTRAL GREECE or STEREA ELLADA PGI
Central Greece

This Regional PGI is for white (dry, semi-dry or semi-sweet, sweet), rosé (dry, semi-dry or semi-sweet, sweet), and red (dry, semi-dry or semi-sweet, sweet) from grapes as authorised for any of the PDOs in the region indicated. There are 29 varieties allowed, including Assyrtiko, Roditis, Savatiano, Agiorgitiko, Mouchtaro, and Cabernet Sauvignon.

CORFU PGI
See Kerkyra (Corfu) PGI

CORINTHOS or CORINTHIA PGI
See Korinthos (Corinth) PGI

CRETE PGI
Crete

This Regional PGI is for white (dry, semi-dry, semi-sweet, or sweet), rosé (dry, semi-dry, semi-sweet, or sweet), and red (dry, semi-dry, semi-sweet, or sweet) and also includes wines from grapes authorised for any of the PDOs in the region indicated. There are 29 varieties allowed, including Assyrtiko, Muscat Blanc, Vidiano, Athiri, Liatiko, Kotsifali, Syrah, and Cabernet Sauvignon.
✓ *Idaia • Estate Michalakis • Rous • Minos Miliarakis • Lyrarakis • Manousakis*

DAFNES PDO
Crete

This PDO covers dry sweet reds. Douloufakis is the best-known producer of these wines made from the Liatiko grape variety.

DODEKANISA (DODECANESE) PGI
Aegean Sea

This PGI District is for white (dry, semi-dry, or semi-sweet), rosé (dry, semi-dry, or semi-sweet), and red (dry, semi-dry, semi-sweet, or sweet), and white and rosé sparkling and semi-sparkling from Dodekanisa. There are 16 varieties allowed, including Athiri, Assyrtiko, Malagousia, Muscat Blanc, Muscat Trani, Chardonnay, Sauvignon Blanc, Mandilaria, Mavrothiriko, Cabernet Sauvignon, Merlot, and Syrah.

DRAMA PGI
Macedonia

This PGI District, a potentially exciting wine-producing region in northern Greece, is worthy of a PDO status. The PGI covers white (dry, semi-dry, semi-sweet, or sweet), rosé (dry, semi-dry, semi-sweet, or sweet), and red (dry, semi-dry, semi-sweet, or sweet) from Drama. There are 20 varieties allowed, including Assyrtiko, Malagousia, Chardonnay, Sauvignon Blanc, Sémillon, Ugni Blanc, Agiorgitiko, Limnio, Cabernet Sauvignon, Cabernet Franc, Merlot, Nebbiolo, Refosco, Syrah, and Tempranillo.
✓ *Oinogenesis • Pavlidis • Wine Art • Costa Lazaridi*

ELASSONA PGI
Thessaly

This PGI Area is for white (dry, semi-dry, semi-sweet, or sweet), rosé (dry, semi-dry, semi-sweet, or sweet) and red (dry, semi-dry, semi-sweet, or sweet) from Thessaly.

Debina, Sauvignon Blanc, Chardonnay, Ugni Blanc, Muscat of Hamburg, Cabernet Sauvignon, Merlot, Syrah

EPANOMI PGI
Macedonia

This PGI Area is for dry white and reds from Epanomi. Oenologist Evangelos Gerovassiliou single-handedly put this PGI on the map, definitively producing some of the country's greatest wines, both red and white.

Assyrtiko, Malagousia, Viognier, Chardonnay, Sauvignon Blanc, Limnio, Xinomavro, Grenache Rouge, Cabernet Sauvignon, Merlot, Syrah
✓ *Gerovassiliou*

EPIRUS PGI
Epirus

This Regional PGI is for white (dry, semi-dry, semi-sweet, or sweet and sparkling), rosé (dry or semi-dry), and red (dry). The whites and the rosés may be semi-sparkling.

Debina, Riesling, Chardonnay, Gewürztraminer, Bekari, Vlahiko, Xinomavro, Grenache Rouge, Cabernet Sauvignon, Merlot, Syrah

EVVOIA PGI
Central Greece

This PGI District is for white (dry, semi-dry, semi-sweet, or sweet), rosé (dry, semi-dry, semi-sweet, or sweet), and red (dry, semi-dry, semi-sweet, or sweet) from Evvoia. There are 20 varieties allowed, including Assyrtiko, Malagousia, Monemvasia, Moschofilero, Roditis, Savatiano, Sauvignon Blanc, Agiorgitiko, Karabraimis, Liatiko, Mandilaria, Cabernet Sauvignon, Merlot, and Syrah.
✓ *Avantis • Vrinioti • Lycos*

EVROS PGI
Thrace

This PGI District is for white (dry, semi-dry, semi-sweet, or sweet), rosé (dry, semi-dry, semi-sweet, or sweet), and red (dry, semi-dry, semi-sweet, or sweet) from Evros. There are 21 varieties allowed, including Assyrtiko, Zoumiatiko, Malagouzia, Mavroudi, Keratsouda, Karnachalades, Limnio, Pamidi, Cabernet Sauvignon, Merlot, and Syrah.

FLORINA PGI
Macedonia

This PGI District is for white (dry, semi-dry, semi-sweet, or sweet), rosé (dry, semi-dry, semi-sweet, or sweet), and red (dry, semi-dry, semi-sweet, or sweet). The white may be semi-sparkling. There are 19 varieties allowed, including Assyrtiko, Malagousia, Riesling, Chardonnay, Sauvignon Blanc, Gewürztraminer, Limnio, Xinomavro, Cabernet Sauvignon, Montepulciano, Pinot Noir, Syrah, and Tannat.
✓ *Alpha Estate*

FTHIOTIDA PGI
Central Greece

This PGI is for white (dry, semi-dry, semi-sweet, or sweet), rosé (dry, semi-dry, semi-sweet, or sweet), and red (dry, semi-dry, semi-sweet, or sweet). There are 25 varieties allowed, including Assyrtiko, Verdicchio, Robola, Savatiano, Chardonnay, Mavroudi, Vradyano, Limnio, and Cabernet Sauvignon.

GERANEIA PGI
Central Greece

This PGI Area is for white (dry), rosé (dry), and red (dry) from Attiki (Attica).

Assyrtiko, Savatiano, Chardonnay, Sauvignon Blanc, Roditis, Agiorgitiko, Grenache Rouge, Carignan, Merlot, Syrah

GOUMENISSA PDO
Macedonia

A red wine from the Goumenissa district, northeast of Naoussa, it is usually of good fruit and has a certain elegance. The best wines undergo a light maturation in cask and can be relatively rich in flavour.

🍷 *Xinomavro requires a minimum of 20% Negoska*
✔ *Boutaris • Chatzivaritis • Aidarinis • Tatsis*

GREVENA PGI
Macedonia

This PGI District is for rosé (dry, semi-dry, or semi-sweet) and red (dry, semi-dry, or semi-sweet) from Grevena.

🍷 *Roditis, Xinomavro, Moschomavro, Merlot, Cabernet Sauvignon, Syrah*

IKARIA PGI
North Aegean

This PGI Area is for white (dry, semi-dry, or semi-sweet), rosé (dry, semi-dry, or semi-sweet), and red (dry, semi-dry, or semi-sweet) from Ikaria.

🍷 *Athiri, Assyrtiko, Begleri, Vaftra, Mandilaria, Fokiano*
✔ *Afianes*

ILEIA PGI
Peloponnese

This PGI District is for white (dry or semi-dry), rosé (dry or semi-dry) , and red (dry or semi-dry) from the Ileia (Elis) region. There are 24 varieties of grape allowed, including Asproudes, Assyrtiko, Robola, Fileri, Roditis, Viognier, Chardonnay, Sauvignon Blanc, Agiorgitiko, Avgoustiatis, Mavrodaphne, Mavroudi, Cabernet Sauvignon, Merlot, Refosco, and Syrah.

ILION PGI
Central Greece

This PGI Area is for dry white wines from Attiki (Attica).

🍷 *Malagousia, Savatiano, Chardonnay, Sauvignon Blanc, Roditis*

IMATHIA PGI
Macedonia

This PGI District is for white (dry, semi-dry, or semi-sweet), rosé (dry, semi-dry, or semi-sweet), and red (dry, semi-dry, or semi-sweet) from Imathia.

🍷 *Assyrtiko, Athiri, Priknadi, Malagouzia, Roditis, Xinomavro, Merlot, Syrah*

IOANNINA PGI
Epirus

This PGI District is for white (dry or semi-dry) , rosé (dry or semi-dry), and red (dry). The white may be sparkling or semi-sparkling. The rosé may be semi-sparkling. There are 15 varieties allowed, including the local Debina, Vlachiko, and Bekari along with grapes such as Roditis, Xinomavro, Riesling, Cabernet Sauvignon, and Syrah.

IRAKLEIO (HERAKLION) PGI
Crete

This PGI District is for white (dry, semi-dry, semi-sweet, or sweet), rosé (dry, semi-dry, semi-sweet, or sweet), and red (dry, semi-dry, semi-sweet, or sweet) from Irakleio on Crete. There are 22 varieties of grapes allowed, including Vilana, Vidiano, Dafni, Thrapsathiri, Muscat Blanc, Plyto, Kotsifali, Liatiko, Mandilaria, Cabernet Sauvignon, Carignan, Merlot, Mourvèdre, and Syrah.
✔ *Boutari*

ISMAROS PGI
Thrace

This PGI Area is for white (dry, semi-dry, or semi-sweet), rosé (dry, semi-dry, or semi-sweet), and red (dry, semi-dry, or semi-sweet) from Maronia/Rodopi.

🍷 *Muscat of Alexandria, Chardonnay, Sauvignon Blanc, Roditis, Limnio, Mavroudi, Grenache Rouge, Cabernet Sauvignon, Merlot, Mourvèdre, Syrah*
✔ *Kikones*

KARDITSA PGI
Thessaly

This PGI District is for white (dry, semi-dry, or semi-sweet), rosé (dry, semi-dry, or semi-sweet), and red (dry, semi-dry, or semi-sweet) wines.

🍷 *Assyrtiko, Malagousia, Batiki, Debina, Chardonnay, Roditis, Limniona, Mavro Mesenikola, Cabernet Sauvignon, Carignan, Cinsault, Merlot, Syrah*

KARYSTOS (KARISTOS) PGI
Central Greece

This PGI Area is for white (dry, semi-dry, semi-sweet, or sweet), rosé (dry, semi-dry, or semi-sweet), and red (dry, semi-dry, semi-sweet, or sweet) from Evvoia. There are 16 varieties allowed, including Assyrtiko, Savatiano, Roditis, Agiorgitiko, Liatiko, Mandilaria, Cabernet Sauvignon, and Syrah.

KASTORIA PGI
Macedonia

This PGI District is for white (dry or sweet) and red (dry or sweet) from Kastoria.

🍷 *Roditis, Sauvignon Blanc, Xinomavro, Sefka, Cabernet Sauvignon, Cinsaut, Merlot, Syrah*

KAVALA PGI
Macedonia

This District PGI is for white (dry, semi-dry, semi-sweet, or sweet), rosé (dry, semi-dry, semi-sweet, or sweet), red (dry, semi-dry, semi-sweet, or sweet) wines from Kavala. There are 19 grape varieties allowed.

🍷 *Assyrtiko, Malagousia, Muscat of Alexandria, Roditis, Viognier, Gewürztraminer, Chardonnay, Semillon, Sauvignon Blanc, Trebbiano, Agiorgitiko, Lemnio, Moschomavro, Pamidi, Grenache Rouge, Cabernet Franc, Cabernet Sauvignon, Merlot, and Syrah.*
✔ *Lalikos*

KERKYRA (CORFU) PGI
Ionian Islands

This PGI District is for white (dry) from Kerkyra (Corfu).

🍷 *Kakotrygis, Petrokoritho*

KISSAMOS PGI
Crete

This PGI Area is for white (dry), rosé (dry), and red (dry) wines from Chania.

🍷 *Athiri, Thrapsathiri, Vilana, Romeiko, Ugni Blanc (a maximum of 20%), Cabernet Sauvignon, Grenache Rouge, Carignan, Syrah*

KLIMENTI PGI
Peloponnese

This PGI Area is for white (dry), rosé (dry), and red (dry) wine from the Korinthos (Corinth region).

🍷 *Malagousia, Chardonnay, Moschofilero, Agiorgitiko, Cabernet Sauvignon, Merlot, Syrah*

KORINTHOS (CORINTH) PGI
Peloponnese

This PGI District is for white (dry), rosé (dry), and red (dry) wines from the Korinthos (Corinth) region.

🍷 *Assyrtiko, Savatiano, Asproudes, Lagorthi, Malagousia, Moschofilero, Roditis, Chardonnay, Sauvignon Blanc, Agiorgitiko, Mavroudi, Cabernet Sauvignon, Merlot, Syrah*
✔ *Boutari*

KOROPI PGI
Central Greece

This PGI Area is for dry white wines from Attiki (Attica).

🍷 *Savvatiano*

KOS PGI
Aegean Sea

This PGI Area is for white (dry, semi-dry, or semi-sweet), rosé (dry, semi-dry, or semi-sweet), and red (dry, semi-dry, or semi-sweet) from the Kos island.

🍷 *Athiri, Assyrtiko, Malagousia, Chardonnay, Sauvignon Blanc, Grenache Rouge, Cabernet Sauvignon, Merlot, Mourvèdre, Cinsaut, Syrah, Tempranillo*
✔ *Triantafillopoulos*

KOZANI PGI
Macedonia

This PGI District is for white (dry, semi-dry, or semi-sweet), rosé (dry, semi-dry, or semi-sweet), and red (dry, semi-dry, or semi-sweet) from Kozani.

🍷 *Malagousia, Batiki, Priknadi, Chardonnay, Ugni Blanc, Gewürztraminer, Roditis, Limnio, Moschomavro, Xinomavro, Cabernet Sauvignon, Cinsaut, Merlot, Syrah*

KRANIA PGI
Thessaly

This PGI Area is for white (dry) and red (dry) from Larissa.

🍷 *Chardonnay, Cabernet Sauvignon, Merlot*
✔ *Katsaros*

KRANNONAS PGI
Thessaly

This PGI Area is for white (dry) and red (dry) from Larissa.

🍷 *Sauvignon Blanc, Cabernet Sauvignon, Merlot, Syrah*
✔ *Karipidis*

KRITI (CRETE) PGI
See Crete PGI

KYKLADES (CYCLADES) PGI
Aegean Sea

This PGI District is for white (dry, semi-dry, semi-sweet, or sweet), rosé (dry, semi-dry, semi-sweet, or sweet), and red (dry, semi-dry, semi-sweet, or sweet) from the Kyklades islands. There are 25 varieties allowed, including Athiri, Aidani White, Asprouda Santorinis, Assyrtiko, Gaidouria, Katsano, Kritiko, Maloukato, Mandilaria White, Muscat White, Savvatiano, Athiri Black, Aidani Black, Avgoustiatis, Vaftra, Voydomatis, Kotsifali, Mandilaria, Mavrotragano, and Fokiano.

LAKONIA (LACONIA) PGI
Peloponnese

This PGI District is for white (dry, semi-dry, semi-sweet, or sweet), rosé (dry, semi-dry, or semi-sweet), and red (dry, semi-dry, semi-sweet, or sweet) wines.

🍷 *Athiri, Assyrtiko, Aidani, Kydonitsa, Malagousia, Monemvasia, Petroulianos, Roditis, Agiorgitiko, Thrapsa, Mandilaria, Mavroudi, Cabernet Sauvignon, Merlot*
✔ *Tsibidis • Theodorakakos • Vatistas*

LASITHI PGI
Crete

This PGI District is for white (dry, semi-dry, or semi-sweet), rosé (dry, semi-dry, or semi-sweet), and red (dry, semi-dry, or semi-sweet) from Lasithi. There are 18 varieties allowed, including Assyrtiko, Vilana, Thrapsathiri, Muscat Blanc, Plyto, Chardonnay, Kotsifali, Liatiko, Mandilaria, Cabernet Sauvignon, Carignan, Merlot, and Syrah.

LEFKADA PGI
Ionian Islands

This PGI District is for white (dry), rosé (dry, semi-dry, semi-sweet, or sweet), and red (dry, semi-dry, or semi-sweet) from Lefkada.

🍷 *Vardea, Vertzami, Merlot*

LESVOS PGI
Lesvos

This PGI District is for white (dry, semi-dry, semi-sweet, or sweet), rosé (dry, semi-dry, semi-sweet, or sweet), and red (dry, semi-dry, semi-sweet, or sweet) wines from Lesvos (Lesbos).

🍇 *Assyrtiko, Savatiano, Athiri, Chiridiotiko, Mandilaria, Vaftra, Fokiano*

LETRINA PGI
Peloponnese

This PGI Area is for dry red wines from Ileia.

🍇 *Agiorgitiko, Mavrodaphne, Merlot, Refosco, Syrah*
✓ Mercouri

LIMNOS PDO
North Aegean

This PDO covers white (dry, semi-dry, semi-sweet, or sweet) and red (dry, sweet). Limnos white wines are usually soft and flowery, with an attractive Muscat aroma, followed by clean fruit. Sweet PDO reds are based on Limnio but can also have 10 per cent of Muscat.

🍇 *Muscat of Alexandria, Limnio*
✓ *Chatzigeorgiou • Limnos Organic Wines*

LILANTIO PEDIO PGI
Central Greece

This PGI is for white (dry, semi-dry, semi-sweet, or sweet), rosé (dry, semi-dry, semi-sweet, or sweet), and red (dry, semi-dry, semi-sweet, or sweet) from Evvoia. There are 16 varieties allowed.

🍇 *Athiri, Assyrtiko, Malagousia, Moschofilero, Roditis, Savvatiano, Sauvignon Blanc, Agiorgitiko, Vradiano, Karabraïmis, Mandilaria (Kountoura Black), Ritino, Grenache Rouge, Cabernet Sauvignon, Merlot, Syrah*

MACEDONIA PGI
Macedonia

This Regional PGI is for white (dry, semi-dry, or semi-sweet), rosé (dry, semi-dry, or semi-sweet), and red (dry, semi-dry, or semi-sweet), and white semi-sparkling (dry or semi-sweet) wines from 44 grape varieties, including all the grapes authorised for any of the PDOs in the region indicated.

MAGNISIA PGI
Thessaly

This PGI District is for white (dry, semi-dry, or semi-sweet), rosé (dry, semi-dry, or semi-sweet), and red (dry, semi-dry, or semi-sweet) wines.

🍇 *Asproudes, Assyrtiko, Savatiano, Ugni Blanc, Roditis, Vradiano, Limnio, Mavroudi, Xinomavro, Sykiotis, Grenache Rouge, Cabernet Sauvignon, Merlot, Syrah*

MALVASIA CANDIA PDO
Crete, Irakleio

This white sweet wine appellation, also known as Malvasia Chandakas-Candia, uses the grapes Assyrtiko, Athiri, Vidiano, Thrapsathiri, Liatiko Muscat Spinas, and Malvasia di Candia Aromatica. The last two cannot constitute more than 15 per cent of the blend. Wines can be fortified as well as naturally sweet.

MALVASIA PAROS PDO
Paros

This white sweet wine appellation requires at least 85 per cent Monemvasia and Assyrtiko. Wines can be fortified as well as naturally sweet.

MALVASIA SITEIA PDO
Crete, Lasithi

This white sweet wine appellation uses the grapes Assyrtiko, Athiri, Thrapsathiri, Liatiko Muscat Spinas, and Malvasia di Candia Aromatica. The last two cannot constitute more than 15 per cent of the blend. Wines can be fortified as well as naturally sweet.

MANTINIA PDO
Peloponnese

This is a dry white wine from mountain vineyards in the centre of the Peloponnese, where the vines grow at an altitude of 650 metres (2,130 feet), producing wines with light palate structure, nice lively fruit, floral nose, and crisp acidity.

🍇 *A minimum of 85% Moschofilero, with Asproudes*
✓ *Semeli • Tselepos • Spiropoulos • Troupis • Boutari • Bossinakis*

MANTZAVINATA PGI
Ionian Islands

This PGI Area is for white (dry, semi-dry, or semi-sweet), rosé (dry, semi-dry, or semi-sweet), and red (dry, semi-dry, or semi-sweet) wines from Kefalonia. There are 42 varieties allowed, including Goustolidi (Vostilidi), Tsaousi, Moschatela, Araklino, Thiako, and Mavrodaphne.

MARKOPOULO PGI
Central Greece

This PGI Area is for dry white wines from Attiki (Attica).

🍇 *A minimum of 80% Savatiano, plus any of the following: Assyrtiko, Athiri, Aidani Aspro, Roditis*

MARTINO PGI
Central Greece

This PGI Area is for white (dry, semi-dry, semi-sweet, or sweet), rosé (dry, semi-dry, or semi-sweet), and red (dry, semi-dry, semi-sweet, or sweet) from Fthiotida.

🍇 *Athiri, Malagousia, Robola, Chardonnay, Sauvignon Blanc, Ugni Blanc, Roditis, Cabernet Sauvignon, Cabernet Franc, Merlot, Syrah*

MAVRODAPHNE OF KEFALONIA PDO
Ionian Islands

The Mavrodaphne of Kefalonia is a sweet, red fortified wine, similar in character to the Mavrodaphne of Patra but not quite in the same class.

🍇 *Mavrodaphne*

MAVRODAPHNE OF PATRA PDO
Peloponnese

A rich, sweet, red liqueur wine with a velvety smooth, sweet-oak finish, initially made in the style of port. One delightful aspect is that this wine can be drunk with equal pleasure either young and fruity or smooth and mature.

🍇 *A minimum of 51% Mavrodaphne and Corinthiaki*
✓ *Achaia Clauss • Parparoussis*

MESENIKOLA PDO
Thessaly

The primary variety of this appellation, Mavro Mesenikola, is grown at an altitude of 250 to 600 metres (820 to 1,970 feet).

🍇 *70% Mavro Messenikola, 30% Carignan and Syrah*
✓ *Monsieur Nicolas*

MESSINIA PGI
Peloponnese

This PGI District is for white (dry) and red (dry) from Messinia.

🍇 *Assyrtiko, Lagorthi, Arintho, Chardonnay, Sauvignon Blanc, Ugni Blanc, Roditis, Fokiano, Grenache Rouge, Cabernet Sauvignon, Carignan, Merlot*

METAXATA PGI
Ionian Islands

This PGI Area is for red (dry, semi-dry, or semi-sweet) from Kefalonia (Cephalonia).

🍇 *Mavrodaphne*

METEORA PGI
Thessaly

This PGI Area is for white (dry, semi-dry, semi-sweet, or sweet), rosé (dry, semi-dry, semi-sweet, or sweet), and red (dry, semi-dry, or semi-sweet), and white sparkling (dry or semi-dry) from Trikala.

🍇 *Assyrtiko, Malagousia, Batiki, Debina, Roditis, Zalovitiko, Xinomavro, Limniona, Vlachiko, Cabernet Sauvignon, Cinsaut, Syrah, Merlot*

METSOVO PGI
Epirus

This PGI Area is for white (dry, semi-dry, semi-sweet, or sweet) and red (dry) from Ioannina.

A VOTIVE SHRINE SITS IN A NEMEA VINEYARD
The Nemea PDO makes richly coloured, spicy-flavoured wines from the Agiorgitiko grape variety (*see* p538).

🍇 *Debina, Gewürztraminer, Vlahiko, Cabernet Sauvignon, Cabernet Franc, Merlot*

✓ *Averoff*

MONEMVASIA MALVASIA PDO

Peloponnese

Both fortified and naturally sweet styles are eligible for this historic sweet wine, produced from sun-dried grapes predominantly from the white grape Monemvasia, which has to be at least 51 per cent of the blend. The rest can be Assyrtiko, Kydonitsa, and Asproudes.

✓ *Monemvasia Winery*

MUSCAT OF KEFALONIA PDO

Ionian Islands

Muscat of Kefalonia is one of the lesser-known, sweet liqueur wines from the Muscat grape.

🍇 *Muscat Blanc*

MUSCAT OF LEMNOS PDO

North Aegean

A superior liqueur Muscat wine, though not reaching the character of Muscat from Samos. A *grand gru* designation can be used for wines from selected vineyards.

🍇 *Muscat of Alexandria*

MUSCAT OF PATRA PDO

Peloponnese

The PDO produces a gold-coloured, sweet liqueur Muscat that can be delicious.

🍇 *Muscat Blanc*

✓ *Parparoussis*

MUSCAT OF RODOS PDO

Aegean Sea

The PDO covers all sweet wines from sun-dried to fortified.

🍇 *Muscat Blanc, Muscat Trani*

MUSCAT OF RIO PATRA PDO

Peloponnese

From a defined area around Rio, this PDO produces a gold-coloured sweet Muscat that is often more floral and less intense than Muscat of Patra.

🍇 *Muscat Blanc*

NAOUSSA PDO

Macedonia

Generally reliable wines are grown west of Thessaloniki at an altitude ranging from 150 to 400 metres (492 to 1,312 feet). Xinomavro in Naoussa is often compared to Nebbiolo due to its tannic structure, the lack of primary fruit, and the high acidity. A Naoussa is one of the great Greek wines that can display a remarkable complexity over a long period of ageing, which can stretch over decades. Semi-dry and semi-sweet reds are also allowed.

🍇 *Xinomavro*

✓ *Boutari • Taralas • Kir-Yianni • Thimiopoulos • Dalamaras • Karidas • Fountis*

NEA MESIMVRIA PGI

Macedonia

This PGI Area is for white (dry or semi-dry) and red (dry or semi-dry) from Thessaloniki.

🍇 *Zoumiatiko, Malagousia, Chardonnay, Sauvignon Blanc, Roditis, Limnio, Grenache Rouge, Cabernet Sauvignon, Merlot, Cinsaut, Syrah*

NEMEA PDO

Peloponnese

Grown in the Korinthos (Corinth) district at an altitude of about 230 to 900 metres (755 to 2,953 feet), the Agiorgitiko grape, meaning St-George, comes into its own, producing deep-coloured red wines, fruity with sweet spices. A Nemea can be light or, with more extract, capable of ageing. Semi-sweet and sweet styles are also permitted.

🍇 *Agiorgitiko*

✓ *Gaia • Mitravelas • Papaïoannou • Pavilos • Repanis • Tselepos • Semeli Estate • Skouras*

OPOUNTIA LOKRIS PGI

Central Greece

This PGI Area is for white (dry, semi-dry, semi-sweet, or sweet), rosé (dry, semi-dry, or semi-sweet), and red (dry, semi-dry, semi-sweet, or sweet) wines from Fthiotida. There are 15 varieties allowed including, Assyrtiko, Malagouzia, Chardonnay, Roditis, Merlot, Limnio, Sauvignon Blanc, Xinomavro, Cabernet Sauvignon, Cabernet Franc, and Syrah.

✓ *Chatzimichalis (Opus ıβ΄)*

PAIANIA (PEANEA) PGI

This PGI Area is for white (dry) wine from Attiki (Attica).

🍇 *Assyrtiko, Savatiano*

PALLINI PGI

Central Greece

This PGI Area is for white (dry) wine from Attiki (Attica).

🍇 *Savatiano, Assyrtiko, Malagouzia, Sauvignon Blanc, Roditis*

PANGEON PGI

Macedonia

This PGI Area is for white (dry, sweet), rosé (dry or semi-dry), and red (dry or semi-dry).This is an exciting regional wine area south of Drama with a sea-influenced mountainous climate, the elevation and weather of which result in a significantly longer growing season. There are 19 varieties allowed, including Assyrtiko, Muscat of Alexandria, Malagouzia, Chardonnay, Sauvignon Blanc, Agiorgitiko, Limnio, Cabernet Sauvignon, Merlot, and Syrah.

✓ *Biblia Chora • Hatzigeorgiou • Ampeloeis*

PARNASSOS PGI

Central Greece

This PGI Area is for white (dry), rosé (dry), and red (dry) from Fokida and Fthiotida. There are 14 varieties allowed, including Assyrtiko, Malagousia, Robola, Savatiano, Roditis, Mavroudi, Agiorgitiko, Cabernet Sauvignon, Merlot, and Syrah.

PAROS PDO

Aegean Sea

The red is a deep-coloured wine of one-third black Mandilaria grape and two-thirds from the white aromatic Monemvassia grape. A pure white Monemvassia wine is also produced.

✓ *Moraitis*

PATRA PDO

Peloponnese

Made exclusively from the Roditis grape, Patra is a white wine that can be dry, semi-dry, semi-sweet, or sweet; it is most often found in the dry style. A Roditis from high altitude can be intense, full of ripe fruit, crisp, and well-balanced.

🍇 *Roditis*

✓ *Rouvalis*

PELLA PGI

Macedonia

This PGI District is for white (dry or semi-dry), rosé (dry or semi-dry), and red (dry or semi-dry) wines.

🍇 *Chardonnay, Sauvignon Blanc, Ugni Blanc, Roditis, Agiorgitiko, Limnio, Moschomavro, Xinomavro, Negoska, Cabernet Sauvignon, Merlot, Cinsaut, Syrah*

✓ *Ligas*

PELOPONNESE PGI

Peloponnese

This Regional PGI is for white (dry, semi-dry, semi-sweet, or sweet), rosé (dry, semi-dry, semi-sweet, or sweet), and red (dry, semi-dry, semi-sweet, or sweet), and white and rosé semi-sparkling wines from 49 grape varieties including grapes authorised for any of the PDOs in the region indicated.

PEZA PDO

Crete

The white wines are made from Vilana and can be aromatic, fresh, and full of ripe fruit, whereas the reds are blends of 75 per cent Kotsifalii and 25 per cent Mandilaria that display elegance and bright, spicy fruit.

PIERIA PGI

Macedonia

This PGI District is for white (dry, semi-dry, semi-sweet, or sweet), rosé (dry, semi-dry, or semi-sweet), and red (dry, semi-dry, semi-sweet, or sweet) from Pieria. There are 16 varieties allowed, including Assyrtiko, Malagousia, Chardonnay, Sauvignon Blanc, Roditis, Agiorgitiko, Xinomavro, Cabernet Sauvignon, Merlot, and Syrah.

PISATIS PGI

Peloponnese

This PGI Area is for white (dry) wine from Ileia.

🍇 *Fileri, Roditis, Chardonnay, Sauvignon Blanc*

PYLIA PGI

Peloponnese

This PGI Area is for white (dry) wine from Messinia.

🍇 *Assyrtiko, Chardonnay, Ugni Blanc, Roditis*

RAPSANI PDO

Thessaly

In Rapsani, vineyards located in the lower foothills of Mount Olympus reach up to 700 metres (2,300 feet) in altitude. The wines are made from equal parts of Xinomavro, Krasato, and Stavroto grapes. Modern Rapsani wines are polished, with a great depth of fruit, capable of ageing in the long term.

✓ *Tsantalis • Dougos • Liapis*

RETHYMNO PGI

Crete

This PGI District is for white (dry, semi-dry, semi-sweet, or sweet), rosé (dry, semi-dry, or semi-sweet), and red (dry, semi-dry, semi-sweet, or sweet) from Rethymno. There are 12 varieties allowed, including Muscat Blanc, Vidiano, Thrapsathiri, Vilana, Romeiko, Mandilaria, Kotsifali, and Syrah.

RETSINA PGI

Although rosé is not unknown, most Retsina is white, with most of the wines coming from the Savatiano or Roditis grapes. Production is mostly in Central Greece, and most of that is from Attiki (Attica). There are 15 Retsina PGI Traditional Appellations: Attiki, Evvoia, Gialtra, Chalkida, Karystos, Koropi, Markopoulo, Megara, Mesogia, Pallini, Paiania, Pikermi, Spata, Thiva, and Voiotia.

Traditionally, Retsina is made by adding pine resin to the wine during fermentation – a practice that dates back to antiquity, when wine was stored in jars and amphorae. As these vessels were not air-tight, the wines rapidly deteriorated. Over the course of time, people learned to seal the jars with a mixture of plaster and resin, and the wines naturally lasted longer. This increased longevity was attributed to the antiseptic effect of resin, the aroma and flavour of which was very noticeable in the best-preserved wines. Within a short time, resin was used as a flavouring agent added directly to the wine and practically masking any off flavours developed during vinification. Over the years, the amount of resin added has been gradually decreased, and more careful use of it resulted in more balanced wines.

Today, according to the legislation, the presence of resin in the wine must be restricted between 0.15 and 1 per cent of the final product, while the minimum acidity required must be 4.5 grams per litre, and the alcohol between 10 per cent and 13.5 per cent ABV. The same standards apply for the rosé Retsina called Kokkineli. Nowadays a number of producers around Greece produce some exciting top-quality Retsinas in an array of different styles such as sparkling or orange, sometimes using oak barrels or amphorae for ageing. Varieties grown are primarily Roditis and/or Savatiano, but numerous other grape varieties are permitted outside Central Greece.

✓ *Gaia • Kechris • Mylonas • Georgas • Tetramythos*

RITSONA PGI
Central Greece

This PGI Area is for white (dry, semi-dry, semi-sweet, or sweet), rosé (dry, semi-dry, or semi-sweet), and red (dry, semi-dry, or semi-sweet) from Evvoia. There are 20 varieties allowed, including Assyrtiko, Malagousia, Savatiano, Moschofilero, Agiorgitiko, Vradiano, Cabernet Sauvignon, Merlot, and Syrah.

ROBOLA OF KEFALONIA PDO
Ionian Islands

This important white wine appellation (dry) is located on the island of Kefalonia. Robola is considered one of the noblest Greek varieties, producing wines of elegance with depth of fruit and flavoursome extract.

🍷 *Robola*

RODOS (RHODES) PDO
Aegean Sea

Reds are blends of at least 70 per cent Mandilaria, locally called Amoriano, and Mavrothiriko, while the island's white appellation includes wines 70 per cent Athiri ,with Malagouzia and Assyrtiko. CAIR is the major wine producing company on Rodos, of great commercial significance nationally and well-known for its sparkling wine production.

SAMOS PDO
North Aegean

This is one of the greatest sweet Muscats in the world. The local Samos Union of Winemaking Cooperatives is the major supplier of Samos wines. Sweet wines are made in distinctive styles that range from the very sweet fortified and without-ageing Samos Vin Doux, Vin Doux Naturel, and Vin Doux Naturel Grand Cru to the sun-dried, naturally sweet "Nectar" that contains about 130 g/L of unfermented sugars and is aged for six years. Samos dry wines are not allowed to bear the PDO designation, yet they can be delightful, with the fresh and floral character that only a Muscat, coming from the island's mountainous vineyards can develop.

🍷 *Muscat Blanc*

✓ *Union of Winemaking Cooperatives • Nopera • Jason Ligas*

SANTORINI PDO
Aegean Sea

The beauty of the wines produced here is indisputable. Its main grape variety, Assyrtiko, is producing wines with intense character, steely acidity, high extract, minerality, and a volcanic *goût de terroir* that can convince even the most challenging palates. PDO Santorini encompasses both dry and sweet styles. Sweet *vin santo* wines can age over a period of many decades, and they can be just majestic. Other traditional styles include Nykteri, a high-alcohol wine made from overripe grapes that undergo oxidative ageing in oak barrels, developing complex aromatics and flavours.

🍷 *A minimum of 75% Assyrtiko, with Athiri (only in dry whites) and Aidani*

✓ *Argyros • Boutari • Hatzidakis • Vassaltis • Karamolegos • Sigalas • Santo wines • Gaia*

SERRES PGI
Macedonia

This PGI District is for white (dry, semi-dry, or semi-sweet), rosé (dry, semi-dry, or semi-sweet), and red (dry, semi-dry, or semi-sweet) wines from Serres. There are 20 varieties allowed, including Zoumiatiko, Assyrtiko, Muscat of Alexandria, Batiki, Roditis, Limnio, Pamidi, Merlot, and Syrah.

SIATISTA PGI
Macedonia

This PGI Area is for white (dry or sweet), rosé (dry or sweet), and red (dry) from Kozani.

🍷 *Batiki, Priknadi, Gewürztraminer, Moschomavro, Xinomavro, Chondromavro, Cabernet Sauvignon, Merlot, Cinsaut*

SITHONIA PGI
Macedonia

This PGI Area is for white (dry) and red (dry) from Chalkidiki. There are 26 varieties allowed; for white wines at least 60 per cent Malagousia is required and for the reds a minimum of 60 per cent Syrah.

SITEIA PDO
Crete

The red wines can be dry or sweet, fortified or not, and must be made from at least 80 per cent Liatiko and Mandilaria. White Siteia wines must be dry from at least 70 per cent Vilana, together with Thrapsathiri. They cannot be fortified. Winemaker Yiannis Economou has been working hard since the mid-1990s to bring elegance and finesse to these wines while creating a great demand from export markets.

✓ *Economou Domaine*

SLOPES OF AIGIALEIA PGI
Peloponnese

This PGI Area is for white (dry), rosé (dry), and red (dry), and white semi-sparkling (dry, semi-dry, or semi-sweet) from the hilly coast of the Achaia province.

🍷 *Lagorthi, Assyrtiko, Robola, Malagouzia, Chardonnay, Riesling, Sauvignon Blanc, Sideritis, Roditis, Agiorgitiko, Mavrodaphne, Volitsa, KalavritinoMavro, Cabernet Sauvignon, Merlot, Syrah*

✓ *Rouvalis*

A VINE DRESSER WEAVES THE CANES OF A GRAPEVINE TO FORM A *KOULOURA* IN A SANTORINI VINEYARD
This pruning method developed on Santorini. The grapes will grow facing inward in the wreath-like basket, which protects them from the brilliant sun and harsh winds of the island.

SLOPES OF AINOS PGI
Ionian Islands

This PGI Area is for white (dry, semi-dry, semi-sweet, or sweet), rosé (dry, semi-dry, semi-sweet, or sweet), and red (dry, semi-dry, semi-sweet, or sweet), and white semi-sparkling from Kefalonia. There are 19 varieties allowed: Goustolidi, Moschatela, Robola, Skiadopoulo, Zakynthino, Roditis, Tsaoussi, Avgoustiatis, and Mavrodaphne are most important.

✓ *Petrakopoulos*

SLOPES OF AMPELOS PGI
Aegean Sea

This PGI Area is for dry white wines.

🍇 *Muscat Blanc*

SLOPES OF KITHERONA PGI
Central Greece

This PGI Area is for white (dry, semi-dry, or semi-sweet), rosé (dry or semi-dry), and red (dry, semi-dry, or semi-sweet) from Attiki (Attica) and Viotia.

🍇 *Athiri, Assyrtiko, Savatiano, Chardonnay, Malagousia, Sauvignon Blanc, Chardonnay, Roditis, Agiorgitiko, Grenache, Cabernet Sauvignon, Carignan, Merlot, Syrah*

SLOPES OF KNIMIDA PGI
Central Greece

This PGI Area is for white (dry, semi-dry, semi-sweet), rosé (dry or semi-dry), and red (dry, semi-dry, or semi-sweet) from Fthiotida. There are 19 varieties allowed, of which Assyrtiko, Savatiano, Malagousia, Chardonnay, Sauvignon Blanc, Roditis, Vradiano, Limnio, Mavroudi, Cabernet Sauvignon, Merlot, and Syrah are the most important.

SLOPES OF MELITON PDO
Macedonia

This appellation covers the red and dry white wines on the central-western coast of Sithonia. The Porto Carras Estate winery was created here and sold to the Stegos family in the 1990s. In the early 1980s, Domaine Porto Carras's flagship Château Carras was one of the first Greek wines of modern times to establish a truly international reputation. At the time the estate was the largest private viticultural enterprise of Europe. Professors Logothetis and Émile Peynaud were hired as consultants with Evangelos Gerovassiliou, who later became the senior oenologist of the estate. Evangelos, while in Porto Carras, started his own estate, Domaine Gerovassiliou (see Epanomi PGI), and is regarded until today as a trend-setter in modern winemaking in Greece, playing a major role in the revival of the almost extinct Malagouzia grape.

🍇 *Athiri, Assyrtiko, Roditis (white); Limnio, Cabernet Franc, and Cabernet Sauvignon (red)*

✓ *Porto Carras*

SLOPES OF PAIKO PGI
Macedonia

This PGI Area is for white (dry, semi-dry, semi-sweet, or sweet), rosé (dry, semi-dry, or semi-sweet), and red (dry, semi-dry, semi-sweet, or sweet) from Kilkis.

🍇 *Assyrtiko, Malagousia, Sauvignon Blanc, Chardonnay, Roditis, Limnio, Xinomavro, Negoska, Cabernet Sauvignon, Merlot, Syrah*

SLOPES OF PARNITHA PGI
Central Greece

This PGI Area is for white, rosé, and red wines and includes sites from Attiki (Attica) and Viotia prefectures.

SLOPES OF PENTELIKO PGI
Central Greece

This PGI Area is for dry white wines from Attiki (Attica).

🍇 *Assyrtiko, Savatiano, Aidani, Athiri, Malagouzia, Riesling, Chardonnay*

SLOPES OF PETROTO PGI
Peloponnese

This PGI Area is for dry red wines from Achaia.

🍇 *Mavrodaphne, Cabernet Sauvignon.*

SLOPES OF VERTISKOS PGI
Macedonia

This PGI Area is for white (dry), rosé (dry), and red (dry) from Thessaloniki.

🍇 *Athiri, Assyrtiko, Malagousia, Ugni Blanc, Chardonnay, Roditis, Xinomavro, Pamidi, Cabernet Sauvignon, Syrah*

SPATA PGI
Central Greece

This PGI Area is for white (dry) from Attiki (Attica).

🍇 *A minimum of 85% Savatiano and/or Assyrtiko, Aidani Aspro, Athiri, Malagouzia, Riesling, Chardonnay, and Roditis.*

✓ *Anastasia Fragou • Markou*

SYROS
Aegean Sea

This PGI Area is for dry white wines from the island of Syros in the Kyklades.

🍇 *Monemvassia, Assyrtiko*

TEGEA PGI
Peloponnese

This PGI Area is for red (dry) from Arcadia.

🍇 *Cabernet Sauvignon, Cabernet Franc, Merlot*

✓ *Tselepos*

THAPSANA PGI
Aegean Sea

This PGI Area is for white (dry) from Paros

THASOS PGI
Macedonia

This PGI District is for white (dry), rosé (dry), and red (dry) wines from the island of Thasos.

🍇 *Muscat of Alexandria, Assyrtiko, Limnio, Cabernet Sauvignon, Merlot*

THESSALY PGI
Thessaly

This Regional PGI is for white (dry, semi-dry, semi-sweet, or sweet), rosé (dry, semi-dry, semi-sweet, or sweet), and red (dry, semi-dry, semi-sweet, or sweet), and rosé semi-sparkling (dry or semi-dry) and white and rosé sparkling (dry or semi-dry) wines from 25 grape varieties and the grapes authorised for any of the PDOs in the region indicated.

✓ *Dougos • Tsililis*

THESSALONIKI PGI
Macedonia

This PGI District is for white (dry), rosé (dry), and red (dry) wines from Thessaloniki. There are 18 varieties allowed.

THIVA (THEBES) PGI
Central Greece

This PGI Area is for white (dry, semi-dry, semi-sweet), rosé (dry, semi-dry, semi-sweet, or sweet), and red (dry, semi-dry, or semi-sweet) wines from Viotia. There are 16 varieties allowed.

THRACE PGI
Thrace

This Regional PGI is for white (dry, semi-dry, semi-sweet, or sweet), rosé (dry, semi-dry, semi-sweet, or sweet), and red (dry, semi-dry, semi-sweet, or sweet) wines from Thrace.

There are 24 varieties allowed, including Assyrtiko, Malagouzia, Chardonnay, Muscat of Alexandria, Limnio, Mavroudi, Pamidi, Syrah, Merlot, and Cabernet Sauvignon.

✓ *Tsantali • Anatolikos • Kikones*

TRIFILIA PGI
Peloponnese

This PGI Area is for white (dry, semi-dry, semi-sweet), rosé (dry, semi-dry, or semi-sweet), and red (dry, semi-dry, or semi-sweet) from Messinia. There are 16 varieties allowed, including Chardonnay, Roditis, Fileri, Agiorgitiko, Cabernet Sauvignon, Carignan, Merlot, Syrah, and Tempranillo.

TYRNAVOS (TIRNAVOS) PGI
Thessaly

This PGI Area is for white (dry, semi-dry, semi-sweet), rosé (dry, semi-dry, or semi-sweet), and red (dry, semi-dry, semi-sweet, or sweet), and rosé semi-sparkling (dry or semi-dry) from Larissa. There are 15 varieties allowed, including Debina, Savatiano, Chardonnay, Sauvignon Blanc, Roditis, Batiki, Macabeu, Limnio, Limniona, Muscat de Hamburg, Grenache, Cabernet Sauvignon, Merlot, Syrah.

✓ *Zafeirakis*

VALLEY OF ATALANTI PGI
Central Greece

This PGI Area is for white (dry, semi-dry, semi-sweet, or sweet), rosé (dry, semi-dry, or semi-sweet), and red (dry, semi-dry, semi-sweet, or sweet) wines from Fthiotida. There are 20 varieties allowed, including Athiri, Assyrtiko, Malagousia, Roditis, Robola, Savatiano, Chardonnay, Sauvignon Blanc, Semillon, Vradyano, Limnio, Mavroudi, Cabernet Franc, Cabernet Sauvignon, Merlot, Refosco, and Syrah.

VELVENTOS PGI
Macedonia

This PGI Area is for white (dry), rosé (dry), and red (dry) from Kozani.

🍇 *Batiki, Assyrtiko, Malagouzia, Chardonnay, Roditis, Xinomavro, Limniona, Cabernet Franc (Tsapournakos), Moschomavro, Cabernet Sauvignon, Merlot, Syrah*

✓ *Voyatzi*

VERDEA OF ZAKYNTHOS PGI
Ionian Islands

This is a Traditional Designation for white (dry) wine from Zakynthos made from minimum of 50 per cent Skiadopoulo, plus other local recommended varieties, including Gourstolidi, Robola, and Pavlos.

VILITSA
Central Greece

This PGI Area is for dry red wines from Attiki (Attica)..

🍇 *Cabernet Sauvignon*

ZAKYNTHOS PGI
Ionian Islands

This PGI District is for white (dry, semi-dry, semi-sweet), rosé (dry, semi-dry, semi-sweet, or sweet), and red (dry, semi-dry, or semi-sweet) from Zakynthos. There are 19 varieties allowed, including Moschatela, Skiadopoulo, Asproudes, Mygdali, Robola, Pavlos, Goustolidi, Mavrodaphne, Skylopnichtis, Katsakoulias, and Avgoustiatis.

ZITSA PDO
Epirus

The grape variety in Zitsa is the white Debina, producing still dry, sparkling dry, and semi-dry. The wines are light bodied with low alcohol and with delicate fruity and floral aromas. The vineyards are located on the slopes of Zitsa, where the vines grow at an altitude of 700 metres (2,300 feet).

🍇 *Debina*

✓ *Glinavos • Zoinos*

The Levant

The wine regions of the Levant bear the most ancient traces of wine production and consumption, yet they have often failed to live up to that reputation in their modern-day offerings. The good news is that the last decade has brought an overall rise in technological standards of winemaking that has elevated many of these regions to world class status.

Lebanon and Israel have the most extensive boutique winery industries (*see* pp544–545); Cyprus is just beginning to show the world that there is more to this island's wines than Commandaria, as wonderfully complex as that dessert wine can be; and Turkey has made the most progress as a wine region, with a growing number of wineries that are starting to master the skills to showcase their rich portfolio of native varieties.

Lagging behind a little are the more embryonic regions of Syria, Jordan, and Egypt, where both political and religious dynamics make the resulting wine culture a niche and restricted output; as a result, it may be a long time before these countries can fulfil the rich winemaking history that have been uncovered on the walls of their archaeological ruins.

CYPRUS

Wines have been made on this beautiful island for at least 4,000 years, and Cyprus's historically famous wine, Commandaria St John, is one of a handful that claim to be the world's oldest wine. It can be traced back to 1191, when Richard the Lionheart, King of England, acquired the island during the Crusades. He subsequently sold it to the Knights Templar. The Knights Templar (and later the Knights of the Order of St John, or Knights Hospitaller) established themselves as Commanderies.

Commandaria St John is a *solera*-matured, sweet dessert wine, which is made from a blend of black and white grapes that have been left in the sun for between 10 and 15 days after the harvest to shrivel and concentrate the grape sugars. In its heyday Commandaria won the first-ever international wine competition called the "Battle of Wines" run by the crusaders in the 13th century.

A STAND IN OMODOS, CYPRUS, DISPLAYS THE LOCAL SWEET WINE
Whether the claim that Commandaria is the world's oldest sweet wine is true or not, this Cypriot speciality does indeed have long history dating to the 12th century.

It is now hard to imagine that Cyprus was the third-largest producer of wine next to France and Spain in the 1930s. Sadly this period did more to damage than elevate the reputation of its wines – its main output was a cheap, sweet sherry beloved of the British, and the industry was dominated by four large cooperatives (Keo, ETKO, SODAP, Leol) whose focus was volume rather than quality production.

It is only in the past two decades that a quality-driven industry has started to emerge, largely due a fall in demand from the Russian market since the 1980s. At the same time subsidies from the EU helped encourage the growth of smaller boutique wineries that have explored sites farther inland, at altitude, and process grapes closer to the vineyards. This in turn roused the larger wineries that still dominate over 80 per cent of the production, to benchmark and improve their quality in line with their new competitors. It is perhaps a shame that the funding for these plantings encouraged the use of international rather than native varieties, because their indigenous grapes are now starting to show real promise.

The red Mavro grape accounts for approximately 50 per cent of plantings, but it is fairly bland and indistinct, whereas the black-skinned Maratheftiko shows the most promise as a structured and age-worthy red capable of great concentration. For the whites, Xinisteri is producing wines that are surprisingly crisp, mineral, and fragrant for the island climate. Other native grapes of interest are whites Spourtiko and Promara, as well as reds Ofthalmo and Yiannoudi. Of the international varieties planted, Syrah is producing the best results with many extremely polished and impressive wines. Despite the island's ancient history of winemaking, all traditional methods such as amphora ageing have been lost, so it will be interesting to see if the future brings any more consideration to reviving these techniques.

Wine districts: *Akama, Commandaria, Laona, Krasohoria Lemesou, Krasohoria Lemesou-Afames, Krasohoria Lemesou-Laona, Pitsilia, Vouni Panayia-Ampelitis*

✓ *Tsiakkas (Xynisteri, Commandaria) Tsangarides (Shiraz, Maratheftiko), Zambartas (Single Vineyard Xynisteri), Vouni Panaiya (Alina Xynisteri, Barba Yiannis Maratheftiko) Ezouza (Eros Rose, Maratheftiko) Kyperounda (Petritis) SODAP (St Barnabas Commandaria)*

EGYPT

Sometime between 3000 BC and 2890 BC, the world's earliest "wine labels" were stamped into clay on sealed wine jars later found in the tomb of King Den at Abydos, certifying that the wine they contained was produced at a vineyard dedicated to Horus. Tomb paintings from the same era show vines grown in raised troughs to avoid wasting precious irrigation resources and trained into arbours for easy harvesting, demonstrating that Egyptian viticulture was already extraordinarily sophisticated almost 1,700 years before Greece reached its peak. Paintings in the tomb of Parennefer (1330 BC) illustrate that the Egyptians had even invented a primitive but effective form of air-conditioning, burying fermentation jars in sand kept wet and fanned, so that the jar and its contents were cooled as the water evaporated. Egyptian winemaking came to an abrupt end in AD 641, however, when it the country was conquered by Muslim Arabs. The first winery in more recent terms was established in 1882 by Nestor Gianaclis, a Greek-born Egyptian cigarette manufacturer. The Gianaclis (pronounced "jaana-cleesse") winery was nationalized in 1963, but the government maintained production for the lucrative tourist industry before privatizing business in 1999 for a swift and healthy return on its investment. At the time of purchase by Al-Ahram Beverages Company (ABC), all noble grape plantings had been lost, and most of its production was made using grapes brought in from Lebanon or imported

BLACK SEA

MEDITERRANEAN SEA

AEGEAN SEA

T U R K E Y

SYRIA

LEBANON
see p547

ISRAEL
see p546

WINE-PRODUCING REGIONS OF THE LEVANT
The Levant is an imprecise geographic term that refers to the eastern shore of the Mediterranean Sea and nearby islands and includes Cyprus, Jordan, Syria, Turkey, and Egypt. Greece, Israel, and Lebanon also fall into the Levant (see individual entries).

concentrated grape must that was diluted to ferment into wine. By 2004 they started planting vines along the fertile Nile delta, and in 2010 they had made an impressive turnaround, and by then only 2 per cent of their wine was made from imported grapes.

For the most part Al-Ahram Beverages Company (ABC) has been able to monopolize the local market, buying its main competitor El Gouna Beverages Company in 2001 and easily beating off potential foreign competition due to imports being taxed at 300 per cent. Although it has had some competition from organic winery Koroum of the Nile set up in 2003 based in the Red Sea resort El Gouna. This winery has also been the first to attempt to cultivate an indigenous grape called Bannati to make its Beausoleil white wine brand. Koroum's Jardin du Nil brand has also gained some international recognition. ABC was bought by Heineken International in 2002.

The future of Egypt's wine industry looks uncertain for climatic rather than political reasons. With less than 12 centimetres (4.7 inches) of rainfall a year and summer temperatures than can easily exceed 40°C (104°F), it is heavily reliant on groundwater, which is becoming increasingly sparse, for irrigation. In quality terms the output is still at a stage of infancy with the Shahrazade and Jardin du Nil brands being the most reliable of those on offer.

Wine districts: *Dr Armanios Vineyard, Karm El Nada, Minya, Sahara Vineyards (Khatatba), Sahara Vineyards (Luxor), Soos Vineyard, Taba*

✓ *Giancalis (Domaine de Giancalis Ayam Syrah Carignan)* • *Jardin du Nile (Grand Vin d'Egypt blends)*

JORDAN

Jordan is a country where winemaking once flourished, and archaeological digs near Petra have found evidence of 82 ancient wine presses. Yet its fast-diminishing vineyards now cover barely more than one-quarter of the area they did in the late 1990s, and only a small percentage of these vines

A VITICULTURIST INSPECTS A BUNCH OF CABERNET SAUVIGNON GRAPES IN THE MENDERES VINEYARD OF SEVILEN
Founded in 1942, Sevilen is a major wine producer in Turkey. Sevilen produces grapes for its wines on two sites: the Aegean area, which has a warm Mediterranean climate, and on the Anatolian Plateau, with its cool Mediterranean climate.

produce wine. Jordanians are not wine drinkers, preferring Arrack, the aniseed-flavoured spirit that is ubiquitous in the Levant. The wine industry in its modern form was first established by the Haddad distillery company in 1975, which now makes Jordan River and Mount Nebo wines. This was followed in the mid-1990s by a formidable character, Omar Zumot, who arrived determined to put this country back on the wine map and began planting his first three vineyards to produce wines under the Saint-George label. Both companies have land in the northern Mafraq region that benefits from higher elevation sites above 600 metres (1,970 feet), a good underground water supply, and basalt rich soils, and together their production now equals approximately a million bottles. Zumot is somewhat visionary in his approach to producing wine; rather than rely on nitrate fertilizers, he constructed an industrial-sized fishpond, fed by an aquifer, and stocked it with 150,000 fish to provide natural nitrate-rich irrigation water. The local ancient varieties appear to have been lost, but the two wineries are experimenting with a surprisingly broad array of international plantings. Zumot introduced 30 noble varieties and continues to experiment today with unusual grapes such as Tocai and Carmenere. In quality terms the Zumot wines are the most reliable, whereas the Haddad wine brands lack consistency.

Wine district: *Amman-Zarqua*

SYRIA

Syria once had a fledgling wine industry with similar roots to the Lebanese. French troops were stationed in the country during World War II and were the catalyst for the wine industry to grow, because the French, being French, demanded wine.

Sadly, the recent decade of civil war has largely put an end to their accomplishments. The only serious operational winery that remains is Domaine Bargylus, which was set up in 2003 by the half Syrian–half Lebanese Saadé family in mountainous coastal range of Latakia. It is an ambitious, quality-minded operation that has employed the renowned Bordelais Stéphane Derenencourt as consultant. Despite the disruption of the war, the Saadé have continued to produce wine, albeit they are forced to do so remotely from a base in Beirut where they also run the Lebanese winery Chateau Marsyas. The family cannot visit Syria, for fear of kidnapping and therefore manage the process with the resident winemaker in a somewhat cloak-and-dagger manner, having grape and wine samples driven over the border for assessment. Hostilities have come within a few metres of their winery on several occasions. All of this has helped give their wine the tagline "the world's most dangerous wine". Considering these obstacles, it is impressive to see that the Saadé family has achieved excellent global distribution for their wine, which can be found in many top restaurant lists in Europe.

Wine districts: *Aleppo, Homs, Damascus*

TURKEY

Turkey's history as a grape grower goes as far back as 9000 BC, recently proven through DNA profiling of grape varieties by Dr Jose Vouillamoz, who concluded the southeast part of Turkey was the origin of grape seeds and grape domestication. It is estimated there are up to 1,200 different local native varietals, which is four times more than its neighbour Greece, although only 60 are produced commercially. Given innumerable areas in Turkey are cut off from the rest of the country, individual varieties have been able to survive unchanged for centuries, possibly millennia. This country has the fourth-largest area under vine in the world (after Spain, Italy, and France), but because its population is predominantly Muslim, most vines produce table grapes, sultanas, or currants. Just 2.5 per cent of Turkish vineyards produce wine. This all suggests the incredible potential for Turkey to produce wine, which is only just being realised.

A commercial wine industry has existed since the 14th century, albeit under the tension of existing within a Muslim state. It was never entirely outlawed under Ottoman rule because it brought in useful taxes for the government. Their infamous leader Mustafa Kemal Atatürk was instrumental in establishing the industry that remains today, setting up a state-owned winery in 1925 that funded research into indigenous varieties and potential regional boundaries with the help two French viticulturists, M

FACTORS AFFECTING TASTE AND QUALITY

LOCATION
Technically, the Levant comprises the countries along the eastern Mediterranean shores, which for practical reasons this encyclopedia has extended northwest to Turkey, southwest to Egypt, east to Jordan, and this necessitates that Cyprus is also included.

CLIMATE
Hot and dry for the most part, with a minority of cooler microclimates due to their proximity to the Mediterranean, or high altitude. Vines in Lebanon's Bekaa Valley receive an amazing 300 days of sunshine per year and no rain during the harvest.

ASPECT
Vines grow on all types of land from coastal plains to higher mountain slopes.

SOIL
Soils vary greatly from volcanic origin on Cyprus, through the alluvial river and sandy coastal plains to the Bekaa Valley's gravel over limestone.

VITICULTURE & VINIFICATION
Better-quality vineyards are being planted on higher mountain slopes, and grapes are being harvested at lower sugar levels. The resulting wines are being fermented at cooler temperatures, with increasing use of new French oak for top-of-the-range wines.

GRAPE VARIETIES
Cyprus: Maratheftiko, Mavro, Xinisteri, Ofthalmo, Yiannoudi, Syrah
Turkey: Narince, Sultaniye, Emir, Sauvignon Blanc, Semillon, Muscat, Okuzgozu, Boğazkere, Kalecik Karasi, Syrah, Cabernet Sauvignon, Merlot
Israel: Cabernet Franc, Cabernet Sauvignon, Carignan, Chardonnay, Colombard, Merlot, Riesling, Sauvignon Blanc
Lebanese varieties: Cabernet Sauvignon, Carignan, Chardonnay, Chasselas, Cinsault, Clairette, Gamay, Grenache, Muscat (various), Pinot Noir, Riesling, Sémillon, Syrah, Ugni Blanc

---

Bouffart and Marcel Biron. This eventually led to the creation of 28 state-owned wineries. Sadly, the output turned to serve the Soviet market, which demanded a deluge of basic and mass-produced wines.

The government closed its state-owned wine monopoly in 2004, which created a turning point for the wine industry. Since that time a new wave of quality-minded wineries has evolved, many with an exciting focus on their indigenous varietals. In 2008 Wines of Turkey was set up by the government, helping to bring global attention to the industry and its producers. A tension between the wine industry and the government remains, however, and in 2013 then prime minister Recep Tayyip Erdoğan imposed strict controls on domestic sales and advertisement, leading to producers increasingly relying on exports and tourism for sales.

There are now more than 100 wineries, with more than half the production on the Aegean coastline. Many of the wineries have emerged from the state-controlled era, but there is also a boutique winery movement emerging as wealthy individuals invest in the industry.

Of the indigenous varietals, there are several showing real promise to compete on the international market. For the whites, Narince provides the most interest, with its gentle orchard fruit spectrum and a natural high acidity that allows it to age with a complex, nutty, and exotic profile. Emir makes lively, crisp, thirst-quenching whites. The local palate favours the strongly tannic red grapes like Okuzgozu and Bogazkere, the latter having particularly dense tannins that means it blends well with Okuzgozu, a grape that offers plush fruit to plump out the mid-palate. But it will perhaps be the more delicate local red Kalecik Karasi that will find international appeal, given that at its best it has the delicacy and fine perfume of a Pinot Noir.

Of the countries largest wineries those that have a quality focus are Sevilen, Kavaklidere, Pammukale, Doluca, and Kayra.

Wine districts: *Marmara, Aegean, Mediterranean, Mid-Northern Anatolia, Mid-Southern Anatolia, Mid-Eastern Anatolia, South-Eastern Anatolia*

✓ *Sevilen (Nativus Kalecik Karasi, Plato Syrah Okuzgozu, Isabey Sauvignon Blanc) • Pammukale (Syrah Kalecik Karasi) • Kavaklidere (Ancyra Narince) • Doluca (Tugra Okuzgozu) • Arcadia (Sauvignon Gris) • Kayra (Alpagut Versus Okuzgozu) • Diren (Narince, Okuzgozu)*

Israel

Israel has developed a quality-driven wine industry, based on a terroir of enormous variety, with young, internationally trained winemakers using advanced technology in both winery and vineyards. Israelis are great experimenters and researchers, and some of the wineries are making innovations of obvious interest to international wineries interested in combating climate change.

A long history of winemaking in ancient times was revived by Baron Edmond de Rothschild, owner of Château Lafite-Rothschild, at the end of the 19th century. He founded a modern wine industry with the historic Carmel Winery in Rishon Le Zion and Zichron Yaacov. It was not until the 1980s that Golan Heights Winery introduced New World technology by using high-altitude vineyards. Their award-winning Yarden wines put Israel on the international wine map.

In the 1990s a small winery revolution began, with many wineries being founded. They made wines of character and individuality, raising the quality bar considerably. Margalit Winery and Domaine du Castel were the pioneers that led the way. As a result, some of the larger wineries responded by investing in quality.

The epicentre of Israel wine moved northwards and eastwards from the coastal regions in search of high elevation, so crucial to making quality wine in Israel. Most of the quality vineyards are about 450 to 1,200 metres (1,475 to 2,940 feet) above sea level.

The initial leap forward was made using varieties such as Cabernet Sauvignon, Merlot, Chardonnay, and Sauvignon Blanc. Cabernet Franc and Petit Verdot also show interesting results in Israel. Mediterranean varieties have come to the fore lately, such as Shiraz, Grenache, and Mourvedre amongst the reds and Viognier and Roussanne the whites. Some existing workhorse varieties like Carignan, Petite Sirah, Chenin Blanc, and Colombard have also been looked at anew. By lowering yields and aiming for quality, some interesting wines are being made using these varieties.

Israel is becoming known for making Southern Rhône–style blends, sometimes blending Bordeaux and Mediterranean varieties together, and there has been a big step forwards in the production of fresh, quality white wines showing good varietal character, representing the local *terroir*.

A recent development is ongoing research on indigenous Holy Land varieties, such as Bittuni, Dabouki, and Marawi (aka Hamdali). The Cremisan Monastery revived these, and Barkan (Israel's largest winery) and Recanati are amongst those making wines from these grapes. The locally developed Argaman variety is also being used more, notably by Barkan.

The main quality regions are the Judean Hills with *terra rossa* on limestone (Castel, Tzora) that rise towards Jerusalem, the volcanic plateau of the Golan Heights with volcanic tuff, (Golan Heights Winery), and the Upper Galilee with basalt and limestone (Margalit, Shvo, and some Flam).

✓ Castel (Grand Vin, C Blanc du Castel)
• Clos du Gat (Syrah, Merlot, Chardonnay)
• Flam (Noble) • Golan Heights
(especially Yarden and prestige Katzrin)
• Margalit (Enigma, Cabernet Franc)
• Recanati (Carignan) • Sphera
(all whites) • Shvo • Tzora (Misty
Hills, Shoresh) • Vitkin (Carignan)
• Yatir (Yatir Forest)

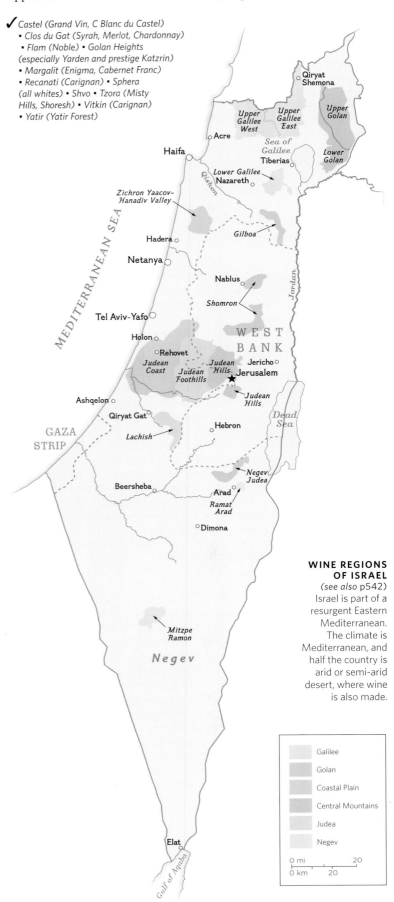

WINE REGIONS OF ISRAEL
(see also p542)
Israel is part of a resurgent Eastern Mediterranean. The climate is Mediterranean, and half the country is arid or semi-arid desert, where wine is also made.

	Galilee
	Golan
	Coastal Plain
	Central Mountains
	Judea
	Negev

0 mi   20
0 km   20

GOLAN HEIGHTS WINERY IN QATSRIN
Israel's third-largest winery focusses on technological innovation. Its sophisticated approach has proved successful, earning its wines international awards and accolades.

Lebanon

The land that is now Lebanon has been making wine for over 7,000 years. It's finest "hour" was arguably in the second millennium BC when Phoenician wine, the best of which was considered the Petrus of its day, was the toast of the Greeks and Romans. Lebanese wine is now in its seventh millennium and still, albeit in a more modest way, going strong, despite Lebanon's current crop of winemakers having to contend with the wars and political instability that have plagued the country since its independence from France in 1943.

The French legacy is profound. The French Mandate may have lasted only 20 years, but it was enough for a Francophone culture to permeate a country that was open to a new identity after centuries of Ottoman rule. And when it comes to wine, France is everywhere: Lebanon's winemakers are mostly French-trained, while the blends, especially the reds, can trace an arc of influence from Bordeaux to the Rhône valley and down to southern France.

Château Musar is still the most celebrated Lebanese producer, but the late Serge Hochar's famous creations – earthy reds and oxidative whites – are an anomaly and are not really like the rest of Lebanon's wines, which

VINEYARDS OF CHÂTEAU KEFRAYA
Located in the western Bekaa Valley, this Lebanese vineyard spreads over terraced slopes above the Mediterranean Sea, on the foothills of Mount Barouk.

WINE REGIONS OF LEBANON
(see also p542)
The vast majority of Lebanese wines still comes from the Bekaa Valley, but the emergence of new wineries has seen viticulture spread westwards and northwards towards more Mediterranean-influenced areas.

Sub-zone named on map: eg, *Bekaa Valley*

0 mi   10
0 km   10

are, by and large, more approachable. Château Musar may have planted Lebanon's flag in the world wine map, but the other 50 or so producers have also shown that this tiny industry – a mere 10 million bottles a year – can compete with the very best in the world.

As you might expect from a country that is hot and dry, the reds are powerful and concentrated and, given the Lebanese penchant for doing things properly, often see a fair amount of wood. The whites, due to the availability of high-altitude vineyards – around 1,000 to 1,800 metres (3,280 to 5,900 feet) above sea level – demonstrate a freshness and complexity that can often belie their hot climate origins.

Lebanon no longer has any indigenous red wine grapes, so producers have embraced Bordeaux and Rhône varieties, especially for the upper-end wines, as well as grapes from areas closer to the Mediterranean, such as Cinsault, Carignan, and Grenache, which have been the backbone of the industry since the Jesuits founded what would become Château Ksara in the mid-19th century. These hardy varieties are often cited as the country's "native" grapes.

The whites, made mainly with Chardonnay, Sauvignon Blanc, Clairette, Semillon, and Viognier (as well as a soupçon of Muscat), offer surprising diversity of style. Producers are also getting intriguing results from the indigenous Obeideh and Merwah, for so long seen as peasant grapes used only for Arak, the local *eau de vie*.

The Bekaa Valley – more a plateau – is still the epicentre of the modern industry, with the majority of the vineyards in the western Bekaa, although there are increased plantings in the eastern Bekaa in the hills above Zahlé; the area around Baalbek, home to the famous Temple of Bacchus, and Hermel farther north. Other wine regions include the northern coastal district of Batroun, home to the new, shiny, and eco-friendly IXSIR winery, whose presence has cemented the area's reputation as Lebanon's second wine region, and Bhamdoun in Mount Lebanon and Jezzine in the south.

✓ *Adyar (Expression Monastique) • Batroun Mountains (Prestige Rouge) • Cave Kouroum (Petit Noir) • Chateau Ka (Source Rouge) • Chateau Kefraya (Comte de M) • Chateau Ksara (Chardonnay) • Chateau Marsyas (B-Qa) • Chateau Heritage (Plaisir du Vin) • Chateau Khoury (Chateau Symphonie) • Chateau St Thomas (Les Emirs) • Coteaux du Liban (Chateau) • Domaine de Baal (Domaine de Baal) • Domaine des Tourelles (Vielle Vigne Cinault) • Domaine Wardy (Les Terroirs) • IXSIR (Altitudes Red) • Karam Wines (St John) • Massaya (Terrasses De Baalbeck)*

The Wines *of* Africa

AFRICAN WINES—AND THEIR APPELLATION SYSTEMS— still bear the influence of earlier European settlers, yet many countries have put their own spin on the industry. Not surprisingly, there is little interest in or excitement over wine in the North African countries of Algeria, Morocco, and Tunisia, where increasing Islamic fundamentalism has stifled production in spite of the area's potential. Recently, however, sales figures are increasing, and some producers have been upgrading their vineyards and wineries. If growers continue this trend—and survive these difficult times—the world will soon have to alter its opinion of North African wines. Madagascar, Ethiopia, Kenya, Namibia, and Tanzania are home to a number of commercial wineries; Ethiopia, in particular, has seen an upswing in sales since 2013. The French can even boast of a *vin de pays* on the island of Réunion. At the tip of the continent lies the jewel in Africa's wine crown—the vineyards of South Africa. The initial surge of sales during the Mandela years was less an indicator of the wines' quality and more a show of international solidarity. Today, after a technological revolution that impacted both quality and diversity, the wines now stand on their own merit in the world market.

THE STELLENBOSCH WINE DISTRICT LIES IN THE SHADOW OF THE PEAKS OF THE CAPE FOLD BELT
Stellenbosch is the second-oldest wine region in South Africa and perhaps the most famous. It is part of Western Cape's Coastal Region.

South Africa

Since the arrival of democracy to South Africa in 1994 – and the consequent entry of its wine industry into the world market – the rate of change has been remarkable. After years of isolation, tight control, and a general orientation to bulk production, the small high-quality element has grown greatly, and the Cape has become one of the New World's consistently most exciting areas.

Accompanying an adjustment to international standards of wine-making, newer, mostly cooler winegrowing areas were developed, especially along the south coast from Elgin via the Hemel-en-Aarde to Agulhas, Africa's southernmost tip. Chardonnay, Sauvignon Blanc, and Pinot Noir showed new promise. Farther inland, well established but largely cooperative-dominated regions like Tulbagh and the Swartland reinvented themselves; the latter was particularly important in allowing a new generation of young winemakers to build their reputations and lead the industry towards fresher, less extracted, and less oaky wines. The country's most planted variety, Chenin Blanc, became in a sense the symbol of the new whites, as the Swartland's older vineyards (many dating to the 1960s, a veritable neglected treasure trove) were rediscovered. They were in Paarl and Stellenbosch too, not to mention such "unknown" areas as Citrusdal in the Olifants River region. Syrah, and blends based on it, became the red equivalent for the Swartland. Another rather despised variety of the past, Cinsaut, has more recently become greatly fashionable as a fragrant, generally lighter-styled varietal wine and also as a blending partner for the Rhône varieties and, increasingly, even for Cabernet Sauvignon in Stellenbosch. In fact, the Cabernet-Cinsaut

WORKERS HAND-HARVESTING GRAPES IN STELLENBOSCH
Much of the wine produced in South Africa is created from grapes grown in the districts and wards around Cape Town, including Stellenbosch.

FACTORS AFFECTING TASTE AND QUALITY

LOCATION
South Africa is located on the southern-most reaches of the African continent, and the majority of its wine is grown in the Western Cape, at latitude 33.22°S.

CLIMATE
The climate is generally mild Mediterranean with hot, dry summers and wet winters. Meso-climates vary considerably, depending on proximity to the oceans (Atlantic and Indian). Fogs and winds accompanying the currents, particularly the cold Benguela that flows northwards along the west coast, moderate the temperature and increase rainfall in coastal areas like Stellenbosch, Walker Bay, and Constantia. It is hotter and drier inland in areas like Paarl, Wellington, Tulbagh, and the Klein Karoo. Other areas, like Elgin and Cederberg, rely on altitude and/or cloud cover to lower the heat. Persistent drought negatively impacted the size and quality of the 2015 to 2018 vintages in some areas; dryland-farmed Swartland and Vredendal, where the Olifants River on which it relies for irrigation all but ran dry, were two of the worst hit.

ASPECT
Many of the vineyard sites cultivated for centuries on the Cape's gently undulating valley floors still exist. Where there is elevation, however, earth movement and erosion have provided myriad orientations. Higher slopes are keenly sought after.

SOIL
Soils range from gravel and heavy loams of sandstone, shale, and granitic origin to deep alluvial, sandy, and lime-rich, red-shale soils. The shales (Bokkeveld group) persist in the valley floors and the erosion-resistant sandstones (Table Mountain sandstone) form the parallel ranges.

VITICULTURE & VINIFICATION
Contrary to popular belief, a Cape vintage is never homogenous. Some vineyards may fail to ripen their grapes fully – Plettenberg Bay, for example – although there are many areas where grapes can quickly over-ripen. Successful cultivation often depends on the availability of water for irrigation, as in Robertson, Worcester, Olifants River, and Vredendal, while night harvests are increasingly common. As a whole, the industry has a history of overproduction, because of too great an area under vine, too little emphasis on high-quality varieties, poor clones, and undesirably high yields. The past 25 years have seen the country work hard to upgrade its vineyards and invest in research in new clones and rootstock, and the area under vine has fallen below 100,000 hectares (247,105 acres). The relatively large, often family-owned properties still play a major role in driving the country's quality agenda, but the importance of a growing number of ambitious winemaker-cum-brand-owners who often don't own vineyards but make wine in contract cellars cannot be denied.

GRAPE VARIETIES
Primary varieties: Cabernet Franc, Cabernet Sauvignon, Chardonnay, Chenin Blanc (Steen), Cinsaut, Colombard, Hanepoot (Muscat d'Alexandrie), Merlot, Pinot Noir, Pinotage, Ruby Cabernet, Sauvignon Blanc, Sémillon, Syrah (Shiraz), White Muscadel (Muscat Blanc à Petits Grains)
Secondary varieties: Albarinho, Barbera, Carignan, Chenel (Chenin Blanc x Ugni Blanc), Crouchen Blanc (Cape Riesling), Fernão Pires, Gamay, Gewürztraminer, Grenache Blanc, Grenache Noir, Malbec, Marsanne, Mourvèdre, Nebbiolo, Petit Verdot, Pinot Gris, Red Muscadel (Muscat Rosé à Petits Grains), Riesling, Roussanne, Sangiovese, Souzào, Tinta Barroca, Tinta Amarela, Touriga Nacional, Ugni Blanc, Viognier, Zinfandel

Klein Karoo Wine region
—— *Swellendam* Wine district

blend was a re-creation of the some of the great red blends of the mid-20th century (Château Libertas, Zonnebloem, and others) and as such also expressed the mood of re-connecting to the traditions of the local past.

This renewed pride in the best elements of the past was part of the way that South African wines were acquiring a distinctiveness of their own. Even more significant during this century has been the increased concentration on finding unique identity in the vineyards themselves – many of which are overseen by the newest additions to the South African scene, winemakers – not winegrowers – who own no vineyards but are making some of the most highly sought-after wines. Although most of South Africa's qualitatively important vineyards are found in the comparatively circumscribed area of the Western Cape, fanning out no farther than 220 kilometres (138 miles) from Cape Town, there is a remarkable variety of soils and meso-climates elsewhere in the region. An indication of this is the fact that this is also the area of the Cape Floristic Region, the smallest and richest of the world's six floral kingdoms. Part of the quest to express both *terroir* and tradition is evident in the enormous growth in interest in older vineyards: now, thanks to the work of the Old Vine Project in conjunction with the industry authorities, wines from vineyards older than 35 years can carry a world-first Certified Heritage Vineyard seal, giving the attested planting date of the vines.

RECENT SOUTH AFRICAN VINTAGES

2019 A very mixed vintage – described as "weird" by many producers in all areas, with unusual patterns of ripening even in those areas less impacted by the preceding drought years. Lower volumes generally. Could prove exceptional for whites; overall possibly longer lived than 2018.

2018 Third year of drought, the coastal areas escaping best; generally challenging and with lower yields; many earlier-peaking wines, but many producers doing well.

2017 Less heat, though dryness persisted, leading to reduced yields in inland areas. But the smaller crop was mostly excellent, plenty of freshness and flavour concentration. Whites quickly showed very well, with reds also promising.

2016 Exceptionally hot, dry and challenging, especially for dryland farming as in the Swartland. In cooler areas, with later-ripening varieties, some excellent results.

2015 Generally an earlier vintage, but there is exceptional quality across the regional and stylistic spectrum. Possibly the finest vintage of the century.

Alongside these developments, however, by far the larger part of the South African wine industry is still given over to bulk production by the cooperatives and a few large merchants, but at ever-lower profit margins. The heavily mechanised area of the hot Northern Cape, dependent on fertiliser and irrigation, falls into this category. Other outposts of the industry are tiny, albeit promising – whether at Plettenberg Bay in the Eastern Cape or in semi-tropical KwaZulu-Natal.

THREE CENTURIES OF WINEMAKING

The first wine harvest in the infant Dutch settlement at the foot of Africa was recorded in 1659. Slowly, inexorably, vineyards (and land-usurping settlers and their slaves) spread down the peninsula and also into the mountainous areas of what were to become Stellenbosch and Paarl. Governor Simon van der Stel helped raise quality, not least by establishing his great estate, Constantia. This was later broken up, but the area became internationally renowned for its sweet wine, beloved of, amongst others, Napoleon and Jane Austen.

European wars and revolutions saw the Cape Colony ceded to Britain in 1814. Preferential trade terms led to greatly increased exports (and plantings of vines), especially as the colonial power was then deprived of French supplies. Many of the iconic gabled manor houses among the vines date from this period. But increased tariffs and renewed commerce with France, allied with the generally poor quality of Cape wines apart from Constantia, led to a market collapse, especially after 1861.

Slavery had been abolished in 1834, and the wine industry was in real trouble, not helped by the arrival of phylloxera in 1886. Colonial wars in Southern Africa worsened the depression; the merging of the British colonies into the Union of South Africa in 1910 made wine far less significant in an economy in thrall to gold and diamonds.

The situation of this under-capitalised and suffering industry led to the formation in 1918 of the KWV (Cooperative Winegrowers' Association). This association later gained enormous power. Wine farmers' livelihoods were saved, and the industry expanded though the 20th century – but tight regulation and a strategy of quantity over quantity prevailed. Merchants made some fine wines at the time, but only with the growing number of individual estates in the 1970s did a quality trajectory begin. For political reasons, though, the rest of the world was shunning South Africa. Isolation prevailed until the liberation of the 1990s arrived.

PINOTAGE, THE "CAPE BLEND", AND CHENIN BLANC

There are those who think that the Pinotage grape, the locally developed cross between Pinot Noir and Cinsaut, or a "Cape Blend" with an obligatory Pinotage component, should be the flagship of South Africa's export drive. Others point to the historical role of Chenin Blanc and its brilliant current showing and argue that this variety, including Chenin-based blends, is the country's "unique selling point". Why not both? Pinotage remains controversial and is still responsible for many big and cumbersome wines with too much alcohol and oak. Yet these are fewer of these wines now, and at the opposite extreme there are some deliciously fresh Pinotages, as well as some classically styled versions, all showing that the grape itself is not the problem, just the way it is sometimes handled.

SOUTH AFRICAN LABEL LANGUAGE

Note: Official South African languages (effectively only Afrikaans and English) may appear on labels, but English is used on the large majority. Only wine certified by the Wine and Spirit Board may indicate variety, vintage, and origin.

Alternative White and Red, Skin Macerated White To enable avant garde–style wines to be certified, some newer, more tolerant categories were introduced. Both of these are for dry wine, with sulphur dioxide content below 40 milligrams per litre.

Blanc de Noir A rosé entirely from black grapes (may otherwise be a blend of white and red wine).

Cape Tawny, Vintage, Vintage Reserve, Late Bottled Vintage Port-style wines may not mention the P-word. The traditional names for Port styles may be used if preceded by the word "Cape".

Certification seal Carried on the bottle neck guaranteeing the veracity of claims about origin, vintage, and variety, with a tracking number by which the wine may be traced back to its precise origin.

Certified Fair Labour Practice A sticker indicating qualification according to the WIETA audit for reasonable working conditions.

Certified Heritage Vineyard A seal from the Old Vine Project, in conjunction with the industry authorities, testifying that the vineyard source for the wine is over 35 years old, and giving the original planting date.

Estate wine The estate used to be the smallest category in the Wine of Origin system until it was abolished in 2004 and allowance was made for single vineyards. Today an "estate wine" comes from a property registered to make wine on the same principle: the wine must be grown, vinified, and bottled on the one property.

Fino, Oloroso, Pale Dry, etc For the few Sherry-style wines now made, the styles may be indicated, but the word Sherry may not.

Grape variety As per the usual international rule, a wine must contain at least 85 per cent of the variety indicated. When more than one variety is named, they must be given in descending order according to volume.

Integrity & Sustainability Certified A version of the certification seal indicating the producer's adherence to the Integrated Production of Wine programme – meaning anything from fully organic to a commitment to sustainable practices.

Jerepigo or **Jerepiko** A very sweet *vin de liqueur* made by fortifying unfermented grape juice. Muscadel/Muskadel is a Jerepigo made from Muscat Blanc à Petits Grains.

Méthode Ancestrale A bottle-fermented sparkling wine but with the bubbles coming from the first fermentation, not a secondary one.

Méthode Cap Classique (MCC) A sparkling wine made according to the traditional methods associated with Champagne. Numerous categories of sweetness are legislated, from Brut Nature (less than 3 grams per litre) to Brut, the most common, (less than 12 grams per litre), to Sweet (over 50 grams per litre).

Natural Sweet/Sweet Natural A frequently used catch-all category for more or less sweet wines, with at least 50 grams per litre of residual sugar.

Noble Late Harvest Essentially a sweet dessert wine showing the character of botrytised grapes.

Single vineyard This strictly legislated category is for a wine from a registered single vineyard. The vineyard cannot exceed 6 hectares (15 acres) and must be monovarietal.

Steen A synonym for Chenin Blanc, now sometimes used to stress "Capeness".

Straw Wine The translation of the French *Vin de Paille;* a dessert wine from ripe grapes allowed to shrivel, usually after being picked. A synonym for Wine from Naturally Dried Grapes.

Sweetness indications For still wines, the official sweetness levels range from Extra-Dry (no more than 2.5 grams per litre residual sugar) to Dry (no more than 5 grams, but higher sugar levels are permitted if the total acidity is within 2 grams per litre of the sugar level) to Special Late Harvest and to Natural Sweet and Noble Late Harvest (mentioned separately).

Vintage Wine must be of at least 85 per cent from the year given (the balance from either the preceding or succeeding vintage).

Wine from Naturally Dried Grapes *See* Straw wine.

Wine of Origin (WO) This system is South Africa's equivalent of the French *Appellation d'Origine Contrôlée.* WO wines must be 100 per cent from the area indicated. The official demarcations range from a geographical unit such as Western Cape to a region (eg Coastal) to a district (eg Stellenbosch) to a ward (eg Banghoek). More than one origin may be mentioned for a blend.

Nieuwoudtville

Koekenaap

*Lutzville
Valley*

○ Vanrhynsdorp

→ *Vredendal*

Spruitdrift

Doring

Bamboes Bay

*Citrusdal
Valley*

Sutherland-Karoo

*Lamberts
Bay*

*Leipoldtville-
Sandveld*

*Citrusdal
Mountain*

Piekenierskloof

*Saint
Helena
Bay*

*St Helena
Bay*

Cederberg

Olifants

Gamka

Saldanha

Swartland

Ceres Plateau

Riebeekberg
Riebeeksrivier

Tulbagh

Calitzdorp

Darling

*Mid-Berg
River*

Ceres

Hex River Valley

Touws

Malmesbury

Limietberg

Worcester

Groenekloof

Wellington *Groenberg*

Nuy *Vinkrivier*
Hoopsrivier

Montagu

Paardeberg
Paardeberg South
Voor-Paardeberg

Breedekloof
Slanghoek
Goudini ○ **Worcester**

Zandrivier
Klaasvoogds

Goudmyn *Tradouw Highlands*

Philadelphia
Agter-Paarl
Durbanville

Paarl

Bovlei
Blouvlei

Eilandia

Goree

Ashton

Tradouw

Langeberg-Garcia

Simonsberg-Paarl
Simonsberg-Stellenbosch

Paarl
Franschhoek

Scherpenheuvel
Stettyn

Goedemoed
Robertson

Herbertsdale

Cape Town ★

Bottelary

Le Chasseur
Agterkliphoogte

Bonnievale

○ Swellendam

Devon Valley

Banghoek

McGregor

Breede

Buffeljags

Gourits

Hout Bay

Cape Town

Papegaaiberg
*Polkadraai
Hills*

*Jonkershoek
Valley*

Theewater

Elandskloof

Greyton

Stormvlei

Swellendam

Malgas

*Lower
Duivenhoks
River*

Still Bay East

Constantia

Stellenbosch

Overberg

Boesmansrivier

Elgin

Bot River

*Upper Hemel-en-
Aarde Valley*

Hemel-en-Aarde Ridge

Klein River

Cape Agulhas

*Hemel-en-
Aarde Valley* ○ Hermanus

Stanford Foothills

Walker Bay

Napier

○ Bredasdorp

Springfontein Rim

Cape Agulhas

Elim

Sunday's Glen

Cape of Good Hope

Cape Agulhas

▨	*Swartland*	Wine district
—	*Stormsvlei*	Wine ward

0 mi ———————— 40
0 km ———————— 40

Inset:

Prince Albert Valley

*Cango
Valley*

→ *Swartberg*

○ Oudtshoorn

→ *Upper Langkloof*

Outeniqua

○ George

○ Knysna

*Plettenberg
Bay*

○ Mossel Bay

SOUTH AFRICAN DISTRICTS AND WARDS

Within the larger wine regions of South Africa (*see* p549) are two other categories: districts (such as Franschhoek, Stellenbosch, and Walker Bay) and wards (such as Hemel-en-Aarde Valley, Wellington, and Elgin). The wards, which lie within the districts, are defined by their *terroir*. The inset at right shows the Plettenberg Bay district, along with the easternmost wards.

THE APPELLATIONS OF
SOUTH AFRICA

The largest WO appellations are the six "geographical units", with Limpopo and Free State recently joining Western Cape (infinitely the most important), Northern Cape, Eastern Cape, and KwaZulu-Natal. The large "regions "come next, with "districts" nested within them, and then "wards" – which is where *terroir* considerations become fully significant. Registered single vineyards are the smallest demarcated WO (Wine of Origin) areas. Wine wards are the most specific type of WO appellation and are listed here under the district name.

EASTERN CAPE WO
St Francis Bay

This vast area is a geographical unit Wine of Origin between the Western Cape and Kwazulu-Natal; it includes the coastal wine ward of St Francis Bay but contains very few vineyards.

FREE STATE WO

An inland area coterminous with the province of that name, it contains the ward of Rietrivier (intimately connected to the Northern Cape WO, however), but little in the way of viticulture.

KWAZULU-NATAL WO

As usual, this geographical unit is coterminous with the coastal, generally tropical province of that name. There are two districts, Central Drakensberg WO and Lions River, and a few wine estates, mostly high lying to attain coolness and avoid excessive humidity, but conditions are not easy for viticulture. The province's tourist trade is an encouragement, however, and there are some estimable wines produced. There are no wine wards here.

LIMPOPO

This is the northernmost of South Africa's six geographical units, with limited hectares under vine.

NORTHERN CAPE
Central Orange River, Hartswater, Prieska

This region now encompasses everything from the newest district – Sutherland-Karoo – in the south to the ward of Hartswater, South Africa's northernmost viticultural area, 80 kilometres (50 miles) north of Kimberley. Along with Sutherland-Karoo, there is the wine-producing district of Douglas and three unattached wards. In the Prieska ward, Lowerland now has a fascinating and ambitious small range of wines, specialising in Colombard and Tannat, made in the Western Cape by different new-wave wine stars.

DOUGLAS
This warmer region has a climate similar to Tuscany.

SUTHERLAND-KAROO
There are definite quality moves afoot in winter-snowy, semi-arid Sutherland-Karoo – the country's smallest and highest-altitude district – with a number of varieties being experimented with. The overwhelming majority of wine from the hot, heavily irrigated vineyards farther north is destined for bulk, but producer-owned wineries make decent fortified wines and an improving quality of good-value, fruity varietals.

WESTERN CAPE WO
Ceres, Cederberg, Prince Albert Valley, Swartberg

This geographical unit Wine of Origin, whose name appears on a great many blends with no pretensions to *terroir* interest, covers the vast majority of South Africa's vineyards. It encompasses the WO regions Breede River Valley, Cape South Coast, Coastal, Klein Karoo and Olifants River, as well as a handful of unattached wine districts and wards. Ceres Plateau district (with Ceres as a ward) is a large, high-lying, inland area between the Tulbagh and Worcester districts, increasingly turned to by producers seeking cooler sites for planting – Pinot Noir and Chardonnay are already showing promise. Cederberg in the Cederberg Mountains has some of the Cape's most remote and highest vineyards (some over 1,000 metres or 3,280 feet above sea level), and just two, highly regarded, producers. Prince Albert Valley ward lies at the foot of the Swartberg Mountain Nature Reserve, north of Klein Karoo, in an area better known for its olives than for wine. Swartberg is another small and minor ward a little farther along the Swartberg Mountain slopes.

BOBERG REGION
This region is invoked only for fortified wine made within the Franschhoek, Paarl, Wellington, and Tulbagh districts. It is rarely used.

BREEDE RIVER VALLEY REGION
This region's vineyards, located east of the Drakenstein Mountains and mostly heavily dependent on irrigation, tend towards white or fortified wines.

BREEDEKLOOF
Goudini, Slanghoek

Breedekloof covers a large proportion of the Breede River Valley and its tributaries, where vineyards are found on alluvial valley soils over well-drained gravel beds. As generally in the Breede River region, the vines can be highly productive and are mostly produced cooperatively: much of the wine goes to brandy and the merchant trade, but also some ambitious bottles of Chenin Blanc, Pinotage, and Chardonnay. Du Toitskloof is perhaps the most dynamic of the "producer cellars" (former co-ops).

ROBERTSON
Agterkliphoogte, Bonnievale,
Boesmansrivier, Eilandia, Hoopsrivier, Klaasvoogds,
Le Chasseur, McGregor, and Vinkrivier

Robertson is known as "the valley of vines and roses", but its hot climate relegated this region to largely bulk-produced wine until the emergence of fine-wine hot spots on its lime-rich soils. It now has a handful of estates with established reputations for white wines, particularly Chardonnay and sparkling varieties. Top producers include De Wetshof and Graham Beck along with rising star, Mont Blois.

WORCESTER
Hex River Valley, Nuy, Scherpenheuvel, Stettyn

Worcester in the Breede River catchment area, is particularly intensively cultivated with vines. The hot climate is tempered in the west by high rainfall, while in the east rainfall is very low, with soils derived from Table Mountain sandstone and the fertile red shale of Little Karoo. Although there are some good-value red wines made here, Worcester is best known for its white and fortified wines. Top producers include Nuy and Alvi's Drift.

CAPE SOUTH COAST REGION
This broad region tidies up the frontier wine areas on the south coast – including some that are definitely southwest.

CAPE AGULHAS
Elim

Africa's most southerly vineyards are in Cape Agulhas, where bracing sea winds seem suited particularly to Sauvignon Blanc and Sémillon, though Syrah and even Pinot Noir show promise.

ELGIN
Elgin has become a district in its own right –an increasingly important one, with a growing reputation for Chardonnay and Sauvignon Blanc, but also for Pinot Noir and now Syrah.

OVERBERG
Elandskloof, Greyton, Klein River, Theewater

Overberg is a large district, with some marked differences. There are some high-altitude vineyards in Elandskloof, best known for Chardonnay and Pinot Noir, and notably cool ones in Greyton, well suited to thrilling Sauvignon Blanc, Chardonnay, and Syrah. Raka in Klein River produces good, bold reds, notably Merlot and Shiraz.

PLETTENBERG BAY
The first vines in Plettenberg Bay were planted in 2000 and were mostly Sauvignon Blanc, still a speciality for this district, as well as Cap Classique.

SWELLENDAM
Swellendam (with few wineries, though including the wine wards of Buffeljags, Malgas, and Stormsvlei) has a Mediterranean-type climate and no particular specialities – with Sijnn in Malgas the undoubted star of the region, having a range of varieties doing well.

WALKER BAY
Hemel-en-Aarde Ridge, Hemel-en-Aarde Valley,
Upper Hemel-en-Aarde Valley, Bot River,
Stanford Foothills, Sunday's Glen, Herbertsdale,
Lower Duivenhoks River, Napier, Stilbaai East

When Hamilton Russell was established in 1975 in Walker Bay district's Hemel-en-Aarde area (now with three wards: Hemel-en-Aarde Ridge, Hemel-en-Aarde Valley, and Upper Hemel-en-Aarde Valley), the conservative industry – conservative in terms of both politics and viticultural approach – thought the comparatively cool location was too much of a gamble. But its suitability for Chardonnay and Pinot Noir became apparent, and now, with a dozen or more producers, it is almost certainly the area in the Cape with the highest average price per bottle. Much the same quality orientation goes for the larger Walker Bay district (also including the wards of Bot River, Stanford Foothills, and Sunday's Glen). Cape South Coast also includes four wine wards not nestled in any district: Herbertsdale, Lower Duivenhoks River (which has already produced some brilliant Chardonnay), Napier (where veteran Jean Daneel has a fine reputation for Chenin Blanc), and Stilbaai East.

COASTAL REGION
One of the most frequently encountered appellations, the arguably too expansive Coastal Region, has seen some important developments in recent years, most notably the creation of a new Cape Town district.

CAPE TOWN
Constantia, Durbanville, Hout Bay, Philadelphia

Cape Town was created to contain four wards within the city or on its fringes. This involved repealing two districts: Cape Peninsula and Tygerberg. Constantia counts as a cool area in South African terms (as does the whole Peninsula, where Cape Point Vineyards is a lonely estate further south). The vineyards of Constantia are situated on the eastern, red-granitic slopes of Constantia Mountain, south of Cape Town. Open to False Bay, the mountain boasts a climate that is moderate, of Mediterranean character, but quite wet, with up to 120 centimetres (47 inches) annual rainfall. Constantia is, of course, the most historic and thus most famous of South Africa's winelands, being the location of Simon van der Stel's original Constantia estate. Sauvignon Blanc and Sémillon dominate plantings in much of the district, used for both varietal wines and some stunning blends; Chardonnay is not far behind – notably successful in the area's sparkling wines (usually partnered by Pinot Noir). Red wines, notably the Bordeaux varieties and Syrah, do well where the vineyards are more exposed to sun and sheltered from the often fierce and cool winds off

False Bay. The small plantings in Hout Bay mostly go towards classic sparkling wine. Sauvignon Blanc dominates Durbanville's vineyards too, which are cooled and dried by sea breezes off Table Bay (there are splendid views of Table Mountain across the Bay) though Merlot from here has some reputation too.

DARLING

Groenekloof

Darling was once part of Swartland, but secession was prompted by the individuality of its cooler vineyards, particularly those in the higher Groenekloof area where the red, decomposed granite hills running parallel to the ocean make attractively aromatic Sauvignon Blanc. Merlot and Shiraz have the best reputations for reds.

FRANSCHHOEK/FRANSCHHOEK VALLEY

Franschhoek is a beautiful district surrounded on three sides by towering mountains, with the charming, hospitable, and well-touristed town of the same name at its centre. Its name (Dutch for "French Corner") recalls the role French Huguenots played here in the 17th century. The Huguenot influence is also seen in the names of many estates. There's a wide range of varieties grown and styles made, with vineyards on the lower slopes generally producing better quality than those on the valley floor. Cap Classique is something of a speciality, as is Sémillon.

PAARL

Simonsberg-Paarl, Voor Paardeberg, Agter-Paarl

Paarl now includes the wards of *Simonsberg-Paarl* (adjacent to Stellenbosch), *Voor Paardeberg* (which touches the south of Swartland and shares with it many characteristics), and *Agter-Paarl* (a flatter valley floor expanse.) Paarl has a wide range of meso-climates, soils, and aspects and so produces an array of successful wine styles. The climate is Mediterranean, and though this was once a white wine area, it is definitely now seen as more red wine country (Cabernet Sauvignon, Pinotage, and the Rhône red varieties, especially), though Chenin Blanc excels, as does Chardonnay in cooler spots.

STELLENBOSCH

Banghoek, Bottelary, Devon Valley, Jonkershoek Valley, Papegaaiberg, Polkadraai Hills, Simonsberg-Stellenbosch

Stellenbosch is located between False Bay to the south and Paarl to the north and is centred around the leafy university town of Stellenbosch itself. This is where the greatest concentration of South Africa's finest wine estates has traditionally been situated. With vines now growing ever higher up the lower slopes of the magnificent mountains, there is a range of aspects and soils, and many varieties perform extremely well here, though the Bordeaux varieties, especially Cabernet Sauvignon, are perhaps the area's chief calling card. There are three main types of soil: granite-based in the east (considered best for red wines), Table Mountain sandstone in the west (favoured for white wines), and alluvial soils around the Eerste River. With warm, dry summers, and cool and moist winters, irrigation can often be kept to a minimum. Lamberts Bay is a ward without a district on the Atlantic coast in the no-man's land between the Olifants River and Swartland districts; it favours Sauvignon Blanc.

SWARTLAND

Paardeberg, Paardeberg South, Riebeekrivier, St Helena Bay, Malmesbury, Riebeekberg

Swartland is situated farther up the West (Atlantic) Coast than Cape Town (with the wards of Paardeberg, Paardeberg South, Riebeekrivier, and St Helena Bay latterly joining Malmesbury and Riebeekberg). It is a vast area, mostly planted to grain, but with vineyards clustering around the granitic or shale mountains in the south. With little water available, most of the vineyards are dry farmed, despite the warmth. As the locus par excellence of the modern Cape wine revolution, the dynamic Swartland has arguably become the country's best-known origin for many international wine enthusiasts, with the modern pioneers (Sadie, Mullineux, and Badenhorst) joined by a number of new producers keen to ride the new wave even further. Old-vine Chenin Blanc is the crucial contribution to white wines here (joined by other suitable varieties in the unique, splendid blends). Syrah, Grenache, Cinsaut, and other Mediterranean varieties are the backbone of the reds, frequently blended but with some superb *terroir*-specific Syrahs especially.

TULBAGH

To the east of Darling and farther inland (far enough to make a legitimate association with the Coastal Region somewhat tenuous) is Tulbagh, whose soil, climate, and topography are, for the most part, similar to those of the Karoo, although it is somewhat more temperate. Tulbagh has long been associated with Cap Classique, but the rediscovery of the area early this century revealed its propensity for bold reds too – and more restrained ones on the cooler slopes where Fable is the notable producer.

WELLINGTON

Blouvlei, Bovlei, Groenberg, Limietberg, Mid-Berg River

Warm-country Wellington, with a cluster of its own wards is increasingly known for its Syrah and other gutsy reds. Like Franschhoek, it used to be incorporated into Paarl.

KLEIN KAROO WO

This appellation is a long, narrow strip that stretches from Montagu in the west to De Rust in the east. The vineyards require irrigation to survive the hot and arid climate. The famous red, shale-based Karoo soil and the deep alluvium closer to the various rivers are very fertile and well suited to the Jerepigo, Muscadel, and other fortified dessert wines for which this area is best known.

CALITZDORP

Calitzdorp has summer temperatures of up to 40°C (104°F) moderated by afternoon sea breezes, and cool nights providing a beneficial diurnal effect of ensuring better acidity levels than would otherwise be found in such a hot climate. Port varieties flourish here, increasingly used for robust table wines as well as the Port-style wines for which a few estates are famous.

LANGEBERG-GARCIA

Cango Valley, Montagu, Outeniqua, Tradouw, Tradouw Highlands, Upper Langkloof

Langeberg-Garcia is located north of the Langeberg mountain range and is only lightly planted. The Klein Karoo, of which this is part, includes a number of wine wards unattached to any district. There are few wineries, with producers often producing grapes for the cooperatives and the brandy industry.

OLIFANTS RIVER

The Olifants River was long associated with intensively farmed, hot, and irrigated vineyards feeding some enormous cooperatives.

CITRUSDAL MOUNTAIN

Piekenierskloof

Intersting wines can come from anywhere it the Cape, but Citrusdal Mountain is where the real revelation came with the discovery of fine old dryland-farmed bushvine plantings of Chenin Blanc by some of the Cape's leading newer winemakers, and grapes from this district find their way into some of the country's most prestigious wines, from Alheit and Sadie, for example. Its wine ward of Piekenierskloof is now increasingly associated with Grenache Noir, though Tierhoek is its only significant winery.

CITRUSDAL VALLEY

Bamboes Bay

Citrusdal Valley is altogether less significant in terms of production and quality. The tiny coastal ward of *Bamboes Bay* is well known for fine Sauvignon Blanc from Fryer's Cove.

LUTZVILLE VALLEY

Koekenaap, Vredendal, Spruitdrift

Lutzville Valley produces large volumes of everyday (and sometimes better) red and white wines from several varieties, including Chardonnay, Sauvignon Blanc, Sémillon, Chenin Blanc, and Colombard for whites, and Cabernet Sauvignon, Merlot, Pinotage, and Shiraz for reds. As everywhere in the Cape, though, there are isolated ambitious producers holding back some of their best grapes from the co-ops and merchants and producing good and interesting wines – such as Cape Rock in Vredendal.

SUN PROTECTION OVER VINEYARDS IN RIEBEEK WEST, SWARTLAND
This district of the Western Cape wine region is notoriously dry. Growers often use plastic sheeting to further protect the vines from too much sun. Most vineyards here dry-farm, meaning that they rely on rainfall rather than irrigation.

THE WINE PRODUCERS OF
SOUTH AFRICA

Notes: The geographical indication given here is where the winery business is situated and not necessarily the wine of origin (source of the grapes) for the wines. This is because so many of South Africa's new (and old) winemakers do not own the land from which they source fruit nor a cellar, and are increasingly sourcing from far-flung sites.

The terms "classic red blend" and "classic white blend" have been reserved for wines from the Bordeaux varieties. Blends containing varieties other than those regarded as Bordeaux white or black grapes have been referred to as "red blend" or "white blend".

AA BADENHORST FAMILY
Swartland
★★☆

This family-run business is known for exciting Syrah-led red and eclectic white blends, as well as several single-vineyard expressions of old vines, including Tinta Barocca. Also offers a decent Vermouth.

 Entire range

ALHEIT VINEYARDS
Hermanus
★★★

Lauded as setting a new standard and expression for South African wines, Chris and Suzaan Alheit's eloquent offerings from widely-sourced fruit (they own no vineyards) includes a handful of the country's best single-vineyard Chenin Blancs.

 Entire range

ALLESVERLOREN
Swartland
★ⓥ

Dating from the turn of the 18th century, this is South Africa's oldest winery. Traditionally one of the Cape's greatest port producers, this Malan family estate focuses on Shiraz and Portuguese grapes, but also makes a charming Chenin Blanc.

 Chenin Blanc • Fortified (Port) • Tinta Barocca

ALVI'S DRIFT
Worcester
★ⓥ

This is the region's standout producer with a much-improved value-for-money range plus a handful of big but not alcoholic special *cuvées*.

 Chenin Blanc (Albertus Viljoen) • Red blends (Albertus Viljoen Bismark, Drift Fusion)

ANTHONIJ RUPERT
Franschhoek
★★

Johan Rupert's wine enterprise, named after his late brother, has been expertly remodelled after nearly a decade of lost form. Its brands include the impressive *terroir*-driven Cape of Good Hope, Jean Roi (rosé), L'Ormarins (now focusing on MCC), Terra del Capo (Italian varieties), the value-for-money Protea, and Anthonij Rupert itself.

 Cabernet Franc (Anthonij Rupert) • Chardonnay (Cape of Good Hope) • Chenin Blanc (Cape of Good Hope) • Sparkling wine (L'Ormarins Brut Rosé)

ARCANGELI
Bot River
★★

Texture, structure, and age-ability are keys to this Italianesque producer.

 Nebbiolo • Verdelho

ARISTEA
Constantia
★☆

Aristea is a new venture from British businessman Martin Krajewski, owner of Bordeaux's Clos Cantenac and Chateau Séraphine, and Matt Krone, who earned his MCC stripes alongside his father, Nicky Krone, at Tulbagh's renowned Tweejongegezellen cellars.

 Sparkling wine (Brut Rosé)

ARTISANAL BOUTIQUE WINERY
Stellenbosch
★★

The more personal project of Stellenrust viticulturist Kobie van der Westhuizen and winemaker Tertius Boshoff. The wines here are more experimental in nature, mostly micro *cuveés* from special sites or rare in South Africa varieties.

 Carignan (SeriesRARE) • Red blend (Artisanal JJ Eight Pillars) • Syrah (SeriesRARE After Eight Shiraz)

ASHBOURNE
Walker Bay
★★

Anthony Hamilton Russell's Pinotage-championing venture also puts out an easy-drinking rosé and white blend.

 Pinotage • Pinotage-Cinsault

ATARAXIA
Walker Bay
★★☆

Kevin Grant was formerly the winemaker at Hamilton Russell, so it's no surprise his masterful Chardonnay leads his Sauvignon Blanc, Pinot Noir, and red blend.

 Chardonnay

AVONDALE
Paarl
★★ⓞ

Terra est vita – "earth is life". One of only a handful of producers with organic certification, this estate smartly marries organic, biodynamic, and scientific principles. A large array of Georgian *qvevri* was added in 2018.

 Classic red blend (La Luna) • White blend (Cyclus)

BARTINNEY
Stellenbosch
★★

This vintner's steadily improving range of wines is produced from hillside vineyard sites on the Helshoogte Pass outside Stellenbosch.

 Entire range

BEAU CONSTANTIA
Constantia
★☆

Beau Constantia bucks the area's trend for cool-climate offerings with rich, sometimes heady red wines. Probably better known for its on-site restaurant, Chef's Warehouse.

 Classic red blend (Lucca) • Classic white blend (Pierre)

BEAUMONT FAMILY
Walker Bay
★★

This producer was established in 1993 by Raoul and Jayne Beaumont, whose son Sebastian crafts top-quality Chenin and impressive red blends, including a Mourvèdre-led offering with Cabernet Franc, Pinotage, Syrah, and Petit Verdot. More recently Jayne has been dabbling in the cellar and has released a set of five vintages of Pinot Noir.

 Chenin Blanc (Hope Marguerite) • Classic red blend (Ariane) • Red blend (Vitruvian)

BEESLAAR
Stellenbosch
★★★

Abrie Beeslaar crafts Kanonkop's stellar wines, as well as a solo Pinotage under his own label.

 Pinotage

BEIN
Stellenbosch
★

Swiss couple Luca and Ingrid Bein produce five wines, all Merlot.

 Merlot (Merlot, Merlot Reserve)

BELLINGHAM
Franschhoek
★

Owned by Douglas Green, Bellingham is a stand-alone operation that seldom disappoints, regardless of the price point.

 Chenin Blanc (The Bernard Series, Homestead Series) • Pinotage (The Bernard Series) • White blend (The Bernard Series)

BENGUELA COVE
Hermanus
★☆

A new lifestyle estate from entrepreneur Penny Streeter OBE offers a range to please both the casual imbiber and serious collector.

 Sémillon (Catalina)

BEYERSKLOOF
Stellenbosch
★★

Partly owned by Beyers Truter, who earned his reputation for Pinotage at Kanonkop, this producer's wines are now made by his son.

 Classic red blend (Field Blend, Synergy) • Pinotage (Diesel, Reserve)

BLACKWATER
Stellenbosch
★★

Francois Haasbroek, an artisan winemaker with a fondness for lighter and fresher reds, continues to add new wines to his quietly impressive range from widely sourced vineyards.

 Carignan (Omerta) • Chenin Blanc (Picquet, The Underdog) • Grenache (Daniel)

BLANKBOTTLE
Somerset West
★★

Owner/winemaker Pieter Wahlser's goal was to create a wine brand with no limitations when it came to style, vintage, area, or variety. The intention was to break down preconceived expectations. The reality has been the opportunity to introduce one-off limited runs of interesting wines, making it hard to recommend specific wines. If you have an open mind, this producer is for you.

 Entire range

BOEKENHOUTSKLOOF
Franschhoek
★★

This star producer continues to make good entry-level wines that are also sold under the Wolftrap and Porcupine Ridge labels, with the latter's Syrah punching well above its weight.

 Cabernet Sauvignon • Classic red blend (The Chocolate Block) • Late Harvest (Noble Late Harvest Sémillon) • Sémillon • Syrah

BOPLAAS
Calitzdorp
★

Lurking in a large range of commercially acceptable products are a number of extremely fine fortified wines, including one, the Vintage Reserve Port, that is considered one of South Africa's very greatest "ports".

 Fortified (Cape Tawny Port, Red Muscadel, Vintage Reserve Port)

BOSCHKLOOF
Stellenbosch
★★

Shiraz is the focus of father and son team, Jacques and Reneen Bornman, while Reenen's "new wave" fascination gets expression in the Kottabos range and his own Ron Burgundy wines.

✓ *Classic red blend • Syrah (Epilogue Shiraz, Syrah)*

BUITENVERWACHTING VINEYARDS, CONSTANTIA VALLEY
Buitenverwachting originally formed part of the Constantia Estate. The Cape Dutch manor house dates back to 1773. This estate is along the Constantia Wine Route, which takes travellers past Groot Constantia, Constantia Glen, Constantia Uitsig, Eagles Nest, High Constantia, Klein Constantia, and Steenberg.

BOSMAN FAMILY
Wellington/Hermanus
★★☆

The Bosman family, owners of one of South Africa's leading vine nurseries, which is located in Wellingotn, relies on the inspirational Corlea Fourie to craft their appealing range, including the country's first Nero d'Avola.
 Entire range

BOTANICA WINES
Stellenbosch
★★

With her outstanding Mary Delany Collection, owner-winemaker Ginny Povall has quietly earned an enviable reputation for Pinot Noir, Chenin Blanc, and Sémillon, but even her "lesser" labels impress.
 Entire range

BOUCHARD-FINLAYSON
Walker Bay
★★

Founded by Hamilton Russell's winemaker Peter Finlayson and Paul Bouchard from Burgundy, this Hemel-en-Aarde stalwart is today owned by Bea and Stanley Tollman.
 Chardonnay (Kaaimansgat, Missionvale)
  • Red blend (Hannibal)

BUITENVERWACHTING
Constantia
★☆ **V**

Buitenverwachting is a reliable producer that can be counted on for consistently producing quality Cabernet Sauvignon, Chardonnay, and Sauvignon Blanc.
 Classic red blend (Christine)

B VINTNERS VINE EXPLORATION CO
Stellenbosch
★★☆

Joint venture between cousins Bruwer Raats (*See* Raats Family) and Gavin Bruwer Slabbert includes a rare-in-SA dry Muscat d'Alexandrie and standout white blend celebrating heirloom varieties Chenin Blanc, Sémillon, and Muscat d'Alexandrie (Harlem to Hope).
 Entire range

CHAMONIX
Franschhoek
★★☆

This winery's mountain slopes site and clay-rich soils are ideal for top-class whites and reds.
 Chardonnay (especially Reserve)
  • Classic red blend (Troika) • Pinotage (Greywacke) • Pinot Noir

CAPENSIS
Stellenbosch
★★

This is a fledgling but ambitious joint venture between American Anthony Beck, whose father started Graham Beck in Robertson, and Barbara Banke, owner of Jackson Family Wines in California.
 Chardonnay

CAPE POINT VINEYARDS
Noordhoek
★★

The only wine farm on the Cape Peninsula's cool, narrow, southern tip concentrates on white Bordeaux varieties.
✓ *Classic white blend (Isliedh)*

CAP MARITIME
Hemel-en-Aarde

This is the latest venture from Marc Kent whose name is synonymous with the high-end Boekenhoutskloof, Chocolate Block, and Porseleinberg labels, as well as the larger volume brands The Wolftrap and Porcupine Ridge. Here the focus will be on Chardonnay and Pinot Noir; the vineyards have yet to be planted, but the intention is that the wine will rival their neighbours' for quality and price.

CARINUS FAMILY
Stellenbosch
★★☆

Swartland vines provide good-quality Chenin Blanc grapes and Stellenbosch provides the Syrah for this quietly confident venture. The wines are made by rising star Lukas van Loggerenberg.
 Chenin Blanc (Carinus, Rooidraai)

CATHERINE MARSHALL
Elgin
★★

The wines from one of South Africa's most respected women winemakers, especially the Pinot Noirs, are vital and long-lived.
 Pinot Noir (Finite Elements, On Clay Soils)

CEDERBERG
Cederberg
★★

South Africa's highest-altitude vineyards produce wines of great length and finesse, rather than heavy in weight and alcohol.
 Entire range

CITY ON A HILL
Swartland
★★

This estate produces impressive whites off Paardeberg granite and reds from shale soils in the Kasteelberg.
 Chenin Blanc • White blend

CHARLES FOX
Elgin
★★☆

This is an exciting young MCC venture in cool-climate Elgin, where the wines are crafted with input from Reims consultant Nicola Follet.
 Entire range

COLMANT
Franschhoek
★★

Colmant celebrates classy, vital MCCs from classic Champagne varieties.
 Sparkling wine (Absolu Zero Dosage, Brut Chardonnay)

CONSTANTIA GLEN
Constantia
★★

This producer creates elegant and refined wines at a magnificent location overlooking False Bay.
✓ *Entire range*

CREATION
Hermanus
★★

Dynamic husband and wife team, Swiss-born Jean-Claude and Carolyn Martin, have the talent to match their ambition. They are creating fine wines at this Walker Bay winery.
 Entire range

CRYSTALLUM
Walker Bay
★★☆

Peter-Allan and Andrew Finlayson, sons of Peter Finlayson of Bouchard-Finlayson fame, are only doing what comes naturally to the Finlaysons – which is making brilliant Pinot Noir and Chardonnay.
 Entire range

DAVID & NADIA
Swartland
★★★

Chenin Blanc and Grenache Noir are the signature grape varieties of winemaker, David Sadie, and viticulturist and soil scientist, Nadia Sadie. This young husband and wife team crafts wines are as understated and elegant as this couple, but they speak volumes.
 Entire range

DE GRENDEL
Tygerberg
★

De Grendel's Koetshuis has, for many years, been regarded as one of the country's best Sauvignon Blancs.
 Entire range

DE KRANS
Calitzdorp
★

Highly-regarded fortified wines from Portuguese varieties lead the otherwise middle-of-the-road pack here.
 Fortifieds (Cape Tawny Limited Release, Cape Vintage Reserve, Muscat de Frontignan)

DELAIRE GRAFF
Stellenbosch
★★☆

Belonging to international jeweller Laurence Graff, this property has leapt upwards in quality since Morné Vrey, who honed his knowledge of international palates at the famous Summertown Wine Café in Oxford, took up the winemaking reins.
 Classic red blends (Laurence Graff Reserve, Botmaskop)

DELHEIM
Stellenbosch
★☆ **V**

Always a good value, in recent years Delheim has also shown a dramatic improvement in quality.
 Cabernet Sauvignon (Grand Reserve) • Late Harvest (Edelspatz) • Pinotage (Vera Cruz) • Shiraz (Vera Cruz)

DEMORGENZON
Stellenbosch
★★★

Wendy and Hylton Appelbaum's hillside estate has gone from strength to strength with GM and cellarmaster Carl van der Merwe at the helm, particularly when it comes to Chenin Blanc and white blends.

✓ *Entire range*

DE TOREN
Stellenbosch
★

This winery achieved an almost immediate cult following for its maiden wine, Fusion V, a "big and bold" blend of the five red Bordeaux varieties.

✓ *Classic red blends (Fusion V, Z)*

DE TRAFFORD
Stellenbosch
★★☆

David Trafford's De Trafford wines, like his classic red blend Elevation 393 are often powerful and alcoholic, but always in balance. There's more restraint in his Sijnn label, made from vineyards close to the lower Breede River some 40 kilometres (25 miles) from Stellenbosch.

✓ *Entire range, both labels*

DE WETSHOF
Robertson
★

Peter de Wet has taken over from his legendary father, Danie de Wet, in the cellar while viticulture-savvy sibling Johann holds the CEO and international marketing reins. Both inherited Danie's passion for Chardonnay which thrives, despite the heat, in the area's lime-rich soils.

✓ *Chardonnay (Bateleur, The Site)*

DIEMERSDAL
Durbanville
★☆

Sixth-generation winemaker Thys Louw lives and breathes Sauvignon Blanc; he makes seven here and a handful of other labels for friends and family. His reds are not too shabby, either.

✓ *Entire range*

DIEMERSFONTEIN
Wellington
★

This producer's full-throttle Pinotage in the "coffee" style is converting many South Africans to wine.

✓ *Pinotage (Carpe Diem)*

DISTELL
Stellenbosch
★

Distell is Africa's largest producer of wines, spirits, ciders, and other ready-to-drink beverages that include go-to wine brands like Durbanville Hills, Fleur du Cap, the House of JC le Roux, Nederburg, and Zonnebloem, among others.

✓ *Cabernet Sauvignon (Nederburg Two Centuries)*

THE DRIFT
Napier
★★

Bruce Jack (of Flagstone Winery fame) has a number of South African-based ventures, including this label from the family farm, Appelsdrift. The Pinot Noir, from wind-swept slopes, is exceptionally characterful while the other red grapes – such as Barbera, Malbec, Syrah, several Tourigas, and Tintas – are co-fermented into interesting blends.

✓ *Pinot Noir • Red blend (The Drift Moveable Feast, Over the Moon)*

EAGLES' NEST
Constantia
★☆

If this vineyard focused exclusively on its Shiraz, its rating would be two and a half stars. Lovers of low-acid Viognier might like to try Eagles' Nest's opulent offering.

✓ *Classic red blend (Verreaux) • Syrah (Shiraz)*

EIKENDAL
Stellenbosch
★

This Swiss-owned producer has dramatically improved its quality in recent years under cellarmaster Nico Grobler.

✓ *Entire range*

ELEMENTAL BOB
Somerset West
★★☆

Hands-off owner-winemaker, Craig Sheard, wants his boutique wines to take you on a sensory and spiritual journey.

✓ *Chenin Blanc (The Rupert, Retro) • Palomino (Farmer Red Beard) • Tinta Barocca (Graveyard)*

ERIKA OBERMEYER
Stellenbosch
❶

This former long-time Graham Beck maker of white wine had a stellar maiden vintage under her own name. If she continues in the same vein with future vintages, she'll soon rival others for rising star status.

✓ *Cabernet Sauvignon (Erika O) • Red blend (Erika O Syrah-Grenache Noir-Cinsault)*

ERNIE ELS
Stellenbosch
★

This range of mostly red wines is as powerful and consistent as the South African golfer himself.

✓ *Entire range*

FABLE MOUNTAIN VINEYARDS
Tulbagh
★★☆

Elevation, aspect, cool nights, and mainly schist/clay soils impart both freshness and longevity to this exciting range.

✓ *Syrah (Fable Mountain, Small Batch Series SYB80) • Red blend (Night Sky) • White blend (Jackal Bird)*

FAIRVIEW
Paarl
★☆

For many years, Charles Back's estate has been one of the most innovative wine producers in South Africa, offering a large and constantly expanding range of wine styles.

✓ *Carignan (Pegleg) • Classic red blend (Caldera) • Red blend (Extraño) • Sémillon (Oom Pagel) • Shiraz (Eenzaamheid, Jakkalsfontein, The Beacon)*

FLEUR DU CAP
Stellenbosch
★ⓥ

Distell-owned brand Fleur du Cap has established a high standard of quality for its basic wines, but also offers a handful of wines that box above their price bracket.

✓ *Cabernet Sauvignon (Unfiltered) • Chardonnay (Unfiltered) • Late Harvest (Bergkelder Noble Viognier)*

THE FOUNDRY
Voor Paardeberg
★★☆

This private label of Chris Williams (winemaker at Meerlust) and his business partner James Reid of Kumala fame includes lean yet flavoursome single-vineyard wines from Voor Paardeberg, neighbouring Swartland, and other *terroirs* around the Cape

✓ *Entire range*

GABRIËLSKLOOF
Bot River
★☆

Cellarmaster Peter-Allan Finlayson (see Crystallum) and his team have been working hard in the vineyards to up their game, and this effort is beginning to show in the wines.

✓ *Cabernet Franc • Syrah (Landscape Series Syrah on Sandstone, Syrah on Shale)*

GLENELLY
Stellenbosch
★★

The range of wines at Glenelly, owned by Madame May-Éliane de Lencquesaing, the former proprietor of the famous Pauillac winery Château Pichon Longueville Comtesse-de-Lalande, is not as large as at other South African wineries, but it includes fine reds as well as two Chardonnays.

✓ *Chardonnay (Estate) • Red blend (Lady May, Estate)*

GLENWOOD
Franschhoek
★☆

The year-on-year improvements, most evident in the Grand Duc Syrah range, are to be admired at this relatively small production estate.

✓ *Chardonnay (Grand Duc, Vigneron's Selection) • Syrah (Grand Duc)*

GRAHAM BECK
Robertson
★★☆

After the death of founder Graham Beck, it was decided that the focus would shift solely to high-end sparkling wines made by long-standing cellarmaster, Pieter "Bubbles" Ferreira.

✓ *Sparkling wine (Blanc de Blancs Brut, Brut Zero, Cuvée Clive)*

GRANGEHURST
Stellenbosch
★☆

Consistent, well-crafted, elegant yet substantive wines have been made by

OAK WINE BARREL OF GROOT CONSTANTIA
A few kilometres from False Bay, Groot Constantia is the oldest wine estate in South Africa, originally cultivated by Simon van der Stel, first Governor of the Cape Colony. Van der Stel brought his viticultural knowledge and winemaking skills from his Muiderbergh vineyards in the Netherlands.

owner-winemaker Jeremy Walker for over a quarter of a century on the slopes of the Helderberg. Profits from the newest addition, The Reward, go to the cellar team, hence the name.

✓ *Cabernet Sauvignon (Reserve, The Reward)*

GROOT CONSTANTIA
Constantia
★☆

This property is part of the original Constantia farm, the oldest and most famous of Cape estates, and is currently being run with passion and great skill.

✓ *Entire range*

HAMILTON RUSSELL
Walker Bay
★★☆

One of South Africa's best-known estates, Unusually in a South African context, it has been planted with just two varieties, Chardonnay and Pinot Noir, since it was established in 1975.

✓ *Entire range*

HARTENBERG
Stellenbosch
★★

"Quality first regardless of style or price point" seems to be the mantra by which Hartenberg's team does business. Here, winemaker for more than 25 years, Carl Schultz, is best known for his Syrah and Syrah blends but lavishes equal care on all the grapes that come into the cellar.

✓ *Cabernet Sauvignon • Chardonnay • Merlot • Syrah (Gravel Hill Shiraz, The Stork, CWG Auction Reserve Shiraz) • Red blend (The Megan)*

HASKELL
Stellenbosch
★★

Owned by American-born Preston Haskell IV, the wines from this highly regarded site in the Helderberg foothills were established and nurtured by Rianie Strydom, but they are now made by Rudolf Steenkamp.

✓ *Entire range*

HERMANUS-PIETERSFONTEIN
Hermanus
★★

Owner of Hermanuspietersfontein and viticulturist-winemaker Bartho Eksteen, best known for his dab hand with Sauvignon Blanc, also produces superbly sleek reds.

✓ *Cabernet Franc (Swartskaap) • Classic red blend (Die Arnoldus, Die Martha) • Sauvignon Blanc (Kat Met Die Houtsbeen)*

HIGHLANDS ROAD
Elgin
★

This small producer in Elgin with a reputation for stellar white wines is producing increasingly good cool-climate Syrah.

✓ *Chardonnay • Sauvignon Blanc (White Reserve) • Syrah*

HOGAN
Stellenbosch
★★

Elegant and quietly eloquent are two ways to describe the wines available from hands-off winemaker Jocelyn Hogan Wilson.

✓ *Entire range*

IONA
Overberg
★★☆

This is another property that built its reputation on Sauvignon Blanc but now offers world-class red and white blends.

✓ *Entire range*

JC WICKENS
Malmesbury
★★

Another young Swartland couple – Jasper and Franziska Wickens, he acting as the winemaker as he also does at AA Badenhorst and she the viticulturist – are turning out a lovely red blend grounded in Tinta Barocca and an understated Chenin Blanc.

✓ *Entire range*

JH MEYER SIGNATURE
Hermon
★★

Another of the Cape's increasing band of non-interventionist labels, this one with a focus on Chardonnay and Pinot Noir and the only solo venture for Johan Meyer. His Mother Rock label – a partnership with a UK importer – offers one of the few ranges of genuinely "natural" wines in the Cape, with grapes from organic vineyards, while his Mount Abora Vineyards offerings are a collaboration with two fellow young guns.

✓ *Chardonnay (Palmiet) • Pinot Noir (Elands River)*

JOOSTENBERG
Stellenbosch
★★

Look for fine, honest expressions of the organic vineyards on cellarmaster-viticulturist Tyrrel Myburgh's family farm in Stellenbosch.

✓ *Chenin Blanc (Die Agteros) • Syrah (Klippe Kou) • Red blend (Bakermat)*

JORDAN
Stellenbosch
★★☆

This well-established winery delivers consistent quality at a high level across a broad range of blended wines as well as single variety bottlings. Owners Kathy and Gary Jordan recently invested in a small country estate an hour south of London where they plan to plant Pinot Noir and Chardonnay for sparkling wine.

✓ *Entire range*

KAAPZICHT
Stellenbosch
★★

Here is another Cape stalwart, with Danie Steytler Junior having a somewhat lighter touch when it comes to alcohol than his father, Danie Steytler Senior.

✓ *Chenin Blanc (The 1947) • Classic red blend (Pentagon) • Red blend (Steytler Vision) • Pinotage (Steytler)*

KANONKOP
Stellenbosch
★★★

Owned by the same family for four generations, Kanonkop is widely regarded as South Africa's leading *"premier cru"*, or "First Growth" and has historically only made red wines from mostly Cabernet Sauvignon and Pinotage. That changed when it launched a dry rosé made from Pinotage under its second label to take advantage of increasing consumer interest in the style.

✓ *Entire range*

KEERMONT
Stellenbosch
★★☆

Quality increases year-on-year at this steep hillside property.

✓ *Chenin Blanc (Riverside, Terrasse) • Classic red blend (Estate Reserve) • Syrah (Steepside, Topside)*

KEN FORRESTER
Stellenbosch
★★

Owner Ken Forrester is widely accredited with inspiring South Africa's Chenin Blanc revolution, so it is therefore not surprising that there are several in his stable, in addition to a handful of impressive reds.

✓ *Chenin Blanc (The FMC, Old Vine Reserve) • Red blend (Gypsy, Three Halves, Renegade)*

KLEIN CONSTANTIA
Constantia
★★☆

This property was part of Simon van der Stel's original Constantia Estate. It is still best known for its vine-dried Vin de Constance, a replica of the sweet wine (*Constantia wyn*) supposedly favoured by Napoleon. It also still produces several excellent Sauvignon Blancs.

✓ *Sauvignon Blanc (Glen Dirk, Metis, Perdeblokke, Block 382) • Dessert (Vin de Constance)*

KLEINE ZALZE
Stellenbosch
★★

One of South Africa's largest privately owned estates expertly balances volume with quality.

✓ *Entire range*

KOTTABOS
Stellenbosch
★★

Kottabos features Boschkloof's Reenan Bornman's more experimental range, which he creates from bought-in grapes. These wines are pure-fruited and fragrant with less reliance on oak.

✓ *Red blend*

KWV
Paarl
★

The privatised former super-sized cooperative can still turn out great-value wines, with a few gems along the way.

✓ *Chenin Blanc (The Mentors) • Petit Verdot (The Mentors) • Classic red blend (Tributum) • Red blend (Canvas, Triptych) • Syrah (The Mentor Shiraz)*

LAMMERSHOEK
Malmesbury
★☆

A German consortium (with soccer legend Franz Beckenbauer being one of the investors) acquired this Swartland estate in 2013 by. Since then there's been considerable investment in the vineyards and cellar facilities, and the range was revamped. It includes a rare in South Africa Hárslevelü, plus a Tinta Barocca from among the oldest vines of that variety.

✓ *Chenin Blanc • Grenache • Hárslevelü • Tinta Barocca*

L'AVENIR
Stellenbosch
★☆

Owned by France-based AdVini, L'Avenir may have a French name but the focus is local with Pinotage and Chenin Blanc leading the range.

✓ *Chenin Blanc (Single Block) • Pinotage (Single Block)*

LEEU PASSANT
Franschhoek
★★★

The wines from this venture between Chris and Andrea Mullineux (*see* Mullineux Family), Indian businessmen Analijt Singh, and Peter Dart honour old vines and *terroir*.

✓ *Entire range*

LEEUWENKUIL FAMILY
Swartland
★☆

At this large estate, the focus was previously on easy-drinking wines. Today, in addition to delivering value for money, there's a desire to make wines that are smaller in quantity but celebrated for their stellar quality.

✓ *Chenin Blanc (Heritage) • Syrah (Heritage) • White blend (Reserve)*

LE LUDE
Franschhoek
★★

This Méthode Cap Classique specialist offers South Africa's first *agrafé* sparkling.

✓ *Entire range*

LE RICHE
Stellenbosch
★★☆

Etienne Le Riche struck out on his own after 20 years at the Rustenberg estate. Today, his legacy is in the hands of son Christo and daughter Yvonne.

✓ *Cabernet Sauvignon (Reserve, CWG Auction Reserve)*

LOURENS FAMILY
Hermanus
 ❶

Owner-winemaker Franco Lourens makes his wines in the Alheit cellar, where he is assistant winemaker to Chris Alheit (*see* Alheit). His early vintages are very promising, auguring well for future ratings.
✓ *Entire range*

LOURENSFORD
Somerset West
★

The range at South African businessman Christo Wiese's Cape Town estate was dramatically reduced after under-performing vineyards were grubbed up, so expectations are that improvements in quality will follow.
✓ *Classic red blend (Chrysalis) • White blend (Chrysalis)*

LUDDITE
Bot River
★☆

Shiraz is the standout variety in this boldly handsome but small range of wines grown on the slopes of the Houw Hoek Mountains.
✓ *Entire range*

MEERLUST
Stellenbosch
★★☆

This is another of South Africa's iconic estates, now with eighth-generation Hannes Myburgh in overall charge and the highly accomplished Chris Williams at the winemaking helm.
✓ *Entire range*

METZER FAMILY
Somerset West
★

The Metzers are producing understated yet characterful wines from old Stellenbosch vines.
✓ *Chenin Blanc (Maritime, Montane) • Shiraz*

MILES MOSSOP
Stellenbosch
★★☆

Previously the winemaker at Tokara, where he made wines under his own label, Miles Mossop is now 100 per cent invested in his brand.
✓ *Entire range*

MONT BLOIS
Robertson
 ❶

Nina-Mari, the young wife of Ernst Bruwer, the owner of this large grape growing concern, recently revived its winemaking tradition with several standout whites. Should she continue on this path, her wines will do much to polish the reputation of this warmer-climate area.
✓ *Chardonnay (Kweekkamp) • Chenin Blanc • Dessert (Harpie Muscadel, Pomphuis Muscadel)*

MÔRESON
Franschhoek
★★☆

This estate produces increasingly impressive wines at both ends of the price spectrum.
✓ *Entire range*

MORGENSTER
Stellenbosch
★★

Late owner Giulio Bertrand had three passions; he indulged two here at this model olive and wine estate with its impeccable olive groves (planted to Italian varieties) and hillside vineyards (planted to Italian varieties, as well as those from Bordeaux).
✓ *Classic red blend (Lourens River, Reserve, Tosca) • Nebbiolo (Nabusco) • Red blend (Tosca)*

MOTHER ROCK
Hermon
★★

Johan Meyer (*see* JH Meyer) and UK importer Ben Henshaw embrace what some call a "radical approach" – early picking, skin contact for whites, and oxidative handling and maturation – in this idiosyncratic natural line-up.
✓ *Chenin Blanc (Kweperfontein, Liquid Skin) • Grenache • Red blend (Holocene)*

MOYA'S VINEYARDS
Hermanus
★★

Look for boutique quantities of high-quality Chardonnay and Pinot Noir from this Hemel-en-Aarde Valley estate.
✓ *Entire range*

MULDERBOSCH
Stellenbosch
★☆

Best known for its lightly oaked Chenin Blanc, Steen Op Hout, there are numerous top-flight wines from superbly sited mountain vineyards from this producer.
✓ *Chenin Blanc (Block S2, Block W) • Cabernet France • Classic red blend (Faithful Hound)*

MULLINEUX
Swartland
★★★

Arguably South Africa's most stellar husband-and-wife wine team, with Chris Mullineux concentrating on the vineyards and Andrea – *Wine Enthusiast*'s Winemaker of the Year in 2016 – on the cellar, but it's probably the teamwork that matters.
✓ *Entire range*

MURATIE
Stellenbosch
★☆

Year-on-year improvement can be found at this historic estate, owned by the Melck family.
✓ *Cabernet Sauvignon (Martin Melck Family Reserve) • Classic red blend (Ansela van de Caab) • Syrah (Ronnie Melck Shiraz)*

NAUDÉ
Stellenbosch
★★

Whether the wines are single-vineyard or multi-*terroir*, Ian Naudé's focus is on venerable old vines in the Swartland and elsewhere. (His wines were previously under the Adoro label.)
✓ *Entire range*

NEDERBURG
Paarl
★

Owned by Distell (*see* Distell), Nederburg continues to build its profile locally, in Africa, and internationally with established wines and new offerings. For example, the maiden Albariño is going exclusively to the Waitrose chain in the UK. The Private Bin range continues to be sold exclusively at the Nederburg Auction, which now accepts wines from other producers.
✓ *Classic red blend (The Brew Master) • Classic white blend (The Young Airhawk) • White blend (Ingenuity White) • Late Harvest (Eminence Noble Late, Edelkeur Noble Late)*

NEIL ELLIS
Stellenbosch
★★☆

Often cited as South Africa's first wine-maker-*négociant*, Neil Ellis has eventually acquired business premises and passed on the winemaking reins to son, Warren.
✓ *Entire range*

NEWTON JOHNSON FAMILY
Hermanus
★★★

The family-owned Newton Johnson Vineyards is one of the brightest stars in the exciting Hemel-en-Aarde wine scene, with no fewer than five outstanding Pinot Noirs, accomplished Chardonnays, a handful of red blends, and the first South African Albariño.
✓ *Entire range*

OAK VALLEY
Elgin
★★

Oak Valley is a massive 1,786-hectare (4,413-acre) mountainside polycultural farm where great quality wine, especially its Groenlandberg label, competes with beef cattle and cut flowers for attention.
✓ *Entire range*

OPSTAL
Rawsonville
★

Renewed energy and a focus on minimum-intervention winemaking, heritage vines and varieties – chiefly Chenin Blanc and Sémillon – have catapulted the flagship Heritage range of this wine estate into a new quality zone. Expect other ranges to follow suit.
✓ *Chenin Blanc (Heritage Carl Everson) • Sémillon (Heritage The Barber) • Red blend (Heritage Carl Everson Cape) • White blend (Heritage Carl Everson Cape White)*

PASERENE
Franschhoek
★★

Hail a new boutique label from former Vilafonté winemaker Martin Smith.
✓ *Entire range*

PAUL CLUVER
Elgin
★★

Apart from Pinot Noir, only white wines are made on this large apple and wine farm with its keen focus on environmental and social sustainability.
✓ *Entire range*

PERDEBERG
Paarl
★ Ⓥ

With thousands of vines to draw upon, mostly unirrigated and many heritage, this dynamic Paarl winery is expected to raise the bar considerably in future. In fact, it has already started with its Dryland Collection.
✓ *Chenin Blanc (Dryland Collection Courageous Barrel Fermented) • Red blend (Dryland Collection Red) • White blend (Dryland Collection Rossouw's Heritage)*

PORSELEINBERG
Swartland
★★★

Just one wine, a Syrah, from stony hillside and hilltop vineyards – a stunner – provides proof that Boekenhoutskloof's Marc Kent is as much a wine businessman as he is a wine lover, collector, and maker.
✓ *Syrah*

RAATS FAMILY
Stellenbosch
★★★

Highly regarded Chenin Blanc and Cabernet Franc specialist Bruwer Raats works alongside his cousin Gavin Bruwer Slabber to produce these justifiably oft-awarded wines.
✓ *Entire range*

RADFORD DALE
Stellenbosch
★★

Barossa-born Ben Radford and British-born Alex Dale partnered in 1998 to form the Winery of Good Hope, in which the Radford Dale label was the flagship label. The name change in 2018 prompted a tweaking of the ranges to Radford Dale, Labeye, Pearce Predhomme, Land of Hope, and Winery of Good Hope.
✓ *Chardonnay (Radford Dale) • Chenin Blanc (Radford Dale The Renaissance of Chenin Blanc, Pearce Prodhomme) • Pinot Noir (Radford Dale AD, Radford Dale Freedom, Labeye) • Red blend (Radford Dale Gravity, Radford Dale Black Rock) • Syrah (Radford Dale Nudity, Radford Dale Syrah)*

RALL
Malmesbury
★★★

Donovan Rall is a vine explorer and experimenter with a fascination for the

Swartland. His newest wine, however, a Cinsault Blanc, comes from Wellington and are the last few vines of the variety in the country.

✓ *Entire range*

RESTLESS RIVER
Hermanus
★★☆

The Restless River wines are produced from boutique vines by Craig and Anne Wessels, whose commitment to site and restraint is admirable.

✓ *Entire range*

REYNEKE
Stellenbosch
★★★

Reyneke offers an excellent range of focused wines from environmental ethics graduate Johan Reyneke. Standout wines are the entire Biodynamic Reserve range and the Syrah from the Biodynamic range.

✓ *Entire range*

RICHARD HILTON
Stellenbosch
★

The boutique vintner's new skin-fermented Viognier is a natural progression of Richard Hilton's wine journey.

✓ *Entire range*

RICHARD KERSHAW
Elgin
★★★

Central to British-born Master of Wine Richard Kershaw's internationally acclaimed offerings are the blends from cool-climate Elgin showcased in his Clonal Selection range. A self-proclaimed "geek", Kershaw also seeks to express the sometimes subtle clonal, site, and soil differences between the grapes he selects for his Deconstructed range of Syrahs and Chardonnays, while his GPS Series features parcels from outside Elgin.

✓ *Entire range*

RICKETY BRIDGE
Franschhoek
★

With a winemaking provenance dating back to the 17th century, history meets creativity at this reliable Franschhoek wine estate.

✓ *Cabernet Sauvignon (The Bridge)*
• *Sémillon (Road to Santiago)*

RIJK'S PRIVATE CELLAR
Tulbagh
★

Many regard winemaker Pierre Wahl as a Pinotage expert, and there are four to try from his cellar – a fruit-driven version from the Touch of Oak range for early drinking, the somewhat more serious Private Cellar and Reserve bottlings, and finally the 888, so-called because only 888 bottles of this wine are produced each vintage and are individually numbered and registered to the buyer's name.

✓ *Entire range*

RON BURGUNDY
Stellenbosch
★★☆

A trio of friends, including Reenan Bornman, winemaker for both these wines and those at his family farm, Boschkloof produce a small range of exceptionally tight and fresh whites, plus a Syrah.

✓ *Entire range*

RUPERT & ROTHSCHILD
Franschhoek
★★☆

A mega-rich partnership delivering world-class wines of various origins, both pure and blended, from Barrydale, Bot River, Darling, Elgin Robertson, Simonsberg-Paarl, and Stellenbosch. The two top wines are Baroness Nadine (a *barrique*-fermented Chardonnay) and Baron Edmond (a Cabernet-Merlot blend)

✓ *Entire range*

RUSTENBERG
Stellenbosch
★★☆

Once heralded as the very best of South Africa's top growths, the quality at Rustenberg faltered in the early 1990s. It picked up from the 1996 vintage but there followed some confusion about the aesthetic to be followed. More recently, with original owner Peter Barlow's grandson Murray at the helm, quality is higher than it has ever been. There are now three ranges – Flagship, Site Specific, and Regional.

✓ *Cabernet Sauvignon (Site Specific Peter Barlow, Regional Stellenbosch)*
• *Chardonnay (Flagship Stellenbosch)*
• *Classic red blend (John X Merriman)*
• *Syrah (Flagship Buzzard Kloof)*

RUST EN VREDE
Stellenbosch
★

Owned by Jean Engelbrecht, son of former Springbok rugby player and legendary winemaker Jannie Engelbrecht, these estate wines, previously "big and dense" in style, are now crafted by long-time winemaker Coenie Snyman to showcase – relatively – greater finesse.

✓ *Entire range*

SADIE FAMILY
Swartland
★★★

There are numerous oenophiles who have had an impact on South Africa's vinous fortunes over the years, but Eben Sadie has to be the best known and most respected, both locally and abroad – and for good reason. His Columella (Syrah with Mourvèdre, Grenache, Cinsault, Tinta Barocca, and Carignan) has, since its debut in 2000, been a standard bearer for the country's reds, while his Palladius (Chenin Blanc, now with 10 other varieties fermented and/or aged on lees in concrete eggs and clay amphorae) has redefined the concept of the South African white blend. Together with independent viticulturist Rosa Kruger, Eben realised the potential of the country's old vineyards – and the threat of their destruction if growers could not sustain them – and launched the impressive Old Vine Series of mostly single-variety bottlings from far-flung, single-vineyard sites. This move firmly put the spotlight on heritage vineyards and it mobilised the industry to protect them, so much so that the country is the first in the world to have a Certified Heritage Vineyard seal.

✓ *Entire range*

SARONSBERG
Tulbagh
★

Since its 2002 acquisition by Nick van Huyssteen and the 2003 appointment of Dewaldt Heyns as winemaker, this estate has been central to the growth in the area's reputation for fine wine.

✓ *Syrah (Shiraz, Full Circle)*

SAVAGE WINES
Cape Town
★★☆

This is an exceptionally small but dynamic one-man venture, the man being winemaker and viticulturist Duncan Savage, previously of Cape Point Vineyards. Their quirky names belie the serious nature of the wines, with each vintage seemingly more elegant than the one before.

✓ *Entire range*

SHANNON
Elgin
★★☆

The grapes for this impressive range are grown by owner-viticulturist James Downes and vinified by the husband and wife team, Gordon and Nadia Newton Johnson (see Newton Johnson). The Mount Bullet Merlot and Capall Bán classic white blend are standouts.

✓ *Entire range*

SIJNN
Malgas
★★☆

David Trafford of De Trafford's project near the mouth of the Breede River yields characterful wines from mostly Mourvèdre, Syrah, Touriga Nacional, and

RICKETY BRIDGE WINERY RAILWAY STATION, FRANSCHHOEK
The Rickety Bridge Winery is one of the stops for a tourist tram ride between vineyards in the Franschhoek Valley.

VINE-COVERED PERGOLA SHADES A WALKWAY AT THE VERGELEGEN WINE ESTATE, STELLENBOSCH
Vergelegen's cellars and winery plant are open to visitors. The property features a Cape Dutch manor house and also includes themed gardens and an upscale restaurant.

Trincadeira made with minimal intervention to reflect the dry and stony landscape.

 Red blend (Sijnn Red, Free Reign) • White blend

SILVERTHORN

Robertson

★★☆

The vineyards for co-owner and winemaker John Loubser's fledgling Methodé Cap Classique range come from the warm Robertson Valley, a far cry from cool-climate Constantia, where he was cellarmaster at Steenberg for many years. The area is renowned for limestone and Chardonnay, however, and the quality of which shows in his line-up.

✓ *Sparkling wine (CWG Auction Reserve Big Dog IV, Jewel Box, The Green Man)*

SIMONSIG

Stellenbosch

★

This winery boasts huge production, and its wines range from the sensational down to pretty standard stuff, with an increasing proportion of the former largely thanks to Debbie Thompson's effort in the cellar.

 Classic red blend (Tiara) • Syrah (Merindol, Heirloom Shiraz)

SPICE ROUTE

Swartland

★★

This estate was originally owned by a consortium of Charles Back, Jabulani Ntshangase, John Platter, and Gyles Webb, but it is now solely owned by Back. Spice Route does best with red wines, and its signature style is typically deep, dark, rich, and spicy, with masses of black fruit.

 Chenin Blanc • Grenache • Red blend (Chakalaka, Malabar)

SPIER

Stellenbosch

★★ **Ⓥ**

This riverside estate – in the grandest and broadest sense of the word – effortlessly combines winegrowing and lifestyle-cum-tourism attractions. Owned by the Enthoven family, its top five ranges all contain seriously smart wines while the remaining three offer amongst the best value-for-money.

 Cabernet Sauvignon (21 Gables) • Chenin Blanc (21 Gables, Farm House Organic) • Classic red blend (CWG Auction Reserve Frans K Smit 20 Year Celebration, Creative Block Five) • Red blend (Creative Block Three) • Classic white blend (Frans K Smit White)

SPIOENKOP

Elgin

★★

Owner and winemaker Koen Roose-Vandenbroucke makes wines with as much character as he can muster, including two rare-in-Elgin Chenin Blancs.

 Chenin Blanc (Johanna Brand, Sarah Raal) • Pinotage (1900, Spioenkop)

SPRINGFIELD

Robertson

★

Considered a bit of an eccentric in his own country, owner-winemaker Abrie Bruwerowner-winemaker, but virtually all of his practices merely represent an organic approach.

 Cabernet Sauvignon (Méthode Ancienne, Whole Berry) • Classic red blend (The Work of Time) • Sauvignon Blanc (Life From Stone)

STARK-CONDÉ

Stellenbosch

★★☆

This picturesque Jonkershoek Valley winery excels with red single varieties and white blends. Three Pines come from almost inaccessible high vineyards, and the Stark-Condé range from lower slopes.

✓ *Entire range*

STEENBERG

Constantia

★★

This estate produces small quantities of gorgeously fresh Sauvignon Blanc as well as a fabulous range of stunning wines in a myriad of styles, including one of the country's best Sémillons.

 Classic red blend (Catharina) • Classic white blend (Magna Carta) • Nebbiolo • Sauvignon Blanc (The Black Swan) • Sémillon

STELLENBOSCH VINEYARDS

Stellenbosch

★

Since 1996 when the cooperatives of Helderberg, Welmoed, Bottelary, and Eersterivier amalgamated into a commercial entity called the Stellenbosch Exchange, this winery has gone by several names including Omnia and The Company of Wine People. Today, its majority shareholder is French wine company AdVini (*see* L'Avenir), potentially heralding that it will look to export more than the 80 per cent of output it currently does. Over the past few years, the range has been culled significantly, and there's a new focus on the fine wine offerings, mostly under the Flagship and Limited Release brands. Of particular interest is the seldom-seen single-variety bottling of Therona, the hybridisation of Crouchen Blanc and Chenin Blanc.

✓ *Cinsault (Limited Release) • Grenache (Limited Release) • Petit Verdot (Flagship) • Verdelho (Limited Release)*

STELLENRUST
Stellenbosch

This family-owned enterprise produces wines with grapes from two different sites in Stellenbosch. One is on the slopes of the Helderberg, and the other is located in the Bottelary hills. Steelenhurst also includes the more personal project of viticulturist Kobie van der Westhuizen and winemaker Tertius Boshoff: Artisanal Boutique Winery.

✓ *Cabernet Franc (Stellenrust) • Carignan (Artisanal Villain Vines) • Chardonnay (Stellenrust Barrel Fermented) • Chenin Blanc (Stellenrust Barrel Fermented, Stellenrust Premium) • Classic red blend (Stellenrust Timeless) • Syrah (Artisanal After Eight Shiraz)*

STONY BROOK
Franschhoek

This estate offers big and bold wines which nevertheless impress with their balance and attention to detail.

✓ *Cabernet Sauvignon (Ghost Gum) • Classic red blend (The Max) • Syrah (Reserve) • Tempranillo (Ovidius)*

STORM
Hermanus

Hannes Storm, who earned a reputation as a pinophile (or "Pinot lover") at Hamilton Russell, experienced success with his own label right out of the starting blocks. There are three Pinot Noirs, plus a Chardonnay in the small range.

✓ *Entire range*

STRANDVELD
Elim

This small but expanding operation in cool-climate Elim excels with both Sauvignon Blanc and Syrah.

✓ *Classic white blend (Adamastor) • Sauvignon Blanc • Syrah (Shiraz)*

STRYDOM VINTNERS
Stellenbosch

Husband and wife team Louis and Rianie Strydom are both respected winemakers – he is MD and cellarmaster at Ernie Els and she recently left Haskell/Dombeya to take over the winemaking reins at their new family venture.

✓ *Cabernet Sauvignon • Classic red blend*

TERRACURA
Swartland

Ryan Mostert is another "hands-off" winemaker. His Terracura range is the more serious and tightly wound, his Silverwis is more exuberant yet still very pure, and the Smiley NV bottlings deliciously idiosyn-

cratic. None are higher than 13.5 per cent alcohol; most less than 12.5per cent.

✓ *Silvervis Cinsault*

TESTALONGA
Swartland

Owned by Craig and Carla Hawkins, Testalonga is widely regarded as South Africa's first "natural wine" brand. Their practice of minimalist winemaking, a considerable use of skin contact and lees-aging for white grapes, and eschewing new oak and often sulphur, allowed the Hawkinses to create wines that rapidly achieved cult status. They were helped – no doubt – by the wines' quirky names and provocative labels. The Chenin Blancs (Baby Bandito and Cortez) are exceptionally good.

✓ *Entire range*

THELEMA
Stellenbosch

Rudi Schultz, who ably assisted owner-winemaker Gyles Webb since 2000, has taken over the helm at this classically inspired South African stalwart, making wines from its Stellenbosch vineyards as well as those in cool climate Elgin (Sutherland).

✓ *Cabernet Sauvignon (Thelema, Thelema The Mint) • Chardonnay (Sutherland Reserve) • Classic red blend (Thelema Rabelais) • Merlot (Thelema Reserve) • Syrah (Sutherland)*

THISTLE & WEED
Stellenbosch

This tiny venture. currently producing small quantities of Chenin Blanc, is set to double production soon with the addition of more single-vineyard site Chenin Blancs, a Verdelho, and a red blend.

✓ *Entire range*

THORNE & DAUGHTERS
Bot River

Non-interventionist John Thorne Seccombe delights in the fresh and pure, with no obvious oak. Although all of his fruit comes from what he describes as "compelling vineyards", one of the more interesting is Tin Soldier Sémillon, from the red-skinned version.

✓ *Cape White Blend (Rocking Horse)*

TOKARA
Stellenbosch

Miles Mossop established the reputation of this mountain-side winery but handed over the reins to Stuart Botha, ex-Eagles' Nest, in 2018. The Director's Reserve blends are particularly fine.

✓ *Entire range*

TRIZANNE SIGNATURE
Cape Town

Trizanne Barnard is passionate about the cool-climate sites in Elim, South Africa's

most southerly winegrowing region, and so their fruit, especially Sauvignon Blanc and Sémillon, dominates her small range.

✓ *Entire range*

UVA MIRA
Stellenbosch

This estate offers dramatic quality from a dramatic setting, especially the compelling Chardonnay and classy red blend.

✓ *Entire range*

VAN LOGGERENBERG
Stellenbosch

Freshness and elegance are the hallmarks of this incredibly good, understated young winemaker, who leads the pack of those adopting a non-interventionist approach while seeking lower alcohols and less extraction. His Cabernet Franc (Breton) could change the way people view the variety.

✓ *Entire range*

VERGELEGEN
Stellenbosch

It's been rather quiet at this immaculately restored, historic wine estate recently; possibly because it is that much harder to raise an already very high bar. Nevertheless, that's what's on the Vergelegen team's mind with its virus-free vineyards and ace consultant Michel Rolland. The entire range continues to impress, particularly the classic red and white blends.

✓ *Entire range*

VILAFONTÉ
Simonsberg-Paarl

This is a joint venture between winemaking guru Zelma Long, viticulturist Phillip Freese, and Mike Ratcliffe, who recently sold his family estate, Warwick, to San Francisco–based Eileses Capital. Vilafonté is taken from the term *vilafontes*, the grey, sandy soil on the property, which its owners believe has a significant impact on the flavour profile and development of their wines. The vineyards are in the Simonsberg-Paarl district, on the northern side of the Simonsberg Mountains, and planted in high density with four Bordeaux varieties (Cabernet Franc, Cabernet Sauvignon, Malbec, and Merlot). Series C is Cabernet Sauvignon-based, whereas Series M is Merlot-based.

✓ *Classic red blend (Series C, Series M)*

VILLIERA
Paarl

Villiera Estate is now a well-established producer by current Cape wine standards, producing a wide range at various price and quality points.

✓ *Classic red blend (The Clan) • Merlot (Monro) • Sparkling wine (Brut Natural Chardonnay, Monro Brut) • Late Harvest (Inspiration Noble Late Harvest)*

VONDELING
Paarl

Considerable emphasis is placed on sustainability of both viticulture and winemaking practices, and management of this property's considerable area of fynbos – the Western Cape's indigenous flora – so it is apt that its Flagship wines are named for rare fynbos species.

✓ *Chenin Blanc (Flagship Babiana) • Pinotage (Limited Releases Philosophie) • Red blend (Flagship Monsonia)*

VUURBERG
Stellenbosch

Stellar winemaker Donovan Rall (*see* RALL) makes the two wines – a classic red blend (although fairly alcoholic) and a white blend led by Chenin Blanc – for Netherlander Sebastian Klaassen.

✓ *Entire range*

WARWICK
Stellenbosch

This family-owned estate was sold in 2018 to San Francisco–based investment company Elises Capital. It also acquired neighbouring Uitkyk from Distell and combined the two under the Warwick brand to yield some 700 hectares (1,729 acres) of prime Simonsberg-Stellenbosch *terroir*. Much of this is intended to be replanted to Cabernet Sauvignon, Cabernet Franc, and Merlot – no doubt to boost the properties' Classic red blend output. The wines are currently in an opulent vein: sweet-fruited with considerable reliance on oak.

✓ *Cabernet Franc • Chardonnay (The White Lady) • Classic red blend (Trilogy)*

WATERFORD
Stellenbosch

A partnership between IT magnate Jeremy Ord and his wife, Leigh, and cellarmaster Kevin Arnold, this quarter-of-a-century-old property on the slopes of the Helderberg has quietly improved its offerings over the past decade. The flagship wine, The Jem, is a blend of the 11 red grapes on the estate, while the Antique Chenin Blanc is a *"solera*-style" NV collector's item.

✓ *Entire range*

WATERKLOOF
Somerset West

This biodynamically farmed estate overlooking False Bay is owned by British wine merchant Paul Boutinot. Quality has been consistently high here since the outset (2005).

✓ *Classic red blend (Circle of Life) • Classic white blend (Circle of Life) • Mourvèdre (Circumstance) • Sauvignon Blanc (Waterkloof) • Syrah (Circumstance)*

North Africa

After decades of fierce political and fundamental religious crackdown on the production and sale of alcohol, a gentle refreshing breeze is wafting across the region, with Algeria, Morocco, and Tunisia all reporting higher sales to Europe. Since 2013 even Ethiopia is showing slow, steady signs of recovery – but it's all relative.

The wine industries of Algeria, Morocco, and Tunisia, and the appellation systems within which they work, are all based on the structure left behind by the colonial French. Some good wines can be found, and a trickle of foreign investment exists, but the development of wine industries in these countries has been held back by their governments, which do not promote alcohol in Islamic states.

Cultivation of the vine flourished in Algeria, Morocco, and Tunisia as far back as 1000 BC, and winemaking was firmly established by at least the sixth century BC. Under French rule, North Africa was one of the most important winemaking regions in the world. Yet today it is difficult to see how the quality of wine can progress in any of these countries, where wine production is at best tolerated, and even that level of forbearance is often undermined by government officials who would personally like to see a total ban.

ALGERIA

In 1830 Algeria became the first North African country to be colonized by the French, which gave it a head start in viticulture over other North African countries. Algeria has always dominated North African viticulture in terms of quantity, if not quality. The French desperately planted vines in Algeria in the second half of the 19th century to offset the loss of their own vineyards to phylloxera. Viticulturally, it was to be the "second France", and for a while it was not only the second-largest wine producer in the world but also the largest exporter of wine. Most of it went to France, where no doubt it was blended into some of the most famous red wines in the world. The demise of the country's wine industry after it gained independence in 1962 made it clear that cynical remarks about the use of its wine in bolstering Burgundies were true, but the ironic fact is that both the Algerian and the Burgundian wines benefited from being blended together. It was not the best Burgundies that were enhanced

with dark Algerian red, but the poorest, thinnest, and least attractive wines. After the blending, they were not Burgundian in character, but were superior to the original thin, low-quality Burgundy. By 1938 the Algerian wine industry had reached a peak of 400,000 hectares (988,000 acres) and 22 million hectolitres (244 million cases). Since independence, these vineyards have shrunk, and the number of wineries has plummeted from more than 3,000 to around 70.

Algeria has never known phylloxera, thus its varieties retain their own rootstock. Of the few reds that can honestly be recommended, the best are from Château Tellagh in Coteaux de Médéa and Cave de Bourkika in Coteaux de Mascara. Other wineries that produce reasonably good wine include: Coteaux de Mascara (Domaine El Bordj, Domaine Mamounia, Domaine Beni Chougrane); Coteaux de Tlemcen (Domaine de Sebra); Coteaux du Zaccar (Chateau Romain); and Dahra (Domaine de Khadra).

An independent wine producer, Grand Crus de l'Ouest Algerie (GCO) founded in 2001 by Rachid Hamamouche, has invested in modernising vineyards and wineries, training *fellahs* (vineyard workers) and building relationships with growers and cooperatives across the region. The company now employs more than 300 people, and their range of wines are exported to Europe and Canada via France, as well as being available to purchase online.

Wine districts *Aïn-Bessem-Bouïra, Coteaux de Mascara, Coteaux de Tlemcen, Coteaux du Zaccar, Dahra, Médéa, Monts du Tessala*

MOROCCO

Another Arab country where the French had vast input into the wine industry during the colonial era, which was then followed by decades of declining fortune, Morocco's wine industry has recently benefited from major international investment and vineyard-lease programmes since the 1990s and is the second biggest producer of wine in the region.

WINE-PRODUCING REGIONS OF NORTH AFRICA

Morocco, Algeria, and Tunisia are all wine-producing countries. Most vineyards are concentrated along the coastal belts.

A WORKER CLEARS A VINEYARD IN MEKNÈS, MOROCCO
A former French colony, Morocco still shows that influence in its winemaking. Today most wineries are owned by French companies and run by French winemakers and viticulturists.

Wines were made here during Roman times, but after more than 1,000 years of Muslim rule, viticulture died out. In 1912, however, most of Morocco came under either French or Spanish control (with an international zone encompassing Tangiers), and it once more became an active winemaking country. When Morocco gained independence in 1956, the new government introduced a quality-control regime similar to the French AOC (*Appellation d'Origine Contrôlée*) system and, in 1973, it nationalized the wine industry. Few wines carry the official Appellation d'Origine Garantie designation. In 1998, the first *Appellation d'Origine Contrôlée* was created, *Les Coteaux de l'Atlas,* encompassing the districts of Sidi-Slimane, Mjat, and Boufekrane on the Atlantic-cooled slopes of the Atlas Mountains, where the lime-rich clay soils over dry, siliceous subsoil are particularly suitable for viticulture.

Here we find Cabernet Sauvignon, Chardonnay, Merlot, and Syrah growing, and the wines produced by Les Celliers de Meknès, particularly the classic red blend, are among the best that Morocco currently has to offer.

French Bordeaux baron Bernard Magrez created Les Deux Domaines at Guerrouane, which produces a Syrah-Grenache blend bottled under the Kahina label. More recent plantings include Chardonnay, Vermentino, and Alicante vines.

The country's more traditional wines are made from Carignan, Cinsault, and Grenache, usually in a pale rosé style, and often sold as *vin gris*. These can be pleasant when served chilled, but the best of even the traditional wines are red, also produced by Les Celliers de Meknès, but sold under the Les Trois Domaines label. Started by one of Morocco's most successful businessman, Brahim (Reda) Zniber, who died in 2016 at the age of 96, Les Celliers de Meknès produces 85 per cent of the country's wine, and most of it is at least drinkable. Or it is when it leaves the cellars. Most is sold on the home market, where poor storage conditions in high temperatures inevitably spoil both local and imported wines. Moroccans themselves do drink wine, although there is a notable discrepancy between international per capita consumption figures (one litre) and Morocco's own religiously sensitive official estimation (zero).

Wine districts *L'Oriental: Beni Sadden AOG, Berkane AOG, Angad AOG; Meknès-Fès: Guerrouane AOG, Beni M'tir AOG, Saiss AOG, Zerhoune AOG, Coteaux de l'Atlas 1er Cru; La Plain du Gharb: Gharb AOG; Rabat-Casablanca: Chellah AOG, Zemmour AOG, Zaër AOG, Zenatta AOG, Sahel AOG; El-Jadida: Doukkala AOG*

✓ *(with care) Cave Viticole de Bou-Argoub • Castel Frères • Les Deux Domaines • Ksar Bahia • Les Celliers de Meknès • Château Roslane • Thalvin • Val d'Argan • Volubilia*

TUNISIA

Wines were first made – mostly likey by the Phoenicians – around the Carthage area in Punic times (814–146 BC), but production of wine was forbidden for 1,000 years under Muslim rule beginning in the seventh century AD. After French colonization in 1881, viticulture resumed, and by Tunisian independence in 1955, the foundations of a thriving wine industry had been laid. At this time two basic appellations had been established – *Vin Supérieur de Tunisie* for table wines and *Appellation Contrôlée Vin Muscat de Tunisie* for Liqueur Muscat.

These designations did not, however, incorporate any controls to safeguard origin, and in 1957 the government introduced a classification system that established four levels. They are as follows: *Vins de Consommation Courante, Vins Supérieurs, Vins de Qualité Supérieure,* and *Appellation d'Origine Contrôlée.*

The best traditional Tunisian wines are those made from the Muscat grape, ranging from the lusciously sweet, rich, and viscous Vin de Muscat de Tunisie, to fresh, delicate, dry Muscats such as the Muscat de Kelibia. There are also a small number of good red wines, such as Château Feriani, Domaine Karim, and Royal Tardi, with Carignan and Cinsault the most important black grape varieties. The finest new-wave wines are made by the partly German-owned Domaine Magon, which is working proficiently with Cabernet Sauvignon, Merlot, Pinot Noir, and Syrah. There is also encouraging news of progress from research scientist Amor Jaziri, who produces organic wines under the Almory label, from his 14-hectare (35-acre) vineyard near Jedeida.

Wine districts *Coteaux de Tébourba, Coteaux d'Utique, Grand Cru Mornag, Kélibia, Mornag, Sidi Salem, Thibar*
✓ *(With care) Ceptunes • Dom Neferis • Kurubis • Vignerons de Carthage*

CHÂTEAU BOU ARGOUB, GROMBALIA, TUNISIA
This Tunisian producer makes its red wines from Carignan and Syrah varieties.

AVERAGE ANNUAL PRODUCTION

COUNTRY	VINEYARDS TOTAL hectares (acres)	VINEYARDS WINE GRAPES hectares (acres)	WINE PRODUCTION		MAJOR GRAPE VARIETIES
			HECTOLITRES	CASES	
Algeria	75,000ha (185,000ac)	No available data	520,000	46,800	Alicante Bouschet, Carignan, Cinsault, Grenache, Muscat d'Alexandria
Morocco	114,000ha (282,000ac)	46,000ha (113,60ac)	359,000	32,300	Alicante Bouschet, Cabernet Sauvignon, Carignan, Cinsault, Grenache
Tunisia	71,600ha (177,000ac)	23,000ha (22,240ac)	273,000	24,600	Alicante Bouschet, Carignan, Cinsault, Grenache, Sangiovese

OTHER WINEMAKING COUNTRIES OF

AFRICA

ETHIOPIA

Grape growing in Ethiopia? In fact, viticulture has existed here for many centuries. This country converted to Christianity in the 4th century, and locally grown Communion wine has been used by the Ethiopian Orthodox Church since at least the 12th century. So entrenched is grape-growing in Ethiopian history that it has entered the language as a specific term. Ethiopians distinguish three altitude-defined climatic zones: *dega, kola,* and *wayna-dega. Dega* means "cold" and indicates areas that are higher than 2,500 metres (8,200 feet) above sea level. *Kola* means "hot place" and applies to areas below 1,800 metres (5,900 feet). *Wayna-dega* literally means "grape elevation", or "place where you grow grapes", and refers to areas with elevations between 1,800 and 2,500 metres (5,900 to 8,200 feet).

The big game changer in Ethiopia in recent years has been the arrival of Castel, one of France's biggest drinks companies, which has set up shop in Ziway.

Ethiopia's largest vineyard is located at Awash, 160 kilometres (100 miles) east of Addis Ababa, in the Harar province, where some 30 hectares (74 acres) of Aleatico, Barbera, Chenin Blanc, Grenache, Petit Sirah, and Ugni Blanc are grown. There are also 75 hectares (185 acres) of so-called "local" varieties, including Debulbul Attere, Key Dubbi, Nech Debulbul, and Tiku Weyn. There are two harvests each year, and the grapes are trucked to the Awash Winery in the outskirts of Addis Ababa, where reasonable-quality red and white wines are produced. The most commonly encountered wines are Axumite (sweetish white), Crystals (dry white), Dukam (red), and Gouder (Ethiopia's best red, from the Gouder vineyard, 135 kilometres, or 84 miles, west of Addis Ababa). The Awash

Winery is located next to the Awash River, however, which irrigates the Awash vineyards 160 kilometres (100 miles) downstream, and in 1998 researchers from the University of Turku found the Awash River to be polluted downstream from the winery's own effluent discharge, but not upstream. This situation will have to be rectified before any serious thoughts of export can be entertained, as will its practice of using rehydrated imported raisins in some of the wines. Currently, the most exported wine is made from honey, not grapes. Called Tej, it is made all over the country and is much more famous than its grape-based counterparts. The Awash Winery exports Tej to Sweden and the USA, but Ethiopia is not really wine country, whether from honey or grapes. It is coffee country, to the point of giving the world the name itself – the province of Kaffa was the etymological origin of the word *coffee.*

KENYA

Lake Naivasha Vineyards are located on volcanic soils beside Lake Naivasha, situated 1,900 metres (6,200 feet) above sea level in the Rift Valley. The Rift Valley is the Flower Garden of Africa, the blooms of which are flown to florists all over the world. In fact, you can grow absolutely anything here, and grow it twice a year, too. You can certainly grow grapes, as John and Elli D'Olier did between 1982 and 1992, with cuttings they brought back from California. They ripped out most of the vines in 1992 when overnight hikes in local taxes made Kenyan wine economically unfeasible, making just the odd wine for home consumption, only to replant in earnest in 2002. Sadly, in 2010 John was murdered in a brutal attack at their private campsite.

A young South African winemaker, James Farquharson, relocated to Kenya in 2008 to take over winemaking at the Rift Valley Winery at Morendat in Naivasha. Cooling systems were installed and in 2015 they produced more than 100,000 bottles under the Leleshwa label. The farm, a division of the Kenya Nut Company, makes and exports a Sauvignon Blanc, Merlot-Shiraz blend, and a rosé. Farquharson has moved on, and the wines at Morendat are now made by Emma Nderitu and Christine Kasimu.

Richard Leakey is most famous for finding a 2.5-million-year-old human skull in Kenya, but he did also make an heroic attempt to grow Pinot Noir in Ngong, and succeeded in bottling some (and Sauvignon Blanc) in 2001 but was ultimately defeated by baboons (and other vineyard interlopers) and an intermittent and unreliable electricity supply that made controlled fermentation impossible.

MADAGASCAR

Affectionately known as "Mad" to Anglo-Saxon inhabitants of the African continent, Madagascans have been growing grapes and making wine since the 19th century. Most of the vines growing on the island today are hybrids. With not a little tongue in cheek, John and Erica Platter write in *Africa Uncorked* (2002) that Stéphane Chan Foa Tong's sparkling Grand Cru d'Antsirabe is "Unquestionably the best sparkling wine in all Madagascar. Probably the finest Couderc Blanc *méthode champenoise* in the world!", and certainly it seems better than his still wines. The clean and decent, but hardly exciting, wines of Clos Malaza in the Ambatomena area are generally agreed by the locals to be the best that Madagascar currently produces, from 30 hectares (74 acres) of well-trained Chambourcin, Couderc Blanc, Petit Bouchet, Villardin, Varousset, and Villard Noir vines. From an outsider's point of view, however, the Chan brothers in Ambalavao produce more promising wine under the Côte de Fiana appellation (their best wine is sold as an exclusivity under the Tsara be label). Domaine Manomisoa is also known as Soavita Estate, and even more confusingly, its best wine is sold as Château Verger. Other wines are produced under the following labels: Clos de la Maromby (from the Masina Maria monastery, near Fiana) and Lazan'i Betsileo (a small cooperative in Fiana). Nothing really stands out among these wines, but with better grape varieties growing at altitude on the drier, eastern side of the island, which are pruned to yield less than half the current average crop, Madagascar does have the potential to produce some very much more interesting wines.

MAURITIUS

Beware Mauritian-looking wine sold under the Oxenham label, which is made from imported grape concentrate. There is nothing wrong with that, but the fact should be made clear.

NAMIBIA

The first vineyards in Namibia were planted by German Roman Catholic priests at the end of the 19th century, just outside the capital, Windhoek. Commercial Namibian wine is a relatively recent phenomenon, dating back to 1994, when Helmuth Kluge established Kristall Kellerei at Omaruru, on the periphery of the Namib desert. All wines are clean and well made, with fresh, zesty Colombard the best.

There are also two small wineries at Otavi where Shiraz, Viognier, Cabernet Sauvignon, Mouvèdre, Tempranillo, and Chardonnay are grown.

RÉUNION

This African island is, believe it or not, the most far-flung of all French wine areas: Vin de Pays de Cilaos. Created in 2004, the *vin de pays* regulations forced the removal of the hybrids that used to grow here. Hybrids are still the mainstay of viticulture on Madagascar, which arguably offers greater potential, but the performance of the classic *vinifera* varieties growing on Réunion (Chenin Blanc, Pinot Noir, and Malbec) could well change the mind of those "Mad" people. There are plans to plant Gros Manseng, Pinotage, Syrah, and Verdelho. The styles produced are red, white, rosé, and, because the tendency is for grapes to overripen rather too easily, Moelleux. The only wine producer currently is the local cooperative, Chais du Cilaos, which harvested just 28 tonnes of grapes in 2004. The vines grow at an altitude of 1,200 metres (3,930 feet) on volcanic soil, cropping in January. The wines are fresh, crisp, and well made, fermented in new, temperature-controlled, stainless-steel vats.

TANZANIA

German missionaries brought vines and vineyards to Tanzania and there are three wineries – Tanganyika Vineyards, Dodoma Wine Company, and the Central Tanzania Wine company – and two harvests a year, producing a decent, but not exciting, Chenin Blanc, Syrah, Cabernet Sauvignon, and a local variety called Makutupora.

ZIMBABWE

Although the first production of wine began in the early 1950s under the labels of Worringham and Lorraine, these semi-sweet wines were made from table grapes. It was not until UDI (the Unilateral Declaration of Independence) that this most beautiful country in Africa gave birth to a truly commercial wine industry. During the mid-1960s, when sanctions bit deep, and anything that could be produced locally was produced locally, Rhodesian farmers grew the grapes to make the wines they could no longer import. In the 1990s, many low-quality grape varieties were ripped up and replaced by classic vinifera, with drip irrigation. The grape varieties grown in Zimbabwe today include Cabernet Sauvignon, Cinsault, Clairette Blanche, Colombard, Chenin Blanc, Cruchen, Gewürztraminer, Merlot, Pinotage, Pinot Noir, Riesling, Ruby Cabernet, Sauvignon Blanc, and Seneca (*Lignan blanc* x *Ontario* hybrid), but many wine farms have been occupied by "war vets". Worringham still exists, both as a wine farm (45 hectares; 111 acres) and as a brand and is owned by African Distillers (also known as Stapleford Wines), but other familiar wine names from the early days, such as Philips and Monis, were absorbed by the Mukuyu Winery, which itself was taken over by Cairns Foods. Export brands such as Flame Lily, which made an ephemeral appearance on the UK market, have also fallen by the wayside. Meadows Estate used to be a shining example of cutting-edge viticulture in the 1980s, growing Gewürztraminer, Riesling, Chardonnay, and Pinot Noir, but it is a rose farm now. The 1990s also saw the installation of cool-fermentation technology and internationally trained winemakers deployed on flying visits, including New Zealander Clive Hartnell (of Hungary's Chapel Hill renown) currently winging his way around the wineries of African Distillers. The late Peter Raynor and his son Humpfrey established Zimbabwe's largest private vineyard near Wedza, but its once impeccably trained vines have been unattended since the Raynor family were evicted by Minister of Youth Brigadier Ambrose Mutinhiri in December 2004. In addition to the human tragedy, this sadly marks the end of the brief promise shown by Zimbabwe's first and, now, only single-vineyard, pure varietal wines – Fighill 2002 Chardonnay and Fighill 2003 Shiraz, which the Raynors made from a super-selected five per cent of their own production. Most of their crop used to be split 50-50 between African Distillers (which bottled their domain wines) and Mukuyu (which at 65 per cent government owned escaped its threatened compulsory acquisition in December 2001). Private farms have been so vulnerable to compulsory acquisition (ie occupation by "war vets") that African Distillers and Mukuyu are now the only producers here, and this looks set to be the case for the foreseeable future.

TEJ, A HONEY WINE, IS BREWED AND CONSUMED IN ETHIOPIA
Grape wines are gaining a foothold in this Horn of Africa country, but Tej is still its most exported type of wine.

The Wines *of the* Americas

FOUR OF THE FIVE MOST FAMOUS WINE REGIONS of the Americas – California, Washington, Oregon, and Chile – are found on or inland from the western coast of the continent. This is because the Eastern Seaboard is generally too humid and prone to severely cold winters. Of the famous regions in the Americas, California is, of course, the most important. It is staggering to think that the 25 largest wineries in this state on their own produce twice the output of Argentina (which is itself the ninth-largest winemaking country in the world), but California is more about quality than quantity, producing some of the world's greatest wines. Although Washington has always been a considerably larger wine-producing state than Oregon, from an international perspective it used to live in the shadow of the latter's excellent but sometimes over-hyped Pinot Noir. Well, no more. Washington is now justly famous for powerful Bordeaux and Rhône varietals, and British Columbia has far more in common with Washington State than it does with Ontario. Chile and Argentina rule in South America.

"THE GRAPE CRUSHER" SITS HIGH ON A HILL OVERLOOKING THE NAPA VALLEY VINEYARDS
This towering bronze monument stands just off Highway 29 in southern Napa. Created by New Mexico artist Gino Miles, it was unveiled during Napa County's sesquicentennial celebration in May 1988 as a tribute to vineyard workers and to honour 200 years of winemaking in the valley.

North America

The United States is the fourth-largest wine-producing country in the world, and the output of California alone is more than that of the fifth-largest wine-producing country (China), yet almost every state in this vast country produces wine. And, of course, North America does not consist of just the United States; it also encompasses the viticultural areas of Ontario and British Columbia in Canada and the Baja California and Sierra Madre in Mexico.

In 1521, within just one year of invading Mexico, the Spanish had planted vines and set about making the first North American wines. Fourteen years later, when French explorer Jacques Cartier sailed down the Saint Lawrence to New France, he discovered a large island overrun by wild vines and decided to call it the Île de Bacchus. He had second thoughts, however, and later renamed it the Île d'Orléans, a calculated move in view of the fact that the then Duke of Orléans was the son of King Francis I of France. It is assumed that, circa 1564, the Jesuit settlers who followed in the wake of Cartier's explorations were the first winemakers in what was to become Canada. The earliest wines made in what is now the United States of America came from Florida. In 1564, French Huguenot settlers produced wines from native Scuppernong grapes on a site that would become Jacksonville.

VITIS LABRUSCA GRAPES
These North American natives were the source of nearly all wines produced in North America — other than California wines — until quite recently.

NATIVE NORTH AMERICAN GRAPE VARIETIES

All classic grape varieties belong to one species, *Vitis vinifera,* but North America's native varieties belong to several different species, not one of which happens to be *Vitis vinifera.* There were plenty of native vines growing wild wherever the early settlers travelled, and so they came to rely on them for their initial wine production. Settlers in Australia, on the other hand, were forced to wait for precious shipments of classic European vines before they could plant vineyards. Although various European varieties were taken across the Atlantic in the 19th century, nearly all North American wines, apart from a few from California, remained products of native varieties until relatively recently.

The most common native North American species, *Vitis labrusca,* has such a distinctive aroma and flavour that it seems truly amazing that those pioneers who were also winemakers did not pester their home countries for supplies of more acceptable vines. The *labrusca* character, commonly referred to as "foxy", has such an exotic, highly-floral and cloying, strawberry Jell-O flavour that it is not appreciated by European and antipodean palates.

PROHIBITION IN THE UNITED STATES

Although total Prohibition in the United States was confined to 1920 to 1933, the first "dry legislation" was passed as early as 1816, and the first state to go completely dry was Maine, in 1846. By the time the 18th Amendment to the Constitution was put into effect in 1920, forbidding "the manufacture, sale, or transport of intoxicating liquors", more than 30 states were already totally dry.

The result of Prohibition was chaos. It denied the government legitimate revenue and encouraged bootleggers to amass fortunes. The number of illicit stills multiplied quicker than the authorities could find and dismantle them, and the speakeasy became a way of life in the cities. Not only did the authorities often realize that it was much easier to turn a blind eye to what was going on, the federal government actually found it useful to open its own speakeasy in New York. Many vineyards were uprooted, but those grapes that were produced were often concentrated, pressed, and sold as "grape bricks". These came complete with a yeast capsule and instructions to dissolve the brick in one gallon of water, but warned against adding the yeast because it would start a fermentation. This would turn the grape juice into wine "and that would be illegal", the warning pointed out.

Prohibition and the wine industry

By the mid- to late-19th century, the Californian wine industry had such a positive reputation that great French wine areas, such as Champagne, began to form syndicates to protect themselves, in part, from the potential of California's marketing threat. The 13 years of Prohibition coincided with a vital point in the evolution of wine, however, and set the Californian wine industry back a hundred years, as other wine regions had just recovered from the effects of phylloxera (*see* p578) and were busily re-establishing their reputations and carving out future markets. In Europe, World War I had robbed every industry of its young, up-and-coming generation, but the rich tradition of the wine industry enabled it to survive until the arrival of a new generation. The early 1900s were also the era of the foundation of the French *Appellation Contrôlée* laws, a quality-control system that many other serious winemaking countries would eventually copy.

The United States also lost much of its young adult generation in World War I, but it had less of a winemaking tradition to fall back on, and, by 1920, there was virtually no wine industry whatsoever to preserve. After Prohibition came one of the worst economic depressions in history, followed by World War II, which took yet another generation of bright young minds. It was, therefore, little wonder that by the late 1940s the wine industry of the United States was so out of date. It had lost touch with European progress and resorted to the production of awful *labrusca* wines

that had been the wine drinker's staple diet in pre-Prohibition days. California produced relatively little *labrusca* compared to the eastern states, but its winemakers also resorted to old-fashioned styles, making heavy, sweet, fortified wines.

The fact that Californian wine today is a match for the best of its European counterparts and that its industry is healthy, growing fast, and looking to compete on foreign markets, clearly proves that in the United States opportunity is boundless.

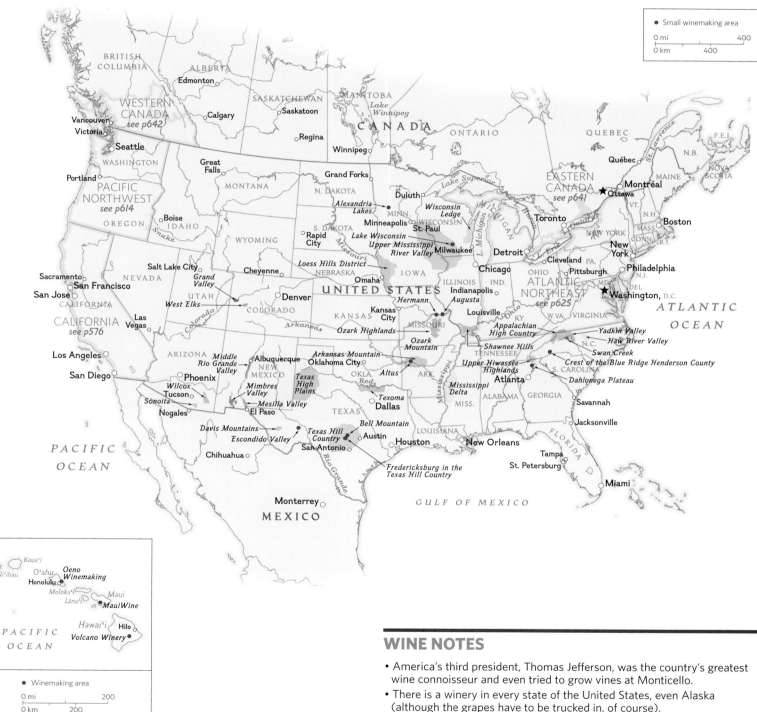

WINE REGIONS OF NORTH AMERICA

Wine is produced across the entire continent of North America, in Canada, the United States, and Mexico, whether from native grapes or *vinifera* varieties. Winemakers work in enormously varied conditions according to climate.

WINE NOTES

- America's third president, Thomas Jefferson, was the country's greatest wine connoisseur and even tried to grow vines at Monticello.
- There is a winery in every state of the United States, even Alaska (although the grapes have to be trucked in, of course).
- Although almost every state in the United States grows wine grapes, California accounts for more than 90 per cent of the wines produced.
- It is legal to produce and sell American Champagne or Burgundy, but not American Bordeaux or Rhône.
- There are still "dry" counties in some US states.
- The oldest winery in the country is the Brotherhood Winery (1839) at Washingtonville in the foothills of the Catskills of New York State.

THE VINEYARDS OF THE MOUNT PLEASANT WINERY FALL IN THE AUGUSTA AVA, THE FIRST FEDERALLY APPROVED AMERICAN VITICULTURAL AREA
The first delimited grape-growing region in the United States was in Missouri. With its first wines produced in 1830, the state's wine industry flourished during the 19th century.
Since the turn of the 21st century, the number of wineries just keeps increasing, skyrocketing from fewer than 40 in 2000 to 165 in 2019.

THE APPELLATION SYSTEM OF THE UNITED STATES

The older generation of appellations of the United States based on political boundaries, such as counties or states, still exists. During the mid-1970s, however, the Department of Treasury's Bureau of Alcohol, Tobacco, and Firearms (BATF, now the TTB) considered the concept of specific controlled appellations based on homogenous geography and climate and in September 1978 BATF published its first laws and regulations designed to introduce a system of American Viticultural Areas (AVAs), to supplement the old system, which is now run by the Alcohol and Tobacco Tax and Trade Bureau (TTB). Every state, from Alaska to Nebraska, and the counties they contain, are recognized in law as their own individual appellations of origin, but other generic appellations are also recognized.

American or United States

This appellation classifies blended and varietal wines from anywhere in the United States of America, including the District of Columbia and the Commonwealth of Puerto Rico. These wines, like *vins de tables* in France, are not allowed to carry a vintage, which is, in my opinion, irrational. It is the only appellation allowed for wines shipped in bulk to other countries.

Multi-State Appellation

This is used to classify a wine from any two or three contiguous states. The percentage from each state must be clearly indicated on the label.

State Appellation

A wine from any state may use this appellation; at least 75 per cent of grapes must come from grapes grown within the one state indicated (except for Texas, which is 85 per cent, and California 100 per cent). Thus wine claiming the appellation of Oregon may contain up to 25 per cent produce of other states. The grapes can be shipped to a neighbouring state and still be given the originating state name, but not to any state that does not share a border. Grapes can therefore be shipped from California to Oregon, where they are turned into wine, and may be called California wine (as long as the wine is at least 75 per cent from those grapes), but the same grapes cannot be shipped to Washington State and be called California wine (although such a wine could be labelled California-Oregon-Washington, if it contained grapes from all three states, *see* Multi-State Appellation). The same principles apply to County Appellations and Multi-County Appellations.

Multi-County Appellation

This classifies a wine from any two or three contiguous counties. The percentage of wine from each county must be clearly indicated on the label.

County Appellation

A wine from any county within any state may use this appellation: at least 75 per cent of the wine must come from grapes grown within the one county indicated.

THE SEMI-GENERIC SAGA

It is not only easy to understand how the misuse of famous European wine names started in America, it is also perfectly understandable. As European immigrants poured into this land, some brought vines with them, and when they made different styles of wine, it was inevitable that they would use the only wine names they knew to distinguish between them. Names like Burgundy, Chablis, Champagne, and the like. There is nothing to forgive. Until now, that is.

Ever since the California wine industry became a significant exporter, opening the door for other wine-producing states, it was incumbent upon the United States to respect all other legitimate geographical appellations, however well entrenched the usage might have become within the United States itself. In 1978, however, well after California wines had stepped onto the international stage, the director of the Department of Treasury's Bureau of Alcohol, Tobacco, and Firearms (BATF) legally enshrined the concept of semi-generic wine names within Federal Regulations, citing, but not restricting itself to, 14 such examples: "Burgundy, Claret, Chablis, Champagne, Chianti, Madeira, Malaga, Marsala, Moselle, Port, Rhine Wine, Sauterne, Sherry, Tokay". The regulations state that these so-called semi-generic names must be qualified by their true place of origin: American Chablis, California Champagne, and so on.

Yet, American regulations are not only unfair, they are decidedly quirky. While it is perfectly legal for US wineries to sell wines labelled as Champagne or Chablis, the name Liebfraumilch is protected by the full might of Federal Law. And whereas Burgundy is permitted, Bordeaux or Rhône are not. Where is the logic? As the most abused wine names were French, it was left initially to that country to redress the situation, but they got nowhere and, frankly, were not helped by their own antics in South America, where the *champenois* were abusing the Champagne appellation. The French were also chronically short-sighted when turning down an under-the-table deal that would have seen "Champagne" dropped in favour of "Champagne Style".

Since then, the production of American wine sold under famous Italian appellations has become a substantial industry in itself, which with the use of other European wine names has seen the European Union take up the fight. Its negotiators have been hammering away at the United States through the World Trade Organization, but they have not been any more successful than the French. What they fail to understand is that they have nothing to bring to the negotiating table that the United States is in the slightest bit bothered about, but when they recognize this, it should provide the European authorities with the obvious solution. If the French, Italians, Spanish, Portuguese, and Hungarians officially created a few new appellations like Napa Valley, Sonoma County Green Valley, and, to catch the attention of the East Coast elite, North Fork of Long Island and Northern Neck George Washington Birthplace, the EU would soon find that it had something to bargain with, and the United States might at least understand why Europeans are so protective of their appellations. The irony is that should such appellations be formally adopted and the wines put into distribution in their country of origin, they would be automatically eligible for sale in the United States, where the law defines an acceptable foreign appellation on an imported wine to be "a delimited place or region the boundaries of which have been recognized and defined by the country of origin for use on labels of wine available for consumption within the country of origin".

Perhaps the wine-producing countries that feel they have been treated unfairly by the United States will realize that sometimes it does take a wrong to right another wrong. Virtually all self-respecting, quality wine producers in the United States will be on their side because they are proud of their own products, which do not need any famous European names as a crutch.

LARGEST WINE GROUPS OF
AMERICA

E & J GALLO
75 million cases
(97 million with takeover of Constellation Brands)

Accounting for one-quarter of all US wine sales, this company was founded by Ernest and Julio Gallo, who were driven by tragedy, division, and a legendary zeal for hard work to create (in their own words) the "Campbell Soup Company of the wine industry". Elder brother Ernest became head of the family in 1933, after his father, a grape-grower, inexplicably killed his mother, and then shot himself. With the end of Prohibition, the brothers began winemaking, selling in bulk until 1940, when the first wines to carry the Gallo label appeared. They achieved their ambition by the 1960s, when E & J Gallo was easily the most popular wine brand in the country, but after the United States, the brothers wanted to conquer the world. The way to achieve that, they decided, was with a profusion of inexpensive brands, and by the 1980s, E & J Gallo (the company, not the brand) was the world's largest wine producer, churning out more wine than famous wine-producing countries like Australia. Ernest and Julio divided everything. Ernest ran the business and marketing, while Julio looked after the production, until he died in a car crash in May 1993.

Ernest's son, Joseph E Gallo, took over in 2001, enabling his father to enjoy six years of retirement before passing away at the ripe old age of 97 in 2007. In 2012 Joseph was named "Wine Person of the Year" by *The Wine Enthusiast*. When asked how he managed to run the largest wine company in the world, with production vineyards and production facilities in 15 countries (including the world's largest at the company's HQ in Modesto) and exporting to 75 different countries, he said "hire the best people and keep out of their way". In recent years, Gallo has been buying up premium brands to raise the average quality and increase margins, with the latest additions coming from Constellation in the form of 30-odd wine and spirit brands for $1.1 billion in 2019. The majority of production, however, remains in the cheap and trying-to-be-cheerful range, which is the nature of the beast for the world's largest wine producer. The real mind-boggling fact here is that such a company is still family owned. *See also* p610.

Brands: *Alamos, André Sparkling, Apothic, Arbor Mist, Ballatore, Barefoot, Bear Flag, Bella Sera, Black Box, Blackstone, Blufeld, Bridlewood, Canyon Road, Capri, Carlo Rossi, Carnivor, Chateau Souverain, Clos du Bois, Columbia Winery, Copper Ridge, Covey Run, Cribari Tables & Desserts, Dancing Bull, Dark Horse, DaVinci, Diseño, Dolcea, Don Miguel Gascón, Ecco Domani, Edna Valley Vineyard, Estancia, Fleur de Mer, Franciscan, Gallo Estate, Gallo Family Vineyards, Gallo of Sonoma, Gallo Signature Series, Ghost Pines, Hidden Crush, Hogue Cellars, J Vineyards & Winery, Le Terre, Liberty Creek, Livingston Cellars, Louis M. Martini, MacMurray Estate Vineyards, Manischewitz, Mark West, Milestone, Matthew Fox Vineyards, Mirassou, Naked Grape, Orin Swift, Pahlmeyer, Peter Vella, Polka Dot, Primal Roots, Proverb, Rancho Zabaco, Ravenswood, Red Bicyclette, Red Rock, Redwood Creek, Rex Goliath, Richards Wild Irish Rose, Simply Naked, Starborough, Talbott Vineyards, Taylor, Thrive, Thunderbird, Tisdale, Toasted Head, Turning Leaf, VNO, Vendange, Vin Vault, Whitehaven, Wild Horse, Wild Vines, William Hill, Wycliff*

THE WINE GROUP
53 million cases

This group's heritage involves Franzia and Coca-Cola origins. Franzia was founded in 1906, when Teresa Carrara and Giuseppe Franzia planted the first vines at Ripon in California's Central Valley. In 1915 they built their first winery, and by 1961 Franzia had become the sixth-largest wine producer in the United States, a position it maintained throughout the 1960s and 1970s by pioneering bag-in-the-box wines. In 1971 Coca-Cola formed Wine Spectrum and purchased Franzia just two years later. Some members of the Franzia family continued in the wine business, creating the JFJ Bronco label (*see below*). In 1981, The Wine Group (TWG) was formed as part of a leveraged buyout headed by Arthur Ciocca, the president and CEO of Franzia and formerly a marketing executive with Gallo. From a slow start, due to a slump in the California wine industry in the early 1980s, TWG climbed to the country's third-largest wine producer by the 1990s.

Brands: *10 Span Vineyards, 7 Deadly, Benziger Family Winery, Big House Wine Co, Chloe Wine Collection, Cocobon, Concannon Vineyard, Cupcake Vineyards, Franzia, Imagery Estate Winery, McManis Family Vineyards, Mogen David, San Francisco Wine Co, Save Me, Stave & Steel*

CONSTELLATION BRANDS
50 million cases
(31 million after sale of brands to E & J Gallo)

Constellation Brands was established by Marvin Sands in 1945 as Canandaigua Industries. This company sold bulk-wine to bottlers on the eastern seaboard and by 1951 had achieved an annual turnover of $1 million (equivalent of $10 million today). Sands attempts to establish his own wine brand were unsuccessful until 1954, when he launched a dessert wine called Richard's Wild Irish Rose. This was named after his son, Richard, and was uniquely franchised to five bottlers across the country. This form of focused bulk-selling would eventually transform the company to brand-led business. By 1963 Canandaigua turned over $10 million a year and in 1973, changed its trading name to Canandaigua Wine Company (CWC), floating it on the New York Stock Exchange. By 1979 son Richard had entered the business, and by 1980 it was the eighth-largest wine producer in the country, marketing 100 wine brands. By 1997 annual sales had reach $1.1 billion, and in 2000 the company changed its

name to Constellation Brands. In 2001 Constellation and BRL Hardy formed a joint venture called Pacific Wine Partners and liked the synergy so much that it took over the Australian giant to become the largest wine producer in the world (but still not the largest US wine producer) just two years later. Its global position was reinforced in 2006, when Constellation purchased Vincor, already the fifth largest producer in the United States. In 2007 billionaire Richard Sands handed over the reins of CEO and president to his billionaire brother, Rob Sands, but retained his position as chairman of the board. In 2011 Constellation sold 80 per cent of BRL Hardy to an equity firm, and in 2016 sold its Canadian wine business (formerly Vincor), but continued to build its international portfolio of premium and super-premium brands. In 2018 Rob Sands stepped back from the roles of CEO and president

(now assumed by Bill Newlands), and in 2019 Constellation sold a raft of its brands to E&J Gallo.

Brands: *7 Moons, Black Box, Cook's, Cooper and Thief, Franciscan Estate, J Roget, Kim Crawford, Mark West, Meiomi, Mount Veeder, Nobilo, Ravage, Robert Mondavi, Ruffino, Schrader Cellars, Simi, Spoken Barrel, The Dreaming Tree, The Prisoner Wine Company, The Snitch, Tom Gore, Woodbridge*

TRINCHERO FAMILY ESTATES

20 million cases

Brothers John and Mario Trinchero moved from New York to Napa Valley, where they purchased Sutter Home Winery in 1947. Sutter Home has always had more brand awareness than Trinchero, even though Trinchero

Family Estates have become one of the largest wine producers in America. In 1975 Sutter Home invented "White Zinfandel" by mistake, when a tank of Zinfandel suffered a stuck fermentation. This left a significant amount of residual sugar and a pink colour, the hallmarks of the much maligned but hugely popular White Zinfandel wine style. Sutter Home White Zinfandel became the best-selling wine in the United States in 1987 and peaked at 2.7 million cases in 1989 (not far short of the sales of all brands and styles of the 10th-largest producer today). Bob and Roger Trinchero are regarded as sustainable-friendly proprietors at this level of production, which sees Trinchero Family Estates brands sold to nearly 50 countries around the world. *See also* Trinchero Winery, p599.

Brands: *Bandit Wines, Barbed Wire, Bieler Père et Fils, Bravium, Charles & Charles, Cloudfall, Complicated, Cucina*

THE AVAS OF THE UNITED STATES

The very first AVA was not, strangely enough, Napa, but Augusta in Missouri, which was established in June 1980. There are now 253 AVAs, not including states and counties, and the number continues to rise, with another 11 waiting in the wings (the AVAs listed and italicised below have been accepted in principle, but await final approval). Although the establishment of these AVAs cannot be taken for granted, few if any are

likely to be rejected. The most famous denial was for a proposed California Coast AVA, which was not only 1,050 kilometres (650 miles) long, but also penetrated inland, encompassing interior basins and valleys, high mountain ranges (up to 1,500-metres, or 5,000 feet, in altitude), as well as shorelines and coastal plains, thus providing no homogeneity of either geography or climate.

- Adelaida District – California
- Alexander Valley – California
- Alexandria Lakes – Minnesota
- Alta Mesa – California
- Altus – Arkansas
- Ancient Lakes of Columbia Valley – Washington
- Anderson Valley – California
- Antelope Valley of the California High Desert – California
- Appalachian High Country – North Carolina, Tennessee, Virginia
- Applegate Valley – Oregon
- Arkansas Mountain – Arkansas
- Arroyo Grande Valley – California
- Arroyo Seco – California
- Atlas Peak – California
- Augusta – Missouri
- Ballard Canyon – California
- Bell Mountain – Texas
- Ben Lomond Mountain – California
- Benmore Valley – California
- Bennett Valley – California
- Big Valley District–Lake County – California
- Borden Ranch – California
- *The Burn of Columbia Valley – Washington*
- California Shenandoah Valley – California
- Calistoga – California
- Capay Valley – California
- Cape May Peninsula – New Jersey
- Carmel Valley – California
- Catoctin – Maryland

- Cayuga Lake – New York
- Central Coast – California
- Central Delaware Valley – New Jersey, Pennsylvania
- Chalk Hill – California
- Chalone – California
- Champlain Valley of New York – New York
- Chehalem Mountains – Oregon
- Chiles Valley – California
- Cienega Valley – California
- Clarksburg – California
- Clear Lake – California
- Clements Hills – California
- Cole Ranch – California
- Columbia Gorge – Oregon, Washington
- Columbia Valley – Oregon, Washington
- Coombsville – California
- Cosumnes River – California
- Covelo – California
- Crest of the Blue Ridge Henderson County – North Carolina
- Creston District – California
- Cucamonga Valley – California
- Cumberland Valley – Maryland, Pennsylvania
- Dahlonega Plateau – Georgia
- Diablo Grande – California
- Diamond Mountain District – California
- Dos Rios – California
- Dry Creek Valley – California
- Dundee Hills – Oregon
- Dunnigan Hills – California

- Eagle Foothills – Idaho
- Eagle Peak Mendocino County – California
- Eastern Connecticut Highlands – Connecticut
- Edna Valley – California
- El Dorado – California
- El Pomar District – California
- Elkton Oregon – Oregon
- Eola–Amity Hills – Oregon
- Escondido Valley – Texas
- Fair Play – California
- Fennville – Michigan
- Fiddletown – California
- Finger Lakes – New York
- Fort Ross-Seaview – California
- Fountaingrove District – California
- Fredericksburg in the Texas Hill Country – Texas
- *Gabilan Mountains – California*
- *Goose Gap – Washington*
- Grand River Valley – Ohio
- Grand Valley – Colorado
- Green Valley of Russian River Valley – California
- Guenoc Valley – California
- Hames Valley – California
- The Hamptons – Long Island, New York
- Happy Canyon of Santa Barbara – California
- Haw River Valley – North Carolina
- Hermann – Missouri
- High Valley – California
- Horse Heaven Hills – Washington
- Howell Mountain – California
- Hudson River Region – New York

- Indiana Uplands – Indiana
- Inwood Valley – California
- Isle St George – Ohio
- Jahant – California
- Kanawha River Valley – West Virginia
- Kelsey Bench–Lake County – California
- Knights Valley – California
- Lake Chelan – Washington
- Lake Erie – New York, Ohio, Pennsylvania
- Lake Michigan Shore – Michigan
- Lake Wisconsin – Wisconsin
- Lamorinda – California
- Lancaster Valley – Pennsylvania
- Leelanau Peninsula – Michigan
- Lehigh Valley – Pennsylvania
- Leona Valley – California
- Lewis–Clark Valley – Idaho, Washington
- Lime Kiln Valley – California
- Linganore – Maryland
- Livermore Valley – California
- Lodi – California
- Loess Hills District – Iowa, Missouri
- Long Island – New York
- *Long Valley-Lake County – California*
- Loramie Creek – Ohio
- Los Carneros – California
- Los Olivos District – California
- *Lower Long Tom – Oregon*
- Madera – California
- Malibu Coast – California

Mista, Echo Bay, Folie à Deux, FRE, Hopes End, Joel Gott Wines, Luminara, Main Street Winery, Mason Cellars, Ménage à Trois, Montevina, Napa Cellars, Newman's Own, Neyers, Pacific Heights, Pomelo, Seaglass Wine Company, Shatter, Sutter Home, Sycamore Lane, Taken, Terra d'Oro, The Show, Three Pears, Three Thieves, Trinchero, Trinity Oaks, Ziata Wines

TREASURY WINE ESTATES
14 million cases

In 2010 Berringer-Blass and Southcorp were renamed and spun-off as a separate company called The Wine Treasury TWT), which in 2015 purchased the US-based Chateau and Estate Wines from Diageo. According to some financial analysts, TWT have been stockpiling luxury brands to increase margins by rationed releases in China and the USA. See also p669.

Brands: *Acacia Winery, Beaulieu Vineyard, Belcreme De Lys, Beringer Vineyards, Blossom Hill, Chateau St Jean, Etude Wines, Hewitt Vineyards, Meridian Vineyards, Provenance Vineyards, Run Riot, Stags' Leap Winery, St Clement Vineyards, Sterling Vineyards, Sledgehammer*

DELICATO FAMILY WINES
13 million cases

Established in 1924 by Gaspare Indelicato, an immigrant from Sicily. Production was 3,451 gallons in 1934, but soared to 15,000 gallons by 1940, 74,107 gallons by 1955, and 403,000 gallons by 1964. After more than a decade of double-digit annual financial growth, this group changed its name from Delicato Family Vineyards to Delicato Family Wines, highlighting that it is in the wine business, not just grape-growing, although it still owns over 4,050 hectares (10,000 acres) of vineyards. It also remains a family-owned company, with Delicato as its primary brand.

Brands: *1924, Belle Ambiance, Black Stallion, Bota Box, Brazin, Diora, Dobbes Family Estates, Diora, Domino, Donati Family Vineyard, Gnarly Head, HandCraft, Irony, Mercer Family Vineyards, Merryvale Vineyards, Noble Vines, Starmont Winery and Vineyards, Three Finger Jack, Toad Hollow Vineyards, Twisted, Wine By Joe, Z Alexander Brown*

BRONCO WINE COMPANY
10 million cases

In 1973, following the acquisition of Franzia by Coca-Cola, the Bronco Wine Company was formed by Fred Franzia, a nephew of legendary Ernest Gallo, his brother Joseph, and their cousin John (Bronco is a contraction of "Brothers" and "Cousin"). The biggest brand in the company is Charles Shaw, which was listed by Trader Joe's in 2002 and sold so cheap that it became known as "Two Buck Chuck". By 2017 Charles Shaw had already sold one billion bottles, averaging 5.5 million cases a year, making that brand alone significantly larger than the 10th-largest group in this list. Bronco is the largest US vineyard owner, with more than 18,200 hectares (45,000 acres) in California. It also claims to be the largest producer of organic wines in America today, with 10 organic wine brands and 15 vegan-friendly wine brands.

Brands: *6° Six Degrees Cellars, Albertoni, Allure Winery, Balletto Vineyards, Bell Wine Cellars, Bella Grace Vineyards, Big Guy Wine Cellars, Brady Vineyard, Buttonwood Farm Winery, Carmenet Reserve, Cass Vineyard & Winery, Cellar Four 79, Charles Shaw, Chateau La Paws, Cottonwood Creek*

- Malibu-Newton Canyon – California
- Manton Valley – California
- Martha's Vineyard – Massachusetts
- McDowell Valley – California
- McMinnville – Oregon
- Mendocino – California
- Mendocino Ridge – California
- Merritt Island – California
- Mesilla Valley – New Mexico, Texas
- Middle Rio Grande Valley – New Mexico
- Middleburg Virginia – Virginia
- Mimbres Valley – New Mexico
- Mississippi Delta – Mississippi, Louisiana, Tennessee
- Mokelumne River – California
- Monterey – California
- Monticello – Virginia
- Moon Mountain District Sonoma County – California
- Mount Harlan – California
- *Mount Pisgah Polk County – Oregon*
- Mount Veeder – California
- Naches Heights – Washington
- Napa Valley – California
- Niagara Escarpment – New York
- North Coast – California
- North Fork of Long Island – New York
- North Fork of Roanoke – Virginia
- North Yuba – California
- Northern Neck George Washington Birthplace – Virginia
- Northern Sonoma – California
- Oak Knoll District of Napa Valley – California
- Oakville – California
- Ohio River Valley – Indiana, Kentucky, Ohio, West Virginia
- Old Mission Peninsula – Michigan
- Outer Coastal Plain – New Jersey

- Ozark Highlands – Missouri
- Ozark Mountain – Arkansas, Missouri, Oklahoma
- Pacheco Pass – California
- Paicines – California
- *Palos Verdes Peninsula – California*
- Paso Robles – California
- Paso Robles Estrella District – California
- Paso Robles Geneseo District – California
- Paso Robles Highlands District – California
- Paso Robles Willow Creek District – California
- *Paulsell Valley – California*
- Petaluma Gap – California
- Pine Mountain-Cloverdale Peak – California
- Potter Valley – California
- Puget Sound – Washington
- Ramona Valley – California
- Rattlesnake Hills – Washington
- Red Hill Douglas – Oregon
- Red Hills Lake County – California
- Red Mountain – Washington
- Redwood Valley – California
- Ribbon Ridge – Oregon
- River Junction – California
- Rockpile – California
- The Rocks District of Milton-Freewater – Oregon
- Rocky Knob – Virginia
- *Rocky Reach – Washington*
- Rogue Valley – Oregon
- Russian River Valley – California
- Rutherford – California
- Saddle Rock-Malibu – California
- Salado Creek – California
- San Antonio Valley – California
- San Benito – California
- San Bernabe – California
- San Francisco Bay – California
- San Juan Creek – California
- San Lucas – California

- *San Luis Obispo Coast (SLO Coast) – California*
- San Miguel District – California
- San Pasqual Valley – California
- San Ysidro District – California
- Santa Clara Valley – California
- Santa Cruz Mountains – California
- Santa Lucia Highlands – California
- Santa Margarita Ranch – California
- Santa Maria Valley – California
- Santa Ynez Valley – California
- Seiad Valley – California
- Seneca Lake – New York
- Shawnee Hills – Illinois
- Shenandoah Valley – Virginia, West Virginia
- Sierra Foothills – California
- Sierra Pelona Valley – California
- Sloughhouse – California
- Snake River Valley – Idaho, Oregon
- Snipes Mountain – Washington
- Solano County Green Valley – California
- Sonoita – Arizona
- Sonoma Coast – California
- Sonoma Mountain – California
- Sonoma Valley – California
- South Coast – California
- Southeastern New England – Connecticut, Rhode Island, Massachusetts
- Southern Oregon – Oregon
- Spring Mountain District – California
- Squaw Valley-Miramonte – California
- St. Helena – California
- Sta. Rita Hills – California
- Stags Leap District – California
- Suisun Valley – California
- Swan Creek – North Carolina
- *Tehachapi – California*

- Temecula Valley – California
- Templeton Gap District – California
- Texas Davis Mountains – Texas
- Texas High Plains – Texas
- Texas Hill Country – Texas
- Texoma – Texas
- Tip of the Mitt – Michigan
- Tracy Hills – California
- Trinity Lakes – California
- *Ulupalakua – Hawaii*
- Umpqua Valley – Oregon
- Upper Hiwassee Highlands – Georgia, North Carolina
- Upper Hudson – New York
- *Upper Lake Valley – California*
- Upper Mississippi River Valley – Illinois, Iowa, Minnesota, Wisconsin
- Van Duzer Corridor – Oregon
- *Verde Valley – Arizona*
- *Virginia Peninsula – Virginia*
- Virginia's Eastern Shore – Virginia
- Wahluke Slope – Washington
- Walla Walla Valley – Oregon, Washington
- *Wanapum Village – Washington*
- Warren Hills – New Jersey
- West Elks – Colorado
- Western Connecticut Highlands – Connecticut
- *White Bluffs – Washington*
- Wild Horse Valley – California
- Willamette Valley – Oregon
- Willcox – Arizona
- Willow Creek – California
- Wisconsin Ledge – Wisconsin
- Yadkin Valley – North Carolina
- Yakima Valley – Washington
- Yamhill-Carlton – Oregon
- York Mountain – California
- Yorkville Highlands – California
- Yountville – California
- *Yucaipa Valley – California*

Cellars, Crane Lake Cellars, Daffodil Winery, De La Costa, Estrella River Winery, Falcone Family Vineyards, Fat Cat Cellars, Garnet Vineyards, Gravel Bar Winery, Green Truck Winery, Haraszthy Family Cellars, Heliotrope, Komodo Dragon Cellars, La Catrina, Lake Sonoma Winery, Masked Rider Winery, Muddy Boot, Picket Fence Vineyards, PINO Cellars, Powder Keg Wine, Rancho Sisquoc Winery, Rare Earth, Red Truck Winery, Richard Grant Estate, Rosenblum Cellars, Rusack Vineyards, Shaw Organic, Stark Raving, Stone Cellars, Summers Estate Wines, Vin Glogg

STE MICHELLE WINE ESTATES
8.2 million cases

The origins of this Washington company date back to the two biggest wineries in the state following the repeal of Prohibition, Pommerelle (fruit wines) and NAWICO (aka National Wine Company, which made wines from the Concord grape). They were both established in 1934 and merged to form American Wine Growers (primarily fortified sweet wines) in 1954. Having followed closely the *vinifera* research projects conducted by Washington State University, American Wine Growers experimented first with Grenache, then planted White Riesling in 1965 in the Yakima Valley, followed later by Cabernet Sauvignon, Pinot Noir, and Sémillon. These wines were launched under the "Ste Michelle Vintners" label, with the collaboration of the legendary Andre Tchelistcheff. In 1976 American Wine Growers changed its name to Ste Michelle Vintners and built its iconic French-style château in Woodinville. It was instrumental in obtaining approval for the Columbia Valley AVA in 1984. Today, Ste Michelle Wine Estates own nearly 1,620 hectares (4,000 acres) of vineyards in Washington and California and formed partnerships with winemakers of global fame, such as its Col Solare Winery (with Tuscany's Piero Antinori), Eroica Wine Riesling (with the Mosel's Ernst Loosen), and Tenet (with Michel Gassier and former French rugby forward and Rhône consultant Philippe Cambie).

Brands: *14 Hands, Altered Dimension, ANEW Wines, Borne of Fire, Chateau Ste Michelle, Columbia Crest, Conn Creek Winery, Erath Winery, Fruit & Flower, Intrinsic, Liquid Light, Merf Wines, Northstar Winery, Patz & Hall, Prayers of Sinners & Saints Wine, Red Diamond Wine, Snoqualmie Vineyards, Stag's Leap Wine Cellars, Stimson Estate Cellars, Tenet Wines, The Cosmic Egg Wine Co, Villa Mt Eden*

JACKSON FAMILY WINES
6 million cases

In 1974 Jess Jackson and his then wife, Jane Kendall Jackson, purchased a pear and walnut orchard in Lakeport and converted it to a vineyard, He sold his grapes to local wineries until 1981, when a slump in the California wine industry led to a surplus of grapes, and, faced with selling his grapes for a loss, Jackson decided to make his own wine. Rather than follow the trend for inexpensive wines, however, he believed there was a shortage of premium-quality wines at affordable prices and decided to fill that gap. In 1983 he released the first vintage of Kendall-Jackson Vintner's Reserve Chardonnay, a wine that later that year became the first-ever

AMERICAN LABEL LANGUAGE

Alcohol It is obligatory to indicate the alcoholic strength of a US wine; this is expressed as a percentage by volume.

Angelica An authentic American fortified wine style of 17 to 24 per cent, traditionally made from Mission grapes and typically blended 50-50 wine and brandy, with the spirit added after fermentation has achieved just 5 per cent alcohol, resulting in excess of 100 grams per litre of residual sugar. If the strength is less than 18 per cent, the wine must be labelled "Light Angelica". Other grapes that are sometimes used include Grenache, Muscat, and Palomino. The quality can be extremely high. Angelica was named after Los Angeles by Spanish missionary winemakers in the 18th century.

Bottle fermented A sparkling wine derived through a second fermentation in "glass containers of not greater than 1 (US) gallon capacity". Usually refers to a wine that, after second fermentation, is decanted and filtered under pressure before rebottling

Burgundy A semi-generic name allowed under US federal law for a red wine of legally undefined "Burgundy-style". Such products today are inevitably of basic quality and made from almost anything, although high-yielding Carignan, Grenache, and Zinfandel are traditional. American Burgundy is typically dark in colour and full in body, but with some residual sweetness to give an impression of softness.

Carbonated wine Still wine made bubbly through the addition of carbon dioxide from a bottle of gas, the method used to produce fizzy drinks such as ginger ale or cola

Champagne A semi-generic name allowed under US federal law for a wine rendered sparkling through a second fermentation in "glass containers of not greater than 1 (US) gallon capacity"

Chianti A semi-generic name allowed under US federal law for a red wine of legally undefined style, American Chianti is generally medium-bodied, falling somewhere between American claret and American Burgundy in colour, character, and residual sweetness, with a more pronounced fruitiness, but is equally basic in quality

Claret A semi-generic name allowed under US federal law for a red wine of legally undefined style, American claret is a basic wine that is generally drier than American Burgundy, as well as being lighter in colour and body.

Crackling wine A sparkling wine derived through a second fermentation in "glass containers of not greater than 1 (US) gallon capacity", thus the same as for American "Champagne", although with a lesser degree of effervescence. It may also be called *pétillant*, *frizzante*, or *crémant*.

Crackling wine — bulk method I have not actually seen "Bulk Method" on a wine label, but it does exist, and it means the same as for "Crackling Wine" but made by the *cuve close* method.

Dessert wine A wine with an alcohol content of 14 to 24 per cent that is usually fortified. If less than 18 per cent, the wine must be labelled "Light Dessert Wine".

Fermented in the bottle A simple and clever way of describing a wine that has been made by the traditional *méthode champenoise*

Fumé blanc Widely used term for oak-aged Sauvignon Blanc

Grape variety If a single grape variety is mentioned on a label, the wine must contain at least 75 per cent of that variety (90 per cent in Oregon). Although the potential of varietal labelling was recognized in Alsace in the 1920s, the international marketing phenomenon these wines represent today was brought about by the widespread application of the concept by California wineries.

Haute Sauterne (*sic*) Rarely encountered, highly questionable, semi-generic name allowed and misspelled under US federal law for medium-sweet and sweet variants of Sauterne (*sic*)

Health warning Warnings are required by federal law, yet none of the proven health advantages are permitted on the label (*see* Micropedia).

Hock A semi-generic name allowed under US federal law for an undefined style of wine that, curiously, is light and dry, rather than vaguely approximating some sort of German Rhine wine. The grapes used are often the same as for American Chablis but are blended to produce something lighter (which is correct) and drier (which is not).

Light Wine A wine with an alcohol level not in excess of 14 per cent (synonymous with table wine)

Madeira A semi-generic name allowed under US federal law for any fortified wine of 17 to 24 per cent, made in a legally undefined "Madeira-style" that is usually sweet to some degree. If it contains less than 18 per cent alcohol, the wine must be labelled "Light Madeira".

Malaga A semi-generic name allowed under US federal law for any fortified wine of 17 to 24 per cent, made in a legally undefined "Malaga-style" that is usually sweet to some degree. If it contains less than 18 per cent alcohol, the wine must be labelled "Light Malaga". Made primarily from Concord grapes, American Malaga is often a kosher product.

Marsala A semi-generic name allowed under US federal law for any fortified wine of 17 to 24 per cent, made in a legally undefined "Marsala-style" that is amber or brown in colour, usually sweet to some degree, and sometimes flavoured with herbs. If it contains less than 18 per cent alcohol, the wine must be labelled "Light Marsala".

Moselle A semi-generic name allowed under US federal law for an undefined style of wine that is traditionally made from Chenin Blanc and Riesling grapes. American Moselle is light in body, pale in colour, medium sweet to taste, and sometimes very lightly *pétillant*. It is softer and lighter than American hock.

winner of a Platinum Award at the American Wine Competition. President Reagan only ever served California wine at the White House, and his wife so loved Kendall-Jackson Vintner's Reserve Chardonnay that it became known as "Nancy's wine" according to Herb Caen, the Pulitzer Prize–winning columnist at the *San Francisco Chronicle*. In a continued effort to upscale its wines, the Jackson Estate label returned to 100 per cent estate-grown in 2005, and for the last six years has been strategically buying up vineyards and smaller premium wineries that specialise in Burgundian styles. *See also* Kendall-Jackson, p613.

Brands: *Anakota, Arrowood Vineyards & Winery, Atalon, Brewer-Clifton, Byron Wines, Cambria Estate Winery, Capture, Cardinale Winery, Carmel Road Winery, Cenyth, Champ de Reves, Copain Wines, Edmeades, Freemark Abbey Winery,*

Galerie, Gran Moraine, Hartford Family Winery, Kendall-Jackson, La Crema, La Jota Vineyard Co., Legacy, Lokoya, Maggy Hawk, Matanzas Creek Winery, Mt Brave, Murphy-Goode, Nielson, Penner-Ash, Siduri, Stonestreet Wines, Vérité, Villa Kenzie, Wind Racer, Zena Crown Vineyard

DEUTSCH FAMILY WINE AND SPIRITS

3.5 million cases

Like all the other largest US wine companies in this Top 10 section, the number of cases does not include exports or imported brands, and Deutsch is the exclusive importer of Yellow Tail, which even in its decline, accounts for a further 4 million cases. *See also* Casella Family Brands, p671.

Brands: *Girard, Josh Cellars, Joseph Carr, Kunde Family Estate, Skyfall, The Calling*

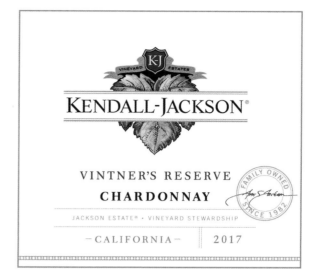

KENDALL-JACKSON VINTNER'S RESERVE CHARDONNAY Once regularly served at the White House, this wine helped raise the image of the Kendall-Jackson label, which is now part of Jackson Family Wines.

Muscatel Generally a more basic quality version of sweet Muscat wine (which is usually sold varietally as Muscat Frontignan or Muscat Canelli). It is usually a blend of two or more of the following: Aleatico, Malvasia Bianca, and Muscat (Alexandria, Canelli, and Frontignan). Black Muscatel is a blend of Aleatico and Black Muscat (also known as Muscat Hamburg).

Natural wine A wine may be called "natural" only if it has not been fortified with grape brandy or alcohol.

Port A semi-generic name allowed under US federal law for any fortified wine of 17 to 24 per cent that has been made in a legally undefined "Port style". If it contains less than 18 per cent alcohol, the wine must be labelled "Light Port". Zinfandel, Petite Sirah, and Cabernet Sauvignon are the most common grapes to be declared on a label, but others widely used include Alicante Bouschet, Grenache, and Mission. Additionally, there is a small but growing category of quality Ports, blended and varietal, that are made from various Portuguese varieties such as Tinta Madeira, Tinta Cão, Alvarelhão, Touriga, and Bastardo or Trousseau. When blended from different Portuguese varieties, the wine is often sold as a Tinta Port. Whichever grapes are used, the wines are usually sweet to some degree, and often qualified by traditional Port terminology, such as Ruby, Tawny, and Vintage. Some examples can be very good, particularly from Port specialists.

Rhine wine Synonymous with American hock

Sauterne (*sic*) A semi-generic name allowed and misspelled under US federal law for a wine that is usually of basic quality, less sweet than genuine Sauternes, and with no hint of botrytis. A few better-quality Sauterne wines are made, usually from Sémillon and Sauvignon Blanc, but most US Sauternes are nothing more than bland blends of Palomino, Sauvignon Vert, Thompson Seedless, and other even less respectable varieties. Amazingly for a semi-generic wine that should by definition be intensely sweet, Sauterne wines range from dry to very sweet, and are often qualified by "Dry" or "Sweet", with "Chateau" weirdly used for the sweetest styles. The cheapest dry Sauterne wines are not dissimilar to American Chablis. Some California producers spell Sauternes correctly, but it is difficult to work out which spelling constitutes the worst insult.

Semi-generic A semi-generic name is a geographical designation that has, in the opinion of the Director of the Department of Treasury's Alcohol and Tobacco Tax and Trade Bureau (TTB), become at least as well known for a specific style of wine, wherever it might happen to be produced, and is thus allowed under US federal law to be used by any wine producer in the United States. It is thus entirely legal within US borders to abuse the appellations of Burgundy, Chablis, Champagne, Chianti, and many others.

Sherry A semi-generic name allowed under US federal law for any fortified wine of 17 to 24 per cent that has been made in a legally undefined "Sherry style", usually sweet to some degree. If the alcohol content is less than 18 per cent, the wine must be labelled "Light Sherry". Palomino is the favoured variety for high-class sherries and is also known as the Golden Chasselas or Napa Golden Chasselas. Palomino is the only varietally labelled Sherry that is likely to be encountered, since the other grapes used are hardly worth shouting about. Although legally undefined by federal law, the best Sherries often involve the use of *flor* and will be blended under a *solera* system. The most common production method is to heat the wines for several months in order to replicate Sherry's oxidative style, and some efforts are worse than others. There are numerous sub-styles produced. Some US Sherries are actually labelled Dry, Medium, or Sweet, in addition to which there are Pale Dry or Cocktail sherries, which are dry; Golden or Amber Sherries, which are medium sweet; and Cream, which is sweet.

Sparkling wine This term may be used to describe "carbonated", *cuve close*, "bottle-fermented", or *méthode champenoise* wine. If the label bears no other information, the consumer should fear the worst (carbonated or *cuve close*); however, the label may indicate any of the following terms: Champagne, Bottle Fermented, Fermented in this Bottle,

Crackling Wine (including Crackling Wine – Bulk Method), and Carbonated Wine.

Sulfites It is obligatory for all American wines to state "Contains sulfites" or "Contains sulfiting agent/s" when sulphur dioxide or a sulphuring agent can be detected at a level of 10 or more parts per million.

Table wine A wine with an alcohol level not in excess of 14 per cent (synonymous with light wine)

Tokay All-American blend of Angelica, Dry Sherry (to reduce the sweetness), and a dash of Port to provide its tawny-pink colour, Tokay is a medium-sweet dessert wine with a slightly nutty flavour, and 17 to 24 per cent in strength. If less than 18 per cent, the wine must be labelled "Light Tokay".

Vintage At least 95 per cent of the wine must be from the vintage indicated. Until the 1970s, the law demanded that the figure was 100 per cent, but wine-makers petitioned for a small margin to enable topping-up, and this was granted.

Volume By law, this must be stated somewhere on the bottle

KORBEL LABEL The label identifies this wine as the semi-generic California "Champagne".

California

California has a vast range of climates, running just over 1,610 kilometres (1,000 miles) north to south, from latitude 32°3N to 42°N. The most significant variation is west to east, however, from the cooling Pacific coastal breezes and famous fog and then crossing the Pacific Coastal Range to the warm, dry, high production volume of the Central Valley. Combined with a focus on technical precision in both the vineyard and winery, the state accounts for about 80 per cent of total American wine production, including a selection of the finest, most highly acclaimed and expensive wines on the international market, as well as some of the most commercial and successful mass-market global brands.

California was first settled by the Spanish in 1769 and formed part of Mexico until 1848, when it was ceded to the United States, becoming a State of the Union in 1850. The first California wine was made in 1782 at San Juan Capistrano, by Fathers Pablo de Mugártegui and Gregorio Amurrió. Mission grapes, from vines brought to California by Don José Camacho on the *San Antonio,* which docked at San Diego on 16 May 1778, were used to make the wine. It was not until 1833, however, that *Bordelais* Jean-Louis Vignes established California's first commercial winery. He was the first Californian winemaker to import European vines and, in 1840, he also became the first to export Californian wines. The discovery of gold in 1848 in the Sierra Nevada foothills sparked the California Gold Rush, attracting immigrants, many hailing from European winemaking regions, and drawing the burgeoning wine industry north to San Francisco.

THE AMAZING HARASZTHY

Eight years before California passed from Mexican to American sovereignty, a certain Hungarian political exile named Ágoston Haraszthy de Mokcsa settled in Wisconsin. Haraszthy was a colourful, flamboyant entrepreneur in the mould of Barnum or Champagne Charlie. Among other things, he founded a town in Wisconsin and modestly called it

BUENA VISTA WINERY
Now a registered landmark, this venerable winery has seen many changes since its inception in 1857. Today the winery sits on its original property and is owned by the Boisset Collection, led by Jean-Charles Boisset, who purchased the winery and the historic property in May 2011.

RECENT CALIFORNIA VINTAGES

2018 Long, moderate, and dry growing season allowed slow ripening with harvest up to three weeks later than 2017. High quality with balance of fruit concentration, alcohol, and acidity.

2017 Higher than average rainfall brought an end to the five-year drought. A heat wave in September led to raisining and reduced yields. North Coast wildfires in October fortunately did not significantly affect vineyards and wineries with 85 to 90 per cent harvests already completed in Napa, Sonoma, and Mendocino. Variable quality with best examples generous and highly concentrated.

2016 Generally cooler than 2015 and 2014m with comparisons to 2011. Winter rains alleviated drought and helped increase yields on 2015. Consistent quality of vibrant, fresh wines.

2015 One of the earliest harvests on record, with small volumes but outstanding quality. Continued drought throughout growing season combined with a cool spring reduced yields below the preceding record years of 2012, 2013, and 2014. Small berries led to concentrated but balanced wines.

2014 The third in a string of excellent vintages and third-largest on record, down from the 2013 peak, despite on-going drought. A mild winter and spring led to early bud-break, followed by excellent weather throughout culminating in an early harvest. Fresher, more elegant and less opulent than 2013.

Haraszthy (it was later renamed Sauk City), ran a Mississippi steamboat, and cultivated the first vineyard in Wisconsin – all within two years of beginning a new life in a strange country.

In 1849 Haraszthy moved to San Diego, leaving his business interests in the hands of a partner, who promptly took advantage of a rumour that he had perished during his transcontinental trek, sold all the business and properties, and vanished with the money. Haraszthy was broke, yet within six months he was farming his own 65-hectare (160-acre) fruit and vegetable ranch. Such rapid success followed by disaster and then an even greater triumph was to become Haraszthy's trademark. Within a few months of acquiring his ranch, he also became the owner of a butcher's shop and a livery stable in Middleton, a part of San Diego that still boasts a Haraszthy Street. In addition, he ran an omnibus company, started a construction business, was elected the first sheriff of San Diego, was made a judge, and became a lieutenant in the volunteer militia. He also began importing cuttings of numerous European vine varieties.

The Buena Vista Winery

Having imported no fewer than 165 different vine varieties from Europe, in 1856 Haraszthy purchased 230 hectares (560 acres) of land near the town of Sonoma, in an area called the Valley of the Moon. Here he built a winery, which he named Buena Vista, and dug six cellars out of the sandstone hill. With northern California's first significant wine estate, Haraszthy won several awards, and he attracted much publicity for both his vineyard and his wine. This venture drew so much attention that, in 1861, the governor of California commissioned Haraszthy to visit Europe and report on its winegrowing areas. Haraszthy's trip took him to every wine region in France, Germany, Italy, Spain, and Switzerland, where he interviewed thousands of winegrowers, took notes, consulted foreign literature, and so accumulated a library of reference material. He returned to the United States with a staggering 100,000 cuttings of 300 different vine varieties, only to have the state senate plead poverty when he presented them with a bill for 12,000 dollars for his trip, although the cuttings alone were worth three times that amount. He was never reimbursed for his trouble, and many of the cuttings, which he had expected to be distributed among the state's other winegrowers, simply rotted away.

Haraszthy was not deterred: within seven years he managed to expand Buena Vista to 2,430 hectares (6,000 acres). In doing so, he totally changed the course of Californian viticulture, transferring the focus of attention from the south of the state to the north. At the height of its fame Buena

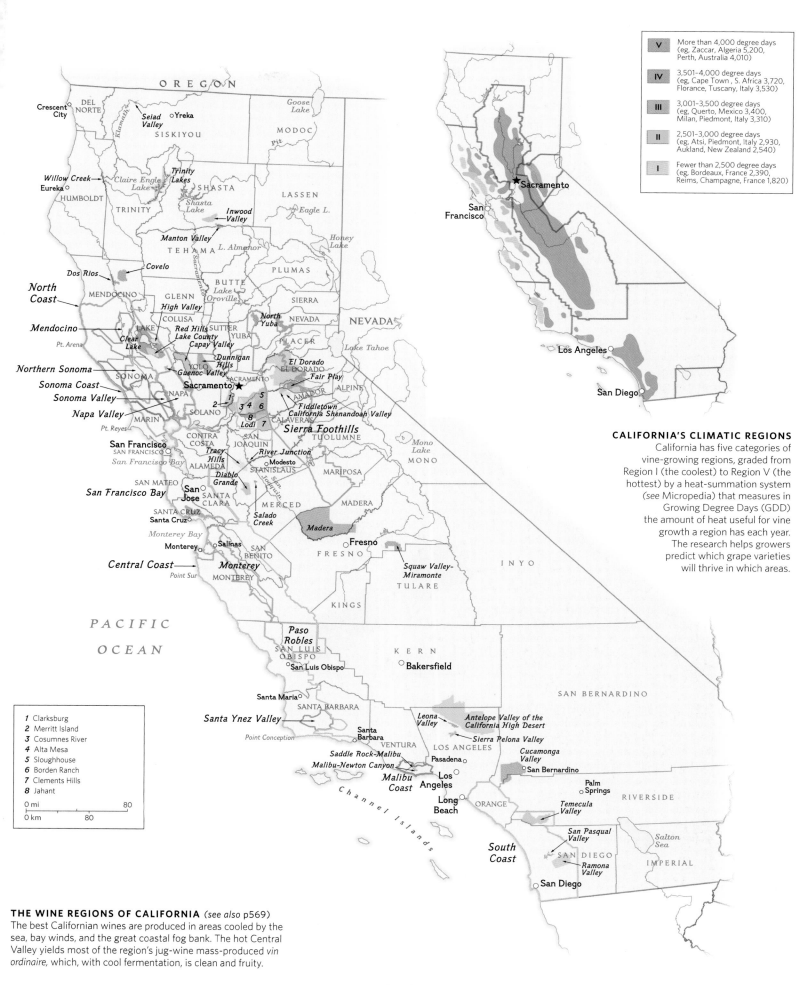

O R E G O N

DEL NORTE
Crescent City
Seiad Valley ○Yreka
SISKIYOU
MODOC
Goose Lake
Pit
Klamath
Willow Creek→
Eureka○
HUMBOLDT
Trinity Lakes
Claire Engle Lake
SHASTA
TRINITY
Shasta Lake
LASSEN
Eagle L.
Inwood Valley
Manton Valley
TEHAMA
L. Almanor
Honey Lake
North Coast
Dos Rios
Covelo
GLENN
PLUMAS
Sacramento
Lake Oroville
SIERRA
Mendocino
MENDOCINO
Pt. Arena
High Valley
COLUSA
BUTTE
Red Hills Lake County
Capay Valley
North Yuba
NEVADA
SUTTER
YUBA
Clear Lake
LAKE
Northern Sonoma
Sonoma Coast
Sonoma Valley
Napa Valley
Dunnigan Hills
Guenoc Valley
YOLO
El Dorado
EL DORADO
Fair Play
NEVADA
PLACER
Lake Tahoe
SONOMA
NAPA
SOLANO
Sacramento ★
ALPINE
AMADOR
Fiddletown
California Shenandoah Valley
CALAVERAS
Sierra Foothills
MARIN
Pt. Reyes
San Francisco
SAN FRANCISCO
San Francisco Bay
CONTRA COSTA
ALAMEDA
Tracy Hills
Diablo Grande
SAN JOAQUIN
River Junction
●Modesto
STANISLAUS
Lodi
TUOLUMNE
Mono Lake
MONO
MARIPOSA
San Francisco Bay
SAN MATEO
San Jose
SANTA CLARA
SANTA CRUZ
Santa Cruz○
Monterey Bay
Monterey○
Salado Creek
MERCED
MADERA
Madera
●Fresno
FRESNO
Squaw Valley-Miramonte
INYO
Central Coast→
Point Sur
Monterey
MONTEREY
○Salinas
SAN BENITO
TULARE
Clarksburg
1 Clarksburg
2 Merritt Island
3 Cosumnes River
4 Alta Mesa
5 Sloughhouse
6 Borden Ranch
7 Clements Hills
8 Jahant

0 mi ———— 80
0 km ———— 80

P A C I F I C

O C E A N

KINGS
Paso Robles
SAN LUIS OBISPO
●San Luis Obispo
K E R N
●Bakersfield
Santa Maria○
SANTA BARBARA
SAN BERNARDINO
Santa Ynez Valley→
Point Conception
Leona Valley
Antelope Valley of the California High Desert
Sierra Pelona Valley
Cucamonga Valley
Santa Barbara
VENTURA
LOS ANGELES
Pasadena○
●San Bernardino
Saddle Rock-Malibu
Malibu-Newton Canyon
Malibu Coast
Los Angeles
ORANGE
Long Beach
Palm Springs○
RIVERSIDE
Temecula Valley
Channel Islands
San Pasqual Valley
SAN DIEGO
Ramona Valley
Salton Sea
IMPERIAL
South Coast
San Diego

V	More than 4,000 degree days (eg, Zaccar, Algeria 5,200, Perth, Australia 4,010)	
IV	3,501–4,000 degree days (eg, Cape Town , S. Africa 3,720, Florance, Tuscany, Italy 3,530)	
III	3,001–3,500 degree days (eg, Querto, Mexico 3,400, Milan, Piedmont, Italy 3,310)	
II	2,501–3,000 degree days (eg, Atsi, Piedmont, Italy 2,930, Aukland, New Zealand 2,540)	
I	Fewer than 2,500 degree days (eg, Bordeaux, France 2,390, Reims, Champagne, France 1,820)	

★ Sacramento
San Francisco
Los Angeles
San Diego

CALIFORNIA'S CLIMATIC REGIONS

California has five categories of vine-growing regions, graded from Region I (the coolest) to Region V (the hottest) by a heat-summation system (*see* Micropedia) that measures in Growing Degree Days (GDD) the amount of heat useful for vine growth a region has each year. The research helps growers predict which grape varieties will thrive in which areas.

THE WINE REGIONS OF CALIFORNIA (see also p569)

The best Californian wines are produced in areas cooled by the sea, bay winds, and the great coastal fog bank. The hot Central Valley yields most of the region's jug-wine mass-produced *vin ordinaire,* which, with cool fermentation, is clean and fruity.

THE SECOND COMING

There is a widespread belief that since the destructive root-feeding insect phylloxera was imported to Europe on plant material from the United States that California's vineyards have always been safe, especially as the only preventative measurement is to graft European vines on to resistant American vine species rootstocks. Nothing could be further from the truth. Phylloxera's home was east of the Rockies, where over aeons native vines, such as *Vitis berlandieri* and *V riparia,* developed a natural resistance to phylloxera. When European *vinifera* vines were introduced to California, phylloxera was bound to follow via the wagon trains across the Rockies or on the vines themselves, and *vinifera* varieties proved to be as vulnerable in California as anywhere else.

THE ANTI-BEAST

Phylloxera was first identified at Sonoma in 1873, coincidentally at the same time as another native American bug, *Tyroglyphus,* was deliberately being shipped to France. The idea of using *Tyroglyphus,* which was harmless to the vine but a deadly enemy of phylloxera, to infect phylloxera-infested vineyards was an imaginative one. Unfortunately, unlike phylloxera, *Tyroglyphus* did not care for the European climate and failed to settle.

In California, the effect was devastating. Phylloxera multiplies at a terrifying rate (one female aphid can be responsible for one billion offspring in 12 generation cycles of 30 days every year). By 1891 Napa's area under vine was reduced from 7,200 hectares (18,000 acres) to a mere 1,200 hectares (3,000 acres). Under the auspices of Professor Eugene Hilgard, Head of Agriculture at the Department of Viticulture and Enology in the University of California at Berkeley (established in 1880, moved to Davis in 1906), California's growers eventually adopted the same method as the Europeans to control the pest, grafting *vinifera* vines on to phylloxera-resistant varieties. Ironically, it was some time before Californians realized these wonder vines originated from eastern American states and initially imported them from France. Only half the vineyards had been grafted when they were hit by another, even more lethal, plague – Prohibition. By the late 1940s, UC Davis had assembled one of the most formidable teams of viticultural and oenological experts in the world. Headed by such legendary figures as Maynard Amerine, Harold Olmo, and Albert Winkler, the university was primarily responsible for making California the wine force it is today. Along the way, however, they made some errors, such as placing too much emphasis on volume and technical correctness, but the biggest mistake of all was to recommend the widespread use of AxR#1 rootstock.

ROOT CAUSE

Despite warnings from various European sources about its susceptibility to phylloxera (all acknowledged but brushed aside by numerous university textbooks), UC Davis recommended AxR#1's use on fertile valley floors, such as Napa, because of its ability to increase yields. As Winkler et al put it in *General Viticulture:* "This is a case where the choice of (root) stock cannot be based entirely on its resistance to phylloxera".

These words came back to haunt the faculty at Davis in the 1980s, when, slowly but inexorably, vineyards grafted on to AxR#1 failed, and the culprit, phylloxera, was identified. Most of California's AxR#1 vineyards were replanted by the turn of the millennium, at an estimated cost of billions. The Napa and Sonoma districts alone cost in excess of 1.5 billion dollars. There have been notable long-term qualitative benefits, however, as growers have taken the opportunity to modernise vineyards, replanting with higher quality vine material and improving matching varieties and rootstocks to site, as well as altering planting density and trellising methods as required.

SHARPSHOOTERS

Just as California's growers were counting the cost of dealing with phylloxera, so their vineyards were being invaded by another, possibly more dangerous insect – *Homalodisca coagulata,* or the Glassy-Winged Sharpshooter. This leaf-hopper is not harmful in itself, but acts as a vector, carrying the bacterium *Xylella fastidiosa* that infects vines with Pierce's disease, attacking and blocking the xylem, thereby preventing the movement of water around the plant. This disease is considerably faster-acting than phylloxera, killing vines in just one to five years, but what really frightens the industry is that there is no known cure. With an estimated cost to the Californian wine industry of 100 million dollars per year and no way to defeat Pierce's disease, most efforts have been focused on control and eradication of the Glassy-Winged Sharpshooter through insecticides and the release of natural predators, including the tiny, stingless wasp *Gonatocerus triguttatus.* On-going research by UC Davis and the California Department of Food and Agriculture (CDFA) Pierce's Disease and Glassy-winged Sharpshooter Board is exploring more sustainable and permanent solutions, including the deactivation of the enzymes that enable the bacteria to infect the vine, as well as treatments to enhance vine immunity and ultimately cure infection.

GLASSY-WINGED SHARPSHOOTER
These leaf-hoppers carry bacterium harmful to grapevines.

Vista had offices in San Francisco, Philadelphia, Chicago, New York, and London. This success was purely superficial, however, for the vineyard was described in 1864 as "the largest wine-rowing estate in the world and the most unprofitable". Haraszthy also suffered a number of losses on the stock exchange and was faced with a new tax on brandy, which resulted in further loss of income. A fire at the winery then destroyed much of his stock and the bank proceeded to cut off his credit.

Enough was enough, even for Ágoston Haraszthy de Mokecsa. He left California for Nicaragua, where he was successful in obtaining a government contract to distil rum from sugar. An enigmatic character to the end, Haraszthy disappeared altogether in 1869, presumed drowned while trying to cross an alligator-infested stream on his plantation.

LEARNING CURVE

The 1930s to 1950s saw slow growth and recovery of the wine industry from the impact of Prohibition, the Great Depression, and two World Wars. The majority of wine at this time remained fortified and sweet. Dry wine did not become the dominant category until the late 1960s, coinciding with the introduction of high-tech wineries and the use of 100 per cent new oak. The precision of style and focus of fruit that resulted were welcomed at the time, as were the supple tannins that replaced harsh ones. The 1970s saw the emergence of a new generation of winemakers, boosting production and quality, highlighted in the 1976 Judgement of Paris, and establishing California's quality reputation on the world stage. The truly magnificent 1985 vintage, the best year since 1974, was another benchmark for California's wine industry. Virtually everyone got it right. From the mid-1980s to the mid-1990s, producers widely pursued a balance of elegance without stripping away their natural generous fruit expression that the Californian climate affords.

TOO MUCH OF A GOOD THING?

Domaine Chandon established the first French-owned sparkling wine venture in 1973, paving the way for number of dedicated sparkling wine operations to follow suit throughout the 1980s: Piper-Sonoma in 1980, Maison Deutz in 1981, Freixenet's Gloria Ferrar in 1982, Roederer Estate in 1982, Mumm Napa in 1985, Taittinger's Domaine Carneros in 1987, and Codorníu in 1991. The combined release of their products onto the market sparked off a fizz war as each tried to grab a slice of the shrinking pie. The only real winner was Domaine Chandon, which had already recouped its capital outlay and was the only one that could afford to play the discount game and still make a profit. Roederer Estate refused to participate and has subsequently built up a reputation as one of the greatest sparkling wine producers outside of Champagne itself. Domaine Carneros and Mumm Napa have both survived by discounting their entry-level wines, while creating awareness for some of their higher-quality cuvées. Piper-Sonoma sold their original winery in the mid-90s to Judy Jordan for the production of her premium-quality "J" sparkling wine, which was subsequently sold to E & J Gallo in 2015. Maison Deutz has been defunct since 1997, and Codorníu was renamed Artesa and now specializes in up-market still wines.

History repeated itself on a grander scale at the turn of the 21st century. In the mid-1990s, California was riding on the crest of the "French Paradox" wave, which started with the *60 Minutes* television programme in 1991 and increased per capita wine consumption in the USA by almost one-third. In 1995 California experienced the largest growth in its wine sales in 20 years, when the industry was hit by a small crop. Sales continued to soar in 1996, and that year's crop was also down. Faced with the prospect of increasing sales and diminishing crops, demand was sky high for all available fruit from the 1997 vintage. The vintage was both high yielding and high quality, with dry warm weather ripening fruit early. In many cases, the sheer volume resulted in full tanks with fruit still in the vineyards. Once the first fermentations had finished in tank and been transferred to barrel, a second harvest produced very ripe wines that were then blended with the first. The results were widely heralded by press and consumers, leading to the birth of "extended hang-time" and the quest for the super-ripe style. Concurrently growers commenced planting new vineyards at such a rate that by 2001 the acreage under vine had increased by more than 60 per cent, but then global economic decline was heightened by September 11. Wine drinking did not slow down, but consumers became more sensitive to price points. This resulted in 10-dollar wines being discounted to 6 dollars, and cheaper wines being sold off anonymously. The ripple effect caused tens of thousands of acres to be grubbed up. This may not have been such a bad thing, as they were mostly fit for little more than blending, but it did seem a bit back to front that while vines were being planted at the fastest rate ever in marginal areas like the Atlantic Northeast, vineyards were actually disappearing in sunny California. There was even talk that California's wine glut was affecting sales of its ultra-premium products, as the minuscule tip of the California wine market had become overcrowded with 100-dollar-plus wines. Their primary patrons, California's new money, nonetheless eventually realized the value of scarcity, with both demand and volume continuing to grow.

A VINEYARD WORKER PLOUGHS THROUGH A BRIGHT YELLOW CARPET OF MUSTARD PLANTS
Mustard plants – whether wild or cultivated – thrive just until the grapevines' bud-break, when vine-growers in Napa County, Sonoma County, and other California wine regions plough it under as a "green manure". The mustard mulch then serves to feed the emerging grape plants with valuable nutrients and phosphorus.

CALIFORNIA'S GRAPE VARIETIES

During the 1980s white varieties accounted for 57 per cent of production. A great period of expansion followed, coinciding with the so-called "French Paradox" (see Micropedia), which aired on American CBS television's 60 Minutes programme, and held largely accountable for a colossal shift in consumer demand away from white to red wine. The eight years from 1995 to 2003 saw the plantings of black varieties increase by 101 per cent, nearly double the 58 per cent growth of white grape varieties. Since the early 2000s modest overall growth in total planted acres has continued with the proportion of white to black varieties remaining relatively stable. During this period of consolidation on total planting, there has been radical change in individual varietal acreages as vineyards are replanted to meet changing consumer demand.

GRAPE COLOUR	2003 ACRES	2010 ACRES	2018 ACRES	INCREASED ACREAGE 2003 TO 2018	PERCENTAGE GROWTH 2003 TO 2018
Black	287,075	294,528	302,837	15,762	5.49%
White	185,373	181,849	176,879	-8,494	-4.58%
TOTAL	472,448	476,377	479,716	7,268	1.54%

CALIFORNIA'S WHITE GRAPE VARIETIES

In 2018 there were 176,879 acres planted to white grape varieties, with the top 10 accounting for 93 per cent of the total. Plantings of Chardonnay have continued the downward trend, with Pinot Gris seeing the largest increase in acreage, as was the case from 2003 to 2010. The greatest proportional rate of growth has been in Grüner Veltliner and Albarino, as growers increasingly experiment with new varieties and explore new locations, both of which reflect the current consumer trend for lighter, fresher wine styles.

LE5	FG5	GRAPE VARIETY	2010 ACREAGE	2018 ACREAGE	INCREASED ACREAGE	PERCENTAGE GROWTH
		Chardonnay	95,271	93,452	-1,819	-1.91%
		French Colombard	24,824	18,907	-5,917	-23.84%
1		Pinot Gris	12,907	16,880	3,973	30.78%
		Sauvignon Blanc	15,407	14,851	-556	-3.61%
		Chenin Blanc	7,223	4,790	-2,433	-33.68%
3		Muscat of Alexandria	3,391	4,620	1,229	36.24%
		Riesling	3,831	3,849	18	0.47%
2	3	Muscat Blanc a Petits Grains	1,908	3,152	1,244	65.20%
		Viognier	2,993	2,710	-283	-9.46%
		Gewürztraminer	1,735	1,636	-99	-5.71%
4	4	Symphony	940	1,492	552	58.72%
		Malvasia Bianca	1,317	1,110	-207	-15.72%
		Triplett Blanc	856	985	129	15.07%
		Burger	1,222	848	-374	-30.61%
		Sémillon	890	578	-312	-35.06%
		Muscat Orange	272	377	105	38.60%
5	2	Albariño	153	347	194	126.80%
		Sauvignon Musque	220	335	115	52.27%
		Grenache Blanc	266	329	63	23.68%
		Roussanne	362	323	-39	-10.77%
		Pinot Blanc	456	299	-157	-34.43%
	1	Grüner Veltliner	61	183	122	200.00%
		Palomino	315	155	-160	-50.79%
	5	Verdelho	94	149	55	58.51%
		Marsanne	115	121	6	5.22%
		Catarratto	187	99	-88	-47.06%
		Vermentino	N/A	91	91	N/A
		Arneis	N/A	55	55	N/A
		Sauvignon Vert	N/A	53	53	N/A
		Tocai Friulano	121	51	-70	-57.85%
		Picpoul Blanc	N/A	45	45	N/A
		Ugni Blanc	197	N/A	-197	-100.00%
		Emerald Riesling	149	N/A	-149	-100.00%
		Other	4,166	4,007	-159	-3.82%
WHITE WINE TOTAL			181,849	176,879	-4,970	-2.73%

CALIFORNIA'S BLACK GRAPE VARIETIES

In 2018 there were 302,837 acres planted to black grape varieties in California with the top 10 accounting for 90 per cent of the total. Cabernet Sauvignon established itself as the number one variety, increasing its share of total planting from 26 per cent in 2010 to 30 per cent in 2018. Pinot Noir continues its surge, reflecting the on-going consumer trend, having now almost doubled in acreage since 2003. In contrast, the most under-the-consumer-radar growth, has been seen by Teroldego, primarily planted in Central Valley and used for blending given its deep colour, high acid, and high yield potential.

LE5	FG5	GRAPE VARIETY	2010 ACRES	2018 ACRES	INCREASED ACREAGE	PERCENTAGE GROWTH
1		Cabernet Sauvignon	77,602	91,834	14,232	18.34%
2		Pinot Noir	37,290	45,304	8,014	21.49%
		Zinfandel	49,136	43,210	-5,926	-12.06%
		Merlot	46,762	39,786	-6,976	-14.92%
		Syrah	19,283	16,448	-2,835	-14.70%
		Rubired	11,844	11,590	-254	-2.14%
3		Petite Sirah	7,999	11,468	3,469	43.37%
		Barbera	6,936	4,889	-2,047	-29.51%
		Ruby Cabernet	5,761	4,701	-1,060	-18.40%
		Grenache	6,170	4,413	-1,757	-28.48%
4	3	Malbec	1,616	3,822	2,206	136.51%
		Cabernet Franc	3,480	3,517	37	1.06%
5	5	Petit Verdot	1,927	3,408	1,481	76.86%
		Carignane	3,393	2,403	-990	-29.18%
		Sangiovese	1,950	1,569	-381	-19.54%
	2	Primitivo	206	1,335	1,129	548.06%
		Mataro (Mourvèdre)	939	1,107	168	17.89%
	1	Teroldego	81	980	899	1109.88%
		Tempranillo	957	942	-15	-1.57%
		Alicante Bouschet	1,065	902	-163	-15.31%
	4	Tannat	248	569	321	129.44%
		Mission	639	409	-230	-35.99%
		Muscat Hamburg	345	301	-44	-12.75%
		Touriga Nacional	227	295	68	29.96%
		Carnelian	782	262	-520	-66.50%
		Gamay	309	251	-58	-18.77%
		Royalty	240	230	-10	-4.17%
		Meunier	163	204	41	25.15%
		Nebbiolo	185	142	-43	-23.24%
		Montepulciano	90	98	8	8.89%
		Souzão	68	86	18	26.47%
		Dolcetto	123	84	-39	-31.71%
		Aglianico	51	80	29	56.86%
		Centurian	84	80	-4	-4.76%
		Cinsaut	118	76	-42	-35.59%
		Charbono	86	75	-11	-12.79%
		Dornfelder	N/A	68	68	N/A
		Carmenère	57	68	11	19.30%
		Lagrein	87	58	-29	-33.33%
		Counoise	52	53	1	1.92%
		Pinotage	55	53	-2	-3.64%
		Graciano	N/A	50	50	N/A
		Salvador	106	N/A	N/A	N/A
		Other	6,016	5,617	-399	-6.63%
RED WINE TOTAL			294,528	302,837	8,309	2.82%

Notes: LE5 = Five largest-expanding varieties by increased acreage.
FG5 = Five fastest-growing varieties by percentage difference.

A CHANGE OF STYLE

The 2008 global financial crisis is yet another turning point in the development of the Californian wine industry. A tightening of disposable income and conspicuous consumption coincided with a new generation of winemakers and consumer preferences, all combined to bring about the birth of a new "restrained" stylistic movement. The focus moved from maximum ripeness and opulence, to balance of structure, with lower alcohol, higher acid, and less new oak, as producers look to express and reflect regional and site-specific qualities. The IPOB (In Pursuit of Balance) movement (2011–2016) founded by Rajat Parr and Jasmine Hirsch, with the explicit aim of globally showcasing the debate around balance and winemaking styles of Californian Pinot Noir and Chardonnay, epitomised this step change within the industry that continues to evolve.

This learning curve has established in California, as it has elsewhere, that the secret of fine wine is merely a matter of balance. The difficulty is in achieving it, for perfect balance is by definition a natural state that can be achieved only in the vineyard, not in the winery.

CALIFORNIA WINERIES

The industry has seen huge growth over the last decade with the number of wineries (as reported by the Wine Institute) almost tripling, from 1,367 in 2006 to 3,900 in 2018. It is important to note that figures for wineries vary by source. The Wine Institute figures quoted are based on physical brick and mortar wineries, which would increase to approximately 4,500 (Wines and Vines Analytics) if "virtual" wineries are included – these are essentially brands working out of shared facilities.

According to industry analysts Wines and Vines, of this total number of wineries (bricks and mortar, plus "virtual"), just over three-quarters are considered "micro" producers with an output of fewer than 5,000 cases per year, responsible for just over 1 per cent of California's total production. At the other end of the scale, "large" wineries producing more than 500,000 cases account for just over 1 per cent of total producers but are responsible for almost 87 per cent of Californian wine output.

THE WINE STYLES OF

CALIFORNIA

BOTRYTISED AND LATE-HARVEST STYLES

Some of the world's most succulent late-harvest and botrytised wines are made in California. The Golden State brings a lovely ripe peachiness to Riesling and mouth-watering tropical fruit freshness to Chenin and Muscat.

Ready when released (but last well in bottle)

Arrowood (Riesling Owl Hoot) • Beringer (Nightingale) • Bonny Doon (Muscat Vin de Glaciere) • Dolce • Domaine de la Terre Rouge (Muscat à Petits Grains) • Château St Jean (Special Late Harvest Johannisberg Riesling) • Martinelli (Jackass Hill) • The Ojai Vineyard (Viognier Ice Wine)

BOTTLE-FERMENTED SPARKLING WINES

California used to be notorious for its cheap sparkling wine until Schramsberg and Domaine Chandon forged a new path in the early 1970s. Roederer Estate then followed in the 1980s, and arguably put Mendocino – and California – on the global wine map as a source of outstanding quality traditional-method sparkling wines.

2–5 years (up to 10 in very exceptional cases)

Caraccioli • Domaine Carneros (Le Rêve) • Gloria Ferrer (Royal Cuvée) • Handley (Vintage Blanc de Blancs, Vintage Brut, Vintage Rosé) • Iron Horse (Classic Vintage Brut) • J (magnums) • Mumm (DVX) • Roederer Estate

CABERNET FRANC

Originally gained popularity in the 1980s primarily for blending with Cabernet Sauvignon to contribute finesse and perfume. Since then, planting of Cabernet Franc vines has doubled, and pure varieties are more commonplace.

1–4 years (up to 10 in exceptional cases)

Arietta • Detert Family Vineyards • La Jota • Nevada City Winery • Paradigm • Peju • Pride • Rancho Sisquoc • Raymond Burr Vineyards • Reverie • Robert Sinskey Vineyards • Gainey Vineyard • Titus

CABERNET SAUVIGNON

California's most widely planted – and potentially its finest – black grape variety. In areas that are too cool, Cabernet Sauvignon wines can have capsicum character, but in most other areas they tend to produce saturated, purple-hued wines with a smooth texture, deliciously ripe blackcurrant fruit, and oak-derived notes of vanilla, mocha, cedar, and spice. Both French and American oak is used with increasing restraint, and some of the most exciting Cabernet wines are blended with Merlot, Cabernet Franc, and/or Petit Verdot, although not to the extent of a classic blend

(*see* Meritage and Red Blends entries on the following page). At their best, these wines are a match for the very greatest vintages of the top Bordeaux *grands crus*. California is so blessed with Cabernet Sauvignon that it is impossible to come up with a definitive top 10, so this represents just a cross-section of some of the best.

3–5 years (inexpensive), 5–12 years (top wineries), 8–25 (or more) years (exceptional wines)

Abreu • Atalon • Caymus (Special Selection) • Colgin (Tychson Hill) • Dalla Valley • Dunn • Joseph Phelps • Lokoya • Robert Mondavi (Reserve)

CHARDONNAY

As California's most widely planted white variety, a vast range of Chardonnay is produced in multiple styles at all quality levels. Although opulent and heavily oaked Chardonnays still abound, there has been an on-going shift to a more restrained and elegant style, achieved both in the vineyard and winery. There could be dozens of the 10 best California Chardonnays, hence, this represents just a selection.

2–8 years (15 or more years in very exceptional cases)

Au Bon Climat (Harmony Nuits-Blanches au Bouge) • Beringer (Distinction Series) • Blackjack Ranch • Brewer-Clifton (Sweeney Canyon) • Cuvaison (Carneros Estate) • Du Mol (Chloe) • Hartford Court (Three Jacks) • Kistler (Vine Hill Road) • Rhys Vineyards • Walter Hansel (Cahill Lane)

CHENIN BLANC

This Loire Valley grape is famous for wines such as the sweet, honey-rich Vouvray and the very dry, searingly flavoured Savennières. It was widely planted in Central Valley in the 1970s and 1980s, primarily for blending due to its naturally high acidity and yield potential. It is currently the fifth-most widely planted white wine grape, yet it is seldom spoken of and acreage has fallen significantly since 2010. Historically Clarksburg AVA has always been considered the home of Chenin Blanc, but it is starting to gain a following amongst boutique producers are exploring the variety's potential in cooler regions. All the best wines below are dry or off-dry. *See* Botrytised and Late-Harvest Wines, at left, for sweeter styles.

1–4 years

Baron Herzog (Clarksburg) • Casa Nuestra • Chalone • Dry Creek (Clarksburg) • Haarmeyer • Leo Steen • Pax Mahle

FORTIFIED STYLES

A small number of California wineries produce fortified wines, with Port emulations the most common. Long before its demise, Inglenook used to be the yardstick for California Palomino Sherry.

Ready when released (but last well in bottle)

Ficklin (Vintage Port) • Geyser Peak Winery (Tawny Port) • Joseph Filippi (Cucamonga Valley Sherry, Cucamonga Valley Port) • Quady • V Sattui (Madeira)

FUMÉ BLANC

This term was first coined by Mondavi in the 1970s, who borrowed two words from the French appellation Pouilly Blanc Fumé. It was very simple, yet very clever, and by fermenting the wines in oak *barriques*, he was able to sell what was then a very unfashionable grape variety under a new and catchy name. Fumé Blanc is now used as a generic name for a Sauvignon Blanc wine that has been fermented and/or aged in oak, with the results ranging from a subtle, *barrique*-fermented influence to a heavily oaked character, however, there is no legal requirement for oak influence.

1–5 years

Chateau St Jean (Lyon Vineyard Fumé Blanc) • Grgich Hills Cellar (Estate Grown) • Larkmead (Lillie) • Robert Mondavi (To Kalon I Block) • Murphy-Goode (Reserve)

GRENACHE

Although cultivation of this variety has been in decline in recent years, it is still the 10th-most-widely planted black grape in California. Historically used to produce high volume, entry-level jug wine from plantings primarily in Central Valley, the variety is experiencing a resurgence in quality for single-varietal expressions, primarily from the Central Coast where it has found its home. The wines are produced in a range of styles from full-bodied, rich, and rich and oaked to earlier harvested, lighter, and fresher unoaked examples.

1–3 years (lighter styles), 5–10 years (more structured, top wineries)

Alban Vineyards • Baker & Brain (Le Mistral) • Bonny Doon (Clos de Gilroy) • Casa Dumetz • Epiphany Cellars • Holly's Hill • Tablas Creek

MALBEC

Previously regarded fundamentally as a blending variety, Malbec has been one of the fastest-growing varieties by acreage since 2010. Most likely this is due to the on-going consumer demand for the variety based on the continued international success of Argentinian Malbec, plus the variety's suitability for warmer climates.

2–5 years

Arrowood • Clos du Bois Devil Proof • Francis Ford Coppola (Diamond Collection) • Ilaria • Rancho Sisquoc (Flood Family Vineyard)

MERITAGE

Concocted from "merit" and "heritage", the term *Meritage* was devised as a designation for upmarket Bordeaux-style blends in 1988 and trademarked the following year. Meritage was first coined by Neil Edgar, who was one of 6,000 entrants in an international contest to come up with a name. There are no rules or regulations governing wines that use this trademark, other than that the wineries must be paid-up members of the Meritage Alliance, and the blends must conform to very simple rules of varietal content. Such blends must consist of two or more of the following varieties: Cabernet Sauvignon, Merlot, Cabernet Franc, Petit Verdot, St Macaire, Gros Verdot, Carmenère, and Malbec for reds; and Sauvignon Blanc, Sémillon, Muscadelle, Sauvignon Vert, and Sauvignon Musque for whites; with no single variety making up more than 90 per cent of the blend. Worth noting that many Meritage blends are not marketed as such.

MERLOT

Merlot became one of California's most fashionable and sought-after varieties in the early 1990s, based on the approachability of its velvety texture and lush fruit. A dramatic fall from grace followed, largely accredited to the 2004 film *Sideways*. Despite having the most acreage removed of any variety in the state from 2010 to 2018 it remains California's fourth-most-planted variety.

3–5 years (inexpensive), 5–15 years (top wineries, exceptional wines)

Beringer (Founders' Estate) • Cosentino (Reserve) • Duckhorn • Gainey Vineyards (Limited Edition) • Hartwell • La Jota • Newton (Epic) • Paloma • Pride (Mountaintop Vineyard)

MUSCAT BLANC À PETITS GRAINS

Muscat Blanc à Petits Grains was the second-most-planted white variety by acreage 2010 to 2018, in line with the growth and popularity of Moscato. It is primarily aimed at the mass market from producers such as Gallo and Sutter Home. The grape produces perfumed, flowery-flavoured wines in off-dry through to very sweet and dessert styles. *See* Botrytised and Late-Harvest Wines on the previous page for even sweeter styles.

Upon purchase

Bonny Doon (Vin de Glaciere) • Eberle • Fresno State Winery (John Diener Vineyard) • Maurice Car'rie • Rosenblum (Muscat de Glacier) • Robert Pecota (Muscato di Andrea) • St Supéry (Moscato)

PETITE SIRAH

Although California's growers had referred to this grape as Petite Sirah since the 19th century, it was unknown to consumers until the 1960s, when varietal wines began to take off. Petite Sirah quickly gained in popularity on the back of the Rhône's Syrah, only for prices to plummet ridiculously in the 1970s, when it was identified as the Durif, a lowly French variety. It was the same wine, but its humble origins obviously lacked cachet for new wine consumers of that era. In 1997 Dr Carole Meredith used DNA "fingerprinting" to determine that only four of the seven Petite Sirahs in the UC Davis collection were in fact Durif, and that one was identified as true Syrah. As consumers began to wonder if their favourite Petite Sirah might not in fact be Syrah, the prestige and price of this varietal began to soar. Yet, though 10 per cent of Petite Sirah is indeed Syrah, the mix is found on a vine-by-vine basis. Rather than find one vineyard of Syrah for every nine of Durif, growers find they have 10 per cent Syrah in their Petite Sirah vineyards. And not just Syrah. Despite increased prices, the purple-coloured wine made by Petite Sirah, with its floral, black-fruit aromas, is not considered the equal of a true Syrah and is far more commonly blended for its deep colour and tannic structure rather than produced as a single varietal.

4–8 years

Biale (Thomann Station) • Delectus • Girard Winery • Orin Swift (Machete) • Quixote • Ridge (Lytton Estate) • Rosenblum (Rockpile) • Switchback Ridge • Turley Cellars (Petite Sirah)

PINOT GRIS

The fastest-growing varietal by acreage from 2010 to 2018, the vast majority of Pinto Gris is grown in Central Valley and produced as crowd-pleasing, ripe, fruity Pinot Grigio. The best expressions of Pinot Gris are found in the cooler regions of Sonoma and Mendocino.

Upon purchase

Cline • Eden Rift (Terraces) • Etude (Carneros Pinot Gris) • Luna (Pinot Grigio) • Monteviña (Pinot Grigio) • Navarro (Pinot Gris) • Mondavi (Private Selection) • Windchaser (Anderson Valley)

PINOT NOIR

This Burgundian grape has found a natural home in parts of California, where it is most widely planted in Sonoma, Monterey, and Santa Barbara. The variety's popularity has surged over the last few years, reflected in its rise to become the second-most-widely-planted black grape in California in 2018, overtaking Zinfandel and Merlot since 2010. A wide range of styles is produced, from generously fruity California regional blends, such as the wildly successful Constellation-owned Meiomi, to the internationally acclaimed, single-site examples thriving in the cooler subregions, most notably Russian River Valley, Carneros, and Santa Rita Hills.

1–3 years (inexpensive), 5–10 years (top wineries), 15+ years (exceptional wines)

Au Bon Climat (La Bauge Au-dessus) • Brewer-Clifton (Melville) • Calera • Cuvaison • Etude • Fiddlehead • Hirsch • Kistler (Occidental Vineyard Cuvée Elizabeth) • Littorai • Marcassin • Rhys • Sandford • Siduri (Pisoni Vineyard) • Williams Selyem (Rochioli Riverblock Vineyard)

RED BLENDS

The current trend for Red Blends sits at the more commercial end of the market, usually classic-based Bordeaux varieties, as well as Zinfandel and Petite Sirah. The wines are mainly brands created by large producers using the generic California denomination that enables them to source fruit throughout the state and create consistent, crowd-pleasing wines. The style is big, ripe, soft, and intensely fruity and usually with a degree of residual sugar to round them out. The global success of Gallo-owned Apothic, Constellation's The Prisoner, and Trinchero's Ménage à Trois epitomise this current market development. At the more premium end of the market the wines are frequently labelled as "Proprietary Blend" or "Meritage" (see Meritage at left). There is no strict legal definition for the use of Proprietary Blend; however, Meritage must consist of Bordeaux varieties. Rhône blends are also increasingly common from the Central Coast.

1–3 years (inexpensive), 5–12 years (top wineries), 8–25 (or more) years (exceptional wines)

Beringer (Alluvium Quantum, Dark Red Blend) • Colgin (Cariad) • Dalla Vale (Maya) • Harlan • L'Aventure (Cuvée) • Peter Michael (Les Pavots) • Pride Mountain Vineyards (Reserve Claret) • Ridge (Three Valleys) • Sanguis • Sean Thackrey (Pleiades) • Tablas Creek

SAUVIGNON BLANC

See Fumé Blanc.

SYRAH

Plantings of Syrah skyrocketed in the late 1990s and early 2000s because many believed the grape could one day rival the popularity of Cabernet Sauvignon. Unfortunately, as acreage increased, the bulk of the wines produced were generic and modest in quality, a reputation that has arguably remained with the general consumer and still tarnishes the grape to this day. Despite falling planted area from growers re-grafting to more popular varieties, a number of top-quality boutique producers remain committed to the grape, which has found its home primarily in the cooler areas of the Central Coast.

3–10 years

Alban (Lorraine Vineyard) • Andremilly • Araujo (Eisel Vineyards • Donelan • Jaffurs • Kongsgaard (Hudson Vineyard) • MacLaren • The Ojai Vineyard (Thompson Vineyard) • Pax (Adler Springs The Terraces)

VIOGNIER

This is an example of another Rhône grape variety that has shown it can excel in California. Viognier was non-existent in California until 1985, when the first vines were planted by Joseph Phelps. At that time, just 4.5 hectares (11 acres) of these vines existed. Now there are more than 1,000 hectares (2,700 acres), making it the ninth-most-widely-planted white variety in the state. Viognier is not over-blessed with acidity, but when it is grown at low yields and harvested physiologically ripe – rather than sugar-ripe – there is a certain acid balance that, together with the proper fruit structure, will provide sufficient grip and length to avoid any connotation of flabbiness. Some of the best examples are fermented and aged in well-managed oak, adding further textural and aromatic complexity.

Upon purchase

Alban (Alban Estate Vineyard) • Arrowood (Saralee's Vineyard) • Calera • Cold Heaven • Eberle (Mill Road Vineyard) • Joseph Phelps • Pride Mountain Vineyards • The Ojai Vineyard

ZINFANDEL

This variety was once thought to be America's only indigenous *Vitis vinifera* grape, but it was later positively identified by Dr Carole Meredith as the Primitivo grape of southern Italy by Isozyme "fingerprinting", a method of recording the unique pattern created by the molecular structure of enzymes found within specific organisms. This led to a conundrum, however, because the earliest records of the Primitivo variety in Italy date from the late 19th century, whereas the Zinfandel name appears in the nursery catalogue, dated 1830, of William Prince of Long Island. One clue was that Italian growers often referred to their Primitivo as a "foreign" variety. This led Meredith farther afield and, with the help of Croatian researchers, she was able to announce in 2002 that the Zinfandel and Primitivo are in fact an obscure variety called Crljenak Kaštelanski (pronounced Sirl-YEN-ak Kastel-ARN-ski). Referred to as Crljenak, for short, this grape was once widely planted throughout Dalmatia and on more than 1,000 islands, but it has mostly been replaced by Croatia's best-known variety, Plavac Mali (the result of a cross-pollination of Crljenak and Dobričić).

While acreage of Zinfandel has declined since the early 2000s, plantings of Primitivo have spurted, increasing five-fold since 2010, as producers explore new clones and vinify the Primitivo grape separately from the Zinfandel. Depending on the way it is vinified, Zinfandel produces many different styles of wine, from rich and dark to light and fruity and from dry to sweet and white to rosé, as well as dessert wines or sparkling wines. One such reason for this flexibility in styles is a fundamental characteristic of uneven grape ripening within bunches. When most of the bunch achieves perfect ripeness, some grapes are green, but if the grower waits for these to ripen, the shoulder clusters quickly raisin. This is also a factor in the sometimes elevated (15 per cent plus) alcohol levels given the high levels of sugar accumulation in the berries. There are several ways to overcome this, but they are all labour-intensive, seemingly wasteful, and inevitably costly, which is why great Zinfandel is never cheap. Great Zinfandel is as rich and deep-coloured as only California could produce, with ripe, peppery-spicy fruit, liquorice intensity, and a chocolate-herb complexity. It is often blended with Petite Sirah, which contributes acid and tannin backbone, and commonly aged in American oak, which contributes a layer of coconut aromatics.

2–5 years (good but inexpensive wines), 5–15 years (expensive, more structured styles), 20+ years (exceptional wines)

Carisle • DeLoach (Heritage Collection) • Edmeades Gary Farrell (Grist Vineyard) • Hartford Court (Hartford Dina's Vineyard) • Heitz • Joseph Swan • Nalle (Reserve) • Ridge Vineyards • Rosenblum • Turley Cellars • Williams-Selyem (Forchini Vineyard)

Mendocino County

The most northerly qualitative region within California, Mendocino was first planted with vines by settlers in the 1850s during the great Gold Rush. This coastal region has a huge variety of climates due to the combination of Pacific influence and the dividing Mendocino Range (part of the Pacific Coastal Range) forming a barrier between the warmer and drier inland areas.

With around 7,600 hectares (18,800 acres) under vine and just over 100 wineries, Mendocino is relatively sparsely planted, yet renowned for its overall quality. Champagne Louis Roederer arguably put Mendocino and the cool-climate AVA of Anderson Valley on the map in the 1980s, when it invested in a 200-hectare (500-acre) vineyard and winery, realising the potential to produce world-class sparkling wines. The 1980s also saw the start of the county's focus on "green" credentials, boasting the first winery, Frey Vineyards, to be certified organic and subsequently biodynamic in 1996. Mendocino remains an industry leader in this field with 20 per cent of wine grapes currently certified organic, representing a third of California's total organic vineyard area. Today red grapes account for around two-thirds of total plantings, led by Pinot Noir, Cabernet Sauvignon, and Zinfandel. Chardonnay is by the far the most popular white variety, with around three-quarters of total white plantings devoted to it, but with aromatic varieties, such as Sauvignon Blanc and Riesling, gaining favour in cooler sub-regions.

FACTORS AFFECTING TASTE AND QUALITY

LOCATION
Mendocino County is the most northerly of the major viticultural coastal counties of California at 160 kilometres (100 miles) northwest of San Francisco.

CLIMATE
The mountain ridges surrounding the Upper Russian and Navarro rivers climb as high as 1,070 metres (3,500 feet), and they form a natural boundary that creates the reputed transitional climate of Mendocino. This climate is unusual in that either coastal or inland influences can dominate for long or short periods, although it generally has relatively warm winters and cool summers. This provides for a growing season with many warm, dry days and cool nights. The Ukiah Valley has the shortest, warmest growing season north of San Francisco.

ASPECT
This region consists of mainly flat ground on valley bottoms or gentle, lower slopes at a height of 76 to 445 metres (250 to 1,460 feet), with some rising to 490 metres (1,600 feet). The vines generally face east, though just south of Ukiah they face west.

SOIL
There are deep, diverse alluvial soils in the flat riverside vineyards of Mendocino County, gravelly-loam in parts of the Russian River Valley, and a thin scree on the surrounding slopes.

VITICULTURE & VINIFICATION
The average growing season in Mendocino is 268 days, compared with 308 in Sonoma County (bud-break is 10 days earlier here), and 223 days in Lake County.

GRAPE VARIETIES
Primary varieties: Cabernet Sauvignon, Chardonnay, Gewürztraminer, Merlot, Pinot Gris, Pinot Noir, Sauvignon Blanc, Syrah, Zinfandel

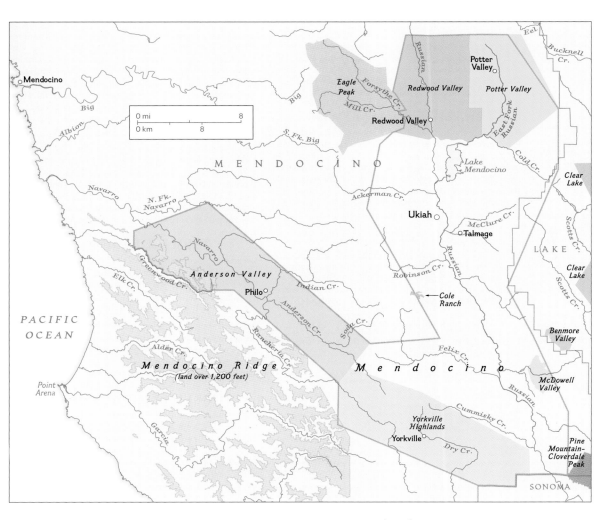

THE VITICULTURAL AREAS OF MENDOCINO COUNTY
(see also p577)
Along with the Mendocino AVA, the county is home to several smaller appellations, including the northerly AVAs of Dos Rios and Covelo (*see Extended Area below*). This area's northerly location need not mean harsher microclimates – Mendocino AVA is known for the cultivation of Mediterranean-climate grapes, such as Carignan, Charbono, Grenache, and others. Inland regions are protected by the mountains, though the Anderson Valley, known for its Pinot Noir, is cooler.

A WINE TASTING AT HANDLEY CELLARS OFFERS A SELECTION OF RED AND WHITE WINES
Located in the town of Philo in the Anderson Valley AVA, this small family-owned winery specializes in Pinot Noir and Alsace white varietals, such as Gewürztraminer and Zinfandel.

THE APPELLATIONS OF
MENDOCINO COUNTY

ANDERSON VALLEY AVA

Anderson Valley's coastal-influenced microclimate is cooler than in the rest of Mendocino and is seen as one of California's coolest winegrowing areas. The Pacific Ocean proximity contributes to high diurnal temperature fluctuations, which are ideally suited to retain acid freshness in cooler-climate varieties, such as the Riesling, Pinot Noir, and Gewürztraminer vines that thrive in the region. The valley's soil consists of more than 20 alluvial soils, providing the diversity required for wineries such as Roederer Estate and Scharffenberger to produce a range of base wines, the building blocks of any fine-quality sparkling wine. Farther inland, there is sufficient warmth to ripen Cabernet Sauvignon and Zinfandel.

COLE RANCH AVA

North America's smallest appellation, consisting of just 25 hectares (62 acres) situated in a small, narrow valley between Russian River and Anderson Valley. In 2019 the entire acreage of Cabernet Sauvignon, Chardonnay, and Riesling went up for sale for 3.3 million dollars. Cole Ranch soils range from gravelly-clay loam to gravelly-silty clay.

COVELO AVA

Located about 72 kilometres (45 miles) north of Ukiah, the appellation comprises Round Valley, Williams Valley, and the surrounding foothills. The bowl shape of the valleys combined with surrounding peaks, which effectively cut off any cooling coastal influence, result in a continental climate and shorter growing season than in most of Mendocino County. The soils are also very deep loam.

DOS RIOS AVA

Soils and climate conditions in Dos Rios are markedly different from other Mendocino AVAs, consisting of steep, rocky slopes and a unique combination of continental and maritime influences. Vin de Tevis is the only winery in the appellation, and its vineyards are almost exclusively planted to red varieties of Cabernet Franc, Cabernet Sauvignon, Merlot, and Zinfandel.

EAGLE PEAK AVA

Situated in California's Coastal Range, just west of Redwood Valley AVA, the region takes its name from nearby Eagle Peak summit. The area is mountainous, with only around 49 hectares (120 acres) of vines planted, divided between five producers, including the sole winery of Masút, owned and run by Jake and Ben Fetzer.

MCDOWELL VALLEY AVA

Overlooking the Russian River to the west, the McDowell Valley perches at around 300 metres (1,000 feet). The microclimate here warms up when other areas are experiencing spring frosts, yet the region is slightly cooler than surrounding areas during the growing season. Rhône varietals are its speciality – Marsanne and Viognier for whites, with Grenache and Syrah the predominant reds. Zinfandel is also successful, with some century-old plantings.

MENDOCINO AVA

This may only be used for wines produced from grapes that are grown in the southernmost third of the county. It encompasses the following AVAs, plus surrounding vineyards: Anderson Valley, Cole Ranch, McDowell Valley, Potter Valley, Redwood Valley, and Yorkville Highlands.

MENDOCINO COUNTY AO

This appellation (not an AVA) covers the wines from anywhere within Mendocino County.

MENDOCINO RIDGE AVA

The only non-contiguous AVA in the US applying exclusively to areas at 365 metres (1,200 feet) or higher, with all lower-altitude areas covered by the Mendocino County appellation or Anderson Valley AVA. Given the altitude, the patchwork of tiny vineyards sit above the fog line, resulting in a combination of long ripening sunshine hours and a high diurnal temperature range that preserves acidity levels. The region produces some of the most highly regarded Zinfandel in California, many from old vine plantings dating back to the late 19th century, when the area was first planted.

POTTER VALLEY AVA

Situated to the east of Redwood Valley, this inland valley is shielded from the cooling coastal influence by the surrounding hills. It benefits from a great diurnal temperature swing, which mitigates the high daytime temperatures. As a result, aromatic varieties of Pinot Noir, Sauvignon Blanc, Gewürztraminer, and Riesling are increasingly finding a home alongside the most widely planted Chardonnay.

REDWOOD VALLEY AVA

This area is where the first Mendocino vines were planted. and has been known as Redwood Valley since it was settled in the mid-1850s. A gap in the Coastal Range funnels the cooling Pacific influence into the valley, leading to an extended, more gradual ripening of the fruit. The resulting wines are structured and refined, with most notable success from Cabernet Sauvignon and Zinfandel.

TALMAGE REGION AVA

Located on the eastern edge of the Ukiah Valley, the region has gravelly loam soils and is known as a source of quality Cabernet Sauvignon and Zinfandel, as well as Chardonnay.

UKIAH VALLEY AVA

Encompassing the area in proximity to the town of Ukiah, the county seat of Mendocino. The Russian River flows the length of the valley, providing both a fertile flood plain and surrounding benchland. Mainly consists of small vineyard plots, many less than 2 hectares (5 acres).

YORKVILLE HIGHLANDS AVA

Geographically, but not topographically, the Yorkville Highlands AVA appears to be an extension of the Anderson Valley AVA. The porous, rocky soil with high gravel content and a distinctive climate led to the AVA approval in 1998. The climate is similar to the middle of the Anderson Valley, with moderate temperatures suiting Sauvignon Blanc, as well as yielding increasingly successful results from Cabernet Sauvignon and Merlot.

THE WINE PRODUCERS OF
MENDOCINO COUNTY

BAXTER WINERY
Philo
★★

Run by Philip Baxter Senior and Philip Baxter Junior, both winemakers with many years' experience in California and Burgundy. Baxter specialises in single-vineyard expressions of Pinot Noir sourced throughout Mendocino.

✔ *Pinot Noir*

BONTERRA
Ukiah
★

Part of the Concha y Toro group, with strong organic/biodynamic philosophy.

✔ *Chardonnay • Merlot*

EDMEADES VINEYARDS
Philo
★

Part of the Kendall-Jackson group.

✔ *Zinfandel*

FETZER VINEYARDS
Redwood Valley
★

Owned by Concha y Toro, Fetzer produces value-for-money wines and is one of the pioneers of sustainable production practices in the industry.

✔ *Cabernet Sauvignon • Gewürztraminer • Merlot • Zinfandel*

FREY
Redwood Valley
☆

The oldest and largest fully organic winery in the United States, Frey was set up in 1980 and was the first US winery to make biodynamic wines.

✔ *Petite Sirah Z • Sangiovese • Syrah*

GOLDENEYE
Philo
★

Established in 1996 by the pioneering owners of Duckhorn Vineyards (Napa Valley) to exclusively explore the quality potential of Pinot Noir in Anderson Valley. Widely considered to produce some of the region's top examples.

✔ *Pinot Noir*

GREENWOOD RIDGE VINEYARDS
Philo
☆

This is an underrated winery best known for its Riesling, particularly its Late Harvest, although other varieties are well worth a look.

✔ *Cabernet Sauvignon • Classic red blend (Home Run Red) • Pinot Noir • White Riesling (Late Harvest)*

HANDLEY CELLARS
Philo
★

This Anderson Valley winery is an excellent, underrated *méthode champenoise* specialist. Handley's crisp Vintage Brut has creamy-vanilla richness, but he also makes flavourful Pinot Noir and Sauvignon Blanc of increasing elegance.

✔ *Pinot Noir (Estate Reserve) • Sauvignon Blanc (Handley Vineyard) • Syrah • Sparkling wine (Vintage Brut Rose)*

HUSCH VINEYARDS
Philo
☆

Founded by Tony Husch, this winery was sold in 1979 to Hugo Oswald, the current owner and winemaker, who also has large vineyard holdings in the Anderson

Valley and Ukiah Valley, most notable for high-quality white varietals.

✔ *Chardonnay • Chenin Blanc (La Ribera) • Muscat Canelli (La Ribera) • Sauvignon Blanc (La Ribera)*

MAPLE CREEK WINERY
Yorkville
★

Artist Tom Rodrigues left Marin County in 2001 to establish this boutique winery in the Yorkville Highlands of Anderson Valley. His emphasis is on French-style winemaking with offerings of Pinot Noir, Chardonnay, Zinfandel, and Merlot

✔ *Chardonnay (Estate) • Pinot Noir (Lost Creek) Zinfandel (Largo Ridge)*

MASÚT
Redwood Valley
★☆

Founded by brothers Ben and Jake Fetzer, grandsons of Barney Fetzer who founded Fetzer Vineyards, to produce estate Pinot Noir. Planting began in the late 1990s with their work eventually leading to the classification of Eagle Peak AVA, where Masút is the only winery.

✔ *Pinot Noir*

NAVARRO VINEYARDS
Philo
★

This winery is owned and run by the Navarro family, who established the estate in 1974. With a dedication to sustainable farming practices, Navarro produces a wide range of good value, varietal-driven wines, including late-harvest and sparkling styles.

✔ *Pinot Gris • Pinot Noir • Sparkling wine (Gewürztraminer)*

PARDUCCI WINE CELLARS
Ukiah
☆

Parducci is a large winery, with more than 142 hectares (351 acres), which was founded in Sonoma in 1918 and moved to Ukiah in 1931, where a new winery was built in preparation for the end of Prohibition. The wines are clean and fruity and possess a good depth of flavour for their price.

✔ *Petite Sirah*

PHILLIPS HILL
Philo
★

Owned and run by Toby Hill, artist-turned-winemaker, who specialises in single-site expressions of Pinot Noir sourced from top growers primarily in Anderson Valley.

✔ *Pinot Noir*

ROEDERER ESTATE
Philo
★★☆

The first New World venture to achieve a sparkling wine comparable to Champagne, highlighted in the flagship Roederer Estate (or Quartet as it is known on export markets). L'Ermitage is the greatest wine here, but the standard non-vintage Brut can be almost as stunning.

✔ *Entire range*

SCHARFFENBERGER CELLARS
Ukiah
★☆

Founded by John Scharffenberger in 1981 and part-owned by Pommery until it was transferred to Veuve Clicquot and the name was changed to Pacific Echo. Following the purchase in July 2004 by the Roederer-owned Maisons Marques & Domaines USA, the name was returned to the original Scharffenberger.

✔ *Sparkling wine*

TOULOUSE VINEYARDS
Philo
★

This producer's 21 acres (hectares) are set out in the Anderson Valley at elevations of about 90 to 150 metres (300 to 500 feet). Making wine from 2002, Toulouse employs organic practices in the vineyard and allow wild yeasts to complete fermentation.

✔ *Pinot Noir*

A SIGN DIRECTS VISITORS TO THE MAPLE CREEK WINERY
Maple Creek releases wines under its Artevino label, which reflects the interests of its founder: art and wine. Owner Tom Rodrigues paints an Art Nouveau–inspired design for each of his varietal wines.

Sonoma County

Situated between the Mayacamas Mountains in the east and Pacific Ocean in the west, six fertile valleys combine with a vast diversity of climates to make Sonoma County one of California's most important and renowned wine-producing areas.

Sonoma has some of the earliest-recorded vine plantings in California, dating back to 1812 with the arrival of Russian colonists. The region's wine production, and reputation, grew throughout the 19th century, including the establishment of Buena Vista winery in 1957 by Ágoston Haraszthy, "the Father of the Californian Wine Industry", in Sonoma Valley. Prohibition saw a switch to entry-level bulk production that took until the late 1960s to recover, as demand for quality wine returned. In 1969 Russell Green, a former oil mogul and owner of a fast-growing vineyard in the Alexander Valley, purchased Simi, a once-famous winery founded in 1876 that was in decline at the time. Green had great plans for Simi, many of which he successfully carried out, but soaring costs forced him to sell up in 1973. During those four brief years, however, he managed to restore the pre-Prohibition reputation of the old winery by creating a new genre of high-quality varietal Sonoma wines. With this achievement, he made other winemakers in the district eager to improve the quality of their own wines.

Green's activity attracted new blood to the area, and by 1985 there was a remarkable total of 93 wineries. By 1989 wine grapes became the county's top agricultural product by revenue, and by 1999 the number of wineries had swollen to 180. Since then the number has increased to in excess of 425 across 18 AVAs.

FACTORS AFFECTING TASTE AND QUALITY

LOCATION
North of San Francisco, between Napa and the Pacific.

CLIMATE
Extremes of climate range from warm in the north of the county (Region III, see p577) to cool in the south (Region I), mainly due to ocean breezes. Fog is prevalent, with a significant influence on temperature fluctuations – drops of up to 5°C (42°F) are common in Russian River Valley – as well as sunshine exposure, all dependant on proximity to the coast, orientation of coastal ranges, and altitude.

ASPECT
The Sonoma Creek drains into the San Francisco Bay, and the Russian River flows directly into the Pacific Ocean. Vines are grown at diverse altitudes, with steeper and higher slopes constantly being explored for the impact of elevation and fog line on temperature and sunshine hours.

SOIL
The soil situation varies greatly, from low-fertile loams in the Sonoma Valley and Santa Rosa areas to highly fertile alluvial soils in the Russian River Valley, with limestone at Cazadera, a gravelly soil in Dry Creek, and vent-based volcanic soils within the fall-out vicinity of Mount St Helena.

VITICULTURE & VINIFICATION
More than 60 grape varieties are grown in Sonoma County, capable due to the diversity of soil and climate. Cabernet Sauvignon is the most widely planted variety in the county, followed by Pinot Noir and Chardonnay, gaining great acclaim for more restrained, elegant styles in cooler climate locations.

GRAPE VARIETIES
Cabernet Sauvignon, Chardonnay, Merlot, Pinot Blanc, Pinot Gris, Pinot Noir, Sauvignon Blanc, Syrah, Viognier, Zinfandel

THE VITICULTURAL AREAS OF SONOMA COUNTY *(see also p577)*
One of California's most important wine regions, Sonoma has a varied climate with different soils, which produces a broad spectrum of wines. It is perhaps aptly named, as the name *Sonoma* is derived from the local Wintun Native American word for "nose", and there are many Sonoma wines that appeal to the sense of smell as well as taste.

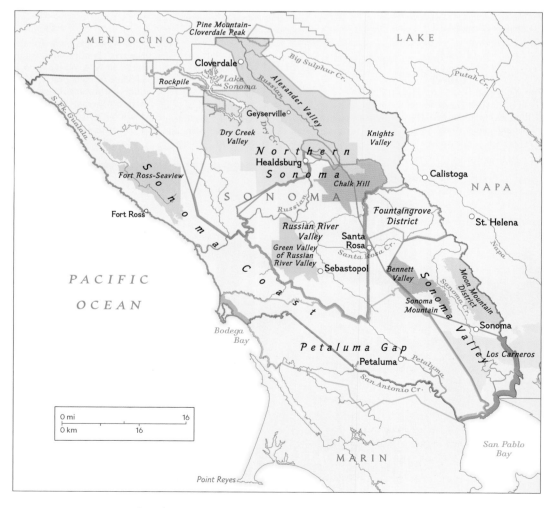

CABERNET SAUVIGNON VINES
One of the primary varieties grown in Sonoma County, these inky black grapes were the favourites of California winemakers and consumers in the 1950s and 1960s.

THE APPELLATIONS OF
SONOMA COUNTY

ALEXANDER VALLEY AVA

Located in the northeast of the county, this appellation extends from the banks of the Russian River into the foothills of the Mayacamas Mountains. In 1986 its boundaries were extended so that it overlapped the Russian River AVA and were extended farther in 1990 into the foothills east of Geyserville. Formerly part of Dry Creek AVA, the Gill Creek watershed area was reclassified as Alexander Valley AVA in 2001. Cabernet Sauvignon is the specialist variety, ideally suited to gravel and alluvial soils the valley. Merlot, Nebbiolo, Sangiovese, and Chardonnay are all highly successful, whilst the area is additionally home to some of the county's best old-vine Zinfandel.

BENNETT VALLEY AVA

Volcanic, clay-based soils are ideally suited to Merlot, which, combined with a moderately cool climate producing an extended ripening period, excels in the region. Chardonnay is also widely planted and benefits from the longer growing season produced by the cooling Pacific winds and fog that make their way into the Valley from the Petaluma Gap.

CARNEROS AVA

Carneros was the first wine region in the United States to be founded purely on climate and topography, rather than political boundaries. The appellation covers an area of low, rolling hills that straddle the counties of Sonoma and Napa. This was originally sheep country, but the cool sea breezes that come off San Pablo Bay to the south provide an excellent fine-wine growing climate, especially for Chardonnay and Pinot Noir, and for both still and sparkling wines. Merlot also stands out on the areas of clay-based soils.

CHALK HILL AVA

Located in the northeast of Russian River AVA, the region has a unique soil, climate, and topography that clearly separates it from other parts of the Russian River Valley. There is no chalk here: the soil's whiteness is in fact derived from volcanic ash with a high quartzite content. Emitted by Mount St Helena over many centuries, the ash has mixed with local sandy and silty loams to provide a deep soil that is not particularly fertile. The area is protected by a thermal belt that promotes a September harvest, compared with October in surrounding areas, whilst altitudes range from 60 to 400 metres (200 to 1,300 feet). Chardonnay and Sauvignon Blanc are considered the most successful whites, with Cabernet Sauvignon the primary red variety.

DRY CREEK VALLEY AVA

This appellation faces that of the Alexander Valley across the Russian River and at 26 kilometres (16 miles) long and 3.2 kilometres (2 miles) wide is one of the smallest enclosed AVAs in the United States. Its climate is generally wetter and warmer than surrounding areas, with a longer growing season than the Russian River appellation to the south. Varietal suitability is very flexible, stretching from Sauvignon Blanc to Zinfandel.

FORT ROSS-SEAVIEW AVA

This appellation sits within the greater Sonoma Coast AVA. Differentiated by the close proximity to the Pacific Ocean and mountainous terrain, it offers an array of aspects, exposures, and elevations, with most vineyards planted at or above 240 metres (800 feet). Pinot Noir and Chardonnay are most highly acclaimed and sought after.

FOUNTAINGROVE DISTRICT AVA

The appellation's eastern location in the county and warmer climate is moderated by elevation – plantings can reach up to 600 metres (2,000 feet) – combined with a gap in the Coastal Range in Santa Rosa that facilitates cooling winds and fog. As such, the region experiences greater coastal influence than Alexander Valley but remains warmer than Russian River Valley. Bordeaux varietals of Cabernet Sauvignon, Merlot, and Cabernet Franc are most highly regarded from the region, along with Syrah and Zinfandel.

GREEN VALLEY OF THE RUSSIAN RIVER VALLEY AVA

This area within the Russian River Valley AVA is one of the smallest and coolest AVAs in Sonoma County, due to most vineyards facing east and the area's persistent fog. The AVA is tied in name to the Russian River Valley to avoid confusion with Green Valley in Solano County. Established around strict soil and climatic conditions, the region produces some of Sonoma's highest quality cool-climate Pinot Noir and Chardonnay.

KNIGHTS VALLEY AVA

Knights Valley sits next to Mount St Helena and is one of the most remote AVAs in the county. The vines grow on rocky and gravelly soil of low fertility at altitudes that are generally higher than those in the adjacent AVAs, making it ideal Cabernet country.

MOON MOUNTAIN AVA

A high elevation region within Mayacamas on the east of the Sonoma Valley AVA, Moon Mountain is recognised for its vineyards planted at altitudes ranging from 120 to 670 metres (400 to 2,200 feet). It possesses ideal growing conditions for top-quality, concentrated, structured Cabernet Sauvignon and other Bordeaux varietals, including Merlot, Cabernet Franc, Petit Verdot, and Malbec. Some of California's oldest Zinfandel and Cabernet Sauvignon vines are located in the iconic Monte Rosso vineyard, originally planted in the late 19th century.

NORTHERN SONOMA AVA

The large appellation of Northern Sonoma completely encapsulates the following AVAs: Alexander Valley, Chalk Hill, Dry Creek Valley, Knights Valley, Russian River Valley, most of Green Valley, and areas of Rockpile and Pine Mountain-Cloverdale Peak. It is separated from the Sonoma Valley appellation to the south by the city of Santa Rosa.

PETALUMA GAP AVA

Only 40 kilometres (25 miles) north of San Francisco, the region is the most southerly in Sonoma, with a proportion crossing into Marin County, as AVAs increasingly focus on climate and topographical classification over political boundaries. Petaluma Gap was designated in 2018, and wineries can choose to label their wines as either Sonoma Coast or Petaluma Gap. The region is the gateway for the cooling winds and fog, and hence is one of the coolest in the county. Pinot Noir, Chardonnay, and increasingly Syrah are most widely planted, with a large proportion of fruit used in sparkling wines.

PINE MOUNTAIN – CLOVERDALE PEAK AVA

The appellation straddles the northern borders of Alexander Valley AVA and Northern Sonoma AVA. It was delineated based on the mountainous soils, high-elevation vineyards at 480 to 910 metres (1,600 to 3,00 feet), and steep topography that clearly differentiate the area from the growing conditions of the Alexander Valley floor. Cabernet Sauvignon is ideally suited to the mountainous climate and is by far the most widely planted.

ROCKPILE AVA

Located between the Mendocino County border and Lake Sonoma, this high-altitude region sits above Dry Creek Valley. Vineyards planted up to elevations of 640 metres (2,100 feet) produce intensely concentrated reds, primarily famed for its "signature clone" of Zinfandel.

A VINTAGE WINE PRESS SITS IN A COLOURFUL SUMMER GARDEN AT CLINE CELLARS
This Sonoma winery is part of the Carneros AVA, the first American wine appellation delineated solely by its unique climate and geographic features.

RUSSIAN RIVER VALLEY AVA

"Russian River" began appearing on wine labels in 1970, although vineyards in this region date from the 19th century. The early-morning coastal fog provides a cooler growing season than that of neighbouring areas, making the Russian River Valley very well suited to Pinot Noir and Chardonnay – and indeed it produces some of California's finest cool-climate examples of both varieties.

SONOMA COAST AVA

An appellation covering 1,940 square kilometres (750 square miles), Sonoma Coast AVA is made up of the area directly inland from the length of Sonoma's Pacific coastline – the AVA's western boundary – running from San Pablo Bay to the Mendocino border. It is significantly cooler than other areas owing to the persistent fog that envelops the Coast Ranges, the mountains that are within sight of the Pacific Ocean. The region is celebrated for high-quality cool-climate Pinot Noir and Chardonnay.

SONOMA MOUNTAIN AVA

This tiny appellation is within the Sonoma Valley AVA. Sonoma Mountain AVA has a thermal belt phenomenon that drains cold air and fog from its steep terrain to the slopes below, creating a climate that is characterized by more moderate temperatures than the surrounding areas. Elegant, structured Cabernet Sauvignon and Chardonnay are the specialist varieties.

SONOMA VALLEY AVA

Sonoma Valley is in the south of the county, enclosed by the Mayacamas to the east and the Sonoma Mountains to the west. As such, rainfall is lower than elsewhere in the county and fog rarely penetrates from the coast. It is red wine country, particularly for Cabernet Sauvignon and Zinfandel.

THE WINE PRODUCERS OF
SONOMA COUNTY

A RAFANELLI WINERY
Healdsburg
★

It was always said that Americo Rafanelli, who died in 1987, made wine "as they used to in the olden days". Well, he would be proud of the blockbusters made by his son, David, and now his daughter team, Rashell and Stacy, from this small, unirrigated, hillside vineyard.

✓ *Zinfandel*

ALEXANDER VALLEY VINEYARDS
Healdsburg
★

An underrated winery built on the original homestead of Cyrus Alexander, who gave his name to the Alexander Valley, now a famed AVA. Most products are well worth buying as value-for-money, everyday drinking wines, but some of tem are much finer than that, despite their modest prices.

✓ *Cabernet Sauvignon • Chardonnay • Classic red blend (Cyrus) • Merlot • Pinot Noir*

ARROWOOD VINEYARDS
Glen Ellen
★★

Richard Arrowood makes very stylish wines, and the quality gets better each year as he hones down his varietals. It has an excellent-value second label (Grand Archer). Arrowood is now part of Jackson Family Wines.

✓ *Cabernet Sauvignon (Réserve Spéciale) • Chardonnay (Cuvée Michel Berthod) • Viognier*

BR COHN
Glen Ellen
☆

Sometimes you have to spit the splinters out of these wines, but you cannot deny their quality and have to admire the producer's enthusiasm.

✓ *Cabernet Sauvignon (Olive Hill) • Merlot (Sonoma Valley)*

BENZIGER FAMILY WINERY
Glen Ellen
★☆

High-profile Mike Benziger has handed over reins to his brother Chris Benziger, and both this winery and the highly successful Imagery winery now sit under The Wine Group umbrella. Benziger concentrates on producing more focused, quality wines from their four estates.

✓ *Cabernet Sauvignon (Sonoma Coast) • Chardonnay (Carneros) • Pinot Noir (Sonoma Coast) • Syrah (California)*

BUENA VISTA WINERY
Sonoma
☆

Haraszthy's original winery (*see* p576) is now owned by Boisset Collection. Buena Vista's Carneros Estate wines are usually good value, particularly the Pinot Noir and Cabernet Sauvignon red wines.

✓ *Cabernet Sauvignon*

CARMENET VINEYARD
Sonoma
★

Owned by Chalone Vineyards of Monterey, the original is now split into labels Carmenet, Moon Mountain Vineyard, and Dynamite and offer some excellent-value wines.

✓ *Cabernet Sauvignon (North Coast Dynamite) • Merlot • Sauvignon Blanc (Reserve)*

CHALK HILL WINERY
Healdsburg
★

This excellent operation was known as Donna Maria Vineyards when the owner only grew and sold grapes but became Chalk Hill Winery when it first began to make and sell wines in 1981. Chardonnay has always been a major focus.

✓ *Cabernet Sauvignon (Estate Bottled) • Chardonnay (Estate Bottled)*

CHÂTEAU SOUVERAIN
Geyserville
☆

This winery offers a good-value basic range. Bought by Gallo in 2015, Souverain now make only single varietal wines: Cabernet Sauvignon, Chardonnay, Merlot, and Sauvignon Blanc.

CHATEAU ST JEAN
Kenwood
★☆

Owned by Treasury Wine Estates, 2017 saw the 25th anniversary of Chateau St Jean's flagship wine, the Cinq Cépages (a classic Bordeaux blend consisting of Cabernet Sauvignon, Merlot, Cabernet Franc, Malbec, and Petit Verdot).

✓ *Cabernet Sauvignon (Sonoma Coast, Sonoma Coast Reserve) • Chardonnay (Robert Young Vineyard Reserve) • Classic red blend (Cinq Cépages) • Merlot (Sonoma Coast) • Riesling (Special Late-Harvest)*

CHRISTOPHER CREEK
Healdsburg
★☆

Formerly known as the Sotoyme Winery, Christopher Creek was snapped up by Englishman John Mitchell, who makes a wonderfully deep, rich, smoky Petite Sirah and a silky, stylish Syrah. The 2019 Kincade fires affected the vineyards

✓ *Petite Sirah (Russian River Estate Bottled) • Syrah (Reserve, Russian River Estate Bottled) • Zinfandel (Dry Creek Valley)*

CLINE CELLARS
Sonoma
★☆

Excellent Rhône-style wines range from the generous Côtes d'Oakly to the serious, oaky Mourvèdre, but it is the superb range of full-throttle, all-American Zinfandels that Cline Cellars is truly famous for.

✓ *Carignane (Ancient Vines) • Mourvèdre (Ancient Vines, Late Harvest) • Syrah (Los Carneros) • Zinfandel (Ancient Vines)*

CLOS DU BOIS
Healdsburg
★★

Owned by Allied-Domecq, Clos du Bois has made fleshy, well-textured, medal-winning wines from its impressive 370 hectares (925 acres) of vineyards under a variety of highly talented winemakers.

✓ *Cabernet Sauvignon • Chardonnay • Classic red blend (Marlstone) • Malbec • Merlot • Pinot Noir (Sonoma Coast) • Syrah (Shiraz) • Zinfandel (Dry Creek Valley Reserve, Russian River)*

H COTURRI & SONS LTD
Glen Ellen
★

A wide range of sometimes excellent varietals, but as is often the case with organic winemaking, the quality ranges between rustic and brilliant. If they are drunk on release, the best examples of the following can be mind-blowing. The Albarello is essentially a Zinfandel and Petite Sirah field blend that includes a mishmash of other varieties.

✓ *Classic red blend (Albarello) • Zinfandel*

DE LOACH VINEYARDS
Santa Rosa
★☆

This winery's signature wine used to be Chardonnay, but its reputation is now firmly embedded in its exceptional range of Zinfandel wines. Top-of-the-range wines are sold as OFS (Our Finest Selection). De Loach was acquired by Burgundy's Jean-Claude Boisset in 2003.

✓ *French Colombard • Pinot Noir (OFS) • Sauvignon Blanc (Russian River Valley Fumé Blanc) • Zinfandel*

DRY CREEK VINEYARD
Healdsburg
★☆

Not all of elegant Dry Creek Vineyard wines actually come from the Dry Creek Valley AVA. Interestingly, in 2016, the winery was awarded a US patent for wine cork closures with sustainable sourcing information.

✓ *Cabernet Sauvignon • Chenin Blanc (Clarksburg) • Fumé Blanc • Zinfandel (Heritage, Old Vines)*

DU MOL
Orinda
★★

Although based in Orinda, this small operation produces some of the classiest Chardonnay and Pinot Noir in the Russian River Valley, not to mention one of California's finest Syrahs.

✓ *Chardonnay • Pinot Noir • Syrah*

DUXOUP
Healdsburg
★☆

The first time I visited Duxoup, I managed to get lost and started to express my apologies for being two hours late, but Andy Cutter waved away my apology with "Don't worry. When your colleagues Robert Joseph and Charles Metcalfe paid a visit, they were two days late," and I am very happy to say the delicious wines are equally laid-back. Grown on half an acre adjacent to the winery Duxoup are making some headway with their Gamay Noir.

✓ *Charbono • Sangiovese (Gennaio) • Syrah*

FERRARI-CARANO
Healdsburg
★☆

The Caranos sold up their Reno casino and hotel to buy 200 hectares (500 acres) of Sonoma vineyards in 1981 becoming in 2015 a Certified California Sustainable Vineyard. Their Italian-styled wines have always been a special feature.

✓ *Classic red blend (Sienna, Tresor) • Dessert style (Eldorado Noir) • Fumé Blanc*

FISHER VINEYARDS
Santa Rosa
★☆

With two hillside vineyards situated in the Mayacamas Mountains and another one on the Napa Valley floor, Fred Fisher has always crafted some high-quality wines but in particular from his Cabernet Sauvignon grapes.

✓ *Cabernet Sauvignon (Lamb Vineyard, Wedding Vineyard) • Chardonnay (Unity, Whitney's Vineyard) • Merlot (RCF Vineyard)*

FRITZ CELLARS
Cloverdale
★

This winery is best known for Chardonnay and Sauvignon, but it is the intense, extraordinarily long-lived Zinfandel that excites most. Particularly the Old Vine, made from a combination of 50-, 60-, and 80-year-old vines, and Rogers Reserve, from 100-year-old vines.

✓ *Cabernet Sauvignon • Zinfandel (Old Vines)*

GARY FARRELL
Healdsburg
★★

Gary Farrell makes truly special Russian River Pinot Noir, full of style and rich in fruit, with amazing finesse and complexity.

✓ *Chardonnay (Redwood Ranch) • Classic red blend (Encounter) • Pinot Noir • Sauvignon Blanc (Redwood Ranch) • Zinfandel (Maple Vineyard)*

GALLO OF SONOMA
Healdsburg
★★

Gina and Matt Gallo literally moved hillsides to create this estate, but they have been unable to bulldoze away the prejudice that exists in some quarters against anything remotely connected with the Gallo empire, while true enthusiasts just let the wine do the talking.

✓ *Barbera • Cabernet Sauvignon • Chardonnay • Pinot Gris • Zinfandel*

GEYSER PEAK WINERY
Geyserville

Hugely underrated because of low price points, the Block Collection range is nonetheless a stunning success, and Australian winemaker Daryl Groom is a magician with even the least expensive, most modest of wines.

✓ *Cabernet Sauvignon • Chardonnay • Classic red blend (Reserve Alexandre Meritage) • Fortified (Tawny Port) • Merlot (Sonoma County) • Sauvignon Blanc • Zinfandel (Sonoma County)*

GLORIA FERRER
Sonoma

Established by Cava giant Freixenet in 1982 and named after the wife of José Ferrer, the Spanish firm's president, Gloria Ferrer. Royal Cuvée might not be the most expensive wine in the range, but it is arguably the best.

✓ *Sparkling wine (Rosé, Royal Cuvée, Brut)*

GUNDLACH-BUNDSCHU WINERY
Sonoma
★

Expect high-quality Cabernet and Zinfandel from the fifth generation of the family of the founder, Jacob Gundlach. This winery also shows promising signs for other varietals.

✓ *Cabernet Sauvignon (Rhinefarm Vineyard) • Tempranillo (Rhinefarm Vineyard) • Zinfandel (Rhinefarm Vineyard)*

HANZELL VINEYARDS
Sonoma
★☆

James Zellerbach, the founder of Hanzell Vineyards, revolutionized the California wine industry in 1957 by ageing a Chardonnay wine in imported Burgundian *barriques*. The emphasis today of the 100 per cent estate-bottled wines is *terroir,* rather than oak.

✓ *Chardonnay • Pinot Noir*

HARTFORD COURT
Sonoma
★☆

Owned by Don Hartford, who is married to the daughter of Jess Jackson of Kendall-Jackson fame, Hartford Court makes extremely classy Pinot Noir, Chardonnay, and Zinfandel.

✓ *Chardonnay • Pinot Noir (Arrendell Vineyard) • Zinfandel (Hartford Vineyard)*

HIRSCH
Fort Ross
★★★

This specialist producer of Pinot Noir and Chardonnay produces world-class wines from single sites on its estate. Founded by David Hirsch in 1980.

✓ *Entire range*

HOP KILN
Healdsburg

The hop-drying barn that houses Hop Kiln winery was built in 1905 and has been declared a national historic landmark. A Thousand Flowers is a very inexpensive, unpretentious, soft, flowery, off-dry blend of Riesling, Chardonnay, and Gewürztraminer. Hop Kiln was bought by Landmark Vineyards in 2016.

✓ *Classic white blend (A Thousand Flowers) • Zinfandel (Old Windmill Vineyard, Primitivo Vineyard)*

IMAGERY ESTATE
Glen Ellen

Mike Benziger's Imagery series now has a dedicated winery and has branched out into single-vineyard varietals.

✓ *Cabernet Franc • Cabernet Sauvignon • Merlot*

IRON HORSE VINEYARDS
Sebastopol
★☆

Once the only railway stop in Sonoma Green Valley (hence its name, as well as that of Tiny Pony, its second wine), Iron Horse is famous for its textbook *méthode champenoise,* but also makes some elegant Pinot Noir wines.

✓ *Pinot Noir • Sparkling wine (Blanc de Blancs, Classic Vintage Brut, Rosé, Russian Cuvée)*

J ROCHIOLI VINEYARDS
Healdsburg
★★

The Rochioli family has been growing grapes at this long-established vineyard since the 1930s but only built a winery in 1984 and made their first wine in 1985. They now offer a large range of all too drinkable estate-bottled wines, including some of the most velvet-textured, varietally pure Pinot Noirs in all of California.

✓ *Pinot Noir • Sauvignon Blanc*

J WINE COMPANY
Healdsburg
★☆

Established in 1986 by Judy Jordan, who rented space in her father's Jordan Winery until 1997, when she purchased Piper Sonoma's specialist sparkling wine facility, renaming it the J Wine Company. Now owned by Gallo.

✓ *Pinot Gris (Russian River Valley) • Pinot Noir (Russian River Valley) • Sparkling wine (Brut)*

JORDAN
Healdsburg
★☆

This winery majors on rich, complex Cabernet Sauvignon and is gradually honing its Chardonnay into a much finer wine than it used to be.

✓ *Cabernet Sauvignon • Chardonnay*

JOSEPH SWAN VINEYARDS
Forestville
☆

For the son of teetotal parents, the late Joe Swan was an extraordinary winemaker in many senses, and his son-in-law, Rod Berglund, has continued his tradition of producing very small batches of high-quality wines, with Zinfandel his most consistent style.

✓ *Zinfandel*

KENWOOD VINEYARDS
Kenwood

Concentrated wines with good attack are consistently produced at these vineyards. Bought by Pernod Ricard in 2014.

✓ *Cabernet Sauvignon (Art Series, Jack London Vineyard) • Chardonnay (Russian River Reserve) • Pinot Noir (Olivet Reserve, Russian River Reserve) • Sauvignon Blanc (Sonoma County)*

KISTLER VINEYARDS
Trenton

Founded by Mark Bixler and Steve Kistler in 1978, all winemaking and vineyard operations are now Jason Kesner's responsibility. Kistler is an outstanding single-vineyard Chardonnay specialist that also turns out masterly Pinot Noir. No apologies for the value rating. Put these Sonoma Coast wines side by side with Burgundies of the same price, and Kistler wins more often than not.

✓ *Entire range*

KUNDE ESTATE
Kenwood
★☆

Kunde Estate winery was set up by a well-established group of Sonoma growers who started selling their own wines in 1990 and rapidly made a huge impact with critics and consumers alike.

✓ *Cabernet Sauvignon (Drummond) • Chardonnay (Wildwood) • Sauvignon Blanc (Magnolia Lane) • Zinfandel*

LA CREMA
Geyserville

Purchased into the Jackson Family Wines stable, La Crema holds a well-deserved reputation for reasonably priced, high-quality Chardonnay and Pinot Noir from the Russian River (9 Barrel being top of the range). La Crema also produces less expensive wines of reliable quality from the Sonoma Coast AVA.

✓ *Chardonnay (9 Barrel, Russian River) • Pinot Noir (9 Barrel, Anderson Valley, Carneros, Russian River)*

LANDMARK VINEYARDS
Windsor
★

These vineyards are best known for Chardonnay of a much fatter, richer, more complex style than they used to be and are now developing a growing reputation for fine Pinot Noir.

✓ *Chardonnay (Lorenzo) • Pinot Noir*

A STATUE OF SAINT JEAN STANDS NEAR THE ENTRANCE TO THE PATIO AND TASTING LAWN AT THE CHATEAU ST JEAN
With a 1920s villa and formal gardens, the grounds of this winery evoke a European feel. This influence is also reflected in the French-influenced wine styles produced here.

WINE BARRELS LINE THE CAVE ENTRANCE AT KENWOOD VINEYARDS
This winery, located in the heart of Sonoma County, concentrates its Kenwood Estate Vineyards on Merlot and Zinfandel vines.

LAUREL GLEN VINEYARD

Glen Ellen
★☆

Bought in 2011 by wine industry veteran Bettina Sichel, Laurel Glen offers a succulent Laurel Glen Cabernet Sauvignon, with Counterpoint as a half-price version.

 Cabernet Sauvignon

LAURIER

Forestville
★ V

Formerly called Domaine Laurier, with production restricted to a 12-hectare (30-acre) vineyard, this is now a brand belonging to the Bronco Wine Company.

 Chardonnay

LITTORAI

Sebastapol
★★★

Globally renowned producer of outstanding Pinot Noir and Chardonnay. Heidi and Ted Lemon, having previously worked at some of the most prestigious domaines in Burgundy, founded it in 1993. The majority of sites in the Sonoma Coast and Anderson Valley are farmed biodynamically but eschew certification.

 Entire range

LYETH WINERY

Geyserville
☆

Pronounced "Leeth", this was an exciting new winery in the 1980s, but when the founder died in a plane crash in 1988, it was sold off and is now owned by Burgundian *négociant* Boisset. Their second label is Christophe.

✓ Classic red blend (Lyeth)

MARCASSIN WINE CO.

Windsor
★★★

Helen Turley, renowned consultant winemaker, and husband Jon Wetlaufer, esteemed viticulturist, produce some of the most sought after – and limited – Pinot Noir and Chardonnay in California, regularly receiving 100-point scores to further increase demand. Produced from their 50-hectare (20-acre) estate on the Sonoma Coast, plus some neighbouring vineyards, the wines are regularly compared in power and complexity of *grand cru* Burgundy with prices to match.

✓ Entire range

MARIETTA CELLARS

Healdsburg
★

Chris Bilbro started up this operation in 1980 after selling off his share in the Bandeira Winery. Chris built his reputation on lush Merlot and great-value wines, a legacy taken on now by son, Scot.

✓ Classic red blend (Old Vine Red) • Petite Sirah • Zinfandel (Cuvée Angeli)

MARIMAR ESTATE

Sebastopol
★☆

Miguel Torres planted his sister's vineyard with Chardonnay and Pinot Noir, at three to four times the normal density for California. These high-density vines work best for Pinot Noir, which has a combination of silky fruit, finesse, and age-worthiness that is uncommon in California.

 Chardonnay • Pinot Noir

MARTINELLI VINEYARDS

Fulton
★★

Experienced grape growers, who turned winemakers in the early 1990s, the Martinelli family turns out limited quantities of lush, stylish wines. The range includes wonderfully classy Pinot Noir, stunning Syrah and Zinfandel, and sumptuous late-harvest Muscat Alexandria.

✓ Entire range

MATANZAS CREEK WINERY

Santa Rosa
★☆

This small, highly reputed winery has been owned by Jess Jackson and Barbara Banke of Kendall-Jackson since 2000, when the founding McIver family retired and sold up. The style is fine, rich, and elegant, with a signature of good fruit-acidity, probably better suited to European palates than American.

✓ Chardonnay • Sauvignon Blanc

MICHEL-SCHLUMBERGER

Healdsburg
☆

Formerly known as Domaine Michel Winery, this producer is now owned by the Adams Wine Group.

 Cabernet Sauvignon • Chardonnay (La Brume) • Merlot (Dry Creek Valley) • Pinot Blanc (Dry Creek Valley)

NALLE

Healdsburg
★★

Doug Nalle was winemaker at Quivira, when he started making Zinfandel under his own label in 1984. His Zinfandel wines are usually augmented with 6 to 12 per cent of Petite Sirah, Syrah, Carignane, Alicante Bouchet, and Gamay and aged exclusively in French oak. They are not the massive monsters that win gold after gold or make national headlines, but the Brits appreciate them.

 Zinfandel

PAX

Sebastopol
★★☆

Pax Mahle established California's fastest-rising Syrah superstar in 2000, and the mesmerizing quality of his low-yield, hands-off, single-vineyard wines has charmed the critics off their perch. Mahle also founded Wind Gap in 2006 but has subsequently now sold 50 per cent of the business. Pax now is experimenting with Chenin Blanc, Gamay, and Carignan.

 Entire range

PEDRONCELLI WINERY

Geyserville
☆ V

Independent winery, owned and run by the Pedroncelli family since 1927. Source of excellent value wines.

 Zinfandel (Bushnell Vineyard, Courage, Mother Clone), Cabernet Sauvignon (Block 007, Three Vineyards, Wisdom Estate), Sauvignon Blanc (East Side Vineyards)

PELLEGRINI FAMILY VINEYARDS

South San Francisco
★ V

These vineyards produce delicious, vividly flavoured red wines with oodles of fresh fruit.

 Pinot Noir (Olivet Lane Estate) • Zinfandel

PETER MICHAEL WINERY

Calistoga
★★

With no expense spared, Englishman Sir Peter Michael applied New World technology to the very best of traditional methods to produce some of the classiest wines in California. The Chardonnays especially can be a match for the greatest examples from all of California, as well as the Côte d'Or.

✓ Chardonnay (Cuvée Indigène, Pointe Rouge for power) • Classic red blend (Les Pavots) • Sauvignon Blanc (L'Après-Midi)

PIPER SONOMA

Healdsburg
★☆ V

Piper Sonoma was founded in Healdsburg in 1980 as a joint venture between Piper-Heidsieck and Renfield Importers.

 Sparkling wine (Brut)

PRESTON VINEYARDS
Healdsburg
★☆

These Dry Creek vineyards are planted with an amazing mix of varieties and are capable of making high-quality wines. Carignane, Cinsault, and Mourvèdre are worth keeping an eye on.

✓ *Barbera • Classic red blend (L Preston Red) • Zinfandel*

PRIDE MOUNTAIN VINEYARDS
Santa Rosa
★★

Established in 1990 by the late Jim Pride, who transformed a rundown, under-performing vineyard into the current immaculate wine estate, which produces equally immaculate red wines and one of California's finest Viogniers.

✓ *Cabernet Franc • Cabernet Sauvignon • Classic red blend (Reserve Claret) • Merlot • Sangiovese • Viognier*

QUIVIRA
Healdsburg
☆

Quivira was named after a legendary American kingdom that Europeans spent 200 years searching for. Not able to locate Quivira, the earliest settlers apparently used the name for this part of northern California before it was called Sonoma.

✓ *Zinfandel*

RADIO COTEAU
Sebastapol
★★

A focus on organic and biodynamic practices in the vineyard combines with a minimal-intervention approach in the winery to produce some of finest Chardonnay, Pinot Noir, and Syrah in California. Following stints at some of finest Burgundy domaines before a period at Bonny Doon, Eric Sussman founded it in 2002. Fruit is sourced from their estate on a ridge above the town of Occidental, purchased in 2012, as well as a selection of top-quality sites throughout Sonoma Coast.

✓ *Chardonnay, Pinot Noir, Syrah*

RAMEY WINE CELLARS
Healdsburg
★★

Founded by pioneering winemaker David Ramey and his wife, Carla, in 1996, utilising David's nearly 20 years of experience at some of the most famous wineries in Napa and Sonoma. Arguably most famous for Chardonnay sourced from Sonoma and Carneros, particularly the single-vineyard expressions Hudson and Hyde, Ramey also produce highly acclaimed Syrah, Cabernet Sauvignon, and Pinot Noir.

✓ *Chardonnay*

RAVENSWOOD
Sonoma
★☆

Ravenswood was founded by Joel Peterson in 1976 and has been known for some quite sensational Zinfandel blends. Zinfandel remains the key attraction now that Gallo purchased Ravenswood in 2019. Since then, however, the brand has seen some repositioning further down the shelf.

✓ *Zinfandel*

RIDGE VINEYARDS LYTTON SPRINGS
Healdsburg
★★★

Ridge Vineyard's history began in the 19th century, and it added the Lytton Springs winery and vineyards in 1991. Sister estate to Monte Bello, Ridge had sourced fruit from the old-vine vineyards that surround the winery since 1972. As with Monte Bello, there is a focus on sustainable practices in the vineyard and winery, combined with a minimal-intervention wine-making ethos. Outstanding, expressive, age-worthy wines.

✓ *Entire range*

ROBERT STEMMLER
Sonoma
★

This winery formerly produced a full range of wines at Robert Stemmler's own winery in Healdsburg, when Pinot Noir was its biggest seller, and that variety is still its best wine, even though Robert Stemmler is now produced at and is marketed by Buena Vista.

✓ *Chardonnay (Three Clone) • Pinot Noir*

RODNEY STRONG VINEYARDS
Windsor
★

Rodney Strong Vineyards produces rich Cabernet Sauvignon and soft-styled Chardonnay, but the fresh, easy-drinking Sauvignon is now emerging as his best wine. Constantly improving.

✓ *Cabernet Sauvignon • Chardonnay (Chalk Hill) • Sauvignon Blanc (Charlotte's Home Vineyard) • Zinfandel (Knotty Vines Estate)*

SEBASTIANI
Sonoma
☆

Founded by Don Sebastiani in 1904, the winery has been owned since 2008 by the Foley Wine group.

✓ *Barbera • Cabernet Sauvignon (Alexander Valley) • Chardonnay • Classic red blend • Pinot Noir (Russian River Valley) • Zinfandel*

SEGHESIO
Healdsburg
★☆

This winery owns a considerable spread of well-established vineyards, the oldest of which date back to 1895, when Edoardo Seghesio planted his first vineyard in the Alexander Valley. Although he made and sold wine in bulk, it was not until 1983 that his grandchildren were persuaded by their own children to sell wines under the Seghesio label. Now owned by Napa-based Crimson Wine Group.

✓ *Zinfandel*

SIDURI
Santa Rosa
★★

Named after the Babylonian goddess of wine by Adam Lee and Diana Novy Lee, a couple of Pinot Noir fanatics who sourced their wines from as far north as Oregon's Willamette Valley to as far south as Santa Barbara's Santa Rita Hills. Siduri was snapped up by Jackson Family Wines in 2015. Siduri stull sources from far and wide, crafting single-vineyard Pinot Noirs from 20 individual sites.

✓ *Pinot Noir*

SIMI WINERY
Healdsburg
☆

Italian brothers Guiseppe and Pietro Simi produced their first wines in 1876, but they both died suddenly in 1904. At that point Giuseppe's 18-year-old daughter, Isabelle, took over the reins, a post she held for 64 years, retiring in 1970 and selling the winery to Alexander Valley grape grower, Russell Green. Moët-Hennessy bought it in 1981, and since 1999 it has been part of Constellation Brands Melissa Stackhouse is now Director of Winemaking.

✓ *Chardonnay*

SONOMA-CUTRER VINEYARDS
Windsor
★

Now owned by Brown-Forman, Sonoma-Cutrer continues to specialize in producing high-quality, single-vineyard Chardonnays.

✓ *Chardonnay (Les Pierres, Founder's Reserve)*

SONOMA-LOEB
Geyserville
☆

Owned originally by John Loeb, a former American ambassador to Denmark, who grew grapes here since the early 1970s, Chappellet Vineyard & Winery bought Sonoma-Loeb in 2011.

✓ *Chardonnay (Private Reserve)*

ST FRANCIS VINEYARD
Kenwood
★

Situated just across the road from Chateau St Jean, this was an underrated winery but is not producing as many cellar wines as it used to.

✓ *Cabernet Sauvignon • Chardonnay (Reserve Behler) • Zinfandel*

STONESTREET
Healdsburg
★☆

One of the wineries in Kendal-Jackson's Artisans & Estates group, Stonestreet produces good Chardonnay, brilliant Sauvignon, and gorgeously sumptuous Cabernet and Merlot but truly excels with its classy Legacy blend.

✓ *Cabernet Sauvignon (Christopher's) • Chardonnay (Upper Barn Vineyard) • Merlot (Sonoma County) • Sauvignon Blanc*

TRENTADUE WINERY
Geyserville
★

The fast-improving Trentadue Winery produces a characterful selection of good-value varietal wines that are rarely, if ever, filtered.

✓ *Petite Sirah • Zinfandel (La Storia)*

UNTI VINEYARDS AND WINERY
Healdsburg
★

Established in 1997 by the Unti family, who remain owners and operators, it specialises in Mediterranean varietals from southern Europe.

✓ *Barbera • Cuvee Foudre • Grenache • Rosé*

VML WINERY
Healdsburg
★★

Specialists in high-quality Pinot Noir and Chardonnay, VMR focusses on Russian River Valley single-vineyard expressions and barrel selections.

✓ *Chardonnay (Cresta Ridge Vineyard, Knowlton Farms) • Pinot Noir (Cresta Ridge Vineyard, Starscape Vineyard)*

WALTER HANSEL
Santa Rosa
★★

A Burgundian specialist who uses wild yeast and does not fine or filter, resulting in some of California's finest Chardonnay and Pinot Noir wines.

✓ *Chardonnay • Pinot Noir*

WILLIAMS SELYEM
Fulton
★★☆

Established in 1981 by Burt Williams and Ed Selyem (since retired), Williams Selyem has carefully carved out a reputation for tiny quantities of some of the most sought-after Pinot Noir in California. The signature note of these wines is their beautiful balance and finesse.

✓ *Chardonnay • Pinot Noir • Zinfandel*

WINDSOR VINEYARDS
Asti
★☆

This producer is grossly underrated because it is corporate-owned and offers so many wines at such inexpensive prices that many think there cannot possibly be anything worth drinking. Yet on the competition circuit, Windsor Vineyards is very well known and highly respected, ranking as one of the three most award-winning wineries in the United States every year in the 1990s and 2000s.

✓ *Cabernet Sauvignon (Private Reserve, Signature Series) • Carignane • Chenin Blanc • Classic red blend (Signature Series Meritage) • Gewürztraminer • Merlot (Signature Series) • Petite Sirah (Mendocino County) • Pinot Noir (Private Reserve, Russian River Valley - Signature Series) • Sauvignon Blanc (Middle Ridge Vineyard) • Syrah • Zinfandel (North Coast, Signature Series)*

Napa County

Napa is the heart and soul of the California wine industry. Producing only 4 per cent of the state's total output by volume, Napa punches well above its weight, accounting for about 25 per cent of the total value. Its vineyards are the state's most concentrated and costly, combined with more wineries than any other county, which produce the greatest number and variety of fine wines in all of North America.

It is hard to believe that Napa County was planted after Sonoma, but it was, by some 13 years. In 1838 a fur trapper from North Carolina, George Yount, acquired a few Mission vines from General Mariano Vallejo's Sonoma vineyard and crossed the Mayacamas Mountains, where he planted them outside his log cabin, 3 kilometres (2 miles) north of present-day Yountville. His intention was simply to make a little wine for himself. Within six years he was harvesting an annual average of 900 litres (200 gallons); by the turn of the decade, other vineyards had sprung up, and in 1859, Samuel Brannan, an ex-Mormon millionaire, bought 8 square kilometres (3 square miles) of valley land and planted cuttings of various European vine varieties.

It was Charles Krug, however, who truly pioneered the Napa's wine industry, establishing the valley's first winery in 1861 just north of St Helena. Krug was a driving influence on the industry's development, with a persuasive and vocal focus on quality and innovation that laid the foundations for future generations: Krug's introduction of a cider press for winemaking being a prime example. Just over 100 years later, in 1966, Robert Mondavi established his eponymous winery, which was another fundamental turning point in the development of the valley. With a similar emphasis on excellence and technology, Mondavi travelled globally, acting as a spokesman and firmly putting Napa on the international wine industry map – a position and reputation that was cemented with the famous 1976 Judgement of Paris. Today, Napa's wines, particularly Chardonnay and Cabernet Sauvignon, are some of the most sought-after and highly prized in the Americas, with prices to match.

DOMAINE CARNEROS OFFERS THANKS TO THE FIRST RESPONDERS WHO PUT THEIR LIVES ON THE LINE DURING THE NORTH BAY FIRES OF 2017
Wildfires in the state have posed increasing threats to California wine country. In the devastating fires of 2017, at least eight wineries were significantly or totally damaged in Napa County, Sonoma County, and Mendocino County. Many others reported damage to their wineries, other buildings, or vineyards.

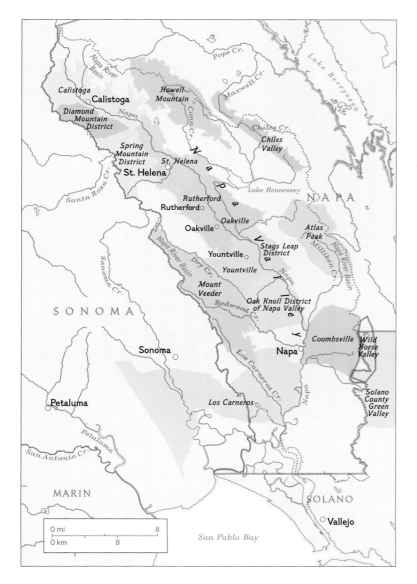

THE VITICULTURAL AREAS OF NAPA COUNTY (see also p577)
The intensive vine-growing areas of this illustrious California district occupy a long, narrow strip running roughly parallel to those of Sonoma County. Many famous names are crowded around Route 29. To the east, Sursun Valley is also included (see Extended Area at right).

FACTORS AFFECTING TASTE AND QUALITY

LOCATION
Starting at San Francisco Bay, Napa runs 54 kilometres (34 miles) north and west to the foothills of Mount St Helena. Flanking it are the Sonoma Valley to the west and Lake Berryessa to the east.

CLIMATE
The Mayacamas in the west prevent the cool Pacific winds reaching the valley, whilst the Vaca Mountains protect against the high temperatures of Central Valley. There is still significant maritime influence and fog rolling in from the south, across San Pablo Bay. The result is a diversity of microclimates, broadly ranging from cool (Region I, see p577) at the southern end in the often foggy Carneros District, to warm (Region III) in the northern section of the Napa Valley and in Pope Valley.

ASPECT
There is great diversity of aspect on Napa County. The altitude of the valley floors ranges from barely 5 metres (17 feet) above sea level at Napa itself to 70 metres (230 feet) at St Helena and 122 metres (400 feet) at Calistoga in the north. The wooded western slopes provide afternoon shade, which adds to the tempering effect of altitude and so favours white grapes, whereas the eastern slopes favour red varieties. Altitude of planting in turn plays a key role, dictating the influence of the moderating fogs by sitting above or below the fog line.

SOIL
Varied soils occur throughout the valley, with fertile clay and silt loams primarily in the south of the region and gravel loams of better drainage and lower fertility in the north. There is further variation east to west, also ranging through elevation, with the most fertile soils generally on the valley floor, rocky well-drained on the hillsides, and alluvial on the benchland plains, located mostly on the western foothills.

VITICULTURE & VINIFICATION
About 75 per cent of the valley is planted with red varieties, of which 70 per cent is Cabernet Sauvignon. Of the white plantings, Chardonnay accounts for around 75 per cent. Fruit costs are some of the most expensive in California, with the most sought-after growers and vineyards commanding huge premiums, in turn reflected in top-end bottle prices. There is a vast range of wineries here, from a small number of large firms employing the latest high-tech methods to an increasing number of small boutique wineries. The latter have limited production, based on traditional methods, although often combined with judicious use of modern techniques.

GRAPE VARIETIES
Cabernet Franc, Cabernet Sauvignon, Chardonnay, Merlot, Pinot Noir, Sauvignon Blanc, Zinfandel

THE APPELLATIONS OF
NAPA COUNTY

ATLAS PEAK AVA
The Atlas Peak appellation is located on and around Atlas Peak, one of the county's highest points, on the western side of the Vaca Mountains above the foothills of the Stags Leap District. The regions' elevation is key, with most planting above 230 metres (760 feet), resulting in the vines benefiting from significantly cooler days and warmer nights compared to the valley floor. With most vineyards sitting above the fog line, they also get the benefit of long sunshine hours. Cabernet Sauvignon and other Bordeaux varietals thrive in the climate and thin, rocky soils.

CALISTOGA AVA
Situated at the northernmost end of the valley, the climate of the Calistoga AVA is warm to hot, and is moderated by diurnal temperature swings of up to 10°C (50°F) and cooling daytime airflows down from Mount St Helena. Black grape varieties dominate, and the focus is on Cabernet Sauvignon and Zinfandel.

CARNEROS AVA
Also known as Los Carneros, this AVA overlaps Napa and Sonoma counties. See also Sonoma.

CHILES VALLEY AVA
This narrow area of vines in the Vaca Mountains is separated from Napa Valley by the first group of hills in the Vaca range. As such, the cooling bay breezes that moderate Napa Valley do not reach Chiles Valley, but its altitude of 240 to 300 metres (800 to 1,000 feet), in combination with the airflow from the Vaca Mountains, tend to cool down the grapes more quickly than in the main areas of the Napa Valley itself. The vines growing here are primarily Zinfandel, Cabernet Sauvignon, Chardonnay, and Sauvignon Blanc.

COOMBSVILLE AVA
Situated in the southeast of Napa Valley in a bowl-shaped depression at the foot of the Vaca Mountains, the region is moderated by the proximity to San Pablo Bay; fogs settle earlier and remain for longer compared to more northerly AVAs. Cabernet Sauvignon, Merlot, and Chardonnay are the predominant plantings.

DIAMOND MOUNTAIN AVA
Diamond Mountain AVA is located in the Mayacamas Mountains in the northeastern corner of Napa Valley, where porous volcanic soils and extended exposure to the sun are the key to its legendary success for Cabernet Sauvignon and, increasingly, Cabernet Franc. The wines are firmly structured and renowned for their ageing potential.

HOWELL MOUNTAIN AVA
The vineyards of the relatively flat table-top of Howell Mountain, located in the east of the valley, are planted above the fog-line at an altitude of between 420 and 890 metres (1,400 and 2,200 feet). Volcanic tufa and red clay are the two main soil types, both nutrient poor. The consistent warm climate and long sunshine hours are best for Cabernet Sauvignon but also capable of producing top-flight Zinfandel and Chardonnay.

MOUNT VEEDER AVA
Situated on the east-facing slopes of the Mayacamas Mountains and named after the volcanic peak that dominates this AVA, Mount Veeder is one of the most rugged mountainous areas in which winemakers went to escape the Napa Valley's fertile valley floor. Cool sea breezes from San Pablo Bay and occasional marine fogs temper the climate, resulting in one of the longest growing seasons in Napa Valley. Chardonnay and Cabernet Sauvignon are the two dominant varieties, the latter being some of the most age-worthy examples from Napa, with quite exceptional varietal intensity and completely different in structure from the lush valley-floor Cabernets.

NAPA COUNTY AO
This is an appellation that covers grapes grown anywhere in the entire county.

NAPA VALLEY AVA
The first AVA to be designated in California – and only the second in the United States – the area includes all of the county with the exception of the area around Putah Creek and Lake Berryessa. The Napa Valley appellation is 40 kilometres (25 miles) long and between 12 and 16 kilometres (8 and 10 miles) wide, sheltered by two parallel mountain ranges. The majority of vineyards occupy the flat valley floor in a continuous strip from Napa to Calistoga.

OAK KNOLL DISTRICT OF NAPA VALLEY AVA
Well-drained, infertile, gravelly alluvial soils are the key characteristic of the AVA, sitting at the southern end of Napa Valley. The impact of morning fogs, which can remain until late morning, and cooling daytime breezes from San Pablo Bay results in a moderate-to-cool climate with a long, slow-ripening period. Diurnal temperatures can swing by 20°C (40°F) in mid-summer, retaining acidity to balance the intense ripeness achieved. Cabernet Sauvignon, Chardonnay, and Merlot are most widely planted.

OAKVILLE AVA
If Rutherford boasts the greatest concentration of famous Cabernet Sauvignon vineyards, then neighbouring Oakville, which is equally Napa's heartland, excels by its very diversity. Warm daytime temperatures are tempered by cool nights and morning fogs, producing ideal conditions for great Cabernet Sauvignon, Merlot, and Chardonnay, and even the Sauvignon Blanc from these vineyards can be extraordinary.

RUTHERFORD AVA
One of the most famous names in Napa, yet one of the last to gain AVA status because many growers feared that dividing the Napa Valley's heartland into smaller, possibly more prestigious sub-appellations would gradually have the effect of diluting the reputation of Napa itself. Average temperatures are slightly warmer than Oakville and Stags Leap District, producing rich, full-bodied, and ripe wines with supple tannins. This AVA encompasses the historic heart of Napa Valley, including some of California's most famous wineries, and many of Napa's greatest Cabernet Sauvignon vineyards, many situated on the alluvial "Rutherford Bench" in the west of the valley.

SPRING MOUNTAIN DISTRICT AVA
This appellation is located just west of St Helena, on the eastern flank of the Mayacamas Mountains, defined by a diversity of soils and mountainous topography. The majority of vineyards sit above the fog line, benefitting from the long

growing season that the cooler days and warmer evenings bring. Cabernet Sauvignon, Cabernet Franc, Chardonnay, Merlot, and Zinfandel are the principal varieties.

ST HELENA AVA

The St Helena AVA lies immediately north of Rutherford in one of the warmest areas of the valley, rarely receiving the cooling winds and fogs from San Pablo Bay in the south. Cabernet Sauvignon, Cabernet Franc, and Merlot from this area are ripe and generous.

STAGS LEAP DISTRICT AVA

Confusingly, Stags Leap is spelled in three slightly different ways: with an apostrophe before the "s" in the famous Stag's Leap Wine Cellar, with an apostrophe after the "s" in the lesser-known Stags' Leap Winery, and without an apostrophe in the AVA name. Stag's Leap Wine Cellar shot to fame in 1976 when its Cabernet Sauvignon trumped top Bordeaux

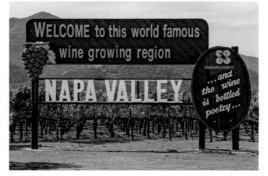

NAPA VALLEY WELCOME SIGN
Since 1950 two of these signs have stood in the Napa landscape – one on Highway 29 between St Helena and Calistoga and the other in Oakville.

wines at the 1976 Judgement of Paris. Stags Leap District vies with Rutherford as the crème of Napa Valley Cabernet Sauvignon, but also produces excellent Merlot and stunning red Bordeaux-style blends. Cooling afternoon marine winds from San Pablo Bay are fundamental to the balance and perfume these wines typically offer.

WILD HORSE VALLEY AVA

Technically attached to Napa, the Wild Horse Valley AVA actually straddles the county line, occupying more of Solano County than Napa County.

YOUNTVILLE AVA

Although less famous than Rutherford or Oakville, this AVA does have one of the Napa Valley's coolest vineyard exposures with high diurnal fluctuations of up to 20°C (40°F), which is why some people rate its Cabernet Sauvignon above those of its neighbours.

THE WINE PRODUCERS OF
NAPA COUNTY

ABREU VINEYARDS
Yountville
★★☆

David Abreu, renowned viticulturist, founded Abreu Vineyards in 1980. He still focusses on single-vineyard expressions of Cabernet Sauvignon–driven blends from a selection of the most exceptional sites in Napa Valley. Very limited production makes them some of the most highly coveted wines produced in Napa Valley.

 Cabernet Sauvignon

ACACIA WINERY
Carneros
★★

Acacia Winery produces wines of increasing finesse, ranging from a fine Chardonnay, which has a touch of spice, to a lovely Pinot Noir that is silky textured and all cherries and vanilla. Diageo sold Acacia to Peju Province Winery of Rutherford in 2016.

 Chardonnay • Pinot Noir (Carneros)

ARTESA
Napa
★☆

Formerly called Codorníu Napa, this is the Californian arm of the famous Catalan Cava producer and originally opened in 1991. Although the quality increased as the years went by, Codorníu found it difficult to sell a sparkling wine at California prices, because consumers obviously wondered why they should pay twice the price for Codorníu with "Product of California", rather than "Product of Spain" on the label. It did not help that Codorníu had always been synonymous with Cava, one of the cheapest sparkling wines on the US market. The decision to pull the plug was taken in 1996 – the winery was totally transformed into one suited for the production of classic varietal wines, and Artesa was born.

 Cabernet Sauvignon • Sauvignon Blanc (Napa Valley Reserve)

BARNETT VINEYARDS
St Helena
★☆

These mountain vineyards make intense, yet delicious, very elegant Cabernet Sauvignon, among other wines.

 Cabernet Sauvignon (Rattlesnake Hill)

BEAULIEU VINEYARD
Rutherford
★

This was the centre of innovation in California under the legendary winemaker André Tchelistcheff but has long since passed into the corporate hands of Treasury Wine Estates. Not the force it was, yet on its own, the legendary Georges de Latour would earn two stars.

 Cabernet Sauvignon (Georges de Latour)

BEHRENS FAMILY WINERY
Oakville
★★

Born in the back-of-beyond (viticulturally speaking) of Arcata, Humboldt County, not far from the Oregon border, where they made 175 cases a year from trucked-in grapes in 1991, Behrens and Hitchcock built their winery in the 1990s. Hitchcock retired in 2005, however, and the Behrens family made the winery their own. They craft some of California's finest Cabernet Sauvignon.

 Cabernet Sauvignon • Petite Sirah

BERINGER VINEYARDS
St Helena
★★ Ⓥ

Another Treasury Wine Estates property (since 2011), Beringer Vineyards wines are among the finest in California. You get proportionately more for your money if you pay extra for the better wines, with the singular exception of the Founders' Estate Merlot, which offers extraordinary finesse for so few dollars. Yet I must take issue with the name of this entry-level range: Beringer may be open about its diverse origins, hence the California AVA, but Founders' Estate does imply that the

vineyards in question are part of a single estate of historic origins, and I'm not so sure that Jacob and Frederick Beringer owned vineyards stretching from Santa Barbara to Mendocino.

 Cabernet Sauvignon • Chardonnay • Classic red blend • Merlot (Bancroft Ranch, Founders' Reserve)

BOND
Oakville
★★

Cult winery founded by Bill Harlan, the focus of BOND is solely on single-site expressions of Cabernet Sauvignon sourced from prized hillside vineyards on both sides of Napa Valley.

 Cabernet Sauvignon

BIALE VINEYARDS
Napa
★★

Amazing, huge, rich, and gorgeously ripe, single-vineyard Zinfandel, and one of California's greatest Petite Sirah.

✓ *Zinfandel (Aldo's Vineyard)*

BOUCHAINE VINEYARDS
Napa
★

Notable by its absence from most US critics' thoughts, Bouchaine's Pinot Noir and Chardonnay are probably too light and elegant to stir up much opinion in America, but have a purity and finesse much appreciated by European palates. The 2018 wildfires saw Bouchaine temporarily shut its doors due to smoke and ash, and their 2019 purchase of a generator shows how California's wineries are adapting to recent climatic issues.

✓ *Pinot Noir*

BRYANT FAMILY VINEYARD
Calistoga
★★

Since their first vintage in 1992 this boutique winery is reputedly turning out masterly Cabernet Sauvignon.

 Cabernet Sauvignon

BUEHLER VINEYARDS
St Helena
★★

Buehler's focus is on Cabernet Sauvignon, which is the main event here. Zinfandel and Chardonnay are also available.

 Cabernet Sauvignon

BURGESS CELLARS
St Helena
★☆ Ⓥ

Founded by Tom Burgess and now run by his sons, Burgess Cellars often produce well-weighted, oaky wines of fine quality.

 Syrah

CAIN CELLARS
St Helena
★☆

These are interesting, innovative, and well-made Cabernet Sauvignon wines.

 Classic red blend (Cain Cuvée, Cain Concept, Cain Five)

CAKEBREAD CELLARS
Rutherford
★☆

Bruce Cakebread has earned his reputation for producing one of the most delicious Sauvignon Blanc wines in the state.

 Cabernet Sauvignon (Vine Hill Ranch) • Sauvignon Blanc • Syrah

CARDINALE
Oakville
☆ Ⓥ

Cardinale was created from the Robert Pepi property by Jackson Family Wines when they acquired it in 1994; they focus on a single red blend wine driven by Cabernet Sauvignon.

 Classic red blend

CASA NUESTRA
St Helena
★☆

Casa Nuestra produces classic styles, such as Merlot and Meritage, but some quirky ones, too, such as one of California's finest Chenin Blanc and a classic red blend (Tinto) made from a field mix of seven varieties

✓ *Chenin Blanc • Classic red blend (Tinto)*

CAYMUS VINEYARDS
Rutherford
★★☆

This extraordinary winery now focusses exclusively on deep, dark, brooding Cabernet Sauvignon of superb quality.

✓ *Cabernet Sauvignon (Napa Valley, Special Selection)*

CHAPPELLET VINEYARDS
St Helena
★☆

Chappellet is an excellent winery with a range of skilfully produced wines that show great finesse.

✓ *Cabernet Sauvignon • Chardonnay • Chenin Blanc • Merlot (Napa Valley)*

CHARLES KRUG
St Helena
☆

Purchased by the Mondavi family in 1943 and run by Bob Mondavi's less-prominent brother Peter, this winery used to be known for its Cabernet Sauvignon, but Krug's reputation has been diluted by its production of jug wine and generics sold under a second label – C K Mondavi.

✓ *Cabernet Sauvignon • Merlot (Reserve)*

CHÂTEAU MONTELENA WINERY
Calistoga
★☆

The style of this small, prestigious winery has changed over the years, but the quality has always been high, especially for Cabernet Sauvignon, although the Chardonnay has in the past been almost as good; it was a Montelena Chardonnay that was entered in the "Judgement of Paris" tasting.

✓ *Cabernet Sauvignon • Chardonnay*

CHÂTEAU POTELLE
Mount Veeder
★

This winery was established in 1985 by a *bordelais* couple sent by the French government to "spy" but who instead decided to stay, and, ironically, were best known in their early days for their slow-evolving Chardonnay, but have since developed a reputation for other age-worthy varietals.

✓ *Cabernet Sauvignon (VGS) • Chardonnay (VGS) • Zinfandel (VGS)*

CHIMNEY ROCK
Napa
★

Part of the Terlato Wine Group (which includes Sandford, Rutherford Hill, and Alder Brook), this was originally Chimney Rock Golf Club, until its nine holes were bulldozed and a vineyard planted.

✓ *Cabernet Sauvignon • Classic red blend (Elevage)*

CLIFF LEDE VINEYARD
Napa
★

Cliff Lede took over the S Anderson site in 2002, but rather than the traditional-method wines that Carol Anderson garnered such a reputation for, Cliff Lede produces still wines in a style as elegant as the labelling, with layers of intense, finely poised fruit.

✓ *Cabernet Sauvignon • Sauvignon Blanc*

CLOS DU VAL
Napa
★

A Clos Du Val Cabernet Sauvignon was entered into the famed 1976 "Judgement of Paris" tasting.

✓ *Cabernet Sauvignon (Stags Leap District) • Chardonnay (Carneros)*

CLOS PEGASE
Calistoga
★

A welcome, understated use of oak coupled with lush fruit produces wines of some elegance. Sold to Vintage Wine Estates in 2013.

✓ *Cabernet Sauvignon • Chardonnay • Classic red blend (Hommage) • Merlot*

COLGIN
Napa
★★☆

Tiny production of exorbitantly priced, high-class Cabernet Sauvignon, with sufficient structure to support the weight. LVMH appreciated the qualities of Colgin and bought a majority stake in the winery in 2017.

✓ *Cabernet Sauvignon • Classic red blend (Cariad)*

CONN CREEK WINERY
St Helena
☆

Owned by Ste Michelle Wine Estates (who also own Château Ste Michelle, Columbia Crest, and Villa Mount Eden), Conn Creek is best known for its supple, fruity Cabernet Sauvignon but does not possess the cachet it once had.

✓ *Cabernet Sauvignon • Classic red blend (Anthology)*

CORISON
St Helena
★★

After 10 years as Chappellet's winemaker, Cathy Corison struck out on her own, and has not looked back since, purchasing small parcels of the finest Napa Valley grapes to produce lush, supple Cabernet wines of surprising longevity.

✓ *Cabernet Sauvignon*

CORNERSTONE CELLARS
Oakville
★★

Owned by two Memphis doctors, Michael Dragutsky and David Sloas. In 1991 Sloas happened to be tasting on Howell Mountain with Randy Dunn, who mentioned that he had received almost five tonnes of grapes more than

he needed from one of his farmers. Sloas telephoned his friend Dragutsky in Memphis, they agreed they had both lost their minds, and purchased the grapes, thus going from wine consumers to wine producers in a few minutes.

✓ *Cabernet Sauvignon • Merlot*

COSENTINO
Yountville
★☆

Founded by Mitch Cosentino (no longer involved), and sold to Vintage Wine Estates in 2011.

✓ *Classic red blend (The Poet) • Merlot (Reserve) • Sangiovese • Zinfandel (Cigar Zin, The Zin)*

CUVAISON
Calistoga
★

This Swiss-owned winery is making better reds than whites.

✓ *Chardonnay (Carneros Estate Selection) • Pinot Noir (Carneros Estate Selection) • Syrah (Carneros)*

DALLA VALLE
Oakville
★★☆

Dalla Valle produces great Cabernet Sauvignon in addition to a truly heroic Cabernet-based blend.

✓ *Cabernet Sauvignon • Classic red blend (Maya)*

DELECTUS
Oakville
★☆

Small, ultra-high-quality operation established in 1989 by Austrian-born Gerhard Reisacher and his wife, Linda. Bought by Vintage Wine Estates in 2016.

✓ *Classic red blend (Cuvée Julia) • Petite Sirah*

DETERT FAMILY VINEYARDS
Oakville
★★

Former Opus One Cabernet Franc vines produce this relatively new winery's best and cheapest wine, but don't let that put you off Detert's beautifully focused Cabernet Sauvignon.

✓ *Cabernet Franc • Cabernet Sauvignon*

DIAMOND CREEK VINEYARDS
Calistoga
★★☆

This small winery specializes in awesomely long-lived, awesomely priced Cabernet Sauvignon.

✓ *Cabernet Sauvignon (Red Rock Terrace, Gravelly Meadow, Volcanic Hill)*

DOLCE
Oakville
★★☆

Established by Far Niente as the only winery in the United States solely devoted to producing a single late-harvest wine, Dolce is a lusciously sweet, heavily botrytised, Sauternes-style dessert wine made

from approximately 90 per cent Sémillon and 10 per cent Sauvignon Blanc. The only pity is that they went for such a kitsch, nouveau riche presentation instead of the understated elegance that such a classy, unique wine deserves.

✓ *Botrytised (Dolce)*

DOMAINE CARNEROS
Napa
★★

The Domaine Carneros winery has the pristine façade of Taittinger's 17th-century Château de la Marquetterie in Champagne, and Le Rêve is one of the greatest 21st-century sparkling wines produced outside of Champagne.

✓ *Pinot Noir (Avant-Garde) • Sparkling wine (Brut, Brut Rosé, Le Rêve)*

DOMAINE CHANDON
Yountville
★

Moët & Chandon's seal of approval kick-started California's sparkling wine industry, which until then had consisted of just one producer, Schramsberg. Although the style of Chandon's sparkling wine has lacked finesse in the past, there has been a noticeably lighter hand in recent years. The same could be said for its Pinot Noir, but that all changed with the 2000 vintage, which was impressive by any yardstick, while the Pinot Meunier has always been a delight.

✓ *Chardonnay (Carneros) • Pinot Meunier • Pinot Noir (Carneros) • Sparkling wine (Mt Veeder, Riche, Vintage)*

DOMINUS
Yountville
★★

Christian Moueix of Château Pétrus joined forces with the daughters of John Daniel to produce Dominus, a massively structured Bordeaux-like blend from the historic Napanook vineyard.

✓ *Classic red blend (Dominus)*

DUCKHORN VINEYARDS
St Helena
★☆

These superb, dark, rich, and tannic wines have a cult following. Also owns the Goldeneye Winery in Mendocino. Other labels include Paraduxx (a Zinfandel–Cabernet Sauvignon blend), Decoy (Bordeaux-style blend), Migration (Pinot Noir), Calera, and Canvasback.

✓ *Cabernet Sauvignon • Merlot (Three Palms) • Sauvignon Blanc*

DUNN VINEYARDS
Angwin
★★

Randy Dunn makes a very small quantity of majestic Cabernet Sauvignon from his Howell Mountain vineyard.

✓ *Cabernet Sauvignon*

EISELE VINEYARD ESTATE
Calistoga
★★

Previously named Araujo Estate, in 2013 Bart and Daphne Araujo sold to the

Pinault family, the French owners of first-growth Château Latour. Located north of Cuvaison, Eisele employs organic and biodynamic practices to make Cabernet Sauvignon of outstanding quality.

✓ Cabernet Sauvignon (Eisele Vineyard) • *Sauvignon Blanc (Eisele Vineyard)* • *Syrah (Eisele Vineyard)*

ELYSE
Napa
★☆

These wines have always had richness of fruit and expressive style, but they increasingly show greater finesse.

✓ Cabernet Sauvignon (Morisoli Vineyard, Tietjen Vineyard) • Zinfandel (Morisoli Vineyard)

ETUDE
Oakville
★★

Some beautifully proportioned, silky-smooth, succulent, and stylish wines at the venture established by Tony Soter and now owned by Beringer Blass.

✓ Cabernet Sauvignon (Napa Valley) • *Pinot Gris (Carneros)* • *Pinot Noir (Heirloom)*

FAILLA
Calistoga
★★

Chardonnay and Syrah are a focus at this winery, with Failla now boasting eight different bottlings of Chardonnay and three of Syrah. Fruit is sourced from such venerable Napa locations as the Hudson Vineyard in Carneros and the Haynes Vineyard in Coombsville.

✓ Chardonnay • *Pinot Noir* • *Syrah*

FAR NIENTE
Oakville
★

A revitalized pre-Prohibition winery that made its mark with fine Chardonnay in the 1980s. Hats off to Far Niente with

their forward thinking on sustainability; they boast a floating solar farm, making them fully self-sufficient for electricity.

✓ Cabernet Sauvignon • *Chardonnay*

FLORA SPRINGS
St Helena
★★

This winery produces an exciting Trilogy, a Bordeaux blend, and Soliloquy, a Sauvignon Blanc-driven blend with Chardonnay, Malvasia, and Pinot Gris.

✓ Cabernet Sauvignon • *Classic red blend (Poggio del Papa, Trilogy)* • *Merlot (Napa Valley)* • *Sauvignon Blanc (Soliloquy)*

FORMAN VINEYARDS
St Helena
★

Rick Forman, formerly of Sterling Vineyard, set up this small winery in 1983. He has a following for classic yet opulent wines from his own vineyard, which sits on a gravel bed as deep as 17 metres (56 feet).

✓ Cabernet Sauvignon • *Chardonnay*

FRANCISCAN VINEYARDS
Rutherford
★☆

Since the mid-1980s Franciscan has been a consistent producer of smooth, stylish, premium-quality wines.

✓ Chardonnay • *Classic red blend (Magnificat Meritage)*

FREEMARK ABBEY
St Helena
★☆

When it comes to its Cabernet Bosché and Chardonnay in general, Freemark Abbey consistently combines high quality and good value. Freemark is now part of Jackson Family Wines.

✓ Cabernet Sauvignon (Bosché) • *Chardonnay*

FROG'S LEAP WINERY
St Helena
★

One of California's best-known organic producers, Frog's Leap was established in 1981 by John Williams. Williams had taken inspiration from his first Napa Valley winemaking post at Stag's Leap Cellars as the name for his new venture.

✓ Chardonnay • *Merlot* • *Sauvignon Blanc* • *Zinfandel*

GALLICA WINE
St. Helena
★☆

Callica wine was established in 2007 by Rosemary Cakebread, who previously managed the estate (vineyard and winemaking) at Spottswoode. The aim was to reflect the unique sites the fruit is sourced from (not just within Napa Valley) with a clear expression of vintage and varietal. The wines are elegant, focussed, and perfumed.

✓ Cabernet Sauvignon • *Albarino* • *Grenache* • *Petit Sirah*

GIRARD WINERY
Oakville
★☆

Top-quality, smooth, and stylish Napa Valley Red boasts seamlessly integrated oak and is the top wine here.

✓ Chardonnay (Russian River) • *Classic ed blend (Artistry)*

GRACE FAMILY VINEYARDS
St Helena
★★

The Grace Family Vineyards produce minuscule amounts of huge, dark, extremely expensive Cabernet Sauvignon wine, with wonderfully multi-layered fruit, oak, and *terroir* flavours.

✓ Cabernet Sauvignon

GRGICH HILLS CELLAR
Rutherford
★☆

Created by Mike Grgich, who was a winemaker at Château Montelena, this winery produces intense, rich, and vibrant styles of high-quality wine.

✓ Cabernet Sauvignon (Napa Valley) • *Merlot (Napa Valley)* • *Sauvignon Blanc (Estate Grown Fumé Blanc)* • *Zinfandel (Miljenko's Old Vine)*

GROTH VINEYARDS
Oakville
★★

Still family owned, Groth produces a Cabernet Sauvignon that can still hack it with the finest in California.

✓ Cabernet Sauvignon (Oakville)

HARLAN ESTATE
Napa
★★☆

Classic quality from the hills above Oakville, but you have to pay for it.

✓ Bordeaux-style red (Napa Valley)

HARTWELL VINEYARDS
Napa
★★

Bob and Blanca Hartwell's vineyard is still operated by the Hartwell family, albeit the new generation. This producer is known for immaculate grapes and a deep, full, lush style, the focus is now on two wines only – Cabernet Sauvignon and Sauvignon Blanc.

✓ Cabernet Sauvignon • *Sauvignon Blanc*

HAVENS
Napa
★★

Mike Havens gave his name to this winery, which was purchased in 2010 by the Smith-Anderson Wine Group. Havens generated a reputation for their supremely rich Merlot Reserve, which is certainly one of California's finest, but they also make a Syrah that is in a class of its own.

✓ Classic red blend • *Merlot* • *Syrah*

HEITZ WINE CELLARS
St Helena
★★

The TCA of the 1990s (or bacterial infection as some have claimed) appears to have been overcome, although too few American critics dared tell the late king that he had no clothes on at the time. It will be a surprise to some, but Heitz is a master of Grignolino, making a delightful, quaffing red and a gorgeous rosé, not to mention a very appealing Port-style fortified wine.

✓ Cabernet Sauvignon (Martha's Vineyard, Trailside Vineyard) • *Fortified (Heitz Cellar Port)* • *Grignolino (Rosé)*

INGLENOOK
Rutherford
★★

Formally Niebaum-Coppola winery and then Rubicon Estate Winery. Hollywood movie director Francis Ford Coppola purchased the old Gustave Niebaum homestead and half of the original

THE WINE IN TASTING ROOM AT INGLENOOK, A HISTORIC NAPA VALLEY WINE ESTATE
Founded in 1879 by Finnish-born Gustave Niebaum, this lovely Napa County property draws scores of tourists each year.

Inglenook vineyards in 1975, producing his first wine, a classic red Bordeaux-style blend called Rubicon, in 1978. He bought almost all of the remaining Inglenook vineyards in 1995, plus the Inglenook winery, but not the brand at that time. It took till 2011 and "an offer they couldn't refuse", to expensively wrestle the Inglenook name from The Wine Group, paying more, he said, than he had for the entire estate. In 2002 Coppola paid out a record 31.5 million dollars for the 24-hectare (60-acre) adjoining J J Cohn vineyard.

✓ Cabernet Sauvignon • Classic red blend (Rubicon) • Merlot (Estate) • Zinfandel (Edizione Pennino)

JOSEPH PHELPS VINEYARDS
St Helena
★★

This prestigious winery continues to make a wide range of top-quality wines, with an increasing emphasis on exciting Rhône-style wines, Insignia remains its very greatest product.

✓ Cabernet Sauvignon (Backus) • Classic red blend (Insignia) • Syrah (Napa Valley)

KENT RASMUSSEN
Napa
★★

As a librarian, Kent Rasmussen loved wine and yearned to know more, so he took a BS in oenology at UC Davis in 1983, picked up some useful experience at Robert Mondavi and Domaine Chandon, took a stage at Stellenbosch Farmers' Winery in South Africa, and then at Saltram's Wine Estate in Australia, before setting up his own winery in 1986. Since then he has created a reputation for producing Carneros Pinot Noir at its most luscious.

✓ Chardonnay (Napa Valley) • Pinot Noir (Carneros)

KONGSGAARD
Oakville
★★☆

Many small wineries claim to make hand-crafted wines, but John and Maggy Kongsgaard's is one of the few that actually does, limiting their production – literally – to the wine they can make with their own hands. This and their nigh fanatical minimalist approach yield wines of exceptional richness, balance, and finesse.

✓ Entire range

LA JOTA
Angwin
★☆

La Jota is best known for lavishly flavoured Cabernet Sauvignon. It is part of Jackson Family Wines.

✓ Cabernet Sauvignon

ROBERT KEENAN WINERY
St Helena
★

This Spring Mountain hillside winery produces admirably restrained Cabernet Sauvignon and Merlot.

✓ Cabernet Sauvignon • Merlot

LAIL
Napa
★★

Established in 1995 by Robin Lail, the daughter of John Daniel (himself the great-nephew of the legendary Gustave Niebaum). Lail trained at Mondavi and was the co-founder of Dominus and Merryvale, before selling her share in those businesses to set up on her own.

✓ Classic red blend (Blueprint, J Daniel Cuvée) • Sauvignon Blanc (Georgia)

LARKIN
St Helena
★★

After representing a number of small, high-quality wineries, Sean Larkin made his first wine in 1999 and achieved instant critical success from all the biggest names in the business. Selling wine also in cans named a "Larkan".

✓ Cabernet Franc

LARKMEAD VINEYARD
Calistoga
★★

This one of the most historic wineries in Napa Valley, originally established in 1895. Under the current family ownership since 1948, the estate's quality and reputation has been meticulously transformed and elevated over the last couple of decades.

✓ Cabernet Sauvignon (The Lark, Solari, Dr Olmo) • Bordeaux blends (Firebelle, LMV Salon) • Tocai Friuliano • Sauvignon Blanc (Lillie)

LEWELLING VINEYARDS
St Helena
★★

The Lewelling family established these vineyards in 1864 and, after the usual Prohibition hiatus, developed a reputation for supplying some of Napa's greatest wineries with grapes. The wines, however, only date from 1992, when the Wight family, descendants of the Lewellings, decided to produce wine under their own vineyard designation, and now produce some of California's greatest Cabernet Sauvignon.

✓ Cabernet Sauvignon

LEWIS CELLARS
Napa
★★

A small, hands-on winery run by ex-race car driver Randy Lewis and his wife and son. The Merlot could be called a California version of top-class Pomerol.

✓ Chardonnay • Classic red blend (Alec's Blend, Cuvée L) • Merlot • Syrah

LOKOYA
Oakville
★★☆

Named after a Native American tribe that lived on Mount Veeder, Lokoya is a Napa Cabernet Sauvignon specialist par excellence. All three styles are outstanding, from the Howell Mountain, which is the most lush, through the Mount Veeder, which packs the most complex fruit, to Diamond Mountain, which is

probably the closest to Bordeaux in style. A Jackson Family Wines brand.

✓ Cabernet Sauvignon

LOUIS M MARTINI
St Helena
☆

The quality of this winery had been slowly turning around for the better when it was purchased by E & J Gallo in 2002, although its reputation is still very patchy from vintage to vintage, even for its most successful wines.

✓ Cabernet Sauvignon (Monte Rosso) • Zinfandel (Monte Rosso)

LUNA VINEYARDS
Napa
★

This is an artisanal winery with an Italian bent, located at the southern tip of the Silverado Trail.

✓ Classic red blend (Canto) • Pinot Grigio • Sangiovese (Riserva)

MARKHAM VINEYARDS
St Helena
★☆

Markham, an underrated Japanese-owned (Mercian Corporation) winery, is best known for Merlot and Cabernet from its Napa Valley vineyards situated at Calistoga, Yountville, and Oakville, but other varietals are beginning to emerge. Good-value wines are also produced under Markham's second label, Glass Mountain.

✓ Cabernet Sauvignon • Merlot • Zinfandel

MAYACAMAS VINEYARDS
Napa
★★

This prestigious small winery is renowned for its tannic, long-lived Cabernet, but Mayacamas actually produces much better, more user-friendly Chardonnay as well.

✓ Chardonnay

MERRYVALE VINEYARDS
St Helena
★☆

These are classy, stylish wines, particularly the Bordeaux-style blends.

✓ Cabernet Sauvignon • Chardonnay (Silhouette) • Classic red blend (Profile)

MONTICELLO CELLARS
Napa
☆

This producer used to be better known for its white wines over its reds, but although Monticello still receives many plaudits for its Chardonnay, it is the red wines that are now established as its most exciting.

✓ Cabernet Sauvignon (Corley Reserve) • Classic red blend (Corley Proprietary Red) • Merlot (Estate Grown)

MUMM NAPA WINERY
Rutherford
★☆

This impeccably run sparkling wine operation produces a Sparkling Pinot

Noir which has an outrageous deep cerise colour, an aroma of strawberries, and intensely perfumed Pinot fruit.

✓ Sparkling wine (Blanc de Blancs, DVX, Sparkling Pinot Noir)

MURPHY-GOODE
Geyserville
★

The top wines are all rich, oaky blockbusters, but I get the most pleasure from Murphy-Goode's Fumé Blanc.

✓ Fumé Blanc

NAPA CELLARS
Oakville
★☆

A brilliant source of great-value, great-quality wines that hit well above their weight, often outclassing famous brands at two or three times the price. Owned by the Trinchero Family Estates.

✓ Cabernet Sauvignon • Chardonnay • Merlot

NEYERS
Rutherford
★★

Bruce (now retired) and Barbara Neyers have been producing classy wines since 1992. Part of the production is from the Neyers' own 20-hectare (50-acre) Conn Ranch vineyard, but most is bought in from selected growers. Another of the Trinchero Family Estates.

✓ Chardonnay • Grenache • Syrah

NEWTON VINEYARDS
St Helena
★★

This top-notch estate has a track record for wines with impeccable oak integration, silky smooth fruit, and supple tannin structure.

✓ Cabernet Sauvignon • Chardonnay

OPUS ONE
Oakville
★☆

The product of collaboration between Robert Mondavi and the late Baron Philippe de Rothschild, Opus One was the first Bordeaux–California joint venture, and there is no denying the quality of this Cabernet-Merlot blend, just as there is no getting away from its outlandish price.

✓ Opus One

PAHLMEYER
Napa
★★

Typical of the boutique wineries established by America's affluent professional class, this operation was founded in 1985 by lawyer Jayson Pahlmeyer, who has hired various consultants, such as Randy Dunn, Helen Turley, and David Abreu, in his pursuit of varietal concentration allied to an expression of terroir. Gallo acquired Pahlmeyer in 2019 to add to their bulging portfolio. The Jayson is a blend of the wines that did not make the cut for Pahlmeyer's Proprietary Red.

✓ Chardonnay • Classic red blend (Jayson, Proprietary Red) • Merlot

PALOMA

Spring Mountain District

★★

This reclaimed 19th-century vineyard is located at a high altitude on Spring Mountain, and the difference between these wines and valley floor wines is that you get more complexity and length for the same weight of wine.

✓ *Cabernet Sauvignon • Merlot • Syrah*

PATZ & HALL

Napa

★☆

This Ste Michelle Wine Estates winery specializes in rich, lush, single-vineyard Burgundian-style varietals.

✓ *Chardonnay • Pinot Noir (Alder Springs, Pisioni Vineyard)*

PEJU PROVINCE

Rutherford

★★

Established by Tony and Herta Peju in 1982, but with 90 per cent of sales going through its tasting room, it has only been in the past 10 years or so that Peju Province has really started to grab wider attention.

✓ *Cabernet Franc • Merlot (Napa Valley Estate) • Sauvignon Blanc*

PHILIP TOGNI VINEYARD

St Helena

★☆

Philip Togni earned great respect at Chalone, Chappellet, Cuvaison, Mayacamas, and Chimney Rock, before establishing this vineyard.

✓ *Cabernet Sauvignon • Classic red blend*

PINE RIDGE WINERY

Napa

★☆

These are interesting wines of exciting quality that are in the mid-price range.

✓ *Cabernet Sauvignon (Stags Leap District) • Classic red blend • Classic white blend (Chenin-Viognier) • Merlot*

PLUMPJACK

Napa

★☆

Owned by the Getty family and Gavin Newsom, Plumpjack is named after the roguish spirit of Shakespeare's Sir John Falstaff, dubbed "Plump Jack" by Queen Elizabeth. There was a whole Plumpjack empire out there, starting with shops and cafés, before the Plumpjack winery was established in 1997. It made headlines in 2000, when it sold its 1997 Reserve Cabernet Sauvignon under screwcap at $135 a bottle, while the regular cork-sealed version retailed for $125. Not only was Newsom making a statement, but he was also years ahead of his time in making it as far as red wines are concerned.

✓ *Cabernet Sauvignon*

RAYMOND

St Helena

☆

The Raymond family remain true to their reputation for vividly fruity Chardonnay and stylish Bordeaux-style reds.

✓ *Cabernet Sauvignon • Chardonnay • Merlot*

ROBERT CRAIG

Mount Veeder

★★

Capable of classy, purple-hearted Cabernet Sauvignon.

✓ *Cabernet Sauvignon • Chardonnay*

ROBERT FOLEY

Angwin

★★

This red wine specialist makes – most notably – a classic blend (essentially Cabernet Sauvignon, plus a little Merlot and Petit Verdot) from his own vineyard on Candlestick Ridge and Charbono. Other varietals such as Petit Sirah, Syrah, and Pinot Noir have been added to the Robert Foley offering.

✓ *Charbono • Red wine blend (Claret)*

ROBERT MONDAVI

Oakville

★☆

Established in 1966 by Bob Mondavi, who hoped to build a wine dynasty and appeared to have succeeded beyond his wildest dreams by the 1980s. The Mondavi brand struggled to maintain its value in the new millennium, however, when profits plunged by 40 per cent, and the company was sold to Constellation Brands in November 2004. The irony is that the top Robert Mondavi wines remain as great as they have ever been. Cabernet Sauvignon is Mondavi's greatest strength, even at the basic Napa Valley appellation, which can be very special in years such as 2000 and 2002. This winery has been rightly accused of turning out large volumes of very ordinary wine, yet it would be rated two and a half stars if judged on Cabernet Sauvignon alone, as its Stags Leap District *cuvée*, and single-vineyard wines from To Kalon are on the same top-quality level as the Reserve.

✓ *Cabernet Sauvignon • Classic red blend (Momentum) • Sauvignon Blanc (To Kalon Block Fumé Blanc)*

ROBERT SINSKEY VINEYARDS

Napa Stags Leap

★☆

Robert Sinskey produces vividly flavoured Pinot, classy Vineyard Reserve, crisp Vin Gris, and the wonderfully balanced Edelzwicker-style blend "Abraxas" all from its own estate vineyards in Stags Leap and Carneros. Eschewing critical acclaim, the wines are created with a food focus in mind.

✓ *Classic red blend • Pinot Noir • Rosé (Vin Gris of Pinot Noir) • Abraxas (Riesling, Pinot Blanc, Pinot Gris, Gewurztraminer)*

SADDLEBACK

Oakville

★☆

Nils Venge, formerly at Groth, is a master of the massive, old-fashioned, port-like style of Zinfandel.

✓ *Zinfandel*

SAINTSBURY

Napa

★☆

David Graves and Dick Ward make plump, juicy Pinot Noir at this increasingly impressive winery.

✓ *Chardonnay • Pinot Noir*

SCHRADER CELLARS

Calistoga

★★

Having been established in 1998, Schrader Cellars quickly rose to cult status, repeatedly receiving 100-point scores from the critics. Sold to Constellation Brands (owner of Robert Mondavi) in 2017 to bolster their premium wine portfolio.

✓ *Cabernet Sauvignon*

SCHRAMSBERG VINEYARDS

Calistoga

★

This prestigious winery started the modern California sparkling wine industry rolling in 1965 and, until Domaine Chandon was set up a few years later, was the only producer of serious-quality bottle-fermented sparkling wine in the United States. The head winemakers at Codorníu, Franciscan, Kristone, Mumm Napa Valley, and Piper-Sonoma all cut their teeth here.

✓ *Sparkling wine (Blanc de Noirs, Crémant, J Schram)*

SCREAMING EAGLE

Oakville

★☆

They make tiny quantities of cult Cabernet Sauvignon, always excruciatingly expensive and mostly of exceptional quality, although I have found the odd under-performer at blind tastings.

✓ *Cabernet Sauvignon*

SEAVEY

St Helena

★

Small quantities of beautifully crafted Cabernet Sauvignon wines are produced here. The Merlot is good, but less intense.

✓ *Cabernet Sauvignon • Merlot*

SEQUOIA GROVE

Napa

★

Jim and Steve Allen have been producing great Cabernet Sauvignon since 1978, and are now sourcing fine Chardonnay from Carneros.

✓ *Cabernet Sauvignon • Chardonnay*

SHAFER VINEYARDS

Napa

★☆

Although the Cabernet Sauvignon Hillside Select is definitely the best wine, Shafer Vineyards also makes very stylish Chardonnay. Relentless is a powerful blend of Syrah and Petite Sirah.

✓ *Cabernet Sauvignon (Hillside Select) • Chardonnay • Classic red blend (Relentless)*

SIGNORELLO VINEYARDS

Napa

★☆

The late Ray "Padrone" Signorello bought a vineyard on the Silverado Trail as long ago as the 1970s but only came to the fore with a bewildering display of fruity yet complex wines in the 1990s. The quality and consistency stepped up a gear under Ray Signorello Jr and is now stewarded by the Birebents. Destroyed in a wildfire in 2017, the winery has subsequently been rebuilt.

✓ *Cabernet Sauvignon (Estate) • Chardonnay (Hope's Cuvée) • Classic red blend (Padrone)*

SILVER OAK CELLARS

Oakville

★

Concentrates solely on one variety, Cabernet Sauvignon, but there are two brands – Napa Valley and Alexander Valley. I prefer the latter because it has the least oak and the most finesse.

✓ *Cabernet Sauvignon (Alexander Valley)*

SILVERADO VINEYARDS

Napa

★☆

Silverado is owned by the widow of the late Walt Disney, but it has never produced a Mickey Mouse wine.

✓ *Cabernet Sauvignon (Limited Reserve) • Sangiovese • Sauvignon Blanc*

SMITH-MADRONE

St Helena

★☆

This winery consists of some 16 hectares (40 acres) of high-altitude vines producing a small, high-quality range of wines.

✓ *Cabernet Sauvignon • Chardonnay • Riesling*

SPOTTSWOODE

St Helena

★★

On benchland in the Mayacamas Mountains, the Novak family produce just two wines – tiny amounts of powerful but stylish Cabernet Sauvignon and a deliciously fresh Sauvignon Blanc.

✓ *Cabernet Sauvignon • Sauvignon Blanc*

SPRING MOUNTAIN VINEYARDS

St Helena

★☆

Known to millions as "Falcon Crest", this winery can produce wines that require ageing and develop finesse.

✓ *Classic red blend (Elivette – formerly sold as Reserve) • Sauvignon Blanc*

ST CLEMENT VINEYARDS

St Helena

★☆

St Clement remains one of the best-performing California wineries in the Beringer-Blass group.

✓ *Cabernet Sauvignon • Classic red blend (Oroppas)*

ST SUPÉRY
Rutherford
★☆

Capable of producing Cabernet of a rare level of finesse, and a deliciously tangy, yet inexpensive Sauvignon Blanc. Owned since 2015 by Alain and Gérard Wertheimer (Chanel).

✓ *Cabernet Sauvignon (Dollarhide Ranch, Napa Valley) • Classic red blend (Elú Meritage, Rutherford Limited Edition) • Merlot • Muscat Canelli (Moscato) • Sauvignon Blanc • Sémillon (Dollarhide Ranch)*

STAG'S LEAP WINE CELLARS
Napa
★★

This is the famous Stag's Leap, with the apostrophe before the "s" and its legendary, ultra-expensive Cabernet Sauvignon Cask 23, although its Fay Vineyard and SLV can sometimes rival it in quality. Artemis is a lighter, easier-drinking, highly recommended, less-expensive blend of the Stag's Leap Arcadia vineyard with the leftovers from the selection of the three top Cabernet Sauvignons. Its second label is Hawk Crest. In 2007 it was scooped up by Washington's Ste Michelle and Tuscan producer Piero Antinori.

✓ *Cabernet Sauvignon • Chardonnay (Arcadia)*

STAGS' LEAP WINERY
Napa
★★

This is the other Stags' Leap winery, with the apostrophe after the "s" and less well known, although, ironically, the first to be established, as long ago as 1893. Stags' Leap Winery makes good Cabernet Sauvignon, particularly the Reserve, but it is not in the same class as the better-known Stag's Leap.

✓ *Cabernet Sauvignon*

STERLING VINEYARDS
Calistoga
★

Providing astonishingly high quality in the mid-1970s, Sterling's standards dropped after Coca-Cola purchased it in 1978, but it started a revival when Seagram took over in 1983 and has continued to improve. It has been a Treasury Wine Estates property since 2016.

✓ *Cabernet Sauvignon (Napa Valley Reserve) • Chardonnay (Napa Valley Reserve, Carneros Winery Lake)*

STONY HILL VINEYARD
St Helena
★

The Chardonnays here are still world class, but fatter than before.

✓ *Chardonnay*

STORYBOOK MOUNTAIN VINEYARDS
Calistoga
★★

Jerry Seps, a former university professor, has created a reputation for one of California's finest Zinfandels on this mountain vineyard. Storybrook was originally established by Adam and Jacob Grimm in 1880.

✓ *Zinfandel*

SUTTER HOME WINERY
St Helena
☆

Zinfandel of one style or another accounts for the majority of this winery's vast production under the state-wide California AVA. Some wines are brilliant one year, but nothing special the next, making recommendations difficult, but when the wines are good, the value is always great.

✓ *Chenin Blanc • Sauvignon Blanc*

SWITCHBACK RIDGE
Calistoga
★★

This converted orchard on the Peterson Ranch at the northern end of the Silverado Trail makes some of Napa's best and biggest reds.

✓ *Cabernet Sauvignon • Merlot • Petite Sirah*

THE HESS COLLECTION
Napa
★☆

The Hess family make wines to be consumed not collected, although the wines do last well. Readers should note that Hess Select is a second label and not a premium *cuvée*.

✓ *Cabernet Sauvignon (Mount Veeder)*

TITUS
St Helena
★★

Lee Titus brought his family from Minnesota to Napa, where he purchased this vineyard in separate lots in 1967. He replanted with classic varieties and began to supply some of the top Napa Valley brands until the next generation decided to make and sell their own wine. With Phillip Titus as consultant winemaker, and his brother Eric running both the business and the vineyard, Titus now threatens to become "the most fashionable new name in California winedom", according to Robert Parker.

✓ *Cabernet Franc*

TREFETHEN VINEYARDS
Napa
☆

This winery combines quality and consistency with good value.

✓ *Cabernet Sauvignon • Chardonnay • Riesling*

TRINCHERO WINERY
Napa
★

Although Sutter Home is the bigger brand and gets all the press, this winery has been owned by the Trinchero family since 1947, and they make sure that some pretty good wines go out under their own label.

✓ *Cabernet Sauvignon (Mario's Vineyard) • Classic red blend • Sauvignon Blanc*

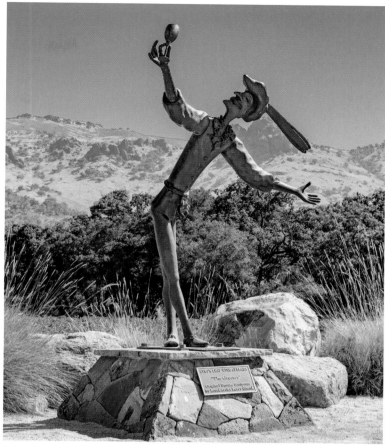

"THE GREETER" WELCOMES VISITORS TO STAG'S LEAP WINE CELLARS
Local artist Larry Shank crafted the whimsical bronze sculpture for this well-regarded winery. The famous Judgement of Paris 1976 awarded highest honours to its 1973 Cabernet Sauvignon, which showed the world that California reds could compete with the best of Bordeaux – and Stag's Leap has maintained its standards to this day.

TURLEY
Napa
★★☆

Talented brother and sister winemakers Larry and Helen Turley produce a raft of spectacular Zinfandel, with Petite Syrah (*sic*) and other interesting odds and ends.

✓ *Petite Syrah • Zinfandel*

V SATTUI
St Helena
★☆

This underrated winery boasts a large range of relatively inexpensive, award-winning wines, but it is let down by its rustic, 1950s labelling and presentation. A change to understated livery would do wonders.

✓ *Cabernet Sauvignon (Morisoli Vineyard) • Chardonnay (Napa Valley) • Classic red blend (Family Red) • Fortified (Madeira) • Merlot (Napa Valley) • Riesling (Dry, Off-Dry) • Rosé (Gamay Rouge) • Zinfandel*

VIADER
St Helena
★☆

Tiny quantities of deeply coloured, concentrated, red Bordeaux-style wine blended from vines grown on the slopes of Howell Mountain. Viader is a blend of 55 to 70 per cent Cabernet Sauvignon with Cabernet Franc, while the firmer V is mostly Petit Verdot with the rest being Cabernet Sauvignon.

✓ *Classic red blend (V, Viader)*

WHITEHALL LANE WINERY
St Helena
★☆

Despite various owners, Whitehall Lane consistently produces inspired wines, as endorsed by the Decanter World Wine Awards, when its 2001 Reserve Cabernet Sauvignon carried off the North American Bordeaux Varietal Trophy.

✓ *Cabernet Sauvignon • Merlot*

WILLIAM HILL WINERY
Napa
★☆

An improving winery in the Gallo stable.

✓ *Cabernet Sauvignon • Chardonnay • Merlot*

ZD WINES
Napa
★☆

This winery used to make variable Cabernet Sauvignon but is now very consistent and this wine now challenges Chardonnay's position as ZD's best varietal. wines ZD sometimes comes out with a killer Pinot Noir (such as the 2000 Carneros Reserve).

✓ *Cabernet Sauvignon • Chardonnay • Pinot Noir (Carneros Reserve)*

The Central Coast (North)

Running from south of San Francisco in the north, down to the southern Monterey County border with San Luis Obispo, the northern Central Coast was originally a district where a few big companies produced a vast quantity of inexpensive wines. As producers have continued to explore and plant interesting new sites, the northern Central Coast now boasts a large number of top-quality, highly individual AVAs and wineries.

Winemaking in the northern Central Coast dates from the 1830s, established in and around Santa Clara County, which remained the focus until the late 1950s and early 1960s, when the growing urban sprawl of San Jose forced the industry to search out new areas for vine growing. This search coincided with the publication of a climatic report based on heat summation (*see* p577) by the University of California, pinpointing cooler areas farther south, particularly in Monterey, that could support fine-wine vineyards.

THE MOVE TO MONTEREY

In 1957 Mirassou and Paul Masson were the first companies to make the move, buying some 530 hectares (1,300 acres) in the Salinas Valley. Some areas, however, were too cool or exposed to excessive coastal winds. The blame for these failures lay with producers who could not conceive that grapes would not ripen in California.

In October 1966 Albert Winkler and Maynard Amerine, the authors of the heat-summation study, were honoured at a luncheon where a toast was made to "the world's first fine-wine district established as the direct result of scientific temperature research". This might have seemed premature but has been substantiated by Monterey's viticultural growth. The pioneering grape-grower Bill Jekel first brought the quality potential of this area to international fame in 1972 with his eponymous winery, founding a reputation that has continued to grow, especially for producing world-class cooler-climate Chardonnay and Pinot Noir.

FACTORS AFFECTING TASTE AND QUALITY

LOCATION
The Central Coast's northern sector stretches from the San Francisco Bay area to the southern Monterey County border, running approximately 100 kilometres (60 miles) inland from the Pacific Coast.

CLIMATE
The region is generally warm (Region III, *see* p277), but with variations such as the cooler (Region I) areas of the Santa Cruz Mountains and the northern part of the Salinas Valley. Mountainous topography has a fundamental influence on the ranging mesoclimates, either blocking or funnelling cooling Pacific winds and fog.

ASPECT
Vines are planted mainly on the flat and sloping lands of the valleys. Variations are found on the steep slopes of the Santa Cruz Mountains and the high benchland of the Pinnacles above Soledad, for example. Altitude is another key factor in determining whether vineyards sit above or below the fog line.

SOIL
A wide variety of gravel loams, often high in stone content and rich in limestone, in the Livermore Valley; clay and gravel loams in Santa Clara; sandy and gravelly loams over granite or limestone in San Benito; and gravelly, well-drained, low-fertility soils in Monterey. There is further variation from the more fertile, alluvial soils generally found in the valley floors, to the shallow, less productive mountainous sites.

VITICULTURE & VINIFICATION
The climatic range results in a large selection of varieties being grown. A small number of big wine companies produce a vast quantity of inexpensive wines utilizing high-tech, production-line methods. This is balanced by the abundance of boutique, quality-focussed wineries, a number of which have garnered international recognition.

GRAPE VARIETIES
Cabernet Sauvignon, Chardonnay, Pinot Noir, Sauvignon Blanc, Zinfandel

---

THE VITICULTURAL AREAS OF THE NORTHERN CENTRAL COAST
(see also p577)
The northern section of the Central Coast includes Alameda, Contra Costa, and San Francisco counties and portions of San Mateo, Santa Clara, and Santa Cruz counties, as well as the Monterey AVA (see Extended Area at right.)

THE APPELLATIONS OF
THE CENTRAL COAST (NORTH)

ALAMEDA COUNTY AO

This appellation covers grapes grown anywhere within Alameda County.

ARROYO SECO AVA
Monterey County

With a name meaning "dry riverbed", this triangular-shaped appellation runs from a narrow gorge opening up to wider, sandy loam soils of the Salinas Valley. Wide variation in topography and microclimates is reflected in the range of varieties planted: Bordeaux and Rhône varieties, Chardonnay, Riesling, and Zinfandel.

BEN LOMOND MOUNTAIN AVA
Santa Cruz County

This AVA in the west of the Santa Cruz Mountains sits above the fog-line, tempered by cool Pacific winds and elevation. Encompassing just 28 hectares (70 acres) of vines, Beauregard Vineyards is currently producing the only wines under the Ben Lomond Mountain AVA designation.

CARMEL VALLEY AVA
Monterey County

Vineyards in this AVA are predominantly planted on mountainous terrain around the Carmel River and Cachagua Creek. A distinctive microclimate is created by the valley's elevation above the fog line and the northeastern Tularcitos Ridge, which curbs the marine fog and provides more sunny days. Dramatic diurnal temperature swings of up to 35°C (70°F) extend the ripening season and retain acidity. Cabernet Sauvignon and Merlot account for more than 70 per cent of all plantings.

CHALONE AVA
Monterey County

High altitude benchland vineyards at 500 metres (1,650 feet) above sea level, with AVA-specific volcanic and granitic soils of high limestone content. The soil here combines with an arid climate to stress the vines, which are offered respite by cool nights. Suited to Chardonnay, Pinot Noir, Pinot Blanc, and Syrah.

CIENEGA VALLEY AVA
San Benito County

The Cienega Valley is located on the western edge of San Benito County at the base of the Gabilan (or Gavilan) Mountain Range where the Pescadero Creek is used artificially to augment the area's rainfall. The valley floor soils are divided by the San Andreas Fault, with granite and sandstone predominantly in the east compared to granite and limestone in the west. Cabernet Sauvignon and Zinfandel are the principle varieties.

CONTRA COSTA COUNTY AO

An appellation covering grape varieties grown anywhere within Contra Costa County.

HAMES VALLEY AVA
Monterey County

Situated north of Lake Nacimiento, at the southern end of Monterey County, the area is sheltered from cooling Pacific winds resulting in a warmer climate compared to appellations farther north. Rhône varietals are ideally suited to the climate.

LAMORINDA AVA
Contra Costa County

Within the greater Central Coast AVA, located east of San Francisco in a suburban area incorporating part of the cities of Lafayette, Moraga, and Orinda. The area is composed of moderate-to-steep slopes and narrow valleys that are exposed to modest maritime influence from the San Francisco and San Pablo Bays. The warm climate and clay-rich, but thin, soils suit Cabernet Sauvignon.

LIME KILN VALLEY AVA
San Benito County

Located within Cienega Valley AVA, this region's annual rainfall ranges from 41 centimetres (16 inches) on the valley floor to 102 centimetres (40 inches) in the mountainous west. Soils are sandy and gravelly loams over limestone, with a high magnesium carbonate content. Home to old-vine Mourvèdre plantings.

LIVERMORE VALLEY AVA
Alameda County

One of the coastal intermountain valleys just 48 kilometres (30 miles) east of San Francisco, the Livermore Valley has a moderate climate, cooled by sea breezes and morning fog drawn from San Francisco Bay through its east-to-west orientation. Soils are primarily well-drained gravel. Chardonnay, Merlot, Italian, Rhône, and Spanish grape varieties are all planted.

MONTEREY AVA
Monterey County

Running almost the entire length of the county, the region has a wide diversity of sub-climates from the cooler north, due to closer coastal exposure, to the warmer more fertile Salinas Valley in the south. Chardonnay accounts for more 50 per cent of total plantings, with Pinot Noir and Riesling suited to the north, moving to Bordeaux and Rhône varieties farther south.

MONTEREY COUNTY AO

An appellation covering grapes that are grown anywhere within Monterey County.

MOUNT HARLAN AVA
San Benito County

Mount Harlan is situated in the far west of San Benito on limestone outcrops in the same range of hills as Chalone, at a slightly higher altitude of 670 metres (2,200 feet). It is one of California's highest, coolest, and driest AVAs. Calera in the only commercial winery in the appellation, predominantly planted with Chardonnay and Pinot Noir.

PACHECO PASS AVA
Santa Clara and San Benito Counties

Straddling Santa Clara and San Benito County lines, it is a small valley with a flat or gently sloping topography that contrasts with the rugged hills of the Diablo Range to the east and west. The climate is moderate and wetter than that of the Hollister Basin to the south, primarily producing mass-market wines for local consumption.

PAICINES AVA
San Benito County

This appellation is a warm sub-region in the centre of San Benito County, still benefitting from cool nights. Originally associated with high volume, commercial wine production, a number of small wineries are now producing high-quality Chardonnay, Merlot, and Cabernet Sauvignon.

SAN ANTONIO VALLEY AVA
Monterey County

Located in a bowl-shaped valley in the far south of the county surrounded by the Santa Lucia range, the region is still influenced by cooling Pacific winds. Primarily gravelly loam and clay soils combined with the warm, dry climate are ideally suited to Rhône and Bordeaux varieties.

SAN BERNABE VALLEY AVA
Monterey County

The average annual temperature in this appellation is similar to Napa Valley; cooler evening temperatures, however, mean that harvest can be up to four weeks later. Irrigation is also essential due to low annual rainfall allied with sandy soils with low water retention. More than 20 varietals are produced including Chardonnay, Merlot, Pinot Noir, Riesling, Sauvignon Blanc, and Syrah.

SAN BENITO AVA
San Benito County

Not to be confused with San Benito County, this AVA encapsulates the smaller AVAs of Paicines, Cienega Valley, and Lime Kiln Valley.

A GOLF COURSE AND VINEYARDS DOT THE HILLS OF DEL VALLE REGIONAL PARK IN LIVERMORE
The park is located in the Livermore Valley AVA, not far from San Francisco. Vineyards in this appellation benefit from afternoon breezes blowing off the San Francisco Bay. Wine grapes were first cultivated here in the 19th century.

SAN BENITO COUNTY AO

This is an appellation covering grapes grown anywhere within San Benito County.

SAN FRANCISCO BAY AVA

This sizable appellation encompasses more than 607,030 hectares (1.5 million acres) across the counties of San Francisco, San Mateo, Santa Clara, Contra Costa, and Alameda, plus parts of San Benito and Santa Cruz. These areas are all, to some degree, affected by coastal fog and winds coming from the San Francisco Bay. It also includes the city of San Francisco, where a number of urban wineries have sprung up.

SAN LUCAS AVA
Monterey County

The San Lucas AVA consists of a 16-kilometre (10-mile) segment of the Salinas Valley in the southern section of Monterey County, primarily on alluvial loam soils. Maritime breezes are minimal, but fog still affects most of the region, which combined with a diurnal range regularly around 20°C (40°F) extend the growing season. Chardonnay, Sauvignon Blanc, Cabernet Sauvignon, and Merlot are most popular.

SANTA LUCIA HIGHLANDS AVA
Monterey County

Cool (Region I), high-altitude area planted on southeast-facing terraces above the Salinas River Valley, which benefits from cooling afternoon breezes. Chardonnay and Pinot excel in the cool, long growing season whilst Syrah has found a home in the warmer, more protected slopes.

SAN MATEO COUNTY AO

This appellation covers grapes grown anywhere within San Mateo County.

SAN YSIDRO AVA
Santa Clara County

The appellation covers a small enclave of vineyards at the southeast end of the Santa Clara AVA in the foothills of the Diablo Range. San Ysidro sits between two hills that channel the sea breezes coming up the Pajaro River, making it one of the coolest areas in the Santa Clara Valley. Particularly noted in the region are Chardonnay and Pinot Noir.

SANTA CLARA COUNTY AO

This appellation covers grapes grown anywhere within Santa Clara County.

SANTA CLARA VALLEY AVA
Santa Clara County

Encompassing the entire municipality of San Jose to the north and the area better known as Silicon Valley in the south, this all include based in the towns of Morgan Hill, Saratoga, San Martin, and Gilroy, meaning this AVA must surely qualify as the most built-up wine appellation in the world. It is home to more than two dozen wineries, ranging from multi-generational operations to newer, garage-based projects.

SANTA CRUZ MOUNTAINS AVA
Santa Clara County

The term "Santa Cruz Mountains" was first recorded in 1838. Characterised by its rugged topography, the region's climate is influenced in its western reaches by ocean breezes and maritime fog, while the eastern portion is moderated by the San Francisco Bay. The soils here are forms of shale that are particular to the area, with vineyards planted up to 790 metres (2,600 feet) on the ridgetops. Cool air coming down from the mountains forces warmer air up, lengthening the growing season to a full 300 days. Chardonnay, Pinot Noir, and Cabernet Sauvignon are the primary varieties and in high demand for their high quality.

THE WINE PRODUCERS OF
THE CENTRAL COAST (NORTH)

ALBATROSS RIDGE
Monterey County
★☆

Father and son, Brad and Garrett Bowlus, planted the vineyards in 2008 in the Carmel Valley to focus exclusively on estate-grown Pinot Noir and Chardonnay vines. Albatross Ridge is fast gaining a following for the taut, precise structure of the wines.

✓ *Pinot Noir • Chardonnay*

BEAUREGARD VINEYARDS
Santa Cruz
★

Now run by fourth-generation Ryan Beauregard, the Beauregard Vineyards estate is one of the most historic wine properties in California, with records of vine plantings dating back to the 1880s. The work exclusively with high-elevation vineyards in the Santa Cruz and Ben Lomond AVAs (which Ryan's father, Jim, was instrumental in establishing), the focus is on pure expressions of Pinot Noir, Chardonnay, Cabernet Sauvignon, and Zinfandel.

✓ *Pinot Noir • Chardonnay • Cabernet Sauvignon • Zinfandel*

BOEKENOOGEN
Monterey County
★☆

The property was established in 1998 when John Boekenoogen saw the potential to convert his fifth-generation cattle farm in the Santa Lucia highlands to a wine estate, producing cool-climate Chardonnay, Pinot Noir, and Syrah. A second estate, Bell Ranch, was founded in 2005 in the Carmel Valley, growing Viognier, Syrah, Petit Sirah, Cabernet Sauvignon, and Zinfandel.

✓ *Pinot Noir • Chardonnay • Syrah*

BONNY DOON VINEYARD
Santa Cruz County
★★

Founded in 1983, this winery has been one of California's brightest, most bizarre, and most innovative stars, with wines that are electrifying in their brilliance and style, often showing surprising finesse. I have got lost trying to find Bonny Doon on two occasions, but it was worth the effort just to meet the wacky winemaker Randall Grahm. In 2001 Grahm officially declared himself "off his rocker" when he macerated three wines on ground-up rocks from three different parcels of his Cigare Volant red wine. This led to an extremely limited run of "The Rock Quartet". In January 2020, after 35 years in business, Grahm sold the Bonny Doon brand to WarRoom Ventures, and the winery closed its tasting room in the coastal town of Davenport in December 2019. Grahm is staying involved, however, but his focus will be on the vineyards and winemaking, particularly Rhône-style wines.

✓ *Entire range*

DAVID BRUCE
Santa Cruz County
★

Having successfully gone through a learning curve, David Bruce now produces far more elegant wines than he used to. Pinot Noir has emerged as his consistently finest style.

✓ *Petite Syrah (sic) (Central Coast, Shell Creek Vineyard) • Pinot Noir • Zinfandel (Paso Robles)*

CALERA WINE COMPANY
San Benito County
★★

One of California's premier pioneers in the continuing quest for perfect Pinot Noir, Calera continues to produce some of the state's greatest examples of this wine, although the focus on varietal purity can sometimes result in overly pretty fruit. Purchased by the Duckhorn Wine Company in 2017.

✓ *Pinot Noir • Viognier*

CARACCIOLI
Monterey County

This small family-owned enterprise has cleverly secured the services of Michel Salgues, who established the reputation of Roederer Estate and is now one of just a few truly world-class sparkling wine consultants. Some would say he is the best. Certainly these *cuvées* demonstrate a degree of class rarely seen in California sparkling wines.

✓ *Sparkling wine (Brut, Brut Rosé)*

CEDAR MOUNTAIN
Alameda County
☆

This boutique winery was established in 1990, and early vintages are promising, if a little too correct and in need of a personal signature.

✓ *Cabernet Sauvignon • Fortified (Chardonnay del Sol) • Zinfandel*

CHALONE VINEYARD
Monterey County
★★

Purchased in 2016 by Foley Family Wines from Diago, this winery's 63-hectare (155-acre) vineyard has its own AVA. The Chenin Blanc is from the Gavilan Mountain ridge, 540 metres (1,800 feet) above the Salinas Valley, where the vines were planted as long ago as 1919. These venerable vines have the capacity to make California's finest Chenin Blanc, and in some years they succeed in doing so, although there is variation in quality.

✓ *Chardonnay • Chenin Blanc • Pinot Blanc • Pinot Noir*

CONCANNON VINEYARD
Alameda County
☆

This winery continues to make fine Petite Sirah wines. Concannon has been owned by the Wine Group since 2002.

✓ *Petite Sirah*

CONUNDRUM
Monterey County
★★

Under the same ownership as Caymus Vineyard, where Conundrum started out, this winery was created by Jon Bolta, who had assisted in the making of every vintage to date of Caymus wines. The idea behind Conundrum was to make a blended white wine to suit the new Pacific cuisine of coastal California. Jon originally experimented with 11 white grapes before deciding on the 5 that have been used ever since: Sauvignon Blanc, Chardonnay, Muscat Canelli, Sémillon, and Viognier.

✓ *Classic white blend (Conundrum)*

EDEN RIFT
San Benito County
★☆

Eden Rift is one of the oldest continually producing wine estates in California, having been first planted in 1849. Today the estate is primarily planted with Pinot Noir and Chardonnay, ideally suited to the limestone soils, but also includes a block of old-vine Zinfandel dating from 1906, as well as a small plot of Pinot Gris. A low intervention philosophy in the vineyard and winery produces wines of purity and precision that are highly acclaimed and sought after.

✓ *Pinot Noir • Chardonnay*

EDMUNDS ST JOHN
Alameda County
★☆

Steven Edmunds and Cornelia St John established this exciting winery in 1985 in what used to be the East Bay Wine Works, where they have honed some of California's most beautifully balanced wines, with a particular penchant for classy Syrah.

✓ *Syrah*

FRICK WINERY
Santa Cruz County
★

This winery is an up-and-coming Rhône-style specialist with a hands-off approach to winemaking.

✓ *Cinsaut • Classic red blend (C2) • Syrah*

HAHN ESTATES
Monterey County
★☆

Spanish missionaries first planted grapevines in Monterey County as early as the 1790s. Nicky Hahn of Smith & Hook launched this label in 1991 and offers flavour-packed bargain wines.

✓ *Merlot (Central Coast)*

MASSA ESTATE VINEYARDS
Monterey County
★☆

Formerly Heller Estate Vineyards, and now under new owner Bill Massa. The vineyards have previously produced heroically dark, rich, densely flavoured Cabernet Sauvignon, and splendidly long-lived Chardonnay, but we have yet, however, to see the new direction the wines will take.

✓ *Cabernet Sauvignon • Chardonnay*

J C CELLARS
Alameda County
★★

The personal label of Jeff "Rhône-in-his-Bones" Cohn specializes in single-vineyard, hillside-grown Syrah, Petite Syrah (one of California's best), Viognier, and Zinfandel.

✓ *Petite Sirah • Syrah • Viognier*

J LOHR
Santa Clara County
★

The Valdigué used to be labelled Gamay until correctly identified by UC Davis. It is made in a fruity Gamay style that is not just a good quaffer, but also a really quite elegant and well-perfumed wine with plenty of refreshing acidity.

✓ *Cabernet Sauvignon • Chardonnay (Arroyo Vista Vineyard) • Classic red blend (Cuvée Pom) • Valdigué (Wildflower)*

MORGAN WINERY
Monterey County
★★

Possibly California's most underrated, age-worthy Pinot Noir and Chardonnay are produced here. Although plenty of Americans love good Burgundy, such consumers are by and large not avid fans of West Coast wines, and regular California wine drinkers are often put off by Morgan's Burgundian-like character. Those of you who fall between these two stools should give these wines a try. The fresh, zesty Sauvignon Blanc is consistently good, but the Rhône-style wines need some work.

✓ *Chardonnay • Pinot Noir • Sauvignon Blanc*

MOUNT EDEN VINEYARDS
Santa Clara County
★

This winery produces just three varietals. The range of Chardonnay excels, but the deep, dark, chocolaty Old Vine Cabernet comes a close second.

✓ *Cabernet Sauvignon (Old Vine Reserve) • Chardonnay*

MURRIETA'S WELL
Alameda County

Philip Wente and Sergio Traverso partnered together to revive the winery in 1990, with Phil's blends then significantly better than his pure varietal wines, in a style generally understated but finer than that of his more prominent brother, Eric Wente of Wente Vineyards. Today Philip joins his brother at Wente with Murrieta's Well under a new winemaker. The red Spur is a classic Bordeaux-style blend, while Zarzuela is interestingly Portuguese Touriga-based.

✓ *Classic red blend (Spur, Zarzuela) • Classic white blend (Spur)*

RHYS VINEYARDS
Santa Cruz County
★★

Founded in 2004 by Silicon Valley entrepreneur and Burgundy fanatic Kevin Harvey, the Santa Cruz estate was established based on extensive research: Harvey was looking for the best sites to produce Pinot Noir and Chardonnay in California. Rhys now produces seven single-vineyard expressions from unique sites in Santa Cruz and Mendocino.

✓ *Pinot Noir • Chardonnay • Syrah*

RIDGE VINEYARDS
Santa Clara County
★★

Those who wonder about the long-term maturation potential of California wines should taste these stunning ones, some of which need between 10 and 20 years in bottle before they are even approachable.

✓ *Cabernet Sauvignon (Monte Bello) • Chardonnay (Santa Cruz Mountains) • Classic red blend (Geyserville) • Petite Sirah (York Creek) • Zinfandel*

ROSENBLUM CELLARS
Alameda County
★★

Established in 1978 by Minnesotan Kent Rosenblum. He produced more than 100,000 cases of over 40 different wines from more than 75 vineyards throughout the state. Acquired by Bronco Wines.

✓ *Cabernet Sauvignon • Classic red blend • Merlot • Marsanne (Dry Creek) • Petite Sirah • Syrah • Viognier • Zinfandel*

TALBOTT VINEYARDS
Monterey County
★

Added to the Gallo portfolio in 2015.

✓ *Chardonnay*

WENTE BROS
Alameda County
☆

Established for more than a century, Wente produces wines that used to range from overtly fruity to distinctly dull, but started to improve around the millennium.

✓ *Chardonnay (Riva Ranch) • Sauvignon Blanc*

THE BOTTLING OPERATION OF THE CALERA WINE COMPANY
Calera is the only commercial winery in the Mount Harlan AVA. The 1990 designation is a result of a petition by Calera owner Josh Jensen, who produced his first vintage in 1978.

The Central Coast (South)

The southern Central Coast has become one of the very best vine-growing areas in the world – outside of Burgundy – itself, for cultivating Pinot Noir grapes. The Chardonnay is equally exciting, whilst the Rhône varieties are some of the most prized in California.

Paso Robles in San Luis Obispo County was originally planted with vines in the late 18th century, the Santa Ynez Valley in Santa Barbara County had a flourishing wine industry in the pre-Prohibition era, and the town of Santa Barbara was once dotted with vineyards. Yet both San Luis Obispo and Santa Barbara counties were virtually devoid of vines in the early 1960s. It was not until Estrella, in Paso Robles, and Firestone, in Santa Ynez Valley, established vineyards in 1972 that others followed suit.

Quite why Santa Barbara, of all areas in the southern Central Coast, has suddenly become the mecca for Pinot Noir specialists is difficult to unravel. In the late 1980s, California seemed the least likely place to be in a position to challenge Burgundy for the Holy Grail of winemaking. The foundations for Santa Barbara's sudden surge of wonderful Pinot Noir wines were innocently laid in the 1970s, when the land was relatively cheap and planted with this variety in order to supply the sparkling wine industry in the north of the state. It is impossible to pinpoint exactly when local winemakers realized the potential of making their own still wine, but much of California's finest, purest, and most consistent Pinot Noir wines now come from this valley.

The Santa Ynez Valley is home to some of the most highly regarded Pinot, yet the area where it excels is restricted: 14 kilometres (9 miles) from the ocean and the valley is too cool to ripen grapes; 26 kilometres (16 miles) and it is ideal for Pinot Noir; but for every 1.5 kilometres (1 mile) farther from the ocean the grapes gain an extra degree (ABV) of ripeness, and by 32 kilometres (20 miles) from the coast, this is Cabernet country. Sandford in the Sta Rita Hills AVA and Bien Nacido in the Santa Maria Valley AVA are widely considered the most outstanding vineyards for Pinot Noir in the region.

THE VITICULTURAL AREAS OF THE SOUTHERN CENTRAL COAST
(see also p577)
Based in San Luis Obispo and Santa Barbara Counties, this area has a fine reputation for top-quality Pinot Noir.

A COPSE OF OAK TREES FORMS A HEART SHAPE ON THE VINEYARD HILLS OF THE NINER WINE ESTATES
This landmark lies in the Paso Robles Willow Creek District on scenic Highway 4 in Paso Robles (which means "Pass of the Oaks") in the vineyard known fittingly as Heart Hill.

FACTORS AFFECTING TASTE AND QUALITY

LOCATION
This region stretches southwards along the coast from Monterey and includes the counties of San Luis Obispo and Santa Barbara.

CLIMATE
The vine-growing region of California's southern Central Coast is generally warm (Region III, see p577), except for areas near the sea, where some of the coolest spots in California are found (Regions I and II) because of the regular incursion of the tail of the great coastal fog bank. Orientation of the coastal ranges is key to the level of influence the Pacific exerts. Annual rainfall ranges from 25 centimetres (10 inches) t o 114 centimetres (45 inches).

ASPECT
Most vines in this region grow on hillsides in San Luis Obispo and southern-facing benchland in Santa Barbara County, at altitudes from 37 to 180 metres (120 to 600 feet) in the Edna Valley to 180 to 300 metres (600 to 1,000 feet) in Paso Robles, and 460 metres (1,500 feet) on York Mountain.

SOIL
Mostly sandy, silty, or clay loams, but soil can be more alkaline, as in the gravelly lime soils on the foothills of the Santa Lucia Mountains.

VITICULTURE & VINIFICATION
More than 20 varieties are produced

throughout the region, reflecting the diversity of climates. By the mid-1980s, the cooler areas had invited experimentation with the Burgundian varieties planted in the early 1970s to supply sparkling wine producers in the north of the state. By the late 1980s, a number of world-class Pinot Noirs had been produced, a trend that continued throughout the 1990s with the planting of better-suited and varied clones, as well as new vineyard practices that allowed the coolest regions to be further explored.

GRAPE VARIETIES
Cabernet Sauvignon, Chardonnay, Gewürztraminer, Merlot, Pinot Gris, Pinot Noir, Riesling, Syrah, Viognier, Zinfandel

THE APPELLATIONS OF
THE CENTRAL COAST (SOUTH)

ADELAIDA DISTRICT AVA
San Luis Obispo County

Situated in the northwest of Paso Robles AVA and spread across the rugged Santa Lucia foothills, this region offers diverse aspects and exposures. The Adelaida District experiences moderate Pacific influence and high average rainfall 64 centimetres (25 inches). Soils are predominantly calcareous limestone. Primarily red and white Bordeaux and Rhône varieties are planted.

ARROYO GRANDE VALLEY AVA
San Luis Obispo County

Located 19 kilometres (12 miles) southeast of the town of San Luis Obispo, the Arroyo Grande Valley enjoys a Region I to II climate (see p577), thanks primarily to its proximity to the ocean and the frequent fog produced by marine air in the mornings and evenings. Planted at between 90 and 300 metres (300 and 1,000 feet), vines grow at much higher altitudes than those in the neighbouring Edna Valley AVA and also receive slightly more rain. The aspect of the hillsides cultivated range from moderate to very steep slopes, with deep, well-drained, sandy-clay and silty-clay loam soils. With a drop of 17°C (30°F) in night-time temperatures, relatively high acidity levels are maintained throughout ripening. Chardonnay and Pinot are the primary varieties planted for production of both still and sparkling wines. Farther inland away from the fog line, Zinfandel and Syrah are successful.

BALLARD CANYON AVA
Santa Barbara County

Designated due its unique soils and climate in the heart of Santa Ynez Valley AVA. Fog and diurnal swings of around 20°C (40°F) are key to countering the short periods of daytime heat and prolonging the ripening season. Ballard Canyon is emerging as the "home" of Syrah in the United States – the variety accounts for over 50 per cent of total plantings, with other Rhône grapes (Grenache, Viognier, and Roussanne) making up a further 30 per cent.

CRESTON DISTRICT AVA
San Luis Obispo County

The Creston District is located on a plateau of fertile alluvial soils at the base of the La Panza Range, which lies in the east of Paso Robles AVA. The region's warmer climate and lower rainfall is tempered by maritime fog and winds through the Templeton Gap. Cabernet Sauvignon and Merlot are the most widely planted varieties.

EDNA VALLEY AVA
San Luis Obispo County

This elongated valley is located just south of Paso Robles and is well defined by the Santa Lucia Mountains to the northeast, the San Luis Range to the southwest, and a low hilly complex to the southeast. In the northwest, the Edna Valley merges with the Los Osos Valley, forming a wide-mouthed funnel that sucks in ocean air from Morro Bay. This marine air flows unobstructed into the valley, where it is captured by the pocket of mountains and hills, providing a moderate summer climate and one of the longest growing seasons in California, which differentiates it from surrounding areas. The vines grow on the valley floor, rising to 120 metres (600 feet) in the Santa Lucia Mountains on soils that are mostly sandy-clay loam, clay loam, and clay. Conditions are ideally suited to producing top-quality Chardonnay and Pinot Noir, which account for the majority of plantings, whilst Syrah is increasingly finding a home in the region.

EL POMAR DISTRICT AVA
San Luis Obispo County

Covering an area of high terraces and hills, the El Pomar District AVA's plantings range in elevation from 225 to 490 metres (740 to 1,600 feet). Located in the central-west of Paso Robles AVA, the area experiences a strong marine influence with heavy, long-lasting fogs due to its proximity to the Templeton Gap. Cabernet Sauvignon and Merlot dominate plantings.

HAPPY CANYON OF SANTA BARBARA AVA
Santa Barbara County

Situated in the far east of the Santa Ynez Valley AVA, the district runs into the rolling foothills of the San Rafael Mountains. Due to its inland location, the climate is significantly warmer, resulting in full-bodied, concentrated wines. Most suited to later ripening Bordeaux varieties, including Cabernet Franc, Cabernet Sauvignon, Merlot, and Petit Verdot.

LOS OLIVOS DISTRICT AVA
Santa Barbara County

The Los Olivos District AVA is located within the Santa Ynez Valley AVA. This wine region covers a plain of uniform alluvial soils running from the Santa Ynez River to an elevation line of 300 metres (1,000 feet). The climate is slightly warmer than Ballard Canyon AVA to the west and

cooler than Happy Canyon of Santa Barbara AVA to the east. Planting is split almost 50-50 between Bordeaux and Rhône varieties.

PASO ROBLES AVA
San Luis Obispo County

This area was given its name of El Paso de Robles, or Paso Robles (meaning "Pass of the Oaks" or "Oak Pass") in the 18th century, when travellers passed through it on their way from the San Miguel to the San Luis Obispo missions. It is one of California's oldest winegrowing regions: grapes have been harvested in this area of rolling hills and valleys since at least 1797. Limited penetration by coastal winds or marine fog means there is the equivalent of an additional 500 to 1,000 degree days appropriate for grape cultivation here compared to viticultural areas to the west and east. Despite an overall continental climate, however, there is a wide variation in sub-climates, dependent on proximity to the Templeton "Gap, which funnels a degree of maritime influence, as well as rainfall, soils, and altitude, ranging from 210 to 730 metres (700 to 2,400 feet). This led to the designation of 11 sub-AVAs in 2014: Adelaida District, Creston District, El Pomar District, Estrella District, Geneseo District, Highlands District, San Juan Creek, San Miguel District, Santa Margarita Ranch, Templeton Gap District, and Willow Creek District.

PASO ROBLES ESTRELLA DISTRICT AVA
San Luis Obispo County

The largest district within Paso Robles AVA, the appellation encompasses rolling plains of fertile alluvial soils in the Estrella River Valley. Its location on the northern limit of Paso Robles AVA results in a warm climate with modest maritime influence. Cabernet Sauvignon is the principle variety grown.

PASO ROBLES GENESEO DISTRICT AVA
San Luis Obispo County

This hilly appellation sits in the centre of Paso Robles AVA on high rolling terraces from 225 to 395 metres (740 to 1,300 feet). It is bordered by Paso Robles Estrella District AVA to the north and El Pomar District AVA to the south and west, and is one of the warmer, drier sub-AVAs with only moderate maritime influence. Cabernet Sauvignon, Merlot, and Syrah are the most planted varieties.

PASO ROBLES HIGHLANDS DISTRICT AVA
San Luis Obispo County

This district sits at elevations of 350 to 630 metres (1,160 to 2,080 feet) in the foothills of the La Panza Range in the southeast of Paso Robles AVA on primarily sandy loam and alluvial soils. It is the warmest district in the region with wide diurnal swings of up to 25°C (50°F), making the climate more continental than maritime. The top varieties are Chardonnay, Merlot, Zinfandel, and Cabernet Sauvignon.

PASO ROBLES WILLOW CREEK DISTRICT AVA
San Luis Obispo County

This mountainous appellation lies on the western boundary of Paso Robles AVA, bordered by Adelaida District AVA to the north and Templeton Gap District AVA to the south. Altitudes of 290 to 485 metres (960 to 1,600 feet) combined with significant maritime influence lead to large diurnal swings, as great as 35°C (70°F). A variety of soils with high proportion of calcareous limestone. Red Rhône varieties are most prized.

SAN JUAN CREEK AVA
San Luis Obispo County

Set on the warmer, drier eastern edge of Paso Robles AVA, to the north of Paso Robles Highlands District AVA, with a similar continental-leaning climate and minimal maritime influence. The area primarily covers the valley floor on alluvial and sandy loam soils, ranging in elevation from 295 to 485 metres (980 to 1,600 feet). Cabernet Sauvignon and Rhône varieties predominate.

SAN LUIS OBISPO AO

This appellation covers grapes grown anywhere within the entire county of San Luis Obispo.

SAN MIGUEL DISTRICT AVA
San Luis Obispo County

Located in the Santa Lucia Range foothills in the north of Paso Robles AVA, bordering Paso Robles Estrella District AVA to the east. Maritime influence is minimal, whilst the terrain is rugged with alluvial terraces from the Salinas and Estrella Rivers. Top planted varieties are Cabernet Sauvignon, Merlot, and Zinfandel, followed by Southern Rhône red grapes, Grenache, Syrah, and Mourvèdre.

SANTA BARBARA AO

An appellation covering grapes grown anywhere within the entire county of Santa Barbara.

SANTA MARGARITA RANCH AVA
San Luis Obispo County

The southernmost appellation within Paso Robles AVA sits only 23 kilometres (14 miles) from the Pacific coast. The coastal proximity results in consistent cooling breezes and fog, plus the highest annual rainfall in the region of 74 centimetres (29 inches) and one of the coolest, latest growing seasons. The terrain consists of steep slopes ranging in elevation from 270 to 425 metres (900 to 1,400 feet) with a range of alluvial soils. A wide range of varieties is planted, with Bordeaux grapes the most popular.

SANTA MARIA VALLEY AVA
Santa Barbara County

The Pacific winds blow along this funnel-shaped valley, causing cooler summers and winters, plus warmer autumns than in the surrounding areas. The terrain climbs from 60 to 240 metres (200 to 800 feet), with most of the vineyards concentrated at 90 metres (300 feet). The soil is sandy and clay loam and is free from the adverse effects of salts. This is top-quality Pinot Noir country, and Bien Nacido (to which many winemakers have access) in the Tepesquet Bench area is far and away its best vineyard. The Santa Maria Valley also grows top-quality Chardonnay and Syrah.

SANTA YNEZ VALLEY AVA
Santa Barbara County

The Santa Ynez Valley is bounded by mountains to the north and south, by Lake Cachuma to the east, and by a series of low hills to the west. The valley's close proximity to the ocean serves to moderate the weather and overall temperatures with maritime fog. The Santa Rita Hills block penetration of the coldest of the sea winds, however, so that temperatures rise moving east along the valley. The vineyards are located at an altitude of between 60 and 120 metres (200 and 400 feet) in the foothills of the San Rafael Mountains, with the vines grown on soils that are mostly well-drained, sandy, silty, clay, and shale loam. Chardonnay and Pinot Noir thrive in the cooler, more westerly sites, with varieties shifting to the Rhône, Bordeaux, and Zinfandel moving east up the valley. Sta Rita Hills AVA, Ballard Canyon AVA, Los Olivos District AVA, and Happy Canyon of Santa Barbara AVA all lie within the appellation's boundaries.

STA RITA HILLS AVA

This appellation is located on the western boundary of the Santa Ynez Valley AVA and includes all the latter's coolest areas. The region experiences a pronounced maritime influence of morning fogs and cooling afternoon ocean breezes, funnelled by the east-west orientation of the surrounding hills, resulting in one of the coolest climates in California. Widely regarded as one of the best sources of Chardonnay and Pinot Noir in California, which are the two primary planted varieties (which are ideally suited to the areas of limestone). The iconic Sandford and Benedict was the first planted vineyard in the region in 1971, and it is one of the most highly sought-after sources of Pinot Noir in the state.

TEMPLETON GAP DISTRICT AVA
San Luis Obispo County

Situated on the western edge of Paso Robles AVA, this region is exposed to significant maritime cooling fogs and winds that are funnelled through gaps in the coastal Santa Lucia Range mountains. The terrain comprises steep slopes and wide alluvial terraces at altitudes of 210 to 545 metres (700 to 1,800 feet). Cabernet Sauvignon, Syrah, Zinfandel, and Viognier are most widely planted.

YORK MOUNTAIN AVA
San Luis Obispo County

This small appellation is just 11 kilometres (7 miles) from the Pacific coast, situated to the west of the Paso Robles AVA at an altitude of 450 metres (1,500 feet) with thin, calcareous limestone soils. The region's elevation and coastal proximity produce a cooler and wetter climate compared to surrounding areas, receiving 114 centimetres (45 inches) of rain per year. The primary varieties are Pinot Noir, Rhône varieties, and Zinfandel, benefitting from large diurnal shifts of up to 25°C (50°F).

THE WINE PRODUCERS OF

THE CENTRAL COAST (SOUTH)

ALBAN VINEYARDS
San Luis Obispo County
★★

John Alban claims that his family-owned vineyard, founded in 1989, was "the first American winery and vineyard established exclusively for Rhône varieties". He certainly produces some of the best California Rhône varietals.

 Classic red blend (Pandora) • Grenache • Syrah • Viognier

ANDREW MURRAY VINEYARDS
Santa Barbara County
★★

Andrew Murray is a Rhône-style specialist. His original hillside family plot was sold in 2006 to a Santa Monica-based Greek real estate agent; the original vineyards are now called Demetria. Andrew is still making his white blend Enchanté, albeit from different fruit sources.

 Classic white blend (Enchanté) • Syrah • Viognier

AU BON CLIMAT
Santa Barbara County
★★

It was Jim Clendenen's wines of the 1980s that convinced me of Santa Barbara's suitability to Pinot Noir; since that time, of course, this area of the Central Coast has established a soaring reputation for Rhône grape varieties, as well as Burgundian varieties. He has been steadily collecting accolades since Au Bon Climat's inception.

 Chardonnay (especially Sanford & Benedict) • Pinot Noir

BECKMEN VINEYARDS
Santa Barbara County
★☆ Ⓥ

The Beckmen family founded their winery on an estate near Los Olivos in 1994 and then purchased what was to become Purisma Mountain Vineyard in 1996 in Ballard Canyon, now widely deemed to be one of the finest sources of Syrah in California. Their estate bottlings showcase the quality of their biodynamically farmed vineyards and the potential for cool-climate Syrah in Ballard Canyon.

 Syrah (PMV Syrah, PMV Block Six Syrah)

BABCOCK VINEYARDS
Santa Barbara County
★☆

Babcock vineyards best wines are the succulent Pinot Noir and, the deep, dark, rewarding Syrah. In 2008 Bryan Babcock reshaped his vineyards using a new "pedestular cane suspension" (PCS) method, which reduced vineyard costs by 25 per cent. Babcock coined the term pedestular due to the use of metal pedestals that support the vine's fruiting canes.

 Chardonnay • Pinot Noir • Sauvignon Blanc • Syrah

BARON HERZOG
Santa Barbara County
★ Ⓥ

A story of riches to rags and back to riches again. Eugene Herzog was born into an aristocratic Jewish family, who were renowned as suppliers of wine since the days of the Austro-Hungarian Empire. Having survived the Nazis only to see his property in the Trnava region of Slovakia confiscated by the Communists, he fled Slovakia in 1948, taking his family to New York City, where he came down to earth with a bump, working as a driver and salesman for the Royal Wine Company. The company had a cash-flow problem and paid part of his salary in shares, but Herzog's selling acumen was such that he became the majority shareholder, purchasing the rest of the company within 10 years of setting foot in the United States. Eugene, with the help of his four sons, opened Kedem Winery, and in 1986 established the Baron Herzog label. Since then it has not been one of the flashy high-flyers, but has steadily produced a number of modestly priced, award-winning wines.

 Chenin Blanc (Clarksburg) • Sauvignon Blanc • Zinfandel (Lodi Old Vines)

BIEN NACIDO ESTATE
Santa Barbara County
★★

Widely considered a pioneer of the Californian wine industry since the establishment of the vineyard in 1973, Bien Nacido is located in Santa Maria Valley AVA. Renowned for the quality of the Pinot Noir, Chardonnay, and Syrah fruit that is coveted by winemakers throughout California and considered one of the preeminent vineyard sites in the country. Since 2011 the estate-produced wines have been made by Trey Fletcher, who previously worked as associate winemaker at Littorai (Sonoma) and garnered much critical acclaim.

✓ *Chardonnay • Pinot Noir • Syrah*

BLACKJACK RANCH
Santa Barbara County
★★☆

In 1989 Roger Wisted invented, patented, copyrighted, and trademarked the casino game California Blackjack, the proceeds of which enabled him to indulge in his lifelong fantasy of planting his own vineyard. His first vintage was 1997, and since then his wines have displayed a structure and acidity balance that sets them apart.

✓ *Chardonnay • Classic red blend (Harmonie) • Merlot • Syrah*

BONACCORSI
Santa Barbara County
★★

In 1999 Master Sommelier Michael Bonaccorsi set up his own winery and swiftly achieved an enviable reputation for the highly polished style of its wines; after his passing in 2004 his family now continues Bonaccorsi.

✓ *Chardonnay • Pinot Noir*

BREWER-CLIFTON
Santa Barbara County
★★

Jackson Family Wines bought this top-class Burgundian varietal specialist in 2017. It sources fruit from four *"mono-pole"* single-vineyards from some of the best areas of Santa Rita Hills AVA.

✓ *Chardonnay • Pinot Noir*

BYRON
Santa Barbara County
★☆

Part of Jackson Family Wines, Byron concentrates on the Pinot Noir and Chardonnay varieties from two appellations, Santa Maria Valley and Sta Rita Hills.

✓ *Chardonnay • Pinot Noir*

CAMBRIA
Santa Barbara County
★☆

Personally owned by Jess Jackson's wife, Barbara Banke, rather than the Kendall-Jackson Company, Cambria's strength is its extraordinary range of high-quality Chardonnay. Also produces interesting single-clone Pinot Noir wines from the same vineyard and vintage.

✓ *Chardonnay • Pinot Noir*

CLAIBORNE & CHURCHILL
San Luis Obispo County
★☆ Ⓥ

Named for its original owners, Alsatian specialists Claiborne Thompson and Fredericka Churchill, this winery produces genuinely dry, classically structured Gewürztraminer and Riesling. I was sceptical at first, but as soon as I tasted them, I was convinced. Drink on purchase. Since 2019 stewarded under a new head winemaker.

✓ *Gewürztraminer (Dry Alsatian Style) • Riesling (Dry Alsatian Style)*

DOMAINE DE LA COTE
Santa Barbara County
★★

Former 16-hectare (40-acre) vineyard project of Evening Land (Oregon), bought in 2013 by Rajat Parr and winemaker Sashi Moorman, who had been involved in the initial planting in 2007. The estate consists of six individual vineyards in the Sta Rita Hills that are vinified separately to express their diverse individual characters. As with their *négociant* project, Sandhi, the wines are showcasing the highest quality of what cool climate Pinot Noir and Chardonnay can achieve in California.

✓ *Pinot Noir • Chardonnay*

EBERLE WINERY
San Luis Obispo County
★☆

Look no further if you want massively built Zinfandel, but the Cabernet is the better-balanced, much classier wine from Eberle, while the Steinbeck Syrah is the best of both worlds.

✓ *Barbera • Cabernet Sauvignon • Muscat Canelli • Sangiovese • Syrah (Steinbeck) • Viognier*

EDNA VALLEY VINEYARD
San Luis Obispo County
★☆

Founded as a joint venture between Chalone and the Niven family, who own the Paragon Vineyard, Edna Valley Vineyard produces rich, toasty Chardonnay and very good Pinot Noir. Became part of Gallo in 2011.

✓ *Chardonnay • Pinot Noir*

EPIPHANY VINEYARDS
Santa Barbara County
★Ⓥ

The personal winery of Eli Parker, son of Fess (*see* below); it is known for its Rhône varietals.

✓ *Grenache (Revelation Rodney's Vineyard) • Pinot Gris*

FESS PARKER WINERY
Santa Barbara County
★☆ Ⓥ

Unless you enjoyed a 1950s childhood, the name of actor Fess Parker might not mean anything to you, but his major role was that of Davy Crockett in the American TV series from Disney, which ran from 1954 to 1955. Both his eponymous winery and Epiphany are led by UC Davis–trained winemaker Blair Fox.

THE FOXEN CANYON WINE TRAIL
This route guides Central Coast tourists along a 48-kilometre (30-mile) stretch of Foxen Canyon Road from Los Olivos to Santa Maria in the Santa Maria Valley AVA. Along the way they can sample the offerings from 14 wineries and tasting rooms.

Like Epiphany, Fess Parker includes Rhône varietals Viognier and Syrah but also those popular California staples Pinot Noir and Chardonnay.

✓ *Chardonnay • Pinot Noir • Syrah • Viognier*

FIRESTONE VINEYARD
Santa Barbara County
★Ⓥ

Although Firestone can occasionally lapse, the wines are usually rich and ripe, and are gradually assuming more finesse. In 2007 Firestone became part of the Foley Food & Wine Society.

✓ *Johannisberg Riesling • Merlot • Sauvignon Blanc • Syrah*

FOXEN VINEYARD
Santa Barbara County
★☆ Ⓥ

Bill Wathen and Dick Doré founded Foxen Winery & Vineyard at the historic Rancho Tinaquaic, named in memory of William Benjamin Foxen, an English sea captain, hence the anchor motif.

✓ *Classic red blend • Pinot Noir • Syrah*

GAINEY VINEYARDS
Santa Barbara County
★★Ⓥ

This winery still makes one of Santa Barbara's best Pinots, but is now best known for its Merlot, and also turns out a number of amazing bargains.

✓ *Chardonnay (Limited Selection) • Merlot (Limited Selection) • Pinot Noir (Limited Selection Santa Rita Hills) • Riesling • Sauvignon Blanc (Limited Selection) • Syrah (Limited Selection)*

THE HITCHING POST
Santa Barbara County
★☆ Ⓥ

Frank Ostini is the soft-spoken, friendly, fun-loving chef-owner of the Hitching Post Restaurant in Santa Maria, which serves tasteful, unpretentious American barbecue food at its best. Frank also enjoys making wines with his fisherman friend Gray Hartley. They started out in a corner of Jim Clendenen's Au Bon Climat winery, but in 2001 moved to Central Coast Wine Services in Santa Maria, and continue to make a number of excellent wines, particularly the polished Pinot Noirs.

✓ *Pinot Noir • Syrah (Purisima Mountain)*

L'AVENTURE
San Luis Obispo County
★★☆

Monstrous wines from *bordelais* Stephan Asseo, whose vision of Bordeaux-meets-Rhône in Pas Robles, has attracted rave reviews. In 2018 the Thienot family (of Champagne fame) took a share in L'Aventure (which is also known as Stephan Vineyards).

✓ *Classic red blend (Côte-à-Côte, Cuvée)*

LONGORIA
Santa Barbara County
★★

Rick Longoria started making tiny quantities under his own label while he was at Gainey Vineyard, which he put on the map as winemaker there for 12 years. Now that he has his own winery, he still produces small batches of wine (ranging between 70 and 850 cases), but the quantities are massive by comparison with his early years. The quality is, however, even better – especially the Pinot Noir.

✓ *Chardonnay • Pinot Noir • Syrah (Alisos Vineyard)*

MOSBY WINERY
Santa Barbara County
★☆ Ⓥ

Bill Mosby runs this interesting, sometimes provocative, Italian-style wine specialist operation.

✓ *Dolcetto • Pinot Grigio • Teroldego*

PALMINA
Santa Barbara County
★★

Palmina offers some of California's greatest Italian-style wines, especially the single-vineyard offerings, from the owners of Brewer-Clifton.

✓ *Barbera • Classic red blend (Savoia) • Nebbiolo • Pinot Grigio*

PIEDRASASSI
Santa Barbara County
★★

Primarily specialising in cool-climate Syrah, Sashi Moorman and his wife, Melissa, produce some of the finest expressions of the variety in the state. The highest-quality fruit is meticulously sourced from vineyards in Santa Maria Valley and Sta Rita Hills, as well as their own estate in the Arroyo Grande Valley, followed by a low-intervention ethos in the winery.

✓ *Syrah • Mourvèdre*

QUPÉ
Santa Barbara County
★☆

Great for Rhône-style reds, particularly the various single-vineyard Syrahs; bought by Vintage Wine Estates in 2018.

✓ *Syrah*

RANCHO SISQUOC
Santa Barbara County
★★

These vividly fruity wines come from one of Santa Barbara County's first vineyards. Rancho Sisquoc covers about 120 hectares (292 acres) of the massive 14,973-hectare (37,000-acre) Flood Ranch – and keeps growing. Its vineyards grow many varietals, including Cabernet Franc, Cabernet Sauvignon, Chardonnay, Johannesburg Riesling, Malbec, Merlot, Nebbiolo, Petit Syrah, Petite Verdot, Pinot Noir, Sangiovese, Sauvignon Blanc, Sylvaner, and Syrah.

✓ *Cabernet Franc • Classic red blend (Sisquoc River Red Meritage) • Chardonnay (Rancho Sisquoc Winery) • Malbec (Rancho Sisquoc Winery) • Riesling (Rancho Sisquoc Winery) • Sylvaner*

SANDHI
Santa Barbara County
★★

This *négociant* business collaboration between Rajat Parr and acclaimed winemaker Sashi Moorman was founded in 2010 to explore, source, and express the finest vineyards of Pinot Noir and Chardonnay throughout Santa Barbara County. A benchmark for top-quality, cool-climate-grown Burgundian varieties in California.

✓ *Pinot Noir • Chardonnay*

SANFORD WINERY & VINEYARDS
Santa Barbara County
★★

With botanist Michael Benedict, Richard Sanford founded the eponymous Sanford & Benedict Vineyard in 1973. Sanford, a modern-day pioneer of Santa Barbara winemaking, released his first wine in 1976. Sanford moved on to another project, bringing on the Terlato family in 2002, and since 2007 the Terlatos have owned and run Sanford Winery.

✓ *Pinot Noir (La Riconada, Sanford & Benedict, Vin Gris Santa Rita Hills) • Sauvignon Blanc*

SANTA BARBARA WINERY
Santa Barbara County
★☆

Although this is the oldest Santa Barbara County winery – dating back to the early 1960s – it did not produce premium-quality wines until a decade later, after the Sanford & Benedict revolution. The Santa Barbara Winery soon began making a number of great-value Burgundian varietals, and it has since moved to Rhône-style wines, making some of the region's most impressive Syrah.

✓ *Chardonnay (Lafond Vineyard, SRH) • Pinot Noir (Lafond Vineyard, SRH) • Sauvignon Blanc • Syrah*

SEA SMOKE ESTATE VINEYARD
Santa Barbara County
★★☆

Established in 1998 in the western Santa Rita Hills, Sea Smoke produces highly acclaimed and coveted cool-climate Pinot Noir and Chardonnay from their biodynamically farmed estate.

✓ *Chardonnay • Pinot Noir*

TABLAS CREEK VINEYARD
San Luis Obispo County
★☆

Spearheading California's Rhône movement, Tablas Creek was founded in 1989 as a joint venture between the Perrin Family of Château de Beaucastel in France and the Haas family of wine importer Vineyard Brands. The estate in the Adelaida District AVA was selected based on similar climate and soils to those of Château de Beaucastel in the Southern Rhône and planted with vine material imported from France. The range of wines is based on traditional Southern Rhône red and white blends, at a range of quality levels, as well as a small proportion of rosé.

✓ *Classic Southern Rhône red blends (Esprit de Tablas, Cotes de Tablas, Patelin de Tablas) • Classic Southern Rhône white blends (Esprit de Tablas Blanc, Cotes de Tablas Blanc, Patelin de Tablas Blanc)*

TALLEY VINEYARDS
San Luis Obispo County
☆

Talley Voneyards is a Burgundian-style specialist with 100 per cent estate-bottled production.

✓ *Chardonnay • Pinot Noir*

VERDAD
Santa Barbara County
★☆

Owned by Bob and Louisa Lindquist (having established and sold Qupé), Verdad is focussed on Spanish-style wines. They primarily source from two vineyards: the Ibarra-Young Vineyard in the Santa Ynez Valley, which is organically farmed, and the biodynamically certified Sawyer Lindquist Vineyard in Edna Valley.

✓ *Albariño • Rosé • Tempranillo (Santa Ynez Valley)*

WILD HORSE
San Luis Obispo County
☆ V

Established in 1982 by Ken Volk, Wild Horse was acquired by Constellation Brands. Kip Lorenzetti is now winemaker.

✓ *Blaufrankisch • Grenache (Equus James Berry Vineyard)*

ZACA MESA WINERY
Santa Barbara County
★ V

The producer is best for inexpensive, refreshing Chardonnay and exciting, low-yield Syrah.

✓ *Chardonnay • Grenache • Roussanne • Syrah (Black Bear Block, Eight Barrel)*

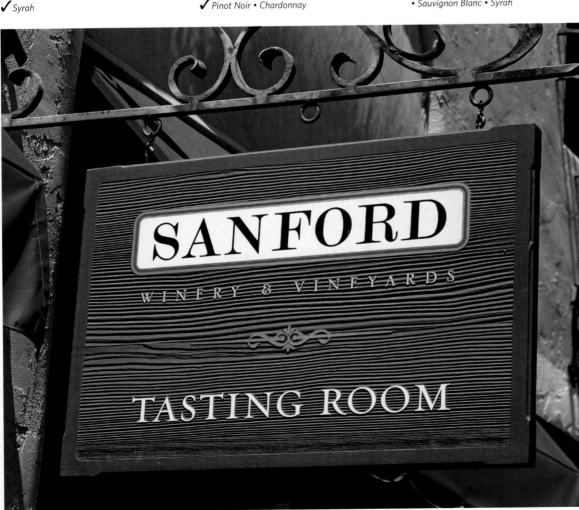

SANFORD WINE TASTING ROOM
This top-quality producer was a vine-growing pioneer, its original owners seeing the potential of the *terroir* in in the Santa Rita Hills.

The Central Valley

Often overlooked in favour of the smaller, more famous coastal regions of the state, hot, flat, dry Central Valley is the real powerhouse, accounting for nearly three-quarters of California's total wine production and acting as the source for some of the world's most successful and well-known brands.

The Central Valley lies inland of California's Coastal Ranges, which limit any significant maritime influence. With the Sierra Nevada as its eastern boundary, the region lies about 160 kilometres (100 miles) inland and meanders southward as it runs parallel with the Pacific coastline, starting north of Sacramento and finishing around 160 kilometres (100 miles) north of Los Angeles. Encompassing the Sacramento and San Joaquin Valleys, the area is extremely fertile, with an ideal climate for growing just about anything. When supported with irrigation, yields can be very high – hence it is the ideal grape source for mass-market, generously ripe, and fruity wines.

Although not an AVA itself, the Central Valley region covers a range of AVAs, including Clarksburg, Diablo Grande, Dunnigan Hills, Lodi, Madera, Merritt Island, River Junction and Salado Creek, a number of which are starting to gain more attention as a source of quality, rather than volume, production. The area well known for producing outstanding Zinfandel, especially in the Lodi AVA, as well as Chenin Blanc and Petite Sirah.

FACTORS AFFECTING TASTE AND QUALITY

LOCATION
The viticultural area of this huge fertile valley stretches 640 kilometres (400 miles) from Redding in the north to Bakersfield in the south, running parallel to the coast approximately 160 kilometres (100 miles) inland between the Coastal Ranges to the west and the Sierra Nevada to the east (see map of California, p577).

CLIMATE
Generally homogenous from end to end, warming steadily from Region IV (see p577) in the north to Region V in the south. The area around Lodi is the only exception, cooled as it is by sea air sweeping up the Sacramento River.

ASPECT
The vines are grown on the vast flat area of the valley floors, intersected and irrigated by a network of levees.

SOIL
Very fertile sandy loam dominates the length and breadth of the valley.

VITICULTURE & VINIFICATION
This region is capable of producing vast quantities of reliable mass-market wine, often employing the latest technologies available in the vineyard and winery to ensure the highest levels of yield, provide consistency of desired quality, and limit production costs. Somewhat higher-quality Zinfandel and sweet dessert wines are produced around Lodi.

GRAPE VARIETIES
Barbera, Cabernet Sauvignon, (Carignan), Chardonnay, Chenin Blanc, Colombard, Grenache, Merlot, Mourvèdre, Muscat, Petite Sirah, Ruby Cabernet, Sauvignon Blanc, Sémillon, Zinfandel

THE APPELLATIONS OF
THE CENTRAL VALLEY

ALTA MESA AVA
Sacramento County

Situated in the north-central region of Lodi AVA, alta mesa (Spanish for "high table") is a reference to the elevated plateau that makes up the appellation. The area's clay and gravel soils, combined with one of the warmest climates within Lodi, are ideally suited to Zinfandel, Syrah, Cabernet Sauvignon, Cabernet Franc, and Merlot.

BORDEN RANCH AVA
Sacramento and San Joaquin counties

This hilly region within the Lodi AVA, sits centrally on the eastern border; it is warmer than those farther west but receives cooling breezes coming down off the Sierra Foothills. Red wines predominate, thriving in the range of exposures and elevations combined with stony, well-drained soils. Cabernet Sauvignon, Merlot, Zinfandel, and Syrah produce structured examples.

CAPAY VALLEY AVA
Yolo County

This appellation is located 130 kilometres (80 miles) northeast of San Francisco, just east of Napa over the Blue Ridge Mountains, bordering Napa, Lake, and Colusa Counties. The Sacramento Delta has a cooling influence on this otherwise warm region. Chardonnay, Tempranillo, Syrah, and Viognier are key varieties.

CLARKSBURG AVA
Sacramento County

A large area south of Sacramento encompassing the AVA of Merritt Island, Clarksburg is cooled by the breezes that roll in off Suisun Bay and extend the growing season. Warm days are balanced by cool evenings, creating ideal conditions to produce good yields of consistent quality fruit, with over 35 varieties planted.

CLEMENTS HILLS AVA
San Joaquin County

One of the seven smaller AVAs within Lodi AVA, the area to the southeast of the larger appellation consists of rounded, rolling hills. Similar to Borden Ranch to the north, the area is warmer than those in the west but the climate is moderated by cool air drainage from the Sierra foothills to the east, producing wide diurnal swings. Home to some of the regions oldest Zinfandel vines, with Bordeaux and Rhône varieties also well suited.

COSUMNES RIVER AVA
Sacramento County

Located in the northwest corner of Lodi AVA, this area is one of the cooler appellations of the region, with moderately fertile, alluvial soils. The cooling influence from the Sacramento/San Joaquin Delta to the west is well suited to white varieties, including Chardonnay, Sauvignon Blanc, Pinot Grigio, and Vermentino.

DIABLO GRANDE AVA
Sacramento County

Located 90 minutes southeast of San Francisco, this AVA is, unusually, owned entirely by the Diablo Grande Resort Community, comprising a private golf resort and residential development. Named after Mount Diablo, the highest peak in the Pacific Coast Range, the appellation lies in its western foothills where, due to the elevation of 300 to 540 metres (1,000 to 1,800 feet), there is greater rainfall and lower average temperatures than in surrounding areas.

DUNNIGAN HILLS AVA
Yolo County

This appellation is located northwest of Sacramento in the rolling Dunnigan Hills, where the soils are primarily well-drained gravelly loams. Cooling breezes from the Sacramento Delta combine with diurnal swings to preserve acidity in a range of varieties, including Chardonnay, Sauvignon Blanc, and Viognier, as well as Cabernet Sauvignon, Merlot, and Syrah.

JAHANT AVA
Sacramento and San Joaquin counties

The smallest appellation within Lodi AVA, situated in the centre of the larger region, it is relatively flat with sandy clay loam soils. Delta fog influence allied with cold air drainage create one of the coolest growing seasons in Lodi. Chardonnay, Viognier, and Sauvignon Blanc are the principal varieties planted.

LODI AVA
Sacramento and San Joaquin counties

This appellation is situated in a large, warm, inland area that comprises alluvial fan, plains that are prone to flood, and terrace lands both above and below the levees. It expanded its southern and western boundaries in August 2002, increasing the viticultural area by 17 per cent, with greater planted acreage than Napa and Sonoma combined. This was followed in 2006 by seven new sub-AVAs being designated to highlight the climatic and soil variation within this larger area. The sub-AVAs are: Alta Mesa, Borden Ranch, Clements Hills, Cosumnes River, Jahant, Mokelumne River, and Sloughhouse.

MADERA AVA
Madera and Fresno Counties

A viticultural area not to be confused with Madera County, Madera AVA is located in the heart of the San Joaquin Valley straddling both Madera and Fresno Counties. The warm, sunny climate of this large, flat area with fertile alluvial soils makes it an ideal source for high-volume, ripe, and fruity mass-market wines.

MERRITT ISLAND AVA
San Joaquin County

This AVA sits on a fertile alluvial island within the Sacramento River Delta, bounded on the west and north by Elk Slough, Sutter Slough on the south, and the Sacramento River on the east. Its climate is tempered by cooling southwesterly breezes from the Carquinez Straits, which reduce the temperature substantially compared with that of Sacramento, located just 10 kilometres (6 miles) north. Plantings are mainly Petite Sirah, Cabernet Sauvignon, and Chardonnay.

MOKELUMNE RIVER AVA
San Joaquin County

Located in the southwest of Lodi AVA, it is the largest and one of the coolest sub-regions, moderated by breezes from the Delta, directly to the west. The flat terrain and fertile sandy loam soils are suited to a vast range of varieties, with around 100 planted. It is also a key source for some of Lodi's oldest vines, many planted on their own rootstocks, including Zinfandel, Carignan, and Alicante Bouschet.

RIVER JUNCTION AVA
San Joaquin County

As the name suggests, River Junction AVA is located at the confluence of two rivers, the San Joaquin and the Stanislaus, south of Lodi. It is an area where cool maritime air collects, causing land surrounding the river junction to be significantly cooler than other Central Valley locations, and it is the only place in which a substantial amount of Grangeville fine sandy loam can be found. The majority of planting in this small appellation is Chardonnay.

SALADO CREEK AVA
Stanislaus County

Named after the river that runs through the AVA from the Diablo Mountains in the west to the San Joaquin River. It is situated in the west of the Central Valley, separated from the Santa Clara Valley AVA by the Diablo Range, which also inhibits any cooling coastal breezes. This warm part of Central Valley is mainly home to contract growers, primarily producing ripe, fruity Cabernet Sauvignon, Syrah, Sauvignon Blanc, and Viognier.

SLOUGHHOUSE AVA
Sacramento County

Located in the northeast in Lodi AVA, this appellation sits in the lower Sierra Nevada foothills at elevations of up to 180 metres (590 feet) with a range of alluvial, sandy loam and rocky soils. Slightly cooler winters leads to later budbreak followed by an intense growing season with a significant diurnal temperature range and little to no fog. Red grapes are predominant, including Bordeaux and Rhône varieties, as well as Petite Sirah and Zinfandel.

SUISUN VALLEY AVA
Solano County

Adjacent to Solano County's Green Valley, and a stone's throw from the Central Valley, Suisun Valley AVA enjoys the same cool winds that blow in from San Pablo Bay from spring until autumn in both of these areas. Situated between the southern ends of the Vaca Mountains in the east and the Mount George Range to the west, the area's soils consist of various forms of clay, along with silty and sandy loams. It has been gaining a reputation for the Petite Sirah (Durif) variety.

TRACY HILLS AVA
San Joaquin and Stanislaus counties

This appellation is approximately 55 miles southeast of San Francisco, sitting in the lower rolling foothills of the Diablo Range, southwest of the city of Tracy. Cooling mountain breezes help moderate the warm temperatures. Italian varieties are most widely planted, including Nero d'Avola, Montepulciano, Sagrantino, and Barbera.

THE WINE PRODUCERS OF
THE CENTRAL VALLEY

BOGLE VINEYARDS
Sacramento County

Chris Smith, a former assistant winemaker at Kendall-Jackson in Lake County, is improving quality in the vineyards (including large vineyards on Merritt Island) while Eric Aafedt leads the winemaking team.

✓ Merlot • Petite Sirah

BRONCO WINE COMPANY
Stanislaus County

Established in 1973 by Fred T John and Joseph S Franzia following the sale of the multi-generational Franzia family winery, the firm is now owned by the Wine Group. Based in Ceres, south of Modesto, Bronco is currently the fifth-largest wine company and largest vineyard owner in the US, producing and selling over 60 brands, both domestic and imported, in more than 90 countries. The primary focus is on delivering value at the mass-market end of the industry. Epitomised by the Charles Shaw varietal range, nicknamed "Two Buck Chuck" with reference to the price, the release of the brand in 2002 was a seminal point for the Californian wine industry, making wine accessible to those who previously viewed it as an expensive, inaccessible luxury. Much debate continues in the industry regarding the price point and quality, yet it remains one of the most successful brands in the country.

✓ Charles Shaw • Allure Moscato

DELICATO
San Joaquin County
★

Their output ranges from penny-pinching jug wines to real quality bargains, with pure fruit and focus presented in smart blue livery. How this Central Valley winery has come up in the world!

✓ Cabernet Sauvignon

E & J GALLO WINERY
Stanislaus County

From the air, Gallo's Modesto premises might look like an oil refinery, but from the ground you see peacocks roaming freely over well-manicured lawns, and inside the building there is a marble-bedecked reception area, with waterfalls cascading into large pools surrounded by lush tropical vegetation and containing carp the size of submarines. Gallo not only makes a lot of cheap plonk, but it also produces inexpensive yet drinkable varietal wines. The plonk comes under a variety of labels (such as Totts, André, Bartles & James, Carlo Rossi), which encompass wine coolers, sweet wines from foxy-flavoured native grapes, such as Concord, and most of the generics, while the varietals bear the Gallo name itself (the non-vintage Cabernet Sauvignon is a real penny-saver). This was how the brothers Ernest and Julio Gallo, two struggling grape-growers who had to borrow money to buy a crusher when Prohibition ended, built up their business to become the most powerful wine producers in the world. Commercial success was not enough, however, because both brothers also wanted the critical acclaim of making some of California's finest wines, which is why they literally moved mountains to terraform their prized Sonoma estate. But if the Gallo family is satisfied, it should not be. To create Gallo Sonoma with the money it threw at a single project was no challenge at all.

I have never been one to condemn the huge wineries because of size. The bigger they are, then the more likely it is that they will come across an outstanding batch of grapes or wine. The choice between losing these little gems in some mega-blend or of giving them their due recognition is what separates the best large wineries from the worst. Instead of creating small-batch, more premium wines themselves, Gallo have taken a different route, and they've bought up other producers with these wines already in their portfolios. This recent strategy saw Gallo purchase lauded Napa-based Pahlmeyer in 2019, who are well known for producing premium Bordeaux-style wines.

FICKLIN VINEYARDS
Madera County
★☆

California's leading estate-bottled port-style specialist. The vineyards are planted with authentic Douro varieties, such as Touriga Nacional and Tinta Cão, plus Sousão, which is used for Dão.

✓ Fortified (Old Vine Tinta Port, Touriga Vintage Port, Vintage Port)

FRESNO STATE WINERY
Fresno County
★

UC Davis has the best-known university wine course in the United States, but Cal State Fresno is the only one with a commercial winery on its campus, and a very successful one it is too, having amassed a large number of medals since its inception in 1997.

✓ Muscat Canelli • Syrah

JEFF RUNQUIST
Amador County
★☆

Standout wines from an underrated producer that majors in Zinfandel from the Shenandoah Valley but also makes stunning Central Valley wines.

✓ Barbera • Cabernet Sauvignon • Petite Sirah • Pinot Noir • Sangiovese • Syrah • Zinfandel ("Z" Masoni Ranch)

JESSIE'S GROVE
San Joaquin County
★

Tiny production of high-quality Lodi varietals, with especially noteworthy Zinfandel and a tip-top Carignane, made from vines over 120 years old.

✓ Carignane (Ancient Vines) • Zinfandel

KLINKER BRICK WINERY
Lodi
★

I came across Steve and Lori Felten's sleek, elegant, beautiful Zinfandel at the IEC wine competition in 2006. These fifth-generation growers and their 100-year-old Zinfandel vines should continue to be treasured.

✓ Zinfandel (Old Vines) • Syrah (Farah)

MARKUS WINE CO.
San Joaquin County
★☆

This company was established in 2014 in partnership with Borra Vineyards, where Markus Niggli was also winemaker; here, small parcels of outstanding fruit are sourced throughout Lodi from ancient vines and eclectic varieties. The wines are vinified at their winery just outside Lodi with minimal intervention to fully express their character.

✓ Entire range

QUADY WINERY
Madera County
★

Quady is an eclectic fortified wine specialist par excellence, although I am not so fond of the Vya vermouth-style offerings. By contrast, Electra is a low-alcohol (4 per cent) Moscato d'Asti-style light wine.

✓ Fortified (Muscat – Elysium, Essensia; Port – Starboard) • Muscat (Electra)

CALIFORNIA

Note: This section contains a round-up of the various California wine areas that are not dealt with in the preceding pages. Each of the AVAs listed below can be found on the regional California map (*see* p577).

ANTELOPE VALLEY OF THE CALIFORNIA HIGH DESERT AVA
Kern and Los Angeles counties

A hot, dry area approximately 110 kilometres (70 miles) inland of Los Angeles in the California High Desert (the western fringe of the Mojave Desert), separated from any coastal influence by the Sierra Pelona Mountains. The continental climate and low rainfall mean that irrigation is essential, with producers planting at altitudes of up to 1,200 metres (4,000 feet) to mitigate the heat.

BENMORE VALLEY AVA
Lake County

Surrounded by the 1,000-metre (2,900-foot) peaks of the Mayacamas Mountains, this small mountaintop AVA was designated in 1991, but as of December 2019 is not yet planted with any vineyards.

BIG VALLEY DISTRICT – LAKE COUNTY AVA
Lake County

Sitting at 425 metres (1,400 feet) in a basin within the Clear Lake AVA in the Mayacamas Mountains, this appellation has gained a reputation for Sauvignon Blanc. Diurnal swings of up to 30°C (50°F) allied with cool mountain breezes moderate the high-UV intensity of the region.

CALIFORNIA SHENANDOAH VALLEY AVA
Amador and El Dorado Counties

The area, preceded by "California" to avoid confusion with the Shenandoah Valley of Virginia, was also delimited in 1983. The California Shenandoah Valley AVA is most westerly of the Sierra Foothills sub-AVAs and as such is set at the lowest altitude and in the warmest location amid the Sierra Foothills. The soil is well drained and moderately deep and consists mostly of coarse sandy loams formed from weathered granitic rock over heavy, often clayey, loam. Given the climate, the wines achieve high levels of ripeness with full-bodied, high-alcohol Zinfandels prominent.

CLEAR LAKE AVA
Lake County

Located between the Mayacamas Mountains and the Mendocino National Forest at elevations of 395 to 455 metres (1,300 to 1,500 feet), Clear Lake's large water mass moderates the AVA's climate. Although this area is best for Chardonnay and Sauvignon Blanc, well-textured Cabernet Sauvignon is also produced, with the UV intensity inducing thick skins and high phenolics. The appellation encompasses the follow sub-AVAs: Big Valley District Lake County, High Valley, Kelsey Bench Lake County, and Red Hill Lake County.

CUCAMONGA VALLEY AVA
Los Angeles and San Bernardino Counties

An area of more than 40,500 hectares (100,000 acres), some 72 kilometres (45 miles) from Los Angeles, the Cucamonga Valley was first planted with Mission grapes in about 1840. Despite the interruption of Prohibition, the vineyards reached their peak in the 1950s, when more than 14,000 hectares (35,000 acres) of land were under vine. As the viticultural emphasis in California moved northwards, these vineyards declined, also facing pressure from urban sprawl. The hot climate supports ripe, full-bodied examples of Zinfandel and other Southern Rhône varieties.

EL DORADO AVA
El Dorado County

Situated between Sacramento and Lake Tahoe at elevations of 365 to 1,065 metres (1,200 to 3,500 feet), diurnal temperature swings and cooling mountain breezes maintain fresh acidities. The soil is mostly decomposed granite, except on Apple Hill, east of Placerville, where an old lava flow exists (*see* Lava Cap Winery). Apple Hill is one of two main winegrowing areas, the other key vineyard is located in the southeast of the AVA, between Quitingdale and Fairplay.

FAIR PLAY AVA
El Dorado County

Named after a former gold-mining camp, Fair Play is the smallest of the Sierra Foothills sub-AVAs. Located in rolling hills at an altitude of 600 to 900 metres (2,000 to 3,000 feet), when averaged, it has the highest elevation of any California AVA. Well-draining decomposed granite soils, allied with ranging exposures and microclimates, suit a host of varieties. Cabernet Sauvignon, Zinfandel and Petite Sirah are the most popular reds, complemented by Chardonnay, Sauvignon Blanc, and Roussanne for the whites.

FIDDLETOWN AVA
Amador County

The area is located in the eastern Sierra Foothills of Amador County. It differs from the neighbouring California Shenandoah Valley because of its higher elevations, colder temperatures at night, and greater rainfall. Grapes are grown without any irrigation, and the vineyards are located on deep, moderately well-drained, sandy loams. Known for the quality of old vine Zinfandel, Syrah, and Grenache.

GUENOC VALLEY AVA
Lake County

South of McCreary Lake and east of Detert Reservoir and situated within the North Coast AVA, the valley has a more extreme climate, lower rainfall, and less severe fog than the nearby Middletown area. Primarily Bordeaux varieties.

INWOOD VALLEY AVA
Shasta County

This sparsely planted region lies in the north of Shasta County, in turn in the north of California, situated in the foothills of the southern end of the Cascade Range. Designated due to the unique soil composition of the area.

KELSEY BENCH – LAKE COUNTY AVA
Lake County

Located to the south of Big Valley District, the vineyards are planted up to 485 metres (1,600 feet), primarily on rich, red volcanic soils. High diurnal swings of up to 30°C (50°F) and cooling mountain winds are key to extending the growing season and preserving acidity levels in intense Sauvignon Blanc, Zinfandel, and Cabernet Franc.

LEONA VALLEY AVA
Los Angeles County

Situated in the Sierra Pelona Mountains northeast of Los Angeles County and 900 metres (3,000 feet) in altitude, the AVA covers the valley floor and surrounding gentle rolling hills. Cut-off from any coastal influence by the San Gabriel and Sant Susana ranges, the fertile alluvial soils and warm, dry climate support a large range of varieties.

MALIBU COAST AVA
Los Angeles and Ventura Counties

First planted in the early 1800s, this region covers the Santa Monica Mountains Recreation Area, just west of the centre of Los Angeles. The climate is classified as Region II or low III (*see* p577), with influences from both the Pacific Coast, as well as cooling breezes from the Santa Monica Mountains behind. There are currently around 50 vineyards in the region, which are being threatened by city zoning issues.

MALIBU-NEWTON CANYON AVA
Los Angeles County

This is a one-vineyard appellation on the south-facing slopes of the Newton Canyon basin within the Santa Monica Mountains and Malibu Coast AVA.

MANTON VALLEY AVA
Shasta and Tehama Counties

A small Northern California AVA named after the town that lies within the AVA set on low-vigour, shallow, volcanic, and sandy loam soils. The moderate-to-warm climate is balanced by diurnal temperature swings, allowing Merlot, Cabernet Sauvignon, and Zinfandel to ripen whilst retaining acidity.

NORTH YUBA AVA
Yuba County

North Yuba is located in the remote middle and upper foothills of Yuba County, immediately west of the Sierra Nevada and north of the Yuba River. The climate is relatively temperate compared to the rest of the AVA, as the area escapes both the early frosts and the snow of higher elevations in the Sierra Nevada, as well as the heat, humidity, and fog common to the Sacramento Valley lowlands.

RAMONA VALLEY AVA
San Diego County

Located northeast of San Diego, the valley sits in the foothills of the coastal ranges at an average altitude of 425 metres (1,400 feet). Morning fogs still influence the valley, which combined with diurnal swings, preserve acidity in the majority rich, full-bodied red wines produced.

RED HILLS LAKE COUNTY AVA
Lake County

Mountainous terrain with free-draining, red volcanic soils and the majority of vineyards planted at altitudes in excess of 610 metres (2,000 feet). Cabernet Franc, Cabernet Sauvignon, and Sauvignon Blanc are intense but balanced.

SADDLE ROCK-MALIBU AVA
Los Angeles County

A high-elevation valley within the Santa Monica Mountains and Malibu Coast AVA that is gaining a reputation for rich but structured Cabernet Sauvignon, Merlot, and Syrah. Vineyards are planted on south-facing slopes at altitudes of up to 610 metres (2,000 feet), resulting in cool nights that preserve acidity.

SAN PASQUAL AVA
San Diego County

A natural valley located in the Santa Ysabel watershed, next to Ramona Valley, it is fed by natural streams that feed the San Dieguito River and is substantially affected by coastal influences. Temperatures are warm in summer, but seldom very hot, and ocean breezes cool the area, especially during the night; well suited to the Rhône varieties that predominate.

SEIAD VALLEY
Siskiyou County

California's most northerly AVA, located just 21 kilometres (15 miles) south of the Oregon border, the Seiad Valley AVA measures 877 hectares (2,165 acres). Granted in 1994, the only winery in the region has since closed, arguably a significant factor in the tightening of AVA classification rules since.

SIERRA FOOTHILLS AVA

This vast region sits to the east of and above Central Valley, covering an area of 1.05 million hectares (2.6 million acres). It attracts rugged specialists up for the challenge of making expressive wine in limited quantities in one of the few areas of California where cultivating grapes is not always possible – here, vineyards generally lie at altitudes of 455 to 915 metres (1,500 to 3,000 feet). Zinfandel is most widely planted, followed by Cabernet Sauvignon, Syrah, and Chardonnay, yet aromatic varieties are also finding a home. There are five sub-AVAs within the AVA – California Shenandoah Valley, El Dorado, Fair Play, Fiddletown, and North Yuba.

SIERRA PELONA VALLEY AVA

Los Angeles County

Located just 48 kilometres (30 miles) from downtown Los Angeles in the heart of the Sierra Pelona Mountains that separate inland LA County from the cooler coastal regions. The Santa Clara River valley funnels maritime winds into the region, ensuring a far more moderate climate than the desert-like conditions of the Antelope Valley farther inland.

SOLANO COUNTY GREEN VALLEY AVA

Solano County

This small valley, approximately 1.5 wide by 6.5 kilometres long (1 mile by 4 miles), is sandwiched between the Napa Valley to the west and the Suisun Valley to the east, bordering Napa's Coombsville AVA. The soil here is a clay loam, and the climate is influenced by the cool, moist winds that blow inland from the Pacific and San Francisco Bay almost continuously from spring through to autumn.

SOUTH COAST AVA

Orange, Riverside, San Bernardino, and San Diego Counties

Established in 1985, this AVA is often assumed to include Los Angeles County; however, as the TTB Final Rule document stated: "Since no grapes come from Los Angeles County and it is very unlikely that any ever will, it was considered confusing to include the County in 'South Coast'."

SQUAW VALLEY–MIRAMONTE AVA

Fresno County

A sub-region of Fresno Valley AVA, it is situated in the steep, southwest-facing foothills of the Sierra Nevada Mountains at elevations from 485 to 1,050 metres (1,600 to 3,500 feet). Cooling mountain air drainage moderates the long sunshine hours and prolongs the growing season, resulting in cooler days and warmer nights than the Central Valley floor below. Petite Sirah and Carmenere are most widely planted.

TEMECULA AVA

Riverside County

Temecula is located in Riverside County in Southern California, between San Diego and Los Angeles, and is gaining a reputation for quality. Situated 35 kilometres (22 miles) from the Pacific coast, the vineyards are cooled to moderate temperatures by marine breezes entering the area through the Deluz and Rainbow Gaps. Merlot, Sauvignon Blanc, and Chardonnay are the most widely planted varieties.

TRINITY LAKES AVA

Trinity County

One of the most northerly and remote AVAs in the state, surrounding two reservoirs, Trinity and Lewiston Lakes, which have a tempering effect on the cool, high-elevation climate. A small area of cool climate varieties is planted on the surrounding steep, narrow valleys on shallow alluvial soils.

WILLOW CREEK AVA

Humboldt County

Not to be confused with the recently classified Paso Robles Willow Creek County AVA, this Northern California appellation was delimited many years earlier, in 1983. The area is influenced primarily by two major climatic forces – namely the Pacific Ocean and the warmer climate of the Sacramento Valley, 160 kilometres (100 miles) to the east. These create easterly winds that give Willow Creek fairly cool temperatures in the summer and infrequent freezes in the winter. The area to the east of Willow Creek experiences colder temperatures in the winter and hotter temperatures in the summer.

OTHER WINE PRODUCERS OF

CALIFORNIA

SIERRA FOOTHILLS

AMADOR FOOTHILL WINERY

Amador County

With its small Shenandoah Valley vineyards, Italian varietals are offered with Zinfandel being the biggest draw.

✓ *Zinfandel*

BOEGER WINERY

El Dorado County

☆

The first winery to re-open the pre-Prohibition Sierra Foothill vineyards, Boeger has established itself as a producer of understated fine wines that take time to flesh out in bottle, more appealing perhaps to European palates than to those in California.

✓ *Barbera • Classic red blend (Meritage Reserve) • Zinfandel*

DOMAINE DE LA TERRE ROUGE

Amador County

★☆

Billed as the winery where the Rhône Valley meets the Sierra Nevada, Domaine de la Terre Rouge produces elegant, stylish Syrah and an exotically sweet late-harvest Muscat à Petits Grains.

✓ *Muscat à Petits Grains • Syrah (Ascent) • Zinfandel*

HOLLY'S HILL VINEYARDS

El Dorado County

★

Tom and Holly Cooper's boutique winery concentrates to a large degree on Rhône-style wines. The Holly's Hill vineyards are planted with the latest French clones from Beaucastel and Rayas.

✓ *Grenache (El Dorado) • Syrah*

IRONSTONE

Calaveras County

★

Opened in 1994, Ironstone, John Kautz's winery in Gold Rush country has already shown a capability for smooth, lush Cabernet Franc.

✓ *Cabernet Franc • Cabernet Sauvignon • Zinfandel*

LAVA CAP

Placerville

★☆

This underrated producer just gets better and better. Brilliant-value wines, including one of California's finest Petite Sirahs, grown on the volcanic soils, some 760 metres (2,500 feet) up Apple Hill in the El Dorado AVA.

✓ *Cabernet Franc • Petite Sirah • Syrah (Reserve) • Viognier • Zinfandel*

MONTEVINA

Amador County

★

Part of Trinchero Family Estates, together with Sutter Home, Montevina is best known for Zinfandel, but has also helped the Latin trend with its Italian varieties planted in the Shenandoah Valley.

✓ *Barbera • Pinot Grigio • Zinfandel*

NEVADA CITY WINERY

Nevada County

The ultimate garage winery, Nevada City Winery is located in the historic Miners Foundry Garage, on Spring Street in downtown Nevada City. It was founded in the early 1980s and has gone from strength to strength, winning several awards. Red, white, rosé, and dessert wines are produced.

✓ *Zinfandel*

SOBON ESTATE

Amador County

★

Family-owned and -run estate specialising in old vine Zinfandel from high-altitude vineyards planted in the early 1900s, offering excellent value.

✓ *Zinfandel (Rocky Top)*

SOUTHERN CALIFORNIA

ANDREMILY

Ventura County

★★☆

Rhône specialist producing some of the most highly sought-after Syrah in California. Founded in 2011 by winemaker Jim Binn, who previously worked at cult-status producer Sine Qua Non. The highest quality Syrah, Mourvèdre, and Viognier fruit is sourced from celebrated Alta Mesa, Larner, and White Hawk vineyards in Santa Barbara. Intensely concentrated, perfumed, and balanced.

✓ *Syrah • Mourvèdre*

CALLAWAY

Riverside County

☆

One of the first wineries to prove that parts of Southern California have, in fact, excellent microclimates for producing fine-wine grapes.

HART FAMILY WINES

Riverside County

☆

Benchmark producer in the moderate-climate Temecula Valley was founded in 1980, producing a range of wines from Bordeaux, Rhône, and Italian varieties.

✓ *Sauvignon Blanc • Petite Sirah*

JOSEPH FILIPPI

San Bernadino County

★

This 35-year-old winery has moved from Fontana to Ranch Cucamonga, where it continues to produce a wide range of wines, including a number of very successful fortified styles.

✓ *Fortified (Alicante Bouschet Port, Oloroso) • Mourvèdre • Muscat*

MAURICE CAR'RIE WINERY

Riverside County

☆

The Van Roekels came to Temecula to retire in 1984, but were sidetracked into just "one more venture".

✓ *Muscat Canelli • Sauvignon Blanc*

MORAGA

Los Angeles County

★★

This winery is owned by media mogul Rupert Murdoch. Viticulturally this tiny sandstone and limestone canyon has a discernible microclimate with 61 centimetres (24 inches) of rain each year, compared with 38 centimetres (15 inches) on nearby properties, and the wines are also truly fine by any standards. Bel Air is a soft, deep-coloured red

Bordeaux-style blend with seamless oak integration and beautifully layered fruit, the flavour building in the mouth. Moraga also makes a tiny amount of fresh, soft, delicately rich Sauvignon Blanc.

✓ *Classic red blend (Moraga Red)* • *Sauvignon Blanc (Moraga White)*

ORFILA VINEYARDS & WINERY
San Diego County

Good value wines, including the odd gem that punches well above its weight, come from the winery of Argentinian Alejandro Orfila, who was twice elected secretary general of the Organization of American States.

✓ *Muscat Canelli* • *Sangiovese (Collina Estate)* • *Syrah (San Pasqual Valley)* • *Viognier (Lotus)*

THE OJAI VINEYARD
Ventura County

Adam Tolmach, a former partner in Au Bon Climat, teamed up with Helen Hardenbergh to create this boutique winery. Stylish, smoky-blackcurrant Syrah stands out, but some great Pinot Noir too; all wines of all designations can be top-notch.

✓ *Chardonnay* • *Pinot Noir* • *Sauvignon Blanc* • *Syrah* • *Viognier* • *Riesling (Dry, Ice Wine)*

THORNTON
San Diego County

Best known for its sparkling wine, particularly its Cuvée Rouge (although not my favourite), yet these wines would be much better if sold younger and fresher. In recent years, however, the quality of Thornton's table wines has far exceeded that of its fizz, and its Nebbiolo can rival the best that California has to offer.

✓ *Classic red blend* • *Muscat Canelli* • *Nebbiolo* • *Rosé (Grenache)* • *Sparkling wine (Brut Rosé)*

LAKE, MARIN, AND SOLANO COUNTIES

GUENOC WINERY
Lake County

This was once the property of actress Lillie Langtry (mistress of Edward VII, when he was the Prince of Wales), and her picture is used to market one of these excellent-value wines. Now part of the Foley portfolio.

✓ *Cabernet Sauvignon* • *Chardonnay* • *Petite Sirah*

KALIN CELLARS
Marin County

After waiting for four years before selling his 1990 Chardonnay, only to attract great acclaim, microbiologist Terry Leighton stretched his late-release philosophy to bizarre lengths. In October 2004, for example, Leighton decided the time was right to release his 1996 Potter Valley Sauvignon Blanc! On average, he puts his whites on the market when they are six years old and his reds when they are eight years old. This works better for some styles than others. Leighton was initially to be applauded, but it is always best to err on the side of youth, as the wine-buying public can always cellar a wine, but cannot turn back the clock.

KENDALL-JACKSON
Lake County

It's amazing to think that when I started researching the first edition of this encyclopedia, the Kendall-Jackson brand was barely a year old. Since then, K-J has become one of California's best-selling brands, although its impact on export markets has been minimal. In the United States this company has become so successful that some critics now love to bash the brand they helped to create, when really it is only the obviousness of the entry-level Vintner's Reserve and cheaper subsidiary brands they object to. Yet it is the obviousness of these wines that make them so popular, and the people who buy Vintner's Reserve seldom read a wine critic's column, so what's the point of putting question marks in the minds of those who do? Better instead to highlight the plus points, and K-J has plenty of those, not least in its higher-level ranges, where the wines pack in a lot of quality at what are relatively inexpensive prices.

K-J is a private company belonging to Jess Jackson, who also owns Cambria (which is run by his wife, Barbara Banke), La Crema, Edmeades, and Stonestreet. Other brands include Camelot and Pepi. K-J is part of Jackson Family Wines.

✓ *Cabernet Sauvignon (Hawkeye Mountain, Stature)* • *Chardonnay (Camelot, Grand Reserve)* • *Classic red blend (Stature Meritage)* • *Pinot Noir* • *Riesling* • *Zinfandel*

SEAN THACKREY AND COMPANY
Marin County

San Francisco art dealer Sean Thackrey loved Burgundy, which led him to establish this winery in 1980 and, naturally enough, try his hand at sculpting Pinot Noir. It was not until the late 1980s that his big, thick, richly flavoured wines began to appear and quickly attracted attention. Since then Thackrey's obsession for fermenting under the stars, along with several other artisanal quirks of ancient winemaking, have intrigued critics and consumers alike.

✓ *Petite Sirah (Sirius)*

SKYWALKER VINEYARDS
Marin County
★☆

Founded by George Lucas in 1991, estate-grown cool-climate Pinot Noir and Chardonnay are released under Skywalker and Sommita labels, the latter being from ridge-top higher-altitude sites.

✓ *Pinot Noir* • *Chardonnay*

STEELE
Lake County
★★

This is the personal label of Jed Steele, who is a former Kendall-Jackson winemaker and a highly regarded wine consultant. His debut vintage was 1991, and his brightly flavoured, brilliantly stylish wines made an instant impact. More than a quarter century later, Jed still produces wines that are excellent all around, but he truly excels in his single-vineyard Pinot Noir range. The second label is Jed Steele's Shooting Star.

✓ *Chardonnay* • *Pinot Blanc (Santa Barbara)* • *Pinot Noir* • *Zinfandel (Catfish Vineyard)*

VEZÉR FAMILY VINEYARD
Salano County
★☆

Frank and Liz Vezér originally founded a vineyard in the Suisun Valley as grape growers but in 2003 released their inaugural estate Zinfandel, which is gaining a following, along with their Petite Sirah. Chardonnay and Sauvignon Blanc are equally receiving critical acclaim.

✓ *Zinfandel* • *Petite Sirah*

THE SKYWALKER RANCH IN NICASIO, MARIN COUNTY
Famed *Star Wars* creator, George Lucus, began cultivating vines at his Central Valley property in 1991, but the first vintage under the Skywalker Vineyards label only came in 2007 with a total production of 60 cases of Pinot Noir and 30 cases of Chardonnay.

The Pacific Northwest

A collection of wine regions located north of California are bundled together as the Pacific Northwest – Washington, Oregon, and Idaho – yet aside from proximity to the ocean, more things separate them than unite them.

Cold Pacific winds and even rain near the Cascade Mountains block more than 95 per cent of Washington's vineyards. In Oregon, the Cascades are to the east of the vineyards, and the Coast Range is an older, lower set of mountains. Protection is sketchy, so timing of the ocean winds and even rains can dramatically affect the harvest. Washington's near-desert conditions require irrigation; harvest timing is often only a matter of ripeness. Meanwhile, its continental climate – which can be extreme – can encumber Idaho's grape production.

The lush rolling hills and teeming forests of Oregon, and the dry scrublands of central and eastern Washington, where vineyards prosper today, were long occupied by only native peoples and fur trappers. In the early years of colonization, both states were valued mostly for their seaports and, though it required a brief military spat with England to cleave them from British Columbia, the two joined the American experiment with little bloodshed. Idaho came along with the rest of Oregon Country, as it was known then, but it was mostly Indian Territory until the latter part of the 19th century. It then became notable for ranchers, small-scale farming, and, eventually, Mormon and Basque enclaves. The vine didn't feature in anyone's plans.

The end of the 19th century saw each state grow its farming capacities, occasionally with grapes, but the first real burst of energy into wine was launched by a teetotaling scientist in the post–World War II years. The state's agricultural sector hired Dr Walter Clore to assess the fortunes of the vast central and eastern regions, more desert than farmland. The mighty Columbia River had been dammed for energy in the post-war boom; it could be relied upon for water as well. But which crops should be prioritised? Clore selected a dozen and, perhaps surprisingly, he included not simply table grapes but a more lucrative and nascent crop, wine grapes.

Clore was quickly joined by another group of scientists, but these were not agronomists. Rather it was the University of Washington and the Hanford Nuclear Plant that supplied some whiz kids, most of them hobbyists, for anything that might provide some activity outside the laboratory. They approached grape-growing and winemaking like scientists: plant a little of everything and measure the results. To this day, a little of everything is what you can find in the state.

Wine writer Leon Adams, preeminent in the 1960s, tasted some wines from a new Washington company called American Wine Growers, a merger between two wineries making mostly Concord wines. Adams introduced that company to legendary winemaker André Tchelistcheff, and they shortly launched Chateau Ste Michelle with a plan to make *vinifera* wines. Around the same time, the University of Washington group established Associated Vintners, soon to become the Columbia Winery (now owned by Gallo).

WINE REGIONS OF THE PACIFIC NORTHWEST

(see also p569)

Encompassing thousands of square kilometres, the Pacific Northwest is defined by its proximity to the cold Pacific; in Oregon, most areas are barely protected by the Coast Range, and the vineyards can be cool and misty. In Washington, most vineyards are far to the east of the coastal mountains; the vineyard climate is arid and challenging. In Oregon, wet harvests can bedevil winemakers. Washington vineyards are, in the main, unlikely to survive without irrigation; Idaho suffers from the same deficits.

OREGON VINEYARDS OVERLOOK THE COLUMBIA RIVER GORGE AT RUTHTON POINT
On the other side of the river lies Washington State. The Columbia Gorge AVA spans both sides of the river, encompassing vineyards in both Oregon and Washington.

While a bunch of straight-laced pocket protector–wearing nerds spent the 1960s and 1970s transforming Washington State wine, the hippies were descending upon Oregon – or so it must have seemed to the farmers there. Some of them were interested in growing Pinot Noir and were convinced it was the Burgundian mecca that America had long promised and never delivered. First Richard Sommer down south, and then David Lett, Dick Erath, and others up in the Red Hills . . . each cracked open abandoned orchards. They liked Burgundy, and because California wasn't doing much with Pinot Noir, they decided to give it a try. The landscape was not particularly helpful: it was too cool for most reds and too wet for most vines. But in the right spots things prospered, and the industry's single-minded focus upon one grape got them some notice, especially after Lett's 1975 Eyrie Pinot Noir South Block bested all but Drouhin Chambolle Musigny in a 1979 Gault-Millau tasting, which was reported worldwide.

WASHINGTON STATE

No such watershed moment catapulted Washington's wines to the world's attention, but for those lucky enough to taste it, the 1975 Chateau Ste Michelle Cold Creek Cabernet Sauvignon was just as jolting an experience – full, refined, and elegant. By the late 1970s and early 1980s, a wave of pioneers was giving chase in numerous corners of the state: Marty Clubb (L'Ecole No. 41), Gary Figgins (Leonetti), and Rick Small (Woodward Canyon) in Walla Walla; Jim Holmes (Ciel du Cheval) and John Williams (Kiona) on Red Mountain; Charles Henderson and Dr William Anderson (Celilo) in the Columbia Gorge; Mike Sauer (Red Willow Vineyard) and Dick Boushey (Boushey Vineyards) in Yakima Valley; Don Mercer (Mercer Estates) and Paul Champoux (Champoux Vineyard) in Horse Heaven Hills; and many others like Mike Wallace (Hinzerling), Bill Preston (Preston Vineyards), Maury Balcom (Quarry Lake, Balcom & Moe), Jerry Bookwalter (Bookwalter Winery), Rob Griffin (Barnard Griffin), and Tom Hedges (Hedges Family).

Initial interest was in Riesling (and later, Chardonnay) but the red wine craze of the 1990s inspired fervour for red Bordeaux varieties, particularly for Merlot. The state's reputation soared on the wings of Merlot- and Cabernet-centric wineries such as Leonetti, Quilceda Creek (far away in Woodinville, outside Seattle), and Andrew Will (even farther away on Vashon Island). Leonetti had some vines in its Walla Walla homelands, but they and the rest were dependent upon grapes from throughout the vast Columbia Valley, which encompasses almost all AVAs. The Yakima Valley was a mainstay; it was soon parcelled into Red Mountain AVA and the Rattlesnake Hills AVA; Snipes Mountain AVA followed in 2009. Horse Heaven Hills AVA (extending into Oregon) and Wahluke Slope AVA were carved off from the greater Columbia Valley in the mid 2000s.

In the meanwhile, trendsetters embraced Rhône varieties; some of the country's most sought-after Syrah is grown around the state. Early adopters such as Doug McCrea (McCrea Cellars) have been eclipsed, along with many wineries throughout the rest of the state, by Christophe Baron of Cayuse Vineyards. Baron was the first to focus upon an orchard-rich region south of the Oregon-Washington border (but still within the Walla Walla AVA) that had long been known as The Rocks. Fruit farmers swore by it, but grape-growers were frightened off by the tendency of the area to suffer from frost and freeze damage. Baron adopted some spots deemed to be slightly higher in elevation, teeming with basalt cobblestones, looking for all the world like Châteauneuf-du-Pape in America. With high-density planting and biodynamic viticulture, Baron has helped change the conversation in Walla Walla and beyond.

Some are critical of the low-acidity, plush, even fat wines typical of the Rocks District. But Washington's rich, plush wines have garnered international acclaim from critics and consumers alike. Newer wineries, such as Gramercy Cellars and Rotie Cellars, seem to be pointing the way to more classically structured wines, with higher acidity and less alcohol. Newer AVAs such as Lake Chelan and Ancient Lakes are decidedly cooler places, so they can show similar character and are planted more to whites than reds.

Today there are more than a thousand wineries in the state, supplied by nearly 24,000 hectares (60,000 acres) of grapes from the state's 14 AVAs. Almost 60 per cent of the output is in red grapes; 41 per cent is white. Nearly 18 million cases are produced annually, and the number is still growing.

OREGON

The first *vinifera* vines were planted in Oregon's Rogue River Valley as early as 1854. These and other *vinifera* vineyards were still in existence at the time of Prohibition but, as in Washington, the wine industry in Oregon relied almost entirely on *labrusca* grapes of the Concord variety until the 1970s. Change began in a small way in the 1960s and 1970s, when California drop-outs established wineries, such as Richard Sommer (Hill Crest, 1961), David Lett (The Eyrie Vineyards, 1965), and Bill Fuller (Tualatin, 1973).

Oregon's overnight fame came in 1979, when Lett entered his 1975 Pinot Noir in a blind wine-tasting competition organized by Robert Drouhin. Drouhin's Chambolle-Musigny 1959 had won, but Lett's The Eyrie Vineyards 1975 Pinot Noir came second, trouncing Drouhin's fabulous 1961 Clos-de-Bèze and many other prestigious Burgundies in the process. Since that eventful day, most critics have believed that it would only be a matter of time before the Pinot Noirs of Oregon would rival those of Burgundy.

It might be that Oregon has morphed its Pinot Noir style; that wouldn't be different from any other evolving New World wine region. Certainly, the early 1990s were the time of the cold soak; the late 1990s and early aughts might be dubbed the era of the barrel salesman. But things have equilibrated since then.

As Drouhin dropped anchor, the French haven't stopped visiting and some have stayed. Jean-Nicolas Meo of the great house of Meo-Camuzet (who was mentored by the late cult Burgundian Henri Jayer) opened Nicholas-Jay with music entrepreneur Jay Boberg. The mind behind the modern ascendance of Louis Jadot, Jacques Lardiere, has moved in as well, though only metaphorically – but his wines at Résonance convey elegance and bearing. Dominique Lafon has been a frequent visitor since the 1980s; he is a partner in a newish project, Lingua Franca, and it includes local hero and master sommelier Larry Stone as well.

RECENT PACIFIC NORTHWEST VINTAGES

2019 For some northwest vintners, 2019 was a complete shock to the system: cool, wet, marred by freezes in Washington and mildew in Oregon. For the old guard, it was merely a return to normalcy. After seven warm-to-hot vintages, most vineyard managers were challenged to alter their now-habitual practices. Quantities were strangely higher than expected, and some inexpensive wines suffered. Still, the more expensive wines may be better balanced and longer-lived than any produced in other years.

2018 This was the eighth-warmest year on record for Oregon and dry overall. Fires were a source of much worry; smoke taint was expected in southern Oregon and central Washington. But the timing seemed to have saved most grapes (though one large, well-known winery used the emergency to drop numerous vineyard contracts); the atmospheric haze also slowed ripening just enough to allow a bit of phenolic maturity that might have been otherwise impaired.

2017 Another hot year overall, though things started off cool and wet in Oregon. But either some cool moments took the edge off, or vintners started to understand how to deal with the heat. Either way, as in Washington, wines are tasty and even, at times, elegant.

2016 Another early, warm vintage in Oregon with bud-break significantly advanced from historical dates, a good two-plus weeks ahead of average. By early summer temperatures swung back towards normal in both Oregon and Washington.

2015 Though 2013 and 2014 were hot and early, 2015 vied to outdo them. The Columbia Valley averaged 3,157 Growing Degree Days (GDDs), compared to a long-term average of 2,628 GDDs, with bud-break, bloom, and harvest occurring two to three weeks ahead of historical averages. Harvest began historically early, with some wineries bringing in fruit in mid-August – a pattern that would last till 2019. Oregon too saw crazy early harvests (mid to late September).

All this attention has led to dramatic boosts in Willamette Valley plantings and, because valley floor plantings are most likely to see frost damage, most new vineyards creep along the hills (the usual rule is that vineyards need to be at least 60 metres, or 197 feet, off the floor). Lack of available land has led to price competition for plantable sites. The big companies have taken notice, too: Jackson Family is the most notable – they have bought properties like Zena Crown Vineyard in Eola-Amity Hills, Solena Estate, and wineries such as Penner-Ash. Yet, even off the record, winemakers at competing wineries don't castigate Jackson's efforts. "They've generally been good actors," as one winemaker said to me.

Although a focus upon individual sites has always been part of Oregon's DNA, the rise in the mid-aughts of the new so-called sub-AVAs is intended to differentiate one spot in the vast Willamette Valley from another, to justify both pricing and marketing alike. At the moment no consumer seems to know why they ought to buy a Yamhill-Carlton over a Chehalem Mountains wine, though the move towards regional specificity is amplified by the genuine soil and climate differences. And Oregon's original love of Burgundy suggests that it was only a matter of time before a series of hyphenated vineyard regions became code for excellence.

Today Oregon's vineyards have increased to a total of 14,564 hectares (35,972 acres) and there are nearly 800 wineries across the state. Interestingly, the highest growth rate in planted acreage last year was in the Umpqua and Rogue Valley AVAs. Year over year, grape acreage increased by 10 per cent as southern Oregon continued its development into a wine region worthy of the world's attention.

IDAHO

Idaho had a pre-Prohibition wine industry of some note, but the modern industry has been slow in growing. Although the fetters have long been off, the state has a sizable Mormon population and so alcohol beverages aren't always celebrated. There are now almost 60 wineries reliant upon 525 hectares (1,300 acres) of vineyards, as well as fruit from Washington State's prolific acreage.

A MECHANICAL TRIMMER SIMULTANEOUSLY PRUNES TWO ROWS OF GRAPEVINES IN A WILLIAMETTE VALLEY VINEYARD
The hillsides of the Williamette Valley AVA are particularly suited to Pinot Noir vines.

FACTORS AFFECTING TASTE AND QUALITY

LOCATION
This arbitrary grouping of three northwest states: Washington, Oregon, and Idaho, includes the large Columbia Valley, which runs from northern Oregon into and including central and eastern Washington. Oregon and Washington sit within the Pacific Ocean's Ring of Fire, a bi-hemispheric band that accounts for 75 per cent of the world's active volcanoes and 90 per cent of the world's earthquakes. Volcanic and plate activity is relatively quiet for now, but the geology and topography of each state is shaped by their fiery and dramatic pasts.

CLIMATE
The temperatures generated by continental air masses that can hammer Idaho and eastern Oregon and Washington are most often moderated by westerly winds from the Pacific Ocean, flowing along the Columbia River or its tributaries or through the Van Duzer Corridor, a break in the Cascades. Oregon is the coolest state, and Washington the wettest (though only on paper). With the exception of Puget Sound, all the viticultural areas of Washington are in the much hotter and drier central and eastern areas, with plentiful sunshine (averaging over 17 hours per day in June) and crisp, cool nights during the critical ripening period. Climatic conditions are more continental in Idaho.

ASPECT
The overbroad Columbia Valley includes long valleys that run alongside the Columbia River perched on slopes both low-lying and elevated. In Oregon's Willamette Valley, vineyards necessarily sit above 60 metres (196 feet) higher to avoid the worst of the weather. In the south, vintners generally find the cooler spots, especially those along the Rogue or Umpqua Rivers, benefiting from at least some cool Pacific Ocean winds.

SOIL
Oregon's vineyards are often roughly delineated as having volcanic or marine soils (or blends of the two). The Pacific Ocean reached Idaho's border 200 million years ago and Oregon's hills are the remaining uplifted marine bedrock, mostly clay soils. The Cascades exploded in dramatic volcanic action 10 to 50 million years ago, and the volcanic soils dominate in southern Oregon and appear in pockets such as the Dundee Hills, McMinnville, and Eola-Amity Hills AVAs. Those soils would appear too in Washington State were it not for the powerful Missoula floods (10,000 to 15,000 years ago) that wiped away topsoils in many places, so soils can be very shallow and poor. Much of the available "soils" are deep pockets of wind-blown loess, devoid of much organic materials. Oregon's Coast Range was pushed up above the ocean 10 to 16 million years ago, and streams flowing down its ridges have provided weathered sedimentary soils for the Willamette Valley. Here, too, the Missoula Floods deposited much of what they scoured from Idaho, Washington, and eastern Oregon. So Idaho, Washington, and Oregon include deeper, fertile, silty, sandy, or clay loams or loess, much of it over volcanic bedrock, and with a bit more clay and volcanic ash in Oregon.

VITICULTURE & VINIFICATION
Washington vines are generally un-grafted, with irrigation broadly necessary. In many frost- and freeze-prone areas, vines will have two trunks, with one of them expected to survive in the worst of winter. Washington declares itself phylloxera-free, but it's not true. The insect has been slow to attack, due to the poor soils and the often intense winters. But climate change may be giving the creature a leg up so, in some areas, grafting has already begun. Oregon and Idaho, like most of the rest of the world, rely upon grafted vines. In all three states, the arguments revolve around row orientation, planting density, and canopy and trellising choices. Producers throughout often maintain sustainable and environmentally friendly practices, as well as organic and biodynamic viticulture and winemaking. Virtually half of Oregon's vineyards are certified by the half dozen or so certifying bodies. Yet the devil is in the details; some insist upon their adherence to stringent strategies but still bend or break the rules as needed. Far fewer are certified organic and biodynamic than claim such philosophies as adherents. In Washington State, in particular, biodiversity is not the consideration that it should be.

GRAPE VARIETIES
Washington –Primary varieties: Cabernet Sauvignon, Merlot, Syrah, Chardonnay, Riesling
Secondary varieties: Sauvignon Blanc, Pinot Gris, Cabernet Franc, Viognier
Oregon – Primary varieties: Pinot Noir, Pinot Gris, Chardonnay
Secondary varieties: Cabernet Sauvignon, Gewürztraminer, Merlot, Pinot Blanc, Riesling, Cabernet Franc, Gamay Noir, Sauvignon Blanc, Syrah, Grenache, Zinfandel
Idaho – Primary varieties: Chardonnay, Riesling, Pinot Gris
Secondary varieties: Cabernet Sauvignon, Merlot, Syrah, Cabernet Franc, Chenin Blanc, Sauvignon Blanc, Gewürztraminer, Lemberger, Pinot Gris, Pinot Noir, Sémillon

---

THE APPELLATIONS OF

THE PACIFIC NORTHWEST

IDAHO

The explosion of wines in Washington has not gone unnoticed, and a few winemakers have wandered across the state line to join an expanding group of creative souls in southeast Idaho. Ste Chapelle reckoned to be Idaho's equivalent of Ste Michelle, but similar success didn't come. The winery continues to make good wines, some of them remarkable values, but they seem to shrink a little bit each year. Idaho now has almost 60 wineries, with most of them in the Snake River AVA in Idaho's Panhandle area, around Coeur d'Alene, the state's largest and deepest lake, just a short drive from Spokane in Washington.

EAGLE FOOTHILLS AVA

Encompassing nearly hectares 20,235 hectares (50,000 acres), there are currently only 27 hectares (67 acres) planted in the AVA, but plans include (at this printing) an additional 202 hectares (500 acre)s to be planted shortly. Contained within the larger Snake River Valley AVA, this is the only AVA located completely within the state of Idaho. Nestled in the foothills, the slope and aspect of the Eagle Foothills play an important role in sunlight reception, cold air drainage, and frost and wind protection.

LEWIS-CLARK VALLEY AVA

Named for explorers Meriwether Lewis and William Clark, who used the area as a refuge for during their expedition from 1804 to 1806, this new AVA (2019) is Idaho's third wine appellation. Lewis-Clark hosts vineyards between 200 and 600 metres (656 and 1,970 feet) in elevation. Growing Degree Days (GDDs) vary between 2,600 and 3,000, as opposed to the areas surrounding it, most of which are under 2,000 GDDs. Currently 40 hectares (100 acre)s are planted to vineyards along the steep canyons of the Clearwater and Snake Rivers and their tributaries.

SNAKE RIVER VALLEY AVA

The Snake River Valley AVA, located in eastern Oregon and southwest Idaho and spanning a total area of about 20,720 square kilometres (8,000 square miles), at elevations of 760 to 915 metres (2,500 to 3,000 feet), has little uniformity; soils vary widely throughout the region. There are at present only about 490 hectares (1205 acres) of grapevines. It's more than 400 miles (645 km) from the Pacific Ocean; the climate is continental and can be extreme, but the elevation offers distinctive cooling temperatures during summer nights. To date, few wines have been produced on the Oregon side but this AVA offers the largest density of vineyards and wineries in Idaho.

OREGON

APPLEGATE VALLEY AVA

Surrounded by the Rogue Valley AVA, but following the contours of the Applegate River, the Applegate Valley AVA offers higher elevations than other areas. The Siskiyu Mountains buffer the cool coastal winds (though the river mitigates that protective effect, particularly in the west), allowing the cultivation of Cabernet Sauvignon, Chardonnay, Merlot, Pinot Noir, Syrah, and Tempranillo. Stream sediments predominate.

COLUMBIA GORGE AVA

Situated in both Oregon and Washington and straddling the Columbia River, the Columbia Gorge AVA contains some of the most picturesque vineyards in either state. Vineyards are tiny in the main, and the small-scale approach has benefited some artisanal projects of quality. One of the Pacific Northwest's most famed Chardonnay vineyards, Celilo, is found on the Washington side of the Gorge. Throughout there are 545 hectares (1,350 acres) planted to everything from Arneis to Zinfandel, on soils that vary from volcanic basalt flows to loess, silt, and sand, but all of them scoured at some point by the mighty

Missoula Floods of the past. The climate transitions from the very wet and cool, marine-influenced western portion to the very dry, continental, high desert in the east, and elevations range from sea level to 610 metres (2,000 feet). Even in the eastern portion, winds along the Columbia River are a significant factor.

COLUMBIA VALLEY AVA

See entry under Washington.

CHEHALEM MOUNTAINS AVA

This elevated AVA (its tallest point, Bald Peak, is at 498 metres, or 1633 feet) is wholly contained within Willamette Valley AVA and has mostly basaltic and marine sedimentary soil on the southern and western slopes. But there is a distinct area of windblown loess on its northeast slope – they call the soil Laurelwood – and this area is proposed as a future AVA called Laurelwood District. The area's 1,075 hectares (2,660 acres) are planted to Pinot Noir, Chardonnay, Pinot Gris, Riesling, Pinot Blanc, Gamay Noir, and Gewürztraminer. The Chelahem Mountains, the highest in the Willamette Valley, shelter the vineyards from cold Columbia Gorge winds. Wineries include some of the pioneers: Erath, Adelsheim, and Ponzi chief amongst them. Most of the vineyards are planted between 60 and 305 metres (200 and 1,000 feet) in elevation. Ribbon Ridge AVA is contained within the Chehalem Mountains AVA.

DUNDEE HILLS AVA

Contained within the Willamette Valley AVA, this is a continuous landmass rising above the low, flat Willamette and Chehalem Valleys and protected from the Pacific breezes as well as the Columbia Gorge winds. Vineyards begin at the level of 60 metres (200 feet) and extend to the AVA's peak of 325 metres (1,067 feet). Volcanic, basaltic Jory soils give the hills their reddish hue, but there are marine sedimentary soils at the lower elevations. Vineyards cover 900 hectares (2,225 acres) and are planted to Pinot Noir, Chardonnay, Pinot Gris, and Pinot Blanc. Oregon's first famed Pinot Noir was planted here in 1966 by David Lett of Eyrie Vineyards; he was followed by Dick Erath, Susan Sokol, and Bill Blosser, among others.

ELKTON OREGON AVA

Elkton Oregon's elevation and ocean exposure (only 40 kilometres, or 25 miles, from the coast) ensure coolness and rainfall that nearly approaches that of Willamette Valley. Clay soils provide some challenges but also allow consistent crops in drier, warmer years. Pinot Noir, Pinot Gris, Riesling, and Gewürztraminer do well here, but with only 40 hectares (100 acres), there's not a lot to go around.

EOLA HILLS–AMITY AVA

Far south of the Columbia Gorge, but nonetheless exposed to cold from the Pacific winds from the Van Duzer Corridor, the Eola Hills–Amity AVA can see significant diurnal temperature shifts throughout the day. The region's 1,220 hectares (3,005 acres) are generally planted to volcanic soils, but the lower elevations have plenty of marine sedimentary soils. The hills run north and south, but there are plenty of lateral ridges; throughout vineyards are found between 75 to 215 metres (250 to 700 feet), where they are generally safe from frosts. The area came slightly later to the quality Pinot Noir efforts amongst its fellow Willamette Valley-contained AVAs, but it has definitely caught up with its neighbours.

LAURELWOOD DISTRICT AVA

(Proposed)

As with Tualatin Hills AVA, the petition for an AVA has not yet been granted, though it appears that will happen at some point in the near future.

McMINNVILLE AVA

Somewhat exposed to ocean winds through the Van Duzer Corridor, this AVA's vineyards cover 305 hectares (750 acres) on shallow, primarily marine sedimentary soils with some basalt and, unusually for the state, alluvium. The wines can be a bit more tannic than their neighbours', but Pinot Noir, Chardonnay, Pinot Gris, Riesling, and Pinot Blanc prosper nonetheless. Contained within the Willamette Valley AVA, this region sits in the Coast Range foothills in the Nestucca Formation, a thick bedrock formation, adding to the shallowness of the soils. The town of McMinnville is the site of the International Pinot Noir Celebration, a three-day event in which winemakers and enthusiasts from all over the world congregate for Pinot Noir tastings, winery tours, and seminars.

RED HILL DOUGLAS COUNTY AVA

If the name reminds the reader of the state's Red Hills of Dundee, it's intentional. Similar volcanic soils cover this thin AVA of less than 200 hectares (just under 500 acres) at elevations between 240 to 350 metres (800 and 1,200 feet). These volcanic soils from seabeds uplifted 40 million years ago furnish structure and aromatics to Pinot Gris and Pinot Noir here, just as they do in the Willamette.

RIBBON RIDGE AVA

This tiny AVA is carved out of Chehalem Mountains AVA. There are about 260 hectares (650 acres) of Pinot Noir, Chardonnay, Pinot Gris, Riesling, and Gamay Noir along this eponymous ridge 200 metres (680 feet) above the Chehalem Valley, floating like an island above the valley floor. It's a relatively sheltered series of vineyards, warmer and drier than the valleys around it, with finer, consistent, and well-drained soils. Harry Peterson-Nedry was the first to established vineyards here (Ridgecrest Vineyards).

THE ROCKS OF MILTON FREEWATER AVA

The gently sloping alluvial fan that is the Rocks area south of the town of Walla Walla was orchard land until Christophe Baron of Cayuse (okay, a few other folks, too) saw its stony ground as ideal vineyard land and within a few years of production had proved it with Bordeaux varieties, but also lofty, powerful Grenache and Syrah. Unlike much of the state, this is not defined by Missoula Flood deposits but by rocks and stones drawn from the Blue Mountains by the Walla Walla River. Elevations range from 240 to 300 metres (800 to 1,000 feet). These cobble soils are planted to 138 hectares (330 acres) of grapes, most of them Syrah, Cabernet Sauvignon, Grenache, and Cabernet Franc.

ROGUE VALLEY AVA

The Rogue Valley AVA straddles Applegate Valley and a small portion butts up against California and enjoys a mixture of elevations and exposures. But protected behind the Siskiyou Mountains, it is both the warmest and driest of Oregon's wine regions (though the region sees fewer Growing Degree Days than Napa Valley). The region's 800 hectares (1,900 acres) rely upon the river for cooling, and the vineyards can be dominated by alluvial materials, as well as loamy clay, granite, and volcanic rocks, especially in the northern portion. There are some successful Pinot Noirs and Chardonnays in the coolest parts, while Merlot, Cabernet Sauvignon, Syrah, Grenache, and Tempranillo provide more frequent plantings.

SNAKE RIVER VALLEY AVA

See entry under Idaho.

SOUTHERN OREGON AVA

This portmanteau appellation encompasses four river valleys and all five AVAs south of the critically important Willamette Valley: Applegate Valley AVA, Elkton Oregon AVA, Red Hill Douglas County AVA, Rogue Valley AVA, and Umpqua Valley AVA. But the Southern Oregon AVA is gaining importance as a Pinot Noir producer less pricey than its northern neighbours. The latitude and climate are not so different from central Italy; temperatures tend to be warm but are mitigated by Pacific winds following the rivers into the interior. The portions that are unprotected by the Klamath Range (Elkton and Red Hill Douglas County) can be very cool and wet, so white varieties fare best. In the rest of the Southern Oregon AVA, Bordeaux and Rhône varieties dominate, though there are other gems to be found, albeit in very small quantities.

TUALATIN HILLS AVA

(Proposed)

As with Laurelwood District AVA, the petition for an AVA has not yet been granted, though it appears that will happen at some point in the near future.

UMPQUA VALLEY AVA

As in other southern Oregon AVAs, there is a wide variety of microclimates, most often determined by proximity to and protection by the Klamath Mountains. The nearly 1,100 hectares (2,660 acres) extend well to the north of the undulating Umpqua River, and the soils can be stream sediments, marine sedimentary bedrock and/or volcanic soils. Cooler sites harbour Pinot Noir, Chardonnay, and Pinot Gris; the warmer places see Bordeaux varieties and Syrah, but with degree days varying by 1000 GDDs, Dolcetto, Albariño, Gewürztraminer, Grenache, Grüner Veltliner, Petit Verdot, and Tempranillo are prospering too.

VAN DUZER CORRIDOR AVA

This new (2019) appellation represents an area called Perrydale Hills, which was required by the Alcohol and Tobacco Tax and Trade Bureau (the agency that grants AVAs) to come up with a historical rather than geographic name. Vintners chose Van Duzer Corridor because that dramatic, mountain opening allows in a stiff, cold ocean wind that makes this the coldest AVA in Oregon. Plantings of Pinot Noir, Chardonnay, Pinot Gris, Pinot Blanc, Riesling, and Sauvignon Blanc cover 405 hectares (1,000 acres) of mostly marine sedimentary soils. But the winds that offer coolness to most of the Willamette Valley are 40 to 50 per cent stronger in the afternoon here compared to the other AVAs.

WALLA WALLA VALLEY AVA

See entry under Washington.

WILLAMETTE VALLEY AVA

The large (3,3450 square kilometres, or 5,390 square miles) Willamette Valley, well known for Pinot Noir, stealthily began in 1970, when David Lett planted the Eyrie Vineyards in the red volcanic soils of the Dundee Hills. After Lett's 1975 Pinot Noir embarrassed Drouhin's 1961 Clos-de-Bèze, and Drouhin became serious about founding a winery in Oregon, it was no surprise that the Burgundian chose the Dundee Hills for the location. The Willamette has since been parcellated into seven smaller AVA's: Chehalem Mountains AVA, Dundee Hills AVA, Eola Hills-Amity AVA, McMinnville AVA, Ribbon Ridge AVA, Van Duzer Corridor AVA, and Yamhill-Carlton District AVA. There are probably more on the way.

YAMHILL-CARLTON DISTRICT AVA

This holds mostly well-drained marine sedimentary materials and the oldest soils in the Willamette. The AVA's 970 hectares (2,405 acres) are bordered by the Coast Range to the west, Chehalem Mountains to the north, and the Dundee Hills to the east, resulting in warmer temperatures and the earliest harvest dates in the area. The top grapes are Pinot Noir, Pinot Gris, and Chardonnay.

WASHINGTON

ANCIENT LAKES AVA

Ancient Lakes, named for a series of about three dozen lakes, is a relatively new and chilly AVA planted more to Riesling than any other single grape. Plantings of that grape, as well as Chardonnay, Pinot Gris, Gewürztraminer, and others, total about 647 hectares (1,600 acres), with a few red grapes thriving as well on these soils left over from the Missoula Floods. Elevations range from 174 metres (570 feet) at the edge of the Columbia River to 583 metres (1,912 feet) in the Frenchman Hills in the southern portion of the AVA. Though there are dozens of soil types throughout, virtually all the landscape is arid and poor in organic matter. The AVA is little known at present, but Evergreen Vineyard is a large and successful project that easily justifies this Riesling-centric AVA's existence.

COLUMBIA VALLEY AVA
Oregon and Washington

A sort of one (giant) size fits all, the Columbia Valley represents one-third of Washington State's land mass, with three and a half million hectares (8,748,949 acres) in Washington and another million hectares (2,559,687 acres) in Oregon. There are almost 23,000 hectares (56,835 acres) of planted grapes within Washington's portion of the AVA, so there are few commonalities to its many landscapes. Contained within its borders are the AVAs of Red Mountain, Yakima Valley, Walla Walla Valley, Wahluke Slope, Rattlesnake Hills, Horse Heaven Hills, Snipes Mountain, Lake Chelan, Naches Heights, and Ancient Lakes. Amongst Washington's other AVAs only Lewis-Clark Valley, Puget Sound, and Columbia Gorge are excluded.

Each of these AVAs offers its own set of flavours, aromatics, and structures to the many different grapes residing there. Yet the climate proffers an over-arching tendency, at least in comparison to California, its erstwhile competitor. Wines tend to exhibit less of the plush and easy personality of the Golden State; instead, the longer sunshine days (courtesy of higher latitudes) and shorter growing season give many of the wines more acidity or at least simply less ripe-to-overripe flavours and aromas. Grape plantings mirror those of Washington in general: Cabernet Sauvignon is king, with Merlot accounting for just over half as much acreage. Chardonnay, Riesling, and Syrah dominate thereafter.

The Missoula Floods, a remarkable series of fast-moving floods, 100 metres high or higher, created the vineyard soils of the Columbia Valley. There are dried-up pools, basalts scraped clean by the raging torrent, or piles of windblown loess, with occasional alluvial fans. The common thread is that all are well drained and are poor in organic materials. As well, all of the Columbia Valley lies in the rain shadow of the Cascade Mountain range. The region has an arid and semi-arid continental climate, receiving an average of 15 to 20 centimetres (6 to 8 inches) of precipitation annually. Irrigation is therefore required to grow *vinifera* grapes. Early- and late-season frosts, along with hard winter freezes, are the main environmental threats. Due to dry temperatures and sandy soils, phylloxera has only a small foothold in the area, and probably 90 per cent of the vines are grown on their own rootstock, in contrast to most other vineyards throughout the world.

The Columbia Valley encompasses numerous microclimates, but most fall between 1,240 and 1,440 degree-days Celsius (2,232 and 2,592 degree-days Fahrenheit), which overlaps Regions I and II on the California heat summation system (*see* p577). Due to its northerly latitude and cloudless climate, the Columbia Valley averages two hours more sunlight during midsummer than the Napa Valley, a state and a half farther south. Astonishingly, there are more than 300 cloud-free days every year, and, although Washington as a whole is the wettest state in the nation, annual rainfall in the Columbia Valley is usually no more than 38 centimetres (15 inches).

LAKE CHELAN AVA

The nearly 90 kilometre-long (55 miles) Lake Chelan would be bucolic were it not for all the tourists. The climate of this AVA has been regarded as too cool for red grapes, but both red and white varieties are planted, and many are thriving. Best of all, there are thirsty visitors waiting to buy what is for sale.

PUGET SOUND AVA

Seattle has a reputation as one of the world's wettest cities and rainfall initially put severe restrictions on what was grown. Bainbridge Island Winery boasts the nearest vines to downtown Seattle, with Müller-Thurgau and Siegerrebe growing just a couple of kilometres away by ferry, while farther south on the mainland, Johnson Creek Winery grows Müller-Thurgau. Bainbridge Island was the first vineyard in Puget Sound to grow Pinot Noir.

RED MOUNTAIN AVA

One of Washington's smallest appellations, Red Mountain overlaps Yakima AVA, encompassing 1,616 hectares (4,040 acres), of which more than 280 (700 acres) are under vine.

A HARVEST OF SIEGERREBE GRAPES FILL A BUCKET ON WHIDBEY ISLAND IN THE PUGET SOUND
In Washington State, the Siegerrebe variety is grown only in the Puget Sound AVA. These low-acid grapes yield a finished wine that has an intense Muscat-like aroma, and they tend to be used for white blends rather than varietals.

This AVA has a full southern exposure, with gentle slopes that are protected from frost by excellent air drainage. The sandy loam soil has a high calcium content, which with the diurnal influence of warm summer days and cool nights, encourages increased acidity levels in the grapes grown (Cabernet Franc, Cabernet Sauvignon, Sangiovese, and Syrah). But the true character of Red Mountain wine is as a red wine with intense muscle and structure.

WALLA WALLA VALLEY AVA

This area has been called the Walla Walla Valley since it was settled in the 1850s, before the creation of either Oregon or Washington. With less than half of one per cent of the state's vineyards, Walla Walla would not perhaps seem to warrant its own AVA, but it sometimes receives up to 50 centimetres (20 inches) of rain, which is more than twice as much as the rest of the Columbia Valley, and this makes Walla Walla a truly distinctive viticultural area, with the potential to produce outstanding non-irrigated wine. For every kilometre you travel east from Walla Walla towards the Blue Mountains, you get another 1.6 centimetres (0.63 inches) of rain.

Viticulture is growing fast in Walla Walla – from just a dozen wineries cultivating 69 hectares (170 acres) of vines in 1998, to nearly 100 and more than 1200 hectares (nearly 3,000 acres) of vineyards in 2019 – but wine has been slowly stealing land from the wheat farmers who still dominate much of the agricultural life.

Walla Walla wine punches well above its weight when it comes to the excitement factor. Wineries from all over the state now line up to get their hands on some of its production. This AVA also encompasses The Rocks of Milton Freewater AVA, which lies wholly within Oregon. Within the AVA's nearly 1m215 hectares (3,000 acres) of vineyard, Cabernet Sauvignon is dominant, with Merlot and Syrah close behind, and a bit of Cabernet Franc and Malbec, among others. At least for the moment, a majority (57 per cent) of the vineyard acreage lies on the Washington side of the Walla Walla Valley, with 43 per cent on the Oregon side, but those ratios are ever in flux.

YAKIMA VALLEY AVA

One of Columbia Valley's many sub-appellations, this AVA contains a third of the state's vineyards; with 7,661 hectares (18,923 acres). It is in every sense the home and historical centre of the Washington wine industry. Many historical vineyards are on the southeast-facing slopes of the Rattlesnake Hills, especially in the mid-valley area from Sunnyside to Prosser, and once intermingled with the apple, cherry, and peach orchards that were found between two old irrigation canals, the Roza and the Sunnyside. The Roza is higher up the slopes and sometimes an odd plot of vines, a windbreak of trees, or an orchard can be seen above this canal, where the lush green of irrigated vegetation stands out in contrast to the yellow-ochre starkness of semi-arid desert. In this way, the Yakima is representative of the agricultural success brought to eastern Washington by irrigation projects dating back to the turn of the 20th century, although real prosperity occurred under the great Columbia irrigation scheme funded by the New Deal. Everything has been grown here: apples, apricots, asparagus, cherries, hops, pears, lentils, mint, peas, plums, potatoes, and raspberries. The apple farmers arrived first and took all the best vineyard sites (apple trees and vines prefer a similar growing environment). But in recent years, vineyards have nearly replaced all the once-prime fruit orchards.

Within its borders Yakima Valley AVA harbors three other AVAs: Red Mountain, Snipes Mountain, and Rattlesnake Hills. Chardonnay is the most widely planted grape, followed by Merlot, Cabernet Sauvignon, Riesling, and then Syrah, among the 40-plus varieties growing there. The landscape is a remnant of the Missoula Floods, with silt, loam, or loess, each allowing excellent drainage for the vines. That's the good news. The bad news is that there generally isn't enough rain, so irrigation is common. Still, the AVA includes Washington's oldest Cabernet Sauvignon vines at Otis (1957) and Harrison Hill (1963) vineyards. And other famed vineyards, such as Red Willow or Cold Creek, express intense character and longevity despite irrigation's questionable reputation amongst some of the world's wine cognoscenti. Red Willow is located 24 kilometres (15 miles) southwest of Yakima itself, on Ahtanum Ridge, which is on the opposite side of the valley to the Roza and Sunnyside canals. It has a steep slope, particularly on the west side, where there is little topsoil, most of it having been dispersed by the winds that blow across the ridge. The west side produces small, thick-skinned berries, whereas the east side, which has less of a slope and a deep topsoil, produces larger berries and softer, less tannic wines. Mike Sauer, who first planted Red Willow in 1973, believes that the complexity of the east-slope soils and the different air movements that occur over the various blocks of the vineyard create eight distinctly different microclimates. Red Willow is the source of great Cabernet Sauvignon and of top-class Syrah and Merlot. Cold Creek is close to being the driest, warmest spot in the state and one of the first vineyards to start harvest, yet when the vineyards around Prosser are covered in an autumnal carpet of leaves, just over the low Rattlesnake Hills, the vines of Cold Ridge are still verdant. The reason for this is that it has one of the longest growing seasons in Washington, so it is not only the first to start harvesting, but the last to finish, whereas the explanation for its name is that it is one of the most bitterly cold places in the state in winter. Cold Creek Vineyard has matured into the consistent source for most of Chateau Ste Michelle's premium varietal.

THE WINE PRODUCERS OF
THE PACIFIC NORTHWEST

IDAHO

CINDER WINERY
Garden City
★★

Owner-winemaker Melanie Krauss has a deft and proven hand.

✓ *Tempranillo • Chardonnay*

CLEARWATER CANYON
Lewiston
★★

Karl and Coco Umister have taken on the burden of ushering this winery to its next level; the wines are big and impressive.

✓ *Cabernet Sauvignon (Louis Delsol)*

COILED
Garden City
★★

Leslie Preston is both owner and vintner, and she has stylish wines on offer.

✓ *Sidewinder (Syrah blend)*

COLTER'S CREEK
Juliaetta
★★

Mike Pearson and Melissa Sanborn have focused upon quality every step of the way. All of the wines are worth consideration but particularly their Syrah.

✓ *Syrah (Estate, Reserve) • Arrow Rim (Grenache-Syrah-Mourvedre)*

HELLS CANYON
Caldwell
★

Founded in 1980 by Steve and Leslie Robertson. Their long experience makes this boutique winery stand out from many of the others in this state.

✓ *Merlot (Middle Fork) • Syrah (Deer Slayer)*

KOENIG
Caldwell
★

Andy and Greg Koenig have been running an award-winning distillery and offering both spirits and wines. Their Riesling Ice Wine is a silly bargain. New owners are taking over but hopes are high.

✓ *Cabernet Sauvignon (Bitner Vineyard) • Riesling (Ice Wine, Late Harvest) • Syrah (Three Vineyard Cuvée), Zinfandel (Sawtooth Vineyard)*

PAR TERRE
Garden City
★

Former ballet dancers have created a tiny winery; one to watch.

✓ *Rose of Cabernet Franc • Semillon (skin-fermented)*

PARMA RIDGE VINEYARDS
Parma
★

Owners Dick and Shirley Dickstein grow Chardonnay, Gewürztraminer, Merlot,

---

Syrah, Viognier, and even Zinfandel at an altitude of 720 metres (2,400 feet) on their 4-hectare (9.5-acre) vineyard overlooking the Boise River.

✓ *Chardonnay (Reserve)*

PEND D'OREILLE WINERY
Parma
★

Julie and Steve Meyer started this winery in 1995 and have just recently sold it to two key employees, Jim Bopp and Kaylie Presta, both with extensive experience in wine industry. The wine should continue to be tasty and to represent very good value.

✓ *Chardonnay (Reserve)*

RIVAURA
Juliaetta
★★

Billo Naravane MW shepherds this brand-new Lewis-Clark winery, where early results have been offering great confidence in the future.

✓ *Rosé (Lewis Clark)*

SAWTOOTH VINEYARDS
Caldwell
★

Though owned by the large concern Precept Brands, Sawtooth has if anything improved with time and new ownership.

✓ *Pinot Gris (Estate) • Riesling (Classic Fly) • Petit Verdot (Classic Fly) • Tempranillo (Classic Fly)*

STE CHAPELLE
Caldwell
★

Let's face it, these wines are not cheap. No, they're practically given away. Established in 1976 and named after the Gothic-style Ste Chapelle in Paris, Idaho's oldest winery has set a particularly fine standard for others to follow. Its Riesling Ice Wine is absurdly delicious for the money.

✓ *Cabernet Sauvignon • Riesling (Ice Wine)*

SNAKE RIVER WINERY
Caldwell
★

This small winery offers reliable wines from a wide variety of grapes.

✓ *Cabernet Franc • Merlot • Barbera*

OREGON

A TO Z
Dundee
☆

If A to Z has any address, it is a post office box in Dundee, but if you wanted to find its winemakers, Sam Tannahill (former Archery Summit winemaker) or his wife, Cheryl Francis (former Chehalem winemaker), you would have to search a number of wineries in Yamhill County. Together with their partners Bill

---

Hatcher, who developed and managed Domaine Drouhin, and his wife, Debra, these two families have established a *négociant*-style business to mop up any excess production from the wineries they work with. Although the individual wines purchased by A to Z are theoretically inferior, Tannahill and Francis blend them into a harmonious product that is greater than the sum of its parts.

✓ *Pinot Noir (Willamette Valley)*

ABACELA VINEYARDS
Roseburg
★

A family-owned and -operated winery, with an interesting gaggle of grapes, including Tempranillo, Syrah, Merlot, Dolcetto, Malbec, Cabernet Franc, Grenache, Viognier, and Albariño, growing on sunny south-sloping rocky hillsides. They virtually set the standard for Tempranillo in the United States.

✓ *Albariño • Tempranillo (South East Block Reserve) • Syrah • Viognier*

ADELSHEIM VINEYARD
Newberg
★★

They offer fine wines across the range, and they are most successful with Pinot Noir, which is consistently among the best dozen in Oregon.

✓ *Chardonnay • Pinot Gris • Merlot (Layne Vineyards Grant's Pass) • Pinot Noir • Sauvignon Blanc*

AIRLIE WINERY
Monmouth
★★

Airlie owner Elizabeth Clark makes tasty wines across the board.

✓ *Pinot Blanc • Pinot Noir • Riesling*

ARCHERY SUMMIT
Dundee
★★

Californian Gary Andrus put Archery Summit on the map with his smoky, toasty, sinewy Pinot Noir, and the winery continues on after him.

✓ *Pinot Noir*

ARGYLE WINERY
Dundee
★★

Argyle's long-time and exceptional sparkling wine performance is due to several people: Allen Holstein, who ran Domaine Drouhin's vineyards as well as Argyle's; Rollin Sole, who was this winery's gifted, laid-back winemaker; and Brian Croser, who was one of the original partners and whose intellectual curiosity into the hows and whys of everything that sparkles was the very catalyst of Argyle's success. Winemaker Nate Klosterman continues the work Sole and Croser began. Some excellent still wines are made, even though the bubbly is the most worthy of your attention.

✓ *Chardonnay (Nuthouse) • Pinot Noir • Riesling • Sparkling wine • Syrah (Nuthouse)*

AYRES VINEYARD & WINERY
Newberg
★★

This winery offers lovely, supple wines from along Ribbon Ridge. Its offerings include a Pinot Blanc and Rosé of Pinot Noir, a Gamay Noir, and a Willamette Valley Pinot Noir.

✓ *Pinot Noir • Gamay Noir*

BEAUX FRÈRES
★★☆
Dundee

Wine critic Robert Parker once owned a share of this vineyard and winery, which is run by his brother-in-law, Mike Etzel. At approximately 6,000 vines per hectare (2,430 vines per acre), only Domaine Drouhin is planted at a higher density. Though there is new ownership, the wines remain delightful. Etzel's son is now the winemaker, and Etzel has another first-rate project called Sequitur.

✓ *Pinot Noir*

BERGSTRÖM
Dundee
★★

Established in 1997 by the Bergström family, who sent son Josh to France for a post-graduate course in viticulture and oenology at Beaune. The standard has been nothing less than exemplary since the first vintage in 1999. Their own winery was built in 2001.

✓ *Pinot Noir*

BETHEL HEIGHTS VINEYARD
Salem
★★☆

One of the top dozen wineries in the state, Bethel Heights offers a basic Pinot Noir that is often as good as the individual block bottlings, and it can sometimes even be even better balanced. It not only ranks as one of Oregon's best-quality Pinot Noirs, it also ranks as one of the best values.

✓ *Pinot Noir*

BIG TABLE FARMS
Carlton
★★☆

This fascinating little project makes wines of character and immediacy, but the wines age well too.

✓ *Entire range*

BRICK HOUSE VINEYARD
Newburg
★★

These organic wines have consistently ranked amongst the tops in elegance and finesse. They are produced with biodynamic practices.

✓ *Gamay • Pinot Noir • Chardonnay*

BRITTAN VINEYARDS
McMinnville
★★

Robert Brittan worked for years in Napa and made some top-flight wines; he and his wife Ellen now have this passion project to focus upon.

✓ *Pinot Noir (Gestalt Block)* • *Chardonnay*

BROOKS WINES
Amity
★★

Established by the late Jimi Brooks, the winery has continued on and even improved of late.

✓ *Entire range*

CAMERON WINERY
Dundee
★★

An underrated winery with a good, easy-drinking, and often elegant, style.

✓ *Chardonnay* • *Pinot Blanc* • *Pinot Noir*

COOPER MOUNTAIN
Beaverton
★☆

This winery produces fine-quality, barrel-fermented Chardonnay from an extinct volcano overlooking the Tualatin Valley. Cooper Mountain was certified organic in 1995, and biodynamic in 1999. Some wines are sulphite free.

✓ *Chardonnay* • *Pinot Gris* • *Pinot Noir*

COWHORN VINEYARDS
Jacksonville
★★

This biodynamic-dedicated winery has looked at the Applegate Valley and concluded that Rhône varieties are the ideal match for this fractured soil, warm wine region. Based on the results, Syrah, Grenache, Viognier, Marsanne, and Roussanne are clearly happy in this place.

✓ *Rhône blends (Spiral 36, Moonraker)* • *Grenache* • *Syrah (Estate, Sentience)*

CRISTOM
Salem
★★

Steve Doerner once made wines at California's Calera Winery; his work at Cristom has improved all Oregon Pinot Noirs with his focus upon the intelligent use of whole cluster and stems.

✓ *Pinot Noir* • *Viognier*

DROUHIN ESTATE
Dundee
★★★

Robert Drouhin was intrigued by this state's potential after David Lett's The Eyrie Vineyard 1975 Pinot Noir came second to Drouhin's 1959 Chambolle-Musigny in a tasting organized in France in 1979. Drouhin purchased a property close to the Eyrie Vineyard in the Dundee Hills, planted his own vineyard and built a winery. Just nine years after that eventful tasting, he produced his first vintage of Oregon Pinot Noir. When it was released, locals and critics alike marvelled at how Drouhin's wine instantly possessed more colour, depth, and com-

plexity than any other Oregon Pinot Noir yet still maintained the grape's varietal purity and finesse. The secret, it was assumed, was the density of Drouhin's vines, which he planted three and a half times closer than the Oregon average, but that first vintage (1988) did not contain a single grape from his own still-too-young vineyard. The real secret lay in the way that Drouhin had handled the grapes, which itself was ironic, because the only reason he made a wine in 1988 with bought-in grapes was because he wanted hands-on experience of Oregon fruit. The first pure Domaine Drouhin was 1992, and all *cuvées* of every Domaine Drouhin vintage have been exceptional since then.

✓ *Entire range*

DOMAINE SERENE
Carlton
★★

This winery made an excellent debut with the 1992 vintage and produces firm wines today.

✓ *Pinot Noir*

ELK COVE VINEYARDS
Gaston
★★

Adam Campbell and family produce deceptively easy-going Pinot Noir, but the wines are excellent and longer-lived than they might at first seem.

✓ *Pinot Gris* • *Pinot Noir (all of them)*

ERATH WINERY
Dundee
☆

Formerly Knudsen Erath, this is an old pioneer with Oregon Pinot Noir, but vintages can lurch between firm-tannic and soft, creamy, and voluptuous.

✓ *Chardonnay* • *Pinot Gris* • *Pinot Noir*

EVESHAM WOOD
Salem
★★

Evesham Wood has long been a rising star, but it's true even now with another winemaker and owner at the helm.

✓ *Pinot Gris* • *Pinot Noir*

HIYU WINE FARM
Hood River
★★

Master Sommelier Nate Ready was raised in Napa and worked the floor at Boulder's famous Frasca Restaurant, and he's chosen to plant a mixed varietal vineyard (dozens and dozens of grapes planted higgledy-piggledy) and to create wines that reflect this small and beautiful place. Biodynamic, but not certified as such.

✓ *Entire range*

J CHRISTOPHER VINEYARDS
Newberg
★★

Oregon winemaker Jay Somers was making a small-production Willamette Valley Pinot when he and famed German winemaker Ernie Loosen met. They hit it off, and we are the beneficiaries.

✓ *Entire range*

KELLEY FOX
Gaston
★★★

This is a small-scale producer of wines with heart, soul, and deliciousness.

✓ *Chardonnay* • *Pinot Noir (all of them)*

KEN WRIGHT CELLARS
McMinnville
★★☆

This is the high-performance winery founded by former Panther Creek owner Ken Wright. He has established a reputation for his sumptuous, stylish, expressive Pinot Noir wines sold under single-vineyard names.

✓ *Entire range*

KING ESTATE
★
Eugene

This vast and still-expanding vineyard surrounds a high-tech, hilltop winery that pushes to place Oregon on the commercial map with its large-scale production capacity.

✓ *Pinot Gris (Vin Glacé)* • *Pinot Noir*

LINGUA FRANCA
McMinnville
★★☆

Co-founded by Larry Stone, David Honig, and Dominique Lafon in 2015. They work with winemaker Thomas Sayre to offer excellent Pinot Noir and Chardonnay.

✓ *Entire range*

McKINLAY VINEYARDS
Newberg
★★

Matt Kinne makes a limited amount of wine, and the customers line up to make it disappear quickly.

✓ *Entire range*

MONTINORE VINEYARDS
Forest Grove
★

A large project by Oregon standards, with all wines 100 per cent estate grown, Montinore is at long last beginning to reflect its true potential.

✓ *Pinot Gris (Entre Deux)* • *Pinot Noir (Parson's Ridge, Winemaker's Reserve)*

NICHOLAS JAY VINEYARDS
Newberg
★★

This joint project includes the great Jean-Nicolas Meo (of Meo-Camuzet fame), continued proof that Burgundy is fascinated by Oregon Pinot Noir.

✓ *Entire range*

OWEN ROE
Newberg
★★

This winery is the result of a collaboration between Jerry Owen, who looks after the vineyards, and David O'Reilly, who makes and sells these Oregon and Washington wines. The top-of-the-range wines have beautiful photogravure labels bearing the inspired pho-

tography of fellow Irishman David Brunn. Entry-level wines are sold under the O'Reilly label. *See Sineann for other David O'Reilly wines.*

✓ *Cabernet Sauvignon (DuBrul Vineyard)* • *Chardonnay (Casa Blanca Vineyard)* • *Merlot (DuBrul Vineyard)* • *Pinot Noir* • *Syrah (DuBrul Vineyard)*

PANTHER CREEK CELLARS
McMinnville
★☆

This tiny winery specializes in opulent Pinot Noir wines. It also makes a delicious white wine from the Muscadet grape, sold under its synonym of Melon.

✓ *Melon* • *Pinot Noir*

PATRICIA GREEN WINERY
Newberg
★★★

The late Patty Green was one of Oregon's greatest winemakers. The ship is now helmed by her right-hand man, Jim Anderson, who creates remarkably consistent, delicious, elegant Pinot Noir.

✓ *Entire range*

PENNER-ASH VINEYARDS
Newberg
★★

Established by one of the great Oregon winemakers, Lynn Penner-Ash, the eponymously named winery continues to generate stylish Pinot Noir.

✓ *Pinot Noir* • *Riesling*

PONZI VINEYARDS
Beaverton
★★

Ponzi Vineyards has offered one of Oregon's top-performing Pinot Noirs for decades; second generation winemaker Luisa Ponzi is making some of the best Chardonnays and Pinot Noirs on the West Coast.

✓ *Pinot Noir* • *Chardonnay*

RAPTOR RIDGE
Cheshire
★★

"Raptor" is a bird of prey, and the name was chosen because of the large number of hawks, kestrels, and owls that hunt within the vicinity of this winery and its vineyards. The bright, fruity wines produced here are made by winemaker Shannon Gustafson in partnership with owners-vintners Scott and Annie Shull.

✓ *Pinot Noir* • *Rosé*

RÉSONANCE VINEYARDS
Carlton
★★☆

Back in April 2013 Thibault Gagey, along with Jadot and Burgundian legend Jacques Lardière, selected their site for Résonance Vineyards just shortly after Jacques' "retirement". His stint at Maison Louis Jadot lasted 42 years; Thibault's family has operated Maison Louis Jadot since 1962. Thus far the wines are very good, perhaps unsurprisingly.

✓ *Chardonnay (Hyland)* • *Pinot Noir (Resonance Vineyard)*

REX HILL VINEYARDS

Newberg

★☆

Best known for its range of gentle, stylish Pinot Noir, the estate wines are biodynamic, but some bought-in grapes used in non-estate wines may not be organic.

✓ *Pinot Gris • Pinot Noir*

SHEA WINE CELLARS

Yamhill

★☆

You could not get two more diametrically opposed worlds than Wall Street and farming in Oregon, but Dick Shea made the transition with ease in 1996 and quickly achieved respect from local winemakers for his classy, complex Pinot Noir.

✓ *Chardonnay • Pinot Noir*

SINEANN

Santa Rosa, CA

★★

Winemaker Peter Rosback and his business partner David O'Reilly source wines for their own label from all over the Northwest: Pinot Noir and Pinot Gris from the Willamette and the Hood River valleys; Cabernet Sauvignon, Merlot, and Zinfandel from the Columbia Valley; and Gewürztraminer from the Willamette Valley and the Columbia Gorge. *See* Owen Roe for other David O'Reilly wines.

✓ *Cabernet Sauvignon ("Block One" Champoux Vineyard, Baby Poux Vineyard) • Fortified (CJ Port) • Merlot (Columbia Valley) • Pinot Noir (all Oregon wines) • Zinfandel (Old Vine Columbia Valley)*

SOKOL BLOSSER WINERY

Dundee

★☆

This second-generation winery produces some lush, seductive, early-drinking Pinot Noirs that regularly rank among some of Oregon's finest. Biodynamic experimentation continues.

✓ *Chardonnay • Pinot Noir*

SOLENA ESTATE

Carlton

★☆

With long careers in the Oregon wine industry, Laurent Montalieu and Danielle Andrus Montalieu purchased a 32-hectare (80 acres) estate as their wedding gift to each other in 2000 and registered with premium nurseries for six different clones of Pinot Noir vines.

✓ *Entire range*

SOTER VINEYARDS

Yamhill

★★

Michelle and Tony Soter's Oregon venture was established in 1997 and abuts properties such as Beaux Frères and Brick House. In 2003 this vineyard was awarded a LIVE Farming Certificate, which is not organic as such, but recognizes the "sustainable farming practices" that have been employed since the Soters purchased the land.

✓ *Chardonnay • Pinot Noir*

THE EYRIE VINEYARDS

McMinnville

★★

The man who started it all, the late David Lett, also made Pinot Noir to age, Chardonnay for medium-term storage, and Pinot Gris that is delightful when drunk young. His son Jason continues in the same elegant style, though, in frankness, the wines are now better than they have ever been.

✓ *Entire range*

TRISAETUM WINERY IN OREGON'S RIBBON RIDGE AVA
This family-owned winery produces Pinot Noir and Riesling from its estate vineyards in two of Oregon's AVAs. The older vineyard, the Coast Range Estate, is located in the Yamhill-Carlton District AVA. The new vineyard plots in the Red Ribbon AVA surround the winery complex, which features a tasting room and art gallery.

TRISAETUM WINERY

Newberg

★★

This excellent Riesling specialist is named as an amalgam of owners James and Andrea Frey's children: Tristen and Tatum.

✓ *Riesling (all)*

TROON VINEYARD

Grants Pass

★★

Troon Vineyard is an historic part of Oregon wine, founded by pioneer Dick Troon in 1972. It is situated on the ancient, higher, second bench above the Applegate River, in the Siskiyou Mountains, where the Applegate Valley opening allows the Pacific Ocean breezes to flow in.

✓ *Vermentino (Kubli Bench, Estate) • Zinfandel (Kubli Bench, Estate) • Grenache*

VALLEY VIEW VINEYARD

Jacksonville

★

One of the few consistent Cabernet Sauvignon in Oregon. Jacksonville in the Rogue Valley is one of those idyllic towns everyone should visit. The Anna Maria *cuvée* is a reserve produced only in the best years and is therefore not available for all varietals in every year.

✓ *Chardonnay (Anna Maria) • Cabernet Sauvignon (Anna Maria) • Reserve (Anna Maria) • Sauvignon Blanc (Anna Maria)*

WILLAKENZIE ESTATE

Yamhill

★☆

Located on the rolling hillsides on the Chehalem Mountains, this estate's vineyards are today maintained by sustainable viticulture.

✓ *Pinot Noir*

YAMHILL VALLEY VINEYARDS

McMinnville

★☆

This is one of the pioneering estates, and though the founder is no longer with us, his work continues in the able hands of winemaker Ariel Eberle. These wines can be tannic and hard, courtesy of the soils and the proximity to the Van Duzer, but they can be seductive too.

✓ *Pinot Noir*

WASHINGTON

AMAVI

Walla Walla

★★

Amongst the luminaries in Walla Walla should be included the Goff family, the McKibben family, and the Pellet family. These three families have come together to create admirable and reliable wines.

✓ *Syrah (Les Collines) • Sémillon*

ANDREW WILL

Vachon

★★★

Owner Chris Camarda named the winery after his nephew Andrew and his son Will, who is now deeply involved in the winery. These wines are superb, especially the range of beautifully crafted classic red blends.

✓ *Cabernet Franc (Sheridan) • Cabernet Sauvignon (Klipsun) • Classic red blend (Champoux, Ciel du Cheval, Seven Hills, Sheridan, Sorella) • Merlot (Klipsun) • Sangiovese (Ciel du Cheval)*

BARNARD GRIFFIN

Kennewick

★☆

Owners Deborah Barnard and Rob Griffin have made a track record for Merlot but have made some splendid Cabernets and some stunning Syrahs.

✓ *Merlot (Red Mountain Ciel du Cheval) • Sauvignon Blanc (Fumé Blanc) • Syrah (Handcrafted Selection)*

BETZ FAMILY VINEYARDS

Woodinville

★★☆

When I first met Bob Betz, he was a director at Chateau Ste Michelle. I thought that here was someone who was good at connecting with visitors, but also a person who, I realized, would be happier with his hands in the soil, getting back to his winemaking roots. Over the years he has become a Master of Wine, and in 1997 he began releasing wine under his own label, while still at Chateau Ste Michelle. At that first meeting, Betz was full of enthusiasm for Red Mountain, and that was long before any Washington wine drinkers had heard of it. He left Chateau Ste Michelle in 2003, and it is no surprise to find that he now sources many of his wines from Red Mountain.

✓ *Cabernet Sauvignon (Père de Famille) • Classic red blend (Clos de Betz) • Syrah (La Côte Rousse, La Serenne)*

BOOKWALTER

Pasco

★★

Jerry Bookwalter graduated from UC Davis in 1962, but did not start up on his own until 1983, after 20 years' experience with Sagemoor Farms and other vineyards in Washington and Oregon. The winery is now run by his son, John.

✓ *Cabernet Sauvignon • Merlot*

BUTY

Walla Walla

★★

Artisanal winery established and run since 2000 by Nina Buty Foster, whose white and red wines are reared in some of Walla Walla's best sites, particularly in the Rocks District.

✓ *Chardonnay • Syrah (Rediviva of the Stones)*

CADENCE

Seattle

★★

Gaye McNutt and Ben Smith produce small lots of classy red blends: Bel Canto (Merlot-based), Coda (Cabernet Franc-based), Ciel du Cheval (Cabernet Sauvignon-based), Klipsun (Merlot-based), and Tapteil (Cabernet Sauvignon-based). I like all five of these wines, but four are a class apart, and of those, Ciel du Cheval and Tapteil stand out.

✓ *Classic red blend (Bel Canto, Ciel du Cheval, Klipsun, Tapteil)*

CAMARADERIE CELLARS

Port Angeles

★☆Ⓥ

Don and Vicki Corson make tasty reds that are more powerful than elegant, but most consumers prefer that.

✓ *Grace (Bordeaux-style blend)*

CANOE RIDGE

Walla Walla

★☆Ⓥ

Purchased by Precept Brands (which owns many good wineries, including the small and fascinating Cavatappi Cellars), this winery continues to do good, solid work.

✓ *Merlot (Reserve)*

CAYUSE

Walla Walla

★★★Ⓑ

This exciting boutique winery is owned by Christophe Baron of Champagne Baron Albert, and his wines are better than they have ever been. He has planted five vineyards totalling 16 hectares (41 acres), all on stony ground, which prompted Baron to name his venture after one of three local Native American tribes, as a play on the French word for stony: *cailloux*. Mostly planted with Syrah, plus a few other Rhône varieties. Look also for his wines made under other labels, such as No Girls and Horsepower. Most of Baron's wines have single-vineyard designations, while those that don't are usually of a humorous ilk; this is a Frenchman who does not mind gently mocking himself, as his top-performing Bionic Frog Syrah demonstrates.

✓ *Syrah*

CHARLES SMITH WINES

★☆Ⓥ

Former rock-and-roll roadie Charles Smith seems to understand how to market wines as well as or better than most of his peers. The style of the wines reflects that same "spider" sense.

✓ *Merlot (Velvet Devil)*

CHATEAU STE MICHELLE

Woodinville

★☆Ⓥ

If rated on the top-of-the-range wines alone, Chateau Ste Michelle would deserve two and a half stars; however, the Altria Group–owned company (owner of the largest tobacco brands) produces a wide range of brilliant wines even at entry level. Col Solare is a Cabernet-Merlot blend with a dash of Syrah and Malbec, made in partnership with Piero Antinori. Eroica is the result of a collaboration with Ernst Loosen, and it has caused such a renaissance for this grape in Washington that today Riesling is grown on over 2,100 hectares (5,300 acres) in the state.

✓ *Cabernet Sauvignon (Canoe Ridge Estate, Cold Creek) • Chardonnay (Canoe Ridge Estate, Indian Wells) • Classic red blend (Artist Series Meritage, Col Solare) • Merlot (Cold Creek) • Riesling (Eroica, Ice Wine, Single Berry Select) • Syrah (Reserve)*

COLUMBIA CREST

Prosser

★☆Ⓥ

Owned by Michelle Wine Estates (part of the Altria Group), this winery continues to impress, but is in the middle of nowhere and will never realize its tourist potential. Because tourism not only provides profit, but also increases the reputation of a wine area, it seems odd that they did not build this wine wonder palace in the Yakima Valley, which has the greatest concentration of wineries and almost half the state's vineyards.

✓ *Cabernet Sauvignon (Reserve) • Classic red blend (Walter Clore Reserve) • Merlot (Grand Estates, Reserve) • Syrah (Shiraz, Two Vines Shiraz)*

COLUMBIA WINERY

Bellevue

★☆Ⓥ

This estate was once helmed by the late and great Master of Wine David Lake, and now Gallo oversees the continuance of Washington's best-value red wines.

✓ *Cabernet Sauvignon (Red Willow Vineyard) • Merlot (Milestone) • Syrah (Red Willow Vineyard)*

CORLISS ESTATE

Walla Walla

★★

Michael and Lauri Corliss have catapulted themselves to the top ranks, courtesy of very deep pockets and very shrewd taste. Winemaker Andrew Trio works with famed and ubiquitous consultant Philippe Melka to make excellent, if expensive, wines across the board.

✓ *Cabernet Sauvignon • Estate Red • Syrah*

DE LILLE CELLARS

Woodinville

★★

If anything is proof of the strength of Washington's classic red wine blends, it is Chris Upchurch, a self-confessed inveterate blender, who has swiftly realized the potential of this state's vineyards with his range of top-performing blends. Even De Lille's second wine, D2, is so good that he had to name it twice (the D stands for *Deuxième*), as if to emphasize its modest aspirations, yet it regularly knocks seven bells out of most other proprietary reds. All percentages vary from vintage to vintage, of course, but the D2 is Merlot-based, while both Harrison Hill and Chaleur Estate are effectively 65 per cent Cabernet Sauvignon, 25 per cent Merlot, and 10 per cent Cabernet Franc. The main difference is that Harrison Hill is a single-vineyard wine, whereas Chaleur Estate is an assemblage from half a dozen different vineyards.

✓ *Classic red blend (D2, Harrison Hill, Chaleur Estate) • Syrah (Doyenne) • Cabernet Sauvignon (Grand Ciel) • Syrah (Grand Ciel) • Chaleur Blanc (Sauvignon Blanc-Semillon)*

DELMAS

Milton-Freewater

★★

Stephen Robertson was one of the first to grasp the significance of the Rocks District.

✓ *Entire range*

DOUBLEBACK WINERY

Walla Walla

★★

During his 14-year NFL career, Walla Walla hometown boy Drew Bledsoe wanted to create a wine estate vineyard in honour of his family's agricultural roots in Walla Walla Valley. The wines are well built and both powerful and supple.

✓ *Cabernet Sauvignon • Estate Reserve*

DUNHAM CELLARS

Walla Walla

★★

This winery was established in 1995 by Eric Dunham, and since his passing it has been ably shepherded forward.

✓ *Cabernet Sauvignon (Columbia Valley, Roman Numeral series) • Classic red blend (Trutina) • Syrah (Columbia Valley, Lewis Vineyard)*

DUSTED VALLEY VINTNERS

Woodinville

★☆

Starting in 2003, and with strong roots in agriculture, brothers-in-law Chad Johnson and Corey Braunel focussed upon sustainable practices, low (or modest) yields, and minimalistic winemaking. Dusted Valley produces very good wines, even if slightly under the radar.

✓ *Cabernet Sauvignon (VR Special)*

FIGGINS FAMILY

Walla Walla

★★☆

While owner-winemaker Chris Figgins continues to helm the work at Leonetti, he is also hard at work building a new and beautiful facility for his own project, Figgins Family. The wines tend to be more robust than those of Leonetti.

✓ *Entire range*

FORCE MAJEUR VINEYARDS

Walla Walla

★★

Grand Reve Vintners was transformed into Force Majeur in 2004, and the wines unquestionably have become more impressive, as well as expensive and harder to procure. Former Bryant Family winemaker Todd Alexander is in charge; he is aided by fellow Napa alum Helen Keplinger. There are prominent vineyards in the winery's control; one is co-owner Ryan Johnson's fascinating and densely planted Red Mountain vineyard. The other is a new estate in the nearly alpine North Fork of the Walla Walla. That likely Rhône-centric project will take years to come to fruition, but it is sure to demand attention as amongst the most beautiful vineyards in the Northwest.

✓ *Rhône blend (Parata) • Syrah (Red Mountain) • Cabernet Sauvignon • Bordeaux blend (Épinette)*

GRAMERCY CELLARS

Walla Walla

★★☆

Master Sommelier Greg Harrington has helped run wine programs for some of America's most important chefs – Emeril Lagasse, Wolfgang Puck, and Danny Meyer – so it should be no surprise that he creates wines that lack the bluster of score chasers and instead feel right at home on the table.

✓ *Syrah (Columbia Valley, John Lewis) • Rhône blend (Third Man) • Cabernet Sauvignon*

HEDGES CELLARS

Benton City

★☆ⓋⒷ

I have never tasted a Hedges wine I would not want to drink. Tom Hedges has the amazing knack of being able to produce premium-quality Bordeaux-style reds that are lush, drinkable, and not lacking complexity and finesse within less than a year of the grapes being picked.

✓ *Classic red blend (CMS, Red Mountain Reserve) • Classic white blend (Fumé Chardonnay)*

HOGUE CELLARS

Prosser

★☆Ⓑ

Hogue Cellars, part of a large family-owned agricultural concern, has at times been an important player in the quality wine game, even if it is no so exciting today.

✓ *Cabernet Sauvignon (Terroir Red Mountain) • Syrah (Genesis) • Malbec (Terroir) • Merlot (Reserve)*

JANUIK

Woodinville

★★

This estate is owned and run by Mike Januik, former Head Winemaker at Chateau Ste Michelle. Both "basic" and single-vineyard wines of each varietal are of equally excellent quality.

✓ *Cabernet Sauvignon • Merlot*

K VINTNERS

Walla Walla

★★

A raunchy range of silky, sensuous, saturated Syrah created by the wildly successful Charles Smith, whose House Wine was sold to Precept for a pretty penny. Some wonderfully eclectic classic red blends are emerging too. All of Smith's wines are interesting and sometimes fascinating.

✓ *Syrah (Morrison Lane)*

KIONA VINEYARDS

Benton City

★☆Ⓥ

The arid backdrop to this lush vineyard illustrates how irrigation can transform a desert. Although Kiona might not always hit the bull's eye in the Kiona wines themselves, the same fruit has consistently shown its potential in the wines of Woodward Canyon and has been one of three main ingredients of Quilceda Creek.

✓ *Cabernet Sauvignon (Red Mountain) • Chenin Blanc • Gewürztraminer (Late Harvest) • Merlot (Red Mountain Reserve) • Riesling • Sangiovese*

L'ECOLE No 41
Lowden
★★

Marty Clubb has carved out quite a reputation from the wines made at this old schoolhouse. L'Ecole No 41 is primarily known for its Sémillon white wine, but, although this is very good (particularly the Fries Vineyard), Marty's red wines are a class apart.

✓ *Cabernet Sauvignon (Ferguson Vineyard, Walla Walla Valley)* • *Classic red blend (Pepperbridge Vineyard Apogee)* • *Merlot (Columbia Valley, Walla Walla Valley)* • *Sémillon* • *Syrah (Seven Hills Vineyard)*

LEONETTI CELLAR
Walla Walla
★★★

Owner-winemaker Chris Figgins has an intense following that is matched only by Quilceda Creek, Reynvaan, or Cayuse. Figgins, a consummate vinegrower, has perfected his handling of oak to match the distinctive style of the wines. Although some of the earlier vintages could be criticized for showing too much oak, any objective observer must acknowledge that for some time now they have been beautifully balanced, highly polished, and finely focused wines of great complexity, finesse, and longevity.

✓ *Entire range*

LONG SHADOW VINTNERS
Walla Walla
★★

Long Shadow is the culmination of all the dreams and fantasies Allen Shoup had during the 20 years he ran Ste Michelle Wine Estates (formerly Stimson Lane). It was his collaboration with Piero Antinori and, particularly, Ernst Loosen (which led to Eroica), that convinced Shoup his idea could work. After retiring from Ste Michelle in 2000, he invited several of the world's most iconic winemakers to the sunny slopes where the Snake River and the Yakima River flow into the mighty Columbia and explained his vision. He wanted them to tour the region, run the soil through their fingers, survey the leafy trellises that spill down the hillsides of the Columbia Valley appellation, and decide where, ideally, they would grow those varieties they had had a lifetime's experience with. The objective was to create seven or eight stand-alone, ultra-premium wineries, producing no more than 5,000 cases or so, with a focus on a single varietal or blend, made under the supervision of world-famous winemakers, comparable in stature to those they crafted in their native wine regions. Long Shadow Vintners is so named in tribute to the reputation of Shoup's select group of winemakers. They are: Armin Diel's Poet's Leap Riesling, Randy Dunn's Feather Cabernet, John Duval's Sequel Syrah, Augustin Huneeus and Philippe Melka's Pirouette (Cabernet-Merlot blend), Allen Shoup and Gilles Nicault's Chester-Kidder (Merlot-Cabernet blend with a classic Washington twist of Syrah), and Michel Rolland's Pedestal (Merlot-Cabernet blend).

✓ *Entire range*

McCREA
Rainier
★☆

Promoted as "Washington State's first winery dedicated to Rhône varietals", McCrea's efforts were initially rather patchy, as he tried to understand a number of varieties, some of which might never have the potential in this state, but he was one of the first to perfect Syrah.

✓ *Classic red blend (Sirocco)* • *Syrah*

MOUNTAIN DOME
Spokane
★☆

The Manz family live in a dome that is cut into the side of a mountain. This is one of half a dozen producers dotted around the world who have the potential to make truly exceptional sparkling wine.

✓ *Entire range*

NORTH BY NORTHWEST
Columbia Valley
★☆

North by Northwest – or NxNW as it is also known – is under the umbrella of Oregon's King Estate Winery and has been busy making red wines from Washington State, If the wines aren't necessarily inspiring, they represent good and consistent value nonetheless.

✓ *Cabernet Sauvignon (Columbia Valley)* • *Syrah (Walla Walla)*

NORTHSTAR
Walla Walla
★☆

Consistently this is one of the prettiest Merlots in the Pacific Northwest.

✓ *Classic red blend (Stella Maris)* • *Merlot*

PEPPER BRIDGE
Walla Walla
★★

Having supplied grapes for some of Washington's greatest wines from its Pepper Bridge and Seven Hills vineyards, Norm McKibben built a classic gravity-fed winery on his property.

✓ *Cabernet Sauvignon* • *Merlot* • *Trine*

PRECEPT BRANDS
Seattle
★☆

Established in 2003, Precept Brands is headed by Andrew Browne, CEO of Corus until it was sold to Canandaigua in 2001. Browne is attempting to repeat the success of Corus, but without the overheads and on an even larger scale. It currently own five wineries and produce 25 brands.

✓ *Cabernet Sauvignon (Waterbrook, Pendulum)* • *Riesling (Washington Hills)* • *Syrah (Washington Hills)*

QUILCEDA CREEK
Seattle
★★☆

Washington's first truly important red wine, Quilceda Creek is still produced in tiny quantities and remains one of this state's very elite.

✓ *Entire range*

RASA VINEYARDS
Walla Walla
★★

Billo Naravane MW started this project in 2007 along with his brother, Pinto, and for many their wines have been nearly cultish in their allure.

✓ *Merlot blend (Fianchetto)* • *Bordeaux blend (Creative Impulse)*

REININGER
Walla Walla
★★

Since Chuck Reininger established this up-and-coming Walla Walla boutique winery in 1997, he has hardly put a foot wrong with his intense reds.

✓ *Cabernet Sauvignon* • *Merlot* • *Syrah*

REYNVAAN ESTATE
Walla Walla
★★☆

A small family winery that produces Rhône-style red and white wines from two vineyard properties in the Walla Walla Valley: one in the Rocks District and one in the much cooler Blue Mountains. The wines are extremely good if rather pricey.

✓ *Cabernet Sauvignon (The Classic)* • *Syrah (Stonessence, In the Hills, The Contender, Foothills Reserve)*

SAGELANDS VINEYARDS
Wapato
★☆

Formerly known as Staton Hills, this winery and its vineyards were taken over by the Chalone Group in 1999 and relaunched as Sagelands in 2000.

✓ *Cabernet Sauvignon (Four Corners)* • *Merlot (Four Corners)*

SEVEN HILLS WINERY
Walla Walla
★☆

A grapegrower since 1980, Casey McClellan has some of Washington's best vineyards at his fingertips, and it shows in the wonderfully expressive wines he produces at Seven Hills.

✓ *Cabernet Sauvignon (Klipsun, Seven Hills, Walla Walla Reserve)* • *Merlot (Seven Hills)*

SLEIGHT OF HAND CELLARS
Walla Walla
★★

Sleight of Hand Cellars was founded in 2007 by Trey Busch and Jerry and Sandy Solomon. Along with Trey's love of music, his knowledge of the diversity of wine styles available to a creative Washington winemaker allows a great deal of imaginative latitude.

✓ *Syrah (Levitation)* • *Cabernet Sauvignon (The Illusionist)* • *Red Blend (The Conjurer, Archimage)*

SPRING VALLEY VINEYARDS
Walla Walla
★★

These delightful wines are made exclusively in Spring Valley's own 16 hectares (40 acres) of vineyards, situated in the midst of over 240 hectares (600 acres) of wheat land that the Derby family has been farming for over a century. This is a spot north of Walla Walla that will surely grow in importance.

✓ *Cabernet Sauvignon (Derby)* • *Classic red blend (Frederick, Uriah)* • *Merlot (Muleskinner)* • *Syrah (Nina Lee)*

TAMARACK CELLARS
Walla Walla
★☆

A highly regarded Walla Walla boutique winery, Tamarack Cellars was established in 1998 by Ron and Jamie Coleman. Their rich, ripe, opulent Cabernet Sauvignon has quickly earned them a reputation. Theses days they are in fact better than ever.

✓ *Cabernet Sauvignon (Columbia Valley)* • *Classic red blend (Firehouse Red)*

VALDEMAR ESTATES
Walla Walla
★★

One of the newest and most striking of Washington wineries, created by Jesus Martinez Bujanda Mora, a fifth-generation Rioja winemaker. They are still working only from purchased grapes, but the size and scope of this estate suggests that it will be a major player in the near future. Ably helmed by long-time Walla Walla winemaker Marie-Eve Gilla.

✓ *Syrah (Klipsun, Blue Mountains, Red Mountain)*

WALLA WALLA VINTNERS
Walla Walla
★★

This long established and remarkably consistent winery has new ownership, and everything looks to continue with excellent wines at great prices.

✓ *Cabernet Sauvignon (Columbia Valley, Walla Walla Valley)* • *Merlot (Columbia Valley, Walla Walla Valley)*

WATERBROOK
Walla Walla
★☆

Founded in 1984 by Eric and Janet Rindal, the winery was named to reflect that of Walla Walla itself, which means "running water". Now part of Precept Brands.

✓ *Cabernet Sauvignon (Columbia Valley)* • *Classic red blend (Mélange, Red Mountain Meritage)* • *Merlot (Columbia Valley)* • *Viognier (Columbia Valley)*

WOODWARD CANYON WINERY
Lowden
★★☆

This Walla Walla winery has hardly put a foot wrong since it was first set up in 1981, making some of the state's most consistent and most elegant, award-winning red wines that show great finesse and focus. Rick Small is not only a pioneer but also one of the great minds in Washington wine.

✓ *Cabernet Sauvignon* • *Classic red blend (Charbonneau)* • *Chardonnay* • *Merlot (WWV)*

The Atlantic Northeast

The wine regions of the eastern United States face hurdles. New York wineries, like those in Michigan or Virginia, are challenged not only by the weather (and those challenges are frequent and many) but by the prevailing wine styles. Many consumers enjoy plush, fat, alcoholic wines and, regardless of the state of origin, the states around the Atlantic Ocean and the Great Lakes will never produce those kinds of wines.

This region is also held back by is its harsh winters. This does not prevent the cultivation of classic grape varieties, but it does make grafting wounds highly vulnerable and as such dictates where such grafted vines are planted. If vineyards could be chosen for their ripening potential, rather than for winter survival, the eastern seaboard could rival California, because it has a much greater variation of soils and microclimates. Since the first transgenic vines were produced, only the identity of two genes has stood in the way of this region achieving its full potential – the gene that makes *Vitis amurensis* immune to Siberian winters, together with the one that enables native American vines to resist phylloxera.

AMERICA'S OLDEST WINE INDUSTRY

Wines have been made in America's Mid-Atlantic since the middle of the 17th century, when vineyards were first established on Manhattan and Long Island. For the first few centuries, vineyards were planted with poorly understood hybrids or even on the notoriously "foxy" *labrusca* varieties. *Vinifera* vines were not cultivated until 1957, although the series of events that culminated in this most important development in the region's quest for quality wines began in 1934.

Immediately after Prohibition, Edwin Underhill, the president of Gold Seal Vineyards, went to Champagne and persuaded Charles Fournier, the *chef de cave* at Veuve Clicquot, to return with him to the United States. But Fournier found the *Vitis labrusca* grape varieties planted in New York State's Finger Lake vineyards far too aromatic. Persuaded by local wine growers that *Vitis vinifera* vines could not survive the harsh winters, he began planting hybrid vines (choosing crosses between French and native American varieties). These were initially shipped from France and then acquired from a winemaker by the name of Philip Wagner, who

had already established a considerable collection of hybrids at his Boordy Vineyard in Maryland.

In 1953 Fournier heard that one Konstantin Frank, a Ukrainian viticulturist who had arrived in the United States in 1951, had been criticizing the industry for not planting European *vinifera* vines. On his arrival in America, Frank, who spoke no English and had no money, had washed dishes to support his wife and three children. Yet as soon as he had learned enough English to get by, he applied for a job at the New York State viticultural research station at Geneva, informing his prospective employers of his studies in viticulture at Odessa and his experience in organizing collective farms in the Ukraine, teaching viticulture and oenology at an agricultural institute, and managing farms in Austria and Bavaria. When told that the winters were too harsh for European vines, he dismissed the idea as absurd. Two years later, when Fournier heard Frank's claims, he employed him, taking the chance that his theory would prove correct. Frank's claims were justified, particularly after the great freeze of February 1957. Later that year, some of the hardiest *labrusca* vines failed to bear a single grape, yet fewer than 10 per cent of the buds on Frank's Riesling and Chardonnay vines were damaged, and they produced a bumper crop of fully ripe grapes.

In the 1980s Frank was still battling with the Geneva viticultural station. *Vinifera* had not taken off in New York State, despite Frank's success. He blamed this on the "Genevians", who maintained that *vinifera* was too risky to be cultivated by anyone other than an expert. Frank had, however, become articulate in his new language: "The poor Italian and Russian peasants with their shovels can do it, but the American farmer with his push-button tools cannot".

The last few decades have been a steady march towards relevance. Not so long ago, New Yorkers were ignorant of the wealth of wines grown only a few hours from the metropolis. Today, any decent store or wine list has at least a reasonable representation from Long Island, the Finger Lakes, and sometimes from the Hudson Valley. Virginia too has gone from being thoroughly ignored by the Beltway imbibers to being celebrated at least by the hipster restaurants and bars, with the more staid establishments allowing one or two Commonwealth wines on their lists. The locavore movement hasn't made local and regional wines a must all around the country, but Michigan, New York, and Virginia can enjoy a modicum of respect and representation.

WINE-PRODUCING REGIONS OF THE ATLANTIC NORTHEAST
(see also p569)

(see also p569)

The Atlantic Northeast covers the Eastern Seaboard states of southern New England down the coast to Virginia and then west to the Great Lakes. New York, the regions' most prominent wine producer, has established this area's reputation for *Vitis vinifera* wines, but Virginia and Michigan are now rivalling this supremacy. The wine industries of many of the other states are dominated almost entirely by native *labrusca*.

FACTORS AFFECTING TASTE AND QUALITY

LOCATION
An arbitrary grouping of states situated around the Atlantic Ocean and Great Lakes.

CLIMATE
Despite severe winters, the moderating effect of large masses of inland water, such as the Great Lakes or the Finger Lakes, creates microclimates that make cultivation of *vinifera* vines possible. Harsh winter temperatures represent the most prevalent factor holding back the wine industry in the region. There are other, more localized drawbacks, such as Virginia's heat and humidity at harvest time that can cause rapid acidity drop and encourage cryptogamic diseases, but these problems can and are managed through sound viticultural practices.

ASPECT
Many of the vineyards is this region are planted on flat ground around the various lakes' shores, and on the nearby lower slopes of the various mountain ranges.

SOIL
In New York the soil is composed of shale, slate, schist, and limestone in the Hudson River Region. Virginia has silty loam and gravel at Rocky Knob and limestone and sandstone at North Fork of Roanoke. Michigan has glacial scree in Fennville and more such deposits on the peninsulas. Ohio has shallow drift soil over fissured limestone bedrock on Isle St George. Pennsylvania has deep limestone-derived soils in the Lancaster Valley.

VITICULTURE & VINIFICATION
In many areas, despite some advantageous microclimates, *vinifera* vines can survive the harsh winters only by being buried under several feet of earth before winter arrives. Sparkling wines are a speciality of New York and of the Finger Lakes area in particular. Through very careful vineyard practices, the use of the latest sprays, and the aid of new vinification technology, the number of *vinifera* varietals produced is increasing, and their reputation is quickly growing.

GRAPE VARIETIES
Cabernet Franc, Cabernet Sauvignon, Chardonnay, Gewürztraminer, Pinot Blanc, Pinot Gris, Pinot Noir, Riesling, Sauvignon Blanc, Syrah, Viognier

---

THE APPELLATIONS OF

THE ATLANTIC NORTHEAST

CONNECTICUT

No documented evidence exists of the first vineyard in this state, but based on the Great Seal of Connecticut it seems certain that its earliest settlers at least attempted to plant vines. The Great Seal today depicts three grapevines bearing fruit, but the original seal, carried over from England in 1639, had 15 grapevines and the motto *Sustinet Qui Transtulit* ("who transplants sustains"). It soon became clear, however, that tobacco was the farm crop of choice. The first Connecticut wines of modern times were made by Ciro Buonocore, who planted various French hybrid varieties at North Haven, but the industry got its start with the passage of the Connecticut Farm Winery Act in 1978, allowing wineries to sell directly to consumers. It had almost 40 wineries in 2019.

EASTERN CONNECTICUT HIGHLANDS AVA

Approved at the end of 2019, this appellation covers approximately 3,227 square kilometres (1,246 square miles) and includes major towns like Hartford and New Haven. It contains 16 vineyards with a total of 46 hectares (115 acres) of vineyards and six wineries. The soil is glacial till over difficult-to-erode metamorphic rock, resulting in the hills and mountains that characterize the terrain. The eastern and western edges are characterized by sharp ridgelines and high elevations, while the central portion is composed of rounded hills, with elevations ranging from 60 to 305 metres (200 to 1,000 feet).

SOUTHEASTERN NEW ENGLAND AVA
Connecticut, Rhode Island, and Massachusetts

A massive area (758,867 hectares, or 1,875,200 acres) distinguished in New England by its moderate climate, caused by its proximity to various coastal bodies of water.

WESTERN CONNECTICUT HIGHLANDS AVA

A vast 3,900-square-kilometre (1,500-square-mile) area of rolling hills that rise to 150 metres (500 feet) above sea level, and the Western Connecticut Highlands, small mountains that reach to 460 metres (1,500 feet).

DELAWARE

The Swedes encouraged their Delaware settlers to plant vines and make wine in 1638, as did the Dutch, following the Swedes, but all found it easier to grow apples. Nassau Valley Vineyards was the first winery in this state, but there isn't much to report at present.

DISTRICT OF COLUMBIA

John Adlum, the author of America's first book on winemaking, tried unsuccessfully to persuade the federal government in the 1820s to support a national experimental vineyard with every variety of native vine in existence "to ascertain their growth, soil, and produce". Jefferson had tried and famously failed. Now Georgetown is all restaurants, hotels, and lawyers' offices.

INDIANA

The first vines were planted by Jean-Jacques Dufour in 1804 at what would be Vevay in Switzerland County, 12 years before Indiana became a state. Dufour was a Swiss immigrant who had been sent by his family from war-torn Europe to search the newly opened lands of America for a location where they could establish a Swiss colony dedicated to winemaking. He landed in 1796, and by 1799 had already planted a vineyard at Big Bend, near Lexington in Kentucky, but it was not a success, so he purchased land north of the Ohio River, in the newly surveyed Indiana Territory. The wines made by this Swiss colony achieved two firsts: they were made from the first American hybrid (commonly known as "Alexander" or "Cape"), and they were the very first American-grown wines sold to the public. Hybrids such as Traminette have found happy grounds in Indiana, and there are nearly 100 wineries today.

INDIANA UPLANDS AVA

There are 17 wineries in the 12,432 square kilometres (4,800 square miles) of this AVA with about 80 hectares (200 acres) under vine. The Indiana Uplands are the remnant of a highly dissected plateau that remained free from repeated glacial advances due to the height of the plateau and the southern latitude. The bedrock, composed of alternating layers of limestone, shale, and sandstone, is exposed in deep valleys and high ridges throughout.

MAINE

With vintners relying on *vinifera* grapes shipped in from as far afield as Washington and only just steeling themselves to make the transition from fruit wines to own-grown French hybrids, there is hardly even an embryonic wine industry in this most northerly of America's Eastern Seaboard states. In 2019, there were 10 wineries, but although native and hybrid vines are grown, most of the production is fruit wine, and Bartlett Estate Winemaker's Reserve Blueberry is better known than any grape-based wines. Wineries like Cellardoor Winery utilize purchased grapes from far away, but there are cold-climate hardy vineyards nonetheless; Cellardoor has a delightful ice wine from their own Frontenac Gris and Frontenac Blanc.

MARYLAND

In 1662 Lord Baltimore instructed his son, Charles Calvert, the governor of Maryland, to plant a vineyard and make wine. Within three years he had planted 136 hectares (340 acres). The vines were native, not European, but the wine was reportedly "as good as the best Burgundy". A century and three-quarters later, Philip Wagner published the definitive text on US winemaking, *American Wines and How to Make Them*, (later revised as *Grapes Into Wine*). In 1945 he opened Boordy Winery; even today, tasters are surprised if not shocked at the quality and strength of the Bordeaux varieties that Boordy and some of its neighbours can craft. There are more than Maryland wineries as of 2019.

CATOCTIN AVA

Situated west of the town of Frederick, this area's specific *terroir* was well known before the AVA was established, due to the fact that it roughly coincides with the Maryland Land Resource Area. This was determined by the US Soil Conservation Service on the basis of identifiable patterns of soil, climate, water availability, land use, and topography.

CUMBERLAND VALLEY AVA
Maryland and Pennsylvania

The Cumberland Valley is situated between the South Mountains and the Allegheny Mountains and is 120 kilometres (80 miles) long, bending in a northeasterly direction. Although this AVA covers approximately 3,100 square kilometres (1,200 square miles), its vines are confined to small areas where the soil, drainage, rainfall, and protection from lethal winter temperatures permit viticulture. Vineyards are found on high terraces along the north bank of the Potomac River, on the hills and ridges in the basin of the valley, and in the upland areas of the South Mountains.

LINGANORE AVA

Maryland's first viticultural area lies east of Frederick. It is generally warmer and wetter than the areas to the east, and slightly cooler and drier than those to the west.

MASSACHUSETTS

Wines were made from native grapes in the very first summer of the Massachusetts Bay Colony in 1630, but its dubious quality was probably the reason why the settlers immediately petitioned the Massachusetts Bay Company for Frenchmen experienced in planting vines. Two years later, as part of an agreement with the Colonial Legislature, Governor John Winthrop planted a vineyard on Governor's Island, from which he was supposed to supply as annual rent "a hogshead of the best wine". There is no evidence that he succeeded and, as his rent was changed to "two bushels of apples" within just a few years, there is every reason to suspect that he had failed. In modern times, the first *vinifera* vines were planted in 1971 by the Mathieson family on Martha's Vineyard. Most wineries are in the south, within the Southeastern New England AVA. Although the coastal location helps moderate the harshness of winter, both non-*vinifera* and *vinifera* are being utilized. As of 2019 there were a bit more than four dozen wineries.

MARTHA'S VINEYARD AVA

This AVA is an island off Massachusetts that is surrounded to the north by Vineyard Sound, to the east by Nantucket Sound, and to the south and west by the Atlantic. The AVA's boundaries include an area known as Chappaquiddick, which is connected to Martha's Vineyard by a sand bar. Ocean winds delay the coming of spring and make for a cooler autumn, extending the growing season to an average of 210 days, compared with 180 days on the mainland.

SOUTHEASTERN NEW ENGLAND AVA

See entry under Connecticut.

MICHIGAN

Michigan was already a mature winemaking region by 1880, when the first national winegrowing census was taken, but the grapes used were native varieties, and they were grown in the southeast of the state, on the shore of Lake Erie, where virtually no vines exist today. Michigan made the transition to French hybrids in the 1950s and 1960s, although it was not in full swing until the mid-1970s. Meanwhile, the first European varieties were planted at Tabor Hill Vineyard by Len Olsen and Carl Banholzer in 1970, although 20 years later 85 to 90 per cent of Tabor Hill wines were still being made from hybrids. That was in the southwest, however, in what is now Lake Michigan Shore AVA. The future as far as quality and diversity of *vinifera* wines was concerned, and a whole raft of new boutique wineries that were about to emerge, lay in the Old Mission Peninsula and Leelanau Peninsula in the northwest of the state. Only one man was crazy enough to plant *vinifera* 225 kilometres (140 miles) north of Canada's Niagara District in 1974, and that was Ed O'Keefe of Château Grand Traverse, where not one single hybrid was ever planted. His adamant arguing of his case did not made him many friends amongst his fellow winegrowers, who were clinging to their hybrids, but he was right, and in the end he did everyone in the Michigan wine industry a big favour. Today Michigan has 5,400 hectares (13,100 acres) of vineyards, making it the fourth-largest grape-growing state, but most of this area is still devoted to juice grapes, such as Concord and Niagara. There are 1,230 hectares (3,050 acres) devoted to wine grapes, making Michigan the sixth-largest state for wine grape production. As of 2019, there were 160 wineries.

FENNVILLE AVA

Lake Michigan moderates this area's climate, providing slightly warmer winters and cooler summers than other areas within a 48-kilometre (30-mile) radius. Fennville covers 310 square kilometres (120 square miles) and has been cultivating various fruits for well over a century, including grapes for wine production. The soil is mostly scree of glacial origin.

LAKE MICHIGAN SHORE AVA

Located in the southwest corner of Michigan, this AVA is a geographically and climatically uniform region, although it does encapsulate smaller, very specific *terroirs*, such as Fennville, which has its own AVA.

LEELANAU PENINSULA AVA

This AVA is on the western shore of Lake Michigan, northwest of Traverse City. The lake delays fruit development beyond the most serious frost period in the spring, and prevents sudden temperature drops in the autumn. Most of Michigan's boutique wineries are located here.

OLD MISSION PENINSULA AVA

This AVA is surrounded on three sides by Grand Traverse Bay and connected to the mainland at Traverse City. The waters, coupled with warm southwesterly winds, provide a unique climate that makes cultivation of *vinifera* vines possible. The longer established of the two peninsulas, Old Mission houses the smallest number of wineries, but there is renewed interest in the area since Black Star Farms has sourced so many award-winning red wines from here, particularly the Leori Vineyard, and the success of numerous white wine producers.

THE TIP OF THE MITT AVA

Established in 2016, this AVA is located in the northern Lower Peninsula of Michigan, with most wineries grouped around Petoskey.

NEW HAMPSHIRE

The first European vines were planted at the mouth of the Piscataqua River by Ambrose Gibbons in 1623, although from his diary we know he suspected that, unlike the native vines that prospered in the wild, they would not survive: "The vines that were planted will come to nothing. They prosper not in the ground where they were set, but them that grow naturally are very good of divers sorts." It would be almost another 250 years before anyone managed to grow wine grapes in New Hampshire. In 1965 John J Canepa planted 800 Maréchal Foch vines and, although some died that first winter, they yielded enough that he gave up his day job and built a winery, bonded in 1969. Today there are still no more than two dozen wine producers in the entire state.

NEW JERSEY

Although far more famous for cider than wine, New Jersey does, in fact, boast the first American-grown wine to win an international award. It was as early as 1767 that London's Royal Society of the Arts recognized two New Jersey vintners for producing the first quality wine derived from colonial agriculture, although they were from wild vines, not *vinifera*. This state also gave birth to Dr Thomas Bramwell Welch, the wine-hating dentist who curiously read Pasteur's studies of fermentation, which he turned on their head to sterilize grape juice. It was "Dr Welch's Grape Juice" that established the grape juice industry, though it was first sold as "Dr Welch's Unfermented Wine". There are more than 60 wineries in the state, and *vinifera* vines are making serious inroads, with more than 100 varieties planted. There are still plenty of native and hybrid vines, however, and, indeed, fruit wines.

CAPE MAY PENINSULA AVA

This sub-AVA, established in 2018, is fully contained within the Outer Coastal Plains AVA, all inside the borders of New Jersey.

CENTRAL DELAWARE VALLEY AVA

See entry under Pennsylvania.

WARREN HILLS AVA

Located in Warren County. Wines made in the eastern half of the Central Delaware Valley AVA (*see* entry, p629), which consists of five narrow valleys rather than one broad one, may use this sub-appellation. The narrow valleys provide hillsides that are exposed and funnel the winds, reducing the risk of frost and rot.

OUTER COASTAL PLAIN AVA

Now covering 911,000 hectares (2,250,000 acres) in southeast New Jersey, this AVA includes all of the coastal counties in the south, as well as portions of the other southern counties. Soils are generally well-drained sandy or sandy loam soils with temperatures moderated by the Atlantic Ocean and Delaware Bay. The AVA contains one subregion, the Cape May Peninsula AVA, established in 2018.

NEW YORK

The first vineyard in what is now New York State was cultivated by the Dutch, when the colony was still New Amsterdam, in 1642. Very little is known about it other than that it failed to survive the winter, so it was probably planted with *vinifera* vines. Long Island was an important nursery for imported vines in the late 18th and early 19th

BLACK STAR FARMS NEAR TRAVERSE CITY ON THE OLD MISSION PENINSULA IN MICHIGAN
Black Star Farms has winery production facilities with tasting rooms on both the Old Mission and Leelanau Wine Trails. The Old Mission Peninsual AVA and the Leelanau Peninsula AVA lie near the northwest corner of the state.

centuries. At one time, New York was second only to California in terms of acreage of vines, but this peaked in 1975 and had dropped to just under 15,000 hectares (37,000 acres) by 2018, while the acreage in trailing states has increased significantly since the late 1990s. About 75 per cent of New York's vineyards are planted to native varieties (three-quarters being Concord), intended for juice or table consumption; the remaining 25 per cent is split between hybrids and *vinifera,* with 1,822 hectares (4500 acres) of *vinifera* and 1,093 hectares (2,700 acres) of hybrids.

Not long ago, New York was number two in terms of wine production, but with its annual production now at 28.5 million gallons, it is well behind Washington State with its output of 35.6 million gallons. Climate change seems to be increasing the success of red wines here, but the most dramatic industry changes are reflected in the newly enthusiastic embrace of local wines by New York restaurateurs – a phenomenon significantly altering the opportunities for many of the state's 400 wineries.

CAYUGA LAKE AVA

This area encompasses vines grown along the shores of Lake Cayuga, making it part of the Finger Lakes AVA. The soil here is predominantly shale, and the growing season approximately one month longer than that of most of the Finger Lakes area.

CHAMPLAIN VALLEY OF NEW YORK

The appellation application described this as "a long, narrow, relatively flat valley on the western shore of Lake Champlain." The AVA is near the Adirondack Mountains and the border with Vermont. The AVA is approximately 132 kilometres (82 miles) long, and 32 kilometres (20 miles) wide at its northernmost and widest point (along the New York–Canadian border). There are only seven wineries at present.

FINGER LAKES AVA

This name is derived from the 11 finger-shaped lakes in west-central New York State. These inland water masses temper the climate, and the topography of the surrounding land creates "air drainage", which moderates extremes of temperature in winter and summer.

HUDSON RIVER REGION AVA

This AVA encompasses all or portions of eight counties, but only has 64 hectares (159 acres) of grapes at present, with 62 wineries. This is the Taconic Province, one of the most complex geological divisions, where the soil is made up of glacial deposits of shale, slate, schist, and limestone.

LAKE ERIE AVA

New York, Pennsylvania, and Ohio

Overlapping three states and encompassing the AVAs of Isle St George and Grand River Valley, Lake Erie moderates the climate and is the fundamental factor that permits viticulture. This is predominately juice and table grapes.

LONG ISLAND AVA

The Long Island AVA covers the entire island, encompassing the two pre-existing AVAs of North Fork of Long Island and The Hamptons, Long Island. Virtually all of the vineyards re-planted solely to *vinifera* varieties.

NIAGARA ESCARPMENT

Approved in 2005, this large 7,284-hectare (18,000-acre) AVA is somewhat underdeveloped, at least in comparison to the Canadian side of the Niagara Escarpment. The *terroirs* are similar – dolomitic limestone soil and gravel silts near the lake shore – along with a moderate climate.

NORTH FORK OF LONG ISLAND AVA

The sea that surrounds this AVA makes it a more temperate region than many other places of the same latitude in the interior of the United States. The growing season is about one to three weeks longer than in the South Fork of the Island and, in general, the sandy soils contain less silt and loam, but are slightly higher in natural fertility.

SENECA LAKE AVA

Seneca Lake is the deepest of New York's 11 finger lakes, so it is in permanent use by the US Navy's sonar testing platform. At its deepest (203 metres, or 678 feet), this long body of water (56 kilometres, or 35 miles) in the heart of New York wine country does not freeze; its stored heat warms the surrounding area, an effect largely responsible for the region's winegrowing success. Best known for Riesling – the steep, slate slopes are ideal for these vines – this variety is often planted on silty, sandy, or loamy soils, while other varieties, such as Pinot Noir, are wasted on slate-rich soils, when they would do much better elsewhere.

THE HAMPTONS, LONG ISLAND AVA

The Hamptons has been a productive agricultural area for 300 years. It lies within Suffolk County, next to the North Fork of Long Island, with the Peconic River and Peconic Bay its northern boundary. The AVA includes Gardiners Island.

UPPER HUDSON AVA

This new (2018) AVA is located to the northwest of Albany, New York, and north of the Hudson River Region AVA. A portion of it forming the border with Vermont.

OHIO

When Nicholas Longworth arrived in Cincinnati in the early 1820s, he studied law for a mere six months before setting up in a practice that made him a millionaire. As a hobby, he planted a vineyard in a part of Cincinnati known as Tusculum, but many of the vines he imported from Europe died. Longworth thus turned to native American varieties and quickly discovered Catawba, the so-called "wonder grape", which he planted in 1825. Three years later he made his first Catawba wine and was so impressed that he retired from law to devote his energy to viticulture and winemaking, planting hundreds of acres. At first he produced still wine, but as soon as he made sparkling Catawba, it took off. In 1854 Henry Longfellow even wrote an "Ode to Catawba Wine" in which he compared it with two of the most famous growths of Champagne, and by 1858 the fame of Catawba had spread as far as Europe, where the *Illustrated London News* reported that "Sparkling Catawba, of the pure, unadulterated juice of the Catawba grape, transcends the Champagne of France". By 1860 one-third of all the vines in America were planted along the banks of the Ohio, which boasted twice the acreage of California's vineyards, making Ohio the wine centre of the New World prior to the Civil War. Between 1980 and 2000, the number of wineries in this state hardly wavered, moving from 44 to an equally sedate 47, but it has more than exploded since, reaching 223 in 2018 and indicating that some people now think that Ohio could regain its past glories.

GRAND RIVER VALLEY AVA

Located within the Lake Erie AVA, the lake protects these vines from frost damage and enables a longer growing season than vineyards situated in inland areas. The river valley increases "air drainage", giving this AVA a sufficiently different microclimate to warrant its distinction from the Lake Erie AVA.

ISLE ST GEORGE AVA

The northernmost of the Bass Islands, vines have been grown here since 1853. Today they cover over half the island. Although tempered by Lake Erie, the climate is cooler in the spring and summer and warmer in the winter than mainland Ohio. The region is frost-free for 206 days a year, which is longer than for any other area in Ohio. The shallow drift soil over fissured limestone bedrock is well suited to viticulture.

RIPE CATAWBA GRAPES
Known for the "foxy" flavour it imparts to wine, this hybrid of *Vitis labrusca* and *V vinifera* was extremely popular with mid-19th-century wine consumers. The Catawba, promoted by winemaker Nicholas Longworth, gave rise to a flourishing wine industry in Ohio, helping to make in the largest in the country in the pre–Civil War era.

LAKE ERIE AVA
See entry under New York.

LORAMIE CREEK AVA

This small AVA covers only 1,460 hectares (3,600 acres) in Shelby County in west-central Ohio. Moderate-to-poor drainage means vines must be grown on slopes and ridges to prevent "wet feet". There is no operating winery there at the moment.

OHIO RIVER VALLEY AVA

Ohio, Indiana, West Virginia, and Kentucky

This is a vast AVA of 67,300 square kilometres (26,000 square miles). For a time, Ohio was the leading wine-producing state. The Cincinnati Wine Company sold massive amounts of sparkling Catwaba (almost a quarter million cases in 1859); Longfellow gave it poetic praise. During the Civil War, however, black rot and powdery mildew took hold and destroyed nearly all its vineyards. Since then newer and less aromatically intense hybrid grapes prevailed until the last few decades as *vinifera* is taking its place alongside them.

PENNSYLVANIA

In 1683, colonist William Penn established a vineyard with French and Spanish varieties he had brought with him, but they failed. Conrad Weiser, on the other hand, who arrived at the Penn colony in 1729 and became the greatest Indian interpreter of his time, was spectacularly successful with his vineyard, planted near Womelsdorf in the Tulpehocken Valley. Chambourcin is widely regarded as the most successful variety in this state today, although some good *vinifera* wines are also produced. In 2019 there are almost 300 wineries in this state, so the industry is clearly expanding.

CENTRAL DELAWARE VALLEY AVA

Pennsylvania and New Jersey

This appellation, hugging both sides of the Delaware River in southeastern Pennsylvania and New Jersey, covers 388 square kilometres (150 square miles), although very little of it is actually planted with vines. The Delaware River modifies the climate.

CUMBERLAND VALLEY AVA
See entry under Maryland.

LAKE ERIE AVA
See entry under New York.

LANCASTER VALLEY AVA

Grapes have been grown in Lancaster County since the early 19th century but have only recently provoked any interest from outside. The vines are grown on a virtually level valley floor, at an average altitude of 120 metres (400 feet), where the deep, limestone-derived soils are well drained. Even so, the soils have good moisture retention, are highly productive, and differ sharply from those in the surrounding hills and uplands. There are presently 162 hectares (400 acres) of grapevines.

LEHIGH VALLEY AVA

The Lehigh Valley includes 93 hectares (230 acres) of vineyards, with both *vinifera* and hybrids. Nearly one-fifth of the wine produced in Pennsylvania is made from Lehigh Valley grapes, although there are fewer than a dozen wineries in the AVA.

RHODE ISLAND

The smallest and most densely populated state, Rhode Island is not, of course, an island, but is surrounded on three sides by two other states, Connecticut and Massachusetts. The first vines were planted here in the 1820s by Huguenot settlers, who successfully made wine. These settlers were driven out by legal difficulties, however, after which viticulture did not continue in the state. There are now fewer than 10 wineries in Rhode Island.

SOUTHEASTERN NEW ENGLAND AVA
See entry under Connecticut.

VERMONT

Although the home of two native grape varieties, Vergennes and Green Mountain, which were commercially significant in the 19th century, Vermont is now more cider country than wine country, which explains why there were no more than 2 wineries until 1990. In 2004 there were 9 wineries, but they mostly sold fruit wines. In 2019, however, there are 36 wineries in the state, most producing wine from their own grapes.

VIRGINIA

This was not the first state to make wine: that honour goes to Florida, where wine was made from wild grapes circa 1563. The first settlers in Virginia made wine from "hedge grapes" in 1609. Virginia was the first state to attempt the cultivation of *vinifera* grapes for wine, however, although not by its first governor, Lord De la Warr (later corrupted to Delaware), as many sources claim. The first documented attempt to transplant European vines to eastern America was recorded by S M Kingbury of the Virginia Company of London, who reported that French vines and eight French *vignerons* from Languedoc had been sent out to Virginia in 1619, one year after De la Warr had died at sea. The Virginia Company caused a law to be enacted requiring every householder to plant 10 vines a year "until they have attained the art and experience of dressing a vineyard", but all efforts to cultivate vineyards failed, even though native grapes were rampant in the wild. So desperate was Virginia to succeed that an Act of Assembly was passed in 1658 offering 10,000 pounds of tobacco to the first person to make "two tunne of wine raised out of a vineyard made in this colony". No one ever claimed the prize and, almost 30 years later, the offer was quietly dropped.

Today Virginia has the ability to be the Washington State of America's East Coast, but the negative influence of its hot and humid weather during the growing season, particularly towards the end of *véraison* and at harvest time, must first be overcome. There are two main difficulties. First, the sugar-ripeness of the grapes tends to soar away from the rate at which physiological ripeness progresses, resulting in the production of wines bearing green tannins and depleted acidity. Second, the humidity encourages cryptogamic diseases. Various oenological practices are employed to combat the harsh tannins, but the goal must ultimately be to produce grapes with ripe tannins and a good sugar-acidity balance, rather than to sift out unripe seeds (*délestage*). The only way to achieve this is by improving canopy management. The state is home 140 wineries.

APPALACHIAN HIGH COUNTRY AVA
See entry under North Carolina.

MIDDLEBURG VIRGINIA AVA

This AVA in the northern Piedmont region was planted by Meredyth Vineyards back in 1972. The application was filed by the VP of the estimable Boxwood Vineyards, as important today as Meredyth was in its day. Only 80 kilometres (50 miles) west of Washington, DC, this is a delightful area of rolling hills, bounded on the north by the Potomac River. Yet excellent wineries, such as Breaux, were drawn just outside the AVA.

MONTICELLO AVA

Monticello is well known as the home of Thomas Jefferson, who is recorded as having planted wine grapes here. Many of Virginia's best wineries are found in this AVA.

NORTH FORK OF ROANOKE AVA

Located on the eastern slopes of the Allegheny Mountains, this AVA lies in a valley protected from excessive rainfall in the growing season by mountain slopes to the west and east. The vines are on the limestone southeast-facing slopes and limestone-with-sandstone north-facing slopes. These soils are very different from those in the surrounding hills and ridges.

VERITAS VINEYARD WINERY, AFTON, VIRGINIA
This vineyard and winery is located in a beautiful site nestled in the foothills of the Blue Ridge Mountains.

NORTHERN NECK
GEORGE WASHINGTON BIRTHPLACE AVA

This AVA lies on a peninsula 160 kilometres (100 miles) long, between the Potomac and Rappahannock Rivers in the Tidewater District of Virginia, which runs from Chesapeake Bay in the east to a few kilometres from the town of Fredericksburg to the west. The vines grow in sandy clay soils on the slopes and hills and in alluvial soils on the river flats. The favourable climate, with excellent air-drainage, is moderated by the surrounding water.

ROCKY KNOB AVA

This AVA is in the Blue Ridge Mountains, and in spring is colder than nearby areas. This means the vines flower later, enabling them to survive the erratic, very cold early spring temperatures. It also causes a late fruit-set, extending the growing season by about a week. The silty-loam and gravel soil provides good drainage.

SHENANDOAH VALLEY AVA

Virginia and West Virginia

The Shenandoah Valley lies between the Blue Ridge Mountains and the Allegheny Mountains. This appellation extends south beyond the Shenandoah Valley almost as far as Roanoke, Virginia.

VIRGINIA'S EASTERN SHORE AVA

Located in Accomack and Northampton counties along the 120-kilometre (75-mile) narrow tip of the Delmarva Peninsula, with the Atlantic to the east and Chesapeake Bay to the west. The climatic influence of these two large bodies of water helps to alleviate the severest winter temperatures but retards the ripening process and can be problematic at harvest time. There are only a few wineries here.

WEST VIRGINIA

Grape growing began here, along the Ohio River, in the 1830s, when this state was still part of Virginia, and continued after West Virginia had been created during the Civil War. In 2010 there were 17 wineries, mostly producing native, hybrid, and fruit wines, with none standing out as yet.

KANAWHA RIVER VALLEY AVA

The AVA area covers 2,600 square kilometres (1,000 square miles) yet contains just 6 hectares (15 acres) of vines and one bonded winery. It sits within the gargantuan Ohio River Valley AVA.

OHIO RIVER VALLEY AVA
See entry under Ohio.

SHENANDOAH VALLEY AVA
See entry under Virginia.

THE ATLANTIC NORTHEAST

CONNECTICUT

CHAMARD VINEYARDS

Clinton

★☆

Founded back 1983, Chamard Vineyards is just 3 kilometres (2 miles) from Long Island Sound and enjoys some of the same climatic benefits of wineries located on Long Island.

✓ *Chardonnay (Estate Reserve)* • *Rosé*

STONINGTON VINEYARDS

Stonington

★☆

Since 1987, this small vineyard and winery has been family owned and has had the same capable winemaker, Mike McAndrew. They have been producing solid wines (particularly Chardonnay) for decades.

✓ *Chardonnay (Oak, Sheer)* • *Cabernet Franc*

SHARPE HILL VINEYARD

Pomfret

★☆

The Vollweiler family first planted vines here in 1992 and opened the winery in 1997. Located in the historic town of Pomfret, Sharpe Hill's wines are under the guidance of winemaker Howard Bursen, who has earned them the claim of being Connecticut's most-awarded winery. Some of their top honours have gone to Chardonnays and Riesling.

✓ *Chardonnay (Vineyard Reserve, Cuvee Ammi Phillips')* • *Dry Riesling*

SALTWATER FARM

Stonington

★☆

New York City lawyer Michael Connery bought this old airport and hanger almost two decades ago and has converted it into an exciting winery project.

✓ *Chardonnay (Estate, Gold Arc)* • *Sauvignon Blanc*

INDIANA

HUBER WINERY

Borden

★☆

This orchard and vineyard estate is being run by the seventh generation of this family; wine production began more than three decades ago, and the winery continues to be a leader in the state.

✓ *Petit Verdot* • *Traminette*

OLIVER WINERY

Borden

★☆

Oliver Winery started from modest roots back in the 1960s as a hobby in the basement of Indiana University law professor William Oliver. His son is CEO.

✓ *Creekbend III (Vignoles, Chardonel, Vidal Blanc)* • *Crimson Cabernet (Creekbend)* • *Chambourcin (Creekbend)*

MARYLAND

BASIGNANI WINERY

Sparks Glencoe

★☆

With wine a family tradition, Bertero Basignani bought 4 hectares (10 acres) of farmland in Butler, Maryland, in 1972 to plant as a vineyard. He was an amateur winemaker for years, selling the remainder of his grapes to Maryland wineries. Basignani Winery is still small but has gained a large following.

✓ *Cabernet Sauvignon* • *Bordeaux blend (Lorenzino)* • *Vidal Blanc*

BIG CORK VINEYARDS

Rohrersville

★☆

Former Breaux (Virginia) winemaker Dave Collins started this winery in 2015, and he relished the chance to create his own legacy, working with a wealth of grapes with legacies of their own.

✓ *Muscat Canelli* • *Muscat Blanc (Russian Kiss)* • *Nebbiolo* • *Vidal Ice Dessert*

BOORDY VINEYARDS,

Hydes

★☆

Establishing a true pioneer winery, Philip and Jocelyn Wagner bonded Boordy as a winery in 1945. Rob Deford III, a UC Davis graduate, joined in 1980, and four generations of the Deford family are working with him today.

✓ *Cabernet Franc (Maryland, Reserve)* • *Chambourcin-Merlot* • *Albariño* • *Bordeaux blend (Landmark Reserve)*

CATOCTIN BREEZE VINEYARDS

Brookeville

★

Catoctin Breeze , overlooking the Catoctin Mountains, is a newish family-owned property. Since its first harvest in 2012, it has been earning a place among the top ranks of Maryland wineries.

✓ *Cabernet Franc (Reserve)* • *Bordeaux blend (Concerto)*

THE VINEYARDS AT DODON

Davidsonville

★☆

Owner, winemaker, and physician Tom Croghan is farming and making wine from vines grown on his wife's family land, in the family since 1725.

ELK RUN VINEYARD

Mount Airy

★☆

The owner-winemaker, Fred Wilson, who started the winery in 1983 and trained with the legendary Dr Konstantin Frank (see pp 625, 631), was an early adopter of Pinot Noir and Syrah in Maryland.

✓ *Pinot Noir* • *Bordeaux blend (The Red Door)*

SUGARLOAF MOUNTAIN VINEYARD

Dickerson

★

Only 30 minutes away from the Beltway, Sugarloaf Mountain Vineyard has a plot specializing in Bordeaux-style wines. It is benefiting from the help and guidance of one of the legends of East Coast viticulture, Lucie Morton.

✓ *Pinot Gris* • *Rosé* • *Evoe!* • *Cabernet Franc Reserve*

OLD WESTMINSTER WINERY

Westminster

★

Farmer Drew Baker, along with his two sisters, winemaker Lisa Hinton and general manager Ashli Johnson, runs the vineyard on the Carroll County farm where they grew up.

✓ *Grüner Veltliner* • *Pét-Nat*

MASSACHUSETTS

WESTPORT RIVERS

Westport

★★

Bob and Carol Russell converted a 17th-century turnip farm into the largest vineyard in New England in 1986, establishing the Westport Rivers winery in 1989. The family tradition was to make sparkling wine, and they do it well today.

✓ *Sparkling RJR* • *Blanc de Blancs*

TRURO VINEYARDS OF CAPE COD

North Truro

★☆

Truro Vineyards was founded in 1992, but the main house dates back two centuries; it was painted by Edward Hopper in 1930. Former wine industry CEO Dave Roberts and his family took over in 2007.

✓ *Cabernet Franc*

MICHIGAN

AMORITAS WINERY

Lake Leelanau

★☆

A very small, new, family-owned winery; this is a labour of love for the tight-knit Goodell family.

✓ *Muscat Ottonel (Fascinator)* • *Riesling (entire range)*

BLACK STAR FARMS

Suttons Bay

★★

Since 1998 Black Star has not only been making delicious wines, it also offers some of the best *eaux de vies* in the United States. Winemaker Lee Lutes is a pro.

✓ *Gamay Noir (Arcturos)* • *Pinot Noir (Arcturos)* • *Pinot Blanc (Arcturos)* • *A Capella Ice Wine* • *Classic red blend (Leorie Vineyard, Merlot-Cabernet Franc)*

BOWERS HARBOR VINEYARDS

Traverse City

★☆

The Stegenga family have steadily built this estate since 1991, and the wines just keep getting better.

✓ *Riesling (Block II, Smokey Hollow)* • *Bordeaux blend (2896 Langley)*

BRYS ESTATE VINEYARD

Traverse City

★☆

What started out in 2004 as a fun retirement project has morphed into a reliable and trustworthy winery. Now the second generation of the Brys family is helping to run it.

✓ *Riesling (Artisan Dry, Reserve)* • *Pinot Noir (Reserve)*

BEL LAGO VINEYARDS

Cedar

★★

This winery was founded all the way back in 1987 by Charlie Edson and Bel Lago, who are old school, but the wines they make are not. They produce a surprisingly good Chardonnay.

✓ *Riesling* • *Chardonnay* • *Auxerrois (Moreno Block)* • *Edelzwicker* • *Cabernet Franc*

CHATEAU CHANTAL WINERY

Traverse City

This husband-and-wife team were once Nadine, a Felician nun, and Robert, a Roman Catholic priest, but in 1974 the Begins married and in 1983 opened a bed and breakfast and a winery. The wines show a steady improvement as the project has grown into a larger-scale wine producer helmed by daughter Marie-Chantal Dalese.

✓ *Pinot Grigio* • *Riesling (Semi-dry, Late Harvest)*

CHATEAU GRAND TRAVERSE

Traverse City

★☆

Chateau Grand Traverse was established in 1974 by Ed O'Keefe, the first person in Michigan exclusively to plant *vinifera* vines, ruffling many feathers among the locals in the process. And a good job it was, too, for Michigan wine would not be where it is if he had not been as forceful as he was. This estate has long carried the banner for delightful Riesling of varying styles (sweet, dry, and everything in between). Ed O'Keefe established it in 1974, and son Eddie runs it today.

✓ *Riesling (entire range)*

DOMAINE BERRIEN CELLARS

Berrien Springs

★

Winemaking hobbyist Tom Fricke bought a cherry orchard and planted grapes in Lake Michigan Shore AVA in 1992 and focused immediately on Rhône varieties, such as Syrah, Marsanne, Roussanne, and Viognier. His whites are particularly enjoyable.

✓ *Marsanne • Lemberger*

FENN VALLEY VINEYARDS

Fenville

★

Approaching its 50-year anniversary and run by fourth-generation vintners, this winery is worthy of celebration. Fenn Valley is particularly skilled with hybrid grapes. It is also the first large-scale winery in Michigan to produce a canned wine (Vino Blanco).

✓ *Traiminette • Riesling (sweet) • Vignoles (Late Harvest)*

LEFT FOOT CHARLEY

Traverse City

★★★

Bryan Ulbrich is one of the top white wine producers in the country. Tom Stevenson has called him a white wine genius. Believe it.

✓ *Entire range*

MARI VINEYARDS

Traverse City

★★

This new winery is owned by the Lagina family, who have wisely joined with veteran Michigan winemaker Sean O'Keefe to found a beautiful Italianate estate.

✓ *Any and all Rieslings*

MAWBY WINES

Suttons Bay

★★☆

Founded in 1973 by Larry Mawby and his partners, Michael and Peter Laing, Mawby is making wines as good or better than any other sparkling wine in the country, utilizing both *vinifera* and hybrid grapes for their delicious wines.

✓ *CA. 2012 (and subsequent vintages) Sparkling • Grace Sparkling Rosé • Talis Sparkling • Blanc Brut Sparkling*

PENINSULA CELLARS

Traverse City

★☆

This same farm has been held by the Kroupa family for more than a century; not particularly notable in the Old World, but in North America, it's a big deal. They diversified decades ago from orchards to vineyards, and they've exhibited brilliance in employees (Black Star Farms' Lee Lutes and Left Foot Charley's Bryan Ulbrich are each former winemakers).

✓ *Riesling (all types)*

SHADY LANE

Suttons Bay

★★

Planted in 1989 with Larry Mawby as their early winemaker and Adam Satchwell (Jed Steele's nephew) as their next winemaker, Shady Lane has had a fairly charmed existence. Under Kasey Wierzba's tutelage, the wines are even tastier.

✓ *Riesling (all types)*

ST JULIAN WINERY

Paw Paw

★☆

Founded in 1921 by Mariano Meconi, this winery was established in Canada, where it was called Border City Wine Cellars for a short while before being renamed Meconi Wine Cellars. It did not become American until 1934, when Meconi moved his business to Detroit. In 1936 he moved the business to its current location, changing its name once again, to the Italian Wine Company. When the US entered World War II, Meconi sought to avoid anti-fascist sentiment by changing his company's name for the third and final time. The St Julian Wine Company is thus the oldest continuously operating winery in Michigan. It is also by far the largest, producing almost as much as all the state's other wineries put together. Although the St Julian Wine Company is rated one and half stars, one of its products would deserve three stars on its own. The perpetually award-winning Solera Cream Sherry is world class, which would be achievement enough for any winery outside of Jerez, but the fact that it is made exclusively from the Niagara grape makes it all the more miraculous.

✓ *Solera Cream Sherry • Traminette (Braganini Reserve) • Vignoles (BR Late Harvest) • Riesling (SJ Reserve Late Harvest)*

TABOR HILL WINERY

Buchanan

★

The Moersch agricultural group that owns this winery, as well as the Round Barn Winery, makes clean, solid wines.

✓ *Albarino • Lemberger*

TWO LADS

Traverse City

★★

A Michigan winemaker and a South African cellar rat have joined up to give chase to the some of the other top-flight Michigan wineries. It's going well.

✓ *Riesling (all types) • Pinot Noir (all types)*

VERTERRA WINERY

Leland

★☆

In 2007 Paul and Marty Hamelin bought a cherry orchard and turned it into a winery. Initially they were all about Chardonnay, but their other white wines are pretty good too.

✓ *Pinot Blanc • Riesling (Late Harvest) • Chardonnay (Unoaked)*

NEW JERSEY

ALBA VINEYARD

Milford

★★

Alba's 38-hectare (93-acre) estate lies in the rolling hills of Warren Hills AVA, with elevations of 76 and 198 metres (250 feet and 650 feet) and with the nearby Musconetcong River creating air drainage and protection from frosts. Gently sloping, south-facing vineyards make this a very pretty and useful site, dating back to the 1700s, but the land has never been commercially cultivated.

✓ *Chambourin • Chardonnay • Cabernet Franc • Riesling*

TOMASELLO

Hammonton

★☆

This winery, which opened up immediately after Prohibition was repealed, has had over 50 years' experience of make decent sparkling wine.

✓ *Petit Verdot (Palmaris Reserve) • Merlot (Outer Coastal Plain Reserve) • Blaufrankisch*

UNIONVILLE VINEYARDS

Ringoes

★☆

This is one of those New Jersey wineries that will change your mind about what is possible in this state. The current winemaking tandem of Zeke Johnson and Conor Quilty have established a very good track record.

✓ *White blend (Hunterdon Mistral Blanc) • Syrah (Pheasant Hill)*

NEW YORK

ANTHONY ROAD WINERY

Penn Yan

★★

John and Joan Martini made their first wine in 1990, and over the years they have been crucial in establishing the lofty reputation of Finger Lakes Riesling. It remains a family enterprise.

✓ *Riesling (entire range)*

BEDELL

Cutchogue

★★

Kip and Susan Bedell planted the vineyard in 1980; in 2000 Bedell was sold to the Lynne Family. Sadly, Michael and his son Jonathan both passed away in 2019 but MIchael's wife, Ninah, and the team continue to advance this remarkable project. Winemaker Richad Olsen-Harbich wrote the application for the North Fork of Long Island AVA and collaborated with the late Paul Pontallier of Château Margaux in France's Médoc region to improve the vineyards and winery.

✓ *Chardonnay • Classic red blend (Taste Red, Musée) • Merlot • Cabernet Franc*

BENMARL WINE COMPANY

Marlboro

★

The state's first farm license winery, this old establishment may include the prettiest B&B in the Hudson Valley. It gets its name Benmarl from the Gaelic word for the vineyard's slate-marl soil.

✓ *Baco Noir • Sauvignon Blanc*

BROTHERHOOD WINERY

Washingtonville

★

The oldest winery in continuous operation in the USA, Brotherhood Winery was established by a shoemaker named Jean Jacques, who initially sold wine to the First Presbyterian Church. The wines used to be very old school but are now fresher and crisper in character.

CHANNING DAUGHTER

Bridgehampton

★☆

This winery has been on the fast track since Walter and Molly Channing and partner Larry Perrine launched their first wines with the 1997 vintage. Perrine's expertise in soil impact has been crucial to the development of this winery, as well as the entire North Fork. Winemaker Chris Tracy is wildly creative with the styles and wines made: orange wines, single-variety rosés, crazy-quilt blends, and unusual varietal bottlings.

✓ *White blend (Silvanus, Mosaico) • Pinot Grigio (Ramato) • Lagrein (Home Farm Vineyard)*

DR KONSTANTIN FRANK

Hammondsport

★★

Back in the 1950s, Konstantin Frank was mocked for holding the view that *vinifera* grapes could survive, much less prosper, in New York. He opened the winery in 1962 and proved them all wrong, and today the old pioneering winery is still running at the front of the pack.

✓ *Riesling (entire range) • Rkatsiteli • Saperavi*

ELEMENT WINERY

Geneva

★★

Master Sommelier Christopher Bates brings a unique perspective and a willingly experimental attitude to Finger Lakes winemaking. Check out his small restaurant, FLX Wienery, when you visit the Finger Lakes.

✓ *Syrah • Riesling • Gamay*

FOX RUN

Penn Yan

★★

Scott Osborn and his family planted this spot in 1984, and his winemaker, Peter Bell, is an old and true hand with reds and whites. His Cabernet Franc is probably the best in the state, and he makes numerous fantastic Rieslings.

✓ *Riesling (all types) • Cabernet Franc (Finger Lakes, Single Barrel)*

GLENORA WINE CELLARS

Dundee

★Ⓥ

Named after the nearby Glenora waterfall, this winery was the first to open on Seneca Lake, way back in 1977. They consistently produce crisp and stylish wines.

✓ *Riesling • Seyval Blanc • Sparkling Wine (Vintage Blanc de Blancs, Vintage Brut)*

GOOSE WATCH WINERY

Romulus

★☆

Under the same ownership as Swedish Hill and Penguin Bay wineries, Goose Watch has lovely Riesling, as well as the best Diamond *(labrusca)* around – not that there is much competition. They also make lovely wines from hybrids such as Noiret, Chambourcin, and Melody, as well as the less common *vinifera* grapes Lemberger and Viognier.

✓ *Lemberger • Viognier • Diamond • Melody*

CASTELLO DI BORGHESE VINEYARDS

Cutchogue

★☆

Purchased as Hargrave Vineyards and re-named in 1999, this is Long Island's first vineyard (1973). Giovanni Borghese's parents, who were running the estate, passed away in 2014, and he has ably taken on the mantle.

✓ *Cabernet Franc (Estate, Reserve)*

HAZLITT'S 1852 VINEYARDS

Hector

★

In 1985 Jerry and Elaine Hazlitt founded this vineyard on land that had been in the Hazlitt family since 1852 (hence the name). They have the clever ability to make both sweet and popular wines, and well-crafted dry and off-dry wines.

✓ *Riesling (entire range)*

HERON HILL WINERY

Hammondsport

★

The name Heron Hill is a flight of fancy on the part of advertising copy writer Peter Johnstone, who so liked the Finger Lakes on a visit in 1968 that he stayed, set up in business with major shareholder John Ingle, and turned winemaker in 1977. Ingle still owns this beautiful winery, tasting room, and café on Keuka Lake, along with two other tasting rooms around Seneca and Canandaigua lakes.

✓ *Pinot Blanc (Reserve) • Grüner Veltliner (Reserve)*

HUNT COUNTRY VINEYARDS

Branchport

★

The Hunts are now seventh-generation farmers on this pretty spot of land overlooking Keuka Lake. They are also the longest continuous producers of ice wine in the state (since 1987); their dessert wines are really stunning.

✓ *Vignoles (Late Harvest, Ice Wine)*

HEART & HANDS WINE COMPANY

Union Springs

★★

Tom and Susan Higgins' small estate offers the best-yet example of a warming climate improving New York's fortunes with the Pinot Noir variety.

✓ *Pinot Noir*

HERMANN J WIEMER VINEYARD

Dundee

★★☆

Fred Merwarth took over at this winery-when Wiemer retired in 2007, and under his watch the wines have gone from very good to superb.

✓ *Cabernet Franc (all) • Riesling (entire range)*

KEMMETER WINES

Penn Yan

★★★

Owner-winemaker Johannes Reinhardt, who hails from a winemaking family in Germany, used to make the delightful Rieslings at neighbouring Anthony Road and now is the brilliant mind behind Kemmeter Rieslings, which we can't recommend highly enough.

✓ *Entire range*

KNAPP VINEYARD

Romulus

★☆

Opened in 1984, Knapp Winery was the first winery in the Finger Lakes to make Cabernet Franc. It was also the first to open a winery restaurant (back in 1992), and it remains skilled at both tasks.

✓ *Cabernet Franc • Riesling • Vignoles (Dry)*

LAKEWOOD WINERY

Watkins Glen

★☆

Chris Stamp's family has been farming in this spot since the 1950s, and in the 1980s they got serious about making *vinifera* wine. They have Riesling down cold.

✓ *Riesling (all types) • Dessert (Glaciavinum)*

LAMOUREAUX LANDING WINE CELLARS

Lodi

★☆

Mark Wagner's family have been in the Finger Lakes grape business since 1949; the next generation is already groomed to take over the winery at the right time.

✓ *Riesling (Yellow Dog, Red Oak, Round Rock)*

LENZ

Peconic

★

While running a restaurant, Patricia and Peter Lenz developed a passion for wine, which led them to establish this winery on a potato farm. Fellow winery owners Peter and Deborah Carroll bought out the Lenzes in 1994.

✓ *Sparkling (Cuvee, Cuvee RD) • Chardonnay • Merlot*

MACARI

Mattituck

★★☆

Growing up, Joseph Macari saw his father and grandfather making wine in their basement in Queens. Years later he bought a 200-hectare (500-acre) former potato farm on Long Island Sound and decades after that, he planted grapevines. In 1995 he moved his family out to the farm to fulfil his dream. He then learned of biodynamic viticulture from California's late, great Alan York and Chile's Alvaro Espinoza, so Macari's approach remains ecological and holistic.

✓ *Merlot (Reserve) • White blend (Dos Aguas White) • Bordeaux blend (Alexandria, Bergen Road, Sette) • Cabernet Franc • Pinot Meunier*

MARTHA CLARA VINEYARDS

Riverhead

The first vintage was 1998, and Martha Clara continues today under the new ownership of the Rivero-González family, wine producers based in Coahuila, Mexico. The wines have been made at a custom-crush facility, so the new owners plan to build their own winery to supplement the attractive tasting room.

✓ *Riesling (Estate Reserve) • Sémillon-Sauvignon Blanc*

McGREGOR VINEYARD

Dundee

★

The McGregors established their *vinifera* vineyard in 1971, making it one of the first vineyards in New York devoted to *Vitis vinifera*. The winery was established in 1980, and remains family owned today.

✓ *Riesling (Dry) • Saperavi • Sparkling wine (Blanc de Noir, Blanc de Blanc)*

PALMER VINEYARDS

Riverhead

★

Bob Palmer converted a potato and pumpkin farm into a vineyard in 1983, and built a winery three years later, quickly producing attention-grabbing wines. One of the North Fork's original wineries, Palmer was acquired in 2018 by the Massoud family, who also own Paumanok, which was established the same year as Palmer. The vineyard has diversified into 20-plus varietals, sustainably farmed in the unique maritime climate of the East End of Long Island

✓ *Cabernet Franc • Albariño • Riesling*

PAUMANOK VINEYARDS

Aquebogue

★★

Founded in 1983, this 51-hectare (127-acre) estate is owned and run by Ursula and Charles Massoud and their three sons.

✓ *Chenin Blanc • Cabernet Franc (Grand Reserve) • Merlot (Grand Reserve)*

PELLEGRINI VINEYARDS

Cutchogue

★

Pellegrini Vineyards has been around for nearly 40 years and has consistently been an innovator both in the vineyard and winery. The current winemaker shares a common history of Long Island wine as well; he is Zanda Hargrave, the son of Louisa and Alex Hargrave, pioneers of Long Island viticulture.

✓ *Cabernet Franc • Sauvignon Blanc*

PINDAR VINEYARDS

Peconic

★

With the intention of growing grapes, Dr Herodotus Damianos converted a potato farm on Long Island into a vineyard. In 1982 he built a winery, named it after the Greek lyric poet, and it is now Long Island's largest producer. It is still family owned and run.

✓ *Sparkling (Cuvee Rare) • Chardonnay • Gamay Noir*

RAPHAEL

Peconic

★☆

From its establishment in 1996, it was obvious that Raphael had the cash and intent to make it one of the top two or three East Coast wineries. For a time, Paul Pontallier of Château Margaux was consulting, but the winery's full potential is still to be realized.

✓ *Cabernet Franc • Classic red blend (La Fontana) • Sauvignon Blanc*

HERON HILL WINERY IN HAMMONDSPORT, NEW YORK
Heron Hill lies along the Keuka Lake Wine Trail, which winds its way through the Finger Lakes wine country, one of New York's best-known wine-producing regions.

RAVINES WINE CELLAR
Geneva
★★

European-born winemaker Morten Hallgren and his chef wife, Lisa, rub a two-decades-long project that offers good red wines but excellent Rieslings.

 Riesling (entire range)

RED NEWT CELLARS
Penn Yan
★★

Founded by David and Debra Whiting in 1998, Red Newt Cellars has since become one of the Finger Lakes region's preeminent Riesling producers.

✓ *Riesling (Finger Lakes, Muse Dry)*

RED TAIL RIDGE WINERY
Penn Yan
★☆

Nancy Irelan and husband, Michael Schnelle, make deft and balanced Rieslings.

✓ *Riesling (Finger Lakes, Muse Dry)*

SHELDRAKE POINT
Ovid
★☆

Named for a prominent point of land jutting into Cayuga Lake, this was orchard land for a century before a small group of wine enthusiasts bought the property and created the winery.

✓ *Gewürztraminer • Riesling (entire range) • Muscat Ottonel*

SILVER THREAD VINEYARD
Lodi
★ Ⓥ

The original owner-winemaker, Richard Figiel, was one of the first organic producers in the Atlantic Northeast. The new owners, Paul and Shannon Brock, have continued in the same vein; since 2015 the winery has been generating 100 per cent of its energy needs from a 28-kilowatt solar power system located on-site.

✓ *Riesling (all types)*

SWEDISH HILL WINERY
Romulus
★☆

Founded in 1986 by Dick and Cindy Peterson and run today by their highly skilled and experienced son and his wife, Swedish Hill has become one of the largest wineries in the region, producing nearly 60,000 cases of wine a year. *See also Goose Watch.*

✓ *Riesling (entire range) • Sparkling wine (Riesling Cuvée, Spumante Blush, Vintage Brut) • Cabernet Franc*

WAGNER VINEYARDS
Lodi
★☆

The Wagners got into the wine grape growing business in 1972; four years later they built a fun, quirky eight-sided winery building. Their consistently high-quality wines are actually better than ever after more than four decades.

✓ *Gewürztraminer (Dry) • Riesling (Dry, Icewine) • Sparkling (Riesling Champagne)*

WÖLFFER ESTATE
Sagaponack
★★☆

Wölffer has proved that Long Island's South Fork has the potential to produce good wines as well as the North Fork. Founded in 1988 by Christian Wölffer, the estate today is owned and operated by his children, Marc and Joey, and winemaker-partner Roman Roth. They cultivate 150-plus hectares (400 acres) on the South Fork, 20-plus hectares (50 acres) on the North Fork, and vineyard acreage in Argentina and Mallorca. Wölffer made history with its 2000 Premier Cru Merlot when it was released as Long Island's first 100-dollar wine. Winemaker Roman Roth also has some first-rate wines under the label Grapes of Roth.

✓ *Merlot • Classic red blend • Rosé • Sparkling (Blanc de Blancs)*

OHIO

DEBONNÉ VINEYARDS
Madison
★☆

The state's largest estate winery, Debonné was first farmed by the current owner's grandfather in 1916. The vineyard was greatly expanded in the 1960s. The winery was created in 1972, and the family continues its reign in the state.

✓ *Pinot Gris • Cabernet Franc*

FERRANTE WINERY
Geneva
★★

The Ferrante Family has been producing wine since 1937, and way back in the 1970s Peter Ferrante built an up-to-date winery in the family vineyard. In 1994, after a fire destroyed it, the third generation built a new facility . . . and still the wines continue to improve.

✓ *Riesling (Golden Bunches, GRV) • Grüner Veltliner • Dolcetto*

GERVASI VINEYARDS
Canton
★

This really well-designed and beautiful facility boasts some solid estate wines from hybrid vines, select Ohio *vinifera*, and out-of-state grape sources.

✓ *Malvasia Bianca • Chardonnay (Bellina)*

HARPERSFIELD VINEYARD
Geneva
★★

Patricia Ribic travelled in Europe to visit her husband's family in Slovenia and was smitten by classic wines. In 2005, she and her husband bought the Harpersfield Winery, which had originally opened in 1979. The winery works only with *vinifera* grapes from Ohio, unlike many others.

✓ *Chardonnay (Fût de Chêne) • Pinot Gris (St. Fiacre) • Pinot Noir (Clos mes Amis)*

HENKE WINERY
Cincinnati
★

This is an urban winery working with both Ohio grapes from local growers and out-of-state grape sources, but everything here is of very good quality.

✓ *Sparkling Chardonnay • Seyval Blanc • Vidal Blanc*

KOSICEK VINEYARD
Geneva
★★

The Kosicek family has been grape farming in northeast Ohio since 1929. In 2012 third-generation farmer Tony Kosicek and his wife, Mauri, established the winery using grapes from areas along the Lake Erie shore.

✓ *Traminette • Sparkling (Rhapsody)*

M CELLARS
Geneva
★★

Tara and Matt Meinke's passion drove them to buy property and plant it in 2008; this small estate has now become one of the standard-bearers for excellent Grand River Valley wines.

✓ *Pinot Gris • Rkatsiteli • Riesling*

MERANDA NIXON WINERY
Ripley
★★

Both Seth Meranda and his late wife, Tina Nixon, came from farming and agricultural backgrounds. They opened the winery in 2007 in a Ohio Valley area that was one of the largest grape-growing regions in the United States during the 19th century. Seth and his second wife, Maura, now operate the winery with help from their four children, producing red, whites, and even sparkling wines.

✓ *Sparkling Catawba • Chardonnay • Cabernet Franc*

SOUTH RIVER VINEYARD
Geneva
★

This "church winery" exists because of a chance encounter: South River Vineyard owner Gene Sigel was travelling when he spotted an abandoned Methodist chapel. He asked a neighbour about it, who told him he could have the building for the trouble of dismantling it. Sigel moved the building, piece by piece, to his place in Geneva.

✓ *Vidal Blanc • Pinot Noir*

ST JOSEPH VINEYARD
Madison
★

Doreen and Art Pietrzyk opened their winery more than 20 years ago, though they were growing wine grapes nearly a decade before that. There may be a third generation in the wings; grandson Kevin's school science project is on the effects of yeast strains on fermentation.

✓ *Vidal Blanc • Pinot Noir*

VALLEY VINEYARDS
Morrow
★☆

The third generation of Schuchters is in charge here now, and they continue to make very good wines.

✓ *Cabernet Franc*

PENNSYLVANIA

ALLEGRO VINEYARDS
Brogue
★☆

Carl Helrich and wife, Kris Miller, purchased Allegro Winery from founders John and Tim Crouch; the estate's vines are among the oldest in Pennsylvania, with the first ones planted in 1973.

✓ *Cabernet Sauvignon-Chambourcin • Geurztraminer-Traminette*

CHADDSFORD WINERY
Chadds Ford
★

Eric and Lee Miller created Chaddsford Winery in 1982, and they were definitely ahead of their time. They were able to convert this Brandywine Valley winery to one of the most important in the Northeast. But they retired a few years ago, and the new owners have wisely chosen to begin turning the winery back to dry wines, much as Eric and Lee had done in the beginning.

✓ *Red blend (Harbinger) • White blend (The White Standard)*

CLOVER HILL VINEYARDS
Breinigsville
★☆

John and Pat Skrip began planting grapes as a hobby at their home in the early 1970s. In 1985 they opened Clover Hill Vineyards and since that time have been joined by two of their children. John Skrip III studied oenology at Fresno State University in California. Daughter Kari studied wine marketing at the University of Adelaide in Australia.

✓ *Chambourcin (Frizzante) • Vidal Blanc (Vidal Verde)*

CROSSINGS VINEYARDS
Newtown
★

Ten-year-old Tom Carroll Jr told his parents that their new historic home on hectares 8 acres (20 acres) in Bucks County "would make a great vineyard". Tom and his parents started Crossing Vineyards and Winery 15 years later. The first vines were planted in April 2002.

✓ *Chambourcin (Reserve) • Vidal Blanc • Pinot Gris (Pinot Grigio)*

GALEN GLEN VINEYARDS
Andreas
★☆

Galen Troxell and his winemaker wife, Sarah Troxell, have been picking up gold medals for some time at this substantial 20,000-case winery.

✓ *Riesling • Grüner Veltliner*

MOUNT NITTANY VINEYARDS
Centre Hall
★

Winery founder Joe Carroll was raised an Indiana farm boy. He and wife, Betty, bought a 26-hectare (65-acre) site on the side of Mount Nittany in 1983, and since he'd been an amateur winemaker for years, they decided to plant a vineyard. By 1987 they had launched a winery. Now run by daughter and son-in-law, Linda and Steve Weaver, Mount Nittany keeps getting better at its craft.

 Rosé (Linden Vale) • Geisenheim • Chardonnay (Reserve)

VA LA VINEYARDS
Avondale
★★

Anthony Vietri is one of the more original winemakers in America. He has more than two dozen French and Italian varietals, and he creates fascinating blends from them, such as La Prima Donna (made from Tocai, Malvasia Bianca, Fiano, Pinot Grigio, and Petit Manseng).

 Entire range

VOX VINETI VINEYARDS
Breinigsville
★★

These are really interesting wines, particularly from Galloping Cat Vineyard, a small 2-hectare (5-acre) plot planted with a variety of grapes. The Discantus blend is typical of the style of the winery: 40 per cent Barbera, 33 per cent Cabernet Franc, 14 per cent Cabernet Sauvignon, and 13 per cent Nebbiolo.

✓ *Entire range*

RHODE ISLAND

CAROLYN'S SAKONNET VINEYARDS
Little Compton
★★

This was the first winery to open in Rhode Island after Prohibition. The winery has changed hands a couple of times since then but has remained a financial and critical success throughout. Current owner is Carolyn Rafaelian, a jewellery chain owner, and she has appended her name to the well-established estate. The viticultural and winemaking team have been working the estate for decades, so things should remain consistent.

 Gewürztraminer • Vidal Blanc

VERMONT

LINCOLN PEAK VINEYARD
New Haven
★☆

Owners Chris and Michaela Granstrom started farming his land in 1981, first with apples, then strawberries, and those fruit crops paid the bills. But after receiving some cold-climate grape varieties from a colleague in Minnesota, they decided to focus upon vine-growing. They've led the way with grapes like the hybrid Marquette, and their daughter, Sara, is now part of the team.

 Marquette • La Crescent

SHELBURNE VINEYARD
Shelburne
★

In the early 1980s, Ken Albert, an IBM engineer, was inspired by the new Quebec wine industry and so he started a small vineyard with the help of Scott Prom, a

Shelburne neighbour. It remains small, but they've had very notable success with cold-hardy varieties such as Marquette.

✓ *Marquette*

VIRGINIA

BARBOURSVILLE VINEYARDS
Barboursville
★★★

Once the property of James Barbour, a former governor of Virginia, this historic estate located between Monticello and Montpelier, was developed by Italian firm Zonin in 1976. They might have been more lucky than wise, but the wines have been good from the beginning. Perhaps it is because winemaker Luca Paschina has been at the winery for so long, but the wines remain consistently entertaining, just like Luca. The Octagon requires ageing after release.

 Barbera (Reserve) • Classic red blend (Octagon) • Nebbiolo (Reserve)

BREAUX VINEYARDS
Purcellville
★★

Originally amateur winemakers owning a small parcel of vines, Paul and Alexis Breaux have developed a small vineyard purchase into a very reliable producer.

✓ *Cabernet Franc • Nebbiolo*

BOXWOOD ESTATE WINERY
Middleburg
★★★

A former NFL franchise owner has built a beautiful estate focusing upon Bordeaux varieties. He is aided by famed consultant Stephane Derenoncourt.

 Cabernet Franc • Nebbiolo

CHATEAU MORISETTE
Floyd
★★

A long-established (1980) winery, but there are still new things happening here: a new Chambourcin and a lovely Petit Manseng are in the fold.

✓ *Chambourcin • Viognier (Reserve) • Petit Manseng*

CHRYSALIS VINEYARDS
Middleburg
★★☆

Chrysalis produces the exemplary Norton, and owner-winemaker Jenny McCloud loves discussing its history. McCloud is one of the country's best practitioners of balanced elegant Norton.

 Norton (Estate, Locksley Reserve, Buttorfleoge) • Tannat

LINDEN VINEYARDS
Linden
★★☆

Founder Jim Law started things back in 1983, and his winery remains one of the stars of East Coast winedom. His dessert wine is stunning.

 Petit Manseng (Late Harvest) • Petit Verdot • Bordeaux blend (Avenius) • Chardonnay (Hardscrabble)

RDV
Delaplane
★★★

Rutger de Vink has created arguably the best winery on the Atlantic Coast. Look for these wines; they're worth the search.

 Bordeaux blend (Rendezvous, Lost Mountain)

HORTON
Gordonsville
★★

Dennis Horton was once known as Virginia's lone Rhône Ranger, but he now performs best with grapes that are more at home in the southwest of France. Even though there is an inconsistency in Horton's Norton, I have to recommend it because the highlights are minor classics, and everyone just loves the rhyme.

 Cabernet Franc • Norton • Petit Manseng • Tannat

JEFFERSON VINEYARDS
Charlottesville
★★

Located just outside Charlottesville, a mile from Monticello on vineyard sites chosen and planted by Thomas Jefferson and Philip Mazzei in 1774, but obviously more successful, as these bear fruit.

 Merlot (Reserve) • Viognier • Pinot Gris (Skin-fermented) • Meritage

KESWICK VINEYARDS
Keswick
★

Al and Cindy Schornberg planted a vineyard at the Edgewood Estate, part of the original 1727 Nicholas Meriwether Crown Grant, the site of important Revolutionary and Civil War battles. South African winemaker Stephen Barnard, who started at top spots like Groot Constantia and Flagstone, has been at the helm for Keswick for 15 years

 Cabernet Franc (Block 7 Reserve) • Viognier (Reserve) • Touriga (Reserve)

KING FAMILY VINEYARDS
Crozet
★☆

Michael Shaps is considered one of Virginia's most talented winemakers, and King Vineyards is one of the most beautiful locations in the scenic state. One barrel of 2002 Merlot was for Tom Stevenson "by leaps and bounds the greatest red wine component I have ever tasted from any Virginia winery". A few years later, Mathieu Finot took over the winemaking, and the family is still in charge of this vast and lovely place.

 Cabernet Franc • Merlot • Meritage • Viognier

PRINCE MICHEL VINEYARDS & WINERY
Culpeper
★

Founded by Jean Leducq, this is now owned by Kristin Easter. Supplemented with grapes from the company's own vineyards in Napa, the wines are also sold under the Rapidan River labels.

✓ *Chardonnay • Bordeaux blend (Symbius)*

RAPPAHANNOCK CELLARS
Huntly
★

John Delmare relocated to the beautiful Blue Ridge Mountains area from California after making wine in the Santa Cruz Mountains. Some of his children are deeply involved in the winery and have been since its founding in 2000.

 Cabernet Franc (Black Label) • Classic red blend (Right Bank Meritage)

ROCKBRIDGE CELLARS
Raphine
★

Owner-winemaker Shepherd Rouse earned a masters degree in oenology at UC Davis while at Schramsberg, Chateau St Jean, and Carneros Creek. Returning to his native Virginia in 1986, he was the winemaker at Montdomaine Cellars when its 1990 Cabernet Sauvignon won the 1993 Virginia Governor's Cup. Rouse established Rockbridge Vineyard in 1988, and he continues to run this worthy winery, along with his wife, Jane.

 Meritage (DeChiel Reserve) • Traminette • Vignoles

TARARA
Leesburg
★☆

In 1985 Whitie Hubert retired on 192 hectares (475 acres) on the Potomac River in rural Loudoun County. The winery he established in 1989 on a 16-hectare (40-acre) vineyard has always been in the in the top ranks.

 Viognier • Petit Manseng (Late Harvest)

VALHALLA VINEYARDS
Roanoke
★☆

Since the Vascick family pulled out the peach trees in this orchard and established Valhalla in 1999, their wines have won hundreds of medals in national and international competitions.

 Classic red blend (Valkyrie) • Alicante Bouschet

VERITAS
Afton
★★

British-born neurologist Andrew Hodson and his wife, Patricia, established a winery, with the help of daughter Emily, who then went on to earn a masters in oenology at Virginia Tech. Their other children are crucial to this stylish project as well.

✓ *Petit Manseng • Petit Verdot • Claret*

WHITE HALL
White Hall
★★

In 1991 Tony and Edie Champ planted a small plot with Chardonnay, Cabernet Sauvignon, Cabernet Franc, and Merlot near the Blue Ridge Mountains. They later added other varieties and now have 19 hectares (48 acres) of vines and a state-of-the-art winery.

 Petit Verdot • Viognier • Petit Manseng (Soliterre)

OTHER WINEMAKING AREAS OF
THE UNITED STATES

SOUTHERN STATES

ALABAMA

The first commercially cultivated vines were planted here in the 1830s, and the state had a flourishing wine industry prior to Prohibition. Yet whether it is due to religious enmity towards alcohol or the climate's hostility toward viticulture, Alabama has malingered as the rest of the United States has joined the global wine market. There are 19 wineries supported by only 60 hectares (150 acres) of total vineyards in the state, much of it Muscadine grapevines. Other varieties are trucked in from out of state.

ARKANSAS

More than 100 wineries sprang up in Arkansas after Prohibition, but lack of winemaking skills meant that very few survived (there are 19 as of 2019). A fortified wine made by Cowie Wine Cellars and called Robert's Port is well respected, as is this winery's Cynthiana.

ALTUS AVA

This plateau lies between the Arkansas River bottomlands and the climatically protective high peaks of the Boston Mountains. This AVA sits in the middle of the Arkansas Mountain AVA; there are five wineries here as of 2019.

ARKANSAS MOUNTAIN AVA

Covering a vast area in the mountainous region of Arkansas, this AVA sits within the even larger Ozark Mountain AVA. The Arkansas Mountains moderate winter temperatures and provide shelter from violent northerly winds and sudden changes in temperature. Hybrid and native varieties are at home here, but like in much of the South, the grapevine has struggled for acceptance.

OZARK MOUNTAIN AVA

See entry under Missouri.

FLORIDA

The very first wine made in what is now the USA was made by French Huguenots at Fort Caroline in 1564, but the only viticulture of any significance occurred after the Civil War, when Florida was settled by dispossessed Northerners and Southerners alike. The Northerners planted their beloved *labrusca* vines, while the Southerners competed with their even more bizarre Muscadine. Some 200 hectares (500 acres) were cultivated towards the end of the 19th century, and it was the Southerners who won this battle of the grapes, as *labrusca* succumbed to Florida's extreme heat and humidity. Amid this North-South viticultural divide, a Frenchman by the name of Emile Dubois produced Norton and Cynthiana (then believed to be two different varieties) of sufficient quality to win a medal at the Paris Exposition of 1900, but virtually all of Florida's vineyards disappeared shortly thereafter, because it was widely acknowledged that this semi-tropical state was better suited to citrus groves. Not surprisingly, a lot of "orange wine" is made today, not to mention the wines made from Key lime, mango, and passionfruit. Odd though these concoctions might be, they are less intense than the wine made from the local Muscadine grapes. Since the 1950s, the University of Florida has released a number of complex Pierce's disease-resistant hybrids created by crossing French hybrids with American ones, with Lake Emerald (developed in 1954), Stover (1968), Conquistador (1983), Suwannee (1983), and Blanc du Bois (1987) being the most widely planted. Set against this backdrop, it is difficult to understand why there has been an explosion of wineries opening up of late. There were just 6 in 2000, yet no fewer than 85 by 2019.

GEORGIA

Georgia has cultivated grapes since 1733 and by 1880 was the sixth-largest wine-growing state. Even in modern times, its wines have been based on native Muscadine grapes until fairly recently, and those are an acquired taste. Since the late 1990s, however, wineries such as Chateau Elan, Creekstone, Frogtown Cellars, Habersham, Tiger Mountain, Three Sisters Vineyard, and Wolf Mountain Vineyards have used *vinifera* grapes planted at an altitude of 540 to 600 metres (1,800 to 2,000 feet) in northern Georgia, where humidity is less of a problem, and cool nights help preserve acidity.

DAHLONEGA PLATEAU AVA

This is defined as a long, narrow plateau leaning northeast-southwest in the northern foothills of the Georgia Piedmont. The gently sloping hills and valley floors are more elevated in the north, east, and southeast portions of the AVA. With about 45 hectares (110 acres) planted out of the 345-square-kilometre (133-square-mile) area, there is plenty of space, but weather considerations play a strong role in site selection. The AVA includes some of the more notable Georgia wineries such as Frogtown, Montaluce, Three Sisters, and Wolf Mountain.

UPPER HIWASSEE HIGHLANDS AVA

Georgia and North Carolina

Where the Blue Ridge Mountains taper from North Caroline into Georgia and, following the Hiwassee River, this AVA spans 1,787 square kilometres (690 square miles) with five wineries on the North Carolina side and four in the Georgia portion of the AVA. At present only 22 hectares (54 acres) of vineyards are planted, somewhat equally split between *vinifera* and hybrids. Yet although the climate is warm and humid, the elevation of 600 to 730 metres (2,000 to 2,400 feet) offers cooling nights.

KENTUCKY

In 1799, John James Dufour set up the Kentucky Vineyard Society, which expected to raise $10,000 in capital through the sale of 200 shares at $50 each. Without waiting for the subscriptions certificates to be printed, let alone sold, Dufour had purchased 253 hectares (633 acres) of land at Big Bend on the banks of the Kentucky River and began planting the optimistically named First Vineyard in 1798. Almost all the vines had failed by 1802, however. Like so many early ventures, the cause was probably Pierce's disease, phylloxera, or both. Another more successful early Kentucky venture was planted by Colonel James Taylor at Newport. This vineyard was described by John Melish, an English traveller, as "the finest that I have yet seen in America". In 2019 there were 75 wineries in Kentucky; the most notable include Chrisman Mill and Equus Run.

OHIO RIVER VALLEY AVA

See entry under Ohio.

LOUISIANA

Jesuit priests made altar wine here as early as 1750, and by 1775 an Englishman we know of only as Colonel Ball produced a wine from the banks of the Mississippi, north of New Orleans, that was considered worthy of sending to George III. Sadly, the vineyard died after the colonel and his family were massacred by Indians, according to the August 1822 and December 1826 issues of *American Farmer*. Louisiana's climate is difficult for any vine other than Scuppernong, so grapes imported from California, or other fruit such as oranges, were relied upon for winemaking. There are more than a dozen wineries in the state today.

MISSISSIPPI DELTA AVA

See entry under Mississippi.

MISSISSIPPI

The pattern here is very similar to that in Kentucky. Heavy and restrictive taxes were reduced in 1984, however, when the Mississippi Delta was granted its own AVA. By 2019, there were three wineries, but none are outstanding at the time of writing.

MISSISSIPPI DELTA AVA

A fertile alluvial plain with loess bluffs that abruptly rise 30 metres (100 feet) along the entire eastern side of the Delta that also covers parts of Louisiana and Tennessee.

THE MOTHERVINE SCUPPERNONG WINE
Duplin Winery, North Carolina's oldest operating winery, produces this sweet white wine from clippings of the "Mother Vine" on Roanoke Island, the oldest-living grapevine in the state at over 400 years old. The Scuppernong grape, which is native to the American South, was once a mainstay variety of the region's wine producers.

NORTH CAROLINA

Both the first and second colonies to settle this area were expected to produce wine, and although these hopes are documented, no evidence of any success exists. The hot climate and viticultural challenges dictated that most wines were produced from the hardy, native Scuppernong vine, which offers the most extraordinary, weirdly exotic, and unusual wine from large, cherry-like grapes. It was these grapes that made North Carolina the largest wine-producing state in 1840. It was also from Scuppernong grapes that "Captain" Paul Garrett first made his best-selling Virginia Dare wine from a string of East Coast wineries in the 1900s. After Prohibition, he created the first singing commercial for wine, and Virginia Dare once again became the country's biggest-selling wine. Scuppernong grapes were scarce, however, and despite successfully encouraging many growers in other Southern states to grow this vine, he had to blend in more California wine, and in the process Virginia Dare lost its distinctive (and, by today's standards, not very attractive) character, so sales dropped. There are 150-plus wineries in North Carolina, and Scuppernong is almost forgotten. North Carolina has grown significantly; it vies with Virginia to be the number-seven-producing state. Like its Atlantic Coast neighbours, it has only recently grown reliable as a *vinifera* producer; there are tasty examples of Cabernet Franc, Nebbiolo, Vignoles, Norton, and even sparkling wine.

APPALACHIAN HIGH COUNTRY AVA
Virginia, North Carolina, and Tennessee

This large and somewhat amorphous AVA wanders into three states and covers 6,216 square kilometres (2,400 square miles) in portions of northeastern Tennessee, northwestern North Carolina, and southwestern Virginia;. Vineyards are planted at elevations between 700 and 1,400 metres (2,290 and 4,630 feet) and more than half of the vineyards are on slopes with angles of 30 per cent or more. There are 10 wineries currently in the AVA.

CREST OF THE BLUE RIDGE HENDERSON COUNTY AVA

This 2019-approved AVA is located in southwest North Carolina. The Crest of the Blue Ridge Mountains covers 557 square kilometres (215 square miles). There are currently 14 vineyards with a total of 38 hectares (70 acres) of vineyards. The name refers to the way in which the Eastern Continental Divide splits two portions – the Blue Ridge Escarpment (the AVA's south and east portions) and the Blue Ridge Plateau (its north and west portions). Almost 80 per cent of the area is planted to *vinifera*.

HAW RIVER VALLEY AVA

Though this has been a traditional area for Muscadine grapes, there is much about the hills of the Haw River Valley AVA that is encouraging for *vinifera* production. The landscape can be wet, but cool too, and the soils are very well drained with low fertility. With only 24 hectares (60 acres) of grapes in this 2,250-square-kilometre (870-square-mile) AVA, there is room for growth and experimentation.

SWAN CREEK AVA

Partially found within the Yadkin Valley AVA, this is a slightly cooler portion of that Valley. This 470-square-kilometre (180-square-mile) appellation is distinguished by its loamy soil with schist and mica. There are currently seven wineries there.

UPPER HIWASSEE HIGHLANDS AVA
See entry under Georgia.

YADKIN VALLEY AVA

North Carolina's first official appellation, Yadkin Valley was recognized as an AVA in 2003. This viticultural area covers 566,800 hectares (1.4 million acres) spanning seven counties along the Yadkin River. Elevations range from 243 to 364 metres (800 to 1,200 feet), with *vinifera* making up roughly 70 percent of the area's grapes (Muscadines comprise the remainder). Chardonnay, Riesling, Cabernet Sauvignon, and Merlot fare best over the long, warm growing season and mild winters on the Yadkin Valley's gently rolling hills of well-drained clay and loam soils. Swan Creek AVA overlaps with the southern portion of Yadkin Valley.

SOUTH CAROLINA

Winemaking began here as long ago as 1748, when Robert Thorpe was paid the handsome prize of £500 by the Commons of South Carolina for being "the first person who shall make the first pipe of good, strong-bodied, merchantable wine of the growth and culture of his own plantation". The state was slow to recover from Prohibition, however, with only a few wineries in existence till the 2000s – today there are 25. The grapes used were once invariably Muscadine, but now hybrid varieties are ascendant; there are many locally grown fruit wines.

TENNESSEE

The best-known episode in this state's wine history took place in the 1850s, when a French, German, and Italian Catholic community called Vineland, or Vinona, established terraced vineyards and made wine, which was sold to the Scottish Presbyterians in the southern Appalachians. Unfortunately, the Civil War put an end to both the activity and the community. In the 1880s, so-called Victorian Utopians established themselves at Rugby, and German settlers did likewise in the Allardt community of Fentress County. Although Tennessee still has some "dry" counties, the state now has 60-plus wineries. Along with hybrid vines, *vinifera* vines have been planted here, but those can struggle under the severe winter conditions. Most producers purchase grapes (*vinifera*, hybrid, and native varieties such as Scuppernong and Concord) from both in-state and out-of-state sources. Fruit wines are commonly encountered, and some mixed grape and fruit wines are also made. Pure Tennessee grape wines are produced, and there are good wines to be found from Arrington, Beachhaven, Beans Creek, Blue Goose, Hillside Winery, and the Winery at Seven Springs Farm, among others.

APPALACHIAN HIGH COUNTRY AVA
See entry under North Carolina.

MISSISSIPPI DELTA AVA
See entry under Mississippi.

CENTRAL STATES

ILLINOIS

By 1847 there were more than 240 hectares (600 acres) of vineyards on the banks of the Mississippi, supplying no fewer than 40 wineries in Nauvo. In 1851 the oldest Concord vineyard in Illinois was planted in Nauvo State Park, where the vineyard is still producing wine today, as is Baxter's, established in 1857. By 1868 Illinois was producing 852,000 litres (225,000 gallons) a year – very nearly as much as New York. Today, there are roughly 150 wineries in the state, and skill and experience are winning the day. Illinois has some Chicagoland wineries utilizing purchased grapes from either coast; of more interest are their vineyards downstate to the south. The usual Midwestern grapes prevail, particularly Chambourcin, and some of the wineries you can trust include Alto, August Hill, Blue Sky, Galena, and Prairie State; the bubblies from the Illinois Sparkling Wine Company are also delightful.

SHAWNEE HILLS AVA

Approved in 2006, this is an elevated series of plateaus between the Mississippi River and the Ohio River in southern Illinois. The wine appellation includes more than 5,500 square kilometres (2,139 square miles) and includes the vast majority of the Shawnee National Forest. Elevation ranges from 400 to 800 feet (122 to 244 metres) higher than the glaciated land to the north and the delta-coastal land to the south, with some areas above 300 metres (1,000 feet). Shawnee Hills AVA enjoys warmer winter temperatures and a longer growing season as a result of its higher elevations. There is limestone bedrock as well as thin loess soils, particularly where the AVA's approximately 120 hectares (300 acres) are planted.

UPPER MISSISSIPPI RIVER VALLEY AVA

This largest of all AVAs covers 77,477 square kilometres (29,914 square miles) or 7,859,093 hectares (19,144,960 acres). It follows the Upper Mississippi River and its tributaries in northwest Illinois, northeast Iowa, southeast Minnesota, and southwest Wisconsin. The climate is continental, at times cold and extreme, and many soils around the rivers are clay and silt loam on limestone bedrock. Perhaps that provides some common connection along with the fact that these areas didn't see glaciation during the last Ice Age.

IOWA

Hiram Barney established 40 hectares (100 acres) of vines at Keokuk in 1869, which he called White Elk Vineyards, and it produced 115,000 litres (30,000 gallons) of wine a year before succumbing to a lethal combination of disease and Prohibition. Although wineries have struggled with pesticide drift from neighbouring farms, Iowa viticulture has expanded rapidly – there are now more than 100 wineries. In the southern part of the state, Norton can perform well, but cold-climate-hardy varieties, particularly the newer ones (LaCrescent, La Crosse, GR7, Marquette, and Frontenacs Gris and Blanc) are gaining favour and are widely planted in the north. Among the 100-plus wineries in the state, Cedar Ridge, Eagle's Landing, Loess Hills, Soldier Creek, and, especially, Fireside Winery are making tasty and worthwhile wines.

LOESS HILLS DISTRICT
Iowa and Missouri

Running north and south on the western side of Iowa and leaning into northwest Missouri, this AVA covers 33,403 square kilometres (12,897 square miles) along the Big Sioux and Missouri Rivers. The eponymous wind-deposited loess soils create some rolling hills and evenly eroded, steep slopes, which can protect against frosts. The climate is cool to cold; hybrids comprise most of the wine grapes; the AVA has just 45 hectares (112 acres) under vine, and there about a dozen wineries.

UPPER MISSISSIPPI RIVER VALLEY AVA
See entry under Illinois.

KANSAS

Kansas has never been a huge winemaking state, but it did produce as many as 110,000 cases of wine a year until it voted to go "dry" in 1880. Although no modern-day wineries existed until 1990, there were 22 wineries by 2010, and Norton and Chambourcin are the best red grapes. The track record among the notable wineries is even better, with white wines from Seyval, Traminette, Valvin Muscat, Vidal Blanc, and Vignoles. Today there are more than three dozen wineries, with notables examples including Aubrey, Bluejacket Crossing, Stone Pillar, Somerset Ridge, and, especially, Holy-Field.

MINNESOTA

As in many northern US and Canadian vineyards, vines have to be buried beneath earth and/or vegetation in order to survive Minnesota's winters. And as with other Upper Plains wineries, hybrids and fruit wines are the norm. The University of Minnesota has been a crucial player in the development of new cold-climate-hardy grapevines; their successes have been as useful in central Canada as they have been for the Upper Plains. At least some of their creations were based upon earlier vines from breeder Elmer Swenson. Still, present or past University of Minnesota professors, such as Peter Hemsted, have continued Swenson's legacy; consider trying Itasca, LaCrescent, Marquette, and others from any of the state's 60-plus wineries, but particularly Alexis Bailly, Cannon River, Carlos Creek, Chankaska Creek, Crow River, Four Daughters, Morgan Creek, Northern Hollow, Parley Lake, Saint Croix, Three Oak, Two Rivers, and WineHaven.

ALEXANDRIA LAKES AVA

This AVA is bounded by six lakes, each of which is amongst the deepest in Minnesota and which provide some moderation to the extreme winters. There is no Lake Alexandria, rather the AVA is near the city of Alexandria, which is surrounded by lakes. While a bit of *vinifera* is grown, virtually all wineries and vineyards are reliant upon hybrid vines or other fruits for their wines. This AVA covers 4,400 hectares (10,880 acres), but there is currently only one major producer, Carlos Creek Winery. The region was scoured by glaciers, and the soils are loamy, deep, and well-drained.

UPPER MISSISSIPPI RIVER VALLEY AVA

See entry under Illinois.

MISSOURI

The first wines were produced in Missouri in the 1830s, and the state had a flourishing wine industry in the mid-19th century. The wine industry has exploded since the turn of the 21st century, with the number of wineries increasing from 37 in 2000 to 165 in 2019. Norton is the official state grape of Missouri but, amongst reds, Chambourcin is easier to love. Sparkling wines are increasingly successful; white wines, whether dry, sweet, or in-between, are too. Seyval Blanc and Chardonel are less popular than a few years ago; vintners are instead increasingly focusing upon Traminette, Valvin Muscat, Vidal Blanc, and Vignoles. The big-name winery is Stone Hill, but Adam Puchta, Amigoni, Augusta, Blumenhof, Fence Stile, Heinrichshaus, Hermannhof, Jowler Creek, KC Wineworks, Les Bourgeois, Montelle, Mount Pleasant, Noboleis, Pirtle, St James, Stonehaus, Van Till, and Vox are reliable too.

AUGUSTA AVA

Grape-growing in Augusta, the country's first established AVA, dates from 1860. The bowl-like ridge of hills from west to east and the Missouri River on the southern edge of the viticultural area provide a microclimate that distinguishes Augusta from the surrounding areas. As of 2019, 15 wineries are located in the AVA.

HERMANN AVA

In 1904 this area furnished 97 per cent of the wine produced in Missouri. The soils are well drained, can carry and store a good amount of water, and provide proper root development. Ten wineries lie in or around Hermann; this sits within the Ozark Mountain AVA.

LOESS HILLS DISTRICT AVA

Iowa and Missouri

Running north and south on the western side of Iowa and leaning into northwest Missouri, this AVA covers 33,403 square kilometres (12,897 square miles) along the Big Sioux and Missouri Rivers. The eponymous wind-deposited loess soils create some rolling hills to steep slopes, which can protect against frosts. The climate is cool to cold; hybrid varieties comprise most of the wine grapes, and the AVA has just 45 hectares (112 acres) under vine; there are about a dozen wineries.

OZARK HIGHLANDS AVA

This AVA is located within the much larger Ozark Mountain AVA. This area's climate is frost free and relatively cool during spring and autumn, compared with surrounding areas. The eight wineries within it include St. James and Heinrichshaus.

OZARK MOUNTAIN AVA

Missouri, Arkansas, and Oklahoma

Five major rivers make up this AVA's boundaries: the Mississippi, the Missouri, the Osage, the Neosho, and the Arkansas. The Ozark Mountain appellation straddles southern Missouri, northwest Arkansas, and northeast Oklahoma and is hilly-to-mountainous. It is the sixth-largest of all AVA in the United States.

NEBRASKA

The earliest-known producer in Nebraska was Peter Pitz, who cultivated 5 hectares (12 acres) of vines near Plattsmouth in the 1890s and claimed to make red, white, and yellow wine. The first vineyard in modern times was planted at Pierce by Ed Swanson of Cuthills Vineyards (now sadly closed). Nebraska has about three dozen wineries in 2019; some are quite reliable, especially Cellar 426, Feather River, Glacial Till, James Arthur, Mac's Creek, Milletta Vista, and Whiskey Run Creek. Hybrids, and, particularly, the cold-climate varieties (Marquette, Frontenac, LaCrescent, La Crosse, and the like) are the most planted.

WISCONSIN

The present-day Wollersheim Winery is the historic site selected by the famous Agoston Haraszthy to plant his Wisconsin vineyard in the 1840s. He later moved on to San Diego, after experiencing several years of winter damage on his vines. The Haraszthy vineyard was taken over by Peter Kehl, a German immigrant, who built the existing winery during the Civil War period. Until relatively recently, this state had a rather more considerable reputation for its cherry, apple, and other fruit wines than for those made from grapes. Fruit wines still dominate in Wisconsin, but grape hybrids are beginning to gain ground, with some *vinifera* varieties, such as Riesling, just starting to emerge. The soils in the region are predominantly clayey loams. There were 80 wineries in the state as of 2019.

LAKE WISCONSIN AVA

Lake Wisconsin and the Wisconsin River moderate the climate locally, providing winter temperatures that are several degrees higher than those found to the north, south, and west of this region, while the good air circulation helps to prevent both frost and rot.

UPPER MISSISSIPPI RIVER VALLEY AVA

See entry under Illinois.

WISCONSIN LEDGE AVA

With low fertility and well-drained glacial soils over limestone bedrock, the soils here are excellent for high-quality wines – the challenge her, as in all northern sites, is the climate. But proximity to Lake Michigan helps dry the grapes in humid summers and warm them in ice-cold winters. It covers 9,850 square kilometres (3,800 square miles), with 14 wineries and just over 120 hectares (300 acres) of vineyard. Despite the Great Lakes' assistance, hybrids are best. The Ledge is named for the westernmost portion of the 1050-kilometre (650-mile) Niagara Escarpment, which also forms the cliff that underpins Niagara Falls.

WESTERN STATES

ALASKA

How do they do it? Well, not with Alaska-grown grapes, that's for certain. There are nine wineries in 2019, mostly making wines from locally grown soft fruits, although the Denali Winery produces a range of *vinifera* wines (Cabernet Sauvignon, Chardonnay, Merlot, Riesling, and the inevitable Alaska Icewine).

ARIZONA

The American Southwest has quietly built a burgeoning industry; there are 120-plus wineries in Arizona, and two AVAs (with another on the way). Most of these areas are an extension of southern California's arid and elevated scrublands. Much of them are desert. But as with Las Vegas or Phoenix-Scottsdale, whole cities can sprout up courtesy of the heavily pilfered underground aquifers that have taken millennia to fill. What will happen when they are depleted is easy to guess. Summer temperatures can easily hit 40°C (104°F) but elevated sites (between 1,200 and 1,500 metres, or 4,000 and 5,000 feet) suffer less of the worst of it, and they cool considerably at nights. Not surprisingly, Rhône and Italian varieties most often succeed. Those who haven't tasted wines from Arizona Stronghold, Bodega Pierce, Callaghan, Caduceus, Dos Cabezas, Garage East, Keeling-Schaefer, Merkin, Oddity, Page Springs, Pillsbury, Sand Reckoner, Southwest Wine Center, and Tumbleweed in Arizona are in for a happy surprise.

SONOITA AVA

The Santa Rita, Huachuca, and Whetstone Mountain ranges isolate this AVA. Geologically, this appellation is an upland basin rather than a valley, because it comprises the headwaters for three distinct drainages: Sonoita Creek to the south, Cienega Creek to the north, and the Babocamari River to the east. The origins of modern Arizona viticulture are here: Dr Gordon Dutt created Sonoita Vineyards in 1983, believing that the *terra rosa* soil, along with elevation, would offer great potential to this region.

VERDE VALLEY AVA

(Proposed)

Proposed in 2017, this AVA in central Arizona covers an area in northeast Yavapai County centred on the confluence of Oak Creek and the Verde River. It covers an area of about 567 square kilometres (219 square miles), with 36 per cent of the AVA privately owned land. Federal and state entities manage the remainder.

WILLCOX AVA

This 212,956-hectare (526,000-acre) area is a relatively flat, closed basin in southeast Arizona, surrounded by the Chiricahua, Dos Cabezas, Pinalenos, Dragoon, Little Dragoon, and Winchester Mountains. A range of grapes is grown in the area, with a major nod to Bordeaux and Mediterranean varieties. There are more than a dozen commercial wineries and a total of 184 hectares (454 acres) planted to vine.

COLORADO

The growing season in Colorado is a challenge; most vineyards are on the western slopes of the Rocky Mountains. There summers are usually not too hot, and the seasons can end gently enough. But spring and autumn frosts are a constant danger. Altitude is the one requirement for the 150-plus wineries there. The higher altitude Colorado wineries of importance include Alfred Eames, Balistreri, Book-Cliff, Carlson, Colterris, Cottonwood, Garfield, Grande River, Infinite Monkey Theorem, Jack Rabbit Hill, Plum Creek, S. Rhodes, Stone Cottage, Sutcliffe, Terror Creek, Two Rivers, and Winery at Holy Cross Abbey.

GRAND VALLEY AVA

Grand Valley is located just west of Grand Junction and incorporates three localities known as Orchard Mesa, Redlands, and Vinelands. It is located roughly 320 kilometres (200 miles) west of Denver and encompasses 30,750 hectares (75,990 acres), though only a fraction of that is planted to grapevines. This high-desert AVA has an average elevation of between 1,220 and 1,524 metres (4,000 and 5,000 feet). More than two dozen wineries are found here.

WEST ELKS AVA

Very high-altitude viticulture, ranging from 1,646 to 1,951 metres (5,400 to 6,400 feet), is practiced in this AVA in Delta County, in and around the towns of Cedaredge, Hotchkiss, and Paonia. It's found within the larger North Fork Valley of the Rocky Mountains in west-central Colorado. There are about a dozen wineries in the West Elks AVA.

HAWAII

Only Alaska seems a less likely wine-producing state, but vines were first planted here in 1814. The first vines planted in modern times were at Tedeschi Vineyard on Maui in 1974, and the first wines made in 1980. Tedeschi went back to California and Paula Hegele took over. For 25 years she and her team have persevered – the Ulupalakua Vineyard includes numerous *vinifera* (Syrah, Malbec, Viognier, and more), but they still make the popular pineapple wine. In 1987 Volcano Vineyard on Hawaii Island was planted with the Symphony cross, but that was eventually lost, and hybrids and fruit wines are now the mainstay. The Hawaiian Islands boast six wineries in total, although most are producing tropical fruit wines.

MONTANA

Although the growing season is generally too short for grapes, and most of this state's 14 wineries produce wines from locally grown soft fruits, or grapes trucked in from Washington, Idaho, and California, vines have grown around Flathead Lake since 1979, when Dr Thomas Campbell established Mission Mountain Winery on Flathead Lake's west shore. There are still fewer pure Montana-grown wines than wines made from out-of-state grapes, but other wineries are worth consideration, particularly Ten Spoon Vineyard and Tongue River Winery.

NEVADA

Even though the growing season in Nevada is theoretically too short for growing vines, the first experimental vines were planted here in 1992, and by 1994 there were three vineyards: two in the Carson Valley of southern Nevada, and one in southern Nevada's Pahrump Valley. There are now 11 wineries, with Nevada Ridge (part of Pahrump Valley Vineyards) and Churchill Vineyards producing Nevada-grown *vinifera* wines but with others growing various *vinifera* and hybrid varieties to supplement their purchases. Now the question is whether or not cannabis will overwhelm any opportunities for vineyards in this desert state.

NEW MEXICO

Only Florida can claim an older winemaking industry than New Mexico's, which dates from the early 1600s but was unknown to Americans until 1821, when the Santa Fe Trail (known then as the Rio del Norte) was opened, and reports of its vineyards filtered back to the East Coast. As with Arizona, although the hot desert environment might seem inhospitable, elevation ameliorates the worst of the weather. Today, there are 52 wineries. Those who haven't tasted wines from Black Mesa, D H Lescombes, Jaramillo, La Chiripada, Milagro, Vivac, and, particularly, Gruet would likely be shocked at the quality.

MESILLA VALLEY AVA

New Mexico and Texas

This AVA spans two states, following the Mesilla Valley along the Rio Grande River from an area just north of Las Cruces, New Mexico (where most of its vineyards are situated) to El Paso, Texas. Soils are alluvial, stratified, deep, and well drained.

MIDDLE RIO GRANDE VALLEY AVA

This narrow valley stretches along the Rio Grande River from Albuquerque to San Antonio. Franciscan missions grew vines here during the 17th century, and winemaking existed until Prohibition. At an altitude of between 1,465 and 1,585 metres (4,800 and 5,200 feet), the climate is characterized by low rainfall and hot summers.

MIMBRES VALLEY AVA

This largest AVA in the state covers 113,500 hectares (636,000 acres) of semi-desert, with 810 hectares (2,000 acres) of it planted to irrigation grapevines. Elevations in this semi-desert area are between 1,200 and 1,800 metres (4,000 and 6,000 feet); soils are well drained, comprised of alluvials from the once much larger Mimbres River.

NORTH DAKOTA

There are fewer than a dozen wineries in this state, but all are dependent upon fruit wines.

OKLAHOMA

The first known Oklahoma wines were made by Edward Fairchild, who in 1890 moved from the viticultural paradise of New York's Finger Lakes to establish a vineyard and orchard near Oklahoma City. He made commercial quantities of Concord and Delaware wine between 1893 and 1907, when Oklahoma joined the US as a dry state. In 2019 there are 60-plus wineries utilizing either purchased out-of-state grapes or crafting wines from hybrid grapes, such as Chambourcin, Norton, Vignoles, Seyval Blanc, and Vidal Blanc.

OZARK MOUNTAIN AVA

See entry under Missouri.

SOUTH DAKOTA

Valiant Vineyards was the first-ever winery in South Dakota, established in 1993, when Eldon and Sherry Nygaard planted a small vineyard overlooking Turkey Ridge Creek in Turner County. At the time it would have been illegal to build and operate a winery, so the Nygaards drew up a farm winery bill and caused it to be introduced by the South Dakota Legislature, which passed it into law in July 1996. By 2019, there were 22 wineries in this state, but few aside from Prairie Berry Winery stand out.

TEXAS

The Franciscan missions were making wines here at least 130 years before the first vines were grown in California, and the first commercial Texan winery, Val Verde, was established before the first one opened in California. In 1880 Thomas Volney Munson saved the phylloxera-infected French wine industry by sending Texas rootstock to Charente and was awarded the French Legion of Honour medal, only the second American recipient at that time (the first being Benjamin Franklin). By 1919 there were 19 wineries in Texas, and a Texas wine won a gold medal at the Paris Exposition that year. Yet Prohibition, mandated the next year, destroyed winemaking in Texas, as it did everywhere else. The revival of Texas wine began by accident in the mid-1950s, when Robert Reed, a professor of viticulture at Texas Tech University, took home some discarded vine cuttings. He planted them in his garden and was astonished at how well they grew. Reed and colleague Clinton "Doc" McPherson planted an experimental vineyard of 75 grape varieties, which led to the founding of Llano Estacado winery in 1975. In 1987 Cordier, the famous Bordeaux *négociant,* invested in a massive 405-hectare (1,000- acre) vineyard called Ste Geneviève, and by 2019 there were nearly 400 Texas wineries. In 1983 Texas was the smallest wine-producing state in the United States; today it is the fifth largest. If a wine states "For Sale in Texas Only", it does not have to state its origin. Texas wineries rely far too often on Chardonnay and Cabernet Sauvignon – it seems Texans will buy anything that says "Texas Grown". Yet there are more reliable wineries, such as McPherson, William Chris, and Perissos that use actual Texas fruit in their worthy wines, and other notables include Becker, Duchman, Eden Hill, Fairhaven, Fall Creek, Flat Creek, Fredericksburg, Haak, Inwood, Lewis, Pedernales, Red Caboose, Sister Creek, Spicewood, and Wedding Oak.

BELL MOUNTAIN AVA

Located in Gillespie County, north of the Fredericksburg AVA (which like Bell Mountain is a sub-appellation of the Texas Hill Country AVA), this is a single-winery denomination consisting of some 22 hectares (55 acres) of vines, growing on the southern and southwestern slopes of Bell Mountain. Over a third of the vines are Cabernet Sauvignon, which has so far proved to be significantly better suited to this area than the Pinot Noir, Sémillon, Riesling, and Chardonnay that also grow here. Bell Mountain is drier than the Pedernales Valley to the south and the Llano Valley to the north and also cooler due to its elevation and its constant breezes. Its soils are sandy-loam, with light, sandy-clay subsoil.

ESCONDIDO VALLEY AVA

Proposed by Cordier, the Escondido Valley AVA in Pecos County encompasses a single producer, the Ste Genevieve Winery, once owned by Cordier. Escondido is an upland valley formed by a series of mesas to the north and south; the vineyards are at an elevation of between 795 and 825 metres (2,600 to 2,700 feet).

FREDERICKSBURG IN THE TEXAS HILL COUNTRY AVA

This AVA's full title is Fredericksburg in the Texas Hill Country, but it is usually shortened to just Frederickburg AVA to distinguish it from the massive Texas Hill Country appellation. Fredericksburg contained 8 vineyards when established, but the area was once primarily known for its peach orchards. Because the peach tree has proved as sensitive to soil and climate as *vinifera* vines, many peach farmers have experimented with grapes, and 19 wineries are now selling their wares. The soil is sandy loam over a mineral-rich reddish clay.

MESILLA VALLEY AVA

See entry under New Mexico.

TEXAS DAVIS MOUNTAIN AVA

Described as a mountain island, this Texas appellation is cooler, twice as wet, and climatically more diverse than the Chihuahua Desert that surrounds it. Located in Jeff Davis County, Davis Mountain AVA covers 108,000 hectares (270,000 acres) in the Trans-Pecos region of western Texas that was formed during the glacially slow tectonic and volcanic catastrophe that formed the front range of the Rockies. This viticultural region ranges in height from 1,350 to 2,490 metres (4,500 to 8,300 feet), with about 20 hectares (50 acres) of vines currently growing on a porous, well-drained soil composed of granitic, porphyritic, and volcanic rocks, as well as limestone.

TEXAS HIGH PLAINS AVA

This massive appellation in the northwest of Texas covers 32,400 square kilometres (12,500 square miles), encompassing 24 counties and 800 hectares (2,000 acres) of vineyards. Soils are generally brown clay loams to the north and fine sandy loam to the south. Rainfall ranges from 36 centimetres (14 inches) in the west to 51 centimetres (20 inches) in the east.

TEXAS HILL COUNTRY AVA

A vast appellation, Texas Hill Country consists of the eastern two-thirds of the Edwards Plateau and encompasses two tiny AVAs (Bell Mountain and Fredericksburg), 400-plus hectares (almost 1,000 acres) of vineyards, and 60-plus wineries. Most of the hillsides are on limestone, sandstone, or granite, while the valleys contain various sandy or clayey loams.

TEXOMA AVA

Another large AVA comprising 94,500 hectares (233,600 acres) in north-central Texas on the south side of Lake Texoma and the Red River border with Oklahoma. Yet the planted acreage is tiny – 22 hectares (55 acres) and a half dozen wineries. Historically, though, this area has importance: it was here that 19th-century viticulturist TV Munson perfected the grafting of *vinifera* vines onto native American varieties to create phylloxera-resistant vines. Munson, after helping to rescue European viticulture, went on to create hundreds of phylloxera-resistant hybrids, and a few vintners are working with those today.

UTAH

Despite the enormous influence wielded by the powerful Mormon Church based in Salt Lake City, winemaking has a history in this state, ironically originating with the arrival of the seemingly abstemious Latter-Day Saints. Brigham Young, who led the Mormons to Utah in 1847, ordered the planting of vineyards and the building of a winery. He required one of his followers, an experienced German winemaker, to make as much wine as he could, and although he permitted Mormons to drink it for Communion (no longer allowed), he recommended that the bulk should be sold. His advice was not taken by the Dixie Mormons, who ran the winery and kept back their best wines for consumption. Winemaking peaked at the end of the 19th century, declining once the Church clamped down on drinking, and died out during Prohibition. Eleven wineries now exist.

WYOMING

Established in 1994, Terry Bison Ranch was this state's first winery, and by 2019, there were just five wineries. Most grapes are trucked in from out of state, though Table Mountain Vineyards uses Wyoming-grown grapes.

OTHER WINE PRODUCERS OF

THE UNITED STATES

SOUTHERN STATES

GEORGIA

CREEKSTONE

Sautee Naucoochee

★☆

Habersham Winery's premium label is reserved for *vinifera* wines grown primarily in its Mossy Creek vineyard.

✓ *Viognier*

FROGTOWN WINERY

Dahlonega

★☆

A sense of fun pervades this entire project, which produces excellent reds.

✓ *Marsanne • Red blend (404)*

TIGER MOUNTAIN VINEYARDS

Tiger

★★

In 1995, John and Martha Ezzard planted Norton and an eclectic bunch of *vinifera* varieties (Cabernet Franc, Malbec, Mourvèdre, Tannat, Tinta Cão, Touriga Nacional, and Viognier) on mineral-rich, well-drained slopes of their 40-hectare (100-acre) farm, at an altitude of 600 metres (2,000 feet) in the southern Blue Ridge Mountains. Martha continues the project, and the wines just keep getting better.

✓ *Cabernet Franc • Touriga • Norton*

NORTH CAROLINA

BILTMORE ESTATE WINERY

Asheville

★

French grape varieties were planted in 1979 by the grandson of George Washington Vanderbilt, who built the country's largest mansion (250 rooms) on this 3,200-hectare (8,000-acre) estate. At a height of 1,400 metres (4,500 feet) in the Blue Ridge Mountains, it is not only cool enough for *vinifera* varieties to grow but can also sometimes even be too cold for them to ripen. Many of the wines are based upon grapes brought in from California, and some of the wines may be the better for it.

✓ *Cabernet Sauvignon (Signature) • Sparkling*

CENTRAL STATES

ILLINOIS

ILLINOIS SPARKLING WINE COMPANY

Utica

★★

Mark Wenzel, who created August Hill Winery, is doing even more important work at his ISC project, making sparkling wines that are improving each year.

✓ *Entire range*

KANSAS

HOLY-FIELD

Basehor

★★☆

This winery sets standards for wine in the middle part of the United States. Holy-Field demonstrates that parts of "fly-over America" are worth a stopover.

✓ *Chambourcin • Norton (Cynthiana) • Seyval Blanc • Vignoles (Dry, Late Harvest)*

MISSOURI

ADAM PUCHTA

Hermann

★★

Originally established as Adam Puchta & Son Wine Company in 1855, this winery was put out of business during the Prohibition years. It was resurrected in 1990 by Adam Puchta's great-great-grandson Timothy. Best known for Norton, but probably best at Port-style wines, particularly the Signature Port.

✓ *Fortified (Signature Port, Vintage Port) • Norton • Vignoles*

AUGUSTA WINERY / MONTELLE WINERY

Augusta

★★☆

Look for impressive Norton, called Cynthiana in this part of the state, but everything Tony Kooyumjian makes at Augusta and at his other winery, Montelle, is absolutely first-rate.

✓ *Chambourcin • Norton (Cynthiana) • Vignoles • Seyval Blanc*

NOBOLEIS WINEYARDS

Augusta

★★ ⓥ

Even a tornado (2011) couldn't stop Bob and Lou Ann Nolan; they have steadily improved upon their wines and are reaching the top ranks.

✓ *Chambourcin • Vignoles (Dry) • Norton*

STONE HILL WINERY

Hermann

★★☆ ⓥ

Established in 1847, Stone Hill is the oldest winery in Missouri. Its cellars were used as a mushroom farm during Prohibition but were revived in 1965 by Jim and Betty Held. The Helds are best known for their Norton, both as a red wine and a Port style. But, truthfully, everything here is excellent.

✓ *Fortified (Cream Sherry, Port) • Norton • Seyval Blanc • Vignoles (Late Harvest)*

TERRAVOX WINERY

Platte City

★★

Jerry Eisterhold is (along with RL Winters in Texas) the only vintner savvy enough to understand that the TV Munson varieties might unlock the key to viticultural success in this part of the United States. He is spreading the word with presentations and with good wine.

✓ *Wetumka (Port, Estate) • Norton • Cloeta*

WISCONSIN

CEDAR CREEK WINERY

Cedarburg

★☆

A sister winery to the 800-pound gorilla in Wisconsin, Wollersheim Winery. Winemaker Philippe Coquard runs a very smart ship at all these admirable Wisconsin wineries.

✓ *Syrah • Vidal Blanc*

WOLLERSHEIM WINERY

Prairie du Sac

★★

The Wollersheim Winery was built in 1858 by the Kehl family, from Nierstein in Germany and then re-established in 1972 by Robert and Joann Wollersheim. Future son-in-law Philippe Coquard arrived in 1984, and the wines morphed into benchmarks for Middle American wine. Both hybrid and *vinifera* grapes are grown here. Wines are also sold under the Domaine du Sac label.

✓ *Seyval Blanc (Prairie Fumé) • Riesling (Dry) • Rosé (Prairie Blush) • Ice Wine*

WESTERN STATES

ARIZONA

CADUCEUS CELLARS

Jerome

★★☆

People find it hard to believe that rock star Maynard James Keenan, formerly of Tool, is the actual winemaker for this project, but they should believe it – and should appreciate the fine, limited-production wines.

✓ *Entire range*

CALLAGHAN VINEYARDS

Sonoita

★★★

Extraordinarily impressive wines have been made here for a good number of years, with Syrah a particularly standout in their range of wines.

✓ *Entire range*

DOS CABEZAS WINEWORKS

Sonoita

★★☆

This project was pioneered by the late visionary Al Buhl. Todd Bostock joined as winemaker in 2002, and a few years later his family took on the winery. They

BANNERS DIRECT VISITORS AT WISCONSIN'S WOLLERSHEIM WINERY
This Midwestern winery in Prairie du Sac produces grape wines alongside the company's distillery, which produces spirits such as apple brandy and vermouth.

have two estate vineyards: Pronghorn Vineyard in Sonoita and Cimarron Vineyard in the Kansas Settlement.

✓ *Red blends (Áquileón, El Norte, El Campo)*

SAND-RECKONER VINEYARDS

Sonoita

★★★

Rob and Sarah Hammelman have created an inspiring, small-scale, high-quality project in southern Arizona.

✓ *Entire range*

COLORADO

ALFRED EAMES WINERY

Paonia

★★

These wines can be big and burly.

✓ *Tempranillo • Syrah • Bordeaux blend (Collage)*

BOOKCLIFF VINEYARDS

Boulder

★★

Ulla Merz and John Garlich have created one of the enduring wineries of the Western Rockies.

✓ *Cabernet Franc • Syrah • Bordeaux blend (Ensemble, Encore)*

COLTERRIS CELLARS

Palisade

★★

Scott and Teresa High have a vineyard-focused, farm-first winery that is both smart and significant.

✓ *Merlot (Estate) • Cabernet Sauvignon (Estate) • Cabernet Franc*

JACK RABBIT HILL FARM

Hotchkiss

★★🅑

A true-to-its-beliefs biodynamique winery that produces soulful wines.

✓ *Riesling (Mitzi's) • Red blend (Meunier/ Pinot Noir) • Chardonnay (Upper West Side) • Petit Verdot*

PLUM CREEK CELLARS

Palisade

★★

Founded in 1984, Plum Creek Cellars is a pioneer and remains one of the most successful wineries in Colorado.

✓ *Cabernet Franc • Cabernet Sauvignon • Riesling*

THE STORM CELLAR

Paonia

★★

Two Denver sommeliers have a great start-up project. One to watch.

✓ *Riesling (Entire range)*

NEW MEXICO

GRUET

Albuquerque

★★

The Gruet family established a successful cooperative in the Sézanne district of

Champagne but are prone to inconsistency in New Mexico. Yet some of their best wines have shown extraordinary acidity.

✓ *Sparkling (Blanc de Blancs, Grande Reserve)*

MILAGRO VINEYARDS

Corrales

The Hobson family planted these vineyards in 1985; after excellent vintages, they just keep making better wines.

✓ *Riesling • Grüner Veltliner • Zinfandel*

TEXAS

BECKER FAMILY WINERY

Lubbock

★★

Starting with a simple log cabin, what was just a country retreat has, three decades onward, turned into 125 hectares (308 acres) of vineyard and winery estate, producing over 100,000 cases a year.

✓ *Cinsault • Counoise • Dolcetto (Reserve)*

CAPROCK

Lubbock

★☆

The winery began in 1988 as Teysha Cellars, changed ownership a few times and now is English Newsom Cellars at CapRock Winery (the new owners are the English and Newsom families). Fresh money doesn't hurt, and all their wines are sourced from their own vineyards.

✓ *Roussanne • Dry Muscat • Tempranillo*

DUCHMAN WINERY

Driftwood

★☆

Founded in 2004 by Drs Lisa and Stan Duchman, who have a passion for Italian grape varieties, and the aid of Texas viticultural legend Bobby Cox.

✓ *Vermentino • Sangiovese • Dolcetto • Aglianico*

EDEN HILL WINERY

Celina

★

Since 2003 this family affair has been creating wines just north of Dallas; Linda and Clark Hornbaker manage the vineyard, with its good soil and limestone base, and son Chris is the winemaker.

✓ *Moscato Gialla • Roussanne (Estate, Reserve)*

FAIRHAVEN WINERY

Hawkins

★☆

RL Winters was an original pioneer of the forgotten Munson clones, and he is one of the few vintners that understands the Texas treasure that is the Black Spanish grape. His Lomanto (another heritage grape) should make a believer of you.

✓ *Lenoir (Black Spanish) • Lomanto (Reserve)*

FALL CREEK VINEYARDS

Driftwood

★☆

Named after the waterfall that feeds Lake Buchanan from an upper ridge of

Ed and Susan Auler's ranch, Fall Creek has been growing wine since 1975. The Aulers are the original Texas Hill Country vintners, making 100 per cent Texas wines.

✓ *Red blend (Terroir Reflection GSM, Meritus) • Sangiovese (Vintner's Selection) • Chenin Blanc*

HAAK WINERY

Santa Fe

★☆

The king of Blanc de Bois, it would be safe to say (though the Floridians who created the grape might argue otherwise); Haak has crafted this and other grapes admirably.

✓ *Blanc de Bois (all types, including Madeira)*

INWOOD WINERY

Fredericksburg

★★

Dan Gatlin started this winery way in 1981: he was one of the first to adopt Tempranillo and other Spanish grapes.

✓ *Tempranillo (Cornelious, Tempranillo-Cabernet) • Palomino-Chardonnay*

LLANO ESTACADO WINERY

Lubbock

★

Llano Estacado was the first Texas winery to get national and even international notice, founded in 1976. The wines continue to grow in sales and quality.

✓ *Red blend (Viviano) • Montepulciano*

MESSINA HOF WINE CELLARS

Bryan

★

Paul Bonnarrigo's family originally came from Messina in Italy, while his wife, Merrill, has a German heritage, hence the name of this winery, the third to be founded in the modern history of Texan wine. Paul's son and daughter-in-law manage the winery today.

✓ *Muscat Canelli (Aggie Network) • Riesling • Red blend (GSM) • Primitivo (Paulo)*

McPHERSON CELLARS WINERY

Lubbock

★★☆

Kim McPherson's father was a true pioneer in Texas wine (Doc McPherson was an original founder of Llano Estacado), and Kim's success as a winemaker offers a fantastic testimonial to his legacy. These are my favourite Texas wines.

✓ *Entire range*

PEDERNALES CELLARS WINERY

Stonewall

★★

Larry and Jeanine Kuhlken planted their first vineyard in the early 1990s and quickly became known for generating high-quality fruit. In 2005 their children David and Julie created Pedernales Cellars, David as winemaker and Julie overseeing hospitality, marketing, and design, with Larry and Jeanine managing the family vineyards.

✓ *Vermentino • Tempranillo (all types) • Mourvèdre • GSM (Melange)*

PERISSOS VINEYARD

Burnet

★★

Just northwest of Austin, this newish spot is in Hoover's Valley. The first vines were planted in 2005, expanding to 16 acres (6.4 hectares), and the tasting room opened in 2009.

✓ *Aglianico • Dolcetto • Montepulciano*

PHEASANT RIDGE WINERY

Lubbock

★

Owner Bobby Cox has been in the vineyard business since 1973; he has always focussed upon Texas grown wines, and his track record as a grower is spotless.

✓ *Chenin Blanc • Viognier*

RED CABOOSE WINERY

Meridian

★☆

This is a well-run winery with sustainability in its DNA. They utilize advanced geothermal cooling, photo-voltaic solar cells, subterranean thermal mass, rainwater harvesting, and shading, to achieve their energy goals.

✓ *Tempranillo • Black Spanish • Syrah • Touriga Nacional*

SISTER CREEK WINERY

Sisterdale

★

Founded in 1988, this winery features a charming old cotton gin receiving house.

✓ *Muscat Canelli • Merlot*

SPICEWOOD WINERY

Spicewood

★☆

Creating this winery as an estate-wine-only project, the original owners planted vines that are now more than a quarter century old. Ron Yates, who has owned and run the estate for more than a decade, has expanded to 12 hectares (30 acres) of *vinifera*.

✓ *Red blend (The Good Guy) • Syrah • Tempranillo*

WEDDING OAK WINERY

San Saba

★☆

A popular and skilled producer of 100 per cent Texas wines, it named for a 400-year-old oak just north of the winery.

✓ *Montepulciano • Sangiovese • Roussanne • Tempranillo*

WILLIAM CHRIS VINEYARDS

Hye

★★

Bill Blackmon and Chris Brundrett have decades of experience in the Texas winery and vineyard industry; their partnership was bound to succeed the minute it started in 2008. They focus upon Texas-grown wine.

✓ *Mourvèdre (all types) • Tannat (Hye) • GSM (Artist Blend)*

Canada

Canada is an exciting and dynamic country in the world of wine. Despite viticultural activities being nestled in its southernmost reaches, the sheer geographic vastness of the country provides a fantastic spectrum in diversity of terroir.

Much like their neighbours to the south, Canada – and its wine-producing industry – was blighted with Prohibition in the early 20th century. Winemaking had existed in Canada for more than 200 years; it was not until 1974, however, when Inniskillin, founded in Niagara-on-the-Lake, marked a turning point in quality wine production. The 1980s ushered in key events that shaped the current wine landscape, including the establishment of the Vintners Quality Alliance (VQA), a major vine-pulling and replacement program; the advent of a number of keen pioneering spirits; as well as a major trade agreement with the USA.

ICEWINE: A CANADIAN DELICACY

Canada's northern longitude creates a cool climate for the majority of the country's wine-growing regions. As such, top-shelf icewine can be reliably produced vintage-to-vintage. Canada is the world's leader in icewine production in terms of quantity. To contextualize the volume, Canada produces more icewine than all other countries combined. Often, the hybrid grape Vidal Blanc is used due to its winter hardiness; a spectrum of other grapes is used, however, from Riesling and Cabernet Franc all the way to Chardonnay and Sangiovese. Icewine (also known as *Eiswein*) is a true vinous luxury and prized around the world by Michelin-starred restaurants and oenophiles alike. Production is both labour-intensive and meticulously regulated by the VQA. Grapes must naturally freeze on the vine and must not be harvested until a sustained -8°C (17°F) is achieved. Typically, whole-cluster picking is conducted in the veil of darkness during night with juice ranging between 35 to 39° Brix. Pressing these frozen grapes enables the water content to be greatly reduced (a vine typically only produces one glass of icewine), leaving the luscious essence of the grape to be fermented.

THE FUTURE OF CANADIAN WINE

The exciting future of Canadian wine lies with its dry table wines. From coast to coast, top-shelf wines are entering the market both domestically and internationally. The diversity of *terroir* in the Oakanagan Valley produces myriad styles, from finessed and full-bodied Merlot, Cabernet Sauvignon, and Syrah to crisp Rieslings. Ontario continues to gain international recognition for its elegant and balanced cool-climate Chardonnay and Pinot Noir. Nova Scotia, in the eastern Maritime Provinces, is proving to be a dynamic and compelling region for traditional sparkling wines.

INNISKILLIN ICEWINE
The Innislillin winery is best known for its production of icewine, which is well-suited to the cool climate of Canada's wine-growing regions. The winter season in both the Niagara Peninsula and the Okanagan Valley produces some of the best examples.

| Niagara Peninsula | Wine region |
| Niagara Escarpment | Wine sub-region |

0 mi — 40
0 km — 40

WINE REGIONS OF EASTERN CANADA
As are many of the cool-climate wine regions in Europe, the wine-growing regions of Canada are situated within the recognized growing zones of 30° and 50°N latitude. The Niagara Peninsula in southern Ontario is the primary wine-growing region of eastern Canada, along with a few other areas of Ontario and smaller ones in Quebec and Nova Scotia. In total, Canada's wine-growing regions cover 11,950 hectares (29,500 acres).

WINE REGIONS OF WESTERN CANADA

British Columbia in Canada's Pacific Northwest is rapidly closing the gap with Ontario. When NAFTA gave cheap California wines free access to the Canadian market, both regions had to upgrade their vineyards to survive. British Columbia's advantage was the smaller size of its vineyards and the tiny amount of *labrusca* planted, because this enabled the province to rapidly replace almost all of its hybrids with *vinifera* varieties.

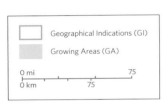

Geographical Indications (GI)

Growing Areas (GA)

0 mi   75
0 km   75

Ontario

Ontario is the heartland of Canadian wine production. This fertile land is blessed with vast natural diversity, from abundant forests to the Great Lakes. In terms of grape and wine production there is notable diversity in soil and a spectrum of microclimates, as well as the enthusiasm of wine-makers, who continue to produce wines that are making headlines internationally. The history of Ontario wine can be traced back to 1811, when Johann Schiller transplanted hybrid and *labrusca* grapes in Cooksville, Ontario. *Vinifera* grapes were not considered winter hardy enough for the often-frigid winters. In 1866 the province's first winery, Vin Villa, was established in its southern reaches at Pelee Island, but it was not until the 1900s that the industry started to evolve. The controversial provincial alcohol monopoly the Liquor Control Board of Ontario (LCBO) was established in 1927. Despite the ability and demand of private enterprise to facilitate distribution and sale of wine, the LCBO still controls all vinous activities in the province. In terms of vineyard development, 1979 brought the first quality *vinifera* plantings with German vintner Herman Weis planting Riesling in what is now the Vineland Estate Winery. But everything changed in 1974 when Donald Ziraldo and Karl Kaiser established Inniskillin Wines in Niagara. Even today when the region is still fresh in terms of its history, quality and quantity output continues to grow. The deeper experience and expertise of the vintners and the character of the terroir will only lend more success in the future.

Nova Scotia

Grape-growing has been present in the region since the 1600s. Nova Scotia has encountered a number of obstacles, however, such as the scourge of Prohibition, an extremely cool, marginal climate, and a culture more attuned to beer and spirits. All of these factors have kept the wine industry at bay until mid 1900s. The founding father of the region, Roger Dial, opened Grand Pré Winery in the late 1970s and is credited with renaming the signature grape of the region from V-53261 to L'Acadie Blanc, in honour of the early French settlers of the region. Today Nova Scotia is bursting with enthusiasm and hospitality and proudly producing top-shelf traditional sparkling wines. With a new appellation, Nova Scotia holds a bright future in the world of wine.

Quebec

Quebec is a newcomer to the Canadian wine scene but is both enthusiastic and focused. The first commercial vineyards were established in the 1980s. Resourcefulness and a strong constitution are required to produce wines in Quebec. Despite the moderating effects of bodies of water such as the Deaux-Montagnes, the Saint-Louis Lakes, and the Ottawa and St Lawrence Rivers, winters can be bitterly cold. As such, protection of vines from frost and damage is paramount. Hybrids are leveraged due to their resistance to the cold and ability to ripen during their brief growing season. Frontenac Noir, Frontenac Blanc, and Vidal are popular choices among growers. In terms of *Vitis vinifera* varieties, Pinot Noir and Chardonnay are the most planted but only represent 6 per cent of varieties planted as of 2018. Putting this production in context, Quebec currently represents 3.8 per cent of Canadian wine. Needless to say this region has room to grow. In November 2018 the Quebec government officially recognized the protected geographical indication (PGI) of "Quebec Wines". With close to 150 producers in the province it is fair to say that Quebec has an exciting future ahead.

British Columbia

Located on Canada's Pacific west coast, British Columbia boasts some of the most breath-taking landscapes in the country. The vast majority of wine is produced in the Okanagan Valley, which pushes the boundaries of growing grapes on the 49th parallel, but wine is grown across the province. Vastly diverse landscapes of varying climates and soils produce more than 80 different grapes. Styles range from light-bodied crisp whites to robust reds. Standouts in terms of quality wines include Merlot, Pinot Noir, and Cabernet Sauvignon, as well as Pinot Gris, Chardonnay, and Riesling. The invigorating spirit of the region sees new wineries consistently opening, along with established producers creating beautiful wines.

VQA SEAL OF QUALITY

The Vintners Quality Alliance (VQA) is a regulatory and appellation system for Canadian wines made in British Columbia and Ontario. The VQA guarantees typicity and quality, as well as origin of grapes and finished wines. Further, the VQA quality standards encompass label integrity for accuracy and clarity. In addition to the VQA, the BC Wine Authority also implements the Wines of Marked Quality Regulation, which is a certification program applicable to wine processed in British Columbia.

FACTORS AFFECTING TASTE AND QUALITY

LOCATION
Canada is a vast landmass with its main wine regions located in the southeast (Ontario, Nova Scotia, and Quebec) and southwest (British Columbia) corners, provinces that are more than 3,200 kilometres (2,000 miles) away from each other.

CLIMATE
Despite Ontario's shared latitude with Burgundy in France, the growing season is distinctly Canadian. Winters can be savagely cold, posing significant viticultural threats. Conversely, the growing season tends to be compressed and often hot – viticulturalists frequently experience significant heat events over 30°C (86°F). Three of the Great Lakes – Lake Huron, Lake Erie, and Lake Ontario – help moderate the temperature extremes. Nova Scotia's growing regions are influenced by three large bodies of water: the Atlantic Ocean, the Gulf of St Lawrence, and the Bay of Fundy. The climate is notably wet, with copious and sustained precipitation and strong westerly winds. The North and South Mountains, as well as the Bay of Fundy, are key geological features that contribute toward a warm micro-climate in the Annapolis and Gaspereau Valleys.

ASPECT
Vineyards are located in myriad locations and aspects across the country. Leveraging slope to maximize sunlight exposure and warm breezes is commonplace in the Niagara Peninsula region. Vineyards are located in ranges from flat former seabeds to hillsides and benches. British Columbia can see top-shelf vineyards nestled in valleys to maximize heat. The diversity of *terroir* and topography occurring coast to coast will ultimately dictate which aspect viticulturalists will leverage.

SOIL
Ontario has both unique topography and soil diversity due to glacial events that created the Niagara Escarpment via years of erosion of rock and reef structures. The exposure of ancient seabed adds to this variety with an extensive concentration of limestone as well as desirable water drainage. Nova Scotia's surface is varied due to numerous episodes of glaciation and the rise and fall of the ocean, as well as volcanic activity and shifting tectonic plates. British Columbia is also diverse in terms of soil and topography, from coastal islands to inland valleys.

VITICULTURE & VINIFICATION
Viticulture and vinification varies greatly throughout the country. Vastly different climates, soil, and topography influence both how grapes are grown as well as how wine is made. Where threats of frost damage and vine death are present, practices such as burying vines in the winter, smudge pots, windmills, and the like are all used to mitigate the cold climate. In high-rainfall areas, such as Nova Scotia, risks from mould and mildew persist from prolonged and ample precipitation. Mindful canopy management is necessary to mitigate the disease pressure. An individual producer's desired styles most often dictate vinification techniques.

GRAPE VARIETIES
Baco Noir, Cabernet Franc, Cabernet Sauvignon, Chardonnay, Gamay, Gewürztraminer, L'Acadie Blanc, Merlot, Pinot Gris, Pinot Noir, Riesling, Sauvignon Blanc, Seyval Blanc, Syrah, Vidal Blanc

RECENT CANADIAN VINTAGES

Although viticulture is clustered in the southern reaches of the country. the sum of almost all of Europe's landmass would fit into Canada. With an area of more than 9.9 million square kilometres (3.8 million square miles) and a width from east to west of 5,514 kilometres (3,426 miles) the country is vast. Therefore, any vintage generalizations coast to coast must be considered with a tempered mindset.

2018 A variable season in Ontario, with top producers making good wines. In British Columbia, the vintage was cooler than expected, producing restrained and fresh styles. Smoke from wildfires and ravenous starlings provided intrigue for growers.

2017 Devastating flooding occurred in Prince Edward County in June. An advantageous summer and fall, however, enabled growers to produce a good vintage. British Columbia suffered a wet and cool spring. This vintage also saw the worst wildfire season on record for the province. The flames avoided the winegrowing regions, however, and created a smoke blanket for sun coverage. This had the fortuitous result of slowing down ripening during the rather hot summer.

2016 A standout vintage for Ontario with near-perfect conditions, including a mild winter and long warm summer and autumn. On the west coast, growers were also fortunate, despite an abnormally hot spring, and enjoyed a long, dry, and moderate summer.

2015 Ontario suffered record-breaking cold during the winter months. Additionally, rain in autumn provided challenges for growers. Despite this, a long and favourable autumn produced classically styled wines from mindful producers. On the west coast, growers experienced an excellent vintage, both warm and dry, producing a top-shelf vintage.

2014 Technique and experience were key factors necessary in this variable vintage across Ontario. In British Columbia the season was moderately warm, seeming even warmer after a buildup in temperatures from cool 2010 and 2011.

AN ENGLISH PHONE BOOTH STANDS IN THE LUCKETT VINEYARDS, NOVA SCOTIA
Located in the Gaspereau Valley, this winery falls in the Tidal Bay wine region. Luckett produces a range of both red and white wines, including those in the fittingly named Phone Box series.

THE APPELLATIONS OF

CANADA

BRITISH COLUMBIA

Below are listed the major wine-producing regions of British Columbia. Other regions include Vancouver Island, Fraser Valley, Gulf Islands, Thompson Valley, Shuswap, Lillooet, and Kootenays.

OAKANAGAN VALLEY

As of 2018 the Province of British Columbia has nine Geographical Indications (GIs). The most important of these, in terms of volume and quality wine production, is the Okanagan Valley. With over 4,050 hectares (10,000 acres) of planted vines, this region boasts more than 80 per cent of the provincial acreage. In addition, four official subregions express unique micro climates and soils. These subregions include Golden Mile Bench, Naramata Bench, Okanagan Falls, and Skaha Bench. Nestled between the Cascades and the Monashee Mountains the region enjoys a continental climate with long daylight hours. The region relishes an additional two hours of sunlight per day during the growing season than that of California. Further, the large diurnal shift from day to night ensures refreshing acidity and balance. In terms of longitude, the region reaches the northern boundaries of wine production at the 49th parallel. Top-shelf Bordeaux varieties are standouts in the region, while equally impressive Chardonnay and Pinot Gris are produced.

SIMILKAMEEN VALLEY

This region, nestled amongst mountains with the Similkameen River running through it, is one to watch. The valley is blessed with notable winds, which keep the valley pest free, and there is a natural heat trap formed by the surrounding mountains and reflective rocks. As such, organic viticulture is common among the small, boutique wineries that are making lovely wines.

NEW BRUNSWICK

Beyond the production of fruit wines and fine meads, the province is home to a number of pioneering spirits in winemaking. Growth in this market has been encouraged by the provincial liquor monopoly introducing a grocery store program for local wines, thus increasing demand. This cool-climate Atlantic region has a bright future.

NEWFOUNDLAND

The most easterly province in Canada is ruggedly picturesque. Newfoundland is making lovely fruit wines and meads. In the future, greater expansion into winemaking is inevitable for this region.

NOVA SCOTIA

TIDAL BAY

Established in 2012, Tidal Bay was Nova Scotia'a first appellation. The wine from this region is a blend of aromatic, white grapes grown in Nova Scotia. Finished wines have moderate alcohol, elevated, crisp acidity, and residual sugar typically seen at 20 grams per litre to create a balanced mouth-feel that pairs well with the local maritime cuisine. In terms of geographic regions, much of Nova Scotia's quality grape-growing occurs in the Annapolis Valley and Gaspereau Valley.

ONTARIO

LAKE ERIE NORTH SHORE

Although small in terms of producers in the region, Lake Erie North Shore is producing some lovely wines from Cabernet Sauvignon, Cabernet Franc, and Riesling. The favourable southerly location combined with the warming effect of the shallow waters of Lake Erie allows this appellation to enjoy a long growing season. These factors ultimately promote ripe fruit with balance between natural sweetness and acidity. This region also includes the South Islands sub-appellation.

NIAGARA PENINSULA

As the heart of Ontario wine production, the Niagara Peninsula is the most planted area in the province. The Niagara Escarpment acts as a backbone to the region, rising 177 metres (580 feet) above sea level and running like a spine down the district. This north-facing cliff has several positive effects, such as advantageous sunlight exposure via slope, as well as elevation. The natural aspect also encourages the circulating flow of air from Lake Ontario, which favourably moderates temperatures throughout the year. This region also includes the Beamsville Bench, Creek Shores, Four Mile Creek, Lincoln Lakeshore, Niagara Lakeshore, Niagara River, Short Hills Bench, St David's Bench, Twenty Mile Bench, and Vinemount Ridge sub-appellations and the Niagara Escarpment and Niagara-on-the-Lake regional appellations.

PRINCE EDWARD COUNTY

Officially identified in 2007 as a VQA appellation of origin, Prince Edward County is now one of Canada's most exciting regions, producing exceptional quality cool-climate wines. Still Pinot Noir and Chardonnay, as well as traditional sparkling wines, are the region's fortes. At a latitude of 44°N, this is Ontario's most northerly appellation and relies on the lake to provide a moderated and productive cool-climate growing season. The region is full of mesoclimates and *terroir* conditions that support the basis for the distinct wines that are produced.

THE WINE PRODUCERS OF

CANADA

BRITISH COLUMBIA

BLASTED CHURCH

Okanagan Valley

With Evelyn Campbell at the helm, viticulturist John Bayley and winemaker Evan Saunders produce a number of varietals in their 17 hectares (42 acres). Beyond eye-catching bottles, of note is the Small Blessing series, aptly named for their small-lot production. These wines are made with minimal intervention and much pride.

✓ *Bible Thumper • Small Blessings Semillon • OMFG • Nectar of the Gods • Small Blessings Be Fruitful*

CEDARCREEK ESTATE WINERY

Okanagan Valley, Kelowna

As a newly certified organic estate CedarCreek produces bright and expressive wines in the northern end of the Okanagan Valley. This cool-climate location is ideal for their single-vineyard expressions of Pinot Noir. Their flagships of this varietal are Platinum Block 2 Pinot Noir and Platinum Block 4 Pinot Noir. Winemaker Taylor Whelan and viticulturist Kurt Simcic work together with their unique estate *terroirs* to create balanced and beautiful wines.

✓ *Block 3 Riesling • Haynes Creek Viognier • Block 4 Pinot Noir • Desert Ridge Meritage*

DESERT HILLS ESTATE WINERY

Okanagan Valley, Black Sage Bench
★★

Local legend Anthony Buchanan is making top-shelf and approachable wines at this family-run estate. While paying homage to tradition with their much-prized Gamay and old vine Syrah, the estate does not shy away from experimenting with new techniques, such as concrete fermentations, skin contact whites, and even ancestral-method sparkling.

✓ *Sauvignon Blanc • Gamay Noir • Cabernet Franc • Syrah • Mirage • Gewurztraminer*

THE HATCH

Okanagan Valley and Similkameen Valley
★★

Irreverent, and bursting with swagger, this estate thumbs its nose at convention. In some vintages the estate will release upwards of 60 different wines, and it prides themselves in all of them, including their self-acknowledged "duds". Winemaker Jason Parkes and his team may liken themselves to punk rockers, keeping their awards in a toilet, but their Cabernet Franc is serious enough for any traditionalist oenophile to appreciate.

✓ *Dynasty Red • Monarch Syrah • Melville Vineyard Chardonnay • Desert Valley Cabernet Franc • Palo Solera Vineyard Pinot Noir*

LITTLE FARM WINERY

Similkameen Valley, Mulberry Tree Vineyard
★★★

Nestled in picturesque Cawston, this boutique estate is owned and operated by Rhys Pender MW and Alishan Driediger. This talented duo is making bone-dry, racy, intense wines that reflect the calcium carbonate soils of the Mulberry Tree Vineyard. Little Farm uses old neutral barrels, wild ferments, and lees ageing to give texture and complexity. This creates balance with the intense, racy minerality they naturally get from the site. Their Pied de Cuve series is a hands-off, dry, intense wine style that reflects the *terroir*. Little Farm is a pioneer as one of the first to use old neutral barrels to ferment and age Riesling. The estate was also an early adopter of making orange wine and a pioneer of low-intervention winemaking in British Columbia

✓ *Pied de Cuve Riesling • Pied de Cuve Chardonnay*

MISSION HILL FAMILY ESTATE

Okanagan Valley
★★★

Mission Hill Family Estate is a leader in Canadian wine in almost every aspect of their operations. At this family-owned estate winemaker Ben Bryant is making meticulously crafted, award-winning wines. The estate is committed to sustainable practices, and even their winery is architecturally iconic. Within their broad portfolio of wines produced, their top "Legacy" collection contains Perpetua Chardonnay, which is complex and balanced. Several red wines in this collection are also of note, from Bordeaux blends to Pinot Noir.

✓ *Vista's Edge Cabernet Franc • Jagged Rock Vineyard Chardonnay • Prospectus • Compendium • Oculus • Quatrain • Silver Ranch Riesling Icewine*

MOON CURSER VINEYARDS
Okanagan Valley, Osoyoos East Bench
★☆

Moon Curser Vineyards is a small estate with a big reputation. Beata and Chris Tolley are growing unusual varieties, such as Tannat, Dolcetto, Touriga Nacional, and Arneis, to name a few. These varieties have not historically been a part of South Okanagan viticulture, but nonetheless thrive here.

 ✓ *Malbec • Touriga Nacional • Syrah • Tannat • Carménère • Dead of Night*

NK'MIP CELLARS
Osoyoos
★☆

According to Nk'Mip Cellars, as the first indigenous-owned winery in North America, they are inspired to express their culture in everything they do. Winemakers Randy Picton and Justin Hall craft award-wining wines at the scenic property overlooking the Osoyoos Lake. Standouts include their Riesling and Bordeaux varieties. Their Meritage blend, both red and white "Mer'r'iym", are complex and elegant.

 ✓ *Riesling • Chardonnay • Merlot • Pinot Noir • Syrah • Mer'r'iym Red & White • Riesling Icewine*

PAINTED ROCK
Okanagan Valley
★★

A true family collaboration, the Skinner family focuses primarily on full-bodied reds including Bordeaux varieties and Syrah at Painted Rock. A mainstay in the region, the estate continues to be focussed on not only top-shelf wine production but also sustainable practices.

✓ *Chardonnay • Red Icon • Cabernet Franc • Cabernet Sauvignon • Merlot • Syrah*

QUAILS' GATE WINERY
Okanagan Valley
★★

A mainstay in the region, this estate is known for their Pinot Noir and Chardonnay. The Stewarts family prides themselves on their balance between innovation and sustainability.

✓ *Rosemary's Block Chardonnay • Pinot Gris • Family Reserve Pinot Noir • Dry Riesling*

ROAD 13 VINEYARDS
Oakanagan Valley, Golden Mile Bench
★★

With winemaker Jeff Del Nin leading the charge, the estate focusses on big, ripe, full-bodied reds, especially Rhône varietals. Their flagship wine – called 5th Element – is both age-worthy and complex. The estate also boasts some of the oldest vines in Canada with Chenin Blanc planted in 1968.

 ✓ *Sparkling Chenin Blanc • Blind Creek Collective Chardonnay • Blind Creek Collective Cabernet Sauvigonon • GSM • Jackpot Petit Verdot*

A VINEYARD OVERLOOKS OKANAGAN LAKE IN NARAMATA, BRITISH COLUMBIA
The Okanagan Valley region is the province's top wine-producing area, with many of its vineyards planted with Bordeaux varieties.

TANTALUS VINEYARDS
Okanagan Valley, Kelowna
★★☆

The site where Tantalus now sits was first planted to table grapes in 1927, and today this estate is one of the oldest continuously producing vineyards in British Columbia. Winemaker David Paterson and owner Eric Savics proudly produce wines which are almost exclusively single-vineyard selections.

 ✓ *Blanc de Blancs • Old Vines Riesling • Chardonnay • Reserve Pinot Noir*

LE VIEUX PIN
Okanagan Valley, Oliver
★★★

Beyond the estate's majestic old-growth pine tree that lent the moniker, Le Vieux Pin is making consistently beautiful wines. Winemaker Severine Pinte is fastidious with minimal intervention, minimal additions, and low sulphites.

 ✓ *"Ava" Viognier Roussanne Marsanne • Syrah Cuvée Violette • Syrah Cuvee Classique • Equinoxe Syrah*

NOVA SCOTIA

BENJAMIN BRIDGE
Gaspereau Valley
★★★

Producing arguably the nation's most stunning traditional method sparkling wines, Benjamin Bridge is a leader in the Nova Scotia winemaking scene. The stylistic pursuit of head winemaker Jean-Benoit Deslauriers is in harmony with the Bay of Fundy's tidal influence, which works in concert with their cool-climate *terroir*. Styles range from their quaffable Pétillant Naturel and aromatic Nova 7 to

their top-shelf Méthode Classique Brut Reserve made from Chardonnay and Pinot Noir. Vintage-dated Blanc de Blancs and Blanc de Noirs are splendidly complex and elegant, spending upward of eight years on the lees. Unsurprisingly, this estate is making waves internationally, recognized for their quality, complexity, and elegance.

 ✓ *NOVA 7 • Cabernet Franc Rosé • Brut Reserve • Blanc de Noirs • Blanc de Blancs*

BLOMIDON ESTATE WINERY
Annapolis Valley
★★

This boutique estate is tucked into the Minas Basin, off the Bay of Fundy. Embracing their cool-climate *terroir*, the Ramey family, along with winemaker Simon Rafuse, is producing exceptional traditional sparkling wines. The crown jewel, gaining international notoriety, is their Blanc de Blancs from 100 per cent Chardonnay, which spends over 80 months on the lees. Also not to miss is their Blanc de Noirs, which lends lovely texture and finesse to the palate.

✓ *Tidal Bay • Blanc de Blancs • Blanc de Noirs • Brut Réserve • Riesling • Pinot Noir*

DOMAINE DE GRAND PRÉ
Annapolis Valley, Grand Pré
★☆

The Stutz Family are the proud stewards of the oldest winery in the province of Nova Scotia, which is nestled in the Annapolis Valley near the Minas Basin. Domaine de Grand Pré produces fresh and aromatic white wines, in particular their Tidal Bay bottling, which contains L'Acadie Blanc, Seyval Blanc, Vidal Blanc, Ortega, and New York Muscat varieties. The estate also embraces the

region's major agricultural product, apples; as such the Stutz' produce dessert apple wine, Pomme d'Or Ice Cider, which is wildly delicious.

 ✓ *Tidal Bay • L'Acadie Blanc • Pomme d'Or Ice Cider*

LIGHTFOOT & WOLFVILLE VINEYARDS
Annapolis Valley
★★☆ B

The Lightfoot Family has been farming in the Annapolis Valley for eight generations and on their current land, where the Home Farm Vineyard is today, they have been farming for four generations. The estate is fully certified biodynamic by Demeter, and the vineyards of the estate are planted along the coast of the Atlantic Ocean, influenced by the Bay of Fundy. This cool-climate estate focusses on Chardonnay and Pinot Noir. From these head winemaker Josh Horton crafts the "Ancienne" series of still wines, along with top-shelf traditional sparkling wines.

 ✓ *Terroir Series Riesling • Blanc de Blanc Brut Nature • Blanc de Blancs Brut • Pinot Rosé • Ancienne Chardonnay • Ancienne Pinot Noir*

LUCKETT VINEYARDS
Gaspereau Valley
★★☆

This vineyard overlooks and is influenced by the Bay of Fundy. Since 2010, it has made wine to showcase the local and unique grape varietals grown in rocky and clay-rich soil of this maritime area.

✓ *Lucie Kuhlman, Leon Millot and Marechal Foch (Buried Red)*

ONTARIO

BACHELDER NIAGARA

Niagara Peninsula

★★★

As a self-styled micro-*negociant*, winemaker Thomas Bachelder, along with co-owner Mary Delaney-Bachelder, works with top growers to produce wines from vineyards that truly express the Ontario *terroir*. They are responsible for some of the country's top cool-climate wines including delicate and complex Pinot Noir and finessed Chardonnay.

✓ *Chardonnay (Wismer-Wingfield)* • *Pinot Noir (Lowrey Old Vines)*

BIG HEAD WINES

Niagara Peninsula, Four Mile Creek

★★☆

The Lipinski family owns and operates this boutique Niagara estate with patriarch Andrzej Lipinski making varieties such as Chenin Blanc, Savagnin, Syrah, Petit Verdot, and Malbec. Big Head uses wild fermentation, aiming for balance and maximizing the flavour potential of each variety. Interestingly, *appassimento* is used for Bordeaux varieties. Although Andrzej does not have formal wine training, he is producing excellent wines.

✓ *Big Red* • *Merlot Select* • *Pinot Noir Select* • *The Biggest Red* • *Petit Verdot Select*

CAVE SPRING VINEYARD

Niagara Peninsula, Niagara Escarpment, Beamsville Bench, and Lincoln Lakeshore

★☆

Sustainable practices in the vineyard and minimal intervention in the cellars ensure that these wines reflect their cool-climate origin. The estate's Riesling CVS is made from top parcels established by the Pennachetti family dating back to the 1970s. Always bottling single varietals, the estate is well known for their Rieslings, but also produces *terroir*-driven Cabernet Franc of note.

✓ *Chardonnay (CVS)* • *Riesling (CVS)* • *Riesling Icewine*

CHARLES BAKER WINES

Niagara Peninsula

★★★

As Ontario's first virtual winery, this estate exclusively produces Riesling from top sites in Niagara. Baker, along with winemaker Jean-Laurent Groux, pride themselves in getting the Riesling from the vineyard to the glass without interference. Through bottlings, such as the Picone and Ivan Vineyard, expressing vivacity, steely tension, and complexity it is no wonder that this estate is arguably making the county's finest dry Riesling.

✓ *Riesling (Picone Vineyard)* • *Riesling (Ivan Vineyard)*

CHÂTEAU DES CHARMES

Niagara-on-the-Lake, St. David's Bench, and Four Mile Creek

★☆

This well-established winery is owned and operated by the Bosc Family, with Amélie Boury making wine. Top wines include the fabled Gamay Noir "Droit". The estate focuses on respecting *terroir*

and fruit with balance and purity, especially their St. David's Bench Vineyard and Pinot Noir.

✓ *Pinot Noir (Paul Bosc Estate Vineyard)* • *Gamay Noir "Droit"*

CLOSSON CHASE VINEYARDS

Hillier, Prince Edward County

★★☆

Since 2015 Keith Tyers has been producing this estate's signature *terroir*-focussed wine. Fractured limestone and a boundary-pushing cool climate create brisk acidity and note-worthy minerality, while great care creates complex and finessed wines that are both elegant and refined.

✓ *Churchside Chardonnay* • *South Clos Pinot Noir*

FLAT ROCK CELLARS

Niagara Peninsula, Twenty Mile Bench

★★★

Edward Madronich is known for his big personality and warm welcomes at this state-of-the-art, five-tier gravity-flow facility. Certified sustainable, this property produces a range of styles. The fruit is ripe, but elegant, with their Pinot Noirs expressing sun-kissed cherries and wild raspberries on the mid-palate.

✓ *Gravity Pinot Noir* • *Rusty Shed Chardonnay* • *Nadja's Vineyard Riesling* • *Riddled Sparkling*

GRANGE OF PRINCE EDWARD VINEYARDS AND ESTATE WINERY

Prince Edward County, Hillier

★★

Caroline Granger goes by many titles: founder, winemaker, mother, and local leader. Her County Crémant Series showcases the best of her wines and a significant style for Prince Edward County. No new oak, limited intervention, health conscious vineyard practices all drive towards one goal – wines that are made in the vineyard. The Grange is run entirely by women who feel that positivity is something you can taste in every bottle – and they are right.

✓ *County Crémant Sparkling (Amber, Citrine, Quartz)* • *Botanical Chardonnay & Pinot Noir*

HIDDEN BENCH ESTATE WINERY

Niagara Peninsula, Beamsville Bench

★★☆

This Niagara flagship focuses on Pinot Noir, Chardonnay, and Riesling. In addition, winemaker Jay Johnston, and proprietor-*vigneron* Harald Thiel produce limited quantities of white and red Meritage, as well as traditional-method sparkling wine. This winery proudly makes organic estate wines and continues to be focussed on producing age-worthy, mineral-driven wines.

✓ *La Brunate* • *Rosomel Vineyard Pinot Noir* • *Felseck Chardonnay* • *Nuit Blance*

HINTERLAND WINE COMPANY

Prince Edward County

★★☆

The enterprising duo of Vicki Samaras and Jonas Newman are making some of Canada's best sparkling wines. Their

range includes many styles of fizz from *charmat* to ancestral, as well as traditional. Their top bottle "Les Etoiles" is made in minute quantities and is a blend of Chardonnay and Pinot Noir with a fine *mousse* and long finish worthy of any top-shelf comparison. It is no wonder Hinterland is making waves internationally for their breath-taking bubbles.

✓ *Les Etoiles* • *Ancestral Rosé*

INNISKILLIN WINES

Niagara-on-the-Lake

★

Austrian native Karl Kaiser stands as a true visionary in the Canadian wine landscape. Kaiser, along with Donald Ziraldo, founded Inniskillin Wines in 1974, which was the first in Ontario since Prohibition. Kaiser's vision of quality Canadian wine was executed via his meticulously high standards and thus started a new era for the region and the country. His 1989 Inniskillin Icewine won the Grand Prix d'Honneur at Vinexpo in Bordeaux. The winery still is known for its icewines.

✓ *Icewine*

KARLO ESTATES

Hillier, Prince Edward County

★☆

At this award-winning estate Sherry Karlo and winemaker Derek Barnett make Old World–style wines that are fruit forward and clean on the palate. It is the first vegan-certified winery in North America. Along with the classic varieties of the region, Pinot Noir and Chardonnay, the estate produces Malbec, Carménère, and Port-styled wines.

✓ *Estate Pinot Noir* • *Lae on the Mountain Pinot Noir* • *Estate Chardonnay*

MALIVOIRE WINE

Niagara Peninsula, Beamsville Bench

★☆

Shiraz Mottiar continues to make expressive single-vineyard Gamay. As Canada's first gravity-flow winery, this estate also produces classically cool-climate and vibrant Chardonnay, rosé, and Pinot Noir.

✓ *Small Lot Gamay* • *Small Lot Pinot Noir* • *The Colleen Cabernet Franc* • *Moira Cat on the Bench Chardonnay*

PEARL MORISSETTE

Niagara Peninsula

★★☆

François Morissette prides himself in his non-interventionist style, which at times flies in the face of the VQA standards. The estate embraces this adversarial relationship so far as to name their flagship Riesling "Cuvée Black Ball". While opinions continue to vary on this style, any oenophile can agree and appreciate that the estate's Cabernet Franc Cuvée Madeline is complex and age-worthy.

✓ *Chardonnay (Cuvée Dix-Neuvième)* • *L'Oublié* • *Cabernet Franc (Cuvée Madeline)* • *Pinot Noir* • *Svet Nat*

SOUTHBROOK VINEYARDS

Niagara Peninsula, Four Mile Creek

★★☆

Southbrook is owned and operated by the Redelmeier family. This estate became

Canada's first organic-certified and Demeter-certified biodynamic winery and vineyard in 2008. Canadian wine icon Ann Sperling has been the estate's winemaker since 2005. Beyond the flagship Chardonnay and Cabernet Sauvignon, Southbrook proudly produces skin-contact white wines.

✓ *Poetica Chardonnay* • *Poetica Red* • *Laundry Vineyard Pinot Noir* • *Vidal Icewine*

STANNERS VINEYARD

Hillier, Prince Edward County

★★★

For such a small family operation, Stanners Vineyard has a big reputation in the region. Although several white wines are produced, this estate in is best known for their wonderful red wine. Pale and elegant, perfumed and complex, their Pinot Noir is a standout. This estate, located in Hillier, captures the beauty of the province's *terroir* and proves that world-class wines are being made in Prince Edward County.

✓ *Pinot Noir*

STRATUS VINEYARDS

Niagara-on-the-Lake, Niagara Lakeshore

★★☆

Always considered a top winery in the Canadian scene Stratus Vineyards was the first winery in the world to be fully LEED (Leadership in Energy and Environmental Design) certified. Winemaker Jean-Laurent (J-L) Groux crafts their flagship Stratus Red, which varies year to year but can contain all of the Bordeaux varieties, in addition to Tannat and Syrah. This 22-hectare (55-acre) estate is also dedicated to sustainability and quality.

✓ *Stratus Red* • *Stratus White* • *Riesling Icewine* • *Semillon* • *Sauvignon Blanc* • *Decant Cabernet Franc*

TAWSE WINERY

★★★

Niagara Peninsula, Vineland

Any discussion of classic Niagara Chardonnay and Pinot Noir cannot be complete without mention of winemaker Paul Pender. He is both an innovator and a mainstay, who is producing wines that are elegant, as well as showing finesse and pure varietal character. This estate is certified organic and biodynamic.

✓ *Robyn's Block Chardonnay* • *Cherry Avenue Pinot Noir*

QUEBEC

VIGNOBLE LES PERVENCHES

Quebec

★☆

This ambitious estate is growing a number of hybrids, such as Seyval Blanc, Marshal Foch, and Frontenac, as well as the Zweigelt and Chardonnay varieties. The estate has also achieved biodynamic and organic certification via Demeter and Ecocert respectively. Their Chardonnay, Le Couchant, is creating buzz among fans of natural wine.

✓ *Chardonnay (Le Couchant, Le Feu, and Les Rosiers)*

Mexico

Just one year after the conquista of Mexico, in 1521, the Spanish were already planting vines and producing wine, making Mexico the oldest wine-producing country in the Americas.

In 1524 Hernán Cortés, then governor of New Spain, decreed that all of the nascent colony's Spanish residents who had been granted land by the Crown would have to plant 1,000 vines for every 100 indigenous workers under their charge.

During the 16th century production increased dramatically and many vineyards, such as Hacienda de San Lorenzo (currently known as Casa Madero), were founded, leading to Mexico becoming self-sufficient in wine production towards the end of the century. However, this angered many Spanish producers, and bowing to domestic pressure, King Phillip II of Spain, halted the planting of new vines all over the New World. Production stopped until the mid 19th Century when some vineyards, such as Bodega Ferriño in Coahuila (1860) and Santo Tomás in Baja California (1888), were rebuilt and began to produce wine on a small scale.

The industry remained relatively dormant until the 1950s to 1970s, when a resurgence in small vineyards took place. Domecq (1958) and L.A. Cetto (1964) in Baja California and La Redonda (1972) in Queretaro were some of the first of these new wineries to take root. Production began to creep up, and many of the current wine-producing regions came into existence. That was until the mid-1990s, however; the signing of the North American Free Trade Agreement and the Mexico–EU Free Trade Agreement changed things. The availability of cheaper imported alternatives to Mexican wine led to an industry-wide upheaval, and many wineries simply went out of business. The inability for many wineries to compete with their international counterparts led to a modernisation of the industry, and the focus shifted to quality and national pride in Mexican products.

FACTORS AFFECTING TASTE AND QUALITY

LOCATION
Grape are grown in Mexico from Baja California in the north to San Juan del Rio, just north of Mexico City in the south.

CLIMATE
Half of Mexico lies south of the Tropic of Cancer, but altitude moderates the temperature of the vineyards. Most are situated on the high central plateau, and some are cooled by the nearby ocean. Principal problems include extreme fluctuation of day and night temperatures, and the fact that most areas have either too little or too much moisture. The dry areas often lack adequate sources of water for irrigation, and the wet districts suffer from too much rain during the growing season.

ASPECT
In the states of Aguascalientes, Querétaro, and Zacatecas, vines are grown on flat plateau lands and the sides of small valleys, at altitudes of 1,600 metres (5,300 feet), rising to nearly 2,100 metres (7,000 feet) in Zacatecas State. In Baja California, vines are located in valley and desert areas at much lower altitudes of between 100 and 335 metres (330 and 1,100 feet).

SOIL
The soils of Mexico can be divided into two wide-ranging categories: slope or valley soils are thin and low in fertility, while plains soils are of variable depth and fertility. In the Baja California, the soils range from a poor, alkaline sandy soil in Mexicali to a thin spread of volcanic soil, which is intermixed with gravel, sand, and limestone to provide excellent drainage. In Sonora, the soils of Caborca are similar to those found in Mexicali, but those in Hermosillo are very silty and of alluvial origin. The high plains of Zacatecas have mostly volcanic and silty-clay soils. In the Aguascalientes, the soil in both the valley and the plains is of a scarce depth with a thin covering of calcium. The volcanic, calcareous sandy-clay soil in Querétaro has a good depth and drainage and is slightly alkaline, while in La Laguna the silty-sandy alluvium is very alkaline.

VITICULTURE & VINIFICATION
Irrigation is widely practised in dry areas such as Baja California and Zacatecas. Most wineries are relatively new and staffed by highly trained oenologists.

GRAPE VARIETIES
Primary varieties: Cabernet Sauvignon, Merlot, Cariñena (Carignan), Chenin Blanc, Grenache, Mission, Nebbiolo, Shiraz, Tempranillo, cabernet Franc, Chardonnay, Sauvignon Blanc, Petite Sirah (Durif), Zinfandel
Secondary varieties: Barbera, Colombard, Malbec, Viognier

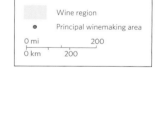

	Wine region
•	Principal winemaking area

0 mi ... 200
0 km ... 200

WINEMAKING AREAS OF MEXICO
(see also p569)
There are three major wine regions in Mexico: Central Mexico includes the areas around the country's capital – Querétaro, Guanajuato, San Luis Potosí, Zacatecas, and Aguascalientes; the La Laguna region includes Coahuila and can boast of Valle de Parras, home of the oldest winery in Mexico; and the North, which includes Baja California, Chihuahua, and Sonora. At just north of the 30° latitude line, Baja California accounts for the vast majority of Mexican wine production (about 85 per cent).

MONTE XANIC VINEYARDS IN THE VALLE DE GUADALUPE, BAJA CALIFORNIA
Baja California produces a vast majority of Mexico's wine, and on the peninsula, the Valle de Guadalupe can truly be considered Mexico's wine country,

MODERN MEXICAN WINE

Since the early 2000s changes in the market meant that wineries had to change their practices in order to survive and thrive. These changes have led Mexico's wine production to blossom into a vibrant countrywide industry. As of 2019, there were approximately 33,000 hectares (81,545 acres) of vines planted in Mexico with 69 per cent produced for table grapes, 11 per cent for raisins, and 20 per cent for wine production, for a total of approximately 6,500 hectares (16,060 acres) producing grapes for wine. These 6,500 hectares grow -between 5 and 8 tonnes of grapes per hectare annually, which produces approximately 25.7 million bottles.

The wine produced in Mexico ranges in both style and quality, but it is generally accepted that the quality is improving year on year. As of 2019 more than 750 labels have won awards nationally, and internationally and this number looks set to increase as participation in international awards becomes the yardstick for wineries in Mexico. A few large producers dominate wine production. The largest two, L.A. Cetto and Casa Madero, account for almost 75 per cent of the entire countrywide production, and if we include Freixenet and Santo Tomás this percentage rises to almost 85 per cent, leaving only 15 per cent for the approximately 255 remaining wineries.

WINE CONSUMPTION IN MEXICO

The consumption of wine in Mexico, excluding sacramental wine, is a relatively new phenomenon. Traditionally drinks like tequila, mescal, and pulque were most widely consumed; however, the explosion of beer production has led to an increasing consumption of domestically produced beer. Although consumption levels of beer are not as high as other Latin American countries, beer has become the de facto alcoholic beverage in the country because it is easily accessible and at an affordable price.

Wine in Mexico has always struggled with high taxes, and therefore the price point has been less accessible than other alcoholic beverages. This has led to wine being viewed as a luxury product, available only to the middle and upper classes. As the economy has grown stronger, however, and the middle classes swelled in number, wine is increasingly being viewed as an aspirational product to many young consumers. This has led to consumption more than doubling since the year 2000.

As of 2018, there are approximately 2.7 million regular wine drinkers in the country, and almost half of these are made up of young people between the ages of 18 and 34. Men account for 58 per cent of consumers with women making up the remaining 42 per cent. Yet more than half of the population consume no wine whatsoever, with only 6.6 per cent of the population considered frequent consumers. Overall Mexico consumed almost 89.5 million litres of wine in 2018, making it the 45th-largest consumer worldwide. Of the wine consumed, only 30 per cent is domestic and 70 per cent is wine imported mainly from Chile, Argentina, and Spain. The average price per bottle is around 80 Mexican pesos, which equates to roughly 4 US dollars. By far the most popular style of wine consumed is red wine, which makes up 71 per cent of consumption. This is followed by white wine (11 per cent), sparkling wine (9 per cent), and other styles making up the remaining 9 per cent.

MEXICAN LABEL LANGUAGE

Many terms found on Mexican wine labels are the same as, or similar to, those seen on Spanish labels (*see* Spanish Label Language p393). Some common terms are listed below.

Bodega Winery

Combinados Indicates a blend of different grape varieties

Contenido neto Contents

Cosechas Seleccionadas Special blend

Hecho en México Made in Mexico

Seco; extra seco Dry; extra dry

Variedad Grape variety

Varietales Single-varietal wines

Viña Vineyard

Vino wine

Vino Blanco White wine

Vino Espumoso Sparkling wine ("foamy wine")

Vino de Mesa Table Wine

Vino Tinto Red wine

MAJOR WINE GROWING REGIONS OF
MEXICO

Note: Within the major regions, there are smaller principal growing areas. These are listed under the regional name.

AGUASCALIENTES

Calvillo, Paredón, Los Romo

Inhabitants of Aguascalientes began cultivating wine as early as the 1790s; however, most of the current wineries were founded in the last 20 years. The area is growing and developing very quickly because there is a decent level of state aid to wineries in order to promote agricultural products and tourism.

Vineyards are planted high up (1,800 to 2,000 metres, or 5,905 to 6,560 feet) in a semi-desert climate, with very high diurnal temperature changes that receive an average of 55 centimetres (22 inches) of rainfall per annum. Well-draining, stony soils of gravel, clay, and sand patches with a chalky subsoil.

🍷 *Cabernet Sauvignon, Malbec, Nebbiolo, Syrah, Carignan, Touriga Nacional, Chardonnay, Macabeo, Moscatel and Chenin Blanc.*

BAJA CALIFORNIA

Valle de Guadalupe, Valle de Calafia, Valle de San Antonio de las Minas, Valle de Tecate, Valle de Santo Tomas, Valle de Ojos Negros, Valle de San Vicente

By far Mexico's largest winegrowing region with more than 120 producers and 4,610 hectares (11,390 acres) under vine, accounting for more than 70 per cent of the total vine-growing area of the country. The region has a history of international ties with many of the original producers being of Russian, Spanish, or Italian descent. Located on the Baja California peninsula, the majority of the vines are grown in valleys at 100 to 800 metres (330 to 2,625 feet) in altitude that are cooled by morning sea breezes and occasional fogs, while warming in the afternoon. The Mediterranean climate with cooling winds arriving from the bay of Alaska causes high diurnal temperature changes, with daytime temperature reaching 42°C (107°F) in summer and then lowering to 12°C (53°F) in winter.

In the north of the state near Mexicali the soil tends to be slightly alkaline and rather thin, consisting predominantly of sand. Farther south towards the Valles de Ensenada the soil consists of a thin volcanic layer mixed with gravel, sand, and limestone. Continuing south there are also some elements of red clay in the soil.

The region itself has been quite susceptible to droughts; however, good winemaking practices and the use of technology stemming from both international investment and education programs mean that the overall quality of the wines has improved markedly and continues to do so. Wines from the region are generally high in alcohol, full bodied with ripe, sun-kissed flavours. As there are no rules governing Mexican winemaking, many winemakers use their creativity to employ unique and unorthodox blends that you might not find anywhere else.

🍷 *Cabernet Sauvignon, Nebbiolo, Tempranillo, Petite Sirah, Merlot, Grenache, Cabernet Franc, Petit Verdot, Carignan, Syrah, Barbera, Sangiovese, Chardonnay, Sauvignon Blanc, Chenin Blanc*

CHIHUAHUA

Delicias, De los Encinos, Bachíniva

Chihuahua has a history of wine production, and it is currently being seen as one of the regions with the greatest potential to produce some excellent wines. Many large wineries, including L.A. Cetto, are investing heavily in the region believing in that high potential for production. This is mainly due to its altitude (above 1,500 metres, or 4,920 feet), fine gravel and clay soil, warm and dry climate (40 to 50 centimetres, or 8 to 20 inches, annual rainfall) and, most important, its access to water.

🍷 *Cabernet Sauvignon, Merlot, Cabernet Franc, Petit Verdot, Syrah, Gewürztraminer and Chardonnay.*

COAHUILA

Parras, Cuatro Ciénagas, General Cepeda, Saltillo.

Coahuila is the oldest winegrowing region in the Americas, with more than 420 years of history. The oldest vines were planted in Parras. With its founding dating to 1597, Casa Madero is the oldest winery that still operates in the region, but several other wineries are some of the most long-lived on the continent.

With 680 hectares (1,680 acres) under vine and more than 30 producers, today Coahuila still plays a major role in Mexico's wine scene. Vineyards are planted at 1,500 to 2,650 metres (4,920 to 8,695 feet) in altitude, with a range of climates depending on the area (dry, semi-desert, and even temperate climates can be found) in nutrient-rich, varied clay-sandy soils. The region typically has cold winters and hot summers with temperatures ranging from highs of 33°C (91°F) in the summer to an average temperature of around 15°C (59°F) in winter. Rain tends to fall over the summer months from April to October with roughly 20 to 40 centimetres (8 to 16 inches) of rainfall per year.

🍷 *Cabernet Sauvignon, Merlot, Syrah, Grenache, Mouvedre, Chardonnay and Chenin Blanc.*

GUANAJUATO

Dolores Hidalgo, San Miguel de Allende, Guanajuato, San Felipe

Mountainous Guanajuato is one of the fastest-growing wine regions in Mexico, with the number of vines having tripled to 350 hectares (865 acres) between 2010 and 2019. Its unique position in Mexico as one of its premier tourist destinations means that international demand is driving increased production. It is also home to the Museo del Vino (Wine Museum), which is helping to highlight the importance of wine historically in the region, while also helping producers with the latest technology and knowledge.

🍷 *Cabernet Sauvignon, Merlot, Tempranillo, Aglianico, Syrah, Merlot, Sauvignon Blanc, Chardonnay and Semillon.*

A CROSS STANDS AT THE ENTRANCE TO CASA MADERO IN PARRAS DE LA FUENTE
Established in the 16th-century, Casa Madero is usually recognized as the oldest winery in Mexico.

QUERÉTARO

Ezequiel Montes, Tequisquiapan, Bernal

Querétaro has become one of the country's most well-known wine-producing regions due to its investment in tourism. The region now welcomes millions of tourists per year, and wine is one of the driving forces behind this growth. Querétaro is located in the centre of the county, and its close proximity to Mexico City and other major urban areas means that it is one of the most accessible wine regions in the country. Querétaro is also the nation's biggest producer of sparkling wine.

Located in the Altiplano in the centre of Mexico, Querétaro is a semi-desert region that is mainly dry and warm, with diurnal temperature changes of up to 15°C (59°F). Rain is scarce, but there are short periods of very intense rain, normally in June and September, which averages out to between 35 to 60 centimetres (14 to 24 inches) of rainfall per year.

The 420 hectares (1,038 acres) of vineyards are shared between 28 producers who planted vines between 1,700 and 2,200 metres (5,575 to 7,220 feet) above sea level. The majority of soil is clay based but, in some areas, this can be mixed with varying levels of sand, the limestone bedrock means that the soil can be slightly alkaline.

🍷 *Malbec, Merlot, Cabernet Sauvignon, Pinot Noir, Gamay, Viura, Parrellada, Xarel-lo, Chardonnay, Sauvignon Blanc, Colombard and Pinot Gris.*

SAN LUIS POTOSÍ

Villa de Arista, Moctezuma, Altiplano Potosino, Soledad de Graciano Sanchez

A Spanish settler fleeing Franco's dictatorship started wine production in San Luis Potosí. He planted vines in order to keep alive the tradition of what was produced back home in Cataluña. More recently San Luis Potosi has been recognised as an area in which white grapes retain their acidity compared with other regions, meaning that some of the best white wines in the country come from this area.

🍷 *Merlot, Syrah, Malbec, Cabernet Sauvignon, Sauvignon Blanc, Semillon and Rosa de Peru.*

SONORA

Sierra de Sonora, Agua Prieta, Cananea

With just 25 hectares (62 acres) of vines, Sonora is a small, high-altitude growing region (ranging in elevation from 1,400 to 1,700 metres, or 4,595 to 5,575 feet) located in the north of Mexico. Sonora has a desert climate, with very high temperatures during the summer and high diurnal temperature changes. The soil here is predominantly sand with some gravel.

🍷 *Malbec, Syrah, Tempranillo, Merlot, Nebbiolo, Chardonnay and Verdejo.*

ZACATECAS

Ojo Caliente, Valle de la Macarena

Zacatecas is one of the largest producers of grapes in the country, accounting for almost 15 per cent of the nation's production, yet only a small quantity of it is used for vinification. Most of the 20 wineries in the region have been founded in the last 10 to 15 years.

Vines are grown on 157 hectares (388 acres) at altitudes of 1,900 to 2,300 metres (6 235 to 7,545 feet), where relatively cool summers with higher levels of rain and humidity lead to high diurnal shifts. The area gets an average of 40 centimetres 16 (inches) of rainfall per year. Hail around harvest time can be an issue, and in 2018 almost 50 per cent of the harvest was lost due to hail. The soil in Zacatecas is mainly mineral rich clay with some sandy areas.

🍷 *Malbec, Cabernet Sauvignon, Syrah, Merlot, Tempranillo, Petite Sirah, Ruby Cabernet, Carignan, Sauvignon Blanc, Macabeo, Moscatel, Colombard and Ugni Blanc*

THE WINE PRODUCERS OF
MEXICO

AGUASCALIENTES

BODEGAS ORIGEN
Aguascalientes

This winery displays a postmodern philosophy, using a mix of traditional and modern tools in the winemaking process.

BODEGAS DE LA PARRA
Aguascalientes

This organic vineyard specialises in Malbec, producing the intense, deep purple Paradoja.

HACIENDA DE LETRAS
Pabellón de Arteaga

One of the largest wineries in the state, it was originally founded in 1854 and has been producing wine since 1978.

SANTA ELENA
Aguascalientes

The most well-known winery in the region produces Entrelineas, the second-most-widely sold label in Mexico. Founded in 2005, the winery uses a mix of traditional and modern methods.

BAJA CALIFORNIA

Along with the producers listed below, other boutique wineries in Baja California worth mentioning include Veramendi, Vena Cava, Villa Montefiori, Decantos Vinicola, Hacienda Guadalupe, Alximia, Epicentro, El Cielo, Las Nubes, Finca Carrodilla, JC Bravo, Casa Baloyan, and Concierto Enologico.

BICHI

Founders Noel and Jair Téllez made their first Bichi (which means "naked") vintage in 2014. They farm their vineyards biodynamically, with a focus on Rosa del Peru, Misión, Tempranillo, and Mystical grape varieties.

BODEGAS DE SANTO TOMÁS
Ensenada

The very first winery to be established in Baja California, Bodegas de Santo Tomás was founded in 1888.

CASA DOMECQ
Ensenada

This was the first commercial winery in Valle de Guadalupe and built a reputation for Mexico's best Cabernet Sauvignons. After changing hands several times, it was closed for a while. It is again open and under the care of new owners from Spain.

CASA MAGONI
Valle de Guadalupe

One of the pioneers of viticulture in the state. As did many current wineries in the region, they began by growing and selling grapes to other wineries, but now produce their own estate wines.

L.A. CETTO
Valle de Guadalupe

By far the largest winery in the region, L.A. Cetto is also the largest exporter of wine in the republic, exporting to more than 17 countries. Along with Santo Tomás, Cavas Valmar, and Casa Domecq, this producer is considered one of the founders of wine in the region.

CHÂTEAU CAMOU
Ensenada

The Favela family began by cultivating Bordeaux varieties like Merlot, Cabernet Sauvignon, and Cabernet Franc, adding others over the years, including Malbec, Nebbiolo, Sauvignon Blanc, and Viognier in 2017. They craft their wines in a state-of-the-art winery in the Valle de Guadalupe.

MONTE XANIC
Valle de Guadalupe

Founded in 1987, Monte Xanic was one of the first of the new generation of boutique wineries in Baja California. Along with others like Baron Balché, Adobe Guadalupe, Casa de Piedra, and Château Camou, it has been producing some of the highest quality "cult" wines in Mexico for the past 30 years and continues to be instrumental in changing public perception of the quality of Mexican wine.

VIÑEDOS DE LA REINA
Ensenada

This estate winery, based in the Valle de San Vincente, focusses on quality. It site selection has enabled it to be one of very few Pinot Noir producers in Mexico.

VINICOLA SIERRA VITA
Valle de Guadalupe

In the heart of the Valle de Guadalupe, this boutique artisanal winery practices minimal intervention in the vineyard. Worth the trip if you are looking for something different.

VINISTERRA VITIVINÍCOLA
Villa de Juárez

This modern, boutique winery typifies the "new Baja", where knowledge and a more scientific approach is used, analysing site and varietal selection in order to improve quality.

CHIHUAHUA

CAVALL 7
Ciudad Delicias

The largest producer in the region with almost 70 hectares (172 acres) of vines in 2019, the winery plans to increase production to 150 hectares (370 acres) over the next few years.

ENCINILLAS
Valle de Encinillas

Based in an old hacienda dating from 1707 and along the historic Camino Real (recently declared a World Heritage Site), Viñedos y Bodegas Encinillas has been making wine since 2004.

COAHUILA

BODEGAS FERRIÑO
Cuatro Ciénegas

Founded in 1860, Ferriño, just like many wineries in Mexico, started by producing brandy but now produces table wine.

CASA MADERO
Parras de la Fuente

Founded in 1597, Casa Madero is the oldest winery in the Americas, as well as Mexico's first certified organic vineyard. It is most famous for its label 3V, which is one of the most consumed wines in all of Mexico. Its highly anticipated label, 1597, was finally released in 2019 after more than 20 years of planning, aiming to be the best wine in Mexico.

VINOS DON LEO
Parras de la Fuente

Don Leo began his first vine planting in the year 2000. This family-owned and -operated winery is one of the very few kosher wine producers in the country.

GUANAJUATO

CAVA EL GARAMBULLO
San Miguel de Allende

This is an artisanal winery focusing on excellent quality.

CUÑA DE TIERRA
Dolores Hidalgo

Cuña de Tierra is one of Mexico's larger producers, and its wines have won international awards. The whites are Sémillon grapes, while the reds are blends.

OCTAGONO
San Felipe

This boutique/artisanal natural winery produces reds, as well golden Sémillons.

VIÑEDO SAN MIGUEL
San Miguel de Allende

The largest winery in the region is based in the beautiful town of San Miguel de Allende. It is managed by Cuña de Tierra (see above entry).

VIÑEDOS PÁJARO AZU
San Felipe

One of the highest-quality wineries in the region, Pájaro Azu has won many international accolades with its Guanamé label.

QUERÉTARO

BODEGAS DE COTE
Tunas Blancas

Founded in 2008, this is now the second-largest wine producer in the region,

with state-of-the-art facilities, heavily focusing on tourism. It relies completely on local grapes, using varieties such Cabernet Sauvignon, Tempranillo, Merlot, and Shiraz for its reds.

FREIXENET MÉXICO
Ezequiel Montes

This is the largest sparkling wine producer in Mexico, with the most-visited cellar in the country, drawing almost 300,000 visitors each year. It exports more than 25 per cent of its production to Japan and the United States.

LA REDONDA
Ezequiel Montes

Founded in 1972, La Redonda was the first grape producer in the area. Its wines are made from Cabernet Sauvignon, Merlot, Malbec, Tempranillo, and Chardonnay varieties. It is also a tourist destination, celebrating four wine festivals every year.

SAN LUIS POTOSÍ

CAVA QUINTANILLA
Villa de Arista

This estate began planting in 2011 and has since become the largest winery in San Luis Potosí. It produces good-quality wine using the latest technology, and it also invests heavily in attracting tourists.

POZO DE LUNA
Soledad

Launched in 2015, this is one of the pioneers of wine in the area, interestingly focussing on high-quality white wines.

VIÑA CORDELIA
Soledad

At more than 70-years-old, Viña Cordelia is one of the oldest wineries in Central Mexico. They produce a mistela (a fortified sweet wine) with grapes grown on the estate.

SONORA

UVAS DE ALTURA
Cananea

This is the largest producer in the area and the first Sonoran wine company located north of the state between Cananea and Agua Prieta. It is known for the 4S label, which refers to the four mountains that surround its vineyards.

ZACATECAS

TIERRA ADENTRO

This is the largest and oldest producer in the region with 40 hectares (100 acres) of vines and more than 35 years as a winery. It has many award-winning wines, the most noteworthy a Syrah.

Central America and the Caribbean

Wine has existed in the Caribbean and Central America for many years. Yet, the majority of this wine is not actually made from grapes. Confusingly many countries in this region refer to any alcohol produced from fruit as wine, and it is common to find pineapple wine in Costa Rica or ginger wine in Jamaica.

The rainy and humid climate in much of the region has, for years, been seen as an obstacle to growing vines and producing wine of quality, and few have even attempted. But there are a few areas where brave winemakers have decided to embark on this noble pursuit and run the risk. Although not traditionally a wine-growing region, it is an important market for many wine producers due to the growth in tourism, both in hotels and restaurants, as well as on cruise ships, making the region one of the highest consumers per capita.

CUBA

Most Cuban wineries started as joint ventures between the Cuban government and other national governments. Many of these still work under this model. Most wineries on the island grow some of the grapes they use, but it is also common practice to import juices from European or South American sources.

Bodegas del Caribe

Founded in the early 2000s, the winery imported 22 grape varieties from Spain in order to begin production. They produce red, white, and rosé wines under their label.

Bodegas San Cristobal

Founded as part of a Cuban-Italian partnership, which still exists today. Cuban grapes are mixed with Italian pulp in order to produce wine.

Bermejales Wine Estate

To the west of Havana, one of the country's better-known wineries producing from 12 hectares of vines.

Orestes Estévez

This 20-year-old Havana winery is based entirely on the city's rooftops. It blends locally grown grapes with fruit juice before fermentation.

WINEMAKER ORESTES ESTÉVEZ
Estévez uses condoms to produce tropical fruit wines. When the condom falls off, it indicates the fermentation process has ended, and the wine is ready to be bottled.

DOMINICAN REPUBLIC

Legend has it that when Hernan Cortez arrived here in 1504, he planted vines in Azua and made the first wine in the Caribbean. The country's only winery is retracing these steps to produce wine in the same location.

Ocoa Bay Winery

Based in Azua Province, this organic vineyard grows Colombard, Black Muscat, and Tempranillo in sandy soils of volcanic origins. They also produce a fruit wine made with mango and passionfruit.

US VIRGIN ISLANDS

Its St John's Winery was founded in 2016 but destroyed by hurricanes Irma and Maria, and the project was abandoned.

CURACAO

The short-lived Curacao Winery's last harvest was in 2016 due to legal issues with local government.

ARUBA

This island's Adore Wines mixes wine production with tourism and vineyard visits. It grows some grapes but also buys juices from abroad.

COSTA RICA

Based in San Jose, Viña el Espavey produces an artisanal wine made from the Isabella grape. The production is small, and they describe the wine as having a sweet-and-sour flavour profile.

Vinos Don Teofilo

Based in Puriscal in San Jose, it produces an artisanal wine.

Vinícola Costa Rica SA (VICOSA)

VICOSA was founded in 1997 as a joint partnership with the Taiwanese government and works with 11 national producers. For its white wines, it blends its estate-produced Ruby Seedless grapes with Sauvignon Blanc, Malvoise, and Chenin Blanc from Chile and Italy. For its reds, it blends its estate Isabella with Cabernet Sauvignon and Merlot. Interestingly, it has two harvests per year, the first in January/February and the second in July/August. It has six wines and three labels.

BAHAMAS

Bahama Barrells, the first winery in the Bahamas, was founded in 2017. Located in Nassau it is housed in a 1937 church built by the Sisters of Charity. It blends juices from Italy and Argentina.

PUERTO RICO

The small (3.2 hectares, or 8 acres) Bodegas Andreu Sole was planted in the early 1980s and produces Tempranillo and Merlot. The winery uses no sulphites and adds some port to the wine to increase its body. Its main label is called Doce Calles.

GUATEMALA

As in most of this region, Guatemala relies on imported grapes.

Vinicola Centroamericana

Started in the 1960s, working in a joint venture with an Italian organization, it uses grapes imported from Italy to make a sweet sparkling wine that is often mixed with fruits to make sweet beverages.

Vinos, Bodegas y Fabricas Carlos Kong

This producer uses imported grapes to produce sweet wines that are often mixed with fruits.

PANAMA

Although no wine is grown or made here, it is worth mentioning because it is may be the part of the world where wine costs the least, as there is almost no tax whatsoever and the country could be considered a hub for premium wines in the region

South America

Viticulture was introduced to the Americas by the Spanish: first to Mexico in 1521, and then farther afield as the conquistadors opened up other areas of the southern continent. Now in the 2020s, the largest producer remains Argentina, with Chile and Peru playing catch-up.

Argentina's Malbec has further cemented its reputation as both a good-value, bang-for-buck wine and also one from which some producers are able to coax very good examples. Chile doubled its planted hectares, and Peru increased output by three times since the last edition of this book. Sparkling wine is surging in most countries here, and many wineries seek to adventurously expand into new areas or reform older ones. The face of South American wine is changing, with more experimentation and greater risks being taken to see what the region is capable of.

South America's wine industries are inextricably linked to Spain's expansionist policies of the 16th century, although the conquistadors were not primarily concerned with the spread of viticulture. They were in South America to plunder gold for Ferdinand of Spain, and when the Indians grew tired of the coloured-glass beads traded for their treasures, the conquistadors took what they wanted by more direct and brutal means. In response, the Indians poured molten gold down the throats of captured soldiers, which no doubt quenched the Spanish thirst for the precious metal, but also served as a sardonic retort to the Christian missionaries who had forced them to drink wine as part of the Sacrament.

In more recent times, it has been the traditional beer-drinking culture of the local populations that has held back the development of South American wines as a whole. This phenomenon has even affected the two major wine-producing countries, Chile and Argentina. Brazil could be leading the way, however, as the switch to wine from traditionally popular drinks, such as spirits and beer, mostly amongst the younger generation, makes this one of the few countries in the world where wine consumption is actually growing.

WINE REGIONS OF SOUTH AMERICA
Climatic conditions and inhospitable terrain prevent much of South America from producing wine, but such is the size of this continent that most countries have some vineyards. Argentina and Chile are the top producers.

Caribbean Sea
Barranquilla
Caracas
Lara
VENEZUELA
San Cristóbal
Medellín
Valle del Saquenzipa
Bogotá
Cali
COLOMBIA
Orinoco
Boa Vista
GUYANA
SURINAME
French Guiana Fr.
PACIFIC OCEAN
Mira
Quito
Yaruquí
ECUADOR
Guayaquil
Negro
Manaus
Amazon
Santarém
Belém
São Luís
Fortaleza
Marañón
Juruá
Madeira
Tapajós
Xingu
Chiclayo
Porto Velho
BRAZIL
Tocantins
Vale do São Francisco
Juàzeiro
Recife
Cerro de Pasco
Lima
Iténez
São Francisco
Salvador
Cusco
Ica
BOLIVIA
Arequipa
La Paz
Brasília
Moquegua
Tacna
Cochabamba
Sucre
Chuquisaca
Campo Grande
Potosí
Belo Horizonte
Tarija
PARAGUAY
São Paulo
Rio de Janeiro
Bermejo
Asunción
Curitiba
ATLANTIC OCEAN
Paraná
Salado
Planalto Catarinense
Campos de Cima da Serrra
CHILE see p655
Campanha Gaúcha
Serra Gaúcha
Porto Alegre
Córdoba
Salto
Rivera
Serra do Sudeste
Paysandú
URUGUAY
Valparaíso
Florida
Santiago
Colonia
Canelones
Buenos Aires
Maldonado
Montevideo
Concepción
ARGENTINA see p661
San José
Smaller winemaking area
0 mi 500
0 km 500

SOUTH AMERICAN COUNTRIES: AREA UNDER VINE AND YIELD

COUNTRY	HECTARES (Acres)	HECTOLITRES	CASES	YIELD (Hectolitres per hectare)
Argentina	224,258 (554,153)	9,447	104,966,667	42 hl/ha
Chile	209,038 (516,543)	10,143	112,700,000	49 hl/ha
Brazil	86,408 (213,518)	1,257	3,966,667	15 hl/ha
Peru*	30,042 (74,235)	750	8,333,333	25 hl/ha
Uruguay	7,251 (17,918)	755	8,388,889	104 hl/ha
Bolivia	3,283 (8,112)	83	922,222	25 hl/ha
Colombia	2,592 (6,405)		-	-
Venezuela	1,247 (3,081)		-	-
Paraguay	372 (919)	16		
Ecuador	82 (203)		-	-

* Figures include production of Pisco

THE ANDES FORM A DRAMATIC BACKDROP FOR THE VIK CHILE'S AVANT-GARDE VINEYARD RETREAT

Its luxury hotel in Millahue, Chile, may have brought this vineyard to the world's attention, but the state-of-the art winery of Viña Vik also produces high-quality red wines. Chile is the South American continent's second-leading producer of wines, doubling its vineyard plantings in a bit more than a decade.

THE WINE PRODUCERS OF

SOUTH AMERICA

BOLIVIA

LA CONCEPCIÓN

La Concepción

☆

La Concepción was Bolivia's first modern winery, established in 1978, and it remains the best.

✓ *Cabernet Sauvignon*

BRAZIL

CAVE MARSON

Cotiporã

★

In 1887, the first members of the Marson family emigrated to Brazil from the Veneto region of Italy with the objective of developing international-quality wines that expressed the local *terroir*. Now in the hands of the grandchildren and great-grandchildren of pioneer Antônio Marson, this producer is known for fabulous sparkling sweet Moscatel from the Serra Gaúcha region.

✓ *Sparkling (Cave Marson Moscatel Espumante)*

COOPERATIVA VINÍCOLA AURORA

Bento Gonçalves

This enterprise began in 1931 when 16 families of grape producers in the municipality of Bento Gonçalves, in the Serra Gaúcha, gathered to form Vinícola Aurora. This wine cooperative now has more than 1,100 members, who amongst them account for more than one-third of the country's vineyards. There are various labels, including Conde Foucolde and Clos de Nobles. This might be Brazil's second-largest winery and, in a commercial sense, very successful, but in terms of quality, with the exception of an excellent sweet Moscatel, Vinícola Aurora could do much better.

✓ *Dessert (Moscatel)*

CAVE GEISSE

Bento Gonçalves

☆

This award-winning winery is under the guidance of Mario Geisse, formerly of Chandon Brazil.

✓ *Cabernet Sauvignon • Merlot*

CHANDON BRASIL

Rio Grande do Sul

☆

Part of LVMH's empire, this is the least impressive of Chandon's far-flung outposts. Yet, at least the sparkling wines are no longer sold shamelessly on the Brazilian market as Champaña.

✓ *Classic red blend (Grand Philippe)*
  • Sparkling (Diamantina)

LOVARA

Rio Grande do Sul

☆ **Ⓥ**

Despite its relatively ancient origins (established by Italian immigrants in the 1870s), Lovara is an up-and-coming winery that has been improving with each year since the 1990s.

✓ *Cabernet Sauvignon • Merlot*

MIOLO

Rio Grande do Sul

★ **Ⓥ**

Grape-growers since 1897, the Miolo family built a winery in 1989. Since then

their best, medal-winning wines have been some of the most exciting wines made in Brazil.

✓ *Classic red blend (Lote 43, Quinta do Seival, RAR, Terranova Cabernet Sauvignon-Shiraz)*

ECUADOR

CHAUPI ESTANCIA WINERY

Yaruqui

☆

A boutique winery with just 6 hectares (15 acres) of vineyard located 40 kilometres (25 miles) northeast of Quito, Chaupi Estancia Winery was established in 1989 to grow only *Vitis vinifera* grapes. Due to the rarefied altitude (2,440 metres, or 8,100 feet) and equatorial latitude of the vineyard plots, viticulturist and winemaker Dick Handall has been experimenting with 32 different varieties and many styles and combinations of wine to find what works best, with Chilean agronomist-winemaker Hector Olivares Madrid consulting.

✓ *Palomino*

TACAMA SELECCIÓN ESPECIAL RED BLEND
Founded in 1540, Tacama is the oldest winery in South America. Tacama produces red and white wines, along with Pisco, a brandy produced in Peru and Chile.

EMERGING MARKETS, NEW TRENDS, AND EXPERIMENTATION

With the population slowly moving away from drinking spirits and beer, Brazil's wine industry is increasing both production and domestic consumption, experiencing a surge in sparkling wines, a trend which is mirrored throughout South America. Mario Geisse was hired by Moet & Chandon to start up Chandon Brasil in the 1970s and is still an important sparkling producer under his eponymous Cave Geisse. Traditional Brazilian producers like Casa Valduga are also getting in on the act.

Sparkling is booming in other countries, not just Brazil: Chandon's operations in Argentina are winning medals, Zuccardi in the Uco Valley also offers a decent *blanc de blancs*, Miguel Torres in Chile produces some good bubbly, and an interesting Tannat Brut Nature is offered by the Pisano brothers in Uruguay.

David Bonomi, the winemaker of large Argentinian producer Norton, produced a Chardonnay that matures for seven years under *flor*, the yeast film more associated with Jerez winemaking and its Sherry. This Volare de Flor from PerSe was a limited-edition release, but the seeds have been sown with other winemakers across the region, who are now looking to their own white wines and what can be achieved using *flor*. Argentinian winemakers are also experimenting in areas such as orange wines, where a variety like Torrontés lends its aromatic nature to produce these structured and complex skin-contact wines.

Speaking of Torrontés, when winemakers got a sniff of the commercial uplift in sales of this varietal, both domestically and abroad, parcels of existing Sémillon vines, which was front and centre of white wine consumption in the country, were eschewed; today there is a resurgence of Sémillon, along with wines being made from new plantings of Verdelho, Fiano, Marsanne, and Rousanne, to name but a few.

Chile continues to push the frontiers by developing the more marginal climates – Lago Ranco for one and even places such as Añihué, which is an island, part of a small archipelago in southern Chile. Climate change is a factor, with northern areas struggling more and more for water, but there is also the drive for more risk taking to see what is possible from a quality perspective, especially the potential for fresher, lower-alcohol, structured wines in which the fruit dials are not pushed up to the max.

PERU

TACAMA
Ica
☆

This is the only significant producer in Peru and, although the wine is exported, only the Malbec is up to international standards. Other reds are short or bitter, the whites are fresh but uninteresting, and the sparkling wines unpleasantly explosive. In 2008, another Bordeaux professor, Pierre-Louis Teissedre, discovered a unique clone of Petit Verdot growing in Tacama's vineyard in the Ica Valley.
✓ Classic red blend • Classic white blend

URUGUAY

ARIANO
Las Piedras
☆

Established in 1927 by Adelio and Amilcar Ariano, this winery has become a shining example since the 1990s.
✓ Tannat

BODEGAS CARRAU
Montevideo
★

Established in the 1970s by Juan Carrau, this winery has since been carving a name for itself with well-crafted Tannat.
✓ Classic red blend • Tannat

CASTILLO VIEJO
San José
☆

The winery of Castillo Viejo makes sparkling and sharp, fruity reds from a private estate, which is 100 kilometres (60 miles) northwest of Montevideo.

DE LUCCA
El Colorado
☆

De Lucca is an up-and-coming winery run by passionate winemaker Reinaldo de Lucca, whose family's tradition of winemaking dates back to the 19th century in Piedmont, Italy. With vineyards situated 30 kilometres (19 miles) inland from the Atlantic coast, De Lucca is able to grow vines that do very well in Uruguay's relatively mild maritime climate.
✓ Classic red blend (Tannat Syrah) • Sauvignon Blanc

H STAGNARI
Canelones
★

This winery produces exclusively own-grown wines, with no bought-in fruit; the Tannat in particular benefits from the extremely high density of the vineyards, which boast 10,000 vines per hectare.
✓ Tannat

JUANICO
Canelones
★

One of Uruguay's fastest-rising wineries, Juanico performs to international standards, thanks to a little help in the past from flying winemaker Peter Bright.
✓ Chardonnay (Reserve) • Classic red blend • Classic white blend (Don Pasqual Chardonnay-Viognier) • Tannat

PISANO
Progreso
★

Established in 1914 by Don Cesare Secundino Pisano, the first generation of his Italian-Basque family to be born in Uruguay, the winery is now run by Eduardo Pisano, who has produced some of Uruguay's best wines in recent years, along with other family members. Around half of all the wine produced by this winery is sold abroad. The latest generation has a range of sparkling wines, which includes a black sparkling Tannat.
✓ Tannat

TRAVERSA
Montevideo
★

Carlos Domingo Traversa started off in 1937 as a grower of Isabela and Moscatel vines, and in 1956 he established a small winery, which has just started to blossom under his grandsons, who have produced some of Uruguay's greatest Tannat.
✓ Tannat

Chile

Chile has always possessed incredible potential as a wine producer, leading Miguel Torres to call it a "viticultural paradise" in the mid-1970s. The emphasis has often been placed on excellent-value red wines, leaving more premium wines and white wines behind. This has changed in recent times, however, and we have seen pioneering of new areas alongside a renaissance of Chile's rich vinous heritage to see it converted into one of the most dynamic countries in the wine world.

The historical heartland of Chile's wine industry lies close to the city of Concepción, in the areas that are now known as the Itata Valley and the southern part of the Maule region. First planted in 1550 this area was originally home to dry-farmed Pais and Moscatel, which thrived without irrigation. Increased centralization and the burgeoning wealth of Santiago saw these regions marginalised with the arrival of Bordeaux varieties in the 1850s. Plantings around Santiago increased, driven by wealthy, illustrious families with a distinct French influence. French winemakers, fleeing from phylloxera-ravaged Europe, flocked to Chile, which was spared (and still remains so) from the pest's devastation.

After a period of high production and extremely high domestic consumption in the early 20th century, the situation in Chile changed dramatically in the 1970s and 1980s with economic uncertainty and political

THE VERANDA OF THE CONCHA Y TORO WINERY IN SANTIAGO DISPLAYS VINTAGE WINEMAKING EQUIPMENT
This largest producer of wine in Latin America owns and operates vinification plants, bottling plants, and a wine distribution network. It has vineyards throughout Chile in Maipo, Maule, Rapel, Colchagua, Curico, and Casablanca.

Viticultural Region
- Atacama
- Coquimbo
- Aconcagua
- Valle Central
- Sur
- Austral

0 mi   50
0 km   50

Caldera
Copiapó   *Valle de Copiapó*
Tierra Amarilla

Huasco
Vallenar
Valle del Huasco
Chanaral

La Serena
Valle del Elqui
Vicuna

Ovalle
Valle del Limarí

PACIFIC
OCEAN

Valle del Choapa
Illapel
Salamanca

Valle del Aconcagua
San Felipe
Quillota
Valparaiso
Aconcagua
Valle de Casablanca
Santiago
Valle de San Antonio — San Antonio
San Bernardo
Valle del Maipo

Rancagua
Pichilemu
Valle del Rapel
San Fernando

Curico
Valle de Curicó
Constitucion

Valle del Maule
Linares
Cauquenes

Quirihue
Valle del Itata   Chillan

Concepción
Valle del Bío-Bío

Angol   Collipulli
Valle del Malleco
Victoria

Valle del Cautín
Temuco

Valdivia
Valle de Osorno —
La Union
Rio Bueno
Osorno

WINE REGIONS OF CHILE
(*see also p652*)
Chile's vineyards are located on the non-humid west coast of South America, stretching 1,300 kilometres (800 miles) from Copiapó Valley in the north, to Malleco Valley in the south.

instability. The industry was reborn in the mid-1980s with free market policies, seeing significant plantings of international varieties and a clear focus towards export markets.

As Chile exported more wine during the 1980s, it became evident that, although it made some really good-value reds, its white wines left much to be desired. In the late 1970s the quality of Chilean wines greatly improved with the introduction of temperature-controlled, stainless steel vats. The early to mid-1980s saw the removal from the winemaking process of raulí wood, a native variety of beech, which left a taint in the wines that the Chileans had got used to but that international consumers found unpleasant.

With the introduction of new oak *barriques,* mostly of French origin, the 1989 vintage saw a transformation in the quality of Chardonnay, and with this change it was widely believed that Chile had at long last cracked the secret to successful white wine production. It soon became evident, though, that even the best producers could not improve their dismal Sauvignon Blanc. Why was this so?

This was all due to the fact that the majority of Chilean Sauvignon Blanc was in fact Sauvignonasse. After Chile's non-existent Sauvignon Blanc was exposed in an article in 1991, a number of Chilean wineries commissioned French experts to go through their vineyards with a fine-tooth comb. They came to the same conclusion, and since then great swathes of vines have been chip-budded over to true Sauvignon Blanc. Today Sauvignon Blanc is Chile's most-planted white variety (more than 15,000 hectares, or 37,065 acres) – most successfully grown in Casablanca and the cooler coastal areas of San Antonio and Leyda. The quality can be very good indeed in these coastal areas and has significantly strengthened Chile's white production.

THREE KEYS TO SUCCESS

There are three key factors that have proved pivotal in Chile's development in recent times. First, the planting of cooler coastal areas, such as Casablanca (1982), San Antonio, and Leyda, as well as a drive to plant farther north and at altitude. Allied to this, there has been a move to more southerly vineyards (see southern regions and Austral) where water availability is greater, and fresher styles can be made. Second, the renaissance of Chile's historical heartland of Maule, Itata, and Bío-Bío has become a factor. For years maligned and marginalised, renewed interest in old vineyards of Pais, Muscat, Cinsault, and Carignan, amongst others, has added diversity, depth, and interest. Some of the País vineyards here are more than 200 years old. Third, the rise of smaller producers and groups of independent producers (MOVI, VIGNO, Chanchos Deslenguados, and Colchagua Singular, for example) has enriched and invigorated the industry greatly. There is still consolidation in the industry – dominated by Concha y Toro, San Pedro, and Santa Rita – but the past decade (since circa 2010) has seen a positive increase in smaller, "artisan" producers that are pushing the boundaries and the status quo.

CARMÉNÈRE'S PROGRESS

Chile has always been red wine country, and with the official identification of Carménère in 1994 it looked as though Chile had its own answer to Argentine Malbec. Carménère is a Bordeaux variety, but it had virtually died out over there when it was discovered in Chile, masquerading as Merlot. Here was an almost extinct, ancient *bordelais* grape that had been preserved in its pure, ungrafted format – a unique grape variety of impeccable breeding with which to fly the flag for Chilean wines, unchallenged by major plantings anywhere else.

Yet Carménère was been a slow burn in Chile because producers needed time to fully understand the grape. Is it a blending grape or a single variety? The answer I believe is both, and only in recent years have producers realised that it is a very site-specific grape. Unlike Malbec, which can perform well in many areas, Carménère requires warm (not hot) sites, and real care is needed with regard to harvest dates to avoid either green or – at the other extreme – soupy, flabby examples. Things are progressing well and will only continue to improve, adding further diversity to Chile's offerings.

FACTORS AFFECTING TASTE AND QUALITY

LOCATION
Vnes are grown along 1,300 kilometres (800 miles) of the Pacific coast and are most concentrated south of Santiago. Much is produced between latitudes 32° to 38°, although there are experimental vineyards outside of these latitudes that show promise.

CLIMATE
Extremely variable conditions prevail in Chile, ranging from arid and extremely hot in the north to very wet in the south. Maipo – the main wine area around Santiago – is dry, with around 33 centimetres (13 inches) of rain per year, no spring frosts, and clear, sunny skies. The temperature drops substantially at night due to the proximity of the snow-covered peaks of the Andes, enabling the grapes' acidity levels to remain high. There is relatively little disease pressure, with no phylloxera and mildew not often a problem. Areas such as the Casablanca Valley are proving suitable for winegrowing, particularly for white wines, but the coastal range of hills is important to Chile's fine wine future, because they receive enough rainfall to allow viticulture at modest yields without irrigation. This is the area most affected by the ice-cold Humboldt Current, which brings a tempering Arctic chill to all Chilean vineyards but is effectively blocked from flowing elsewhere in Chile by the coastal hills. Rainfall increases from around 8 centimetres (3 inches) per year in Elqui to around 120 centimetres (47 inches) per year in Bío-Bío.

The need to irrigate therefore diminishes farther south and, given climate change has seen drought farther north, this is a key issue. Those who think Chile doesn't experience much vintage variation need to read the recent vintage reports (see below) – climatic variations are happening; El Niño and La Niña can play a role, as does climate change.

ASPECT
Chile functions both on a west-east axis and a north-south axis. The cooler coastal areas (Humboldt influenced) are separated from the warmer inland plains by the coastal mountain range. On the eastern side the Andes provide a cooling influence from altitude (diurnal temperature variations) as well as winds. This has led to Chile's viticultural zoning to recognise three distinct areas that appear on labels: Costa (from the coast), Entre Cordilleras (between the mountain ranges), and Andes. The north-south axis is important with respect to rainfall and this irrigation, but the mitigating coastal breeze means that vines (with irrigation) can be cultivated even in Atacama.

SOIL
Vines are grown on a vast variety of soils, from alluvial soils in Aconcagua to limited areas of limestone in Limarí. Maipo has a mixture of loam, clay, and gravel. Coastal areas and Maule down to Bío-Bío have large quantities of volcanic soils, such as decomposed granite with a high proportion of quartz.

VITICULTURE & VINIFICATION
With no phylloxera, little downy mildew, and almost no summer rain, Chile is suited to organic or biodynamic viticulture. Over half of Chile's vineyard area is estimated to be less than 15 years old. In Itata and Bío-Bío, however, there are plenty of vineyards (País, Moscatel, and even Malbec) that are certainly well over 100 years old and anecdotally up to 300 years old. There are approximately 138,000 hectares (341,005 acres) planted – 36,000 are white and 102,000 are red. Around 120,000 hectares (296,525 acres) are irrigated with the balance dry-farmed. There has been a welcome move away from the overuse of oak in premium red wines over the past decade. Chile's wineries are now typically high-tech, a far cry from the 1970s and before, with foreign investment and the targeting of export markets to thank. The nation's winemakers are now very well travelled and have a deeper global understanding of trends and styles than at the turn of the century.

GRAPE VARIETIES
Primary varieties: Cabernet Sauvignon, Sauvignon Blanc, Merlot, Chardonnay, Carménère, Syrah, Pinot Noir
Secondary varieties: País, Carignan, Cinsault, Cabernet Franc, Gewürztraminer, Malbec, Petit Verdot, Mourvèdre, Riesling, Viognier, Sémillon

RECENT CHILEAN VINTAGES

2019 A good vintage with some fine results. A very dry winter was followed by a cool spring and then some real heat spikes in the summer. Yields were generally lower than average.

2018 An excellent vintage, no real problems. Slightly above average winter rainfall and a warm but not hot end to the season. Worth seeking out.

2017 Raging forest fires burnt almost 500,000 hectares (1,235,525 acres) of land in late January and early February. Smoke-tainted grapes were a huge problem and can be tasted in many wines.

2016 Extremely heavy rainfall at harvest time wreaked havoc, with up to 30 per cent of Chile's production lost. Some of the early-ripening varieties performed well (pre-deluge) including Sauvignon Blanc, Chardonnay, and Pinot Noir.

2015 A largely warm year. Some full and generous wines were made – when harvested at the correct time (not too late) there are some superb reds and good quality whites.

2014 Severe frosts decimated vineyards in and around Casablanca. Some fine reds were made; however, yields were generally low, and the season ended dry and hot.

ELQUÍ VALLEY VINEYARDS IN THE COQUIMBO VITICULTURAL REGION
The green of the vines stand out against the chalky mountains that surround this desert-like region. Varieties such as Syrah have done best here, and many Muscat varieties grown here are used to make pisco, an amber Chilean brandy.

THE APPELLATIONS OF
CHILE

Note: Chile is divided into six wine regions, with smaller districts within them. In the list below, the names of the wine districts appear after the regional designations.

ATACAMA VITICULTURAL REGION
Copiapó Valley, Huasco Valley

The Atacama Viticultural Region is the appellation with the most northern DO in Chile. There is no real significant winemaking here, but there is the production of *pajarete*, a traditional *passito*-style sweet wine made from dried pisco (brandy) grapes. Viña Ventisquero has the exciting premium project Tara in Huasco.

COQUIMBO VITICULTURAL REGION
Elquí Valley, Limarí Valley, Choapa Valley

The Coquimbo Viticultural Region is traditionally a pisco and table-grape growing area. Initially the coastal areas showed most potential. More recently there has been a push to altitude Andean sites over 1,600 metres (5,249 feet) above sea level. Elquí experiences some of the clearest skies in the world, meaning high UV exposure (producing deep coloured intense wines, but with freshness.) Frosts and snowfall can be a problem, but there is real potential here. Still wine production started in Limarí in 1995. Syrah and Chardonnay have shown excellent results. Water shortage in the area is a major problem, due to restricted water rights. For this reason, only premium grapes are grown here. Choapa is the farthest south of the sub regions.

ACONCAGUA VITICULTURAL REGION
Aconcagua Valley, Casablanca Valley, San Antonio Valley (Leyda Valley)

The interior area (around San Felipe) of the Aconcagua Viticultural Region is hot and dry. The renowned Viña Errázuriz winery is based in nearby Panquehue. Annual rainfall is low, at 25 centimetres (10 inches), and is best for late-ripening varieties such as Cabernet, Syrah, Carménère, and Petit Verdot. Recently they have planted closer to the coast, and the results are extremely encouraging for Chardonnay, Pinot Noir, and Sauvignon Blanc.

Casablanca Valley has been a success story and is often referred to as a cool-climate region. The coastal influence is important, with cooling winds and frequent cloud cover

that delays ripening. Spring frosts are an issue here, and early-ripening varieties (Sauvignon Blanc, Chardonnay, and Pinot Noir) fare best. There has also been an increase in quality sparkling wines from the Casablanca Valley in recent years.

The San Antonio Valley District contains the DO of Leyda Valley, where Sauvignon Blanc is the star player. This cool, coastal area has vines planted as close as 2 kilometres (a bit over 1 mile) from the Pacific coast. Weathered granitic soils dominate here, and Syrah and Pinot Noir can also fare very well. Water access can be an issue and has restricted plantings.

AUSTRAL VITICULTURAL REGION
Cautín and Osorno

Still relatively few plantings, but serious players such as de Martino are planting here. Shows excellent potential for Chardonnay, Sauvignon Blanc, and Pinot Noir.

CENTRAL VALLEY VITICULTURAL REGION
Curicó Valley (Lontue Valley), Maipo Valley, Maule Valley (Claro Valley, Teno Valley), Rapel Valley (Cachapoal Valley, Colchagua Valley (Apalta Valley)

The oldest, most central, and most traditional wine region, the Central Valley Viticultural Region contains four wine districts encompassing seven wine areas. The Maipo Valley around Santiago itself is one of the warmest growing districts in the country but is tempered by cooler nights. The alluvial and gravel soils are capable of regularly producing very good wines (including Cabernet and Carménère), especially around Puente Alto.

The Rapel Valley comprises Colchagua and Cachapoal, and together they provide the largest vineyard area in Chile. Cachapoal (especially Peumo) has become one of the highest-quality areas for Carménère. Colchagua is an increasingly important area. The DO of Apalta is arguably the most significant (and most expensive) where Montes, Lapostolle, Ventisquero and others own significant land. Colchagua contains other important areas such as Marchigue (high-class Syrah) and farther to the east in the Andean foothills Los Lingues, where the cooling effect of the mountains allow for longer ripening and excellent quality reds especially Carménère.

Curicó Valley was chosen as the base by the first foreign investor in Chile, Miguel Torres, which arrived in district in 1979. The area is also home to San Pedro (founded in 1865) and Valdivieso (dating back to 1879). The warm DO Sagrada Familia is slowly improving in quality, and the Curicó Valley generally can produce almost Old World (semi-rustic) reds, specifically Cabernet Sauvignon.

Maule Valley was for years the heartland of bulk wine production and was dominated by País plantings. Average annual rainfall here is around 80 centimetres (31 inches). Recently Cabernet has been widely planted, yet there are still significant plantings of old-vine, dry-farmed País and pockets of old-vine Sémillon, as well as Carignan, which truly shines around Cauquenes in the southern part of the region. The self-regulating appellation of VIGNO (*Vignadores de Carignan*) has focussed on old-vine, dry-farmed Carignan with excellent results. The Maule revival has been a welcome progression.

SOUTHERN VITICULTURAL REGION
Itata Valley, Bio Bio Valley, Malleco Valley

Possibly the biggest development in the past decade has been the reinvigoration and revival of the Southern Viticultural Region. Itata Valley is the only region that is not dominated by French varieties; here old-vine Pais, Moscatel, and Cinsault lead the way. Annual rainfall is around 100 centimetres (39 inches), and therefore many vineyards are unirrigated. The area is dominated by small holdings (the average holding is only about 1 hectare (2.47 acres), unlike in the rest of Chile. Larger players, such as de Martino, Torres, and Montes, to name a few, are making wine here, with Torres and Montes having bought properties. There is also potential for other varieties, and Pandolfi Price shows the potential for Chardonnay in the eastern part near Chillan. Bío-Bío is a similar story, with old vines but also some larger areas that are now planted to varieties such as Riesling and Pinot Noir by producers including Cono Sur. Expect to see more investment in these areas in the coming years.

The Malleco Valley wine district is small, but it is producing some stellar mineral and racy Chardonnays from producers such as Altos Las Gredas, Aquitania, and William Févre.

THE WINE PRODUCERS OF

CHILE

ALMAVIVA
Maipo Valley
★★★

A joint venture between Domaines Baron Philippe de Rothschild and Concha y Toro, Almaviva is a classy single-vineyard, Bordeaux-style red wine. Rarely fails to shine.

✓ *Classic red blend (Almaviva)*

ALTO LAS GREDAS
Traiguen
★★

Alto las Gredas produces very small-scale, high-quality Chardonnay from the southerly area of Traiguen. The vineyard was planted in 2001 by the effusive owner, María Victoria Petermann. Production rarely exceeds 4,000 bottles per year. The wine is racy and steely with superb concentration.

✓ *Gran Reserva Chardonnay*

ANTIYAL
Maipo
★★

Viña Antiyal is the boutique biodynamic, Maipo Valley–based project of Alvaro Espinoza, one of Chile's finest wine-makers and consultants, and his wife, Marina. There is real purity in these wines, based largely on Cabernet Sauvignon, Carmènere, Syrah, Petit Verdot, and Garnacha.

✓ *Antiyal, Kuyen • Pura Fe*

AQUITANIA
Peñalolen
★★☆

After a shaky start, this international collaboration between Bruno Prats (ex-Château Cos d'Estournel), Ghislain de Montgolfier (Bollinger), the late Paul Pontallier (Château Margaux), and Felipe de Solminihac (Santa Monica) has become one of Chile's best wineries. The SOLdeSOL is from Solminihac's own property at Traiguén, the most southerly vineyard in Chile, 650 kilometres (400 miles) south of Santiago.

✓ *Chardonnay (SOLdeSOL)*

BOUCHON
Maule
★★☆

Reinvigourated under the guidance of Julio Bouchon Jr and winemaker Cristián Sepúlveda, this Maule stalwart is producing some of Chile's most exciting wines. A mixture of classics and also a focus on low-intervention wines from older vines of País and Sémillon.

✓ *Granito Sémillon • Pais Viejo • Carignan (VIGNO)*

BODEGAS RE
Casablanca
★★

This project, led by Pablo Morandé Sr and Jr, is based in Casablanca and offers plenty of fascinating, experimental wines (many amphora based). There is great vibrancy throughout the wines.

✓ *Syragnan (Syrah/Carignan blend) • Chardonnoir (Chardonnay/Pinot Noir blend)*

CARMEN
Alto Jahuel
★☆

Established in 1850, this is the oldest brand in Chile, now part of the same group as Santa Rita. The winery is overseen by the very talented Emily Faulconer, who is really driving this winery forward. New and innovative styles (*flor*-aged Sémillon) are appearing along with Itata Cinsault and great field blends from Maule.

✓ *Florillon • Quijada Sémillon*

CASA LAPOSTOLLE
Apalta
★★

Owned by the Marnier-Lapostolle family of Grand Marnier renown (The iconic Clos Apalta used to be the top wine but is now considered a separate estate). Winemaker Andrea Leon has really brought an elegance and refinement to these wines that have traditionally been full-throttle in style.

✓ *Cuvée Alexandre Carmènere • Collection Cinsault (Itata)*

CASA MARIN
San Antonio
★★☆

Tour de force Maria Luz Marín was the pioneer of the cool-climate San Antonio Valley, and is joined by her son, Felipe, as head winemaker. They produce some of Chile's finest Syrah and Sauvignon Blanc.

CASA SILVA
Los Lingues
★★

Casa Silva is based in Los Lingues, in the foothills of the Andes in Colchagua Valley. Here they produce some hefty and fine Carmènere, yet they have also pioneered projects in the cool areas of Paredones (coastal Colchagua) and in the extreme south of Chile at Lago Ranco – these are showing very exciting results.

✓ *MicroTerroir Los Lingues Carmènere • Lago Ranco Sauvignon Blanc and Riesling*

CONCHA Y TORO
Pirque
★★☆

Concha y Toro is the largest producer in Chile, yet still remains a serious winery. Concha y Toro also owns Cono Sur and is involved with Domaines Baron Philippe de Rothschild in their Almaviva joint venture. Arguably the best value wines are the Marques de Casa Concha, overseen by the magisterial Marcelo Papa.

✓ *Cabernet Sauvignon (Don Melchor, Marques de Casa Concha, Terrunyo) • Carmènere (Terrunyo) • Chardonnay (Amelia)*

CONO SUR
Chimbarongo
★★

This fast-track winery is owned by Concha y Toro and was driven forward by Chief winemaker Adolfo Hurtado, who has very recently left. Pinot Noir is a focus and producing it was a brave move that is paying off in the long run.

✓ *Cabernet Sauvignon (20 Barrels) • Pinot Noir (20 Barrels) • Ocio Pinot Noir*

COUSIÑO MACUL
Buin
★★

One of Chile's stalwarts. The estate has been through ups and downs, yet it is back in sparkling form. The Finis Terrae rarely disappoints.

✓ *Cabernet Sauvignon (Antigua Reserva) • Classic red blend (Finis Terrae) • Riesling (Doña Isidora)*

DE MARTINO
Maipo

Established in 1934, de Martino is now led by the young brothers Marco and Sebastián de Martino alongside the supremely talented winemaker Marcelo Retamal, who has actively pursued lighter, more elegant wines. They work on organic vineyards throughout Chile from Elqui to Itata, where they have been one of the leading lights in its renaissance and the use of Tinajas (amphora).

✓ *Viejas Tinajas Cinsault • Viejas Tinajas Muscat*

ERRÁZURIZ
Aconcagua
★★☆

Another of the pioneers of the Chilean industry (and the first to plant Syrah). Led by Eduardo Chardwick and the talented winemaker Francisco Baettig, they are producing some superb wines. A recent venture in coastal Acongaua, Las Pizarras, is fast becoming an icon.

✓ *Las Pizarras Chardonnay • Don Maximiliano Founders Blend*

GARAGE WINE COMPANY
Maule
★★

Founded in 2001 by Canadian Derek Mossman Knapp, and run with his winemaker wife Pilar. They work closely with small growers and with a "natural/low-intervention" approach, often with very old parcels of vines. Some beguiling and fascinating wines and very much a handcrafted operation.

✓ *VIGNO Carignan*

LEYDA
Leyda
★★☆

This winery just keeps getting better and is proving that the Leyda Valley can produce some of Chile's greatest Pinot Noir (amongst others). Winemaker Viviana Navarrete is one of Chile's (many) highly talented female winemakers.

✓ *Pinot Noir Las Brisas • Neblina Riesling • Lot 4 Sauvignon Blanc*

MATETIC
Casablanca
★★

A modern, biodynamic-organic estate based in San Antonio and Casablanca. With over 150 hectares (370 acres) set in their own estate, Matetic's standout is their superb Syrah, which ranks amongst Chile's finest examples.

✓ *Matetic Syrah • EQ Syrah*

MIGUEL TORRES
Curicó
★★

Miguel Torres was the first foreign investor in Chile in 1979. Choosing Curicó as his base, he also makes wins from other areas. There is much to admire about this enterprise: its philosophy, environmental stance, and most recently it's pivotal role in the renaissance of the Chilean viticultural patrimony –specifically the revival of Pais.

✓ *Empedrado Pinot Noir • Manso de Velasco • Reserva del Pueblo Pais (excellent value)*

MONTES
Apalta
★★

The brainchild of Aurelio Montes, these wines have traditionally been full in style and with plenty of oak. Recently there has been more finesse, especially with the Outer Limits range. They produce a number of Chile's best-known wines: Montes Folly, Purple Angel, and Montes Alpha M.

✓ *Syrah (Folly) • Alpha M*

PANDOLFI PRICE
Itata
★★

A winery that shows Itata is not just about old-heritage varieties but is capable of producing world-class Chardonnay as well, led by the energetic Enzo Pandolfi alongside winemaker Francois Massoc. The coup de Coeur is the Los Patricios Chardonnay, a Meursault-esque beauty.

✓ *Los Patricios Chardonnay • Larkun Chardonnay*

ROGUE VINE
Itata
★★

Leo Erazo is one of Itata's leading lights, and this is one of his projects with co-founder Justin Decker. They produce elegant and refined wines from old vines on the steep, granite slopes of Guarilihue in Itata.

✓ *Los Amigos Grand Itata (red blend) • Super Itata Sémillon*

SAN PEDRO

Molina

★★

San Pedro is one of the pioneers that earned respect for Chilean wines on export markets. As well as plenty of good-value wines, the 1865 range are very smart as are top-of-the-range Altair and Cabo de Hornos. Recent years has seen more restraint in the wines.

✓ *1865 Carménère • Cabo de Hornos • Kankana de Elqui*

SANTA RITA

Buin

★★

Well known for the "120" range (so-called because Bernardo O'Higgins, the liberator of Chile, and his 120 men hid in these cellars after the battle of Rancagua in 1810) that has remained reliable throughout. The winery is in an exciting stage with the excellent Sebastian Labbé at the reins in the cellar. The Medalla Real range is consistent and offers great value, but it is the Floresta range that is really exciting.

✓ *Floresta Carménère • Medalla Real Chardonnay*

SEÑA

Panquehue

★★★

Initially a joint venture between Errázuriz and Robert Mondavi, Seña had established itself as one of Chile's grands crus when Errázuriz purchased Mondavi's share in 2004. One of Chile's finest.

✓ *Classic red blend (Seña)*

UNDURRAGA

Talagante

★★☆

One of Chile's oldest producers, yet also one of the most innovative under talented winemaker Rafael Urrejola. They are based in Talagante, Maipo, but have vineyards over the country. The Terroir Hunter (TH) series of wines are some of the finest in the portfolio.

✓ *TH Syrah, Altazor • TH Sauvignon Blanc*

VENTISQUERO

Lo Miranda

★★☆

Viña Ventisquero is another winery producing wines over the length of Chile

since the turn of the millennium. Now under Cheif Winemaker Felipe Tosso, this winery's finest are possibly the Tara range of wines that come from Huasco Valley in Atacama Viticultural Region, and the wines from their Apalta estate in the Colchagua Valley District.

✓ *Vertice • Tara Chardonnay*

VIK

Cachapoal Valley

★★☆

VIK is a newcomer to the Chilean wine scene. It was founded in 2006 by Norewgian Alexander Vik, who purchased 11,000 acres in Cachapoal Valley. No expenses have been spared in building the winery, which is also a luxury hotel. Three lines of red are made, based largely on the Cabernet Sauvignon and Carménère varieties.

✓ *Vik • Milla Calla*

VIÑEDOS DE ALCOHUAZ

Elqui Valley

★★★

Located at 1,800 metres above sea level (5,905 feet) in the Elqui Valley, Viñedos de Alcohuaz is one of Chile's rising superstar estates, and their wines grow under some of the clearest skies in the world. Sandwiched in a corridor between the cool Pacific coast of northern Chile and the towering Andes Mountains in the east, this estate's vineyards are planted with several varieties led by Syrah, Malbec, Grenache, and others. The wines show intense purity as well as superb ageing potential.

✓ *Rhu de Alcohuaz (red blend) • Tococo (Syrah)*

VIU MANENT

Santa Cruz

★☆

Founded in 1935, Viu Manent is an old, established family-owned winery with a vineyard estate that dates back to the 1800s. In recent years, it has moved away from bulk wines into more expressive, high-quality wines under its own label. Viu Manent wines are made with grapes from their own vineyards in the Colchagua Valley in the Central Valley Viticultural Region, with Malbec as their signature variety. The winery property also boasts a wonderful restaurant to attract tourists. Consistently good.

✓ *Vibo Vinedo Centenario • Malbec (Single Vineyard San Carlos)*

SANTA RITA 120 RESERVA ESPECIAL CARMÉNÈRE, VINTAGE 2016
Santa Rita, one of Chile's prominent wine producers, sources it grapes from its vineyards scattered throughout the country's viticultural regions, including the Aconcagua and Casablanca Valleys in the Aconcagua VR; the Colchagua and Rapel Valleys in the Central Valley VR; and the Limarí Valley in the Coquimbo VR. Cabernet Sauvignon and Carménère, with some Merlot, account for much of their reds, and for whites they grow Sauvignon Blanc and Chardonnay.

Argentina

Argentina is the largest wine-producing country in South America and the fifth biggest in the world, with 218,233 hectares (539,265 acres) under vine. Consumption is in decline and now stands at 18 litres per capita. Yet wine still plays an important part in everyday life and was declared the national beverage of choice by presidential decree in 2010. Argentina's Spanish and Italian heritage makes it feel distinctly "European" at times.

G rapes are grown in 18 of Argentina's 23 provinces, and the industry makes an important contribution to the economy, accounting for 0.4 per cent of GDP. Just as significantly, Argentina's 23,931 vineyards and 895 wineries employ 385,000 people. The most important of those provinces by far is Mendoza, which accounts for 70.1 per cent of plantings, followed by San Juan (21.4 per cent). The other 16, led by La Rioja (3.6 per cent), Salta (1.5 per cent), and Catamarca (1.3 per cent) are tiny by comparison, though certainly important in terms of quality.

Argentina's viticultural roots date back to 1551, but the modern wine industry is barely 30 years old and is still evolving at a rapid pace. Two of its best vineyards – Estancia Uspallata and Per Se – are less than a decade old. These changes have come about partly because of the influence of international consultants and external investment (rarer these days because of the parlous state of the economy) but mostly because of local winemakers and viticulturists. The country still makes a lot of basic cheap wine, mostly for the domestic market, yet there is increasing focus on reds and whites with a sense of place.

MALBEC GRAPES RIPEN IN A VINEYARD IN THE LUJÁN DE CUYO SUBREGION OF MENDOZA
Malbec is the flagship grape of Argentina, but the country produces wine from more than 100 other varieties.

Malbec is Argentina's most famous variety, even if the country is considerably more diverse than that one type. Roughly 80 per cent of its vineyards are planted with 122 other French, Italian, Spanish, and "native" grapes, making Argentina one of the most diverse wine-producing countries in the New World.

FACTORS AFFECTING TASTE AND QUALITY

LOCATION
Vines grow mainly in the provinces of Mendoza and San Juan, east of the Andes, but the country's increasingly diverse vineyards stretch from Jujuy, close to Bolivia, in the north (23° latitude), to Chubut in the south (43° latitude) and Chapadmalal, on the Atlantic Coast, in the east.

CLIMATE
In Argentina's intensively cultivated Mendoza district, the climate is officially described as continental-semi-desertic, with scarce rainfall, except in atypical vintages like 2015 and 2016. Argentina has way less rainfall than most other Pacific Ocean–influenced countries at a mere 20 to 25 centimetres (8 to 10 inches) per year, although this is spread over the growing season in most vintages. Temperatures can be very hot in summer, with 40°C (104°F) not unusual, especially if the Zonda wind is blowing, but there is often a compensating amount of diurnal variation, most notably closer to the Andes. Argentina is often perceived as sweltering, yet it's much more varied than people think. In the space of 128 kilometres (80 miles), you can travel from the heat of Rivadivia (a V on the Winkler Scale) to the distinctly chilly conditions of the higher parts of the Uco Valley (Winkler I and suitable for sparkling wine base).

ASPECT
Argentina contains the greatest concentration of high-altitude vineyards in the world, with Jujuy holding the record at 3,329 metres (10,920 feet), but not all of its plantings are so extreme. The farther south and east you go, away from the peaks of the Andes, the lower they get. Most of them are indeed located within sight of South America's largest mountain range, but the lowest are at 39 metres (128 feet) in Chapadmalal on the east coast. The majority of Argentina's vineyards are flat and mechanised, although there has been a move to slopes in recent

years with the development (or redevelopment) of areas in Barreal, Mendoza (El Challao, El Espinillo, Gualtallary, La Carrera, Los Chacayes, and Uspallata) ,and San Juan (the Pedernal Valley).

SOIL
Vines are grown on a vast variety of soils in Argentina, ranging from sand to clay, with a predominance of deep, loose soils of alluvial and Aeolian origin. Regions closer to the Andes, particularly the Uco Valley, often contain calcium carbonate deposits, washed down from the mountains.

VITICULTURE & VINIFICATION
Partly because of their Italian and Spanish heritage, Argentines have always been good grape growers, even if their yields were necessarily high in bulk wine areas to satisfy domestic price points. The biggest game changer was the arrival of drip irrigation in the mid-1990s, making it possible to plant on slopes, rather than relying on flood-irrigated vineyards on the plain. The last decade has witnessed a further shift towards greater precision, based on scientific rigour rather than empiricism. Soil pits are common in the best vineyards to match variety or wine style to *terroir* and meso-climate. And more and more *bodegas* are picking early to retain freshness. There's been an equally important transformation in the country's wineries with many of the best producers opting to use little if any new oak, preferring fermentation and ageing in concrete or larger wooden *foudres*. Extraction levels have declined; balance has largely improved.

GRAPE VARIETIES
Primary varieties: Balbariño, Barbera, Caladoc, Cereza, Chenin Blanc, Grenache, Marsanne, Merlot, Moscatel Rosado, Pinot Gris, Riesling, Roussanne, Sangiovese, Sauvignon Gris, Syrah, Tannat, Tempranillo, Viognier

THE SNOWY PEAK OF ACONCAGUA RISES ABOVE OTHER ANDEAN MOUNTAINS THAT OVERLOOK VINEYARDS IN MENDOZA
Not to be confused with the Aconcagua Viticultural Region in Chile, this Andean stratovolcano is the highest mountain in both the Southern and Western Hemispheres. The Andean valleys in the Mendoza Province have become the most exciting wine regions to watch in Argentina. The excellent climate and soil produces some of the region's most famous wines.

WINE REGIONS OF ARGENTINA
(see also p652)
From the low-lying Río Negro, to the seriously high-altitude mountainside sites of Salta and Jujuy, Argentina's vineyards get higher the farther north and closer to the equator they go.

RECENT ARGENTINIAN VINTAGES

2019 Early, dry harvest with cooler temperatures than usual. Early days to say for certain how good the resulting wines will be, but several producers describe it as exceptional.

2018 Cool, dry, sunny vintage with low rainfall, seen as a typically Argentinian growing season. At its best for whites, but there are some impressive reds too.

2017 mall, cool vintage that one producer called "Bordeaux-like". Yields were very low, grape prices high, and quality mixed, depending on whether people picked pre- or post-rains.

2016 Cool, wet, "European-style" vintage marked by the El Niño phenomenon in Mendoza and San Juan. The best wines are elegant and refined, the worst dilute and unlikely to age well.

2015 An El Niño vintage marked by heavy rainfall and late frosts (in some zones), with disease pressure high in many vineyards because of marked humidity. Pick and choose to find the best wines of this year.

2014 Marked by heavy rains in Mendoza in February, this was a vintage that required a lot of work in the vineyards. Earlier picked varieties, especially the whites, tended to fare best.

2013 A very good vintage that benefited from dry, sunny conditions and cooler nights close to harvest. There was some hail damage in some regions in March, but the best reds are impressively long-lived.

2012 Yields were down on 2011 because of poor flowering, producing red wines that are rich and intensely coloured at their best, with balancing acidity and good ageing potential.

THE WINE REGIONS OF

ARGENTINA

CATAMARCA

Isolated, mountainous and less famous than neighbouring Salta, Catamarca is still worth keeping an eye on. El Esteco's first reds from Chañar Punco, an area that has a high concentration of limestone soils, are promising.

CHAPADMALAL

An exciting frontier for Argentina, Atlantic-influenced Chapadmalal has more in common with Uruguay than Mendoza. Tiny for now, with just one producer (Costa & Pampa), but there's very good potential for aromatic whites and Pinot Noir.

CHUBUT

The southern frontier of Argentinean viticulture, located 595 kilometres (370 miles) from Neuquén, and prone to frost risk. Against the odds, brave pioneers are making increasingly good Chardonnay and Pinot Noir.

JUJUY

Spectacularly beautiful region in northern Argentina that's home to the world's highest vineyard, planted at 3,329 metres (10,920 feet) by Claudio Zucchino of Uraqui. Full-bodied reds are the order of the day, with some surprisingly good Sauvignon.

LA PAMPA

Mostly known for the production of one winery (Bodega del Desierto), this is a new, isolated area that has been making wine since 2004. The results are increasingly promising, made with help from American consultant, Paul Hobbs.

LA RIOJA

The site of Argentina's first vineyards, this under-rated area is home to La Riojana, one of the country's best co-ops and a Fairtrade stalwart. Two areas to look out for are Valle de la Puerta and more isolated, high-altitude Chañarmuyo.

MENDOZA

Mendoza Province is the hub of the Argentinian wine industry. It's a large and varied area that makes most of the country's best reds and whites, but also a lot of its everyday fare. Hotter Maipú, East Mendoza, and San Rafael tend to produce cheaper styles than the best sub-regions: traditional Luján de Cuyo (which includes Agrelo, Chacras de Coria, Las Compuertas, Lunlunta, Perdriel, Ugarteche, and Vistalba) and the cooler, higher altitude Uco Valley (see separate entry), which has risen to prominence in the last decade.

NEUQUÉN

Neuquén is a comparatively new, low-lying area created thanks to water from a large irrigation dam. Quality is good rather than spectacular, as this is quite a warm area. But the wines are improving with every year, as the vines get older. Añelo is the most exciting sub-zone, especially for Pinot Noir.

RÍO NEGRO

Patagonia's original vine-growing area and the source of most of its top wines. Latitude is more important than altitude, as the vineyards are all below 350 metres (1,150 feet). Diurnal variation and old vines make it an area to watch for Malbec, Merlot, Pinot Noir, Riesling, Sémillon, and Trousseau.

SALTA

An arid, cactus-strewn region that's named after the sub-tropical city of the same name, even if most of its vineyards are located three hours south in the cooler and more elevated Calchaquí Valleys. Best known for the local white speciality, Torrontés Riojano, and for rich, spicy, high-octane reds made from Malbec, it is starting to make its name with Tannat, Cabernet Sauvignon, and red blends too.

SAN JUAN

A large, mostly hot area that is second only to Mendoza in size, San Juan produces a lot of basic, quaffing wine, much of it destined for domestic consumption. More recent plantings in the higher, cooler Pedernal and Zonda Valleys and the old-vine reds and whites from rediscovered Calingasta are more interesting. Makes some of Argentina's best Syrahs.

SAN RAFAEL

Administratively, San Rafael is part of Mendoza Province, even though it lies 240 kilometres (787 miles) to the south of the city. It's a little cooler than Luján de Cuyo, despite its lower altitude. It grows many different grapes well, but is not readily associated with any of them, with the possible exception of Cabernet Sauvignon.

TUCUMÁN

Tucumán is best known for its share of the Calchaquí Valleys (with neighbouring Catamarca and Salta). Plantings are small and are mostly located in the Tafí del Valle department. High-altitude vineyards (starting at 1,800 metres, or 5,905 feet) are promising.

UCO VALLEY

Unquestionably the most exciting region in Argentina, thanks to a combination of foreign investment, high altitudes, the proximity of the Andes, old vines (especially in La Consulta), and new plantings in the higher northern part of the valley. The potential of Chardonnay, Cabernet Franc, and Malbec (particularly if they are grown on limestone soils) is enormous, especially as some of the young vineyards mature, but the Uco Valley can make almost any style well. Look out for sub-regional wines from recently delimited Paraje Altamira, as well as El Peral, Gualtallary, La Carrera, La Consulta, Los Chacayes, and San Pablo.

THE WINE PRODUCERS OF

ARGENTINA

ALEJANDRO SEJANOVICH

Patagonia, Mendoza, Salta, and Uco Valley

★★

"Colo" Sejanovich runs a number of impressive projects in different parts of the country with his business partner, Jeff Musbach: Teho, Buscado Vivo o Muerto, Estancia Uspallata, Tinto Negro, and Estancia Los Cardones. The emphasis is on *terroir* expression rather than oak or fruit power.

 Estancia Uspallata Malbec • Bodega Teho Corte Tomal Vineyard • Buscado Vivo o Muerto El Indio

ALTOS LAS HORMIGAS

Maipú and the Uco Valley

★★

Based in Maipú but drawing most of its grapes from the Uco Valley, this is Italian consultant Alberto Antonini's home base, where he works with Chileans Perdo Parra and Leo Erazo. Elegant, age-worthy Malbecs are the highlights, but the Bonarda is fun too.

 Malbec Appellation Gualtallary

ACHÁVAL-FERRER

Luján de Cuyo, Maipú, and the Uco Valley

★★☆

Now under Russian ownership (the SPI Group that also owns Stolichnaya vodka) and without the input of joint founder, Santiago Achával, Achával-Ferrer is still among the country's top *bodegas*. Winemaker-oenolgist Gustavo Rearte travelled to New Zealand and California after graduation to hone his skills before joining Achával-Ferrer. Single-vineyard Malbecs from different parts of Mendoza are the focus here, as well as a blend of the three, called Quimera.

 Finca Altamira Malbec • Quimera

BODEGA ALEANNA

Uco Valley

★★☆

Co-owned by Adrianna Catena and Alejandro Vigil, this is where Catena's chief winemaker gets to let his hair down a bit, with Malbec/Cabernet Franc blends, a *flor*-aged Chardonnay, and a series of old-vine Bonardas.

 Gran Enemigo El Cepillo • El Enemigo Cabernet Franc • El Barranco SV Bonarda

CATENA ZAPATA

Mendoza

★★★ⓥ

Nicolás Catena, now aided by his daughter Laura, is one of the key figures in the development of Argentina's modern wine industry, as much for the people he has employed and trained over the years as for his drive and vision. The characterful Alejando Vigil is the winemaker here. Catena makes a wide range of wines at different price points, including some of the country's best whites and reds. He also has a joint venture, Caro, with the Rothschilds of Château Lafite Rothschild, while Laura has Domaine Nico. The research and development carried out by the Catena Institute is also of a very high quality.

 Argentino Malbec • Nicolás Catena Zapata (blend) • White Stones Chardonnay; White Bones Chardonnay

CHACRA

Río Negro

★★

Italian Piero Inchisa della Rocchetta is the grandson of the creator of Tuscany's Sassicaia, which makes Merlot, a range of Pinot Noirs, and now, in partnership with Jean-Marc Roulot of Domaine Roulot in Meursault, a Chardonnay. The top wine here comes from a vineyard planted in 1932.

 Treinta y Dos Pinot Noir

CHEVAL DES ANDES

Luján de Cuyo

★★☆

Owned by French luxury goods company LVMH, this property benefits from the input of Pierre Lurton and his team from Château Cheval Blanc in Saint-Émilion. This Mendoza Province *bodega* is now producing one of Argentina's finest red blends, made entirely with its own grapes since 2016.

 Cheval des Andes

CLOS DE LOS SIETE

Uco Valley

★ ⓥ

Still blended by globetrotting oenologist, Michel Rolland, this pioneering project in Campo Los Andes is one of Argentina's most reliable branded red blends, with a stylish label to match.

✓ *Clos de los Siete*

COLOMÉ
Molinos
★★☆

Founded in 1831, Colomé is now part of the Hess group (which has other estates in Napa Valley, South Africa, and Australia), this isolated Salta province winery boasts some of the highest vineyards in the world. Amalaya is its entry-point brand, but the real excitement comes from the estate's 19th-century parcels and its dramatic Altura Máxima vines at over 3,000 metres (9,840 feet).

 Auténtico Malbec • 1831 Malbec • Altura Máxima Malbec

DE ANGELES
Vistalba
★★

Malbec, Cabernet Sauvignon, and a blend of the two, all from one of Vistalba's oldest and most highly prized vineyards, are the highlights of this traditional *bodega's* range. Old-vine concentration at its best.

 Viña 1924 Single Vineyard Gran Corte

DOÑA PAULA
Mendoza
★☆

Overseen by top viticulturist Martín Kaiser, Doña Paula makes a large range of impressive wines, including Riesling and Sauvignon Blanc (two specialities) and modern-style Uco Valley Malbecs.

✓ *Estate Black Edition • Selección de Bodega Malbec*

FABRE MONTMAYOU
Patagonia, the Uco Valley, and Vistalba
★☆

Frenchman Hervé Joyaux Fabre was one of the first overseas winemakers to see the potential of Malbec as a varietal wine in Argentina. He now makes a widely distributed range under the Viñalba, HJ Fabre, and Fabre Montmayou labels in Patagonia as well as Mendoza.

✓ *Fabre Montmayou Grand Vin • Viñalba Diane*

MATERVINI
Mendoza and Salta
★★☆

After leaving Achával-Ferrer, Santiago Achával set up a new winery close by, producing reds and one white to the same exacting standards. Mostly made with purchased grapes from both Mendoza and Salta, these are some of Argentina's most exciting new wines.

✓ *Piedras Viejas El Challao • Antes Andes Calchaqui Valleys Malbec*

MENDEL
Mendoza
★★

Roberto de la Mota comes from a family of winemakers – his father was the legendary Don Raúl – and has continued its fine tradition at Mendel as well as with his own brand, Revancha. Alongside an iconic old-vine Sémillon, the *bodega* produces Finca Remota, one of Paraje Altamira's best Malbecs.

 Sémillon • Finca Remota Malbec

NOEMÍA
Río Negro
★★☆

Danish winemaker Hans Vinding-Diers makes his eponymous Malbec from a remarkable dry-farmed vineyard planted in 1932 in Mainque. He also produces more approachable wines under the A Lisa label. One of the south's top names.

✓ *Noemía Malbec*

NORTON
Mendoza
★★

Although Norton dates back to 1895, it was not until 1989, when Austrian Gernot Langes-Swarovski purchased the winery, that it started to show its true potential. Now better than ever under winemaker David Bonomi, the highlights are its two blended reds, Lote Negro and Gernot Langes.

✓ *Gernot Langes • Lote Negro • Privada Malbec*

PEÑAFLOR
Argentina
★★

Peñaflor is Argentina's biggest producer, with several different brands under the talented baton of chief winemaker Daniel Pi and his team, including Costa & Pampa, El Esteco, Finca Las Moras, La Mascota, Navarro Correas, and (best of all) Trapiche. Quality is generally very good at all prices levels.

✓ *Costa & Pampa Albariño • El Esteco Old Vines 1945 Torrontés • Trapiche Terroir Series Finca Ambrosía*

PER SE
Gualtallary
★★★

In the space of less than a decade, viticulturist Edy del Pópolo and oenologist David Bonomi have taken this small, 4.2-hectare (10.3 acres) estate sited on limestone soils in high-altitude Gualtallary to the first rank of Argentina's *bodegas*. The vines are still young, but this is a *grand cru* site in the making. Look out for blended La Craie, pure Malbec

Uní del Bonnesant, and a *solera*-aged Chardonnay called Volare de Flor.

✓ *La Craie • Uní del Bonnesant • Volare de Flor*

RICCITELLI WINES
Mendoza and Patagonia
★☆

Matías Riccitelli is one of the most exciting young producers in Argentina and is the son of Jorge, the long-time winemaker at Norton. Using striking labels and good marketing, he has brought style as well as humour to the industry with wines like Hey! Malbec. The whites from La Carerra and Patagonia are exceptional.

✓ *Old Vines Sémillon • República del Malbec • Vino de la Carrera Sauvignon Blanc*

SALENTEIN
Tunuyán
★★☆

Salentein is one of the most beautiful properties in the Uco Valley, producing a large range of high-end wines under the hand of experienced winemaker Pepe Galante as well as a cheaper El Portillo label. Almost everything they make, from Malbec to Pinot Noir to Chardonnay is worth buying.

✓ *Las Sequoias SV Chardonnay • Los Cerezos SV Malbec • Los Jabiles SV Pinot Noir*

SUSANA BALBO WINES
Mendoza
★★

Susana Balbo was the first high-profile female winemaker and remains a very important figure in the industry. Working alongside her children and Edy del Pópolo, she is making increasingly refined whites and reds and has backed off on her use of oak in recent years. The Crios range supplies great value for money.

✓ *Benmarco Expresivo • Signature White Blend*

TERRAZAS DE LOS ANDES
Mendoza
★★☆

Part of the same group that makes Cheval des Andes and Chandon sparkling wines, this *bodega* has grown in stature over the last five years, with increased focus on single-site reds and whites. Terrazas also produces one of the country's best sweet wines from Petit Manseng.

✓ *Grand Chardonnay • Grand Terroir Petit Manseng • Los Castaños Malbec*

THE MICHELINI FAMILY
Uco Valley
★★

It's hard to keep up with the various projects that the Michelini family (four brothers, one son, and Andrea Mufatto, who is the wife of Gerardo Michelini), so creative is the clan. Michelini I Mufatto, Passionate Wine, Plop, SuperUco, and Zorzal all involve one or more of this Gualtallary-based group, with the emphasis on low-alcohol and low-intervention reds and whites.

✓ *Michelini I Mufatto Propósitos Chenin Blanc • Passionate Wine Agua de Roca • Zorzal Eggo Franco*

VIÑA COBOS
Agrelo and Uco Valley
★★★

Viña Cobos is superstar American winemaker Paul Hobbs' home base in Argentina, although he also consults for other wineries. The style has shifted here in recent vintages towards less oak and more freshness, making the wines even more appealing. Ambitiously priced, but these are some of the country's best reds.

✓ *Bramare Zingaretti Malbec • Bramare Rebon Vineyard Malbec • Cobos Marchiori Estate Malbec*

ZUCCARDI
Mendoza
★★★★

Sebastián Zuccardi, who works alongside his father, José, has switched the focus of this dynamic *bodega* in the last decade from the flatlands of eastern Mendoza to the best *terroirs* of the Uco Valley. The Q and Santa Julia range still deliver value for money, while the top wines, made with little or no oak, are world class.

✓ *Fósil Chardonnay • Piedra Infinita Malbec • Polígonos Cabernet Franc*

OAK BARRELS BEAR THE STAMP OF ZUCCARDI, A FAMILY-OWNED WINERY IN MENDOZA
Since 2008 Zuccardi has had a Research and Development arm that studies *terroir* and its different variables that affect wine quality.

The Wines *of* Australia, New Zealand, *and* Asia

IN THE REGION OF ASIA AND THE SOUTH PACIFIC, Australia had long been the stand-out as far as winemaking went, especially with their famed spicy Shiraz and the phenomenal growth of boutique wineries, along with quality bulk producers like Yellow Tail. (Even when the world market seemed glutted, the grape-growers battled back by focusing on site-specific wines, re-establishing their dominance.) Yet New Zealand's wine industry began in earnest barely 30 years after their Aussie cousins, around 1819, and their wines, lead by crisp Sauvignon Blanc, assumed classic status in the 1990s, only a decade after Australia was so honoured. Sustainability has become their battle cry for the new millennium. Recent floods and fires in both countries, sadly, may have imperilled a number of established wineries and taken a toll on current vintages. Asia, meanwhile, has been eagerly playing catch-up, with hundreds of wineries found in more than 15 countries. China, India, and Japan may be the current leaders in terms of quality and production, but many other Asian countries have an ancient heritage of growing grapes and producing distinctive wines, including Bali, Bhutan, Cambodia, Myanmar, Nepal, the Philippines, Sri Lanka, South Korea, Taiwan, and Thailand.

KANGAROOS INVESTIGATE THE GRAPE VINES IN A HUNTER VALLEY VINEYARD IN NEW SOUTH WALES, AUSTRALIA
Kangaroos will feed on tender buds and shoots and can be occasional vineyard pests. Far more of a threat are the great number of fruit-eating birds, such as rosellas, starlings, and thrushes, that feast on ripening grapes.

Australia

Having built its name on the consistent quality of its high-volume brands, which have been aggressively marketed through multi-bottle discount deals, "Brand Australia" grew into the biggest success story in modern wine history. In 2007, this upward growth reached an abrupt stop, however, and Australian winegrowers were faced with the reality of a market that did not want the quantity of wines they were producing. Fortunately, this trend has once again reversed towards growth, and since 2013, Brand Australia has been successfully demonstrating the lessons learned in the five-year slump.

It is no accident that the public perception of Australian wine is so dependent on the concept of branded wine. Australian brands have always been strong, but in 1996 their position was formally promoted by the industry's 30-year plan to become the "world's most influential and profitable supplier of branded wines". Following the drop in growth post 2007, Australian branded wines found themselves competing with a range of alternatives from places such as Chile and Argentina, making domination of this sector a much greater challenge. A renewed focus on site-specific wines at a higher quality level has been instrumental in helping bring the industry back to its projected pathway.

THE BRANDED APPROACH

Australian brands have been criticized as overripe, over-alcoholic, formulaic wines that have no soul. Perhaps. Yet the fact that so many might taste very similar is in itself an indication of their very consistent quality. They might not exude individuality, and wine lovers might wonder why this commercial category of wine has been singled out for official blessing in the 30-year plan, but Australia's branded wines are made in a soft, rich, consumer-friendly, fruit-driven style. They are nothing more than the equivalent of branded generic wines from, say, Bordeaux or Beaujolais, but infinitely more enjoyable. Certainly consumers think so. By 2003 Australia had become the second-largest exporter of wine to the United States and had even pushed out France from the number one position on the UK market. Australian wines remain first in the UK off-trade, in large part due to the still-ubiquitous nature of "sunshine in a bottle" wines in every corner off-licence. While it might not seem so surprising that an up-and-coming wine country like Australia should appeal more to consumers, particularly younger ones, than an old and traditional producer such as France, it is an extraordinary feat considering the contrasting size of these two wine industries.

Did Australia achieve its export victories because its brands were sold at economically unsustainable prices, as some now fear? No. How could this be true when the country's largest wine groups have made handsome profits in the process? There is nothing wrong with multi-state blends, such as the all-encompassing Southeastern Australian appellation, which straddles 95 per cent of the country's vineyards; however, brand owners have started concentrating less on such wines and more on expressing some level of individuality.

THE INFLUENCE OF CHINA

China is now Australia's largest export market, with a dramatic increase following the 2015 Fair Trade Agreement. Australia's proximity to China gives it an advantage over its competitors, and indeed there is an increasing interest from Chinese companies in investing in Australian wineries. The vast majority of wines exported to China are reds, due in large part to the cultural connection between red and good fortune – and indeed it is Australia's good fortune that the majority of its production is red wine.

RIPE GRAPES AWAIT HARVEST IN AN ADELAIDE HILLS VINEYARD
Although the Adelaide Hills wine region is believed to be one of the areas worst hit by the bushfires of 2019–20, the loss of vineyard acres is a lesser concern than other possible aftereffects. It will take some time to truly evaluate the fires' impacts, but with an estimate of only 1 per cent of Australia's vineyard land located in fire zones, issues such as smoke taint to grapes might come to the fore.

RECENT AUSTRALIAN VINTAGES

In general, Australia is known better for its consistency across vintages, unlike the fickler climates of many European regions. When the weather strikes out, however, it can have a devastating impact; fire and flood provide a particularly contrasting influence in comparison with the usually temperate norm.

2019 A warm year, though not quite as much as 2018. Somewhat varied in yield levels and with early harvest in many areas.

2018 Perhaps a textbook year? This is considered by many to be universally excellent.

2017 Another year with good quality grapes at harvest, which was slightly later than 2018 due to a cooler season.

2015 The impact of bushfires can be seen in some wines from the area around Adelaide, though not as widespread as in some earlier years.

2011 This was an exceptionally challenging year, due mainly to rain. After almost 10 years of drought, broken in 2010 with much lower rainfall, wineries were not prepared to deal with the disease pressure of this unusually cold and wet vintage.

2009 Generally, a very good to excellent year, with great Hunter Valley Sémillon the standout. Bushfires in Victoria led to smoke-tainted wines, and in a few cases the loss of vineyards.

2002 The 2000s were an example of exactly how consistent Australian vintages can be – however, small variations become more apparent with age, and the 2002s have stood the test of time particularly well.

Unlike the wines that helped bring Australia to prominence in the early 2000s, those shipped to China are relying more on prestige than low prices. By taking a more individually brand-specific and site-specific approach, it seems that Wine Australia and wineries themselves are building this growing market on a more stable and sustainable base. China's domestic market is also growing, however, and time will only tell how this affects the status of imports such as those from Australia.

Mainland China is not the only location helping to build the Australian exports; Australia is second only to France in imports to Hong Kong, a market where quality perception is particularly important. This achievement is as impressive as the French/Australian comparison in the UK market; Australia contributes a far smaller percentage of global production yet is able to compete effectively at the top level.

THE GROWTH OF AUSTRALIA'S WINE TRADE

The first Australian vineyard was planted at Farm Cove in New South Wales in 1788, with vines originating not from France but from Rio de Janeiro and the Cape of Good Hope. These were collected by the first governor, Captain Arthur Phillip, en route to Sydney. The rich soil of Farm Cove and its humid climate proved fine for growing vines, but not for making wine. Phillip persevered, however, and planted another vineyard at Parramatta, just north of Sydney. Soil and climate were more suitable, and the success of this new venture encouraged Phillip's official requests for technical assistance. England responded by sending out two French prisoners of war, who were offered their freedom in exchange for three years' service in New South Wales, in the belief that all Frenchmen knew something about making wine. This did not prove true: one was so bad at it that he was transported back to England; the other could make only cider, and mistakenly used peaches instead of apples. From these shaky beginnings an industry grew, but with no thanks to the British. At first, Australia's wine trade was monopolized by the needs of the British Empire, later the Commonwealth. It thus gained a reputation for cheap fortified wines – not because these were the only ones it could make, but because Britain wanted them. Unfortunately, Australians also acquired

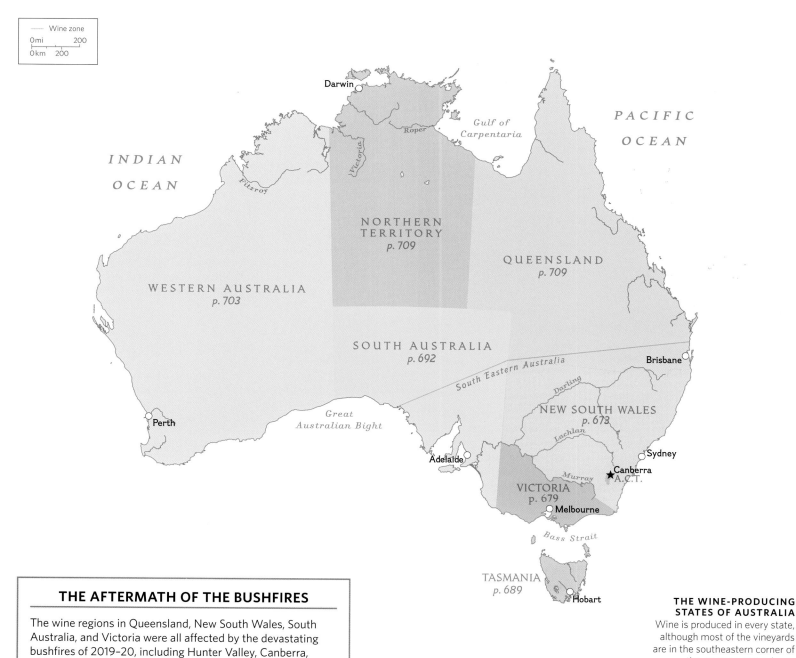

THE AFTERMATH OF THE BUSHFIRES

The wine regions in Queensland, New South Wales, South Australia, and Victoria were all affected by the devastating bushfires of 2019–20, including Hunter Valley, Canberra, Rutherglen, Gippsland, and Adelaide Hills. How smoke taint and heat will affect the 2020 vintage is uncertain.

THE WINE-PRODUCING STATES OF AUSTRALIA
Wine is produced in every state, although most of the vineyards are in the southeastern corner of the country, centred on a semi-circular band running from Sydney in New South Wales to Adelaide in South Australia.

MADE OF RECYCLED CORKS, THE CHURCH BLOCK BOTTLE OF WINE WAS BUILT TO CELEBRATE THE BRAND'S 35TH ANNIVERSARY
First released in 1999, Church Block is the Wirra Wirra winery's flagship label. The 10-metre (33-foot) installation also marked the beginning of this South Australian producer's phasing out of cork closure and its move to screw caps.

the taste. Although Australia made (and still makes) some of the world's finest dessert wines (botrytised as well as fortified), most wines were heavy, very sweet, and sluggish at a time when the rest of the world was drinking lighter and finer wines. Small quantities of truly fine table wine were produced but, until the 1960s, most of those exported were remarkably uniform in style.

THE RENAISSANCE OF AUSTRALIAN WINES

Over the past 40 years or so, the technology used in Australian winemaking has made huge advances, taking with it the quality of the wines. Until the early 1980s, as much as 99 per cent of all Australian wine was consumed domestically. When output dramatically increased, producers had to concentrate on exports, which increased tenfold in the 1980s, and a further tenfold in the 1990s. Today, they represent more than 60 per cent of the country's total production. There are around 2,500 wineries in Australia, employing nearly 200,000 people.

RESEARCH AND DEVELOPMENT

The Australian Wine Research Institute (AWRI) was established in 1955 and is one of the world's leading sources of research and development in the wine industry. Much of the Institute's work focusses on assisting Australian wineries with problem-solving investigations and strategic research. The commercial arm of the organization provides advanced consulting and analysis, and many university viticultural projects are overseen by the AWRI.

The institute recently set a new RDE plan 2017–2025, building on previous work from 2013. Fifty projects have been selected, covering everything from consumers and markets to microbial diversity and wine faults.

Another organization with particular relevance to the industry is the Commonwealth Scientific and Industrial Research Organisation (CSIRO). Globally recognised for their research into yeast strains, vine training, and climate change impact, this organization often works with the AWRI and other international institutions in order to further the progress of scientific research.

If oenological universities had an Ivy League, Australia would most certainly be able to claim one of the members; Adelaide University's Roseworthy College has seen many of the country's best-known winemakers, and indeed some journalists and researchers, pass through its doors. Time spent studying in Australia and working the harvest is seen as something of a rite of passage for the up-and-coming generation in many well-established European regions.

GOING GREEN?

Taking the organic route in the vineyard is harder in some countries than others. Essentially, the more rain or humidity there is, the more rot affects the grapes and the more difficult it becomes to control this by natural means. Some growers are game for the challenge, even in rain-soaked countries like England, whereas fewer than expected have converted in sunny Australia, despite the fact that Australia has the largest amount of certified-organic land in the world. It should be pointed out that many good growers (and all the best) are "almost" organic the world over, since anyone who is interested in producing the best wine they possibly can will realize that it is in their own interest to ensure that their own viticulture is sustainable for the long term. The market has showed a markedly increased interest in organic and biodynamic viticulture in recent years; certification can be expensive and is a bureaucratic nightmare, but the mark on the label pays for itself in many ways. Not only does a sensitivity towards the environment ensure sustainability of the industry for future generations, it's a useful selling point in any setting, from supermarket shelf to Michelin-starred restaurant.

As this category grows in importance, the relevance of the certifying body with regards to the export market adds a further complication. Not all the wineries listed below are certified biodynamic, but all are known for their adherence to the principles of this philosophy.

- Avonmore Estate (Avonmore, Victoria)
- Castagna (Beechworth, Victoria)
- Cowbaw Ridge (Macedon, Victoria)
- Cullen Wines (Margaret River, Western Australia)
- Goulburn Terrace Wines (Nagambie, Victoria)
- Hochkirch (Henty, Victoria)
- Kalleske Wines (Barossa Valley, South Australia)
- Krinklewood (Hunter Valley, New South Wales)
- Lark Hill Winery (Canberra, New South Wales)
- Maverick Wines (Vine Vale, South Australia)
- Ngeringa (Mount Barker, South Australia)
- Pennyeight (Beechworth, Victoria)
- Robinvale Wines (Murray Darling, Vicoria)
- Stefano Lubiana (Derwent Valley, Tasmania)
- Temple Breuer (Langhorne Creek, South Australia)

THE LARGEST WINE GROUPS OF
AUSTRALIA

It is quite extraordinary that, in 2005, when this feature last appeared in *Sotheby's Wine Encyclopedia,* the largest wine group in Australia was Southcorp, churning out 30 per cent of the country's wine and ranking seventh-largest wine producer in the world.

The five largest wine groups account for 75 per cent of all Australian wines and nearly 90 per cent of Australian wine exports. The following wine groups are now the largest in terms sales of wines by volume.

TREASURY WINE ESTATES

(1st by value of sales)

Vineyards: 8,609 hectares (1st)

The remnants of Southcorp are buried in the relatively recent origins of Treasury Wine Estates. In 1990 the Penfolds Wine Group was formed when Penfolds, which had already acquired the Barossa Co-op Winery (Kaiser Stuhl) and Allied Vintners wineries (Glenloth, Seaview, Tulloch, and Wynns), took over Lindemans (including Leo Buring and Rouge Homme). Later that year, the Penfolds Wine Group was acquired by SA Brewing (aka South Australia Brewing), which owned only one wine company at the time (Seppelt). SA Brewing sold off its brewing interests, thus requiring a new name for its newly formed wine-only group, and so in 1993 Southcorp was born. In 1996 the group purchased Coldstream Hills in the Yarra Valley, and Devil's Lair Winery and Vineyards in Margaret River. Its strategy to focus on premium-quality wines and increase the standard of its less-expensive wines was the underlying factor for its success. The proficient synergy went awry in 2001, however, when Southcorp Wines acquired the privately owned Rosemount Estates from the Oatley family. There were just too many similarities in the Lindemans and Rosemount products, particularly on export markets, and totally opposing management styles, not to mention the radically different winemaking regimes. When Penfolds' chief winemaker John Duval left after 28 years with the company amid rumours of a rift with Rosemount's Philip Shaw, and Shaw was appointed chief winemaker for the entire Southcorp group, the deal that saw Southcorp pay AU$1.5 billion (US$730 million) for the merger was billed by the press as a reverse takeover by the smaller Rosemount. In 2002, when Southcorp was still blissfully ignorant of the massive loss it would soon endure, the company sold the Tulloch winery and brand, the Hungerford Hill brand, and the Rouge Homme winery. In January 2003 Southcorp's accounts showed profits had plummeted 97 per cent and in June 2004, following the shockwave of the loss of almost a one billion Australian dollars, it came up with various plans to rationalize its structure and operation, but nothing could turn the downward plunge. In 2005 Southcorp was taken over by the brewing conglomerate Foster's, which already owned Berringer-Blass and tried to run Southcorp like a beer business, but that did not work either. In 2010 Berringer-Blass and Southcorp were renamed and spun-off as a separate company called The Wine Treasury, which in 2015 purchased the US-based Chateau and Estate Wines from Diageo. In addition to this, The Wine Treasury owns the Squealing Pig brand, which sources from a wide range of origins (Australia, NZ, France and Italy).

Brands: *19 Crimes, Annie's Lane, Coldstream Hills, Devil's Lair, Fifth Leg, Heemskerk, Ingoldby, Jamieson's Run, Killawarra, Leo Buring, Lindeman's, Metala, Penfolds, Pepperjack, Rawson's Retreat, Rosemount Estate, Saltram, Samuel Wynn & Co, Seppelt, St Huberts, T'Gallant, Wolf Blass, Wynn's Coonawarra Estate, Yellowglen*

PERNOD RICARD

(2nd by value of sales)

Vineyards: 1,662 hectares (6th)

In 2019 Jacob's Creek was voted the eighth-strongest brand in Australia. Not the eighth-strongest wine brand, but the eighth-strongest brand of anything, with other winners from the world of airlines to banking and everything else between. No other wine brand made the top 10. This pretty much justified the bizarre and some would say sacrilegious strategy of its owners, Pernod Ricard, who rebranded all of its premium Orlando-Wyndham wines as Jacob's Creek. Some people thought, not unreasonably, that this was madness, because the group owned many famous brands, including Wyndham Estate, of course. It was not only a great shame to lose such historic names, but downright perverse to rebrand everything under its cheapest brand. Surely it would make sense to rebrand under its most premium brand? Well, apparently not. It seems that Pernod Ricard saw Jacob's Creek not as its cheapest brand, but as its most significant brand, and therefore developed the strategy to diversify its the range and raise the quality. To valorise the brand, as the French would say. And no one can argue with the result.

Brands: *Jacob's Creek*

ACCOLADE

(3rd by value of sales)

Vineyards: 1,073 hectares (10th)

Formerly BRL Hardy, Accolade was purchased by the giant American-owned multinational Carlyle Group in 2018 and its core brand Hardy's remains top-selling wine band in Australia and a number of important export markets.

VISITORS TO THE PENFOLDS TASTING BUILDING CAN PERUSE DISPLAYS DETAILING THE ESTATE'S HISTORY
Penfolds, one of Australia's most distinguished names in wine, is part of Treasury Wine Estate, a global winemaking and distribution conglomerate headquartered in Melbourne.

ROWS OF YELLOW TAIL SHIRAZ LINE A WINE SHOP SHELF IN CANADA
Produced by Casella Family Brands, the Yellow Tail label of wines proved to be export phenomenon. At the height of its popularity in 2004, the Shiraz was Australia's top seller.

The country's largest wine group was originally established by Thomas Hardy, a farmworker who arrived in Adelaide on board the *British Empire* in 1850. He was 20 years of age and had travelled with two cousins, the sisters Joanna and Mary Anna Hardy. They were also accompanied by a friend and fellow farmworker, John Holbrook. Hardy went to work for John Reynhill, who owned a cattle station in Normanville, and within 18 months he had made enough money driving cattle to feed miners at the Victorian goldfields to purchase 7 hectares (16.5 acres) of land next to the Torrens River, just outside Adelaide. He married Joanna, and John Holbrook married Mary Anna. Hardy called their homestead Bankside and planted grapes, making his first wine in 1857 and exporting two hogsheads to London in 1859. By 1865 the Bankside winery was producing 53,000 litres (14,000 gallons). Hardy's big break came in 1873, when he was the sole bidder for the Tintara Vineyards Company, which included a winery, cellars, vast stocks of wine, and 280 hectares (700 acres) of land a few miles northeast of McLaren Vale, where the vineyards had been planted in the 1850s and were thus mature. By selling the stocks, he recouped the entire cost of the bid and set about extending the cellars and planting even more vineyards. In 1878 he converted a disused flour mill in McLaren Vale into a brand-new winery, selling off the mill machinery to fund a large part of the construction. He might have been a farmworker in England, but he was evidently a natural-born business-

man, and in Victorian Australia's land of opportunity, he had the freedom to express his talents, not least of which was grape-growing and winemaking, although no one knows to this day how and where he learned those skills. By 1895 Thomas Hardy was the largest wine producer in South Australia, a position Hardy's has maintained ever since. In 1976, Thomas Hardy and Sons purchased the Emu Wine Company, which was known primarily for exporting Kangaroo Red, but also owned a little gem called Houghton Wines. Chateau Reynella and Rhine Castle Wines were taken over in 1982. Perhaps its most innovative decision was made in 1990, when the company extended its wine holdings into France, where it acquired Domaine de La Baume on the outskirts of Béziers (sold off in 2003). The company's rationale for this move was that it believed the French had not exploited varietal wines. In 1992 Thomas Hardy and Sons merged with Berri-Renmano, becoming BRL Hardy, which listed itself on the Australian stock exchange. The funding that resulted helped the company to purchase 1,800 hectares (4,450 acres) – only 285 hectares s (705 acre) of which was plantable – at Banrock Station in 1994. Over the next three years BRL Hardy purchased Yarra Burn, a one-third share in National Liquor Distributors (New Zealand), the Hunter Ridge brand (from McGuigan Wines), 50 per cent of Brookland Valley Vineyards (which is certified organic) in Western Australia, and announced plans for its ambitious Kamberra winery and vineyards in the Canberra region. In

1998 BRL Hardy increased its shareholding in National Liquor Distributors, which had by then merged with Nobilo Vintners, effectively taking control by 2000. Expansion continued as the company took a 50 per cent shareholding in Barossa Valley Estates and, in 2001, entered a joint venture with the US-based Constellation Brands. This latter arrangement developed into a full merger in 2003 to form Constellation Wines, the largest wine group in the world, and the Australian group was renamed The Hardy Group. In 2016 it acquired the Australian premium wine portfolio (Fine Wine Partners) of beverage giant Lion Nathan, but in 2017, The Hardy Group was itself taken over by CHAMP Private Equity (Constellation Wines retained a 20 per cent holding), who changed its name to Accolade. One year later, the Carlyle Group bought Accolade, including the 20 per cent held by Constellation, and is now sole owner. The current strategy appears to be one of moving from owning vineyards, where it can, to buying in grapes. Accolade also produces wine Chile, Italy, New Zealand, North America, and South Africa.

Brands: *Amberley, Banrock Station, Bay of Fires, Berri Estates, Brookland Valley, The Busselton Boys, Croser, Days of Rosé, Eddystone Point, Goundrey, Grant Burge Wines, Hardys, Houghton, House of Arras, Jam Shed, Knappstein, Leasingham, Moondah Brook, Omni, Petaluma, Renmano, Reynella, Stanley Wines, Starve Dog Lane, St Hallett, Stonier, Tatachilla, Yarra Burn*

CASELLA FAMILY BRANDS

(4th by value of sales)

Vineyards: 5,300 hectares (2nd)

Filippo Casella emigrated to Australia from Sicily in the 1950s, purchased a farm in 1965, planted vines, and made his first wine in 1969, but he did so as a bulk supplier to other wineries. Consequently, the company he established went unnoticed until it started the transition to marketing wines under its own brand. In more recent years, Casella became a legend, thanks to the "Yellow Tail" phenomenon, which took Australia's biggest wine players by surprise. Yellow Tail went from zero to 5.9 million cases in just four years. In 2002, midway through this extraordinary growth, Yellow Tail had become the top-selling Australian wine in the United States, and by 2004 Yellow Tail Shiraz was that country's best-selling bottled red wine, as well as being officially ranked as the fastest-growing export brand in the history of the Australian wine industry. That segment of the market has been declining, however, particularly in the United States, where Casella has spent millions on Yellow Tail ads for the Super Bowl three years running (2017–19) just to cling on at the bottom end of market where margins are already squeezed. Fortunately, being family-owned has helped when taking over other family-owned wineries that have found themselves floundering, such as Morris Wines, which has fared extremely well in this group. Casella's quarter-million-tonne facility at Yenda is the largest winery in Australia.

Brands: *Baileys of Glenrowan, Brands Laira, Casella Family Brands, Magic Box, Morris Wines, Peter Lehman Wines, Shaw Family Vintners, Yellow Tail*

AUSTRALIAN VINTAGE

(5th by value of sales)

Vineyards: 2,950 hectares (3rd)

The origins of this group technically date back to 1992, when McGuigan Wines in the Hunter Valley was established by the legendary Brian McGuigan, but the full story starts a little earlier than that. It was Brian McGuigan who founded Wyndham Estate (which became Jacob's Creek; *see* Pernod Ricard) on a property purchased in 1830 by George Wyndham, one of the very earliest pioneers of viticulture in the Hunter Valley. This property had been sold by Penfolds to its former cellar master, McGuigan's father, in the early 1960s. With the help of his wife, Fay, who was the export manager for Wyndham Estate under Penfolds in the 1980s, McGuigan Wines began exporting to 20 different countries, growing the company quickly. In 1994 the company purchased the Buonga (the third-largest winery in the country) from Orlando Wyndham and in 1998 took over Yaldara Barossa Valley. In 2002 McGuigan Wines merged with Simeon Wines (which included the Australian Vintage, Coldridge, Loxton, and Manton's Run brands) and changed the trading name to McGuigan Simeon, with Brian as CEO. In so doing, it had become the fifth-largest wine producer in Australia. In 2003 McGuigan Simeon acquired Miranda Wines and in 2007 took over Nepenthe Wines. In 2008, due to shareholders' demand, the name of the company was changed to Australian Vintage. This group is super keen on renewable energy and all other aspects of sustainability, but although it is extremely vineyard-rich, it is unfortunately premium-brand poor. Being the fifth-largest wine group in Australia is a great achievement, but with a successful range of premium and super-premium wines, Australian Vintage could be in a league of its own.

Brands: *McGuigan, Miranda, Nepenthe, Passion Pop, Tempus Two*

McWILLIAMS WINES

(6th by value of sales)

Vineyards: 980 hectares (11th)

This company was established in 1877 by Samuel McWilliam, who planted his first vineyard at Corowa. In 1941, McWilliam's purchased the Mount Pleasant Estate, which was founded at Pokolbin by the legendary Maurice O'Shea. In 1989 McWilliam's took over Barwang Vineyard, which coincidentally had been established using cuttings from McWilliam's some 20 years earlier. Between 1990 and 1994, McWilliam's bought Brands Winery in Coonawarra, and Lillydale vineyards in the Yarra Valley. In 2000 this company won 13 trophies and 611 medals at Australian wine shows. It had also firmly established itself as the 5th-largest exporter of Australian wine by volume and 6th by value. McWilliams was swimming in waters inhabited by corporate sharks; however, in 2001 McWilliam's entered into an international distribution agreement with Gallo Wines (US), and by 2002 had dropped to 15th-largest exporter of Australian wine by volume, and 20th by value. This downward tumble probably saved McWilliams from being taken over and having its asset stripped. Despite the drop in sales, McWilliams remained financially strong and in 2007 took over Evans & Tate. Resisting growth for growth's sake, it became a founding member of Australia's First Families in 2009 and has been the 7th-, 8th- or 9th-largest Australian wine producer ever since.

Brands: *Evans & Tate, McWilliams, Mount Pleasant*

DE BORTOLI WINES

(7th by value of sales)

Vineyards: 1,172 hectares (8th)

Established in 1928 by Vittorio De Bortoli, who had emigrated from Treviso in Italy in 1924. To conserve money, Vittorio had literally lived under a rainwater tank until his fiancée, Giuseppina, arrived in Australia. Together they purchased a 135-hecatre (55-acre) mixed-fruit farm in 1928, a year that saw a glut of Shiraz grapes. Many farmers had decided it was cheaper to leave the grapes rot on the vines and did not object when Vittorio asked if he could harvest the grapes for free. He made wine from 15 tonnes of free grapes and sold it to local Italian immigrants and so began a business that initially focussed on selling bulk wines to immigrant families. The company expanded in the 1970s and received international acclaim for its Noble One botrytised dessert wine in the 1980s (still considered to be Australia's best-known Sauternes-style wine), enabling it to continue its shift away from bulk to premium.

Brands: *De Bortoli, Down the Lane, Este, La Bohème, Melba, Noble One, Palette Series, PHI, Riorret, Sheep Shape, The Accomplice, Rutherglen Estates, Vinoque*

WARBURN ESTATE

(8th by value of sales)

Vineyards: 1,2017 hectares (7th)

In 1968 Giuseppe and Maria Sergi bought a farm in the Riverina area and planted their first vineyard. They had worked hard and saved as much as they could since emigrating from Calabria in Italy in 1952. In 1969 they started cellar-door sales of The House of Sergi's in two-litre flagons. The wines were mostly fortified, which was the preferred style back then, and they quickly became known for their Cream Cherry, Blackberry Marsala and Choc Mint Marsala, to name but a few specialties. Giuseppe's son Antonio would make an annual trip to Northern Queensland to sell their wine in 200-litre barrels to the Italian community for the festive season. By the 1990s the company had followed the demand for varietal wines, growing more Shiraz, Cabernet, Sauvignon Blanc, and Chardonnay. The Sergi family then moved to a property by the name of 1164, which they changed to Warburn Estate, and planted it with Italian varietals. In 1995 Antonio moved into his own 650-hectare (1,600-acre) property. Still family-owned, Warburn Estate is currently run by the third and fourth generation.

Brands: *1164, AC/DC The Wine, Bushman's Gully, Coolabah Cask, Gossips, Koobah Estate Cask, Rumours, Warburn Estate, Wine Gang*

ANDREW PEACE

(17th by value of sales)

Vineyards: 684 hectares (16th)

This family-owned business has seen a meteoric rise from 1981, when Jim and Pam Peace bought a vineyard, to 1995, when the winery was built by son Andrew and the first harvest processed, to 1999, when the first exports were made, to 2015, when Andrew Peace became the ninth-largest-selling Australian producer, and 2018, when it climbed to eighth place.

Brands: *Australia Felix, Empress, Heart & Soul, Tall Poppy, Masterpeace, The Unexpected, Wrattonbully Wines*

IDYLL WINE CO.

(18th by value of sales)

The 20-hectare (50-acre) Idyll Vineyard in the Moorabool Valley was planted in 1966 with Shiraz, Cabernet Sauvignon, and Gewürztraminer by Darryl Sefton, a veterinarian turned self-taught winemaker and pioneer of the Geelong area. Daryl and his wife, Nini, whose artistic expertise shone through on the labels, had something of a cult following in Australia, Denmark, Ireland, and Switzerland. The Seftons retired in the late 1980s and in the 1990s the vineyard and winery were purchased by Vince and David Littore, who merged it with their Jindalee Estate operation. This morphed into Littore Family Wines, which got into financial difficulties and was taken over in 2016 by the Costa Group, which invested heavily in Idyll's production facilities and controversially released a blue-coloured wine called Blue Bird.

Brands: *Arcadian, Eden Grove, Idyll, Jindalee, Punters Corner, St Andrews, Trails End, Wispers, Winton Rd*

OTHER NOTABLY LARGE PRODUCERS

Brown Brothers (*see* p683) is the 9th-largest producer in sales by value; Kingston Estate is the 4th largest owner of vineyards and 6th largest wine producer (including 50 million litres sold in bulk); Qualia is 8th-largest wine producer (including bulk-wine); Yalumba (*see* p702) is the 10th-largest in sales by value and 11th by volume of branded wines; Zilzie is 9th-largest wine producer (including bulk-wine).

A VINTAGE GRAPE PRESS DECORATES THE CELLAR DOORS OF THE YALUMBA WINERY
With vineyards planted in the mid-19th century, this Barossa valley winery has flourished to become one of Australia's top wineries in both sales by value and volume of branded wines in the 21st century.

New South Wales

From the Hunter Valley's weighty Shiraz, through its honeyed, bottle-aged Sémillon, and the easy-drinking Riverina wines, to the fast-rising wines from areas such as Orange, Tumbarumba, and Hilltops – the wines of New South Wales are better in quality and more varied in style than they have ever been.

Although vines arrived with the British First Fleet, and Australia's first governor, Captain Arthur Phillip, planted the country's first vineyard in 1788 (at Farm Cove, known by the Aborigines as Woccanmagully and now the site of Sydney's Royal Botanic Gardens), it was not until the next century that Gregory Blaxland made Australia's first wine. Blaxland planted his vineyard at Ermington, a few miles down river from Phillip's second vineyard at Parramatta. In 1822 Blaxland also became the first person to export Australian wine, when he brandied a quarter pipe (137.5 litres, or 26.3 US gallons) of red wine so that it might withstand the rigours of being shipped to London, where it won a Silver Medal from the Society for Encouragement of Arts, Manufactures and Commerce (now the Royal Society of Arts).

HUNTER SHIRAZ

The reputation of this state's wine industry has historically revolved around the Hunter Valley, particularly the Upper Hunter. Because Australia has claimed Shiraz as its own, Hunter Shiraz will always play a pivotal role. But our perception of this wine has changed; it is hard to imagine that the Hunter was once famous for a huge, beefy style of Shiraz that gave off a strong, gamey, sweat-and-leather odour and possessed an earthy, almost muddy taste that was "chewed" rather than swallowed. Its "sweaty saddle" smell was supposed to derive from the Hunter Valley's volcanic basalt soil, until it was widely declared to be a defect. For quite some time, it was thought to be a mercaptan fault, but the sweaty saddle odour (also described as "horsey" or "stables") is now known to be ethyl-4-phenol, a specific volatile phenol defect caused by the Brettanomyces yeast.

A VIEW FROM ABOVE SHOWS A PATCHWORK OF VINEYARDS IN THE HUNTER VALLEY WINE REGION
Known for its earthy Shiraz, Hunter Valley was one of the first regions planted with vines in the early 19th century and has since become familiar to wine drinkers worldwide. The disastrous bushfires of 2019–20 might affect its immediate future, however, with the thick smoke of backburning in some areas leaving the grapes smoke tainted.

WINE REGIONS OF NEW SOUTH WALES

(see also p667)

North of Sydney, the Lower and Upper Hunter Valley and the Mudgee area excel, while the Murrumbidgee Irrigation Area to the southwest proves that quantity can, to a certain extent, co-exist with quality.

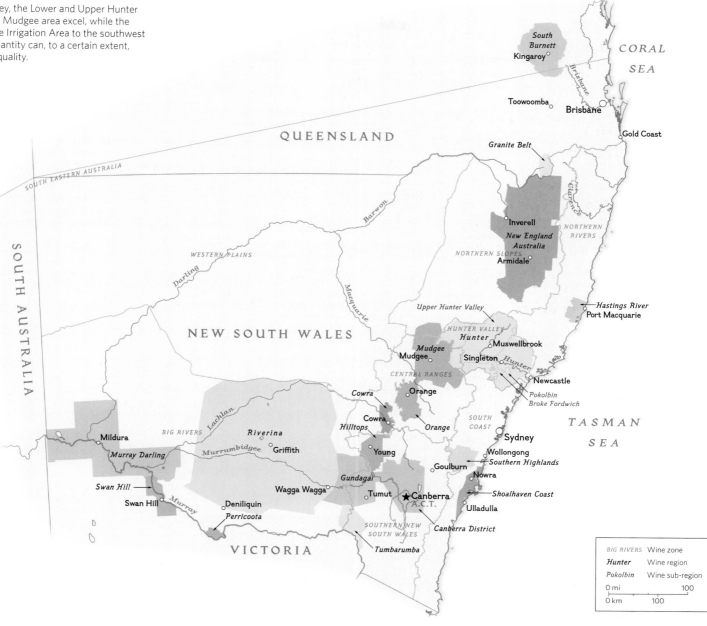

BIG RIVERS	Wine zone
Hunter	Wine region
Pokolbin	Wine sub-region
0 mi	100
0 km	100

FACTORS AFFECTING TASTE AND QUALITY

LOCATION
The southern part of Australia's east coast, between Victoria and Queensland.

CLIMATE
Growing-season temperatures are similar to those of the Languedoc in southern France. Cloud cover can temper the heat in the Hunter Valley, but the accompanying rains often promote rot. The growing season is later, and the climate sunnier, in Mudgee, Orange, and Cowra, while it is hotter and drier in Riverina, and significantly cooler in Tumbarumba.

ASPECT
Vines are grown on generally low-lying, flat, or undulating sites, but also on steeper slopes such as the fringes of the Brokenback Range in the Lower Hunter Valley, where vines are grown

at altitudes of up to 500 metres (1,600 feet); on the western slopes of the Great Dividing Range, where some vineyards can be found at an altitude of 800 metres (2,600 feet); and in Canberra at 820 metres (2,706 feet).

SOIL
Soils are varied, with sandy and clay loams of varying fertility found in all areas. Various other types of soil, such as the red-brown volcanic loams, are scattered about the Lower Hunter region and the fertile, but well-drained, alluvial sands and silts of the flat valley floors. Hilltops has a deep, gravelly, red soil, mixed with basalt and sandy granite, while Tumbarumba much farther south also has basalt and granite soils.

VITICULTURE & VINIFICATION
Irrigation is practised throughout the state,

particularly in the mainly bulk-wine-producing inland area of Riverina. The range of grape varieties is increasing, and grapes are harvested several days earlier than they used to be, for a crisper style. Temperature-controlled fermentation in stainless-steel vats is common, but new oak is used judiciously.

GRAPE VARIETIES
Primary varieties: Cabernet Sauvignon, Chardonnay, Pinot Noir, Riesling, Sémillon, Shiraz (Syrah)

Secondary varieties: Cabernet Franc, Colombard, Grenache, Malbec, Merlot, Muscat Gordo Blanco (Muscat d'Alexandrie), Petit Verdot, Ruby Cabernet, Sauvignon Blanc, Trebbiano (Ugni Blanc), Verdelho

THE APPELLATIONS OF
NEW SOUTH WALES

BIG RIVERS GI

This zonal GI encompasses the regional GIs of Murray Darling, Perricoota, Riverina, and Swan Hill (Murray Darling, and Swan Hill GIs are shared with North-West Victoria Zone in Victoria).

BROKE FORDWICH GI

This sub-region of Hunter centres in and around Broke, but also extends south and southwest, encompassing the catchment area of Wollombi Brook, and going north beyond Bulge. Chardonnay and Cabernet Sauvignon seem to be the most successful varieties grown here.

CANBERRA DISTRICT GI

One of the quirkiest GI demarcations must be that of Canberra District, because most of the appellation is not even located within the Australian Capital Territory (ACT), but in New South Wales. The reason for this, however, has nothing to do with any lack of viticultural potential within ACT. The absence of viticulture here is due solely to the fact that the concept of freehold property does not exist within ACT, deterring anyone in the long-term business of establishing a vineyard from setting up within the bounds of the ACT itself. More than 30 wineries are within a 35-minute drive of the city, however, and Canberra District has become a well-established, ever-growing region. The altitude of most vineyards is 600 to 820 metres (2,000 to 2,700 feet), but most face north or northeast, so the sunshine hours are very high at 7.2 hours per day, enabling grapes to ripen on nearly all aspects. The climate is cool in Australian terms, which equates to something much closer to moderate in a European setting. Summers here are warm and dry, with three out of every four years yielding degree-day equivalents between those of the Médoc and Hermitage. The vineyards are dependent on irrigation, and water is a declining resource. All the usual grape varieties are grown, but snappy Rieslings, elegant Pinot Noir, and Rhône-style Shiraz excel.

CENTRAL RANGES GI

A zonal GI that encompasses the regional GIs of Cowra, Mudgee, and Orange.

COWRA GI

This is a small but growing viticultural area, which is situated some 180 kilometres (120 miles) inland and west from Sydney, and nearly the same distance north of Canberra. As in most of Australia's lesser-known wine areas, viticulture flourished here in the 19th century, and was reignited in the 1970s. But lesser known does not necessarily mean lesser quality, as demonstrated by the earliest vintages of Brian Croser's legendary Petaluma Chardonnay, which were sourced from Cowra.

GUNDAGAI GI

It is surprising how many emerging wine regions in New South Wales happen to be conveniently located for tourist appeal, and Gundagai is no different, since it is touted as a near-perfect stopover between Sydney and Melbourne. Protected by the Snowy Mountains to the southeast, most vineyards are located at 500 to 1,000 metres (1,640 to 3,280 feet) in altitude, with a small handful of wineries producing whites and reds from Chardonnay, Cabernet Sauvignon, and Shiraz.

HASTINGS RIVER GI

This small region is located on the north coast of New South Wales, where the climate is maritime and subtropical, with high humidity and rainfall. The dominant grape variety is Chambourcin, a French hybrid that survives everything that the local squally weather can throw at it. Chambourcin was pioneered by John Cassegrain, formerly of Tyrrell's, who established a vineyard here in 1980, the first to be planted in the locality since the 1860s. Chardonnay and Sémillon are also supposed to perform well, but it is worth noting that Cassegrain now sources most of his grapes from elsewhere.

HILLTOPS GI

Cool climatic conditions produce an elegant style of Chardonnay, and distinctive Sémillon, with red wines capable of a slow-building complexity, particularly Cabernet Sauvignon and Shiraz. This is essentially cherry orchard country, but the first vines were planted as long ago as the 1860s by Nichole Jaspprizza, a Croatian immigrant who made his money selling the wine he made to diggers in the surrounding goldfields. Peter Robertson planted the first vines of the modern era at Barwang in 1969. Other varieties growing here include Merlot, Pinot Noir, Riesling, Sauvignon Blanc, and even Zinfandel. Vines grow at an elevation of 450 to 600 metres (1,475 to 2,000 feet) on a deep, gravelly, red soil, mixed with basalt and sandy granite. The climate is continental, cool, and dry, with snow in winter and most rainfall occurring from mid-spring to mid-autumn.

HUNTER GI

The Hunter sounds as if it should encompass the Hunter Valley, not the other way round. The Hunter covers the GIs Broke Fordwich, Pokolbin, and Upper Hunter Valley.

HUNTER VALLEY GI

This zonal GI encompasses the regional GI of Hunter.

MUDGEE GI

With a continuous history of grape-growing since 1856, this is the oldest district on the western side of the Great Dividing Range. Mudgee produces rich and succulent red wines, particularly from Shiraz and Cabernet Sauvignon, and is building a reputation as an excellent source for Sémillon.

MURRAY DARLING GI

The Murray Darling GI straddles New South Wales and Victoria and is essentially an industrialized wine-growing area. It has a hot and dry continental climate, with little rain, making it ideal for risk-free, large-scale, irrigated production of bag-in-the-box wines, although exceptional wines do exist, with some producers focusing on alternative grape varieties such as Nero d'Avola and Sangiovese. Almost every grape variety found in Australia grows here, but the most significant in terms of bulk production are Chardonnay and Cabernet Sauvignon, with Muscat Gordo Blanco, and clones of Sémillon and Colombard best suited to fortified styles.

NEW ENGLAND AUSTRALIA GI

An upland plateau with cold winters and cool-to-warm summers, New England Australia is completely contained within the Northern Slopes GI. The main grape varieties are Chardonnay, Riesling, Sauvignon Blanc, and Shiraz.

NORTHERN RIVERS GI

Northern Rivers is a zonal GI consisting of New South Wales' northern coastal strip and encompasses the regional GI of Hastings River.

NORTHERN SLOPES GI

Sandwiched between the Western Plains and the Northern Rivers, and almost as barren as the former.

ORANGE GI

In the era of orange wines more commonly referring to white wines with skin contact, Orange GI has become something of a controversial region. The question has arisen as to whether "Orange" should be considered a PGI, requiring orange wine to change its name to something else. Although Orange is not yet a large enough region on a global scale to warrant such dramatic changes, it is, however, rapidly gaining national recognition for its excellent sparklings and cool-climate reds. The elevation is from 600 to 1,200 metres above sea level (2,000 to 4,000 feet),

ROWS OF GRAPEVINES GROW NEAR SUTTON FOREST IN THE SOUTHERN HIGHLANDS GI
The Southern Highlands form part of the Great Dividing Range, Australia's largest mountain range. This cool, moist, and slightly humid region saw very little viticulture until the 1980s, when the Joadja Estate was established.

with wineries in the region taking full advantage of the variety of wines this range in altitude allows, from Cabernet blends to high-quality traditional-method sparkling.

Orange is located between Cowra and Mudgee and, like most famous regions, it has expanded beyond its place of origin. Its most recent and most important plantings are found at Little Boomey, northeast of Molong. This extended region is best known for Cabernet Sauvignon and Chardonnay, but also grows Sauvignon Blanc and increasingly impressive Shiraz, although it could also succeed with Pinot Noir.

PERRICOOTA GI

The Perricoota region is located in the Big Rivers zone on the Victoria border, with most wine production near or around the town of Moama, and only two significant producers. Four wineries are listed under this GI, which has remained small since it gained GI status in 1999. St Anne is owned by the McLean family, whose vines are up to 40 years old, with Chardonnay, Cabernet Sauvignon, Merlot, and Shiraz the main varieties, alongside fortified wines.

POKOLBIN GI

This GI was established in 2010, but it is one of the most well-established Shiraz, Sémillon, and (later) Cabernet Sauvignon areas of the Hunter Valley, having been exploited long ago by names such as Lindemans, McWilliam's Mount Pleasant, and Tyrrell's.

RIVERINA GI

The centre of this region was formerly known as Murrumbidgee Irrigation Area, which was always a bit of a mouthful, or MIA, which sounded like an American military acronym. Riverina is a much more romantic name to see on a bottle of wine, although Murrumbidgee Irrigation Area accurately described a wine region that was made possible by the flooding and pumping of the Murrumbidgee River. This previously infertile land now cultivates rice and many fruits, including 15 per cent of all of Australia's wine grapes. The most commonly grown varieties are Chardonnay and Semillon; much of this is used in blends labelled under the "South-East Australia" catch-all GI, but conversely some of the country's best premium botrytized Semillons also come from here.

SHOALHAVEN COAST GI

An emerging region on the south coast of New South Wales, where the distinct maritime climate is responsible for Chambourcin being the primary grape, as it is farther up the coast at Hastings. Other varieties include Chardonnay, Gewürztraminer, Sauvignon Blanc, Sémillon, Verdelho, Cabernet Sauvignon, Merlot, and Shiraz. The first vines were planted in the 1820s, by Alexander Berry, at Coolangatta Estate, which was revived in 1988, encouraging other vineyards to be established locally, although surprisingly it was not until 1996 that the first Chambourcin was planted (by Humphries Wines). A few wineries are focusing on making Semillon in a style similar to that found in Hunter Valley, with mixed success.

SOUTH COAST GI

This zonal GI encompasses the regional GIs of Shoalhaven Coast and the Southern Highlands.

SOUTHERN HIGHLANDS GI

The Southern Highlands is a little farther inland than Shoalhaven Coast, but it is still subject to a maritime climate. With its proximity to both Canberra and Sydney, one eye was obviously on the potential tourist trade when it was decided to plant vineyards in a region that had previously been devoted to sheep, dairy farming, and horse studs. The grapes grown here are primarily Chardonnay, Riesling, Sauvignon Blanc, Cabernet Sauvignon, Pinot Noir, and Shiraz.

SOUTHERN NEW SOUTH WALES GI

This zonal GI encompasses the regional GIs of Canberra District, Gundagai, Hilltops, and Tumbarumba.

DIRECTIONAL SIGNS POINT THE WAY TO AUSTRALIAN GRAPE VARIETIES IN HUNTER VALLEY
Black grapes, particularly Shiraz (or Syrah as it is known elsewhere), are most associated with Aussie wines.

SWAN HILL GI

This regional GI overlaps New South Wales and Victoria, with most producers located in the latter.

TUMBARUMBA GI

One of Australia's coolest and most picturesque wine regions, Tumbarumba has an alpine climate that ripens grapes significantly later than in neighbouring Gundagai region (late March to mid-April, as opposed to late February), making it one of the most exciting sources for producers of premium-quality sparkling wine. Still varietal wine successes include Chardonnay, Sauvignon Blanc, Cabernet Sauvignon, and Pinot Noir. The vineyards are in the foothills of the Australian Alps, growing at elevation of 500 to 800 metres (1,640 to 2,600 feet), on basalt or granite soils. Viticulture is relatively new in Tumbarumba, with the very first vines planted in 1983, but in keeping with much New

World wine history, it was originally orchard country and remains known today for its famous Batlow apples.

UPPER HUNTER VALLEY GI

The Upper Hunter Valley was pioneered in the 1960s by Penfolds and put on the map internationally by the sensational performance of Rosemount Estate, whose Show Reserve Chardonnay took export markets by storm in the early 1980s. Although this district is just as hot as the Lower Hunter Valley, and the growing season very similar, the climate is drier and the vineyards need irrigation. Despite fertile alluvial soils and high yields – two factors that do not augur well for quality – some very fine wines are made.

WESTERN PLAINS GI

A virtual wilderness as far as wine today is concerned, this zonal GI covers more than one-third of the entire state.

THE WINE PRODUCERS OF
NEW SOUTH WALES

ALLANDALE
Lower Hunter Valley
★☆

Amazingly good quality is found at this Lower Hunter Valley estate, which also uses grapes from Hilltops.

✓ *Cabernet Sauvignon • Chardonnay • Sémillon • Shiraz • Verdelho*

BLOODWOOD
Orange
★☆

Owned by Stephen Doyle, a pioneer of Orange, Bloodwood's wines are made at the Reynolds Yarraman Estate in the Upper Hunter Valley.

✓ *Cabernet • Merlot*

BRANGAYNE OF ORANGE
Orange
★☆

Don and Pamela Hoskins make supremely stylish wines, with the ubiquitous dynamic duo of Richard Smart and Simon Gilbert consulting.

✓ *Chardonnay • Classic red blend (The Tristan)*

BRIAR RIDGE
Lower Hunter Valley
★☆

Each year, this winery plays host to guest winemaker Karl Stockhausen, one of the Hunter's legendary Sémillon stylists.

✓ *Chardonnay (Hand Picked) • Sémillon (Signature Stockhausen)*

BRINDABELLA HILLS
Canberra District
★

Owner Dr Roger Harris produces some of Canberra District's most elegant wines. The Shiraz is fleshy without being heavy, and a 1994 Riesling tasted in 2001 ranked as the best mature Australian Riesling I had ever tasted, with wonderful zesty-fresh aromas mingling with petrolly mature fruit.

✓ *Chardonnay • Cabernet Sauvignon • Riesling • Shiraz*

BROKENWOOD
Lower Hunter Valley
★★☆

This winery was established in 1970 by three partners, one of whom was none other than Australia's very own wine guru, James Halliday, but he has long since moved on. Iain Riggs has been winemaker here since 1983, and he is a driven man when it comes to Sémillon. The basic release is stunning, but bottle-age puts his ILR Aged Reserve Sémillon in a class of its own, demonstrating the potentially huge longevity and sheer quality of Hunter Sémillon.

✓ *Cabernet Sauvignon (Graveyard Vineyard) • Sauvignon Blanc/Sémillon (Cricket Pitch) • Sémillon • Shiraz (Graveyard Vineyard, Rayner Vineyard)*

CANOBOLAS-SMITH
Orange
★

The hard-to-find, intensely ripe, complex wines of Canobolas-Smith show the great potential of Orange.

✓ *Chardonnay • Classic red blend (Alchemy)*

CASELLA FAMILY BRANDS
See p671.

CASSEGRAIN VINEYARDS
Hastings River
★

The Cassegrain family established the winery the bears its name back in 1980. Some believe the maritime climate is far too marginal for viticulture, but despite the bracing sea breezes and lashing rain – which compelled John Cassegrain to plant the rot-resistant Chambourcin grape variety amongst others – some exciting wines are to be found here amidst the disappointments.

✓ *Chambourcin • Fortified (Old Yarras Tawny Port) • Fromenteau-Chardonnay (Reserve) • Sémillon (Reserve) • Verdelho*

CHALKERS CROSSING
Hilltops
★★

Established in 2000, Chalkers Crossing draws on grapes sourced from Tumbarumba and Gundagai to increase the range of wines available from its own 10-hectare (25-acre) vineyard. Winemaker Xanthe Freeman produces regionally defined wines that combine ripe fruit with graceful lines, including a Sémillon to die for.

✓ *Cabernet Sauvignon (Hilltops) • Chardonnay (Tumbarumba) • Sémillon (Hilltops)*

CHARLES STURT UNIVERSITY
Wagga Wagga
★

Wagga Wagga is the oenological campus of Charles Sturt University, which has always sold the "students' wines" from its cellar door but built a brand-new commercial-scale winery in 2002.

✓ *Chardonnay (Orange)*

CHÂTEAU PÂTO
Lower Hunter Valley
★★

This tiny, family-owned estate produces one of the Hunter Valley's finest Shiraz wines, rich and intense.

✓ *Shiraz*

CLONAKILLA
Canberra District
★★☆

Since taking over from his father, Tim Kirk has not only put his own stamp on Clonakilla, but he has also achieved fame above, beyond, and despite Canberra District's low profile. His Shiraz-Viognier has been recognized for more than a decade within Australia itself as one of the country's best reds, and it will be known by many Australians who do not realize that Canberra District does, in fact, produce any wine. Any emerging region seeking recognition needs to be famous for something, and if other producers in this region had been quick enough, they might have been able to make Shiraz-Viognier Canberra District's signature wine.

✓ *Chardonnay • Classic red blend (Shiraz-Viognier) • Riesling • Shiraz • Viognier*

CRAIGMOOR WINES
Mudgee
★☆

Previously known as Poet's Corner, then Robert Oatley Vineyard, and now Craigmoor Wines.

✓ *Shiraz (Montrose Black)*

CRANSWICK ESTATE
Riverina
☆

Cash-strapped when acquired by Evans & Tate (*see* Western Australia) for A$100 million in March 2003, Cranswick Estate nevertheless increased its new parent company's profits by more than 40 per cent, due to rolling its operations into the fold, and by in excess of 70 per cent the second year. It is currently known for a broad line of elegant, approachable varietal wines produced under the direction of winemaker Sam Trimboli.

✓ *Sparkling wine (Sparkling Shiraz)*

DALWOOD ESTATE
Lower Hunter Valley
★

Formerly Wyndham Estate Wines, this historic property is owned by developer and pub group Iris Capital. Today Dalwood estate is more a wedding venue than a winery, but it still produces decent Chardonnay and Semillon.

✓ *Chardonnay • Semillon*

DAVID HOOK WINES
Lower Hunter Valley
★★

Previously called Pothana this is the family-owned brand of well-travelled, former Lake's Folly winemaker David Hook.

✓ *Chardonnay • Sémillon • Shiraz*

DE BORTOLI
See p671.

EDEN ROAD
Canberra District
★★

Formerly Doonkuna Estate, the property has been renowned for its consistently excellent white wines. Celebrated French-born winemaker Celine Rousseau runs this operation since 2017, having moved from Chalkers Crossing.

✓ *Riesling • Chardonnay*

HELM WINES
Canberra District
★

Run by the irrepressible aspiring politician Ken Helms (he ran for the senate in 2001), Helm Wines is a microcosm of Canberra District itself, hallmarked by inconsistency, yet always capable of shining here and there. Among his best examples is a soft, smooth, easy-drinking Reserve Merlot, but Ken's hobby-horse of the past few years is his Riesling Classic Dry. This is picked at lower ripeness to avoid the typically high alcohol levels of most Australian Rieslings in an attempt to achieve a more European balance, with its light-bodied, refreshingly elegant, lime-free fruit.

✓ *Riesling (Classic Dry, Premium)*

HUNGERFORD HILL
Lower Hunter Valley
★

Hungerford Hill was established in 1969 by John Parker, who quickly expanded his vineyards to more than 200 hectares (500 acres), before shrinking the estate to a fifth of its former size. In 1980 a vineyard in Coonawarra was acquired, and some of the wines were blended from these two regions, which was the geographical equivalent of blending Rioja in Spain with a wine from the Great Hungarian plains! Sales increased to 200,000 cases by the late 1980s, attracting the attention of Seppelt, which took over the company. Seppelt was part of SA Brewing, which acquired Penfolds the same year, thus Hungerford Hill soon became just another Southcorp brand, and wines ended up being sourced from even more diverse regions (Hilltops, Cowra, Gundagai, and Tumbarumba), although the results were equally good. In 2002 the Hungerford Hill brand was sold to the Kirby family, who also retained the services of Phillip John, the winemaker under Southcorp ownership. In 2016 the winery was sold again, this time to Sydney-based developer and pub group Iris Capital, and Bryan Currie (formerly of McWilliams Wines) was appointed head winemaker.

✓ *Chardonnay (Tumbarumba) • Merlot (Orange) • Pinot Noir (Tumbarumba) • Riesling (Clare Valley)*

HUNTINGTON ESTATE
Mudgee
★☆

Bob Roberts entered the business with a law degree and learned winemaking through a British-based correspondence course. Not the strongest base for success, perhaps, but his clean, fleshy, well-balanced, stylish wines regularly outshine wines made by highly qualified professionals. Some vintages of the Cabernet Sauvignon and Sémillon possess remarkable longevity.

✓ *Cabernet Sauvignon (Special Reserve) • Sémillon • Shiraz*

WITH A BARREL-SHAPED TASTING ROOM, HUNGERFORD HILL'S WINERY HAS BECOME A LANDMARK OF THE HUNTER VALLEY WINE REGION
This property has changed hands a few times, but the boutique estate's Muse restaurant, with stunning views across the vineyards to the Brokenback Ranges, as well as its. tasting room and underground working cellar draws wine tourists to include Hungerford Hill on their vineyard stops.

KEITH TULLOCH
Lower Hunter Valley
★★

This eponymous brand was established in 1998 by a former Lindemans and Rothbury Estate winemaker with a portfolio that covers both reds and whites.

 Chardonnay • Classic red blend (Forres Blend) • Sémillon • Shiraz (Kester)

KYEEMA ESTATE
Canberra District
★

Kyeema has won nearly 30 competition trophies and is renowned for its reds, particularly Shiraz and Merlot. It was sold to Four Winds Vineyard in 2019, and the label ceased to be produced.

✔ *Chardonnay • Merlot • Shiraz*

LAKE GEORGE
Canberra District
☆

In its original incarnation as Cullarin Vineyard, this was the first winery in Canberra District, and you can still visit the wooden shed where Edgar Rieck made his wines. Rieck not only established this operation in 1971 but was also the driving force behind Canberra's National Wine Show. Lake George is now owned by Anthony and Sarah McDougall, and much has changed since Rieck's days, not just the vineyards and winery, but also the lake that this property is named after and overlooks. Lake George is mostly dry (they even play cricket on it), yet it was once full, highlighting the dire shortage of water this region faces.

✔ *Merlot • Pinot Noir*

LAKE'S FOLLY
Lower Hunter Valley
★★☆

Sydney surgeon Max Lake had no winemaking experience when he founded this property in 1963, hence the name. Yet he introduced the use of new oak to the region and consistently produced vivid, vibrant, and stylish wines. Lake's Folly was sold for A$8 million in 2000 to Peter Fogarty, a Perth businessman, who installed ex-McGuigan's winemaker Rodney Kempe, since when the quality has, if anything, become even better with each vintage.

 Chardonnay • Classic red blend (Cabernet Blend)

LARK HILL
Canberra District
★★Ⓑ

The highest vineyard in the Canberra District is one of its best, and where Pinot Noir is concerned, it is the very best. I am also impressed by the richness, complexity, finesse, and length of Lark Hill's Shiraz – not quite in the Clonakilla league, but certainly the best pure Shiraz I have tasted from Canberra District.

✔ *Chardonnay • Pinot Noir (Exaltation) • Riesling • Shiraz*

LAMBERT
Canberra District
☆

Owner-winemakers Steve and Ruth Lambert planted their first vines in 1991 and studied winemaking and viticulture at Charles Sturt University in Wagga before they had their first crop. An impressive-looking winery that can also boast of a fine restaurant.

✔ *Pinot Noir • Shiraz*

LINDEMANS WINES
Lower Hunter Valley

This brand is part of Treasury Wine Estates, and its historic Hunter Valley home is now no more than a cellar-door operation, bizarrely selling Lindemans wines from other areas. It's a shame that Lindemans' once-legendary Hunter Valley Sémillon wine is no longer produced. *See* Lindemans Wines, Victoria.

MARGAN FAMILY
Lower Hunter Valley
★

Although the vineyard was established in 1989, the winery was built in 1998, and the quality of their output is very good.

 Botrytis (Sémillon) • Cabernet Sauvignon

A CANBERRA VINEYARD WORKER PRUNES THE BARE GRAPEVINES
Pruning is essential for controlling canes and producing high-quality yields. It is best to trim in late winter when vines are dormant. As they grow, they will be trained onto the wires, and all shoots between the wires will be removed.

McGUIGAN WINES
Lower Hunter Valley
★★

Now part of Australian Vintage, this Hunter Valley winery was established by Brian McGuigan in 1992, yet was the precursor to the fifth-largest wine company. A two-star rating might appear controversially high to some critics, but McGuigan produces some superb wines at giveaway prices, performing at least as well as the two-star McWilliam's winery.

✓ *Cabernet Sauvignon • Chardonnay (Bin 7000) • Sémillon (Bin 9000) • Shiraz*

McWILLIAM'S MOUNT PLEASANT
Lower Hunter Valley
★★

Established in 1921 by Maurice O'Shea. McWilliam's Mount Pleasant bottle-aged Elizabeth Sémillon is one of Australia's finest examples of this unique wine style, and one of its greatest bargains too. Look out for this award-winning winery's museum releases. In 2018 Adrian Sparks was appointed Chief Winemaker.

✓ *Chardonnay (Maurice O'Shea) • Classic red blend ("1877" Cabernet Sauvignon-Shiraz) • Sémillon (Elizabeth, Lovedale) • Riesling (Collection) • Shiraz (Maurice O'Shea, OP & OH)*

MOUNT MAJURA
Canberra District
★

The best wine offered by Mount Majura has consistently been its rich, soft, and creamy Chardonnay, but I believe the vineyard will eventually produce greater Pinot Noir. It is also worth keeping an eye on the Cabernet Franc-Merlot blend and the Tempranillo.

✓ *Chardonnay • Pinot Noir • Riesling • Tempranillo*

MOUNT VIEW ESTATE
Lower Hunter Valley
★☆

This property was established in 1971 by Harry and Anne Tulloch.

✓ *Cabernet Sauvignon • Chardonnay (Reserve) • Sémillon (Reserve) • Shiraz (Reserve)*

MURRUMBATEMAN WINERY
Canberra District
☆

The creamy-rich, black-pepper fruit in Murrumbateman's Shiraz just pips the menthol-laden Cabernet-Merlot for this winery's best wine.

✓ *Shiraz*

PANKHURST
Canberra District
★

Since establishing this vineyard in 1986, Allan and Christine Pankhurst have built up a reputation for Pinot Noir.

✓ *Chardonnay • Classic red blend (Cabernet-Merlot) • Pinot Noir*

PETERSONS WINES
Lower Hunter Valley
★★

Petersons Wines is a family business, with their first vintage debuting in 1981. Petersons has two cellar doors in the Hunter Valley: one at Mount View and another on Broke Road.

✓ *Sémillon • Cabernet Sauvignon • Chardonnay • Sémillon • Shiraz*

PHILIP SHAW WINES
Orange
★★

Former Rosemount winemaker Philip Shaw planted his first vineyards in 1988 in the cool climate, high altitude Orange. His first commercial vintage was 2004 and the winery has gone from strength to strength. Today Pilip's two sons run the company.

✓ *Merlot • Cabernet Franc (No 17) • Chardonnay (No 11)*

RAVENSWORTH
Canberra District
★

The personal brand of Bryan Martin, winemaker at Clonakilla, which, incidentally, is where these wines are made.

TULLOCH
Upper Hunter Valley
★★

Once one of the most traditional wineries in the Hunter Valley, when Southcorp bought it in 1969 it became little more than a brand. The Tulloch family eventually regained control in 2001. These wines have gone from good to great under Jay Tulloch, who was given a free hand as winemaker. Jay's youngest daughter, Christina, is now CEO.

✓ *Verdelho*

TYRRELL'S VINEYARDS
Lower Hunter Valley
★★

This long-established winery is still family-owned (2018 saw the 160th anniversary) and produces some excellent Sémillon and Chardonnay.

✓ *Chardonnay (Vat 47, Vat 91, Vat 63) • Sémillon (Belford, Steven's Reserve, Vat 1) • Shiraz (Old Winery, Vat 7)*

Victoria

Victoria is the smallest of Australia's mainland states, but it is second only to South Australia for the volume of wine produced and is by far the most diverse in terms of cool-climate terroirs and the styles of premium-quality wine made.

The wines of Victoria range from deep-coloured, cassis-flavoured Cabernet Sauvignon to some surprisingly classy Pinot Noir and from light and delicate aromatic whites to rich, oaky, yet finely structured Chardonnay and Sémillon, as well as fortified and sparkling wines of the highest reputation. Victoria is one of Australia's oldest winemaking states – and the most famous when it comes to the country's classic fortified specialities, especially its rich and sticky Rutherglen fortifieds.

John Batman established the city of Melbourne in 1834, and within four years, William Ryrie, a sheep farmer, planted the first Yarra Valley vineyard in a place that became known as Yering. The most important sequence of events in the viticultural history of the state began with the appointment of Swiss-born Charles La Trobe as Superintendent of Melbourne in 1839 and culminated in the arrival of 11 fellow Swiss vignerons from his home canton of Neufchâtel in 1846. They laid the foundation of Victoria's future wine industry when they settled in the Geelong district and planted vineyards around their homes.

STICKY STUFF: VICTORIA'S FORTIFIED WINES

Victoria's fortified wines are world class, with Liqueur Muscat and Topaque its supreme classics: these are great wines that know no peers. The older these wines are, the sweeter they are likely to be, in addition to being richer and more complex. Fortified wines such as these were for a long period the most widely exported products to Europe, up until the 1970s. At that time, they were named after their European equivalents; fortified reds were "Port", fortified whites were "Sherry", "Tokay", or "Liqueur Muscat" depending on their style. However, an agreement between Australia and the EU in 1994, and a further agreement in 2010 following Hungary's entry to the EU, resulted in these names being phased out of use. By 2020, all geographic references were replaced with new terms.

Liqueur Muscat

Liqueur Muscat has retained its name, as it does not reference a geographic location in the EU. The most luscious of Australia's fortified wines, the best Liqueur Muscat can be like liquid Christmas cake. This wine is usually made from Brown Muscat (a red-skinned mutation of Muscat Blanc à Petits Grains, sometimes referred to as Muscat à Petits Grains Rouges). While it is well worth paying a premium for a step or two up on a producer's entry-level Liqueur Muscat, there is no need to be seduced by ultra-rare, super-expensive, extremely ancient bottlings, unless you just want a mind-blowing, tiny sip, as they are seldom drinkable in the ordinary sense. Whilst there are no EU restrictions on the naming of this product, "liqueur" is not permitted for any wine in the US, where this term is reserved for spirits, so Liqueur Muscat should be labelled as Muscat Australian Dessert Wine. In the US, the sweeter versions may be labelled as Australian Dessert Wine. *See also* Classifications, right.

Vintage/Ruby/Tawny

Australian fortified wines made in the Port style are labelled "Australian Vintage", "Australian Ruby", or "Australian Tawny". The word "Australian" can be substituted for the relevant GI, as with "Liqueur" in "Liqueur Muscat". Traditionally, Australian grape varieties used for these styles include Shiraz, Grenache, and Mataro (Mourvèdre), with more recent plantings of grape varieties such as Touriga Nacional, Tinta Roriz (Tempranillo), and Touriga Franca bringing a distinctively Portuguese influence. It is more common to find rancio Tawny styles than ageworthy Vintage, but regions like Rutherglen retain a tradition of bottling their best wines as vintages designed to age. "Ruby" is a term used to describe the

WINE REGIONS OF VICTORIA
(see also p667)
Victoria lies beneath New South Wales and neighbouring South Australia. Although it is the smallest of Australia's mainland states, it is home to more individual wineries than any other in the country.

GIPPSLAND	Wine zone
Sunbury	Wine region
Nagambie Lakes	Wine sub-region

0 mi — 75
0 km — 75

youthful style made with a view to retaining primary aroma and colour. Both Ruby and Vintage styles require a minimum ageing of four months in barrel. Tawny wines are usually blended from more than one vintage. In the United States, the sweeter versions may be labelled as Australian Dessert Wine.

Apera

Australian Sherry-style wines are referred to as "Australian Aperitif Wine" or "Apera". Muscat, Palomino, and Pedro Ximenez are the main grape varieties used for these wines, and a further classification into different styles is also required: Dry (for *fino* style with under 15 grams per litre of residual sugar), medium dry (for *amontillado* style with under 27 g/L sugar), medium sweet (for *oloroso* style with 28 to 73 g/L sugar), sweet (over 73 g/L), and cream (over 90 g/L). In the United States, the sweeter versions may be labelled as Australian Dessert Wine. Note that the classifications at right do not apply to these "Sherry" styles.

Topaque

Australia's very own, unique fortified wine style, Topaque is made exclusively from Muscadelle grapes and is lusciously sweet. The grapes are semi-raisined on the vine before being partly fermented, fortified with grape spirit, and aged in large barrels. This style is not only unique to Australia, it is specific to the region of Rutherglen in North-East Victoria, and is often considered to be a hidden gem of Australia's industry. Some producers have library stock going back more than 100 years.

THE CLASSIFICATIONS OF FORTIFIED WINE

The following classifications apply only to Australian fortified wines and are based on organoleptic (relating to qualities such as taste, colour, odour, and feel) guidelines. Ageing requirements apply only to Tawny styles.

AUSTRALIAN MUSCAT/TOPAQUE/TAWNY

The wines should show fresh fruit character and be the most youthful in style. ("Australian" can be replaced by the GI name, eg, "Rutherglen").

"TOPAQUE" OR "BAROSSA MUSCAT" CLASSIC

Theses wines are just beginning to show *rancio* and cask-ageing characters, maturing but still retaining some youthful notes. Tawny wines must have an average age of greater than five years.

GRAND

The wine should be fully mature, with concentrated *rancio* and cask-aged character, as well as an aged fruit flavour. Tawny wines must have an average age of greater than 10 years.

RARE

These are also fully mature, but dominated by the *rancio* character acquired through prolonged oxidative ageing, displaying intensely pronounced flavours and structural elements. Tawny wines must have an average age of greater than 15 years.

A EUCALYPTUS TREE SOARS ABOVE VINEYARDS IN THE BEECHWORTH WINE REGION
Within the North East Victoria GI, Beechworth has a range of altitudes and topographical aspects, which allows it to produce high-quality wines from a variety of grapes.

FACTORS AFFECTING TASTE AND QUALITY

LOCATION
Victoria is the smallest of Australia's mainland states, situated in the very southeastern corner of the continent.

CLIMATE
Climates are extremely varied, ranging from the hot continental conditions of northwest Victoria around Mildura, to the temperate coastal climes of the Mornington Peninsula.

ASPECT
Vines are grown on all types of land; the topography of Victoria includes flat plains suitable for bulk production, undulating hills, and some vineyards planted on higher sites up to 800 metres (2,600 feet). above sea level.

SOIL
A wide variety of soils, ranging from the red loam of northeast Victoria that produces its famous fortified wines; through the sandy alluvial soils of the Murray Basin that produce mainly bulk wines; and the gravelly soil mixed with quartz and shale on a clay sub-soil found in the premium wine-producing Pyrenees area; to the rich, poorly drained, volcanic soils in Geelong.

VITICULTURE & VINIFICATION
Victoria is the only Australian state to have been totally devastated by phylloxera (1875–81). So, unlike neighbouring South Australia, its vines all have to be grafted on to American rootstock. It generally utilizes high-quality wine-producing techniques. The northeast is traditionally dessert-wine country, but here, as elsewhere in Australia, winemakers have searched for cool-climate areas to expand the state's premium varietal industry since the 1970s. The famous traditional-method sparkling wines of Victoria's Great Western region have suffered with competition from the new up-and-coming sparkling-wine areas such as the Yarra Valley and King Valley in the centre of the state.

GRAPE VARIETIES
Primary varieties: Brown Muscat, Cabernet Sauvignon, Chardonnay, Marsanne, Merlot, Muscadelle, Pinot Noir, Riesling, Sauvignon Blanc, Sémillon, Shiraz (Syrah)
Secondary varieties: Arneis, Barbera, Cabernet Franc, Carignan, Chasselas, Cinsault, Cortese, Dolcetto, Durif, Flora, Folle Blanche, Gamay, Gewürztraminer, Glera, Graciano, Grenache, Lagrein, Malbec, Mondeuse, Mourvèdre, Muscat Gordo Blanco (Muscat d'Alexandrie), Nebbiolo, Orange Muscat (Muscat Fleur d'Oranger), Petit Verdot, Pinot Gris, Pinot Meunier, Roussanne, Sangiovese, Tarrango, Tempranillo, Verduzzo, Viognier

THE APPELLATIONS OF

VICTORIA

ALPINE VALLEYS

Most of the wineries and vineyards in the Alpine Valleys region are located in and around the upper reaches of the Ovens River, although vines grow as far north as the Kiewa Valley. This is a cool-climate region, with primarily Chardonnay, Sauvignon Blanc, Cabernet Sauvignon, Merlot, Pinot Noir, and Shiraz grapes growing on alluvial soils. There is also increasing success with plantings of emerging varieties such as Sangiovese and Vermentino,

BEECHWORTH

This small GI region within the North East Victoria zone is former gold-digging country, where Ned Kelly once plied his trade as Australia's most infamous "bushranger". Beechworth is located in the foothills of the Victorian Alps, at the northeastern edge of the Alpine Valleys, of which it is geographically a part, but the concentration of truly outstanding producers in this small area led to its designation as a more specific region in 2000. The region is dominated by a number of small, independent wineries with a focus on high quality wines that tend to show both a refinement reflective of the frosty, sub-alpine climate and a surprising weight, a power that makes it clear why this region deserves its reputation. The soils vary from fertile sandy alluvium on the Ovens Valley flood plain to granitic loam over decomposed gravel and clay at higher elevations, and it is the latter that produces some of Australia's finest Chardonnay, as well as some excellent Pinot Noir and Shiraz. Fortified wine is a small but strong feature of this region.

BENDIGO

Situated about 160 kilometres (100 miles) northwest of Melbourne, this part of the Central Victoria zonal GI has a dry climate, although only some of its vines are irrigated. Bendigo is gold-mining country, and viticulture began here during the 1850s Gold Rush. The vineyards tend to be small and scattered over three main growing areas. The Granite Slopes lies to the southeast, Loddon Valley to the northwest, and Marong and Golden Waters to the southwest. Bendigo is primarily known for its menthol-eucalyptus-tasting red wines. The sandy loam over red clay, quartz, and ironstone soil is well suited to producing top-quality Shiraz, but Cabernet Sauvignon and Chardonnay can also excel. More recent plantings have even included such contrasting varietals as Sauvignon Blanc and Viognier. And, as

Domaine Chandon (known as Green Point on export markets) has shown, the potential for premium-quality sparkling wines also exists.

CENTRAL VICTORIA

Also known as Victorian High Country, this zonal GI encompasses the regional GIs of Bendigo, Goulburn Valley, Heathcote, Strathbogie Ranges, and Upper Goulburn.

GEELONG

South-east of Ballarat, facing Melbourne across Port Phillip Bay, the vineyards of this district fan beyond the town of Geelong itself. Viticulture was established by Swiss immigrants in the mid-1800s, but came to a halt in 1875, when phylloxera struck, and the Victorian government ordered the removal of all vines. Geelong began its viticultural revival in 1966, somewhat earlier than other rediscovered areas of Australia, when Nini and Daryl Sefton planted the Idyll vineyard. The cool climate and volcanic soil produce wines of fine acidity and varietal character, particularly from Chardonnay and Pinot Noir, which tend to show a more savoury character than many of their counterparts from other regions. Cabernet Sauvignon and Shiraz are also widely planted and can produce long-lived wines with moderate ripeness. Geelong is another region dominated by family-owned businesses, and its proximity to a major population centre has helped to cement the region's thriving cellar-door culture.

GIPPSLAND

Sometimes referred to as Coastal Victoria, this vast zonal GI encompasses a large and diverse area that is generally rather vaguely divided into East and West Gippsland. Beyond just wine production, the area of Gippsland includes some of the best agricultural farming sites in Victoria, with lush, undulating green hills and a distinctive coastal influence. The number of wineries in operation has increased in the last ten years, as the region represents a large, relatively untapped resource where new winemakers are able to experiment without the burden of tradition. It is also quite common for producers based in Mornington or Yarra to include a Gippsland Pinot Noir in their portfolio; the style tends to be quite distinctive and provides an interesting comparison with the other regions. Producers such as William Downie, Circe, and Jane Eyre are leading the way with this practice.

GLENROWAN

Part of the North East zone, this region was once better known outside Australia as Milawa, thanks to Brown Brothers, which used the name on its widely exported wines. Glenrowan has always, however, been more widely used by other winemakers, and is another Gold Rush town with a vibrant passing tourist-trade. This is a classic dessert-wine area, but individual vineyards produce excellent Cabernet Sauvignon and other premium varietals, even Graciano.

GOULBURN VALLEY

The Goulburn Valley has long been known as a fruit-growing area, due to the readily available water source of the Goulburn River, and the warm, inland climate. The majority of top-quality producers in the Goulburn are located within the subregion of Nagambie Lakes (see Nagambie Lakes). Beyond this subregion, most Goulburn Valley vines are relatively new, and viticulture is far overshadowed by other mixed farming.

The region's climate is warm and dry in summer, making it dependent on irrigation from the Goulburn River and aquifers. The wines are generally full-bodied, with Cabernet Sauvignon and Shiraz the best varieties, whether in their pure form or, more traditionally, blended together. Marsanne has a certain hold here, mainly due to the famous 1927 vines of Tahbilk, claimed as the oldest in the world. Many of the newer wineries have chosen to experiment with more alternative varieties ranging from Zinfandel to the first Australian plantings of the Turkish varietal Boğazkere.

GRAMPIANS

The famous old sparkling wine district of Great Western was renamed Grampians under the GI scheme, but Great Western has retained its use as a subregional GI with conditions of use that are so specific they dictate how certain producers from the region are able to use the term. The dropping of Great Western has allowed for the region's finest varietal, Shiraz, to shine without being lumbered by a name that is linked historically and indelibly to the connotation of fizz. Cabernet Sauvignon, Chardonnay, and Riesling can also be excellent. Top-performing wineries are Mount Langi Ghiran, Seppelt, and Best's, the latter known for producing one of the few age-worthy versions of a 100 per cent Meunier wine.

HEATHCOTE

Formerly considered by many to be part of Bendigo, Heathcote has been recognized by Australia's GI system as a wine region in its own right since 2002. This is due, in part, to its reputation as a premium Shiraz-producing area, which is well deserved, and has been built up slowly over more than 25 years of often outstanding vintages from the likes of Jasper Hill. It is perhaps no coincidence, however, that the influential critic Robert Parker awarded Wild Duck Creek Estate 1997 Duck Muck Shiraz a perfect 100-point score, which not only propelled its owner, David "Duck" Anderson, from obscurity to international fame overnight, but also reflected well on the potential of the entire locality. This assessment was responsible for the US discovering Australian Shiraz in general, and a GI of its own was the least that Heathcote deserved in the circumstances. The region sits on the north side of the Great Dividing Range at elevations between 160 to 320 metres (525 to 1,050 feet). The soil is a rich, red volcanic type known as Cambrian, the mineral content of which is said to favour the production of red wine varietals, particularly Shiraz.

HENTY

Formerly known as Drumborg, then Far South West, and now a GI by the name of Henty, this remote region might have a changing identity, but is still best known for Seppelt's Drumborg winery. Henty takes its name from the Henty brothers, who in 1834 were the first to settle this part of the Portland Bay area, where they established a sheep station. In addition to the famous Merino sheep, the Henty brothers also brought vines with them, plus "13 heifers, 5 pigs, 4 working bullocks, 2 turkeys, and 6 dogs". Grapes were more of an afterthought, and it seems that the idea of a vineyard literally withered on the vine. The earliest truly commercial viticultural operation was not set up until much later, when Seppelt established its Drumborg winery in the 1960s. The company had been looking for new supplies of grapes to furnish its growing production of sparkling wines, and the grapes from this area have been very useful components in many an Australian fizz blend ever since. The soil is volcanic and the climate so cool that some varieties did not ripen. It is surprising, therefore, that Cabernet Sauvignon and Merlot should be so successful here, but they can be excellent; Riesling and Pinot Noir are also very good. Whilst the number of wineries in this area is still relatively small, the region is well recognised in its local market.

KING VALLEY

Until 1989 Brown Brothers were the sole buyers of grapes in King Valley, but the fruit has since been sold far and wide, and other wineries have been built here. It has now become a full-fledged GI, known for its strong Italian farming heritage that has translated more recently to a successful range of wines made from varietals such as Pinot Grigio, Barbera, Nebbiolo, and Sangiovese. King Valley is also the home of Australia's Prosecco production; the name "Prosecco" is under debate, but still able to be used at the time of writing, as the varietal was already established in King Valley prior to its 2009 renaming as Glera.

MACEDON RANGES

Macedon Ranges is Australia's coolest mainland wine region, with vineyards planted at elevations of 300 to 800 metres (980 to 2,600 feet). It often snows here in winter, which is particularly unusual by Australian standards. The proximity of this region to Melbourne (it can be reached in less than an hour by car) makes it quite appealing to the cellar-door trade, whilst the challenges of growing in this frosty climate have kept most wineries to a boutique size. Sparkling wine, Chardonnay, and Pinot are the stars here, with producers such as Bindi and Cobaw Ridge setting the benchmark.

MORNINGTON PENINSULA

Winemaking in this region took off in the 1970s, and it has now established itself as an important local industry, with more than 200 vineyards planted. The area has long been known as a weekend playground for city-dwellers from Melbourne, and the rapid expansion of the winegrowing industry has been well supported by these visitors. Provided the vineyards are adequately protected from strong sea winds that whip up from Western Port Bay to the east and Port Phillip Bay to the west, Mornington Peninsula's cool, often wet, maritime climate shows great potential for first-class winemaking. Chardonnay, Pinot Noir, and Pinot Gris show the most potential, with additional plantings of premium Shiraz, Cabernet Sauvignon, Malbec, and Merlot.

MURRAY DARLING

This regional GI overlaps two zonal GIs: North West Victoria and Big Rivers, New South Wales. The region takes its name from the Murray and Darling Rivers, two of Australia's great wine rivers. Farther upstream is the Rutherglen area, while just downstream is Riverland. The Murray Darling region brings together several areas of mainly irrigated vineyards dotted along the banks of the middle section of the river. Whilst the majority of plantings are destined for bulk production, there have been a number of boutique wineries emerging in the last 10 years, with a focus on alternative varietals.

NAGAMBIE LAKES

The Nagambie Lakes subregion GI is part of the Goulburn Valley, where the very best wines of the region have traditionally been made. The most well-recognised names include the historic Tahbilk, (established in 1860) and Mitchelton winery (established in 1969). The region's boundaries were determined on the basis of the Nagambie Lakes system, a large body of water made up of lakes, billabongs, lagoons, and streams of, and linked by, the Goulburn River. It is the only wine region in Australia where the climate is dramatically influenced by an inland water mass, which enables the vines to enjoy a milder, cooler climate than would otherwise be expected. Another factor in determining the Nagambie Lakes GI boundary was the rare soil type found in the area. Known as duplex 2.2, this soil is a red, sandy loam with a very high iron oxide content (hence the colour), over a gravelly-sandy alluvium deposited by the river system, and this is said to give a certain distinctive regional character to the wines.

NORTH EAST VICTORIA

This zonal GI encompasses the regional GIs of Alpine Valleys, Beechworth, Glenrowan, and Rutherglen.

NORTH WEST VICTORIA

This zonal GI encompasses the Victorian parts of the regional GIs of Murray Darling and Swan Hill.

PORT PHILLIP

This zonal GI encompasses the regional GIs of Geelong, Macedon Ranges, Mornington Peninsula, Sunbury, and Yarra Valley.

RUTHERGLEN'S WINE BOTTLE WATER TOWER
The iconic water tower was completed in 1900. It has since become a symbol of this wine region.

PYRENEES

Formerly known as the Avoca district, the vineyards around Mount Avoca, Redbank, and Moonambel are now called Pyrenees after a nearby mountain range. The region is more suited to the production of reds, making wines of a distinctive and attractive minty character. Shiraz, Cabernet Sauvignon, and Merlot are the outstanding wines, with Chardonnay, and perhaps Pinot Noir, on the next rung of quality. Interestingly, this is where France's Rhône Valley producer, Michel Chapoutier, decided to set up his Australian venture.

RUTHERGLEN

Rutherglen was once considered by many to include also Glenrowan (or Milawa), but it has always been the heart of the North East region and the soul of Victoria's wine industry, and since 1997 its own boundaries have been determined by the GI process. Due to the region's long, unbroken history of winemaking, with many fifth- and sixth-generation families continuing the work of their ancestors, it remains the hub around which the entire state's most traditional viticultural activities revolve. Rutherglen comprises a collection of wineries and vineyards on the south bank of the Murray River, around the town of Rutherglen, and farther south into the hinterland. This is Australia's greatest dessert wine country, producing fortified Muscats and Topaque wines that know no peers. In addition to these world-famous fortified wines, however, the cooler areas of Rutherglen have emerged as an innovative light wine-producing region. Chardonnay and Sémillon are clean, fresh, and vibrant; Gewürztraminer is performing relatively well; and Durif, Carignan, Shiraz, and Cabernet Sauvignon all show promise to one degree or another.

STRATHBOGIE RANGES

Better known in Australia than abroad, this region was first planted in 1968 by Alan Plunkett, whose natural inclination was to grow cool-climate varieties, such as Riesling and Gewürztraminer. Others followed, and in recent years the Strathbogie Ranges has developed a growing reputation as a source for sparkling-wine components and for crisp, fresh white wines of excellent fruit intensity from a number of boutique wineries. The vineyards are located at varying altitudes, ranging from 150 to 650 metres (500 to 2,130 feet). This, plus the diversity of soil types throughout the region, suggests there may be an untapped potential for numerous styles of wine beyond the grape varieties already cultivated, which include Chardonnay, Pinot Gris, Riesling, Sauvignon Blanc, Cabernet Franc, Cabernet Sauvignon, Merlot, and Pinot Noir.

SUNBURY

Sunbury is lower in altitude and much warmer in climate than the Macedon Ranges, which abut this region along the 400-metre (1,310-foot) contour line. It has alluvial soils on the plain and basalt-based loams on the slopes. Sunbury was lumped together with Macedon as part of the same region until it achieved its own GI in 1998. This status was deserved for historical reasons alone, Sunbury having been productive as a viticultural region earlier than most in Victoria, as long ago as the 1860s. The most distinctive wines in this region are from Chardonnay and Shiraz, as exemplified by the region's pioneering Craiglee winery. Other varieties grown include Sauvignon Blanc, Sémillon, Cabernet Sauvignon, and Pinot Noir, alongside a number of Italian varietals such as Fiano and Pinot Grigio championed by Galli Estate. The number of hectares under vine is still quite small, and in the near future, much of the once-rural area is likely to be repurposed by the encroaching suburban sprawl of Melbourne's inner west.

SWAN HILL

This regional GI overlaps New South Wales, extending to Kyalite, but most producers are located in Victoria. Swan Hill has a Mediterranean climate, with high temperatures and low summer rainfall, making viticulture dependent on irrigation from the Murray River. The region encompasses Brown Brothers' Mystic Park vineyard to the south, and a couple of other large wineries (Andrew Peace and RL Buller), but mainly comprises smaller boutique wineries.

There are some very good red wines from Cabernet Franc, Cabernet Sauvignon, Shiraz, Durif, and Sangiovese; whites are less exciting, although Chardonnay and Riesling fare quite well. The Golden Mile Wine Trail runs from Goodnight in the north to Beverford in the south, with many of the wineries along the way offering lunch, as well as the inevitable cellar doors sales.

UPPER GOULBURN

South of the more famous Goulburn Valley region, this GI covers the upper reaches of the river. Upper Goulburn encompasses the Lake Eidon area, between Alexander and Mansfield, and is set against the backdrop of the snow-capped Australian Alps, where the river begins its journey. This region makes crisp Chardonnay, which is better used as a component for sparkling-wine production than as a still varietal wine, the best varieties for which are the more aromatic types, such as Riesling or Sauvignon Blanc. Some Gewürztraminer is also grown. Perhaps the best-known winery in this region is Delatite, just outside Mansfield.

WESTERN VICTORIA

This zonal GI encompasses the regional GIs of Grampians, Henty, and Pyrenees.

YARRA VALLEY

Considered by many as Australia's answer to Burgundy, this recently rediscovered wine region benefits not only from the enriching after-effect of earlier sheep farming, but also from a cool climate, a long growing season, relatively light yields, and some very talented winemakers. The vines grow on the grey-brown loams (to the south) and red volcanic soils (to the north). Viticulture began in 1837, when the three Ryrie brothers planted vines on their cattle ranch, Yering Station, and with the help of Swiss and German immigrants, Yering and two other wineries, St Huberts and Yeringberg, achieved an international reputation for the Yarra Valley. This region declined in the depression of the 1890s and the situation was exacerbated by the consumer shift to fortified wines, so that by the 1920s the only surviving winery was forced to close down.

Viticulture in the Yarra Valley was revived in the 1960s by medical doctors who anticipated that consumers would return to table wines. Dr John Middleton pioneered this revival when he established the Mount Mary vineyard. This was followed by a wave of interest in the late 1980s, in large part due to wine critic James Halliday after he established Coldstream Hills winery in 1985. It was Halliday more than anyone who made consumers believe in Yarra Valley Pinot Noir. It also helped that Moët & Chandon set up its Domaine Chandon winery a few years later. The region is quite versatile, able to produce some fine Chardonnay, Cabernet Sauvignon, Merlot, Shiraz, and Sauvignon Blanc, and of course, with the option for winemakers to plant at increasingly high elevations in response to warming growing seasons, the sparkling wine section of the industry is booming. Pinot Noir remains the most renowned varietal, however, and with its propensity to show in the finished wine even the smallest variations in site, it is paving the way for a future delineation of this large region into further subregions.

THE WINE PRODUCERS OF
VICTORIA

ALL SAINTS

Rutherglen

★☆

Established in 1864, All Saints was owned personally by the late Peter Brown of Brown Brothers fame, as was the nearby St Leonards winery.

 Durif (Estate Family Cellar) • Fortified (All Liqueur Muscats, all Muscadelle, Vintage Port, Old Tawny Port) • Shiraz (Estate Family Cellar)

ARMSTRONG VINEYARDS

Grampians

★★

The limited quantity, top-performing Shiraz is from Tim Armstrong with Erica Orr acting as a consultant.

 Shiraz

AVANI

Mornington Peninsula

★★🅱

The "good earth" is the translation of Avani is Sanskrit. Shashi (who formerly work for eight years at Bass Philip) and husband Devendra Singh acquired the winery in 1998. Their north-facing vineyard was originally planted with Pinot Noir and Chardonnay. She felt that the site was too warm and on the advice of Philip Jones, decided to plant Syrah only. The wines always show great purity and have never had more than 13 per cent abv. The second wines are sold under the Amrit label, the grapes are sourced from organic vineyards.

 Syrah (Avani) • Chardonnay (Amrit) • Sauvignon Blanc (Amrit)

BALGOWNIE

Bendigo

★☆

Balgownie went through a patchy period in the early 1990s, but this winery now seems to have pulled out of the doldrums as far as its reds are concerned. The whites are disappointing.

✓ Cabernet Sauvignon (Estate) • Shiraz (Estate)

BANNOCKBURN VINEYARDS

Geelong

★★

Winemaker Gary Farr's experience at Domaine Dujac in Burgundy shows through in the finesse of these wines, which are as appreciated in Europe as they are in Australia. The SRH Chardonnay and Serre Pinot Noir are ultra-premium limited releases.

✓ Chardonnay • Classic red blend (Cabernet-Merlot) • Pinot Noir • Shiraz

BASS PHILLIP

Gippsland

★★🅱

Ex-research engineer Philip Jones converted his estate to biodynamic practices in 2002. He produces some of the most stunning Pinot Noir of Australia in minuscule quantities.

✓ Pinot Noir • Gamay

BERRYS BRIDGE

Pyrenees

★☆

Established in 1990, this estate sold grapes initially, with 1997 the first vintage. The intense Shiraz may be the best and most consistent wine so far.

✓ Shiraz • Cabernet

BEST'S WINES

Great Western

★★

This historic winery boasts a large, superb range that includes many gems, such as the stunning Thomson Family Shiraz and the lovely zesty-fresh, lime-scented Riesling.

✓ Cabernet Sauvignon • Chardonnay • Merlot • Riesling • Shiraz

BINDI WINE GROWERS

Macedon Ranges

★★

Expect spectacular Chardonnay and Pinot Noir of stunning quality and individuality made by Michael Dillon at Bindi. I have had a few back vintages of his reds which were truly memorable.

✓ Chardonnay • Pinot Noir

BIRD ON A WIRE

Yarra Valley

★★

This tiny property has attracted a lot of eyes in recent years. Caroline Mooney has done a great job at making complex and balance wines. The Syrah is a must.

✓ Marsanne • Syrah • Pinot Noir

BLACKJACK VINEYARD

Bendigo

★

Blackjack Vineyard was named after an American sailor who caught gold fever in the 1850s and jumped ship, but this family-owned venture was established in 1987. Everyday drinking Shiraz is sold under the Chortle's Edge label.

 Classic red blend (Cabernet Merlot) • Shiraz

BLUE PYRENEES

Pyrenees

★

The original winemaker here was Frenchman Vincent Gere, who tended to make his sparkling wines on the sweet side until the 1990 vintage, when they became more classic in style. He was succeeded by Kim Hart, who placed most emphasis on still wines. Today, the estate is owned by Andrew Koerner who is also the chief winemaker.

 Cabernet Sauvignon (Richardson Reserve) • Viognier • Shiraz • Sparkling wine (Midnight Cuvée)

BROWN BROTHERS

King Valley

★☆

The size of this family-owned firm is deceptive until you visit Brown Brothers' so-called micro-vinification winery, which is bigger and better equipped than 70 per cent of the wineries I regularly visit, with so-called micro-vinification tanks that are like a battery of boutique wineries. Brown Brothers' experimental wines are thus as polished as any commercially produced wine, which explains why this is one of Australia's most innovative wine producers. One of Brown Brothers' experiments has been with the Spanish Graciano grape, and that has demonstrated incredible longevity, with a 1974 vintage blowing away every example of the "genuine thing" put up by Rioja wineries at the London wine trade fair in 2004. The Tarrango might not be a fine wine, but if you like Beaujolais, it is made in that style, only better than 90 per cent of wines bearing that appellation.

✓ Botrytis (Noble Riesling) • Cabernet Sauvignon (Patricia) • Chardonnay (Patricia) • Classic red blend (Shiraz Mondeuse Cabernet) • Fortified (Reserve Muscat, Very Old Tokay) • Graciano • Shiraz (Banksdale) • Sparkling (Patricia, Pinot Noir, and Chardonnay, Vintaged Pinot Chardonnay)

RL BULLER & SON

Rutherglen

★☆

This firm dates back to 1921, and also has a winery at Beverford in Swan Hill. In 2013, the estate was purchased by the Judd family who have invested heavily in the facility. They have six different brands. Best advice is to concentrate on the beautiful fortified wines. Judged on fortified alone, RL Buller would deserve two and a half stars.

✓ Fortified (Calliope Rare Liqueur Muscat, Calliope Rare Liqueur Tokay, Fine Old Muscat, Fine Old Tawny)

BY FARR

Geelong

★★☆

Previous winemaker on the other side of the road at Bannockburn, the very talented Gary Farr opened his own winery in 1994. Nowadays, his son Nick runs the daily operation and carries on the legacy while refining the style of By Farr.

 Chardonnay • Pinot Noir (Sangreal, Farrside, Tout Prêt) • Shiraz • Viognier

CAMPBELLS WINERY
Rutherglen
★★

This is one fortified wine specialist that has expanded into varietal wines with brilliant success.

 ✓ *Durif (The Barkly)* • *Fortified (Allen's Port, Isabella Liqueur Tokay, Liquid Gold Tokay, Merchant Prince Liqueur Muscat)* • *Shiraz (Bobbie Burns)*

CHAMBERS ROSEWOOD
Rutherglen
★★

Stephen Chambers is now the sixth generation running this winery and represents one of the all-time great producers of fortified wine. The Old Rare Muscat and Tokay have no peers.

 ✓ *Fortified (All Liqueur Muscat, all Liqueur Tokay)*

COFIELD
Rutherglen
★

Cofield specializes in small quantities of high-quality red fizz and produces a good Durif for Rutherglen.

 ✓ *Durif (Rutherglen)* • *Shiraz* • *Sparkling (Shiraz, Durif)*

COLDSTREAM HILLS
Yarra Valley
★★

James Halliday, the doyen of Australia's wine writers, has sold this property to Treasury Wine Estates, but is still involved with the winemaking on a consultancy basis. Coldstream is well known for beautiful Pinot Noir and classic Chardonnay, the Reserves of which are well worth the price for those with deep enough pockets. The Cabernet Sauvignon, however, and, more recently, the Merlot have been the surprises, reflecting Halliday's penchant for length, elegance, and finesse in these Bordeaux varieties, rather than mere weight or punch.

 ✓ *Cabernet Sauvignon* • *Chardonnay* • *Merlot* • *Pinot Noir*

CRAIGLEE
Sunbury
★☆

This famous 19th-century, four-storey, bluestone winery was re-established in 1976 by Pat Carmody, who released his first wine, a 1980 Shiraz, in 1982. It is still a family-run vineyard where they grow cool-climate wines and host a cellar door once a month.

✓ *Shiraz*

CRAWFORD RIVER
Western District
★★

This property was established in 1982 by John Thomson, who quickly built a reputation for Riesling, particularly in the botrytis style, but is also very consistent these days with Cabernet.

 ✓ *Cabernet Sauvignon* • *Classic red blend (Cabernet Merlot)* • *Classic white blend (Sémillon Sauvignon Blanc)* • *Riesling*

CRITTENDEN AT DROMANA
Mornington Peninsula
★☆ Ⓥ

Garry Crittenden was the first to plant grapes in the Mornington Peninsula back in 1982. He also introduced Italian grapes on the peninsula. Second-generation Rollo and Zoe have now enlarged the grape collection by planting Spanish varieties. It is a sure bet to recommend Crittenden's wines in general.

CURLEWIS WINERY
Geelong
★★

Established in 1998 by Rainer Brett, who has very quickly established an enviable reputation for great Pinot Noir. He sold the winery to Stefano Marasco and Leesa Freyer in 2008 and still gives them advice as they remain good friends. Their Shiraz is damn good too.

✓ *Pinot Noir* • *Shiraz* • *Chardonnay*

CURLY FLAT
Macedon Ranges
★☆

Owner Jeni Kolkka's production is relatively small, but specializes in Burgundian varietals, and the quality is outstanding. A new winery was built in 2002, and the old one was turned over to cellar sales. Second wines are sold under the Williams Crossing label.

✓ *Chardonnay* • *Pinot Noir*

DALWHINNIE
Pyrenees
★★

Intensely rich and well-structured wines are consistently produced from these vineyards adjacent to Taltarni.

✓ *Cabernet Sauvignon (Moonambel)* • *Chardonnay* • *Pinot Noir* • *Shiraz (Moonambel)*

DAL ZOTTO
King Valley
★

The Dal Zottos (originally from Valdobbiadene) started planting vines on a former tobacco farm in 1987. Today they grow mainly Italian varieties on nearly 500 hectares (200 acres) and were the first to plant commercial acreage of Glera. They launched their own "Prosecco" in 2004, which continues to enjoy a huge success.

 ✓ *Fiano* • *Arneis* • *Nebbiolo* • *Prosecco*

DE BORTOLI
See p671.

DELATITE WINERY
Central Victorian High Country
★ Ⓥ

Situated on its own in the Great Divide, Delatite has produced some very good value wines that always show rich, tangy fruit. This winery has a reputation for its Deadman's Hill Gewürztraminer.

 ✓ *Chardonnay* • *Classic red blend (RJ Cabernet-Merlot)* • *Pinot Noir* • *Riesling (VS)*

DIAMOND VALLEY VINEYARDS
Yarra Valley
★★

David Lance is well known for Pinot Noir of great intensity yet elegance, but his other wines should not be overlooked.

 ✓ *Cabernet Sauvignon* • *Chardonnay* • *Classic red blend (Cabernet-Merlot)* • *Pinot Noir* • *Sémillon-Sauvignon*

DOMAINE CHANDON
Yarra Valley
★☆ Ⓥ

Moët & Chandon's Australian winery was set up later than its California venture, but quickly overtook it in quality. Much of the credit should go to Dr Tony Jordan, who ran the enterprise at that time, but even he will admit that, with sources stretching across an entire continent, he had a wider choice of quality components to draw on. Jordan left to take over at Wirra Wirra, but has returned to head all of LVMH's winemaking operations in Australia and New Zealand, including Western Australia's Cape Mentelle. The rosé and *blanc de noirs* are both very good. In 2004 Chandon became the first major producer to market a sparkling wine sealed by a crown-cap: Brut ZD (for Zero Dosage). A pure Chardonnay blend, the inaugural 2000 Brut ZD vintage was one of the few non-*dosage* wines that could be consumed easily upon purchase. The wines are known on export markets as Green Point. Currently, they host a cellar door at the winery, in part to promote their cocktail range, which includes French Australian Fizz.

 ✓ *Chardonnay (Reserve)* • *Pinot Noir (Reserve)* • *Sparkling wine (Blanc de Blancs, Cuvée Riche, Pinot Shiraz, Vintage Brut)*

DOMINIQUE PORTET
Yarra Valley
★☆ Ⓥ

After 22 years as CEO at Taltarni, Dominique Portet left to set up his own venture in the Yarra Valley. So far the Heathcote Shiraz is the standout.

✓ *Cabernet Sauvignon (Heathcote)* • *Sauvignon Blanc* • *Shiraz (Heathcote)*

FIGHTING GULLY ROAD
Alpine Valleys

Mark Valpole has grown Italian and international varieties at his Beechworth property since 1997. His Sangiovese is worth looking out for.

✓ *Sangiovese (Bolck 2)* • *Aglianico* • *Chardonnay (Smiths Vineyard)*

KANGOROOS IN THE DOMAINE CHANDON VINEYARD MAKE IT CLEAR THAT THIS BRANCH OF THE FRENCH FINE WINERY MOËT & CHANDON IS IN AUSTRALIA

In 1986 Domaine Chandon set up shop at Green Point, an old dairy farm in Victoria's Yarra Valley, producing *méthode-traditionnelle* sparkling wines. The Green Point name was given to wines intended for the export market,

GALLI ESTATE

Sunbury

★☆V

This estate was established in 1997 by Pam and Lorenzo Galli, who have built up a formidable reputation.

✓ *Cabernet Sauvignon • Pinot Grigio • Shiraz*

GIACONDA

Beechworth

★★☆

Exquisitely crafted, highly sought-after wines, which only those on the mailing list have any chance of acquiring. The Pinot Noir is by far the best of an amazing range of wines.

✓ *Cabernet Sauvignon • Chardonnay • Classic white blend (Les Deux) • Pinot Noir • Roussanne (Aeolia)*

GIANT STEPS

Yarra Valley

The story of this multi award-winning winery started in 1997 when brewer-turned-winemaker Phil Sexton found the cool-climate spot he was searching for to plant Chardonnay and Pinot Noir. The wines are still excellent, as are the cellar door facilities.

✓ *Entire range*

HANGING ROCK

Macedon Ranges

★

This was once the largest winery in the Macedon area, and specialized in bottle-fermented sparkling wine, but Shiraz is the best wine today. Although the Heathcote Shiraz is top of the range and certainly the best, the basic Victoria Shiraz can be very nearly as good at almost half the price.

✓ *Shiraz*

JASPER HILL

Heathcote

★★B

Not exactly known as the epitome of elegance, but there is no doubting the quality of Jasper Hill's biodynamic red wines in the big, muscular, complex style.

✓ *Classic red blend (Emily's Paddock Shiraz-Cabernet Franc) • Shiraz (Georgia's Paddock)*

KARA KARA

Pyrenees

★V

At one time this winery seemed to be doing good things with Sauvignon Blanc, pure and blended, but excels today with its red wines.

✓ *Cabernet Sauvignon • Shiraz • Sauvignon Blanc*

KOOYONG ESTATE

Mornington Peninsula

Some of the best Chardonnays and Pinot Noirs in Mornington Peninsula come from this property, owned by the Gjergja family, who also own Port Philip Estate, a fabulous architectural feat of a winery.

✓ *Chardonnay (Faultline, Farrango) • Pinot Noir (Haven, Meres, Ferrous)*

LETHBRIDGE

Geelong

Maree Collis and Jay Nadeson gave up their careers in science for their vinous passions. Established in 1996, the winery produces a range of Pinot Noirs and Chardonnays, as well as wines from Italian varieties.

✓ *Chardonnay (Allegra) • Primarone*

LEVANTHINE HILL

Yarra Valley

★☆

Founder Elias Jreissati has attracted ex-Yarra Yering winemaker Paul Bridgeman to make the wines at his modern property. My favourite is the Samantha's Paddock, a classy Bordeaux blend capable of aging well.

✓ *Sauvignon Blanc • Chardonnay • Classic red blend (Samantha's Paddock) • Shiraz (Melissa's Paddock)*

LILLYDALE VINEYARDS

Yarra Valley

★☆V

Established in 1975, Lillydale has been owned by McWilliams Wines since 1994, with Max McWilliams in charge of winemaking. I always thought of Lillydale as being best for white wines, but under McWilliams' hand, the reds have outclassed everything except the elegantly rich and stylish Yarra Valley Chardonnay.

✓ *Chardonnay • Classic red blend (Cabernet-Merlot) • Shiraz*

LINDEMANS WINES

Murray Darling

★☆V

This vast winery is now part of Treasury Wine Estates and produces wine from almost everywhere, although its historic home in the Hunter Valley has been reduced to cellar-door sales only. The Karadoc winery in the Murray Darling region churns out copious quantities of very respectable, sub-A$10 wines, including the good-value Cawarra range and umpteen millions of cases of Bin 65 Chardonnay, and would deserve a rating of no more than one star, and that would be for miraculous consistency, rather than quality per se. Lindemans' Coonawarra winery, which also makes the Padthaway wines, would, however, easily rate two stars, hence the compromise of one and a half. Even at Coonawarra, regionality has gradually been replaced by state-wide blends, ripping the heart and soul out of the Lindemans heritage. The real pity is that under Treasury Wine Estates, Lindemans no longer produces a modern equivalent of its legendary Hunter Riesling, some 70-year-old examples of which are still in amazing condition. The message to corporate Treasury Wine Estates must be to reverse the decline of regionality in general and reintroduce some form of ultra-premium bottle-aged Hunter Sémillon in particular.

✓ *Cabernet Sauvignon (St George) • Chardonnay (Padthaway) • Classic red blend (Limestone Ridge Shiraz-Cabernet, Pyrus-Cabernet-Merlot-Cabernet Franc) • Fortified (Macquarie Tawny Port) • Pinot Noir (Padthaway) • Shiraz (Padthaway)*

MERRICKS CREEK

Mornington Peninsula

★★☆

Established in 1998 by Peter and Georgina Parker, who are reputed to have a ridiculously high-density, low-canopy vineyard planted with a single variety, Pinot Noir, but this couple are not completely mad. They are indeed Pinot purists, but most of the 2-hectare (3-acre) vineyard is planted with vines 1 metre apart, and rows a very generous 2.5 metres (8 feet) apart, with the canopy a good 2 metres (6.5 feet) high. However, a separate plot is planted with vines just 600 millimetres (2 feet) apart and row spacing at 1 metre (3.3 feet), with the fruiting wire just 20 centimetres (8 inches) above the ground – this plot must be frost-free. And only half barking. The results are very impressive, with a helping hand from contract winemaker Nick Farr, who is the son of Pinot genius Gary (ex-Bannockburn and By Farr), who acted as consultant viticulturist.

✓ *Pinot Noir • Sparkling wine (Pinot Noir)*

MERRICKS ESTATE

Mornington Peninsula

★☆

Expect consistently stunning Shiraz from this estate owned by the Kefford family since 1977, with no connection to Merricks Creek above.

✓ *Shiraz*

MITCHELTON VINTNERS

Goulburn Valley

★☆V

This estate is now part of the Lion Nathan wine division since its takeover of Petaluma. Innovative Rhône-style wines have added to Mitchelton's reputation as one of Australia's greatest value-for-money brands, although I find the whites such as Viognier and Airstrip (Marsanne-Roussanne-Viognier) too fat and soft.

✓ *Botrytis (Blackwood Park Botrytis Riesling) • Chardonnay (Vineyard Series) • Classic red blend (Crescent Shiraz-Mourvèdre-Grenache) • Merlot (Chinaman's Ridge) • Riesling (Blackwood Park) • Shiraz (Print)*

MONTALTO VINEYARDS

Mornington Peninsula

★☆

Pinot Noir is definitely the standout at Montalto, which was established in 1998. Although the Pennon Hill range is effectively Montalto's entry-level label, some vintages of the Pennon Hill Pinot Noir are of such superb quality that they rank as one of the world's best bargains for this sought-after variety.

✓ *Chardonnay • Pinot Noir • Riesling*

MOOROODUC ESTATE

Mornington Peninsula

★★

Richard McIntyre's well-established estate has long been reputed for its Chardonnay, but if anything, the Pinot Noir is even better.

✓ *Chardonnay • Pinot Noir • Riesling*

MORRIS WINES

Rutherglen

★★V

The table wines are decent enough, but no one visits Morris Wines to taste anything other than the fortifieds, the older and rarer blends of which are arguably the best of their type in the world. In fact, on one occasion I was so eager not to miss an appointment to taste such nectar here that I received an on-the-spot speeding fine of A$350 from Victoria's finest as I sped across the state line from New South Wales. The legendary Mick Morris was amazed at the size of the fine, asking what on earth I had done to deserve it, and I shall never forget that twinkle in his eye and the characteristic chuckle as I told him that my mind was so intent on getting to the winery that I hadn't realized that I had overtaken a police car in my haste!

✓ *Fortified (Entire range)*

MOUNT AVOCA VINEYARD

Pyrenees

★

This Pyrenees winery was purchased by Barrington Estate in the Hunter Valley in 2002 and is currently performing better with red wines than white.

✓ *Cabernet Sauvignon • Classic red blend (Arda's Choice) • Merlot • Shiraz*

MOUNT BECKWORTH

Pyrenees

★☆V

The estate was established in 1984, yet I first came across these wines some time later, perhaps because the grapes were sold in bulk, and wines under the Mount Beckworth label did not appear until much more recently.

✓ *Chardonnay • Pinot Noir • Shiraz*

MOUNT LANGI GHIRAN

Grampians

★★V

This winery and vineyards were purchased by the Rathbone Family Group in 2002, but former owner Trevor Mast has stayed on as winemaker. Mast produces rich, ripe, complex reds (particularly Shiraz), and increasingly exciting Riesling.

✓ *Cabernet Sauvignon (Joanna) • Classic red blend (Billi Billi, Cabernet Sauvignon-Merlot) • Riesling • Shiraz (Cliff Edge)*

MOUNT MARY VINEYARD

Yarra Valley

★★☆

Only small quantities of reds are made at Mount Mary, but these are outstanding with a well-deserved cult following.

✓ *Chardonnay • Classic red blend (Quintet) • Classic white blend (Triolet) • Pinot Noir*

NICHOLSON RIVER

Gippsland

★★

There has been great improvement over recent years from this well-established Gippsland producer. Second wines sold under the Mountview label.

✓ *Chardonnay • Classic red blend (The Nicholson) • Pinot Noir*

OAKRIDGE
Yarra Yarra
★★

Stunning quality has been the watchword since the arrival of winemaker David Bicknell in 2001. The 864 Riesling is a manufactured icewine style, produced by freezing the grapes in the winery, then pressing them in a refrigerated press!

✓ *Chardonnay (864, Yarra Valley) • Riesling (864) • Shiraz (864, Yarra Valley)*

PARINGA ESTATE
Mornington Peninsula
★★

Established in 1985, Lindsay McCall's Paringa Estate is well entrenched as one of the *"grands crus"* vineyards of the Mornington Peninsula region.

✓ *Pinot Noir • Shiraz • Chardonnay*

PASSING CLOUDS
Macedon Ranges
★☆

Originally established in Bendigo, the winery is now located in the Macedon Ranges due to several years of drought which put the business at risk. Cameron Leith continues to produce classy, unirrigated red wines from Bendigo but has now expanded into making Pinot Noir and Chardonnay from Macedon Ranges too.

✓ *Classic red blend (Graeme's Blend, Angel's Bland) • Shiraz (Reserve)*

PHILLIP ISLAND VINEYARD
Port Phillip
★☆

Established in 1994 by David Lance of Diamond Valley fame, Phillip Island must be the only vineyard within spitting distance of penguins and a seal colony.

✓ *Chardonnay • Pinot Noir • Riesling*

PIZZINI
King Valley
★☆

This family winery was founded in 1978 and has continued to retain its Italian heritage: they grow varieties such as Nebbiolo, Sangiovese, Sagrantino, and Verduzzo to great critical success.

✓ *Nebbiolo • Sangiovese • Verduzzo*

PONDALOWIE VINEYARDS
Bendigo
★★

Dominic and Krystina Morris have established Pondalowie Vineyards not only as the *"grand cru"* of Bendigo, but also as one of the best wineries in Australia. Look out for the annually changing "Special Release" – I loved the delicious Sparkling Shiraz.

✓ *Classic red blend (Shiraz-Viognier, Vineyard Blend) • Fortified (Vintage Port) • Tempranillo (Mt Unwooded)*

PORT PHILLIP ESTATE
Mornington Peninsula
★☆

Established in 1987, Port Phillip Estate was sold in 2000 to current owners Giorgio and Dianne Gjergja. Along with running a dining room and cellar door,

THE BRANDY NOOK IN THE UNDERGROUND CELLARS OF SEPPELT GREAT WESTERN IS SET UP FOR A TASTING
This heritage-listed labyrinth, known as The Drives, wind beneath the historic Great Western home. Excavation of the tunnels and cellars began in 1868 by out-of-work gold miners, and the digging continued for more than 60 years. At 3 kilometres (almost 2 miles), The Drives forms the largest underground cellars complex in the Southern Hemisphere.

the Gjergjas manage to produce some truly stunning red wines.

✓ *Pinot Noir (Reserve) • Shiraz (Reserve)*

PRINCE ALBERT
Geelong
★☆

Bruce Hyett produces tiny quantities of highly acclaimed Pinot Noir.

✓ *Pinot Noir*

PUNCH
Yarra Valley
★★

Instantly established as iconic Pinot Noir, Punch is produced in minuscule volumes by James and Claire Lance from exactly the same fruit as Diamond Valley White Label Pinot Noir (produced by David Lance).

✓ *Pinot Noir (Close Planted)*

RED EDGE
Heathcote
★★

Named after the red volcanic soil of the vineyard, this estate was purchased in 1994 by Peter and Judy Dredge – and I don't think it is a coincidence that it is also almost an anagram of their surname. The vineyard was planted in 1971 by Vert Vietman, who originally named it Red Hill. Although the Dredges have expanded the vineyard, they still have quite a lot of mature vines to work with – and this shows through in the intensity of some of their wines.

✓ *Cabernet Sauvignon • Shiraz (Heathcote, Degree)*

RED HILL ESTATE
Mornington Peninsula
★★

Described by the president of the Australian Winemakers Federation as

the "best view of any vineyard in the world", it's no wonder Red Hill's restaurant is packed. They also offer one of the peninsula's very best Pinot Noirs.

✓ *Chardonnay • Pinot Noir*

ROCHFORD WINES
Yarra Valley
★☆

Part of this enterprise was formerly known as Eyton on Yarra. It was taken over in 2002 by Rochford Wines, whose winemaker, David Creed, coincidentally was a contract winemaker for Eyton on Yarra in 1993. The Rochford and Eyton wines have evolved into a three-tiered range, all under the Rochford label. The Rochford R Range is from the Macedon Ranges and represents those wines formerly sold as Rochford; the Rochford E Range is from Yarra Valley and is effectively the old Eyton wines; while the Rochford V Range is sourced from both wineries plus bought-in grapes and consequently sold under the broader Victoria GI.

✓ *Chardonnay (R Range)* • *Pinot Noir (R Range)*

SCOTCHMAN'S HILL
Geelong
★

This winery is owned by David and Vivienne Browne, whose Burgundy-style specialist vineyard and winery overlook the Bellarine Peninsula and benefit from its cool maritime breezes. Second wines are sold under the Spray Farm label.

✓ *Chardonnay* • *Pinot Noir*

SEPPELT GREAT WESTERN
Great Western
★★ V

Part of the Treasury Wine Estates group, this legendary sparkling wine producer makes everything from cheap-but-expertly-made lime-and-lavender fizz under its most basic Great Western Brut label, through numerous exceptional-value, relatively inexpensive, genuinely premium cuvées, to the truly fine-quality, upmarket Salinger *cuvée*. And then there is the Show Sparkling Shiraz, which is one of Australia's icons: the biggest, brashest, and most brilliant of sparkling "Burgundies", even though its massive, concentrated blackcurrant-syrup fruit is too much for many to swallow more than half a glass. But it is definitely a "show" wine in the biggest of senses. And just because this winery has a well-deserved reputation for sparkling wine, it would be wrong to overlook some of Seppelt's exceptionally fine table wines. A two-star classification is not too high, because some individual wines, both with and without bubbles, would rate two and a half stars on their own.

✓ *Chardonnay (Drumborg, Jaluka Henty)* • *Shiraz (One Mile Drive)* • *Riesling (Drumborg)* • *Sparkling (Original Sparkling Shiraz, The Great Entertainer Pinot Noir-Chardonnay, Salinger, Show Sparkling Shiraz, Sparkling Shiraz)*

SEVILLE ESTATE
Yarra Valley
★☆ V

This winery made its name on minuscule releases of its superb botrytised Riesling, but has been owned by Brokenwood since 1997, and now concentrates on the more marketable premium varietals.

✓ *Chardonnay* • *Pinot Noir* • *Shiraz*

SHADOWFAX
Geelong
★★☆

Shadowfax: is this name making some pun on email? No, it comes from *The Lord of the Rings*, in which Shadowfax was "chief among horses". Deep and meaningful, then? No, I suspect it was just named by some Tolkien nutter. Shadowfax is not run by nutters, though. This ultra-modern and stylish winery, which is located in Werribee Park, has its own award-winning luxury hotel and restaurant, The Mansion Hotel, and makes wines that are as lush and as manicured as its surroundings.

✓ *Chardonnay* • *Pinot Noir* • *Shiraz* • *Sauvignon Blanc* • *Mataro*

SHANTELL
Yarra Valley
★☆ V

An underrated producer of fine white wines, Shantell has recently started to make excellent reds.

✓ *Cabernet Sauvignon* • *Chardonnay* • *Sémillon* • *Shiraz*

STANTON & KILLEEN
Rutherglen
★★ V

Affectionately known as "Stomp It and Kill It", this famous old fortified wine producer also manages to make some excellent table wines.

✓ *Classic red blend (Shiraz Durif)* • *Fortified (Entire Classic range, Grand Rutherglen Muscat)* • *Shiraz (Moodemere)*

STONIER WINES
Mornington Peninsula
★★

Stonier has been owned by Lion Nathan since the takeover of Petaluma, but its Burgundy-style wines get even more stunning by the vintage under winemaker Mike Symons, who has started specializing in the expression of single vineyards.

✓ *Entire range*

TAHBILK
Goulburn Valley
★☆ V

The oldest winery in Victoria was established in 1860, when it was known as Chateau Tahbilk, but it dropped the French pretension at the turn of the millennium. It still makes very traditional wines that always improve in bottle. Second wines are sold under the Dalfarras, Everyday Drinking, and Republic labels.

✓ *Cabernet Sauvignon* • *Marsanne* • *Riesling* • *Sémillon* • *Shiraz*

TALTARNI VINEYARDS
Pyrenees
★☆

Taltarni is Aboriginal for "red earth", the soil in this vineyard being an iron-rich siliceous clay and thus red in colour. The wines have always been noticeably rich in extract, but they are no longer as austere, lean or – for reds – as tannic as they used to be under Dominique Portet (who now has his own label). The style has become more relaxed and easier to access at a much earlier age. Clover Hill and Lalla Gulley are premium single-vineyard wines, while second wines are sold under the Fiddleback label.

✓ *Cabernet Sauvignon (Dynamic)* • *Chardonnay (Dynamic)* • *Sauvignon Blanc (Fumé Blanc)* • *Shiraz (Taltarni)* • *Sparkling (Essence, Brut Taché)*

TARRINGTON VINEYARDS
Henty
★☆

Look for beautiful Burgundy-style Pinot Noir and refreshing unoaked Chardonnay.

✓ *Chardonnay* • *Pinot Noir*

TEN MINUTES BY TRACTOR
Yarra Valley
★★

You have to admire the ingenuity of such a name for a winery in a country that thinks nothing of trucking grapes across several state lines. But it's not a gimmick. This venture is the combination of three separate family-owned vineyards (hence the three single-vineyard labels: Judd, McCutcheon, and Wallis), each of which is just 10 minutes by tractor from the other two. The wines are brilliant.

✓ *Chardonnay* • *Pinot Noir*

TIMO MAYER
Yarra Valley
★☆ V

German-born Timo Mayer made his first vintage from his steep Bloody Hill vineyard in 2002. A man of presence whose minimal intervention philosophy set the standard for great wines, he is also involved in the wines of Gembrook Hill.

✓ *Pinot Noir* • *Nebbiolo* • *Riesling*

VIRGIN HILLS VINEYARDS
Macedon
★★

Virgin Hills is the creation of Hungarian-born Melbourne restaurateur Tom Lazar. This "one-wine-winery" has been one of Australia's hidden jewels for more than three decades. Lazar did not learn wine-making; he simply read a book, planted a vineyard, and made his wine. Virgin Hills has always consisted primarily of Cabernet Sauvignon, plus a dash of Syrah, Merlot, and Malbec. When he retired, Mark Sheppard took over the winemaking, and continued in the Lazar fashion, making it one of the Australia's least-known, greatest wines. Sheppard left to join Vincorp but returned when Vincorp owned Virgin Hills for a short while in 1998, after which it was purchased by Michael Hope, the current proprietor, who had been successful producing premium wines in the Hunter Valley. Now, two decades later, Lazar's vision lives on.

✓ *Virgin Hills*

A HIGHLAND COW WANDERS ALONGSIDE A MORNINGTON PENINSULA VINEYARD
The Mornington Peninsula, south of Melbourne, is home to more than 200 vineyards, with those such as Paringa Estates and Red Hill Estate receiving praise as some of the best in Australia. With its cool-climate conditions, Chardonnay, Pinot Noir, and Pinot Gris grape varieties fare particularly well here, producing some praise-worthy wines.

WARRABILLA
Rutherglen
★☆

Expect a great quality-price ratio from former All Saints owner-winemaker Alan Sutherland Smith.

✓ *Classic red blend (Reserve Shiraz Durif) • Durif (Reserve) • Fortified (Reserve Muscat) • Merlot (Reserve) • Shiraz (Parola's Limited Release, Reserve)*

WARRENMANG VINEYARD
Pyrenees
★

The Grand Pyrenees does not indicate the nature of its blend on the label, but contains Merlot, Cabernet Franc, and Shiraz.

✓ *Cabernet Sauvignon • Chardonnay • Shiraz*

WATER WHEEL
Bendigo
★

In 1989 owner-winemaker (and local lad) Peter Cumming purchased this vineyard, which had been established in 1970 and named after a nearby flour mill. To this day, the family-run concern still makes well-crafted wines at very reasonable prices, deserving its ★ rating on value alone.

✓ *Cabernet Sauvignon • Chardonnay*

WEDGETAIL ESTATE
Yarra Valley
★☆

Established back in 1994, this winery continues to offer superb Burgundian varietals, and although the quality of the Cabernet (Cabernet Sauvignon-Shiraz blend) is vulnerable to change of vintage, it does hit the spot when the weather is right.

✓ *Chardonnay • Classic white blend (Cabernet) • Riesling*

YARRA BURN
Yarra Junction
★★

Established in 1975, but owned by BRL Hardy since 1995, the Yarra Burn range has been revamped. The quality is up and is still improving. I used to favour the Bastard Hill *cuvées* of Chardonnay and Pinot Noir, but the straight Yarra Burn label is often as good nowadays, and sometimes even a little better!

✓ *Chardonnay • Pinot Noir • Shiraz • Sparkling (Pinot Noir-Chardonnay-Pinot Meunier)*

YARRA YARRA
Yarra Valley
★★

Ian McLean makes a tiny amount of superbly expressive wines from a vineyard that is so good, they named it twice!

✓ *Classic red blend (Cabernets) • Classic white blend (Sémillon-Sauvignon Blanc) • Cabernet Sauvignon (Reserve) • Merlot • Classic red blend (Syrah-Viognier)*

YARRA YERING
Yarra Valley
★★

The first in a number of Yarra-named wineries, Yarra Yering was founded in 1969 by New Zealand botanist Dr Bailey Carrodus, whose blended reds are legendary. Remarkably, Carrodus remained as the day-to-day winemaker until his death in 2008. Nowadays, the 2017 winemaker of the year Sarah Crowe is in charge and has done a great job at maintaining the high quality of the wines intact. Dry Red No.1 is Cabernet-based; Dry Red No.2 is Shiraz-based; Dry Red No.3 is a Portuguese red blend.

✓ *Cabernet Sauvignon (Carrodus) • Classic red blend (Dry Red No.1, Dry Red No.2) • Merlot • Shiraz (Underhill)*

YELLOWGLEN VINEYARDS
Ballarat
★

This operation was started by Australian Ian Home and *champenois* Dominique Landragin and his wife, Anna, but it is now owned by Beringer Blass. The wines of Yellowglen Vineyards have always shown excellent potential, and they did improve enormously, especially starting in early 1990s. The non-vintage Pinot Noir-Chardonnay is often superior to the vintaged Brut. Wines are also sold under Botanics Series and Colours Collection label.

✓ *Sparkling*

YERING STATION
Yarra Valley
★★

Part of the Rathbone Family Group since 1996, Yering Station sells table wines under its own name and sparkling wines under the Yarrabank label. Their Late Disgorged Cuvee is a very impressive sparkling wine.

✓ *Chardonnay (Reserve) • Classic red blend (Reserve Shiraz-Viognier) • Pinot Noir (Scarlett) • Sparkling (Yarrabank)*

YERINGBERG
Yarra Valley
★★

The oldest winery in the Yarra Valley, Yeringberg was established in 1862 by Swiss-born Frédéric Guillaume, Baron de Pury, and re-established in 1969 by his grandson, Guillaume de Pury, who has consistently produced fine, sometimes exceptionally fine, wines. The Dry Red is a Bordeaux-style blend.

✓ *Chardonnay • Classic red blend (Dry Red) • Classic white blend (Marsanne-Roussanne) • Pinot Noir*

THE ROOF OVERHANG OF THE YERING STATION WINERY FORMS A GOLDEN SWEEP OVER THE GROUNDS
This Yarra Valley site is Victoria's oldest vineyard. In 1838 the Scottish-born Ryrie brothers had planted two grape varieties here. By the late 19th century, Yering Station wines were winning awards, such as a Grand Prix at the Universal Exhibition in Paris in 1889.

Tasmania

With its cool, maritime climate, increasingly well-recognised wine styles, and a burgeoning interest from external investors, it's not surprising that Tasmania is considered by many to be leading the way into Australia's wine future.

Whilst Tasmania's output represents only a very small portion of Australia's overall production, the wines themselves are skewed disproportionately in the direction of the premium end of the market. This is a region that has built its industry on quality and a willingness to experiment, and the surge in both production and interest levels since the late 2000s has shown the rewards of this approach. Although this is really a new wine region, nine vines were planted here by William Bligh of HMS *Bounty* as early as 1788. They were planted at the eastern end of Adventure Bay on Bruny Island, which is much farther south than Tasmanian vines grow today, and not surprisingly, Bligh found they had not survived when he returned in 1791. The first true vineyard was established in 1823 at Prospect Farm by Bartholomew Broughton, an extraordinary convict who had been granted a pardon and eventually came to own a substantial amount of property. In 1834, settler William Henty took vine cuttings from Tasmania to Victoria, and these cuttings have been credited as the source of the first vineyards in Victoria and South Australia.

The history of wine in Tasmania follows a familiar trajectory, along the same path as the majority of Australian regions: a non-existence from the late 1800s through to the mid 20th century, followed by a resurgence and eventual success. Tasmania's renaissance of the 1950s was led by a Frenchman called Jean Miguet, who came to Tasmania to work on the construction of a hydro-electric plant. His new home at La Provence, north of Launceston, reminded him of his native Haute-Savoie and, since it was protected from ocean winds by trees, he believed that grapes might ripen there. In 1956 he cleared his bramble-strewn land and planted vines that grew successfully enough to encourage another European, Claudio Alcorso, to establish the now-revered Moorilla Estate vineyard on the Derwent River in 1958. The growth of the industry from this point onwards was slow but steady, and there remains an exciting potential for further development.

A COOL CLIMATE IN A WARMING WORLD

Tasmania's reputation, post-20th century resurgence, was initially built on its sparkling wines; The industry has since expanded well beyond just one style, however. Pinot Noir and Chardonnay have established a unique expression when grown on Tasmanian soils, and, due to the marginality of the climate, may be able to withstand a greater change in overall temperature than many of their mainland counterparts. The Tasmanian industry has been heavily influenced by the ideas of Richard Smart with regards to canopy management, constantly seeking more efficient and suitable ways of growing vines. At a time when other regions are beginning to struggle with moderating ripeness levels, and indeed with the increased risk of bushfires wiping out vineyards on the mainland, Tasmania is well-placed to capitalise on its unique environment and climate.

THE ENTRANCE TO DEVIL'S CORNER CELLAR DOOR AND LOOKOUT
The property can boast of its panoramic view over the Freycinet Peninsula. The underlying structure of the buildings is made from repurposed shipping containers.

WINE REGIONS OF TASMANIA *(see also p667)*
The island of Tasmania lies due south, on the same latitude as New Zealand's South Island. This area covers a wide variety of climates, terrains, and wines.

East Coast Wine sub-region (unofficial)

0 mi   50
0 km   50

FACTORS AFFECTING TASTE AND QUALITY

LOCATION
The island of Tasmania lies 160 kilometres (100 miles) due south of Victoria.

CLIMATE
Tasmania is significantly cooler than Victoria, with very similar conditions to Marlborough in New Zealand. The Tamar Valley and even the Coal River in the south, however, are warmer and drier than parts of Victoria. The north is warmer and sunnier than the south of the island, but this is slightly compensated for by the lower-lying vineyards in the south. Tasmania does not have the propensity for drought that many wine regions on the mainland suffer from.

ASPECT
Vines are mostly grown on gentle slopes up to 210 metres (690 feet) high in the north, but they are planted in lower-lying sites in the south of the island.

SOIL
The soil of Tasmania is famously clayey. In the north is it clay over gravel, but it is more sandy and sandstone in the south.

VITICULTURE & VINIFICATION
The harvest is approximately one week later in the south of Tasmania compared to the north of the island. The influence of Richard Smart's viticultural theories is very much evident in the vineyards of this region. High-tech equipment, such as pressing in an inert-gas atmosphere, can be found, but the deployment of technology is dependent on the size of the operation, and many Tasmanian wineries are very small.

GRAPE VARIETIES
Primary varieties: Chardonnay, Pinot Noir
Secondary varieties: Cabernet Franc, Cabernet Sauvignon, Gamay, Merlot, Pinot Gris, Pinot Meunier, Riesling, Sauvignon Blanc, Sémillon, Shiraz, Traminer

THE WINE PRODUCERS OF

TASMANIA

Note: At least a third of the island's most successful medal-winning wines are made by or under the guidance of Tasmanian Vintners (formerly Winemaking Tasmania). Frogmore Creek (formerly Hood winemaking services) also contract for 20 different labels, including many award-winning wines.

APOGEE
★
Lebrina

Andrew Pirie has gone back to the original single-vineyard concept that produced one of the world's greatest sparkling wines in 1995. Pirie continues to research wine climates to more precisely predict the effects of *terroir*.

ARRAS

Not a winery as such, so no location, but one of Australia's greatest sparklers. *See* Bay of Fires.

BAY OF FIRES
Pipers River
★★ V

This outpost of the Hardy Wine Company produces classy boutique wines, thanks to the team led by Penny Jones.
✓ *Entire range*

BREAM CREEK
Bream Creek
★★

Fred Peacock's wines are a surprise and a delight, including cool-climate Tasmanian red, white, and sparkling wines.
✓ *Entire range*

CHARTLEY ESTATE
Rowella
★☆

Tiny even by boutique standards, Chartley Estate is a family-owned and -run vineyard. It produces typically cool-climate wines from hand-picked varieties like Pinot Gris, Pinot Noir, Riesling, and Sauvignon Blanc.
✓ *Riesling*

CLEMENS HILL WINES
Cambridge
★★

Clemens Hill's cool-climate vineyards in the picturesque Coal River Valley deliver delicious, fruit-driven wine styles.
✓ *Entire range*

CLOVER HILL
Lebrina
★☆

Established in an old dairy farm, Clover Hill specialises in sparkling wines of elegance and finesse.
✓ *Sparkling wine*

CRAIGOW VINEYARD
Cambridge
★☆ V

One of the first vineyards established in southern Tasmania's Coal River Valley, which has worked well for classic cool-climate grape varieties such as Riesling, Chardonnay, Gerwürztraminer, and Pinot Noir. Wines of exciting quality are produced here by Hobart surgeon Barry Edwards.
✓ *Pinot Noir • Riesling*

DALRYMPLE VINEYARDS
Piper's Brook
★

Established in 1987, Dalrymple's is run by Peter Caldwell, who crafts balanced, elegant wines from Pinot Noir, Chardonnay, and Sauvignon Blanc.
✓ *Pinot Noir (Single Site Coal River Valley, Swansea)*

DELAMERE
Piper's Brook
★

Shane Holloway and Fran Austin are making structured, ripe Pinot Noirs and promising single-vineyard sparkling wines, Chardonnay, and Pinot Noir at their small, family-run establishment in northeast Tasmania.
✓ *Pinot Noir • Chardonnay (Block 3)*

DEVIL'S CORNER WINERY
Apslawn

Winemaker Tom Wallace focusses on crafting Pinot Noir from the wild environment of the east coast.

DOMAINE A
Campania
★★

Peter Althaus maintains the highest standards and is unremitting in his aim for a northern-European style from Southern Hemisphere grapes. It is owned by Moorilla Estate since 2018.
✓ *Entire range*

ELSEWHERE VINEYARD
Glaziers Bay
★☆ V

Terry and Mara Doyle produce superb Pinot Noir from their Huon Valley vineyard, as well as very good Riesling.
✓ *Pinot Noir • Riesling*

FREYCINET
Swansea
★★

Lindy Bull, daughter of founders Geoff and Susan Bull, and her partner, Claudio Radenti, craft surprisingly big and beautiful blends of Cabernet Sauvignon and Merlot and various other fine wines, including the stand-out Radenti sparkling wine, Riesling, and Louis Chardonnay. It is, however, the exquisitely crafted Pinot Noir that surpasses everything else here.
✓ *Entire range*

FROGMORE CREEK
Cambridge
★★ O

Owned by Tony Scherer of Tasmania and Jack Kidwiler of California, who have made some weird and wonderful wines over the years, from the only true wild ferment I ever saw. (Normally the yeast resident in the winery itself dominates the fermentation after the first 2 to 3 per cent of alcohol, but Frogmore puts the fermenting vessel in the vineyard itself!) Alain Rousseau is now Chief Winemaker, and John Brown is Winemaker. The Riesling FGR (Forty Grams Residual) is stunning. The range includes the excellent 42°S label.
✓ *Entire range*

GLAETZER-DIXON
Coal River Valley
★★

Originally from the Barossa Valley, Nick Glaetzer moved to Tasmania in 2005 and eventually transformed an old ice factory and urban winery and tasting room. The entire range is terrific, from the succulent Uberblanc Riesling to the excellent but expensive La Judith Pinot Noir.
✓ *Riesling • Pinot Noir*

HOLM OAK
Rowella
★ V

Owners winemaker Rebecca and her viticulturist husband Tim Duffy make first-rate Riesling at their rustic vineyard in the Tamar Valley.
✓ *Riesling*

JANSZ
Pipers Brook
★★

With its climate almost identical to France's Champagne region, it is no surprise that the Jansz vineyard turned to producing Tasmania's first sparkling wine made using the traditional *Méthode Champenoise*. Now owned by Hill-Smith Family Vineyards and under the careful eye of Jennifer Doyle, these sparklers have more finesse than ever and are produced in what they call *Méthode Tasmanoise*
✓ *Entire range*

JOSEF CHROMY CELLARS
Relbia
★★

Immaculate wines are made in this technologically advanced artisanal winery. A high-tech German press, for example, allows low-pressure juice extraction, which yields more delicate aromatic juices.

✓ *Entire range*

MEADOWBANK VINEYARD

Cambridge

★☆

Run by the Ellis family since 1976, Meadowbrook Vineyard produces classic black-cherry Pinot Noir from just north of Hobart, including the brilliant Henry James Pinot Noir.

✓ *Syrah • Chardonnay • Pinot Noir*

MILTON VINEYARD

Cranbrook

★

This boutique winery, sited in a dream location on the Freycinet Coast, has a unique micro-climate that suits the ancient grape Pinot Noir. They also grow Pinot Gris, Riesling, and Gewürztraminer and focus on producing small quantities of handmade wine.

✓ *Pinot Noir • Riesling*

MOORILLA ESTATE

Berriedale

★★

Established in 1962, this pioneering Tasmanian winery produces some brilliant cool-climate wines from estate-grown fruit. They have three different ranges: Praxis, Muse, and Cloth Label.

✓ *Entire range*

PIPERS BROOK VINEYARD

Pipers Brook

★

This Tamar region winery is now owned by the Belgian-owned Kreglinger Wine Estates (along with Ninth Island and Kreglinger Wine). They have yet to convince me about their sparkling wines.

✓ *Chardonnay • Pinot Noir (Reserve) • Riesling*

PIRIE TASMANIA

Rosevears

★★

Internationally acclaimed Andrew Pirie, Australia's first PhD in viticulture and one of the country's most respected wine-makers, makes only sparkling wine at his estate. Its site in the Tamar Valley region, with its cool slopes, gives it great potential.

✓ *Entire range*

POOLEY WINES

Campania

★★

Previously called Cooinda, this winery in the heart of the Coal River Valley has always produced award-winning Riesling, but its Pinot Noir may be even better. Pooley was Tasmania's first and only fully accredited Environmentally Certified Sustainable Vineyard.

✓ *Pinot Noir • Riesling*

PRESSING MATTERS

Hobart

★☆

Owned by the irrepressible Greg Mellick, who is Tasmania's highest-ranking barrister, a major-general in the Australian Defence Force Reserves, and Cricket Australia's special investigator into corruption. He is also the only senior wine judge I've caught asleep during a live competition panel discussion! The winery produces Pinot Noir and Mosel-inspired Riesling.

✓ *Riesling*

PROVIDENCE VINEYARDS

Lalla

★★

These vineyards were established in 1956, with vines illegally removed from France and smuggled into Tasmania.

✓ *Chardonnay • Pinot Noir*

SINAPIUS

Piper's Brook

★☆

High-density plantings, a hands-off approach, and small batches of characterful wines define this producer.

STEFANO LUBIANA WINES

Granton

★★

Probably best known for excellent sparkling wine, this boutique winery produces brilliant Pinot Noir, Chardonnay, and a whole lot more. Aims for finesse and minerality. Stefano Lubiana is Tasmania's first certified biodynamic vineyard.

✓ *Entire range*

STONEY RISE

Gravelly Beach

★★

Joe and Lou Holyman make great Pinot Noir, even in off-years.

✓ *Entire range*

TAMAR RIDGE

Kayena

★★

Purchased by Brown Brothers in 2010, Tamar Ridge is the largest vineyard owner and wine producer in Tasmania. With the winemaking foundations laid down by Andrew Pirie (since left) and Richard Smart as the on-site viticultural consultant (still on contract), the best-practice regime here is as high as it gets.

✓ *Entire range*

TOLPUDDLE VINEYARD

Coal River Valley

★☆

A joint venture of Shaw + Smith owners Martin Shaw and Michael Hill-Smith MW, it produces Chardonnay and Pinot Noir that is sold on allocation.

✓ *Chardonnay, Pinot Noir*

TWO BUD SPUR VINEYARD

Gardners Bay

★★

The team at Two Bud Spur produces great Pinot Noir at one of Tasmania's most southerly vineyards.

✓ *Pinot Noir*

VÉLO WINES

Legana

★☆

This Tasmanian producer is best by far for classic Chardonnay.

✓ *Chardonnay*

A PATIO AT THE FROGMORE CREEK WINERY INVITES VISITORS TO ENJOY VIEWS OF THE VINEYARDS
Frogmore Creek Winery, located in in the Coal River Valley of southern Tasmania, is one of Tasmania's most-awarded wineries.

South Australia

This is Australia's most productive and contradictory wine region, accounting for nearly half of its total wine output, ranging from its cheapest to its most expensive wines.

Vines were first planted in South Australia by John Barton Hack in 1837, and from there the industry began to take off. Winemaking flourished until the end of the 19th century, although during this time Victoria produced well over double the quantity of South Australia's output. The impact of phylloxera on South Australia's vineyards was devastating, but since recovering and replanting, the region achieved and maintained its position as the largest producer by volume in Australia. Despite a vine-pull scheme in the late 1980s that saw a large number of older vines ripped up and lost, this is still the go-to region for old vines in Australia, and what was left is now carefully preserved.

MULTIREGIONAL – REGIONAL – SUBREGIONAL

The paradox is that if consumers are asked what South Australia means to them as a winemaking state, they will probably say that it has a reputation for cheap wine, but ask them about Coonawarra, and they will say it is Australia's most famous, premium-quality wine region. Ask them about the Barossa Valley, and they will say that is where Australia's greatest Shiraz comes from. Since the drop in sales of Australian wine in the mid-2000s, "Brand Australia" has refocused on the individuality of each regional GI. This has been of particular benefit to South Australian regions, which in recent years have been able to more easily differentiate themselves from the more generic "Southeast Australia" blends that do not necessarily equal the quality level of the boutique wines of the smaller Geographical Indications (GIs). This is not to say that all cross-regional blends lack quality; famously, Penfold's Grange falls within this category, and yet it has the highest price on release of any Australian wine. For the majority of fine wine producers, however, the ability to draw on the strength of their regional branding can only be considered a benefit. Sub-regionality is being explored in earnest in regions like the Clare and Barossa Valleys, where there is enough recognition of regional style to warrant further differentiation within the boundaries. Whether or not this translates to new GIs has yet to be seen, but the future of South Australian wines is definitely in focus.

WINE REGIONS OF SOUTH AUSTRALIA
(*see also p667*)
The state perches over the eastern half of the Great Australian Bight, a bay whose climatic influence decreases farther inland.

FACTORS AFFECTING TASTE AND QUALITY

LOCATION
This is the southern central part of the country, with Australia's five other mainland states to the east, north, and west, and half of the Great Australian Bight forming the coastline to the south.

CLIMATE
The climate varies greatly, from the intensely hot continental conditions of the largely cask-wine producing Riverland area, through the less extreme but still hot and dry Barossa Valley, to the cooler but still dry Coonawarra region. Sea breezes reduce humidity in the plains around Adelaide, which receives low annual rainfall, as does the whole region.

ASPECT
Vines are grown on all types of land, from the flat coastal plain around Adelaide and flat interior Riverland district to the varied locations of the Barossa Valley, where vines are grown from the valley floor at 250 metres (820 feet), up to the slopes to a maximum of 600 metres (1,970 feet) at Pewsey Vale.

SOIL
Soils are varied, ranging from sandy loam over red earth (*terra rossa*) on a limestone-marl sub-soil in the Adelaide and Riverland areas (the latter having suffered for some time from excess salinity), through variable sand, loam, and clay topsoils, over red-brown loam and clay sub-soils in the Barossa Valley, to the thin layer of weathered limestone, stained red by organic and mineral matter, over a thick limestone sub-soil in the Coonawarra area.

VITICULTURE & VINIFICATION
This varies enormously, from the bulk-production methods of the large modern wineries that churn out vast quantities of clean, well-made, inexpensive wine from grapes grown in Riverland's high-yielding irrigated vineyards to the use of new oak on restricted yields of premium-quality varietals by top estate wineries in areas such as Coonawarra, the Barossa Valley, and the up-and-coming Padthaway or Wrattonbully districts, which produce some of Australia's greatest wines.

GRAPE VARIETIES
Primary varieties: Cabernet Franc, Cabernet Sauvignon, Chardonnay, Malbec, Merlot, Petit Verdot, Pinot Noir, Riesling, Sauvignon Blanc, Sémillon, Shiraz (Syrah)
Secondary varieties: Brown Muscat (Muscat Blanc à Petits Grains), Cinsault, Clare Riesling (Crouchen), Durif, Gewürztraminer, Grenache, Mourvèdre, Muscadelle, Muscat Gordo Blanco (Muscat d'Alexandrie), Palomino, Pedro Ximénez, Ruby Cabernet, Sangiovese, Portugal (Tinto Amarella), Tempranillo, Touriga (Touriga Nacional), Ugni Blanc

THE APPELLATIONS OF
SOUTH AUSTRALIA

ADELAIDE GI

A superzonal GI that encompasses the regions of Barossa, Fleurieu, and the Mount Lofty Ranges.

ADELAIDE HILLS GI

Subregion: Lenswood, Piccadilly Valley

This hilly region is just 15 kilometres (9 miles) from the coast, overlooking Adelaide, the Adelaide Plains, and McLaren Vale. It has two sub-regions of its own, Lenswood and Piccadilly Valley, which are determined by differing temperatures and soil types. Along with Clare Valley and Adelaide Plains, this region is part of the much larger Mount Lofty Ranges GI. Although the Adelaide Hills have a Mediterranean climate, this is by far the coolest of all South Australian wine regions, being strongly influenced by elevation ranging from 350 to 700 metres (1,150 to 2,300 feet), and the winds that sweep across St Vincents Gulf. Undulating hills are interspersed with steep slopes, which drop into frost-prone gullies, creating many different kinds of *terroir* according to height, aspect, and the various soils, which range from (but are not restricted to) sandy alluvium on the valley floor, to yellow and red loams on the slopes. The first vines were planted here in the 1840s, but viticulture struggled for economic survival, finally succumbing to the Great Depression. Its revival finally came in the 1970s, when boutique wineries started to satisfy the demand of consumers who were switching from fortified to table wine, most especially with the arrival of Petaluma Winery. More recently, this region has served as an outpost for a new generation of winemakers utilising experimental and natural approaches. Chardonnay, Riesling, Pinot Noir, and bottle-fermented sparkling wines are now the classics here, with Cabernet Franc and Merlot faring better than Cabernet Sauvignon. Sauvignon Blanc and Sémillon also do well, with Sangiovese and the inevitable Shiraz showing promise. Brushfires in 2019 damaged a third of vines growing here, specifically in the Tumbarumba region.

ADELAIDE PLAINS GI

The Adelaide Plains is situated within a half-hour drive of the city of Adelaide, and therefore receives a disproportionately large number of visitors each year, given the small number of operational wineries actually located there. It is an arid region with an extremely low annual rainfall and very high summer temperatures. Irrigation is essential, but there is minimal disease pressure, and red-brown loamy soils on the flat plain that allow for high yields and easy mechanisation. A number of operations based in other regions such as Barossa use this area to boost their production levels, but this is also the region in which Penfolds' legendary Magill Vineyard is located; a small agricultural outpost surrounded by Adelaide's suburban sprawl.

BAROSSA GI

A zonal GI encompassing the Barossa Valley and Eden Valley regional GIs. The Barossa Old Vine Charter was instituted in 2009 to recognise the value of the incredible resource of vineyards with significant age in the region. A Barossa Old Vine is equal or greater than 35 years, Barossa Survivor Vine is equal or greater than 70 years, Barossa Centenarian Vine is equal or greater than 100 years, and Barossa Ancestor Vine is equal or greater than 125 years.

BAROSSA VALLEY GI

This district is the oldest and most important of South Australia's premium varietal areas. With a hot, dry climate, the vines mostly grow on flatlands at an altitude of 240 to 300 metres (800 to 1,000 feet), although in some areas the altitude rises to 550 metres (1,800 feet), and the vines are cooled by ocean breezes. The Barossa Valley is heavily dependent on irrigation, but boreholes are metered, and the amount of water that may be drawn is strictly regulated. Historically – and still today – there were far more grape-growers than winemakers in the region, with certain highly reputed sites staying in the same family for generations and appearing on the labels of some of the best producers in much the same manner as many Californian vineyards. Big, brash Shiraz is king in this region, although

ROWS OF WIRE STAKES IN A BAROSSA VALLEY VINEYARD AWAIT THE GROWTH OF NEW VINES
Founded by German settlers, the Barossa Valley is one of Australia's oldest viticultral regions and produces some of its premier wines, often made from the Rhône varietal Shiraz.

JABON'S CREEK CABERNET SAUVIGNON, VINTAGE 2014
The Australian wine industry's focus on cultivating huge brands like Jacob's Creek, a label first released in 1976 by Orlando Wines, sparked the popularity of Aussie wines in the 1990s. Orlando first sourced its grapes for Jacob's Creeks from vineyards in the Langhorne Creek GI. The label is now owned by Pernod Ricard, who have elevated the label to signature status. It is now exported to more than 60 countries.

a significant volume of white wine is also produced. Chardonnay, Sémillon, and Riesling dominate, and depending on the climate and the soil (which may be limestone, clay, or sand), these white wines can range from full-bodied to surprisingly delicate.

CLARE VALLEY GI

Clare Valley is the most northerly winegrowing district in South Australia, and its climate is correspondingly hotter and drier. Many of the vineyards are not irrigated, however, and the result is a low yield of very intensely flavoured, big-bodied, often strapping wines. There are a number of unofficial sub-regions, designated by soil type and geographic location, the most well-known of which include Watervale, Clare, and Sevenhill. Riesling is the valley's most important variety and botrytis-affected wines are rich, fine, and mellifluous. Other good wines are Cabernet Sauvignon (often blended with Malbec or Shiraz), Sémillon, and Shiraz. Clare Valley is also credited with influencing the majority of the Australian wine industry to accept screwcap closures as the norm. In 2000 a joint decision to switch over was made by the Riesling producers of this region; others followed, and these days it can be difficult to find an Australian wine under cork.

COONAWARRA GI

This famous district is the most southerly in South Australia. *Coonawarra* is Aboriginal for "wild honeysuckle"; it

also happens to be easy on Anglo-Saxon tongues, and this has been useful when it comes to marketing wine from the district in English-speaking countries. Consequently, this region achieved worldwide fame, which paradoxically held it back during the GI process, while other regions that were not as well known were demarcated relatively quickly. Everyone with the slightest chance of being considered part of Coonawarra wanted to be included within the boundaries, while the bureaucrats operating the GI scheme demonstrated little or no imagination. It thus took eight years to settle all the disputes. The crux of the matter revolved around Coonawarra's red earth, or *terra rossa*. Black soil areas are interspersed among the *terra rossa*. and these soils produce quite different wines, even though they share the same limestone subsoil. The Coonawarra ridge hardly stands out topographically, rising just 59 metres (194 feet) above sea level, but it does stand out viticulturally, because the surrounding land is flat, frosty, and poorly drained. This and the Mediterranean climate, which enjoys the cooling maritime influences off the Southern Ocean, make height-challenged Coonawarra superbly suited to viticulture, whether on black or red soils. It is just that those vineyards on *terra rossa* soils have historically been responsible for some of Australia's outstanding wines, particularly Cabernet Sauvignon. Vineyards on black soils can produce some very good wines, but those on *terra rossa* are demonstrably more expressive of their *terroir*. Cabernet Sauvignon might be king of the *terra rossa*, but

other varieties that perform well throughout Coonawarra include Shiraz, Petit Verdot, Pinot Noir, Malbec, and Merlot. Although Coonawarra is red wine country par excellence, successful white grapes include the Chardonnay, Riesling, Sauvignon Blanc, and Sémillon varieties.

CURRENCY CREEK GI

This regional GI is located on the Fleurieu Peninsula, extending from Port Elliot in the west to the shores of Lake Alexandrina in the east, and includes three islands: Hindmarsh, Mundoo, and Long Islands. The proximity to Lake Alexandrina and the mouth of the Murray River moderates the hot Mediterranean climate, making the area suitable for viticulture. Grapes growing here include Cabernet Sauvignon, Chardonnay, Sauvignon Blanc, and Shiraz.

EDEN VALLEY GI

Subregion: High Eden

The Eden Valley has a rockier, more acid soil than that found in the neighbouring Barossa Valley, with a cooler, much wetter climate. Vineyards are located at an altitude of 400 to 600 metres (1,300 to 2,000 feet) in the Mount Lofty Ranges, which form part of the Barossa Range. The Eden Valley GI is not a valley as such but rather takes its name from the township of Eden Valley. While the pre-eminence of Henschke might suggest that the Eden Valley is world-class Shiraz country, that is very much due to one vineyard, Hill of Grace, and the 130-year-old Shiraz vines planted there. Eden Valley is best known for its quintessentially Australian (that is, bone-dry, racy) Riesling, and deservedly so. Chardonnay also fares well.

FAR NORTH GI

Apart from the Southern Flinders Ranges GI in the far south, the vast Far North zonal GI is very much a viticultural wilderness.

FLEURIEU GI

This zonal GI consists of the peninsula jutting out towards Kangaroo Island, and encompasses the regional GIs of Currency Creek, Kangaroo Island, Langhorne Creek, McLaren Vale, and Southern Fleurieu.

HIGH EDEN GI

So named because, unsurprisingly, this sub-regional GI is located at the highest point in Eden Valley, which also makes it one of the coolest and most windy areas. The terrain is hilly and rugged, with poor sandy soils that are well suited to Chardonnay, Riesling, and Shiraz, as is the rest of the Eden Valley. Cabernet Sauvignon is also grown.

KANGAROO ISLAND GI

Located off the coast of South Australia, Kangaroo Island is renowned for its natural beauty and an abundance of wildlife, including colonies of Australian sea lions, New Zealand fur seals, penguins, and the once-endangered Cape Barren goose. In recent years, Kangaroo Island has also been making a name for its Bordeaux-style blends. Most vines are grown on ruddy-coloured ironstone or sandy loam soils on the north side of the island, near Kingscote, where the maritime climate is marked by strong winds and moderate humidity. In addition to Bordeaux varieties, Chardonnay and Shiraz are also grown.

LANGHORNE CREEK GI

This once-tiny, traditional area, located southeast of Adelaide on the Bremer River, has expanded rapidly since 1995, when Orlando Wines started sourcing grapes from here for huge brands such as Jacob's Creek. This district was named after Alfred Langhorne, a cattle herder from New South Wales, who arrived in 1841. The meagre annual rainfall of 35 centimetres (14 inches) means that the vineyards require irrigation. This area is well regarded for its full-bodied, approachable reds and for its dessert wines.

LENSWOOD GI

This sub-regional GI is located north of Piccadilly, the other sub-region in the Adelaide Hills. The vines grow on steep slopes and benefit from a cool climate, but at 400 to 560 metres (1,300 to 1,840 feet), thus lower than those in Piccadilly (up to 710 metres, 2,330 feet), and do not experience quite the same snap of coldness. The grape varieties that excel are the same – Chardonnay and Pinot Noir – and the quality can be equally exciting.

LIMESTONE COAST GI

This zonal GI contains the regional GIs of Coonawarra, Mount Benson, Mount Gambier, Padthaway, Robe, and Wrattonbully within its borders. The name is particularly apt as it refers to the one geological factor that links these regional GIs.

LOWER MURRAY GI

A zonal GI that encompasses the regional GI of Riverland.

MCLAREN VALE GI

The rolling green hills of McLaren Vale's vineyards and orchards begin south of Adelaide and extend to south of Morphett Vale. With 56 centimetres (22 inches) of rain and a complex range of soils, including sand, sandy loam, limestone, red clay, and many forms of rich alluvium, there is great potential for quality in a range of styles from various grapes. This is perhaps why it is the most volatile wine district, attracting a lot of new talent, but also seeing wineries close or change hands. This is another region that has been divided into unofficial sub-regions by the local producers. Blewitt Springs is probably the best know of these, appearing on numerous labels, and distinctive for its sandy soils. McLaren Vale produces big red wines of excellent quality from Grenache and Shiraz (sometimes blended together with Mourvedre), and Cabernet Sauvignon (often blended with Merlot), with some increasingly fresh and vital white wines from Chardonnay, Sémillon, and Sauvignon Blanc. Fine dessert wines are also produced, alongside a number of other varieties such as Tempranillo and Nebbiolo.

MOUNT BENSON GI

Prior to 1989 this region had no history of viticulture, and it is therefore one of the newest regions in Australia. Located right on the coast, to the west of Coonawarra, this isolated region has remained relatively small. The climate is maritime, and much cooler than the inland regions of the Limestone Coast, allowing for greater success with white varieties, including Grüner Veltliner, alongside the ubiquitous Chardonnay and Sauvignon Blanc. The main red grapes of the area include Cabernet Sauvignon, Merlot, and Shiraz, although there has been some recent experimentation with Italian and Spanish varieties.

MOUNT GAMBIER GI

With a much cooler GI than others in the Limestone Coast Zone, the vineyards of Mount Gambier are able to produce Pinot Noirs, sparkling wines, and aromatic whites. The region encompasses the area from south of Coonawarra all the way to the coast. As yet, it has not made a significant impression on the local or international wine scene, but there is certainly room for development.

MOUNT LOFTY RANGES GI

This zonal GI encompasses the regional GIs of Adelaide Hills, Adelaide Plains, and Clare Valley.

PADTHAWAY GI

Established in 1963, this area has been developed almost exclusively by the larger companies (Hardy's built Australia's largest winery in 20 years here in 1998). A majority of the harvest is blended away, rather than labelling with the specific regional GI, but a number of smaller producers in the region are helping to increase the awareness and quality perception of wines from this area. Cabernet Sauvignon, Chardonnay, and Shiraz are the main varieties planted.

THE PENINSULAS GI

If the Far North zonal GI is a massive viticultural wilderness above an imaginary line, then the Peninsulas is that part of South Australia to the south of the same line, where hardly any wineries exist. It is not as if there is no scope for viticulture here. There are plenty of gently undulating hills, with the right aspect and altitude. The sub-soil is limestone, with everything from sandy-clayey loams to gravel for topsoil. There is even *terra rossa*. The area remains untapped, however.

PICCADILLY GI

Sub-region situated in the west of the Adelaide Hills GI, between Ashton and Bridgewater. The vines grow at altitudes of 400 to 710 metres (1,300 to 2,300 feet), benefiting from a very cool, wet climate, which stretches the ripening process, yielding grapes with naturally high tartaric acid levels. Chardonnay and Pinot Noir excel.

RIVERLAND GI

The wine areas that cluster around the Murray River in Victoria continue into Riverland, South Australia's irrigated answer to Riverina in New South Wales and Murray Darling in Victoria. Although a lot of cheap cask-wine is made in the Riverland district, rarely is a bad one encountered. Cabernet Sauvignon, Cabernet Sauvignon-Malbec, Chardonnay, and Riesling all fare well, and some relatively inexpensive wines are produced in the area. In contrast to the generic cask blends most commonly associated with this region, a few pioneering producers, Unico Zelo at their head, are working to cement the reputation of Riverland as the home of the alternative varietal. "From Arneis to Zinfandel", the output of boutique, quirky wines is steadily growing.

ROBE GI

Robe lies directly south of Mount Benson on the east coast of the Great Australian Bight. Like Mount Benson, this region is relatively new, and has remained small, with just a couple of listed producers. The strong influence of the Southern Ocean helps to moderate the climate, allowing Sauvignon Blanc to flourish alongside Cabernet Sauvignon, Shiraz, and Chardonnay.

SOUTHERN FLEURIEU GI

This regional GI is located at the southern end of the Fleurieu Peninsula, where its gentle, undulating coastal country benefits from cooling winds coming from both the west and south, and vines grow in sandy loam and gravelly-ironstone soils. An up-and-coming area for small, boutique wineries, where Shiraz grows best, but other varieties include Cabernet Sauvignon, Malbec, Riesling, and Viognier.

SOUTHERN FLINDERS RANGES GI

Although the first grapes were grown here in the 1890s, it is only in the past 15 years that viticulture has become viable on a commercial scale, hence its regional GI status. This has been due to three factors: the so-called "Goyder's line of rainfall", which reduces demand on water supplies; an early harvest; and a ready market in the neighbouring Barossa and Clare Valleys. The Southern Flinders Ranges are divided by "Goyder's line of rainfall", which was named after the surveyor George W Goyder, who first determined the "northernmost limits of feasible agriculture". Irrigation comes from underground water, but with an annual rainfall of 450–650 millimetres (17–25 inches), it is possible to grow grapes without it, which explains the focus on dry-grown, premium-quality grapes. The elevation extends from just 20 metres (65 feet) on the coast to 718 metres (2,355 feet) at Frypan Hill, but most vineyards are located on slopes at a height of 350 to 550 metres (1,150 to 1,800 feet), where the region's warm to hot climate is tempered by the cooling effects of altitude and winds crossing the gulf. The soils range from deep sandy loam on the coastal plains to the west of the Flinders Ranges, to deep loam over red clay to the east, around the Wild Dog Creek Land System, and more shallow, stony loam on the slopes. The best-suited grapes are Cabernet Sauvignon, Merlot, and Shiraz.

WRATTONBULLY GI

Despite the peculiar name, it is surprisingly easy to pronounce for Anglo-Saxon tongues, perhaps because it is itself an Anglo-Saxon interpretation of the Aboriginal expression for "a place of rising smoke signals" Wrattonbully was not the original name for this region, sandwiched between Coonawarra and Padthaway. It was formerly known as Koppamurra, but there is a vineyard of that name in the area, and there were trademark problems in using it for a GI, thus it was decided to rename the region Wrattonbully, presumably to send out a signal. The first commercial vineyard was established in 1968, and by the 1990s most of Australia's biggest wine groups were sourcing fruit from here. The climate is very similar to that of Coonawarra, but its slightly higher aspect provides better air drainage, therefore a reduced risk of frost, and lower humidity, thus fewer cryptogamic disease problems. The sub-soil is limestone, of course, with various topsoils consisting of red-brown sandy and clayey loams, including some of the purest *terra rossa* soils, making this ideal *terroir* for Cabernet Sauvignon and Shiraz. Potentially, the reputation of Wrattonbully will be at least as great as Coonawarra's. We have barely begun to hear about these wines.

THE WINE PRODUCERS OF

SOUTH AUSTRALIA

ANGOVE'S
Riverland
★

Most wines are fairly cheap and usually rack up a bronze medal at wine competitions, thus the odds are that even those not specifically recommended below can be relied on for above-average quality and very good value.

✓ *Cabernet Sauvignon (Studio Series)*
• *Chardonnay (Long Row)* • *Shiraz (Long Row)* • *Fortified (Tawny Grand)*

ANNIE'S LANE AT QUELLTALER ESTATE
Clare Valley
★★ Ⓥ

The winemaker Alex MacKenzie has been working since 2002 for this precious part of the Beringer Blass empire.

✓ *Classic red blend (Cabernet-Merlot, Shiraz-Grenache-Mourvèdre)* • *Riesling* • *Shiraz*

ASHTON HILLS
Adelaide Hills
★★

Owner-winemaker Stephen George has honed his style into one of great elegance. Wines are also sold under the Galah label.

✓ *Chardonnay* • *Pinot Noir (Picadilly Valley)* • *Riesling* • *Sparkling (Pinot Noir)*

BALNAVES OF COONAWARRA
Coonawarra
★★

Established in 1975 by Doug Balnaves, who sold his entire harvest as grapes until 1990, when he hired a contract winemaker to launch his own label. Quality and consistency surged upwards in 1996, when Balnaves of Coonawarra got its own brand-new winery, with Wynns' former assistant winemaker Peter Bissell in charge.

✓ *Cabernet Sauvignon* • *Chardonnay* • *Classic red blend (The Blend)* • *Shiraz*

BAROSSA VALLEY ESTATE
Barossa Valley
★

Owned by BRL Hardy, Barossa Valley Estate makes Shiraz from the low-yielding Barossa vineyards of Elmor Roehr and Elmore Schulz. Their GSM blend is also a great wine.

✓ *Shiraz* • *Classic red blend (Grenache-Shiraz-Mourvèdre)* • *Cabernet*

BERRI ESTATES
Riverland

The merger between Thomas Hardy and Sons and the huge Berri-Renmano winery formed BRL Hardy (today's Accolade Wines), and this cooperative winery, servicing some 850 private grape-growers, was absolutely essential for the growth of that group. Despite this pivotal role, however, and although Berri Estates is Australia's largest winery, processing up to 220,000 tonnes, neither its Berri Estates nor its Renmano brand contribute much to the wealth and grandeur of Accolade today. Most wines are bag-in-the-box and very cheap, although well-made for what they are.

BIRD IN HAND
Adelaide Hills
★☆

Founded in 1997 by the Nugent family, Bird in Hand produces clean, classy wines with chief winemaker Kym Milne MW (formerly of Villa Maria) at the helm. Some of their vines suffered damage from the 2019 bushfires.

✓ *Chardonnay (Bird in Hand)*

BRASH HIGGINS
McLaren Vale
★

Chicago-born Brad Hickey founded the brand after a long and varied career in hospitality, brewing, and distilling, together with McLaren Vale winemaker Nicole Thorpe. The fresh labels and the minimal-intervention wines speak of their love for European styles.

✓ *Nero d'Avola* • *Grüner Veltliner/Riesling*

BOWEN ESTATE
Coonawarra
★★

Doug Bowen continues to produce some truly exceptional wines.

✓ *Cabernet Sauvignon* • *Chardonnay* • *Shiraz*

BRAND'S OF COONAWARRA
Coonawarra
★★

This winery is sometimes known as Brand's Laira, because of its famous Laira vineyard in Coonawarra's *terra rossa* heartland, the original part of which is planted with 110-year-old Shiraz vines. The Brand family did not purchase the Laira vineyard until 1946, and the first wine sold under the Brand's Laira label was not until the 1966 vintage. In 1990, McWilliams purchased 50 per cent of what was then known as Brand's Estate, buying up the other 50 per cent in 1994, retaining Jim and Bill Brand as winemakers. Jim Brand is still one of this winery's two winemakers. The old Laira Vineyard Shiraz is sold today as Tall Vines. Other fruit from the extended Laira vineyard is also found in the Blockers Cabernet Sauvignon.

✓ *Cabernet Sauvignon (Blockers)* • *Riesling* • *Shiraz (Tall Vines)*

BREMERTON WINES
Langhorne Creek
★☆

Bremerton was stablished in 1988, but the fact that rapidly expanding production in recent years has been matched with either constant or improved quality is something that the Wilson family should be proud of.

✓ *Cabernet Sauvignon (Coulthard)* • *Classic red blend (Tamblyn)* • *Shiraz (Old Adam)*

LEO BURING
Eden Valley and Clare Valley
★★

Orlando Wines took over the Leo Buring Winery (which now operates as Richmond Grove), but Treasury Wine Estates bought the Leo Buring name. Although just one of Treasury Wine Estates' many brands, Leo Buring remains one of Australia's foremost producers of fine Rieslings. Indeed, the range is now reduced to this one varietal, and just three versions of that. On the one hand, this is a tragic end to a once-great winery, but on the other, as an exclusive Riesling brand, Leo Buring must rate a good two stars.

✓ *Entire range*

CHAIN OF PONDS
Adelaide Hills
★☆

Zar Brooks, the young man who put a smile on the face of d'Arenberg, was hired as CEO to put his impish ideas to work here, before he moved onto stranger things (Stranger & Stranger actually, a brand consultancy firm).

✓ *Chardonnay (Corkscrew Road)* • *Pinot Noir (Mornington Star)* • *Sémillon (Adelaide Hills)* • *Shiraz (Forreston Reserve)*

CHAPEL HILL WINERY
McLaren Vale
★★

This Chapel Hill has no connection with the Australian-made wines of the same name from Hungary, but instead derives its name from the deconsecrated, 19th-century hilltop chapel which is used as its winery. In December 2000, Chapel Hill was purchased by the Swiss Schmidheiny family, which also owns Cuvaison Winery in California, as well as vineyards in Argentina and Switzerland.

✓ *Cabernet Sauvignon (The Prophet)* • *Chardonnay* • *Shiraz (The Prophet)* • *Verdelho*

M CHAPOUTIER
Mount Benson
★★

This famous Rhône Valley *négociant* Michel Chapoutier established his presence in Australia in 1997, when he purchased land at Mount Benson, and began planting. He now has vineyards in Pyrenees and Heathcote which are bottled under Domaine Tournon, plus two joint ventures in Victoria: one with Jasper Hill at Heathcote called La Pleiade, the other with Anthony Terlato in the Pyrenees region.

✓ *Entire range*

CHARLES MELTON
Barossa Valley
★★

Most famous for its Châteauneuf-du-Pape-type blend, cheekily named Nine Popes, but all the reds produced here are high-flyers, and Melton's sweet, seductive, sparkling Shiraz is regularly one of Australia's two or three finest. Rose of Virginia is a Grenache-Cabernet blend.

✓ *Cabernet Sauvignon* • *Classic red blend (Nine Popes)* • *Rosé (Rose of Virginia)* • *Shiraz*

CLARENDON HILLS
McLaren Vale
★☆

Roman Bratasiuk strives for the biggest, most concentrated wines he can possibly extract from his venerable old vines.

✓ *Grenache (Romas Vineyard)* • *Shiraz (Astralis)*

CORIOLE
McLaren Vale
★★

Expect full-throttle reds, including top-flight Shiraz and what was once the New World's best Sangiovese (still the same brilliant quality, but lots of top producers are now growing this and other Italian varieties). Redstone is a Shiraz-Cabernet blend, while Mary Kathleen is a Bordeaux-style blend.

✓ *Classic red blend (Mary Kathleen)* • *Chenin Blanc* • *Nebbiolo* • *Sangiovese* • *Shiraz (En Bonne Sante)*

CROSER
See Petaluma

DANDELION VINEYARDS
McLaren Vale

The hugely talented Elena Brooks makes an interesting range of wines from grapes sourced from Adelaide Hills, McLaren Vale, Barossa, Fleurieu Peninsula, and Eden Valley.

✓ *Riesling (Eden Valley)* • *Shiraz*

D'ARENBERG WINES
McLaren Vale
★★☆

The original vineyards were planted in the 1890s and purchased in 1912 by teetotaller John Osborn, whose family has owned the business ever since. I have been tasting these wines since the 1970s, when the style of the reds was sort of dusty-dry, but could they age! The wines became more rounded in the 1980s, but the real revolution came in the 1990s, when the wines became more voluptuous. The 1990s also saw the arrival of Zar Brooks, whose marketing flair, and ability to dream up catchy names, woke this slumbering giant. Today, all of d'Arenberg's Shiraz wines are stunning, but none more so than Dead Arm, which is a true Australian classic. Old Vine Shiraz and Footbolt Shiraz are almost in the same league, as is Sticks & Stones, the Rioja-style blend, and The Ironstone Pressings blend of Grenache, Mourvèdre, and Shiraz.

✓ *Botrytis (Noble Riesling)* • *Cabernet Sauvignon (The Coppermine Road)* • *Chardonnay* • *Classic red blend (Bonsai Vine GSM, Sticks & Stones Tempranillo-Grenache, The Ironstone Pressings)* • *Shiraz (The Dead Arm)*

JOHN DUVAL
Barossa Valley
★★

The former chief winemaker at Penfolds started his own label in 2003. He uses Rhône varieties for his inspired blends and varietal bottlings and has built up a cult following worldwide.

✓ *Entire range*

ELDERTON
Barossa Valley
★★ V

Established in 1984, Elderton has gained an enviable reputation for its high-quality, often highly oaked, wines since the early 1990s, with Cabernet Sauvignon king of this particular castle.

✓ *Cabernet Sauvignon (Ashmead) • Classic red blend (CSM) • Merlot • Shiraz (Command) • Sparkling (Pinot Pressings)*

ELDRIDGE
Clare Valley
★☆ V

The quality from Leigh and Karen Eldridge gets better with each vintage. Surprisingly good Gamay.

✓ *Cabernet Sauvignon • Classic white blend (Sémillon-Sauvignon Blanc) • Gamay • Riesling (Blue Chip) • Shiraz (Blue Chip)*

EPEROSA
Barossa Valley

A born-and-bred Barossan, Brett Grocke continues the family tradition of six generations of winegrowing with some stellar results.

✓ *Shiraz (especially Stonegarden)*

GLAETZER WINES
Barossa Valley
★★

This winery offers brilliant Shiraz, both still and sparkling.

✓ *Shiraz • Sparkling (Shiraz)*

GEOFF MERRILL
McLaren Vale
★☆ V

While Geoff Merrill was senior winemaker at Chateau Reynella (from 1977 to 1985), he started selling wines under his own name. He was appointed consultant red winemaker for Reynella's parent company, Thomas Hardy and Son, in 1985, but was actively searching for somewhere to make his own wines. That same year, Merrill came across the rundown Mount Hurtle Winery, which he purchased and restored, before leaving Hardy's in 1988 to concentrate on his own operation. At first, the emphasis was on the Mount Hurtle brand, but since 1993, when he took up the position as a consultant winemaker in Italy to UK supermarket Sainsbury and Gruppo Italiano Vini in Italy, the international fame of the moustachioed Merrill began to outshine that of Mount Hurtle, thus it was inevitable that his eponymous brand would one day take centre stage. Merrill's style is for elegance and length over concentration and weight, with the occasional over-oaking the only significant chink in his armour. Virtually all the wines offer value for money to one degree or another. The one obvious exception being the stunningly beautiful, but horrendously expensive Henley Shiraz. Mount Hurtle survives, but as an exclusive brand for Liquorland.

✓ *Chardonnay • Classic red blend (Shiraz-Grenache-Mourvèdre) • Classic white blend (Sémillon-Chardonnay) • Shiraz (Henley, Reserve)*

GEOFF WEAVER
Adelaide Hills
★★ V

For four years Geoff Weaver was chief winemaker for the entire Hardys wine group, which effectively made him personally responsible for 10 per cent of Australia's entire wine production. Not the sort of pressure that many of us would want to endure for too long, so little surprise that he left to pursue his own small wine operation in 1992. He had in fact planted his vineyard at Lenswood 10 years earlier, so it was nicely matured, and wines had been made on a part-time basis since 1986. Weaver's vineyard was also history in the making, as it was a pioneering venture into an untried region. Today his flavourful wines remain expressive of their *terroir* and are made with little need for human intervention.

✓ *Chardonnay • Classic red blend (Cabernet-Merlot) • Pinot Noir • Riesling • Sauvignon Blanc*

GRANT BURGE
Barossa Valley
★★☆ V

Owner-winemaker Grant Burge launched this venture in 1988. Since then, it has built up a reputation for richness, quality, and consistency, and he has not let his concentration lapse once. Grant Burge is probably best known for his iconic Shiraz, Meshach, but the Filsell is probably just as outstanding at one-third the price, and there are numerous other wines, red and white, that are in a similar class. The Holy Trinity is a Grenache-Shiraz-Mourvèdre blend; MSJ2 is Shiraz-Cabernet; Balthasar is Shiraz-Viognier. The Lily Farm Frontignac is not a fortified wine, but is made in a wonderfully fresh, sweetish, late-harvest style.

✓ *Cabernet Sauvignon (Cameron Vale) • Chardonnay (Summers) • Classic red blend (MSJ2, The Holy Trinity) • Fortified (20-Year-Old Tawny) • Frontignac (Lily Farm) • Riesling (Thorn) • Sémillon (Barossa Vines, RBS2, Zerk) • Shiraz (Filsell, Meshach, Miamba)*

GROSSET
Clare Valley
★★☆

Owner-winemaker Jeffrey Grosset is widely regarded as the king of Australian Riesling, but everything he produces is special. Gaia is a Bordeaux-style blend; Mesh is a Yalumba-Grosset collaboration.

✓ *Chardonnay (Piccadilly) • Classic red blend (Gaia) • Classic white blend (Sémillon Sauvignon Blanc) • Pinot Noir • Riesling (Alea, Polish Hill, Watervale)*

THE D'ARENBERG CUBE SITS WITHIN THE WINERY'S VINEYARDS
Designed by Chief Winemaker Chester Osborn, the d'Arenberg Cube is a five-storey building situated within the d'Arenberg vineyards on Osborn Road that was completed in 2017. The Cube houses a restaurant, a wine sensory room, a virtual fermenter, a 360-degree video room, and the Alternate Realities Museum. Commanding impressive views of McLaren Vale, the Willunga Hills, and Gulf St Vincent, the avant-garde Cube attracted 1,000 people a day in its first month of operation.

HAMILTON
McLaren Vale
★☆

Richard Hamilton makes stunning wines, particularly reds. Hut Block Cabernets is a blend of Cabernet Sauvignon and Cabernet Franc. Hamilton also owns his late uncle's vineyard in Coonawarra, Leconfield.

✓ *Cabernet Sauvignon (Hut Block) • Classic red blend (Hut Block Cabernets) • Merlot (Lot 148) • Shiraz (Centurion 125 Year Old Vines)*

HARDY'S
McLaren Vale
★★

This large brand was the basis of the BRL Hardy group (today Accolade Wines), which continues to produce a number of wines under various labels bearing the Hardy's name. This diverse selection ranges from penny-pinchers through penny-savers to premium-quality wines. Eileen Hardy Chardonnay (a wine named after the late, much-loved matriarch of the family for 40 years) consistently retains its reputation as one of Australia's very greatest wines.

✓ *Botrytis (Padthaway Noble Riesling) • Cabernet Sauvignon (Padthaway, Thomas Hardy, Tintara) • Chardonnay (Eileen Hardy, Tintara) • Fortified (Ports: Show, Tall Ships, Tawny, Vintage, Whiskers Blake Tawny) • Riesling (Siegersdorf) • Shiraz*

THE CELLARS AT HENSCHKE ARE OPEN TO VISITORS
One of the oldest and best wine producers in the country, Henschke has been operating since 1868. It is an inaugural member of Australia's First Families of Wine.

(Eileen Hardy, Oomoo, Tintara) • Sparkling (Arras as from 1997 vintage, Banrock Station Sparkling Shiraz, Sir James, Nottage Hill, Omni)

HEGGIES VINEYARD
Eden Valley
★★

Part of Yalumba, this winery has always been known for its Riesling, but has started to produce a fine Chardonnay in recent years.

✓ *Chardonnay • Riesling*

HENSCHKE
Eden Valley
★★☆

Founded in 1868, this is one of Australia's oldest and greatest wine-producing icons, with Hill of Grace Shiraz, and Cyril Henschke Cabernet Sauvignon lording it over the rest of the highly acclaimed range. Other offerings are Henry's, a Shiraz-Grenache-Viognier blend, and Johann's Garden, which is a Grenache-Mourvèdre-Shiraz.

✓ *Cabernet Sauvignon (Cyril Henschke) • Chardonnay (Crane's Eden Valley, Lenswood Croft) • Classic red blend (Johann's Garden, Henry's Seven) • Merlot (Abbot's Prayer) • Pinot Noir (Giles) • Riesling (Julius) • Sémillon (Hill of Peace, Louis Eden Valley) • Shiraz (Mount Edelstone, Hill of Grace)*

HEWITSON
Adelaide
★★☆

Dean Hewitson was a winemaker at Petaluma for 10 years, where he achieved the impossible by shining even in the bright light of Brian Croser. During his time at Petaluma, he completed a vintage at that company's subsidiary in Oregon, the Argyle Winery, three harvests in France, and picked up a master's degree at UC Davis in California. Hewitson's speciality is to source grapes from very old, superbly sited vineyards in various regions. The Mourvèdre is the extreme example, coming from 160-year-old vines in Barossa.

✓ *Entire range*

HOLLICK WINES
Coonawarra
★

Owner-viticulturist Ian Hollick set up his venture in 1983 and has produced some fresh, clean, well-focused wines, including a red Bordeaux-style blend that is simply sold by its Coonawarra appellation.

✓ *Cabernet Sauvignon (Ravenswood) • Chardonnay (Reserve) • Classic red blend (Coonawarra Cabernet Sauvignon-Merlot)*

IRVINE
Eden Valley
★☆

Established in 1980 by the indefatigable James Irvine, whose name has become synonymous with Merlot. Cheaper wines sold under the Eden Crest range.

✓ *Classic red blend (Eden Crest Merlot-Cabernet) • Merlot (Grand Merlot)*

JAMIESONS RUN
Coonawarra
★

At one time, this was the most underrated brand in the Beringer Blass stable. It is not quite the treasure trove of inexpensive goodies it used to be but can still come up with exceptional-value Chardonnay.

✓ *Shiraz • Chardonnay*

JIM BARRY WINES
Clare Valley
★★

Run by third generation Tom, Sam and Olivia, grandchildren of the eponymous Jim Barry. Their top Shiraz, The Armagh, is one of Australia's greatest wines. It is also one of the country's most expensive, and, at one-third of the price and almost as good quality as The Armagh, the McRae Wood Shiraz is better value. Virtually all other wines produced by Jim Barry are sold at bargain prices.

✓ *Cabernet Sauvignon (McRae Wood) • Chardonnay • Riesling (Florita, Watervale) • Shiraz (Lodge Hill, McRae Wood, The Armagh)*

KAESLER WINES
Barossa Valley
★★

Toby and Treena Hauppauff have shone with Shiraz since purchasing the Kaesler vineyard in 1990, helped no doubt by a

core of 120-year-old Shiraz vines. Now they are making top Cabernet Sauvignon.

✓ *Cabernet Sauvignon • Shiraz (Old Vine, Old Bastard)*

KATNOOK ESTATE
Coonawarra
★★

Part of the Wingara Group, which is majority-owned by Freixenet of Spain. With 330 hectares (815 acres) of vineyards in prime *terra rossa* soil, Katnook is an upmarket producer of intensely flavoured wines that display remarkable finesse, the very best being Cabernet Sauvignon, Shiraz, and Sauvignon Blanc. Between 1979 and 2005, Katnook wines won a remarkable 14 trophies, as well as 110 gold, 243 silver, and 645 bronze medals; they then stopped counting.

✓ *Cabernet Sauvignon (Odyssey) • Merlot • Riesling • Sauvignon Blanc • Shiraz (Prodigy)*

KILIKANOON
Clare Valley
★☆

Kilikanoon has been the talk of the town since 2002, when its Riesling, Shiraz, and Cabernet Sauvignon walked away with six out of seven trophies at the Clare Valley Wine Show.

✓ *Cabernet Sauvignon (Killerman's Run) • Grenache (Duke Reserve) • Riesling (Mort's Block) • Shiraz (Covenant, Oracle)*

KNAPPSTEIN
Clare Valley
★★

Today this winery is part of Yinmore Wines group, which bought the property from Accolade in 2019, with Mike Farmilo at the helm.

✓ *Cabernet Sauvignon (Enterprise) • Classic white blend (Sauvignon Blanc–Sémillon) • Riesling (Hand Picked) • Shiraz (Enterprise) • Sparkling (Chainsaw Sparkling Shiraz)*

LEASINGHAM
Clare Valley
★☆

Part of the Accolade group, Leasingham makes wines that offer outstanding quality and value. Bastion (sold on some export markets as Magnus) is a Shiraz-Cabernet blend, while Bin 56 is Cabernet-Malbec.

✓ *Cabernet Sauvignon (Classic Clare) • Classic red blend (Bastion Shiraz Cabernet, Bin 56) • Riesling (Bin 7, Classic Clare) • Shiraz (Bin 61, Classic Clare, Domaine)*

LECONFIELD
Coonawarra
★☆

Owned by Richard Hamilton (see Hamilton), whose uncle, Sydney Hamilton, established Leconfield in 1974.

✓ *Cabernet Sauvignon • Merlot • Shiraz*

LINDEMANS WINES
Coonawarra

Although this is the best-performing part of Lindemans' current operations, it is impossible to subdivide the brand according to location, particularly when Treasury Wine Estates is progressively dropping Lindemans' regionality in favour of state blends. Consequently, all this brand's best wines can be found in the Victoria section, where the overall score is ★☆, although the wines produced in Lindemans Coonawarra winery would easily rate ★★ at the moment. *See* Lindemans Wines, Victoria.

MAJELLA
Coonawarra
★★

The Lynn family had kept Merino sheep for four generations, when in 1968 George and Pat Lynn decided to invest in vineyards. The first 10 years were less than successful, with 30 tonnes or more of grapes left to rot on the vines in some years due to low demand. In 1980, however, they entered into an agreement to supply Wynns, a relationship that lasted for over 20 years, giving the Lynns confidence to strike out on their own. Cabernet Sauvignon and Shiraz are the standout wines here. The Malleea is a Shiraz-Cabernet Sauvignon blend.

✓ *Cabernet Sauvignon • Classic red blend (The Malleea) • Shiraz • Sparkling (Sparkling Shiraz)*

MITCHELL
Clare Valley
★☆

A small winery, Mitchell produces one of Australia's classic examples of Riesling.

✓ *Riesling (Watervale) • Shiraz (Pepper Tree Vineyard)*

MITOLO
Adelaide Plains
★★

Fast-rising star with award-winning wine made by Ben Glaetzer. The wines are mostly sourced from McLaren Vale, with (so far) one from the Barossa Valley.

✓ *Cabernet Sauvignon (Serpico) • Shiraz (G.A.M., Savitar)*

MOUNT HORROCKS
Clare Valley
★★

The quality and definition in style of these wines have really shone since the late 1990s, especially in the Riesling wines, which are world class. All Rieslings are highly recommended, but one – the Cordon Cut – should be explained. Its beautifully balanced sweet style is achieved by cutting, but leaving in place the cordon, which disconnects the grape clusters from the vine's metabolism, artificially inducing passerillage.

✓ *Classic red blend (Cabernet-Merlot) • Riesling • Sémillon*

MOUNTADAM VINEYARDS
Eden Valley
★☆

In 2006 David and Jenni Brown purchased Mountadam Vineyards from LVMH. This estate produces wines that are rich, yet lean, and of excellent quality. They come from vineyards that once belonged to Adam Wynn, son of the founder of the almost legendary Wynns

Coonawarra estate. The Red is a Cabernet Sauvignon-Merlot blend. Other brands include David Wynn and Eden Ridge.

✓ *Classic red blend (The Red) • Pinot Noir • Riesling*

NEPENTHE VINEYARDS
Adelaide Hills
★☆

In 1996 the Tweddall family became the second-ever vineyard owner to receive permission to build a winery in the Adelaide Hills (Petaluma being the first, almost 20 years earlier).

✓ *Chardonnay • Pinot Noir • Riesling*

NOON WINERY
McLaren Vale
★★

I don't know whether it's a trick of the mind, or if it really happened, but I have this memory of Drew Noon in 1991 at Cassegrain, where he was assistant winemaker, and he is telling me that he dropped the "e" from the end of his name because he didn't want to be "no one"! Since then, he has passed his Master of Wine examination, taken over his parents' winery, and caused enough of a stir with his massive, super-ripe wine oddities to catch the attention of critic Robert Parker. Eclipse is a Grenache-Shiraz blend, and VP is a vintage port style.

✓ *Cabernet Sauvignon (Reserve) • Classic red wine (Eclipse) • Fortified (VP) • Shiraz (Reserve)*

OCHOTA BARRELS
Adelaide Hills
★★☆

This husband-and-wife team started their own venture after many years making wine (and surfing) around the world. Their hands-off, minimalist approach and the characterful wines make this producer a wine list favourite.

✓ *Grenache • Chardonnay*

O'LEARY WALKER
Clare Valley
★☆

David O'Leary and Nick Walker started this enterprise in 2001, and the Rieslings have continued to be splendid.

✓ *Classic red blend (Cabernet Sauvignon Merlot) • Riesling (Watervale) • Sémillon (Watervale)*

PARACOMBE
Adelaide Hills
★

Owner-viticulturist-winemaker Paul Drogemuller produces one of Australia's best-value Cabernet Franc wines.

✓ *Cabernet Franc • Chardonnay • Sauvignon Blanc*

PARKER ESTATE
Coonawarra
★★☆

Sold by the Rathbone Family Group to the Hesketh family in 2013, the wines are now made by Andrew Hardy who used to work at Petaluma and Knappstein. The estate's small, low-yielding vineyard

produces super-premium wines, headed by First Growth, a blend of Cabernet Sauvignon and Merlot.

✓ *Cabernet Sauvignon (Terra Rossa) • Classic red blend (Terra Rossa First Growth) • Merlot (Terra Rossa)*

PAULETT'S
Clare Valley
★☆

I have always enjoyed Neil Paulett's beautifully focused white wines and am now rapidly becoming impressed with some of his reds.

✓ *Classic red blend (Cabernet-Merlot) • Riesling • Sauvignon Blanc • Shiraz*

PENFOLDS WINES
Barossa Valley
★★☆

The superb Grange (formerly Grange Hermitage) from Penfolds is Australia's most famous wine, created by the late and legendary Max Schubert. And through this single masterpiece Penfolds itself has become a living legend. Naturally enough, Grange is a very expensive product – expect to pay the price of a Bordeaux *premier cru* – but other far-less-expensive Penfolds wines are miniature masterpieces in their own right.

Some of their wines, however, such as Bin 707 and The Magill, have become very pricey (though nowhere near as expensive as Grange), and The Magill is, in terms of its origin, more Grange than Grange has ever been, being sourced entirely from the vineyard that many people once believed to be the sole source of Grange. It is now part of Treasury Wine Estate. RWT is a Shiraz of incredible finesse for its size. "RWT" stands for "Red Wine Trial", only the trial has long since stopped, and the name has stuck. Between Grange and Magill, Bin 707 is positioned as the Cabernet Sauvignon equivalent of RWT. Bin 138 is a Barossa Valley Shiraz-Mourvèdre-Grenache. Bin 389 is a Cabernet-Shiraz. Rawson's Retreat is the penny-pinching, decent-value label, but for just a little more, Koonunga Hill offers far better value for money, and even has the occasional real standout. Penfolds' Clare Valley operation is certified organic.

✓ *Cabernet Sauvignon (Bin 407, Bin 707) • Chardonnay (Bin 00A, Bin 98A, Yattarna) • Classic red blend (Bin 138, Bin 389) • Fortified (Grandfather Port) • Riesling (Eden Valley Reserve) • Sémillon (Adelaide Hills) • Shiraz (Grange, Kalimna Bin 28, Magill Estate, RWT)*

PENLEY ESTATE
Coonawarra
★★

Established in 1988 by Kym Tolley (whose mother was a Penfold and father a Tolley, hence the cheekily named Penley). Penley Estates leans towards Penfolds in style, as Kym's penchant is clearly for red wines, but he has also produced an excellent, rich, lush, toasty Chardonnay. Reserve and Phoenix Cabernet Sauvignons, however, are the absolute standouts here.

✓ *Cabernet Sauvignon • Chardonnay • Classic red blend (Shiraz-Cabernet) • Sparkling (Pinot Noir-Chardonnay)*

PENFOLDS GRANGE HERMITAGE, BIN 95, VINTAGE 1988, BOTTLED 1989 IN THE BAROSSA VALLEY
The Grange wines from Penfolds are made for ageing; do not drink them until at least 12 to 15 years after vintage. The best vintages can age for 20 to 50-plus years.

PETALUMA
Adelaide Hills
★★☆

When run by Brian Croser, Petaluma became a master of beautifully ripe, classically dry, top-quality Riesling that ages slowly and gracefully and one of the best exponents of sparkling wine in the New World (*see* Argyle Winery, Oregon). Not to mention one of Australia's best Merlots and a highly reputed Chardonnay. Nowadays, Mike Mudge is running the operation, carrying on Croser's legacy.

✓ *Entire range*

PETER LEHMANN
Barossa Valley
★★

Often underrated in Australia itself, Peter Lehmann's range of always excellent-value, mostly brilliant-quality wines is highly regarded in international markets. The Mentor is a Cabernet Sauvignon-Malbec-Shiraz blend, while the Seven Surveys is a Mourvèdre-Shiraz-Grenache blend.

✓ *Cabernet Sauvignon • Classic red blend (Art'n'Soul) • Fortified (The King) • Riesling (District Eden Valley) • Sémillon (Masters Margaret) • Shiraz (Masters Eight Songs)*

PEWSEY VALE
Eden Valley
★★

Forget the Cabernet Sauvignon, there is only one reason to buy Pewsey Vale, and that is for its absolutely classic Eden Valley Riesling. Even the "basic" Riesling is superb, but The Contour is magnificent.

✓ *Riesling*

PIKES
Clare Valley
★☆

Created by Neil Pike, formerly a winemaker at Mitchell, these excellent-quality

WORKING VATS STAND IN THE YARD OF OUTBUILDINGS AT ROCKFORD WINES IN THE BAROSSA VALLEY
This Barossa Valley producer founded its winery on the grounds of a 19th-century farm, which fits its traditional approach to winemaking. The original 1850s barn now houses the Cellar Door tasting room, and the winery buildings were built in the same style.

red wines lean less on weight than subtlety and elegance to achieve complexity. The Luccio Red is a Super Tuscan-type blend of Sangiovese, Merlot, and Cabernet Sauvignon.

 Classic red blend (Luccio Red) • Riesling • Shiraz

POWELL & SON
Barossa Valley
★★

Some strong relationships remained between Dave Powell and vineyard owners from the Barossa and Eden Valley after his departure from Torbreck. In 2014, when the time came to open his winery along with his son, Callum, that obviously was an advantage. The Powells manage vineyards organically, though not certified. They have two ranges of wines: the regional series and the single-vineyard series. The first one represents the good-value label with good wines at very accessible prices, while the second one is regarded as the top bracket, coming from six selected vineyards of old vines composed of Shiraz, Garnacha, and Mataro. Quality is very high and on par with Torbreck; prices too.

 Riesling • Shiraz (Kraehe Marananga, Loechel Eden Valley) • Grenache (Brennecke Seppeltsfield) • Mataro (Kleinig Greenock)

PRIMO
Adelaide Plains
★★ V

I cannot think why I was not wildly enthusiastic about these wines in previous editions. Owner-winemaker Joe Grilli produces some truly excellent wines, including the exquisitely rich, creamy-cedary Joseph sparkling red, its cult following being easily justified by the fact that it is regularly one of Australia's top two or three sparkling red wines. Although the best wines are sold under the Joseph label, don't ignore the basic Primo Estate white (La Biondina) and red (Il Briccone), which represent tremendous, everyday drinking value.

 Botrytis (Joseph La Magia Riesling) • Classic red blend (Il Briccone Shiraz-Sangiovese, Joseph Cabernet-Merlot) • Colombard (Il Biondina) • Sparkling (Joseph Red)

PUNTERS CORNER
Coonawarra
★★

This venture started life quite modestly in 1988, with 16 hectares (40 acres) of vineyards in Victoria and Coonawarra but now amounts to more than 150 hectares (370 acres), virtually all in Coonawarra, including 12 hectares (30 acres) surrounding the Punters Corner cellar sales (the wines are made under contract elsewhere). The basic Shiraz is stunning, but pales in comparison with The Track. The Cabernet is almost entirely Cabernet Sauvignon, with just 9 per cent (or thereabouts) of Merlot.

 Cabernet Sauvignon • Shiraz (The Track)

RESCHKE WINES
Coonawarra
★★

The launch of the first wine, the 1998 Empyrean, caused something of a stir. First, because the asking price was A$100 for a wine with no track record. Second, because its big, porty character was praised to heaven and back by those who believe big is best. The next two vintages, however, saw a substantial drop down in alcohol level: virtually 1 per cent per year, so that by 2000 Empyrean was a Bordeaux-like 12.9 per cent alcohol. Currently, the wine is more elegant, and possibly longer-lived, depending what you expect in a fully mature wine. The Vitulus is 100 per cent Cabernet Sauvignon, whereas the Empyrean has always included 10 per cent Merlot, but has also had 3 per cent Cabernet Franc since 1999 (thus remains within the limit of a so-called pure varietal, which may include up to 15 per cent of other grape varieties).

 Cabernet Sauvignon (Empyrean, Vitulus) • Shiraz

ROCKFORD WINES
Barossa Valley
★★

The very traditional reds from low-yielding vines might be a bit soupy for some, but the Black Shiraz will be admired by anyone who enjoys Australia's wonderfully eccentric, show-style, sparkling Shiraz. Winemaker Ben Radford took up the reins from owner Robert O'Callaghan in 2012 and introduced minimal intervention practices, which creates even more personality in the wines.

 Cabernet Sauvignon (Rifle Range) • Grenache (Moppa Springs) • Sémillon (Local Growers) • Shiraz (Basket Press) • Sparkling (Black Shiraz)

ROSEMOUNT ESTATE
McLaren Vale
★ V

This winery had produced incredibly successful wines in the past, but it seems under Treasury Wine Estates' (formerly Southcorp) direction volumes skyrocketed and the brand is now reduced to a supermarket label.

RUSDEN
Barossa Valley

In 2017 Christian and Amy Canute took over from Christian's parents and Rusden founders, Christine and Dennis Canute, where they make a range of red, white, and rosé.

 Shiraz (Black Guts)

RYMILL
Coonawarra
★☆

This property is owned and run by the descendants of John Riddoch, who planted the first Coonawarra vineyard in 1861. The MC2 is so named because it is a blend of Merlot and two Cabernets (Sauvignon and Franc), but it is also half of Einstein's famous formula, E = mc2, which demonstrates that energy equals mass and, as nuclear fission scientists would discover 30-odd years later, a minuscule mass can equal fabulously large amount of energy in some circumstances, so maybe there is also a more mystical meaning behind this label?

✓ *Cabernet Sauvignon • Classic red blend (MC2) • Shiraz*

ST HALLETT
Barossa Valley
★★

Established in 1944, the winery makes splendidly rich reds from old vines. Indeed, the quality has deepened over the time. SGT is a blend of Shiraz, Grenache, and Touriga.

✓ *Cabernet Sauvignon (Gamekeeper's) • Classic red blend (SGT) • Merlot • Riesling (Eden Valley) • Shiraz (Blackwell, Old Block)*

SALOMON
Fleurieu Peninsula
★☆

The Salomon family makes wine in the Finniss River Valley area and in Kremstal, Austria. Their super-premium Cabernet and Shiraz are made with the help of industry legend Mike Farmilo.

✓ *Shiraz*

SALTRAM
Barossa Valley
★★

Saltram has refocused on its red wines, for which it was once justifiably famous, and is again today.

✓ *Cabernet Sauvignon (Mamre Brook) • Chardonnay (Mamre Brook) • Classic red blend (Barossa Cabernet-Merlot) • Shiraz (Mamre Brook, No.1 Reserve, Graded McLaren Vale)*

SC PANNELL
McLaren Vale
★★

Stephen and Fiona Pannell started their own winery in 2004, after Stephen's long and hugely successful career making wine for others. Their large range of wines are all competently made.

✓ *Entire range*

SEPPELT
Barossa Valley
★★☆

Seppelt Great Western in Victoria is Treasury Wine Estates' sparkling wine specialist, while this Seppelt, in Seppeltsfield of the Barossa Valley, is its fortified wine specialist. Even though part of a corporate empire, Seppelt remains as priceless a piece of Australia's wine history as it has ever been, and Treasury Wine Estates has emphasized that by restricting this winery and cellars exclusively to fortified wine. Table wines used to be made here, but no longer – they are all sourced from Victoria. The entire range of Seppelt's fortified wine is recommended, none more so than the Para range of liqueur ports. Yet it is with these wines that I have my only quibble. Or, more accurately, with the bottles that contain the wines. Worst of all is the Para 100 Year Old Liqueur Port, vintages of which easily fetch between A\$1,000 and A\$2,000 a bottle, and yet is contained in a bottle that is a cross between a pretentious Cognac bottle and Mateus! What is wrong with a top-quality classic bottle shape?

✓ *Entire range*

SHAW + SMITH
Adelaide Hills
★★

This venture was set up in 1989 by Martin Shaw, who is well known for his flying winemaker activities, and Michael Hill-Smith, the first Australian to become a Master of Wine. After squatting at Petaluma for a decade, Shaw + Smith built their own high-tech winery at Balhannah in 2000. They have extended their activities to Tasmania with the successful Tolpuddle Vineyard brand established in the Coal River Valley.

✓ *Chardonnay (M3) • Sauvignon Blanc • Shiraz*

SKILLOGALEE
Clare Valley
★☆

Owner-winemaker David Palmer produces increasingly stylish wines from his 60-hectare (150-acre) property, particularly the Riesling.

✓ *Classic red blend (The Cabernets) • Riesling • Shiraz*

STANDISH
Barossa Valley
★★☆

This mailing list-only winery produces only a few different labels of Shiraz, but all of them are world class.

✓ *Entire range*

TAPANAPPA WINES
Wrattonbully

This must be one to watch: a partnership formed between Brian Croser, Champagne Bollinger (which was a major shareholder in Petaluma before the takeover by Lion Nathan), and Jean-Michel Cazes of Château Lynch-Bages in Bordeaux. This high-powered trio has purchased Koppamurra Vineyard in Wrattonbully. Tapanappa is named after a sandstone formation near Croser's beach house. Apparently, it is an Aboriginal word meaning "stick to the path". Their Chardonnay is succulent.

✓ *Chardonnay (Tiers Vineyard) • Pinot Noir • Classic red blend (Cabernet Sauvignon-Shiraz)*

TAYLORS
Clare Valley
★★☆

The basic Cabernet Sauvignon, Shiraz, and Riesling would probably be singled out at any other winery. They're not just some of the best bargains in Australia, but some of the country's greatest wines too. The St Andrews range is, however, in a different league, as is part of the relatively new Jaraman range. The best classic red blend is the Promised Land Shiraz-Cabernet, but it too cannot live in the company of those recommended below. Wines also sold under the Wakefield label (the Cabernet Sauvignon and Merlot can be brilliant).

✓ *Cabernet Sauvignon (Jaraman, St Andrews) • Chardonnay (Jaraman) • Riesling (Jaraman, Clare, St Andrews) • Shiraz (St Andrews)*

TIM ADAMS
Clare Valley
★★

The Tim Adams style is very pure, deeply flavoured, and bursting with fruit, no wine more so than the fabulous, pure lime-juice Sémillon.

✓ *Classic red blend (The Fergus) • Sémillon • Shiraz (Aberfeldy)*

TORBRECK
Barossa Valley
★★☆

Dave Powell made his first wine at Torbreck in 1994. A few years later, American critic Robert Parker's gushing accolades helped building the reputation of those iconic wines. Less than 10 years after his first vintage and due to financial issues, Powell sold a majority stake to Jack Cowin who owned Hungry Jack's. In 2008 American billionaire Pete Knight bought Torbreck from Cowin and included Powell's minority stake, giving him a chance to raise the funds to buy Torbreck back. Unfortunately for Powell, having failed to do that, the decision to force him out was made in 2013. He recovered from what he considered an unfair decision and started his own eponymous estate with his son (see Powell & Son).

The range and style of the wines at Torbeck have remained the same since Powell's departure. Both RunRig and Descendant are Shiraz-Viognier blends, but the former will set you back A\$300, whereas the latter costs a mere A\$125 to \$150 – so what is the difference? Well, RunRig contains just 3 per cent of Viognier, compared to 8 per cent in Descendant. Furthermore, the Viognier is vinified separately for RunRig, whereas both grapes are crushed and fermented together for Descendant, which makes the cheaper wine (if it is possible to describe Descendant as "cheaper") more classic in French terms. They are each from different vineyards, the Descendant vineyard having been planted from cuttings from RunRig, hence the name. Descendant spends 18 months in two-year-old to three-year-old French oak, whereas RunRig is given 30 months in French oak, 60 per cent of which is new. The Steading is Grenache, Mataro (Mourvèdre), and Shiraz, while Cuvée Juveniles is a full-throttle, fruit-driven, unoaked blend of Grenache, Shiraz, and Mataro. Juveniles is all old vines, but best drunk as young and as fresh as possible. RVM is a Roussanne, Viognier, Marsanne blend, with winning acidity balance. The Bothie is Torbeck's interpretation of a Muscat de Beaumes-de-Venise.

✓ *Classic red blend (Cuvée Juveniles, Descendant, RunRig, The Steading) • Classic white blend (RVM) • Grenache (Les Amis) • Fortified (The Bothie) • Shiraz (The Factor, The Struie)*

TURKEY FLAT
Barossa Valley
★★

Turkey Flat is a high-flying winery producing several exceptionally rich, lush, complex red wines from superbly sited vineyards, the core of which is more

than 150 years old. Butcher's Block is a blend of Mataro (Mourvèdre), Shiraz, and Grenache.

✓ *Cabernet Sauvignon • Classic red blend (Butcher's Block) • Classic white blend (Marsanne Sémillon) • Fortified (Pedro Ximénez) • Grenache Noir • Rosé • Shiraz*

UNICO ZELO
Riverland
★

This funky, dynamic winery takes their socio-economical responsibility very seriously. Unico Zelo grows mainly Italian varieties, with respect for the land and the people.

✓ *Truffle Hound (Barbera, Nebbiolo) • Esoterico (Zibbibo, Fiano, Moscato Giallo)*

WENDOUREE
Clare Valley
★★☆

This has been one of Australia's iconic wineries since the late 19th century. Since taking over in 1974, Wendouree's current CEO-winemaker, Tony Brady (aka "the arms and legs"), alongside Ashton Hill founder Steve George (aka "the palate"), has continued to produce some of this country's longest-lived reds. Shiraz is the jewel in Wendouree's crown, but given some bottle-age, Brady's wonderfully individual and expressive blended reds are near equals.

✓ *Cabernet Sauvignon • Classic red blend (Shiraz-Malbec, Shiraz-Mataro, Cabernet Sauvignon-Malbec) • Shiraz*

WIRRA WIRRA
McLaren Vale
★★

When Greg and Roger Trott took over Wirra Wirra in 1969, it was a wreck, literally. But the Trotts built it into a stylish winery making stylish wines, helped in the early days by that dynamic flying winemaker duo, Croser & Jordan (or should that be Jordan & Croser?). Jordan returned for a stint as managing director, when he grew restless, having achieved everything he had set out to do at Domaine Chandon, but returned to that fold when LVMH made him an offer he couldn't refuse. Whomever Greg Trott has had around the place, Wirra Wirra has always had an excellent reputation, but over the last few years the quality has become more exciting and deepened across the range.

✓ *Cabernet Sauvignon (The Angelus) • Chardonnay • Classic red blend (Grenache-Shiraz) • Fortified (VP Vintage Fortified Shiraz) • Grenache (McLaren Vale) • Shiraz (Church Block, McLaren Vale, RSW)*

WOLF BLASS
Barossa Valley
★

This company is now part of Beringer Blass, but its range is still huge, and the wines are marketed as aggressively as ever. Wolf Blass wines are mostly graded according to the colour of their label, which works well enough, although the Yellow Label tends to be more cheap than cheerful these days (the last exception being the 1999 Yellow Label Cabernet

Sauvignon, which really did punch above its weight). The Red Label is cheaper than the Yellow, and Eaglehawk cheaper still (but try the Riesling!). Contrary to what most people intuitively think, the pecking order after Yellow Label is Gold, Grey, Black, and, logically, at the very top comes Platinum. It is this role reversal of what the colours really mean that has been the genius of Wolf Blass marketing, and no Australian winery is more marketing-driven than Wolf Blass, but its best wines have to be among the country's most prolific award-winners for this to work, and that has always been so. By 1990, the total num-

ber of awards that Wolf Blass had garnered stood at 2,575, including an amazing 135 trophies and no fewer than 712 gold medals. That was 30 years ago. Who knows what the total is nowadays, but I can safely say that at the upper end of the Wolf Blass range, the quality has never been better. The Grey Label, which has been around since 1967, has replaced the Brown Label, but I've retained the Brown Label recommendations as the wines will be around for some time to come.

✓ *Cabernet Sauvignon-Shiraz (Black Label, Grey Label, Platinum Label, President's Selection)* • *Classic red blend (Black Label*

Cabernet Sauvignon-Shiraz, Grey Label Cabernet Sauvignon-Shiraz) • *Riesling (Clare Valley, Eaglehawk, Gold Label, South Australia)* • *Shiraz (Brown Label, McLaren Vale Reserve, Platinum Label, President's Selection)*

WOODSTOCK
McLaren Vale
★☆

Doug and Mary Collett acquired a 10-hectare (25-acre), derelict vineyard and an old cottage named Woodstock in 1973. Apparently, the Townsend family had settled at the site back in 1859,

emigrating from Woodstock in Oxfordshire, England, which was the origin of the name. They are probably all distantly related to the original settlers of this property. Five Feet is a blend of Cabernet Sauvignon, Shiraz, Merlot, and Petit Verdot. The name supposedly comes from the wooden stocks found in English Woodstock, which had five holes for five feet – pull the other one!

✓ *Red blend (Five Feet)* • *Shiraz (The Stocks)*

WYNNS
Coonawarra
★☆

John Riddoch is supposed to be the top wine here, but, superb as it is, the straight Black Label Cabernet Sauvignon, although disappointingly thin during the 1980s, has been the epitome of elegance ever since, and, at half the price of John Riddoch, represents much better value. The Michael Shiraz in on a par with the John Riddoch, in terms of quality and price, but the basic Shiraz is not the equivalent of the Black Label Cabernet Sauvignon, neither is it meant to be, as its much cheaper price indicates. Wynns is part of Treasury Wine Estate.

✓ *Cabernet Sauvignon* • *Chardonnay* • *Classic red blend (Cabernet Shiraz-Merlot)* • *Shiraz* • *Riesling*

YALUMBA
Barossa Valley
★★☆

This winery produces a vast range of styles and qualities, both still and sparkling. The Signature is a blend of Cabernet Sauvignon and Shiraz. Mesh is a Yalumba-Grosset collaboration. Oxford Landing is its penny-pinching, entry-level brand, offering very good value for money. Yalumba also owns Heggies Vineyard in Eden Valley and Nautilus in New Zealand.

✓ *Cabernet Sauvignon (The Menzies)* • *Chardonnay (Organic, Y Series)* • *Classic red blend (Barossa Valley Shiraz-Viognier, Hand Picked Mourvèdre-Grenache-Shiraz, The Reserve, The Signature)* • *Fortified (all Antique wines)* • *Riesling (Mesh, Block 44, Y Series)* • *Shiraz (The Octavius)* • *Viognier (Eden Valley, Y Series)*

YANGARRA
McLaren Vale
★★Ⓑ

Winemaker Peter Fraser focuses on bush vine Grenache on this biodynamic estate. Their Shiraz wines are just as good.

✓ *Old Vine Grenache*

ZEMA ESTATE
Coonawarra
★☆

This estate was established in 1982 by the Zema family, who currently cultivate 60 hectares (150 acres) of prime *terra rossa* land in Coonawarra. Cluny is a blend of Cabernet Sauvignon, Merlot, Cabernet Franc, and Malbec.

✓ *Cabernet Sauvignon (Family Selection)* • *Classic red blend (Cluny)* • *Shiraz (Family Selection)*

A VINTAGE GRAPHIC GIVES CHARACTER TO THE WINE ROOM SIGN AT THE YALUMBA WINERY
Located near the town of Angaston in the Barossa Valley, Yalumba was established by Samuel Smith, a brewer who emigrated from England. In 1849 Samuel and his son, Sidney, planted their first vineyards. Yalumba is a founding member of Australia's First Families of Wine (AFFW), an initiative to raise the global profile of Australian wines.

Western Australia

Western Australia possesses less than 10 percent of Australia's vineyard lands, and the wines produced show that this region is all about quality over quantity.

The state of Western Australia is the country's largest, though the vast majority of it is sparsely populated. Both winemaking and population are concentrated in the southwest corner, more than 3,000 kilometres (2,000 miles) from Sydney, the country's largest city on the east coast. This isolation ensured that the industry remained small throughout the boom of the late 1800s, and production levels rose slowly throughout the 20th century. The region has always been suited to agriculture, but the majority of its wine regions are relatively new.

Thomas Waters, a botanist who had learned his winemaking skills from Boers in South Africa, established the first winemaking vineyards in 1829 in the region of Swan Valley. This area remained the hub of the state's wine industry for the following 150 years, and, with an influx of Europeans, all experienced in viticulture and winemaking, the expertise of this industry grew. The focus, as in many other parts of Australia, was on fortified wine, which well-suited the hot climate of the established regions. In the latter half of the 20th century, however, the focus shifted towards table wines, and the cooler regions of Mount Barker, then Frankland, and last, but certainly not least, Margaret River emerged. Swan Valley remains a recognised GI, but as an already-hot region in an increasingly hot climate, the viability of the area is slowly decreasing.

LOOKING FORWARDS

Remarkably, Margaret River has made itself the most well-recognised region in Western Australia in under 50 years. Isolation remained a key factor in the success of this region; until recently it was a three-hour drive from the nearest major airport, meaning that it remained a well-kept local secret for some time, with only smaller boutique producers committing to production. In 2019 Lonely Planet named Margaret River as its No.1 destination to visit in the Asia-Pacific region. As a place that has built much of its reputation on a slow-but-steady growth, it may soon face new challenges relating to the likely influx of international interest.

WINE REGIONS OF WESTERN AUSTRALIA (see also p667) Situated at Australia's southwestern tip, this state's winemaking areas are remote from those to the east. The Margaret River region produces some of Australia's finest wines.

GREATER PERTH	Wine zone
Peel	Wine region
Denmark	Wine sub-region

0 mi — 50
0 km — 50

FACTORS AFFECTING TASTE AND QUALITY

LOCATION
Vine-growing areas sweep around the southwestern corner of Australia from Perth to Albany.

CLIMATE
The climate here is very variable, from the long, very hot, dry summers and short, wet winters of the Swan Valley, one of the hottest winegrowing areas of the world, through the Mediterranean-type conditions of the Margaret River, with a higher rainfall and summer heat tempered by ocean breezes such as the "Fremantle Doctor" and "Albany Doctor", to the even cooler Lower Great Southern Area, which also has some light rainfall in summer. Ocean winds can exacerbate salinity problems and high coastal humidity helps with the development of botrytis.

ASPECT
Most vines are planted on the relatively flat coastal plain and river valley basins, but also on some rather more undulating, hilly areas, such as those around Denmark and Mount Barker near Albany in the south and east of the region. Vineyards generally grow at an altitude of between 90 and 250 metres (295 and 820 feet) but can be as high as 340 metres (1,115 feet) in the Blackwood Valley.

SOIL
Soils are fairly homogenous, being mainly deep, free-draining, alluvial, sandy, gravelly, or clay loams over clay subsoils. Loam soils such as karri and marri are named after the gum trees that used to grow on them and have significantly different properties. The Geographe area has a fine, white-grey topsoil called tuart sand, over a base of limestone with gravel in parts of the Margaret River.

VITICULTURE & VINIFICATION
Drip irrigation is widespread in Western Australia because of the regions's general lack of summer rain and the free-draining nature of the soil, although, ironically, winter waterlogging due to clay subsoils is also a problem. Wide planting, mechanized harvesting, and the use of the most modern vinification techniques typify the vine-growing areas of Western Australia, which has generally concentrated on developing the cooler regions away from the Swan Valley in recent years. Well-equipped boutique wineries dominate the Margaret River wine region.

GRAPE VARIETIES
Primary varieties: Cabernet Sauvignon, Chardonnay, Merlot, Riesling, Sauvignon Blanc, Sémillon, Shiraz (Syrah)
Secondary varieties: Cabernet Franc, Chenin Blanc, Malbec, Muscat Gordo Blanco (Muscat d'Alexandrie), Muscadelle, Pinot Noir, Verdelho, Zinfandel

THE APPELLATIONS OF

WESTERN AUSTRALIA

ALBANY GI

One of five subregional GIs within the Great Southern region, Albany has a Mediterranean climate that is moderated by sea breezes known locally as the "Albany Doctor". Soils suitable for viticulture are quite patchy, and mostly confined to the slopes. Pinot Noir has something of a reputation thanks, primarily, to Bill Wignall (Wignall Wines), but Cabernet Sauvignon, Chardonnay, and Sauvignon Blanc probably perform at least as well.

BLACKWOOD VALLEY GI

This GI is the least known and one of the newest of Western Australia's wine regions, sandwiched between Manjimup and Geographe, to the east of Margaret River. The first vineyard was Blackwood Crest, planted by Max Fairbrass in 1976, and by 2004 there were more than 50 vineyards and 10 wineries. The climate is Mediterranean, with dry summers and wet winters. Vines grow at an elevation of 100 to 340 metres (330 to 1,115 feet) on gravelly loam soils, which are thinner and more gravelly, with some red earths, on the steeper slopes. Cabernet Sauvignon is the most widely planted variety and the most successful, with Shiraz on the increase. Other grapes include Chardonnay, Riesling, Sauvignon Blanc, and Sémillon.

CENTRAL WESTERN AUSTRALIA GI

This zonal GI covers a vast area of the Darling Scarp, from the Great Southern GI to north of the Swan District, but size is the only impressive aspect of what is currently a viticultural wasteland.

DENMARK GI

The newest subregion within the Great Southern GI, Denmark is situated some 60 kilometres (37 miles) along the coast from Albany. With its wet winters and warm to hot summers, the ocean's moderating influence, and a mix of marri and karri loam soils, the region's focus has rightly been on Chardonnay and Pinot Noir. Other varieties grown include Cabernet Franc, Cabernet Sauvignon, Sauvignon Blanc, Sémillon, and Shiraz. The etymological origin of this Denmark had nothing to do with the Scandinavian country, but was named after Dr Alexander Denmark, an English naval surgeon.

EASTERN PLAINS, INLAND AND NORTH OF WESTERN AUSTRALIA GI

Hardly rolls off the tongue, does it? I sincerely doubt that any wines will be sold under this zonal GI, which is the ultimate expression of bureaucratic logic, referring to all areas of Western Australia not already covered.

FRANKLAND RIVER GI

A small area on the western edge of the Great Southern zone, Frankland River was put on the map by Houghton. This GI is, however, something of a misnomer, because four rivers converge in this subregion: the Frankland, Gordon, Kent, and Tone. The climate is more Continental than Mediterranean, and, unlike the other subregions of the Great Southern zone, Frankland River has virtually no maritime influence. Surprisingly crisp Rieslings are made here. It is also good for Cabernet Sauvignon. Other grapes grown include Chardonnay, Sauvignon Blanc, and Shiraz.

GEOGRAPHE GI

Formerly known as the Southwest Coastal Plains, this regional GI takes in the curve of land on Geographe Bay, where the somewhat Mediterranean climate is cooled and humidified by the Indian Ocean. The region extends inland, where it creeps eastward into the foothills of the Darling Range. Geographe has a limestone subsoil, and is dissected by four rivers, the Capel, Collie, Ferguson, and Harvey, which have deposited alluvial soils that have a much higher nutrient content than the deep sandy soils in the coastal areas. Chardonnay is perhaps the most successful variety, followed by Cabernet Sauvignon and Merlot, with Sémillon and Shiraz on the rise.

GREAT SOUTHERN GI

This regional GI is the coolest of Western Australia's viticultural areas, and, although it has similar climatic influences to the Margaret River, it has a lower rainfall. The vineyards are scattered throughout a vast area and mostly consist of Riesling, Cabernet Sauvignon, Shiraz, Malbec, Pinot Noir, and Chardonnay. The subregions are Albany, Denmark, Frankland River, Mount Barker, and Porongurup.

GREATER PERTH GI

This zonal GI encompasses the regional GIs of the Peel, Perth Hills, and Swan District.

MANJIMUP GI

This GI is immediately north of Pemberton, on the same latitude as the Margaret River. Manjimup GI finalization was delayed until 2005 because some producers wanted to be part of Pemberton, but the two regions differ in both soils and climate. Manjimup is warmer, with more sunshine and less humidity, while the marri loam is less fertile, having more sand and gravel than Pemberton's karri loam. In addition to wines produced by local boutique wineries, the region also sells a lot of its fruit to other wineries in the state. Chardonnay is the most widely planted variety, but Cabernet Sauvignon and Merlot fare best. There is more hope than promise for Pinot Noir, even though Picardy, the region's top producer, makes excellent wines from this grape. Other grapes grown include Sauvignon Blanc and Verdelho.

MARGARET RIVER GI

This regional GI remains Australia's premier region for wine lovers who seek class and finesse, rather than weight and glory. Situated south of Perth, the Margaret River district attracted much attention in 1978 when it established Australia's first Appellation of Origin system. Like similar schemes, it was unsuccessful; however, the region has a number of well-recognised unofficial subregions that are perhaps, of all Australia's unofficial regions, the most likely to next gain GI status. The first vineyard was planted in the Margaret River area at Bunbury as long ago as 1890. Yet it was a vineyard planted by Dr Tom Cullity at Vasse Felix in 1967 that was the first step in the Margaret River's journey to success. The region follows a ridge that runs along the coast from Cape Naturaliste to Cape Leeuwin in the south, with Margaret River flowing westward through its centre, while the Blackwood River flows southwest to Augusta. The land is undulating, with vines planted up to a maximum altitude of 90 metres (295 feet) on gravelly, sandy loams, with the majority of vineyards and wineries located in and around Wilyabrup. The warm maritime climate is cooled by ocean breezes, with most rain falling in autumn and winter. Relatively minor problems do exist in the area, notably powdery mildew, parrots, wind, and, most serious, dry summers. The powdery mildew seems to be under control, and the vineyard workers plant sunflowers to distract the parrots from the vines, while rye grass acts as a windbreak. A lot of vines experience water-stress, not a heat-related problem, but dry-summer induced, and one that is exacerbated by the wind factor. The greatness of Margaret

A BOWER OF RAMBLING ROSES OFFERS SCENTED SHADE TO VISITORS OF THE LEEUWIN ESTATE IN MARGARET RIVER WHILE THEY LOOK OUT OVER THE VINEYARDS AND GROUNDS
The family-owned Leeuwin Estate is one of the founding wineries in the Margaret River GI, producing its first commercial vintage in 1979. Surrounded on three sides by the Indian Ocean, this Western Australian wine region benefits from the resulting maritime climate.

A VINTAGE LORRY LOADED WITH BARRELS IS ONE OF THE DISPLAYS AT THE SWANBROOK WINERY & CAFE IN THE SWAN VALLEY GI
In this hottest of Australia's wine regions, the number of wineries is decreasing; nonetheless, the Australian sense of humour and whimsey still shows in the grounds of the survivors.

River wines cannot be disputed. This quality of fruit is not bettered by any other Australian wine region. The best varieties are Cabernet Sauvignon, Chardonnay, Sauvignon Blanc, Sémillon, and Shiraz.

MOUNT BARKER GI

A subregion of the Great Southern zonal GI, where vines enjoy a Mediterranean climate and grow at altitudes of 180 to 250 metres (590 to 820 feet) on gravelly-sandy loams over gently undulating hills. Growers avoid the valley floors due to the salinity of their soils. Mount Barker has a reputation for its lime-laden Riesling and elegant red wines from Cabernet Sauvignon and Shiraz. Pinot Noir has also won awards. Other varieties grown include Cabernet Franc, Chardonnay, Malbec, Merlot, Sauvignon Blanc, Sémillon, and Shiraz.

PEEL GI

This large region takes in the Peel Inlet just south of Mandurah, where the vines enjoy a Mediterranean climate that is cooled and moderated by this large body of water, backed up by the Indian Ocean, although this effect diminishes farther inland. The soils are extremely diverse and range from deep, free-draining tuart sand on the coast to sandy-alluvium on the plains, with marri-jarrah loams, yellow duplex soils, and gravel farther inland. Although vines were planted at Pinjarra as early as 1857, and remained productive for 40 years, it was not until the 1970s that viticulture in this region underwent sustained development. Chenin Blanc and Shiraz are the primary varieties. Other grapes grown include Cabernet Sauvignon, Chardonnay, Merlot, Sémillon, Shiraz, and Verdelho.

PEMBERTON GI

This GI south of Manjimup is one of Western Australia's less consistent wine regions, but with vineyards established in the 1980s and 1990s, it is still a relatively young area and the locals are still confident. Pemberton is commonly known as "Karri country" after its magnificent forests of Karri gum trees, which give their name to the soil in which they grow. Karri loam is a deep, red, fertile soil that is too fertile for vines, causing excessive vigour problems, although growers overcome this by deliberate water-stressing and hard pruning to control the canopy. Pemberton was originally thought to be a Pinot Noir area, but Cabernet Sauvignon, Merlot, and Shiraz have been far more successful. Other varieties grown include Chardonnay, Sauvignon Blanc, Sémillon, and Verdelho.

PERTH HILLS GI

This regional GI is adjacent to the Swan Valley and consists of a strip of the lower slopes of the Darling Scarp. Although this area has a hot climate, it is higher and cooler than the Swan Valley, with grapes ripening some two weeks later. Cabernet Sauvignon and Chardonnay are the most successful grapes, with Chenin Blanc, Shiraz, and Pinot Noir the other primary varieties grown.

PORONGURUP GI

Set against a backdrop of the Porongurup Range, this picturesque region is one of five subregional GIs encompassed by the Great Southern region. The vines benefit from a Mediterranean climate and grow on granite-based karri loam soils. The primary varieties are Cabernet Franc, Cabernet Sauvignon, Chardonnay, Merlot, Pinot Noir, Riesling, Sémillon, Shiraz, and Verdelho.

SOUTH WEST AUSTRALIA GI

This zonal GI encompasses the majority of Western Australia's most exciting wine areas, including the regional GIs of Blackwood Valley, Geographe, Great Southern, Manjimup, Margaret River, and Pemberton.

SWAN DISTRICT GI

This regional GI contains the Swan Valley subregion itself, but it also extends northwards from Perth, encompassing former unclassified areas such as Gingin and Moondah Brook.

SWAN VALLEY GI

This subregional GI is located northeast of Perth and represents the heart and soul of the Swan District zone. The Swan Valley has the dubious distinction of being one of the hottest viticultural regions in the world. Partly because of this, and partly as a reaction to the phenomenal success of the Margaret River, several producers have deserted the area, and the number of Swan Valley vineyards is shrinking. What was once the traditional centre of Western Australia's wine industry is now a waning force, although the best areas are cooled by the so-called Fremantle Doctor wind, allowing the old-fashioned, foursquare wines to be replaced by lighter and fresher styles. If the Swan Valley can claim to make any classic wine whatsoever, it has to be the fortified wines made from the Muscat Gordo Blanco and Muscadelle grapes. Other important varieties from this region include Cabernet Sauvignon, Chardonnay, Merlot, Sémillon, and Shiraz.

WEST AUSTRALIAN SOUTH EAST COASTAL GI

This zonal GI covers a large swathe of coastal area that lies immediately to the east of the Great Southern zone.

WESTERN AUSTRALIA

ALKOOMI WINES
Frankland
★★

Sheep farmers Mervyn and Judy Lange established this venture in 1971, making their first wines five years later. They always used to produce fresh, fruity, and rich-flavoured wines that could be recommended with confidence but were seldom special, but since the mid-to-late 1990s the quality, style, and expressiveness have become ever more impressive. Sandy and Rod Hallett (Merv and Judy's daughter and son-in-law) have been behind the success of the winery since 2010. Blackbutt is a blend of Malbec, Cabernet Sauvignon, Cabernet Franc, and Merlot. Young, easy-drinking wines are sold under the Sutherlands label.

✓ *Cabernet Sauvignon (Black Label) • Cabernet Franc (Black Label) • Chardonnay (Victrix) • Sémillon (Wandoo)*

ASHBROOK ESTATE
Margaret River
★★

Great quality and even better value.

✓ *Chardonnay • Cabernet Sauvignon (Reserve) • Sémillon • Verdelho*

BLIND CORNER
Margaret River
★★

Ben and Naomi Gould have put everything they have to open their winery in 2005. Tired of seeing chemicals use in vineyards and during the wine-making process from other wineries, they have adopted a hands-off philosophy. It's straightforward, there is no additions of any kind, just a little sulphur.

✓ *Chenin Blanc (Quindalup) • Cabernet Sauvignon (Quindalup, Bernard) • Aligoté*

BROOKLAND VALLEY
Margaret River
★★

Malcolm and Deirdre Jones established Brookland Valley in 1984, with BRL Hardy purchasing a 50 per cent shareholding in 1997. The quality then went from good to excellent. Hardys moved to full ownership in 2004, and Brookland Valley is now part of Accolade Wines.

✓ *Chardonnay (Reserve) • Cabernet Sauvignon (Reserve) • Classic white blend (Verse 1 Sémillon-Sauvignon Blanc) • Merlot • Sémillon (Verse 1)*

CAPE MENTELLE
Margaret River
★★☆

Referred to by locals as the "Mentelle asylum", this is where David Hohnen started his cult-driven empire, which included New Zealand's Cloudy Bay phenomenon. He retired in 2003, to be succeeded by Dr Tony Jordan, who now oversees all of LVMH's operations in Australia and New Zealand. The quality remains exemplary, showing an uncanny ability to straddle both Old and New World styles, respecting tradition, and understanding restraint and finesse, yet revealing the best of Margaret River's ripe fruit flavours. This is the legacy of Hohnen's own philosophy, and it even applies to the Zinfandel, which under Cape Mentelle boasts infinitely more finesse than many a California icon of this variety. Marmaduke is primarily Shiraz-Grenache-Mataro (Mourvèdre), with a little Merlot and Pinot Noir. Trinders is Cabernet Sauvignon and Merlot, with a little Cabernet Franc and Petit Verdot. Georgina is Sauvignon Blanc and Chardonnay, with a little Sémillon and Chenin Blanc. Wallcliffe is a blend of Sauvignon Blanc and Sémillon.

✓ *Cabernet Sauvignon • Chardonnay • Petit Verdot (Wallcliffe) • Classic white blend (Wallcliffe) • Shiraz (Two Vineyards) • Zinfandel*

CAPEL VALE WINES
Geographe
★☆

Owner Peter Pratten, a former radiologist, was the driving force behind this expanding brand, which is known for its delicious, vibrantly fruity wines. It is now guided by son Simon Pratten.

✓ *Cabernet Sauvignon • Classic white blend (Sauvignon Blanc-Sémillon) • Merlot (Howecroft) • Riesling (Whispering Hill) • Shiraz*

CHERUBINO
Margaret River
★

Larry Cherubino and his wife, Edwina, started their own winemaking business in 2005. The venture now covers a wide range of wines, but his Rieslings and red and white blends always shine.

✓ *Laissez Faire range*

CLAIRAULT
Margaret River
★☆

Clairault became part of John H Streicker portfolio in 2012, making important vineyard management changes and taking the quality to new heights.

✓ *Cabernet Sauvignon (Estate) • Chardonnay (Estate)*

CULLEN WINES
Margaret River
★★

Di Cullen sadly passed away in 2003, leaving the winery that she and her late husband, Kevin, established in 1966 in the hands of her youngest daughter, Vanya. She could not have left the business in safer hands. Vanya Cullen has been the winemaker since 1989, and makes an outstanding range of wines, which get better with every vintage. If winemakers of the stature of Dr Andrew Pirie are eager to subject themselves to a masterclass with her, it is little wonder that she was voted Australia's Winemaker of the Year in 2000. Cullen is a biodynamic producer and, in 2006, became Australia's first carbon-neutral winery.

✓ *Entire range*

DEVIL'S LAIR
Margaret River
★★

Planted in 1981, and purchased by Southcorp in 1996, Devil's Lair takes its name from the nearby Devil's Lair cave, where the fossil remains of the so-called Tasmanian devil have been discovered. The Cabernets blend of Cabernet Sauvignon, Cabernet Franc, and Merlot is deep, dark, and yet finely structured, showing great potential. The Fifth Leg wines are more for ready drinking, but they are fine, elegant wines nonetheless.

✓ *Classic red blend (Cabernet-Merlot) • Chardonnay*

EDWARDS WINES
Margaret River
★★

Sensational wines, particularly reds, from brothers Michael and Christo Edwards.

✓ *Entire range*

EVANS & TATE
Margaret River
★☆

This winery once owned vineyards in the Swan Valley, as well as the Margaret River, but sold off its Swan Valley holdings. In 2003 Evans & Tate purchased the cash-strapped Cranswick Estates, but despite the latter's precarious financial situation, it fuelled Evans & Tate's profits by more than 40 per cent in the first year and in excess of 70 per cent the second year. The best wines at Evans & Tate combine richness with finesse.

✓ *Cabernet-Merlot (Hullabaloo) • Chardonnay (Broadway) • Merlot • Sémillon • Sémillon-Sauvignon • Shiraz (Hullabaloo)*

FERMOY ESTATE
Margaret River
★

The most consistent wines produced at the Fermoy Estate now are Chardonnay and Merlot.

✓ *Chardonnay • Cabernet (Estate Reserve)*

FERNGROVE VINEYARDS
Great Southern
★☆

Murray Burton ordered the first vines to be planted in 1996 and now has more than 400 hectares (1,000 acres) of vineyards, plus a winery built in 1999. Since then these wines have really taken off, reaching orbit with the 2002 Cossack Riesling, which won a record seven trophies and four golds. The winery has been controlled since 2011 by Chinese manufacturing and food company Pegasus.

✓ *Chardonnay (Orchid) • Cabernet Sauvignon (The Stirlings)*

FLAMETREE
Margaret River
★★☆

Every wine this family-owned winery has produced since its start-up in 2007 has been jaw-droppingly fabulous.

✓ *Entire range*

FLINDERS BAY
Margaret River
★

Established in 1995 as a joint venture between two families, the grape-growing Gillespies and wine-retailing Irelands.

✓ *Classic white blend (Sauvignon-Sémillon Blanc) • Merlot • Shiraz*

FRANKLAND ESTATE
Frankland
★

Frankland Estate is best for its blended red, which consists of Cabernets, Merlot, and Malbec. It is dedicated to the late Professor Olmo, California's famous grape breeder, not for creating varieties such as Ruby Cabernet, but because he was instrumental in selecting areas of Frankland best suited to viticulture in the early days of this wine region. Olmo's Reward is a blend of Cabernet Franc and Merlot primarily, supported by Malbec, Cabernet Sauvignon, and Petit Verdot. Poison Hill refers to the heartleaf plant growing there, which is poisonous to sheep and dogs, and thus was a threat to early settlers.

✓ *Classic red blend (Olmo's Reward) • Riesling (Isolation Bridge, Poison Hill)*

GALAFREY
Great Southern
★

Anyone who names a winery after the mythical planet of the Time Lords in Doctor Who should be locked up in a dark place – preferably with a good supply of Galafrey wines.

✓ *Cabernet Sauvignon • Chardonnay (Reserve) • Muller-Thurgau • Riesling • Shiraz*

GILBERTS
Great Southern
★★☆

The Gilberts have their wines made by contract at Plantagenet, which does a better job for them than it does with its own wines, which are good, but just not in the same class. This is a classic demonstration of the benefits of superior *terroir*. Best known to an informed few for their Riesling, the class of these wines can also be seen throughout the range.

✓ *Entire range*

GRALYN ESTATE
Margaret River
★★

The Graylyn Estate style leans more towards size than finesse, but you cannot help being seduced by the lush quality of these wines.

Cabernet Sauvignon (Reserve) • Fortified (Vintage Port) • Shiraz (Reserve)

HACKERSLEY

Geographe

★★

These wines have come a long way since Hackersley was established as a joint venture between three families. The quality is high, and the production is tiny.

 Cabernet Sauvignon • Merlot • Shiraz

HAPPS VINEYARD

Margaret River

★

Erl Happ makes several wines, all of which are eminently drinkable, but you can tell that he puts his heart into his reds, of which Merlot is the most highly regarded. Charles Andreas is a blend of Cabernet Sauvignon, Merlot, Cabernet Franc, Malbec, and Petit Verdot.

 Shiraz • Classic red blend (Cabernet Sauvignon-Merlot) • Merlot

HIGHER PLANE

Margaret River

★☆

Purchased by Roger Hill and his wife, Gillian Anderson, in 2006, their hands-off philosophy is proving them right.

 Chardonnay • Malbec • Classic red blend (Cabernet-Merlot) • Merlot

HOUGHTON WINES

Swan Valley

★★

Houghton wines are generally underestimated because of the winery's location in the underperforming Swan Valley, its huge production, and the fact that it is part of the even more gigantic Constellation, but this winery is an exemplary instance of how big can be beautiful. The entry-level wines offer great value for money. The range is divided between the Stripe, Reserve, Crofters, Small Batch, Wisdom and Icon Wines. At the top of the Houghton hierarchy is Jack Mann, which is a super-premium wine named after the late and legendary winemaker Jack Mann MBE. Jack Mann is usually a blend of Cabernet Sauvignon with maybe Malbec one year, Shiraz another. At one time it was a blend that included both of those ancillary varieties, and occasionally it can be pure Cabernet Sauvignon. The Crofters wines are a red blend made from Cabernet Sauvignon and Merlot.

 Cabernet Sauvignon (Gladstones) • Chardonnay (Wisdom) • Sauvignon Blanc (Reserve) • Shiraz (Thomas Yule) • Verdelho (Reserve)

HOWARD PARK

Denmark

★★

Howard Park is now owned by the Burch family, who also have vineyards in Margaret River, where they have built a new winery that has taken over some of the production. And lest the high rating mislead anyone, the production is very substantial indeed. Entry-level wines are sold under the Maimup label,

of which the Chardonnay, Sémillon-Sauvignon Blanc, and Shiraz are all recommended as strongly as the Howard Park wines listed below.

 Cabernet Sauvignon • Chardonnay • Classic red blend (Cabernet Sauvignon-Merlot) • Pinot Noir (Scotsdale) • Riesling • Shiraz (Scotsdale)

KARRIVIEW

Great Southern

★★

Karriview produces a tiny production of expressive Burgundian varietals.

✓ *Chardonnay • Pinot Noir*

KILLERBY

Margaret River

★☆

This winery's move to the Margaret River has coincided with a noticeable increase in quality.

✓ *Cabernet Sauvignon • Chardonnay • Shiraz*

LA VIOLETTA

Great Southern

★☆

Winemaker Andrew Hoadley makes tiny quantities of superb Shiraz and Riesling near Denmark since 2008.

✓ *Riesling*

LEEUWIN ESTATE

Margaret River

★★☆

Leeuwin Estate is one of the founding fives wineries of Margaret River (along with Cape Mentelle, Cullen, Moss Wood and Vasse Felix). For many years, winemaker Phil Hutchison produced lush, stylish, and very classy wines that belong with food. The current senior winemaker is Tim Lovett.

 Cabernet Sauvignon (Prelude, Art Series) • Chardonnay (Art Series) • Sauvignon Blanc (Sibblings) • Pinot Noir (Art Series) • Riesling (Art Series) • Sauvignon Blanc (Art Series)

LENTON BRAE

Margaret River

★☆

Established in 1983 by Bruce and Jeanette Tomlinson, whose son, Edward, is the winemaker today.

 Cabernet Sauvignon • Chardonnay • Classic red blend (Cabernet-Merlot)

MERUM

Margaret River

★★

Tiny production of stunning Sémillon, and a beautiful Shiraz.

✓ *Sémillon • Shiraz*

MOONDAH BROOK

Swan District

★☆

This brand belongs to Constellation and its good value, fruit-driven wines are made at the Houghton winery.

✓ *Cabernet Sauvignon • Chardonnay • Chenin Blanc • Shiraz • Verdelho*

RIPE CABERNET SAUVIGNON GRAPES AWAIT HARVEST IN SWAN VALLEY
Western Australia sees its harvest in March, which is autumn in the Southern Hemisphere.

MOSS BROTHERS

Margaret River

★☆

Established in 1984 by Jeff Moss, who had been a viticulturist at Houghton and Moss Wood, and Peter Moss, who actually built boutique wineries for a living. They were later joined by Jeff's daughter, Jane (graduate of winemaking and viticulture at Roseworthy) and son David, who had worked in various wineries in the Margaret River. With such experience, how could they fail? Well, they didn't, and Sémillon is this winery's flagship. The wines are made under three labels: Moss Brothers, Moses Rock (named after a wild stallion that roamed the area in the 1920s), and Jane Moss.

 Classic red blend (Cabernet Sauvignon-Merlot) • Sémillon • Shiraz • Verdelho

MOSS WOOD

Margaret River

★★

One of the very best Margaret River wineries, Moss Wood was the first in the area to perfect Pinot Noir, and arguably the best Sémillon outside the Hunter Valley.

✓ *Cabernet Sauvignon • Chardonnay • Merlot (Ribbon Vale) • Pinot Noir • Sémillon*

PICARDY

Pemberton

★★

Owners Bill and Sandra Pannell established Picardy in 1993 and pride themselves in producing excellent Pinot Noir, Chardonnay, and Shiraz on their sustainable vineyard.

✓ *Pinot Noir, Chardonnay, Shiraz*

PIERRO

Margaret River

★☆

Dr Michael Peterkin, who is a genius with white wines, makes one of the best examples of Chardonnay in Margaret River. His LTC is a blend of Sémillon and Sauvignon Blanc.

 Chardonnay • Classic red blend (Cabernet-Merlot) • Classic white blend (LTC)

PLANTAGENET WINES

Great Southern

★☆

A winery named after the shire in which it is situated, Plantagenet was the first to cultivate Mount Barker, and is the leading winery in the area today.

 Cabernet Sauvignon • Chardonnay • Classic red blend (Omrah Merlot-Cabernet) • Pinot Noir • Riesling • Shiraz

SANDALFORD
Margaret River
★☆

A historic winery dating back to 1840, Sandalford was purchased in 1991 by Peter and Debra Prendiville, who have successfully resurrected this great old name.

✓ *Cabernet Sauvignon (Premium) • Classic white blend (Premium Sémillon-Sauvignon Blanc) • Fortified (Sandalera) • Riesling • Shiraz (Premium) • Verdelho*

SETTLERS RIDGE
Margaret River
★

Wayne and Kaye Nobbs purchased this 40-hectare (100-acre) property in 1994. Originally part of a dairy farm in the heart of the Margaret River region, its gravelly loam soils were immediately planted with vines, which yielded a first crop in 1997. Since then, Settlers Ridge organic wines have won more than 50 medals at wine shows.

✓ *Cabernet Sauvignon • Shiraz*

SUCKFIZZLE
Margaret River
★☆

The creation of two well-known Margaret River winemakers, Stuart Pym and Janice McDonald. The seemingly bizarre name was inspired by the Great Lord Suckfizzle from Rabelais's Gargantua, whereas the wines themselves are simply inspired.

✓ *Cabernet Sauvignon • Classic white blend (Sauvignon Blanc-Sémillon) • Muscat (Stella Bella Pink) • Sauvignon Blanc (Stella Bella)*

VASSE FELIX
Margaret River
★☆

One of the Margaret River's pioneering wineries of the modern era, Vasse Felix's speciality has always been its long-lived Cabernet Sauvignons. It remains so today, although since the change of ownership, and the move to a new winery in 1999, they are softer and easier to drink at a younger age than they used to be. Paul Holmes à Court is the second generation of the family to own and operate Vasse Felix since 1987, and since 2008 the winery has been in good hands under winemaker Virgini Willcok. Heytesbury is a Cabernet-Shiraz blend. Classic Dry White is a Sémillon-Sauvignon Blanc blend.

✓ *Cabernet Sauvignon • Chardonnay (Heytesbury) • Classic red blend (Heytesbury) • Classic white blend (Classic Dry White)*

VOYAGER ESTATE
Margaret River
★★

The Voyager Estate property was purchased in 1991 by the late mining magnate Michael Wright. Wright's daughter Alex Burt, took over, and she continues making some excellent wines here.

✓ *Cabernet Sauvignon (Tom Price) • Chardonnay • Classic red blend (Cabernet Sauvignon-Merlot) • Classic white blend (Sauvignon Blanc-Sémillon, Tom Price Sémillon-Sauvignon Blanc) • Shiraz*

WILLESPIE
Margaret River
★

The Squance family continues to produce fine-quality Verdelho, but the expressiveness of Willespie's red wines, especially the Shiraz, has improved dramatically, which is a revelation.

✓ *Cabernet Sauvignon (Reserve) • Classic red blend (Cabernets) • Shiraz • Verdelho*

WOODLANDS
Margaret River
★★

The Watson family has been making wine at Woodlands since 1973. The business is now run by the second generation, but it still excels in the Cabernet Sauvignons and Bordeaux-style blends that made them famous.

✓ *Entire Range*

WOODY NOOK
Margaret River
☆

Neil Gallagher, who is the son of Woody Nook's original owner, the late Jeff Gallagher, has always made a lovely Cabernet Sauvignon, as well as a fine Sauvignon Blanc. In 2000, when the elder Gallagher retired, Peter and Jane Bailey became the proud new owners of Wood Nook, and the couple wisely kept on Neil as the estate's winemaker and viticulturist.

✓ *Cabernet Sauvignon • Sauvignon Blanc*

XANADU
Margaret River
★☆

Formerly Chateau Xanadu, this ex-boutique winery was bought by Rathbone in 2005. Glenn Goodall has been their chief winemaker since then.

✓ *Cabernet Sauvignon (Exmoor, Stevens Road, Museum)*

VASSE FELIX FROM THE AIR DISPLAYS ITS VINEYARD AND WINERY COMPLEX
Vasse Felix was the first vineyard and winery to be established in the Margaret River wine region. With its welcoming grounds, it has become one Australia's top winery tour destinations, offering a restaurant, cellar door, and art gallery.

Queensland and Northern Territory

Queensland and the Northern Territory both make wine, but neither state can claim conditions ripe for developing a truly thriving grape wine scene.

These two states cover nearly half of Australia but contain less than a quarter of its population. The Northern Territory has the lowest population density, and correspondingly low winery density, but Queensland has shown that it has more to offer than just surf and sun.

QUEENSLAND

For a place that is globally recognised for its tropical rain forests and white-sand beaches, Queensland, surprisingly, has over 100 wineries. Like the majority of Australian wine regions, the first vineyards were planted in the late 1800s, and fortified wine was the main style produced. Following the national trend of decline and revival, still wine production began to take off in the early 1980s, with two regions emerging as the main areas for high-quality production. Granite Belt and South Burnett both received GI status in the early 2000s, although the former is certainly considered to be the more suited to winemaking (*see* map of New South Wales, which includes Queensland, p673).

Summers are much cooler in Queensland's high-altitude Granite Belt than in many other, far more famous Australian regions, whilst South Burnett experiences less seasonal variation in temperature. In addition to these official GIs, there are a number of other emerging regions along the eastern coast which are recognised by Wine Australia as unofficial sub-regions of Queensland. This is particularly apparent in the traditionally ocean-facing tourist destinations of the Sunshine Coast and Gold Coast, where a number of cellar doors have sprung up in the hinterlands of the Great Dividing Range.

NORTHERN TERRITORY

An even less likely location for winemaking than Queensland is hard to imagine, but its western neighbour, Northern Territory, is such a place. Much of the state is arid desert or tropical rainforest, and commercial winemaking in the region has not been a success.

THE APPELLATIONS OF
QUEENSLAND

GRANITE BELT GI

Italian immigrants and their families pioneered viticulture and winemaking in the 1920s, selling to relations and friends in the Italian cane-growing communities to the north, but the current boutique-based industry started in the 1960s, when Shiraz was planted. This is the largest of Queensland's regions in terms of production, with the vines grown in an area that surrounds the town of Ballandean, on an elevated granite plateau 240 kilometres (150 miles) west of Brisbane. It is the altitude of this district, between 790 and 1,000 metres (2,500 and 3,280 feet), that provides a sufficiently cool climate. Indeed, at Felsberg, located at an altitude of 850 metres (2,750 feet), the grapes sometimes struggle to ripen. As can be imagined, the soil is granitic, with Riesling the most suitable variety, although there are a surprisingly large number of alternative varietals grown here, ranging from Fiano to Saperavi. Other varieties that do well include Cabernet Sauvignon, Chardonnay, Sémillon, and Shiraz.

GOLD COAST HINTERLANDS

Most wineries in this emerging region (not an official GI) are based around Mount Tamborine, around 45 minutes' drive inland from the famous seaside resort town of Gold Coast. The climate is cooler than at the coast, due to the elevation of the vineyards, with plantings mainly focused on Chardonnay, Sauvignon Blanc, Cabernet Sauvignon. and Shiraz. This is an area catering very specifically to the local tourist trade, with cellar doors being the main sales route.

DARLING DOWNS/TOOWOOMBA

Located in the southeast of the state, below South Burnett, this unofficial sub-region is centred around the town of Toowoomba and has a sub-tropical climate. Most wineries are of boutique size.

ROMA

Roma is not a GI wine region, and viticulture in this hot and arid area of Queensland is improbable. Historically it was known for the fortified wines produced by is solitary winery, Romavilla, which remained in continuous operation from 1866 until the mid-2010s. At the time of writing there are no operational wineries in the region, although it is still acknowledged by Wine Australia as an unofficial sub-region of Queensland.

SCENIC RIM

Another unofficial but emerging region; this area in the volcanic basin around the city of Brisbane was originally planted in the 1870s. As in many other parts of the country, the industry declined, and was non-existent for most of the 20th century. Winemaking in the area was revived in the 1980s and continues to be an added attraction for tourists.

SOUTH BURNETT GI

For anyone who has followed the tiny, emerging wine industry in Queensland, it might come as a surprise to discover that South Burnett was the first official wine region in this state. Observers might be forgiven for expecting that honour would go to the Granite Belt, which is Queensland's oldest and most famous wine region, yet it received GI status in 2002, almost two years after South Burnett. Vineyards in this regional GI were established in the early 1990s and currently stretch from Goomeri in the north, through the Murgon region to the Bunya Mountains, Kingaroy, and south to Nanango. Not surprisingly, the climate is hot, humid, and subtropical, with high rainfall during the growing season. There are various soil types – from red granitic-sandy soils on the plains and terraces, to basalt-based soils, light sandy soils, red-coloured, light clayey soils, and heavier brown and black clayey soils elsewhere. Grape varieties include Cabernet Sauvignon, Chardonnay, Merlot, Riesling, Sauvignon Blanc, Sémillon, and Shiraz.

SUNSHINE COAST HINTERLANDS

This new, unofficial region, at a latitude similar to South Burnett, but close to the coast, is another that has developed in the wake of a tourist trade hungry for winery visits to break up their beach-side holiday. The majority of wineries operate as combined cellar door and event spaces, and grow mostly Merlot, Pinot Gris, and Semillon.

THE WINE PRODUCERS OF
QUEENSLAND

BALLANDEAN
Granite Belt
★☆

This family-owned winery is justly famous for its cane-cut late-harvest Sylvaner, but its other wines are among the best in the Granite Belt, with the vintages assuming an even greater consistency since 2000, when Dylan Rhymer took on the day-to-day winemaking responsibility.

✓ *Chardonnay (Family Reserve)* • *Classic red blend (Generation 3)* • *Fortified (Liqueur Muscat)* • *Shiraz (Family Reserve)* • *Sylvaner (Late Harvest)*

CLOVELY ESTATE
South Burnett
★★

This ambitious 180-hectare (445-acre) project was established in 1997, and since then it has produced some of South Burnett's finest wines. It is owned by the Heading family, who wisely steered clear of hands-on management, electing instead to head-hunt a highly professional team to put the winery firmly on the map, which it has certainly done.

✓ *Botrytis (Left Field Sémillon)* • *Cabernet Sauvignon (Reserve)* • *Chardonnay (Left Field, Reserve)* • *Classic red blend (Shiraz-Merlot-Cabernet)* • *Petit Verdot (Left Field)* • *Sémillon (Left Field)* • *Shiraz (Double Pruned)*

GOLDEN GROVE ESTATE
Granite Belt
★★

The Costanzo family have lived here since the 1940s, when they established an orchard and fruit farm. The first vines, Shiraz, went in the ground in 1972, but a winery was not established until 1992, when Sam and Grace Costanzo quickly made a name for themselves. Their son Ray has made even greater strides with a wide range of wines, however, including many curveballs, all terrifyingly good. In addition to excellent classics, such as the Shiraz from the first vines planted on this estate, Golden Grove produces some outstanding, more eclectic wines, such as Barbera, Durif, Malbec, Mourvèdre, and Tempranillo, plus an intriguing classic red blend of Tempranillo, Durif, and Barbera. As the range widens, the quality just gets better and better.

✓ *Entire range*

JESTER HILL
Granite Belt
★☆

This winery was founded in 1993 by John and Genevieve Ashwell, who achieved a good reputation, particularly for Cabernet Sauvignon and Shiraz. Then, one day in March 2010, two motorcyclists, Michael and Ann Bourke, set off for the Granite Belt and came across Jester Hill. They loved the place but had no idea that they would be the new own-

ers nine months down the line. John Ashwell remains as winemaker, and two main labels are the same: Touchstone for the premium wines, and Two Fools for a bit of fun; apparently John and Genevieve used this name because you have to be fool to own a vineyard.

✓ *Cabernet Sauvignon (Touchstone) • Shiraz (Touchstone) • Sparkling wine (Sparkling Shiraz)*

MASON WINES
Granite Belt
★☆

The Mason family consistently produces some excellent wines from two former stone-fruit orchards.

✓ *Classic white blend (Sémillon-Sauvignon Blanc) • Petit Verdot*

PRESTON PEAK
Toowoomba
★

One of the more successful Queensland operations, Preston Peak draws on fruit from two different vineyards, the largest being at Devil's Elbow in the Granite Belt, where the winery is also located, but all other operations, including the

VINES SHOW AUTUMN COLOURS IN THE GRANITE BELT OF QUEENSLAND
Despite the image of Queensland as a tropical paradise, the Granite Belt region is cooler than a number of much more southern regions and even experiences snow in winter.

cellar door, are at the Preston Vineyard near Toowoomba, with glorious views of Table Top Mountain, Lockyer Valley, and the Darling Downs.

✓ *Cabernet Sauvignon (Leaf Series) • Chardonnay (Leaf Series) • Petit Verdot (Leaf Series) • Sauvignon Blanc (Leaf Series) • Sémillon (Leaf Series) • Shiraz (Reserve)*

ROBERT CHANNON
Granite Belt
★

Yet another winery that benefits from the services of consultant winemaker Peter Scudamore-Smith MW, Robert Channon goes to great lengths to ensure best practices throughout its vineyard, such as the installation of a permanent bird netting. The Verdelho still stands out.

✓ *Chardonnay • Pinot Gris • Verdelho*

SETTLERS RISE
Sunshine Coast Hinterland
★

Established in 1998, this winery has most of its vineyards in the Granite Belt, and operates under the guidance of Peter Scudamore-Smith MW.

✓ *Shiraz • Chardonnay • Sauvignon Blanc*

SIRROMET WINES
Granite Belt
★★

Established by the Morris family in 1998, Sirromet's avant-garde 200-seat restaurant was recognized in 2004 with numerous awards, including Brisbane's Best Tourism Restaurant, Brisbane's Best Restaurant in a Winery, and Brisbane's Restaurant of the Year. Part of this success is due, no doubt, to its Brisbane-friendly location, just 30 minutes southeast of the city. But not only has the restaurant maintained its incredible reputation, but so have the wines, which have gone from excellent to world-class. Australia's leading wine authority James Halliday attributes this, at least in part, to Adam Chapman, whom he believes to be "the most skilled winemaker practising in Queensland". A good part must also be down to the location of Sirromet's 144 hectares (356 acres) of vineyards in and around the Ballandean area of the Granite

Belt:, the St Jude's vineyard has 22 hectares (54 acres) growing Shiraz, Chardonnay, Verdelho, and Cabernet Sauvignon; The Night Sky has 23 hectares (57 acres) of Pinot Gris, Verdelho, Chardonnay, and Viognier), and Seven Scenes has 101 hectares (250 acres) in 40 separate blocks growing 17 grape varieties. Sirromet claims that Seven Scenes is Australia's highest-altitude vineyard and the source of many of its medal-winning wines.

✓ *Entire range*

SYMPHONY HILL WINES
Granite Belt
★★

Owned by the Macpherson family, who are obviously perfectionists, based on the finesse of these wines. Top quality across the board.

✓ *Entire range*

TOBIN WINES
Granite Belt
★☆

This family-owned eponymous winery was established in 1964, and the Shiraz and Sémillon that went in over the first three years are the oldest surviving vines in the Granite Belt.

✓ *Cabernet Sauvignon (Luella) • Chardonnay (Lily) • Dessert wine (Liqueur Muscat) • Merlot (Elliott) • Shiraz (Max)*

WITCHES FALLS WINERY
North Tamborine
★☆

Jon Heslop produces some elegantly rich wines from selected Granite Belt fruit in his winery just northwest of Gold Coast.

✓ *Cabernet Sauvignon (Prophecy Unfiltered) • Chardonnay (Prophecy Wild Ferment) • Grenache • Marsanne • Merlot (Prophecy) • Shiraz (Syrah)*

SETTLERS RISE
Sunshine Coast Hinterland
★

Established in 1998, this winery has most of its vineyards in the Granite Belt and operates under the guidance of Peter Scudamore-Smith MW.

✓ *Shiraz • Chardonnay • Sauvignon Blanc*

FACTORS AFFECTING TASTE AND QUALITY

LOCATION
Queensland is situated in the northeastern corner of Australia, with the Coral Sea to the east. The Northern Territory, the north-central state, is to the west.

CLIMATE
Annual rainfall at inland Roma is only 51 centimetres (20 inches), but the Granite Belt receives 79 centimetres (31 inches). Much of it tends to fall at vintage time and this can be a problem, but frost and hail pose a greater danger. Temperatures are high (similar to the Margaret River in Western Australia), but not unduly so, due to the tempering effect of altitude.

ASPECT
The altitude of Stanthorpe in the Granite Belt, where the surrounding vineyards lie between 750 and 940 metres (2,500 and 3,100 feet) above sea level, helps to temper the otherwise scorching summer heat in both areas. The vines here are generally grown in the hilly area, on sloping sites, whereas at South Burnett vines grow at 300–442 metres (990–1,459 feet) on more rolling countryside.

SOIL
The soils of Queensland are phylloxera-free. As the name suggests, they are granitic in the Granite Belt around Stanthorpe.

South Burnett varies from red granitic-sand with basalt on the plains and terraces, to basalt-based soils, light sandy soils, red-coloured light, clayey soils, and heavier brown and black clayey soils elsewhere.

VITICULTURE & VINIFICATION
Winegrowing in Queensland tends to be on a small-to-boutique scale, with a large number of producers focusing on providing a "winery experience" to the passing tourist trade, rather than large-scale production. Major seasonal flooding is an occasional but devastating risk to vineyards; several wineries were driven out of business by the 2011 and 2012 floods, including the well-recognised Romavilla Winery.

GRAPE VARIETIES
Primary varieties: Cabernet Sauvignon, Chardonnay, Riesling, Sauvignon Blanc, Sémillon, Shiraz (Syrah), Verdelho
Secondary varieties: Barbera, Cabernet Franc, Chambourcin, Chenin Blanc, Gamay, Malbec, Marsanne, Merlot, Mourvèdre, Muscat (Muscat Blanc à Petits Grains), Nebbiolo, Petit Verdot, Pinot Gris, Pinot Noir, Ruby Cabernet, Sangiovese, Saperavi, Sylvaner, Tarrango, Tempranillo, Touriga (Touriga Nacional), Viognier, Zinfandel

New Zealand

New Zealand's isolation and cool, maritime climate make it one of the most exciting places on earth for wine. For years, the country's incredibly successful industry has built itself on the backbone of Marlborough Sauvignon Blanc, somewhat eclipsing the many other regions and styles. There is a movement, however, even from a number of producers of "Marlborough Sav", towards sustainability as the true driver of the industry, rather than reliance on one single grape and region.

Whilst Sauvignon Blanc may not be considered the classiest of wines by all producers, it is the ideal choice to open up wider markets and has led the way for the rest of New Zealand's vinous bounty to make its presence known internationally. With a rising interest in Chardonnay, Pinot Noir, various Bordeaux varieties, Riesling, sparkling wine, Syrah, and alternative varieties, the breadth and depth of New Zealand's reputation will only continue to grow. The easy, quaffable fruitiness of Sauvignon Blanc provides an entry point for new wine drinkers, and for most it quickly becomes very specifically *Marlborough* or *New Zealand* Sauvignon. Producers no longer need to strive simply to have their country recognised as a wine-growing nation but can capitalise on the work already done by the Sauvignon Blanc and its fame, enticing drinkers to try something that is new but also familiar.

SUSTAINABILITY IN FOCUS

Despite the success of the Marlborough Sauvignon Blanc brand, there is some validity in the concerns raised by producers regarding the industry's "putting all its eggs in one basket". To this end, it would certainly be wise for producers to at least consider ensuring some security through diversity of production, and indeed this seems to be where the industry is headed.

SWNZ (Sustainable Winegrowing New Zealand) was set up in the 1990s in order to provide "a framework of efficient and economical viticultural and winemaking practices that encourage environmental stewardship". At the time of writing, around 98 per cent of all 700 wineries in New Zealand are certified by SWNZ, and the majority of those who are not members chose not to join because they are more stringent with their own sustainability practices, not less, and do not see the value in the additional paperwork required to qualify. New Zealand benefits from having a relatively cohesive industry, allowing an organisation like SWNZ to roll out changes and policies with relative ease. The organisation focuses not only on environmental concerns such as biodiversity and water use, but also extends its standards to cover people and business. It remains to be seen how well the industry fares when truly tested, but it is certainly reassuring to see some advance planning in place.

PAST, PRESENT, AND FUTURE

Up to the 1980s, the (now defunct) large companies Corbans and Montana reigned over the New Zealand wine scene, and Sauvignon Blanc was a rarity. It may be hard to believe now, but Müller-Thurgau dominated plantings until the 1990s; a well-timed vine-pull scheme instituted by the government in 1986 allowed most of the vines to be removed and replaced by Sauvignon Blanc, paving the way for its inexorable rise. Now Pernod Ricard NZ's Brancott Estate has subsumed both Corbans and Montana, and after Brancott Estate and Nobilo, the largest wine producer is Villa Maria. There exist a few other fairly big, well-known wineries, but the majority are very small operations, the number of which has rapidly increased, along with foreign ownership (around one-third of the industry at the time of writing), since the 1990s. The range of varietals has blossomed in their wake, as boutique producers have sought something

WINE-PRODUCING REGIONS OF NEW ZEALAND
Vine-growing in New Zealand spans both islands of the country. Although South Island has a younger wine industry, with its lighter rainfall, it boasts a slightly more favourable growing climate than North Island.

different with which to gain an edge. Nowhere has this been more evident than in red wine varieties. Bordeaux varieties, Pinot Noir, and Syrah are all excelling, and there is space for even more experimentation with varieties such as Tempranillo, Montepulciano, and Barbera.

The large range of New Zealand brands available even on export markets like the UK shows that ultimately small wineries can benefit from the larger companies, if the system is managed correctly. It is a role model that would certainly help small producers in other regions around the world, if only the large companies could be induced to follow suit.

FACTORS AFFECTING TASTE AND QUALITY

LOCATION
With the exception of Waiheke Island, all of New Zealand's grape-growing areas are on its two principal islands, North Island and South Island, stretched over 1,600 kilometres (1,000 miles) between latitudes 36°S and 45°S. Central Otago is the most southerly location.

CLIMATE
North Island generally has a cool maritime climate, similar to that of Bordeaux in temperature but with much higher rainfall. The crucial autumn periods are rarely dry; heavy rains and high humidity lead to problems of grape damage and rot. South Island is significantly cooler, but sunnier and drier. Marlborough is South Island's warmest area, and often has the country's most hours of sunshine. Rainfall is variable. Using the California heat summation system, the most important viticultural areas in both islands are all Region 1.

ASPECT
Most vines are planted on flat or gently sloping land and are easy to work. Some north-facing slopes have been planted in Auckland and Te Kauwhata; these provide better drainage and longer hours of intensive sunlight. Some steep vineyards are found in South Island's Central Otago district, and in general there is a move within regions towards exploring steeper hillside sites.

SOIL
Soils are varied, mostly clay or loam-based, often sandy or gravelly, with schistous loess over gravel in Central Otago, and volcanic subsoils in parts of Northland (Cottle Hill) and around Canterbury (Banks Peninsula).

VITICULTURE & VINIFICATION
Harvests begin in March and April, six months ahead of wine regions in the Northern Hemisphere. Canopy management is particularly important in many regions, both to control excessive vigour, and to protect the grapes from the strong UV light. Most winemakers have studied oenology in Australia and done at least one stage in Europe. This straddling of Old and New World tradition and technique has helped shape New Zealand's wine reputation. Over 90 per cent of all New Zealand wine is bottled under screwcap.

GRAPE VARIETIES
Primary varieties: Cabernet Sauvignon, Chardonnay, Merlot, Pinot Gris, Pinot Noir, Riesling, Sauvignon Blanc, Syrah
Secondary varieties: CCabernet Franc, Chenin Blanc, Gewürztraminer, Grüner Veltiner, Malbec, Müller-Thurgau, Muscat (Muscat Blanc à Petits Grains), Pinotage, Sémillon

THE WEKA PASS RAILWAY CHUGS TOWARDS THE WAIPARA WINE REGION
Vintage steam and diesel-electric locomotives take tourists on a journey through the through scenic limestone outcrops and past vineyards on their route through the Weka Pass.

THE APPELLATIONS OF
NEW ZEALAND

New Zealand instituted its first GI (Geographic Indication) regulations in 2017; prior to this, regional designations were generally recognised, but not official. As this is a relatively new initiative, there are still many GIs pending approval, especially for the numerous sub-regions within each larger GI. The regions listed below are official GIs, unless stated otherwise.

A note on names: A large number of regions and wineries utilise Māori words (one of New Zealand's three official languages, that of the indigenous population), and it is particularly common to come across the syllable *wai* in many names. It is much easier to recognise its relevance to wine once the meaning is known: "water." Of secondary note is that the letters "wh" are pronounced as "f" in Māori words.

NEW ZEALAND

The New Zealand appellation is usually found on exported cask wines (bag-in-the-box) and inexpensive blended wines sold on the domestic market. This is one of the three "enduring" GIs under the new registration scheme, meaning that it cannot be removed from the register, unlike other regional appellations that may be subject to change or review in future. The other two enduring GIs are North Island and South Island.

NORTH ISLAND

Supporting more than two-thirds of the population, North Island is where New Zealand's wine industry began and was confined to until 1973. North Island is technically an appellation in its own right, but seldom seen as such.

AUCKLAND

Registered sub-regions: Kumeu,
Matakana, Waiheke Island

Unofficial sub-regions: Clevedon, Henderson, Huapai,
Mahurangi, South Auckland

In the 1960s, a decade before the first vines were planted in Marlborough, Auckland possessed more than half of New Zealand's vineyards. Even though the area under vine has continued to increase, its percentage of the country's overall production has dwindled rapidly against the explosion of winegrowing on the South Island. Despite this minuscule contribution in terms of vineyard land, Auckland remains an important centre for the local wine industry, primarily because it houses the headquarters of the largest wine companies: Brancott Estate and Villa Maria both bottle wines here from all over the country.

Auckland's various wine districts produce some of the country's very finest wines. Clevedon's vineyards are sited on steep, north-facing hillsides, where Cabernet Sauvignon, Malbec, Merlot, Syrah, and even Montepulciano vines are low-yielding and mostly hand-picked. There is also a growing reputation for Chardonnay and Pinot Gris in dessert wine styles. In the 1960s the Henderson district in West Auckland was second only to Hawke's Bay as the country's largest wine producing region, and it is still the centre of New Zealand's fortified wine industry. The Kumeu district escapes some of the rain that falls on West Auckland and has long been established as a source of premium varietal wines, including Chardonnay, Merlot, and Cabernet Sauvignon, although some wineries source their fruit from farther south. Matakana is one of the three official sub-regions, located on the east coast to the north of Auckland. A small, but fast-growing area, it is known for its pinkish-red, iron-rich, granulated clay soil, upon which the local wineries have built a certain red wine reputation and are beginning to show promise for full, fruity white wines. Waiheke Island, also an official GI is much drier and sunnier than mainland Auckland and is one of New Zealand's most exciting districts for Bordeaux-style reds. Tiny quantities of Cabernet-Merlot have been produced on Great Barrier Island, which is way out in the Pacific, well beyond Waiheke Island. Transport costs have prevented this island from developing a wine industry with any aspirations for a national reputation.

GISBORNE

Unofficial sub-regions: Central Valley, Golden Slope,
Manutuke, Ormond, Ormond Valley, Patutahi,
Patutahi Plateau, Riverpoint, Waipaoa

Chardonnay is this region's major grape variety, accounting for over half of all plantings, much of this owned by Pernod Ricard. Most vineyards are located on the plains on either side of the Waipaoa River, and although the grape varieties grown have improved, this area is primarily flat and extremely fertile, thus it is inclined to churn out even the most classic of grapes at relatively high yields. Little wonder then that Gisborne regularly produces a higher tonnage of grapes than Hawke's Bay, despite having half the area under vine. There is no sense of place to most of the wines produced, since Gisborne Chardonnay averages over 100 hectolitres per hectare. It is, however, ideal for bulking out young, easy-drinking multi-regional blends, and this is one reason why only a small proportion ends up under the Gisborne appellation.

The number of smaller, boutique wineries is beginning to grow, however, and in the hands of quality-conscious growers, even vineyards on the flats can produce expressive Chardonnay. With moves into the hills overlooking the plains, we can expect to see more individual, higher quality Gisborne Chardonnay in future years.

More than one-third of Gisborne's vines are in flat-as-a-pancake Patutahi. To the south is Manutuke, the region's oldest winegrowing district (1894), where some of Gisborne's rare red wine varietals can be found, including Malbec and Pinot Noir. White grapes vary from Chenin Blanc (probably Gisborne's best) to Chardonnay, Muscat, and Riesling. On the other side of the plain is the area referred to as the Golden Slope, with a reputation for producing some of the best Chardonnays. Ormond, further to the north, is likewise known for Chardonnay. The two coastal regions of Riverpoint and Manutuke benefit from sea breezes that keep the temperature down.

Gisborne is essentially white wine country, where loose-cluster grapes perform best. Varieties with tightly packed clusters run the risk of rot, while most red varieties (less than 10 per cent of all Gisborne vines) are difficult to ripen. This may change as more early-ripening clones of classic red wine varieties become available, and if growers plant a thirst-quenching cover-crop between the rows to mop up as much of the diluting rain as possible.

HAWKE'S BAY

Registered sub-regions: Central Hawke's Bay

Unofficial sub-regions: Bridge Pa Tirangle, Crownthorpe
Terraces, Esk River Valley, Gimblett Gravels, Havelock
Hills, Korokipo, Mangatahi, Ngaruroro River Valley,
Ohiti, Te Awanga, Tukituki River Valley, Dartmoor
Valley/Tutuaekuri River Valley

Hawke's Bay is the driest wine region in the country and has the most diverse range of soils, including but not restricted to fine alluvium over greywacke gravel, hardpan clays, heavy silts, sandy loams, sandy loams over clay, sandy silt loams over gravel, stones over gravel, and stony gravels. Parts of the region were underwater less than 100 years ago, and only appeared as dry land after a major earthquake in 1931, adding further to the soil diversity. Hawke's Bay has always offered more potential for Cabernet Sauvignon and its derivative blends than Marlborough or any other New Zealand region, but it is wrong to class this region as only red wine country, let alone the sole province of Bordeaux varieties. It is at least as exciting for Chardonnay, and a handful of producers compete even with Marlborough's best Sauvignon Blancs, albeit in a richer, fuller style.

The two most important grape varieties are Chardonnay and Merlot, but the complex matrix of soils, the differences in aspect, and local variations in climate open up Hawke's Bay to myriad different grape varieties, some of which are yet to be planted. Roy's Hill on the northern side of the area known as the Bridge Pa Triangle is beginning to

establish its reputation for Syrah that is neither Australian nor Rhône in style, but a balance of both. Gimblett Gravels, just east of Bridge Pa, has a much rockier soil, as the name suggests, and internationally is the most well-recognised sub-region of Hawke's Bay – due in part to the excellent marketing of the local winemakers but also because of the quality of wines coming from this area. "Gimblett Gravels" is a trademarked name, initiated in 2001 by the producers in the area to bring greater recognition to their particular *terroir*, and has the unique attribute of being defined by a particular gravel soil type within an 800-hectare (1,976-acre) area, rather than a delineated block of land. Although not yet an official GI, this area would seem likely to be registered at some point in the future.

To the west of the Bridge Pa Triangle, inland and further upriver along the Ngaruroro is the up-and-coming area of Mangatahi. Growers believe that Mangatahi's greater diurnal extremes of temperature and cooler climate will enable it to establish a reputation as Burgundy country, in both red and white formats. On the opposite, northern bank is the even newer Crownthorpe area, best suited to early-ripening varieties.

The Tutaekuri River Valley has its vineyards laid out along the banks and hinterland of the Tuteukuri River. This is indeed Bordeaux country, favouring Cabernet Sauvignon on the lower reaches and Merlot farther upriver. Chardonnay also fares well, particularly between the Ngaruroro and Tutaekuri Rivers, at Ohiti and Korokipo. The Esk River Valley is north of Napier, and the climate is cooler than the more southern coastal sub-regions due to sea breezes drawn up the valley, although nights can be warmer. Soils are relatively fertile, although more gravelly towards the coast, where the wines take on more finesse. Big, bold Bordeaux-type reds are the hallmark of the area.

Sheltered by Te Mata Peak, the north-facing slopes of Te Mata's famous Coleraine vineyard at Havelock North, southeast of Hastings, are suntraps that have established over 30 years their ability to produce top-class Cabernet Sauvignon. The Te Mata Special Character Zone was established by the Hawke's Bay Regional Council in the mid 1990s, prior to the implementation of a national GI register, and was the first winegrowing area in New Zealand to be legally protected at any governmental level. The southernmost coastal district on the bay is Te Awanga, a coastal block of vines where some surprising big, bold Chardonnay and Bordeaux-style blends are grown. Central Hawke's Bay is the only registered sub-regional GI so far, but the name is slightly misleading; it may be central in terms of geography, but in relation to the rest of the viticultural areas, it is very much the most southern area. The hillsides offer an elevation of up to 300 metres (984 feet) for vines, and this region is best suited to Sauvignon Blanc, Pinot Gris, and Pinot Noir.

NORTHLAND

Unofficial sub-regions: Kaitaia, Kerikeri, Whangarei

It was in Kerikeri, Northland, that Samuel Marsden planted New Zealand's first vines in 1819. Shortly after, in 1833, James Busby, the so-called father of Australian viticulture, settled at Waitangi, where two years later he established his own vineyard. Despite these auspicious beginnings, local winemaking dwindled to almost nothing until quite recently. This is by far the smallest region in New Zealand, with only a handful of small producers catering almost exclusively to the local market of Auckland-based holidaymakers. Currently most vines are located in three districts: Kaitaia on the far northwest, the Bay of Islands (which contains the appellation Kerikeri) in the northeast, and around Whangarei, Northland's largest city. These areas boast the country's warmest climate, hence, Cabernet Sauvignon, Chardonnay, and Merlot are the three most widely planted varieties. They also have the highest rainfall and humidity, however, particularly on the west coast, and thus suffer fungal diseases and variations in yield. Sémillon, Pinot Gris, and Syrah are on the rise.

WAIKATO & BAY OF PLENTY (UNOFFICIAL)

Unofficial sub-regions: Coromandel Peninsula, Hamilton, Lake Taupo, Te Awamutu, Te Kauwhata

These two regions span the area south of Auckland, along the east coast and inland. Known more for dairy products than for wine, these create a large region containing relatively few wineries, and even fewer vineyards, as most of the producers source their fruit from Hawke's Bay or elsewhere. Of note is the Te Kauwhata region, where humidity levels are such that the conditions allow for the production of botrytised wine. The most famous names here are Morton Estate and Mills Reef.

WAIRARAPA

Registered sub-regions: Gladstone, Martinborough

Unofficial sub-regions: Masterton

This small region is located in the far south of the North Island, near the city of Wellington. Whilst Wairarapa is the official regional GI, it is the sub-region of Martinborough (GI) that is the most well-recognised. Martinborough rapidly achieved a name in the early 1990s as a potentially exciting area for Pinot Noir; and over the last few decades it has cemented its place as a rival to Central Otago for top Pinot Noir region in New Zealand. Central Otago certainly still wins on this front, but Martinborough has more to offer than just Pinot Noir. It is also one of the best places to grow Syrah and Sauvignon Blanc on the North Island, and in addition it is a fantastic location for Alsatian varieties such as Riesling, Pinot Gris, and Gewürztraminer.

Wairarapa contains one other official GI, Gladstone, and an unofficial one, Masterton. Gladstone is southwest of Masterton, north of Martinborough, where some of the first Wairarapa vines were planted. Gladstone is well suited to Riesling and Sauvignon Blanc in the main, whereas Pinot Noir is grown on the sunniest, north-facing slopes. This area is attracting a lot of interest; so too is Masterton itself, where the same varieties are planted, though many of the vineyards are really quite young.

SOUTH ISLAND

South Island has a much lower population than North Island and, due to transport difficulties, was not cultivated as a winegrowing region until 1973. It has quickly demonstrated exciting potential for premium varietals. Technically an appellation in its own right, it is seldom seen as such.

CANTERBURY

Registered sub-regions:
North Canterbury (under examination), Waipara Valley

This large region stretches along the eastern coast of the South Island, divided up by a series of west-east river valleys. The variation in soil type, climate, and elevation is quite significant throughout this region, although most vineyards are located in the plains surrounding Christchurch, New Zealand's third-largest city, or at Waipara, a coastal area to the north. Although summer here can be as warm as in Marlborough, autumn is cooler, and overall temperatures are lower. Canterbury receives some of the lowest rainfall in New Zealand, and irrigation is necessary in most sites. The flat Canterbury Plains are seen as somewhat lesser in comparison with the slightly warmer and more topographically diverse vineyards around Waipara, and the majority of the most exciting vineyards are located in this northern area.

Winegrowing in the region began as early as 1840 on the Banks Peninsula, but it has been stop-start here ever since. Waipara is the one classic winemaking area in the Canterbury region, with a variety of soils, including clays, gravels, and, in the promising Waikari-Weka Pass area to the north, brilliant-white, hard limestone. This encourages a number of different grapes to be planted. Currently, Pinot Noir, Chardonnay, Sauvignon Blanc, and Riesling fare best, but other varieties include Pinot Gris, Sémillon, and Gewürztraminer. Even Cabernet Sauvignon and Merlot are grown.

CENTRAL OTAGO

Unofficial sub-regions: Alexandra Basin, Bannockburn, Bendigo, Cromwell/Lowburn/Pisa, Gibbston, Wanaka

Home to the only continental climate in New Zealand, Otago is ringed by mountains on all sides. Even here, the ocean is a maximum of 100 kilometres (62 miles) away to the east and west, but the vineyards are protected from the ocean by the surrounding mountainous terrain, allowing winemaking to prosper in the southernmost region of the world. Central Otago is the most renowned quality wine region in New Zealand and is the most sought-after place in the country for vineyard investors. The cause of all this activity is the grape variety being grown. The region is still very much at the beginning of its existence, with the pioneers still making wine today. When they first started, no one knew what they should be growing in Central Otago. In fact, so confused was the issue that local growers would lobby every passing wine journalist for his or her opinion. With heat summation well below the supposed minimum for ripening grapes of 1,000°C degree-days, the growers assumed that it must be white wine country; but what variety, and whether to plump for sparkling, nobody was quite sure. If anything, there was a tendency to think of playing it safe with German varieties. No one ever imagined Central Otago could be red wine country, let alone a potential home for Pinot Noir, the Holy Grail of all red wine grapes. But since Central Otago Pinot Noir took off in the late 1990s, there has been no stopping it. A lesson, perhaps, for all those "fruit salad" regions throughout the world: find the right grape and create a reputation, rather than trying to be all things to all consumers.

So, what makes Pinot Noir work in heat summation-challenged Central Otago? Well, heat is not everything, and summation does not take account of the beneficial effect of diurnal temperature difference. Warm daytimes ripen the fruit, but the cooler it is at night, the more acidity is preserved, yielding brighter, more vibrant fruit. In Central Otago, the difference between the highest daytime and lowest night-time temperatures can be as much as 30°C (86°F). Furthermore, the days this far south are long, with little cloud and low rainfall, thus exceptionally sunny and dry. The schistous loess is a heavy-textured soil, which the Pinot likes, but schist tends to powder, rather than clay up, and thus provides excellent drainage. The fact that there is very little rot, due to low humidity (30 to 40 per cent), is a further benefit, keeping the vineyards clean and free from the chemicals that are often used to combat cryptogamic diseases and allowing a high proportion of organic/biodynamic viticulture.

This tight-knit region is mostly made up of small, independent wineries, as the majority of vineyards are restricted in size by the topography of the land and thus less appealing to a large conglomerate looking for a large plot. Cloudy Bay opened their own cellar door in Otago in 2018, however, a few years after investing in the region, and outside investment with a view to premium production seems to be a more viable entry point. Almost a dozen producers vinify their wines at the Central Otago Wine Company (COWCO), allowing these smaller growers to benefit from the equipment and experience of the larger facility whilst maintaining their unique labels.

Although trial blocks of vines had been planted as far south as Alexandra in the 1950s, it was not until 1976 that Central Otago's first commercial vineyard, Rippon, was established in Wanaka. And it took another five years for the first commercial vineyards to be planted in what would become Pinot country, between Gibbston (Gibbston Valley, 1981) and Alexandra (Black Ridge, 1981). Alexandra is the most southerly established wine area in New Zealand; yet, counterintuitively, it is one of the warmest wine areas in Central Otago. Some claim it to be the very hottest, but it is an area of extremes. Whereas the sunny, north-facing sites are indeed among Central Otago's warmest, other parts are too cool for any sort of viticulture. By contrast, Gibbston is

SHEEP GRAZE IN A MARLBOROUGH AREA VINEYARD
Vines were first brought to this region in the early 1970s, and since then Marlborough wines have gained a reputation as some of New Zealand's best, especially the zesty and flavourful Sauvignon Blancs.

A SELECTION OF WELL-REVIEWED NEW ZEALAND WHITES DEMONSTRATES THIS COUNTRY'S RICH HERITAGE OF WINEMAKING
From left to right: Cloudy Bay Sauvignon Blanc from Marlborough, which brought New Zealand's Sauvignon Blanc to international attention in the 1980s; Chardonnays from Martinborough (Wairarapa) and Kumeu River (Auckland); a single-vineyard Viognier from Gisborne; a Felton Road Riesling from Central Otago; and Wild Earth Pinot Gris from Otago.

one of the coolest districts in the region, and its Pinot Noir wines are just beginning to shine. More than two-thirds of all the vines planted in Central Otago are found between Lowburn and the Cromwell Basin, including Bannockburn, the most intensively planted area in the region. The regions are still at the stage of defining their styles, but based on the wines produced so far, there is enough diversity to warrant official recognition of these sub-regional divisions. Bendigo, located at the northern end of Lake Dunstan in the Cromwell Basin, is another warmer region that has only just begun to show its potential.

Despite the tendency to focus only on Pinot Noir when considering Central Otago, there are a number of white grapes that flourish and have shown a unique expression from this region. Riesling, Pinot Gris, and Chardonnay all do well, and there is an opportunity for some excellent sparkling from cooler sites.

MARLBOROUGH

Unofficial sub-regions: Awatere Valley,
Southern Valleys, Wairau Valley

Allan Scott planted the first vines here in 1973, and Marlborough quickly became New Zealand's most famous wine region, with Sauvignon Blanc its greatest asset. The part of Marlborough planted with vines is primarily the Wairau Plains between the town of Blenheim and the Waihopai River. Protected by the Inland Kaikoura Range from cold, southerly winds and by North Island from northeasterly winds, the vines also benefit from the combination of free-draining soils, abundant sunshine (the sunniest region in New Zealand!), cool nights, and a long growing season, resulting in the ideal environment to produce wines that are marked by electrifyingly bright fruit flavours. The flat plains of Wairau are almost entirely planted

to vines, as the area is easily mechanised and churns out the bulk of Marlborough's production. There is a move by smaller producers to planting vines on hillside sites; more labour intensive, perhaps, but able to provide wines of much greater complexity.

Awatere Valley is located to the southeast, just over the Wither Hills, where grapes often take one or two weeks longer to ripen and show intense flavour. The Southern Valleys region includes the Brancott, Omaka, Fairhall, Ben Morvan, and Waihopai Valleys, between Wairau and the Wither Hills. Soils here are slightly heavier than Wairau, with Pinot Noir succeeding in this area.

With such a rapid expansion and rise to fame, many Marlborough winemakers have sought ways to ensure the integrity of their brand is not compromised. Marlborough is the first region in New Zealand to have its own appellation "Appellation Marlborough Wine", that is not only a GI, but also requires adherence to certain criteria such as cropping parameters in order to be used on a label. Of note is the requirement for wine to be bottled in New Zealand, helping local businesses distinguish themselves from the many wines shipped in bulk and bottled at market. Another group known as MANA (Marlborough Natural) Winegrowers represents a slightly more exclusive club of producers that must, among other requirements, be certified organic and be estate bottled and owned. Clearly not all can agree on exactly the best way to go about protecting the Marlborough brand, but there is a certainly a general move in the same direction.

NELSON

Unofficial sub-regions: Moutere Hills, Waimea Plains

Nelson is overshadowed by the fame of Marlborough, on the other side of Mount Richmond, but not because of any lack

of quality potential. Nelson is known for its long, warm summers, cool autumn nights, and abundance of sunshine. The only negative factor is abundant rainfall, although that is nowhere near as high as in Auckland. The low profile has been due simply to there being no parcels of land of sufficient size available to attract any of New Zealand's larger wineries. Nelson has thus never benefited from any significant marketing, and the high price of land has not helped either. The region is slowly but steadily increasing in size, however, and, thanks to the Neudorf winery, is becoming especially famed for its Riesling and Chardonnay.

Many of Nelson's wineries, including most of the oldest established producers, are located on the Waimea Plains to the east of Richmond, where rich, fruity wines are made. The Moutere Hills are located to the west of the city of Nelson, and this is where the region's best Chardonnay is produced, alongside some aromatic Sauvignon Blanc and Pinot Noir.

WAITAKI VALLEY, NORTH OTAGO

This small region is characterised by its limestone soils, unlike any in the surrounding areas, and by its isolation, which is enough to discourage any significant investment. The region, which remains New Zealand's smallest, was first planted in 1998, and has remained boutique in size despite the promising quality of the wines produced. It is known especially for Pinot Noir, although Pinot Gris, Gewürztraminer, and Riesling also do well. Frost risk is a real danger to vineyards in this area, yields are low, and the difficulty of bringing a vineyard to ripeness is another reason why large-scale investment has not taken off. Ostler is the best-known producer from this region, with Central Otago's Valli Vineyards bringing the most significant outside influence.

NEW ZEALAND

BOTRYTISED WINES

Botrytis, or noble rot, appears on an unpredictable basis throughout all of New Zealand's wine regions. As such, it is very much up to the vintage and producer as to whether the wine is made in a botrytis style or not. Riesling is, without doubt, the premier grape for this style in New Zealand, producing wines of the most vivid flavour enhanced by a scintillating acidity to deliver a razor-sharp sweetness; a number of producers follow the German classification system (*Spätlese* TBA) when naming the sweetness of their wines, making it relatively simple to discern the style – for those who are familiar. Others find success with Chenin Blanc, Pinot Gris, Gewürztraminer, Sauvignon Blanc, Sémillon, even Viognier and Chardonnay.

🍷 *1–15 years*

👑 *Ata Rangi (Kahu Botrytis Riesling) • Dry River (Selection Riesling) • Fromm (Riesling Auslese) • Framingham (TBA) • Giesen (Marlborough Late Harvest Sauvignon Blanc) • Greywacke (Botrytis Pinot Gris) • Loveblock (Noble Chenin Blanc) • Villa Maria (Marlborough Noble Sémillon Botrytis Selection)*

CABERNET FRANC

Increasingly popular as a pure varietal wine, Cabernet Franc has always been considered a useful component in classic red wine blends. Hawke's Bay is the best region and Gimblett Gravels the best sub-region.

🍷 *2–5 years (up to 15 for exceptional wines)*

👑 *Clearview Estate • Man o' War (Ironclad Bordeaux Blend) • Pyramid Valley • Sileni (Pacemaker)*

CABERNET SAUVIGNON
Pure and Blended

Not so long ago, New Zealand suffered from a reputation for green, aggressive red wines. It was white-wine country, and its cool climate was thought to be too cool to ripen such thick-skinned grapes as Cabernet Sauvignon. Today, however, almost every wine region in New Zealand has become a gold mine for rich, ripe red wines of the most serious and sensual quality. Even volume-selling brands are big, full, and spicy, with more quality and character than you can find in Bordeaux for twice the price. The most exciting Cabernet Sauvignon wines are produced by

strict selection from top vineyards or crafted by smaller boutique wineries. Hawke's Bay is the best region, with a smaller area planted around Auckland.

🍷 *2–5 years (up to 15 for exceptional wines)*

👑 *Craggy Range (The Quarry) • Paritua (21.12) • Stonyridge (Larose) • Te Mata (Coleraine) • Trinity Hill (Prison Block) • Vidal (Legacy) • Villa Maria (Reserve Gimblett Gravels)*

CHARDONNAY

The world's most widespread classic grape really does produce something special in New Zealand, and whilst its plantings have been far outstripped by Sauvignon Blanc, many consider Chardonnay to carry the title for best-quality white variety. It is generally less predictable, more expressive, and capable of a slower, more classic rate of maturation than its Australian equivalent. This is due not only to cooler climatic conditions, but also to hand-picking, whole-bunch pressing, a wider use of natural (local) yeast fermentation, and less obvious oak. There is more weight and complexity in Hawke's Bay, greater finesse in Marlborough, lusher qualities in Nelson, and more upfront fruit in Gisborne. Central Otago still has to establish its style.

🍷 *1–5 years*

👑 *Ata Rangi (Craighall) • Dog Point (Chardonnay) • Esk Valley (Winemakers Reserve) • Bell Hill (Chardonnay) • Kumeu River (Maté's Vineyard) • Felton Road (Block 2) • Neudorf (Moutere) • Te Mata Estate (Elston) • Villa Maria (Reserve Barrique Fermented Gisborne)*

CHENIN BLANC

Despite promising wines from a small number of producers, the variety has never been widely planted here, and what's there has been declining of late. The style tends to be on the richer, more honeyed side, and can reward the adventurous consumer.

🍷 *1–3 years*

👑 *The Millton Vineyard (Te Arai)*

GEWÜRZTRAMINER

This is one of New Zealand's potentially exciting niche wines. The style has yet to take off, however, and is still only exported in miniscule quantities.

🍷 *1–3 years*

👑 *Dry River • Lawson's Dry Hills • Margrain • Brancott Estate ("P")*

MALBEC
Pure and Blended

Increasing at a rate similar to Syrah, Malbec does not have the same single-varietal cachet. It is judged more useful for its ability to add some mid-palate smoothness to classic red wine blends. A number of producers are beginning to vinify the grape separately, resulting in vibrant but as-yet undefined wines. Hawke's Bay is the most extensively planted to this variety.

🍷 *2–8 years*

👑 *Decibel (Gimblett Gravels) • Esk Valley (The Terraces) • Fromm (Fromm Vineyard) • Pask (Declaration)*

MERLOT
Pure and Blended

Merlot far outweighs Cabernet Sauvignon in plantings, with the vast majority planted in the Hawke's Bay area. Merlot is much lusher than any of the Cabernet varieties, but the New Zealand version shows more European structure than California Merlot. It is more common to find this grape making up the majority of a blend, rather than bottled as a single varietal. Hawke's Bay winemakers note that

BUNCHES OF PINOT GRIS RIPEN IN A CENTRAL OTAGO VINEYARD
Pinot Gris is a white wine grape variety, usually with greyish blue fruit, a hue that accounts for the "Gris" (French for "grey") in its name There is a wide colour range, though, and the fruit can be brownish pink to black and even white.

in hot years it is likely a blend will have a higher percentage of Cabernet rather than Merlot, so we may yet see Merlot plantings matched by Cabernet Sauvignon.

🍷 *2–8 years*

👑 *Alpha Domus (The Collection) • Craggy Range (Sophia) • Esk Valley (Winemaker's Reserve) • Pask (Declaration) • Pegasus Bay (Maestro) • Sileni (Exceptional Vintage) • Unison Vineyard (Unison Selection)*

PINOT GRIS

A variety that has seen a great rise in popularity over the last few years, this is now New Zealand's third-most-planted white grape. Many producers tend towards a more Alsatian style – rich and weighty, often with a touch of residual sugar – whilst others, generally aiming at larger volume, follow the more neutral Italian Pinot Grigio approach. There are plantings springing up all over the country, although the majority remains in Marlborough, and some of the best wines come from Central Otago and Martinborough.

🍷 *1–3 years*

👑 *Amisfield • Ata Rangi (Lismore) • Chard Farm • Greywacke • Dry River • Neudorf (Moutere) • Palliser Estate*

PINOT NOIR

The Holy Grail of all grapes was first thought to excel in Martinborough. Some excellent Pinot Noir wines have been and still are being made there, but the South Island has produced most of the best Pinot Noir wines on a consistent basis. First it was Nelson that showed the most promise; then it was Canterbury, followed by Marlborough (with around half of all Pinot Noir plantings) and, most exciting of all, Central Otago. Pinot Noir is by far the most widely planted red grape in the country, and interest seems to be only increasing.

🍷 *2–8 years*

👑 *Bell Hill • Ata Rangi • Craggy Range (Te Muna) • Felton Road (Block 3, Block 5) • Prophet's Rock • Burn Cottage • Neudorf (Moutere) • Rippon (Mature Vine) • Two Paddocks (The Last Chance) • Valli (Gibbston Vineyard)*

RIESLING

Truly dry Riesling is still a minority product in New Zealand, where most renditions of this variety have some residual sugar, even if the result is not a distinctive sweetness. New Zealand Riesling seems to be generally crisper than its Australian counterpart, with more citrus finesse. It does not have the same simplistic lime fruit and is not inclined to go petrolly very quickly either. In structure and style it leans closer to northern Europe, although not as close as the Riesling produced in Michigan or the Finger Lakes of the United States.

🍷 *1–5 years*

👑 *Domaine Rewa • Dry River (Craighall) • Felton Road • Framingham • Fromm • Mount Edward • Neudorf (Dry Moutere) • Palliser Estate • Pegasus Bay (Aria)*

SAUVIGNON BLANC

The fate of New Zealand's industry has long ridden on the unique properties of its locally grown Sauvignon Blanc. Certainly this country in general, but Marlborough in particular, has the rare capability to ripen Sauvignon Blanc slowly to a near-perfect state, enabling a large number of wineries to produce very fresh wines with ripe gooseberry fruit and such an electrifying balance of mouth-watering acidity that the flavours can intensify as grapefruit or passion fruit. Asparagus and peas, especially canned peas, are unwanted characteristics. The particularities of why exactly these characteristics are produced are still the subject of research, but essentially it comes down to a perfect combination of soil, climate, and latitude. Whilst Marlborough is overwhelmingly the most extensively planted area, Hawke's Bay, Nelson, Canterbury, and Wairarapa all have enough plantings to be able to claim their own regional styles.

These days there are two distinctive styles produced: the classic fruit-driven, early-drinking style, and a *"fumé blanc"* style matured in oak barrels and often using indigenous yeasts; the most famous of these is Te Koko, pioneered by Cloudy Bay in 1996 and leaving the way open for a more diverse use of Sauvignon Blanc. There is also an extensive amount of research going into low-alcohol styles, with the alcohol reduction enacted in the vineyard rather than the winery. New Zealand is very much at the forefront of this research, managing to produce their classic Sauvignon Blanc (usually around 12.5 per cent) at just 9 to 10 per cent ABV through canopy management techniques pioneered by Dr John Forrest of Forrest Wines.

🍷 *1–2 years (not older, except in a few particular wines made with the intention of aging)*

👑 *Clos Henri • Cloudy Bay • Craggy Range • Dog Point • Forrest (The Doctors') • Framingham (F-Series) • Greywacke (Wild) • Giesen • Seresin • Te Mata (Cape Crest) • Villa Maria*

SPARKLING WINES

At one time, Lindauer Brut and Rosé exemplified New Zealand's sparkling wine industry. They are very good for creamy, easy-drinking fizzes made by the transfer method, but it took *champenois* Daniel Le Brun to demonstrate Marlborough's true potential for serious sparkling wine. Meanwhile, Cloudy Bay's Pelorus is slowly developing a cult following to match that of its Sauvignon Blanc. When Hunter's released its deliciously fresh and easy-drinking Miru Miru, the international wine press was willing to concede the potential of Kiwi fizz. Quartz Reef now promises to put Central Otago on the sparkling-wine map, Kumeu River has demonstrated the potential of Auckland, but Marlborough still reigns supreme.

🍷 *1–2 years (from purchase)*

👑 *Kumeu River (Crémant) • Hunter's (Hunter's Brut, Miru Miru) • Deutz Marlborough (Cuvée Blanc de Blancs) • Cloudy Bay (Pelorus) • Quartz Reef (Méthode Traditionnelle) Brut)*

SYRAH

This could be one of New Zealand's greatest wines. The quality of this grape has come a long way since the first commercial release of Stonecroft Syrah, the earliest vintages of which were tight and green. Hawke's Bay is definitely the right place for this variety, where the Syrah grows dark and deep, developing a truly lush style that is set off by the classic cracked-peppercorn character. The best examples are neither Rhône nor Australian in style. The Waiheke Island sub-region of Auckland is another area well suited to this grape, again producing a vibrant style that is neither jammy nor lean.

🍷 *2–5 years (up to 15 for exceptional wines)*

👑 *Bilancia • Craggy Range (Le Sol) • Man o' War (Dreadnnought) • Schubert • Te Mata (Bullnose) • Trinity Hill (Gimblett Gravels) • Vidal (Legacy)*

THE WINE PRODUCERS OF

NEW ZEALAND

AKARUA
Central Otago
★★ Ⓥ

This Bannockburn winery produces beautiful Pinot Noir. The entry-level Pinot Noir is often better than most producers' top-of-the-line Pinot Noir.

✓ *Pinot Noir*

ALLAN SCOTT FAMILY WINEMAKERS
Marlborough
★☆

Allan Scott established Marlborough's very first vineyard in 1973, while working for Montana. He later became Corbans' chief viticulturist but left to set up the Allan Scott Family Winemakers label in 1990. Allan Scott produces a good, but not special, Pinot Noir, and has been persevering with a sparkling wine that frankly needs more finesse. But it truly excels at still white wines, which seem to have more fruit than wines from surrounding vineyards, particularly his Riesling.

 Chardonnay • Riesling • Sauvignon Blanc

ALPHA DOMUS
Hawke's Bay
★☆

Sometimes the wines have too much of a fruit-accentuated style, both reds and whites. They are never unfriendly; they just try too hard. Reds have become quite big in recent vintages. Now is the time for this producer to start thinking more about finesse rather than power.

 Classic red blend (The Aviator, The Navigator) • Chardonnay • Pinot Noir • Sauvignon Blanc • Sémillon (Leonarda)

AMISFIELD
Central Otago
☆

Amisfield makes wines of impressive acidity from vineyards in the Pisa Range foothills, just north of Lowburn. Only their Riesling disappoints.

✓ *Pinot Gris • Pinot Noir*

ASTROLABE WINES
Marlborough
★

Simon Waghorn is the winemaker of this husband-and-wife team, making a wide range of wines, including some delicious Albariños and Rieslings.

 Kerkerengu Coast Sauvignon Blanc • Albariño

ATA RANGI
Martinborough
★★ Ⓥ

Clive Paton consistently demonstrates that Martinborough can hack it with Central Otago's very best Pinot Noirs Ata Rangi also makes one of Martinborough's best Chardonnays.

✓ *Chardonnay (Craighall) • Pinot Gris (Lismore) • Pinot Noir (McCrone)*

BABICH WINES
Henderson
★ Ⓥ

Babaich is one of New Zealand's largest family-owned wine companies. The entry-level varietal wines of this large producer have never been exciting, but the top-of-the-range wines, such as Irongate and The Patriarch, always excel (they are two-star wines in their own right), and Babich's Marlborough wines have seen a recent climb in quality.

 Classic red blend (Irongate) • Cabernet Sauvignon (The Patriarch) • Chardonnay (Irongate, The Patriarch) • Pinot Gris (Marlborough) • Sauvignon Blanc (Cowslip Valley Marborough, Marlborough, Winemakers Reserve)

BELL HILL
North Canterbury
★★

The Bell Hill winery was established in 1997 by Marcel Giesen and Sherwyn Veldhuisen, who have planted 2 hectares (4.9 acres) of Chardonnay and Pinot Noir on limestone-derived soil in North Canterbury. The yields are low, the wines are stunning, but because the production is so tiny, prices tend to be high. Bell Hill applie biodynamic- and organic-growing practices in the vineyards and does not use additives in the winery.

✓ *Chardonnay • Pinot Noir*

BILANCIA
Hawke's Bay
★

Bilancia is Italian for "balance". This is the personal label of Trinity Hill's winemaker, Warren Gibson, and his partner and Bilancia's winemaker, Lorraine Leheny. They make some exceptional Pinot Gris, Chardonnay, Viognier, and Syrah.

✓ *Syrah*

BLACK RIDGE VINEYARD
Central Otago
★

In 2014 Verdun Burgess retired and handed over the winery to Joss Purbrick and Belinda Green. You can rely on good-quality Riesling and Gewürztraminer from his Alexandra vineyard.

✓ *Riesling • Gewürztraminer*

BRANCOTT ESTATE
Auckland, Gisborne, Hawke's Bay, and Marlborough
★★

Formerly Montana Wineries – Pernod Ricard NZ, this property was purchased by Allied-Domecq in 2001 and technically renamed Allied-Domecq Wines New Zealand in 2004. It was bought by Pernod Ricard in 2005 and again renamed, this time as Brancott Estate, five years later. The company owns 3,000 hectares (7,400 acres) of vines, and accounts for 55 per cent of all New Zealand's wines, making it the largest wine producer in the country, yet remarkable quality and consistency exist at all price points.

BROOKFIELDS VINEYARDS
Napier
★☆

Impressive quality red wines, particularly in recent vintages.

✓ *Classic red blend (Highland) • Syrah (Hillside)*

BURN COTTAGE
Central Otago
★★

Marquis Sauvage, a wine importer from Chicago, fell in love with Central Otago and decided to buy the property in 2002. He asked the help of Ted Lemon, who was the first-ever American winemaker and vineyard manager of a Burgundian estate and founder of Littorai on the Sonoma Coast. The site is truly spectacular, and the wines are too.

✓ *Pinot Noir*

CABLE BAY VINEYARDS
Waiheke Island
★

It might seem strange that the best wine from this Waiheke Island producer is a very fine Marlborough Sauvignon Blanc, but even its Waiheke Syrah is blended from various vineyards spread over the island.

✓ *Sauvignon Blanc*

CAIRNBRAE
Marlborough
★☆

This producer was purchased by Sacred Hill in 2001. The quality here remains high, and the value is even better. Cairnbrae concentrates on the white Sauvignon Blanc, Chardonnay, and Pinot Gris, along with red Pinot Noir.

CARRICK
Central Otago
★

With an organic vineyard located in Bannockburn, Carrick makes mainly Pinot Noir in a clean, rich, fruity, easy-drinking style. It also grows and makes Chardonnay, Riesling, Pinot Gris, and Sauvignon Blanc whites.

✓ *Pinot Noir*

CHARD FARM
Central Otago
★

One of the earliest viticultural pioneers this far south, Chard Farm enjoys a dramatic location on a spectacular terrace above the Kawarau River Gorge. It is approached along a rough-hewn track that bends around a mountainside, with the steep gorge on one side and barely enough room to drive a car. Some wines tend to rapidly develop a canned-pea character, but Mason Vineyard Pinot Noir is a banker.

✓ *Pinot Gris • Pinot Noir (Mason Vineyard) • Riesling*

CHARLES WIFFEN
Marlborough
★

Charles Wiffen is based at Cheviot in North Canterbury, his vineyards are in Marlborough, and his wines are made in Auckland by Anthony Ivicevich of West Brook. Rieslings have been nothing less than stunning, especially the Late Harvest. The Chardonnay and Sauvignon Blanc are also very good.

✓ *Chardonnay • Riesling (Late Harvest) • Sauvignon Blanc*

CLEARVIEW ESTATE WINERY
Hawke's Bay
★☆

Clearview Estate delivers unashamedly bold-flavoured wines crafted by founder, owner, and winemaker Tim Turvey, along with chief winemaker Matt Kirby and assistant winemaker Rob Bregmen. The red wines have always benefited most from Turvey's bold approach, and some would claim that Clearview's Chardonnay is one of New Zealand's very best.

✓ *Cabernet Franc • Chardonnay • Classic red blend (Old Olive Block)*

CLIFFORD BAY
Marlborough
★

An instant success from its very first vintage (1997), Clifford Bay's white wines have seduced critics and customers alike.

✓ *Chardonnay • Pinot Noir • Sauvignon Blanc*

CLOUDY BAY VINEYARDS
Marlborough
★★☆

Cloudy Bay is part of the French LVMH group. With 140 hectares (350 acres) supplying a fraction of its needs, even when newly planted vineyards come on stream, as well as acquiring Calvert and Northburn vineyards as its own in 2013 and 2014, this is a big business. Yet, it maintains an aura of a boutique winery, which has been essential to holding on to its perceived cult following and thus its relatively high prices (particularly on export markets). The Chardonnay is at least as good as the Sauvignon Blanc, but it is the latter that has made Cloudy Bay synonymous with Marlborough's marketing success. This is why the winery released its *barrique*-fermented Te Koko Sauvignon Blanc, rather than mess with the winning formula of its entry-level Sauvignon Blanc. Amongst its other offerings is Pelorus, a *méthode-tradition-nelle*) sparkling wine.

✓ *Chardonnay • Pinot Noir • Sauvignon Blanc • Sparkling wine (Pelorus)*

COOPERS CREEK VINEYARD
Huapai
★☆

Despite their penchant for quirky names in the Select Vineyards range (Cat's Pee on a Gooseberry Bush Sauvignon or Guido in Velvet Pants Montepulicano, for example), winemaking here is very serious, and the result has been award winning.

✓ *Chardonnay (Swamp Reserve) • Classic red blend (Merlot–Cabernet Franc) • Pinot Noir (Glamour Puss, Marlborough) • Riesling (Hawke's Bay, Marlborough Late Harvest, Nelson) • Sauvignon Blanc (Marlborough)*

CRAGGY RANGE
Hawke's Bay
★★☆

Craggy Range is owned by American billionaire and immigrant to Australia Terry Peabody. This might be one of New Zealand's most ambitious wine projects, with 285 hectares (700 acres) of vineyards and a winery that is a mix of technology and tradition and finely tuned to produce single-vineyard wines. All the wines show great finesse, from Sauvignon Blanc to Syrah, and everything between.

✓ *Entire range*

CROSSROADS WINES
Hawke's Bay
★☆

The flagship wine here is a classic red blend called Talisman. Crossroads is happy to confirm that it is composed of six different black grape varieties, but still refuses to divulge what they are, which has piqued many a New Zealand critic. The Destination Series is an entry-level range of 10 up-front, easy-drinking varietal wines. There also a Milestone Series and reserve-level Winemakers Collection. Their varieties include Sauvignon Blanc, Merlot-Cabernet, Syrah, Riesling, Gewürztraminer, Pinot Gris, Pinot Noir, and Chardonnay.

✓ *Chardonnay (Milestone Collection) • Classic red blend (Talisman)*

DELEGAT'S
Henderson
★☆

The style is generally quite restrained, with more elegance than richness, but seldom lacking depth or length. Marlborough wines are sold under the Delegat and Oyster Bay label

✓ *Merlot (Crownthorpe Terraces) • Sauvignon Blanc (Awatere Valley)*

DRY RIVER WINES
Martinborough
★★

Neil McCallum sold Dry River to New York investment manager Julian Robertson and Napa Valley vineyard owner Reg Oliver (see Te Awa) in February 2003 but stayed on as winemaker until he retired in 2011. Wilco Lam continues the values and philosophy established by Neil. Dry River is known for his Gewürztraminer and Pinot Gris, and it is producing one of New Zealand's most delicately perfumed dry Rieslings; the Pinot Noirs evolve beautifully, and the decadently rich botrytised wines are the stuff of legend.

✓ *Chardonnay (Botrytised, Craighall) • Gewürztraminer • Pinot Gris • Pinot Noir • Riesling (Botrytised, Craighall) • Syrah*

ESCARPMENT
Martinborough
★☆

If anyone put Martinborough Pinot Noir on the map, it was Larry McKenna, when he was the winemaker at Martinborough Vineyard. This venture (in partnership with the Kirby family) that so tantalized for Martinborough Pinot Noir lovers is living up to its promise.

✓ *Pinot Noir*

ESK VALLEY ESTATE
Hawke's Bay
★☆

Esk Valley red wines became increasingly impressive under the ownership of Villa Maria. In 2018 production shifted to a new winery in Gimblett Gravels, adding technological advancements to its artisanal approach. Its Malbec-Merlot-Cabernet Franc is one of New Zealand's most outstanding classic red blends.

✓ *Classic red blend (Merlot-Malbec-Cabernet Sauvignon, Merlot-Cabernet Sauvignon) • Chardonnay • Chenin Blanc • Verdelho*

FAIRHILL DOWNS
Marlborough
★

This vineyard has been planted since 1982, but the Small family did not sell their own wine until 1996. Since then they have developed a reputation for their bold fruit, balanced by crisp, ripe acidity. There are plans in place to convert Fairhall Downs to organics, using biodynamic principles.

✓ *Pinot Gris • Sauvignon Blanc*

FELTON ROAD

Central Otago

★★☆

This is the producer of New Zealand's greatest Pinot Noir, and a stunning medium-sweet Riesling. Other labels include Cornish Point (good value, easy-drinking Drystone Pinot Noir). The difference between Felton Road's top-performing Block 3 and Block 5 Pinot Noirs has more to do with the clones than the soil. These vineyards are biodynamic since 2003.

✓ *Chardonnay (Block 2, Block 6) • Pinot Noir (Block 3, Block 5) • RieslingWaikato (Block 1)*

FORREST WINES

Marlborough

★☆

An interesting range of fine wines is produced here, especially the soft, classic Sauvignon Blanc.

✓ *Pinot Noir (John Forrest Collection) • Cabernet Sauvignon (John Forrest Collection) • Riesling (Botrytised, Late Harvest) • Sauvignon Blanc*

FOXES ISLAND

Marlborough

★

Founded in 1992 by the great winemaker John Belsham, Foxes specialises in the two classic Burgundian grapes and makes a tremendous Sauvignon Blanc.

✓ *Chardonnay • Pinot Noir • Sauvignon Blanc*

FRAMINGHAM

Marlborough

★☆

Although Framingham produces a large and fascinating range of varietal wines, including a rare Montepulciano, its true speciality is Riesling in all its styles.

✓ *Riesling (F-Series) • Sauvignon Blanc*

FROMM WINERY

Marlborough

★☆

The vineyards here are organic, and they strongly believe in minimum intervention, regularly relying on wild yeasts and never filtering the red wines. Of course, the Sauvignon Blanc is very good, but the reds and the Riesling should not be overlooked, particularly the *Auslese*.

✓ *Malbec • Pinot Noir • Riesling • Syrah • Sauvignon Blanc*

GEORGES MICHEL WINE ESTATE

Hawke's Bay

★

Fine quality Chardonnay from a Beaujolais grower who now lives in New Zealand. There is also good Sauvignon under the Domaine Georges Michel label and fairish Pinot Noir. Since 2005 Georges' daughter, Swan, has headed the winemaking team.

✓ *Chardonnay*

GIBBSTON VALLEY

Central Otago

★

Grant Taylor's Pinot Noirs are becoming increasingly fleshy, and this stylistic progression has continued after the warmer vineyards in Alexandra and Bendigo came into production.

✓ *Pinot Noir (Reserve)*

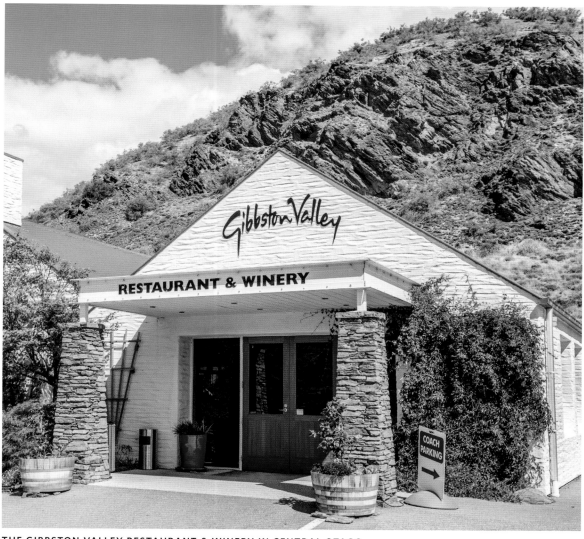

THE GIBBSTON VALLEY RESTAURANT & WINERY IN CENTRAL OTAGO
Like many wineries throughout the world, Gibbston Valley offers more than cave tours and wine tastings to attract visitors. Wine tourism is a lucrative business, and Gibbston Valley offers not only a restaurant, but also a lodge, spa, and even a bike centre for visitors who might want to cycle through the vineyard-filled surrounding area.

GIESEN WINES

Canterbury and Marlborough

★☆

Once the owners of the most important winery in Canterbury, the Giesen brothers have invested so heavily in a Marlborough winery and vineyards that this is now their primary location.

✓ *Pinot Noir (Glenlee) • Riesling (GV Collection, Red Shed)*

GOLDWATER

Waiheke Island

★★

When Kim and Jeanette Goldwater sold their winery to New Zealand Wine Fund in 2006, the production of one of New Zealand's greatest red wines, the Goldwater Cabernet-Merlot stopped. They are now focused on Burgundy grapes along with Sauvignon Blanc and Pinot Gris.

✓ *Pinot Noir • Sauvignon Blanc*

GREENHOUGH

Nelson

★☆

One of the Nelson area's longest-established cellars, Greenhough was established in 1991 by Andrew Greenhough and Jenny Wheeler. It is a small organic vineyard that produces wines just brimming with fruit. As well as being managed organically, it gained full BioGro certification in 2011.

✓ *Chardonnay (Hope Vineyard) • Pinot Noir (Hope Vineyard) • Riesling (Hope Vineyard) • Sauvignon Blanc*

GREYWACKE

Marlborough

★★

This top-class venture is by Kevin Judd, who established the label in 2009 with his wife, Kimberley. The grapes are sourced from mature vineyards, and Dog Point winery provides space for vinification. They specialise in Sauvignon Blanc and Pinot Noir, but their small production of aromatic varietals is also excellent.

✓ *Wild Sauvignon Blanc • Botrytis Pinot Gris*

GROVE MILL

Marlborough

★

In 1988 a group of local grape growers and wine enthusiasts purchased the Malt house (The Mill), a historic landmark in Grovetown, Blenheim, to establish a winery. In 2006 Grove Mill became New Zealand's (and the world's) first carbon-neutral winery, thanks in part to its "smart cellar". This involves energy-efficient technology to draw cold night air into the winery and high-performance insulation to keep temperatures low during the day. There is also a heat-exchanger to supply the entire complex with hot water.

✓ *Chardonnay • Riesling • Sauvignon Blanc*

HERON'S FLIGHT

Matakana

★

David Hoskins and Mary Evans used to make splendid Cabernet Sauvignon, but they caught the Italian bug. By 2004 they had replaced all the French varieties with Italian vines, specialising in Sangiovese and Dolcetto.

✓ *Sangiovese*

HERZOG
Marlborough
★☆

These wines might seem expensive, but if a Michelin-starred restaurant in a classic European wine region sold the wine, its customers would not blink at such prices. And that's not an unreasonable comparison, for although Hans Herzog makes the wines, his wife, Therese, runs the winery's restaurant, drawing on her Michelin-starred culinary experience in Switzerland. They have made a name for their artisan winery, handcrafting wines from their single-estate organic vineyard.

✓ *Classic red blend (Spirit of Marlborough)* • *Montepulciano*

HIGHFIELD
Marlborough
★

In 2015 Highfield and TerraVins merged, and those two wineries are run by three formers employees. The Highfield TerraVins winery is brightly coloured Tuscan-styled, and the restaurant is overlooking the Omaka Valley. The rich and smooth Pinot Noir is the wine of most consistent style here.

✓ *Pinot Noir* • *Sauvignon Blanc*

HUIA VINEYARDS
Marlborough
★☆

Claire and Mike Allan are white wine specialists, making only one red, a Pinot Noir. The white wines are all hand-picked and whole-bunch pressed, which works particularly well for Sauvignon Blanc and an improving sparkling wine.

✓ *Sauvignon Blanc* • *Riesling*

HUNTER'S WINES
Marlborough
★★

This large, established, and expanding winery is run by Jane Hunter, a gifted viticulturist whose medal-winning standards have earned her an OBE, the respect of her neighbours, and the very first IWS Women in Wine Award. Gary Duke is her longstanding winemaker.

✓ *Chardonnay* • *Riesling* • *Sauvignon Blanc (Kahi Roa)* • *Sparkling wine (Miru Miru)*

HYPERION WINES
Northland
☆

In Greek mythology, Hyperion was one of the 12 Titans – the Titan of light. I'm not quite sure what light is shining out of John Crone's winery, a former cowshed north of Auckland, but of all his wines, which are named after one Titan or another, his Asteria Syrah is the one that most deserves to be set free.

✓ *Syrah (Asteria)*

ISABEL VINEYARD
Marlborough
★☆

For 40 years, this family-owned winery, nestled into the heart of the Wairau Valley in Marlborough, has been consistently producing one of New Zealand's most impressive Sauvignon Blancs. Watch out for the occasional release of Noble Sauvage, a Sauternes-like botrytised version of Marlborough's famous Sauvignon Blanc varietal.

✓ *Pinot Gris* • *Pinot Noir* • *Sauvignon Blanc*

JACKSON ESTATE
Marlborough
★☆

At one time, Jackson Estate consistently produced Marlborough's greatest Sauvignon Blanc. The quality is still just as high, but a number of wineries now make Sauvignon Blanc that is at least as fine, and which of them produces the best of the crop varies from year to year. But Jackson Estate excels in quality and value throughout the range.

✓ *Chardonnay* • *Pinot Noir* • *Riesling* • *Sauvignon Blanc*

KAHURANGI ESTATE
Nelson
★

This is the original Seifried winery and 10.5-hectare (26-acre) vineyard. It was sold in 1998 to the current owners, Amanda and Greg Day, who renamed it Kahurangi, which is Māori for "treasured possession". Kahurangi might be a recent name, but it boasts the oldest commercial vineyard in the region. And with a new 12-hectare (30-acre) vineyard just a kilometre away (on the northwest-

THE WINE TASTING ROOM AT THE MISSION ESTATE WINERY AWAITS THE ARRIVAL OF VISITORS
French Catholic Marist missionaries established this estate in 1851 for the production of sacramental wine. It is now New Zealand's oldest-surviving winemaking company.

ern side of Upper Moutere, planted in 2001), production is expanding fast.

✓ *Chardonnay • Riesling • Sauvignon Blanc*

KIM CRAWFORD
Hawke's Bay
★★

One of New Zealand's most experienced, talented, and award-winning winemakers, Kim Crawford has had his own winery since 1996. It was eventually bought by Constellation Brands in 2006. Since then, the production has more than doubled and the range has been simplified.

✓ *Sauvignon Blanc • Chardonnay • Pinot Noir*

KUMEU RIVER
Auckland
★★

Master of Wine Michael Brajkovich regularly produces one of New Zealand's greatest Chardonnays, but his reds are just as good. Excellent value, easy-drinking wines are sold under the Kumeu River Village label.

✓ *Chardonnay (Maté's Vineyard) • Classic red blend (Melba) • Pinot Gris • Pinot Noir*

LAKE CHALICE WINES
Marlborough
★

In 2016, the Lake Chalice wine brand, including vineyards, was been purchased by the family-owned Marlborough winery, Saint Clair Family Estate. Top-of-the-range wines are sold under the Platinum label.

✓ *Chardonnay (Platinum F V) • Sauvignon Blanc (Marlborough)*

LAWSON'S DRY HILLS WINERY
Marlborough
★☆

This Marlborough winery can produce intensely flavoured white wines. The Gewürztraminer may be a bit too pretty-pretty, but it is highly regarded by New World standards. The Sauvignon Blanc is my favourite.

✓ *Chardonnay • Gewürztraminer • Riesling • Sauvignon Blanc*

LINCOLN
Henderson
☆

Most critics would put Lincoln's Chardonnay first, but I prefer the richer reds, with their minty finesse and fruit-cake complexity. The Chardonnay Heritage Patricia is alarmingly overwhelmed by coconutty American oak.

✓ *Classic red blend (Home Vineyards, Vintage Selection)*

MAMMOTH
Nelson
★★

Michael Glover spent nine years at Bannockburn Vineyards following. Before coming back to his native Nelson, he decided to spend some time in Europe around vineyards in Italy, France, and Germany. Charismatic and uncompromising, Michael is one of

those free-spirited individuals who will challenge the conventional thinking. His single-vineyard Pinot Noir and Sauvignon Blanc are sensational.

✓ *Sauvignon Blanc • Pinot Noir*

MAN O' WAR VINEYARDS
Waiheke Island
★☆

Although 60 hectares (150 acres) of vineyards were planted amid 4,500 hectares (11,120 acres) of rugged clifftop farmland on Waiheke in 1993, this family venture only came to international attention around 2005. Its vineyards are scattered, with a variety of *terroirs,* that allow winemaker Duncan Mctavish to vary his blends. Man O' War is the only winery with a beachfront tasting room on Waiheke

✓ *Classic red blend (Dreadnought, Ironclad)*

MARGRAIN VINEYARD
Martinborough
★

This vineyard was planted in 1992 by Graham and Daryl Margrain, who made their first wine in 1995. Out of the Margrain's range, the Pinot Noir is the better wine.

✓ *Chardonnay • Pinot Noir • Riesling (Botrytis Selection)*

MARTINBOROUGH VINEYARDS
Martinborough
★★

This winery's reputation was established by winemaker Larry McKenna, followed by Claire Mulholland (ex-Gibbston Valley); nowadays Paul Mason is in charge and has proved more than a match.

✓ *Chardonnay • Pinot Noir (Reserve) • Riesling*

MATAKANA ESTATE
Auckland
☆

Matakana's most successful winery keeps steadily improving, focussing on the best varieties for the *terroir* in which they grow.

✓ *Chardonnay • Pinot Gris*

MATARIKI WINES
Hawke's Bay
★☆

Although owners John and Rosemary O'Connor planted this vineyard as long ago as 1981, they did not start making and selling wines until 1997. It is now owned by the Taurus Group, whose mission statement is to produce "fruit-driven wines of elegance and complexity" using a "hands-off" approach to viticulture, and a "hands-on" approach in the winery. It has proved to be a very successful winemaking philosophy.

✓ *Classic red blend (Quintology) • Sauvignon Blanc • Syrah*

MATAWHERO WINES
Gisborne
★☆

Owners Kirsten and Richard Searle bought this legendary property in 2008. Kim

Crawford acts as a consultant and has helped them to revive this dormant beauty.

✓ *Gewürztraminer • Sauvignon Blanc • Chardonnay • Merlot*

MATUA VALLEY
Waimauku
★

Matua Valley is located northwest of Auckland but produces wine mainly from three other regions: Hawke's Bay, Gisborne, and Marlborough. This is a very large producer, with a high-tech winery that regularly produces a range of fine, expressive wines that are a pleasure to drink. The Lands & Legends label is reserved for Matua Valley's very best wines. At the other end of the price spectrum, the Regional Range concentrates on exceptional value, fruit-driven, entry-level wines. Excellent Marlborough wines are sold under the Taste Range label.

✓ *Chardonnay • Sauvignon Blanc • Syrah*

MILLS REEF WINERY
Hawke's Bay
★☆

Father-and-son team Paddy and Tim Preston produce deliciously rich, fruity whites, and some stunning reds from Bordeaux varietal reds and Syrah.

✓ *Classic red blend (Elspeth) • Malbec (Elspeth) • Merlot (Elspeth) • Syrah (Elspeth)*

MILLTON VINEYARDS & WINERY
Gisborne
★

This biodynamic producer is justly famous for its award-winning, lightly botrytised, medium-sweet Opou Riesling. It also produces a stylish, barrel-fermented Chardonnay and is appreciated locally for Chenin Blanc.

✓ *Chardonnay (Essencia, Opou Vineyards) • Chenin Blanc (Te Arai) • Pinot Noir (Clos de Ste Anne)*

MISSION ESTATE WINERY
Taradale
★

From a wine perspective, the Society of Mary fell into a rut for a while, which is understandable considering it has been producing wine here for more than 150 years (making it New Zealand's oldest winery). Now things are back on track, with good-quality and value wines.

✓ *Cabernet Sauvignon (Reserve) • Chardonnay (Reserve) • Riesling*

Bay of Plenty

MOUNT EDWARD
Central Otago
★

Owner-winemaker Alan Brady was the original proprietor of Gibbston Valley, where he was the first person to plant grapes in what is now the viticultural boom area of southern Central Otago. John Buchanan is now the proprietor of the unique winery, and more grapes have been planted while the winemaking remains hands-off.

✓ *Pinot Noir • Riesling • Grüner Veltliner*

MOUNT RILEY
Marlborough
★☆

Owned and operated by John Buchanan, his daughter Amy Murphy and her husband Matt Murphy, Mount Riley boasts of a "fiercely independent spirit and a commitment to making wines of excellent value". This award-winning winery has 100 hectares (250 acres) of vines spread over six vineyards.

✓ *Chardonnay • Pinot Noir • Sauvignon Blanc*

MT DIFFICULTY
Central Otago
★☆

With no fewer than six vineyards and a heap of experience in South Africa, Italy, and Oregon, a lot was expected from Matt Dicey. His talent and vision have served the wines well, and the Pinot Noirs are the highlight here. And their sustainability program is inspiring. The winery was sold to Foley Family Wines in 2019, which is slowly building an empire. Let's hope that the quality will remain high.

✓ *Pinot Noir*

MUDDY WATER WINES
Waipara
★☆

Waipara is Maori for "muddy water", but there is nothing murky about the pure fruit in these wines. Muddy Water was the first certified organic winery in Waipara. The Thomas family (owner of Greystone) took over for Michael East in 2011. Their wines include a surprising Pinotage.

✓ *Chardonnay • Pinot Noir (Hare's Breath, Slowhand) • Pinotage*

NAUTILUS ESTATE
Marlborough
★

Nautilus invested heavily in developing its Pinot Noir (with the first dedicated Pinot Noir winery in the Southern Hemisphere), but the rich, ripe, Sauvignon Blanc is still its number-one wine. Wines are also sold under the Twin Islands second label.

✓ *Pinot Gris • Pinot Noir • Sauvignon Blanc • Sparkling wine (Marlborough Brut)*

NEUDORF VINEYARDS
Nelson
★★

I have always enjoyed Tim and Judy Finn's creamy, Burgundian-style wines – particularly the Pinot Noir, which benefited from its own winery.

✓ *Chardonnay (Moutere) • Pinot Gris (Moutere) • Pinot Noir (Moutere) • Riesling (Dry Moutere, Late Harvest Moutere) • Sauvignon Blanc*

NGATARAWA WINES
Hawke's Bay
★★

The winery was founded by the Glazebrook family and Alwyn Corban in 1981. The Glazebrooks sold up in 1999, and Alwyn's cousin Brian Corban joined the company. In 2017 the brand was acquired by Mission Estate. This is a quality-conscious Hawke's Bay winery making its mark with distinctive red wines. But with the exception of Alwyn Riesling and Glazebrook Chardonnay, its whites are so understated that their subtlety is lost on me. Second wines are sold under the Stables brand.

✓ *Classic red blend (Alwyn)* • *Merlot (Glazebrook)* • *Chardonnay (Glazebrook)*

NO1 FAMILY ESTATE
Marlborough

This brand is the result of Daniel Le Brun (known as the "grandfather" of sparkling wine in New Zealand) being unable to market wine under his own name for three years following the severing, in 1996, of his ties with Cellier Le Brun. In 1999, three years to the day after the split, Daniel cheekily opened up Le Brun Family Estate (since renamed No1 Family Estate) around the corner. He produces three sparkling wines: Cuvée No.1 (non-vintage *blanc de blancs*, rosé, and reserve), Cuvée Number Eight (non-vintage classic blend), Cuvée Remy (vintage blend dominated by Pinot Noir), and Cuvée Virginie (vintage blend dominated by Chardonnay).

✓ *Entire range*

NOBILO
Auckland
★

So much change in so little time. This was already a large wine producer before it purchased Selaks in 1998. That deal brought with it substantial additional vineyards, and the new Drylands Winery in Marlborough, allowing Nobilo both to increase production and to develop its premium wine range. A new publicly listed company was formed, attracting investment from Australian giant BRL Hardy. The Aussies liked what they saw, invested more in the company, then effectively took it over in 2000, when the Nobilo family exchanged its remaining shareholding in the old family firm for shares in BRL Hardy. Nobilo is now the second-largest wine producer in New Zealand. There is an exciting Icon Marlborough Sauvignon Blanc, which is a welcome addition to the basic, but consistently excellent, Marlborough Sauvignon Blanc. Nobilo needs to improve its red wines, but the premium Icon range is obviously the one to watch.

✓ *Pinot Noir (Icon)* • *Sauvignon Blanc (Marlborough, Icon)*

OKAHU ESTATE
Northland
☆

Established in 1984 Okahu Estate now consistently produces award winners, so the Knight family must be doing something right. Located 3.5 kilometres (2 miles) from Kaitaia on the road to Ahipara, at the southern end of the Ninety Mile Beach, this is the most northerly vineyard in New Zealand.

✓ *Chambourcin* • *Syrah* • *Viognier*

OMAKA SPRINGS ESTATES
Marlborough

This estate is owned by Geoff and Robina Jensen, who were pioneers of the New Zealand olive industry, before venturing into the wine industry. They produce inexpensive, unpretentious wines from some 60 hectares (150 acres) of vineyards.

✓ *Sauvignon Blanc*

OSTLER
Waitaki Valley

Founders Jeff Sinnott and Jim Jerram were pioneers in the Waitaki Valley (North Otago) in 2002, when they planted their first Pinot Noir vineyard. They also produce notable Riesling and Pinot Gris.

✓ *Pinot Noir*

PALLISER ESTATE WINES
Martinborough

This is a substantial producer of consistently classy wines. Even those under second label Pencarrow show some finesse.

✓ *Chardonnay (including Pencarrow)* • *Pinot Gris* • *Pinot Noir* • *Riesling* • *Sauvignon Blanc (including Pencarrow)*

PASK WINERY
Hawke's Bay
★★

Paul Smith continues to churn out a wide range of classic-quality, richly flavoured wines. There are also good value, inexpensive wines sold under the Roy's Hill label.

✓ *Classic red blend (Declaration, Gimblett Road Cabernet-Merlot)* • *Chardonnay (Gimblett Road)* • *Merlot (Gimblett Road Reserve)*

PEGASUS BAY
Waipara
★★

The Donaldson family operate a very fine winery and restaurant south of Waipara. The style of wine produced is richly flavoured with a classic structure. Second wines are sold under the Vergence label.

✓ *Chardonnay (Virtuoso)* • *Classic red blend (Cabernet-Merlot)* • *Pinot Noir (Prima Donna)* • *Riesling (Bel Canto, Aria)* • *Classic white blend (Sauvignon-Sémillon)*

PEREGRINE WINES
Central Otago

Peregrine sources fruits from three different locations: Gibbston, Pisa, and Bendigo. Only grapes from organic vineyards are used. The Pinot Noir is a must but, the range of whites contributes to the solid reputation of the winery.

✓ *Pinot Noir* • *Riesling* • *Sauvignon Blanc* • *Chardonnay*

PROVIDENCE
Matakana

This is a small, export-oriented, red wine specialist, whose wine often costs an arm and a leg, but the price can vary considerably from country to country.

✓ *Classic red blend (Cabernet Franc-Merlot-Malbec)*

PROPHET'S ROCK
Central Otago

This boutique winery specialises in Pinot Noir and aromatic whites. Winemaker Paul Pujol left his native New Zealand to hone his skills in Alsace, Burgundy, and Oregon, amongst many others. The wines enjoy a hands-off approach and are intense but elegant.

✓ *Home Vineyard Pinot Noir* • *Cuvée Aux Antipodes Pinot Noir*

PYRAMID VALLEY VINEYARDS
Canterbury

In 1996 Mike and Claudia Elze Weersing moved to New Zealand, and Mike became the winemaker of Neudorf. In 2000, they felt it was the time to open their own winery and found a spot in North Canterbury, near Waikari. They planted four small parcels which were farmed biodynamically since the beginning. Fast forward, and 2017 saw the pair Steve Smith MW and Brian Sheth purchase the winery, injecting a new vision, starting with some renovations and expansion of the estate holdings.

✓ *Pinot Noir* • *Chardonnay*

QUARTZ REEF
Central Otago

This estate is owned by Rudi Bauer, who is from Austria. He fell in love with Central Otago, and when the opportunity to stay came his way, he didn't hesitate twice. Quartz Reef specialises in *méthode-traditionnelle* sparkling wine. Other ranges include some very fine Pinot Noir, along with a few white varieties.

✓ *Pinot Noir (Bendigo Estate, Franz Ferdinand)* • *Sparkling wine* • *Pinot Gris*

RICHMOND PLAINS
Nelson

Also known as Holmes Brothers, this was the first vineyard on South Island to go organic and biodynamic,

✓ *Sauvignon Blanc (Marlborough)*

RIPPON
Wanaka

In 1974, against all advice, Lois and Rolfe Mills chose one of the most beautiful areas in the world to grow vines. They have certainly succeeded, although it took until 1989 to make their first commercial wine. Rippon is certified biodynamic.

✓ *Pinot Noir (Mature Vines, Emma's Block)* • *Riesling* • *Gamay*

ROCKBURN WINES
Central Otago
☆

As well as Pinot Noir, Rockburn produce a range of white wines. Former assistant

winemaker at Felton Road, Malcom Rees-Francis has been the winemaker since 2005.

✓ *Pinot Noir*

RONGOPAI WINES
Te Kauwhata
☆

The original Rongopai winery was established in 1932 by the Gordon family, and this continued until the death of Lou Gordon in 1954. The winery fell into disrepair until 1982, when it was resurrected by Dr Rainer Eschenbruch and Tom Van Dam from the nearby Te Kauwhata Viticultural Research Station. They produced the first new Rongopai wines in 1985, attracting instant acclaim for their botrytised wines. But Eschenbruch left in 1993, and Van Dam retired in 2001. Scottish businessman Derek Reid had invested in Rongopai since 1991, and in 1995 he purchased the old viticultural research station, which has since become Rongopai's winery. In 2007, Babich Wines Ltd took over the brand, moving the winemaking to the Henderson site in West Auckland. They produce six wines as of 2019.

✓ *Chardonnay* • *Sauvignon Blanc* • *Pinot Noir*

SACRED HILL
Hawke's Bay
★☆

Since Sacred Hill produced its first wine in 1986, its reputation has been built on its vineyards in the Dartmoor Valley hills of Hawke's Bay. In 2001, the owners purchased Cairnbrae in Marlborough, and in addition to maintaining that brand, its vineyards now contribute to Sacred Hill's growing prestige, particularly its vibrantly fresh Sauvignon Blanc. Less expensive, early-drinking wines are sold under the Whitecliff label.

✓ *Chardonnay (Rifleman's)* • *Classic red blend (Helmsman)* • *Merlot (Brokenstone)* • *Sauvignon Blanc (Marlborough)*

SAINT CLAIR FAMILY ESTATE
Marlborough

This large vineyard, owned by Neal and Judy Ibbotso, consistently produces fresh, crisp, expressive whites, including one of New Zealand's best Sauvignon Blanc. There is also a rich, deliciously fresh, deep-flavoured Merlot.

✓ *Chardonnay (Omaka)* • *Merlot (Origin Gimblett Gravels)* • *Riesling* • *Sauvignon Blanc*

SEIFRIED ESTATE
Nelson

Hermann Seifried's original winery and 10.5-hectare (26-acre) vineyard now belong to Amanda and Greg Day, who purchased and renamed it Kahurangi Estate in 1998. Two years earlier, the Seifried family had opened a brand-new winery in closer proximity to most of its other vineyards, which now total 160 hectares (395 acres). Seifried always was Nelson's biggest wine producer, and still is. Consumers outside New Zealand will

THE VINES OF THE RIPPON VINEYARD SPILL DOWN THE SLOPE TO LAKE WANAKA ON SOUTH ISLAND
Located in the Central Otago wine region, Rippon's spectacular views make it one of the most beautiful winery sites in New Zealand and one that is often photographed.

know these wines as Redwood Valley. Seifried is best for white wines, especially the superb botrytised Riesling and one of this country's better dry Rieslings.

✓ *Chardonnay (Old Coach Road) • Pinot Noir • Riesling (Botrytis Dry Riesling)*

SELAKS
Auckland
★★ **V**

This has been the award-winning division of the Nobilo group since it was taken over by the group in 1998. Selaks continues to produce fine quality, stylish wines by the barrel-load, with its superb, fruity red wines now notching up as many successes as its crisp, refreshing whites.

✓ *Chardonnay (Reserve, The Taste Collection) • Classic red blend (Reserve) • Sauvignon Blanc (The Taste Collection) • Pinot Noir (The Taste Collection)*

SERESIN ESTATE
Marlborough
★☆ **O**

Michael Seresin sold the winery in 2018 to two New Zealand winemakers: Ben Glover and Rhyan Wardman. The vines are all hand-tended, and some wines are

fermented with wild yeasts. Winemaker Tamra Kelly-Washington grew up in Marlborough and is a flying winemaker who worked in the Unoted States, Italy, and Tunisia. The wines from Seresin rank among the very best from Marlborough. The vineyard is certified organic. The Tatou and Raupo Creek vineyards have been converted to biodynamic. Future impetus will be towards polishing up the already impressive Pinot Noir.

✓ *Chardonnay (Reserve) • Pinot Gris • Pinot Noir (Tatou) • Riesling (Memento) • Sauvignon Blanc (Marama)*

SHERWOOD ESTATE WINES
Canterbury
☆

Sherwood Estate's wines are sourced from vineyards around Waipara Valley. Other labels include Stratum, an entry-level range.

✓ *Sauvignon Blanc*

SILENI
Hawke's Bay
★☆ **V**

After crushing its first crop in 1998, this winery quickly built up a reputation for

wines of style and finesse. Wines in the EV range are made only in exceptional vintages, while those in the Estate Selection are premium wines from Sileni's own vineyards. Entry-level wines are sold under the Cellar Selection label, and often represent exceptional value.

✓ *Chardonnay (Estate Selection) • Classic red blend (Estate Selection Merlot-Cabernets) • Merlot (EV) • Sémillon (Estate Selection)*

SMITH & SHETH
Hawke's Bay, Marlborough, Central Otago
★★

Craggy Range founder Steve Smith MW teamed up with Texan investor and wildlife conservationist Brian Sheth in this boutique winery venture. They are sourcing premium quality fruit from all over the country to create elegant, balanced Sauvignon Blancs, Chardonnays, and Syrahs.

✓ *Entire cru range*

SOLJANS ESTATE WINERY
Auckland
☆

Three generations of the Soljan family are involved in the winery business.

Soljans can sometimes go over the top with the amount of VA lift he gives to the fruit in his wines, but it works well when it is understated.

✓ *Cabernet-Merlot (Estate) • Port-style wine (10-Year-Old Tawny)*

SPENCER HILL ESTATE
Nelson
☆

The Spencer Hill Estate was established in the Upper Moutere area in 1992 by Philip and Sheryl Jones, whose first release was in 1994. Production was moved to its current location in the Coastal Ridge, north of Nelson, in 2000. Other labels include Mariner Vineyards (sourced from their own vineyards, plus from purchased fruit), and Latitude 41.

✓ *Chardonnay • Pinot Noir • Sauvignon Blanc*

SPY VALLEY WINES
Marlborough
★ **V**

Owned by the Johnson family, Spy Valley is a name to watch out for. The name has its origins in the cold war: Spy Valley is what the locals have called the Waihopai Valley since the

RECENT NEW ZEALAND VINTAGES

2019 A hot summer in all parts of the country, with rainfall at harvest in some areas. Hawke's Bay wines are a particular standout in terms of quality. Overall a very well-received vintage.

2018 Harvest was interrupted by rain, making this a vintage defined by how well grapes were sorted before processing.

2017 Overall this was a somewhat cooler year, with wet conditions and a reduced yield in comparison with 2016.

2015 A dry summer with concentrated harvests and high ripeness levels. Especially notable in Marlborough and Central Otago.

2013 A very well-regarded vintage; yields were up 28 per cent on 2012, with "near perfect conditions" over the summer.

2012 An exceptionally cool year, especially on the North Island. Central Otago and Martinborough Pinots show a lighter, more delicate fruit character.

2010 One of the greatest vintages on record, similar to 2013 but with lower-than-average yields.

Americans built a listening station there. The grapes come from the 145-hectare (360-acre) Johnson Estate, which might sound vast for a relatively new winery, but the Johnsons have been growing for other wineries since 1992. Fresh, crisp, vibrantly fruity Sauvignon Blanc; sweet, ripe Pinot Noir; and classy Riesling are the best so far.

✓ *Chardonnay • Pinot Noir • Riesling • Sauvignon Blanc*

STAETE LANDT VINEYARD
Marlborough
☆

From the beginning (in 1998), Ruud Maasdam and Dorien Vermaas have always aimed to produce single-vineyard wines. All of their offerings do in fact come from the same 21-hectare (52-acre) vineyard but they have been selected on a parcel-by-parcel basis according to which combination of variety and rootstock grows best. If there are differences, they are effectively what the Burgundians would call *climats* (individual block or plots), and it would be a good idea to name them.

✓ *Pinot Noir • Sauvignon Blanc*

STONECROFT
Hawke's Bay
☆

Stonecroft is a small, family-run winery making organic Gimblett Gravels wines. Owner Dermott McCollum is also the viticulturalist and winemaker.

✓ *Chardonnay • Classic red blend (Ruhani)*

STONYRIDGE VINEYARD
Waiheke Island
★★☆

Since 1985 Stephen White has produced one of New Zealand's truly great red wines, Larose, which sells out within hours of being offered *en primeur* each year. Its indicative blend is Cabernet Sauvignon, Merlot, Cabernet Franc, Malbec, Carménère, and Petit Verdot.

✓ *Classic red blend (Larose)*

TEAWA WINERY
Hawke's Bay
★

Sold in December 2002 to New York investment manager Julian Robertson and Napa Valley vineyard owner Reg Oliver (*see* Dry River), TeAwa remains under the same day-to-day control, with viticulturist John van der Linden in overall charge. Winemaker Jenny Dobson worked for a year at Neudorf before moving to Owen Roe in Oregon. Other labels include Left Field and Kidnapper Cliffs.

✓ *Chardonnay • Cabernet Sauvignon • Syrah*

TE KAIRANGA
Martinborough
★☆

With 100 hectares (250 acres) under vine, Te Kairanga is no boutique winery, but its rich, creamy Pinot Noir is the sort of impressive quality expected from the smallest, top-performing producer.

✓ *Chardonnay • Pinot Noir • Riesling*

TE MATA ESTATE
Hawke's Bay
★★

First established back in 1896, Te Mata Estate was purchased in 1978 by John Buck, who, with the help of winemaker Peter Cowley, fashioned two of the country's greatest red wines, Awatea and Coleraine. In some years, however, Awatea and/or Coleraine are too attenuated to deserve top billing. Phil Brodie has led the winemaking at Te Mata since 1992.

✓ *Chardonnay (Elston) • Classic red blend (Awatea, Coleraine) • Sauvignon Blanc (Cape Crest) • Syrah (Bullnose) • Viognier (Zara)*

TE MOTU
Waiheke Island
★☆

The Te Motu vineyard is just 60 metres (196 feet) from Stonyridge Vineyard, where Larose, one of New Zealand's true

iconic wines is produced. Currently three styles are being marketed: Kokoro, Tipua, and Te Motu. The wines are excellent but are on the expensive side.

✓ *Classic red blend (Kokoro, Te Motu) • Cabernet Franc (Tipua)*

TE WHARE RA WINES
Marlborough
★

Established by Allen and Joyce Hogan in 1979, Te Whare Ra (Māori for "the house in the sun") became the first Marlborough vineyard to be planted after Montana had paved the way. The Hogans sold Te Whare Ra to Roger and Christine Smith in 1998, and the Smiths sold it on to the present owners, Jason and Anna Flowerday. The wines hit their stride with the 2009 and 2010 vintages, especially the Rieslings. Te Whare Ra is certified organic.

✓ *Pinot Noir • Riesling (Dry, medium) • Sauvignon Blanc • Syrah*

TOHU WINES
Marlborough
☆

Established in 1998, Tohu is New Zealand's first Māori-owned winery. Their wide range of wines, from Marlborough and Nelson vineyards, are made with respect for the environment.

✓ *Sauvignon Blanc*

TORLESSE WINES
Waipara
☆

Established in 1987 by 20-odd grower-suppliers, Torlesse soon went into receivership. It re-emerged in 1990 as a new company owned by several shareholders, also with vineyards dotted over Waipara. The spread of the vineyards gave winemakers Kym Rayner and Paul Hewett the opportunity to create a range of individual *terroir*-based wines. Today, Torlesse is 100-per cent sustainable as a winery and vineyard, crafting estate-grown, -made, and -bottled wines.

✓ *Sauvignon Blanc*

TRINITY HILL
Hawke's Bay
★★

The trinity here consists of managing director John Hancock, who helped establish Morton Estate's reputation before setting up Trinity Hill. Robyn and Robert Wilsonare are now co-owners, with Warren Gibson as chief winemaker and Damian Fischer winemaker. Trinity Hill's most outstanding wines so far, however, are Merlot and Syrah. The only thing I could fault it on was its straightforward, upfront, easy-drinking quality.

✓ *Cabernet Sauvignon (Gimblett Gravels) • Classic red blend (The Trinity) • Tempranillo (Gimblett Gravels) • Syrah (Homage)*

TWO PADDOCKS
Gibbston Valley
★☆

This estate is owned by actor Sam Neill, who lives in Central Otago and loves

Pinot Noir. The original 2-hectare (5-acre) Two Paddocks on the Gibbston Back Road is so called because it was one of two neighbouring paddocks planted by Neill and friend Roger Donaldson in 1993. Donaldson went on to own his own label (Sleeping Dogs), while Neill retained the name for the overall business, although this vineyard is now known as One Paddock. In 1998 Neill purchased Alex Paddocks in the Earnscleugh Valley, Alexandra, which is a warmer site, where 3 hectares (7 acres) have been planted. Alex Paddocks yielded its first crop in 2001, enabling the first truly *two-paddock* Two Paddocks to be produced that year. The latest addition has been Redbanks Paddock, also in the Earnscleugh Valley, but significantly warmer still. With potentially 25 hectares (60 acres) of vineyards, this became Neill's most ambitious project so far. Since 2002, a soft, easy-drinking blend of all three paddocks has been produced under the Picnic Pinot Noir label. Today, the vineyards and wine are certified organic and revolve around a holistic sustainable farming model.

✓ *Pinot Noir (Proprietor's Reserve)*

UNISON VINEYARD
Hawke's Bay
★★

Unison's high-density Gimblett Gravels vineyard used to produce just two reds, both of which blended Merlot, Cabernet Sauvignon, and Syrah. One called Unison, the other Classic Blend. Both are brilliant. Now there is a refreshing rosé from the same varieties and also three whites being made.

✓ *Classic red blend (Classic Blend, Unison)*

VALLI VINEYARDS
Central Otago
★

This is the own label of Gibbston Valley's winemaker, Grant Taylor. It is named after a distant relative who emigrated to New Zealand in the 19th century. Taylor has two small vineyards at his disposal, one in Gibbston Valley, which is supposed to be better in cooler vintages, the other in Bannockburn, which excels in warmer years. In 2015, Grant received the assistance of Jen Parr to make his wines.

✓ *Pinot Noir (Bannockburn Vineyard)*

VAVASOUR
Marlborough
★☆

This vineyard is situated south of Blenheim, in the Awatere Valley, where the intense sun and dry winds are responsible for the intense fruit flavour of these well-structured wines. Excellent value wines are also sold under the Dashwood label.

✓ *Chardonnay • Pinot Noir • Sauvignon Blanc*

VIDAL
Hawke's Bay

Vidal Estate closed the doors of its historic winery in 2018, and it relocated to a new winery in Gimblett Gravels.

VILLA MARIA
Auckland
★★

New Zealand's best all-round wine producer is the country's third-largest (after Brancott Estate and Nobilo), and the only one of the three that is still New Zealand-owned. Villa Maria has a vast range of wines that are all beautifully crafted to enhance fruit and finesse. Where oak is used, the touch is light and impeccably integrated. For such a large company to offer the level of quality and consistency that Villa Maria does is pretty remarkable. But some of the kudos must go to Brancott Estate for chasing Villa Maria on quality and value at all price points, despite its production being considerably larger still. The very best Villa Maria wines deserve two and a half stars. Full marks to proprietor George Fistonich, who converted the entire production to screwcap bottling.

 Entire range

VIN ALTO WINERY
Auckland

This boutique winery has received high praise from critics, such as Bob Campbell MW. Anyone curious about how Italian varieties and styles might work in New Zealand should start at Vino Alto, where Margaret and Enzo Bettio grow, among others varieties, Arneis, Barbera, Corvina, Montepulciano, Nebbiolo, and Sangiovese. The Retico is, apparently, well worth seeking out as an example of a rare Amarone-style wine, made from grapes that have been dried on racks for two months prior to fermentation.

WAIPARA HILLS
Canterbury
★

An ambitious, young venture that makes and sells wines from Canterbury and Marlborough, as well as Waipara.

✓ *Riesling*

WAIPARA SPRINGS
Waipara
★

Owned by the Moore family, who tend the vineyard, and the Grants, who make and sell the wines, Waipara Springs produces consistently fine-quality wines.

✓ *Chardonnay (Springs Premo) • Pinot Noir (Springs Premo) • Riesling (Springs)*

WAIPARA WEST
Canterbury
★

This winery and vineyard is owned by a partnership called Tutton Sienko Hill. Paul Tutton is a London wine merchant, Olga Sienko is his wife, and Lindsay Hill is his sister-in-law and the viticulturist who manages the Waipara West vineyard.

✓ *Chardonnay • Classic red blend (Ram Paddock Red) • Pinot Noir*

WAIRAU RIVER
Marlborough
★

In 1991 Phil and Chris Rose launched their own label and now take the pick of the crop from their 125 hectares (310 acres) of vines spread over three vineyards. Sam Rose is now responsible for all wines along with winemaker Nick Entwistle.

✓ *Pinot Gris • Riesling • Sauvignon Blanc*

WEST BROOK
Auckland
★

Founded by Mick Ivicevich in 1937 in what was rural Henderson. The estate was surrounded on all sides by urban sprawl by 2000, when grandson Anthony Ivicevich moved West Brook to its present location in the Ararimu Valley.

✓ *Chardonnay (Waimauku) • Sauvignon Blanc*

WHITEHAVEN WINES
Marlborough
★☆

Whitehaven Wines produces sensational Riesling and Sauvignon Blanc, particularly the Greg collection.

✓ *Chardonnay • Riesling • Sauvignon Blanc*

WITHER HILLS
Marlborough
★☆

In 2002 Wither Hills winery was purchased by Lion Nathan, the brewing giant that owns Australia's Petaluma. The vineyards are in South Island's famed Marlborough region; the winery used to be at Henderson, Auckland, until a brand-new winery was build next to the vineyards in 2015

✓ *Chardonnay • Pinot Noir • Sauvignon Blanc*

THE FAMILY HOME ACROSS FROM THE BUCK'S TE MATA ESTATE SHOWS ART DECO AND SPANISH MISSION INFLUENCES
The Bucks have owned Te Mata since 1974. Established in the late 19th century, the winery and original vineyards were the first to be heritage-protected in New Zealand. Netting is placed over the vines here in the Coleraine Vineyard, as it is in many vineyards around the country, to protect the fruit from hungry birds.

Asia

It is no surprise that the world's largest continent is home to hundreds of wineries spread over 15 countries. China, India, and Japan are by far the most important winemaking nations of Asia—though their wine industries are of different degrees of maturity and different stages of sophistication.

As the most prolific of the Asian wine-producing nations, China, India, and Japan may overshadow the rest of the continent, but there are many other Asian nations that have a tradition of growing wine grapes or are launching new efforts to gain traction in the winemaking world. The following countries all produce wines of varying degrees of quality.

BHUTAN

The first planting of wine grapes was in the early 1990s at Paro, near the capital Thimphu, at an altitude of 2,300 metres (7,500 feet). An Australian winery, Taltarni, provided technical support, with trial plantings of a range of classic varieties that were selected because they were expected to cope best in the high alpine conditions. There are now at least two local wine labels, but there is no evidence that the wines are made solely from locally grown grapes.

CAMBODIA

Prasat Phnom Banan Winery is Cambodia's only winery. It is located in the northwest, just outside the country's second largest city, Battambang. The family-owned and -managed winery planted vines brought in from Thailand in 2000 and had its first vintage in 2004. It makes red wines and rosés from just under 3 hectares (7.5 acres) of Shiraz, Cabernet Sauvignon, Black Queen, and Kyoho vines.

INDONESIA (BALI)

Indonesia's pioneer winery, Hatten Wines, began making wine on the Island of Bali in 1994. Its first wine was a rosé made from the descendants of an obscure French variety, Alphonse Lavallée, that was brought to the island a couple of centuries ago to grow table grapes for colonial settlers. Hatten has expanded very creatively in subsequent years to add a sparkling version of the rosé and a red from this variety, white wines from another obscure variety, Belgia (from the Alexandria family, similar to Muscat St Vallier), a dessert wine from a combination of both these varieties, and a white sparkling wine from a mystery grape known locally as Probolinggo Biru.

Hatten is still, by far, the nation's largest winery and, following its success, there are now four other producers making wine in commercial quantities. Two of them, like Hatten, are using locally grown grapes: Sababay, now the

RED MOUNTAIN ESTATE NEAR LAKE INLE, MYANMAR
This winery near Nyaungshwe follows French practices to produce their wines and grow mostly Sauvignon Blanc grapes, along with Pinot Noir and Chardonnay.

second-largest winery, started a decade ago, and Isola, the most recent entrant, is a much smaller producer. The other two Balinese wineries, Plaga Wines and Cape Discovery Wines, are using only imported grapes or must for their wines. Hatten and Sababay also make wines from imported grapes or must as separate operations and released under separate labels.

There are now around 400 hectares (990 acres) of grapevines on the island, mainly in the northwest, at 8 degrees below the equator, and around a third are used for winemaking. The vines are mostly grown on pergolas to help cope with the tropical climate and intense humidity, and they produce grapes in 120-day cycles. Growers stagger the cycles with prunings across their vineyards to permit a virtually constant harvest.

A range of classic wine grape varieties have been trialled, but few have progressed to commercial quantities. Shiraz and Chenin Blanc seem to be the most comfortable in the challenging local growing conditions. Hatten now has expanding plantings of both varieties. They are double-pruned for enforced dormancy, allowing just one harvest per year. Hatten was awarded Winery of the Year in the 2017 edition of *Asian Wine Review* and won the prize for Best Sparkling Wine with its NV Tunjung Brut (Probolinggo Biru).

KYRGYZSTAN

Situated between China and Kazakhstan, this country's vineyards total just over 6,000 hectares (14,825 acres), and only a small proportion is used for wine: in 2015 wine production was 15,000 hectolitres.

MYANMAR (BURMA)

The country's first wine venture was Myanmar Vineyard Estate, launched in 1999 by German entrepreneur Bert Morsbach. He planted vines on the slopes of Taunggyi Mountain at an altitude of 1,200 metres (3,940 feet), near the village of Aythaya in Shan State, 30 minutes by car from the state's main tourist attraction, Lake Inle. The winery is now sourcing a large and growing proportion of its grapes from contract farmers in three other locations, in contrasting climates and altitudes, and the total area of vineyards is just over 26 hectares (64 acres). After lots of experimentation with many varieties, those that are now the core of the winery's Aythaya label are Shiraz, Dornfelder, and Tempranillo for red wines; Sauvignon Blanc, Chenin Blanc, and Sémillon for white wines; and Red Muscat for a dessert wine. The annual crush is around 200 tonnes.

Myanmar's second winery, Red Mountain Estate, was established in 2003. It is also located in Shan State, on a hill overlooking Lake Inle. Reflecting its French connections it began with trial plantings of a range of classic French varieties and now gets the best results from Sauvignon Blanc and Syrah, which represent almost 70 per cent of its 75 hectares (185 acres) of vineyards. Minor plantings of Muscat Blanc à Petits Grains and Pinot Noir are doing well, but Chardonnay and Cabernet Sauvignon have struggled.

NEPAL

Isolated small grape vineyards have been spotted at up to 2,750 metres (9,000 feet) near the town of Jomsom in midwestern Nepal. Local drinks labelled as wine may include some grapes, but they are primarily made from berries, apples, or other fruits.

PHILLIPINES

Wine grapes have been grown at Ilcos and Cebu in the past, but there is no commercial wine production in the Phillipines. Rice wine has been made locally since before the Spanish conquistadors, according to local sources. Palm, coconut, and fruit wines also exist.

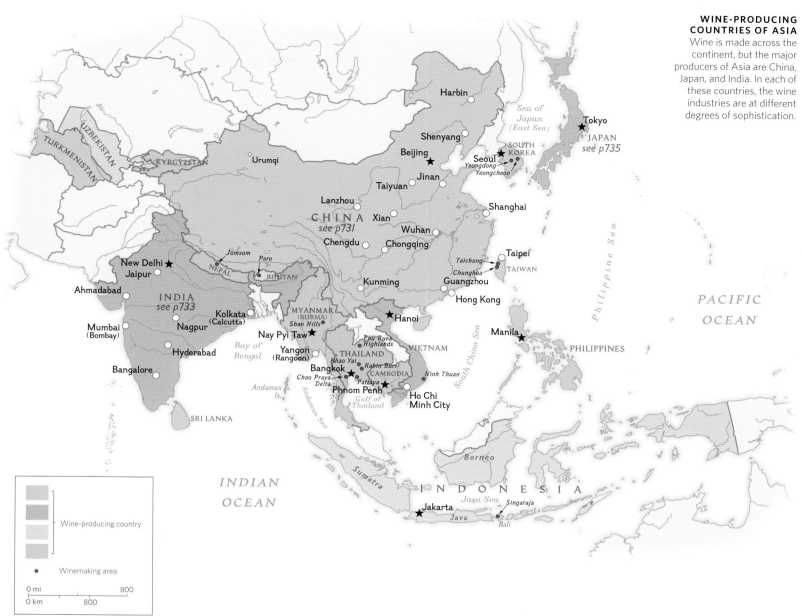

Wine-producing country

• Winemaking area

0 mi   800
0 km   800

SRI LANKA

In this country where table grapes are commonly grown – principally, Cardinal and Black Muscat – there have been some fledgling ventures making wine on a tiny scale. There are now two companies licensed to make wine from imported grape must and concentrates as well as local fruit, including grapes, but are still in their early stages.

SOUTH KOREA

South Korea has a long history of growing grapes, primarily for table grapes and dried fruit, but making wine with grapes did not begin until the early 1970s. There are 17,000 hectares (42,000 acres) of vines in 14 regions, and a growing number of wineries draw on a small proportion of these vines. There are now 60 wineries, mostly very small-scale, farm-based operations, and most of them are focused primarily on producing fruit wine, rather than grape wine. The two largest wine-producing regions are Yeongcheon and Yeongdong and both are in the south of the country.

The Japanese hybrid grapes Campbell Early and Muscat Bailey A and the local wild mountain *Vitis amurensis* grape, known locally as Sanmeo-ru, are the varieties most commonly used for winemaking. However, a local hybrid called Cheongsoo, developed recently by the National Institute of Horticultural and Herbal Science, is becoming popular with wine producers because of its tolerance of the extreme-cold growing conditions and the attractive attributes of the resulting wines. There are tiny plantings of *Vitis vinifera* varieties – Riesling, Chardonnay, Sauvignon Blanc, Muscat, Cabernet Sauvignon, and Merlot – but very little wine is made from any of these varieties at present.

TAIWAN

A nascent winemaking industry has emerged in Taiwan following the termination in 2002 of the state-run monopoly that had controlled all alcohol production and sales and had been the exclusive buyer of grapes for wine production. Now the Government's Council of Agriculture is working with farmers to promote viticultural development and wine production. It formally recognises 12 privately owned grape wineries among a much larger number of fruit wine producers. Official statistics show that domestic wine accounted for just under 9 per cent of wine sold in 2017.

Grape-wine production is centred in two counties in the central-west: Taichung County and Chunghua County. Black Queen and Golden Muscat have been the most widely grown varieties, but there has been recent experimentation to find other varieties that might do well in local growing conditions. A new hybrid white grape variety developed by the Taichung District Agricultural Research and Extension Station is promising. Its technical name is Taichung No.3 and it has been trademarked by the leading local winery, Weightstone, as Musann Blanc for commercial release. Major style choices for most local producers with all varieties have been, and are still, sweet wine styles.

TAJIKISTAN

There are 34,000 hectares (84,015 acres) of vineyards in this central Asian state, most of which are found in Leninabad, Ghissar, and Vakhsh. Wine production has been shrinking in recent years and the surviving wineries produce just 2 hectolitres of wine.

THAILAND

Thailand has 4,838 hectares (11,954 acres) of vineyards, mostly of the Pokdum and Malaga Blanc varieties that have been traditionally grown in the Chao Praya Delta, west of Krung Thep (Bangkok), for table grapes. The wine industry didn't get underway until the early 1990s, following a 12-year viticultural experimental study in the mid-1970s and a ruling by the Thai government, in 1992, that officially allowed wine production.

The country's first winery, Château de Loei, was established in 1993 in the Phu Ruea Highlands by the late Dr Chaijudh Karnasuta. Prior to this, grapes from Malaga Blanc and Pokdum vines in the Chao Praya Delta were used to make what became a very popular wine cooler called Spy. The maker, Siam Winery, now Thailand's largest wine producer, then progressed to table wines made from Shiraz, Colombard, and Chenin Blanc planted in the Khao Yai region, 175 kilometres (110 miles) northeast of Krung Thep. Its main vineyards and winemaking operation for its Monsoon Valley label are on the east of the Thai Gulf, near the city of Hua Hin.

There are now eight active wineries in Thailand, and most of them are in the Khao Yai region and nearby Kabin Buri, at altitudes up to 550 metres (1,800 feet). Here GranMonte Vineyard, another industry pioneer, has demonstrated the industry's creativity, trialling 30 different grape varieties and currently using 12 of them in its commercial wine releases – including such rare varieties as Durif and Verdelho. Other wineries in the region are Village Farm (selling wine under Château de Brumes and Village Thai labels), Alcidini Winery, Moonlight Valley, J&J Vineyard, and PB Valley Winery. Silverlake Winery is on the west side of the Thai Gulf, near the tourist resort of Pattaya.

The Thai Wine Association, with six members in three regions (Khao Yai, Kabin Buri, and Pattaya), has been a cohesive force in setting and maintaining standards for the industry. It has been common practice throughout Asia to supplement locally grown grapes with imported grapes, grape concentrate, or bulk wine. But the Thai Wine Association requires that its members declare any imported component clearly on their labels if it is above 10 per cent – welcome rigour in wine's new territory.

Given the tropical climate it is quite possible to pick five crops every two years, but most Thai wineries harvest just once every 12 months.

UZBEKISTAN

The total vineyard area is 130,182 hectares (321,686 acres), of which around one-quarter is used to produce grape juice, wine, and spirits. In a 2006 wine sector reorganisation the government set up a holding company, Uzvinsanoat, to oversee more than 80 domestic wine-producing operations across the country. Most production is in the southeast, with the largest concentrations in Bukhara, Samarkand, and Tashkent. The wines are mostly high-strength reds and dessert and sparkling wines, and the varieties used include Soyaki, Saperavi, Rkatsiteli, Aleatico, Bayanshirei, and Hindogni. Wine production in 2014 was 303,680 hectolitres.

VIETNAM

There was some early viticulture and basic winemaking during the French colonial period in the Central Highlands and around the capital, Hanoi. Its first modern winery was the ill-fated joint venture established in 1995 by Allied-Domecq in the Province of Ninh Thuan, drawing on fruit from long-established table grape vineyards, mostly of the Cardinal variety. It lasted only a few years, but wine production started again not long after this and has been growing steadily over the past decade.

Ladora Winery, owned by the large food processing company, Lam Dong (Lado) Food Co, is the nation's largest wine producer and has been leading the revival. Its winery is in the Central Highlands, near the city of Dalat but, surprisingly, its vineyards are 100 kilometres (62 miles) away on the coastal plains in Ninh Thuan Province. It has moved beyond the traditional model of, basically, low-priced fruit wine, including some made from grapes and now has added a range of premium grape wines under its Chateau Dalat label, based on plantings of Shiraz, Cabernet Sauvignon, Sauvignon Blanc, Merlot, and Chardonnay, and in its new Ninh Thuan vineyards.

There are also some small-scale wineries that are progressively modernising, moving on from fruit wines and now making grape wines from revitalised vineyards. Some of them are also moving on from Cardinal, and the popular choices are Sauvignon Blanc, Shiraz, and Cabernet Sauvignon. The Ba Moi Winery Farm is also working with two varieties known locally as NH.01.48 and NH.01.152 that are similar to Malaga Blanc and Black Queen that are commonly grown in similar conditions in Thailand.

FACTORS AFFECTING TASTE AND QUALITY

LOCATION
The Asian continent accounts for nearly one-third of the globe's land mass. Of its major wine-producing regions, China alone takes up 9,600,000 square kilometres (3,700,000 square miles). Japan lies off the eastern coast, and India, the seventh-largest country by area, lies in the southern region in the Indian subcontinent.

CLIMATE
China: Vineyards are classified as "humid micro-thermal cool", similar to those of Austria and Hungary, where continental conditions are influenced by great water masses, creating hot, damp summers and very cold, dry winters.
India: Hot with no real winter. Two harvests a year: one dry, one humid monsoon (usually avoided). Altitude of Maharashtra and Dodballapur provides a relatively cool climate.
Japan: Extremes of climate: freezing winds in the winter, monsoon rains in spring and autumn, and typhoons in the summer. The temperature drops northwards; humidity rises southwards. The climate of Miyazaki Prefecture on Kyushu has been likened to that of Florida.

ASPECT
China: Recent plantings have been on well-drained, south-facing slopes to overcome the earlier problem of high water-tables on the flatter sites.
India: Gentle east- and south-facing slopes at an altitude of 750m (2,460ft) in Narayangaon, and 600m (1,980ft) in Nasik and Dodballapur
Japan: On Honshu Island, the best vineyards are on the south-facing valley slopes around Kofu.

SOIL
China: Generally alluvial.
India: Lime-rich soils at Narayangaon, sandy-clay loam at Nasik, and red-sandy loam at Dodballapur.
Japan: Predominantly acidic soils that are unsuitable for viticulture, except around Kofu, where the soil is gravelly and of volcanic origin.

VITICULTURE & VINIFICATION
In tropical regions, the grape vines never winter, and up to 15 vintages can be harvested.
China: Big, new wineries in the western provinces of Xinjiang, Gansu, and Ningxia are responsible for most of this expansion. Export-orientated wineries are driving quality standards upwards and have had some remarkable success with super-premium wines, but attempts to raise standards across the board are held back by high-yielding varieties that crop up to 150 hl/ha.
India: Vine training is moving from the Lenz Moser high system to more sophisticated high-trellis systems.
Japan: Recent developments have concentrated on the growing of top European varietals, and although this continues at an increasing pace, attention is also shifting to the Koshu grape. The biggest problem is still rain.

GRAPE VARIETIES
China: Beichun, Cabernet Franc, Cabernet Sauvignon, Carignan, Chardonnay, Chasan, Chenin Blanc, Crystal, Gamay, Gewürztraminer, Grenache, Italian Riesling, Maru (Mare's Nipple), Marsanne, Merlot, Muscat d'Hambourg, Muscat à Petits Grains, Merlot, Pinot Noir, Rkatsiteli, Rose Honey, Saperavi, Sauvignon Blanc, Sémillon, Sylvaner, Syrah, Welschriesling
India: Bangalore Blue (Isabelle), Cabernet Sauvignon, Chardonnay, Karachi Gulabi (Muscat Hambourg), Pinot Noir, Ruby Red, Thompson Seedless, Ugni Blanc
Japan: Cabernet Sauvignon, Campbell's Early, Chardonnay, Delaware, Koshu, Merlot, Müller-Thurgau, Muscat Bailey, Riesling, Sémillon

China

There is recent archaeological work that attests to the use of local wild grapes for winemaking as far back as circa 7000 BC and the arrival of Vitis vinifera *grape vines in far western Xinjiang Province in the fourth century BC. Vine-growing then seems to have moved progressively across the country. In 126 BC,* Vitis vinifera *vines were planted at the Imperial Palace in Chang An (now Xi'an) 1,000 kilometres (620 miles) southwest of the present-day capital, Beijing.*

It was during the Tang Dynasty (AD 618 to 907) that grapes and wine became a feature of Chinese culture. Documentation from that time refers to wines made from an elongated grape called Manaizi (Mare's Nipple). But the significant shift in the growing and use of grapes for winemaking came when links to European wine specialists began to be forged with a view to development and expansion. In 1892 Chinese businessman Zhang Bishi brought cuttings of many *vinifera* grape varieties from Europe and built the Zhang Yu (now Changyu) Winery at Yantai in eastern Shandong province. In 1910 a French priest opened a winery in Beijing called Shangyi (later known as Beijing Friendship Winery) and, in 1930 a German company set up the Shandong Melco Winery, now called Qingdao Huadong Winery, at Tsingtao (Qingdao) on the Shandong Peninsula.

The current expansion phase began in the early 1980s when the Chinese Government launched its Four Modernisation Policy to revitalise the domestic economy and encourage practical international linkages. Rémy Martin was one of the first foreign parties to establish a China base when it joined forces with the city of Tianjin to apply modern viticulture and winemaking techniques in a new venture to make contemporary-style wines that became Dynasty, now one of China's top 3 producers. In 1987 the Pernod Ricard Group teamed up with the Beijing Friendship Winery to create the Dragon Seal brand for locally produced European-style wines. Pernod Ricard pulled out of the venture in 2001, but Dragon Seal remains a major force in the industry. A wide range of other international parties have committed to ventures in China in increasing numbers as the momentum in the domestic industry gains substance and diversity.

Progressing on from these relatively humble beginnings, the wine industry has grown over three decades to become Asia's wine powerhouse: coming from nowhere to rank among the world's top 10 wine-producing countries. The International Organisation of Vine and Wine (OIV) has ranked China as high as the world's fifth-largest producer – when it recorded annual production in 2012 at 13.5 million hectolitres. But there is controversy regarding the official statistics that the OIV and other sources are relying on, with some critics claiming that they overstate actual domestic production. It is noteworthy that production as recorded by the OIV has fallen marginally each year since the high point in 2012, to 10.8 million hectolitres in 2017, and there are industry observers who believe it could be even lower than this. Some suspect that a proportion of the substantial quantity of imported bulk wine is counted in official figures as *domestic* wine when used by winemakers in locally bottled wines. Others say it is not clear that wine made in one region and bottled in another region is not sometimes double counted, especially some of the wine from the large-scale operations in the Western Regions that is shipped across to East Coast wineries. It *is* clear that actual domestic production has fallen in recent years: wineries have closed down or merged and vineyards have been reduced. Acknowledging the double-counting possibilities and actual production contractions, some informed sources in the industry put total production as low as 6.7 million hectolitres in 2018. That would rank China just outside the top 10 producers worldwide, but still with a high ranking, at number 11.

CHÂTEAU CHANGYU-CASTEL IN YANTAI, SHANDONG PROVINCE, RISES ABOVE ITS STATELY GARDENS
The winery of the Chinese wine brand Changyu, the largest producer in China, features a French-style château and formal gardens.

INKY DARK MARSELAN GRAPES DISPLAY A WHITISH BLOOM
This red French wine grape variety is a cross between Cabernet Sauvignon and Grenache. The Chinese and French worked together to bring the Marselan to China, where it has thrived, and it is now heading toward signature status.

OPENING UP CHINA'S WINE REGIONS

The area under vine has trebled in the last two decades and was recorded by the OIV at 870,000 hectares (2.15 million acres) in 2017, the second-largest vineyard area in the world, after Spain. However, almost 80 per cent of China's vines are table grapes, and a further 10 per cent are for raisins. Only around 10 per cent are for wine.

Wine is now produced in 25 of China's 28 Provinces and Autonomous Regions and the number of active grape wineries stood at 625 in 2017, as measured by the long-standing wine data gatherer, China Wine Online. The industry is concentrated in four broad regions – the Shandong Peninsula, the far North-East (Liaoning, Jilin, and Heilongjiang Provinces), the Mid-North-East (Beijing, Tianjin, Hebei Province, and Shanxi Province), and the far North-West (Xinjiang and Ningxia Autonomous Regions and Gansu Province).

The Shandong Peninsula is both the historical and ongoing hub of China's modern wine industry, where the production of wine with imported European varieties began 130 years ago and where the first wave of the new industry development took off from the 1980s. It is now home to 160 wineries, responsible for 39 per cent of national production, with several of them among the nation's largest producers: Changyu is China's largest and Weilong (Grand Dragon) is also in the top 10. The major wineries are located around the provincial city of Yantai, and the vineyards are now concentrated in Penglai County. There are also a few wineries in Qingdao, on the south side of the Shandong Peninsula.

The rapid development in Shandong over the past two decades has featured some standout joint ventures, including Domaine de Penglai, set up in 2008 by Château Lafite and China's largest state-owned investment company, CITIC Group. Its initial plantings were more than half Cabernet Sauvignon, then some Cabernet Franc and Merlot, with 10 per cent each of Syrah and the emerging local choice, Marselan. Several of the region's major wineries have also expanded through large-scale vineyard plantings in the far North-West Region, providing some relief from the persistent problem of rot in the East and offering vineyard scale that could never be achieved in the closely settled East Coast.

The second-largest producing region now is in the North-West, comprising the Autonomous Region of Xinjiang, where grape-growing and winemaking in China originated, and, in more recent years, the Autonomous Region of Ningxia and Gansu Province. In these regions it is possible to grow grapes on a "New World" scale, and the region now accounts for almost 20 per cent of national wine production. There are 76 wineries in Xinjiang, 52 in Ningxia, and 12 in Gansu.

In Xinjiang there are around 100,000 hectares (240,000 acres) of vineyards, of which just under 17,000 hectares (42,000 acres) are used for winemaking. The venture that set the scale for winemaking in this region was Suntime (Xintian), set up by a provincial state-owned enterprise and the national investment company, CITIC. Launched in 1998 it quickly became one of the largest wineries in China, with most of its wine output sold to producers in Eastern Regions looking to expand volumes under their labels. It restructured and now operates as CITIC Guoan Wine Company and is still one of the top 10 producers. More recent arrivals include Tiansai, quickly winning recognition for its quality wines, and Zhongfei. There are also operations specifically set up to add volume capacity for wineries in the East.

Ningxia has rapidly earned a reputation for its quality wines led, in the early stages, by Pernod Ricard's Helan Mountain winery, now with 400 hectares (990 acres) of classic varieties. The region overall has earned a national and international reputation for quality wines. A stand-out winery in this regard is Silver Heights. Other wineries working to define Ningxia's styles and win global respect include Xixia King, Leirenshou, and Helan Qingxue. LVMH combined with a state-owned enterprise to produce sparkling wines here and launched Chandon China in 2013. In Gansu, where there are now eight wineries, Mogao Winery has been the region's pioneer winery, particularly with Pinot Noir. They have also collaborated with a Greek winery, Kir-Yianni, to plant the first Xinomavro vines in China. Shandong-based Weilong (Grand Dragon) has committed to large-scale plantings in Gansu, including organic vineyards.

The third major region, with 132 wineries, is an expansive geographic area that incorporates the national capital, Beijing, coastal Tianjin, as well as Hebei and Shanxi Provinces. It is home to two of the nation's top 10 producers – Great Wall and Dynasty, respectively located in Beijing and Tianjin. Interesting ventures in Hebei include the 100 per cent Austrian-owned Bodega Langes and Changyu's AFIP Global (Disney meets Bordeaux). Hebei is also home to the Sino-French Demonstration Vineyard which has pioneered the development and widespread uptake of the Marselan variety, which may one day become China's signature variety. On a new frontier, in central Shanxi Province, Grace Vineyards has demonstrated the premium-quality potential of this Province and has grown rapidly to now rank amongst the largest producers in China which, in turn, has attracted new ventures to the region. A more recent quality-focused arrival is Chateau Rongxi. Both are working predominantly with Cabernet Sauvignon

In the North-East region there are 102 wineries. Wine was traditionally made here using the local wild mountain grape varieties (literally, *shan-putao*) and indigenous *Vitis vinifera* varieties – the former being mostly a *Vitis amurensis* variety called Amur and the latter being Longyan (Dragon's Eye) and Niu-nai (Cow's Nipple). They continue to be the major resource for wine producers in this region, but there has also been ongoing breeding research crossing *Vitis amurensis* and *vinifera* varieties to create hybrids that cope better with the extreme-cold climate. Liaoning Province has developed a reputation for its ice wines, and national industry leader, Changyu, added real substance to this reputation when it set up the Changyu Golden Ice Wine Valley project with 330 hectares (815 acres) of Vidal. The company now claims to be the world's largest planting of this variety, exclusively to make ice wine.

There are smaller numbers of wineries in other provinces, some of them very recent arrivals. The latest rush to plant has been in southern Yunnan Province, on the Chinese side of the Himalayas, bordering Vietnam, Laos, and Myanmar. There are three varieties that have been growing here since the late 19th century, having been brought in by French Christian missionaries: two red grapes, Honey Rose and Wild French, and a white grape, Crystal (possibly *labrusca*). Patient experimental work has been done in recent years, at extreme altitudes for viticulture, with new plantings of classic European varieties, with very encouraging results. The Shangri-la Wine Company, established in 2001, was one of the pioneers, working with Cabernet Sauvignon and Merlot and some Grenache and Chardonnay. Moet Hennessy joined with a Yunnan-based Chinese liquor producer VATS to produce a super-premium Bordeaux blend with grapes drawn from vineyards as high as 2,200 to 2,600 metres (6,560 to 8,330 feet) above sea level Its first release under the Ao Yun label was in 2013, a blend of 90 per cent Cabernet Sauvignon and 10 percent Cabernet Franc, and subsequent releases have attracted high rankings from some of the most respected critics in the region, and internationally.

The majority of China's wine, around 80 per cent, is red wine. This is because red is the consumers' preference, not because white vines do not do well. The most widely planted red grape variety, by a long margin, is Cabernet Sauvignon. There are also substantial plantings of Merlot and Cabernet Franc, and some Syrah and Grenache. Of growing interest to local winemakers is Marselan, a crossing of Cabernet Sauvignon and Grenache. There has also been an expanding uptake of Cabernet Gernischt, a variety that was brought to Shandong from France in the late 19th century with the collection to establish the Changyu winery – and now known to be genetically very close, if not identical, to Chile's Carménère. White wines are predominantly of Welschriesling (known in China as Italian Riesling), Chardonnay, Muscat, and the native *vinifera* grapes, Longyan and Nui-nai.

STACKS OF OAK BARRELS LINE A WINE CELLAR IN HEBEI PROVINCE
Hebei is an expansive wine region with 132 vineyards, as well as the Sino-French Demonstration Vineyard, a Chinese-French government-joint venture project.

✓*Dragon Seal* (Huailai Cabernet Sauvignon, Huailai Syrah) • *Shanxi Grace Vineyard* (Cabernet Franc)

Wine-producing province labeled on map

Wine region

0 mi   400
0 km   400

WINE REGIONS OF CHINA
(*see also* p727)
The wine-grape vineyards of China have grown massively and rapidly since the 1990s, when most commercially grown vines were centred on the Shandong and Beijing regions.

India

Grapevines have been grown in India for more than 6,000 years, and the country now has 131,000 hectares (323,700 acres) of vineyards, of which around 5,000 hectares (12,360 acres) are wine grapes, concentrated in the southeastern states of Maharashtra and Karnataka. There are now 60 wineries, producing over 2 million cases of wine per year. Most of them are located in Maharashtra State.

The famous Indian medical treatises *Sasruta Samhita* and *Charaka Samhita,* written between 1356 BC and 1220 BC, mention the medicinal properties of grapes already growing in India, and specific grape varieties are named in *Arthashastra,* a treatise on political economy, once thought to have been written in the fourth century BC but now considered to date from the second century AD. The first large-scale cultivation of grapes was by Persian invaders in about AD 1300, and they quickly spread the practice to the Aurangabad district of Maharashtra so that, by 1430, a Moorish trader by the name of Ibn Batuta reported having seen vineyards flourishing in the south of India.

So, viticulture is not new to this country, but contemporary winemaking did not begin until the early 1980s after self-made multi-millionaire Sham Chougule convinced Piper-Heidsieck to assist him in his quest to locally create a world-class sparkling wine. And so the Indage Vintners (formerly Champagne Indage) empire was born, making sparkling wines at Pune, in Maharashtra State, initially from local table grapes and progressing boldly to Chardonnay and Ugni Blanc. Aggressive expansion, however, including substantial international wine-related investments just as the global financial crisis began, brought it to an end and the company was wound up in 2011.

The other early industry pioneer was Kanwal Grover, who founded India's second winery, Grover Zampa Vineyards, which is now the country's second-largest producer. After assessing potential sites in several states in the mid to late 1980s with technical support from French associates, he settled on the Nandi Foothills in Karnataka State as the best location for his vineyard and winery. Grover Vineyards began with a 40-acre site and its first commercial vintage was in 1992. Through careful experimentation and ongoing technical inputs from French oenologist Michel Rolland, the Karnataka operation has expanded progressively with plantings of Chenin Blanc, Sauvignon Blanc, Viognier, Shiraz, Cabernet Sauvignon, Merlot, Zinfandel, and, more recently, Tempranillo.

Because of regulations in Maharashtra State restricting sales of domestic wines to wines made only in that state, Grover decided to commit to production there in order to expand its market base with access to what is the nation's largest market for wine. To achieve this, it joined forces with a Maharashtra winery, Valleé de Vin, adding that winery's Zampa range to its portfolio and re-naming the winery as Grover Zampa Vineyards in 2012. In December 2018 it further expanded its Maharashtra presence in a major move that added two other large local operations to its wine portfolio – Four Seasons Winery and Charosa Vineyards. This brings the total vineyard area that Grover Zampa is now drawing its fruit from to 243 hectares (600 acres). With the most recent additions to its portfolio its annual production, from both states, is expected to grow from 220,000 cases per year to beyond 300,000 cases.

A TRAPEZOID OF PAINTED WINE BARRELS ADDS COLOUR TO THE GROUNDS OF SULA VINEYARDS
Sula Vineyards, India's leading wine producer, started off in 1998 as a small family estate of just 12 hectares (30 acres). Today Sula covers approximately 730 hectares (1,800 acres) across Nashik and the state of Maharashtra. Founder Rajeev Samant took what he learned in California wine country and applied it in his homeland to build a thriving business.

A VINEYARD IN THE NASHIK VALLEY, MAHARASHTRA, GROWS WINE GRAPES
Long a site of vineyards growing table grapes, the region around the ancient city of Nashik has become the centre of India's winemaking industry.

The wine industry in Maharashtra had its origins in Pune, but the industry's growth has been concentrated in Nashik, which has become, by far, the nation's largest producing region, now accounting for more than 60 per cent of national production. The winery that has led this growth from the outset is Sula Vineyards. It was started in the mid-1990s when Rajeev Samant, a local who had been studying and working in California, returned to India with a vision to create locally what had impressed him during his years in California. In collaboration with seasoned Californian winemaker Kerry Damskey, he planted Sula's first vineyards in 1996 in soils of volcanic origin at an altitude of 600 metres (1,970 feet). The first Sula wines hit the market in 1999 and the company quickly grew to become India's largest wine producer, accounting now for around half of the Indian wine industry's total production. Chenin Blanc, Sauvignon Blanc, Viognier, and Riesling are the principal white grape varieties and Shiraz, Cabernet Sauvignon, Zinfandel, Merlot, and Grenache are the main red varieties. Sula is the first Asian winery outside China to reach an annual sales level of 1 million cases and, although the major share of its wine is consumed domestically, it is now exporting to 20 countries. Sula has also pioneered wine tourism, with more than 356,000 visitors visiting its cellar door and associated hospitality facilities in 2018.

Undoubtedly reflecting the outstanding achievements and contributions to the local communities by the pioneering wineries, the restrictive government regulations controlling alcohol production and consumption, which had severely constricted growth in the wine industry in the past, were changed: in Maharashtra in 2001 and Karnataka in 2007. The changes have spurred a quite spectacular growth of the wine industry. There have been around 100 wine ventures over the years since the beginning of the industry in Maharashtra and Karnataka, but not all have survived. There have also been mergers and integrations. The total number of fully operational wineries in India now stands at 56.

Among the next rank of producers are York Winery, Fratelli Vineyards, Reveilo Vineyards, Myra Vineyards, Soma Vineyards, Vallonne Vineyards, Vintage Wines, and Good Drop Cellars. Grover Zampa Vineyards was awarded Winery of the Year in the 2018 edition of *Asian Wine Review*.

WINE REGIONS OF INDIA (*see also* p727)
Although the vineyards of India cover 2,100 kilometres (1,300 miles) of the subcontinent, most vineyards and wineries are in the Maharashtra province, where Indage Vintners, then Champagne Indage, started the modern Indian wine industry in 1982.

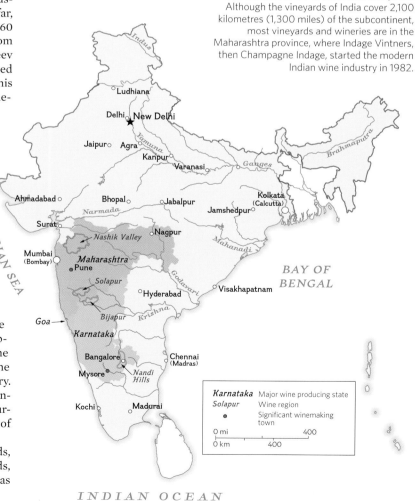

Karnataka	Major wine producing state
Solapur	Wine region
●	Significant winemaking town

0 mi 400
0 km 400

Japan

Wine production in Japan was first documented at the time of the Portuguese settlement in the southern island of Kyushu in the 16th century, but it may date back to much earlier times. There are indigenous wild mountain grapes and the arrival of the Koshu variety, the core of the wine industry's identity, has been traced back at least a thousand years.

In the 17th century, under the all-powerful Tokugawa shogunate, anything perceived as Christian or Western was condemned, so wine production would not have been officially allowed. But after military and civil powers were handed back to the emperor in 1867, winemaking was possible, and the first commercial winery was established in the Yamanashi district in 1875, followed over the next decade by a small number of other wine ventures. It was the Koshu variety, which had been growing there for centuries, that was the resource for these pioneer wineries. It originated in Asia Minor and was carried to Japan via China along the Silk Route by Buddhist monks, arriving as long ago as the eighth century. It is still the most widely grown variety nationally: it accounts for 40 per cent of Japanese vineyards overall and a higher proportion in Yamanashi.

It was not until the 1980s that wine drinking became an aspect of social life, and it is still only a minor share of alcohol consumption, with the overwhelming share of wine consumed coming from imported sources. The local industry has, in recent decades, however, expanded, modernised, and evolved distinct wine styles that are, not surprisingly, well suited to Japanese cuisine and, for that reason, are starting to win customers in international markets. In 2017 exports totalled 2,389 hectalitres and the major destinations were Hong Kong, China, Taiwan, Spain, and the UK.

Reflecting this, the National Tax Agency in 2018 introduced a mandatory industry labelling code to ensure the authenticity of all wine produced in Japan. If using imported ingredients, a wine can still be labelled as "Domestically Produced Wine", but it cannot state a local production region, varietal names or the harvest year – and must declare if it uses concentrated juice (from the country of origin) or imported bulk wine (from the country of origin). A wine which uses only grapes harvested in Japan is now labelled as "Japan Wine". If the producer wants to show more detailed origin it is permitted to add a region – if at least 85 per cent of the grapes were harvested in that region – and the winery site must be physically located in the declared region. A Geographical Indication (GI) system introduced at the same time adds further authenticity: Yamanashi Prefecture was the first to be formally assigned under the GI regime and Hokkaido, the second-largest producing region, was the second GI formalised. Other wine-producing prefectures will follow progressively. To add further authenticity, a wine that is labelled as a single varietal must be made from at least 85 per cent of the declared variety: if the wine is a blend the varietal names must be listed in order from major to minor varietal.

The contemporary wine industry initially switched from a dependence on local varieties to classic European grape varieties, mostly French (particularly Cabernet Sauvignon, Chardonnay, and Merlot). However, more recently, it has gone back the other way with a renewed commitment to Koshu, with its growing international recognition as Japan's flagship wine. And the Koshu wines have progressively moved on from chaptalised,

TOKACHI-IKEDA RESEARCH INSTITUTE FOR VITICULTURE AND ENOLOGY, IKEDA, NAKAGAWA DISTRICT, HOKKAIDO
This research institute is known colloquially as the Ikeda Wine Castle. Set on a hillside overlooking the town, the fortress-like structure is also the home of the Tokachi Winery, which hosts a wine and food festival in Ikeda every year in October.

KOSHU GRAPES RIPEN IN A YAMANASHI PREFECTURE VINEYARD
Bunches of grapes are wrapped in paper bags to protect them from bad weather and hungry birds and wasps.

oaked, *sur lie* styles to purer, natural, dry fruit styles. A second unique Japanese variety that is quickly becoming a signature wine is Muscat Bailey A, a cross of Muscat Hamburg and Bailey bred in the 1920s by one of the industry's icons, Kawakami Zenbei: it delivers a range of styles from rosé to fuller-bodied reds. Entries in the Japan Wine Competition provide a measure of its growing popularity. In 2018 there were 105 entries (77 as single variety and 28 in blends): in 2008 it was less than half that, just 42 entries (28 single variety and 14 in blends). Sustained experimental vine breeding efforts to develop varieties that will do well in the some-times-challenging growing conditions has led to an increasing number of unique local varieties, and these are also showing up in larger numbers at the Japan Wine Competition. In 2018 among the red and rosé entries alone there were 20 unique hybrid varieties – principally Campbell Early, Bailey Alicante, Kai Noir, Yamasauvignon, Shokoshi, and Black Queen. The number of varieties now used in commercially available wines in Japan totals at least 110.

Geographically, the wine industry was historically concentrated in Yamanashi Prefecture, with smaller bits in neighbouring Nagano and Yamagata Prefectures. But the most rapid growth in recent years has been in Hokkaido, the large island in the far north of Japan. In 2000 there were just 8 wineries there and now there are 34. The biggest producer in Hokkaido, Hokkaido-Wine, is now the largest wine producer in the country. In contrast to the rest of Japan, this producer has very large-scale viticulture, owning and operating about 150 hectares (370 acres) as well as drawing on about 40 contract grape growers. A large-scale vineyard on the main island, Honshu, is about 10 hectares (25 acres). The varieties that are doing best here are Kerner, Pinot Blanc, Müller Thurgau, Bacchus, Gewürztraminer, Zweigelt, Lemberger, and Seibel.

There are now 283 commercial wineries, in 42 of Japan's 47 Prefectures – with growing conditions stretching for extreme cold in Hokkaido to semi-tropical conditions in the southern island of Kyushu. Yamanashi has the largest number of wineries, with 81, followed by Hokkaido and Nagano with 34 each. The nation's largest producers are Chateau Mercian, Sapporo, and Suntory.

✓Ajima Budoshu (Chardonnay) • Château Mercian (Hokushin Reserve Chardonnay, Private Reserve Kikyogahara Merlot) • Domaine Sogga (Chardonnay) • Grace (Misawa Private Reserve Cabernet-Merlot, Misawa

Private Reserve Koshu – for Koshu lovers only!) • Hayashi Farm (Kifugo, Kikyogahara Mer lot) • Kumamoto (Night Harvest Chardonnay) • Manns (Solaris Chikumagawa Nagano Merlot) • Okuizumo (Shimane Chardonnay) • Sapporo (Grande Polaire Furusato Cabernet Sauvignon, Grande Polaire Nagano Furusato Vineyard Cabernet Sauvignon, Grande Polaire Yoichi Kifu) • Suntory (Tomi no Oka Cabernet Merlot, Tsutaishizaka Chardonnay) • Tsuno (Campbell's Early Rose)

Guide to Good Vintages

This chart serves as a useful fingertip reference guide to the comparative performance of more than 700 vintages, covering 24 different categories of wine.

As with any vintage chart, the ratings should be seen merely as "betting odds". They express the likelihood of what might reasonably be expected from a wine of a given year and should not be used as a guide to buying specific wines. No blanket rating can highlight the many exceptions that exist in every vintage, although the higher the rating, the fewer the exceptions; quality and consistency do to some extent go hand in hand.

LOW-RATED VINTAGES

The low-rated vintages should be treated with extreme caution, but you need not ignore them. Although wine investors buy only great vintages of blue-chip wines, the clever wine drinker makes a beeline for unfashionable vintages at a tasting and searches for the exceptions. This is because no matter how successful the wine, if it comes from a modest year, it will be relatively inexpensive.

WINE CATEGORY	2018	2017	2016	2015	2014	2013	2012	2011	2010	2009	2008	2007	2006	2005	2004	2003	2002	2001
BORDEAUX – MÉDOC and GRAVES	92	87	97	90	88	82	88	91	96	99	89	80	85	96	94	93	91	90
BORDEAUX – ST-ÉMILION and POMEROL	94	89	95	96	87	84	90	89	96	98	90	83	87	98	94	89	85	91
BORDEAUX – SAUTERNES and BARSAC	88	91	88	93	95	91	77	95	95	98	87	98	87	95	90	92	95	98
BURGUNDY – CÔTE D'OR – red	85-100	91	97	98	89	86	93	88	75	97	87	85	86	95	84	75	90	84
BURGUNDY – CÔTE D'OR – white	80-95	96	87	94	97	88	92	91	80	91	89	90	89	90	86	65	90	90
BURGUNDY – BEAUJOLAIS – red	90	94	92	88	90	82	86	88	88	97	84	85	88	92	80	90	85	80
CHAMPAGNE	50-85	65	62	89	65	90	98	60	70	90	98	85	80	75	90	50-85	95	35
ALSACE	85	87	90	92	89	89	83	85	89	90	70-85	87	70	89	87	65-90	88	90
LOIRE – sweet white	93	92	94	92	86	84	85	88	94	95	88	88	75	95	80	93	85	95
LOIRE – red	92	91	94	90	87	86	89	88	90	95	80	85	85	95	85	95	85	80
RHÔNE – NORTHERN RHÔNE	90	92	88	97	82	84	86	88	93	96	80	90	92	96	86	96	70	85
RHÔNE – SOUTHERN RHÔNE	82	90	97	92	82	83	87	83	93	92	85	95	90	96	90	93	55	89
GERMANY – MOSEL	87	89	89	91	68	70	88	91	89	91	90	92	85	96	91	85-97	92	90
GERMANY – RHEIN	92	90	87	89	72	74	85	90	89	90	92	90	80	90	90	85-94	90	92
ITALY – BAROLO	92	90	96	92	85	94	85	84	89	90	92	92	95	86	96	90	75	96
ITALY – CHIANTI	88	87	97	95	88	93	89	90	89	90	90	91	94	85	93	89	75	91
SPAIN – RIOJA	87	86	92	89	82	82	89	83	92	90	87	85	87	95	95	80	70	93
PORTUGAL – VINTAGE PORT	90	98	96	90	88	–	–	99	89	85	87	95	70	80	88	94	70	86
US – CALIFORNIA – red	92	86	96	94	88	95	92	79	87	89	90	88	90	96	95	90	92	94
US – CALIFORNIA – white	93	90	90	92	90	92	90	84	88	90	87	88	90	96	90	90	91	90
US – PACIFIC NORTHWEST – red	88	89	89	88	90	88	91	88	89	80-90	94	88	90	85	89	88	93	89
US – PACIFIC NORTHWEST – white	87	89	95	91	92	88	95	78	84	90	94	80-88	90	82	84	85	90	90
AUSTRALIA – SOUTH-EASTERN AUSTRALIA	94	79	93	95	89	92	94	68	89	89	78	88	90	95	82	98	70	70
AUSTRALIA – WESTERN AUSTRALIA	94	87	89	92	91	91	89	92	90	90	90	92	85-95	88	89	70	90	93

CLOSE-SCORING WINES

The smaller the gap between two scores, the less difference one might expect in the quality of the wines, but many readers may think that to discriminate by as little as one point is to split hairs. Certainly, a one-point difference reveals no great divide but, on balance, it indicates which vintage has the edge.

UNRATED VINTAGES

It is impossible to rate generally undeclared vintages of Port and Champagne: the few wines released often prove to be truly exceptional anomalies. Take the Champagne vintage of 1951: this was one of the worst in Champagne's history. To almost everyone who remembers the harvest, no pure vintaged Champagnes were produced, yet my research uncovered three. If I were to rank 1951 on the two I tasted (Clos des Goisses and Salon), it would be one of the finest vintages of the century, which is clearly nonsense. Port poses a similar problem: should 1992 be judged on the basis of the superb Taylor's, by the handful of other good, but distinctly lesser, 1992s, or by the majority of shippers, who did not produce good enough wine to declare a vintage?

KEY TO VINTAGE RATINGS	
Excellent to superb	90–100*
Good to very good	80–89
Average to good	70–79
Disappointing	0–69
Very bad	40–59
Disastrous	0 – 39

Remember that some wines are not particularly enjoyable, nor even superior, in so-called great years. Such wines, which are normally light-bodied, aromatic whites that are best drunk while young, fresh, and grapey (eg, Muscat d'Alsace, German QbA, or Kabinett), favour vintages with ratings of between 70 and 85, or lower.

*No vintage can be accurately described as perfect, but those achieving a maximum score are truly great vintages.

2000	1999	1998	1997	1996	1995	1994	1993	1992	1991	1990	1989	1988	PRE-1988 GREAT VINTAGES
98	85	85	83	95	90	85	82	78	78	90	95	88	1986, 1985, 1982, 1970, 1961, 1953, 1949, 1945, 1929, 1928, 1900
97	88	90	80	90	90	85	85	78	75	92	95	88	1985, 1982, 1961, 1953, 1949, 1945, 1929, 1928, 1900
87	87	89	95	90	95	79	60	65	60	92	95	95	1986, 1983, 1975, 1962, 1959, 1955, 1949, 1947, 1945, 1937
85	91	89	88	97	90	70	87	85	82	95	92	98	1986, 1985, 1971, 1961, 1959, 1949, 1945, 1929, 1919, 1915
87	89	90	92	97	92	80	87	95	70	98	95	92	1986, 1985, 1962, 1928, 1921
87	88	83	85	90	85	90	85	92	85	90	90	85	1985, 1978, 1961, 1959, 1957, 1949, 1945, 1929
85	86	92	85	93	90	83	87	87	83	93	89	90	1985, 1982, 1979, 1975, 1964, 1959, 1947, 1945, 1928, 1921, 1914
85	85	85	90	60	90	87	85	80	75	95	90	90	1983, 1975, 1971, 1969, 1961, 1959, 1953, 1949 1945, 1937, 1921
60	60	60	91	90	95	90	55	55	50	90	90	90	1975, 1975, 1971, 1969, 1961, 1959, 1949, 1945, 1921
85	80	75	90	92	90	60	60	40	40	90	90	85	1985, 1976, 1975, 1969, 1961
90	91	89	85	85	92	75	65	75	85	96	95	95	1985, 1982, 1975, 1972, 1970, 1961
92	88	95	70	85	90	80	80	75	70	90	95	92	1983, 1978, 1972, 1970, 1967, 1961
80	87	75	90	85	92	92	94	92	88	100	98	95	1986, 1985, 1983, 1976, 1975, 1973, 1971, 1969, 1959, 1953, 1949, 1945, 1921
80	87	85	95	84	90	85	90	90	80	95	95	90	1986, 1985, 1983, 1976, 1975, 1971, 1969, 1959, 1953, 1949, 1945, 1921
90	95	94	92	96	90	88	85	75	80	93	95	95	1985, 1982, 1978, 1971, 1958, 1947, 1931, 1922
87	92	85	93	80	88	90	85	75	80	90	65	90	1985, 1971, 1947, 1931, 1928, 1911
85	84	80	85	75	87	95	87	85	85	78	85	80	1982, 1970, 1968, 1964, 1962, 1942, 1934, 1924, 1920, 1916
95	75	80	90	–	88	95	–	85	95	–	–	–	1977, 1969, 1963, 1945, 1935, 1931, 1927, 1908
89	94	89	87	88	92	94	88	91	93	80	80	85	1987, 1985, 1974, 1970, 1968, 1951, 1946
89	90	89	88	88	92	88	89	92	85	80	82	88	1986, 1984, 1974, 1970, 1967
90	95	92	85	90	88	88	87	89	85	85	85	85	1985, 1975
90	90	88	85	90	85	75	69	85	85	88	80	80	1985, 1983
88	80	90	65	95	85	70	90	65	95	80	85	80	1985, 1984, 1980, 1979, 1975, 1967, 1965
89	88	86	88	92	94	94	85	89	86-98	87	86	87	

Micropedia

The following is an ever-expanding compendium of terms that not even the most obsessed wine geek will ever need to read page-by-page yet will enjoy the occasional journey from one definition to another. The primary reason for the Micropedia is to answer 99 per cent of the most basic, as well as the most technical, questions that readers of every capability could possibly ask about wine.

Terms that are explained more comprehensively within the main body of the book are accompanied by a cross-reference to the appropriate page. Terms that appear within a Micropedia entry and have their own separate entry appear set in **bold type**. For more tasting terms, *see* "Tastes and Aromas", pp96–101.

Key to abbreviations:

Fr = French
Ger = German
Gr = Greek
It = Italian
Port = Portuguese
S Afr = South African
Sp = Spanish

ABC An acronym for "Anything but Cabernet" or "Anything but Chardonnay", ABC was a more-than-acceptable term when originally conceived by Randall Grahm of Bonny Doon. Grahm was selling Cabernet at the time, but saw it as a rut that every California winery was trapped in. He wanted to explore the quality potential of other grapes, particularly the Rhône varieties, but was severely restricted by the public demand for Cabernet and Chardonnay. While Cabernet walked off the shelf, Grahm had to work hard at selling the virtues of anything more exotic. Compelled to sell Cabernet to fund other activities, he came up with the ABC term. Everyone loved it when Grahm invented it. It has since been hijacked by inverted snobs and myopic critics, however, who have been zealots in their crusade to rid the world of two great wine grapes.

Abscisic acid A plant hormone associated with *véraison*.

ABV Abbreviation for **alcohol by volume**, usually expressed as a percentage

AC (Port, Gr) This is short for *Adega Cooperativa* in Portugal and Agricultural Cooperative in Greece or other titles denoting a local or regional cooperative in these countries.

Accessible Literally, wine that is easy to approach, with no great barriers of **tannin, acidity,** or undeveloped extract to prevent drinking enjoyment. This term is often used for young, fine-quality wine that will undoubtedly improve with age but whose tannins are supple and thus approachable.

Acetaldehyde The principal **aldehyde** in all wines but found in much greater quantities in Sherry. In light, unfortified **table wines**, a small amount of acetaldehyde enhances the **bouquet**, but an excess is undesirable because it is unstable, halfway to complete **oxidation**, and evokes a **Sherry-like smell**.

Acetic When a wine is acetic, the **acetaldehyde** has been converted by bacterial action to **acetic acid**, and the **aroma** has moved to vinegary.

Acetic acid The most important **volatile acid** found in wine, apart from **carbonic acid**. Small amounts of acetic acid contribute positively to the attractive flavour of a wine, but large quantities produce a taste of **vinegar**.

Acetification The production of **acetic acid** in a wine.

Acetobacter The **vinegar** bacillus (rod-shaped bacterium), which can cause **acetification**.

Acid line Can apply to many wines, particularly white and **rosé**, but most especially to sparkling, where the acid line ideally carries the wine to a long and hopefully lingering or tapering **finish**, along which any *dosage* should ride as close to weightless as possible.

Acidity Essential for the life and vitality of all wines. Just enough adds a lip-smacking tang, but too much will make wine too sharp (not sour – that's a fault), whereas not enough will make it taste flat, dull, and short, and the flavour will not last the month. *See also* pH; TOTAL ACIDITY.

Active acidity Acids contain positively charged hydrogen ions, the concentration of which determines the total **acidity** of a wine. The **pH** is the measure of the electrical charge of a given solution (positive acidity hydrogen buffered by negative alkalinity hydrogen ions). Thus the pH of a wine is a measure of its active acidity.

Adega (Port) Cellar or winery. Often used as part of a firm's title.

Aerobic In the presence of air. Often applied to **fermentation**.

Aftertaste A term that describes the flavour and **aroma** left in the mouth after the wine has been swallowed. When the aftertaste is attractive, it could be the reason why you prefer one wine to a similar wine with no particular aftertaste.

Ages gracefully Describes wine that retains **finesse** as it **matures** and that sometimes may even increase in finesse.

Agglomerated cork Made of granulated **cork**, usually produced from the leftovers of the manufacture of natural **cork closures.** The granulated cork is washed and dried and mixed with a food-grade glue before being either individually injected into moulds or produced by extrusion (ie a continuous "sausage" that is cut into individual corks).

Aggressive The opposite of **soft** and **smooth**.

Agrafe A wire mesh fastening used on a bottle of **Champagne** or other effervescing wine to hold the cork in place during the final **fermentation**. Also known as a filling bracket, wire basket, wire cage, shipping bracket, or French Muselet.

Alban *or* **Albanum** An **ancient Roman wine** that rose to fame after the demise of **Falernian**, Alban was first mentioned by Dionysius of Halicarnassus and was usually consumed when 10 to 20 years old. Some sources claim it to be very sweet, while others believe it to be sharp, even sour. Sweetness was a mark of a great wine in those days, however, so its fame would suggest that it was indeed very sweet.

Albany Doctor One of Western Australia's beneficially cooling sea breezes, the Albany Doctor provides similar relief to that of the famous Fremantle Doctor, only farther south and closer to the coast. *See also* CANBERRA DOCTOR; FREMANTLE DOCTOR.

Albariza (Sp) A white-surfaced soil formed by diatomaceous (decomposed deep-sea algae) deposits, which is found in the Sherry-producing area of Spain. *See also* "Sherry Country", pp406–409.

Alcohol In wine terms, this is **ethyl alcohol**; a colourless flammable liquid. Alcohol is essential to the flavour and **body** of alcoholic products; thus, a de-alcoholised wine is intrinsically difficult to perfect.

Alcoholic This term is usually employed in a pejorative rather than a literal sense and implies that a wine has too much **alcohol** to be in **balance**.

Aldehyde The midway stage between an **alcohol** and an **acid**, formed during the **oxidation** of an alcohol. **acetaldehyde** is the most important of the common wine aldehydes, and forms as wine alcohol oxidises to become **acetic acid** (**vinegar**). Small amounts of acetaldehyde add to the complexity of a wine, but too much will make a table wine smell like Sherry.

Aldehydic An excessive form of **oxidative**. *See* ACETALDEHYDE; ALDEHYDE.

Allier A famous type of **oak** used for barrelmaking from **Tronçais**, Grosbois, Civrais, Dreuille, or any other forest in the Allier *département* in the centre of France.

Alte Reben (Ger) Old vines.

Altec The first technical **microagglomerate** cork, Altec, was launched by Sabaté in 1995 as a response to unacceptable incidence of **TCA** (trichloroanisole) found in natural **corks** and perceived consumer resistance to synthetic corks and **screw caps** on the French market. It was literally ahead of its time. By removing lenticels and the woody material surrounding the lenticels in the sifting stage of microagglomerate production, most TCA and its precursors were eliminated. Most TCA was not all TCA for every batch, however, and the remaining TCA was evenly distributed at a very low level throughout tainted batches due its production method of mixing and baking the closures. The TCA might have been very low, indeed too low to detect on the nose in most cases, but it was enough to **scalp** the fruit in a wine. This led to legal action taken (from 2001) by various wine producers, particularly in the United States, and the negative publicity that followed in the wake of these court cases prompted Sabaté Diosos to change its name to Oeneo Group, Marc Sabaté to step down as CEO, and Altec relaunched as **DIAM**, produced by the **Diamant® Process**.

Amino acids **Proteins** formed by a combination of fruit **esters**, amino acids are found naturally in grapes and are both created and consumed during **fermentation** and **autolysis**. They are essential precursors to the complexity and **finesse** of a sparkling wine. *See also* MAILLARD REACTIONS.

Amino sugars Amine-containing **sugars** found naturally in grapes and wine. There are some 60 known amino sugars, but only one of any consequence in wine and that is **galactosamine** (derived from **galactose**).

Amorin This Portuguese company is the world's largest **cork** manufacturer. *See* NDTech and ROSA Evolution.

Ampelographer An expert who studies, records, and identifies grapevines.

Amylic The peardrop, banana, or bubblegum **aromas** of amyl or isoamyl acetate are aromas created by a first **fermentation** that is too cool. Small amounts can "blow away" after a few months post-**disgorgement** ageing, but excessive amounts remain and can have a banalising effect. In its pure form, amyl acetate is used as an aromatising flavourant and is commercially known as "banana oil" or "pear oil". *See also* "Tastes and Aromas", pp96–101.

Anaerobic Occurring in the absence of oxygen. Most maturation processes that take place in a sealed bottle are considered to be anaerobic.

Anbaugebiet A wine region in Germany, such as Pfalz or Mosel, that is divided into districts (**Bereiche**). All **QbA** and **QmP** wines must show their Anbaugebiet of origin on the label.

Ancient Greek wines The Greeks preferred sweet wines that were very high in **alcohol**, but they always diluted their wine, usually with three or four parts water, the intention of such a mix being to enjoy the aesthetic pleasure of the wine and to be intoxicated just enough to have the mind released from inhibition. At banquets, it was the *symposiarchos* (master of drinking) who decided the ratio of water to wine. The wine was always added to the water, never the water to the wine, a practice that developed from the earliest times when it was necessary to add wine to water, albeit in smaller proportions, as a "disinfectant" to protect against disease and to disguise any stagnant smell. This was particularly necessary on long voyages. In Homer's *Odyssey*, for example, a ratio of 1 part wine to 20 parts water is mentioned. *See* specific ancient Greek wines: ANTHOSMIAS, ARIUSIAN; CHIAN; COAN; DENTHIS; ERESOS; LESBIAN; PRAMNIAN; THASIAN.

Ancient Roman wines The Romans added herbs, spices, and resin to perfume their wines, a practice that was copied from the Egyptians. Like the Greeks, the Romans diluted their wine with water, although with just one or two parts water, which was considered barbaric by the Greeks, just as the Gauls, who drank wine undiluted, were regarded as barbaric by the Romans. *See* specific ancient Roman wines: ALBAN; CAECUBAN; CAUCINIAN; CONDITUM; FALERNIAN; FALERNUM; FAUSTIAN; FORMIAN; GAURAN; LORA; MAMERTINE; MARSIC; MASSICUM; MULSUM; PASSUM; PRAENESTINE; PRIVERNATINUM; RHAETIC; RHEGIUM; SORRENTUM; SETINE; STATAN; TIBURTINE; TRIFOLIAN.

Année (Fr) Year.

Ano (Port, Sp) Year.

Anthocyanins Various colouring pigments in the skins of black grapes. Of the several factors that affect the colour these anthocyanins produce, **pH** is the most notable. Grapes with a high pH (low acidity) are dominated by blue and purple tints, while grapes with a low pH (high acidity) are simply red and any purple hues that may exist will drop out quickly.

Anthosmias This ancient Greek wine was made "stronger with the fruit of new vines rather than of old" and cut with 1 part seawater to 50 parts wine.

Antioxidant Any chemical that prevents grapes, **must**, or wine from **oxidising**, such as **ascorbic acid** or SO_2 (sulphur dioxide).

AOC (Fr) *Appellation d'Origine Contrôlée* is the top rung in the French wine-quality system, although in practice it includes everything from the greatest French wines to the worst; thus, it is almost always better to buy an expensive *vin de pays* than a cheap AOC wine.

AOP (Fr) Common abbreviation for *Appellation d'Origine Protégée*, the French-language version of the EU umbrella classification (formerly VQPRD) for all national quality-wine regimes, such as **AOC** in France, **DOC** in Italy, **DO** in Spain etc. The English equivalent is **PDO** (Protected Denomination of Origin).

Aperitif Originally used exclusively to describe a beverage prescribed purely for laxative purposes, the term *aperitif* now describes any drink that is taken before a meal in order to stimulate the appetite.

Appellation Literally means "a name", this term is usually used to refer to an official geographically based designation for a wine. An appellation does not necessarily infer a whole bunch of rules dictating what must be grown and how, or what sort of wine must be made. That level of bureaucratic interference is a matter for each individual country; thus, countries such as France, Italy, Spain, and Portugal dictate these details, whereas the likes of Germany, the United States, and Australia do not.

Appellation d'origine contrôlée *See* AOC.

Aquifer A water-retaining geological formation into which rainfall from the surrounding area drains.

Arabinose A **monosaccharide**, arabinose is a **pentose** that is often found as one of the **residual sugars** even in fully fermented, bone-dry wines. Although of little significance in wine, it is also a **wood sugar**, thus relatively more abundant in oak-aged wines. Furthermore, it is a constituent of gum arabic, which may be added to the *dosage*, thus any arabinose in a sparkling wine could come from three different sources. Arabinose is an unfermentable reducing sugar.

Arabinitol *or* **Arabitol** A **sugar alcohol** commonly found in small amounts in wine, arabinitol can be converted from **arabinose** by **yeast** or during the **malolactic** process.

Are (Fr) One-hundredth of a **hectare**.

Argon An expensive noble inert gas, superior to both nitrogen and **carbonic gas** (CO_2) for winemaking purposes, such as blanketing wines inside vats to prevent **oxidation**.

Argonne A well-known type of **oak** used for barrel-making from a forest in the Ardennes *département* of northern France.

Ariusian This ancient Greek wine was supposedly the best **Chian** wine.

Aroma In an ideal world, this should really be confined to the **fresh** and **fruity** smells reminiscent of grapes, rather than the more winey or bottle-mature complexities of bouquet, but a book is not an ideal world, and such correctness would merely result in a repetitiveness that would interrupt the flow of the text. It is thus inevitable that "aroma" and "bouquet" will become synonymous, and the technically correct, yet contradictory term "bottle aroma" is evidence of this. *See* BOUQUET; BOTTLE AROMAS.

Aromatic grape varieties The most aromatic classic varieties are Gewürztraminer, Muscat, and Riesling, and they are defined as such because when ripe they possess high levels of various **terpenes** in their skins. Although attractive when young, these terpenes take a few years in bottle to develop their full **varietal** potential (except for the Muscat or Moscato, which should be consumed as young and as fresh as possible).

Aromatised wine Usually **fortified**, these wines are flavoured by as few as 1, or as many as 50, aromatic substances and range from bitter-sweet vermouth to Retsina. The various herbs, fruits, flowers, and other sometimes less-appetising ingredients used include strawberries, orange peel, elderflowers, wormwood, quinine, and pine resin.

Ascorbic acid Otherwise known as Vitamin C, ascorbic acid is a powerful and speedy **antioxidant** that is often used in conjunction with sulphur, which is primarily antibacterial with lesser, slower antioxidant capabilities. It has a more freshening effect than SO_2, which tends to dampen the aromatics in wine. It also enables less SO_2 to be used in the **vinification** process, but the reduction is not proportionate. The function of ascorbic acid is more of a quick sweep than a replacement, as it mops up oxygen that SO_2 would take some time getting around to and, in the meantime, without ascorbic acid, oxidation would begin. Ascorbic acid is not a magic potion, however. It has its downside; if the *dosage* of both SO_2 and ascorbic acid is not correctly calculated, and there is still ascorbic acid available after all of the free SO_2 has been used up, ascorbic acid becomes the bad guy, actively promoting **oxidation**. Ascorbic acid should not be confused with **sorbic acid**.

Aseptic A substance with an antimicrobial capacity that can kill bacteria, such as **sorbic acid** or SO_2.

Aspect The topography of a vineyard, including its altitude, the direction in which the vines face, and the angle of any slope.

Aspersion A method of protecting vines from frost by spraying with water.

Assemblage (Fr) A blend of base wines that creates the final *cuvée*.

Asti (It) A town in northern Italy that gives its name to the world's finest sweet sparkling wines, Asti and Moscato d'Asti.

ATA *See* Atypical ageing.

Atmosphere A measure of atmospheric pressure: 1 atmosphere = 15 pounds per square inch. The average internal pressure of a bottle of **Champagne** is 6 atmospheres.

Atoms Composed of electrons, neutrons, and protons, atoms join together to form **molecules**, which in turn form the chemical and biochemical **compounds** found in wine (and are the basic building blocks of ordinary matter all around you). *See also* ELECTRONS; NEUTRONS; PROTONS.

Attack A wine with good attack suggests one that is complete and readily presents its full armament of taste characteristics to the **palate**. The wine is likely to be youthful rather than **mature**, and its attack augurs well for its future.

Atypical ageing *or* **ATA** Although first noticed in Germany in 1977, ATA was not widely known in the global wine industry until about 1995, when viticultural researchers from Germany, Austria, and Switzerland began comparing notes. Wines affected by ATA are thin or green or unripe, with atypical **aromas** and a loss of **varietal** flavour. The worst cases develop a dirty dishcloth, dirty floorcloth, or mothball aroma. ATA is a phenomenon associated with vine stress during drought conditions, which restricts nitrogen uptake, increasing a plant hormone called indole acetic acid (IAA), which breaks down into aminoacetophenone and unpleasant smelling indoles (skatole being the worst offender). High levels of UV irradiation in the vineyard can have a similar effect. ATA can also have a naphthalene-like (mothball) smell.

Auslese Category of German **QmP** wine (above **Spätlese** but below **Beerenauslese**) that is very sweet, made from late-picked grapes. May contain some **botrytised grapes**.

Austere This term is used to describe wine that lacks **fruit** and is dominated by harsh **acidity** and/or **tannin**.

Autoclave Generically, this is a hermetically sealed tank capable of withstanding high internal pressure. In the wine world, the most common use for an autoclave is the tank in which the Charmat method or Metodo Martinotti is carried out. *See* CUVE CLOSE.

Autolysis The enzymatic breakdown of **yeast** cells that increases the possibility of bacterial spoilage; the autolytic effect of ageing a wine on its **lees** is therefore undesirable in most wines, exceptions being those bottled *sur lie* (principally Muscadet) and sparkling wines.

Autolytic With the **aroma** of a freshly disgorged *brut*-style sparkling wine, which is not **yeast**y at all but has a flowery, often acacia-like freshness.

Autophagic Self-digesting.

AVA This stands for American Viticultural Area, the American equivalent of **AOC**. It assures geographical integrity, without any bureaucratic regulations to control the grape varieties, how they are to be grown or the style of wine that may be produced.

AWRI Common abbreviation for the Australian Wine Research Institute.

Back-blend To blend fresh, unfermented grape juice into a fully fermented wine, with the aim of adding a certain fresh, grapey sweetness commonly associated with German wines. Synonymous with the German practice of adding *Süssreserve*.

Backwards Describes a wine that is slow to develop (the opposite of **precocious**).

Baked Applies to wines of high alcoholic content that give a sensory perception of grapes harvested in great heat – either from a hot country or from a classic wine area in a swelteringly hot year. This characteristic can be controlled to some extent by the following methods: early harvesting, night harvesting, rapid transport to the winery, and modern cool-**fermentation** techniques.

Balance, balanced Refers to the harmonious relationship between acids, **alcohol**, **fruit**, **tannin**, and other natural elements. If you have two similar wines but you definitely prefer one of them, its balance is likely to be one of the two determining factors (length being the other).

Balthazar Large bottle equivalent to 16 normal-sized 75cl bottles.

Ban de vendange (Fr) Official regional start of grape-picking for the latest **vintage**.

Barrel-fermented Some white wines are still traditionally fermented in **oak** barrels – new for top-quality Bordeaux, Burgundy, and premium **varietal** wines; old for middle-quality wines and top-quality **Champagnes**. New barrels impart oaky characteristics; the older the barrels, the less oaky and more **oxidative** the influence. Barrel-fermented wines have more complex **aromas** than wines that have simply been **mature**d in wood. See also "The Cost of New Oak", p63.

Barrique (Fr) Generic term for "barrel", now widely used throughout English-speaking countries for any small cask, often denoting the use of new **oak**.

Basic A marketing term for a quality category. See PREMIUM.

Basic taste There are only five known basic tastes: **bitterness**, sourness (or **acidity**), sweetness, saltiness, and **umami**. These are the only tastes we actually experience; all others are tastes that we smell through the olfactory bulb. See OLFACTORY BULB; TASTE BUD; TASTE RECEPTOR.

Bâtonnage (Fr) **Lees**-stirring. This practice is often carried out on fine wines made from relatively neutral grape varieties, such as Chardonnay, to instil some additional fullness, but it can easily be overdone. *Bâtonnage* should be "felt, not smelt" – you should never be able to pick up a wine and smell a leesy-*bâtonnage* **aroma**.

Baumé (Fr) A scale of measurement used to indicate the amount of **sugar** in grape **must**.

Beerenauslese A category of German **QmP** wine that comes above *Auslese* but beneath *Trockenbeerenauslese* (TbA) and is made from **botrytised grapes**. It has more **finesse** and **elegance** than any other intensely sweet wine, with the possible exception of **Eiswein**.

Bellême Lesser-known type of **oak** used for barrelmaking, from a forest in the Orne *département* of Normandy.

Bench or **benchland** The flat land between two slopes, this term describes a form of natural, rather than artificial, terrace.

Bentonite This is a fine clay containing a volcanic ash derivative called montromillonite, which is a hydrated silicate of magnesium that activates a precipitation in wine when used as a **fining** agent.

Berce Lesser-known type of **oak** used for barrelmaking, from a forest in the Sarthe *département* of the Loire Valley.

Bereich A wine district in Germany, which contains smaller *Großlagen* (or *Grosslagen*) and is itself part of a larger *Anbaugebiet.*

Bertrange A well-known type of **oak** used for barrelmaking, from a small forest in the Nievre *département* in the centre of France.

Bianco (It) White.

Big vintage, Big year These terms are usually applied to great years, because the exceptional weather conditions produce bigger (ie fuller and richer) wines than normal. They may also be used literally to describe a year with a big crop.

Big wine This term describes a full-bodied wine with an exceptionally rich flavour.

Biochemical compound Any carbon-based compound that is found in living things. There are four classes: **carbohydrates**, **proteins**, **lipids**, and **nucleic acids**. See also COMPOUND.

Biodynamic A spiritual-scientific agricultural concept based on Rudolf Steiner's philosophy of biodiversity and respect for the rhythm of the earth and the universe. Wines produced biodynamically are made without the aid of chemical or synthetic sprays or **fertilizers** and are vinified with natural **yeast** and the minimum use of **filtration**, SO2, and **chaptalisation**. There are several competing organizations that certify biodynamic farming.

Biomethylation See METHYLATION.

Biscuity A desirable aspect of **bouquet** that is found in some **Champagnes** – particularly in well-matured blends dominated by Pinot Noir (Chardonnay-dominated Champagnes tend to go **toasty**, although some top-quality Chardonnay Champagnes can slowly acquire a creamy biscuitiness).

Bite A very definite qualification of **grip**. Bite is usually a desirable characteristic, although an unpleasant bite is possible.

Bitterness One of the five basic tastes, bitterness may be an unpleasant aspect of a poorly made wine or an expected characteristic of an as-yet-undeveloped concentration of flavours that, with maturity, become rich and delicious. The term is often applied to **tannin**.

Blackstrap A derogatory term that originated when Port was an unsophisticated product, coloured by elderberries and very coarse.

Blanc de blancs (Fr) Literally "white of whites", describing a white wine made from white grapes. It is a term often, but not exclusively, used for sparkling wines.

Blanc de noirs (Fr) Literally "white of blacks", describing a white wine made from black grapes. It is a term that is often, but not exclusively, used for sparkling wines. In the New World, such wines usually have a tinge of pink (often no different from a fully fledged rosé), but a classic *blanc de noirs* should be as white as possible without artificial means.

Blanco (Sp) White.

Blind, blind tasting A winetasting at which the identity of the wines is unknown to the taster until after he or she has made notes and given scores. All competitive tastings are made blind.

Bloom Bacteria and **yeast** adhering to the grape's **pruina** creating a coating on all grapes, but it is most noticeable on black varieties.

Blowzy An overblown and exaggerated **fruity aroma**, such as fruit jam, which may be attractive in a cheap wine, but would indicate a lack of **finesse** in a more expensive product.

Blush wine A **rosé** wine that is probably cheap.

BOB An acronym for "buyer's own brand", under which many retailers and restaurants sell wine of increasingly good value – particularly in the supermarket sector, in which the selection process has been increasingly honed to a fine art since the early 1980s.

Bodega (Sp) Commercial winery. The Spanish equivalent of the Portuguese *adega* (ie a cellar or winery).

Body The impression of weight in the mouth, which is brought about by a combination of the fruit extract and alcoholic strength.

Bota (Sp) A Sherry butt (cask) with a capacity of between 600 and 650 litres (158 and 171 US gallons).

Botrytis A generic term for rot, but also often used as an abbreviation of *Botrytis cinerea.*

Botrytis cinerea The technically correct name for **noble rot**, the only rot that is welcomed by winemakers, particularly in sweet wine areas, as it is responsible for the world's greatest sweet wines. See also "The noble rot", pp176–177.

Botrytised grapes, botrytised Literally "rotten grapes", but commonly used for grapes that have been affected by *Botrytis cinerea.*

Bottle-age The length of time a wine spends in bottle before it is consumed. A wine that has good bottle-age is one that has sufficient time to **mature** properly. Bottle-ageing has a mellowing effect.

Bottle aromas Synonymous with **tertiary aromas**. In sparkling wine, bottle aromas are the mellowing complex aromas created after **disgorgement**, such as **toasty** and **biscuity**.

Bottle-fermented In sparkling wine production, **secondary fermentation** often takes places in the wine bottle that the wine will be sold in. See MÉTHODE CHAMPENOISE; TRADITIONAL METHOD; TRANSFER METHOD.

Bottle variation This can occur in all wines with time, when the most minuscule differences in each bottle have had time to take effect. **corks** commonly have a variation in oxygen permeability (the rate at which oxygen enters a bottle) of five-fold, which means that once the oxygen inside a bottle has been consumed, one bottle could age five times faster than another bottle due its different rate of **oxygen ingress**.

Bouchonné (Fr) Means "corked", which usually infers **TCA** taint.

Bouillage (Fr) French for the initial, extremely vigorous, stage of **fermentation**, literally and aptly translated as "boiling".

Bound All sorts of **compounds** in wine are chemically or biochemically bound (also known as fixed), but the most important bound (or fixed) compound is SO2.

Bound SO2 or **Bound sulphur** See SO2, BOUND.

Bouquet This should really be applied to the combination of smells directly attributable to a wine's maturity in bottle – thus "aroma" for grape-related smells and "bouquet" for maturation-related smells. But it is not always possible to use these words in their purest form – hence, aroma and bouquet may be considered synonymous in a literary, but not literal, sense. See AROMA.

Bourbes The heavy sediment, or *gros lies*, of formerly suspended matter in the juice, such as grapeskin, stalks, and pips removed during *débourbage*.

Bourgeois (Fr) *Cru bourgeois* is a Bordeaux château classification beneath *cru classé*.

Bourgogne 1. The wine of Burgundy. 2. A type of **oak** used for barrelmaking, from Châtillon or any other forest in the Burgundy region.

Branco (Port) White.

Breathing A term used to describe the interaction between a wine and the air after a bottle has been opened and before it is drunk.

Breed The **finesse** of a wine that is due to the intrinsic quality of grape and *terroir* combined with the irrefutable skill and experience of a great winemaker.

Brett or **Brettanomyces** A genus of **yeast** that can inhabit a winery or barrels (which may be sold from one winery to another, spreading the infection). Brett (as it is commonly referred to) is a common spoilage organism that mainly, although not exclusively, affects red wines. It creates various **volatile phenols**, including ethyl-4-phenol, which is responsible for the Brett off odours, such as the so-called sweaty saddle, barnyard, stables, and generally horsey smells.

Brut Normally reserved for sparkling wines, *brut* literally means "raw" or "bone dry". Even the driest wines, however, contain a little **residual sugar** (formerly

0–15g/l residual sugar; now 0–12g/l). *See also* "Levels of Champagne Sweetness", p256.

Brut Nature A style of sparkling wine that receives no *dosage*. Good, even great Brut Nature is possible to make in **Champagne**, but it is climatically difficult. It is also difficult in England and the mountain-vineyards of Trentodoc, but relatively easy in Franciacorta and for **Cava**. Most Brut Nature *cuvée*s should be enjoyed as soon as they are released, as they do not age well and are likely to become **oxidative**, possibly excessively oxidative (*See* SO2, THE MYTH OF BOUND). Furthermore, no sparkling wine without **residual sugar** can benefit from the potentially complex post-**disgorgement aromas** created by **Maillard reactions**. *See also* "Levels of Champagne Sweetness", p256.

Burnt Synonymous with baked and marginally uncomplimentary.

Burnt match The **aroma** of molecular **SO2**, burnt match is more acceptable, desirable even, in sparkling wine than in any other type of wine, where it might be construed as a fault. There is more molecular **sulphur** in low-**pH** sparkling wines than most other styles of wine. *See* SO2, FREE; SO2, MOLECULAR.

Butt *See* BOTA.

Butterscotch An excessive form of buttery. Can come from time in **oak** but is usually from diacetyl produced by **malolactic**.

Buttery This is normally a rich, fat, and positively delicious character found in white wines, particularly those that have undergone **malolactic fermentation**.

Buyer's Own Brand Often seen as the acronym BOB, this is a brand that belongs to the wine buyer, which could be a wine merchant, supermarket, or restaurant (the buyer is the seller as far as the consumer is concerned).

CA (Sp) Short for *Cooperativa Agrícola* and other titles denoting a local or regional cooperative.

Caecuban An **ancient Roman wine** (Lazio coast) considered to be a First Growth around 70 BC, thus on a par with **Falernian** and **Setine**, but had disappeared by the time of Pliny the Elder (AD 23–79). Smoother than **Falernian**, Caecuban was a sweet, strong, intoxicating white wine that turned "fire-coloured" as it aged.

Calenum This **ancient Roman wine** was made from the Molle Calenum, a large grape that gave a light wine that was, according to Pliny, kinder on the stomach than **Falernian**.

Canberra Doctor An easterly evening wind from the coast that helps to cool Canberra, Australia, although it virtually blows itself out by the time it reaches Braidwood or Bungendore. *See also* ALBANY DOCTOR; FREMANTLE DOCTOR.

Canopy The leafy area surrounding the vine.

Canopy management The vine's canopy is the collective arrangement of its shoots, leaves, and fruit. Ideally, a canopy will have most of its leaves well-exposed to sunlight, as this promotes fruit ripening through photosynthesis, and air circulation will be good, which provides the least-favourable environmental conditions for the development of fungal diseases. References to an excessive or too-vigorous canopy imply that the ratio of leaves to fruit is too high, causing herbaceous flavours in the wines produced from such vines.

Canopy microclimate The climatic environment immediately surrounding the canopy of a vine.

Cantina (It) Winery.

Cantina sociale (It) A grower's cooperative.

Cap The manta, or layer of skins that rises to the top of the vat during *cuvaison*.

Caramel An excessive form of buttery. Can come from time in **oak** but is usually from diacetyl produced by malolactic. *See also* "Caramel" in "Guide to Tastes and Aromas", p98.

Carbohydrate One of four classes of biochemical

compound. The most important carbohydrates are **sugars**. Grapes also contain cellulose.

Carbon Used by some to remove excess colour (due either to hot years like 2003 or to the style, such as *blanc de noirs)*, active carbon is frowned upon by some, including some of those who use it but pretend not to.

Carbon dioxide *See* CARBONIC GAS.

Carbonate A salt or **ester** of **carbonic acid**. Active or free carbonates increase the alkalinity of soil and are thus found in limestone soils, such as **chalk**.

Carbonic acid The correct term for carbon dioxide (CO2) when it dissolves in the water content of wine (to become H2CO3). Although sometimes referred to as a **volatile acid**, carbonic acid is held in equilibrium with the gas in its dissolved state and cannot be isolated in its pure form.

Carbonic gas Synonymous with carbon dioxide (**CO2**), this gas is naturally produced during the **fermentation** process (when the sugar is converted into almost equal parts of **alcohol** and carbonic gas). It is normally allowed to escape during fermentation, although a tiny amount will always be present in its dissolved form (**carbonic acid**) in any wine, even a still one; otherwise, it would taste dull, flat, and lifeless. If the gas is prevented from escaping, the wine becomes sparkling.

Carbonic maceration A generic term covering several methods of vinifying wine under the pressure of **carbonic gas**. Such wines, Beaujolais Nouveau being the archetypal example, are characterised by **amylic aromas** (peardrops, bubblegum, nail varnish). If this method is used for just a small part of a blend, however, it can lift the **fruit** and **soften** a wine without leaving such telltale aromas. *See also* "Carbonic Maceration", p68.

Casein A milk **protein** sometimes used for **fining**.

Cask-fermented *See* BARREL-FERMENTED.

Cassis (Fr) Literally "blackcurrant". If cassis is used by wine tasters in preference to "blackcurrant", it probably implies a richer, more concentrated, and viscous character.

Caucinian An individual-named site on the top slopes of the famous ancient Roman vineyard of **Falernian**.

Cava The generic **appellation** for **traditional-method** wines produced in delimited areas of Spain, although most are located in Cataluña.

Cava de Paraje Calificado A new designation for Cava. It is reserved for Cavas that come from a smaller area – *Paraje* ("place", meaning a single-vineyard or a group of vineyards) – of DO Cava, labelled as such due to its unique *terroir*. In addition, both the **viticulture** and **vinification** must comply with the strict and specific list of rules managed by the Cava Regulatory Board.

Cave, caves (Fr) French for "cellar", "cellars".

Caviste (Fr) A cellar-worker.

Cedarwood A purely subjective word applied to a particular **bouquet** associated with the bottle-maturity of a wine previously stored or fermented in wood, usually **oak**.

Cellar door A sales point at a winery. This is often a sophisticated retail operation in the New World, with sales of wine accessories, books, and T-shirts as well as the wines. For small growers in the Old World, however, purchases are more-often-than-not conducted in the producer's kitchen.

Cellar palate When the winemaker and other people within a single winery taste their own product so often, usually in isolation from the wines of other producers, that they cannot see any faults. The classic case is when a cellar gradually becomes infested with a mould, which when absorbed through the casks into a wine may be harmless to humans, but which most seasoned consumers would find objectionable. It can creep up so slowly that no one inside the firm notices, yet to a visitor it is so obvious.

Cellier (Fr) The French term for above-ground storage or warehousing premises.

Cendres noires (Fr) French for lignites.

Centrifugal filtration Not filtration in the pure sense, but a process in which unwanted matter is separated from wine or grape juice by so-called centrifugal force.

Cépage (Fr) Literally "grape variety", this is sometimes used on the label immediately prior to the variety, while in the plural format, *cépages*, it is used to refer to the **varietal** recipe of a particular *cuvée*.

Ceramic filtration An ultra-fine depth filtration that utilises **perlite**.

Chai, chais (Fr) Building(s) used for wine storage.

Chalk Commonly misused as a synonym for limestone, but whereas all chalk is limestone, not all limestone is chalk. Limestone applies to any sedimentary rock consisting essentially of carbonates, but chalk is geologically unique: the soft, white, fine-grained limestone of the Upper Cretaceous period. Seas covered most of what is now Europe 95 million years ago, and for 30 million years, while the earth experienced great calm, the microscopic calcareous material secreted by unicellular planktonic algae built up a lime-mud on the sea floor. During this monumental length of time, numerous forms of sea-life evolved, flourished, and became extinct, their calcareous bodies forming thin layers in the ever-thickening lime-mud. Some 65 million years ago this seemingly endless sedate marine activity was abruptly halted when earth movements lifted up part of the sea-bed to form the continental land mass. The lime-mud dried up to form chalk up to 300-metres (984-feet) thick, creating, for example, the South Downs of England, parts of the Loire and, of course, **Champagne**. The geological conditions under which the chalk formed are unique, and, as a result, so is the chalk.

Champagne Specifically a sparkling wine produced in a delimited area of northern France, the Champagne name is protected within the EU and in various other countries. It is, however, abused elsewhere, especially in the United States, where it is legal to sell domestically produced "Champagne", although the *champenois* did not do themselves any favours by abusing their own **appellation** on the fizz they produced in South America for 30-odd years.

Chaptalisation The addition of sugar to fresh grape juice in order to raise a wine's alcoholic potential. Theoretically it takes 1.7 kilograms of sugar per hectolitre of wine to raise its alcoholic strength by 1 per cent, but red wines actually require 2 kilograms to allow for evaporation during the **remontage**. The term is named after Antoine Chaptal, a brilliant chemist and technocrat who served under Napoleon as minister of the interior from 1800 to 1805 and instructed winegrowers on the advantages of adding sugar at pressing time.

Charm This is a subjective term; if a wine charms, it appeals without attracting in an obvious fashion.

Charmat method *See* CUVE CLOSE.

Château (Fr) Literally "castle" or stately home. While many château-bottled wines do actually come from magnificent edifices that could truly be described as châteaux, many may be modest one-storey villas, some are no more than purpose-built *cuveries,* while a few are merely tin sheds. The legal connotation is the same as for any domaine-bottled wine.

Châtillon A well-known type of **oak** used for barrelmaking, from any forest in the Burgundy region of France.

Cheesy This is a characteristic element in the **bouquet** of a very old **Champagne**, although other wines that have an extended contact with their **lees** – possibly those that have not been **racked** or **filtered** – may also possess it. It is probably caused by the production during **fermentation** of a very small amount of butyric acid that may later develop into an ester called ethyl butyrate.

Chef de caves (Fr) The cellarmaster or chief winemaker in **Champagne**.

Chemical compound *See* COMPOUND; BIOCHEMICAL COMPOUND.

Chestnut barrels Although chestnut is usually considered too porous and tannic for winemaking purposes and can colour a wine, barrels are made on a commercial basis from this wood. Most are produced by cooperages in Burgundy (by Billon in Beaune, Dargaud & Jaegle in Romanèche-Thorins, and Seguin Moreau in Chagny, for example) but seldom for Burgundian vintners. Some chestnut barrels are made in Portugal and Italy. Several wines specifically vinified in chestnut are made in Sardinia, principally from Vernaccia di Oristano and Cannonau (aka Grenache) grapes. A sparkling wine that undergoes its first **fermentation** in chestnut barrels is made from organically grown Xarel-lo near Sitges in Spain and sold under the Ancestral label. *See also* OAK.

Chewy An extreme qualification of meaty.

Chian A "most delicious" ancient Greek wine, according to Virgil and Pliny, with the best Chian coming from Ariusium on the island of Chios. Probably the greatest Greek wine of antiquity; the Romans certainly believed this to be the case. According to mythology, it was Oinopion, son of Dionysus and founder of Chios, who had personally taught the Chians the art of making "black wine", which meant red wine but was described as black because a red wine looked black in the earthenware drinking vessels of the time.

Chip-budding A method of propagating vines in which a vine bud with a tiny wedge-shape of phloem (live bark) and xylem (inner wood) is inserted into a **rootstock** in an existing root system.

Chlorosis A vine disorder caused by a mineral imbalance (too much active lime; not enough iron or magnesium) that is often called "green sickness".

Chocolaty, chocolate-box This is a subjective term often used to describe the odour and flavour of Cabernet Sauvignon or Pinot Noir wines. Sometimes the term "chocolate-box" is used to describe the **bouquet** of fairly **mature** Bordeaux. The **fruity** character of a wine may also be described as chocolaty in wines with a **pH** above 3.6.

Christmas cake A more intense version of the tasting term "fruitcake". Commonly found in maturing **Champagne**, usually from a high percentage of Pinot Noir.

Cigar-box A subjective term often applied to a certain complex **bouquet** in wines that have been **mature**d in **oak** and have received good **bottle-age** (usually used in relation to red Bordeaux).

Citric acid An organic acid, tiny amounts of which may be found in grapes (typically 0.25g/l in **Champagne**), often concentrated in the skin membrane. Citric acid is also sometimes added for acidification purposes. Citric acid is susceptible to microbial degradation by lactic acid bacteria, however, which could lead to the production of acetic acid and undesirably high levels of diacetyl, thus only **tartaric acid** is allowed for acidification in Champagne.

Citrus, citrussy This describes **aromas** and flavours of far greater complexity than mere lemony implies.

Civrais Lesser-known type of **oak** used for barrelmaking, from a forest in the Allier *département* in the centre of France.

Clairet (Fr) A wine that falls somewhere between a dark **rosé** and a light red wine, which is not the most endearing colour for any rosé, so if one scores highly, it must be seriously good to overcome the visual.

Claret An English term for a red Bordeaux wine. Etymologically, it has the same roots as the French term *clairet.*

Clarification, Clarify To remove any suspended matter that may cloud a wine (by **fining** or **filtration**) or grape juice (by *débourbage* or centrifuge).

Classic, classy These are both subjective words to convey an obvious impression of quality. These terms are applied to wines that not only portray the correct characteristics for their type and origin, but also possess the **finesse** and style indicative of top-quality wines.

Classico (It) This term may be used only for wines produced in the historic, or classic, area of an **appellation** – usually a small, hilly area at the centre of a **DOC**.

Clean A straightforward term applied to any wine devoid of unwanted or unnatural undertones of **aroma** and flavour.

Climat (Fr) A single plot of land with its own name, located within a specific vineyard. *See also* LIEU-DIT.

Clone A variety of vine that has developed differently due to a process of selection, either natural, as in the case of a vine adapting to local conditions, or by human intervention. Clones may be developed for specific reasons, such as to increase yield, resist certain diseases, or to suit local conditions etc. Identical clones of this one vine can then be replicated an infinite number of times by micro-biogenetic techniques. *See also* LOCALISED CLONE and "Clones and cloning", p39.

Clos (Fr) Synonymous with *climat*, except that this plot of land is, or was, enclosed by walls.

Closed Refers to the **nose** or **palate** of a wine that fails to show much character (or "open"). It implies the wine has some qualities – even if "hidden"– that should open up as the wine develops in bottle.

Closure Nothing to do with coming to terms, this is, as far as winespeak goes, a **cork**, **screw cap**, **crown cap**, or, indeed, anything used to seal a bottle.

Cloves Often part of the complex **bouquet** found on a wine fermented or **mature**d in **oak**, the **aroma** of cloves is actually caused by eugenic acid, which is created during the **toasting** of oak barrels.

Cloying Describes the sickly and sticky character of a poor-quality sweet wine, where the **finish** is heavy and often unclean.

CO2 *See* CARBONIC GAS.

Coan An ancient Greek wine from the island of Kos, Coan was mixed with seawater, and the salty wine became all the rage in the 4th century BC, leading to the process being copied in various parts of Greece.

Coarse Applies to a "rough and ready" wine, not necessarily unpleasant but certainly not fine.

Coconut Commonly found in very **mature**, impeccably preserved, fine quality **Champagne**.

Coconutty-oak Coconutty **aromas** are produced by various wood lactones that are most commonly found in American **oak**.

Coffee A hint of this luxuriantly rich, profoundly deep, yet often fleeting **aroma** is found in only the finest **Champagnes** of significant maturity.

Col Fondo (It) The traditional but newly fashionable style of Prosecco re-fermented in bottle and sold hazy with **yeast lees** still in the bottle. Due to difficulties in registering the name it is officially called *rifermentazione in bottiglia* ("re-fermented in bottle").

Colmated corks A type of natural **cork closure** that have has their lenticels sealed with cork dust and food-grade glue.

Commercial A commercial wine is blended to a widely acceptable formula. At its worst it may be bland and inoffensive, at its best it will be **fruity**, quaffable, and uncomplicated.

Compact fruit This term suggests a good weight of fruit with a correct **balance** of **tannin** (if red) and **acidity** that is presented on the **nose** and **palate** in a distinct manner that is opposite to **open-knit**.

Compiegne Lesser-known type of **oak** used for barrelmaking, from a forest in the Seine-et-Marne *département* of the Île de France.

Complete Refers to a wine that has everything (**fruit**, **tannin**, **acidity**, **depth**, **length**, and so on) and thus feels satisfying in the mouth.

Complexity An overworked word that refers to many different nuances of smell or taste. Great wines in their youth may have a certain complexity, but it is only with maturity in bottle that a wine will achieve its full potential in terms of complexity.

Compound A chemical or biochemical compound is a molecule that contains at least two different elements, thus all compounds are **molecules**, but not all molecules are compounds. *See also* BIOCHEMICAL COMPOUND; MOLECULE

Concoction Usually a derogatory term, but it can be used in a positive sense for a medley of flavours in an inexpensive wine.

Conditum Spiced **ancient Roman wine** that was sweetened with honey and contained laurel, mastic, saffron, date seeds, and dates soaked in wine.

Cooked Similar to **baked**, but may also imply the addition of grape concentrate to the wine during **fermentation**.

Cool A cool sensation in the mouth is usually attributed to menthol (as in peppermint). *See* HEAT.

Cool-fermented An obviously cool-fermented wine is very fresh, with simple **aromas** of apples, pears, and bananas.

Cork This material has been around since Roman times – or since the ancient Greek and Egyptian civilization s if we consider the sealing of amphorae with cork stoppers. The earliest post-Roman use of cork as a bottle stopper was in England in the 16th century (Shakespeare's *As You Like It*, 1598), although the English did not start binning bottles until the early 17th century, following the development of strong bottle glass. Prior to this, glass bottles stood upright and were little more than glorified carafes. Since the time bottles have been cellared on their side, which keeps the cork in contact with the wine, maintaining its elasticity and ensuring its seal, the humble cork has been the catalyst for our understanding of how wine **matures**.

Cork and agrafe Although originally conceived as a temporary closure for the second **fermentation** and ageing on **yeast**, the cork and agrafe is commonly found as a commercial seal for Coteaux Champenois, Ratafia, and Lambrusco, plus a few special *cuvée*s by **Champagne** producers such as Henri Bazin, Le Brun de Neuville, Henri Giraud, and René Jolly. Other sparkling wine producers using a cork and agrafe for the final presentation include Le Lude Agrafe Brut (Cap Classique). *See also* AGRAFE; CORK, FIRST.

Cork closures Corks can contribute **tannins** and **oak** characteristics to a wine, which is not surprising because cork is the bark of an oak tree (*Quercus suber*) and contains many of the same phenolic **compounds** and **volatile phenols**, including **vanillin**. Some critics view this as a natural part of a wine's maturation, while others point out that because its effect cannot be predicted, this phenomenon blemishes the wine in a way the winemaker never intended and must therefore be seen as a fault. Even if that is put to one side, the **oxygen ingress** of a cork can vary by more 300-fold (up to 1,000-fold, according to one study, but that result has never been repeated). This variability can be realistically reduced to 5-fold if narrowed down the same wine sealed with the same same grade of high-quality cork, but even a 5-fold difference in the rate of maturation is unacceptable. As an AWRI study stated, if you put the same wine into 12 different bottles, sealed each one with a cork that is **TCA**-free, and left them for 10 years, you would end up with 12 different wines. Although the incidence of **cork taint** has dramatically reduced from 8 per cent in the early 1990s to maybe 2 to 3 per cent today, the more we know about cork, the more issues we find. Unfortunately, unlike wines, which have feasible alternative closures, such as **screw caps**, sparkling wines are stuck with the cork for the time being. This has encouraged the rise of **microagglomerate technical corks**, however, the most successful to date being **DIAM MytiK**. *See also* AGGLOMERATED CORK; CORK HARVESTING; CORK MIRROR, CROWN CAP; DIAM; MICROAGGLOMERATE, SCREW CAP; SYNTHETIC CLOSURES

Cork, First The original practice of using a c**ork and agrafe** to seal the bottle for **second fermentation** and thence ageing on **yeast lees** (*sur lattes* and *sur point*) is seldom encountered these days, although the tradition is preserved by a few prestige *cuvées* in **Champagne** (Laurent-Perrier's Grand Siècle, and Taittinger's Comte de Champagne, for example) and various prestige producers elsewhere (most notably Gramona in Spain) and can be identified by the neck of the bottle, which needs a *bague carré* to secure the **agrafe**. **Champagne** seems well-equipped to handle the aromatic contribution of cork, whereas other more delicate sparkling wines can be so dominated by years of contact with cork that they taste decidedly oaky, and when this becomes excessive, it is a far more crude and aggressive oak **aroma** than that provided by **oak** barrels, even new oak barrels. Even when ideally suited wines are closed with **TCA**-free corks for second fermentation and yeast-ageing, however, there will still be a huge oxygen-ingress variation of up to 300 times to consider. If producers do not want to introduce, from such an early stage in production, the radically different rates of development that are intrinsic to natural cork, they should consider **DIAM MytiK** for temporary as well as final closures. *See also* CORK AND AGRAFE.

Cork flour The raw material used to make a **microagglomerate** closure. It looks nothing like flour and is not cork dust (an unwanted material in the production of any cork, technical or natural) and looks more like fine granules (far finer than **agglomerate**). After grinding cork into tiny particles, it is sifted to remove cork dust at one extreme, and woody parts, lenticels, and detritus at the other.

Cork dust This is not cork "flour" but it does look flour-like and it is unwanted. If the surface of a cork (any cork, technical or natural) is not sealed or polished, a lot of dust can and does accumulate in bags when corks are emptied into a hopper and can eventually find its way into a wine.

Cork harvesting Cork comes from the bark of the *Quercus suber*, an evergreen species of **oak** that can live for 200 to 250 years. It should be 20 or 25 years old before its first cork is harvested. Harvesting is conducted by removing the bark from the longest, straightest part of the trunk. This bark grows back and will be ready for harvesting every 7 to 12 years. The time between harvests may be affected by climatic conditions (eg in hotter climes, it is possible to harvest at shorter intervals than in cooler climes), but essentially, the longer a tree is left between harvests, the thicker the bark; and the thicker the bark, the higher quality cork it yields. The vast majority of cork trees are not part of a natural environment but have, like vines, been cultivated, with Portugal supplying half of the world's cork. The next major producer is Spain, after which, in order of importance (albeit on a much lower scale of production) come Algeria, Italy, Morocco, Tunisia, and France. In China, cork is produced to a small extent from *Quercus variabilis,* a deciduous oak found throughout Asia, but its potential yield is significantly lower than that of *Quercus suber*.

Cork mirror Known as a *rondelle* in French, a cork mirror is a disc of pure **cork** that is glued onto the wine-contact end of a regular sparkling wine cork. There can be one, two, or even three mirrors, depending on quality and cost. Mirrors are usually cut at a right angle to how full-sized, regular corks are cut, which gives a more attractive surface to the disc but exposes the wine from mirror to mirror by a "snakes and ladders" network of lenticels, fast-tracking the wine to the main body of **agglomerate cork**, negating any benefit of the mirrors. This could be resolved by ensuring the lenticels are colmated with **TCA**-free

cork dust, since TCA can migrate from the agglomerate to mirrors during the gluing process.

Cork taint *or* **Corked wine** This fault has been a recognised since the late 17th century, but until the second half of the 20th century, wine was the preserve of the rich and privileged, so the problem was contained. Since the 1970s, however, the consumption of wine has been enjoyed by a much wider and larger customer base, and consequently the incidence of "corked wine" has become common knowledge. The cause of cork taint was first identified in 1981 by a Swiss research scientist, and by the 1990s its incidence had been quantified at such an alarmingly high percentage (8 per cent) that a significant proportion of winemakers started looking for alternative closures. It was originally believed that a penicillin or aspergillus mould in the cork might be the cause of this taint, but such infections have always been extremely rare, which is why, in the late 1970s, Hans Tanner at the Wädenswil Research Institute (now merged with Changins research centre to become Agroscope) was commissioned by Swiss winegrowers to discover the true cause. After several years of little progress, Tanner was helped by a colleague who was doing pioneering work with gas-chromatography and mass-spectrometry, and eventually he became the first person to determine that the cork taint was due to extremely low levels of various chloroanisoles, with 2,4,6-trichloroanisole (commonly referred to as **TCA**) the main culprit. Initially thought to be exclusively the unwanted by-product of sterilizing corks with chlorine, TCA has since been identified at source in cork oak trees, in oak barrels, wooden pallets, and wooden roofs. Other **compounds** that may also be responsible include **geosmin**, which gives beetroots (beets) their earthy taste and can be found in reservoir water.

Correct This word describes a wine with all the correct characteristics for its type and origin, but not necessarily an exciting wine.

Côte, côtes (Fr) Slope(s) or hillside(s) of one contiguous slope or hill.

Coteaux (Fr) Slopes and hillsides in a hilly area that are not contiguous.

Coulure (Fr) A physiological disorder of the vine that occurs as a result of alternating periods of warm and cold, dry and wet conditions after bud-break. If this culminates in a flowering during which the weather is too sunny, the sap rushes past the embryo bunches to the shoot tips, causing a vigorous growth of foliage, but denying the clusters an adequate supply of essential nutrients. The barely formed berries thus dry up and drop to the ground.

Coupage (Fr) The process of blending by cutting one wine with another.

Creamy A subjective term used to convey the impression of a creamy flavour that may be indicative of the variety of grape or method of **vinification**. I tend to use this word in connection with the **fruitiness** or oakiness of a wine. Dr Tony Jordan believes that creaminess in a sparkling wine is probably a combination of the **finesse** of the *mousse* (created by the most minuscule of bubbles and their slow release) and an understated **malolactic** influence, the combined effect of which is picked up at the back of the throat on the **finish** of the wine, and this is most apparent in Chardonnay-based wines.

Creamy-oak A more subtle, lower-key version of the **vanilla-oak** character that is most probably derived from wood lactones during maturation in small **oak** barrels.

Crémant (Fr) Although traditionally ascribed to a **Champagne** with a low-pressure and a soft, creamy *mousse*, this term has now been phased out in Champagne as part of the bargain struck with other producers of French sparkling wines who have agreed to drop the term *méthode champenoise*. In return they have been exclusively permitted to use this old Champagne term to create their own **appellation**s, such as Crémant de Bourgogne and Crémant d'Alsace.

Crisp A clean wine, with good **acidity** showing on the **finish**, yielding a refreshing, clean taste.

Cross A vine that has been propagated by crossing two or more varieties within the same species (within *Vitis vinifera*, for example). In contrast, a **hybrid** is a cross between two or more varieties from more than one species. *See also* " Crosses and Hybrids", p39.

Crossflow filtration A relatively new, high-speed form of micro-filtration in which the wine flows across (not through), a membrane filter, thus avoiding build-up.

Crown cap The *capsule couronne* in French or common beer-bottle cap, was first patented in the United States in 1892. It is now widely used as the temporary closure for sparkling wine when it undergoes its **second fermentation** and its subsequent ageing on **yeast**. Crown caps have been used as the final closure following **disgorgement** by Domaine Chandon in Australia but have failed to catch on. Crown caps constructed from standard steel are for short-term storage only, and for cellars with less than 70 per cent humidity; those made from aluminium are for short- to mid-term storage; while stainless steel crown caps are for long-term storage and cellars with more than 70 per cent humidity.

Crown cap, First generation Crown caps were first introduced into **Champagne** in the mid 1960s, when they were constructed of standard steel and had cork liners.

Crown cap, Second generation This generation saw the phasing out of cork liners, which were prone to **TCA**, and replaced by synthetic liners in the late 1980s. *See* CROWN-CAP LINERS.

Crown-cap liners This is the lining that sits between the **crown cap** and the glass, cushioning the seal. Since the demise of cork as a crown-cap liner, there are three primary types of lining available:

- Expanded polyethylene: For aging up to 24 months ($0.64cc$ CO_2 loss per 24 hours).

- Injected polyethylene: For aging up to 36 months ($0.25cc$ CO_2 loss per 24 hours).

- Saranex®: For aging up to 48 months ($0.14cc$ CO_2 loss per 24 hours), this is an expanded polyethylene liner with Saranex both sides.

- Polyethylene-tin-Saranex and Polyethylene-aluminium-Saranex: These liners exist for **screw caps,** but apparently not for **crown caps,** which is strange because the CO_2 loss would amount to about a tenth of Saranex-only liners, and their construction is even better suited to the reductive environment of **yeast**-ageing than it is for protecting still wine. *See also* SARANEX.

Cru or **crû** (Fr) Literally means "growth", as in *cru bourgeois* or *cru classé*.

Cru bourgeois (Fr) A non-classified growth of the Médoc.

Cru classé (Fr) An officially classified French vineyard.

Crush Grapes are often crushed so that the juice can macerate in the skins prior to and during **fermentation** – to obtain colour for red wines and aromatic qualities for white wines. In the United States and Australia, "the crush" is synonymous with the harvest in general and with the crushing/pressing in particular.

Cryptogamic Refers to a fungus-based disease such as grey rot.

CS (It) Short for *Cantina Sociale* and other titles denoting a local or regional cooperative.

Cultivar A term used mainly in South Africa for a cultivated variety of wine grape.

Cut 1. In blending, a wine of a specific character may be used to cut (mix with) a wine dominated by an opposite quality. This can range from a bland wine that is cut by a small quantity of very acidic wine, to a white wine that is cut with a little red wine to make a rosé, as in pink **Champagne**. The most severe form of cutting is also called stretching and involves diluting wine with water, an illegal practice. 2. A cut in pressing terms is a point at which the quality of juice changes, the term deriving from the days of old vertical presses when the lid of the press would be lifted and workers would cut up the compacted mass with sharp spades, piling it in the middle so that more juice may be extracted. 3. In matching food and wine, a wine with high **acidity** may be used to cut (**balance**) the organoleptic effect of grease from a grilled or fried dish or an oily fish, just as the effervescence of a fine sparkling wine cuts the creamy texture of certain soups and sauces.

Cuvaison (Fr) The **fermentation** period in red wine production, during which the juice is kept in contact with its skins.

Cuve (Fr) Vat; a *cuve* should not be confused with *cuvée*.

Cuve close (Fr) The tank-fermented method of producing sparkling wine was first described by Francesco Scacchi in De Salubri Potu Dissertatio in 1622, then invented by Federico Martinotti in 1895, before being adapted on a large-scale commercial basis by Eugène Charmat in 1907. This bulk-production method of making sparkling wine by a second **fermentation** inside a sealed vat often attracts low-quality raw materials, thus has a low reputation, but for aromatic grape varieties it is the best method of sparkling wine production, bar none. Also known as the tank method, Charmat method, and Metodo Martinotti.

Cuvée (Fr) This originally meant the wine of one *cuve*, or vat, but now refers to a specific blend or product that, in current commercial terms, will be from several vats.

Cuverie, Cuvier (Fr) The room or building housing the fermenting vats (*cuves*).

CV (Fr) Short for *Coopérative de Vignerons* and various other titles that denote a local or regional cooperative.

Cytoplasm The liquid (80 per cent water) found inside a **yeast** cell.

Cytoplasmic ribosomes Ribosomes are found within the intracellular membrane of a **yeast** cell. They are effectively molecular machines that knit amino acids together to form polypeptide chains.

Darnay Lesser-known type of **oak** used for barrel-making, from a forest in the Vosges *département* of northeastern France.

Deacidification The only methods of reducing **acidity** that fit well with the sparkling wine structure are tartrate precipitation and **malolactic**. Some producers in certain countries have resorted to chemical deacidification (adding calcium carbonate), but this should be avoided like the plague.

Definition A wine with good definition is one that is not just clean with a correct **balance** but that also has a positive expression of its grape variety or origin.

Dégorgement (Fr) *See* DISGORGEMENT.

Dégorgement Tardive (Fr) Late-disgorged. *See* LD.

Degree days *See* HEAT SUMMATION.

Délestage (Fr) Commonly known to anglophile winemakers as "rack and return". *Délestage* is a process designed to produce softer red wines by reducing harsh **tannins** from the grape seeds and is particularly successful in areas where unripe seeds are common due to uneven ripening. The basic *délestage* procedure starts after a cold soak of juice and skins. The juice is drawn off into a separate tank, allowing the **cap**, or manta, to fall to the bottom of the first tank, where it is left to drain for several hours. During this time, many of the seeds are loosened from the pulp and can be caught by a filter that allows free passage to the draining juice, which also goes to the second tank. Once the cap has drained out, the drain is closed off, and the juice from the second tank is pumped back into the first tank, where it is mixed with cap. This process is repeated on a daily basis until all the seeds and their harsh **tannins** are removed.

Delicate Describes the quieter characteristics of quality that give a wine **charm**.

Demi-muid (Fr) A large oval barrel with a capacity of 300 litres (600 litres in **Champagne**).

Demi-sec (Fr) This literally means "semi-dry" but such wines actually taste quite sweet (formerly 33–50g/l **residual sugar** for sparkling wines in the EU; now 32–50g/l). *See also* "Levels of Champagne Sweetness", p256.

Dendometer A very accurate device that measures the minute swelling and shrinkage of the vine trunk in response to water use. It can be used to control the amount of irrigation water taken up by the vine, rather than the amount that goes into the ground.

Denominação de Origem Controlada (Port) *See* DOC.

Denominación de Origen (Sp) *See* DO.

Denominación de Origen Calificada (Sp) *See* DOCa.

Denominazione di Origine Controllata (It) *See* DOC.

Denominazione di Origine Controllata e garantita (It) *See* DOCG.

Denthis An "unfired" and "unboiled" ancient Greek wine from the western foothills of the Mount Taygetus region of Messinia that was described by the Greek poet Alkman as having *anthosmias,* or "the **bouquet** of flowers".

Département (Fr) A French geopolitical region, geographically similar to a UK county but more geopolitically similar to a US state.

Depth This refers primarily to a wine's depth of flavour and secondarily to its depth of interest.

Depth filtration The separation of solids from a liquid solely inside a **filtration** medium such as **kieselguhr** (*See* DIATOMACEOUS EARTH). A rotary drum vacuum or plate-and-frame filter is commonly used.

Deutscher Qualitatsschaumwein (Ger) Same as *Deutscher Sekt.*

Deutscher Sekt A sparkling wine made by any method (though probably **tank method**), exclusively comprising wine from German-grown grapes. It may indicate a maximum of two grape names and should be at least 10 months old when sold.

Dextrose *See* GLUCOSE

Diacylglycerols These **molecules** consist of **glycerol** and two fatty acids. *See* FATTY ACID; GLYCEROL.

DIAM The **technical cork** arm of the Oeno Group and the generic brand of **TCA**-free still wine corks.

DIAM MytiK DIAM's sparkling wine brand of technical **microagglomerate** cork is currently produced by two processes (Diamant® and Revtech) and in two formats: Classic and Access. The **Diamant® Process** removes all detectable **TCA**, but the **Revtech** removes only 75 to 85 per cent of TCA. All MytiK closures offer a uniformity of post-**disgorgement** maturation due to their consistency of **oxygen ingress** (and thus development in bottle), which regular corks, with up to 300 times variation, cannot hope to rival. *See* DIAMANT® PROCESS, REVTECH; SUPERCRITICAL CO2.

DIAM MytiK Access This slightly larger-grained closure releases fractionally more oxygen into the bottle on insertion and has a slightly higher **oxygen ingress**. It was developed for those sparkling wine producers who felt that DIAM MytiK was so effective that it left their wines too tight and closed. They wanted its **TCA**-free guarantee and its consistent oxygen ingress qualities, but they also wanted a closure that allowed the wine to open up after **disgorgement**, and so Access was born.

DIAM MytiK Classic The original **DIAM MytiK** closure has the finest granules, lowest oxygen release when inserted, and the lowest **oxygen ingress** throughout its life.

Diamant® Process Since 1973 **supercritical CO2 extraction** has been widely used in the perfume and food industries, but **DIAM** and **DIAM MytiK**, the first and only **TCA**-free corks produced by this process, have only been available since as recently 2005. Cork flour is placed in an autoclave, which is then injected with compressed supercritical **CO2**. When in a super-critical state, CO2 combines the penetrative behaviour of gas with the extractive property of liquid, removing TCA, TCA precursors, and all other volatile **compounds** to yield cork granules with no smell whatsoever. The contaminated CO2 is piped into a sealed system where it is cooled, reverting the CO2 to gas form, forcing the TCA and all other volatile contaminants to drop out. The CO2 is then filtered back to a pure state so that it can be re-used endlessly in the Diamant Process. Meanwhile, the odourless cork granules are mixed with a food-grade binding agent and food-grade polymer microspheres, moulded, and baked. When mixed, the microspheres have the consistency of talcum powder, but when baked they swell to fill the gaps between the cork granules. There are two basic, commercial DIAM MytiK formats, each visibly recognisable by its size of granule: Classic for the finest and Access for slightly larger (although nowhere near as large as found in **agglomerate corks**), but is another invisible difference, the proportion of microspheres used. The most widely known quality of microspheres are their elasticity, but they are also impervious to oxygen, thus the proportion of microspheres-to-cork affects oxygen-ingress, and OTR is the primary criterion for using either Classic or Access. In addition to these two commercial offerings, DIAM can produce bespoke MytiK closures to suit the specific requirements of any sparkling wine producer by varying the proportion of microspheres, playing with the size of cork granules, and tweaking the food-grade glue. *See* DIAM; DIAM MYTIK; DIAM MYTIK ACCESS; DIAM MYTIK CLASSIC; SUPERCRITICAL CO2.

Diatomaceous earth Also known as **kieselguhr,** this is a fine, powdered, silaceous earth evolved from decomposed deep-sea algae called diatoms. *See also* PERLITE; CERAMIC FILTRATION; POLISHING.

Dimethyldisulphide *See* LIGHT-STRUCK.

Diose A 2-carbon **monosaccharide**, only one of which exists and that can be found in wine: **glycolaldehyde**.

Dipeptide One of the more common groups of **peptide** found in sparkling wine, a dipeptide contains two amino acids. *See* Peptides.

Direct-producing hybrid Also "direct producer" and "hybrid direct producer", this term dates back to 1880, when *Vitis vinifera* arieties were crossed with various native American species in an attempt to produce grapes with the character of the former from vines with the resistance to **phylloxera** of the latter, so that the phylloxera-infested vineyards of France could be replanted. This was one of the two approaches to overcome phylloxera and, ultimately, the least successful, since no direct-producing hybrid – then or now – has ever matched the quality of the classic varieties that have established the specific reputation of each famous winemaking area, whereas the alternative strategy of grafting those classic *vinifera* vines to American **rootstock** did.

Dirty This applies to any wine with an unpleasant off-taste or off-smell and is probably the result of poor **vinification** or bad bottling.

Disaccharide A sugar containing two monosaccharides. **Sucrose** (common table sugar), for example, is comprised of **glucose** and **fructose**. There are some 25 disaccharides, but the most common are sucrose, **lactose**, and **maltose** (only the first and last have any relevance for wine).

Disgorgement This is part of the process of making a **bottle-fermented** sparkling wine such as **Champagne**. After **fermentation**, the **yeast** forms a deposit, which must be removed. To allow for this removal, the bottles are inverted in a freezing brine for just long enough for the sediment to form a semi-frozen slush that adheres to the neck of the bottle. This enables the bottle to be re-inverted without disturbing the wine. The temporary cap used to seal the bottle is removed and the internal pressure is sufficient to eject or "disgorge" the slush of sediment without losing very much wine at all. The wine is then topped up and a traditional Champagne cork is used to seal the bottle.

Dissolved oxygen This refers to oxygen **molecules** in the wine itself. *See also* HEADSPACE OXYGEN; OXYGEN TRANSMISSION RATE.

Distinctive Describes a wine with a positive character. All fine wines are distinctive to some degree or other, but not all distinctive wines are necessarily fine.

Diurnal difference In **viticulture**, any reference to a diurnal or daily difference will invariably be a reference to temperature, comparing the highest daytime temperature with the lowest nighttime temperature – the greater the difference, the better the grape's **acidity** retention. There can be a wide diurnal difference in relatively cool wine areas, such as **Champagne**, as well as in essentially hot ones, such as Idaho or mountain-influenced viticultural areas. The month of harvest is vital for diurnal difference. In Champagne, for example, there is a diurnal difference of 10° to 15°C (50° to 59°F) in September, the usual month of harvest, whereas it is just 5°C (41°F) in August, when the acid level in any ripe grapes would plummet. There has always been the occasional August harvest in Champagne, but the first decade of the 21st century uniquely produced three August harvests within a 10-year span.

DMDS Dimethyldisulphide. *See* LIGHT-STRUCK.

DO (Sp) This stands for Spain's *Denominación de Origen,* which is theoretically the equivalent of the French **AOC**.

Doble pasta (Sp) Red wines macerated with double the normal proportion of grape skins to juice during **fermentation**. *See also* "Spanish Label Language", p393.

DOC (It, Port) Short for Italy's *Denominazione di Origine Controllata* and Portugal's *Denominação de Origem Controlada,* which are theoretically the equivalent of the French **AOC**.

DOCa (Sp) Abbreviation for Spain's *Denominación de Origen Calificada,* which is the equivalent of the Italian **DOCG**.

DOCG (It) Italy's *Denominazione di Origine Controllata e Garantita* is theoretically one step above the French **AOC**. Ideally, it should be similar to, say, a *premier cru* or *grand cru* in Burgundy or a *cru classé* in Bordeaux, but in reality, it is almost as big a catch-all as Italy's *Denominazione di Origine Controllata* itself.

Doppelstück (Ger) A very large oval cask with a capacity of 2,400 litres (634 US gallons).

Dosage Sugar added to sparkling wine after **disgorgement**, prior to shipping, via the *liqueur d'expédition,* the amount of which is controlled by the terminology used on the label (**Brut Nature, Extra Brut, Brut, Sec, Extra Sec, Demi-Sec,** and **Doux**). Since the sugar is used to **balance** the **acidity**, which is necessary for any sparkling wine, to keep it as fresh as possible throughout its lengthy production process, and the acidity slowly rounds out with age, the older the wine, the less dosage required. Sometimes producers declare this either as the dosage added or the total **residual sugar** (ie the dosage plus any residual sugar that may happen to be in the wine prior to adding the dosage, which could be up to two grams per litre). They do not usually clarify which, however, thus, although the sweetness may appear to be precisely defined in grams per litre, more often than not there will be no way of telling whether, say, 5g/l is actually 5g/l of residual sugar or perhaps 5g/l of dosage and maybe 1 or 2g/l of unfermentable sugar, thus one 5g "dosage" could be as sweet or sweeter than someone else's 6g "dosage", regardless of acidity balance.

Doux (Fr) "Sweet", as applied to wines (50 g/l or more of **residual sugar** for sparkling wines in the EU). Rarely encountered, although there are more examples in **Champagne** today than there were 10 years ago. The last commercially available *doux* made by a *grande marque* was produced by Roederer in 1983 and sold under its famous Carte Blanche label. That wine had 60 grams of residual sugar, not really sweet by *doux* standards, although ten years earlier Roederer's Carte Blanche had 80 g/l of residual sugar, and 100 years ago it was 180 grams. By 1998, Roederer's Carte Blanche *dosage* had dropped to 45 grams (i.e., just a demi-sec), while today it is just 38 grams. *See also* "Levels of Champagne Sweetness", p256.

Dreuille Lesser-known type of **oak** used for barrelmaking, from a forest in the Allier *département* in the centre of France.

Drip irrigation Various forms exist, but at its most sophisticated, this is a computer-controlled watering system programmed with the vine's general water requirement and constantly amended by a continuous flow of data from soil sensors. The water is supplied literally drip-by-drip through a complex system of pipes with metered valves.

Drying up Describes a wine that has dried up and lost some of its **freshness** and **fruit** through ageing in the bottle. It may still be enjoyable, but remaining bottles should not be kept long.

Duplex soils So-called when two contrasting soil textures are found layered, with a sharp divide between the two. Duplex soils usually consist of a coarse soil over a fine-grained soil, and are commonly found in Western Australia, where they are invariably sand over clay. They are categorised by colour (red, yellow, brown, dark, and grey duplex soils) based on the colour of the subsoil, not the topsoil.

Dusty Akin to "peppery" in a red wine; a blurring of **varietal** definition in a white wine (in which case, it might be due to **TCA**).

Earth filtration This term can be synonymous with **depth filtration**. *See* DEPTH FILTRATION.

Earthy Describes a drying impression in the mouth. Some wines can be enjoyably earthy, but the finest-quality wines should be as clean as a whistle. When a wine is very earthy, it is usually due to a preponderance of **geosmin**, which can occur naturally in grapes, but in excess can give a wine a corked taste.

Easy This term is to a certain extent synonymous with accessible, but probably implies a cheaper, value-for-money wine, whereas "accessible" often applies to finer wines.

Easy to drink *or* **Easy drinking** This is not the opposite of hard to drink; it is the opposite of a wine that you do not want to rush and similar, therefore, to a magazine or a page-turner of a book being "easy to read".

Eau-de-vie (Fr) Literally, "water of life"; specifically, a grape-derived spirit.

Edelfäule (Ger) The term for **noble rot**. *See* BOTRYTIS CINEREA.

Edelkeur (S Afr) The term for **noble rot**. *See* BOTRYTIS CINEREA.

Edge Almost, but not quite, synonymous with **grip**; wine can have an edge of **bitterness** or **tannin**. Edge usually implies that a wine has the capacity to develop, while grip may be applied to a wine in various stages of development, including fully **mature** wine.

Edgy Synonymous with nervy or "nervous".

Eggs, rotten *See* ROTTEN EGGS.

Egg white A traditional **fining** agent that fines out negatively charged matter.

Einzellage (Ger) A single-vineyard wine area; the smallest geographical unit allowed under German wine law.

Eiswein (Ger) An *Eiswein* occurs through extremely unusual circumstances, whereby grapes left on the vine to be affected by **noble rot** are frozen by frost or snow. They are harvested and rushed to the winery where they must be pressed in their frozen state so that the ice rises to the surface and can be skimmed off. This results in a fantastic concentration of juice, sugar, **acidity**, extract, and minerals. Some so-called Icewines produced in other less-fastidious countries are made from grapes that have been frozen, but not pressed in this state, therefore no ice is skimmed off and the wines are no more concentrated than they would have been if picked before they were frozen.

Electrons Negatively charged particles that surround the atom's nucleus.

Elegant; elegance A subjective term applied to wines that may also be termed "**stylish**" or "possessing **finesse**".

Elevated fruit Synonymous with **VA lift**.

Élevé en fûts de chêne (Fr) Aged in **oak** barrels.

Eleveur, élevage (Fr) Literally "bringing up" or "raising" the wine. Both terms refer to the traditional function of a *négociant*: namely to buy ready-made wines after the harvest and take care of them until they are ready to be bottled and sold. The task involves **racking** the wines and blending them into a marketable product as each house sees fit.

Embryo bunches In spring, the vine develops little clusters of miniature green berries that will form a **bloom** a few weeks later. If a berry successfully flowers, it is capable of developing into a grape. The embryo bunch is, thus, an indication of the potential size of the crop.

Encépagement (Fr) The relative proportions of the grape varieties in a blend.

En foule (Fr) Meaning "in a crowd", *en foule* is the rather haphazard effect created when ungrafted vines are cultivated by various methods of layering.

Enologist, enology The American spelling of oenologist, oenology. *See* OENOLOGIST, OENOLOGY.

En primeur (Fr) Classic wines such as Bordeaux are offered for sale *en primeur,* which is to say within a year of the harvest, before the final blending and bottling has taken place. For experienced buyers given the opportunity to taste, this is a calculated risk, and the price should reflect this element of chance.

Entry-level wine From the producer's point of view, this will be his cheapest, most basic quality of wine. From a critic's point of view, this will be the cheapest wine worth buying.

Enzymes These are **proteins** produced by living organisms, which can be anything from human beings down to the most basic life forms, such as **yeast** cells. Enzymes function as catalysts for specific biochemical reactions, breaking down **molecules**, such as the yeast enzymes that break down molecules of **sugar** into molecules of **carbonic gas** and **alcohol** during **fermentation**. Or the enzymatic breakdown of the yeast cells themselves in the biochemical process of **autolysis**. In fact, enzymes play so many important and varying roles in the **vinification** process that some have been isolated and developed into commercial products to assist in basic tasks such as pressing and settling or to tweak a wine this way or that. Most commercial enzyme products are based on the following enzymes: pectinase (primarily assists **maceration** and **clarification**; used specifically for macerating or pressing of white juice, clarification of white juice, yeast autolysis, red colour extraction, red colour stability, and **filtration**); cellulase (primarily assists maceration and colour stability; used specifically for macerating or pressing of white juice, red colour extraction, and red colour stability); hemicellulase (primarily assists maceration and colour stability; used specifically for macerating or pressing of white juice, red colour extraction, and red colour stability); glucanase (primarily assists clarification;

used specifically for macerating or pressing of white juice, red colour stability, and filtration, including filtration of **botrytised wines**); glycosidase (primarily assists maceration and colour stability; used specifically for macerating or pressing of white juice); polygalacturonase (used specifically for macerating or pressing of white juice, and filtration); ß-glucosidase (primarily assists clarification; used specifically to release bound **terpenes** in aromatic wines); rhamnosidase (used specifically to release bound terpenes in aromatic wines); apiosidase (used specifically to release bound terpenes in aromatic wines); arabinofuransidase (used specifically to release bound terpenes in aromatic wines); and lysozyme (used specifically to kill or control lactic bacteria. Some enzyme preparations (notably those extracted from moulds such as *Aspergillus niger* or *Trichoderma harzianum*) are so aggressive that they come with warnings not to use before pressing whites or fermenting reds. The enzymes in questions destroy grapeskin, creating a haze of solids so fine that they are either very difficult or impossible to remove. Some winemakers swear by commercial enzyme products; others view them as unnecessary tinkering.

Enzymes, reducing Also known as a reductase, a reducing enzyme removes **oxygen** by catalysing a chemical reduction reaction.

Eresos This ancient Greek wine was reputedly the best wine from Lesbos.

Erythritol A tetrose (4-carbon **monosaccharide**) formed by **yeast** during **fermentation,** found in very small quantities. Not fermentable. First isolated by the French in rhubarb in 1849, the anti-cancer properties of erythritol have been recognised in the United States since 2010.

Esters Sweet-smelling **compounds,** formed during **fermentation** and throughout maturation, that contribute to a wine's **aroma** and **bouquet.**

Estufagem (Port) The process whereby Madeira is heated in ovens called *estufas,* then cooled.

Ethanoic acid Synonymous with **acetic acid.**

Ethanol Synonymous with **ethyl alcohol.**

Ethyl alcohol This main **alcohol** in wine is so important in quantitative terms that to speak of a wine's alcohol is to refer purely to its ethyl alcohol content.

EU Lot Number Proposed by an EC directive in 1989 and implemented by all member states of the Community by 1992, this Lot Number must be indicated on every bottle of wine produced in or sold to the EU. Should a wine have to be removed from general distribution for any reason, this code can save unnecessary waste by pinpointing the shipment involved.

Everyday wines Inexpensive, **easy-drinking** wines.

Ex-cellars Wines offered *en primeur* are usually purchased "ex-cellars"; the cost of shipping the wine to the importer's cellars is extra, on top of which any duty and taxes will be added.

Expansive Describes a wine that is big, but open and accessible.

Expressive A wine that is expressive is true to its grape variety and area of origin.

Extra Brut A very dry style of **Champagne** (0–6g/l), of which only those with the maximum permitted *dosage* have any realistic chance of ageing smoothly, so it is best consumed on purchase. *See also* "Levels of Champagne Sweetness", p256.

Extra Sec Literally "extra dry", but with between 12 and 17g/l of **residual sugar** (formerly 12–20g/l for sparkling wines in the EU), an extra sec **Champagne** or sparkling wine can have a distinct hint of sweetness. *See also* "Levels of Champagne Sweetness", p256.

Extract Sugar-free soluble solids that give body to a wine. The term covers everything from **proteins** and vitamins to **tannins,** calcium, and iron.

Falernian An individual-named site encompassing the best lower slopes of the famous ancient Roman vineyard of **Falernian**. This site produced sweet white wines that were very high in **alcohol** content, according to Pliny, who claimed that it is "the only wine that ignites when a flame is applied to it". *See* ANCIENT ROMAN WINES; *see also* 121 BC in "A Chronology of Wine", p111.

Falernum Possibly the most well-known ancient Roman vineyard, Falernum was produced from Aminean grapes (known as Greco today) grown on the slopes of Mount Falernus in the Campanium hills. These slopes were subdivided into three individually named sites: **Caucinian** (on the top slopes), **Falernian** (from the best mid-slopes), and **Falernian** (from the lower slopes). Falernum was a late-harvested white wine that was usually drunk at between 10 and 20 years of age, when it was amber-coloured. The most fashionable of all **ancient Roman wines**, Falernum was probably the first wine to experience such a heavy demand that the quality was noted to decrease as the volume produced increased. *See also* 121 BC in "A Chronology of Wine", p111.

Fall bright A liquid that becomes limpid after cloudy matter drops as sediment to the bottom of the vessel is said to fall bright.

Fall over A wine that goes past its peak and starts to decline at a relatively young age, and at a faster than normal rate, is said to fall over.

Farmyardy In 1982, Anthony Hanson wrote "great Burgundy smells of shit", but whether just a touch of farmyard or a full-blown manure **aroma**, these **mercaptan**-derived aromas are no longer acceptable. *See* "Manure" in "Guide to Tastes and Aromas", p99.

Fat A wine full in **body** and **extract.** It is good for any wine to have some fat, but fat in an unqualified sense can be derogatory, and no wine should be too fat, as it will be **flabby** or too **blowzy.**

Fatty acids A term sometimes used for **volatile acid**s.

Faustian An individual-named site encompassing the best mid-slopes of the famous ancient Roman vineyard of **Falernian**. These mid-slopes were part on the estate of belonging to Faustus Cornelius Sulla, son of the dictator Sulla, the only man in history who successfully attacked and occupied both Athens and Rome. Many regarded Faustus to be the best Falernum. *See* ANCIENT ROMAN WINES.

Feminine A subjective term used to describe a wine with a preponderance of delicately attractive qualities, rather than weight or strength. Descibes a wine of striking beauty, grace, and **finesse**, with a silky texture and exquisite style.

Fermentation The biochemical process by which **enzymes** secreted by **yeast** cells convert sugar **molecules** into almost equal parts of **alcohol** and **carbonic gas.** *See also* "Fermentation", pp64–66.

Fermentazione naturale (It) Literally "naturally fermented" in Italian, which should apply to every wine ever made – even a carbonated fizz must have been naturally fermented in the first place. What it is supposed to imply, however, is that a wine has been rendered sparkling by natural refermentation in a tank or bottle (usually the former).

Fertilizer A chemical product used to enrich the soil with one or more of the three basic requirements for all plant life: potassium (for fruit development and general plant metabolism), phosphorus (for root development), and nitrogen (for leaf development). Technically the term also refers to manure, compost, and other natural means of soil enrichment.

Feuillette (Fr) A small Burgundian barrel with a capacity of 114 litres, or 30 US gallons (132 litres, or 34 US gallons in Chablis).

Ficelage Securing the permanent **cork** and placque with a wire cage or muzzle.

Ficelè a l'ancìenne The precursor to the wire cage, this 18th-century method of hand-tying a cork to the bottle with twine was reprised by Audoin de Dampierre for his Family Reserve and Cuvèe de Prestige, since when it has been followed by a number of other **Champagnes**, including Michel Boilleau Descendance, Bovière Doria L'Équilibre, Wolf's Doria L'Équilibre, Dom Bacchus Cuvée Antique, and Sanger Les Oubliés.

Field blend, field mix The best description I have seen for this is "a wine recipe planted in the ground". It is not a homogenous vineyard planted to a single grape variety (of which there may be several different clones), but a vineyard planted with a collection of grape varieties that reflect traditional Old World practices of several generations ago. The advantage is that if a disease or disorder affected one variety, the others would probably pull through unscathed. The disadvantage, however, is that the different varieties do not ripen at the same time; this was not a problem in the old days, however, since it was common practice to make several *tries*, or sweeps, through the vineyards, picking only the ripe grapes and cutting out any rotten ones.

Filter, filtration The removal of suspended matter by one of four basic methods: **depth filtration** (also known as **earth filtration**); **pad filtration** (also known as sheet filtration), **membrane filtration** (also known as micro-porous filtration), and **crossflow filtration**. There is also **centrifugal filtration**, which is not filtration in the pure sense but achieves the same objective of removing unwanted particles suspended in wine or grape juice. *See also* "Filtration", pp66–67.

Fine wines Quality wines, representing only a small percentage of all wines produced.

Finesse That elusive, indescribable quality that separates a fine wine from those of lesser quality.

Fining The clarification of fresh grape juice or wine is often sped up by the use of various fining agents that operate by an electrolytic reaction to fine out oppositely charged matter. *See also* "Fining", p66.

Finish The quality, and a person's enjoyment, of a wine's **aftertaste.**

Firm Refers to a certain amount of **grip**. A firm wine is a wine of good constitution, held up with a certain amount of **tannin** and **acidity**.

First pressing The first pressing yields the sweetest, cleanest, clearest juice.

Fixed acidity The total **acidity** less the **volatile acidity.**

Fixed sulphur The principal reason why SO_2 (sulphur dioxide) is added to grape juice and wine is to prevent **oxidation**, but only **free sulphur** can do this. Upon contact with wine, some SO_2 immediately combines with oxygen and other elements, such as **sugars** and acids, and is known as fixed or bound sulphur. What remains is free sulphur, capable of combining with **molecules** of oxygen at some future date.

Flabby The opposite of **crisp**, referring to a wine lacking in **acidity** and consequently dull, weak, and short.

Flash pasteurization A **sterilization** technique that should not be confused with full **pasteurization** . It involves subjecting the wine to a temperature of about 80°C (176°F) for between 30 and 60 seconds.

Flat 1. A sparkling wine that has lost all of its *mousse*. 2. A term that is interchangeable with **flabby**, especially when referring to a lack of **acidity** on the **finish**.

Fleshy This term refers to a wine with plenty of **fruit** and **extract** and implies a certain underlying firmness.

Flood-irrigated The crudest form of irrigation and the cheapest where water is readily available (such as from the Andes snow-melt in parts of Chile and Argentina). Sluice-gates are opened up, allowing water to flow into irrigation channels around and between rows of vines.

Flor (Sp) A scum-like **yeast** film that naturally occurs and floats on the surface of some Sherries as they

mature in part-filled wooden butts. It is the *flor* that gives Fino Sherry its inimitable character.

Flurbereinigung (Ger) A modern viticultural method of growing vines in rows that run vertically up and down slopes, rather than across in terraces.

Flying winemaker The concept of the flying wine-maker (a consultant who gets his hands dirty) was born in Australia, where, due to the size of that continent and the staggered picking dates, highly sought-after consultants would hop by plane from harvest to harvest. The term was first applied to Brian Croser and Tony Jordan.

Focus Can apply to both **aroma** and flavour, focus indicates a clarity, preciseness, and purity that can range from broad to linear.

Foliar feeds Plant nutrients that are sprayed directly onto, and are absorbed by, the foliage.

Fontainbleau Lesser-known type of **oak** used for barrelmaking, from a forest in the Seine-et-Marne *département* of the Île de France.

Formian *or* **Formianum** An ancient **Roman wine** from the Gulf of Caieta, Formian was compared by Aelius Galenus (also known as Galen of Pergamon) to **Privernatinum** and **Rhegium** but was richer and earlier-developing. Athenaeus of Naucratis also compared it to those two wines, claiming Formian was smoother and that it **matured** more quickly.

Fortified Fortification with pure **alcohol** (usually very strong grape spirit of 77 to 98 per cent) can take place either before **fermentation** (as in Ratafia de Champagne and Pineau des Charentes), during fermentation (as in Port and Muscat de Beaumes de Venise), or after fermentation (as in Sherry).

Foudre (Fr) A large wooden cask or vat.

Foxy The very distinctive, highly perfumed character of certain indigenous American grape varieties that can be sickly sweet and cloying to unconditioned **palates**.

Free-run juice *See* VIN DE GOUTTE.

Free SO2 *or* **Free sulphur** The active element of **SO2** (sulphur dioxide) in wine, produced by free sulphur combining with intruding **molecules** of **oxygen**. *See* SO2, FREE.

Fremantle Doctor Also known as the "Freo Doctor", this afternoon sea breeze brings a cooling relief to better parts of the Swan Valley in Western Australia. *See also* ALBANY DOCTOR; CANBERRA DOCTOR.

French paradox In 1991, Morley Safer, host of the CBS show *60 Minutes*, screened a programme about the so-called French paradox. This described how the high-cholesterol-consuming, high-alcohol-drinking, low-exercising French have a very low mortality rate from heart disease compared to health-conscious Americans, who have low-cholesterol diets, exercise frequently, and drink relatively little alcohol. Part of the explanation was attributed to the Mediterranean diet, in which milk plays a negligible role and wine – particularly red wine – a very important one. Although it is a complete food for the young, milk is unnatural for adults, who cannot digest it properly. The more milk an adult drinks (and Americans are particularly high consumers of milk), the greater the risk of cardiovascular disease, while three glasses of wine a day have a proven protective effect against cardiovascular disease. *See also* HEALTH BENEFITS OF WINE.

Fresh Describes wines that are **clean** and still vital with youth.

Friable Term used to describe a soil structure that is crumbly or easily broken up.

Frizzante (It) Semi-sparkling.

Frizzantino (It) Very lightly sparkling, between still and semi-sparkling (ie perlant).

Fructose One of the two most important **monosaccharides** found naturally in grapes, fructose is a fermentable **hexose** (6-carbon) **sugar**. Also known as levulose and fruit sugar.

Fruit, fruity Wine is made from grapes and must therefore be 100 per cent fruit, yet a fruity flavour depends on the grapes used having the correct combination of ripeness and **acidity**.

Fruit-bomb The etymological origin of this term provides its own definition. The **fruit** in any wine described as a fruit-bomb will be super-rich, super-lush, and super-concentrated. California was probably the first to produce such wines, but the term was not coined at that juncture. The concept of a fruit-bomb first emerged on the Australian wine-competition circuit – for what variety originally, no one is sure, but it was soon applied to Pinot Noir more than any other wine. Having failed so miserably to achieve the basic **varietal character** of Pinot Noir, Australian wine producers and judges were happy for a short while in the early 1990s, when more and more Pinot Noir wines demonstrated this varietal purity. They soon got fed up with simple varietal Pinot Noir, however. Both the producers and judges wanted the right structure on which to hang that fruit and more **finesse** and potential **complexity**, but all they got was more fruit. There were successes, of course, but the harder most tried, the more of a caricature their Pinot Noir became. As an example of a fruit-bomb, somebody once wrote "think Jim Carrey rather than Jeremy Irons", and that's about as close as any analogy can get.

Fruitcake This is a subjective term for a wine that tastes, smells, or has the complexity of the mixed dried-fruit richness and spices found in fruitcake. Commonly found in maturing **Champagne**, usually from a high percentage of Pinot Noir.

Fuder (Ger) A large oval cask with a capacity of 1,000 litres (264 US gallons), more prevalent in Mosel areas than in those of the Rhine.

Full This term usually refers to body, as in "full-bodied". A wine can be light in **body** yet full in flavour, however.

Fully fermented A wine that is allowed to complete its natural course of **fermentation** and so yield a totally dry wine.

Fût (Fr) A wooden cask, usually made of **oak**, in which wines are aged, or fermented and aged.

Galactosamine A 6-carbon **amino sugar** found naturally in grapes and wine.

Galactose An unfermentable **hexose** (6-carbon **monosaccharide**) found naturally in grapes. Although of little significance in wine, it is also a **wood sugar**, thus relatively more abundant in oak-aged wines. Furthermore, it is also a constituent of gum arabic, which may be added to the *dosage*, thus any galactose in a sparkling wine could come from one or more of three different sources.

Ganau Revolution A custom-designed high-pressure autoclave with a replenishable water- and steam-cleansing system that is said to dramatically decrease **TCA** levels in cork.

Garrigue (Fr) A type of moorland found in Languedoc-Roussillon.

Gauran *or* **Gauranum** An **ancient Roman wine** produced a few kilometres west of Naples in the hills above Puteoli. It was the source of "Purpŭra", the finest and most expensive dye of the ancient world. According to Athenaeus of Naucratis, "Gauran is both rare and excellent, besides being vigorous and rich." He also believed it to be smoother than either **Praenestine** or **Tiburtine**.

Gelatine A positively charged **fining** agent used for removing negatively charged suspended matter in wines, especially an excess of **tannin**.

Generic Describes a wine, usually blended, of a general **appellation**.

Generous A generous wine gives its **fruit** freely on the **palate**; an ungenerous wine is likely to have little or no fruit and, probably, excess **tannin**. All wines should have some degree of generosity.

Genus The botanical family Ampelidaceae has 10 genera, one of which, *Vitis*, through the sub-genus *Euvites*, contains the species **Vitis vinifera**, to which all the famous winemaking grape varieties belong.

Geosmin A chemical **compound** sometimes found in wine; responsible for the characteristic earthiness of beetroot and the **earthy** taste of some potatoes.

Glass closures *See* VINO-LOK.

Glucans **Polysaccharides** that are found naturally in **yeast**, where they account for some 60 per cent of the yeast's walls, the membrane through which **autolysis** is carried out.

Gluconic acid Recording gluconic acid levels started as recently as 1996. As a by-product of grey rot, the presence of gluconic acid can be used as an indicator of how clean a crop is or how much rot it contains.

Glucose One of the two most important **monosaccharides** found naturally in grapes. Glucose is a fermentable **hexose** (6-carbon) **sugar**. Also known as dextrose.

Gluggy Easy to guzzle.

Glycerol A sugar alcohol. After **alcohol** and **carbon dioxide**, glycerol is the most abundant product of **fermentation**.

Glycogen During the second phase of accelerated **yeast** activity, the cells start to grow and divide, storing **sugar** in the form of glycogen, which is later broken down and used to continue yeast growth, eventually leading to **autolysis**.

Glycolaldehyde This is not only a **diose**, it the only possible 2-carbon **monosaccharide**, although not strictly a saccharide nor indeed a true **sugar**. Glycolaldehyde can be found in wine, when formed by bacterial activity during **malolactic**.

Good grip A healthy structure of **tannin** supporting the **fruit** in a wine.

Goût de terroir (Fr) Literally translates as "taste of earth" but does not infer any sort of **earthy** taste. It denotes a combination of good **typicity** and "sense of place" – the essence of *terroir*.

Graft The joint between the **rootstock** and the scion of the **producer vine**.

Grand cru (Fr) Literally "great growth". In regions such as Burgundy, where the term's use is strictly controlled, it has real meaning (in other words, the wine should be great relative to the quality of the year), but in other winemaking areas where there are no controls, it will mean little.

Grand vin (Fr) Normally used in Bordeaux, this term applies to the main wine sold under the château's famous name, and it will have been produced from only the finest barrels. Wines excluded during this process go into second, third, and sometimes fourth wines that are sold under different labels.

Grande marque (Fr) Literally a "great or famous brand". In the world of wine, the term *grande marque* is specific to **Champagne** and applies to members of the *Syndicat de Grandes Marques,* which include, of course, all the famous names.

Granvas (Sp) Spanish term for tank method. *See* CUVE CLOSE.

Grape acids The most important acids in grapes are tartaric and malic. These are organic acids and as such they are fixed acids. Other organic acids found in grapes include citric, gluconic, succinic, and, possibly, lactic. Also found are volatile acids (acetic, sulphurous); phenolic acids, amino acids, and fatty acids. See Fixed acidity; Titratable acidity; Total acidity; and individual named acids.

Grapey This term may be applied to an **aroma** or flavour that is reminiscent of grapes rather than wine, and is a particular characteristic of German wines and wines made from various Muscat or Muscat-like grapes.

Grappa (It) A rough spirit distilled from the grapeskins and stalks that are left after pressing, which are mixed with water and fermented.

Grassy Often used to describe certain wines (commonly Colombard, Scheurebe, or Sauvignon Blanc) portraying a grassy type of **fruitiness**, usually through low ripeness (as opposed to under-ripeness, which would be green).

Green Young and tart, as in Vinho Verde. It can be either a derogatory term or simply a description of a youthful wine that might well improve.

Green pruning Pruning is a bit of a misnomer, as this is really a method of reducing yields by thinning out the potential crop when the grapes are green (unripe) by cutting off a certain percentage of the bunches, so that what remains achieves a quicker, greater, and more even ripening. Also called summer pruning.

Grip This term applies to a firm wine with a positive **finish**. A wine showing grip on the finish indicates a certain bite of **acidity** in white wines and of **tannin** in red wines.

Grippy Good grippy **tannins** imply ripe tannins that have a nice tactile effect without seeming in the least **firm**, **harsh**, or **austere**.

Grosbois Lesser-known type of **oak** used for barrel-making, from a forest in the Allier *département* in the centre of France.

Großlage, Grosslage (Ger) A wine area in Germany that is part of a larger district or Bereich.

Growth *See* Cru.

Gunpowder A complex, sulphidic **aroma** that precedes **toast**.

Gutsy A wine full in **body**, **fruit**, **extract**, and – usually – **alcohol**. The term is normally applied to wines of fairly ordinary quality.

Guzzly This term is synonymous with **gluggy**.

Halbfüder (Ger) An oval cask with a capacity of 500 litres (132 US gallons), more prevalent in Mosel areas than in those of the Rhine.

Halbstück (Ger) An oval 600-litre (158-US gallon) cask.

Hard The opposite of ample, it suggests a lack of **fruit**.

Harsh A more derogatory term than **coarse**.

Headspace oxygen The amount of oxygen in the headspace of a bottle.

Health benefits of wine The consumption of wine in moderation flushes out the cholesterol and fatty substances that can build up inside the body's artery walls. It does this through the powerful **antioxidant** properties of various chemical **compounds** found naturally in wine (through contact with grapeskins), the most important of which are polyphenols such as procyanidins and rytoalexins such as resveratrol. Most cholesterol in the body is carried around the body on LDLs (low-density lipoproteins), which clog up the arteries. By contrast, HDLs (high-density lipoproteins) do not clog up the arteries, but take the cholesterol straight to the liver, where it is processed out of the system. The antioxidants convert LDL into HDL, literally flushing away the cholesterol and other fatty substances. Together with alcohol itself, these antioxidants also act as an anti-coagulant on the blood, diminishing its clotting ability, which reduces the chances of a stroke by 50 per cent in contrast with non-drinkers. (I would, however, be equally as dishonest as the neo-Prohibitionists who make phoney health-danger claims if I did not also point out the one true health danger of moderate drinking that came to light in 2002, when the British *Journal of Cancer* published a study demonstrating that a woman's risk of contracting breast cancer increases by 6 per cent if she consumes just one drink per day, and this rises to 32 per cent if she has three or four drinks per day. The report concludes that 4 per cent of all breast cancers are attributable to alcohol. This should be put into context, however, as Dr Isabel dos Santos Silva of the International Agency for Research on Cancer wrote, "Alcohol intake . . . is likely to account, at present, for a small proportion of breast cancer cases in developed countries, but for women who drink moderately, its lifetime cardioprotective effects probably outweigh its health hazards." And as Dr Philip Norrie pointed out, 10 times the number of women die from vascular disease than from breast cancer.)

Heat The hot and cool sensations in the mouth are physiological experiences created when receptors on the tongue and in the throat are activated by compounds such as capsaicin (the hot sensation of chillies) and menthol (peppermint). Although these receptors have a direct neural link to the brain and are located close to receptors on the taste buds, they do not activate the taste receptors and are not therefore classified as tastes. They work independently from any **aromas** picked up by olfaction. (Menthol triggers its receptors even when applied directly to the tongue with the olfactory bulb fully occluded.)

Heat summation A system of measuring the growth potential of vines in a specific area in terms of the environmental temperature, expressed in degree days. A vine's vegetative cycle is activated only above a temperature of 10°C (50°F). The time during which these temperatures persist equates to the vine's growing season. To calculate the number of degree days, the proportion of the daily mean temperature significant to the vine's growth – the daily mean minus the inactive 10°C (50°F) – is multiplied by the number of days of the growing season. For example, a growing season of 200 days with a daily mean temperature of 15°C (59°F) gives a heat summation of 1,000 degree days Celsius (1,800 degree days Fahrenheit) based on the following calculation: $(15 – 10) \times 200 = 1,000$.

Hectare A measurement of area; 1 hectare is equal to 10,000 square metres or 2.471 acres.

Hectolitre Equivalent of 100 litres (26 US gallons).

Heptose A 7-carbon **sugar**, very few of which exist and none of which can be found in wine.

Herbaceous A green-leaf or white-currant characteristic that is usually associated with too much vigour in the vine's **canopy**, which can cause under-ripeness, resulting in excessive pyrazine content. A herbaceous quality can also be the result of aggressive extraction techniques employed for red wines fermented in stainless steel.

Herbal, herbal-oak These terms apply to wines **mature**d in cask, but unlike **vanilla-oak**, **creamy-oak**, **smoky-oak**, and **spicy-oak**, their origin is unknown. A herbal character devoid of oak is usually derived from the **varietal character** of a grape and is common to many varieties.

Herbicide A weed-killer that is usually, but not necessarily, a highly toxic concoction of chemicals.

Heurige (Austrian) The name given to the new wine that, from 11 November each year, may be sold in the small taverns of the same name that are owned by *vignerons* who made the wine. Vine branches are hung over the entrance to the *Heurige* when the new wine is available and should be removed during the hours when the premises are not actually open.

Hexose A 6-carbon **monosaccharide**. All fermentable **sugars** are hexoses, but not all hexoses are fermentable. Those found in grapes or wine include: **fructose** (important), **galactose**, **glucose** (important), and **mannose**.

High-density vines Vines planted close together compete with each other to yield higher-quality fruit, but less of it per vine, than vines planted farther apart. Initial planting costs are higher and more labour is required for pruning and other activities, but if the vineyard is in **balance**, the greater number of vines should produce the same overall volume per hectare, even though the output per vine is reduced. Quantity can therefore be maintained while significantly raising quality, although there is a threshold density which vineyards must reach before real benefits appear. For example, more than half the vineyards in the New World are planted at less than 2,000 vines per hectare (800 per acre) and 1,200 to 1,500 per hectare (485 to 600 per acre) is very common, whereas in **Champagne**, 6,666 vines per hectare (2,699 per acre) is the minimum allowed by law, 7,000 to 8,000 (2,830 to 3,240 per acre) the average, and 11,000 (4,450 per acre) possible. In **pre-phylloxera** times, it was something like 25,000 vines per hectare (10,000 per acre). Indeed, before California's vineyards were mechanised, the average density of vines was twice what it is now because every other row has been ripped up to allow entry for tractors. When Joseph Drouhin planted his vineyard in Oregon, he planted 7,450 vines per hectare and brought over French tractors that straddled the rows of vines, rather than going between them. All of a sudden high-density vineyards entered the American vocabulary, although Drouhin did not consider them to be high density – merely a matter of course.

High-tone A term used in this book to describe elements of the **bouquet** that aspire to **elegance**, but that can become too exaggerated and be slightly reminiscent of vermouth.

Higher alcohols Whereas regular **alcohol** (ie **ethyl alcohol** aka ethanol) has only two carbons, higher alcohols (aka fusel oils) have more than two carbons, thus they have greater molecular weight and a higher boiling point. The most important higher alcohols in wine are amyl or isoamyl alcohol, isobutyl alcohol, and propyl alcohol. These and other less-common higher alcohols can have an aromatic effect on wines that may be positive or negative. The type and quantity of higher alcohol found in wine is primarily a **yeast** function.

Hogshead A barrel with a capacity of between 300 and 315 litres (79 and 83 US gallons), commonly found in Australia and New Zealand.

Hollow A wine that appears to lack any real flavour in the mouth compared to the promise shown on the **nose**. Usually due to a lack of **body**, **fruit**, or **acidity**.

Honed Skilfully crafted.

Honest Applied to any wine, but usually to one that is of a fairly basic quality, honest implies it is true in character and typical of its type and origin. It also implies that the wine does not give any indication of being souped-up or mucked around with in any unlawful way. The use of the word "honest" is, however, a way of damning with faint praise, for it does not suggest a wine of any special or truly memorable quality.

Honeyed Many wines develop a honeyed character through **bottle-age**, particularly sweet wines and more especially those with some **botrytis** character. Some dry wines can also become honeyed, however, a **mature** Riesling being the classic example.

Horizontal tasting A tasting of different wines of the same style or **vintage**. A vertical tasting consists of different vintages of the same wine.

Hot 1. A hot sensation in the mouth is usually attributed to an imbalance between **fruit** and **alcohol** so that the latter dominates, or the searing effect (if the wine is dry) of spice-laden **terpene** aromatics on the **finish** of a Gewürztraminer. *See* HEAT. 2. Synonym for **baked**.

House claret An unpretentious and not too expensive everyday-drinking red Bordeaux.

Hybrid A cross between two or more grape varieties from more than one species.

Hydrogen sulphide When hydrogen combines with SO_2 (sulphur dioxide), the result is a smell of **rotten eggs**. If this occurs prior to bottling and is dealt with immediately, it can be rectified. If allowed to progress, the hydrogen sulphide can develop into **mercaptans** and ruin the wine.

Hydrolysis The chemical breakdown of a **compound** by water, which is split it into its component parts. The most common hydrolysis in winemaking is the hydrolysis of **sucrose**, which is split into its component parts of **glucose** and **fructose**. All forms of hydrolysis may be assisted and speeded-up by **enzymes**, which in the case of sucrose would be either sucrase or invertase.

Hyperoxidation Also known as hyperoxygenation, a deliberate browning of non-sulphured grape juice by exposing it to air prior to **fermentation**. This might seem radical and counter-intuitive as an anti-oxidation technique, but it works by oxidising polyphenols, which drop out during clarification, taking the oxygen **molecules** away. The theory and practice of hyperoxygenation was established by Müller-Späth (1977) and Guerzoni (1981). It was used to reduce the colour and oxidation potential of *vins de tailles* in Champagne in 1989. The benefits of hyperoxygenation are, however, disputed and certainly the results vary according to grape variety and style. Nebbione is the only wine to this author's knowledge that must be made by hyperoxygenation.

Hyperoxygenation *See* Hyperoxidation.

Icewine *See Eiswein.*

Icon wine This category is above ultra-premium and will be priced in hundreds or thousands of dollars.

IGP (Fr) Common abbreviation for *Indication Géographique Protégée,* the French language version of the EU umbrella classification encompassing *vins de pays* and other national equivalents. The English equivalent is PGI (Protected Geographical Indication).

Indicação de proveniência regulamentada (Port) *See* IPR.

Inky Can refer either to a wine's opacity of colour or to an inkiness of character indicating a deep flavour with plenty of **supple tannin**.

IPR (Port) Short for *Indicação de Proveniência Regulamentada,* a Portuguese quality designation that falls between **DOC** and **VR**.

Iron This is found as a trace element in fresh grapes that have been grown in soils in which relatively substantial ferrous deposits are located. Wines from such sites may naturally contain a tiny amount of iron, which is barely perceptible on the **palate**. If there is too much iron, the flavour becomes medicinal. Above 7 milligrams per litre for white and 10 milligrams per litre for red, there is a danger of the wine going cloudy. But wines of such high iron levels should have been blue-fined prior to bottling. *See* Fining.

Isinglass A gelatinous **fining** agent obtained from the swim-bladder of freshwater fish and used to clear hazy, low-**tannin** wines.

Jammy Commonly used to describe a **fat** and eminently drinkable red wine rich in **fruit**, if perhaps a bit contrived and lacking **elegance**.

Jug wine California's mass-produced *vin de table*, synonymous with carafe wine.

Jupille A lesser-known type of **oak** used for barrelmaking, from a small forest in the Sarthe *département* of the Loire Valley.

Kabinett (Ger) The first rung of predication in Germany's **QmP** range, one below **Spätlese**, and often drier than a **QbA**.

Ketonic acids A classification that includse alpha-ketoglutaric acid, pyruvic acid, glutaric acid, gluconic acid, and galacturonic acid. *See* Gluconic acid.

Kieselguhr A form of **diatomaceous earth**.

Labrusca Native American grape, not to be confused with Lambrusco (although it has the same etymological origins). *Vitus labrusca* is a species, not a variety, and is included in the Micropedia for clarification.

Lactic acid The acid of sour milk, this is also created during the **malolactic fermentation**, which converts harsh **malic acid** into soft lactic acid and **carbonic gas**. Insignificant amounts may also be found in grapes. *See* Malolactic.

Lactose The so-called **milk sugar**, lactose is a **disaccharide** that consists of **galactose** and **glucose**. Included for reassurance only, lactose is never found in wine, not even wine's fined with casein (a milk **protein**).

Lagar (Port) A rectangular concrete receptacle in which people tread grapes.

Laid-back A term that has come into use since the arrival of California wines on the international scene in the early 1980s. To call a wine laid-back usually implies that it is very relaxed, **easy to drink**, and confident of its own quality.

Landwein German equivalent of *vin de pays*.

Late disgorged *See* LD.

LD A sparkling-wine term that stands for "late disgorged" and, paradoxically, means the same as "recently disgorged". The use of LD implies that the wine in question is of a **mature vintage** that has been kept on its **yeast** deposit for an extended period. *See also* RD.

Leaching A term that may be used to refer to the deliberate removal of **tannin** from new oak by steaming – or when discussing certain aspects of soil, such as **pH**, that can be affected when carbonates are leached (removed) by rainwater.

Lees Deposits of dead **yeast** or residual yeast and other particles that precipitate, or are carried by the action of **fining**, to the bottom of a vat of wine after **fermentation** and ageing. *See also* Lees, Gros or heavy; Lees, Fine or light.

Lees, Fine *or* **light** After the first **racking**, when the wine is removed from its **heavy lees** (*gros lie*), the sediment builds up more slowly and is much finer and silkier. When producers talk about keeping a wine on its lees and/or stirring the lees (*bâttonage*), they are invariably referring to the light lees (*fine lie*). *See also* Lees, Gros or heavy.

Lees, Gros *or* **heavy** The first **lees**, which drop out prior to or during **fermentation**. The lees from the settling are called the *bourbes*. The *gros lie* will contain pips, bits of skin, stalk, and other debris and should always be removed, although where *gros lie* ends and *fine lie* begins is always open to discussion. *See also* Lees, Fine or light.

Lemony Many dry and medium-sweet wines have a tangy, **fruity acidity** that is suggestive of lemons.

Length A wine that has length is one whose flavour lingers in the mouth a long time after swallowing. If two wines taste the same, yet you definitely prefer one, but do not understand why, it is probably because the one you prefer has a greater length. *See also* Balance.

Lesbian The sweet ancient Greek wine of Lesbos. *See also* Eresos.

Lie (Fr) The French for **lees**. *See also* Sur lie.

Liège et agrafe (Fr) *See* Cork and Agrafe.

Lieu-dit (Fr) A named site (plural: *lieux-dits*). This term is commonly used for wines of specific growths that do not have grand cru status.

Light-struck When wine is exposed to light, it develops an unpleasant **compound** called dimethyldisulphide (**DMDS**). At it's very worst, DMDS smells of stagnant water, old drains, and sewage, as anyone who attended the International Sparkling Wine Symposium in 2013 and witnessed my demonstration can testify. At lower thresholds, DMDS merely inflicts an otherwise fresh wine **aroma** with the barest hint of rotten cabbage. Although DMDS can be formed in other ways, this compound is most commonly found in wine when a sulphur-bearing **amino acid** called methionine is broken down by exposure to various wavelengths of light: 341nm, 370nm, 375nm, 380nm, 440nm, and 442nm are the most dangerous wavelengths often mentioned in scientific papers. The wavelengths responsible for the light-strike fault are typically referred to as ultraviolet (UV). UV light extends from 100nm to 400nm and although this range harbours more harmful wavelengths than visible light, the 440nm and 442nm peaks are definitively within the visible spectrum. Moreover, they are peaks and all 400–523nm wavelengths pose a danger of some measure. It's just that they take longer to wreak their havoc. According to studies (Vierra), peak wavelengths can degrade methionine into DMDS after just 60 minutes exposure to fluorescent light, while other research has shown that carbonic gas elevates the detection of this compound. Clear glass is commonly used for **rosé** and **blanc de blancs Champagnes**, including some of the very highest quality **cuvées**. Whatever the producer does to safeguard his Champagne from DMDS while it is within his own production facilities, he has no control over its exposure to light once it has left. Readers are advised never to purchase a clear bottle of any wine straight from the shelf, particularly sparkling wine. Always request a bottle that has never been removed from its gift box, and when you take it home, always keep it in that gift box unopened until you want to drink it, storing it in a cool and – most important – dark place, until that moment. If a sparkling wine is not sold with its own gift box with tamper-proof cellophane wrapping, ask for one to be removed from an as yet unopened carton of six bottles. If for any reason the store owner cannot do this or refuses, you should not buy the wine.

Light vintage A light **vintage** or year produces relatively light wines. Not a great vintage, but not necessarily a bad one either.

Lime This is the classic character shared by both the Sémillon and Riesling grape varieties when grown in many areas of Australia – which explains why Sémillon from the Hunter Valley used to be sold as Hunter Riesling.

Limousin A famous type of **oak** used for barrelmaking, from any forest in the Haute-Vienne, Creuse, and Correze *départements* in the centre of France.

Linalool A **compound** found in some grapes, particularly the Muscat and Riesling varieties. It contributes to the peachy-flowery fragrance that is characteristic of Muscat wines.

Lingering Normally applied to the **finish** of a wine – an **aftertaste** that literally lingers.

Lipids One of four classes of **carbohydrate** found in wine, lipid content (monoacylglycerols, diacylglycerols, and triacylglycerols) increases during the **second fermentation** and **autolysis**. These lipids represent a significant source of **aroma** and flavour **compounds** in the finished product and are generally believed to improve foam stability, although there have been some contradictory findings.

Liqueur d'expédition Sugar dissolved in wine that is added to **Champagne** and other sparkling wines after **disgorgement** to produce various styles (**brut nature, extra brut, brut, extra sec, sec, demi-sec,** and **doux**), each depending on the amount of **residual sugar** in the final product. **Sulphur** is usually also added to prevent premature **oxidation** (*See* Low sulphur regimes). As the oxidative effect of **disgorgement** tends to magnify whatever is added at this stage of the process, the choice of the *dosage* wine is important, whether it is to maintain a neutral effect (ie by using exactly the same as the rest of the bottle) or to inject any colour, freshness, crispness, weight, **complexity**, a **varietal** touch, or a hint of **oak** that might be deemed lacking.

Liqueur de tirage The bottling liqueur, which consists of wine, **yeast**, **sugar**, and yeast nutrients, is added to still **Champagne** or other traditional sparkling wine to induce the *mousse*. A little bentonite (**fining** agent) will also be included at this juncture, to promote full sedimentation prior to *remuage*. The amount of sugar used determines how fizzy the wine will be. In theory between 4 and 4.3

grams of sugar per litre of wine will produce one atmosphere of pressure, the fluctuation being due to the alcoholic degree of the base wine (the more **alcohol**, the less efficient the yeast becomes). Between 24 and 25.8 grams of sugar are therefore required in theory to produce 6 atmospheres, but account must be taken of the potential loss of pressure at **disgorgement**, thus 27 grams is generally regarded as the rule-of-thumb *dosage* for a fully sparkling wine. Most Champagnes utilise 22 to 24 grams of sugar and are around 5 atmospheres, whereas most New World sparkling wines use 18 to 22 grams and are less fizzy at some 4.5 atmospheres.

Liquoreux (Fr) Literally "liqueur-like", this term is often applied to dessert wines of an unctuous quality. (Sometimes also "liquorous".)

Liquorice A quality often detected in Monbazillac, but may be found in any rich, sweet wine. The term refers to the concentration of flavours from heat-shrivelled grapes, rather than **botrytised grapes**.

Liveliness, lively A term that usually implies a certain youthful **freshness** of **fruit** due to good **acidity** and a touch of **carbonic gas**.

Localised clone A variant of vine variety that is peculiar to a specific *terroir*. With time, some localised clones can evolve into a genetically distinct variety. *See also* MASSAL SELECTION.

Longevity Potentially long-lived wines may owe their longevity to a significant content of **tannin, acidity, alcohol,** and/or **sugar**.

Lora This **ancient Roman wine** was effectively an undistilled form of *grappa*, having been produced from the cake of grapeskins left after pressing, which was then mixed with water. Bitter and tannic, Lora was further cut with water and given to slaves.

Low-sulphur regimes Some producers have switched to low-sulphur regimes, and a number of those have got it wrong, ending up with overly **oxidative, aldehydic** wines. To produce wine with less **sulphur** is admirable, but to produce it and maintain the same smoothly maturing style requires adjustments in the whole winemaking process. Of these adjustments, the most important is to shift the emphasis of when to first add **SO2** and to ensure that just before final corking is the one stage that absolutely must receive SO2, as the oxidative shock of **disgorgement** must be compensated. It is one thing if a producer is deliberately making a distinctly oxidative style from the very start (and all the more reason for that producer to add SO2 after disgorgement), but it is a completely different matter if the producer makes a classic, reductive style sparkling wine, only to destroy that wine by not adding SO2 with the **dosage**, turning it prematurely oxidative.

Luscious, lusciousness This term is almost synonymous with **voluptuous**, although more frequently used to describe an unctuous, sweet white wine than a succulently rich red.

Maceration A term that is usually applied to the period during the **vinification** process when the fermenting juice is in contact with its skins. This process is traditionally used in red winemaking, but it is on the increase for white wines utilising pre-**fermentation** maceration techniques.

Macération carbonique (Fr) The French term for **carbonic maceration**.

Macroclimate Regional climate.

Macromolecules Very large molecules (such as **nucleic acids, proteins, sugars,** and **lipids**) containing 100 **atoms** or more.

Maderised All Madeiras are maderised by the *estufagem*, in which the wines are slowly heated in specially constructed ovens, and then by cooling them. This is undesirable in all wines except for certain Mediterranean wines that are deliberately made in a *rancio* style. Any ordinary, light, **table wine** that is maderised will often be erroneously diagnosed as

oxidized, but there is a significant difference in the symptoms: maderised wines have a duller **nose**, have rarely any hint of the Sherry-like character of **acetaldehyde**, and are flatter on the **palate**. All colours and styles of wine are capable of maderising and the likely cause is storage in bright sunlight or too much warmth.

Magnum A large bottle format that is equivalent to two normal-sized 75cl.

Magnum effect Magnums are better at preserving wine due to the greater ratio of wine to oxygen in the headspace, but the so-called "magnum effect" is far more noticeable in **bottle-fermented** wines than for any other style. The basic ratio of wine to oxygen has been well understood by sparkling wine consumers for a number of years, with well-informed collectors preferring this larger bottle format for laying down. At the time of **disgorgement**, all the oxygen in a bottle-fermented sparkling wine will have been scavenged by the **yeast**, thus, as the 75cl bottle and 150cl magnum both have the same diameter neck and are sealed with the same size **cork**, the ratio of oxygen to wine is almost exactly half in the magnum compared to the bottle, thus the potential for oxidation in a magnum is half that of a bottle. What few have realised until recently, however, is that the magnum effect begins from the moment the wine is bottled, as the second **fermentation** in magnum starts on average two days later than the same wine in 75cl bottles and takes approximately a week longer. This is due to the same-sized headspace, which is the primary source of oxygen for the voracious yeast, which have twice the work to do in magnums, and (almost) literally gasp for air in the process, thereby delaying the second fermentation, effectively producing a slightly different sparkling wine from exactly the same base wine. It is this variation in end product from the same raw material, plus the slower evolution in bottle due to twice the raitio of wine to oxygen, that produces the magnum effect, and it is why the magnum effect is more effective for sparkling wines than still.

Maillard reactions Chemical interactions between amino acids created during **autolysis** and **residual sugar** added by *dosage*, which are responsible for part of the mellow, complex post-**disgorgement aromas** adored by drinkers of **mature Champagne**. They also play an important role in the raisining of grapes, and they occur in cooking as part of the caramelising process that happens during the sealing of meat. Due to the lower temperatures involved, Maillard reactions in wine take weeks, months, and years, whereas in cooking they take seconds or minutes. Furthermore, low temperatures require catalysts, such as pressure.

Malic Tasting term that describes the green apple **aroma** and flavour found in some young wines due to the presence of **malic acid**, the dominant acid found in apples.

Malic acid A very strong-tasting acid that diminishes during the fruit's ripening process, but still persists in ripe grapes and, although reduced by **fermentation**, in wine too. The quantity of malic acid present in a wine may sometimes be considered too much, particularly in a red wine, and the smoothing effect of replacing it with just two-thirds the quantity of the much weaker **lactic acid** is often desirable. *See also* "Malolactic Fermentation", p64.

Malolactic The **malolactic fermentation** is often termed a secondary fermentation, but it is not an alcoholic fermentation. It is an entirely different biochemical process that converts the hard **malic acid** of unripe grapes into soft **lactic acid** and **carbonic gas**. The level of malolactic contribution will vary according to the style of wine in question. It is essential for red wine but not for white, rosé, or sparkling – and malolactic for some white-wine **varietals** should be avoided altogether. Whatever the grape variety or style of wine, malolactic should never be immediately discernible. There is nothing so vulgar and clumsy in a wine as a nostril-full of diacetyl, whether it is **caramel, butterscotch,** or simply **buttery**. *See also* "Malolactic Fermentation", p64.

Malolactic "cocktails" Mixtures of malolactic bacteria specifically prepared to produce low diacetyl, proprietary brands of which include BL01, IOC IB (InoBacter), Lalvin MT01, and Lalvin VP41.

Maltose A (**glucose + glucose**) **disaccharide**, tiny amounts of which may be found naturally in wine.

Mamertine or **Mamertinum** This **ancient Roman wine** from the northeastern tip of Sicily (roughly the same area as Faro DOC today) is said to have been the favourite of Julius Caesar.

Mannitol A sugar alcohol that is also considered to be a fault because it is usually the result of bacterial infection and is often accompanied by high **volatile acid**, with an unpleasant slimy texture on the **finish**. Not directly fermentable.

Mannose A fermentable **hexose** (6-carbon) **monosaccharide**, tiny amounts of mannose may be formed from **glucose** or by **oxidation** of **mannitol**. This is a curiously anomalous **sugar** as its sweetness does not increase in ratio to its concentration.

Manure A very extreme form of **farmyardy**.

Mannoprotein Nitrogenous matter secreted from **yeast** during **autolysis**.

Manta *See* CAP.

Marc 1. The residue of skins, pips, and stalks after pressing. 2. The name given to a four-tonne load of grapes in **Champagne**. 3. A rough brandy made from the residue of skins, pips, and stalks after pressing.

Marque (Fr) A brand or make.

Marsic An **ancient Roman wine** that was described as "very dry and wholesome".

Massal selection What plant nurseries used to do before genetically identical clones were produced. Grafts are taken from the very best old vines.

Martinotti, Federico Inventor who patented the Metodo Martinotti.

Martinotti Lungo Literally "Long Martinotti", this involves leaving the wine on **yeast** in the tank for at least 12 months. EC legislation for Metodo Martinotti (or Tank method) is very lax, requiring just 30 days between the start of refermentation and marketing the wine. For VSQ and VSQPRD, a minimum of 80 days on **lees** is required (although this may be reduced to 30, if the tank is equipped with an agitator to stir up the lees) and not less than 6 months between the beginning of the refermentation and the marketing the wine. Also curiously referred to in Italy as Charmat Lungo. *See* CUVE CLOSE

Massicum An **ancient Roman wine** from the Naples area that was known for its firm structure.

Mature, maturity Refers to a wine's development in bottle, as opposed to **ripe**, which describes the maturity of the grape itself.

MD Code found on **DIAM MytiK** corks. *See* DIAM MYTIK; DIAMANT® PROCESS

MDA Code found on **DIAM MytiK Access** corks. *See* DIAM MYTIK; DIAM MYTIK ACCESS; DIAMANT® PROCESS

MDC Code found on **DIAM MytiK Classic** corks. *See* DIAM MYTIK, DIAM MYTIK CLASSIC AND DIAMANT® PROCESS.

Mean An extreme qualification of ungenerous.

Meaty This term suggests a wine so rich in body and extract that the drinker feels almost able to chew it. Wines with a high **tannin** content are often meaty.

Melitose or **melitriose** *See* RAFFINOSE.

Mellow Describes a wine that is round and nearing its peak of maturity.

Membrane filtration Use of a thin screen of biologically inert material, perforated with micro-sized pores that occupy 80 per cent of the membrane, to filter wine. Anything larger than these holes is denied passage when the wine is pumped through during **filtration**.

Mercaptans Methyl and ethyl **alcohol**s can react with **hydrogen sulphide** to form mercaptans (sometimes referred to as thiols), foul-smelling **compounds** that are often impossible to remove and can ruin a wine. Mercaptans can smell of garlic, onion, burnt rubber, or stale cabbage.

Mesoclimate In strict scientific terms, mesoclimate is site climate, while microclimate is more specific.

Metal, metallic Some *terroir* can give a wine a distinctly metallic **finish**, particularly when produced from relatively neutral grape varieties, such as Chardonnay growing in the *lieu-dit* of Les Bionnes in Avize on the Côte des Blancs of **Champagne**.

Metayage The cultivation of land for a proprietor whereby the tenant receives a proportion of the crop produced in payment for his labours. This form of sharecropping dates back to Roman times and was recommended by Oliver de Serres in Le Theatre d'Agriculture, written in 1600, as a means of sharing the risk of winemaking.

Méthode champenoise (Fr) The process by which effervescence is produced through a **secondary fermentation** in the same bottle in which the wine is sold (in other words, not by transvasage). This procedure is used for **Champagne** and other quality sparkling wines. In Europe, the term is forbidden on the label of any wine other than Champagne, which of course never uses it. The first description of what became the so-called *méthode champenoise* was penned by Dr Christopher Merret in London in 1662.

Méthode gaillaçoise (Fr) A variant of *méthode rurale* involving disgorgement.

Méthode rurale (Fr) The precursor of *méthode champenoise*, this method involves no **secondary fermentation**. The wine is bottled before the first alcoholic fermentation has finished, and **carbonic gas** is produced during the continuation of fermentation in the bottle. There is also no **disgorgement**. *See also* MÉTHODE RURALE.

Methuselah Large-format equivalent to eight normal-sized 75cl bottles.

Methylation A chemical reaction in which a hydrogen atom is replaced by a methyl ion. In winemaking this can have both positive and negative effect. Positives include the methylation of polyphenols, which can improve colour stability, but negatives most famously include the biomethylation of **TCP**, **TeCP**, and **TBP** to **TCA**, **TeCA**, and **TBA** respectively. The catalyst for this biomethylation is the enzyme chlorophenol O-methyltransferase, which can be, but is not always, found in fungus residing in wood, particularly **oak**, **cork** (which is oak bark, of course), and, by transfer, in the fungal growth of some cellars.

Metodo Classico (It) *See* TRANSFER METHOD.

Metodo Martinotti *See* CUVE CLOSE.

Metodo Tradizionale (It) Italian for "transfer method". *See* TRANSFER METHOD.

Microagglomerate The visually distinguishable, fine-textured composition of a **technical cork**. Many manufacturers produce technical corks, but for a brief explanation of how microagglomerate closures are produced. *See* DIAMANT® PROCESS.

Microclimate Due to a combination of shelter, exposure, proximity to mountains and/or water mass, and other topographical features unique to a given area, a vineyard can enjoy (or be prone to) a specific microclimate that differs from the standard climate of the region as a whole. Technically, this is a **mesoclimate**, not a microclimate. Scientists refer to the conditions around just one or a handful of vines as a microclimate, but the variations recorded on such microscale are meaningless to non-scientists and, indeed, to any wine, no matter how tiny the vineyard might be. This is why "microclimate" is still in common usage for the environment of a single site.

Micro-oxygenation This process involves the ultra-slow diffusion of oxygen throughout a wine from a device that is little more than a sophisticated version of those oxygenators you see bubbling away in aquariums, only the bubbles are virtually microscopic. It was lampooned by Jonathan Nossiter's film *Mondovino*, which gave the impression that world-famous consultant Michel Rolland advised most of his clients to use it. Although used primarily to soften harsh **tannins**, micro-oxygenation can also intensify colour by as much as 30 per cent by fixing colour **molecules** to tannin molecules that would otherwise be unfixed and drop out. Furthermore, it can remove stinky, reductive tank **aromas** and reduce **herbaceous** notes. Micro-oxygenation was devised in 1991 by Patrick Ducournau of Domaine Mouréou and Chapelle L'Enclos in Madiran, the home the Tannat grape, which is legendary for its harsh tannins.

Micro-porous filtration Synonymous with **membrane filtration.**

Micro-vinification This technique involves **fermentation** in small, specialised vats, which are seldom bigger than a washing machine. The process is often used to make experimental wines. There are certain dynamics involved in fermentation that determine a minimum optimum size of vat, which is why home-brewers seldom make a polished product and why most wines made in research stations are dull.

Mid-palate 1. The centre-top of your tongue. 2. A subjective term to describe the middle of the taste sensation when taking a mouthful of wine. It may be hollow if the wine is **thin** and lacking or **full** if it is rich and satisfying.

Millerandage (Fr) A physiological disorder of the vine that occurs after cold or wet weather at the time of the flowering. This makes fertilization very difficult, and consequently many berries fail to develop, remaining small and seedless even when the rest of the bunch is full-sized and ripe.

Millésime (Fr) **Vintage** year. *See also* RÉCOLTE

Millésimé (Fr) **Vintaged** wine.

Minerality This tasting term causes controversy because minerals are not taken up from the soil into grape juice and it is impossible to taste whatever minerals there might be in any wine. The perception of minerality in a wine can be likened to a certain nervosity of **fruit** and always shows at its best and at its most in wines that are leaner, less alcoholic, and more acidic. The richer and fatter a wine is, the less likely it will show any minerality. **Malolactic** is the enemy of minerality. There are all sorts of minerality, from the saltiness of **Champagne**'s chalk-inspired *blanc de blancs*, to the sulphidic pungency of Santorini's volcanic Assyritiko. A fresh, lean Assyritiko from Macedonia rather than the island of Santorini has a simple yet evident minerality, just as a fresh Albariño from Gallicia has. Certain grapes have a tendency towards minerality, especially if grown in cooler climes. Those who describe wines as mineral are often labelled snobs, particularly by the pedants who look only at the science and cannot see any justification for the term, but most experienced tasters understand what is meant by minerality, just as they understand what is meant by the the **petrol aroma** of Riesling, even though we all know that there cannot be any petrol whatsoever in a wine, and everyone using that term knows perfectly well that petrolly Riesling does not smell or taste anything like petrol. All specialist subjects have their own vocabulary as a necessity to convey ideas and opinions quickly. These vocabularies often contain words that seem contradictory or misleading to other people, but it is only when the subject is wine or some other cultural topic that the users of such words are considered to be snobs.

Mistelle (Fr) Fresh grape juice that has been muted with **alcohol** before any **fermentation** can take place (ie a **VdL**).

Mitochondria In each living **yeast** cell there are about 50 mitochondria located in the periphery of the cytoplasm, either spherical or tubular, and they are effectively the respiratory organs.

Mitophagy A selective form of autophagy that specifically degrades the **mitochondria.**

MLF Short for "malolactic fermentation". *See* MALOLACTIC.

Moelleux (Fr) Literally soft or smooth, this term implies a rich, medium-sweet style in most areas of France. In the Loire, however, it is used to indicate a truly rich, sweet **botrytis** wine, thereby distinguishing it from *demi-sec*.

Molasses Also known as "black treacle" or "blackstrap", this is effectively the second pressing in sugar production: dark, rich, and heady, containing only 50 to 75 per cent sugar and a high vitamin and mineral content not found in refined **sugars** and syrups.

Molecule A molecule forms when two or more **atoms** are bonded together chemically. All **compounds** are molecules, but not all molecules are compounds. *See also* ATOM.

Mono-cru (Fr) French for a wine made from the grapes of one village.

Mono-parcelle (Fr) French for a wine made from the grapes, not just of one vineyard, but of one parcel within a single vineyard.

Monopole (Fr) Denotes the single ownership of one vineyard.

Monosaccharide The most basic form of **sugar**, also known as a "simple sugar". The two most common monosaccharides in nature are also the two most important sugars found in grapes: **glucose** and **fructose**. Only **hexoses** and, to a lesser extent, **pentoses** are important as far as wine and winemaking are concerned.

Mousse (Fr) The effervescence of a sparkling wine, which is best judged in the mouth because a wine may appear to be flat in one glass and vigorous in another due to the different surfaces. The bubbles of a good *mousse* should be small and persistent; the strength of effervescence depends on the style of wine.

Mousseux (Fr) Literally "sparkling".

Mouth-feel The texture, overall feel and experience, of the wine in the mouth.

Muid (Fr) A large oval barrel with a capacity of 600 litres (158 US gallons).

MR Code found on MytiK **Revtech** corks. *See* DIAM MYTIK; REVTECH.

MRA Code found on MytiK **Revtech** Access corks. *See* DIAM MYTIK; DIAM MYTIK ACCESS; REVTECH.

MRC Code found on MytiK **Revtech** Classic corks. *See* DIAM MYTIK; DIAM MYTIK CLASSIC; REVTECH.

Mulsum An ancient Roman mixed drink consisting of wine (usually **Massicum** or **Falernian**) sweetened with honey, which was mixed in just before serving as an aperitif. It was often freely dispensed to lower classes at public events to solicit their political support. It may also, according to taste, have included myrrh, cassia, costum, malobathrum, nard, or pepper. *See* ANCIENT ROMAN WINES.

Multi-vintage A misused and often misleading term that originated in **Champagne** by those who claim **non-vintage** is a negative term, an argument that is grammatically flawed. Very few hyphenated terms that are qualified by a "non" prefix are negative. Indeed, many are intrinsically positive (such as non-addictive, non-abusive, or non-violent), but most are neutral (such as non-aligned and, of course, non-vintage). Non-vintage merely means "without **vintage**" and "without year"; it is the literal translation of the equivalent French term of *sans année*. This issue is more important than semantics, however, because selling a wine as "multi-vintage" is deliberately misleading if it does not contain exclusively wines from years that have been or will be released by that producer as pure vintages. I am sure that many producers have not thought this through and are using the terms innocently, but they should think this through because they are not being truthful.

Mushroom Not a mustiness found in wine, but rather the **aroma** of freshly peeled mushrooms. Quite common in older **vintages** of **Champagne**, the mushroomy aroma can affect some bottles, yet not others, thus not a naturally occurring **tertiary aroma.** Not unpleasant, but the bottles without this penetrating aroma are preferable. Thought to be due to overfilling in a less technological era, when the wine overflows in the corking operation and some gets trapped between the **cork** and the neck of the bottle, where the theory is that it picks up fungus from the walls of the cellars and eventually transmits it, perhaps through the lenticels of the cork, to the wine inside the bottle.

Mushroom-shaped corks According to Article 69, paragraph 1(a) of EU Commission Regulation 538/2011 "a mushroom-shaped stopper made of cork or other material permitted to come into contact with foodstuffs, held in place by a fastening, covered, if necessary, by a cap and sheathed in foil completely covering the stopper and all or part of the neck of the bottle". As Jamie Goode pointed out at the Ontario Sparkling Wine Technical Symposium in 2014, "If you go by the letter of the law, alternatives such as Zork, **crown caps,** and **screw caps** are not allowed for fizzy wines produced or sold in the EU!" Article 69, however, covers wine produced, not sold, in the EU, and the bureaucrats who wrote the law were inept in their choice of the word "mushroom", which grows in far too many shapes to act as any sort of legal definition.

Must Unfermented or partly fermenting grape juice.

Must weight The amount of **sugar** in ripe grapes or grape **must.**

Mustum Ancient Roman concoction containing partially fermented grape juice and **vinegar.** *See* ANCIENT ROMAN WINES.

Mutage (Fr) The addition of pure **alcohol** to a wine or to fresh grape juice either before **fermentation** can take place, as in the case of a **VdL** (*vin de liqueur*), or during fermentation, as in the case of a **VDN** (*vin doux naturel*).

MytiK Origine® by DIAM Made with a plant-based binder and beeswax filler to replace food-grade glue and plastic microspheres, and aimed at organic, **biodynamic,** and "green"-minded producers.

NDtech Amorim's TCA-screening Non Destructive Gas Chromatography detection technology can be applied in conjunction with its proprietary ROSA Evolution treatment to any **cork**, natural, **agglomerate**, or **microagglomerate.** Claims that this offers an unprecedented quality control level, however, effectively eliminating all risk of **cork taint** in natural corks by ensuring that any remaining TCA is below the detection threshold of 0.5 nanograms of TCA per litre (parts per trillion, the equivalent of one drop of water in 800 Olympic swimming pools), is impossible if applied to the entire cork. Non Destructive Gas Chromatography detection effectively measures surface-released volatiles and cannot penetrate deep inside the cork. *See* ROSA EVOLUTION

Nanoclimate The environmental climate in the air cavities through the **canopy** and immediately above the leaves.

Natural wine According to at least one authority, a natural wine is farmed organically or biodynamically and made (or rather transformed) without adding or removing anything in the cellar. This involves no additives or processing aids and any intervention in the naturally occurring **fermentation** process is kept to a minimum, with neither **fining** nor **filtration.** If this results in a fresh, bright, and clean wine, who could possibly object? Most, however, do not.

Négociant (Fr) "Trader" or "merchant". The name is derived from the traditional practice of negotiating with growers (to buy wine) and wholesalers or customers (to sell it).

Négociant-éleveur (Fr) "Trader" or "merchant" and "farmer" or "breeder") A wine firm that buys ready-made wines for *élevage.* The wines are aged, possibly blended, and then bottled under the *négociant's* label.

Nervy, nervous A subjective term usually applied to a dry white wine that is **firm** and vigorous, but not quite settled down.

Neutral grape varieties Such grapes include virtually all the minor, nondescript varieties that produce bland-tasting, low-quality wines, but also encompass better known varieties such as the Melon de Bourgogne, Aligoté, Pinot Blanc, Pinot Meunier, and even classics such as Chardonnay and Sémillon. The opposite of aromatic grape varieties, these are ideal for oak-maturation, bottling *sur lie*, and turning into fine sparkling wines because their characteristics are enhanced rather than hidden by these processes.

Neutrons Uncharged particles found within an atom's nucleus.

Nevers A famous type of **oak** used for barrelmaking, from a forest in the Nievre ***département*** in the centre of France.

Nievre A type of **oak** used for barrelmaking, from Nevers, Bertrange, or any other forest in the Nievre ***département*** in the centre of France.

Nitrogen flushing Flushing bottles with nitrogen immediately prior to filling was first applied to sparkling wine production by Cherie Spriggs and Brad Greatrix at Nyetimber, who have been nitrogen-flushing across their entire range and in all bottle formats since the 2011-based bottling in spring of 2012. According to Greaterix "his was the simplest way to eliminate as much **headspace oxygen** as possible for fresher results". Other producers have since experimented with the process in an attempt to replicate the "**magnum effect**" in 75cl bottles. According to Virginia Tech's Bruce Zoecklein, to reduce oxygen from atmospheric levels to less than 1 per cent requires flushing with 3.25 times the volume of nitrogen, thus it takes in excess of 243.75cl of nitrogen (75 x 3.25) to flush a 75cl bottle thoroughly because the internal capacity of a 75cl bottle with its headspace and the volume of neck that will be occupied by a cork is, of course, greater than 75cl.

Noble rot The (frequently beneficial) shrivelling or decaying of grapes caused by the fungus *Botrytis cinerea* under certain conditions.

Non-vintage A blend of at least two different years.

Nonose An 8-carbon **sugar** that can be synthetically produced, but does not exist in nature.

Nose The smell or odour of a wine, encompassing both **aroma** and **bouquet.**

Nucleic acids One of four classes of **carbohydrate** found in wine, nucleic acids are biological macromolecules that are essential for all known forms of life. Nucleic acids include DNA and RNA and are comprised of nucleotides.

Nucleosides Effectively a nucleotide without phosphate, nucleosides are produced during **autolysis** from the degeneration of DNA and are essentially flavouring agents.

Nucleotides When **sugar** and a nitrogenous base combine they form a nucleotide. Nucleotides are produced during **autolysis** from the degeneration of DNA and are essentially flavouring agents.

O2 The chemical formula for oxygen.

Oak Many wines are fermented or aged in casks and most commonly the wood used is oak. There are two main categories of oak, French (*Quercus sessilis* and *Quercus robur*) and American (*Quercus alba*), and they are both used the world over. *See also* "Regional Oak Varieties", p78.

Oak grain The tightness of the oak grain is a measure of its quality for winemaking purposes. Strangely enough, the tighter the grain is, the more porous its wood will be. The rationale is simple enough, however, because the more porous the wood, the greater the micro-oxygenation of the wine – hence, the softer the **tannins.** A compilation of various standards from around the world can be distilled into the following classification: very fine (less than 1.5mm), fine (1.5–2mm or 2.5mm), medium-fine (2.5–3mm or 3.5mm), and wide (greater than 3mm or 3.5mm).

Octose An 8-carbon **sugar** that can be synthetically produced, but does not exist in nature.

Oechsle level A system of measuring the **sugar** content in grapes for wine categories in Germany and Austria. *See also* "Quality Requirements and Harvest Percentages" p439.

Oeneo This is the world's largest **oak** barrel producer (owning Seguin Moreau et al) and the world's second-largest **cork** producer, whose specialist manufacturer, **DIAM**, was the only supplier of supercritical CO_2 technical **cork** until the patent ran out in 2020).

Oenologist, oenology Pronounced "enologist" and "enology" (and spelled this way in the United States). Oenology is the scientific study of wine. It is a branch of chemistry, but with practical consequences, hands-on production experience, and an understanding of **viticulture.**

Off vintage An off **vintage** or year is one in which many poor wines are produced due to adverse climatic conditions, such as very little sunshine during the summer, which can result in unripe grapes, and rain or humid heat at the harvest, which can result in rot. Generally an off vintage is a vintage to be avoided, but approach any opportunity to taste the wines with an open mind because there are always good wines made in every vintage, and they have to be sold at bargain prices if a vintage has a bad reputation.

Oidium A fungal disease of the vine that turns leaves powdery grey and dehydrates grapes.

Oily A subjective term meaning **fat** and viscous, and often also **flat** and **flabby.**

OIR *See* OXYGEN INITIAL RELEASE.

Olfaction The sense of smell, much of which is perceived as taste. *See* "How We Taste Smells", p90.

Olfactory bulb A sensory organ situated above the nose and between the eyes, the olfactory bulb provides us with not only the sense of smell but also much of what we interpret as taste.

Oligopeptide As *oligos* in Greek means "few" and *poly* "many" this group encompasses all **peptides** containing between two and either 10 or 20 **amino acids** (definition differs according to source). *See* PEPTIDES.

Oligosaccharide *Oligos* in Greek means "few", but in this case it refers not to few, as in hardly any, but to "a few" as in "quite a few". This group of **saccharides** therefore refers to saccharides comprised of more than one **sugar**, but there are also **polysaccharides** (from the Greek *poly* meaning "many"), so where do oligosaccharides begin and end? The devil is in the details because some authorities include **disaccharides** within the oligosaccharide group and some do not. Furthermore, the highest number of **sugars** in an oligossacharide can be anything between 9 and 15! If nothing else, the definition of oligossacharides is as imprecise as the Greek meaning of *oligos,* which is at least fitting. *See also* POLYSACCHARIDE.

Oloroso (Sp) A Sherry style, naturally dry but usually sweetened for export markets.

Open-knit An open and enjoyable **nose** or **palate**, usually found in a modest wine that is not capable of much development.

Opulent Suggestive of a rather luxurious **varietal aroma**; very rich, but not quite **blowzy**.

Orange wine The orange wine discipline has invaded the sparkling wine category in recent years. It started out fine, as nothing more than an extended prefermentation soak on the skins, to emphasise the aromatics, but was quickly owned by natural wine producers. This has rendered most wines cloudy and **oxidative**. If "orange" sparkling wines are fresh and star-bright, who could possibly object? Unfortunately, most are not. *See* Natural wine.

Organic wines A generic term for wines made using the minimum amount of **SO2** (sulphur dioxide), from grapes grown without the use of chemical fertilizers, **pesticides**, or **herbicides**.

Organoleptic Affecting a bodily organ or sense, usually that of taste or smell.

Osmotic pressure When two solutions are separated by a semi-permeable membrane, water will leave the weaker solution for the more concentrated one in an endeavour to equalise the differing solution strengths. In winemaking, this is most commonly seen when **yeast** cells are put to work in grape juice with an exceptionally high **sugar** content. Because water accounts for 65 per cent of a yeast cell, osmotic pressure causes the water to escape through the semi-permeable cell membrane. The cell caves in (a phenomenon called plasmolysis), and the yeast dries up and eventually dies.

OTR *See* Oxygen Transfer Rate.

Ouillage (Fr) Topping-up of casks or vats.

Overtone A dominating element of **nose** and **palate**; often one that is not directly attributable to the grape or wine.

Oxidation, oxidized These terms are ambiguous: as soon as grapes are pressed or crushed, oxidation sets in and the juice or wine will become oxidized to a certain and increasing extent. Oxidation is also an unavoidable part of **fermentation** and essential to the maturation process. In this case, however, in order not to mislead it is best to speak of a **mature** or, at the extreme, **oxidative** wine. This is because when the word "oxidized" is used, even among experts, it will invariably be in an extremely derogatory manner, to highlight the **Sherry-like** odour of a wine that is in a prematurely advanced stage of oxidation. In a totally oxidized wine, the **acetaldehyde** is converted to **acetic acid**, and the **aroma** turns to **vinegar**.

Oxidation enzymes Enzymatic catalysts of **oxidation** such as tyrosinase and laccase are inhibited by **SO2**, which eventually destroys them with time.

Oxidation-reduction *or* **Redox** Also known as redox potential, this is the summation of all the **oxidative** and reductive reactions existing in a wine at any given moment and indicates whether the wine is more oxidative or more reductive. Together with a gain or loss of oxygen, oxidation and reduction involve the transfer of **electrons** from one **compound** to another. Oxygen can take a pair of electrons from another compound and cause this compound to be oxidized, while the oxygen **atom** with its newly acquired electrons is itself reduced. Oxidation involves a loss of electrons and an increase in redox potential, while reduction involves a gain of electrons and a decrease in redox potential. To remember this, wine schools use the mnemonic OIL-RIG (Oxidation Is Loss of electrons – Reduction Is Gain of electrons). Do not forget that oxidation and reduction are irrevocably coupled: there is no oxidation without a corresponding reduction and vice-versa. An oxidant (eg oxygen) causes oxidation and to do this will itself be reduced; reductants (eg **SO2**, **ascorbic acid**, **alcohol**, **phenols**) cause reduction and to do this they are themselves oxidized. White wines are said to have less of a buffering capacity than red wines. This is because reductants form a buffer

against oxidation and as red wines have more types and higher volumes of phenolic compounds, red wines naturally have a greater buffering capacity. It is worth noting that the loss of electrons can occur under **anaerobic** conditions, thus oxidation can take place in the absence of oxygen! *See* Redox.

Oxidative A wine that openly demonstrates the characteristic **aroma** or flavour of browning apple, raisins, or hazelnuts. This can develop into an overtly **aldehydic** character when excessively oxidative. A degree of oxidative aroma is legitimate even for **bottle-fermented** sparkling wines, despite the fact that they are intrinsically reductive (because the **second fermentation** removes all oxygen **molecules** and the time spent on **yeast** following the second fermentation it is held a completely **anaerobic** environment). However, some bottle-fermented sparkling wines are not made in an oxidative style, but become oxidative due to low or no **SO2** after **disgorgement**.

Oxidative or acetic? If it is aldehydic, it is oxidative, if it is acetic, it's **VA**. Oxidative **aromas** include bruised apples, dried-fruits, or hazelnuts on the **nose** or **palate**, which can develop into an overtly **aldehydic** character, particularly the **Sherry-like aroma** of **acetaldehyde**, when excessively oxidative. When a wine is acetic, the acetaldehyde has been converted by bacterial action to acetic acid, and the aroma has moved from Sherry-like to a vinegary, **volatile acid** aroma.

Oxygen ingress Molecules of oxygen penetrate **corks** against the internal pressure because the internal pressure is **CO2**-pressure, not oxygen-pressure. Because there is only an infinitesimal amount of oxygen in any sparkling wine, its oxygen pressure is, for all intents and purposes, zero, whereas oxygen represents 21 per cent of the air, thus the oxygen pressure outside the bottle is 0.21 atmospheres at sea level. As far as any oxygen is concerned, this means that the pressure is reversed, forcing oxygen molecules through the cork into the bottle, and this will continue until 21 per cent of the bottle's volume is occupied by oxygen, which at the rate of ingress of even the most porous cork would be long after the wine had **oxidized** and died.

Oxygen Initial Release Also known as OIR, this term was recently coined by **DIAM**'s research team for the release of oxygen into the headspace of a wine when a **cork**, natural or technical, is compressed and inserted into the bottle. The volume of oxygen released varies enormously between one natural cork and another, but also varies between DIAM's Classic (less) and Access (more) lines, albeit by a minuscule amount and totally consistent for all closures of each respective genre. OIR has long been overlooked, but for those who remember when corks would be soaked prior to insertion, and the "curtain of brown water" that ran down the inside of the neck of the bottle, it will be obvious that air would be expelled in the same process using dry corks. Even if the soaked corks were **TCA**-free, it was not taint-free. That "curtain of brown water" introduced **aromas**, **phenolics**, and a multitude of other **compounds** that varied from cork to cork. The influence of those compounds on each individual bottle was not knowable and could not be planned for by the winemaker. Dry insertion has removed this particular source of taint, but it has introduced a new source of oxygen at the time of bottling for still wines and after **disgorgement** for sparkling wines.

Oxygen Transmission Rate Known simply as OTR, this would be more accurately termed "Closure Oxygen Transmission Rate". The amount of oxygen that passes through a closure differs according to the composition of the closure as follows:

- Screw cap with tin-saran liner: 0.0001cc O2 per 24hrs (constant)
- Screw cap with saranex-only liner: 0.001 (constant)
- Natural cork: 0.0005 cc O2 per 24hrs (variable)

- Technical cork: 0.0005 (constant)
- Synthetic cork: 0.002 - 0.005 (constant)
- Vino-Lok: 0.003 (constant)

In a sparkling wine, there is also an exchange of gases, with a loss of CO2 (**carbonic gas**) occurring simultaneously with the ingress of oxygen. *See also* Dissolved oxygen; Headspace oxygen.

Pad filtration A **filtration** system utilising a plate-and-frame filter with a series of cellulose, asbestos, or paper sheets through which wine is passed.

Palate The flavour or taste of a wine.

Partial rootzone drying *See* PRD.

Passerillage (Fr) Grapes without **noble rot** that are left on the vine become cut off from the plant's metabolic system as its sap withdraws into its roots. The warmth of the day, followed by the cold of the night, causes the grapes to dehydrate and concentrate in a process known as *passerillage*. The sweet wine produced from these grapes is prised in certain areas. A *passerillage* wine from a hot autumn will be totally different to one from a cold autumn. Roasted is an excessive and quick form of *passerillage* caused by great heat, often at the height or end of summer, rather than the normal, drawn-out autumnal *passerillage*.

Passito (It) The Italian equivalent of *passerillage*. *Passito* grapes are semi-dried, either outside – on the vine or on mats – or inside a warm building. This concentrates the pulp and produces strong, often sweet wines.

Passum An **ancient Roman wine** made from grapes left to raisin on the vine, the concept having been passed to the Romans from Carthage by the Phoenicians.

Pasteurization A generic term for various methods of **stabilization** and **sterilization**.

PDO *See* AOP.

Peak The ideal maturity of a wine. Those liking **fresher**, **crisper** wines will perceive an earlier peak in the same wine than drinkers who prefer **mature** wines. As a rule of thumb that applies to all extremes of taste, a wine will remain at its peak for as long as it took to reach it.

Peardrop *See* Amylic; Carbonic maceration.

Pentose A 5-carbon **monosaccharide** that is not fermentable by any conventional means but can be converted by wine spoilage bacteria into **lactic** and **acetic acids**. Pentoses often found in minute volumes in the **residual sugars** of healthy bone-dry wines include: **arabinose**, **ribose**, **rhamnose**, and **xylose**.

Peppery A term applied to young wines whose components are raw and not yet in harmony, sometimes quite fierce and **prickly** on the nose. It also describes the characteristic odour and flavour of southern French wines, particularly Grenache-based ones. Syrah can smell of freshly crushed black pepper, while white pepper is the character of great Grüner Veltliner. Young Ports and light red Riojas can also be very peppery.

Peptides Nitrogenous, organic **compounds** consisting of at least two **amino acids**. Peptides are produced during **fermentation** and, especially, **autolysis**. The production of peptides during autolysis exceeds that of individual amino acids, thus the volume of peptides in a sparkling wine is widely regarded as the best indicator for the dynamics of autolysis. Peptides are classified by the number of peptides: dipeptides (two amino acids), tripeptides (three amino acids) etc., then oligopeptides (2–10 or 2–20 amino acids, definitions vary according to source), and polypeptides (10 or 20 upwards, with definitions again varying according to source, polypeptides can contain as many as 100–300 amino acids). Most peptides in sparkling wine are dipeptides and tripeptides. Peptides are involved in **Maillard reactions** and have a positive effect on the quality of the *mousse*.

Perfume An agreeable scented quality of a wine's **bouquet**.

Perlant (Fr) Very slightly sparkling, less so than *crémant* and *pétillant*.

Perlite A fine, powdery, light, lustrous substance of volcanic origin with **diatomaceous earth**-like properties. When perlite is used for **filtration**, it is sometimes referred to as **ceramic filtration**.

Pesticide Literally a pest-killer, but more accurately a parasite-killer, the term "pesticide" implies a highly toxic concoction of chemicals capable of eradicating parasitic insects that attack the vine, including larvae, flies, moths, and spiders.

Pétillance, pétillant (Fr) This term describes a wine with sufficient **carbonic gas** to create a light sparkle.

Petit château (Fr) Literally "small castle", this term is applied to any wine château that is neither a *cru classé* nor a *cru bourgeois*.

PetNat or **Pét-nat** A recently fashionable style of **pétillant** so-called "natural wine", PetNat is bottled before the first and only **fermentation** has finished. It is not disgorged and sold hazy with **yeast lees** remaining in the wine. *See* COL FONDO; NATURAL WINE.

Petrol, petrolly With some **bottle-age**, the finest Rieslings have a vivid **bouquet** that some call petrolly. This petrolly character has an affinity with various zesty and citrussy odours, but many **lemony**, **citrussy**, **zesty** smells are totally different from one another and the Riesling's petrolly character is both singular and unmistakable. As great Riesling **matures**, so it also develops a **honeyed** character, bringing a classic, honeyed-petrol richness to the wine.

pH A commonly used chemical abbreviation of "potential hydrogen-ion concentration", a measure of the active **acidity** or alkalinity of a liquid. It does not give any indication of the total acidity in a wine, but neither does the human **palate**. When we perceive the acidity in wine through taste, it is more closely associated with the pH than with the total acidity. A seemingly small difference in pH is in fact quite meaningful, as it is logarithmic and, for example, a pH of 3 is 10 times as acidic as a pH of 4. Numerically, pH is the opposite of acidity: the lower the pH, the higher the acidity, thus in the pH scale of 0–14, 0 is extremely acid, 7 is neutral, and 14 is extremely alkaline. The pH rises slightly during **fermentation** as the final product becomes less acidic than the freshly squeezed juice it started from. Wine normally ends up with a pH between just under 3 (white) and 4 (red). For comparative purposes, it is interesting to note that battery acid is 0, gastric acid in the stomach is 1, water ranges between 6 and 8.5 (chalkstreams are 7.5 to 8), bleach is 13, and caustic soda is 14.

Phenols, phenolic compounds Compounds found in the skin, seeds, and stalks of grapes, the most common being **tannin** and **anthocyanins**.

Photosynthesis The process by which light energy is trapped by chorophyll, a green chemical in the leaves, and is converted into chemical energy in the form of **glucose**. This is then carried around the plant in special tubes called phloem to grow shoots, leaves, flowers, and fruit.

Phylloxera The vine louse *Phylloxera vastatrix,* which devastated the vineyards of Europe in the late 19th century, still infests the soils of nearly all the world's winegrowing regions. At the time, it was considered the greatest disaster in the history of wine, but with hindsight, it was a blessing in disguise. Before phylloxera arrived, many of Europe's greatest wine regions had gradually been devalued because of increased demand for their wines. This led to bulk-producing, inferior varieties being planted, and vineyards being extended into unsuitable lands. As phylloxera spread, it became apparent that every vine had to be grafted on to phylloxera-resistant American **rootstock**. This forced a much-needed rationalization, in which only the best sites in the classic regions were replanted and only noble vines were cultivated, a costly operation that vineyard owners in lesser areas could not afford.

The grafting took France 50 years and enabled the **AOC** system to be set up. It is hard to imagine what regional or **varietal** identities might now exist if phylloxera had not occurred.

Pièce standard (Fr) A **Champagne** measure of 205 litres (54 US gallons). When a Champagne cask is empty it is called a *fût*, when it is full it is a *pièce*.

Pipe (Port) The most famous Portuguese barrel, a Douro *pipe* has a capacity of 550 litres (145 US gallons).

Piquant 1. This term refers to a perfect knife-edge balance between sweetness and **acidity** that could well describe that found in the greatest Mosel *Kabinett*. 2. (Fr) Usually applied to a pleasing white wine with positive underlying **fruit** and *acidity*.

Plafond limité de classement. *See* PLC.

PLC Plafond Limité de Classement A legalised form of cheating whereby producers of AOC wines are allowed to exceed the official maximum limit by as much as 20 per cent.

Plum pudding A subjective term for a rich and spicy red wine; a more intense term than Christmas cake.

Plummy An **elegant**, juicy flavour and texture that resembles the fleshiness of plums.

Polished Describes a wine that has been skilfully crafted, leaving no rough edges. It is smooth and refined to drink.

Polishing The very last, ultra-fine **filtration** of a wine, usually with **kieselguhr** (*See* DIATOMACEOUS EARTH) or **perlite**. It is so called because it leaves the wine bright. Many high-quality wines are not polished because the process can wash out natural flavours.

Polypeptide Literally many **peptides**, these are peptides containing at least 10 or 20 **amino acids**, but which of these it is depends on the definition of oligopeptides and that varies between 2–10 and 2–20 amino acids according to different scientific sources. A polypeptide can be part or of a **protein** and all proteins consist of one or more polypeptides, but not all polypeptides are proteins. *See also* PROTEIN.

Polyol Often used synonymously for **sugar alcohols**, but although all sugar alcohols are polyols, not all polyols are necessarily sugar alcohols.

Polysaccharide *Poly* in Greek means "many", and as such this grouping is seen to cover all **saccharides** above and beyond **oligossacharides**, but as *oligos* means "few" and no one can agree exactly how many that is (somewhere between and including 9 and 15 according to various sources), the starting point of polysaccharides remains unknown. *See also* OLIGOSACCHARIDE.

Posca A wine deliberately made to go sour, then mixed with water and given to soldiers and lower classes. Posca was occasionally sweetened with honey, and Roman legionnaires considered it refreshing. *See* ANCIENT ROMAN WINES.

Post-disgorgement ageing The period between **disgorgement** and when the wine is consumed. With the sudden exposure to air after an extended period of ageing under **anaerobic** conditions, the development of a sparkling wine after disgorgement is very different from its development before.

Pourriture noble (Fr) **Noble rot**, which is caused by the fungus *Botrytis cinerea* under certain conditions.

Praenestine An **ancient Roman wine** (red) produced in Praeneste (now Palestrina) in Lazio.

Pramnian Ancient Greek wine produced in several regions. Athenaeus of Naucratis suggested that Pramnian could be a generic name for a dark red wine of good quality and ageing potential. *See* ANCIENT GREEK WINES.

PRD "Partial rootzone drying", a clever way of fooling the vine into thinking that it is not being irrigated, when, in fact, it is. This is achieved by alternating irrigation between two separate parts of the root system. Part of

the vine receives a carefully metered out **drip irrigation**, but the rest of the plant system is unaware of this and, not sensing the irrigation, believes that it is, in fact, experiencing a mild water stress. The vine thus diverts its metabolism (energy) from the leaves to the grape clusters, improving the quality of the fruit. When the water is drawn into the part of the vine that has shut down the metabolism of its leaves, this part of the vine reverses the metabolic process. This is the very time that the drip irrigation is switched to that side of the vine, as it has already accepted the water's presence. By turning off the irrigation to the other side of the vine, however, that side now believes it is experiencing a mild water stress, and it is its turn to divert the vine's metabolism from the leaves to the grape clusters. And so it goes on, drip-feeding either side of a vine that perpetually experiences a mild state of water stress. This conserves water, and whilst it does not increase yields per se, it does produce better quality grapes at normal yields.

Precocious A wine that develops early.

Pre-fermentation maceration The practice of **maceration** of juice in grape skins prior to **fermentation**, to enhance the **varietal character** of the wine. This maceration is usually carried out cold and is normally employed for **aromatic** white varieties but can be undertaken warm – or even quite hot for red wines.

Premier cru (Fr) Literally "first growth", this term is of relevance only in those areas where it is controlled, such as in Burgundy and Champagne.

Premium wine A marketing term for a quality category. So-called premium or premium-quality wine is not as expensive as you might think and certainly not the top category of wine. *See also* SUPER-PREMIUM WINE; ULTRA-PREMIUM WINE; ICON WINE.

Premox Short for "premature oxidation", this phenomenon is sometimes further shortened to the alarmingly apt "pox". This term encapsulates various stages of **oxidation** that would not normally be expected in a wine of any given age (ie compared to its counterparts of the same age). The first incidence of premox came to light in white Burgundy **vintages** of the 1990s, but due to denial (by collectors and producers alike) and the fact that it generally takes five to six years to emerge in bottle, it was not widely discussed until the 2000s. Premox has often been attributed to various factors, including vineyard practices (which reduce a natural **antioxidant** called glutathione), the trend towards richer wines with lower **pHs**, introduction of pressing techniques (which reduce the levels of antioxidant **lees**), **corks,** and other factors, but this is overcomplicating the problem. If we apply Occam's razor we are left with just one common factor: all premox wines have exhausted their supply of free sulphur. Even if any or all of the possible reasons postulated for premox wines are verified, the ultimate cause has to be the failure to calculate the correct final addition of SO_2 immediately prior to bottling or (for **bottle-fermented** sparkling wines) following **disgorgement**.

Pre-phylloxera vine So called because the vines have been planted *en foule* prior to **phylloxera**. *See* EN FOULE; PHYLLOXERA.

Press Wine *See* VIN DE PRESSE.

Prickle, prickly This term describes a wine with residual **carbonic gas**, but with less than the light sparkle of a **pétillant** wine. This characteristic can be desirable in some **fresh** white and **rosé** wines, but it is usually taken as a sign of an undesirable **secondary fermentation** in red wines, although it is deliberately created in certain South African examples.

Primary aromas These aromas are attributed to the grape. *See also* SECONDARY AROMAS; TERTIARY AROMAS.

Primat Large-bottle equivalent to 36 normal-sized 75cl bottles, the Primat format is exclusively produced by Champagne Drappier.

Prise de mousse (Fr) 1. During the **second fermentation**, the *liqueur de tirage* is converted into almost equal quantities of **alcohol** and **carbonic gas**. When a bottle of **Champagne** is opened, the **carbonic gas** escapes from the wine in the form of bubbles creating the *mousse*, thus this whole process is encapsulated in the French term *prise de mousse* or "capturing the sparkle". 2. *Prise de mousse* is also synonymous with Lalvin EC1118, a brand of dried **yeast**.

Privernatinum or **Privernian** An ancient Roman wine that was light and pleasant, from the Volscian Hills in the south of Lazio. Athenaeus of Naucratis found Privernatinum to be thinner than **Rhegium**.

Producer vine Vines are usually grafted on to **root-stock** resistant to **phylloxera**, but the grapes produced are characteristic of the above-ground producer vine, or scion, which is normally a variety of *Vitis vinifera*.

ProMalic A proprietary brand of encapsulated *Schizos-accharomyces pombe*, a **yeast** that metabolises malic acid into **alcohol**. This yeast would be a good alternative to **malolactic fermentation** (MLF), but it has been ignored until recently because it can produce off-odours if still active after fermentation. Proenol, a Portuguese biotech company, has encapsulated *Schizo-saccharomyces pombe* in double-layered alginate beads, which are added to the juice at the beginning of the alcoholic fermentation and removed once the desired malic level is achieved. This might be particularly useful for producers faced with high levels of **malic acid**, when full MLF can dominate the aromatics undesirably, even when using low-diacetyl bacteria. Converting some of the malic acid to alcohol will also reduce the amount, therefore the cost, of **chaptalisation**, albeit by not very much (a yield of 0.1 per cent ABV per 2.33g/l of malic acid), but the simplicity of removing the alginate beads should allow producers to tweak the malic acid content and thus calibrate the precise MLF required, if any, for the house style.

Protein One of four classes of **carbohydrate**, small amounts of which are found in wine, originating from either grapes or **yeast**. They include **amino acids**, **enzymes**, and glycoproteins (**mannoproteins**). A protein is a chain of amino acids. In other words it consists of one or more **polypeptides**, but these are polypeptides that fold into a fixed three-dimensional structure. Most polypeptides of fewer than 40 amino acids in length do not fold, which is why many sources classify proteins as polypeptides of a certain minimum length, but because protein's determining factor is one of structure, not length, such definitions inevitably vary (40, 50, or 100 amino acids) and none can possibly be correct. Indeed, the lowest number of amino acids in a naturally forming protein is TRP-Cage, which consists of just 20 amino acids, and as recently as 2004, Honda et al designed Chignolin, the first synthetic protein with as few as 10 amino acids. Neither have anything to do with wine, but these examples are given to emphasise that proteins cannot be defined by the number of their constituent amino acids. Indeed, although all proteins are polypeptides, the point at which proteins emerge from polypeptides cannot be usefully or universally defined.

Protein haze Protein is present in all wines. Too much **protein** can react with **tannin** to cause a haze, in which case **bentonite** is usually used as a **fining** agent to remove it.

Protons Positively charged particles that surround the atom's nucleus. Protons are made from quarks.

Pruina Waxy substance on the surface of grapes to which **yeast** and bacteria adhere to form a **bloom.**

Puncheon A 450-litre (119-US gallon) barrel commonly found in Australia and New Zealand.

PVPP Abbreviation for "polyvinylpolypyrrolidone", a **fining** agent used to remove **compounds** sensitive to browning from white wines.

Pyrazines One of the most important groups of aromatic **compounds** found in grapes (especially methoxypyrazines), pyrazines typically have green, leafy, grassy characteristics through to bell pepper, green pea, and asparagus. The more herbaceous pyrazine **aromas** are symptomatic of an excessively vigorous vine **canopy**, particularly in red wines. Although pyrazines become less abundant as grapes ripen, they are considered a vital element in the **varietal character** of Sauvignon Blanc.

QbA (Ger) Germany's *Qualitätswein bestimmter Anbaugebiete* is the theoretical equivalent of the French **AOC**.

QmP (Ger) The abbreviation for *Qualitätswein mit Prädikat*. Literally a "quality wine with predication", this term is used for any German wine above **QbA**, from *Kabinett* upwards. The predication carried by a QmP wine depends upon the level of ripeness of the grapes used in the wine.

Quaffable, quaffing wine An unpretentious wine that is enjoyable and **easy to drink.**

Qualitätsschaumwein (Ger) A so-called "quality sparkling wine", this can be produced by any member state of the EU, but the term should be qualified by the country of origin (of the wine), thus only *Deutscher Qualitätsschaumwein* will necessarily be from Germany.

Qualitätswein bestimmter anbaugebiete See QbA.

Qualitätswein mit prädikat See QmP.

Quercus The Latin name for "oak" or "oak tree", of which there are more than 600 species. Those of particular interest for wine include *Quercus alba* (American oak or white oak), *Q gariana* (sometimes called *Q garryana*, a tighter-grained American oak found in Oregon), *Q ilex* (Holm oak), *Q robur* (solitary oak trees, also known as *Q pdeunculata*, or common oak, English oak, pedunculate oak, and truffle oak), *Q sessilis* (forest oak, with a longer and straighter trunk, also known as *Q petraea* or *Q sessiflora*, or durmast oak, roble oak, sessile oak, steineiche oak, and Welsh oak), *Q suber* (evergreen species of oak commonly known as cork oak, the bark of which is harvested for the cork industry), and *Q variabilis* (deciduous species of oak commonly known as Chinese cork oak). *See also* CORK.

Quinta (Port) A wine estate.

Quintal The equivalent of 100 kilograms. Argentina, Chile, and other Spanish-speaking countries often talk about harvest yields in terms of *quintales* per hectare.

R2 A **yeast** strain (*Saccharomyces cerevisiae* race *bayanus*) discovered by Danish-born winemaker Peter Vinding-Diers.

Rack, racking Draining the wine off its **lees** into a fresh cask or vat. The term derives from the different levels (or racks) from which the wine is run from one container to another. In some **Champagnes**, particularly the fuller, potentially more complex styles, this is desirable, but in lighter Champagnes that have a less **oxidative**, more of a **crisp**, **elegant** style it is not and can be avoided or reduced by using gravity or by employing various enclosed pumping systems that either utilise an inert gas, such as nitrogen, or a physical, snakelike action that requires no gas at all.

Racy Often applied to wines of the Riesling grape. The term "racy" accurately suggests the liveliness, vitality, and **acidity** of this grape.

Raffinose A (**glucose** + **galactose** + **fructose**) trisac-**charide**, raffinose is more commonly found in vegetables, but also occurs naturally in very small amounts in wine. Also known as melitiose and melitriose, but should not be confused with melibiose, even though melibiose can be hydrolysed from raffinose.

Raisin 1. A dried Muscat grape (as opposed to a sultana, which is a dried Thompson Seedless grape, or a currant, which is a dried Zante grape). 2. Often used generically in winemaking for a grape that has shrivelled on the vine due to *passerillage*. 3. (Fr) French for grape.

Rancio Wines that have followed a process of ageing, noticeably rusted, with abrupt changes of temperature in presence of air, or in wood packages or crystal packages. Description of a **VDN** (*vin doux naturel*) stored in **oak** casks for at least two years, often with the barrels exposed to direct sunlight.

Ratafia A liqueur made by combining marc with grape juice, Ratafia de Champagne being the best known.

RD A sparkling-wine term that stands for "recently disgorged", the initials RD are the trademark of Champagne Bollinger. *See also* LD.

Recioto (It) A strong, sweet wine made in Italy from *passito* grapes.

Récoltant (Fr) Vineyard owner.

Récolte (Fr) Harvest. *See also* MILLÉSIME.

Rectified grape must Sometimes concentrated, sometimes not, rectified grape must is a clear, odourless, and tasteless **sugar** solution, which has had all non-sugar components removed. Neutral in **acidity**, with very little SO_2, RGM can be used for both **chaptalisation** and *dosage*. The result is invert sugar (ie 50-50 **glucose** and **fructose**). Often preferred to cane or beet sugar by producers who want to work exclusively with grape-based products, **organic** and **biodynamic** versions exist. Some producers dislike RGM because, they claim, it is not as neutral as it proponents claim, citing distinct and negative aromatic and textural taints.

Redox The ageing process of wine was originally conceived as purely **oxidative**, but it was then discovered that when one substance in wine is oxidized (gains oxygen), another is reduced (loses oxygen). This is known as a reductive-oxidative, or redox reaction. Organoleptically, however, wines reveal either oxidative or **reductive** characters. In the presence of air, wine is prone to an oxidative character, but shut off from a supply of oxygen, reductive characteristics begin to dominate; thus, the **bouquet** of **bottle-age** is a reductive one and the **aroma** of a fresh, young wine is more oxidative than reductive.

Redox potential Although "redox" is a concatenation of "**reduction**" and "**oxidation**", the potential it measures is defined the other way around, as "oxidation-reduction", but why it was not termed "oxred" has been lost in the mists of time. Redox potential is measured in millivolts (mV) and the higher it is the more **oxidative** a wine, while the lower it is the more reductive the wine. A deliberately **oxidized** wine such as Sherry or *vin jaune* has the highest redox potential. A red wine generally has a higher redox potential than a white wine, which itself is higher than that of a sparkling wine, although it is much more complex than that. When the juice is pressed and exposed to the air, the redox potential will be at its highest, but will drop very low at the height of **fermentation**, when the consumption of oxygen by the **yeast**s is greatest. An aerated red wine will have a redox potential of 400 to 450mV, but this will fall to 200 to 250mV after a while in bottle. The amount of oxygen in a wine has the most dramatic effect on the redox potential (eg a wine with 0.1mg/l of oxygen will typically have a redox potential of 263mV, compared to 340mV for a wine with 2.5mg/l of oxygen), but other variables include **alcohol**, **acidity**, **pH**, **phenolics**, and, of course, temperature. *See also* OXIDATION-REDUCTION.

Reduced Although **reductive** is not necessarily a negative term, reduced is, because it is at the extreme end of reduction, bringing with it reduced **aromas**, such as those of **fixed sulphur compounds** and **mercaptans**. *See* FIXED SULPHUR; MERCAPTAN.

Reductive The less exposure it has to air, the more reductive a wine will be. Different as they are in basic character, **Champagne** (although there are some Champagnes that are deliberately made less reductive than others), Muscadet *sur lie*, and Beaujolais Nouveau are all examples of reductive, as opposed to **oxidative** wines – from the vividly autolytic Champagne, through Muscadet *sur lie* with its barest hint of autolytic character, to the **amylic aroma** of Beaujolais

Nouveau. A good contrast is Madeira, which is reductive, while Sherry is **oxidative**. The term is, however, abused, as many tasters use it to describe a fault, where the wine is heavily reduced.

Refractometer An optical device used to measure the **sugar** content of grapes when out in the field.

Remontage (Fr) Pumping wine over the **cap** (or manta) of skins during the *cuvaison* of red wine.

Remuage (Fr) An intrinsic part of the *méthode champenoise*; deposits thrown off during **secondary fermentation** are eased down to the neck of the bottle and are then removed at **disgorgement**.

Reserve wines Still wines from previous **vintages** that are blended with the wines of one principal year to produce a **balanced non-vintage Champagne**.

Residual sugar The sweetness of a wine expressed in grams per litre (thus a 75cl bottle will contain only three-quarters of the grams expressed). Even dry wines can have up to 2 grams of unfermentable residual sugar per litre.

Residual sweetness In one sense, this is synonymous with **residual sugar**, but in another it infers the basic taste of sweetness, rather than grams of sweetness per litre – and a greater number of grams per litre does not necessarily translate as sweeter. With no more than 2g/l of unfermentable sugar, the **sugars** present in all but so-called completely dry wine will be primarily **glucose** and **fructose**. Fructose is approximately twice as sweet as glucose, but as the residual sugar found in most wines is predominantly fructose, this is of little relevance. The only significant exception is the relative sweetness of sparkling wine made by different methods. For most sparkling wines the residual sugar is added via a *dosage*, which inverts to equal proportions fructose and glucose, whereas any residual sugar in a sparkling made by the rural method, or *méthode rurale*, will be exclusively fructose because it is literally the residue of the first **fermentation**. It follows, therefore, that a **traditional-method** sparkling wine with 8g/l residual sugar (4g fructose plus 4g glucose) will be 25 per cent less sweet than a rural-method sparkling wine with 8g of residual sugar (all fructose), providing everything else is equal in terms of both the wines and the taster.

Resveratrol A beneficial **antioxidant** found naturally in some grapes and wines. *See* HEALTH BENEFITS OF WINE.

Reticent A wine that is holding back on its **nose** or **palate**, perhaps through youth, and may well develop with a little more maturity.

Retroussage (Fr) Breaking-up the *marc* after pressing and shovelling it into the centre of the *maie* for re-pressing. This process takes place after grapes are pressed for the first time in a traditional champagne press.

Revtech This is not **DIAM**'s patented Diamant supercritical C02 process, which offers a **TCA**-free guarantee. Revtech is an older, cheaper, steam-cleaning technology that is used by various producers of **cork closures**, including DIAM, but removes only 75 to 85 per cent of TCA. Any remaining TCA will be equally divided amongst all the corks in any one batch, and it was precisely that, the spread of low levels of TCA, that was at the heart of the Altec debacle of **scalped** wines. Revtech corks are rarely seen in **Champagne**, but commonly found in sparkling wines of the Loire and **Cava**.

Rhaetic *or* **Rhaeticum** Sweet **ancient Roman wine** that was made from grapes grown near Verona. Suetonius claims that this wine was the favourite of Augustus, the first Roman emperor, while other writers attribute that honour to **Setine**.

Rhamnose A 5-carbon **sugar** that is sometimes found in tiny volumes as one of the unfermentable sugars in even the most bone-dry wines. Rhamnose is also a **monosaccharide** and a constituent of gum arabic, which is sometimes added with the *dosage*, thus its presence can be from two sources, one natural, the other added. *See also* PENTOSE.

Rhegium A light and pleasant **ancient Roman wine** from Rhegium (now Reggio di Calabria), smoother than that of **Sorrentum** and "fit to use after 15 years".

Ribonucleotide A nucleotide that contains **ribose** and is a precursor of nucleic acids. Ribonucleotides are produced in the latter stages of **autolysis** and contribute to the **umami** taste of extended **lees** ageing in more mature Champagnes.

Ribose Another 5-carbon **monosaccharide** that is sometimes found in tiny volumes as one of the unfermentable **sugars** in even the most bone-dry wines. Ribose is also an important component of RNA. *See also* PENTOSE.

Ribosomes *See* CYTOPLASMIC RIBOSOMES.

Ribotide A concatenation of and synonymous with ribonucleotide. *See* RIBONUCLEOTIDE.

Rich, richness 1. A term used by **Champagne** producers for a sweeter style, such as **sec** or *demi-sec*. 2. A wine with **balanced** wealth of **fruit** and **depth** on the **palate**, and a good **finish**.

Ripasso (It) Re-**fermentation** of wine on the **lees**.

Ripe Grapes ripen; wines **mature**. The **fruit** and even the **acidity** in wine can be referred to as ripe, however. Tasters should be careful not to mistake a certain **residual sweetness** for ripeness.

Ripe acidity The main acidic component in ripe grapes (**tartaric acid**) tastes refreshing and **fruity**, even in large proportions, whereas the main **acidity** in unripe grapes (**malic acid**) tastes **hard** and unpleasant.

Ripe grapes Grapes are ripe as soon as birds start eating them, but the ideal ripeness level depends on the style of wine to be produced. Generally, still white wines require riper grapes than sparkling wines, and red wines need to be even riper. When harvesting, winemakers used to speak about **sugar** ripeness and more recently so-called physiological ripeness, but all forms of fruit ripeness are physiological, so as a term it is as meaningless and as tautological as "chronological time". There is also acid ripeness, which might be a bit simplistic, but still works, and the rule-of-thumb acid ripeness is 50-50 **tartaric/malic**.

Roasted Describes the character of grapes subjected to the shrivelling or roasting on the vine (*passerillage*), due to excessive heat, rather than **noble rot**.

Robust A milder form of **aggressive**, which may frequently be applied to a **mature** product. A wine is robust by nature, rather than aggressive through youth.

Rondelle (Fr) The equivalent of a **cork mirror**, the disk of pure cork found on **agglomerate** sparkling-wine corks. *See* CORK MIRROR.

Rootstock Since the spread of *Phylloxera vastatrix* in the 19th century, the majority of vines in most commercial vineyards have been grafted onto **phylloxera-resistant rootstock**. Not detracting from this basic fact, the development and selection of rootstock varieties (which are crosses between two or more, usually American, species of vine) has become very sophisticated owing to the fact that a rootstock should not only induce a level of resistance to phylloxera, but the choice made must suit the prevailing conditions of soil and climate and can affect when the vine flowers, when its grapes ripen, the quality and quantity of those grapes, and other characteristics. *See* "Rootstock", p38.

Rooty Usually refers to a certain rooty richness found in Pinot Noir. Although a connotation of root vegetables, "rooty" is a positive term, albeit a very distinctive one that people will love or hate, whereas "vegetal" is a negative term.

ROPP Common abbreviation for "roll-on pilfer-proof", a type of **screw cap**. *See also* ROTE.

ROSA Evolution Amorim's proprietary steam-cleaning process is used for all of its **cork** production, reducing **TCA** levels by approximately 80 per cent. *See also* REVTECH; NDTECH.

Rosado (Sp) Spanish term for a rosé.

Rosato (It) Italian term for a pink wine.

Rosé (Fr) This term for "pink wine" has become as anglicised as rendezvous has for "appointment". In most cases, a rosé is made by crushing black grapes and keeping the juice in contact with the grapeskins for a short while prior to pressing or by running off coloured juice (*saignée*). It will have no discernible **tannin** content. Champagne rosé is a rare case where the wine may be made by blending a little red wine into a white wine.

ROTE Common abbreviation for "roll-on tamper-evident", a type of **screw cap**. *See also* ROPP.

Rotten eggs If a wine has a rotten-egg **aroma**, the **sulphur** has been reduced to **hydrogen sulphide** and the wine may well have formed **mercaptans** that are impossible to remove.

Round A wine that has rounded off all its edges of **tannin**, **acidity**, extract, and so on through **maturity** in bottle.

Route du vin (Fr) "Wine road". A clearly sign-posted *Route du Vin* that takes in all the most important wine centres, plus ancillary sites of historic interest and can be followed using a map obtained from the local tourist office.

Rural method The first commercial method of producing a sparkling wine entailed the bottling of a wine before its first and only **fermentation** had finished, thus creating a naturally effervescent wine. Without a *liqueur de tirage* the strength of *mousse* was inconsistent until the role of **yeast** was discovered and a means of measuring **sugar** content devised. The true rural method does not employ *dégorgement*, leaving the sediment to cloud the wine. **Champagnes** were produced by this method until the late-18th century and some producers were not adding sugar for either **second fermentation** or *dosage* until the late-19th century.

Sabaté (Sabaté Diosos from 2003) This producer of **oak** barrels and **corks** suffered a setback in reputation and sales due to the **Altec fiasco**, after which the group changed its name to Oeneo and launched a **TCA**-free version of Altec renamed **DIAM** in 2005.

Saccharide Effectively synonymous with "**sugar**", the most important for wine are the **monosaccharides** glucose and fructose. *See also* DISACCHARIDE; MONOSACCHARIDE; OLIGOSACCHARIDE; POLYSACCHARIDE; TETRASACCHARIDE; TRISACCHARIDE.

Saccharometer A lab device used for measuring the **sugar** content of grape juice, based on specific gravity.

Saccharomyces cerevisiae The *Saccharomyces* genus is usually tolerant to high **sugar** concentrations and resists increasing levels of **alcohol**, thus *S cerevisiae* represents the principle winemaking species of **yeast** for the **fermentation** of all types of wine, including sparkling wine.

Saignée (Fr) A term that literally means "bleeding" and is used for the process of drawing off surplus liquid that has bled from the grapes under their own weight in the press. This free-run juice is used to produce a **rosé** wine. In cooler wine regions, this process may be used to produce a darker red wine than would normally be possible from the remaining mass of grape pulp because the greater ratio of solids to liquid provides more colouring pigment.

Saint-Germain Lesser-known type of **oak** used for barrelmaking, from a forest in the Seine-et-Marne *département* of the Île de France.

Saint Hélène Lesser-known type of **oak** used for barrelmaking, from a forest in the Vosges *département* of northeastern France.

Salmanazar Large bottle equivalent to 12 normal-sized 75cl bottles.

Saranex Used as a liner for **screw caps** and **crown caps**, this is a five-layer film consisting of LDPE-EVA-PVDC-EVA-LDPE, which is to say two outer layers of "low-density polyethylene", and two inner layers of "ethylene vinyl acetate", sandwiching a core of actual Saran ("polyvinylidene chloride"). It is very thin and clear and is in fact the same material that is sold as Cling Film in Britain (aka Saran Wrap in the United States and Glad Wrap in Australia).

Sassy Should be a less-cringing version of the cheeky, audacious character found in a wine with bold, brash but not necessarily big flavour.

Scalp, scalped, scalping This refers to a flavour or **fruit** loss that could be due to **TCA**, which might be present at a level undetectable by the taster, but whose scalping effect is noticeable. If in doubt, open a second bottle to see if it does have more flavour or fruit.

Schirmeck Lesser-known type of **oak** used for barrelmaking, from a forest in the Vosges *département* of northeastern France.

Scion See PRODUCER VINE.

Screw cap Hated by some but loved by many, including the author of this encyclopedia, especially for wines to drink upon purchase and for wines of short- and medium-term ageing. It's a bit of a generalization, but the French do not like screw caps, yet the long-skirted Stelvin, which was the world's first wine-specific screw cap and has thus become synonymous with the generic term "screw cap" in the wine industry, was invented by the French. *See also* CROWN CAP; SYNTHETIC CLOSURES.

Sec (Fr) "Dry". When applied to wine, this means without any sweetness, but it does not mean there is no **fruit**. Dry wines with plenty of very ripe fruit can sometimes seem so rich they may appear to have some sweetness. *See also* "Levels of Champagne Sweetness", p256.

Second or **secondary fermentation** The fermentation that occurs in bottle during the *méthode champenoise*. The term is sometimes also used, mistakenly, to refer to **malolactic fermentation**.

Secondary aromas Not so much "fermentation aromas" as some sources state, but more the aromas of a freshly fermented wine. **Fermentation** aromas suggest the smell of a fermentation, which would be considered a fault in a finished wine, as both primary and secondary aromas are found in commercially available wines. *See also* PRIMARY AROMAS; TERTIARY AROMAS.

Sekt (Ger) German term for "sparkling wine", which (unless qualified as *Deutscher Sekt*) may be made from grapes grown in any EU country and by almost any method, Charmat being most important.

Selection de Grains Nobles (Fr) Literally "selection of noble berries." In Alsace, a rare, intensely sweet, **botrytised** wine. Often abbreviated to SGN.

Semi-carbonic maceration An adaption of the traditional **carbonic maceration** method of **fermentation**, in which whole bunches of grapes are placed in a vat that is then sealed while its air is displaced with **carbonic gas** (CO_2).

Setine or **Setinum** This **ancient Roman wine** from grapes grown in the hills of Sitia is generally considered one of the great wines of antiquity because it was favoured by Augustus and most of his courtiers, although the Roman historian Gaius Suetonius Tranquillus claimed **Rhaetic** wine was the first Roman emperor's favourite wine.

Sharp This term applies to **acidity**, whereas **bitterness** applies to **tannin** and, sometimes, other natural solids. Immature wines may be sharp. If used by professional tasters, however, the term is usually a derogatory one. The opposite to sharp acidity is usually described as ripe acidity, which can make the **fruit** refreshingly tangy.

Sheet filtration See PAD FILTRATION.

Sherry-like This term refers to the odour of a wine in an advanced state of **oxidation**, which is undesirable in low-strength or unfortified wines. It is caused by excessive **acetaldehyde**.

Short Refers to a wine that may have a good **nose** and initial flavour, but falls short on the **finish**, its taste quickly disappearing after the wine has been swallowed.

Sin Cosecha (Sp) Spanish term for "non-vintage".

Skin-contact The **maceration** of grape skins in **must** or fermenting wine can extract varying amounts of colouring pigments, **tannin**, and aromatic **compounds**.

Slavonian oak A type of **oak** used for barrelmaking, from forests in eastern Croatia.

Sleek Synonymous with smart and **stylish**, sleek infers a focused, fault-free modern style.

Smart Synonymous with **sleek** and **stylish**, smart implies a focused, fault-free modern style.

Smokiness, smoky, smoky complexity, smoky-oak Some grapes have an inherent smoky character (particularly Syrah and Sauvignon Blanc). This charcter can also come from well-**toasted oak** casks, but may also indicate an unfiltered wine. Some talented winemakers do not **rack** their wines and sometimes do not filter them in a passionate bid to retain maximum character and create an individual and expressive wine.

Smooth The opposite of **aggressive** and more extreme than **round**.

SO2 The commonly used chemical formula for sulphur dioxide, an **antioxidant** with aseptic (antibacterial) qualities that may be added in the production of wine. SO_2 is also a natural by-product of **yeast**, with as much as 41mg/l found in **fermentation**s where no SO_2 has been added.

SO2 allergies, intolerance, and headaches While every responsible winemaker tries to minimise the amount of **sulphur** added, it should be remembered that sulphur is not intrinsically harmful, and wine is not necessarily the prime source. Sulphur is a vital mineral for health that is found naturally in a vast number of common foodstuffs, helping us to resist bacteria and protecting the body against toxic substances. Without sulphur bonds we could not metabolise **proteins**, and we would die. Our bodies contain more sulphur than can be found in over 2,000 bottles of **Champagne**. Of course a small number of individuals are allergic to or intolerant of sulphur. Although they also need it to survive, they have to be very careful about how much they consume, and know more than anyone else how frustratingly long the list is of sulphite-containing food and drinks they need to avoid. Wine is indeed on that list, but it is not at the top. If any non-allergic or non-intolerant person is worried about sulphur levels in wine, he or she should be even more worried about the rest of their diet because there are more naturally occurring sulphites in a four-ounce portion of healthy broccoli than in one *flûte* of Champagne, and broccoli is nowhere near as sulphite-rich as french fries, popcorn, and certain dried fruits.

SO2 aromas Although it is often said that **sulphur** should not be noticeable in a finished wine, it forms the basis of many sulphidic **aromas** that are key characteristics to classic *brut*-style sparkling wines, together with **yeast**-complexed and post-*disgorgement* aromas. These typically commence with "**gunpowder**" and ultimately end up as "**toasty**", sometimes even "burnt toast", but whatever variant the toastiness might be, the longer it takes for to emerge, the more **finesse** it will have.

SO2, Bound Upon contact with grapes, juice or wine, some SO_2 immediately combines with oxygen and various **compounds**, and these combined forms are known as bound (or fixed) SO_2. This binding action is not immediate, but is fastest during the first 24 hours of SO_2 addition. After this, it takes four to five days before full binding is complete. Over this period, a slow decrease in free SO_2 and a corresponding increase in bound SO_2 occurs. After this, an equilibrium exists until the free SO_2 decreases due to **oxidation**. It should be noted that **sulphur** forms a tight or

stable binding only with **aldehyde**s. Much looser (unstable) bindings are made with **sugars**, ketonic acids, uronic acids, and, in red wines, **anthocyanins** and quinones etc. In its aseptic role, SO_2 also binds to **yeast**, bacteria, **protein**, and other cellular particulates. It is therefore more effective to add SO_2 after clarification of the juice, otherwise it will be wasted on compounds that will be removed from the wine. The later SO_2 is added, the better.

SO2, Free The principal reason SO_2 is added to wine is to prevent **oxidation**, even though it is a more powerful antiseptic (combatting bacteria). Most **sulphur** in wine is **bound SO2**, however, and it is only free SO_2 that is effective as an antioxidant. Free SO_2 consists of sulphite, bisulphite, and molecular (or active) **sulphur** *See* SO2, MOLECULAR.

SO2, Molecular Existing as either a gas or as single **molecules** in juice and wine, molecular SO_2 is the quickest form of **sulphur** to bind with chemical or biochemical **compounds**, thus it represents the most important form of SO_2 in wine.

SO2, The myth of bound Total SO_2 found in wine consists of **free** and (mostly) **bound SO2**. When SO_2 is added, much of it quickly binds with **sugars**, pigments, and other **compounds** to become bound SO_2. Outside of academic circles, it is often said that once bound, SO_2 is no longer available to protect a wine against **oxidation**, but in legal parlance, this is not the whole truth. How this myth was born is perfectly understandable, as the phrase "once SO_2 becomes bound it is no longer available as free SO_2" or words to that effect, can be found in a mountain of texts, including scientific papers, and this notion is then repeated unqualified elsewhere. There are various degrees at which SO_2 binds, however, depending on the properties of the compound to which it is bound and the degree to which SO_2 is bound to a compound affects its availability to bind elsewhere. If **acetaldehyde** is present and there is sufficient free SO_2 available, no "loosely" bound SO_2 is required, but if the amount of free SO_2 has become depleted, then it will be topped up through the release of SO_2 that is "loosely" bound to unstable compounds. In a classically dosaged sparkling wine, the most common unstable compounds are sugars. SO_2 will only "loosely" bind to **glucose** and hardly bind at all to **fructose** (whereas in red wine more common unstable compounds are pigments and **anthocyanins**). The *dosage* in a sparkling wine therefore acts as potential reinforcement for the free SO_2 and is therefore an additional line of defence against oxidation. The equilibrium between free and bound SO_2 is therefore subject to a certain flux, and the amount of that flux is dependent on freeing up SO_2 from that which is bound to unstable compounds, primarily sugars. A *brut nature* is thus defenceless in this respect, which is why it is always prudent to consume, rather than cellar, any good or great *brut nature*. The only compound that SO_2 is "tightly" bound to is **aldehyde** (primarily but not exclusively acetaldehyde). Although SO_2 can directly bind oxygen **molecules**, it more frequently binds acetaldehyde, and it is the tight binding of acetaldehyde that eliminates the characteristic **aroma** of oxidation, covering up the oxidation, rather than stopping it. *See also* SO2, FREE; SO2, TOTAL

SO2, The role of To paraphrase Dr Roger Boulton of UC Davis, there is a general misconception that SO_2 will protect against **oxidation**. Only a fraction of the SO_2 interacts directly with O_2 molecules. The problem is that the speed at which **sulphur** reacts is glacial compared to that of oxygen, thus its main **antioxidant** role is to bind with aldehyde formed by oxidation. It is not stopping the oxidation, just removing the smell of the oxidation. A bit like an oenological air-freshener (that's my description, don't blame Dr Boulton!).

SO2, Total The total SO_2 is the sum of its **free** and **bound** components. *See also* SO2, BOUND; SO2, FREE.

Soft A term interchangeable with **smooth**, although it usually refers to the **fruit** on the **palate**, whereas smooth is more often applied to the **finish**. Softness is a very desirable quality, but "extremely soft" may be derogatory, implying a weak and **flabby** wine.

Solera (Sp) A system of continually refreshing an established blend with a small amount of new wine (equivalent in proportion to the amount of the blend that has been extracted from the *solera*) to produce a wine of consistent quality and character. Some existing *soleras* were laid down in the 19th century, and whereas it would be true to say that every bottle of that *solera* sold today contains a little of that first **vintage**, it would not even be in a teaspoon. You would have to measure it in **molecules**, but there would be infinitesimal amounts of each and every vintage from the date of its inception to the year before bottling. *See also* "The solera blending system", p408.

Solid This term is interchangeable with **firm**.

Solomon Large bottle equivalent to 24 normal-sized 75cl bottles.

Solumological The science of soil and, in the context of wine, the relationship between specific soil types and vine varieties.

Sorbitol A **sugar alcohol** commonly found in small amounts in wine, sorbitol is not a constituent of grapes (although it is of other fruits, particularly orchard fruits) unless affected by rot, noble or otherwise. In normal circumstances dry or sparkling white wines would usually contain the lowest level of sorbitol, all of which would be produced by **yeast** during the **fermentation** process. Not fermentable.

Sorbic acid A **yeast**-inhibiting **compound** found in the berries of mountain ash, sorbic acid is sometimes added to sweet wines to prevent re-fermentation, but it can give a powerful geranium odour if the wine subsequently undergoes **malolactic fermentation**.

Sorrentum, Surrentine, *or* **Surrentinum** The Pavie/ Parker/Robinson of antiquity, Sorrentum is a wine from the promontory forming the southern horn of the Bay of Naples. It divided Roman opinion 2,000 years ago, when the poet Publius Papinius Statius ranked it as the equal of **Falernian**, yet Tiberius described it as "only generous vinegar", while his successor, Caligula, thought it "worthless"!

Souped up, soupy Implies a wine has been blended with something **richer** or more **robust**. A wine may well be legitimately souped up, or use of the term could mean that the wine has been played around with. The wine might not be correct, but it could still be very enjoyable.

Sour In purely scientific terms of basic taste, this is synonymous with **acidity**, but in wine-tasting terms it is definitely a negative, indicating something off, with **acetic acid** dominating.

Sous marque (Fr) A *marque* under which wines, usually second-rate wines, are offloaded.

Southern-style This term describes the obvious characteristics of a wine from the sunny south of France. For reds, it may be complimentary at an honest basic level, indicating a full-bodied, full-flavoured wine with a peppery character. For whites, it will probably be derogatory, implying a **flabby** wine with too much **alcohol** and too little **acidity** and freshness.

Soutirage (Fr) Synonymous with **racking**.

Sparging A process in which **carbonic gas** is introduced into a wine before bottling, often simply achieved through a valve in the pipe between the vat and the bottling line.

Spätlese (Ger) A **QmP** wine that is one step above Kabinett, but one below *Auslese*. It is fairly sweet and made from late-picked grapes.

Spicy 1. A **varietal characteristic** of some grapes, such as Gewürztraminer. 2. An aspect of a complex **bouquet** or **palate**, probably derived from **bottle-age** after time spent in wood.

Spicy-oak A subjective term describing complex **aromas** derived from **fermentation** or maturation in **oak** that can give the impression of various spices – usually "creamy" ones such as cinnamon or nutmeg – and that are enhanced by **bottle-age**.

Spritz, spritzig (Ger) Synonymous with **pétillant**.

Spumante (It) Fully sparkling.

Stachyose A (**galactose** + galactose + **fructose** + **glucose**) tetrasaccharide that is sometimes found in tiny volumes in wine, but more commonly in vegetables.

Stabilization The process by which a heaving broth of biochemical activity becomes firmly fixed and not easily changed. Most wines are stablised by **tartrate** precipitation, **filtration**, **fining**, and the addition of **SO2** (sulphur dioxide).

Stage (Fr) A period of practical experience. It has long been traditional for vineyard owners to send their sons on a *stage* (internship) to a great château in Bordeaux. Now the *bordelais* send their sons and daughters on similar *stages* to California and Australia.

Stalky 1. The herbaceous-tannic **varietal characteristic** of Cabernet grapes. 2. Applies literally to wines made from grapes which were pressed with their stalks. 3. Could be indicative of **cork taint** or **corked wine**.

Statan According to Athenaeus of Naucratis, this was one of the best of the **ancient Roman wines**, resembling **Falernian** but "lighter and innocuous", leading us to suspect that inoffensive wines were preferable in those times.

Stelvin This brand of **screw cap** was developed specifically for wine in 1959 by La Bouchage Mecanique (a French firm later taken over by Pechiney), which effectively put a longer skirt on its existing Stelcap, a general-purpose closure. The product was initially called Stelcap-vin but was soon shortened to Stelvin.

Sterilization The ultimate sterilization of a very cheap, commercial wine may be **pasteurization** or flash pasteurization.

Stickies Common parlance for very sweet wines, usually **fortified** or **botrytised**.

Strain Refers to a selected **yeast** and oenological bacteria, usually sold under a proprietary brand.

Straw Straw-like **aromas** often blight sparkling wines. Producers claim it is part of the **complexity**, but if true, it is a very dull sort of complexity and one that is not completely clean. Perhaps it comes from the **yeast**, or maybe rotten grapes, or even the reaction of yeast contact to wine made from a certain percentage of rotten grapes. In any case, this is not a positive attribute, although where it appears in this guide the wines obviously have sufficient going for them to overcome these straw-like aromas, otherwise they would not be recommended.

Stretched This term describes a wine that has been diluted or cut with water (or a significantly inferior wine), which is usually illegal in an official **appellation**. It can also refer to wine that has been produced from vines that have been "stretched" to yield a high volume of attenuated **fruit**.

Structure The structure of a wine is literally composed of its solids (**tannin**, **acidity**, **sugar**, and **extract** or density of **fruit** flavour) in **balance** with the **alcohol**, and how positively they form and feel in the mouth.

Stück (Ger) A 1,200-litre (317-US gallon) oval cask.

Stuck fermentation A stuck, literally "halted", fermentation is always difficult to rekindle and, even when done successfully, the resultant wine can taste strangely bitter. The most common causes for a stuck fermentation are: 1. temperatures of 35°C (95°F) or above; 2. nutrient deficiency, which can cause **yeast** cells to die; 3. high **sugar** content, which results in high osmotic pressure, which can cause yeast cells to die.

Stylish Describes wines possessing all the subjective qualities of **charm**, **elegance**, and **finesse**. A wine might have the "style" of a certain region or type, but this does not mean it is stylish. A wine is either stylish or it is not – it defies definition.

Subtle Although this description should mean a significant yet understated characteristic, it is often employed by wine snobs and frauds who taste a wine with a famous label and know that it should be special, but cannot detect anything exceptional. They need an ambiguous word to get out of the hole they have dug for themselves.

Succinic acid Although the second-most-common organic acid found in Muscadine grapes (*Vitis rotundifolia*), succinic acid is insignificant in all **Vitis vinifera** wine grapes, and even though still insignificant in *vinifera* wines, its presence in wine will be higher than that found in the grapes used to make that wine because more succinic acid is produced during **fermentation** than through grape metabolism. Its bitter, salty taste precludes its use for wine acidification.

Sucrose This **disaccharide** is a natural plant **sugar** that is produced by the vine via photosynthesis, but is not found in grapes, as it is broken down into its component parts (**glucose** + **fructose**) by **hydrolysis** as soon as it has moved from the vine into the fruit. Elsewhere, sucrose is most commonly found in sugar-cane and sugar-beet, both of which are refined into ordinary table sugar. Sucrose is a non-reducing sugar. It is not a **hexose**, but its component parts – glucose and fructose – are. Technically, sucrose does not contain 50-50 per cent glucose and fructose, but for all practical purposes that is what it breaks down into.

Sugar All sugars are **carbohydrates** and sweet to one degree or another (*See* SUGARS, THE SWEETNESS OF). Sugar is effectively synonymous with **saccharide**. There are two primary methods of classifying sugars and one of these is by the number of saccharides involved. The simplest saccharide is a **monosaccharide**, and a monosaccharide is composed of just one sugar, whereas a **disaccharide** is composed of two sugars, a trisaccharide of three sugars, a tetrasaccharide of four, and so on. The other method of classifying sugars is by size and this is expressed by the number of carbon **molecules**: **diose** for a 2-carbon molecule sugar; **triose** for a 3-carbon; tetrose for 4-carbon; **pentose** for 5; **hexose** for 6; **heptose** for 7; **octose** for 8, and **nonose** for 9.

Sugar acids *See* URONIC ACIDS.

Sugar alcohols These **compounds** are so-called because their molecular structure is a **hybrid** between a sugar **molecule** and an **alcohol** molecule, but they are misleadingly named, as they are not sugars, nor do they contain alcohol. They do, however, possess sweetness, thus must be taken into account when considering the **residual sweetness** of a wine. Sugar alcohols are not completely absorbed into the blood stream from the small intestines, thus they have less (but not zero) impact on blood **glucose** than regular sugar. They also provide fewer calories per gram, making them popular sweeteners in the food industry. Sugar alcohols are not fermentable. For winemaking considerations, they mostly include **arabinitol**, **erythritol**, **glycerol**, **mannitol**, and **sorbitol**.

Sugars, Fermentable The only readily fermentable sugars are **hexoses** (6-carbon **monosaccharides**) and by far the most important of these are **glucose** and **fructose**. In fact, glucose is the most common form of sugar found in nature. **Sucrose** (regular table sugar) is not naturally found in grapes, but it is produced by photosynthesis in vines and stored in its leaves until the ripening process begins, when it is hydrolysed into its constituent parts of glucose and fructose, which are translocated to the grapes. The two most common fermentable sugars are therefore readily available as soon as grapes are crushed or pressed. Although sucrose is not naturally found in grapes, it is commonly added to a fermenting wine via **chaptalisation** and will quickly breakdown into fermentable sugars of glucose and fructose by **hydrolysis**. It should be noted that while all fermentable sugars are reducing sugars, not all reducing sugars are fermentable (eg **pentoses**).

Sugars, Invert A mixture of equal parts of **glucose** and **fructose**, formed through inversion of **sucrose** by **hydrolysis**.

Sugars, The sweetness of We tend to think of **sugar** in terms of the white, refined sugar as purchased from grocery stores, the slightly sticky Demerara, or other more exotic sugars, such as Muscovado, with its wonderfully pungent **aroma** of molasses. They are all essentially **sucrose**: Muscovado is 80 per cent sucrose, for example, while refined white sugar is 95 per cent sucrose (*see* SUCROSE). The problem with sucrose is that we all have a rough idea of how sweet it is, not just through our everyday consumption, but also from the echo of childhood memories, which in many ways has a stronger influence over our perception of sweetness. There is no sucrose in wine, however, the primary sugars of which are **glucose** and **fructose** (*see* SUGARS, FERMENTABLE) and they are not the same sweetness. Glucose is only 65 per cent as sweet as sucrose, whereas fructose is at least twice as sweet as glucose (*see* SUGARS, RELATIVE SWEETNESS OF DIFFERENT). In wines with a natural **residual sugar** (either wholly or partially), the proportions of these two sugars may vary wildly, thus the **residual sweetness** expressed in grams per litre can be extremely misleading. Without even taking into account the **acidity, alcohol,** and other constituents of a wine that affect how we perceive its sweetness, a wine with a natural residual sweetness of, say, 20g/l, could conceivably taste significantly sweeter than another wine with a natural residual sweetness of, say, 25g/l (assuming the acidity, alcohol, and all other constituents being equal, of course). It would make much more sense if sweetness were to be in grams per litre expressed as sucrose, just as acidity is expressed as **tartaric** or **sulphuric acid**, but although technically possible, the cost of identifying and quantifying the individual sugars to formulate such an expression would be far too much a test applied universally to every wine produced in the world. In sparkling wines, however, which are fermented dry and sweetened with a *dosage*, the proportion of glucose and fructose should be equal. Even when the dosage added is pure sucrose, it quickly breaks down into equal parts glucose and fructose by **hydrolysis**. Some producers use rectified grape **must** (which might or might not be concentrated) for their *dosage*, and that is already in the form of invert sugar (ie 50-50 glucose and fructose). The relative sweetness of most sparkling wines should thus be comparable at any given residual sugar, albeit not with wines containing natural residual sugar. The only significant exception would be those Prosecco that are sweetened with grape juice. Where variation in seemingly the same residual sweetness of sparkling does exist, it is in the different ways in which producers determine that sweetness. Most simply state the *dosage* added, whereas some refer to residual sweetness, combining both the *dosage* and any unfermentable sugar, which could be up to two grams per litre. Even at the upper end of their possible existence, unfermentable sugars (**arabinose, rhamnose, xylose,** and **ribose**) do not bring any sweetness to a so-called bone-dry wine, but when tasted in conjunction with a *dosage*, they do make a perceptible difference to the overall sweetness of a sparkling wine. Sometimes producers are very clear in the way they determine the sweetness, either as the *dosage* added or the total residual sugar (ie the *dosage* plus any residual sugar that may happen to be in the wine prior to adding the *dosage*). They do not usually clarify which, however, although the sweetness may appear to be precisely defined in grams per litre, more often than not there will be no way of telling whether one producer's 5g/l has 5g/l of residual sugar or actually has 7g/l and could thus be sweeter than another producer's 6g/l.

Sugars, Reducing A saccharide (sugar) that can reduce other **compounds** and in turn be **oxidized** itself. For wine these are primarily **fructose** and **glucose** but also include other sugars of minimal significance, such as **arabinose, galactose, maltose, mannose,** melibiose, **ribose, rhamnose, stachyose,** and **xylose. Sucrose** is not a reducing sugar, but its two component parts, glucose and fructose, are.

Sugar, Simple A monosaccharide.

Sugars, Unfermentable Even fully fermented bone-dry wines have a tiny amount of **residual sugar**, due to the presence of unfermentable sugars. **Yeast**s are unable to convert these sugars in any practical way, so there will be up to 2g/l of residual sugar in the driest wines. The most important unfermentable sugars commonly found in wine are all **pentoses** or 5-carbon **monosaccharides**, primarily **arabinose** and to a lesser extent **rhamnose**, then **xylose**, and **ribose**.

Sulphur dioxide *See* SO2.

Sulphuric acid Whether **sulphurous acid** exists or not (*see* SULPHUROUS ACID), it is a weak acid whereas sulphuric acid is a strong acid and is never found in wine, although it is often used to express total **acidity**, particularly by the French.

Sulphurous acid Often used synonymously with SO_2, but sulphurous acid is H_2SO_3 and thus hydrated (H_2O + SO_2 = H_2SO_3). Some sources will state that sulphurous acid does not exist, but that is more an argument of semantics than chemistry. Suffice to say that sulphurous acid is SO_2 in an aqueous state and that for all practical purposes it can be used as synonymous with SO_2.

Summer pruning Synonymous with **green pruning**.

Super-premium wine This category comes between **premium** and **ultra-premium** wines and may be priced at, say, US$10 to 20.

Super-second A term that evolved when Second-Growth (*deuxième cru*) châteaux, such as Palmer and Cos d'Estournel, started making wines that came close to First-Growth (***premier cru***) quality at a time when certain First Growths were not always performing well. The first super-second was Palmer 1961, although the term did not evolve until some time during the early 1980s.

Supercritical CO2 In thermodynamics, there is a critical state at the end point of a pressure-temperature environment when a gas stops being a gas, and a beginning point when it becomes a liquid, or vice-versa. Between these two critical points, the gas or liquid is said to be in a supercritical state, where it is neither gas nor liquid, yet possesses some – not all – characteristics of both. Although Charles Baron Cagniard de la Tour observed the supercritical state in 1822, he was unaware of its significance and it would be left to Thomas Andrews of Queen's University, Belfast, to coin the term "critical state" in 1869 and Johannes Diderik van der Waals to develop the equation defining near-critical, critical, and supercritical states in 1873, eventually earning him a Nobel Prize in 1910. The first description of a practical application for supercritical CO2 was by Kurt Zosel of Max Planck Institute in 1963. Zosel patented the supercritical CO2 decaffeination process in 1973, since when it has been applied to many other extraction processes in the perfume and food industries. In collaboration with Sabaté (now Oeneo Group) Guy Lumia, Christian Perre, and Jean-Marie Aracil of *Commissariat à l'Energie Atomique* patented the Diamant® Supercritical CO2 process to remove all releasable **TCA** (and its precursors) from **technical cork closures** in 1999, but it was not until 2005 that the first **DIAM** Diamant® closures were launched. *See* DIAMANT® PROCESS.

Supertaster A term that applies to those people who have more taste buds than the rest of us. Contrary to what the name implies, however, being a so-called supertaster is more of a burden than a blessing when it comes to tasting wine because a supertaster is super-sensitive to bitterness and, consequently, finds

even the softest **tannins** in the greatest red wines too bitter to enjoy. This super-sensitivity is thought to be a survival trait from our hunter-gatherer days, when the only instrument we had to avoid toxic foods was our **palate**, since the majority of the most poisonous substances are bitter to the taste. *See* "What Is a Supertaster?", p90.

Super Tuscan This term was coined in Italy in the 1980s for the Cabernet-boosted *vini da tavola* blends that were infinitely better and far more expensive than Tuscany's traditional Sangiovese-based wines. *See* "Birth of the Super-Tuscans", p367.

Supervin A screw cap developed by Auscap in Australia.

Supple Describes a wine that is **easy to drink**, not necessarily **soft,** but the term suggests more ease than **round** does. With age, the **tannin** in wine is said to become supple.

Supple tannin Tannins are generally perceived to be **harsh** and mouth-puckering, but the tannins in a ripe grape are **supple,** whereas those in an unripe grape are not.

Sur lie (Fr) Describes wines, usually Muscadet, that have been kept on their **lees** and have not been **racked** or **filtered** prior to bottling. Although this practice increases the possibility of bacterial infection, the risk is worth taking for those wines made from neutral grape varieties. In the wines of Muscadet, for example, this practice enhances the fruit of the normally bland Melon de Bourgogne grape and adds a **yeast**y dimension of depth that can give the flavour of a modest white Burgundy. It also avoids aeration and retains more of the carbonic gas created during **fermentation**, thereby imparting a certain **liveliness** and **freshness**.

SurePure A South African process that uses light energy to purify wine. First deployed in the wine industry at the Franschhoek wine estate of L'Ormarins in 2009, SurePure uses patented Turbulator technology to transmit ultraviolet energy (UV-C at 250–270 nm) to deactivate microbes, reducing the need to add sulphur.

Süßreserve (Ger) Unfermented, fresh grape juice commonly used to sweeten German wines up to and including *Spätlese* level. It is also added to cheaper *Auslesen*. Use of *Süßreserve* is far superior to the traditional French method of sweetening wines, which utilises grape concentrate instead of grape juice. *Süßreserve* provides a fresh and grapey character that is desirable in inexpensive medium-sweet wines.

Synthetic closures The first commercially available synthetic **corks** were produced by companies such as Novembal (France) and Metal Box (UK), using an injection-moulded ethylene vinyl acetate polymer, which was insufficiently flexible to provide what would be regarded today as an acceptable quality of wine closure, due to an unacceptable level of **oxygen ingress**. The next level of technology, achieved in the early 1990s by companies such as Supremecorq and Integra, provided closures of greater flexibility and thus improved performance, yet they were still not a satisfactory alternative to cork, particularly for longer than a year or two in bottle. Finally, by dropping the injection-moulded process, two-part closures such as Nomacorc and Neocork, which are comprised of an inner core produced by an extrusion process and sleeved in a softer and more flexible gasket or outer sleeve, managed to achieve a very satisfactory performance as an alternative to cork. The outer sleeve is so soft and flexible that, if used as a one-piece closure, it would be practically impossible to insert into a bottle and even harder to extract – but as an outer skin to the extruded inner core, it provides an almost perfect seal with the inside of the bottle. *See also* CROWN CAP; SCREW CAP.

Table wine A term that originally distinguished between a light, unfortified wine (which was generally the preserve of the dining table) and a **fortified** wine (which was served at other times, such as Sherry in the morning, Madeira at teatime, and even Port, with which the men would retire after dining). The term ***vin de table*** was then usurped by the French wine regime, which cast it as the lowest of the low. Now the term has been replaced by ***vin de France***.

Tafelwein (Ger) "**Table wine**" or ***vin de table***.

Talento (It) Since March 1996, producers of Italian ***méthode champenoise*** wines may use the new term *"Talento"*, which has been registered as a trademark by the *Instituto Talento Metodo Classico* – established in 1975 and formerly called the *Instituto Spumante Classico Italiano. Talento* is almost synonymous with the Spanish term ***Cava***, although to be fully compatible it would have to assume the mantle of a **DOC** and to achieve that would require the mapping of all the areas of production. It will take all the *talento* they can muster, however, to turn most Italian ***spumante brut*** into an international class of sparkling wine.

Tank Method *See* CUVE CLOSE.

Tannic, tannin Tannins are various ***phenolic*** substances found naturally in wine that come from the skin, seeds, and stalks of grapes. They can also be picked up from **oak** casks, particularly new ones. Grape tannins can be divided into ripe and unripe, the former being most desirable. In a proper **balance**, however, both types are essential to the structure of red wines, in order to knit the many flavours together. Unripe tannins are not water-soluble and will remain harsh no matter how old the wine is, whereas ripe tannins are water-soluble, have a suppleness or, at most, a **grippy** feel from an early age, and will drop out as the wine **matures**. Ripe grape tannin softens with age, is vital to the structure of a serious red wine, and is useful in wines chosen to accompany food.

Tart Refers to a noticeable **acidity** that is much more than **piquant** and very close to **sharp**.

Tartaric acid The ripe acid of grapes that increases slightly when the grapes increase in sugar during the ***véraison.***

Tartrates, tartrate crystals Deposits of **tartaric acid** look very much like sugar crystals at the bottom of a bottle and may be precipitated when a wine experiences low temperatures. Tartrates are also deposited simply through the process of time, although seldom in a still or sparkling wine that has spent several months in contact with its **lees**, as this produces a **mannoprotein** called MP32, which prevents the precipitation of tartrates. A fine deposit of glittering crystals can also be deposited on the base of a **cork** if it has been soaked in a sterilizing solution of metabisulphite prior to bottling. All are harmless. *See also* "Cold stabilization", p66.

Taste bud Humans have approximately 10,000 taste buds, mostly on the tongue but also on the soft or upper **palate**, the insides of the cheeks, around the back of the throat, the upper throat, and even on the lips. As the name implies, these are bud-like protrusions, and they contain taste receptors, which detect only the basic tastes of **bitterness**, sourness (or **acidity**), sweetness, saltiness, and **umami**.

Taste receptor In each taste bud, there are approximately 100 taste receptors; thus, we each have around one million taste receptors. There are many different types of taste receptor, including a number whose purpose is unknown or not fully understood, but essentially they all transmit the detection of a basic taste directly to the brain without any input from the olfactory bulb. Some taste receptors detect only **bitterness**, while others detect only sourness (**acidity**) or sweetness or saltiness or **umami**, but with 100 or so taste receptors in each taste bud, each taste bud has the ability to detect all basic tastes.

Tastevin (Fr) A shallow, dimpled, silver cup used for tasting, primarily in Burgundy.

TBA Short for tribromoanisole, TBA was first identified as one of the **TCA**-like cork-taint culprits by Pascal Chatonnet et al as recently as 2004.

TBP Short for tribromophenol, TBP is the precursor to **TBA** and can be found in wooden structures treated with preservatives and/or flame retardants. It can also be found in various plastics and other synthetic polymers. It is methylated to TBA.

TCA Short for trichloroanisole, the prime (but by no means only) culprit responsible for **corked wines**, TCA was first identified by Hans Tanner in 1981. **TeCA** and **TBA** were later identified (by Tanner and Chatonnet respectively) as sole or contributing **compounds** responsible for this taint. TCA can be found in or methylated from **TCP** found in **oak** staves, wooden pallets, or even winery beams and is so volatile that it may migrate to the smallest wooden part, inert plastic, and even into an open packet of **bentonite**, the use of which should ironically clean a wine, but could in such circumstances simultaneously contaminate it. It can also get into water supplies that are stored in wooden or plastic tanks (*see also* GEOSMIN). TCA is not therefore restricted to cork, so it is possible to find a "corked" wine in a bottle sealed with a **screw cap** and, although rare, I have personally encountered two such examples. Experienced tasters can detect levels of just one ppt TCA. That is one part in a trillion. We have heard the word "trillion" so much that we have forgotten how large a number it is (and thus how microscopic one in a trillion is). A trillion is an old-fashioned billion: that is a million-million. A million seconds lasts for less than 12 days, but a trillion seconds is more than 31,700 years. Check it out. Imagine trying to isolate a single, specific second within a timespan of almost 32 millennia, but as George Taber, author of *To Cork or Not to Cork,* writes "that's enough to ruin a bottle of wine". Even most non-experienced consumers can detect TCA at 5ppt. Human detection levels for TCA in sparkling wine are just half that for still wine, because **carbonic gas** accentuates the presence of this compound, and even subliminal levels can **scalp** the fruit in a wine. *See also* CORK TAINT OR CORKED WINE; SCALP, SCALPED, SCALPING.

TeCA Short for tetrachloroanisole, one of the **TCA**-like **cork-taint** culprits. TeCA was first identified as one of the TCA-like cork-taint culprits by Hans Tanner in 1983.

TeCP Short for tetrachlorophenol, TeCP is the precursor to **TeCA** and can be found in wooden structures treated with pentachlorophenol-based preservatives. It is methylated to TeCA.

TDN Common abbreviation for trimethyldihydronaphthalene, the active chemical **compound** responsible for the so-called **petrol** (kerosene, gasoline, or paraffin) **aroma** of bottle-**matured** Riesling. TDN develops through the degradation of beta-carotene, an **antioxidant** that is itself derived from another antioxidant – lutein. The ratio of beta-carotene to lutein is higher in Riesling than in any other white grape variety. Studies show that the lower the **pH** of a wine, the higher its potential for developing TDN, thus its propensity to develop in warmer climes. The longer it takes for the petrol aromas to emerge, the more **finesse** they have. Interestingly, **cork** absorbs 40 per cent of TDN, so **screw caps** effectively preserve petrol aromas.

Technical cork These corks consist of a dense **agglomerated cork** body with natural cork disks glued on one or both ends to provide a closure that is chemically very stable and mechanically very strong; used for wines that are destined to be consumed within a period of two to three years. *See* CORK CLOSURES, MICROAGGLOMERATE

Teinturier (Fr) A grape variety with coloured (red), as opposed to clear, juice.

Terpenes These **compounds** are responsible for some of the most powerful spice-laden and full-blown floral aromatics found in ***Vitis vinifera*** grapes, particularly varieties such as Gewürztraminer, Muscat, and Riesling. **terpenes** and their derivatives (terpenoids) are present in wine in five forms: **terpenoid acids**, **terpenoid alcohols**, **terpenoid aldehydes, terpenoid esters**, and **terpenoid oxides.** All these terpenoids are derived from isopentenyl pyrophosphate (IPP), which is a five-carbon isoprenic; thus, terpenes are always multiples of five units (5, 10, 15, 20 etc), up to caratenoids (40), after which they are lumped together as polyisoprenoids. Although there are more than 400 naturally occurring terpenoid compounds in the plant world, only 44 have so far been found in grapes or wine, and of these, 18 or more possess some significant degree of influence over wine **aromas**. The six most important terpenoids are: citronellol, geraniol, hotrienol, **linalool**, nerol, and alpha-terpineol. As only free terpenes are odiferous and all free terpenes have to be present at above-threshold levels to have any aromatic effect, there is an enormous untapped aromatic potential in almost every wine produced, since a large proportion of terpenes are found in their bound form. Bound terpenes can be freed by hydrolytic action, however, via either acids or **enzymes**, and to a certain extent this happens naturally in the ***véraison*** of the grapes and **vinification** of the wine. To capitalise on this, wine laboratories all over the world have formulated various specialised enzymatic preparations to achieve the same ends but to a much greater extent. Some of the most experienced wine judges in Australia believe that enzyme treatments on Rieslings produce an unnatural, overly floral character. No studies have been made, as yet, to determine the veracity of such claims, and no similar accuzations have been made against the use of enzymes for either Gewürztraminer or Muscat, but the practice of using enzyme preparations for this and other purposes remains controversial, not least because the purpose of enzymes is to break down **molecules**, and sometimes it is impossible to predict what they might do next. When referencing any terpenoid compound, you will find that many are found in the **aromas** of various flowers, fruits, spices, and herbs. This does not mean that each terpene will contribute all the aromas indicated to any specific wine; it might be just one or two aromas, or none of the above, or they might together create entirely different aromas. The range of aromas for which these compounds are responsible beyond the confines of grape and glass should, however, give readers some idea of the aromatic pathway down which each terpene can be found.

Terpenoid acids The only terpenoid acid of interest as far as wine **terpenes** are concerned is geranic acid (found in Muscat and contributing to the aroma of cardamom and peppermint).

Terpenoid alcohols By far the most prolific, influential, and important category of terpenoids found in wine, terpenoid alcohols are present in increasing quantities in grapes as they ripen: citronellol (found in garden rose, geranium, ginger, black pepper, basil, peppermint, and cardamom; also plays a supporting role to citronellal in the **aroma** of lemon eucalyptus); diendol (not aromatic in itself but breaks down into hotrienol and nerol oxide); eugenol (the most herbal-influenced, spice-laden aroma of all terpenoid alcohols; found in bay leaves, cloves, and allspice); farnesol (the only terpenoid important to grape and wine aroma that is not a monoterpene or single-carbon terpene; farnesol is a sesquiterpene [15 carbon **atoms**] alcohol that is found in linden oil and is a constituent of garden-rose aroma); geraniol (although found in nutmeg, ginger, basil, rosemary, sage, cardamom, and grapefruit, geraniol can bring elements of peach and orange to the Muscat aroma, for which it is one of three main aromatic constituents); hotrienol (the aroma of linden or lime tree, but at levels higher than 30ug/l it is regarded as an indicator of premature ageing, probably due to poor storage conditions); **linalool** (or linalol; found in lavender, bergamot, jasmine, basil, rosemary, sage, star anise, cinnamon, cloves, nutmeg, coriander, cardamom, ginger, black pepper, and mandarin; one of the three terpene compounds principally responsible for Muscat aroma); nerol (found in orange blossom, ginger, basil, cardamom, mint, and mandarin; also one of the three terpene compounds principally responsible for the Muscat aroma); and a-terpineol (lilac and lime).

Terpenoid aldehydes Just two of any importance as far as wine **terpenes** are concerned: citronellol (although found in ginger, black pepper, geranium, and peppermint, citronellol is overwhelmingly lemony-resinous in character, representing a minimum of 82 per cent of lemon eucalyptus oil); and geranial (cinnamon, clove, ginger, basil, and peppermint).

Terpenoid esters Only two of any real importance as far as wine terpenes are concerned: geranyl acetate (found in lemongrass, coriander, nutmeg, cinnamon, peppermint, and, of course, geranium); and linalyl acetate (lavender, bergamot, jasmine, cinnamon, cardamom, bell pepper, basil, rosemary, sage, and peppermint).

Terpenoid oxides The three of most interest as far as wine **terpenes** are concerned are: **linalool** oxide (hay and pine; found mostly in white Muscat varieties); nerol oxide (an oxidation of nerol alcohol; sweet-**fruity aroma** found in Bulgarian rose); and rose oxide (found in Bulgarian rose but can also be green and geranium-like).

Terroir This French term literally means "soil", but in a viticultural sense *terroir* refers more generally to a vineyard's whole growing environment, which also includes altitude, aspect, climate, and any other significant factors that may affect the life of a vine, and thereby the quality of the grapes it produces. It used to be said that *terroir* had no direct equivalent in English, but that has not been true since 16 September 2006, when it was defined in the OED. Like so many other native French words, *terroir* is now officially part of the English language.

Tertiary aromas Synonymous with **bottle aromas**, these are the more mellow aromas that build up with age in the bottle. *See also* PRIMARY AROMAS; SECONDARY AROMAS.

Tête de cuvée (Fr) The first flow of juice during the pressing of the grapes, and the cream of the *cuvée*. It is the easiest juice to extract and the highest in quality, with the best **balance** of acids, **sugars**, and minerals.

Tetrasaccharide A sugar containing four **monosaccharides**. **Stachyose** is the only tetrasaccharide of interest as far wine is concerned.

Thasian *or* **Thasos** The wines of Thasos were mixed with honeyed dough and stored before straining and serving. Thasian was famous until the 2nd century BC, when production slowly deteriorated due to higher-volume, lower-quality wines produced in Kos, Rodos (Rhodes), Knedos, and various Ionian coastal cities. *See* ANCIENT GREEK WINES.

Thin A term used to describe a wine that is lacking in **body**, **fruit**, and other properties.

Third leaf A term derived from the earliest French **AOC** laws that determined that wine should be made only from vines that have given their *troisième feuille*, which is to say their third vegetative cycle, or third year. Many vines provide a small crop in their second year, and it is debatable whether any such crop should be entirely and immediately **green pruned**.

Tiburtine A thin **ancient Roman wine**, Tiburtine "easily evaporates" and "matures in 10 years, but it is better when aged", according to Athenaeus of Naucratis. This statement, combined with the extremely old age at which the best Roman wines were generally consumed, implies that a 10-year-old wine was not considered "aged" in this period.

Tight A **firm** wine of good extract and possibly significant **tannin** that seems to be under tension, like a wound spring waiting to be released. Its potential is far more obvious than that of **reticent** or **closed** wines.

Tirage (Fr) Bottling.

Titratable acidity Like **total acidity**, titratable acidity is often referred to as TA. As the eminently down-to-earth Dr Roger Boulton of UC Davis, states "Many people use titratable acidity and total acidity as synonyms, but they are not. The titratable acidity is always less than the total acidity, because not all of the hydrogen ions expected from the acids are found during the determination of titratable acidity. However, titratable acidity is easier to measure." So technically titratable acidity is not the same as total acidity, but as everyone uses the former to express the latter, for all practical purposes readers may take the two as synonymous. *See also* TOTAL ACIDITY.

Toast 1. A slow-developing, bottle-induced **aroma** commonly associated with Chardonnay, but that can develop in wines made from other grapes (including red wines). Toasty **bottle aromas** are initially noticeable on the **aftertaste**, often with no indication on the **nose**. 2. A fast-developing **oak**-induced aroma. 3. Barrels are toasted during their construction to one of three grades: light or low, medium, and heavy or high.

Tobacco A subjective **bouquet**/tasting term often applied to **oak**-**matured** wines, usually Bordeaux.

Tonnelier A cooper or barrelmaker.

Tonnellerie A cooperage or barrel works.

Torrefaction, torrefied A roasted, **toasted**, or charred **aroma**.

Torula The shortened version of *Torulaspora delbrueckii*, a cultured wild **yeast** that can reduce SO_2 levels. *Torula* prevents the development of unwanted micro-organisms that are traditionally eliminated by SO_2, thus it has been developed as a tool for assisting safe **low-sulphur regimes**. It also reduces VA levels by about two-thirds, reduces **acetaldehyde volatile phenols**, and can enhance certain **fruity esters**. Proprietary brands include Biodiva TD921 and Oenoferm Wild & Pure.

Total acidity Often referred to as TA, this is the total amount of both fixed and volatile **acidity** in a wine and is usually measured in grams per litre, because each acid is of a different strength, expressed either in terms of **sulphuric acid** (a 19th-century habit still utilised in France) or **tartaric acid**. To determine total acidity is difficult, as it entails measuring all the hydrogen ions of both the **fixed** and **volatile acids** present, including potential hydrogen ions able to be released as well as those already released, existing as free hydrogen ions in solution. For this reason the method most commonly used is one of titration, even though it will always be slightly lower than the true total acidity. *See also* TITRATABLE ACIDITY.

Total sulphur *See* SO_2, TOTAL.

Total package oxygen Known simply as TPO, total package oxygen is the total oxygen found in the headspace (HO) and dissolved in the wine (DO). Typically, 60 to 70 per cent of the total package oxygen in a wine will be contained in the headspace. *See also* DISSOLVED OXYGEN; HEADSPACE OXYGEN.

Traditional method The method of producing a sparkling wine by a **second fermentation** in the same bottle in which it is sold.

Transfer method Non-*méthode champenoise* sparkling wines undergo a **second fermentation** in bottle, and are then decanted, **filtered**, and re-bottled under pressure to maintain the *mousse*.

Transvasage (Fr) Synonymous with **transfer method**, although most often used on a small scale to fill very large-format bottles from 75cl bottles.

Trichloroanisole *See* TCA.

Trie (Fr) This term usually refers to the harvesting of selected over-ripe or **botrytised grapes** by numerous sweeps (*tries*) through the vineyard.

Trifolian According to Athenaeus of Naucratis, this

ancient Roman wine apparently **matured** more slowly than **Sorrentum**, and critics of modern wine-speak should note that he considered it to be "more **earthy**".

Triose A 3-carbon **monosaccharide**. Trioses are important in winemaking for the production of **glycerol**, but not for **fermentation** per se or indeed for **residual sweetness**.

Tripeptide One of the more common groups of **peptide** found in sparkling wine, a tripeptide contains three **amino acids**. *See* PEPTIDES.

Trisaccharide A **sugar** containing three **monosaccharides**. The only trisaccharides of any relevance for wine are **raffinose** and, to an even lesser extent, melezitose. *See also* OLIGOSACCHARIDE

Trockenbeerenauslese, TBA (Ger) A **QmP** for wines produced from individually picked, **botrytised grapes** that have been left on the vine to shrivel. The wine is golden-amber to amber in colour, intensely sweet, viscous, very complex, and as different from **Beerenauslese** as that wine is from **Kabinett**.

Tronçais A famous type of **oak** used for barrelmaking, from a small forest in the Allier *département* in the centre of France.

Typical Over-used, less-than-honest form of the term "honest".

Typicity A wine that shows good typicity is one that accurately reflects its grape variety or varieties, but it will be dependent on the person using this term to indicate whether it also encompasses a "sense of place". Typicity does not have to be confined to **varietal character**, but its meaning is not precise without further qualification.

UC (Fr) Short for *union coopérative* or other titles denoting a local or regional cooperative.

UC Davis The University of California's oenology department at Davis.

Ullage (Fr) 1. The space between the top level of the wine and the head of the bottle or cask. An old bottle of wine with an *ullage* beneath the shoulder of the bottle is unlikely to be any good. 2. The practice of topping up wine in a barrel to keep it full and thereby prevent excessive **oxidation**.

Ultra-premium wine This category comes between **super-premium** and **icon wines** and may be priced at US$10 up to, say, $150.

Umami The fifth basic taste (after sweetness, sourness, **bitterness**, and saltiness – *see* "The Taste, or 'Palate', of a Wine", p90–91) is the Eastern concept of umami (meaning "deliciousness" and implying a satisfying taste of completeness), which is triggered by the **amino acid** glutamate, hence the use of monosodium glutamate as a taste enhancer in Chinese cuisine. The concept of umami as a basic taste is contested by some scientists, who argue that it is artificial and created out of a complex continuum of perceptions. Other scientists claim that this explains all basic tastes, including sweetness, sourness, bitterness, and saltiness. Yet others propose further basic taste sensations, such as the taste of free **fatty acids** and a **metallic** sensation. *See* "What Is Umami?", p93.

Undertone A subtle and supporting characteristic that does not dominate like an overtone. In a fine wine, a strong and simple overtone of youth can evolve into a delicate undertone with maturity, adding to a vast array of other nuances that give the wine **complexity**.

Unfermentable residual sugar Even fully fermented dry wines can and usually do have a tiny amount of **residual sugar**, due to the presence of unfermentable sugars such as **arabinose, rhamnose, ribose,** and **xylose**. **Yeast**s are unable to convert these sugars, so there will be up to 2g/l of residual sugar in the driest wines.

Ungenerous A wine that lacks generosity has little or no **fruit** and also far too much **tannin** (if red) or **acidity** for a correct and harmonious **balance**.

Ungrafted vines This refers to vines grown on their own roots, rather than most *Vitis vinifera* vines, which are grafted onto American **rootstock**.

Unripe acid malic acid, as opposed to **tartaric acid** or ripe acid.

Upfront This term suggests a wine with an attractive, simple, immediately recognizable quality that says it all. Such a wine may initially be interesting, but it will not develop further and the last glass would say nothing more about its characteristics than the first.

Uronic acids So-called sugar acids such as gluconic, glucuronic and galacturonic acids are created from the **oxidation** of **sugars** and are all indicators of rot-infected grapes. Uronic acids are also found in urine, hence the name.

UTA (Ger) Acronym for *Untypischer Alterungs,* or atypical ageing (ATA).

Uvaggio (It) Wine that has been blended from various grape varieties.

VA The abbreviation for **volatile acidity**.

VA lift A winemaking "trick" whereby the **volatile acidity** is elevated to enhance the **fruitiness** of wine, but it is never allowed to rise anywhere near the level where the wine becomes unstable. Acceptable only in wines that are ready to drink, as this phenomenon does not improve with age.

Value for money This is the difference between penny-saving and penny-pinching. True value for money exists at £50, $50, or whatever, just as much as it does at £5, $5, or whatever, and the decision whether to buy will depend on how deep your pocket is, not whether it is 10 times better.

Vanilla, vanilla-oak Often used to describe the **nose** and sometimes the **palate** of an **oak**-aged wine, especially Rioja. It is the most basic and obvious of oak-induced characteristics.

Vanillin An aldehyde with a **vanilla aroma** that is found naturally in **oak** to one degree or another.

Varietal, varietal aroma, varietal character The unique and distinctive character of a single grape variety as expressed in the wine it produces.

VC (Sp) Short for *vino comarcal,* which literally means a "local wine" and can be compared to the *vin de pays* of France.

VdL A common abbreviation of *vin de liqueur,* a **fortified** wine that is normally muted with **alcohol** before **fermentation** can begin.

VdlT (Sp) Short for *vino de la tierra,* which literally means a "country wine", but is closer to the **VDQS** of France than its *vin de pays*.

VDN A common abbreviation for *vin doux naturel*. This is, in fact, a **fortified** wine, such as Muscat de Beaumes de Venise, that has been muted during the **fermentation** process, after it has achieved a level of between 5 and 8 per cent **alcohol**.

VDQS A common abbreviation for *vin délimité de qualité supérieure,* a lapsed quality-control system below **AOC** but above *vin de pays*. VDQS is no longer in existence, but the definition will be retained in the Micropedia for historical reference.

VdT (It) Short for *vino da tavola. See* VIN DE TABLE.

Vegetal A tasting term that can range from relatively neutral root vegetables, often cooked, through the more aromatic celeriac, to certain **herbaceous** notes, such as asparagus, bell pepper, and so on.

Vendange (Fr) The harvest.

Vendange tardive (Fr) Late harvest.

Vendangeur (Fr) Harvester or picker.

Vendangoir (Fr) A purpose-built structure to house and feed the *vendangeurs* at harvest time.

Vendemia (Sp) "Harvest", often used to indicate **vintage**.

Véraison (Fr) The ripening period, during which the grapes do not actually change very much in size, but do gain in colour (if black) and increase in **sugar** and **tartaric acid**, while at the same time decreasing in unripe **malic acid**.

Vermouth An aromatised wine. The name "vermouth" originates from *Wermut,* the German word for "wormwood", which is this wine's principal ingredient. The earliest examples of this wine were made in Germany in the 16th century and were for local consumption only. The first commercial vermouth was Punt-é-Mes, created by Antonio Carpano of Turin in 1786. Italian vermouth is traditionally red and sweet, while French vermouth is white and dry, but both countries make both styles. Vermouth is made by blending very bland base wines (they are two or three years old and come from Apulia and Sicily in Italy and Languedoc-Roussillon in France) with an extract of aromatic ingredients, then sweetening the blend with **sugar** and fortifying it with pure **alcohol**. Chambéry, a pale and delicately aromatic wine made in the Savoie, France, is the only vermouth with an official **appellation**.

Vertical tasting *See* HORIZONTAL TASTING.

Viertelstück (Ger) A 300-litre (79-US gallon) oval cask.

Vigneron (Fr) Vineyard worker; often a vineyard owner.

Vignoble (Fr) Vineyard.

Vigour Although this term could easily apply to wine, it is invariably used when discussing the growth of a vine, and particularly of its **canopy**. In order to ripen grapes properly, a vine needs about 50 square centimetres (7.75 square inches) of leaf surface to every gram of fruit, but if a vine is too vigorous (termed "high vigour"), the grapes will have an over-**herbaceous** character, even when they are theoretically ripe.

Vin clair (Fr) A still base wine used for sparkling wine-making, before the blending and **secondary fermentation**.

Vin de l'année (Fr) This term is synonymous with *vin primeur*.

Vin de café (Fr) This category of French wine is sold by the carafe in cafés, bistros, and so on.

Vin de cuvée (Fr) Wine made from the first (and best) pressing only.

Vin délimité de qualité supérieur *See* VDQS.

Vin doux naturel *See* VDN.

Vin de France This term has replaced *vin de table* as the lowest-quality wine in the French wine regime, but because these wines are now allowed to indicate grape variety and **vintage**, some producers have started selling interesting wines at this level. The effect has been to raise standards from the bottom up.

Vin de garde (Fr) Wine that is capable of significant improvement if it is allowed to age.

Vin de glace (Fr) French equivalent of **Eiswein**.

Vin de goutte (Fr) Also called the **free-run juice**. In the case of white wine, this is the juice that runs free from the press before the actual pressing operation begins. With red wine, it is fermented wine drained off from the manta, or **cap**, before this is pressed.

Vin gris (Fr) A delicate, pale version of rosé.

Vin jaune (Fr) This is the famous "yellow wine" of the Jura that derives its name from its honey-gold colour that results from a deliberate **oxidation** beneath a Sherry-like *flor*. The result is similar to an aged Fino Sherry, although it is not **fortified**. *See also* "The Jura, Bugey, and Savoie", pp307–310.

Vin de liqueur *See* VdL.

Vin mousseux (Fr) This literally means "sparkling wine" without any particular connotation of quality one way or the other. But because all fine sparkling wines in France utilise other terms, for all practical purposes it implies a cheap, low-quality product.

Vin nouveau (Fr) This term is synonymous with *vin primeur*.

Vin ordinaire (Fr) Literally an "ordinary wine", this term is most often applied to a French *vin de table,* although it can be used in a rather derogatory way to describe any wine from any country.

Vin de paille (Fr) Literally "straw wine", a complex sweet wine produced by leaving late-picked grapes to dry and shrivel in the sun on straw mats. *See also* "The Jura, Bugey, and Savoie", pp307–310.

Vin de pays (Fr) A rustic style of country wine that is one step above *vin de table,* but one beneath VDQS. *See also* "Vins de Pays and Vins de France", p331–341.

Vin de presse (Fr) Very dark, **tannic**, red wine pressed out of the manta, or **cap**, after the *vin de goutte* has been drained off.

Vin primeur Young wine made to be drunk within the year in which it is produced. Beaujolais Primeur is the official designation of the most famous *vin primeur,* but export markets see it labelled as Beaujolais Nouveau most of the time.

Vin de table (Fr) Literally "table wine", although it is not necessarily a direct translation of this term. It is used to describe the lowest level of wine in France and is not allowed to give either the grape variety or the area of origin on the label. In practice, it is likely to consist of various varieties from numerous areas that have been blended in bulk in order to produce a wine of consistent character, or lack thereof, as the case may be.

Vin d'une nuit (Fr) A **rosé** wine or very pale red wine that is allowed contact with the manta, or **cap**, for one night only.

Vinegar Alcohol that has been **oxidized** into **acetic acid**.

Vinegar fly A synonym for the *Drosophila melanogaster,* or common fruit fly.

Vinho regional (Port) *See* VR.

Vinifera *See* VITIS VINIFERA.

Vinification Far more than simply describing **fermentation**, vinification involves the entire process of making wine, from the moment the grapes are picked to the point at which the wine is finally bottled.

Vinimatic This is an enclosed, rotating **fermentation** tank with blades fixed to the inner surface, that works on the same principle as a cement-mixer. Used initially to extract the maximum colour from the grape skins with the minimum **oxidation**, it is now being utilised for pre-fermentation **maceration**.

Vino comarcal (Sp) *See* VC.

Vino-Lok A commercial glass-stopper alternative to **cork**, Vino-Lok utilises a sealing ring made of DuPont Elvax, a registered eythylene vinyl acetate polymer. It is the click-on, click-off quality of this seal that sets it apart, making it not only tremendously easy to bottle wine but also effortless to reseal not just the bottle it comes with but many other bottles, too – so don't throw these closures away!

Vino de mesa (Sp) The same as *vin de table* or **table wine**.

Vino novello (It) The same as *vin primeur*.

Vino da tavola (It) The same as *vin de table* or **table wine**.

Vino de la tierra (Sp) *See* VdLT.

Vinous Of, or relating to, a characteristic of wine. When used to describe a wine, this term implies basic qualities only.

Vintage 1. A wine of one year. 2. Synonymous with harvest: a vintage wine is the wine of one year's harvest only (or at least 85 per cent according to EU regulations) and the year may be anything from poor to exceptional. It is, for this reason, a misnomer to use the term vintage for the purpose of indicating a wine of special quality.

Viscosity Higher viscosity as expressed by rivulets gathering on the inside of a glass after swirling it (known as "legs" or "tears") is a crude indicator of higher **alcohol** levels in a wine and, contrary to popular belief, has no bearing on the relative quality of a wine.

Viticulture Cultivation of the vine. Viticulture is to grapes what horticulture is to flowers.

Vitis International Varietal Catalogue, The This is a database that lists the accepted international standard grape variety names and their known synonyms.

Vitis vinifera A species covering all varieties of vines that provide classic winemaking grapes. *See also* The Wine Vine Tree, p38.

Vitispirane An aromatic **compound** that forms during bottle **maturation**, vitispirane can have a chrysanthemum **bouquet**, while trans-vitispirane is heavier, more exotic, and more **fruity**, even camphor-like.

Vivid The **fruit** in some wines can be so fresh, ripe, clean-cut, and expressive that it quickly gives a vivid impression of complete character in the mouth.

Volatile acids These acids, sometimes called **fatty acids**, are capable of evaporating at low temperatures, hence the volatility. Too much volatile acidity is always a sign of instability, but small amounts actually play a significant role in the taste and **aroma** of a wine. Formic, butyric, and proprionic are all volatile acids that may be found in wine, but **acetic acid** and **carbonic acid** are the most important, and acetic acid is the one referred to when an experienced taster makes the comment "VA" or "too much VA". The volatile aroma of acetic acid will be easily recognised by anyone who has ever made a sweet-and-sour sauce in Chinese cuisine or a sauce diable in French cuisine and has caught a whiff of **vinegar** evaporating over the heat.

Volatile phenols Almost one-third of all French wines tested have volatile phenols above the level of perception, so they are clearly not always bad. Some volatile phenols such as ethyl-4-guaiacol (**smoky-spicy aroma**) and, to a lesser degree, vinyl-4-guaiacol (carnation aroma) can actually contribute attractive elements to a wine's **bouquet**. Volatile phenols are generally considered to be faults, however, and the amount of ethyl and vinyl phenols present in a wine is increased by harsh methods of pressing (particularly the use of continuous presses), insufficient settling, use of particular strains of **yeast**, and, to a lesser extent, increased skin-contact. Ethyl-4-phenol is responsible for the so-called **Brett** off-aromas (stables, horsey, sweaty-saddles), while vinyl-4-phenol has a Band-Aid (sticking plaster) off-aroma. *See* "Tastes and Aromas", pp96–101.

Voluptuous A term used to describe a succulently rich wine, often a red wine, which has a seductive, mouth-filling flavour. *See also* Luscious.

Vosges 1. A mountain range bordering and sheltering the Alsace region of France. 2. A well-known type of **oak** used for barrelmaking, from Darnay, Saint Hélène, Schirmeck, or any other forest in the Vosges *département* of eastern France.

VQPRD (Fr) Common abbreviation for *vin de qualité produit dans une région délimité*, the French language version of the lapsed EU umbrella classification encompassing national quality wine regimes, such as **AOC**, **DOC** etc. QWPSR, or "quality wine produced in a specified region", is the English equivalent. VQPRD has now been replaced by **AOP**.

VR (Port) The abbreviation for *vinho regional*, the lowest rung in Portugal's **appellation** system. A VR can be compared to the regional *vin de pays* category in France.

Warm, warmth Terms suggestive of a good-flavoured red wine with a high alcoholic content; if these terms are used with an accompanying description of cedary or **creamy**, they can mean well-matured in **oak**.

Water immersion A fairly recently developed barrel-making process whereby the **oak** staves are rinsed prior to cooperage (whereas soaking barrels in water following cooperage is a long-established practice). According to a review of the chemical and sensory profile of water-immersion barrels published by *Practical Vineyard & Winery* in the November/December 2010 issue, water immersion is said to have a significant impact on refining the flavour profile, making the oak influence more subtle, resulting in more **elegant** wines.

Watershed A term used for an area where water drains into a river system, lake, or some other body of water.

Watery An extreme qualification of **thin**.

Weißherbst, Weissherbst (Ger) A single-variety **rosé** wine produced from black grapes only.

Wet feet *Vitis vinifera* vines famously do not like "wet feet" – that is, they do not respond well to being grown in areas where water accumulates. Some varieties, even closely related varieties, are even more sensitive to this than others, such as Sauvignon Gris, which is not as resilient to water accumulation as Sauvignon Blanc.

Wine The fermented juice of grapes, as opposed to "fruit wine", which may be made from other fruits.

Wine acids The most important acids in wine are **tartaric** and **lactic**, unless the wine has not undergone **malolactic** (MLF), in which case the most important will be tartaric and **malic**. These are organic acids and as such they are fixed acids. Other organic acids found in wine include **citric**, **gluconic**, **succinic**, and, even in non-MLF wines, traces of lactic. Also found are **volatile acids** (acetic, carbonic, sulphurous); **phenolic acids**, **amino acids**; and **fatty acids**. *See* Fixed acidity, Titratable acidity, Total acidity, and individual named acids.

Wine diamonds A euphemistic synonym for **tartrate crystals**.

Wine lake A common term for the EU surplus of low-quality **table wine** .

Winkler scale A term synonymous with the **heat summation** system.

Winzersekt (Ger) Literally a "grower *Sekt*", this can either be the product of a single grower or a cooperative of growers but must be a Sekt bA.

Wood lactones These are various **esters** that are picked up from new **oak**; they may be the source of certain **creamy-oak** and **coconutty** characteristics.

Wood-matured This term normally refers to a wine that has been aged in new **oak**.

Xylose A **monosaccharide**, xylose is sometimes found as one of the **unfermentable sugars** in even the most full-fermented, bone-dry wines. Although of little significance in wine, it is also a **wood sugar**, thus relatively more abundant in oak-aged wines.

Yeast A kind of fungus that is absolutely vital in all winemaking. Yeast cells excrete a number of yeast **enzymes**, some 22 of which are necessary to complete the biochemical chain reaction known as **fermentation**. *See also* "Yeast: The Fermenter", p63.

Yeast, killer Some species of **yeast**, known as "killer strains" or yeasts with a "killer factor" release **protein** factors that are selectively toxic to other species of yeast. Providing that at least 2.5 per cent of the yeast population is a killer strain, it will take over and exclusively complete the **fermentation**. Although a useful attribute, yeasts are not and should not be selected on the basis of their killer factor. Less than 8 per cent of *Saccharomyces cerevisiae* yeast varieties are killer strains.

Yeast, First and second fermentation For sparkling wine, first and second fermentation **yeasts** have two diametrically opposed purposes and therefore require two totally different yeasts. The first fermentation is short and conducted at a relatively higher temperature, with the aim of producing a somewhat simple, lower- **alcohol** end product that is not supposed to be consumed. The second fermentation should be long and conducted at a comparatively lower temperature, to produce a more complex, higher alcohol end product that must be conducive to yeast-ageing. Many sparkling winemakers outside of **Champagne** make the basic error of using the same yeast for both and commercial laboratories unwisely market a *prise de mousse* yeast as equally suitable for both.

Yeast enzymes Each **yeast** enzyme acts as a catalyst for one particular activity in the **fermentation** process and is specific for that one task only.

Yeast nutrient Just as we need various nutrients to function, so does **yeast**. For winemaking purposes yeast nutrients are required in the most minute of quantities. These nutrients are: **amino acids**, vitamins (thiamine), and minerals (potassium, sulphate, phosphate, and magnesium). Trace elements of iron, copper, zinc, boron, and manganese are also responsible for stimulating yeast growth, but calcium, although it activates **enzymes**, is not essential.

Yeasty This is not a complimentary term for most wines, although a yeasty **bouquet** can sometimes be desirable in a good-quality sparkling wine, especially if it is young. A simple yeastiness, however, does not convey the **finesse** that the **aromas** of certain yeast products can, such as bread dough, bready, or brioche.

Yield 1. The quantity of grapes produced from a given area of land. 2. How much juice is pressed from this quantity of grapes. Wine people in Europe measure yield in hl/ha (hectolitres per hectare – a hectolitre equals 1,000 litres), referring to how much juice has been extracted from the grapes harvested from a specific area of land. This is fine when the amount of juice that can be pressed from grapes is controlled by European-type **appellation** systems, but in the New World, where this seldom happens, they tend to talk in terms of tons per acre. It can be difficult trying to make exact conversions in the field, particularly after a heavy tasting session, when even the size of a ton or gallon can become quite elusive. This is why, as a rough guide, I multiply the tons or divide the hectolitres by 20 to convert one to the other. This is based on the average extraction rates for both California and Australia, which makes it a good rule-of-thumb. Be aware that white wines can benefit from higher yields than reds (although sweet wines should have the lowest yields of all) and that sparkling wines can get away with relatively high yields. For example, Sauternes averages 25hl/ha, Bordeaux 50hl/ha, and **Champagne** 80hl/ha.

Zesty A lively characteristic that is suggestive of a zippy, tactile impression combined, maybe, with a distinctive hint of **citrus aroma**.

Zing, zingy, zip, zippy Terms that are all indicative of something that is notable for being **refreshing, lively**, and vital in character, resulting from a high **balance** of ripe **fruit acidity** in the wine.

Zygosaccharomyces bailii A so-called spoilage **yeast** that can generate a flocculant deposit in wine.

Zymase An **enzyme** involved in the conversion of sugar into **alcohol**.

Zymology The science of **fermentation**.

About the Authors

TOM STEVENSON has been writing about wine for more than 40 years. He has won 33 literary awards, including Wine Writer of the Year three times and the Wine Literary Award, as well as America's only lifetime achievement award for wine writers. In 1998 he was the first person to publish a 17th-century document proving that the English invented sparkling Champagne six years before Dom Pérignon set foot in Hautvillers. This ensured his *Christie's World Encyclopedia of Champagne & Sparkling Wine* also made history as the only wine book to warrant a leader in any UK national newspaper (*The Guardian*, 14 October 1998). In 2011 he was inducted into the New York Wine Media Guild's Hall of Fame. He has written 28 books, the most important of which have been published internationally by more than 50 publishers and translated into more than 25 languages. Additionally, he conceived and was the contributing editor of *Wine Report*. His *Sotheby's Wine Encyclopedia* (now *The New Sotheby's Wine Encyclopedia*) has sold over 700,000 copies worldwide.

DR O S SZENTKIRALYI is a Hungarian-born wine professional based in London. She is a Doctor of Law, specialising in protected origin of wine. Her vinous career started in hospitality: she worked as a sommelier in top establishments around the world before moving to luxury wine retail and eventually starting her own consulting company, Liquid Talent. She is a judge and ambassador at the International Wine Challenge, in charge of tasting quality control at the Champagne and Sparkling Wine World Championships, and Head of Tasting at the wine app Wotwine. She is currently a candidate for both the Master of Wine and Master Sommelier exams.

Contributors

France – Bordeaux
WENDY NARBY
Based in Bordeaux for more than 30 years, Wendy Narby is a senior lecturer at the Ecole du Vin de Bordeaux, specialised in training international Bordeaux wine tutors for the Bordeaux Wine Council. She is also an international trade educator for the Médoc Wine Council. She published *Bordeaux Bootcamp: The Insider Tasting Guide to Bordeaux Basics*, in 2015 and *The Drinking Woman's Diet: A Liver Friendly Lifestyle Guide* in 2018.

France – Burgundy
GYÖRGY MÁRKUS
György Márkus has written about wine since the late-1990s, having started his career as deputy editor of *Borbarát*, Hungary's former leading wine magazine. He has specialized in Burgundy since 2001 and visits the region several times a year to research the highly respected *Burgundia & Champagne* magazine, which he has published since 2010.

France – Burgundy – Beaujolais
BILL NANSON
Bill has been writing about Burgundy for 17 years both online and in print. His 2012 book *The Finest Wines of Burgundy* enjoyed much critical success, and his regular *Burgundy Reports* are a reference point in the industry. He also runs tailor-made wine tours in the region and contributes to *Decanter* magazine.

France – Loire Valley
CHRIS HARDY AND ALEX MEUNIER
British-born Chris has spent 22 years as Senior Buyer for UK retailer Majestic Wine before purchasing Loire specialist Charles Sydney Wines in 2016. He was joined by Bordeaux Business School graduate Alex Meunier, who leads the sales department of the company. Their hands-on approach and excellent relationship with most of the top producers allow them unrivalled insight into the region's life.

France – Jura, Bugey, and Savoie
WINK LORCH
Wink Lorch has worked as a wine writer and educator since the 1980s, based between London and the French Alps. She has become the leading source of knowledge on the wines of Jura, Savoie, and Bugey. Wink has written and self-published two widely acclaimed books: *Jura Wine* (2014) and *Wines of the French Alps: Savoie, Bugey and Beyond* (2019).

France – Rhône Valley
DOMINIC BUCKWELL
Dominic has 25 years experience as an English barrister/solicitor in international trade and containerised transport. He qualified as a sommelier in 2016, judging wines and writing about *terroir* and appellations. He conducts tasting and consumer events, with a focus on the Rhône Valley. He is a non-executive director of WineGB responsible for lobbying on UK wine regulation and helping to formulate policy for government in matters concerning wine regulation.

France – Languedoc-Roussillon
ROSEMARY GEORGE MW
Rosemary George was one of the first women to become a Master of Wine, back in 1979. She has worked as a freelance wine writer for nearly 40 years, with Languedoc-Roussillon one of her fields of particular expertise, with her most recent publications *The Wines of Faugères* and *The Wines of the Languedoc, Infinite Ideas*, 2017 and 2018 respectively. She also writes a regular blog.

France – Southwest France, Provence, and Corsica
PAUL SCHOFIELD
London-based Paul Schofield is a WSET Diploma student. He is a judge at international wine competitions and has worked vintages in Italy, the United States, Canada, and Hungary.

Italy
PIETRO RUSSO, GABRIELE GORELLI, AND ANDREA LONARDI
Pietro graduated in viticulture and winemaking at the University of Conegliano Veneto and received his Masters degree between Enita of Bordeaux and Supagro Montpellier. His career includes working periods spent in France, Italy, Spain, and New Zealand, where he has developed first-hand experience and an insatiable passion for wine. Today, he makes wines in the most compelling appellations of Sicily and studies to become a Master of Wine.

Montalcino born and raised, Gabriele developed his passion for wine in the land of Brunello. After a degree in foreign languages, he soon started working as a graphic designer while nurturing his knowledge and talent for wine. Today he operates as a wine export and communication consultancy and studies in London to become a Master of Wine.

Valpolicella-born Andrea has a degree in Agriculture and a Masters in control and management at the Grande Ecole of Montpellier. He spent several years working in different stages of the wine field's value chain: from sales and marketing to production and viticulture. In 2012 he was appointed as Chief Operating Officer of Bertani Domaines. He is currently studying for the Master of Wine title.

Spain
PEDRO BALLESTEROS TORRES MW
Brussels-based Pedro Ballesteros Torres is a Master of Wine with a degree in Agrofood Engineering and a Masters in Viticulture and Oenology. He is a columnist at several magazines in Spain and Belgium and writes regularly for magazines in the UK, the United States, and Italy. He is chair/senior jury at DWWA (UK), 5StarsWines (Italy), Bacchus (Spain), Concours Mondial Bruxelles (Belgium), Mundus Vini (Germany), IWC Merchants Awards Spain, and Silk Road (China). He is also actively engaged in promotion, education, and advisory roles.

Portugal
CARLOS SIMOES MS
Melbourne-based Carlos imparts his wine knowledge with passion and erudition. His fascination with wine began at his family's vineyard in Portugal, which developed into a life-long journey exploring the world of wine. He is a Master Sommelier and has been working in some of the world's best restuarants around the world.

Germany
MICHAEL SCHMIDT
Michael has been involved with the *Sotheby's Wine Encyclopedia* since it was first published in 1988. Over the years he has worked with Tom on several projects, most notably the *Wine Report* from 2004 to 2009. He is the German correspondent of Jancis Robinson's *Purple Pages* (since 2008) and the German consultant for Hugh Johnson's and Jancis Robinson's *World Atlas of Wine*. In his home country his professional tasting, writing, and translating skills are much in demand and have made him a regular contributor to German wine magazines and books. He lives in the Ahr Valley with his wife.

Austria
KRISTIAN KIELMAYER
Economist with a degree in winemaking, Kristian is a WSET Diploma graduate and a lecturer of the Austrian Wine Academy (*Weinakademie Österreich*). He has worked many years for the largest wine merchant in Vienna, Austria. Currently based in Budapest, Hungary, working as a consultant, he specialises in training, education, sensory evaluation, protection of origin, presentation, and tourism.

Switzerland
CHANDRA KURT MW
Award-winning writer, Chandra Kurt is one of Switzerland's foremost wine authors, having written

more than 20 books on wine, including *Wine Tales*, the yearly guide *Weinseller*, while also having launched the wine magazine *Weinseller Journal*. Chandra is an international wine consultant, a member of the Circle of Wine Writers in London, the Ordre des Coteaux de Champagne, and the Confrérie du Guillon. Chandra spent her childhood in Asia. She now lives in Zurich, Switzerland.

Other Winemaking Countries of Northwestern Europe
KATHRINE LARSEN-ROBERT MS
Kathrine is a Danish-born wine professional living in London. She is an award-winning sommelier, currently working for a major UK distributor. She is a Master Sommelier since 2013, and a Master of Wine candidate.

Southeastern Europe
CAROLINE GILBY MW
Dr Caroline Gilby MW is an author, writer, and consultant, specialising in Central and Eastern Europe. In 2018 she published her first book, *The Wines of Bulgaria, Romania and Moldova,* and she has also contributed to Hugh Johnson's *Pocket Wine Book, The Oxford Companion to Wine, The World Atlas of Wine,* and Tom Stevenson's *Wine Report,* as well as several magazines and websites. She holds a doctorate in plant biology and has been a Master of Wine since 1992.

Greece
KONSTANTINOS LAZARAKIS MW
Piraeus-born Konstantinos studied mechanical engineering as well as jazz before becoming a Master of Wine at the age of 32, in 2002. He was awarded with the Bollinger Medal for outstanding performance in wine tasting and the Villa Maria Award for the viticulture paper during his MW exams. Being the first Master of Wine from Greece and the broader region of Southeastern Europe, he has been an active advocate for Greek wine around the world. A charismatic communicator and educator, he has been a source of inspiration for both the last and the coming generations of wine professionals in Greece and beyond and has contributed decisively to the shaping of a sustainable wine culture and the establishment of world-acclaimed wine education in his birth land. Konstantinos' book *The Wines of Greece,* updated in 2018, is considered a textbook in English for anyone interested in the development of the Greek wine industry.

Levant
EMMA DAWSON MW
Emma Dawson is a Master of Wine and senior wine buyer for the UK wine merchant Berkmann Wine Cellars who specialises in the on-trade. Previously she worked for the high-end retailer Marks and Spencer for nine years. Over this time she has covered most of the wine world within her buying areas. She currently manages regional France, Southern Italy, Germany, Australia, South America, and the Eastern Mediterranean. She has also taken a special interest in exploring emerging wine regions across the world, travelling broadly to areas including Greece, Lebanon, Georgia, Turkey, Israel, Croatia, Brazil, Bolivia, Uruguay, Mexico, and Japan.

Israel
ADAM S MONTEFIORE
Israeli-based Adam S Montefiore is the wine writer for the *Jerusalem Post* and the member of the Circle of Wine Writers. He is also an educator and consultant, specialising in the wines of Israel and the Eastern Mediterranean. He contributes to Hugh Johnson's *Pocket Wine Book, The World Atlas of Wine,* and Jancis Robinson MW's *The Oxford Companion To Wine*. He has written or contributed to the books *Wine Route of Israel, The New Book of Israeli Food,* and *Wines of Israel.*

Lebanon
MICHAEL KARAM
London-based Michael Karam is the author of *Wines of Lebanon,* which won the Gourmand Award for the Best New World Wine Book in 2005, *Arak and Mezze: The Taste of Lebanon,* and *Lebanese Wine: A Complete Guide to Its History and Winemakers.* He is also the contributing editor to *Tears of Bacchus: A History of Wine in the Arab World.*

South Africa
CATHY VAN ZYL MW AND TIM JAMES
Cathy van Zyl is an associate editor of the *Platter's South African Wine Guide.* A Master of Wine, she is also a member of its council and travels widely to present South African wines in both existing and new markets.

Tim James is a wine journalist based in Cape Town and the author of *Wines of the New South Africa: Tradition and Revolution* (California University Press, 2013). He is also an associate editor of the annual *Platter Guide,* for which he has tasted for many years.

North Africa and Other Winemaking Countries
LINDA GALLOWAY
Chef-consultant Linda Galloway trained and worked as a journalist for 20 years before switching her focus to food and wine, studying at Leiths School of Food and Wine and the Wine and Spirit Education Trust. Her love of wine took hold while living on the doorstep of the Cape Winelands in South Africa. Writing combines all three of her interests.

The United States – California
JAMES CLARKE
James is currently an MW student with over 10 years' experience in the wine industry. His interest in California peaked and then cemented during both personal and business trips throughout the States, as well as working directly with a number of top-end Napa wineries during his time with UK agent Pol Roger Portfolio.

The United States – Pacific Northwest, Atlantic Northeast, and Other Winemaking Areas
DOUG FROST MS MW
Doug is a Kansas City author who is one of only four people in the world to have achieved the remarkable distinctions of Master Sommelier and Master of Wine. He has written three books: *Uncorking Wine* (1996), *On Wine* (2001), and the *Far From Ordinary: The Spanish Wine Buying Guide* (third edition, 2011). He is the global wine and spirits consultant for United Airlines and writes about wine and spirits for many publications, including the *Oxford Companion to Spirits and Cocktails* (due in 2020). Frost is the director of the Jefferson Cup Invitational Wine Competition, the Mid-American Wine Competition, the host of the Emmy Award winning PBS-TV show *FermentNation,* and is a founding partner of Beverage Alcohol Resource, an educational and consulting company.

Canada
EMILY PEARCE
Emily is a Toronto-based wine professional who is both a Master Sommelier candidate with the Court of Master Sommeliers, as well as a Master of Wine student. As a contributing writer for *Sotheby's Wine Encyclopedia* and *Decanter Magazine Online,* Emily leverages her knowledge to promote the Canadian wine scene to the broader world. She is a respected international wine judge, an award-winning sommelier, and the founder and president of Femmes du Vin, which is a not-for-profit focused on mentoring women in the hospitality community while creating a network of collaboration and respect.

Mexico and the Caribbean
SHYNTIA PEREZ
Mexican-born Shyntia is a WSET Diploma holder and educator and founder of import and distribution company Exclusive Terroirs Co and the Wine Education Centre Mexico. She has more than 14 years experience in the wine industry, having previously worked in the UK and Hong Kong. She is the first and so far only Mexican MW candidate.

Chile
ALISTAIR COOPER MW
Alistair holds a degree in Modern Languages (Spanish, Portuguese, and Latin American Studies) and has spent several years in South America as export manager for various wine producers. He is a Master of Wine since 2017 and currently works as a freelance writer, wine judge, and consultant.

Argentina
TIM ATKIN MW
Tim is a Master of Wine since 2001, an award-winning wine writer, judge, and presenter. He regularly contributes to *Decanter, The World of Fine Wine, Harper's,* and the *Drinks Business* amongst many other publications. He is co-chair of the International Wine Challenge and one of the Three Wine Men. His annual *Argentina Special Report* is an important reference point in the industry.

Australia and New Zealand
JULIA SEWELL AND DORIAN GUILLON MS
Julia Sewell began her wine career in the Melbourne and Sydney restaurant scene, working her way through various sommelier roles. This opportunity for immersion in the wines of her Australian homeland has ensured a continued connection even after leaving the country. She now lives in London, where she is involved in miscellaneous wine endeavours, including education, writing, and sales.

Dorian grew up in France's Loire Valley, but didn't fall in love with wine until a backpacking adventure through South America, where he was beguiled by an elegant sommelier in Mendoza, Argentina. He has since been working at some of the best restaurants in London and Melbourne. He considers his greatest achievement to be earning the prestigious title of Master Sommelier in 2018, making him one of only 269 people worldwide to have ever received that accreditation.

Asia
DENIS GASTIN
Denis Gastin has performed a wide range of roles in the wine industry over the past 30 years in Australia and Asia, including as a wine writer, wine judge, and wine educator. In 2016 he was awarded the national Order of Australia Medal (OAM) for his contribution to the wine industry. He has been a long-standing contributor of the Asia content for some of the world's most authoritative wine reference books.

Credits

MAPS

NATIONAL GEOGRAPHIC

Debbie Gibbons

Jerome N Cookson

Greg Ugiansky

Maureen Flynn

CONTRACTORS

Martin Darlison, Encompass Graphics

Ed Merritt

Martin von Wyss, vW Maps

Map data

General: OpenStreetMap contributors. Map data is available under the Open Database License; EuroGeographics EuroGlobalMap; Natural Earth

Argentina: Wines of Argentina

Australia: Commonwealth of Australia; Wine Australia

Canada/British Columbia: British Columbia Wine Authority

Canada/Ontario: VQA Wines of Ontario

Chile: Límite Comunas Chile; Servicio Agricola Y Ganadero / División Protección Agrícola Y Forestal, Subdepartamento De Viñas Y Vinos, Inocuidad Y Biotecnología

Croatia: Croatian Chamber of Commerce

France: datagouvfr; https://www.inao.gouv.fr/

Georgia: https://www.georgianwine.uk/news/pdf-downloads/

Hungary: https://mtu.gov.hu/documents/prod/borte--rke--p.pdf

Israel: IPEVO (Israel Professional Enology and Viticulture Organization)

Italy: Marcello Leder www.quattrocalici.it; Regione Veneto - Servizio qualificazione delle produzioni agroalimentari, Regione Lombardia; Regione Lombardia - tutti i diritti riservati; https://idt2.regione.veneto.it/

Luxembourg: Fonds de Solidarité Viticole

Moldova: http://www.wineofmoldova.com/en/

New Zealand: New Zealand Winegrowers; Intellectual Property Office of New Zealand, Ministry of Business, Innovation and Employment

Romania: https://www.onvpv.ro/en

Slovenia: Republic of Slovenia, Ministry of Agriculture, Forestry and Food

South Africa: Jackie Cupido, S A Wine Industry Information & Systems NPC / S A Wynbedryf-Inligting & -Stelsels MSW www.sawis.co.za

Spain: ICEX Spain Trade and Investment, Zonas de Calidad Diferenciada: Vinos; Consejos Reguladores, Consejerías CC.AA. y Ayuntamientos

Switzerland: swisswine.ch 2019

United Kingdom: British Geological Survey

United States: TTBGov - American Viticultural Area (AVA) Map Explorer https://www.ttb.gov/wine/ava-map-explorer

PHOTOGRAPHS

KEY TO ABBREVIATIONS

l = left   r = right   t = top   m = middle   b = bottom   s = sidebar box

SS = Shutterstock.com   DT = Dreamstime.com   ASP = Alamy Stock Photo   PD = Public Domain

CC = Creative Commons (Creative Common licenses can be found at https://creativecommons.org/licenses/.)

Frontispiece: NewFabrika/SS

Contents page: Sotheby's, Inc.

Foreword: Sotheby's, Inc.

A New Edition for a New Decade:
8 Tom Stevenson; 9 Csaba Villanyi

Understanding the Fine Wine Market:
All photos Sotheby's, Inc.

Part One: TASTE AND QUALITY

Factors Affecting Taste and Quality

14–15 LightField Studios/SS; 16–17 Peek Creative Collective/SS; 20 Rostislav Glinsky/SS; 21 PIXEL to the PEOPLE/SS; 22 Climber 1959/SS; 23 Dmitrii Postnov/SS; 24 Juan Aunion/SS; 25t Colleen Ashley/SS; 25b robertonencini/SS; 26 Photo-Graphia/SS; 27 bepsy/SS; 28 David Prado PeruchaSS; 29 Martha Almeyda/SS; 30 FreeProd33/SS; 36tl Ganka Trendafilova/SS; 36tr nnattalli/SS; 36bl nnattalli/SS; 36br nnattalli/SS; 37tl LAFS/SS; 37tr Lucky Business/SS; 37mr FreeProd33/SS; 37bl B.E. Lewis/SS; 37br Horst Lieber/SS; 40 alfredo ravanetti/SS; 41t gab90/SS; 41b Chiyacat/SS; 42 Rosenzweig/CC BY-SA 2.5; 43t Sherri R. Camp/SS; 43b Prayut Piangbunta/SS; 44 S. Kuelcue/SS; 45 Chiyacat/SS; 46t TP Gronlund/SS; 46b Ralf Geithe/SS; 47 PosiNote/SS; 48 Buntan/SS; 49t AdryPhoto1/SS; 49b Vbecart/CC BY-SA 3.0; 50t wjarek/SS; 50b Rosenzweig/CC BY-SA 3.0; 51 Alessandro Cristiano/SS; 52t TFoxFoto/SS; 52b PD; 53 James Zandecki/SS; 54 sonsart/SS; 55t Buffy1982/SS; 55b patjo/SS; 56t Chateau La Tour de Chollet/CC BY 2.0; 56b Ralf Geithe/SS; 57t sylv1rob1/SS; 57b Josep M Penalver Rufas/SS; 58 Arnaud25/CC BY-SA 3.0; 59t Nalidsa/SS; 59b Deborah Kolb/SS; 60 Josef Mohyla/SS; 61 Petra Nowack/SS; 62 Craig_Camp/CC BY 2.0; 63 goran cakmazovic/SS; 64 Ivan Kovbasniuk/SS; 66 Sergey Ryzhov/SS; 69 peizais/SS; 70 Kondor83/SS; 71 Thomas Dutour/SS; 72 lunamarina/SS; 75 Victor Grigas/CC BY-SA 4.0; 77 Bondarenco Vladimir/SS; 78 Archimëa/CC BY-SA 3.0; 81 Kiev.Victor/SS; 82 FJAH/SS; 83t Ugis Riba/SS; 83b KH-Pictures/SS; 85 Sean Moore/Karen Prince; 86l Sigur/SS; 86r UfaBizPhoto/SS; 87 Iurii Korolev/SS; 88 Riedel; 89t Zalto; 89s Peter Hermes Furian/SS; 91 limpido/SS; 92–93 Sean Moore/Karen Prince; 94–95 dien/SS; 97l BestPhotoPlus/SS; 97m Vanebtyn75/DT; 97tr exopixel/SS; 97br Natali Zakharova/SS; 98tl Nerss/DT; 98ml horiyan/SS; 98bl Sergio33/SS; 98tm Snowbelle/SS; 98bm Cozine/SS; 98tr Max Lashcheuski/DT; 98mr Enlightened Media/SS; 98br Kovaleva_Ka/SS; 99tl Lev Kropotov/SS; 99bl Inna Kyselova/DT; 99tm Rudmer Zwerver/SS; 99bm gresei/SS; 99tr Hekla/SS; 99br New Africa/SS; 100tl Hekla/SS; 100ml Melica/DT; 10bl Food Impressions/ SS; 100tm Eric Isselee/SS; 100bm Discovod/SS; 100tl Anna1311/DT; 100bl bonchan/SS; 102 Plateresca/SS; 103 Eduard Zhukov/SS; 105 Ekaterina Pokrovsky/SS

Part Two: WINE THROUGH THE AGES

A Chronology of Wine

106–107 Alessandro Giordano 1981/SS; 108–109 PD; 111 PD; 114 PD; 116 PD; 119 www_visitaland_com

Part Three: A WORLD OF WINE

The Wines of the World

120–121 Africa Studio/SS; 122–123 Suse Schulz/DT; 124 EQRoy/SS; 127 PD

The Wines of France

128–120 Lee Jorgensen/SS; 130–131 flydragon/SS; 133 Petr Kovalenkov/SS; 135 HUANG Zheng/SS; 137 wjarek/SS; 138 Iryna Savina/SS; 140 vovidzha/SS; 142 Arnieby/SS; 146 Arnieby/SS; 149 Douethe; 150 Renaud Camus/CC BY 2.0; 151 Isabelle Albucher/CC BY-SA 4.0; 152 Meandering Trail Media/SS; 154 Alex Brown/CC BY 2.0; 155 PA/CC BY-SA 3.0; 156 el-meister/CC BY 2.0; 158 Jim Budd/CBY-NC-ND 2.0; 159 Pascal Gaudette/CC_BY-NC-SA 2.0; 160 Château La Tour-Haut-Caussan; 161 Marie-Pierre Samel/SS; 164 Hedley Lamarr/SS; 165 FreeProd33/SS; 167 Marzolino/SS; 168 Bernd Zillich/SS; 169 Arnieby/SS; 170 Jordi Muray/SS; 171 Marzolino/SS; 174 wjarek/SS; 176 Brum/SS; 177 FreeProd33/SS; 179 Dreamer Company/SS; 182 FreeProd33/SS; 183 SpiritProd33/SS; 186 Chateau Quinault; 187 Relaisfrancmayne/CC BY-SA 3.0; 188 SpiritProd33/SS; 189 wjarek/SS; 191 ThoDut/SS; 193 Antoine Bertier/CC- BY 2.0; 195 wjarek/SS; 198 HUANG Zheng/SS; 199 Elena Dijour/SS; 201 kid315/SS; 203 Château de la Rivière; 204 O_Voltz/SS; 205 SpiritProd33/SS; 206 michael Clarke/CC BY-SA 2.0; 207 hebdromadaires/CC BY-SA 2.0; 209 FreeProd33/SS; 210 SpiritProd33/SS; 211 Olaf Schulz/SS; 212 RossHelen/SS; 214 Iryna Savina/SS; 215 PD; 216 Claude PIARD/CC- BY 4.0; 217 Meandering Trail Media/SS; 221 RnDmS/SS; 222 jorisvo/SS; 223 Mpmpmp/CC BY-SA 3.0; 224 Matt Ragen/SS; 227 andre quinou/SS; 228 Massimo Santi/SS; 229 Ghischeforever/SS; 232 Martin M303/SS; 234 RICIfoto/SS; 236 Nevskii Dmitrii/SS; 239 Agnes27/CC BY-SA 3.0; 240 Richard Semik/SS; 241 Richard Semik/SS; 242 michusa/SS; 244 Alec Issigonis/SS; 245 Gaelfphoto/SS; 246 robert paul van beets/SS; 251 Daan Kloeg/SS; 253 Massimo Santi/SS; 254 Tomas er/CC BY-SA 3.0; 256 Ekaterina_Minaeva/SS; 258 Luca Querzoli/SS; 260 Daan Kloeg/SS; 263 Daan Kloeg/SS; 264 RICIfoto/SS; 265 Daan Kloeg/SS; 266 StudioPortoSabbia/SS; 268 FreeProd33/SS; 271 Jef Wodniack/SS; 273 Jakapong Paoprapat/SS; 275 Pawel

Kazmierczak/SS; 276 Sergey Dzyuba/SS; 279 nnattalli/SS; 281 leoks/SS; 283 Sebastien DURAND/SS; 285 Pack-Shot/SS; 287 Heliosphile/SS; 288 leoks/SS; 290 barmalini/SS; 291 sylv1rob1/SS; 293 RYSAN/SS; 294 Peter Titmuss/SS; 295 Omaly Darcia/SS; 298 wjarek/SS; 301 mrstam/SS; 305 Krisztian Juhasz/SS; 307 Jorge Alves/SS; 308 Agnes27/CC BY-SA 3.0; 310 Deborde/SS; 311 Alison Cornford-Matheson/SS; 312 Nagarazoku/CC BY 2.0; 314 Paul Cullen Photography/SS; 315 syvwich/CC BY 2.0; 316 Richard Semik/SS; 317 Jibi44/CC BY-SA 3.0; 319 gabriel12/SS; 321 Stephen M Brooks/SS; 322 barmalini/SS; 324 PAUL ATKINSON/SS; 328 Jon Ingall/SS; 329 Marianne Casamance/CC BY-SA 4.0; 333 Richard Semik/SS; 336 ConnieMa/CC BY-SA 2.0; 341 barmalini/SS

The Wines of Italy
342–343 iacomino FRiMAGES/SS; 344 Mirabelle Pictures/SS; 346 Lasterday/SS; 349 Olga Zinovskaya/SS; 352 Poogie/SS; 357 Richard Semik/SS; 359 Lenscap Photography/SS; 362 zakaz86/SS; 365 Salvador Maniquiz/SS; 367 Dmitriy Yakovlev/SS; 370 Simona Bottone/SS; 373 Giuseppe Sanfilippo; 376 wjarek/SS; 387 Sara Babusci/SS; 388 REDMASON/SS

The Wines of Spain
390 Anatolijs Gizenko/SS; 393 ZRyzner/SS; 398 Alberto Loyo/SS; 399 Ander Dylan/SS; 401 Unai Huizi/SS; 402 F. JIMENEZ MECA/SS; 403 Photosite/SS; 406 barmalini/SS; 409 Uwe Aranas/SS; 410 Lobroart/SS; 411 berni0004/SS; 411 Fernando Garcia Esteban/SS; 414 Noradoa/SS

The Wines of Portugal
416–417 Lmarcal/SS; 418 maloff/SS; 419 Ricard_Bernardo/CC BY 2.0; 423 Roman Tiraspolsky/SS; 426 Marafona/SS; 427 arcangela/SS; 429 ribeiroantonio/SS; 431 NICOLA MESSANA PHOTOS/SS; 433 Cicero Castro/SS

The Wines of Germany, Austria, and Switzerland
434-435 maloff/SS; 438 wsf-s/SS; 439 Donnhoff; 440 www.germanwines.de; 442 gallofilm/SS; 444 Marcel Werner/SS; 447 M. Volk/SS; 448 Alice-D/SS; 449 trabantos/SS; 451 Richard Semik/SS; 453 travelview/SS; 457 travelview/SS; 458 mese.berg/SS; 459 Blue Nun; 461 Christian Mueller/SS; 463 nnattalli/SS; 465 EWY Media/SS; 467 Sina Ettmer Photography/SS; 469 Moritz Klingenstein/SS; Rolf Krahl/CC BY 4.0; 472 Timo Haberl/SS; 474 www.germanwines.de; 477 2199_de/SS; 479 Sina Ettmer Photography/SS; 480 TFoxFoto/SS; 481 Bildagentur Zoonar GmbH/SS; 483 Four season lover/SS; 485 DeepGreen/SS; 489 Mikadun/SS

The Wines of Northwestern Europe
490–491 ingridagrants/SS; 492 Matthew Dixon/SS; 494 SubstanceTproductions/SS; 497 Heiko Kalweit/SS; 500 Brian Clifford/SS; 502 Martine WALLENBORN/SS; 504 Guido Vermeulen-Perdaen/SS; 505 Matt Lewis/CC BY-SA 2.0; 506 Radoslaw Maciejewski/SS

The Wines of Southeastern Europe
508–509 posztos/SS; 511 k_samurkas/SS; 512 hdesislava/SS; 513 Szilard Csaki/SS; 516 BalkansCat/SS; 520 Marek Vostik/SS; 521 irakite/SS; 522 Petr Pohudka/SS; 524 ollirg/SS; 528 PicProductionMD/SS; 529 Zhuravlev Andrey/SS

The Wines of the Eastern Mediterranean
530–531 WR7/SS; 532 Peter Titmuss/SS; 534 editorial_head/SS; 537 ollirg/SS; 539 Lydia Vero/SS; 541 Chrispictures/SS; 542 Scott Woodham Photography/SS; 544 GreSiStudio/SS; 545 Areej Khaddaj/SS

The Wines of Africa
546–547 T. Wilbertz/SS; 548 LongJon/SS; 553 Peter Titmuss/SS; 555 David Steele/SS; 556 Marisa Estivill/SS; 559 Peter Titmuss/SS; 560 PhotoSky/SS; 563 Hemis/ASP; 564 Alastair Wallace/SS; 565 Iulian Ursachi/SS

The Wines of the Americas
566–567 tishomir/SS; 568 Ljubov/DT; 570 Mount Pleasant; 575t KJ; 575b california-champagne-social; 576 Buena Vista Winery; 578 USDA; 579 Alexandra Latypova/SS; 584 Handley; 585 melissamn/SS; 586 KarenWibbs/SS; 587 Gail Johnson/SS; 589 Stephen Beaumont/SS; 590 Kent Sorensen/SS; 592 Sheila Fitzgerald/SS; 594 Andrew Zarivny/SS; 596 Leonard Zhukovsky/SS; 599 Arnieby/SS; 601 Hank Shiffman/SS; 603 Noatake_Murayama/CC BY 2.0; 604 Carlos Gandiaga/SS; 607 randy andy/SS; 608Ken Wolter/SS; 613 Mike McCune/CC BY 2.0; 615 dorinser/SS; 616 Michael Warwick/SS; 619 Danita Delmont/SS; 622 Trisaetum Wine/CC BY 3.0; 627 Kirk Hewlett/SS; 628 indigolotos/SS; 629 Justin Hollinger/SS; 632 Bruce Goerlitz Photo/SS; 635 Southern Foodways Alliance//CC BY 2.0; 639 Helene Woodbine/SS; 641 Atomazul/SS; 643 Gareth Janzen/SS; 645 Nalidsa/SS; 648 Moises Bissu/SS; 649 Monica Garza 73/SS; 651 dpa picture alliance/ASP; 653 Agent Wolf/SS; 654 H.J. Herrera/SS; 655 Larisa Blinova/SS; 657 Ksenia Ragozina/SS; 659 CEXM/SS; 660 Juan Cruz de Frias/SS; 661 Ksenia Ragozina/SS; 663 Guadalupe Polito/SS

The Wines of Australia, New Zealand, and Asia
664–665 Mick Fitzsimmons/SS; 666 amophoto_au/SS; 668 Bilby/CC BY 3.0; 669 Milleflore Images/SS; 670 Atomazul/SS; 671 lkonya/SS; 672 Elias Bitar/SS; 674 Phillip Minnis/SS; 675 Maurizio Callari/SS; 677 will Perrett/ASP; 678 AJ Nitz Images/SS; 680 Steven J Scott/SS; 682 Ian Crocker/SS; 684 Oliver Foerstner/SS; 686 FiledIMAGE/SS; 687 Neil Cox/SS; 688 designium/SS; 689 ChameleonsEye/SS; 691 Benny Marty/SS; 693 Bernhard Richter/SS; 694 SUKIYASHI/SS; 697 Greyboots40/SS; 698 Milleflore Images/SS; 699 Milleflore Images/SS; 700 Keith 316/SS; 702 Keith 316/SS; 704 lkonya/SS; 705 lkonya/SS; 707 Ian Geraint Jones/SS; 708 Agent Wolf/SS; 710 Silken Photography/SS; 712 Sheryl Watson/SS; 714 patjo/SS; 715 Jonathan Harker/CC BY_SA 4.0; 716 Billybytheglass/SS; 719 Rolf_52/SS; 720 Peter Titmuss/SS; 723 Andrew Korson/SS; 725 Andrew George Barton/SS; 726 Wojtek Chmielewski/SS; 729 Luoxi/SS; 730 Vbecart/CC BY 3.0; 731 junrong/SS; 732 Dhananjay Bhagat/SS; 733 Pikoso.kz/SS; 734 Chatama/CC BY 3.0; 735 Princess_Anmitsu/SS

ILLUSTRATIONS

All illustrations by Remie Geoffroi (remgeo.com)

Acknowledgements

The authors wish to acknowledge the staff of National Geographic, Sotheby's Inc, and Moseley Road Inc, who made this book possible.

Tom would also like to thank Léna Martin (CIVA), Olivier Sohler (Fédération Nationale des Producteurs et Elaborateurs de Crémant), Foulques Aulagnon (CIVA), Thierry Fritsch (CIVA), Claire Sertnig (IVV), Romain Mondloch (IVV), Serge Fischer (IVV), and Francoise Peretti (Champagne Bureau) for their help in creating this book. Orsi would like to thank Sherry Stolar for her help in researching producers and creating content for the general part of the book and Paul Schofield for his help with researching California producers, apart from his role as contributor for Southwest France and Provence and Corsica.

NATIONAL GEOGRAPHIC

PUBLISHER AND EDITORIAL DIRECTOR Lisa Thomas

MANAGING EDITOR Jennifer Thornton

SENIOR EDITORIAL PROJECT MANAGER Allyson Johnson

CREATIVE DIRECTOR Melissa Farris

DIRECTOR OF PHOTOGRAPHY Susan Blair

SENIOR PHOTO EDITOR Adrian Coakley

PHOTO EDITOR Jill Foley

SENIOR PRODUCTION EDITOR Judith Klein

SOTHEBY'S INC.

WORLDWIDE HEAD, SOTHEBY'S WINE Jamie Ritchie

VICE PRESIDENT, SOTHEBY'S WINE ADVISORY Julia Gilbert

WINE AUCTION DEPARTMENT, SOTHEBY'S WINE Jackie Perrotta

MOSELEY ROAD INC.

PRESIDENT Sean Moore

GENERAL MANAGER Karen Prince

ART AND EDITORIAL DIRECTOR Lisa Purcell

COPY EDITOR Nancy J Hajeski

INDEXER Vikas Makkar, WPS Service

Index

National Geographic Partners
1145 17th Street NW
Washington, DC 20036-4688 USA

Get closer to National Geographic explorers and photographers, and connect with our global community. Join us today at nationalgeographic.com/join

For rights or permissions inquiries, please contact
National Geographic Books Subsidiary Rights: bookrights@natgeo.com

ISBN: 978-1-4262-2141-5

Printed in Hong Kong

20/PPHK/2

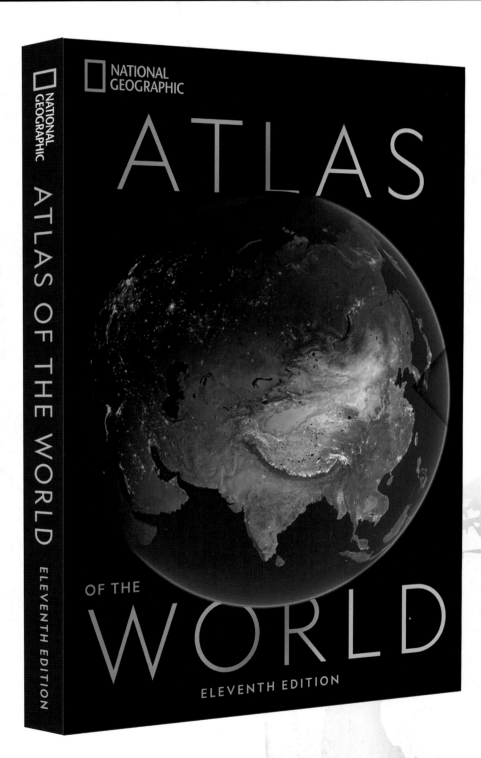